Interest Boxes

Student Aids

Mousa Nashad

ORGANIC CHEMISTRY

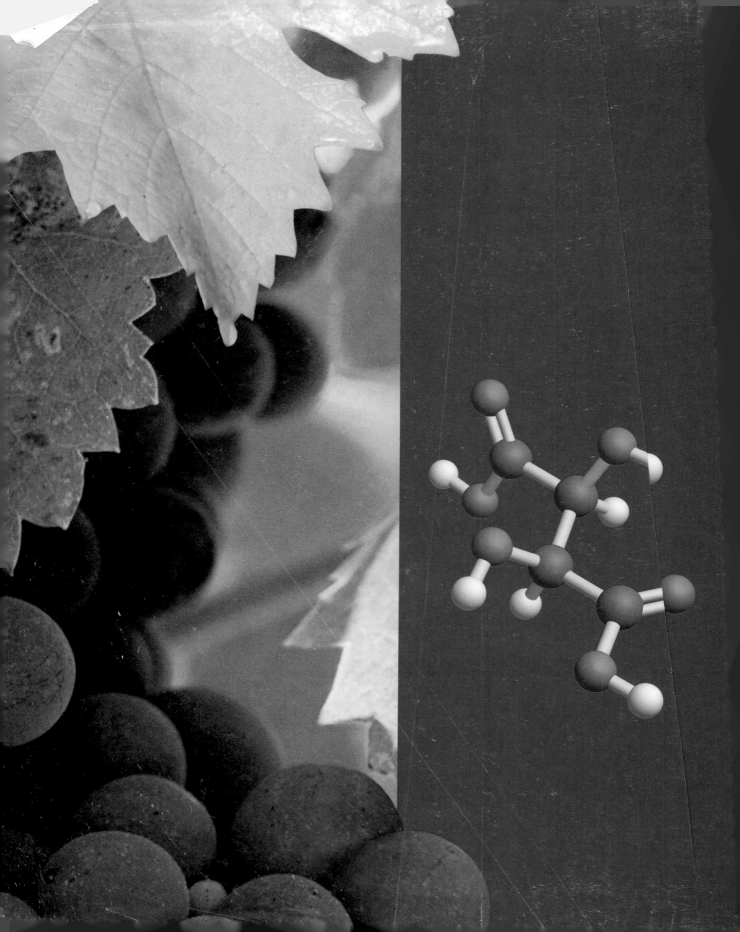

ORGANIC CHEMISTRY

SECOND EDITION

Paula Yurkanis Bruice

University of California, Santa Barbara

PRENTICE HALL
Upper Saddle River, New Jersey 07458

Library of Congress Cataloging-in-Publication Data

Bruice, Paula Yurkanis
    Organic chemistry / Paula Yurkanis Bruice.—2nd ed.
      p.   cm.
    Includes index.
    ISBN 0-13-841925-6
    I. Chemistry, Organic   I. Title.
QD251.2.B784 1998
   547-dc21   97-29107
                                CIP

Senior Editor: *John Challice*
Editor in Chief: *Paul F. Corey*
Associate Editor in Chief, Development: *Carol Trueheart*
Executive Managing Editor: *Kathleen Schiaparelli*
Assistant Vice President of Production and Manufacturing: *David W. Riccardi*
Marketing Manager: *Linda Taft MacKinnon*
Editorial Assistant: *Betsy Williams*
Associate Editor: *Mary Hornby*
Manufacturing Manager: *Trudy Pisciotti*
Creative Director: *Paula Maylahn*
Art Manager: *Gus Vibal*
Art Editor: *Karen Branson*
Interior Designer: *Judy Matz-Coniglio*
Art Director/Cover Designer: *Joseph Sengotta*
Cover Photo: *Christi Carter/Grant Heilman Photography, Inc.*
Cover Art: *Kenneth Eward/BioGrafx*
Cover Digital Imaging: *Rolando Corujo*
Photo Researcher: *Yvonne Gerin*
Photo Research Administrator: *Melinda Reo*
Art Studio: *Academy ArtWorks, Inc.* and *Thompson Steele Production Services, Inc.*
Copyediting, Text Composition, and Electronic Page Makeup: *Thompson Steele Production Services, Inc.*

Spectra reproduced by permission of Aldrich Chemical Co.
Chapter opener renderings © 1998 Kenneth Eward/BioGrafx.

© 1998, 1995 by Prentice-Hall, Inc.
Simon & Schuster /A Viacom Company
Upper Saddle River, New Jersey 07458

Printed in the United States of America
10 9 8 7 6 5 4 3 2

ISBN 0-13-841925-6

Prentice-Hall International (UK) Limited, *London*
Prentice-Hall of Australia Pty. Limited, *Sydney*
Prentice-Hall Canada Inc., *Toronto*
Prentice-Hall Hispanoamericana, S.A., *Mexico*
Prentice-Hall of India Private Limited, *New Delhi*
Prentice-Hall of Japan, Inc., *Tokyo*
Simon & Schuster Asia Pte. Ltd., *Singapore*

To Meghan, Kenton, and Alec
with love and immense respect
and to Tom, my best friend

BRIEF CONTENTS

CONTENTS

PART I

An Introduction to the Study of Organic Chemistry 1

1 Electronic Structure and Bonding. Acids and Bases 3

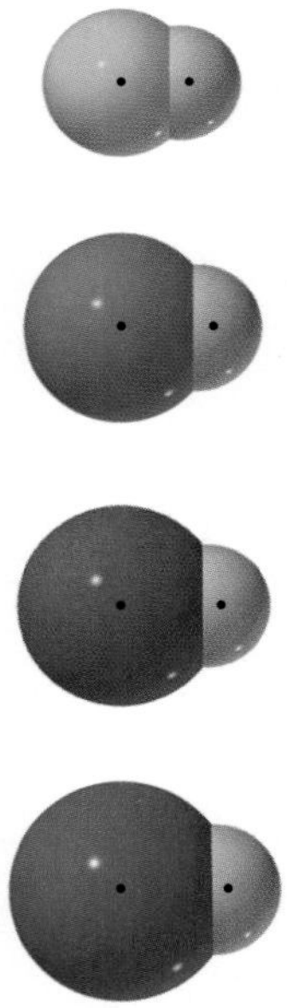

Hydrogen halides

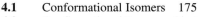

4 Stereochemistry:
The Arrangement of Atoms in Space;
The Stereochemistry of Addition Reactions   174

5 Reactions of Alkynes.
Introduction to Multistep Synthesis   235

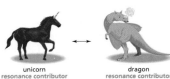

unicorn
resonance contributor

dragon
resonance contributor

rhinoceros
resonance hybrid

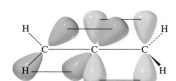

PART IV Identification of Organic Compounds   479

12 Mass Spectrometry and Infrared Spectroscopy   480

13 NMR Spectroscopy and Ultraviolet/Visible Spectroscopy   522

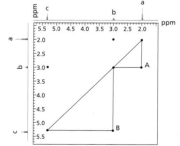

PART V

Aromatic Compounds   599

14 Reactions of Benzene and Substituted Benzenes   601

PART VI Carbonyl Compounds 667

15 Carbonyl Compounds I: Reactions of Carboxylic Acids and Their Derivatives with Oxygen and Nitrogen Nucleophiles 668

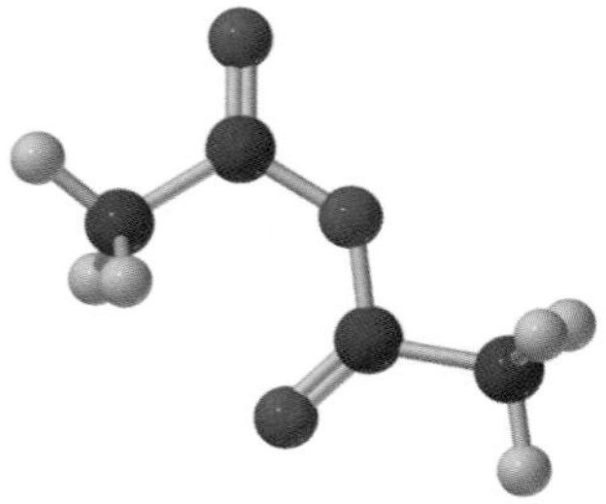
acetic anhydride

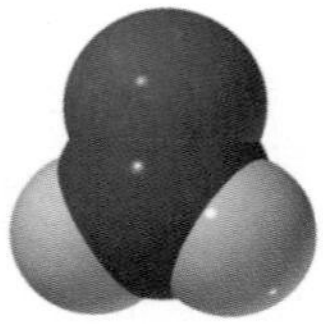
formaldehyde

16 Carbonyl Compounds II: Reactions of Carbonyl Compounds with Carbon and Hydrogen Nucleophiles; Reactions of Aldehydes and Ketones with Oxygen and Nitrogen Nucleophiles; Reactions of α,β-Unsaturated Carbonyl Compounds 725

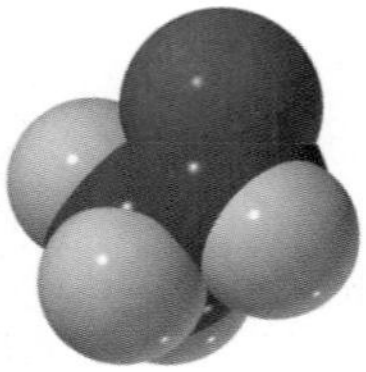
acetaldehyde

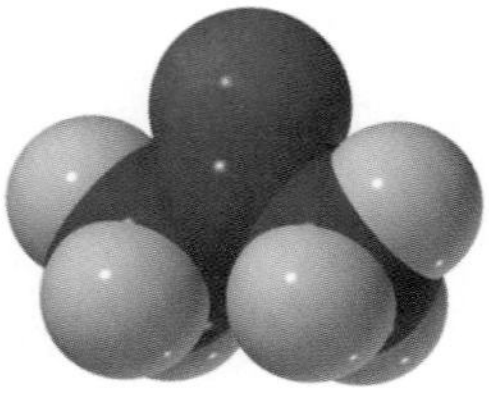
acetone

17 Oxidation–Reduction Reactions 777

18 Carbonyl Compounds III:
Reactions at the α-Carbon 816

PART VII

Bioorganic Compounds   869

19 Carbohydrates   870

20 Amino Acids, Peptides, and Proteins   907

21　Catalysis　953

22　The Organic Mechanisms of the Coenzymes　987

23　Lipids　1029

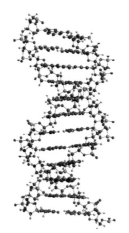

24 Nucleosides, Nucleotides, and Nucleic Acids 1061

PART VIII Special Topics in Organic Chemistry 1109

25 Synthetic Polymers 1110

26 Heterocyclic Compounds 1141

27 Pericyclic Reactions 1173

28 More About Multistep Organic Synthesis 1204

29 The Organic Chemistry of Drugs:
Discovery and Design 1236

PREFACE

TO THE INSTRUCTOR

Our students come to us because they want to make organic chemistry a part of their lives. It then becomes our job to make the subject come alive for them; to imbue in them the same kind of excitement that caused us to make this our life's work. This has been my guiding principle in the creation of this book. It is written for the student encountering the subject for the first time, not for the person who is already comfortable with the discipline.

My first resolve in creating a book for the student is that accessibility need not come at the expense of subject matter. Making a book "student-friendly" by sacrificing rigor or by exposing the student to less than the whole story is not fair to the discipline. The beauty of organic chemistry is not that something occurs, but rather *why* it occurs. Understanding the "why" causes a subject to come alive. Hiding "why" hurts the discipline and, in the long run, hurts the student.

I'd like to draw your attention to the following features of the text:

A Mechanistic Organization that Separates Synthesis and Reactivity

This book is organized so as to discourage the student from rote memorization. The subject has, in fact, been organized around mechanistic similarities, thus allowing a lot of material to be understood based on unifying principles of reactivity. As students master these principles, they will see that many of them reappear. The students soon get to the point where they can predict (perhaps with a little help) much of what will come.

In addition, I have separated the reactions of the functional groups from the synthesis of those groups. For example, in the alkene chapter, students learn about the reactions of alkenes. Students do *not* learn here about the synthesis of alkenes. Instead, they learn about the synthesis of alkyl halides, alcohols, ethers, and alkanes—the compounds formed when alkenes react. Because alkenes are synthesized as a result of the reactions of alkyl halides and alcohols, the synthesis of alkenes is covered when the reactions of alkyl halides and alcohols are discussed. Separating the synthesis of a functional group from its reactivity leaves the student without lists of unrelated reactions to memorize. But because such lists are useful in designing syntheses, a compilation of how each of the functional groups can be prepared appears in Appendix IV.

A Detailed Look at the Organization of the Text

The book has been divided into eight sections. Each section starts with a one-page overview so that students can understand where they are "going." Chapter 1 provides a summary of the material that students need to recall from General Chemistry. Acids and bases are covered (they are reviewed even more extensively in the Study Guide/Solutions Manual), and the student is told why this information will be important in his or her study of organic chemistry. In Chapter 2, students learn how to name five classes of organic compounds—those that will be the products of the reactions in the chapters that immediately follow. Chapter 2 also covers topics that are necessary before the study of reactions can begin (structures, conformations, and physical properties). Chapter 3 covers the reactions of alkenes. In this chapter, students are first introduced to the concept of "curved arrows." (The Study Guide/Solutions Manual contains an extensive exercise on "electron pushing." I have found this exercise to be very successful in making my students comfortable with a topic that should be easy but somehow perplexes even the best of them unless they have sufficient practice.) Chapter 3 also contains a discussion of thermodynamics and kinetics. Rate equations are derived in an appendix for those who wish to give their students a more mathematical treatment, and the Study Guide/Solutions Manual contains a section on calculating kinetic parameters.

I chose to lead off with the reactions of alkenes because of their simplicity. There is a wide variety of reactions in Chapter 3, but as soon as the student sees the similarity in their mechanisms (the electrophile adds to the sp^2 carbon bonded to the most hydrogens, and the nucleophile adds to the other sp^2 carbon), they understand that a lot can be learned relatively easily if there is a common thread. The reactions in Chapter 3 are discussed without regard to stereochemistry because I have found that students do quite well as long as only one new concept is introduced at a time. Chapter 4 reviews the isomers the students were introduced to in Chapters 2 and 3 (conformers and cis–trans isomers) and then discusses isomers that result from having a chirality center (the most recently IUPAC-approved term for a carbon bonded to four different groups). There is an extensive model-building exercise in the Solutions Manual to help the students get used to building molecular models.

Now that the students are comfortable with both isomers and electrophilic addition reactions, the two topics are considered together at the end of Chapter 4 when the stereochemistry of the reactions covered in Chapter 3 is presented. Chapter 5 covers alkynes. This chapter should build the students' confidence because of the similarity of the material to what was presented in Chapter 3.

Because of the importance of electron delocalization in organic chemistry, it has its own chapter (Chapter 6). Chapter 7 covers the reactions of dienes, allowing students to apply the concept of electron delocalization just learned to the electrophilic addition reactions that were mastered in Chapter 3. Chapter 8 covers the reactions of alkanes; the lack of a functional group in alkanes further demonstrates the importance of a functional group to chemical reactivity.

The next three chapters cover substitution and elimination reactions at an sp^3-hybridized carbon. Chapter 9 covers substitution reactions of alkyl halides. Chapter 10 covers elimination reactions of alkyl halides and then goes on to consider competition between substitution and elimination. Chapter 11 covers substitution and elimination reactions at an sp^3-hybridized carbon when a group other than a halogen is the leaving group (for example, alcohols, ethers, epoxides, and quaternary ammonium ions).

Next are two spectroscopy chapters (mass spectrometry and IR, followed by NMR and UV/Vis). Each spectral technique is written as a "stand-alone topic" so it can be covered independently at any time during the course. Chapter 12 opens with a table of functional groups for those who want to cover spectroscopy before students have been introduced to all the functional groups. Chapter 14 covers the reactions of benzene and substituted benzenes. Because not all organic courses cover the same amount of material in a semester, Chapters 12–14 have been strategically placed to fall around the end of the first semester. Ending a semester before Chapter 12 or after Chapter 14 or somewhere in between won't interfere with the flow of information.

The order of the topics dealing with the chemistry of carbonyl compounds in Chapters 15 and 16, where carboxylic acid derivatives are treated before aldehydes and ketones, was met with skepticism by some reviewers. However, when the opinions of those who had used the first edition of the book were sought, there was strong agreement that this is a preferred order of treatment. Chapter 17 covers redox reactions. Many of these reactions appear in earlier chapters, but this is an opportunity for students to extend their knowledge within the framework of what they already know. Chapter 18 deals with reactions of carbonyl compounds at the α-carbon.

Organic Chemistry with a Bioorganic Flavor

Bioorganic material has been introduced throughout the text to encourage students to recognize that organic chemistry and biochemistry are not separate entities, but two parts of a continuum of knowledge. For example, students can learn not just how carboxylic acids are activated for reaction in the laboratory, but also how they are activated for reaction in biological systems and why the mode of activation differs in the two situations. Once students learn how such things as electron delocalization, leaving-group tendency, electrophilicity, and nucleophilicity affect the reactions of simple organic compounds, they can appreciate how these factors are involved in the reactions of more complicated organic molecules such as enzymes, nucleic acids, and vitamins.

In the first part of the book, the bioorganic material is limited primarily to interest boxes and end-of-chapter sections, so the material is available to the curious student but the instructor is not compelled to introduce bioorganic topics into the course. At the end of the book there are several chapters that focus heavily on bioorganic topics, including the mechanisms of coenzymes, the chemistry of drug design, and the chemistry of biological macromolecules. Each instructor may choose which, if any, of these chapters to cover.

The chapters on bioorganic chemistry, unlike the coverage of such topics in most biochemistry texts, stress the chemistry associated with the topic. For example, in the lipids chapter, the mechanisms of prostaglandin formation (allowing students to understand how aspirin works), fat breakdown (allowing students to understand the odor associated with rancidity), and terpene biosynthesis are covered. The chapter on coenzymes stresses the role of vitamin B_1 as an electron delocalizer, vitamin K as a strong base, vitamin B_{12} as a radical initiator, biotin as a compound that can transfer a carboxyl group, and how the many different reactions of vitamin B_6 are controlled by the overlap of p orbitals. The chapter on nucleic acids explains mechanistically such things as the function of ATP and why DNA contains "T" instead of "U." The chapter on catalysis explains the various modes of catalysis that occur in organic reactions and then shows that these are identical to the modes of catalysis that occur in enzymatic reactions—all presented in a way that allows students to understand the lightening-fast rates of enzymatic reactions. Thus, these chapters are not a repetition of what will be covered in a biochemistry course, but are designed to serve as a bridge between the two disciplines.

An Early Emphasis on Organic Synthesis

Students are introduced to synthetic chemistry and retrosynthetic analysis early in this book (Chapters 3 and 5, respectively) so that they can use this technique throughout the course as they design multistep syntheses. In addition, there are six sections spread throughout the book on synthesis design, each with a different focus. For example, one emphasizes the proper choice of reagents and reaction conditions to maximize the yield of the target molecule (Chapter 10), and another focuses on the synthesis of cyclic compounds (Chapter 16). There is also a chapter at the end of the text titled "More About Multistep Organic Synthesis." This chapter discusses such things as protecting groups, control of stereochemistry during synthesis, and retrosynthetic analysis at a more advanced level. The use of combinatorial methods in organic synthesis is described in the chapter that focuses on drug design (Chapter 29).

A Modular Organization in the Last Third of the Book

I anticipate that the first twenty chapters will be covered by most instructors during the course of a year. An instructor can then choose among the remaining chapters depending on his or her preference and the interests of the majority of the students enrolled in the course. Those teaching students whose interests are primarily in the biological and health sciences might be more inclined to cover Chapter 21 (Catalysis), Chapter 22 (The Organic Mechanisms of the Coenzymes), Chapter 23 (Lipids), and Chapter 24 (Nucleosides, Nucleotides, and Nucleic Acids). Those teaching courses designed for chemistry majors may decide to include Chapter 25 (Heterocyclic Chemistry), Chapter 26 (Synthetic Polymers), Chapter 27 (Pericyclic Reactions), and Chapter 28 (More About Multistep Organic Synthesis). The book ends with a chapter on drug discovery and design—a topic that in my experience interests students enough that they will choose to read it on their own, even if it is not covered in the course.

Margin Notes and Boxed Material to Engage the Student

Two kinds of material are presented in the margin: margin notes that remind students of important principles, and biographical sketches that give students some appreciation of the history of chemistry and the people who contributed to that history. The boxed material I included to bring life to the topic being discussed (for example: Alkyl Halides as Survival Compounds, Chimney Sweeps and Cancer, Ultraviolet Light and Sunscreens, Penicillin and Drug Resistance) or to give the student extra help (for example: Calculating Kinetic Parameters, A Few Words About Curved Arrows, Incipient Primary Carbocations).

Problems, Solved Problems, and Problem-Solving Strategies

The book contains more than 1500 problems. The problems within each chapter are primarily drill problems. They are designed so the students can test themselves on the material just covered before they go on to the next section. Solutions to selected problems are shown to give the student an understanding of how to solve a problem. Each chapter also contains at least one Problem-Solving Strategy that teaches the student how to approach certain kinds of problems. For example, the Problem-Solving Strategy in Chapter 9 teaches students how to determine whether a reaction will be more apt to take place by an S_N1 or an S_N2 pathway. Each Problem-Solving Strategy is followed by an exercise that gives the student an opportunity to use the problem-solving skill just learned.

The end-of-chapter problems are of varying difficulty. The initial problems are drill problems that cover material from the whole chapter. This, I feel, provides a greater challenge by requiring the student to think about all the material in the chapter rather than individual sections. The end-of-chapter problems become more challenging as the student proceeds through them in order. However, I suspect the problems will not appear more difficult to the student because as the student works through the problems, his or her ability increases and so does his or her confidence (that is why the more difficult problems are not marked as such). I made a concerted effort to make none of the problems so difficult that the student loses confidence. A few problems contain references to the journal in which the original work was published; this enables curious students to find out more about the work that led to the conclusion shown in the problem.

Three-Dimensional, Computer-Generated Structures

In this edition, energy-minimized, three-dimensional structures appear throughout the text to give the students an appreciation of the three-dimensional shapes of organic molecules.

Pedagogical, Efficient Use of Color

Organic chemistry is sufficiently challenging without forcing students to use a road map to understand the use of color in a textbook. Therefore, this book does not have such things as blue incoming groups and green leaving groups. Students will not find themselves asking "Why did the nucleophile change its color?" Color is used to make the book visually more appealing and to make learning easier by highlighting points of focus. An attempt has been made to make the color consistent (mechanism arrows are always red, for example), but there is no need in this text for a student to memorize a color palette.

Enhanced Integration of Spectroscopy Problems
In response to user feedback, in this edition I've added more problems that involve interpreting spectra. As in the last edition, most ^{1}H-NMR spectra I show are 300 MHz FT-NMR spectra, which are more like the NMR spectra students are likely to see in their labs and after they graduate.

Instructional Aids
There are a number of products for students and faculty that are designed to complete this learning package:

FOR THE STUDENT

Study Guide/Solutions Manual (ISBN 0-13-889387-X)   The Study Guide/ Solutions Manual contains not just answers to the problems, but complete explanations of how the answers were obtained. The Study Guide/Solutions Manual also contains:

- a section on acid/base chemistry at a more advanced level than what is covered in the text, with a set of problems
- an 18-page exercise on "pushing electrons"
- an exercise on building molecular models
- an exercise on calculating kinetic parameters
- 20 practice tests

For the sake of making the path to mastery of the subject as smooth as possible, I thought it was important that I also write the Study Guide/Solutions Manual.

ChemCentral Organic Website (http://www.prenhall.com/~chem)   This companion website for students features additional problems for each chapter, regular updates of articles in the popular and lay scientific press that relate to organic chemistry, and a gallery of organic molecules that students can manipulate in real-time on their computers. And for students who are not familiar with the world wide web, we have:

Chemistry on the Internet   This free, 55-page, easy-to-understand book features a description of the internet (including the world wide web) and suggestions for students who are planning to use ChemCentral. This book can be packaged with this text; please see your Prentice Hall representative for details.

Prentice Hall Molecular Model Kit (ISBN 0-205-508136-3)   This best-selling model kit allows students to build space-filling and ball-and-stick models of common organic molecules. It allows accurate depiction of double and triple bonds, including heteroatomic molecules.

Prentice Hall Framework Molecular Model Kit (ISBN 0-13-330076-5)   This model kit allows students to build scale models that show the mutual relations of atoms in organic molecules, including precise interatomic distances and bond angles. This is the most accurate model kit available.

Chemistry: Themes of the Times   This mini-newspaper is produced by *The New York Times* for students of chemistry. It features selected articles related to chemistry from the pages of *The New York Times* and is available free to adopters of the text.

FOR THE INSTRUCTOR

Organic Matter: A Presentational CD-ROM   This CD-ROM includes almost all images from the book, animations of selected organic reactions, and videos of selected organic lab demos. In addition, it includes Presentation Manager 3.0, an easy-to-use software tool that allows instructors to organize material from the CD, material from the internet, and material from any other networked source into a cohesive presentation. Available for Windows (ISBN 0-13-919283-2) and Macintosh (ISBN 0-13-889379-9).

Transparency package (ISBN 0-13-889338-1)   This set features 150 full-color images from the text.

Test Item File (ISBN 0-13-889320-9)   Written by Gary Hollis of Roanoke College, this test bank features over 2000 multiple-choice questions. For instructors who prefer to customize their own tests, we also feature the test item file electronically along with test management software. Available for Windows (ISBN 0-13-889346-2) and Macintosh (0-13-889353-5).

ACKNOWLEDGMENTS

It gives me great pleasure to acknowledge the dedicated efforts of many good friends that made this book a reality: E. Paul Papadopoulos, whose work was truly invaluable in the creation of this edition; Robert Bly of the University of South Carolina, Ronald Starkey of the University of Wisconsin, Green Bay, James Lane of the University of Wisconsin, Melvin Rueppel of Rueppel Consulting, Leonora Gray Rueppel of St. Louis Community College, Meramec Campus, and Homer Smith, Jr. of the University of Richmond all of whom checked every inch of the book for accuracy; Edward Skibo of Arizona State University, who read and critiqued each chapter of the first edition as it was developed and whose many contributions persist in this edition; Alfred Burger, for the historical perspective he supplied for Chapter 29; Wayne Schnatter of Yeshiva University, for his contribution to Chapter 28; David L. Yerzley, M.D., for his assistance with the section on MRI; Andrei Blasko, for his contribution to the discussion of 2-D NMR; John Perona, University of California, Santa Barbara, who provided several of the computer-generated molecular images used in this edition, Helgi Adalsteinsson and Felice Chu Lightstone, for immediate response to any computer problems; and Chuck Huber, Chemistry Librarian at UCSB, who was always able to find what I needed. And I am deeply indebted to Paul Gilbert, who created the original index.

I also want to acknowledge the efforts of the many users of the first edition who generously gave their time to help me make this a better book. My particular thanks go to William Daly, Louisiana State University; Gordon Hambly, CEGEP John Abbott College; Jack Kirsch, University of California, Berkeley; Stanley Kudzin, State University of New York, New Paltz, and especially Kazusato Shibata, University of Birmingham. I am also very grateful to my students, who pointed out sections that needed clarification, worked the problems, and searched for errors.

I am greatly indebted to the following colleagues whose suggestions and encouragement were invaluable:

Mahamed Asgar Ali, *Howard University*
Shelby R. Anderson, *Trinity College*
John Barbas, *Valdosta State University*
Rick Bolesta, *Mt. Hood Community College*
Joyce C. Brockwell, *Northwestern University*
Thomas A. Bryson, *University of South Carolina*

George B. Clemans, *Bowling Green State University*
Barry A. Coddens, *Northwestern University*
John Cullen, *Monroe Community College*
Jan M. Fleischer, *Earlham College*
Warren P. Giering, *Boston University*
Michael M. Haley, *University of Oregon*
James E. Hanson, *Seton Hall University*
David Harpp, *McGill University*
John Isidor, *Montclair State University*
Dennis Lehman, *Harold Washington College*
James W. Long, *University of Oregon*
Jerry Manion, *University of Central Arkansas*
Amanda Martin-Esker, *Northwestern University*
John N. Marx, *Texas Tech University*
Anthony Masulaitis, *Jersey City State College*
Barbara J. Mayer, *California State University, Fresno*
Robert McClelland, *University of Toronto*
Gary W. Morrow, *University of Dayton*
Thomas W. Nalli, *Winona State University*
Lawrence Principe, *Johns Hopkins University*
Michael Rathke, *Michigan State University*
J. Ty Redd, *Southern Utah University*
Todd Richmond, *The Claremont Colleges*
Homer A. Smith, Jr., *University of Richmond*
Andrew B. Turner, *Saint Vincent College*
T. K. Vinod, *Western Illinois University*
Maria Vogt, *Bloomfield College*

I particularly want to thank the many wonderful and talented students I have had over the years who taught me how to be a teacher. And I want to thank my children, from whom I may have learned the most.

I also want to thank the talented and dedicated people at Prentice Hall: John Challice, not only for his truly superior editorial ability but also for keeping me happy and sane and on track, no matter what, and Carol Trueheart, for her magical ability to fix my prose and who miraculously kept track of everything I sent her and turned it into a book.

I want to make this text as user friendly as possible, and I would appreciate any comments that would help me achieve this goal in future editions. If you find sections that could be clarified or expanded, or examples that could be added, please let me know. Finally, this edition has been painstakingly combed for typographical errors. Any that remain are my responsibility; if you find any, please let me know so that they can be corrected in future printings.

Paula Yurkanis Bruice
University of California, Santa Barbara
pybruice@bioorganic.ucsb.edu

TO THE STUDENT

Welcome to organic chemistry. This book has been written for you, someone who is encountering the subject for the first time. The first thing you should do is familiarize yourself with the book. The material on the inside of the front and back covers is information that you may want to refer to many times during the course. The Key Terms and Reaction Summaries at the end of the chapters, and the Glossary at the end of the book can be useful study aids. Also look at the Appendices. The pictures of molecular models throughout the book are there to give you an appreciation of what molecules look like in three dimensions.

Be sure to work all the problems within each chapter. Short answers are given in the back of the book for all problems marked with a diamond. These are drill problems that will enable you to check whether you have mastered the material. Work as many end-of-chapter problems as you can. The more problems you work, the more comfortable you will be with the subject. Do not let any problem frustrate you. If you cannot figure out the answer in a reasonable amount of time, turn to the Solutions Manual to learn how you should have approached the problem. Later on go back and try to work the problem again without consulting the Solutions Manual.

The most important thing for you to remember in organic chemistry is: **DO NOT GET BEHIND.** Organic chemistry consists of a lot of simple steps, so it is not a difficult subject. But the subject can become overwhelming if you don't keep up.

Before many of the theories and mechanisms were worked out, organic chemistry was a discipline that could be mastered only through memorization. That is no longer true. You will find many common threads that will allow you to use what you have learned in one situation to predict what will happen in other situations. So as you read the book and study your notes, always try to understand *why* each thing happens. If the reasons behind the behavior are understood, most reactions can be predicted. If you approach the class with the misconception that you must memorize hundreds of unrelated reactions, it could be your downfall. There is simply too much material to memorize, and if all your knowledge is based on memorization, you won't have the necessary foundation on which to lay subsequent material. From time to time, some memorization will be required. Some fundamental rules have to be memorized, and you will have to memorize the common names of some organic compounds. But the latter should not be a problem. All your friends have common names that you have been able to learn.

Students who take organic chemistry to gain entrance into medical school sometimes wonder why medical schools pay so much attention to how they do in organic chemistry. The importance of organic chemistry is not in the subject matter alone. Mastering organic chemistry requires a thorough understanding of fundamentals and the ability to use these fundamentals to analyze, classify, and predict. This parallels the study of medicine; a physician uses an understanding of fundamentals to analyze, classify, and diagnose.

Good luck in your study. I hope that you will enjoy your course in organic chemistry and learn to appreciate the logic of the discipline. If you have any comments about the book or any suggestions about how it can be improved for students who will follow you, I would be happy to hear from you.

Paula Yurkanis Bruice
pybruice@bioorganic.ucsb.edu

A GUIDE TO USING THIS TEXT

The following two pages walk you through some of the key features of this text. This book was designed with you, the student, in mind—to help you succeed in organic chemistry. My main goal in writing this text was to make the material you study as accessible and appealing as possible. I hope you enjoy your study of organic chemistry.

—Paula Bruice

PROBLEM SOLVING

The key to success in this course lies in lots of practice with problem solving.

PROBLEM-SOLVING STRATEGY

Propose a mechanism for the following reaction.

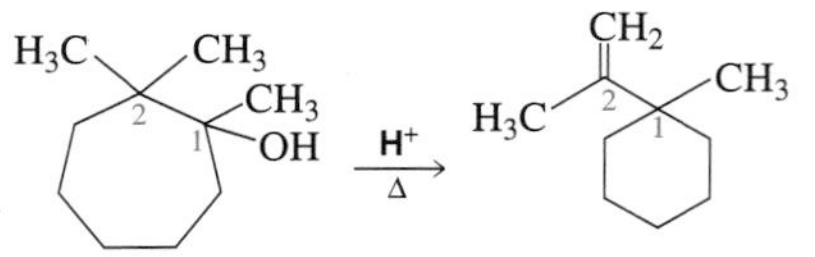

$$H_3C \quad CH_3 \quad CH_3 \quad OH \quad \xrightarrow[\Delta]{H^+} \quad H_3C \quad \overset{CH_2}{\underset{1}{\longleftarrow}} \quad CH_3$$

Even the most complicated-looking mechanism can be worked out if you proceed one step at a time, keeping in mind the structure of the final product. It can be helpful to label the equivalent carbons in the reactant and product. The OH group is the only base in the starting material, so it must react with the proton. Loss of water forms a tertiary carbocation.

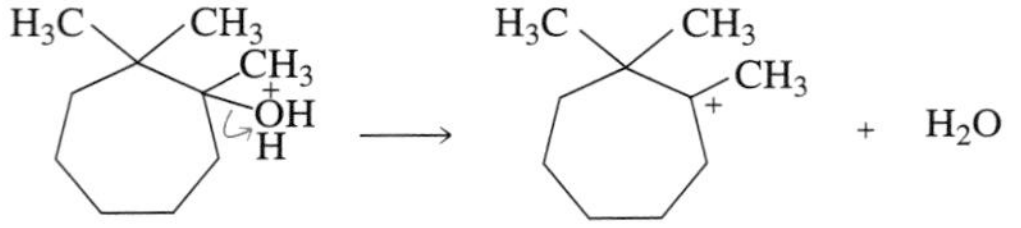

$$H_3C \quad CH_3 \quad \overset{+}{CH_3} \quad \overset{+}{OH} \quad H \quad \longrightarrow \quad H_3C \quad CH_3 \quad \overset{+}{CH_3} \quad + \quad H_2O$$

Since the starting material contains a seven-membered ring and the final product has

PROBLEM-SOLVING STRATEGIES

At least one problem-solving strategy in each chapter teaches you how to approach certain kinds of problems, organize your thoughts, and improve your problem-solving ability. Every strategy is followed by an exercise that allows you to immediately practice the strategy just discussed.

PROBLEM 27 / SOLVED

HCl is a weaker acid than HBr. Why then is ClCH$_2$COOH a stronger acid than BrCH$_2$COOH?

SOLUTION When comparing the acidities of HCl and HBr, we are comparing the ease of breaking two different kinds of bonds, an H—Cl bond and an H—Br bond. The electron density of a $3p$ orbital (of Cl) is greater than the electron density of a $4p$ orbital (of Br), which means that the attraction of Cl$^-$ for a proton is greater than the attraction of Br$^-$ for a proton. Size is more important than electronegativity in determining acidity, so H—Cl is less acidic even though chlorine is more electronegative.

In both carboxylic acids, the same kind of bond is broken (an O—H bond). Therefore, the only factor to be considered is the difference in the electronegativity of the

SOLVED PROBLEMS

Throughout the text, you will find solved problems that walk you through how to solve a particular type of problem. These step-by-step solutions help you learn how to approach a given problem and help you succeed at organic chemistry.

PROBLEM 2◆

Determine ($\pi + r$) for each of the following molecular formulas.

a. C$_{10}$H$_{16}$ c. C$_8$H$_{16}$
b. C$_{20}$H$_{34}$ d. C$_{12}$H$_{20}$

PROBLEM 3◆

Draw possible structures for compounds with the following molecular formulas.

a. C$_3$H$_6$ b. C$_3$H$_4$ c. C$_4$H$_6$

This diamond indicates a problem whose answer is found at the end of the text. Full solutions to these and all other problems in the book are in the accompanying Study Guide and Solutions Manual.

STUDY AIDS

A FEW WORDS ABOUT CURVED ARROWS

1. Make certain that the arrows are drawn in the direction of the electron flow and never against the flow. This means that an arrow will always be drawn away from a negative charge and/or toward a positive charge.

correct

$$CH_3-\overset{\overset{\displaystyle :\ddot{O}:}{|}}{\underset{\underset{\displaystyle CH_3}{|}}{C}}-\ddot{B}\ddot{r}: \longrightarrow CH_3-\overset{\overset{\displaystyle :\ddot{O}}{\|}}{\underset{\underset{\displaystyle CH_3}{|}}{C}} + :\ddot{B}\ddot{r}:^-$$

$$CH_3-\overset{+}{\underset{\underset{\displaystyle H}{|}}{\ddot{O}}}-H \longrightarrow CH_3-\ddot{O}-H + H^+$$

incorrect

$$CH_3-\overset{\overset{\displaystyle :\ddot{O}:}{|}}{\underset{\underset{\displaystyle CH_3}{|}}{C}}-\ddot{B}\ddot{r}: \longrightarrow CH_3-\overset{\overset{\displaystyle :\ddot{O}}{\|}}{\underset{\underset{\displaystyle CH_3}{|}}{C}} + :\ddot{B}\ddot{r}:^-$$

$$CH_3-\overset{+}{\underset{\underset{\displaystyle H}{|}}{\ddot{O}}}-H \longrightarrow CH_3-\ddot{O}-H + H^+$$

2. Curved arrows are drawn to indicate the movement of electrons. Never use a curved arrow to indicate the movement of an atom (in this case the movement of a proton).

continued

BOXES FOR STUDENT AID

This boxed material provides information that will further your understanding of organic chemistry. For example, here I discuss the right and wrong ways to draw curved arrows.

When you are asked to design a synthesis, one way to approach the problem is to look at the given starting material to see if there is an obvious series of reactions that can get you started on the pathway to the **target molecule** (the desired product). This is often the best way to approach a simple synthesis. The following examples will give you practice in designing a successful synthesis.

**10.11
DESIGNING A
SYNTHESIS II**

Example 1. How could you prepare 1,3-cyclohexadiene from cyclohexane?

Since the only reaction an alkane can undergo is halogenation, deciding what the first reaction should be is easy. An E2 reaction using a high concentration of a strong and bulky base to encourage elimination over substitution will form cyclohexene; therefore, *tert*-butoxide ion is used as the base and *tert*-butyl alcohol is used as the solvent. Bromination of cyclohexene will give an allylic bromide, which will form the desired target molecule by undergoing another E2 reaction.

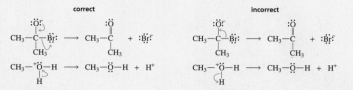

Example 2. Starting with methylcyclohexane, how could you prepare the following vicinal *trans*-dihalide?

Again, since the starting material is an alkane, the first reaction must be a radical substitution; the bromine will selectively substitute for the tertiary hydrogen. Under E2 conditions, tertiary alkyl halides undergo only elimination, so there will be no

DESIGNING A SYNTHESIS

Retrosynthetic analysis is introduced in Chapter 5 so that you can use this technique throughout your course to design multistep syntheses. In addition, you will find six sections spread throughout your book on synthesis design, each with a different focus. Look for this icon.

SUMMARY OF REACTIONS

1. Electrophilic addition reactions

   a. Addition of hydrogen halides (Markovnikov orientation) (Section 3.9)

$$RCH{=}CH_2 + HX \longrightarrow RCHCH_3$$
$$\underset{\displaystyle X}{|}$$

$$HX = HF, HCl, HBr, HI$$

   b. Addition of hydrogen bromide (apparent anti-Markovnikov orientation) Section 3.18)

$$RCH{=}CH_2 + HBr \xrightarrow{\text{peroxide}} RCH_2CH_2Br$$

REACTION SUMMARIES

At the end of any chapter that discusses organic reactions, you will find a summary of these reactions. Read these summaries when you finish the chapter, and make sure you can explain *how* each reaction occurs.

ABOUT THE AUTHOR

Paula Bruice and Orangathorpe.

Paula Yurkanis Bruice was raised primarily in Massachusetts, Germany, and Switzerland and was graduated from the Girls' Latin School in Boston. She received an A.B. from Mount Holyoke College in 1963 and a Ph.D. in chemistry from the University of Virginia in 1968. She received an NIH postdoctoral fellowship for postdoctoral study in biochemistry at the University of Virginia Medical School, and she held a postdoctoral appointment in the Department of Pharmacology at Yale Medical School.

Since 1971, she has been a member of the faculty at the University of California, Santa Barbara where she has received The Associated Students Teacher of the Year Award, the Academic Senate Distinguished Teaching Award, and two Mortar Board Professor of the Year Awards. Her research interests are in the application of organic reaction mechanisms to biological problems. Paula has a daughter and a son who are physicians and a son who is a lawyer. Her main hobbies are reading mystery/suspense novels and her pets (a dog, a parrot, and a couple of cats).

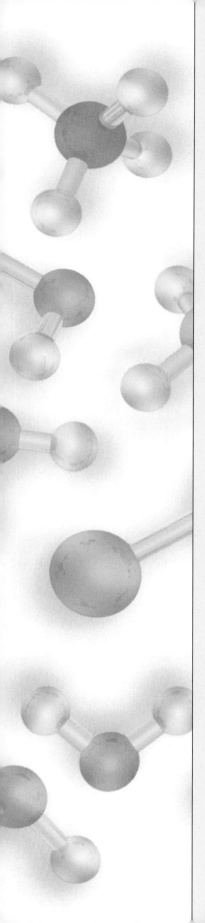

AN INTRODUCTION TO THE STUDY OF ORGANIC CHEMISTRY

The first two chapters cover a variety of topics that are designed to lay the foundation for your study of organic reactions.

Chapter 1 reviews the topics from general chemistry that will be important in your study of organic chemistry. The chapter starts with a description of the structure of atoms and then proceeds to a description of the structure of molecules. Acid-base chemistry is central to an understanding of organic reactions, so the concepts of acid-base chemistry are reviewed. You will see how the structure of a molecule affects its acidity, and how the acidity of a solution affects molecular structure.

To discuss organic compounds, it is necessary to be able to name them. In **Chapter 2,** you will learn how to name five different classes of organic compounds, and you will start to understand the rules used in naming compounds. Because the compounds examined in this chapter are the products of the reactions presented in the following ten chapters, you will have the opportunity to review the nomenclature of these compounds as you proceed. You will also compare and contrast the structures and physical properties of these compounds, which makes learning about them a little easier than if each compound were presented separately. Because organic chemistry is a study of compounds that contain carbon, the last part of Chapter 2 discusses the spatial arrangement of the atoms in chains of carbon compounds and in rings of carbon compounds.

ELECTRONIC STRUCTURE AND BONDING. ACIDS AND BASES

CH_4, H_2O, NH_3, $\overset{+}{N}H_4$, Cl_2, $HOOH$, H_2NNH_2

To stay alive, early humans must have been able to tell the difference between two kinds of material in their world. "You can live on roots and grubs," they might have said, "but you can't live on dirt. You can burn branches, but you can't burn rocks."

By the eighteenth century, scientists thought they had grasped the nature of that difference. Compounds from living organisms, they reasoned, must contain an unmeasurable vital force, the essence of life. Because chemists could not create life in the laboratory, neither could they create compounds with this vital force. In 1807, Jöns Jakob Berzelius gave names to the two kinds of materials. Those derived from living organisms—those containing the vital force—he called "organic." Compounds derived from minerals—those lacking that force—were "inorganic."

With this mind-set, you can imagine how surprised chemists were in 1828 when Friedrich Wöhler produced an organic compound called urea by heating ammonium cyanate, a compound that did not come from a living organism.

$$\overset{+}{N}H_4 \ \overset{-}{O}CN \xrightarrow{\text{heat}} H_2N-\overset{\overset{\displaystyle O}{\|}}{C}-NH_2$$

ammonium cyanate
an inorganic compound

urea
an organic compound

The urea Wöhler prepared was identical to natural urea, a chemical synthesized by mammals. But he had prepared it by heating an inorganic salt. For the first time, an "organic" compound had been obtained from something other than a living system and certainly without the aid of any kind of vital force. Clearly, chemists needed a new definition for "organic compounds." **Organic compounds** are now defined as compounds that contain carbon.

Why is an entire branch of chemistry devoted to the study of just carbon-containing compounds? Essentially all chemical reactions that take place in living

Jöns Jakob Berzelius (1779–1848) *not only coined the terms "organic" and "inorganic," but also invented the system of chemical symbols still used today. He published the first list of accurate atomic weights and proposed the idea that atoms carry an electric charge. He purified or discovered the elements cerium, lithium, silicon, thorium, titanium, and zirconium.*

Generally credited with the first synthesis of an organic compound from an inorganic one, German chemist **Friedrich Wöhler (1800–1882)** *began his professional life as a physician and later became a professor of chemistry at the University of Göttingen. Wöhler codiscovered the fact that two different chemicals could have the same molecular formula. He also developed methods of purifying aluminum—at the time, the most expensive metal on Earth—and beryllium.*

systems, including our own bodies, are organic reactions, because the molecules of life—proteins, enzymes, vitamins, lipids, carbohydrates, and nucleic acids—all contain carbon. We are dependent on organic compounds that occur in nature for our food, for many of our medicines, for our clothing (cotton, wool, silk), and for energy (natural gas, petroleum). In addition to naturally occurring compounds, chemists have learned to synthesize millions of organic compounds never found in nature, including synthetic fabrics, plastics, synthetic rubber, medicines, and preservatives. Many of these synthetic compounds prevent shortages of naturally occurring products. For example, it has been estimated that if synthetic materials were not available for clothing, all of the arable land in the United States would have to be used for the production of cotton and wool for clothing.

What makes carbon so special? Why are there so many carbon-containing compounds? The answer lies in carbon's position in the periodic table. Carbon is in the center of the second row of elements. The atoms to the left of carbon have a tendency to give up electrons, whereas the atoms to the right have a tendency to accept electrons.

the second row of the periodic table

Carbon, being in the middle, neither readily gives up nor accepts electrons. Instead, it shares electrons. Not only can it share electrons with several different kinds of atoms, but it can share electrons with other carbon atoms, allowing the formation of a wide variety of chainlike molecules. The fact that carbon can also share electrons with atoms of different electronegativity adds to the range of stable carbon-containing compounds. Consequently, carbon is able to form millions of stable compounds with a wide range of chemical properties simply by sharing electrons.

When we study organic chemistry, we study how organic compounds react. When a compound reacts, old bonds break and new bonds form. Bonds form when two atoms share electrons, and bonds break when two atoms no longer share their electrons. How readily a bond breaks and how easily it forms depend on the particular electrons that are shared, which, in turn, depends on the atom to which the electrons belong. So if we are going to start our study of organic chemistry at the beginning, we must start with an understanding of the structure of an atom and how the electrons in that atom are distributed.

1.1
THE STRUCTURE OF AN ATOM

An atom consists of a dense nucleus that contains positively charged protons and neutral neutrons. The nucleus, therefore, is positively charged. The space around the nucleus contains negatively charged electrons. The amount of positive charge on a proton is the same as the amount of negative charge on an electron, so a neutral atom has an equal number of protons and electrons. Protons and neutrons have approximately the same mass (1.67×10^{-27} kg) and are 1833 times as massive as electrons (9.11×10^{-31} kg). In other words, the mass of a proton is equal to the combined mass of 1833 electrons. This means that most of the mass of an atom is in the nucleus. Atoms can become charged as a result of gaining or losing electrons. However, the number of protons in an atom does not change.

The **atomic number** of an atom is equal to the number of protons in its nucleus. For a neutral atom, the atomic number is also equal to the number of electrons surrounding the nucleus. For example, the atomic number of carbon is 6, which means that carbon has six protons and six electrons. The atomic number of a particular element is always the same; all carbon atoms have six protons in their nuclei and an atomic number of 6.

The **mass number** of an atom is the sum of the number of protons and the number of neutrons it has. Not all carbon atoms have the same mass number because not all of them have the same number of neutrons. For example, 98.89% of naturally occurring carbon atoms have six neutrons (a mass number of 12), and 1.11% have seven neutrons (a mass number of 13). These two different kinds of carbon atoms (^{12}C and ^{13}C) are called isotopes. **Isotopes** have the same number of protons but different mass numbers because they have different numbers of neutrons. The chemical properties of isotopes of a given element are nearly identical.

The **atomic weight** of an element is the average mass of the atoms in the naturally occurring element. Because an atomic mass unit (amu) is defined as exactly 1/12 of the mass of an atom of ^{12}C, the atomic weight of ^{12}C is 12.0000 amu; the atomic weight of ^{13}C is 13.0034 amu. Therefore, the atomic weight of carbon is 12.011 amu—since $(0.9889 \times 12.000) + (0.0111 \times 13.0034) = 12.011$.   Notice that the atomic weight of an element is not an integer because it includes the contributions of all the isotopes.

**1 amu =
1.66054 × 10^{-27} kg**

PROBLEM 1 ◆

Oxygen is 99.759% ^{16}O (atomic weight = 15.9949 amu), 0.037% ^{17}O (atomic weight = 16.9991 amu), and 0.204% ^{18}O (atomic weight = 17.9992 amu).

a. What are the mass numbers of the three isotopes of oxygen?

b. What is the atomic weight of oxygen?

Electrons are continuously moving. Like anything that is moving, electrons have kinetic energy, and this energy is what defies the attractive force of the positively charged protons that would otherwise pull the negatively charged electrons into the nucleus. For a long time, electrons were thought of as particles—infinitesimal planets swinging around the nucleus. But in 1924, de Broglie, a French physicist, postulated that electrons also have wavelike properties. This startling idea spurred physicists to propose a mathematical concept known as quantum mechanics. **Quantum mechanics** uses the same mathematical equations that describe the wave motion of a guitar string to characterize the motion of an electron around a nucleus. The version of quantum mechanics most useful to chemists was proposed by Erwin Schrödinger. According to Schrödinger, the behavior of each electron in an atom or a molecule can be described by a **wave equation.** A wave equation has a series of solutions that are known as **wave functions.** Solving the wave equation for a given electron tells us the volume of space around the nucleus where the electron is most likely to be found. This volume of space is called an **orbital.**

According to quantum mechanics, the electrons in an atom can be thought of as occupying a set of concentric shells that surround the nucleus. The first shell is the one closest to the nucleus. The second shell lies farther from the nucleus, and even farther out lie the third and higher-numbered shells. Each shell contains subshells known as **atomic orbitals.** Each atomic orbital has a characteristic energy, which is predicted by the Schrödinger equation. The closer the orbital is to the nucleus, the lower its energy.

The first shell contains only an *s* atomic orbital; the second shell contains *s* and *p* atomic orbitals; the third shell contains *s, p,* and *d* atomic orbitals; and the fourth and higher shells contain *s, p, d,* and *f* atomic orbitals (Table 1.1).

1.2 DISTRIBUTION OF ELECTRONS IN AN ATOM

Prince Louis Victor Pierre Raymond de Broglie (1892–1987) *studied history at the Sorbonne. During World War I he was stationed in the Eiffel Tower as a radio engineer. Intrigued by his exposure to radio communications, he returned to school after the war and earned a Ph.D. in physics. He received the Nobel Prize in physics in 1929, five years after obtaining his degree, for his work that showed electrons to have properties of both particles and waves.*

TABLE 1.1	Distribution of Electrons in the First Four Shells That Surround the Nucleus			
	First shell	Second shell	Third shell	Fourth shell
Atomic orbitals	s	s, p	s, p, d	s, p, d, f
Number of atomic orbitals	1	1, 3	1, 3, 5	1, 3, 5, 7
Number of electrons	2	8	18	32

Erwin Schrödinger (1887–1961) *was teaching physics at the University of Berlin when Hitler rose to power. Although not Jewish, Schrödinger left Germany to return to his native Austria, only to see it later taken over by the Nazis. He then moved to the School for Advanced Studies in Dublin and then to Oxford University. He shared the Nobel Prize in physics in 1933 with Paul Dirac, a professor of physics at Cambridge University, for mathematical work on quantum mechanics.*

There is only one *s* orbital in each shell. The second and higher shells each contain three degenerate *p* orbitals. **Degenerate orbitals** are orbitals that have the same energy. The third and higher shells also contain five degenerate *d* orbitals, and the fourth and higher shells contain seven degenerate *f* orbitals. Since a maximum of two electrons can coexist in a single orbital (see the Pauli exclusion principle, below), the first shell, with only one orbital, can contain no more than two electrons. The second shell, with four orbitals—one *s* and three *p*—can have a total of eight electrons. Eighteen electrons can occupy the nine orbitals of the third shell, and 32 electrons can occupy the 16 orbitals of the fourth shell. In studying organic chemistry, we will be concerned primarily with atoms that have electrons only in the first and second shells.

The **ground-state electronic configuration** of an atom is a description of the orbitals occupied by an atom's electrons when they are all in the lowest available energy orbitals. If energy is applied to an atom in the ground state, one or more electrons can be made to jump into a higher energy orbital. The atom then would be in an **excited-state electronic configuration.** The ground-state electronic configurations of the 11 smallest atoms are shown in Table 1.2. The following principles are used to determine which orbitals electrons occupy.

1. The **aufbau principle** (*aufbau* is German for "building up") tells you the first thing you must know to assign electrons to the various atomic orbitals. According to this principle, an electron always goes into the available orbital with the lowest energy. The relative energies of the atomic orbitals are:

$$1s < 2s < 2p < 3s < 3p < 4s < 3d < 4p < 5s < 4d < 5p < 6s < 4f < 5d < 6p < 7s < 5f$$

A 1*s* orbital is of lower energy and thus is closer to the nucleus than a 2*s* orbital, which is of lower energy (and is closer to the nucleus) than a 3*s* orbital. Comparing orbitals in the same shell, an *s* orbital is of lower energy than a *p* orbital, and a *p* orbital is of lower energy than a *d* orbital.

2. The **Pauli exclusion principle** states that (a) no more than two electrons can occupy each orbital and (b) the two electrons must be of opposite spin ($+\frac{1}{2}$ or $-\frac{1}{2}$).[1] This is called an "exclusion" principle because it states that only so many electrons can occupy any particular shell. Notice in Table 1.2 that arrows indicate the opposite directions of electron spin.

As a teenager, Austrian **Wolfgang Pauli (1900–1958)** *wrote articles on relativity that caught the attention of Albert Einstein. Pauli went on to teach physics at the University of Hamburg and at the Zurich Institute of Technology. When World War II broke out, he emigrated to the United States, where he joined the Institute for Advanced Study at Princeton.*

---

[1]A helpful conceptual picture of electron spin (although not strictly correct since an electron has both particle-like and wave-like properties) is to imagine an electron as a spinning, charged sphere for which $+\frac{1}{2}$ and $-\frac{1}{2}$ indicate the two opposite directions in which it can rotate.

TABLE 1.2 The Ground-State Electronic Configurations of the Smallest Atoms

Atom	Name of element	Atomic number	$1s$	$2s$	$2p_x$	$2p_y$	$2p_z$	$3s$
H	Hydrogen	1	↑					
He	Helium	2	↑↓					
Li	Lithium	3	↑↓	↑				
Be	Beryllium	4	↑↓	↑↓				
B	Boron	5	↑↓	↑↓	↑			
C	Carbon	6	↑↓	↑↓	↑	↑		
N	Nitrogen	7	↑↓	↑↓	↑	↑	↑	
O	Oxygen	8	↑↓	↑↓	↑↓	↑	↑	
F	Fluorine	9	↑↓	↑↓	↑↓	↑↓	↑	
Ne	Neon	10	↑↓	↑↓	↑↓	↑↓	↑↓	
Na	Sodium	11	↑↓	↑↓	↑↓	↑↓	↑↓	↑

From these first two rules, we can assign electrons to orbitals for atoms that contain one, two, three, four, or five electrons. The single electron of a hydrogen atom occupies a $1s$ orbital; the second electron of a helium atom fills the $1s$ orbital; the third electron of a lithium atom occupies a $2s$ orbital; the fourth electron of a beryllium atom fills the $2s$ orbital; the fifth electron of a boron atom occupies one of the $2p$ orbitals.

3. **Hund's rule** states that when there are degenerate orbitals—two or more orbitals of equal energy—an electron will occupy an empty orbital before it will pair up with another electron. That is because energy is required to pair up electrons. The sixth electron of a carbon atom, therefore, goes into an empty $2p$ orbital rather than pairing up with the electron already occupying a $2p$ orbital.

Using these three rules, the location of the electrons in the remaining elements can be assigned.

Friedrich Hermann Hund *was born in Germany in 1896. He was a professor of physics at several German universities, the last one being the University of Göttingen. He spent a year as a visiting professor at Harvard University. In February 1996, the University of Göttingen held a symposium to honor Hund on the occasion of his 100th birthday.*

PROBLEM 2 ◆

Potassium has an atomic number of 19 and one unpaired electron. What orbital does the unpaired electron occupy?

PROBLEM 3 ◆

Write electronic configurations for chlorine (atomic number 17), bromine (atomic number 35), and iodine (atomic number 53).

In trying to explain why atoms form bonds, G. N. Lewis proposed the theory that *an atom is most stable if it has a filled shell or an outer shell of eight electrons and no electrons of higher energy.* According to the theory, an atom will give up, accept, or share electrons in order to achieve a filled shell or an outer shell that contains eight electrons. This theory has come to be called the **octet rule.**

**1.3
IONIC, COVALENT,
AND POLAR BONDS**

Lithium (Li) has a single electron in its $2s$ orbital. If it loses this electron, a lithium atom ends up with a filled first shell and no other electrons—a stable configuration. Removing an electron from an atom takes energy. The energy required to remove an electron from an atom is called the **ionization energy** of that atom. Lithium has a relatively low ionization energy because it loses an electron relatively easily. Sodium (Na) has a single electron in its $3s$ orbital. Consequently, sodium easily loses one electron, which leaves the atom with an outer shell of eight electrons. Elements (such as lithium and sodium) that have low ionization energies are called **electropositive** elements (they readily lose an electron and thereby become positively charged). The elements in the first column of the periodic table are electropositive; each readily loses an electron because each has a single electron in an s atomic orbital. We can see, then, that the chemical behavior of an element depends on its electronic configuration.

When we draw the electrons around an atom, as in the following equations, the electrons in any completely filled shells are not shown; only the electrons in shells that are not filled are shown. These outer electrons are known as **valence electrons;** each valence electron is shown as a dot. Lithium and sodium each bear a single valence electron. When the electron is removed, the resulting atom—now called an ion—carries a positive charge.

$$Li\cdot \longrightarrow Li^+ + e^-$$

$$Na\cdot \longrightarrow Na^+ + e^-$$

Fluorine has seven valence electrons (Table 1.2). Consequently, fluorine readily acquires an electron in order to have an outer shell of eight electrons. When an atom such as fluorine or chlorine acquires an electron, energy is released. The energy released is called the **electron affinity.** The other elements in the same column as fluorine and chlorine (bromine, iodine) also need only one electron to have an outer shell of eight electrons, so they also readily acquire an electron. Elements that readily acquire an electron are called **electronegative** elements (they easily acquire an electron and thereby become negatively charged).

$$:\!\ddot{F}\!\cdot + e^- \longrightarrow :\!\ddot{F}\!:^-$$

$$:\!\ddot{C}\!l\!\cdot + e^- \longrightarrow :\!\ddot{C}\!l\!:^-$$

Ionic Bonds

Because sodium easily gives up an electron and chlorine readily acquires an electron, will sodium donate an electron to chlorine to form sodium chloride (Na^+Cl^-) when sodium metal and chlorine gas are combined? Because the energy given off when chlorine accepts an electron (83.3 kcal/mol; 349 kJ/mol)[2] is not as great as the energy required to remove an electron from sodium (118 kcal/mol; 494 kJ/mol), one might be tempted to predict that the reaction will not happen. However, when sodium metal and chlorine gas are mixed, we observe the formation of crystalline sodium chloride (table salt), so we know the reaction occurs. Where does the extra energy come from? The sodium ions and chloride ions in sodium chloride are held together because the opposite charges attract each other (Figure 1.1). These forces of attraction are what provide the energy required to transfer the electrons. Attractive forces between opposite charges are called **electrostatic attractions.** A bond is an attractive force between two atoms. A bond that

---

[2]1 kcal = 4.184 kJ. Joules are the Système International units for energy, although many chemists use calories. We will use both in this book.

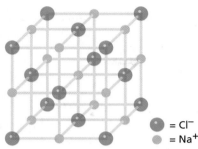

= Cl⁻
= Na⁺

◀ **Figure 1.1**
(a) Crystalline sodium chloride. (b) You can see that, in the crystal lattice, each chloride ion is surrounded by six sodium ions and each sodium ion is surrounded by six chloride ions.

is the result of electrostatic attractions is called an ionic bond. An **ionic bond** is formed as the result of a *transfer of electrons.*

$$Na\cdot \; + \; \cdot\ddot{\underset{..}{C}}l\colon \; \longrightarrow \; Na^+ \; :\ddot{\underset{..}{C}}l\colon^-$$

an ionic bond

sodium chloride

Sodium chloride is an example of an ionic compound. **Ionic compounds** are formed when an element on the left side of the periodic table (an electropositive element) reacts with an element on the right side of the periodic table (an electronegative element).

Covalent Bonds

Instead of giving up or acquiring electrons, an atom can also achieve an outer shell of eight electrons by sharing electrons. Two fluorine atoms can each obtain a filled shell by sharing their unpaired valence electrons. A bond formed as a result of *sharing electrons* is called a **covalent bond.**

$$:\ddot{F}\cdot \; + \; \cdot\ddot{F}\colon \; \longrightarrow \; :\ddot{F}\colon\ddot{F}\colon$$

a covalent bond

Two hydrogen atoms can react to form a covalent bond. By sharing electrons, each hydrogen acquires a completely filled first shell (two electrons).

$$H\cdot \; + \; \cdot H \; \longrightarrow \; H\colon H$$

Hydrogen and chlorine can form a covalent bond by sharing electrons. In doing so, hydrogen fills its first shell and chlorine achieves an outer shell of eight electrons.

$$H\cdot \; + \; \cdot\ddot{\underset{..}{C}}l\colon \; \longrightarrow \; H\colon\ddot{\underset{..}{C}}l\colon$$

A hydrogen atom can also achieve a completely empty shell by losing an electron, forming a positively charged hydrogen ion, which is called a **hydrogen ion** (or a **proton**). A hydrogen atom can achieve a completely filled first shell by gaining an electron, thereby forming a negatively charged hydrogen ion. A negatively charged hydrogen ion is called a **hydride ion.**

$$H\cdot \; \xrightarrow{\;-e^-\;} \; H^+$$

a hydrogen ion
a proton

$$H\cdot \; \xrightarrow{\;+e^-\;} \; H\colon^-$$

a hydride ion

Oxygen needs to form two covalent bonds in order to fill its second shell. Nitrogen needs to form three covalent bonds, and carbon must form four covalent bonds. Notice that all of the atoms in water, ammonia, and methane have complete shells of electrons.

$$2 \text{ H}\cdot \ + \ \cdot\ddot{\text{O}}\text{:} \ \longrightarrow \ \text{H:}\ddot{\text{O}}\text{:}$$
$$\overset{\textstyle}{\underset{\textstyle\text{H}}{}}$$
water

$$3 \text{ H}\cdot \ + \ \cdot\ddot{\text{N}}\cdot \ \longrightarrow \ \text{H:}\ddot{\text{N}}\text{:H}$$
$$\underset{\textstyle\text{H}}{}$$
ammonia

$$\overset{\textstyle\text{H}}{}$$
$$4 \text{ H}\cdot \ + \ \cdot\dot{\text{C}}\cdot \ \longrightarrow \ \text{H:}\dot{\text{C}}\text{:H}$$
$$\underset{\textstyle\text{H}}{}$$
methane

Polar Covalent Bonds

In the fluorine–fluorine and hydrogen–hydrogen covalent bonds shown previously, the atoms that share the bonding electrons are identical. Therefore, they share the electrons equally; each electron spends as much time in the vicinity of one atom as in the other. An even, or nonpolar, distribution of charge results. Such a bond is a **nonpolar covalent bond.**

In hydrogen chloride, water, and ammonia, in contrast, the bonding electrons are more attracted to one atom than another because the atoms that share the electrons in these compounds have different electronegativities. **Electronegativity** is the tendency of an atom to pull bonding electrons toward itself. The bonding electrons in these molecules are more attracted to the atom with the greater electronegativity. This results in a polar distribution of charge. A **polar covalent bond** is a covalent bond between atoms of different electronegativities. The electronegativities of some of the elements are shown in Table 1.3. Notice that the electronegativity

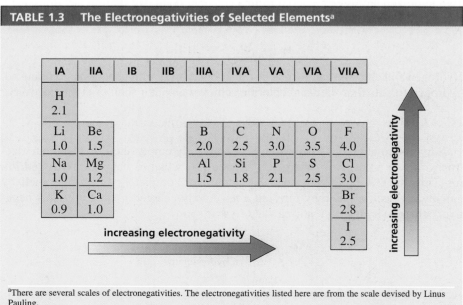

TABLE 1.3   The Electronegativities of Selected Elements[a]

IA	IIA	IB	IIB	IIIA	IVA	VA	VIA	VIIA
H 2.1								
Li 1.0	Be 1.5			B 2.0	C 2.5	N 3.0	O 3.5	F 4.0
Na 1.0	Mg 1.2			Al 1.5	.Si 1.8	P 2.1	S 2.5	Cl 3.0
K 0.9	Ca 1.0							Br 2.8
								I 2.5

increasing electronegativity

increasing electronegativity

[a]There are several scales of electronegativities. The electronegativities listed here are from the scale devised by Linus Pauling.

increases as you go across a row of the periodic table to the right and increases as you go up any of the columns.

A polar covalent bond has a slight positive charge on one end and a slight negative charge on the other. Polarity in a covalent bond is indicated by the symbols $\delta+$ and $\delta-$, which indicate partial positive and partial negative charges. The negative end of the bond is the end that has the more electronegative atom. The greater the electronegativity difference between the bonded atoms, the more polar the bond will be. The carbon–hydrogen bonds in methane are relatively nonpolar because carbon and hydrogen have similar electronegativities.

$$\overset{\delta+ \quad \delta-}{H-Cl} \qquad \overset{\delta+ \quad \delta-}{H-\underset{\underset{\delta+}{H}}{O}} \qquad \overset{\delta+ \quad \delta- \quad \delta+}{H-\underset{\underset{\delta+}{H}}{N}-H}$$

The direction of bond polarity can be indicated using an arrow. By convention, the electrons are pulled in the direction of the arrow, so the head of the arrow is at the negative end of the bond; a short, perpendicular line is drawn near the tail of the arrow to show partial positive character.

$$\overset{\longrightarrow}{H-Cl}$$

You can think of ionic bonds and nonpolar covalent bonds as being at the opposite ends of a continuum of bond types. An ionic bond involves no sharing of electrons. A nonpolar covalent bond involves exactly equal sharing. Polar covalent bonds are in the middle, and the greater the electronegativity difference between the atoms forming the bond, the closer the bond is to the ionic end of the continuum.

continuum of bond types

ionic bond	polar covalent bond	nonpolar covalent bond

Because a polar bond has a negative end and a positive end, it has a **dipole.** The size of the dipole is indicated by the dipole moment (μ). The dipole moment of a bond is defined as the size of the charge (e) on the atom (either the positive charge or the negative charge because they are both the same size) times the distance between the two charges (d). A dipole moment is measured in a unit called the debye (D) (pronounced de-bye).

$$\text{dipole moment} = \mu = e \times d$$

Because the charge on an electron is 4.80×10^{-10} electrostatic units (esu) and the distance between charges in a polar bond is on the order of 10^{-8} cm, the product of charge and distance is on the order of 10^{-18} esu·cm. A dipole moment of 1.5×10^{-18} esu·cm is stated as 1.5 D. The dipole moments of some bonds commonly found in organic compounds are listed in Table 1.4.

In a molecule with only one covalent bond, the dipole moment of the whole molecule is identical to the dipole moment of the bond. For example, the dipole moment of hydrogen chloride (HCl) is 1.1 D because the dipole moment of the single hydrogen–chlorine bond is 1.1 D. The dipole moment of a molecule with more than one covalent bond depends on the geometry of the molecule as well as on the dipole moments of all the bonds in the molecule. We will examine the dipole moments of such molecules in Section 1.14 after we learn about the geometry of molecules.

Peter Debye (1884–1966) *was born in the Netherlands. He taught at the universities of Zürich (succeeding Einstein), Leipzig, and Berlin, but returned to his homeland in 1939 when the Nazis ordered him to become a German citizen. Arriving at Cornell to give a lecture, he decided to stay, becoming a U.S. citizen in 1946. He received the Nobel Prize in chemistry in 1936 for his work on dipole moments and the properties of solutions.*

TABLE 1.4   The Dipole Moments of Some Commonly Encountered Bonds

Bond	Dipole moment (D)	Bond	Dipole moment (D)
H—C	0.4	C—C	0
H—N	1.3	C—N	0.2
H—O	1.5	C—O	0.7
H—F	1.7	C—F	1.4
H—Cl	1.1	C—Cl	1.5
H—Br	0.8	C—Br	1.4
H—I	0.4	C—I	1.2

PROBLEM 4◆

Which of the following compounds would you expect to be ionic?

a. CCl_4 c. NaOH   e. $NaNH_2$ g. $MgCl_2$

b. KBr   d. BF_3 f. LiCl   h. NaI

PROBLEM 5◆

Use the symbols $\delta+$ and $\delta-$ to show the direction of polarity of the indicated bond in each of the following compounds (for example, $\overset{\delta-}{HO}-\overset{\delta+}{H}$).

a. H_3C-Cl c. H_3C-NH_2 e. $HO-Br$

b. $F-Br$ d. H_3C-OH f. $H_3C-MgBr$

1.4
LEWIS STRUCTURES

*American chemist **Gilbert Newton Lewis (1875–1946)** was the first to prepare "heavy water," which has deuterium atoms in place of the usual hydrogen atoms (D_2O versus H_2O). Because heavy water can be used as a moderator of neutrons, it became important for the development of the atomic bomb. Lewis started his career as a professor at the Massachusetts Institute of Technology and joined the faculty at the University of California, Berkeley, in 1912.*

The chemical symbols we have been using, in which the valence electrons in molecules are represented as dots, are called **Lewis structures.** The Lewis structures for H_2O, H_3O^+, HO^-, and H_2O_2 are shown below. Lewis structures are useful because they show us what atoms are bonded together and tell us whether any atoms possess a charge. When you draw a Lewis structure you must make sure that hydrogen atoms are surrounded by two electrons, and C, O, N, and halogen (F, Cl, Br, I) atoms are surrounded by eight electrons. In other words, they must obey the octet rule. Valence electrons not used in bonding are called **nonbonding electrons** or **lone pair electrons.**

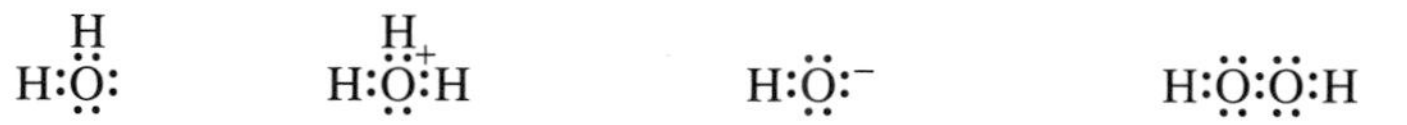

$$\begin{array}{cccc} \text{H} & \text{H} & & \\ \text{H:}\ddot{\text{O}}\text{:} & \text{H:}\overset{+}{\ddot{\text{O}}}\text{:H} & \text{H:}\ddot{\text{O}}\text{:}^- & \text{H:}\ddot{\text{O}}\text{:}\ddot{\text{O}}\text{:H} \\ \text{water} & \text{hydronium ion} & \text{hydroxide ion} & \text{hydrogen peroxide} \end{array}$$

Once the atoms and the electrons are in place, you must determine whether a charge should be assigned to any of the atoms. A positive or a negative charge assigned to an atom is called a **formal charge;** the oxygen atom in the hydronium ion has a formal charge of $+1$; the oxygen atom in the hydroxide ion has a formal charge of -1. It is called a formal charge because the entire charge may not necessarily belong to the atom that bears the formal charge. But it does indicate which atom in the molecule bears *most* of the charge. A formal charge is the difference between the number of valence electrons an atom has when it is not bonded to any other atoms and the number of electrons it actually "owns" when it is bonded. An atom "owns" all of its nonbonding electrons and half of its bonding electrons.

formal charge =
   number of valence electrons − (number of nonbonding electrons +
                          $\frac{1}{2}$ number of bonding electrons)

For example, oxygen has six valence electrons when not bonded (Table 1.2). In water (H_2O), oxygen "owns" six electrons (four nonbonding electrons and two (half of four) bonding electrons). Because the number of electrons it "owns" is equal to the number of its valence electrons, the oxygen atom in water has no formal charge. The oxygen atom in the hydronium ion (H_3O^+) "owns" five electrons: two nonbonding plus three (half of six) bonding. Because the number of electrons it "owns" is one fewer than the number of its valence electrons, it is positively charged; its formal charge is +1. The oxygen atom in hydroxide ion (HO^-) "owns" seven electrons: six nonbonding plus one (half of two) bonding. Because it "owns" one more electron than the number of its valence electrons, it is negatively charged; its formal charge is −1.

Knowing that nitrogen has five valence electrons (Table 1.2), convince yourself that the appropriate formal charges have been assigned to the nitrogen atoms in the following molecules.

ammonia	ammonium ion	amide anion	hydrazine

Since carbon has four valence electrons, you should be able to understand why the carbon atoms in the following compounds have the indicated formal charges. A compound containing a positively charged carbon atom is called a **carbocation,** and a compound containing a negatively charged carbon is called a **carbanion.** (A cation is a positively charged ion, and an anion is a negatively charged ion.) Carbocations were formerly called carbonium ions, so you will see this term in older literature. A compound containing an atom with a single unpaired electron is called a **radical.**

methane	methyl cation	methyl anion	methyl radical	ethane
	a carbocation	a carbanion		

Hydrogen has one valence electron, and the halogens have seven valence electrons, so the following species have the indicated formal charges.

hydrogen ion	hydride ion	hydrogen radical	bromide ion	bromine radical	bromine	chlorine

In studying the molecules in this section, notice that oxygen has two covalent bonds when it is neutral; when nitrogen is neutral it has three covalent bonds; when carbon is neutral it has four covalent bonds. Neutral hydrogen and neutral halogens each have one covalent bond. If the atoms have more bonds or fewer bonds than the number required for a neutral molecule, they will have a formal charge or an unpaired electron. These numbers are very important to remember when you are first drawing structures of organic compounds because they provide a quick way to recognize when you have made a mistake.

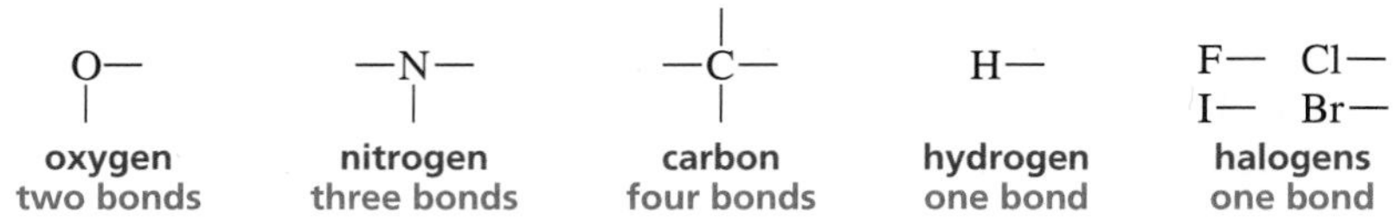

O—	—N—	—C—	H—	F— Cl—
				I— Br—
oxygen	**nitrogen**	**carbon**	**hydrogen**	**halogens**
two bonds	**three bonds**	**four bonds**	**one bond**	**one bond**

In the Lewis structures for CH_2O_2, HNO_3, CH_2O, and CO_3^{2-}, shown next, notice that each atom has a complete octet (except hydrogen, which has a filled first shell) and that each atom has the appropriate formal charge. (In drawing a compound that has two or more oxygen atoms, avoid oxygen–oxygen single bonds. These are weak bonds, and few compounds have such bonds.)

Lewis structures

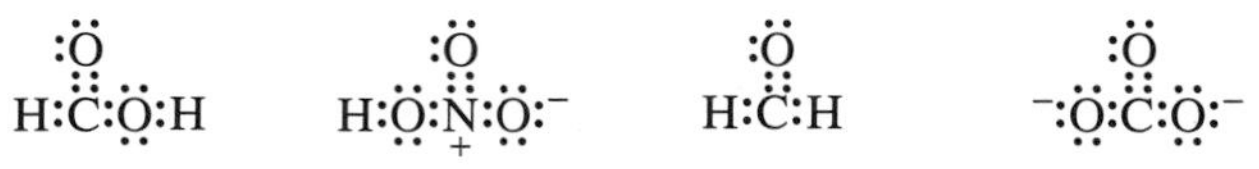

$$H\!:\!\overset{\ddot{\text{O}}}{\underset{\ddots}{C}}\!:\!\ddot{\text{O}}\!:\!H \qquad H\!:\!\ddot{\text{O}}\!:\!\overset{\ddot{\text{O}}}{\underset{+}{N}}\!:\!\ddot{\text{O}}\!:^{-} \qquad H\!:\!\overset{\ddot{\text{O}}}{C}\!:\!H \qquad {}^{-}\!:\!\ddot{\text{O}}\!:\!\overset{\ddot{\text{O}}}{C}\!:\!\ddot{\text{O}}\!:^{-}$$

If you are asked to draw the Lewis structure for HNO_2, once you give all the atoms complete octets of electrons you are left with a molecule that has two negative charges. Because HNO_2 is a neutral molecule, it is clear that you have given the molecule two too many electrons. You can erase two electrons and still maintain complete octets if you connect two of the atoms with a double bond.

$$H\!:\!\ddot{\text{O}}\!:\!\ddot{\text{N}}\!:\!\ddot{\text{O}}\!:^{-} \qquad\qquad H\!:\!\ddot{\text{O}}\!:\!\ddot{\text{N}}\!:\!:\!\ddot{\text{O}}\!:$$

$\qquad\qquad$ **incorrect Lewis structure** $\qquad$ **correct Lewis structure**
$\qquad\qquad$ **for HNO₂** $\qquad\qquad\qquad$ **for HNO₂**
$\qquad\qquad\qquad\qquad\qquad\qquad\qquad$ double bond

A simpler way of drawing compounds—already used in this chapter—is the Kekulé structure. In **Kekulé structures,** the bonding electrons are drawn as lines and the nonbonding electrons are left out entirely unless they are needed to draw attention to some chemical property of the molecule. (However, even though they may not be shown, you should remember that neutral oxygen and nitrogen atoms always have nonbonding electrons: two pairs in the case of oxygen; one pair in the case of nitrogen.)

Kekulé structures

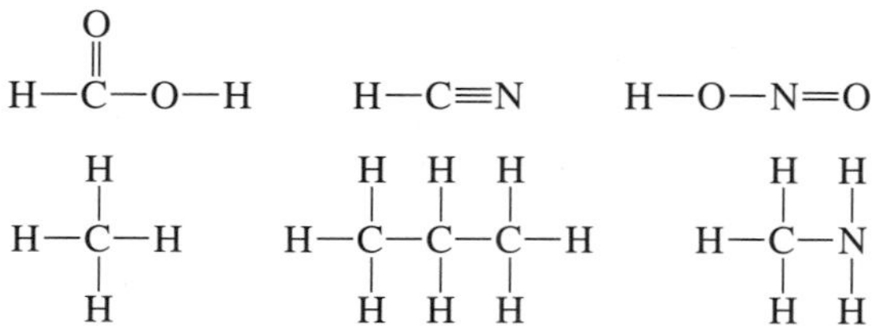

$$\begin{array}{c} \text{O} \\ \| \\ \text{H}-\text{C}-\text{O}-\text{H} \end{array} \qquad \text{H}-\text{C}\!\equiv\!\text{N} \qquad \text{H}-\text{O}-\text{N}\!=\!\text{O}$$

$$\begin{array}{c} \text{H} \\ | \\ \text{H}-\text{C}-\text{H} \\ | \\ \text{H} \end{array} \qquad \begin{array}{c} \text{H} \;\; \text{H} \;\; \text{H} \\ | \;\;\; | \;\;\; | \\ \text{H}-\text{C}-\text{C}-\text{C}-\text{H} \\ | \;\;\; | \;\;\; | \\ \text{H} \;\; \text{H} \;\; \text{H} \end{array} \qquad \begin{array}{c} \text{H} \;\; \text{H} \\ | \;\;\; | \\ \text{H}-\text{C}-\text{N} \\ | \;\;\; | \\ \text{H} \;\; \text{H} \end{array}$$

Structures are often further simplified by omitting some (or all) of the covalent bonds and listing atoms bonded to a particular carbon (or nitrogen, or oxygen) next to it with a subscript to indicate the number of such atoms. Such structures are called **condensed structures.**

condensed structures

$$HCO_2H \qquad HCN \qquad HNO_2 \qquad CH_4 \qquad CH_3CH_2CH_3 \qquad CH_3NH_2$$

PROBLEM 6 / SOLVED

Draw the Lewis structure for each of the following:

a. NO_3^- d. CO_2 g. $CH_3NH_3^+$

b. NO_2^- e. HCO_3^- h. $^+C_2H_5$

c. NO_2^+ f. N_2 i. $^-CH_3$

SOLUTION TO 6a The only way you can arrange one N and three O's and avoid O—O bonds is to arrange the three O's around the N. Doing this and giving each of the atoms a complete octet results in an ion with three negative charges, two more than it should have.

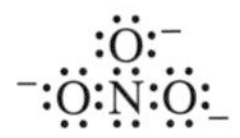

You can erase two electrons and still maintain all the atoms' octets if a double bond is formed between two atoms. Doing this results in an ion with two negative charges and one positive charge, so overall it has one negative charge. It is the desired ion, NO_3^-.

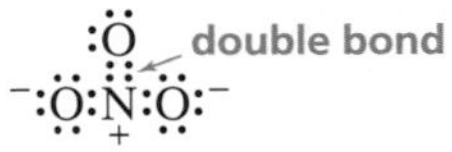

PROBLEM 7 ◆

Draw two Lewis structures for C_2H_6O.

PROBLEM 8

Expand the condensed structures to show the covalent bonds and nonbonding pairs of electrons.

a. $CH_3NHCH_2CH_3$ c. $(CH_3)_2CHCHO$

b. $(CH_3)_2CHCl$ d. $(CH_3)_3C(CH_2)_3CH(CH_3)_2$

We have seen that electrons are distributed into different atomic orbitals (Table 1.2). An orbital is a three-dimensional region around the nucleus where there is a high probability of finding an electron. But what does an orbital look like? Mathematical calculations indicate that the s atomic orbital is a sphere with the nucleus at its center, and experimental evidence supports this theory. When we say that an electron occupies a $1s$ orbital, we mean that there is a greater than 90% probability that the electron is in the space defined by the sphere. The **Heisenberg uncertainty principle** states that both the precise location and the momentum of an atomic particle cannot be simultaneously determined. This means we can never say precisely where an electron is; we can only describe its probable location. A $2s$ atomic orbital is represented by a larger sphere, since the average distance from the nucleus is greater for an electron in a $2s$ orbital than for an electron in a $1s$ orbital. Consequently, the average electron density in a $2s$ orbital is less than the average electron density in a $1s$ orbital.

1.5
ATOMIC ORBITALS

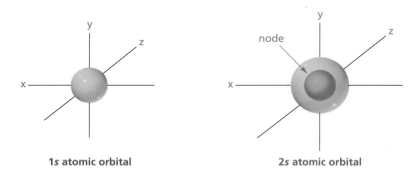

1s atomic orbital   **2s atomic orbital**

The wave-like properties of electrons help to elaborate our picture of orbitals. An electron in a 1s orbital can be anywhere within the 1s sphere. But a 2s atomic orbital is more complex. There are parts of the orbital where the probability of finding an electron falls to zero. This area is called a node.

To understand why nodes occur, remember that electrons have both particle-like and wave-like properties. A node results because of the wave-like properties of an electron. To further understand, consider the two types of waves: traveling waves and standing waves. Traveling waves move through space; light is an example of a traveling wave (it travels through space at 3×10^8 m/s). A standing wave is confined to a limited space. A vibrating string of a guitar is an example of a standing wave; the string moves up and down, but the wave itself does not travel through space. A **node** is simply a region where a standing wave has an amplitude of zero; the string rests motionless with no transverse displacement. An electron is also a standing wave, but—unlike the wave created by a vibrating string—it is three-dimensional. This means that the node of a 2s orbital is actually a surface; it is a spherical surface within the 2s orbital. Because the electron wave has zero amplitude at the node, there is zero probability of finding an electron at the node.

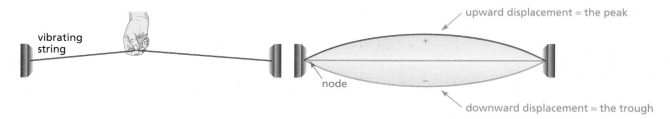

Although 1s and 2s orbitals resemble spheres, p atomic orbitals have two lobes and are dumbbell-shaped. The two lobes are of opposite phase, which can be designated by a plus (+) and a minus (−) or by two different colors. (Notice that in this context + and − do not indicate charge; they indicate the phase of the orbital.) A nodal plane passes through the center of the nucleus, bisecting the two lobes of the p orbital. Because the standing wave has zero amplitude at the node, there is zero probability of finding an electron in the nodal plane of the p orbital.

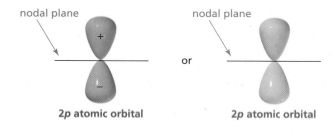

2p atomic orbital   or   **2p atomic orbital**

There are three degenerate $2p$ atomic orbitals: the p_x orbital is symmetrical about the x-axis, the p_y orbital is symmetrical about the y-axis, and the p_z orbital is symmetrical about the z-axis.

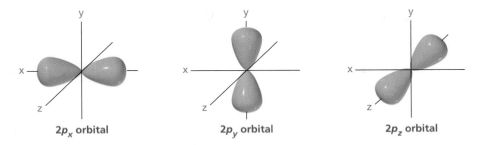

How do atoms form covalent bonds in order to form molecules? The Lewis model, in which atoms attain a complete octet simply by sharing electrons, represents only an approximation of what really happens in covalent bond formation. It is only an approximate representation of chemical reality. As chemists learn more, theories become more sophisticated and more closely fit all the observations of real atoms and molecules. One drawback of the Lewis model is that it treats electrons like particles and does not take into account their wave-like properties. De Broglie was the first to show that an electron has both particle-like and wave-like properties. He did this by combining a formula developed by Einstein that relates mass and energy with a formula developed by Planck relating frequency and energy.

Molecular orbital (MO) theory combines the tendency of electrons to fill their octets (the Lewis model) with their wave-like properties. According to MO theory, covalent bonds form when atomic orbitals overlap, forming molecular orbitals— orbitals that belong to a whole molecule instead of to a single atom. We will see later that the geometry of a molecule depends on the particular orbitals that participate in the overlap.

Let's look first at the bonding in a hydrogen molecule (H_2). The $1s$ atomic orbital of one hydrogen atom overlaps with the $1s$ atomic orbital of a second hydrogen atom. As a result, a molecular orbital forms. It is called a **molecular orbital** because it is an orbital of an entire molecule rather than an orbital of a single atom. The electron density of a molecular orbital is greatest in the region of overlap. The hydrogen–hydrogen bond that forms when the two s orbitals overlap is called a **sigma (σ) bond.** The electrons in a σ bond are symmetrically distributed about the internuclear axis (an imaginary line between the two nuclei).

**1.6
MOLECULAR
ORBITALS AND
BONDING**

During bond formation, the two orbitals approach each other and start to overlap. Energy is released as the electron in each atom becomes attracted to the positively charged nucleus of the other atom as well as to its own nucleus (Figure 1.2). The more the orbitals overlap, the more the energy decreases until the atoms approach each other so closely that their nuclei start to repel each other electrostatically. This causes a large increase in energy. This means that maximum stability (minimum

Figure 1.2 ▶
The change in energy that occurs as two 1s atomic orbitals approach each other. The internuclear distance at minimum energy is the length of the hydrogen–hydrogen covalent bond.

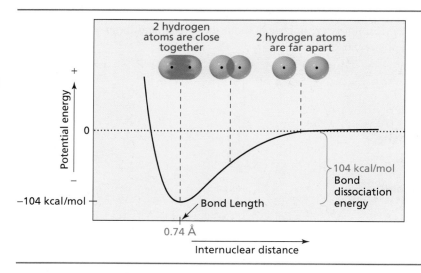

Albert Einstein (1879–1955)
was born in Germany. His only interest in high school was mathematics, and he was expelled for poor performance in Latin and Greek. Einstein was visiting the United States when Hitler came to power, so he accepted a position at the Institute for Advanced Study in Princeton, becoming a U.S. citizen in 1940. Although a lifelong pacifist, he wrote a letter to President Roosevelt warning of ominous advances in German nuclear research. This led to creation of the Manhattan Project, which developed the atomic bomb and tested it in New Mexico in 1945.

energy) is achieved when the nuclei are a certain, finite distance apart. This distance corresponds to the **bond length** of the new covalent bond. The bond length of the hydrogen–hydrogen bond is 0.74 Å.[3] As Figure 1.2 shows, there is a net release of energy when a covalent bond forms. When the hydrogen–hydrogen bond forms, 104 kcal/mol of energy is released. Breaking the bond requires precisely the same amount of energy. This is the dissociation energy of the hydrogen–hydrogen bond. Every covalent bond has a characteristic bond length and **bond dissociation energy.**

Orbitals are conserved—when two atomic orbitals combine, two molecular orbitals result. But in discussing the formation of a hydrogen–hydrogen bond, we combined two atomic orbitals and discussed only one molecular orbital. Where is the other molecular orbital? It is there, but it contains no electrons.

A molecular orbital that holds atoms together results when two orbitals with the same phase interact. This is called a **bonding molecular orbital.** When two in-phase atomic orbitals overlap, there is increased electron density between the nuclei because each of the electrons in this region is electrostatically attracted to both nuclei, making each electron more stable than it was in the 1s orbital of the individual hydrogen atom. This is similar to two light waves or two sound waves that reinforce, or enhance, each other (Figures 1.3, 1.4). When two out-of-phase atomic orbitals overlap, they cancel each other and produce a node between the nuclei. This is called an **antibonding molecular orbital** because any electrons that occupy this orbital detract from, rather than aid, formation of a bond between the atoms. The cancellation is similar to the darkness that occurs when two light waves cancel each other or to the silence that occurs when two sound waves cancel each other.

The aufbau principle and the Pauli exclusion principle, which are followed when electrons occupy atomic orbitals, are also followed when electrons occupy molecular orbitals. So the two electrons of the covalent bond occupy the lower-energy bonding molecular orbital (Figure 1.4). Notice that, in the bonding orbital, electron density is greatest between the nuclei, while in the antibonding orbital, electron density is greatest away from the area between the nuclei.

The electrons in the bonding molecular orbital are attracted to both positively charged nuclei. It is this increase in electrostatic attraction that gives a covalent bond its strength. The greater the overlap of the atomic orbitals, the stronger the

[3]The angstrom (Å) is not a Système International unit. Those who opt to adhere strictly to SI units can convert it into picometers: 1 picometer (pm) = 10^{-12} m; 1 Å = 100 pm. Because the angstrom continues to be used by many organic chemists, we will use angstroms in this book.

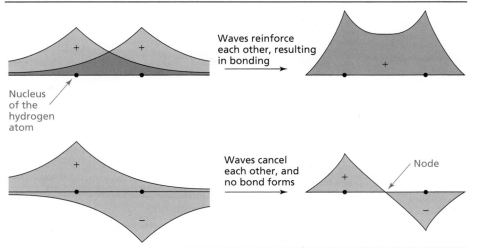

The wave functions of two hydrogen atoms can interact to reinforce, or enhance, each other (top) or can interact to cancel each other (bottom).

covalent bond. The more nuclei an electron "sees," the more stable it is. Consequently, the bonding molecular orbital is more stable (of lower energy) than the individual atomic orbitals (Figure 1.4). In an antibonding orbital, there are no electrons between the nuclei to weaken the internuclear repulsion. The antibonding molecular orbital, therefore, is less stable (of higher energy) than the atomic orbitals.

So a hydrogen molecule *does* have two molecular orbitals: the bonding molecular orbital with two electrons, and the antibonding molecular orbital with no electrons. Electrons in a bonding orbital favor bonding, whereas electrons in an antibonding orbital favor the separated atoms (disfavor bonding).

Using molecular orbital theory (and Figure 1.4), we can readily understand why H_2^+ is not as stable as H_2 since the former has only one electron in the bonding orbital. We can also predict that He_2 does not exist. Because He_2 has four electrons, two electrons will fill the lower-energy bonding molecular orbital and the remaining two will fill the higher-energy antibonding molecular orbital. The two electrons in the antibonding molecular orbital will cancel the advantage to bonding gained by the two electrons in the bonding molecular orbital.

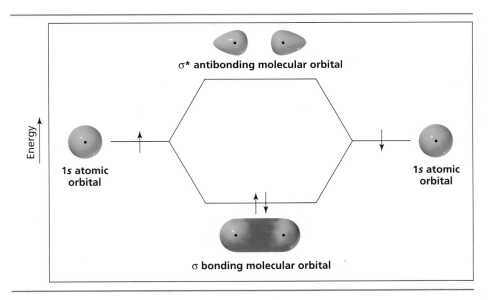

◀ **Figure 1.4**
Atomic and molecular orbitals of H_2. Before covalent bond formation, each electron is in an atomic orbital. After covalent bond formation, both electrons are in the bonding molecular orbital. The antibonding molecular orbital is empty.

The covalent bond in F_2 results from end-on overlap of a $2p$ atomic orbital of one fluorine atom with a $2p$ atomic orbital of another fluorine atom.

2p atomic
orbital   2p atomic
orbital   σ bonding molecular orbital

The greater the electron density in the region of orbital overlap, the stronger the bond.

Because a $2p$ atomic orbital encompasses a greater volume than a $1s$ atomic orbital, the average electron density in a $2p$ atomic orbital is less than the average electron density in a $1s$ atomic orbital. So when the two $2p$ orbitals of F_2 overlap, the average electron density of the resulting bonding molecular orbital is less than the average electron density of the bonding molecular orbital of H_2. Therefore, the electrons cannot pull the electrostatically repulsed fluorine nuclei as close together. Consequently, the F_2 bond is a longer and weaker bond than the H_2 bond: F—F bond length = 1.42 Å, H—H bond length = 0.74 Å: F—F bond strength = 38 kcal/mol (159 kJ/mol), H—H bond strength = 104 kcal/mol (435 kJ/mol).

There are two ways in which p orbitals can overlap. When a p orbital of one atom overlaps end-on with a p orbital of another atom, as it does in the fluorine–fluorine bond, a sigma (σ) bond forms (Figure 1.5). If the overlapping lobes of the p orbitals are in phase (are the same color), a constructive bonding molecular orbital forms. If the lobes are out of phase (are different colors), a destructive antibonding molecular orbital forms. The bonding molecular orbital is a σ bonding molecular orbital because it results from end-on overlap of p atomic orbitals. The antibonding molecular orbital is a σ^* antibonding molecular orbital (an antibonding orbital is indicated by an asterisk). The electron density of the bonding molecular orbital is concentrated between the nuclei, which causes the back lobes of the molecular orbital to be quite small.

Instead of overlapping end-on, two p atomic orbitals could also overlap side to side (Figure 1.6). A bond formed as a result of side-to-side overlap of p atomic orbitals is called a **pi (π) bond.** Side-to-side overlap of two in-phase p atomic orbitals forms a π bonding molecular orbital, while side-to-side overlap of out-of-phase p orbitals forms a π^* antibonding molecular orbital. The maximum electron density in the bonding orbital is on either side of the internuclear axis.

Figure 1.5 ▶
End-on overlap of two p orbitals to form a σ bonding molecular orbital and a σ^* antibonding molecular orbital.

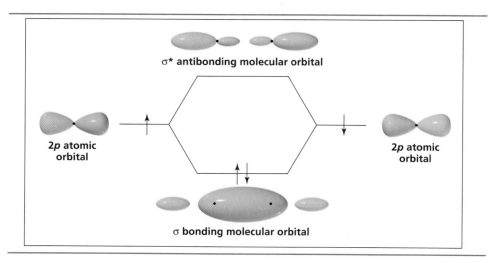

σ* antibonding molecular orbital

2p atomic
orbital   2p atomic
orbital

σ bonding molecular orbital

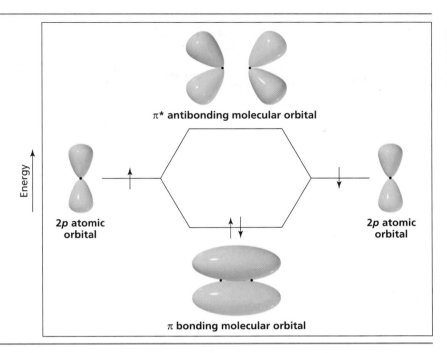

◀ **Figure 1.6**
Side-to-side overlap of two
parallel *p* orbitals to form a π
bonding molecular orbital
and a π* antibonding
molecular orbital.

Organic chemists find that the information obtained from MO theory, where valence electrons occupy bonding and antibonding molecular orbitals, does not always yield the most useful information about a covalent bond. The valence bond model combines the concept of atomic orbitals with the Lewis concept of shared electron pairs and nonbonding electron pairs, so it really is an electron pair covalent bond model. Adding a third principle—minimization of electron repulsions—results in the **valence shell electron pair repulsion (VSEPR) model.** In this model, because electron pairs repel each other, the pairs of bonding electrons and pairs of nonbonding electrons around an atom are positioned as far apart as possible.

Because organic chemists generally think of chemical reactions in terms of the change that occurs in the bonds in a molecule, the VSEPR model often provides the easiest way of visualizing chemical change. The model is inadequate for molecules in excited states because it does not allow for antibonding orbitals. (A molecule is in an excited state when one or more of its electrons are not in the lowest available energy level.) We will use both the MO model and the VSEPR model in this book. Our choice will be determined by which model provides the best description of the molecule under discussion.

PROBLEM 9 ◆

a. Predict the relative lengths and strengths of the bonds in Cl_2 and Br_2.

b. Predict the relative lengths and strengths of the bonds in HF, HCl, and HBr.

PROBLEM 10 ◆

a. How many nodes are there in the π bonding molecular orbital in Figure 1.6?

b. How many nodes are there in the $π^*$ antibonding molecular orbital in Figure 1.6?

1.7
BONDING IN METHANE AND ETHANE. SINGLE BONDS

We will start the discussion of bonding in organic compounds by looking at the bonding in methane, a compound with only one carbon atom. We will then examine the bonding in ethane (a compound with a carbon–carbon single bond), ethylene (a compound with a carbon–carbon double bond), and acetylene (a compound with a carbon–carbon triple bond).

Next, we will look at bonds formed by atoms other than carbon that are commonly found in organic compounds—bonds formed by oxygen, nitrogen, and the halogens. Because the distribution of electrons in the atoms that make up a molecule determines the orbitals used in bond formation and because the orbitals used in bond formation determine the bond angles in a molecule, you will see that, if you know the distribution of electrons in the atoms and the bond angles in the molecule, you can determine the kinds of orbitals that are involved in bond formation.

Bonding in Methane

Methane (CH_4) has four covalent carbon–hydrogen bonds. Since all four bonds have the same length and all the bond angles are the same (109.5°), one can conclude that the four covalent bonds in methane are identical.

Three different ways to represent the methane molecule are shown here. In a wedge-and-dash model, bonds protruding out from the plane of the paper are drawn as solid wedges, those protruding into the paper are drawn as dashed lines, and solid lines are used for bonds in the plane of the paper.

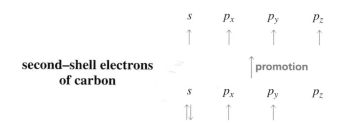

wedge-and-dash model of methane   ball-and-stick model of methane   space-filling model of methane

Initially it may seem surprising that carbon forms four covalent bonds. In its ground state, carbon has only two unpaired electrons (Table 1.2), so one would expect it to form only two covalent bonds. However, if carbon formed only two covalent bonds, it would not complete its octet. In order to form four covalent bonds and complete its octet, carbon must promote an electron from the $2s$ orbital into the empty $2p$ orbital. After promotion, there are four unpaired electrons in the new electronic configuration so four covalent bonds can be formed.

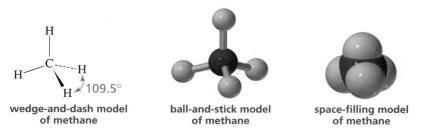

A p orbital is of higher energy than an s orbital, and so it takes energy to promote an electron from an s orbital to a p orbital. The amount of energy required for promotion is 96 kcal/mol (402 kJ/mol). The dissociation energy of a carbon–hydrogen bond is ~100 kcal/mol (~418 kJ/mol), which means that, when four carbon–

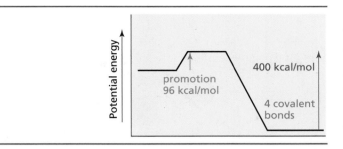

◀ **Figure 1.7**
As a result of electron promotion, carbon forms four covalent bonds with a release of 400 kcal/mol of energy. Without promotion, carbon would form two covalent bonds with a release of 200 kcal/mol of energy. Because it requires 96 kcal/mol to promote an electron, the overall energy advantage of promotion is more than 100 kcal/mol.

hydrogen bonds are formed, 400 kcal/mol is released. If promotion did not occur, carbon could form only two covalent bonds for a net energy change of 200 kcal/mol. By spending 96 kcal/mol to promote an electron, an extra 200 kcal/mol is released as a result of the formation of two additional covalent bonds (Figure 1.7).

The four covalent carbon–hydrogen bonds in methane are identical. Each has a bond length of 1.10 Å and a bond dissociation energy of 104 kcal/mol (435 kJ/mol). If carbon used an *s* orbital and three *p* orbitals in covalent bond formation, the bond formed with the *s* orbital would differ from the three bonds formed with *p* orbitals. How can carbon form four identical bonds with one *s* and three *p* orbitals?

The fact that all four bonds are identical means that carbon must use hybridized orbitals. **Hybridized orbitals** are mixed orbitals. The concept of mixing different orbitals, called **orbital hybridization,** was first proposed by Linus Pauling in 1931. If the four atomic orbitals in the second shell (s, p_x, p_y, p_z) are mixed together and then divided into four equal orbitals, each of the four resulting orbitals will be $\frac{1}{4}s$ and $\frac{3}{4}p$. Such mixed orbitals are called sp^3 orbitals. Each sp^3 orbital has 25% *s* character and 75% *p* character. The four sp^3 orbitals are degenerate. In other words, they have the same energy.

A friend's home chemistry laboratory sparked in **Linus Carl Pauling (1901–1994)** *an early interest in science. Pauling was born in Portland, Oregon, in 1901. He received a Ph.D. from the California Institute of Technology and remained there for his entire academic career. He received the Nobel Prize in chemistry in 1954 for his work on molecular structure. Like Einstein, Pauling was a pacifist, winning the 1962 Nobel Peace Prize for his work on behalf of nuclear disarmament.*

Because an sp^3 orbital has 25% of the characteristics of an *s* orbital and 75% of the characteristics of a *p* orbital, its shape is a mixture of both kinds of orbitals. It has two lobes like a *p* orbital, but the lobes differ in size since the sp^3 orbital has some of the spherical shape of an *s* orbital. This means that the electron density is greater in one lobe than in the other. The larger lobe is used in covalent bond formation.

The four sp^3 orbitals arrange themselves in space in a way that allows them to get as far away from each other as possible (Figure 1.8a). This occurs because electron pairs repel each other and getting as far from each other as possible minimizes the mutual repulsion of the electrons (Section 1.6). When four orbitals spread themselves into space as far from each other as possible, they point toward the corners of a regular tetrahedron (a pyramid with four faces, each an equilateral triangle). Each of the four carbon–hydrogen bonds in methane forms as the result of overlap of an *s* orbital of hydrogen with an sp^3 orbital of carbon (Figure 1.8b). Now you can understand why the four carbon–hydrogen bonds are identical.

The angle formed between any two bonds of methane is 109.5°. This bond angle is called the **tetrahedral bond angle.**

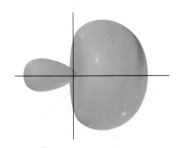

An sp^3 hybridized orbital. It has 25% *s* character and 75% *p* character.

Figure 1.8 ▶
(a) The four sp^3 orbitals are directed toward the corners of a tetrahedron, causing each bond angle to be 109.5°.
(b) An orbital picture of methane, showing the overlap of the sp^3 orbitals of carbon with the s orbital of hydrogen. (For clarity, the back lobes of the sp^3 orbitals are not shown.)

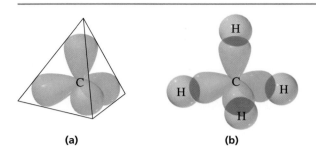

(a)   (b)

Electron pairs spread themselves into space as far from each other as possible.

The fact that covalent bonds arrange themselves in space to minimize mutual repulsion explains why the geometry of a molecule depends on the kinds of covalent bonds in the molecule. A carbon such as the one in methane, which forms covalent bonds using four equivalent sp^3 hybridized orbitals, is called a **tetrahedral carbon.**

The postulation of hybridized orbitals may seem to be a theory contrived just to make things fit. And that is exactly what it is. But it is a theory that gives us a very good picture of how chemical compounds behave.

Bonding in Ethane

The two carbons in ethane are tetrahedral. Each carbon uses four sp^3 atomic orbitals to form four covalent bonds.

$$\begin{array}{ccc} & H & H \\ & | & | \\ H - & C - C & - H \\ & | & | \\ & H & H \end{array}$$
ethane

One sp^3 orbital of one carbon overlaps an sp^3 orbital of the other carbon to form the carbon–carbon bond. The remaining three sp^3 orbitals of each carbon overlap an s orbital of hydrogen to form carbon–hydrogen bonds. So the carbon–carbon bond forms by sp^3–sp^3 overlap and each carbon–hydrogen bond forms by sp^3–s overlap (Figure 1.9). Each of the bond angles in ethane is nearly the tetrahedral bond angle of 109.5°, and the length of the carbon–carbon bond is 1.54 Å.

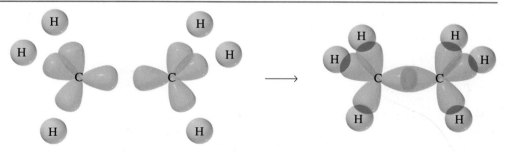

▲ Figure 1.9
An orbital picture of ethane. The carbon–carbon bond is formed by sp^3–sp^3 overlap, and each carbon–hydrogen bond is formed by sp^3–s overlap. (The back lobes of the sp^3 orbitals are not shown.)

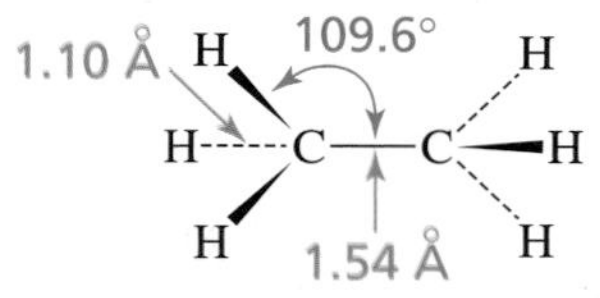

1.10 Å H   109.6°   H

H----C——C—H

H   1.54 Å   H

wedge-and-dash model
of ethane

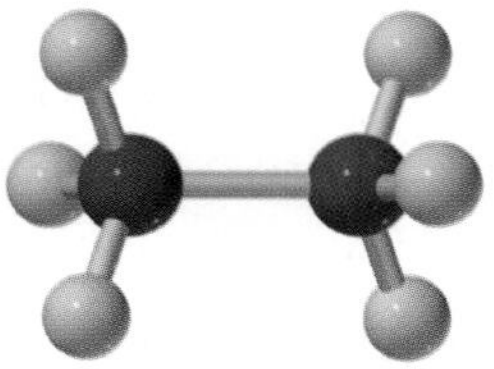

ball-and-stick model
of ethane

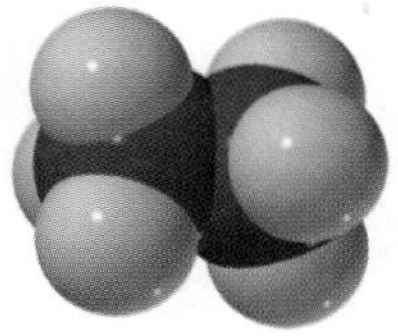

space-filling model
of ethane

All the bonds in methane and ethane are sigma (σ) bonds because they all are formed by end-on overlap of atomic orbitals. (All **single bonds** found in organic compounds are sigma bonds.)

PROBLEM 11◆

Explain what orbitals are involved in formation of the covalent bonds in propane ($CH_3CH_2CH_3$).

Each of the carbon atoms in ethene forms four bonds, but each is bonded to only three atoms.

$$H_2C{=}CH_2$$

ethene
ethylene

**1.8
BONDING IN ETHENE.
DOUBLE BONDS**

To bond to three atoms, each carbon hybridizes three orbitals. Because three orbitals are hybridized (an s orbital and two of the p orbitals), three hybridized orbitals are obtained. These are called sp^2 orbitals. After hybridization, each carbon atom has three degenerate sp^2 orbitals and one p orbital.

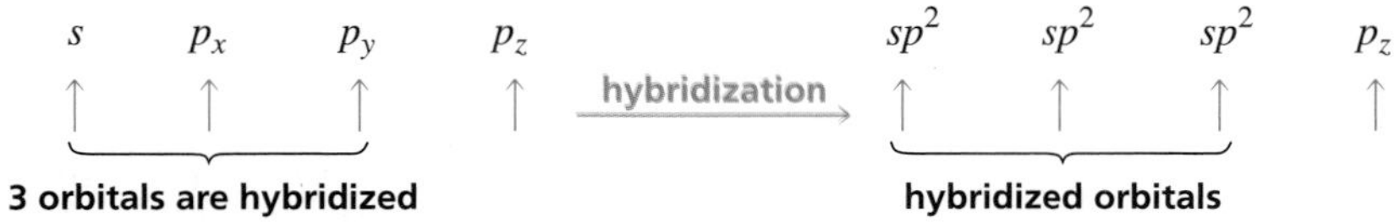

s	p_x	p_y	p_z	hybridization	sp^2	sp^2	sp^2	p_z
↑	↑	↑	↑		↑	↑	↑	↑

3 orbitals are hybridized   **hybridized orbitals**

Each carbon uses its three sp^2 orbitals to bond to two hydrogens and a carbon. The carbon–hydrogen bonds result from overlap of an sp^2 orbital of carbon and an s orbital of hydrogen. One of the carbon–carbon bonds in ethene results from overlap of an sp^2 orbital of one carbon with an sp^2 orbital of the other carbon; this is a sigma (σ) bond because it is formed by end-on overlap (Figure 1.10a). However, the carbons in ethene form two bonds with each other. This is called a **double bond.** The two carbon–carbon bonds are not identical. The second

Figure 1.10 ▶
(a) The carbon–carbon σ bond in ethene is formed by sp^2–sp^2 overlap, and the carbon–hydrogen bonds are formed by sp^2–s overlap. (b) The carbon–carbon π bond is formed by side-to-side overlap of a p orbital of one carbon with a p orbital of the other carbon.

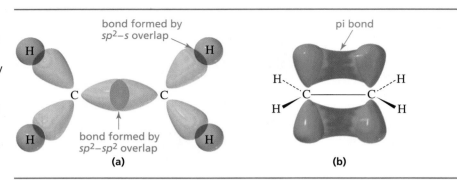

(a)

(b)

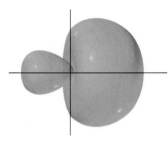

An sp^2 hybridized orbital. It has 33.3% s character and 66.7% p character.

carbon–carbon bond results from overlap of the two unhybridized p orbitals. They overlap in a side-to-side manner (Figure 1.10b) and therefore form a pi (π) bond. So one of the bonds in a double bond is a σ bond and the other is a π bond. The carbon–hydrogen bonds are all σ bonds.

In order to get as far from each other as possible, the axes of the three sp^2 orbitals lie in a plane, directed to the corners of an equilateral triangle with the carbon nucleus in the center. This means that the bond angles are all close to 120°. Because the sp^2 hybridized carbon atom is bonded to three atoms that define a plane, it is called a **trigonal planar carbon.** The unhybridized p orbital is perpendicular to the triangular plane defined by the axes of the sp^2 orbitals. The two p orbitals that overlap to form the π bond must be parallel to each other in order for maximum overlap to occur. This forces the second triangle formed by the other carbon and two hydrogens to lie in the same plane as the first. So all six atoms in ethene lie in the same plane, and the electrons in the p orbitals occupy a volume of space above and below this plane.

Both bonds in the double bond of ethene contribute to its strength. In a carbon–carbon double bond, four electrons hold the carbons together; in a carbon–carbon single bond, only two electrons bind the atoms. This means that a carbon–carbon double bond in ethene is stronger (152 kcal/mol, 636 kJ/mol) and shorter (1.33 Å) than the carbon–carbon single bond in ethane (88 kcal/mol, 368 kJ/mol, and 1.54 Å).

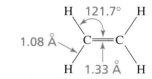

a double bond consists of
one sigma bond and one pi bond

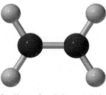

ball-and-stick model
of ethene

space-filling model
of ethene

DIAMOND, GRAPHITE, AND BUCKMINSTERFULLERENE: SUBSTANCES CONTAINING ONLY CARBON ATOMS

Diamond is the hardest of all substances. Graphite, in contrast, is a slippery, soft solid most familiar to us as the "lead" in pencils. Both materials, in spite of their very different physical properties, contain only carbon atoms. The two substances differ only in the nature of the carbon–carbon bonds holding them together. Diamond consists of an extended three-dimensional network of atoms, with each carbon bonding to four other carbons using sp^3 orbitals. The carbon atoms in graphite are sp^2 hybridized carbons. Each carbon bonds to only three other carbon atoms. This planar trigonal arrangement causes atoms in graphite to lie in flat sheets that can shear off of neighboring sheets. As you rub a pencil across a page, sheets of carbon atoms shear off, leaving a thin trail of graphite. There is a third substance found in nature that contains only carbon atoms; it is called buckminsterfullerene. Like graphite, buckminsterfullerene contains only sp^2 hybridized carbons but, instead of forming planar sheets, the sp^2 carbons in buckminsterfullerene form spherical structures. (Buckminsterfullerene is discussed in more detail in Section 6.11.)

The carbon atoms in ethyne are each bonded to two atoms—a hydrogen and another carbon.

$$HC{\equiv}CH$$
ethyne
acetylene

1.9 BONDING IN ETHYNE. TRIPLE BONDS

Since each carbon forms covalent bonds with only two atoms, two orbitals (an s and a p) are hybridized. Two degenerate sp hybridized orbitals result. Each carbon atom in ethyne, therefore, has two sp hybridized orbitals and two unhybridized p orbitals.

One of the sp orbitals of one carbon in ethyne overlaps with an sp orbital of the other carbon to form a carbon–carbon σ bond. The other sp orbital of each carbon overlaps with the s orbital of a hydrogen to form a carbon–hydrogen σ bond (Figure 1.11). In order to minimize electron repulsions, the two sp orbitals point away from each other, and so the bond angles are 180°.

Each of the unhybridized p orbitals engages in side-to-side overlap with a parallel p orbital on the other carbon, with the result that two π bonds are formed. The overall result is a triple bond. A **triple bond** consists of one σ bond and two π bonds. Because the two unhybridized p orbitals on each carbon are perpendicular to each other, there is a region of electron density above and below *and* in front of and in back of the internuclear axis of the molecule. You can think of a triple bond as consisting of a σ bond surrounded by a cylinder of electrons (Figure 1.11).

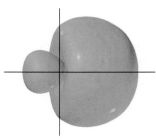

An sp hybridized orbital. It has 50% s character and 50% p character.

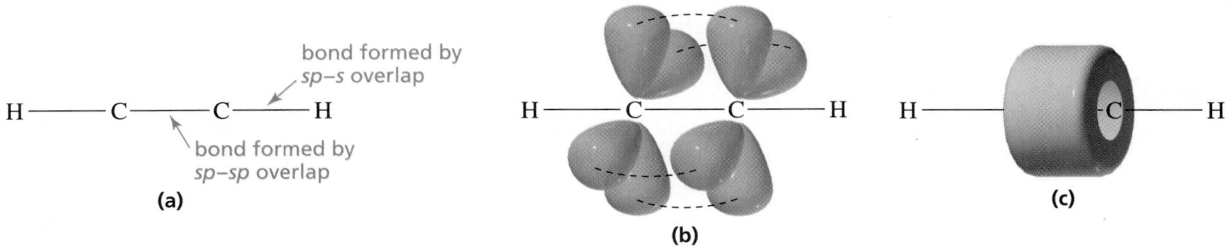

▲ **Figure 1.11**
(a) The carbon–carbon σ bond of ethyne is formed by *sp–sp* overlap, and the carbon–hydrogen bonds are formed by *sp–s* overlap. (b) The two carbon–carbon π bonds are formed by side-to-side overlap of the *p* orbitals of one carbon with the *p* orbitals of the other carbon. (c) The carbon atoms in a triple bond and the atoms bonded to them are in a straight line. The triple bond can be represented by a cylinder of electrons.

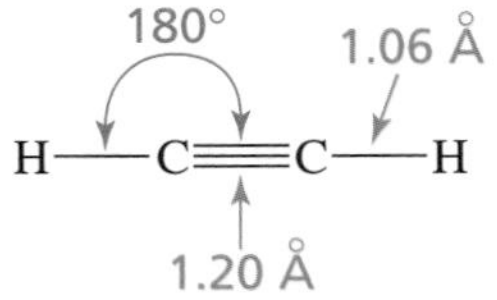

a triple bond consists of one sigma bond and two pi bonds

ball-and-stick model of ethyne   space-filling model of ethyne

Because the two carbon atoms in a triple bond are held together by six electrons, a triple bond is stronger (200 kcal/mol, 837 kJ/mol) and shorter (1.20 Å) than a double bond.

1.10 BONDING IN THE METHYL CATION, THE METHYL RADICAL, AND THE METHYL ANION

Not all carbon atoms form four bonds. A carbon atom with a positive charge, a negative charge, or an unpaired electron forms only three bonds. We will now take a look at the bonding in carbon atoms that form three bonds.

The Methyl Cation ($^+CH_3$)

A positively charged carbon atom forms three covalent bonds, so it hybridizes three orbitals—an *s* orbital and two *p* orbitals. Therefore, it forms its three covalent bonds using three sp^2 orbitals. Its unhybridized *p* orbital remains empty. Because the three sp^2 orbitals lie in a plane, a carbocation is flat. The *p* orbital stands perpendicular to the plane defined by the three atoms covalently bonded to the positively charged carbon.

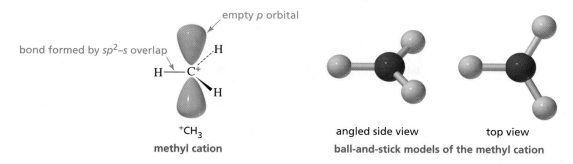

empty *p* orbital

bond formed by *sp²–s* overlap

+CH₃
methyl cation

angled side view

top view

ball-and-stick models of the methyl cation

The Methyl Radical (·CH₃)

The carbon atom in the methyl radical is also sp^2 hybridized. The methyl radical differs by one unpaired electron from the methyl cation. That electron occupies the *p* orbital. Notice the similarity in the ball-and-stick models for the methyl cation and the methyl radical.

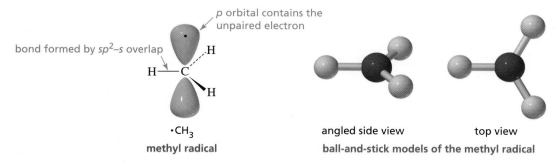

p orbital contains the unpaired electron

bond formed by *sp²–s* overlap

·CH₃
methyl radical

angled side view

top view

ball-and-stick models of the methyl radical

The Methyl Anion (⁻:CH₃)

A negatively charged carbon atom has three pairs of bonding electrons and one pair of nonbonding electrons. Because pairs of electrons are positioned as far apart as possible (Section 1.6), the four orbitals containing the bonding and nonbonding electrons point toward the corners of a tetrahedron. In other words, a negatively charged carbon is sp^3 hybridized. In the methyl anion, three of carbon's sp^3 orbitals each overlap with an *s* orbital of hydrogen and the fourth sp^3 orbital holds the nonbonding pair of electrons.

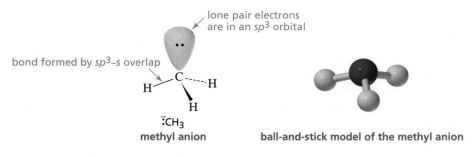

lone pair electrons are in an *sp³* orbital

bond formed by *sp³–s* overlap

⁻:CH₃
methyl anion

ball-and-stick model of the methyl anion

1.11
BONDING IN WATER

The oxygen atom in water (H_2O) forms two covalent bonds. Because oxygen has two unpaired electrons in its ground-state electronic configuration (Table 1.2), it does not need to promote an electron to form the required number of covalent bonds. But this simple picture presents a problem. Oxygen has two nonbonding pairs of electrons. These two nonbonding pairs of electrons are chemically identical, which means that they must be in identical orbitals. However, the ground-state electronic configuration of oxygen suggests that one of the nonbonding pairs of electrons occupies an s orbital and the other pair occupies a p orbital, which would make them nonidentical. Furthermore, if oxygen uses p orbitals to form the two covalent bonds as predicted by the ground-state electronic configuration, the oxygen–hydrogen bonds should have a bond angle of about 90° since the two p orbitals are at right angles to each other. But the experimentally observed bond angle is 104.5°. In order to explain the observed bond angle and the fact that the two pairs of nonbonding electrons are identical, one must assume that oxygen, like carbon, uses hybridized orbitals to form covalent bonds. The s orbital and the three p orbitals must hybridize to produce four sp^3 orbitals.

second-shell electrons of oxygen

Each of the two oxygen–hydrogen bonds is formed by the overlap of an sp^3 orbital of oxygen with an s orbital of hydrogen. A nonbonding pair of electrons occupies each of the two remaining sp^3 orbitals.

The bond angle in water is a little smaller (104.5°) than the tetrahedral bond angle (109.5°) in methane—presumably because each nonbonding electron pair "sees" only one nucleus, which makes the nonbonding pair more diffuse than the bonding pair that "sees" two nuclei and is therefore relatively confined between the two nuclei. These more diffuse orbitals take up more room, causing the oxygen–hydrogen bonds to squeeze closer together, decreasing the bond angle.

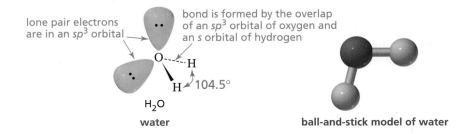

WATER—A UNIQUE COMPOUND

Water is the most abundant compound found in biological organisms. It has unique properties that allowed life to originate and evolve. Its high heat of fusion protects organisms from freezing at low temperatures. The high heat capacity of water minimizes temperature changes in organisms, and its high heat of vaporization allows animals to cool themselves with a minimal loss of body fluid. Because liquid water is more dense than ice, ice formed on the surface of water floats and insulates the water below. This is why oceans and lakes don't freeze from the bottom up.

The experimentally observed bond angles in NH_3 are 107.3°. The bond angles indicate that nitrogen also uses hybridized orbitals when it forms covalent bonds. Like carbon and oxygen, the one s and three p orbitals of the second shell of nitrogen hybridize to form four degenerate sp^3 orbitals.

**1.12
BONDING IN
AMMONIA AND THE
AMMONIUM ION**

**second-shell electrons
of nitrogen**

s	p_x	p_y	p_z	hybridization	sp^3	sp^3	sp^3	sp^3
↑↓	↑	↑	↑	→	↑↓	↑	↑	↑

The nitrogen–hydrogen covalent bonds in NH_3 form as a result of the overlap of an sp^3 orbital of nitrogen and the s orbital of hydrogen. The single nonbonding pair of electrons occupies an sp^3 orbital. The bond angle (107.3°) is smaller than the tetrahedral bond angle (109.5°) because the nonbonding pair of electrons takes up more space than the bonding electrons. But the NH_3 bond angle is larger than the H_2O bond angle (104.5°) because nitrogen has only one nonbonding pair of electrons while oxygen has two such pairs.

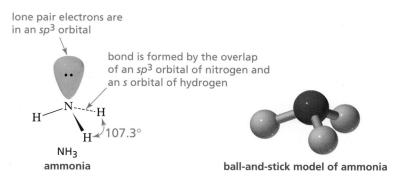

lone pair electrons are
in an sp^3 orbital

bond is formed by the overlap
of an sp^3 orbital of nitrogen and
an s orbital of hydrogen

107.3°

NH_3
ammonia

ball-and-stick model of ammonia

Because the ammonium ion ($^+NH_4$) has four identical nitrogen–hydrogen bonds and has no nonbonding pairs of electrons, all the bond angles are 109.5°.

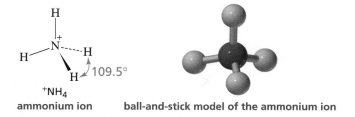

109.5°

$^+NH_4$
ammonium ion

ball-and-stick model of the ammonium ion

1.13 BONDING IN THE HYDROGEN HALIDES

Fluorine, chlorine, bromine, and iodine are collectively known as the halogens. HF, HCl, HBr, and HI are called hydrogen halides. The hydrogen–halogen bond in the hydrogen halides is formed by the overlap of a p orbital of the halogen with the s orbital of hydrogen.

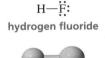

H—Ḟ:

hydrogen fluoride

In the case of fluorine, the p orbital used in bond formation belongs to the second shell of electrons. Chlorine uses a p orbital that belongs to the third shell of electrons. Because the average distance from the nucleus is greater for an electron in the third shell than for an electron in the second shell, the hydrogen–halogen bond becomes longer and weaker as the atomic weight of the halogen increases (Table 1.5).

Halogen halides

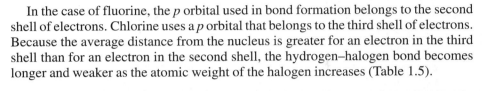

TABLE 1.5   Hydrogen–Halogen Bond Lengths and Bond Strengths

Hydrogen halide			Bond length (A)	Bond strength ($DH°$)[a]	
				kcal/mol	kJ/mol
H—F	H :F	H F	0.917	136	568
H—Cl	H :Cl	H Cl	1.2746	103	431
H—Br	H :Br	H Br	1.4145	88	366
H—I	H :I	H I	1.6090	71	297

[a]$DH°$ is the bond dissociation energy.

1.14 SUMMARY OF ORBITAL HYBRIDIZATION, BOND LENGTHS, BOND STRENGTHS, AND BOND ANGLES

All single bonds are σ bonds. All double bonds are composed of one σ bond and one π bond. All triple bonds are composed of one σ bond and two π bonds. *The hybridization of a carbon, oxygen, or nitrogen atom can be determined by the number of π bonds it forms: If it forms no π bonds, it is sp^3 hybridized; if it forms one π bond, it is sp^2 hybridized; if it forms two π bonds, it is sp hybridized.* The exceptions are carbocations and carbon radicals, which are sp^2 hybridized—not because they form a π bond, but because they have an unfilled or partially filled p orbital (Section 1.10).

$$CH_3-NH_2$$
$$\underset{sp^3}{\uparrow} \quad \underset{sp^3}{\uparrow}$$

$$\underset{sp^2}{\overset{CH_3}{\underset{CH_3}{>}}} C=\underset{sp^2}{N}-\underset{sp^3}{NH_2}$$

$$CH_3-C\equiv N$$
$$\underset{sp^3}{\uparrow} \quad \underset{sp}{\uparrow} \quad \underset{sp}{\uparrow}$$

$$CH_3-OH$$
$$\underset{sp^3}{\uparrow} \quad \underset{sp^3}{\uparrow}$$

$$\overset{O \leftarrow sp^2}{\underset{sp^3 \; sp^2 \; sp^3}{CH_3-\overset{\|}{C}-OH}}$$

$$O=C=O$$
$$\underset{sp^2}{\uparrow} \quad \underset{sp}{\uparrow} \quad \underset{sp^2}{\uparrow}$$

Comparing the lengths and strengths of carbon–carbon single, double, and triple bonds, we see that the more bonds holding two carbon atoms together, the shorter and stronger is the carbon–carbon bond (Table 1.6). Triple bonds are shorter and stronger than double bonds, which are shorter and stronger than single bonds. Because a double bond (a σ bond plus a π bond) is less than twice as strong as a single bond (a σ bond), we can conclude that a π bond is weaker than a σ bond. If we subtract the σ bond energy from the total bond energy of the double bond, we find that the π bond has a dissociation energy of 64 kcal/mol ($152 - 88 = 64$) or 268 kJ/mol ($626 - 368 = 268$). There is some evidence that a σ bond formed from sp^2–sp^2 overlap is stronger (91 kcal/mol, 381 kJ/mol) than that formed from sp^3–sp^3 overlap (88 kcal/mol, 368 kJ/mol). In that case, the π bond of ethylene could be as weak as 61 kcal/mol ($152 - 91 = 61$) or 255 kJ/mol ($636 - 381 = 255$).

A π bond is weaker than a σ bond.

The data in Table 1.6 indicate that a carbon–hydrogen σ bond is shorter and stronger than a carbon–carbon σ bond. The s orbital of hydrogen is closer to the nucleus than is the sp^3 orbital of carbon, so the carbon–hydrogen bond is shorter because the nuclei are closer together in sp^3–s overlap than in sp^3–sp^3 overlap. The carbon–hydrogen bond is stronger because there is greater electron density in the region of overlap of an sp^3 orbital with an s orbital (of hydrogen) than in the region of overlap of an sp^3 orbital with an sp^3 orbital (of carbon).

The more s character, the shorter and stronger the bond.

It is apparent that the length and strength of a carbon–hydrogen bond is dependent on the hybridization of the carbon atom to which the hydrogen is attached. The more s character in the orbital used by carbon to form the bond, the shorter and stronger the bond—again, because an s orbital is closer to the nucleus than a p orbital. So a carbon–hydrogen bond formed by an sp hybridized carbon (50% s) is shorter and stronger than a carbon–hydrogen bond formed by an sp^2 hybridized carbon (33.3% s), which in turn is shorter and stronger than a carbon–hydrogen bond formed by an sp^3 hybridized carbon (25% s).

The more s character, the larger the bond angle.

The bond angle is also dependent on the orbital used by carbon in bond formation. The greater the amount of s character in the orbital, the larger the bond angle. For example, sp hybridized carbons have bond angles of 180°, sp^2 hybridized carbons have bond angles of 120°, and sp^3 hybridized carbons have bond angles of 109.5°.

You may wonder how an electron knows what orbital it should go into. Clearly, electrons do not know anything about orbitals. They simply arrange themselves around atoms in the most stable manner possible. It is chemists who use the concept of orbitals to explain this arrangement.

TABLE 1.6   Comparison of the Bond Angles and the Lengths and Strengths of the Carbon–Carbon and Carbon–Hydrogen Bonds in Ethane, Ethene, and Ethyne

Molecule	Hybridization of carbon	Bond angles	Length of C—C bond (Å)	Strength of C—C bond (kcal/mol)	Strength of C—C bond (kJ/mol)	Length of C—H bond (Å)	Strength of C—H bond (kcal/mol)	Strength of C—H bond (kJ/mol)
H—C—C—H ethane (with H H on top and H H on bottom)	sp^3	109.5°	1.54	88	368	1.10	101	423
C=C ethene (H's top and bottom)	sp^2	120°	1.33	152	636	1.08	107	448
H—C≡C—H ethyne	sp	180°	1.20	200	837	1.06	131	548

PROBLEM 14

a. Indicate the hybridization of each of the carbon atoms in the following compound.

$$CH_3CHCH=CHCH_2C\equiv CCH_3$$
$$\underset{\displaystyle CH_3}{|}$$

b. Indicate the hybridization of each of the carbon, oxygen, and nitrogen atoms in the following compound.

$$\overset{\displaystyle O}{\underset{\displaystyle \parallel}{}}$$
$$CH_3CCH_2OCH_2NHCH_2C\equiv N$$

PROBLEM 15

Describe the orbitals used in bonding and the bond angles in the following compounds. (Hint: see Table 1.6.)

    a. BeH_2 **b.** BH_3 **c.** CCl_4 **d.** CO_2

1.15
DIPOLE MOMENTS
OF MOLECULES

We looked at the dipole moments of some commonly encountered single covalent bonds in Section 1.3. For molecules that contain more than one covalent bond, the geometry of the molecule and therefore the **vector sum**[4] of all the individual bond dipole moments have to be taken into account when determining the overall dipole moment of the molecule. For example, since the carbon atom in carbon dioxide is bonded to two atoms, it uses sp hybridized orbitals to form the carbon–oxygen σ bonds. The remaining two p orbitals on carbon form the two carbon–oxygen π bonds. Because sp hybridized orbitals form a bond angle of 180°, the individual carbon–oxygen bond dipole moments cancel each other, giving carbon dioxide a dipole moment of 0 D.

$$O=C=O$$
carbon dioxide
$\mu = 0$ D

carbon tetrachloride
$\mu = 0$ D

The carbon atom in carbon tetrachloride uses sp^3 hybridized orbitals. Because the four atoms bonded to carbon are identical and project symmetrically out from the central atom, the four bond dipole moments cancel and carbon tetrachloride has no dipole moment.

The dipole moment of chloromethane (CH_3Cl) is greater (1.87 D) than the dipole moment of the carbon–chlorine bond (1.5 D). This is because the C—Cl dipole is reinforced by the C—H dipoles, since the direction of the vector sum of the three C—H dipoles is the same as the direction of the dipole of the C—Cl bond. The dipole moment of water (1.85 D) is greater than the dipole moment of a single oxygen–hydrogen bond (1.5 D) because the dipoles of the two O—H bonds of water reinforce each other. Similarly, the dipole moment of ammonia (1.47 D) is greater than the dipole moment of a single nitrogen–hydrogen bond (1.3 D).

---

[4]The vector sum takes into account both the magnitudes and the directions of the bond dipoles.

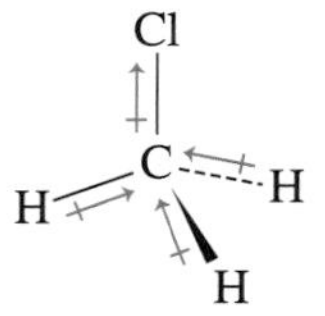

chloromethane
$\mu = 1.87$ D

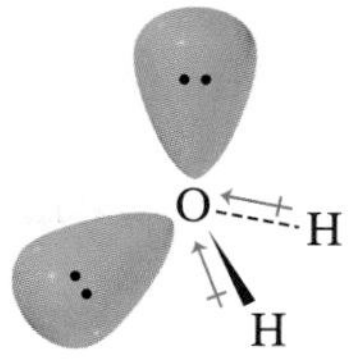

water
$\mu = 1.85$ D

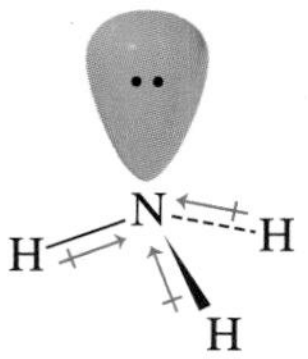

ammonia
$\mu = 1.47$ D

PROBLEM 16◆

Which of the following molecules would you expect to have a dipole moment of 0?

a. CH_3CH_3

b. $H_2C=O$

c. CH_2Cl_2

d. $BeCl_2$

e. $H_2C=CH_2$

f. $H_2C=CHBr$

g. NH_3

h. BF_3

Early chemists called any compound that tasted sour an acid (from *acidus,* Latin for "sour"). Some familiar acids are citric acid (found in lemons and other citrus fruits), acetic acid (found in vinegar), and hydrochloric acid (found in stomach acid—the sour taste associated with vomiting). Compounds that neutralize acids, such as wood ashes and other plant ashes, are called bases, or alkaline compounds ("ash" in Arabic is *al kalai*). Glass cleaners and solutions designed to unclog drains are basic solutions.

The terms "acid" and "base" were more precisely defined by Svante Arrhenius. He defined an acid as a substance that ionizes in solution to release a hydrogen ion (H^+) and a base as a substance that ionizes in solution to release a hydroxide ion (HO^-). His definition of an acid was acceptable, but his definition of a base applied to only one kind of base. New definitions were provided by Brønsted and Lowry in 1923. They defined acids and bases as compounds that are involved in the transfer of a proton. (Hydrogen ions are also called protons.)

In the Brønsted–Lowry definitions, an **acid** is a substance that donates a proton, and a **base** is a substance that accepts a proton. In the following reaction, hydrochloric acid (HCl) meets the Brønsted–Lowry definition of an acid because it donates a proton to water. Water meets the definition of a base because it accepts a proton from HCl. Water can accept a proton because it has two nonbonding pairs of electrons. Either lone pair can form a covalent bond with a proton. In the reverse reaction, H_3O^+ is an acid because it donates a proton to Cl^-, and Cl^- is a base because it accepts a proton from H_3O^+. Such proton-donating and proton-accepting species are commonly called **Brønsted acids** and **Brønsted bases,** respectively.

$$H\ddot{\underset{\cdot\cdot}{C}}l\colon \; + \; H_2\ddot{\underset{\cdot\cdot}{O}}\colon \; \rightleftharpoons \; \colon\!\ddot{\underset{\cdot\cdot}{C}}l\colon^- \; + \; H_3\overset{+}{\ddot{O}}$$

an acid   a base   a base   an acid

According to the Brønsted–Lowry definitions, any compound that contains a hydrogen can potentially act as an acid and any compound that contains a lone pair of electrons can act as a base. Both an acid and a base must be present in a proton-transfer reaction because an acid cannot donate a proton unless a base is present to accept it. **Proton-transfer reactions** are often called **acid–base reactions.**

1.16
AN INTRODUCTION TO THE CONCEPT OF ACID AND BASE

Swedish chemist **Svante August Arrhenius (1859–1927)** *received a Ph.D. from the University of Uppsala. Threatened with a low passing grade on his dissertation because his examiners did not understand his thesis on ionic dissociation, he sent his dissertation to several scientists who defended it. This dissertation earned Arrhenius the 1903 Nobel Prize in chemistry. He was the first to describe the "greenhouse" effect, predicting that as concentrations of atmospheric carbon dioxide (CO$_2$) increase, so will Earth's surface temperature (Section 12.10).*

Born in Denmark, **Johannes Nicolaus Brønsted (1879–1947)** *studied engineering before he switched to chemistry. He was a professor of chemistry at the University of Copenhagen. During World War II, he became known for his anti-Nazi position and in 1947 was elected to the Danish parliament—but died before he could take his seat.*

Thomas M. Lowry (1874–1936) *became the first professor of chemistry in a British medical school.*

In a reaction involving ammonia and water, ammonia (NH_3) is defined as a base because it accepts a proton, and water is an acid because it donates a proton. In the reverse reaction, the ammonium ion ($^+NH_4$) is an acid because it donates a proton and hydroxide ion (HO^-) is a base because it accepts a proton.

$$\ddot{N}H_3 \;+\; H_2\ddot{O}: \;\rightleftharpoons\; ^+NH_4 \;+\; H\ddot{O}:^-$$

a base   an acid   an acid   a base

Notice that water can behave as an acid or as a base. It can behave as an acid because it has a proton that it can donate, but it can also behave as a base because it has a lone pair of electrons that can accept a proton. A compound that can behave either as an acid or as a base is called an **amphoteric compound.**

When a compound loses a proton, the resulting species is called its **conjugate base:** Cl^- is the conjugate base of HCl; HO^- is the conjugate base of H_2O; H_2O is the conjugate base of H_3O^+. When a compound accepts a proton, the resulting species is called its **conjugate acid:** $^+NH_4$ is the conjugate acid of NH_3; HCl is the conjugate acid of Cl^-; H_2O is the conjugate acid of HO^-. An important relationship exists between an acid and its conjugate base: *The stronger the acid, the weaker its conjugate base.* For example, we will see that HBr is a stronger acid than HCl. This means that Br^- is a weaker base than Cl^-. **Basicity** is a measure of how well a compound shares its electrons with a proton.

The stronger the acid, the weaker its conjugate base.

PROBLEM 17◆

a. Draw the conjugate acid of each of the following:
   1. NH_3 2. Cl^- 3. HO^- 4. H_2O

b. Draw the conjugate base of each of the following:
   1. NH_3 2. HBr   3. HNO_3 4. H_2O

PROBLEM 18

CH_3OH and NH_3 are amphoteric compounds. For each compound, write an equation showing its reaction as an acid and an equation showing its reaction as a base.

1.17 ACID STRENGTH

The strength of an acid depends on its ability to donate a proton. Hydrochloric acid is a strong acid (it readily donates a proton), while acetic acid is a relatively weak acid (a weak proton donor). When a strong acid such as hydrochloric acid is put into water, it dissociates almost completely, which means that products are favored when the reaction has reached equilibrium. When a much weaker acid such as acetic acid is put into water, it dissociates only to a small extent, so reactants are favored at equilibrium. Two half-headed arrows are used to designate equilibrium. A longer arrow is drawn toward the species favored at equilibrium.

$$HCl \;+\; H_2O \;\rightleftharpoons\; H_3O^+ \;+\; Cl^-$$
hydrochloric acid

$$CH_3-\overset{O}{\overset{\|}{C}}-OH \;+\; H_2O \;\rightleftharpoons\; H_3O^+ \;+\; CH_3-\overset{O}{\overset{\|}{C}}-O^-$$
acetic acid

Whether a reversible reaction favors reactants or products at equilibrium is indicated by the **equilibrium constant** of the reaction (K_{eq}). Remember that brackets are used to indicate concentration in moles/liter = molarity (M).

$$HA + H_2O \rightleftharpoons H_3O^+ + A^-$$

$$K_{eq} = \frac{[H_3O^+][A^-]}{[H_2O][HA]}$$

The degree to which an acid (HA) dissociates is described by its **acid dissociation constant (K_a).** The acid dissociation constant is obtained by multiplying the equilibrium constant by the concentration of the solvent in which the reaction takes place.

$$K_a = K_{eq}[H_2O] = \frac{[H_3O^+][A^-]}{[HA]}$$

The larger the acid dissociation constant, the stronger the acid (the more readily the compound gives up a proton). Hydrochloric acid has an acid dissociation constant of 10^7, whereas acetic acid has a much smaller acid dissociation constant of 1.74×10^{-5}. For convenience, the strength of an acid is generally indicated by its **pK_a** value rather than its K_a value.

$$pK_a = -\log K_a$$

The pK_a of hydrochloric acid is -7, and the pK_a of acetic acid is 4.76. Notice that the smaller the pK_a, the stronger the acid: strong acids have pK_a values less than 1; acids with pK_a values between 1 and 5 are considered to be moderately strong acids; weak acids have pK_a values between 5 and 15; extremely weak acids have pK_a values greater than 15. Unless otherwise stated, the pK_a values given in this text indicate the strength of the acid *in water.* Later (Section 9.10) you will see how the pK_a of an acid is affected by a change in the solvent.

The **pH** of a solution indicates the concentration of hydrogen ions in the solution. The concentration of hydrogen ions can be indicated as $[H^+]$ or, because a hydrogen ion in water is solvated, $[H_3O^+]$. The lower the pH, the more acidic the solution.

$$pH = -\log [H_3O^+]$$

The pH of a solution can be changed simply by adding acid or base to the solution. Do not confuse pH and pK_a. The pH scale is used to describe the acidity of a *solution.* The pK_a is characteristic of a particular *compound,* much like a melting point or a boiling point; it tells how readily the compound gives up a proton.

PROBLEM 19 ◆

a. Which is a stronger acid, one with a pK_a of 5.2 or one with a pK_a of 5.8?

b. Which is a stronger acid, one with an acid dissociation constant of 3.4×10^{-3} or one with an acid dissociation constant of 2.1×10^{-4}?

PROBLEM 20 ◆

An acid has a K_a of 4.53×10^{-6}. What is its K_{eq}? ($[H_2O] = 55.5$ M)

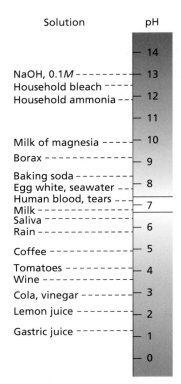

Solution	pH
	— 14
NaOH, 0.1M --------	— 13
Household bleach -----	
Household ammonia --	— 12
	— 11
Milk of magnesia -----	— 10
Borax ------------	— 9
Baking soda --------	
Egg white, seawater --	— 8
Human blood, tears ---	— 7
Milk --------------	
Saliva ------------	
Rain -------------	— 6
Coffee -----------	— 5
Tomatoes ---------	— 4
Wine -------------	
Cola, vinegar -------	— 3
Lemon juice --------	— 2
Gastric juice -------	— 1
	— 0

Organic acids and bases will be very important in our discussions of how and why organic compounds react. The most common organic acids are carboxylic acids. Acetic acid and formic acid are examples of carboxylic acids. Carboxylic acids have pK_a values ranging from about 3 to 5 (they are moderately strong acids). The pK_a values of a variety of organic compounds are given in Appendix II.

$$CH_3\overset{\displaystyle O}{\overset{\|}{C}}OH \qquad H\overset{\displaystyle O}{\overset{\|}{C}}OH$$

acetic acid   formic acid
pK_a = 4.76   pK_a = 3.75

Carboxylic acids are frequently written using condensed structures.

$$CH_3COOH \qquad HCOOH$$

acetic acid   formic acid

Alcohols are much weaker organic acids, with pK_a values close to 16. Methanol and ethanol are examples of alcohols.

$$CH_3OH \qquad CH_3CH_2OH$$

methanol   ethanol
pK_a = 15.5   pK_a = 15.9

Alcohols are amphoteric compounds. An alcohol can behave as an acid and donate a proton, or as a base and accept a proton.

$$CH_3OH + HO^- \rightleftharpoons CH_3O^- + H_2O$$
an acid

$$CH_3OH + H_3O^+ \rightleftharpoons CH_3\overset{+}{O}H + H_2O$$
a base   H

Carboxylic acids are also amphoteric compounds. A carboxylic acid can behave as an acid and donate a proton, or as a base and accept a proton.

$$CH_3\overset{\displaystyle O}{\overset{\|}{C}}OH + HO^- \rightleftharpoons CH_3\overset{\displaystyle O}{\overset{\|}{C}}O^- + H_2O$$
an acid

$$CH_3\overset{\displaystyle O}{\overset{\|}{C}}OH + H_3O^+ \rightleftharpoons CH_3\overset{\displaystyle +OH}{\overset{\|}{C}}OH + H_2O$$
a base

A protonated alcohol or a protonated carboxylic acid is a very strong acid. For example, protonated methanol has a pK_a of -2.5, protonated ethanol has a pK_a of -2.4, and protonated acetic acid has a pK_a of -6.1.

$$CH_3\overset{+}{O}H \qquad CH_3CH_2\overset{+}{O}H \qquad CH_3\overset{\displaystyle \overset{+}{O}H}{\overset{\|}{C}}OH$$
      H   H

protonated methanol   protonated ethanol   protonated acetic acid
pK_a = -2.5 pK_a = -2.4 pK_a = -6.1

Amines are also amphoteric.

$$CH_3NH_2 + HO^- \rightleftharpoons CH_3\overset{-}{N}H + H_2O$$
an acid

$$CH_3NH_2 + H_3O^+ \rightleftharpoons CH_3\overset{+}{N}H_3 + H_2O$$
a base

Amines, however, have such high pK_a values that they rarely show acidic behavior.

$$CH_3NH_2$$
methylamine
pK_a = 40

Amines are much more likely to act as bases. In fact, amines are the most common organic bases. Instead of talking about the strength of a base in terms of its pK_b (the basic equivalent of a pK_a), it is much easier to talk about the pK_a of its conjugate acid, remembering that the stronger the acid, the weaker its conjugate base. The pK_a values of protonated amines are about 10 to 11. Protonated methylamine is a stronger acid than protonated ethylamine, which means that methylamine is a weaker base than ethylamine.

The stronger the acid, the weaker its conjugate base.

$$CH_3\overset{+}{N}H_3 \qquad CH_3CH_2\overset{+}{N}H_3$$
protonated methylamine   **protonated ethylamine**
pK_a = 10.7   **pK_a = 11.0**

It is important to remember the approximate pK_a values of the various classes of compounds we have discussed. An easy way to remember these is in units of five (Table 1.7). Protonated alcohols, protonated carboxylic acids, and protonated water have pK_a's less than 0, carboxylic acids have pK_a's of about 5, protonated amines have pK_a's of about 10, and alcohols and water have pK_a's of about 15. These are listed inside the back cover for easy reference.

In determining the position of equilibrium (whether reactants or products are favored at equilibrium) for an acid–base reaction, remember that the equilibrium favors *reaction* of the strong acid and strong base and *formation* of the weak acid and weak base. In other words, *strong reacts to give weak*. So the equilibrium lies away from the strong and toward the weak.

TABLE 1.7   Approximate pK_a Values

pK_a < 0	pK_a ~ 5	pK_a ~ 10	pK_a ~ 15
$CH_3\overset{+}{O}H_2$	$CH_3\overset{O}{\overset{\|}{C}}OH$	$CH_3\overset{+}{N}H_3$	CH_3OH
$\overset{+}{O}H$			H_2O
$CH_3\overset{\|}{C}OH$			
H_3O^+			

$$CH_3-\overset{\overset{\displaystyle O}{\|}}{C}-OH \quad + \quad NH_3 \quad \rightleftharpoons \quad CH_3-\overset{\overset{\displaystyle O}{\|}}{C}-O^- \quad + \quad \overset{+}{N}H_4$$

stronger acid	**stronger base**	**weaker base**	**weaker acid**
$pK_a = 4.8$			$pK_a = 9.4$

$$CH_3CH_2OH \quad + \quad CH_3NH_2 \quad \rightleftharpoons \quad CH_3CH_2O^- \quad + \quad CH_3\overset{+}{N}H_3$$

weaker acid	**weaker base**	**stronger base**	**stronger acid**
$pK_a = 15.9$			$pK_a = 10.7$

The precise value of the equilibrium constant (K_{eq}) can be calculated by dividing the K_a of the reactant acid by the K_a of the product acid. The equilibrium constant for the reaction of acetic acid with ammonia (top reaction) is 4.0×10^4. The equilibrium constant for the reaction of ethanol with methylamine (bottom reaction) is 6.3×10^{-6}.

$$K_{eq} = \frac{K_a \text{ reactant acid}}{K_a \text{ product acid}}$$

$$K_{eq} = \frac{10^{-4.8}}{10^{-9.4}} = 10^{4.6} = 4.0 \times 10^4$$

$$K_{eq} = \frac{10^{-15.9}}{10^{-10.7}} = 10^{-5.2} = 6.3 \times 10^{-6}$$

PROBLEM 21

Show that

$$K_{eq} = \frac{K_a \text{ reactant acid}}{K_a \text{ product acid}} = \frac{[\text{products}]}{[\text{reactants}]}$$

PROBLEM 22◆

a. Which is a stronger base, CH_3COO^- or $HCOO^-$? (The pK_a of CH_3COOH is 4.8; the pK_a of $HCOOH$ is 3.8.)

b. Which is a stronger base, HO^- or $^-NH_2$? (The pK_a of H_2O is 15.7; the pK_a of NH_3 is 36.)

c. Which is a stronger base, H_2O or CH_3OH? (The pK_a of H_3O^+ is -1.7; the pK_a of $CH_3OH_2^+$ is -2.5.)

PROBLEM 23

a. For each of the acid–base reactions in Section 1.17, by comparing the pK_a values of the acids on either side of the equilibrium arrows, convince yourself that the position of equilibrium is in the direction indicated. (The pK_a values you need can be found in Section 1.17 or in Problem 22.)

b. Do the same thing for the equilibria in Section 1.16 (the pK_a of $^+NH_4$ is 9.4).

PROBLEM 24◆

Calculate the equilibrium constants for acid-base reactions between the following pairs of reactants.

a. $HCl + H_2O$

b. $CH_3COOH + H_2O$

c. $CH_3NH_2 + H_2O$

d. $CH_3\overset{+}{N}H_3 + H_2O$

What determines the strength of an acid? In other words, what determines how readily an acid gives up a proton? One factor is the *electronegativity* of the atom to which the hydrogen is bonded. A second factor is the *size* of the atom to which the hydrogen is bonded.

The elements in the second row of the periodic table are all about the same size, but they have very different electronegativities, which increase across the row from left to right.

1.18
THE EFFECT OF STRUCTURE ON pK_a

relative electronegativities: $C < N < O < F$

If we look at the acidities of hydrogens attached to some of the elements in the second row, we see that the acid strength increases as we move to the right (Table 1.8). Ammonia is a stronger acid than methane, water is a stronger acid than ammonia, and hydrofluoric acid is a stronger acid than water. We can conclude, therefore, that when the atoms are similar in size, the more electronegative the atom to which the hydrogen is attached, the more acidic the compound. Fluorine is more electronegative than oxygen, so a hydrogen bonded to a fluorine is more acidic than a hydrogen bonded to an oxygen.

The effect that the electronegativity of the atom bonded to a hydrogen has on the acidity of that hydrogen can be appreciated when the pK_a's of alcohols and amines are compared. Because oxygen is more electronegative than nitrogen, an alcohol is about 24 pK_a units more acidic than an amine.

$$CH_3OH \qquad\qquad CH_3NH_2$$
$$pK_a = 15.5 \qquad\qquad pK_a = 40$$

A protonated alcohol is about 13 pK_a units more acidic than a protonated amine.

$$CH_3\overset{+}{O}H_2 \qquad\qquad CH_3\overset{+}{N}H_3$$
$$pK_a = -2.5 \qquad\qquad pK_a = 10.7$$

As we proceed down a column in the periodic table, the electronegativity of the elements decreases but the acidity of the hydrogens attached to them increases. That occurs because the atoms get larger as we move down a column, and the size of the atom bonded to hydrogen is more important than its electronegativity in determining acidity.

relative electronegativities $I < Br < Cl < F$

relative acid strengths $HI > HBr > HCl > HF$

TABLE 1.8 The pK_a Values of Some Simple Acids			
CH_4	NH_3	H_2O	HF
$pK_a = 50$	$pK_a = 36$	$pK_a = 15.7$	$pK_a = 3.2$
		H_2S	HCl
		$pK_a = 7.0$	$pK_a = -7$
			HBr
			$pK_a = -9$
			HI
			$pK_a = -10$

Why does the size of an atom have such a significant effect on the acidity of the hydrogen attached to it? It is because, as we move down a column, the orbitals used by the elements to form covalent bonds increase in volume. For example, the covalent bond in HF (second row) is formed by the overlap of a $2p$ orbital of fluorine with the s orbital of hydrogen, whereas the covalent bond in HCl (third row) is formed by the overlap of a $3p$ orbital of chlorine with the s orbital of hydrogen. In the case of HBr (fourth row), the p orbital involved in bonding is a $4p$ orbital, and in HI (fifth row) it is a $5p$ orbital.

The volume of space occupied by a $3p$ orbital is significantly greater than the volume of space occupied by a $2p$ orbital because a $3p$ orbital extends out farther from the nucleus. The increased volume of a $3p$ orbital causes its average electron density to be less than that of a $2p$ orbital. In other words, as the halogen increases in size, the electron density of the p orbital involved in covalent bond formation decreases because of the orbital's increased volume. Thus the attraction of a proton for Cl^- is much less than the attraction of a proton for F^- because the negative charge of Cl^- is less concentrated since it occupies a larger volume. This means that HI is the strongest acid of the hydrogen halides even though iodine is the least electronegative of the halogens (Table 1.8).

In summary, moving across a row of the periodic table, the orbitals have approximately the same volume so the only variable affecting the ease with which a proton is lost from a second-row element is the electronegativity of the element. Moving down a row of the periodic table, the orbitals have very different volumes and, therefore, very different electron densities, and the electron density of the orbital is more important than electronegativity in determining the ease with which a proton is lost; the less the electron density, the weaker the bond and the stronger the acid.

The acidic proton of each of the five carboxylic acids shown below is attached to an oxygen atom, but the five compounds have different acidities. Therefore, there must be a factor—other than the nature of the atom to which the hydrogen is bonded—that affects acidity. From the structures of the five carboxylic acids, you can see that changing one of the atoms of the carboxylic acid affects the acidity of the compound. (Chemists call this *substitution,* and the new atom is called a *substituent.*) An electronegative substituent such as iodine increases the acidity of a carboxylic acid. This is because an electronegative substituent pulls the bonding electrons of the O—H bond away from the proton, which makes it easier for the proton to depart. Pulling electrons away through sigma (σ) bonds is called **inductive electron withdrawal.** As the pK_a values of the carboxylic acids show, the acidity increases as the electronegativity (the electron-pulling ability) of the substituent increases.

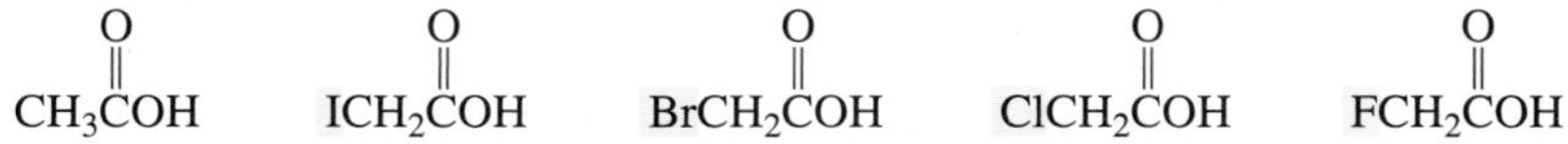

O‖	O‖	O‖	O‖	O‖
CH_3COH	ICH_2COH	$BrCH_2COH$	$ClCH_2COH$	FCH_2COH
pK_a = 4.76	pK_a = 3.15	pK_a = 2.86	pK_a = 2.81	pK_a = 2.66

The effect of a substituent on the acidity of a compound decreases as it is moved farther away from the O—H bond.

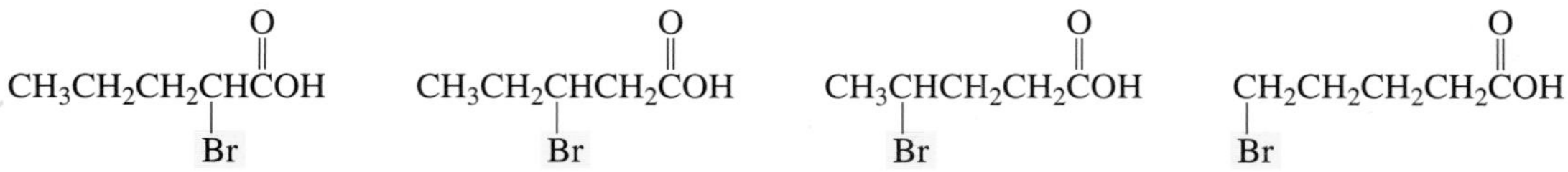

O‖	O‖	O‖	O‖
$CH_3CH_2CH_2CHCOH$	$CH_3CH_2CHCH_2COH$	$CH_3CHCH_2CH_2COH$	$CH_2CH_2CH_2CH_2COH$
Br	Br	Br	Br
pK_a = 2.97	pK_a = 4.01	pK_a = 4.59	pK_a = 4.71

Now we can understand why a carboxylic acid is more acidic than an alcohol. One factor that contributes to the greater acidity of a carboxylic acid is the doubly bonded oxygen that is in place of the two hydrogens of the alcohol. This oxygen pulls the bonding electrons of the O—H bond away from the proton, making it easier for the proton to depart.

$$\overset{\overset{\displaystyle O}{\|}}{CH_3COH} \qquad CH_3CH_2OH$$

$$\text{p}K_a = 4.76 \qquad\quad \text{p}K_a = 15.9$$

Organic acids and bases will be covered in greater depth in later chapters, including a more comprehensive discussion of the factors responsible for the greater acidity of a carboxylic acid compared with an alcohol.

PROBLEM-SOLVING STRATEGY

a. Which is a stronger acid:

$$\underset{\displaystyle F}{CH_3CHCH_2OH} \quad \text{or} \quad \underset{\displaystyle Br}{CH_3CHCH_2OH}$$

When you are asked to compare two items, such as two compounds, pay attention to how they differ; ignore where they stay the same. These two compounds differ only in the halogen atom that is attached to the middle carbon of the molecule. Because fluorine is more electronegative than bromine, there is greater electron withdrawal from the O—H bond in the fluorinated compound, causing it to be the stronger acid.

b. Which is a stronger acid:

$$\underset{\displaystyle Cl}{\overset{\displaystyle Cl}{CH_3CCH_2OH}} \quad \text{or} \quad \underset{\displaystyle Cl}{\overset{\displaystyle Cl}{CH_2CHCH_2OH}}$$

These two compounds differ in the location of one of the chlorine atoms. Because that chlorine in the compound on the left is closer to the O—H bond than is the chlorine in the compound on the right, it is more effective at withdrawing electrons from the O—H bond, causing the compound on the left to be the stronger acid.

Now continue on to Problem 25.

PROBLEM 25◆

For each of the following pairs of compounds, indicate which is the stronger acid.

a. CH_3OCH_2OH or $CH_3CH_2CH_2OH$?

b. $CH_3CH_2CH_2\overset{+}{N}H_3$ or $CH_3CH_2CH_2\overset{+}{O}H_2$?

c. $CH_3OCH_2CH_2OH$ or $CH_3CH_2OCH_2OH$?

d. $\underset{}{\overset{\overset{\displaystyle O}{\|}}{CH_3CCH_2OH}}$ or $\overset{\overset{\displaystyle O}{\|}}{CH_3CH_2COH}$?

PROBLEM 26◆

List the following compounds in order of decreasing acidity.

$$\underset{\displaystyle F}{CH_2CH_2OH} \qquad CH_3CH_2OH \qquad \underset{\displaystyle F}{CH_3CHOH} \qquad \underset{\displaystyle Cl}{CH_2CH_2OH}$$

> **PROBLEM 27 / SOLVED**
>
> HCl is a weaker acid than HBr. Why then is ClCH$_2$COOH a stronger acid than BrCH$_2$COOH?
>
> **SOLUTION**   When comparing the acidities of HCl and HBr, we are comparing the ease of breaking two different kinds of bonds, an H—Cl bond and an H—Br bond. The electron density of a 3p orbital (of Cl) is greater than the electron density of a 4p orbital (of Br), which means that the attraction of Cl$^-$ for a proton is greater than the attraction of Br$^-$ for a proton. Size is more important than electronegativity in determining acidity, so H—Cl is less acidic even though chlorine is more electronegative.
>
> In both carboxylic acids, the same kind of bond is broken (an O—H bond). Therefore, the only factor to be considered is the difference in the electronegativity of the atoms that are pulling electrons away from the O—H bond. Because Cl is more electronegative than Br, the O—H bond in the Cl-containing carboxylic acid is easier to break.

> **PROBLEM 28◆**
>
> **a.** Which of the halide ions (F$^-$, Cl$^-$, Br$^-$, I$^-$) is the strongest base?
>
> **b.** Which is the weakest base?
>
> **PROBLEM 29◆**
>
> **a.** Which is more electronegative, oxygen or sulfur?
>
> **b.** Which is a stronger acid, H$_2$O or H$_2$S?
>
> **c.** Which is a stronger acid, CH$_3$OH or CH$_3$SH?
>
> **PROBLEM 30**
>
> Using the table of pK_a values given in Appendix II, answer the following:
>
> **a.** Which is the most acidic organic compound in the table?
>
> **b.** Which is the least acidic organic compound in the table?
>
> **c.** Which is the most acidic carboxylic acid in the table?
>
> **d.** Which is more electronegative, an sp^3 hybridized oxygen or an sp^2 hybridized oxygen?
>
> **e.** What are the relative electronegativities of sp^3, sp^2, and sp hybridized nitrogen atoms?
>
> **f.** What are the relative electronegativities of sp^3, sp^2, and sp hybridized carbon atoms?
>
> **g.** Which is more acidic, HNO$_3$ or HNO$_2$? Explain the relative acidities.

**1.19
THE EFFECT OF pH
ON THE STRUCTURE
OF AN ORGANIC
COMPOUND**

Whether or not a given acid will lose a proton in an aqueous solution depends on the pK_a of the acid *and* on the pH of the solution. The relationship between the two is given by the **Henderson–Hasselbalch equation,** which is a very useful equation because it tells us whether a compound will exist in its acidic form (with its proton retained) or in its basic form (with its proton removed) at a particular pH.

the Henderson–Hasselbalch equation

$$pK_a = pH + \log \frac{[HA]}{[A^-]}$$

If the pH is greater than the pK_a, the compound will exist primarily in its basic form.

The Henderson–Hasselbalch equation tells us that, when the pH of a solution equals the pK_a of the compound that undergoes dissociation, the concentration of the compound in its acidic form [HA] will equal the concentration of the compound in its basic form [A$^-$]. If the pH of the solution is less than the pK_a of the compound, the compound will exist primarily in its acidic form. If the pH of the solution is greater than the pK_a of the compound, the compound will exist primarily in its basic form. In other words, *compounds exist primarily in their acidic forms in solutions that are more acidic than their pK_a values and they exist primarily in their basic forms in solutions that are more basic than their pK_a values.*

If we know the pH of the solution and the pK_a of the compound, the Henderson–Hasselbalch equation allows us to calculate precisely how much of the compound will be in its acidic form and how much will be in its basic form. For example, when a compound with a pK_a of 5.2 is in a solution of pH 5.2, half the compound will be in the acidic form and the other half will be in the basic form (Figure 1.12). If the pH is one unit less than the pK_a of the compound (pH = 4.2), there will be ten times more compound present in the acidic form than in the basic form (because log 10 = 1). If the pH is two units less than the pK_a of the compound (pH = 3.2), there will be 100 times more compound present in the acidic form than in the basic form (because log 100 = 2). If the pH is changed to 6.2, there will be ten times more compound in the basic form than in the acidic form, and at pH = 7.2 there will be 100 times more compound present in the basic form than in the acidic form.

If the pH is less than the pK_a, the compound will exist primarily in its acidic form.

The Henderson–Hasselbalch equation can be very useful in the laboratory when you must separate compounds from each other. Water and diethyl ether are immiscible liquids and, therefore, when combined will form two layers. The ether layer will lie over the more dense water layer. Charged compounds are more soluble in water, whereas neutral compounds are more soluble in diethyl ether. Two compounds, such as a carboxylic acid (RCOOH) with a pK_a of 4.0 and a protonated amine (R$\overset{+}{N}$H$_3$) with a pK_a of 10.0, dissolved in a mixture of water and diethyl ether, can be separated by adjusting the pH of the water layer. (R means that the particular carboxylic acid or amine is not specified.) For example, if the pH of the water layer is 2, the carboxylic acid and the amine will both be in their acidic forms because the

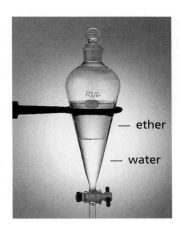
— ether
— water

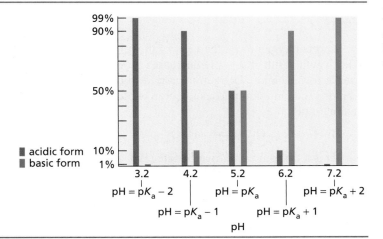

◀ **Figure 1.12**
The relative amounts of a compound with a pK_a of 5.2 in the acidic and basic forms at different pH values.

pH of the water is less than the pK_a's of both compounds. The acidic form of a carboxylic acid is neutral, whereas the acidic form of an amine is charged. Therefore, the carboxylic acid will be more soluble in the ether layer, and the amine will dissolve in the water layer.

acidic form   basic form

$RCOOH \rightleftharpoons RCOO^- + H^+$

$R\overset{+}{N}H_3 \rightleftharpoons RNH_2 + H^+$

For the most effective separation, it is best if the pH of the water layer is at least two units away from the pK_a's of the compounds being separated so the relative amounts in the acidic and basic forms will be at least 100:1 (Figure 1.12).

The Henderson–Hasselbalch equation is useful in dealing with buffer solutions. A **buffer solution** is a solution that maintains a nearly constant pH when small amounts of H^+ or HO^- are added to it. The Henderson–Hasselbalch equation and buffer solutions are discussed in the Solutions Manual/Study Guide.

DERIVATION OF THE HENDERSON–HASSELBALCH EQUATION

The Henderson–Hasselbalch equation can be derived from the expression defining the acid dissociation constant:

$$K_a = \frac{[H_3O^+][A^-]}{[HA]}$$

Taking the logarithms of both sides of the equation and then multiplying both sides of the equation by -1, we obtain

$$\log K_a = \log [H_3O^+] + \log \frac{[A^-]}{[HA]}$$

$$-\log K_a = -\log [H_3O^+] - \log \frac{[A^-]}{[HA]}$$

Substituting and remembering that a positive log is the reciprocal of a negative log, we have

$$pK_a = pH + \log \frac{[HA]}{[A^-]}$$

BLOOD: A BUFFERED SOLUTION

Blood is the fluid that transports oxygen to all the cells of the human body. The normal pH of human blood is 7.35 to 7.45. Death will result if this pH decreases to less than ~6.8 or increases to a value greater than ~8.0 for even a few seconds. Oxygen is carried to cells by a protein in the blood called hemoglobin. When hemoglobin binds O_2, hemoglobin loses a proton, which would make the blood more acidic if it did not contain a buffer to maintain pH.

$$HbH^+ + O_2 \rightleftharpoons HbO_2 + H^+$$

The buffer that controls the blood pH is a carbonic acid/bicarbonate (H_2CO_3/HCO_3^-) buffer. An important feature of this buffer is that carbonic acid decomposes to CO_2 and H_2O.

$$CO_2 + H_2O \rightleftharpoons \underset{\text{carbonic acid}}{H_2CO_3} \rightleftharpoons \underset{\text{bicarbonate}}{HCO_3^-} + H^+$$

It is necessary for cells to be provided with a constant supply of O_2, especially during periods of strenuous exercise when very high levels of O_2 are needed. When O_2 is consumed, the hemoglobin equilibrium shifts to the left, which means that the concentration of H^+ decreases. However, increased metabolism during exercise causes large amounts of CO_2 to be produced. This shifts the carbonate/bicarbonate equilibrium to the right, which increases the concentration of H^+. Significant amounts of lactic acid are also produced and this adds to the increased concentration of H^+. Receptors in the brain respond to the increased concentration of H^+ and trigger a reflex that increases the rate of breathing. This increases the concentration of O_2 available to cells. It also causes CO_2 to be eliminated by exhalation, which decreases the concentration of H^+.

PROBLEM 31◆

a. At what pH will 99% of a compound with a pK_a of 8.4 be in its basic form?

b. At what pH will 90% of a compound with a pK_a of 3.7 be in its acidic form?

c. At what pH will 10% of a compound with a pK_a of 5.9 be in its basic form?

d. At what pH will 50% of a compound with a pK_a of 7.3 be in its basic form?

e. At what pH will 1% of a compound with a pK_a of 7.3 be in its acidic form?

PROBLEM 32◆

a. Indicate whether a carboxylic acid with a pK_a of 5 will be mostly charged or mostly neutral at each of the following pH values.

1. pH = 1	**3.** pH = 5	**5.** pH = 9	**7.** pH = 13
2. pH = 3	**4.** pH = 7	**6.** pH = 11	

b. Answer the same question for a protonated amine with a pK_a of 9.

c. Answer the same question for an alcohol with a pK_a of 15.

PROBLEM 33◆

For each compound, indicate the pH at which:

a. 50% of the compound will be in a form that possesses a charge.

b. > 99% of the compound will be in a form that possesses a charge.

  1. CH_3CH_2COOH (pK_a = 4.9)

  2. $CH_3\overset{+}{N}H_3$ (pK_a = 10.7)

PROBLEM 34◆

The following compounds are all shown in their acidic forms. Draw the form in which each will predominantly exist at pH = 7.

a. CH_3COOH (pK_a = 4.76)

b. $CH_3CH_2\overset{+}{N}H_3$ (pK_a = 11.0)

c. H_3O^+ (pK_a = −1.7)

d. CH_3CH_2OH (pK_a = 15.9)

e. $CH_3CH_2\overset{+}{O}H_2$ (pK_a = −2.5)

f. $\overset{+}{N}H_4$ (pK_a = 9.4)

g. $HC{\equiv}N$ (pK_a = 9.1)

h. HNO_2 (pK_a = 3.4)

i. HNO_3 (pK_a = −1.3)

j. HBr (pK_a = −9)

1.20
LEWIS ACIDS AND BASES

In 1923, G. N. Lewis offered new definitions for the terms "acid" and "base." He defined an acid as a substance that accepts an electron pair and a base as a substance that donates an electron pair. All Brønsted acids fit the Lewis definition because all Brønsted acids lose a proton and the proton accepts an electron pair.

$$H^+ \quad + \quad :NH_3 \quad \rightleftharpoons \quad H-\overset{+}{N}H_3$$

acid **base**
accepts a shares a pair
share in a pair of electrons
of electrons

The Lewis definition of an acid is much broader than the Brønsted–Lowry definition because it is not limited to compounds that lose protons. According to the Lewis definition, compounds such as aluminum chloride and boron trifluoride are acids because they have unfilled valence orbitals and thus can accept a pair of electrons. They react with a compound that has a lone pair of electrons just like a proton reacts with ammonia. But aluminum chloride and boron trifluoride are not Brønsted acids because they are not proton donors.

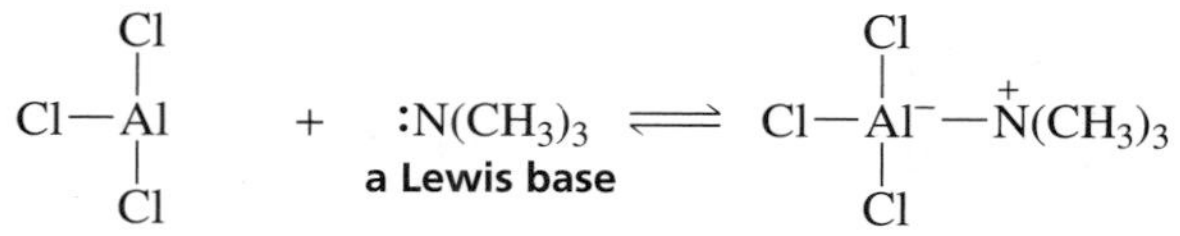

$$Cl-\underset{\underset{Cl}{|}}{\overset{\overset{Cl}{|}}{Al}} \quad + \quad :N(CH_3)_3 \quad \rightleftharpoons \quad Cl-\underset{\underset{Cl}{|}}{\overset{\overset{Cl}{|}}{Al}}^- -\overset{+}{N}(CH_3)_3$$

aluminum chloride   **a Lewis base**
a Lewis acid

$$F-\underset{\underset{F}{|}}{\overset{\overset{F}{|}}{B}} \quad + \quad CH_3\ddot{O}CH_3 \quad \rightleftharpoons \quad F-\underset{\underset{F}{|}}{\overset{\overset{F}{|}}{B}}^- -\overset{+}{\ddot{O}}\underset{CH_3}{\overset{CH_3}{}}$$

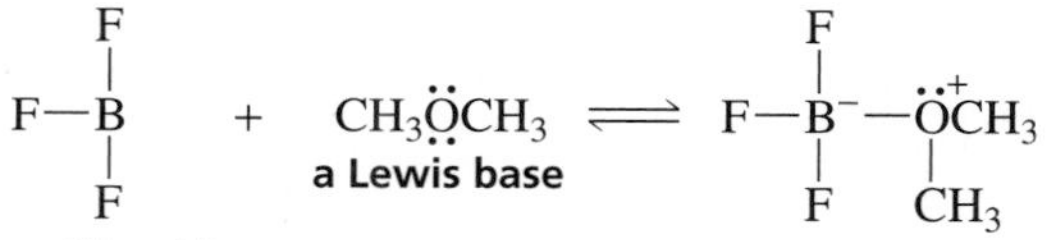

boron trifluoride   **a Lewis base**
a Lewis acid

Lewis base: Have pair, will share.

So the Lewis definition of an acid includes all Brønsted acids and some additional acids that are not Brønsted acids. All **Lewis bases** are Brønsted bases and *vice versa;* they have a pair of electrons that they can share. We will generally use the terms "acid" and "base" to mean a Brønsted acid or a Brønsted base and use the term "**Lewis acid**" to refer to non-Brønsted acids such as $AlCl_3$ or BF_3.

Lewis acid: Need two, from you.

PROBLEM 35

Zinc chloride ($ZnCl_2$) and ferric bromide ($FeBr_3$) are Lewis acids that are commonly used in organic reactions. Give the product of each of the following reactions.

a. $ZnCl_2$ + CH_3OH $\rightleftharpoons$

b. $FeBr_3$ + Br^- $\rightleftharpoons$

c. $AlCl_3$ + Cl^- $\rightleftharpoons$

d. BF_3 + $H\overset{O}{\overset{\|}{C}}H$ $\rightleftharpoons$

PROBLEM 36

Show how each of the following compounds reacts with HO^-.

a. CH_3OH
b. $\overset{+}{N}H_4$
c. $CH_3\overset{+}{N}H_3$
d. BF_3
e. $\overset{+}{C}H_3$
f. $FeBr_3$
g. $AlCl_3$
h. CH_3COOH

PROBLEMS

37. Draw a Lewis structure for each of the following compounds.
   a. H_2CO_3 **d.** N_2H_4 **g.** CO_2
   b. CO_3^{2-} **e.** CH_3NH_2 **h.** NO^+
   c. H_2CO **f.** $CH_3N_2^+$ **i.** H_2NO^-

38. Give the hybridization of the central atom of each of the following species and tell whether the compound is linear, trigonal planar, or tetrahedral.
   a. NH_3 **d.** $\cdot CH_3$ **g.** HCN
   b. BH_3 **e.** $^+NH_4$ **h.** $C(CH_3)_4$
   c. $^-CH_3$ **f.** $^+CH_3$ **i.** H_3O^+

39. Draw a compound that contains only carbon atoms and hydrogen atoms and that has:
   a. three sp^3 hybridized carbons.
   b. one sp^3 hybridized carbon and two sp^2 hybridized carbons.
   c. two sp^3 hybridized carbons and two sp hybridized carbons.

40. Predict the indicated bond angles:
   a. the C—N—H bond angle in $(CH_3)_2NH$
   b. the C—N—C bond angle in $(CH_3)_2NH$
   c. the C—N—C bond angle in $(CH_3)_2\overset{+}{N}H_2$
   d. the C—O—C bond angle in CH_3OCH_3
   e. the C—O—H bond angle in CH_3OH
   f. the H—C—H bond angle in $H_2C{=}O$

  g. the F—B—F bond angle in $^-BF_4$
  h. the C—C—N bond angle in $CH_3C\equiv N$

41. Give each atom the appropriate formal charge.

  a. $H:\ddot{O}:$

  b. $H:\ddot{O}\cdot$

  c. $CH_3-\underset{\underset{CH_3}{|}}{\overset{\overset{CH_3}{|}}{N}}-CH_3$

  d. $H-\ddot{N}-H$

  e. $H-\ddot{C}-H$

  f. $H-\underset{\underset{H}{|}}{\overset{\overset{H}{|}}{N}}-\underset{\underset{H}{|}}{\overset{\overset{H}{|}}{B}}-H$

  g. $H-\underset{\underset{H}{|}}{\overset{}{\ddot{C}}}-H$

  h. $CH_3-\underset{\underset{H}{|}}{\ddot{O}}-CH_3$

42. Draw the ground-state electronic configuration for:
  a. Ca
  b. Ca^{2+}
  c. Ar
  d. Mg^{2+}

43. Write the Kekulé structure for each of the following compounds.
  a. CH_3CHO
  b. CH_3OCH_3
  c. CH_3COOH
  d. $(CH_3)_3COH$
  e. $CH_3CH(OH)CH_2CN$
  f. $(CH_3)_2CHCH(CH_3)CH_2C(CH_3)_3$

44. Show the direction of the dipole moment in each of the following bonds. (Use the electronegativities given in Table 1.3.)
  a. CH_3—Br
  b. CH_3—Li
  c. HO—NH_2
  d. I—Br
  e. CH_3—OH
  f. $(CH_3)_2N$—H

45. **a.** For each compound, tell which of the indicated bonds is the shorter bond.
  b. Indicate the hybridization of the C, O, and N atoms in each of the molecules.

  1. $CH_3CH{=}CHC{\equiv}CH$

  2. $CH_3\overset{\overset{O}{\|}}{C}CH_2{-}OH$

  3. $CH_3NH{-}CH_2CH_2N{=}CHCH_3$

  4. $\underset{H}{\overset{H}{>}}C{=}CHC{\equiv}C{-}H$

  5. $\underset{H}{\overset{H}{>}}C{=}CHC{\equiv}C{-}\underset{\underset{CH_3}{|}}{\overset{\overset{CH_3}{|}}{C}}{-}H$

46. For each of the following compounds, draw the form in which it will predominantly exist at pH = 3, pH = 6, pH = 10, and pH = 14.
  a. CH_3COOH
    $pK_a = 4.8$
  b. $CH_3CH_2\overset{+}{N}H_3$
    $pK_a = 11.0$
  c. CF_3CH_2OH
    $pK_a = 12.4$

47. Indicate the hybridization of the indicated atom in each of the following compounds.

  a. $CH_3CH{=}CH_2$
  b. $CH_3\overset{\overset{O\leftarrow}{\|}}{C}CH_3$
  c. CH_3CH_2OH
  d. $CH_3C{\equiv}N$
  e. $CH_3CH{=}NCH_3$

48. The following compound has two isomers. One isomer has a dipole moment of 0 D, and the other has a dipole moment of 2.95 D. Propose structures for the two isomers that are consistent with these data.

  $ClCH{=}CHCl$

49. Give the products of the following acid–base reactions and indicate the direction of the equilibrium. (Use the pK_a values that are given in Section 1.17.)

a.
$$CH_3\overset{O}{\overset{\|}{C}}OH \ + \ CH_3O^- \ \rightleftharpoons$$

b. $CH_3CH_2OH \ + \ {}^-NH_2 \ \rightleftharpoons$

c.
$$CH_3\overset{O}{\overset{\|}{C}}OH \ + \ CH_3NH_2 \ \rightleftharpoons$$

d. $CH_3CH_2OH \ + \ HCl \ \rightleftharpoons$

50. Explain why NF_3 has a smaller dipole moment than does NH_3.

51. a. Estimate the pK_a value of each of the following acids without using a calculator (i.e., between 3 and 4, between 9 and 10, etc.).
b. Determine the pK_a values using a calculator.
c. Which is the strongest acid?
  1. nitrous acid (HNO_2), $K_a = 4.0 \times 10^{-4}$
  2. nitric acid (HNO_3), $K_a = 2.2 \times 10$
  3. bicarbonate (HCO_3^-), $K_a = 6.3 \times 10^{-11}$
  4. hydrogen cyanide (HCN), $K_a = 7.9 \times 10^{-10}$
  5. formic acid (HCOOH), $K_a = 2.0 \times 10^{-4}$

52. a. List the following carboxylic acids in order of decreasing acidity.

1. $CH_3CH_2CH_2COOH$
$K_a = 1.52 \times 10^{-5}$

3. $ClCH_2CH_2CH_2COOH$
$K_a = 2.96 \times 10^{-5}$

2. $CH_3CH_2\underset{Cl}{\overset{|}{C}}HCOOH$
$K_a = 1.39 \times 10^{-3}$

4. $CH_3\underset{Cl}{\overset{|}{C}}HCH_2COOH$
$K_a = 8.9 \times 10^{-5}$

b. How does the presence of an electronegative substituent such as Cl affect the acidity of a carboxylic acid?
c. How does the location of the substituent affect the acidity of a carboxylic acid?

53. a. For each of the following pairs of reactions, indicate which one has the more favorable equilibrium constant (that is, which one most favors products).
b. Which of the four reactions has the most favorable equilibrium constant?

1. $CH_3CH_2OH + NH_3 \rightleftharpoons CH_3CH_2O^- + \overset{+}{N}H_4$
or
$CH_3OH + NH_3 \rightleftharpoons CH_3O^- + \overset{+}{N}H_4$

2. $CH_3CH_2OH + NH_3 \rightleftharpoons CH_3CH_2O^- + \overset{+}{N}H_4$
or
$CH_3CH_2OH + CH_3NH_2 \rightleftharpoons CH_3CH_2O^- + CH_3\overset{+}{N}H_3$

54. Water and diethyl ether are immiscible liquids. Charged compounds dissolve in water and uncharged compounds dissolve in ether. $C_6H_{11}COOH$ has a pK_a of 4.8 and $C_6H_{11}NH_3{}^+$ has a pK_a of 10.7.
a. What pH would you make the water layer to cause both compounds to dissolve in the water layer?
b. What pH would you make the water layer to cause the acid to dissolve in the water layer and the amine to dissolve in the ether layer?
c. What pH would you make the water layer to cause the acid to dissolve in the ether layer and the amine to dissolve in the water layer?

55. How could you separate a mixture of the following compounds? The reagents available to you are water, ether, 1.0 M HCl, and 1.0 M NaOH. (*Hint:* see Problem 54.)

$pK_a = 4.17$ $pK_a = 10.00$ $pK_a = 10.66$

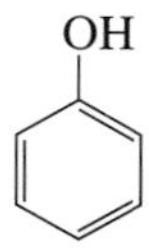

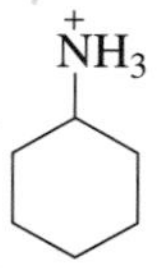

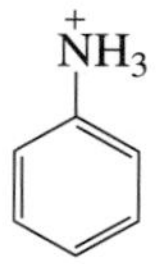

$pK_a = 4.60$

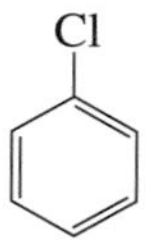

For help in answering
Problems 56, 57, and 58,
see Special Topic I in the
Solutions Manual/Study
Guide.

56. Carbonic acid has a pK_a of 6.1 at physiological temperature. Is the carbonic acid/bicarbonate buffer system that maintains the pH of the blood (at pH = 7.4) better at neutralizing excess acid or excess base?

57. a. If an acid with a pK_a of 5.3 is in an aqueous solution of pH 5.7, what percent of the acid will be present in the acidic form?
  b. At what pH will 80% of the acid exist in the acidic form?

58. Calculate the pH values of the following solutions.
  a. a 1.0 M solution of acetic acid ($pK_a = 4.76$)
  b. a 0.1 M solution of protonated methylamine ($pK_a = 10.7$)
  c. a solution containing 0.3 M HCOOH and 0.1 M HCOO$^-$ (pK_a of HCOOH = 3.76)

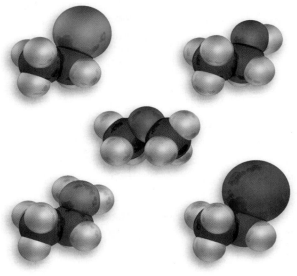

AN INTRODUCTION TO ORGANIC COMPOUNDS:
Nomenclature, Physical Properties, and Representation of Structure

CH_3CH_2Cl, CH_3CH_2OH, $CH_3CH_2NH_2$, CH_3CH_2Br, CH_3OCH_3

This book organizes organic chemistry according to how organic compounds react. But we must not forget that, whenever an organic compound undergoes a reaction, a new organic compound is synthesized. In other words, while we are learning how organic compounds react, we will also be learning how to synthesize new compounds.

$$Y \longrightarrow Z$$

Y is reacting　　　**Z is being synthesized**

The main classes of compounds that are synthesized by the reactions that we will study in Chapters 3 to 11 are alkanes, alkyl halides, ethers, alcohols, and amines. The first of these classes of compounds, alkanes, takes on great importance because the names of alkanes form the basis for the names of all organic compounds. So we will start our study of organic chemistry by learning how to name alkanes. We will build on this knowledge to learn how to name alkyl halides, ethers, alcohols, and amines. We will then be able to name all the compounds that are synthesized in the next nine chapters.

Hydrocarbons are compounds that contain only carbon atoms and hydrogen atoms. **Alkanes** are hydrocarbons that contain only single bonds. Alkanes in which the carbons form a continuous chain with no branches are called **straight-chain alkanes.** The names of several straight-chain alkanes are given in Table 2.1. It is important that you learn the names of at least the first ten.

The family of alkanes shown in Table 2.1 is an example of a homologous series. A **homologous series** (*homos* is Greek for "the same as") is a family of compounds each member of which differs from the next by one **methylene (CH_2) group.** The members of a homologous series are called **homologs.** Propane ($CH_3CH_2CH_3$) and butane ($CH_3CH_2CH_2CH_3$) are homologs.

By examining the relative numbers of carbon and hydrogen atoms in the alkanes listed in Table 2.1, you can conclude that the general molecular formula for an

TABLE 2.1	Nomenclature and Physical Properties of Straight-Chain Alkanes					
Number of carbons	Molecular formula	Name	Condensed structure	Boiling point (°C)	Melting point (°C)	Density^a (g/ml)
1	CH_4	methane	CH_4	−167.7	−182.5	
2	C_2H_6	ethane	CH_3CH_3	−88.6	−183.3	
3	C_3H_8	propane	$CH_3CH_2CH_3$	−42.1	−187.7	
4	C_4H_{10}	butane	$CH_3CH_2CH_2CH_3$	−0.5	−138.3	
5	C_5H_{12}	pentane	$CH_3(CH_2)_3CH_3$	36.1	−129.8	0.5572
6	C_6H_{14}	hexane	$CH_3(CH_2)_4CH_3$	68.7	−95.3	0.6603
7	C_7H_{16}	heptane	$CH_3(CH_2)_5CH_3$	98.4	−90.6	0.6837
8	C_8H_{18}	octane	$CH_3(CH_2)_6CH_3$	127.7	−56.8	0.7026
9	C_9H_{20}	nonane	$CH_3(CH_2)_7CH_3$	150.8	−53.5	0.7177
10	$C_{10}H_{22}$	decane	$CH_3(CH_2)_8CH_3$	174.0	−29.7	0.7299
11	$C_{11}H_{24}$	undecane	$CH_3(CH_2)_9CH_3$	195.8	−25.6	0.7402
12	$C_{12}H_{26}$	dodecane	$CH_3(CH_2)_{10}CH_3$	216.3	−9.6	0.7487
13	$C_{13}H_{28}$	tridecane	$CH_3(CH_2)_{11}CH_3$	235.4	−5.5	0.7546
20	$C_{20}H_{42}$	eicosane	$CH_3(CH_2)_{18}CH_3$	343.0	36.8	0.7886
21	$C_{21}H_{44}$	heneicosane	$CH_3(CH_2)_{19}CH_3$	356.5	40.5	0.7917
30	$C_{30}H_{62}$	triacontane	$CH_3(CH_2)_{28}CH_3$	449.7	65.8	0.8097

^{a}Density is temperature dependent. The densities given are those determined at 20 °C ($d^{20°}$).

alkane is C_nH_{2n+2}, where n is any integer. So, if an alkane has one carbon atom, it must have four hydrogen atoms. Carbon forms four covalent bonds and hydrogen forms only one covalent bond (Section 1.4). This means that there is only one possible structure for an alkane with molecular formula CH_4 (methane) and only one structure for an alkane with molecular formula C_2H_6 (ethane). We have examined the structures of these compounds in Section 1.7. There is also only one structure for an alkane with molecular formula C_3H_8 (propane).

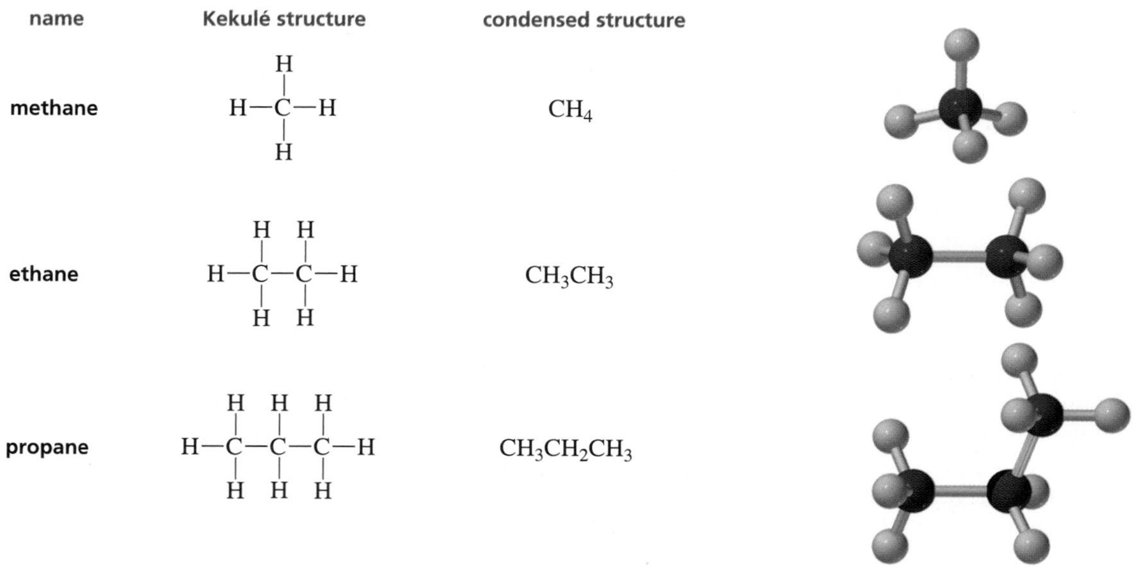

butane   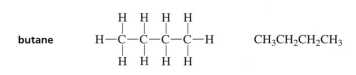 $CH_3CH_2CH_2CH_3$

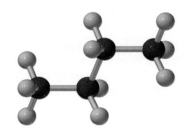

However, as the number of carbons in the alkane increases, the number of possible structures increases. There are two possible structures for an alkane with molecular formula C_4H_{10}. In addition to butane—a straight-chain alkane—there is a branched butane called isobutane. Both of these structures fulfill the requirement that each of the sp^3 hybridized carbons bonds to four atoms and each of the hydrogens forms only one bond.

Molecules such as butane and isobutane that have the same molecular formula but differ in the order in which the atoms are connected are called **constitutional isomers** (the molecules have different constitutions). Constitutional isomers are also known as **structural isomers** because the isomers have different structures. This is how isobutane got its name: it is an "iso"mer of butane. The structural unit with two methyl groups and a hydrogen bonded to a carbon that occurs in isobutane has come to be called "iso." So the name isobutane tells you that it is a four-carbon alkane with an iso structural unit.

$$CH_3CH_2CH_2CH_3$$
butane

$$CH_3CHCH_3$$
$$|$$
$$CH_3$$
isobutane

$$CH_3CH-$$
$$|$$
$$CH_3$$
**an "iso"
structural unit**

There are three alkanes with molecular formula C_5H_{12}. Pentane is the straight-chain alkane. Isopentane, as its name indicates, has an iso structural unit and a total of five carbon atoms. The third isomer is called neopentane. The structural unit with a carbon surrounded by four other carbons is called "neo."

$$CH_3CH_2CH_2CH_2CH_3$$
pentane

$$CH_3CHCH_2CH_3$$
$$|$$
$$CH_3$$
isopentane

$$CH_3$$
$$|$$
$$CH_3CCH_3$$
$$|$$
$$CH_3$$
neopentane

$$CH_3$$
$$|$$
$$CH_3CCH_2-$$
$$|$$
$$CH_3$$
**a "neo"
structural unit**

There are five isomers with molecular formula C_6H_{14}. Three of them (hexane, isohexane, and neohexane) can be named using "iso" and "neo" structural units. But we cannot name the other two without defining names for new structural units.

common name:
IUPAC name:

$$CH_3CH_2CH_2CH_2CH_2CH_3$$
hexane
hexane

$$CH_3CHCH_2CH_2CH_3$$
$$|$$
$$CH_3$$
isohexane
2-methylpentane

$$CH_3$$
$$|$$
$$CH_3CCH_2CH_3$$
$$|$$
$$CH_3$$
neohexane
2,2-dimethylbutane

$$CH_3CH_2CHCH_2CH_3$$
$$|$$
$$CH_3$$
3-methylpentane

$$CH_3CH-CHCH_3$$
$$|\quad\ |$$
$$CH_3\ \ CH_3$$
2,3-dimethylbutane

There are nine isomers with molecular formula C_7H_{16}. We can name only two of them (heptane and isoheptane) without defining new structural units. (At this point, ignore the names written in blue.) Notice that neoheptane cannot be used as a name because three different heptanes have a carbon that is bonded to four other carbons and *a name has to be specific to only one compound.*

A compound can have more than one name, but a name must specify only one compound.

$$CH_3CH_2CH_2CH_2CH_2CH_2CH_3$$

common name: heptane
IUPAC name: heptane

$$CH_3CHCH_2CH_2CH_2CH_3$$
$$CH_3$$
isoheptane
2-methylhexane

$$CH_3CH_2CHCH_2CH_2CH_3$$
$$CH_3$$
3-methylhexane

$$CH_3CH-CHCH_2CH_3$$
$$CH_3\ CH_3$$
2,3-dimethylpentane

$$CH_3CHCH_2CHCH_3$$
$$CH_3\ \ CH_3$$
2,4-dimethylpentane

$$CH_3$$
$$CH_3CCH_2CH_2CH_3$$
$$CH_3$$
2,2-dimethylpentane

$$CH_3$$
$$CH_3CH_2CCH_2CH_3$$
$$CH_3$$
3,3-dimethylpentane

$$CH_3CH_2CHCH_2CH_3$$
$$CH_2CH_3$$
3-ethylpentane

$$CH_3\ \ CH_3$$
$$CH_3C-CHCH_3$$
$$CH_3$$
2,2,3-trimethylbutane

You can see that the number of different structural units will need to increase rapidly as the number of carbons in an alkane increases. (There are 75 alkanes with molecular formula $C_{12}H_{22}$, and 4347 alkanes with molecular formula $C_{15}H_{32}$.) This means that naming compounds would require memorizing the names of many different structural units. To avoid this problem, chemists devised rules that allow compounds to be named on the basis of their structures. This way, only the rules have to be learned. Because the name is based on the structure, these rules also allow the structure of a compound to be deduced from its name.

This method of nomenclature is called **systematic nomenclature.** It is also called **IUPAC nomenclature** since it was designed by a commission of the International Union of Pure and Applied Chemistry (abbreviated IUPAC and pronounced "Eye-You-Pack") at a meeting in Geneva, Switzerland in 1892. The IUPAC rules have been continually revised by the commission since that time. Names such as isobutane and neopentane—nonsystematic names—are called **common names** and are shown in red. The IUPAC names are shown in blue for all of the alkanes with six and seven carbon atoms. Before we can understand how an IUPAC name for an alkane is constructed, we must learn how to name alkyl substituents.

2.1 NOMENCLATURE OF ALKYL SUBSTITUENTS

An **alkyl substituent** (or an alkyl group) is an alkane from which a single hydrogen has been removed. Alkyl substituents are named by replacing the "ane" ending of the alkane with "yl." The letter "R" is used to indicate any alkyl group.

$$CH_3—$$
a methyl group

$$CH_3CH_2—$$
an ethyl group

$$CH_3CH_2CH_2—$$
a propyl group

$$CH_3CH_2CH_2CH_2—$$
a butyl group

$$R—$$
any alkyl group

If a hydrogen of an alkane is replaced by an OH, the compound becomes an **alcohol;** if it is replaced by an NH_2, the compound becomes an **amine;** and if it is replaced by a halogen, the compound becomes an **alkyl halide.**

$$R—OH$$
an alcohol

$$R—NH_2$$
an amine

$$R—X$$
an alkyl halide

X = F, Cl, Br, or I

An alkyl group name followed by the name of the class of the compound (alcohol, amine, etc.) yields the common name of the compound. The following examples show how alkyl group names are used to build common names. Notice that there is a space between the name of the alkyl group and the name of the class of compound, except in the case of amines.

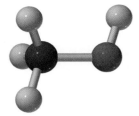

methyl alcohol

CH_3OH
methyl alcohol

$CH_3CH_2NH_2$
ethylamine

$CH_3CH_2CH_2Br$
propyl bromide

$CH_3CH_2CH_2CH_2Cl$
butyl chloride

CH_3I
methyl iodide

CH_3CH_2OH
ethyl alcohol

$CH_3CH_2CH_2NH_2$
propylamine

$CH_3CH_2CH_2CH_2OH$
butyl alcohol

Two alkyl groups—a propyl and an isopropyl—contain three carbon atoms. A propyl group results when a hydrogen is removed from a primary carbon of propane. A **primary carbon** is one that is bonded to only one other carbon. An isopropyl group is obtained when a hydrogen is removed from the secondary carbon of propane. A **secondary carbon** is one that is bonded to two other carbons. Notice that an isopropyl group, as its name indicates, has three carbon atoms arranged as an iso structural unit.

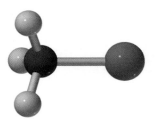

methyl chloride

a primary carbon

$$CH_3CH_2CH_2—$$

a propyl group

a secondary carbon

$$CH_3CHCH_3$$

an isopropyl group

Molecular structures can be drawn in different ways. Isopropyl chloride, for example, is drawn here in two ways. Both indicate the same compound: a central tetrahedral carbon bonded to a hydrogen, a chlorine, and two methyl groups. Studying the models might help convince you that the two molecular structures represent the same compound.

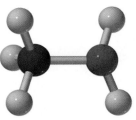

methylamine

two different ways to draw isopropyl chloride

$$CH_3CH_2CH_2Cl$$

$$CH_3CHCH_3$$
$$|$$
$$Cl$$

$$CH_3CHCl$$
$$|$$
$$CH_3$$

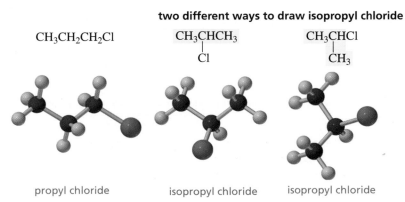

propyl chloride   isopropyl chloride   isopropyl chloride

There are four alkyl groups that contain four carbons. Butyl and isobutyl groups have both had a hydrogen removed from a primary (1°) carbon. A *sec*-butyl group has had a hydrogen removed from a secondary (2°) carbon (*sec-,* often abbreviated *s-,* stands for secondary); and a *tert*-butyl group has had a hydrogen removed from a tertiary (3°) carbon (*tert-,* sometimes abbreviated *t-,* stands for tertiary). A **tertiary carbon** is a carbon that is bonded to three other carbons.

$CH_3CH_2CH_2CH_2$—

a butyl group
an *n*-butyl group

CH_3CHCH_2—
  |
  CH_3

an isobutyl group

CH_3CH_2CH—
       |
       CH_3

a *sec*-butyl group

a tertiary carbon CH_3
                      |
            CH_3C—
                  |
                  CH_3

a *tert*-butyl group

A straight-chain alkyl group, such as butyl, is often called *n*-butyl (*n* for normal) to emphasize that its four carbon atoms are in a continuous chain. However, if the name does not have a prefix such as "*n*" or "iso," it is assumed that the carbons are in a continuous chain.

The hydrogens in a molecule are also referred to as primary, secondary, and tertiary. **Primary hydrogens** are attached to primary carbons, **secondary hydrogens** are attached to secondary carbons, and **tertiary hydrogens** are attached to tertiary carbons.

primary hydrogens
$CH_3CH_2CH_2CH_2OH$

tertiary hydrogen
CH_3CHCH_2OH
      |
      CH_3

secondary hydrogens
CH_3CH_2CHOH
          |
          CH_3

        CH_3
         |
CH_3COH
         |
        CH_3

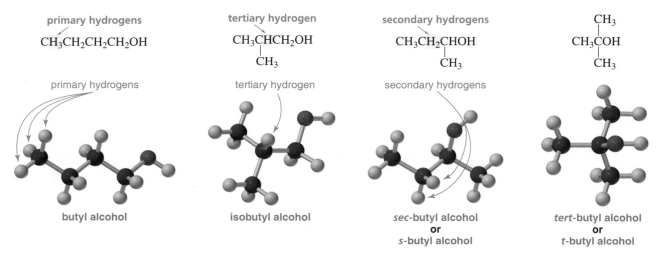

primary hydrogens

butyl alcohol

tertiary hydrogen

isobutyl alcohol

secondary hydrogens

sec-butyl alcohol
or
s-butyl alcohol

tert-butyl alcohol
or
t-butyl alcohol

Because a chemical name must apply to only one compound, the only time that you will see the prefix "*sec*" is in *sec*-butyl. The name "*sec*-pentyl" cannot be used because pentane has two different secondary carbon atoms, which means that there are two different alkyl groups that result from the removal of a hydrogen from a secondary carbon of pentane.

$CH_3CHCH_2CH_2CH_3$
      |
      Cl

$CH_3CH_2CHCH_2CH_3$
          |
          Cl

***sec*-pentyl chloride cannot be used as a name because both of these compounds would be *sec*-pentyl chlorides**

The prefix "*tert*" is found in *tert*-butyl and *tert*-pentyl because each of these substituent names describes only one alkyl group. The name "*tert*-hexyl" cannot be used because it describes two different alkyl groups. Amyl is sometimes used instead of pentyl to designate five carbons.

$$CH_3\overset{\overset{\displaystyle CH_3}{|}}{\underset{\underset{\displaystyle CH_3}{|}}{C}}\!-\!Br$$

tert-butyl bromide

$$CH_3\overset{\overset{\displaystyle CH_3}{|}}{\underset{\underset{\displaystyle CH_2CH_3}{|}}{C}}\!-\!Br$$

tert-pentyl bromide
or
tert-amyl bromide

$$CH_3CH_2\overset{\overset{\displaystyle CH_2CH_3}{|}}{\underset{\underset{\displaystyle CH_3}{|}}{C}}\!-\!Br$$

$$CH_3CH_2CH_2\overset{\overset{\displaystyle CH_3}{|}}{\underset{\underset{\displaystyle CH_3}{|}}{C}}\!-\!Br$$

Both alkyl bromides have six carbon atoms with a bromine attached to a tertiary carbon, so both compounds would be *tert*-hexyl bromides

The prefix "iso" can be used regardless of the number of carbons in the alkyl group. This prefix means that there is an iso structural unit (a carbon bonded to two methyl groups and a hydrogen) at one end of the molecule and a group replacing a hydrogen at the other end.

$$CH_3\overset{\overset{}{\underset{\underset{\displaystyle CH_3}{|}}{C}}H}{}CH_2CH_2OH$$

isopentyl alcohol
or
isoamyl alcohol

$$CH_3\underset{\underset{\displaystyle CH_3}{|}}{C}HCH_2CH_2CH_2Cl$$

isohexyl chloride

$$CH_3\underset{\underset{\displaystyle CH_3}{|}}{C}HCH_2CH_2CH_2CH_2NH_2$$

isoheptylamine

$$CH_3\underset{\underset{\displaystyle CH_3}{|}}{C}HCH_2Br$$

isobutyl bromide

$$CH_3\underset{\underset{\displaystyle CH_3}{|}}{C}HCH_2CH_2CH_2CH_2CH_2OH$$

isooctyl alcohol

$$CH_3\underset{\underset{\displaystyle CH_3}{|}}{C}HBr$$

isopropyl bromide

Notice that *all* isoalkyl compounds have the substituent (OH, Cl, NH$_2$, etc.) on a primary carbon except for isopropyl, which has the substituent on a secondary carbon. The isopropyl group could have been called a *sec*-propyl group. Either name would have been appropriate, since the group has an iso structural unit and a hydrogen has been removed from a secondary carbon. Chemists decided to call it isopropyl, which means that "*sec*" is used only for *sec*-butyl.

Alkyl group names are used so frequently that you should learn them. Some of the most common alkyl group names are compiled in Table 2.2 to make learning them a little easier.

TABLE 2.2   Names of Some Alkyl Groups

methyl	CH_3-	*sec*-butyl	$CH_3CH_2\underset{\underset{\displaystyle CH_3}{|}}{C}H-$	neopentyl	$CH_3\underset{\underset{\displaystyle CH_3}{|}}{\overset{\overset{\displaystyle CH_3}{|}}{C}}CH_2-$
ethyl	CH_3CH_2-				
propyl	$CH_3CH_2CH_2-$	*tert*-butyl	$CH_3\underset{\underset{\displaystyle CH_3}{|}}{\overset{\overset{\displaystyle CH_3}{|}}{C}}-$	hexyl	$CH_3CH_2CH_2CH_2CH_2CH_2-$
isopropyl	$CH_3\underset{\underset{\displaystyle CH_3}{|}}{C}H-$			isohexyl	$CH_3\underset{\underset{\displaystyle CH_3}{|}}{C}HCH_2CH_2CH_2-$
butyl	$CH_3CH_2CH_2CH_2-$	pentyl	$CH_3CH_2CH_2CH_2CH_2-$	heptyl	$CH_3CH_2CH_2CH_2CH_2CH_2CH_2-$
isobutyl	$CH_3\underset{\underset{\displaystyle CH_3}{|}}{C}HCH_2-$	isopentyl	$CH_3\underset{\underset{\displaystyle CH_3}{|}}{C}HCH_2CH_2-$	isoheptyl	$CH_3\underset{\underset{\displaystyle CH_3}{|}}{C}HCH_2CH_2CH_2-$

PROBLEM 1◆

Give a structure for each of the following compounds.

   a. isopropyl alcohol   **d.** neopentyl chloride

   b. isopentyl fluoride   **e.** *tert*-butylamine

   c. *sec*-butyl iodide   **f.** isooctyl bromide

2.2 NOMENCLATURE OF ALKANES

In order to name an alkane by the IUPAC system of nomenclature, the following rules must be followed.

1. Determine the number of carbons in the longest continuous carbon chain. This chain is called the **parent hydrocarbon.** The alkane name derived from the number of carbons in the parent hydrocarbon becomes the alkane's "last name." Notice that the longest continuous chain is chosen regardless of how the molecule is written. Sometimes you have to "turn a corner" in defining the longest continuous chain.

Determine the number of carbons in the longest continuous chain.

$$\overset{8}{C}H_3\overset{7}{C}H_2\overset{6}{C}H_2\overset{5}{C}H_2\overset{4}{C}H\overset{3}{C}H_2\overset{2}{C}H_2\overset{1}{C}H_3 \qquad \overset{8}{C}H_3\overset{7}{C}H_2\overset{6}{C}H_2\overset{5}{C}H_2\overset{4}{C}HCH_3$$

$$\underset{}{|} \qquad\qquad\qquad\qquad\qquad\qquad\qquad\qquad |$$

$$CH_3 \qquad\qquad\qquad\qquad\qquad\qquad\qquad \underset{3}{C}H_2\underset{2}{C}H_2\underset{1}{C}H_3$$

three different ways to draw 4-methyloctane

$$\overset{4}{C}H_3\overset{3}{C}HCH_2\overset{2}{C}H_2\overset{1}{C}H_3$$
$$|$$
$$\underset{5}{C}H_2\underset{6}{C}H_2\underset{7}{C}H_2\underset{8}{C}H_3$$

2. The name of the alkyl substituent that hangs off the parent hydrocarbon is cited before the last name with a number to designate the carbon to which it is attached. The chain is numbered in the direction that gives the substituent the lowest possible number. The substituent name and the "last name" are joined in one word, and there is a hyphen between the number and the substituent name.

Number the chain so that the substituent gets the lowest possible number.

$$\overset{1}{C}H_3\overset{2}{C}HCH_2\overset{3}{C}H_2\overset{4}{C}H_2\overset{5}{C}H_3 \qquad \overset{6}{C}H_3\overset{5}{C}H_2\overset{4}{C}H_2\overset{3}{C}HCH_2\overset{2}{C}H_2\overset{1}{C}H_3$$
$$|\qquad\qquad\qquad\qquad\qquad\qquad |$$
$$CH_3 \qquad\qquad\qquad\qquad\qquad CH_2CH_3$$

      2-methylpentane   3-ethylhexane

$$\overset{1}{C}H_3\overset{2}{C}H_2\overset{3}{C}H_2\overset{4}{C}HCH_2\overset{5}{C}H_2\overset{6}{C}H_2\overset{7}{C}H_2\overset{8}{C}H_3$$
$$|$$
$$CHCH_3$$
$$|$$
$$CH_3$$

          4-isopropyloctane

Notice that only IUPAC names have numbers; common names never contain numbers.

$$CH_3$$
$$|$$
$$CH_3CHCH_2CH_2CH_3$$

   common name: isohexane
   IUPAC name: 2-methylpentane

3. If more than one substituent is attached to the longest continuous chain, the chain is numbered in the direction that will result in the lowest possible number in the name of the compound. The substituents are listed in alphabetical order (not numerical order), with each substituent getting its own number. In the following example, the correct name (5-ethyl-3-methyloctane) contains a 3 as its lowest number, while the incorrect name (4-ethyl-6-methyloctane) contains a 4 as its lowest number.

List substituents in alphabetical order.

CH₃CH₂CHCH₂CHCH₂CH₂CH₃
| |
CH₃   CH₂CH₃
5-ethyl-3-methyloctane
not
4-ethyl-6-methyloctane
because 3 < 4

If two or more substituents are the same, the prefixes di, tri, and tetra are used with the substituent names. The numbers that indicate the locations of the identical substituents are cited together, separated by commas.

CH₃CH₂CHCH₂CHCH₃
| |
CH₃   CH₃
2,4-dimethylhexane

CH₃ CH₃
| |
CH₃CH₂CH₂C—CCH₂CH₃
| |
CH₃ CH₃
3,3,4,4-tetramethylheptane

The prefixes di, tri, tetra, *sec,* and *tert* are ignored in alphabetizing substituent groups, but the prefixes iso, neo, and cyclo are not ignored.

di, tri, tetra, *sec,* and *tert* are ignored in alphabetizing.

CH₂CH₃   CH₃
| |
CH₃CH₂CCH₂CH₂CHCHCH₂CH₂CH₃
| |
CH₂CH₃ CH₂CH₃
3,3,6-triethyl-7-methyldecane

CH₃
|
CH₃CH₂CH₂CHCH₂CH₂CHCH₃
|
CHCH₃
|
CH₃
5-isopropyl-2-methyloctane

iso, neo, and cyclo are not ignored in alphabetizing.

4. When both directions lead to the same lowest number for one of the substituents, the direction is chosen that gives the lowest number to one of the remaining substituents.

CH₃
|
CH₃CCH₂CHCH₃
| |
CH₃ CH₃
2,2,4-trimethylpentane
not
2,4,4-trimethylpentane
because 2 < 4

CH₂CH₃
|
CH₃CHCHCH₂CH₂CHCH₃
| |
CH₃   CH₃
3-ethyl-2,6-dimethylheptane
not
5-ethyl-2,6-dimethylheptane
because 3 < 5

5. If the same substituent numbers are obtained in both directions, the first cited group receives the lower number.

Only if the same set of numbers is obtained in both directions does the first cited group get the lowest number.

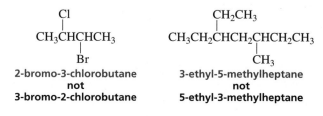

Cl
|
CH₃CHCHCH₃
|
Br
2-bromo-3-chlorobutane
not
3-bromo-2-chlorobutane

CH₂CH₃
|
CH₃CH₂CHCH₂CHCH₂CH₃
|
CH₃
3-ethyl-5-methylheptane
not
5-ethyl-3-methylheptane

6. If a compound has two or more chains of the same length, the parent hydrocarbon is the chain with the greatest number of substituents.

In the case of two hydrocarbon chains with the same number of carbons, choose the one with the most substituents.

$$\overset{3}{C}H_3CH_2\overset{4}{C}HCH_2\overset{5}{C}H_2\overset{6}{C}H_3$$
$$|$$
$$\overset{2}{C}HCH_3$$
$$|$$
$$\overset{1}{C}H_3$$

3-ethyl-2-methylhexane **(two substituents)**

$$\overset{1}{C}H_3\overset{2}{C}H_2\overset{3}{C}HCH_2\overset{4}{C}H_2\overset{5}{C}H_2\overset{6}{C}H_3$$
$$|$$
$$CHCH_3$$
$$|$$
$$CH_3$$

not
3-isopropylhexane **(one substituent)**

7. Names such as isopropyl, *sec*-butyl, and *tert*-butyl are acceptable substituent names in the IUPAC system of nomenclature, but systematic substituent names are preferable. Systematic substituent names are obtained by numbering the substituent starting at the carbon that is attached to the parent hydrocarbon. This means that the carbon that is attached to the parent hydrocarbon is always the number 1 carbon of the substituent. In a compound such as 4-(1-methylethyl)-octane, the substituent is in parentheses; the number inside the parentheses indicates a position on the substituent, whereas the number outside the parentheses indicates a position on the parent hydrocarbon.

$$CH_3CH_2CH_2CH_2CHCH_2CH_2CH_3$$
$$|$$
$$\overset{1}{C}H\overset{2}{C}H_3$$
$$|$$
$$CH_3$$

4-isopropyloctane
or
4-(1-methylethyl)octane

$$CH_3CH_2CH_2CH_2CHCH_2CH_2CH_2CH_3$$
$$|$$
$$\overset{1}{C}H_2\overset{2}{C}H\overset{3}{C}H_3$$
$$|$$
$$CH_3$$

5-isobutyldecane
or
5-(2-methylpropyl)decane

Some substituents have only a systematic name.

$$CH_2CH_2CH_3$$
$$|$$
$$CH_3CH_2CH_2CH_2CCH_2CH_2CH_2CH_3$$
$$|$$
$$CH_3\overset{1}{C}H\overset{2}{C}H\overset{3}{C}H_3$$
$$|$$
$$CH_3$$

5-(1,2-dimethylpropyl)-5-propyldecane

Numbers are used only for IUPAC names and never for common names.

These rules will allow you to name thousands of alkanes. Since all compounds can be named by an IUPAC name, you may wonder why you must learn common names. Common names have been in existence for so long and are so well entrenched in chemists' vocabulary that they are widely used in scientific conversation and are often found in the literature. Of course, the IUPAC names must be learned because the rules enable you to name thousands of compounds; if you ever want to look up a compound in the scientific literature, it will usually be listed by its IUPAC name. Now you should study the IUPAC names shown for the isomeric hexanes and isomeric heptanes at the beginning of this chapter.

A number and a word are separated by a hyphen; numbers are separated by a comma.

PROBLEM 2

Give the structure for each of the following compounds.

a. 2,3-dimethylhexane

b. 4-isopropyl-2,4,5-trimethylheptane

c. 4,4-diethyldecane

d. 2,2-dimethyl-4-propyloctane

e. 4-isobutyl-2,5-dimethyloctane

f. 4-(1,1-dimethylethyl)octane

PROBLEM 3 / SOLVED

  a. Draw the eighteen isomeric octanes.

  b. Name each isomer by the IUPAC system.

  c. How many isomers have common names?

  d. Which isomers contain an isopropyl group?

  e. Which isomers contain a *sec*-butyl group?

  f. Which isomers contain a *tert*-butyl group?

SOLUTION TO 3a   Start with the isomer with an 8-carbon continuous chain, then list isomers with a 7-carbon continuous chain plus one methyl group, then isomers with a 6-carbon continuous chain plus two methyl groups or one ethyl group, then isomers with a 5-carbon continuous chain plus three methyl groups or one methyl group and one ethyl group, then a 4-carbon continuous chain with four methyl groups. (You will be able to tell if you list duplicate structures by your answers to **3b,** because if two structures have the same IUPAC name they are the same compound.)

PROBLEM 4

The IUPAC name for isobutane is methylpropane. Why does this name not contain a number?

PROBLEM 5◆

Give the systematic name for each of the following compounds.

  a. 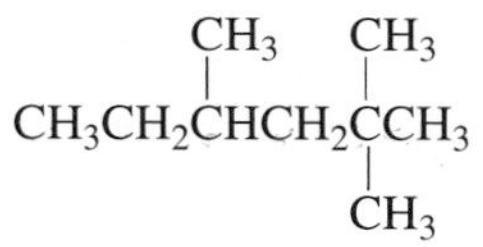
$$CH_3CH_2CHCH_2CCH_3$$
with CH_3 and CH_3 on the CH and C, and CH_3 below.

  b. $CH_3CH_2C(CH_3)_3$

  c. 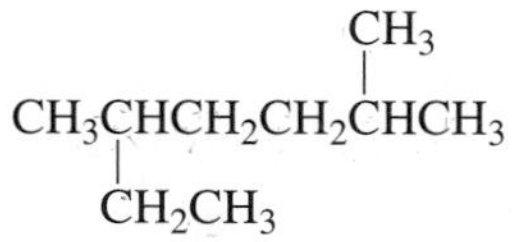 $CH_3CHCH_2CH_2CHCH_3$ with CH_3 above and CH_2CH_3 below

  d. $CH_3CH_2C(CH_2CH_3)_2CH_2CH_2CH_3$

  e. $CH_3CH_2C(CH_2CH_3)_2CH(CH_3)CH(CH_2CH_2CH_3)_2$

  f. 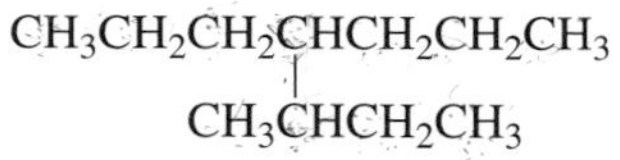 $CH_3CH_2CH_2CHCH_2CH_2CH_3$ with $CH_3CHCH_2CH_3$ below

  g. $CH_3C-CHCH_2CH_3$ with $CH_3CH_2CH_2CH_3$ above and $CH_2CH_2CH_3$ below

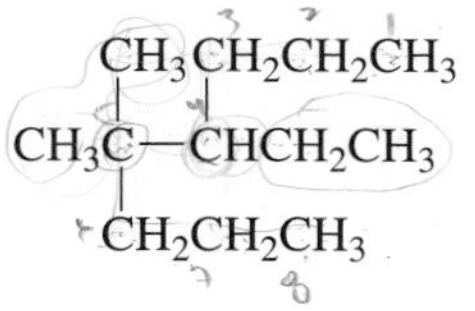

  h. $CH_3CH_2CH_2CH_2CHCH_2CH_2CH_3$ with $CH(CH_3)_2$ below

---

Cycloalkanes are alkanes with their carbons arranged in a ring. Because of the ring, a cycloalkane has two fewer hydrogens than a noncyclic alkane with the same number of carbons. This means that the general molecular formula for a cycloalkane is C_nH_{2n}. Cycloalkanes are named by adding the prefix "cyclo" to the alkane name that signifies the number of carbons in the ring.

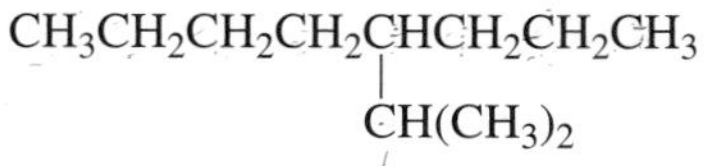

**2.3
NOMENCLATURE OF
CYCLOALKANES**

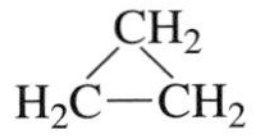

cyclopropane 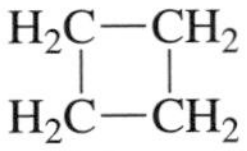   cyclobutane 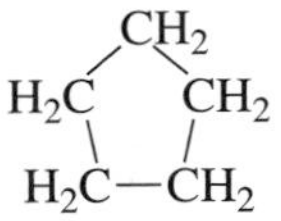   cyclopentane 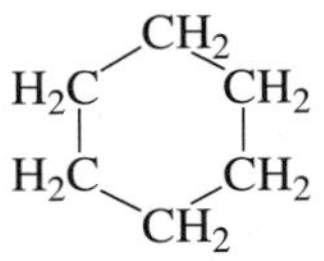   cyclohexane

Cycloalkanes are almost always written using **skeletal structures.** Skeletal structures show the carbon–carbon bonds as lines but do not show the carbon–hydrogen bonds. Each vertex in a skeletal structure represents a carbon. It is understood that each carbon is bonded to the appropriate number of hydrogens to give it four bonds.

cyclopropane 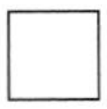   cyclobutane 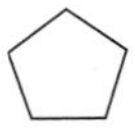   cyclopentane 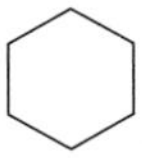   cyclohexane

Noncyclic molecules can also be represented by skeletal structures. In a skeletal structure, the carbon chains are represented by zigzag lines. Again, each vertex represents a carbon, and carbons occupy the ends of the chain.

octane 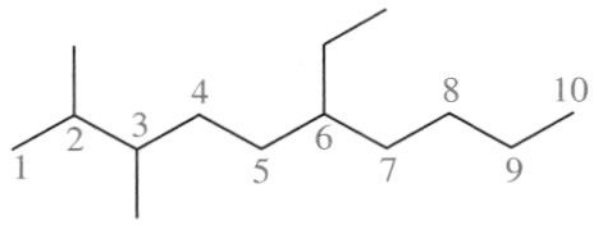   6-ethyl-2,3-dimethyldecane 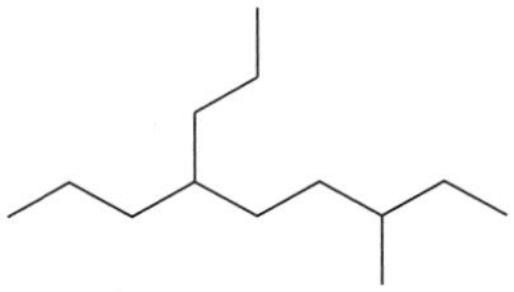   3-methyl-6-propylnonane

Rules for naming cycloalkanes resemble the rules for naming noncyclic alkanes.

1. In the case of a cycloalkane with an attached alkyl substituent, the ring is the parent hydrocarbon unless the substituent has more carbons than the ring. In that case, the substituent is the parent hydrocarbon and the ring is named as a substituent.

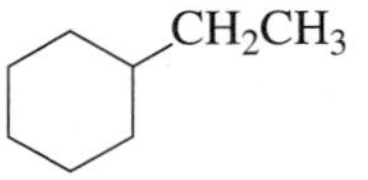

ethylcyclohexane 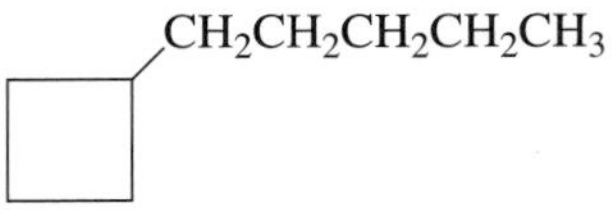   1-cyclobutylpentane

2. If there is more than one substituent on the ring, the substituents are cited in alphabetical order. One of the substituents is given the number 1 position and the ring is numbered from that position in a direction (either clockwise or counterclockwise) that gives a second substituent the lowest possible number.

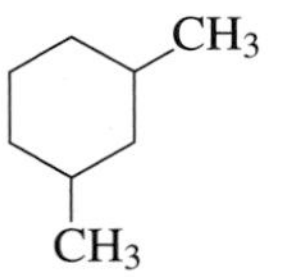

1,3-dimethylcyclohexane 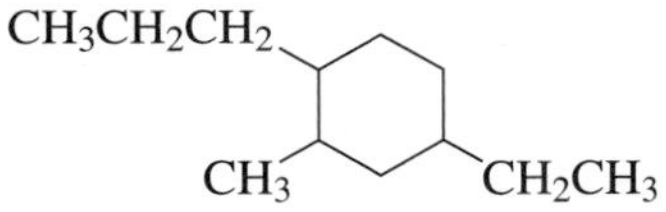   4-ethyl-2-methyl-1-propylcyclohextane   1,1,2-trimethylcyclopentane
    not   not
    1-ethyl-3-methyl-4-propylcyclohexane   1,2,2-trimethylcyclopentane
    because 2 < 3   because 1 < 2

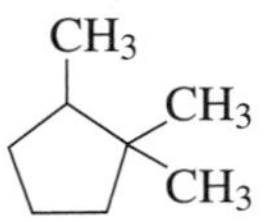

3. If the ring has only two substituents and they are different, the substituents are cited in alphabetical order and the number 1 position is given to the first cited substituent.

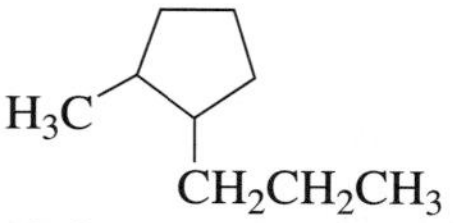

1-methyl-2-propylcyclopentane

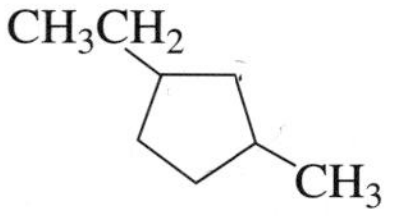

1-ethyl-3-methylcyclopentane

PROBLEM 6

Convert the following condensed structures into skeletal structures. (Remember that condensed structures show atoms but few, if any, bonds; skeletal structures show bonds but few, if any, atoms.)

a. $CH_3CH_2CH_2CH_2CH_2CH_2OH$

c. 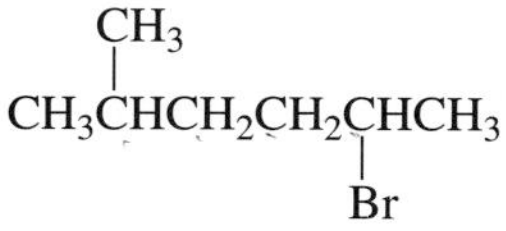
$$CH_3$$
$$CH_3CHCH_2CH_2CHCH_3$$
$$Br$$

b. 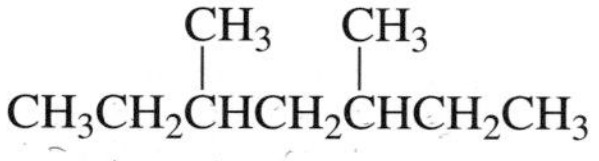
$$CH_3 \quad CH_3$$
$$CH_3CH_2CHCH_2CHCH_2CH_3$$

d. $CH_3CH_2CH_2CH_2OCH_3$

PROBLEM 7 ◆

Give the IUPAC name for each of the following compounds.

a.

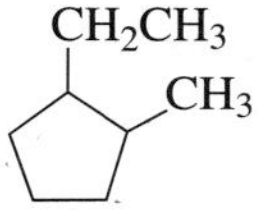

e.  $CH_3CHCH_2CH_2CH_3$

b.

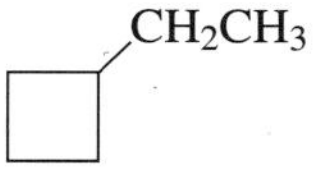

f.

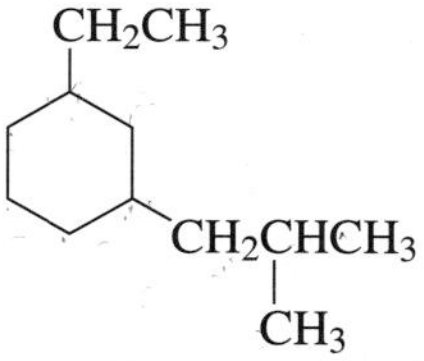

c.

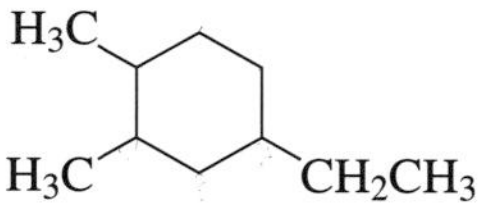

g.

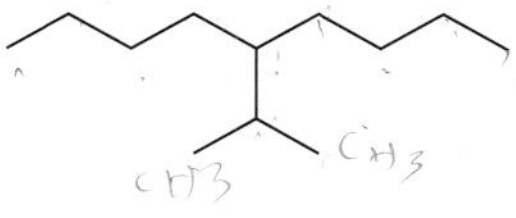

d.

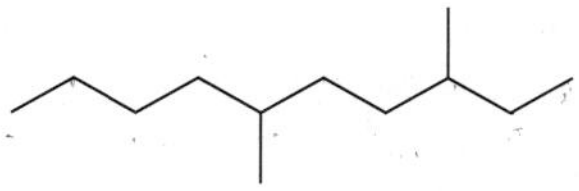

h.

Alkyl halides are compounds in which a hydrogen of an alkane has been replaced by a halogen. Alkyl halides are classified as primary, secondary, or tertiary depending on the carbon to which the halogen is attached. The halogen is bonded to a primary carbon (a carbon bonded to one carbon) in a **primary alkyl halide,** to a secondary carbon (a carbon bonded to two carbons) in a **secondary alkyl halide,** and to a tertiary carbon (a carbon bonded to three carbons) in a **tertiary alkyl**

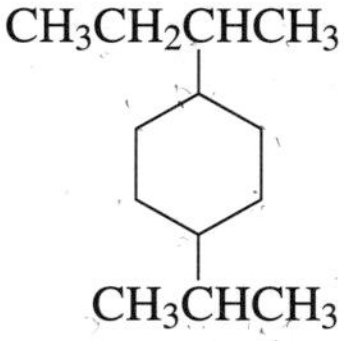

2.4

NOMENCLATURE OF ALKYL HALIDES

halide. Nonbonding electrons are generally not shown unless they are needed to draw your attention to some chemical property of the atom.

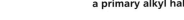

a primary carbon   a secondary carbon   a tertiary carbon

$$R-CH_2-Br \qquad\qquad R-\underset{\underset{Br}{|}}{CH}-R \qquad\qquad R-\underset{\underset{Br}{|}}{\overset{\overset{R}{|}}{C}}-R$$

a primary alkyl halide   a secondary alkyl halide   a tertiary alkyl halide

The common names of the alkyl halides are obtained by citing the name of the alkyl group followed by the name of the halogen.

$$CH_3Cl \qquad CH_3CH_2F \qquad CH_3\underset{\underset{CH_3}{|}}{CHI} \qquad CH_3CH_2\underset{\underset{CH_3}{|}}{CHBr}$$

methyl chloride   ethyl fluoride

isopropyl iodide   *sec*-butyl bromide

In the IUPAC system, alkyl halides are called haloalkanes. They are named as substituted alkanes. The substituent prefix names for the halogens end in "o" (fluoro, chloro, bromo, iodo).

$$CH_3CH_2\underset{\underset{Br}{|}}{\overset{\overset{CH_3}{|}}{CH}}CH_2CH_2CHCH_3 \qquad CH_3\underset{\underset{CH_3}{|}}{\overset{\overset{CH_3}{|}}{C}}CH_2CH_2CH_2CH_2Cl$$

2-bromo-5-methylheptane   1-chloro-5,5-dimethylhexane

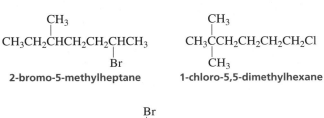

4-bromo-2-chloro-1-methylcyclohexane

CH₃F
methyl fluoride

CH₃Cl
methyl chloride

CH₃Br
methyl bromide

CH₃I
methyl iodide

PROBLEM 8

Draw **a–c,** by substituting a chlorine for a hydrogen of methylcyclohexane.

a. a primary alkyl halide   **c.** a tertiary alkyl halide

b. three different secondary alkyl halides   **d.** name each of the alkyl halides

PROBLEM 9 ◆

Give two names for each of the following compounds, and tell whether each alkyl halide is primary, secondary, or tertiary.

a. $CH_3CH_2\underset{\underset{Cl}{|}}{CH}CH_3$ **c.** $\overset{\overset{\displaystyle Br}{|}}{\bigcirc}$

b. $CH_3\underset{\underset{CH_3}{|}}{CH}CH_2CH_2CH_2CH_2Cl$ **d.** $CH_3\underset{\underset{F}{|}}{CH}CH_3$

PROBLEM 10◆

Give the structure and the IUPAC name of a compound with a molecular formula C_5H_{12} that has (**a**) only primary and secondary hydrogens, (**b**) only primary hydrogens, and (**c**) one tertiary hydrogen.

2.5 NOMENCLATURE OF ETHERS

Ethers are compounds in which an oxygen is bonded to two alkyl substituents. If the alkyl substituents are identical, the ether is a **symmetrical ether.** If they are not identical, the ether is an **asymmetrical ether.**

R—O—R   R—O—R′
a symmetrical ether   **an asymmetrical ether**

The common name of an ether is obtained by citing the names of the two alkyl substituents (in alphabetical order) followed by the word "ether." The smallest ethers are almost always named by their common names.

Chemists sometimes neglect the prefix "di" when they name symmetrical ethers. Try not to do this.

$CH_3OCH_2CH_3$
ethyl methyl ether

$CH_3CH_2OCH_2CH_3$
diethyl ether
often called ethyl ether

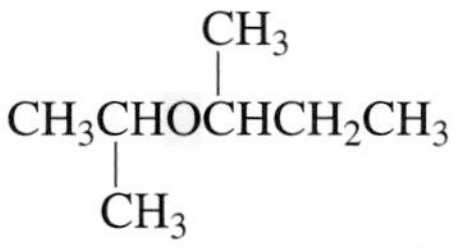

CH₃
|
$CH_3CHOCHCH_2CH_3$
|
CH₃
sec-butyl isopropyl ether

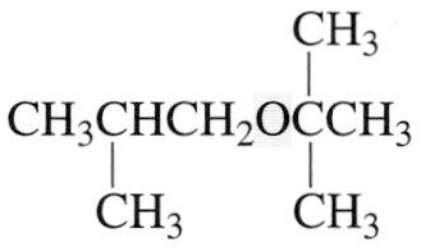

CH₃
|
$CH_3CHCH_2OCCH_3$
| |
CH₃   CH₃
tert-butyl isobutyl ether

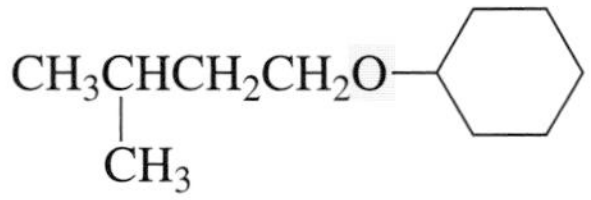

$CH_3CHCH_2CH_2O$—⬡
|
CH₃
cyclohexyl isopentyl ether

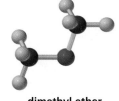

dimethyl ether

The IUPAC system names an ether as an alkane that bears an alkoxy substituent. Alkoxy substituents are named by removing the "yl" from the name of the alkyl substituent and adding "oxy."

CH_3O—
methoxy

CH_3CH_2O—
ethoxy

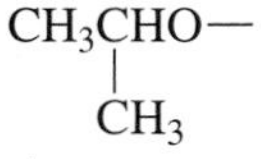

CH_3CHO—
|
CH₃
isopropoxy

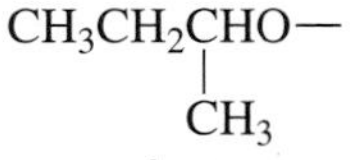

CH_3CH_2CHO—
|
CH₃
sec-butoxy

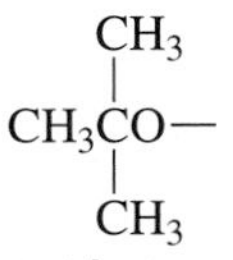

CH₃
|
CH_3CO—
|
CH₃
tert-butoxy

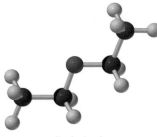

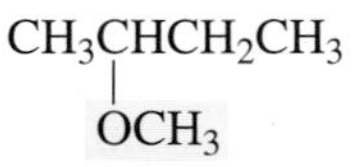

diethyl ether

$CH_3CHCH_2CH_3$
|
OCH₃
2-methoxybutane

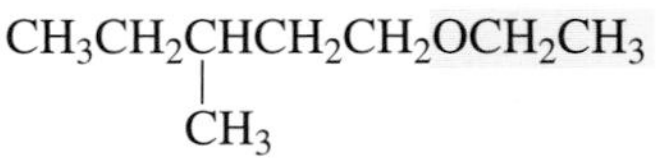

$CH_3CH_2CHCH_2CH_2OCH_2CH_3$
|
CH₃
1-ethoxy-3-methylpentane

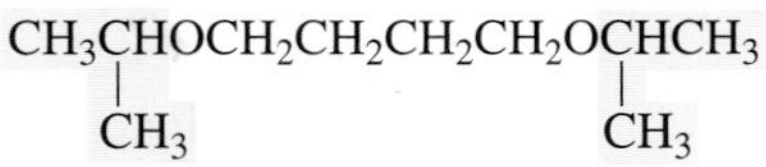

$CH_3CHOCH_2CH_2CH_2CH_2OCHCH_3$
| |
CH₃   CH₃
1,4-diisopropoxybutane

PROBLEM 11

(1) Name the following ethers by the IUPAC system. (2) Do all of these ethers have common names? (3) What are their common names?

a. $CH_3OCH_2CH_3$

b. $CH_3CH_2OCH_2CH_3$

c. $CH_3CH_2CH_2CH_2CHCH_2CH_2CH_3$
$\qquad\qquad\qquad\qquad |$
$\qquad\qquad\qquad\quad\ \, OCH_3$

d. $CH_3CH_2CH_2OCH_2CH_2CH_2CH_3$

e. $CH_3CHOCHCH_2CH_2CH_3$
$\qquad\quad |\qquad\ \ |$
$\qquad\ CH_3\qquad CH_3$
with CH_3 above

f. $CH_3CHOCH_2CH_2CHCH_3$
$\qquad\ |\qquad\qquad\quad |$
$\qquad CH_3\qquad\qquad\ CH_3$

2.6
NOMENCLATURE OF ALCOHOLS

Alcohols are compounds in which a hydrogen of an alkane has been replaced by an OH. Like alkyl halides, alcohols are classified as **primary alcohols, secondary alcohols,** or **tertiary alcohols** depending on whether the OH substituent is bonded to a primary, secondary, or tertiary carbon.

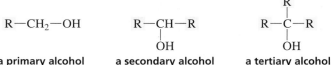

$$R-CH_2-OH \qquad R-\underset{\underset{OH}{|}}{CH}-R \qquad R-\underset{\underset{OH}{|}}{\overset{\overset{R}{|}}{C}}-R$$

a primary alcohol a secondary alcohol a tertiary alcohol

The common name of an alcohol is obtained by citing the name of the alkyl group to which the OH group is attached, followed by the word "alcohol."

methyl alcohol

ethyl alcohol

$$CH_3CH_2OH \qquad CH_3CH_2CH_2OH \qquad CH_3\underset{\underset{CH_3}{|}}{CH}OH \qquad CH_3\underset{\underset{CH_3}{|}}{\overset{\overset{CH_3}{|}}{C}}CH_2OH$$

ethyl alcohol   propyl alcohol   isopropyl alcohol   neopentyl alcohol

The **functional group** is the center of reactivity in a molecule. In an alcohol molecule, the OH is the functional group. The IUPAC system uses a suffix to denote a functional group. The IUPAC name of an alcohol, for example, is obtained by removing the "e" from the name of the alkane and adding the suffix "ol."

$$CH_3OH \qquad CH_3CH_2OH$$
methanol   ethanol

When necessary, the position of the functional group is indicated by a number immediately preceding the name of the parent alcohol. In naming a compound that has a functional group suffix, two points are important.

1. The parent hydrocarbon is the longest continuous chain *containing the functional group.*

propyl alcohol

2. The chain is numbered in the direction that gives the *functional group the lowest possible number.*

1 2 3 4
$CH_3CH_2CHCH_3$
|
OH
2-butanol

5 4 3 2 1
$CH_3CH_2CH_2CHCH_2OH$
|
CH_2CH_3
2-ethyl-1-pentanol

3 2 1
$CH_3CH_2CH_2CH_2OCH_2CH_2CH_2OH$
3-butoxy-1-propanol

↑

the longest continuous chain has six carbon atoms, but the longest continuous chain containing the OH functional group has five carbon atoms, so the compound is named as a pentanol

the longest continuous chain has four carbon atoms, but the longest continuous chain containing the OH functional group has three carbon atoms, so the compound is named as a propanol

3. If there is a functional group and a substituent, the functional group gets the lowest possible number. A number is not needed to designate the position of a functional group in a cyclic compound since it is assumed to be at the 1-position.

1 2 3 4 5
$HOCH_2CH_2CH_2CH_2CH_2Br$
5-bromo-1-pentanol

4 3 2 1
$ClCH_2CH_2CHCH_3$
|
OH
4-chloro-2-butanol

CH_3

3-methylcyclohexanol

4. If the same number for the functional group is obtained in both directions, the chain is numbered in the direction that gives a substituent the lowest possible number.

$CH_3CHCHCH_2CH_3$
| |
Cl OH

2-chloro-3-pentanol
not
4-chloro-3-pentanol

$CH_3CH_2CH_2CHCH_2CHCH_3$
| |
OH CH_3

2-methyl-4-heptanol
not
6-methyl-4-heptanol

When there is only a substituent, the substituent gets the lowest number.

When there is only a functional group, the functional group gets the lowest number.

5. If there is more than one substituent, the substituents are cited in alphabetical order.

CH_2CH_3
|
$CH_3CHCH_2CHCH_2CHCH_3$
| |
Br OH
6-bromo-4-ethyl-2-heptanol

CH_2CH_3

H_3C OH
2-ethyl-5-methylcyclohexanol

HO CH_3

CH_3
3,4-dimethylcyclopentanol

When there is a functional group and a substituent, the functional group gets the lowest number.

Remember that the name of a substituent is stated *before* the name of the parent hydrocarbon and the functional group ending is stated *after* the name of the parent hydrocarbon.

[Substituent]   [Parent Hydrocarbon]   [Functional Group]

PROBLEM 12◆

Give the IUPAC name for each of the following compounds, and tell whether each of the alcohols is primary, secondary, or tertiary.

PROBLEM 12◆

Give the IUPAC name for each of the following compounds, and tell whether each of the alcohols is primary, secondary, or tertiary.

a. $CH_3CH_2CH_2CH_2CH_2CH_2OH$

b. $CH_3CHCH_2CHCH_2CH_3$
$\quad\quad\ \ |\quad\quad\ \ |$
$\quad\quad CH_3\quad OH$

c. $CH_3CCH_2CH_2CH_2Cl$
with CH_3 above the central C and OH below it

d. $CH_3CHCH_2CHCH_2CH_2CH_3$
$\quad\quad\ \ |\quad\quad\ \ |$
$\quad\quad CH_3\quad OH$

e.
A cyclohexane ring with OH at the top, CH_2CH_3 on the lower right, and Cl at the bottom.

PROBLEM 13◆

Give the structures of all the tertiary alcohols with molecular formula $C_6H_{14}O$. Name each one by the IUPAC system.

2.7 NOMENCLATURE OF AMINES

Amines are compounds in which one or more hydrogens of ammonia have been replaced by alkyl groups. Smaller amines are characterized by their fishy odors. For example, fermented shark, a traditional dish of Iceland, smells exactly like triethylamine. There are **primary amines, secondary amines,** and **tertiary amines.** The classification depends on how many alkyl groups are bonded to the nitrogen. Primary amines have one alkyl group bonded to the nitrogen, secondary amines have two, and tertiary amines have three.

$$R—NH_2 \quad\quad\quad R—NH—R \quad\quad\quad R—\underset{\underset{R}{|}}{\overset{\overset{R}{|}}{N}}—R$$

a primary amine   **a secondary amine**   **a tertiary amine**

Notice the difference between primary, secondary, and tertiary amines and primary, secondary, and tertiary alkyl halides or alcohols. The number of alkyl groups attached to the nitrogen determines the classification of an amine; the number of alkyl groups attached to the carbon to which the halogen or the OH is bonded determines the classification of an alkyl halide or an alcohol.

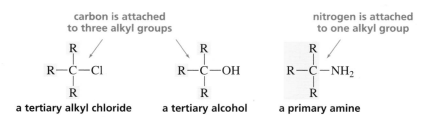

carbon is attached to three alkyl groups   nitrogen is attached to one alkyl group

a tertiary alkyl chloride   **a tertiary alcohol**   **a primary amine**

Remember that the prefixes di, tri, tetra, *sec*, and *tert* are ignored in alphabetizing but the prefixes iso, neo, and cyclo are not ignored.

The common name of an amine is obtained by citing the names of the alkyl groups bonded to the nitrogen in alphabetical order followed by "amine." The entire name is written as one word (unlike the names of alcohols, ethers, and alkyl halides, in which "alcohol," "ether," or "halide" are separate words).

CH_3NH_2 $CH_3NHCH_2CH_2CH_3$ $CH_3CH_2NHCH_2CH_3$
methylamine   methylpropylamine   diethylamine

$\overset{\displaystyle CH_3}{\underset{\displaystyle |}{CH_3NCH_3}}$ $\overset{\displaystyle CH_3}{\underset{\displaystyle |}{CH_3NCH_2CH_2CH_2CH_3}}$ $\overset{\displaystyle CH_3}{\underset{\displaystyle |}{CH_3CH_2NCH_2CH_2CH_3}}$
trimethylamine   butyldimethylamine   ethylmethylpropylamine

The IUPAC system uses a suffix to denote the amine functional group. This method is similar to the way in which alcohols are named. The "e" at the end of the alkane name for the longest continuous carbon chain in the amine is replaced by "amine." A number is used to signify the carbon in the longest continuous chain to which the nitrogen is attached. The name of any other alkyl group bonded to nitrogen is preceded by an "*N*" (in italic) to indicate that the group is bonded to a nitrogen rather than to a carbon.

$\overset{4\quad3\quad2\quad1}{CH_3CH_2CH_2CH_2NH_2}$ $\overset{1\quad2\quad3\quad4\quad5\quad6}{CH_3CH_2CHCH_2CH_2CH_3}$ $\overset{3\quad2\quad1}{CH_3CH_2CH_2NCH_2CH_3}$
1-butanamine $\underset{NHCH_2CH_3}{|}$ $\underset{CH_3}{|}$
          N-ethyl-3-hexanamine   *N*-ethyl-*N*-methyl-1-propanamine

All substituents, whether they are attached to the nitrogen or to the parent hydrocarbon, are listed in alphabetical order. The chain is numbered such that the functional group gets the lowest number.

$\overset{4\quad3\quad2\quad1}{CH_3CHCH_2CH_2NHCH_3}$ $\overset{1\quad2\quad3\quad4\quad5\,|\,6}{CH_3CH_2CHCH_2CHCH_3}$
$\underset{Cl}{|}$ $\underset{NHCH_2CH_3}{|}$
3-chloro-*N*-methyl-1-butanamine   *N*-ethyl-5-methyl-3-hexanamine

$\overset{Br}{\overset{5\ \ 4|\ \ 3\ \ 2\ \ 1}{CH_3CHCH_2CHCH_3}}$ $-NHCH_2CH_2CH_3$
$\underset{CH_3NCH_3}{|}$
4-bromo-*N,N*-dimethyl-2-pentanamine   2-ethyl-*N*-propylcyclohexanamine

Nitrogen compounds with four alkyl groups bonded to the nitrogen—thereby placing a positive charge on the nitrogen—are called **quaternary ammonium salts.** Their names are obtained by citing the names of the alkyl groups in alphabetical order as a prefix to "ammonium," followed by the name of the counter-ion.

$CH_3-\overset{\displaystyle CH_3}{\underset{\displaystyle CH_3}{\overset{|}{\underset{|}{N^+}}}}-CH_3\ \ HO^-$ $CH_3CH_2CH_2-\overset{\displaystyle CH_3}{\underset{\displaystyle CH_2CH_3}{\overset{|}{\underset{|}{N^+}}}}-CH_3\ \ Cl^-$
tetramethylammonium hydroxide   ethyldimethylpropylammonium chloride

Table 2.3 summarizes the ways in which alkyl halides, ethers, alcohols, and amines are named.

You can find a discussion of elemental analysis— how the relative proportions of the elements present in a compound are determined—in the **Solutions Manual/Study Guide.**

	IUPAC Name	Common Name
TABLE 2.3	**Nomenclature Summary**	
alkyl halide	substituted alkane	alkyl group to which halogen is attached plus halide
	CH_3Br bromomethane	CH_3Br methyl bromide
	CH_3CH_2Cl chloroethane	CH_3CH_2Cl ethyl chloride
ether	substituted alkane	alkyl groups attached to oxygen plus ether
	CH_3OCH_3 methoxymethane	CH_3OCH_3 dimethyl ether
	$CH_3CH_2OCH_3$ methoxyethane	$CH_3CH_2OCH_3$ ethyl methyl ether
alcohol	functional group suffix is ol	alkyl group to which OH is attached plus alcohol
	CH_3OH methanol	CH_3OH methyl alcohol
	CH_3CH_2OH ethanol	CH_3CH_2OH ethyl alcohol
amine	functional group suffix is amine	alkyl groups attached to N plus amine
	$CH_3CH_2NH_2$ ethanamine	$CH_3CH_2NH_2$ ethylamine
	$CH_3CH_2CH_2NHCH_3$ *N*-methyl-1-propanamine	$CH_3CH_2CH_2NHCH_3$ methylpropylamine

PROBLEM 14◆

Give common and IUPAC names for each of the following compounds.

a. $CH_3CH_2CH_2CH_2CH_2CH_2NH_2$

b. $CH_3CH_2CH_2NHCH_2CH_2CH_2CH_3$

c. $CH_3CHCH_2NHCHCH_2CH_3$
$\quad\quad\ \ |\quad\quad\quad\ |$
$\quad\quad\ CH_3\quad\quad CH_3$

d. $CH_3CH_2CH_2NCH_2CH_3$
$\quad\quad\quad\quad\ \ \ |$
$\quad\quad\quad\quad\ \ CH_2CH_3$

e. 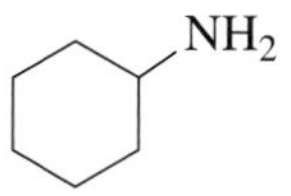 NH_2

PROBLEM 15

Give the structure for each of the following compounds.

a. 2-methyl-*N*-propyl-1-propanamine

b. *N*-ethylethanamine

c. 5-methyl-1-hexanamine

d. methyldipropylamine

e. *N,N*-dimethyl-3-pentanamine

f. cyclohexylethylmethylamine

PROBLEM 16◆

Give the IUPAC name and the common name (for those that have common names) for each of the following compounds. Indicate whether the amines are primary, secondary, or tertiary.

a. $CH_3CHCH_2CH_2CH_2CH_2CH_2NH_2$
$\quad\quad\ \ |$
$\quad\quad\ CH_3$

b. $CH_3CH_2CH_2NHCH_2CH_2CHCH_3$
$\quad\quad\quad\quad\quad\quad\quad\quad\quad |$
$\quad\quad\quad\quad\quad\quad\quad\quad CH_3$

c. $(CH_3CH_2)_2NCH_3$

d.

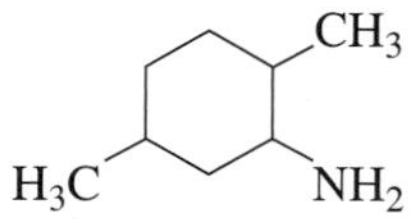

The carbon–halogen bond in an alkyl halide is formed as a result of the overlap of an sp^3 orbital of carbon with a p orbital of the halogen (Section 1.13). Fluorine uses a $2p$ orbital, chlorine uses a $3p$ orbital, bromine uses a $4p$ orbital, and iodine uses a $5p$ orbital. Consequently, the carbon–halogen bond becomes longer and weaker as the atomic weight of the halogen increases (Table 2.4). Notice that this is the same trend shown by the hydrogen–halogen bond (Table 1.5).

An alcohol has the same geometry as water (Section 1.12). An alcohol molecule can be thought of as a water molecule with an alkyl group in place of one of the hydrogens. As in water, the oxygen atom in an alcohol is sp^3 hybridized. One of the sp^3 orbitals of oxygen overlaps with an sp^3 orbital of a carbon, one sp^3 orbital overlaps with an s orbital of hydrogen, and the other two sp^3 orbitals each contain a pair of nonbonding electrons.

2.8
STRUCTURES OF ALKYL HALIDES, ALCOHOLS, ETHERS, AND AMINES

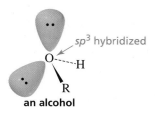

*sp*³ hybridized

an alcohol

TABLE 2.4	Carbon–Halogen Bond Lengths and Bond Strengths				
	Orbital interactions	Bond lengths	Bond length Å	Bond strength ($DH°$) kcal/mol	kJ/mol
$H_3C—F$	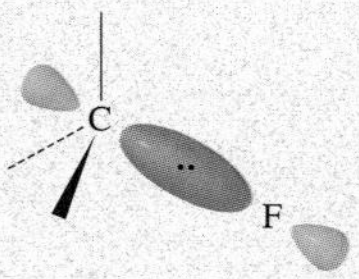	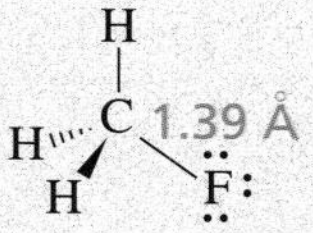 1.39 Å	1.39	108	451
$H_3C—Cl$		1.78 Å	1.78	84	349
$H_3C—Br$	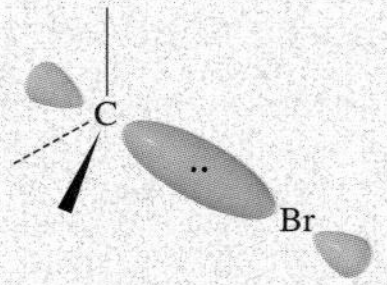	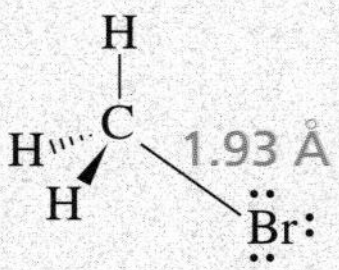 1.93 Å	1.93	70	293
$H_3C—I$		2.14 Å	2.14	56	234

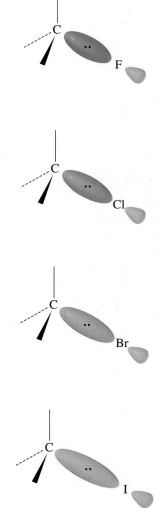

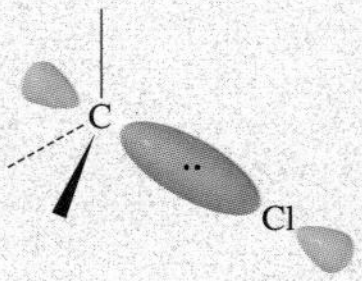

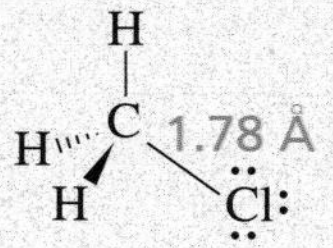

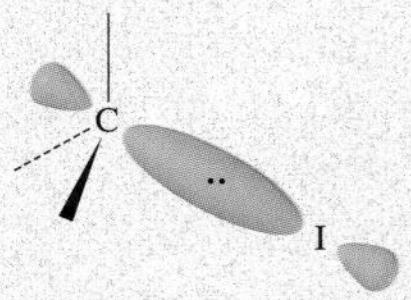

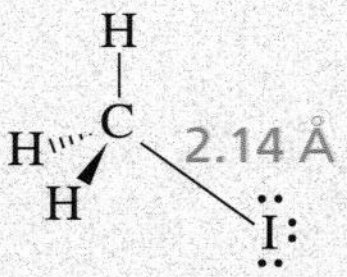

Ethers also have the same geometry as water. An ether molecule can be thought of as a water molecule with alkyl groups in place of both of the hydrogens.

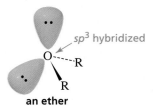

an ether

An amine has the same geometry as ammonia (Section 1.13). One, two, or three hydrogens may be replaced by alkyl groups. The number of hydrogens replaced by alkyl groups determines whether the amine is called primary, secondary, or tertiary (Section 2.7).

a primary amine   **a secondary amine**   **a tertiary amine**

PROBLEM 17◆

Predict the size of each of the following angles.

a. the C—O—C bond angle in an ether

b. the C—N—C bond angle in a secondary amine

c. the C—O—H bond angle in an alcohol.

d. the C—N—C bond angle in a quaternary ammonium salt

2.9
PHYSICAL PROPERTIES OF ALKANES, ETHERS, ALCOHOLS, AMINES, AND ALKYL HALIDES

Boiling Points

The **boiling point (bp)** of a compound is the temperature at which the liquid form of the compound can become a gas (vaporize). In other words, it is the temperature at which the compound's vapor pressure equals the surrounding pressure. In order for a compound to vaporize, the forces that hold the molecules close to each other must be disrupted. This means that the boiling point of a compound depends on the attractive forces between the individual molecules. If the molecules are held together by strong forces, it will take a lot of energy to pull the molecules away from each other, and the compound will have a high boiling point. If, however, the molecules are held together by weak forces, only a little energy will be required to pull the molecules away from each other, and the compound will have a low boiling point.

Relatively weak forces hold alkane molecules together. An alkane contains only carbon and hydrogen atoms. Because the electronegativities of carbon and hydrogen are similar, the bonds in an alkane are nonpolar. Consequently, there are no partial charges on any of the atoms in an alkane. Alkanes are neutral molecules.

However, it is only the average charge distribution over the molecule that is neutral. Because electrons are continuously moving, at any instant one side of the molecule can have slightly more electron density than the other side. This means that at any instant, one end of the molecule will have a slight negative charge and the other end will have a slight positive charge. This gives the molecule a temporary dipole.

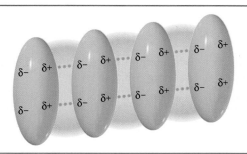

◀ **Figure 2.1**
Van der Waals forces are
induced dipole–induced
dipole interactions.

The temporary dipole in one molecule can induce a temporary dipole in a nearby molecule. As a result, the negative side of the first molecule is lined up adjacent to the positive side of the second molecule (Figure 2.1). Since the dipoles in the molecules are induced, the interactions between the molecules are called **induced dipole–induced dipole interactions.** The molecules of an alkane are held together by these induced dipole–induced dipole interactions known as **van der Waals forces** or **London forces.** In order for an alkane to boil, these van der Waals forces must be disrupted.

Van der Waals forces are the weakest of all the intermolecular attractions. The size of the van der Waals forces that hold alkane molecules together depends on the area of contact between the molecules. The greater the area of contact, the greater the van der Waals forces and the greater the amount of heat needed to overcome these forces. If you look at the homologous series of alkanes in Table 2.1, you will see that, as alkanes increase in size, their boiling points also increase. This occurs because each additional methylene group increases the area of contact between the molecules. With each additional methylene group added to the molecule, the energy required to boil the compound increases by 1.0 to 1.5 kcal/mol (4.2 to 6.3 kJ/mol). This amounts to an increase in boiling point of about 20 to 35 °C per methylene group. Because the four smallest alkanes have boiling points below room temperature (room temperature is about 25 °C), they exist as gases at room temperature. Pentane is the smallest alkane that is a liquid at room temperature.

Branching in a compound lowers its boiling point. A branched compound has a more compact, more nearly spherical shape. If you think of the unbranched alkane pentane as a cigar and branched neopentane as a tennis ball, you can see that branching decreases the contact area between molecules. Two cigars have contact over a greater area than do two tennis balls. So if two alkanes have the same molecular weight, the more highly branched alkane will have a lower boiling point.

$$CH_3CH_2CH_2CH_2CH_3 \qquad CH_3CHCH_2CH_3 \qquad \overset{\displaystyle CH_3}{\underset{\displaystyle CH_3}{\vert \;\; CH_3CCH_3 \;\; \vert}}$$

pentane	isopentane	neopentane
bp = 36.1 °C	bp = 27.9 °C	bp = 9.5 °C

Any homologous series of compounds will show increasing boiling points with increasing molecular weight because of the increase in van der Waals forces. So the boiling points of a homologous series of ethers, alkyl halides, alcohols, and amines increase with increasing molecular weight (Appendix I). The boiling points of these compounds, however, are affected by a property that alkanes do not have. All of these molecules have some polar character, because nitrogen, oxygen, and halogens are more electronegative than the carbon to which they are bonded.

$$\overset{\displaystyle|}{\underset{\displaystyle|}{R-\overset{\delta+\ \ \delta-}{C-Z}}} \qquad Z = N, O, F, Cl, Br, or\ I$$

The dipole moment of a bond is equal to the size of the charge on one of the bonding atoms times the distance between the bonding atoms.

The magnitude of the charge differential between the two bonding atoms is indicated by the dipole moment (Sections 1.3 and 1.17).

H_3C-NH_2	$H_3C-O-CH_3$	H_3C-I	H_3C-OH
1.30 D	1.31 D	1.64 D	1.71 D

H_3C-F	H_3C-Br	H_3C-Cl
1.82 D	1.79 D	1.94 D

Alcohols and amines also have a polar bond between the oxygen or nitrogen atom and the less electronegative hydrogen.

$$\overset{\displaystyle|}{\underset{\displaystyle|}{R-\overset{\delta+\ \ \delta-\ \ \delta+}{C-O-H}}} \qquad \overset{\displaystyle|}{\underset{\displaystyle|}{R-\overset{\delta+\ \ \delta-\ \ \delta+}{C-N-H}}}$$

Molecules with dipole moments are attracted to one another because they can align themselves in such a way that the positive end of one dipole is close to the negative end of another dipole. These electrostatic attractive forces, called **dipole–dipole interactions,** are stronger than van der Waals forces but not as strong as ionic or covalent bonds.

More extensive tables of physical properties can be found in Appendix I.

Ethers generally have higher boiling points than alkanes of comparable molecular weight because both the van der Waals forces and the dipole–dipole interactions in an ether must be overcome for the ether to boil (Table 2.5). Notice, however, that diethyl ether, an ether with a molecular weight comparable to that of pentane, has a lower boiling point than pentane. This is because the oxygen atom interferes with the van der Waals contact area of the floppy ethyl groups. Tetrahydrofuran, a cyclic ether with approximately the same molecular weight as diethyl ether, has a higher boiling point than cyclopentane because it has a better van der Waals contact area.

$CH_3CH_2CH_2CH_2CH_3$	$CH_3CH_2OCH_2CH_3$	cyclopentane	tetrahydrofuran
pentane	diethyl ether		
bp = 36.1 °C	bp = 34.5 °C	bp = 49.3 °C	bp = 65 °C

TABLE 2.5 Comparative Boiling Points of Alkanes, Ethers, Alcohols, and Amines (°C)			
$CH_3CH_2CH_3$	CH_3OCH_3	CH_3CH_2OH	$CH_3CH_2NH_2$
−42.1	−23.7	78	16.6
$CH_3CH_2CH_2CH_3$	$CH_3OCH_2CH_3$	$CH_3CH_2CH_2OH$	$CH_3CH_2CH_2NH_2$
−0.5	10.8	97.4	47.8
$CH_3CH_2CH_2CH_2CH_3$	$CH_3CH_2OCH_2CH_3$	$CH_3CH_2CH_2CH_2OH$	$CH_3CH_2CH_2CH_2NH_2$
36.1	34.5	117.3	77.8

Alcohols have much higher boiling points than alkanes or ethers of comparable molecular weight (Table 2.5) because, in addition to van der Waals forces and the dipole–dipole interactions of the carbon–oxygen bond, a special kind of dipole–dipole interaction can occur between alcohols. A hydrogen bonded to an oxygen, a nitrogen, or a fluorine can form a weak association with a second oxygen, nitrogen, or fluorine. This association is known as a **hydrogen bond.** The length of the covalent bond between oxygen and hydrogen within an alcohol molecule is 0.96 Å. The hydrogen bond between an oxygen of one molecule and a hydrogen of another molecule is almost twice as long (1.69–1.79 Å). So a hydrogen bond is not nearly as strong as an oxygen–hydrogen covalent bond, but a hydrogen bond is stronger than other dipole–dipole interactions. The strongest hydrogen bonds are linear, where the two electronegative atoms and the hydrogen between them lie in a straight line.

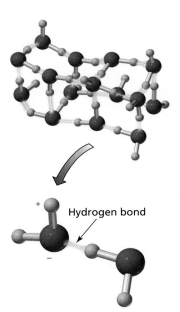

Hydrogen bonding in water

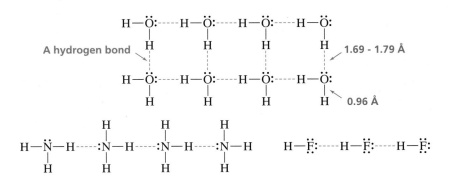

Although each individual hydrogen bond is weak (requiring about 5 kcal/mol to break), there are many such bonds holding alcohol molecules together. The extra energy required to break these hydrogen bonds is the reason alcohols have much higher boiling points than either alkanes or ethers with similar molecular weights.

The boiling point of water illustrates the dramatic effect of hydrogen bonding on boiling points. Water has a molecular weight of 18 and a boiling point of 100 °C. The alkane nearest in size is methane, with a molecular weight of 16. Methane boils at −167.7 °C.

Primary and secondary amines also form hydrogen bonds, so these amines have higher boiling points than alkanes with similar molecular weights. Nitrogen is not as electronegative as oxygen, however, which means that the hydrogen bonds between amine molecules are weaker than the hydrogen bonds between alcohol molecules. (Remember that hydrogen bonds are a special kind of dipole–dipole interaction.) Amines, therefore, have lower boiling points than alcohols with similar molecular weights.

Because primary amines have two N—H bonds, hydrogen bonding is more important for primary amines than it is for secondary amines. Tertiary amines cannot form hydrogen bonds with each other because they do not have a hydrogen attached to the nitrogen. Consequently, if you compare amines with the same molecular weight, primary amines have higher boiling points than secondary amines, which have higher boiling points than tertiary amines.

$CH_3CH_2CH_2CH_2CH_2NH_2$
pentylamine
a primary amine
bp = 104 °C

$CH_3CH_2NHCH_2CH_2CH_3$
ethylpropylamine
a secondary amine
bp = 90 °C

$CH_3CH_2NCH_2CH_3$ with CH_3 on N
diethylmethylamine
a tertiary amine
bp = 66 °C

PROBLEM-SOLVING STRATEGY

a. Which of the following compounds (1–3) will form hydrogen bonds between its molecules?

b. Which will form hydrogen bonds with a solvent such as ethanol?

1. $CH_3CH_2CH_2CH_2OH$ 3. $CH_3CH_2OCH_2CH_3$

2. $CH_3CH_2CH_2CH_2SH$

In solving this type of question, start by defining the kind of compound that will do what is being asked.

a. A hydrogen bond forms when a hydrogen that is attached to an O, N, or F forms a bond with a different O, N, or F. Therefore, a compound that will form hydrogen bonds with itself must have a hydrogen bonded to an O, N, or F. Only compound **1** will be able to form hydrogen bonds with itself.

b. Ethanol has an H bonded to an O, so as long as the compound has an O, N, or F, hydrogen bonding with ethanol can occur. Compounds **1** and **3** will be able to form hydrogen bonds with ethanol.

Now continue on to Problem 18.

PROBLEM 18 ◆

a. Which of the following compounds (1–6) will form hydrogen bonds between its molecules?

b. Which will form hydrogen bonds with a solvent such as ethanol?

1. $CH_3CH_2CH_2COOH$ 4. $CH_3CH_2CH_2NHCH_3$

2. $CH_3CH_2N(CH_3)_2$ 5. $CH_3CH_2OCH_2OH$

3. $CH_3CH_2CH_2CH_2Br$ 6. $CH_3CH_2CH_2CH_2F$

PROBLEM 19

Explain why:

a. H_2O has a higher boiling point than CH_3OH (65 °C).

b. H_2O has a higher boiling point than NH_3 (−33 °C).

c. H_2O has a higher boiling point than HF (20 °C).

PROBLEM 20

Explain why the dipole moment of methyl fluoride is less than the dipole moment of methyl chloride even though fluorine is more electronegative than chlorine.

PROBLEM 21 ◆

List the following compounds in order of decreasing boiling point.

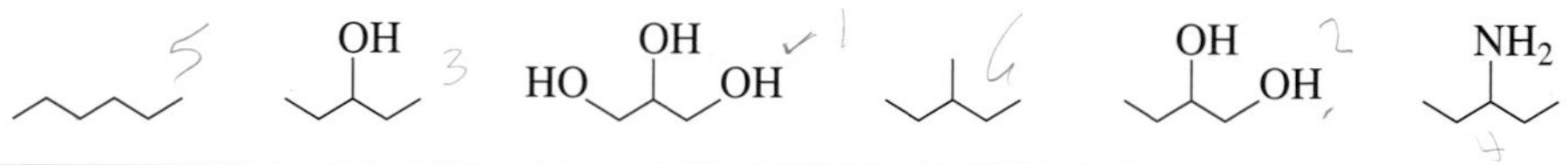

Both van der Waals forces and dipole–dipole interactions must be overcome for an alkyl halide to boil. An alkyl fluoride has a lower boiling point than an alkyl

TABLE 2.6	Comparative Boiling Points of Alkanes and Alkyl Halides (°C)				
			Y		
	H	F	Cl	Br	I
CH_3-Y	−161.7	−78.4	−24.2	3.6	42.4
CH_3CH_2-Y	−88.6	−37.7	12.3	38.4	72.3
$CH_3CH_2CH_2-Y$	−42.1	−2.5	46.6	71.0	102.5
$CH_3CH_2CH_2CH_2-Y$	−0.5	32.5	78.4	101.6	130.5
$CH_3CH_2CH_2CH_2CH_2-Y$	36.1	62.8	107.8	129.6	157.0

chloride with the same alkyl group. Similarly, alkyl chlorides have lower boiling points than alkyl bromides, which have lower boiling points than alkyl iodides (Table 2.6), because of the difference in size of the halogen atoms. As the halogen atom increases in size, the van der Waals contact area increases.

It is difficult to compare the boiling points of alkyl halides and alkanes. If we compare alkyl halides with alkanes that have the same number of carbon atoms, clearly alkyl halides will have higher boiling points because they have dipole–dipole interactions in addition to their van der Waals interactions. However, this is not a fair comparison because an alkyl halide has a much greater molecular weight than an alkane with the same number of carbons. If we compare alkanes and alkyl halides with similar molecular weights, it is still not a fair comparison because the alkanes are much larger molecules than the alkyl halides and therefore have much greater van der Waals contact areas. For example, decane (bp = 142 °C) has the same molecular weight as methyl iodide (bp = 42.5 °C).

PROBLEM 22

List the following compounds in order of decreasing boiling point.

a. $CH_3CH_2CH_2CH_2CH_2CH_2Br$ $CH_3CH_2CH_2CH_2Br$ $CH_3CH_2CH_2CH_2CH_2Br$

b.
$$CH_3\overset{\displaystyle CH_3}{\underset{\displaystyle CH_3}{\overset{|}{\underset{|}{C}}}}-\overset{\displaystyle CH_3}{\underset{\displaystyle CH_3}{\overset{|}{\underset{|}{C}}}}CH_3 \qquad CH_3CH_2CH_2CH_2CH_2CH_2CH_2CH_3$$

$$\overset{}{\underset{\displaystyle CH_3}{\overset{|}{CH_3CHCH_2CH_2CH_2CH_2CH_3}}}$$

c. $CH_3CH_2CH_2CH_2CH_3$ $CH_3CH_2CH_2CH_2OH$ $CH_3CH_2CH_2CH_2Cl$

$CH_3CH_2CH_2CH_2CH_2OH$

Melting Points

The **melting point (mp)** is the temperature at which a solid is converted into a liquid. If you examine the melting points of the alkanes in Table 2.1, you will see that the melting points increase (with a few exceptions) in a homologous series as the molecular weight increases. But the increase in melting point is less regular than the increase in boiling point because of an important factor that influences the melting point of a compound—**packing.** Packing is a property that determines how

Figure 2.2 ▶
Melting points of alkanes. Alkanes with even numbers of carbon atoms fall on a melting point curve that is higher than the melting point curve for alkanes with odd numbers of carbon atoms.

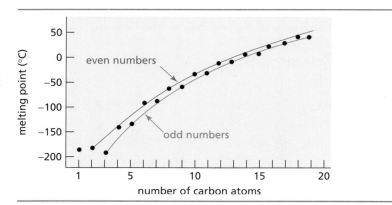

well the individual molecules in a solid fit together in the crystal lattice. The tighter the fit, the more energy is required to break the lattice and melt the compound. In Figure 2.2, you can see that the melting points of alkanes with even numbers of carbon atoms fall on a smooth curve. The melting points of alkanes with odd numbers of carbon atoms also fall on a smooth curve. But the two curves do not coincide because alkanes with even numbers of carbon atoms fit together better in a crystal lattice, and this increases their chances for intramolecular attraction, which increases their melting points.

Solubility

Oil and water don't mix. The general rule that governs **solubility** with regard to the polarity of molecules is "like dissolves like." In other words, polar compounds dissolve in polar solvents and nonpolar compounds dissolve in nonpolar solvents. This is because a polar solvent such as water has partial charges that can interact with the partial charges on a polar compound. The negative poles of the solvent molecules surround the positive pole of the polar solute, and the positive poles of the solvent molecules surround the negative pole of the polar solute. Clustering of the solvent molecules around the solute molecules separates solute molecules from each other, which is what makes them soluble. The interaction between a solvent and a molecule or an ion that is dissolved in that solvent is called **solvation.**

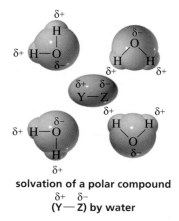

solvation of a polar compound
$\delta+$ $\delta-$
(Y — Z) by water

Because nonpolar compounds have no charge, polar solvents are not attracted to them. In order for a nonpolar molecule to dissolve in a polar solvent such as water, the nonpolar molecule would have to push the water molecules apart, disrupting their hydrogen bonding. Hydrogen bonding is so strong that it excludes the nonpolar compound. In contrast, nonpolar solutes dissolve in nonpolar solvents because of the van der Waals interactions between solvent and solute molecules.

Alkanes are nonpolar molecules, which causes them to be soluble in nonpolar solvents and insoluble in polar solvents such as water. The densities of alkanes (Table 2.1) increase with increasing molecular weight. But even a 30-carbon alkane is less dense than water (density of H_2O at 20 °C = 0.9982 g/ml). This means that a mixture of an alkane and water will separate into two distinct layers with the less dense alkane floating on top. The Alaskan oil spill of 1989 and the Persian Gulf spill of 1991 are large-scale examples of this phenomenon. (Crude oil is a mixture of alkanes.)

An alcohol has both a nonpolar alkyl group and a polar OH group. So, is the molecule as a whole nonpolar or polar? Is it soluble in a nonpolar solvent or is it soluble in water? The answer depends on the size of the alkyl group. As the alkyl group increases in size, it becomes a more significant fraction of the alcohol molecule and the compound becomes less and less soluble in water. In other words, the molecule becomes more and more like an alkane. Four carbons tend to be the dividing line. Alcohols with fewer than four carbons are soluble in water, but alcohols with more than four carbons are insoluble in water. In other words, an OH group can drag about three or four carbons into solution in water.

The four-carbon dividing line is only an approximate guide because the solubility of an alcohol also depends on the structure of the alkyl group. Branched alkyl groups are more water-soluble than nonbranched alkyl groups with the same number of carbons, because branching minimizes the contact surface of the nonpolar portion of the molecule. For example, *tert*-butyl alcohol is more soluble than *n*-butyl alcohol in water.

The oxygen atom of an ether can also drag only about three carbons into solution in water (Table 2.7).

Oil from the Alaskan oil spill on the shores of Prince William Sound.

TABLE 2.7	Solubilities of Ethers in Water	
2 C's	CH_3OCH_3	soluble
3 C's	$CH_3OCH_2CH_3$	soluble
4 C's	$CH_3CH_2OCH_2CH_3$	slightly soluble (10 g/100 g H_2O)
5 C's	$CH_3CH_2OCH_2CH_2CH_3$	minimally soluble (1.9 g/100 g H_2O)
6 C's	$CH_3CH_2CH_2OCH_2CH_2CH_3$	insoluble (0.25 g/100 g H_2O)

Low-molecular-weight amines are soluble in water because an amine can form a hydrogen bond with water. Comparing amines with the same number of carbon atoms, primary amines are more soluble than secondary amines because the former have two hydrogens that can engage in hydrogen bonding. Because tertiary amines can accept a hydrogen bond but do not have a hydrogen to donate to form a hydrogen bond, a tertiary amine is less soluble in water than a secondary amine with the same number of carbon atoms.

Alkyl halides have some polar character, but only the alkyl fluorides have an atom that can form a hydrogen bond. This means that alkyl fluorides are the most water soluble of the alkyl halides. The other alkyl halides are considerably less soluble than ethers or alcohols with the same number of carbons. The solubilities of alkyl halides in water are shown in Table 2.8. Alkyl halides with three carbons are only slightly soluble in water, and alkyl halides with four or more carbons are insoluble.

TABLE 2.8 Solubilities of Alkyl Halides in Water			
CH_3F very soluble	CH_3Cl soluble	CH_3Br slightly soluble	CH_3I slightly soluble
CH_3CH_2F soluble	CH_3CH_2Cl slightly soluble	CH_3CH_2Br slightly soluble	CH_3CH_2I slightly soluble
$CH_3CH_2CH_2F$ slightly soluble	$CH_3CH_2CH_2Cl$ slightly soluble	$CH_3CH_2CH_2Br$ slightly soluble	$CH_3CH_2CH_2I$ slightly soluble
$CH_3CH_2CH_2CH_2F$ insoluble	$CH_3CH_2CH_2CH_2Cl$ insoluble	$CH_3CH_2CH_2CH_2Br$ insoluble	$CH_3CH_2CH_2CH_2I$ insoluble

PROBLEM 23

Rank the following groups of compounds in order of decreasing solubility in water.

a. $CH_3CH_2CH_2OH$ $CH_3CH_2CH_2CH_2OH$ $CH_3CH_2CH_2CH_2Cl$

$HOCH_2CH_2CH_2OH$

b.   CH₃   NH₂   OH

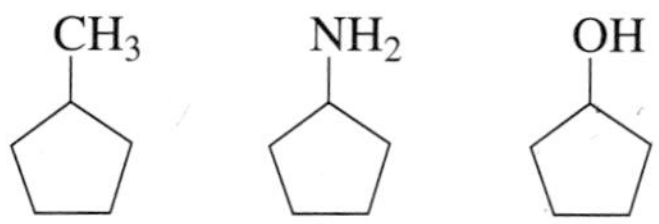

PROBLEM 24 ◆

In which of the following solvents would cyclohexane have the lowest solubility: pentanol, diethyl ether, ethanol, or tetrahydrofuran?

2.10
HEAT OF COMBUSTION AND HEAT OF FORMATION

The standard **heat of combustion** ($\Delta H°$) is the amount of heat released when a carbon-containing compound burns—reacts completely with oxygen—under standard conditions (298 K; 1 atm of pressure). The more stable the compound, the less heat it gives off on burning. Figure 2.3 shows that isobutane gives off less heat than butane, which means that isobutane is more stable.

If we compare the heats of combustion of four isomeric octanes, one with no branches, one with one branch, one with two branches, and one with four branches, we see that branching causes a compound to give off less heat when it burns. Therefore we can conclude that branching increases the stability of the alkane.

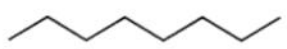

$\Delta H° = -1307.5$ kcal/mol −5470.6 kJ/mol	$\Delta H° = -1306.3$ kcal/mol −5465.6 kJ/mol	$\Delta H° = -1304.6$ kcal/mol −5458.4 kJ/mol	$\Delta H° = -1303.0$ kcal/mol −5451.8 kJ/mol

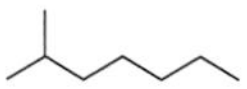

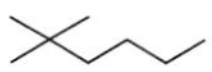

Branching increases the stability of an alkane because, although it *decreases* the van der Waals forces of attraction *between* molecules, it *increases* the van der Waals forces of attraction *within* the molecule. These intramolecular van der Waals attractions stabilize the molecule.

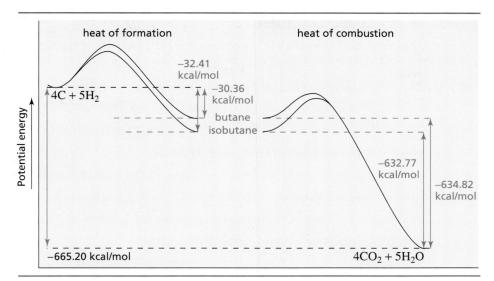

◀ **Figure 2.3**
The more stable compound (left) has the more negative heat of formation. The less stable compound (right) has the more negative heat of combustion.

The **heat of formation** ($\Delta H°_f$) of a compound is the heat given off or consumed when the compound is formed from its elements under standard conditions. For example, the heat of formation of butane from carbon (graphite) and hydrogen gas is -30.36 kcal/mol (-127.0 kJ/mol). The heat of formation of isobutane is -32.41 kcal/mol (-135.6 kJ/mol). The observation that isobutane formation releases more heat than butane formation releases tells us that isobutane is more stable (Figure 2.3).

$$4C + 5H_2 \longrightarrow CH_3CH_2CH_2CH_3 \quad \Delta H°_f = -30.36 \text{ kcal/mol}$$
graphite   gas -127.0 kJ/mol

$$4C + 5H_2 \longrightarrow CH_3CHCH_3 \quad \Delta H°_f = -32.41 \text{ kcal/mol}$$
graphite   gas $\underset{CH_3}{|}$ -135.6 kJ/mol

We have seen that *the more stable isomer has a less negative heat of combustion and a more negative heat of formation.* Figure 2.3 shows the relationship between the heat of combustion and the heat of formation. The sum of the heat of combustion and the heat of formation of a particular alkane is equal to the heats of combustion of graphite (-94.05 kcal/mol carbon, -393.5 kJ/mol carbon) and hydrogen (-57.80 kcal/mol H_2, -241.8 kJ/mol H_2) for that alkane. This means that, if we know either the heat of combustion or the heat of formation of a compound, we can calculate the other value. Figure 2.3 shows that the heats of combustion of graphite and hydrogen for C_4H_{10} is -665.20 kcal/mol [$4(-94.05) + 5(-57.80) = -665.20$].

PROBLEM 25◆

a. Which of the following compounds has the most negative heat of formation?

b. Which of the following compounds has the most negative heat of combustion?

$$\underset{\underset{CH_3}{|}}{CH_3CHCH_2CH_3} \qquad CH_3CH_2CH_2CH_2CH_3 \qquad \underset{\underset{CH_3}{|}}{\overset{\overset{CH_3}{|}}{CH_3CCH_3}}$$

PROBLEM 26 / SOLVED

The heats of combustion of pentane and isopentane are -782.2 and -780.4 kcal/mol, respectively. Calculate their heats of formation.

SOLUTION   The heats of combustion of graphite and hydrogen for C_5H_{12}

$$(5C \text{ and } 6H_2) = 5(-94.05) + 6(-57.80)$$

$$= -817.1 \text{ kcal/mol}$$

Heat of formation of pentane + heat of combustion of pentane

$$= -817.1 \text{ kcal/mol}$$

Heat of formation of pentane $- 782.2$ kcal/mol $= -817.1$ kcal/mol

Heat of formation of pentane $= -34.9$ kcal/mol

Heat of formation of isopentane + heat of combustion of isopentane

$$= -817.1 \text{ kcal/mol}$$

Heat of formation of isopentane $- 780.4$ kcal/mol $= -817.1$ kcal/mol

Heat of formation of isopentane $= -36.7$ kcal/mol

2.11 CONFORMATIONS OF ALKANES: ROTATION ABOUT CARBON–CARBON SINGLE BONDS

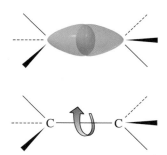

▲ **Figure 2.4**
A carbon–carbon bond is formed by the overlap of cylindrically symmetrical sp^3 orbitals. Therefore, the bond can rotate without changing the amount of orbital overlap.

A carbon–carbon single bond is formed as the result of the overlap of an sp^3 orbital of one carbon with an sp^3 orbital of a second carbon. Because sp^3 orbitals are cylindrically symmetrical (they are symmetrical about an imaginary line connecting the centers of the two atoms joined by the sigma bond), rotation about a carbon–carbon single bond is possible without any change in the amount of orbital overlap (Figure 2.4). The different spatial arrangements of the atoms that result from rotation about a single bond are called **conformations**. Different conformations are also called **conformational isomers** or **conformers.**

When an ethane molecule rotates about its carbon–carbon bond, two extreme conformations can result: the staggered conformation and the eclipsed conformation. An infinite number of conformations between these two extreme conformations is also possible. Compounds are three-dimensional, but we are limited to a two-dimensional sheet of paper when showing their structures. There are several ways to represent on paper the three-dimensional spatial arrangements of the atoms that occur as a result of rotation about a sigma (σ) bond. Wedge-and-dash structures, sawhorse projections, and Newman projections are all commonly used methods. In a wedge-and-dash structure, bonds protruding out from the plane of the paper are drawn as solid wedges, those protruding into the paper are drawn as dashed lines, and solid lines are used for bonds that lie in the plane of the paper. In a sawhorse projection, you are looking at the carbon–carbon bond from an oblique angle. In a Newman projection, you are looking down the length of a particular carbon–carbon bond. The carbon that is in front is represented by the point at which three bonds intersect, and the carbon that is in back is represented by a circle. The three lines emanating from each of the carbons represent the carbon's other three bonds. We will use Newman projections because they do the best job of representing the conformational isomers.

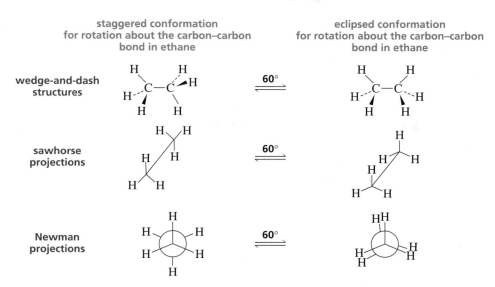

The electrons in a carbon–hydrogen bond will repel the electrons in another carbon–hydrogen bond if the bonds get too close to each other. The **staggered conformation,** therefore, is the most stable conformation because the carbon–hydrogen bonds are as far away from each other as possible. The **eclipsed conformation** is the least stable conformation, because in no other conformation are the carbon–hydrogen bonds that close to one another. The extra energy of the eclipsed conformation is called torsional strain. **Torsional strain** is the name given to the repulsion felt by the bonding electrons of one substituent as they pass close to the bonding electrons of another substituent. In the staggered conformation, the distance between the hydrogen nuclei is 2.55 Å, but they are only 2.29 Å apart in the eclipsed conformation.

Rotation about a carbon–carbon single bond is not completely free because of the energy difference between the staggered and eclipsed conformations. The eclipsed conformation is of higher energy, so an energy barrier must be overcome when rotation about the carbon–carbon bond occurs (Figure 2.5). This energy barrier is only 2.9 kcal/mol (12 kJ/mol), which means that at room temperature there is enough thermal energy in the surroundings to allow an easy transition over the barrier. So even though there is an energy barrier to rotation, the barrier is small enough to allow the conformational isomers to interconvert millions of times per second at room temperature.

Butane has three carbon–carbon single bonds, and the molecule can rotate about each of them.

the C-2—C-3 bond

$$\overset{1}{C}H_3-\overset{2}{C}H_2-\overset{3}{C}H_2-\overset{4}{C}H_3$$

butane

the C-1—C-2 bond   the C-3—C-4 bond

For example, a staggered and an eclipsed conformer can be drawn for rotation about the C-3—C-4 bond. The carbon in the foreground in a Newman projection has the lower number.

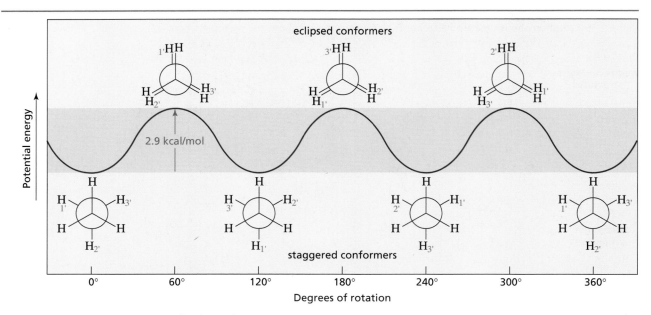

▲ Figure 2.5
Potential energy of ethane as a function of the angle of rotation about the carbon–carbon bond.

staggered conformation for rotation about the C-3—C-4 bond in butane

eclipsed conformation for rotation about the C-3—C-4 bond in butane

Although the staggered conformers resulting from rotation about the C-3—C-4 bond in butane all have the same energy, the staggered conformers resulting from rotation about the C-2—C-3 bond do not have the same energy. The staggered conformers for rotation about the C-2—C-3 bond in butane are shown below. Conformer C, in which the two methyl groups are as far apart as possible, is more stable than the other two staggered conformers (A and E). The most stable of the staggered conformers is called the **anti conformer,** and the other two staggered conformers are called **gauche conformers** (*anti* is Greek for "opposite of "; *gauche* is French for "left"). In the anti conformer, the largest substituents are opposite each other; in the gauche conformer, they are adjacent. The two gauche conformers have the same energy, but each is 0.9 kcal/mol (3.8 kJ/mol) less stable than the anti conformer.

staggered
gauche

eclipsed

staggered
anti

eclipsed

staggered
gauche

eclipsed

A   B   C   D   E   F

Anti and gauche conformers do not have the same energy because of steric strain. **Steric strain** (or **steric hindrance**) is the strain put on a molecule when atoms or groups are too close to each other, which causes repulsion between the electron clouds of the atoms or groups. The **van der Waals radius** is defined as half the distance between the nuclei of two equivalent atoms or groups when they are at an energy minimum. The van der Waals radius of a methyl group is 2.0 Å, whereas the van der Waals radius of a hydrogen atom is 1.2 Å. Two methyl groups, therefore, must be at least 4.0 Å apart (the sum of their van der Waals radii), and a methyl and a hydrogen must be at least 3.2 Å apart, to avoid steric strain. There is more steric strain in the gauche conformer than in the anti conformer, because the two methyl groups are closer together in the gauche conformer. This is called a **gauche interaction.**

The eclipsed conformers resulting from rotation about the C-2—C-3 bond in butane also have different energies. The eclipsed conformer in which the two methyl groups are closest to each other (F) is less stable than the other eclipsed conformers (B and D). The energy diagram for rotation about the C-2—C-3 bond of butane is shown in Figure 2.6. All of the eclipsed conformers have both torsional and steric strain—torsional strain because of bond–bond repulsion, and steric strain because of the closeness of the eclipsing groups. In general, steric strain in molecules increases as the size of the eclipsing groups increases.

Because there is continuous rotation about all of the carbon–carbon single bonds in a molecule, organic molecules with carbon–carbon single bonds are not static balls and sticks. Molecules with carbon–carbon single bonds have many interconvertible conformers. Conformers cannot be separated because they rapidly interconvert.

The number of molecules in a particular conformation at any one time depends on the stability of the conformation: the more stable the conformation, the greater the number of molecules that will be in that conformation. So most molecules are in staggered conformations, and more molecules are in an anti conformation than in a gauche conformation. Calculations reveal that at room temperature about 72% of the molecules of butane are in the anti conformation and about 28% are in a gauche conformation. The preference for a staggered conformation causes carbon chains to orient themselves in a zigzag fashion, as shown by the structure of decane.

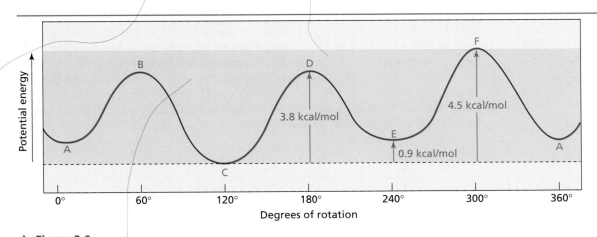

▲ **Figure 2.6**
Potential energy of butane as a function of the degree of rotation about the C-2—C-3 bond. Green letters refer to the conformers (A–F) shown on page 86.

a computer generated picture of decane

PROBLEM 27

  a. Draw all the staggered and eclipsed conformers for rotation about the C-2—C-3 bond of pentane.

  b. Draw a potential energy diagram for rotation of the C-2—C-3 bond of pentane through 360°.

2.12 CYCLOALKANES: RING STRAIN

Early chemists noted that cyclic compounds found in nature generally had five- or six-membered rings. Compounds with three- and four-membered rings were found much less frequently. This finding indicated that compounds with three- and four-membered rings are not as stable as compounds with five- or six-membered rings.

The German chemist Baeyer was the first to suggest that the instability of these small rings was due to angle strain. We know that an sp^3 hybridized carbon has bond angles of 109.5° (Section 1.7). In 1885, Baeyer suggested that the stability of a cyclic compound could be predicted by determining how close the bond angle of a planar cyclic compound is to the desired tetrahedral bond angle of 109.5°. The angles in a regular triangle are 60°. The bond angles in cyclopropane, therefore, are compressed from the desired tetrahedral angle of 109.5° to 60°. The deviation of the bond angle from the desired bond angle causes the strain known as **angle strain.**

The angle strain in a three-membered ring can be appreciated by looking at the orbitals that overlap to form the sigma (σ) bonds in cyclopropane (Figure 2.7). Normal σ bonds are formed by the overlap of two sp^3 orbitals that point directly toward each other. In cyclopropane, overlapping orbitals cannot point directly toward each other. Therefore, orbital overlap is less effective than in a normal carbon–carbon bond. The less effective orbital overlap causes the carbon–carbon bond to be weaker than a normal carbon–carbon bond. Because the bonds in cyclopropane have shapes that resemble bananas, they are sometimes called **banana bonds.**

In addition to angle strain, three-membered rings also have torsional strain, because all of the adjacent carbon–hydrogen bonds are eclipsed.

banana bonds

Figure 2.7 ▶
(a) Overlap of sp^3 orbitals in a normal σ bond. (b) Overlap of sp^3 orbitals in cyclopropane.

good overlap strong bond
(a)

poor overlap weak bond
(b)

The bond angles in planar cyclobutane would have to be compressed from 109.5° to 90°, the bond angle associated with a planar four-membered ring. Planar cyclobutane would be expected to have less angle strain than cyclopropane because the bond angles in cyclobutane are only 19.5° away from the normal tetrahedral bond angle.

Baeyer predicted that a five-membered ring compound would be the most stable of the cyclic compounds because its bond angles (108°) are the closest to the tetrahedral bond angle. He predicted that six-membered ring compounds with bond angles of 120° would be less stable, and as cyclic compounds became larger than six-membered rings they would be less and less stable.

cyclopropane

"planar" cyclopentane
bond angles = 108°

"planar" cyclohexane
bond angles = 120°

"planar" cycloheptane
bond angles = 128.6°

Contrary to what Baeyer predicted, however, cyclohexane is more stable than cyclopentane. Furthermore, cyclic compounds do not become less and less stable as the number of sides increases. The mistake that Baeyer made was to assume that all cyclic compounds are planar. Because three points define a plane, the carbons of cyclopropane must lie in a plane. But the other cycloalkanes are not planar. Cyclic compounds twist and bend in order to achieve a final structure that minimizes the three different kinds of strain that can destabilize a cyclic compound.

1. *Angle strain* is the strain that results when the bond angles are different from the desired tetrahedral bond angle of 109.5°.

2. *Torsional strain* is caused by repulsion of the bonding electrons of one substituent with the bonding electrons of a nearby substituent.

3. *Steric strain* is caused by atoms or groups of atoms approaching each other too closely.

cyclobutane

Although planar cyclobutane has less angle strain than cyclopropane, it has more torsional strain because it has eight pairs of eclipsed hydrogens compared with six pairs in cyclopropane. In addition, the hydrogens in cyclobutane are closer together. So cyclobutane is not planar—it is bent. One of its methylene groups is bent at an angle of about 25° from the plane of the other three methylene groups. This increases the angle strain, but the increase is more than compensated for by the decreased torsional strain that results from the adjacent hydrogens not being as eclipsed as they would be in a planar structure.

Unlike what Baeyer predicted, cyclopentane has some angle strain. If cyclopentane were flat, there would be essentially no angle strain but there would be ten pairs of adjacent hydrogens that would experience considerable torsional strain. Cyclopentane puckers so that the hydrogens become nearly staggered. In the process, cyclopentane acquires some angle strain. This puckered form of cyclopentane is called the envelope conformation, because the shape resembles a squarish envelope with the flap up.

PROBLEM 28◆

The bond angles of a regular polygon with *n* sides are equal to

$$180° - \frac{360°}{n}$$

a. What are the bond angles in a regular octagon?   **b.** In a regular nonagon?

cyclopentane

BAEYER AND BARBITURIC ACID

Johann Friedrich Wilhelm Adolf von Baeyer (1835–1917) was born in Germany. He discovered barbituric acid in 1864 and it is agreed that he named it after a woman named Barbara. Who Barbara was is not certain. Some sources say that she was his girlfriend, others say that she was a waitress. Since Baeyer discovered barbituric acid in 1864, the same year that Prussia defeated Denmark, there are some who believe that he named the acid after Saint Barbara, the patron saint of artillerymen. Baeyer was the first to synthesize indigo, the dye that is used in the manufacture of blue jeans. He was a professor of chemistry at the University of Strasbourg and later at the University of Munich. He received the Nobel Prize in chemistry in 1905 for his work in synthetic organic chemistry.

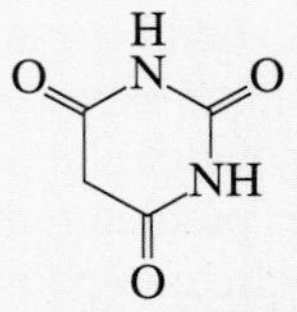

barbituric acid

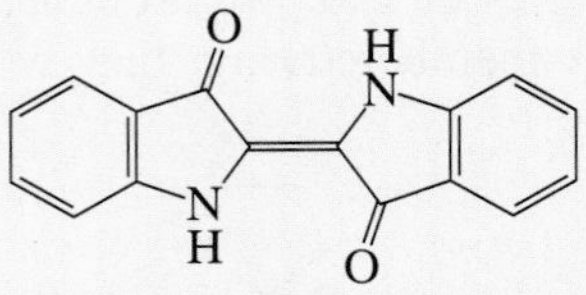

indigo

HIGHLY STRAINED HYDROCARBONS

Organic chemists have been able to synthesize some highly strained cyclic hydrocarbons such as bicyclo[1.1.0]butane, cubane, and prismane.[1]

bicyclo[1.1.0]butane

cubane

prismane

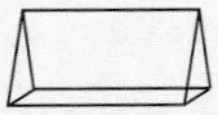

[1]Bicyclo[1.1.0]butane was synthesized by David Lemal, Fredric Menger, and George Clark at the University of Wisconsin (*Journal of the American Chemical Society* 1963, *85*, 2529). Cubane was synthesized by Philip Eaton and Thomas Cole, Jr. at the University of Chicago (*Journal of the American Chemical Society* 1964, *86*, 3157). Prismane was synthesized by Thomas Katz and Nancy Acton at Columbia University (*Journal of the American Chemical Society* 1973, *95*, 2738).

2.13 CONFORMATIONS OF CYCLOHEXANE

Despite Baeyer's prediction that five-membered ring compounds would be the most stable, the cyclic compounds most commonly found in nature contain six-membered rings. Six-membered ring compounds are so prevalent because they can exist in a conformation that is almost completely free of strain. This conformation is called the **chair conformation** (Figure 2.8). In the chair conformation of cyclohexane, for example, all the bond angles are 111°, which is very close to the tetrahedral bond angle (109.5°), and all the adjacent carbon–hydrogen bonds are staggered. The chair conformation is such an important conformation that you should learn how to draw it.

1. Draw two parallel lines of the same length, slanted upward. Both lines should start at the same height.

2. Connect the tops of the lines with a V. The left-hand side of the V should be slightly longer than the right-hand side. Connect the bottoms of the lines with an inverted V. The lines of the V and the inverted V are parallel. This completes the framework of the six-membered ring.

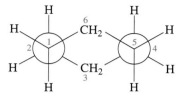

chair conformation of
cyclohexane

Newman projection of
the chair conformation

◄ **Figure 2.8**
The strain-free chair conformation of cyclohexane and
the Newman projection of
the chair conformation,
showing that all of the
carbon–hydrogen bonds are
staggered.

ball-and-stick model of the
chair conformer of cyclohexane

3. Each carbon has an axial and an equatorial bond. The **axial bonds** are parallel
   to the edge of the page and alternate their directions. The one on the uppermost carbon is up, the next is down, the next is up, and so on.

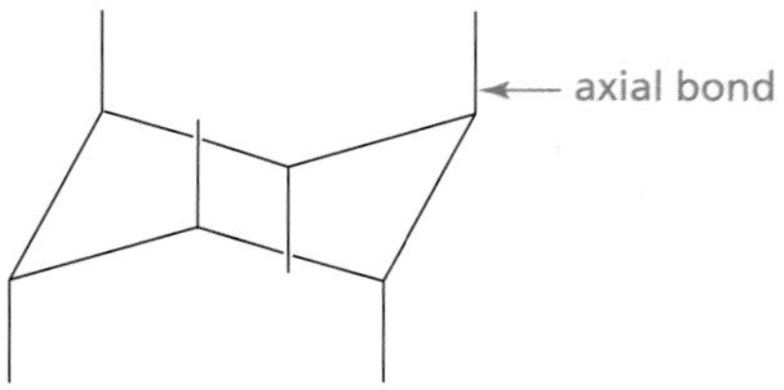

←— axial bond

4. The **equatorial bonds** (blue circles) point outward from the ring. If the axial
   bond is up, the equatorial bond on the carbon is below the plane perpendicular
   to the bottom of the axial bond. If the axial bond is down, the equatorial bond
   is above the plane perpendicular to the top of the axial bond.

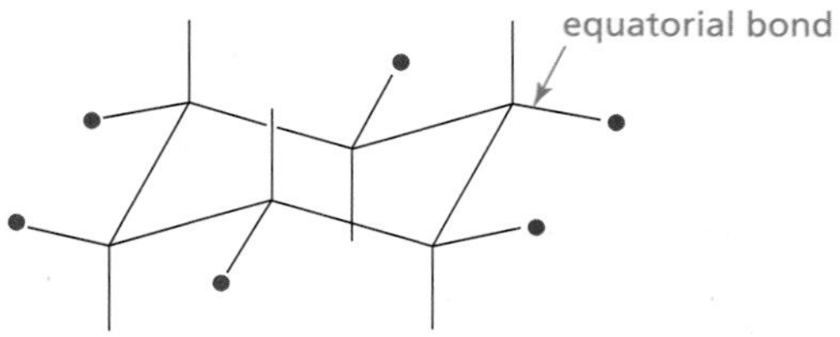

equatorial bond

5. Notice that each equatorial bond is parallel to a ring bond "two carbons over."

6. Remember that cyclohexane is viewed on edge. The lower bonds of the ring are in front, and the upper bonds of the ring are in back.

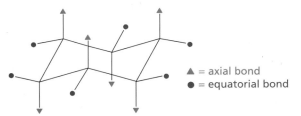

▲ = axial bond
● = equatorial bond

If we assume that cyclohexane is completely free of strain, we can calculate the amount of total strain energy (angle strain plus torsional strain plus steric strain) that each of the other cycloalkanes experiences. Taking the heat of formation of cyclohexane (Table 2.9) and dividing by 6 for the six CH_2 groups in cyclohexane gives us a value of −4.92 kcal/mol (−20.6 kJ/mol) for a "strainless" CH_2 group (−29.5/6 = −4.92). We can now calculate the heat of formation of a "strainless" cycloalkane by multiplying the number of CH_2 groups in the ring by −4.92 kcal/mol. The total strain in the compound is the difference between the "strainless" heat of formation and the actual heat of formation (Table 2.8). For example, cyclopentane has a "strainless" heat of formation of 5 × −4.92 = −24.6 kcal/mol. Because its actual heat of formation is −18.4 kcal/mol, it has a strain energy of 6.2 kcal/mol [−18.4 − (−24.6) = 6.2].

As a result of rotation about its carbon–carbon single bonds, cyclohexane rapidly interconverts between two stable chair conformations. This interconversion is known as **ring-flip** (Figure 2.9). When the two chair conformers interconvert, bonds that are equatorial in one chair conformer become axial in the other chair conformer.

Cyclohexane can also exist in a **boat conformation** (Figure 2.10). Like the chair conformation, the boat conformation is free of angle strain. However, the boat con-

TABLE 2.9	Heats of Formation and Total Strain Energies of Cycloalkanes					
	Heat of formation		**"Strainless" heat of formation**		**Total strain energy**	
	(kcal/mol)	(kJ/mol)	(kcal/mol)	(kJ/mol)	(kcal/mol)	(kJ/mol)
cyclopropane	+12.7	53.1	−14.6	−61.1	27.3	114.22
cyclobutane	+6.8	24.5	−19.7	−82.4	26.5	110.9
cyclopentane	−18.4	−77.0	−24.6	−102.9	6.2	25.9
cyclohexane	−29.5	−123.4	−29.5	−123.4	0	0
cycloheptane	−28.2	−118.0	−34.4	−143.9	6.2	25.9
cyclooctane	−29.7	−124.3	−39.4	−164.8	9.7	40.6
cyclononane	−31.7	−132.6	−44.3	−185.4	12.6	52.7
cyclodecane	−36.9	−154.4	−49.2	−205.9	12.3	51.5
cycloundecane	−42.9	−179.5	−54.1	−226.4	11.2	46.9

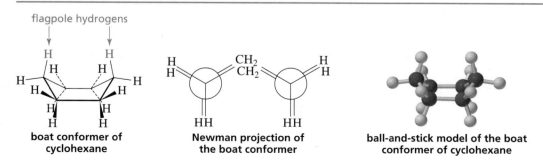

The bonds that are axial in one chair conformation are equatorial in the other chair conformation. The bonds that are equatorial in one chair conformation are axial in the other chair conformation.

formation is not as stable as the chair conformation because some of the carbon–hydrogen bonds in the boat conformation are eclipsed. The boat conformation is further destabilized by the close proximity of the **flagpole hydrogens.** These hydrogens (at the "bow" and "stern" of the boat) are 1.8 Å apart, but the sum of their van der Waals radii is 2.4 Å. The flagpole hydrogens are also called **transannular hydrogens** because they are across the ring from each other.

The conformations that cyclohexane assumes when interconverting from one chair conformer to the other are shown in Figure 2.11. To convert from the boat conformation to one of the chair conformations, one of the uppermost carbons of the boat conformation must be pulled down so it becomes the bottommost carbon. When the uppermost carbon is pulled down just a little, the **twist-boat (or skew-boat) conformation** is obtained. The twist-boat conformation is more stable than the boat conformation because the flagpole hydrogens have moved away from each other. When the uppermost carbon is pulled down to the point where it is in the same plane as the sides of the boat, the very unstable **half-chair conformation** is obtained. Pulling the uppermost carbon down farther produces the *chair conformation*. Figure 2.11 shows the energy of a cyclohexane molecule as it interconverts from one chair conformer to the other. The energy barrier for interconversion from one chair conformer to the other is 10.8 kcal/mol (45.2 kJ/mol). From this value, it can be calculated that cyclohexane undergoes 10^5 flips per second at room temperature. In other words, the two chair conformers are in rapid equilibrium.

flagpole hydrogens

| boat conformer of cyclohexane | Newman projection of the boat conformer | ball-and-stick model of the boat conformer of cyclohexane |

▲ **Figure 2.10**
The boat conformation of cyclohexane and the Newman projection of the boat conformation, showing that some of the carbon–hydrogen bonds are eclipsed.

Figure 2.11 ▶
Diagram showing the conformations of cyclohexane (and their relative energies) as one chair conformation interconverts to the other chair conformation.

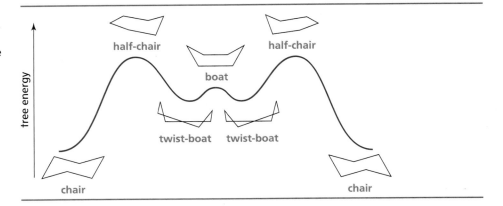

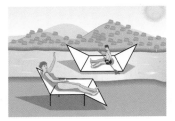

Because the chair conformers are the most stable of the conformers, at any instant more molecules of cyclohexane are in chair conformations than in any other conformation. It has been calculated that, for every thousand molecules of cyclohexane in the chair conformation, no more than two molecules are in the next most stable conformation—the twist-boat.

2.14
CONFORMATIONS OF MONOSUBSTITUTED CYCLOHEXANES

Unlike cyclohexane, which has two identical chair conformers, the two chair conformers of a monosubstituted cyclohexane such as methylcyclohexane are not identical. Because the methyl substituent is in an equatorial position in one conformer and in an axial position in the other conformer (Figure 2.12), the two chair conformers do not have the same stability.

The chair conformer with the substituent in the equatorial position is more stable than the chair conformer with the substituent in the axial position. The reason is the same as the reason that an anti conformer is more stable than a gauche conformer (Section 2.11). We saw that an anti conformer is more stable than a gauche conformer because the gauche interaction causes steric strain. Gauche interactions also make a cyclohexane with a substituent in the axial position less stable than a cyclohexane with a substituent in the equatorial position. When the methyl group is in the equatorial position, it is anti to the C-3 and C-5 carbons, and it extends into space away from the rest of the molecule (Figure 2.13).

When the methyl group is in the axial position, however, it is gauche to both the C-3 and the C-5 carbons (Figure 2.14). If you take a few minutes to build

Figure 2.12 ▶
A substituent is in the equatorial position in one chair conformation and in the axial position in the other chair conformation. The conformation with the substituent in the equatorial position is more stable.

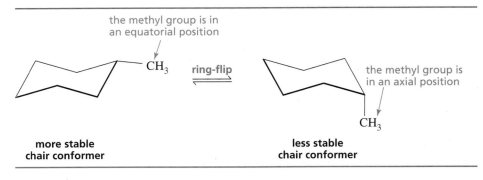

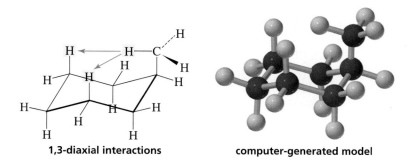

◀ Figure 2.13
An equatorial substituent on the C-1 carbon is anti to C-3 and C-5 carbons.

◀ Figure 2.14
An axial substituent on the C-1 carbon is gauche to C-3 and C-5 carbons.

methyl is anti to C-3   methyl is anti to C-5

methyl is gauche to C-3   methyl is gauche to C-5

models, you can easily see that a substituent has more room if it is in an equatorial position.

In addition to the steric strain resulting from the gauche relationship of the axial substituent and the C-3 and C-5 carbons, the axial substituent also has unfavorable steric interactions with the two other axial substituents (in this case, hydrogens) that are on the same side of the ring. Notice that the three axial bonds on the same side of the ring are parallel to each other. The unfavorable steric interaction between an axial substituent and a substituent on one of these parallel bonds results from the distance between them being less than the sum of their van der Waals radii. Because the interacting substituents are on 1,3-positions relative to each other, these unfavorable steric interactions are called **1,3-diaxial interactions.**

Build a model of cyclohexane and interconvert it from one chair conformer to the other. To do this, pull the topmost carbon down and push the bottommost carbon up. Now do the same for methylcyclohexane.

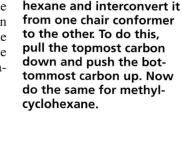

1,3-diaxial interactions   computer-generated model

Since a substituent has more freedom from repulsive interactions with other atoms if it is in an equatorial position, at any instant more monosubstituted cyclohexane molecules are in the chair conformer with the substituent in the equatorial position than in the chair conformer with the substituent in the axial position. The relative amounts of the conformers with the substituent in the equatorial and axial positions depend on the substituent (Table 2.10). For example, the equilibrium constant (K_{eq}) for the conformers of methylcyclohexane indicates that 95% of methylcyclohexane molecules have the methyl group in the equatorial position at 25 °C (see the calculation below). In the case of *tert*-butylcyclohexane, where the steric interactions are even more destabilizing because a *tert*-butyl group is larger than a methyl group, more than 99.9% of the molecules have the *tert*-butyl group in the equatorial position. The investigation of the various conformations of a compound and their relative stabilities is called **conformational analysis.**

TABLE 2.10   Equilibrium Constants for Several Monosubstituted Cyclohexanes at 25 °C

Substituent	Axial $\xrightleftharpoons{K_{eq}}$ Equatorial	Substituent	Axial $\xrightleftharpoons{K_{eq}}$ Equatorial
H	1	F	1.5
CH_3	18	Cl	2.4
CH_3CH_2	23		
$CH_3\overset{\displaystyle CH_3}{\underset{}{CH}}$	38	Br	2.2
		I	2.2
$CH_3\overset{\displaystyle CH_3}{\underset{\displaystyle CH_3}{C}}$	4000	HO	5.4

$$K_{eq} = \frac{[\text{equatorial conformer}]}{[\text{axial conformer}]} = \frac{18}{1}$$

$$\% \text{ of equatorial conformer} = \frac{[\text{equatorial conformer}]}{[\text{equatorial conformer}] + [\text{axial conformer}]} \times 100$$

$$\% \text{ of equatorial conformer} = \frac{18}{18 + 1} \times 100 = 95\%$$

PROBLEM 29

Bromine has a larger van der Waals radius than chlorine, but the equilibrium constants in Table 2.10 indicate that a chloro substituent has a greater preference for the equatorial position. Suggest an explanation.

KEY TERMS

alcohol (page 57)
alkane (page 53)
alkyl halide (page 57)
alkyl substituent (page 56)
amine (page 57)
angle strain (page 88)

anti conformer (page 86)
asymmetrical ether (page 67)
axial bond (page 91)
banana bond (page 88)
boat conformation (page 92)
boiling point (bp) (page 74)

chair conformation (page 90)
common name (page 56)
conformational analysis (page 95)
conformational isomer (page 84)
conformation (page 84)
conformer (page 84)

PROBLEMS

30. Give a structural formula for each of the following compounds.

a. *sec*-butyl *tert*-butyl ether
b. isoheptyl alcohol
c. *sec*-butylamine
d. neopentyl bromide
e. 1,1-dimethylcyclohexane
f. 4,5-diisopropylnonane
g. triethylamine
h. cyclopentylcyclohexane
i. 4-*tert*-butylheptane
j. 5,5-dibromo-2-methyloctane
k. 1-methylcyclopentanol
l. 3-ethoxy-2-methylhexane
m. 5-(1,2-dimethylpropyl)nonane
n. 3,4-dimethyloctane

31. Give the IUPAC name for each of the following compounds.

a.
$$CH_3CHCH_2CH_2CHCH_2CH_2CH_3$$
with Br above and CH_3 below

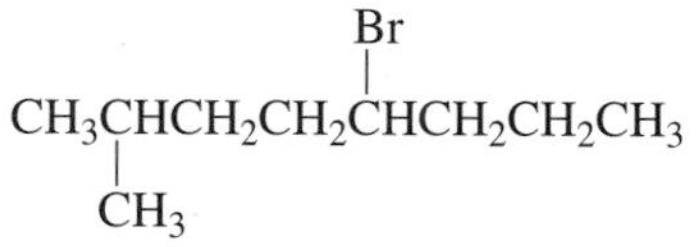

b. $(CH_3)_3CCH_2CH_2CH_2CH(CH_3)_2$

c.
$$CH_3CHCH_2CHCHCH_3$$
with CH_3 above, and CH_3, CH_3 below

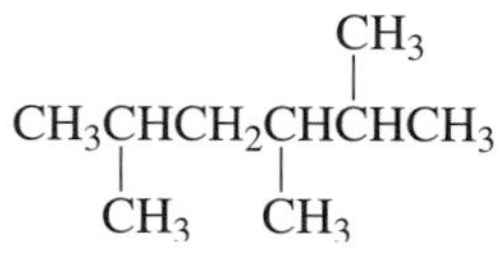

d. $(CH_3CH_2)_4C$

e.
$$CH_3CHCH_2CH_2CHCH_2CH_3$$
with CH_3 and OH below

f.
$$CH_3CH_2CHOCH_2CH_3$$
with $CH_2CH_2CH_2CH_3$ below

g. cyclohexane ring with CH_3 at top and Br at bottom

h. cyclohexane ring with NCH_3 and CH_3

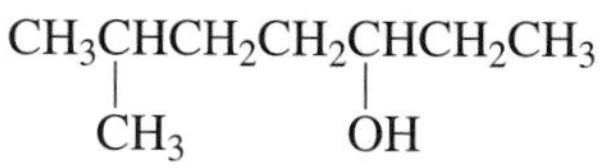

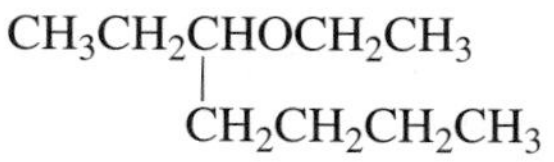

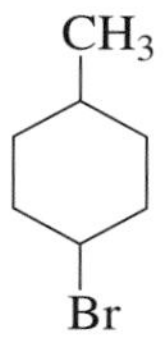

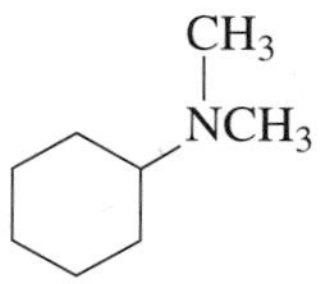

i.

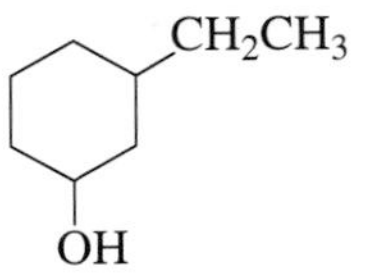

j. $CH_3OCH_2CH_2CH_2OCH_3$

32. Answer the questions for the structure shown below.
   a. How many primary carbons does it have?
   b. How many secondary carbons does it have?
   c. How many tertiary carbons does it have?

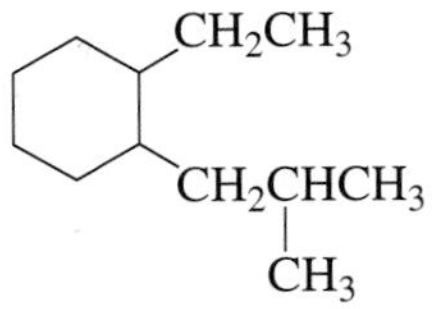

33. Draw the structural formula of an alkane that has
   a. six carbons, all secondary carbons.
   b. eight carbons and only primary hydrogens.
   c. seven carbons with two isopropyl groups.

34. Give two names for each of the following compounds.

   a. $CH_3CH_2CH_2OCH_2CH_3$

   b. $CH_3CHCH_2CH_2CH_2OH$
      $|$
      CH_3

   c. $CH_3CH_2CHCH_3$
      $|$
      NH_2

   d. $CH_3CH_2CHCH_3$
      $|$
      Cl

   e. $CH_3CHCH_2CH_2CH_3$
      $|$
      CH_3

   f. $CH_3\overset{\displaystyle CH_3}{\underset{\displaystyle CH_2CH_3}{\overset{|}{\underset{|}{C}}}}Br$

   g.

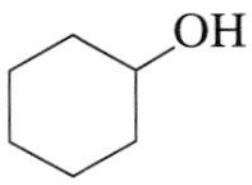

   h.

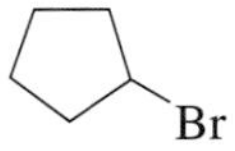

   i. CH_3CHNH_2
      $|$
      CH_3

35. Which of the following pairs of compounds has
   a. the higher boiling point: 1-bromopentane or 1-bromohexane?
   b. the higher boiling point: pentyl chloride or isopentyl chloride?
   c. the greater solubility in water: 1-butanol or 1-pentanol?
   d. the higher boiling point: 1-hexanol or 1-methoxypentane?
   e. the higher melting point: hexane or isohexane?
   f. the higher boiling point: 1-chloropentane or 1-pentanol?
   g. the more negative heat of combustion: methylpropane or butane?
   h. the higher boiling point: 1-bromopentane or 1-chloropentane?
   i. the higher boiling point: diethyl ether or butyl alcohol?
   j. the greater density: heptane or octane?
   k. the more negative heat of combustion: methylcyclobutane or cyclopentane?
   l. the higher boiling point: isopentyl alcohol or isopentylamine?
   m. the higher boiling point: hexylamine or dipropylamine?
   n. the more negative heat of formation: methylcyclobutane or cyclopentane?

36. Fred Leadhead was given the structural formulas of several compounds and was asked to name them by the IUPAC system. How many did Fred name correctly? Correct the ones that are misnamed.

a. 4-bromo-3-pentanol
b. 2,2-dimethyl-4-ethylheptane
c. 5-methylcyclohexanol
d. 1,1-dimethyl-2-cyclohexanol
e. 5-(2,2-dimethylethyl)nonane
f. isopentyl bromide

g. 3,3-dichlorooctane
h. 5-ethyl-2-methylhexane
i. 1-bromo-4-pentanol
j. 3-isopropyloctane
k. 2-methyl-2-isopropylheptane
l. 2-methyl-*N*,*N*-dimethyl-4-hexanamine

37. Give IUPAC names for all the alkanes with molecular formula C_7H_{16} that do not have any secondary hydrogens.

38. Draw the structures of the following compounds using skeletal structures.

a. 5-ethyl-2-methyloctane
b. 1,3-dimethylcyclohexane
c. 2,3,3,4-tetramethylheptane

d. propylcyclopentane
e. 2-methyl-4-(1-methylethyl)octane
f. 2,6-dimethyl-4-(2-methylpropyl) decane

39. Considering rotation about the C-3—C-4 bond of 2-methylhexane,

a. draw the Newman projection of the most stable conformer.
b. draw the Newman projection of the least stable conformer.
c. how many other carbon–carbon bonds in the compound can rotate?
d. how many of the carbon–carbon bonds in the compound have staggered conformers that are all equally stable?

40. Draw all the isomers that have molecular formula $C_5H_{11}Br$.

a. Give the IUPAC name for each of the isomers.
b. Give a common name for each isomer that has a common name.
c. How many isomers do not have common names?
d. How many of the isomers are primary alkyl halides?
e. How many of the isomers are secondary alkyl halides?
f. How many of the isomers are tertiary alkyl halides?

41. Give the IUPAC name for each of the following compounds.

a. 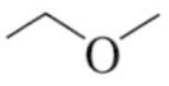
b. 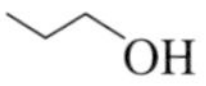
c. 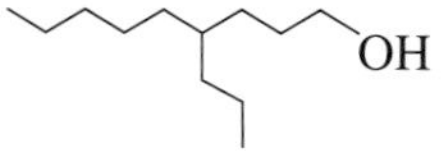
d.

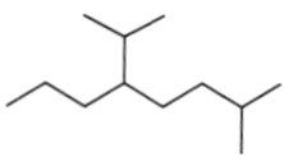

e. 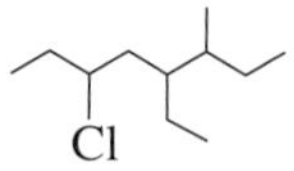
f. 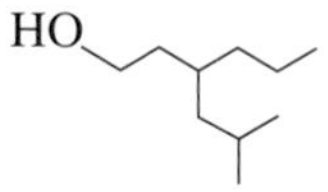
g.

42. Why are alcohols of lower molecular weight more water soluble than those of higher molecular weight?

43. The most stable conformer of *N*-methylpiperidine is shown at the right.

a. Draw the other chair conformer.
b. Which takes up more space, the nonbonded pair of electrons or the methyl group?

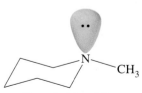

N-methylpiperidine

44. How many ethers have molecular formula $C_5H_{12}O$? Give the structural formula and IUPAC name for each. What are their common names?

45. Give the IUPAC name for each of the following compounds.

a. 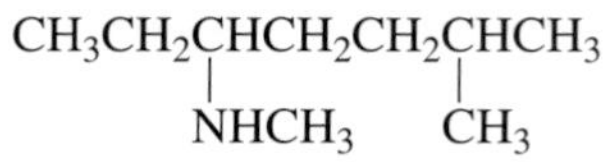 $CH_3CH_2CHCH_2CH_2CHCH_3$
     $\underset{NHCH_3}{|}\underset{CH_3}{|}$

b. 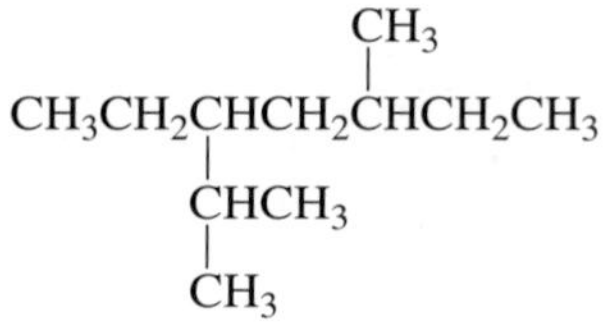 $CH_3CH_2CHCH_2CHCH_2CH_3$
     $\overset{CH_3}{|}$
     $\underset{CHCH_3}{|}$
     $\underset{CH_3}{|}$

c. 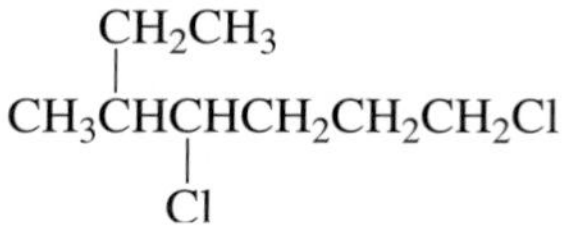 $CH_3CHCHCH_2CH_2CH_2Cl$
     $\overset{CH_2CH_3}{|}$
     $\underset{Cl}{|}$

d. 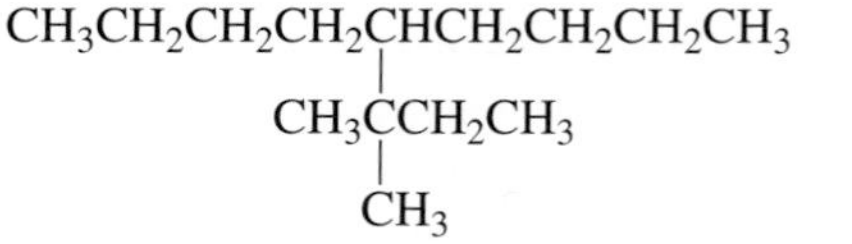 $CH_3CH_2CH_2CH_2CHCH_2CH_2CH_2CH_3$
     $\underset{CH_3CCH_2CH_3}{|}$
     $\underset{CH_3}{|}$

e. 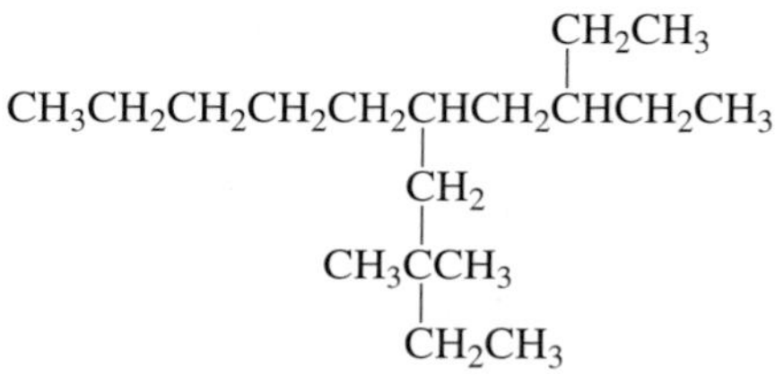 $CH_3CH_2CH_2CH_2CH_2CHCH_2CHCH_2CH_3$
     $\overset{CH_2CH_3}{|}$
     $\underset{CH_2}{|}$
     $\underset{CH_3CCH_3}{|}$
     $\underset{CH_2CH_3}{|}$

46. Al Kane went to buy heating fuel and found that both propane and cyclopropane fuels were on sale at the same price per pound. Which fuel should Al have bought? Why?

47. Explain the following:
a. 1-Hexanol has a higher boiling point than 3-hexanol.
b. Diethyl ether has only very limited solubility in water but tetrahydrofuran is essentially completely soluble.

HYDRO-CARBONS, STEREOCHEMISTRY, AND RESONANCE

Four of these six chapters cover the reactions of hydrocarbons—compounds that contain only carbon and hydrogen. The other two chapters cover two topics that are so important to the study of organic reactions that each deserves its own chapter: the first is stereochemistry and the second is electron delocalization and resonance.

In **Chapter 3,** you will study the reactions of alkenes—*hydrocarbons that contain carbon–carbon double bonds.* You will learn how alkenes react and what kinds of products are formed from the reactions. Although there are many different reactions in this chapter, you will see that they all take place by a similar pathway. This chapter also discusses the principles of thermodynamics and kinetics—principles that are central to an understanding of how and why organic reactions take place.

Chapter 4 is all about stereochemistry. Here you will learn about the different kinds of isomers that organic compounds form. Then we will revisit the reactions from Chapter 3 to determine whether the products of these reactions can exist as isomers and, if so, which isomers are formed.

Chapter 5 covers the reactions of alkynes—*hydrocarbons that contain carbon–carbon triple bonds.* Because alkenes and alkynes both have reactive carbon–carbon π bonds, you will discover that their reactions share many similarities. Finally, this chapter will introduce you to some of the techniques chemists use to design syntheses of organic compounds, and you will then have your first opportunity to design some multistep syntheses.

All the organic compounds in Chapters 1 to 5 have only localized electrons. In **Chapter 6,** you will learn about delocalized electrons and the important concepts known as resonance and aromaticity.

In **Chapter 7,** you will take a look at the reactions of dienes—*hydrocarbons that have two carbon–carbon double bonds.* You will see that if the two double bonds in a diene are sufficiently separated from one another, the reactions that occur are identical to the reactions of alkenes. If, however, the double bonds are separated by only one single bond, resonance plays an important role in the reactions of that compound. So Chapter 7 combines many of the theories you learned in Chapter 3 with those you studied in Chapter 6.

Chapter 8 covers the reactions of alkanes—*hydrocarbons that contain only single bonds.* In the previous chapters you will have discovered that when an organic compound reacts, the weakest bond in the molecule is the one that breaks. Alkanes, however, have only strong bonds. Therefore, you can correctly predict that alkanes undergo reactions only under extreme conditions.

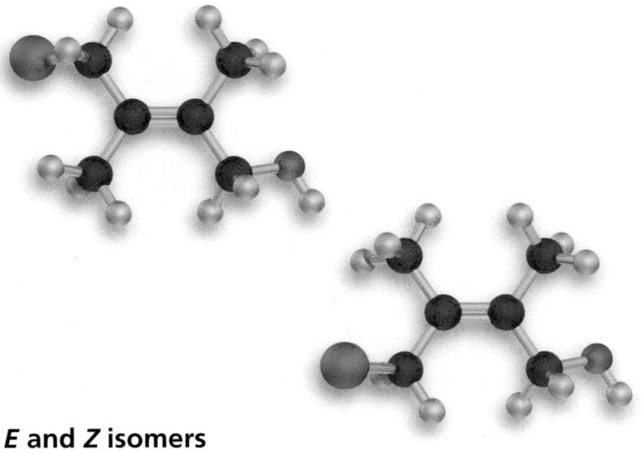

3

E and Z isomers

REACTIONS OF ALKENES. THERMODYNAMICS AND KINETICS

Yǒu learned in Chapter 2 that alkanes are hydrocarbons that contain only carbon–carbon single bonds. Hydrocarbons that contain a carbon–carbon double bond are called **alkenes.** Early chemists noted that when ethene ($H_2C\!=\!CH_2$), the smallest alkene, reacted with chlorine, an oily substance was formed. This observation caused early organic chemists to call alkenes **olefins** (oil forming).

Alkenes play many important roles in biology. For example, ethene (ethylene) is a plant hormone. A **hormone** is a compound that controls growth and other changes in tissues. Ethene plays an important role in seed germination, flower maturation, and fruit ripening.

Insects communicate by releasing **pheromones,** chemical substances that other insects of the same species detect with their antennae. There are sex, alarm, and trail pheromones, and many of these are alkenes. Interfering with an insect's ability to send or receive chemical signals is an environmentally safe way to control insect populations. For example, traps with synthetic sex attractants have been used to destroy certain crop-destroying insects.

Many of the flavors and fragrances produced by certain plants also belong to the alkene family.

Ethene is the hormone that causes tomatoes to ripen.

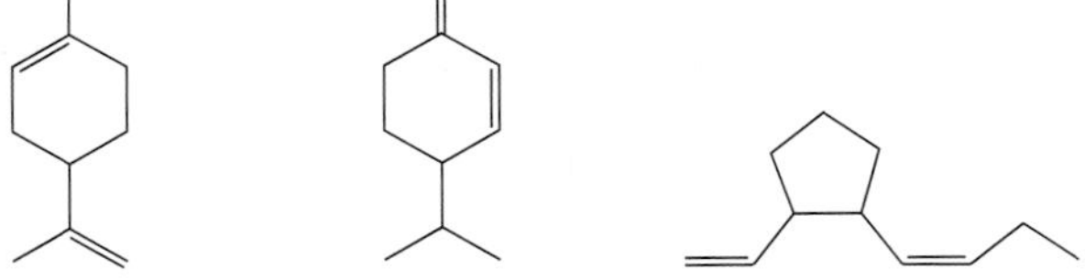

**limonene
from lemon and
orange oils**

**β-phellandrene
oil of eucalyptus**

**multifidene
sex attractant of
brown algae**

muscalure
sex attractant of the house fly

α-farnesene
found in the waxy coating
on apple skins

The double bond is the **functional group**—the center of reactivity—of an alkene. A double bond consists of a strong sigma (σ) bond and a weaker pi (π) bond. The relatively loosely held π electrons are attracted to a variety of reagents, with the result that the π bond breaks and the carbons form new σ bonds with the reagents. As you study the many reactions presented in this chapter, notice that the products that are formed differ only in the nature of Y and Z.

the double bond is composed
of a σ bond and a π bond

the π bond has broken and
new σ bonds have formed

This reactivity makes alkenes an important class of organic compounds because they can be used to synthesize a wide variety of products. In this chapter we will look at the kinds of reagents that react with alkenes and the nature of the products that are formed.

3.1 GENERAL MOLECULAR FORMULA FOR A HYDROCARBON

In Chapter 2 you learned that the general molecular formula for a *noncyclic alkane* is C_nH_{2n+2}. The general molecular formula for a *cyclic alkane* is C_nH_{2n} because the cyclic structure causes the compound to have two fewer hydrogens. Noncyclic compounds are also called **acyclic** compounds.

The general molecular formula for a *noncyclic alkene* is also C_nH_{2n} and the general molecular formula for a *cyclic alkene* is C_nH_{2n-2} because, as a result of the carbon–carbon double bond, an alkene has two fewer hydrogens than an alkane with the same number of carbon atoms. Therefore, we can make the following general statement: *The general molecular formula for a hydrocarbon is C_nH_{2n+2} minus two hydrogens for every π bond and/or ring present in the molecule.*

> The general molecular formula for a hydrocarbon is C_nH_{2n+2} minus two hydrogens for every π bond and/or ring present in the molecule.

$CH_3CH_2CH_2CH_2CH_3$ $CH_3CH_2CH_2CH{=}CH_2$

an alkane
C_5H_{12}
C_nH_{2n+2}

an alkene
C_5H_{10}
C_nH_{2n}

a cyclic alkane
C_5H_{10}
C_nH_{2n}

a cyclic alkene
C_5H_8
C_nH_{2n-2}

Thus, for every *two* hydrogens that are missing from the general molecular formula, C_nH_{2n+2}, a hydrocarbon has either a π bond or a ring. For example, a compound with a molecular formula of C_8H_{14} needs four more hydrogens to become $C_8H_{2\times8+2}$ (C_8H_{18}). Therefore, the compound has either two double bonds, a ring and a double bond, two rings, or a triple bond. [Remember that a triple bond consists of 2π bonds and a σ bond (Section 1.9).]

compounds with molecular formula C_8H_{14}

$CH_3CH{=}CH(CH_2)_3CH{=}CH_2$

$CH_3(CH_2)_5C{\equiv}CH$

Because alkanes contain the maximum possible number of carbon–hydrogen bonds—that is, they are saturated with hydrogen—they are called **saturated hydrocarbons.** Alkenes, because they have fewer hydrogens, are called **unsaturated hydrocarbons.**

$$CH_3CH_2CH_2CH_3 \qquad\qquad CH_3CH{=}CHCH_3$$
a saturated hydrocarbon   an unsaturated hydrocarbon

PROBLEM 1 / SOLVED

Knowing the number of π bonds and rings $(\pi + r)$, you can calculate how many hydrogens are missing from the general molecular formula of an alkane (C_nH_{2n+2}).

a. What is the value of $(\pi + r)$ for each of the compounds below?

b. Give the molecular formula for each of the following compounds without actually counting the number of hydrogens.

1. $CH_3CH_2C{\equiv}CH$

2.

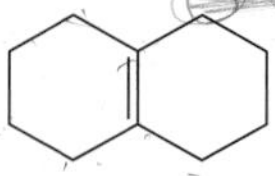

3. $CH_3CH{=}CHCH_2CH_2CH{=}CH_2$

4. 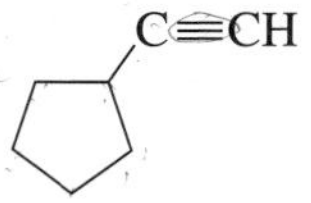 $C{\equiv}CH$

SOLUTION TO 1a(1)   This compound has two π bonds and no rings. Therefore, $\pi + r = 2 + 0 = 2$.

SOLUTION TO 1b(1)   Because this compound has four carbons, $C_nH_{2n+2} = C_4H_{10}$; since $\pi + r = 2$, the compound has four fewer hydrogens than C_4H_{10}. The molecular formula, therefore, is C_4H_6.

PROBLEM 2 ◆

Determine $(\pi + r)$ for each of the following molecular formulas.

a. $C_{10}H_{16}$

b. $C_{20}H_{34}$

c. C_8H_{16}

d. $C_{12}H_{20}$

PROBLEM 3 ◆

Draw possible structures for compounds with the following molecular formulas.

a. C_3H_6 **b.** C_3H_4 **c.** C_4H_6

The suffix for the double bond functional group is "ene." Thus the IUPAC name of an alkene is obtained by replacing the "ane" ending of the alkane with "ene." For example, a two-carbon alkene is called ethene and a three-carbon alkene is called propene. The two-carbon alkene is frequently referred to by its common name (ethylene).

**3.2
NOMENCLATURE
OF ALKENES**

$$H_2C{=}CH_2 \qquad CH_3CH{=}CH_2 \qquad \overset{\displaystyle CH_3}{\underset{\displaystyle}{CH_3C{=}CH_2}}$$

IUPAC name:	ethene	propene	methylpropene
common name:	ethylene	propylene	isobutylene

cyclohexene

Most alkene names need a number to indicate the position of the double bond. (The names above do not because there is no ambiguity.) The same IUPAC rules that we learned in Chapter 2 are followed in naming alkenes.

1. The longest continuous chain containing the functional group (in this case, the carbon–carbon double bond) is numbered in a direction that gives the functional group the lowest possible number. For example, 1-butene signifies that the double bond is between the first and second carbons of butene; 2-hexene signifies that the double bond is between the second and third carbons of hexene.

$$\overset{4}{C}H_3\overset{3}{C}H_2\overset{2}{C}H=\overset{1}{C}H_2 \qquad \overset{1}{C}H_3\overset{2}{C}H=\overset{3}{C}H\overset{4}{C}H_3 \qquad \overset{1}{C}H_3\overset{2}{C}H=\overset{3}{C}H\overset{4}{C}H_2\overset{5}{C}H_2\overset{6}{C}H_3$$

1-butene   **2-butene**   **2-hexene**

$$\overset{6}{C}H_3\overset{5}{C}H_2\overset{4}{C}H_2\overset{3}{C}H_2\overset{2}{C}CH_2CH_2CH_3$$
$$\|$$
$$_1CH_2$$

2-propyl-1-hexene

the longest continuous chain has eight carbons but the longest continuous chain containing the functional group has six carbons, so the parent name of the compound is hexene

Notice that 1-butene does not have a common name. You might be tempted to call it "butylene," but this is not appropriate because a name must be unambiguous and "butylene" could signify either 1-butene or 2-butene. Isobutylene is an appropriate common name for methylpropene because there is only one compound that has an iso structural unit, four carbons, and a double bond.

2. If the chain has substituents, it is still numbered in the direction that gives the functional group the lowest possible number.

$$\begin{array}{c} CH_3 \\ | \\ \overset{1}{C}H_3\overset{2}{C}H=\overset{3}{C}H\overset{4}{C}H\overset{5}{C}H_3 \end{array} \qquad \begin{array}{c} \overset{2}{C}H_2\overset{1}{C}H_3 \\ | \\ \overset{3}{C}H_3\overset{}{C}=\overset{4}{C}H\overset{5}{C}H_2\overset{6}{C}H_2\overset{7}{C}H_3 \end{array}$$

4-methyl-2-pentene   **3-methyl-3-heptene**

$$\overset{}{C}H_3\overset{}{C}H_2\overset{}{C}H_2\overset{}{C}H_2\overset{}{C}H_2\overset{4}{O}CH_2\overset{3}{C}H_2\overset{2}{C}H=\overset{1}{C}H_2$$

4-pentoxy-1-butene

3. If a chain has more than one substituent, the substituents are cited in alphabetical (not numerical) order, using the same rules for alphabetizing that we learned in Section 2.2 (the prefixes di, tri, *sec,* and *tert* are ignored in alphabetizing, but iso, neo, and cyclo are not ignored).

3,6-dimethyl-3-octene   **5-bromo-4-chloro-1-heptene**

$$\begin{array}{c} Br\ \ Cl \\ |\ \ \ | \\ \overset{7}{C}H_3\overset{6}{C}H_2\overset{5}{C}H\overset{4}{C}H\overset{3}{C}H_2\overset{2}{C}H=\overset{1}{C}H_2 \end{array}$$

4. In cyclic compounds, a number is not needed to denote the position of the functional group since the ring is always numbered so that the double bond is between carbons 1 and 2.

4,5-dimethylcyclohexene

5. If the same number for the alkene functional group is obtained in both directions, the correct name is the one that contains the lowest substituent number. For example, in 2,5-dimethyl-4-octene, the compound is a 4-octene whether the longest continuous chain is numbered from left to right or from right to left. If you number from left to right, the substituents are at positions 4 and 7, but if you number from right to left they are at positions 2 and 5. Of those four substituent numbers, 2 is the lowest number, so the compound is named 2,5-dimethyl-4-octene and not 4,7-dimethyl-4-octene.

$$CH_3CH_2CH_2C\!\!=\!\!CHCH_2CHCH_3$$
$$\qquad\qquad\quad | \qquad\qquad |$$
$$\qquad\qquad\ CH_3 \qquad\ CH_3$$

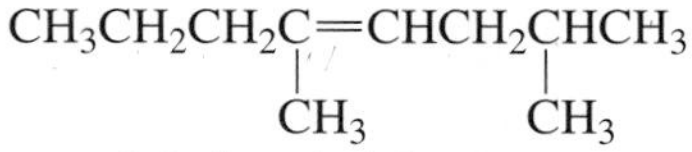

2,5-dimethyl-4-octene
not
4,7-dimethyl-4-octene
because 2 < 4

In the following cyclohexenes, the double bond is between C-1 and C-2 regardless of whether you move around the ring clockwise or counterclockwise. Therefore, you move around the ring in the direction that puts the lowest substituent number into the name. Notice that it is *not* the lowest sum of the substituents that is important: For example, 1,6-dichlorocyclohexene is not called 2,3-dichlorocyclohexene even though the latter has the lowest sum of the substituents ($1 + 6 = 7$ versus $2 + 3 = 5$); 1,6-dichlorocyclohexene is the correct name because it has the lowest substituent number (1).

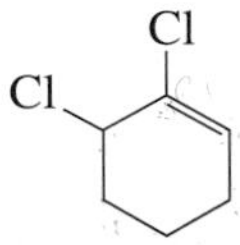

1,6-dichlorocyclohexene
not
2,3-dichlorocyclohexene
because 1 < 2

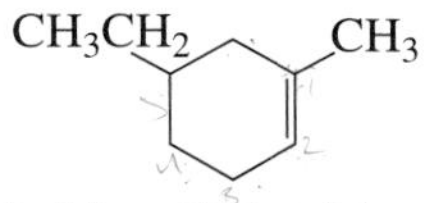

5-ethyl-1-methylcyclohexene
not
4-ethyl-2-methylcyclohexene
because 1 < 2

6. If both directions lead to the same number for the alkene functional group and the same low number(s) for one or more of the substituents, then those substituents are ignored and the direction is chosen that gives the lowest number to one of the remaining substituents.

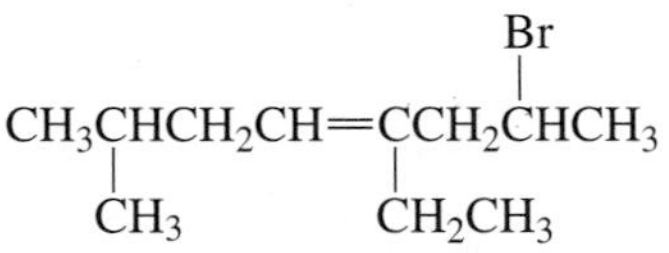

2-bromo-4-ethyl-7-methyl-4-octene
not
7-bromo-5-ethyl-2-methyl-4-octene
because 4 < 5

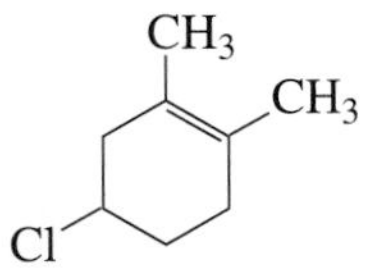

4-chloro-1,2-dimethylcyclohexene
not
5-chloro-1,2-dimethylcyclohexene
because 4 < 5

The sp^2 carbons of an alkene are called **vinylic carbons.** An sp^3 carbon that is adjacent to a vinylic carbon is called an **allylic carbon.**

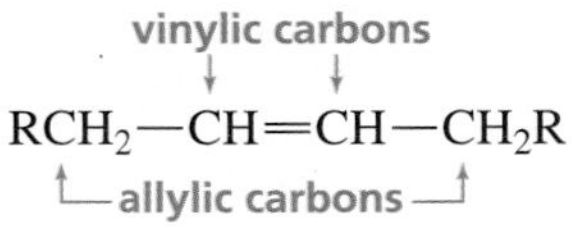

vinylic carbons
$$\downarrow \qquad \downarrow$$
$$RCH_2\!-\!CH\!\!=\!\!CH\!-\!CH_2R$$
└─ allylic carbons ─┘

There are two groups that contain a carbon–carbon double bond that are used as substituent groups in common names, the **vinyl group** and the **allyl group.** The vinyl group is the smallest possible group that contains a vinylic carbon, and the allyl group is the smallest possible group that contains an allylic carbon. When allyl is used in nomenclature, the substituent must be attached to the allylic carbon.

$$H_2C=CH— \qquad H_2C=CHCH_2—$$
$$\textbf{the vinyl group} \qquad \textbf{the allyl group}$$

	$H_2C=CHCl$	$H_2C=CHCH_2Br$
IUPAC name:	chloroethene	3-bromopropene
common name:	vinyl chloride	allyl bromide

PROBLEM 4 ◆

Draw the structure for each of the following compounds.

  a. 3,3-dimethylcyclopentene   **c.** vinylcyclohexane

  b. 6-bromo-2,3-dimethyl-2-hexene   **d.** allyl alcohol

PROBLEM 5 ◆

Give the IUPAC name for each of the following compounds.

  a. $CH_3CHCH=CHCH_3$
$\qquad\quad |$
$\qquad\;\; CH_3$

  b. $CH_3CH_2C=CCHCH_3$
$\qquad\qquad\; | \quad\; |$
$\qquad\quad\; CH_3 \; Cl$

  c. Br

  d. $BrCH_2CH_2CH=CCH_3$
$\qquad\qquad\qquad\qquad |$
$\qquad\qquad\qquad\quad CH_2$
$\qquad\qquad\qquad\qquad |$
$\qquad\qquad\qquad\quad CH_3$

  e. $H_3C \qquad CH_3$

  f. $CH_3CH=CHOCH_2CH_2CH_2CH_3$

3.3
THE STRUCTURE
OF ALKENES

The structure of the smallest alkene (ethene) was described in Section 1.8. Other alkenes have similar structures. Each double-bonded carbon of an alkene has three sp^2 orbitals that lie in a plane with angles of 120°. Each of these sp^2 orbitals overlaps with an orbital of another atom to form a σ bond. So one of the carbon–carbon bonds is a σ bond, formed by the overlap of an sp^2 orbital of one carbon with an sp^2 orbital of the other carbon. The second carbon–carbon bond (the π bond) is formed as a result of side-to-side overlap of the remaining p orbitals on the sp^2 carbons. Because three points determine a plane, each sp^2 hybridized carbon and the two atoms bonded to it are in a plane. The two p orbitals, in order to achieve maximum orbital–orbital overlap, must be parallel to each other with the result that all six atoms are in the same plane. A molecular orbital description of a carbon–carbon double bond is shown in Figure 1.6.

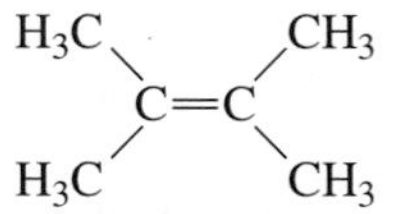

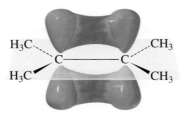

the six carbon atoms in the molecule
are in the same plane

It is important to remember that the π bond represents the cloud of electrons that is above and below the plane defined by the two sp^2 carbons and the four atoms bonded to them.

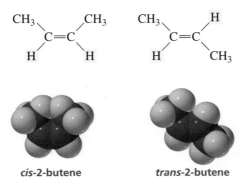

p orbitals overlap to form
a π bond

Because the two *p* orbitals that form the π bond must be parallel to one another to achieve maximum overlap, carbon–carbon double bonds, unlike carbon–carbon single bonds, are not free to rotate. If rotation around a double bond did occur, the two *p* orbitals would cease to overlap and the π bond would be destroyed (Figure 3.1).

Since there is no rotation around a carbon–carbon double bond, an alkene such as 2-butene can exist in two distinct forms: the hydrogens bonded to the sp^2 carbons can be on the same side of the double bond or on opposite sides of the double bond. The isomer with the hydrogens on the same side of the double bond is called the **cis isomer** (*cis* is Latin for "on this side"), and the isomer with the hydrogens on opposite sides of the double bond is called the **trans isomer** (*trans* is Latin for "across").

**3.4
CIS–TRANS
ISOMERISM**

cis-2-butene   trans-2-butene

A pair of isomers such as *cis*-2-butene and *trans*-2-butene are called **cis–trans stereoisomers** or **geometric isomers.** They have the same atoms and the same order of linkage of the atoms, but they differ in the spatial (stereo) arrangement of the atoms.

If one of the sp^2 carbons of the double bond is attached to two identical substituents, there is only one possible structure for the alkene. Cis and trans stereoisomers are not possible for such a compound.

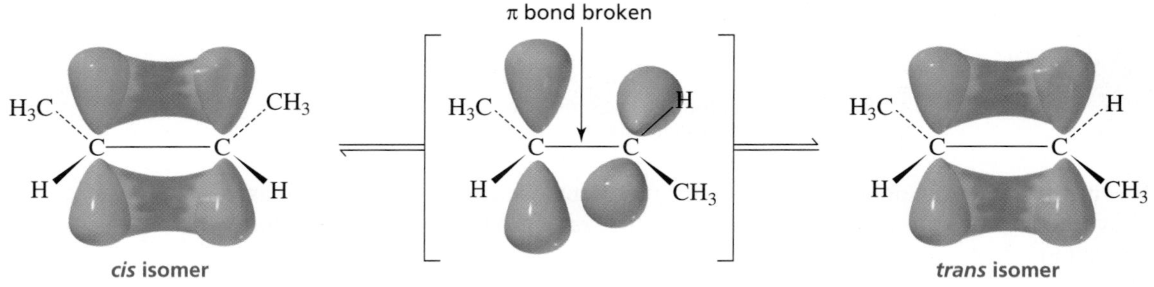

▲ **Figure 3.1**
Rotation about the carbon–carbon double bond destroys the π bond.

**cis and trans isomers are not possible for these compounds because
two substituents on one of the sp² carbons are the same**

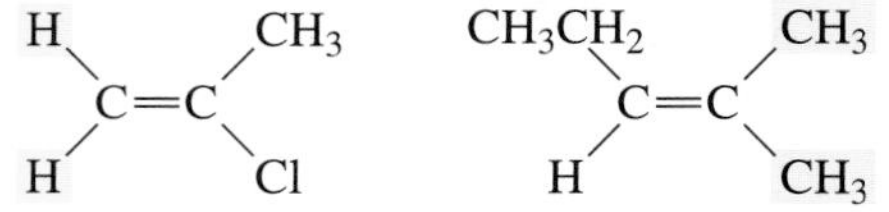

Since double bonds do not rotate, the cis and trans isomers cannot interconvert (except under extreme conditions). This means that they can be separated from each other. The two isomers are different compounds with different physical properties, such as different boiling points and different dipole moments. Notice that *trans*-2-butene and *trans*-1,2-dichloroethene have resultant dipole moments (μ) of zero because the bond dipole moments cancel.

Cis and trans stereoisomers can be interconverted under extreme conditions. Because the conditions are so extreme, cis–trans interconversion is not a practical laboratory process. Interconversion occurs when the molecule absorbs sufficient heat or light energy to cause the π bond to break. Once the π bond is broken, the remaining σ bond is free to rotate.

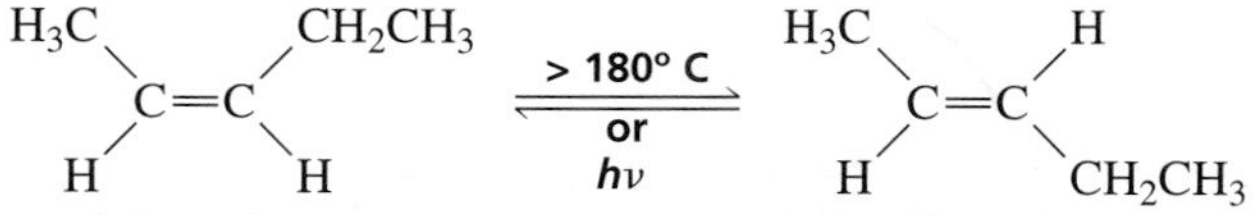

PROBLEM 6◆

a. Which of the following compounds can exist as cis–trans stereoisomers?

b. For those compounds, draw and label the cis and trans stereoisomers.

1. $CH_3CH{=}CHCH_2CH_3$ 3. $CH_3CH{=}CHCH_3$

2. $CH_3C{=}CHCH_3$ 4. $CH_3CH_2CH{=}CH_2$
    $|$
    CH_3

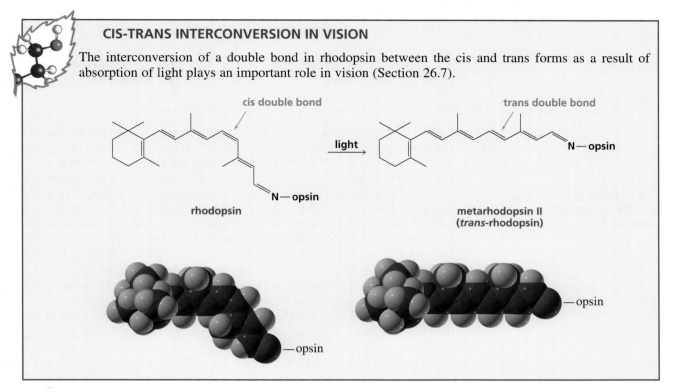

CIS-TRANS INTERCONVERSION IN VISION

The interconversion of a double bond in rhodopsin between the cis and trans forms as a result of absorption of light plays an important role in vision (Section 26.7).

cis double bond

trans double bond

light

N—opsin

rhodopsin

metarhodopsin II
(*trans*-rhodopsin)

—opsin

—opsin

As long as each of the sp^2 carbons of an alkene is bonded to only one substituent, we can use the terms cis and trans to designate the structure of the alkene: if the hydrogens are on the same side of the double bond, it is the **cis isomer;** if they are on opposite sides of the double bond, it is the **trans isomer.** But how would you determine cis and trans isomers for a compound such as 1-bromo-2-chloropropene that has three substituents bonded to the sp^2 carbons?

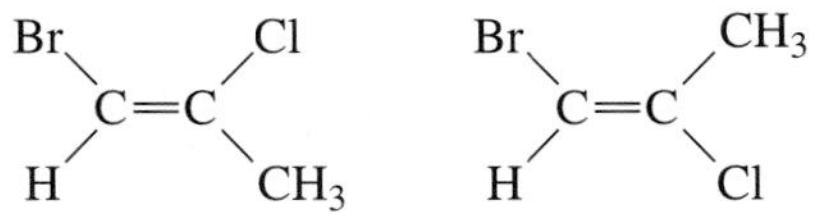

**3.5
THE *E, Z* SYSTEM OF
NOMENCLATURE**

Which isomer is cis and which is trans?

For such compounds, the cis–trans system of nomenclature cannot be used. The *E, Z* system of nomenclature was devised for such situations.[1]

In order to name an isomer by the *E, Z* system, first determine the relative priorities of the two groups bonded to one of the sp^2 carbons and then the relative priorities of the two groups bonded to the other sp^2 carbon. (Rules for assigning relative priorities are explained next.) If the high-priority groups are on the same side of the double bond, the isomer is said to have the *Z* configuration (*Z* is for *zusammen*, German for "together"). If the high-priority groups are on opposite sides of the double bond, the isomer has the *E* configuration (*E* is for *entgegen*, German for "opposite").

---

[1]Actually, since the *E, Z* system of nomenclature can be used for all alkene stereoisomers, the IUPAC prefers the *E* and *Z* designations. However, many chemists continue to use the cis and trans designations for simple molecules.

high priority   high priority   high priority   low priority

C=C   C=C

low priority   low priority   low priority   high priority

the Z isomer   **the E isomer**

The following priority rules are used to determine the E and Z isomers.

The Z isomer has the high-priority groups on the same side.

• Rule 1. The relative priorities of the groups depend on the atomic numbers of the atoms bonded directly to a particular sp^2 carbon. The greater the atomic number, the higher the priority. For example, in naming the isomers of 1-bromo-2-chloropropene, one of the sp^2 carbons is bonded to a bromine and to a hydrogen. Bromine has a greater atomic number than hydrogen, so bromine has the higher priority. The other sp^2 carbon is bonded to a chlorine and to a carbon. Chlorine has a greater atomic number than carbon, so chlorine has the higher priority. (Notice that you use the atomic number of carbon, not the formula weight of the methyl group, because the priorities are based on the atomic numbers of atoms and not on the formula weights of groups.) The isomer on the left has the high-priority groups (Br and Cl) on the same side of the double bond, so it is the **Z isomer** (Zee Groups are on Zee Zame Zide). The isomer on the right has the high-priority atoms on opposite sides of the double bond, so it is the **E isomer.**

Br   Cl   Br   CH₃

C=C   C=C

H   CH₃   H   Cl

(Z)-1-bromo-2-chloropropene   **(E)-1-bromo-2-chloropropene**

• Rule 2. In the case of a tie in the atomic numbers, the atoms that are attached to the "tied" atoms must be considered. In the following pair of isomers, chlorine has a greater atomic number than carbon, so the chloro group has a higher priority than the chloromethyl group. Both atoms bonded to the other sp^2 carbon are carbon, so there is a tie at this point. One of the carbons is bonded to C, C, and H; the other is bonded to C, H, and H. A carbon can be canceled in each of the two groups. Of the remaining atoms, carbon has a greater atomic number than hydrogen, so the isopropyl group has a higher priority than the ethyl group. The E and Z isomers are as shown.

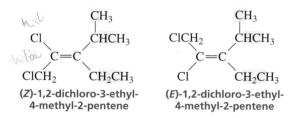

CH₃   CH₃

Cl   CHCH₃   ClCH₂   CHCH₃

C=C   C=C

ClCH₂   CH₂CH₃   Cl   CH₂CH₃

(Z)-1,2-dichloro-3-ethyl-   **(E)-1,2-dichloro-3-ethyl-**
4-methyl-2-pentene   **4-methyl-2-pentene**

• Rule 3. If an atom is doubly bonded to another atom, the priority system treats it as if it were bonded to two of those atoms. One of the sp^2 carbons in the following pair of isomers is bonded to an ethyl group and to a vinyl group. Because the atoms immediately bonded to the sp^2 carbon are both carbons, there is a tie. The carbon of the ethyl group is bonded to C, H, and H. The carbon of the vinyl group is bonded to an H and doubly bonded to a C. Therefore, the carbon of the vinyl group is considered to be bonded to C, C, and H. Consequently, the vinyl group has a higher priority than the ethyl group. Both atoms that are bonded to the other sp^2 carbon are carbons, so there is a tie. The carbon of the hydroxymethyl group is bonded to O, H, and H, and the carbon of the isopropyl group to C, C, and H. Of these six atoms, oxygen has the greatest atomic number, so the hydroxymethyl group has the higher priority. (Notice that you do not add the atomic numbers; you take the single atom with the greatest atomic number.)

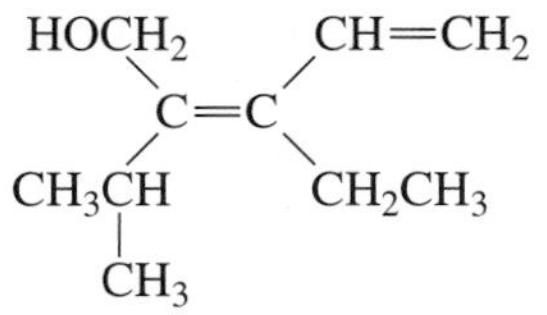

the *Z* isomer 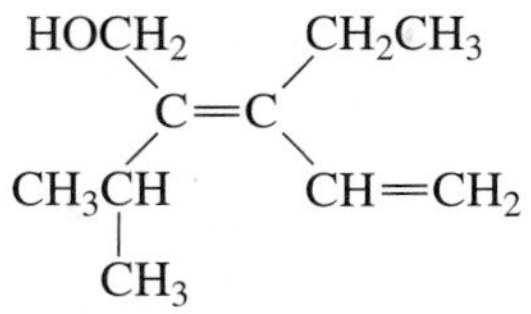   the *E* isomer

- Rule 4. In the case of isotopes (atoms with the same atomic number but different mass numbers), the mass number is used to determine their relative priorities. Deuterium and hydrogen have the same atomic number, but deuterium has a greater mass number, so it has the higher priority. The carbons that are bonded to the other sp^2 carbon are both bonded to C, H, and H, so you must go out to the next set of atoms to break the tie. The second carbon of the isobutyl group is bonded to C, C, and H, while the second carbon of the pentyl group is bonded to C, H, and H, so the isobutyl group has the higher priority.

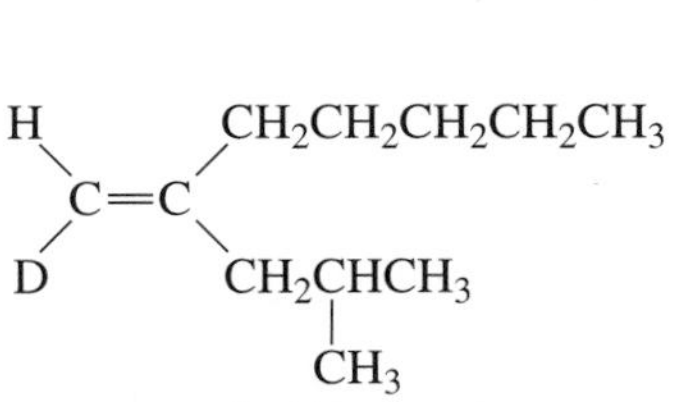

the *Z* isomer 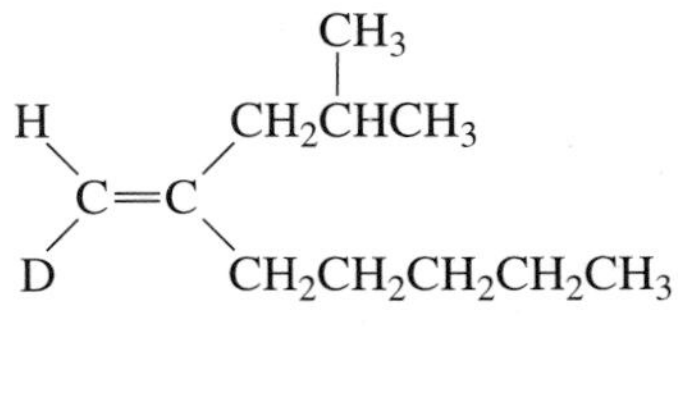   the *E* isomer

---

PROBLEM 7

a. Draw and label the *E* and *Z* isomers for each of the following compounds.

1. $CH_3CH_2CH{=}CHCH_3$

3. 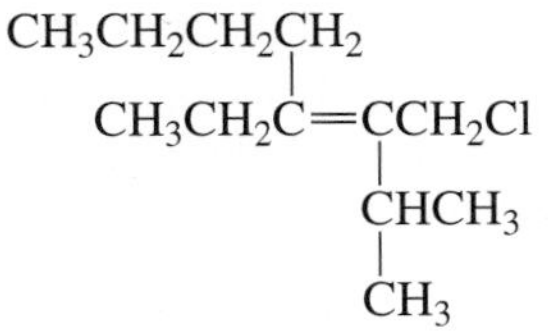
$$CH_3CH_2\overset{\overset{\displaystyle CH_3CH_2CH_2CH_2}{|}}{C}{=}\overset{\overset{\displaystyle }{}}{C}CH_2Cl$$
$$\underset{\underset{\displaystyle CH_3}{|}}{\overset{|}{C}}HCH_3$$

2. $CH_3CH_2\underset{\underset{\displaystyle Cl}{|}}{C}{=}CHCH_2CH_3$

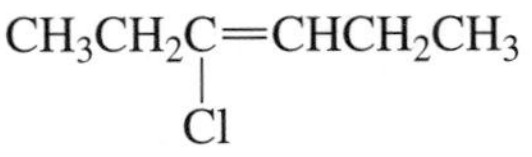

4. 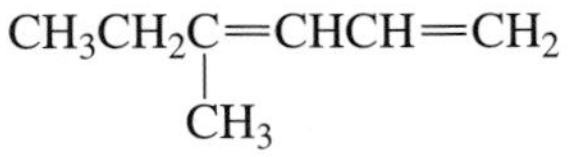 $CH_3CH_2\underset{\underset{\displaystyle CH_3}{|}}{C}{=}CHCH{=}CH_2$

b. Draw the structure of (Z)-3-isopropyl-2-heptene.

We have seen that the *functional group* is the site of reactivity in a molecule. In each of the chapters in this book where we are concerned with the reactivity of a particular functional group, we will look at the structure of the functional group to see if its structure allows us to predict the kind of reactions that it will undergo.

**3.6
REACTIVITY
CONSIDERATIONS**

It is important to be able to figure out why molecules react the way they do. If you understand why a functional group reacts the way it does, you will not need to memorize hundreds of organic reactions. When you are confronted with a reaction you have never seen before, if you understand how the structure of a molecule affects its reactivity, you will likely be able to predict the products of the reaction.

Chemistry is all about the interaction between positive and negative charges. It is these forces of attraction that make chemical reactions happen. Therefore, a very important rule determines the reactivity of organic compounds: *Electron-rich atoms or molecules are attracted to electron-deficient atoms or molecules.*

Let's look at this rule a little more closely. An electron-deficient ion or molecule is called an **electrophile**. An electrophile can accept a pair of electrons or it can have a single electron in need of a partner in order to complete its octet. An electrophile, therefore, looks for electrons. Literally, electrophile means electron loving (*phile* is the Greek suffix for "loving").

a few examples of electrophiles

$$H^+ \quad CH_3\overset{+}{C}H_2 \quad BH_3 \qquad\qquad :\ddot{B}r\cdot \quad R\ddot{O}\cdot$$

these are electophiles because they can accept a pair of electrons   these are electophiles because they are seeking an electron

Nucleophiles react with electrophiles.

An electron-rich atom or molecule is called a **nucleophile.** A nucleophile has a pair of electrons it is willing to share. Notice that some nucleophiles are neutral and some are negatively charged. Since a nucleophile can share electrons and an electrophile can accept electrons, it is not surprising that they get together. Here, then, is another way of stating the foregoing rule: *A nucleophile and an electrophile react with each other.*

a few examples of nucleophiles

$$H\ddot{O}:^- \quad :\ddot{C}l:^- \quad CH_3\ddot{N}H_2 \quad H_2\ddot{O}:$$

these are nucleophiles because they have a pair of electrons to share

You know that a π bond is weaker than a σ bond (Section 1.14). The π bond, therefore, is the weakest bond in an alkene, which means that it is the bond most likely to break in a reaction. You also know that the π bond of an alkene consists of a cloud of electrons above and below the σ bond (Section 3.3). As a result of this cloud of electrons, an alkene is an electron-rich molecule; it is a nucleophile. You can, therefore, predict that an alkene will react with an electrophile. If a reagent such as hydrogen bromide is added to an alkene, the alkene (a nucleophile) will react with the partially positively charged hydrogen (an electrophile) of hydrogen bromide and a carbocation will be formed. In the second step of the reaction, the positively charged carbocation (an electrophile) will react with the negatively charged bromide ion (a nucleophile) to form an alkyl halide.

$$CH_3CH{=}CHCH_3 + \overset{\delta+}{H}{-}\overset{\delta-}{Br} \longrightarrow CH_3CH\underset{\underset{H}{|}}{\overset{+}{-}}CHCH_3 + Br^- \longrightarrow CH_3CH\underset{\underset{Br}{|}}{-}\underset{\underset{H}{|}}{C}HCH_3$$

a carbocation   2-bromobutane an alkyl halide

Such a description of the step-by-step process by which reactants are changed into products is called the **mechanism of the reaction.** The mechanism can be seen more clearly if "curved arrows" are used to show the movement of the electrons. The curved arrows are drawn to show how the electrons move as new covalent bonds are formed and existing covalent bonds are broken. Remember that the arrows show how the electrons move, so *the arrows are drawn from an electron-rich center to an electron-deficient center.* A two-sided arrowhead (⤵) represents the simultaneous movement of two electrons. We will see later that a one-sided arrowhead (⤴) represents the movement of one electron. (Notice that we have described these arrows as "curved" arrows in order to distinguish them from the "straight" arrows that link reactants on the left with products on the right.)

For the reaction of 2-butene with HBr, an arrow is drawn to show that the two electrons of the π bond are attracted to the partially positively charged hydrogen of HBr. The hydrogen, however, is not free to accept this pair of electrons because it is already bonded to a bromine, and hydrogen can be bonded to only one atom at a time. Therefore, as the π electrons move toward the hydrogen, the H—Br bond breaks, with bromine keeping the bonding electrons. Notice that the π electrons are pulled away from one carbon but remain attached to the other. The product of the first step of the reaction is a carbocation; the two electrons that formerly formed the π bond of the alkene now form a σ bond between carbon and hydrogen. The sp^2 carbon that did not form the new bond with hydrogen is positively charged because it has lost a share in a pair of electrons.

$$CH_3CH = CHCH_3 \;+\; H - \overset{\cdot\cdot}{\underset{\cdot\cdot}{Br}} : \;\longrightarrow\; CH_3\overset{+}{CH} - \underset{\underset{H}{|}}{CH}CH_3 \;+\; : \overset{\cdot\cdot}{\underset{\cdot\cdot}{Br}} :^{\bar{}}$$

In the second step of the reaction, a pair of nonbonded electrons on the negatively charged bromide ion forms a bond with the positively charged carbon of the carbocation. Notice that both steps of the reaction involve the reaction of an electrophile with a nucleophile.

$$CH_3\overset{+}{CH} - \underset{\underset{H}{|}}{CH}CH_3 \;+\; : \overset{\cdot\cdot}{\underset{\cdot\cdot}{Br}} :^{\bar{}} \;\longrightarrow\; CH_3\underset{\underset{:Br:}{|}}{CH} - \underset{\underset{H}{|}}{CH}CH_3$$

A FEW WORDS ABOUT CURVED ARROWS

1. Make certain that the arrows are drawn in the direction of the electron flow and never against the flow. This means that an arrow will always be drawn away from a negative charge and/or toward a positive charge.

correct	incorrect

$$CH_3 - \underset{\underset{CH_3}{|}}{\overset{\overset{:\ddot{O}:^{\bar{}}}{|}}{C}} - \overset{\cdot\cdot}{\underset{\cdot\cdot}{Br}} : \;\longrightarrow\; CH_3 - \underset{\underset{CH_3}{|}}{\overset{\overset{:\ddot{O}}{||}}{C}} \;+\; : \overset{\cdot\cdot}{\underset{\cdot\cdot}{Br}} :^{\bar{}} \qquad CH_3 - \underset{\underset{CH_3}{|}}{\overset{\overset{:\ddot{O}:^{\bar{}}}{|}}{C}} - \overset{\cdot\cdot}{\underset{\cdot\cdot}{Br}} : \;\longrightarrow\; CH_3 - \underset{\underset{CH_3}{|}}{\overset{\overset{:\ddot{O}}{||}}{C}} \;+\; : \overset{\cdot\cdot}{\underset{\cdot\cdot}{Br}} :^{\bar{}}$$

$$CH_3 - \underset{\underset{H}{|}}{\overset{+}{\ddot{O}}} - H \;\longrightarrow\; CH_3 - \ddot{O} - H \;+\; H^+ \qquad CH_3 - \overset{+}{\ddot{O}} - H \;\longrightarrow\; CH_3 - \ddot{O} - H \;+\; H^+ $$

2. Curved arrows are drawn to indicate the movement of electrons. Never use a curved arrow to indicate the movement of an atom (in this case the movement of a proton).

correct

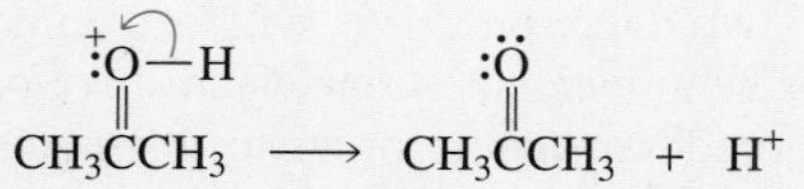

$$\overset{+}{:}\!O\!-\!H \qquad\qquad :\overset{..}{O}:$$
$$CH_3\overset{\|}{C}CH_3 \longrightarrow CH_3\overset{\|}{C}CH_3 + H^+$$

incorrect

$$\overset{+}{:}\!O\!-\!H \qquad\qquad :\overset{..}{O}:$$
$$CH_3\overset{\|}{C}CH_3 \longrightarrow CH_3\overset{\|}{C}CH_3 + H^+$$

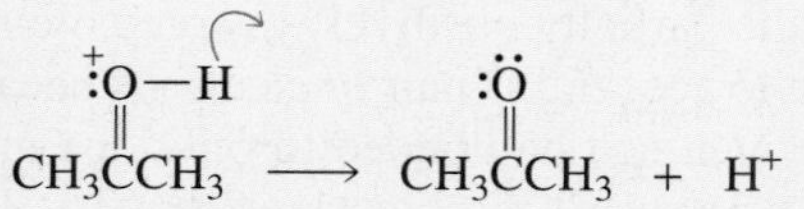

3. The arrow starts at the electron source. It does not start at an atom (in this case at a carbon atom).

correct

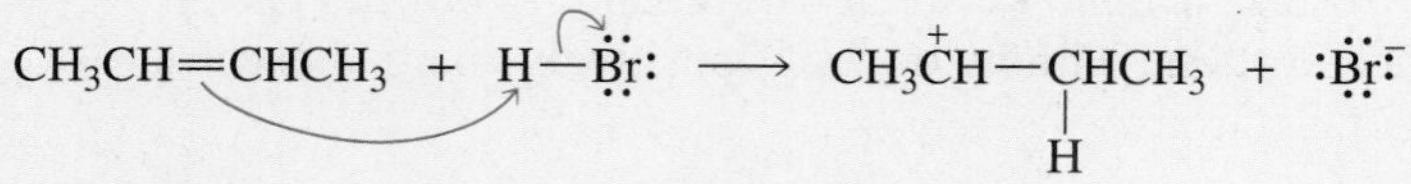

$$CH_3CH\!=\!CHCH_3 + H\!-\!\overset{..}{\underset{..}{Br}}: \longrightarrow CH_3\overset{+}{C}H\!-\!CHCH_3 + :\overset{..}{\underset{..}{Br}}:^-$$
$$\qquad\qquad\qquad\qquad\qquad\qquad\qquad\qquad\qquad\quad |$$
$$\qquad\qquad\qquad\qquad\qquad\qquad\qquad\qquad\qquad\quad H$$

incorrect

$$CH_3CH\!=\!CHCH_3 + H\!-\!\overset{..}{\underset{..}{Br}}: \longrightarrow CH_3\overset{+}{C}H\!-\!CHCH_3 + :\overset{..}{\underset{..}{Br}}:^-$$
$$\qquad\qquad\qquad\qquad\qquad\qquad\qquad\qquad\qquad\quad |$$
$$\qquad\qquad\qquad\qquad\qquad\qquad\qquad\qquad\qquad\quad H$$

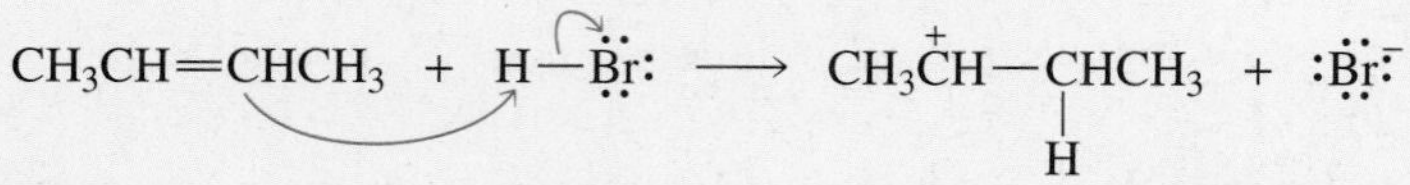

You will find it helpful to do the exercise on drawing curved arrows that is in the Study Guide/Solution Manual (Special Topic II).

Solely from the knowledge that an electrophile reacts with a nucleophile, you have been able to predict that the product of the reaction of 2-butene and HBr is 2-bromobutane. Overall, the reaction involves the addition of 1 mole of HBr to 1 mole of the alkene. The reaction, therefore, is called an **addition reaction.** Since the first step of the reaction involves the addition of an electrophile (H⁺) to the alkene, the reaction is more precisely called an **electrophilic addition reaction.** Electrophilic addition reactions are characteristic reactions of alkenes.

Since the reaction of 2-butene with HBr is the first organic reaction that you have encountered, you may think that it would be easier just to memorize that 2-bromobutane is the product of the reaction without trying to understand the mechanism of the reaction that explains why 2-bromobutane is the product. However, as the number of reactions that you encounter increases, you will find that, if you have a thorough understanding of the mechanism of each of the reactions, you will become aware of the unifying features behind the discipline known as organic chemistry, and mastering organic chemistry will be much easier and a lot more fun.

PROBLEM 8 ◆

Which of the following are electrophiles and which are nucleophiles?

$$H^- \qquad AlCl_3 \qquad CH_3O^- \qquad CH_3C\!\equiv\!CH \qquad CH_3\overset{+}{C}HCH_3 \qquad NH_3$$

PROBLEM 9

Draw arrows to show the movement of electrons in the three reactions in Section 1.20.

PROBLEM 10

Use curved arrows to show the movement of electrons in each of the following reaction steps.

a. $CH_3\overset{\overset{O}{\|}}{C}-O-H \ + \ H\ddot{\underset{..}{O}}{:}^{-} \ \longrightarrow \ CH_3\overset{\overset{O}{\|}}{C}-O^- \ + \ H_2\ddot{\underset{..}{O}}{:}$

b. $+ \ Br^+ \ \longrightarrow$

c. $CH_3\overset{\overset{\ddot{O}:}{\|}}{C}OH \ + \ H-\overset{+}{\underset{\underset{H}{|}}{O}}-H \ \longrightarrow \ CH_3\overset{\overset{+\ddot{O}H}{\|}}{C}OH \ + \ H_2O$

d. $CH_3-\overset{\overset{\displaystyle CH_3}{|}}{\underset{\underset{\displaystyle CH_3}{|}}{C}}-Cl \ \longrightarrow \ CH_3-\overset{\overset{\displaystyle CH_3}{|}}{\underset{\underset{\displaystyle CH_3}{|}}{C^+}} \ + \ Cl^-$

The mechanism of a reaction shows the various steps that are involved in the conversion of a reactant into a product. A **reaction coordinate diagram** describes the energy changes that take place in each of the steps.

**3.7
THERMODYNAMICS
AND KINETICS**

Before you can understand the energy changes that take place in a reaction such as the addition of HBr to an alkene, you must have an understanding of *thermodynamics,* which describes a reaction at equilibrium, and an appreciation of *kinetics,* which deals with the rates of chemical reactions.

If we consider a reaction in which A is converted to B, the thermodynamics of the reaction tells us the relative amounts of A and B that are present when the reaction has reached equilibrium, whereas the kinetics of the reaction tells us how fast A is converted into B.

$$A \ \rightleftharpoons \ B$$

Reaction Coordinate Diagrams

The energy changes that take place during the course of a reaction, as reactants are converted into products, can be visualized using a reaction coordinate diagram. In a reaction coordinate diagram, the energy of the species involved in the reaction is plotted against the progress of the reaction. A reaction progresses from left to right, starting with the reactants and ending with the products. The energy of the reactants is plotted on the left-hand side of the x-axis, and the energy of the products is plotted on the right-hand side of the x-axis. The more stable the species, the lower its energy.

As the reactants are converted into products, the reaction passes through a *maximum* energy state called the **transition state.** The structure of the transition state is between the structure of the reactants and the structure of the products. Bonds that break and bonds that form in the conversion of reactants to products are partially broken and partially formed in the transition state. A typical reaction coordinate diagram is shown in Figure 3.2.

$$A-B \ + \ C \ \rightleftharpoons \ A \ + \ B-C$$
Reactants   **Products**

Figure 3.2 ▶
A reaction coordinate diagram. The dashed lines in the
transition state indicate
bonds that are partially
formed or partially broken.

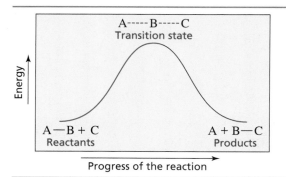

Thermodynamics

The field of chemistry that describes the properties of a system at equilibrium is
called **thermodynamics.** When reactants form products, an equilibrium results that
depends on the concentrations of reactants and products. This equilibrium can be
expressed numerically as an equilibrium constant, K_{eq}. For example, in a reaction in
which m moles of A react with n moles of B to form s moles of C and t moles of D,
K_{eq} is equal to the relative concentrations of products and reactants at equilibrium.

$$m\,A \;+\; n\,B \;\rightleftharpoons\; s\,C \;+\; t\,D$$

$$K_{eq} = \frac{[\text{products}]}{[\text{reactants}]} = \frac{[C]^s\,[D]^t}{[A]^m\,[B]^n}$$

The relative concentrations of products and reactants at equilibrium depend on
the relative stabilities of the reactants and products; the more stable the compound,
the greater its concentration at equilibrium. Thus, if the products are more stable
(have a lower free energy) than the reactants (Figure 3.3a), there will be a higher
concentration of products than reactants at equilibrium. Therefore, K_{eq} will be
greater than 1. In contrast, if the reactants are more stable than the products (Figure
3.3b), there will be a higher concentration of reactants than products at equilibrium,
and K_{eq} will be less than 1.

Several thermodynamic parameters are used to describe a reaction at equilibrium. The difference between the free energy content of the products and the free
energy content of the reactants at equilibrium under standard conditions (all species
at 1 M, temperature at 25 °C, pressure at 1 atm) is called the **Gibbs standard free
energy change** ($\Delta G°$).

$$\Delta G° = (\text{free energy of the products}) - (\text{free energy of the reactants})$$

From this equation you can see that, if the products have a lower free energy (are
more stable) than the reactants, $\Delta G°$ will be negative (Figure 3.3a). The reaction
will release more energy into the system than is consumed from the system—it will
be an **exergonic reaction.** If the products have a higher free energy (are less stable)
than the reactants, $\Delta G°$ will be positive and the reaction will consume more energy
from the system than is released into the system—it will be an **endergonic reaction**
(Figure 3.3b). (Notice that the terms *exergonic* and *endergonic* refer to whether the
reaction has a negative $\Delta G°$ or a positive $\Delta G°$, respectively. Do not confuse these
terms with *exothermic* and *endothermic,* which are defined later.)

As we have already mentioned, the relative concentrations of reactants and products at equilibrium (K_{eq}) depend on the relative free energies (stabilities) of those
reactants and products at equilibrium ($\Delta G°$). In other words, the difference in free

**Josiah Willard Gibbs
(1839–1903)** *was born in New
Haven, Connecticut, the son of
a Yale professor. In 1863 he
obtained the first Ph.D.
awarded by Yale in engi-
neering. After studying in
France and Germany, he
returned to Yale to become a
professor of mathematical
physics. His work on free
energy received little attention
for more than 20 years because
few chemists could understand
his mathematical treatment
and because Gibbs published it
in* Transactions of the
Connecticut Academy of
Sciences, *a relatively obscure
journal. In 1950 he was elected
to the Hall of Fame for Great
Americans.*

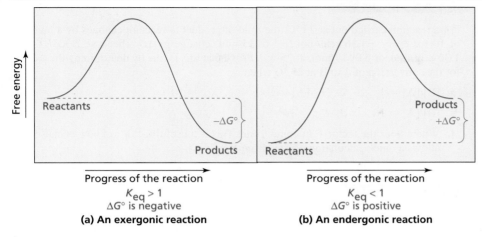

$K_{eq} > 1$
$\Delta G°$ is negative
(a) An exergonic reaction

$K_{eq} < 1$
$\Delta G°$ is positive
(b) An endergonic reaction

▲ **Figure 3.3**
Reaction coordinate diagrams for (a) a reaction in which the products are more stable than the reactants (an exergonic reaction) and (b) a reaction in which the products are less stable than the reactants (an endergonic reaction).

energy determines the position of equilibrium. These two quantities are related by the equation

$$\Delta G° = -RT \ln K_{eq}$$

where R is the gas constant (1.986×10^{-3} kcal K^{-1} mol^{-1}, or 8.314×10^{-3} kJ K^{-1} mol^{-1}; 1 kcal = 4.184 kJ); T is the temperature in Kelvin (K = °C + 273; therefore, 25 °C = 298 K).

Base 10 logarithms (log) are more commonly used than natural logarithms (ln). The two are related by the equation

$$\ln x = 2.303 \log x$$

Substitution results in the following relationship between the Gibbs standard free energy change ($\Delta G°$) and the equilibrium constant (K_{eq}). This is the equation that is commonly used to relate the difference in free energy and the position of equilibrium. By solving Problem 11, you will see that a small difference in $\Delta G°$ gives rise to a large difference in the ratio of product to reactant.

$$\Delta G° = -2.303 \, RT \log K_{eq}$$

The gas constant R *is thought to be named after* **Henri Victor Regnault,** *who was commissioned by the French Minister of Public Works in 1842 to redetermine all the physical constants involved in the design and operation of the steam engine. Regnault was known for his work on the thermal properties of gases. Later, van't Hoff (p. 186) determined, while studying the thermodynamics of dilute solutions, that* R *could be used for all chemical equilibria.*

PROBLEM 11

a. Which of the monosubstituted cyclohexanes in Table 2.10 has a negative $\Delta G°$ for interconversion of the axial and equatorial conformers?

b. Which one has the most negative $\Delta G°$ value?

c. Which one has the greatest preference for the equatorial position?

d. Calculate $\Delta G°$ for interconversion of the axial and equatorial conformers of methylcyclohexane.

PROBLEM 12

For a reaction carried out at 25 °C, the ratio of product to reactant changes by a factor of 10 for every −1.36 kcal/mol (−5.69 kJ/mol) change in $\Delta G°$ (because $2.303RT =$ 1.36 kcal/mol, or 5.09 kJ/mol, at 25 °C). Prove that this is true by determining the $\Delta G°$ for the conversion of A to B at 25 °C when

a. [B]/[A] = 1   **c.** [B]/[A] = 100   **e.** $K_{eq} = 10^4$

b. [B]/[A] = 10   **d.** $K_{eq} = 10^3$

f. Where does the factor 1.36 come from? This is a useful factor to know because it allows a quick calculation of $\Delta G°$ from K_{eq}.

PROBLEM 13 / SOLVED

a. The $\Delta G°$ for conversion of "axial" fluorocyclohexane to "equatorial" fluorocyclohexane at 25 °C is −0.25 kcal/mol. Calculate the percentage of fluorocyclohexane molecules that have the fluoro substituent in the equatorial position.

b. Do the same calculation for isopropylcyclohexane (the $\Delta G°$ at 25 °C is −2.1 kcal/mol).

c. Why does isopropylcyclohexane have a greater percentage of conformers with the substituent in the equatorial position?

SOLUTION TO 13a

$$\text{fluorocyclohexane} \rightleftharpoons \text{fluorocyclohexane}$$
$$\text{axial} \qquad\qquad\qquad \text{equatorial}$$

$$\Delta G° = -0.25 \text{ kcal/mol at 25 °C}$$

$$\Delta G° = -2.303RT \log K_{eq}$$

$$-0.25 \ \frac{\text{kcal}}{\text{mol}} = -2.303 \times 1.986 \times 10^{-3} \ \frac{\text{kcal}}{\text{mol K}} \times 298 \text{ K} \times \log K_{eq}$$

$$\log K_{eq} = 0.183$$

$$K_{eq} = 1.53 = \frac{[\text{fluorocyclohexane}]_{\text{equatorial}}}{[\text{fluorocyclohexane}]_{\text{axial}}} = \frac{1.53}{1}$$

Now we must determine the percentage of the total that is equatorial:

$$\frac{[\text{fluorocyclohexane}]_{\text{equatorial}}}{[\text{fluorocyclohexane}]_{\text{equatorial}} + [\text{fluorocyclohexane}]_{\text{axial}}} = \frac{1.53}{1.53 + 1} = \frac{1.53}{2.53} = 60\%$$

The Gibbs standard free energy change ($\Delta G°$) has an enthalpy ($\Delta H°$) component and an entropy ($\Delta S°$) component.

$$\Delta G° = \Delta H° - T\Delta S°$$

The **enthalpy** term ($\Delta H°$), which is sometimes called the heat of the reaction, is the heat given off or the heat absorbed during the course of a reaction. Heat is given off when bonds are formed, and heat is consumed when bonds are broken. So $\Delta H°$ is a measure of the bond-making and bond-breaking processes that occur as reactants are converted into products.

If the bonds that are formed in a reaction are stronger than the bonds that are broken, more energy will be released as a result of bond formation than will be consumed in the bond-breaking process, and $\Delta H°$ will be negative. A reaction with a negative $\Delta H°$ is called an **exothermic reaction.** If the bonds that are broken are stronger than those that are formed, $\Delta H°$ will be positive. A reaction with a positive $\Delta H°$ is called an **endothermic reaction.**

$$\Delta H° = \text{(energy of bonds being broken)} - \text{(energy of bonds being formed)}$$

Entropy $(\Delta S°)$ is defined as the degree of disorder. It is a measure of the freedom of motion in a system. Restricting the freedom of motion of a molecule causes a decrease in entropy. For example, in a reaction in which two molecules come together to form a single molecule, the entropy in the product will be less than the entropy in the reactants, because two individual molecules can move in ways that are not possible when the two are bound together in a single molecule. In such a reaction, $\Delta S°$ will be negative. But in a reaction in which a single molecule is cleaved into two separate molecules, the products will have greater freedom of motion than the reactant, and $\Delta S°$ will be positive.

$$\Delta S° = \text{(freedom of motion of products)} - \text{(freedom of motion of reactants)}$$

A reaction with a negative $\Delta G°$ is said to have a favorable driving force, or a favorable equilibrium constant, since the products are more stable than the reactants. If you examine the expression for the Gibbs standard free energy change, you will find that negative values of $\Delta H°$ and positive values of $\Delta S°$ contribute to make $\Delta G°$ negative. In other words, *the formation of products with stronger bonds and with greater freedom of motion causes $\Delta G°$ to be negative.*

Since values of $\Delta H°$ are relatively easy to calculate, organic chemists frequently think of reactions only in terms of $\Delta H°$. This is all right if the reaction involves only small changes in entropy, because in such cases the $T\Delta S°$ term is small and therefore the value of $\Delta H°$ is very close to the value of $\Delta G°$. Ignoring the entropy term can be a dangerous practice, however, because many organic reactions occur with a significant change in entropy and so have significant $T\Delta S°$ terms. (Organic reactions that occur in biological systems commonly have significant $T\Delta S°$ terms.) It is permissible to use values of $\Delta H°$ to approximate whether a reaction occurs with a favorable equilibrium constant, but if a precise answer is needed, values of $\Delta G°$ must be used. When $\Delta G°$ values are used to construct reaction coordinate diagrams, the y-axis is free energy; when $\Delta H°$ values are used, the y-axis is potential energy.

The formation of products with stronger bonds and with greater freedom of motion causes $\Delta G°$ to be negative.

PROBLEM 14◆

For which reaction will $\Delta S°$ be more significant?

a. A $\rightleftharpoons$ B   or   A + B $\rightleftharpoons$ C

b. A + B $\rightleftharpoons$ C   or   A + B $\rightleftharpoons$ C + D

PROBLEM 15◆

For a reaction with $\Delta H° = -12 \text{ kcal mole}^{-1}$ and $\Delta S° = 0.01 \text{ kcal deg}^{-1} \text{ mol}^{-1}$, calculate the $\Delta G°$ and the equilibrium constant:

a. at 30 °C   **b.** at 150 °C

c. What can you conclude about the relationship between $\Delta G°$ and T?

d. What can you conclude about the relationship between K_{eq} and T?

Values of $\Delta H°$ can be calculated from bond dissociation energies (Table 3.1). For example, the $\Delta H°$ for the addition of HBr to ethene is calculated as shown below. The value of $\Delta H°$ for breaking a bond (the bond dissociation energy) is indicated by the special term $DH°$. Notice that the value for the bond dissociation energy of the π bond is obtained by taking the bond dissociation energy of the double bond ($\sigma + \pi = 152$ kcal/mol) and subtracting the bond dissociation energy of the σ bond (88 kcal/mol).

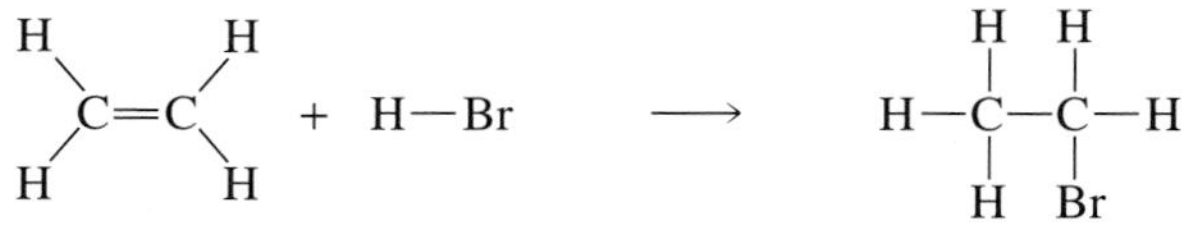

$$
\begin{array}{ccc}
\underset{\substack{\text{H} \\ \diagdown}}{\text{H}}C=C\underset{\substack{\text{H} \\ \diagup}}{\overset{\text{H}}{}} + \text{H}-\text{Br} & \longrightarrow & \text{H}-\underset{\substack{| \\ \text{H}}}{\overset{| \\ \text{H}}{\text{C}}}-\underset{\substack{| \\ \text{Br}}}{\overset{| \\ \text{H}}{\text{C}}}-\text{H}
\end{array}
$$

bonds being broken	**bonds being formed**

π bond of ethene	$DH° = 64$ kcal/mol	C—H	$DH° = 101$ kcal/mol
H—Br	$DH° = \underline{\ 88}$ kcal/mol	C—Br	$DH° = \underline{\ 69}$ kcal/mol
	$DH°_{total} = 152$ kcal/mol		$DH°_{total} = 170$ kcal/mol

$\Delta H°$ for the reaction = $DH°$ for bonds being broken − $DH°$ for bonds being formed

$$= 152 \text{ kcal/mol} - 170 \text{ kcal/mol}$$

$$= -18 \text{ kcal/mol}$$

J. Berkowitz, G.B. Ellison, D. Gutman, J. Phys. Chem. **98**, 2744 (1994).

TABLE 3.1 Homolytic Bond Dissociation Energies ($DH°$)

Y—Z $\longrightarrow$ Y· ·Z

Bond	$DH°$ kcal/mol	$DH°$ kJ/mol	Bond	$DH°$ kcal/mol	$DH°$ kJ/mol
CH_3—H	105	439	H—H	104	435
CH_3CH_2—H	101	423	F—F	38	159
$CH_3CH_2CH_2$—H	101	423	Cl—Cl	58	242
$(CH_3)_2CH$—H	99	414	Br—Br	46	192
$(CH_3)_3C$—H	97	406	I—I	36	150
			H—F	138	568
CH_3—CH_3	88	368	H—Cl	103	431
CH_3CH_2—CH_3	85	355	H—Br	88	366
$(CH_3)_2CH$—CH_3	84	351	H—I	71	297
$(CH_3)_3C$—CH_3	80	334			
			CH_3—F	108	451
H_2C=CH_2	152	635	CH_3—Cl	84	349
HC≡CH	200	836	CH_3CH_2—Cl	82	343
			$(CH_3)_2CH$—Cl	81	338
			$(CH_3)_3C$—Cl	79	330
HO—H	119	497	CH_3—Br	70	293
CH_3O—H	102	426	CH_3CH_2—Br	69	289
CH_3—OH	91	380	$(CH_3)_2CH$—Br	68	285
			$(CH_3)_3C$—Br	63	264
			CH_3—I	56	234
			CH_3CH_2—I	55	230

The value of -18 kcal/mol for $\Delta H°$—calculated by subtracting the $DH°$ for the bonds being formed from the $DH°$ for the bonds being broken—indicates that the addition of HBr to ethene is an exothermic reaction. But does this mean that the $\Delta G°$ for the reaction is also negative? Is the reaction exergonic as well as exothermic? Because $\Delta H°$ has a significant negative value (-18 kcal/mol), you can assume that $\Delta G°$ is also negative. If the value of $\Delta H°$ were close to 0, you could no longer assume that both $\Delta H°$ and $\Delta G°$ have the same sign.

You must remember that two assumptions are being made when you use $\Delta H°$ values to predict values of $\Delta G°$: the first is that the entropy changes in the reaction are small, with the result that the value of $\Delta H°$ is very close to the value of $\Delta G°$; the second is that the reaction is taking place in the gas phase.

When reactions are carried out in solution, which is the case for the vast majority of organic reactions, the solvent molecules can interact with the reagents and with the products. Polar solvent molecules cluster around a charge (either a full charge or a partial charge) on the reactant or product so that the negative poles of the solvent molecules surround the positive charge and the positive poles of the solvent molecules surround the negative charge. The interaction between a solvent and a molecule (or ion) in solution is called **solvation.**

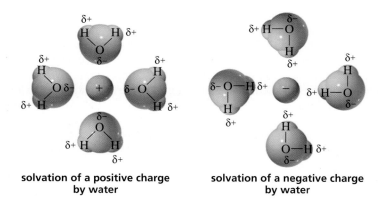

solvation of a positive charge   solvation of a negative charge
            by water   by water

Solvation can have a large effect on the $\Delta H°$ of a reaction, and it can also affect the $\Delta S°$ of a reaction. For example, in a reaction in which a polar reagent is solvated, the $\Delta H°$ for breaking the dipole–dipole interactions between the solvent and the reagent has to be taken into account. In addition, solvation of a polar reagent by a polar solvent can greatly reduce the freedom of motion of the solvent molecules, thereby making $\Delta S°$ more negative (which makes $\Delta G°$ more positive).

PROBLEM 16◆

a. Using the bond dissociation energies in Table 3.1, calculate the $\Delta H°$ for the addition of HCl to ethene.

b. Calculate $\Delta H°$ for the addition of H_2 to ethene.

c. Are the reactions exothermic or endothermic?

d. Do you expect the reactions to be exergonic or endergonic?

Kinetics

Knowing whether a given reaction is exergonic or endergonic will not tell you how fast the reaction occurs, because the $\Delta G°$ of a reaction tells you only the relative difference between the stability of the reactants and the stability of the products. It does not tell you anything about the energy barrier of the reaction, which is the energy hill that must be climbed for the reactants to be converted into products. The higher the energy barrier, the slower the reaction. **Kinetics** is the field of chemistry that deals with the rates of chemical reactions and the factors that affect those rates.

This energy barrier, indicated by $\Delta G^{\ddagger}$, is called the **free energy of activation.** It is the difference between the free energy of the transition state and the free energy of the reactants (Figure 3.4).

$$\Delta G^{\ddagger} = \text{(free energy of the transition state)} - \text{(free energy of the reactants)}$$

The greater the $\Delta G^{\ddagger}$, the slower the reaction. Like $\Delta G°$, $\Delta G^{\ddagger}$ has both an enthalpy component and an entropy component. Notice that any quantity that relates to the transition state is represented by the "double dagger" symbol (‡).

$$\Delta G^{\ddagger} = \Delta H^{\ddagger} - T\Delta S^{\ddagger}$$

$$\Delta H^{\ddagger} = \text{(enthalpy of the transition state)} - \text{(enthalpy of the reactants)}$$

$$\Delta S^{\ddagger} = \text{(entropy of the transition state)} - \text{(entropy of the reactants)}$$

Some exergonic reactions have small free energies of activation, and the reactions take place spontaneously (Figure 3.4a). Some exergonic reactions have free energies of activation that are so large that the reaction cannot take place without the addition of energy (Figure 3.4b). Endergonic reactions can also have either small free energies of activation (Figure 3.4c) or large free energies of activation (Figure 3.4d). If a reaction has a free energy of activation less than about 20 kcal/mol, there is sufficient energy at room temperature to enable the reacting molecules to get over the energy barrier. If, however, the $\Delta G^{\ddagger}$ is greater than 20 kcal/mol, energy in the form of heat has to be supplied to give the molecules sufficient energy to overcome the energy barrier in a reasonable amount of time.

Notice that $\Delta G°$ relates to the equilibrium constant of the reaction, whereas $\Delta G^{\ddagger}$ relates to the rate of the reaction. The **thermodynamic stability** of a compound is

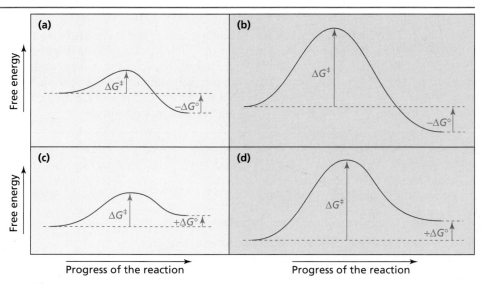

▲ **Figure 3.4**
Reaction coordinate diagrams for (a) a fast exergonic reaction; (b) a slow exergonic reaction; (c) a fast endergonic reaction; (d) a slow endergonic reaction. (The four reaction coordinates are drawn on the same scale.)

indicated by $\Delta G°$. For example, if $\Delta G°$ is negative, the product is *thermodynamically stable* compared with the reactant. But if $\Delta G°$ is positive, the product is *thermodynamically unstable* compared with the reactant. The **kinetic stability** of a compound is indicated by $\Delta G^{\ddagger}$. If $\Delta G^{\ddagger}$ is large, the compound is *kinetically stable*. In other words, it is not very reactive. But if $\Delta G^{\ddagger}$ is small, the compound is *kinetically unstable* (it is reactive).

PROBLEM 17◆

Which of the reactions in Figure 3.4 are most likely to take place spontaneously?

PROBLEM 18

Draw a reaction coordinate diagram for a reaction in which:

a. the product is thermodynamically unstable and kinetically unstable.

b. the product is thermodynamically unstable and kinetically stable.

The rate of a chemical reaction is the speed at which reacting substances are used up or the speed at which products are formed. The rate of a reaction is dependent on the following three factors.

1. *The number of collisions that take place between the reacting molecules in a given period of time.* The greater the number of collisions, the faster the reaction.

2. *The fraction of the collisions that occur with sufficient energy to get the reacting molecules over the energy barrier.* If the free energy of activation is small, more of the collisions will lead to reaction than if the free energy of activation is large.

3. *The fraction of the collisions that occur with the proper orientation.* For example, in the reaction of 2-butene with HBr, reaction will occur only if the molecules collide with the hydrogen of HBr approaching the π bond of 2-butene. If collision occurs with the hydrogen approaching the methyl group of 2-butene, no reaction will take place regardless of the energy of the collision.

$$\text{rate of a reaction} = \left(\begin{array}{c} \text{number of collisions} \\ \text{per unit of time} \end{array} \right) \times \left(\begin{array}{c} \text{fraction with} \\ \text{sufficient energy} \end{array} \right) \times \left(\begin{array}{c} \text{fraction with} \\ \text{proper orientation} \end{array} \right)$$

Increasing the temperature speeds up the motion of the molecules. This increases the rate of a reaction because it increases the number of collisions that occur in a given period of time and it increases the number of collisions that have sufficient energy to get the reacting molecules over the energy barrier. Increasing the concentration of the reactants increases the rate of the reaction because it increases the number of collisions that occur in a given period of time.

For a reaction in which a single reactant molecule (A) is converted into a product molecule (B), the rate of the reaction is proportional to the concentration of A: If the concentration of A is doubled, the rate of the reaction will double; if the concentration of A is tripled, the rate of the reaction will triple. Since the rate of this reaction is proportional to the concentration of only one reactant, it is a **first-order reaction.**

$$A \longrightarrow B$$

$$\text{rate} \propto [A]$$

We can replace the "proportional to" sign ($\propto$) with an "equal to" sign if we use a proportionality constant, k, which is called a **rate constant.** The rate constant of a first-order reaction is called a **first-order rate constant.**

$$\text{rate} = k[A]$$

A reaction whose rate is dependent on the concentrations of two reactants is called a **second-order reaction.** If the concentration of either A or B is doubled, the rate of the reaction will double. If the concentrations of both A and B are doubled, the rate of the reaction will increase by a factor of 4. The rate constant, k, is a **second-order rate constant.**

$$A + B \longrightarrow C + D$$
$$\text{rate} = k[A][B]$$

A reaction in which two molecules of A combine to form a molecule of B is also a second-order reaction. If the concentration of A is doubled, the rate of the reaction will quadruple.

$$A + A \longrightarrow B$$
$$\text{rate} = k[A]^2$$

RATE CONSTANTS

The rate of a reaction measures the change that has occurred in the concentration of some species in a period of time. Rate is generally expressed in moles per liter (molarity) per second ($M\ s^{-1}$). Therefore, the rate constant of a first-order reaction has units of s^{-1}, so when it is multiplied by the molar concentration of the reactant (M), the rate will have the appropriate units ($M\ s^{-1}$). Likewise, the rate constant of a second-order reaction has units of $M^{-1}\ s^{-1}$.

First-order: rate $(M\ s^{-1}) = k(s^{-1})M$

Second-order: rate $(M\ s^{-1}) = k(M^{-1}\ s^{-1})MM$

It is important to understand the difference between rate and rate constant. The reaction "rate" is a measure of the amount of product that is formed per unit of time, and this is what we actually measure in the laboratory. The "rate constant" tells us how easy it is to reach the transition state (how easy it is to get over the energy barrier). *The smaller the free energy of activation ($\Delta G^{\ddagger}$), the greater the rate constant.* In other words, low energy barriers are associated with large rate constants, whereas high energy barriers have small rate constants. From the reaction coordinate diagrams in Figure 3.5, we can predict that the rate constant for conversion of A to B is greater than the rate constant for conversion of C to D.

Now that we know that the rate constant for formation of B is greater than the rate constant for formation of D, does this also mean that the rate of formation of B is greater than the rate of formation of D? In other words, is more B formed than D in a given period of time? If we go into the lab and set up the reactions so that the concentration of A that we use in the first experiment is identical to the concentration of C that we use in the second experiment, we will find that both the rate constant for the reaction and the rate of the reaction (amount of product formed per unit of time) are greater for the conversion of A to B. However, if we were to use a small concentration of A for one experiment and a large concentration of C for the other

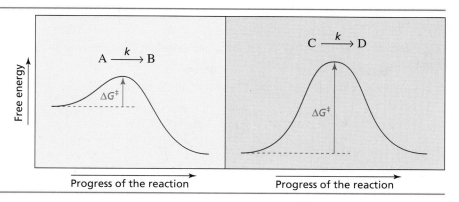

◀ **Figure 3.5**
Reaction coordinate diagrams
for (a) a reaction with a large
rate constant and (b) a reaction with a small rate
constant.

experiment, we might find that the rate of formation of B is lower than the rate of formation of D, but we would get the same values for the rate constants that we obtained previously. *Reaction rates are dependent on concentration, while rate constants are independent of concentration.* So when we compare two reactions to see which one occurs more easily, we must compare their rate constants and not their concentration-dependent rates of reaction. (Appendix III explains how rate constants are determined.)

THE DIFFERENCE BETWEEN $\Delta G^{\ddagger}$ AND E_a

It is important to distinguish between the *free energy of activation,* $\Delta G^{\ddagger}$, and the **experimental energy of activation,** E_a. The free energy of activation ($\Delta G^{\ddagger} = \Delta H^{\ddagger} - T\Delta S^{\ddagger}$) has both an enthalpy component and an entropy component. The experimental energy of activation ($E_a = \Delta H^{\ddagger} + RT$) has only an enthalpy component. The energy of activation is an approximate energy barrier to a reaction. The true energy barrier to a reaction is given by $\Delta G^{\ddagger}$, since some reactions are driven by a change in enthalpy, some by a change in entropy, but most by a change in both enthalpy and entropy.

Although rate constants are independent of concentration, they are dependent on temperature. Increasing the temperature increases the rate constant of a reaction by increasing the average energy of each collision so that a greater fraction of the collisions boost the reactants over the energy barrier. The **Arrhenius equation** relates the rate constant of a reaction to the energy of activation and to the temperature at which the reaction is carried out. A good rule of thumb is that an increase of 10 °C in temperature will double the rate constant for a reaction.

the Arrhenius equation

$$k = Ae^{-E_a/RT}$$

where k is the rate constant, E_a is the experimental energy of activation, R is the gas constant = 1.986×10^{-3} kcal mol^{-1} deg^{-1}, and T is the temperature (K). In this equation, A is the frequency factor; it accounts for the fraction of collisions that occur with the proper orientation for reaction. Take the logarithms of both sides and then change to base 10 logarithms:

$$\ln k = \ln A - E_a/RT$$

$$2.303 \log k = 2.303 \log A - E_a/RT$$

$$\log k = \log A - E_a/2.303RT$$

CALCULATING KINETIC PARAMETERS

In order to calculate E_a, $\Delta H^{\ddagger}$, $\Delta G^{\ddagger}$, and $\Delta S^{\ddagger}$ for a reaction, rate constants for the reaction must be obtained at several temperatures.

- E_a can be determined from the Arrhenius equation (the slope of a plot of log k versus $1/T$) since

$$\log k_2 - \log k_1 = -E_a/2.303R \left(\frac{1}{T_2} - \frac{1}{T_1} \right)$$

- At a given temperature, $\Delta H^{\ddagger}$ can be determined from E_a since $\Delta H^{\ddagger} = E_a - RT$.
- $\Delta G^{\ddagger}$, in kcal/mol, can be determined from the following equation, which relates $\Delta G^{\ddagger}$ to the rate constant at a given temperature,

$$-\Delta G^{\ddagger} = RT\, 2.303 \log\ kh/Tk_B$$

where h is Planck's constant (6.625×10^{-27} erg s) and k_B is Boltzmann's constant (1.3806×10^{-16} erg/kcal).
- The entropy of activation can be determined from a knowledge of the other two kinetic parameters $\Delta S^{\ddagger} = (\Delta H^{\ddagger} - \Delta G^{\ddagger})/T$.

MAX KARL ERNST LUDWIG PLANCK

Planck (1858–1947) was born in Germany, the son of a professor of civil law. He was a professor at the Universities of Munich (1880–1889) and Berlin (1889–1926). During World War I, one of his sons was killed in action, and two of his daughters died in childbirth. In 1918, Planck received the Nobel Prize in physics for his development of quantum theory. He became president of the Kaiser Wilhelm Society of Berlin (later renamed the Max Planck Society) in 1930. Planck felt it was his duty to remain in Germany during the Nazi era, but never supported the Nazi regime. He unsuccessfully interceded with Hitler on behalf of his Jewish colleagues and, as a consequence, was forced to resign from the presidency of the Max Planck Society in 1937. A second son was accused of taking part in the plot to kill Hitler and was executed. Planck lost his home to Allied bombings. He was rescued by Allied forces during the final days of the war.

PROBLEM 19◆

The rate constant for a reaction can be increased by _____ the stability of the reactant or by _____ the stability of the transition state.

PROBLEM 20◆

At 30 °C, the second-order rate constant for the reaction of methyl chloride and HO^- is $1.0 \times 10^{-5}\ M^{-1}\,s^{-1}$.

a. What is the rate of the reaction when $[CH_3Cl] = 0.10$ M and $[HO^-] = 0.10$ M?

b. If the concentration of methyl chloride is decreased to 0.01 M, what effect will this have on the *rate* of the reaction?

c. If the concentration of methyl chloride is decreased to 0.01 M, what effect will this have on the *rate constant* of the reaction?

How are the rate constants for a reaction related to the equilibrium constant? At equilibrium, the rate of the forward reaction must be equal to the rate of the reverse reaction because the amounts of reactants and products are not changing.

$$A \underset{k_{-1}}{\overset{k_1}{\rightleftharpoons}} B$$

forward rate = reverse rate
$$k_1[A] = k_{-1}[B]$$

Therefore,

$$K_{eq} = \frac{k_1}{k_{-1}} = \frac{[B]}{[A]}$$

From this equation, you can see that the equilibrium constant for a reaction can be determined from the relative concentrations of the reactant and product or from the relative rate constants of the forward and reverse reactions. The reaction shown in Figure 3.5(a) has a large equilibrium constant because B is much more stable than A. You could also say that it has a large equilibrium constant because the rate constant of the forward reaction is much greater than the rate constant of the reverse reaction.

Ludwig Edward Boltzmann (1844–1906), *the son of a civil servant, was born in Vienna and became a professor at the University of Vienna. He shares the credit for the kinetic theory of gases with Maxwell, although their work was done independently. He was the founder of statistical mechanics, a branch of physics that deals with the macroscopic behavior of a system based on the microscopic interactions of its constituents.*

Reaction Coordinate Diagram for the Addition of HBr to 2-Butene

You have already seen that the addition of HBr to 2-butene is a two-step reaction. The structure of the transition state for each of the steps is shown here in brackets. Notice that bonds that break and bonds that form during the reaction are partially broken and partially formed in the transition state; atoms that become charged or lose their charge during the reaction are partially charged in the transition state. A reaction coordinate diagram can be drawn for each of the steps (Figure 3.6a and b).

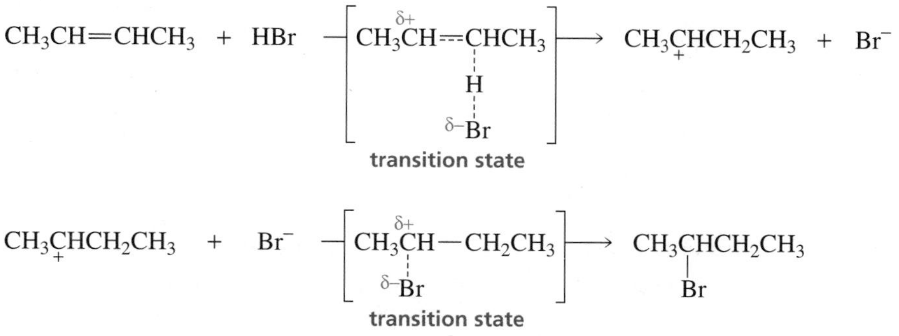

$$CH_3CH{=}CHCH_3 + HBr \longrightarrow \left[\begin{array}{c} \overset{\delta+}{CH_3CH}{=\!\!=}CHCH_3 \\ | \\ H \\ | \\ \overset{\delta-}{Br} \end{array} \right] \longrightarrow CH_3\overset{+}{C}HCH_2CH_3 + Br^-$$

transition state

$$CH_3\overset{+}{C}HCH_2CH_3 + Br^- \longrightarrow \left[\begin{array}{c} \overset{\delta+}{CH_3CH}{-}CH_2CH_3 \\ | \\ \overset{\delta-}{Br} \end{array} \right] \longrightarrow \begin{array}{c} CH_3CHCH_2CH_3 \\ | \\ Br \end{array}$$

transition state

The first step of the reaction involves the conversion of an alkene into a carbocation that is less stable than the alkene. The first step of the reaction, therefore, is endergonic ($\Delta G°$ is positive). In the second step of the reaction, the carbocation reacts with a nucleophile to form a product that is more stable than the carbocation reactant. This step, therefore, is exergonic ($\Delta G°$ is negative).

Since the product of the first step is the reactant for the second step, we can hook the two reaction coordinate diagrams together to obtain the reaction coordinate diagram for the overall reaction (Figure 3.7). The $\Delta G°$ for the overall reaction is the difference between the free energy of the final products and the free energy of the initial reactants. Figure 3.7 shows that $\Delta G°$ for the overall reaction is negative. The overall reaction is exergonic.

A chemical species that is the product of one step and the reactant for the next step is called an **intermediate.** The carbocation intermediate in this reaction is too unstable to be isolated, but some reactions have more stable intermediates that can

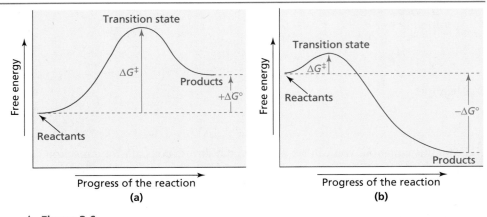

▲ **Figure 3.6**
Reaction coordinate diagrams for the two steps in the addition of HBr to 2-butene:
(a) the first step; (b) the second step.

Figure 3.7 ▶
Reaction coordinate diagram
for the addition of HBr to 2-
butene.

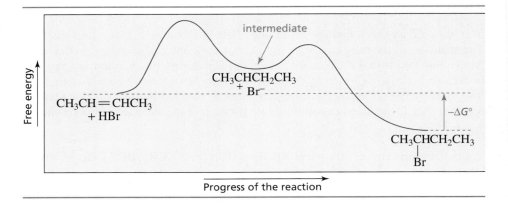

be isolated. Transition states, in contrast, represent the highest-energy structures that are involved in the reaction. They exist only fleetingly and can never be isolated. Do not confuse transition states and intermediates: *Transition states have partially formed bonds whereas intermediates have fully formed bonds.*

You can see from the reaction coordinate diagram that the free energy of activation for the first step of the reaction is greater than the free energy of activation for the second step. In other words, the rate constant for the first step is smaller than the rate constant for the second step. This is what you would expect, since in the first step of this reaction, the molecules must collide with sufficient energy to break covalent bonds, while no bonds are broken in the second step. The reaction step that has the transition state with the highest energy is called the **rate-determining step** or **rate-limiting step.** The rate-determining step controls the overall rate of the reaction. From Figure 3.7 it is apparent that when an electrophilic reagent such as HBr adds to an alkene, the first step of the addition reaction is the rate-determining step.

Being able to construct reaction coordinate diagrams will be useful in explaining why a given reaction forms a particular product and not others. We will see the first example of this in Section 3.11.

**Transition states have
partially formed bonds.
Intermediates have fully
formed bonds.**

PROBLEM 21◆

Which step in the following reaction sequence is the rate-determining step?

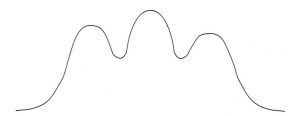

PROBLEM 22

Draw a reaction coordinate diagram for the following reaction, in which C is the most stable and B the least stable of the three species and the transition state going from A to B is more stable than the transition state going from B to C.

$$A \underset{k_{-1}}{\overset{k_1}{\rightleftharpoons}} B \underset{k_{-2}}{\overset{k_2}{\rightleftharpoons}} C$$

a. How many intermediates are there?

b. How many transition states are there?

c. Which is the faster step in the forward direction?

d. Which is the faster step in the reverse direction?

e. Of the four steps, which is the fastest?

f. Which is the rate-determining step in the forward direction?

g. Which is the rate-determining step in the reverse direction?

You have seen that an alkene such as 2-butene undergoes an electrophilic addition reaction with HBr. The first step in the reaction is the slow addition of the electrophilic proton to the alkene to form a carbocation intermediate. The intermediate reacts rapidly with the nucleophilic Br^- in the second step of the reaction.

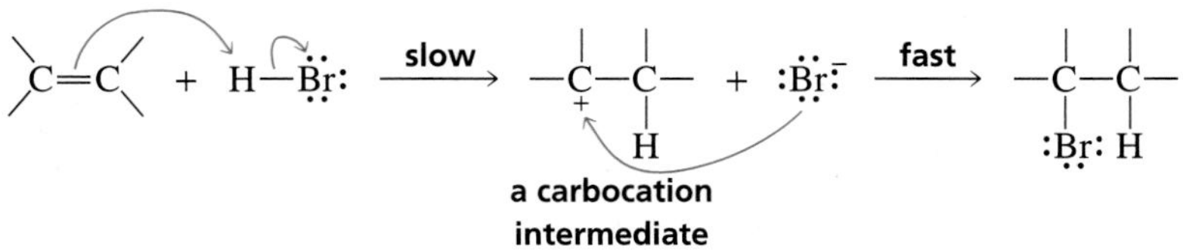

**3.8
GENERAL
MECHANISM FOR
ELECTROPHILIC
ADDITION
REACTIONS**

$$\overset{}{C}=\overset{}{C} \quad + \quad H-\overset{..}{\underset{..}{Br}}: \quad \xrightarrow{\text{slow}} \quad -\overset{|}{\underset{+}{C}}-\overset{|}{\underset{|}{C}}- \quad + \quad :\overset{..}{\underset{..}{Br}}: \quad \xrightarrow{\text{fast}} \quad -\overset{|}{\underset{|}{C}}-\overset{|}{\underset{|}{C}}-$$

$$\overset{}{\underset{H}{|}} \qquad\qquad\qquad\qquad\qquad :\overset{..}{\underset{..}{Br}}: H$$

**a carbocation
intermediate**

In the following sections, we will look at a wide variety of alkene reactions. You will see that many of the reactions form a carbocation intermediate like the carbocation intermediate formed when HBr reacts with an alkene. Some of the reactions form other kinds of intermediates and some form no intermediate at all. As you study each reaction, notice the feature that all alkene reactions have in common: *They are all addition reactions and each reaction starts with the addition of an electrophile to one of the sp^2 carbons of the alkene and concludes with the addition of a nucleophile to the other sp^2 carbon.*

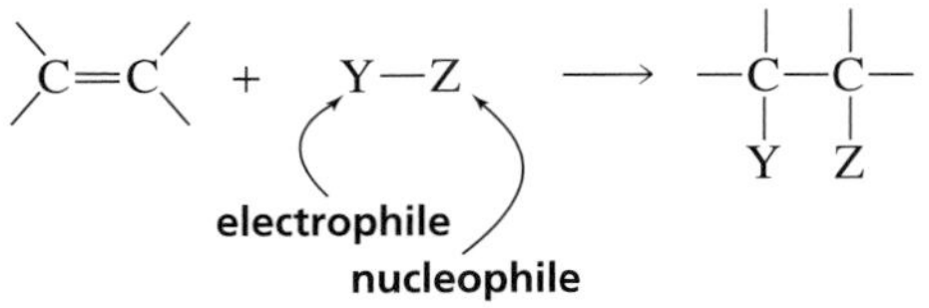

electrophile
nucleophile

The reaction of an alkene can lead to the formation of a wide variety of products (alkyl halides, alcohols, ethers, etc.). The particular product obtained from the reaction of an alkene depends only on the reagent that is used in the addition reaction (Y–Z).

3.9
ADDITION OF
HYDROGEN HALIDES

If the electrophilic reagent that adds to an alkene is a hydrogen halide (HCl, HBr, HI), the product of the reaction will be an alkyl halide.

$$CH_2=CH_2 \ + \ HCl \longrightarrow CH_3CH_2Cl$$
ethene   ethyl chloride

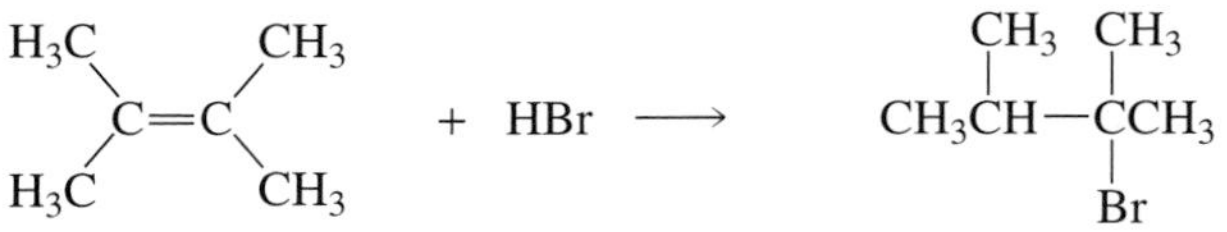

2,3-dimethyl-2-butene   2-bromo-2,3-dimethylbutane

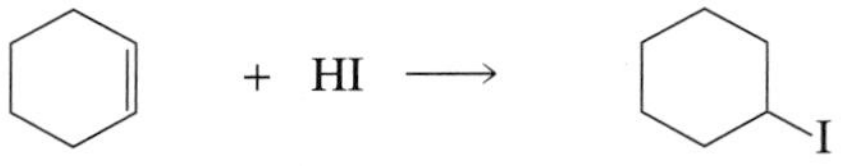

cyclohexene   iodocyclohexane

Because the alkenes in these reactions are symmetrical, it is easy to determine the product of the reaction: the electrophile (H^+) adds to one of the sp^2 carbons, and the nucleophile (X^-) adds to the other sp^2 carbon. But what happens if the alkene is not symmetrical? Which sp^2 carbon gets the hydrogen? For example, does the addition of HCl to methylpropene produce *tert*-butyl chloride or isobutyl chloride?

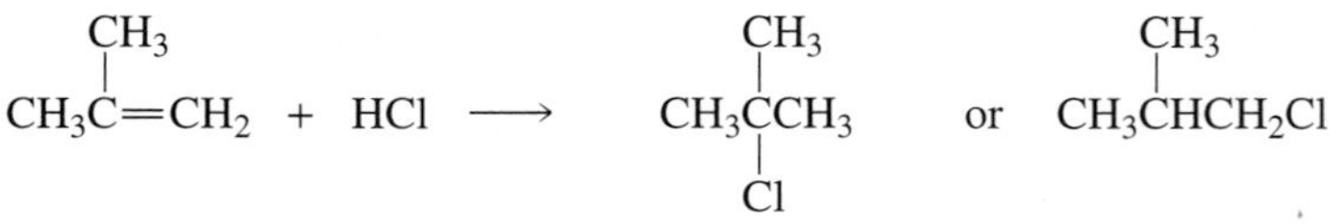

methylpropene   *tert*-butyl chloride   isobutyl chloride

To answer this question we need to look at the mechanism of the reaction. The first step of the reaction, the addition of H^+ to an sp^2 carbon to form either the *tert*-butyl cation or the isobutyl cation, is the rate-determining step. If there is any difference in the rate of formation of these two carbocations, the one that is formed faster will be the preferred product of the first step. The particular carbocation that is formed in the first step determines the final product of the reaction because, if the *tert*-butyl cation is formed, it will react with Cl^- to form *tert*-butyl chloride; if the isobutyl cation is formed, isobutyl chloride will be the product. If we were actually to carry out this reaction in the laboratory, we would find that the only product of the reaction is *tert*-butyl chloride. This means that the *tert*-butyl cation is formed faster than the isobutyl cation. Let's take a look at the factors that affect the stability of carbocations and therefore the ease with which they are formed.

$$\underset{\substack{\text{CH}_3 \\ |}}{\text{CH}_3\text{C}}=\text{CH}_2 \; + \; \text{HCl}$$

$$\underset{\text{\textit{tert}-butyl cation}}{\overset{\substack{\text{CH}_3 \\ |}}{\text{CH}_3\overset{+}{\text{C}}\text{CH}_3}} \xrightarrow{\text{Cl}^-} \underset{\substack{\text{\textit{tert}-butyl chloride} \\ \text{only product formed}}}{\overset{\substack{\text{CH}_3 \\ |}}{\underset{\substack{| \\ \text{Cl}}}{\text{CH}_3\text{C}\text{CH}_3}}}$$

$$\underset{\text{isobutyl cation}}{\overset{\substack{\text{CH}_3 \\ |}}{\text{CH}_3\text{CH}\overset{+}{\text{C}}\text{H}_2}} \xrightarrow{\text{Cl}^-} \underset{\substack{\text{isobutyl chloride} \\ \text{not formed}}}{\overset{\substack{\text{CH}_3 \\ |}}{\text{CH}_3\text{CHCH}_2\text{Cl}}}$$

3.10 CARBOCATION STABILITY

Carbocations are classified according to the number of alkyl substituents that are bonded to the positively charged carbon: a **primary carbocation** has one alkyl substituent bonded to the positively charged carbon; a **secondary carbocation** has two; a **tertiary carbocation** has three. The stability of a carbocation depends on the number of alkyl substituents that are bonded to the positively charged carbon atom: *The greater the number of alkyl substituents bonded to the positively charged carbon, the more stable the carbocation will be.* In other words, tertiary carbocations are more stable than secondary carbocations, which are more stable than primary carbocations.

> The greater the number of alkyl substituents bonded to the positively charged carbon, the more stable the carbocation.

relative stabilities of carbocations

$$\underset{\substack{\text{a tertiary} \\ \text{carbocation}}}{\overset{\substack{\text{R} \\ |}}{\underset{\substack{| \\ \text{R}}}{\text{R}-\overset{+}{\text{C}}}}} \; > \; \underset{\substack{\text{a secondary} \\ \text{carbocation}}}{\overset{\substack{\text{R} \\ |}}{\underset{\substack{| \\ \text{H}}}{\text{R}-\overset{+}{\text{C}}}}} \; > \; \underset{\substack{\text{a primary} \\ \text{carbocation}}}{\overset{\substack{\text{H} \\ |}}{\underset{\substack{| \\ \text{H}}}{\text{R}-\overset{+}{\text{C}}}}} \; > \; \underset{\text{methyl cation}}{\overset{\substack{\text{H} \\ |}}{\underset{\substack{| \\ \text{H}}}{\text{H}-\overset{+}{\text{C}}}}}$$

← increasing stability

Why does the stability of the carbocation increase if the number of alkyl substituents increases? Alkyl groups decrease the concentration of positive charge on the carbon—and decreasing the concentration of charge increases the stability of the carbocation.

Alkyl groups decrease the concentration of positive charge because they are more readily polarized than are hydrogens. In other words, the electrons in a carbon–alkyl group σ bond can move toward the positively charged carbon more readily than can the electrons in a carbon–hydrogen σ bond. The greater the electron density that can move toward the positive charge, the more that charge will be dispersed—and the more the charge is dispersed, the more the carbocation is stabilized.

$$\underset{\text{greater electron flow}}{\text{R}\xrightarrow{\quad}\overset{+}{\text{C}}} \qquad\qquad \underset{\text{less electron flow}}{\text{H}\xrightarrow{\quad}\overset{+}{\text{C}}}$$

electron flow toward the positively charged carbon is greater from an alkyl group than from a hydrogen atom

Alkyl substituents also stabilize carbocations by hyperconjugation. A charged species is more stable if its charge is spread out (delocalized) over more than one

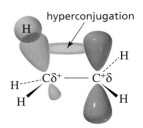

▲ **Figure 3.8**
Stabilization of a carbocation by hyperconjugation: the σ electrons in the adjacent carbon–hydrogen bond spread into the empty *p* orbital.

atom. The positive charge on a carbon signifies an empty *p* orbital (Section 1.10). The σ orbital of an adjacent carbon–hydrogen or carbon–carbon bond can overlap with the empty *p* orbital (Figure 3.8). The movement of the electrons of the σ orbital toward the vacant *p* orbital causes a partial positive charge to develop on the σ bond orbital. Therefore, the positive charge is no longer localized solely on one atom but is spread out over a greater volume of space, thereby stabilizing the carbocation. Delocalization of electrons by overlap of a carbon–hydrogen or carbon–carbon σ bond orbital with an empty *p* orbital is called **hyperconjugation.**

Hyperconjugation occurs only if the σ bond orbital and the empty *p* orbital have the proper orientation. Since the carbon–carbon bond is free to rotate, achieving the proper orientation is not a problem. In the case of the *tert*-butyl cation, there are nine carbon–hydrogen bond orbitals that can potentially overlap with the empty *p* orbital on the positively charged carbon. The isopropyl cation has six such orbitals and the ethyl cation has three carbon–hydrogen bond orbitals that can overlap with the empty *p* orbital. Therefore, there is greater stabilization through hyperconjugation in the tertiary *tert*-butyl cation than in the secondary isopropyl cation and greater stabilization in the secondary isopropyl cation than in the primary ethyl cation.

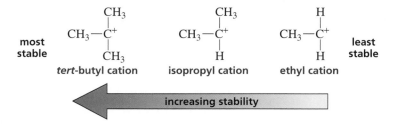

PROBLEM 23◆

List the carbocations in order of decreasing stability.

a. CH₃CH₂C⁺CH₃ (with CH₃ substituent)   CH₃CH₂C⁺HCH₃   CH₃C⁺HCH₂CH₂⁺ (with CH₃ substituent)

b. CH₃CHCH₂CH₂⁺ (with Cl substituent)   CH₃CH₂CH₂CH₂⁺   CH₃CH₂CHCH₂⁺ (with Cl substituent)

PROBLEM 24◆

a. How many carbon–hydrogen bond orbitals are available for overlap with the vacant *p* orbital in the methyl cation?

b. Which is more stable, a methyl cation or an ethyl cation?

3.11
THE STRUCTURE OF THE TRANSITION STATE

A knowledge of the structure of a transition state is important when you are trying to predict the products of a reaction. In Section 3.7 we saw that the structure of the transition state lies between the structure of the reactants and the structure of the products. But what do we mean by "between"? Does the structure of the transition state lie exactly halfway between the structures of the reactants and products (as in

II, below), or does it resemble the reactants more closely than it resembles the products (as in I), or is it more similar to the products than to the reactants (as in III)?

$$A\text{—}B \ + \ C \quad \longrightarrow \quad \begin{bmatrix} \text{(I)} & A\text{----}B\cdots\cdots\cdots C \\ \text{(II)} & A\text{-------}B\text{-------}C \\ \text{(III)} & A\cdots\cdots\cdots B\text{----}C \end{bmatrix} \quad \longrightarrow \quad A \ + \ B\text{—}C$$

reactants   transition state   products

According to the **Hammond postulate,** *the transition state will be more similar in structure to the species that it is closer to energetically.* In the case of an exergonic reaction, the transition state is energetically closer to the reactant than it is to the product (Figure 3.9, curve I). Therefore, the structure of the transition state will more closely resemble the structure of the reactant than the structure of the product. In an endergonic reaction (Figure 3.9, curve III), the transition state is energetically closer to the product, so the structure of the transition state will be closer to the structure of the product. The only time that we would expect the structure of the transition state to be halfway between the structure of the reactant and that of the product is if the reactant and the product have similar energies (Figure 3.9, curve II).

Now we can understand why, when methylpropene reacts with HCl, the *tert*-butyl cation is formed faster than the isobutyl cation. The addition of an electrophile to an alkene to form a carbocation is an endergonic reaction (Figure 3.10). Knowing this, you can predict that the structure of the transition state for this step will be close to the structure of the carbocation product. This means that the transition state will have a significant amount of positive charge on a carbon. Of the two possible products, the *tert*-butyl cation with three alkyl groups bonded to its positively charged carbon is more stable (has a lower free energy) than the isobutyl cation, which has only one alkyl substituent bonded to its positively charged carbon. Because the transition states are close to the structures of the products, *the same factors that stabilize the products also stabilize the transition states.* Therefore, the transition state leading to the *tert*-butyl cation is more stable than the transition state leading to the isobutyl cation. Since the amount of positive charge in the transition state is not as great as the amount of positive charge in the product, the difference in the stabilities of the two transition states will not be as great as the difference in the stabilities of the two products.

You have seen that the rate of a reaction is determined by the free energy of activation, which is the difference between the free energy of the transition state and the free energy of the reactant. Since the free energy of activation for formation of

George Simms Hammond *was born in Maine in 1921. He received a B.S. from Bates College in 1943 and a Ph.D. from Harvard University in 1947. He was a professor of chemistry at Iowa State University and at the California Institute of Technology and a scientist at Allied Chemical Co.*

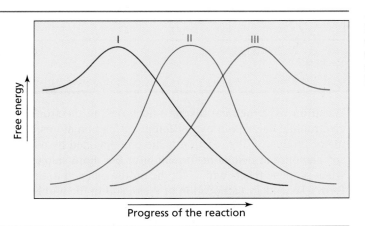

◀ **Figure 3.9**
Reaction coordinate diagrams for reactions with (I) an early transition state; (II) a mid-way transition state; (III) a late transition state.

Figure 3.10 ▶
Reaction coordinate diagram
for the addition of H$^+$ to
methylpropene to form the
primary isobutyl cation and
the tertiary *tert*-butyl cation.

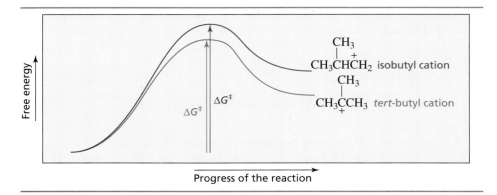

the *tert*-butyl cation is less than the free energy of activation for formation of the isobutyl cation, the *tert*-butyl cation will be formed faster. So, in an electrophilic addition reaction, the more stable carbocation will be the carbocation formed more rapidly.

The difference in the rate of formation of the two carbocations determines the relative amounts of the products that are formed. If the difference in the rates is small, both products will form, but the more stable product will be the major product. If the difference in the rates is sufficiently large, the more stable product will be the only product of the reaction. In the addition of HCl to methylpropene, for example, the rates of formation of the two carbocations (one is primary and the other is tertiary) are sufficiently different to cause *tert*-butyl chloride to be the only product of the reaction.

$$
\underset{\substack{| \\ CH_3}}{CH_3C=CH_2} + HCl \longrightarrow \underset{\substack{| \\ Cl \\ \textbf{only product formed}}}{\overset{\substack{CH_3 \\ |}}{CH_3CCH_3}} \quad \left[\underset{\substack{| \\ CH_3}}{CH_3CHCH_2Cl}\right] \\ \textbf{not formed}
$$

PROBLEM 25◆

Which compound will react faster with HCl? (*Hint:* the two alkenes have about the same stability.)

$$
CH_3CH=CHCH_3 \quad \text{or} \quad \underset{\substack{| \\ CH_3}}{CH_3C=CH_2}
$$

**3.12
REGIOSELECTIVITY
OF ELECTROPHILIC
ADDITION
REACTIONS:
MARKOVNIKOV'S
RULE**

When an asymmetrical alkene undergoes an electrophilic addition reaction, the electrophile can add to two different sp^2 carbon atoms. You have just seen that the major product of the reaction is the one obtained by adding the electrophile to the sp^2 carbon that results in formation of the more stable carbocation. For example, when propene reacts with HCl, the proton can add to the number 1 carbon (C-1) to form a secondary carbocation or it can add to the number 2 carbon (C-2) to form a primary carbocation. The secondary carbocation is formed more rapidly because it is more stable than the primary carbocation. (Primary carbocations are so unstable that they form only with great difficulty.) The product of the reaction, therefore, is isopropyl chloride.

The *sp²* carbon that does *not* become attached to the proton is the one that is positively charged in the carbocation.

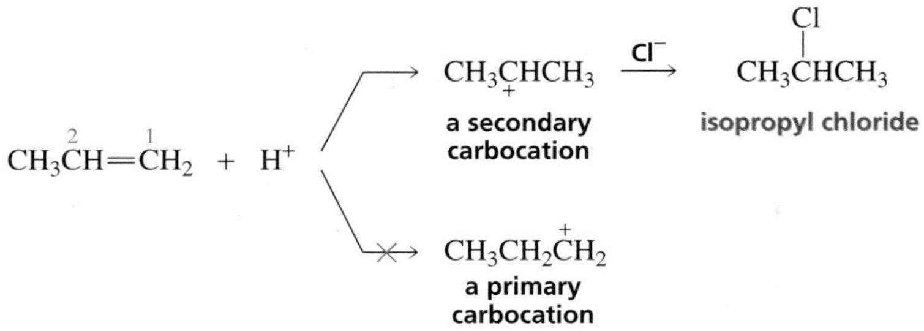

$$CH_3\overset{2}{CH}=\overset{1}{CH_2} \ + \ H^+$$

$$CH_3\overset{+}{CH}CH_3 \quad \xrightarrow{\ Cl^-\ } \quad CH_3\overset{Cl}{\underset{|}{CH}}CH_3$$

a secondary carbocation

isopropyl chloride

$$\xrightarrow{\ \times\ } \ CH_3CH_2\overset{+}{CH_2}$$

a primary carbocation

The major product obtained from the addition of HI to 2-methyl-2-butene is 2-iodo-2-methylbutane; only a small amount of 2-iodo-3-methylbutane is obtained. The major product obtained from addition of HBr to 1-methylcyclohexene is 1-bromo-1-methylcyclohexane. In both cases, the more stable tertiary carbocation is formed more rapidly than the less stable secondary carbocation, so the major product of each reaction is the one that results from forming the tertiary carbocation.

$$CH_3CH=\overset{CH_3}{\underset{|}{C}}CH_3 \ + \ HI \ \longrightarrow \ CH_3CH_2\overset{CH_3}{\underset{|}{\underset{I}{C}}}CH_3 \ + \ CH_3\overset{CH_3}{\underset{|}{CH}}\overset{}{\underset{I}{CH}}CH_3$$

2-methyl-2-butene

2-iodo-2-methylbutane
major product

2-iodo-3-methylbutane
minor product

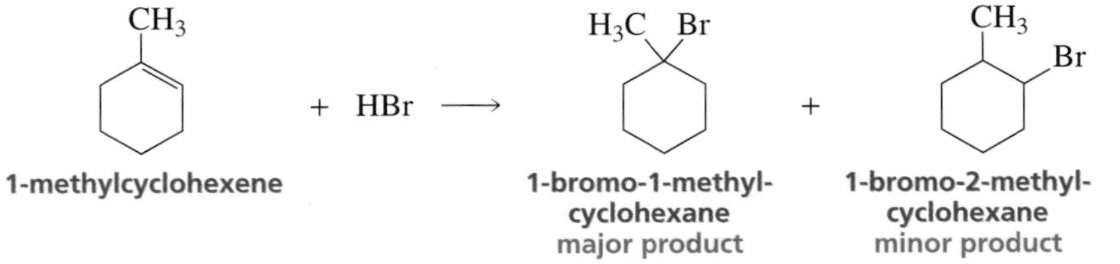

1-methylcyclohexene   + HBr ⟶   1-bromo-1-methyl-cyclohexane
major product   +   1-bromo-2-methyl-cyclohexane
minor product

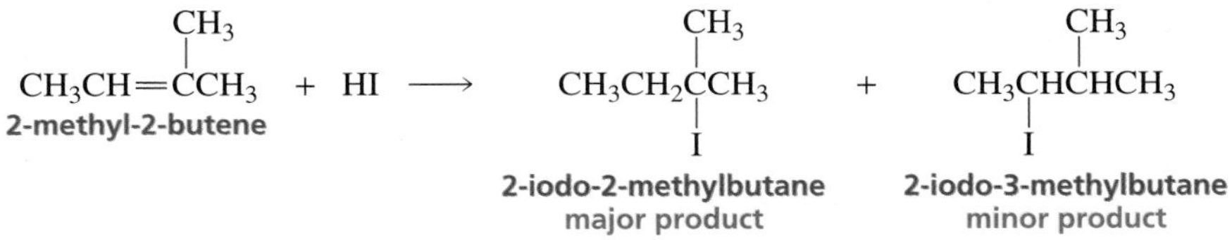

The two different products in each of these reactions are constitutional isomers. **Constitutional isomers** differ only in how the atoms are linked. A reaction (such as those shown above) in which two or more constitutional isomers could be obtained as products but one of them predominates is called a **regioselective reaction.**

There are different degrees of regioselectivity. A reaction can be *moderately regioselective, highly regioselective,* or *completely regioselective.* In a completely regioselective reaction, one of the possible products is not formed at all. The addition of a hydrogen halide to methylpropene (where the choice is between formation of a tertiary or a primary carbocation) is more highly regioselective than the addition of a hydrogen halide to 2-methyl-2-butene (where the choice is between formation of a tertiary or a secondary carbocation) because, in the latter, the two carbocations are closer in stability.

The addition of HBr to 2-pentene is not regioselective. Both isomers are obtained because the addition of the proton to either of the *sp²* carbons produces a secondary carbocation. Both carbocation intermediates have the same stability, so both will form equally easily, resulting in the formation of approximately equal amounts of the two alkyl halides.

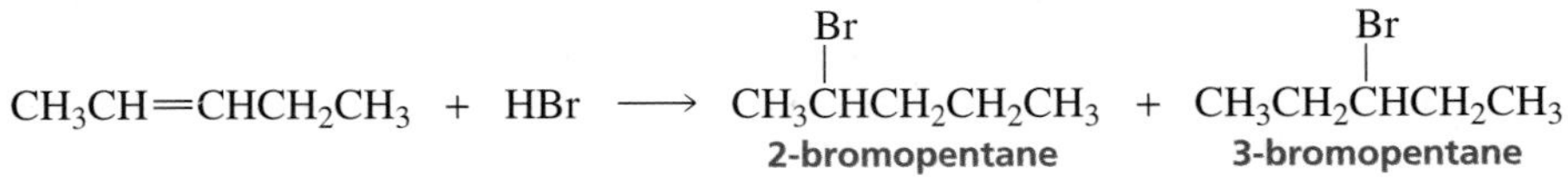

$$CH_3CH=CHCH_2CH_3 \ + \ HBr \ \longrightarrow \ CH_3\overset{Br}{\underset{|}{CH}}CH_2CH_2CH_3 \ + \ CH_3CH_2\overset{Br}{\underset{|}{CH}}CH_2CH_3$$

2-bromopentane   3-bromopentane

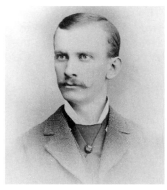

Vladimir Vasilevich Markovnikov

Vladimir Vasilevich Markovnikov (1837–1904) *was born in Russia, the son of an army officer. His family name was transliterated Markownikoff in his original paper in* Annalen der Chemie, *but it is now most often transliterated Markovnikov. He was a professor of chemistry at Kazan, Odessa, and Moscow Universities. By synthesizing rings containing four carbons and seven carbons, he disproved the notion that carbon could form only five- and six-membered rings.*

The electrophile adds to the *sp*² carbon that is bonded to the greater number of hydrogens.

You have seen that, if you want to predict the major product of an electrophilic addition reaction, you must first determine the relative stabilities of the two possible cations. This tells you the sp^2 carbon to which the electrophile will add. In 1870, Vladimir Markovnikov published a paper in *Annalen der Chemie* that described a faster way to predict the major product resulting from the addition of a hydrogen halide to an asymmetrical alkene. His shortcut is known as **Markovnikov's rule:** "When a hydrogen halide adds to an asymmetrical alkene, the addition occurs such that the halogen attaches itself to the carbon atom of the alkene bearing the least number of hydrogen atoms." Since H^+ is the first species that adds to the alkene, most chemists state the rule in the following way: "The H^+ adds to the sp^2 carbon that is bonded to the greater number of hydrogens." Because electrophiles other than H^+ can add to alkenes, the following is a more all-encompassing statement of Markovnikov's rule and it is the one that you should learn. *The electrophile adds to the sp^2 carbon that is bonded to the greater number of hydrogens.*

Whether you determine the major product of an electrophilic addition reaction by determining relative carbocation stabilities or by using Markovnikov's rule, you will get the same answer. In the reaction shown below, the H^+ adds preferentially to C-1 because that results in the formation of a secondary carbocation that is more stable than the primary carbocation that would be formed if H^+ were to add to C-2 (choice made using relative carbocation stabilities). Alternatively, the electrophile (H^+) adds preferentially to C-1, which is bonded to two hydrogens, rather than to C-2, which is bonded to only one hydrogen (choice made using Markovnikov's rule).

$$CH_3CH_2\overset{2}{C}H{=}\overset{1}{C}H_2 + HCl \longrightarrow CH_3CH_2\underset{}{\overset{\overset{\displaystyle Cl}{|}}{C}}HCH_3$$

PROBLEM 26◆

What would be the major product obtained from the addition of HBr to each of the following compounds?

a. $CH_3CH_2CH{=}CH_2$

b. $CH_3CH{=}\overset{\overset{\displaystyle CH_3}{|}}{C}CH_3$

c.

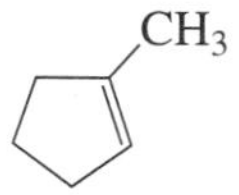

d. $CH_3CH{=}CHCH_3$

PROBLEM-SOLVING STRATEGY

a. What alkene should be used to synthesize 3-bromohexane?

$$? + HBr \longrightarrow CH_3CH_2\underset{\underset{\displaystyle Br}{|}}{C}HCH_2CH_2CH_3$$

3-bromohexane

In answering this kind of question, start by listing all the alkenes that could be used. Since you want to synthesize an alkyl halide that has a bromo substituent at the C-3 position, the alkene should have an sp^2 carbon at the C-3 position. Therefore, there are two alkenes that can be used to form 3-bromohexane: 2-hexene and 3-hexene.

$$CH_3CH=CHCH_2CH_2CH_3 \qquad CH_3CH_2CH=CHCH_2CH_3$$

2-hexene   **3-hexene**

Since there are two possibilities, look to see if there is any advantage to using one over the other. The addition of H^+ to 2-hexene can form two different carbocations. Since they are both secondary carbocations, they have the same stability and, therefore, approximately equal amounts of each will form. As a result, half of the product will be 3-bromohexane and half will be 2-bromohexane.

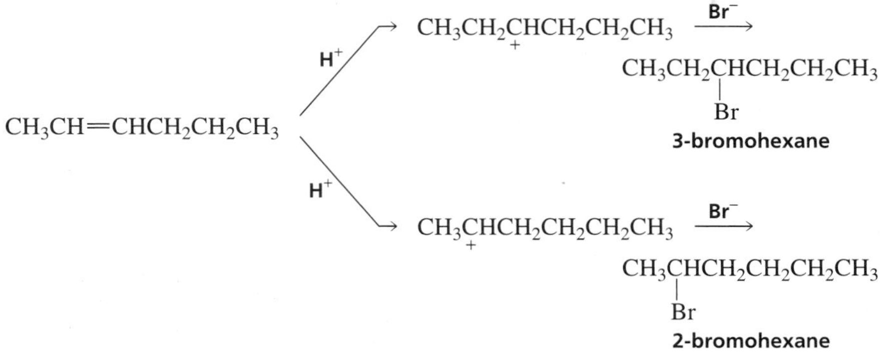

The addition of H^+ to either of the sp^2 carbons of 3-hexene forms the same carbocation. Therefore, all of the product will be the desired 3-bromohexane.

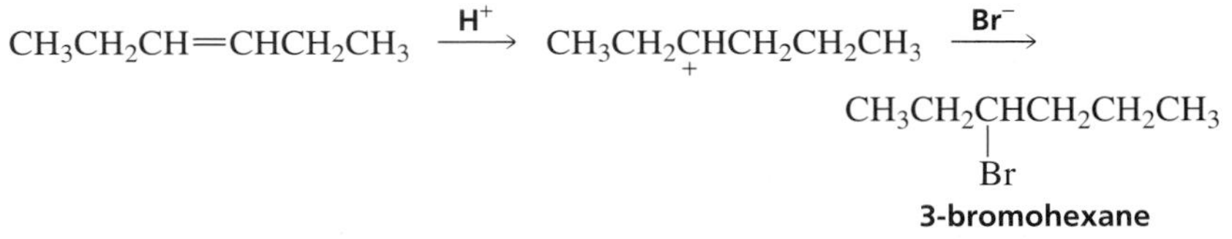

Since all the alkyl halide formed from 3-hexene is 3-bromohexane, but only half the alkyl halide formed from 2-hexene is 3-bromohexane, 3-hexene is the best alkene to use to prepare 3-bromohexane.

b. What alkene should be used to synthesize 2-bromopentane? Since you want to synthesize an alkyl halide with a bromo substituent at the C-2 position, the alkene should have an sp^2 carbon at the C-2 position. Therefore, there are two alkenes that can be used to prepare 2-bromopentane: 1-pentene and 2-pentene.

$$CH_2=CHCH_2CH_2CH_3 \qquad CH_3CH=CHCH_2CH_3$$

1-pentene   **2-pentene**

When H^+ adds to 1-pentene, one of the carbocations that could be formed is secondary and the other is primary. A secondary carbocation is more stable than a primary carbocation, which is so unstable that little, if any, will form. So 2-bromopentane will be the only product of the reaction.

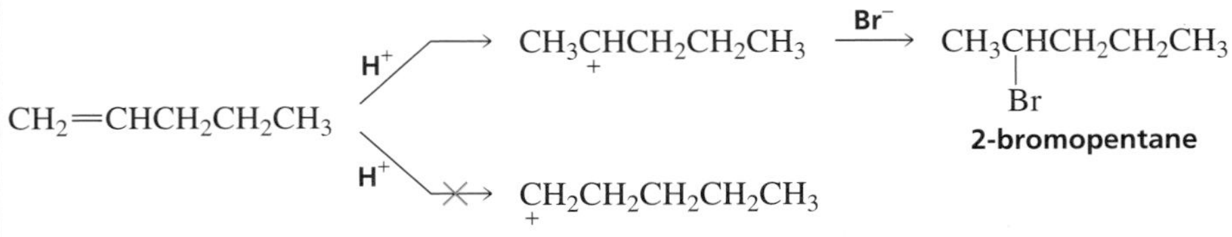

When H^+ adds to 2-pentene, the two carbocations that can form are both secondary. They are equally stable, so they will form in equal amounts. Half of the product of the reaction will be 2-bromopentane and half will be 3-bromopentane.

$$CH_3CH=CHCH_2CH_3 \quad \overset{H^+}{\longrightarrow} \quad CH_3\overset{+}{C}HCH_2CH_2CH_3 \quad \overset{Br^-}{\longrightarrow} \quad \underset{\underset{\text{2-bromopentane}}{|}}{CH_3CHCH_2CH_2CH_3} \\ Br$$

$$\overset{H^+}{\longrightarrow} \quad CH_3CH_2\overset{+}{C}HCH_2CH_3 \quad \overset{Br^-}{\longrightarrow} \quad \underset{\underset{\text{3-bromopentane}}{|}}{CH_3CH_2CHCH_2CH_3} \\ Br$$

Since all the alkyl halide formed from 1-pentene is 2-bromopentane, but only half the alkyl halide formed from 2-pentene is 2-bromopentane, 1-pentene is the best alkene to use to prepare 2-bromopentane.

Now continue on to answer the questions in Problem 27.

PROBLEM 27◆

What alkene should be used to synthesize each of the following alkyl bromides?

a. $\underset{\underset{Br}{|}}{CH_3\overset{\overset{CH_3}{|}}{C}CH_3}$

c. (cyclohexyl)—$\underset{\underset{Br}{|}}{\overset{\overset{CH_3}{|}}{C}CH_3}$

b. (cyclohexyl)—$\underset{\underset{Br}{|}}{CH_2CHCH_3}$

d. (cyclohexyl with CH_2CH_3 and Br)

PROBLEM 28◆

The addition of HBr to which of the following alkenes is more highly regioselective?

(cyclohexane with $=CH_2$)   or   (cyclohexene with CH_3)

3.13
ADDITION OF WATER AND ALCOHOLS

If you add water to an alkene, no reaction will take place, because the oxygen–hydrogen bonds of water are too strong (water is too weakly acidic) to allow the hydrogen to act as an electrophile. Consequently, there is no electrophile present to add to the nucleophilic alkene to start the reaction.

$$RCH=CH_2 \quad + \quad H_2O \quad \longrightarrow \quad \text{no reaction}$$

If, however, you add acid (for example, H_2SO_4) to the solution, water will add to the alkene, forming an alcohol. When acid is added to the aqueous solution, H_3O^+ is formed and its proton can act as an electrophile. It will react with the alkene to form a carbocation, and the carbocation will then react with the nucleophilic water. Addition of water to a molecule is called **hydration,** so we can say that in the presence of water and acid an alkene will be **hydrated.**

The mechanism for the addition of water is exactly the same as the mechanism for the addition of a hydrogen halide. The addition also follows Markovnikov's rule: the electrophile (H^+) adds to the sp^2 carbon that is bonded to the most hydrogens, and the nucleophile (H_2O) adds to the other sp^2 carbon.

mechanism for the acid–catalyzed addition of water

$$CH_3CH\!\!=\!\!CH_2 \;+\; H\!\!-\!\!\overset{..+}{O}H \xrightarrow{\;slow\;} CH_3\overset{+}{C}HCH_3 \;+\; H_2\overset{..}{O}: \xrightarrow{\;fast\;} CH_3CHCH_3$$

with H below the $\overset{..+}{O}H$

$$CH_3CHCH_3$$
$$\overset{+}{:}OH$$
$$\overset{\,}{H}$$

fast | $H_2\overset{..}{O}$:

$$\downarrow$$

$$CH_3CHCH_3 \;+\; H_3\overset{+}{O}:$$
$$:\overset{..}{O}H$$

As in the addition of a hydrogen halide to an alkene, the addition of the electrophile to the alkene is slow and the subsequent addition of the nucleophile to the carbocation occurs rapidly. The reaction of the carbocation with a nucleophile is so fast that the carbocation combines with whatever nucleophile it collides with first. In the foregoing reaction there are two nucleophiles in solution: water and the counterion of the acid (HSO_4^-) that was used to start the reaction. (Notice that HO^- is not a nucleophile in this reaction because there is no HO^- in an acidic solution.) Since the concentration of water is much greater than the concentration of the counterion, the carbocation will be more likely to collide with water. The product of the collision is a protonated alcohol. Protonated alcohols are strong acids (Sections 1.17 and 1.19). Therefore, the protonated alcohol readily loses a proton to the solvent (H_2O), and the final product of the addition reaction is an alcohol.

The proton adds to the alkene in the first step but is returned to the reaction mixture in the last step. Overall, the proton is not consumed. A species that increases the rate of a reaction and is not consumed during the course of the reaction is called a **catalyst.** Catalysts increase the reaction rate by decreasing the activation energy of the reaction (p. 939). Catalysts do not affect the equilibrium constant of the reaction. In other words, a catalyst increases the *rate* at which a product is formed but does not affect the amount of product formed. In the preceding example, the catalyst is an acid (H^+), so the reaction is an **acid-catalyzed reaction.**

PROBLEM 29 ◆

Give the structure of the alcohol that is formed from the acid-catalyzed hydration of each of the following alkenes.

a. $CH_3CH_2CH_2CH\!\!=\!\!CH_2$ c. $CH_3CH_2CH_2CH\!\!=\!\!CHCH_3$

b. (hexagon)   d. (hexagon)$=\!\!CH_2$

Alcohols react with alkenes in the same way that water does. Like the addition of water, the addition of an alcohol requires an acid catalyst. The product of the reaction is an ether.

$$CH_3CH_2\!\!=\!\!CH_2 \;+\; CH_3OH \xrightarrow{\;H^+\;} CH_3CHCH_3$$
$$\overset{\,}{O}CH_3$$

The mechanism for the addition of an alcohol is exactly the same as the mechanism for the addition of water.

mechanism for the acid-catalyzed addition of an alcohol

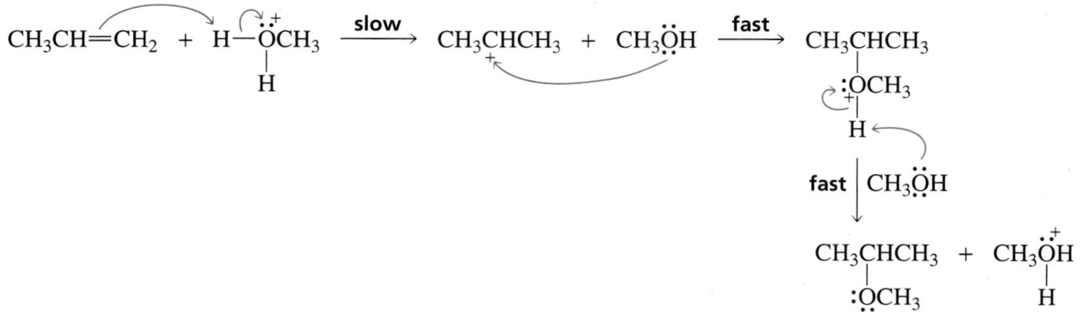

$$CH_3CH{=}CH_2 \ + \ H{-}\overset{+}{\underset{H}{\ddot{O}}}CH_3 \ \xrightarrow{\textbf{slow}} \ CH_3\overset{+}{C}HCH_3 \ + \ CH_3\ddot{O}H \ \xrightarrow{\textbf{fast}} \ CH_3CHCH_3$$

$$\underset{\underset{H}{\overset{|}{\underset{}{}}}}{\overset{\overset{+}{\ddot{O}}CH_3}{|}}$$

$$\textbf{fast} \ \Big| \ CH_3\ddot{O}H$$

$$\downarrow$$

$$CH_3CHCH_3 \ + \ CH_3\overset{\cdot+}{\ddot{O}}H$$

$$\underset{:\ddot{O}CH_3}{|} \qquad \underset{H}{|}$$

PROBLEM 30

Give the mechanism for the following reaction. (Remember to use curved arrows when showing a mechanism.)

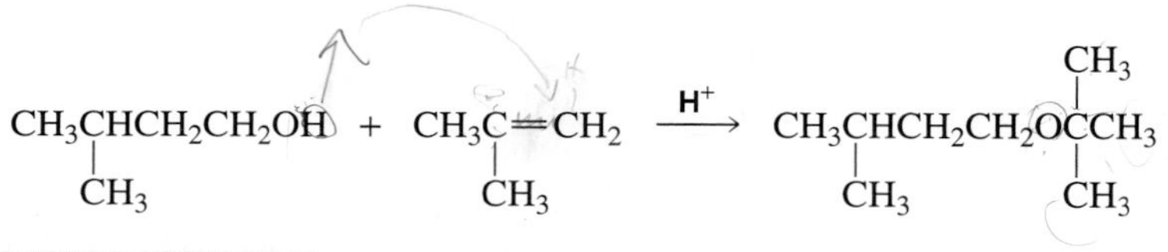

$$\underset{\underset{CH_3}{|}}{CH_3CHCH_2CH_2OH} \ + \ \underset{\underset{CH_3}{|}}{CH_3C{=}CH_2} \ \xrightarrow{H^+} \ \underset{\underset{CH_3}{|}}{CH_3CHCH_2CH_2O}\underset{\underset{CH_3}{|}}{\overset{\overset{CH_3}{|}}{C}CH_3}$$

PROBLEM 31

a. Give the products for each of the following reactions.

1. 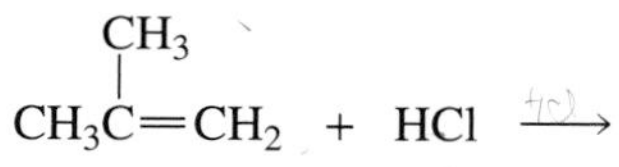 $\underset{\underset{CH_3}{|}}{CH_3C{=}CH_2} \ + \ HCl \ \xrightarrow{\ \ }$
   3. $\underset{\underset{CH_3}{|}}{CH_3C{=}CH_2} \ + \ H_2O \ \xrightarrow{H^+}$

2. $\underset{\underset{CH_3}{|}}{CH_3C{=}CH_2} \ + \ HBr \ \longrightarrow$
   4. 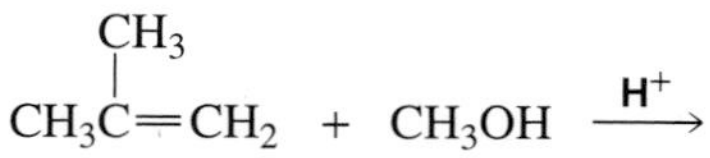 $\underset{\underset{CH_3}{|}}{CH_3C{=}CH_2} \ + \ CH_3OH \ \xrightarrow{H^+}$

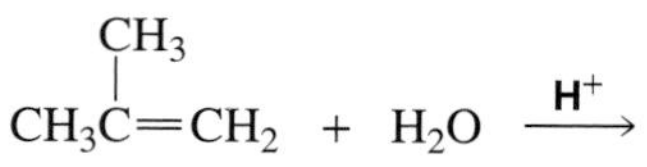

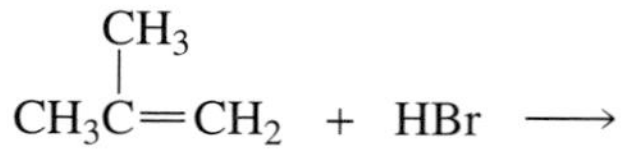

b. What do all the reactions have in common?

c. How do all the reactions differ?

PROBLEM 32◆

How could the following compounds be prepared using an alkene as one of the starting materials?

a.

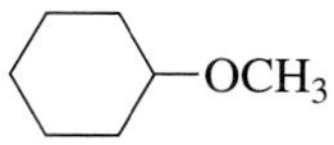

   (cyclohexane)—OCH_3

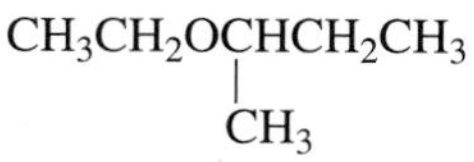

b. 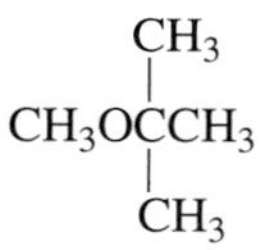 $CH_3O\underset{\underset{CH_3}{|}}{\overset{\overset{CH_3}{|}}{C}}CH_3$

c. $CH_3CH_2OCHCH_2CH_3$
   $\qquad\qquad\quad \underset{CH_3}{|}$

d. 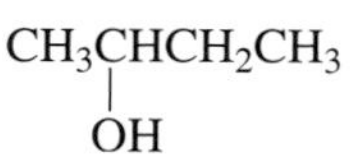 $CH_3CHCH_2CH_3$
   $\qquad \underset{OH}{|}$

Some electrophilic addition reactions give products that are clearly not the result of the addition of an electrophile to the sp^2 carbon bonded to the most hydrogens and the addition of a nucleophile to the other sp^2 carbon. For example, the addition of HBr to 3-methyl-1-butene results in the formation of two products: 2-bromo-3-methylbutane, the product expected as a result of adding H^+ to the sp^2 carbon that is bonded to the most hydrogens and adding Br^- to the other sp^2 carbon, and 2-bromo-2-methylbutane, an "unexpected" product. Furthermore, the "unexpected" product is the major product of the reaction.

**3.14
REARRANGEMENT
OF CARBOCATIONS**

$$CH_3CHCH=CH_2 + HBr \longrightarrow CH_3CHCHCH_3 + CH_3CCH_2CH_3$$

3-methyl-1-butene

2-bromo-3-methylbutane
minor product

2-bromo-2-methylbutane
major product

In another case, the addition of HCl to 3,3-dimethyl-1-butene results in the formation of both an "expected" product, 3-chloro-2,2-dimethylbutane, and an "unexpected product," 2-chloro-2,3-dimethylbutane. Again, the "unexpected" product is obtained in greater yield.

$$CH_3C-CH=CH_2 + HCl \longrightarrow CH_3C-CHCH_3 + CH_3C-CHCH_3$$

3,3-dimethyl-1-butene

3-chloro-2,2-dimethylbutane
minor product

2-chloro-2,3-dimethylbutane
major product

F. C. Whitmore was the first to suggest that the "unexpected" product results from a rearrangement of the carbocation intermediate. Only certain kinds of carbocations rearrange—none of the carbocations that we have seen up to this point rearrange. Carbocations rearrange only if they become more stable as a result of the rearrangement. For example, when an electrophile adds to 3-methyl-1-butene, the carbocation that is initially formed is a *secondary* carbocation. The carbocation has a hydrogen that can shift, with its pair of electrons, to the adjacent positively charged carbon, creating a more stable *tertiary* carbocation.

$$CH_3CH-CH=CH_2 + H-Br \longrightarrow CH_3C-CHCH_3 \xrightarrow{\text{a 1,2-hydride shift}} CH_3C-CH_2CH_3$$

3-methyl-1-butene

a secondary carbocation

a tertiary carbocation

attack on unrearranged carbocation | Br⁻

$$CH_3CH-CHCH_3$$
Br
minor product

attack on rearranged carbocation | Br⁻

$$CH_3C-CH_2CH_3$$
Br
major product

As a result of this **carbocation rearrangement,** two alkyl halides form—one from attack of the nucleophile on the unrearranged carbocation and one from attack on the rearranged carbocation. The major product is the rearranged product. Because the rearrangement involves the shift of a hydrogen with its pair of electrons, it is called a hydride shift ($H:^-$ is a hydride ion). More specifically it is called a **1,2-hydride shift** because the hydride ion moves from one carbon to an adjacent carbon. (Note that this does not mean that it moves from C-1 to C-2.)

3,3-Dimethyl-1-butene adds an electrophile to form a *secondary* carbocation. In this case there is a methyl group that can shift with its pair of electrons to the adjacent positively charged carbon to form a more stable *tertiary* carbocation. This is called a **1,2-methyl shift.**

3,3-dimethyl-1-butene   **a secondary carbocation**   **a tertiary carbocation**

attack on unrearranged carbocation | Cl^- attack on rearranged carbocation | Cl^-

minor product   **major product**

Frank (Rocky) Clifford Whitmore (1887–1947) *was born in Massachusetts. He received a Ph.D. from Harvard University and was a professor of chemistry at Minnesota, Northwestern, and Pennsylvania State Universities. Whitmore never slept a full night; when he got tired, he took a one-hour nap. Consequently, he had the reputation of being an indefatigable worker; 20-hour work days were common. He generally had 30 graduate students working in his lab at a time, and he wrote an advanced textbook that was considered a milestone in the field of organic chemistry.*

Hydride ions and methyl groups are not the only groups that can shift. We will see that a carbon can shift to increase the number of carbons in a ring, and phenyl groups can shift as well. Other groups do not shift.

A shift involves only the movement of a species from one carbon to an adjacent carbon; 1,3-shifts normally do not occur. If the rearrangement does not lead to a more stable carbocation, it can be assumed that carbocation rearrangement does not occur. For example, when a proton adds to 4-methyl-1-pentene, a secondary carbocation is formed. A 1,2-hydride shift would cause the formation of a different secondary carbocation. Since there is no energetic advantage to the shift, rearrangement does not occur and only one alkyl halide is formed.

4-methyl-1-pentene   the carbocation does not rearrange

Carbocation rearrangements also can occur by ring expansion, which is another type of 1,2-shift. In the following example, the carbocation formed initially is a secondary carbocation. Ring expansion leads to a tertiary carbocation, which is more stable and, because of the five-membered ring, has less angle strain.

In any reaction that forms a carbocation intermediate, check to see if the carbocation will rearrange.

In subsequent chapters you will study other reactions that involve formation of carbocation intermediates. It is important to remember that *whenever a reaction leads to the formation of a carbocation, you must check its structure for the possibility of rearrangement.*

PROBLEM 33 / SOLVED

Which of the following carbocations would you expect to rearrange?

a. 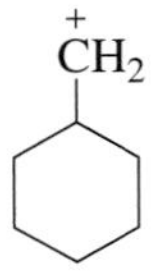 $\overset{+}{C}H_2$ (cyclohexane ring)

c. CH_3 (cyclohexane ring with +)

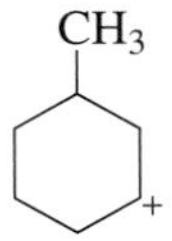

e. $CH_3CH_2\overset{+}{C}HCH_3$

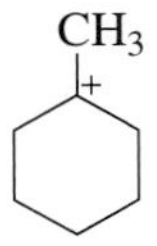

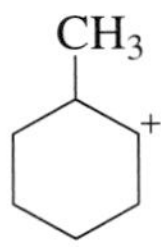

b. CH_3 (cyclohexane ring with +)

d. CH_3 (cyclohexane ring with +)

f. 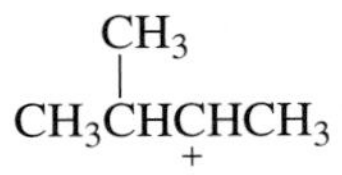 $\underset{\overset{|}{C}H_3}{CH_3}CH CH\overset{+}{C}HCH_3$ $CH_3\overset{|}{C}H\overset{+}{C}HCH_3$

SOLUTION

a. This carbocation will rearrange because a 1,2-hydride shift will convert a primary carbocation into a tertiary carbocation.

$$H \curvearrowright \overset{+}{C}H_2 \quad CH_3$$

(structures with arrow →)

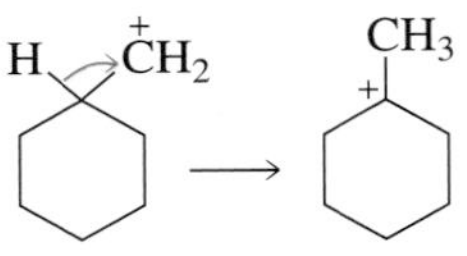

b. This carbocation will not rearrange because it is tertiary and its stability cannot be improved by a carbocation rearrangement.

c. This carbocation will rearrange because a 1,2-hydride shift will convert a secondary carbocation into a tertiary carbocation.

$$CH_3 \curvearrowright H \quad CH_3$$

(structures with arrow →)

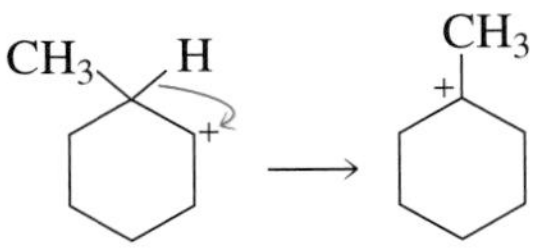

d. This carbocation will not rearrange because it is a secondary carbocation and a carbocation rearrangement would yield another secondary carbocation.

e. This carbocation will not rearrange because it is a secondary carbocation and a carbocation rearrangement would yield another secondary carbocation.

f. This carbocation will rearrange because a 1,2-hydride shift will convert a secondary carbocation into a tertiary carbocation.

$$\underset{\overset{|}{H}}{\overset{\overset{\displaystyle CH_3}{|}}{CH_3\overset{+}{C}-\overset{+}{C}HCH_3}} \longrightarrow \underset{}{\overset{\overset{\displaystyle CH_3}{|}}{CH_3\overset{+}{C}CH_2CH_3}}$$

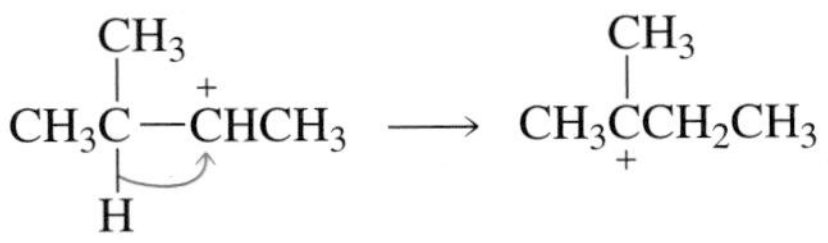

PROBLEM 34

Give the major product obtained from the reaction of each of the following with HBr.

a. 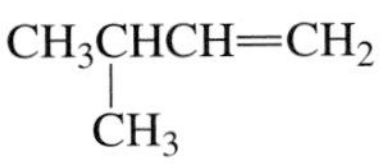 $CH_3CHCH=CH_2$
       $\overset{|}{C}H_3$

b. CH_2 (cyclohexane ring with =CH_2)

c. CH_3 (cyclohexene ring)

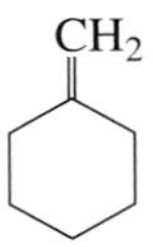

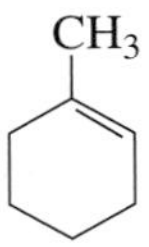

d. 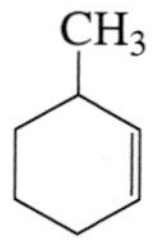 [structure: methylcyclohexene]

e. 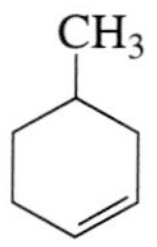 [structure: methylcyclohexene]

f. CH_2=$CHCCH_3$ with CH_3 groups

3.15 ADDITION OF HALOGENS

The halogens, Br_2 and Cl_2, add to alkenes. At first glance this may be surprising because it is not immediately apparent that an electrophile, necessary to start an electrophilic addition reaction, is present.

Halogens are electronegative atoms, and electronegative atoms are electron-withdrawing. Since both atoms in Br_2 or Cl_2 are strongly electron-withdrawing, the bond joining the two halogen atoms is weak (examine the bond dissociation energies in Table 3.1) and therefore easily broken. When the π electrons of the alkene approach a molecule of Br_2 or Cl_2, one of the halogens accepts them and releases the shared electrons to the other halogen. So, in an electrophilic addition reaction, Br_2 behaves as if it were Br^+ and Br^-, and Cl_2 behaves as if it were Cl^+ and Cl^-.

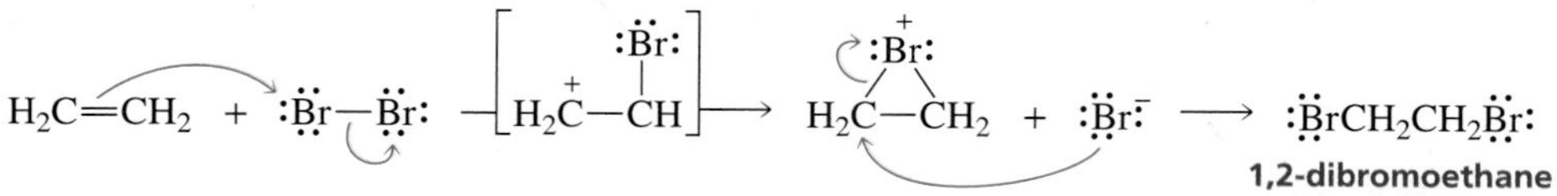

a bromonium ion

1,2-dibromoethane
a vicinal dibromide

The carbocation product of the first step is shown in brackets because it is never actually formed. Instead, a nonbonded pair of electrons on bromine forms a bond with the positively charged carbon, thereby forming a cyclic bromonium ion. The bromonium ion is more stable than the carbocation because all the atoms (except H) in the bromonium ion have complete octets, whereas the positively charged carbon of the carbocation does not have a complete octet. (Remember the octet rule; Section 1.3.) In the second step of the reaction, Br^- attacks a carbon atom of the bromonium ion. This releases the strain in the three-membered ring and forms a vicinal dibromide. **Vicinal** indicates that the two bromines are on adjacent carbons (*vicinus* is the Latin word for "near"). Vicinal dichlorides form from the reaction of alkenes with Cl_2.

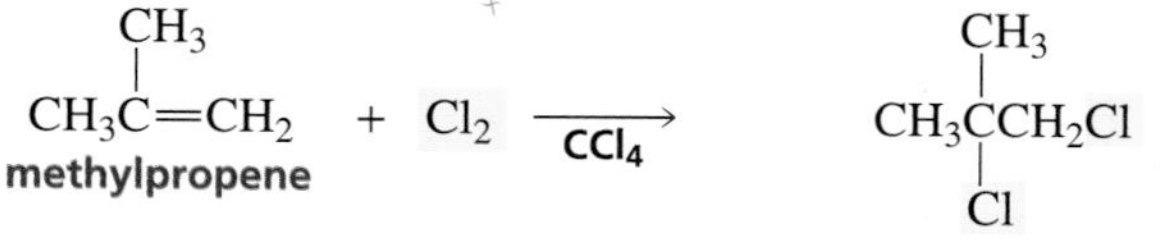

1,2-dichloro-2-methylpropane
a vicinal dichloride

Since a carbocation is not formed when a Br_2 or Cl_2 adds to an alkene, carbocation rearrangements do not occur.

CH₃
|
CH₃CHCH=CH₂ + Br₂ $\xrightarrow{CCl_4}$ CH₃CHCHCH₂Br
3-methyl-1-butene |
                                                  Br

**1,2-dibromo-3-methylbutane
a vicinal dibromide**

PROBLEM 35

a. How does the first step in the reaction of ethene with Br₂ differ from the first step in the reaction of ethene with HBr?

b. To understand why Br⁻ attacks a carbon atom of the bromonium ion and does not attack the positively charged bromine atom, draw the product that would be obtained if Br⁻ *did* attack the bromine atom.

Reactions of alkenes with Br₂ or Cl₂ are generally carried out by mixing the alkene and the halogen in an inert solvent such as carbon tetrachloride (CCl₄) that readily dissolves both reactants but does not participate in the reaction. The preceding reactions illustrate the way in which organic reactions are typically written. The reagents are placed to the left of the reaction arrow, the products are placed to the right of the arrow, and the reaction conditions, such as solvent, temperature, or any required catalyst, are indicated above or below the arrow. Sometimes reactions are written by placing only the organic (carbon-containing) reagent on the left-hand side of the arrow and placing any inorganic reagent above the arrow.

CH₃CH=CHCH₃ $\xrightarrow[CCl_4]{Cl_2}$ CH₃CHCHCH₃
                                          | |
                                          Cl Cl

F₂ and I₂ are not employed as reagents in electrophilic addition reactions. Fluorine reacts explosively with alkenes, so electrophilic addition of F₂ is not a synthetically useful reaction. Vicinal diiodides are unstable at room temperature, decomposing back to the alkene and I₂.

CH₃CH=CHCH₃ + I₂ $\underset{CCl_4}{\overrightarrow{}}$ CH₃CHCHCH₃
                                                                     | |
                                                                     I I

If H₂O is used as the solvent rather than CCl₄, the major product of the reaction will be a vicinal halohydrin. A **halohydrin** is an organic molecule that contains both an OH group and a halogen. In a vicinal halohydrin, the two groups are bonded to adjacent carbons.

CH₃CH=CH₂ + Br₂ $\xrightarrow{H_2O}$ CH₃CHCH₂Br + CH₃CHCH₂Br + HBr
propene | |
                                          OH   Br

a halohydrin   **minor product**
major product

CH₃   CH₃   CH₃
| | |
CH₃CH=CCH₃ + Cl₂ $\xrightarrow{H_2O}$ CH₃CHCCH₃ + CH₃CH—CCH₃ + HCl
2-methyl-2-butene | | | |
                                        Cl OH   Cl   Cl

a halohydrin   **minor product**
major product

The mechanism for halohydrin formation involves formation of a cyclic bromonium ion (or chloronium ion) in the first step of the reaction, since Br^+ is the only electrophile potential in the reaction mixture. The second step is fast; the bromonium ion rapidly reacts with whatever nucleophile it bumps into. There are two nucleophiles present in solution, H_2O and Br^-. Since H_2O is the solvent, its concentration far exceeds the concentration of Br^-. Consequently, the cyclic bromonium ion is more likely to collide with a molecule of water than with Br^-. The protonated halohydrin that is formed is a strong acid (Section 1.19) and immediately loses a proton.

mechanism for halohydrin formation

In the preceding addition reaction, the electrophile (Br^+) ends up on the sp^2 carbon that is bonded to the most hydrogens. This is because, in the transition state for the second step of the reaction, breaking of the carbon–bromine bond has occurred to a greater extent than has formation of the carbon–oxygen bond. As a result, there is a partial positive charge on the carbon that is attacked by the nucleophile.

more stable transition state   less stable transition state

The most stable transition state is achieved by adding the nucleophile to the sp^2 carbon that is bonded to the fewest hydrogens, because the partial positive charge will be on a secondary carbon rather than on a primary carbon. So this reaction also follows Markovnikov's rule. In this case, the electrophile is Br^+ rather than H^+.

Nucleophiles other than H_2O can be added to the reaction mixture and, just as water changed the product of the Br_2 addition from a vicinal dibromide to a vicinal bromohydrin, the added nucleophile will change the product of the reaction. The concentration of the added nucleophile will be greater than the concentration of the halide ion generated from Br_2 or Cl_2, so it will be the nucleophile that will be most likely to participate in the second step of the reaction. Markovnikov's rule will be followed in order to achieve the most stable transition state for the second step of the reaction. (Since positive ions such as Na^+ and K^+ cannot form covalent bonds, they do not react with organic compounds but serve only as counterions to negatively charged species. Therefore, their presence generally is ignored in writing equations.)

PROBLEM 36

Why are Na^+ and K^+ unable to form covalent bonds?

PROBLEM 37

Two products result from each of the following reactions. Identify the products. Which will be the major product of each reaction?

a. $CH_2=\overset{\underset{\displaystyle |}{CH_3}}{C}-CH_3$ + Cl_2 $\xrightarrow{\ CH_3OH\ }$

b. $CH_2=CHCH_3$ + $2\ NaI$ + HBr $\longrightarrow$

c. $CH_3CH=CHCH_3$ + HCl $\xrightarrow{\ H_2O\ }$

d. $CH_3CH=CHCH_3$ + HBr $\xrightarrow{\ CH_3OH\ }$

PROBLEM 38 ◆

What will be the product of the addition of I—Cl to 1-butene? (*Hint:* chlorine is more electronegative than iodine.)

PROBLEM 39 ◆

What would be the major product obtained from the reaction of Br_2 with 1-butene if the reaction were carried out in:

a. carbon tetrachloride?   c. ethyl alcohol?

b. water?   d. methyl alcohol?

In Section 3.13 you saw that water can add to an alkene if an acid catalyst is present. This is the way alkenes are converted into alcohols industrially. However, under normal laboratory conditions, water is generally added to an alkene by a procedure known as **oxymercuration–demercuration.** Addition of water by oxymercuration–demercuration has two advantages over the acid-catalyzed addition of water: it does not require acidic conditions that are harmful to many organic molecules, and carbocation rearrangements do not occur.

In oxymercuration, the alkene is treated with mercuric acetate in aqueous tetrahydrofuran (THF). When reaction with that reagent is complete, sodium borohydride and hydroxide ion are added to the reaction mixture. (The numbers 1 and 2 in front of the reagents in the following equation indicate that the second reagent is not added until reaction with the first reagent is complete.)

3.16 OXYMERCURATION AND ALKOXY-MERCURATION

$$R-CH=CH_2 \xrightarrow[\text{2. NaBH}_4,\ \text{HO}^-]{\text{1. Hg(OAc)}_2,\ \text{H}_2\text{O, THF}} R-\underset{\underset{\displaystyle OH}{|}}{CH}-CH_3$$

In the first step of the oxymercuration mechanism, the electrophilic mercury of mercuric acetate adds to the alkene. Since it is known that carbocation rearrangements do not occur, we can conclude that the product of the addition reaction is a cyclic mercurinium ion rather than a carbocation. The reaction is analogous to the addition of Br_2 to an alkene to form a cyclic bromonium ion.

mechanism for oxymercuration

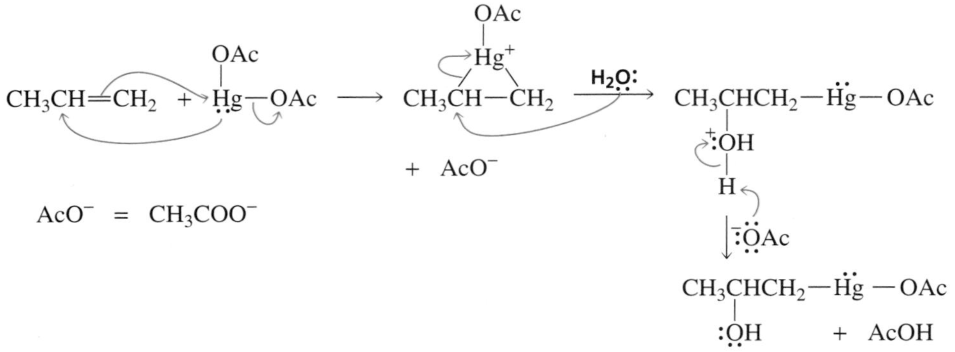

$AcO^- = CH_3COO^-$

In the second step of the reaction, water attacks the most substituted carbon (the one bonded to the *fewest* hydrogens) of the cyclic mercurinium ion for the same reason that it attacks the most substituted carbon of the bromonium ion (Section 3.15): attack at the most substituted carbon leads to the most stable transition state.

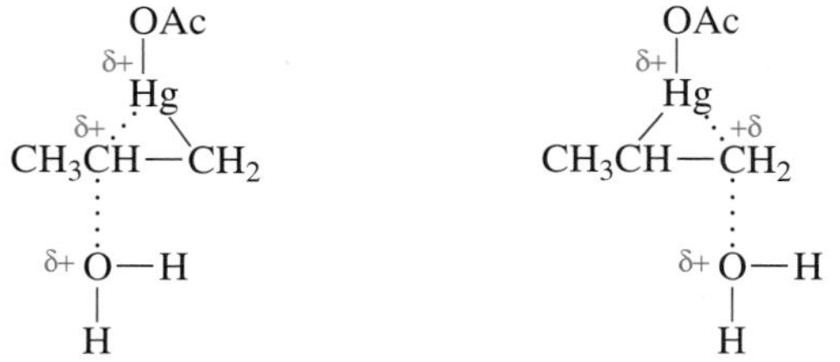

more stable transition state   less stable transition state

Sodium borohydride ($NaBH_4$) converts the carbon–mercury bond into a carbon–hydrogen bond. Because the reaction results in the loss of mercury, it is called *demercuration.*

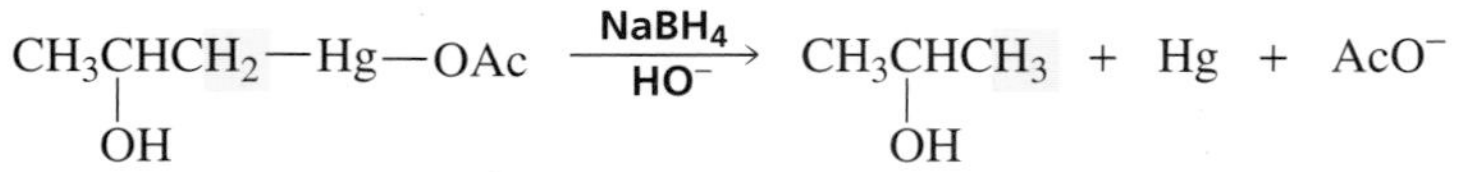

The overall reaction (oxymercuration–demercuration) results in the addition of water to the double bond in a manner that would be predicted by Markovnikov's rule: hydrogen adds to the sp^2 carbon bonded to the most hydrogens, and OH adds to the other sp^2 carbon.

You have seen that, in the presence of an acid catalyst, alkenes react with alcohols to form ethers (Section 3.13). Just as addition of water works better in the presence of mercuric acetate than in the presence of a strong acid, addition of an alcohol works better in the presence of mercuric acetate. [Mercuric trifluoroacetate, $Hg(O_2CCF_3)_2$, works even better.] This reaction is called **alkoxymercuration.**

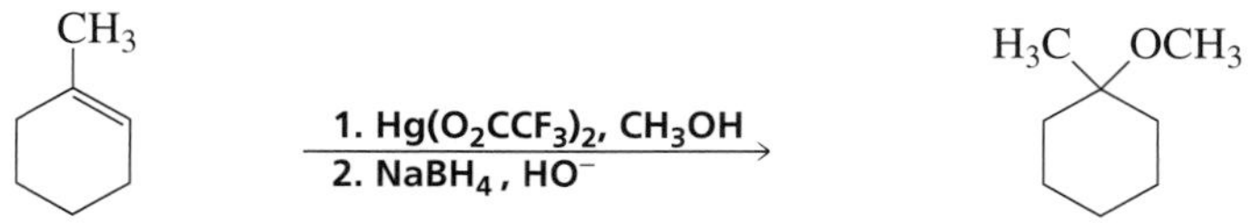

1-methylcyclohexene

1-methoxy-1-methylcyclohexane
an ether

The mechanisms for oxymercuration and alkoxymercuration are essentially identical. The only difference is that water is the nucleophile in oxymercuration and an alcohol is the nucleophile in alkoxymercuration. Therefore, the product of oxymercuration is an alcohol while the product of alkoxymercuration is an ether.

PROBLEM 40

How could each of the following compounds be synthesized from an alkene?

a. (structure: cyclopentane ring with OH substituent)

b. (structure: cyclopentane ring with OCH_2CH_3 substituent)

c. $CH_3\overset{\overset{\displaystyle CH_3}{|}}{\underset{\underset{\displaystyle OH}{|}}{C}}CH_2CH_3$

d. $CH_3\overset{\overset{\displaystyle CH_3}{|}}{\underset{\underset{\displaystyle OCH_3}{|}}{C}}CH_2CH_3$

3.17 ADDITION OF BORANE: HYDROBORATION–OXIDATION

An atom or molecule does not have to be positively charged to be an electrophile. Borane (BH_3), a neutral molecule, is an electrophile because boron has an empty orbital and therefore readily accepts a pair of electrons in order to complete its octet. So alkenes undergo electrophilic addition reactions with borane serving as the electrophile. When the addition reaction is over, an aqueous solution of sodium hydroxide and hydrogen peroxide is added to the reaction mixture and the resulting product is an alcohol. The addition of borane to an alkene followed by reaction with hydroxide ion and hydrogen peroxide is called **hydroboration–oxidation.** Hydroboration–oxidation was first reported by the chemist H. C. Brown in 1959.

$$CH_2{=}CH_2 \quad \xrightarrow[\text{2. HO}^-,\ \text{H}_2\text{O}_2,\ \text{H}_2\text{O}]{\text{1. BH}_3} \quad CH_3CH_2OH$$

The alcohol that is formed from hydroboration–oxidation of an alkene has the H and OH groups on opposite carbons compared with the alcohol that is formed from the acid–catalyzed addition of water (Section 3.13). In other words, the product looks as if water has added to the alkene in a way that violates Markovnikov's rule. However, you will see that the all-encompassing statement of Markovnikov's rule is not violated: the electrophile adds to the sp^2 carbon that is bonded to the greater number of hydrogens. The product appears to violate Markovnikov's original statement of the rule because the original statement was created for a reaction in which H^+ is the electrophile. In hydroboration–oxidation, H^+ is not the electrophile; H^- is the nucleophile.

Herbert Charles Brown (his original family name was Brovarnik) was born in London in 1921 and was brought to the United States by his parents at the age of two. He received his Ph.D. from the University of Chicago and has been a professor of chemistry at Purdue University since 1947. For his studies on boron-containing organic compounds, he shared the 1979 Nobel Prize in chemistry with G. Wittig (Chapter 16, p. 754).

$$CH_3CH{=}CH_2 \quad \xrightarrow[\text{2. HO}^-,\ \text{H}_2\text{O}_2,\ \text{H}_2\text{O}]{\text{1. BH}_3} \quad CH_3CH_2CH_2OH$$
$$\text{propene} \qquad\qquad\qquad\qquad\qquad \text{1-propanol}$$

$$CH_3CH{=}CH_2 \quad \xrightarrow[\text{H}_2\text{O}]{\text{H}^+} \quad CH_3\underset{\underset{\displaystyle OH}{|}}{C}HCH_3$$
$$\text{propene} \qquad\qquad\qquad\qquad\qquad \text{2-propanol}$$

Because borane is a flammable, toxic, and explosive gas, a solution of borane in an ether such as tetrahydrofuran (THF) is a more convenient and less dangerous reagent. A pair of nonbonded electrons on the ether oxygen atom satisfies boron's immediate requirement for electrons. So the reagent that is actually used for the first step of the hydroboration–oxidation reaction is a borane–THF complex that reacts in a similar manner as BH_3.

BORANE AND DIBORANE

In the gaseous state, borane exists primarily as diborane because, in diborane, the two electrons in the hydrogen–boron bond can be shared by two boron atoms, giving each boron a share in eight electrons and satisfying boron's requirement for additional electrons. Diborane is a **dimer**—a molecule formed by the joining of two identical molecules. The hydrogen bridges in diborane are shown as dotted lines to indicate that the bond is made up of fewer than the normal two electrons.

$$2 \quad \underset{H}{\overset{H}{>}}B{-}H \quad \rightleftharpoons \quad \text{diborane}$$

borane   diborane

tetrahydrofuran
THF

$$+ \ BH_3 \ \rightleftharpoons$$

To understand why hydroboration–oxidation of propene forms 1-propanol, the mechanism of the reaction must be examined. Borane, being electron deficient, is an electrophile and consequently reacts with the nucleophilic alkene. As boron accepts the π electrons from the alkene, it eliminates a hydride ion that also adds to the alkene. In all the addition reactions that we have seen up to this point, the electrophile adds to the alkene in the first step and the nucleophile adds to the positively charged intermediate in the second step. In the addition of borane to an alkene, the addition of the electrophilic boron and the nucleophilic hydride ion take place in one step. Therefore, no intermediate is formed.

$$CH_3CH{=}CH_2 \longrightarrow CH_3CH_2CH_2$$
$$H{-}BH_2 \qquad\qquad BH_2$$

an alkylborane

Addition of borane to an alkene is an example of a concerted reaction. A **concerted reaction** is a reaction in which all the bond-making and bond-breaking processes occur in a single step. Addition of borane to an alkene is also an example of a pericyclic reaction. (Pericyclic means "around the circle.") A **pericyclic reaction** is a concerted reaction that takes place as the result of a cyclic rearrangement of electrons.

The electrophilic boron adds to the sp^2 carbon that is bonded to the most hydrogens. The electrophiles that we have looked at previously also added to the sp^2 carbon bonded to the most hydrogens in order to form the most stable carbocation intermediate. But hydroboration is a concerted reaction, so no carbocation intermediate is formed. Why, then, does boron show the same preference in its addition?

To answer this question, we need to examine the transition state for the addition of borane. In the transition state, the sp^2 carbon that does not become attached to boron has a partial positive charge. The addition of borane to the sp^2 carbon of propene that is bonded to the most hydrogens results in the formation of a more stable transition state because the partial positive charge is on a secondary carbon.

If boron had added to the other sp^2 carbon, the partial positive charge would have been on a primary carbon. So, even though a carbocation intermediate is not formed, a carbocation-like transition state is formed. Thus, both the addition of borane and the addition of an electrophile such as H^+ occur at the same sp^2 carbon and for the same reason—to form the more stable carbocation-like transition state.

The alkylborane formed in the first step of the reaction reacts with another molecule of alkene to form a dialkylborane, which then reacts with yet another molecule of alkene to form a trialkylborane. In each of these reactions, boron adds to the sp^2 carbon bonded to the most hydrogens and the nucleophilic hydride ion adds to the other sp^2 carbon.

$$CH_3CH{=}CH_2 \ + \ \underset{\text{an alkylborane}}{RBH_2} \ \longrightarrow \ \underset{\text{a dialkylborane}}{CH_3CH_2CH_2BHR}$$

$$CH_3CH{=}CH_2 \ + \ \underset{\text{a dialkylborane}}{R_2BH} \ \longrightarrow \ \underset{\text{a trialkylborane}}{CH_3CH_2CH_2BR_2}$$

The alkylborane, BH_2R, is a bulkier molecule than BH_3 because R is a larger substituent than H. The dialkylborane with two R groups, BHR_2, is even bulkier than the alkylborane. The alkylborane and the dialkylborane add to the sp^2 carbon that is bonded to the most hydrogens not only to achieve the most stable carbocation-like transition state but also because there is more room at this carbon for the bulky group to attach itself. **Steric effects** are space-filling effects. **Steric hindrance** refers to bulky groups at the site of the reaction that make it difficult for the reactants to approach each other. Steric hindrance associated with the alkylborane and particularly with the dialkylborane causes the addition to occur at the sp^2 carbon that is bonded to the most hydrogens because that is the least sterically hindered of the two sp^2 carbons. Therefore, in each of the three successive additions to the alkene, boron adds to the sp^2 carbon that is bonded to the most hydrogens and H^- adds to the other sp^2 carbon.

When the hydroboration addition reaction is over, aqueous sodium hydroxide and hydrogen peroxide are added to the reaction mixture. Notice that both the hydroxide ion and the hydroperoxide ion are reagents in the reaction.

$$HOOH \ + \ HO^- \ \rightleftharpoons \ HOO^- \ + \ H_2O$$

The end result is replacement of boron by an OH group. Because the replacement of boron by an OH group is an oxidation reaction, the overall reaction is called hydroboration–oxidation.

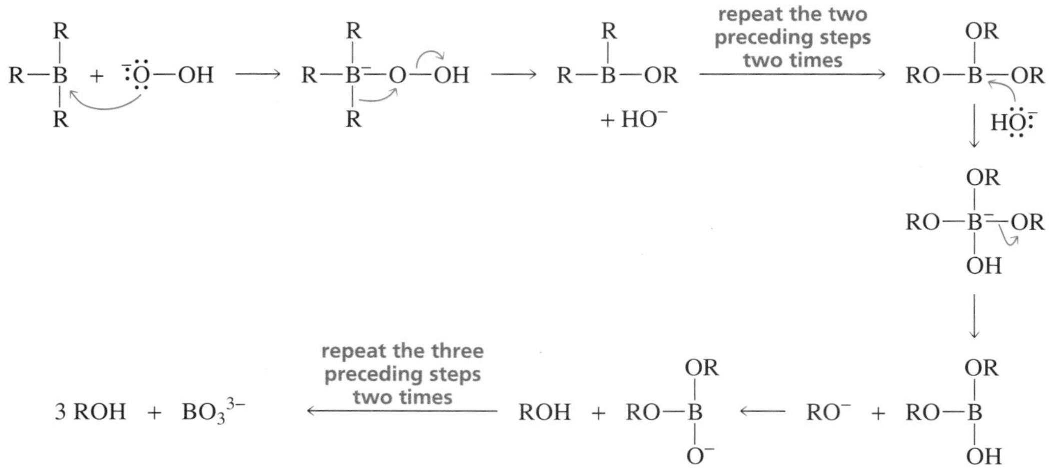

$$ \underset{\underset{R}{|}}{R-B} + \ddot{:}\ddot{O}-OH \longrightarrow R-\underset{\underset{R}{|}}{B}-O-\overset{\frown}{OH} \longrightarrow \underset{+\ HO^-}{R-B-OR} \xrightarrow[\text{two times}]{\substack{\text{repeat the two}\\\text{preceding steps}}} \underset{\downarrow HO\ddot{:}^-}{RO-B=OR} $$

$$ \underset{\underset{OH}{|}}{RO-B=OR} $$

$$ 3\ ROH + BO_3{}^{3-} \xleftarrow[\text{two times}]{\substack{\text{repeat the three}\\\text{preceding steps}}} ROH + \underset{\underset{O^-}{|}}{RO-B} \longleftarrow RO^- + \underset{\underset{OH}{|}}{RO-B} $$

In the overall hydroboration–oxidation reaction, 3 moles of alkene react with 1 mole of BH_3 to form 3 moles of alcohol. The OH ends up on the sp^2 carbon that was bonded to the greater number of hydrogens because it replaces boron, which was the original electrophile in the reaction.

$$ 3\ CH_3CH{=}CH_2 + BH_3 \xrightarrow{\text{THF}} (CH_3CH_2CH_2)_3B \xrightarrow[\text{H}_2\text{O}]{\text{HO}^-,\ \text{H}_2\text{O}_2} 3\ CH_3CH_2CH_2OH + BO^{3-} $$

Since carbocation intermediates are not formed in the hydroboration reaction, carbocation rearrangements do not occur.

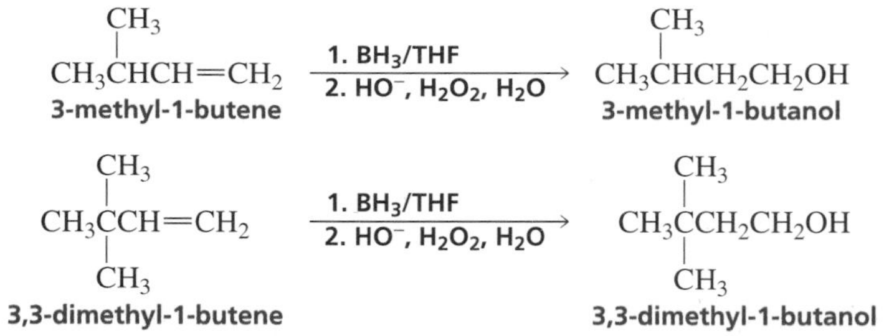

$$ \underset{\textbf{3-methyl-1-butene}}{\overset{\overset{\displaystyle CH_3}{|}}{CH_3CHCH{=}CH_2}} \xrightarrow[\textbf{2. HO}^-,\ \textbf{H}_2\textbf{O}_2,\ \textbf{H}_2\textbf{O}]{\textbf{1. BH}_3/\textbf{THF}} \underset{\textbf{3-methyl-1-butanol}}{\overset{\overset{\displaystyle CH_3}{|}}{CH_3CHCH_2CH_2OH}} $$

$$ \underset{\textbf{3,3-dimethyl-1-butene}}{\overset{\overset{\displaystyle CH_3}{|}}{\underset{\underset{CH_3}{|}}{CH_3CCH{=}CH_2}}} \xrightarrow[\textbf{2. HO}^-,\ \textbf{H}_2\textbf{O}_2,\ \textbf{H}_2\textbf{O}]{\textbf{1. BH}_3/\textbf{THF}} \underset{\textbf{3,3-dimethyl-1-butanol}}{\overset{\overset{\displaystyle CH_3}{|}}{\underset{\underset{CH_3}{|}}{CH_3CCH_2CH_2OH}}} $$

PROBLEM 41◆

How many moles of BH_3 are needed to react with 2 moles of 1-pentene?

PROBLEM 42◆

What product would be obtained from the hydroboration–oxidation of the following alkenes?

a. 2-methyl-2-butene

b. 1-methylcyclohexene

PROBLEM-SOLVING STRATEGY

A **carbene** is an unusual carbon-containing chemical species. It has a carbon with a nonbonded pair of electrons and an empty orbital. The empty orbital makes a carbene very reactive. Methylene (H_2C:) is the simplest carbene. It can be generated by heating diazomethane. Propose a mechanism for the following reaction.

$$:\bar{C}H_2—\overset{+}{N}\equiv N \xrightarrow[\text{H}_2\text{C}=\text{CH}_2]{\Delta} \triangle$$

diazomethane

The information provided in the question is all you need to write a mechanism. Since you know the structure of methylene, you can see that it can be generated simply by breaking the carbon–nitrogen bond of diazomethane. The fact that methylene has an empty orbital means that it is an electrophile and, therefore, will react with ethene (a nucleophile). Now the question is, what is the nucleophile that reacts with the other sp^2 carbon of the alkene? Since you know that cyclopropane is the product of the reaction, the nucleophile must be the nonbonded electrons of methylene.

$$:\bar{C}H_2—\overset{+}{N}\equiv N \longrightarrow N_2 + \underset{\underset{H_2C=CH_2}{}}{:CH_2} \longrightarrow \triangle$$

Note: Diazomethane must be handled with great care because it is both explosive and toxic.

The addition of HBr to 1-butene produces 2-bromobutane. But what if you wanted to synthesize 1-bromobutane? Formation of 1-bromobutane requires the addition of HBr in a manner that would violate Markovnikov's rule. The addition of HBr to 1-butene will produce 1-bromobutane if peroxide is added to the reaction mixture. Either hydrogen peroxide (HOOH) or an alkyl peroxide (ROOR) can be used. The presence of peroxide in the reaction mixture will cause bromine to add to the sp^2 carbon that is bonded to the most hydrogens and hydrogen to add to the other sp^2 carbon. Peroxide reverses the order of addition of the H and Br because it changes the mechanism of the reaction in a way that causes Br• to be the electrophile rather than H^+.

3.18
ADDITION OF RADICALS. THE RELATIVE STABILITIES OF RADICALS

$$CH_3CH_2CH=CH_2 + HBr \longrightarrow CH_3CH_2\overset{\overset{\displaystyle Br}{|}}{C}HCH_3$$

1-butene   **2-bromobutane**

$$CH_3CH_2CH=CH_2 + HBr \xrightarrow{\text{peroxide}} CH_3CH_2CH_2CH_2Br$$

1-butene   **1-bromobutane**

Breaking a bond with the result that both electrons in the bond stay with one of the atoms is called **heterolytic bond cleavage** or **heterolysis**. Breaking a bond with the result that each of the atoms retains one of the bonding electrons is called **homolytic bond cleavage** or **homolysis**. Recall that a double-sided arrowhead signifies the movement of two electrons whereas a single-sided arrowhead, sometimes called a fishhook, signifies the movement of a single electron.

A double-sided arrowhead signifies the movement of two electrons.

heterolytic bond cleavage

$$H—\ddot{\underset{..}{Br}}: \longrightarrow H^+ + :\ddot{\underset{..}{Br}}:^-$$

homolytic bond cleavage

A single-sided arrowhead signifies the movement of one electron.

$$H—\ddot{\underset{..}{Br}}: \longrightarrow H• + •\ddot{\underset{..}{Br}}:$$

Either hydrogen peroxide or an alkyl peroxide can be used to reverse the order of HBr addition. Both contain a weak oxygen–oxygen single bond that is readily broken homolytically in the presence of light or heat to form a radical. An atom or a group of atoms with an unpaired electron is called a **radical.**

$$ HO{-}OH \xrightarrow{\text{light}} 2\ HO\cdot $$
hydrogen peroxide   **hydroxyl radicals**

$$ RO{-}OR \xrightarrow{\text{light}} 2\ RO\cdot $$
an alkyl peroxide   **alkoxyl radicals**

A radical is very reactive because it seeks an electron with which to pair so it can complete its octet. To complete its octet, the hydroxyl (or alkoxyl) radical can remove a hydrogen atom from a molecule of HBr. This results in the formation of a bromine radical.

$$ R{-}\ddot{O}\cdot \ +\ H{-}\ddot{B}r\!: \longrightarrow R{-}\ddot{O}{-}H\ +\ \cdot\ddot{B}r\!: $$
                                                        a bromine radical

The bromine radical now seeks an electron with which to pair to complete its octet. Since the double bond of an alkene is an electron-rich site, the bromine radical completes its octet by combining with one of the electrons in the π bond of the alkene to form a bromine–carbon bond. The second electron of the π bond is the unpaired electron in the resulting alkyl radical. If the bromine radical adds to the sp^2 carbon in 1-butene that is bonded to the most hydrogens, a secondary alkyl radical is formed. Addition of the bromine radical to the other sp^2 carbon would result in the formation of a primary alkyl radical. **Primary, secondary,** and **tertiary alkyl radicals** have the same order of stability as primary, secondary, and tertiary carbocations (tertiary is more stable than secondary, which is more stable than primary). The bromine radical, therefore, adds to the sp^2 carbon that is bonded to the most hydrogens in order to form the more stable secondary radical. The alkyl radical that is formed removes a hydrogen atom from another molecule of HBr to produce a molecule of the alkyl halide product and another bromine radical. Since the first species that adds to the alkene is a radical, the addition of HBr in the presence of peroxides is called a **radical addition reaction.**

$$ \cdot\ddot{B}r\cdot\ +\ CH_2{=}CHCH_2CH_3 \longrightarrow\ :\!\ddot{B}rCH_2\dot{C}HCH_2CH_3 $$
                                                an alkyl radical

$$ BrCH_2\dot{C}HCH_2CH_3\ +\ H{-}\ddot{B}r\!: \longrightarrow\ :\!\ddot{B}rCH_2CH_2CH_2CH_3\ +\ \cdot\ddot{B}r\!: $$

When HBr reacts with an alkene in the absence of peroxide, the electrophile (the first species to add to the alkene) is H^+. In the presence of peroxide, the electrophile is $Br\cdot$. In both cases, the electrophile adds to the sp^2 carbon that is bonded to the most hydrogens, so both reactions follow the all-encompassing statement of Markovnikov's rule (the electrophile adds to the sp^2 carbon that is bonded to the greater number of hydrogens). The addition of peroxide to the reaction mixture changes the order of the addition of the reagents to the double bond, making the reaction appear to "violate" Markovnikov's rule. This is why the presence of peroxide in the reaction mixture is said to cause the *apparent anti-Markovnikov addition* of HBr.

Since the addition of HBr in the presence of peroxide involves the formation of a radical intermediate rather than a carbocation intermediate, rearrangement of the intermediate does not occur. Radicals do not rearrange as readily as carbocations.

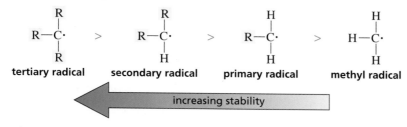

CH$_3$CHCH=CH$_2$ + HBr $\xrightarrow{\text{peroxide}}$ CH$_3$CHCH$_2$CH$_2$Br
3-methyl-1-butene   **1-bromo-3-methylbutane**

As just mentioned, the relative stabilities of primary, secondary, and tertiary alkyl radicals are in the same order as the relative stabilities of primary, secondary, and tertiary carbocations. However, the difference in energy between the radicals is quite a bit smaller than the difference in energy between the carbocations.

relative stabilities of alkyl radicals

$$\underset{\text{tertiary radical}}{R-\overset{\overset{\displaystyle R}{|}}{\underset{\underset{\displaystyle R}{|}}{C}}\cdot} \quad > \quad \underset{\text{secondary radical}}{R-\overset{\overset{\displaystyle R}{|}}{\underset{\underset{\displaystyle H}{|}}{C}}\cdot} \quad > \quad \underset{\text{primary radical}}{R-\overset{\overset{\displaystyle H}{|}}{\underset{\underset{\displaystyle H}{|}}{C}}\cdot} \quad > \quad \underset{\text{methyl radical}}{H-\overset{\overset{\displaystyle H}{|}}{\underset{\underset{\displaystyle H}{|}}{C}}\cdot}$$

$\Longleftarrow$ **increasing stability**

The relative stabilities of the radicals can be determined by comparing the $\Delta H°$ values for formation of the radicals (Table 3.1).

CH$_4$ $\longrightarrow$ $\dot{\text{C}}$H$_3$ + H· $\Delta H° = 105$ kcal/mol

CH$_3$CH$_2$CH$_3$ $\longrightarrow$ CH$_3$CH$_2$$\dot{\text{C}}H_2$ + H· $\Delta H° = 101$ kcal/mol

CH$_3$CH$_2$CH$_3$ $\longrightarrow$ CH$_3$$\dot{\text{C}}HCH_3$ + H· $\Delta H° = 99$ kcal/mol

CH$_3$CHCH$_3$ $\longrightarrow$ CH$_3$$\dot{\text{C}}CH_3$ + H· $\Delta H° = 97$ kcal/mol
   | |
   CH$_3$ CH$_3$

The reaction coordinate diagrams for the four reactions are plotted in Figure 3.11. Since the alkane reactants have similar enthalpies (branching stabilizes an alkane by less than 2 kcal/mol; Section 2.10), the fact that abstraction of a hydrogen atom leads to significantly different values of $\Delta H°$ for each of the four reactions

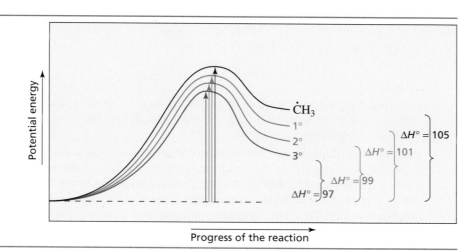

◀ **Figure 3.11**
Relative stabilities of alkyl radicals.

means that the enthalpies of the radical products are different. The tertiary radical is the most stable of the radicals, since it is associated with the smallest enthalpy change. The methyl radical is the least stable of the radicals since it is associated with the greatest change in enthalpy.

Since radical formation is an endothermic reaction, the Hammond postulate (Section 3.11) predicts that the structure of the transition state will be more similar to the structure of the radical product than to the structure of the alkane reactant. This means that the stability differences in the radical products will be reflected in the transition states leading to their formation. Consequently, the more stable the radical, the less energy will be required to make it. Now we can better understand why the bromine radical adds to the sp^2 carbon of 1-butene that is bonded to the most hydrogens to form a secondary alkyl radical rather than adding to the other sp^2 carbon to form a primary alkyl radical. The secondary radical is more stable than the primary radical and therefore is easier to form—the energy barrier is lower for its formation.

The following mechanism for the addition of HBr to an alkene in the presence of peroxide involves seven steps. The steps in a radical chain reaction can be divided into three types: initiation steps, propagation steps, and termination steps.

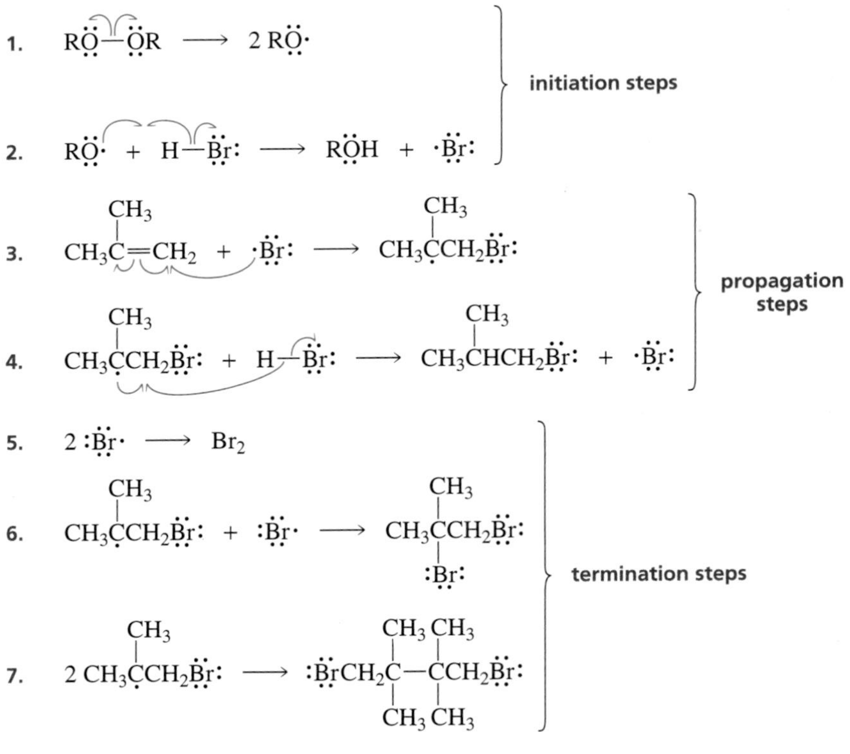

- **Initiation Steps.** The first step is an **initiation step;** it involves the creation of radicals. The second step is also an initiation step; it involves the formation of the chain-propagating bromine radical.
- **Propagation Steps.** Steps 3 and 4 are **propagation steps.** In the first of a pair of propagation steps (step 3), a radical reacts to produce another radical. In the second propagation step (step 4), the radical produced in the first propagation step reacts to form the radical (Br·) that was the reactant in the first propagation step. The two propagation steps are repeated over and over. Hence, the reaction is called a **radical chain reaction;** these two steps propagate the chain.

- **Termination Steps.** Steps 5, 6, and 7 are **termination steps.** In a termination step, two radicals combine to produce a molecule in which all the electrons are paired. Any two radicals present in the reaction mixture can combine in a termination step. Consequently, radical reactions produce a mixture of products.

PROBLEM 43

Write out the propagation steps that occur in the addition of HBr to 1-methylcyclo-hexene in the presence of peroxide.

Peroxide is a **radical initiator.** If peroxide were excluded from the reaction mixture, the preceding radical reaction would not occur. Any reaction that occurs in the presence of a radical initiator and does not occur in its absence must take place by a mechanism that involves radicals as intermediates. Any compound that can readily undergo homolysis (dissociate to form radicals) can act as a radical initiator.

While radical initiators cause radical reactions to occur, **radical inhibitors** have the opposite effect. They trap radicals as they are formed, preventing reactions that take place by mechanisms involving radicals. The reason that radical inhibitors are able to trap radicals is discussed in Section 8.8.

Peroxide has no effect on the addition of HCl or HI to an alkene. In the presence of peroxide, the hydrogen–halogen bond breaks heterolytically and addition occurs just as it does in the absence of peroxide.

$$CH_3CH{=}CH_2 \ + \ HCl \ \xrightarrow{\textbf{peroxide}} \ \underset{\underset{Cl}{|}}{CH_3CHCH_3}$$

$$\underset{\underset{CH_3}{|}}{CH_3C}{=}CH_2 \ + \ HI \ \xrightarrow{\textbf{peroxide}} \ \underset{\underset{I}{|}}{\overset{\overset{CH_3}{|}}{CH_3CCH_3}}$$

Why is the **peroxide effect** observed for the addition of HBr but not for the addition of HCl or HI? This question can be answered by calculating the $\Delta H°$ for the two propagation steps in the radical chain reaction (using the bond dissociation energies in Table 3.1).

$$Cl\cdot \ + \ CH_2{=}CH_2 \ \longrightarrow \ ClCH_2\dot{C}H_2 \qquad \Delta H° = 64 - 82 = -18 \text{ kcal/mol}$$

$$ClCH_2\dot{C}H_2 \ + \ HCl \ \longrightarrow \ ClCH_2CH_3 \ + \ Cl\cdot \qquad \Delta H° = 103 - 101 = +2 \text{ kcal/mol}$$

$$Br\cdot \ + \ CH_2{=}CH_2 \ \longrightarrow \ BrCH_2\dot{C}H_2 \qquad \Delta H° = 64 - 69 = -5 \text{ kcal/mol}$$

$$BrCH_2\dot{C}H_2 \ + \ HBr \ \longrightarrow \ BrCH_2CH_3 \ + \ Br\cdot \qquad \Delta H° = 88 - 101 = -13 \text{ kcal/mol}$$

$$I\cdot \ + \ CH_2{=}CH_2 \ \longrightarrow \ ICH_2\dot{C}H_2 \qquad \Delta H° = 64 - 55 = +9 \text{ kcal/mol}$$

$$ICH_2\dot{C}H_2 \ + \ HI \ \longrightarrow \ ICH_2CH_3 \ + \ I\cdot \qquad \Delta H° = 71 - 101 = -30 \text{ kcal/mol}$$

For the radical addition of HCl, the first propagation step is exothermic and the second one is endothermic. For the radical addition of HI, the first propagation step is endothermic and the second one is exothermic. Only for the radical addition of

HBr are both propagation steps exothermic. In a radical reaction, the steps that propagate the chain reaction compete with the steps that terminate it. Termination steps are always exothermic since only bond making (and no bond breaking) occurs. Therefore, only when both propagation steps are exothermic can propagation compete with termination. When HCl or HI adds to an alkene in the presence of peroxides, any chain reaction that is initiated is terminated rather than propagated, since propagation cannot compete with termination. Consequently, the radical chain reaction does not occur. Only heterolytic addition occurs.

3.19
ADDITION OF HYDROGEN. THE RELATIVE STABILITIES OF ALKENES

In the presence of a metal catalyst such as platinum, palladium, or nickel, hydrogen (H_2) adds to the double bond of an alkene to form an alkane. Without the catalyst, the energy barrier to the reaction would be enormous, since the hydrogen–hydrogen bond is so strong (Table 3.1). The catalyst decreases the energy of activation by breaking the hydrogen–hydrogen bond. Platinum and palladium are used in a finely divided state adsorbed on charcoal (Pt/C, Pd/C). The platinum catalyst is also frequently used in the form of PtO_2, which is known as Adams's catalyst.

$$CH_3CH=CHCH_3 \ + \ H_2 \ \xrightarrow{\text{Pt/C}} \ CH_3CH_2CH_2CH_3$$
2-butene   butane

       CH₃   CH₃

$$CH_3C=CH_2 \ + \ H_2 \ \xrightarrow{\text{Pd/C}} \ CH_3CHCH_3$$
methylpropene   methylpropane

$$\text{cyclohexene} \ + \ H_2 \ \xrightarrow{\text{Ni}} \ \text{cyclohexane}$$

The addition of hydrogen is called **hydrogenation.** Since the reactions above require a catalyst, they are examples of **catalytic hydrogenations.** The metal catalysts are insoluble in the reaction mixture and therefore are classified as **heterogeneous catalysts.** Catalysts that are soluble in a reaction mixture, such as the acid catalyst used in the hydration of alkenes in Section 3.13, are called **homogeneous catalysts.** A heterogeneous catalyst can easily be removed from the reaction mixture by filtration. It can then be recycled. This is an important feature because metal catalysts tend to be expensive.

Roger Adams (1889–1971)
was born in Boston. He received a Ph.D. from Harvard University and was a professor of chemistry at the University of Illinois. He and Sir Alexander Todd (Section 24.1) clarified the structure of tetrahydrocannabinol (THC), the active ingredient of the marijuana plant. Adams's research showed that the test commonly used by the Federal Bureau of Narcotics at that time for marijuana detection was actually detecting an innocuous companion compound.

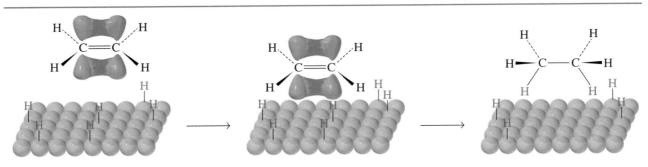

▲ **Figure 3.12**
Catalytic hydrogenation of an alkene.

The details of the mechanism of catalytic hydrogenation are not completely understood. It is known that hydrogen is adsorbed on the surface of the metal and that the alkene complexes with the metal by overlapping its p orbitals with vacant orbitals of the metal. Breaking the π bond of the alkene and the σ bond of H_2 and forming the carbon–hydrogen σ bonds all occur on the surface of the metal. The alkane product diffuses away from the metal surface as it is formed (Figure 3.12).

Hydrogenation reactions are exothermic: they have negative $\Delta H°$ values. The heat released in a hydrogenation reaction $(-\Delta H°)$ is called the **heat of hydrogenation.**

CH₃
|
CH₃C=CHCH₃ + H₂ →(Pt/C)→ CH₃CHCH₂CH₃ $\Delta H° = -26.9$ kcal/mol
2-methyl-2-butene

CH₃
|
CH₂=CCH₂CH₃ + H₂ →(Pt/C)→ CH₃CHCH₂CH₃ $\Delta H° = -28.5$ kcal/mol
2-methyl-1-butene

CH₃
|
CH₃CHCH=CH₂ + H₂ →(Pt/C)→ CH₃CHCH₂CH₃ $\Delta H° = -30.3$ kcal/mol
3-methyl-1-butene

The three catalytic hydrogenation reactions above all result in formation of the same alkane product. Consequently, the energy of the product is the same for each of the three reactions. This must mean that, since the three reactions have different heats of hydrogenation, the three reactants have different energies. The reaction coordinate diagrams for these three reactions are shown in Figure 3.13. The reaction associated with the *least* negative $\Delta H°$ (-26.9 kcal/mol) must have the *most* stable reactant, and the reaction with the most negative $\Delta H°$ (-30.3 kcal/mol) must have the *least* stable reactant. Since potential energy, rather than free energy, is plotted, the difference between the energies of the products and the reactants represents $\Delta H°$ rather than $\Delta G°$.

Examination of the structures of the three alkene reactants shows that the most stable alkene has two alkyl substituents bonded to one of the sp^2 carbons and one alkyl substituent bonded to the other sp^2 carbon, for a total of three alkyl substituents (three methyl groups) bonded to the two sp^2 carbons. The alkene of

Platinum and palladium are expensive metals, so the accidental finding by **Paul Sabatier (1854–1941)** *that nickel, a much cheaper metal, can catalyze hydrogenation reactions made hydrogenation a feasible large-scale industrial process. The conversion of plant oils to margarine is one such hydrogenation reaction. Sabatier was born in France and was a professor at the University of Toulouse. He shared the 1912 Nobel Prize in chemistry with Victor Grignard (Chapter 11, page 449).*

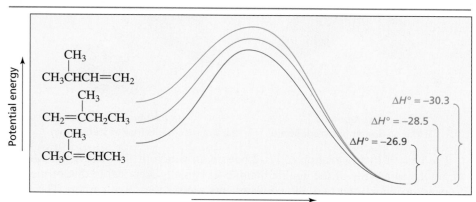

◀ Figure 3.13
Reaction coordinate diagrams for the catalytic hydrogenation of three alkenes that form the same alkane.

The fewer hydrogens bonded to the sp^2 carbons of an alkene, the more stable it is.

intermediate stability has a total of two alkyl substituents (a methyl group and an ethyl group) bonded to the sp^2 carbons, and the least stable of the three alkenes has only one alkyl substituent (an isopropyl group) bonded to the sp^2 carbons. It is apparent that alkyl substituents on the sp^2 carbons of an alkene have a stabilizing effect on the alkene. We can therefore make the following statement: *The more alkyl substituents that are bonded to the sp^2 carbons of an alkene, the more stable it is.* (Some students find it easier to look at the number of hydrogens bonded to the sp^2 carbons. Stated in terms of hydrogens: *The fewer hydrogens that are bonded to the sp^2 carbons of an alkene, the more stable it will be.*)

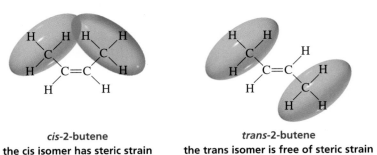

Both *trans*-2-butene and *cis*-2-butene have two alkyl groups bonded to the sp^2 carbons, but *trans*-2-butene has a less negative heat of hydrogenation. This means that the trans isomer, in which the large substituents are farther apart, is more stable than the cis isomer, in which the large substituents are closer together.

When the large substituents are on the same side of the molecule, their electron clouds can interfere with each other. This causes strain in the molecule and makes the molecule less stable. You saw in Section 2.11 that this kind of strain is called *steric strain*. When the large substituents are on opposite sides of the molecule, their electron clouds cannot interact, and the molecule has less steric strain.

cis-2-butene
the cis isomer has steric strain

trans-2-butene
the trans isomer is free of steric strain

The heat of hydrogenation of *cis*-2-butene, in which the two alkyl substituents are on the same side of the double bond, is essentially identical to that of methylpropene, in which the two alkyl substituents are on the same carbon. The three dialkyl-substituted alkenes are all less stable than a trialkyl-substituted alkene and are all more stable than a monoalkyl-substituted alkene.

relative stabilities of dialkyl-substituted alkenes

| alkyl substituents are trans | alkyl substituents are cis | alkyl substituents are on the same sp^2 carbon |

> $>$ $=$

PROBLEM 44 ◆

The same alkane is obtained from the catalytic hydrogenation of either alkene A or alkene B. The heat of hydrogenation of alkene A is -29.8 kcal/mol, and the heat of hydrogenation of alkene B is -31.4 kcal/mol. Which alkene is more stable?

The chemistry associated with living organisms is called **biochemistry** or **biological chemistry.** When you study biochemistry, you study the structures and functions of the molecules found in the biological world and the reactions involved in the synthesis and degradation of these molecules. Since most of the molecules in living organisms are organic molecules, you should expect that many of the reactions that you encounter in organic chemistry will also be found in the chemistry of biological systems. Living cells do not contain molecules such as Cl_2, HBr, or BH_3, so you would not expect to find the addition of such reagents to alkenes in biological systems. But living cells contain water, so hydration reactions (Section 3.13) do occur in biological systems.

Instead of acid or metal catalysts, living organisms use protein catalysts called *enzymes.* (Enzymes will be discussed in Section 21.8.) As an example, let's look at one step in a metabolic pathway called the Krebs cycle. The Krebs cycle (also known as the citric acid cycle) is a series of reactions carried out by cells in which the acetyl group of acetyl-CoA is oxidized to CO_2 and water (see Appendix VI). One step in this cycle involves the hydration of fumarate to produce malate. This hydration is catalyzed by the enzyme fumarase. The rate of the reaction is 10^4 times greater than it would be in the absence of enzyme.

One of the steps in the biosynthesis of fatty acids involves the hydrogenation of a trans alkene. Since cells do not contain H_2 or metal catalysts, the source of hydrogen for this reaction is NADPH + H^+, and the catalyst is an enzyme (enoyl-ACP reductase). At this point, don't worry about what the acronyms NADPH and ACP represent. We will consider these and other biological reactions in further detail in later chapters. What is important now is that you realize that alkenes in biological systems undergo the same kinds of reactions that alkenes undergo in an organic laboratory.

3.20 SOME BIOLOGICAL ADDITION REACTIONS OF ALKENES

Sir Hans Adolf Krebs (1900–1981) *was born in Germany and received a medical degree from the University of Hamburg in 1925. He left Germany in 1933 to escape Hitler. In 1934 he became a member of the faculty at the University of Sheffield and then moved to Oxford University in 1954. He shared the 1953 Nobel Prize in physiology and medicine with Fritz Lipmann (Chapter 15, page 697).*

3.21
REACTIONS AND
SYNTHESIS

This chapter has been concerned with the reactions of alkenes. We have looked at why alkenes react, the kinds of reagents with which they react, the mechanisms by which the reactions occur, and the products that are formed. It is important to remember that when you are studying reactions, you are simultaneously studying synthesis. When you learn that compound A reacts with a certain reagent to form compound B, you are learning not only about the reactivity of A, but also about one way in which compound B can be synthesized.

$$A \longrightarrow B$$

For example, you have seen that alkenes can add many different reagents and that, as a result of adding these reagents, compounds such as alkyl halides, vicinal dihalides, halohydrins, alcohols, ethers, and alkanes are synthesized.

Although you have seen how alkenes react and the kinds of compounds that can be synthesized by the reactions of alkenes, you have not seen how alkenes are synthesized. Reactions of alkenes involve the addition of atoms (or groups of atoms) to the two sp^2 carbons of the double bond. Reactions that lead to the synthesis of alkenes are exactly the opposite. They involve the elimination of atoms (or groups of atoms) from two adjacent sp^3 carbons.

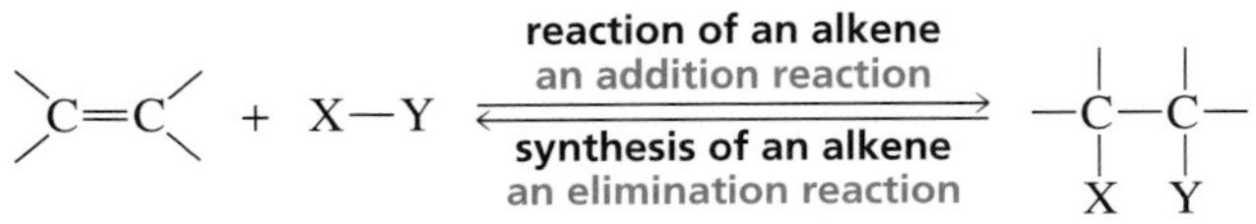

You will learn how alkenes are synthesized when you study compounds that undergo elimination reactions. The various reactions that result in the synthesis of alkenes are listed in Appendix IV.

PROBLEM 45 / SOLVED

Starting with an alkene, indicate how each of the following compounds can be synthesized.

a. 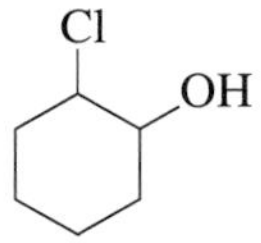   b. 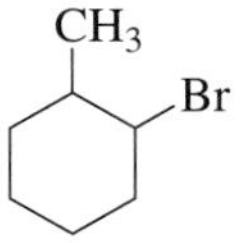   c.

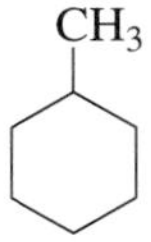

SOLUTION

a. The only alkene that can be used is cyclohexene. To get the desired substituents on the ring, cyclohexene must react with Cl_2 in an aqueous solution so that water will be the nucleophile.

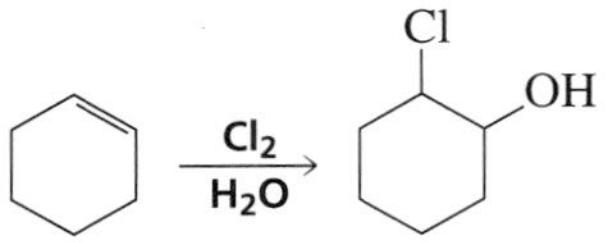

b. The alkene that should be used is 1-methylcyclohexene. To get the substituents in the desired locations, the electrophile in the reaction must be bromine. Therefore, the reagents required to react with 1-methylcyclohexene are HBr and peroxide.

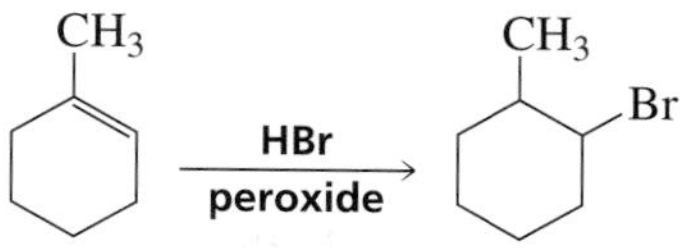

c. In order to synthesize an alkane from an alkene, the alkene must undergo catalytic hydrogenation. Several alkenes could be used for the synthesis.

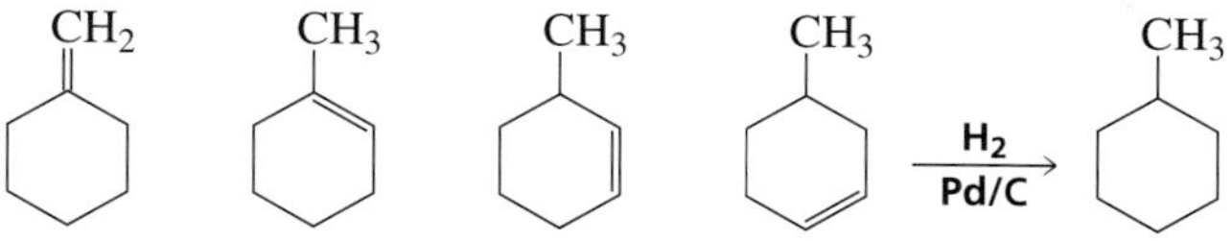

PROBLEM 46

Why should you not use 3-methylcyclohexene as the starting material in Problem 45b?

PROBLEM 47

Starting with an alkene, indicate how each of the following compounds can be synthesized.

a. CH₃CHOCH₃
   |
   CH₃

$$\text{a. } CH_3CHOCH_3 \atop \quad\; CH_3$$

d.

b. CH₃CH₂CHCHCH₃
      | |
      Br   Br

$$\text{b. } CH_3CH_2\underset{Br}{\overset{Br}{\text{CHCH}}}CH_3$$

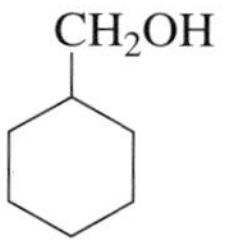

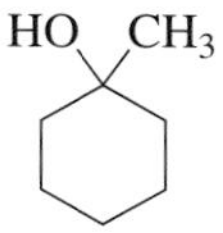

e.

c.

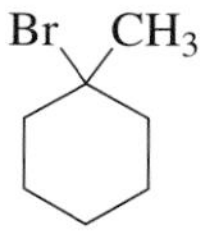

f.

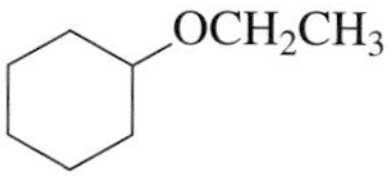

SUMMARY OF REACTIONS

1. Electrophilic addition reactions

   a. Addition of hydrogen halides (Markovnikov orientation) (Section 3.9)

$$RCH{=}CH_2 \;+\; HX \;\longrightarrow\; RCHCH_3 \atop \qquad\qquad\qquad\qquad\;\; X$$

$$HX = HF, HCl, HBr, HI$$

   b. Addition of hydrogen bromide (apparent anti-Markovnikov orientation) Section 3.18)

$$RCH{=}CH_2 \;+\; HBr \;\xrightarrow{\;peroxide\;}\; RCH_2CH_2Br$$

c. Addition of halogen (Section 3.15)

$$RCH=CH_2 \ + \ Cl_2 \ \xrightarrow{CCl_4} \ \underset{\underset{Cl}{|}}{RCHCH_2Cl}$$

$$RCH=CH_2 \ + \ Br_2 \ \xrightarrow{CCl_4} \ \underset{\underset{Br}{|}}{RCHCH_2Br}$$

d. Addition of water and alcohols: acid-catalyzed addition (Markovnikov orientation) (Section 3.13)

$$RCH=CH_2 \ + \ H_2O \ \xrightarrow{H^+} \ \underset{\underset{OH}{|}}{RCHCH_3}$$

$$RCH=CH_2 \ + \ CH_3OH \ \xrightarrow{H^+} \ \underset{\underset{OCH_3}{|}}{RCHCH_3}$$

e. Addition of water and alcohols: oxymercuration–demercuration and alkoxy-mercuration–demercuration (Markovnikov orientation) (Section 3.16)

$$RCH=CH_2 \ \xrightarrow[\text{2. NaBH}_4, \text{ HO}^-]{\text{1. Hg(OAc)}_2, \text{ H}_2\text{O,THF}} \ \underset{\underset{OH}{|}}{RCHCH_3}$$

$$RCH=CH_2 \ \xrightarrow[\text{2. NaBH}_4, \text{ HO}^-]{\text{1. Hg(O}_2\text{CCF}_3)_2, \text{ CH}_3\text{OH}} \ \underset{\underset{OCH_3}{|}}{RCHCH_3}$$

f. Addition of water: hydroboration–oxidation (apparent anti-Markovnikov orientation) (Section 3.17)

$$RCH=CH_2 \ \xrightarrow[\text{2. HO}^-, \text{ H}_2\text{O}_2, \text{ H}_2\text{O}]{\text{1. BH}_3/\text{THF}} \ RCH_2CH_2OH$$

2. Addition of hydrogen (Section 3.19)

$$RCH=CH_2 \ + \ H_2 \ \xrightarrow{\text{Pd/C, Pt/C, or Ni}} \ RCH_2CH_3$$

KEY TERMS

acid-catalyzed reaction (page 141)
acyclic (page 104)
addition reaction (page 116)
alkene (page 103)
alkoxymercuration (page 150)
allyl group (page 108)
allylic carbon (page 107)
Arrhenius equation (page 127)
biochemistry (page 163)
biological chemistry (page 163)
carbene (page 154)
carbocation rearrangement (page 143)
catalyst (page 141)
catalytic hydrogenation (page 160)

cis isomer (page 109)
cis–trans stereoisomers (page 109)
concerted reaction (page 152)
constitutional isomers (page 137)
dimer (page 152)
E isomer (page 112)
electrophile (page 114)
electrophilic addition reaction
   (page 116)
endergonic reaction (page 118)
endothermic reaction (page 121)
enthalpy (page 120)
entropy (page 121)
exergonic reaction (page 118)

exothermic reaction (page 121)
experimental energy of activation
   (page 127)
first-order rate constant (page 126)
first-order reaction (page 125)
free energy of activation (page 124)
functional group (page 104)
geometric isomers (page 109)
Gibbs standard free energy change
   (page 118)
halohydrin (page 147)
Hammond postulate (page 135)
heat of hydrogenation (page 161)
heterogeneous catalyst (page 160)

PROBLEMS

48. Give the IUPAC name for each of the following compounds.

a. $CH_3CH_2CHCH=CHCH_2CH_2CHCH_3$
           | |
          Br   Br

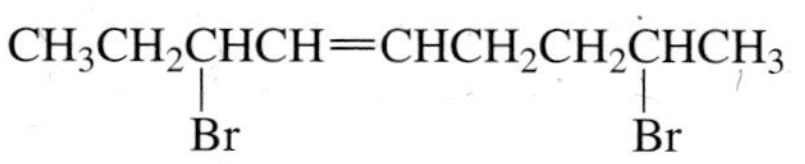

b.
$$H_3C \quad\quad CH_2CH_3$$
$$C=C$$
$$CH_3CH_2 \quad\quad CH_2CH_2CHCH_3$$
$$ CH_3$$

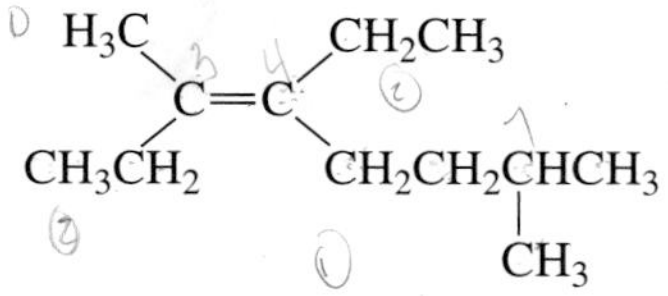

c.
$$CH_3$$
(cyclopentene ring with CH_3 at position 1 and CH_3 substituent)

d.
$$H_3C \quad\quad CH_2CH_3$$
$$C=C$$
$$H_3C \quad\quad CH_2CH_2CH_2CH_3$$

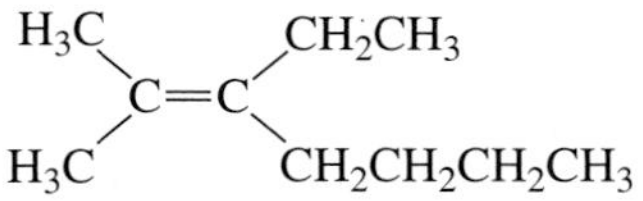

49. Give the structure of a hydrocarbon that has six carbon atoms and:
  a. three vinylic hydrogens and two allylic hydrogens.
  b. three vinylic hydrogens and one allylic hydrogen.
  c. three vinylic hydrogens and no allylic hydrogens.

50. Draw the structure for each of the following.
  a. (Z)-1,3,5-tribromo-2-pentene
  b. (Z)-3-methyl-2-heptene
  c. (E)-1,2-dibromo-3-isopropyl-2-hexene
  d. vinyl bromide
  e. 1,2-dimethylcyclopentene
  f. diallylamine

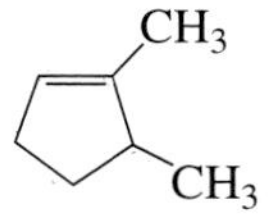

51. Give the structures and the IUPAC names for all alkenes with the molecular formula
    C_6H_{12}.

52. What will be the major product of the reaction of 2-methyl-2-butene with each of the following reagents?

a. HBr
b. HBr + peroxide
c. HI
d. HI + peroxide
e. ICl
f. H_2/Pd
g. HBr + excess NaCl

h. $Hg(OAc)_2$, H_2O followed by $NaBH_4$/HO^-
i. H_2O + trace HCl
j. Br_2/CCl_4
k. Br_2/H_2O
l. Br_2/CH_3OH
m. BH_3/THF followed by H_2O_2/HO^-
n. $Hg(O_2CCF_3)_2$, CH_3OH followed by $NaBH_4$/HO^-

53. Name the following compounds.

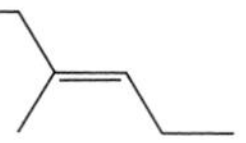

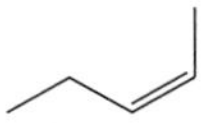

54. Draw curved arrows to show the flow of electrons responsible for the conversion of the reactants into the products.

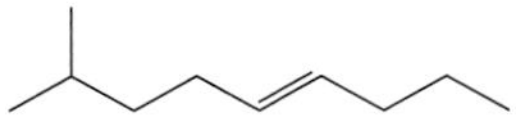

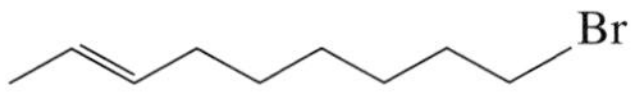

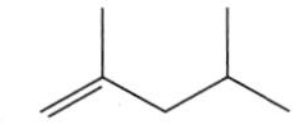

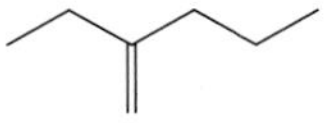

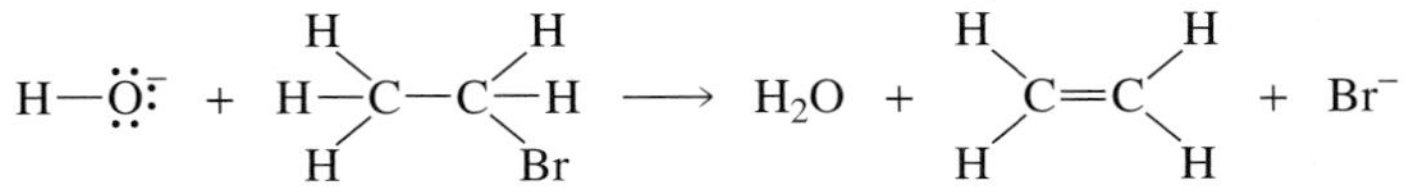

55. When 3-methyl-1-butene reacts with HBr, two alkyl halides are formed, 2-bromo-3-methylbutane and 2-bromo-2-methylbutane. Give a mechanism that explains the formation of these products.

56. In a reaction in which reactant A is in equilibrium with product B at 25 °C, what are the relative amounts of A and B present at equilibrium if $\Delta G°$ at 25 °C is:

a. 2.72 kcal/mol?
b. 0.65 kcal/mol?

c. −2.72 kcal/mol?
d. −0.65 kcal/mol?

57. Give the reagents that would be required to carry out the following syntheses.

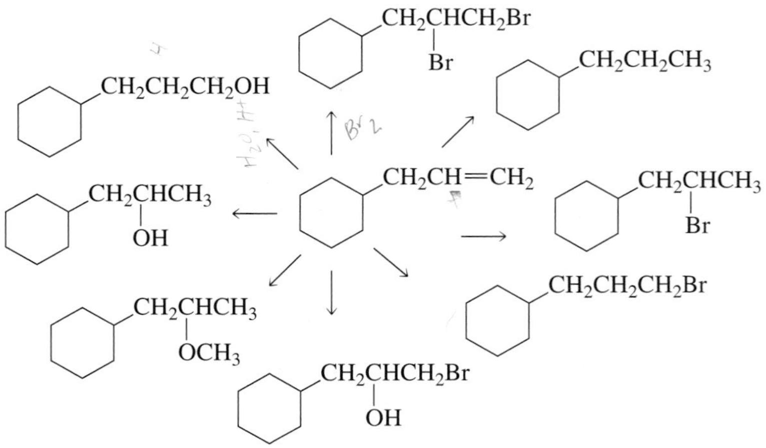

58. Several studies have shown that β-carotene, a precursor of vitamin A, may play a role in preventing cancer. β-Carotene has a molecular formula of $C_{40}H_{56}$, and it contains two rings and no triple bonds. How many double bonds does it have?

59. Which of the indicated bonds has the greater bond strength? Briefly explain why.

    a. 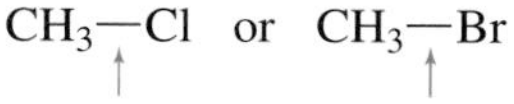 CH_3—Cl   or CH_3—Br

    b. 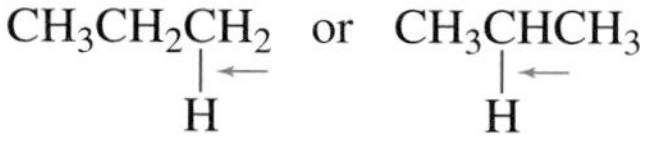 $CH_3CH_2CH_2$ or CH_3CHCH_3
             | |
             H   H

    c. 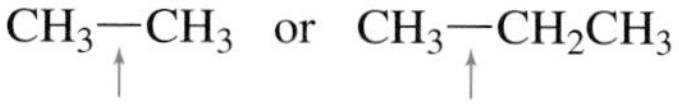 CH_3—CH_3 or CH_3—CH_2CH_3

    d. 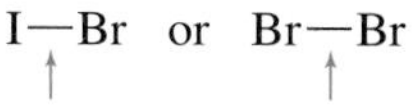 I—Br   or   Br—Br

60. How could you prepare the following compounds? You can use any alkene and any other reagents.

    a.  ⬡

    b. 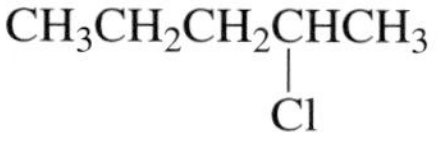 $CH_3CH_2CH_2CHCH_3$
                           |
                           Cl

    c.  ⬡ with CH_2Br

    d. 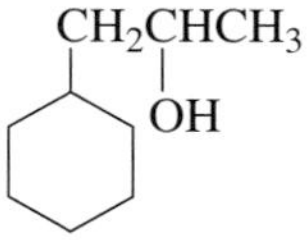 ⬡ with CH_2CHCH_3 and OH

    e. 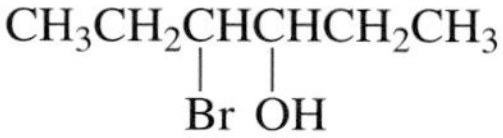 $CH_3CH_2CHCHCH_2CH_3$
                          | |
                        Br OH

    f. 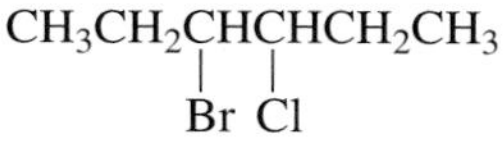 $CH_3CH_2CHCHCH_2CH_3$
                          | |
                        Br Cl

61. Which of the following compounds is the most stable? Which would you expect to have the most negative heat of hydrogenation? Which would you expect to have the least negative heat of hydrogenation?

    3,4-dimethyl-2-hexene; 2,3-dimethyl-2-hexene; 4,5-dimethyl-2-hexene

62. Tell whether each of the following compounds has the E or the Z configuration.

    a. 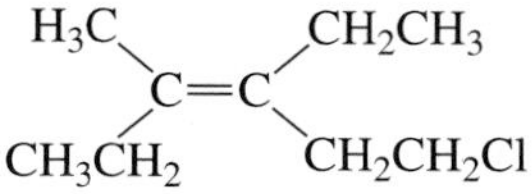 H_3C and CH_2CH_3 / $C=C$ / CH_3CH_2 and CH_2CH_2Cl

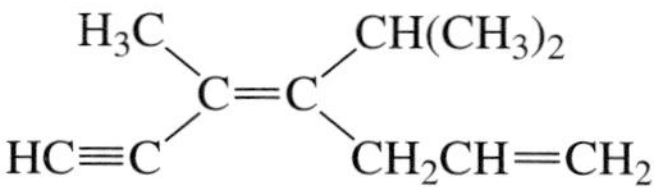

    b. H_3C and $CH(CH_3)_2$ / $C=C$ / $HC\equiv C$ and $CH_2CH=CH_2$

    c. H_3C and CH_2Br / $C=C$ / Br and $CH_2CH_2CH_2CH_3$

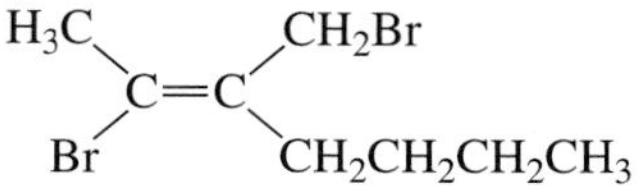

    d. 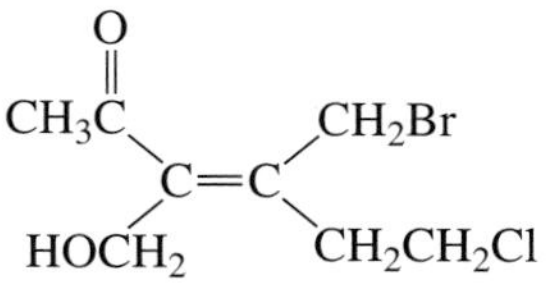 $CH_3\overset{O}{\overset{\|}{C}}$ and CH_2Br / $C=C$ / $HOCH_2$ and CH_2CH_2Cl

63. Squalene is a hydrocarbon with molecular formula $C_{30}H_{50}$ that is obtained from shark liver (*squalus* is Latin for "shark"). Knowing that squalene is not a cyclic compound, determine how many π bonds it has.

64. Assign priorities to each set of substituents.
    a. —Br, —I, —OH, —CH_3
    b. —CH_2CH_2OH, —OH, —CH_2Cl, —CH=CH_2
    c. —$CH_2CH_2CH_3$, —$CH(CH_3)_2$, —CH=CH_2, —CH_3
    d. —CH_2NH_2, —NH_2, —OH, —CH_2OH
    e. —$COCH_3$, —CH=CH_2, —Cl, —CN

65. There are two alkenes that react with HBr to give 1-bromo-1-methylcyclohexane.
    a. Identify the alkenes.

b. Will both alkenes give the same product when they react with HBr/peroxide?
c. With HCl?
d. With HCl/peroxide?

66. Indicate which member of each of the following pairs is the more stable.

a. $\overset{CH_3}{\underset{|}{CH_3\overset{+}{C}CH_3}}$ or $CH_3\overset{+}{C}HCH_2CH_3$

b. $\underset{\underset{CH_3}{|}}{\overset{CH_3}{\underset{|}{CH_3\overset{\cdot}{C}CH_2}}}$ or $CH_3\overset{\cdot}{C}HCH_2CH_3$

c. $CH_3\overset{+}{C}HCH_3$ or $CH_3\overset{+}{C}HCH_2Cl$

d. $CH_3CH_2CH_2\overset{\cdot}{C}HCH_3$ or $CH_3CH_2CH_2CH_2\overset{\cdot}{C}H_2$

e. $\overset{CH_3}{\underset{|}{CH_3C}}=CHCH_2CH_3$ or $\overset{CH_3}{\underset{|}{CH_3CH}}=CHCHCH_3$

f. 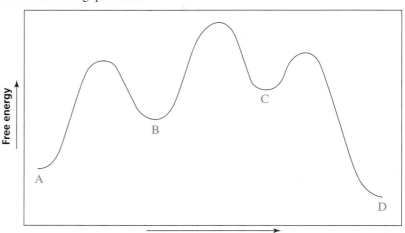 or

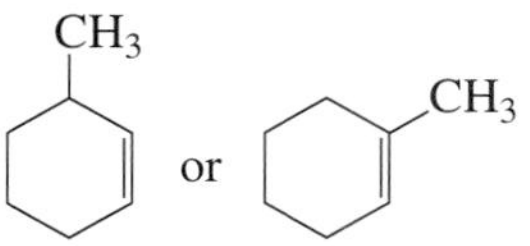

g. $\underset{\underset{Cl}{|}}{CH_3\overset{+}{C}HCHCH_3}$ or $\underset{\underset{Cl}{|}}{CH_3\overset{+}{C}HCH_2CH_2}$

67. Jim Alkeme was a lab technician who was asked by his supervisor to add names to the labels on a collection of alkenes that had only structures on the labels. How many did Jim get right? Correct the incorrect names.
a. 3-pentene
b. 2-octene
c. 2-vinylpentane
d. 1-ethyl-1-pentene
e. 5-ethylcyclohexene
f. 5-chloro-3-hexene
g. 5-bromo-2-pentene
h. (E)-2-methyl-1-hexene
i. 2-methylcyclopentene
j. 2-ethyl-2-butene

68. Given the following reaction coordinate diagram for the reaction of A to give D, answer the following questions.

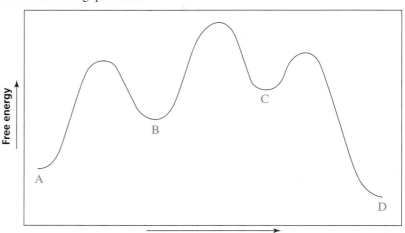

a. How many intermediates are there in the reaction?
b. How many transition states are there?
c. What is the fastest step in the reaction?
d. Which is more stable, A or D?
e. What is the reactant of the rate-determining step?
f. Is the first step of the reaction exergonic or endergonic?
g. Is the overall reaction exergonic or endergonic?

69. The second-order rate constant ($M^{-1} s^{-1}$) for acid-catalyzed hydration at 25 °C is given for the following alkenes.

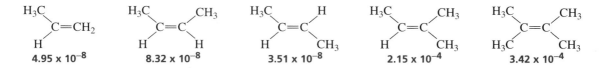

4.95 x 10⁻⁸   8.32 x 10⁻⁸   3.51 x 10⁻⁸   2.15 x 10⁻⁴   3.42 x 10⁻⁴

a. Calculate their relative rates of hydration.
b. Why does (Z)-2-butene react faster than (E)-2-butene?
c. Why does 2-methyl-2-butene react faster than (Z)-2-butene?
d. Why does 2,3-dimethyl-2-butene react faster than 2-methyl-2-butene?

70. Which compound has the greater dipole moment?

a.

b.

c.

71. a. How many alkenes could you treat with H_2/Pt in order to prepare methylcy-clopentane?
 b. Which alkene is the most stable?
 c. Which alkene has the least negative heat of hydrogenation?

72. a. What alkene that contains five carbon atoms will form the same product when it reacts with HBr in the presence of peroxides as it forms when it reacts with HBr in the absence of peroxides?
 b. Give three alkenes containing six carbon atoms that form the same product when they react with HBr in the presence of peroxides as they form when they react with HBr in the absence of peroxides.

73. a. What piece of information do you need to know before you can predict whether methylpropene or 1-butene will react faster with HCl?
 b. Knowing that methylpropene reacts faster than 1-butene with HCl, predict which of the following will react faster with HCl.
   1. ethylene or propylene? Explain.
   2. 2-methyl-1-butene or 3-methyl-1-butene? Explain.

74. a. What is the equilibrium constant of a reaction that is carried out at 25 °C (298 K) and that has $\Delta H° = -20$ kcal/mol and $\Delta S° = -25$ eu (cal deg^{-1} mol^{-1})?
 b. What is the equilibrium constant of the same reaction carried out at 125 °C?

75. Mark Onikoff was about to turn in the products he had obtained from the reaction of HI with 3,3,3-trifluoropropene when he realized that the labels had fallen off his

flasks. He didn't know which label belonged to which flask. The student at the next lab bench told him that since the product obtained by following Markovnikov's rule was 1,1,1-trifluoro-2-iodopropane, he should put that label on the flask that contained the most product and label the flask with the least product 1,1,1-trifluoro-3-iodopropane, the anti-Markovnikov product. Should Mark follow the student's advice?

76. a. For a reaction that is carried out at 25 °C, how much must $\Delta G°$ change in order to increase the equilibrium constant by a factor of 10?
   b. How much must $\Delta H°$ change if $\Delta S° = 0$ eu?
   c. How much must $\Delta S°$ change if $\Delta H° = 0$ kcal?

77. a. Propose a mechanism for the following reaction. (Show all arrows.)

$$CH_3CH_2CH{=}CH_2 \ + \ CH_3OH \ \xrightarrow{\ H^+\ } \ CH_3CH_2CHCH_3$$
$$\underset{OCH_3}{|}$$

   b. Which step is the rate-determining step?
   c. What is the electrophile in the first step?
   d. What is the nucleophile in the first step?
   e. What is the electrophile in the second step?
   f. What is the nucleophile in the second step?

78. a. What product is obtained from the reaction of HCl with 1-butene? With 2-butene?
   b. Which of the two reactions has the greater free energy of activation?
   c. Which of the two alkenes reacts more rapidly with HCl?
   d. Which compound reacts more rapidly with HCl, (Z)-2-butene or (E)-2-butene?

79. a. Propose a mechanism for the following reaction.

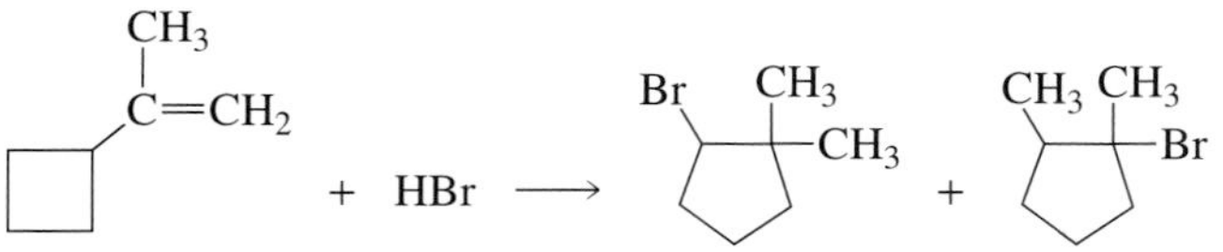

   b. Is the initially formed carbocation primary, secondary, or tertiary?
   c. Is the rearranged carbocation primary, secondary, or tertiary?
   d. Why does the rearrangement occur?

80. When the following compound is hydrated in the presence of acid, the unreacted alkene is found to have retained the deuterium atoms. What does this tell you about the mechanism of hydration?

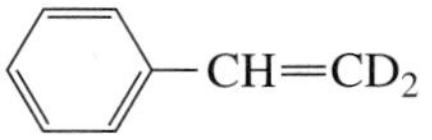

81. Propose a mechanism for each of the following reactions.

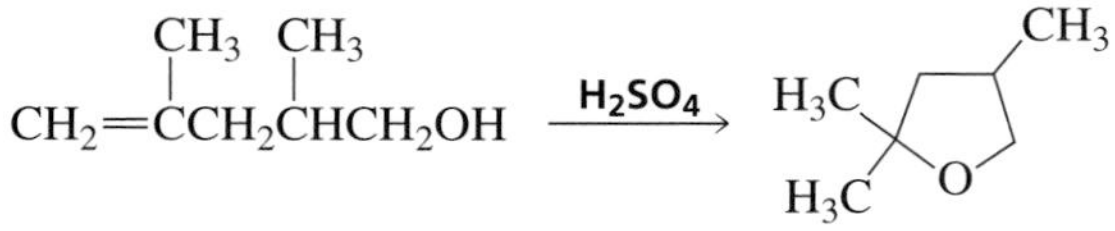

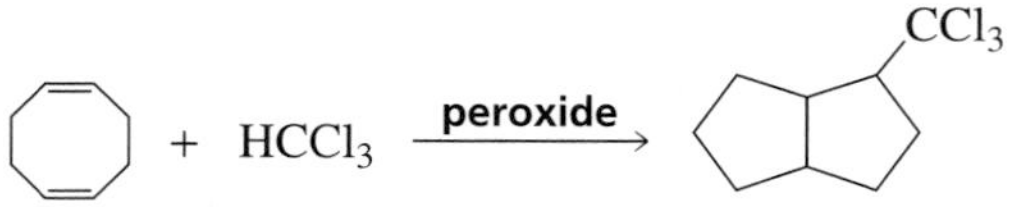

c.

d. $CH_3CH_2\overset{\underset{\textstyle |}{CH_3}}{C}{=}CH_2$ + $CH_2{=}\overset{+}{N}{=}\overset{-}{N}$ $\longrightarrow$

82. Rate constants for a reaction were determined at five temperatures. From the following data, calculate the experimental energy of activation. Then calculate $\Delta G^{\ddagger}$, $\Delta H^{\ddagger}$, and $\Delta S^{\ddagger}$ for the reaction at 30 °C.

Temperature	Observed rate constant
31.0 °C	$2.11 \times 10^{-5}\ s^{-1}$
40.0 °C	$4.44 \times 10^{-5}\ s^{-1}$
51.5 °C	$1.16 \times 10^{-4}\ s^{-1}$
59.8 °C	$2.10 \times 10^{-4}\ s^{-1}$
69.2 °C	$4.34 \times 10^{-4}\ s^{-1}$

83. a. Dichlorocarbene can be generated by heating chloroform with HO^{-}. Propose a mechanism for the reaction.

$$CHCl_3 + HO^{-} \xrightarrow{\Delta} Cl_2C: + H_2O + Cl^{-}$$
chloroform   dichlorocarbene

b. Dichlorocarbene can also be generated by heating sodium trichloroacetate. Propose a mechanism for the reaction.

$$Cl_3C\overset{\overset{\textstyle O}{\|}}{C}O^{-}\ Na^{+} \xrightarrow{\Delta} Cl_2C: + CO_2 + Na^{+}\ Cl^{-}$$
sodium trichloroacetate

4

STEREOCHEMISTRY:
The Arrangement of Atoms in Space; The Stereochemistry of Addition Reactions

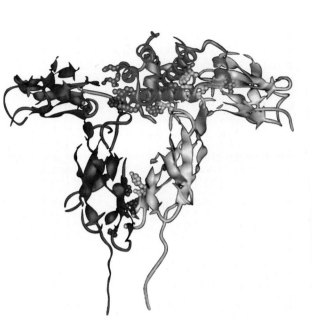

a receptor and its guest

Compounds that have the same molecular formula but are not identical are called **isomers.** Isomers can be divided into two main classes: constitutional isomers and stereoisomers. In Chapter 2, we learned that **constitutional isomers** (also called **structural isomers**) differ in the way their atoms are connected. For example, ethanol and dimethyl ether are isomers because they have the same molecular formula, C_2H_6O. More specifically, they are constitutional isomers because the atoms in each compound are connected differently. The oxygen in ethanol is bonded to a carbon and to a hydrogen, while the oxygen in dimethyl ether is bonded to two carbons.

constitutional isomers

CH_3CH_2OH and CH_3OCH_3
ethanol · · · dimethyl ether

$CH_3\overset{\displaystyle O}{\overset{\|}{C}}CH_3$ and $CH_3CH_2\overset{\displaystyle O}{\overset{\|}{C}}H$
acetone · · · propionaldehyde

$CH_3CH_2CH_2CH_2CH_3$ and $CH_3\overset{\displaystyle CH_3}{\overset{|}{C}H}CH_2CH_3$
pentane · · · isopentane

$CH_3CH_2CH_2CH_2Cl$ and $CH_3CH_2\overset{\displaystyle Cl}{\overset{|}{C}H}CH_3$
1-chlorobutane · · · 2-chlorobutane

Unlike the atoms in constitutional isomers, the atoms in stereoisomers are connected in the same way. **Stereoisomers** differ in the way their atoms are arranged in space. There are two kinds of stereoisomers: conformational isomers and configurational isomers.

Conformational isomers can interconvert rapidly at room temperature. Because they interconvert, conformational isomers cannot be separated. Conformational isomers are also called **conformers.**

Configurational isomers cannot interconvert (unless a covalent bond breaks, which can happen only under vigorous conditions; Section 3.4). Because they cannot interconvert, configurational isomers can be separated.

There are two kinds of conformational isomers: those resulting from rotation about single bonds, and those resulting from amine inversion. There are also two kinds of configurational isomers, **cis–trans isomers** and **isomers that contain chirality** (ky-ral-ity) **centers**. In the following sections, we will meet the members of the stereoisomer family tree (shown here).

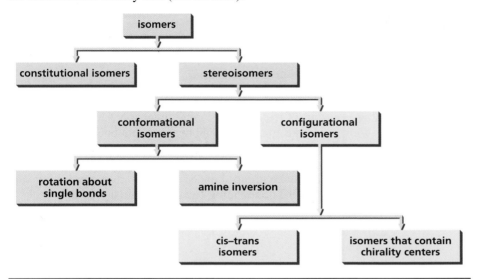

PROBLEM 1♦

a. Draw three constitutional isomers with molecular formula C_3H_8O.

b. How many constitutional isomers can you draw for $C_4H_{10}O$?

One kind of conformational isomer results from the fact that there is free rotation about a carbon–carbon single bond. Since each of the carbon–carbon single bonds in a molecule rotates continually, compounds that contain carbon–carbon single bonds have many interconvertible conformational isomers (Section 2.11).

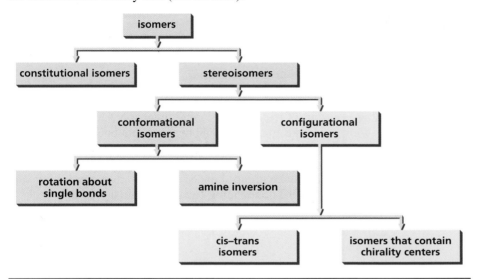

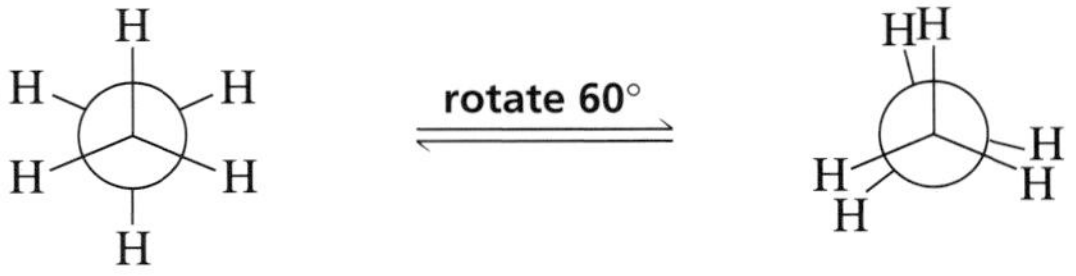

4.1 CONFORMATIONAL ISOMERS

staggered conformation of ethane   **eclipsed conformation of ethane**

As a result of rotation about its carbon–carbon single bonds, methylcyclohexane rapidly interconverts between two chair conformers (Section 2.13).

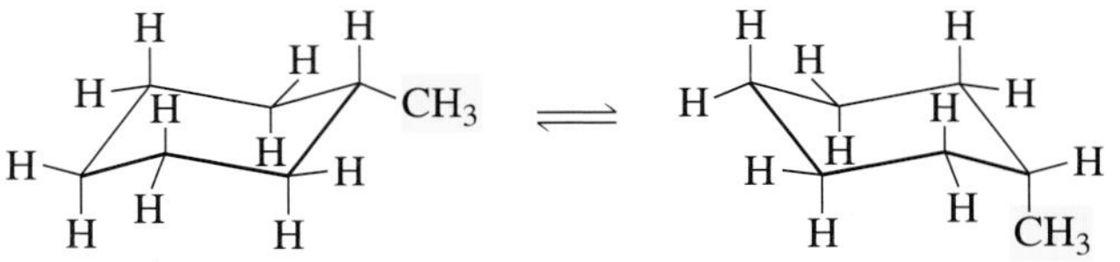

The second kind of conformational isomer results from the fact that, because nitrogen has a pair of nonbonded electrons, it rapidly turns inside out at room temperature. This is called **amine inversion.** The best way to picture amine inversion is to compare it to an umbrella that turns inside out in a windstorm.

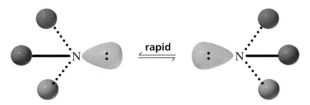

amine inversion

Because of rapid inversion, the individual stereoisomers cannot be isolated. The pair of nonbonding electrons is required for amine inversion. Quaternary ammonium ions—ions with four bonds to nitrogen and hence no nonbonding electrons—do not invert (Section 4.15).

Amine inversion takes place through a transition state in which the sp^3 nitrogen becomes an sp^2 nitrogen. The three groups bonded to the sp^2 nitrogen are coplanar with bond angles of 120°, and the nonbonding electrons are in a p orbital.

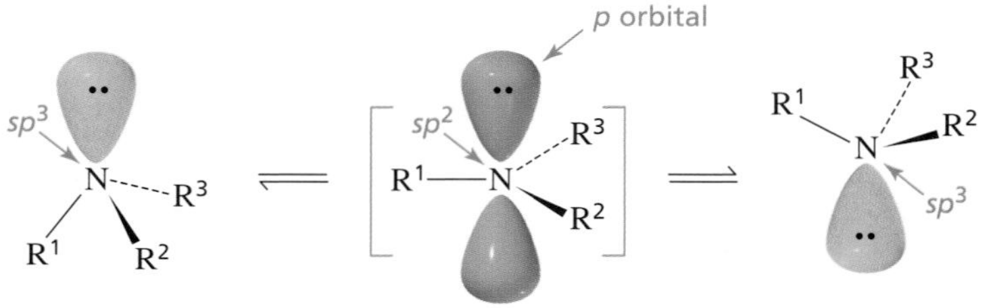

transition state

The energy required for amine inversion is approximately 6 kcal/mol, about twice the amount of energy required for rotation about a carbon–carbon single bond but still less than the thermal energy available from the surroundings at room temperature.

4.2 CONFIGURATIONAL ISOMERS: CIS–TRANS ISOMERS

Cis–trans isomers result from restricted rotation (Section 3.4). Restricted rotation can be caused either by a double bond or by a cyclic structure. Because there is no rotation about a carbon–carbon double bond, an alkene such as 2-pentene can exist as cis and trans isomers. Because the double bond cannot rotate, the cis and trans isomers cannot interconvert. Thus, they are configurational isomers.

cis-2-pentene

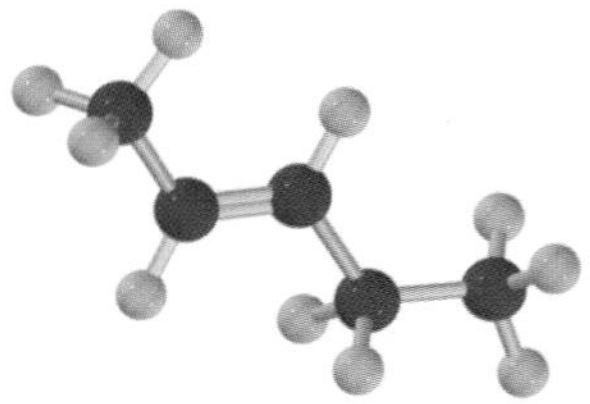

trans-2-pentene

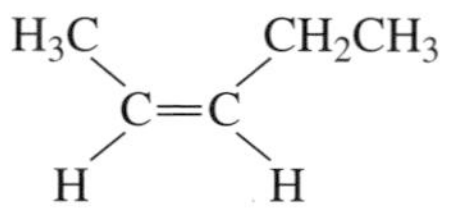

cis-2-pentene

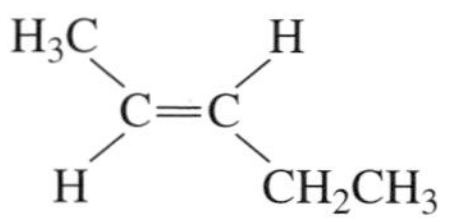

trans-2-pentene

Cyclic compounds can also have cis and trans isomers because the cyclic system prevents free rotation about the single bonds. The **cis isomer** has substituents on the same side of the cyclic structure. The **trans isomer** has substituents on opposite sides of the cyclic structure.

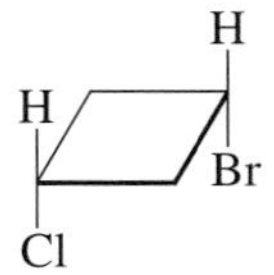

cis-1-bromo-3-chlorocyclobutane

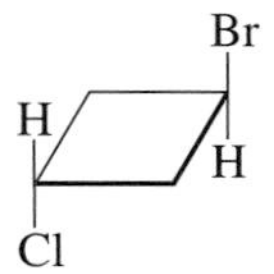

trans-1-bromo-3-chlorocyclobutane

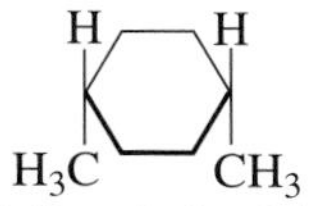

cis-1,4-dimethylcyclohexane

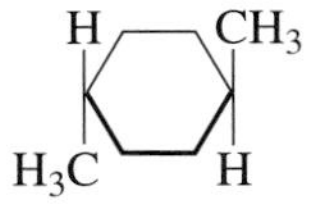

trans-1,4-dimethylcyclohexane

PROBLEM 2◆

Draw the cis and trans isomers for the following compounds.

a. 1-ethyl-3-methylcyclobutane

b. 3,4-dimethyl-3-heptene

c. 1-bromo-4-chlorocyclohexane

d. 1,3-dibromocyclobutane

4.3 CONFIGURATIONAL ISOMERS: ISOMERS WITH ONE CHIRALITY CENTER

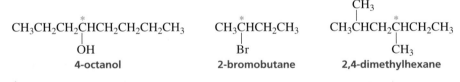

A carbon bonded to four different groups is called a **chirality center.** The chirality center in each of the following compounds is indicated by an asterisk. For example, the fourth carbon in 4-octanol is a chirality center because it is bonded to four different groups (H, OH, $CH_2CH_2CH_3$, and $CH_2CH_2CH_2CH_3$). Notice that the difference in the groups bonded to the chirality center is not necessarily right next to the chirality center; the propyl and butyl groups are different groups. The starred carbon in 2,4-dimethylhexane is a chirality center because it is bonded to four different groups—methyl, ethyl, isobutyl, and hydrogen.

$CH_3CH_2CH_2\underset{\displaystyle OH}{\overset{*}{C}}HCH_2CH_2CH_2CH_3$
4-octanol

$CH_3\underset{\displaystyle Br}{\overset{*}{C}}HCH_2CH_3$
2-bromobutane

$CH_3\overset{\displaystyle CH_3}{C}HCH_2\underset{\displaystyle CH_3}{\overset{*}{C}}HCH_2CH_3$
2,4-dimethylhexane

Notice that the only carbons that can be chirality centers are sp^3 hybridized carbons; sp^2 and sp hybridized carbons cannot be chirality centers because they cannot have four groups attached to them.

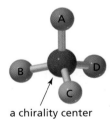

a chirality center

PROBLEM 3◆

Which of the following compounds have chirality centers?

a. $CH_3CH_2\underset{\displaystyle Cl}{C}HCH_3$

b. $CH_3CH_2\underset{\displaystyle CH_3}{C}HCH_3$

c. $CH_3CH_2\underset{\displaystyle Br}{\overset{\displaystyle CH_3}{C}}CH_2CH_2CH_3$

d. CH_3CH_2OH

e. $CH_3CH_2\underset{\displaystyle Br}{C}HCH_2CH_3$

f. $CH_2{=}CH\underset{\displaystyle NH_2}{C}HCH_3$

The term "a chirality center" is the most recently IUPAC-approved name. At various times, it has been called a stereogenic center, a stereo center, a chiral center, or an asymmetric carbon.

left hand

right hand

The solid wedges represent bonds that point out of the plane of the paper toward the viewer.
The dashed lines represent bonds that point back from the plane of the paper away from the viewer.

A compound with a chirality center can exist as two different isomers. Recall that the four groups attached to a carbon do not lie in a plane. The two isomers of 2-bromobutane shown below may look similar on paper, but if you mentally try to put one of the isomers on top of the other—keeping their three-dimensional shape in mind—you cannot superimpose them.

Even turning one molecule over does not help. Because you cannot superimpose them, they are not identical molecules. So there are two different 2-bromobutanes. The two molecules are analogous to a left and a right hand. You cannot superimpose your left hand on your right hand. When you try to superimpose them, either the thumb of one hand lies on top of the little finger of the other hand or the palms and backs face opposite directions.

Take a break and convince yourself that the two 2-bromobutane isomers are not identical by building ball-and-stick models using four different-colored balls to represent the four different groups bonded to the chirality center. Compare your models with those in the margin.

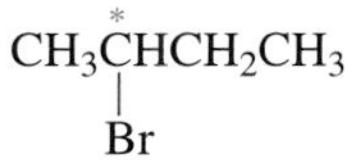

$$CH_3\overset{*}{C}HCH_2CH_3$$
$$|$$
$$Br$$

2-bromobutane

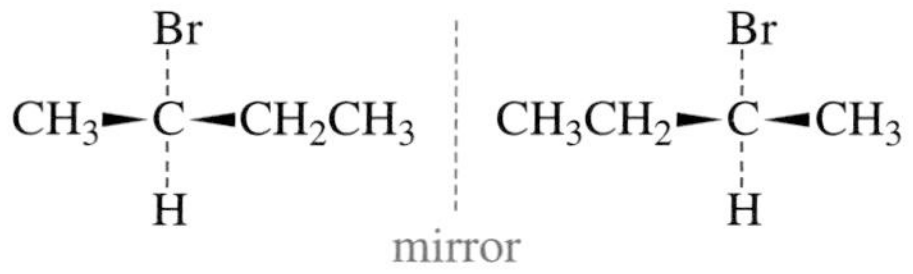

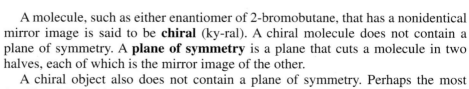

a pair of enantiomers

If you imagine a mirror between the two isomers, you can see that they are mirror images of each other. A nonidentical (nonsuperimposable) mirror image is called an **enantiomer** of the original molecule. "Enantiomer" comes from the Greek *enantion,* which means "opposite." Each of the structures shown above is the enantiomer of the other. A pair of nonidentical mirror images is called a **pair of enantiomers.** The two isomers of 2-bromobutane are a pair of enantiomers.

PROBLEM 4◆

Which of the compounds in Problem 3 can exist as a pair of enantiomers?

A molecule, such as either enantiomer of 2-bromobutane, that has a nonidentical mirror image is said to be **chiral** (ky-ral). A chiral molecule does not contain a plane of symmetry. A **plane of symmetry** is a plane that cuts a molecule in two halves, each of which is the mirror image of the other.

A chiral object also does not contain a plane of symmetry. Perhaps the most familiar chiral objects are your hands. If you imagine a mirror between your two hands, you can see that they are mirror images of each other. Because a hand has a nonidentical mirror image, it is chiral. So a pair of enantiomers, like a pair of hands, are nonidentical mirror images of each other. Interestingly, the term "chiral" comes from the Greek word *cheir,* which means "hand." Other familiar chiral objects are feet, shoes, and ears.

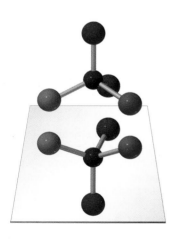

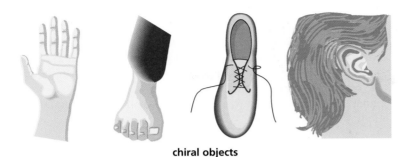

chiral objects

A chiral molecule has at least one chirality center and does not have a plane of symmetry.

A molecule or object that does contain a plane of symmetry is **achiral** (ai-ky-ral). If you cut the object in two halves along the plane of symmetry, the left half is the mirror image of the right half. A fork and a table each have a plane of symmetry; they are achiral. An achiral molecule and its mirror image are identical (they can be superimposed). Therefore, an achiral compound cannot have an enantiomer.

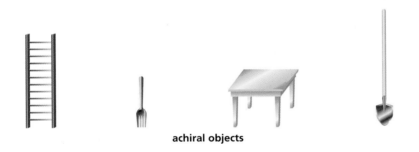

achiral objects

An achiral molecule has a plane of symmetry.

Notice that chirality is a property of an entire object or an entire molecule. The chirality center is what causes a molecule to be chiral.

PROBLEM 5 ◆

Which of the following are chiral?

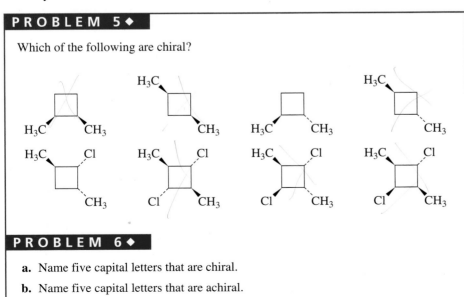

PROBLEM 6 ◆

a. Name five capital letters that are chiral.

b. Name five capital letters that are achiral.

4.4
DRAWING ENANTIOMERS

Emil Fischer

Emil Fischer (1852–1919)
was born in a village near Cologne, Germany. He became a chemist against the wishes of his father, a successful merchant, who wanted him to enter the family business. He was a professor of chemistry at the Universities of Erlangen, Würzburg, and Berlin. In 1902 he received the Nobel Prize in chemistry for his work on sugars. During World War I, he organized German chemical production. Two of his three sons died in that war.

Viktor Meyer (1848–1897)
was born in Germany. To prevent him from becoming an actor, his parents persuaded him to enter the University of Heidelberg, where he earned a Ph.D. at the age of 19. He was a professor of chemistry at the Universities of Stuttgart and Heidelberg. He coined the term "stereochemistry" for the study of molecular shapes, and was the first to describe the effect of steric hindrance on a reaction.

Once you have determined that a compound has a chirality center, there are several ways to draw the enantiomers. **Perspective formulas** show two of the bonds to the chirality center in the plane of the paper, one bond as a solid wedge coming out of the paper, and the fourth bond as a dashed line projecting back from the paper. You can draw the first enantiomer by putting the four groups on the four bonds to the chirality center in any order. Draw the second enantiomer by making a mirror image of the first enantiomer.

perspective formulas of the enantiomers of 2-bromobutane

You can also draw enantiomers using **wedge-and-dash structures.** Draw the first enantiomer by arranging the four groups bonded to the chirality center in any order. Draw the second enantiomer by interchanging two of the groups. It does not make any difference which two groups you interchange. (It is a good idea to make models to convince yourself that this is true.) Many organic chemists prefer to interchange the horizontal groups, because the two enantiomers then look like mirror images on paper.

wedge-and-dash representations of the
enantiomers of 2-bromobutane

A shortcut for showing the three-dimensional arrangement of groups bonded to a chirality center was devised in the late 1800s by Emil Fischer. Fischer used a single dot to represent a bond that was slanted away from the viewer and a solid line for a bond that was directed toward the viewer. Several years later, Viktor Meyer modified Fischer's method. He used the point of intersection of two perpendicular lines to represent the chirality center; horizontal lines represent the bonds that project out of the plane of the paper toward the viewer, and vertical lines represent the bonds that project back from the plane of the paper away from the viewer. This method of representing the arrangement of groups bonded to a chirality center is known as a *Fischer projection formula* or, more simply, a **Fischer projection.**

original Fischer projections

Fischer projections used now

PROBLEM 7◆

Draw enantiomers for each of the following compounds using:

a. perspective formulas   **b.** Fischer projections

$$
\underset{\text{Br}}{\overset{}{|}}
$$

(1) $CH_3\overset{\overset{\displaystyle Br}{|}}{C}HCH_2OH$ (2) $ClCH_2CH_2\overset{\overset{\displaystyle CH_3}{|}}{C}HCH_2CH_3$ (3) $CH_3\overset{\overset{\displaystyle CH_3}{|}}{C}H\underset{\underset{\displaystyle OH}{|}}{C}HCH_3$

Now that we know that a compound such as 2-bromobutane has two stereoisomers, we need a way to name the individual stereoisomers so we can distinguish them. In other words, we need a system of nomenclature that will indicate the configuration (arrangement) of the groups about the chirality center. The *R, S* system of nomenclature indicates the **configuration** of a chirality center. For any pair of enantiomers with a chirality center, one will have the *R* **configuration** and the other will have the *S* **configuration**. The *R, S* system of nomenclature was devised by Cahn, Ingold, and Prelog.

Let's take a look at the enantiomers of 2-bromobutane and determine which has the *R* configuration and which has the *S* configuration.

the enantiomers of 2-bromobutane

1. Rank the groups that are bonded to the chirality center in order of priority. The priority depends on the atomic numbers of the atoms directly attached to the chirality center. The greater the atomic number, the higher the priority. In 2-bromobutane, bromine has the highest priority (1) and the hydrogen has the lowest priority (4). This should remind you of the way that relative priorities are determined in the *E, Z* system of nomenclature because the Cahn–Ingold–Prelog system of priorities was originally developed for the *R, S* system and was later borrowed for the *E, Z* system. You might want to review how relative priorities are determined (Section 3.5).

(S)-2-bromobutane   **(R)-2-bromobutane**

2. If the group with the lowest priority is bonded by a dashed line, draw a curved arrow from the group with the highest priority (1) to the group with the second-highest priority (2). If the arrow points in a clockwise direction, the chirality center has the *R* configuration (*R* is for *rectus,* which is Latin for "right"). If the arrow points in a counterclockwise direction, the chirality center has the *S* configuration (*S* is for *sinister,* which is Latin for "left").

4.5
NAMING ENANTIOMERS: THE *R, S* SYSTEM OF NOMENCLATURE

Robert Sidney Cahn (1899–1981), *a British scientist, received an M.A. from Cambridge University and a doctorate in natural philosophy in France. He edited the* Journal of the Chemical Society (London).

Sir Christopher Ingold (1893–1970) *was born in Ilford, England and was knighted by Queen Elizabeth II. He was a professor of chemistry at Leeds University (1924–1930) and at University College, London (1930–1970).*

Vladimir Prelog *was born in Sarajevo, Bosnia in 1906. He taught at the University of Zagreb from 1935 until 1941, when he fled to Switzerland just ahead of the invading German army. For his work that contributed to an understanding of how living organisms carry out chemical reactions, he shared the 1975 Nobel Prize in chemistry with John Cornforth (page 214).*

left turn

right turn

(S)-2-bromobutane   (R)-2-bromobutane

If you forget which is which, you can imagine driving a car and turning the steering wheel to make a right or a left turn.

3. If the group with the lowest priority is not bonded by a dashed line, first, switch a pair of groups so that the group with the lowest priority is bonded by a dashed line. Then draw a curved arrow from the group with the highest priority (1) to the group with the second-highest priority (2). Because you have switched a pair of groups, you are now determining the configuration of *the enantiomer of the original molecule.* So if the arrow points in a clockwise direction, the "enantiomer" (with the switched pair) has the *R* configuration, which means that the original molecule has the *S* configuration. If the arrow points in a counterclockwise direction, the "enantiomer" (with the switched pair) has the *S* configuration, which means that the original molecule has the *R* configuration.

what is its configuration?

switch
CH_3 and H

this compound has the *R* configuration, which means that the compound before the pair was switched had the *S* configuration

4. In drawing the arrow from group 1 to group 2, you can draw past the group with the lowest priority (4), but never draw past the group with the next-lowest priority (3).

(R)-1-bromo-3-pentanol

If you were to determine the configuration of a compound for which you had a three-dimensional model, you would orient the molecule so that the group with the lowest priority was directed away from you and then draw an imaginary arrow from the group with the highest priority to the group with the next-highest priority. If the arrow points in a clockwise direction, the compound has the *R* configuration; if it points in a counterclockwise direction, the compound has the *S* configuration. These rules allow you to determine quickly and easily the configurations of compounds for which you do not have a three-dimensional model but whose structures are written on a two-dimensional piece of paper.

It is also easy to determine the configurations of compounds drawn in Fischer projections.

1. Draw an arrow from the group with the highest priority (1) to the group with the next highest priority (2). If the arrow points in a clockwise direction, the enantiomer has the *R* configuration; if it points in a counterclockwise direction, the enantiomer has the *S* configuration *provided that the group with the lowest priority (4) is on a vertical bond.*

(*R*)-3-chlorohexane   (*S*)-3-chlorohexane

2. If the group with the lowest priority is on a *horizontal* bond, the answer you get by determining the direction of the arrow will be the opposite of the correct answer. For example, if the arrow points in a clockwise direction, suggesting that the chirality center has the *R* configuration, it actually has the *S* configuration; if the arrow points in a counterclockwise direction, suggesting that the chirality center has the *S* configuration, it actually has the *R* configuration. (Notice that, if you assume that a clockwise arrow specifies an *R* configuration and a counterclockwise arrow specifies an *S* configuration, the answer you get is **V**ery true if the lowest-priority substituent is on a **V**ertical bond and **H**orribly wrong if the lowest-priority substituent is on a **H**orizontal bond.) In the following example, because the group with the lowest priority is on a horizontal bond, clockwise signifies the *S* configuration, not the *R* configuration.

(*S*)-2-butanol   (*R*)-2-butanol

3. In drawing the arrow from group 1 to group 2, you can draw past the group with the lowest priority (4), but never draw past the group with the next-lowest priority (3).

(*S*)-lactic acid   (*R*)-lactic acid

   If you are working with three-dimensional molecular models, you can easily tell whether two molecules are enantiomers (nonidentical mirror images) or identical molecules. But if you are working with structures on a two-dimensional piece of paper, the easiest way to determine whether two molecules with one chirality center and the same sets of substitutents are enantiomers or identical molecules is by determining their configurations. If one has the *R* configuration and the other has the *S* configuration, they are enantiomers. If they both have the *R* configuration or both have the *S* configuration, they are identical.

   When comparing two Fischer projections to see if they are the same or different, never rotate one 90° or turn one over, because this is a quick way to get a wrong answer. A Fischer projection can be rotated 180° in the plane of the paper, but this is the only way to move it without risking an incorrect answer.

184 CHAPTER 4 Stereochemistry

PROBLEM 8 ◆

Indicate whether each of the following structures has the *R* configuration or the *S* configuration.

a. 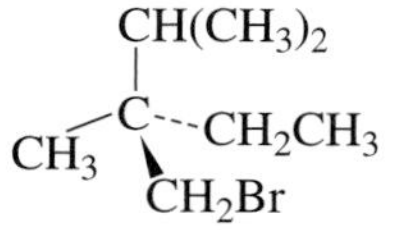
$$CH(CH_3)_2$$
$$CH_3 — C ⋯ CH_2CH_3$$
$$CH_2Br$$

b. 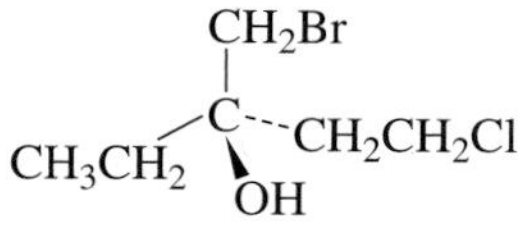
$$CH_2Br$$
$$CH_3CH_2 — C ⋯ CH_2CH_2Cl$$
$$OH$$

c. (structure with HO)

d. 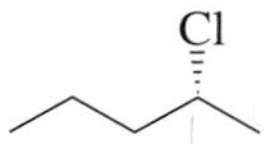 (structure with Cl)

PROBLEM 9 / SOLVED

Do the following structures represent identical molecules or a pair of enantiomers?

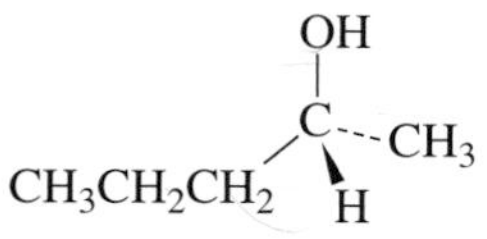

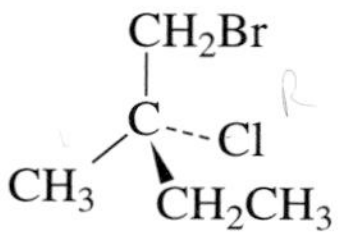

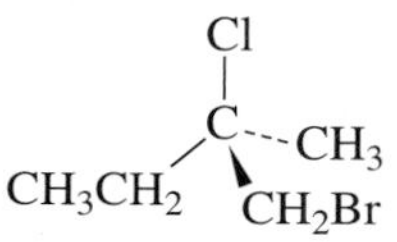

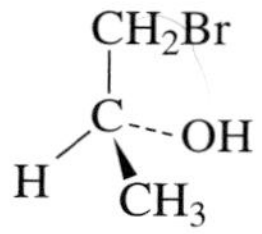

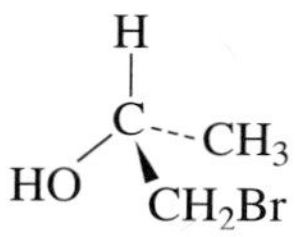

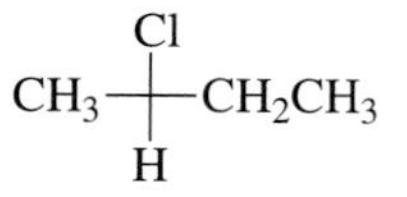

a. 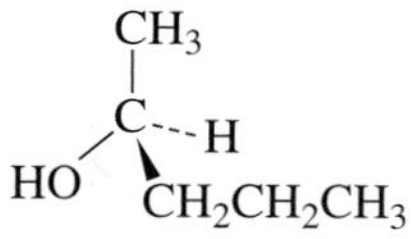
$$CH_3$$
$$HO — C ⋯ H$$
$$CH_2CH_2CH_3$$
and
$$OH$$
$$CH_3CH_2CH_2 — C ⋯ CH_3$$
$$H$$

b.
$$CH_2Br$$
$$CH_3 — C ⋯ Cl$$
$$CH_2CH_3$$
and
$$Cl$$
$$CH_3CH_2 — C ⋯ CH_3$$
$$CH_2Br$$

c.
$$CH_2Br$$
$$H — C ⋯ OH$$
$$CH_3$$
and
$$H$$
$$HO — C ⋯ CH_3$$
$$CH_2Br$$

d. CH_3—|—CH_2CH_3 with Cl top, H bottom   and   H—|—Cl with CH_3 top, CH_2CH_3 bottom

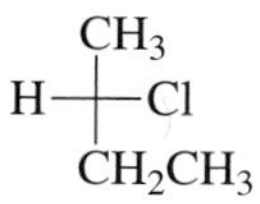

SOLUTION TO 9a The easiest way to determine whether two structures represent identical molecules or a pair of enantiomers is to determine their configurations. If they have the same configuration, they are identical; if they have opposite configurations, they are enantiomers. The first structure shown in part (a) has the *S* configuration and the second structure has the *R* configuration. Therefore, the structures represent a pair of enantiomers.

PROBLEM 10

Assign priority numbers to the following groups.

a. —CH_2OH   —CH_3   —CH_2CH_2OH   —H

b. —CH=O   —OH   —CH_3   —CH_2OH

c. —CH(CH_3)_2   —CH_2CH_2Br   —Cl   —CH_2CH_2CH_2Br

d. —CH=CH_2   —CH_2CH_3   (phenyl)   —CH_3

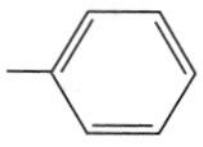

PROBLEM 11◆

Indicate whether each of the following structures has the *R* configuration or the *S* configuration.

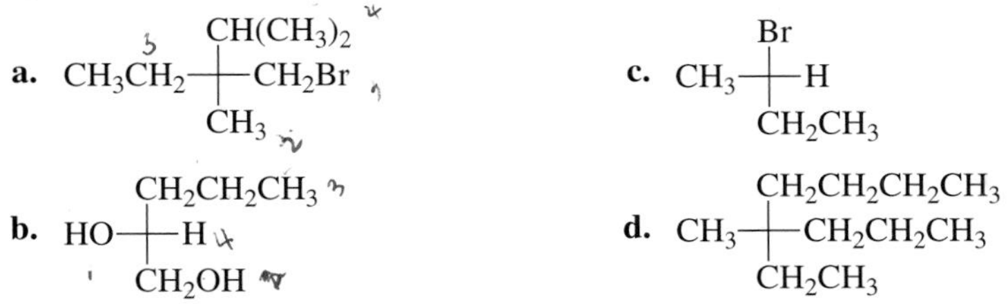

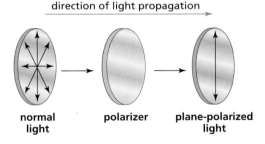

a. $CH_3CH_2 \quad \begin{array}{c} CH(CH_3)_2 \\ \hline \\ CH_3 \end{array} \quad CH_2Br$

b. $HO \quad \begin{array}{c} CH_2CH_2CH_3 \\ \hline \\ CH_2OH \end{array} \quad H$

c. $CH_3 \quad \begin{array}{c} Br \\ \hline \\ CH_2CH_3 \end{array} \quad H$

d. $CH_3 \quad \begin{array}{c} CH_2CH_2CH_2CH_3 \\ \hline \\ CH_2CH_3 \end{array} \quad CH_2CH_2CH_3$

Enantiomers share many of the same properties; they have the same boiling points, the same melting points, and the same solubilities. In fact, all of the physical properties of enantiomers are the same except those properties that depend on how groups bonded to the chirality center are arranged in space. One of the properties that enantiomers do not share is the way an enantiomer interacts with polarized light.

What is polarized light? Normal light consists of electromagnetic waves that oscillate in all planes perpendicular to the direction the light travels. **Plane-polarized light** oscillates only in a single plane. Plane-polarized light is produced by passing normal light through a polarizer such as a polarized lens or a Nicol prism.

4.6
OPTICAL ROTATION

direction of light propagation

normal
light

polarizer

plane-polarized
light

Born in Scotland, **William Nicol (1768–1851)** *was a professor at the University of Edinburgh. He developed the first prism that produced plane-polarized light. He also developed methods to produce thin slices of materials for use in microscopic studies.*

You have experienced the effect of a polarized lens if you own polarized sunglasses. Because polarized sunglasses allow only light oscillating in a single plane to pass through them, they block reflections more effectively than do nonpolarized sunglasses.

In 1815, the physicist Jean-Baptiste Biot discovered that certain natural organic substances such as camphor and oil of turpentine are able to rotate the plane of polarized light. He noted that some compounds rotated the plane of polarized light in a clockwise direction and some in a counterclockwise direction, while others had no effect on the light. He predicted that the ability to rotate the plane of polarized light was attributable to some asymmetry that existed in the molecule. Van't Hoff and Le Bel later determined that the molecular asymmetry was the result of having four different groups attached to a carbon (a chirality center).

When plane-polarized light passes through a solution of achiral molecules, the light emerges from the solution with its direction of polarization unchanged, because there is no asymmetry in the molecules. *An achiral compound does not rotate the plane of polarized light. It is optically inactive.*

Born in France, **Jean-Baptiste Biot (1774–1862)** *was imprisoned for taking part in a street riot during the French Revolution. He became a professor of mathematics at the University of Beauvais and later a professor of physics at the Collège de France. He was awarded the Legion of Honor by Louis XVIII.*

Jacobus Hendricus van't Hoff (1852–1911), *a Dutch chemist, was a professor of chemistry at the University of Amsterdam and later at the University of Berlin. He received the first Nobel Prize in chemistry (1901) for his work on solutions.*

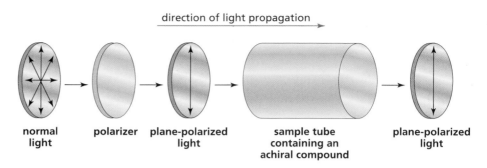

However, when plane-polarized light passes through a solution of a chiral compound, the light emerges with its direction of polarization changed, because the molecules are asymmetric. Thus, *a chiral compound rotates the plane of polarized light.* A chiral compound will rotate the plane of polarized light in either a clockwise or a counterclockwise direction. If one enantiomer rotates the plane of polarized light in a clockwise direction, its mirror image will rotate the plane of polarized light exactly the same amount in a counterclockwise direction.

Joseph Achille Le Bel (1847–1930), *a French chemist, inherited his family's fortune and established his own laboratory. He and van't Hoff independently arrived at the reason for the optical activity of certain molecules. Although van't Hoff's explanation was more precise, they are both given credit.*

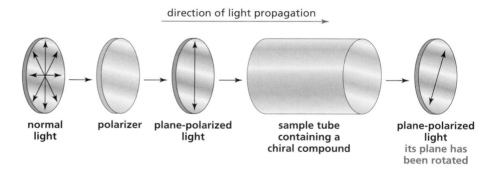

A compound that rotates the plane of polarized light is said to be **optically active.** In other words, chiral compounds are optically active and achiral compounds are **optically inactive.**

If an optically active compound rotates the plane of polarized light in a clockwise direction, it is called **dextrorotatory,** indicated by (+); if an optically active compound rotates the plane of polarized light in a counterclockwise direction, it is called **levorotatory,** indicated by (−). *Dextro* and *levo* are Latin prefixes for "to the right" and "to the left," respectively.

Do not confuse (+) and (−) with *R* and *S*. The (+) and (−) symbols indicate the direction in which an optically active compound rotates polarized light, whereas *R* and *S* indicate the arrangement of the groups about the chirality center. Some compounds with the *R* configuration are (+) and some are (−).

The amount that an optically active compound rotates the plane of polarized light can be measured with an instrument called a **polarimeter** (Figure 4.1). Because the amount of rotation depends on the wavelength of the light used, the light source for a polarimeter must produce light with a single wavelength (monochromatic light). Most polarimeters use light from a sodium arc (called the sodium D-line; wavelength = 589 nm). In a polarimeter, monochromatic light passes through a polarizer and emerges as plane-polarized light. The polarized light then

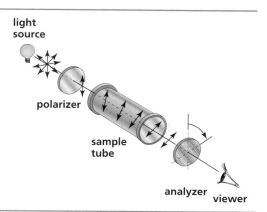

passes through an empty sample tube (or one filled with an optically inactive solvent) and emerges with its direction of polarization unchanged. The light finally passes through an analyzer. The analyzer is a second polarizer mounted on an eyepiece with a dial marked in degrees. The user rotates the analyzer until he or she sees total darkness. At this point the analyzer is at a right angle to the first polarizer, so no light passes through. This setting of the analyzer corresponds to zero rotation.

The sample to be measured is then placed in the sample tube. If the sample is optically active, it will rotate the plane of polarized light. The analyzer will no longer block all the light, and light reaches the user's eye. The user then rotates the analyzer again until no light passes through. The amount the analyzer is rotated can be read from the dial and represents the difference between an inactive sample and the active sample. This is called the **observed rotation** (α), and it is measured in degrees. The amount of rotation caused by the sample depends on the number of optically active molecules the light encounters in the sample. This, in turn, depends on the concentration of the sample and the length of the sample tube. The observed rotation also depends on the temperature and the wavelength of the light source.

Each optically active compound has a characteristic specific rotation. The **specific rotation** is the number of degrees of rotation caused by a solution of 1.0 g of the compound per mL of solution in a sample tube 1.0 dm long at a specified temperature and wavelength. The specific rotation can be calculated from the observed rotation using the following formula:

When polarized lenses are at a 90° angle to each other, there is no transmission of light through them.

$$[\alpha]_{\lambda}^{T} = \frac{\alpha}{l \times c}$$

where $[\alpha]$ is the specific rotation; T is temperature in °C; λ is the wavelength of the incident light (when the sodium D-line is used, λ is indicated as D); α is the observed rotation; l is the length of the sample tube in decimeters; and c is the concentration of the sample in grams per milliliter of solution.

For example, one enantiomer of 2-methyl-1-butanol has been found to have a specific rotation of +5.75°. Because its mirror image rotates the plane of polarized light the same amount but in the opposite direction, the specific rotation of the other enantiomer must be −5.75°.

$$CH_2OH$$

CH$_3$ ---C---H CH$_2$CH$_3$

(R)-2-methyl-1-butanol

$$[\alpha]_D^{20\,°C} = +5.75\,°$$

$$CH_2OH$$

H---C---CH$_3$ CH$_2$CH$_3$

(S)-2-methyl-1-butanol

$$[\alpha]_D^{20\,°C} = -5.75\,°$$

PROBLEM 12 ◆

The observed rotation of 2.0 g of a compound in 10 mL of solution in a polarimeter tube 10 cm long is +134°. What is the specific rotation of the compound?

Knowing whether a chiral molecule has the R configuration or the S configuration does not tell us the direction in which the molecule rotates polarized light. Some compounds with the R configuration rotate light to the right ($+$), and some rotate light to the left ($-$). We can tell by looking at the structure of a compound whether it has the R configuration or the S configuration. But the only way we can tell whether a compound is dextrorotatory ($+$) or levorotatory ($-$) is to put the compound in a polarimeter. For example, (S)-lactic acid and (S)-sodium lactate have very similar structures, but one is dextrorotatory and the other is levorotatory. When we know the direction in which an optically active compound rotates the plane of polarized light, we can incorporate ($+$) or ($-$) into its name.

$$CH_2CH_3$$

HO ---C---H COOH

(S)-(+)-lactic acid

$$CH_2CH_3$$

HO ---C---H COO$^-$Na$^+$

(S)-(–)-sodium lactate

A mixture of equal amounts of a pair of enantiomers is called a **racemic mixture,** a **racemic modification,** or a **racemate.** Racemic mixtures do not rotate polarized light; they are optically inactive. For every molecule in a racemic mixture that rotates the plane of polarized light in one direction, there is a mirror-image molecule that rotates the plane in the opposite direction. As a result, the plane of polarized light emerges from a racemic mixture with its direction unchanged. The symbol ($\pm$) is used to specify a racemic mixture. Thus, ($\pm$)-2-bromobutane indicates a mixture of ($+$)-2-bromobutane and an equal amount of ($-$)-2-bromobutane.

PROBLEM 13 ◆

(S)-($+$)-Monosodium glutamate (MSG) is a flavor enhancer used in many foods. Some people have an allergic reaction to MSG (headache, chest pains, and an overall feeling of weakness). "Fast food" often contains substantial amounts of MSG, and it is widely used in Chinese food as well. MSG has a specific rotation of +24°.

$$COO^-\ Na^+$$

HOOCCH$_2$CH$_2$ ---C---H +NH$_3$

(S)-(+)-monosodium glutamate

a. What is the specific rotation of (R)-(−)-monosodium glutamate?

b. What is the specific rotation of a racemic mixture of MSG?

Whether a particular compound consists of a single enantiomer or a mixture of enantiomers can be determined by its observed specific rotation. For example, an **enantiomerically pure** sample of (S)-(+)-2-bromobutane will have an observed specific rotation of +23.1° because the specific rotation of (S)-(+)-2-bromobutane is +23.1°. If, however, the sample of 2-bromobutane has an observed specific rotation of 0°, we will know that the compound is a racemic mixture. If the observed specific rotation is positive but less than +23.1°, we will know that we have a mixture of enantiomers and the mixture contains more of the enantiomer with the S configuration than the enantiomer with the R configuration.

From the observed specific rotation, we can calculate the amount of each enantiomer present in the mixture. First we must calculate the optical purity using the following formula.

$$\text{optical purity} = \frac{\text{observed specific rotation}}{\text{specific rotation of the pure enantiomer}}$$

For example, if a sample of 2-bromobutane has an observed specific rotation of +9.2°, its optical purity is 0.40. In other words, it is 40% optically pure.

$$\text{optical purity} = \frac{+9.2°}{+23.1°} = 0.40 \text{ or } 40\%$$

Because the observed specific rotation is positive, we know that the solution contains excess (S)-(+)-2-bromobutane. The **optical purity** tells us how much excess (S)-(+)-2-bromobutane is in the mixture. The mixture is 40% optically pure, which means that 40% of the mixture is excess S enantiomer and 60% is a racemic mixture. Half of the racemic mixture plus the amount of excess S enantiomer equals the amount of the S enantiomer present in the mixture. Thus, 70% of the mixture is the S enantiomer (1/2 × 60 + 40) and 30% is the R enantiomer.

PROBLEM 14◆

(+)-Mandelic acid has a specific rotation of +158°. What would be the observed specific rotation of each of the following mixtures?

a. 25% (−)-mandelic acid and 75% (+)-mandelic acid

b. 50% (−)-mandelic acid and 50% (+)-mandelic acid

c. 75% (−)-mandelic acid and 25% (+)-mandelic acid

PROBLEM 15◆

The specific rotation of (R)-(+)-glyceraldehyde is +8.7°. If the observed specific rotation of a mixture of (R)-glyceraldehyde and (S)-glyceraldehyde is +1.4°, what percentage of the glyceraldehyde is present as the R enantiomer?

PROBLEM 16 / SOLVED

A solution prepared by mixing 10 mL of a 0.10 M solution of the R enantiomer and 30 mL of a 0.10 M solution of the S enantiomer was found to have an observed specific rotation of $+4.8°$. What is the specific rotation of each of the enantiomers?

SOLUTION.   One millimole (mmol; 10 mL $\times$ 0.10 M) of the R enantiomer is mixed with 3 mmol of the S enantiomer; 1 mmol of the R enantiomer plus 1 mmol of the S enantiomer will form 2 mmol of a racemic mixture. There will be 2 mmol of S enantiomer left over. Therefore, 2 mmol out of 4 mmol is excess S enantiomer (2/4 = 0.50). The solution is 50% optically pure.

$$\text{optical purity} = 0.50 = \frac{\text{observed specific rotation}}{\text{specific rotation of the pure enantiomer}}$$

$$0.50 = \frac{+4.8°}{x}$$

$$x = +9.6°$$

The S enantiomer has a specific rotation of $+9.6°$; the R enantiomer has a specific rotation of $-9.6°$.

4.8
ISOMERS WITH MORE THAN ONE CHIRALITY CENTER

Many organic compounds have more than one chirality center. The more chirality centers a compound has, the more stereoisomers it has. If we know how many chirality centers a compound has, we can calculate the maximum number of stereoisomers for that compound. *A compound can have a maximum of 2^n stereoisomers, where n equals the number of chirality centers.* For example, 3-chloro-2-butanol has two chirality centers. Therefore, it can have as many as four ($2^2 = 4$) stereoisomers. In each Fischer projection shown below, all four horizontal bonds point out of the paper toward the viewer and the two vertical bonds point behind the paper away from the viewer. Groups can rotate freely about the carbon–carbon single bonds, but the Fischer projections show the stereoisomers in their eclipsed conformations.

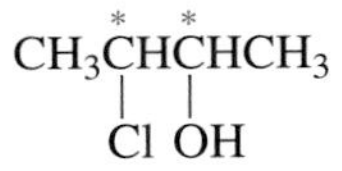

$$CH_3\overset{*}{C}H\overset{*}{C}HCH_3$$
$$\underset{Cl\ \ OH}{|\ \ \ |}$$

3-chloro-2-butanol

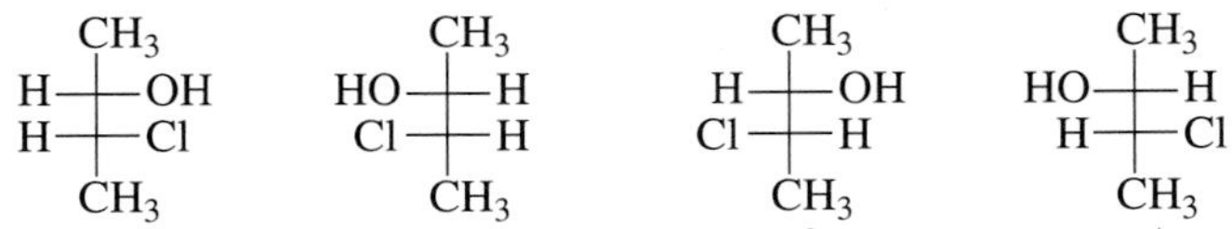

erythro enantiomers   threo enantiomers

Stereoisomers **1** and **2** are a pair of nonidentical mirror images. Therefore, stereoisomers **1** and **2** are enantiomers. Stereoisomers **3** and **4** are also enantiomers. The four stereoisomers of 3-chloro-2-butanol consist of two pairs of enantiomers. When Fischer projections are drawn (with the carbon chain aligned vertically) for stereoisomers with two adjacent chirality centers, the pair of enantiomers with similar groups on the same side of the carbon chain is called the **erythro enantiomers.** The pair of enantiomers with similar groups on opposite sides is called the **threo enan-**

tiomers. Stereoisomers **1** and **2** are the erythro enantiomers of 3-chloro-2-butanol (the hydrogens are on the same side). Stereoisomers **3** and **4** are the threo enantiomers.

Stereoisomers **1** and **3** are not identical and they are not mirror images. Such stereoisomers are called diastereomers. **Diastereomers** are configurational isomers that are not enantiomers. Stereoisomers **1** and **4**, **2** and **3**, and **2** and **4** are also diastereomers. (Notice that cis–trans isomers are also considered to be diastereomers because they are configurational isomers that are not enantiomers.)

Although enantiomers have identical physical properties (except for the direction in which they rotate the plane of polarized light), the physical properties of diastereomers are different. Diastereomers have different melting points, different solubilities, and different specific rotations. Enantiomers also have the same chemical properties, which means they react at the same rate with a given achiral reagent. Diastereomers have different chemical properties; they react with the same reagents at different rates.

Fischer projections are useful because they allow us readily to represent stereoisomers in two dimensions. Unfortunately, they represent the molecule in its relatively unstable, eclipsed conformation. Perspective formulas are harder for some people to visualize, but they can represent the molecule in its more stable, staggered conformation, so they provide a more accurate representation of its structure. We will use both kinds of structures to depict the three-dimensional arrangement of groups bonded to a chirality center.

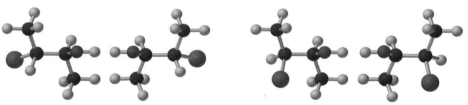

erythro enantiomers   threo enantiomers
perspective formulas of the stereoisomers of 3-chloro-2-butanol

stereoisomers of 3-chloro-2-butanol

PROBLEM 17◆

a. Configurational isomers with two chirality centers are called _____ if the configuration of both of the chirality centers in one isomer is the opposite of the configuration of the chirality centers in the other isomer.

b. Configurational isomers with two chirality centers are called _____ if the configuration of both of the chirality centers in one isomer is the same as the configuration of the chirality centers in the other isomer.

c. Configurational isomers with two chirality centers are called _____ if one of the chirality centers has the same configuration in both isomers and the other chirality center has the opposite configuration in the two isomers.

PROBLEM 18 / SOLVED

Tetracycline is called a broad-spectrum antibiotic because it is active against a wide variety of bacteria. How many chirality centers does tetracycline have?

$$H_3C \overset{1}{\underset{N}{\,}} \overset{2}{CH_3}$$

tetracycline

First, locate all the sp^3 hybridized carbons in tetracycline. (They are numbered in red.) Because a chirality center has four different groups attached to it, sp^3 hybridized carbons can be chirality centers, but sp^2 and sp hybridized carbons cannot. Tetracycline has nine sp^3 hybridized carbons. Four of them (#1, #2, #5, and #8) are not chirality centers because they are not bonded to four different groups. Tetracycline, therefore, has five chirality centers.

PROBLEM 19

The following compound has only one chirality center. Why then does it have four stereoisomers?

$$CH_3CH_2\overset{*}{C}HCH_2CH=CHCH_3$$
$$\underset{Br}{|}$$

2,4-Dichlorohexane is another example of a compound with two chirality centers and four stereoisomers. Stereoisomers **1** and **2** are enantiomers; **3** and **4** are also enantiomers. Stereoisomers **1** and **3**, **1** and **4**, **2** and **3**, and **2** and **4** are diastereomers.

$$CH_3\overset{*}{C}HCH_2\overset{*}{C}HCH_2CH_3$$
$$\underset{Cl}{|}\quad\underset{Cl}{|}$$
2,4-dichlorohexane

CH₃	CH₃	CH₃	CH₃
H——Cl	Cl——H	H——Cl	Cl——H
CH₂	CH₂	CH₂	CH₂
H——Cl	Cl——H	Cl——H	H——Cl
CH₂CH₃	CH₂CH₃	CH₂CH₃	CH₂CH₃
1	2	3	4

1-Bromo-2-methylcyclopentane also has two chirality centers and four stereoisomers. Because the compound is cyclic, the substituents can be in either the cis or the trans configuration. The cis isomer exists as a pair of enantiomers, and the trans isomer exists as a pair of enantiomers.

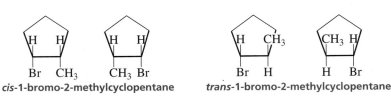

cis-1-bromo-2-methylcyclopentane   *trans*-1-bromo-2-methylcyclopentane

1-Bromo-3-methylcyclobutane does not have any chirality centers. C-1 has a bromine and a hydrogen attached to it, but its other two groups are identical; C-3 has a methyl group and a hydrogen attached to it, but its other two groups are identical. Because the compound does not have a carbon with four different groups attached to it, it has only two stereoisomers, the cis isomer and the trans isomer. The cis and trans isomers do not have enantiomers. Notice that each of them has a plane of symmetry. (In Section 4.3 we saw that, if a compound has a plane of symmetry, it cannot have an enantiomer.)

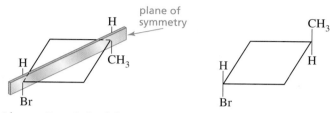

cis-**1-bromo-3-methylcyclobutane**   *trans*-**1-bromo-3-methylcyclobutane**

1-Bromo-3-methylcyclohexane has two chirality centers. The carbon that is bonded to a hydrogen and a bromine is also bonded to two different carbon-containing groups, so it is a chirality center. The carbon that is bonded to a hydrogen and a methyl group is also bonded to two different carbon-containing groups, so it is also a chirality center.

Because the compound has two chirality centers, it has four stereoisomers. The cis isomer exists as a pair of enantiomers, and the trans isomer exists as a pair of enantiomers.

cis-**1-bromo-3-methylcyclohexane**   *trans*-**1-bromo-3-methylcyclohexane**

1-Bromo-4-methylcyclohexane has no chirality centers. Therefore, the compound has only one cis isomer and one trans isomer.

cis-**1-bromo-4-methylcyclohexane**   *trans*-**1-bromo-4-methylcyclohexane**

PROBLEM 20◆

Draw all possible stereoisomers for each of the following compounds.

a. 2-chloro-3-hexanol   **c.** 2,3-dichloropentane

b. 2-bromo-4-chlorohexane   **d.** 1,3-dibromopentane

PROBLEM 21◆

Draw the stereoisomers of 1-bromo-3-chlorocyclohexane.

PROBLEM 22

Of all the possible cyclooctanes that have one chloro substituent and one methyl substituent, which ones do not have any chirality centers?

PROBLEM 23

Draw a diastereomer for each of the following.

a. 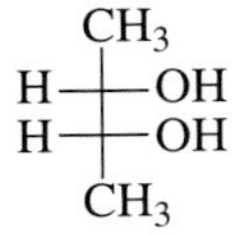
$$
\begin{array}{c}
CH_3 \\
H \!-\!\!|\!-\! OH \\
H \!-\!\!|\!-\! OH \\
CH_3
\end{array}
$$

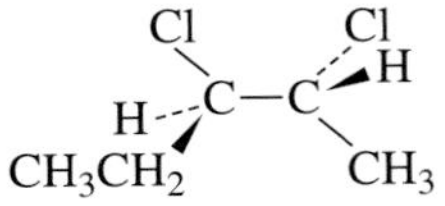

c.
$$
\begin{array}{c}
CH_3 \quad CH_3 \\
\diagdown \; / \\
C\!=\!C \\
/ \quad \diagdown \\
H \qquad H
\end{array}
$$

b.
$$
\begin{array}{c}
Cl \qquad Cl \\
| \qquad | \\
H\text{-}\text{-}C\!-\!C\text{-}\!H \\
| \qquad | \\
CH_3CH_2 \quad CH_3
\end{array}
$$

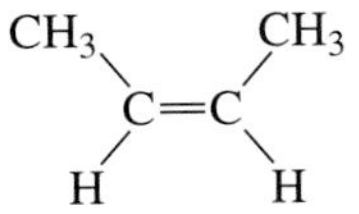

d. 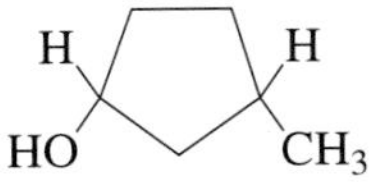
$$
\begin{array}{c}
H \qquad H \\
\\
HO \qquad CH_3
\end{array}
$$

4.9
MESO COMPOUNDS

In the examples that we have just seen, each compound with two chirality centers has four stereoisomers. However, some compounds with two chirality centers have only three stereoisomers. This is why we emphasized in Section 4.8 that the *maximum* number of stereoisomers a compound with n chirality centers can have is 2^n, rather than stating that a compound with n chirality centers has 2^n stereoisomers.

An example of a compound with two chirality centers that has only three stereoisomers is 2,3-dibromobutane.

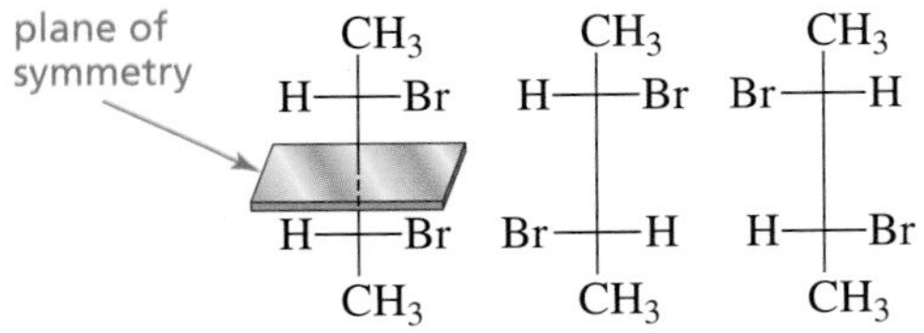

The "missing" stereoisomer is the mirror image of stereoisomer **1**. Stereoisomer **1** has a plane of symmetry, which means that it does not have a nonidentical mirror image. If we draw the mirror image of stereoisomer **1**, we find that it and stereoisomer **1** are identical. To convince yourself that the two structures are identical, rotate the mirror image of stereoisomer **1** that you just drew 180° and you will see that it is identical to stereoisomer **1**. (*Remember, you can move Fischer projections only by rotating them 180° in the plane of the paper.*)

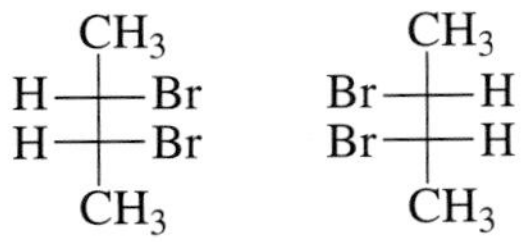

superimposable mirror images

Stereoisomer **1** is called a meso compound. Even though a meso compound has chirality centers, it is an achiral molecule because it has a plane of symmetry. *Mesos* is the Greek word for "middle." Because of the plane of symmetry, half of the molecule is the mirror image of the other half. Consequently, one chirality center will rotate the plane of polarized light in one direction and the other chirality center will cancel that rotation by rotating the plane of polarized light by the same amount but in the opposite direction. As a result, meso compounds do not rotate the plane of polarized light. They are optically inactive. **Meso compounds** can be recognized by the fact that they have two or more chirality centers and a plane of symmetry. *If a compound has a plane of symmetry, it will not be optically active even though it possesses chirality centers.*

> **A compound with one or more chirality centers will be optically active except if it is a racemic mixture or a meso compound.**

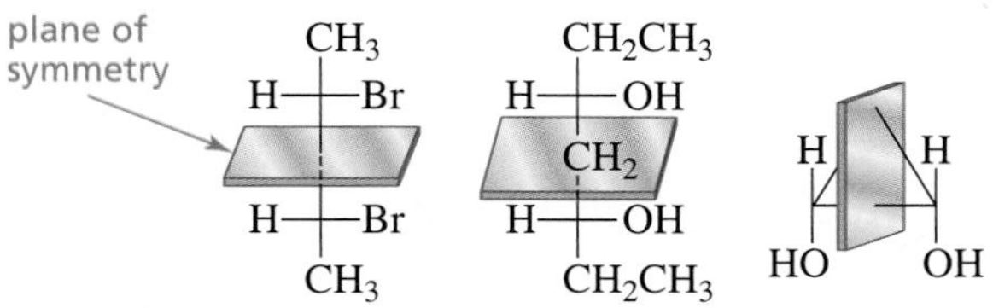

meso compounds

Whenever a compound has two chirality centers and the four groups bonded to one chirality center are identical to the four groups bonded to the other chirality center, the compound will have three stereoisomers. One stereoisomer will be a meso compound, and the other two will be a pair of enantiomers. In the case of cyclic compounds, the cis isomer will be the meso compound.

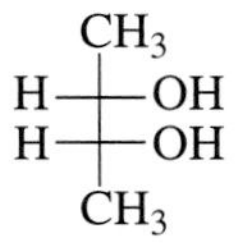

a meso compound 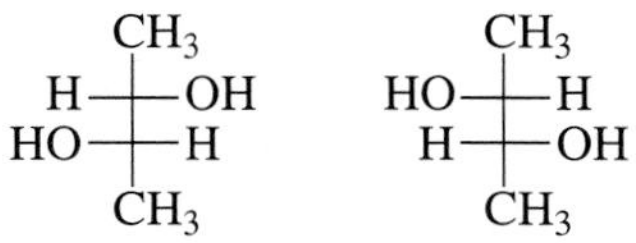   a pair of enantiomers

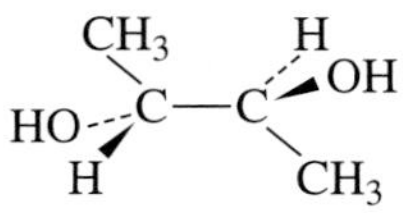

a meso compound 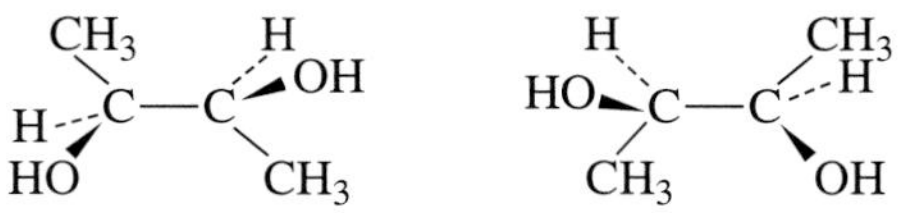   a pair of enantiomers

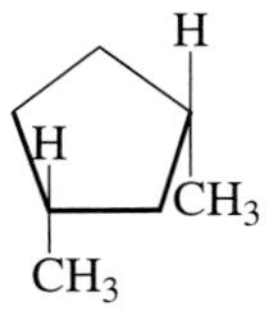

a meso compound 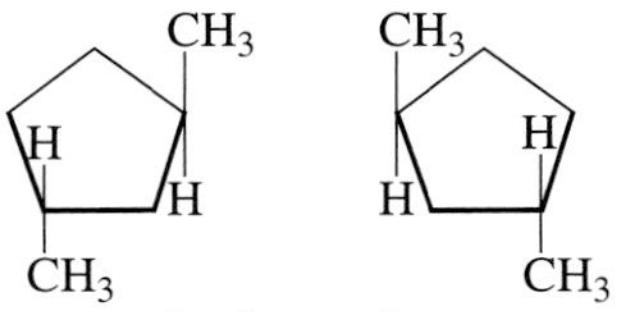   a pair of enantiomers

---

PROBLEM 24 ◆

Which of the following compounds has a stereoisomer that is a meso compound?

a. 2,4-dibromohexane   d. 1,3-dichlorocyclohexane

b. 2,4-dibromopentane   e. 1,4-dichlorocyclohexane

c. 2,4-dimethylpentane   f. 1,2-dichlorocyclobutane

PROBLEM 25

Draw all the stereoisomers for each of the following compounds.

a. 1-bromo-2-methylbutane	**h.** 2,4-dichloropentane
b. 1-chloro-3-methylpentane	**i.** 2,4-dichloroheptane
c. 2-methyl-1-propanol	**j.** 1,2-dichlorocyclobutane
d. 2-bromo-1-butanol	**k.** 1,3-dichlorocyclohexane
e. 3-chloro-3-methylpentane	**l.** 1,4-dichlorocyclohexane
f. 3-bromo-2-butanol	**m.** 1-bromo-2-chlorocyclobutane
g. 3,4-dichlorohexane	**n.** 1-bromo-3-chlorocyclobutane

4.10
THE *R, S* SYSTEM OF NOMENCLATURE FOR ISOMERS WITH MORE THAN ONE CHIRALITY CENTER

If a compound has more than one chirality center, the steps used to determine whether a chirality center has the *R* configuration or the *S* configuration must be applied to each of the chirality centers individually. Let's name one of the stereoisomers of 3-bromo-2-butanol.

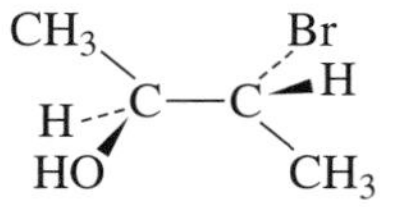

a stereoisomer of 3-bromo-2-butanol

First we will determine the configuration at C-2. The OH group has the number 1 priority, the C-3 carbon (C is attached to Br, C, and H) is number 2, CH_3 is number 3, and H is number 4. Since the group with the lowest priority is bonded by a dashed line, we can immediately draw an arrow from the group with the highest priority to the group with the next-highest priority. Because that arrow points in a counterclockwise direction, the configuration at C-2 is *S*.

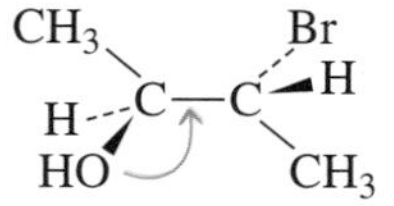

Now we can determine the configuration at C-3. Since the group with the lowest priority is not bonded by a dashed line, we must put it there by temporarily switching a pair of groups.

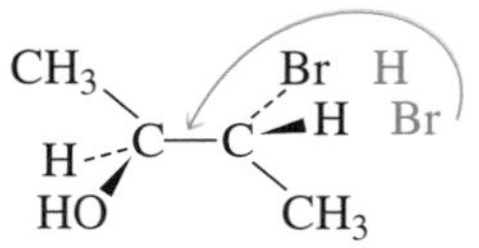

The arrow going from the highest priority to the next-highest priority points in a counterclockwise direction, so our answer is "*S*"; however, because we switched a pair of groups before we drew the arrow, C-3 has the *R* configuration.

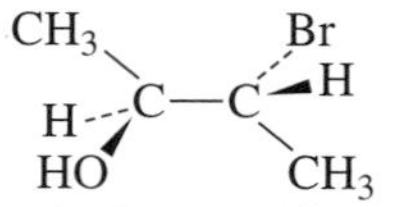

(2*S*,3*R*)-3-bromo-2-butanol

Fischer projections with two chirality centers can be named in a similar manner. Just apply the steps to each chirality center that you learned for a Fischer projection with one chirality center. For C-2, the arrow from the group with the highest priority to the group with the next-highest priority points in a clockwise direction, so our answer is "*R*." But because the group with the lowest priority is on a **h**orizontal bond, that answer is "**H**orribly wrong." You can conclude, then, that C-2 has the *S* configuration.

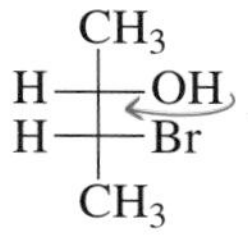

$$
\begin{array}{c}
CH_3 \\
H \!-\!\!-\! OH \\
H \!-\!\!-\! Br \\
CH_3
\end{array}
$$

By repeating these steps for C-3, you will find that it has the *R* configuration. So the isomer is named (2*S*,3*R*)-3-bromo-2-butanol.

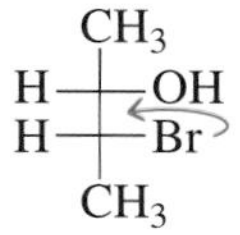

$$
\begin{array}{c}
CH_3 \\
H \!-\!\!-\! OH \\
H \!-\!\!-\! Br \\
CH_3
\end{array}
$$

(2*S*,3*R*)-3-bromo-2-butanol

The four stereoisomers of 3-bromo-2-butanol are named as shown here. Take a few minutes to verify their names.

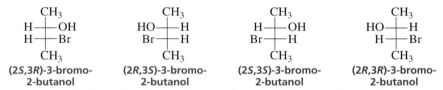

| (2*S*,3*R*)-3-bromo-2-butanol | (2*R*,3*S*)-3-bromo-2-butanol | (2*S*,3*S*)-3-bromo-2-butanol | (2*R*,3*R*)-3-bromo-2-butanol |

Fischer projections of the stereoisomers of 3-bromo-2-butanol

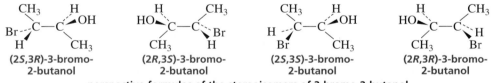

| (2*S*,3*R*)-3-bromo-2-butanol | (2*R*,3*S*)-3-bromo-2-butanol | (2*S*,3*S*)-3-bromo-2-butanol | (2*R*,3*R*)-3-bromo-2-butanol |

perspective formulas of the stereoisomers of 3-bromo-2-butanol

Notice that pairs of enantiomers have the opposite configuration at both chirality centers, while pairs of diastereomers have the opposite configuration at one chirality center and the same configuration at the other chirality center.

PROBLEM 26

Draw and name the four stereoisomers of 1,3-dichloro-2-butanol using:

a. perspective formulas.

b. Fischer projections.

Tartaric acid has three stereoisomers because its two chirality centers have the same set of four substituents. The meso compound and the pair of enantiomers are named as shown.

TABLE 4.1	Physical Properties of the Stereoisomers of Tartaric Acid		
	Melting Point, °C	$[\alpha]^{25°C}_D$	Solubility, g/100 g H_2O at 15 °C
(2R,3R)-(+)-tartaric acid	170	+11.98°	139
(2S,3S)-(−)-tartaric acid	170	−11.98°	139
(2R,3S)-tartaric acid	140	0°	125
(±)-tartaric acid	206	0°	20.6

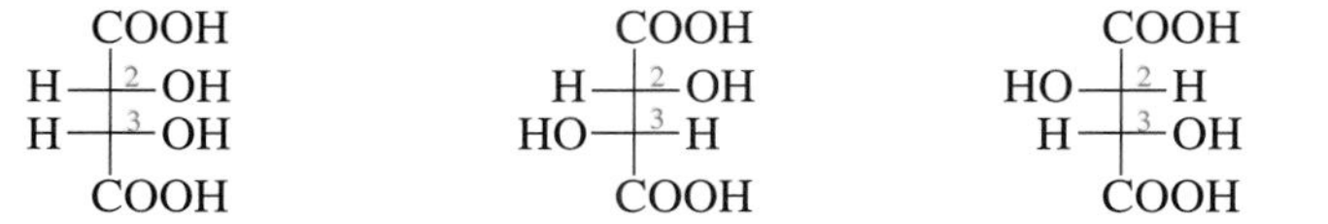

Fischer projections of the stereoisomers of tartaric acid

(2R,3S)-tartaric acid
a meso compound

(2R,3R)-tartaric acid
a pair of enantiomers

(2S,3S)-tartaric acid

Fischer projections of the stereoisomers of tartaric acid

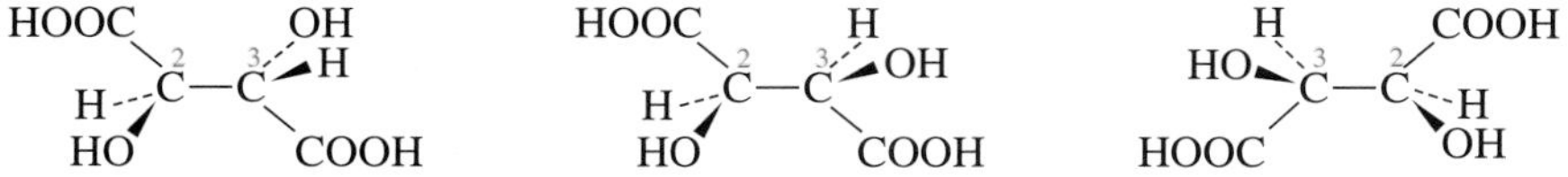

(2R,3S)-tartaric acid

(2R,3R)-tartaric acid

(2S,3S)-tartaric acid

perspective formulas of the stereoisomers of tartaric acid

The physical properties of the three stereoisomers of tartaric acid are listed in Table 4.1. The meso compound and either one of the enantiomers constitute a pair of diastereomers. Notice that the physical properties of the enantiomers are identical, while the physical properties of the diastereomers are different. Notice also that the physical properties of the racemic mixture differ from the physical properties of either of the enantiomers.

PROBLEM-SOLVING STRATEGY

Draw perspective formulas for the following compounds.

a. (R)-2-butanol   **b.** (2S,3R)-3-chloro-2-pentanol

a. First draw the compound so that you know what groups are bonded to the chirality center.

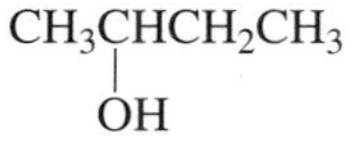

$$CH_3CHCH_2CH_3$$
$$|$$
$$OH$$
2-butanol

Draw the bonds about the chirality center.

Put the group with the lowest priority on the dotted bond. Put the group with the highest priority on any remaining bond.

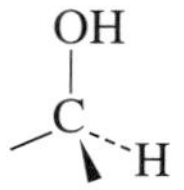

Since you have been asked to draw the *R* enantiomer, draw an arrow clockwise from the group with the highest priority to the next available bond and put the group with the next-highest priority on that bond.

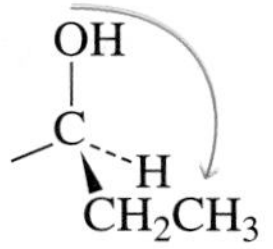

Put the remaining substituent on the last available bond.

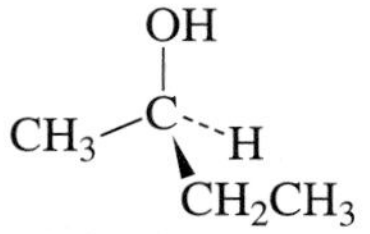

(*R*)-2-butanol

b. First draw the compound.

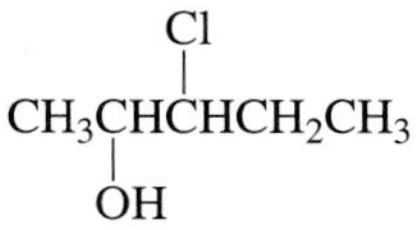

3-chloro-2-pentanol

Draw the bonds about the chirality centers.

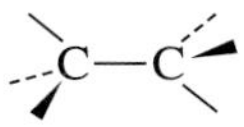

For each chirality center, put the group with the lowest priority on the dotted bond.

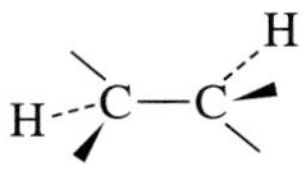

For each chirality center, put the group with the highest priority on a bond such that an arrow points clockwise (if you want the *R* configuration) or counterclockwise (if you want the *S* configuration) to the group with the next-highest priority.

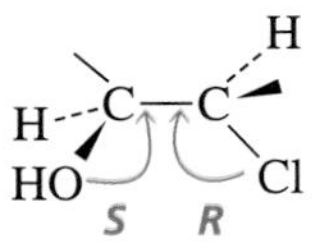

Put the remaining substituents on the last available bonds.

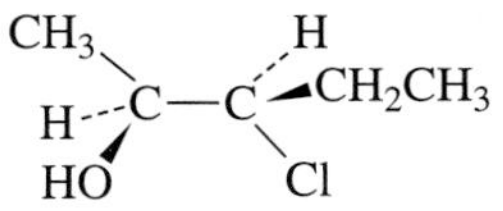

(2*S*,3*R*)-3-chloro-2-pentanol

Now continue on to solve Problem 27.

Louis Pasteur

*The great French chemist and microbiologist **Louis Pasteur** (1822–1895) was the first to prove that microbes cause specific diseases. He developed an antitoxin against rabies. Asked by the French wine industry to find out why wine often went sour while aging, he showed that microorganisms cause grape juice to ferment, producing wine, and cause wine to slowly become sour. Gently heating the wine after fermentation, a process called pasteurization, kills the organisms so they cannot sour the wine.*

PROBLEM 27

Draw perspective formulas for the following compounds.

a. (*S*)-3-chloro-1-pentanol

b. (2*R*,3*R*)-2,3-dibromopentane

c. (2*S*,3*R*)-3-methyl-2-pentanol

d. (*R*)-2-butanol

PROBLEM 28 ◆

For many centuries, the Chinese have used extracts from a family of herbs, known as Ephedra, to treat asthma. Chemists have been able to isolate a compound from these herbs, which they named ephedrine, that is a potent dilator of air passages in the lungs.

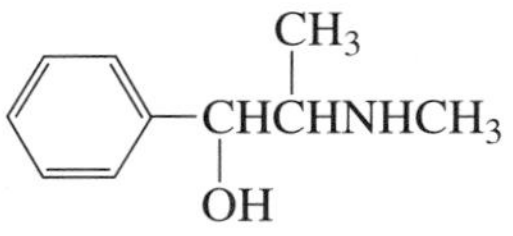

$$\text{C}_6\text{H}_5\text{-CHCHNHCH}_3$$
$$\overset{|}{\text{OH}} \quad \overset{|}{\text{CH}_3}$$

ephedrine

a. How many stereoisomers are possible for ephedrine?

b. The stereoisomer shown below is the one that is pharmacologically active. What is the configuration of each of the chirality centers?

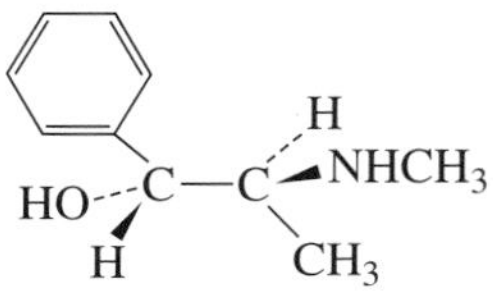

PROBLEM 29

Name the following compounds.

a.

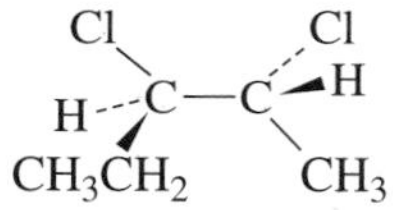

b.

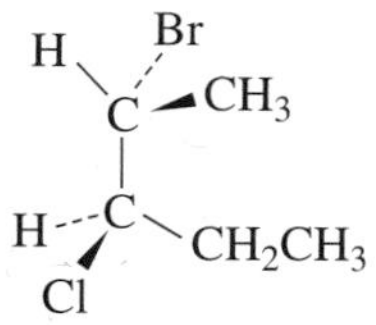

c.

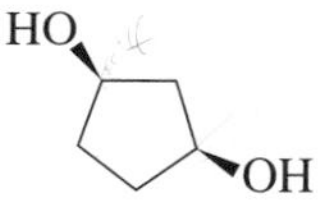

d.

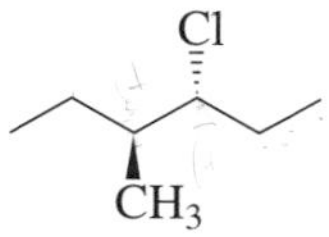

4.11 SEPARATION OF ENANTIOMERS

Enantiomers cannot be separated by the usual separation techniques such as fractional distillation or crystallization. Their identical boiling points and solubilities cause them to distill or crystallize together. Louis Pasteur was the first to separate a pair of enantiomers successfully. While working with crystals of the sodium ammonium double salt of tartaric acid, he noted that the crystals were not identical. Some of the crystals were "right-handed" and some were "left-handed," so he painstakingly separated the two kinds of crystals with a pair of tweezers. He found that a solution of the "right-handed" crystals rotated the plane of polarized light in a clockwise direction, while a solution of the "left-handed" crystals rotated the plane in a counterclockwise direction.

Pasteur was only 26 years old at the time and was unknown in scientific circles. He was concerned about the accuracy of his observations because, a few years earlier, the well-known German organic chemist Eilhardt Mitscherlich had reported that crystals of the same salt were all identical. Pasteur immediately reported his findings to Jean-Baptiste Biot and repeated the experiment with Biot present. Biot was convinced that Pasteur had successfully separated sodium ammonium tartrate into a pair of enantiomers. Pasteur's experiment also created a new chemical term. Tartaric acid is obtained from grapes and thus tartaric acid was also called racemic acid because the term *racemus* is Latin for "a bunch of grapes." When Pasteur found that tartaric acid was actually a mixture of enantiomers, it was termed a "racemic mixture." Separation of enantiomers is called **resolution of a racemic mixture.**

Later, chemists recognized how lucky Pasteur had been. The sodium ammonium salt of tartaric acid forms asymmetric crystals only under certain conditions—precisely the conditions that Pasteur had employed. Under other conditions, the symmetrical crystals that had fooled Mitscherlich are formed. But to quote Pasteur, "Chance favors the prepared mind."

Separating enantiomers by hand, as Pasteur did, is not a universally useful method of resolving a racemic mixture because few compounds form asymmetric crystals. A more commonly used method is to convert the pair of enantiomers into a pair of diastereomers. Diastereomers can be separated because they have different physical properties. After separation, the individual diastereomers are converted back into the original enantiomers.

For example, because an acid reacts with a base to form a salt, a racemic mixture of a carboxylic acid reacts with a naturally occurring chiral base to form two diastereomeric salts. Morphine, strychnine, and brucine are naturally occurring chiral bases commonly used for this purpose. The chiral base exists as a single enantiomer because, when a chiral compound is synthesized in a living system, generally only one enantiomer is formed (Section 4.20). When an *R* acid reacts with an *S* base, an *R,S* salt will form; when an *S* acid reacts with an *S* base, an *S,S* salt will form.

Eilhardt Mitscherlich (1794–1863), *a German chemist, first studied medicine so he could travel to Asia—a way to satisfy his interest in Oriental languages. But he became fascinated by chemistry. He was a professor of chemistry at the University of Berlin and wrote a successful chemistry textbook that was published in 1829.*

$$
\begin{array}{c}
COO^- \ Na^+ \\
H \!-\!\!-\! OH \\
HO \!-\!\!-\! H \\
COO^- \ NH_4^+
\end{array}
$$

sodium ammonium tartrate
right-handed crystals

$$
\begin{array}{c}
COO^- \ Na^+ \\
HO \!-\!\!-\! H \\
H \!-\!\!-\! OH \\
COO^- \ NH_4^+
\end{array}
$$

sodium ammonium tartrate
left-handed crystals

Crystals of potassium hydrogen tartrate that have precipitated on the inner surface of the cork from a bottle of red wine.

COOH   COOH $\xrightarrow{\textbf{\textit{S} base}}$ COO⁻ *S* baseH⁺   COO⁻ *S* baseH⁺

HO—C···H   H—C—OH   HO—C···H   H—C—OH

CH₃   CH₃   CH₃   CH₃

R acid   *S* acid   *R,S* salt   *S,S* salt

a pair of enantiomers   **a pair of diastereomers**

separate

COO⁻ *S* baseH⁺   COO⁻ *S* baseH⁺

HO—C···H   H—C—OH

CH₃   CH₃

R,S salt   *S,S* salt

HCl   HCl

COOH   COOH

S baseH⁺ +   HO—C···H   H—C—OH   + *S* baseH⁺

CH₃   CH₃

R acid   *S* acid

One of the chirality centers in the *R,S* salt is identical to a chirality center in the *S,S* salt, and the other chirality center in the *R,S* salt is the mirror image of a chirality center in the *S,S* salt. Therefore, the salts are diastereomers and have different physical properties. They can be separated by fractional crystallization. After separation, they can be converted back into the carboxylic acids by adding an acid such as HCl. The chiral base then can be separated from the acid and used again.

Enantiomers can also be separated by a technique called **chromatography.** With this method, the mixture to be separated is dissolved in a solvent and the solvent is then passed through a column packed with an adsorbent stationary phase. If the chromatographic column is packed with a chiral stationary phase, the two enantiomers can be expected to move through the column at different rates because they will have different affinities for the chiral stationary phase. Columns packed with chiral materials are commercially available and have been used successfully to separate certain kinds of enantiomers.

THE ENANTIOMERS OF THALIDOMIDE

About half of the commercially available drugs have one or more chirality centers. Most of the drugs obtained from natural sources are single enantiomers. When drugs are synthesized in the laboratory, however, they are usually obtained as racemic mixtures. With few exceptions, these drugs have been marketed as racemic mixtures because of the high cost of separating the enantiomers. Enantiomers can have the same physiological activities, unequal degrees of the same activity, or very different activities. Thalidomide was approved as a sedative for use in Europe in 1956. It was not used in the United States because some neurological side effects had been noted. The dextrorotatory isomer had stronger sedative properties, but the commercial drug was a racemic mixture. However, it wasn't recognized that the levorotatory isomer was highly teratogenic, until it was noticed that women who were given the drug during the first 3 months of pregnancy gave birth to babies with a wide variety of birth defects. It was eventually determined that the dextrorotatory isomer also had mild teratogenic activity and that the enantiomers racemized in vivo. Thus it is not clear whether giving those women only the dextrorotatory isomer would have decreased the severity of the birth defects.

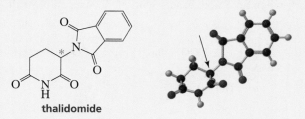

thalidomide

CHIRAL DRUGS

In 1992, the Federal Drug Administration (FDA) issued a policy statement that had the effect of encouraging drug companies to use recent advances in synthetic and separation techniques to develop single enantiomer drugs. Such drugs are becoming more common. Ibuprofen, the popular analgesic (marketed as Advil, Nuprin, and Motrin) will soon be available as (*S*)-(+)-ibuprofen because it exerts its effect in 12 min, while the racemic mixture requires 30 min. The *S* enantiomer of ketoprofen is a nonsteroidal antiinflammatory agent, but the *R* enantiomer prevents bone loss in periodontal disease. There are plans to market both enantiomers: (*S*)-(+)-ketoprofen as an antiinflammatory drug and (*R*)-(−)-ketoprofen as a toothpaste ingredient. The AIDS drug indinavir sulfate, which has five chirality centers, is currently in production as a single enantiomer. The FDA may require drug companies that develop a drug as a single enantiomer rather than as the already approved racemic mixture to show that the missing enantiomer does not provide some type of beneficial effect.

When a compound that contains a chirality center undergoes a reaction, what happens to the configuration of the chirality center depends on the specific reaction. If the reaction *does not break* any of the four bonds to the chirality center, it is obvious that the relative positions of the groups bonded to the chirality center will not change. For example, when (S)-1-chloro-3-methylhexane reacts with hydroxide ion, the relative positions of the groups bonded to the chirality center remain the same because the reaction does not involve breaking a bond to the chirality center.

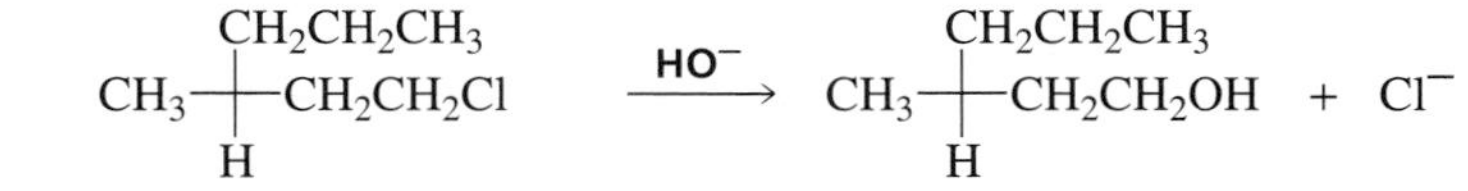

**4.12
REACTIONS OF
COMPOUNDS THAT
CONTAIN A
CHIRALITY CENTER**

$$CH_3 \underset{H}{\overset{CH_2CH_2CH_3}{\rule{0pt}{0pt}}} CH_2CH_2Cl \xrightarrow{\text{HO}^-} CH_3 \underset{H}{\overset{CH_2CH_2CH_3}{\rule{0pt}{0pt}}} CH_2CH_2OH \;+\; Cl^-$$

(S)-1-chloro-3-methylhexane   (S)-3-methyl-1-hexanol

If the four groups bonded to the chirality center maintain their relative positions, it does not necessarily mean that an S reactant will always yield an S product. For example, in the following reaction, the relative positions of the groups bonded to the chirality center are maintained in the product but the S reactant forms an R product (LiAlH$_4$ is a reagent that substitutes an H for a Cl). In other words, the groups maintain their relative positions during the reaction but their relative priorities as defined by the Cahn–Ingold–Prelog rules can change (Section 3.5). The change in priorities—not the change in positions of the groups—is what causes the S reactant to become an R product.

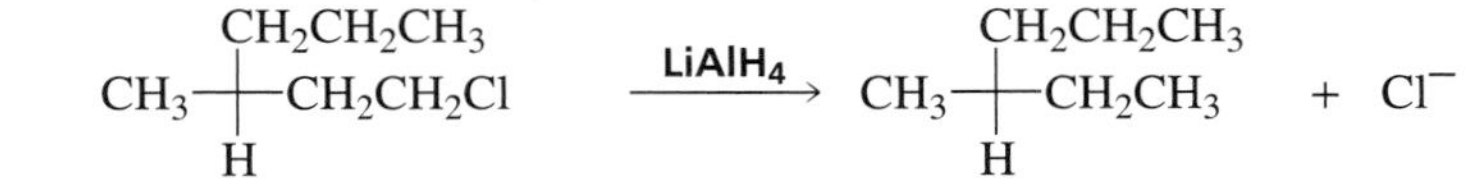

$$CH_3 \underset{H}{\overset{CH_2CH_2CH_3}{\rule{0pt}{0pt}}} CH_2CH_2Cl \xrightarrow{\text{LiAlH}_4} CH_3 \underset{H}{\overset{CH_2CH_2CH_3}{\rule{0pt}{0pt}}} CH_2CH_3 \;+\; Cl^-$$

(S)-1-chloro-3-methylhexane   (R)-3-methylhexane

If the reaction *does break* a bond to the chirality center, the product can have the same configuration as the reactant or it can have the opposite configuration. Which of the products is actually obtained depends on the mechanism of the reaction. Therefore, we cannot predict what the configuration of the product will be unless we know the mechanism of the reaction.

$$CH_3 \underset{H}{\overset{CH_2CH_3}{\rule{0pt}{0pt}}} Y \longrightarrow CH_3 \underset{H}{\overset{CH_2CH_3}{\rule{0pt}{0pt}}} Z \;+\; Z \underset{H}{\overset{CH_2CH_3}{\rule{0pt}{0pt}}} CH_3$$

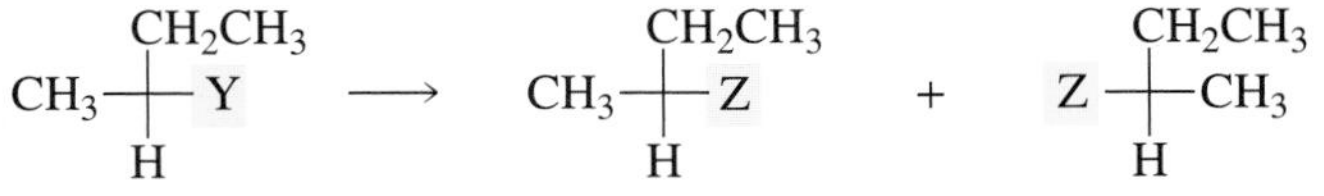

**this product has the
same configuration
as the reactant**　　　**this product has the
opposite configuration
as the reactant**

Enantiomers have the same chemical properties. In other words, they react with *achiral* reagents at the same rate. For example, hydroxide ion is an achiral reagent. It reacts with (R)-2-bromobutane at the same rate that it reacts with (S)-2-bromobutane. If, however, the reagent is *chiral,* the enantiomers react at different rates. One example of a chiral reagent is an enzyme. If you imagine an enzyme to be a right-hand glove and the enantiomers to be a pair of hands, the enzyme typically reacts with only one enantiomer because only the right hand fits into the right-hand glove. For example, the enzyme D-amino acid oxidase reacts only with the R enantiomer. The S enantiomer is unchanged.

$$R \overset{NH_2}{\underset{H}{\mid}} COO^- + \ ^-OOC \overset{NH_2}{\underset{H}{\mid}} R \xrightarrow[\text{oxidase}]{\textbf{D-amino acid}} \underset{R \quad COO^-}{\overset{NH}{\underset{\parallel}{C}}} + \ ^-OOC \overset{NH_2}{\underset{H}{\mid}} R$$

R-enantiomer   S-enantiomer   oxidized   unreacted
                                        R-enantiomer   S-enantiomer

Receptors are proteins that bind particular molecules. Because a receptor is chiral, it will bind one enantiomer better than the other. In Figure 4.2, the receptor binds the *R*-enantiomer but it does not bind the *S*-enantiomer.

Because a receptor will recognize only one of a pair of enantiomers, enantiomers can have different physiological properties. Receptors located on the exterior of nerve cells in the nose are able to perceive and differentiate the estimated 10,000 smells to which they are exposed. (*R*)-(−)-carvone is found in spearmint oil, and (*S*)-(+)-carvone is the main constituent of caraway seed oil. The reason that these two enantiomers have such different odors is that each fits into a different receptor.

(R)-(–)-carvone
spearmint oil

$[\alpha]_D^{20°} = -62.5°$

(S)-(+)-carvone
caraway seed oil

$[\alpha]_D^{20°} = +62.5°$

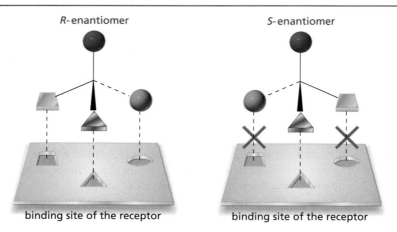

R-enantiomer   S-enantiomer

binding site of the receptor   binding site of the receptor

▲ **Figure 4.2**
Schematic diagram showing why only one enantiomer is bound by a receptor. One enantiomer fits into the binding site and one does not.

Suppose we had a sample of (+)-1-bromo-3-phenyl-2-propanamine and we wanted to know its configuration. Remember, the configuration is the arrangement of the substituents attached to the chirality center. We can draw the R enantiomer and the S enantiomer, but we cannot tell which is (+) by looking at their structures.

**4.13
RELATIVE AND
ABSOLUTE
CONFIGURATIONS**

Which enantiomer is dextrorotatory?

(R)-1-bromo-
3-phenyl-2-propanamine

(S)-1-bromo-
3-phenyl-2-propanamine

The easiest way to determine the configuration of an optically active compound is to find a compound whose configuration is known and that can be converted to the compound whose configuration you want to know by reactions that do not break any bonds to the chirality center. For example, the configuration of amphetamine is known: The levorotatory (−) isomer has the R configuration, and the dextrorotatory (+) isomer has the S configuration.

(R)-(–)-amphetamine

(S)-(+)-amphetamine

If we allow our sample of (+)-1-bromo-3-phenyl-2-propanamine to react with a reagent that will substitute H for Br, we will obtain one of the enantiomers of amphetamine.

(+)-1-bromo-
3-phenyl-2-propanamine
the configuration is not known

amphetamine

If we now place the amphetamine formed in this reaction into a polarimeter to determine in which direction it rotates the plane of polarized light, we can determine whether it has the R or the S configuration. For example, if it is dextrorotatory, we will know that it has the S configuration. Because it was formed in a reaction that did not break any bonds to the chirality center, the product will have to have the same **relative configuration** as the reactant. In other words, the relative positions of the groups bonded to the chirality center must be the same in both compounds. Thus, we would know that (+)-1-bromo-3-phenyl-2-propanamine has the R configuration.

(R)-(+)-1-bromo-
3-phenyl-2-propanamine

(S)-(+)-amphetamine

In contrast, if the product were levorotatory, we would know that (+)-1-bromo-3-phenyl-2-propanamine has the S configuration.

We have determined the configuration of (+)-1-bromo-3-phenyl-2-propanamine by relating it to amphetamine, a compound whose configuration was already known. In other words, the relative configurations of two different compounds can be determined through a reaction that changes one compound into the other. A knowledge of the actual configuration (whether it is *R* or *S*) of either one of them allows us to determine the actual configuration of the other.

The actual configuration is frequently called the **absolute configuration** to indicate that the configuration is known in an absolute sense rather than in a relative sense.

PROBLEM 30 / SOLVED

(*S*)-(−)-2-methyl-1-butanol can be oxidized to (+)-2-methylbutanoic acid without breaking any of the bonds to the chirality center. What is the configuration of (−)-2-methylbutanoic acid?

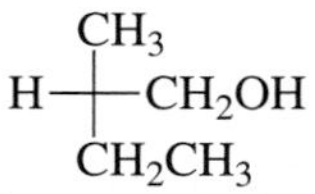

$$CH_3$$
$$H{-}\!|{-}CH_2OH$$
$$CH_2CH_3$$
(S)-(–)-2-methyl-1-butanol

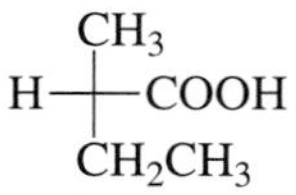

$$CH_3$$
$$H{-}\!|{-}COOH$$
$$CH_2CH_3$$
(+)-2-methylbutanoic acid

SOLUTION.   We know that (+)-2-methylbutanoic acid has the configuration shown above because it was formed from (*S*)-(−)-2-methyl-1-butanol without breaking any bonds to the chirality center. Therefore, we know that (+)-2-methylbutanoic acid has the *S* configuration. We can conclude then that (−)-2-methylbutanoic acid has the *R* configuration.

PROBLEM 31◆

The stereoisomer of 1-iodo-2-methylbutane with the *S* configuration rotates plane-polarized light in a counterclockwise direction. The following reaction results in an alcohol that rotates plane-polarized light in a clockwise direction. What is the configuration of (−)-2-methyl-1-butanol?

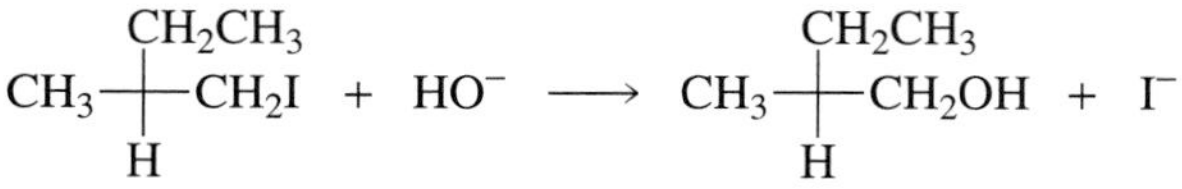

$$CH_2CH_3$$
$$CH_3{-}\!|{-}CH_2I \;+\; HO^- \longrightarrow$$
$$H$$
$$CH_2CH_3$$
$$CH_3{-}\!|{-}CH_2OH \;+\; I^-$$
$$H$$

In the late nineteenth century, Emil Fischer (of Fischer projection fame) assumed that the dextrorotatory isomer of glyceraldehyde had the configuration we now know as *R* and the levorotatory isomer had the configuration we now know as *S*. Assigning *R* to (+)-glyceraldehyde was completely arbitrary, and he had a 50–50 chance of being correct. It would be 60 years before chemists found out whether Fischer was right or wrong.

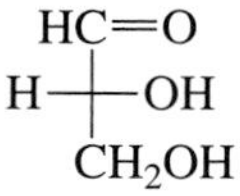

$$HC{=}O$$
$$H{-}\!|{-}OH$$
$$CH_2OH$$
(R)-(+)-glyceraldehyde

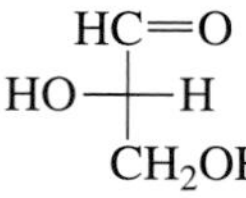

$$HC{=}O$$
$$HO{-}\!|{-}H$$
$$CH_2OH$$
(S)-(–)-glyceraldehyde

For the next half century, the configurations of many organic compounds were "determined" by synthesizing them from (+)-glyceraldehyde or by converting them to (+)-glyceraldehyde, always using reactions that did not break any of the bonds to the chirality center. For example, (−)-lactic acid was assumed to have the configuration shown page 207 because it could be related to (+)-glyceraldehyde through the reactions indicated. The configurations assigned to these molecules were relative configurations, not absolute configurations. They were relative to (+)-glyceraldehyde because they all related back to the *assumption* that (+)-glyceraldehyde was the *R* enantiomer.

HC=O
H——OH $\xrightarrow{\text{HgO}}$ COOH
H——OH   **(+)-glyceraldehyde**   CH$_2$OH

COOH
H——OH $\xleftarrow[\text{H}_2\text{O}]{\text{HNO}_2}$ **(−)-glyceric acid**   CH$_2$OH

COOH
H——OH $\xrightarrow[\text{HBr}]{\text{NaNO}_2}$ **(+)-isoserine**   CH$_2$NH$_2$

COOH
H——OH $\xrightarrow{\text{Zn, H}^+}$ **(−)-3-bromo-2-hydroxypropanoic acid**   CH$_2$Br

COOH
H——OH   **(−)-lactic acid**   CH$_3$

In 1951, the Dutch chemists J. M. Bijvoet, A. F. Peedeman, and A. J. van Bommel, using X-ray crystallography and a new technique known as anomalous dispersion, determined that (+)-tartaric acid has the *R,R* configuration. Because (+)-tartaric acid could be synthesized from (−)-glyceraldehyde, (−)-glyceraldehyde had to be the *S* enantiomer. Fischer had guessed right! The work of these chemists immediately provided absolute configurations for all those compounds whose relative configurations had been determined by relating them to (+)-glyceraldehyde.

HC=O
HO—•—H $\xrightarrow{\text{several steps}}$ COOH H——OH HO—•—H COOH
CH$_2$OH
(−)-glyceraldehyde   **(+)-tartaric acid**

PROBLEM 32◆

What is the absolute configuration of:

a. (−)-glyceric acid?

c. (−)-glyceraldehyde?

b. (+)-isoserine?

If a carbon is bonded to two hydrogens and to two different groups, the two hydrogens are called **enantiotopic hydrogens.** For example, the two hydrogens (H$_a$ and H$_b$) in ethanol are enantiotopic hydrogens because the other two groups bonded to the carbon (CH$_3$ and OH) are not identical. The two hydrogens (H$_a$ and H$_b$) in propane, however, are not enantiotopic hydrogens because the other two groups bonded to the carbon (CH$_3$ and CH$_3$) are identical. The H$_a$ and H$_b$ hydrogens of propane are called **homotopic hydrogens.**

4.14 ENANTIOTOPIC HYDROGENS, DIASTEREOTOPIC HYDROGENS, AND PROCHIRALITY CENTERS

H$_a$
CH$_3$—C—OH
H$_b$
H$_a$ and H$_b$ are enantiotopic hydrogens

H$_a$
CH$_3$—C—CH$_3$
H$_b$
H$_a$ and H$_b$ are homotopic hydrogens

If one of the enantiotopic hydrogens in ethanol were replaced by a deuterium, the carbon to which the enantiotopic hydrogens are attached would become a chirality center. If the H$_a$ hydrogen were replaced by a deuterium, the chirality center would have the *R* configuration. Thus, the H$_a$ hydrogen is called the **pro-*R*-hydrogen.**

The H_b hydrogen is called the **pro-S-hydrogen** because, if it were replaced by a deuterium, the chirality center would have the S configuration.

$$CH_3 \overset{H_a \; \longleftarrow \; \text{pro-}R\text{-hydrogen}}{\underset{H_b \; \longleftarrow \; \text{pro-}S\text{-hydrogen}}{\text{—}OH}}$$

The carbon to which the enantiotopic hydrogens are attached is called a **prochirality center,** because it would become a chirality center if one of the hydrogens were replaced by a deuterium (or any group other than CH_3 or OH) since four different groups would then be bonded to the carbon. The molecule containing the prochirality center is called a prochiral molecule because it would become a chiral molecule if one of the hydrogens were replaced.

The pro-R- and pro-S-hydrogens are chemically equivalent. In other words, they have the same chemical reactivity and cannot be distinguished by achiral chemical reagents. For example, when ethyl alcohol is oxidized (by CrO_3) to acetaldehyde, one of the enantiotopic hydrogens is removed. Because the two hydrogens are chemically equivalent, half of the product will form as a result of removing the H_a hydrogen and the other half as a result of removing the H_b hydrogen.

$$CH_3-\underset{H_b}{\overset{H_a}{\underset{|}{\overset{|}{C}}}}-OH \xrightarrow[\text{pyridine}]{CrO_3} CH_3-\overset{O}{\overset{||}{C}}-H_b \underset{50\%}{} + CH_3-\overset{O}{\overset{||}{C}}-H_a \underset{50\%}{}$$

Enantiotopic hydrogens, however, are not chemically equivalent in enzyme-catalyzed reactions. The enzyme can distinguish between them because an enzyme is chiral (Section 4.12). For example, when the oxidation of ethyl alcohol to acetaldehyde is catalyzed by the enzyme alcohol dehydrogenase, only the H_a hydrogen is removed.

$$CH_3-\underset{H_b}{\overset{H_a}{\underset{|}{\overset{|}{C}}}}-OH \xrightarrow[\text{dehydrogenase}]{\text{alcohol}} CH_3-\overset{O}{\overset{||}{C}}-H_b \underset{100\%}{}$$

If a carbon is bonded to two hydrogens and replacing each of them in turn with deuterium (or other group) creates a pair of diastereomers, the hydrogens are called **diastereotopic hydrogens.**

H_a and H_b are
diastereotopic hydrogens

replace H_a with a D

replace H_b with a D

a pair of diastereomers

Unlike enantiotopic hydrogens, diastereotopic hydrogens do not have the same reactivity with achiral reagents. For example, in Chapter 10 we will see why removal of H_b and Br to form *trans*-2-butene occurs faster than removal of H_a and Br to form *cis*-2-butene.

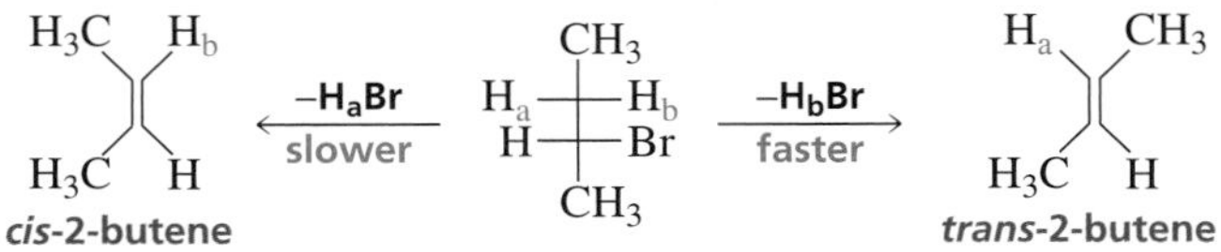

cis-2-butene   trans-2-butene

PROBLEM 33 ◆

Tell whether the H_a and H_b hydrogens in each of the following compounds are homotopic, enantiotopic, or diastereotopic.

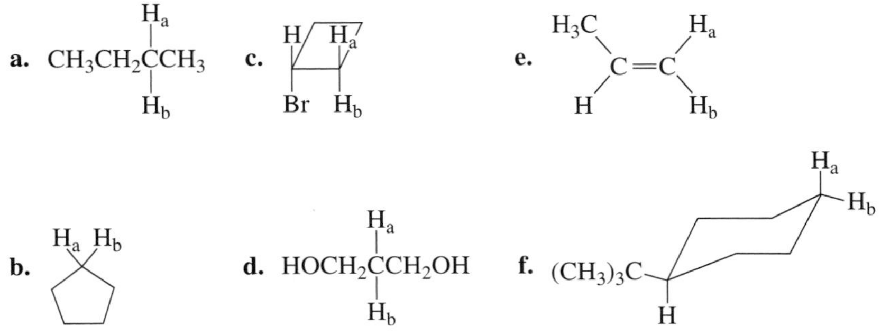

a. $CH_3CH_2\overset{\overset{\displaystyle H_a}{|}}{\underset{\underset{\displaystyle H_b}{|}}{C}}CH_3$

c.

e.

b.

d. $HOCH_2\overset{\overset{\displaystyle H_a}{|}}{\underset{\underset{\displaystyle H_b}{|}}{C}}CH_2OH$

f.

Atoms other than carbon can be chirality centers. When an atom such as nitrogen or phosphorus has four different groups or atoms attached to it, it is a chirality center. Such compounds can exist as a pair of enantiomers, and the enantiomers can be separated.

4.15
NITROGEN AND PHOSPHORUS CHIRALITY CENTERS

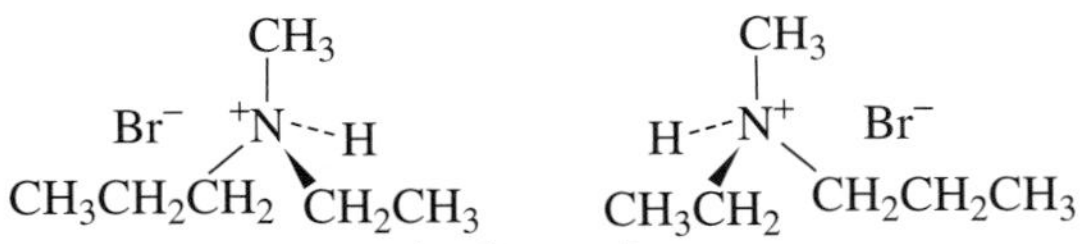

a pair of enantiomers 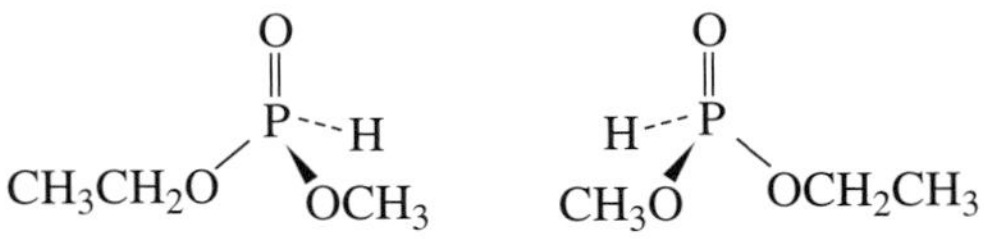   a pair of enantiomers

If one of the four "groups" attached to nitrogen is a pair of nonbonding electrons, the enantiomers cannot be separated because rapid inversion between the two enantiomers takes place at room temperature (Section 4.1).

Nitrogen is sp^2 hybridized in the transition state for amine inversion. If, in a particular compound, the nitrogen atom cannot achieve the 120° bond angles required for sp^2 hybridization, inversion will not take place and the enantiomers can be separated. For example, 1,2-dimethylaziridine cannot undergo inversion at room temperature because the nitrogen is in a three-membered ring. A nitrogen in a three-membered ring cannot achieve a 120° bond angle. The two enantiomers, therefore, can be separated.

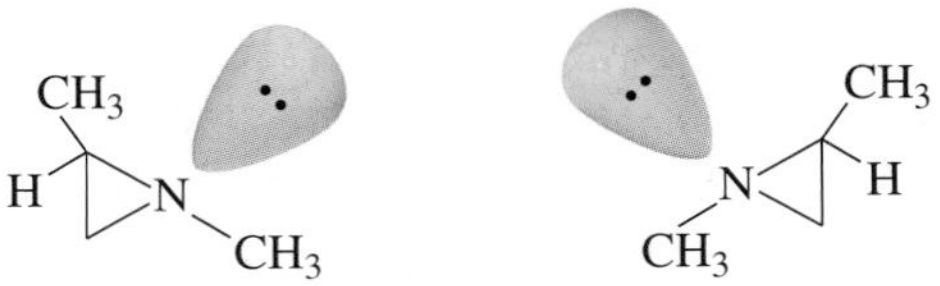

enantiomers of 1,2-dimethylaziridine

PROBLEM 34

Compound I has two stereoisomers, but compounds II and III exist as single compounds. Explain.

$$
\begin{array}{ccc}
\overset{\displaystyle CH=CH_2}{\underset{\displaystyle CH_2CH_3}{CH_3-\overset{+}{N}-H}}\ \ Cl^- & \overset{\displaystyle CH=CH_2}{\underset{\displaystyle CH_3}{CH_3-\overset{+}{N}-H}}\ \ Cl^- & \overset{\displaystyle CH=CH_2}{CH_3-\overset{\displaystyle ..}{N}-H} \\
\mathbf{I} & \mathbf{II} & \mathbf{III}
\end{array}
$$

4.16
CONFORMATIONS OF DISUBSTITUTED CYCLOHEXANES

Two things that we have already learned about monosubstituted cyclohexanes are important to our understanding of the conformations of disubstituted cyclohexanes:

1. The two chair conformations rapidly interconvert, and bonds that are equatorial in one conformation are axial in the other conformation (Section 2.13).

2. There is more room for a substituent in an equatorial position than in an axial position (Section 2.14).

Conformations of 1,4-Disubstituted Cyclohexanes

1,4-Disubstituted cyclohexanes do not have any chirality centers; therefore, they do not have enantiomers. A 1,4-disubstituted cyclohexane has only one cis isomer and one trans isomer.

cis-1-*tert*-butyl-
4-methylcyclohexane

trans-1-*tert*-butyl-
4-methylcyclohexane

Each of the isomers has two chair conformations. In drawing the conformations for a particular disubstituted cyclohexane, do you place the substituents in the axial or in the equatorial positions? If you are working with three-dimensional molecular models, it is easy to determine where the substituents for a given conformation should be placed. If, however, you are working on a two-dimensional paper surface, it is a little more difficult unless you can easily visualize objects in three dimensions. A way around this difficulty is to look first at the axial bonds. It is readily apparent that substituents in the 1,2-diaxial positions and in the 1,4-diaxial positions are trans to each other, while substituents in the 1,3-diaxial positions are cis to each other. For example, we can easily see that one of the chair conformations of *trans*-1-*tert*-butyl-4-methylcyclohexane has both substituents in axial positions. It is harder to see that the other chair conformation has both substituents in equatorial positions, but we know that it does because substituents that are axial in one conformation are equatorial in the other. The more stable conformation is the one with both substituents in equatorial positions, where there is less chance of unfavorable steric interactions (Section 2.14).

TABLE 4.2	Locations of Substituents in Disubstituted Cyclohexanes	
	cis	*trans*
1,2-disubstituted	one axial and one equatorial	both equatorial or both axial
1,3-disubstituted	both equatorial or both axial	one axial and one equatorial
1,4-disubstituted	one axial and one equatorial	both equatorial or both axial

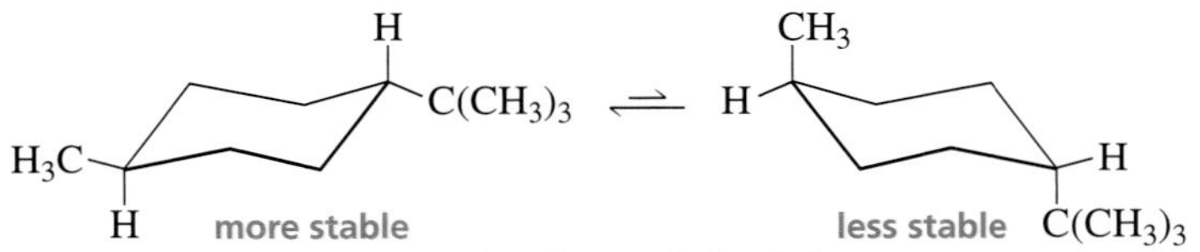

***trans*-1-*tert*-butyl-4-methylcyclohexane**

Because *trans*-1-*tert*-butyl-4-methylcyclohexane either has both substituents in equatorial positions or both substituents in axial positions, the cis isomer must have one substituent in the equatorial position and one substituent in the axial position (Table 4.2). One of the chair conformations of *cis*-1-*tert*-butyl-4-methylcyclohexane has the *tert*-butyl group in the equatorial position and the methyl group in the axial position, while the other chair conformation has the *tert*-butyl group in the axial position and the methyl group in the equatorial position. A *tert*-butyl group is larger than a methyl group, so the more stable conformation is the one with the *tert*-butyl group in the equatorial position, where there is more room for the larger substituent.

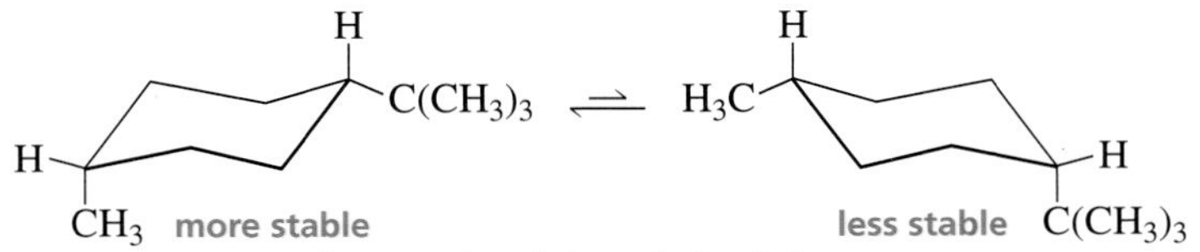

***cis*-1-*tert*-butyl-4-methylcyclohexane**

Conformations of 1,3-Disubstituted Cyclohexanes

1,3-Disubstituted cyclohexanes have two chirality centers. Therefore, a compound such as 1-*tert*-butyl-3-methylcyclohexane has four stereoisomers. The cis isomer exists as a pair of enantiomers, and the trans isomer exists as a pair of enantiomers.

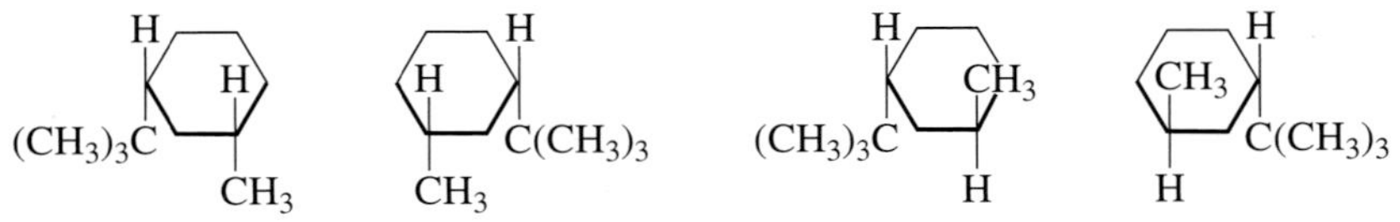

We know that the 1,3-diaxial positions are cis to each other. So each cis enantiomer has a chair conformation in which both substituents are in axial positions and a chair conformation in which both substituents are in equatorial positions. The most stable conformation for each cis isomer is the one in which both substituents are in equatorial positions. Because the two enantiomers of *cis*-1-*tert*-butyl-4-methylcyclohexane are mirror images, the more stable conformation of one enantiomer is the mirror image of the more stable conformation of the other enantiomer.

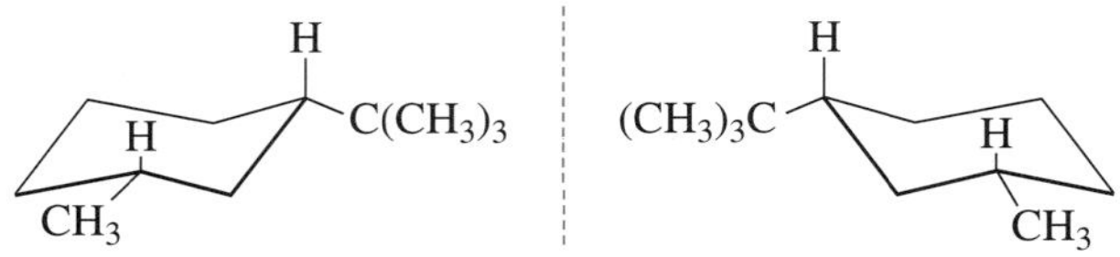

the more stable chair conformation for each of the two enantiomers
of *cis*-1-*tert*-butyl-3-methylcyclohexane

Each enantiomer of *trans*-1-*tert*-butyl-3-methylcyclohexane has two chair conformations. In each conformation, one of the substituents is in the equatorial position
and one is in the axial position. The more stable conformation is the one in which
the larger substituent is in the equatorial position.

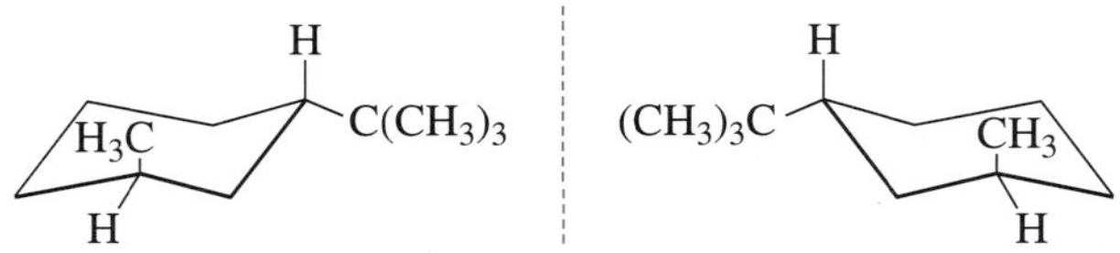

the more stable chair conformation for each of the two enantiomers
of *trans*-1-*tert*-butyl-3-methylcyclohexane

PROBLEM 35

a. Draw the two chair conformations for each of the stereoisomers of *trans*-1-*tert*-
butyl-3-methylcyclohexane.

b. For each pair, indicate which conformation is more stable.

Conformations of 1,2-Disubstituted Cyclohexanes

1,2-Disubstituted cyclohexanes, like 1,3-disubstituted cyclohexanes, have two chirality centers and four stereoisomers.

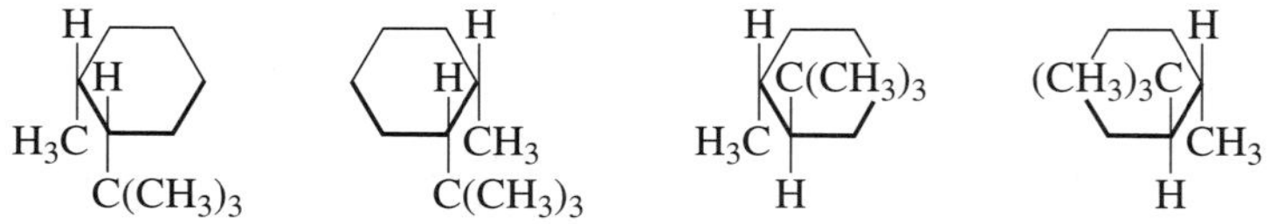

In *cis*-1-*tert*-butyl-2-methylcyclohexane, one substituent is in the equatorial position and the other is in the axial position. For each of the enantiomers, the conformation with the larger *tert*-butyl group in the equatorial position is the more stable
conformation.

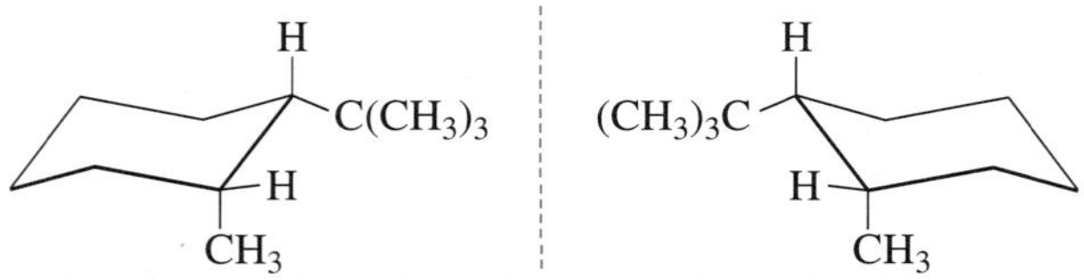

the more stable chair conformation for each of the two
enantiomers of *cis*-1-*tert*-butyl-2-methylcyclohexane

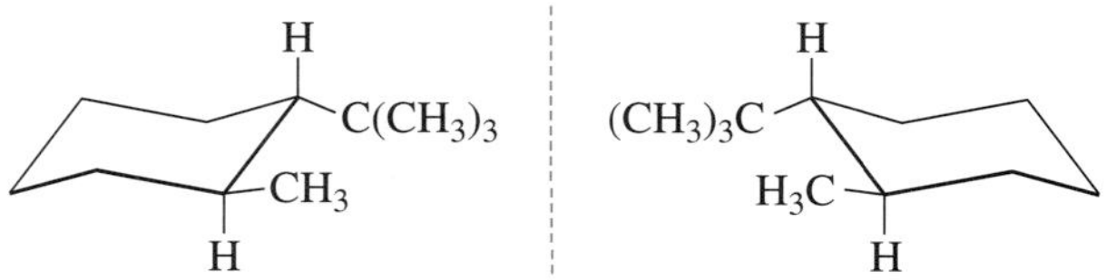

the more stable chair conformation for each of the two
enantiomers of trans-1-*tert*-butyl-2-methylcyclohexane

Because the 1,2-diaxial positions are trans to one another, the substituents in
trans-1-*tert*-butyl-2-methylcyclohexane are both in axial positions in one conformation and both in equatorial positions in the other conformation. The conformation with both substituents in equatorial positions is the more stable conformation.

Meso Compounds

If the two substituents in a 1,3-disubstituted cyclohexane are identical, the cis isomer will be a meso compound (because it has a plane of symmetry) and the trans isomer will exist as a pair of enantiomers.

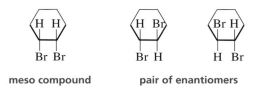

meso compound   pair of enantiomers

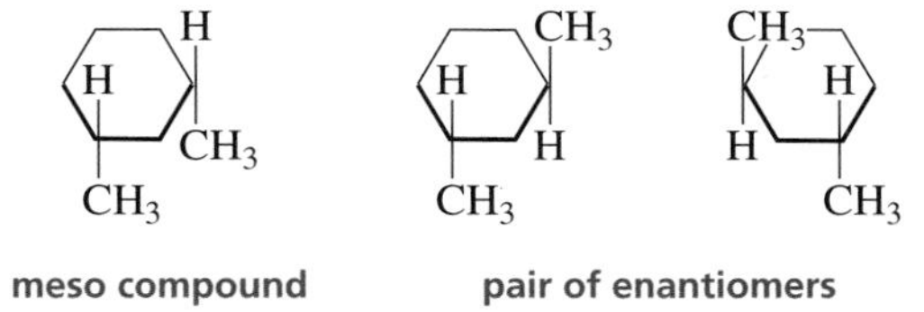

Similarly, if the two substituents in a 1,2-disubstituted cyclohexane are identical, the cis compound will be a meso compound and the trans compound will exist as a pair of enantiomers.

meso compound   pair of enantiomers

So both 1,2-dibromocyclohexane and 1,3-dimethylcyclohexane have three non-interconvertible isomers. The *cis* isomers are meso compounds; the *trans* isomers exist as a pair of enantiomers.

It is apparent from looking at the above structure for *cis*-1,2-dibromocyclo-hexane that the compound has a plane of symmetry. Cyclohexane, however, is not a flat hexagon; it exists preferentially in a chair conformer, and the chair con-former of *cis*-1, 2-dibromocyclohexane does not have a plane of symmetry. Only the much less stable boat conformer of *cis*-1,2-dibromocyclohexane has a plane of symmetry. So is *cis*-1,2-dibromocyclohexane a meso compound? The answer is yes. As long as any one conformer of a compound has a plane of symmetry the compound is achiral, and if the compound has two chirality centers it can be con-sidered to be a meso compound.

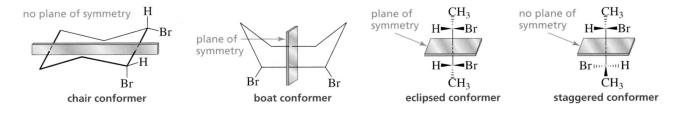

chair conformer   boat conformer   eclipsed conformer   staggered conformer

This holds for acyclic compounds as well. We have seen that 2,3-dibromobutane is an achiral meso compound because it has a plane of symmetry (Section 4.9). But when we made that determination, we were looking at a relatively unstable eclipsed conformer. The more stable staggered conformer does not have a plane of sym-metry. But 2,3-dibromobutane is a meso compound because it has a conformer that has a plane of symmetry.

PROBLEM 36 ◆

Which is more stable:

a. a *cis*-1,2-disubstituted cyclohexane or a *trans*-1,2-disubstituted cyclohexane?

b. a *cis*-1,3-disubstituted cyclohexane or a *trans*-1,3-disubstituted cyclohexane?

c. a *cis*-1,4-disubstituted cyclohexane or a *trans*-1,4-disubstituted cyclohexane?

PROBLEM 37

a. Draw the most stable conformation for each of the following compounds. (If the compound can exist as a pair of enantiomers, answer the question for only one of the enantiomers.)

 1. *cis*-1-ethyl-2-isopropylcyclohexane   **4.** *cis*-1-ethyl-4-isopropylcyclohexane

 2. *trans*-1,4-dibromocyclohexane   **5.** *cis*-1,2-dimethylcyclohexane

 3. *cis*-1-*tert*-butyl-3-methylcyclohexane **6.** *cis*-1,3-dichlorocyclohexane

b. Which of the above compounds are optically active?

4.17
CONFORMATIONS OF FUSED RINGS

When two cyclohexane rings are fused together, the second ring can be considered to be a pair of substituents bonded to the first ring. The two substituents can be either cis or trans. If the cyclohexane rings are drawn in their chair conformations, the trans isomer will have both substituents in the equatorial position. The cis isomer will have one substituent in the equatorial position and one substituent in the axial position. **Trans fused** rings, therefore, are more stable than **cis fused** rings.

For studies on the stereochemistry of enzyme-catalyzed reactions, **Sir John Cornforth** *received the Nobel Prize in chemistry in 1975 (sharing it with Vladimir Prelog, page 177). Born in Australia in 1917, he studied at the University of Sydney, and received a Ph.D. from Oxford. His major research was carried out in laboratories at Britain's Medical Research Council and in laboratories at Shell Research Ltd. He was knighted in 1977.*

trans-decalin
***trans*-fused rings**
more stable

cis-decalin
***cis*-fused rings**
less stable

Trans fused rings cannot undergo chair–chair interconversion (Section 2.13). Cis fused rings are much more flexible and undergo chair–chair interconversion almost as fast as cyclohexane.

Steroids are examples of compounds that have fused rings. They have four rings—three six-membered rings and one five-membered ring (Section 23.9). Cholesterol is an example of a naturally occurring steroid.

cholesterol

PROBLEM 38

a. How many chirality centers does cholesterol have?

b. What is the maximum number of stereoisomers that cholesterol can have? (Only one of these stereoisomers is found in nature.)

In Chapter 3, we learned that alkenes undergo addition reactions. We looked at the different kinds of reagents that add to alkenes, we examined the step-by-step process by which the reactions occur, and we determined what products are formed. But we did not consider the stereochemistry of the reactions. In other words, we did not determine which stereoisomers are formed.

Stereochemistry is the field of chemistry that deals with the structure of molecules in three dimensions. When we study the stereochemistry of a reaction, we are concerned with the following questions:

1. If stereoisomers are possible for a reaction product, does the reaction produce a single stereoisomer, a set of particular stereoisomers, or all possible stereoisomers?

2. If stereoisomers are possible for the reactant, do all stereoisomers react to form the same stereoisomeric product, or does each reactant form a different stereoisomer or a different set of stereoisomers?

Before we examine the stereochemistry of addition reactions, let's become familiar with some terms used in describing the stereochemistry of a reaction.

We learned in Section 3.12 that a **regioselective** reaction is one in which two *constitutional isomers* can be obtained as products but more of one is obtained than of the other. A regioselective reaction selects for a particular constitutional isomer. Recall that a reaction can be *moderately regioselective, highly regioselective,* or *completely regioselective* depending on the relative amounts of the constitutional isomers formed in the reaction.

Stereoselective is a similar term, but it refers to the preferential formation of a *stereoisomer* rather than a *constitutional isomer.* If a reaction that generates a carbon–carbon double bond or a chirality center in a product leads to the preferential formation of one stereoisomer over another, it is a stereoselective reaction. In other words, it selects for a particular stereoisomer. Depending on the degree of preference for a particular stereoisomer, a reaction can be described as being *moderately stereoselective, highly stereoselective,* or *completely stereoselective.*

4.18
STEREOCHEMISTRY OF REACTIONS: REGIOSELECTIVE, STEREOSELECTIVE, AND STEREOSPECIFIC REACTIONS

A regioselective reaction forms more of one constitutional isomer than of another.

a regioselective reaction

$$A \longrightarrow B + C$$

more B is formed than C where B and C are constitutional isomers

a stereoselective reaction

$$A \longrightarrow B + C$$

more B is formed than C where B and C are stereoisomers

A stereoselective reaction forms more of one stereoisomer than of another.

A reaction is **stereospecific** if the reactant can exist as stereoisomers and each stereoisomeric reactant leads to a different stereoisomeric product.

a stereospecific reaction

$$A \longrightarrow B$$
$$C \longrightarrow D$$

A and C are stereoisomers
B and D are stereoisomers

In a stereospecific reaction, each stereoisomeric reactant forms a different stereoisomeric product.

Notice that, in the preceding reaction, because stereoisomer A forms stereoisomer B but does not form D, the reaction is stereoselective in addition to being stereospecific. *All stereospecific reactions, therefore, are also stereoselective. All stereoselective reactions are not stereospecific,* however, because there are stereoselective reactions in which the reactant does not have a carbon–carbon double bond or a chirality center, so it cannot exist as stereoisomers.

All stereospecific reactions are also stereoselective. All stereoselective reactions are not necessarily stereospecific.

4.19 STEREOCHEMISTRY OF ALKENE ADDITION REACTIONS

Now that we are familiar with addition reactions and with stereoisomers, we will combine the two topics and take a look at the stereoisomers that are formed in an addition reaction.

In Chapter 3, we saw that when an alkene reacts with an electrophilic reagent such as HBr, the major product of the addition reaction can be predicted by Markovnikov's rule: the electrophile (H^+) adds to the sp^2 carbon bonded to the most hydrogens. For example, the major product obtained from the reaction of propene with HBr is 2-bromopropane. This particular product does not have stereoisomers. Therefore, we do not have to be concerned with the stereochemistry of the product.

$$CH_3CH{=}CH_2 \xrightarrow{H^+} CH_3\overset{+}{C}HCH_3 \xrightarrow{Br^-} CH_3CHCH_3$$

propene

$$\underset{\text{2-bromopropane}}{\underset{\text{major product}}{|}}$$
Br

If, however, the reaction creates a product with a chirality center, we need to know which stereoisomers are formed. For example, the reaction of HBr with 1-butene forms 2-bromobutane, a compound with a chirality center. What is the configuration of the product? Do we get the *R* enantiomer, the *S* enantiomer, or both?

chirality center

$$CH_3CH_2CH{=}CH_2 \xrightarrow{H^+} CH_3CH_2\overset{+}{C}HCH_3 \xrightarrow{Br^-} CH_3CH_2\overset{*}{C}HCH_3$$

1-butene

Br

2-bromobutane

In our discussion of the stereochemistry of addition reactions, we will look first at reactions that form a product with one chirality center. Then we will look at reactions that form a product with two chirality centers.

Addition Reactions That Form One Chirality Center

When a reactant that does not have a chirality center undergoes an addition reaction that forms a product with one chirality center, the product will be a racemic mixture. For example, the reaction of 1-butene with HBr forms equal amounts of (R)-2-bromobutane and (S)-2-bromobutane. In other words, an addition reaction that forms a compound with a single chirality center from a reactant without any chirality centers is not stereoselective; it does not select for a particular stereoisomer.

We can understand why a racemic mixture is obtained if we examine the structure of the carbocation formed in the first step of the reaction. The positively charged carbon is sp^2 hybridized, and the three atoms to which it is bonded lie in a plane (Section 1.10). When the bromide ion attacks the carbocation from above the plane, one enantiomer is formed, but when it attacks the carbocation from below the plane, the other enantiomer is formed. Because the bromide ion can attack the planar carbocation intermediate from above just as well as it can attack it from below, equal amounts of the R and S enantiomers are obtained from the addition reaction.

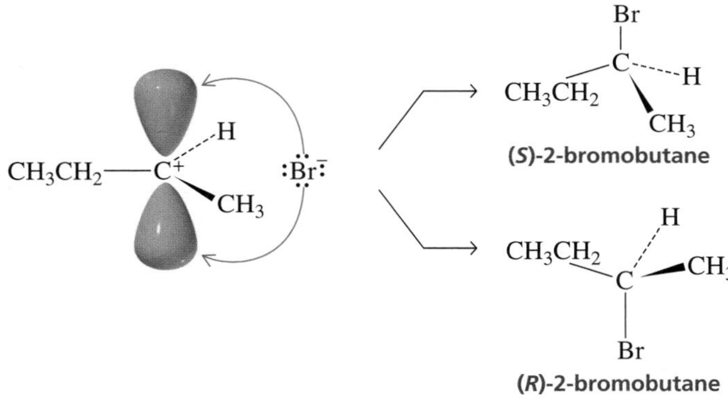

(S)-2-bromobutane

(R)-2-bromobutane

PROBLEM 39◆

a. Is the reaction of 2-butene with HBr stereoselective?

b. Is it stereospecific?

When HBr adds to 2-methyl-1-butene in the presence of peroxide, the product has one chirality center. Therefore, we can predict that equal amounts of the R and S enantiomers are obtained.

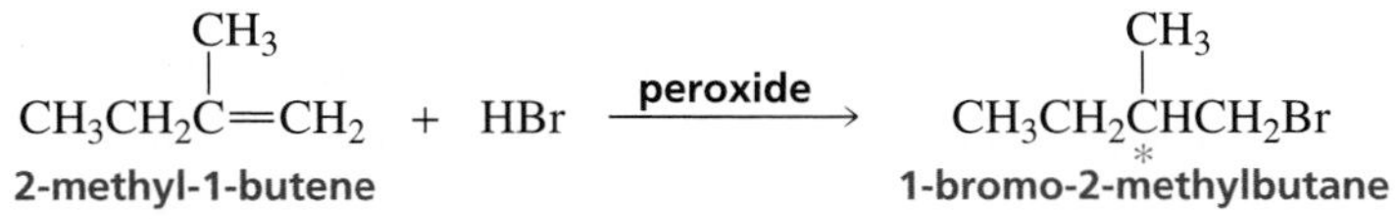

The product is a racemic mixture because the carbon in the radical intermediate that bears the unpaired electron is sp^2 hybridized. This means that the three atoms bonded to it are all in the same plane (Section 1.10). Consequently, the R enantiomer and the S enantiomer will be obtained in equal amounts because HBr has equal access to both sides of the radical. (Although the unpaired electron is shown to be in the top lobe of the *p* orbital, it really is spread out equally through both lobes.)

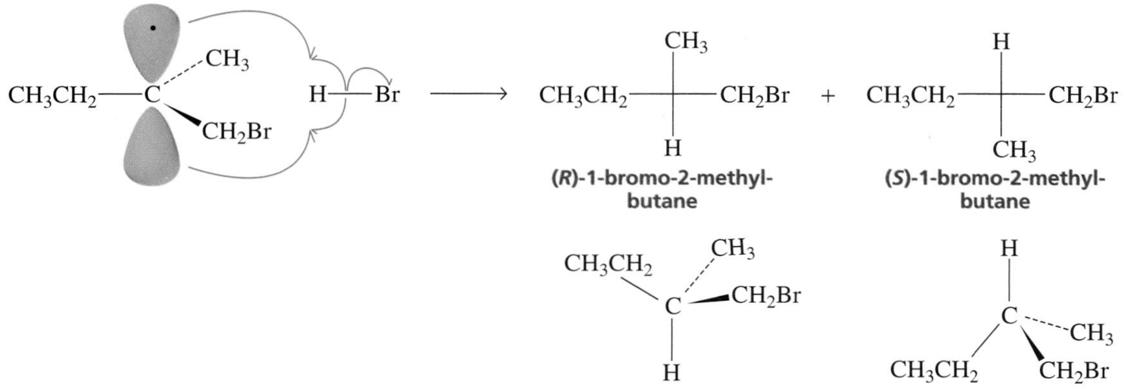

(R)-1-bromo-2-methyl-butane

(S)-1-bromo-2-methyl-butane

PROBLEM 40◆

What stereoisomers are obtained in each of the following reactions?

a. $CH_3CH_2CH_2CH=CH_2$ + HCl ⟶

b. + H_2O $\xrightarrow{H^+}$

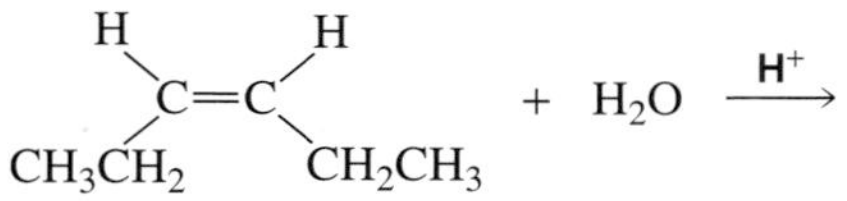

c. + HBr ⟶

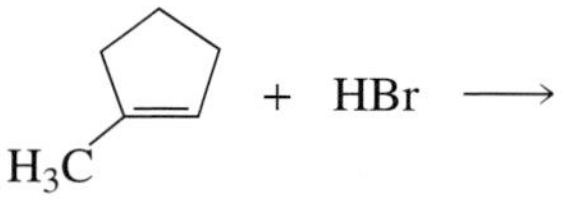

d. + HBr ⟶

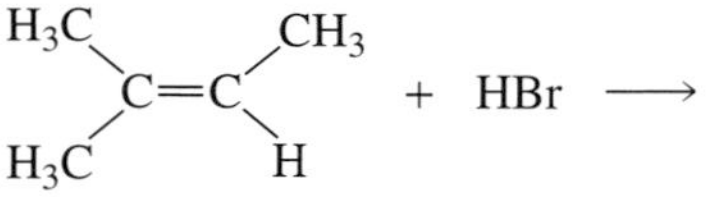

e. + HBr $\xrightarrow{\text{peroxide}}$

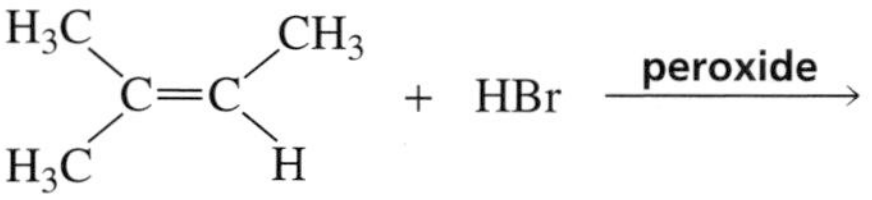

f. + H_2 $\xrightarrow{\text{Pt}}$

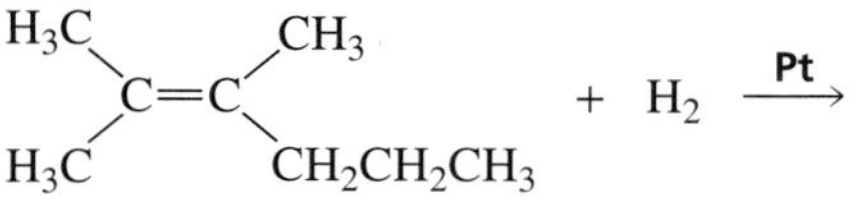

If an addition reaction creates a chirality center in a compound that already has a chirality center, a pair of diastereomers will be formed. Because none of the bonds to the chirality center in (*R*)-3-chloro-1-butene is broken during the addition of HBr, the configuration of this chirality center does not change. Because the bromide ion can attack the planar carbocation intermediate from either the top or the bottom in the process of creating the new chirality center, two stereoisomers result. The stereoisomers are diastereomers because one of the chirality centers has the same configuration in both isomers and the other has opposite configurations in the two isomers.

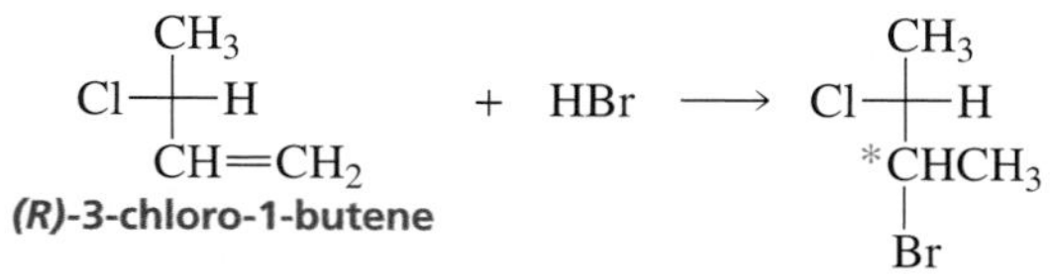

$$\underset{\substack{\text{(R)-3-chloro-1-butene}}}{\overset{\displaystyle CH_3}{\underset{\displaystyle CH=CH_2}{Cl\!-\!\!\!\!\!-\!\!\!\!\!-H}}} \quad + \quad HBr \quad \longrightarrow \quad \underset{\substack{|\\Br}}{\overset{\displaystyle CH_3}{\underset{\displaystyle {}^*CHCH_3}{Cl\!-\!\!\!\!\!-\!\!\!\!\!-H}}}$$

stereochemistry of the product

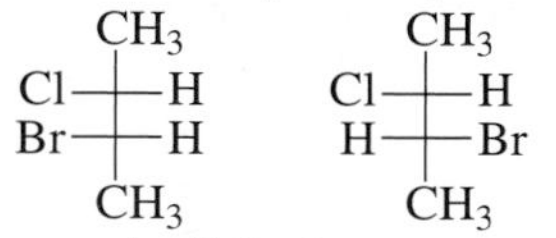

$$\underset{\displaystyle CH_3}{\overset{\displaystyle CH_3}{\underset{\displaystyle Br\!-\!\!\!\!\!-H}{Cl\!-\!\!\!\!\!-H}}} \qquad \underset{\displaystyle CH_3}{\overset{\displaystyle CH_3}{\underset{\displaystyle H\!-\!\!\!\!\!-Br}{Cl\!-\!\!\!\!\!-H}}}$$

a pair of diastereomers

Because there is a chirality center in the carbocation intermediate, it is a chiral ion and therefore does not have a plane of symmetry. If the chirality center is near the positively charged carbon, one face of the carbocation will be more sterically hindered than the other. The incoming bromide ion will have greater access to the less sterically hindered face. The diastereomers, therefore, will be formed in unequal amounts. The reaction is stereoselective; more of one stereoisomer is formed than of the other.

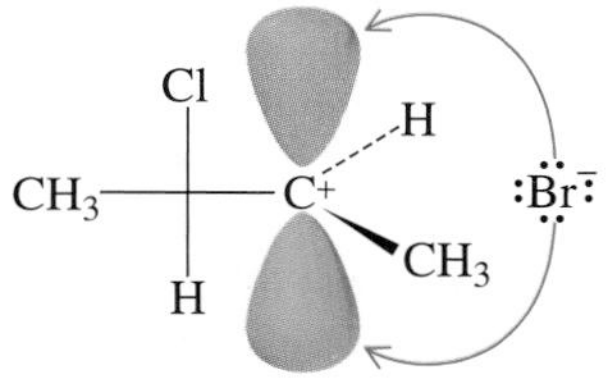

PROBLEM 41

What stereoisomers are formed in the following reactions?

a.

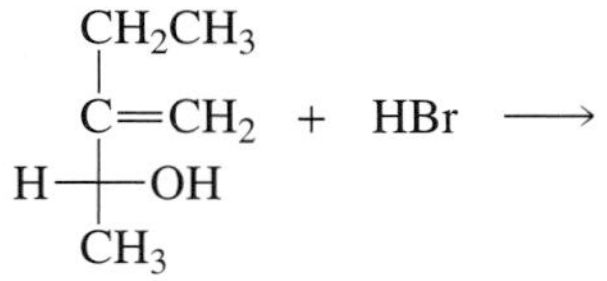

$$\underset{\displaystyle CH_3}{\overset{\displaystyle CH_2CH_3}{\underset{\displaystyle H\!-\!\!\!\!\!-OH}{C=CH_2}}} \quad + \quad HBr \quad \longrightarrow$$

b. (+)-3-methyl-1-pentene   +   HBr $\longrightarrow$

Addition Reactions That Form Products with Two Chirality Centers

When a reactant that does not have a chirality center undergoes an addition reaction that forms a product with two chirality centers, the stereoisomers that are formed depend on the mechanism of the addition reaction.

1. Addition Reactions That Involve a Carbocation Intermediate or a Radical Intermediate

If two chirality centers are created as the result of an addition reaction that forms a carbocation intermediate, four stereoisomers can be obtained as products.

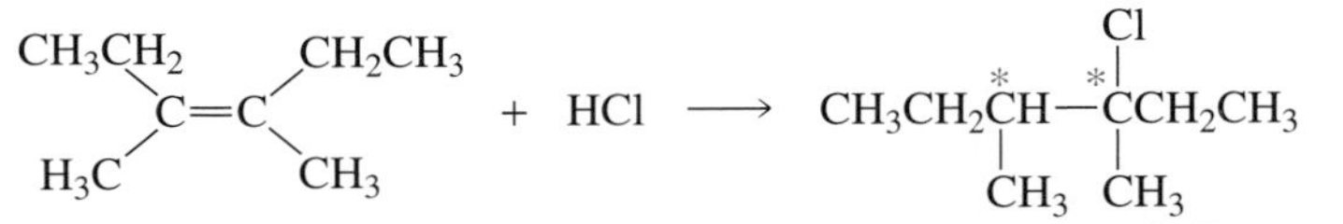

$$CH_3CH_2 \quad CH_2CH_3 \qquad \qquad \qquad \qquad Cl$$

cis-3,4-dimethyl-3-hexene + HCl ⟶ 3-chloro-3,4-dimethylhexane

cis-3,4-dimethyl-3-hexene   3-chloro-3,4-dimethylhexane

stereochemistry of the product

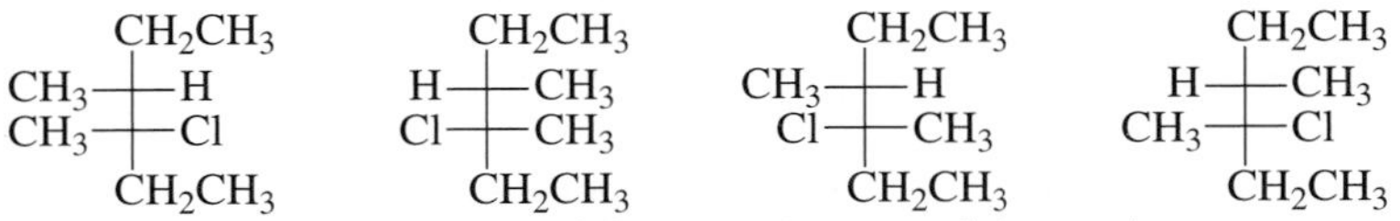

Fischer projections of the stereoisomers of the product

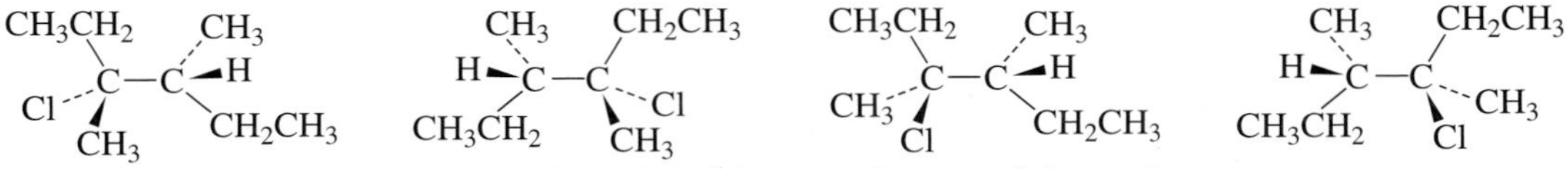

perspective formulas of the stereoisomers of the product

The proton can approach the plane containing the doubly bonded carbons of the alkene from above or below to form the carbocation in the first step of the reaction. Once the carbocation is formed, the chloride ion can attack from above or below. As a result, four stereoisomers are obtained as products: the two substituents can add from above-above, above-below, below-below, or from below-above. When the two substituents add to the same side of the double bond, the addition is called **syn addition.** When the two substituents add to opposite sides of the double bond, the addition is called **anti addition.** Both syn and anti addition occur in alkene addition reactions that take place by way of a carbocation intermediate. Because the four stereoisomers formed by the cis alkene are identical to the four stereoisomers formed by the trans alkene, the reaction is not stereospecific.

Similarly, if two chirality centers are created as the result of an addition reaction that forms a radical intermediate, four stereoisomers can be formed because both syn and anti addition are possible. And since the stereoisomers formed by the cis isomer are identical to those formed by the trans isomer; the reaction is not stereospecific.

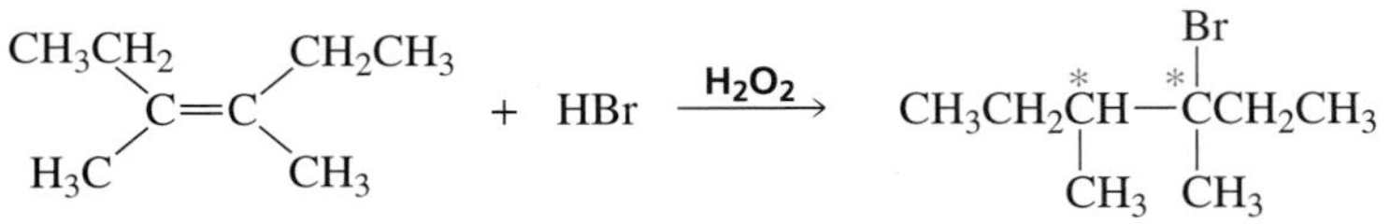

cis-3,4-dimethyl-3-hexene + HBr $\xrightarrow{H_2O_2}$ 3-bromo-3,4-dimethylhexane

stereochemistry of the product

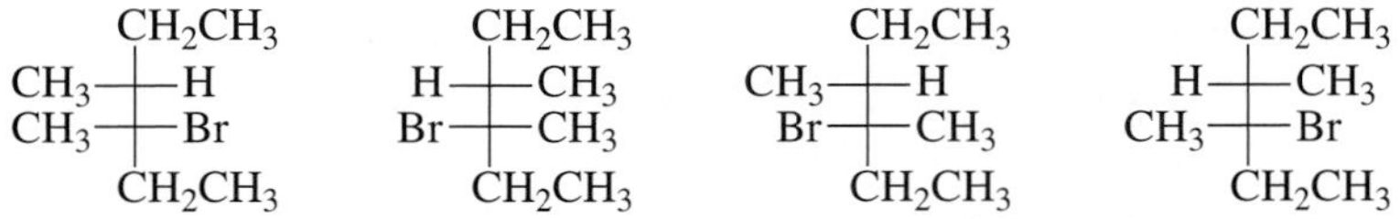

Fischer projections of the stereoisomers of the product

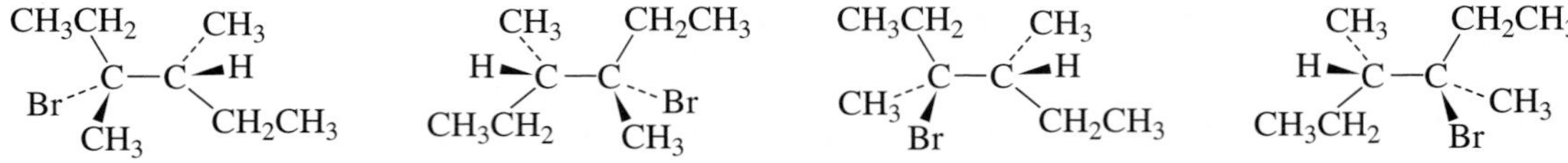

perspective formulas of the stereoisomers of the product

2. Stereochemistry of Hydrogen Addition

When hydrogen (H_2) adds to an alkene, both hydrogen atoms add to the same side of the double bond (Section 3.19), which makes the addition of H_2 a syn addition reaction.

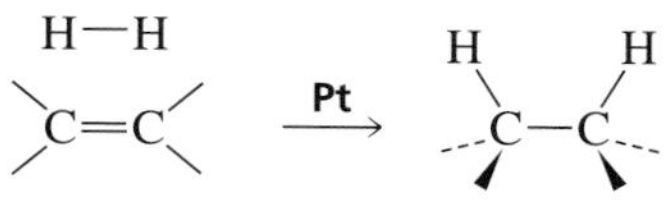

addition of H_2 is a syn addition

If addition of hydrogen to an alkene forms a product with two chirality centers, only two of the four possible stereoisomers are obtained, since only syn addition can occur. (The other two stereoisomers would have to come from anti addition.) Because the doubly bonded carbons of the alkene and the atoms attached to them all lie in a plane, one stereoisomer results from addition of the hydrogens from above the plane, and the other stereoisomer results from addition of the hydrogens from below the plane. The particular pair of stereoisomers that is formed depends on whether the reactant is a cis alkene or a trans alkene. Syn addition of H_2 to a cis alkene forms only the erythro pair of enantiomers. (In Section 4.8, we saw that the erythro pair of enantiomers is the pair with identical groups on the same side of the carbon chain in the Fischer projections.)

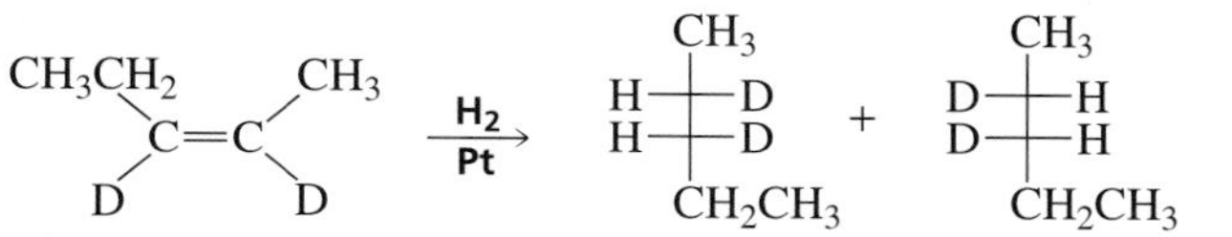

cis-2,3-dideuterio-
2-pentene

Fischer projections of the products
erythro enantiomers

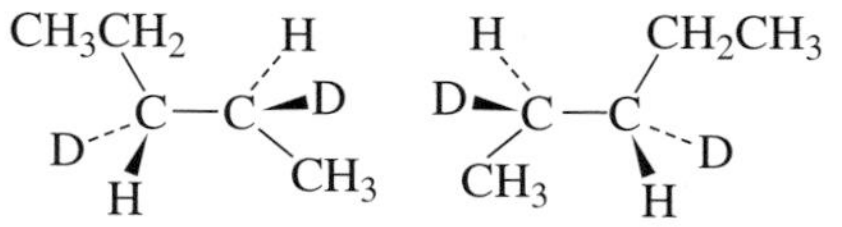

perspective formulas of the products

CIS-SYN-ERYTHRO

If the two chirality centers have the same set of substituents, a meso compound will be obtained instead of the pair of erythro enantiomers.

Syn addition of H_2 to a trans alkene forms only the threo pair of enantiomers. Thus the addition of hydrogen is a stereospecific reaction; the product obtained from addition to the cis isomer is different from the product obtained from addition to the trans isomer.

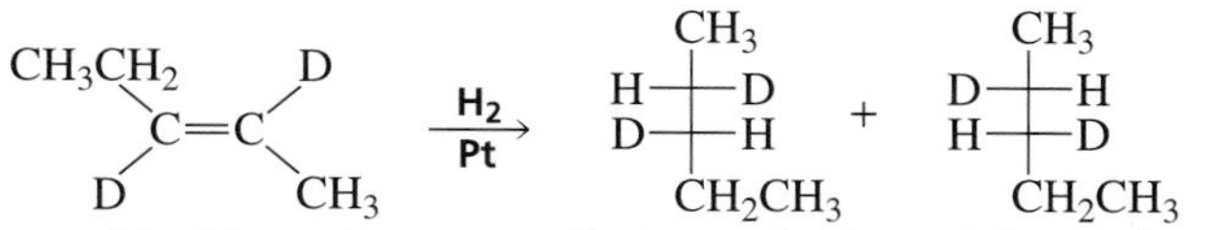

trans-2,3-dideuterio-
2-pentene

Fischer projections of the products
threo enantiomers

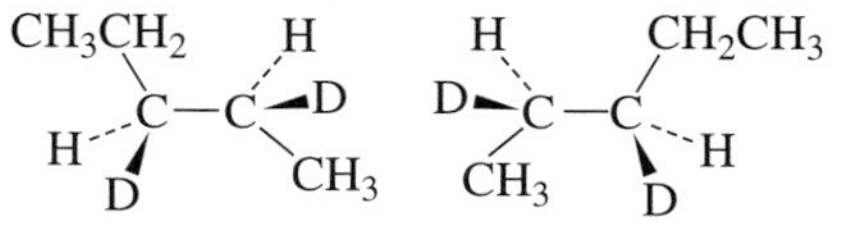

perspective formulas of the products

TRANS-SYN-THREO

It is a little easier to see the outcome of syn addition with cyclic compounds. Because addition of H_2 is syn, only the cis stereoisomers are formed.

1-isopropyl-2-methyl-cyclopentene

Syn addition of H_2 to 1,2-dideuteriocyclopentene forms only one product. Addition from above the plane of the double bond forms the same product as addition from below the plane. The product is a meso compound.

1,2-dideuterio-cyclopentene

CONFIGURATION OF CYCLOALKENES

Cyclic alkenes with fewer than eight carbon atoms, such as cyclopentene and cyclohexene, can exist only in the cis configuration because they do not have enough carbons to incorporate a trans double bond. Therefore, it is not necessary to use the "cis" designation with their names. Both cis and trans isomers are possible for rings containing eight or more carbons, so the configuration of the compound must be specified in its name.

cis-cyclooctene   *trans*-cyclooctene

3. Stereochemistry of Hydroboration–Oxidation

The addition of borane to an alkene is a concerted reaction. The boron and the hydride ion add to the two sp^2 carbons of the double bond at the same time (Section 3.17). Because the two species add simultaneously, they must add to the same side of the double bond. So the addition of borane to an alkene, like the addition of hydrogen, is a syn addition.

syn addition of borane

When the alkylborane is oxidized by reaction with hydrogen peroxide and hydroxide ion, the OH group ends up in the same position as the boron group that it replaces. Consequently, the overall hydroboration–oxidation reaction amounts to a syn addition of water to a carbon–carbon double bond.

an alkyl borane   **an alcohol**

Because only syn addition occurs, hydroboration–oxidation is stereospecific. As we saw when we looked at the addition of H_2 to a cycloalkene, syn addition results in the formation of only the pair of enantiomers that has the added groups on the same side of the ring.

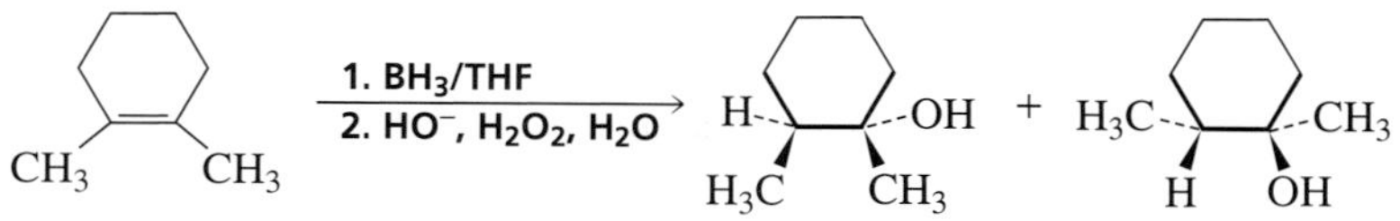

PROBLEM 42

Give the products that would be obtained from hydroboration–oxidation of the following compounds:

a. cyclohexene

b. 1-methylcyclohexene

c. 1,2-dimethylcyclopentene

d. *cis*-2-butene

PROBLEM 43

Reaction of 2-ethyl-1-pentene with Br_2, with H_2/Pt, or with BH_3 followed by HO^- + H_2O_2 leads to a racemic mixture. Explain why a racemic mixture is obtained in each case.

4. Addition Reactions That Involve a Bromonium Ion Intermediate

If two chirality centers are created as the result of an addition reaction that forms a bromonium ion intermediate, only one pair of enantiomers will be formed. In other words, the addition of Br_2 is a stereospecific reaction. Addition of Br_2 to the cis alkene forms only the threo pair of enantiomers.

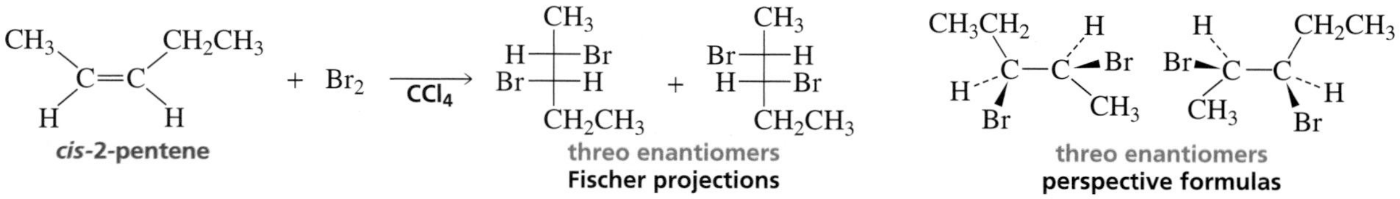

Addition of Br_2 to the trans alkene forms only the erythro pair of enantiomers.

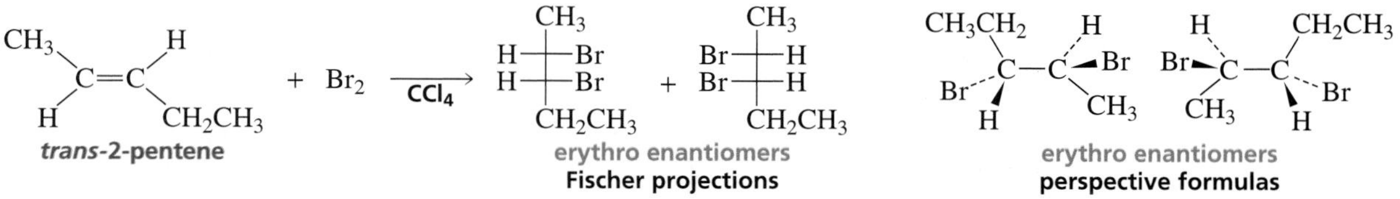

Since addition of Br_2 to the cis alkene forms the threo pair of enantiomers, we know that addition of Br_2 is an example of anti addition because, as we have just seen, syn addition to the cis alkene forms the erythro pair of enantiomers. The addition of Br_2 is anti because the reaction intermediate is a cyclic bromonium ion (Section 3.15). Once the cyclic bromonium ion forms, the bridged bromine atom blocks that side of the ion from further attack. So the negatively charged bromide ion must attack from the opposite side of the ion (following either the green arrows *or* the red arrows). Thus, the two bromine atoms add to opposite sides of the double bond. Because syn addition of Br_2 cannot occur, only two of the four possible stereoisomers are obtained.

CIS-ANTI-THREO

TRANS-ANTI-ERYTHRO

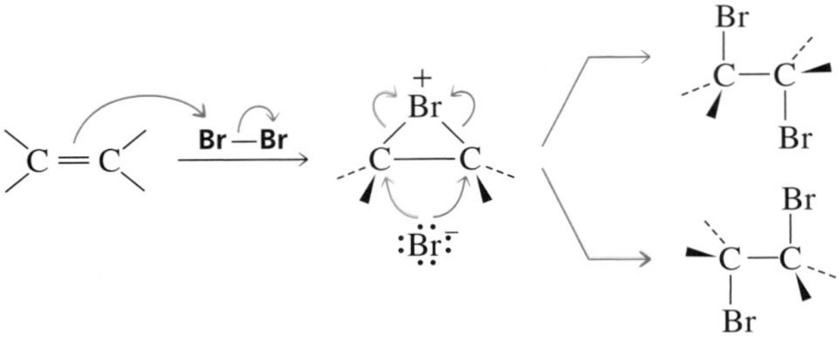

addition of Br_2 is an anti addition

If the two chirality centers in the product each have the same four substituents, the erythro isomers are identical and constitute a meso compound. Therefore, addition of Br_2 to *trans*-2-butene forms a meso compound.

***trans*-butene**

a meso compound
Fischer projection

a meso compound
perspective formula

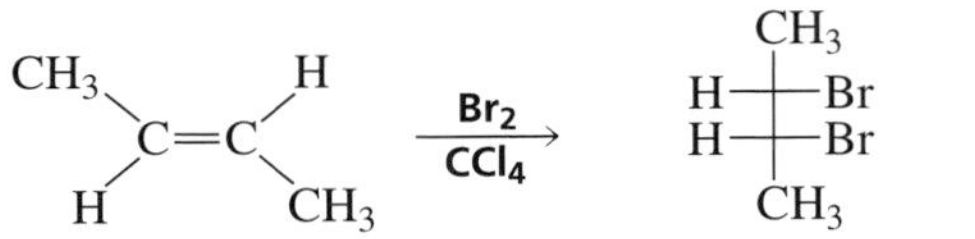

Because only anti addition occurs, addition of Br_2 to a cyclic alkene results in the formation of only the pair of enantiomers that has the added bromine atoms on opposite sides of the ring.

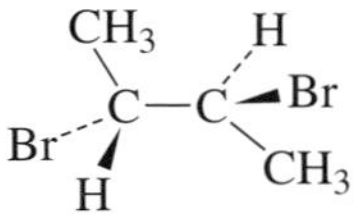

TABLE 4.3 Stereochemistry of Alkene Addition Reactions

Reaction	Type of addition	Products formed
Addition reactions that create one chirality center in the product		1. If the reactant does not have a chirality center; a pair of enantiomers will be obtained (equal amounts of the *R* and *S* isomers). 2. If the reactant has a chirality center, unequal amounts of a pair of diastereomers will be obtained.
Addition reactions that create two chirality centers in the product		
Addition of electrophilic reagents (for example, HCl, H_3O^+, HBr + peroxide	syn and anti	Four stereoisomers can be obtained[a] (the *cis* and *trans* isomers form the same products)
Addition of H_2 Addition of borane	syn	cis ⟶ erythro enantiomers[a] trans ⟶ threo enantiomers
Addition of Br_2	anti	cis ⟶ threo enantiomers trans ⟶ erythro enantiomers[a]

[a]If the two chirality centers have the same substituents, a meso compound will be obtained instead of the pair of erythro enantiomers.

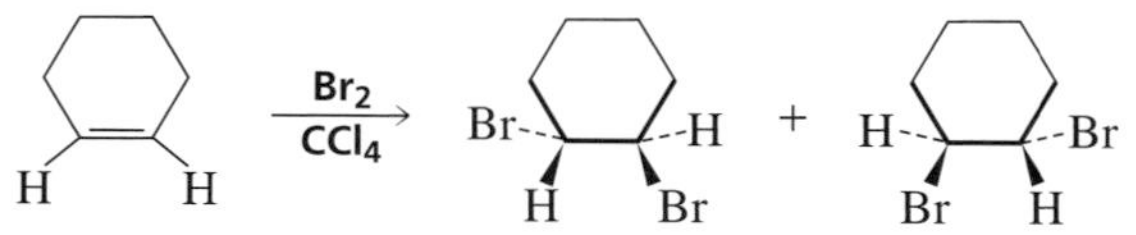

One way to determine what stereoisomers are obtained from a reaction that creates a product with two chirality centers is **"CIS-SYN-ERYTHRO,"** which is easy to remember since all three terms mean "on the same side." You can change any two of the terms, but you cannot change just one. (For example, **TRANS-ANTI-ERYTHRO, TRANS-SYN-THREO,** and **CIS-ANTI-THREO** are allowed, but TRANS-SYN-ERYTHRO is not allowed.)

A summary of the stereochemistry of the products obtained from addition reactions to alkenes is given in Table 4.3.

PROBLEM-SOLVING STRATEGY

Give the configuration of the products obtained from the following reactions.

a. 1-butene + HCl

b. 1-butene + HBr + peroxide

c. 3-methyl-3-hexene + HBr + peroxide

d. *cis*-3-heptene + Br$_2$

e. *trans*-3-heptene + Br$_2$

f. *trans*-3-hexene + Br$_2$

Start by drawing the product without regard to stereochemistry to check if the reaction has created any chirality centers. Then determine the configuration of the products, paying attention to the configuration (if any) of the reactant, how many chirality centers are formed, and the mechanism of the reaction. Let's start with **a.**

a. 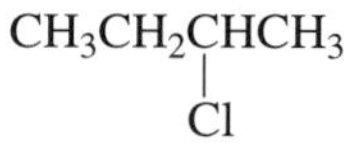 CH$_3$CH$_2$CHCH$_3$
           |
           Cl

The product has one chirality center, so equal amounts of the R and S enantiomers will be formed.

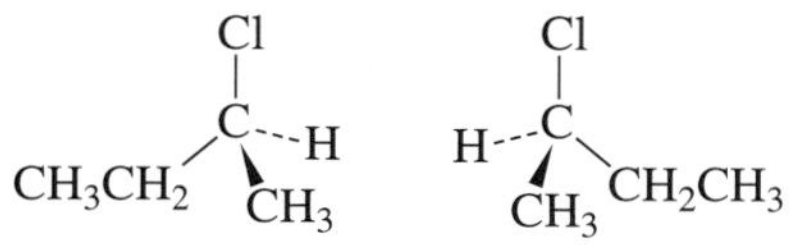

b. CH$_3$CH$_2$CH$_2$CH$_2$Br

The product does not have a chirality center, so there are no stereoisomers.

c. 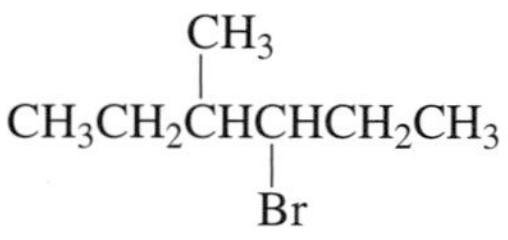 CH$_3$CH$_2$CHCHCH$_2$CH$_3$
               | |
              CH$_3$ Br

Two chirality centers have been created in the product. Because the reaction forms a radical intermediate, two pairs of enantiomers are formed.

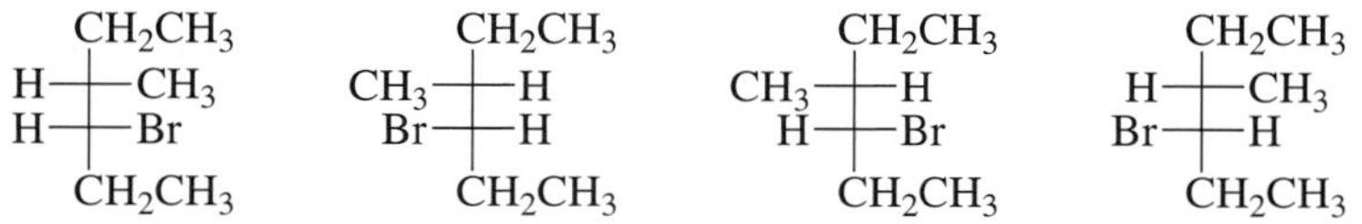

d. 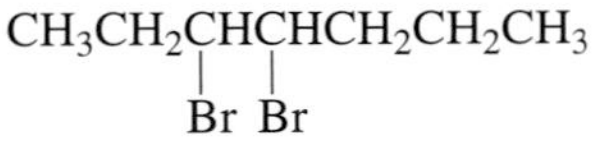 CH₃CH₂CHCHCH₂CH₂CH₃
　　　　　　|　|
　　　　　Br　Br

Two chirality centers have been created in the product. Because the reactant is cis and addition of Br₂ is anti, the threo pair of enantiomers is formed.

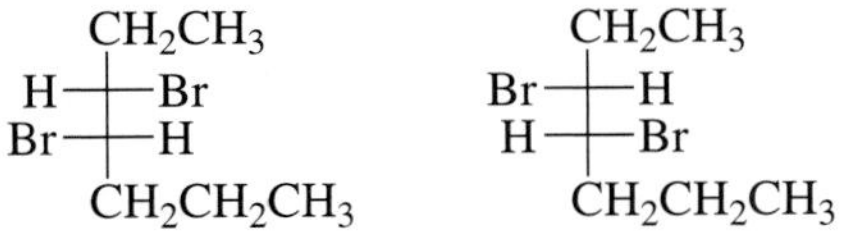

```
      CH₂CH₃              CH₂CH₃
   H──┬──Br            Br──┬──H
  Br──┴──H             H──┴──Br
    CH₂CH₂CH₃           CH₂CH₂CH₃
```

e. 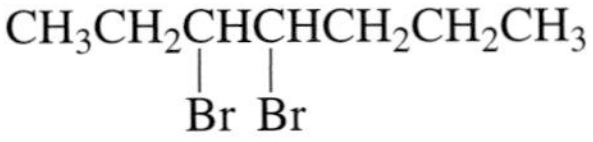 CH₃CH₂CHCHCH₂CH₂CH₃
　　　　　　|　|
　　　　　Br　Br

Two chirality centers have been created in the product. Because the reactant is trans and addition of Br₂ is anti, the erythro pair of enantiomers is formed.

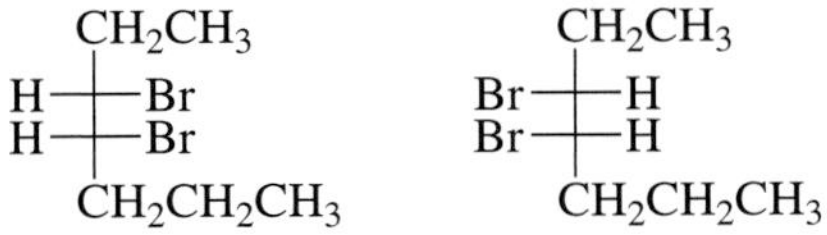

```
      CH₂CH₃              CH₂CH₃
   H──┬──Br            Br──┬──H
   H──┴──Br            Br──┴──H
    CH₂CH₂CH₃           CH₂CH₂CH₃
```

f. 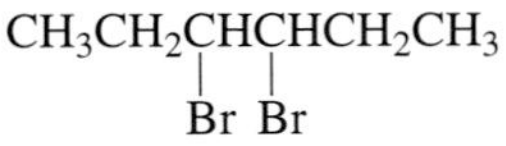 CH₃CH₂CHCHCH₂CH₃
　　　　　　|　|
　　　　　Br　Br

Two chirality centers have been created in the product. Because the reactant is trans and addition of Br₂ is anti, one would expect the erythro pair of enantiomers, but because the two chirality centers are bonded to the same four groups, the erythro product is a meso compound. Thus, only one stereoisomer is formed.

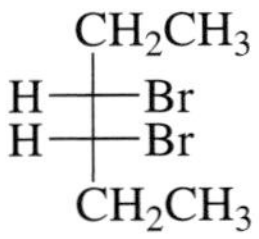

```
      CH₂CH₃
   H──┬──Br
   H──┴──Br
    CH₂CH₃
```

Now continue on to solve Problem 44.

PROBLEM 44

Give the stereochemistry of the products obtained from the following reactions.

a. *trans*-2-butene　+　HBr　+　peroxide

b. *cis*-3-hexene　+　HBr

c. (Z)-3-methyl-2-pentene　+　HBr

d. (Z)-3-methyl-2-pentene　+　HBr　+　peroxide

e. *cis*-2-pentene　+　Br₂

f. 1-hexene　+　Br₂

PROBLEM 45

What products would result from the addition of Br₂ to *cis*-2-pentene if the reaction occurred by way of a carbocation intermediate instead of by way of a bromonium ion intermediate?

PROBLEM 46

When Br_2 adds to an asymmetric alkene such as *cis*-2-heptene, equal amounts of the two threo enantiomers are obtained even though Br^- is more likely to attack the less sterically hindered carbon atom of the bromonium ion. Explain why equal amounts of the stereoisomers are obtained.

PROBLEM 47

What products would be obtained from the addition of Br_2 to cyclohexene if the solvent were H_2O instead of CCl_4? Give the mechanism of the reaction.

PROBLEM 48

What stereoisomers would you expect to obtain from each of the following reactions?

a.
$$\begin{array}{c} CH_3CH_2 \quad\quad CH_3 \\ C=C \\ CH_3 \quad\quad CH_2CH_3 \end{array} \xrightarrow[CCl_4]{Br_2}$$

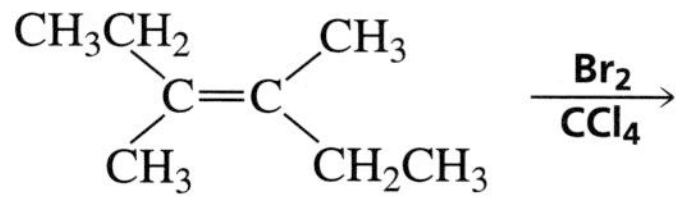

d.
$$\xrightarrow[CCl_4]{Br_2}$$
$CH_3 \quad CH_2CH_3$

b.
$$\begin{array}{c} CH_3CH_2 \quad\quad CH_3 \\ C=C \\ CH_3 \quad\quad CH_2CH_3 \end{array} \xrightarrow[Pt]{H_2}$$

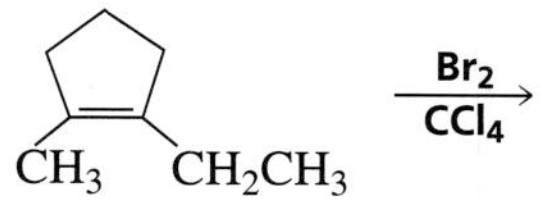

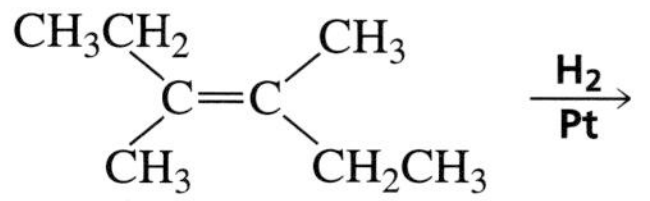

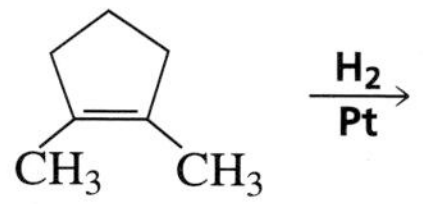

e.
$$\xrightarrow[Pt]{H_2}$$
$CH_3 \quad CH_3$

c.
$$\xrightarrow[CCl_4]{Br_2}$$
$CH_3 \quad CH_3$

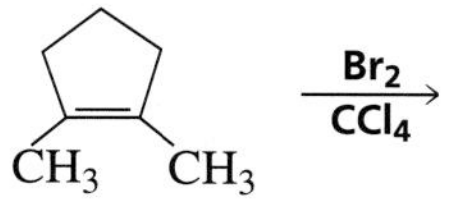

f.
$$\xrightarrow[Pt]{H_2}$$
$CH_3 \quad CH_2CH_3$

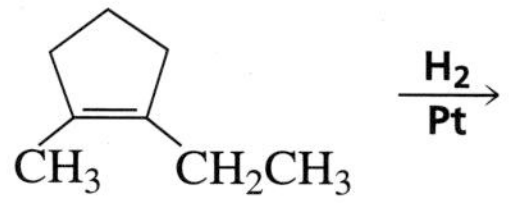

PROBLEM 49

a. What is the major product obtained from the reaction of propene and Br_2 plus excess Cl^-?

b. Indicate the relative amounts of the stereoisomers obtained.

Almost all the organic reactions that occur in biological systems are catalyzed by **enzymes.** Enzyme-catalyzed reactions are completely stereoselective: enzymes catalyze reactions that lead to the formation of only a single stereoisomer. For example, the enzyme-catalyzed addition of water to fumarate forms only (*S*)-malate. The *R* enantiomer is not formed.

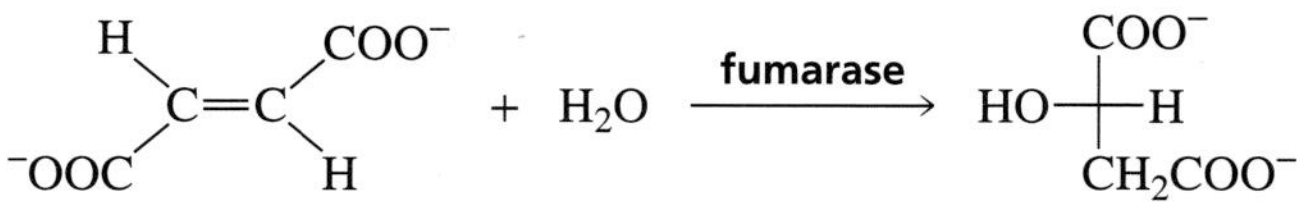

**4.20
STEREOCHEMISTRY
OF ENZYME-
CATALYZED
REACTIONS**

$$\begin{array}{c} H \quad\quad\quad COO^- \\ C=C \\ {}^-OOC \quad\quad H \end{array} + H_2O \xrightarrow{\text{fumarase}} \begin{array}{c} COO^- \\ HO-\!\!\!-H \\ CH_2COO^- \end{array}$$

fumarate   (S)-malate

Enzymes form only one stereoisomer because they possess a chiral binding site. The chiral binding site allows reagents to be delivered to only one side of the functional group of the compound. Consequently, only one stereoisomer is formed.

Enzyme-catalyzed reactions are also stereospecific. Enzymes typically react with only one stereoisomer. For example, fumarase catalyzes the addition of water to fumarate (the trans isomer) but not to maleate (the cis isomer).

Fundamental work on the stereochemistry of enzyme-catalyzed reactions was done by **Frank H. Westheimer.** *Westheimer was born in Baltimore in 1912, and received his graduate training at Harvard University. He served on the faculty of the University of Chicago and subsequently returned to Harvard as Professor of Chemistry. He also was a pioneer in the field of molecular mechanics.*

$$\begin{array}{c} ^-\text{OOC} \qquad\qquad \text{COO}^- \\ \diagdown \quad\quad / \\ \text{C}=\text{C} \\ / \qquad\quad \diagdown \\ \text{H} \qquad\qquad \text{H} \end{array} \quad + \quad H_2O \quad \xrightarrow{\textbf{fumarase}} \quad \text{no reaction}$$

maleate

Enzymes are able to differentiate between stereoisomeric reactants because of the chiral binding site (Section 4.12). A particular stereoisomer will bind to this site because it has substituents in the correct positions to interact with substituents in the binding site. Other stereoisomers do not have substituents in the proper positions. An enzyme's stereospecificity can be thought of as a right-handed glove being able to differentiate between a right hand and a left hand.

PROBLEM 50◆

a. What would be the product of the reaction of fumarate and H_2O if H^+ were used as a catalyst instead of fumarase?

b. What would be the product of the reaction of maleate and H_2O if H^+ were used as a catalyst instead of fumarase?

KEY TERMS

absolute configuration
   (page 206)
achiral (page 179)
amine inversion (page 176)
anti addition (page 220)
chiral (page 178)
chirality center (page 177)
cis fused (page 214)
cis isomer (page 176)
cis–trans isomers (page 175)
chromatography (page 202)
configuration (page 181)
configurational isomers
   (page 174)
conformational isomers (page 174)
conformer (page 194)
constitutional isomers (page 174)
dextrorotatory (page 186)
diastereomer (page 191)
diastereotopic hydrogens (page 208)
enantiomer (page 178)

enantiotopic hydrogens (page 207)
enzyme (page 227)
erythro enantiomers (page 190)
Fischer projection (page 180)
homotopic hydrogens (page 207)
isomers (page 174)
isomers that contain chirality centers
   (page 175)
levorotatory (page 186)
meso compound (page 195)
observed rotation (page 187)
optical purity (page 189)
optically active (page 186)
optically inactive (page 186)
pair of enantiomers (page 178)
perspective formula (page 180)
plane of symmetry (page 178)
polarimeter (page 186)
polarized light (page 185)
pro-*R*-hydrogen (page 207)
pro-*S*-hydrogen (page 208)

prochirality center (page 208)
racemate (page 188)
racemic mixture (page 188)
racemic modification (page 188)
R configuration (page 181)
regioselective (page 215)
relative configuration (page 205)
resolution of a racemic mixture
   (page 201)
S configuration (page 181)
specific rotation (page 187)
stereochemistry (page 215)
stereoisomers (page 174)
stereoselective (page 215)
stereospecific (page 216)
structural isomers (page 174)
syn addition (page 220)
threo enantiomers (page 190)
trans fused (page 214)
trans isomer (page 176)
wedge-and-dash structure (page 180)

PROBLEMS

51. Neglecting stereoisomers, give the structures of all compounds with molecular formula $C_5H_{11}Cl$. Which ones can exist as stereoisomers?

52. Neglecting stereoisomers, give the structures of all compounds with molecular formula C_5H_{10}. Which ones can exist as stereoisomers?

53. Draw all possible stereoisomers for each of the following compounds. If no stereoisomers are possible, so state.

a. 1-bromo-2-chlorocyclohexane
b. 2-bromo-4-methylpentane
c. 1,2-dichlorocyclohexane
d. 2-bromo-4-chloropentane
e. 3-heptene
f. 1-bromo-4-chlorocyclohexane

g. 1,2-dimethycyclopropane
h. 4-bromo-2-pentene
i. 3,3-dimethylpentane
j. 3-chloro-1-butene
k. 1-bromo-2-chlorocyclobutane
l. 1-bromo-3-chlorocyclobutane

54. a. Draw the staggered and eclipsed conformers for rotation about the C-2—C-3 bond of 1-chloropentane.
b. Which conformer is the most stable?
c. Which conformer is the least stable?

55. Name each of the following compounds using *R, S* and *E, Z* designations where necessary.

a.

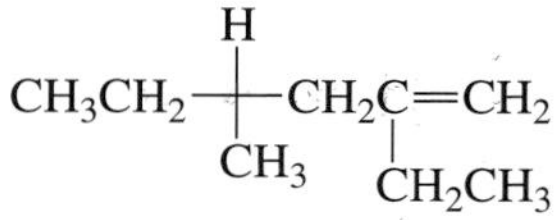

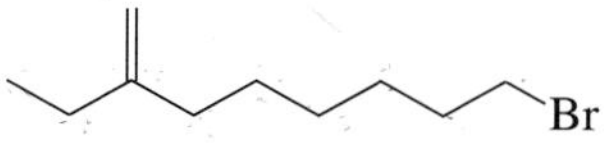

b.

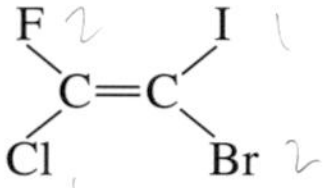

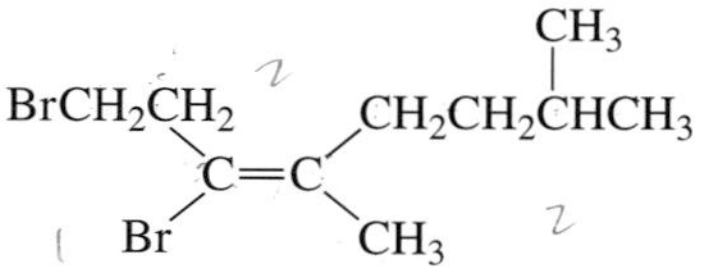

c.

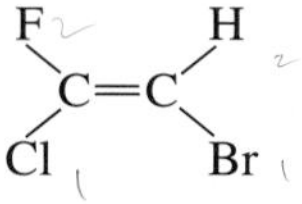

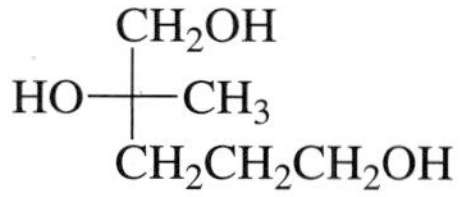

d.

e.

f.

g.

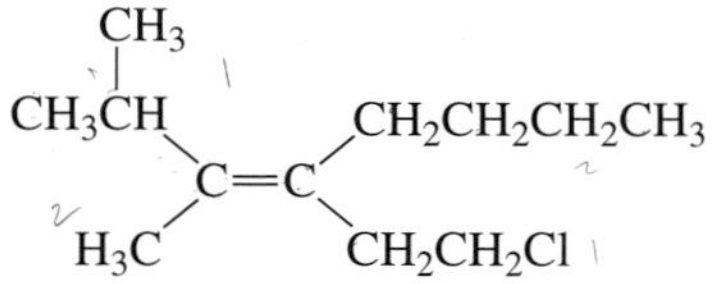

h.

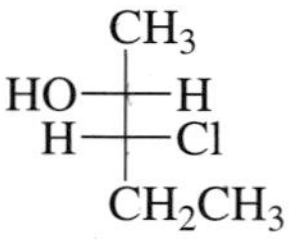

56. Indicate whether each of the following pairs of compounds are identical or are enantiomers, diastereomers, or constitutional isomers.

a.

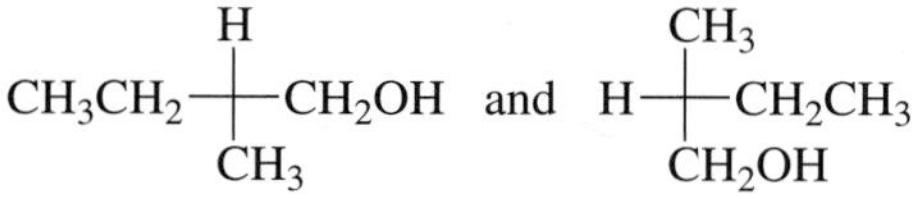

b.

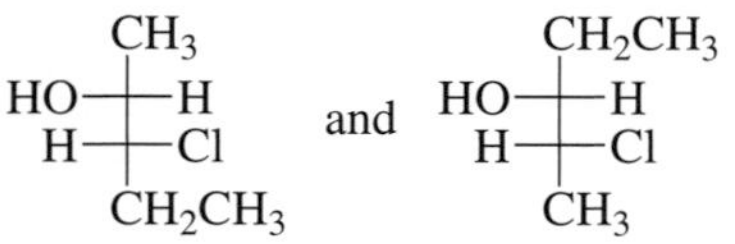

c.

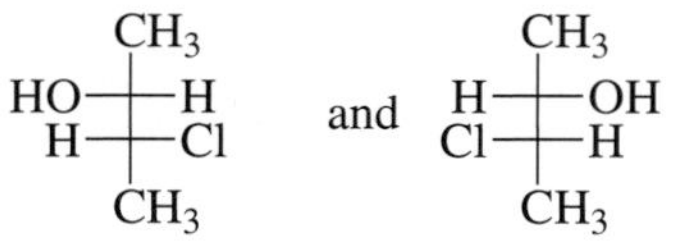

d.

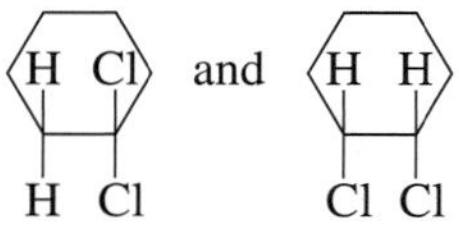

e. 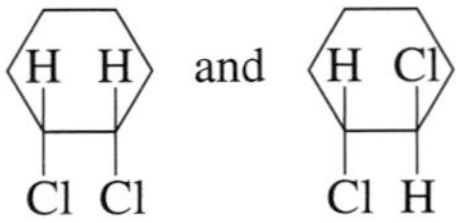 and

f. 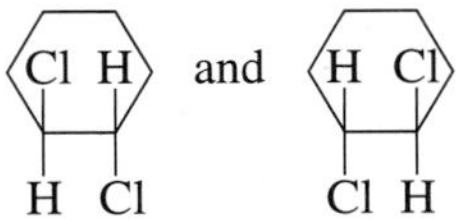 and

g.

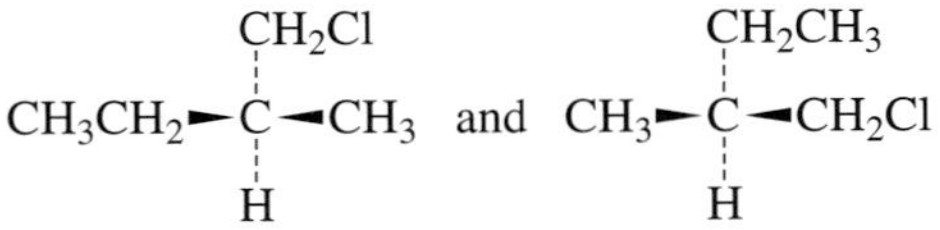

$$CH_3CH_2-\overset{\overset{\displaystyle CH_2Cl}{|}}{\underset{\underset{\displaystyle H}{|}}{C}}-CH_3 \quad \text{and} \quad CH_3-\overset{\overset{\displaystyle CH_2CH_3}{|}}{\underset{\underset{\displaystyle H}{|}}{C}}-CH_2Cl$$

h. and

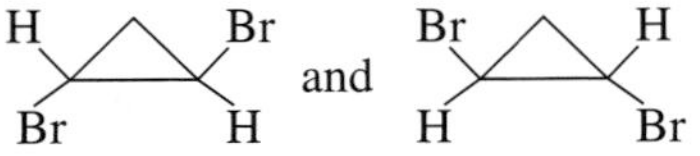

i. 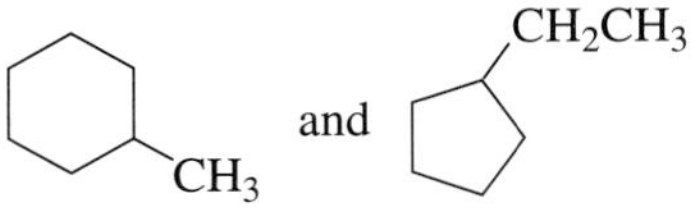 and

j. 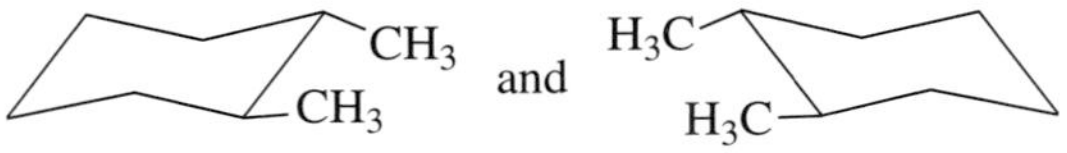 and

k. 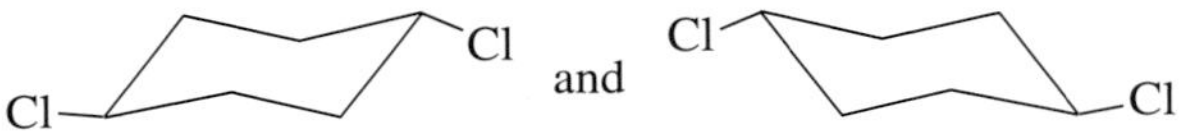 and

l. 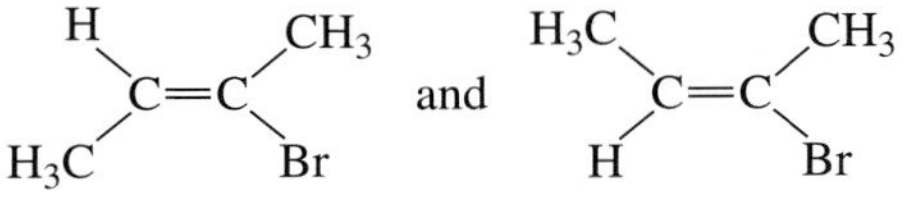 and

m. 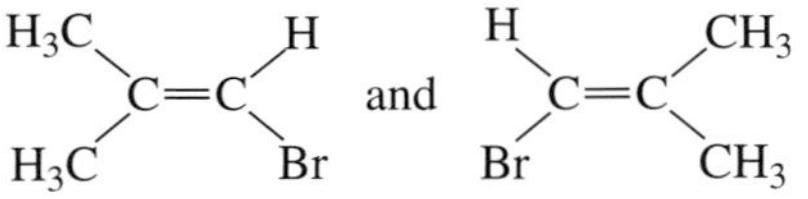 and

n.

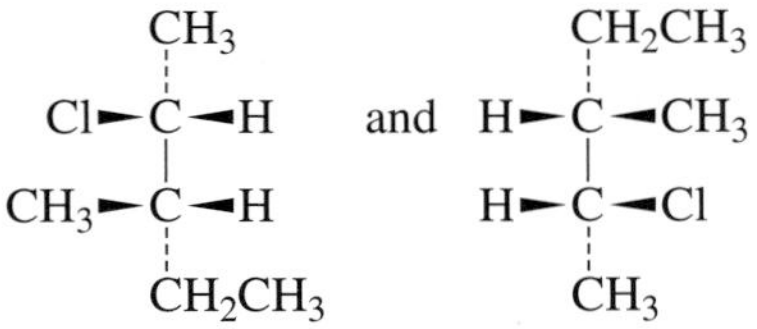

$$\overset{\overset{\displaystyle CH_3}{}}{Cl-\overset{|}{C}-H} \qquad \overset{\overset{\displaystyle CH_2CH_3}{}}{H-\overset{|}{C}-CH_3}$$
$$CH_3-\overset{|}{C}-H \qquad \text{and} \qquad H-\overset{|}{C}-Cl$$
$$\underset{\displaystyle CH_2CH_3}{} \qquad \underset{\displaystyle CH_3}{}$$

o. 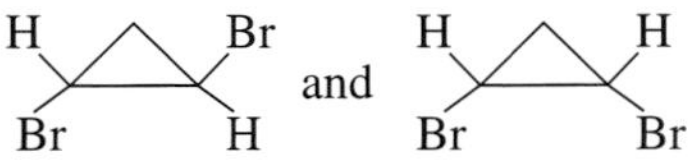 and

p. 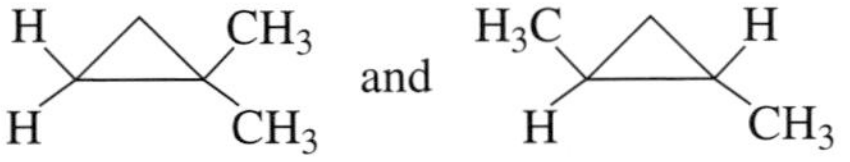 and

57. **a.** Give the product(s) that would be obtained from the reaction of *cis*-2-butene and *trans*-2-butene with each of the following reagents. If the products can exist as stereoisomers, show what stereoisomers are obtained.

   1. HCl
   2. BH$_3$ followed by HO$^-$, H$_2$O$_2$
   3. HBr + peroxide
   4. Br$_2$ in CCl$_4$
   5. Br$_2$ + H$_2$O
   6. H$_2$/Pt
   7. HCl + H$_2$O
   8. HCl + CH$_3$OH

   b. With which reagents do the two alkenes react to give different products?

58. Which of the following compounds have a stereoisomer that is achiral?
a. 2,3-dichlorobutane
b. 2,3-dichloropentane
c. 2,3-dichloro-2,3-dimethylbutane
d. 1,3-dichlorocyclopentane
e. 1,3-dibromocyclobutane
f. 2,4-dibromopentane
g. 2,3-dibromopentane
h. 1,4-dimethylcyclohexane
i. 1,2-dimethylcyclopentane
j. 1,2-dimethylcyclobutane

59. Citrate synthase, one of the enzymes in the series of enzyme-catalyzed reactions known as the Krebs cycle, catalyzes the synthesis of citric acid from oxaloacetic acid and acetyl-CoA. If the synthesis is carried out with acetyl-CoA that has radioactive carbon (^{14}C) in the indicated position, the isomer shown below is obtained.

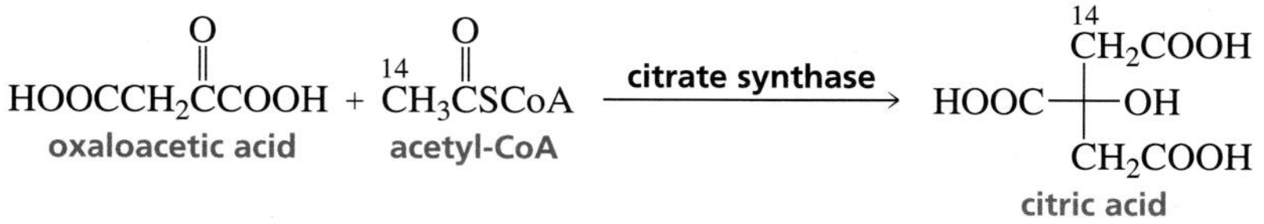

$$\underset{\text{oxaloacetic acid}}{\text{HOOCCH}_2\overset{\text{O}}{\overset{\|}{\text{C}}}\text{COOH}} + \underset{\text{acetyl-CoA}}{\overset{14}{\text{CH}_3}\overset{\text{O}}{\overset{\|}{\text{C}}}\text{SCoA}} \xrightarrow{\text{citrate synthase}} \underset{\text{citric acid}}{\text{HOOC}\overset{\overset{14}{\text{CH}_2\text{COOH}}}{\underset{\text{CH}_2\text{COOH}}{\mid}}\text{OH}}$$

a. Which stereoisomer of citric acid is synthesized?
b. Why is the other stereoisomer not obtained?
c. If the acetyl-CoA used in the synthesis does not contain ^{14}C, will the product of the reaction be chiral or achiral?

60. Give the products of the following reactions. If the products can exist as stereoisomers, show what stereoisomers are obtained.
a. *cis*-2-pentene + HCl
b. *trans*-2-pentene + HCl
c. 1-ethylcyclohexene + H_3O^+
d. 1,2-diethylcyclohexene + H_3O^+
e. 1,2-dideuteriocyclohexene + H_2/Pt
f. 1,2-dimethylcyclohexene + HCl
g. 3,3-dimethyl-1-pentene + Br_2/CCl_4
h. (*E*)-3,4-dimethyl-3-heptene + H_2/Pt
i. (*Z*)-3,4-dimethyl-3-heptene + H_2/Pt
j. 1-chloro-2-ethylcyclohexene + H_2/Pt
k. *cis*-2-pentene + Br_2/CCl_4
l. *trans*-2-pentene + Br_2/CCl_4
m. 1-butene + HCl
n. 1-butene + HBr + peroxide
o. *trans*-3-hexene + Br_2/CCl_4
p. *cis*-3-hexene + Br_2/CCl_4
q. 3,3-dimethyl-1-pentene + HBr
r. *cis*-2-butene + HBr + peroxide
s. (*Z*)-2,3-dichloro-2-butene + H_2/Pt
t. (*E*)-2,3-dichloro-2-butene + H_2/Pt
u. (*Z*)-3,4-dimethyl-3-hexene + H_2/Pt
v. (*E*)-3,4-dimethyl-3-hexene + H_2/Pt

61. Indicate whether each of the following structures is (*R*)-2-chlorobutane or (*S*)-2-chlorobutane. (Use models, if necessary.)

a.

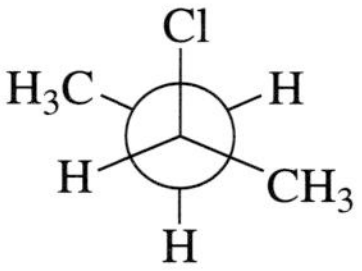

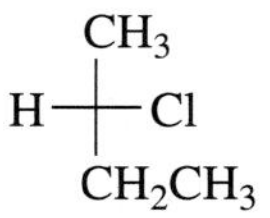

b.

c.

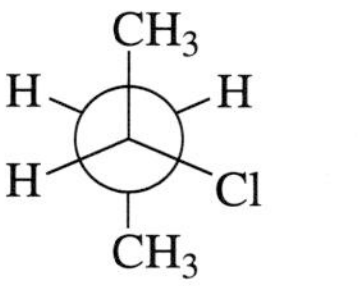

d.

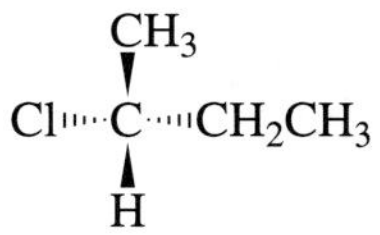

e. 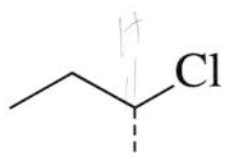   f.

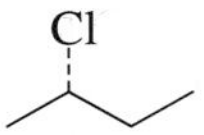

62. A solution of an unknown compound (3.0 g of X per 20 mL of solution), when placed in a polarimeter tube 2.0 dm long, was found to rotate the plane of polarized light 1.8° in a counterclockwise direction. What is the specific rotation of X?

63. Mevacor is used clinically to lower serum cholesterol levels. How many chirality centers does Mevacor have?

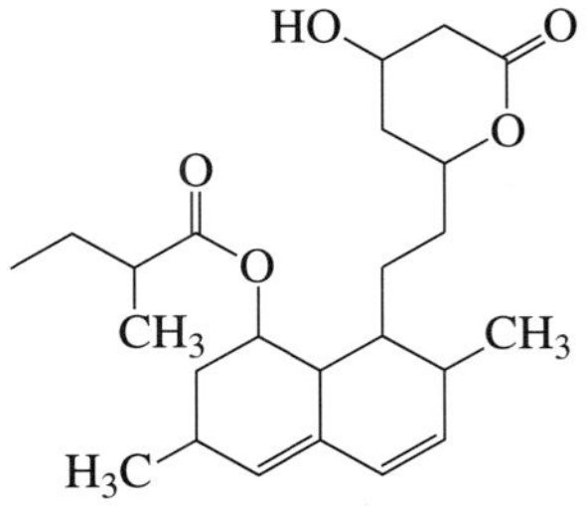

Mevacor

64. Which of the following objects are chiral?
    a. a mug with DAD written on one side
    b. a mug with MOM written on one side
    c. a mug with DAD written opposite the handle
    d. a mug with MOM written opposite the handle
    e. a nail
    f. a wine glass
    g. a screw

65. Explain how *R* and *S* are related to (+) and (−).

66. a. Draw all possible isomers for the following compound.

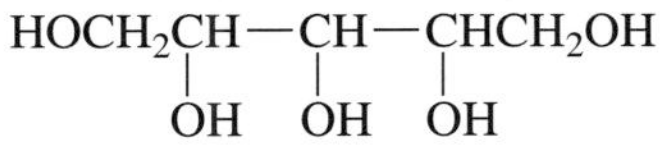

$$HOCH_2CH\!-\!CH\!-\!CHCH_2OH$$
$$\quad\quad\;\; OH \quad OH \quad OH$$

  b. Which isomers are optically inactive (will not rotate plane-polarized light)?

67. Butaclamol is a potent antipsychotic that has been used clinically in the treatment of schizophrenia. Although patients are given a racemic mixture of the drug, only the (+)-enantiomer has pharmacological activity. How many chirality centers does Butaclamol have?

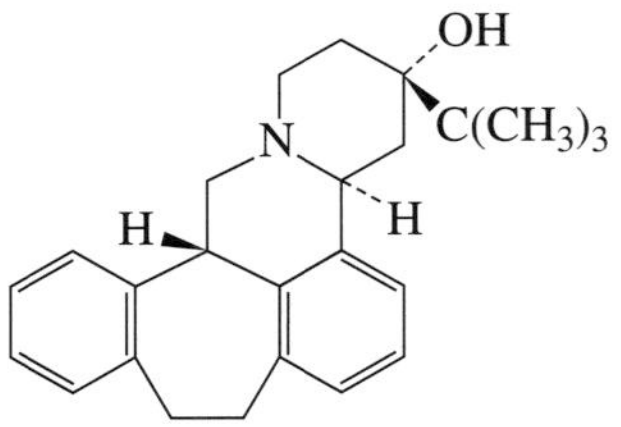

Butaclamol

68. Indicate the configuration of the chirality centers in the following molecules.

a. 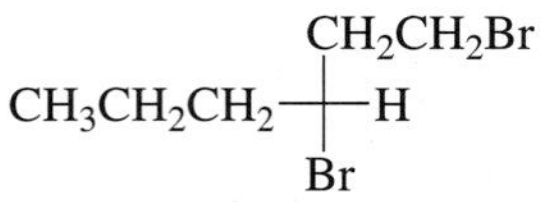 $CH_3CH_2CH_2\!-\!\overset{\displaystyle CH_2CH_2Br}{\underset{\displaystyle Br}{|}}\!-\!H$

b. 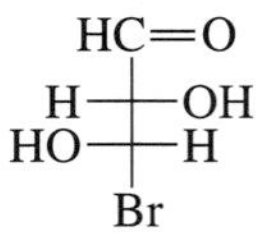 $\overset{\displaystyle HC=O}{\underset{\displaystyle Br}{\underset{\displaystyle |}{HO\!-\!\overset{H\,-\,OH}{|}\!-\!H}}}$

c. 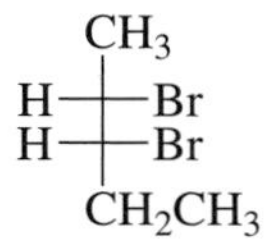

69. **a.** Draw all the isomers with molecular formula C_6H_{12} that contain a cyclobutane ring. (*Hint:* There are seven.)
b. Name the compounds without specifying the configuration of any chirality centers.
c. Identify:
1. constitutional isomers
2. stereoisomers
3. cis–trans isomers
4. chiral compounds
5. achiral compounds
6. meso compounds
7. enantiomers
8. diastereomers

70. Cholesterol has a specific rotation of −39.0°. A solution of a cholesterol sample (0.187 g/mL) has an observed rotation of −6.52° when placed in a polarimeter tube 10 cm long. What is the percent optical purity of cholesterol sample in the sample?

71. Draw structures for each of the following molecules.
a. (*S*)-1-bromo-1-chlorobutane
b. (2*R*,3*R*)-2,3-dichloropentane
c. an achiral isomer of 1,2-dimethylcyclohexane
d. a chiral isomer of 1,2-dibromocyclobutane
e. two achiral isomers of 3,4,5-trimethylheptane

72. Of the possible monobromination products shown for the following reaction, circle any that would not be formed.

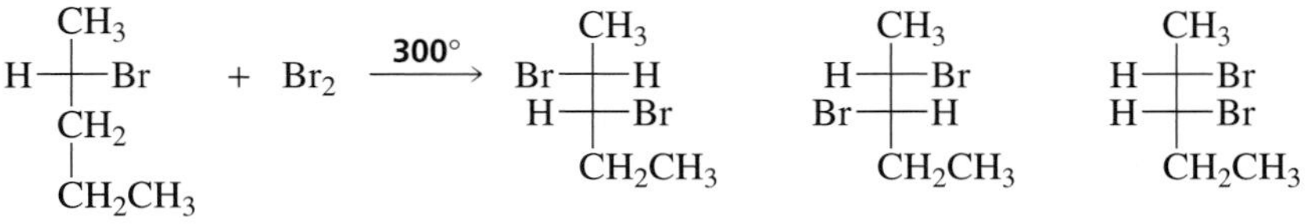

73. A sample of (*S*)-(+)-lactic acid was found to have an optical purity of 72%. What percentage of the *R* isomer is present in the sample?

74. Give the products and their configuration:

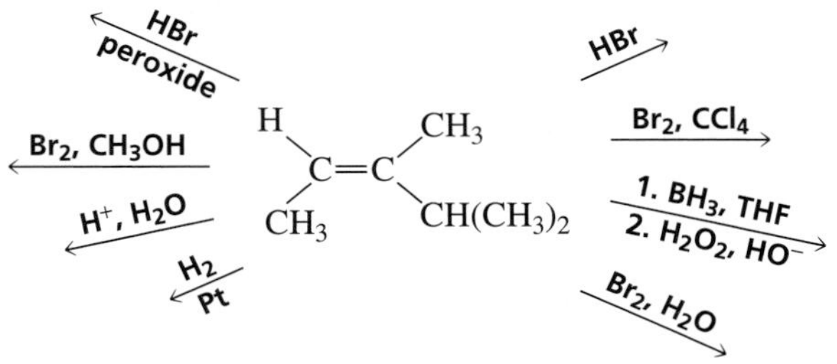

75. **a.** Using the wedge-and-dash notation, draw the nine configurational isomers of 1,2,3,4,5,6-hexachlorocyclohexane.
b. From the nine configurational isomers, identify one pair of enantiomers.
c. Draw the most stable conformation of the most stable configurational isomer.

76. Sherry O. Eismer decided that the configuration of the chirality centers in a sugar such as D-glucose could be determined rapidly by simply assigning the *R* configuration to a chirality center with an OH group on the right and the *S* configuration to a chirality center with an OH group on the left. Is she correct?

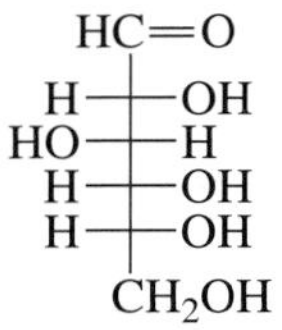

$$
\begin{array}{c}
HC=O \\
H \!\!-\!\!|\!\!-\!\! OH \\
HO \!\!-\!\!|\!\!-\!\! H \\
H \!\!-\!\!|\!\!-\!\! OH \\
H \!\!-\!\!|\!\!-\!\! OH \\
CH_2OH
\end{array}
$$

D-glucose

77. Cyclohexene exists only in the cis form, while cyclodecene exists in both the cis and trans forms. Explain. (Molecular models are helpful for this problem.)

78. When fumarate reacts with D_2O in the presence of the enzyme fumarase, only one isomer of the product is formed. Its structure is shown below. Is the enzyme catalyzing a syn or an anti addition of D_2O?

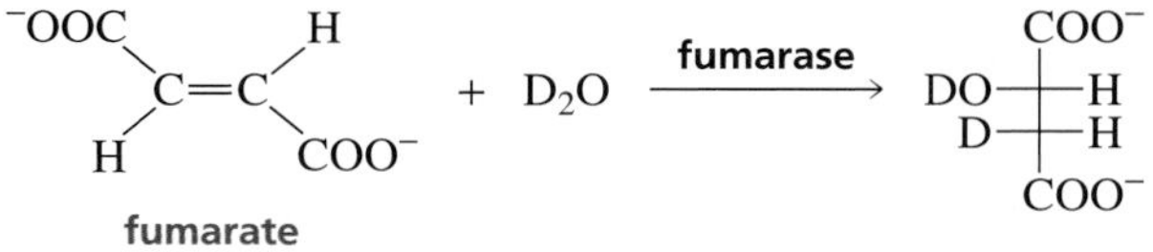

fumarate

79. Two stereoisomers are obtained from the reaction of HBr with (3*S*,4*S*)-4-bromo-3-methyl-1-pentene. One of the stereoisomers is optically active, and the other is not. Give the structures of the stereoisomers, indicating their absolute configurations, and explain the difference in optical properties.

80. When (*S*)-(+)-1-chloro-2-methylbutane reacts with chlorine, one of the products that is formed is (−)-1,4-dichloro-2-methylbutane. Does this product have the *R* or the *S* configuration?

81. Indicate the configuration of the chirality centers in the following molecules.

a.

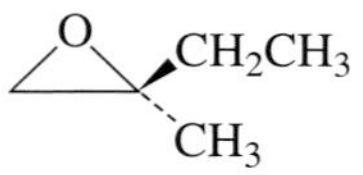

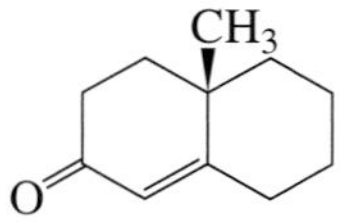

c.

b.

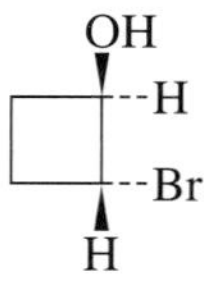

82. **a.** Does the following compound have any chirality centers?

$$CH_3CH=C=CHCH_3$$

b. Is it chiral? (*Hint:* Make models.)

REACTIONS OF ALKYNES. INTRODUCTION TO MULTISTEP SYNTHESIS

1-butyne + 2HCl $\longrightarrow$ 2,2-dichlorobutane

A **lkynes** are hydrocarbons that contain a carbon–carbon triple bond. Because of its triple bond, an alkyne has four fewer hydrogens than does a corresponding alkane. Therefore, the general molecular formula for a noncyclic alkyne is C_nH_{2n-2}.

There are not many naturally occurring alkynes. Examples include capillin, which has fungicidal activity, and ichthyothereol, a convulsant used by the Amazon Indians for poisoned arrowheads. A naturally occurring alkyne generally has at least one other functional group in addition to one or more triple bonds.

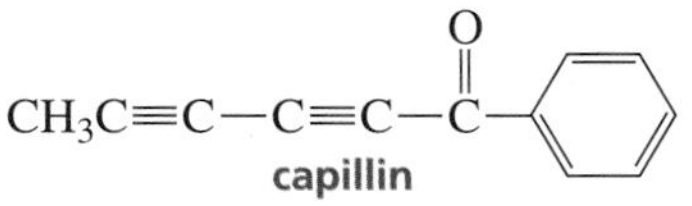

$CH_3C{\equiv}C{-}C{\equiv}C{-}\overset{\displaystyle O}{\overset{\|}{C}}{-}\bigcirc$
capillin

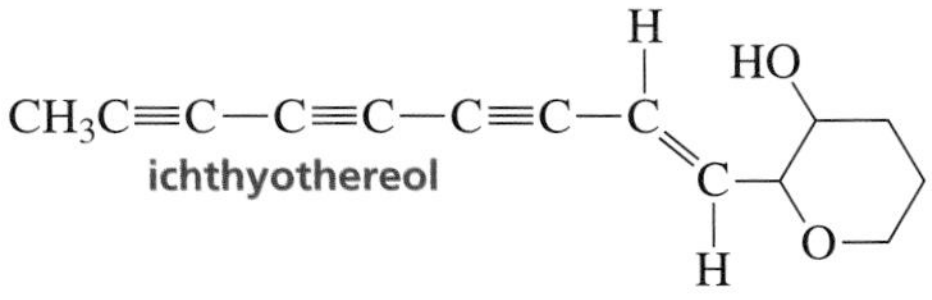

$CH_3C{\equiv}C{-}C{\equiv}C{-}C{\equiv}C{-}\overset{\displaystyle H}{\underset{\displaystyle C}{C}}$ ...
ichthyothereol

There are a few drugs that contain alkyne functional groups; these drugs, however, are synthetic compounds. They exist only because chemists have been able to synthesize them. Their trade names are shown in green in the diagrams below. Trade names are always capitalized and can be used for commercial purposes only by the owner of the registered trademark (Section 29.1).

Parsal
Sinovial

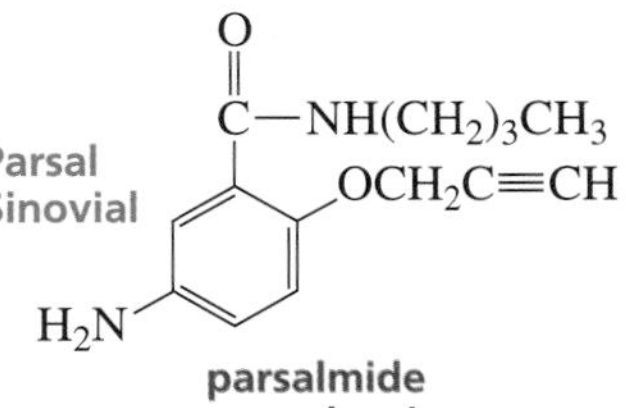

parsalmide
an analgesic

Eudatin
Supirdyl

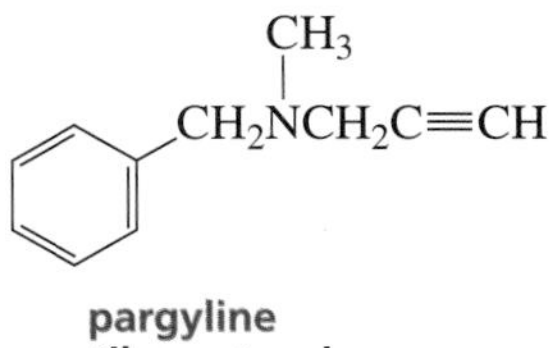

pargyline
an antihypertensive

Norquen
Ovastol

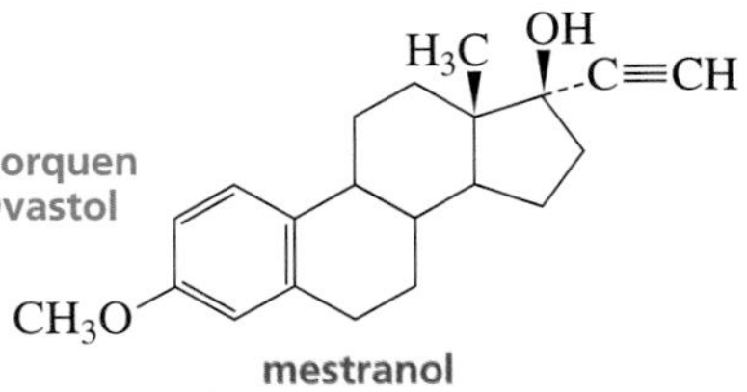

mestranol
a component in oral contraceptives

Acetylene (HC≡CH), the common name for the smallest alkyne, is probably a familiar word to you because of the oxyacetylene torches used in welding. Acetylene is supplied to the torch from one high-pressure gas tank, and oxygen is supplied from another. Burning acetylene produces a high-temperature flame capable of melting and vaporizing iron and steel.

PROBLEM 1◆

What is the molecular formula for a cyclic alkyne with 14 carbons and 2 triple bonds?

5.1 NOMENCLATURE OF ALKYNES

To name an alkyne by the IUPAC system of nomenclature, the "ane" ending of the alkane name is replaced with the functional group suffix "yne." Analogous to naming compounds with other functional groups, the longest continuous chain containing the carbon–carbon triple bond is numbered in a direction that gives the alkyne functional group the lowest possible number. If the triple bond is at the end of the chain, the alkyne is classified as a **terminal alkyne.** Alkynes with triple bonds located elsewhere along the chain are called **internal alkynes.** For example, 1-butyne is a terminal alkyne; 2-pentyne is an internal alkyne.

IUPAC: HC≡CH   ethyne   CH₃CH₂C≡CH 1-butyne   CH₃C≡CCH₂CH₃ 2-pentyne   CH₃CHC≡CCH₃ 4-methyl-2-hexyne
Common: acetylene   ethylacetylene / a terminal alkyne   ethylmethylacetylene / an internal alkyne   sec-butylmethyl-acetylene

1-hexyne

3-hexyne

In common nomenclature, alkynes are named as substituted acetylenes. The common name is obtained by naming the alkyl groups, in alphabetical order, that have replaced the hydrogens of acetylene. Acetylene is an unfortunate common name for the smallest alkyne because its "ene" ending is characteristic of a double bond rather than a triple bond.

If the same number for the alkyne functional group is obtained in both directions along the carbon chain, the correct IUPAC name is the one that contains the lowest substituent number. If the compound contains more than one substituent, the substituents are listed in alphabetical order.

CH₃CHCHC≡CCH₂CH₂CH₃
3-bromo-2-chloro-4-octyne
not 6-bromo-7-chloro-4-octyne
because 2 < 6

CH₃CHC≡CCH₂CH₂Br
1-bromo-5-methyl-3-hexyne
not 6-bromo-2-methyl-3-hexyne
because 1 < 2

The triple-bond-containing propargyl group is used in common nomenclature. It is analogous to the double-bond-containing allyl group that we saw in Section 3.2.

HC≡CCH₂—   **propargyl group**   H₂C=CHCH₂—   **allyl group**

HC≡CCH₂Br   **propargyl bromide**   H₂C=CHCH₂OH   **allyl alcohol**

A substituent receives the lowest possible number only if there is no functional group or if the same number for the functional group is obtained in both directions.

PROBLEM 2

Draw the structure for each of the following compounds.

a. 1-chloro-3-hexyne

b. cyclooctyne

c. isopropylacetylene

d. propargyl chloride

e. 4,4-dimethyl-1-pentyne

f. dimethylacetylene

PROBLEM 3 ◆

Give the IUPAC name for each of the following compounds.

$$CH_2CH_2CH_3$$

a. $BrCH_2CH_2C{\equiv}CCH_3$

b. $CH_3CH_2CHC{\equiv}CCH_2CHCH_3$
       | |
       Br   Cl

c. $CH_3CH_2CHC{\equiv}CH$

d. $CH_3OCH_2C{\equiv}CCH_2CH_3$

PROBLEM 4

Draw the structures and give the common and IUPAC names for the seven alkynes with molecular formula C_6H_{10}.

PROBLEM 5 ◆

Which would you expect to be more stable, an internal alkyne or a terminal alkyne? Why?

5.2 PHYSICAL PROPERTIES OF UNSATURATED HYDROCARBONS

All hydrocarbons have similar physical properties. In other words, alkenes and alkynes have physical properties similar to those of alkanes (Section 2.9). All are insoluble in water and all are soluble in solvents with low polarity such as benzene and ether. They are less dense than water and, like other homologous series, have boiling points that increase with increasing molecular weight (Table 5.1). Alkynes are more linear than alkenes. This gives them stronger van der Waals interactions and therefore higher boiling points.

Internal alkenes have higher boiling points than terminal alkenes. Similarly, internal alkynes have higher boiling points than terminal alkynes. The boiling point of *cis*-2-butene is a slightly higher than that of *trans*-2-butene, because the cis isomer has a small dipole moment while the dipole moment of the trans isomer is zero (Section 3.4).

5.3 THE STRUCTURE OF ALKYNES

The structure of ethyne was discussed in Section 1.9. Each of the carbons in ethyne has two *sp* orbitals. One overlaps with the *s* orbital of a hydrogen, and the other overlaps with an *sp* orbital of the other carbon (Figure 5.1). Because the *sp* orbitals are oriented as far from each other as possible to minimize electron repulsion, ethyne is a linear molecule with bond angles of 180°.

Figure 5.1 ▶
(a) Each of the carbon–carbon π bonds is formed by side-to-side overlap of a p orbital of one carbon with a p orbital of the adjacent carbon.
(b) The triple bond can be represented by a cylinder of electrons.

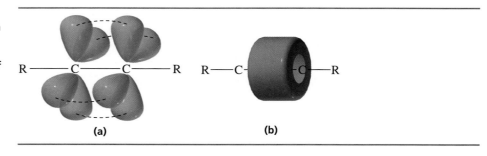

(a)   (b)

Each carbon has two remaining p orbitals that stand at right angles to one another and to the sp orbitals. The two p orbitals on one carbon overlap with the parallel p orbitals on the other carbon to form two π bonds. Consequently, a triple bond consists of a σ bond (formed by sp–sp orbital overlap) and two π bonds (formed by side-to-side p–p orbital overlap). One pair of overlapping p orbitals results in a cloud of electrons above and below the σ bond, and the other pair results in a cloud of electrons in front of and behind the σ bond. The end result can be thought of as a cylinder of electrons wrapped around the σ bond (Section 1.9).

If we subtract the σ bond energy from the total bond energy of the triple bond ($200 - 88 = 112$ kcal/mol), we find that the average bond strength of each of the π bonds in the triple bond is 56 kcal/mol (234 J/mol). The π bonds, therefore, are much weaker than the σ bond. These weak π bonds allow alkynes to react easily.

TABLE 5.1	Boiling Points of the Smallest Hydrocarbons				
	bp (°C)		**bp (°C)**		**bp (°C)**
CH_3CH_3 ethane	−88.6	$H_2C{=}CH_2$ ethene	−104	$HC{\equiv}CH$ ethyne	−84
$CH_3CH_2CH_3$ propane	−42.1	$CH_3CH{=}CH_2$ propene	−47	$CH_3C{\equiv}CH$ propyne	−23
$CH_3CH_2CH_2CH_3$ butane	−0.5	$CH_3CH_2CH{=}CH_2$ 1-butene	−6.5	$CH_3CH_2C{\equiv}CH$ 1-butyne	8
$CH_3(CH_2)_3CH_3$ pentane	36.1	$CH_3CH_2CH_2CH{=}CH_2$ 1-pentene	30	$CH_3CH_2CH_2C{\equiv}CH$ 1-pentyne	39
$CH_3(CH_2)_4CH_3$ hexane	68.7	$CH_3CH_2CH_2CH_2CH{=}CH_2$ 1-hexene	63.5	$CH_3CH_2CH_2CH_2C{\equiv}CH$ 1-hexyne	71
		$CH_3CH{=}CHCH_3$ cis-2-butene	3.7	$CH_3C{\equiv}CCH_3$ 2-butyne	27
		$CH_3CH{=}CHCH_3$ trans-2-butene	0.9	$CH_3CH_2C{\equiv}CCH_3$ 2-pentyne	55
		$CH_3CH_2CH{=}CHCH_3$ cis-2-pentene	36.9	$CH_3CH_2CH_2C{\equiv}CCH_3$ 2-hexyne	84
		$CH_3CH_2CH{=}CHCH_3$ trans-2-pentene	36.4	$CH_3CH_2C{\equiv}CCH_2CH_3$ 3-hexyne	81

With a cylindrical cloud of electrons surrounding the σ bond, an alkyne is an electron-rich molecule. In other words, it is a nucleophile. Because it is a nucleophile, we expect it to react with an electrophile. We just saw that an alkyne's π bonds are weak compared with its σ bond. Consequently, if an electrophilic reagent such as hydrogen chloride is added to an alkyne, a π bond will break because the electrons are attracted to the electrophilic hydrogen. We might, therefore, expect a vinylic cation to be formed, which would then react rapidly with the nucleophilic chloride ion. A **vinylic cation** has a positive charge on a vinylic carbon.

**5.4
REACTIVITY
CONSIDERATIONS**

$$HC\equiv CH \ + \ H-Cl \xrightarrow{\text{slow}} \left[HC\overset{+}{=}CH_2 \right] \ + \ :\overset{..}{\underset{..}{Cl}}:^- \xrightarrow{\text{fast}} \underset{\displaystyle HC=CH_2}{\overset{\displaystyle Cl}{|}}$$

a vinylic cation

However, we learned in Section 3.12 that primary carbocations are so unstable that they form only with great difficulty. Primary vinylic cations are even more unstable than primary alkyl carbocations. Thus it seems unlikely that a vinylic cation is ever formed. A more reasonable mechanism involves the formation of a complex between the electrophile and the triple bond. This complex is called a **pi-complex.** The pi-complex is similar to the cyclic bromonium ion that is formed when Br_2 reacts with an alkene (Section 3.15). Attack on the pi-complex by the nucleophilic chloride ion forms the expected addition product.

$$HC\equiv CH \ + \ H-Cl \longrightarrow \underset{\textbf{a pi-complex}}{HC\overset{\overset{\displaystyle \overset{-\delta}{Cl}}{\vdots}\atop{\overset{\displaystyle H^{+\delta}}{\vdots}}}{\equiv}CH} \xrightarrow[:\overset{..}{\underset{..}{Cl}}:^-]{} \underset{\displaystyle Cl}{\overset{\displaystyle HC=CH_2}{|}}$$

Why is a vinylic cation less stable than a similarly substituted alkyl cation? In other words, why is a secondary vinylic cation less stable than a secondary alkyl cation, and a primary vinylic cation less stable than a primary alkyl cation?

relative stabilities of carbocations

The decreased stability of a vinylic cation results from the fact that the positive charge is on a more electronegative carbon. The positive charge in a vinylic cation is on an sp carbon, while the positive charge on an alkyl cation is on an sp^2 carbon; sp carbons are more electronegative than are sp^2 carbons (Section 5.9). In addition, hyperconjugation is less effective in stabilizing the charge on a vinylic cation.

That a pi-complex rather than a carbocation is formed as an intermediate when an electrophilic reagent such as HCl adds to an alkyne is substantiated by two observations: carbocation rearrangements do not occur, and the reaction is stereoselective. For example, the addition of HCl to 2-butyne forms only (Z)-2-chloro-2-butene, which means that only anti addition of H and Cl occurs.

$$CH_3C\equiv CCH_3 \ + \ HCl \ \longrightarrow$$

2-butyne

(Z)-2-chloro-2-butene

Thus alkynes, like alkenes, undergo electrophilic addition reactions. We will see that the same electrophilic reagents that add to alkenes also add to alkynes and that—again like alkenes—electrophilic addition to alkynes is regioselective, following Markovnikov's rule. There is, however, an added complication in the reactions of alkynes. Because the product of the addition of an electrophilic reagent to an alkyne is an alkene, a second electrophilic addition reaction can occur if there is an excess of the electrophilic reagent.

$$CH_3C\equiv CCH_3 \ \xrightarrow{\textbf{HCl}} \ CH_3\overset{Cl}{\underset{}{C}}=CHCH_3 \ \xrightarrow{\textbf{HCl}} \ CH_3\overset{Cl}{\underset{\underset{Cl}{|}}{C}}CH_2CH_3$$

5.5 ADDITION OF HYDROGEN HALIDES AND ADDITION OF HALOGENS

The addition of a hydrogen halide to a terminal alkyne follows Markovnikov's rule. Presumably this is because the electrophile in the pi-complex binds more strongly to the sp carbon that is bonded to the hydrogen.

$$CH_3C\equiv CH \ + \ H\text{—}Cl \ \longrightarrow \ CH_3C\overset{\overset{\delta^-}{Cl}}{\underset{\underset{:\ddot{C}l:}{\underset{\overset{+\delta}{H}}{\equiv}}}{}}CH \ \longrightarrow \ CH_3\overset{}{\underset{\underset{Cl}{|}}{C}}=CH_2$$

The addition of a hydrogen halide to an alkyne stops after the addition of one equivalent[1] of hydrogen halide—unless excess hydrogen halide is present. This is because the rate of addition of the hydrogen halide to the alkyne is greater than the rate of addition of the hydrogen halide to the halo-substituted alkene that is formed in the first addition reaction.

Both π bonds of an alkyne will react only if there is an excess of the electrophilic reagent.

$$CH_3CH_2CH_2C\equiv CH \ + \ HBr \ \longrightarrow \ CH_3CH_2CH_2\overset{\overset{Br}{|}}{C}=CH_2$$

1-pentyne

2-bromo-1-pentene
a halo-substituted alkene

In the presence of excess hydrogen halide, however, a second addition reaction takes place. Addition of the second equivalent of hydrogen halide also follows Markovnikov's rule. The product of the second addition reaction is a **geminal dihalide,** a molecule with two halogens on the same carbon. "Geminal" comes from *geminus,* which is Latin for "twin."

$$CH_3CH_2CH_2\overset{\overset{Br}{|}}{C}=CH_2 \ + \ HBr \ \longrightarrow \ CH_3CH_2CH_2\overset{\overset{Br}{|}}{\underset{\underset{Br}{|}}{C}}CH_3$$

2-bromo-1-pentene

2,2-dibromopentane
a geminal dihalide

[1]One equivalent means the same number of moles as the other reactant.

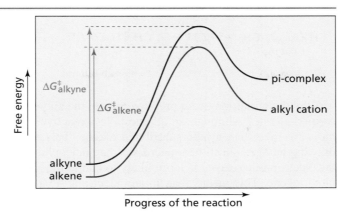

◀ **Figure 5.2**
Comparison of the free energies of activation for the addition of an electrophile to an alkyne and to an alkene. Because the energy difference between the transition states is greater than the energy difference between the reactants, $\Delta G^{\ddagger}$ for the reaction of an alkyne is greater than the $\Delta G^{\ddagger}$ for the reaction of an alkene.

An alkyne is less stable than an alkene because the π bond of an alkyne is weaker (56 kcal/mol) than the π bond of an alkene (64 kcal/mol). Consequently, one might expect an alkyne to be more reactive than an alkene in an electrophilic addition reaction. Actually, the reverse is true; the alkene has the greater reactivity. We cannot assume that the less stable compound will be the more reactive compound because the rate of a reaction depends on the free energy of activation—the difference between the free energy of the transition state and the free energy of the reactant (Section 3.7). Reactivity, therefore, depends on both the stability of the reactant *and* the stability of the transition state. Consequently, we must look at the relative stabilities of the transition states formed by alkenes and alkynes.

The Hammond postulate predicts that the structure of the transition state for electrophilic addition will resemble the structure of the intermediate (Section 3.11). The intermediate formed by addition of an electrophile (such as H^{+}) to an alkene is an alkyl cation, whereas the intermediate formed by addition of an electrophile to an alkyne is a pi-complex. For an alkene to be both more stable and more reactive than an alkyne requires that the alkyl cation be more stable than the pi-complex. It also requires that the difference in stability between the transition states be greater than the difference in stability between the reactants (Figure 5.2). Thus the free energy of activation for electrophilic addition to an alkene is less than the free energy of activation for electrophilic addition to an alkyne ($\Delta G^{\ddagger}_{alkene} < \Delta G^{\ddagger}_{alkyne}$ in Figure 5.2). Consequently, even though alkenes are more stable than alkynes, alkenes are more reactive than alkynes in electrophilic addition reactions.

Alkenes are more reactive than alkynes.

PROBLEM 6 ◆

Under what conditions can you assume that the less stable reactant will be the more reactive reactant?

If an alkene is more reactive than an alkyne in an electrophilic addition reaction, how can a halo-substituted alkene product result from the addition of one equivalent of HBr or HCl to the alkyne? Shouldn't we expect a second molecule of electrophilic reagent to add preferentially to the alkene formed in the first addition reaction rather than to the alkyne starting material? Shouldn't we get a mixture of geminal dihalide and unreacted alkyne?

$$\text{CH}_3\text{CH}_2\text{C}\equiv\text{CH} \;+\; \text{HCl} \;\longrightarrow\; \underset{\substack{| \\ \text{Cl} \\ \textbf{2-chloro-1-butene}}}{\text{CH}_3\text{CH}_2\text{C}=\text{CH}_2} \;\xrightarrow{\textbf{excess HCl}}\; \underset{\substack{| \\ \text{Cl} \\ \textbf{2,2-dichlorobutane}}}{\overset{\overset{\text{Cl}}{|}}{\text{CH}_3\text{CH}_2\text{CCH}_3}}$$

1-butyne

Why does addition occur preferentially to the alkyne rather than to the halo-substituted alkene? Although an alkyne is less reactive than an alkene, an alkyne is more reactive than a halo-substituted alkene. This is because the electron-withdrawing halogen substituent decreases the nucleophilic character of the alkene. So the electrophilic reagent is more likely to add to the alkyne than to the halo-substituted alkene. This means that, in any electrophilic addition reaction of an alkyne that forms an alkene with an electron-withdrawing substituent on an sp^2 carbon, the alkyne will be completely converted to alkene before the alkene reacts with a second equivalent of electrophilic reagent.

Hydrogen peroxide (or an alkyl peroxide) has the same effect on the addition of HBr to an alkyne that it has on the addition of HBr to an alkene (Section 3.18). In the presence of peroxide, HBr adds to a terminal alkyne in an apparent anti-Markovnikov manner, because Br· rather than H^+ is the electrophile.

$$\underset{\textbf{1-butyne}}{\text{CH}_3\text{CH}_2\text{C}\equiv\text{CH}} \;+\; \text{HBr} \;\xrightarrow{\textbf{peroxide}}\; \underset{\textbf{1-bromo-1-butene}}{\text{CH}_3\text{CH}_2\text{CH}=\text{CHBr}}$$

The mechanism of the reaction is the same as that for the addition of HBr to an alkene in the presence of peroxide. Peroxide is a radical initiator and creates a bromine radical that adds to the sp carbon bonded to the hydrogen. The vinylic radical that is formed abstracts a hydrogen atom from HBr and regenerates the bromine radical.

mechanism for the addition of HBr in the presence of peroxide

$$\text{RÖ}\!-\!\text{ÖR} \;\longrightarrow\; 2\,\text{RÖ}\cdot$$

$$\text{RÖ}\cdot \;+\; \text{H}\!-\!\ddot{\text{B}}\text{r:} \;\longrightarrow\; \text{RÖH} \;+\; \cdot\ddot{\text{B}}\text{r:}$$

$$\text{CH}_3\text{CH}_2\text{C}\equiv\text{CH} \;+\; \cdot\ddot{\text{B}}\text{r:} \;\longrightarrow\; \underset{\substack{| \\ :\ddot{\text{B}}\text{r:}}}{\text{CH}_3\text{CH}_2\dot{\text{C}}=\text{CH}}$$

$$\underset{\substack{| \\ \text{Br}}}{\text{CH}_3\text{CH}_2\text{C}=\text{CH}} \;+\; \text{H}\!-\!\ddot{\text{B}}\text{r:} \;\longrightarrow\; \text{CH}_3\text{CH}_2\text{CH}=\text{CHBr} \;+\; \cdot\ddot{\text{B}}\text{r:}$$

The halogens Cl_2 and Br_2 add to alkynes. If an excess of halogen is present, a second addition reaction occurs.

$$\text{CH}_3\text{CH}_2\text{C}\equiv\text{CCH}_3 \;\xrightarrow[\textbf{CCl}_4]{\textbf{Cl}_2}\; \underset{\substack{| \\ \text{Cl}}}{\overset{\overset{\text{Cl}}{|}}{\text{CH}_3\text{CH}_2\text{C}=\text{CCH}_3}} \;\xrightarrow[\textbf{CCl}_4]{\textbf{Cl}_2}\; \underset{\substack{| \quad | \\ \text{Cl} \;\; \text{Cl}}}{\overset{\overset{\text{Cl} \;\; \text{Cl}}{| \quad |}}{\text{CH}_3\text{CH}_2\text{C}-\text{CCH}_3}}$$

$$\text{CH}_3\text{C}\equiv\text{CH} \;\xrightarrow[\textbf{CCl}_4]{\textbf{Br}_2}\; \underset{\substack{| \\ \text{Br}}}{\overset{\overset{\text{Br}}{|}}{\text{CH}_3\text{C}=\text{CH}}} \;\xrightarrow[\textbf{CCl}_4]{\textbf{Br}_2}\; \underset{\substack{| \quad | \\ \text{Br} \;\; \text{Br}}}{\overset{\overset{\text{Br} \;\; \text{Br}}{| \quad |}}{\text{CH}_3\text{C}-\text{CH}}}$$

In Section 3.13, we saw that alkenes undergo acid-catalyzed addition of water. The product of the reaction is an alcohol.

**5.6
ADDITION OF WATER**

$$CH_3CH_2CH=CH_2 + H_2O \xrightarrow{H_2SO_4} CH_3CH_2CHCH_3$$

1-butene   OH

2-butanol

Alkynes also undergo acid-catalyzed addition of water. The product of the reaction is an enol. (The ending "ene" signifies the double bond, and "ol" the OH. When the two endings are joined, the final *e* of "ene" is dropped to avoid two consecutive vowels, but it is pronounced as if the *e* were there, "ene-ol.")

$$CH_3C\equiv CCH_3 + H_2O \xrightarrow{H_2SO_4} CH_3C=CHCH_3 \rightleftharpoons CH_3CCH_2CH_3$$

an enol   **a ketone**

The enol immediately rearranges to a ketone. A carbon doubly bonded to an oxygen is called a carbonyl ("car-bo-neel") group. A ketone is a compound that has two alkyl groups bonded to a carbonyl group; an aldehyde is a compound that has one alkyl group and a hydrogen bonded to a carbonyl group.

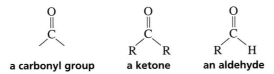

a carbonyl group   **a ketone**   **an aldehyde**

Terminal alkynes are less reactive than internal alkynes toward addition of water. Terminal alkynes will add water if mercuric ion (Hg^{2+}) is added to the acidic mixture. The mercuric ion acts as a catalyst to increase the rate of the addition reaction.

$$CH_3CH_2C\equiv CH \;+\; H_2O \xrightarrow[\text{HgSO}_4]{\text{H}_2\text{SO}_4} \underset{\text{an enol}}{CH_3CH_2\overset{\displaystyle OH}{\underset{}{C}}=CH_2} \;\rightleftharpoons\; \underset{\text{a ketone}}{CH_3CH_2\overset{\displaystyle O}{\overset{\|}{C}}CH_3}$$

The first step in the mercuric-ion-catalyzed hydration of an alkyne is formation of a cyclic mercurinium ion. In the second step, water attacks the most substituted carbon of the cyclic intermediate (Section 3.16). Oxygen loses a proton to form a mercuric enol, which immediately rearranges to a mercuric ketone. Loss of the mercuric ion results in formation of an enol, which rearranges to a ketone. Notice that the overall addition of water follows Markovnikov's rule.

mechanism for the mercuric-ion-catalyzed hydration of an alkyne

A ketone and an enol differ from each other only in the location of a double bond and in the location of a hydrogen. Such isomers are called **tautomers.** The ketone and enol are called **keto-enol tautomers.** The conversion of one tautomer into the other tautomer is called **tautomerization.** We will examine the mechanism of keto–enol interconversion in Chapter 18. For now, the important thing to remember is that the keto and enol tautomers come to equilibrium in solution, and the keto tautomer, because it is usually much more stable than the enol tautomer, predominates at equilibrium.

$$\underset{\text{keto tautomer}}{RCH_2-\overset{\displaystyle O}{\overset{\|}{C}}-R} \;\rightleftharpoons\; \underset{\text{enol tautomer}}{RCH=\overset{\displaystyle OH}{\underset{}{C}}-R}$$

Addition of water to a *symmetrical* internal alkyne forms a single ketone as a product. But if the alkyne is asymmetrical, two ketones are formed, because the initial addition of the proton can occur equally easily to either of the *sp* carbons.

$$CH_3CH_2C\equiv CCH_2CH_3 \;+\; H_2O \xrightarrow{\text{H}_2\text{SO}_4} CH_3CH_2\overset{\displaystyle O}{\overset{\|}{C}}CH_2CH_2CH_3$$

$$CH_3C\equiv CCH_2CH_3 \;+\; H_2O \xrightarrow{\text{H}_2\text{SO}_4} CH_3\overset{\displaystyle O}{\overset{\|}{C}}CH_2CH_2CH_3 \;+\; CH_3CH_2\overset{\displaystyle O}{\overset{\|}{C}}CH_2CH_3$$

PROBLEM 9◆

What ketones would be formed from the acid-catalyzed hydration of 3-heptyne?

Borane adds to alkynes in the same way that it adds to alkenes: one mole of BH_3 reacts with three moles of alkyne to form one mole of boron-substituted alkene (Section 3.17). When the addition reaction is over, aqueous sodium hydroxide and hydrogen peroxide are added to the reaction mixture. The end result, as in the case of alkenes, is the replacement of the boron by an OH group. The enol product immediately rearranges to a ketone.

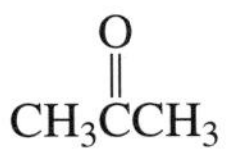

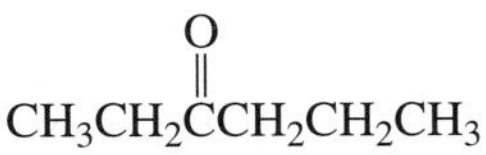

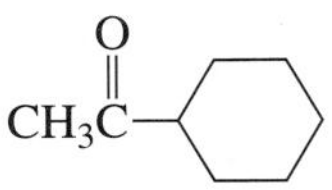

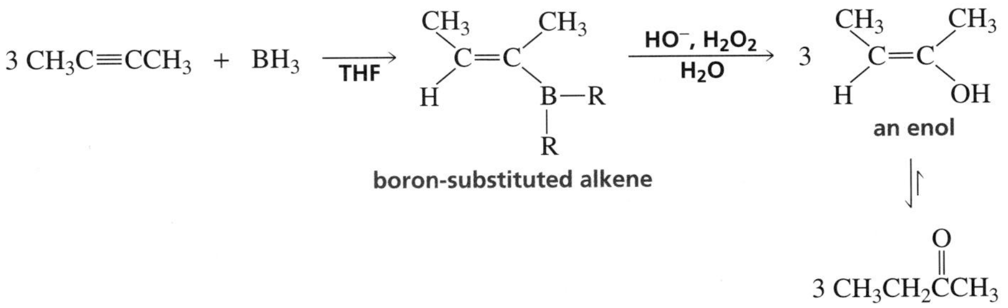

5.7 ADDITION OF BORANE: HYDROBORATION–OXIDATION

$$3\ CH_3C{\equiv}CCH_3\ +\ BH_3\ \xrightarrow{\text{THF}}$$

boron-substituted alkene

an enol

$$3\ CH_3CH_2\overset{\overset{\displaystyle O}{\|}}{C}CH_3$$

In order for the enol to be the product of the addition reaction, the boron-substituted alkene must not undergo a further addition reaction. In other words, the reaction must stop at the alkene stage. In the case of internal alkynes, the bulky groups on the boron-substituted alkene prevent the second addition from occurring. Because there is less steric hindrance in a terminal alkyne, it is harder to stop the addition reaction at the alkene stage. A special reagent called disiamylborane has been developed for use with terminal alkynes. The bulky alkyl groups of disiamylborane prevent a second addition to the boron-substituted alkene. So borane can be used to hydrate internal alkynes, but disiamylborane is preferred for the hydration of terminal alkynes.

$$CH_3CH_2C\equiv CH \ + \ \left(CH_3CH-CH-\right)_2 BH \longrightarrow$$

$$\overset{CH_3\ CH_3}{}$$

disiamylborane
bis(1,2-dimethylpropyl)borane

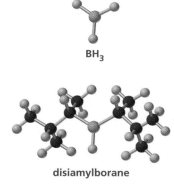

BH₃

disiamylborane

The addition of borane (or disiamylborane) to a terminal alkyne exhibits the same regioselectivity seen in borane addition to an alkene. Boron, with its electron-seeking empty orbital, adds preferentially to the *sp* carbon bonded to the hydrogen. Because the second step of the hydroboration–oxidation reaction results in the replacement of boron by an OH group, the overall reaction is an apparent anti-Markovnikov addition of water to the alkyne. The product appears to violate Markovnikov's original statement of the rule, because the original statement was created for a reaction in which H^+ is the electrophile. In hydroboration–oxidation, H^+ is not the electrophile; $H:^-$ is the nucleophile. Consequently, mercuric-ion-catalyzed addition of water to a terminal alkyne produces a ketone (the carbonyl group is *not* on the terminal carbon), whereas hydroboration–oxidation of the alkyne produces an aldehyde (the carbonyl group *is* on the terminal carbon).

$$CH_3C\equiv CH$$

$$\xrightarrow[\text{HgSO}_4]{\text{H}_2\text{O, H}_2\text{SO}_4} \quad CH_3\overset{OH}{\underset{|}{C}}=CH_2 \ \rightleftharpoons \ CH_3\overset{O}{\overset{\|}{C}}CH_3$$
a ketone

$$\xrightarrow[\text{2. HO}^-,\text{ H}_2\text{O}_2,\text{ H}_2\text{O}]{\text{1. disiamylborane}} \quad CH_3CH=\overset{OH}{\underset{|}{C}H} \ \rightleftharpoons \ CH_3CH_2\overset{O}{\overset{\|}{C}}H$$
an aldehyde

PROBLEM 12◆

Give the products of (1) mercuric-ion-catalyzed addition of water and (2) hydrobora-tion–oxidation for:

a. 1-butyne   **b.** 2-butyne   **c.** 2-pentyne

PROBLEM 13◆

There is only one alkyne that forms an aldehyde as a result of acid- (or mercuric ion-) catalyzed addition of water. Identify this alkyne.

5.8
ADDITION OF HYDROGEN

Hydrogen adds to an alkyne in the presence of a metal catalyst such as palladium, platinum, or nickel in the same manner in which it adds to alkenes. Since hydrogen also readily adds to alkenes, the reaction cannot be stopped at the alkene stage. The product of the hydrogenation reaction, therefore, is an alkane.

$$CH_3CH_2C\equiv CH \ \xrightarrow{\text{H}_2}{\text{Pt/C}} \ CH_3CH_2CH=CH_2 \ \xrightarrow{\text{H}_2}{\text{Pt/C}} \ CH_3CH_2CH_2CH_3$$

The relative reactivities of double and triple bonds toward the addition of hydrogen depend on the metal catalyst. In the presence of a partially deactivated metal catalyst, addition of hydrogen to an alkene is much slower than addition of hydrogen to an alkyne. So if the alkene is the desired product of the reaction, such a catalyst must be used. The most common partially deactivated metal catalyst is Lindlar's catalyst. Because the alkyne sits on the surface of the metal catalyst and the hydrogens are delivered to the triple bond from the surface of the catalyst, only syn addition of hydrogen occurs (Section 3.19). Syn addition of hydrogen to an internal alkyne results in the formation of a cis alkene.

Herbert H. M. Lindlar was born in Switzerland in 1909 and received a Ph.D. from the University of Bern. He worked at Hoffmann–La Roche and Co. in Basel, Switzerland, and he authored many patents. His last patent was a procedure for isolating the carbohydrate xylose from the waste produced in paper mills.

$$CH_3CH_2C{\equiv}CCH_3 + H_2 \xrightarrow[\text{catalyst}]{\text{Lindlar's}}$$

2-pentyne

$$\underset{CH_3CH_2}{\overset{H}{\diagdown}}C{=}C\underset{CH_3}{\overset{H}{\diagup}}$$

cis-2-pentene

LINDLAR'S CATALYST

Lindlar's catalyst can be prepared by precipitating palladium on calcium carbonate and treating it with lead acetate and quinoline. This treatment partially deactivates the catalyst. In the deactivated state, it is much more effective at catalyzing the addition of hydrogen to a triple bond than to a double bond.

$$(CH_3COO^-)_2Pb^{2+}$$
lead acetate

quinoline

Because sodium reacts more rapidly with triple bonds than with double bonds, alkynes can be converted into alkenes using an equivalent of sodium (or lithium) in liquid ammonia. This reaction forms a trans alkene. Since ammonia (bp = −33 °C) is a gas at room temperature, it is kept in the liquid state by using dry ice to cool the reaction flask.

$$CH_3CH_2C{\equiv}CCH_3 \xrightarrow[\text{NH}_3\text{ (liq)}]{\text{Na or Li}}$$

2-pentyne

$$\underset{H}{\overset{CH_3CH_2}{\diagdown}}C{=}C\underset{CH_3}{\overset{H}{\diagup}}$$

trans-2-pentene

Sodium and lithium have a strong tendency to lose the single electron in the outer-shell *s* orbital (Section 1.3). The first step in the mechanism of this reaction is transfer of the *s* orbital electron from sodium (or lithium) to the π framework of the triple bond to form a **radical anion** (a species with a negative charge and an unpaired electron). The radical anion abstracts a proton from ammonia. This results in the formation of a **vinylic radical** (the unpaired electron is on a vinylic carbon). Another single electron transfer from sodium (or lithium) to the vinylic radical forms a vinylic anion, which abstracts a proton from another molecule of ammonia. The product is the trans alkene.

$$R{-}C{\equiv}C{-}R + Na{\cdot} \longrightarrow R{-}\overset{\cdot}{C}{=}\overset{\cdot\cdot}{\overset{-}{C}}{-}R \xrightarrow{H{-}NH_2} R{-}\overset{}{C}{=}C\overset{R}{\underset{H}{\diagup}} \xrightarrow{Na{\cdot}} R{-}\overset{\cdot\cdot}{\overset{-}{C}}{=}C\overset{R}{\underset{H}{\diagup}} \xrightarrow{H{-}NH_2} \underset{R}{\overset{H}{\diagdown}}C{=}C\overset{R}{\underset{H}{\diagup}}$$

a radical anion   **a vinylic radical**   **a vinylic anion**   **a trans alkene**

$$+ Na^+ \qquad\qquad + {^-}NH_2 \qquad\qquad + Na^+ \qquad\qquad + {^-}NH_2$$

The vinylic anion intermediate can be in either a cis or a trans form. The cis and trans forms are in equilibrium, but the equilibrium favors the more stable trans form. The reaction leads to the formation of the trans alkene, because the rate-determining step occurs prior to the step in which the vinylic anion reacts. If reaction of the vinylic anion were the rate-determining step, the cis vinylic anion, being less stable, could react faster than the trans vinylic anion, and the cis product would be formed.

trans vinylic anion
more stable

cis vinylic anion
less stable

PROBLEM 14◆

What alkyne would you start with and what reagents would you use if you wanted to synthesize:

a. pentane?　　　　　　　　　　**c.** *trans*-2-pentene?

b. *cis*-2-butene?　　　　　　　　**d.** 1-hexene?

**5.9
ACIDITY OF A HYDROGEN BONDED TO AN *sp* HYBRIDIZED CARBON**

Carbon forms very strong covalent bonds with hydrogen because their similar electronegativities cause them to share the bonding electrons almost equally. We have seen, however, that not all carbon atoms have the same electronegativity (Section 5.4). An *sp* hybridized carbon is more electronegative than an sp^2 hybridized carbon, which is more electronegative than an sp^3 hybridized carbon, which is just slightly more electronegative than a hydrogen.

relative electronegativities of carbon atoms

$$sp > sp^2 > sp^3$$

increasing electronegativity

　　Why does the type of hybridization affect the electronegativity of the carbon atom? Electronegativity is a measure of an atom's ability to hold electrons close to its nucleus. The closer the electrons are to the nucleus, the more electronegative is the atom. The average distance of a 2s electron from the nucleus is less than the average distance of a 2p electron from the nucleus. Therefore, the electrons in an *sp* hybrid orbital (50% s character) are closer, on average, to the nucleus than those in an sp^2 hybrid orbital (33.3% s character). In turn, sp^2 electrons are closer to the nucleus than sp^3 electrons (25% s character). The *sp* hybridized carbon, then, is the most electronegative.

　　In Section 1.18, we saw that the acidity of a hydrogen attached to some of the second-row elements depends on the electronegativity of the atom to which the hydrogen is attached. The greater the electronegativity of the atom to which the hydrogen is attached, the easier it is for the proton to be released. The easier the proton is released, the stronger the acid. (Remember: The stronger the acid, the lower its pK_a.)

relative electronegativities	N	<	O	<	F

relative acid strengths NH_3 < H_2O < HF
$pK_a = 36$ $pK_a = 15.7$ $pK_a = 3.2$

increasing acid strength and electronegativity →

Because the electronegativity of carbon atoms follows the order $sp > sp^2 > sp^3$, acetylene is a stronger acid than ethylene, which is a stronger acid than ethane.

$HC\equiv CH$ $H_2C=CH_2$ CH_3CH_3
$pK_a = 25$ $pK_a = 44$ $pK_a = 50$

We can compare the acidities of these compounds with the acidities of hydrogens attached to other second row elements.

relative acid strengths

CH_3CH_3 < $H_2C=CH_2$ < NH_3 < $HC\equiv CH$ < H_2O < HF
$pK_a = 50$ $pK_a = 44$ $pK_a = 36$ $pK_a = 25$ $pK_a = 15.7$ $pK_a = 3.2$

increasing acid strength →

The conjugate bases have the following relative base strengths *because the stronger the acid, the weaker its conjugate base.*

relative base strengths

$CH_3CH_2^-$ > $H_2C=CH^-$ > H_2N^- > $HC\equiv C^-$ > HO^- > F^-

← **increasing base strength**

In order to remove a proton from an acid, the base that is removing the proton must be stronger than the base generated as a result of proton removal (Section 1.17). In other words, you must start with a stronger base than the base you are making. The amide ion ($^-NH_2$) can remove a hydrogen bonded to an *sp* carbon because it is a stronger base than the acetylide ion that is formed. The reagent actually used to remove the hydrogen is sodium amide ($Na^+{}^-NH_2$) in liquid ammonia.

The stronger the acid, the weaker its conjugate base.

$RC\equiv CH$ + $^-NH_2$ ⇌ $RC\equiv C^-$ + NH_3
stronger acid   amide anion / stronger base   acetylide anion / weaker base   weaker acid

Hydroxide ion is not a strong enough base to remove a hydrogen bonded to an *sp* carbon because it is a weaker base than the acetylide ion that would be formed.

$RC\equiv CH$ + HO^- ⇌ $RC\equiv C^-$ + H_2O
weaker acid   hydroxide anion / weaker base   acetylide anion / stronger base   stronger acid

The amide ion cannot remove a hydrogen bonded to an sp^2 or an sp^3 carbon. Only a hydrogen bonded to an *sp* carbon is sufficiently acidic to be removed by the amide ion. As a result, a hydrogen bonded to an *sp* carbon is referred to as an "acidic" hydrogen. The fact that terminal alkynes are "acidic" is one way in which their chemistry differs from that of alkenes. We must be careful not to misinterpret what we mean when we say that a hydrogen bonded to an *sp* carbon is acidic. What we mean is that it is more acidic than most other carbon-bound hydrogens. But it is much less acidic than a hydrogen of a water molecule, and we normally do not consider water to be an acidic compound.

SODIUM AMIDE AND SODIUM

Take care not to confuse sodium amide ($Na^{+\ -}NH_2$) in liquid ammonia with sodium (Na^0) in liquid ammonia. Sodium amide is the strong base used to remove a proton from a terminal alkyne. Sodium is used as a source of electrons in the reduction of an alkyne to a trans alkene (Section 5.8).

PROBLEM 15 ◆

Any base whose conjugate acid has a pK_a greater than _____ can remove a proton from a terminal alkyne to form an acetylide anion.

PROBLEM 16 ◆

Which carbocation in each of the following pairs is more stable?

a. $CH_3\overset{+}{C}H_2$ or $H_2C=\overset{+}{C}H$ **b.** $H_2C=\overset{+}{C}H$ or $HC\equiv\overset{+}{C}$

PROBLEM-SOLVING STRATEGY

a. List the following compounds in order of decreasing acidity.

$$CH_3CH_2\overset{+}{N}H_3 \qquad CH_3CH=\overset{+}{N}H_2 \qquad CH_3C\equiv\overset{+}{N}H$$

To compare the acidities of a group of compounds, first look at how they differ. These three compounds differ in the hybridization of the nitrogen to which the acidic hydrogen is attached. Now, recall what you know about hybridization and acidity. You know that hybridization of an atom affects its electronegativity (*sp* is more electronegative than sp^2, which is more electronegative than sp^3), and you know that the more electronegative the atom to which a hydrogen is attached, the more acidic the hydrogen. Now you can answer the question.

relative acidities $CH_3C\equiv\overset{+}{N}H > CH_3CH=\overset{+}{N}H_2 > CH_3CH_2\overset{+}{N}H_3$

increasing acidity

b. Draw the conjugate bases and list them in order of decreasing basicity.

Simply remove a proton from each acid to get the structures of the conjugate bases, and recall that the stronger the acid, the weaker its conjugate base.

relative basicities $CH_3CH_2NH_2 > CH_3CH=NH > CH_3C\equiv N$

increasing basicity

5.10 SYNTHESES USING ACETYLIDE IONS

Reactions that result in the formation of new carbon–carbon bonds are important in the synthesis of organic compounds. Without such reactions, we could not convert molecules with small carbon skeletons into larger ones; the product of a reaction would always have the same number of carbons as the starting material.

One reaction that results in the formation of a new carbon–carbon bond is the reaction of an acetylide ion with an alkyl halide. Only primary alkyl halides or methyl halides should be used in this reaction.

$$CH_3CH_2C \equiv C^- \ + \ CH_3CH_2CH_2Br \ \longrightarrow \ CH_3CH_2C \equiv CCH_2CH_2CH_3 \ + \ Br^-$$
$$\textbf{3-heptyne}$$

The mechanism of this reaction is well understood. Bromine is more electronegative than carbon and, as a result, the electrons in the carbon–bromine bond are not shared equally by the two atoms. This means that the carbon–bromine bond is polar, with a partial negative charge on bromine and a partial positive charge on carbon. The negative charge on the acetylide ion is attracted to the partial positive charge on the carbon of the alkyl halide. As the electrons of the acetylide ion approach the carbon to form a new carbon–carbon bond, they push out the bromine and its bonding electrons because carbon can bond to no more than four atoms at a time. Because an alkyl group has been added to the acetylide ion, this is called an **alkylation reaction.**

$$CH_3CH_2C \equiv \ddot{C}^- \ + \ CH_3CH_2CH_2 \overset{\delta+}{-} \overset{\delta-}{Br} \ \longrightarrow \ CH_3CH_2C \equiv CCH_2CH_2CH_3 \ + \ Br^-$$

The mechanism of this and similar reactions will be discussed in greater detail in Chapter 9. At that time, we will also see why the reaction works best with primary alkyl halides and methyl halides.

Simply by choosing an alkyl halide of the appropriate structure, terminal alkynes can be converted into internal alkynes of any desired chain length.

$$CH_3CH_2CH_2C \equiv CH \xrightarrow[\text{2. } CH_3CH_2CH_2CH_2CH_2Cl]{\text{1. } NaNH_2} CH_3CH_2CH_2C \equiv CCH_2CH_2CH_2CH_2CH_3$$
$$\textbf{1-pentyne} \qquad\qquad\qquad\qquad\qquad\qquad\qquad \textbf{4-decyne}$$

PROBLEM 17 / SOLVED

A chemist wants to synthesize 4-decyne but cannot find any 1-pentyne, the starting material used in the synthesis described above. How else can 4-decyne be synthesized?

SOLUTION   The *sp* carbons of 4-decyne are bonded to a propyl group and to a pentyl group. Therefore to obtain 4-decyne, the acetylide ion of 1-pentyne can react with a pentyl halide (above) or the acetylide ion of 1-heptyne can react with a propyl halide.

$$CH_3CH_2CH_2CH_2CH_2C \equiv CH \xrightarrow[\text{2. } CH_3CH_2CH_2Cl]{\text{1. } NaNH_2} CH_3CH_2CH_2C \equiv CCH_2CH_2CH_2CH_2CH_3$$
$$\textbf{1-heptyne} \qquad\qquad\qquad\qquad\qquad\qquad\qquad \textbf{4-decyne}$$

So far in our study of organic chemistry, we have paid attention to *why* a reaction occurs, *how* it occurs, and the *products* that are formed. A good way to review these reactions is to design syntheses, because, in so doing, you must be able to recall many of the reactions that you have learned.

Synthetic chemists consider time, cost, and yield in designing syntheses. In the interest of time, a well-designed synthesis should require as few steps (sequential reactions) as possible, and those steps should each involve a reaction that is easy to

**5.11
DESIGNING A
SYNTHESIS I:
AN INTRODUCTION
TO MULTISTEP
SYNTHESIS**

carry out. If two chemists in a pharmaceutical company were each asked to prepare a new drug, and one synthesized the drug in three simple steps while the other used 20 difficult steps, which chemist would be out looking for a new job? In addition, each step in the synthesis should result in the greatest possible yield of the desired product. The more reactant needed to synthesize 1 g of product, the more expensive it is to make that product. Sometimes it is preferable to design a synthesis involving several steps if the starting materials are inexpensive, the reactions are easy to carry out, and the yield of each step is high. The alternative would be to design a synthesis with fewer steps, requiring expensive starting materials, and reactions that are more difficult or give lower yields.

The following examples will give you an idea of the type of thinking required for the design of a successful synthesis. This kind of problem will appear repeatedly throughout this book, because working such problems is a good way to learn organic chemistry.

Example 1. Starting with 1-butyne, how could you make the following ketone? You can use any other organic reagents and any inorganic reagents.

$$CH_3CH_2C{\equiv}CH \xrightarrow{\ ?\ } CH_3CH_2\overset{\displaystyle O}{\overset{\|}{C}}CH_2CH_2CH_3$$

Many students find that the easiest way to design such a synthesis is to work backward. Instead of looking at the starting material and deciding what to do first, look at the product and decide what to do last. The product is a ketone. A ketone can be formed by adding water and an acid catalyst to an alkyne. If the alkyne used in the reaction is symmetrical, only one ketone will be obtained as a product. Thus, 3-hexyne is the best alkyne to use for the synthesis of the desired ketone.

$$CH_3CH_2C{\equiv}CCH_2CH_3 \xrightarrow[\text{H}_2\text{SO}_4]{\text{H}_2\text{O}} CH_3CH_2\overset{\displaystyle OH}{\overset{|}{C}}{=}CHCH_2CH_3 \ \rightleftharpoons \ CH_3CH_2\overset{\displaystyle O}{\overset{\|}{C}}CH_2CH_2CH_3$$

3-Hexyne can be obtained from the starting material by removing the hydrogen bonded to the *sp* carbon followed by alkylation. To obtain the desired product, a two-carbon alkyl halide must be used in the alkylation reaction.

$$CH_3CH_2C{\equiv}CH \xrightarrow[\text{2. CH}_3\text{CH}_2\text{Br}]{\text{1. NaNH}_2} CH_3CH_2C{\equiv}CCH_2CH_3$$

Elias James Corey *coined the term "retrosynthetic analysis" (Section 28.2). He was born in Massachusetts in 1928 and is a professor of chemistry at Harvard University.*

Designing a synthesis by working backward from product to reactant is not simply a technique taught to organic chemistry students. It is done so frequently by experienced synthetic chemists that it has been given a name: **retrosynthetic analysis.** Chemists show that they are working backward by using open arrows. For example, the route to the synthesis of the ketone above can be arrived at by the following retrosynthetic analysis.

retrosynthetic analysis

$$CH_3CH_2\overset{\displaystyle O}{\overset{\|}{C}}CH_2CH_2CH_3 \ \Longrightarrow \ CH_3CH_2C{\equiv}CCH_2CH_3 \ \Longrightarrow \ CH_3CH_2C{\equiv}CH$$

Once the sequence of reactions has been worked out by retrosynthetic analysis, the synthetic scheme can be shown by reversing the steps and including the reagents required to carry out each step.

synthesis

$$CH_3CH_2C\equiv CH \xrightarrow[\text{2. } CH_3CH_2Br]{\text{1. } NaNH_2} CH_3CH_2C\equiv CCH_2CH_3 \xrightarrow[H_2SO_4]{H_2O} CH_3CH_2\overset{\overset{\displaystyle O}{\|}}{C}CH_2CH_2CH_3$$

Retrosynthetic analysis will be discussed again in other chapters and will be covered in much greater detail in Chapter 28.

Example 2. Starting with acetylene, how could you make 1-bromopentane?

$$HC\equiv CH \xrightarrow{?} CH_3CH_2CH_2CH_2CH_2Br$$

A primary alkyl halide can be prepared from a terminal alkene (using HBr in the presence of peroxide). A terminal alkene can be prepared from a terminal alkyne. And a terminal alkyne of the appropriate length can be prepared from acetylene and an alkyl halide.

retrosynthetic analysis

$$CH_3CH_2CH_2CH_2CH_2Br \implies CH_3CH_2CH_2CH=CH_2 \implies CH_3CH_2CH_2C\equiv CH \implies HC\equiv CH$$

Now we can write the synthetic scheme.

synthesis

$$HC\equiv CH \xrightarrow[\text{2. } CH_3CH_2CH_2Br]{\text{1. } NaNH_2} CH_3CH_2CH_2C\equiv CH \xrightarrow[\text{Lindlar's catalyst}]{H_2} CH_3CH_2CH_2CH=CH_2$$
$$\xrightarrow[\text{peroxide}]{HBr} CH_3CH_2CH_2CH_2CH_2Br$$

Example 3. How could 2,6-dimethylheptane be prepared from an alkyne and an alkyl halide?

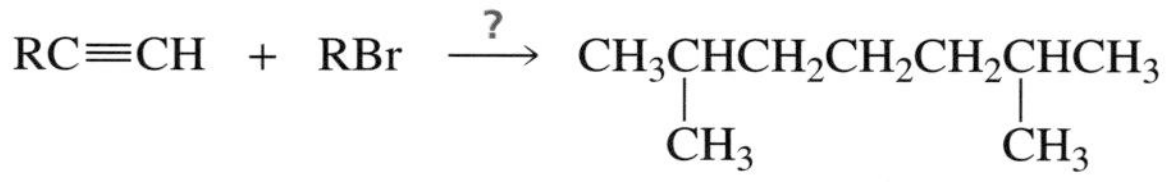

$$RC\equiv CH + RBr \xrightarrow{?} CH_3\underset{CH_3}{CH}CH_2CH_2CH_2\underset{CH_3}{CH}CH_3$$

Only one alkyne will form the desired hydrocarbon upon hydrogenation. This alkyne can be dissected in two different ways. In one case, the alkyne could be prepared from the reaction of an acetylide ion with a primary alkyl halide; in the other case, the acetylide ion would have to react with a secondary alkyl halide.

retrosynthetic analysis

$$CH_3\underset{CH_3}{CH}CH_2CH_2CH_2\underset{CH_3}{CH}CH_3 \implies CH_3\underset{CH_3}{CH}CH_2C\equiv C\underset{CH_3}{CH}CH_3$$

$$CH_3\underset{CH_3}{CH}CH_2Br + CH_3\underset{CH_3}{CH}C\equiv CH \quad\text{or}\quad CH_3\underset{CH_3}{CH}Br + CH_3\underset{CH_3}{CH}CH_2C\equiv CH$$

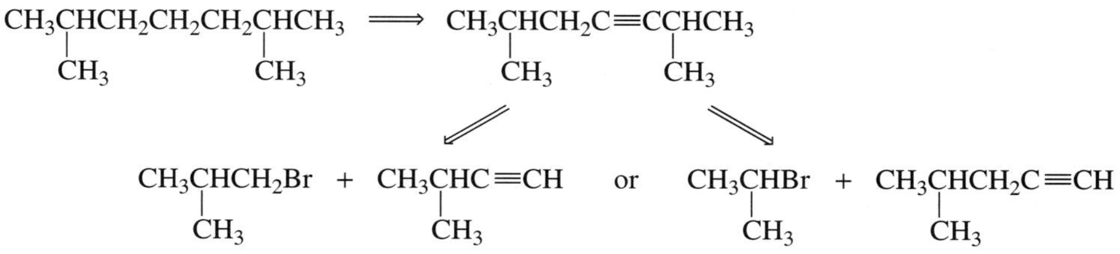

Since we know that the reaction of an acetylide ion with an alkyl halide works best with primary alkyl halides and methyl halides, we know how to proceed.

synthesis

$$CH_3CHC{\equiv}CH \xrightarrow[\text{2. CH}_3\text{CHCH}_2\text{Br}]{\text{1. NaNH}_2} CH_3CHCH_2C{\equiv}CCHCH_3 \xrightarrow[\text{Pd/C}]{\text{H}_2} CH_3CHCH_2CH_2CH_2CHCH_3$$

(with CH_3 substituents as drawn)

Example 4. How could you carry out the following synthesis using the given starting material?

$$\bigcirc{-}C{\equiv}CH \xrightarrow{?} \bigcirc{-}CH_2CH_2OH$$

An alcohol can be prepared from an alkene, and an alkene can be prepared from an alkyne.

retrosynthetic analysis

$$\bigcirc{-}CH_2CH_2OH \Longrightarrow \bigcirc{-}CH{=}CH_2 \Longrightarrow \bigcirc{-}C{\equiv}CH$$

You can use either of the two methods you know to convert an alkyne into an alkene. Hydroboration–oxidation must be used to convert the alkene into the desired alcohol.

synthesis

$$\bigcirc{-}C{\equiv}CH \xrightarrow[\text{or Na/NH}_3\text{ (liq)}]{\text{H}_2 \text{ / Lindlar's catalyst}} \bigcirc{-}CH{=}CH_2 \xrightarrow[\text{2. HO}^-,\text{ H}_2\text{O}_2,\text{ H}_2\text{O}]{\text{1. BH}_3} \bigcirc{-}CH_2CH_2OH$$

Example 5. How could you prepare (*E*)-2-pentyne from acetylene?

$$HC{\equiv}CH \xrightarrow{?} \begin{array}{c} CH_3CH_2 \quad\quad H \\ \diagdown C{=}C \diagup \\ H \quad\quad CH_3 \end{array}$$

A trans alkene can be prepared from an alkyne. The alkyne needed to synthesize the desired alkene can be prepared from 1-butyne and a methyl halide; 1-butyne can be prepared from acetylene and an ethyl halide.

retrosynthetic analysis

$$\begin{array}{c} CH_3CH_2 \quad\quad H \\ \diagdown C{=}C \diagup \\ H \quad\quad CH_3 \end{array} \Longrightarrow CH_3CH_2C{\equiv}CCH_3 \Longrightarrow CH_3CH_2C{\equiv}CH \Longrightarrow HC{\equiv}CH$$

synthesis

$$HC{\equiv}CH \xrightarrow[\text{2. CH}_3\text{CH}_2\text{Br}]{\text{1. NaNH}_2} CH_3CH_2C{\equiv}CH \xrightarrow[\text{2. CH}_3\text{Br}]{\text{1. NaNH}_2} CH_3CH_2C{\equiv}CCH_3 \xrightarrow[\text{Na}]{\text{NH}_3\text{ (liq)}} \begin{array}{c} CH_3CH_2 \quad\quad H \\ \diagdown C{=}C \diagup \\ H \quad\quad CH_3 \end{array}$$

Example 6. How could you prepare 3,3-dibromohexane from reagents that contain no more than two carbon atoms?

$$\text{reagents with no more than 2 carbon atoms} \xrightarrow{?} \underset{\underset{\displaystyle Br}{|}}{\overset{\overset{\displaystyle Br}{|}}{CH_3CH_2CCH_2CH_2CH_3}}$$

A geminal dibromide can be prepared from an alkyne. 3-Hexyne is the desired alkyne because it will form one geminal bromide, whereas 2-hexyne would form two different geminal dibromides. 3-Hexyne can be prepared from 1-butyne and ethyl bromide; 1-butyne can be prepared from acetylene and ethyl bromide.

retrosynthetic analysis

$$CH_3CH_2\overset{\overset{\displaystyle Br}{|}}{\underset{\underset{\displaystyle Br}{|}}{C}}CH_2CH_2CH_3 \implies CH_3CH_2C \equiv CCH_2CH_3 \implies CH_3CH_2C \equiv CH \implies HC \equiv CH$$

synthesis

$$HC \equiv CH \xrightarrow[\text{2. CH}_3\text{CH}_2\text{Br}]{\text{1. NaNH}_2} CH_3CH_2C \equiv CH \xrightarrow[\text{2. CH}_3\text{CH}_2\text{Br}]{\text{1. NaNH}_2} CH_3CH_2C \equiv CCH_2CH_3 \xrightarrow{\textbf{excess HBr}} CH_3CH_2\overset{\overset{\displaystyle Br}{|}}{\underset{\underset{\displaystyle Br}{|}}{C}}CH_2CH_2CH_3$$

PROBLEM 18

Starting with acetylene, how could the following compounds be synthesized?

a. $CH_3CH_2CH_2C \equiv CH$

b. $CH_3CH = CH_2$

c. $CH_3CH_2CH_2CH_2\overset{\overset{\displaystyle O}{||}}{C}H$

d. $CH_3\overset{\overset{\displaystyle Cl}{|}}{\underset{\underset{\displaystyle Cl}{|}}{C}}CH_3$

e. $CH_3\overset{\underset{\underset{\displaystyle Br}{|}}{}}{C}HCH_3$

f. $\underset{H}{\overset{CH_3}{}}C = C\underset{H}{\overset{CH_3}{}}$

There are two sources of acetylene. Heating ethylene to a temperature greater than 1200 °C produces acetylene and hydrogen. This is called **thermal cracking.**

**5.12
INDUSTRIAL USE
OF ACETYLENE**

$$H_2C = CH_2 \xrightarrow{> \textbf{1200 °C}} HC \equiv CH + H_2$$

Acetylene can also be obtained by heating carbon obtained from coal (coke) with calcium oxide in an electric furnace. Addition of water to the calcium carbide that is formed produces acetylene.

$$3C + CaO \xrightarrow{\textbf{heat}} \underset{\textbf{calcium carbide}}{Ca^{2+}(\bar{C} \equiv \bar{C})} + \underset{\textbf{carbon monoxide}}{CO}$$

$$Ca^{2+}(\bar{C} \equiv \bar{C}) + 2H_2O \longrightarrow HC \equiv CH + Ca(OH)_2$$

Most of the acetylene produced commercially is used as a starting material for polymers that you encounter daily, such as vinyl flooring, plastic piping, Teflon, and acrylics. Polymers are large molecules that are made by linking together small molecules known as monomers. The mechanisms by which monomers are converted into polymers will be discussed in Chapter 25.

Poly(vinyl chloride), the polymer produced from the polymerization of vinyl chloride, is known as PVC. Poly(vinyl chloride) is hard and rather brittle, but it can be softened by adding certain compounds known as plasticizers. It then becomes the soft, elastic vinyl commonly used both as a substitute for leather and in the manufacture of such things as garbage bags and plastic covers. Poly(acrylonitrile) looks like wool when it is made into fibers. It is marketed under the trade names Orlon (Du Pont), Creslan (American Cyanamid), and Acrilan (Monsanto).

$$HC\equiv CH \ + \ HCl \ \longrightarrow \ \underset{\textbf{vinyl chloride}}{H_2C=CHCl} \ \xrightarrow{\textbf{polymerization}} \ \begin{bmatrix} CH_2-CH \\ | \\ Cl \end{bmatrix}_n$$

poly(vinyl chloride)
PVC

$$HC\equiv CH \ + \ HCN \ \longrightarrow \ \underset{\textbf{acrylonitrile}}{H_2C=CHCN} \ \xrightarrow{\textbf{polymerization}} \ \begin{bmatrix} CH_2-CH \\ | \\ CN \end{bmatrix}_n$$

poly(acrylonitrile)
Orlon

ACETYLENE CHEMISTRY OR THE FORWARD PASS?

Fr. Julius Arthur Nieuwland (1878–1936) did much of the early work on the polymerization of acetylene. He was born in Belgium and settled with his parents in South Bend, Indiana, two years later. He became a priest and professor of botany and chemistry at the University of Notre Dame, where Knute Rockne (the inventor of the forward pass) worked for him as a research assistant. Rockne also taught chemistry at Notre Dame, but when he received the offer to coach the football team he switched fields, in spite of Fr. Nieuwland's attempts to convince him to continue his work as a scientist.

Knute Rockne in his uniform during the year that he was captain of the Notre Dame football team.

SUMMARY OF REACTIONS

1. Electrophilic addition reactions
   a. Addition of hydrogen halides (Markovnikov orientation) (Section 5.5)

$$RC\equiv CH \ \xrightarrow{\textbf{HX}} \ \underset{X}{RC=CH_2} \ \xrightarrow{\textbf{excess HX}} \ \underset{X}{\overset{X}{RC-CH_3}}$$

$$HX \ = \ HF, \ HCl, \ HBr, \ HI$$

   b. Addition of hydrogen bromide (apparent anti-Markovnikov orientation) (Section 5.5)

$$RC\equiv CH\ +\ HBr\ \xrightarrow{\text{peroxide}}\ RCH=CHBr$$

c. Addition of halogens (Section 5.5)

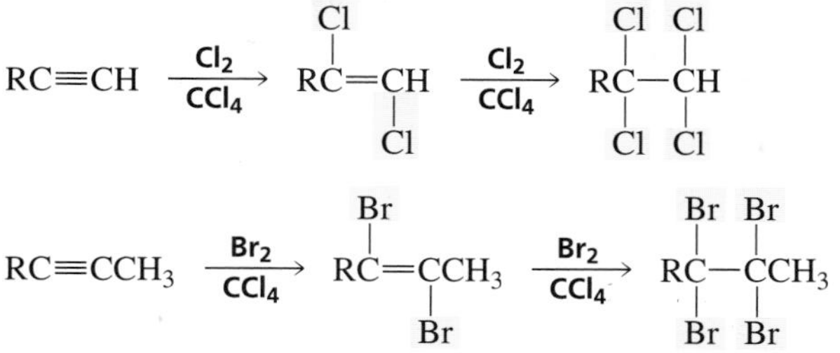

$$RC\equiv CH\ \xrightarrow[\text{CCl}_4]{\text{Cl}_2}\ \underset{\text{Cl}}{\overset{\text{Cl}}{RC=CH}}\ \xrightarrow[\text{CCl}_4]{\text{Cl}_2}\ \underset{\text{Cl Cl}}{\overset{\text{Cl Cl}}{RC-CH}}$$

$$RC\equiv CCH_3\ \xrightarrow[\text{CCl}_4]{\text{Br}_2}\ \underset{\text{Br}}{\overset{\text{Br}}{RC=CCH_3}}\ \xrightarrow[\text{CCl}_4]{\text{Br}_2}\ \underset{\text{Br Br}}{\overset{\text{Br Br}}{RC-CCH_3}}$$

d. Addition of water/hydroboration–oxidation (Sections 5.6 and 5.7)

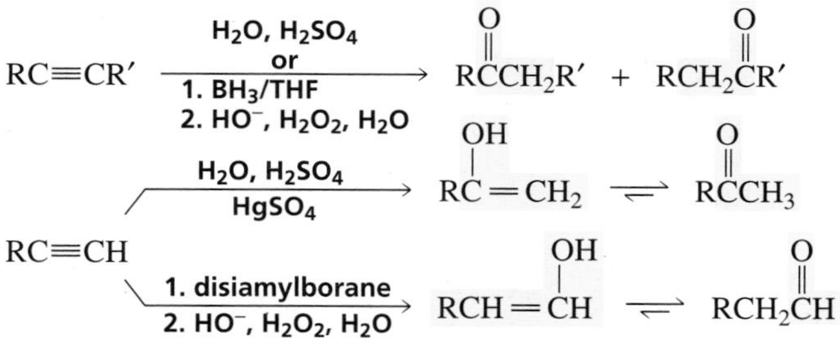

$$RC\equiv CR'\ \xrightarrow[\substack{\text{1. BH}_3/\text{THF}\\ \text{2. HO}^-,\ \text{H}_2\text{O}_2,\ \text{H}_2\text{O}}]{\substack{\text{H}_2\text{O, H}_2\text{SO}_4\\ \text{or}}}\ \overset{O}{\overset{\|}{RCCH_2R'}}\ +\ \overset{O}{\overset{\|}{RCH_2CR'}}$$

$$RC\equiv CH$$

$$\xrightarrow[\text{HgSO}_4]{\text{H}_2\text{O, H}_2\text{SO}_4}\ \underset{}{\overset{\text{OH}}{RC=CH_2}}\ \rightleftharpoons\ \overset{O}{\overset{\|}{RCCH_3}}$$

$$\xrightarrow[\text{2. HO}^-,\ \text{H}_2\text{O}_2,\ \text{H}_2\text{O}]{\text{1. disiamylborane}}\ \underset{}{\overset{\text{OH}}{RCH=CH}}\ \rightleftharpoons\ \overset{O}{\overset{\|}{RCH_2CH}}$$

2. Addition of hydrogen (Section 5.8)

$$RC\equiv CR'\ +\ 2\,H_2\ \xrightarrow[\text{or Ni}]{\text{Pd/C, Pt/C,}}\ RCH_2CH_2R'$$

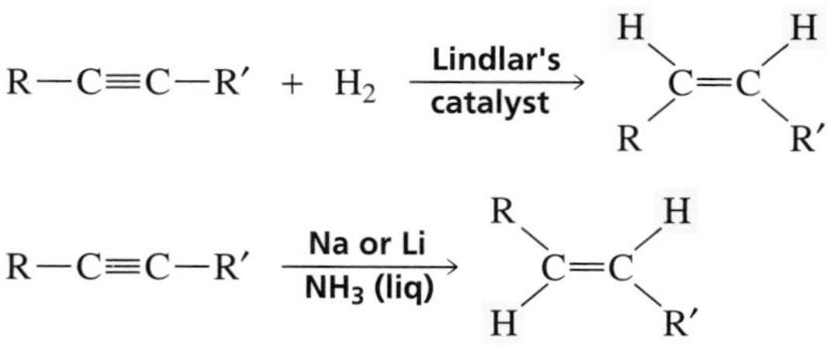

$$R-C\equiv C-R'\ +\ H_2\ \xrightarrow[\text{catalyst}]{\text{Lindlar's}}\ \underset{R}{\overset{H}{C}}=\underset{R'}{\overset{H}{C}}$$

$$R-C\equiv C-R'\ \xrightarrow[\text{NH}_3\text{ (liq)}]{\text{Na or Li}}\ \underset{H}{\overset{R}{C}}=\underset{R'}{\overset{H}{C}}$$

3. Removal of a proton, followed by alkylation (Sections 5.9 and 5.10)

$$RC\equiv CH\ \xrightarrow{\text{NaNH}_2}\ RC\equiv C^-\ \xrightarrow{\text{R'CH}_2\text{Br}}\ RC\equiv CCH_2R'$$

KEY TERMS

alkylation reaction (page 251)
alkyne (page 235)
geminal dihalide (page 240)
internal alkyne (page 236)
keto-enol tautomers (page 244)

pi-complex (page 239)
radical anion (page 247)
retrosynthetic analysis
 (page 252)
tautomerization (page 244)

tautomers (page 244)
terminal alkyne (page 236)
thermal cracking (page 255)
vinylic cation (page 239)
vinylic radical (page 247)

PROBLEMS

19. Give a structure for each of the following.
  a. 2-hexyne
  b. 5-ethyl-3-octyne
  c. methylacetylene
  d. vinylacetylene
  e. methoxyethyne
  f. *tert*-butyl-*sec*-butylacetylene
  g. 1-bromo-1-pentyne
  h. propargyl bromide
  i. diethylacetylene
  j. di-*tert*-butylacetylene
  k. cyclopentylacetylene
  l. 5,6-dimethyl-2-heptyne

20. Give the IUPAC name for each of the following.

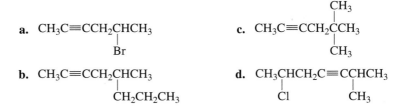

21. What reagents would be required to carry out the following syntheses?

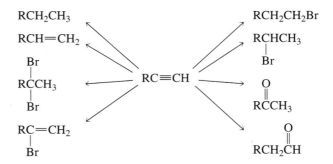

22. How could the following compounds be synthesized, starting with a hydrocarbon that has the same number of carbon atoms as the desired product?

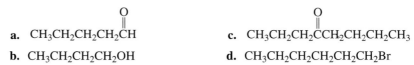

23. What reagents would you use for the following syntheses?
  a. (*Z*)-3-hexene from 3-hexyne
  b. (*E*)-3-hexene from 3-hexyne
  c. hexane from 3-hexyne

24. What is the molecular formula of a hydrocarbon that has 1 triple bond, 2 double bonds, 1 ring, and 32 carbons?

25. What will be the major product of the reaction of 1 mol of propyne with each of the following reagents?
  a. HBr (1 mol)
  b. HBr (2 mol)
  c. Br$_2$ (1 mol)/CCl$_4$
  d. Br$_2$ (2 mol)/CCl$_4$
  e. aqueous H$_2$SO$_4$, HgSO$_4$
  f. disiamylborane followed by H$_2$O$_2$/HO$^-$
  g. HBr + H$_2$O$_2$
  h. excess H$_2$/Pt
  i. H$_2$ (1 mol)/Lindlar's catalyst
  j. sodium in liquid ammonia
  k. sodium amide in liquid ammonia
  l. product of **k** followed by 1-chloropentane

26. Answer Problem 25 using 2-butyne as the reagent in place of propyne.

27. a. Starting with isopropylacetylene, how could you prepare the following alcohols?
    1. 2-methyl-2-pentanol
    2. 4-methyl-2-pentanol
  b. In each case, a second alcohol would also be obtained. What alcohol would it be?

28. How many of the following names are correct? Correct the incorrect names.
  a. 4-heptyne   **d.** 2,3-dimethyl-5-octyne
  b. 2-ethyl-3-hexyne   **e.** 4,4-dimethyl-2-pentyne
  c. 4-chloro-2-pentyne   **f.** 2,5-dimethyl-3-hexyne

29. Which of the following pairs are keto-enol tautomers?

  a. $CH_3CH_2CH{=}CHCH_2OH$ and $CH_3CH_2CH_2CH_2\overset{\overset{\displaystyle O}{\|}}{C}H$

  b. $CH_3\underset{\underset{\displaystyle OH}{|}}{C}HCH_3$ and $CH_3\overset{\overset{\displaystyle O}{\|}}{C}CH_3$

  c. $CH_3CH_2CH{=}CHOH$ and $CH_3CH_2CH_2\overset{\overset{\displaystyle O}{\|}}{C}H$

  d. $CH_3CH_2CH_2CH{=}CHOH$ and $CH_3CH_2CH_2\overset{\overset{\displaystyle O}{\|}}{C}CH_3$

  e. $CH_3CH_2CH_2\underset{\underset{\displaystyle OH}{|}}{C}{=}CH_2$ and $CH_3CH_2CH_2\overset{\overset{\displaystyle O}{\|}}{C}CH_3$

30. Using acetylene as the starting material, how can the following compounds be prepared?

  a. $CH_3CH_2\underset{\underset{\displaystyle Br}{|}}{C}HCH_2Br$ **d.**

  b. $CH_3\overset{\overset{\displaystyle O}{\|}}{C}H$ **e.**

  c. $CH_3\overset{\overset{\displaystyle O}{\|}}{C}CH_3$ **f.**

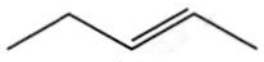

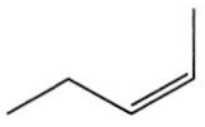

31. Give the stereoisomers obtained from the reaction of 2-butyne with the following reagents.
  a. 1. H_2/Lindlar's catalyst   **2.** Br_2/CCl_4
  b. 1. Na/NH_3 (liq)   **2.** Br_2/CCl_4
  c. 1. Cl_2/CCl_4 **2.** Br_2/CCl_4

32. Draw the keto tautomer for each of the following.

  a. $CH_3CH{=}\underset{\underset{\displaystyle OH}{|}}{C}CH_3$ **c.**

  b. $CH_3CH_2CH_2\underset{\underset{\displaystyle OH}{|}}{C}{=}CH_2$ **d.**

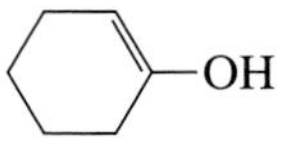

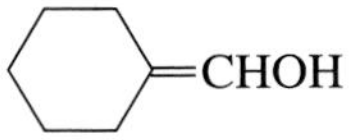

33. Show how each of the following compounds could be prepared using the given starting material, any necessary inorganic reagents, and any necessary organic compound that has no more than four carbon atoms.

a. $HC\equiv CH \longrightarrow CH_3CH_2CH_2CH_2\overset{\displaystyle O}{\overset{\|}{C}}CH_3$

b. $HC\equiv CH \longrightarrow CH_3CH_2\underset{\underset{\displaystyle Br}{|}}{C}HCH_3$

c. $HC\equiv CH \longrightarrow CH_3CH_2CH_2\underset{\underset{\displaystyle OH}{|}}{C}HCH_3$

d. 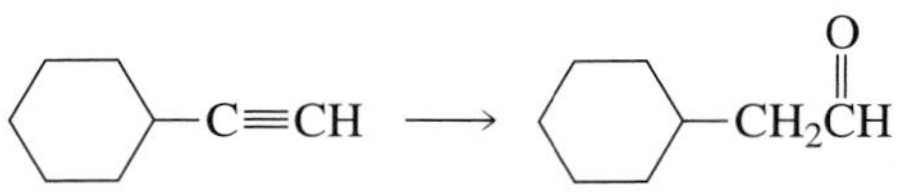 ⬡—C≡CH ⟶ ⬡—$CH_2\overset{\displaystyle O}{\overset{\|}{C}}H$

e. 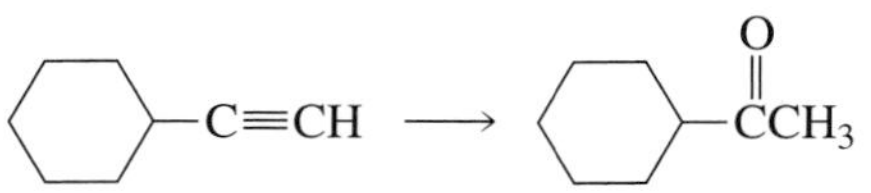 ⬡—C≡CH ⟶ ⬡—$\overset{\displaystyle O}{\overset{\|}{C}}CH_3$

f. 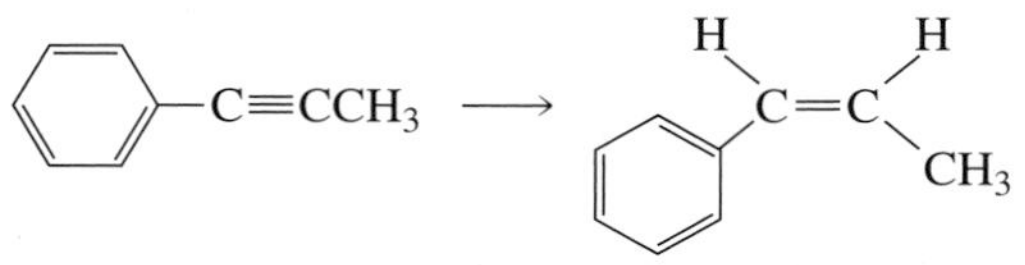 ⬡—C≡CCH₃ ⟶ (structure with C=C, H substituents, phenyl, and CH₃)

34. Dr. Iona Ford was planning to synthesize 3-octyne by adding 1-bromobutane to the product obtained from the reaction of 1-butyne with sodium amide. However, she had forgotten to order 1-butyne. How else can she prepare 3-octyne?

35. The female of a certain species of worms produces a sex attractant, called spodoptol, in low concentrations, which attracts the male worm. The population of these worms can be controlled by putting synthetic spodoptol in traps. Spodoptol can be synthesized by the following series of reactions.

$$HOCH_2CH_2CH_2CH_2CH_2CH_2CH_2CH_2C\equiv CH \xrightarrow[\text{NaNH}_2]{\text{excess}} \textbf{A} \xrightarrow{CH_3CH_2CH_2CH_2Br} \textbf{B} \xrightarrow{EtOH} \textbf{C} \xrightarrow[\substack{\text{Lindlar's} \\ \text{catalyst}}]{H_2} \text{spodoptol}$$

a. Draw the structure of spodoptol.
b. Why is excess $NaNH_2$ used in the first step?

36. a. Explain why a single pure product is obtained from hydroboration–oxidation of 2-butyne whereas two products are obtained from hydroboration–oxidation of 2-pentyne.
b. Name two other internal alkynes that will yield only one product on hydroboration–oxidation.

ELECTRON DELOCALIZATION, RESONANCE, AND AROMATICITY

aniline   ferrocene   thiophene

E lectrons that are restricted to a particular region are called **localized electrons.** They either belong to a single atom or are confined to a bond between two atoms.

$$CH_3 - \ddot{N}H_2 \qquad CH_3 - CH = CH_2$$

localized electrons   **localized electrons**

Many organic compounds contain delocalized electrons. These electrons do not belong to a single atom nor are they confined to a bond between two atoms; **delocalized electrons** are electrons that are shared by more than two atoms. In this chapter you will learn to recognize compounds that contain delocalized electrons, you will learn to draw structures that represent the electron distribution in molecules with delocalized electrons, and you will be introduced to some of the special characteristics of compounds that have delocalized electrons. You will then be able to understand the effects that delocalized electrons have on the reactivity of organic compounds.

We will start by taking a look at benzene, a compound whose structure chemists could not determine until they recognized that electrons in organic molecules could be delocalized.

The structure of benzene puzzled early organic chemists. They knew that benzene had a molecular formula of C_6H_6 and that it was an unusually stable compound that did not undergo the addition reactions characteristic of alkenes (Section 3.8). They also knew the following facts:

**6.1
DELOCALIZED
ELECTRONS:
THE STRUCTURE
OF BENZENE**

1. When a different atom is substituted for one of the hydrogen atoms of benzene, only one product is obtained.
2. When the substituted product undergoes a second substitution, three products are obtained: A, B, and C.

$$C_6H_6 \xrightarrow[\text{with an x}]{\text{replace a hydrogen}} C_6H_5X \xrightarrow[\text{with an x}]{\text{replace a hydrogen}} \underset{A}{C_6H_4X_2} + \underset{B}{C_6H_4X_2} + \underset{C}{C_6H_4X_2}$$

What kind of structure would we arrive at for benzene if we knew only what the early chemists knew? We can tell from its molecular formula (C_6H_6) that benzene has eight fewer hydrogens than a noncyclic alkane with six carbons (C_nH_{2n+2} = C_6H_{14}). Therefore, benzene is either an acyclic (noncyclic) compound with four π bonds, a cyclic compound with three π bonds, a bicyclic compound with two π bonds, a tricyclic compound with one π bond, or a tetracyclic compound (Section 3.1). All the hydrogens in benzene must be identical because only one product is obtained if any one hydrogen is replaced with another atom. Two structures that fit these requirements are shown here.

$$CH_3C{\equiv}C{-}C{\equiv}CCH_3$$

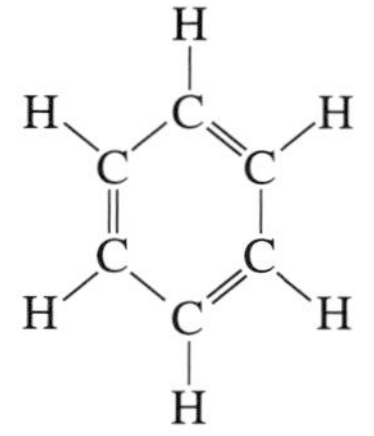

Neither of these structures is consistent with the observation that three compounds are obtained if a second hydrogen is replaced. The noncyclic structure yields two disubstituted products.

$$CH_3C{\equiv}C{-}C{\equiv}CCH_3 \xrightarrow[\text{with Br's}]{\text{replace 2 H's}} CH_3C{\equiv}C{-}C{\equiv}CCHBr$$
$$|$$
$$Br$$

and $\quad BrCH_2C{\equiv}C{-}C{\equiv}CCH_2Br$

The cyclic structure yields four disubstituted products: a 1,3-disubstituted product, a 1,4-disubstituted product, and two 1,2-disubstituted products—because the two substituents can be placed either on two adjacent carbons joined by a single bond or on two adjacent carbons joined by a double bond.

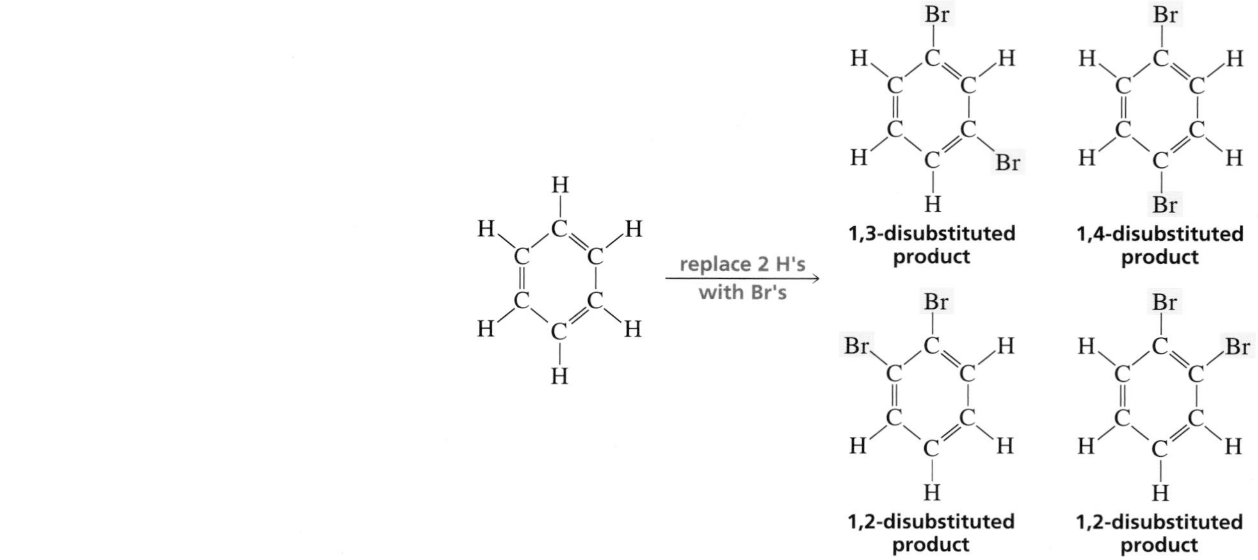

replace 2 H's with Br's

1,3-disubstituted product

1,4-disubstituted product

1,2-disubstituted product

1,2-disubstituted product

In 1865, the German chemist Kekulé suggested an answer to this dilemma. He proposed that benzene was not a single compound but a mixture of two compounds in rapid equilibrium.

Kekulé structures of benzene

This would explain why only three disubstituted products are obtained when a substituted benzene undergoes a second substitution. According to Kekulé, there actually *are* four disubstituted products, but the two 1,2-disubstituted products interconvert too rapidly to be distinguished and separated from each other.

The Kekulé structures of benzene account for the number of isomers obtained as a result of substitution. However, they fail to account for the fact that the double bonds of benzene do not undergo the addition reactions characteristic of alkenes, and they do not explain the unusual stability of benzene. Confirmation that benzene had a six-membered ring came in 1901, when Paul Sabatier (Section 3.19) found that hydrogenation of benzene produced cyclohexane. This, however, did not solve the puzzle of benzene's structure.

benzene   cyclohexane

The controversy over the structure of benzene continued until the 1930s, when the new techniques of X-ray and electron diffraction produced a surprising result. *These methods showed that benzene is a planar molecule in which all six carbon–carbon bonds have the same length.* The length of each of the carbon–carbon bonds in benzene is 1.39 Å, which is shorter than a carbon–carbon single bond (1.54 Å) but a little longer than a carbon–carbon double bond (1.33 Å) (Section 1.14). If all of the carbon–carbon bonds have the same length, they must have the same electron density. The X-ray and electron diffraction observations can be explained only by postulating that the π electrons in benzene are not localized between two pairs of carbon atoms; rather, they are delocalized around the ring. To gain a clearer understanding of delocalized electrons, let's take a close look at the bonding in benzene.

KEKULÉ'S DREAM

Friedrich August Kekulé von Stradonitz (1829–1896) was born in Germany. He entered the University of Giessen to study architecture but switched to chemistry after taking a chemistry course. He was a professor of chemistry at the University of Heidelberg, at the University of Ghent in Belgium, and then at the University of Bonn. In 1890, he gave an extemporaneous speech at the twenty-fifth anniversary celebration of his first paper on the cyclic structure of benzene. In this speech he claimed that he had arrived at the

Kekulé structure as a result of dozing off in front of a fire while working on a textbook. He dreamed of chains of carbon atoms twisting and turning in a snakelike motion, when suddenly the head of one snake seized hold of its own tail and formed a spinning ring. Recently, the veracity of his snake story has been questioned by those who point out that there is no written record of the dream from the time he experienced it in 1861 until the time he related it in 1890. Others counter that dreams are not the kind of evidence one publishes in scientific papers. But it is not uncommon for scientists to report moments of creativity through the unconscious, when they were not thinking about science. After relating his dream, Kekulé said, "Let us learn to dream, and perhaps then we shall learn the truth. But let us also beware not to publish our dreams until they have been examined by the wakened mind." In 1895, he was made a nobleman by Emperor William II. This allowed him to add "von Stradonitz" to his name. Kekulé's students received three of the first five Nobel Prizes in chemistry: van't Hoff in 1901, Fischer in 1902, and Baeyer in 1905.

Friedrich August Kekulé von Stradonitz

Sir James Dewar (1842–1923) *was born in Scotland, the son of an innkeeper. After studying under Kekulé, he became a professor at Cambridge University and then at the Royal Institution in London. Dewar's most important work was in the field of low-temperature chemistry. He used double-wall flasks with evacuated space between the walls in order to reduce heat transmission. These flasks are now called Dewar flasks—better known to nonchemists as thermos bottles.*

Albert Ladenburg (1842–1911) *was born in Germany. He was a professor of chemistry at the University of Kiel.*

PROBLEM 1 ◆

a. How many monosubstituted products would each of the following compounds have? (Notice that each compound has the same molecular formula as that of benzene.)

    1. $HC{\equiv}CC{\equiv}CCH_2CH_3$ **2.** $CH_2{=}CHC{\equiv}CCH{=}CH_2$

b. How many disubstituted products would each of these compounds have? (Do not include stereoisomers.)

PROBLEM 2

Between 1865 and 1890, other possible structures were proposed for benzene. Two of these are shown here. Considering what nineteenth-century chemists knew about benzene, which is a better proposal for the structure of benzene, Dewar benzene or Ladenburg benzene? Why?

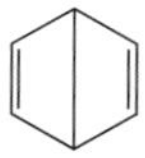

Dewar benzene 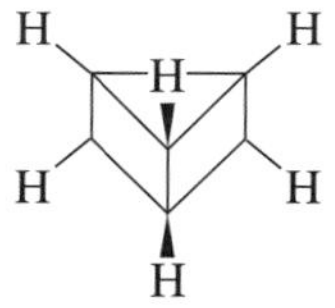   **Ladenburg benzene**

6.2 BONDING IN BENZENE

Benzene is a planar molecule. Each of the six carbons in benzene is sp^2 hybridized. An sp^2 hybridized carbon forms bond angles of 120°, identical to the angles of a planar hexagon. Each of the carbons in benzene uses two sp^2 orbitals to bond to two other carbons. Its third sp^2 orbital overlaps with the s orbital of a hydrogen (Figure 6.1a). Each carbon also has a p orbital at right angles to the sp^2 orbitals. Since benzene is planar, the six p orbitals are parallel (Figure 6.1b). The p orbitals are close enough for side-to-side overlap; each p orbital overlaps the p orbitals on

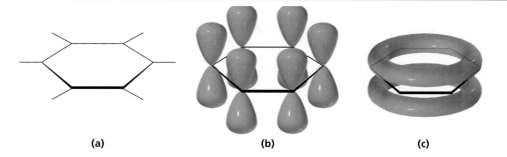

▲ Figure 6.1
(a) The carbon–carbon and carbon–hydrogen σ bonds in benzene. (b) The *p* orbital on each carbon of benzene overlaps with adjacent *p* orbitals. (c) The clouds of π electrons above and below the plane of the benzene ring.

the adjacent carbons. As a result, the overlapping *p* orbitals form a continuous doughnut-shaped cloud of π electrons above, and another doughnut-shaped cloud of π electrons below, the plane of the benzene ring (Figure 6.1c).

Each π electron is, therefore, not localized on a single carbon, nor is it localized in a bond between two carbons (as in an alkene). Instead, each π electron is shared by all six carbons. The six π electrons are delocalized; they roam freely within the doughnut-shaped clouds that lie over and under all six carbons. Consequently, benzene can be represented by a hexagon containing a circle that symbolizes the six delocalized π electrons.

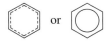

From this type of representation, it is clear that benzene does not contain any double bonds, which explains why benzene does not exhibit the addition reactions that are characteristic of alkenes. We see now that Kekulé's structure for benzene was pretty close to the correct structure. The actual structure of benzene is a Kekulé structure with delocalized electrons.

6.3 DELOCALIZED ELECTRONS AND RESONANCE

A significant disadvantage to using structures that represent electrons as delocalized is that such structures do not tell us how many electrons are present in the molecule. For example, the circle inside the hexagon in the second diagram of the structure of benzene illustrated above shows that the π electrons are shared equally by all six carbons and that all the carbon–carbon bonds have the same length, but it does not show how many π electrons are in the ring. Consequently, chemists prefer to use structures containing localized electrons to approximate the appearance of the actual structure with delocalized electrons. A compound with delocalized electrons is said to have **resonance.** The approximate structure using localized electrons is called a **resonance contributor, a resonance structure,** or a **contributing resonance structure.** The actual structure, drawn using delocalized electrons, is called a **resonance hybrid.** Each resonance contributor of benzene clearly indicates that there are six π electrons in the ring.

resonance contributor   ⟷   resonance contributor

resonance hybrid

Resonance contributors are shown with a double-headed arrow separating them. The double-headed arrow does *not* mean that the structures are in equilibrium with one another; rather, it indicates that the actual structure lies somewhere between the structures of the resonance contributors. The resonance contributors are only a convenient way to show the π electrons; they do not depict any real electron distribution. For example, the bond between C-1 and C-2 in benzene is not a double bond, as shown in the resonance contributor on the left, nor is it a single bond, as shown in the resonance contributor on the right. It is midway between what are represented by the two resonance contributors. Neither of the contributing resonance structures accurately represents the structure of benzene. The actual structure of benzene is given by the average of the two resonance contributors (the resonance hybrid.).

The following analogy illustrates the difference between resonance contributors and the resonance hybrid. Imagine that you are trying to describe to a friend what a rhinoceros looks like. You might tell your friend that a rhinoceros looks like a cross between a unicorn and a dragon. The unicorn and the dragon do not exist; they are like the resonance contributors. They are not in equilibrium; a rhinoceros does not jump back and forth between the two resonance contributors, looking like a unicorn one minute and a dragon the next minute. The rhinoceros always looks like a rhinoceros; it is like the resonance hybrid. The unicorn and the dragon are simply ways to represent what the actual structure (the rhinoceros) looks like. *Resonance contributors, like unicorns and dragons, are imaginary, not real. Only the resonance hybrid, like the rhinoceros, is real.*

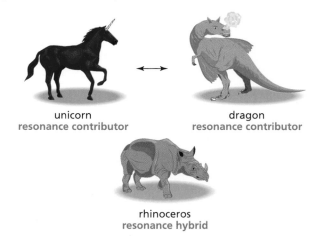

unicorn
resonance contributor

dragon
resonance contributor

rhinoceros
resonance hybrid

**6.4
HOW TO DRAW
RESONANCE
CONTRIBUTORS**

An organic compound with delocalized electrons is generally represented as a structure with localized electrons. For example, nitroethane is represented as having a nitrogen–oxygen double bond and a nitrogen–oxygen single bond.

$$CH_3CH_2-\overset{+}{N}\overset{\nearrow O}{\underset{\searrow O^-}{}}$$

nitroethane

However, the two nitrogen–oxygen bonds in nitroethane are identical (they have the same bond length). A more accurate description of the molecule's structure is obtained by drawing two resonance contributors. Both resonance contributors show the compound with a nitrogen-oxygen double bond and a nitrogen–oxygen single bond, but the double bond in one contributor is the single bond in the other contributor, and vice versa. In other words, the electrons are delocalized.

$$CH_3CH_2-\overset{+}{N}\overset{\nearrow O}{\underset{\searrow O^-}{}} \qquad \longleftrightarrow \qquad CH_3CH_2-\overset{+}{N}\overset{\nearrow O^-}{\underset{\searrow O}{}}$$

resonance contributor   **resonance contributor**

The resonance hybrid shows that the two nitrogen–oxygen bonds are identical and that the negative charge is shared by both oxygens. The resonance hybrid also shows that the *p* orbital of nitrogen overlaps with the *p* orbital of each oxygen. In other words, the two π electrons are shared by three atoms.

Notice that delocalized electrons result from a *p* orbital overlapping with the *p* orbitals of more than one adjacent atom.

$$CH_3CH_2-\overset{+}{N}\overset{\overset{-\delta}{\ddot{O}:}}{\underset{\underset{-\delta}{\ddot{O}:}}{}}$$

resonance hybrid

Rules for Drawing Resonance Contributors

Resonance contributors tell us where the formal charges reside in a molecule, and they also tell us the approximate bond orders. Only by visualizing all the resonance contributors can we appreciate what the actual molecule—the resonance hybrid—looks like.

In order to draw resonance contributors, the electrons in one resonance contributor are moved to generate the next resonance contributor. As you draw contributing resonance structures, note that only electrons move. The nuclei of the atoms never move. Further, the only electrons that can move are π electrons and nonbonding electrons. Also notice that the total number of electrons in the molecule does not change, nor do the numbers of paired and unpaired electrons.

The electrons can be moved in one of the following ways.

1. Move π electrons toward a positive charge or toward a π bond (Figures 6.2 and 6.3).

2. Move a nonbonding pair of electrons toward a π bond (Figure 6.4).

3. Move a single nonbonding electron toward a π bond (Figure 6.5).

Notice that, in all cases, the electrons are moved toward an sp^2 hybridized atom. (Remember that an sp^2 hybridized carbon is either a double-bonded carbon or a carbon that has a positive charge or an unpaired electron.) Electrons cannot be moved toward an sp^3 hybridized carbon because an sp^3 hybridized carbon has four σ bonds and cannot accommodate any more electrons.

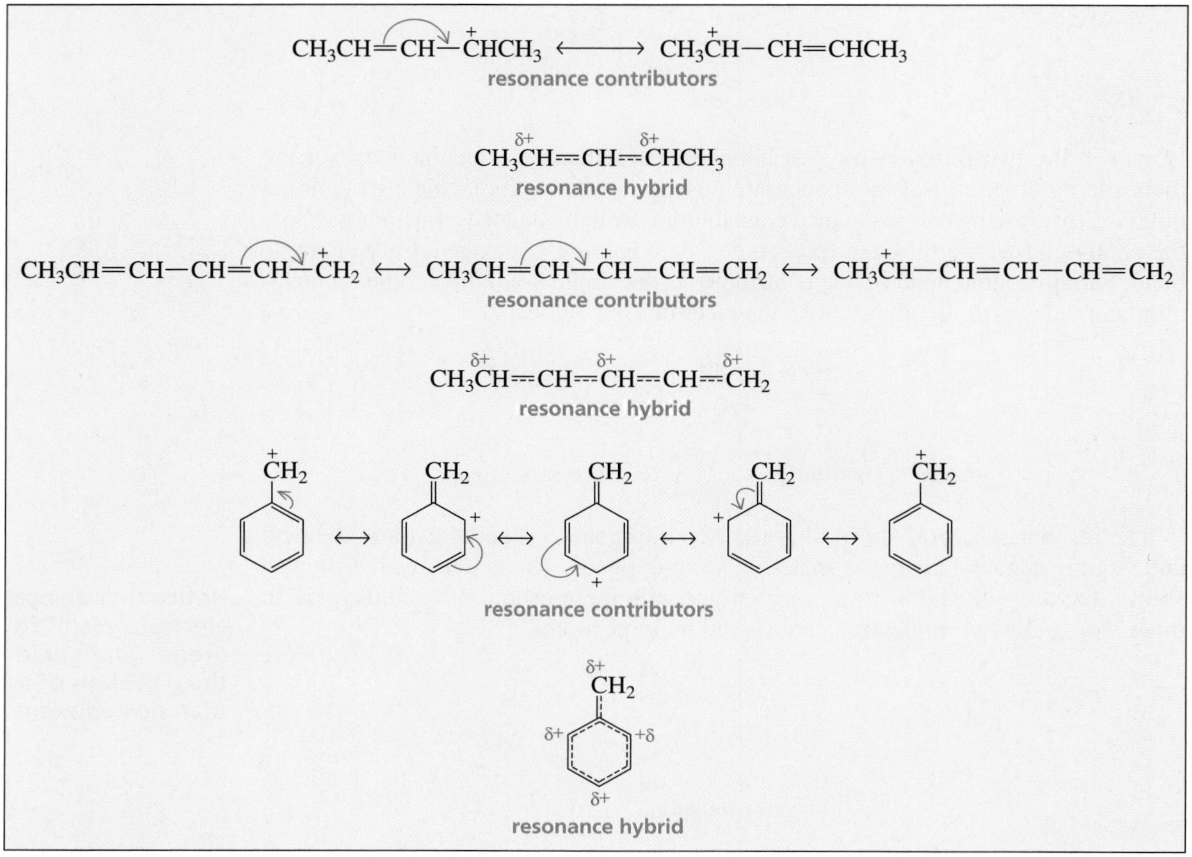

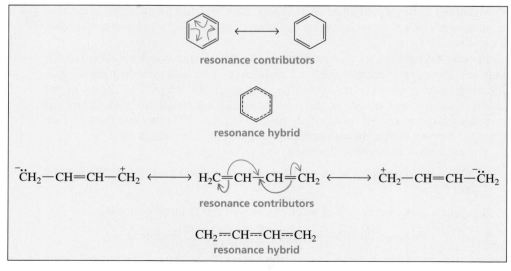

▲ **Figure 6.2**
Resonance contributors are obtained by moving π electrons toward a positive charge.

▲ **Figure 6.3**
Resonance contributors are obtained by moving π electrons toward a π bond. (In the second example, the red arrows lead to the resonance contributor on the right, and the blue arrows lead to the resonance contributor on the left.)

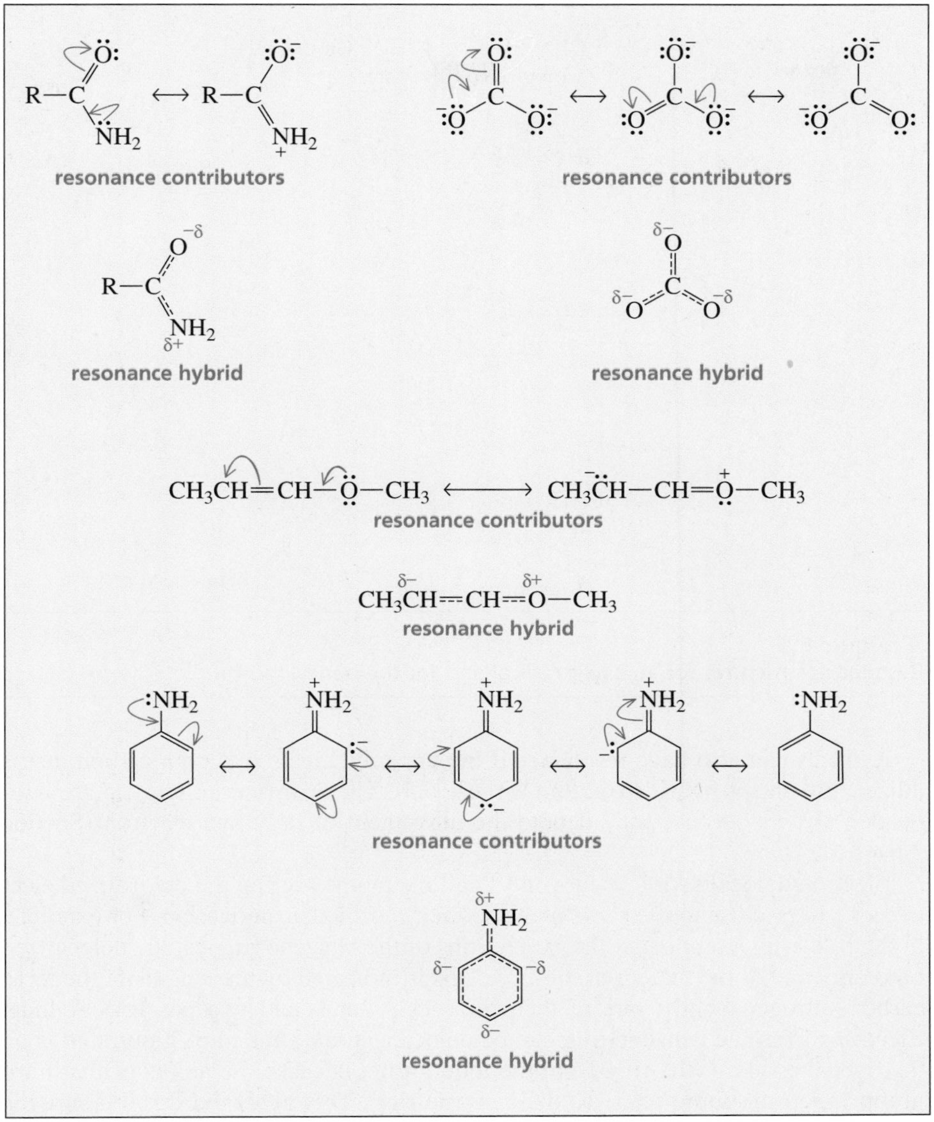

▲ **Figure 6.4**
Resonance contributors are obtained by moving a nonbonding pair of electrons toward a π bond.

Since electrons are neither added to nor removed from the compound when resonance contributors are drawn, each of the resonance contributors for a particular compound must have the same net charge. If one resonance structure has a net charge of −1, all the others must also have net charges of −1; if one has a net charge of 0, all the others must also have net charges of 0. (A net charge of 0 does not necessarily mean that there is no charge on any of the atoms; a molecule with a positive charge on one atom and a negative charge on another atom has a net charge of 0.)

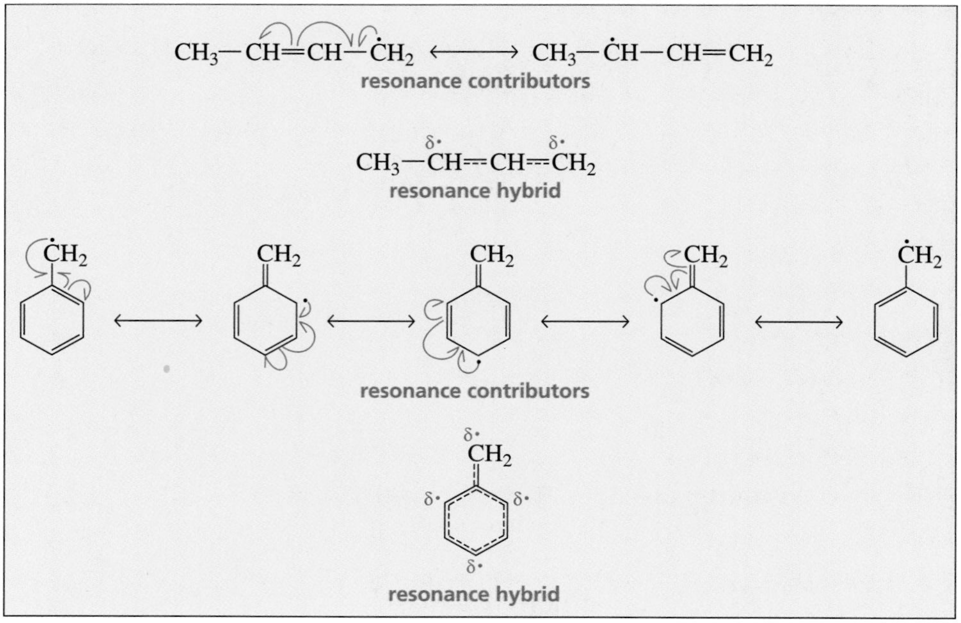

▲ **Figure 6.5**
Resonance structures for an allylic radical and for the benzyl radical.

Radicals can also have resonance if the unpaired electron is on a carbon that is adjacent to an sp^2 hybridized atom (Figure 6.5). The arrows in Figure 6.5 are half-headed arrows because they denote the movement of only one electron (Section 3.6)

Electron delocalization occurs only if all the atoms sharing the delocalized electrons lie in or close to the same plane so their p orbitals can overlap. For example, in the following compound, the π electrons of the benzene ring can be delocalized onto only one of the three nitro groups. This is indicated by the lengths of the three carbon–nitrogen bonds: one of them is 1.35 Å long, and two are 1.45 Å long. Electrons from the benzene ring can be delocalized onto the nitro group that is on the opposite side of the ring from the iodo group, because the atoms of that nitro group are in the same plane as the benzene ring. This gives the carbon–nitrogen bond a partial double-bond character and shortens it. The large iodine atom causes the other nitro groups to twist out of the plane of the benzene ring, and thus electron delocalization onto these nitro groups cannot occur.

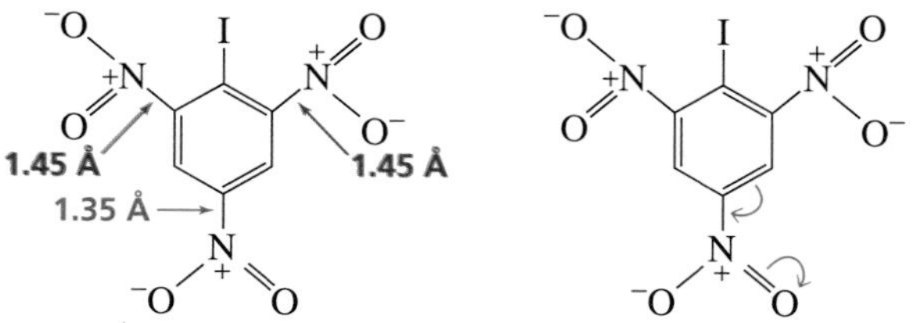

One way to recognize compounds with delocalized electrons is to compare them with similar compounds in which all the electrons are localized. The compound on the left in the first example has delocalized electrons because the nonbonding pair of electrons on nitrogen can be shared with the adjacent sp^2 carbon since the carbon–carbon π bond can be broken. In contrast, all the electrons in the compound on the right are localized. The nonbonding pair of electrons on nitrogen cannot be shared with the adjacent sp^3 carbon because carbon cannot form five bonds. (The octet rule requires that second-row elements be surrounded by no more than eight electrons.) In other words, sp^3 hybridized carbons cannot accept electrons. Because an sp^2 hybridized carbon has a π bond that can break, a positive charge, or an unpaired electron, it can accept electrons without violating the octet rule.

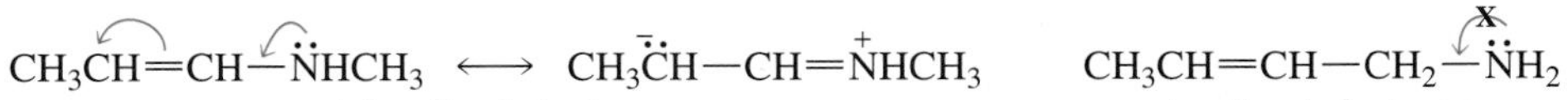

$$CH_3CH{=}CH{-}\ddot{N}HCH_3 \longleftrightarrow CH_3\ddot{C}H{-}CH{=}\overset{+}{N}HCH_3 \qquad CH_3CH{=}CH{-}CH_2{-}\ddot{N}H_2$$

delocalized electrons   localized electrons

The carbocation shown on the left below has delocalized electrons because the π electrons can move into the empty p orbital of the adjacent sp^2 carbon. We know that this carbon has an empty p orbital because it has a positive charge. The electrons in the carbocation on the right are localized because the π electrons cannot move. The carbon that they would move to is sp^3 hybridized, and sp^3 hybridized carbons cannot accept electrons.

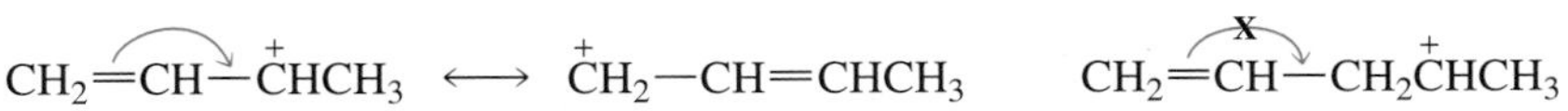

$$CH_2{=}CH{-}\overset{+}{C}HCH_3 \longleftrightarrow \overset{+}{C}H_2{-}CH{=}CHCH_3 \qquad CH_2{=}CH{-}CH_2\overset{+}{C}HCH_3$$

delocalized electrons   localized electrons

The next example shows a ketone with delocalized electrons (left) and a ketone with only localized electrons (right).

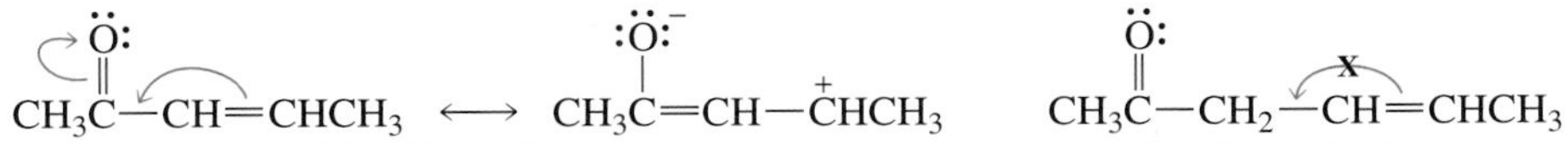

$$CH_3\overset{\ddot{O}:}{\overset{\|}{C}}{-}CH{=}CHCH_3 \longleftrightarrow CH_3\overset{:\ddot{O}:^{-}}{\overset{|}{C}}{=}CH{-}\overset{+}{C}HCH_3 \qquad CH_3\overset{\ddot{O}:}{\overset{\|}{C}}{-}CH_2{-}CH{=}CHCH_3$$

delocalized electrons   localized electrons

PROBLEM 3 ◆

a. Predict the relative bond lengths of the three carbon–oxygen bonds in the carbonate ion ($CO_3{}^{2-}$).

b. What would you expect the charge to be on each oxygen atom?

PROBLEM 4 ◆

a. Which of the following compounds have delocalized electrons?

1. $CH_3CH_2NHCH_2CH{=}CH_2$ **4.** $CH_2{=}CHCH_2CH{=}CH_2$

2. (benzene ring with NH_2 substituent)

5. 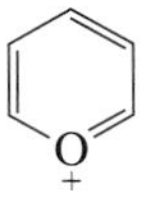 (six-membered ring with O at bottom, positively charged)

3. (benzene ring with CH_2NH_2 substituent)

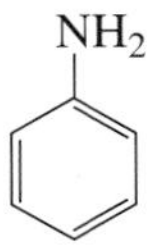

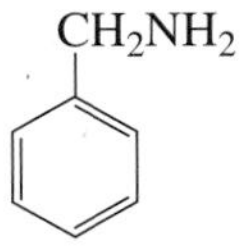

6. $CH_3CH{=}CHCH{=}CH\overset{+}{C}H_2$

b. Draw the contributing resonance structures for these compounds.

6.5
THE RESONANCE HYBRID

All resonance contributors do not contribute equally to the resonance hybrid. The degree to which each resonance contributor contributes to the hybrid depends on its predicted stability. Because the resonance contributors are not real, their stabilities cannot be measured. Therefore, stabilities of resonance contributors have to be predicted on the basis of molecular features that are found in real molecules. *The greater the predicted stability of the resonance contributor, the more it contributes to the hybrid.* And the more it contributes to the hybrid, the more similar the contributor is to the real molecule. The following examples illustrate these points.

Two resonance contributors for a carboxylic acid are shown below. Structure B has separated charges. To say that a structure has **separated charges** means that it has a positive charge and a negative charge that can be neutralized by the movement of electrons. We can predict that resonance contributors with separated charges are relatively unstable because it takes energy to keep the charges separated. Because structure A does not have separated charges, it has a considerably greater predicted stability. Because structure A is predicted to be more stable than structure B, structure A makes a greater contribution to the hybrid.

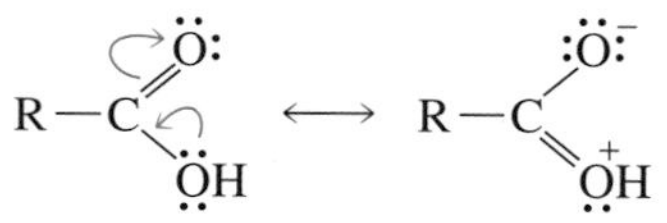

a carboxylic acid

The two resonance contributors for a carboxylate ion are shown next. Structures C and D are predicted to be equally stable and therefore are expected to contribute equally to the hybrid.

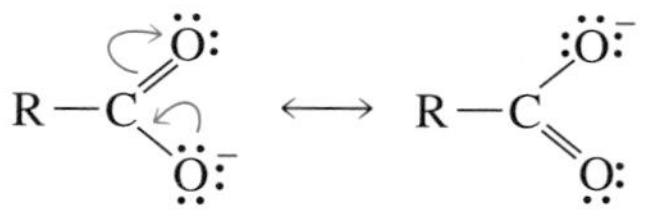

a carboxylate ion

When there are two possible directions in which electrons can be moved, they are always moved toward the more electronegative atom. For example, structure G in the next example results from moving π electrons toward oxygen, the more electronegative of the atoms. Structure E results from moving π electrons toward carbon, the less electronegative atom.

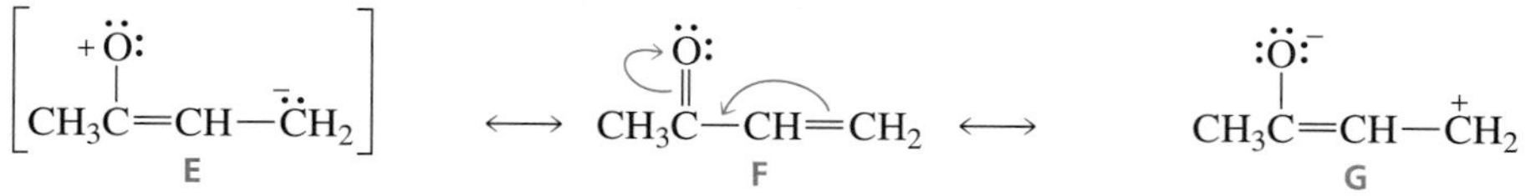

resonance contributor obtained by moving π electrons away from the more electronegative atom

resonance contributor obtained by moving π electrons toward the more electronegative atom

We can predict that structure G will make only a small contribution to the hybrid because it has separated charges as well as an atom with an incomplete octet. Structure E also has separated charges and an atom with an incomplete octet, but its predicted stability is even less than that of structure G because it has a positive charge on the electronegative oxygen. Its contribution to the hybrid is so insignificant that we do not need to include it as one of the resonance contributors. The hybrid, therefore, looks very much like structure F.

The only time that resonance contributors obtained by moving electrons away from the more electronegative atom should be shown is when this is the only way the electrons can be moved. In other words, movement of the electrons away from the more electronegative atom is better than no movement at all, because electron delocalization makes the molecule more stable (Section 6.6). For example, the only resonance contributor that can be drawn for the following compound requires movement of the electrons away from oxygen. Structure I is predicted to be relatively unstable because it has separated charges and its most electronegative atom is the one with the positive charge. Therefore, the structure of the hybrid is very similar to structure H, with only a small contribution from structure I.

The greater the predicted stability of the resonance contributor, the more it contributes to the structure of the resonance hybrid.

$$CH_2=CH-\ddot{O}CH_3 \longleftrightarrow \bar{\ddot{C}}H_2-CH=\overset{+}{\ddot{O}}CH_3$$
$$\phantom{CH_2=CH-\ddot{O}CH_3}H \phantom{\longleftrightarrow \bar{\ddot{C}}H_2-CH=\overset{+}{\ddot{O}}CH}I$$

Of the two contributing resonance structures for the enolate anion, structure J has a negative charge on carbon and structure K has a negative charge on oxygen. Oxygen is more electronegative than carbon and therefore can better accommodate the negative charge. Consequently, structure K has the greater predicted stability of the contributing resonance structures. The resonance hybrid most closely resembles structure K; the hybrid has a greater concentration of negative charge on oxygen than on carbon.

$$R-\overset{\overset{\displaystyle :\ddot{O}:}{\|}}{C}-\overset{\cdot\cdot}{C}HCH_3 \longleftrightarrow R-\overset{\overset{\displaystyle :\ddot{O}:^-}{|}}{C}=CHCH_3$$
$$J \phantom{-\overset{\cdot\cdot}{C}HCH_3 \longleftrightarrow R-\overset{\overset{\displaystyle :\ddot{O}:^-}{|}}{C}=C}K$$

an enolate anion

We can summarize the features that decrease the predicted stability of a contributing resonance structure as follows:

1. an atom with an incomplete octet

2. a negative charge that is not on the most electronegative atom or a positive charge that is not on the least electronegative (most electropositive) atom

3. charge separation

When we compare the relative stabilities of structures each of which has only one of these features, an atom with an incomplete octet (feature 1) generally makes a structure more unstable than does an atom with either feature 2 or feature 3.

PROBLEM 5 / SOLVED

Draw contributing resonance structures for each of the following species. Rank the structures in order of decreasing contribution to the hybrid.

a. $CH_3\overset{+}{\underset{\underset{\textstyle CH_3}{|}}{C}}-CH=CHCH_3$

b. $CH_3\overset{\overset{\textstyle O}{\|}}{C}OCH_3$

c. =O

d. =O

e. $CH_3-\overset{\overset{\textstyle \overset{+}{O}H}{\|}}{\underset{\underset{\textstyle CH_3}{|}}{C}}-NCH_3$

f. $CH_3\overset{+}{C}H-CH=CHCH_3$

SOLUTION TO 5a

$$CH_3\overset{+}{C}-CH=CHCH_3 \longleftrightarrow CH_3C=CH-\overset{+}{C}HCH_3$$
$$\underset{CH_3}{|} \qquad\qquad \underset{CH_3}{|}$$

A   B

Structure A is more stable than structure B because the positive charge is on a tertiary carbon in A and on a secondary carbon in B.

6.6
RESONANCE ENERGY

A compound with delocalized electrons is more stable than that compound would be if all its electrons were localized. The extra stability a compound gains as a result of having delocalized electrons is called **delocalization energy** or **resonance energy.** Let's take a look at the resonance energy of benzene. In other words, let's determine how much more stable benzene (with three pairs of delocalized π electrons) is than a hypothetical "cyclohexatriene" (with three pairs of localized π electrons).

The heat of hydrogenation of cyclohexene, a compound with one localized double bond, has been experimentally determined to be -28.6 kcal/mol. We would expect "cyclohexatriene," a hypothetical compound with three localized double bonds, to have a heat of hydrogenation three times that of the compound with one double bond. Therefore, we can calculate the heat of hydrogenation of "cyclohexatriene" to be $3(-28.6) = -85.8$ kcal/mol.

cyclohexene

$+ \ H_2 \longrightarrow \qquad \Delta H° = \text{-28.6 kcal/mol}$

cyclohexatriene
hypothetical

$+ \ 3 \ H_2 \longrightarrow \qquad \Delta H° = \text{-85.8 kcal/mol}$
calculated

When the heat of hydrogenation was experimentally determined for benzene, it was found to be -49.8 kcal/mol, much smaller than that calculated for the hypothetical "cyclohexatriene" molecule.

benzene

$+ \ 3 \ H_2 \longrightarrow \qquad \Delta H° = \text{-49.8 kcal/mol}$
experimental

If reaction coordinate diagrams are drawn for the addition of hydrogen to "cyclohexatriene" and for the addition of hydrogen to benzene, the energy of the product will be the same in both reaction coordinate diagrams because both reactions produce the same product, cyclohexane (Figure 6.6).

The only way to account for the difference in the heats of hydrogenation is for the reactants to have different stabilities. Since the experimental heat of hydrogenation for benzene is 36 kcal/mol less than the heat of hydrogenation calculated for

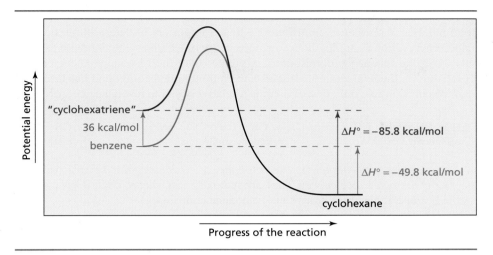

◀ **Figure 6.6**
Reaction coordinate diagrams
for the hydrogenation of
"cyclohexatriene" and
benzene.

"cyclohexatriene" (85.8 − 49.8 = 36), benzene must be 36 kcal/mol more stable than "cyclohexatriene."

Because benzene and "cyclohexatriene" have different stabilities, they must be different compounds. Benzene has six delocalized π electrons, whereas hypothetical "cyclohexatriene" has six localized π electrons. The difference in energy between them is the resonance energy of benzene. *The resonance energy tells us how much more stable a compound with delocalized electrons is than it would be if its electrons were localized.* Benzene, with six delocalized π electrons, is 36 kcal/mol more stable than hypothetical "cyclohexatriene" with six localized π electrons. Now we can understand why nineteenth-century chemists, who didn't know about delocalized electrons, were puzzled by benzene's unusual stability (Section 6.1).

Since the ability to delocalize electrons increases the stability of a molecule, we can conclude that *a resonance hybrid is more stable than the predicted stability of any of its resonance contributors.* The resonance energy associated with a compound that has delocalized electrons depends on the number *and* predicted stability of the resonance contributors. *The greater the number of relatively stable resonance contributors, the greater the resonance energy.* For example, the resonance energy of a carboxylate ion with two relatively stable resonance contributors is significantly greater than the resonance energy of a carboxylic acid that has only one relatively stable resonance contributor.

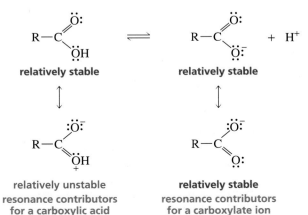

The greater the number of relatively stable resonance contributors, the greater the resonance energy.

Note that it is the number of *relatively stable* resonance contributors—not the total number of resonance contributors—that is important in determining the resonance energy. For example, the resonance energy of a carboxylate ion with two relatively stable resonance contributors is greater than the resonance energy of the compound in the next example because, even though this compound has three resonance contributors, only one of them is a relatively stable resonance contributor.

$$\bar{C}H_2-CH=CH-\overset{+}{C}H_2 \quad \longleftrightarrow \quad CH_2=CH-CH=CH_2 \quad \longleftrightarrow \quad \overset{+}{C}H_2-CH=CH-\bar{C}H_2$$

relatively unstable   **relatively stable**   relatively unstable

The more nearly equivalent the resonance contributors are in structure, the greater is the resonance energy. The carbonate dianion, shown next, is particularly stable because it has three equivalent resonance contributors.

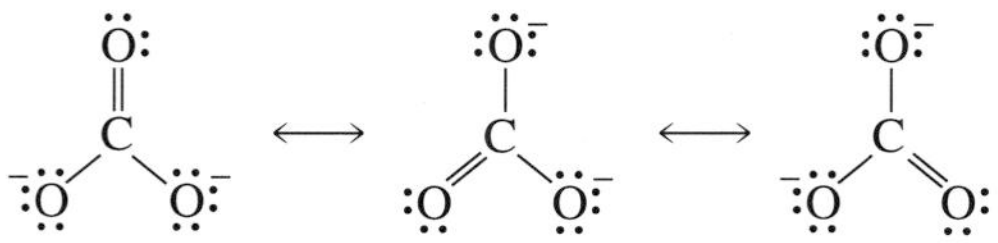

We can now summarize what we know about contributing resonance structures.

1. The greater the predicted stability of a resonance contributor, the more it contributes to the hybrid.

2. The greater the number of relatively stable resonance contributors, the greater the resonance energy.

3. The more nearly equivalent the resonance contributors, the greater the resonance energy.

6.7 STABILITY OF ALLYLIC AND BENZYLIC CATIONS

Allylic and benzylic cations have delocalized electrons. This makes them more stable than similarly substituted carbocations with localized electrons. An allylic carbon is an sp^3 carbon that is adjacent to an sp^2 carbon of an alkene. An **allylic cation** is a carbocation with the positive charge on an **allylic carbon.** The *allyl cation* is an unsubstituted allylic cation. A **benzylic carbon** is an sp^3 carbon that is bonded to a benzene ring. A **benzylic cation** is a carbocation with the positive charge on a benzylic carbon. The *benzyl cation* is an unsubstituted benzylic cation.

$CH_2=CH\overset{+}{C}HR$ $CH_2=CH\overset{+}{C}H_2$

an allylic cation   **the allyl cation**

a benzylic cation   **the benzyl cation**

An allylic cation has two resonance contributors. The positive charge is not localized on a single carbon but is shared by two carbons.

$RCH=CH-\overset{+}{C}H_2 \quad \longleftrightarrow \quad R\overset{+}{C}H-CH=CH_2$

an allylic cation

A benzylic cation has five resonance contributors. The positive charge is shared by four carbons.

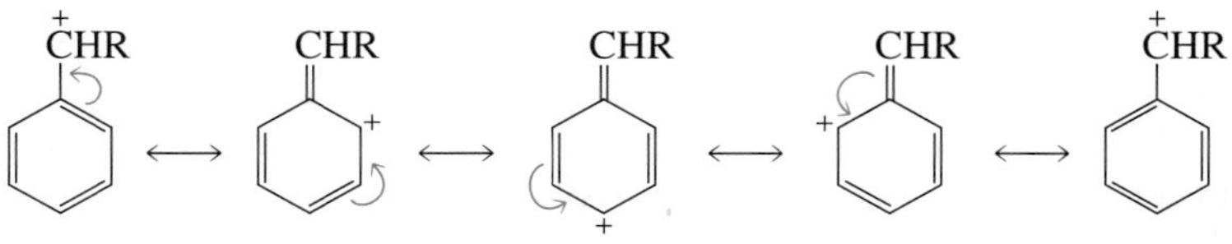

a benzylic cation

Because allyl and benzyl cations have delocalized electrons, they are more stable than other primary carbocations; their stability is approximately the same as that of a secondary carbocation. We can add the benzyl and allyl cations to the group of carbocations whose relative stabilities were shown in Sections 3.10 and 5.4.

relative stabilities of carbocations

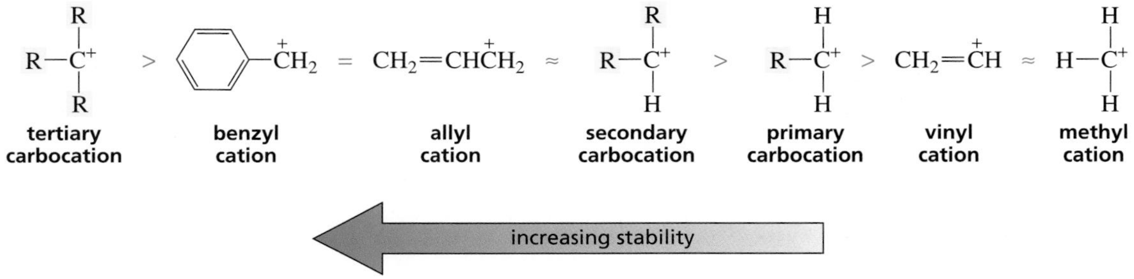

| tertiary carbocation | benzyl cation | allyl cation | secondary carbocation | primary carbocation | vinyl cation | methyl cation |

increasing stability

Not all benzylic and allylic cations have the same stability. Just as a tertiary carbocation is more stable than a secondary carbocation, a tertiary allylic cation is more stable than a secondary allylic cation, which in turn is more stable than the allyl cation. Similarly, a tertiary benzylic cation is more stable than a secondary benzylic cation, which is more stable than the benzyl cation. In this comparison of carbocation stabilities, notice that it is the allyl and benzyl cations that have the same stability as secondary carbocations. Secondary and tertiary allylic and benzylic cations are even more stable.

relative stabilities

| tertiary allylic cation | secondary allylic cation | allyl cation |

| tertiary benzylic cation | secondary benzylic cation | benzyl cation |

increasing stability

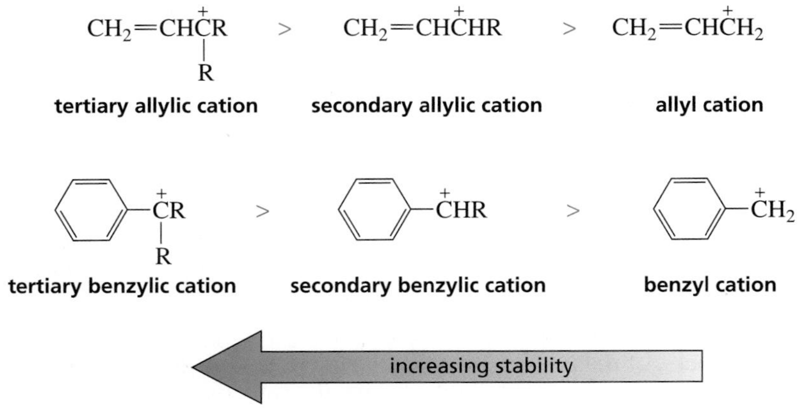

PROBLEM 6♦

Which carbocation in each of the following pairs is more stable?

a. $CH_3\overset{+}{O}CH_2$ or $CH_3\overset{+}{N}HCH_2$

b. 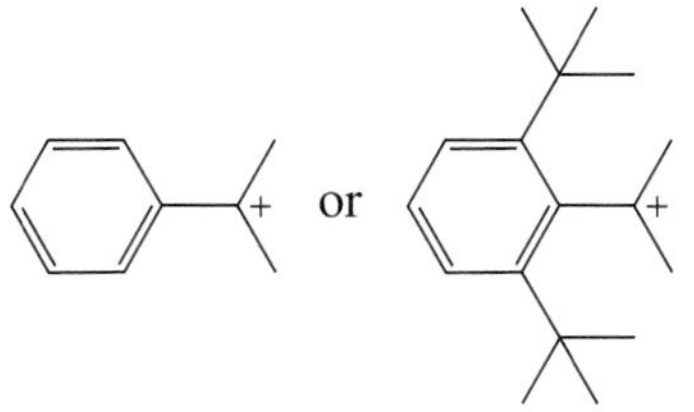 or

---

**6.8
STABILITY OF
ALLYLIC AND
BENZYLIC RADICALS**

An allylic radical has an unpaired electron on an allylic carbon and, like an allylic cation, has two contributing resonance structures. A benzylic radical has an unpaired electron on a benzylic carbon and, like a benzylic cation, has five contributing resonance structures.

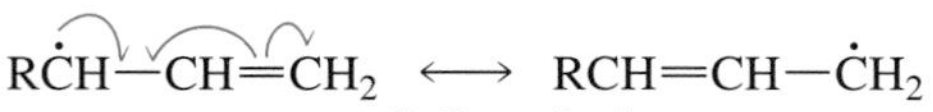

$$R\overset{\cdot}{C}H\!-\!CH\!=\!CH_2 \longleftrightarrow RCH\!=\!CH\!-\!\overset{\cdot}{C}H_2$$
an allylic radical

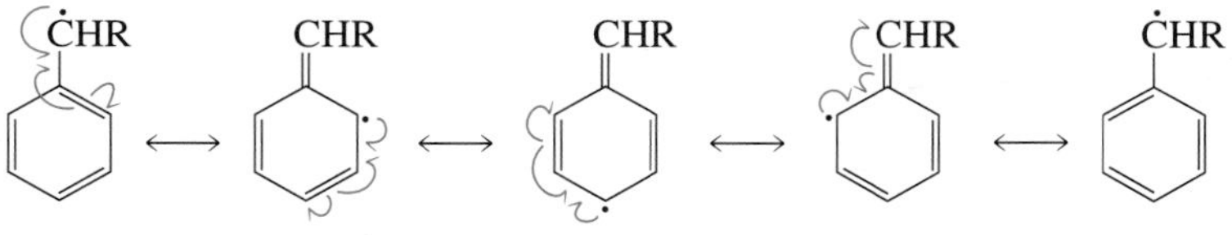

a benzylic radical

Because allylic and benzylic radicals have delocalized electrons, they are more stable than other radicals. The allyl and benzyl radicals are even more stable than tertiary radicals.

relative stabilities of radicals

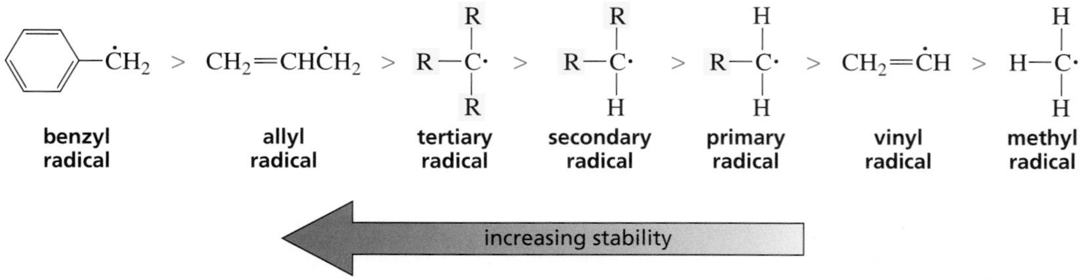

| benzyl radical | allyl radical | tertiary radical | secondary radical | primary radical | vinyl radical | methyl radical |

increasing stability

**6.9
SOME CHEMICAL
CONSEQUENCES OF
ELECTRON
DELOCALIZATION**

Our ability to predict the correct product of an organic reaction often depends on our ability to recognize when organic molecules have delocalized electrons. For example, in the following reaction, both sp^2 carbons of the alkene are bonded to the same number of hydrogens. Therefore Markovnikov's rule predicts that approximately equal amounts of the two addition products will be formed. When the reaction is carried out, however, only one of the products is obtained.

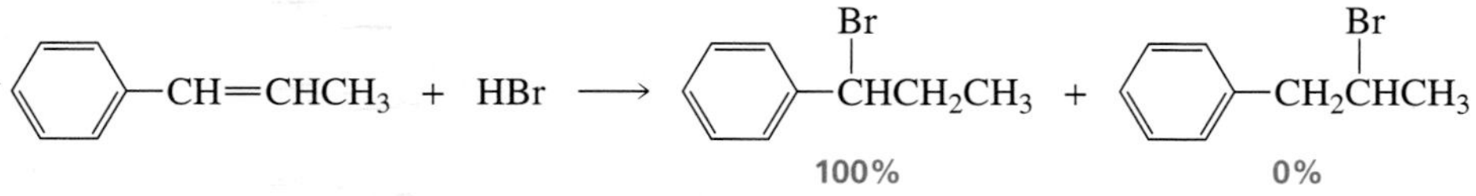

Markovnikov's rule leads us to an incorrect prediction of the product of the reaction shown above because it does not take resonance (electron delocalization) into consideration. It assumes that both carbocation intermediates are equally stable since they are both secondary carbocations. The rule does not recognize that one intermediate is an ordinary secondary carbocation while the other is a secondary benzylic carbocation. Because the secondary benzylic carbocation is stabilized by resonance, it is formed more readily, and therefore only one addition product is obtained.

a secondary benzylic cation   **a secondary carbocation**

This example serves as a warning. Do not use Markovnikov's rule in reactions in which the carbocations can be stabilized by resonance. In such cases, you must look at the stability of the individual carbocations in order to predict the product of the reaction.

The following reaction also demonstrates the importance of recognizing the presence of delocalized electrons in predicting the outcome of a reaction. The addition of a proton to the alkene yields a secondary carbocation. A carbocation rearrangement occurs because a more stable secondary benzylic carbocation is formed as a result of a 1,2-hydride shift. It is electron delocalization that causes the benzylic secondary carbocation to the more stable than the initially formed carbocation. If we had neglected electron delocalization, we would not have expected the carbocation rearrangement, and we would not have correctly predicted the product of the reaction.

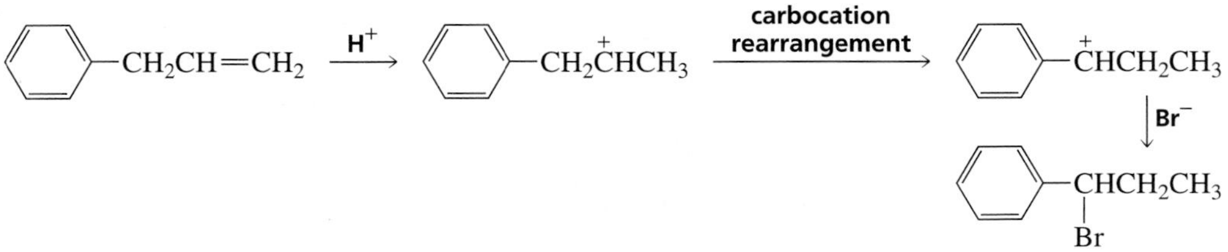

PROBLEM 7 / SOLVED

Predict the sites, on each of the following compounds, where the reaction can occur.

a. $CH_3CH=CHOCH_3$ + H^+

c. 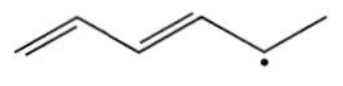 + $Br^{\cdot}$

b. 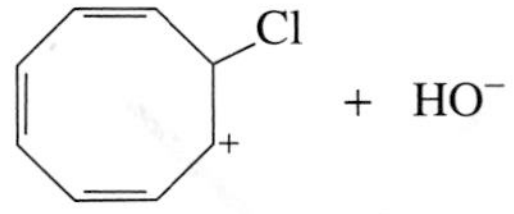 + HO^-

d. 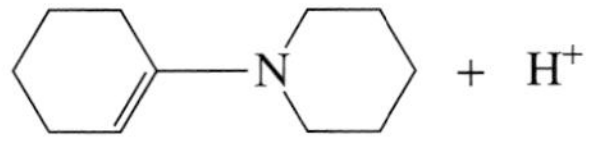 + H^+

SOLUTION TO 7a.   The contributing resonance structures show that there are two sites that can be protonated, the nonbonded electrons on oxygen and the nonbonded electrons on carbon.

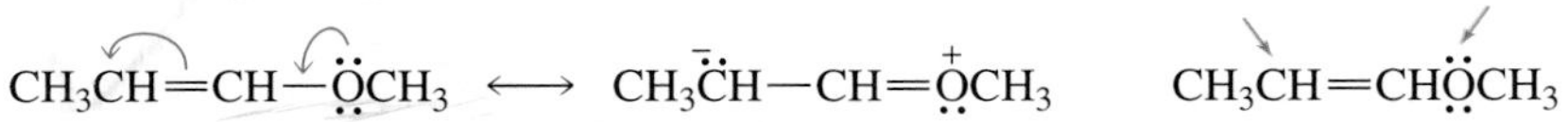

contributing resonance structures   sites of reactivity

**6.10
EFFECT OF
DELOCALIZED
ELECTRONS ON pK_a**

We have seen that a carboxylic acid is a much stronger acid than an alcohol. For example, the pK_a of acetic acid is 4.76, whereas the pK_a of ethyl alcohol is 15.9 (Section 1.17). Now we can explain more fully why a carboxylic acid is a stronger acid than an alcohol.

$$CH_3\overset{O}{\overset{\|}{C}}OH \qquad CH_3CH_2OH$$

acetic acid
pK_a = 4.76

ethyl alcohol
pK_a = 15.9

Substituents that withdraw electrons from the oxygen of the OH group increase the acidity of a compound by making the O—H bond easier to break (Section 1.18). An alcohol such as ethyl alcohol has no electron-withdrawing substituents, but both resonance contributors of a carboxylic acid have an electron-withdrawing group close to the OH group (Section 6.6): one contributor has an electron-withdrawing oxygen two atoms away from the OH group, and the other has an electron-withdrawing positively charged oxygen directly attached to the acidic hydrogen.

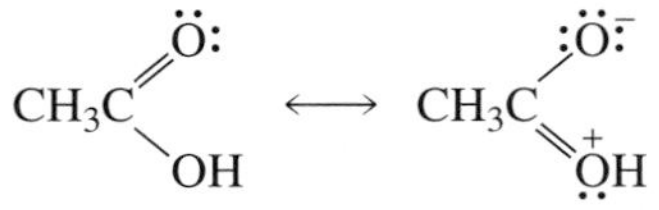

Another reason why the acidity of a carboxylic acid is greater than that of an alcohol is the resonance energy of the carboxylate ion. The carboxylate ion has greater resonance energy than the carboxylic acid because the anion has two equivalent resonance contributors that are predicted to be relatively stable, whereas the carboxylic acid has only one resonance contributor that is predicted to be relatively stable (Section 6.6). Therefore, loss of a proton from a carboxylic acid is accompanied by an increase in resonance energy.

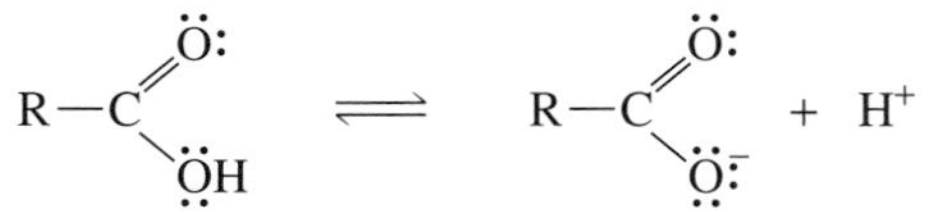

a carboxylic acid

a carboxylate ion

In contrast, all the electrons in an alcohol, such as ethanol, and its conjugate base are localized, so loss of a proton from an alcohol is not accompanied by resonance stabilization.

$$CH_3CH_2\ddot{O}H \rightleftharpoons CH_3CH_2\ddot{O}^- + H^+$$

ethanol

Phenol, a compound in which an OH group is bonded to an sp^2 carbon of a benzene ring, is a stronger acid than an alcohol such as ethanol or cyclohexanol, compounds in which an OH group is bonded to an sp^3 carbon.

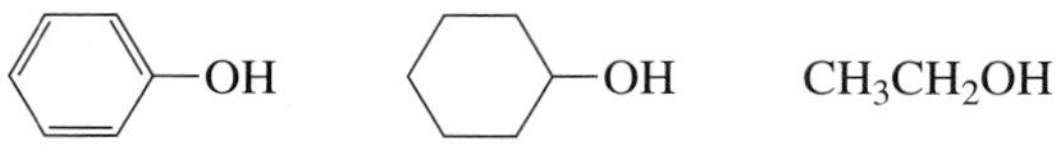

phenol
pK_a = 10

cyclohexanol
pK_a = 16

ethanol
pK_a = 16

One reason for the increased acidity of phenol is that the electronegativity of an

sp^2 carbon is greater than that of an sp^3 carbon. Also, the O—H bond of phenol is weakened by resonance contributors that have electron-withdrawing, positively charged oxygens. Finally, there is electron delocalization in both phenol and the phenolate ion, but the resonance energy of the phenolate ion is greater than the resonance energy of phenol since phenol has resonance contributors with separated charges. The greater resonance energy of the phenolate ion also contributes to the increased acidity of phenol compared with that of an alcohol such as cyclohexanol.

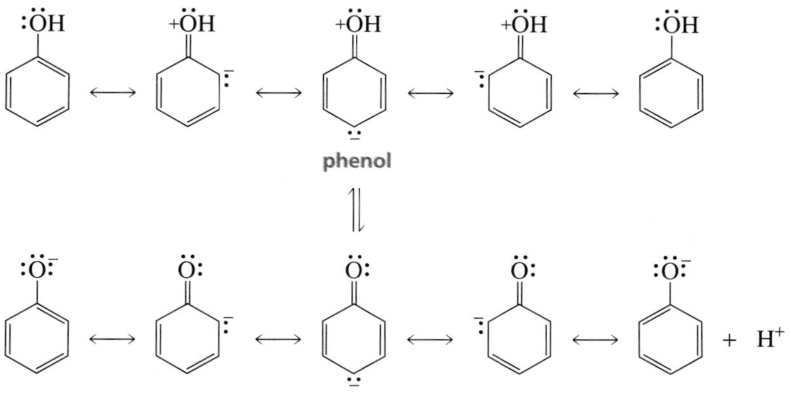

phenol

phenolate ion

The electron withdrawal in phenol is not as great as the electron withdrawal in a carboxylic acid, and the resonance energy of a phenolate ion is not as great as the resonance energy of a carboxylate ion where the negative charge is shared equally by two oxygens. The reduced electron withdrawal in phenol compared with a carboxylic acid and the reduced resonance energy in the phenolate ion compared with a carboxylate ion cause phenol to be a weaker acid than a carboxylic acid while being a stronger acid than an alcohol.

A protonated amine such as protonated aniline, in which the nitrogen atom is attached to an sp^2 carbon of a benzene ring, is a stronger acid than a protonated amine such as protonated cyclohexylamine, in which the nitrogen atom is attached to an sp^3 carbon.

protonated aniline
$pK_a = 4.60$

protonated cyclohexylamine
$pK_a = 11.2$

Part of the increased acidity of protonated aniline is attributable to the greater electronegativity of an sp^2 carbon compared with an sp^3 carbon. But electron delocalization also plays an important role. An amine such as cyclohexylamine does not have any delocalized electrons either in the protonated form or in the unprotonated form.

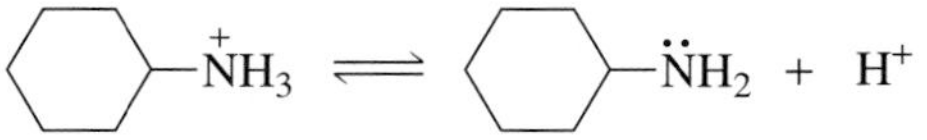

The nitrogen of protonated aniline also does not have any nonbonding electrons that can be delocalized. When it loses a proton, however, the nonbonding electron pair that held the proton can be delocalized. The increased stability achieved as a result of this electron delocalization contributes to the increased acidity of protonated aniline compared with a protonated amine in which loss of a proton does not result in any electron delocalization.

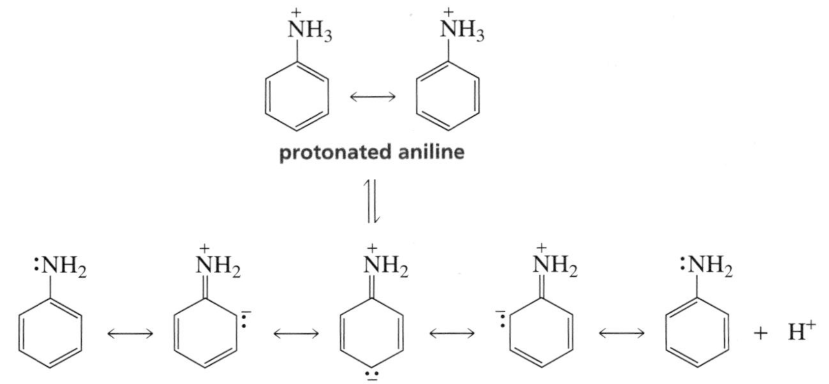

protonated aniline

aniline

The approximate pK_a values of carboxylic acids, alcohols, and protonated amines are listed in Table 1.7. We will add phenol and protonated aniline to the classes of organic compounds whose approximate pK_a values you should know (Table 6.1).

TABLE 6.1   Approximate pK_a Values

pK_a < 0	pK_a ≈ 5	pK_a ≈ 10	pK_a ≈ 15
$\overset{+}{R}\overset{}{O}\overset{}{H}$H	$\overset{O}{\underset{}{RCOH}}$	$R\overset{+}{N}H_3$	ROH
$\overset{+OH}{\underset{RCOH}{\parallel}}$	⬡—$\overset{+}{N}H_3$	⬡—OH	H_2O
H_3O^+			

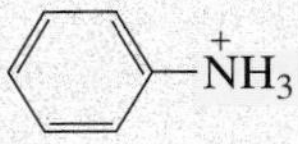

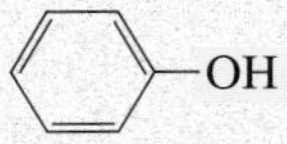

PROBLEM 8◆

Which is a stronger acid?

a. $CH_3CH_2CH_2OH$ or $CH_3CH=CHOH$

b. $CH_3CH=CHCH_2OH$ or $CH_3CH=CHOH$

PROBLEM 9◆

Which is a stronger base?

a. ethylamine or aniline

b. ethylamine or ethoxide ion ($CH_3CH_2O^-$)

c. phenolate ion or ethoxide ion

PROBLEM 10◆

Rank the following compounds in order of decreasing acid strength.

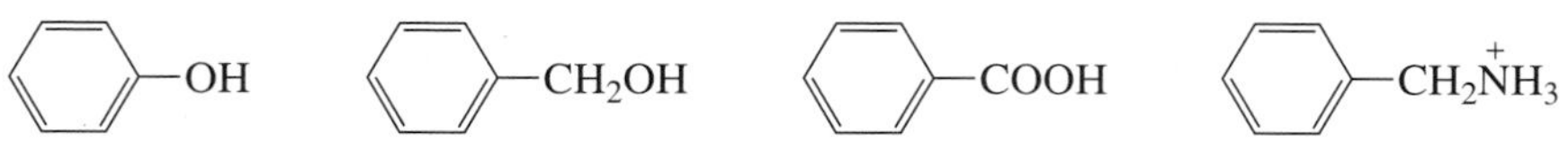

⬡—OH   ⬡—CH_2OH ⬡—COOH   ⬡—$CH_2\overset{+}{N}H_3$

In Section 6.6, we saw that benzene has an unusually large resonance energy of 36 kcal/mol. Most compounds with delocalized electrons have much smaller resonance energies. Compounds, such as benzene, with unusually large resonance energies are called **aromatic** compounds and are particularly stable. *For a compound to be classified as aromatic, it must fulfill both of the following criteria.*

Erich Hückel (1896–1980) *was born in Germany. He was a professor of chemistry at the University of Stuttgart and at the University of Marburg.*

1. It must have an uninterrupted ring of *p* orbital–bearing atoms so that it has an uninterrupted cloud of delocalized π electrons above and below the plane of the molecule (often called a π cloud). This implies that the molecule must be cyclic and planar.

2. The π cloud must contain an odd number of pairs of π electrons.

Benzene is an aromatic compound because it has an uninterrupted ring of *p* orbital–bearing atoms, and the π cloud contains three pairs of π electrons.

The German chemist Erich Hückel was the first to recognize that an aromatic compound must have an odd number of pairs of π electrons. In 1931, he described this requirement by what has come to be known as **Hückel's rule,** or the **4n + 2 rule.** The rule states that, for a planar, cyclic compound to be aromatic, it must have $(4n + 2)$ π electrons, where *n* is any whole number. According to Hückel's rule, then, an aromatic compound must have 2 ($n = 0$), 6 ($n = 1$), 10 ($n = 2$), 14 ($n = 3$), 18 ($n = 4$), etc. π electrons. Because there are two electrons in a pair, Hückel's rule requires that an aromatic compound must have 1, 3, 5, 7, 9, etc. pairs of π electrons. Thus, Hückel's rule is just an odd way to state that an aromatic compound must have an odd number of pairs of π electrons.

Richard E. Smalley *was born in 1943 in Akron, Ohio. He received a B.S. from the University of Michigan and a Ph.D. from Princeton University. He is a professor of chemistry at Rice University.*

Robert F. Curl, Jr. *was born in Texas in 1933. He received a B.A. from Rice University and a Ph.D. from the University of California, Berkeley. He is a professor of chemistry at Rice University.*

PROBLEM 11

Why must an aromatic compound be planar?

PROBLEM 12 ◆

a. What is the value of *n* in Hückel's rule when a compound has 11 pairs of π electrons?

b. Is such a compound aromatic?

Sir Harold W. Kroto *was born in 1939 in England and is a professor of chemistry at the University of Sussex.*

BUCKYBALLS AND AIDS

In addition to diamond and graphite, a third form of pure carbon was discovered while scientists were conducting experiments designed to understand how long-chain molecules are formed in outer space. It is the most symmetrical large molecule known, consisting of a hollow cluster of 60 carbons; each molecule has 32 interlocking rings (20 hexagons and 12 pentagons). It looks like a soccer ball. Its discoverers (R. E. Smalley, R. F. Curl, Jr., and H. W. Kroto, who shared the 1996 Nobel Prize in chemistry for their discovery) named it buckminsterfullerene (often shortened to fullerene) because it reminded them of the geodesic domes popularized by R. Buckminster Fuller, an American architect and philosopher. It is now nicknamed "buckyball." At first glance, the buckyball would appear to be aromatic because of its benzene-like rings. However, it undergoes electrophilic addition reactions like an alkene. Its lack of aromaticity is caused by the curvature of the ball, which prevents the molecule from fulfilling the first criterion for aromaticity (that it must be cyclic *and* planar).

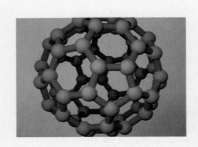

C₆₀
buckminsterfullerene
"buckyball"

Buckyballs have extraordinary chemical and physical properties. They are exceedingly rugged and are capable of surviving the extreme temperatures of outer space. Because they are essentially hollow cages, they can be manipulated to make materials never before known. For example, when a buckyball is "doped" by inserting potassium or cesium into its cavity, it becomes the best organic superconductor known. These molecules are presently being studied for use in many other applications, such as new polymers and catalysts and new drug delivery systems. The discovery of buckyballs is a strong reminder of the technological advances that can be achieved as a result of conducting basic research.

Scientists have even turned their attention to buckyballs in their quest for a cure for AIDS. An enzyme that is required for the HIV virus to reproduce exhibits a nonpolar pocket in its three-dimensional structure. If this pocket is blocked, the production of virus ceases. Because buckyballs are nonpolar and have approximately the same diameter as the pocket of the enzyme, they are being considered as possible blockers. The first step in pursuing this possibility was to equip the buckyball with polar arms to make it water-soluble so that it could flow through the bloodstream. Scientists have now modified the arms so that they bind to the enzyme. It's still a long way from a cure for AIDS, but this represents one example of the many and varied approaches that scientists are taking to find a cure for this disease.

6.12 AROMATICITY

Double bonds separated by one single bond are called **conjugated double bonds.** Cyclobutadiene and cyclooctatetraene have conjugated double bonds. (The ending "diene" signifies that the compound has two double bonds; "triene," three double bonds; "tetraene," four double bonds.) Monocyclic hydrocarbons with alternating single and double bonds are called **annulenes.** Cyclobutadiene, benzene, and cyclooctatetraene are annulenes. A prefix in brackets denotes the number of carbons in the ring.

cyclobutadiene   **benzene**   **cyclooctatetraene**
[4]-annulene   **[6]-annulene**   **[8]-annulene**

Cyclobutadiene has two pairs of π electrons, and cyclooctatetraene has four pairs of π electrons. Therefore, these compounds are not aromatic because they have an even number of pairs of π electrons. There is an additional reason why cylooctatetraene is not aromatic: it is not planar. Because these compounds are not aromatic, they do not have the unusual stability of aromatic compounds.

Which of the following three-membered ring structures is aromatic? Cyclopropene is not aromatic: it does not have an uninterrupted ring of p orbital–bearing atoms since one of its ring atoms is sp^3 hybridized. Therefore, it does not fulfill the first criterion for aromaticity.

cyclopropene   cyclopropenyl   cyclopropenyl
                           cation   anion

The cyclopropenyl cation is aromatic because it does have an uninterrupted ring of
p orbital–bearing atoms and the π cloud contains one pair of delocalized π electrons. The cyclopropenyl anion is not aromatic because it has two pairs of π electrons that could circle around the ring.

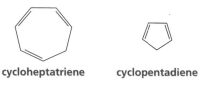

resonance contributors for the cyclopropenyl cation

resonance hybrid

Cycloheptatriene is not aromatic. It has the correct number of pairs of π electrons to be aromatic (three pairs), but it does not have an uninterrupted ring of p
orbital–bearing atoms, since one of the ring atoms is sp^3 hybridized. Cyclopentadiene is also nonaromatic. It has an even number of pairs of π electrons (two
pairs) and does not have an uninterrupted ring of p orbital–bearing atoms.

cycloheptatriene   cyclopentadiene

The criteria for determining whether a monocyclic compound is aromatic can
also be used to determine whether a polycyclic compound is aromatic. Azulene
(five pairs of π electrons), naphthalene (five pairs of π electrons), phenanthrene
(seven pairs of π electrons), and chrysene (nine pairs of π electrons) are
aromatic.

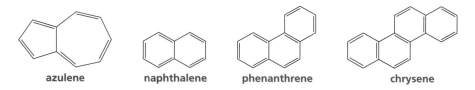

azulene   naphthalene   phenanthrene   chrysene

A compound does not have to be a hydrocarbon to be aromatic. Pyridine, pyrrole, furan, and thiophene are all aromatic compounds.

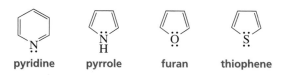

pyridine   pyrrole   furan   thiophene

Each of the six ring atoms of pyridine has a p orbital, and the molecule contains
three pairs of π electrons. (The nonbonding pair of electrons on nitrogen occupies
an sp^2 orbital.)

From the structure of pyrrole, it is not immediately obvious whether the nitrogen atom is sp^3 or sp^2 hybridized. If it is sp^3 hybridized, the nonbonding pair of electrons will occupy an sp^3 orbital and thus pyrrole will have only two pairs of π electrons and will not be aromatic. In contrast, if the nitrogen atom is sp^2 hybridized, the nonbonding pair will occupy a p orbital and thus pyrrole will have three pairs of π electrons and will be aromatic. Because the aromatic compound is more stable, the nitrogen atom is sp^2 hybridized, making pyrrole aromatic.

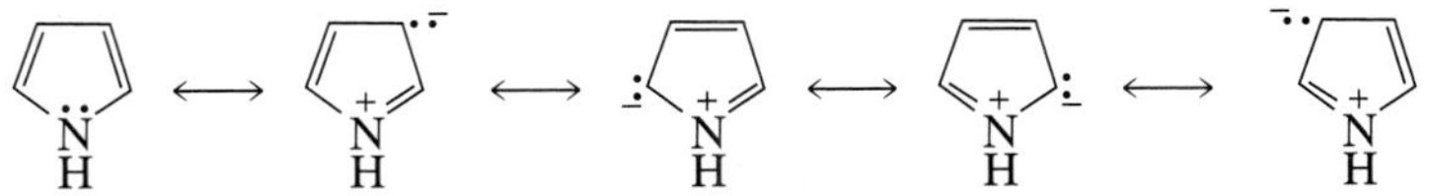

resonance contributors of pyrrole

Similarly, the oxygen atom in furan and the sulfur atom in thiophene are sp^2 hybridized, and thus furan and thiophene are stable aromatic compounds.

PROBLEM 13◆

Which of the following are aromatic?

a.

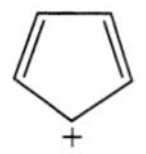

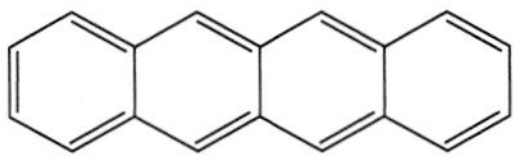

b.

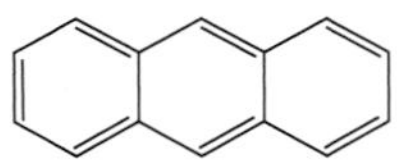

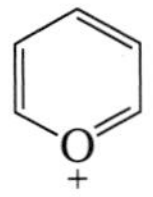

c. cycloheptatrienyl anion

d.

e.

f.

g. cyclononatetraenyl cation

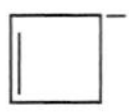

h. $CH_2{=}CHCH{=}CHCH{=}CH_2$

PROBLEM 14 / SOLVED

a. How many monobromonaphthalenes are there?

b. How many monobromophenanthrenes are there?

c. How many monobromoazulenes are there?

SOLUTION TO 14a. There are two monobromonaphthalenes.

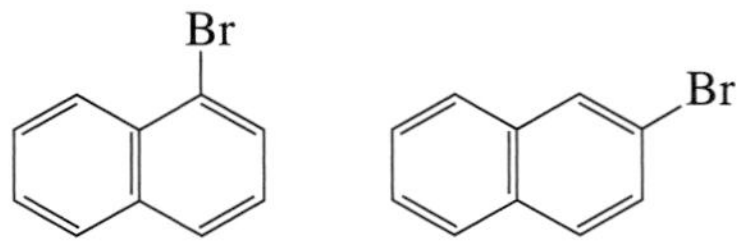

PROBLEM 15

The [10]- and [12]-annulenes have been synthesized, and neither has been found to be aromatic. Explain.

In Section 5.9, we saw that a hydrogen bonded to an *sp* hybridized carbon is more acidic than a hydrogen bonded to an *sp³* hybridized carbon. For example, acetylene has a pK_a of 25, whereas ethane has a pK_a of 50. The pK_a of cyclopentadiene is 15, which is surprisingly acidic for a hydrogen bonded to an *sp³* hybridized carbon. Indeed, it would even be unusually acidic for a hydrogen bonded to an *sp* hybridized carbon.

6.13
SOME CHEMICAL CONSEQUENCES OF AROMATICITY

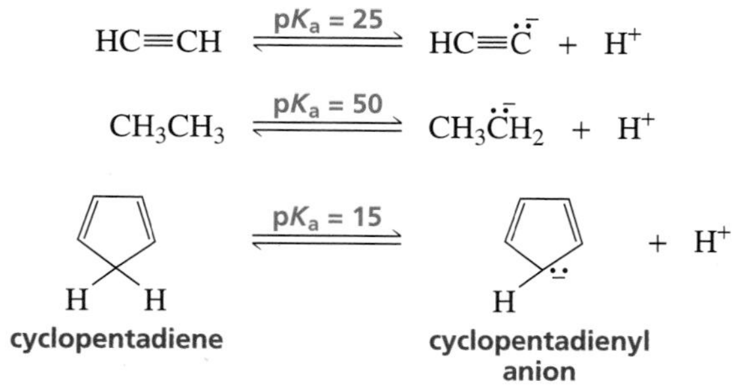

Why does cyclopentadiene have such a low pK_a? To answer this question, we must look at the anion that is formed when cyclopentadiene loses a proton. The cyclopentadienyl anion fulfills the two requirements for aromaticity: each atom in the ring has a *p* orbital, and the π cloud has three pairs of π electrons. Notice that the negatively charged carbon in the cyclopentadienyl anion is *sp²* hybridized; if it were *sp³* hybridized, the ion would not be aromatic. The resonance hybrid shows that all the carbons in the cyclopentadienyl anion are equivalent. Each carbon has exactly one-fifth of the negative charge associated with the anion.

resonance contributors

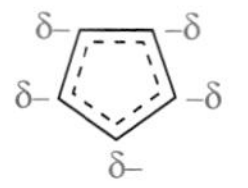

resonance hybrid

As a result of its aromaticity, the cyclopentadienyl anion is an unusually stable carbanion and therefore is easier to form than most carbanions (Figure 6.7). This is what gives cyclopentadiene its unusually low pK_a. In other words, the stability associated with the aromaticity of the cyclopentadienyl anion causes the hydrogen to be lost much more readily than hydrogens bonded to other *sp³* carbons.

PROBLEM 16

a. Draw the arrows showing the movement of electrons in going from one contributing resonance structure to the next in:
  1. the cyclopentadienyl anion   2. pyrrole

b. How many ring atoms share the negative charge in the following?
  1. the cyclopentadienyl anion   2. pyrrole

Figure 6.7 ►
Difference in the ease of formation of cyclopentyl anion and cyclopentadienyl anion.

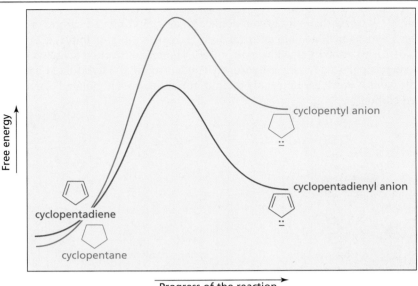

PROBLEM 17◆

Predict the relative pK_a values of cyclopentadiene and cycloheptatriene.

Another example of the influence of aromaticity on chemical reactivity is the unusual chemical behavior exhibited by cycloheptatrienyl bromide. Typically, alkyl halides are relatively nonpolar covalent compounds and therefore are soluble in nonpolar solvents and insoluble in water (Section 2.9). Cycloheptatrienyl bromide is an unusual alkyl halide in that its behavior is typical of an ionic compound; it is insoluble in nonpolar solvents but readily soluble in water. Investigation has shown that cycloheptatrienyl bromide is indeed an ionic compound because its ring is not aromatic in the covalent form, but is aromatic in the ionic form. In the covalent form, it has three pairs of π electrons but it does not have an uninterrupted ring of *p* orbital–bearing atoms. In the ionic form, however, the compound is composed of an aromatic cycloheptatrienyl cation and a bromide ion.

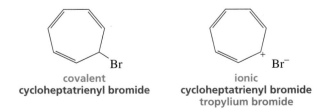

The cycloheptatrienyl cation is aromatic because it is a flat cyclic ion with three pairs of delocalized π electrons. The stability associated with the aromatic cation causes the alkyl halide to exist in the ionic form.

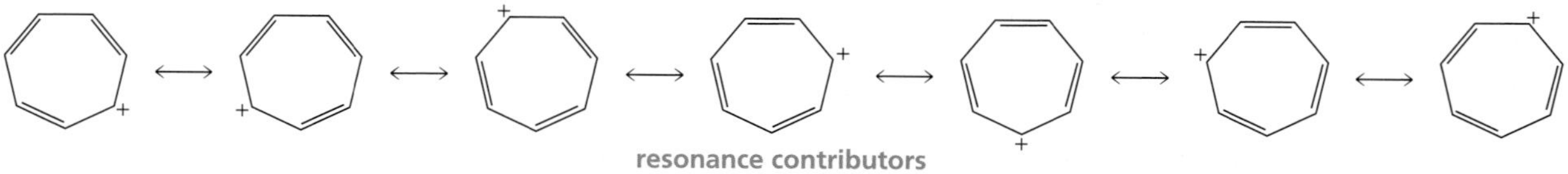

resonance contributors

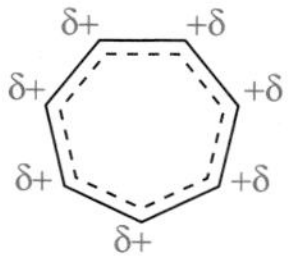

resonance hybrid

The reaction of cyclooctatetraene with potassium metal is yet another example of how the drive for a compound to become aromatic causes unusual chemical behavior. Each equivalent of potassium donates an equivalent of electrons to cyclooctatetraene.

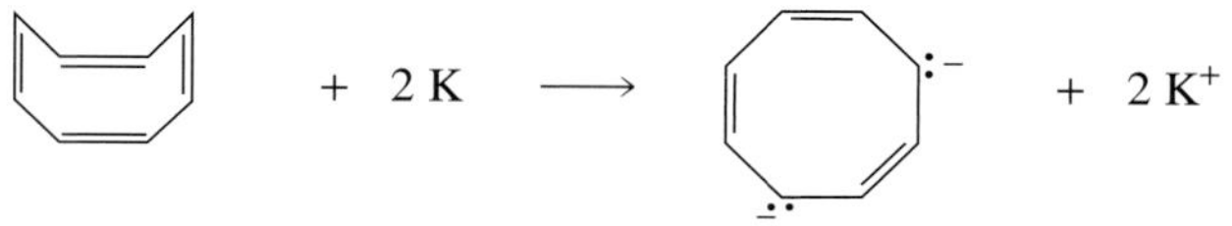

$$+ \ 2 \ K \ \longrightarrow \qquad + \ 2 \ K^+$$

cyclooctatetraene   **cyclooctatrienyl dianion**

This reaction occurs because the nonaromatic cyclooctatetraene molecule becomes an aromatic dianion by accepting two electrons. Cyclooctatetraene is not aromatic because it is not planar and it has four pairs of π electrons. The cyclooctatrienyl dianion is aromatic because it is planar and it has five pairs of delocalized π electrons.

SANDWICH COMPOUNDS

Some interesting compounds containing carbon–metal bonds (**organometallic compounds**) result from the aromaticity of the cyclopentadienyl anion. For example, the organometallic compound ferrocene is formed from the reaction of two equivalents of cyclopentadienyl anion with an equivalent of ferrous ion.

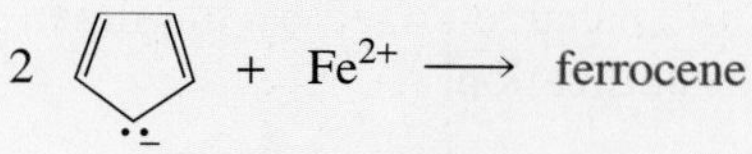

$$2 \quad + \ Fe^{2+} \ \longrightarrow \quad ferrocene$$

Ferrocene is a stable compound with an unusual "sandwich" structure. The ferrous ion is the "meat" of a sandwich in which two cyclopentadienyl anions act as the "bread." Since all the carbons in the cyclopentadienyl anion are equivalent, each is bonded equally to the ferrous ion.

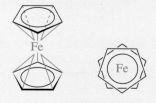

PROBLEM-SOLVING STRATEGY

Which of the following compounds has the greater dipole moment?

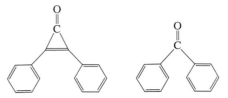

In answering this kind of question, you must first carefully determine what the question is asking. The dipole moment results from the unequal sharing of electrons by carbon and oxygen, and the more unequal the sharing, the greater the dipole moment. Draw the structures with separated charges and study them to determine the relative stabilities of the two structures with separated charges. In the case of the compound on the left, when the electrons of the π bond are placed on oxygen, the three-membered ring becomes aromatic. The aromaticity will stabilize the structure with separated charges. There is no additional aromaticity when the charges are separated in the compound on the right. Thus the compound on the left has the greater dipole moment.

PROBLEM 18

Draw the resonance contributors of the cyclooctatrienyl dianion.

a. Which of the resonance contributors is the least stable?

b. Which of the resonance contributors makes the smallest contribution to the hybrid?

6.14 ANTIAROMATICITY

An aromatic compound is *more stable* than the analogous acyclic compound. An **antiaromatic** compound is *less stable* than the analogous acyclic compound. Aromaticity is characterized by stabilization, whereas antiaromaticity is characterized by destabilization.

relative stabilities

aromatic compound > acyclic compound > antiaromatic compound

⬅ increasing stability

A compound is classified as being antiaromatic if it fulfills the first criterion for aromaticity but violates the second. Stated another way, it must be a planar cyclic compound with an uninterrupted ring of p orbital–bearing atoms and the π cloud must contain an *even* number of pairs of π electrons. Cyclobutadiene is a planar molecule with two pairs of π electrons. Hence it is antiaromatic and very unstable. The

cyclopentadienyl cation is also antiaromatic. Now we can understand why cyclooctatetraene is not a planar molecule; by not being planar, it avoids being antiaromatic.

cyclobutadiene   **cyclopentadienyl
cation**

PROBLEM 19◆

a. Predict the relative pK_a values of cyclopropene and cyclopropane.

b. Which is more soluble in water, 3-bromocyclopropene or bromocyclopropane?

PROBLEM 20◆

Which of the compounds in Problem 13 are antiaromatic?

Resonance has been used to describe the stabilities of compounds such as benzene
and an allylic cation. The stabilities of these compounds can also be described
using molecular orbital (MO) theory.

Molecular orbital (MO) theory was introduced in Section 1.6, where you saw
that the two lobes of a *p* orbital are of opposite phases. You also learned that when
two in-phase *p* orbitals overlap, a covalent bond is formed, and that when two out-
of-phase *p* orbitals overlap, they cancel each other and produce a node between the
two nuclei. A *node* is a region (in this case a plane) in which there is zero probability of finding an electron.

A molecular orbital description of ethene is shown in Figure 6.8. The two *p*
orbitals in ethene can be either in phase or out of phase. (The different phases are indicated by different colors.) Notice that the number of orbitals is conserved; two atomic
orbitals overlap to produce two molecular orbitals. Side-to-side overlapping of in-
phase *p* orbitals (lobes of the same color) produces a **bonding π molecular orbital**
designated ψ_1 (the Greek letter psi). The bonding π molecular orbital encompasses
both carbons. In other words, each electron in the π molecular orbital spreads over
both carbons. The bonding π molecular orbital is lower in energy than the *p* atomic
orbitals. Side-to-side overlapping of out-of-phase *p* orbitals produces an **antibonding
π molecular orbital,** ψ_2, which is higher in energy than the *p* atomic orbitals. The
antibonding orbital has a node between the lobes of opposite phases. The bonding
molecular orbital arises from additive interaction of the atomic orbitals, whereas the
antibonding molecular orbital arises from subtractive interaction.

Molecular orbitals that result from side-to-side overlapping of *p* orbitals contain
only π electrons. The π electrons are placed in molecular orbitals following the
same rules used for placing electrons in atomic orbitals: the aufbau principle
(orbitals are filled in order of increasing energy), the Pauli principle (each orbital
can hold two electrons of opposite spin), and Hund's rule (an electron will occupy
an empty degenerate orbital before it will pair up with an electron already present in
an orbital) apply (Section 1.2). Each π electron is associated with all the atoms
encompassed by the π molecular orbital.

Let's take a look at the molecular orbitals of the allyl cation, the allyl radical, and
the allyl anion. In all three cases, the π electrons are delocalized over three carbon
atoms.

6.15
A MOLECULAR
ORBITAL
DESCRIPTION OF
STABILITY

**Take a few minutes to
review Section 1.6.**

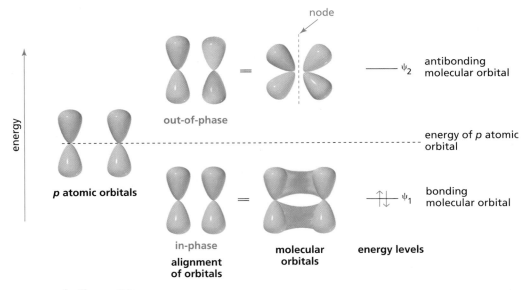

▲ **Figure 6.8**
The distribution of electrons in ethene. Overlapping of in-phase *p* orbitals produces a bonding π molecular orbital that is lower in energy than the *p* atomic orbitals. Overlapping of out-of-phase *p* orbitals produces an antibonding π molecular orbital that is higher in energy than the *p* atomic orbitals.

$$\overset{+}{\overbrace{CH_2 {=\!=\!=} CH {=\!=\!=} CH_2}} \qquad \overset{\bullet}{\overbrace{CH_2 {=\!=\!=} CH {=\!=\!=} CH_2}} \qquad \overset{\bullet\,\overset{-}{}}{\overbrace{CH_2 {=\!=\!=} CH {=\!=\!=} CH_2}}$$

  the allyl cation   **the allyl radical**   **the allyl anion**

Since each of the three carbon atoms contributes one *p* atomic orbital, the allyl group has three π molecular orbitals ψ_1, ψ_2, and ψ_3 (Figure 6.9). The bonding molecular orbital (ψ_1) encompasses all the carbons in the π system. In a noncyclic system, the number of bonding molecular orbitals always equals the number of antibonding molecular orbitals. This means that when there is an odd number of molecular orbitals, one of them is a nonbonding molecular orbital. In an allyl system, ψ_2 is a nonbonding molecular orbital. The nonbonding molecular orbital encompasses only the end atoms, because there is a node that passes through the middle carbon. In other words, the end orbitals are so far apart that there is no overlap between them—hence the term **nonbonding molecular orbital.** Notice that a nonbonding molecular orbital has the same energy as the isolated atomic *p* orbitals. The third molecular orbital (ψ_3) is an antibonding molecular orbital.

The allyl cation has two π electrons. They are both in the bonding π molecular orbital. Consequently, the two carbon–carbon bonds in the allyl cation are identical, with each having some double-bond character. This is another way of showing that the stability of the allyl cation is a result of electron delocalization.

Like the allyl cation, the allyl radical and the allyl anion also have two π electrons in the bonding π molecular orbital. The third π electron of the allyl radical and the third and fourth π electrons of the allyl anion are in the nonbonding molecular orbital.

Benzene, with six carbon atoms, contributes six *p* orbitals. The six *p* orbitals combine to produce six π molecular orbitals: three bonding molecular orbitals (ψ_1,

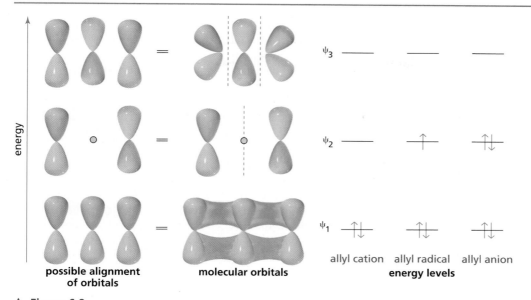

▲ Figure 6.9
Three p atomic orbitals overlap to produce three π molecular orbitals. The distribution of π electrons in the molecular orbitals of the allyl cation, the allyl radical, and the allyl anion.

ψ_2, and ψ_3) and three antibonding molecular orbitals (ψ_4, ψ_5, and ψ_6) (Figure 6.10). Benzene's six π electrons occupy the three bonding molecular orbitals and therefore are delocalized over the six carbon atoms. This cyclic cloud of delocalized π electrons accounts for benzene's large resonance (delocalization) energy. Compare the molecular orbital picture of six atomic p orbitals combining to produce a cyclic compound with a cyclic π electron cloud (in Figure 6.10) with the molecular orbital picture of six p orbitals combining to produce a acyclic compound with a linear π electron cloud (Section 27.3).

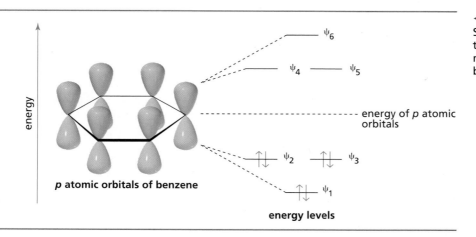

◀ Figure 6.10
Six p atomic orbitals overlap to produce the six π molecular orbitals of benzene.

6.16
A MOLECULAR ORBITAL DESCRIPTION OF AROMATICITY AND ANTIAROMATICITY

Why are planar molecules with continuous π electron clouds very stable (aromatic) if they have an odd number of pairs of π electrons, and why are they very unstable (antiaromatic) if they have an even number of pairs of π electrons? The answer comes from MO theory. Benzene has six carbon atoms, contributes six p atomic orbitals, and therefore has six π molecular orbitals. Cyclobutadiene has four carbon atoms, contributes four p atomic orbitals, and has four π molecular orbitals. The cyclopentadienyl anion has five carbon atoms, contributes five p atomic orbitals, and has five π molecular orbitals. The number of molecular orbitals in a cyclic con-jugated compound is equal to the number of carbon atoms in the ring. The relative energies of the molecular orbitals can be determined by drawing the cyclic compound with one of its vertices pointed down; the relative energies of the molecular orbitals correspond to the relative levels of the vertices (Figure 6.11).

Molecular orbitals that are below the midpoint of the cyclic structure are bonding molecular orbitals, those that are above the midpoint are antibonding molecular orbitals, and those that are at the midpoint are nonbonding molecular orbitals.

The six π electrons of benzene occupy its three bonding π molecular orbitals, and the six π electrons of the cyclopentadienyl anion occupy its three bonding π molecular orbitals. Notice that there is always an odd number of bonding orbitals since one corresponds to the lowest vertex and the others come in degenerate pairs. This means that aromatic compounds (such as benzene and the cyclopentadienyl anion) with an odd number of pairs of π electrons have completely filled bonding orbitals and no electrons in either nonbonding or antibonding orbitals. This is what gives aromatic molecules their stability.

Antiaromatic compounds have an even number of pairs of π electrons. Therefore, they either are unable to fill their bonding orbitals (cylopentadienyl cation) or have a pair of π electrons left over after the bonding orbitals are filled (cyclobutadiene). Hund's rule requires that these two electrons go either into two degenerate nonbonding orbitals or into two degenerate antibonding molecular orbitals. The unfilled bonding orbitals or the unpaired electrons in nonbonding or antibonding orbitals are responsible for the destabilization of antiaromatic molecules.

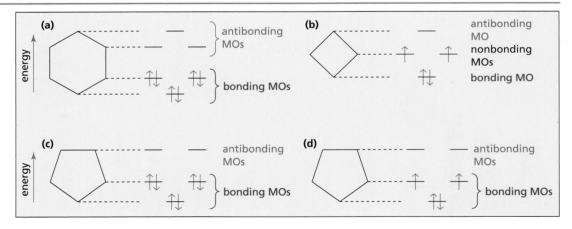

▲ Figure 6.11
The distribution of π electrons in the π molecular orbitals of (a) benzene, (b) cyclobuta-diene, (c) the cyclopentadienyl anion, and (d) the cyclopentadienyl cation.

PROBLEM 21

Why is a cyclic compound with an odd number of pairs of delocalized π electrons more stable than one with an even number of pairs of delocalized π electrons?

PROBLEM 22

Following the instructions given for drawing the π molecular orbital energy levels for the compounds shown in Figure 6.11, draw the π molecular orbital energy levels for the cycloheptatrienyl cation, the cycloheptatrienyl anion, and pyrrole. For each compound, show the distribution of the π electrons. Which of the compounds are aromatic? Which are antiaromatic?

KEY TERMS

allylic carbon (page 276)
allylic cation (page 276)
annulene (page 284)
antiaromatic (page 290)
antibonding π molecular orbital
  (page 291)
aromatic (page 283)
benzylic carbon (page 276)
benzylic cation (page 276)

bonding π molecular orbital (page 291)
conjugated double bonds (page 284)
contributing resonance structure
  (page 265)
delocalization energy (page 274)
delocalized electrons (page 261)
Hückel's rule, or the $4n + 2$ rule
  (page 283)
localized electrons (page 261)

nonbonding molecular orbital
  (page 292)
organometallic compound (page 289)
resonance (page 265)
resonance contributor (page 265)
resonance energy (page 274)
resonance hybrid (page 265)
resonance structure (page 265)
separated charges (page 272)

PROBLEMS

23. Which of the following compounds have delocalized electrons?

a. $CH_2\!=\!CHCCH_3$ (with $\overset{O}{\overset{\|}{C}}$)

g.

b. (cyclohexadiene ring structure)

h. $CH_3CH_2NHCH_2CH\!=\!CHCH_3$

c. (cyclopentadienyl structure)

i. $CH_3CH_2NHCH\!=\!CHCH_3$

d. (bicyclic fused ring structure)

j. (bicyclic fused ring structure)

e. $CH_2\!=\!CHCH_2CH\!=\!CH_2$

k. $CH_3\overset{+}{C}CH_2CH\!=\!CH_2$ with CH_3 substituent

f. (cyclohexene ring structure)

l. $CH_3CH_2\overset{+}{C}HCH\!=\!CH_2$

m. $CH_3CH\!=\!CHOCH_2CH_3$

24. a. Draw resonance contributors for the following species, showing all the nonbonded pairs of electrons.
   b. For each species, indicate the most stable resonance contributor.
   1. CH_2N_2
   2. N_2O
   3. NO_2^-

25. Draw resonance contributors for the following ions.

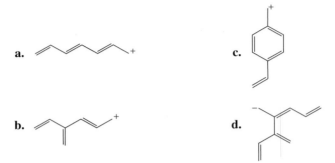

a.

b.

c.

d.

26. Are the following pairs of structures resonance contributors or different compounds?

a. 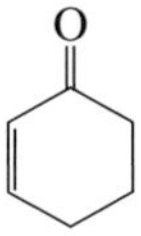 [cyclohexenone structures] and

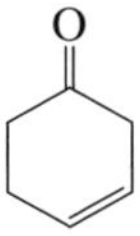

b. $CH_3CH=CH\dot{C}HCH=CH_2$ and $CH_3\dot{C}HCH=CHCH=CH_2$

c. $CH_3\overset{\text{O}}{\overset{\|}{C}}CH_2CH_3$ and $CH_3\overset{\text{OH}}{\overset{|}{C}}=CHCH_3$

d.  [cyclohexene cation] and [cyclohexenyl cation]

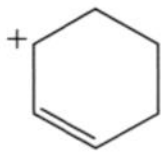

e. $CH_3\overset{+}{C}HCH=CHCH_3$ and $CH_3CH=CHCH_2\overset{+}{C}H_2$

27. Draw resonance contributors for the following species.
   a. Indicate which are major contributors and which are minor contributors to the resonance hybrid. Do not include structures that are so unstable that their contributions to the hybrid would be negligible.
   b. Do any of the species have resonance contributors that all contribute equally to the hybrid?

1. $CH_3CH=CHOCH_3$ **6.** $H\overset{\text{O}}{\overset{\|}{C}}NHCH_3$

2. $CH_3CH_2\overset{\text{O}}{\overset{\|}{C}}OCH_2CH_3$ **7.** [cyclopentadienyl radical]

3. $CH_3CH=\overset{+}{C}HCH_2$ **8.** $CH_2=CH\overset{-}{C}H_2$

4. $CH_3CH=CHCH=CH\dot{C}H_2$ **9.** [benzene ring with OCH_3]

5. $CH_3\overset{-}{C}HC\equiv N$ **10.** CO_3^{2-}

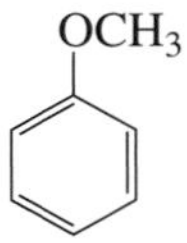

11. 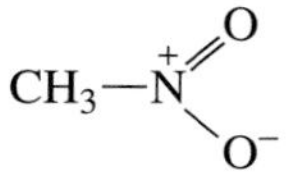 $CH_3-\overset{+}{N}\overset{O}{\underset{O^-}{\diagup}}$

15.

$\underset{\text{(benzene ring)}}{\bigcirc}$ with $CH=CH_2$

12. $\bar{C}H_2\overset{O}{\overset{\|}{C}}OCH_2CH_3$

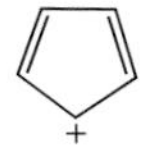

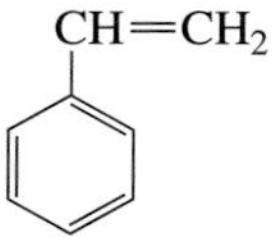

16.

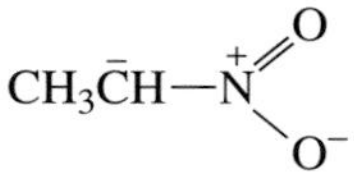

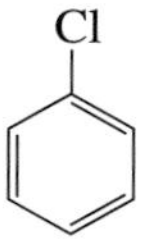

$\underset{\text{(benzene ring)}}{\bigcirc}$ with Cl

13. $\underset{+}{\bigcirc}$ (cyclopentadienyl cation)

17. $CH_3\overset{O}{\overset{\|}{C}}\bar{C}H\overset{O}{\overset{\|}{C}}CH_3$

14. $CH_3\bar{C}H-\overset{+}{N}\overset{O}{\underset{O^-}{\diagup}}$

18. $H\overset{O}{\overset{\|}{C}}CH=CH\bar{C}H_2$

28. Which resonance contributor makes the greater contribution to the hybrid?

a. $CH_3\overset{+}{C}HCH=CH_2$ or $CH_3CH=CH\overset{+}{C}H_2$

b. 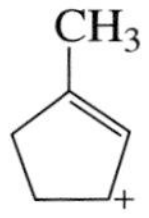 (cyclopentene with CH_3 and + on ring) or (cyclopentene with CH_3 and + on ring)

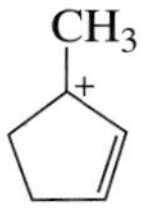

c. 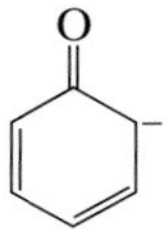 (cyclohexadienone anion) or (phenolate, O^-)

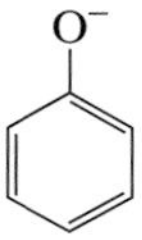

29. Rank the indicated carbon–hydrogen bonds in order of decreasing ease of homolytic cleavage.

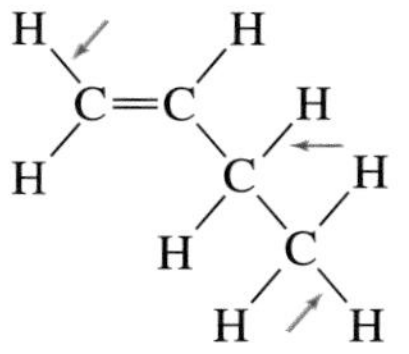

(structure showing $C=C$ and C chain with indicated H atoms)

30. Which of the following compounds are aromatic? Are any antiaromatic? (*Hint:* If possible, a ring will be nonplanar, to avoid being antiaromatic.)

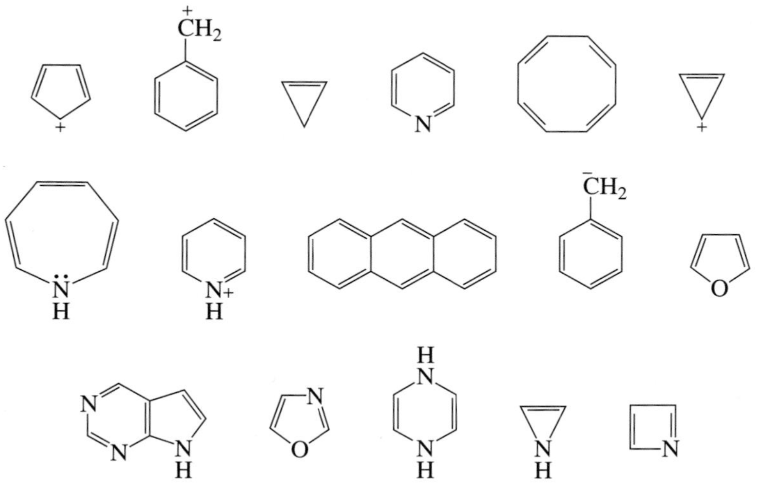

(various structures: cyclopentadienyl cation; $\overset{+}{C}H_2$ on benzene; cyclopropene cation; pyridine; cyclooctatetraene; cyclopropenyl cation; azocine (N–H); pyridinium (N^+–H); anthracene; $\bar{C}H_2$ on benzene; furan; azonine (N–H); oxazole; pyrazine (N–H); aziridine (N–H); azetidine (N); 7-azaindole type (N, N–H))

31. a. Which compound has the greater electron density on its nitrogen atom?

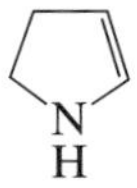

or

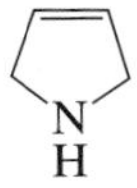

b. Which compound has the greater electron density on its oxygen atom?

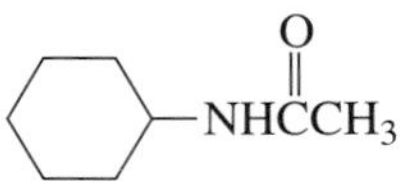

$$\text{cyclohexyl—NHCCH}_3 \quad \text{or} \quad \text{phenyl—NHCCH}_3$$

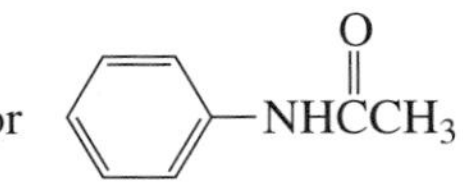

32. Chamazulene is a blue oil with antiinflammatory properties that is obtained from wormwood. Is chamazulene aromatic?

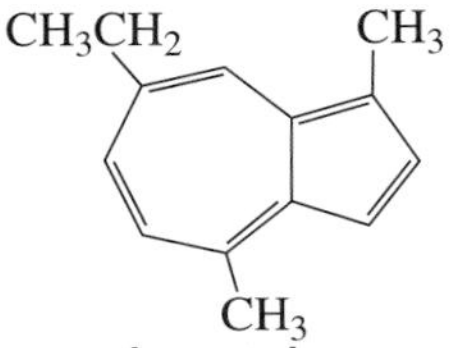

chamazulene

33. In each of the following pairs, which ion is more stable?

a.  or

b. 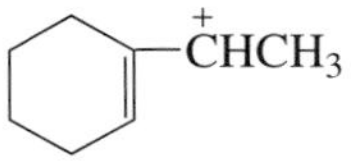 or

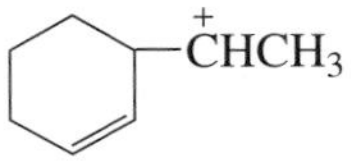

c. 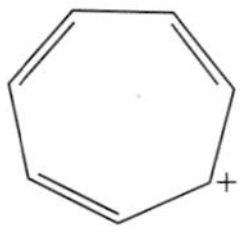 or

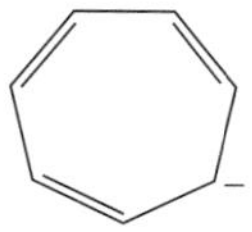

d. 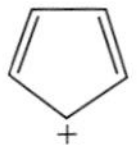 or

e. $CH_3\bar{C}HCH_3$ or $CH_3\bar{C}HCH=CH_2$

f. $CH_3\bar{C}HCH_2\overset{O}{\overset{\|}{C}}CH_3$ or $CH_3\bar{C}H\overset{O}{\overset{\|}{C}}CH_3$

34. Which can lose a proton more readily, a methyl group bonded to cyclohexane or a methyl group bonded to benzene?

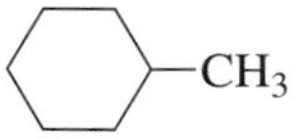

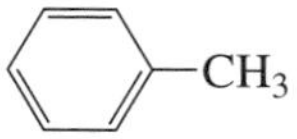

—CH_3 —CH_3

35. Rank the following compounds in order of decreasing acidity.

36. When this compound ionizes, is Cl^- or Cl^+ formed?

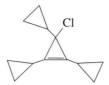

37. We saw in Chapter 5 that acetylene reacts with an equivalent amount of HCl to form vinyl chloride. In the presence of excess HCl, the final reaction product is 1,1-dichloroethane. Why is 1,1-dichloroethane formed in preference to 1,2-dichloroethane?

$$HC{\equiv}CH \xrightarrow{\text{HCl}} H_2C{=}\underset{\underset{Cl}{|}}{CH} \xrightarrow{\text{HCl}} CH_3CHCl_2$$

38. **a.** What is the direction of the dipole moment in fulvene? Explain.
b. What is the direction of the dipole moment in calicene? Explain.

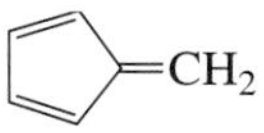

fulvene 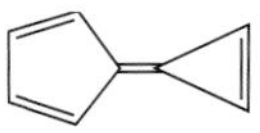   **calicene**

39. Furan has a resonance energy of 16 kcal/mol. Pyrrole has a resonance energy of 21 kcal/mol. Why is the resonance energy of pyrrole greater than the resonance energy of furan?

furan    **pyrrole**

40. Rank the following compounds in order of decreasing boiling point.

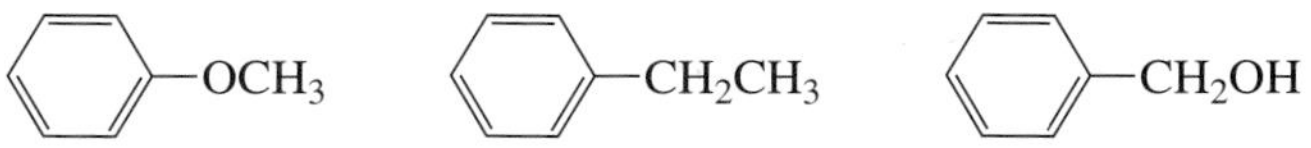

$-OCH_3$ $-CH_2CH_3$ $-CH_2OH$

41. Rank the following compounds in order of decreasing reactivity with HBr.

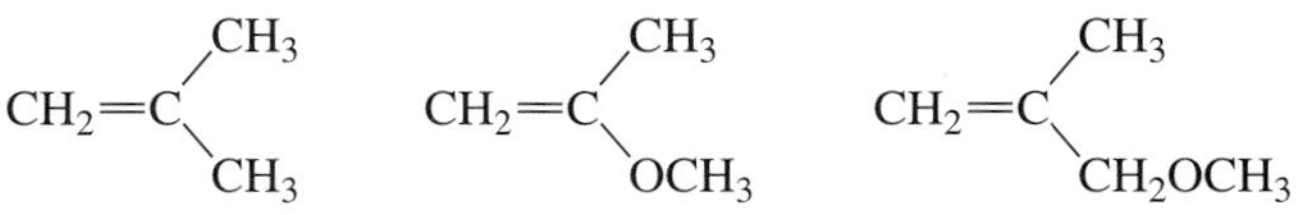

$$CH_2{=}\underset{\underset{CH_3}{|}}{\overset{\overset{CH_3}{|}}{C}} \qquad CH_2{=}\underset{\underset{OCH_3}{|}}{\overset{\overset{CH_3}{|}}{C}} \qquad CH_2{=}\underset{\underset{CH_2OCH_3}{|}}{\overset{\overset{CH_3}{|}}{C}}$$

42. In each of the following pairs, which compound is a stronger base? Why?

a. 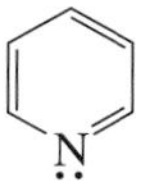   or 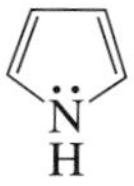   **b.** $CH_3\overset{\overset{NH_2}{|}}{C}HCH_3$ or $CH_3\overset{\overset{NH}{\|}}{C}NH_2$

43. Explain why Markovnikov's rule is followed in reaction **a** but is not followed in reaction **b**.
a. $CH_2{=}CHF + HF \longrightarrow CH_3CHF_2$
b. $CH_2{=}CHCF_3 + HF \longrightarrow FCH_2CH_2CF_3$

44. The acid dissociation constant (K_a) for loss of a proton from cyclohexanol is 1×10^{-16}.

   a. Draw an energy diagram for loss of a proton from cyclohexanol.

$$\bigcirc\!\!-\!OH \;\xrightleftharpoons[\;\;\;\;\;]{K_a = 1 \times 10^{-16}}\; \bigcirc\!\!-\!O^- \;+\; H^+$$

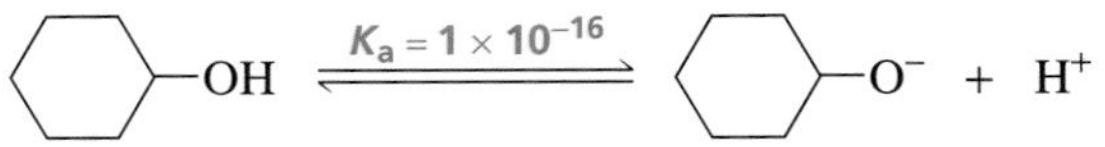

   b. Draw the contributing resonance structures for phenol.
   c. Draw the contributing resonance structures for the phenolate ion.
   d. Draw an energy diagram for loss of a proton from phenol on the same plot with the energy diagram for loss of a proton from cyclohexanol.

$$\bigcirc\!\!-\!OH \;\rightleftharpoons\; \bigcirc\!\!-\!O^- \;+\; H^+$$

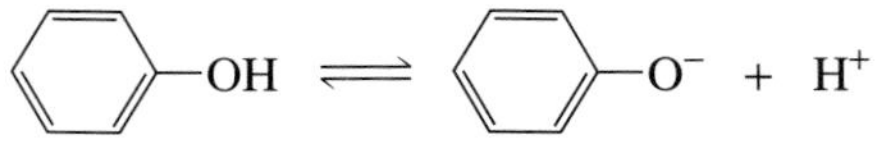

   e. Which has a greater K_a, cyclohexanol or phenol?
   f. Which is a stronger acid, cyclohexanol or phenol?

45. Protonated cyclohexylamine has a K_a of 1×10^{-11}. Using the same sequence of steps as in Problem 44, determine which is a stronger base, cyclohexylamine or aniline.

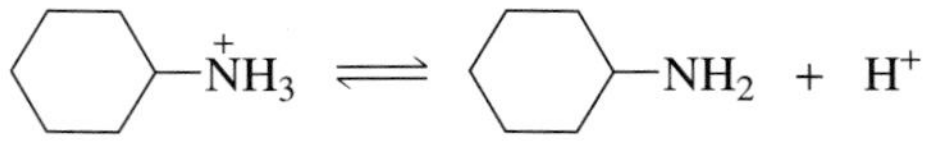

$$\bigcirc\!\!-\!\overset{+}{N}H_3 \;\rightleftharpoons\; \bigcirc\!\!-\!NH_2 \;+\; H^+$$

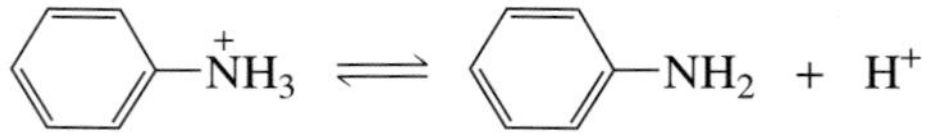

$$\bigcirc\!\!-\!\overset{+}{N}H_3 \;\rightleftharpoons\; \bigcirc\!\!-\!NH_2 \;+\; H^+$$

REACTIONS OF DIENES

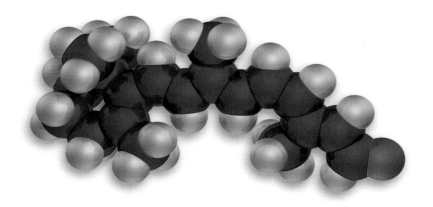

(*Z*)-11-retinal

M any organic compounds contain more than one functional group. When these functional groups are sufficiently separated in a molecule, each undergoes its characteristic reactions. But when functional groups are close enough to allow electron delocalization, the reactivity of each functional group can be affected by the presence of another.

Hydrocarbons with two double bonds are called **dienes,** and hydrocarbons with three double bonds are called **trienes. Tetraenes** have four double bonds, and **polyenes** have several double bonds. Here, we will be concerned mainly with the reactions of dienes, but the same considerations apply to hydrocarbons that contain more than two double bonds.

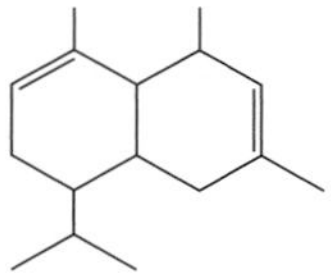

α-cardinene
oil of citronella
a diene

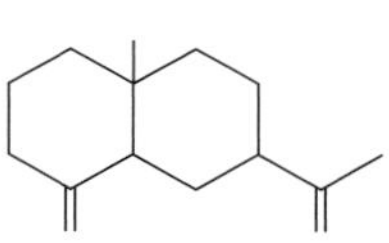

β-selinene
oil of celery
a diene

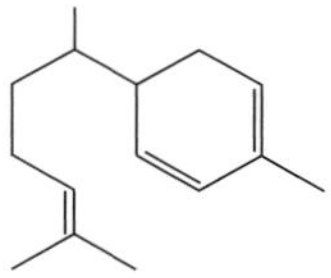

zingiberene
oil of ginger
a triene

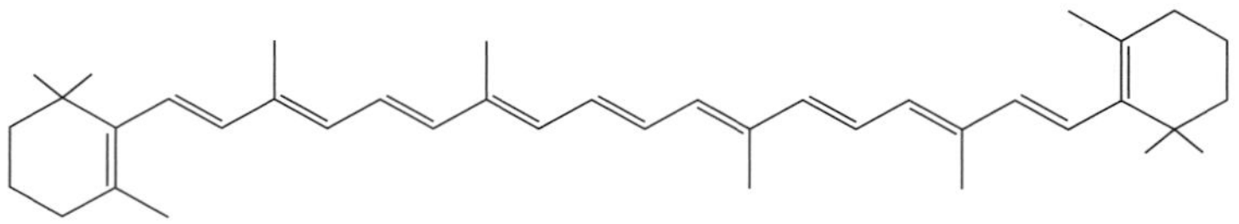

β-carotene
a polyene

Double bonds can be conjugated, isolated, or cumulated. **Conjugated double bonds** are separated by one single bond. **Isolated double bonds** are separated by more than one single bond. In other words, the double bonds are isolated from each other. **Cumulated double bonds** are adjacent to each other. Compounds with cumulated double bonds are called **allenes.**

$$CH_3CH=CH-CH=CHCH_3 \qquad CH_2=CH-CH_2-CH=CH_2$$

a conjugated diene **an isolated diene**

$$CH_3-CH=C=CH-CH_3$$

a cumulated diene
an allene

7.1 NOMENCLATURE OF COMPOUNDS WITH MORE THAN ONE FUNCTIONAL GROUP

To name a diene by the IUPAC system of nomenclature, the longest continuous chain that contains both double bonds is designated by its alkane name with the "ne" ending replaced by "diene." The chain is numbered in the direction that gives the double bonds the lowest possible numbers. The numbers indicating the locations of the double bonds are cited either immediately before the name of the parent compound or immediately before "diene." Substituents are cited in alphabetical order. Allene, the smallest member of the class of compounds known as allenes, is an acceptable IUPAC name for propadiene.

$$CH_2=C=CH_2 \qquad CH_2=\overset{\overset{\displaystyle CH_3}{|}}{C}-CH=CH_2$$

IUPAC: propadiene

2-methyl-1,3-butadiene
2-methylbuta-1,3-diene

common: allene
isoprene

5-bromo-1,3-cyclohexadiene
5-bromocyclohexa-1,3-diene

$$CH_3CH=CHCH_2\overset{\overset{\displaystyle CH_3}{|}}{C}=CH_2$$

2-methyl-1,4-hexadiene
2-methylhexa-1,4-diene

$$CH_3\overset{\overset{\displaystyle CH_3}{|}}{C}=CHCH=\overset{\overset{\displaystyle CH_2CH_3}{|}}{C}CH_2CH_3$$

5-ethyl-2-methyl-2,4-heptadiene
5-ethyl-2-methylhepta-2,4-diene

To name an alkene in which the second functional group is not another double bond, the longest continuous chain containing both functional groups is chosen, and both functional group designations are cited at the end of the name. The "ene" ending is cited first, with the terminal "e" omitted in order to avoid two adjacent vowels.

The location of the first cited functional group is stated immediately before the name of the parent chain. The location of the second functional group is stated immediately before its suffix name. If the functional groups are a double bond and a triple bond, the chain is numbered in the direction that yields the lowest number in the name of the compound.

$$CH_3CH=CHCH_2CH_2C\equiv CH \qquad CH_2=CHCH_2CH_2C\equiv CCH_3$$

5-hepten-1-yne
not 2-hepten-6-yne
since 1 < 2

1-hepten-5-yne
not 6-hepten-2-yne
since 1 < 2

$$CH_2=CHCH\overset{\overset{\displaystyle CH_2CH_2CH_2CH_3}{|}}{C}\equiv CCH_3$$

3-butyl-1-hexen-4-yne

the longest continuous chain has 8 carbons but the 8-carbon chain does not contain both functional groups; the compound is named as a hexenyne since the longest continuous chain containing both functional groups has 6 carbons

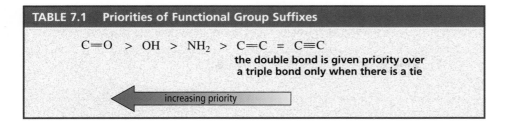

TABLE 7.1   Priorities of Functional Group Suffixes

$$C{=}O \; > \; OH \; > \; NH_2 \; > \; C{=}C \; = \; C{\equiv}C$$

the double bond is given priority over
a triple bond only when there is a tie

← increasing priority

If the same low number is obtained in both directions, the chain is numbered in the direction that gives the double bond the lower number.

$$\overset{1}{C}H_3\overset{2}{C}H{=}\overset{3}{C}H\overset{4}{C}{\equiv}\overset{5}{C}\overset{6}{C}H_3$$
2-hexen-4-yne
not **4-hexen-2-yne**

$$\overset{6}{H}C{\equiv}\overset{5}{C}\overset{4}{C}H_2\overset{3}{C}H_2\overset{2}{C}H{=}\overset{1}{C}H_2$$
1-hexen-5-yne
not **5-hexen-1-yne**

Compounds with functional groups other than double and triple bonds are named using the functional group priorities listed in Table 7.1. The chain is numbered in the direction that assigns the lowest number to the highest-priority functional group.

$$CH_3CH_2CH{=}CHCH_2OH$$
2-penten-1-ol

$$CH_2{=}CHCH_2CH_2CH_2\overset{\overset{\displaystyle NH_2}{|}}{C}HCH_3$$
6-hepten-2-amine

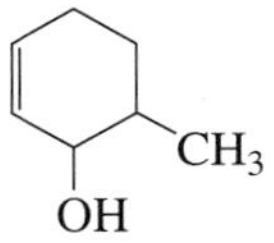

6-methyl-2-cyclohexenol

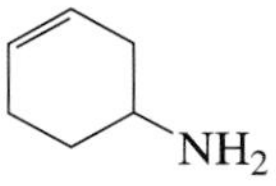

3-cyclohexenamine

PROBLEM 1◆

Give the IUPAC name for each of the following compounds.

a.

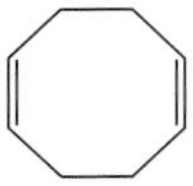

b.

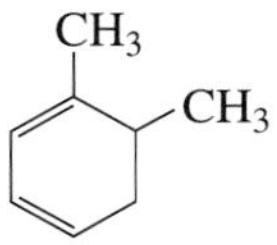

c. $CH_2{=}CHCH_2C{\equiv}CCH_2CH_3$

d. $HOCH_2CH_2C{\equiv}CH$

e. $CH_3CH{=}\overset{\overset{\displaystyle CH_3}{|}}{C}CH_2CH{=}CH_2$

f. $CH_3CH{=}CHCH{=}CHCH{=}CH_2$

g. $CH_3CH{=}\overset{\overset{\displaystyle CH_3}{|}}{C}CH_2\overset{\overset{\displaystyle CH_3}{|}}{C}HCH_2OH$

h. $CH_3CH_2CH=CCH_2CH_2C\equiv CH$
$\qquad\qquad\quad |$
$\qquad\qquad\quad CH=CH_2$

7.2
CONFIGURATIONAL ISOMERS OF DIENES

A diene such as 1-chloro-2,4-heptadiene has four configurational isomers because each of the double bonds can be in either the *E* configuration or the *Z* configuration. Thus there are *E-E*, *Z-Z*, *E-Z*, and *Z-E* isomers. The rules for determining the *E* and *Z* configurations are given in Section 3.5.

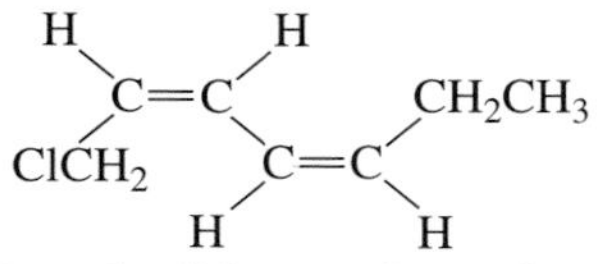

(2Z,4Z)-1-chloro-2,4-heptadiene   (2Z,4E)-1-chloro-2,4-heptadiene

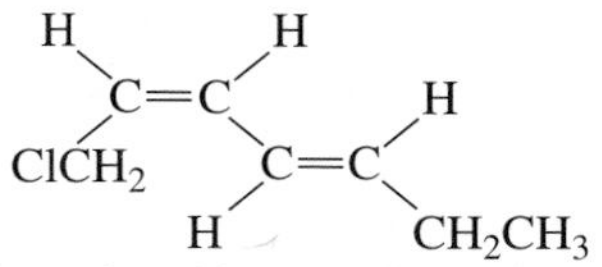

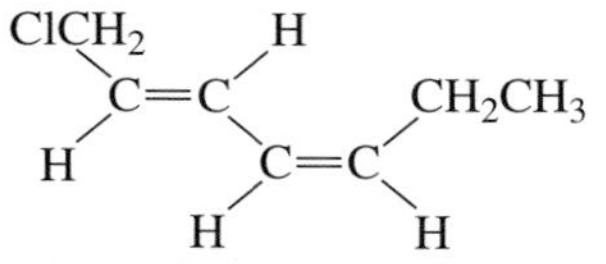

(2E,4Z)-1-chloro-2,4-heptadiene   (2E,4E)-1-chloro-2,4-heptadiene

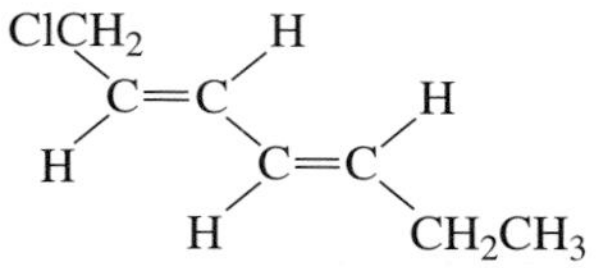

PROBLEM 2

Draw the configurational isomers for the following compounds. Name each one.

a. 2-methyl-2,4-hexadiene   c. 1,3-pentadiene

b. 2,4-heptadiene

7.3
RELATIVE STABILITIES OF DIENES

In Section 3.19, we saw that the relative stabilities of substituted alkenes can be determined by their relative heats of hydrogenation: the least stable alkene gives off the most heat when hydrogenated because it contains more energy to begin with. The heat of hydrogenation of 2,3-pentadiene (a cumulated diene) is more negative than the heat of hydrogenation of 1,4-pentadiene (an isolated diene), which in turn is more negative than the heat of hydrogenation of 1,3-pentadiene (a conjugated diene).

$CH_3CH=C=CHCH_3$ + 2 H_2 $\xrightarrow{\text{Pt}}$ $CH_3CH_2CH_2CH_2CH_3$ $\Delta H° = -70.5$ kcal/mol
2,3-pentadiene

$CH_2=CHCH_2CH=CH_2$ + 2 H_2 $\xrightarrow{\text{Pt}}$ $CH_3CH_2CH_2CH_2CH_3$ $\Delta H° = -60.2$ kcal/mol
1,4-pentadiene

$CH_2=CHCH=CHCH_3$ + 2 H_2 $\xrightarrow{\text{Pt}}$ $CH_3CH_2CH_2CH_2CH_3$ $\Delta H° = -54.1$ kcal/mol
1,3-pentadiene

From the relative heats of hydrogenation of the three pentadienes, we can conclude that conjugated dienes are more stable than isolated dienes, which are more stable than cumulated dienes.

relative stabilities of dienes

conjugated diene   >   isolated diene   >   cumulated diene

increasing stability

Why is a conjugated diene more stable than an isolated diene? Two factors contribute to the stability of conjugated dienes. One is the hybridization of the orbitals forming the carbon–carbon single bonds. The carbon–carbon single bond in 1,3-butadiene is formed as the result of the overlap of an sp^2 orbital with another sp^2 orbital, whereas the carbon–carbon single bonds in 1,4-pentadiene result from the overlap of an sp^3 orbital with an sp^2 orbital.

single bond formed by
sp^2–sp^2 overlap

single bonds formed by
sp^3–sp^2 overlap

$CH_2{=}CH{-}CH{=}CH_2$ $CH_2{=}CH{-}CH_2{-}CH{=}CH_2$
1,3-butadiene   **1,4-pentadiene**

Because a 2s electron is closer, on average, to the nucleus than a 2p electron (Section 5.9), the electrons in an sp^2 orbital, with 33.3% s character, are closer to the nucleus than the electrons in an sp^3 orbital, with 25% s character. The length of a bond depends on how close the electrons in the bonding orbital are to the nucleus. Consequently, the more s character in the orbitals forming the σ bond, the shorter the bond (Table 7.2). This means that a σ bond formed by an sp^2–sp^2 overlap is shorter than a σ bond formed by an sp^3–sp^2 overlap. Shorter bonds are stronger, and hence the molecule with conjugated double bonds is more stable.

A change in hybridization also affects the length of a carbon–hydrogen bond but not to the same extent that it affects the length of a carbon–carbon bond (Table 1.6).

TABLE 7.2   Dependence of the Length of a Carbon–Carbon Single Bond on the Hybridization of the Orbitals Used in Its Formation

Compound	Hybridization	Bond length (Å)
$H_3C{-}CH_3$	sp^3–sp^3	1.54
$H_3C{-}\overset{H}{C}{=}CH_2$	sp^3–sp^2	1.50
$H_2C{=}\overset{H}{C}{-}\overset{H}{C}{=}CH_2$	sp^2–sp^2	1.47
$H_3C{-}C{\equiv}CH$	sp^3–sp	1.46
$H_2C{=}\overset{H}{C}{-}C{\equiv}CH$	sp^2–sp	1.43
$HC{\equiv}C{-}C{\equiv}CH$	sp–sp	1.37

The second factor that causes a conjugated diene to be more stable than an isolated diene is resonance, which means that the compound has delocalized electrons. The π electrons in a conjugated double bond are not localized between two carbons but are delocalized over four carbons. As we discovered in Section 6.6, electron delocalization increases a molecule's stability.

$$\bar{C}H_2-CH=CH-\overset{+}{C}H_2 \longleftrightarrow CH_2=CH-CH=CH_2 \longleftrightarrow \overset{+}{C}H_2-CH=CH-\bar{C}H_2$$

resonance structures

$$CH_2\text{---}CH\text{---}CH\text{---}CH_2$$

resonance hybrid

The resonance hybrid shows that the single bond in 1,3-butadiene is not a pure single bond but has partial double-bond character. This supports the observation that an sp^2–sp^2 bond is shorter and stronger than an sp^2–sp^3 bond.

A molecular orbital description of 1,3-butadiene is shown in Figure 7.1. The four p atomic orbitals of 1,3-butadiene combine to produce four π molecular orbitals, ψ_1, ψ_2, ψ_3, and ψ_4. Half of the molecular orbitals are bonding π molecular orbitals (ψ_1 and ψ_2), and the other half are antibonding π molecular orbitals (ψ_3 and ψ_4). A more complete description of molecular orbitals is presented in Section 27.2.

The four π electrons of 1,3-butadiene are all in bonding molecular orbitals. Figure 7.1 shows that the four π electrons are delocalized over the four p orbitals. Resonance and molecular orbital diagrams are two different ways of showing that the π electrons in 1,3-butadiene are delocalized.

Figure 7.1 ▶
Four p atomic orbitals overlap to produce the four π molecular orbitals of 1,3-butadiene.

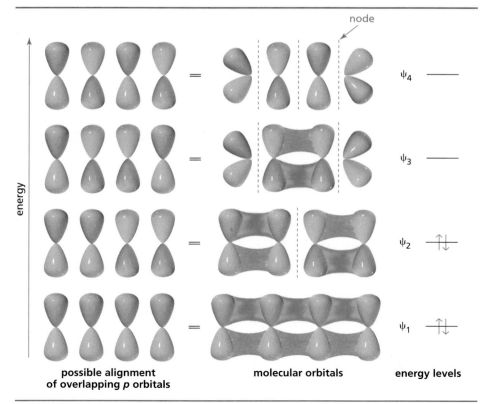

possible alignment of overlapping p orbitals · molecular orbitals · energy levels

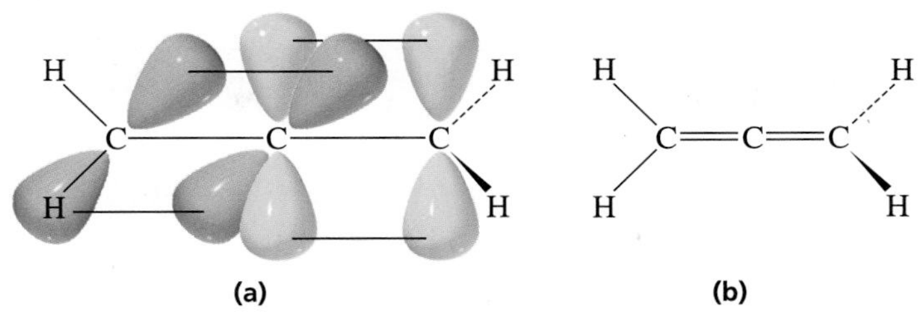

(a)   (b)

PROBLEM 3 ◆

What is the total number of nodes in the ψ_3 and ψ_4 molecular orbitals of 1,3-butadiene?

While a conjugated diene is more stable than an isolated diene, a cumulated diene is less stable than an isolated diene. The doubly-bonded carbon atoms of isolated dienes and conjugated dienes are all sp^2 hybridized. Cumulated dienes are unlike other dienes in that the central carbon is sp hybridized because it has two π bonds. This sp hybridization gives the molecule unique properties. The heat of hydrogenation of allene is very similar to the heat of hydrogenation of 1-butyne, a compound with sp hybridized carbons.

$$CH_2{=}C{=}CH_2 \ + \ 2\,H_2 \ \xrightarrow{\text{Pt}} \ CH_3CH_2CH_3 \qquad \Delta H° = -70.5 \text{ kcal/mol}$$
allene

$$CH_3CH_2C{\equiv}CH \ + \ 2\,H_2 \ \xrightarrow{\text{Pt}} \ CH_3CH_2CH_2CH_3 \qquad \Delta H° = -69.9 \text{ kcal/mol}$$
1-butyne

One of the p orbitals of the central carbon of allene overlaps with a p orbital of an adjacent sp^2 carbon. The second p orbital of the central carbon overlaps with a p orbital of the other sp^2 carbon (Figure 7.2). The two p orbitals of the central carbon are perpendicular to each other. Therefore, the plane containing one H—C—H group is perpendicular to the plane containing the other H—C—H group. We will not consider the reactions of cumulated dienes, because they are rather specialized and are more appropriately covered in an advanced course in organic chemistry.

7.4 REACTIVITY CONSIDERATIONS

Dienes, like alkenes and alkynes, are nucleophiles because of the electron density of their π bonds. This means that they will react with electrophilic reagents. Like alkenes and alkynes, dienes undergo electrophilic addition reactions.

Until now, we have been concerned with the reactions of compounds that have only one functional group. Compounds with two or more functional groups exhibit reactions characteristic of the individual functional groups if the functional groups are sufficiently separated from each other. If the functional groups are close enough to allow electron delocalization, one functional group can affect the reactivity of the other. In other words, we will see that the reactions of isolated dienes are the same as the reactions of alkenes, whereas the reactions of conjugated dienes are a little different because of the proximity of the two double bonds.

**7.5
ELECTROPHILIC
ADDITION
REACTIONS OF
ISOLATED DIENES**

The reactions of isolated dienes are the same as the reactions of alkenes. If an excess of the electrophilic reagent is present, two independent addition reactions will occur, each following Markovnikov's rule.

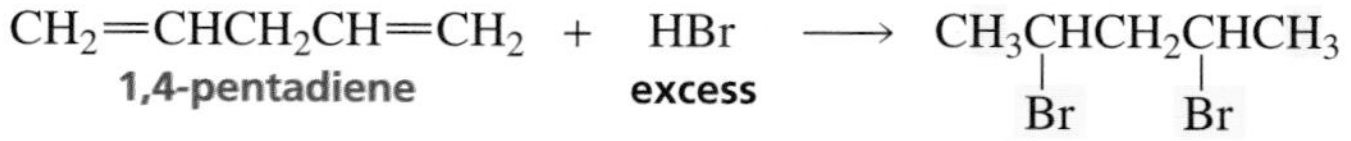

$$CH_2=CHCH_2CH=CH_2 \ + \ HBr \ \longrightarrow \ CH_3CHCH_2CHCH_3$$

1,4-pentadiene   **excess**   Br   Br

The reaction proceeds exactly as we would predict from our knowledge of the mechanism for the reaction of alkenes with electrophilic reagents. The electrophile (H^+) adds to the electron-rich double bond in a manner that produces the more stable carbocation (Section 3.12). The bromide ion then adds to the carbocation. Because there is an excess of the electrophilic reagent, both double bonds undergo addition.

mechanism for the reaction of 1,4-pentadiene with excess HBr

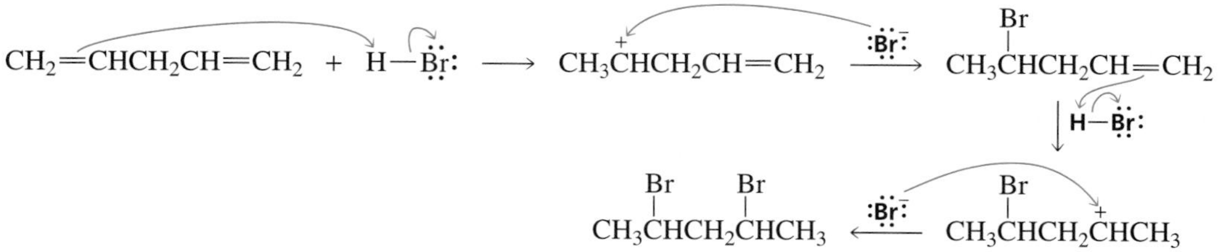

$$CH_2=CHCH_2CH=CH_2 \ + \ H-\overset{..}{\underset{..}{Br}}: \ \longrightarrow \ CH_3\overset{+}{C}HCH_2CH=CH_2 \ \xrightarrow{\ :\overset{..}{Br}:^-\ } \ \overset{Br}{\underset{|}{CH_3CHCH_2CH=CH_2}}$$

$$\downarrow H-\overset{..}{\underset{..}{Br}}:$$

$$\underset{Br \quad\quad Br}{CH_3CHCH_2CHCH_3} \ \xleftarrow{\ :\overset{..}{Br}:^-\ } \ \overset{Br}{\underset{|}{CH_3CHCH_2\overset{+}{C}HCH_3}}$$

If there is only enough electrophilic reagent to add to one of the double bonds, it will add preferentially to the more reactive double bond. For example, in the reaction of 2-methyl-1,4-pentadiene with HCl, addition of HCl to the double bond on the left leads to formation of a secondary carbocation, while addition to the double bond on the right results in formation of a tertiary carbocation. Because the transition state leading to formation of a tertiary carbocation is more stable than the transition state leading to a secondary carbocation, the tertiary carbocation forms faster (Section 3.11). So in the presence of a limited amount of electrophilic reagent, the major product of the reaction will be 4-chloro-4-methyl-1-pentene.

$$\underset{\substack{\textbf{2-methyl-1,4-pentadiene}\\\textbf{1 mol}}}{CH_2=CHCH_2\overset{\overset{\displaystyle CH_3}{|}}{C}=CH_2} \ + \ \underset{\textbf{1 mol}}{HCl} \ \longrightarrow \ \underset{\substack{\textbf{4-chloro-4-methyl-1-pentene}\\\textbf{major product}}}{CH_2=CHCH_2\overset{\overset{\displaystyle CH_3}{|}}{\underset{\underset{\displaystyle Cl}{|}}{C}}CH_3} \ + \ \underset{\substack{\textbf{4-chloro-2-methyl-1-pentene}\\\textbf{minor product}}}{CH_3\overset{\overset{\displaystyle CH_3}{|}}{\underset{\underset{\displaystyle Cl}{|}}{CHCH_2C}}=CH_2}$$

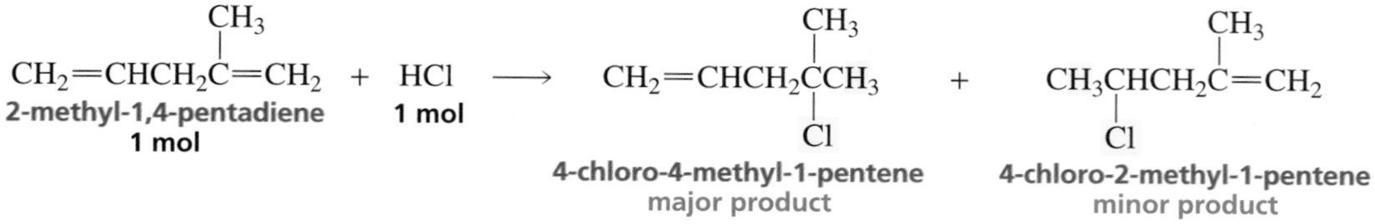

PROBLEM 4 ◆

Give the major product of each of the following reactions. Equivalent amounts of reagents are used in each case.

a. 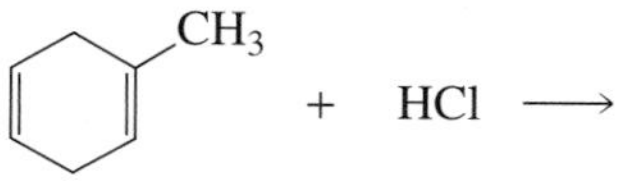 (cyclohexene ring with CH_3 substituent) $+$ HCl $\longrightarrow$

b. 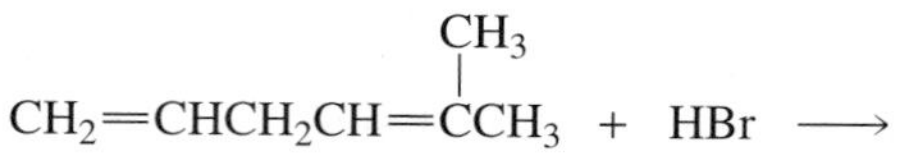 $CH_2=CHCH_2CH=\overset{\overset{\displaystyle CH_3}{|}}{C}CH_3 \ + \ HBr \ \longrightarrow$

$$
\text{c. } CH_2\!=\!CHCH_2CH_2\overset{\overset{\displaystyle CH_3}{\displaystyle |}}{C}\!=\!CH_2 \ + \ HBr \xrightarrow{\textbf{peroxide}}
$$

$$
\text{d. } HC\!\equiv\!CCH_2CH\!=\!CH_2 \ + \ Cl_2 \xrightarrow{CCl_4}
$$

PROBLEM 5

Which of the double bonds in zingiberene (its structure is given on the first page of this chapter) is the most reactive in an electrophilic addition reaction?

If a conjugated diene, such as 1,3-butadiene, reacts with a limited amount of electrophilic reagent so that addition can occur at only one of the double bonds, two addition products are formed. One is a 1,2-addition product, which is a result of addition at the 1- and 2-positions. The other is a 1,4-addition product—the result of addition at the 1- and 4-positions.

**7.6
ELECTROPHILIC
ADDITION
REACTIONS OF
CONJUGATED DIENES**

$$
\underset{\substack{\textbf{1,3-butadiene}\\\textbf{1 mol}}}{CH_2\!=\!CH\!-\!CH\!=\!CH_2} \ + \ \underset{\textbf{1 mol}}{Cl_2} \xrightarrow{CCl_4} \underset{\substack{\textbf{3,4-dichloro-1-butene}\\\text{1,2-addition product}}}{\overset{\overset{\displaystyle Cl\ \ Cl}{\displaystyle |\ \ \ |}}{CH_2CHCH\!=\!CH_2}} \ + \ \underset{\substack{\textbf{1,4-dichloro-2-butene}\\\text{1,4-addition product}}}{\overset{\overset{\displaystyle Cl\ \ \ \ \ \ \ \ Cl}{\displaystyle |\ \ \ \ \ \ \ \ \ |}}{CH_2CH\!=\!CHCH_2}}
$$

$$
\underset{\substack{\textbf{1,3-butadiene}\\\textbf{1 mol}}}{CH_2\!=\!CH\!-\!CH\!=\!CH_2} \ + \ \underset{\textbf{1 mol}}{HBr} \longrightarrow \underset{\substack{\textbf{3-bromo-1-butene}\\\text{1,2-addition product}}}{\overset{\overset{\displaystyle Br}{\displaystyle |}}{CH_3CHCH\!=\!CH_2}} \ + \ \underset{\substack{\textbf{1-bromo-2-butene}\\\text{1,4-addition product}}}{\overset{\overset{\displaystyle Br}{\displaystyle |}}{CH_3CH\!=\!CHCH_2}}
$$

Addition at the 1- and 2-positions is called **1,2-addition** or **direct addition.** Addition at the 1- and 4-positions is called **1,4-addition** or **conjugate addition.** We expect the formation of the 1,2-addition product from our knowledge of how electrophilic reagents add to double bonds. The 1,4-addition product, however, may be surprising. First, the reagent did not add to adjacent carbons; second, a double bond moved. The double bond in the product is between the 2- and 3-positions, whereas the reactant has a single bond in this position.

When we refer to addition at the 1- and 2-positions or at the 1- and 4-positions, we mean that addition occurs at the 1- and 2- or 1- and 4-positions of the four-carbon conjugated system. The 1-position is one of the sp^2 carbons at the end of the conjugated system; it is not necessarily the first carbon in the molecule.

$$
R\!-\!\overset{1}{C}H\!=\!\overset{2}{C}H\!-\!\overset{3}{C}H\!=\!\overset{4}{C}H\!-\!R
$$
the conjugated system

$$
\underset{\textbf{2,4-hexadiene}}{CH_3CH\!=\!CHCH\!=\!CHCH_3} \xrightarrow[CCl_4]{\textbf{Br}_2} \underset{\substack{\textbf{4,5-dibromo-2-hexene}\\\text{1,2-addition product}}}{\overset{\overset{\displaystyle Br\ \ Br}{\displaystyle |\ \ \ |}}{CH_3CHCHCH\!=\!CHCH_3}} \ + \ \underset{\substack{\textbf{2,5-dibromo-3-hexene}\\\text{1,4-addition product}}}{\overset{\overset{\displaystyle Br\ \ \ \ \ \ \ \ Br}{\displaystyle |\ \ \ \ \ \ \ \ \ \ |}}{CH_3CHCH\!=\!CHCHCH_3}}
$$

To understand why both 1,2-addition and 1,4-addition products are obtained from the reaction of a conjugated diene with a limited amount of electrophilic reagent, we must look at the mechanism of the reaction. In the first step of the addition of HBr to 1,3-butadiene, the electrophilic proton adds to C-1, which results in the formation of an allylic cation. Recall that an allylic cation has a positive charge on a carbon that is next to a doubly-bonded carbon. Allylic means "next to a carbon–carbon double bond." Notice that, because the molecule is symmetrical, the proton could just as well have added to C-4. The proton does not add to C-2 or C-3 because this would result in the formation of a primary carbocation, which is less stable than an allylic cation.

The allylic cation has two contributing resonance structures. The positive charge on the carbocation is not localized on C-2 but is shared by C-2 and C-4 as a result of delocalization of the π electrons. Consequently, in the second step of the reaction, the bromide ion can attack either C-2 (direct addition) or C-4 (conjugate addition) to form the 1,2-addition product or the 1,4-addition product, respectively.

mechanism for the reaction of 1,3-butadiene with HBr

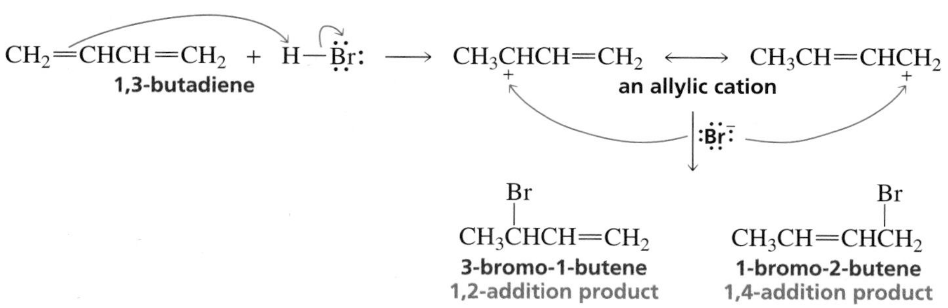

$$CH_2\!=\!CHCH\!=\!CH_2 \;+\; H\!-\!\ddot{B}r\!: \;\longrightarrow\; CH_3CHCH\!=\!CH_2 \;\longleftrightarrow\; CH_3CH\!=\!CHCH_2$$

1,3-butadiene   **an allylic cation**

$$:\!\ddot{B}r\!:\!^-$$

Br	Br
$CH_3CHCH\!=\!CH_2$	$CH_3CH\!=\!CHCH_2$
3-bromo-1-butene	**1-bromo-2-butene**
1,2-addition product	**1,4-addition product**

The two contributing resonance structures and the actual structure of the allylic cation are shown below.

$$CH_3\overset{+}{C}H\!-\!CH\!=\!CH_2 \;\longleftrightarrow\; CH_3CH\!=\!CH\!-\!\overset{+}{C}H_2$$
resonance contributors

$$CH_3CH\overset{\delta+}{=\!=\!=}CH\overset{\delta+}{=\!=\!=}CH_2$$
resonance hybrid

As we look at more examples, notice that the first step in all electrophilic additions to conjugated dienes is addition of the electrophile to one of the sp^2 carbons at the end of the conjugated system. This is the only way to obtain a carbocation that is stabilized by resonance. If the electrophile were to add to one of the internal sp^2 carbons, the resulting carbocation would not be stabilized by resonance.

Now we can compare addition to an isolated diene with addition to a conjugated diene. The carbocation formed by addition of an electrophile to an isolated diene is not stabilized by resonance. The positive charge is localized on a single carbon, so only direct (1,2-) addition occurs.

addition to an isolated diene

$$CH_2\!=\!CHCH_2CH\!=\!CH_2 \xrightarrow{H^+} CH_3\overset{+}{C}HCH_2CH\!=\!CH_2 \xrightarrow{Br^-} CH_3CHCH_2CH\!=\!CH_2$$

1,4-pentadiene   Br | **4-bromo-1-pentene**

The carbocation formed by addition of an electrophile to a conjugated diene, in contrast, is stabilized by resonance. The positive charge is spread over two carbons and, as a result, both direct (1,2-) and conjugate (1,4-) addition occur.

addition to a conjugated diene

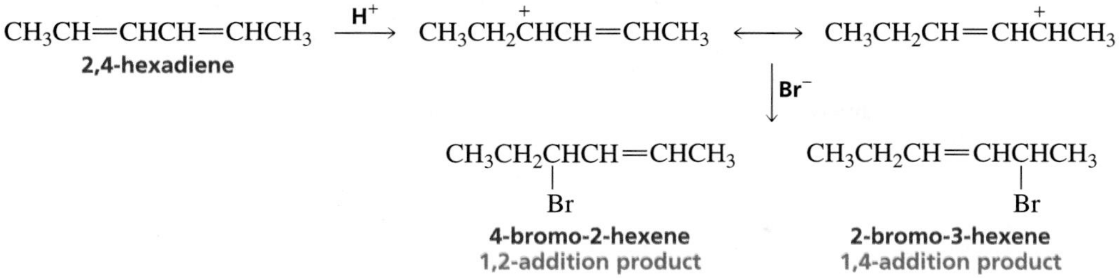

$$CH_3CH=CHCH=CHCH_3 \xrightarrow{\ H^+\ } CH_3CH_2\overset{+}{C}HCH=CHCH_3 \longleftrightarrow CH_3CH_2CH=CH\overset{+}{C}HCH_3$$

2,4-hexadiene

$$\Big\downarrow Br^-$$

$$CH_3CH_2CHCH=CHCH_3 \qquad\qquad CH_3CH_2CH=CHCHCH_3$$
$$\quad\ \ \ |\qquad\qquad\qquad\qquad\qquad\qquad\qquad\qquad\ \ |$$
$$\quad\ \ \ Br\qquad\qquad\qquad\qquad\qquad\qquad\qquad\qquad\ Br$$

4-bromo-2-hexene   **2-bromo-3-hexene**
1,2-addition product   **1,4-addition product**

If the diene is asymmetrical, the major products of the reaction are those obtained by adding the electrophile to whichever terminal sp^2 carbon results in formation of the more stable carbocation. For example, in the reaction of 2-methyl-1,3-butadiene, the proton adds preferentially to C-1 because the positive charge on the resulting carbocation is shared by a tertiary allylic and a primary allylic carbon. Adding the proton to C-4 would form a carbocation with the positive charge shared by a secondary allylic and a primary allylic carbon. Therefore, addition to C-1 forms the more stable carbocation and leads to 3-bromo-3-methyl-1-butene and 1-bromo-3-methyl-2-butene, the major products of the reaction.

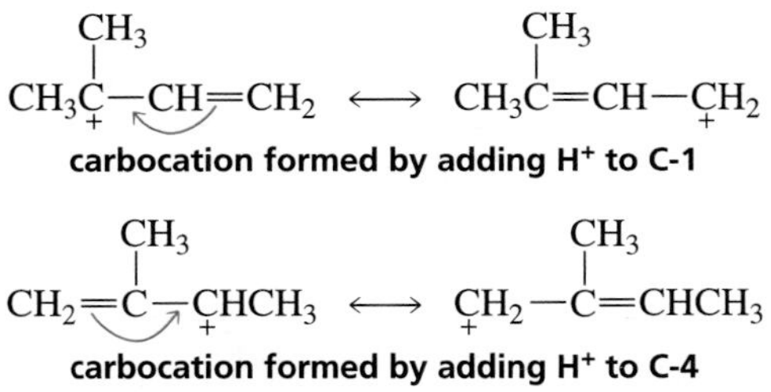

$$\overset{\displaystyle CH_3}{\underset{\displaystyle }{\overset{|}{\underset{1}{CH_2}}=\underset{}{\overset{}{CCH}}=\overset{4}{CH_2}}} + HBr \longrightarrow CH_3CCH=CH_2 + CH_3C=CHCH_2$$

2-methyl-1,3-butadiene   Br   Br

3-bromo-3-methyl-   **1-bromo-3-methyl-**
1-butene   **2-butene**

$$CH_3\overset{+}{C}-CH=CH_2 \longleftrightarrow CH_3C=CH-\overset{+}{C}H_2$$

(both bearing CH_3 substituent)

carbocation formed by adding H⁺ to C-1

$$CH_2=\overset{+}{C}-CHCH_3 \longleftrightarrow CH_2-C=CHCH_3$$

(both bearing CH_3 substituent)

carbocation formed by adding H⁺ to C-4

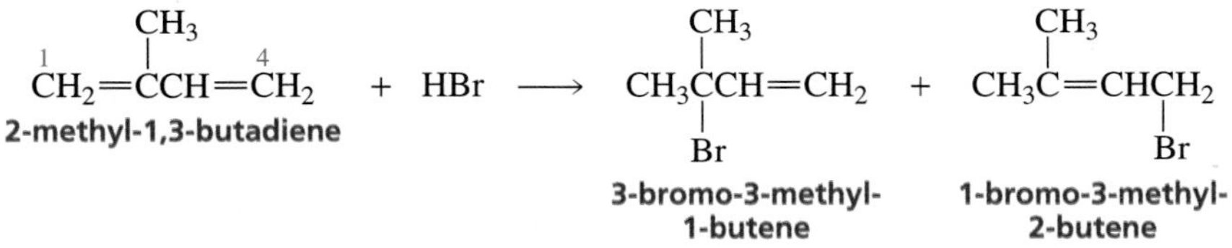

PROBLEM 6 ◆

Give the products of the following reactions. Equivalent amounts of reagents are used in each case.

a. $CH_3CH=CH-CH=CHCH_3 + Cl_2 \xrightarrow{\ CCl_4\ }$

b. 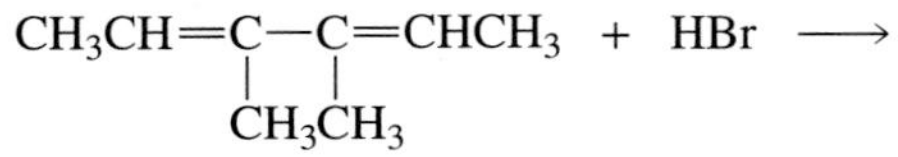 $CH_3CH=C-C=CHCH_3 + HBr \longrightarrow$
$\qquad\qquad\quad |\ \ \ |$
$\qquad\qquad\ CH_3CH_3$

c. $CH_3CH=CH-C=CHCH_3 + HBr \longrightarrow$
$\qquad\qquad\qquad\ |$
$\qquad\qquad\qquad CH_3$

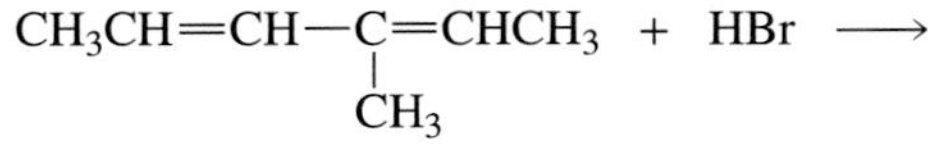

d. 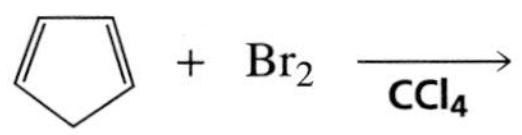 (cyclopentene) $+ Br_2 \xrightarrow{\ CCl_4\ }$

7.7
THERMODYNAMIC VERSUS KINETIC CONTROL OF REACTIONS

When a conjugated diene undergoes an electrophilic addition reaction, which product is formed in the greater amount—the 1,2-addition product or the 1,4-addition product? The two factors that determine this are the structure of the reactant and the temperature at which the reaction is carried out.

When a reaction produces more than one product, the product that is formed the most rapidly is called the **kinetic product.** The most stable product is called the **thermodynamic product.** Reactions that produce the kinetic product as the major product are said to be kinetically controlled. Reactions that produce the thermodynamic product as the major product are said to be thermodynamically controlled. When a reaction is under **kinetic control,** the relative amounts of the products *depend on the rates* at which they are formed. When a reaction is under **thermodynamic control,** the relative amounts of the products *depend on their stabilities.*

For many organic reactions, the kinetic product and the thermodynamic product are one and the same. Electrophilic addition to 1,3-butadiene is an example of a reaction in which the kinetic product and the thermodynamic product are not the same.

$$CH_2{=}CHCH{=}CH_2 \ + \ HBr \ \longrightarrow \ \overset{\displaystyle Br}{\underset{\displaystyle |}{CH_3CHCH{=}CH_2}} \ + \ \overset{\displaystyle Br}{\underset{\displaystyle |}{CH_3CH{=}CHCH_2}}$$

1,3-butadiene   **1,2-addition product**   **1,4-addition product**

Let's first determine which product in the preceding reaction is more stable—the thermodynamic product. The relative stability of an alkene can be determined by the number of alkyl groups that are bonded to the sp^2 carbons; the greater the number of alkyl groups bonded to the sp^2 carbons, the more stable is the alkene (Section 3.19). The two products formed by addition of HBr to 1,3-butadiene have different stabilities: the 1,2-addition product has one alkyl group bonded to the sp^2 carbons; the 1,4-addition product has two alkyl groups bonded to the sp^2 carbons. So the 1,4-addition product is the more stable product. It is, therefore, the thermodynamic product.

Now let's determine which of the products is formed faster—the kinetic product. The first step of the addition reaction, addition of a proton to C-1, is the same whether you are forming the 1,2-product or the 1,4-product. It is the second step that is different, because in the second step the bromide ion attacks either C-2 or C-4. Therefore, to determine which of the products is formed faster, we must look at the relative stabilities of the transition states for the second step of the reaction.

The reactant for the second step is an allylic cation whose structure is represented by two contributing resonance structures. As the bromide ion approaches the allylic

$$CH_2{=}CHCH{=}CH_2 \ \xrightarrow{\textbf{HBr}} \ CH_3\overset{+}{C}HCH{=}CH_2 \ \longleftrightarrow \ CH_3CH{=}CH\overset{+}{C}H_2$$

$$\Big\downarrow \textbf{Br}^-$$

$$\left[\overset{\delta+}{CH_3CHCH{=}CH_2} \atop \underset{\delta-}{:\!\ddot{B}\!r\!:} \right] \qquad\qquad \left[CH_3CH{=}\overset{\delta+}{CH}CH_2 \atop \underset{\delta-}{:\!\ddot{B}\!r\!:} \right]$$

transition state for   **transition state for**
formation of the 1,2-product   **formation of the 1,4-product**

$$\Big\downarrow \qquad\qquad\qquad\qquad\qquad \Big\downarrow$$

$$\underset{\displaystyle Br}{\underset{\displaystyle |}{CH_3CHCH{=}CH_2}} \qquad\qquad \underset{\displaystyle Br}{\underset{\displaystyle |}{CH_3CH{=}CHCH_2}}$$

1,2-product   **1,4-product**
kinetic product   **thermodynamic product**

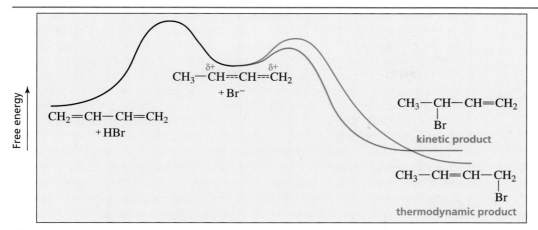

▲ **Figure 7.3**
Reaction coordinate diagram for addition of HBr to 1,3-butadiene.

cation, its positive charge becomes concentrated on the carbon that the bromide ion approaches. This means that the transition state for formation of the 1,2-product resembles the resonance structure in which the positive charge is on a secondary allylic carbon, while the transition state for formation of the 1,4-product resembles the resonance structure in which the positive charge is on a primary allylic carbon. Because a secondary allylic cation is more stable than a primary allylic cation, the transition state for formation of the 1,2-addition product is the more stable transition state (Figure 7.3). As a result, the 1,2-product is formed faster than the 1,4-product. The 1,2-product, therefore, is the kinetic product.

> **The thermodynamic product is the more stable product.**

For a reaction in which the kinetic and thermodynamic products are not the same, the product that predominates depends on the conditions under which the reaction is carried out. If the reaction is carried out under mild (low-temperature) conditions, the major product will be the kinetic product. If the reaction is carried out under more vigorous (high-temperature) conditions, the major product will be the thermodynamic product.

> **The kinetic product is formed faster.**

$$A + B \quad \begin{cases} \xrightarrow{\text{mild conditions}} \text{kinetic product} \\ \xrightarrow{\text{vigorous conditions}} \text{thermodynamic product} \end{cases}$$

For example, when the addition of HBr to 1,3-butadiene is carried out at $-80\ °C$, the major product is the 1,2-addition product.

$$CH_2{=}CHCH{=}CH_2 + HBr \xrightarrow{-80\ °C}$$

$$\underset{\substack{\text{1,2-addition product} \\ \text{80\%}}}{CH_3\overset{\displaystyle Br}{\underset{\displaystyle |}{C}}HCH{=}CH_2} + \underset{\substack{\text{1,4-addition product} \\ \text{20\%}}}{CH_3CH{=}CH\overset{\displaystyle Br}{\underset{\displaystyle |}{C}}H_2}$$

In contrast, when the reaction is carried out at 45 °C, the major product is the 1,4-addition product.

$$CH_2{=}CHCH{=}CH_2 + HBr \xrightarrow{45\ °C}$$

$$\underset{\substack{\text{1,2-addition product} \\ \text{15\%}}}{CH_3\overset{\displaystyle Br}{\underset{\displaystyle |}{C}}HCH{=}CH_2} + \underset{\substack{\text{1,4-addition product} \\ \text{85\%}}}{CH_3CH{=}CH\overset{\displaystyle Br}{\underset{\displaystyle |}{C}}H_2}$$

The reaction coordinate diagram in Figure 7.3 for the addition of HBr to 1,3-buta-diene explains why the major product of the reaction depends on the temperature at which the reaction is carried out. Under mild conditions, the reaction is irreversible. At −80 °C, there is sufficient energy for the reactants to overcome the energy barrier for the first step of the reaction to form the intermediate carbocation. There is also sufficient energy for the carbocation to overcome the energy barrier to form the two addition products. However, there is not sufficient energy for the reverse reaction to occur. The products cannot overcome the large energy barriers separating them from the intermediate carbocation. Consequently, at −80 °C, the relative amounts of the two products formed reflect the relative energy barriers to the second step of the reaction. The energy barrier to formation of the 1,2-product is lower than the energy barrier to formation of the 1,4-product, so the major product is the 1,2-product.

Under vigorous conditions, however, the reaction is reversible. At 45 °C, there is sufficient energy for the reactants to get over the energy barriers to form the prod-ucts and sufficient energy for one or more of the products to go back to the carbo-cation. The carbocation is a **common intermediate,** an intermediate that both products have in common. The ability to return to a common intermediate allows the products to interconvert. Because the products can interconvert, the relative amounts of the two products at equilibrium will depend on their relative stabilities. The thermodynamic product reverses less readily because it has a higher energy barrier to the common intermediate, so it gradually comes to predominate in the product mixture. Thermodynamic control is also called **equilibrium control.**

more stable product ⇌ common intermediate ⇌ less stable product

Thus, a reaction will be under thermodynamic control when there is sufficient energy to allow it to be reversible. When a reaction is under thermodynamic con-trol, the major product will be the more stable product. When the reaction is irre-versible under the conditions used in the experiment, it will be under kinetic control. When a reaction is under kinetic control, the major product will be the one that is formed more rapidly.

The temperature at which a reaction changes from being kinetically controlled to being thermodynamically controlled depends on the reaction. Each reaction that is irreversible under mild conditions and reversible under more vigorous conditions has a temperature at which the change-over happens.

The particular temperature required to shift the reaction of a conjugated diene with an electrophilic reagent from kinetic control to thermodynamic control depends on both the diene and the electrophilic reagent. For example, in the case of addition of HCl to 1,3-butadiene, the reaction remains under kinetic control at 45 °C even though the addition of HBr to that diene is under thermodynamic con-trol at that temperature. Because a carbon–chlorine bond is stronger than a carbon–bromine bond, a higher temperature is required for the products to undergo the reverse reaction. Thermodynamic control is achieved only when there is suffi-cient energy to allow the reaction to be reversible.

Do not assume, however, that the 1,2-product is the kinetic product and the 1,4-product is the thermodynamic product for all conjugated dienes. As the following examples show, the structure of the conjugated diene ultimately determines which of the addition products is under kinetic control and which is under thermodynamic control.

In the reaction of 4-methyl-1,3-pentadiene, the 1,2-product is the thermody-namic product since it has three alkyl groups bonded to the sp^2 carbons, while the 1,4-product has only two alkyl groups bonded to its sp^2 carbons. The 1,4-product is the kinetic product because the positive charge in the transition state leading to its

The thermodynamic product predominates when the reaction is reversible.

The kinetic product predominates when the reaction is not reversible.

formation is concentrated on a tertiary allylic carbon rather than on a secondary
allylic carbon as it is in the transition state leading to the 1,2-product.

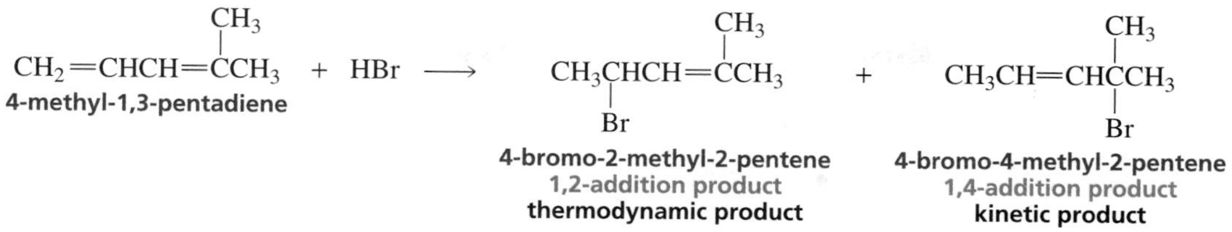

$$CH_2=CHCH=\overset{\overset{\displaystyle CH_3}{|}}{C}CH_3 \;+\; HBr \;\longrightarrow\; CH_3\overset{\overset{\displaystyle CH_3}{|}}{\underset{\underset{\displaystyle Br}{|}}{C}}HCH=CCH_3 \;+\; CH_3CH=CH\overset{\overset{\displaystyle CH_3}{|}}{\underset{\underset{\displaystyle Br}{|}}{C}}CH_3$$

4-methyl-1,3-pentadiene   **4-bromo-2-methyl-2-pentene**   **4-bromo-4-methyl-2-pentene**
                                             1,2-addition product   **1,4-addition product**
                                             thermodynamic product   **kinetic product**

   The two products obtained from addition of HCl to 2,4-hexadiene have the same
stability. Therefore, approximately equal amounts of the two products are obtained at
temperatures high enough for the reaction to be reversible. The two transition states
also have the same stability since, in both cases, the positive charge is shared by two
secondary allylic carbons. Therefore, we also expect equal amounts of the two products when the reaction is carried out at low temperatures. However, more of the 1,2-
product is obtained. Even though the carbocations leading to the 1,2- and
1,4-products have the same stability, the 1,2-product is formed a little faster because
after the proton adds to the C-1, the chloride ion is closer to C-2 than to C-4.

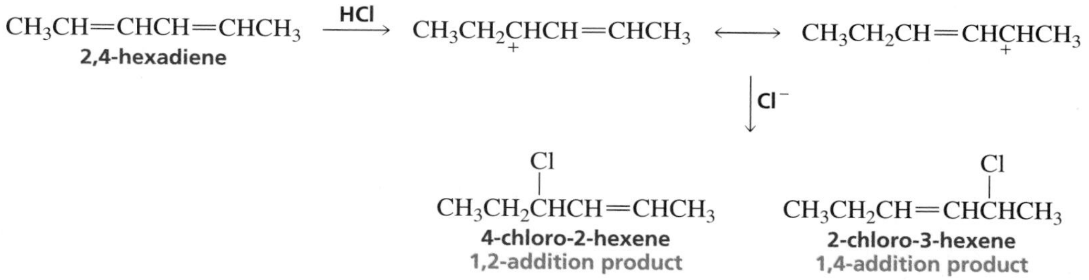

$$CH_3CH=CHCH=CHCH_3 \;\xrightarrow{\;HCl\;}\; CH_3CH_2\overset{+}{C}HCH=CHCH_3 \;\longleftrightarrow\; CH_3CH_2CH=CH\overset{+}{C}HCH_3$$

2,4-hexadiene

$$\Big\downarrow Cl^-$$

$$CH_3CH_2\overset{\overset{\displaystyle Cl}{|}}{C}HCH=CHCH_3 \qquad\qquad CH_3CH_2CH=CH\overset{\overset{\displaystyle Cl}{|}}{C}HCH_3$$

4-chloro-2-hexene   **2-chloro-3-hexene**
1,2-addition product   **1,4-addition product**

PROBLEM 7◆

a. How many products can be obtained from the reaction of one mole of 1,3-penta-
diene with one mole of HBr?

b. What is the major product?

PROBLEM 8 / SOLVED

For each of the following reactions, identify the kinetic product and the thermodynamic product.

a. + HCl $\longrightarrow$

c. + HCl $\longrightarrow$

b. $CH_3CH=CH\overset{\overset{\displaystyle CH_3}{|}}{C}=CH_2$ + HCl $\longrightarrow$

d. + HCl $\longrightarrow$

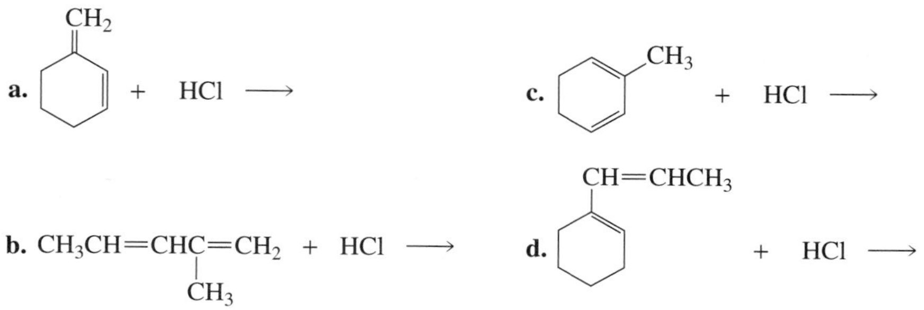

SOLUTION TO 8a   The proton adds to the indicated sp^2 carbon because an intermediate is formed with the positive charge shared by a tertiary allylic and a secondary
allylic carbon. If the proton were to add to the other end of the conjugated system, the
intermediate would be less stable because the positive charge would be shared by a

primary allylic and a secondary allylic carbon. 3-Chloro-3-methylcyclohexene is the kinetic product because, in the transition state for its formation, the positive charge is concentrated on a tertiary carbon. 3-Chloro-1-methylcyclohexene is the thermodynamic product because it is more stable as a result of its more highly substituted double bond.

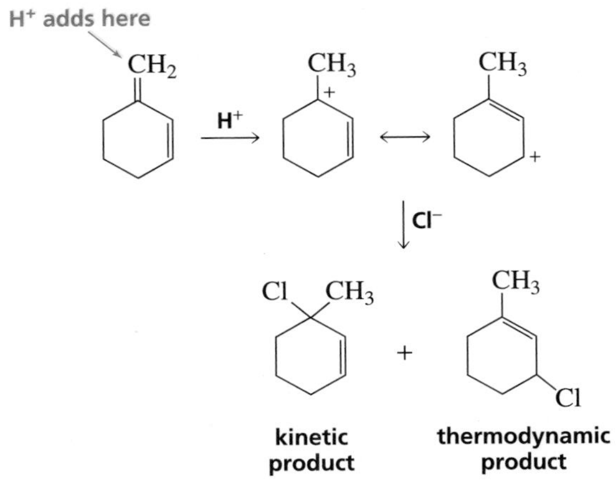

7.8 ADDITION OF A DIENOPHILE TO A CONJUGATED DIENE: THE DIELS–ALDER REACTION

Reactions that create new carbon–carbon bonds are very important in organic synthesis because it is only through such reactions that small carbon skeletons can be converted into larger ones (Section 5.10). The **Diels–Alder reaction** is a particularly important reaction because it creates two new carbon–carbon bonds in a manner that results in the formation of a cyclic molecule. In recognition of the importance of this reaction to synthetic organic chemistry, Otto Diels and Kurt Alder received the Nobel Prize in chemistry in 1950.

In the Diels–Alder reaction, a conjugated diene reacts with a compound containing a carbon–carbon double bond. This compound is called a **dienophile** because it "loves a diene." (Δ signifies heat.)

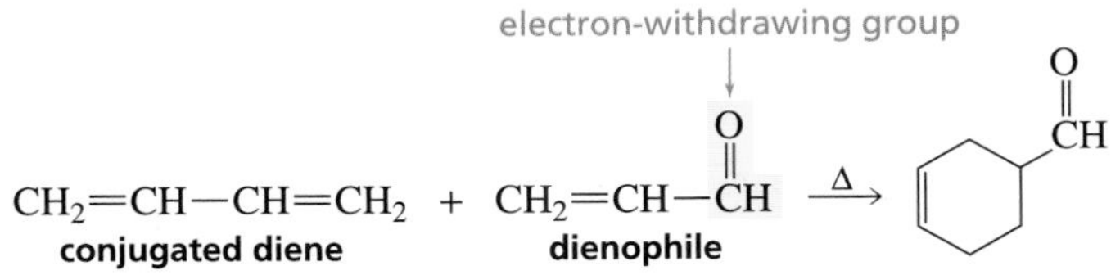

Dienophiles with an electron-withdrawing group bonded to one of the sp^2 carbons are more reactive in Diels–Alder reactions. The electron-withdrawing group polarizes the double bond, making the reaction easier to initiate.

contributing resonance structures of the dienophile

resonance hybrid

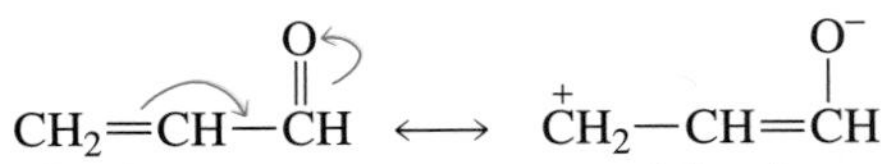

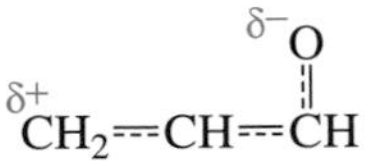

The partially positively charged sp^2 carbon of the dienophile can be thought of as the electrophile that is attacked by electrons from C-1 of the conjugated diene. The other sp^2 carbon of the dienophile is the nucleophile that adds to C-4 of the diene. Dienophiles with an electron-withdrawing group bonded to one of the sp^2 carbons are more reactive in Diels–Alder reactions because the electron-withdrawing group makes the double bond of the dienophile more electrophilic.

Professor Otto Paul Hermann Diels receiving his Nobel Prize from King Gustaf VI.

a 1,4-addition reaction to 1,3-butadiene

Although at first glance a Diels–Alder reaction does not look like any reaction that you have seen before, it is simply a 1,4-addition to a conjugated diene. As with other 1,4-addition reactions, the double bond in the product is between C-2 and C-3 of what was the conjugated diene.

There is, however, a significant difference between a Diels–Alder reaction and the 1,4-addition reactions that we have seen previously. The Diels–Alder reaction is a concerted reaction; the addition occurs in a single step. The other 1,4-addition reactions occur in two successive steps: the electrophile adds to the diene in the first step, and the nucleophile adds to the carbocation in the second step.

The Diels–Alder reaction is another example of a **pericyclic reaction,** a reaction that takes place in one step by a cyclic shift of electrons (Section 3.17). It is also a cycloaddition reaction. A **cycloaddition reaction** is one in which two reactants combine to yield a cyclic product. More precisely, the Diels–Alder reaction is a **[4 + 2] cycloaddition reaction** because, of the six π electrons involved in the cyclic transition state, four come from the conjugated diene and two come from the dienophile.

Otto Paul Hermann Diels (1876–1954) *was born in Germany, the son of a professor of classical philology at the University of Berlin. He became a professor of chemistry at that university and later at the University of Kiel. Two of his sons died in World War II. He retired in 1945 after his home and laboratory were destroyed in bombing raids.*

new bond

| diene
four π electrons | dienophile
two π electrons | six π electron
transition state | new
double bond   new bond |

A wide variety of cyclic compounds can be obtained by varying the structures of the conjugated diene and the dienophile.

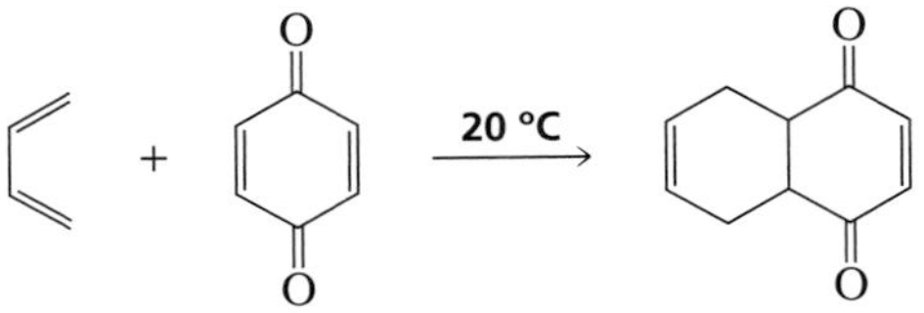

Compounds containing carbon–carbon triple bonds can also be used as dienophiles in Diels–Alder reactions.

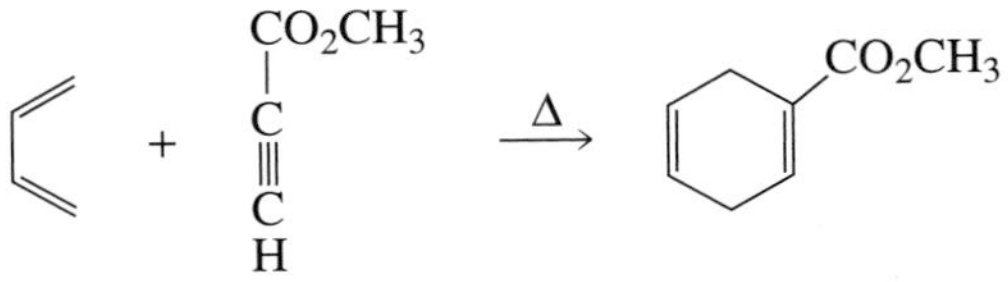

PROBLEM 9 ◆

Give the product of each of the following reactions.

a. CH_2=CH—CH=CH_2 + HC≡CH $\xrightarrow{\Delta}$

b. CH_2=CH—CH=CH—CH_3 + HC≡CH $\xrightarrow{\Delta}$

c. CH_2=CH—C=CH_2 + HC≡CH $\xrightarrow{\Delta}$
　　　　　　　|
　　　　　　CH_3

d. CH_2=CH—CH=CH_2 + HC≡CCH_3 $\xrightarrow{\Delta}$

The Diels–Alder reaction is stereospecific because the configuration of the reactants is maintained during the course of the reaction. If the substituents in the dienophile are cis, they will be cis in the product; if the substituents in the dienophile are trans, they will be trans in the product. The configuration is maintained because the reaction is concerted. The stereochemistry of the reaction will be discussed in terms of MO theory in Section 27.4.

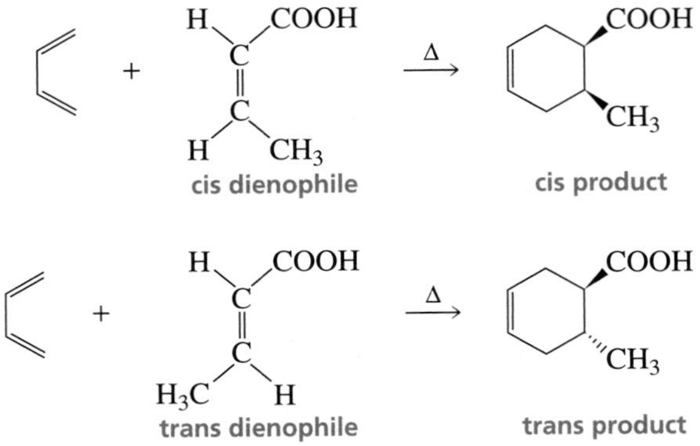

cis dienophile → cis product

trans dienophile → trans product

In each of the preceding Diels–Alder reactions, only one product is formed because in each of these reactions at least one of the reacting molecules is symmetrical. If both the diene and the dienophile are asymmetrical, two products can be formed.

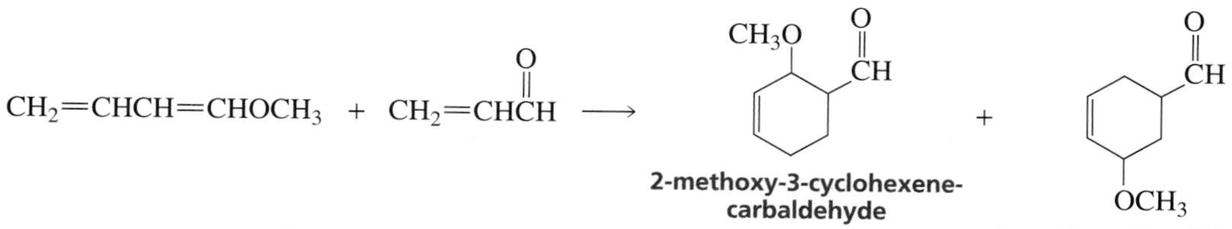

$$CH_2=CHCH=CHOCH_3 \; + \; CH_2=CHCH \longrightarrow$$

2-methoxy-3-cyclohexene-carbaldehyde

+

5-methoxy-3-cyclohexene-carbaldehyde

The two products result from the reactants being aligned relative to each other in two different ways.

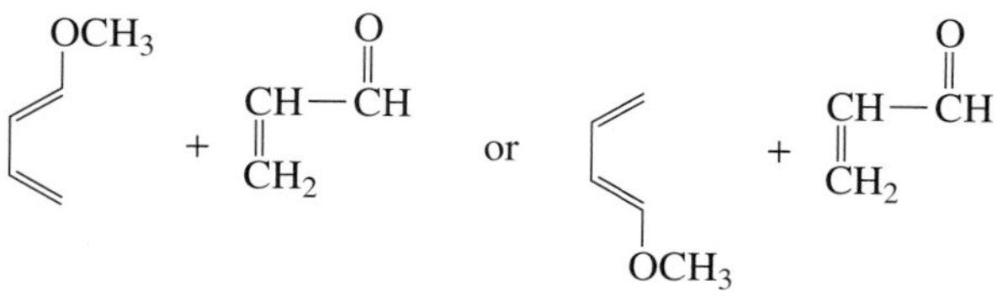

Which of the products will be obtained in greater yield depends on the charge distribution in each of the reactants. It is necessary to draw contributing resonance structures to determine the charge distribution. In the preceding example, the methoxy group of the diene is capable of electron donation by resonance. As a result, the terminal carbon atom bears a partial negative charge.

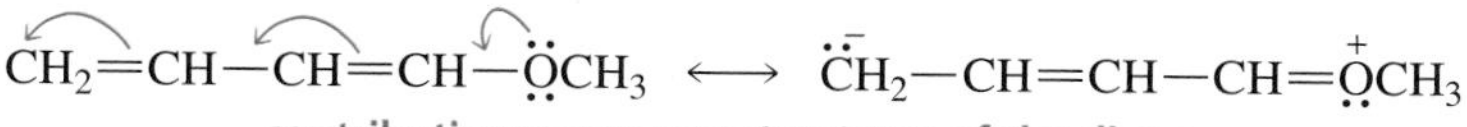

contributing resonance structures of the diene

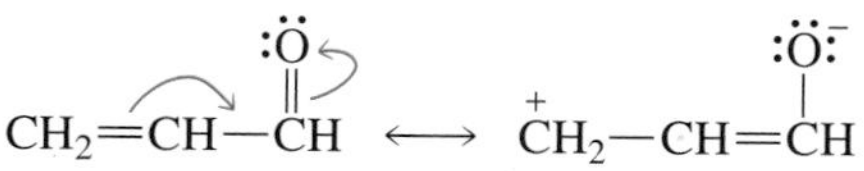

contributing resonance structures of the dienophile

The partially positively charged carbon atom of the dienophile will preferentially attack the most negatively charged carbon in the diene. In this example, 2-methoxy-3-cyclohexenecarbaldehyde is the major product.

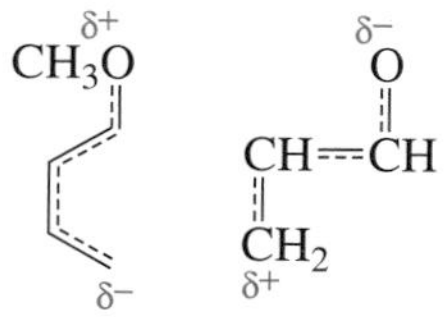

PROBLEM 10◆

What would be the major product if the methoxy substituent in the preceding reaction were bonded to C-2 of the diene rather than to C-1?

A conjugated diene can exist in two different conformations, an **s-cis conformation** and an **s-trans conformation**. By *s*-cis we mean that the double bonds are cis about the single bond (*s* = single). In order to participate in a Diels–Alder reaction, the conjugated diene must be in the *s*-cis conformation. In the *s*-trans conformation, the number 1 and number 4 carbons are too far apart to react with the dienophile.

The rotational barrier between the *s*-cis and *s*-trans conformations is slightly higher than that for a normal single bond.

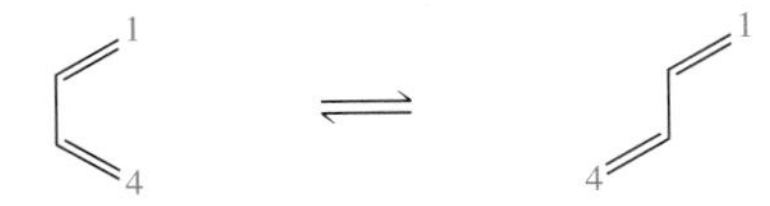

s-cis conformation   s-trans conformation

A conjugated diene that is locked in the *s*-trans conformation cannot undergo a Diels–Alder reaction.

s-trans conformation

A conjugated diene that is locked in the *s*-cis conformation, such as 1,3-cyclopenta-diene or 1,3-cyclohexadiene, is very reactive in a Diels–Alder reaction. When the diene is a cyclic compound, the product of the Diels–Alder reaction is a bridged bicyclic compound. A bridged bicyclic compound contains two rings that share two nonadjacent carbons (Section 7.9).

both rings share these carbons

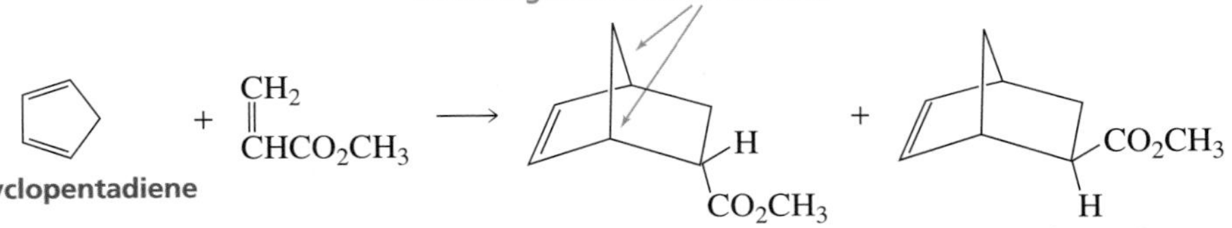

1,3-cyclopentadiene

endo configuration   exo configuration

bridged bicyclic compounds

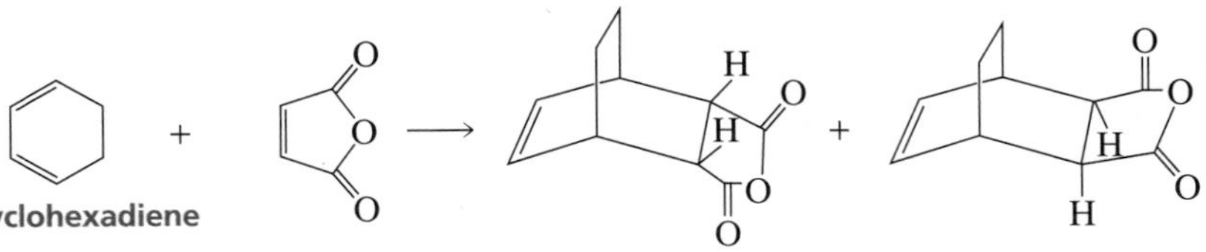

1,3-cyclohexadiene

endo configuration   exo configuration

bridged bicyclic compounds

There are two possible configurations for such compounds. The substituent is **endo** if it is closer to the longer (or more unsaturated) of the two unsubstituted bridges; it is **exo** if it closer to the shorter (or more saturated) bridge.

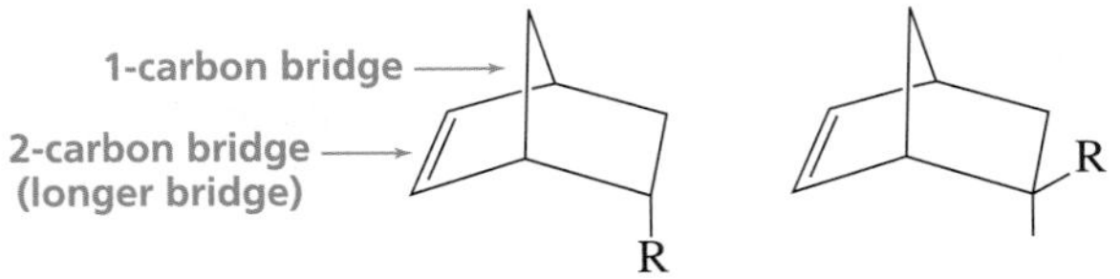

1-carbon bridge ⟶

2-carbon bridge ⟶
(longer bridge)

endo   *exo*

PROBLEM 11◆

Which of the following conjugated dienes would not react with a dienophile in a Diels–Alder reaction?

a. 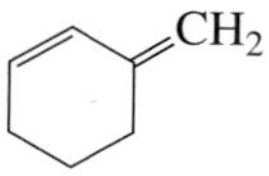 [structure: cyclohexene with =CH₂ substituent]

c. 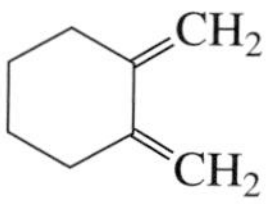 [structure: cyclohexane ring with two =CH₂ groups]

e.  [structure: furan]

b. 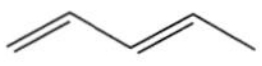 [structure: 2,4-hexadiene]

d. 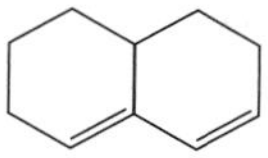 [structure: decalin-type fused bicyclic with two double bonds]

f. 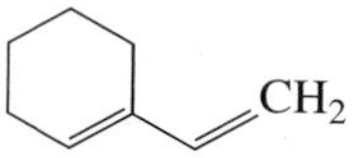 [structure: cyclohexene with vinyl (—CH=CH₂) substituent]

PROBLEM 12 / SOLVED

List the following dienes in order of decreasing activity in a Diels–Alder reaction.

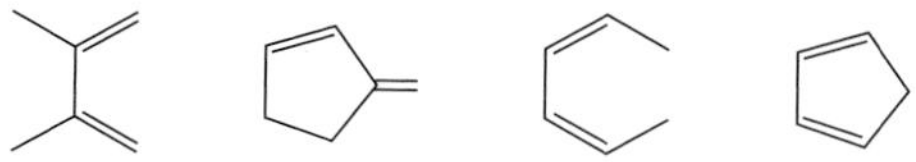

[structures: 2,3-dimethyl-1,3-butadiene; methylenecyclopentane-type ring; 2,4-hexadiene (s-cis drawn); methylenecyclopentene]

SOLUTION   The most reactive diene has the double bonds locked in the *s*-cis conformation, whereas the least reactive diene cannot achieve the required *s*-cis conformation because it is locked in the *s*-trans conformation.

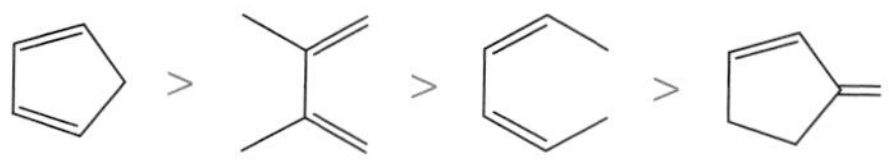

[structures: cyclopentadiene > 2,3-dimethyl-1,3-butadiene > 2,4-hexadiene > methylenecyclopentene]

2,3-Dimethyl-1,3-butadiene and 2,4-hexadiene are of intermediate reactivity because they can exist both in the *s*-cis and *s*-trans conformations. 2,4-Hexadiene is less apt to be in the required *s*-cis conformation because of steric interference between the terminal methyl groups. Consequently, it is less reactive than 2,3-dimethyl-1,3-butadiene.

[structures showing equilibria]

    s-cis   **s-trans**   **s-cis**   **s-trans**

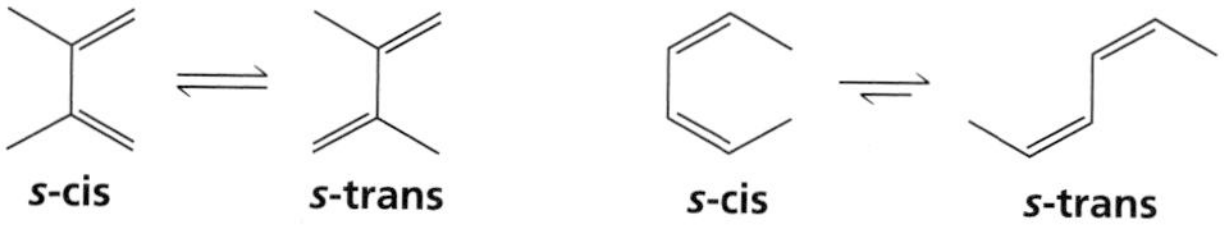

PROBLEM-SOLVING STRATEGY

What diene and what dienophile were used to synthesize the following compound?

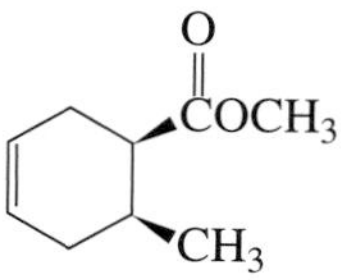

[structure: cyclohexene ring with —COCH₃ and —CH₃ substituents]

$$\overset{O}{\overset{\|}{C}}OCH_3$$

The diene that was used to form the cyclic product had double bonds on either side of the double bond in the product, so put in the π bonds and remove the π bond between them.

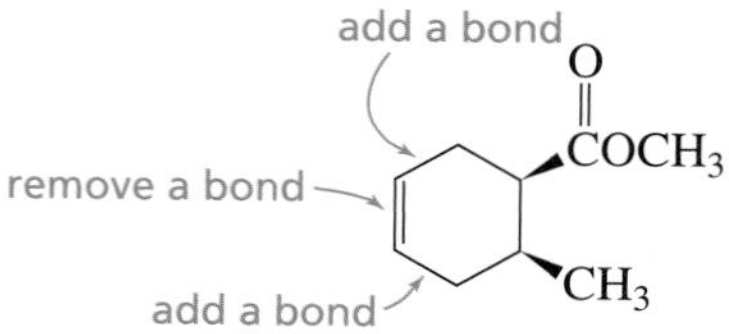

add a bond

remove a bond —

add a bond

[structure: cyclohexene ring with —COCH₃ and —CH₃ substituents, annotated with arrows]

The new σ bonds are now on either side of the double bonds.

Erasing these σ bonds and putting a π bond between the two carbons that were attached by the σ bonds gives the diene and the dienophile.

Now continue on to solve Problem 13.

PROBLEM 13◆

What diene and what dienophile should be used to synthesize the following compounds?

a.

b.

c.

d.

7.9
NOMENCLATURE OF BICYCLIC COMPOUNDS

Bicyclic compounds are compounds that contain two rings. If the two rings share one carbon, the compound is a **spirocyclic compound.** If the two rings share two adjacent carbons, the compound is a **fused bicyclic compound.** Finally, if the rings share two nonadjacent carbons, the compound is a **bridged bicyclic compound.**

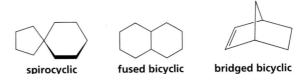

spirocyclic   fused bicyclic   bridged bicyclic

Bicyclic compounds are named by using the alkane name to designate the total number of carbons and the prefix "bicyclo" or "spiro" to indicate the number of shared carbons. "Spiro" indicates one shared carbon and "bicyclo" indicates two shared carbons. After the prefix come brackets that contain numbers indicating the number of carbons in each bridge. These are listed in order of decreasing bridge length. In the molecule shown below, the green bridge has four carbons, the blue bridge has three carbons, and the red bridge has zero carbons.

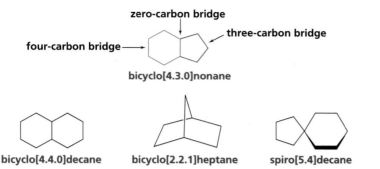

bicyclo[4.3.0]nonane

bicyclo[4.4.0]decane   bicyclo[2.2.1]heptane   spiro[5.4]decane

To determine the position of a substituent, the carbons of the rings are numbered, starting at a shared carbon (a bridgehead carbon) and numbering around the longest bridge first.

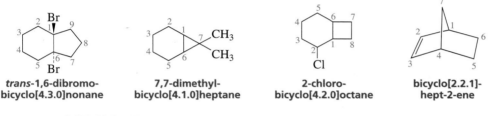

trans-1,6-dibromo-bicyclo[4.3.0]nonane   7,7-dimethyl-bicyclo[4.1.0]heptane   2-chloro-bicyclo[4.2.0]octane   bicyclo[2.2.1]-hept-2-ene

PROBLEM 14◆

Name the following compounds.

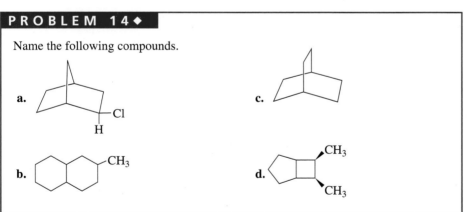

SUMMARY OF REACTIONS

1. If there is excess reagent, both double bonds of an *isolated diene* will undergo electrophilic addition.

$$CH_2=CHCH_2\underset{CH_3}{C}=CH_2 + \underset{\text{excess}}{HBr} \longrightarrow CH_3CHCH_2CCH_3 \quad (CH_3, Br, Br)$$

But if there is only one equivalent of reagent, only the most reactive of the double bonds of an *isolated diene* will undergo electrophilic addition (Section 7.5).

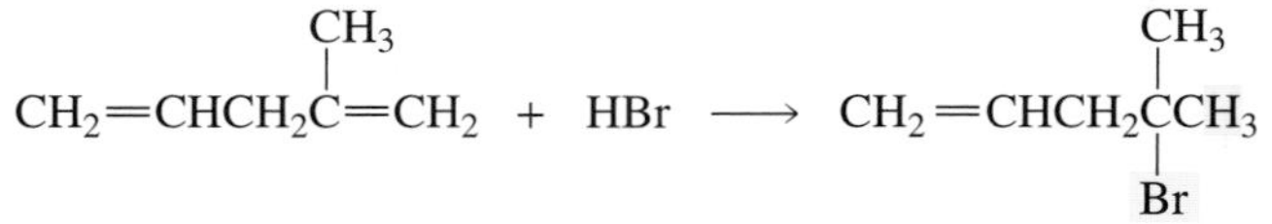

$$CH_2{=}CHCH_2\overset{\displaystyle CH_3}{\underset{}{C}}{=}CH_2 \ + \ HBr \ \longrightarrow \ CH_2{=}CHCH_2\overset{\displaystyle CH_3}{\underset{\displaystyle Br}{C}}CH_3$$

2. *Conjugated dienes* undergo 1,2- and 1,4-addition with one equivalent of an electrophilic reagent (Section 7.6).

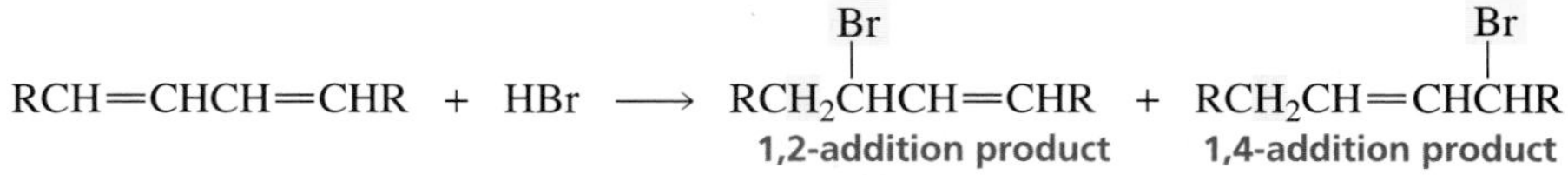

$$RCH{=}CHCH{=}CHR \ + \ HBr \ \longrightarrow \ RCH_2\overset{\displaystyle Br}{\underset{}{C}}HCH{=}CHR \ + \ RCH_2CH{=}CH\overset{\displaystyle Br}{\underset{}{C}}HR$$

1,2-addition product **1,4-addition product**

3. *Conjugated dienes* undergo 1,4-cycloaddition with a dienophile (Section 7.8).

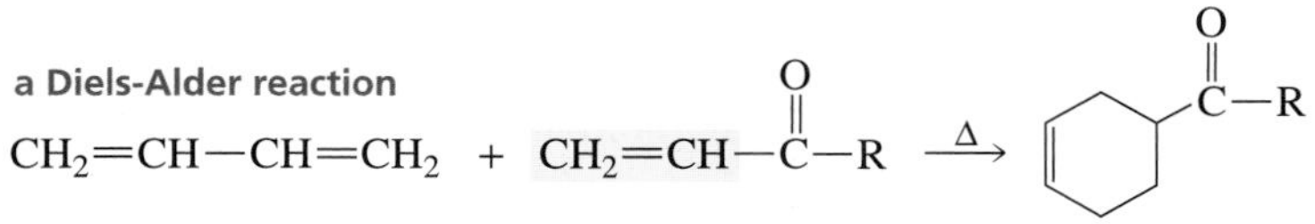

a Diels-Alder reaction

$$CH_2{=}CH{-}CH{=}CH_2 \ + \ CH_2{=}CH{-}\overset{\displaystyle O}{\overset{\|}{C}}{-}R \ \xrightarrow{\ \Delta\ } \ $$

KEY TERMS

1,2-addition (page 309)
1,4-addition (page 309)
allene (page 302)
bicyclic compound (page 322)
bridged bicyclic compound
 (page 322)
common intermediate (page 314)
conjugate addition (page 309)
conjugated double bonds (page 301)
cumulated double bonds (page 302)
cycloaddition reaction (page 317)

[4 + 2] cycloaddition reaction
 (page 317)
Diels–Alder reaction (page 316)
diene (page 301)
dienophile (page 316)
direct addition (page 309)
endo (page 320)
equilibrium control (page 314)
exo (page 320)
fused bicyclic compound (page 322)
isolated double bonds (page 301)

kinetic control (page 312)
kinetic product (page 312)
pericyclic reaction (page 317)
polyene (page 301)
s-cis conformation (page 319)
s-trans conformation (page 319)
spirocyclic compound (page 322)
tetraene (page 301)
thermodynamic control (page 312)
thermodynamic product (page 312)
triene (page 301)

PROBLEMS

15. Give the IUPAC name for each of the following compounds.

a. $CH_3C{\equiv}CCH_2CH_2CH_2CH{=}CH_2$

b. 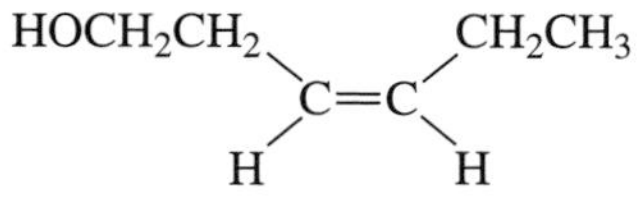
$$\underset{H}{\overset{HOCH_2CH_2}{\diagdown}}C{=}C\underset{H}{\overset{CH_2CH_3}{\diagup}}$$

c. $CH_3CH_2C{\equiv}CCH_2CH_2C{\equiv}CH$

d.

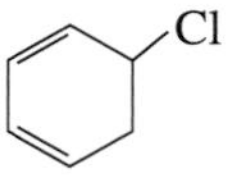

e.

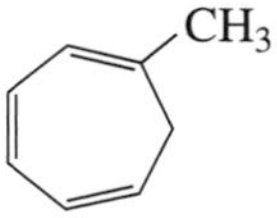

16. Give structures for the following compounds.
  a. (2*E*,4*E*)-1-chloro-3-methyl-2,4-hexadiene
  b. (3*Z*,5*E*)-4-methyl-3,5-nonadiene
  c. (3*Z*,5*Z*)-4,5-dimethyl-3,5-nonadiene
  d. (3*E*,5*E*)-2,5-dibromo-3,5-octadiene

17. **a.** How many linear dienes are there with molecular formula C_6H_{10}? (Ignore cis–trans isomers.)
  b. How many of these dienes are conjugated dienes?

18. Which compound would you expect to have the more negative heat of hydrogenation: 1,2-pentadiene or 1,4-pentadiene?

19. In Chapter 3, we learned that α-farnesene is found in the waxy coatings of apple skins. What is its IUPAC name?

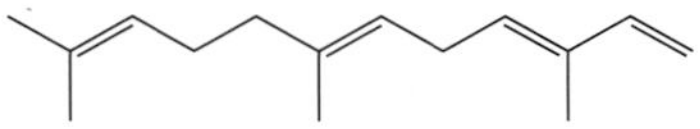

20. Diane O'File treated 1,3-cylohexadiene with Br_2/CCl_4 and obtained two products. Her lab partner treated 1,3-cyclohexadiene with HBr and was surprised to find that only one product was formed. Account for these results.

21. Give the major product of each of the following reactions; one equivalent of each reagent is used.

a. 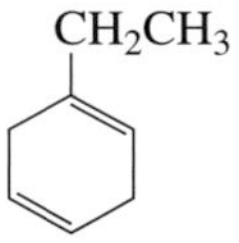 + HBr ⟶

c. 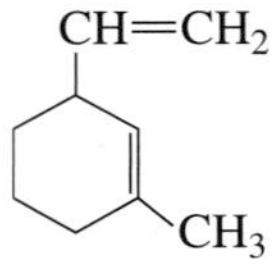 + HBr ⟶

b. 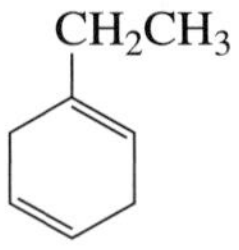 + HBr $\xrightarrow{\text{peroxide}}$

d. 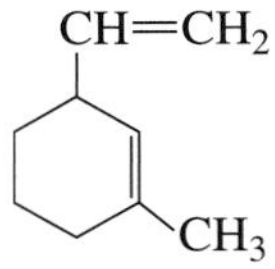 + HBr $\xrightarrow{\text{peroxide}}$

22. **a.** Give the products obtained from the reaction of one mole of HBr with one mole of 1,3,5-hexatriene.
  b. Which product(s) will predominate if the reaction is under kinetic control?
  c. Which product(s) will predominate if the reaction is under thermodynamic control?

23. How could the following compounds be synthesized using a Diels–Alder reaction?

a. 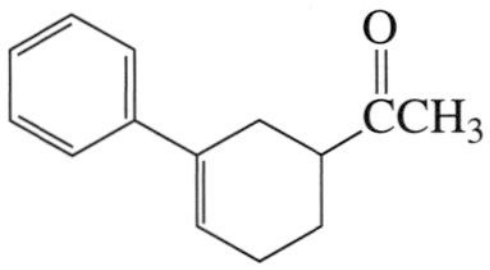

c. 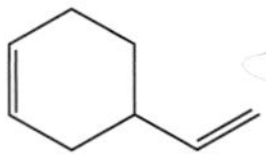

b. 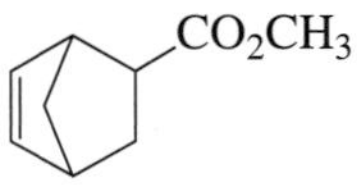

d. 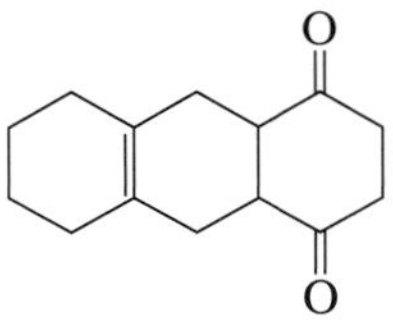

24. **a.** How could each of the following compounds be prepared from a hydrocarbon in a single step?
  b. What other organic compound would be obtained from each synthesis?

1.

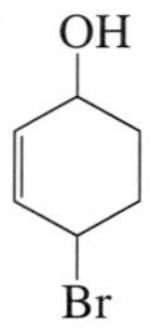

2.

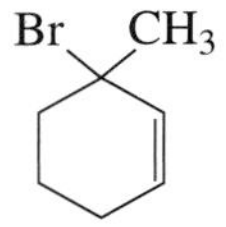

3.

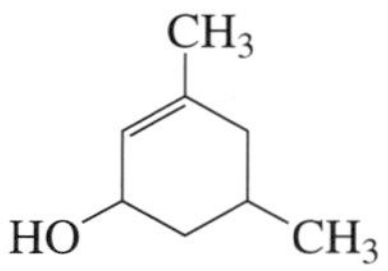

25. How would the following substituents affect the rate of a Diels–Alder reaction?
   a. an electron-donating substituent in the diene
   b. an electron-donating substituent in the dienophile
   c. an electron-withdrawing substituent in the dienophile

26. Give the major products obtained from the reaction of one equivalent of HCl with:
   a. 2,3-dimethyl-1,3-pentadiene
   b. 2,4-dimethyl-1,3-pentadiene
   For each reaction, indicate the kinetic and thermodynamic products.

27. Give the product or products that would be obtained from each of the following reactions.

   a.

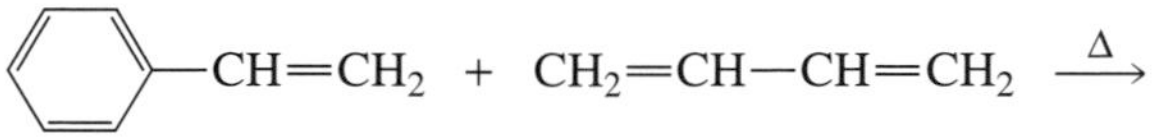

   b.

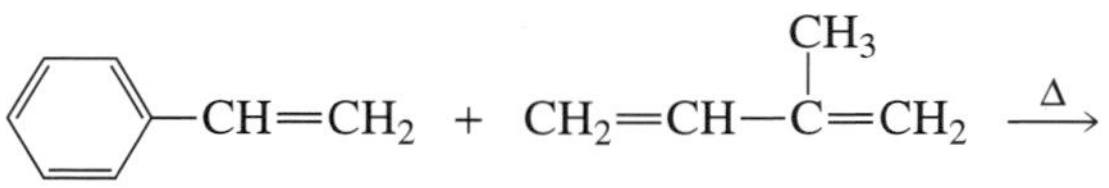

   c.

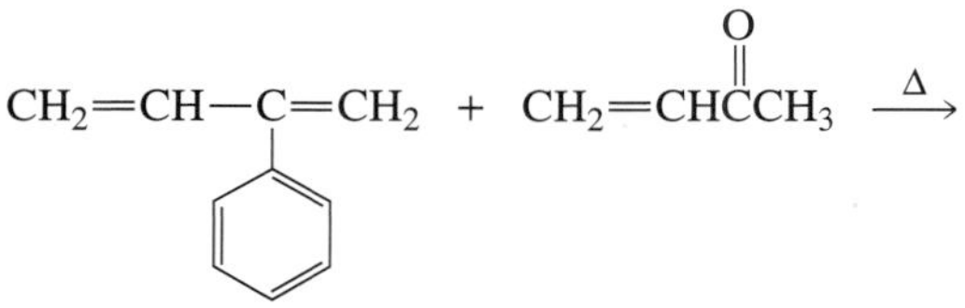

   d.

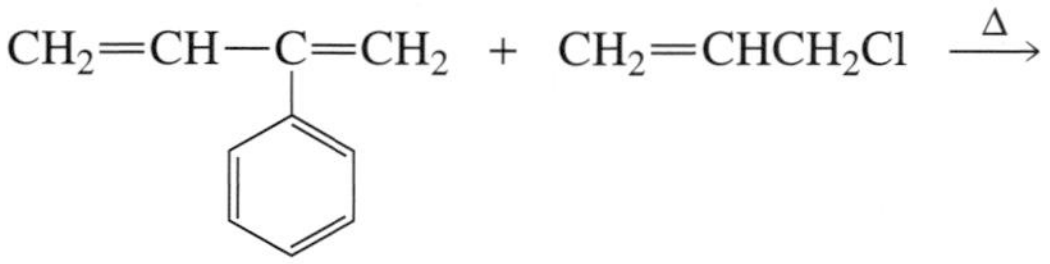

28. On the same graph, draw the reaction coordinate for the addition of one equivalent of HBr to 2-methyl-1,3-pentadiene and one equivalent of HBr to 2-methyl-1,4-penta-diene. Which reaction is faster?

29. While attempting to recrystallize maleic anhydride, Professor Nots O. Kareful dis-solved it in cyclopentadiene rather than in cyclopentane. Was his recrystallization suc-cessful?

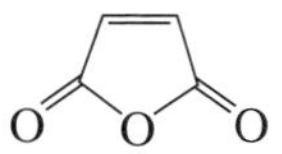

maleic anhydride

30. The following equilibrium is driven to the right if the reaction is carried out in the presence of maleic anhydride. What is the function of maleic anhydride?

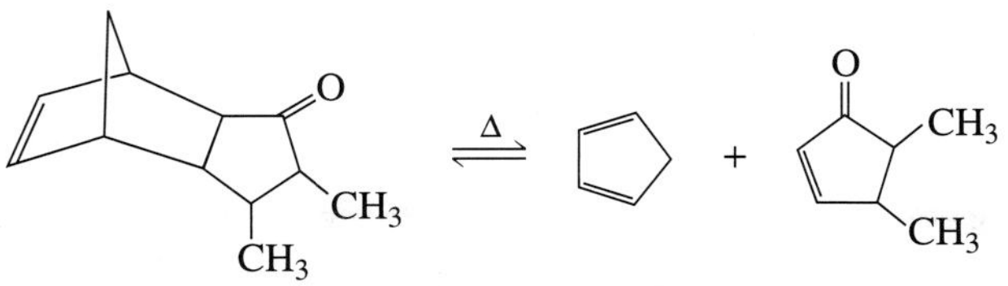

31. Give two sets of a conjugated diene and a dienophile that could be used to prepare the following compound.

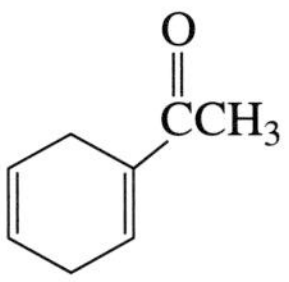

32. a. Which dienophile is more reactive in a Diels–Alder reaction?

1. $CH_2{=}CHCH$ or $CH_2{=}CHCH_2CH$

2. $CH_2{=}CHCH$ or $CH_2{=}CHCH_3$

   b. Which diene is more reactive in a Diels–Alder reaction?

$CH_2{=}CHCH{=}CHOCH_3$ or $CH_2{=}CHCH{=}CHCH_2OCH_3$

33. Cyclopentadiene can react with itself in a Diels–Alder reaction. Draw the endo and exo products.

34. Which diene and which dienophile could be used to prepare each of the following compounds?

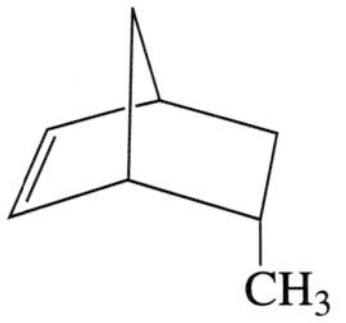

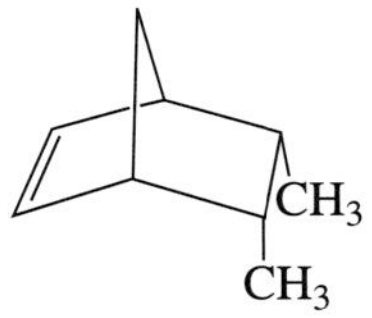

35. As many as nine different products can be obtained by heating a mixture of 1,3-butadiene and 2-methyl-1,3-butadiene. Identify the products.

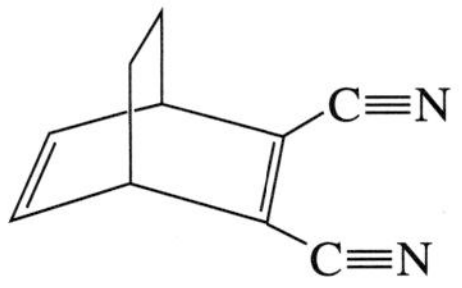

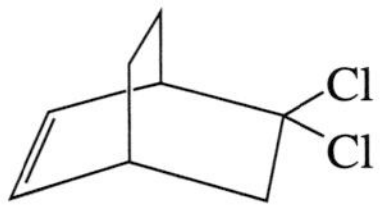

36. a. Give the products of the following reaction.

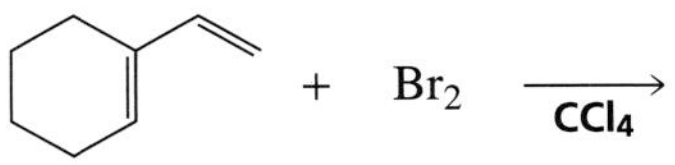

   b. How many stereoisomers of each product could be obtained?

8

REACTIONS OF ALKANES. RADICALS

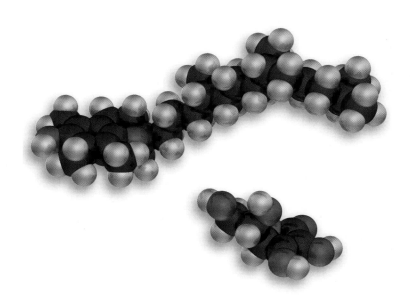

vitamin E and vitamin C

You have learned that there are three classes of hydrocarbons: *alkanes* contain only single bonds; *alkenes* contain double bonds; and *alkynes* contain triple bonds. We have studied the chemistry of alkenes (Chapter 3) and alkynes (Chapter 5). Now we will take a look at the chemistry of alkanes.

Alkanes are called **saturated hydrocarbons** because they are saturated with hydrogen. In other words, they do not contain any double or triple bonds. A few examples of alkanes are shown here; their nomenclature is discussed in Section 2.2.

$CH_3CH_2CH_2CH_3$
butane

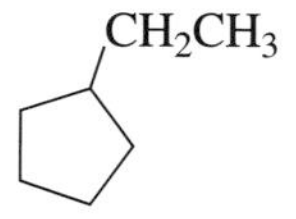

CH_2CH_3 (on cyclopentane)
ethylcyclopentane

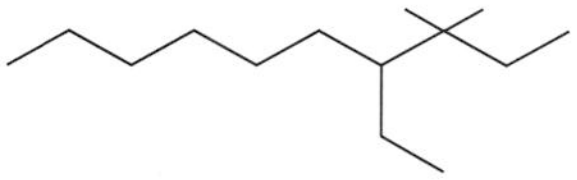

4-ethyl-3,3-dimethyldecane

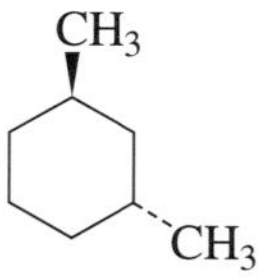

CH_3 ... CH_3
***trans*-1,3-dimethyl-cyclohexane**

Alkanes are widely distributed both on Earth and on other planets. The atmospheres of Saturn and Jupiter contain large quantities of methane (CH_4), the smallest alkane, which is an odorless and flammable gas. Alkanes on Earth are found in natural gas and petroleum. Natural gas and petroleum are formed by the decomposition of plant and animal material that has been buried for long periods of time in Earth's crust, an environment with little oxygen. Natural gas and petroleum, therefore, are known as *fossil fuels*.

Natural gas consists of about 75% methane. The remaining 25% comprises small alkanes such as ethane, propane, and butane. In the 1950s, natural gas replaced coal as the main energy source for domestic and industrial heating in many parts of the United States.

Petroleum is a complex mixture of alkanes and cycloalkanes which can be separated by distillation into fractions. The fraction that boils at the lowest temperature (hydrocarbons containing three and four carbons) is a gas that can be liquefied under pressure. This gas is used as a fuel for cigarette lighters, camp stoves, and barbecues. The fraction that boils at the next-higher temperature (hydrocarbons containing 5 to 11 carbons) is gasoline; the next fraction (9 to 16 carbons) includes kerosene and jet fuel. The fraction with 15 to 25 carbons is used for heating oil and diesel oil, and the highest-boiling fraction is used for lubricants and greases. The nonpolar nature of these compounds is what gives them their oily feel. After distillation, a nonvolatile residue is left behind, which we call asphalt or tar.

The 5- to 11-carbon fraction that is used for gasoline is, in fact, a poor fuel for internal combustion engines and requires a process known as catalytic cracking to become a high-performance gasoline. Catalytic cracking converts straight-chain hydrocarbons that are poor fuels into branched-chain compounds that are high-performance fuels. Originally, cracking (also called pyrolysis) involved heating gasoline to 400 to 500 °C in order to obtain hydrocarbons with three to five carbons. Modern cracking methods use catalysts to accomplish the same thing at much lower temperatures. The small hydrocarbons are then catalytically recombined to form highly branched hydrocarbons.

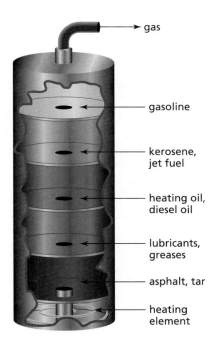

When poor fuels are used in an engine, combustion can be initiated before the spark plug fires. This produces a "pinging" or "knocking" in the running engine. As the quality of a fuel improves, the engine is less likely to knock. The quality of a fuel is indicated by its octane number.

Straight-chain hydrocarbons have low octane numbers and make poor fuels. Heptane, for example, with an arbitrarily assigned octane number of 0, causes an engine to knock badly. Branched-chain fuels do not cause knocking because they burn more slowly; they have high octane numbers. 2,2,4-Trimethylpentane, for example, does not cause knocking and has arbitrarily been assigned an octane number of 100.

$$CH_3CH_2CH_2CH_2CH_2CH_2CH_3$$
heptane
octane number = 0

$$CH_3\overset{\overset{\displaystyle CH_3}{|}}{C}CH_2\overset{\overset{\displaystyle CH_3}{|}}{C}HCH_3$$
$$|$$
$$CH_3$$
2,2,4-trimethylpentane
octane number = 100

The octane number of a gasoline is determined by comparing its knocking tendency to the knocking tendency of mixtures of heptane and 2,2,4-trimethylpentane. The octane number given to the gasoline corresponds to the percent of 2,2,4-trimethylpentane in the matching mixture. The fact that 2,2,4-trimethylpentane contains eight carbons explains the origin of the term "octane number."

FOSSIL FUELS: A PROBLEMATIC ENERGY SOURCE

Three major problems face humans as a consequence of our reliance on fossil fuels for energy. First, fossil fuels are a nonrenewable resource and the world supply is continually decreasing. Second, a group of Middle Eastern and South American countries control a large portion of the world's supply of petroleum. These countries have formed a cartel known as the Organization of Petroleum Exporting Countries (OPEC), which controls both the supply and the price of crude oil. Any political instability in these countries, such as the instability that occurred during the Persian Gulf War in 1991, can seriously affect the world oil supply. Third, burning fossil fuels increases the concentrations of CO_2 and SO_2 in the atmosphere. Scientists have established experimentally that atmospheric SO_2 causes "acid rain," a threat to Earth's plants and therefore to our food and oxygen supplies. Some scientists predict that the CO_2 currently being released into the atmosphere will cause an increase in Earth's temperature as a result of the absorption of infrared radiation by the CO_2. This is called the "greenhouse effect." A steady increase in Earth's temperature can have devastating consequences, including the creation of new deserts, melting of polar ice caps, a rise in sea level, and starvation as a result of crop failure. Clearly, what we need is a renewable, nonpolitical, nonpolluting, and economically affordable source of energy.

8.1 REACTIVITY CONSIDERATIONS

The double and triple bonds of alkenes and alkynes are composed of strong σ bonds and relatively weak π bonds. We have seen that the reactivity of alkenes and alkynes is the result of an electrophile being attracted to the cloud of electrons that constitutes the π bond.

Alkanes contain only strong σ bonds. Because the carbon and hydrogen atoms of an alkane have approximately the same electronegativity, the electrons in the carbon–hydrogen and carbon–carbon σ bonds are shared equally by the bonding atoms. Consequently, none of the atoms in an alkane have any significant charge. This means that neither nucleophiles nor electrophiles are attracted to them. As a result of having only strong σ bonds and having atoms without any partial charge, alkanes are very unreactive compounds. Because of their lack of reactivity, early organic chemists called them **paraffins** from the Latin *parum affinis,* which means "little affinity" (for other compounds).

8.2 CHLORINATION AND BROMINATION OF ALKANES

Alkanes react with molecular chlorine (Cl_2) or bromine (Br_2) to form alkyl chlorides or alkyl bromides. These reactions take place only at high temperatures or in the presence of light (symbolized by *hν*). These are the only reactions that alkanes undergo—with the exception of combustion (Section 2.10).

$$CH_4 + Cl_2 \xrightarrow[\substack{\text{or} \\ h\nu}]{400\ ^\circ C} \underset{\textbf{methyl chloride}}{CH_3Cl} + HCl$$

$$CH_3CH_3 + Br_2 \xrightarrow[\substack{\text{or} \\ h\nu}]{400\ ^\circ C} \underset{\textbf{ethyl bromide}}{CH_3CH_2Br} + HBr$$

The mechanism for the halogenation of an alkane is well understood. The high temperature (or light) supplies the energy required to break the chlorine–chlorine or bromine–bromine bond homolytically. **Homolytic bond cleavage** results in the formation of radicals. This is called the **initiation step** of the reaction because it creates the radical that is used in the first propagation step (Section 3.18). Recall that a one-sided arrowhead signifies the movement of one electron (Section 3.6).

homolytic cleavage

$$:\!\overset{..}{\underset{..}{Cl}}\!-\!\overset{..}{\underset{..}{Cl}}\!: \xrightarrow[\substack{\text{or} \\ h\nu}]{400\ ^\circ C} 2\ :\!\overset{..}{\underset{..}{Cl}}\!\cdot$$

$$:\!\overset{..}{\underset{..}{Br}}\!-\!\overset{..}{\underset{..}{Br}}\!: \xrightarrow[\substack{\text{or} \\ h\nu}]{400\ ^\circ C} 2\ :\!\overset{..}{\underset{..}{Br}}\!\cdot$$

A radical is highly reactive because it wants to form a bond and thereby complete its octet. In the mechanism for the monochlorination of methane, the chlorine radical formed in the initiation step abstracts a hydrogen atom from methane, forming HCl and a methyl radical. The methyl radical abstracts a chlorine atom from Cl_2, forming methyl chloride and another chlorine radical, which can abstract a hydrogen atom from another molecule of methane. These two steps are called **propagation steps** because the radical created in the first propagation step reacts in the second propagation step to produce a radical that reacts in a repetition of the first propagation step. These two steps are repeated over and over. Because the reaction has repeating propagation steps and radical intermediates, it is called a **radical chain reaction.**

mechanism for the monochlorination of methane

$$:\!\overset{..}{\underset{..}{Cl}}\!-\!\overset{..}{\underset{..}{Cl}}\!: \xrightarrow[\substack{\text{or} \\ h\nu}]{400\ ^\circ C} 2\ :\!\overset{..}{\underset{..}{Cl}}\!\cdot \quad \text{initiation step}$$

$$:\!\overset{..}{\underset{..}{Cl}}\!\cdot + H\!-\!CH_3 \longrightarrow H\overset{..}{\underset{..}{Cl}}\!: + \underset{\textbf{a methyl radical}}{\cdot CH_3} \left. \phantom{\begin{matrix}1\\2\end{matrix}} \right\} \text{propagation steps}$$

$$\cdot CH_3 + :\!\overset{..}{\underset{..}{Cl}}\!-\!\overset{..}{\underset{..}{Cl}}\!: \longrightarrow CH_3Cl + :\!\overset{..}{\underset{..}{Cl}}\!\cdot$$

$$:\!\overset{..}{\underset{..}{Cl}}\!\cdot + :\!\overset{..}{\underset{..}{Cl}}\!\cdot \longrightarrow Cl_2$$

$$\cdot CH_3 + \cdot CH_3 \longrightarrow CH_3CH_3 \left. \phantom{\begin{matrix}1\\2\\3\end{matrix}} \right\} \text{termination steps}$$

$$:\!\overset{..}{\underset{..}{Cl}}\!\cdot + \cdot CH_3 \longrightarrow CH_3Cl$$

Any two radicals in the reaction mixture can combine to form a molecule in which all the electrons are paired. The combination of two radicals, called a **termination step,** brings the reaction to an end by decreasing the number of radicals available to propagate the reaction. Radical chlorination of alkanes other than methane follows the same mechanism. A radical chain reaction, with its characteristic initiation, propagation, and termination steps, was first described in Section 3.18.

The reaction of an alkane with chlorine or bromine to form an alkyl halide is called a **radical substitution reaction** because radicals are involved as intermediates, and the end result is the substitution of a halogen atom for one of the hydrogen atoms of the alkane.

PROBLEM 1

Show the initiation, propagation, and termination steps for the monochlorination of cyclohexane.

Using the bond dissociation energies in Table 3.1 and recalling that the $\Delta H°$ of a reaction is equal to the sum of the energies of the bonds broken minus the sum of the energies of the bonds formed (Section 3.7), the $\Delta H°$ for the monochlorination of methane is calculated to be −24 kcal/mol.

$$CH_4 + Cl_2 \xrightarrow[\substack{\text{or} \\ h\nu}]{400\ °C} CH_3Cl + HCl$$

Bonds broken		Bonds formed	
C—H	$DH° = 105$ kcal/mol	C—Cl	$DH° = 84$ kcal/mol
Cl—Cl	$DH° = 58$ kcal/mol	H—Cl	$DH° = 103$ kcal/mol
	$DH°_{total} = 163$ kcal/mol		$DH°_{total} = 187$ kcal/mol

$\Delta H°$ for the reaction $= DH°$ for bonds broken $- DH°$ for bonds formed
$= 163$ kcal/mol $- 187$ kcal/mol
$= -24$ kcal/mol

The $\Delta H°$ for each step of the reaction can also be calculated. Only one of the termination steps (the one needed to balance the overall reaction) is shown in the scheme below. Adding the individual steps and crossing out molecules that appear on both sides of the equation yield the overall reaction. The value of $\Delta H°$ obtained by adding the individual $\Delta H°$ values is identical to the value of $\Delta H°$ that was calculated for the overall reaction. Notice that the overall $\Delta H°$ for the chain reaction equals the sum of the $\Delta H°$ values for the two propagation steps, because the initiation step and the termination step cancel each other.

	Reaction	$\Delta H°$ (kcal/mol)	E_a(kcal/mol)
(i)	$Cl_2 \longrightarrow 2\ Cl\cdot$	$58 - 0 = 58$	58
(ii)	$Cl\cdot + CH_4 \longrightarrow \cdot CH_3 + HCl$	$105 - 103 = 2$	4
(iii)	$\cdot CH_3 + Cl_2 \longrightarrow CH_3Cl + Cl\cdot$	$58 - 84 = -26$	1
(iv)	$Cl\cdot + Cl\cdot \longrightarrow Cl_2$	$0 - 58 = -58$	0

A reaction coordinate diagram can be constructed for the monochlorination of methane from the values of $\Delta H°$ calculated for each of the individual steps and the experimental activation energies (Figure 8.1). The diagram shows that the rate-determining step—the step with the highest energy transition state—is abstraction of a hydrogen atom from the alkane by the chlorine radical.

In order to maximize the amount of monohalogenated product obtained, a radical substitution reaction should be carried out in the presence of excess alkane. The presence of excess alkane in the reaction mixture ensures that there is a greater

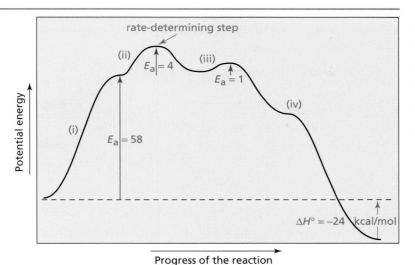

◄ Figure 8.1
Reaction coordinate diagram for the monochlorination of methane. The reaction coordinate diagrams for the individual steps are numbered to correspond to the numbers assigned to the reaction steps in the mechanism (shown in the text).

probability of the halogen radical colliding with a molecule of alkane than with a molecule of alkyl halide. This is true even toward the end of the reaction, by which time a considerable amount of alkyl halide will have been formed. If the halogen radical abstracts a hydrogen from a molecule of alkyl halide rather than from a molecule of unreacted alkane, a dihalogenated product will be obtained.

$$Cl\cdot + CH_3Cl \longrightarrow \cdot CH_2Cl + HCl$$
$$\cdot CH_2Cl + Cl_2 \longrightarrow CH_2Cl_2 + Cl\cdot$$

PROBLEM 2

Give the mechanism for the formation of carbon tetrachloride, CCl_4, from the reaction of methane with $Cl_2 + h\nu$.

PROBLEM 3 / SOLVED

If cyclopentane reacts with excess Cl_2 at a high temperature, how many dichlorocyclopentanes would you expect to obtain as products?

SOLUTION   Seven dichlorocyclopentanes would be obtained as products. Only one isomer is possible for the 1,1-dichloro compound. The 1,2- and 1,3-dichloro compounds have two chirality centers. For each, the cis isomer is a meso compound and the trans isomer is a pair of enantiomers.

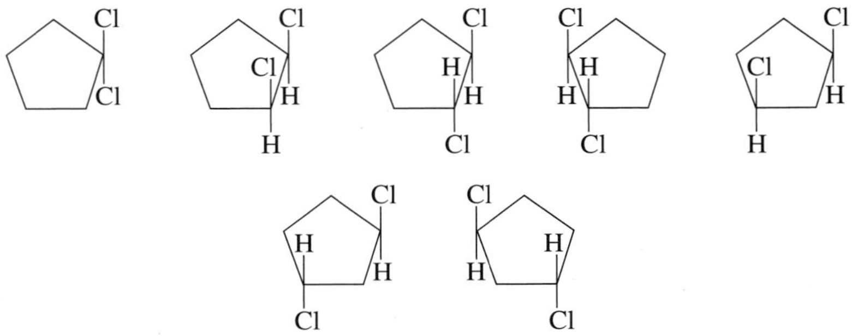

Chlorination and bromination of alkanes follow the same mechanism. The only difference is that chlorination produces alkyl chlorides whereas bromination forms alkyl bromides.

mechanism for the monobromation of ethane

$$Br_2 \xrightarrow{\Delta} 2\ Br\cdot \qquad \text{initiation step}$$

$$Br\cdot + CH_3CH_3 \longrightarrow CH_3\dot{C}H_2 + HBr$$
$$CH_3\dot{C}H_2 + Br_2 \longrightarrow CH_3CH_2Br + Br\cdot \qquad \text{propagation steps}$$

$$Br\cdot + Br\cdot \longrightarrow Br_2$$
$$CH_3\dot{C}H_2 + CH_3\dot{C}H_2 \longrightarrow CH_3CH_2CH_2CH_3 \qquad \text{termination steps}$$
$$CH_3\dot{C}H_2 + Br\cdot \longrightarrow CH_3CH_2Br$$

In a process called **disproportionation,** one alkyl radical transfers a hydrogen atom to another alkyl radical, resulting in formation of an alkane and an alkene. Consequently, small amounts of ethane and ethene are formed in the monobromination of ethane. Because disproportionation destroys radicals, it is a termination step.

disproportionation of ethyl radicals

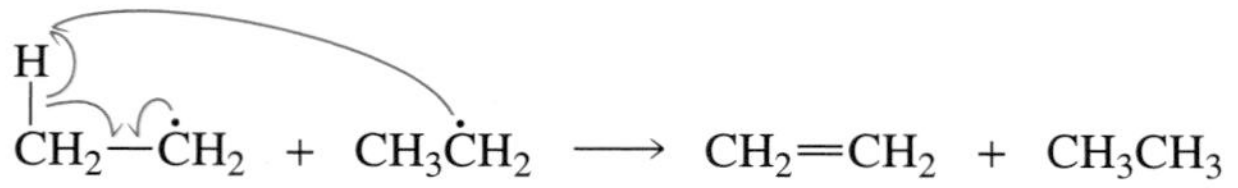

$$\overset{H}{\underset{}{CH_2}}-\dot{C}H_2 + CH_3\dot{C}H_2 \longrightarrow CH_2=CH_2 + CH_3CH_3$$

PROBLEM 4 ◆

What products would be obtained from disproportionation of:

a. an isopropyl radical?   **b.** a *tert*-butyl radical?

8.3
FACTORS THAT DETERMINE PRODUCT DISTRIBUTION

Two different alkyl halides are obtained from the monochlorination of butane. Substitution of a hydrogen bonded to one of the terminal carbons produces 1-chlorobutane, while substitution of a hydrogen bonded to one of the internal carbons leads to 2-chlorobutane.

$$CH_3CH_2CH_2CH_3 + Cl_2 \xrightarrow{h\nu} CH_3CH_2CH_2CH_2Cl + CH_3CH_2\overset{Cl}{\underset{|}{C}}HCH_3 + HCl$$

butane   1-chlorobutane   2-chlorobutane
                      expected = 60%   expected = 40%
                      experimental = 29% experimental = 71%

Butane contains ten hydrogens. Six of them can be substituted to form 1-chlorobutane, and four can be substituted to form 2-chlorobutane. If all of the carbon–hydrogen bonds in an alkane were equally easy to break, probability factors alone would control the relative amounts of the two products. Because the ratio of primary to secondary hydrogens in butane is 6:4, the probability of a chlorine radical colliding with a primary hydrogen compared with a secondary hydrogen is 6 : 4. Therefore, we would expect 60% of the product to be 1-chlorobutane and 40% to be 2-chlorobutane. When we carry out the reaction in the laboratory and analyze the product, however, we find that the product distribution is different than what we predicted: 29% of the product is 1-chlorobutane and 71% is 2-chlorobutane.

The amounts of 1-chlorobutane and 2-chlorobutane obtained from monochlorination of butane clearly indicate that probability alone does not determine the relative amounts of the two products. Because the rate-determining step in a radical chlorination reaction is abstraction of a hydrogen atom by a chlorine radical, the ease with which the hydrogen atom can be removed affects the amounts of products that are obtained. Since more 2-chlorobutane is obtained than expected, we can conclude that it is easier to abstract a hydrogen atom from a secondary carbon than from a primary carbon.

Alkyl radicals have different stabilities (Section 3.18), and the more stable the radical, the more easily it is formed. Thus it is easier to remove a hydrogen atom from a secondary carbon, thereby forming a secondary radical, than it is to remove a hydrogen atom from a primary carbon to form a primary radical.

relative stabilities of alkyl radicals

When a chlorine radical reacts with butane, it can abstract a hydrogen atom from a terminal carbon, forming a primary alkyl radical, or it can abstract a hydrogen atom from an internal carbon, forming a secondary alkyl radical. Since it is easier to form the more stable secondary alkyl radical, 2-chlorobutane is formed faster than 1-chlorobutane.

After experimentally determining the amount of each chlorination product obtained from various hydrocarbons, chemists were able to calculate that *at room temperature* it is 5.0 times easier for a chlorine radical to abstract a hydrogen atom from a tertiary carbon than from a primary carbon, and it is 3.8 times easier to abstract a hydrogen atom from a secondary carbon than from a primary carbon. The precise ratios differ at different temperatures.

relative rates of alkyl radical formation by a chlorine radical

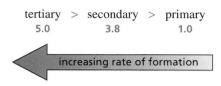

$$\text{tertiary} \quad > \quad \text{secondary} \quad > \quad \text{primary}$$
$$5.0 \qquad\qquad 3.8 \qquad\qquad 1.0$$

increasing rate of formation

To determine the relative amounts of products obtained from radical chlorination of an alkane, both the *probability* factor (the number of hydrogens that can be abstracted that will lead to the formation of the particular product) and the *reactivity* factor (the relative rate at which a particular hydrogen is abstracted) must be taken into account. When both factors are considered, the calculated amounts of 1-chlorobutane and 2-chlorobutane agree with the amounts obtained experimentally.

relative amount of 1-chlorobutane	**relative amount of 2-chlorobutane**
number of hydrogens × reactivity	number of hydrogens × reactivity
$6 \times 1.0 = 6.0$	$4 \times 3.8 = 15$
percent yield $= \dfrac{6.0}{21} = 29\%$	percent yield $= \dfrac{15}{21} = 71\%$

The percent yield of each product is calculated by dividing the relative amount of the particular product by the sum of the relative amounts of all products $(6 + 15 = 21)$.

Radical monochlorination of 2,2,5-trimethylhexane results in the formation of five monochlorination products. Because the relative amounts of the five products total 35 $(9.0 + 7.6 + 7.6 + 5.0 + 6.0 = 35)$, the percent of each product obtained can be calculated as shown here.

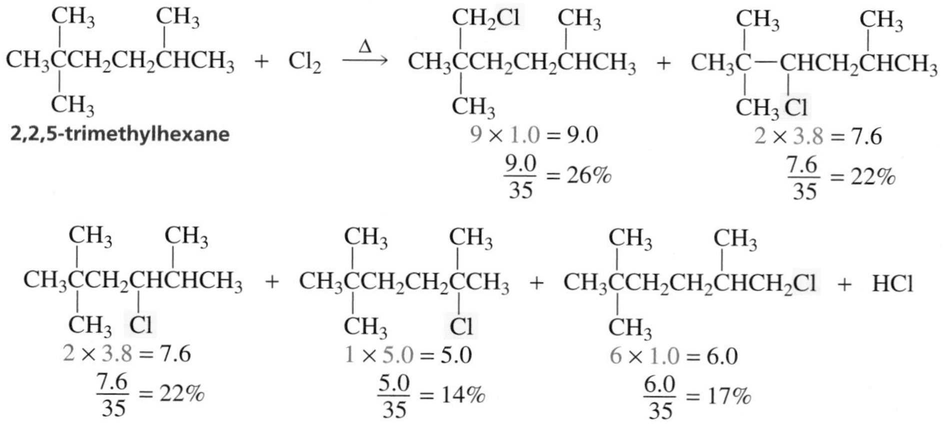

$$
\underset{\textbf{2,2,5-trimethylhexane}}{\text{CH}_3\text{CCH}_2\text{CH}_2\text{CHCH}_3} + \text{Cl}_2 \xrightarrow{\Delta} \text{CH}_3\text{CCH}_2\text{CH}_2\text{CHCH}_3 + \text{CH}_3\text{C}-\text{CHCH}_2\text{CHCH}_3
$$

$$9 \times 1.0 = 9.0 \qquad\qquad 2 \times 3.8 = 7.6$$
$$\frac{9.0}{35} = 26\% \qquad\qquad \frac{7.6}{35} = 22\%$$

$$
\text{CH}_3\text{CCH}_2\text{CHCHCH}_3 + \text{CH}_3\text{CCH}_2\text{CH}_2\text{CCH}_3 + \text{CH}_3\text{CCH}_2\text{CH}_2\text{CHCH}_2\text{Cl} + \text{HCl}
$$

$$2 \times 3.8 = 7.6 \qquad 1 \times 5.0 = 5.0 \qquad 6 \times 1.0 = 6.0$$
$$\frac{7.6}{35} = 22\% \qquad \frac{5.0}{35} = 14\% \qquad \frac{6.0}{35} = 17\%$$

Because radical halogenation of an alkane can yield several different monosubstitution products as well as products that contain more than one halogen, it is not a good method to use for synthesis of alkyl halides. Addition of HCl or HBr to an alkene is a much better way to make an alkyl halide (Section 3.9). However, radical halogenation of an alkane is still a useful reaction because it is the only way that an inert alkane can be converted to a reactive compound. We will see that once the halogen is introduced into the alkane, it can be replaced by a variety of other substituents (Chapter 9).

PROBLEM 5◆

How many alkyl halides can be obtained from monochlorination of each of the following alkanes? Neglect stereoisomers.

a. $CH_3CH_2CH_2CH_2CH_3$

e. 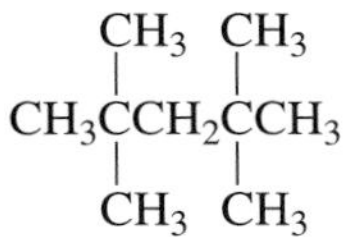
$$
\begin{array}{c}
CH_3\ \ CH_3 \\
|\ \ \ \ \ | \\
CH_3CCH_2CCH_3 \\
|\ \ \ \ \ | \\
CH_3\ \ CH_3
\end{array}
$$

b. 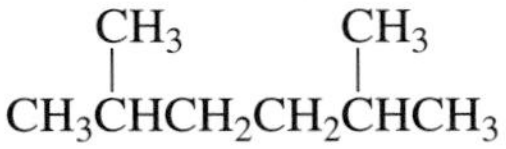
$$
\begin{array}{c}
CH_3\ \ \ \ \ \ \ \ CH_3 \\
|\ \ \ \ \ \ \ \ \ \ \ | \\
CH_3CHCH_2CH_2CHCH_3
\end{array}
$$

f. 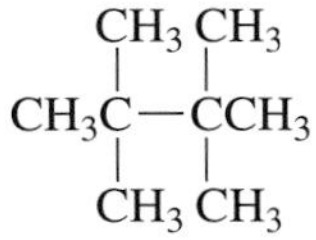
$$
\begin{array}{c}
CH_3\ CH_3 \\
|\ \ \ \ \ | \\
CH_3C\!-\!CCH_3 \\
|\ \ \ \ \ | \\
CH_3\ CH_3
\end{array}
$$

c.  ⬡

g. 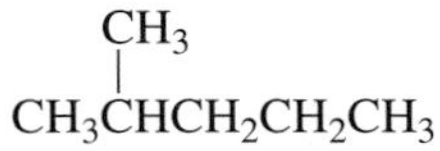
$$
\begin{array}{c}
CH_3 \\
| \\
CH_3CHCH_2CH_2CH_3
\end{array}
$$

d. 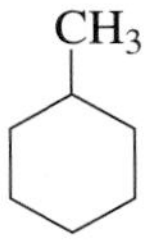
$$
\begin{array}{c}
CH_3 \\
| \\
⬡
\end{array}
$$

h. 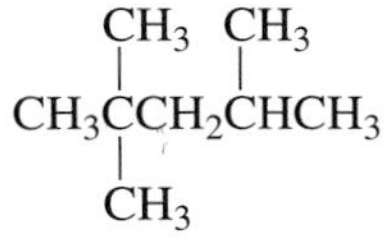
$$
\begin{array}{c}
CH_3\ \ CH_3 \\
|\ \ \ \ \ | \\
CH_3CCH_2CHCH_3 \\
| \\
CH_3
\end{array}
$$

PROBLEM 6◆

Calculate the percent of each product that you would expect to obtain in Problem 5a, b, and g if chlorination were carried out in the presence of light at room temperature.

PROBLEM 7◆

When methylpropane is monochlorinated in the presence of light at room temperature, 36% of the product is 2-chloro-2-methylpropane and 64% is 1-chloro-2-methylpropane. From these data, calculate how much easier it is to abstract a hydrogen atom from a tertiary carbon than from a primary carbon under these conditions.

The relative rates of radical formation when a bromine radical is used to abstract a hydrogen atom are very different from the relative rates of radical formation when a chlorine radical is used. At 125 °C, a bromine radical removes a hydrogen atom from a tertiary carbon 1600 times faster than from a primary carbon, and removes a hydrogen atom from a secondary carbon 82 times faster than from a primary carbon.

**8.4
THE REACTIVITY–
SELECTIVITY
PRINCIPLE**

relative rates of radical formation by a bromine radical

tertiary > secondary > primary
   1600   82   1

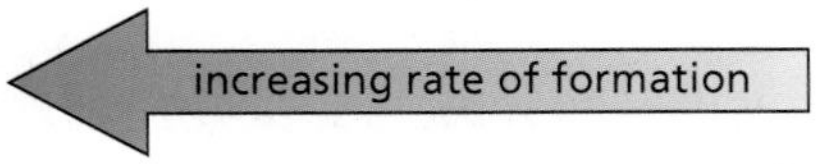

◀ increasing rate of formation

The rate-determining step in bromination of an alkane (like the rate-determining step in chlorination of an alkane) is abstraction of a hydrogen atom by the halogen radical. When a bromine radical is the hydrogen-abstracting agent, the differences in reactivity are so great that the reactivity factor is vastly more important than the probability factor. For example, radical bromination of butane leads to a 98% yield of 2-bromobutane compared with the 71% yield of 2-chlorobutane obtained when butane is chlorinated (Section 8.3).

$$CH_3CH_2CH_2CH_3 + Br_2 \xrightarrow{h\nu} CH_3CH_2CH_2CH_2Br + CH_3CH_2\overset{Br}{\underset{|}{C}}HCH_3 + HBr$$

1-bromobutane   **2-bromobutane**
2%   98%

Similarly, bromination of 2,2,5-trimethylhexane results in an 82% yield of the product in which bromine replaces the tertiary hydrogen. Chlorination of the same alkane results in a 14% yield of the tertiary alkyl halide (Section 8.3).

2,2,5-trimethylhexane   2-bromo-2,5,5-trimethylhexane
                         82%

PROBLEM 8

Perform the calculations that predict that:

a. 2-bromobutane will be obtained in 98% yield.

b. 2-bromo-2,5,5,-trimethylhexane will be obtained in 82% yield.

Why are the relative rates of radical formation so much different when a bromine radical is used as the hydrogen-abstracting reagent rather than a chlorine radical? To answer this question, we must look at the $\Delta H°$ values for formation of primary, secondary, and tertiary radicals as a result of reaction with a chlorine radical and as a result of reaction with a bromine radical. These $\Delta H°$ values can be calculated using the bond dissociation energies in Table 3.1.

		$\Delta H°$ (kcal/mol)
Cl· + $CH_3CH_2CH_3$ ⟶ $CH_3CH_2\dot{C}H_2$ + HCl		$101 - 103 = -2$
Cl· + $CH_3CH_2CH_3$ ⟶ $CH_3\dot{C}HCH_3$ + HCl		$99 - 103 = -4$
Cl· + CH_3CHCH_3(CH₃) ⟶ $CH_3\dot{C}CH_3$(CH₃) + HCl		$97 - 103 = -6$

		$\Delta H°$ (kcal/mol)
Br· + $CH_3CH_2CH_3$ ⟶ $CH_3CH_2\dot{C}H_2$ + HBr		$101 - 88 = 13$
Br· + $CH_3CH_2CH_3$ ⟶ $CH_3\dot{C}HCH_3$ + HBr		$99 - 88 = 11$
Br· + CH_3CHCH_3(CH₃) ⟶ $CH_3\dot{C}CH_3$(CH₃) + HBr		$97 - 88 = 9$

We must also be aware that the activation energy for abstraction of a hydrogen atom by a bromine radical has been found experimentally to be about 4.5 times greater than the activation energy for abstraction of a hydrogen atom by a chlorine radical. Using the calculated $\Delta H°$ values and the experimental activation energies, we can draw reaction coordinate diagrams for the formation of primary, secondary,

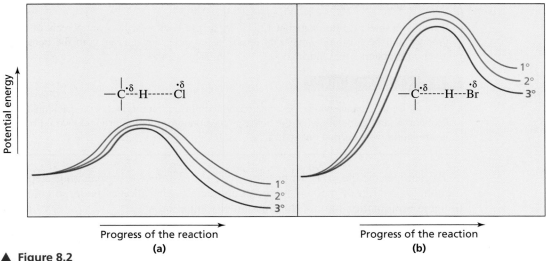

▲ Figure 8.2
Reaction coordinate diagrams for formation of primary, secondary, and tertiary alkyl radicals as a result of abstraction of a hydrogen atom (a) by a chlorine radical and (b) by a bromine radical. The transition states in (a) resemble the reactant; they have relatively little radical character. The transition states in (b) resemble the product; they have a higher degree of radical character.

and tertiary radicals by means of abstraction of a hydrogen atom by a chlorine radical and by a bromine radical. These diagrams are shown in Figure 8.2.

Because the reaction of a chlorine radical with an alkane to form a primary, secondary, or tertiary radical is exothermic, the transition state resembles the reactant more than the product (the Hammond postulate; Section 3.11). Therefore, there will be only a small difference in the activation energies for removal of a hydrogen atom from a primary, secondary, or tertiary carbon. In contrast, the reaction of a bromine radical with an alkane is endothermic, so the transition state resembles the product more than the reactant. Because there is a significant difference in the energies of the product radicals depending on whether they are primary, secondary, or tertiary, there is a significant difference in the activation energies. This means that the transition state for hydrogen atom abstraction by a bromine radical has a higher degree of radical character and is, therefore, more sensitive to the stability of the resulting radical compared with the transition state for hydrogen atom abstraction by a chlorine radical.

A chlorine radical is so reactive that it makes primary, secondary, and tertiary radicals with almost equal ease. A bromine radical is so unreactive that it has a clear preference for formation of the easier-to-form tertiary radical. In other words, because it is unreactive, a bromine radical is very selective as to what hydrogen atom it abstracts. In contrast, the much more reactive chlorine radical is much less selective. These observations illustrate the **reactivity–selectivity principle,** which states that *the greater the reactivity of a species, the less selective it will be.*

Because chlorination is relatively nonselective, it is a useful reaction only when there is one kind of hydrogen in the molecule.

The more reactive a species is, the less selective it will be.

If methylpropane is brominated at 125 °C in the presence of light, what percent of the product will be 2-bromo-2-methylpropane? Compare your answer with the percent given in Problem 7 for chlorination.

Take the same alkanes whose percents of monochlorination products you calculated in Problem 6, and calculate what the percents of monobromination products would be if bromination were carried out at 125 °C.

By comparing the $\Delta H°$ values for the sum of the two propagating steps for monohalogenation of methane, we can appreciate why alkanes undergo chlorination and bromination but not fluorination or iodination.

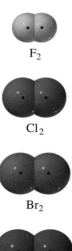

F_2

Cl_2

Br_2

I_2
Halogens

$$F\cdot + CH_4 \longrightarrow \cdot CH_3 + HF \qquad 105 - 136 = -31$$
$$\cdot CH_3 + F_2 \longrightarrow CH_3F + F\cdot \qquad 38 - 108 = -70$$
$$\Delta H° = -101$$

$$Cl\cdot + CH_4 \longrightarrow \cdot CH_3 + HCl \qquad 105 - 103 = \ \ 2$$
$$\cdot CH_3 + Cl_2 \longrightarrow CH_3Cl + Cl\cdot \qquad 58 - 84 = -26$$
$$\Delta H° = -24$$

$$Br\cdot + CH_4 \longrightarrow \cdot CH_3 + HBr \qquad 105 - 88 = \ \ 17$$
$$\cdot CH_3 + Br_2 \longrightarrow CH_3Br + Br\cdot \qquad 46 - 70 = -24$$
$$\Delta H° = -7$$

$$I\cdot + CH_4 \longrightarrow \cdot CH_3 + HI \qquad 105 - 71 = \ \ 34$$
$$\cdot CH_3 + I_2 \longrightarrow CH_3I + I\cdot \qquad 36 - 56 = -20$$
$$\Delta H° = 14$$

A fluorine radical is the most reactive of the halogen radicals, and it reacts violently with alkanes ($\Delta H° = -31$ kcal/mol). Because fluorination is so difficult to control, it is not a synthetically useful reaction. In contrast, the iodine radical is the least reactive of the halogen radicals. It is so unreactive ($\Delta H° = 34$ kcal/mol) that it is unable to abstract a hydrogen atom from an alkane. Consequently, it reacts with another iodine radical and reforms I_2.

Will chlorination or bromination produce a better yield of 1-halo-1-methylcyclohexane?

In solving this kind of problem, first draw the structure of the compound being discussed.

Now that you recognize that the compound is a tertiary alkyl halide, the question becomes "will bromination or chlorination produce a better yield of a tertiary alkyl halide?" Since bromination is more selective, it will produce a greater yield of the desired compound. Chlorination will form some of the tertiary alkyl halide, but it will also form significant amounts of primary and secondary alkyl halides.

Now continue on to solve Problem 11.

PROBLEM 11◆

a. Will chlorination or bromination produce a better yield of 1-halo-2,3-dimethylbutane?

b. Will chlorination or bromination produce a better yield of 2-halo-2,3-dimethylbutane?

c. Which would be a better way to make 1-halo-2,2-dimethylpropane: bromination or chlorination?

Benzyl and allyl radicals are more stable than other primary alkyl radicals. They are even more stable than tertiary radicals.

8.5 RADICAL SUBSTITUTION OF BENZYLIC AND ALLYLIC HYDROGENS

relative stabilities of radicals

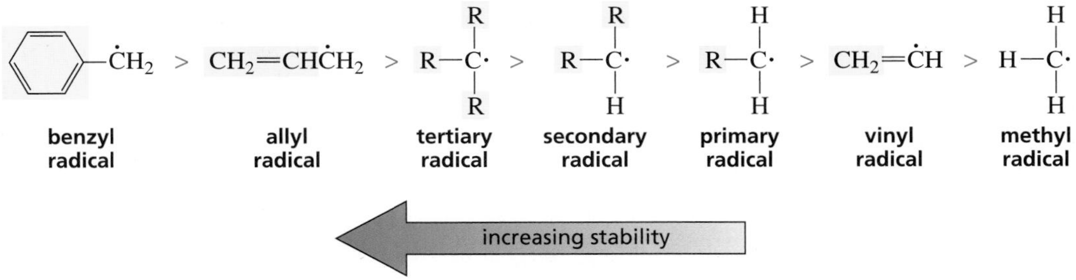

increasing stability

The increased stability of benzyl and allyl radicals arises from delocalization of the unpaired electron. We have already seen that electron delocalization increases the stability of a molecule (Sections 6.6 and 6.8).

We know that the more stable a radical is, the faster it can be formed. This means that halogenation of a compound that has a hydrogen bonded to either a benzylic carbon or an allylic carbon will result in the preferential substitution of that hydrogen. The percent of substitution at the benzylic or allylic carbon is greater in the case of bromination than in the case of chlorination because of the greater selectivity of the bromine radical.

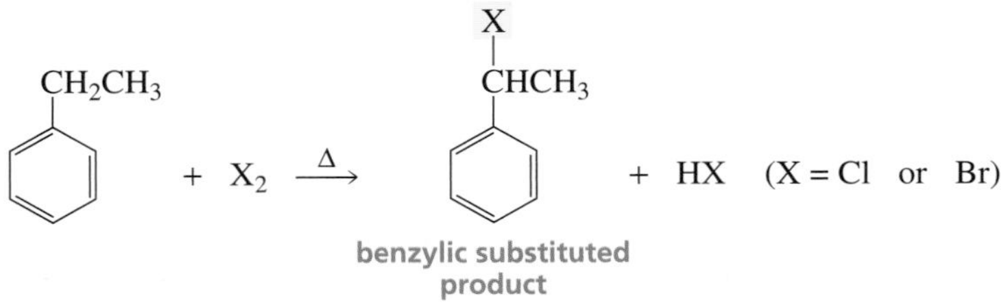

benzylic substituted product

$$CH_3CH{=}CH_2 \ + \ X_2 \ \xrightarrow{\Delta} \ \overset{\overset{\displaystyle X}{|}}{CH_2}CH{=}CH_2 \ + \ HX$$

allylic substituted
product

N-bromosuccinimide (NBS) is frequently used as a source of bromine radicals in radical substitution reactions at benzylic or allylic positions. The bromine radical is obtained as a result of homolytic cleavage of the N—Br bond. NBS allows a radical substitution reaction to be carried out under relatively mild conditions. The solvent used is CCl_4 or CH_2Cl_2. Light or heat can be used to promote the reaction.

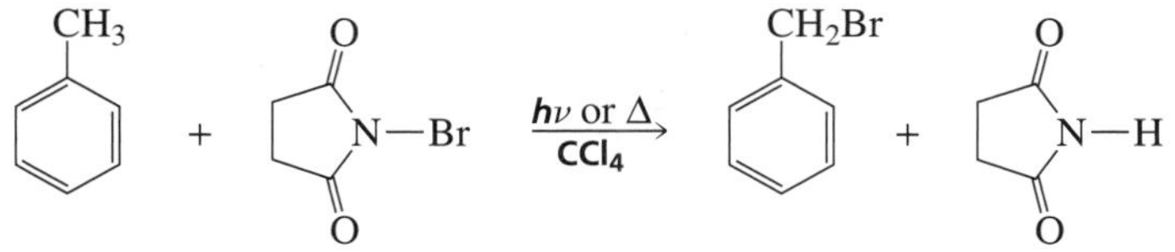

N-bromosuccinimide
NBS

succinimide

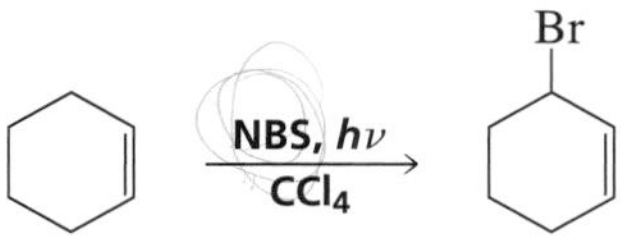

NBS: A CONSTANT SUPPLY OF Br₂ AT A LOW CONCENTRATION

Using NBS rather than Br_2 for the source of bromine radicals for allylic bromination solves the problem of the alkene reacting with Br_2 to form a vicinal dibromide. The first step in allylic bromination by NBS is removal of an allylic hydrogen by a bromine radical. The HBr that is formed in the first step reacts with NBS to form succinimide and Br_2. The allylic radical then reacts with the newly formed Br_2.

$$CH_3CH{=}CH_2 \ + \ Br{\cdot} \ \longrightarrow \ \dot{C}H_2CH{=}CH_2 \ + \ HBr$$

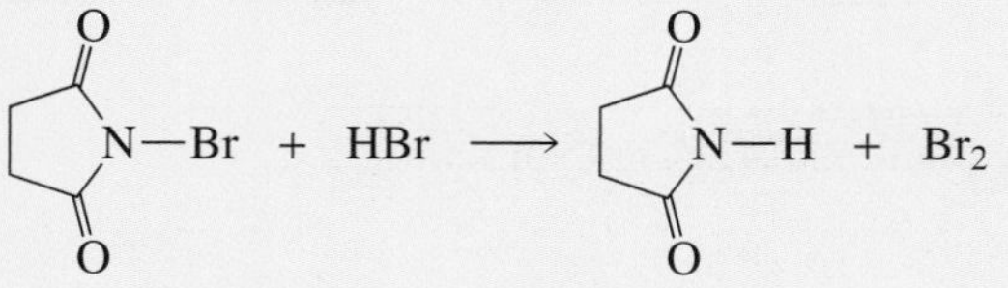

$$\dot{C}H_2CH{=}CH_2 \ + \ Br_2 \ \longrightarrow \ \overset{\overset{\displaystyle Br}{|}}{CH_2}CH{=}CH_2 \ + \ Br{\cdot}$$

Since the Br_2 that is used in the reaction is generated during the course of the reaction, it is always present in a low concentration. This means that the vicinal dibromide—which forms when a high concentration of Br_2 is employed—is not formed.

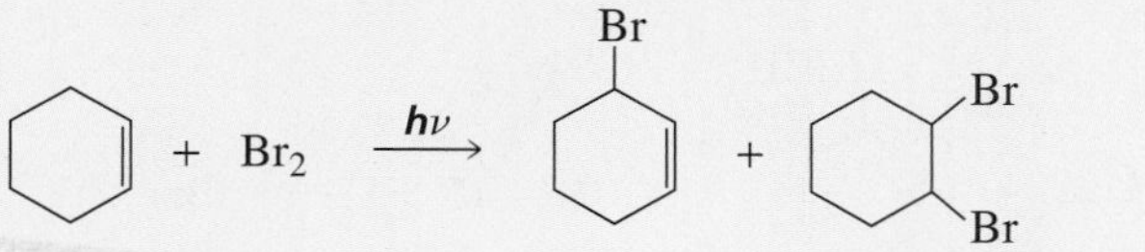

a vicinal dibromide

When a radical abstracts a hydrogen atom from an allylic carbon, the allylic radical that is formed is stabilized by resonance: it has two contributing resonance structures. Because the resonance hybrid of the allyl radical obtained from abstraction of a hydrogen atom from propene is symmetrical, only one substitution product is formed.

$$Br\cdot + CH_3CH=CH_2 \longrightarrow \dot{C}H_2CH=CH_2 \longleftrightarrow CH_2=CH\dot{C}H_2$$

$$\downarrow Br_2$$

$$BrCH_2CH=CH_2 + Br\cdot$$
3-bromopropene

However, if the resonance hybrid of the allylic radical is asymmetrical, two substitution products are formed.

$$Br\cdot + CH_3CH_2CH=CH_2 \longrightarrow CH_3\dot{C}HCH=CH_2 \longleftrightarrow CH_3CH=CH\dot{C}H_2$$

$$\downarrow Br_2$$

$$\underset{\underset{\text{Br}}{|}}{CH_3CHCH=CH_2} + CH_3CH=CHCH_2Br + Br\cdot$$

3-bromo-1-butene   **1-bromo-2-butene**

PROBLEM 12◆

In the preceding reaction, which substitution product would you expect to be formed in greater yield, 3-bromo-1-butene or 1-bromo-2-butene? Why?

PROBLEM 13

Two products are formed when methylenecyclohexane reacts with NBS. Explain how each is formed.

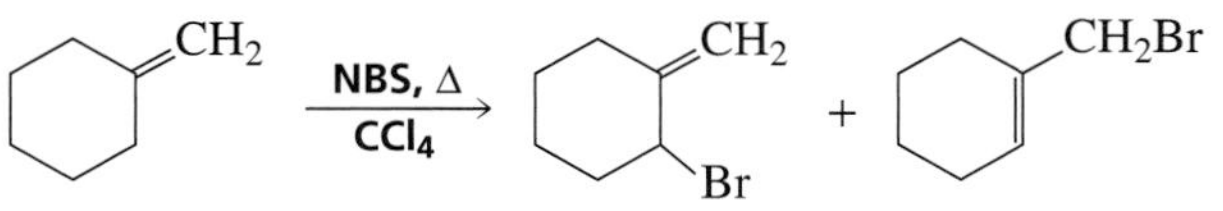

PROBLEM 14 / SOLVED

How many allylic substituted bromoalkenes are formed from the reaction of 2-pentene with NBS? Ignore stereoisomers.

SOLUTION   The bromine radical will abstract a secondary allylic hydrogen from C-4 of 2-pentene in preference to a primary allylic hydrogen from C-1. The resonance hybrid of the resulting radical intermediate is symmetrical, so only one bromoalkene is formed. Because of the high selectivity of the bromine radical, an insignificant amount of radical will be formed by abstraction of a hydrogen from the less reactive position.

$$CH_3CH=CHCH_2CH_3 \xrightarrow[\text{CCl}_4]{\text{NBS}, \Delta} CH_3CH=CH\dot{C}HCH_3 \longleftrightarrow CH_3\dot{C}HCH=CHCH_3$$

$$\downarrow$$

$$\underset{\underset{\text{Br}}{|}}{CH_3CH=CHCHCH_3}$$

8.6 STEREOCHEMISTRY OF RADICAL SUBSTITUTION REACTIONS

If a radical substitution reaction creates a chirality center in the product, both the *R* and *S* enantiomers will be formed.

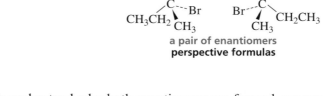

$$CH_3CH_2CH_2CH_3 \ + \ Br_2 \ \xrightarrow{h\nu} \ CH_3CH_2\underset{Br}{CH}CH_3 \ + \ HBr$$

chirality center

configuration of the product

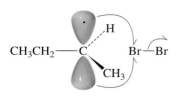

$$CH_3CH_2-\overset{H}{\underset{Br}{|}}-CH_3 \ + \ CH_3CH_2-\overset{Br}{\underset{H}{|}}-CH_3$$

a pair of enantiomers
Fischer projections

a pair of enantiomers
perspective formulas

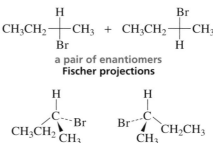

a radical intermediate

To understand why both enantiomers are formed, we must look at the propagation steps in the radical substitution reaction. In the first propagation step, the bromine radical removes a hydrogen atom from the alkane, creating a radical intermediate. The carbon bearing the unpaired electron is sp^2 hybridized and, therefore, the three groups to which it is bonded lie in a plane. In the second propagation step, the incoming halogen has equal access to both sides of the plane. As a result, both the *R* and *S* enantiomers are formed, and they are formed in equal amounts.

Notice that the stereochemical outcome of a radical substitution reaction is identical to the stereochemical outcome of a radical addition reaction (Section 4.19). This is because both reactions involve the formation of a radical intermediate, and it is the reaction of the intermediate that determines the configuration of the products.

What happens if the reactant already has a chirality center and the radical substitution reaction creates a second chirality center? In such a case, a pair of diastereomers will be formed, and they will be formed in unequal amounts. Diastereomers are formed because the new chirality center created in the product can have either the *R* or the *S* configuration, but the configuration of the chirality center in the reactant will be unchanged in the product because none of the bonds to that chirality center is broken during the course of the reaction.

stereochemistry of the product

a pair of diastereomers
Fischer projections

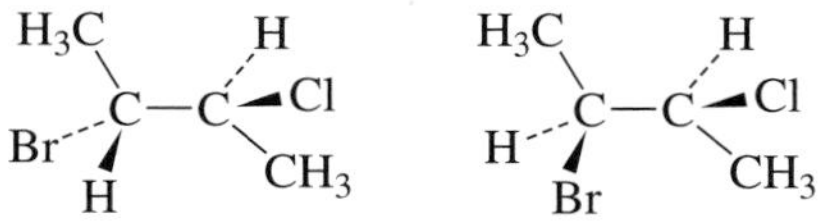

a pair of diastereomers
perspective formulas

More of one diastereomer will be formed than the other because the incoming Br_2 will have greater access to one side of the radical intermediate than to the other due to the presence of the original chirality center.

Br$_2$ has greater access
to this side of the radical

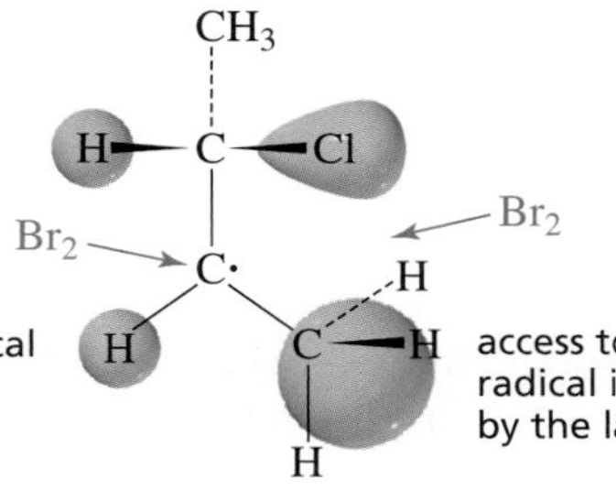

access to this side of the
radical is partially blocked
by the large substituents

PROBLEM 15

Using models, predict which of the diastereomers will be formed in greater yield in the reaction above.

Cyclic compounds undergo the same reactions as do noncyclic compounds. For example, cyclic alkanes, like noncyclic alkanes, undergo radical substitution reactions with chlorine or bromine.

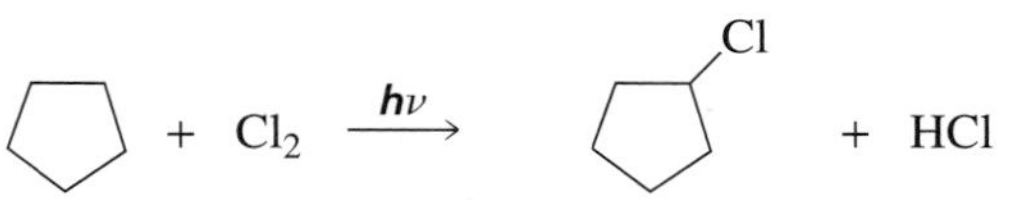

**8.7
REACTIONS OF
CYCLIC COMPOUNDS**

chlorocyclopentane

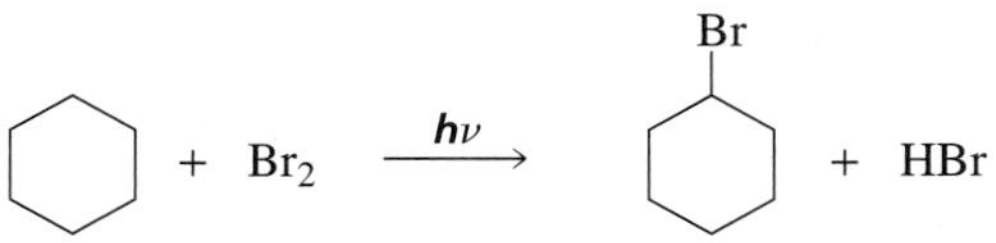

bromocyclohexane

Cyclic alkenes undergo the same reactions as noncyclic alkenes.

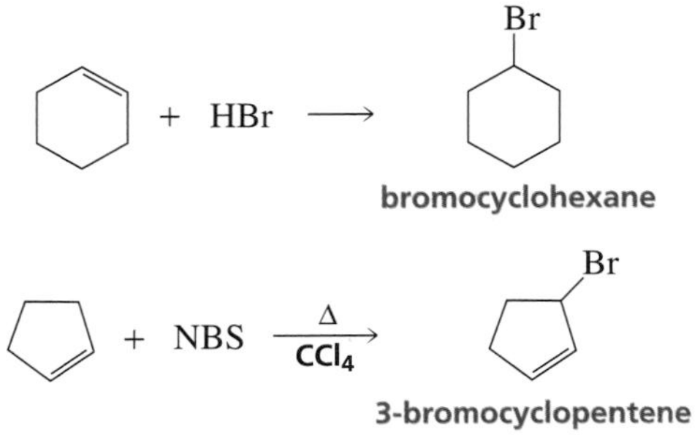

bromocyclohexane

3-bromocyclopentene

In other words, the fact that the compound is cyclic usually has no effect on the chemistry that is characteristic of the functional group.

> ### PROBLEM 16◆
>
> **a.** Give the major product of the reaction of 1-methylcyclohexene with the following reagents, ignoring stereochemistry.
>
>   **1.** NBS/Δ/CCl$_4$ **3.** HBr
>
>   **2.** Br$_2$/CCl$_4$ **4.** HBr/peroxide
>
> **b.** Give the configuration of the products.

Cyclopropane is one notable exception to the generalization that cyclic and noncyclic compounds undergo the same reactions. Although it is an alkane, cyclopropane undergoes electrophilic addition reactions as if it were an alkene.

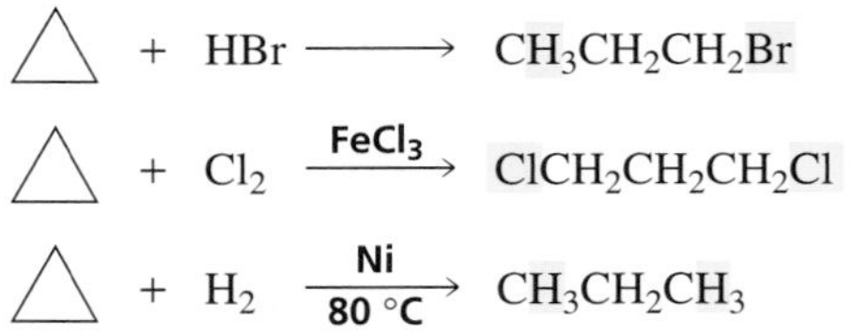

Cyclopropane is more reactive than propene toward addition of acids such as HBr and HCl but is less reactive toward addition of Cl$_2$ and Br$_2$, so a Lewis acid, FeCl$_3$, is needed to catalyze halogen addition (Section 1.20).

Cyclopropane undergoes electrophilic addition reactions as a result of the strain in the small ring (Section 2.12). Because of the 60° bond angles in the three-membered ring, the sp^3 orbitals cannot overlap head-on. Not being able to overlap head-on decreases the effectiveness of the orbital overlap, causing the bonds to be considerably weaker than normal carbon–carbon σ bonds (Figure 2.7). Consequently, three-membered rings undergo ring-opening reactions with electrophilic reagents.

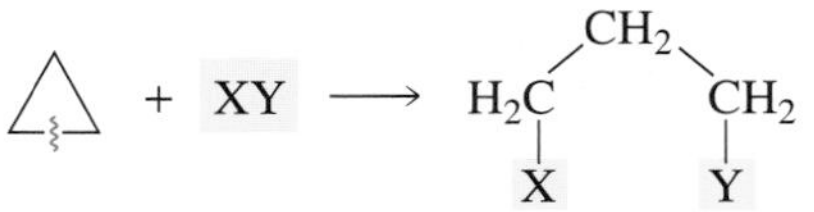

There is less angle strain in cyclobutane than in cyclopropane (Section 2.12). In other words, the sp^3 orbitals in cyclobutane overlap more effectively, which means that the σ bonds are stronger. Cyclobutane, therefore, does not undergo electrophilic addition reactions. It behaves as a normal cyclic alkane except for the fact that it will undergo a ring-opening reaction with hydrogen and a metal catalyst. However, higher temperatures are needed for the hydrogenation of cyclobutane than for the hydrogenation of cyclopropane.

$$\square \quad + \quad H_2 \quad \xrightarrow[\text{200°C}]{\text{Ni}} \quad CH_3CH_2CH_2CH_3$$

cyclobutane   butane

Cyclic alkanes with more than four carbons in the ring do not undergo electrophilic addition reactions because they have considerably less angle strain than four-membered rings (Table 2.9 in Section 2.13).

For a long time, scientists assumed that radical reactions were not important in biological systems because of the large amount of energy required to initiate a radical reaction and because of the problem of controlling the repeating propagation steps of the chain reaction once initiation occurs. However, it is now widely recognized that there are biological reactions that involve radicals. Instead of using heat or light to generate radicals, they are formed by the interaction of an organic molecule with a metal ion. These radical reactions take place in the active site of an enzyme. Containing the reaction in a specific site prevents chain reactions from occurring.

8.8 RADICAL REACTIONS IN BIOLOGICAL SYSTEMS

One radical reaction that takes place in a biological system is the conversion of toxic hydrocarbons to nontoxic alcohols. The hydroxylation of the hydrocarbon is carried out in the liver, catalyzed by an iron-porphyrin-containing enzyme called cytochrome P_{450} (Section 11.7). An alkyl radical intermediate is created as a result of abstraction of a hydrogen atom from an alkane by Fe^VO. In the next step, $Fe^{IV}OH$ dissociates homolytically into Fe^{III} and $HO\cdot$ and the $HO\cdot$ immediately combines with the radical intermediate to form the alcohol. This hydroxylation reaction is also important in the biosynthesis of several important biological compounds.

$$Fe^V{=}O \ + \ H{-}\overset{|}{\underset{|}{C}}{-} \ \longrightarrow \ Fe^{IV}{-}OH \ + \ \cdot\overset{|}{\underset{|}{C}}{-} \ \longrightarrow \ Fe^{III} \ + \ HO{-}\overset{|}{\underset{|}{C}}{-}$$

an alkane   a radical   an alcohol
intermediate

This reaction can also have the opposite toxicological effect. In other words, instead of converting a toxic hydrocarbon into a nontoxic alcohol, substituting an OH for an H in some compounds changes a nontoxic compound into a toxic compound. Therefore, compounds that are nontoxic *in vitro* are not necessarily nontoxic *in vivo*. For example, substituting an OH for an H is thought to be responsible for causing methylene chloride (CH_2Cl_2) to become a carcinogen when it is inhaled.

DECAFFEINATED COFFEE AND THE CANCER SCARE

The finding that methylene chloride becomes a carcinogen on inhalation caused some concern, because methylene chloride is the solvent used to extract caffeine from coffee beans in the manufacture of decaffeinated coffee. However, when methylene chloride was added to drinking water fed to laboratory rats and mice, researchers found no toxic effects. They observed no toxicological responses of any kind either in rats that had consumed an amount of methylene chloride equivalent to the amount that would be ingested by drinking 120,000 cups of decaffeinated coffee per day or in mice that had consumed an amount equivalent to drinking 4.4 million cups of decaffeinated coffee per day. Because of the initial concern, however, researchers sought alternative methods for extraction of caffeine from coffee beans. Extraction by CO_2 at supercritical temperature and pressure turned out to be not only safer but better. This method extracts caffeine without simultaneously extracting some of the flavor compounds that are lost when methylene chloride extraction is used. When coffee is decaffeinated by CO_2 extraction, there is essentially no difference in flavor between regular and decaffeinated coffee.

Another important biological reaction shown to involve a radical intermediate is the conversion of a ribonucleotide into a deoxyribonucleotide. The biosynthesis of ribonucleic acid (RNA) requires ribonucleotides, whereas the biosynthesis of deoxyribonucleic acid (DNA) requires deoxyribonucleotides (Section 24.1). The first step in the conversion of a ribonucleotide to the deoxyribonucleotide needed for DNA biosynthesis involves abstraction of a hydrogen atom from the ribonucleotide to form a radical intermediate.

It is important that unwanted radicals in biological systems be destroyed before they have an opportunity to cause any damage. For example, cell membranes are susceptible to the same kind of radical reactions that cause butter to become rancid as a result of reacting with oxygen (Section 23.3). Imagine the state of your cell membranes if radical reactions could occur readily. Radical reactions in biological systems also have been implicated in the aging process. Unwanted radical reactions are prevented by **radical inhibitors,** compounds that destroy reactive radicals by creating unreactive ones. Hydroquinone is an example of a radical inhibitor. If a reactive radical is formed, hydroquinone can trap it. The semiquinone radical that results is stabilized by resonance and is therefore unreactive in comparison with other radicals.

Two examples of radical inhibitors that are present in biological systems are vitamin C and vitamin E. Like hydroquinone, they form relatively stable radicals. Vitamin C (also known as ascorbic acid) is a water-soluble compound that traps radicals formed in the aqueous environment of the cell and in blood plasma. Vitamin E (also known as α-tocopherol) is a water-insoluble compound that traps radicals formed in nonpolar membranes. Why one vitamin functions in aqueous environments and the other in nonaqueous environments is apparent from their structures (also see structures on page 328).

vitamin C
ascorbic acid

vitamin E
α-tocopherol

FOOD PRESERVATIVES

Radical inhibitors that are present in food are known as preservatives or antioxidants. They preserve food by preventing unwanted radical reactions. Vitamin E is a naturally occurring food preservative found in vegetable oil. BHA and BHT are synthetic radical inhibitors that are added to many packaged foods.

butylated hydroxyanisole
BHA

butylated hydroxytoluene
BHT

food preservatives

A layer of ozone surrounding Earth shields it from harmful solar radiation. The greatest concentration of ozone occurs between 12 and 15 miles above Earth's surface in a part of the atmosphere called the stratosphere. Ozone is formed in the atmosphere as a result of the interaction of molecular oxygen with very short-wavelength ultraviolet light.

$$3\ O_2 \xrightarrow{\ h\nu\ } 2\ O_3$$

molecular   ozone
oxygen

This stratospheric ozone layer acts as a filter for biologically harmful ultraviolet radiation that otherwise would reach Earth's surface. Among other effects, this high-energy short-wavelength ultraviolet light can damage DNA in skin cells, causing mutations that trigger skin cancer (Section 27.6). We owe our very existence to this protective ozone layer. According to current theories of evolution, life could not have developed on land in the absence of this ozone layer; instead, life would have had to remain in the ocean, where water screens out the harmful ultraviolet radiation.

Since about 1985, scientists have noted a precipitous drop in stratospheric ozone over Antarctica. This area of ozone depletion, known as the "ozone hole," is unprecedented in the history of ozone observations. Scientists subsequently noted a similar decrease in ozone over Arctic regions, and in 1988 they detected depletion

8.9
RADICALS AND STRATOSPHERIC OZONE

In 1995, the Nobel Prize in chemistry was awarded to **Sherwood Rowland, Mario Molina,** *and* **Paul Crutzen** *for their pioneering work in explaining the chemical processes responsible for depletion of the ozone layer that serves as Earth's protective shield. Their work demonstrated, for the first time, that human activities could interfere with global processes that support life. This was the first time that a Nobel Prize was presented for work in the environmental sciences.*

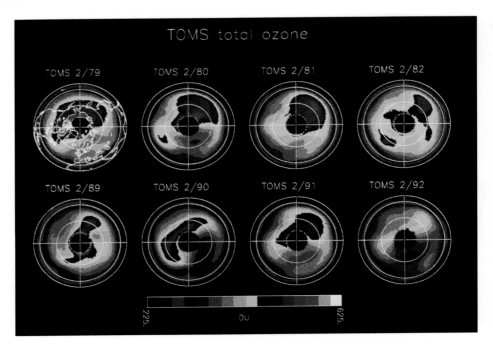

Ozone density over the Northern Hemisphere in February 1979–1982 and ten years later in February 1989–1992. Lowest ozone densities are represented by purples and blues, highest ozone densities by reds and greens. *(NASA)*

F. Sherwood Rowland *was born in Ohio in 1927. He received a Ph.D. from the University of Chicago and is a professor of chemistry at the University of California, Irvine.*

Mario Molina *was born in Mexico in 1943 and subsequently became a U.S. citizen. He received a Ph.D. from the University of California, Berkeley and then became a postdoctoral fellow in Rowland's laboratory. He is currently a professor of earth, atmospheric, and planetary sciences at the Massachusetts Institute of Technology.*

Paul Crutzen *was born in Amsterdam in 1933. He was trained as a meteorologist and taught himself chemistry. He is a professor at the Max Planck Institute for Chemistry in Mainz, Germany.*

of ozone over the United States for the first time. Three years later, scientists determined that the rate of ozone depletion was two to three times faster than originally anticipated. The recently observed increases in cataracts and genetic damage and reduced plant growth are being blamed on the ultraviolet radiation that is able to penetrate the reduced ozone layer. It has been predicted that erosion of the protective ozone layer will cause an additional 200,000 deaths from skin cancer over the next 50 years.

Strong circumstantial evidence implicates synthetic chlorofluorocarbons (CFCs)—alkanes in which all the hydrogens have been replaced by fluorine and chlorine, such as $CFCl_3$ and CF_2Cl_2—as a major cause of ozone depletion. These gases, known commercially as Freons, have been extensively used as cooling fluids in refrigerators and in home and automobile air conditioners, in industrial cleaning solvents, and in the manufacture of some plastic foams. They were once widely used as propellants in aerosol spray cans (deodorant, hair spray, etc.) because of their odorless, nontoxic, and nonflammable properties and because they are chemically inert and thus do not react with the contents of the can. Such use has now been banned in many countries.

Chlorofluorocarbons remain very stable in the atmosphere until they reach the stratosphere. There, they encounter wavelengths of ultraviolet light that cause their homolytic cleavage, generating chlorine radicals.

$$F-\underset{\underset{F}{|}}{\overset{\overset{Cl}{|}}{C}}-Cl \xrightarrow{h\nu} F-\underset{\underset{F}{|}}{\overset{\overset{Cl}{|}}{C}}\cdot \ + \ Cl\cdot$$

The chlorine radicals are the ozone-removing agents. They react with ozone to form chlorine monoxide radicals and molecular oxygen. The chlorine monoxide radicals react with ozone to regenerate chlorine radicals. These two propagating steps are repeated over and over, destroying a molecule of ozone in each step. It has been calculated that each chlorine atom destroys 100,000 ozone molecules.

Polar stratospheric clouds increase the rate of ozone destruction. These clouds form over Antarctica during the cold winter months. Ozone depletion in the Arctic is less severe because it generally does not get cold enough for the polar stratospheric clouds to form.

$$Cl\cdot \; + \; O_3 \; \longrightarrow \; ClO\cdot \; + \; O_2$$
$$ClO\cdot \; + \; O_3 \; \longrightarrow \; Cl\cdot \; + \; 2\,O_2$$

Because of their remarkable chemical stability, chlorofluorocarbons have half-lives of 70 to 120 years. A frightening consequence of such long half-lives is that, once all CFC production is stopped, it will take more than 100 years to recover from the damage caused by CFCs already in the atmosphere.

An international agreement known as the Montreal Protocol went into effect in 1989, which mandated a reduction in the production of CFCs and total elimination of production by the year 2000. When scientists found that the ozone layer was eroding faster than anticipated in the Northern Hemisphere, the deadline for total elimination was moved up to December 31, 1995. By 1993, scientists noted a marked decline in the rate of increase of CFCs in the atmosphere, but the protective ozone layer over the United States was still thinning at a substantial rate. Recent studies have led to the development of environmentally sound CFC substitutes that will not destroy the ozone layer.

THE CONCORDE AND OZONE DEPLETION

Supersonic aircraft cruise in the lower stratosphere, and their jet engines convert molecular oxygen and nitrogen into nitrogen oxides such as NO and NO_2. Like CFCs, nitrogen oxides react with stratospheric ozone. Fortunately, the supersonic Concorde, built jointly by England and France, makes only a limited number of flights each week.

SUMMARY OF REACTIONS

1. Alkanes undergo radical substitution reactions with Cl_2 and Br_2 in the presence of heat or light (Sections 8.2, 8.3, and 8.4).

$$CH_3CH_3 + Cl_2 \xrightarrow{\Delta \text{ or } h\nu} CH_3CH_2Cl + HCl$$
excess

$$CH_3CH_3 + Br_2 \xrightarrow{\Delta \text{ or } h\nu} CH_3CH_2Br + HBr$$
excess
bromination is more selective than chlorination

2. *Alkenes* undergo radical halogenation at the allylic position (Section 8.5).

$$RCH_2CH{=}CH_2 + Br_2 \xrightarrow{h\nu} \underset{\overset{|}{Br}}{R}CHCH{=}CH_2 + RCH{=}CHCH_2{\underset{\overset{|}{Br}}{}} + HBr$$

Alkyl-substituted benzenes undergo radical halogenation at the benzylic position (Section 8.5).

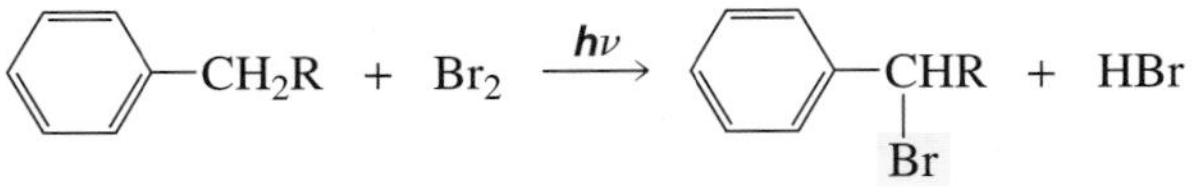

NBS can be used for radical bromination at the allylic and benzylic positions (Section 8.5).

3. *Cyclopropane* undergoes electrophilic addition reactions (Section 8.7).

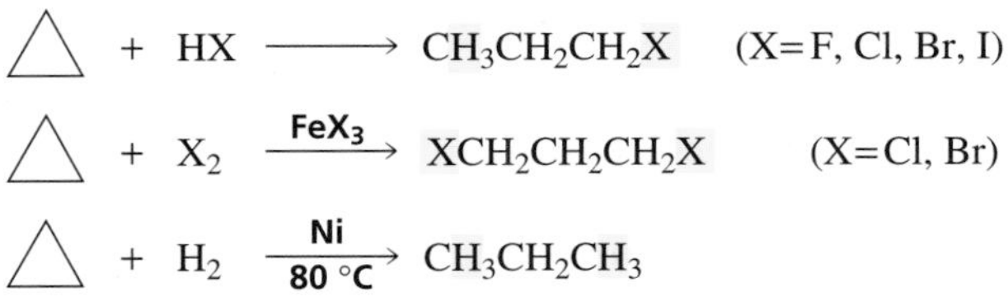

$+ HX \longrightarrow CH_3CH_2CH_2X$ (X=F, Cl, Br, I)

$+ X_2 \xrightarrow{FeX_3} XCH_2CH_2CH_2X$ (X=Cl, Br)

$+ H_2 \xrightarrow[80\,°C]{Ni} CH_3CH_2CH_3$

KEY TERMS

alkane (page 328)
disproportionation (page 334)
homolytic bond cleavage (page 331)
initiation step (page 331)
paraffin (page 330)

propagation step (page 331)
radical chain reaction (page 331)
radical inhibitor (348)
radical substitution reaction
  (page 332)

reactivity–selectivity principle
  (page 339)
saturated hydrocarbon
  (page 328)
termination step (page 331)

PROBLEMS

17. Give the product(s) of each of the following reactions.

    a. $CH_2{=}CHCH_2CH_2CH_3 + Br_2 \xrightarrow{h\nu}$

             $\overset{\displaystyle CH_3}{\overset{|}{}}$
    b. $CH_3C{=}CHCH_3 + NBS \xrightarrow[CCl_4]{\Delta}$

c. 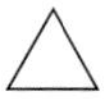 △ + HCl ⟶

   CH$_3$
   |
d. CH$_3$CH$_2$CHCH$_2$CH$_2$CH$_3$ + Br$_2$ $\xrightarrow{h\nu}$

e. 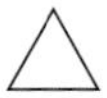 △ + Br$_2$ $\xrightarrow{\text{FeBr}_3}$

f. 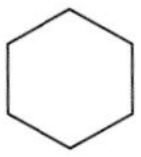 ⬡ + Cl$_2$ $\xrightarrow{h\nu}$

g. 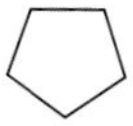 ⬠ + Cl$_2$ $\xrightarrow{\text{FeCl}_3}$

       CH$_3$
h. 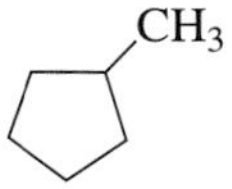 ⬠ + Cl$_2$ $\xrightarrow{h\nu}$

18. **a.** An alkane with molecular formula C$_5$H$_{12}$ forms only one monochlorinated
   product when heated with Cl$_2$. Give the IUPAC name of this alkane.
   b. An alkane with molecular formula C$_7$H$_{16}$ forms seven monochlorinated products
   when heated with Cl$_2$. Give the IUPAC name of this alkane.

19. Dr. Throck Morton wanted to determine experimentally the relative ease of removal of
a hydrogen atom from a tertiary, a secondary, and a primary carbon by a chlorine radical. He allowed 2-methylbutane to undergo chlorination at 300 °C and obtained as
products 29% 1-chloro-2-methylbutane, 24% 2-chloro-2-methylbutane, 33% 2-chloro-
3-methylbutane, and 14% 1-chloro-3-methylbutane. What values did he obtain for the
relative ease of removal of tertiary, secondary, and primary hydrogen atoms by a chlorine radical under the conditions of his experiment?

20. At 600 °C, the ratio of the relative rates of formation of a tertiary, a secondary, and a
primary radical is 2.6 : 2.1 : 1. Explain the change in the degree of regioselectivity
compared with what Dr. Throck Morton found in Problem 19.

21. Iodine (I$_2$) does not react with ethane even though I$_2$ is more easily cleaved homolytically than the other halogens. Explain.

22. **a.** How many monobromination products would be obtained from the radical bromination of methylcyclohexane?
   b. Which one would be obtained in greatest yield? Explain.

23. Give the major product of each of the following reactions.

a. 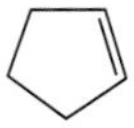 ⬠ + NBS $\xrightarrow[\text{CCl}_4]{\Delta}$

b. CH$_2$=CHCH$_2$CH$_2$CH$_3$ + NBS $\xrightarrow[\text{CCl}_4]{\Delta}$

        CH$_3$
c. 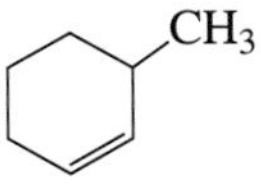 ⬡ + NBS $\xrightarrow[\text{CCl}_4]{\Delta}$

     CH$_3$
     |
d. CH$_3$CHCH$_3$ + Cl$_2$ $\xrightarrow{h\nu}$

     CH$_3$
     |
e. CH$_3$CHCH$_3$ + Br$_2$ $\xrightarrow{h\nu}$

f. 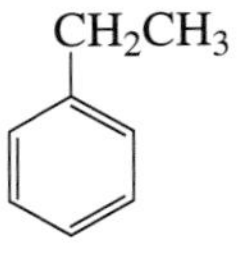 + NBS $\xrightarrow[\text{CCl}_4]{\Delta}$

g.  + NBS $\xrightarrow[\text{CCl}_4]{\Delta}$

h. 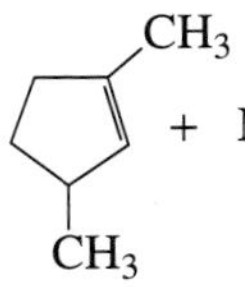 + NBS $\xrightarrow[\text{CCl}_4]{\Delta}$

24. a. Calculate the $\Delta H°$ of the following reaction.

$$CH_3CH_2CH_3 + Br_2 \xrightarrow{h\nu} CH_3\underset{\underset{Br}{|}}{C}HCH_3 + HBr$$

    b. Give the mechanism of the reaction.
    c. Label the initiation, propagation, and termination steps.
    d. Calculate the $\Delta H°$ of each step.
    e. Show how the $\Delta H°$ value for the reaction can be obtained from the $\Delta H°$ values for the individual steps of the reaction.

25. The deuterium kinetic isotope effect for chlorination of an alkane is defined below. Predict which would have a greater deuterium kinetic isotope effect: chlorination or bromination.

$$\begin{array}{l}\text{deuterium kinetic} \\ \text{isotope effect}\end{array} = \frac{\text{rate of homolytic cleavage of a C—H bond by Cl·}}{\text{rate of homolytic cleavage of a C—D bond by Cl·}}$$

26. The rate of bromination of methane is decreased if HBr is added to the reaction mixture. Explain.

27. a. Propose a mechanism for the following reaction.

$$CH_3CH_3 + CH_3{-}\underset{\underset{CH_3}{|}}{\overset{\overset{CH_3}{|}}{C}}{-}OCl \xrightarrow{\Delta} CH_3CH_2Cl + CH_3{-}\underset{\underset{CH_3}{|}}{\overset{\overset{CH_3}{|}}{C}}{-}OH$$

    b. Given that the $\Delta H°$ of the reaction is -42 kcal/mol and that the bond dissociation energies for the C—H, C—Cl, and O—H bonds are 101, 82, and 102 kcal/mol, respectively, calculate the bond dissociation energy of the O—Cl bond.

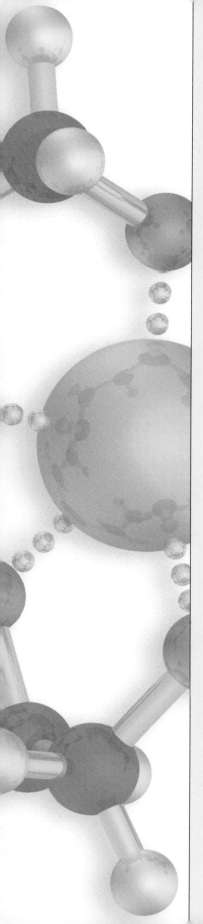

SUBSTITUTION AND ELIMINATION REACTIONS

Chapters 9 to 11 discuss the reactions of compounds that have an electron-withdrawing group (frequently called a leaving group) bonded to an sp^3 hybridized carbon. These compounds undergo two different kinds of reactions: substitution reactions and elimination reactions.

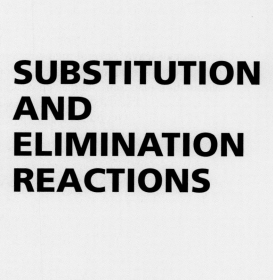

an sp^3 hybridized carbon

Chapter 9 discusses the substitution reactions of alkyl halides. Of the different compounds that undergo substitution and elimination reactions, alkyl halides are discussed first because they have relatively good leaving groups.

Chapter 10 covers the elimination reactions of alkyl halides. Because alkyl halides can undergo both substitution and elimination reactions, Chapter 10 also discusses the factors that determine whether a given alkyl halide will undergo a substitution reaction, an elimination reaction, or both substitution and elimination reactions.

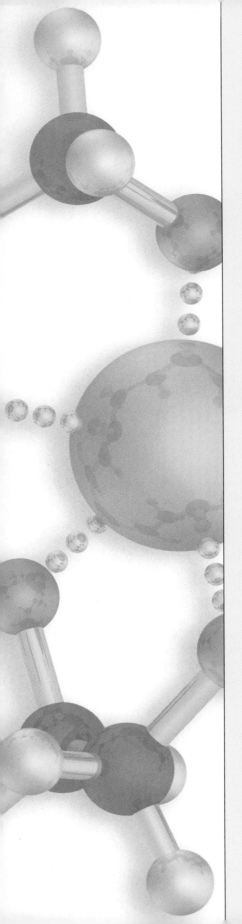

Chapter 11 discusses compounds, other than alkyl halides, that undergo substitution and elimination reactions. You will see that, because alcohols and ethers have relatively poor leaving groups compared with the leaving groups of alkyl halides, alcohols and ethers must be activated before the groups will leave. Several methods commonly used to activate leaving groups will be examined. Other compounds that undergo substitution and/or elimination reactions, such as epoxides, thiols, and quaternary ammonium salts, are also discussed. Finally, this chapter will introduce you to organometallic compounds.

a sea hare

REACTIONS AT AN *sp*³ HYBRIDIZED CARBON I:
Substitution Reactions of Alkyl Halides

Organic compounds in which an sp^3 carbon is bonded to an electronegative atom or group can undergo two types of reactions. They can undergo **substitution reactions,** in which the electronegative atom or group is substituted by another atom or group. They can also undergo **elimination reactions,** in which the electronegative atom or group is eliminated along with a hydrogen from an adjacent carbon. The electronegative atom or group that is *substituted* or *eliminated* is called the **leaving group.**

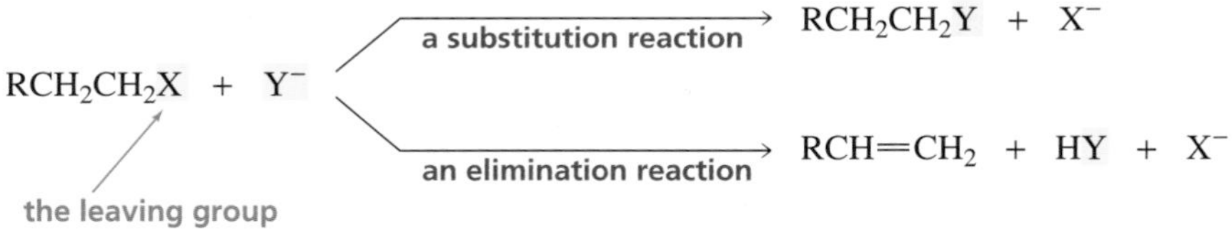

$$RCH_2CH_2X + Y^- \longrightarrow \begin{array}{l} \xrightarrow{\text{a substitution reaction}} RCH_2CH_2Y + X^- \\ \xrightarrow{\text{an elimination reaction}} RCH=CH_2 + HY + X^- \end{array}$$

the leaving group

This chapter will focus on the substitution reactions of alkyl halides, compounds in which the leaving group is a halide ion (F^-, Cl^-, Br^-, I^-). The nomenclature of alkyl halides was discussed in Section 2.4.

alkyl halides

R—F	R—Cl	R—Br	R—I
an alkyl fluoride	an alkyl chloride	an alkyl bromide	an alkyl iodide

In Chapter 10 we will discuss the elimination reactions of alkyl halides and the factors that determine whether substitution or elimination will prevail when an alkyl halide undergoes a reaction.

Because halide ions are relatively good leaving groups (easily displaced), alkyl halides are a good family of compounds with which to start our study of these two very important types of reactions—substitution and elimination. We will then be prepared to discuss the substitution and elimination reactions of compounds with leaving groups other than halide ions in Chapter 11.

Substitution reactions are important reactions in organic chemistry. Through the use of these reactions, readily available alkyl halides can be converted into a wide variety of other compounds. Substitution reactions are also important in the cells of plants and animals. Because cells exist in predominantly aqueous environments and alkyl halides are not water-soluble, biological systems use compounds in which the group that is substituted is more polar than a halogen and therefore more water-soluble. The reactions of some of these biological compounds are discussed in this chapter.

ALKYL HALIDES AS SURVIVAL COMPOUNDS

The few alkyl halides that exist in nature are synthesized by marine organisms that live in salt water (sponges, coral, algae) and are surrounded by a high concentration of chloride ion. Because alkyl halides tend to be toxic, they are synthesized by these organisms for self defense. For example, red algae synthesize a foul-tasting alkyl halide that deters predators from eating them. The sea hare, however, is not deterred. After consuming red algae, it converts the original alkyl halide into a structurally similar alkyl halide that it uses for its own defense. Unlike other mollusks, the sea hare does not have a shell. Therefore it surrounds itself with a slimy material that contains the alkyl halide and in this way protects itself from carnivorous fish.

synthesized by red algae   **synthesized by the sea hare**

A sea hare.

9.1 REACTIVITY CONSIDERATIONS

A halogen is more electronegative than carbon. Consequently, the two atoms do not share their bonding electrons equally. Because the more electronegative halogen has a larger share of the electrons, it has a partial negative charge and the carbon to which it is bonded has a partial positive charge.

$$\overset{\delta+}{R}CH_2 \overset{\delta-}{-X}$$

$$X = F, Cl, Br, I$$

This unequal sharing of electrons causes alkyl halides to undergo substitution and elimination reactions. There are two important mechanisms for the substitution reaction.

1. A nucleophile is attracted to the partially positively charged carbon. As the nucleophile approaches the carbon, it causes the carbon–halogen bond to break heterolytically (the halogen keeps both of the bonded electrons).

$$\left(\overset{..}{\underset{}{Nu}} \; + \; \overset{\delta+}{\underset{}{-C}} \overset{\delta-}{\underset{}{-X}} \; \longrightarrow \; -\overset{|}{\underset{|}{C}} -Nu \; + \; X^- \right)$$

2. The carbon–halogen bond breaks heterolytically without any assistance from the nucleophile, forming a carbocation. The carbocation then reacts with the nucleophile to form the substitution product.

$$\overset{\delta+}{\underset{|}{-C}} \overset{\delta-}{\underset{}{-X}} \; \longrightarrow \; -\overset{|}{\underset{|}{C}}{}^+ \; + \; X^-$$

$$-\overset{|}{\underset{|}{C}}{}^+ \; + \; \overset{..}{\underset{}{Nu}} \; \longrightarrow \; -\overset{|}{\underset{|}{C}} -Nu$$

Regardless of the mechanism by which such a substitution reaction occurs, it is called a **nucleophilic substitution reaction** because a nucleophile is substituted for the halogen. *The mechanism that predominates depends on:*

- *the structure of the alkyl halide,*
- *the reactivity and structure of the nucleophile,*
- *the concentration of the nucleophile, and*
- *the solvent in which the reaction is carried out.*

How is the mechanism of a reaction determined? We can learn a great deal about the mechanism of a reaction by determining the factors that affect the rate of the reaction. These factors are called the **kinetics** of the reaction. The rate of a nucleophilic substitution reaction such as the reaction of methyl bromide with hydroxide ion depends on the concentrations of both reagents. If the concentration of methyl bromide in the reaction mixture is doubled, the rate of the nucleophilic substitution reaction doubles. If the concentration of hydroxide ion is doubled, the rate of the reaction also doubles. If the concentrations of both reactants are doubled, the rate of the reaction quadruples.

**9.2
THE MECHANISM OF
S_N2 REACTIONS**

$$\underset{\text{methyl bromide}}{CH_3Br} \; + \; HO^- \; \longrightarrow \; \underset{\text{methyl alcohol}}{CH_3OH} \; + \; Br^-$$

When you know the relationship between the rate of a reaction and the concentration of the reactants, you can write a **rate law** for the reaction.

rate $\propto$ [alkyl halide][nucleophile]

As we noted earlier (see Section 3.7), the "proportional to" sign ($\propto$) can be replaced by an "equal" sign and a proportionality constant (k).

rate $= k$[alkyl halide][nucleophile]

Because the rate of the reaction depends on the concentration of two reactants, it is a **second-order reaction** (Section 3.7).

The rate law tells us what molecules are involved in the transition state of the rate-determining step of the reaction. From the rate law for the reaction of methyl bromide with hydroxide ion, we know that both methyl bromide and hydroxide ion are involved in the rate-determining transition state. The transition state is **bimolecular;** that is, it involves two molecules. As we saw in Section 3.7, the proportionality constant (k) is the rate constant. The rate constant describes how difficult it is to overcome the energy barrier of the reaction (how hard it is to reach the transition state). The larger the rate constant, the easier it is to reach the transition state.

The reaction of methyl bromide with hydroxide ion is an example of an **S_N2 reaction:** S for substitution, $_N$ for nucleophilic, and 2 for bimolecular. In 1937, Hughes and Ingold[1] proposed a mechanism for an S_N2 reaction. Recall that a mechanism describes the step-by-step process by which reactants are converted into products; it is a theory that fits the experimental evidence that has been accumulated concerning the reaction. Hughes and Ingold based the mechanism for an S_N2 reaction on the following three pieces of experimental evidence.

Edward Davies Hughes (1906–1963) *was born in North Wales. He received a Ph.D. from the University of Wales and earned a D.Sc. from the University of London, working under Sir Christopher Ingold. He was a professor of chemistry at University College, London.*

1. The rate of the reaction depends on the concentration of the alkyl halide *and* on the concentration of the nucleophile. This means that both reactants are involved in the transition state of the rate-determining step.
2. When the hydrogens of methyl bromide are successively replaced with methyl groups, the rate of the reaction with a given nucleophile becomes progressively slower (Table 9.1).
3. The reaction of an alkyl halide in which the halogen is bonded to a chirality center leads to the formation of only one stereoisomer, and its configuration is inverted compared with the configuration of the reacting alkyl halide.

The mechanism proposed by Hughes and Ingold for an S_N2 reaction has one step. The nucleophile attacks the carbon bearing the leaving group and displaces the leaving group. Because the nucleophile hits the carbon on the side opposite to the side bonded to the leaving group, the carbon is said to undergo **back side attack.** Back side attack occurs because the orbital of the nucleophile that contains its nonbonding electrons interacts with the empty σ^* molecular orbital associated with the C—Br bond. This orbital has its larger lobe on the side of the carbon directed away from the C—Br bond. Consequently, the best overlap of the interacting orbitals is achieved through back side attack. An S_N2 reaction is also called a **direct displacement reaction** because the nucleophile displaces the leaving group

TABLE 9.1 Relative Rates of S_N2 Reactions for Several Alkyl Bromides

$$R—Br + Cl^- \xrightarrow{S_N2} R—Cl + Br^-$$

Alkyl bromide	Class	Relative rate
CH_3—Br	methyl	1200
CH_3CH_2—Br	primary	40
$CH_3CH_2CH_2$—Br	primary	16
$CH_3\overset{\mid}{C}H$—Br (CH_3)	secondary	1
$CH_3\overset{\overset{CH_3}{\mid}}{\underset{\underset{CH_3}{\mid}}{C}}$—Br	tertiary	too slow to measure

[1]Sir Christopher Ingold: See Chapter 4, p. 186.

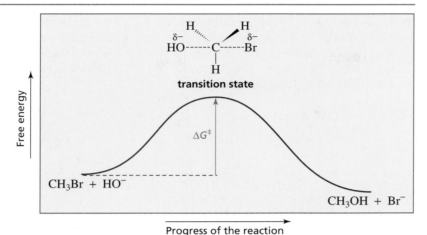

Reaction coordinate diagram for the S_N2 reaction of methyl bromide with hydroxide ion.

in a single step. A reaction coordinate diagram for an S_N2 reaction is shown in Figure 9.1.

mechanism of the S_N2 reaction

$$HO^- \ + \ CH_3{-}Br: \ \longrightarrow \ CH_3{-}OH \ + \ :Br:^-$$

How does this mechanism account for the three observed pieces of experimental evidence? The mechanism shows the alkyl halide and the nucleophile coming together in the transition state of the one-step reaction. Therefore, increasing the concentration of either of them makes their coming together more probable. In other words, the reaction will follow second-order kinetics, exactly as observed.

Because the nucleophile attacks the back side of the carbon that is bonded to the halogen, bulky substituents attached to this carbon will make it harder for the nucleophile to get to the back side and therefore will decrease the rate of the reaction (Figure 9.2). Thus the mechanism explains why substituting methyl groups for the hydrogens in methyl bromide progressively slows the rate of the substitution reaction (Table 9.1).

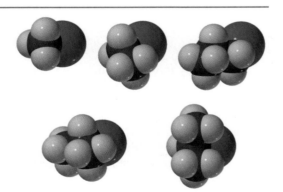

◀ **Figure 9.2**
Three-dimensional structures of the alkyl halides shown in Table 9.1, showing that increasing the bulk of the substituents decreases access to the back side of the carbon that is bonded to the leaving group and therefore decreases the rate of an S_N2 reaction.

Paul Walden

Paul Walden (1863–1957)
was born in Riga, Latvia, the son of a farmer. His parents died when he was still a child, and he supported himself at Riga University and St. Petersburg University by working as a tutor. He moved to Germany after the Russian Revolution, where he became a professor at the University of Rostock and later at the University of Tübingen.

Effects due to groups occupying a certain volume of space are called **steric effects.** A steric effect that decreases reactivity is called **steric hindrance.** Steric hindrance results from groups getting in the way at the reaction site. As a consequence of steric hindrance, alkyl halides have the following relative reactivities in an S_N2 reaction. The steric crowding in tertiary alkyl halides is so great that they are unable to undergo S_N2 reactions.

relative reactivities of alkyl halides in an S_N2 reaction

methyl halide > 1° alkyl halide > 2° alkyl halide > 3° alkyl halide

increasing reactivity in an S_N2 reaction

It is not just the *number* of alkyl groups attached to the carbon undergoing nucleophilic attack that determines the rate of an S_N2 reaction; the *size* of the alkyl groups is also important. For example, ethyl bromide and propyl bromide are both primary alkyl halides, but ethyl bromide is more than twice as reactive in an S_N2 reaction because the methyl group of ethyl bromide provides less steric hindrance to back side attack than does the ethyl group of propyl bromide (Table 9.1).

As the nucleophile approaches the back side of the carbon of methyl bromide, the carbon–hydrogen bonds begin to move away from the nucleophile and its attacking electrons. By the time the transition state is reached, the carbon–hydrogen bonds are all in the same plane and the carbon is pentacoordinate (fully bonded to three atoms and partially bonded to two) rather than tetrahedral. As the nucleophile gets closer to the carbon and the bromine moves farther away from it, the carbon–hydrogen bonds continue to move in the same direction. Eventually, the bond between the carbon and the nucleophile is fully formed and the bond between the carbon and bromine is completely broken, and thus the carbon is once again tetrahedral.

$$HO^- + \overset{\cdots}{\underset{\cdots}{C}}\!-Br \longrightarrow \left[HO\cdots\overset{\delta-}{\underset{|}{C}}\cdots\overset{\delta-}{Br} \right] \longrightarrow HO-C\overset{\cdots}{\underset{\cdots}{}} + Br^-$$

transition state

The best way to visualize the movement of the groups bonded to the carbon at which substitution occurs is to picture an umbrella that turns inside out. This is called **inversion of configuration.** The carbon at which substitution occurs has inverted its configuration during the course of the reaction just as an umbrella has a tendency to invert in a windstorm. The inversion is known as a *Walden inversion,* since Paul Walden was the first to discover that compounds could invert their configurations as a result of substitution reactions.

Because an S_N2 reaction takes place with inversion of configuration, only one product is formed when an alkyl halide—that has the halogen leaving group bonded to a chirality center—undergoes an S_N2 reaction. The configuration of that product is inverted compared with the configuration of the alkyl halide. Therefore, the proposed mechanism accounts for the observed configuration of the product.

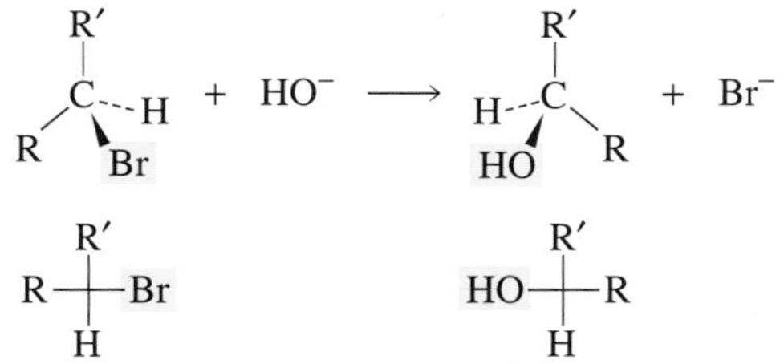

the configuation of the product is inverted compared with the configuration of the reactant

---

PROBLEM 1 ◆

Does increasing the height of the energy barrier to an S_N2 reaction increase or decrease the magnitude of the rate constant for the reaction?

PROBLEM 2 ◆

Arrange the following alkyl chlorides in order of decreasing reactivity in an S_N2 reaction: 1-chloro-2-methylbutane, 1-chloro-3-methylbutane, 2-chloro-2-methylbutane, and 1-chloropentane.

PROBLEM 3 ◆

What product would be formed from the S_N2 reaction of:

a. 2-bromobutane and hydroxide ion?

b. (R)-2-bromobutane and hydroxide ion?

c. (S)-3-chlorohexane and hydroxide ion?

d. 3-iodopentane and hydroxide ion?

The Leaving Group

If an alkyl iodide, an alkyl bromide, an alkyl chloride, and an alkyl fluoride (all having the same alkyl group) were allowed to react with the same nucleophile under the same conditions, we would find that the alkyl iodide is the most reactive and the alkyl fluoride is the least reactive.

9.3
THE S_N2 REACTION

	relative rates of reaction
$HO^- + RCH_2I \longrightarrow RCH_2OH + I^-$	30,000
$HO^- + RCH_2Br \longrightarrow RCH_2OH + Br^-$	10,000
$HO^- + RCH_2Cl \longrightarrow RCH_2OH + Cl^-$	200
$HO^- + RCH_2F \longrightarrow RCH_2OH + F^-$	1

The only difference among these four reactions is the nature of the leaving group. Apparently, the iodide ion is the best leaving group and the fluoride ion is the worst. This brings us to an important rule in organic chemistry that will keep reappearing in this text: *The weaker the basicity of a group, the better its leaving ability.* Leaving ability depends on basicity, because a weak base does not share its electrons as well as a strong base does. Consequently, a weak base is not bonded as strongly to the carbon as a strong base would be. Thus a bond between a carbon and a weak base is more easily broken than a bond between a carbon and a strong base.

The weaker the base, the better it is as a leaving group.

We have seen that the hydrogen halides have the following relative acidities (see Section 1.18).

relative acidities of the hydrogen halides

$$HI > HBr > HCl > HF$$

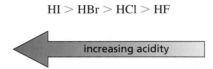

increasing acidity

Because we know that the stronger the acid the weaker its conjugate base, the halide ions have the following relative basicities.

relative basicities of the halide ions

$$I^- < Br^- < Cl^- < F^-$$

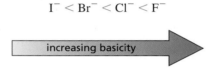

increasing basicity

Further, because weaker bases are better leaving groups, the halide ions have the following relative leaving abilities.

relative leaving abilities of the halide ions

$$I^- > Br^- > Cl^- > F^-$$

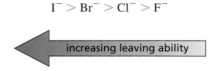

increasing leaving ability

As a consequence of the relative leaving abilities of the halide ions, alkyl halides have the following relative reactivities in an S$_N$2 reaction. (In Section 1.13, we learned why the bond formed by a halogen becomes weaker as the atomic weight of the halogen increases.)

relative reactivities of alkyl halides in an S$_N$2 reaction

$$RI > RBr > RCl > RF$$

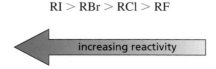

increasing reactivity

The Nucleophile

When we talk about atoms or molecules that have lone pair electrons, sometimes we call them bases and sometimes we call them nucleophiles. What is the difference between a base and a nucleophile?

A **base** shares its lone pair electrons with a proton. **Basicity** is a measure of how strongly the base shares those electrons with a proton. The stronger the base, the better it shares its electrons. Basicity is measured by an acid dissociation constant (K_a) that indicates the tendency of the conjugate acid of the base to lose a proton (Section 1.17).

A **nucleophile** uses its lone pair electrons to attack an electron-deficient atom other than a proton. **Nucleophilicity** is a measure of how readily the nucleophile is able to attack such an atom. It is measured by a rate constant (k). In the case of an S$_N$2 reaction, nucleophilicity is a measure of how readily the nucleophile attacks an sp^3 hybridized carbon bonded to a leaving group.

In comparing molecules *with the same attacking atom,* there is generally a direct relationship between basicity and nucleophilicity. Stronger bases are better nucleophiles. For example, a compound with a negatively charged oxygen is a stronger base *and* a better nucleophile than a compound with a neutral oxygen.

stronger base, better nucleophile		weaker base, poorer nucleophile
HO^-	>	H_2O
CH_3O^-	>	CH_3OH
$^-NH_2$	>	NH_3
$CH_3CH_2\overline{N}H$	>	$CH_3CH_2NH_2$

In comparing molecules *with attacking atoms of approximately the same size,* the stronger bases are again the better nucleophiles. The atoms across the second row of the periodic table have approximately the same size. If hydrogens are attached to the second-row elements, the resulting compounds have the following relative acidities (Section 1.18).

relative acid strengths

$$CH_4 < NH_3 < H_2O < HF$$

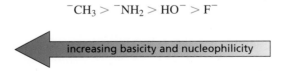

increasing acidity

Consequently, the conjugate bases have the following relative base strengths and nucleophilicities. For example, the methyl anion is the strongest base as well as the best nucleophile.

relative base strengths and relative nucleophilicities

$$^-CH_3 > {}^-NH_2 > HO^- > F^-$$

increasing basicity and nucleophilicity

In comparing molecules *with attacking atoms that are very different in size,* the direct relationship between nucleophilicity and basicity is maintained if the reaction occurs in the gas phase. If, however, the reaction occurs in a solvent—as most organic reactions do—the relationship between nucleophilicity and basicity depends on the solvent.

If the solvent is aprotic (is not a hydrogen bond donor, Section 9.12), the direct relationship between nucleophilicity and basicity holds. For example, both the nucleophilicities *and* basicities of the halogens decrease with increasing size in an aprotic solvent such as dimethylformamide (Table 9.7).

If the solvent is protic (is a hydrogen bond donor such as water, alcohols, etc.), the relationship between basicity and nucleophilicity becomes inverted: as basicity decreases, nucleophilicity increases. Thus, iodide ion, which is the weakest base of the halogen family, is the poorest nucleophile of the family in an aprotic solvent and the best nucleophile in a protic solvent.

Aprotic solvents do not contain a hydrogen bonded to an oxygen or a nitrogen; they are not hydrogen bond donors.

Protic solvents contain a hydrogen that is bonded to an an oxygen or a nitrogen; they are hydrogen bond donors.

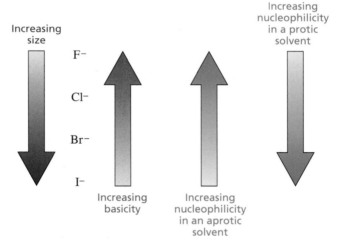

How does the solvent's ability to be a hydrogen bond donor affect the relationship between nucleophilicity and basicity? When a negatively charged species is placed in a protic solvent, the solvent molecules arrange themselves so that their partially positively charged hydrogens point toward the negatively charged species. An aprotic solvent does not have a partially positively charged hydrogen.

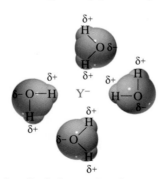

**ion-dipole interactions between
a nucleophile and water**

The interaction between the ion and the dipole of the protic solvent is called an **ion–dipole interaction.** The change from a *direct relationship* between basicity and nucleophilicity in an *aprotic solvent* to an *inverse relationship* in a *protic solvent* results from the ion–dipole interactions between the nucleophile and the protic solvent. This occurs because at least one of the ion–dipole interactions must be broken before the nucleophile can participate in an S_N2 reaction. Weak bases interact weakly with protic solvents; strong bases interact more strongly because they are better at sharing their electrons. It is, therefore, easier to break the ion–dipole interactions between an iodide ion and the solvent than between the more basic fluoride ion and the solvent, because the latter is a stronger base. As a result, iodide ion is a better nucleophile in a protic solvent (Table 9.2).

Stronger bases are better nucleophiles, except if the attacking atoms are very different in size *and* the reaction is carried out in a protic solvent.

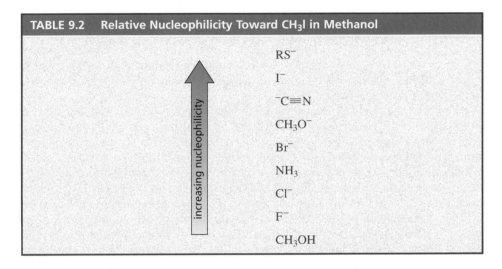

TABLE 9.2   Relative Nucleophilicity Toward CH$_3$I in Methanol

increasing nucleophilicity ↑

RS$^-$

I$^-$

$^-$C≡N

CH$_3$O$^-$

Br$^-$

NH$_3$

Cl$^-$

F$^-$

CH$_3$OH

P R O B L E M   4 ◆

a. Which is a stronger base, RO$^-$ or RS$^-$?

b. Which is a better nucleophile in an aqueous solution?

c. Which is a better nucleophile in dimethylformamide?

P R O B L E M   5 ◆

List the following compounds in order of *decreasing* nucleophilicity in an aqueous solution.

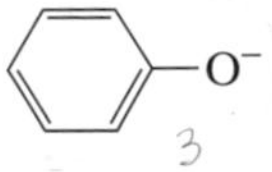

$\langle\bigcirc\rangle$—O$^-$ CH$_3$OH   HO$^-$ CH$_3\overset{\overset{\displaystyle O}{\|}}{C}O^-$ CH$_3$S$^-$ $\checkmark$

Because there are many different kinds of nucleophiles, a wide variety of organic compounds can be synthesized by means of S$_N$2 reactions. The following reactions show just a few of the many kinds of organic compounds that can be synthesized in this way.

$$CH_3CH_2Cl + HO^- \longrightarrow CH_3CH_2OH + Cl^-$$
an alcohol

$$CH_3CH_2Br + HS^- \longrightarrow CH_3CH_2SH + Br^-$$
a thiol

$$CH_3CH_2I + RO^- \longrightarrow CH_3CH_2OR + I^-$$
an ether

$$CH_3CH_2Br + RS^- \longrightarrow CH_3CH_2SR + Br^-$$
a thioether

$$CH_3CH_2F + {}^-NH_2 \longrightarrow CH_3CH_2NH_2 + F^-$$
a primary amine

$$CH_3CH_2Br + {}^-C≡CR \longrightarrow CH_3CH_2C≡CR + Br^-$$
an alkyne

$$CH_3CH_2I + {}^-C≡N \longrightarrow CH_3CH_2C≡N + I^-$$
a nitrile

The reverse of each of these reactions can also be visualized as a nucleophilic substitution reaction. In the first reaction, for example, ethyl chloride reacts with hydroxide ion to form ethyl alcohol and a chloride ion. The reverse reaction appears to satisfy the requirements for a nucleophilic substitution reaction, because the chloride ion is a nucleophile and ethyl alcohol has an HO^- leaving group. But ethyl alcohol and chloride ion do not react.

Why does the nucleophilic substitution reaction take place in one direction but not in the other? We can answer this question by comparing the leaving tendency of Cl^- in the forward direction and the leaving tendency of HO^- in the reverse direction. In order to compare leaving tendencies, we must compare basicities. Most people find it easier to compare the acid strengths of the conjugate acids (Table 9.3), so that is what we will do. HCl is a much stronger acid than H_2O, which means that Cl^- is a much weaker base than HO^- (because the stronger the acid, the weaker its conjugate base). Because it is a weaker base, Cl^- is a better leaving group. Consequently, HO^- can displace Cl^- in the forward reaction, but Cl^- cannot displace HO^- in the reverse reaction. The reaction proceeds in the direction that allows the stronger base to displace the weaker base.

An S$_N$2 reaction proceeds in the direction that allows the stronger base to displace the weaker base.

TABLE 9.3 The Acidities of the Conjugate Acids of Some Leaving Groups

Acid	pK_a	Conjugate base (leaving group)
HI	−10.0	I^-
HBr	−9.0	Br^-
HCl	−7.0	Cl^-
H_2SO_4	−5.0	$^-OSO_3H$
$CH_3\overset{+}{O}H_2$	−2.5	CH_3OH
H_3O^+	−1.7	H_2O
$\langle\!\!\bigcirc\!\!\rangle$—$SO_3H$	−0.6	$\langle\!\!\bigcirc\!\!\rangle$—$SO_3^-$
HF	3.2	F^-
$CH_3\overset{O}{\overset{\|}{C}}OH$	4.8	$CH_3\overset{O}{\overset{\|}{C}}O^-$
H_2S	7.0	HS^-
$HC\equiv N$	9.1	$^-C\equiv N$
$\overset{+}{N}H_4$	9.4	NH_3
CH_3CH_2SH	10.5	$CH_3CH_2S^-$
$(CH_3)_3\overset{+}{N}H$	10.8	$(CH_3)_3N$
CH_3OH	15.5	CH_3O^-
H_2O	15.7	HO^-
$HC\equiv CH$	25	$HC\equiv C^-$
NH_3	36	$^-NH_2$
H_2	36	H^-

PROBLEM 6

Using the pK_a values in Table 9.3, convince yourself that each of the reactions on page 367 proceeds in the direction shown.

If the difference between the basicities of the nucleophile and the leaving group is not very large, the reaction will be reversible. For example, in the reaction of ethyl bromide with iodide ion, Br$^-$ is the leaving group in one direction and I$^-$ is the leaving group in the other direction. Because the pK_a values of the conjugate acids of the two leaving groups are not very different (pK_a of HBr $= -9$; pK_a of HI $= -10$), the reaction is reversible.

$$CH_3CH_2Br \; + \; I^- \; \rightleftharpoons \; CH_3CH_2I \; + \; Br^-$$

You can drive a reversible reaction toward the desired products by removing one of the products as it forms. **Le Châtelier's principle** states that if an equilibrium is disturbed, the components of the equilibrium will adjust to offset the disturbance. In other words, if the concentration of product C is decreased, A and B will react to form more C and D so that the equilibrium constant maintains its value.

$$A \; + \; B \; \rightleftharpoons \; C \; + \; D$$

$$K_{eq} = \frac{[C][D]}{[A][B]}$$

For example, the reaction of ethyl chloride with methanol is reversible because the difference between the basicities of the nucleophile and the leaving group is not very large. If the reaction is carried out in a neutral solution, the protonated product will lose a proton. This disturbs the equilibrium and drives the reaction toward the products.

$$CH_3CH_2Cl \; + \; CH_3OH \; \rightleftharpoons \; CH_3CH_2\overset{+}{O}CH_3 \; \xrightarrow[\text{fast}]{-H^+} \; CH_3CH_2OCH_3$$
$$\underset{H \, + \, Cl^-}{|}$$

Henri Louis Le Châtelier (1850–1936) *was born in France. He studied mining engineering and was particularly interested in learning how to prevent explosions. Because his father was France's inspector general of mines, Henri's interest in mine safety is readily explained. Le Châtelier's research into preventing explosions led him to study heat and its measurement, which in turn led him to study thermodynamics.*

WHY CARBON INSTEAD OF SILICON?

There are two reasons why living organisms are composed primarily of carbon, oxygen, hydrogen, and nitrogen: fitness for specific roles in life processes and availability in the environment. That carbon became the fundamental building block of living organisms instead of silicon—even though silicon is more than 40 times more abundant than carbon in Earth's crust—implies that fitness was more important than availability.

Element	Abundance (atoms/100 atoms)	
	Living organisms	Earth's crust
H	49	0.22
C	25	0.19
O	25	47
N	0.3	0.1
Si	0.03	28

Why are hydrogen, carbon, oxygen, and nitrogen so fit for the roles they play in living organisms? Mainly because they are among the smallest atoms that form covalent bonds; carbon, oxygen, and nitrogen also form multiple bonds. Because the atoms are small, they form the strongest bonds and, therefore, form the most stable molecules. Clearly, the compounds that make up living organisms must be stable and slow to react.

Silicon has almost twice the diameter of carbon and therefore forms longer and weaker bonds. Consequently, an S_N2 reaction at silicon would occur much more rapidly than an S_N2 reaction at carbon. Silicon also has empty d orbitals that are available to accept electrons, so silicon compounds would not exist very long in the presence of any compound with lone pair electrons, including water. This alone would eliminate silicon from consideration as the fundamental building block of living systems. Silicon has yet another problem: it can form only single bonds. The end product of carbon metabolism is CO_2, a compound with two double bonds. The product of silicon metabolism would be SiO_2. Because silicon is only singly bonded to oxygen in SiO_2, silicon dioxide molecules polymerize to form quartz (sea sand). Clearly, life based on animals exhaling CO_2 is preferable to life based on animals exhaling sea sand.

PROBLEM 7◆

What is the product of the reaction of ethyl bromide with each of the following nucleophiles?

a. CH_3OH

b. NH_3

c. $(CH_3)_3N$

PROBLEM 8

The reaction of an alkyl chloride with potassium iodide is generally carried out in acetone in order to maximize the amount of alkyl iodide that is formed. Why does the solvent increase the amount of alkyl iodide? (*Hint:* Potassium iodide is soluble in acetone, but potassium chloride is not.)

9.4 INTERMOLECULAR VERSUS INTRAMOLECULAR REACTIONS

A molecule with two functional groups is called a **bifunctional molecule.** If the two functional groups are able to react with each other, two kinds of reactions can occur. In the case of a molecule that contains both a nucleophile and a leaving group, the nucleophile of one molecule can displace the leaving group of a second molecule of the compound. Such a reaction is called an intermolecular reaction. (*Inter* is Latin for "between.") An **intermolecular reaction** takes place between two molecules.

$$BrCH_2(CH_2)_nCH_2\ddot{\underset{..}{O}}{:}^- \qquad Br-CH_2(CH_2)_nCH_2\ddot{\underset{..}{O}}{:}^-$$

an intermolecular reaction

$$BrCH_2(CH_2)_nCH_2\ddot{\underset{..}{O}}CH_2(CH_2)_nCH_2\ddot{\underset{..}{O}}{:}^- \;+\; Br^-$$

Alternatively, the nucleophile of a molecule can displace the leaving group of that same molecule, thereby forming a cyclic compound. Such a reaction is called an intramolecular reaction. (*Intra* is Latin for "within.") An **intramolecular reaction** takes place within a single molecule.

$$\text{Br}-\text{CH}_2(\text{CH}_2)_n\text{CH}_2\ddot{\text{O}}\text{:}^- \xrightarrow{\textbf{an intramolecular reaction}} \text{H}_2\overset{\displaystyle (\text{CH}_2)_n}{\text{C}} \quad \text{CH}_2 + \text{Br}^-$$

Which reaction is more likely to occur when the nucleophile and the leaving group are parts of the same molecule—an intermolecular reaction or an intramolecular reaction? The answer depends on the concentration of the bifunctional molecule and the size of the ring that will be formed in the intramolecular reaction. The intramolecular reaction has an advantage in that the reacting groups are tethered close together and therefore do not have to wander through the solvent to find a group with which to react. As a result, a low concentration of reactant favors an intramolecular reaction because the two functional groups have a better chance of finding one another if they are in the same molecule. A high concentration of reactant helps make up for the advantage gained by tethering.

How much of an advantage an intramolecular reaction has over an intermolecular reaction depends on the length of the tether and therefore on the size of the ring that is formed. If the intramolecular reaction would form a five- or six-membered ring, it would be highly favored over the intermolecular reaction because of the stability and, therefore, ease of formation of five- and six-membered rings.

Three- and four-membered rings are strained. Consequently, they are less stable than five- and six-membered rings, which means that the transition state leading to their formation is less stable than the transition state leading to the formation of five- and six-membered rings. The higher activation energy for formation of three- and four-membered rings cancels some of the advantage gained by tethering.

The likelihood that the reacting groups can find each other decreases sharply for the formation of seven-membered and larger rings, so the intramolecular reaction becomes less favored as the ring size increases beyond six members.

PROBLEM 9◆

For each of the following pairs of S_N2 reactions, indicate which occurs with the larger rate constant.

a. $CH_3CH_2Br + H_2O$ or $CH_3CH_2Br + HO^-$

b. $CH_3CHCH_2Br + HO^-$ or $CH_3CH_2CHBr + HO^-$
  | |
  CH_3 CH_3

c. $CH_3CH_2Cl + CH_3O^-$ or $CH_3CH_2Cl + CH_3S^-$
  (in ethanol)

d. $CH_3Cl + I^-$ or $CH_3Br + I^-$

e. $BrCH_2CH_2CH_2NH_2$ or $BrCH_2CH_2CH_2CH_2NH_2$

f. Br⌒⌒⌒⌒OH   or   Br⌒⌒⌒⌒⌒OH

(in the presence of KOH)

9.5
THE MECHANISM OF S$_N$1 REACTIONS

Given our understanding of the S_N2 reaction, if we were to measure the rate of reaction of *tert*-butyl bromide with water, we would expect a relatively slow substitution reaction since water is a poor nucleophile and *tert*-butyl bromide is sterically hindered to attack by a nucleophile. However, we would actually discover that the reaction is surprisingly fast. In fact, it is over one million times faster than the reaction of methyl bromide—a compound with no steric hindrance—with water (Table 9.4). Clearly, the reaction must be taking place by a mechanism different from that of the S_N2 reaction.

As we have seen, a study of the kinetics of a reaction is one of the first steps undertaken when investigating the mechanism of a reaction. If we were to investigate the kinetics of the reaction of *tert*-butyl bromide with water, we would find that doubling the concentration of the alkyl halide doubles the rate of the reaction. We would also find that changing the concentration of the water has no effect on the rate of the reaction. Knowing that the rate of this nucleophilic substitution reaction depends only on the concentration of the alkyl halide, we can write a rate law for the reaction.

$$CH_3-\underset{\underset{CH_3}{|}}{\overset{\overset{CH_3}{|}}{C}}-Br + H_2O \longrightarrow CH_3-\underset{\underset{CH_3}{|}}{\overset{\overset{CH_3}{|}}{C}}-OH + HBr$$

tert-butyl bromide   *tert*-butyl alcohol

rate = k[alkyl halide]

Because the rate of the reaction depends on the concentration of only one reactant, it is **a first-order reaction.**

Since the rate law for the reaction of *tert*-butyl bromide with water clearly differs from the rate law for the reaction of methyl bromide with hydroxide ion, the two reactions must follow different mechanisms. We have seen that the reaction between methyl bromide and hydroxide ion is an S_N2 reaction (Section 9.2). The reaction between *tert*-butyl bromide and water is an **S$_N$1 reaction:** S for substitution, $_N$ for nucleophilic, and 1 for unimolecular. The mechanism of an S_N1 reaction is based on the following experimental evidence.

1. The rate law shows that the rate of the reaction depends only on the concentration of the alkyl halide. This means that we must be observing a reaction whose rate-determining transition state involves only the alkyl halide. The rate-determining transition state is **unimolecular;** it involves only one molecule.

2. When the methyl groups of *tert*-butyl bromide are successively replaced by hydrogens, the rate of the S_N1 reaction decreases progressively (Table 9.4). This is opposite to the order of reactivity exhibited by alkyl halides in S_N2 reactions (Table 9.1).

3. The reaction of an alkyl halide in which the halogen is bonded to a chirality center leads to the formation of two stereoisomers: one with the same relative configuration as the reacting alkyl halide and the other with the inverted configuration.

An S_N1 reaction has two steps. In the first step, the carbon–halogen bond breaks heterolytically, with the halogen retaining the previously shared pair of electrons. In the second step, the nucleophile reacts rapidly with the carbocation that was formed in the first step.

mechanism of the S_N1 reaction

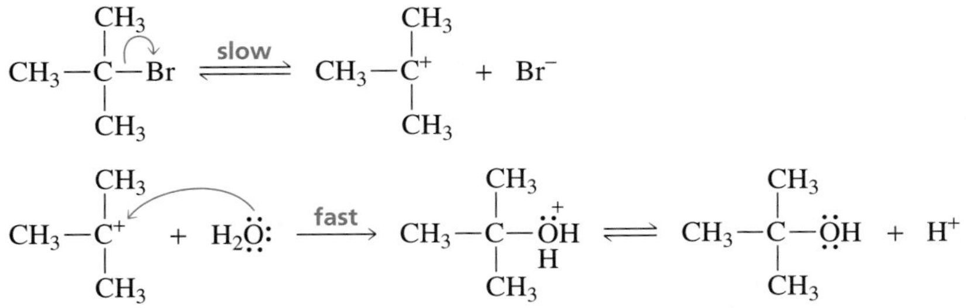

From the observation that the rate of an S_N1 reaction depends only on the concentration of the alkyl halide, we know that the first step is the slow and rate-determining step. Because the nucleophile is not involved in the rate-determining step, its concentration has no effect on the rate of the reaction. If you look at the reaction coordinate diagram in Figure 9.3, you will be able to see why increasing the rate of the second step will not make an S_N1 reaction go any faster.

How does the reaction mechanism that we have just seen account for the three pieces of experimental evidence? First, because the alkyl halide is the only species that participates in the rate-limiting step, the mechanism agrees with the observation that the rate of the reaction depends on the concentration of the alkyl halide and does not depend on the concentration of the nucleophile.

Second, a carbocation is formed in the slow step of an S_N1 reaction. Because a tertiary carbocation is more stable and therefore easier to form than a secondary

TABLE 9.4 Relative Rates of S_N1 Reactions for Several Alkyl Bromides (Solvent Is H_2O, Nucleophile is H_2O)		
Alkyl bromide	**Class**	**Relative rate**
CH_3C-Br (with CH_3 groups above and below)	tertiary	1,200,000
CH_3CH-Br (with CH_3 below)	secondary	11.6
CH_3CH_2-Br	primary	1.00*
CH_3-Br	methyl	1.05*

*Although the rate of the S_N1 reaction of this compound with water is 0, a small rate is observed as a result of an S_N2 reaction.

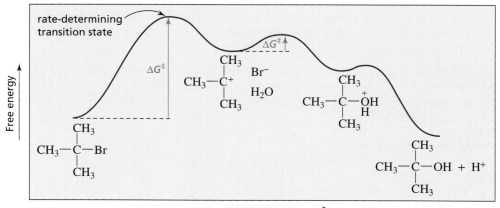

▲ **Figure 9.3**
Reaction coordinate diagram for the S_N1 reaction of *tert*-butyl bromide with water.

carbocation, which in turn is more stable and easier to form than a primary carbo-cation (Section 3.10), tertiary alkyl halides are more reactive than secondary alkyl halides which are more reactive than primary alkyl halides in an S_N1 reaction. Thus the reactivity order agrees with the observation that the rate of an S_N1 reaction decreases as the methyl groups of *tert*-butyl bromide are successively replaced by hydrogens (Table 9.4).

relative reactivities of alkyl halides in an S_N1 reaction

3° alkyl halide > 2° alkyl halide > 1° alkyl halide

Actually, primary carbocations and methyl cations are so unstable that primary alkyl halides and methyl halides do not undergo S_N1 reactions. (The very slow reactions reported for ethyl bromide and methyl bromide in Table 9.4 are S_N2 reactions.)

The positively charged carbon of the carbocation intermediate is sp^2 hybridized, and the three bonds connected to an sp^2 hybridized carbon are in the same plane. In the second step of the S_N1 reaction, the nucleophile can approach the carbocation from either side of the plane.

If the nucleophile attacks the side of the carbon from which the leaving group departed, the product will have the same relative configuration as the reacting alkyl halide. If, however, the nucleophile attacks the opposite side of the carbon, the product will have the inverted configuration compared with the alkyl halide. We can now understand the third piece of experimental evidence for the mechanism of the S$_N$1 reaction. The S$_N$1 reaction of an alkyl halide in which the leaving group is attached to a chirality center leads to formation of two stereoisomers: attack of the nucleophile on one side of the planar carbocation forms one stereoisomer, and attack on the other side produces the other stereoisomer.

If the leaving group in an S$_N$1 reaction is attached to a chirality center, a pair of enantiomers will be formed as products

---

PROBLEM 10♦

Arrange the following alkyl bromides in order of decreasing reactivity in an S$_N$1 reaction: isopropyl bromide, propyl bromide, *tert*-butyl bromide, methyl bromide.

The Leaving Group

Because the rate-determining step of an S$_N$1 reaction is dissociation of the alkyl halide to form a carbocation, two factors affect the rate of an S$_N$1 reaction: the ease with which the leaving group dissociates from the carbon and the stability of the carbocation that is formed. In the preceding section, we saw that tertiary alkyl halides are more reactive than secondary alkyl halides, which are more reactive than primary alkyl halides. This is because the more substituted the carbocation is, the more stable it is and therefore the easier it is to form. But what about a series of alkyl halides with different leaving groups that dissociate to form the same carbocation? The answer is the same for the S$_N$1 reaction as for the S$_N$2 reaction. The weaker the base, the less tightly it is bonded to the carbon and the easier it is to break the carbon–halogen bond. So an alkyl iodide is the most reactive and an alkyl fluoride is the least reactive of the alkyl halides in both S$_N$1 and S$_N$2 reactions.

relative reactivities of alkyl halides in an S$_N$1 reaction

$$RI > RBr > RCl > RF$$

increasing reactivity

The Nucleophile

The nucleophile traps the carbocation that is formed in the rate-determining step of the S$_N$1 reaction. Because the nucleophile comes into play after the rate-

9.6
THE S$_N$1 REACTION

determining step, the reactivity of the nucleophile has no effect on the rate of the S_N1 reaction.

In some S_N1 reactions, the solvent is the nucleophile. For example, the relative rates given in Table 9.4 are for the reactions of alkyl halides with water. Water serves as both the nucleophile and the solvent. When the solvent is the nucleophile, the reaction is called a **solvolysis reaction.** So the relative rates in Table 9.4 are for the solvolysis reactions of the indicated alkyl bromides in water.

Carbocation Rearrangements

A carbocation intermediate is formed in an S_N1 reaction. In Section 3.14 we saw that a carbocation will rearrange if it becomes more stable as a result of the rearrangement. If the carbocation formed in an S_N1 reaction can rearrange, S_N1 and S_N2 reactions can produce different constitutional isomers as products, because carbocations are not formed in an S_N2 reaction and thus rearrangement of the carbon skeleton cannot occur. For example, the product obtained when 2-bromo-3-methylbutane undergoes an S_N1 reaction is different from the product obtained when it undergoes an S_N2 reaction. When the reaction is carried out under conditions that favor an S_N1 reaction, the initially formed secondary carbocation undergoes a 1,2-hydride shift, rearranging to a more stable tertiary carbocation.

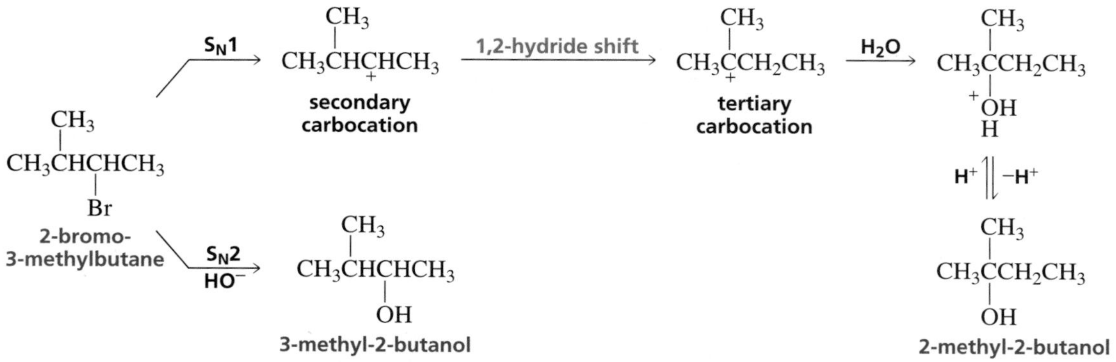

The product obtained from the reaction of 3-bromo-2,2-dimethylbutane with a nucleophile also depends on the conditions under which the reaction is carried out. The carbocation formed under conditions that favor an S_N1 reaction undergoes a 1,2-methyl shift. Because a carbocation is not formed under conditions that favor an S_N2 reaction, the carbon skeleton does not rearrange.

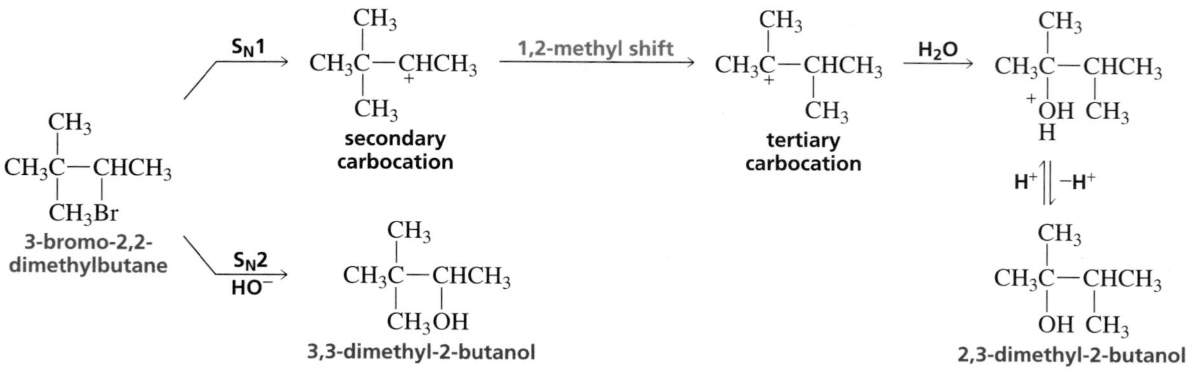

We will see in Sections 9.9 and 9.10 that we can exercise some control over whether an S_N1 or an S_N2 reaction takes place by choosing appropriate reaction conditions.

PROBLEM 11◆

Arrange the following alkyl halides in order of decreasing reactivity in an S_N1 reaction: 2-bromopentane, 2-chloropentane, 1-chloropentane, 3-bromo-3-methylpentane.

PROBLEM 12◆

Which of the following alkyl halides form a substitution product as a result of an S_N1 reaction that is different from the substitution product formed as a result of an S_N2 reaction?

a. 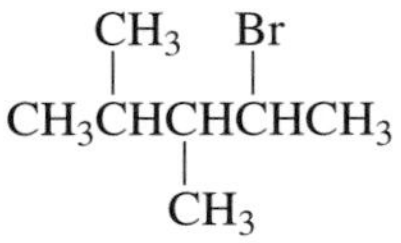
$$\begin{array}{cc} CH_3 & Br \\ | & | \\ CH_3CHCHCHCH_3 \\ | \\ CH_3 \end{array}$$

c. 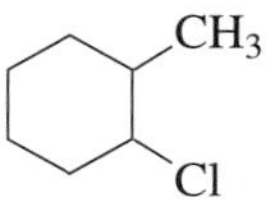
(cyclohexane ring with CH₃ and Cl substituents)

b. 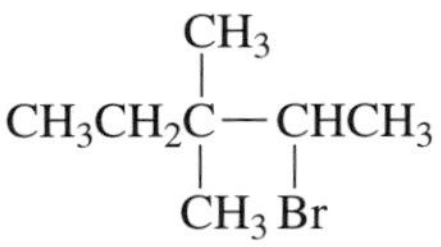
$$\begin{array}{c} CH_3 \\ | \\ CH_3CH_2C-CHCH_3 \\ | \ \ | \\ CH_3 \ Br \end{array}$$

d. 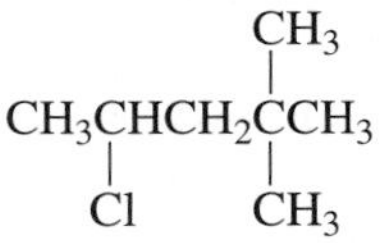
$$\begin{array}{c} CH_3 \\ | \\ CH_3CHCH_2CCH_3 \\ | \ \ \ \ | \\ Cl \ \ \ \ CH_3 \end{array}$$

PROBLEM 13◆

Two substitution products result from the reaction of 3-chloro-3-methyl-1-butene with sodium acetate ($CH_3COO^-Na^+$) in acetic acid under S_N1 conditions. Identify the products. Indicate the thermodynamic product and the kinetic product.

The substitution product obtained from the reaction of 2-bromopropane with hydroxide ion does not have a chirality center. Therefore, stereoisomers are not possible for this product.

$$\begin{array}{cccc} CH_3CHCH_3 & + \ HO^- & \longrightarrow & CH_3CHCH_3 & + \ Br^- \\ | & & & | \\ Br & & & OH \\ \text{2-bromopropane} & & & \text{2-propanol} \end{array}$$

The substitution product obtained from the reaction of 2-bromobutane with hydroxide ion has a chirality center. Therefore, it can exist as a pair of enantiomers.

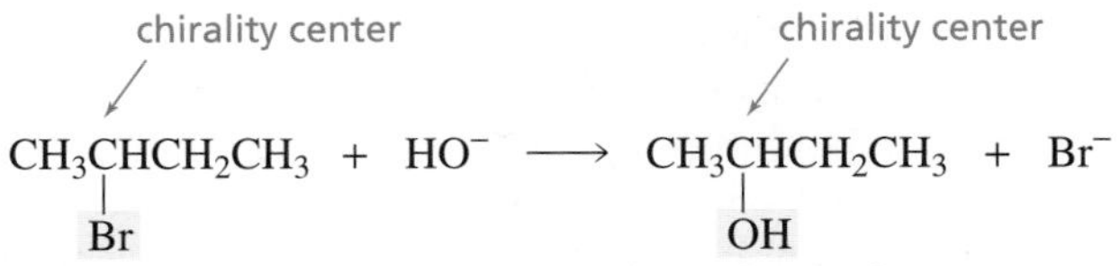

chirality center   chirality center

$$\begin{array}{cccc} CH_3CHCH_2CH_3 & + \ HO^- & \longrightarrow & CH_3CHCH_2CH_3 & + \ Br^- \\ | & & & | \\ Br & & & OH \\ \text{2-bromobutane} & & & \text{2-butanol} \end{array}$$

We cannot specify the configuration of the product formed from the reaction of 2-bromobutane with hydroxide ion unless we know the configuration of the alkyl halide and we know whether the reaction is an S_N2 or an S_N1 reaction. For example, in the S_N2 reaction of (*S*)-2-bromobutane with hydroxide ion, the incoming hydroxide ion attacks the chirality center on the side opposite to where the bromine is bonded. This results in a product whose configuration is inverted compared with the configuration of the reactant. An S_N2 reaction takes place with *inversion of configuration.*

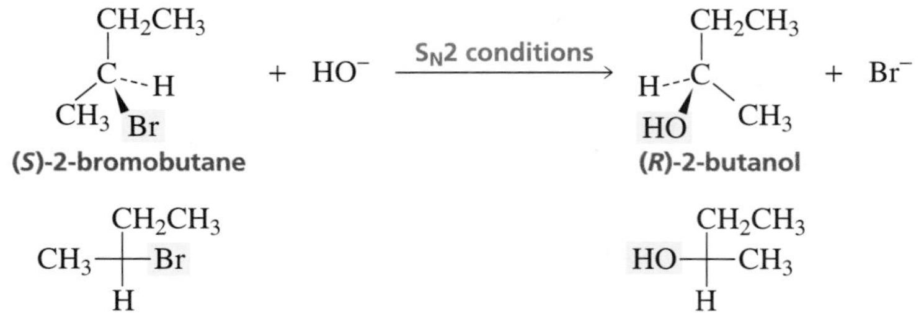

In the S_N1 reaction of (S)-2-bromobutane with water, two substitution products are formed: one product has the same relative configuration as the reactant, and the other has the inverted configuration. In an S_N1 reaction, the leaving group leaves before the nucleophile attacks. This means that the nucleophile is free to attack either side of the planar carbocation. If it attacks the side from which the bromide ion left, the product will have the same relative configuration as the reactant. If it attacks the opposite side, the product will have the inverted configuration.

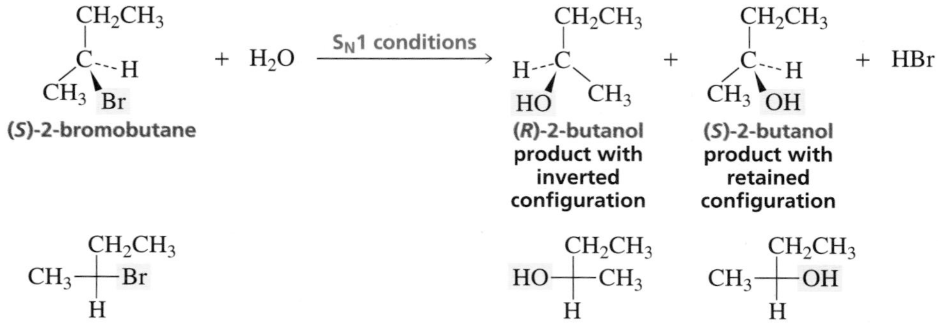

Although you might expect that equal amounts of both products should be formed in an S_N1 reaction, a greater amount of the product with the inverted configuration is obtained in most cases. Typically, 50 to 70% of the product of an S_N1 reaction is the inverted product. If the reaction leads to equal amounts of the two stereoisomers, the reaction is said to take place with **complete racemization.** When more of the inverted product is formed, the reaction is said to take place with **partial racemization.**

Saul Winstein (1912–1969) *was born in Montreal, Canada. He received a Ph.D. from the California Institute of Technology and was a professor of chemistry at University of California, Los Angeles, from 1942 until his death.*

Saul Winstein was the first to explain why extra inverted product is formed. He postulated that dissociation of the alkyl halide initially results in the formation of an **intimate ion pair.** In an intimate ion pair, the bond between the carbon and the leaving group has broken but the cation and anion remain next to each other. This species then forms an ion pair in which one or more solvent molecules have come between the cation and the anion. This is called a **solvent-separated ion pair.** Further separation between the two results in the dissociated ions.

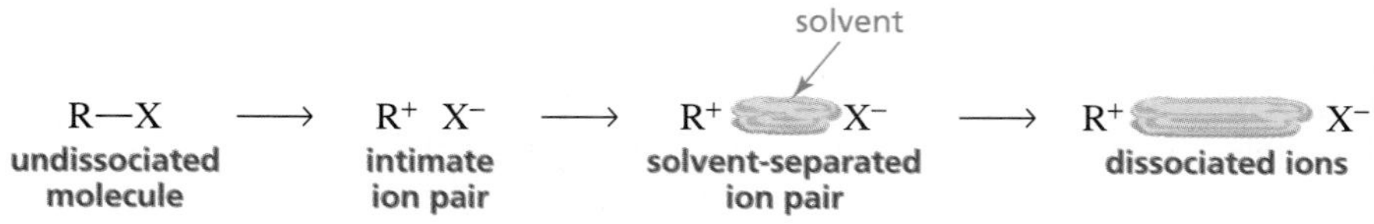

The nucleophile can attack any of these four species. If the nucleophile attacks only the completely dissociated carbocation, the product will be completely racemized. If the nucleophile attacks the carbocation of either the intimate ion pair or the solvent-separated ion pair, the leaving group will be in position to partially block the approach of the nucleophile to that side of the carbocation. As a result, more of the product with the inverted configuration will be obtained. (If the nucleophile attacks the undissociated molecule, it will be an S$_N$2 reaction and all of the product will have the inverted configuration.)

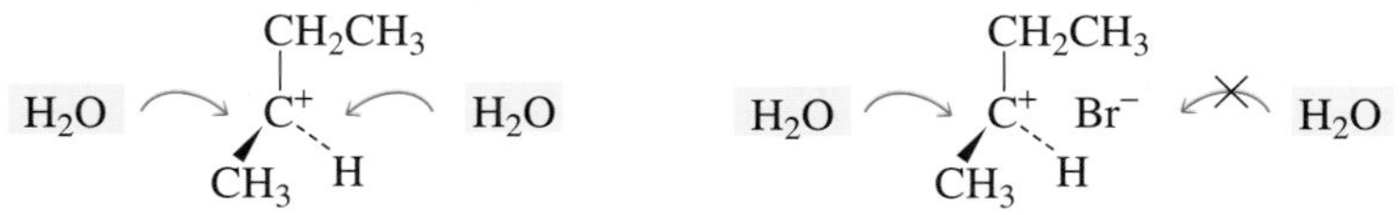

Br$^-$ has diffused away, giving H$_2$O equal access to both sides of the carbocation.

Br$^-$ has not diffused away, so it blocks the approach of H$_2$O to one side of the carbocation.

The difference between the products obtained from an S$_N$1 reaction and from an S$_N$2 reaction is a little easier to visualize in the case of cyclic compounds. When *cis*-1-bromo-4-methylcyclohexane undergoes an S$_N$2 reaction, only the trans product is obtained because the carbon bonded to the leaving group is attacked by the nucleophile only on its back side.

cis-1-bromo-4-methylcyclohexane

trans-4-methylcyclohexanol

When, however, *cis*-1-bromo-4-methylcyclohexane undergoes an S$_N$1 reaction, both the cis and the trans products are formed because the nucleophile can approach the carbocation intermediate from either side.

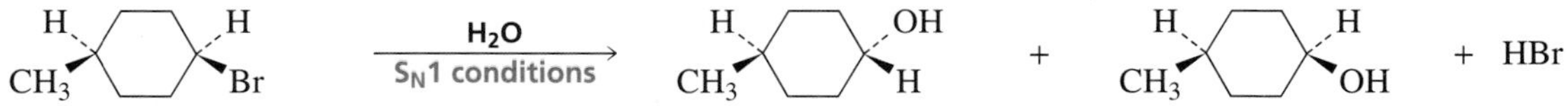

cis-1-bromo-4-methylcyclohexane

trans-4-methylcyclohexanol

cis-4-methylcyclohexanol

PROBLEM 14◆

Which product in the reaction above will be formed in greater yield?

PROBLEM 15

Give the products that will be obtained from the following reactions if:

a. the reaction is carried out under conditions that favor an S$_N$2 reaction.

b. the reaction is carried out under conditions that favor an S$_N$1 reaction.

    (1) *trans*-1-bromo-4-methylcyclohexane + sodium methoxide/methanol

    (2) *cis*-1-chloro-2-methylcyclobutane + sodium hydroxide/water

PROBLEM 16◆

Which of the following reactions will go faster if the concentration of the nucleophile is increased?

a. 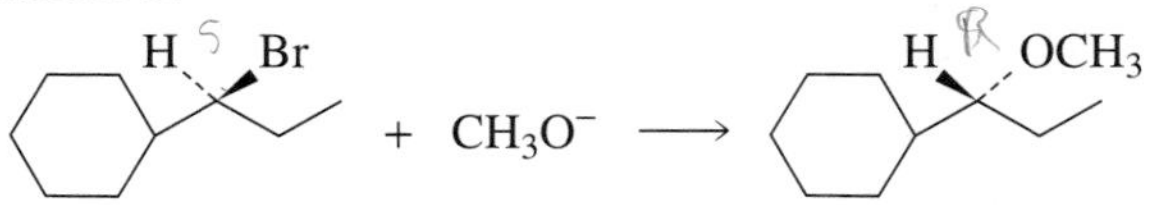   + CH$_3$O$^-$ $\longrightarrow$

b.

$\diagdown\diagup\diagdown\diagup^{Br}$ + CH_3S^- $\longrightarrow$ $\diagdown\diagup\diagdown\diagup^{SCH_3}$

c.

9.8
BENZYLIC HALIDES, ALLYLIC HALIDES, VINYLIC HALIDES, AND ARYL HALIDES

To this point, we have limited our discussion of substitution reactions to methyl halides and primary, secondary, and tertiary alkyl halides. But what about benzylic, allylic, vinylic, and aryl halides? Let's first consider benzylic and allylic halides. Benzylic and allylic halides readily undergo S_N2 reactions unless they are tertiary. Tertiary benzylic and allylic halides, like other tertiary halides, are unreactive in S_N2 reactions because of steric hindrance.

$\bigcirc$—CH_2Cl + CH_3O^- $\xrightarrow{\text{$S_N$2 conditions}}$ $\bigcirc$—CH_2OCH_3 + Cl^-

benzyl chloride   **benzyl methyl ether**

$CH_3CH{=}CHCH_2Br$ + HO^- $\xrightarrow{\text{$S_N$2 conditions}}$ $CH_3CH{=}CHCH_2OH$ + Br^-

1-bromo-2-butene   **2-buten-1-ol**
an allylic halide

Benzylic and allylic halides also undergo S_N1 reactions because of the stability of their carbocations. We have seen that primary alkyl halides (such as CH_3CH_2Br and $CH_3CH_2CH_2Br$) cannot undergo S_N1 reactions because their carbocations are too unstable. Primary benzylic and primary allylic halides, however, readily undergo S_N1 reactions because their carbocations have about the same stability as secondary carbocations, since they are stabilized by resonance (Section 6.7).

$\bigcirc$—CH_2Cl $\underset{}{\overset{S_N1}{\rightleftharpoons}}$ $\bigcirc$—$\overset{+}{C}H_2$ + Cl^- $\xrightarrow{CH_3OH}$ $\bigcirc$—CH_2OCH_3 + H^+

$CH_2{=}CHCH_2Br$ $\underset{}{\overset{S_N1}{\rightleftharpoons}}$ $CH_2{=}CH\overset{+}{C}H_2$ + Br^- $\xrightarrow{H_2O}$ $CH_2{=}CHCH_2OH$ + H^+

Vinylic halides and aryl halides do not undergo S_N2 or S_N1 reactions. They do not undergo an S_N2 reaction because as the nucleophile approaches the back side of the sp^2 carbon, it is repelled by the π electron cloud of the double bond or the aromatic ring.

$RCH{=}CHCl$ $\bigcirc$—Br

       $\times$ nucleophile $\times$ nucleophile

a vinylic halide   **an aryl halide**

There are two reasons why vinylic halides and aryl halides do not undergo S_N1 reactions. First, vinylic and aryl cations are even more unstable than primary carbocations, and we saw that primary alkyl halides do not undergo S_N1 reactions because of the instability of their carbocations (Section 9.5). The instability of vinylic and aryl cations is the result of the positive charge being on an *sp* carbon. Because *sp* carbons are more electronegative than sp^2 carbons, which carry the positive charge of alkyl carbocations, *sp* carbons are more resistant to becoming positively charged. Second, we have seen that sp^2 carbons form stronger bonds than do sp^3 carbons (Section 1.14). As a result, it is harder to break the carbon–halogen bond when the halogen is bonded to an sp^2 carbon.

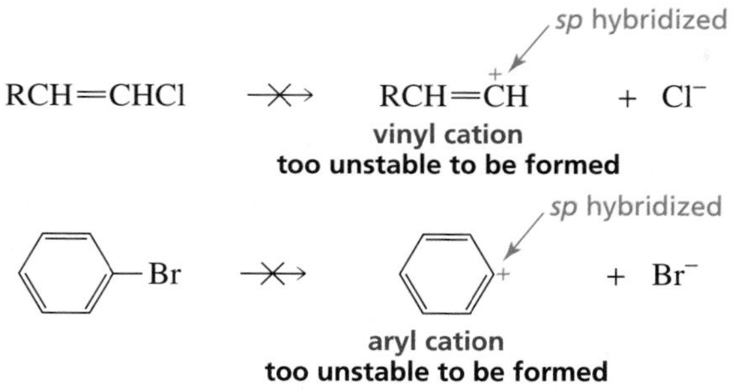

PROBLEM 17 ◆

Which alkyl halide would you expect to be more reactive in an S_N2 reaction with a given nucleophile?

a. $CH_3CH_2CH_2Br$ or $CH_3CH_2CH_2I$

b. $CH_3CH_2CH_2Cl$ or CH_3OCH_2Cl

c.

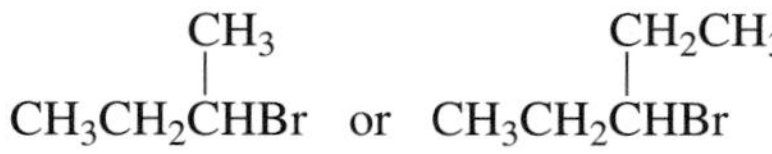

d.

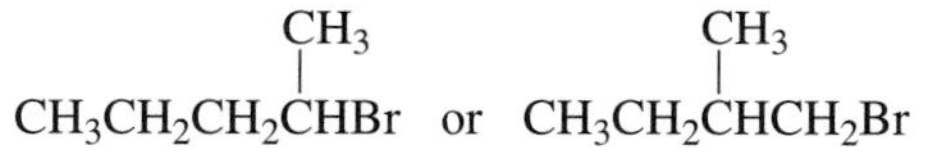

e.

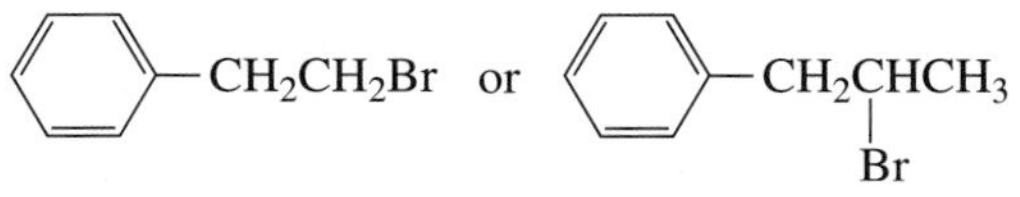

f.

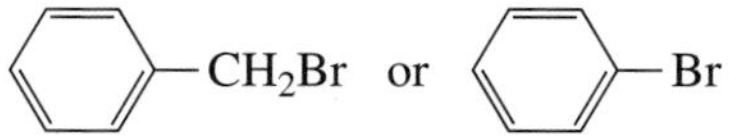

g.

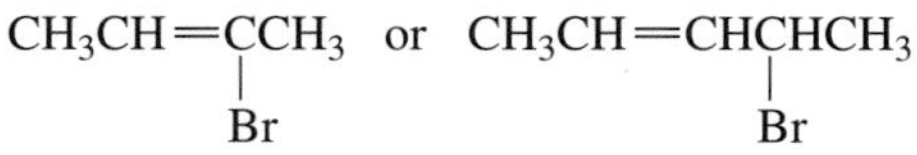

PROBLEM 18 ◆

For each of the pairs in Problem 17, which compound would be more reactive in an S_N1 reaction?

PROBLEM 19

Give the products of the following reaction:

a. under conditions that favor an S_N2 reaction.

b. under conditions that favor an S_N1 reaction.

$$CH_3CH=CHCH_2Br + CH_3O^- \xrightarrow{\ \textbf{CH}_3\textbf{OH}\ }$$

9.9
COMPETITION BETWEEN S_N2 AND S_N1 REACTIONS

The characteristics of the S_N2 and S_N1 reactions are summarized in Table 9.5. It is important to remember that the "2" in S_N2 and the "1" in S_N1 refer to the molecularity of the rate-determining step; the rate-determining step of an S_N2 reaction is bimolecular; the rate-determining step of an S_N1 reaction is unimolecular. These numbers do not refer to the number of steps in the mechanism: the S_N2 reaction proceeds by a one-step concerted mechanism, whereas the S_N1 reaction proceeds by a two-step mechanism with a carbocation intermediate.

We have seen that methyl halides and primary alkyl halides undergo only S_N2 reactions because methyl cations and primary carbocations, which would be formed in an S_N1 reaction, are too unstable to be formed. Tertiary alkyl halides undergo only S_N1 reactions because steric hindrance makes them very unreactive in an S_N2 reaction. Secondary alkyl halides as well as benzylic and allylic halides (unless they are tertiary) can undergo both S_N1 and S_N2 reactions because they form relatively stable carbocations and the amount of steric hindrance associated with these alkyl halides is not very great. Vinylic and aryl halides do not undergo either S_N1 or S_N2 reactions. These results are summarized in Table 9.6.

When an alkyl halide can undergo both S_N1 and S_N2 reactions, both reactions take place simultaneously. The conditions under which the reaction is carried out determine which of the reactions predominates.

What conditions favor an S_N1 reaction? What conditions favor an S_N2 reaction? These are important questions to synthetic chemists because an S_N2 reaction results in the formation of a single substitution product whereas an S_N1 reaction can form two substitution products if the leaving group is bonded to a chirality center. An S_N1 reaction is further complicated by carbocation rearrangements. In other words, an S_N2 reaction is a synthetic chemist's friend, but an S_N1 reaction can be a synthetic chemist's nightmare.

The conditions that determine whether the predominant reaction will be an S_N2 reaction or an S_N1 reaction are the concentration of the nucleophile, the reactivity of the nucleophile, and the solvent in which the reaction is carried out. To understand how the concentration of the nucleophile and the reactivity of the nucleophile affect whether an S_N2 or an S_N1 reaction predominates, we must first examine the rate laws for the two reactions. The rate constants have been given subscripts that indicate the reaction order.

Rate law for the S_N2 reaction: rate $= k_2$ [alkyl halide][nucleophile]

Rate law for the S_N1 reaction: rate $= k_1$ [alkyl halide]

TABLE 9.5 Comparison of the S_N2 and S_N1 Reactions	
S_N2	**S_N1**
A one-step mechanism	A two-step mechanism
A bimolecular rate-determining transition state	A unimolecular rate-determining transition state
No carbocation rearrangements	Carbocation rearrangements
Product has inverted configuration compared with reactant	Products have both identical and inverted configurations compared with reactant
Reactivity order: methyl $>$ 1° $>$ 2° $>$ 3°	Reactivity order: 3° $>$ 2° $>$ 1° $>$ methyl

TABLE 9.6 Summary of the Reactivity of Alkyl Halides in Nucleophilic Substitution Reactions			
methyl and primary alkyl halides	S_N2 only	benzylic and allylic halides	S_N1 and S_N2
secondary alkyl halides	S_N1 and S_N2	vinylic and aryl halides	neither S_N1 nor S_N2
tertiary alkyl halides	S_N1 only		

The rate law for the reaction of an alkyl halide that can undergo both S_N2 and S_N1 reactions simultaneously is the sum of the individual rate laws.

$$\text{rate} \;=\; k_2[\text{alkyl halide}][\text{nucleophile}] \;+\; k_1[\text{alkyl halide}]$$

contribution to the rate
by an S_N2 reaction

contribution to the rate
by an S_N1 reaction

It is apparent that an increase in the concentration of the nucleophile increases the rate of an S_N2 reaction but has no effect on the rate of an S_N1 reaction. Therefore, when both reactions occur simultaneously, increasing the concentration of the nucleophile increases the fraction of the reaction that takes place by an S_N2 pathway. In contrast, decreasing the concentration of the nucleophile decreases the fraction of the reaction that takes place by an S_N2 pathway.

The slow (and only) step of an S_N2 reaction is attack of the nucleophile on the alkyl halide. Increasing the reactivity of the nucleophile increases the rate of an S_N2 reaction by increasing the value of the rate constant (k_2), because more reactive nucleophiles are better able to displace the leaving group. The slow step of an S_N1 reaction is the dissociation of the alkyl halide. The carbocation formed in the slow step rapidly reacts in a second step with any nucleophile present in the reaction mixture. Increasing the rate of the fast step does not affect the rate of the prior slow, carbocation-forming step. This means that increasing the reactivity of the nucleophile has no effect on the rate of an S_N1 reaction. A good nucleophile, therefore, favors an S_N2 reaction over an S_N1 reaction. A poor nucleophile favors an S_N1 reaction, not by increasing the rate of the S_N1 reaction itself but by decreasing the rate of the competing S_N2 reaction.

- An S_N2 reaction is favored by a high concentration of a good nucleophile.

- An S_N1 reaction is favored by a low concentration of a good nucleophile or by a poor nucleophile.

Notice that, when S_N1 reactions are described, poor nucleophiles are generally used (H_2O, CH_3OH); when S_N2 reactions are described, good nucleophiles are used (HO^-, CH_3O^-). In other words, a poor nucleophile is used to encourage an S_N1 reaction, and a good nucleophile is used to encourage an S_N2 reaction.

PROBLEM-SOLVING STRATEGY

This problem will give you practice in determining whether a substitution reaction will take place by an S_N1 or an S_N2 pathway.

Give the configuration(s) of the substitution product(s) that will be obtained from the reactions of the following secondary alkyl halides with the indicated nucleophile.

Because a high concentration of a good nucleophile is used, we can predict that the reaction is an S_N2 reaction. Thus, the product will have the inverted configuration compared with the reactant. (An easy way to draw the inverted product is to draw the mirror image of the reacting alkyl halide, putting the nucleophile in the same location as the leaving group.)

b. 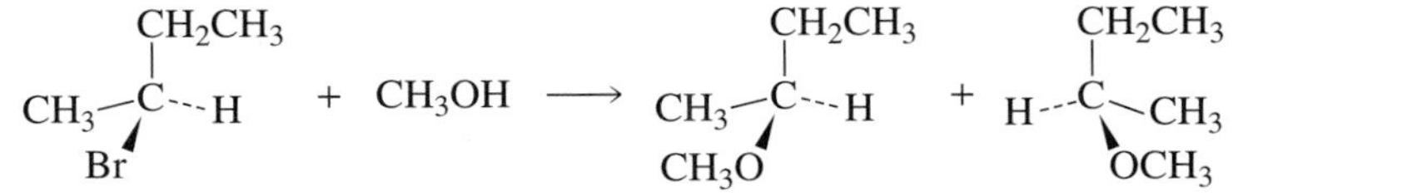   CH₃—C(CH₂CH₃)(H)—Br + CH₃OH ⟶ CH₃—C(CH₂CH₃)(H)—OCH₃ + H—C(CH₂CH₃)(CH₃)—OCH₃

Because a poor nucleophile is used, we can predict that the reaction is an S_N1 reaction. Thus, we will obtain two substitution products, one with the retained configuration and one with the inverted configuration compared with the reactant.

c. 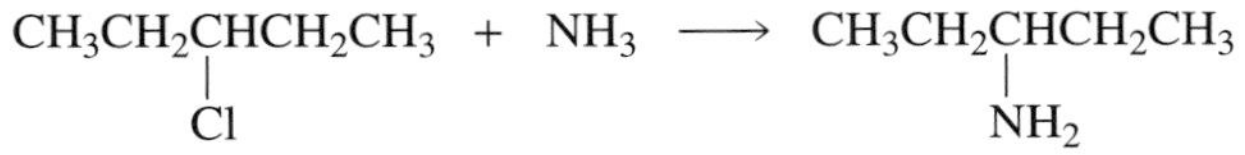   CH₃CH₂CHCH₂CH₃ (Cl) + NH₃ ⟶ CH₃CH₂CHCH₂CH₃ (NH₂)

The poor nucleophile indicates that the reaction is an S_N1 reaction. However, the product does not have a chirality center and so does not have stereoisomers. (The same substitution product would have been obtained if the reaction had been an S_N2 reaction.)

d.    CH₃CH₂CHCH₃ (I) + HO⁻ **high concentration** ⟶ CH₃CH₂CHCH₃ (OH)

Because a high concentration of a good nucleophile is employed, we can predict that the reaction is an S_N2 reaction. Therefore, the product will have the inverted configuration compared with the reactant. But since the configuration of the reactant is not indicated, we do not know the configuration of the product.

e. 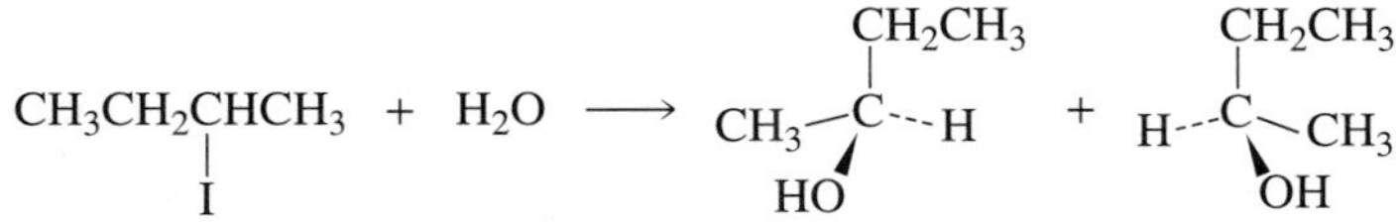   CH₃CH₂CHCH₃ (I) + H₂O ⟶ CH₃—C(CH₂CH₃)(H)—HO + H—C(CH₂CH₃)(CH₃)—OH

Because a poor nucleophile is employed, we can predict that the reaction is an S_N1 reaction. Therefore, both stereoisomers will be formed regardless of the configuration of the reactant.

Now continue on to solve Problem 20.

PROBLEM 20

Give the configuration(s) of the substitution product(s) that will be obtained from the reactions of the following secondary alkyl halides with the indicated nucleophile.

a. 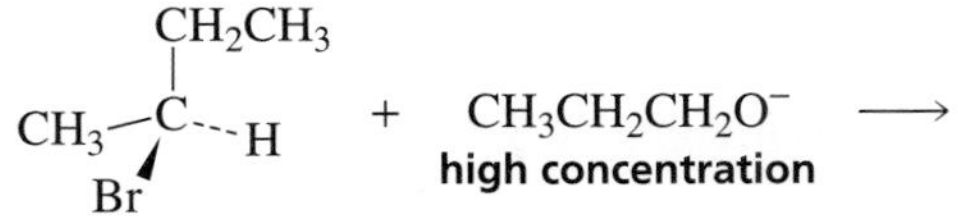   CH₃—C(CH₂CH₃)(H)—Br + CH₃CH₂CH₂O⁻ **high concentration** ⟶

b. 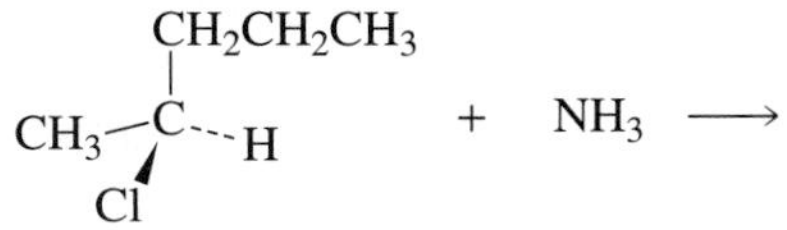   CH₃—C(CH₂CH₂CH₃)(H)—Cl + NH₃ ⟶

c.   (cyclohexane ring with H, CH₃, Cl, H) + CH₃O⁻ **high concentration** ⟶

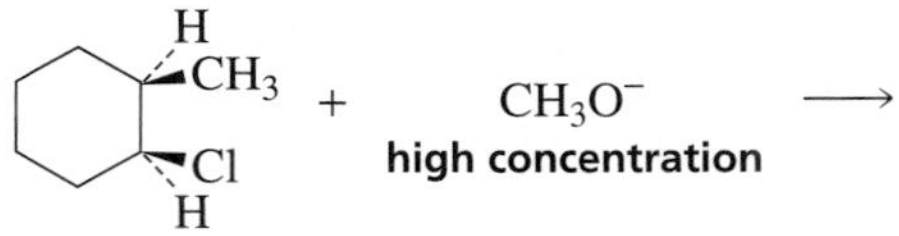

d.

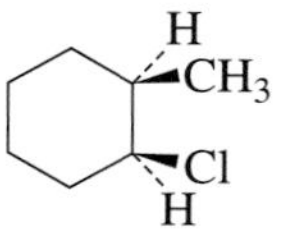

$+ \; CH_3OH \longrightarrow$

e.

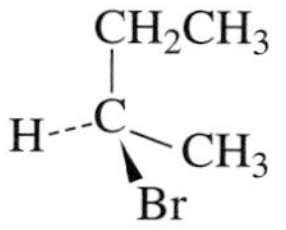

$+ \quad CH_3O^-$ **high concentration** $\longrightarrow$

f.

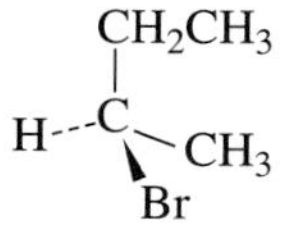

$+ \; CH_3OH \longrightarrow$

PROBLEM 21 / SOLVED

The rate law for the substitution reaction of 2-bromobutane and HO^- in 75% ethanol/25% water at 30 °C is

rate $= \; 3.20 \times 10^{-5}$ [2-bromobutane][HO^-] $+ \; 1.5 \times 10^{-6}$ [2-bromobutane]

a. What percent of the reaction takes place by the S_N2 mechanism when $[HO^-] = 1.00$ M?

b. When $[HO^-] = 0.001$ M?

SOLUTION TO 21a

percentage by S_N2

$$= \frac{S_N2}{S_N2 + S_N1} \times 100$$

$$= \frac{3.20 \times 10^{-5}[\text{2-bromobutane}] \times 1.00 \times 100}{3.20 \times 10^{-5}[\text{2-bromobutane}] \times 1.00 + 1.5 \times 10^{-6}[\text{2-bromobutane}]}$$

$$= \frac{3.20 \times 10^{-5}}{3.20 \times 10^{-5} + 0.15 \times 10^{-5}} \times 100 = \frac{3.20 \times 10^{-5}}{3.35 \times 10^{-5}} \times 100$$

$$= 96\%$$

The solvent in which the nucleophilic substitution reaction is carried out also has an influence on whether an S_N2 or an S_N1 reaction will predominate. Before we can understand how a particular solvent favors one reaction over another, we must understand how solvents stabilize organic molecules.

The **dielectric constant** of a solvent is a measure of how well the solvent can insulate opposite charges from one another. Solvent molecules insulate charges by clustering around a charge so that the negative poles of the solvent molecules surround positive charges while the positive poles of the solvent molecules surround negative charges. Remember that the interaction between a solvent and an ion or molecule dissolved in that solvent is called *solvation* (Section 2.9). When an ion interacts with a polar solvent, the charge is no longer localized solely on the ion but is spread out to the surrounding solvent molecules. Spreading out the charge stabilizes the charged species.

9.10
THE ROLE OF THE SOLVENT IN S_N2 AND S_N1 REACTIONS

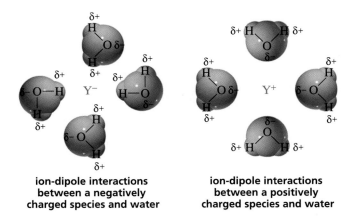

ion-dipole interactions
between a negatively
charged species and water

ion-dipole interactions
between a positively
charged species and water

Polar solvents have high dielectric constants and thus are very good at insulating (solvating) charges. Nonpolar solvents have low dielectric constants and are poor insulators. The dielectric constants of some common solvents are listed in Table 9.7. In this table, solvents are divided into two groups: protic solvents and aprotic solvents. Recall that **protic solvents** contain a hydrogen that is bonded to an oxygen or a nitrogen; in other words, they are hydrogen bond donors. **Aprotic**

TABLE 9.7   The Dielectric Constants of Some Common Solvents

Solvent	Structure	Dielectric constant (ϵ, at 25 °C)
Protic solvents		
Water	H_2O	79
Formic acid	HCOOH	59
Methanol	CH_3OH	33
Ethanol	CH_3CH_2OH	25
tert-Butyl alcohol	$(CH_3)_3COH$	11
Acetic acid	CH_3COOH	6
Aprotic solvents		
Dimethyl sulfoxide	$(CH_3)_2SO$	47
Acetonitrile	CH_3CN	38
Dimethylformamide	$(CH_3)_2NCHO$	37
Acetone	$(CH_3)_2CO$	21
Ethyl acetate	$CH_3COOCH_2CH_3$	6
Diethyl ether	$CH_3CH_2OCH_2CH_3$	4.3
Benzene	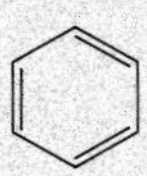	2.3
Carbon tetrachloride	CCl_4	2.2
Hexane	$CH_3(CH_2)_4CH_3$	1.9

solvents do not have a hydrogen bonded to an oxygen or a nitrogen; they are not hydrogen bond donors.

Stabilization of charges by solvent interaction plays an important role in organic reactions. For example, the first step in an S_N1 reaction is dissociation of the carbon–halogen bond of the alkyl halide to form a carbocation and a halide ion. Energy is required to break the bond; but, with no bonds being formed, where does this energy come from? If the reaction is carried out in a polar solvent, the ions that are produced as products are solvated. The energy associated with a single ion–dipole interaction is small, but the additive effect of all the ion–dipole interactions involved in stabilizing a charged species by the solvent represents a great deal of energy. These ion–dipole interactions provide much of the energy necessary for dissociation of the carbon–halogen bond. So, in an S_N1 reaction, the alkyl halide does not fall apart spontaneously; solvent molecules pull it apart. An S_N1 reaction, therefore, can take place in a polar solvent but not in a nonpolar solvent. This reaction also cannot take place in the gas phase, where there are no solvent molecules and consequently no solvation effects.

SOLVATION EFFECTS

The tremendous amount of energy that is provided by solvation can be appreciated by considering the energy required to break the crystal lattice of sodium chloride (Figure 1.1). In the absence of a solvent, sodium chloride must be heated to more than 800 °C to overcome the forces that hold the oppositely charged ions together. However, we are all aware of the fact that sodium chloride (table salt) readily dissolves in water at room temperature. Solvation of the Na^+ and Cl^- ions by water provides the energy necessary to pull the ions apart.

The Effect of the Solvent on the Rate of a Reaction

The rate of a reaction depends on the difference between the energy of the reactants and the energy of the transition state in the rate-limiting step of the reaction. We can therefore predict how changing the polarity of the solvent will affect the rate of a reaction. We do this simply by looking at the charge on the reactant(s) and the charge on the transition state of the rate-limiting step to determine which of these species will be more stabilized by a polar solvent. The greater the charge on the solvated molecule, the stronger it will interact with a polar solvent and the more the charge will be stabilized.

Therefore, if the charge on the reactants is greater than the charge on the rate-determining transition state, a polar solvent will interact more strongly with the reactants than with the transition state. As a result, a polar solvent will stabilize the reactants more than it will stabilize the transition state, increasing the difference in energy ($\Delta G^{\ddagger}$) between the reactants and the transition state. Therefore, increasing the polarity of the solvent will decrease the rate of the reaction (Figure 9.4).

In contrast, if the charge on the rate-determining transition state is greater than the charge on the reactants, a polar solvent will interact more strongly with the transition state than with the reactants, stabilizing the transition state more than it stabilizes the reactants. Therefore, increasing the polarity of the solvent will decrease the difference in energy ($\Delta G^{\ddagger}$) between the reactants and the transition state, which will increase the rate of the reaction (Figure 9.5).

Figure 9.4 ▶
Reaction coordinate diagram
for a reaction in which the
charge on the reactants is
greater than the charge on
the rate-determining transition state.

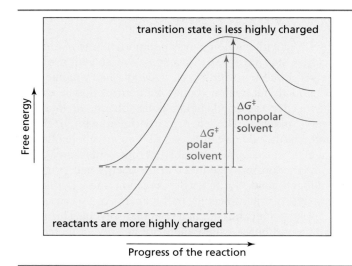

Figure 9.5 ▶
Reaction coordinate diagram
for a reaction in which
the charge on the rate-
determining transition state is
greater than the charge on
the reactants.

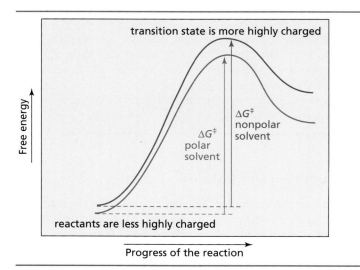

The Effect of the Solvent on the Rate of an S$_N$1 Reaction

Now let's see how increasing the polarity of the solvent affects the rate of an S$_N$1 reaction of an alkyl halide. The alkyl halide is the only reactant in the rate-determining step. It is a neutral molecule with a small dipole moment. The rate-determining transition state has a greater charge because, as the carbon–halogen bond breaks, the carbon becomes more positive and the halogen becomes more negative. Since the charge is greater in the rate-determining transition state, increasing the polarity of the solvent will increase the rate of the S$_N$1 reaction (Figure 9.5 and Table 9.8).

rate-determining step of an S$_N$1 reaction

TABLE 9.8   The Effect of the Polarity of the Solvent on the Rate of Reaction of *tert*-Butyl Bromide in an S$_N$1 Reaction

Solvent	Relative rate
100% water	1200
80% water / 20% ethanol	400
50% water / 50% ethanol	60
20% water / 80% ethanol	10
100% ethanol	1

In Chapter 11, we will see that compounds other than alkyl halides undergo S$_N$1 reactions. As long as the compound undergoing an S$_N$1 reaction is neutral, increasing the polarity of the solvent will increase the rate of the S$_N$1 reaction because the polar solvent will stabilize the dispersed charge on the transition state more than it will stabilize the relatively neutral reactant (Figure 9.5). If, however, the compound undergoing an S$_N$1 reaction is charged, increasing the polarity of the solvent will decrease the rate of the reaction because the more polar solvent will stabilize the full charge on the reactant to a greater extent than it will stabilize the dispersed charge on the transition state (Figure 9.4).

The Effect of the Solvent on the Rate of an S$_N$2 Reaction

The way in which an increase in the polarity of the solvent affects the rate of an S$_N$2 reaction depends on whether the reactants are charged or neutral, just as in an S$_N$1 reaction.

Most S$_N$2 reactions of alkyl halides involve a neutral alkyl halide and a charged nucleophile. Increasing the polarity of a solvent will have a strong stabilizing effect on the negatively charged nucleophile. The transition state also has a negative charge, but the charge is dispersed over two atoms. This means that the interactions between the solvent and the transition state are not as strong as the interactions between the solvent and the fully charged nucleophile. Consequently, a polar solvent stabilizes the nucleophile more than it stabilizes the transition state, so increasing the polarity of the solvent will decrease the rate of the reaction (Figure 9.4).

If, however, the S$_N$2 reaction involves an alkyl halide and a neutral nucleophile, the charge on the transition state will be larger than the charge on the neutral reactants, so increasing the polarity of the solvent will increase the rate of the substitution reaction (Figure 9.5).

We can now summarize the effect that changing the polarity of a solvent has on the rate of a substitution reaction—regardless of the mechanism of the reaction. *If the reactant(s) in the rate-limiting step is (are) charged, increasing the polarity of the solvent will decrease the rate of the reaction. If the reactant(s) is (are) not charged, increasing the polarity of the solvent will increase the rate of the reaction.*

In considering the solvation of charged species by a polar solvent, we have been discussing polar solvents, such as water or alcohols, that are hydrogen bond donors (polar protic solvents). There are also polar solvents (Table 9.7) such as *N,N*-dimethylformamide (DMF) and dimethyl sulfoxide (DMSO) that are not hydrogen bond donors (polar aprotic solvents).

CH$_3$ S CH$_3$:Ö:

CH$_3$:S=Ö: K$^+$:Ö=S: CH$_3$
CH$_3$ CH$_3$
Ö:
S
CH$_3$ CH$_3$

DMSO can solvate a cation better than it can solvate an anion.

$$
\begin{array}{cc}
\text{O CH}_3 & \text{O} \\
\| \ \ | & \| \\
\text{H—C—N—CH}_3 & \text{CH}_3\text{—S—CH}_3
\end{array}
$$

***N,N*-dimethylformamide**
DMF

dimethyl sulfoxide
DMSO

Ideally, one would like to carry out an S_N2 reaction with a negatively charged nucleophile in a nonpolar solvent, since the charge on the reactants is greater than the charge on the transition state. However, a negatively charged nucleophile will not dissolve in a nonpolar solvent. Therefore, a polar aprotic solvent is used. Polar aprotic solvents have atoms with nonbonding electrons that can stabilize positive charges; they are Lewis bases. But, because aprotic solvents are not hydrogen bond donors, they are less effective than polar protic solvents in stabilizing negative charges. Thus the rate of an S_N2 reaction involving a negatively charged nucleophile will be greater in a polar aprotic solvent than in a polar protic solvent. Thus, a polar aprotic solvent is the solvent of choice for an S_N2 reaction in which the nucleophile is negatively charged, whereas a polar protic solvent is used if the nucleophile is a neutral molecule.

We have now seen that when an alkyl halide can undergo both S_N2 and S_N1 reactions, the S_N2 reaction will be favored by a high concentration of a good (negatively charged) nucleophile in a polar aprotic solvent, whereas the S_N1 reaction will be favored by a poor (neutral) nucleophile in a polar protic solvent.

PROBLEM 22 ◆

Indicate whether each of the following solvents is protic or aprotic.

a. chloroform ($CHCl_3$)

b. diethyl ether ($CH_3CH_2OCH_2CH_3$)

c. acetic acid (CH_3COOH)

d. hexane [$CH_3(CH_2)_4CH_3$]

PROBLEM 23 ◆

How will the rate of each of the following S_N2 reactions change if the polarity of the solvent is increased?

a. $CH_3CH_2CH_2CH_2Br + HO^- \longrightarrow CH_3CH_2CH_2CH_2OH + Br^-$

b. $CH_3\overset{+}{S}CH_3 + CH_3O^- \longrightarrow CH_3OCH_3 + CH_3SCH_3$
$\ \ \ \ |$
$\ \ \ \ CH_3$

c. $CH_3CH_2I + NH_3 \longrightarrow CH_3CH_2\overset{+}{N}H_3 + I^-$

PROBLEM 24 ◆

Which reaction in each of the following pairs will take place more rapidly?

a. $CH_3Br + HO^- \longrightarrow CH_3OH + Br^-$

$CH_3Br + H_2O \longrightarrow CH_3OH + HBr$

b. $CH_3I + HO^- \longrightarrow CH_3OH + I^-$

$CH_3Cl + HO^- \longrightarrow CH_3OH + Cl^-$

c. $CH_3Br + NH_3 \longrightarrow CH_3\overset{+}{N}H_3 + Br^-$

$CH_3Br + H_2O \longrightarrow CH_3OH + HBr$

d. $CH_3Br + HO^- \xrightarrow{\text{DMSO}} CH_3OH + Br^-$

$CH_3Br + HO^- \xrightarrow{\text{EtOH}} CH_3OH + Br^-$

e. $CH_3Br + NH_3 \xrightarrow{Et_2O} CH_3\overset{+}{N}H_3 + Br^-$

$ CH_3Br + NH_3 \xrightarrow{EtOH} CH_3\overset{+}{N}H_3 + Br^-$

PROBLEM 25 / SOLVED

The pK_a values given throughout this text are values determined in water. How would the pK_a values of the following classes of compounds change if they were determined in a solvent less polar than water: carboxylic acids, alcohols, phenols, ammonium ions (RNH_3^+), and anilinium ions ($C_6H_5NH_3^+$)?

SOLUTION A pK_a is equal to the negative log of an equilibrium constant, K_a. Because we are determining how changing the polarity of a solvent affects an equilibrium constant, we must look at how changing the polarity of the solvent affects the stability of the reactants and products.

$$K_a = \frac{[B^-][H^+]}{[HB]} \quad \text{or} \quad K_a = \frac{[B][H^+]}{[HB^+]}$$

Carboxylic acids, alcohols, and phenols are neutral in their acidic forms (HB) and charged in their basic forms (B^-). A protic polar solvent will stabilize B^- and H^+ more than it stabilizes HB, thereby increasing K_a. Therefore, K_a will be larger in water than in a less polar solvent. This means that the pK_a values of carboxylic acids, alcohols, and phenols determined in a less polar solvent will be higher than those determined in water.

Ammonium ions and anilinium ions are charged in their acidic forms (HB^+) and neutral in their basic forms (B). A polar solvent will stabilize HB^+ and H^+ more than it will stabilize B. Because the amount of stabilization is slightly greater for HB^+ than for H^+, the pK_a values of ammonium ions and anilinium ions determined in a less polar solvent will be slightly lower than those determined in water.

PROBLEM 26◆

Would you expect acetate ion ($CH_3CO_2^-$) to be a more reactive nucleophile in an S_N2 reaction in methanol or in dimethyl sulfoxide?

PROBLEM 27◆

Under which of the following reaction conditions would (R)-2-chlorobutane form the most (R)-2-butanol: HO^- in 50% water/50% ethanol or HO^- in 100% ethanol?

If an organic chemist wanted to put a methyl group on a nucleophile, methyl iodide would most likely be the methylating agent used. Of the methyl halides, methyl iodide has the most easily displaced leaving group because I^- is the weakest base of the halide ions. The reaction would be a simple S_N2 reaction.

9.11 BIOLOGICAL METHYLATING REAGENTS

$$\overset{..}{Nu} + CH_3-I \longrightarrow CH_3-Nu + I^-$$

If, however, a methyl group were to be put on a nucleophile within a cell, methyl iodide could not be used. Alkyl halides are insoluble in water and therefore are not generally found in biological systems, which have predominantly aqueous environments. Instead, biological systems use S-adenosylmethionine (SAM) and N^5-methyltetrahydrofolate as methylating agents, which are soluble in water. Although they look much more complicated than methyl iodide, they perform the same function: they transfer a methyl group to a nucleophile. Notice that the methyl group in each of these methylating agents is attached to a positively charged atom. In other words, the methyl groups are attached to very good leaving groups; therefore, biological methylation can take place at a reasonable rate.

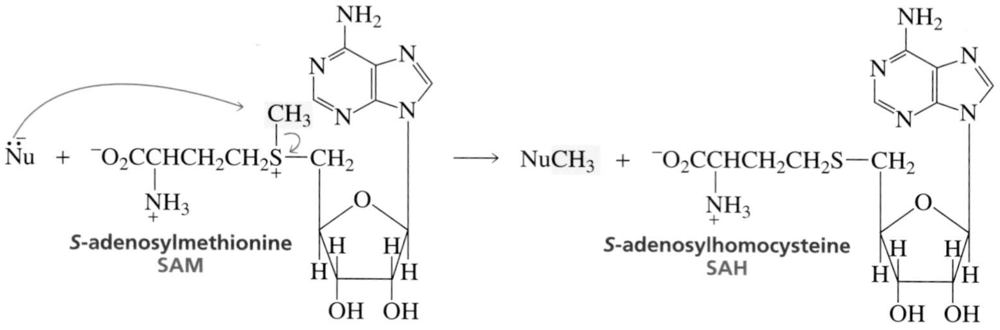

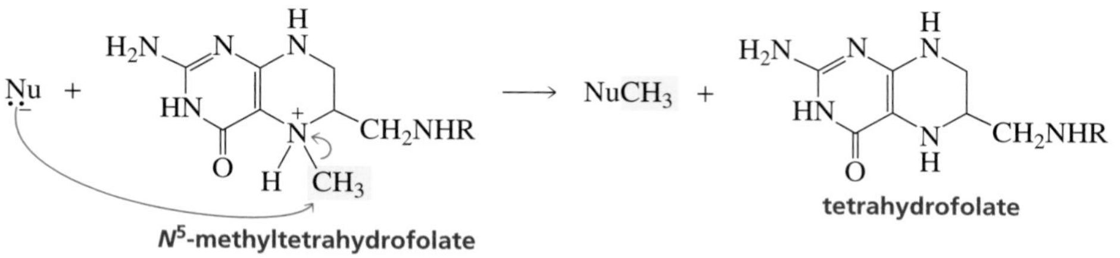

SAM is used in biological systems to convert norepinephrine (noradrenaline) into epinephrine (adrenaline). This is a simple methylation reaction. Norepinephrine and epinephrine are hormones that are released into the bloodstream in response to stress. Epinephrine is the more potent hormone.

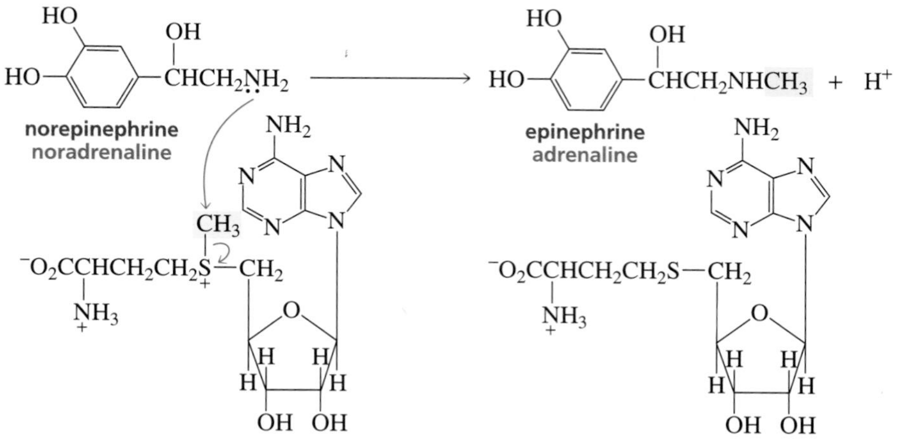

The conversion of the cell membrane phospholipid component phosphatidylethanolamine into phosphatidylcholine requires three methylations by three equivalents of SAM. Biological cell membranes will be discussed in Section 23.4. The use of N^5-methyltetrahydrofolate as a biological methylating agent is discussed in more detail in Section 22.8.

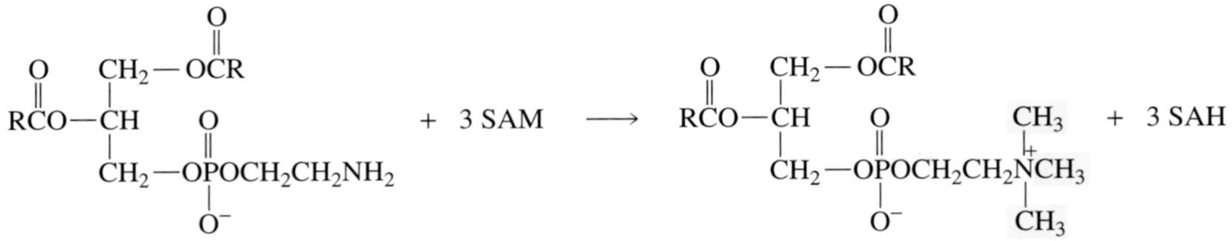

O
‖
O   CH₂—OCR
‖ |
RCO—CH   O
| ‖
CH₂—OPOCH₂CH₂NH₂
|
O⁻

phosphatidylethanolamine

+ 3 SAM ⟶

O
‖
O   CH₂—OCR
‖ |
RCO—CH   O   CH₃
| ‖   +|
CH₂—OPOCH₂CH₂NCH₃   + 3 SAH
| |
O⁻   CH₃

phosphatidylcholine

SUMMARY OF REACTIONS

1. S$_N$2 reaction: a one-step mechanism

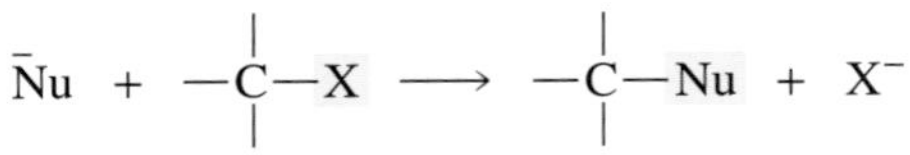

$$\bar{Nu} + -\overset{|}{\underset{|}{C}}-X \longrightarrow -\overset{|}{\underset{|}{C}}-Nu + X^-$$

Relative reactivities of alkyl halides: $CH_3X > 1° > 2° > 3°$.

Only the inverted product is formed.

2. S$_N$1 reaction: a two-step mechanism with a carbocation intermediate

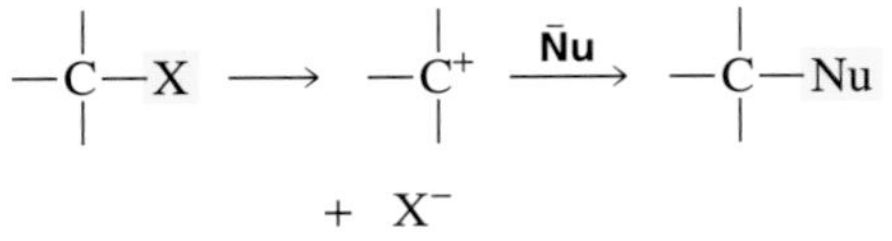

$$-\overset{|}{\underset{|}{C}}-X \longrightarrow -\overset{|}{\underset{|}{C}}{}^+ \xrightarrow{\bar{Nu}} -\overset{|}{\underset{|}{C}}-Nu$$
$$+ X^-$$

Relative reactivities of alkyl halides: $3° > 2° > 1° > CH_3X$.

Both the inverted and noninverted products are formed.

KEY TERMS

aprotic solvent (page 386)
back side attack (page 360)
base (page 365)
basicity (page 365)
bifunctional molecule (page 370)
bimolecular (page 360)
complete racemization (page 378)
dielectric constant (page 385)
direct displacement reaction
    (page 360)
elimination reaction (page 357)
first-order reaction (page 372)

intermolecular reaction (page 370)
intimate ion pair (page 378)
intramolecular reaction (page 370)
inversion of configuration (page 362)
ion–dipole interaction (page 366)
kinetics (page 359)
leaving group (page 357)
Le Châtelier's principle (page 369)
nucleophile (page 365)
nucleophilicity (page 365)
nucleophilic substitution reaction
    (page 359)

partial racemization (page 378)
protic solvent (page 386)
rate law (page 359)
second-order reaction (page 359)
S$_N$1 reaction (page 372)
S$_N$2 reaction (page 360)
solvent-separated ion pair (page 378)
solvolysis reaction (page 376)
steric effects (page 362)
steric hindrance (page 362)
substitution reaction (page 357)
unimolecular (page 372)

PROBLEMS

28. a. Indicate how each of the following factors affects an S$_N$1 reaction.
   b. Indicate how each of the following factors affects an S$_N$2 reaction.
       1. the structure of the alkyl halide   **3.** the concentration of the nucleophile
       2. the reactivity of the nucleophile   **4.** the solvent

29. Which is a better nucleophile in methanol?
a. H_2O or HO^- c. H_2O or H_2S e. I^- or Br^-
b. NH_3 or NH_2^- d. HO^- or HS^- f. Cl^- or Br^-

30. For each of the pairs in Problem 29, indicate which is a better leaving group.

31. What nucleophiles could be used to react with butyl bromide in order to prepare the following compounds?
a. $CH_3CH_2CH_2CH_2OH$
b. $CH_3CH_2CH_2CH_2OCH_3$
c. $CH_3CH_2CH_2CH_2SH$
d. $CH_3CH_2CH_2CH_2SCH_2CH_3$
e. $CH_3CH_2CH_2CH_2NHCH_3$
f. $CH_3CH_2CH_2CH_2C{\equiv}N$

g. $CH_3CH_2CH_2CH_2O\overset{\overset{\displaystyle O}{\|}}{C}CH_3$

h. $CH_3CH_2CH_2CH_2C{\equiv}CCH_3$

32. Rank the following compounds in order of *decreasing* nucleophilicity.

a. $CH_3\overset{\overset{\displaystyle O}{\|}}{C}O^-$, $CH_3CH_2S^-$, $CH_3CH_2O^-$ in methanol

b. 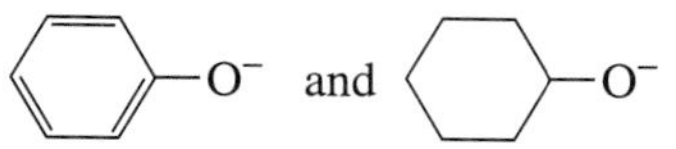 ⬡—O^- and ⬡—O^- in DMSO

c. H_2O, NH_3 in methanol
d. Br^-, Cl^-, I^- in methanol
e. Br^-, Cl^-, I^- in acetone

33. The pK_a of acetic acid in water is 4.76 (Section 1.17). What effect would a decrease in the polarity of the solvent have on the pK_a? Why?

34. Give the substitution products of the following reactions. If the products can exist as stereoisomers, show what stereoisomers are obtained.
a. (R)-2-bromohexane + high concentration of HO^-
b. (R)-2-bromohexane + H_2O
c. *trans*-1-chloro-2-methylcyclohexane + high concentration of CH_3O^-
d. *trans*-1-chloro-2-methylcyclohexane + CH_3OH
e. 3-bromo-2-methylpentane + H_2O
f. 3-bromo-3-methylpentane + H_2O
g. (2S,3S)-2-chloro-3-methylpentane + high concentration of CH_3O^-
h. (2S,3R)-2-chloro-3-methylpentane + high concentration of CH_3O^-
i. (2R,3S)-2-chloro-3-methylpentane + high concentration of CH_3O^-
j. (2R,3R)-2-chloro-3-methylpentane + high concentration of CH_3O^-
k. 3-chloro-2,2-dimethylpentane + CH_3CH_2OH
l. benzyl bromide + CH_3CH_2OH

35. Would you expect methoxide ion to be a better nucleophile if it were dissolved in CH_3OH or if it were dissolved in dimethyl sulfoxide (DMSO)? Why?

36. Which reaction in each of the following pairs will take place more rapidly?

a. 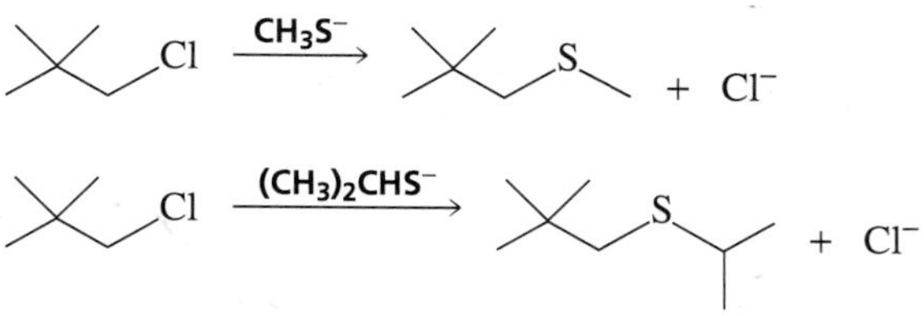 (structure) $\xrightarrow{CH_3S^-}$ (structure) $+ Cl^-$

(structure) $\xrightarrow{(CH_3)_2CHS^-}$ (structure) $+ Cl^-$

b. HO⁻ → OH + Cl⁻

HO⁻ → OH + Cl⁻

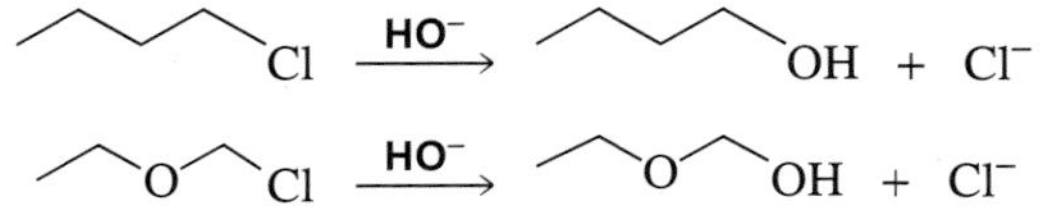

c. Cl $\xrightarrow{H_2O}$ OH + HCl

Cl $\xrightarrow{H_2O}$ OH + HCl

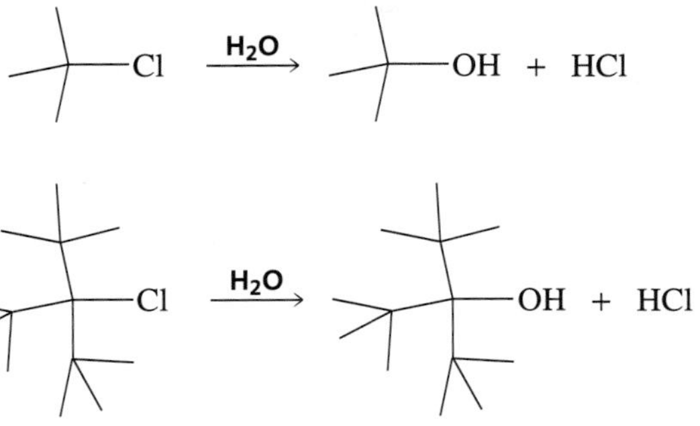

d. $(CH_3)_3CBr \xrightarrow{H_2O} (CH_3)_3COH + HBr$

$(CH_3)_3CBr \xrightarrow{CH_3CH_2OH} (CH_3)_3COCH_2CH_3 + HBr$

37. Which of the following compounds would you expect to be more reactive in an S_N2 reaction?

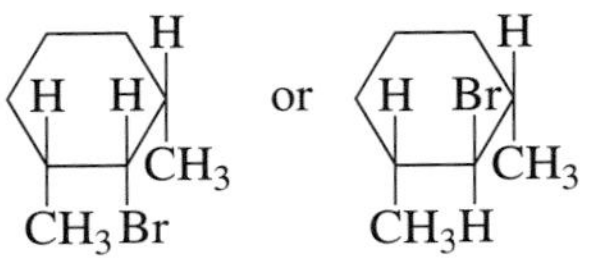

or

38. We saw that *S*-adenosylmethionine (SAM) methylates the nitrogen atom of noradrenaline to form adrenaline, a more potent hormone (Section 9.11). If SAM methylates an OH group on the benzene ring instead, it completely destroys noradrenaline's activity. Give the mechanism for the methylation of the OH group by SAM.

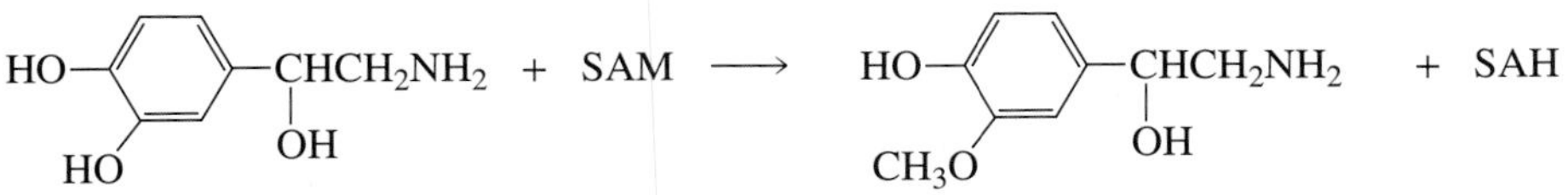

norepinephrine + SAM ⟶ **a biologically inactive compound** + SAH

39. 3-Bromo-3-methyl-1-butene forms two substitution products when it is added to a solution of sodium acetate in acetic acid.
   a. Give the structures of the substitution products.
   b. Which is the kinetically controlled product?
   c. Which is the thermodynamically controlled product?

40. The rate of reaction of methyl iodide with quinuclidine was measured in nitrobenzene, and then the rate of reaction of methyl iodide with triethylamine was measured in the same solvent.
   a. Which reaction had the larger rate constant?
   b. The same experiment was done using isopropyl iodide instead of methyl iodide. Which reaction had the larger rate constant?
   c. Which alkyl halide has the larger $k_{quinuclidine}/k_{triethylamine}$ ratio?

quinuclidine

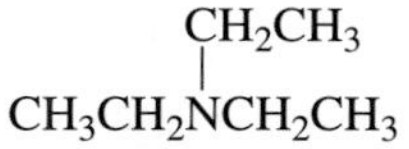

$$CH_2CH_3$$
$$|$$
$$CH_3CH_2NCH_2CH_3$$

triethylamine

41. Starting with cyclohexane, how could the following compounds be prepared?
  a. cyclohexyl bromide
  b. methoxycyclohexane
  c. cyclohexanol

42. Give the substitution products of the following reactions, assuming that all the reactions are carried out under S_N2 conditions. If the products can exist as stereoisomers, show what stereoisomers are formed.
  a. (3S,4S)-3-bromo-4-methylhexane + CH_3O^-
  b. (3S,4R)-3-bromo-4-methylhexane + CH_3O^-
  c. (3R,4R)-3-bromo-4-methylhexane + CH_3O^-
  d. (3R,4S)-3-bromo-4-methylhexane + CH_3O^-

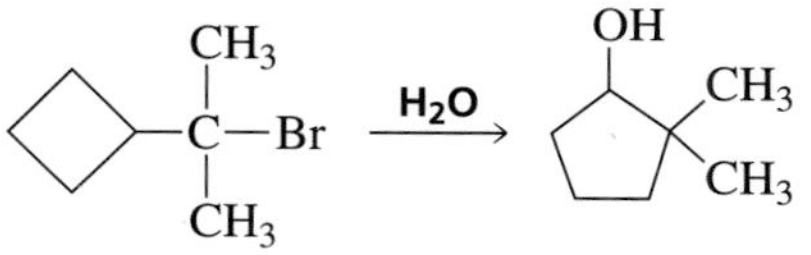

43. Tetrahydrofuran can solvate a positively charged species better than can diethyl ether. Explain.

tetrahydrofuran

$$CH_3CH_2OCH_2CH_3$$

diethyl ether

44. Propose a mechanism for each of the following reactions.

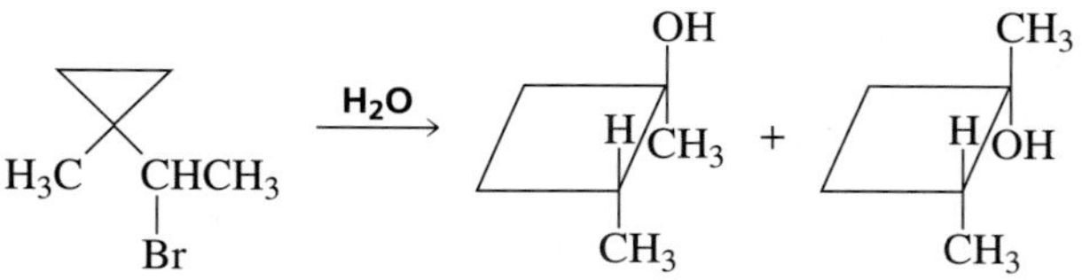

45. Which of the following will react faster in an S_N1 reaction?

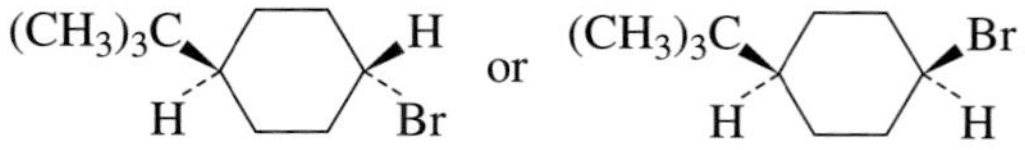

46. The alkyl halide shown below does not undergo a substitution reaction regardless of the conditions under which the reaction is run. Explain.

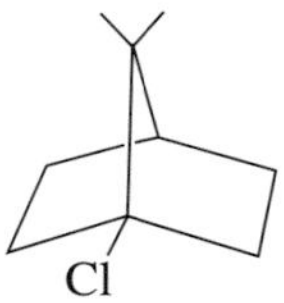

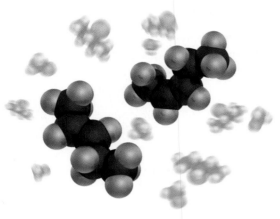

3-chloropentane + methoxide ion

↓

(*E*)-2-pentene + (*Z*)-2-pentene

10

REACTIONS AT AN sp³ HYBRIDIZED CARBON II:
Elimination Reactions of Alkyl Halides; Competition Between Substitution and Elimination

In Chapter 9 you learned that a compound with an electronegative atom or group attached to an sp^3 hybridized carbon can undergo a substitution reaction. Such compounds can also undergo elimination reactions. In an **elimination reaction,** groups are eliminated from the reactant: the electronegative atom or group is removed from the carbon to which it is attached, and a proton is removed from an adjacent carbon. As a result, a double bond is formed between the two carbons from which the atoms or groups are eliminated. Thus the product of an elimination reaction is an alkene.

$$CH_3CH_2CH_2X \ + \ Y^-$$

substitution → $CH_3CH_2CH_2Y \ + \ X^-$

elimination → $CH_3CH{=}CH_2 \ + \ HY \ + \ X^-$

In this chapter we will discuss elimination reactions of alkyl halides, compounds in which the electronegative atom (X) that is eliminated is a halide ion. We will also examine the factors that determine whether a given alkyl halide will undergo a substitution reaction, an elimination reaction, or both substitution and elimination reactions. In the next chapter, we will look at substitution and elimination reactions of compounds with leaving groups other than halide ions.

10.1 THE E2 REACTION

There are two important mechanisms for an elimination reaction. The reaction of ethyl bromide with hydroxide ion is an example of an **E2 reaction:** E for elimination and 2 for bimolecular. It is a second-order reaction because the rate of the reaction depends on the concentrations of both ethyl bromide and hydroxide ion.

$$\underset{\textbf{ethyl bromide}}{CH_3CH_2Br} + HO^- \longrightarrow \underset{\textbf{ethene}}{CH_2{=}CH_2} + H_2O + Br^-$$

rate = *k*[alkyl halide][base]

The rate law tells us that ethyl bromide and hydroxide ion are both involved in the transition state of the rate-determining step of the reaction. The following mechanism agrees with the observed second-order kinetics.

mechanism of the E2 reaction

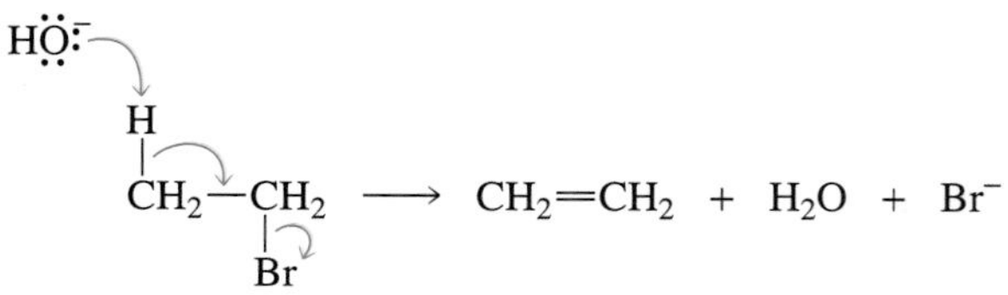

The E2 reaction is a concerted, one-step reaction. The proton and the bromide ion are removed in the same step.

In an E2 reaction, a base removes a proton from a carbon adjacent to the carbon bonded to the halogen. As the proton is removed, the electrons the hydrogen shared with carbon move toward the carbon bonded to the halogen. As these electrons move in, the halogen leaves, taking its bonding electrons with it. The electrons that were bonded to the hydrogen in the reactant have formed the π bond of the double bond in the product. Removal of a proton and a halide ion is called **dehydrohalogenation.**

The carbon to which the halogen is attached is called the α-carbon. An adjacent carbon is called a β-carbon. Because the elimination reaction is initiated by removing a proton from a β-carbon, an E2 reaction is sometimes called a **β-elimination reaction.** It is also called a **1,2-elimination reaction** because the atoms being removed are on adjacent carbons. (B⁚ is any base.)

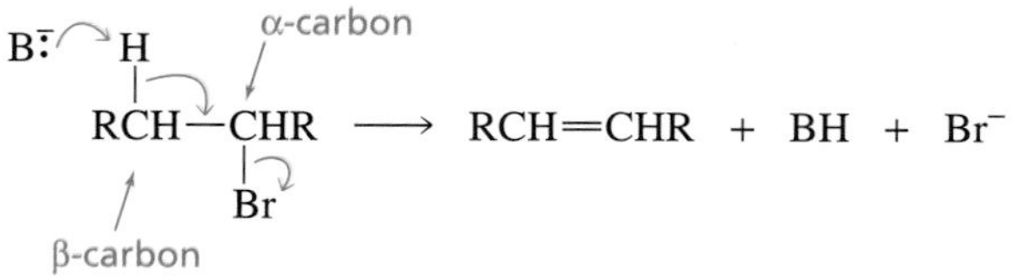

2-Bromopropane has two β-carbons from which a proton can be removed in an E2 reaction. Because the two β-carbons are identical, the proton can be removed from either one. The product of this elimination reaction is propene.

$$\underset{\textbf{2-bromopropane}}{CH_3CH(Br)CH_3} + CH_3O^- \longrightarrow \underset{\textbf{propene}}{CH_3CH{=}CH_2} + CH_3OH + Br^-$$

2-Bromobutane has two structurally different β-carbons from which a proton can be removed. So when 2-bromobutane reacts with a base, two elimination products are formed, 2-butene and 1-butene.

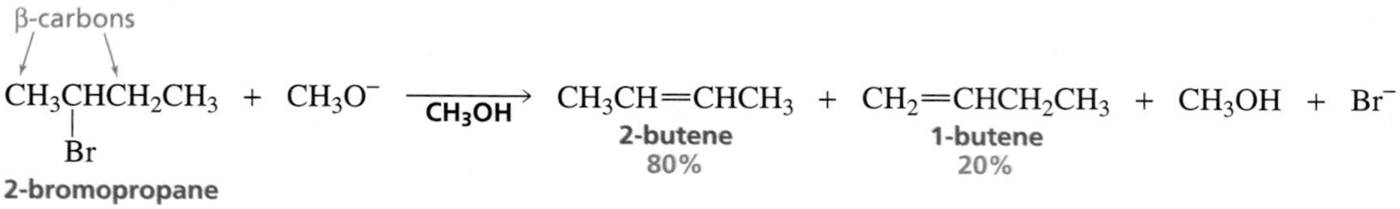

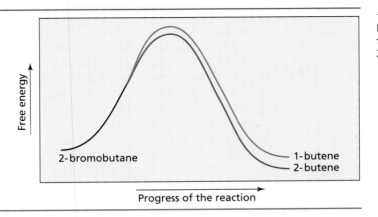

◀ **Figure 10.1**
Reaction coordinate diagram
for the E2 reaction of
2-bromobutane.

How do you know which product will be formed in greater yield? You must determine which of the products is formed more easily—that is, which product is formed faster. The reaction coordinate diagram for the reaction is shown in Figure 10.1.

You know that alkene stability depends on the number of alkyl substituents bonded to the sp^2 carbons; the greater that number, the more stable the alkene (Section 3.19). For example, 2-butene, with a total of two methyl substituents bonded to its sp^2 carbons, is more stable than 1-butene, with one ethyl substituent. In the transition state leading to an alkene, the C—H and C—Br bonds are partially broken and the double bond is partially formed, giving the transition state an "alkene-like" structure. Because the transition state has an alkene-like structure, the transition state leading to 2-butene is more stable than the transition state leading to 1-butene for the same reason that 2-butene is more stable than 1-butene. The more stable transition state allows 2-butene to be formed faster than 1-butene.

$$\overset{\delta^-}{OCH_3}$$
$$H$$
$$CH_3CH{=}{=}{=}CHCH_3$$
$$\underset{\delta^-}{Br}$$

**transition state leading to
2-butene**
more stable

$$\overset{\delta^-}{OCH_3}$$
$$H$$
$$CH_2{=}{=}{=}CHCH_2CH_3$$
$$\underset{\delta^-}{Br}$$

**transition state leading to
1-butene**
less stable

The difference in the rate of formation of the two alkenes is not very great (Figure 10.1). Consequently, both products are formed, but the more stable alkene is the major product of the elimination reaction. Thus an E2 reaction is *regioselective;* more of one constitutional isomer is formed than the other.

The reaction of 2-bromo-2-methylbutane with hydroxide ion produces both 2-methyl-2-butene and 2-methyl-1-butene. Because 2-methyl-2-butene has a greater number of alkyl substituents bonded to its sp^2 carbons, it is the more stable alkene and is the major product of the elimination reaction.

$$\underset{\underset{\displaystyle Br}{|}}{\overset{\overset{\displaystyle CH_3}{|}}{CH_3CCH_2CH_3}} \ + \ HO^- \ \xrightarrow{H_2O} \ \overset{\overset{\displaystyle CH_3}{|}}{CH_3C}{=}CHCH_3 \ + \ \overset{\overset{\displaystyle CH_3}{|}}{CH_2}{=}CCH_2CH_3 \ + \ H_2O \ + \ Br^-$$

2-bromo-2-methylbutane

2-methyl-2-butene
70%

2-methyl-1-butene
30%

10.2 ZAITSEV'S RULE

Alexander M. Zaitsev (1841–1910) *was born in Kazan, Russia. The German transliteration, Saytzeff, is frequently used for his family name. He studied at the University of Paris and became a professor of chemistry at the University of Kazan.*

Alexander M. Zaitsev, a nineteenth-century Russian chemist, devised a shortcut to predict the more substituted alkene product. He pointed out that *the more substituted alkene product is obtained when a proton is removed from the β-carbon that is bonded to the fewest hydrogens.* This is called Zaitsev's rule. 2-Bromo-2-methylbutane, the reactant in the preceding reaction, has three hydrogens bonded to both the β-carbon to the left and the β-carbon on top of the carbon bonded to the halogen, and it has two hydrogens on the β-carbon to the right. **Zaitsev's rule** says that the more substituted alkene will be the one formed by removing a proton from the β-carbon bonded to two hydrogens rather than from one of the β-carbons bonded to three hydrogens.

Alkyl halides have the following relative reactivities in an E2 reaction, because elimination from a tertiary alkyl halide typically leads to a more highly substituted alkene than elimination from a secondary alkyl halide, and elimination from a secondary alkyl halide leads to a more highly substituted alkene than elimination from a primary alkyl halide.

relative reactivities of alkyl halides in an E2 reaction

a tertiary alkyl halide > a secondary alkyl halide > a primary alkyl halide

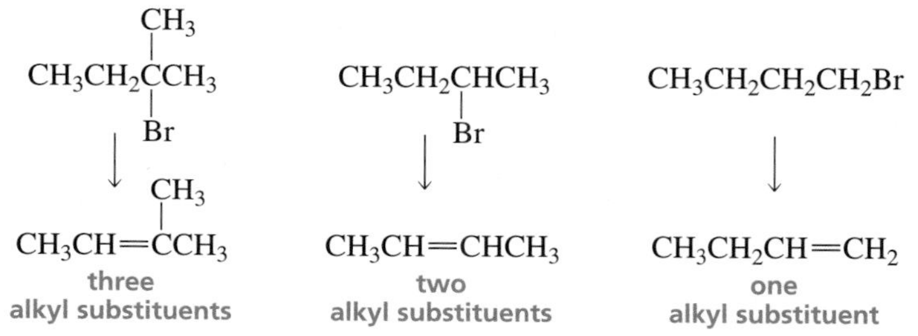

In most elimination reactions, the most substituted alkene is the one that is the easiest to form because it is the most stable of the possible alkene products. Zaitsev's rule is a shortcut to determine which of the possible alkene products is the most substituted alkene. But you must exercise some care in using Zaitsev's rule to predict the major product of an elimination reaction because the most substituted alkene is not always the one that is easiest to form.

For example, if the base in an E2 reaction is sterically bulky, it will preferentially remove the most accessible hydrogen. In the following reaction, it is easier for the bulky *tert*-butoxide ion to remove one of the more exposed terminal hydrogens, which leads to the less substituted alkene. Because the less substituted alkene is the alkene that is the more easily formed, it is the major product of the reaction.

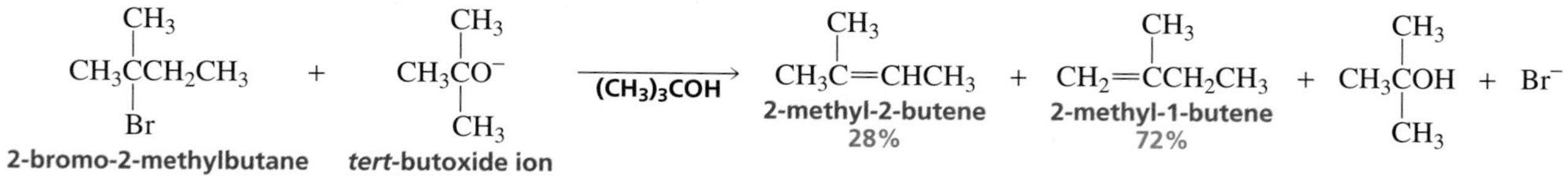

The data in Table 10.1 show that when an alkyl halide undergoes an E2 reaction with a variety of alkoxide ions, as the size of the base increases, the percent of the less substituted alkene increases.

TABLE 10.1   Effect of the Steric Properties of the Base on the Distribution of Products in an E2 Reaction

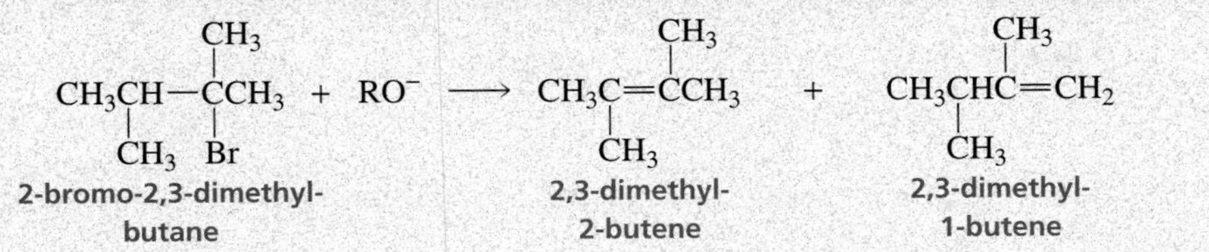

$$CH_3CH-\underset{\underset{Br}{|}}{\overset{\overset{CH_3}{|}}{C}}CH_3 + RO^- \longrightarrow CH_3C=\underset{\underset{CH_3}{|}}{\overset{\overset{CH_3}{|}}{C}}CH_3 + CH_3CH\underset{\underset{CH_3}{|}}{C}=CH_2$$

2-bromo-2,3-dimethyl-butane

2,3-dimethyl-2-butene

2,3-dimethyl-1-butene

Base	More substituted product	Less substituted product
$CH_3CH_2O^-$	79%	21%
$CH_3\underset{\underset{CH_3}{|}}{\overset{\overset{CH_3}{|}}{C}}O^-$	27%	73%
$CH_3\underset{\underset{CH_2CH_3}{|}}{\overset{\overset{CH_3}{|}}{C}}O^-$	19%	81%
$CH_3CH_2\underset{\underset{CH_2CH_3}{|}}{\overset{\overset{CH_2CH_3}{|}}{C}}O^-$	8%	92%

PROBLEM 1

Draw a reaction coordinate diagram for the E2 reaction of 2-bromo-2,3-dimethylbutane with sodium *tert*-butoxide.

Although the major product of the E2 dehydrohalogenation of alkyl chlorides, alkyl bromides, and alkyl iodides is normally the most substituted alkene, the major product of the E2 dehydrohalogenation of alkyl fluorides is the least substituted alkene (Table 10.2).

$$CH_3\underset{\underset{F}{|}}{CH}CH_2CH_2CH_3 + CH_3O^- \xrightarrow{CH_3OH} CH_2=CHCH_2CH_2CH_3 + CH_3CH=CHCH_2CH_3 + CH_3OH + F^-$$

2-fluoropentane   methoxide ion   1-pentene 70%   2-pentene 30%

When a hydrogen and a chlorine, bromine, or iodine are eliminated from an alkyl halide, you have seen that the halogen starts to leave as soon as the base begins to remove the proton. Departure of the halogen and its bonding electrons prevents the buildup of negative charge on the carbon that is losing a proton, giving the transition state alkene character rather than carbanion character (Section 10.1). Of the halogen ions, the fluoride ion is the strongest base and therefore the poorest

TABLE 10.2	Products Obtained from the E2 Reaction of CH_3O^- and 2-Halohexanes			
			More substituted product	Less substituted product

$$CH_3\overset{\overset{\textstyle X}{\textstyle |}}{C}HCH_2CH_2CH_2CH_3 + CH_3O^- \longrightarrow CH_3CH=CHCH_2CH_2CH_3 + CH_2=CHCH_2CH_2CH_2CH_3$$

2-hexene   1-hexene

Leaving group	Conjugate acid	pK_a		
X = I	HI	−10	81%	19%
X = Br	HBr	−9	72%	28%
X = Cl	HCl	−7	67%	33%
X = F	HF	3.2	30%	70%

leaving group. So when a base begins to remove a proton from an alkyl fluoride, the fluoride ion has less tendency to leave than do the other halide ions. As a result, negative charge develops on the carbon that is losing the proton, giving the transition state a carbanion character rather than an alkene character. Normally, carbanion transition states are unstable, but in this case the carbanion transition state is stabilized by the strongly electron-withdrawing fluorine.

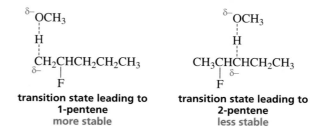

Positively charged carbocations are stabilized by electron-donating groups. Because alkyl groups are more electron donating than is hydrogen, *the more alkyl groups* that are bonded to a positively charged carbon, the more stable is the *carbocation* (Section 3.10).

relative stabilities of carbocations

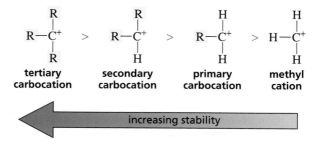

Negatively charged carbanions are stabilized by electron-withdrawing groups. Therefore, *the fewer alkyl groups* that are bonded to a negatively charged carbon, the more stable is the *carbanion*.

relative stabilities of carbanions

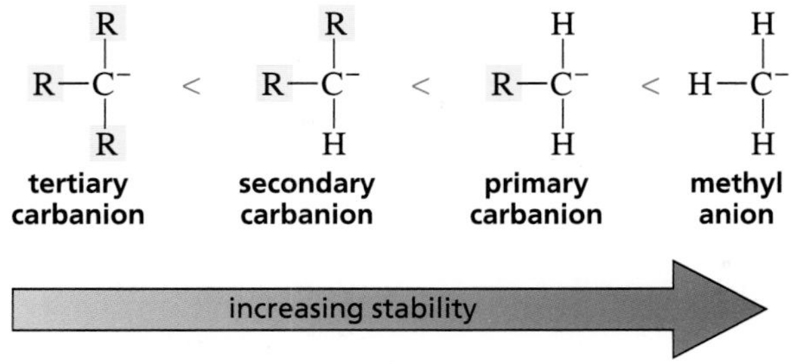

The transition state leading to 1-pentene has the developing negative charge on a primary carbon. This is more stable than the transition state leading to 2-pentene, which has the developing negative charge on a secondary carbon. Because the transition state leading to 1-pentene is more stable, 1-pentene is formed more rapidly and is the major product of the elimination reaction.

The data in Table 10.2 show that, as the halide ion increases in basicity (decreases in leaving ability), there is a decrease in the amount of the more substituted alkene product. However, the more substituted alkene remains the major elimination product in all cases except when the halogen is fluorine.

In the following reactions, Zaitsev's rule does not lead to the more stable alkene because the rule does not take into account the fact that conjugated double bonds are more stable than isolated double bonds (Section 7.3). Since the conjugated alkene is the more stable alkene, it is the one that is most easily formed and is, therefore, the major product of the reaction. So, if there is a double bond or a benzene ring in the alkyl halide, do not use Zaitsev's rule to predict the major product of the elimination reaction.

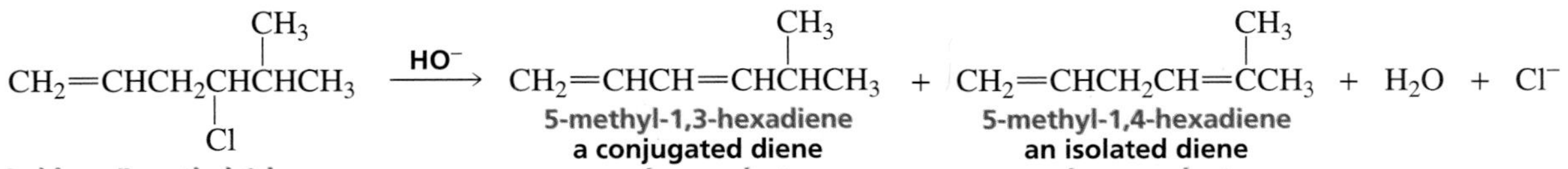

4-chloro-5-methyl-1-hexene → **5-methyl-1,3-hexadiene, a conjugated diene** (major product) + **5-methyl-1,4-hexadiene, an isolated diene** (minor product)

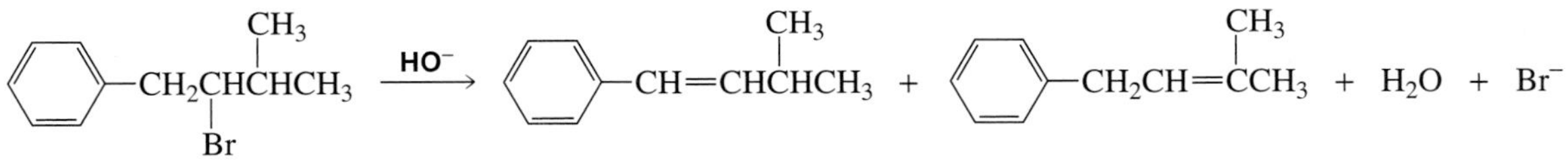

2-bromo-3-methyl-1-phenylbutane → **3-methyl-1-phenyl-1-butene, the double bond is conjugated with the benzene ring** (major product) + **3-methyl-1-phenyl-2-butene, the double bond is not conjugated with the benzene ring** (minor product)

We can summarize by saying that the major product of an E2 elimination reaction is the most substituted alkene *unless one of the following applies:*

- the base is large;
- the alkyl halide is an alkyl fluoride;
- the alkyl halide contains one or more double bonds.

Zaitsev's rule cannot be used to predict the major product of an elimination reaction in these situations. In Section 10.6 you will see that Zaitsev's rule also cannot be used with certain cyclic compounds.

PROBLEM 2 ◆

Give the major elimination product obtained from an E2 reaction of each of the following alkyl halides with hydroxide ion.

a. $CH_3CHCH_2CH_3$
  |
  Cl

b. $CH_3CHCH_2CH_3$
  |
  F

c. $\overset{\displaystyle CH_3}{\underset{}{CH_3CHCHCH_2CH_3}}$
            |
            Cl

d. $\overset{\displaystyle CH_3}{CH_3CHCHCH_2CH_3}$
            |
            F

e. 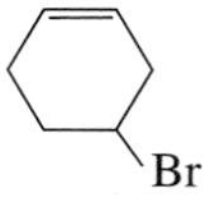 (cyclohexene with Br)

f. $CH_3CHCH_2CH=CH_2$
        |
        Cl

PROBLEM 3 ◆

Which alkyl halide would you expect to be more reactive in an E2 reaction?

a. $CH_3CH_2CH_2CHCH_3$ or $CH_3CH_2CHCH_2CH_3$
                | |
                Br Br

b. 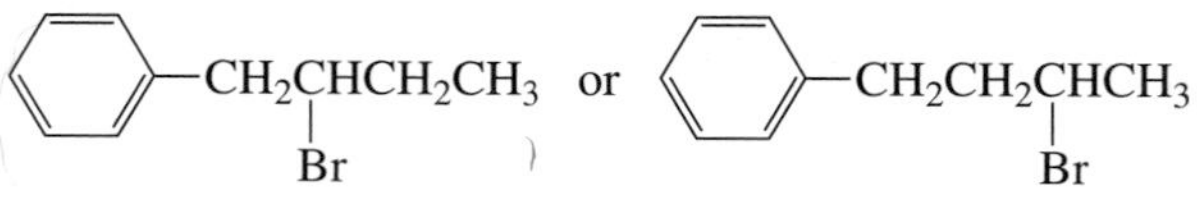 ⟨benzene⟩$-CH_2CHCH_2CH_3$ or ⟨benzene⟩$-CH_2CH_2CHCH_3$
              | |
              Br Br

c. $\overset{\displaystyle CH_3}{CH_3CHCHCH_2CH_3}$ or $\overset{\displaystyle CH_3}{CH_3CHCH_2CHCH_3}$
              | |
              Br Br

d. 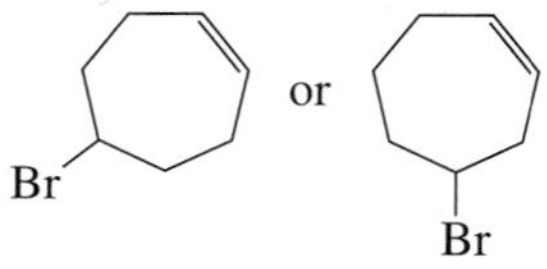 or

---

The reaction of *tert*-butyl bromide with water is a first-order elimination reaction because the rate of the reaction depends only on the concentration of the alkyl halide. It is called an **E1 reaction:** E for elimination, and 1 for unimolecular.

$$CH_3-\underset{\underset{CH_3}{|}}{\overset{\overset{CH_3}{|}}{C}}-Br \ + \ H_2O \ \longrightarrow \ CH_3-\underset{}{\overset{\overset{CH_3}{|}}{C}}=CH_2 \ + \ H_3O^+ \ + \ Br^-$$

tert-butyl bromide   methylpropene

rate = k[alkyl halide]

We know, then, that only the alkyl halide is involved in the transition state of the rate-determining step of the reaction. Therefore, there must be at least two steps in the reaction. The following mechanism agrees with the observed first-order kinetics.

An E1 reaction is a two-step reaction. In the first step, the alkyl halide dissociates heterolytically. This is the rate-determining step of the reaction. In the second step of the reaction, the base forms an elimination product by removing a proton from a carbon adjacent to the positively charged carbon. Because the first step of the reaction is the rate-determining step, increasing the concentration of the base—which comes into play only in the second step of the reaction—has no effect on the rate of the reaction.

mechanism of the E1 reaction

$$CH_3-\underset{\underset{CH_3}{|}}{\overset{\overset{CH_3}{|}}{C}}-Br \underset{slow}{\rightleftharpoons} CH_3-\underset{\underset{CH_3}{|}}{\overset{\overset{CH_3}{|}}{C}}{}^+ + Br^-$$

$$\underset{H_2\ddot{O}:\quad H}{CH_3-\underset{\underset{CH_2}{|}}{\overset{\overset{CH_3}{|}}{C}}{}^+} \xrightarrow{fast} CH_3-\underset{\underset{CH_2}{\|}}{\overset{\overset{CH_3}{|}}{C}} + H_3O^+$$

When two elimination products can be formed, the major product is generally the one obtained by following Zaitsev's rule (the hydrogen is removed from the β-carbon bonded to the fewest hydrogens).

$$\underset{\text{2-chloro-2-methylbutane}}{CH_3CH_2\underset{\underset{Cl}{|}}{\overset{\overset{CH_3}{|}}{C}}CH_3} + H_2O \longrightarrow \underset{\substack{\text{2-methyl-2-butene}\\\text{major product}}}{CH_3CH=\overset{\overset{CH_3}{|}}{C}CH_3} + \underset{\substack{\text{2-methyl-1-butene}\\\text{minor product}}}{CH_3CH_2\overset{\overset{CH_3}{|}}{C}=CH_2} + H_3O^+ + Cl^-$$

You can see why this is so by looking at the reaction coordinate diagram in Figure 10.2. Zaitsev's rule leads to the alkene with the more stable "alkene-like" transition state. Because it has the more stable transition state, it is formed more rapidly.

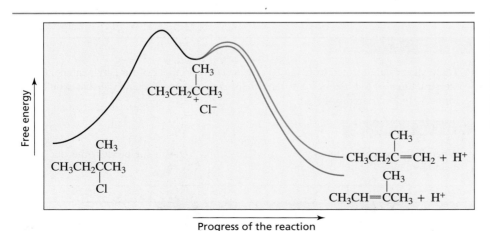

◀ **Figure 10.2**
Reaction coordinate diagram for an E1 reaction of 2-chloro-2-methylbutane. The major product is the more substituted alkene, because its greater stability causes the transition state leading to its formation to be more stable.

Because the first step is the rate-determining step, the rate of an E1 reaction depends on both the ease with which the leaving group leaves and the stability of the carbocation that is formed. Thus the relative reactivities of a series of alkyl halides with the same leaving group parallel the relative stabilities of carbocations. A tertiary benzylic alkyl halide is the most reactive alkyl halide because a tertiary benzylic cation—the most stable carbocation—is the easiest to form (Sections 6.7 and 9.8).

relative reactivities of alkyl halides in an E1 reaction = relative stabilities of carbocations

3° benzylic ≈ 3° allylic > 2° benzylic ≈ 2° allylic ≈ 3° > 1° benzylic ≈ 1° allylic ≈ 2° > 1° > vinyl

⟵ increasing reactivity and stability

Because the E1 reaction involves the formation of a carbocation intermediate, rearrangement of the carbon skeleton can occur before the proton is lost. For example, the secondary carbocation that is formed when a chloride ion dissociates from 3-chloro-2-methyl-2-phenylbutane undergoes a 1,2-methyl shift to form a more stable tertiary benzylic cation.

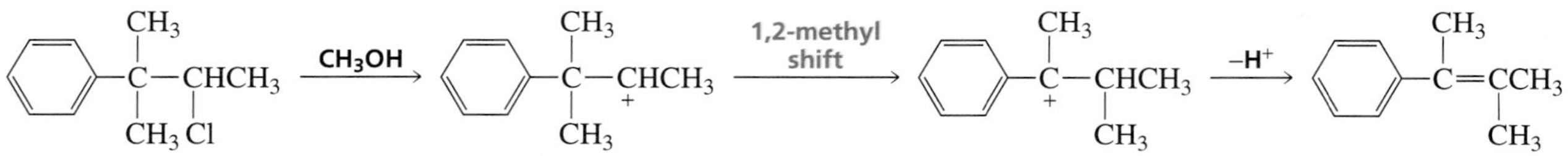

3-chloro-2-methyl-2-phenylbutane **secondary carbocation** **tertiary benzylic cation** **2-methyl-3-phenyl-2-butene**

In the following example, the secondary carbocation undergoes a 1,2-hydride shift to form a more stable secondary allylic cation.

$CH_3CH{=}CHCH_2CH(Br)CH_2CH_3$ (**5-bromo-2-heptene**) $\xrightleftharpoons{CH_3OH}$ $CH_3CH{=}CHCH_2\overset{+}{C}HCH_2CH_3$ (**secondary carbocation**)

↓ 1,2-hydride shift

$CH_3CH{=}CH\overset{+}{C}HCH_2CH_2CH_3$ (**secondary allylic cation**) $\xrightarrow{-H^+}$ $CH_3CH{=}CHCH{=}CHCH_2CH_3$ (**2,4-heptadiene**)

PROBLEM 4◆

Three alkenes are formed from the E1 reaction of 3-bromo-2,3-dimethylpentane. Give the structures of the alkenes and rank them according to the amount that would be formed. (Ignore stereoisomers.)

PROBLEM 5

When 2-fluoropentane undergoes an E1 reaction, is the major product the one predicted by Zaitsev's rule? Explain.

PROBLEM-SOLVING STRATEGY

Propose a mechanism for the following reaction.

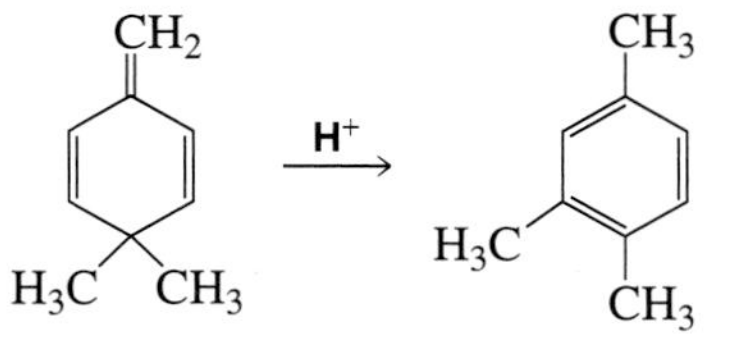

Because the given reagent is a proton, start by protonating the molecule at the position that forms the most stable carbocation; by protonating the CH_2 group, a tertiary allylic carbocation is formed in which the positive charge is delocalized over two other carbons. Then move the π electrons so that the 1,2-methyl shift required to obtain the product can take place. Loss of a proton gives the final product.

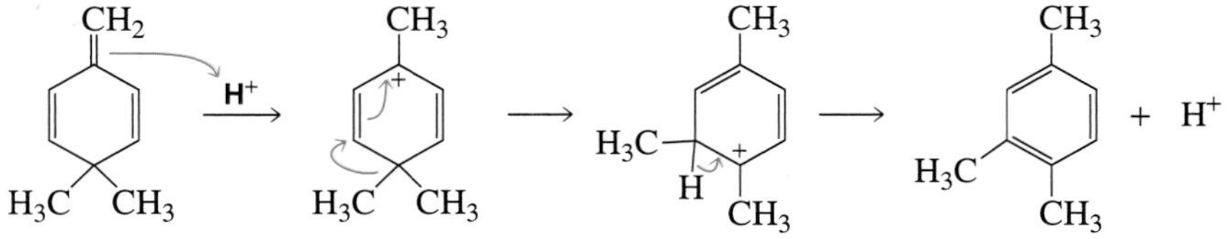

Now continue on to Problem 6.

PROBLEM 6

Propose a mechanism for the following reaction.

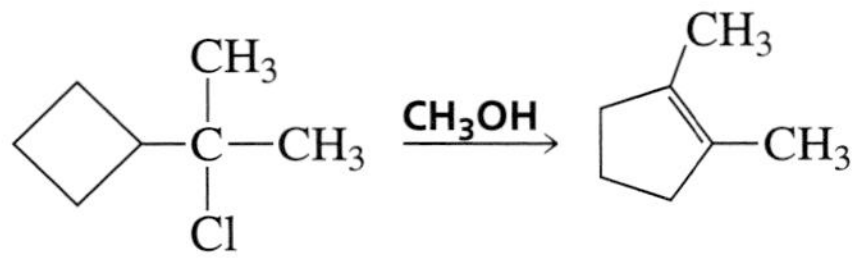

Primary alkyl halides undergo only E2 elimination reactions. They cannot undergo E1 reactions because of the difficulty encountered in forming primary carbocations. *Secondary* and *tertiary* alkyl halides undergo both E2 and E1 reactions (Table 10.3).

For those alkyl halides that can undergo both E2 and E1 reactions, the E2 reaction is favored by the same factors that favor an S_N2 reaction, and the E1 reaction is favored by the same factors that favor an S_N1 reaction. *An E2 reaction is favored by a high concentration of a strong base and an aprotic polar solvent (DMSO, DMF). An E1 reaction is favored by a weak base and a protic polar solvent (H_2O, ROH).* How the solvent affects the mechanism of the reaction was discussed in Section 9.10.

10.4 COMPETITION BETWEEN E2 AND E1 REACTIONS

TABLE 10.3   Summary of the Reactivity of Alkyl Halides in Elimination Reactions

primary alkyl halide	E2 only
secondary alkyl halide	E1 and E2
tertiary alkyl halide	E1 and E2

PROBLEM 7 ◆

Predict whether each of the following reactions is an E2 or an E1 reaction. Give the major product of each reaction.

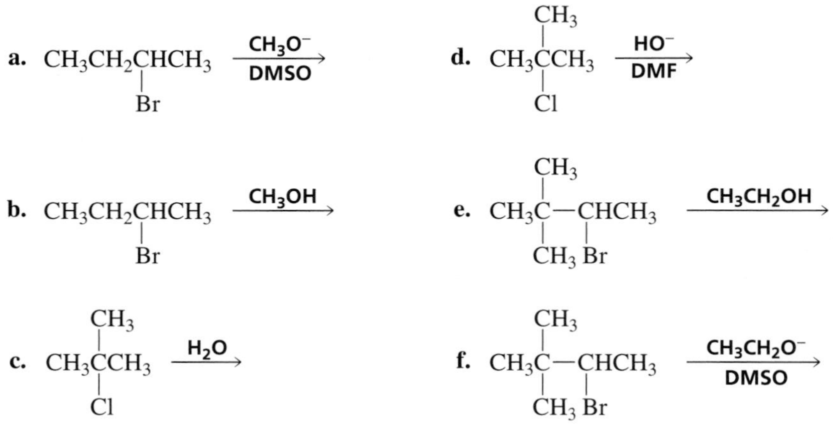

a. $CH_3CH_2CHCH_3$ (with Br), $\xrightarrow[\text{DMSO}]{CH_3O^-}$

b. $CH_3CH_2CHCH_3$ (with Br), $\xrightarrow{CH_3OH}$

c. CH_3CCH_3 (with CH₃ and Cl), $\xrightarrow{H_2O}$

d. CH_3CCH_3 (with CH₃ and Cl), $\xrightarrow[\text{DMF}]{HO^-}$

e. $CH_3C-CHCH_3$ (with CH₃, CH₃, Br), $\xrightarrow{CH_3CH_2OH}$

f. $CH_3C-CHCH_3$ (with CH₃, CH₃, Br), $\xrightarrow[\text{DMSO}]{CH_3CH_2O^-}$

PROBLEM 8 ◆

The rate law for the reaction of HO^- with *tert*-butyl bromide to form an elimination product in 75% ethanol/25% water at 30 °C is:

$$\text{rate} = 7.1 \times 10^{-5}[\textit{tert-butyl bromide}][HO^-] + 1.5 \times 10^{-5}[\textit{tert-butyl bromide}]$$

a. What percent of the reaction takes place by the E2 pathway when $[HO^-] = 5.0$ M?

b. When $[HO^-] = 0.0025$ M?

10.5
STEREOCHEMISTRY OF E2 AND E1 REACTIONS

Stereochemistry of the E2 Reaction

An E2 reaction involves the removal of two substituents from adjacent carbons. It is a concerted reaction because the two substituents are eliminated in the same step. The bonds to the eliminated substituents (H and X) must be in the same plane, because the sp^3 orbital of the carbon bonded to H and the sp^3 orbital of the carbon bonded to X become overlapping p orbitals in the alkene product. Therefore, the orbitals must overlap in the transition state. This overlap can occur only if the orbitals are parallel, and to be parallel they must be in the same plane.

There are two ways in which the C—H and C—X bonds can be in the same plane: they can be parallel to one another either on the same side of the molecule (**syn-periplanar**) or on opposite sides of the molecule (**anti-periplanar**).

substituents are syn-periplanar
an eclipsed conformer

substituents are anti-periplanar
a staggered conformer

If an elimination reaction removes two substituents from the same side of the molecule, the reaction is a **syn elimination.** If the substituents are removed from opposite sides of the molecule, the reaction is an **anti elimination.** Anti elimination is favored in an E2 reaction; syn elimination can occur, but it is a much slower reaction. There are several reasons for this. First, it is apparent from the preceding structures that syn elimination requires the molecule to be in an eclipsed conformation, while anti elimination requires it to be in a staggered conformation. Because the staggered conformer is more stable (Section 2.11), the transition state leading to elimination from the staggered conformer is more stable than the transition state leading to elimination from the eclipsed conformer. Consequently, anti elimination occurs more rapidly.

You can understand the other reasons that anti elimination is preferred by drawing sawhorse projections for the two conformers. In syn elimination, the electrons of the departing hydrogen move to the front side of the carbon bonded to X. In anti elimination, the electrons move to the back side of the carbon bonded to X. You have seen that displacement reactions involve back side attack (Section 9.2). Lastly, anti elimination avoids the repulsion experienced by the electron-rich base when it is on the same side of the molecule as the electron-rich halogen.

syn elimination   anti elimination

In Section 10.1 you saw that an E2 reaction is *regioselective:* more of one constitutional isomer is formed than the other. For example, the major product formed from E2 elimination of 2-bromopentane is 2-pentene.

$CH_3CH_2CH_2\overset{\underset{|}{Br}}{C}HCH_3 \xrightarrow[CH_3CH_2OH]{CH_3CH_2O^-} CH_3CH_2CH{=}CHCH_3 + CH_3CH_2CH_2CH{=}CH_2$

2-bromopentane   2-pentene   1-pentene
                                           55%   25%
                                       (41% *E*, 14% *Z*)

The E2 reaction is also *stereoselective:* more of one stereoisomer is formed than the other. For example, the 2-pentene obtained as the major product from the elimination reaction of 2-bromopentane can exist as a pair of stereoisomers, and more (*E*)-2-pentene (41%) is formed than (*Z*)-2-pentene (14%).

We can make the following general statement about the stereoselectivity of E2 reactions: if the reactant has two (and only two) hydrogens on the carbon from which a hydrogen is to be removed, both the *E* and *Z* products will be formed, but the one with the bulkiest groups on opposite sides of the double bond will be formed in greater yield. More *E*-2-pentene is formed in the preceding reaction because the bulkiest groups (the methyl and ethyl groups) lie opposite each other.

(*E*)-2-pentene   (*Z*)-2-pentene
     41%   14%

The reason that the alkene with the *bulkiest groups* on *opposite sides of the double bond* is formed in greater yield is because it is more stable than the alkene with the bulkiest groups on the same side of the double bond. We saw that the decreased stability of the alkene with the bulkiest groups on the same side of the

Figure 10.3 ► Reaction coordinate diagram for the E2 reaction of 2-bromopentane.

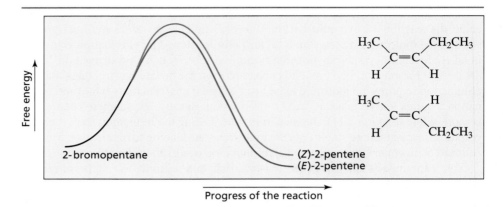

double bond is due to the steric strain that isomer experiences as a result of the interacting electron clouds of the large substitutents (Section 3.19). Because the alkene with the largest groups on opposite sides of the double bond is more stable, it has the more stable transition state and therefore is formed more rapidly (Figure 10.3).

Elimination of HBr from 3-bromo-2,2,3-trimethylpentane leads predominantly to the *E* isomer because this stereoisomer has the methyl group, the bulkiest group on one sp^2 carbon, opposite the *tert*-butyl group, the bulkiest group on the other sp^2 carbon.

3-bromo-2,2,3-trimethylpentane **(E)-3,4,4-trimethyl-2-pentene** major product **(Z)-3,4,4-trimethyl-2-pentene** minor product

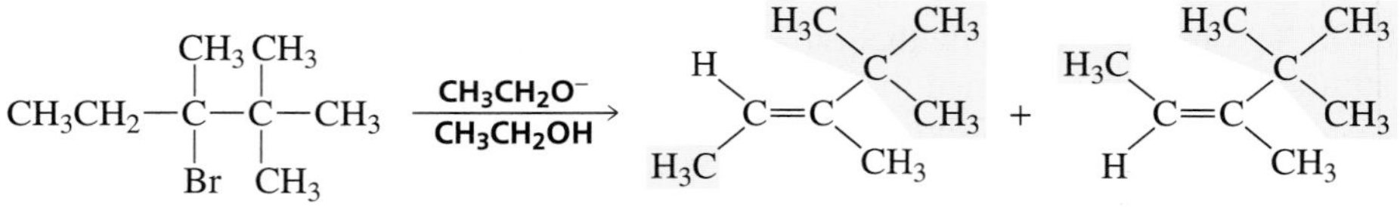

If the β-carbon is bonded to only one hydrogen, only one alkene product can be formed as a result of anti elimination because there is only one conformer in which the groups to be eliminated are anti. The particular isomer obtained depends on the structure of the reactant. For example, anti elimination of HBr from (2*S*,3*S*)-2-bromo-3-phenylbutane forms the *E* isomer, whereas anti elimination of HBr from (2*S*,3*R*)-2-bromo-3-phenylbutane forms the *Z* isomer.

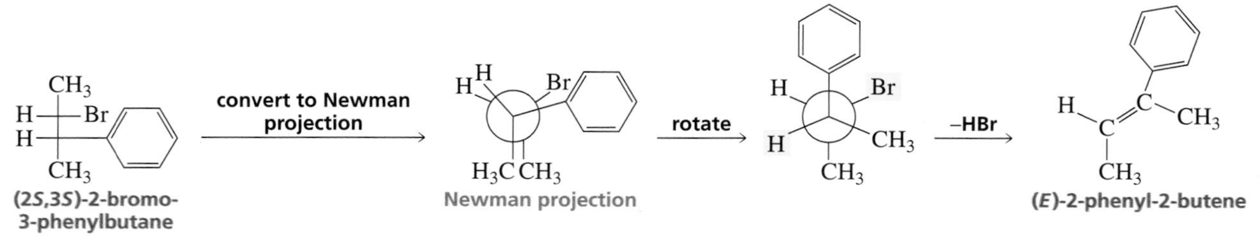

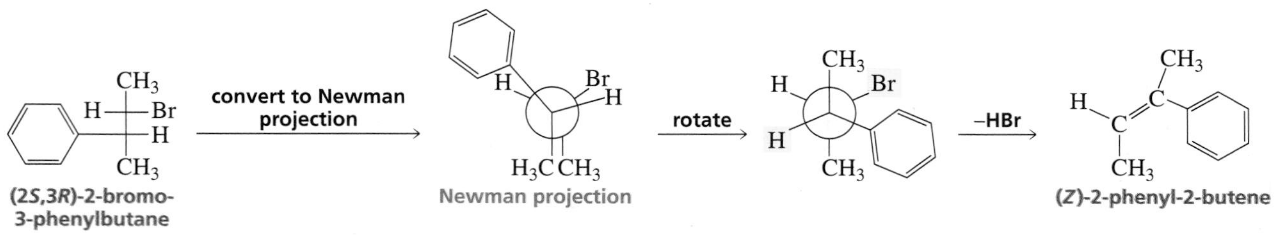

The structure of the elimination product can be determined by following these three steps.

1. Redraw the Fischer projection (or wedge-and-dash structure) as a Newman projection.
2. Rotate the Newman projection to obtain the staggered conformer in which the groups to be eliminated are anti.
3. Replace the eliminated substituents with a double bond.

PROBLEM 9

a. Determine the major product that would be obtained from an E2 reaction of each of the following alkyl halides. In each case, indicate the configuration of the product.

b. Was the product obtained dependent on whether you started with the R or S enantiomer of the reactant?

$$\underset{\underset{Br}{|}\quad\underset{CH_3}{|}}{}$$

1.   CH$_3$CH$_2$CHCHCH$_3$
             Br CH$_3$

2.   CH$_3$CH$_2$CHCH$_2$—⟨phenyl⟩
             |
             Cl

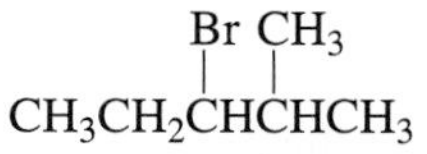

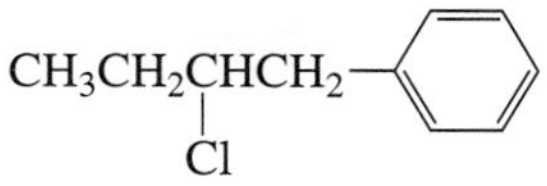

3.   CH$_3$CH$_2$CHCH$_2$CH=CH$_2$
             |
             Cl

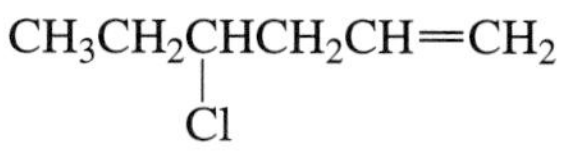

PROBLEM 10◆

What alkene will be formed in an E2 reaction of each of the following compounds? Indicate its configuration.

a. (1S,2S)-1-bromo-1,2-diphenylpropane

b. (1S,2R)-1-bromo-1,2-diphenylpropane

Stereochemistry of the E1 Reaction

An E1 elimination reaction takes place in two steps: the leaving group leaves in the first step, and a proton is lost from an adjacent carbon in the second step, following Zaitsev's rule in order to form the most stable alkene. If the β-carbon from which the proton is removed is bonded to two hydrogens, the products of an E1 reaction will be similar to those of an E2 reaction: the major product will be the one with the bulkiest groups on opposite sides of the double bond.

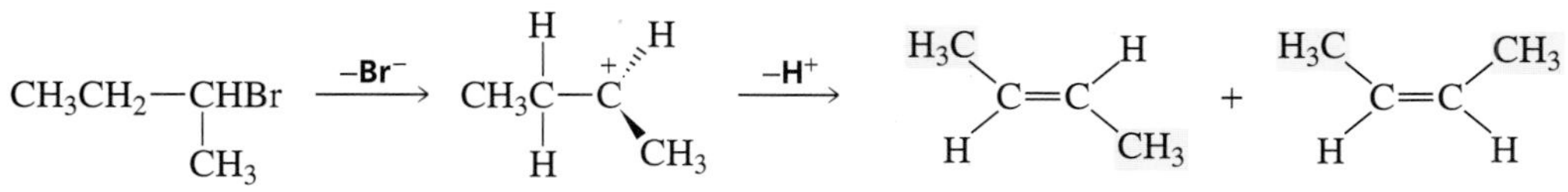

(E)-2-butene
major product

(Z)-2-butene
minor product

The carbocation formed in the first step of an E1 reaction is planar. This means that the electrons from a departing hydrogen can move toward the positively charged carbon from either side. Therefore, both syn and anti elimination can occur.

If the β-carbon is bonded to only one hydrogen, the E1 and E2 reactions will not necessarily form the same products. Because both syn and anti elimination can occur in an E1 reaction, both the *E* and *Z* isomers will be formed, with the major product being the one with the bulkiest groups on opposite sides of the double bond. We just saw that only one of the isomers would be formed in an E2 reaction since only anti elimination can occur.

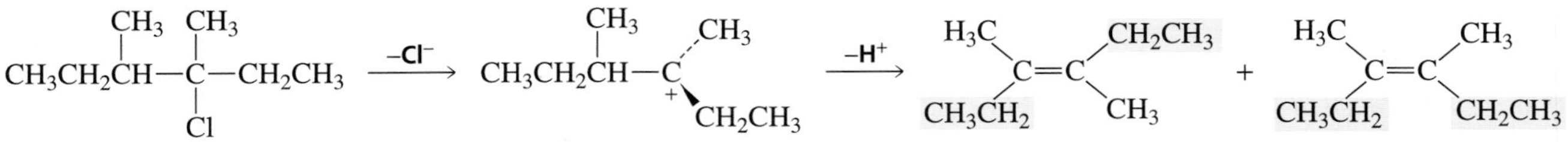

CH3 CH3
CH3CH2CH—C—CH2CH3
Cl
−Cl−
CH3
CH3CH2CH—C
+
CH3
CH2CH3
−H+
H3C
CH2CH3
C=C
CH3CH2
CH3
+
H3C
CH3
C=C
CH3CH2
CH2CH3

(*E*)-3,4-dimethyl-3-hexene major product

(*Z*)-3,4-dimethyl-3-hexene minor product

In summary, except for a reactant that has only one hydrogen on its β-carbon, the products of E2 and E1 reactions are the same: both *E* and *Z* stereoisomers are formed, and the stereoisomer with the bulkiest groups on the opposite sides of the double bond will be the major product. If a reactant has only one hydrogen on its β-carbon, both the *E* and *Z* stereoisomers are formed in an E1 reaction, but only one stereoisomer is formed in an E2 reaction. The particular stereoisomer that is formed depends on the structure of the reactant.

PROBLEM 11◆

Determine the major product that would be formed when each of the following alkyl halides undergoes an E1 reaction.

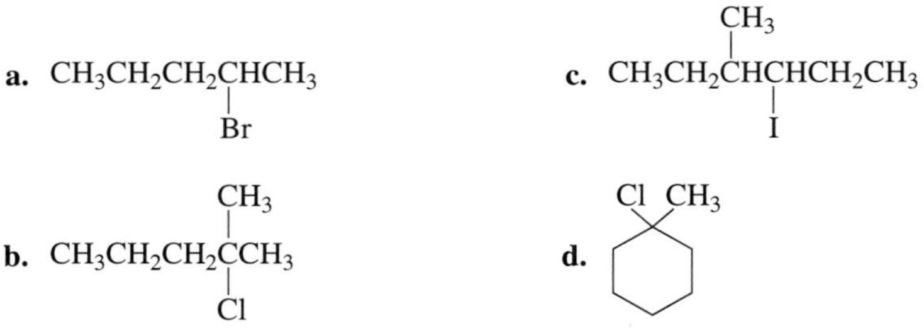

a. CH3CH2CH2CHCH3
Br
b. CH3CH2CH2CCH3
CH3
Cl
c. CH3CH2CHCHCH2CH3
CH3
I
d.
Cl CH3

10.6 ELIMINATION FROM CYCLIC COMPOUNDS

E2 Elimination from Cyclic Compounds

Elimination from cyclic compounds follows the same stereochemical rules as does elimination from open-chain compounds. To achieve the anti-periplanar geometry that is preferred for an E2 reaction, the two substituents that are being eliminated must both be in axial positions.

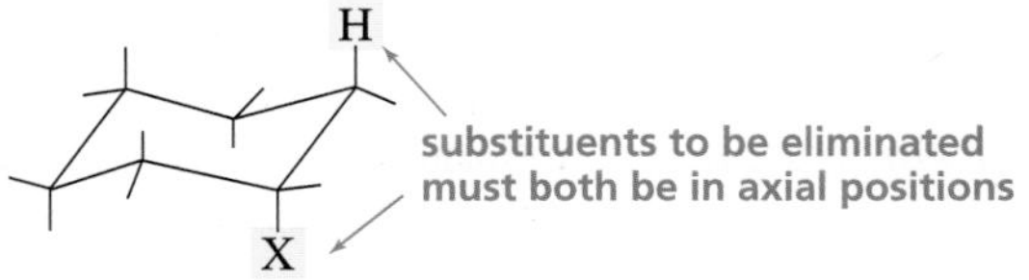

H
X
substituents to be eliminated must both be in axial positions

Recall that the more stable conformation of a monosubstituted cyclohexane is the one in which the substituent is in an equatorial position, since there is more room for a substituent in that position (Section 2.14). Because the chloro substituent is in an equatorial position, the more stable conformation of chlorocyclohexane does not undergo an E2 reaction. The less stable conformation, with the chloro substituent in the axial position, readily undergoes an E2 reaction.

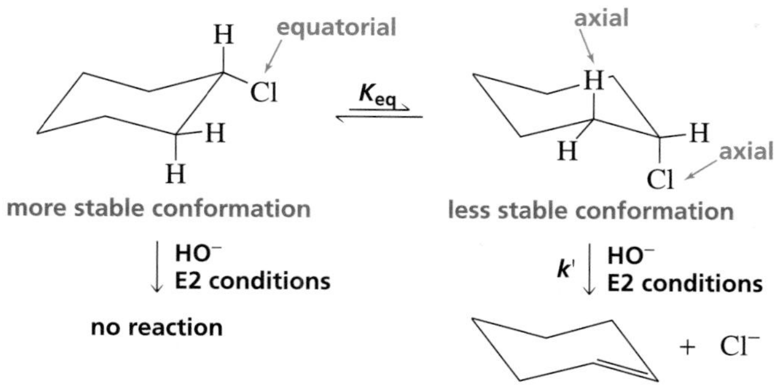

How does the stability of the conformer affect its rate of reaction? Because the rate constant of the reaction is given by $k'K_{eq}$, the reaction is faster if the equilibrium is favorable (K_{eq} is large). Because of the unfavorable equilibrium constant when the reaction has to take place by way of the less stable conformer, the rate of elimination is slower than it would be if elimination could take place by way of the more stable conformer that has a favorable equilibrium constant.

trans-1-Chloro-2-methylcyclohexane also undergoes an E2 reaction through the less stable conformer. Notice that the hydrogen that is eliminated is not removed from the β-carbon bonded to the fewest hydrogens. Remember that Zaitsev's rule states that, *when there is a choice,* a hydrogen is removed from the β-carbon bonded to the fewest hydrogens. Because elimination can occur only if the two substituents that are eliminated are in axial positions, the axial hydrogen is removed even though it is not bonded to the β-carbon with the fewest hydrogens.

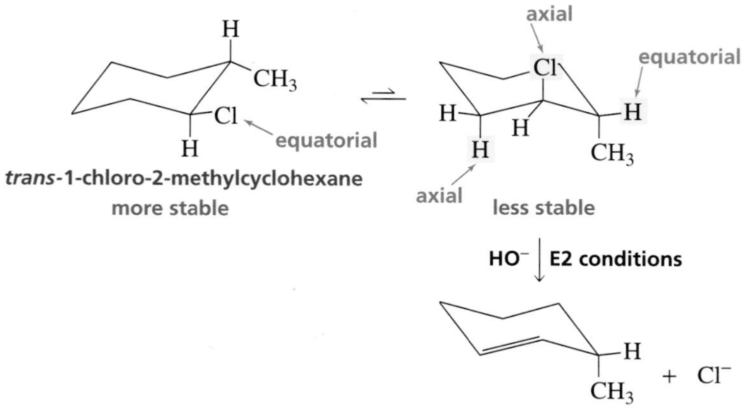

In the following compound, the conformation with the chloro substituent in an axial position has an axial hydrogen on both β-carbon atoms. Therefore, either hydrogen can be eliminated along with the chlorine in an E2 reaction. Because there is a choice of hydrogens, the major product is the one obtained by removing a hydrogen from the β-carbon that is bonded to the fewest hydrogens—the one predicted by Zaitsev's rule.

(a)

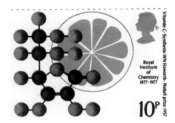

(b)

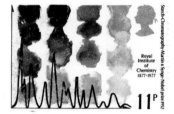

(c)

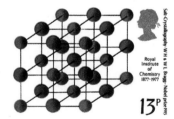

(d)

Stamps issued in honor of English Nobel Laureates:
(a) Sir Derek Barton for conformational analysis, 1969;
(b) Sir Walter Haworth for the synthesis of vitamin C, 1937;
(c) A. J. P. Martin and Richard L. M. Singe for chromatography, 1952;
(d) William H. Bragg and William L. Bragg for crystallography, 1915 (the only father and son to receive a Nobel prize).

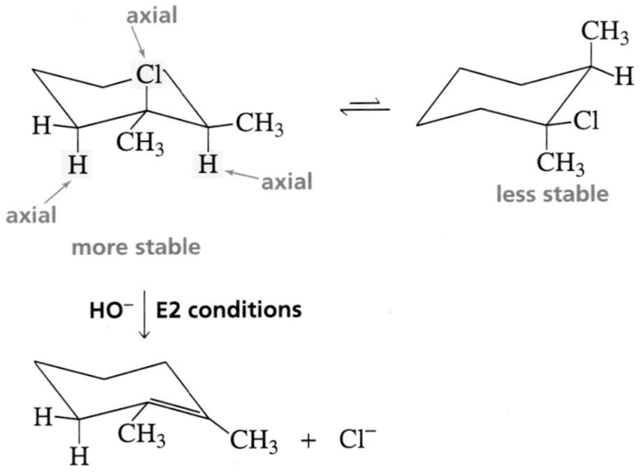

E1 Elimination from Cyclic Compounds

When a substituted cyclohexane undergoes an E1 reaction, the two groups that are eliminated do not have to be diaxial, because the elimination reaction is not concerted. In the following reaction, a carbocation is formed in the first step. It then loses a proton, following Zaitsev's rule.

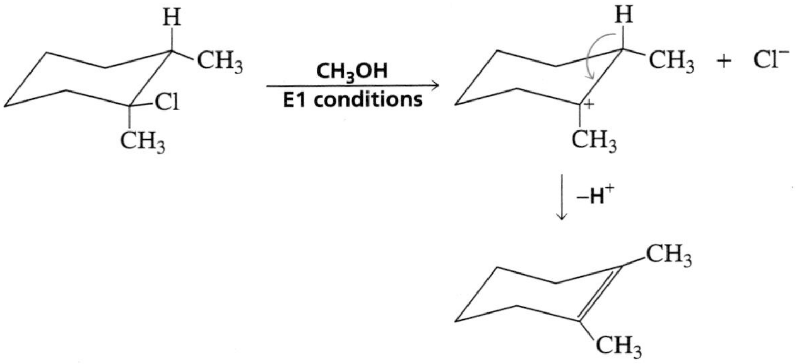

Because a carbocation is formed in an E1 reaction, you must check for the possibility of a carbocation rearrangement before you use Zaitsev's rule to determine the elimination product. In the following reaction, the secondary carbocation undergoes a 1,2-hydride shift, forming a more stable tertiary carbocation.

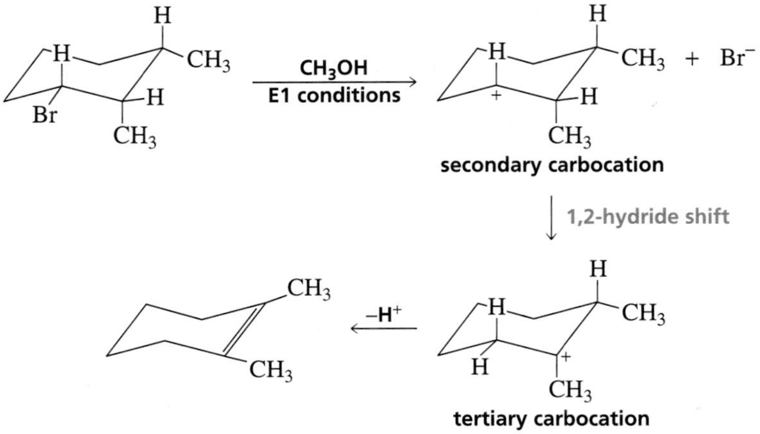

Table 10.4 summarizes the stereochemical outcome of substitution and elimination reactions.

TABLE 10.4	Stereochemistry of Substitution and Elimination Reactions
Mechanism	**Products**
S_N1	Both enantiomers are formed. (more inverted than retained)
E1	Both Z and E stereoisomers are formed (more of the stereoisomer with the bulkiest groups on opposite sides of the double bond).
S_N2	Only the inverted product is formed.
E2	Both E and Z stereoisomers are formed (more of the stereoisomer with the bulkiest groups on opposite sides of the double bond), unless the β-carbon has only one hydrogen, in which case only one stereoisomer is formed; its structure depends on the structure of the reactant.

PROBLEM 12

Which isomer reacts more rapidly in an E2 reaction, *cis*-1-bromo-4-*tert*-butylcyclo-hexane or *trans*-1-bromo-4-*tert*-butylcyclohexane? Explain.

PROBLEM 13

Give the substitution and elimination products for the following reactions, showing the configuration of each product.

a. (S)-2-chlorohexane $\xrightarrow[\text{S}_N2/\text{E2 conditions}]{\text{CH}_3\text{O}^-}$

b. (S)-2-chlorohexane $\xrightarrow[\text{S}_N1/\text{E1 conditions}]{\text{CH}_3\text{OH}}$

c. *trans*-1-chloro-2-methylcyclohexane $\xrightarrow[\text{S}_N2/\text{E2 conditions}]{\text{CH}_3\text{O}^-}$

d. *trans*-1-chloro-2-methylcyclohexane $\xrightarrow[\text{S}_N1/\text{E1 conditions}]{\text{CH}_3\text{OH}}$

e. 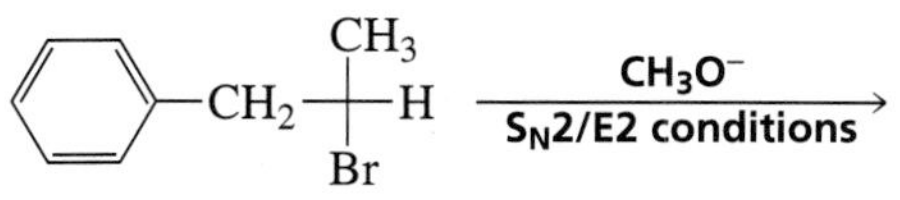 C6H5—CH2—C(CH3)H—Br $\xrightarrow[\text{S}_N2/\text{E2 conditions}]{\text{CH}_3\text{O}^-}$

f. 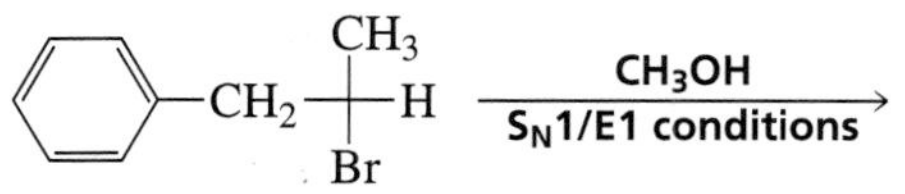 C6H5—CH2—C(CH3)H—Br $\xrightarrow[\text{S}_N1/\text{E1 conditions}]{\text{CH}_3\text{OH}}$

Sir Derek H. R. Barton *was the first to point out that the chemical reactivity of substituted cyclohexanes was controlled by their conformation. Barton was born in Gravesend, Kent, England in 1918. He received a Ph.D. from Imperial College, London in 1942 and became a faculty member there three years later. Currently, he is a professor of chemistry at Texas A & M University. For his work on the relationship of the three-dimensional structure of organic compounds and their chemical reactivity, he received the 1969 Nobel Prize in chemistry. Barton was knighted in 1972.*

10.7 A KINETIC ISOTOPE EFFECT

A mechanism is a theory that accounts for all the experimental evidence known about a reaction. For example, the mechanisms of the S_N1 and S_N2 reactions are based on a knowledge of the order of the reaction, the relative reactivities of the reactants, and the structures of the products.

Another piece of experimental evidence that is helpful in determining the mechanism of a reaction is a kinetic isotope effect. A **kinetic isotope effect** is a comparison of the rate of reaction of a compound with the rate of reaction of the same compound in which one of the atoms has been replaced by an isotope. Deuterium is an isotope of hydrogen. The nucleus of a deuterium contains a proton and a neutron; the nucleus of a hydrogen contains only a proton (Section 1.1).

A **deuterium kinetic isotope effect** is the ratio of the rate constant observed for a compound containing hydrogen to the rate constant observed for an identical compound in which one or more of the hydrogens have been replaced by deuteriums.

$$\text{deuterium kinetic isotope effect} = \frac{k_H}{k_D} = \frac{\text{rate constant for H-containing reactant}}{\text{rate constant for D-containing reactant}}$$

The chemical properties of deuterium and hydrogen are similar. However, a carbon–deuterium bond is harder to break than a carbon–hydrogen bond. When the rate constant (k_H) for elimination of HBr from 1-bromo-2-phenylethane at 30 °C is compared with the rate constant (k_D) for elimination of DBr from 2-bromo-1,1-dideuterio-1-phenylethane determined at the same temperature, k_H is found to be 7.1 times k_D. So the deuterium kinetic isotope effect is 7.1.

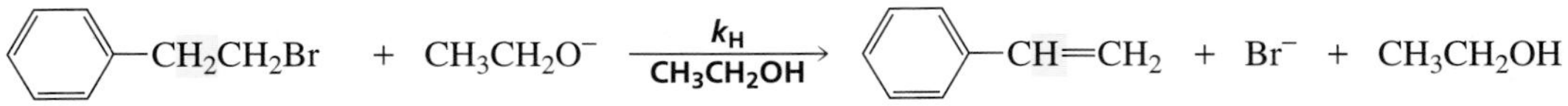

1-bromo-2-phenylethane

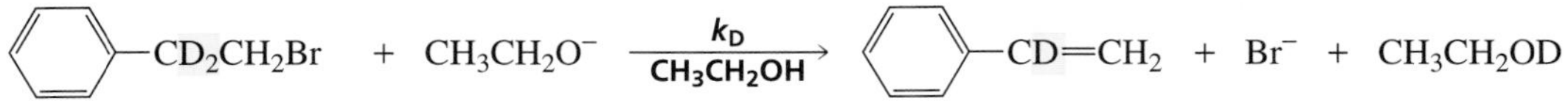

2-bromo-1,1-dideuterio-
1-phenylethane

The fact that the hydrogen-containing compound undergoes elimination faster than the deuterium-containing compound tells us that the carbon–hydrogen (or carbon–deuterium) bond is broken in the rate-determining step. The difference in the reaction rates is due to the difference in the ease of breaking a C—H bond compared with a C—D bond. The fact that the C—H bond is broken in the rate-determining step agrees with the mechanism proposed for an E2 reaction.

PROBLEM 14 ◆

If the preceding reactions took place by an E1 mechanism, what value would you expect to obtain for the deuterium kinetic isotope effect?

10.8 COMPETITION BETWEEN SUBSTITUTION AND ELIMINATION

You have seen that alkyl halides can undergo four types of reactions: S_N2, S_N1, E2, and E1. At this point, it may seem a bit overwhelming to be given an alkyl halide and a nucleophile/base and asked to predict the products of the reaction. So let's now attempt to organize what you have learned about the reactions of alkyl halides to make it a little easier to predict the products of any given reaction.

The first thing you must decide is whether the reaction conditions favor S_N2/E2 or S_N1/E1 reactions. Fortunately, the conditions that favor an S_N2 reaction also favor an E2 reaction, and the conditions that favor an S_N1 reaction also favor an E1 reaction. This means that you do not need to worry about S_N2/E1 or S_N1/E2 combinations; you just decide whether the combination is going to be S_N2/E2 or S_N1/E1.

Your decision is easy if the reactant is a *primary* alkyl halide because, without the ability to form carbocations, primary alkyl halides undergo only S_N2/E2 reactions.

If the reactant is a *secondary* or a *tertiary* alkyl halide, whether it undergoes S_N2/E2 or S_N1/E1 reactions depends on the reaction conditions. S_N2/E2 reactions are favored by a high concentration of a good nucleophile/strong base, whereas S_N1/E1 reactions are favored by a poor nucleophile/weak base (Sections 9.9 and 10.4). In addition, the solvent in which the reaction is carried out can affect the choice of mechanism (Section 9.10).

Once you have decided whether the conditions will lead to $S_N2/E2$ reactions or to $S_N1/E1$ reactions, you must decide how much of the product will be the substitution product and how much will be the elimination product. The relative amounts of substitution and elimination products depend on whether the alkyl halide is primary, secondary, or tertiary and on the nature of the nucleophile/base. This is discussed below and summarized in Table 10.6.

$S_N2/E2$ Conditions

Let's first consider the situation in which you have decided that the conditions lead to $S_N2/E2$ reactions (a high concentration of a good nucleophile/strong base). The negatively charged species can act as a nucleophile and hit the back side of the α-carbon to form the substitution product, or it can act as a base and remove a β-hydrogen, leading to the elimination product. Notice that both reactions occur for the same reason—the electron-withdrawing halogen, which causes the carbon to which it is bonded to have a partial positive charge.

$$CH_3\!-\!CH_2\!-\!\overset{\frown}{Br} \longrightarrow CH_3CH_2OH + Br^-$$
$$\underset{H\ddot{O}\!:^-}{}\qquad\qquad\qquad \text{substitution product}$$

$$CH_2\!-\!CH_2\!-\!\overset{\frown}{Br} \longrightarrow CH_2\!=\!CH_2 + H_2O + Br^-$$
$$\underset{\underset{H\ddot{O}\!:^-}{H}}{|}\qquad\qquad\qquad \text{elimination product}$$

The relative reactivities of alkyl halides in S_N2 and E2 reactions are shown in Table 10.5. Because a *primary* alkyl halide is the most reactive of the alkyl halides in an S_N2 reaction and the least reactive in an E2 reaction, you can predict that a primary alkyl halide will primarily form the substitution product in a reaction carried out under conditions that favor $S_N2/E2$ reactions.

$$CH_3CH_2CH_2Br + CH_3O^- \xrightarrow{CH_3OH} CH_3CH_2CH_2OCH_3 + CH_3CH\!=\!CH_2 + CH_3OH + Br^-$$

propyl bromide　　　　　　　　　　　　　methyl propyl ether　　propene
　　　　　　　　　　　　　　　　　　　　　　90%　　　　　　　　10%

However, if either the primary alkyl halide or the nucleophile/base is sterically hindered, the nucleophile will have difficulty getting to the back side of the α-carbon and, as a result, the elimination product will predominate.

the primary alkyl halide is sterically hindered

$$\underset{\text{propane}}{\underset{\text{1-bromo-2-methyl-}}{\overset{\overset{\displaystyle CH_3}{|}}{CH_3CHCH_2Br}}} + CH_3O^- \xrightarrow{CH_3OH} \underset{\underset{40\%}{\text{isobutyl methyl ether}}}{\overset{\overset{\displaystyle CH_3}{|}}{CH_3CHCH_2OCH_3}} + \underset{\underset{60\%}{\text{methylpropene}}}{\overset{\overset{\displaystyle CH_3}{|}}{CH_3C\!=\!CH_2}} + CH_3OH + Br^-$$

TABLE 10.5	Relative Reactivities of Alkyl Halides		
in an S_N2 reaction:	$1° > 2° > 3°$	in an S_N1 reaction:	$3° > 2° > 1°$
in an E2 reaction:	$3° > 2° > 1°$	in an E1 reaction:	$3° > 2° > 1°$

the nucleophile is sterically hindered

$$CH_3CH_2CH_2CH_2CH_2Br \ + \ CH_3\overset{\underset{|}{CH_3}}{\underset{\underset{|}{CH_3}}{C}}O^- \ \xrightarrow{\text{(CH}_3)_3\text{COH}} \ CH_3CH_2CH_2CH_2CH_2O\overset{\underset{|}{CH_3}}{\underset{\underset{|}{CH_3}}{C}}CH_3$$

tert-butyl pentyl ether
15%

$$+ \ CH_3CH_2CH_2CH{=}CH_2 \ + \ CH_3\overset{\underset{|}{CH_3}}{\underset{\underset{|}{CH_3}}{C}}OH \ + \ Br^-$$

1-pentene
85%

A *secondary* alkyl halide can form both substitution and elimination products under S_N2/E2 conditions. The relative amounts of the two products depend on the base strength (and the bulk) of the nucleophile. The stronger and bulkier the base, the greater the percent of the elimination product. For example, acetic acid is a stronger acid ($pK_a = 4.76$) than ethanol ($pK_a = 15.9$), which means that acetate ion is a weaker base than ethoxide ion. The elimination product is the main product formed from the reaction of 2-chloropropane with the strongly basic ethoxide ion, whereas no elimination product is formed with the weakly basic acetate ion.

$$CH_3\overset{\underset{|}{Cl}}{C}HCH_3 \ + \ CH_3CH_2O^- \ \xrightarrow{\text{CH}_3\text{CH}_2\text{OH}} \ CH_3\overset{\underset{|}{OCH_2CH_3}}{C}HCH_3 \ + \ CH_3CH{=}CH_2 \ + \ CH_3CH_2OH \ + \ Br^-$$

2-chloropropane   ethoxide ion   2-ethoxypropane   propene
                                                                              25%   75%

$$CH_3\overset{\underset{|}{Cl}}{C}HCH_3 \ + \ CH_3\overset{\overset{O}{\|}}{C}O^- \ \xrightarrow{\text{acetic acid}} \ CH_3\overset{\underset{|}{O\overset{\overset{O}{\|}}{C}CH_3}}{C}HCH_3 \ + \ Cl^-$$

2-chloropropane   acetate ion   ispropyl acetate
                                                                      100%

Increasing the temperature at which the reaction is carried out increases the rates of both the substitution and elimination reactions, but the rate of the elimination reaction increases more. Therefore, if the substitution product is the desired product, the reaction should be carried out at a low temperature. High temperatures promote formation of the elimination product.

A *tertiary* alkyl halide is the least reactive of the alkyl halides in an S_N2 reaction and the most reactive in an E2 reaction (Table 10.5). Consequently, only the elimination product is formed when a tertiary alkyl halide reacts with a nucleophile under S_N2/E2 conditions.

$$CH_3\overset{\underset{|}{CH_3}}{\underset{\underset{|}{CH_3}}{C}}Br \ + \ CH_3CH_2O^- \ \xrightarrow{\text{CH}_3\text{CH}_2\text{OH}} \ CH_3\overset{\underset{|}{CH_3}}{C}{=}CH_2 \ + \ CH_3CH_2OH \ + \ Br^-$$

2-bromo-2-methyl-
propane
methylpropene

PROBLEM 15◆

How would you expect the ratio of substitution product to elimination product formed from the reaction of propyl bromide with CH_3O^- in methanol to change when the nucleophile is changed to CH_3S^-?

S$_N$1/E1 Conditions

Let's now look at what happens when conditions lead to S$_N$1/E1 reactions (a poor nucleophile/weak base is used). In S$_N$1/E1 reactions, the alkyl halide dissociates to form a carbocation. The carbocation can either combine with the nucleophile to form the substitution product or lose a proton to form the elimination product.

Alkyl halides have the same order of reactivity in S$_N$1 and E1 reactions because they have the same rate-determining step—ionization of the alkyl halide (Table 10.5). This means that all alkyl halides that react under S$_N$1/E1 conditions will give both substitution and elimination products. Substitution is favored over elimination at lower temperatures; increasing the temperature increases the percentage of elimination product. Remember that primary alkyl halides do not undergo S$_N$1/E1 reactions because primary carbocations are too unstable to be formed.

Table 10.6 summarizes the products obtained when alkyl halides react with nucleophiles under S$_N$2/E2 and S$_N$1/E1 conditions.

PROBLEM 16◆

Indicate whether the following alkyl halides will give mostly substitution products, mostly elimination products, or about equivalent amounts of substitution and elimination products when they react with:

a. methanol under S$_N$1/E1 conditions.

b. sodium methoxide under S$_N$2/E2 conditions.

1. 1-bromobutane   **3.** 2-bromobutane

2. 1-bromo-2-methylpropane   **4.** 2-bromo-2-methylpropane

TABLE 10.6 Summary of the Products Expected in Substitution/Elimination Reactions		
Class of alkyl halide	**S$_N$2 vs. E2**	**S$_N$1 vs. E1**
Primary alkyl halide	Primarily substitution unless there is steric hindrance in the alkyl halide or nucleophile, in which case elimination is favored	Cannot undergo S$_N$1/E1 reactions
Secondary alkyl halide	Both substitution and elimination; the stronger and bulkier the base and the higher the temperature, the greater the percentage of elimination	Both substitution and elimination; the higher the temperature, the greater the percentage of elimination
Tertiary alkyl halide	Only elimination	Both substitution and elimination; the higher the temperature, the greater the percentage of elimination

---

PROBLEM 17

1-Bromo-2,2-dimethylpropane has difficulty undergoing either S_N2 or S_N1 reactions.

a. Explain why this is so.   **b.** Can it undergo E2 and E1 reactions?

10.9
SUBSTITUTION AND ELIMINATION REACTIONS IN SYNTHESIS

When substitution or elimination reactions are used in synthesis, care must be taken to choose reactants and reaction conditions that will give the maximum possible yield of the desired product. In Section 9.3 you saw that a wide variety of organic compounds can be prepared by substitution reactions of alkyl halides. For example, ethers are synthesized by the reaction of an alkyl halide with an alkoxide ion. This reaction was discovered by Alexander Williamson in 1850 and is still considered to be one of the best ways to synthesize an ether.

Williamson ether synthesis:

$$R{-}Br \; + \; R{-}O^- \; \longrightarrow \; R{-}O{-}R \; + \; Br^-$$

alkyl halide   alkoxide ion   ether

The alkoxide ion for the **Williamson ether synthesis** is prepared by using sodium metal (or NaH) to remove a proton from an alcohol.

$$ROH \; + \; Na \; \longrightarrow \; RO^- \; + \; Na^+ \; + \; \frac{1}{2}H_2$$

The Williamson ether synthesis is a substitution reaction and, because it requires a high concentration of a good nucleophile (the alkoxide ion), it is an S_N2 reaction. If you want to synthesize an ether such as butyl propyl ether, you have a choice of starting materials. You can use either a propyl halide and butoxide ion or a butyl halide and propoxide ion.

Alexander W. Williamson

Alexander William Williamson (1824–1904) *was born in London of Scottish parents. As a child he lost an arm and the use of an eye. He was midway through his medical education when he changed his mind and decided to study chemistry. In 1849 he became a professor of chemistry at University College, London.*

$$CH_3CH_2CH_2Br \; + \; CH_3CH_2CH_2CH_2O^- \; \longrightarrow \; CH_3CH_2CH_2OCH_2CH_2CH_2CH_3 \; + \; Br^-$$
propyl bromide   butoxide ion   butyl propyl ether

$$CH_3CH_2CH_2CH_2Br \; + \; CH_3CH_2CH_2O^- \; \longrightarrow \; CH_3CH_2CH_2OCH_2CH_2CH_2CH_3 \; + \; Br^-$$
butyl bromide   propoxide ion   butyl propyl ether

If, however, you want to synthesize *tert*-butyl ethyl ether, the starting materials must be an ethyl halide and *tert*-butoxide ion. If you tried to use a *tert*-butyl halide and ethoxide ion as reactants, you would obtain the elimination product and little or no ether, because reaction of a tertiary alkyl halide under S_N2/E2 conditions primarily forms the elimination product.

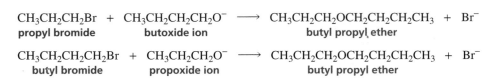

ethyl bromide   *tert*-butoxide ion   *tert*-butyl ethyl ether   ethene

tert-butyl bromide   methylpropene

PROBLEM 18◆

What other organic product will be formed in the synthesis of butyl propyl ether when the alkyl halide used in the synthesis is:

  a. propyl bromide?   **b.** butyl bromide?

PROBLEM 19◆

How could the following ethers be prepared using an alkyl halide and an alcohol?

$$\overset{\displaystyle CH_3}{\underset{\displaystyle |}{}}$$

  a. $CH_3CH_2CHOCH_2CH_2CH_3$ **c.** $CH_3CH_2OCH_2CH_2CH_2CH_2CH_3$

  b. ⟨cyclohexyl⟩$-O\overset{\overset{\displaystyle CH_3}{|}}{\underset{\underset{\displaystyle CH_3}{|}}{C}}CH_3$ **d.** ⟨phenyl⟩$-CH_2O-$⟨phenyl⟩

In Section 5.10 you saw that alkynes can be synthesized by the reaction of an acetylide anion with an alkyl halide.

$$CH_3CH_2C\equiv C^- \ + \ CH_3CH_2CH_2Br \ \longrightarrow \ CH_3CH_2C\equiv CCH_2CH_2CH_3 \ + \ Br^-$$

Now that you know that this reaction is an S_N2 reaction (the alkyl halide reacts with a high concentration of a good nucleophile), you can understand why it is best to use primary alkyl halides and methyl halides in the reaction. These alkyl halides are the only ones that primarily form the desired substitution product. Tertiary alkyl halides give only the elimination product, and secondary alkyl halides give mainly the elimination product, because the acetylide ion is a very strong base.

If you wanted to synthesize propene, you would start with 2-bromopropane rather than 1-bromopropane, because the secondary alkyl halide would give a higher yield of the desired elimination product.

$$\underset{\textbf{2-bromopropane}}{\overset{\overset{\displaystyle Br}{|}}{CH_3CHCH_3}} \ + \ HO^- \ \longrightarrow \ \underset{\text{major product}}{CH_3CH=CH_2} \ + \ \underset{\text{minor product}}{\overset{\overset{\displaystyle OH}{|}}{CH_3CHCH_3}} \ + \ H_2O \ + \ Br^-$$

$$\underset{\textbf{1-bromopropane}}{CH_3CH_2CH_2Br} \ + \ HO^- \ \longrightarrow \ \underset{\text{minor product}}{CH_3CH=CH_2} \ + \ \underset{\text{major product}}{CH_3CH_2CH_2OH} \ + \ H_2O \ + \ Br^-$$

To synthesize 2-methyl-2-butene from 2-bromo-2-methylbutane, you would use $S_N2/E2$ conditions (a high concentration of HO^- in an aprotic polar solvent) because a tertiary alkyl halide under these conditions gives only the elimination product. If $S_N1/E1$ conditions were used (a low concentration of HO^- in a protic polar solvent), both the elimination and substitution products would be obtained.

$$\underset{\textbf{2-bromo-2-methylbutane}}{\underset{\overset{\displaystyle |}{\underset{\displaystyle Br}{}}}{\overset{\overset{\displaystyle CH_3}{|}}{CH_3CCH_2CH_3}}} \ + \ HO^-$$

$\xrightarrow[\text{conditions}]{S_N2/E2}$ $\underset{}{\overset{\overset{\displaystyle CH_3}{|}}{CH_3C=CHCH_3}} \ + \ H_2O \ + \ Br^-$

$\xrightarrow[\text{conditions}]{S_N1/E1}$ $\underset{\textbf{2-methyl-2-butene}}{\overset{\overset{\displaystyle CH_3}{|}}{CH_3C=CHCH_3}} \ + \ \underset{\textbf{2-methyl-2-butanol}}{\underset{\overset{\displaystyle |}{\underset{\displaystyle OH}{}}}{\overset{\overset{\displaystyle CH_3}{|}}{CH_3CCH_2CH_3}}} \ + \ H_2O \ + \ Br^-$

PROBLEM 20◆

Identify the three products that are formed when 2-bromo-2-methylpropane is dissolved in a mixture of 80% ethanol and 20% water.

PROBLEM 21

a. What products (including stereoisomers, if applicable) would be formed from the reaction of 3-bromo-2-methylpentane with HO^-: under $S_N2/E2$ conditions; under $S_N1/E1$ conditions?

b. Answer the same question for 3-bromo-3-methylpentane.

10.10 CONSECUTIVE E2 ELIMINATION REACTIONS

Alkyl dihalides can undergo two consecutive dehydrohalogenations, giving products that contain two double bonds. In the following example, Zaitsev's rule predicts the product of the first dehydrohalogenation but not the product of the second. This is because the conjugated diene is formed more readily since the transition state leading to its formation is more stable than the transition state leading to formation of the isolated diene.

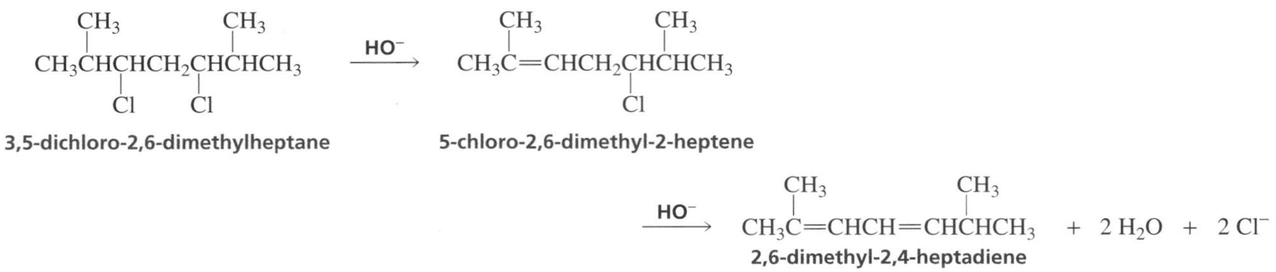

3,5-dichloro-2,6-dimethylheptane

5-chloro-2,6-dimethyl-2-heptene

2,6-dimethyl-2,4-heptadiene

If the two halogens are on the same carbon (geminal dihalides) or on adjacent carbons (vicinal dihalides), the two consecutive E2 dehydrohalogenations result in the formation of a triple bond. This is how alkynes are commonly synthesized.

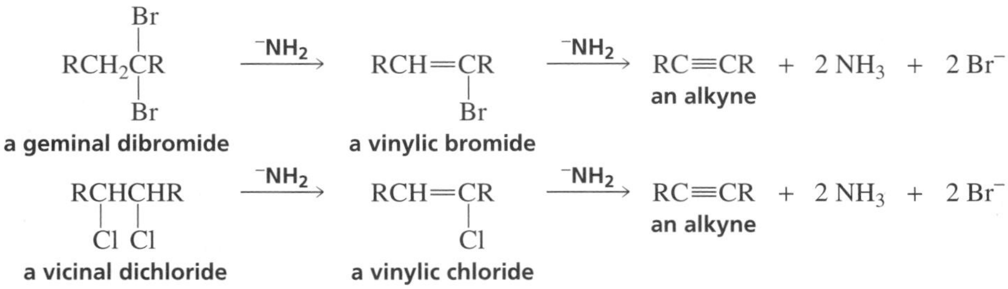

a geminal dibromide **a vinylic bromide** **an alkyne**

a vicinal dichloride **a vinylic chloride** **an alkyne**

The vinylic halide intermediates in the preceding reaction are relatively unreactive. Consequently, a very strong base such as $^-NH_2$ is needed for the second elimination. If a weaker base such as HO^- is used at room temperature, the reaction will stop at the vinylic halide and no alkyne will be formed.

Since a vicinal dihalide is formed from the reaction of an alkene with Br_2 or Cl_2, you have just learned how to convert a double bond into a triple bond.

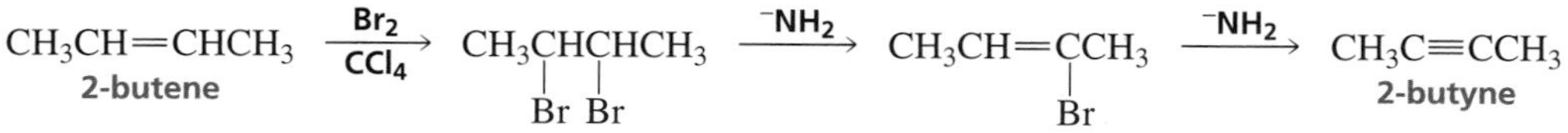

$$CH_3CH=CHCH_3 \xrightarrow[\text{CCl}_4]{\text{Br}_2} \underset{\underset{Br\ Br}{| \ |}}{CH_3CHCHCH_3} \xrightarrow{^-NH_2} \underset{\underset{Br}{|}}{CH_3CH=CCH_3} \xrightarrow{^-NH_2} CH_3C\equiv CCH_3$$

2-butene   2-butyne

PROBLEM 22

Why is a cumulated diene not formed in the preceding reaction?

When you are asked to design a synthesis, one way to approach the problem is to look at the given starting material to see if there is an obvious series of reactions that can get you started on the pathway to the **target molecule** (the desired product). This is often the best way to approach a simple synthesis. The following examples will give you practice in designing a successful synthesis.

**10.11
DESIGNING A
SYNTHESIS II**

Example 1.   How could you prepare 1,3-cyclohexadiene from cyclohexane?

Since the only reaction an alkane can undergo is halogenation, deciding what the first reaction should be is easy. An E2 reaction using a high concentration of a strong and bulky base to encourage elimination over substitution will form cyclohexene; therefore, *tert*-butoxide ion is used as the base and *tert*-butyl alcohol is used as the solvent. Bromination of cyclohexene will give an allylic bromide, which will form the desired target molecule by undergoing another E2 reaction.

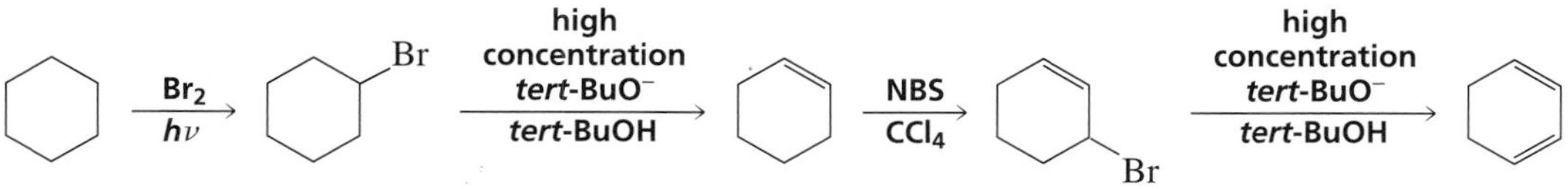

Example 2.   Starting with methylcyclohexane, how could you prepare the following vicinal *trans*-dihalide?

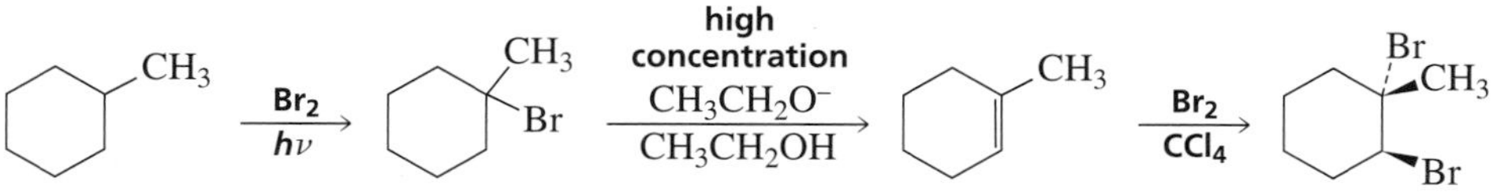

Again, since the starting material is an alkane, the first reaction must be a radical substitution; the bromine will selectively substitute for the tertiary hydrogen. Under E2 conditions, tertiary alkyl halides undergo only elimination, so there will be no competing substitution product in the next reaction. Because addition of Br_2 involves only anti addition, the target molecule is easily obtained.

As you saw in Section 5.11, a useful technique in designing a synthesis—particularly when it is not clear from the starting material how to proceed—is to work backward. Look at the immediate precursor of the target molecule and ask yourself how it could be prepared. Once you have an answer, work backward to the next compound, asking yourself how it could be prepared. Keep working backward one step at a time until you get to a readily available starting material. This technique is called *retrosynthetic analysis*.

Example 3. How could you prepare ethyl methyl ketone from 1-bromobutane?

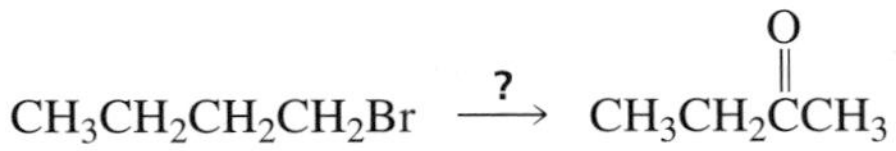

$$CH_3CH_2CH_2CH_2Br \xrightarrow{?} CH_3CH_2\overset{\displaystyle O}{\overset{\displaystyle \|}{C}}CH_3$$

The only method for synthesizing a ketone that you know is to prepare it from an alkyne. The alkyne can be prepared from two successive E2 reactions of a vicinal dihalide. The dihalide is synthesized from an alkene. The desired alkene can be prepared from the given starting material by an elimination reaction. A bulky base is used in the elimination reaction in order to maximize the amount of elimination product.

retrosynthetic analysis

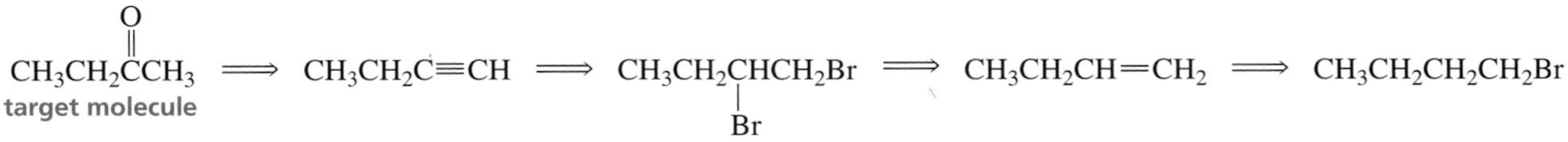

$$CH_3CH_2\overset{\displaystyle O}{\overset{\displaystyle \|}{C}}CH_3 \implies CH_3CH_2C\equiv CH \implies CH_3CH_2\underset{\underset{\displaystyle Br}{|}}{C}HCH_2Br \implies CH_3CH_2CH=CH_2 \implies CH_3CH_2CH_2CH_2Br$$
target molecule

synthesis

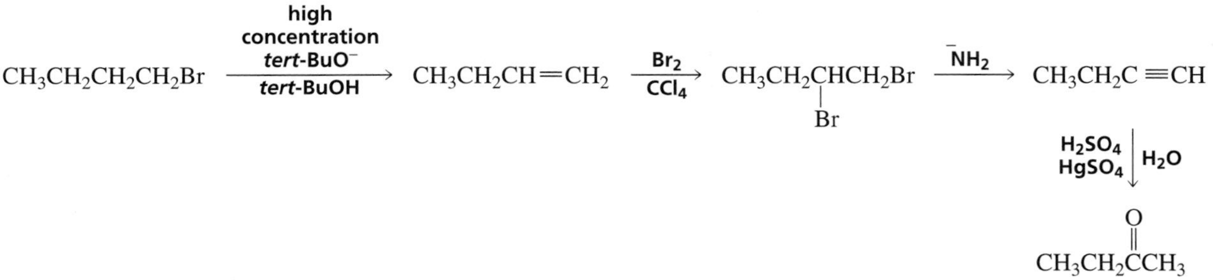

$$CH_3CH_2CH_2CH_2Br \xrightarrow[\textbf{\textit{tert}-BuOH}]{\substack{\textbf{high}\\ \textbf{concentration}\\ \textbf{\textit{tert}-BuO}^-}} CH_3CH_2CH=CH_2 \xrightarrow[\textbf{CCl}_4]{\textbf{Br}_2} CH_3CH_2\underset{\underset{\displaystyle Br}{|}}{C}HCH_2Br \xrightarrow{^-NH_2} CH_3CH_2C\equiv CH$$

$$\downarrow{\substack{\textbf{H}_2\textbf{SO}_4 \\ \textbf{HgSO}_4}}{\Big|}\textbf{H}_2\textbf{O}$$

$$CH_3CH_2\overset{\displaystyle O}{\overset{\displaystyle \|}{C}}CH_3$$

Example 4. How could the following cyclic ether be prepared from the given starting material?

$$BrCH_2CH_2CH_2CH=CH_2 \xrightarrow{?} \text{(cyclic ether)} CH_3$$

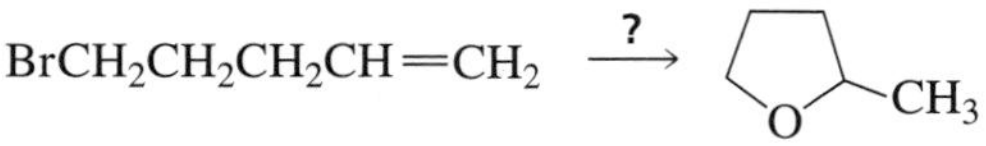

In order to obtain a cyclic ether, the two groups required for ether synthesis (the alkyl halide and the alcohol) must be in the same molecule. To obtain a five-membered ring, the carbons bearing the two groups must be separated by two additional carbons. Addition of water to the given starting material gives the required bifunctional compound.

retrosynthetic analysis

$$\text{(cyclic ether)}CH_3 \implies BrCH_2CH_2CH_2\underset{\underset{\displaystyle OH}{|}}{C}HCH_3 \implies BrCH_2CH_2CH_2CH=CH_2$$
target molecule

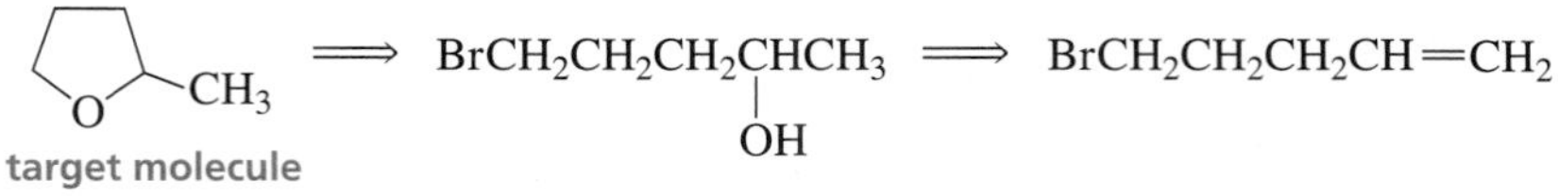

synthesis

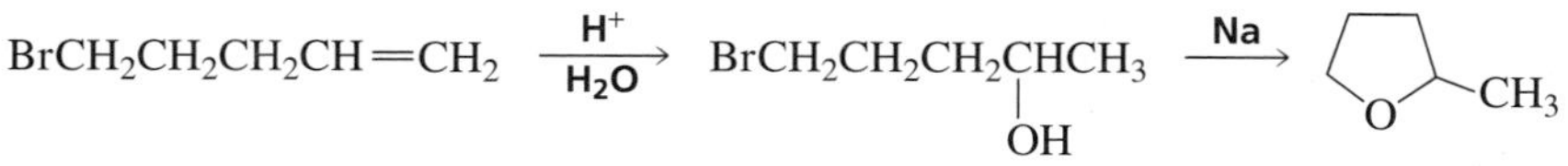

$$BrCH_2CH_2CH_2CH=CH_2 \xrightarrow[H_2O]{H^+} BrCH_2CH_2CH_2CHCH_3 \xrightarrow{Na}$$

with OH below the CHCH₃, leading to a cyclic ether (tetrahydrofuran ring) with CH₃ substituent.

PROBLEM 23

Could you just as well have used a precursor with the OH and Br reversed for the synthesis above? Explain.

PROBLEM 24

Describe how each of the following compounds could be prepared from the given starting material.

a. 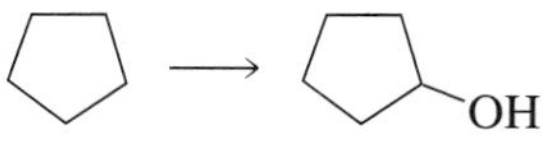 cyclopentane $\longrightarrow$ cyclopentane with OH substituent

b. $CH_3CH_2CH_2CH_3 \longrightarrow CH_3CHCH_2CH_3$ with OCH_2CH_3 below the CH

c. cyclohexane with $CH=CH_2 \longrightarrow$ cyclohexane with CH_2CH bearing a carbonyl ($=O$)

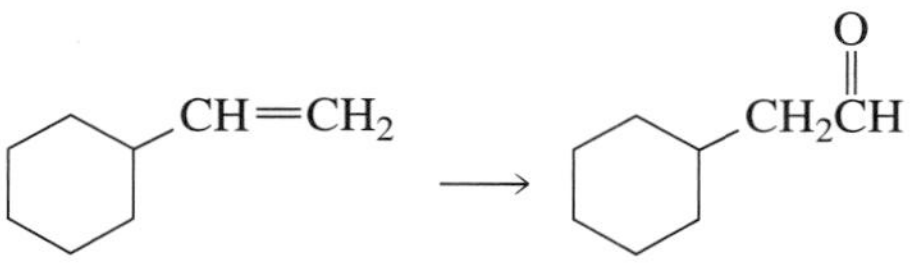

d. $CH_3CH_2CH_2CH_2Br \longrightarrow CH_3CH_2CCH_2CH_2CH_3$ with a carbonyl ($=O$) on the C

e. $BrCH_2CH_2CH_2CH_2Br \longrightarrow$ cyclohexane with CH_2CH_3 substituent

---

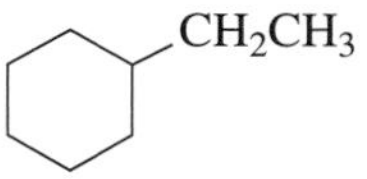

SUMMARY OF REACTIONS

1. E2 reaction: a one-step mechanism

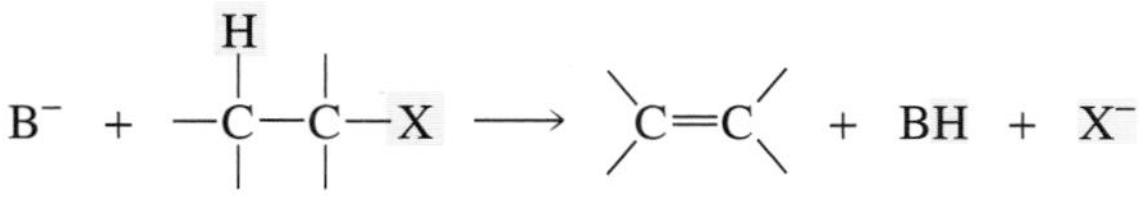

$$B^- + \quad \overset{\overset{\displaystyle H}{|}}{-\underset{|}{C}}-\overset{|}{\underset{|}{C}}-X \longrightarrow C=C + BH + X^-$$

Relative reactivities of alkyl halides: $3° > 2° > 1°$
Anti elimination; both E and Z stereoisomers are formed unless the β-carbon in the reactant has only one hydrogen

2. E1 reaction: a two-step mechanism with a carbocation intermediate

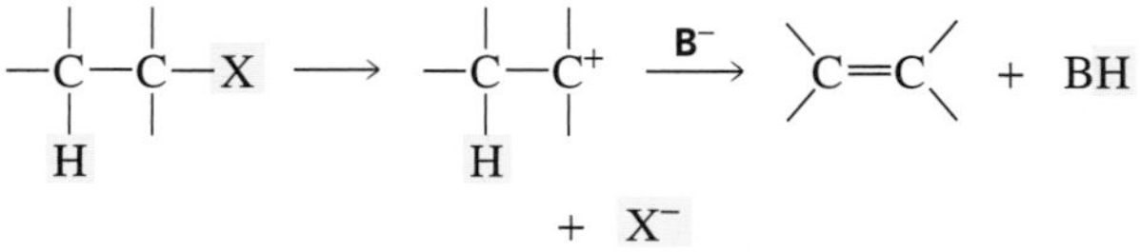

$$-\overset{|}{\underset{\underset{\displaystyle H}{|}}{C}}-\overset{|}{\underset{|}{C}}-X \longrightarrow -\overset{|}{\underset{\underset{\displaystyle H}{|}}{C}}-\overset{|}{C^+} \xrightarrow{B^-} C=C + BH$$

$$+ \; X^-$$

Relative reactivities of alkyl halides: $3° > 2° > 1°$
Anti and syn elimination; both E and Z stereoisomers are formed

Competing S_N2 and E2 Reactions
  Primary alkyl halides: primarily substitution
  Secondary alkyl halides: substitution and elimination
  Tertiary alkyl halides: only elimination

Competing S_N1 and E1 Reactions
  Primary alkyl halides: cannot undergo S_N1 or E1 reactions
  Secondary alkyl halides: substitution and elimination
  Tertiary alkyl halides: substitution and elimination

KEY TERMS

anti elimination (page 409)
anti-periplanar (page 408)
dehydrohalogenation (page 398)
deuterium kinetic isotope effect
  (page 415)
elimination reaction (page 397)

β-elimination reaction (page 398)
1,2-elimination reaction (page 398)
E1 reaction (page 404)
E2 reaction (page 397)
kinetic isotope effect
  (page 415)

syn elimination (page 409)
syn-periplanar (page 408)
target molecule (page 423)
Williamson ether synthesis
  (page 420)
Zaitsev's rule (page 400)

PROBLEMS

25. a. Indicate how each of the following factors affects an E1 reaction.
   b. Indicate how each of the following factors affects an E2 reaction.
       1. the structure of the alkyl halide   **3.** the concentration of the base
       2. the strength of the base   **4.** the solvent

26. Dr. Don T. Doit wanted to synthesize the anesthetic, 2-ethoxy-2-methylpropane. He used ethoxide ion and 2-chloro-2-methylpropane for his synthesis and ended up with very little ether. What was the predominant product of his synthesis? What reagents should he have used?

27. Which reaction in each of the following pairs will take place more rapidly? Explain your choice.

a. $(CH_3)_3CCl$ $\xrightarrow[\text{H}_2\text{O}]{\text{HO}^-}$ $\begin{array}{c} H_3C \\ \diagup \\ H_3C \end{array} C{=}CH_2 \ + \ Cl^-$

$(CH_3)_3CI$ $\xrightarrow[\text{H}_2\text{O}]{\text{HO}^-}$ $\begin{array}{c} H_3C \\ \diagup \\ H_3C \end{array} C{=}CH_2 \ + \ I^-$

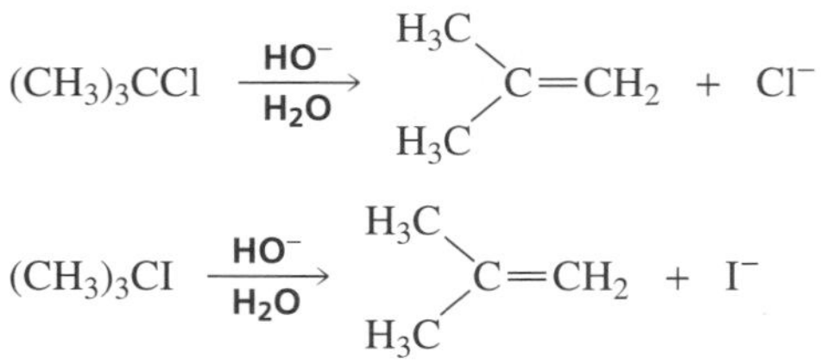

b.

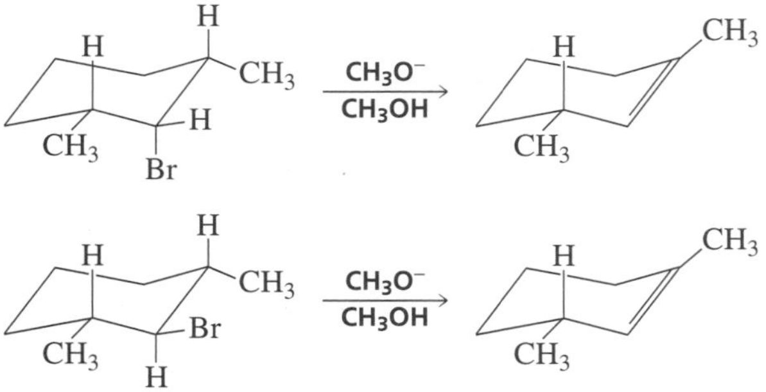

28. Give the elimination products of the following reactions. If the products can exist as stereoisomers, show what stereoisomers are obtained.

a. (*R*)-2-bromohexane + high concentration of HO⁻

b. (*R*)-2-bromohexane + H₂O

c. *trans*-1-chloro-2-methylcyclohexane + high concentration of CH₃O⁻

d. *trans*-1-chloro-2-methylcyclohexane + CH₃OH

e. 3-bromo-3-methylpentane + high concentration of HO⁻

f. 3-bromo-3-methylpentane + H₂O

g. (2*S*,3*S*)-2-chloro-3-methylpentane + high concentration of CH₃O⁻

h. (2*S*,3*R*)-2-chloro-3-methylpentane + high concentration of CH₃O⁻

i. (2*R*,3*S*)-2-chloro-3-methylpentane + high concentration of CH₃O⁻

j. (2*R*,3*R*)-2-chloro-3-methylpentane + high concentration of CH₃O⁻

k. 3-chloro-3-ethyl-2,2-dimethylpentane + high concentration of CH₃CH₂O⁻

29. Using the given starting material and any necessary organic or inorganic reagents, indicate how the desired compounds could be synthesized.

a.

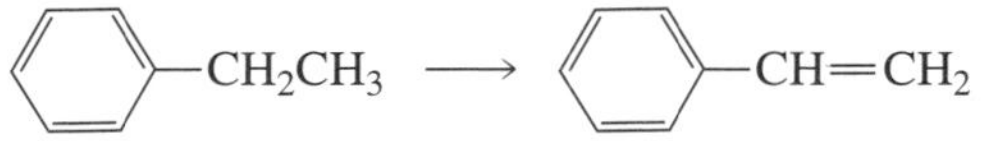

b. CH₃CH₂CH=CH₂ ⟶ CH₃CH₂CH₂CH₂NH₂

c. HOCH₂CH₂CH=CH₂ ⟶

d.

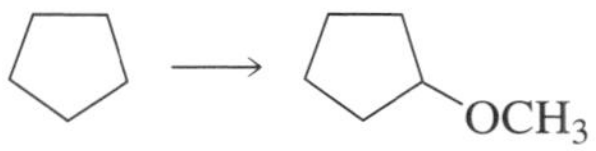

e. CH₃CH₂CH=CH₂ ⟶ CH₂=CHCH=CH₂

30. When 2-bromo-2,3-dimethylbutane reacts with a base under E2 conditions, two alkenes are formed: 2,3-dimethyl-1-butene and 2,3-dimethyl-2-butene.

a. Which of the following bases would give the highest percentage of the 1-alkene?

b. Which would give the highest percentage of the 2-alkene?

31. **a.** Give the structure of the products obtained from the reaction of both enantiomers of *cis*-1-chloro-2-isopropylcyclopentane with a high concentration of sodium methoxide in methanol.

b. Are all the products chiral?

c. How would the products differ if the starting material were the trans isomer?

d. Which pair of enantiomers forms substitution products more rapidly, cis or trans?

e. Which pair of enantiomers forms elimination products more rapidly, cis or trans?

32. Starting with cyclohexane, how could the following compounds be prepared?

a. *trans*-1,2-dichlorocyclohexane　　　**b.** 2-cyclohexenol

33. *cis*-1-Bromo-4-*tert*-butylcyclohexane and *trans*-1-bromo-4-*tert*-butylcyclohexane both react with sodium ethoxide in ethanol to give 4-*tert*-butylcyclohexene. The cis isomer reacts much more rapidly than the trans isomer. Explain.

34. The rate of the reaction of 1-bromo-2-butene with ethanol is increased if silver nitrate is added to the reaction mixture. Explain.

35. Give the products of the following reactions, assuming that all the reactions are carried out under S_N2/E2 conditions. If the products can exist as stereoisomers, show what stereoisomers are formed.

a. (3*S*,4*S*)-3-bromo-4-methylhexane + CH₃O⁻

b. (3*S*,4*R*)-3-bromo-4-methylhexane + CH₃O⁻

    c. (3R,4R)-3-bromo-4-methylhexane + CH_3O^-
    d. (3R,4S)-3-bromo-4-methylhexane + CH_3O^-

36. Two elimination products are obtained from the following E2 reaction.
    a. What are the elimination products?
    b. Which is formed in greater yield? Explain.

$$CH_3CH_2CHDCH_2Br \xrightarrow{\text{HO}^-}$$

37. Three substitution products and three elimination products are obtained from the following reaction. Account for the formation of these products.

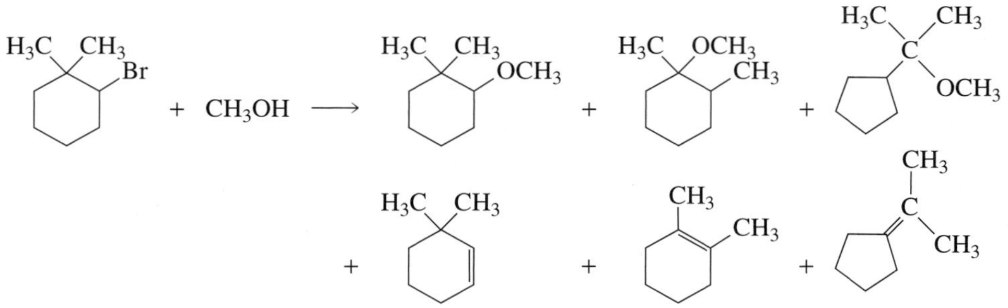

38. When the following stereoisomer of 2-chloro-1,3-dimethylcyclohexane reacts with methoxide ion in a solvent that encourages $S_N2/E2$ reactions, only one product is formed. When the same compound reacts with methoxide ion in a solvent that favors $S_N1/E1$ reactions, eight products are formed. Identify the products that are formed under the two sets of conditions.

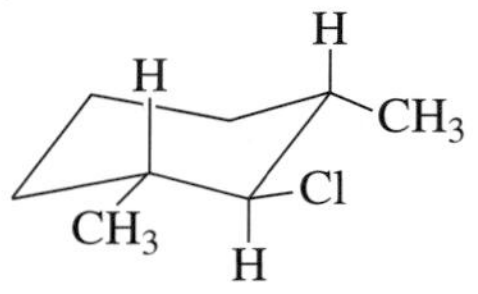

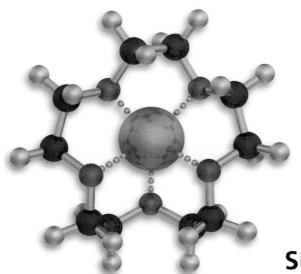

a crown ether

REACTIONS AT AN *sp*^{3} HYBRIDIZED CARBON III:
Substitution and Elimination Reactions of Compounds with Leaving Groups Other than Halogen. Organometallic Compounds

In Chapters 9 and 10 we saw that alkyl halides undergo substitution and elimination reactions because they have an electron-withdrawing halogen bonded to an sp^3 hybridized carbon. Compounds with an electron-withdrawing group other than a halogen bonded to an sp^3 hybridized carbon also undergo substitution and elimination reactions. The reactivity of these compounds compared with alkyl halides depends on the nature of the electron-withdrawing leaving group.

Alcohols, ethers, and quaternary ammonium ions have leaving groups that are stronger bases than halide ions. Therefore, they are less reactive than alkyl halides in substitution and elimination reactions. You will see that alcohols and ethers, because of their strongly basic (poor) leaving groups, have to be "activated" before they can undergo a substitution or elimination reaction. Quaternary ammonium ions, with somewhat less basic leaving groups, can undergo an elimination reaction if they are heated in the presence of a strong base. (See Table 9.3, remembering that the weaker the acid, the stronger its conjugate base, and the stronger the base, the poorer it is as a leaving group.)

$$R—X \qquad R—O—H \qquad R—O—R \qquad R—\overset{\overset{\displaystyle R}{\displaystyle |}}{\underset{\underset{\displaystyle R}{\displaystyle |}}{\overset{+}{N}}}—R$$

an alkyl halide	an alcohol	an ether	a quaternary
X = F, Cl, Br, I			ammonium ion

An alcohol cannot undergo a nucleophilic substitution reaction with a halide ion, because a weakly basic halide ion cannot displace the strongly basic HO⁻ group. [Hydrogen halides (HI, HBr, HCl, HF) are much stronger acids than H_2O; therefore, halide ions (I⁻, Br⁻, Cl⁻, F⁻) are much weaker bases than HO⁻.]

**11.1
SUBSTITUTION
REACTIONS OF
ALCOHOLS**

$$CH_3OH \; + \; \underset{\text{weak base}}{Br^-} \; \xrightarrow{\quad\times\quad} \; CH_3Br \; + \; \underset{\text{strong base}}{HO^-}$$

If, however, a hydrogen halide is used as the reagent in the reaction instead of a halide ion, the substitution reaction will take place. The reaction is slow and requires heat in order to reach completion in a reasonable period of time.

$$CH_3OH \; + \; HBr \; \xrightarrow{\;\Delta\;} \; CH_3Br \; + \; H_2O$$

In the first step of this reaction, H^+ protonates the oxygen atom of the alcohol. Protonation changes the leaving group from HO^- (a strong base) to H_2O (a weak base). The weakly basic H_2O can be substituted by a halide ion.

$$CH_3OH \; + \; H^+ \; \rightleftharpoons \; CH_3\overset{H}{\underset{+}{O}}H \; \xrightarrow{Br^-} \; CH_3Br \; + \; H_2O$$

Primary, secondary, and tertiary alcohols can all undergo nucleophilic substitution reactions with hydrogen halides.

$$\underset{\substack{\text{1-propanol}\\\text{a primary alcohol}}}{CH_3CH_2CH_2OH} \; + \; HI \; \xrightarrow{\;\Delta\;} \; \underset{\text{1-iodopropane}}{CH_3CH_2CH_2I} \; + \; H_2O$$

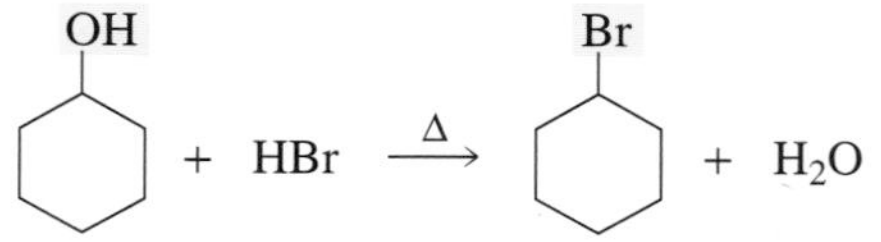

cyclohexanol
a secondary alcohol   bromocyclohexane

$$\underset{\substack{\text{2-methyl-2-butanol}\\\text{a tertiary alcohol}}}{CH_3CH_2\overset{CH_3}{\underset{CH_3}{C}}OH} \; + \; HCl \; \longrightarrow \; \underset{\text{2-chloro-2-methylbutane}}{CH_3CH_2\overset{CH_3}{\underset{CH_3}{C}}Cl} \; + \; H_2O$$

The mechanism of the substitution reaction depends on the structure of the alcohol. Tertiary and secondary alcohols undergo S_N1 reactions.

mechanism of the S_N1 reaction

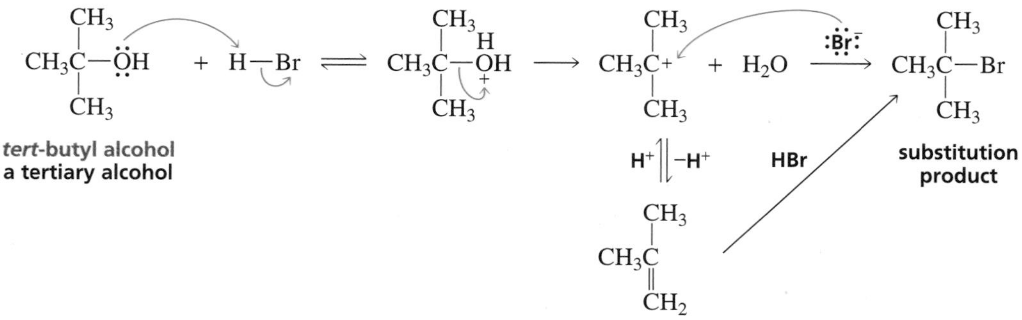

tert-butyl alcohol
a tertiary alcohol

$H^+ \parallel -H^+$ HBr   substitution product

elimination product

The carbocation intermediate formed in an S_N1 reaction with a hydrogen halide has two possible fates: it can combine with a nucleophile and form a substitution product, or it can lose a proton and form an elimination product. However, since the alkene formed in the elimination reaction can undergo an addition reaction with HX, only the substitution product is actually obtained.

Because tertiary carbocations are easier to form than secondary carbocations, the reaction of a hydrogen halide with a tertiary alcohol is faster than the reaction with a secondary alcohol. The reaction of a hydrogen halide with a tertiary alcohol proceeds readily at room temperature; the reaction of a hydrogen halide with a secondary alcohol requires heat.

$$CH_3CHCH_2CH_3 \ + \ HBr \ \xrightarrow{\Delta} \ CH_3CHCH_2CH_3 \ + \ H_2O$$
$$\quad\ | \qquad\qquad\qquad\qquad\qquad\quad |$$
$$\quad OH \qquad\qquad\qquad\qquad\qquad Br$$

Primary alcohols cannot undergo S_N1 reactions because primary carbocations are unstable. Primary alcohols, therefore, have to undergo S_N2 reactions.

mechanism of the S_N2 reaction

Only the substitution product is formed. No elimination product is formed, because the halide ion, although a good nucleophile, is a weak base. A base is needed in an elimination reaction to remove a proton from a β-carbon. (Remember that a β-carbon is adjacent to the carbon bonded to the leaving group.)

When HCl is used instead of HBr or HI, the S_N2 reaction is slower, because Cl^- is a poorer nucleophile than Br^- or I^-. The rate of the reaction can be increased by using $ZnCl_2$ as a catalyst, because the Zn^{2+} cation is a Lewis acid and complexes strongly with the nonbonding electrons on oxygen. This weakens the carbon–oxygen bond and creates a better leaving group.

Tertiary and secondary alcohols undergo S_N1 reactions with hydrogen halides. Primary alcohols undergo S_N2 reactions with hydrogen halides.

$$CH_3CH_2CH_2OH \ + \ HCl \ \xrightarrow[\Delta]{ZnCl_2} \ CH_3CH_2CH_2Cl \ + \ H_2O$$

PROBLEM 1

The observed relative reactivities of primary, secondary, and tertiary alcohols toward reaction with a hydrogen halide are as follows: tertiary alcohols are more reactive than secondary alcohols, which are more reactive than primary alcohols. If secondary alcohols underwent an S_N2 reaction with a hydrogen halide rather than an S_N1 reaction, what would be the relative reactivities of the three classes of alcohols? Explain.

THE LUCAS TEST

Whether an alcohol is primary, secondary, or tertiary can be determined by taking advantage of the relative rates at which the three classes of alcohols react with $HCl/ZnCl_2$. This is known as the *Lucas test.* The alcohol is added to a mixture of concentrated HCl and $ZnCl_2$ (the Lucas reagent). Low-molecular-weight alcohols are soluble in the Lucas reagent, but the alkyl halide products are not. So, as the alkyl halide is formed, the solution turns cloudy. When the test is carried out at room temperature, the solution turns cloudy immediately if the alcohol is tertiary, it turns cloudy in about 5 min if the alcohol is secondary, and it remains clear if the alcohol is primary. Since the test relies on the complete solubility of the alcohol in the Lucas reagent, it is limited to alcohols with fewer than six carbons.

Howard J. Lucas (1885–1963) *was born in Ohio and earned B.S. and M.S. degrees from Ohio State University. He published a description of the Lucas test in 1930. He was a professor of chemistry at the California Institute of Technology.*

Because the reaction of a secondary or a tertiary alcohol with a hydrogen halide is an S_N1 reaction and therefore involves a carbocation intermediate, we must check for the possibility of a carbocation rearrangement when determining the product of a substitution reaction. Remember that a carbocation rearrangement will occur if it leads to formation of a more stable carbocation. For example, the major product of the reaction of 3-methyl-2-butanol with HBr is 2-bromo-2-methylbutane, because a 1,2-hydride shift converts the initially formed secondary carbocation into a more stable tertiary carbocation.

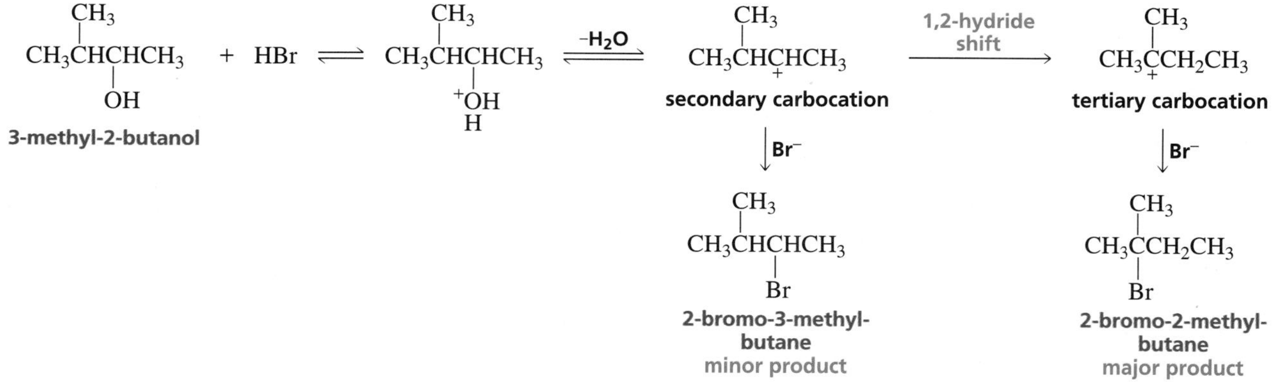

PROBLEM 2◆

Give the major product of each of the following reactions.

a. $CH_3CH_2CH(OH)CH_3 + HBr \xrightarrow{\Delta}$

b. 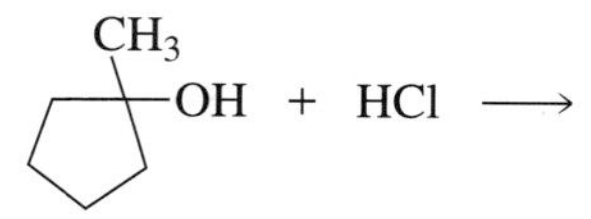

c. $CH_3C(CH_3)_2{-}CH(OH)CH_3 + HBr \xrightarrow{\Delta}$

d. 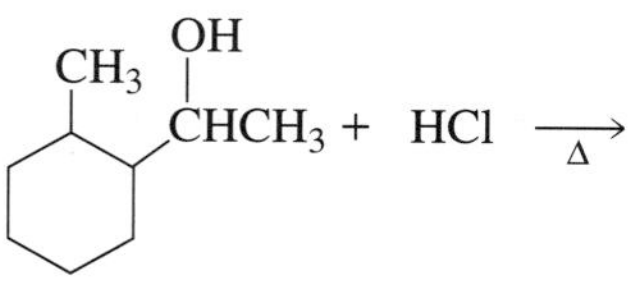

GRAIN ALCOHOL AND WOOD ALCOHOL

When ethanol is ingested, it acts on the central nervous system. Ingestion of moderate amounts affects judgment and lowers inhibitions. Higher concentrations interfere with motor coordination and cause slurred speech and amnesia. Even higher concentrations cause nausea and loss of consciousness. Ingestion of very large amounts interferes with spontaneous respiration and can be fatal.

The ethanol in alcoholic beverages is produced by fermentation of glucose in grapes, corn, malt (sprouted barley), rye, etc., which is why ethanol is also known as grain alcohol.

$$C_6H_{12}O_6 \xrightarrow{\text{yeast enzymes}} 2\ CH_3CH_2OH + 2\ CO_2$$
$$\text{glucose} \qquad\qquad\qquad \text{ethanol}$$

The kind of beverage produced (white or red wine, beer, scotch, bourbon, champagne) depends on the plant species being fermented, whether the CO_2 that is formed is allowed to escape, whether other substances are added, and how the beverage is purified (by sedimentation for wines; by distillation for scotch and brandy).

Because ethanol is used as a starting material in a wide variety of commercial processes, laboratory alcohol is carefully regulated by the federal government to make certain it is not used for the illegal preparation of alcoholic beverages. Denatured alcohol is ethanol that has been made undrinkable by adding a denaturant such as benzene or methanol.

Methanol, also known as wood alcohol because it can be obtained from the distillation of wood in the absence of oxygen, is toxic. Ingestion of even very small amounts can cause blindness. Ingestion of as little as an ounce has been known to be fatal. The antidote to methanol poisoning is discussed in Section 17.11.

Alcohols are readily available, inexpensive compounds. But because they have poor leaving groups, they are not very reactive toward substitution. Alkyl halides, in contrast, have good leaving groups and are very useful to synthetic chemists because the halide can be displaced by a wide variety of nucleophiles, leading to a wide variety of products (Section 9.3). By converting alcohols into reactive alkyl halides, chemists are able to use alcohols as starting materials for the preparation of many organic compounds.

11.2 ADDITIONAL METHODS FOR CONVERTING ALCOHOLS INTO ALKYL HALIDES

$$\begin{array}{ccccc} R\text{—OH} & \longrightarrow & R\text{—X} & \xrightarrow{\text{nucleophile}} & R\text{—Nu} \\ \text{alcohol} & & \text{alkyl halide} & & \\ & & X = \text{Cl, Br, I} & & \end{array}$$

We have seen that an alcohol can be converted into an alkyl halide by treating it with a hydrogen halide. Alcohols can also be converted into alkyl halides using phosphorus tribromide (PBr_3), phosphorus trichloride (PCl_3), or thionyl chloride ($SOCl_2$). These reagents convert the alcohol into an intermediate with a leaving group that is readily displaced by a halide ion. For example, a phosphorus trihalide converts the strongly basic leaving group of an alcohol (^-OH) into a weakly basic halophosphite group.

$$CH_3CHCH_2\ddot{O}H + \underset{\substack{\text{phosphorus} \\ \text{tribromide}}}{\overset{Br}{\underset{Br}{P\text{—Br}}}} \xrightarrow{\text{pyridine}} \underset{\substack{CH_3 \\ + Br^-}}{CH_3CHCH_2\overset{+}{O}PBr_2} \longrightarrow \underset{\substack{CH_3 \\ \underbrace{\quad}_{\substack{a \\ \text{bromophosphite} \\ \text{group}}}}}{CH_3CHCH_2OPBr_2} + H^+$$

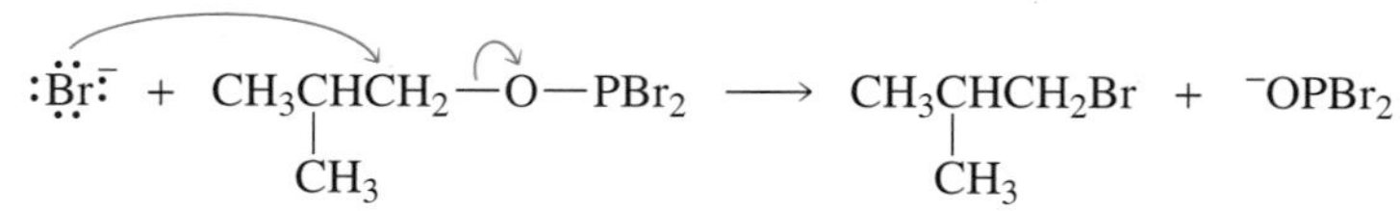

$$:\ddot{Br}\!:^- + CH_3CHCH_2\!-\!O\!-\!PBr_2 \longrightarrow CH_3CHCH_2Br + {}^-OPBr_2$$
$$\quad\quad\quad\quad CH_3 \quad\quad\quad\quad\quad\quad\quad\quad CH_3$$

Thionyl chloride converts an OH group into a chlorosulfite leaving group, which is displaced by Cl⁻.

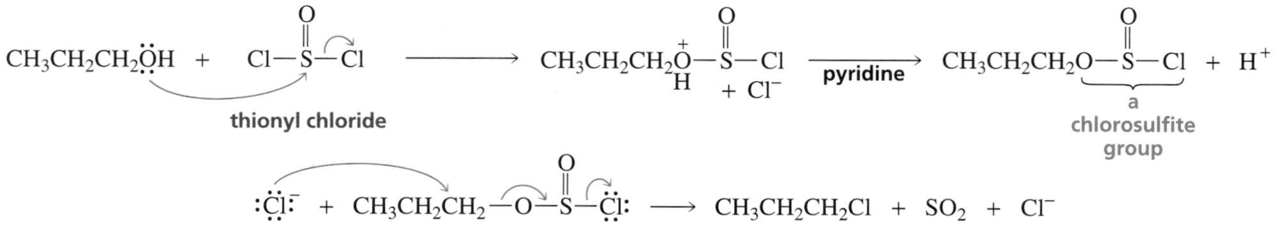

$$CH_3CH_2CH_2\ddot{O}H + Cl\!-\!\overset{O}{\underset{}{S}}\!-\!Cl \longrightarrow CH_3CH_2CH_2\overset{+}{O}\!-\!\overset{O}{\underset{}{S}}\!-\!Cl \xrightarrow{\text{pyridine}} CH_3CH_2CH_2O\!-\!\overset{O}{\underset{}{S}}\!-\!Cl + H^+$$

thionyl chloride   H + Cl⁻   a **chlorosulfite group**

$$:\ddot{Cl}\!:^- + CH_3CH_2CH_2\!-\!O\!-\!\overset{O}{\underset{}{S}}\!-\!\ddot{Cl}\!: \longrightarrow CH_3CH_2CH_2Cl + SO_2 + Cl^-$$

Pyridine is generally used as the solvent in these reactions to neutralize the HCl that is formed.

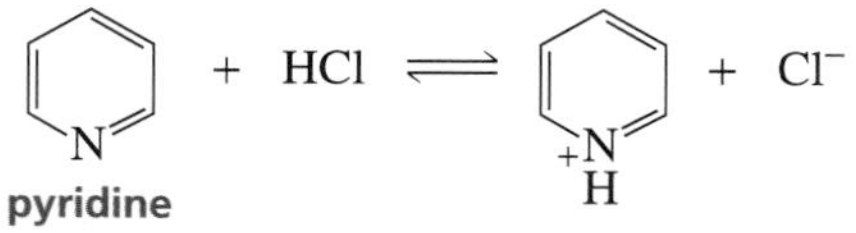

pyridine + HCl ⇌ +N–H + Cl⁻

pyridine

Table 11.1 shows some of the methods that can be used to convert alcohols into alkyl halides.

TABLE 11.1 Commonly Used Methods for Converting Alcohols into Alkyl Halides

ROH + HBr	$\xrightarrow{\Delta}$	RBr
ROH + HI	$\xrightarrow{\Delta}$	RI
ROH + HCl	$\xrightarrow{\Delta}$	RCl
ROH + PBr₃	$\xrightarrow{\text{pyridine}}$	RBr
ROH + PCl₃	$\xrightarrow{\text{pyridine}}$	RCl
ROH + SOCl₂	$\xrightarrow{\text{pyridine}}$	RCl

11.3 CONVERSION OF ALCOHOLS INTO SULFONATE ESTERS

Instead of converting an alcohol into an alkyl halide, another way to activate an alcohol for subsequent reaction with a nucleophile is to convert it into a sulfonate ester. A **sulfonate ester** is formed when an alcohol reacts with a sulfonyl chloride.

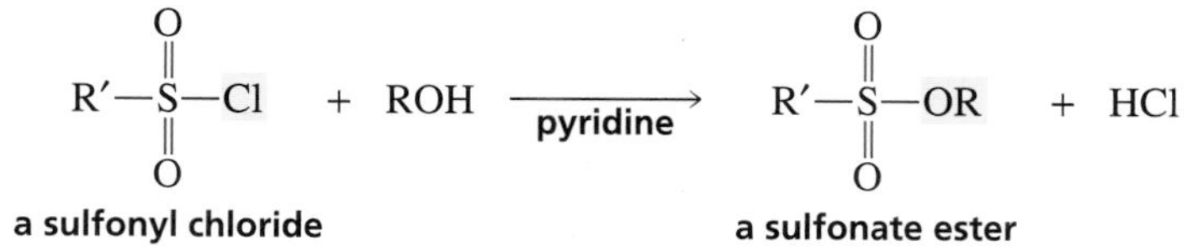

$$R'\!-\!\overset{O}{\underset{O}{S}}\!-\!Cl + ROH \xrightarrow{\text{pyridine}} R'\!-\!\overset{O}{\underset{O}{S}}\!-\!OR + HCl$$

a sulfonyl chloride   **a sulfonate ester**

Sulfonic acids are strong acids ($pK_a \sim -1$) owing to the combination of the electron-withdrawing SO_3 group and the fact that the electrons holding the proton become delocalized over three oxygen atoms when they are no longer bonded to the proton.

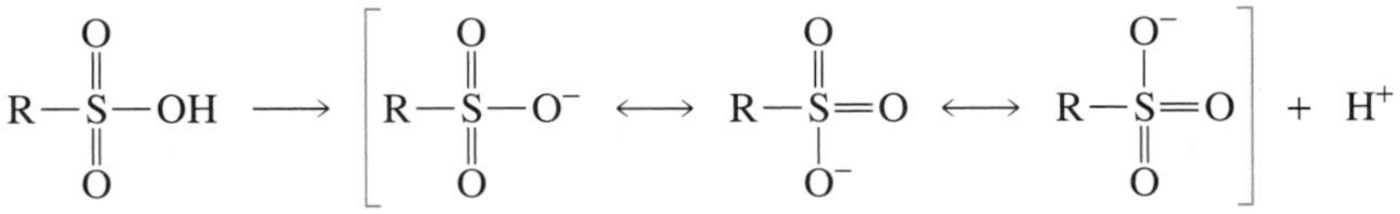

contributing resonance structures

Because a sulfonic acid is a strong acid, its conjugate base is weak. The weak conjugate base gives the sulfonate ester an excellent leaving group.

Many sulfonyl chlorides are available to activate OH groups. The most common one is *para*-toluenesulfonyl chloride.

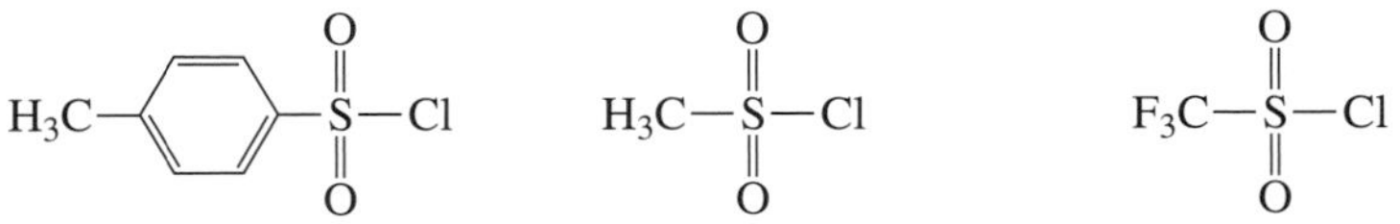

para-toluenesulfonyl
chloride

methanesulfonyl
chloride

trifluoromethanesulfonyl
chloride

Once the alcohol has been activated by being converted into a sulfonate ester, the appropriate nucleophile is added, generally under conditions that favor S_N2/E2 reactions. The reactions take place readily at room temperature because of the weakly basic sulfonate ion. A *para*-toluenesulfonate ion is about 100 times better than a chloride ion as a leaving group. The products are as varied as the nucleophiles chosen to react with the sulfonate ester.

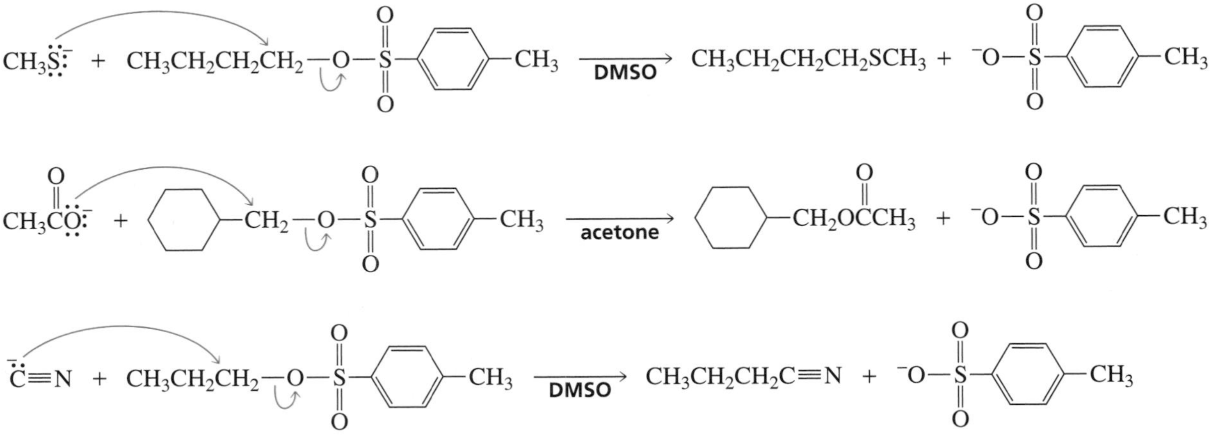

−OTs

para-Toluenesulfonyl chloride is abbreviated to tosyl chloride (TsCl); esters of *para*-toluenesulfonic acid are called **alkyl tosylates** and are abbreviated ROTs. The leaving group, therefore, is ⁻OTs.

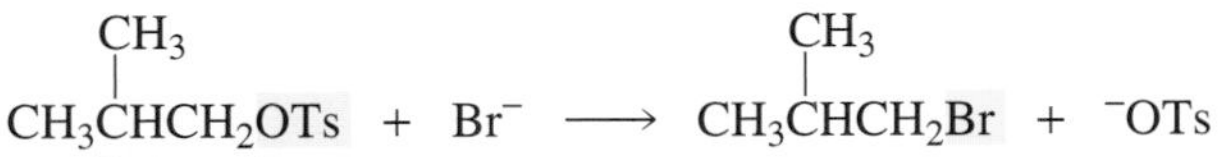

an alkyl tosylate

If the nucleophile that reacts with a sulfonate ester is a halide ion, we have yet another method for converting an alcohol into an alkyl halide (Table 11.1).

$$CH_3CH_2CHCH_2CH_3 \xrightarrow[\text{pyridine}]{\text{TsCl}} CH_3CH_2CHCH_2CH_3 \xrightarrow{Cl^-} CH_3CH_2CHCH_2CH_3 + {}^-OTs$$

with OH, OTs, Cl groups labeled; 3-pentanol, 3-chloropentane

The configuration of the substitution product obtained from an alcohol depends on the method used to activate the alcohol. When the alcohol is activated by being converted into a sulfonate ester, the carbon–oxygen bond of the alcohol is not broken. As a result, the sulfonate ester has the same configuration as the alcohol. When the nucleophile subsequently attacks the sulfonate ester in an S_N2 reaction, the carbon–oxygen bond breaks and a Walden inversion occurs. The configuration of the product is, therefore, *inverted* compared with the configuration of the alcohol. (Recall that a Walden inversion involves nucleophilic back side attack and inversion of configuration; Section 9.2.)

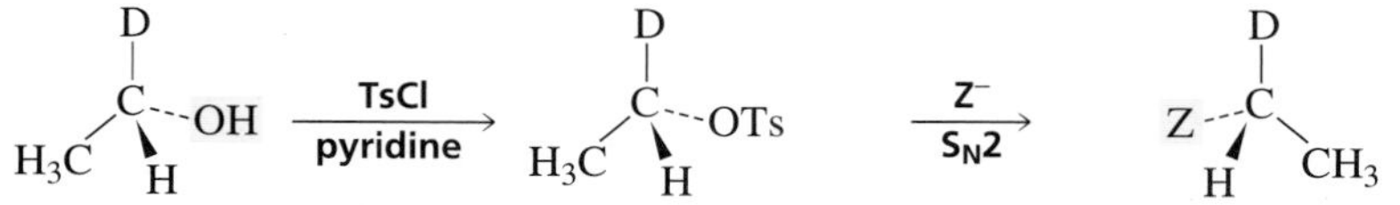

the configuration of the product
is inverted compared with the
configuration of the starting material

When the alcohol is activated by being converted into an alkyl halide, and the halide ion is then displaced by another nucleophile, the product will have the same configuration as the starting material. The configuration of the alcohol is retained because attack of the halide ion on the protonated alcohol is an S_N2 reaction, so it causes a Walden inversion. Subsequent attack of the nucleophile on the alkyl halide causes a second Walden inversion. Because two Walden inversions take place, the final product has the *same* configuration as the alcohol.

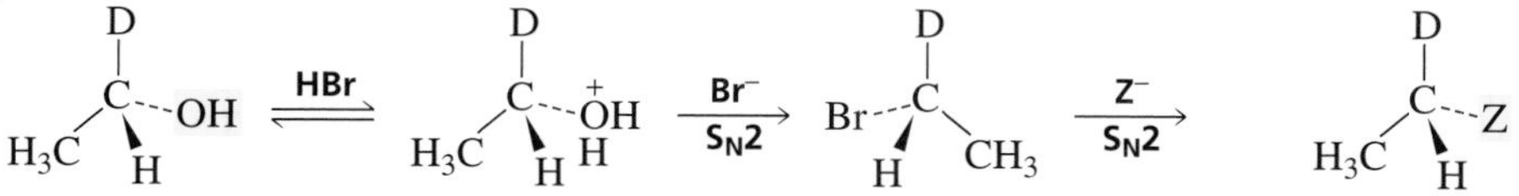

the product and the starting
material have the same
configuration

PROBLEM 3 ◆

Give the structures of X and Z.

a. $\xrightarrow[\text{pyridine}]{\text{TsCl}}$ W $\xrightarrow[\text{DMF}]{\text{NaC}\equiv\text{N}}$ X

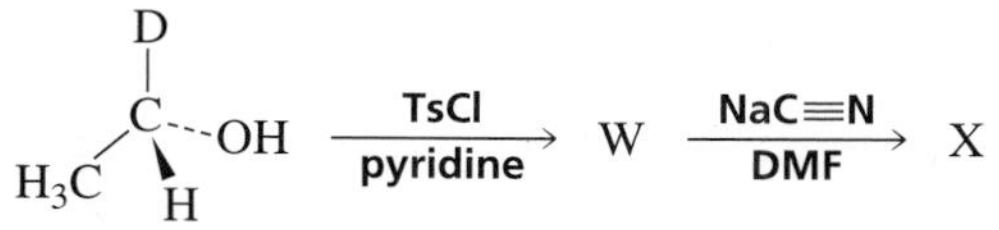

b. $\xrightarrow[\Delta]{\text{HBr}}$ Y $\xrightarrow[\text{DMF}]{\text{NaC}\equiv\text{N}}$ Z

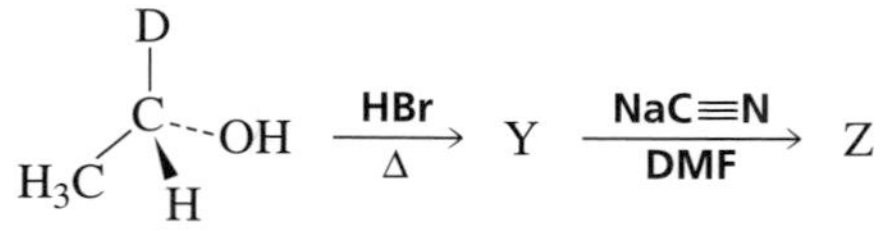

PROBLEM 4

Show how 1-butanol can be converted into the following compounds.

a. $CH_3CH_2CH_2CH_2I$

d. $CH_3CH_2CH_2CH_2NHCH_2CH_3$

b. $CH_3CH_2CH_2CH_2O\overset{\displaystyle O}{\overset{\|}{C}}CH_2CH_3$

e. $CH_3CH_2CH_2CH_2SH$

c. $CH_3CH_2CH_2CH_2OCH_3$

f. $CH_3CH_2CH_2CH_2C{\equiv}N$

An alcohol can undergo an elimination reaction, losing an OH from one carbon and an H from an adjacent carbon. Overall, this amounts to the elimination of a molecule of water. Loss of water from a molecule is called **dehydration.** The dehydration of alcohols requires an acid catalyst and heat. Sulfuric acid (H_2SO_4) and phosphoric acid (H_3PO_4) are the most commonly used acid catalysts.

11.4 DEHYDRATION OF ALCOHOLS

$$CH_3CH_2\underset{\underset{\displaystyle OH}{|}}{C}HCH_3 \xrightarrow[\Delta]{H_2SO_4} CH_3CH{=}CHCH_3 + H_2O$$

In the first step of the dehydration reaction, the acid protonates the oxygen of the alcohol. As we saw earlier, protonation converts a poor leaving group (HO^-) into a good leaving group (H_2O). In the next step, water departs and a carbocation is formed. A base in the reaction mixture (either hydrogen sulfate or water) removes a proton from a carbon adjacent to the positively charged carbon, forming the alkene product and regenerating the acid catalyst.

mechanism of dehydration (E1)

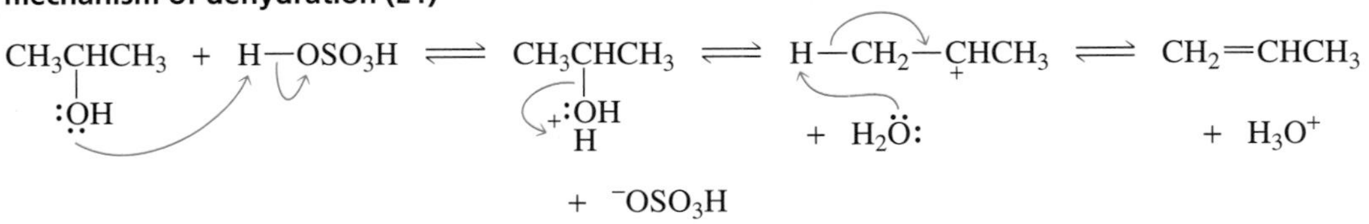

$$CH_3\underset{\underset{\displaystyle :\ddot{O}H}{|}}{C}HCH_3 + H{-}OSO_3H \rightleftharpoons CH_3\underset{\underset{\displaystyle \overset{+}{O}H}{|}{\underset{\displaystyle H}{}}}{C}HCH_3 \rightleftharpoons H{-}CH_2{-}\overset{+}{C}HCH_3 \rightleftharpoons CH_2{=}CHCH_3$$

$$+ \ ^-OSO_3H \qquad\qquad + H_2\ddot{O}: \qquad\qquad + H_3O^+$$

Dehydration is an E1 reaction of a protonated alcohol. The difference between dehydration and an E1 reaction of an alkyl halide is that H_2O is the leaving group in the dehydration of an alcohol whereas a halide ion is the leaving group in the dehydrohalogenation of an alkyl halide.

E1 dehydration of an alcohol

$$H_2\ddot{O}:$$

$$RCH_2\underset{\underset{\displaystyle \overset{+}{O}H}{|}{\underset{\displaystyle H}{}}}{C}HR \rightleftharpoons RCH{-}\overset{+}{C}HR \rightleftharpoons RCH{=}CHR + H_3O^+$$

$$+ H_2O$$

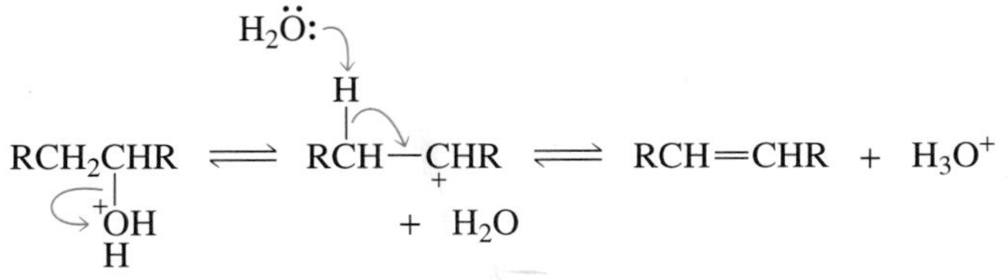

E1 dehydrohalogenation of an alkyl halide

$$RCH_2CHR \rightleftharpoons RCH\!-\!CHR \rightleftharpoons RCH\!=\!CHR + H_3O^+$$
$$\underset{Br}{|} \qquad \overset{+}{} \qquad + Br^-$$

with $H_2\ddot{O}:$ and H shown approaching.

When more than one elimination product can be formed, the major product is the one obtained by following *Zaitsev's rule:* the hydrogen is removed from the β-carbon that is bonded to the fewest hydrogens (Section 10.12). Zaitsev's rule leads to the most substituted of the possible alkene products. The more substituted alkene is more stable and therefore the transition state leading to its formation is more stable, allowing it to be formed more rapidly (Figure 11.1).

$$\underset{OH}{\overset{CH_3}{CH_3CCH_2CH_3}} \underset{\Delta}{\overset{H_3PO_4}{\rightleftharpoons}} \underset{84\%}{CH_3C\!=\!CHCH_3} + \underset{16\%}{CH_2\!=\!CCH_2CH_3} + H_2O$$

$$\underset{}{\overset{H_3C\ \ OH}{\bigcirc}} \underset{\Delta}{\overset{H_2SO_4}{\rightleftharpoons}} \underset{93\%}{\overset{CH_3}{\bigcirc}} + \underset{7\%}{\overset{CH_2}{\bigcirc}} + H_2O$$

In Section 3.13, we saw that an alkene is hydrated (adds water) in the presence of an acid catalyst, thereby forming an alcohol. Hydration of an alkene is the exact reverse of the acid-catalyzed dehydration of an alcohol.

$$\underset{OH}{\overset{}{RCH_2CHR}} + H^+ \underset{hydration}{\overset{dehydration}{\rightleftharpoons}} RCH\!=\!CHR + H_2O + H^+$$

To prevent the alkene formed in the dehydration reaction from adding water and reverting back to the alcohol, the alkene can be removed by distillation as it is formed, since it has a much lower boiling point than the alcohol. Removing products displaces the equilibrium to the right (Le Châtelier's principle; Section 9.3).

Figure 11.1 ▶
The reaction coordinate diagram for dehydration of a protonated alcohol. The major product is the more substituted alkene, because its greater stability causes the transition state leading to its formation to be more stable.

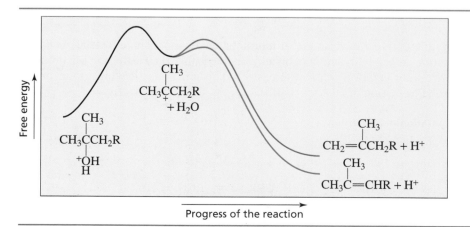

PROBLEM 5

Why is the acid-catalyzed dehydration of an alcohol a reversible reaction whereas the base-promoted dehydrohalogenation of an alkyl halide is not reversible?

Because the rate-determining step in the dehydration of an alcohol is the formation of a carbocation intermediate, the rate of dehydration parallels the ease with which the carbocation is formed. Tertiary alcohols are the easiest to dehydrate because tertiary carbocations are more stable and therefore are easier to form than secondary and primary carbocations (Section 3.10). In order to undergo dehydration, tertiary alcohols must be heated to about 50 °C in 5% H_2SO_4; secondary alcohols must be heated to about 100 °C in 75% H_2SO_4; primary alcohols can be dehydrated only under extreme conditions (170 °C in 95% H_2SO_4).

relative ease of dehydration

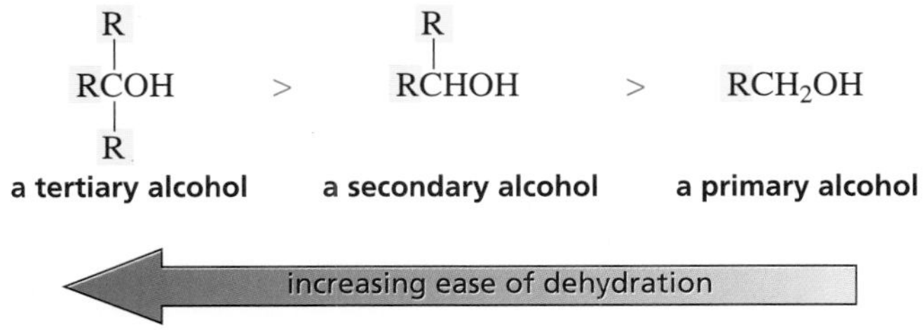

Because dehydration of an alcohol involves the formation of a carbocation intermediate, we must remember to check the structure of the carbocation for the possibility of rearrangement. The carbocation will rearrange if rearrangement produces a more stable carbocation.

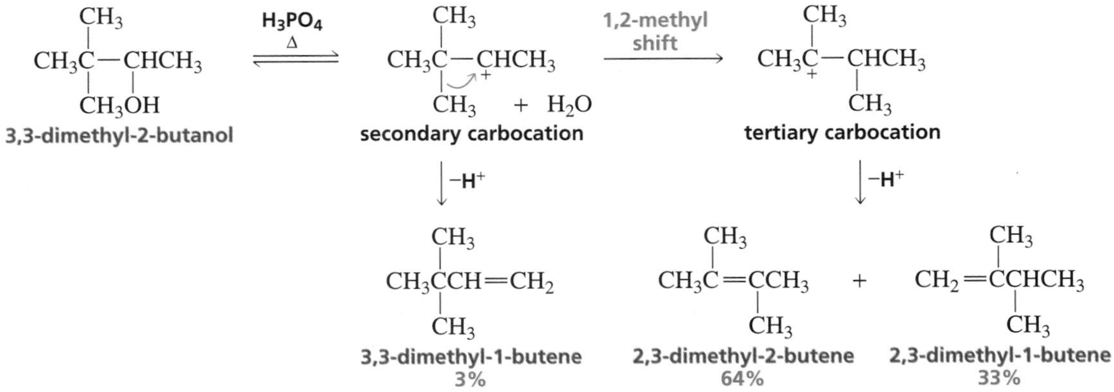

The following reaction is an example of a **ring-expansion rearrangement** (Section 3.14). Both the initially formed carbocation and the carbocation it rearranges to are secondary carbocations, but the initially formed carbocation is less stable because of the strain in the four-membered ring (Section 2.12). Rearrangement relieves this strain. The secondary carbocation then rearranges by a 1,2-hydride shift to an even more stable tertiary carbocation.

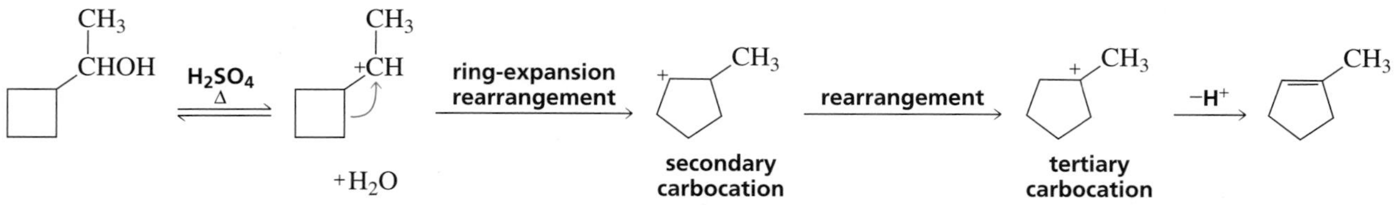

PROBLEM 6◆

What product would be formed if the alcohol above were heated with an equivalent amount of HBr rather than with a catalytic amount of H_2SO_4?

Tertiary and secondary alcohols undergo dehydration by an E1 pathway.

While dehydration of secondary and tertiary alcohols is an E1 reaction, dehydration of primary alcohols is an E2 reaction because of the difficulty encountered in forming primary carbocations. An ether is also obtained as the product of a competing S_N2 reaction.

mechanism of dehydration (E2)

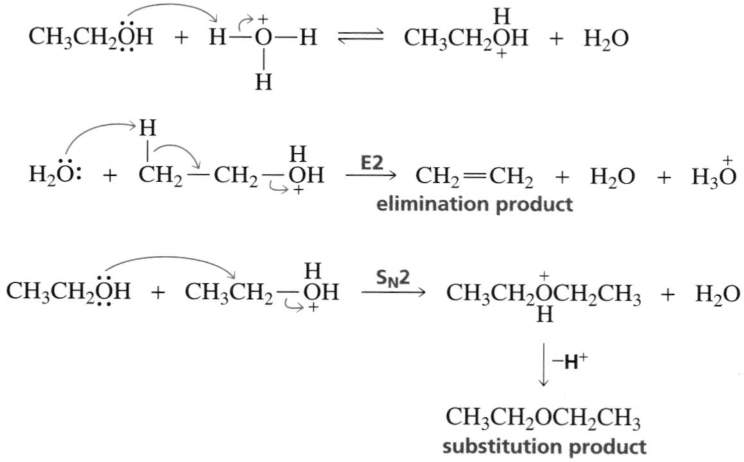

Primary alcohols undergo dehydration by an E2 pathway.

PROBLEM 7

Heating an alcohol with sulfuric acid is a good method for preparing a symmetrical ether such as diethyl ether. It is not a good method for preparing an asymmetrical ether such as ethyl propyl ether.

a. Explain.

b. How would you synthesize ethyl propyl ether?

Although the dehydration of a primary alcohol is an E2 reaction and therefore does not form a carbocation intermediate, the product obtained is identical to the product that would be obtained if a carbocation had been formed in an E1 reaction and then had rearranged. For example, we would expect 1-butene to be the product of the E2 dehydration of 1-butanol. However, we find that the product is actually 2-butene, which would be the product of an E1 reaction in which the carbocation intermediate had rearranged. This occurs because, after the E2 product (1-butene) is formed, a proton from the acidic solution adds to the double bond, following Markovnikov's rule, forming a carbocation. Loss of a proton from the carbocation, following Zaitsev's rule, gives the final product of the reaction. This product is identical to the product that would have been formed if the primary alcohol had undergone an E1 dehydration.

$$CH_3CH_2CH_2CH_2OH \xrightarrow[\Delta]{H_2SO_4} CH_3CH_2CH=CH_2 \xrightarrow{H^+} CH_3CH_2\overset{+}{C}HCH_3 \xrightarrow{-H^+} CH_3CH=CHCH_3$$

1-butanol   1-butene   2-butene

$$+ \quad H_2O$$

The stereochemical outcome of the E1 dehydration of an alcohol is identical to the stereochemical outcome of the E1 dehydrohalogenation of an alkyl halide: both the *E* and *Z* isomers are obtained as products. More of the *E* isomer is formed because it is more stable than the *Z* isomer and therefore the transition state leading to its formation is more stable, causing it to be formed more rapidly (Section 10.5). In the dehydration of 2-butanol, both isomers of 2-butene are more stable than 1-butene, which is formed only in trace amounts.

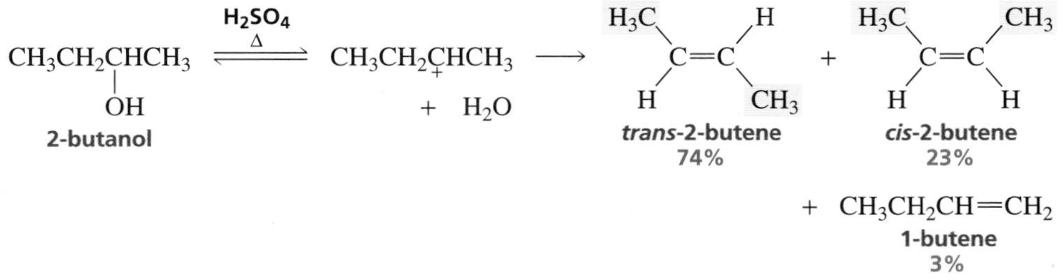

$$CH_3CH_2CHCH_3 \xrightarrow[\Delta]{H_2SO_4} CH_3CH_2\overset{+}{C}HCH_3 \longrightarrow$$

OH $+ \quad H_2O$

2-butanol

trans-2-butene   cis-2-butene
74%   23%

$$+ \quad CH_3CH_2CH=CH_2$$

1-butene
3%

The relatively harsh conditions (acid and heat) required for alcohol dehydration and the structural changes resulting from carbocation rearrangements can cause low yields of a desired alkene to be obtained from the dehydration of an alcohol that has any structural complexity. The rigorous conditions required for dehydration of an alcohol can be avoided, however, if the reaction is carried out in the presence of phosphorus oxychloride ($POCl_3$) and pyridine.

$$CH_3CH_2CHCH_3 \xrightleftharpoons[\text{pyridine, 0 °C}]{POCl_3} CH_3CH=CHCH_3$$

OH

Reaction with $POCl_3$ converts the OH group of the alcohol into a good leaving group. Because the reaction conditions favor an E2 reaction, a carbocation is not formed—so we do not have to be concerned with carbocation rearrangements. Pyridine has two functions in this reaction: it acts as a base to remove a proton in the elimination reaction, and it also serves as the solvent.

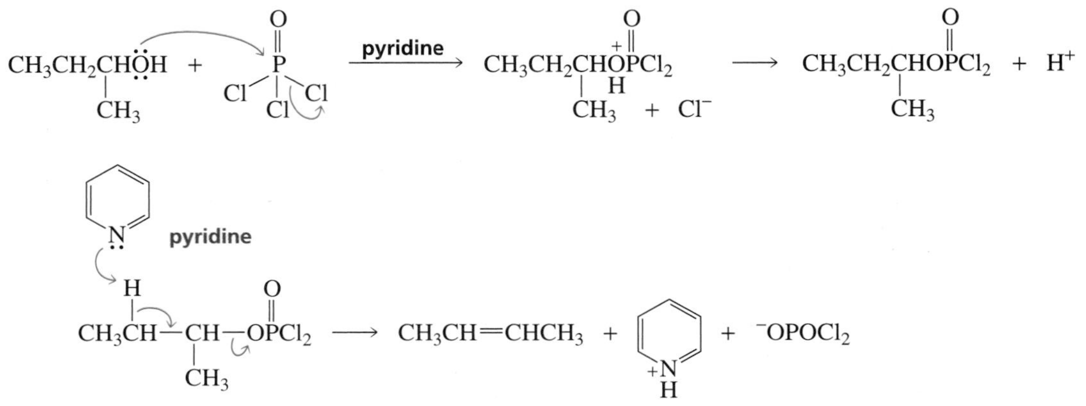

pyridine

PROBLEM-SOLVING STRATEGY

Propose a mechanism for the following reaction.

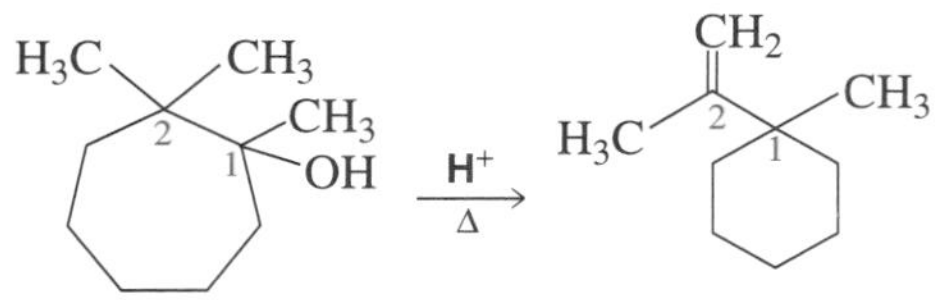

Even the most complicated-looking mechanism can be worked out if you proceed one step at a time, keeping in mind the structure of the final product. It can be helpful to label the equivalent carbons in the reactant and product. The OH group is the only base in the starting material, so it must react with the proton. Loss of water forms a tertiary carbocation.

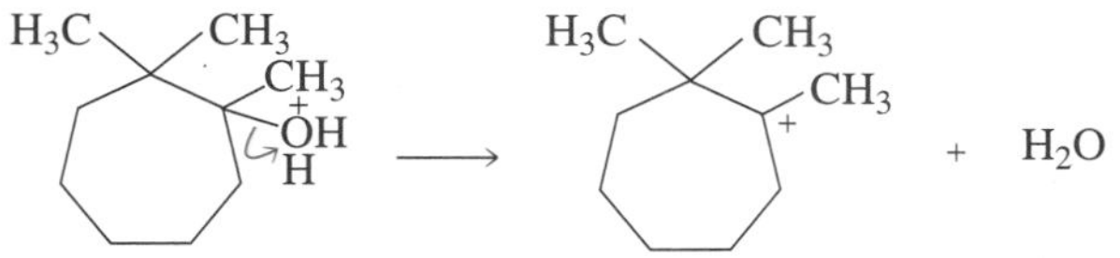

Since the starting material contains a seven-membered ring and the final product has a six-membered ring, a ring-contraction rearrangement must occur. There are two possible pathways for ring contraction: one leads to a tertiary carbocation, and the other leads to a primary carbocation. Since a primary carbocation is unstable and this particular primary carbocation has the wrong arrangement of carbons with regard to the product, the correct pathway must be the one that leads to the tertiary carbocation.

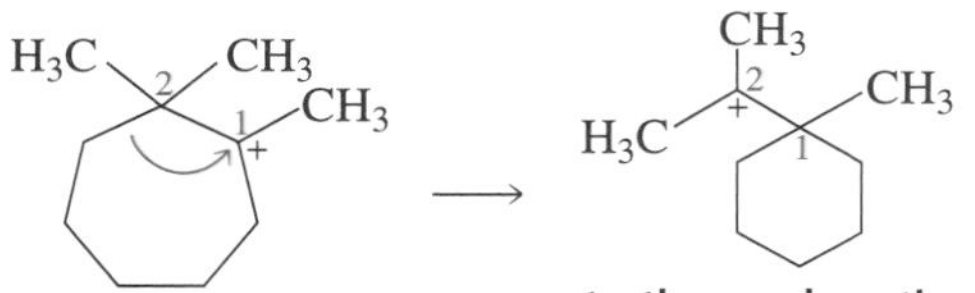

tertiary carbocation

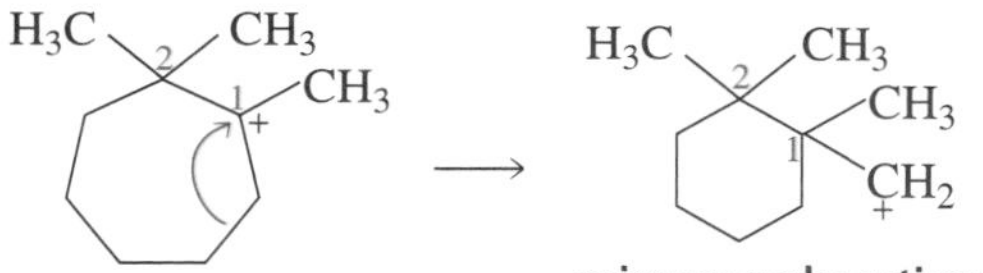

primary carbocation

Now we are close to the product, which we reach by eliminating a proton from the rearranged carbocation.

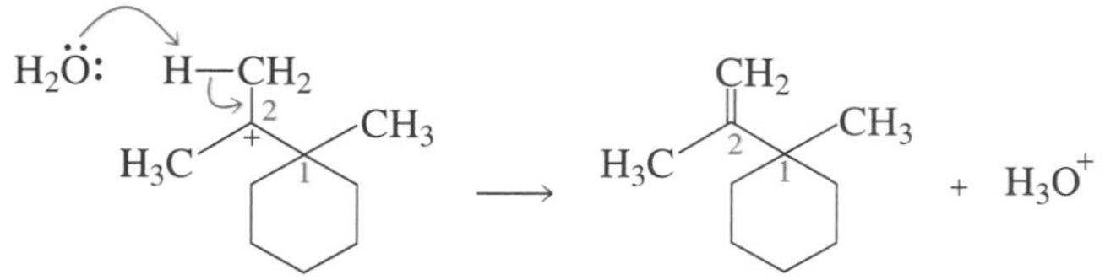

Now continue on to do Problem 8.

PROBLEM 8

Propose a mechanism for each of the following reactions.

a.

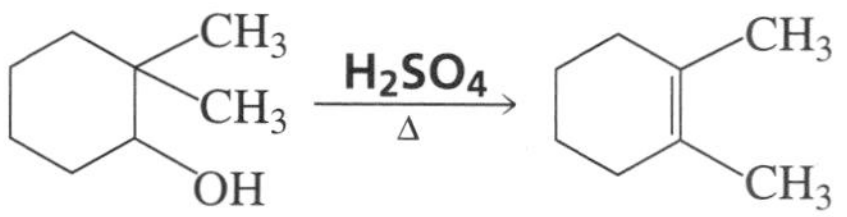

b.

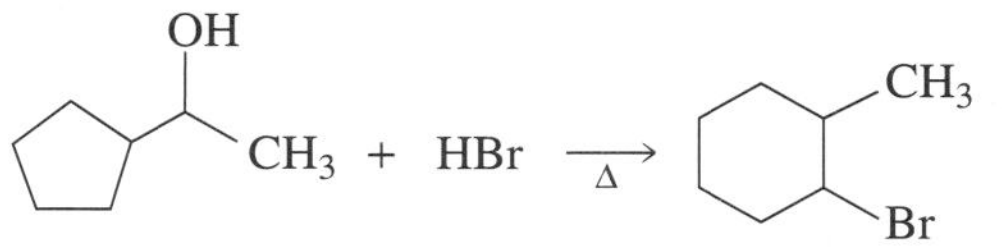

$\text{CH}_3 + \text{HBr} \xrightarrow{\Delta}$

c.

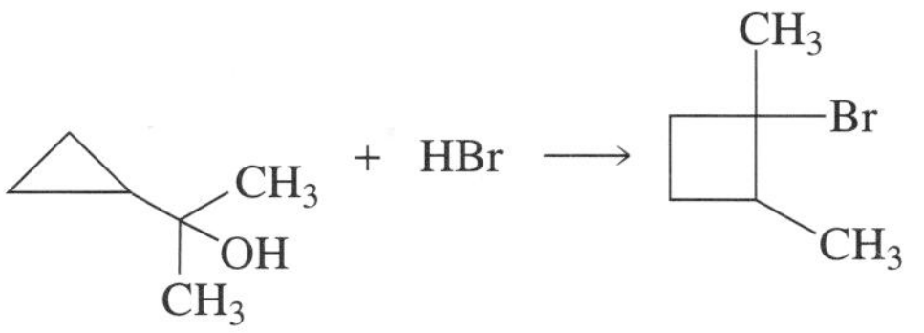

$+ \text{HBr} \longrightarrow$

PROBLEM 9 ◆

Give the major product that is formed when each of the following alcohols is heated in the presence of H_2SO_4.

a. 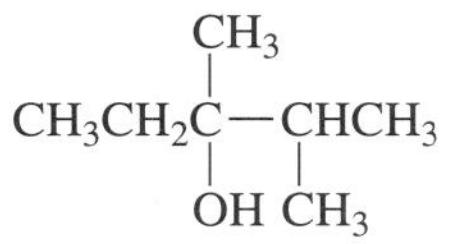 $\underset{\displaystyle \overset{\displaystyle \text{CH}_3}{\underset{\displaystyle \text{OH} \;\; \text{CH}_3}{|}}}{\text{CH}_3\text{CH}_2\text{C}-\text{CHCH}_3}$

b. 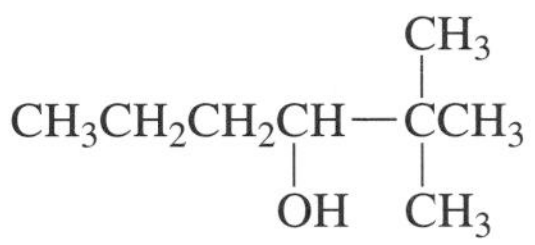 $\underset{\displaystyle \overset{\displaystyle \text{CH}_3}{\underset{\displaystyle \text{OH} \quad \text{CH}_3}{|}}}{\text{CH}_3\text{CH}_2\text{CH}_2\text{CH}-\text{CCH}_3}$

c.

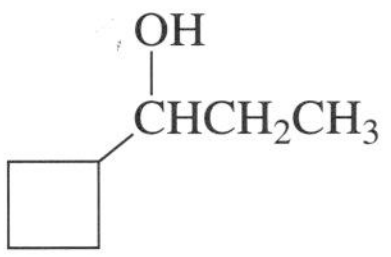

d.

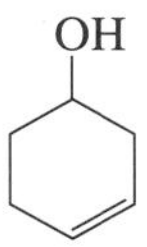

e.

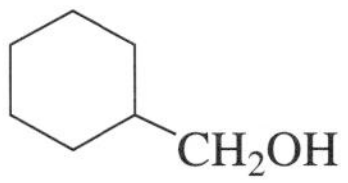

f. $\text{CH}_3\text{CH}_2\text{CH}_2\text{CH}_2\text{CH}_2\text{OH}$

BIOLOGICAL DEHYDRATIONS

Dehydration reactions are known to occur in many important biological processes. Fumarase, for example, is an enzyme that catalyzes the dehydration of malate in the Krebs cycle (Appendix VI). The Krebs cycle is a series of reactions in the terminal oxidation of carbohydrates, fatty acids, and amino acids.

malate   fumarate

Enolase, another enzyme, catalyzes the dehydration of α-phosphoglycerate in glycolysis (Appendix VI). Glycolysis is a series of reactions that prepare glucose for entry into the Krebs cycle.

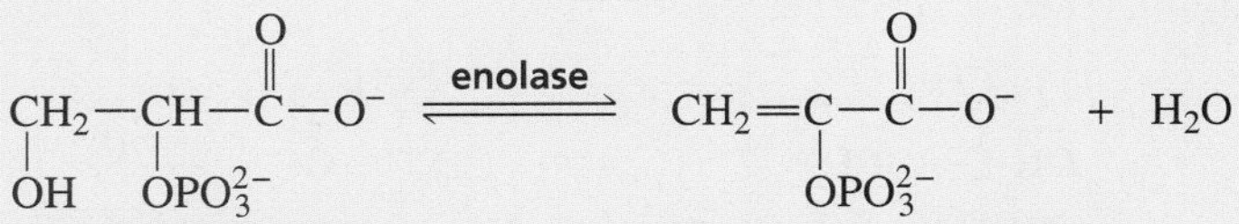

α-phosphoglycerate   phosphoenolpyruvate

11.5 SUBSTITUTION REACTIONS OF ETHERS

Ethers differ from alcohols in that the leaving group of an ether is $^-$OR while the leaving group of an **alcohol** is $^-$OH.

$$\underset{\textbf{an alcohol}}{R-O-H} \qquad \underset{\textbf{an ether}}{R-O-R}$$

The leaving groups of ethers and alcohols have nearly the same basicity and therefore the same leaving tendency. (The pK_a of CH_3OH is 15.5, and the pK_a of H_2O is 15.7.) Consequently, alcohols and ethers are equally unreactive toward nucleophilic substitution.

Many of the reagents that are used to activate alcohols toward nucleophilic substitution cannot be used to activate ethers. For example, when an alcohol reacts with an activating agent such as a sulfonyl chloride, a proton is lost in the second step of the reaction and a stable sulfonate ester results.

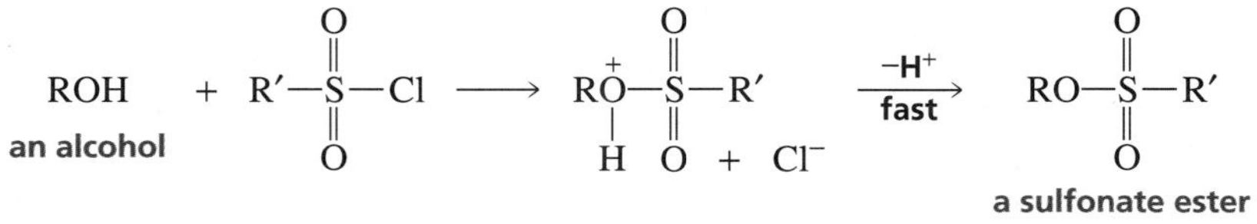

But when an ether reacts with a sulfonyl chloride, the R group cannot be lost, so a stable sulfonate ester cannot be formed. The intermediate returns to the starting materials.

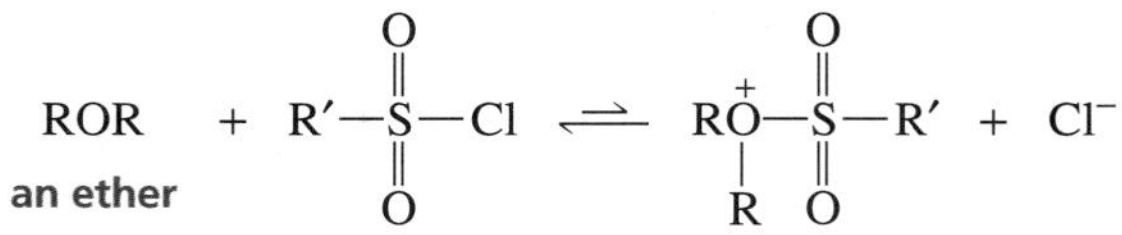

However, like alcohols, ethers can be activated by protonation. Therefore, in the presence of a high concentration of HI or HBr, ethers undergo a nucleophilic substitution reaction. Similar to what is found for alcohols, the reaction of ethers with hydrogen halides is slow and the reaction mixture must be heated in order for the reaction to occur at a reasonable rate.

$$ROR' + HI \xrightarrow{\Delta} ROH + R'I$$

The first step in the cleavage of an ether by a hydrogen halide is protonation of the ether oxygen. This converts the very basic RO^- leaving group into the less basic ROH leaving group. What happens next in the mechanism depends on the structure of the ether.

If departure of the leaving group creates a relatively stable carbocation (say, a tertiary carbocation), an S_N1 reaction occurs: the leaving group departs, and the halide ion combines with the carbocation.

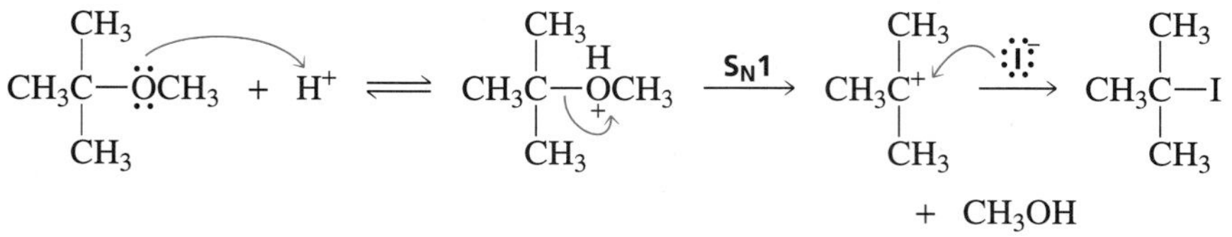

If departure of the leaving group would create an unstable carbocation (say, a methyl carbocation or a primary carbocation), the leaving group cannot depart. It has to be displaced by the halide ion. In other words, an S_N2 reaction occurs. The halide ion preferentially attacks the less sterically hindered of the two alkyl groups.

$$CH_3\overset{\frown}{\ddot{O}}CH_2CH_2CH_3 \ + \ H^+ \ \rightleftharpoons \ CH_3 \overset{\underset{+}{|}}{\underset{\underset{:\ddot{I}:^-}{\frown}}{O}}CH_2CH_2CH_3 \ \xrightarrow{\ S_N2\ } \ CH_3I \ + \ CH_3CH_2CH_2OH$$

If excess hydrogen halide is used, the product alcohol will be converted into an alkyl halide.

$$CH_3CH_2\underset{\underset{CH_3}{|}}{O}CHCH_3 \ \xrightarrow[\Delta]{HI} \ CH_3CH_2I \ + \ CH_3\underset{\underset{CH_3}{|}}{CHOH} \ \xrightarrow[\Delta]{HI} \ CH_3\underset{\underset{CH_3}{|}}{CHI} \ + \ H_2O$$

Cleavage of ethers by HBr or HI occurs more rapidly if the reaction can take place by an S_N1 pathway. If the lack of stability of the carbocation causes the reaction to follow an S_N2 pathway, cleavage will be more rapid if the hydrogen halide is HI than if it is HBr because I^- is a better nucleophile than Br^-. An S_N2 cleavage reaction does not occur at all if the hydrogen halide is HCl, because Cl^- is too weak a nucleophile. The reaction of ethers with hydrogen halides is comparable to the reaction of alcohols with hydrogen halides. The only difference is the leaving group, which is H_2O for alcohols and ROH for ethers. In both cases, only a substitution product is obtained, because the halide ions, being very weak bases, cannot abstract a proton to form an elimination product.

PROBLEM 10

Can HF be used to cleave ethers? Explain.

ANESTHETICS

Because diethyl ether ("ether") is a short-lived muscle relaxant, it has been widely used as an inhalation anesthetic. However, because diethyl ether takes effect slowly and because it has a slow and unpleasant recovery period (ask anyone who has experienced it), other compounds such as ethrane, isoflurane, and halothane have replaced it as an anesthetic. Diethyl ether is still used where there is a lack of trained anesthesiologists, because it is the safest anesthetic to administer by untrained hands. Anesthetics interact with the nonpolar molecules of cell membranes, causing the membranes to swell, which interferes with their permeability.

$CH_3CH_2OCH_2CH_3$	$CF_3CHClOCHF_2$	$CHClFCF_2OCHF_2$	$CF_3CHClBr$
"ether"	isoflurane	ethrane	halothane

Amputation of a leg in 1528 without anesthetic.

Sodium pentothal (also known as thiopental sodium) is commonly used as an intravenous anesthetic. The onset of anesthesia and the loss of consciousness occur within seconds of its administration. It cannot be used as the sole anesthetic; it is generally used to induce anesthesia before an inhalation anesthetic is administered. Care must be taken when administering sodium pentothal because the dose for effective anesthesia is 75% of the lethal dose.

Propofol is a new anesthetic that has all of the properties of the "perfect anesthetic": it can be used as the sole anesthetic by intravenous drip; it has a rapid and pleasant induction period; it has a wide margin of safety; and recovery is rapid and pleasant.

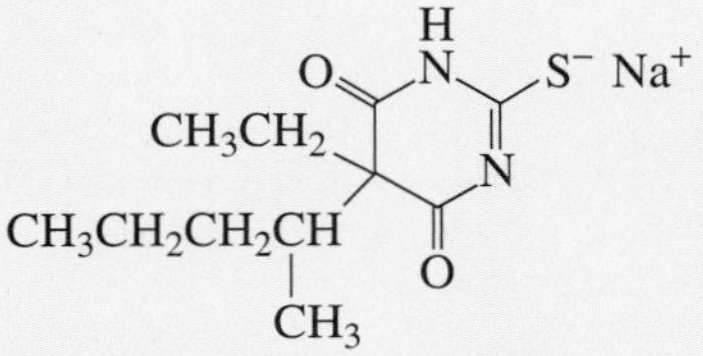

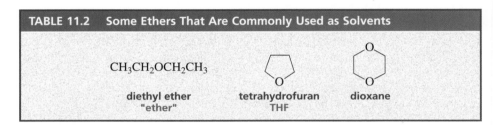

sodium pentothal 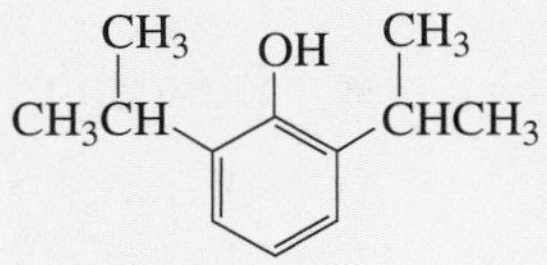   propofol

Ethers are relatively unreactive compounds. Because they do not react with most reagents, they are frequently used as solvents. Some common ether solvents are shown in Table 11.2.

TABLE 11.2	**Some Ethers That Are Commonly Used as Solvents**	

$CH_3CH_2OCH_2CH_3$

diethyl ether
"ether"

tetrahydrofuran
THF

dioxane

PROBLEM 11◆

Give the major products that would be obtained from heating each of the following ethers with HI.

a. $CH_3CH_2\overset{\underset{\textstyle |}{CH_3}}{CH}CH_2OCH_2CH_3$

c. $CH_3CH{=}CHOCH_2CH_3$

b. $CH_3CH_2CH_2O\overset{\underset{\textstyle |}{CH_3}}{\underset{\textstyle |}{C}}CH_2CH_3$ with CH₃ below

d. 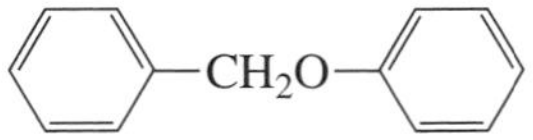 ⬡—CH_2O—⬡

11.6 REACTIONS OF EPOXIDES

Ethers in which the oxygen atom is incorporated into a three-membered ring are called **epoxides** or **oxiranes.** The common name of an epoxide assumes that the oxygen atom is in place of the π bond of an alkene and uses the common name of the alkene followed by "oxide." The simplest epoxide is ethylene oxide.

$H_2C{=}CH_2$ $H_2C{-}CH_2$ (with O bridge) $H_2C{=}CHCH_3$ $H_2C{-}CHCH_3$ (with O bridge)

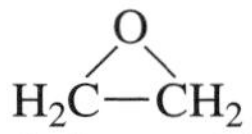

ethylene   ethylene oxide   propylene   propylene oxide

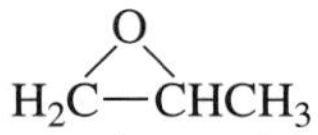

There are two IUPAC-approved ways to name epoxides. One method calls the three-membered oxygen-containing ring "oxirane," with the oxygen occupying

the 1-position. Thus, 2-ethyloxirane has an ethyl substituent at the 2-position of the oxirane ring. Alternatively, an epoxide can be named as an alkane, with "epoxy" and the numbers of the carbons to which the oxygen is attached immediately preceding the alkane name.

$$\underset{\substack{\textbf{2-ethyloxirane}\\\textbf{1,2-epoxybutane}}}{\overset{O}{H_2C-CHCH_2CH_3}} \qquad \underset{\substack{\textbf{2,3-dimethyloxirane}\\\textbf{2,3-epoxybutane}}}{\overset{O}{CH_3CH-CHCH_3}} \qquad \underset{\substack{\textbf{2,2-dimethyloxirane}\\\textbf{1,2-epoxy-2-methylpropane}}}{\overset{O\quad CH_3}{H_2C-C}\underset{CH_3}{}}$$

PROBLEM 12 ◆

Give a structure for each of the following.

a. 2-propyloxirane

b. cyclohexene oxide

c. 2,2,3,3-tetramethyloxirane

d. 2,3-epoxy-2-methylpropane

Although an epoxide and an ether have the same leaving group, epoxides are much more reactive than ethers because of the strain in the three-membered ring (Figure 11.2). Epoxides, therefore, readily undergo ring-opening reactions with a wide variety of nucleophiles.

Epoxides, like other ethers, react with hydrogen halides. In the first step of the reaction, the oxygen is protonated. The protonated epoxide is then attacked by the halide ion. Because epoxides are so much more reactive than ethers, the reaction takes place readily at room temperature.

$$\overset{:\ddot{O}}{\underset{}{H_2C-CH_2}} + H-Br \rightleftharpoons \overset{\overset{H}{\underset{+}{O}}}{\underset{}{H_2C-CH_2}} + :\ddot{Br}:^- \longrightarrow HOCH_2CH_2Br$$

Protonated epoxides can also be opened by poor nucleophiles such as H_2O and alcohols.

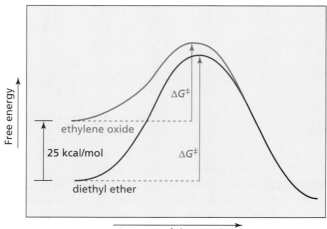

◀ **Figure 11.2**
The reaction coordinate diagrams for nucleophilic attack of hydroxide ion on ethylene oxide and on diethyl ether. The greater reactivity of the epoxide is a result of the strain in the three-membered ring (ring strain and torsional strain), which increases its free energy.

$$H_2\overset{O}{C}\text{—}CH_2 \rightleftharpoons{\scriptstyle H^+} H_2\overset{\overset{H}{\underset{\uparrow}{O^+}}}{C}\text{—}CH_2 \xrightarrow[H_2\ddot{O}:]{} \underset{\underset{H}{\overset{+}{O}H}}{CH_2CH_2OH} \rightleftharpoons{\scriptstyle -H^+} HOCH_2CH_2OH$$

1,2-ethanediol
ethylene glycol

$$CH_3\overset{O}{CH}\text{—}CHCH_3 \rightleftharpoons{\scriptstyle H^+} CH_3\overset{\overset{H}{\underset{\uparrow}{O^+}}}{CH}\text{—}CHCH_3 \xrightarrow[CH_3\ddot{O}H]{} \underset{\underset{H}{\overset{+}{O}CH_3}}{\overset{OH}{CH_3CHCHCH_3}} \rightleftharpoons{\scriptstyle -H^+} \underset{\overset{|}{O}CH_3}{\overset{OH}{CH_3CHCHCH_3}}$$

3-methoxy-2-butanol

If the protonated epoxide is asymmetrical (and the nucleophile is something other than H_2O), the product obtained from nucleophilic attack on the 2-position of the oxirane ring will be different from the product obtained from nucleophilic attack on the 3-position. The major product is the one resulting from attack of the nucleophile on the *more substituted* carbon.

$$CH_3\overset{O}{CH}\text{—}CH_2 \rightleftharpoons{\scriptstyle H^+} CH_3\overset{\overset{H}{\underset{\uparrow}{O^+}}}{CH}\text{—}CH_2 \xrightarrow[Cl^-]{} \underset{\text{2-chloro-1-propanol}}{\overset{Cl}{CH_3CHCH_2OH}} + \underset{\text{1-chloro-2-propanol}}{\overset{OH}{CH_3CHCH_2Cl}}$$

2-chloro-1-propanol 90% 1-chloro-2-propanol 10%

The more substituted carbon is preferentially attacked because, after the epoxide is protonated, it is so reactive that the carbon–oxygen bond begins to break before the nucleophile has a chance to attack. As the carbon–oxygen bond starts to break, a partial positive charge develops on the carbon that is losing its share of the oxygen's electrons. The protonated epoxide breaks preferentially in the direction that puts the partial positive charge on the more substituted carbon, because the more substituted carbocation is more stable. (Recall that tertiary carbocations are more stable than secondary carbocations, which are more stable than primary carbocations.)

partial secondary carbocation

partial primary carbocation

The best way to describe the reaction is to say that it occurs by a pathway that is partially S_N1 and partially S_N2. It is not a pure S_N1 reaction, because a carbocation intermediate is not fully formed. It is not a pure S_N2 reaction because the leaving group begins to depart before the compound is attacked by the nucleophile.

In Section 11.5 we saw that an ether does not undergo a nucleophilic substitution reaction unless the very basic ⁻OR leaving group is converted by protonation into a less basic ROH leaving group. Because of the strain in the three-membered ring, epoxides are reactive enough to open without first being protonated. When a nucleophile attacks an unprotonated epoxide, the reaction is a pure S_N2 reaction: the unprotonated epoxide does not begin to open until it is attacked by the nucleophile. The nucleophile, therefore, is more likely to attack the *less substituted* carbon; the less substituted carbon is more accessible to attack because it is less sterically hindered. So the site of nucleophilic attack on an asymmetrical epoxide under basic conditions (when the epoxide is not protonated) is different from the site of nucleophilic attack under acidic conditions (when the epoxide is protonated).

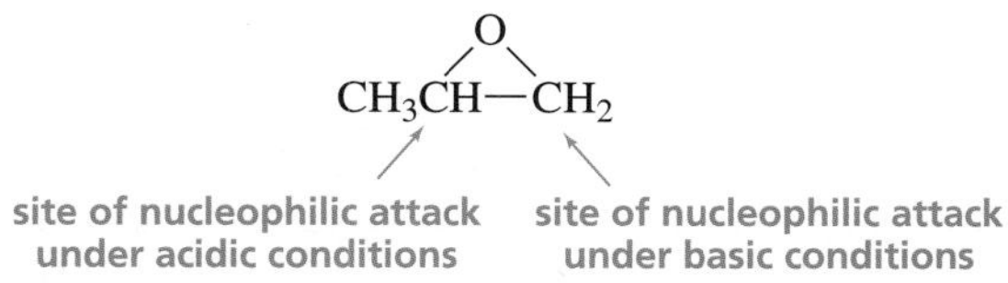

$$CH_3CH-CH_2$$

site of nucleophilic attack   site of nucleophilic attack
under acidic conditions   under basic conditions

After the nucleophile has attacked the epoxide, the alkoxide ion can pick up a proton from the solvent or from an acid added after the reaction is over.

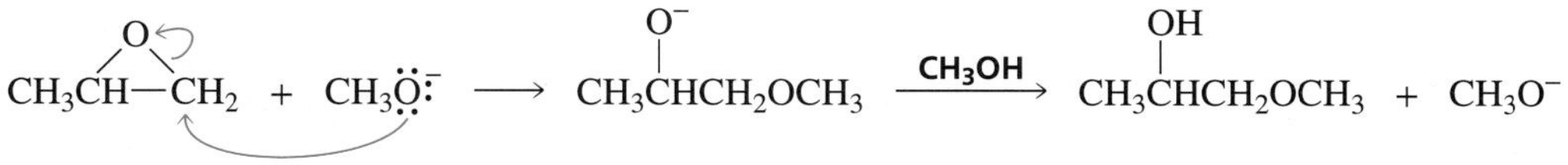

$$CH_3CH-CH_2 + CH_3\ddot{O}^- \longrightarrow CH_3\overset{O^-}{\underset{}{CH}}CH_2OCH_3 \xrightarrow{CH_3OH} CH_3\overset{OH}{\underset{}{CH}}CH_2OCH_3 + CH_3O^-$$

Epoxides are synthetically useful reagents because they react with a wide variety of nucleophiles, leading to the formation of a wide variety of products. They are important in biological processes because they are reactive enough to be attacked by nucleophiles under conditions found in living systems.

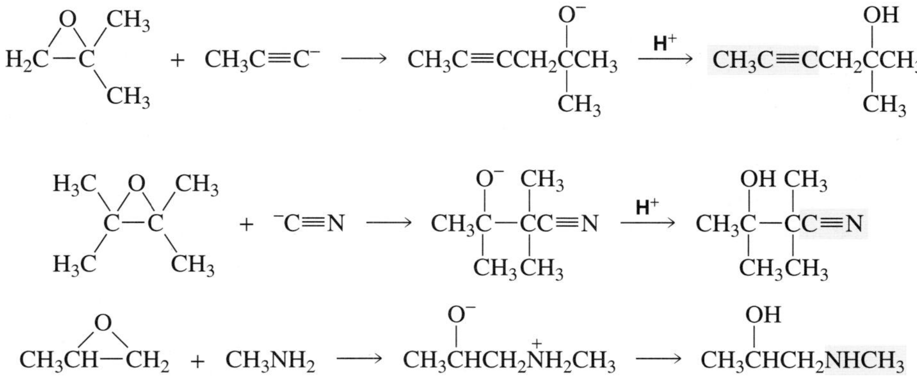

$$H_2C-\overset{CH_3}{\underset{CH_3}{C}} + CH_3C\equiv C^- \longrightarrow CH_3C\equiv CCH_2\overset{O^-}{\underset{CH_3}{C}}CH_3 \xrightarrow{H^+} CH_3C\equiv CCH_2\overset{OH}{\underset{CH_3}{C}}CH_3$$

$$\overset{H_3C}{\underset{H_3C}{}}\overset{O}{C-C}\overset{CH_3}{\underset{CH_3}{}} + {}^-C\equiv N \longrightarrow CH_3\overset{O^-}{\underset{CH_3}{C}}-\overset{CH_3}{\underset{CH_3}{C}}C\equiv N \xrightarrow{H^+} CH_3\overset{OH}{\underset{CH_3}{C}}-\overset{CH_3}{\underset{CH_3}{C}}C\equiv N$$

$$CH_3CH-CH_2 + CH_3NH_2 \longrightarrow CH_3\overset{O^-}{\underset{}{CH}}CH_2\overset{+}{N}H_2CH_3 \longrightarrow CH_3\overset{OH}{\underset{}{CH}}CH_2NHCH_3$$

PROBLEM 13◆

Give the major product of each of the following reactions.

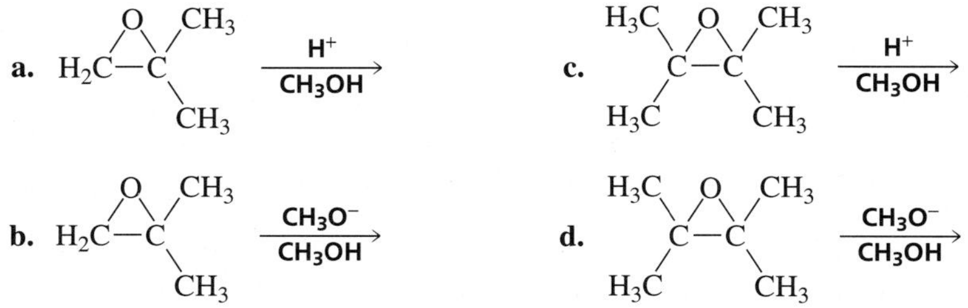

a. $H_2C-\overset{CH_3}{\underset{CH_3}{C}}$ $\xrightarrow[CH_3OH]{H^+}$

b. $H_2C-\overset{CH_3}{\underset{CH_3}{C}}$ $\xrightarrow[CH_3OH]{CH_3O^-}$

c. $\overset{H_3C}{\underset{H_3C}{}}\overset{O}{C-C}\overset{CH_3}{\underset{CH_3}{}}$ $\xrightarrow[CH_3OH]{H^+}$

d. $\overset{H_3C}{\underset{H_3C}{}}\overset{O}{C-C}\overset{CH_3}{\underset{CH_3}{}}$ $\xrightarrow[CH_3OH]{CH_3O^-}$

PROBLEM 14 ◆

Would you expect the reactivity of a five-membered ring ether such as tetrahydrofuran (Table 11.2) to be more similar to an epoxide or to a noncyclic ether?

**11.7
ARENE OXIDES**

When aromatic hydrocarbons are ingested or inhaled, they are enzymatically converted into arene oxides. An **arene oxide** is a compound in which one of the double bonds of the aromatic ring has been converted into an epoxide. Formation of an arene oxide is the first step in changing an aromatic compound that enters the body as a foreign substance (cigarette smoke, drugs, automobile exhaust) into a more water-soluble compound that can eventually be eliminated. Arene oxides are also important intermediates in the biosynthesis of biochemically important phenols such as tyrosine and serotonin.

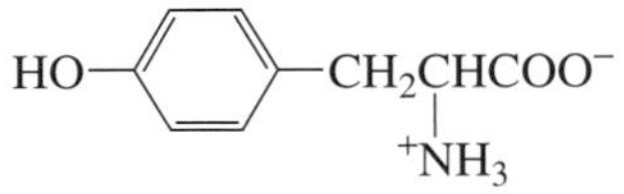

**tyrosine
an amino acid**

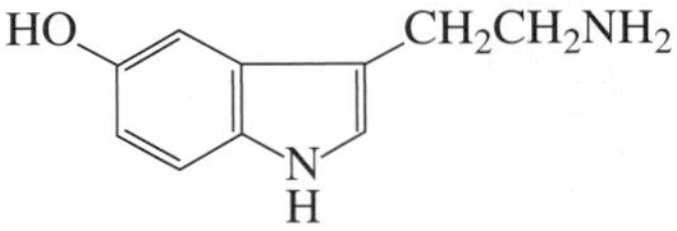

**serotonin
a vasoconstrictor**

The enzyme that detoxifies aromatic hydrocarbons by converting them into arene oxides is known as cytochrome P_{450}.

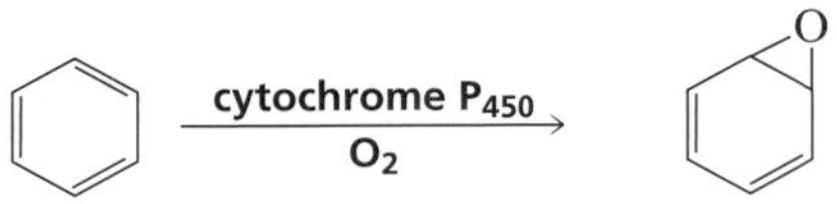

**benzene oxide
an arene oxide**

An arene oxide can react in two different ways. It can react as a typical epoxide, undergoing attack by nucleophiles to form an addition product (Section 11.6). Because of its potential aromatic ring, an arene oxide can also react in a way that epoxides such as ethylene oxide cannot react: it can rearrange to form a phenol.

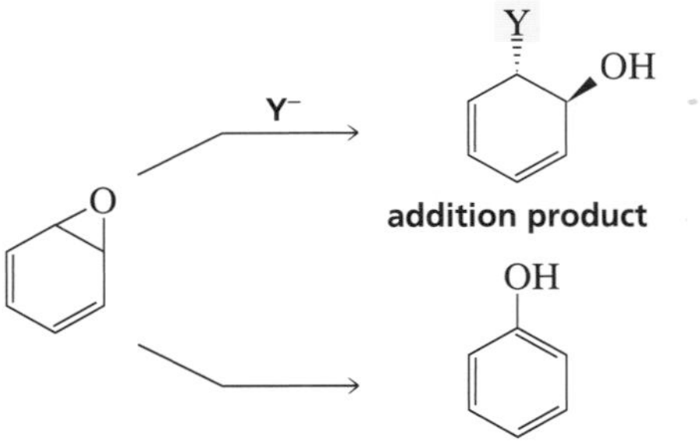

addition product

rearranged product

When an arene oxide reacts with a nucleophile, the three-membered ring undergoes back side nucleophilic attack, forming a trans addition product.

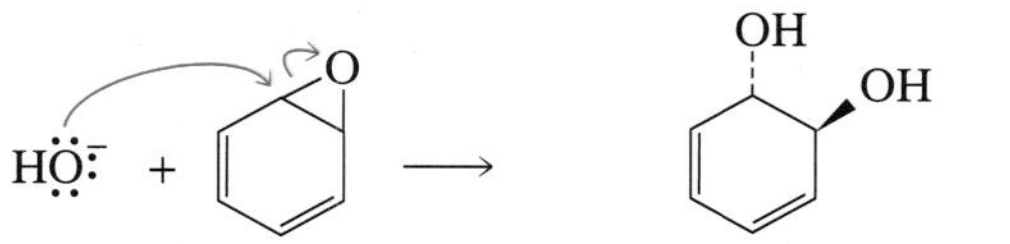

a trans addition product

In the first step of the mechanism for arene oxide rearrangement, the epoxide opens to form a carbocation. An enone is then formed as a result of a 1,2-hydride shift. This is called an *NIH shift* because it was first observed in a laboratory at the National Institutes of Health. Enolization of the enone gives the phenol (Section 5.6).

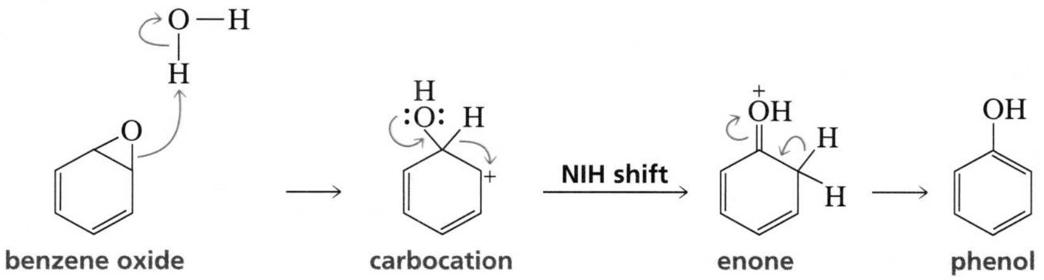

benzene oxide   carbocation   **NIH shift**   enone   phenol

The rate-limiting step in this rearrangement is opening of the epoxide ring to form the carbocation. Therefore, the more stable the carbocation, the easier it will be to open the epoxide ring. This means that the rate of formation of phenol depends on the stability of the carbocation.

Only one arene oxide can be formed from naphthalene, because epoxidation at the 2,3-position or at the bond shared by the two rings would destroy the aromaticity of both rings. Naphthalene oxide can rearrange to form either 1-naphthol or 2-naphthol. The carbocation leading to 1-naphthol is more stable because the carbocation leading to 2-naphthol can be stabilized by resonance only if the aromaticity of the benzene ring on the left of the structure is destroyed. Consequently, rearrangement leads predominantly to 1-naphthol.

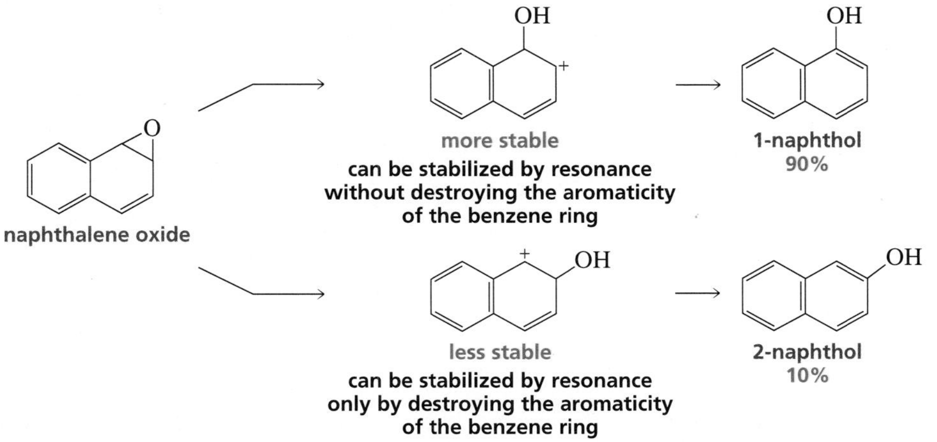

naphthalene oxide

more stable
can be stabilized by resonance without destroying the aromaticity of the benzene ring

1-naphthol
90%

less stable
can be stabilized by resonance only by destroying the aromaticity of the benzene ring

2-naphthol
10%

PROBLEM 15

Draw the resonance contributor for the carbocation leading to 1-naphthol that does not destroy the benzene ring. Compare this with the resonance contributors for the carbocation that leads to 2-naphthol.

PROBLEM 16◆

The existence of the NIH shift was established by the nature of the major product obtained from rearrangement of the following deuterated arene oxide. What would be the major product of the rearrangement if the NIH shift did not occur? (*Hint:* A C—H bond is easier to break than a C—D bond; Section 10.7).

PROBLEM 17◆

How would the major products obtained from rearrangement of the following arene oxides differ?

Some aromatic hydrocarbons are known to cause cancer. Investigation has revealed that it is not the hydrocarbon itself that is carcinogenic; rather, the arene oxide into which the hydrocarbon is converted is the cancer-causing agent. How do arene oxides cause cancer? We have seen that nucleophiles react with epoxides to form addition products. 2′-Deoxyguanosine, a component of DNA (Section 24.1), is a nucleophile and reacts with certain arene oxides. Once 2′-deoxyguanosine becomes covalently attached to the arene oxide, it can no longer fit into the DNA double helix. Thus, the genetic code cannot be properly transcribed. This can lead to mutations that cause cancer. Cancer results when cells lose their ability to control their growth and reproduction.

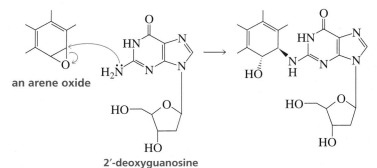

2′-deoxyguanosine

Not all arene oxides are carcinogenic, because not all of them react with nucleophiles. Because an arene oxide has two possible pathways for reaction, whether or not a particular arene oxide will be carcinogenic depends on the relative rates of the two reaction pathways. Arene oxide rearrangement leads to phenolic products that are not carcinogenic, while formation of addition products as a result of nucleophilic attack by DNA leads to cancer-causing products. So if the rate of arene oxide rearrangement is greater than the rate of nucleophilic attack by DNA, the arene oxide will be harmless. If, however, the rate of nucleophilic attack is greater than the rate of rearrangement, the arene oxide will be a carcinogen.

A computer-generated model of DNA.

   Because the rate of arene oxide rearrangement depends on the stability of the carbocation formed in the first step of the rearrangement, an arene oxide's cancer-causing potential depends on the stability of this carbocation. If the carbocation is relatively stable, rearrangement will be relatively fast and the arene oxide will tend to be noncarcinogenic. If the carbocation is unstable, rearrangement will be slow and the arene oxide will be more likely to be carcinogenic. This means that the more reactive the arene oxide (the more easily it opens to form a carbocation), the less likely it is to be carcinogenic.

BENZO[a]PYRENE AND CANCER

   Benzo[a]pyrene is one of the most carcinogenic of the aromatic hydrocarbons. This hydrocarbon is formed whenever an organic compound undergoes incomplete combustion. For example, benzo[a]pyrene is found in cigarette smoke, in automobile exhaust, and in charcoal-broiled meat. Several arene oxides can be formed from benzo[a]pyrene. The two most harmful are the 4,5-oxide and the 7,8-oxide. It has been suggested that people who develop lung cancer as a result of smoking may have a higher-than-normal incidence of cytochrome P_{450} in their lung tissue.

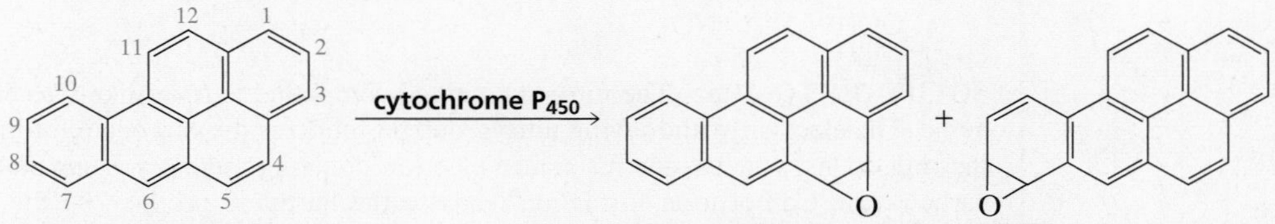

benzo[a]pyrene   4,5-benzo[a]pyrene oxide   7,8-benzo[a]pyrene oxide

   The 4,5-oxide is harmful because it forms a carbocation that cannot be stabilized by resonance without destroying the aromaticity of an adjacent ring. Because the carbocation is relatively unstable, the epoxide will tend not to open until it is attacked by a nucleophile. The 7,8-oxide is harmful because it reacts with water to form a diol, which then forms a diol epoxide. The diol epoxide does not readily undergo rearrangement (the harmless pathway) because it opens to a carbocation that is destabilized by the electron-withdrawing OH groups. Therefore, it can exist long enough to be attacked by nucleophiles (the carcinogenic pathway).

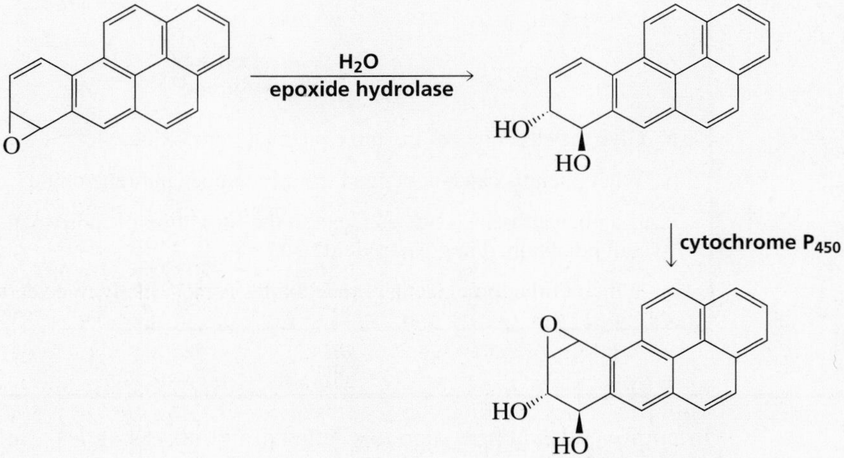

a diol epoxide

CHIMNEY SWEEPS AND CANCER

That environmental factors can cause cancer was first recognized in 1775 by Percival Potts, a British physician, when he became aware that chimney sweeps had a higher incidence of scrotum cancer than the male population as a whole. He realized that something in the chimney soot was apparently causing cancer. We now know that it was benzo[*a*]pyrene.

Titch Cox, the chimney sweep responsible for cleaning the 800 chimneys in Buckingham Palace.

PROBLEM 18 / SOLVED

For each of the following pairs of compounds, indicate the one that is more likely to be carcinogenic.

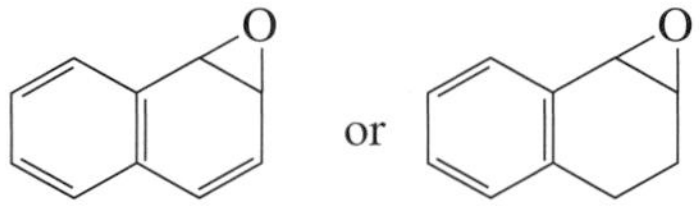

SOLUTION TO 19a   The nitro-substituted compound is more likely to be carcinogenic. The electron-withdrawing nitro group destabilizes the carbocation formed when the ring opens, whereas the resonance electron-donating methoxy group stabilizes the carbocation. Carbocation formation leads to the harmless product, so the methoxy-substituted compound with a more stable (easier-to-form) carbocation will be more likely to form a harmless product. In addition, the electron-withdrawing nitro group increases the arene oxide's susceptibility to nucleophilic attack, the cancer-causing pathway.

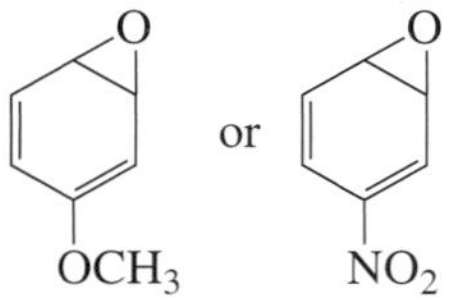

PROBLEM 19

Three arene oxides can be obtained from phenanthrene.

phenanthrene

a. Give the structures of the three phenanthrene oxides.

b. What phenols can be obtained from each phenanthrene oxide?

c. If a phenanthrene oxide can lead to the formation of more than one phenol, which will be obtained in greater yield?

d. Which of the three phenanthrene oxides is most likely to be carcinogenic?

11.8
ORGANOMETALLIC COMPOUNDS

In compounds such as alcohols, ethers, and alkyl halides, carbon is bonded to a more electronegative atom. Carbon, therefore, is electrophilic and the alkyl group reacts with a nucleophile.

$$\overset{\delta+}{CH_3}\overset{\delta-}{CH_2}-Z \;+\; Y^- \longrightarrow CH_3CH_2-Y \;+\; Z^-$$

electrophile   nucleophile

But what if you wanted an alkyl group to react with an electrophile? You would need a compound in which carbon is nucleophilic. To be nucleophilic, carbon would have to be bonded to a less electronegative atom.

$$\overset{\delta-}{CH_3}\overset{\delta+}{CH_2}-Z \;+\; Y^+ \longrightarrow CH_3CH_2-Y \;+\; Z^+$$

nucleophile   electrophile

An **organometallic compound** is a compound that contains a carbon–metal bond. Since most metals are less electronegative than carbon, the carbon bonded to the metal is nucleophilic. Organolithium compounds and organomagnesium compounds are the most common organometallic compounds used in synthesis.

Organolithium compounds are prepared by adding lithium to an alkyl halide in a nonpolar solvent such as hexane.

$$CH_3CH_2CH_2CH_2Br \;+\; 2\,Li \xrightarrow{\text{hexane}} CH_3CH_2CH_2CH_2Li \;+\; LiBr$$
1-bromobutane   **butyllithium**

chlorobenzene   phenyllithium

Organomagnesium compounds, frequently called **Grignard reagents** after their discoverer, are prepared by adding an alkyl halide to magnesium shavings being stirred in diethyl ether or THF. The magnesium is inserted between the carbon and the halogen.

cyclohexyl bromide   cyclohexylmagnesium bromide

$$CH_2=CHBr \;+\; Mg \xrightarrow{\text{THF}} CH_2=CHMgBr$$
vinyl bromide   **vinylmagnesium bromide**

François Auguste Victor Grignard (1871–1935) *was born in France, the son of a sailmaker. His synthesis of the first Grignard reagent was announced in 1900; during the next 5 years, some 200 papers were published about Grignard reagents. He was a professor of chemistry at the University of Nancy and later at the University of Lyon. He shared the Nobel Prize in chemistry in 1912 with P. Sabatier (Chapter 3, p. 157).*

Alkyl halides, vinyl halides, and aryl halides can all be used to form organolithium and organomagnesium compounds. Alkyl bromides are the alkyl halides most often used in the formation of organometallic compounds because they react more readily than alkyl chlorides and are less expensive than alkyl iodides.

The solvent (usually diethyl ether or tetrahydrofuran) plays a crucial role in the formation of a Grignard reagent. Because the magnesium atom of a Grignard reagent is surrounded by only four electrons, it needs two more pairs of electrons to form an octet. Solvent molecules provide these electrons by coordinating with (supplying electron pairs to) the metal. Coordination allows the Grignard reagent to dissolve in the solvent and prevents the Grignard reagent from coating the magnesium shavings, which would make them unreactive.

As a result of the introduction of a metal, an alkyl halide is converted into a compound that will react with electrophiles instead of with nucleophiles. Therefore, organolithium and organomagnesium compounds react as if they were carbanions.

$$CH_3CH_2MgBr \qquad \text{reacts as if it were} \qquad CH_3\overset{-}{C}H_2 \ \overset{+}{M}gBr$$
ethylmagnesium bromide

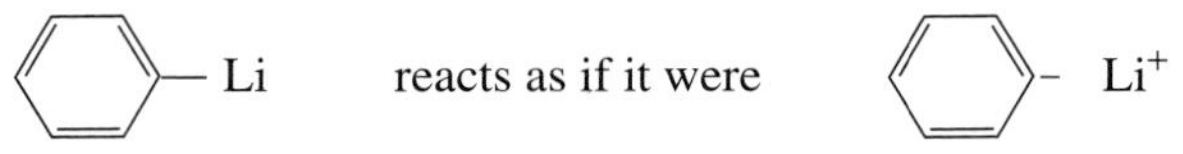

$$\text{phenyllithium} \quad \bigcirc\!\!-\!\text{Li} \qquad \text{reacts as if it were} \qquad \bigcirc\!\!-\!\text{Li}^+$$
phenyllithium

When a Grignard reagent reacts with ethylene oxide, a primary alcohol containing two more carbons than the Grignard reagent is formed.

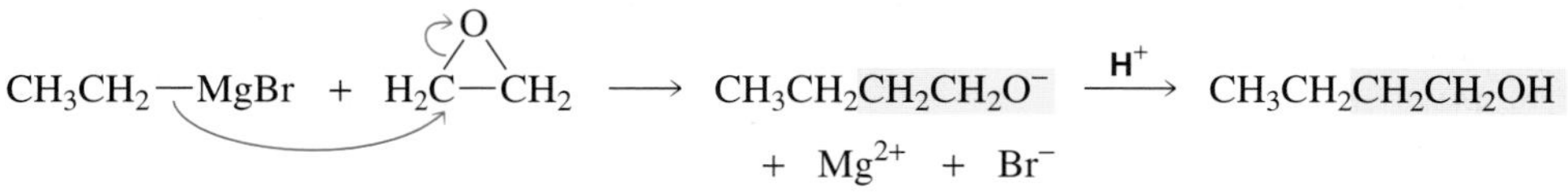

$$CH_3CH_2\!-\!MgBr \ + \ H_2C\!-\!CH_2 \ \longrightarrow \ CH_3CH_2CH_2CH_2O^- \ \xrightarrow{H^+} \ CH_3CH_2CH_2CH_2OH$$
$$+ \ Mg^{2+} \ + \ Br^-$$

PROBLEM 20◆

What alcohols would be formed from the reaction of ethylene oxide with the following Grignard reagents?

a. $CH_3CH_2CH_2MgBr$

c. 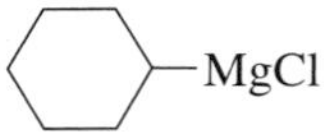 $\bigcirc\!\!-\!MgCl$

b. 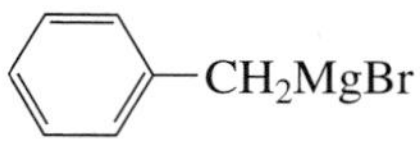 $\bigcirc\!\!-\!CH_2MgBr$

PROBLEM 21 / SOLVED

Show, using any necessary reagents, how the following compounds could be prepared using ethylene oxide as one of the reactants.

a. $CH_3CH_2CH_2CH_2OH$ **c.** $CH_3CH_2CH_2CH_2D$

b. $CH_3CH_2CH_2CH_2Br$ **d.** $CH_3CH_2CH_2CH_2CH_2CH_2OH$

SOLUTION

a.  $CH_3CH_2Br \ \xrightarrow[\text{Et}_2\text{O}]{\text{Mg}} \ CH_3CH_2MgBr \ \xrightarrow[\text{2. H}^+]{\text{1. }\overset{O}{\triangle}} \ CH_3CH_2CH_2CH_2OH$

b. product of **a** $\xrightarrow{\text{PBr}_3} CH_3CH_2CH_2CH_2Br$

c. product of **b** $\xrightarrow[\text{Et}_2\text{O}]{\text{Mg}} CH_3CH_2CH_2CH_2MgBr \xrightarrow{\text{D}_2\text{O}} CH_3CH_2CH_2CH_2D$

d. the same reaction sequence as in **a,** using butyl bromide in the first step

Organomagnesium and organolithium compounds are such strong bases that they will react immediately with even weak acids such as water and alcohols (Problem 21c). When this happens, the organometallic compound is converted into an alkane. Because even trace amounts of moisture can destroy an organometallic compound, it is important that all the reagents are dry when organometallic compounds are being synthesized and when they react with other reagents.

PROBLEM 22 ◆

Which of the following reactions will occur? For the pK_a values necessary to do this problem, see Appendix II.

CH_3MgBr + H_2O $\longrightarrow$ CH_4 + $HOMgBr$

CH_3MgBr + CH_3OH $\longrightarrow$ CH_4 + CH_3OMgBr

CH_3MgBr + NH_3 $\longrightarrow$ CH_4 + H_2NMgBr

CH_3MgBr + CH_3NH_2 $\longrightarrow$ CH_4 + $CH_3NHMgBr$

CH_3MgBr + $HC{\equiv}CH$ $\longrightarrow$ CH_4 + $HC{\equiv}CMgBr$

There are many different organometallic compounds. As long as the metal is less electronegative than carbon, the carbon bonded to the metal will be nucleophilic.

$$\overset{\delta-}{C}{-}\overset{\delta+}{Mg} \qquad \overset{\delta-}{C}{-}\overset{\delta+}{Li} \qquad \overset{\delta-}{C}{-}\overset{\delta+}{Cu} \qquad \overset{\delta-}{C}{-}\overset{\delta+}{Cd} \qquad \overset{\delta-}{C}{-}\overset{\delta+}{Si}$$

$$\overset{\delta-}{C}{-}\overset{\delta+}{Zn} \qquad \overset{\delta-}{C}{-}\overset{\delta+}{Al} \qquad \overset{\delta-}{C}{-}\overset{\delta+}{Pb} \qquad \overset{\delta-}{C}{-}\overset{\delta+}{Hg} \qquad \overset{\delta-}{C}{-}\overset{\delta+}{Sn}$$

The relative reactivity of an organometallic compound toward an electrophile depends on the degree of ionic character in the carbon–metal bond. The degree of ionic character depends in turn on the electronegativity of the metal (Table 11.3). For example, magnesium has an electronegativity of 1.2, compared with 2.5 for carbon. This difference in electronegativity makes the carbon–magnesium bond about 52% ionic. Lithium (1.0) is less electronegative than magnesium (1.2). The carbon–lithium bond, therefore, has an even higher degree of ionic character, which make an organolithium reagent a more reactive nucleophilic reagent.

TABLE 11.3   The Electronegativities of Some of the Elements[a]

IA	IIA	IB	IIB	IIIA	IVA	VA	VIA	VIIA
H 2.1								
Li 1.0	Be 1.5			B 2.0	C 2.5	N 3.0	O 3.5	F 4.0
Na 1.0	Mg 1.2			Al 1.5	Si 1.8	P 2.1	S 2.5	Cl 3.0
K 0.9	Ca 1.0	Cu 1.8	Zn 1.7	Ga 1.8	Ge 2.0			Br 2.8
		Ag 1.4	Cd 1.5		Sn 1.7			I 2.5
			Hg 1.5		Pb 1.6			

[a]From the scale devised by Linus Pauling.

The names of organometallic compounds usually begin with the name of the alkyl group followed by the name of the metal.

CH₃CH₂MgBr	CH₃CH₂CH₂CH₂Li	(CH₃CH₂CH₂)₂Cd	(CH₃CH₂)₄Pb
ethylmagnesium bromide	**butyllithium**	**dipropylcadmium**	**tetraethyllead**

A Grignard reagent will undergo **transmetallation** (metal exchange) if it is added to a metal halide with a more electronegative metal than magnesium. In other words, metal exchange will occur if the result is a carbon–metal bond with greater covalent character. For example, cadmium is more electronegative than magnesium. Consequently, a carbon–cadmium bond has greater covalent character than a carbon–magnesium bond, so metal exchange occurs.

$$2 \text{ CH}_3\text{CH}_2\text{MgCl} + \text{CdCl}_2 \longrightarrow (\text{CH}_3\text{CH}_2)_2\text{Cd} + 2 \text{ MgCl}_2$$

ethylmagnesium chloride · **diethylcadmium**

PROBLEM 23 ◆

What organometallic compound will be formed from the reaction of methylmagnesium chloride and SiCl₄? (*Hint:* see Table 11.3.)

Henry Gilman (1893–1986) *was born in Boston. He received his A.B and Ph.D. degrees from Harvard University. He joined the faculty at Iowa State University in 1919, where he remained for his entire career. He published more than 1000 research papers. More than half of these papers were published after he lost almost all of his sight as the result of a detached retina and glaucoma in 1947. His wife, Ruth, acted as his eyes for 40 years.*

Organolithium compounds are used to make **Gilman reagents.** These lithium dialkylcopper reagents are very useful to synthetic chemists. Gilman reagents are prepared by the reaction of an organolithium reagent with cuprous iodide in ether (diethyl ether or THF).

$$2 \text{ CH}_3\text{Li} + \text{CuI} \xrightarrow{\text{ether}} (\text{CH}_3)_2\text{CuLi} + \text{LiI}$$

organolithium reagent · **Gilman reagent**

When a Gilman reagent reacts with an alkyl halide (with the exception of alkyl fluorides, which do not undergo this reaction), one of the alkyl groups of the Gilman reagent replaces the halogen. The reaction is called a **coupling reaction,** because two alkyl groups are joined (coupled together). The precise mechanism of the reaction is unknown but is thought to involve radicals.

Gilman reagents can even replace halogens in compounds that contain other functional groups.

PROBLEM 24

Muscalure is the sex attractant of the common housefly. Flies are lured to traps by filling them with fly bait containing both muscalure and an insecticide. Eating the bait is fatal.

$$CH_3(CH_2)_7 \quad (CH_2)_{12}CH_3$$
$$C=C$$
$$H \qquad\qquad H$$
muscalure

How could you synthesize muscalure using the following reagents?

$$CH_3(CH_2)_7 \quad (CH_2)_7CH_2Br$$
$$C=C \qquad\qquad \text{and } CH_3(CH_2)_4Br$$
$$H \qquad\qquad H$$

Crown ethers are cyclic compounds that have several ether linkages. A crown ether specifically binds certain metal ions or organic molecules depending on the size of its cavity. The crown ether is called the "host" and the species it binds is called the "guest." Because the ether linkages are chemically inert, the crown ether can bind the "guest" without reacting with it. **The crown–guest complex is called an inclusion compound.** Crown ethers are named [X]-crown-Y, where X is the total number of atoms in the ring and Y is the number of oxygen atoms in the ring. [15]-Crown-5 specifically binds Na^+ because the crown ether has a cavity diameter of 1.7 to 2.2 Å and Na^+ has an ionic diameter of 1.80 Å. Binding occurs as a result of interaction of the positively charged ion with the nonbonding electrons of the oxygens that point into the cavity of the crown ether.

**11.9
CROWN ETHERS**

For their work in the field of crown ethers, Donald J. Cram, Charles J. Pedersen, and Jean-Marie Lehn shared the 1987 Nobel Prize in chemistry.

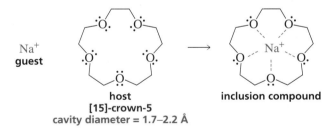

Na^+
guest

**host
[15]-crown-5
cavity diameter = 1.7–2.2 Å**

inclusion compound

Notice that Li^+ (ionic diameter, 1.20 Å) would not be bound by [15]-crown-5 but binds neatly in [12]-crown-4; K^+ (ionic diameter, 2.66 Å) is too large to fit into [15]-crown-5 but is bound specifically by [18]-crown-6.

**[12]-crown-4
cavity diameter = 1.2–1.5 Å**

**[18]-crown-6
cavity diameter = 2.6–3.2 Å**

A remarkable property of crown ethers is that they allow inorganic salts to be dissolved in nonpolar organic solvents, thus permitting many reactions to be carried

Donald J. Cram *was born in Vermont in 1919. He received a B.S. from Rollins College, an M.S. from the University of Nebraska, and a Ph.D. from Harvard University. He is a professor of chemistry at the University of California, Los Angeles.*

Charles J. Pedersen (1904–1989) *was born in Korea as a Norwegian citizen. He received a B.S. in chemical engineering from the University of Dayton and an M.S. in organic chemistry from MIT. He joined DuPont in 1927 and retired from there in 1969.*

Jean-Marie Lehn *was born in France in 1939. He initially studied philosophy and then switched to chemistry. He received a Ph.D. from the University of Strasbourgh. As a postdoctoral fellow, he worked with R. B. Woodward at Harvard on the total synthesis of vitamin B*$_{12}$ *(see pages 1175 and 1229). He is a professor of chemistry at the Université Louis Pasteur in Strasbourg, France and at the College de France in Paris.*

out in nonpolar solvents that otherwise would be able to take place only in polar protic solvents. For example, potassium permanganate ($K^+MnO_4^-$), a common oxidizing agent (Section 17.2), is not soluble in a nonpolar solvent such as benzene because it is an ionic compound. How, then, can potassium permanganate oxidize compounds that are soluble only in nonpolar solvents? Potassium permanganate can be dissolved in benzene if [18]-crown-6 is added to the solution. The crown ether binds potassium in its cavity, and the nonpolar crown ether–potassium complex dissolves in benzene. In order to maintain electrical neutrality, the permanganate ion accompanies the complexed potassium ion into the benzene layer, permitting permanganate to act as an oxidizing agent in the nonpolar solvent.

"purple benzene"
the [18]-crown-6-KMnO₄ complex is soluble in benzene

In this example, the crown ether is acting as a phase transfer catalyst. A **phase transfer catalyst** is a compound that catalyzes a reaction by transferring a reagent (usually an inorganic ion) into the phase in which it is needed (Section 11.12).

AN IONOPHOROUS ANTIBIOTIC

An **antibiotic** is a compound that interferes with the growth of microorganisms. Nonactin is a naturally occurring antibiotic that owes its biological activity to the fact that it is able to disrupt the carefully maintained electrolyte balance between the inside and outside of a cell. It does this by acting like a crown ether. Nonactin's diameter is such that it specifically binds potassium ions. It then transports the potassium ions out of the cell. For proper cell function, the cell must maintain a higher concentration of K^+ inside the cell and a higher concentration of Na^+ outside the cell. The decreased concentration of K^+ within the bacterial cell causes the bacterium to die. Nonactin is an example of an ionophorous antibiotic. An **ionophore** is a compound that transports metal ions by binding them tightly.

nonactin

The ability of a host to bind only certain guests is an example of **molecular recognition.** The very precise reactions of enzymes and other biological molecules are made possible by molecular recognition. Only recently have chemists been able to design and synthesize organic molecules that have molecular recognition, although the specificity of these synthetic compounds for their guests is generally not as highly developed as the specificity exhibited by many biological molecules. The hexaammonium ion shown on page 461 was designed to bind adenosine triphosphate (ATP), a molecule with four negatively charged oxygens. ATP plays a role in many cellular reactions (Section 24.2).

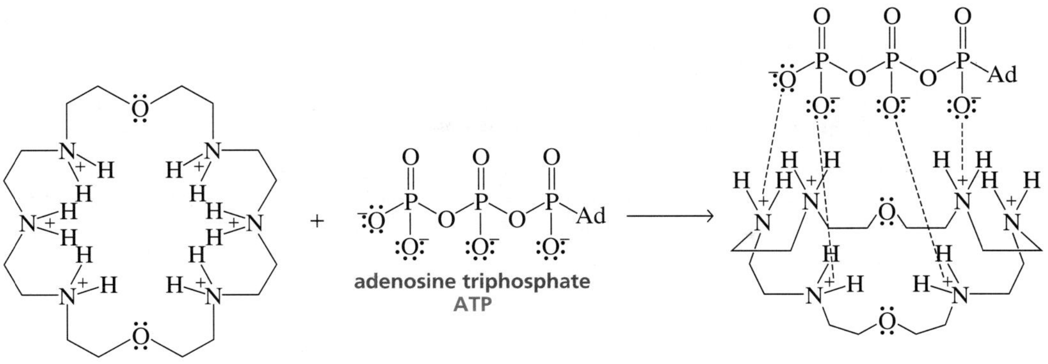

Polycyclic compounds are now being designed that are highly specific in their binding. These compounds use the nonbonding electrons of oxygen and nitrogen atoms to bind "guests." The three-dimensional compounds are called **cryptands** (*kryptos* is the Greek word for "hidden"). The complex that results when the cryptand binds its substrate is called a **cryptate.** The diaminohexaether shown below has been specifically designed to bind a potassium ion. It is better at binding K^+ than [18]-crown-6; the latter is a floppy molecule that becomes rigid when it binds potassium. As a result, some entropy is lost in the binding process. The three-dimensional cryptand is more rigid than the crown ether, and therefore less entropy is lost on binding.

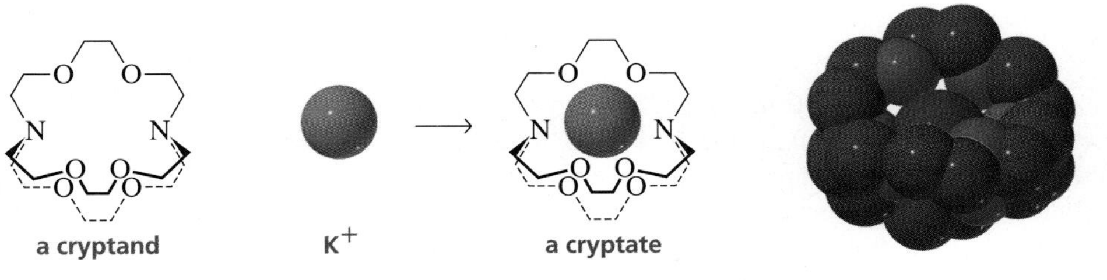

a cryptand K^+ a cryptate

**11.10
THIOLS, SULFIDES, AND SULFONIUM SALTS**

Thiols are sulfur analogs of alcohols. Thiols were formerly called mercaptans because they form strong complexes with heavy metal cations such as mercury and arsenic (they capture mercury).

$$2\ CH_3CH_2SH\ +\ Hg^{2+}\ \longrightarrow\ CH_3CH_2S\!-\!Hg\!-\!SCH_2CH_3\ +\ 2\ H^+$$
$$\text{mercury ion}$$

The IUPAC name for a thiol is obtained by adding the suffix "thiol" to the parent hydrocarbon. This is similar to the way in which alcohols are named except that "thiol" is used in place of "ol." Low-molecular-weight thiols are noted for their strong and pungent odors, such as the odors associated with onions, garlic, and skunks. Natural gas is completely odorless. A small amount of a thiol is added to natural gas to make gas leaks detectable.

			CH_3
			$\mid$
CH_3CH_2SH	$CH_3CH_2CH_2SH$	$CH_2\!=\!CHCH_2SH$	$CH_3CHCH_2CH_2SH$
ethanethiol	**propane-1-thiol**	**2-propene-1-thiol**	**3-methylbutane-1-thiol**

Because sulfur is larger than oxygen, thiols are stronger acids ($pK_a = 10$) than alcohols ($pK_a = 16$). Consequently, thiolate ions are weaker bases than alkoxide ions but better nucleophiles in protic solvents (Section 9.3).

$$CH_3\ddot{\underset{\cdot\cdot}{S}}{:}^- \ + \ CH_3CH_2{-}Br \ \longrightarrow \ CH_3SCH_2CH_3 \ + \ Br^-$$

Because sulfur is not as electronegative as oxygen, thiols are not good at hydrogen bonding. Therefore, they have considerably lower boiling points than alcohols (the boiling point of CH_3CH_2SH is 37 °C; the boiling point of CH_3CH_2OH is 78 °C).

The sulfur analogs of ethers are called **sulfides** or **thioethers.** As a consequence of the strong nucleophilicity of sulfur, sulfides react readily with alkyl halides to form **sulfonium salts.**

$$CH_3\ddot{S}CH_3 \ + \ CH_3{-}I \ \longrightarrow \ CH_3\underset{+}{\overset{CH_3}{S}}CH_3 \ \ I^-$$

dimethyl   trimethylsulfonium iodide
sulfide   a sulfonium salt

Because of its weakly basic leaving group, a sulfonium salt readily undergoes nucleophilic substitution reactions. As with other S_N2 reactions, the reaction works best if the group undergoing nucleophilic attack is a methyl or a primary alkyl group.

$$HO{:}^- \ + \ CH_3\underset{+}{\overset{CH_3}{S}}CH_3 \ \longrightarrow \ CH_3\ddot{O}H \ + \ CH_3\ddot{S}CH_3$$

MUSTARD GAS

Chemical warfare was used for the first time in 1915 when Germany released chlorine gas against French and British forces in the battle of Ypres. For the remainder of World War I, both sides used a variety of chemical agents. One of the more common war gases was mustard gas, a reagent that produces blisters all over the surface of the body. It reacts rapidly because the highly nucleophilic sulfur displaces a chloride ion by an intramolecular S_N2 reaction, forming a cyclic sulfonium salt. The toxicity of mustard gas is due to nucleophiles of cell components becoming alkylated as a result of reacting with the sulfonium salt. The sulfonium salt is particularly reactive because of the strained three-membered ring.

$$ClCH_2CH_2\ddot{S}CH_2CH_2{-}Cl \ \longrightarrow \ ClCH_2CH_2\underset{+}{\overset{\displaystyle CH_2}{S}}\!\!\Big\langle{\overset{|}{\underset{CH_2}{\,}}} \ \ \overset{..}{Nu}{}^- \ \ ClCH_2CH_2SCH_2CH_2{-}Nu$$

mustard gas   sulfonium salt Cl^- alkylated nucleophile

By the onset of World War II, many countries had large quantities of mustard gas and other chemical agents. Such quantities probably served as a deterrent to their use in the war.

ANTIDOTE TO A WAR GAS

Lewisite is a war gas developed in 1917 by W. Lee Lewis, an American scientist. It rapidly penetrates clothing and skin and is poisonous because it contains arsenic, which combines with thiol groups on enzymes. During World War II, the British developed an antidote to lewisite called British anti-lewisite (BAL). BAL contains thiol groups and reacts with lewisite, preventing it from reacting with enzymes.

$$\begin{matrix} CH_2SH \\ | \\ CHSH \\ | \\ CH_2OH \end{matrix} \ + \ \overset{Cl}{\underset{Cl}{\diagdown\,\diagup}}AsCH{=}CHCl \ \longrightarrow \ \begin{matrix} CH_2S \\ | \\ CHS \\ | \\ CH_2OH \end{matrix}\!\!\!\overset{\diagdown}{\underset{\diagup}{\,}}AsCH{=}CHCl \ + \ 2\ HCl$$

BAL   lewisite

PROBLEM 25

How would you prepare the following compounds using an alkyl halide and a thiol as starting materials?

a. $CH_3CH_2SCH_2CH_3$

$$CH_3$$
c. $CH_2{=}CHCHSCH_2CH_3$

$$CH_3$$
b. CH_3SCCH_3
$$CH_3$$

d.  —SCH₂—

The amide ion ($^-NH_2$) is an exceedingly poor leaving group. The pK_a of NH_3 is 36, indicating that an amide ion is a much stronger base and consequently a much poorer leaving group than ^-OH or ^-OR (the pK_a values of H_2O and ROH are about 16). Amines therefore do not undergo substitution or elimination reactions.

$$RNH_2$$
an amine

$$R\overset{+}{N}H_3$$
a protonated amine
an ammonium ion

$$\underset{R}{\overset{R}{R{-}\overset{+}{N}{-}R}}$$
a quaternary ammonium ion

Protonation of the amino group makes it a better leaving group (pK_a of $^+NH_4$ is 9.4), but not good enough to enable the group to dissociate from a carbon to form a carbocation or to be replaced by a halide ion. The protonated amino group cannot be displaced by a strong base such as HO^- because the strong base would first react with the acidic hydrogen, which would turn the protonated amino group back into an unprotonated amino group.

PROBLEM 26

Why is it that a halide ion such as Br^- can react with a protonated primary alcohol but cannot react with a protonated primary amine?

The Hofmann Elimination

Because the leaving group of a **quaternary ammonium ion** has about the same leaving tendency as a protonated amino group but does not have an acidic hydrogen, a quaternary ammonium ion can undergo reaction with a strong base. The reaction of a quaternary ammonium ion with hydroxide ion is known as a **Hofmann elimination reaction.** The leaving group in a Hofmann elimination reaction is a tertiary amine. Because a tertiary amine is only a moderately good leaving group, the reaction requires heat.

$$CH_3CH_2CH_2\overset{CH_3}{\underset{CH_3\ \ HO^-}{\overset{|}{\underset{|}{N}}CH_3}} \xrightarrow{\Delta} CH_3CH{=}CH_2 + \underset{CH_3}{\overset{CH_3}{\overset{|}{\underset{|}{N}}CH_3}} + H_2O$$

**11.11
REACTIONS OF
QUATERNARY
AMMONIUM
COMPOUNDS**

August von Hofmann

August Wilhelm von Hofmann (1818–1892) *was born in Germany. He first studied law, then changed to chemistry. He founded the German Chemical Society. Hofmann taught at the Royal College of Chemistry in London for 20 years, then returned to Germany to teach at the University of Berlin. He was married four times— having three times been left a widower—and had 11 children.*

This reaction is an E2 reaction. The only difference between the Hofmann elimination reaction and an E2 reaction of an alkyl halide is the leaving group. With alkyl halides, the leaving group is a halide ion instead of a tertiary amine. Very little substitution product is formed.

mechanism of the Hofmann elimination

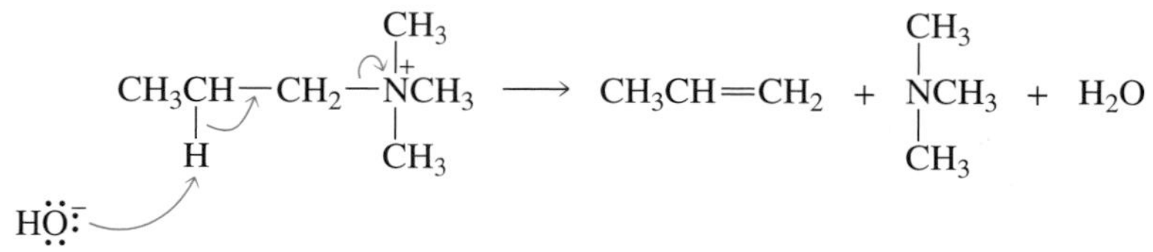

PROBLEM 27◆

What would be the substitution product in the preceding reaction?

Because the carbon to which the tertiary amine is attached is designated as the α-*carbon,* the adjacent carbon, from which the proton is removed, is called the β-*carbon.* If the quaternary ammonium ion has more than one β-carbon, the major alkene product is the one obtained by removing a proton from the β-carbon bonded to the most hydrogens. In the following reaction, the major alkene product is obtained by removing a hydrogen from the β-carbon bonded to three hydrogens, and the minor alkene product results from removing a hydrogen from the β-carbon bonded to two hydrogens.

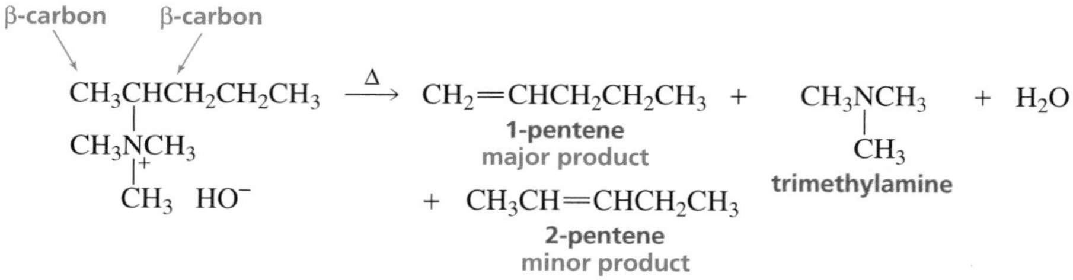

In the following reaction, the major products come from removing a hydrogen from the β-carbon bonded to two hydrogens, because the other β-carbon is bonded to only one hydrogen.

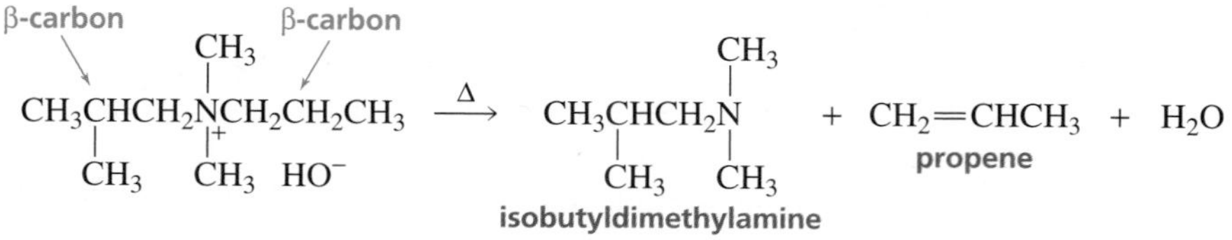

PROBLEM 28◆

What are the minor products in the preceding Hofmann elimination reaction?

Recall that Zaitsev's rule says that the hydrogen is removed from the β-carbon bonded to the fewest hydrogens.

The Hofmann elimination reaction involves **anti-Zaitsev elimination.** The hydrogen is removed from the β-carbon bonded to the most hydrogens.

Electronic factors are responsible for the anti-Zaitsev elimination. When hydroxide ion starts to remove a proton in the elimination reaction of an alkyl halide, the leaving group immediately starts to depart and a transition state with an "alkene-like" structure results. The proton is removed from the β-carbon bonded to the fewest hydrogens in order to achieve the most stable "alkene-like" transition state. However, when hydroxide ion starts to remove a proton in the Hofmann elimination reaction, the leaving group does not immediately start to leave because a tertiary amine is not as good a leaving group as Cl^-, Br^-, or I^-. As a result, a partial negative charge builds up on the carbon from which the proton is leaving. This gives the transition state a "carbanion-like" rather than an "alkene-like" structure. By removing a proton from the β-carbon with the most hydrogens, the most stable "carbanion-like" transition state is achieved. Recall that primary carbanions are more stable than secondary carbanions, which are more stable than tertiary carbanions (Section 10.2).

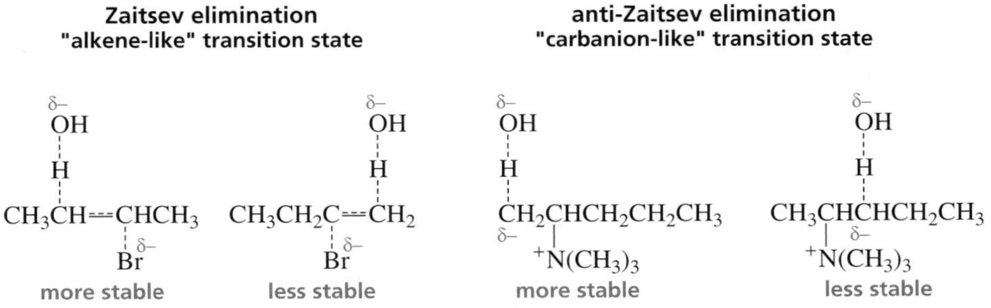

| Zaitsev elimination "alkene-like" transition state | anti-Zaitsev elimination "carbanion-like" transition state |

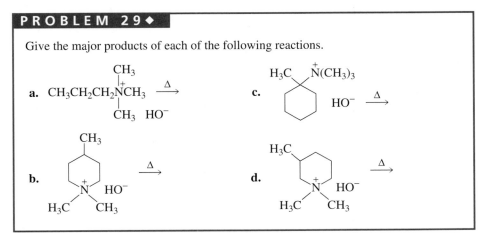

Because the Hofmann elimination reaction occurs in an anti-Zaitsev manner, anti-Zaitsev elimination is also referred to as **Hofmann elimination.** We have previously seen anti-Zaitsev elimination in the E2 reactions of alkyl fluorides as a result of fluoride ion being a poorer leaving group than chloride, bromide, or iodide ions, which causes the transition state to have a "carbanion-like" rather than an "alkene-like" structure (Section 10.2).

PROBLEM 29◆

Give the major products of each of the following reactions.

a. $CH_3CH_2CH_2\overset{\overset{\displaystyle CH_3}{|+}}{\underset{\underset{\displaystyle CH_3\ \ HO^-}{|}}{N}}CH_3 \quad \xrightarrow{\Delta}$

b. (structure) $\xrightarrow{\Delta}$

c. (structure) $\xrightarrow{\Delta}$

d. (structure) $\xrightarrow{\Delta}$

In order for a quaternary ammonium ion to undergo an elimination reaction, the counterion must be hydroxide ion because a strong base is needed to start the reaction by removing a proton from a β-carbon. Because halide ions are weak bases, quaternary ammonium halides cannot undergo Hofmann eliminations. However, a quaternary ammonium halide can be converted into a quaternary ammonium hydroxide by treating it with silver oxide and water. The silver halide precipitates, and the halide ion thus removed from solution is replaced by hydroxide ion. The compound can now undergo elimination.

$$2 \; \underset{\underset{R \;\; I^-}{|}}{\overset{\overset{R}{|}}{RNR}} \; + \; Ag_2O \; + \; H_2O \; \longrightarrow \; 2 \; \underset{\underset{R \;\; HO^-}{|}}{\overset{\overset{R}{|}}{RNR}} \; + \; 2 \; AgI \downarrow$$

The reaction of an amine with sufficient methyl iodide to convert the amine into a quaternary ammonium iodide is called **exhaustive methylation.**

exhaustive methylation

$$CH_3CH_2CH_2NH_2 \; + \; \underset{\text{excess}}{CH_3I} \; \longrightarrow \; \underset{\underset{CH_3 \;\; I^-}{|}}{\overset{\overset{CH_3}{|}}{CH_3CH_2CH_2NCH_3}}$$

PROBLEM 30

Propose a mechanism for the following reaction.

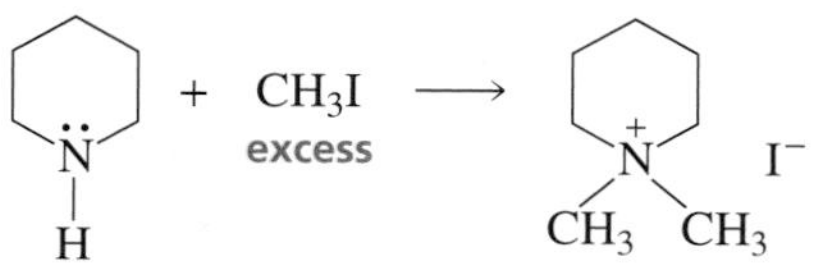

PROBLEM 31

Describe a synthesis for each of the following compounds, using the given starting material and any necessary reagents.

a. $CH_3CH_2CH_2CH_2NH_2 \longrightarrow CH_3CH_2CH{=}CH_2$

b. $CH_3CH_2\underset{\underset{Br}{|}}{\overset{\overset{CH_3}{|}}{CH}}CHCH_3 \longrightarrow CH_3CH_2\overset{\overset{CH_3}{|}}{CH}CH{=}CH_2$

The Cope Elimination

Arthur C. Cope (1909–1966) *was born in Indiana. He received a Ph.D. from the University of Wisconsin and was a professor of chemistry at Bryn Mawr College, Columbia University, and the Massachusetts Institute of Technology.*

The **Cope elimination reaction** is similar to the Hofmann elimination reaction. But in the Cope elimination, a tertiary amine oxide rather than a quaternary ammonium ion undergoes elimination.

$$\underset{\underset{O^-}{\overset{|}{|+}}}{\overset{\overset{CH_3}{|}}{CH_3CH_2CH_2NCH_3}} \; \overset{\Delta}{\longrightarrow} \; CH_3CH{=}CH_2 \; + \; \underset{\underset{OH}{|}}{\overset{\overset{CH_3}{|}}{NCH_3}}$$

a tertiary amine oxide   **a hydroxylamine**

The negatively charged oxygen of the amine oxide is the base that removes a proton from the β-carbon. Because the nucleophile and the leaving group are in the same molecule, the Cope elimination is an intramolecular E2 reaction. The reaction involves syn elimination.

mechanism of the Cope elimination

$$\underset{\substack{|\\H}}{\overset{\substack{CH_3\\|\\+}}{CH_3CH-CH_2-NCH_3}} \quad \underset{:\overset{..}{\underset{..}{O}}{:}^-}{} \quad \xrightarrow{\Delta} \quad CH_3CH{=}CH_2 \;+\; \underset{\substack{|\\OH}}{\overset{\substack{CH_3\\|}}{NCH_3}}$$

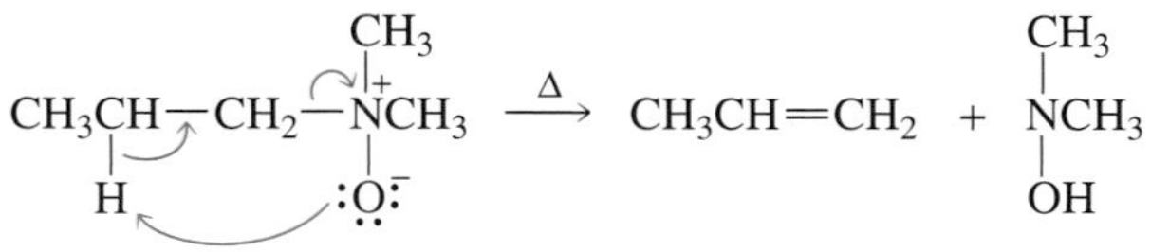

The major product of the Cope elimination, like that of the Hofmann elimination, is the one obtained by removing a hydrogen from the β-carbon bonded to the most hydrogens.

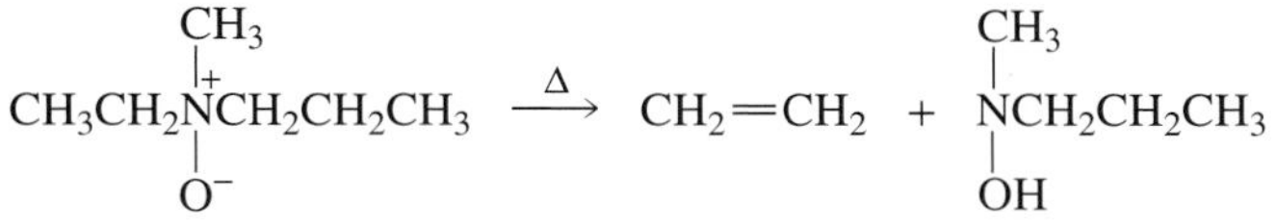

$$\underset{\substack{|\\O^-}}{\overset{\substack{CH_3\\|\\+}}{CH_3CH_2NCH_2CH_2CH_3}} \quad \xrightarrow{\Delta} \quad CH_2{=}CH_2 \;+\; \underset{\substack{|\\OH}}{\overset{\substack{CH_3\\|}}{NCH_2CH_2CH_3}}$$

An amine oxide can be synthesized by treating a tertiary amine with hydrogen peroxide.

$$\underset{\substack{|\\R}}{\overset{\substack{R\\|}}{R\overset{..}{N}R}} + \; HO{-}OH \;\longrightarrow\; \underset{\substack{|\\OH}}{\overset{\substack{R\\|\\+}}{RNR}} + HO^- \;\longrightarrow\; \underset{\substack{|\\O^-}}{\overset{\substack{R\\|\\+}}{RNR}} + H_2O$$

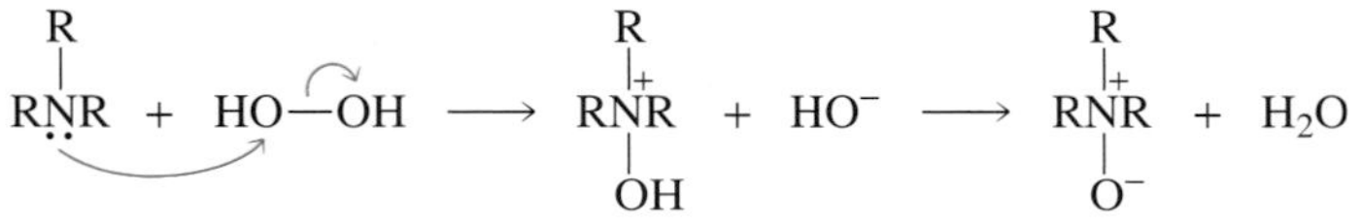

| a tertiary amine | hydrogen peroxide | | an amine oxide | |

PROBLEM 32◆

Does the Cope elimination have an "alkene-like" transition state or a "carbanion-like" transition state?

PROBLEM 33◆

Give the products that would be obtained by treating the following tertiary amines with hydrogen peroxide followed by heat.

a. 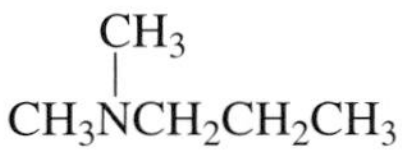 $\underset{\substack{|\\CH_3}}{\overset{\substack{CH_3\\|}}{CH_3NCH_2CH_2CH_3}}$

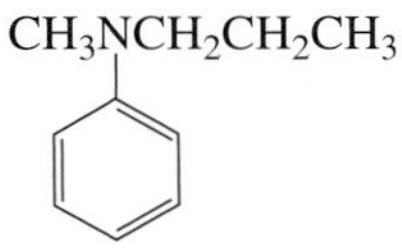

c. 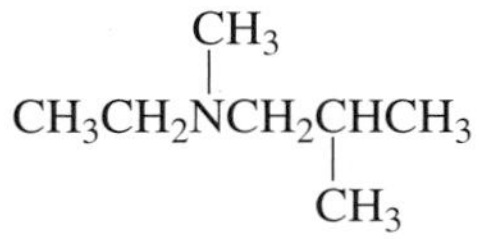 $\underset{\substack{\\ \qquad |\\ \qquad CH_3}}{\overset{\substack{CH_3\\|}}{CH_3CH_2NCH_2CHCH_3}}$

b. $\underset{\substack{|\\ \bigcirc}}{CH_3NCH_2CH_2CH_3}$

d.

A problem that faces organic chemists in the laboratory is finding a solvent that will dissolve all the reagents. For example, if we want 1-bromohexane to react with cyanide ion, we encounter a problem because the alkyl halide is insoluble in water but sodium cyanide, an ionic compound, is soluble only in water.

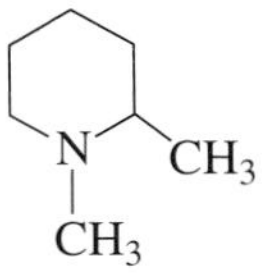

**11.12
PHASE TRANSFER
CATALYSIS**

$$\underset{\text{1-bromohexane}}{CH_3CH_2CH_2CH_2CH_2CH_2Br} \;+\; {}^-C{\equiv}N \;\xrightarrow{?}\; CH_3CH_2CH_2CH_2CH_2CH_2C{\equiv}N \;+\; Br^-$$

If we mix a solution of 1-bromohexane in a nonpolar solvent with an aqueous solution of sodium cyanide, the two solutions will form two layers because they are immiscible. How, then, can sodium cyanide react with the alkyl halide? The two compounds will be able to react with each other if a catalytic amount of a phase transfer catalyst is added to the reaction mixture.

$$CH_3CH_2CH_2CH_2CH_2CH_2Br \ + \ ^-C\equiv N \ \xrightarrow[\text{phase transfer catalyst}]{R_4\overset{+}{N} \ HSO_4^-} \ CH_3CH_2CH_2CH_2CH_2CH_2C\equiv N \ + \ Br^-$$

In Section 11.9 we saw that crown ethers can be used as phase transfer catalysts. Quaternary ammonium salts are the most common phase transfer catalysts.

phase transfer catalysts

$$\underset{\substack{\text{tetrabutylammonium}\\\text{bisulfate}}}{CH_3CH_2CH_2CH_2\overset{\displaystyle \overset{CH_2CH_2CH_2CH_3}{|}}{\underset{\underset{CH_2CH_2CH_2CH_3}{|}}{\overset{+}{N}}}CH_2CH_2CH_2CH_3} \qquad \underset{\substack{\text{hexadecyltrimethylammonium}\\\text{bisulfate}}}{CH_3(CH_2)_{14}CH_2\overset{\displaystyle \overset{CH_3}{|}}{\underset{\underset{CH_3}{|}}{\overset{+}{N}}}CH_3}$$
$$HSO_4^- \qquad\qquad HSO_4^-$$

benzyltriethylammonium bisulfate

How does addition of a phase transfer catalyst allow the reaction of 1-bromo-hexane with cyanide ion to take place? The quaternary ammonium salt, because of its nonpolar alkyl groups, is soluble in nonpolar solvents, but it is also soluble in water because of its charge. This means that it can act as a mediator between two immiscible solvents. When a phase transfer catalyst such as tetrabutylammonium bisulfate passes into the nonpolar layer, it must carry a counterion with it in order to balance its positive charge. The counterion can be either the original counterion (bisulfate) or another ion present in the solution (in the reaction under discussion, it will be cyanide ion). Because there is more cyanide ion than bisulfate ion in the aqueous layer, cyanide ion will more often be the accompanying ion. Once it is in the nonpolar layer, cyanide ion can react with the alkyl halide. (When bisulfate is transported into the nonpolar layer, it is unreactive because it is both a weak base and a poor nucleophile.) The quaternary ammonium ion will pass back into the aqueous layer carrying with it, as a counterion, either bisulfate ion or bromide ion. The reaction continues with the phase transfer catalyst shuttling back and forth between the two phases. **Phase transfer catalysis** has been successfully used in a wide variety of organic reactions.

nonpolar layer $\quad R_4\overset{+}{N} \ ^-C\equiv N \ + \ \underset{\text{starting material}}{R-Br} \ \longrightarrow \ \underset{\text{target compound}}{R-C\equiv N} \ + \ R_4\overset{+}{N} \ Br^-$

water layer $\quad R_4\overset{+}{N} \ ^-C\equiv N \quad \overset{+}{Na} \ HSO_4^- \qquad\qquad\qquad\qquad R_4\overset{+}{N} \ Br^-$

SUMMARY OF REACTIONS

1. Conversion of an *alcohol* to an *alkyl halide* (Sections 11.1 and 11.2).

$$ROH + HBr \xrightarrow{\Delta} RBr$$

$$ROH + HI \xrightarrow{\Delta} RI$$

$$ROH + HCl \xrightarrow[\Delta]{ZnCl_2} RCl$$

$$ROH + PBr_3 \xrightarrow{pyridine} RBr$$

$$ROH + PCl_3 \xrightarrow{pyridine} RCl$$

$$ROH + SOCl_2 \xrightarrow{pyridine} RCl$$

2. Conversion of an *alcohol* to a *sulfonate ester* (Section 11.3).

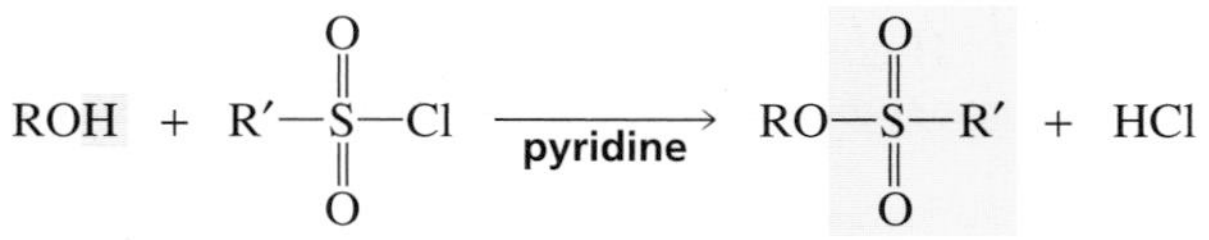

$$ROH + R'\!-\!\overset{\displaystyle O}{\underset{\displaystyle O}{\overset{\|}{\underset{\|}{S}}}}\!-\!Cl \xrightarrow{pyridine} RO\!-\!\overset{\displaystyle O}{\underset{\displaystyle O}{\overset{\|}{\underset{\|}{S}}}}\!-\!R' + HCl$$

3. Conversion of an *activated alcohol* (an alkyl halide or a sulfonate ester) to a *compound with a new group bonded to the* sp^3 *carbon* (Section 11.3).

$$RBr + Y^- \longrightarrow RY + Br^-$$

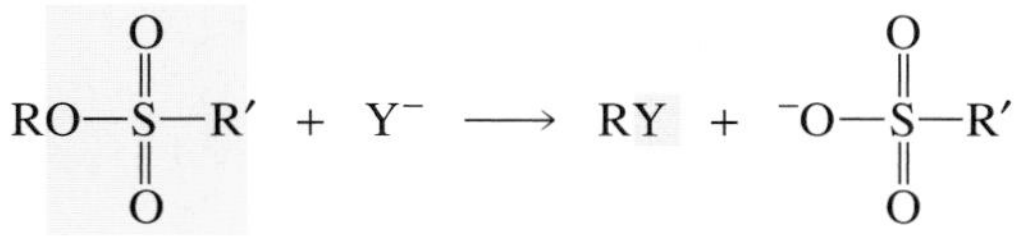

$$RO\!-\!\overset{\displaystyle O}{\underset{\displaystyle O}{\overset{\|}{\underset{\|}{S}}}}\!-\!R' + Y^- \longrightarrow RY + {}^-O\!-\!\overset{\displaystyle O}{\underset{\displaystyle O}{\overset{\|}{\underset{\|}{S}}}}\!-\!R'$$

4. Dehydration of *alcohols* (Section 11.4).

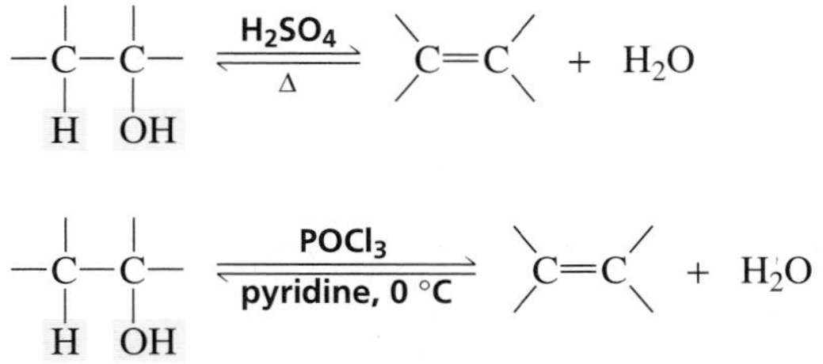

$$-\overset{|}{\underset{|}{\overset{}{C}}}\!-\!\overset{|}{\underset{|}{\overset{}{C}}}- \underset{\Delta}{\overset{H_2SO_4}{\rightleftharpoons}} \,\,\,C\!=\!C \,\,+\, H_2O$$
$$H\;\;OH$$

$$-\overset{|}{\underset{|}{\overset{}{C}}}\!-\!\overset{|}{\underset{|}{\overset{}{C}}}- \underset{pyridine,\,0\,°C}{\overset{POCl_3}{\rightleftharpoons}} \,\,\,C\!=\!C \,\,+\, H_2O$$
$$H\;\;OH$$

relative rate: tertiary > secondary > primary

5. Cleavage of *ethers* (Section 11.5).

$$ROR' + HX \xrightarrow{\Delta} ROH + R'X$$

HX = HBr or HI

6. Ring-opening reactions of *epoxides* (Section 11.6).

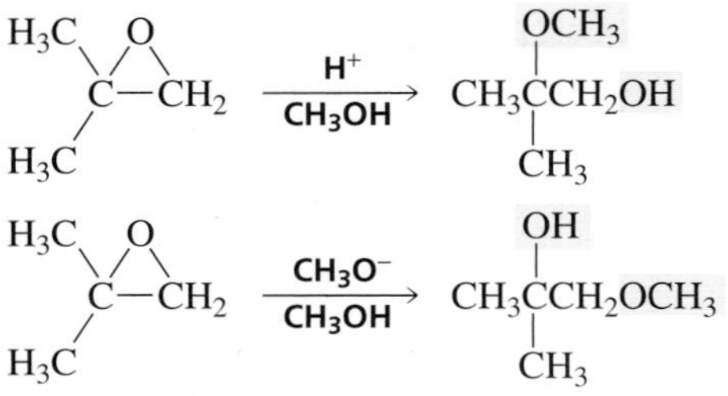

under acidic conditions, the nucleophile attacks the most substituted carbon

under basic conditions, the nucleophile attacks the least sterically hindered carbon

7. Reactions of *arene oxides:* ring opening and rearrangement (Section 11.7).

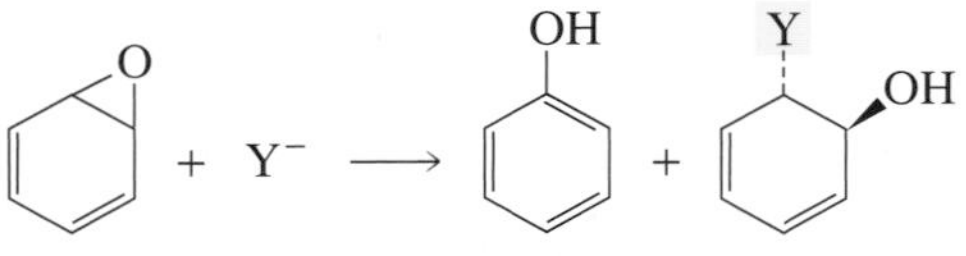

8. Reaction of a *Grignard reagent* with an *epoxide* (Section 11.8).

$$\text{RBr} \xrightarrow[\textbf{ether}]{\textbf{Mg}} \text{RMgBr}$$

a Grignard reagent

$$\text{RMgBr} + \text{H}_2\overset{\text{O}}{\text{C}-\text{C}}\text{H}_2 \longrightarrow \text{RCH}_2\text{CH}_2\text{O}^- \xrightarrow{\textbf{H}^+} \text{RCH}_2\text{CH}_2\text{OH}$$

product alcohol contains two more carbons than the Grignard reagent

9. Reaction of a *Gilman reagent* with an alkyl halide (Section 11.8).

$$2\ \text{RLi} + \text{CuI} \xrightarrow{\textbf{ether}} \underset{\textbf{Gilman reagent}}{\text{R}_2\text{CuLi}} + \text{LiI}$$

$$\text{CH}_3\text{CH}_2\text{CH}_2\text{X} + \text{R}_2\text{CuLi} \xrightarrow{\textbf{ether}} \text{CH}_3\text{CH}_2\text{CH}_2\text{R} + \text{RCu} + \text{LiX}$$

X = Cl, Br, I

10. Reactions of *thiols, sulfides,* and *sulfonium salts* (Section 11.10).

$$2\ \text{RSH} + \text{Hg}^{2+} \longrightarrow \text{RS}-\text{Hg}-\text{SR} + 2\ \text{H}^+$$

$$\text{RS}^- + \text{R}'-\text{Br} \longrightarrow \text{RSR}' + \text{Br}^-$$

$$\text{RSR} + \text{R}'\text{I} \longrightarrow \text{R}\overset{\text{R}'}{\underset{+}{\text{S}}}\text{R}\ \ \text{I}^-$$

$$\text{R}\overset{\text{R}}{\underset{+}{\text{S}}}\text{R} + \text{Y}^- \longrightarrow \text{RY} + \text{RSR}$$

11. Elimination reactions of *quaternary ammonium hydroxides* or *tertiary amine oxides* (Section 11.11).

$$\underset{\text{HO}^-}{\text{RCH}_2\text{CH}_2\overset{\text{CH}_3}{\underset{\text{CH}_3}{\overset{+}{\text{N}}}}\text{CH}_3} \xrightarrow[\textbf{Hofmann elimination}]{\Delta} \text{RCH}=\text{CH}_2 + \overset{\text{CH}_3}{\underset{\text{CH}_3}{\text{N}}}\text{CH}_3 + \text{H}_2\text{O}$$

$$\text{RCH}_2\text{CH}_2\overset{\text{CH}_3}{\text{N}}\text{CH}_3 \xrightarrow{\textbf{H}_2\textbf{O}_2} \text{RCH}_2\text{CH}_2\overset{\text{CH}_3}{\underset{\text{O}^-}{\overset{+}{\text{N}}}}\text{CH}_3 \xrightarrow[\textbf{Cope elimination}]{\Delta} \text{RCH}=\text{CH}_2 + \overset{\text{CH}_3}{\underset{\text{OH}}{\text{N}}}\text{CH}_3$$

in both eliminations the proton is removed from the β-carbon bonded to the most hydrogens

KEY TERMS

alcohol (page 444)
alkyl tosylate (page 435)
antibiotic (page 460)
anti-Zaitsev elimination (page 464)
arene oxide (page 450)
Cope elimination reaction (page 466)

coupling reaction (page 458)
crown ether (page 459)
crown–guest complex (page 459)
cryptand (page 461)
cryptate (page 461)
dehydration (page 437)
epoxide (page 446)
ether (page 444)
exhaustive methylation (page 466)
Gilman reagent (page 458)
Grignard reagent (page 455)
Hofmann elimination (page 465)

Hofmann elimination reaction
 (page 463)
inclusion compound (page 459)
ionophore (page 460)
molecular recognition (page 460)
organolithium compound (page 455)
organomagnesium compound
 (page 455)
organometallic compound (455)
oxirane (page 446)
phase transfer catalysis (page 468)
phase transfer catalyst (page 460)

pinacol rearrangement (page 477)
quaternary ammonium ion
 (page 463)
ring-expansion rearrangement
 (page 439)
sulfide (page 462)
sulfonate ester (page 434)
sulfonium salt (page 462)
thioether (page 462)
thiol (page 461)
transmetallation (page 458)
vicinal diol (page 477)

PROBLEMS

34. Give the product of each of the following reactions.

a. $CH_3CH_2CH_2OH$ $\xrightarrow{\text{1. methanesulfonyl chloride}}$
  2. $CH_3\overset{O}{\overset{\|}{C}}O^-$

b. $CH_3CH_2CH_2CH_2OH + PBr_3$ $\xrightarrow{\text{pyridine}}$

c. 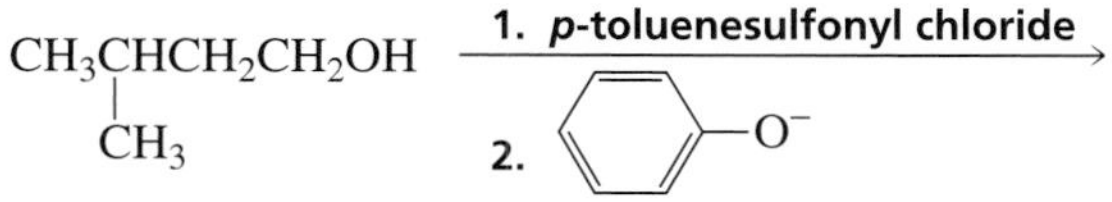 $CH_3\underset{\underset{CH_3}{|}}{CH}CH_2CH_2OH$ $\xrightarrow{\text{1. }p\text{-toluenesulfonyl chloride}}$
  2. ⟨benzene ring⟩—O^-

d. 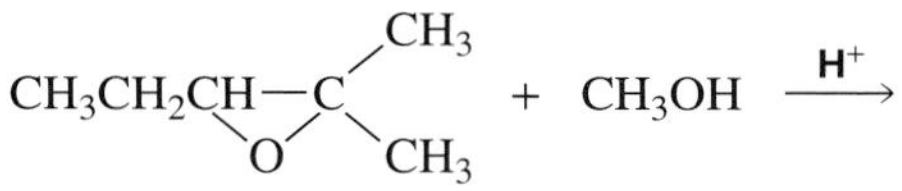 $CH_3CH_2CH—\overset{\overset{CH_3}{|}}{\underset{\underset{O}{\diagdown}}{C}}\diagup^{CH_3}$ $+ CH_3OH$ $\xrightarrow{H^+}$

e. 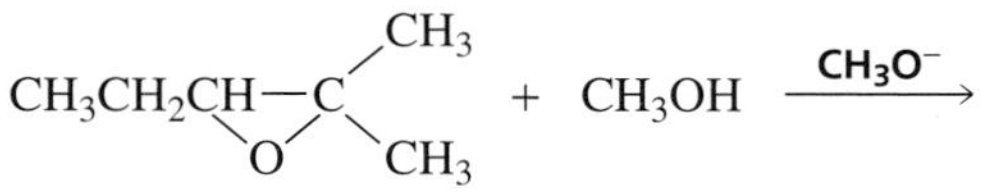 $CH_3CH_2CH—\overset{\overset{CH_3}{|}}{\underset{\underset{O}{\diagdown}}{C}}\diagup^{CH_3}$ $+ CH_3OH$ $\xrightarrow{CH_3O^-}$

f. $CH_3CH_2CH_2CH_2OH$ $\xrightarrow[\Delta]{H_2SO_4}$

g. $CH_3\underset{\underset{CH_3}{|}}{CH}CH_2CH_2OH$ $\xrightarrow[\text{pyridine}]{SOCl_2}$

h. 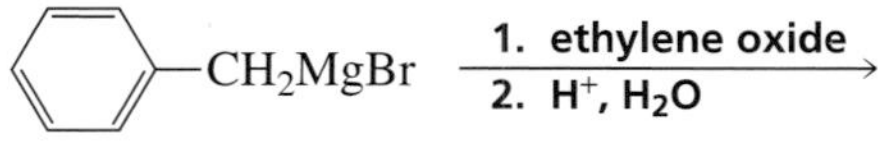 ⟨benzene ring⟩—CH_2MgBr $\xrightarrow[\text{2. H}^+, \text{H}_2\text{O}]{\text{1. ethylene oxide}}$

i. $CH_3\underset{\underset{CH_3}{|}}{\overset{\overset{CH_3}{|}}{C}}OCH_2CH_3 + HBr$ $\xrightarrow{\Delta}$

j. $CH_3\underset{\underset{CH_3}{|}}{CH}CH_2OCH_3 + HI$ $\xrightarrow{\Delta}$

k. $CH_3\underset{\underset{CH_3}{|}}{CH}CH_2\overset{\overset{CH_3}{|}}{\underset{\underset{CH_3}{|}}{\overset{+}{N}}}CH_2CH_3$ $\quad HO^-$ $\xrightarrow{\Delta}$

l. 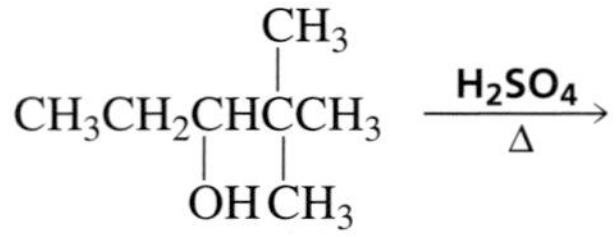 $CH_3CH_2CHCCH_3$ $\xrightarrow[\Delta]{H_2SO_4}$
(with CH_3 substituent above and $OHCH_3$ below)

m. 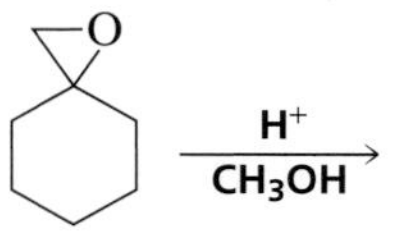 $\xrightarrow[CH_3OH]{H^+}$

n. 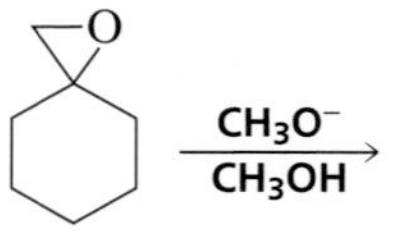 $\xrightarrow[CH_3OH]{CH_3O^-}$

35. Indicate which alcohol, when heated with H_2SO_4, will undergo dehydration more rapidly.

a. 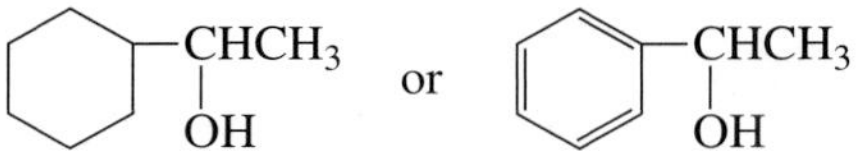   —CHCH₃ (with OH) or —CHCH₃ (with OH)

b. 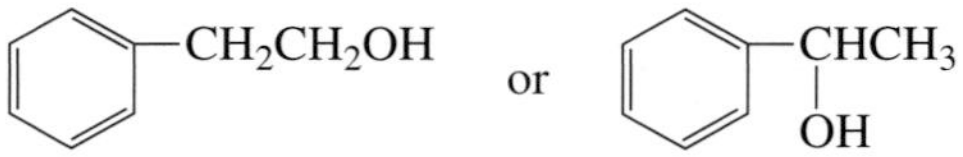   —CH₂CH₂OH or —CHCH₃ (with OH)

c. 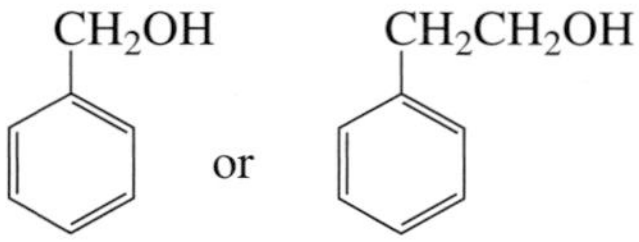   CH₂OH or CH₂CH₂OH

d. 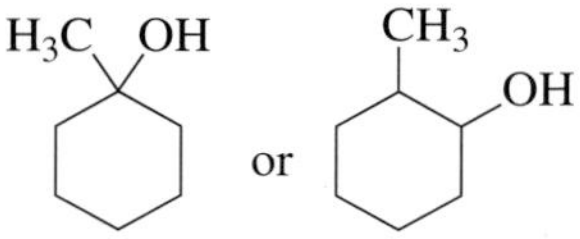   H₃C OH or CH₃ OH

e. 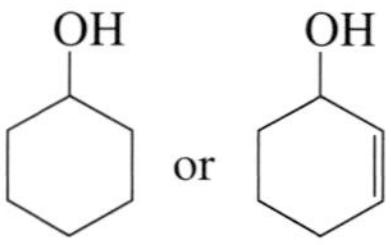   OH or OH

f. $CH_3CH_2CHCH_3$ or $CH_3CCH_2CH_3$
(first with OH, second with CH_3 above and OH below)

36. Which of the following alkyl halides could be successfully used to form a Grignard reagent?
   a. $HOCH_2CH_2CH_2CH_2Br$
   b. $BrCH_2CH_2CH_2\overset{\overset{O}{\|}}{C}OH$
   c. $CH_3NCH_2CH_2CH_2Br$ (with CH_3 below N)
   d. $H_2NCH_2CH_2CH_2Br$

37. Starting with (*R*)-1-deuterio-1-propanol, how could you prepare
   a. (*S*)-1-deuterio-1-propanol?   **c.** (*R*)-1-deuterio-1-methoxypropane?
   b. (*S*)-1-deuterio-1-methoxypropane?

38. What alkenes would you expect to be obtained from the acid-catalyzed dehydration of 1-hexanol?

39. When heated with H_2SO_4, both 3,3-dimethyl-2-butanol and 2,3-dimethyl-2-butanol are dehydrated to form 2,3-dimethyl-2-butene. Which alcohol dehydrates more rapidly?

40. Give the product of each of the following reactions.

a.

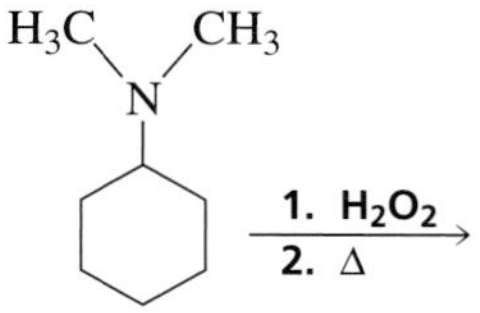

$$\xrightarrow[\text{2. } \Delta]{\text{1. } H_2O_2}$$

b.

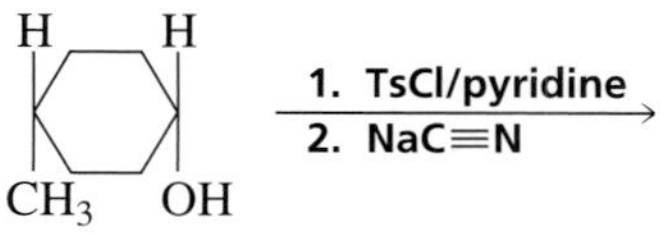

$$\xrightarrow[\text{2. NaC}\equiv\text{N}]{\text{1. TsCl/pyridine}}$$

c.

$$\xrightarrow{\text{(CH}_3\text{CH}_2\text{CH}_2\text{)}_2\text{CuLi}}$$

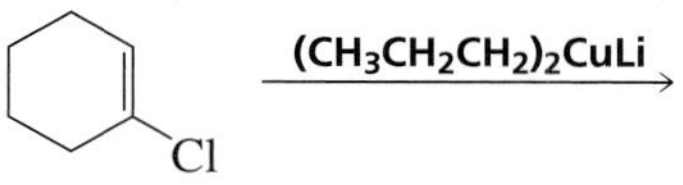

d. $\text{CH}_3\text{CH}-\overset{\overset{\displaystyle \text{CH}_3}{|}}{\underset{\underset{\displaystyle \text{OH}}{|}}{\text{C}}}\text{CH}_3 \xrightarrow[\Delta]{H_2SO_4}$
　　　$\overset{}{\underset{\underset{\displaystyle \text{CH}_3}{|}}{}}$

41. Using the given starting material, any necessary inorganic reagents, and any carbon-containing compounds with no more than two carbon atoms, indicate how the following syntheses could be carried out.

a.

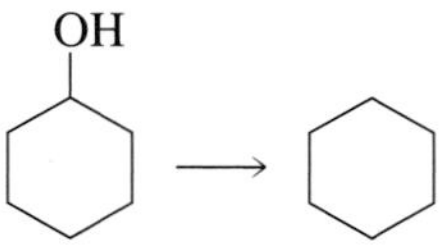

$$\longrightarrow$$

b.

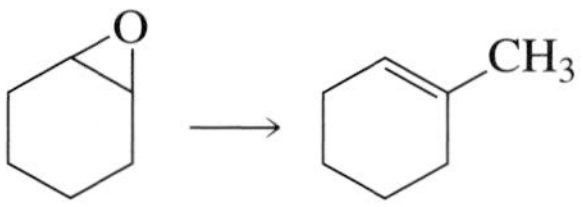

$$\longrightarrow$$

c. $\text{CH}_3\text{CH}_2\text{C}\equiv\text{CH} \longrightarrow \text{CH}_3\text{CH}_2\text{C}\equiv\text{CCH}_2\text{CH}_2\text{OH}$

d. $\text{CH}_3\overset{\overset{\displaystyle }{|}}{\underset{\underset{\displaystyle \text{CH}_3}{|}}{\text{C}}}\text{HCH}_2\text{OH} \longrightarrow \text{CH}_3\overset{}{\underset{\underset{\displaystyle \text{CH}_3}{|}}{\text{C}}}\text{HCH}_2\text{CH}_2\text{CH}_2\text{OH}$

e.

$$\longrightarrow$$

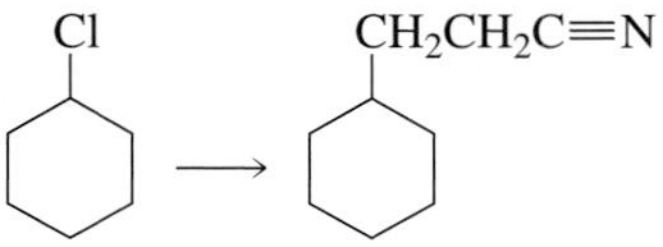

42. Propose a mechanism for the following reaction.

$$\text{CH}_3\overset{\overset{\displaystyle }{|}}{\underset{\underset{\displaystyle \text{Cl}}{|}}{\text{C}}}\text{HCH}-\text{CH}_2 + \text{CH}_3\text{O}^- \xrightarrow{\text{CH}_3\text{OH}} \text{CH}_3\text{CH}-\text{CHCH}_2\text{OCH}_3 + \text{Cl}^-$$

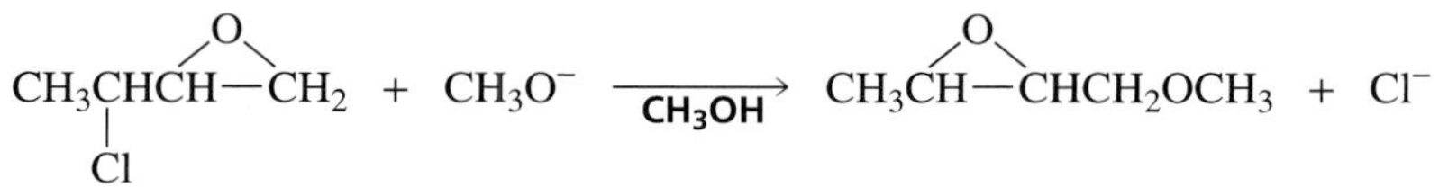

43. When deuterated phenanthrene oxide undergoes an epoxide rearrangement in water, 81% of the deuterium is retained in the product (*J. Am. Chem. Soc.*, 1976, *98*, 2965).

    a. What percentage of the deuterium will be retained if an NIH shift occurs?
    b. What percentage of the deuterium will be retained if an NIH shift does not occur?

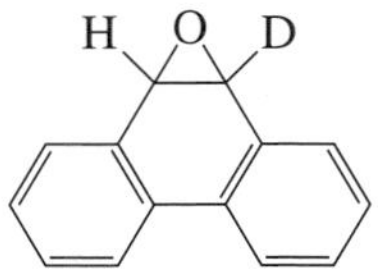

44. When 3-methyl-2-butanol is heated with concentrated HBr, a product with a rearranged carbon skeleton is obtained. When 2-methyl-1-propanol reacts under the same conditions, the product obtained does not have a rearranged carbon skeleton. Explain.

45. When the following seven-membered ring alcohol is dehydrated, three alkenes are formed. Propose a mechanism for their formation.

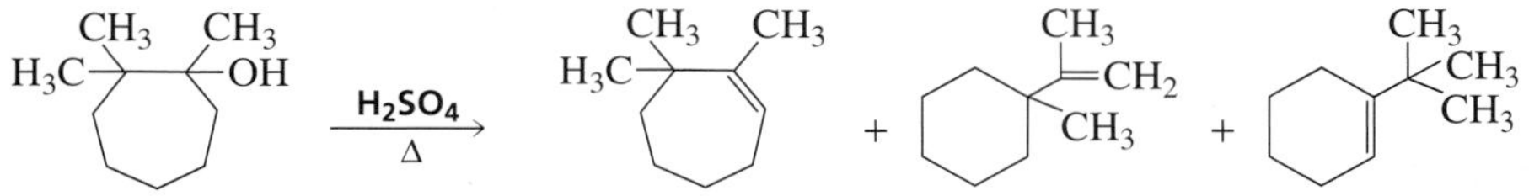

46. How could you synthesize isopropyl propyl ether using isopropyl alcohol as the only carbon-containing reagent?

47. When piperidine undergoes the indicated series of reactions, 1,4-pentadiene is obtained as the product. When the four different methyl-substituted piperidines undergo the same series of reactions, each forms a different diene. The dienes obtained are 1,5-hexadiene, 1,4-pentadiene, 2-methyl-1,4-pentadiene, and 3-methyl-1,4-pentadiene. Which methyl-substituted piperidine yields which diene?

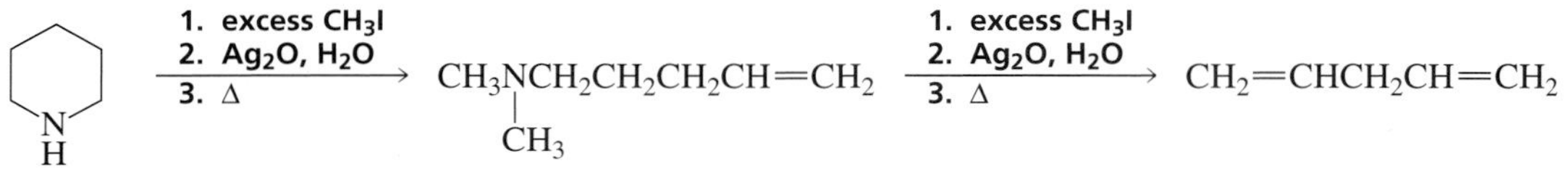

48. Ethylene oxide reacts with HO^- because of the strain in the three-membered ring. Cyclopropane has approximately the same amount of strain, but it does not react with HO^-. Explain.

49. Which of the following ethers would be obtained in greatest yield directly from alcohols?

$$CH_3OCH_2CH_2CH_3 \qquad CH_3CH_2OCH_2CH_2CH_3 \qquad CH_3CH_2OCH_2CH_3 \qquad CH_3OCCH_3$$

50. Propose a mechanism for each of the following reactions.

    a. 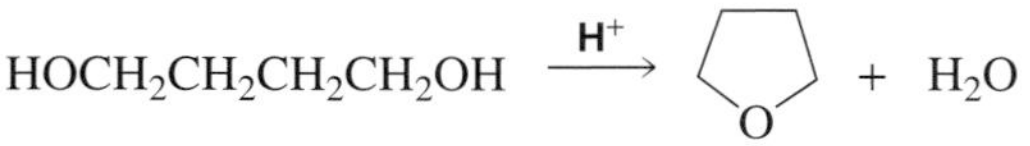 $HOCH_2CH_2CH_2CH_2OH \xrightarrow{H^+}$ $+ H_2O$

    b.  $\xrightarrow[\Delta]{HBr}$ $BrCH_2CH_2CH_2CH_2CH_2Br + H_2O$

51. Indicate how each of the following compounds could be prepared using the given starting material.

a.

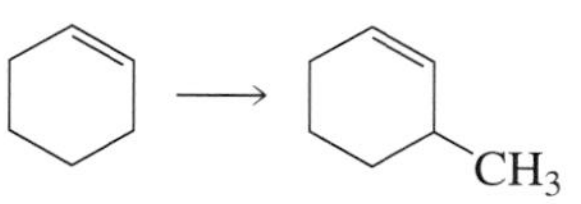

b.

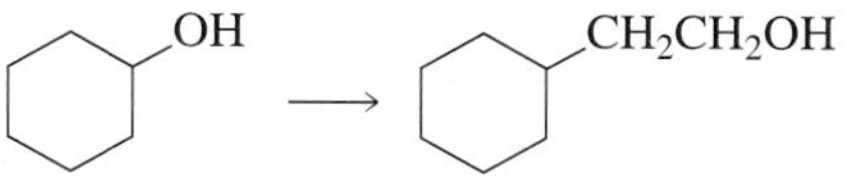

c. 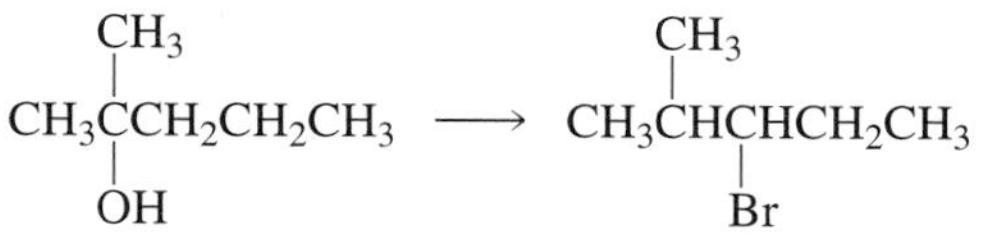 $CH_3\underset{\underset{OH}{|}}{\overset{\overset{CH_3}{|}}{C}}CH_2CH_2CH_3 \longrightarrow CH_3\underset{\underset{Br}{|}}{\overset{\overset{CH_3}{|}}{CH}}CHCH_2CH_3$

d.

e. $CH_3CH_2CH{=}CH_2 \longrightarrow CH_3CH_2CH_2CH_2CH_2CH_2CH_2CH_3$

52. Triethylene glycol is one of the products obtained from the reaction of ethylene oxide and hydroxide ion. Propose a mechanism for its formation.

$$H_2\overset{\overset{O}{\diagdown\!\diagup}}{C}{-}CH_2 \ + \ HO^- \longrightarrow HOCH_2CH_2OCH_2CH_2OCH_2CH_2OH$$

triethylene glycol

53. Give the major product expected from the reaction of 2-ethyloxirane with each of the following reagents.
  a. 0.1 M HCl
  b. 0.1 M NaOH
  c. CH_3OH/H^+
  d. CH_3OH/CH_3O^-
  e. ethyl magnesium bromide in ether followed by 0.1 M HCl

54. When ethyl ether is heated with excess HI for several hours, the only organic product obtained is ethyl iodide. Explain why ethyl alcohol is not obtained as a product.

55. a. Propose a mechanism for the following reaction.

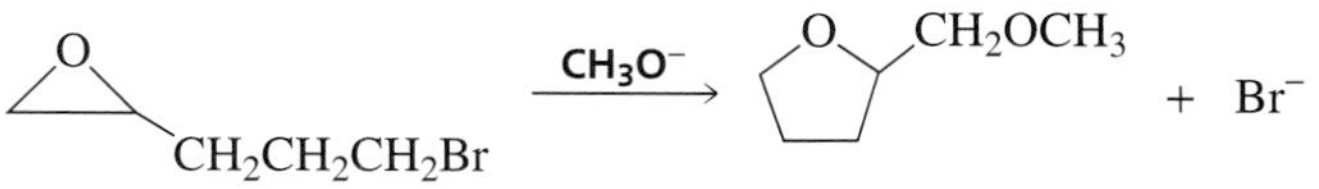

  b. A small amount of a product containing a six-membered ring is also formed. Give the structure of that product.
  c. Why is so little six-membered ring product formed?

56. Identify A through H.

$$CH_3Br \ \xrightarrow[\textbf{B}]{\textbf{A}} \ C \ \xrightarrow[\textbf{2. E}]{\textbf{1. D}} \ CH_3CH_2CH_2OH \ \xrightarrow[\substack{\textbf{2. G}\\ \textbf{3. H}}]{\textbf{1. F}} \ CH_3CH_2CH_2OCH_2CH_2OH$$

57. Greg Nard added 3,4-epoxy-4-methylcyclohexanol to an ether solution of methyl magnesium bromide and then added an aqueous solution of dilute HCl. He expected that the product would be a diol. He did not get any of the expected product. What product did he get?

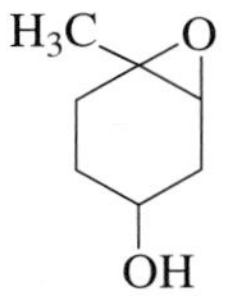

3,4-epoxy-4-methyl-cyclohexanol

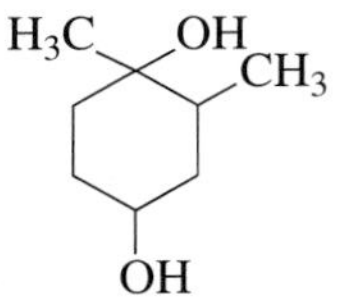

1,2-dimethyl-1,4-cyclohexanediol

58. An ion with a positively charged nitrogen atom in a three-membered ring is called an aziridinium ion. The following aziridinium ion reacts with sodium methoxide to form A and B. If a small amount of aqueous Br$_2$ is added to A, the reddish color of Br$_2$ persists, but the color disappears when Br$_2$ is added to B. When the aziridinium ion reacts with methanol, only A is formed. Identify A and B.

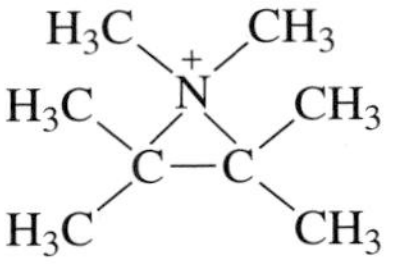

an aziridinium ion

59. Dimerization is a side reaction that occurs during the preparation of a Grignard reagent. Propose a mechanism that accounts for the formation of the dimer.

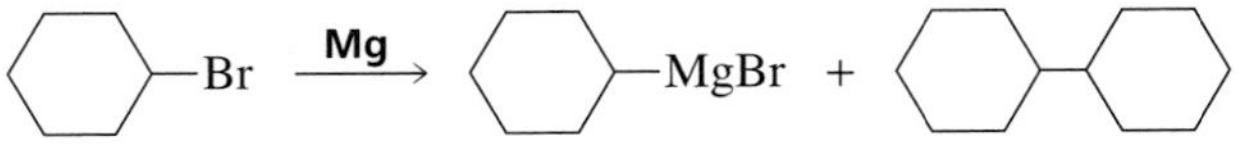

a dimer

60. Propose a mechanism for each of the following reactions.

a.

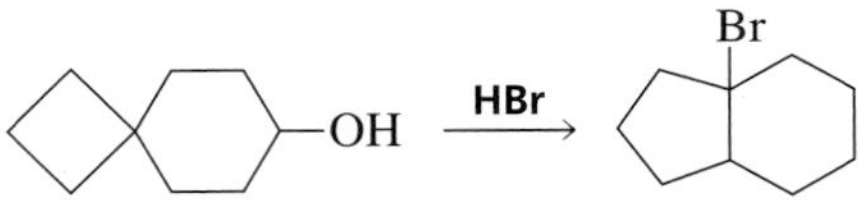

b.

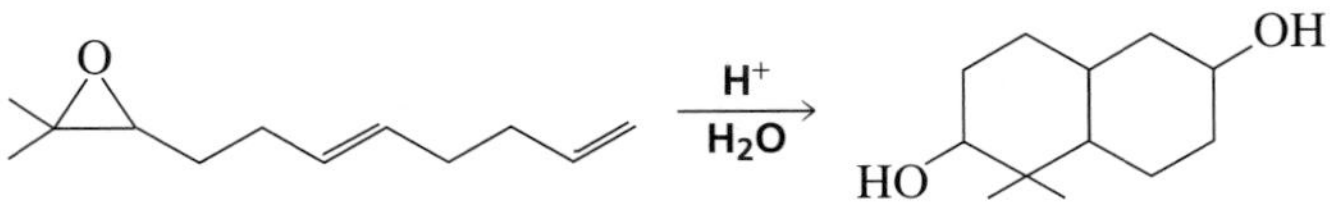

c.

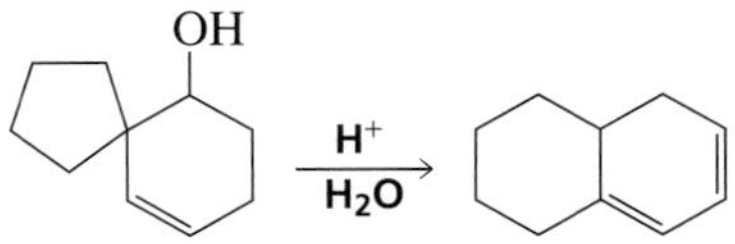

61. One method used for preparing an epoxide is to treat an alkene with an aqueous solution of Br$_2$, followed by an aqueous solution of sodium hydroxide.
    a. Propose a mechanism for the conversion of cyclohexene into cyclohexene oxide by this method.
    b. What is the configuration of the product that is formed when cyclohexene oxide reacts with methoxide ion in methanol?

62. Which of the following reactions is faster? Why?

a.

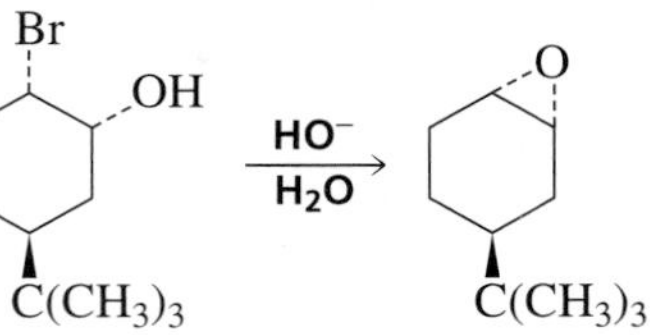

b.

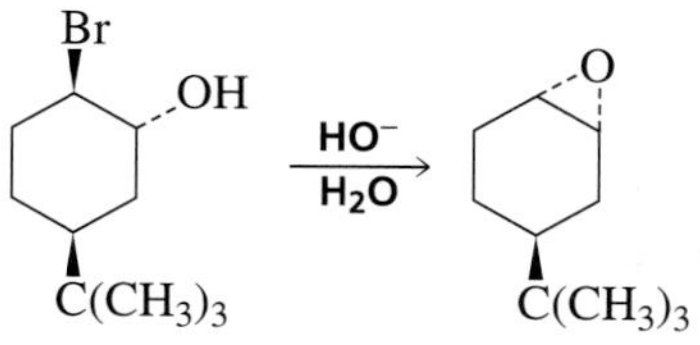

Br
OH
$\xrightarrow[H_2O]{HO^-}$
O

C(CH₃)₃ → C(CH₃)₃

c.

Br
OH
$\xrightarrow[H_2O]{HO^-}$
O

C(CH₃)₃ → C(CH₃)₃

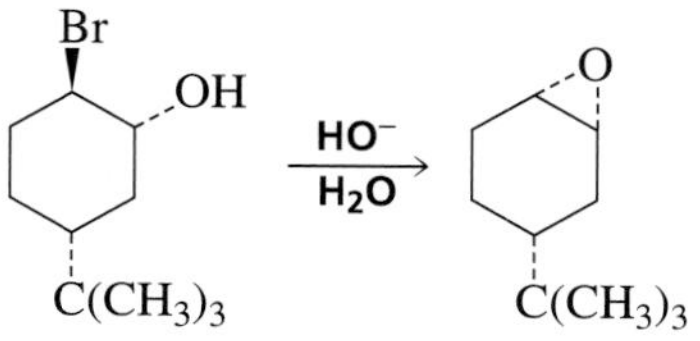

63. A vicinal diol has OH groups on adjacent carbons. The dehydration of a **vicinal diol** is accompanied by a rearrangement called the **pinacol rearrangement.** Propose a mechanism for this reaction.

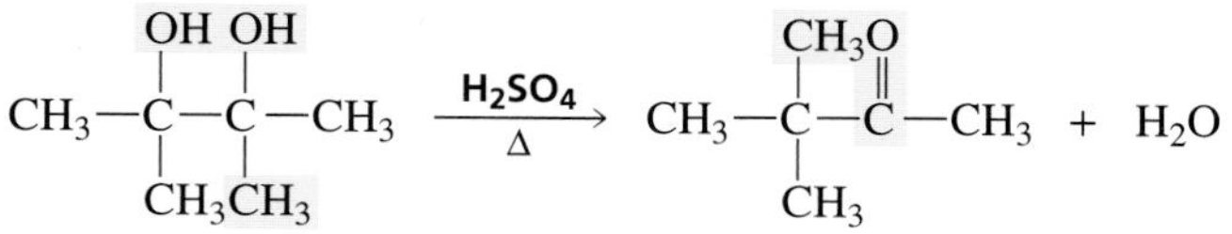

$$CH_3-\underset{\underset{CH_3}{|}}{\overset{\overset{OH}{|}}{C}}-\underset{\underset{CH_3}{|}}{\overset{\overset{OH}{|}}{C}}-CH_3 \xrightarrow[\Delta]{H_2SO_4} CH_3-\underset{\underset{CH_3}{|}}{\overset{\overset{CH_3}{|}}{C}}-\overset{\overset{O}{\|}}{C}-CH_3 + H_2O$$

64. Although 2-methyl-1,2-propanediol is an asymmetrical vicinal diol, only one product is obtained when it is dehydrated in the presence of acid.
   a. What is this product?   **b.** Why is only one product formed?

65. What product will be obtained when the vicinal diol shown below is heated in an acidic solution?

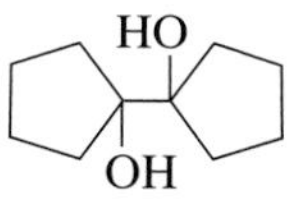

HO
OH

PART IV

IDENTIFICATION OF ORGANIC COMPOUNDS

You have now worked through many problems that required you to design syntheses for given organic compounds. But if you were actually to go into the laboratory to carry out the syntheses you designed, how would you know that the compounds you obtained were the ones you set out to prepare? Clearly, organic chemists need to be able to identify organic compounds. In **Chapters 12 and 13** you will learn about four instrumental techniques that chemists use to identify compounds.

Mass spectrometry is used to determine the molecular weight and the molecular formula of an organic compound as well as certain structural features of the compound. Infrared spectroscopy is used to determine the kinds of functional groups in an organic compound. Nuclear magnetic resonance helps to identify the carbon–hydrogen framework of an organic compound. UV/Vis spectroscopy gives information about organic compounds with conjugated bonds.

12

MASS SPECTROMETRY AND INFRARED SPECTROSCOPY

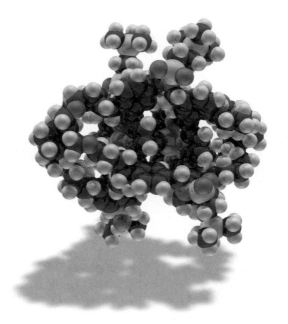

"football porphyrin"

In previous chapters you studied various methods used to synthesize organic compounds. For example, you learned that an aldehyde is formed when a terminal alkyne undergoes hydroboration-oxidation (Section 5.7). But how do you know that the product of that reaction is an aldehyde rather than an alcohol or a ketone? You also learned to predict the product of many different reactions. For example, you know that Markovnikov's rule is followed when an alkene undergoes an addition reaction (Section 3.12). But how do you know what compound is sitting in the reaction flask? Scientists search the world for new compounds with physiological activity. If a promising compound is found, its structure needs to be determined before chemists can design ways to synthesize it or studies can be undertaken to provide insights into its biological behavior.

Clearly, it is important for chemists to be able to determine the structures of the compounds they work with. Before a compound can be identified, it must be isolated. For example, if the product of a reaction is to be identified, it must first be isolated from the solvent and any unreacted starting materials as well as from any side products that might have formed. Any compound found in nature must be isolated from the source from which it was obtained.

At one time, isolating and analyzing products were daunting tasks. The only tools chemists had to isolate products were distillation (for liquids) and fractional recrystallization (for solids). Now a variety of chromatographic techniques allow compounds to be isolated relatively easily.

The only methods organic chemists had to analyze products were simple color and solubility tests. For example, when an aldehyde is added to a test tube containing a solution of silver oxide in ammonia, a silver mirror is formed on the inside of the test tube. Only aldehydes do this. If a mirror forms, you can conclude that the unknown compound is an aldehyde; if a mirror does not form, you know that the compound is not an aldehyde. An example of a solubility test is the Lucas test, which distinguishes primary, secondary, and tertiary alcohols by how rapidly they turn cloudy after the addition of Lucas' reagent (Section 11.1).

TABLE 12.1 Classes of Organic Compounds

Alkane	$-\overset{\displaystyle |}{\underset{\displaystyle |}{C}}-$	contains only C—C and C—H bonds	Aldehyde	$\overset{\displaystyle O}{\overset{\displaystyle \|}{RCH}}$
Alkene	$\overset{\diagup}{\underset{\diagdown}{C}}=\overset{\diagdown}{\underset{\diagup}{C}}$		Ketone	$\overset{\displaystyle O}{\overset{\displaystyle \|}{RCR}}$
Alkyne	$-C\equiv C-$		Carboxylic acid	$\overset{\displaystyle O}{\overset{\displaystyle \|}{RCOH}}$
Nitrile	$-C\equiv N$		Ester	$\overset{\displaystyle O}{\overset{\displaystyle \|}{RCOR}}$
Alkyl halide	RX	where X = F, Cl, Br, or I	Amide (primary)	$\overset{\displaystyle O}{\overset{\displaystyle \|}{RCNH_2}}$
Ether	ROR		Amide (secondary)	$\overset{\displaystyle O}{\overset{\displaystyle \|}{RCNHR}}$
Alcohol	ROH		Amide (tertiary)	$\overset{\displaystyle O}{\overset{\displaystyle \|}{RCNR_2}}$
Phenol	(benzene ring)—OH		Amine (primary)	RNH_2
			Amine (secondary)	R_2NH
Aniline	(benzene ring)—NH_2		Amine (tertiary)	R_3N

Now, however, compounds are commonly identified by means of instrumental techniques. These techniques can be performed quickly on small amounts of a compound and can provide much more information about the compound's structure than either color or solubility tests can provide. In this chapter, we will look at two of these techniques—mass spectrometry and infrared (IR) spectroscopy. **Mass spectrometry** allows us to determine the *molecular weight* and the *molecular formula* of a compound as well as certain *structural features*. **Infrared spectroscopy** allows us to determine the *kinds of functional groups* in a compound. We will examine two additional kinds of spectrophotometric techniques in Chapter 13.

We will refer to different classes of organic compounds as we discuss the various instrumental techniques; these are listed in Table 12.1 for your convenience.

12.1 MASS SPECTROMETRY

In mass spectrometry, a sample of a compound is injected into an instrument called a mass spectrometer, where it is vaporized and then bombarded with a beam of high-energy electrons. The energy of the electron beam can be varied, but a beam of about 70 eV (electron volts) is commonly used. When the electron beam hits a molecule, it knocks out an electron. This produces a *molecular ion*. A molecular ion is a **radical cation,** a species with an unpaired electron and a positive charge. The symbol $^{+}$ indicates the loss of the electron from the molecule because, as a consequence of losing an electron, the compound has an unpaired electron and is positively charged.

$$\text{M} \xrightarrow{\text{70 eV}} \text{M}^{\stackrel{+}{\cdot}} \quad + \quad \text{e}^{-}$$

molecule   **molecular ion**   **electron**
a radical cation

Because the energy of the electron beam is greater than the energy required to break most of the bonds in an organic compound, the electron beam causes many of the molecular ions to break apart into cations, radicals, neutral molecules, and other radical cations. All the *positively charged fragments* pass between two negatively charged plates, which accelerate the fragments into an analyzer tube (Figure 12.1). Neutral fragments are not attracted to the negatively charged plates and therefore are not accelerated. They are eventually pumped out of the spectrometer.

The analyzer tube is surrounded by a magnet, and the magnetic field causes the positively charged fragments to travel in a circular path. At a given magnetic field strength, the radius of the path traveled by a fragment depends on its mass-to-charge ratio (m/z). The smaller the m/z of the fragment, the more the magnet deflects it. If a fragment's path matches the curvature of the analyzer tube, the fragment will pass through the tube and out the ion exit slit. A collector records the relative number of fragments with a particular m/z passing through the slit. By slowly increasing the strength of the magnetic field, fragments with progressively larger values of m/z can be guided through the tube and out the exit slit. A **mass spectrum** consists of a plot of the relative abundance of each fragment versus its m/z value. Because the charge (z) on essentially all of the fragments that reach the collector plate is $+1$, m/z is the molecular weight (m) of the fragment.

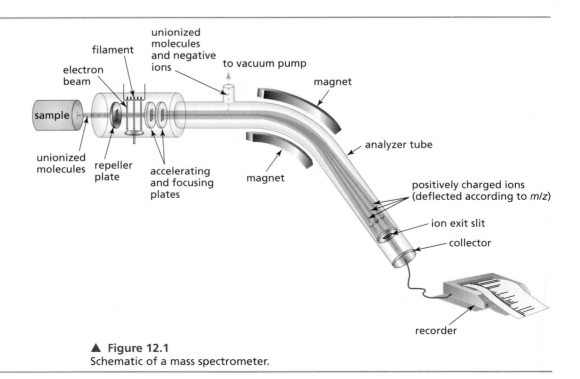

▲ **Figure 12.1**
Schematic of a mass spectrometer.

12.2 THE MASS SPECTRUM. FRAGMENTATION

The mass spectrum of pentane is shown in Figure 12.2. The peak in the spectrum at $m/z = 72$ is due to the fragment that results when an electron is ejected from pentane. It is the **molecular ion** (M) of pentane, also sometimes called the **parent ion.** The m/z value of the molecular ion gives the molecular weight of the compound. Unless all the molecular ions break into fragments before they reach the collector plate, the peak in the spectrum with the largest m/z value is the molecular ion. (For now, ignore the tiny peak at $m/z = 73$. Its significance will be discussed in the next section.) Peaks with smaller values of m/z represent fragments of the molecule.

$$CH_3CH_2CH_2CH_2CH_3 \xrightarrow{\text{70 eV}} [CH_3CH_2CH_2CH_2CH_3]^{+\cdot} + e^-$$

molecular ion
$m/z = 72$

The **base peak** is the one with the greatest intensity. The base peak is assigned a relative intensity of 100%, and the relative intensity of each of the other peaks is reported as a percentage of the base peak. Mass spectra can be shown either as a bar graph or in tabular form.

The way a particular molecular ion fragments depends on the strength of its bonds and the stability of the fragments. Weak bonds break in preference to strong bonds, and bonds that break to form more stable fragments break in preference to those that form less stable fragments.

The carbon–carbon bonds in the molecular ion formed from pentane have about the same strength. However, the C-2—C-3 bond is more likely to be cleaved than the C-1—C-2 bond, because C-2—C-3 fragmentation leads to a primary carbocation and a primary radical, which are more stable than the primary carbocation and methyl radical (or primary radical and methyl cation) obtained from C-1—C-2 fragmentation; C-2—C-3 fragmentation forms ions with $m/z = 43$ or 29, and C-1—C-2 fragmentation forms ions with $m/z = 57$ or 15. The base peak of 43 in the mass spectrum of pentane indicates the preference for C-2—C-3 fragmentation.

$$[\overset{1}{C}H_3-\overset{2}{C}H_2-\overset{3}{C}H_2-\overset{4}{C}H_2-\overset{5}{C}H_3]^{+\cdot}$$

molecular ion of pentane

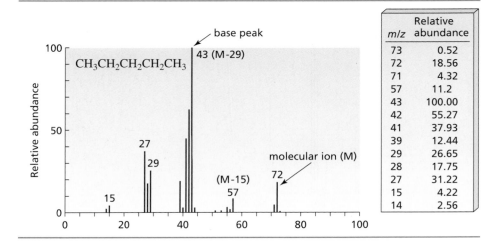

m/z	Relative abundance
73	0.52
72	18.56
71	4.32
57	11.2
43	100.00
42	55.27
41	37.93
39	12.44
29	26.65
28	17.75
27	31.22
15	4.22
14	2.56

◀ **Figure 12.2**
The mass spectrum of pentane shown as a bar graph and in tabular form.

$$\text{CH}_3\text{CH}_2\overset{+}{\text{C}}\text{H}_2 \ + \ \text{CH}_3\overset{\cdot}{\text{C}}\text{H}_2$$
$$m/z = 43$$

$$[\text{CH}_3\text{CH}_2\text{CH}_2\text{CH}_2\text{CH}_3]^{+\cdot} \longrightarrow \text{CH}_3\text{CH}_2\overset{\cdot}{\text{C}}\text{H}_2 \ + \ \text{CH}_3\overset{+}{\text{C}}\text{H}_2$$

molecular ion
$$m/z = 72$$
$$m/z = 29$$

$$\text{CH}_3\text{CH}_2\text{CH}_2\overset{+}{\text{C}}\text{H}_2 \ + \ \overset{\cdot}{\text{C}}\text{H}_3$$
$$m/z = 57$$

$$\text{CH}_3\text{CH}_2\text{CH}_2\overset{\cdot}{\text{C}}\text{H}_2 \ + \ \overset{+}{\text{C}}\text{H}_3$$
$$m/z = 15$$

One method commonly used to identify fragment ions is to determine the difference between the m/z value of a given ion and that of the molecular ion. For example, the ion with $m/z = 43$ in the mass spectrum of pentane is 29 units smaller than the molecular ion (M − 29 = 43). An ethyl group has a molecular weight of 29, so the peak at 43 can be attributed to the molecular ion minus an ethyl group. Similarly, the peak at $m/z = 57$ (M − 15) can be attributed to the molecular ion minus a methyl group. Peaks at $m/z = 15$ and $m/z = 29$ are readily recognizable as due to methyl and ethyl cations, respectively. Appendix V contains a table of common fragment ions and a table of common fragments lost.

Peaks are commonly observed at m/z values one and two units less than the m/z values of the carbocations, because further fragmentation of the carbocation causes the loss of one or two hydrogen atoms.

$$\text{CH}_3\text{CH}_2\overset{+}{\text{C}}\text{H}_2 \ \xrightarrow{\ -\text{H}^\cdot\ } \ [\text{CH}_3\text{CHCH}_2]^{+\cdot} \ \xrightarrow{\ -\text{H}^\cdot\ } \ \overset{+}{\text{C}}\text{H}_2\text{CH}=\text{CH}_2$$
$$m/z = 43 \qquad\qquad m/z = 42 \qquad\qquad m/z = 41$$

2-Methylbutane has the same molecular formula as pentane, and therefore it also has a molecular ion with $m/z = 72$ (Figure 12.3). Its mass spectrum is very similar to that of pentane, with one notable exception: the peak at $m/z = 57$ is much more intense. This occurs because loss of a methyl group from 2-methylbutane forms a secondary carbocation, whereas loss of the same group from pentane forms a less stable primary carbocation.

Figure 12.3 ▶
The mass spectrum of
2-methylbutane.

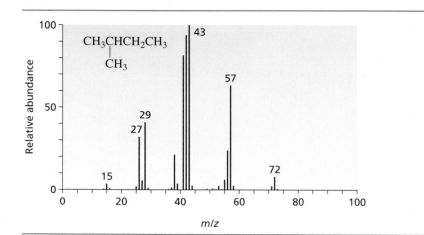

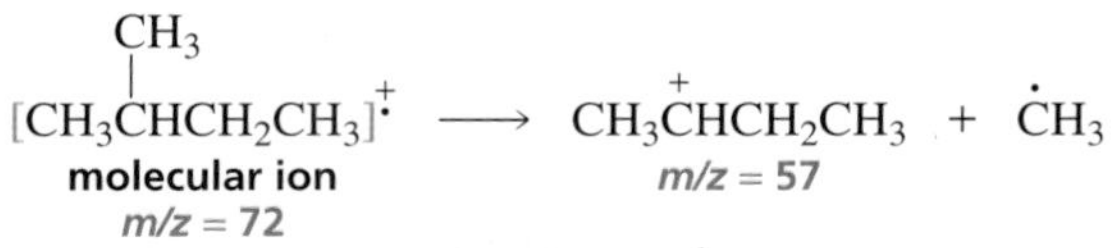

$$\underset{\substack{\text{molecular ion} \\ m/z = 72}}{[CH_3\overset{\displaystyle CH_3}{\overset{|}{C}}HCH_2CH_3]^{\overset{+}{\cdot}}} \longrightarrow \underset{m/z\,=\,57}{CH_3\overset{+}{C}HCH_2CH_3} + \overset{\displaystyle \cdot}{C}H_3$$

PROBLEM 1

How could you distinguish the mass spectrum of 2,2-dimethylpropane from those of pentane and 2-methylbutane?

PROBLEM 2◆

What m/z value would you predict for the base peak in the mass spectrum of 3-methylpentane?

PROBLEM 3 / SOLVED

The mass spectra of two different cycloalkanes both show a molecular ion peak at $m/z = 98$. One spectrum shows a base peak at $m/z = 69$, and the other shows a base peak at $m/z = 83$. Identify the cycloalkanes.

SOLUTION.   The molecular formula for a cycloalkane is C_nH_{2n}. Because the molecular weight of both cycloalkanes is 98, their molecular formulas must be C_7H_{14}. A base peak of 69 means loss of an ethyl substituent ($98 - 69 = 29$), whereas a base peak of 83 means loss of a methyl substituent ($98 - 83 = 15$). A seven-carbon cycloalkane with a base peak signifying loss of an ethyl substituent must be ethylcyclopentane. A seven-carbon cycloalkane with a base peak signifying loss of a methyl substituent must be methylcyclohexane.

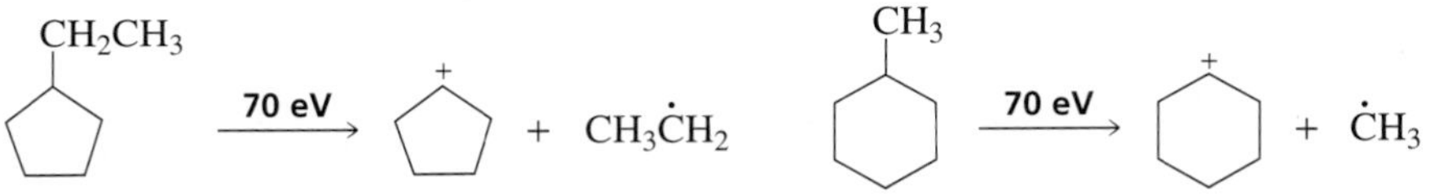

PROBLEM 4

The "nitrogen rule" states that if a compound has an odd-mass molecular ion, the compound contains an odd number of nitrogen atoms—unless the compound contains atoms other than C, H, O, and N.

   a. Explain why the rule holds.

   b. State this rule in terms of an even-mass molecular ion.

Although the molecular ions of pentane and 2-methylbutane both have m/z values of 72, each shows a very small peak at $m/z = 73$ (Figures 12.2 and 12.3). This is called an M + 1 peak, because the ion responsible for this peak is one unit heavier than the molecular ion. The M + 1 peak occurs because there are two naturally occurring isotopes of carbon: 98.89% of natural carbon is ^{12}C and 1.11% is ^{13}C (Section 1.1). The M + 1 fragment results from molecular ions that contain a ^{13}C instead of a ^{12}C.

12.3
ISOTOPES IN MASS SPECTROMETRY

Peaks attributable to isotopes can be useful in identifying the compound responsible for the mass spectrum. For example, if a compound contains five carbon atoms, the relative abundance of the M + 1 ion should be 5(1.1%) = 5.5%, multiplied by the relative abundance of the molecular ion. This means that the number of carbon atoms in a compound can be calculated if the relative intensities of both the M and M + 1 peaks are known.

$$\text{number of carbon atoms} = \frac{\text{relative intensity of M + 1 peak}}{0.011 \times (\text{relative intensity of M peak})}$$

The isotopic distributions of several elements commonly found in organic compounds are shown in Table 12.2. From the information in this table, we can see that the reason the M + 1 peak can be used to predict the number of carbon atoms in a compound is because the contributions to the M + 1 peak by H and O are very small. This formula does not work as well in predicting the number of carbon atoms in a nitrogen-containing compound because the natural abundance of ^{15}N is relatively high.

Mass spectra can show M + 2 peaks as a result of a contribution from ^{18}O, or from having two heavy isotopes in the same molecule (say, ^{13}C and ^{2}H). Most of the time, the M + 2 peak is smaller than the M + 1 peak. The presence of a large M + 2 peak is evidence of a compound containing either chlorine or bromine, because each of these elements has a high percentage of a naturally occurring isotope two units heavier than the most abundant isotope. From the natural abundance of the isotopes of chlorine and bromine in Table 12.2, you can conclude that, if the M + 2 peak is one-third the height of the molecular ion peak, the compound contains one chlorine atom. If the M and M + 2 peaks are about the same height, the compound contains one bromine atom.

In calculating the molecular weights of molecular ions and fragments, the monoisotopic atomic weights of the atoms must be used (Cl = 35 or 37, etc.); the

TABLE 12.2 The Natural Abundance of Isotopes Commonly Found in Organic Compounds

Element		Natural abundance	
carbon	^{12}C 98.89%	^{13}C 1.11%	
oxygen	^{16}O 99.76%	^{17}O 0.37%	^{18}O 0.204%
nitrogen	^{14}N 99.63%	^{15}N 0.37%	
hydrogen	^{1}H 99.99%	^{2}H 0.01%	
fluorine	^{19}F 100%		
chlorine	^{35}Cl 75.77%		^{37}Cl 24.23%
bromine	^{79}Br 50.69%		^{81}Br 49.31%
iodine	^{127}I 100%		

atomic weights in the periodic table (Cl = 35.453) cannot be used, because they are *averages* of naturally occurring isotopes.

PROBLEM 5◆

The mass spectrum of an unknown compound has a molecular ion peak with a relative intensity of 43.27% and an M + 1 peak with a relative intensity of 3.81%. How many carbon atoms are in the compound?

All the mass spectra shown in this text were determined using a low-resolution mass spectrometer. Such spectrometers determine the *nominal mass* of a fragment. The **nominal mass** is the mass to the nearest whole number.

High-resolution mass spectrometers can determine the *exact mass* of a fragment to an accuracy of one part in 10,000. From the exact mass, the molecular formula of the compound can be determined. For example, many compounds have a nominal mass of 122, but each of them has a different exact mass. The exact masses of some common isotopes are listed in Table 12.3). Computer programs exist that allow you to determine the molecular formula of a compound from its exact mass.

12.4 DETERMINATION OF MOLECULAR FORMULAS: HIGH-RESOLUTION MASS SPECTROMETRY

Some Compounds with a Nominal Mass of 122 and Their Exact Masses

Molecular formula	C_9H_{14}	$C_7H_{10}N_2$	$C_8H_{10}O$	$C_7H_6O_2$	$C_4H_{10}O_4$	$C_4H_{10}S_2$
Exact mass	122.1096	122.0845	122.0732	122.0368	122.0579	122.0225

TABLE 12.3 The Exact Masses of Some Common Isotopes

^1H	1.007825	^{32}S	31.9721
^{12}C	12.00000	^{35}Cl	34.9689
^{14}N	14.0031	^{79}Br	78.9183
^{16}O	15.9949		

PROBLEM 6◆

Which molecular formula has an exact mass of 86.1096: C_6H_{14}, $C_4H_{10}N_2$, or $C_4H_6O_2$?

There are characteristic fragmentation patterns associated with specific functional groups that help in the identification of a substance based on its mass spectrum. These patterns were worked out after studying the mass spectra of many compounds that contain a particular functional group. We will look at the fragmentation patterns shown by alkyl halides, ethers, alcohols, and ketones as examples.

12.5 FRAGMENTATION AT FUNCTIONAL GROUPS

Alkyl Halides

From the relative heights of the M and M + 2 peaks in Figure 12.4, we can conclude that the compound contains a bromine atom. If a molecule contains nonbonding electrons, electron bombardment is most likely to dislodge one of the nonbonding electrons, because a molecule does not hold onto its nonbonding electrons as tightly

Figure 12.4 ▶
The mass spectrum of
1-bromopropane.

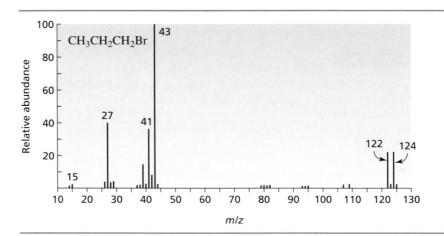

as it holds onto its bonding electrons. Therefore, the electron initially ejected from an alkyl halide is one of the nonbonding electrons of the halogen.

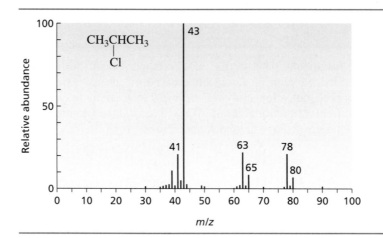

$$CH_3CH_2CH_2Br \xrightarrow{\ -e^-\ } [CH_3CH_2CH_2\overset{+}{\overset{\frown}{-}}\dot{B}r^{79}] + [CH_3CH_2CH_2\overset{+}{\overset{\frown}{-}}\dot{B}r^{81}] \longrightarrow CH_3CH_2\overset{+}{C}H_2 + \dot{B}r$$

1-bromopropane $m/z = 122$ $m/z = 124$ $m/z = 43$

The easiest bond to break in the resulting molecular ion is the carbon–bromine bond, because it is the weakest bond. The bond breaks heterolytically, with both electrons going to the more electronegative of the atoms that were joined by the bond, to give a propyl cation and a bromine atom; consequently, the base peak in the mass spectrum of 1-bromopropane is at $m/z = 43$ [M − 79 or (M + 2) − 81]. The propyl cation shows the same fragmentation pattern that it showed when it was formed from the cleavage of pentane (Figure 12.2).

The mass spectrum of 2-chloropropane is shown in Figure 12.5. That the compound contains a chlorine atom is shown by the M + 2 peak, which is one-third the height of the molecular ion peak. The base peak at $m/z = 43$ results from heterolytic cleavage of the carbon–chlorine bond. The peaks at $m/z = 63$ and $m/z = 65$ have a 3:1 ratio, indicating that these fragments contain a chlorine atom. They result from homolytic cleavage of a carbon–carbon bond at the α-carbon (the carbon bonded to

Figure 12.5 ▶
The mass spectrum of
2-chloropropane.

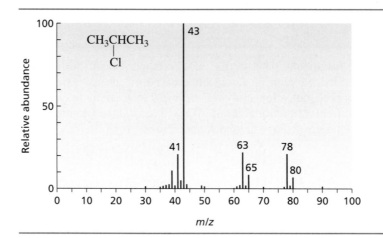

the chlorine). This is known as α-**cleavage.** α-Cleavage occurs because it leads to a particularly stable cation in which all the atoms have complete octets.

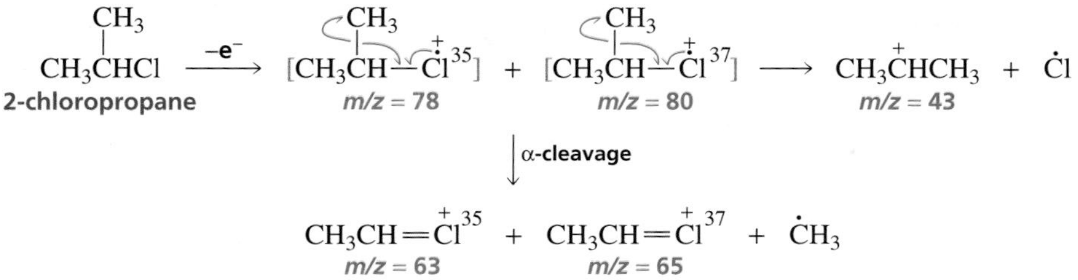

$$CH_3CHCl \xrightarrow{-e^-} [CH_3CH \overset{+}{\cdots} Cl^{35}] + [CH_3CH \overset{+}{\cdots} Cl^{37}] \longrightarrow CH_3\overset{+}{C}HCH_3 + \dot{C}l$$

2-chloropropane $m/z = 78$ $m/z = 80$ $m/z = 43$

$\downarrow$ α-cleavage

$$CH_3CH{=}\overset{+}{C}l^{35} + CH_3CH{=}\overset{+}{C}l^{37} + \dot{C}H_3$$

$m/z = 63$ $m/z = 65$

Recall that a curved arrow with a one-sided arrow-head represents the movement of one electron.

PROBLEM 7

Sketch the mass spectrum of 1-chloropropane.

Ethers

The mass spectrum of *sec*-butyl isopropyl ether is shown in Figure 12.6. Notice that the fragmentation pattern of an ether has many of the characteristics of the fragmentation pattern of an alkyl halide.

1. Electron bombardment dislodges a nonbonding electron from oxygen.
2. Fragmentation of the resulting molecular ion occurs in two principal ways:

a. A carbon–oxygen bond is cleaved heterolytically with the electrons going to the more electronegative oxygen atom.

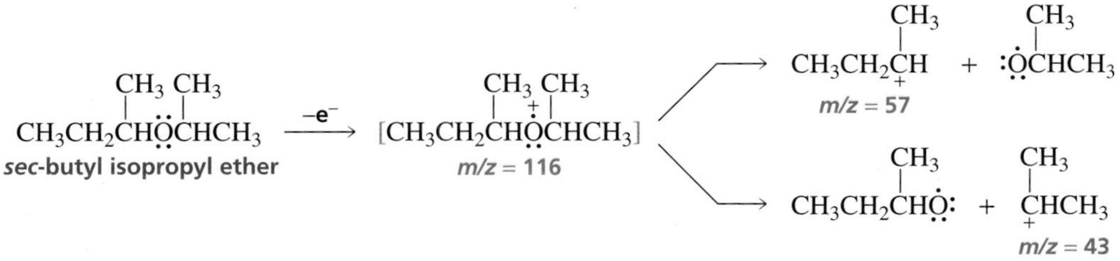

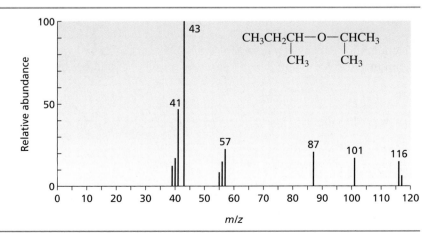

◄ **Figure 12.6**
The mass spectrum of *sec*-butyl isopropyl ether.

b. A carbon–carbon bond is cleaved homolytically at the α-position. The largest α-substituent is the one most readily cleaved.

$$CH_3CH_2-CH-\overset{+}{\underset{..}{O}}-CHCH_3 \xrightarrow{\alpha\text{-cleavage}} CH=\overset{+}{\underset{..}{O}}-CHCH_3 + CH_3\dot{C}H_2$$
$$m/z = 87$$

$$CH_3CH_2CH-\overset{+}{\underset{..}{O}}-CHCH_3 \xrightarrow{\alpha\text{-cleavage}} CH_3CH_2CH=\overset{+}{\underset{..}{O}}-CHCH_3 + \dot{C}H_3$$
$$m/z = 101$$

$$CH_3CH_2CH-\overset{+}{\underset{..}{O}}-CHCH_3 \xrightarrow{\alpha\text{-cleavage}} CH_3CH_2CH-\overset{+}{\underset{..}{O}}=CHCH_3 + \dot{C}H_3$$
$$m/z = 101$$

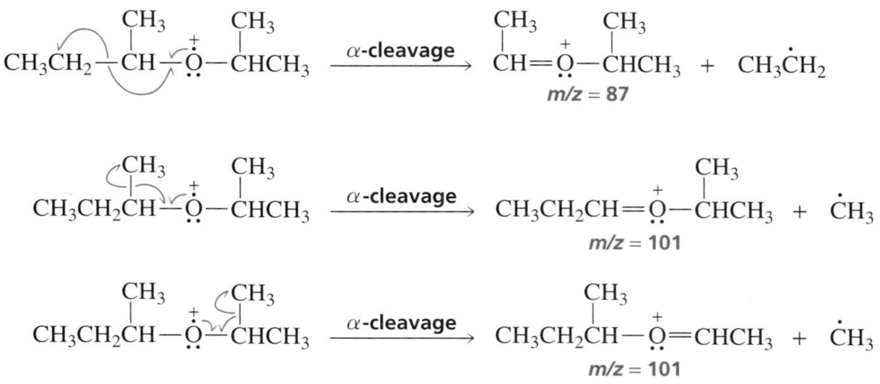

PROBLEM 8◆

The mass spectra of 1-methoxybutane, 2-methoxybutane, and 2-methoxy-2-methyl-propane are shown in Figure 12.7. Match the compounds with the spectra.

Alcohols

The molecular ions obtained from alcohols fragment so readily that few of them survive to reach the collector. As a result, the mass spectra of primary and secondary alcohols show weak molecular ion peaks, and the molecular ions from tertiary alcohols are not detectable.

Alcohols, like alkyl halides and ethers, undergo α-cleavage, with the largest α-substituent being the one most readily cleaved. Consequently, the mass spectrum of 2-hexanol (Figure 12.8) shows a base peak at $m/z = 45$ (α-cleavage of a butyl group) and a smaller peak at $m/z = 87$ (α-cleavage of a methyl group).

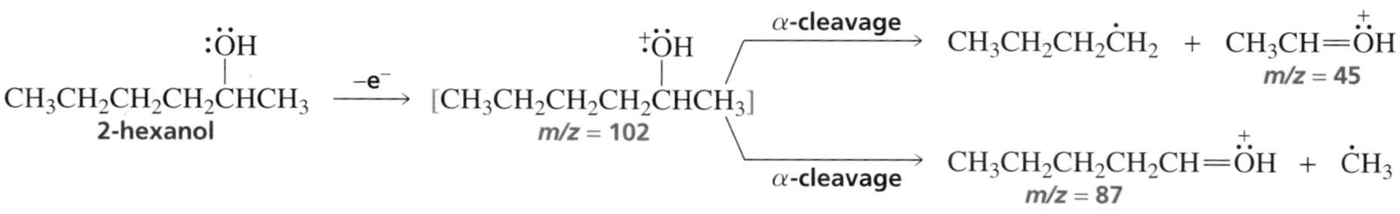

$$CH_3CH_2CH_2CH_2CHCH_3 \xrightarrow{-e^-} [CH_3CH_2CH_2CH_2CHCH_3]$$

2-hexanol, $m/z = 102$

$$\xrightarrow{\alpha\text{-cleavage}} CH_3CH_2CH_2\dot{C}H_2 + CH_3CH=\overset{+}{\underset{..}{O}}H$$
$$m/z = 45$$

$$\xrightarrow{\alpha\text{-cleavage}} CH_3CH_2CH_2CH_2CH=\overset{+}{\underset{..}{O}}H + \dot{C}H_3$$
$$m/z = 87$$

Alcohols show a peak at $m/z = M - 18$ because of the loss of water. The water that is eliminated comes from the OH group of the alcohol and a γ-hydrogen.

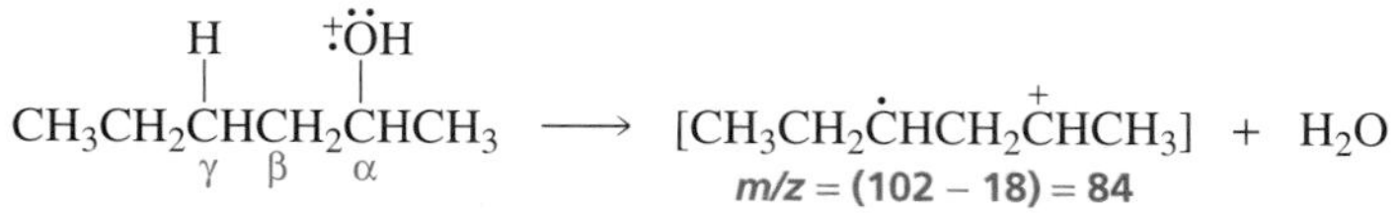

$$CH_3CH_2CHCH_2CHCH_3 \longrightarrow [CH_3CH_2\dot{C}HCH_2\overset{+}{C}HCH_3] + H_2O$$
$$\gamma \quad \beta \quad \alpha$$
$$m/z = (102 - 18) = 84$$

PROBLEM 9◆

Primary alcohols can be readily identified by the presence of a strong peak at $m/z = 31$. What fragment is responsible for this peak?

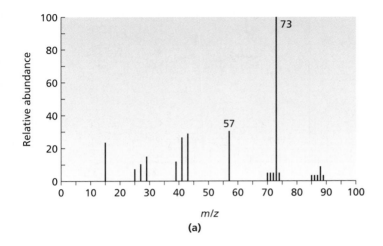

(a)

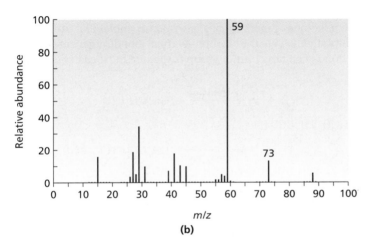

(b)

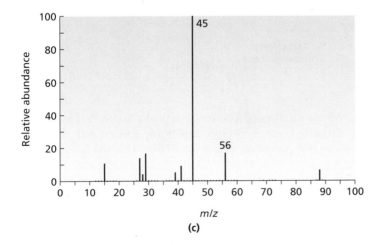

(c)

▲ **Figure 12.7**
The mass spectra for Problem 8.

Figure 12.8 ▶
The mass spectrum of
2-hexanol.

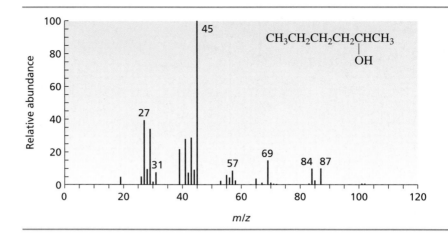

Ketones

The mass spectrum of a ketone generally has an intense molecular ion peak. Ketones fragment homolytically at the carbon–carbon bond adjacent to the carbon–oxygen double bond. The larger alkyl group is the one most easily cleaved.

$$CH_3CH_2CH_2CCH_3 \xrightarrow{-e^-} [CH_3CH_2CH_2CCH_3] $$

2-pentanone   m/z = 86

α-cleavage $CH_3CH_2\dot{C}H_2 + CH_3C{\equiv}\overset{+}{O}{:}$ m/z = 43

α-cleavage $CH_3CH_2CH_2C{\equiv}\overset{+}{O}{:} + \dot{C}H_3$ m/z = 71

Fred Warren McLafferty *was born in Evanston, Illinois in 1923. He received a B.S. and an M.S. from the University of Nebraska and a Ph.D. from the University of Illinois. From 1964 to 1968, he was a professor of chemistry at Purdue University, and he is currently a professor of chemistry at Cornell University.*

If one of the alkyl groups attached to the carbonyl carbon has a γ-hydrogen, a cleavage known as a **McLafferty rearrangement** may occur. In this rearrangement, the bond between the α-carbon and the β-carbon breaks homolytically and a hydrogen atom from the γ-carbon migrates to the oxygen atom.

m/z = 86   **McLafferty rearrangement** $H_2C{=}CH_2 + \cdot CH_2{-}\overset{+}{C}{-}CH_3$ m/z = 58

Notice that the mass spectra of alkanes tend to have peaks at $m/z = 15, 29, 43$, and so on, as a result of methyl, ethyl, propyl, etc. ions. Ethers and alcohols, because of the presence of an oxygen atom, tend to form ions 16 units heavier than alkanes: $m/z = (15 + 16) = 31$, $(29 + 16) = 45$, $(43 + 16) = 59$, etc. Ketones also have an oxygen atom but, because of the double bond, form ions two units lighter than those of alcohols and ethers. So, instead of having peaks at $m/z = 31$, 45, and 59, the mass spectra of ketones have peaks at $m/z = 29, 43$, and 57, which means that they have peaks with m/z values similar to those of alkanes.

PROBLEM 10◆

Identify the ketones that are responsible for the mass spectra shown in Figure 12.9.

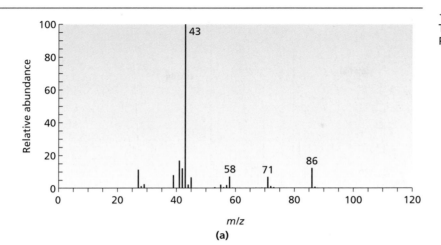

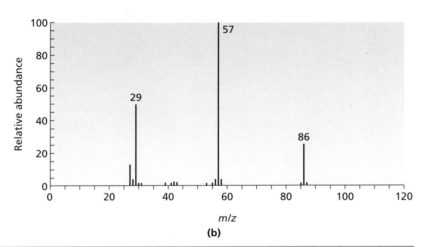

PROBLEM 11

Using curved half-arrows, show the principal fragments that would be observed in the mass spectrum of each of the following compounds.

a. $CH_3CH_2CH_2CH_2CH_2OH$

b.
$$CH_3\overset{\overset{\displaystyle O}{\|}}{C}CH_2CH_2CH_2CH_3$$

c.
$$CH_3-\overset{\overset{\displaystyle CH_3}{|}}{\underset{\underset{\displaystyle CH_3}{|}}{C}}-Br$$

d.
$$CH_3CH_2O\overset{\overset{\displaystyle CH_2CH_3}{|}}{\underset{\underset{\displaystyle CH_3}{|}}{C}}CH_2CH_2CH_3$$

e.
$$CH_3CH_2\overset{}{\underset{\underset{\displaystyle CH_3}{|}}{C}}HCl$$

f.
$$CH_3CH_2\overset{}{\underset{\underset{\displaystyle OH}{|}}{C}}HCH_2CH_2CH_2CH_3$$

Figure 12.10 ▶
The mass spectra for
Problem 12.

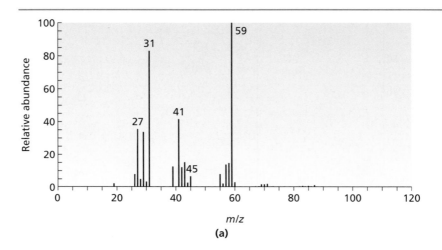

(a)

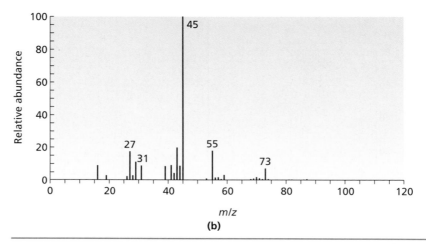

(b)

PROBLEM 12 ◆

Two products are obtained from the reaction of (Z)-2-pentene with water and a trace of
H_2SO_4. The mass spectra of these products are shown above in Figure 12.10. Identify
the compounds responsible for the spectra.

**12.6
SPECTROSCOPY
AND THE
ELECTROMAGNETIC
SPECTRUM**

Spectroscopy is the study of the interaction of matter and electromagnetic radiation. **Electromagnetic radiation** is radiant energy that displays both the properties
of particles and the properties of waves. The different types of electromagnetic
radiation constitute the electromagnetic spectrum (Figure 12.11). Visible light is
the type with which we are most familiar because we can see it. We will examine
three different spectrophotometric techniques used to identify compounds. Each
uses a different type of electromagnetic radiation. In this chapter we will look at
infrared spectroscopy, and in the next chapter we will see how compounds can be
identified using nuclear magnetic resonance (NMR) and ultraviolet/visible
(UV/Vis) spectroscopy.

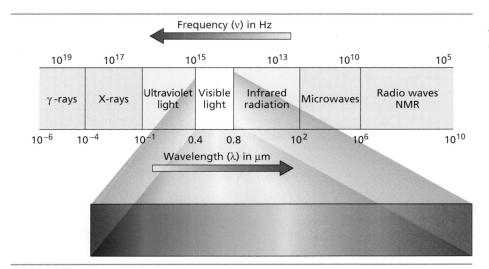

A particle of electromagnetic radiation is called a photon. The relationship between the energy of a photon and the frequency (ν) of the electromagnetic radiation is described by the following equation. Frequency has units of hertz (Hz) or cycles per second (cps). Planck's constant, h, is the proportionality constant linking the radiation frequency and the energy of a photon.

$$E = h\nu$$

- Electromagnetic radiation with the greatest energy (highest frequency) is known as γ-*rays* (gamma rays). These rays are emitted from the nuclei of certain radioactive elements and, because of their high energy, can severely damage biological organisms.
- *X-rays*, somewhat lower in energy than γ-rays, are less harmful except in high doses. Low-dose X-rays are used to examine the internal structure of organisms. The denser the tissue, the more it blocks X-rays.
- *Ultraviolet (UV) light* is responsible for sunburns, and repeated exposure can cause skin cancer by damaging DNA molecules in skin cells (Section 27.6).
- *Visible light* is the electromagnetic radiation we see.
- We feel *infrared radiation* as heat.
- We cook with *microwaves* and use them in radar.
- *Radio waves* have the lowest energy (lowest frequency). We use them for radio and television communication. Radio waves are also used in NMR spectroscopy and in magnetic resonance imaging (MRI) (Chapter 13).

Because electromagnetic radiation has wavelike properties as well as particle-like properties, it can be described by its wavelength (λ). The **wavelength** is the distance from any point on one wave to the corresponding point on the next wave. Wavelength is measured in *micrometers*. One micrometer (μm) equals 10^{-6} m. In older literature, micrometers are called *microns* (μ).

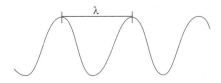

The **frequency** (ν) of the radiation (the number of cycles per second) is equal to the velocity (c) of the wave divided by its wavelength.

$$\nu = \frac{c}{\lambda} \qquad c = 3 \times 10^{10} \text{ cm/s}$$

Thus, electromagnetic radiation can be described either by its frequency or by its wavelength. The energy of electromagnetic radiation is proportional to frequency and inversely proportional to wavelength.

$$E = h\nu = \frac{hc}{\lambda}$$

Wavenumber ($\tilde{\nu}$) is a third unit used to describe electromagnetic radiation, and the one most often used in infrared spectroscopy. It is the number of waves in 1 cm (centimeter) and therefore has units of cm^{-1}. The use of wavenumbers is preferable to wavelengths because, like frequencies, wavenumbers are proportional to energy.

$$\tilde{\nu}(\text{cm}^{-1}) = \frac{1}{\lambda(\text{cm})}$$

The relationship between wavenumber and wavelength is given by the following equation.

$$\tilde{\nu}(\text{cm}^{-1}) = \frac{10^4}{\lambda(\mu\text{m})} \qquad\qquad \text{because 1 } \mu\text{m} = 10^{-4} \text{ cm}$$

So high frequencies, large wavenumbers, and short wavelengths are associated with high energy.

High frequencies, large wavenumbers, and short wavelengths are associated with high energy.

> ### PROBLEM 13 ◆
>
> **a.** Which is higher in energy per photon, electromagnetic radiation with wavenumber 100 cm^{-1} or with wavenumber 2000 cm^{-1}?
>
> **b.** Which is higher in energy per photon, electromagnetic radiation with wavelength 9 mm or with wavelength 8 μm?
>
> **c.** Which is higher in energy per photon, electromagnetic radiation with wavenumber 3000 cm^{-1} or with wavelength 2 μm?
>
> ### PROBLEM 14 ◆
>
> **a.** Light of what wavenumber has a wavelength of 4 μm?
>
> **b.** Light of what wavelength has a wavenumber of 200 cm^{-1}?

12.7 INFRARED SPECTROSCOPY

The bonds in an organic molecule are constantly vibrating. So when we say that a bond between two atoms has a certain length, we are citing an average because the bond behaves as if it were a vibrating spring connecting two atoms. The bond vibrates with both stretching and bending motions. A *stretch* is a vibration occurring along the line of the bond. A *bend* is a vibration that does not occur along the line of the bond. A diatomic molecule such as H—Cl can undergo only a **stretching vibration.**

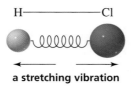

H————————Cl

a stretching vibration

Three bonded atoms can undergo a variety of vibrations These are shown in Figure 12.12. Notice that there are symmetric and asymmetric stretches and bends, and bending vibrations can be either in-plane or out-of-plane vibrations. The **bending vibrations** are often referred to by the descriptive terms *rock, scissor, wag,* and *twist.*

Each stretching and bending vibration of a bond occurs with a certain frequency. When a compound is bombarded with radiation of a frequency that exactly matches the frequency of the vibration of one of its bonds, the vibrating bond will absorb energy. This allows the bond to stretch and bend a bit more. In other words, the absorption of energy increases the amplitude of the vibration but does not change its frequency. By experimentally determining the wavelengths of light that are absorbed by a particular compound, we can determine what kinds of bonds it has. For example, the stretching vibration of a carbon–oxygen double bond absorbs light with wavenumber ~1700 cm^{-1}, whereas the stretching vibration of an oxygen–hydrogen bond absorbs light with wavenumber ~3450 cm^{-1} (Figure 12.13).

$$\text{C}=\text{O} \qquad \text{O}-\text{H}$$

~1700 cm^{-1} ~3450 cm^{-1}

An **infrared spectrum** is obtained by passing a beam of infrared radiation through a sample of the compound. A detector generates a plot of percent transmission of

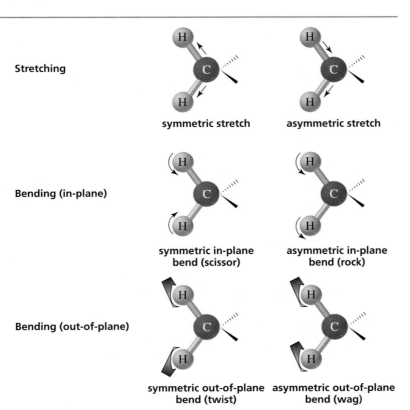

Stretching

symmetric stretch   asymmetric stretch

Bending (in-plane)

symmetric in-plane bend (scissor)   asymmetric in-plane bend (rock)

Bending (out-of-plane)

symmetric out-of-plane bend (twist)   asymmetric out-of-plane bend (wag)

◀ **Figure 12.12** Stretching and bending vibrations of bonds in organic molecules.

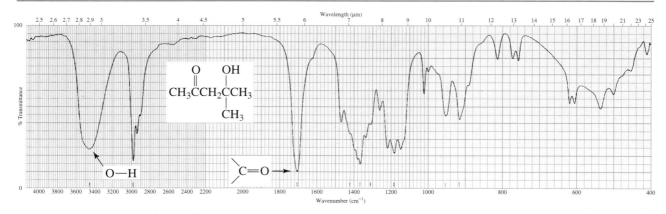

▲ **Figure 12.13**
The infrared spectrum of 4-methyl-4-pentanol-2-one.

radiation versus the wavenumber or wavelength of the radiation that is transmitted. At 100% transmission, all the energy of the radiation is passing through the molecule. At lower values of percent transmission, some of the energy is being absorbed by the compound. Each spike in the infrared (IR) spectrum represents absorption of energy. These spikes are called **absorption bands.** Most chemists report absorption bands using wavenumbers.

An *IR spectrum* can be taken of a gas, a solid, or a liquid sample. Gases are expanded into an evacuated cell (a small container). Solids can be compressed with anhydrous KBr into a disc that is placed in the light beam. A solid can also be examined as a mull. A mull is prepared by mixing a few milligrams of the solid with a drop or two of mineral oil. In the case of liquid samples, a spectrum can be obtained of the neat (undiluted) liquid by placing a few drops of the liquid between two optically polished plates of sodium chloride that are placed in the light beam. A small container—a cell—with optically polished NaCl or KBr windows is used to hold solid or liquid samples dissolved in solvents. KBr discs and NaCl or KBr cells are used because, unlike glass or quartz, these materials do not absorb IR radiation because they have no covalent bonds.

When solvents are used, they must be solvents with few absorption bands in the region of interest. Commonly used solvents are CCl_4, $CHCl_3$, and CS_2. In a double-beam spectrophotometer, the IR radiation is split into two beams: one beam passes through the sample cell, and the other passes through a cell that contains only the solvent. Any absorptions of the solvent are cancelled out, so the absorption spectrum is that of the solute alone.[1]

Electromagnetic radiation with wavenumbers from 4000 to 400 cm^{-1} has just the right energy to correspond to stretching and bending vibrations in an organic molecule. Electromagnetic radiation with this energy is known as **infrared radiation.** An IR spectrum can be divided into two areas. The left-hand two-thirds of the IR spectrum (4000–1000 cm^{-1}) is where most of the functional groups show absorption bands. This is called the **functional group region.** The right-hand third (about 1000–400 cm^{-1}) is called the **fingerprint region.** The overall pattern in the fingerprint region is characteristic of the compound as a whole, because each compound shows a unique pattern in this region. For example, 2-pentanol and 3-

[1]The spectra in this text are FT-IR spectra: the information is digitized and Fourier-transformed by a computer to produce the FT-IR spectrum. FT-IR techniques allow a spectrum to be taken in as little as 10 seconds.

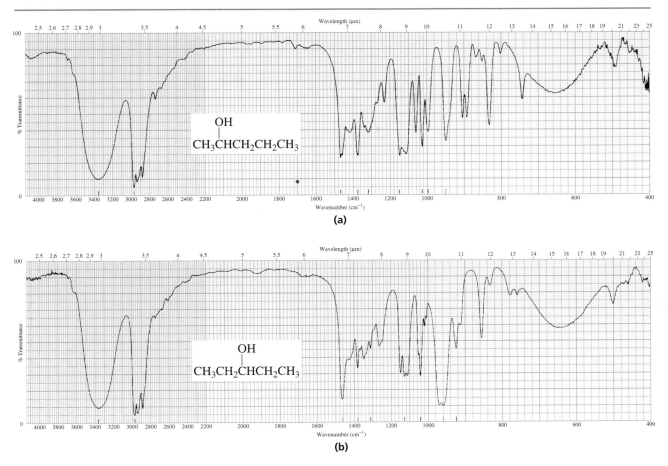

▲ **Figure 12.14**
The IR spectra of (a) 2-pentanol and (b) 3-pentanol.

pentanol have the same functional groups, so they show the same absorption bands in the 4000–1000 cm^{-1} region. Their fingerprint regions, however, are not the same, because the compounds are not the same (Figure 12.14). So a compound can be positively identified by comparing its fingerprint region with the fingerprint region of a known spectrum of the compound.

Each of the different kinds of stretching and bending vibrations shown in Figure 12.12 gives rise to an absorption band, so IR spectra can be quite complex. Because of this complexity, chemists generally do not try to identify all the absorption bands in an IR spectrum. We will look at some characteristic bands so you will be able to tell something about the structure of the compound that gives a particular IR spectrum, but there is a lot more to infrared spectroscopy than we will be able to cover. An extensive table of characteristic group frequencies appears in Appendix V. When identifying an unknown compound, IR spectroscopy is often used in conjunction with information obtained from other spectrophotometric techniques (Chapter 13).

Because it takes more energy to stretch a bond than to bend it, absorption bands for stretching vibrations are found in the 4000–1000 cm^{-1} region, whereas

**12.8
POSITION OF
ABSORPTION BANDS
IN INFRARED
SPECTRA**

TABLE 12.4	Important IR Stretching Frequencies	
Type of bond	$\bar{v}\,(cm^{-1})$	**Intensity**
C≡N	2260–2220	medium
C≡C	2260–2100	medium to weak
C=C	1680–1600	medium
⬡	~1600 and ~1500	strong
C=O	1780–1650	strong
C—O	1250–1050	strong
C—N	1230–1020	medium
O—H (alcohol)	3650–3200	strong, broad
O—H (acid)	3300–2500	strong, very broad
N—H	3500–3300	medium, broad
C—H	3300–2700	medium

absorption bands for bending vibrations are typically found in the fingerprint region. Stretching vibrations are therefore the most useful in determining what kinds of bonds a molecule has. The IR **stretching frequencies** associated with different types of bonds are shown in Table 12.4.

The amount of energy required to stretch a bond depends on the strength of the bond and the masses of the bonded atoms. The approximate wavenumber of an absorption can be calculated using the following equation derived from **Hooke's law,** which describes the motion of a vibrating spring. The stronger the bond, the greater the energy required to stretch it. A stronger bond corresponds to a tighter spring. The heavier the atoms attached to the spring, the more slowly the spring vibrates. So vibrations of heavier atoms occur at smaller wavenumbers. The equation relates the wavenumber of the stretching vibration to the force constant of the bond (f) and the masses of the atoms (in grams) joined by the bond (m_1 and m_2); c is the speed of light (Section 12.6). The force constant is a measure of the strength of the bond. Hooke's law explains why stronger bonds and smaller atoms give rise to greater wavenumbers. For typographic reasons, we will use $[\;]^{1/2}$ to signify the more familiar square root sign ($\sqrt{\;}$).

Stronger bonds and smaller atoms give rise to higher wavenumbers.

$$\tilde{v} = \frac{1}{2\pi c}\left[\frac{f(m_1 + m_2)}{m_1 m_2}\right]^{1/2}$$

One factor affecting bond strength is bond order. A carbon–carbon triple bond is stronger than a carbon–carbon double bond, so a triple bond stretches at a greater wavenumber (C≡C, ~2100 cm^{-1}) than a carbon–carbon double bond (C=C, ~1650 cm^{-1}); likewise, a carbon–oxygen double bond stretches at a greater wavenumber (C=O, ~1700 cm^{-1}) than a carbon–oxygen single bond (C—O, ~1100 cm^{-1}). Carbon–carbon single bonds show stretching vibrations in the region 1200 to 800 cm^{-1}, but they are weak and of little value in identifying compounds.

THE ORIGINATOR OF HOOKE'S LAW

Robert Hooke (1635–1703) was born on the Isle of Wight off the southern coast of England. A brilliant scientist, he contributed to almost every field of science. He was the first to suggest that light had wavelike properties. He discovered that Gamma Arietis was a double star and he discovered Jupiter's Great Red Spot. In a lecture published posthumously, he suggested that earthquakes are caused by the cooling and contracting of Earth. He examined cork under a microscope and coined the term "cell" to describe what he saw. He wrote about evolutionary development based on his studies of microscopic fossils, and his studies of insects were highly regarded as well. Hooke also invented the balance spring for watches and the universal joint currently used in cars.

Robert Hooke's drawing of a "blue fly," which appeared in *Micrographia,* the first book on microscopy, published by Hooke in 1665.

PROBLEM 15◆

a. Which will occur at a greater wavenumber?

   1. C—O stretch or C=O stretch

   2. C—H stretch or C—H bend

   3. ≡C—H stretch or =C—H stretch

b. Assuming the force constants are the same, which will occur at a greater wavenumber?

   1. C—O stretch or C—Cl stretch

   2. C—O stretch or C—C stretch

Table 12.4 shows a range of wavenumbers for each stretch, because the actual position of the stretch depends on other structural features of the molecule. Important details about the structure of a compound can be revealed by the exact position of the absorption band. For example, a carbon–oxygen single bond shows a stretch between 1250 and 1050 cm^{-1}. If the carbon–oxygen bond is in an alcohol, the stretch will occur toward the lower end of the range ($\sim$1100 cm^{-1}). If, however, the carbon–oxygen single bond is in a carboxylic acid, the stretch will occur at the higher end of the range ($\sim$1250 cm^{-1}). This is because the carbon–oxygen bond in an alcohol is a pure single bond, whereas the carbon–oxygen single bond in a carboxylic acid has partial double-bond character as a result of resonance. Because it is harder to stretch a double bond than a single bond, it will be harder to stretch the carbon–oxygen bond with partial double-bond character than a carbon–oxygen bond that is a pure single bond. Consequently, the carbon–oxygen single bond stretch of a carboxylic acid will occur at a greater wavenumber.

CH_3CH_2—OH
carbon–oxygen stretch
$\sim$1100 cm^{-1}

CH_3—$\overset{\overset{\displaystyle O}{\|}}{C}$—OH
carbon–oxygen stretch
$\sim$1250 cm^{-1}

CH_3—$\overset{\overset{\displaystyle O}{\|}}{C}$—O—$CH_3$
carbon–oxygen stretches
at 1100 cm^{-1} and 1250 cm^{-1}

Esters show carbon–oxygen stretches at both 1100 and 1250 cm^{-1}, because esters have two carbon–oxygen single bonds—one that is a pure single bond and one that has partial double-bond character.

TABLE 12.5   The Effects of Electron-Withdrawing Substituents on the Positions of the Carbon–Oxygen Double-Bond Stretch and the Carbon–Oxygen Single-Bond Stretch

$$\text{R}-\overset{\overset{\displaystyle O}{\|}}{\text{C}}-\text{OCH}_2\text{CH}_3$$

R	C=O (cm^{-1})	C—O (cm^{-1})
CH_3	1740	1236
$ClCH_2$	1753	1288
FCH_2	1778	1290
F_2CH	1789	1319

The position of an absorption band can shift if electron-withdrawing or electron-donating substituents are close to the functional group responsible for the absorption band. The data in Table 12.5 indicate how the positions of the carbon–oxygen double-bond stretch and the carbon–oxygen single-bond stretch of a series of ethyl esters are affected by the nature of the substituent bonded to the carbonyl carbon. The more electron-withdrawing the substituent (R), the smaller the contribution of the resonance structure with the positive charge on the carbonyl carbon. This increases the double-bond character of both the carbon–oxygen bonds, which increases the frequency of the absorption.

PROBLEM 16 ◆

Which will occur at a greater wavenumber?

a. The carbon–nitrogen stretch of an amine or the carbon–nitrogen stretch of an amide

b. The carbon–oxygen stretch of phenol or the carbon–oxygen stretch of cyclohexanol

c. The carbon–oxygen double bond stretch of a ketone or the carbon–oxygen double bond stretch of an ester

d. The stretch or the bend of the carbon–oxygen bond in ethanol

The position of the stretch of the oxygen–hydrogen bond in an alcohol depends on the concentration of the alcohol. The more concentrated the alcohol, the more likely it is for the alcohol molecules to form intermolecular hydrogen bonds. It is easier to stretch the oxygen–hydrogen bond of a hydrogen-bonded OH group, because the hydrogen is attracted to the oxygen of a neighboring molecule. So the oxygen–hydrogen stretch of a concentrated (hydrogen-bonded) solution of an alcohol occurs at 3550 to 3200 cm^{-1}, whereas the oxygen–hydrogen stretch of a dilute solution (little or no hydrogen bonding) occurs at a greater wavenumber, 3650 to 3590 cm^{-1}. Hydrogen-bonded OH groups show more intense absorption than non-hydrogen-bonded OH groups. Hydrogen-bonded OH groups also have broader absorption bands, because the hydrogen bonds vary in strength.

$$R—O—H------\overset{\overset{\displaystyle H}{|}}{O}—R \qquad\qquad R—O—H$$

concentrated solution
a hydrogen-bonded O—H bond
$3550–3200$ cm^{-1}

dilute solution
a non-hydrogen-bonded O—H bond
$3650–3590$ cm^{-1}

PROBLEM 17◆

List the following compounds in order of decreasing frequency of the carbon–oxygen absorption band.

$$\underset{CH_3—C—CH_3}{\overset{\overset{\displaystyle O}{\|}}{}} \qquad \underset{H—C—H}{\overset{\overset{\displaystyle O}{\|}}{}} \qquad \underset{CH_3—C—H}{\overset{\overset{\displaystyle O}{\|}}{}}$$

PROBLEM 18◆

An oxygen-containing compound shows an absorption band at 1100 cm^{-1} and no absorption bands either at 1700 cm^{-1} or from 3600 to 3200 $^{-1}$. What class of compound is it?

PROBLEM 19◆

Which will show an oxygen–hydrogen stretch at a greater wavenumber: ethanol dissolved in carbon disulfide or an undiluted sample of ethanol?

Recall that the strength of a carbon–hydrogen bond depends on the hybridization of the carbon. An sp carbon–hydrogen bond is stronger than an sp^2 carbon–hydrogen bond, which is stronger than an sp^3 carbon–hydrogen bond (Section 1.14). In other words, the more s character the bond has, the stronger it is. Because it requires more energy to stretch a stronger bond, the position of a carbon–hydrogen stretch depends on the hybridization of the carbon. If the carbon is sp hybridized, the stretch occurs at ~3300 cm^{-1}; if it is sp^2 hybridized, the stretch occurs at ~3100 cm^{-1}; and if it is sp^3 hybridized, the stretch occurs at ~2900 cm^{-1} (Table 12.6).

A comparison of the IR spectra for methylcyclohexane, cyclohexene, and ethylbenzene shows that a useful first step in the analysis of a spectrum entails looking at the absorption bands in the vicinity of 3000 cm^{-1} (Figures 12.15 to 12.17). The only absorption band in the vicinity of 3000 cm^{-1} in Figure 12.15 is slightly below 3000 cm^{-1}. We know, therefore, that the compound has hydrogens bonded to sp^3 carbons and no hydrogens bonded to sp^2 or sp carbons. The spectrum in Figure 12.16 shows absorption bands slightly above and slightly below 3000 cm^{-1}, which means that the compound responsible for the spectrum contains hydrogens bonded to sp^3 and sp^2 carbons. Once we know that the compound has hydrogens bonded to sp^2 carbons, we can then ask whether they are sp^2 carbons of an alkene or sp^2 carbons of a benzene ring. Because absorption bands at both ~1600 cm^{-1} and ~1500 cm^{-1} indicate a benzene ring while a band only at ~1600 cm^{-1} indicates an alkene (Table 12.3), we know that the compound with the spectrum shown in Figure 12.16 is an alkene and the compound with the spectrum shown in Figure 12.17 has a benzene ring. You should be aware that nitrogen–hydrogen bending vibrations also occur at ~1600 cm^{-1} (Figure 12.18), so absorption at that wavelength does not always indicate the presence of a carbon–carbon double bond. However, absorption bands resulting from nitrogen–hydrogen bends tend to be broader and more intense than those resulting from carbon–carbon double bond stretches.

TABLE 12.6 IR Absorptions of Carbon–Hydrogen Bonds	
Carbon–Hydrogen Stretching Vibrations	$\tilde{\nu}\,(cm^{-1})$
C≡C—H	~3300
C=C—H	3100–3020
C—C—H	2960–2850
C—C̈—H (with O double bonded to middle C)	~2700
Carbon–Hydrogen Bending Vibrations	$\tilde{\nu}\,(cm^{-1})$
CH_3- $-CH_2-$ $-CH-$	1450–1420
CH_3-	1385–1365
H,R / R,H C=C trans	980–960
R,R / H,H C=C cis	730–675
R,R / R,H C=C trisubstituted	840–800
R,R / H,H C=C terminal alkene	890
R,H / H,H C=C terminal alkene	990 and 910

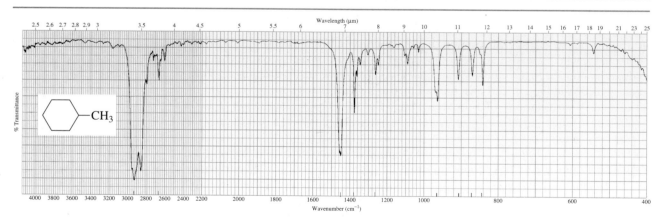

▲ **Figure 12.15**
The IR spectrum of methylcyclohexane.

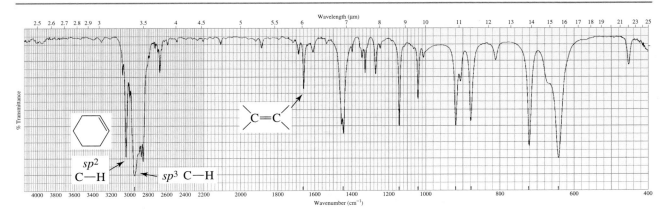

▲ **Figure 12.16**
The IR spectrum of cyclohexene.

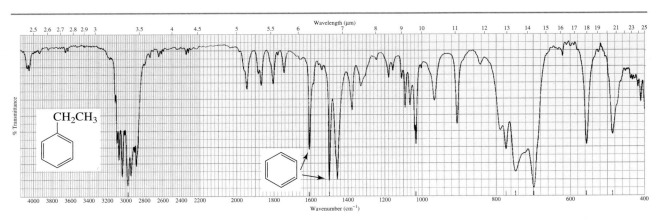

▲ **Figure 12.17**
The IR spectrum of ethylbenzene.

If the compound has sp^3 carbons, a look at ~1400 cm^{-1} will tell you whether the compound has a methyl group. All hydrogens bonded to sp^3 hybridized carbons show a carbon–hydrogen bending vibration at slightly greater than 1400 cm^{-1}. Only methyl groups show a bending vibration at slightly less than 1400 cm^{-1}. So, if the compound has a methyl group, absorption bands will appear at both ~1440 cm^{-1} and ~1380 cm^{-1}. If it does not have a methyl group, only the band at ~1440 cm^{-1} will be present. We see evidence for a methyl group in Figure 12.15 (methylcyclohexane) and in Figure 12.17 (ethylbenzene), but no evidence for a methyl group in Figure 12.16 (cyclohexene). The presence of an isopropyl group in a compound can sometimes be detected by a split in the methyl peak at ~1380 cm^{-1}. This is known as an **isopropyl split** (Figure 12.18).

Carbon–hydrogen bending vibrations, for hydrogens bonded to sp^2 carbons, give rise to absorption bands in the 1000–600 cm^{-1} region (Table 12.6). These bands are useful for identifying the nature of the double bond in an alkene. Terminal alkenes show two absorptions, one at ~910 cm^{-1} and the other at ~990 cm^{-1} if there is only one substituent on C-2, and show a single absorption at 890 cm^{-1} if there are two substituents on C-2. Trans-disubstituted alkenes show an absorption between 980 and 960 cm^{-1}; cis-disubstituted alkenes show an absorption between 675 and

730 cm^{-1}; trisubstituted alkenes show absorptions between 840 and 800 cm^{-1} (Table 12.5). It is important to realize that these absorption bands can be shifted out of the characteristic regions if strongly electron-withdrawing or electron-donating substituents are close to the double bond (Table 12.4). Open-chain compounds with more than four adjacent methylene (CH$_2$) groups show a characteristic absorption band at 720 cm^{-1} due to rocking of all the methylene groups in phase (Figure 12.19).

The shape of an absorption band can be helpful in identifying the compound responsible for an IR spectrum. For example, both oxygen–hydrogen and nitrogen–hydrogen bonds stretch at wavenumbers above 3100 cm^{-1}. It is apparent from the IR spectra of isopentylamine, 1-hexanol, and pentanoic acid (Figures 12.18–12.20) that a nitrogen–hydrogen stretching vibration (~3300 cm^{-1}) is narrower and not as strong as an oxygen–hydrogen stretching vibration (~3300 cm^{-1}), and that the oxygen–hydrogen stretching vibration of a carboxylic acid (~3300–2500 cm^{-1}) is broader than the oxygen–hydrogen stretching vibration of an alcohol. Notice that two absorption bands are detectable in Figure 12.18 for the nitrogen–hydrogen stretch because there are two nitrogen–hydrogen bonds in the compound.

The absence of an absorption band can be as useful as the presence of a band in identifying a compound by IR spectroscopy. For example, a strong absorption at ~1700 cm^{-1} indicates the presence of a carbon–oxygen double bond (Figure

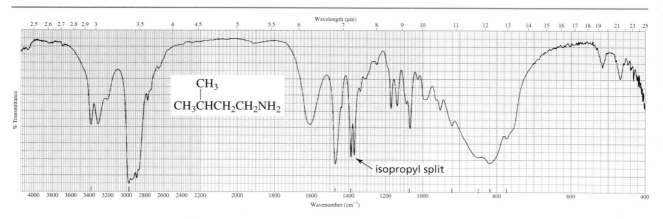

▲ **Figure 12.18**
The IR spectrum of isopentylamine.

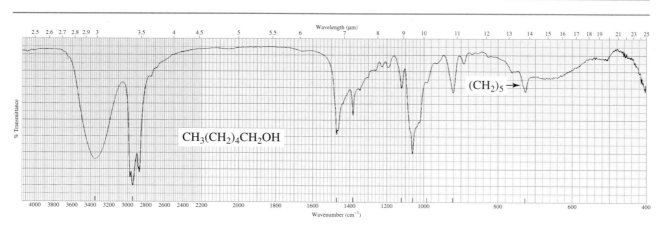

▲ **Figure 12.19**
The IR spectrum of 1-hexanol.

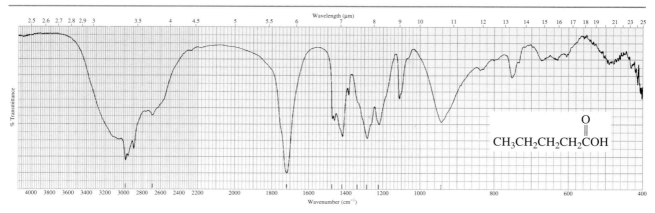

▲ **Figure 12.20**
The IR spectrum of pentanoic acid.

12.21). The fact that it is the carbon–oxygen double bond of a ketone is suggested by the absence of strong absorption at ~3100 cm^{-1}, which would indicate a carboxylic acid (Figure 12.20); by the absence of an absorption at ~2700 cm^{-1}, which would indicate the C—H stretch of an aldehyde (Figure 12.22); and the absence of an absorption at ~1200 cm^{-1}, which would indicate the C—O stretch of an ester or the C—N stretch of an amide (Figure 12.23).

PROBLEM 20◆

Which would show an absorption band at a higher frequency: a carbonyl group bonded to an sp^3 hybridized carbon or a carbonyl group bonded to an sp^2 hybridized carbon?

PROBLEM 21

How could you use IR spectroscopy to distinguish between:

a. a primary amine and a tertiary amine?

b. a cyclic ketone and an open-chain ketone?

c. *cis*-2-hexene and *trans*-2-hexene?

d. cyclohexene and cyclohexane?

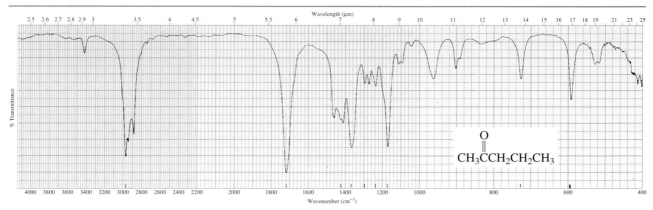

▲ **Figure 12.21**
The IR spectrum of 2-pentanone.

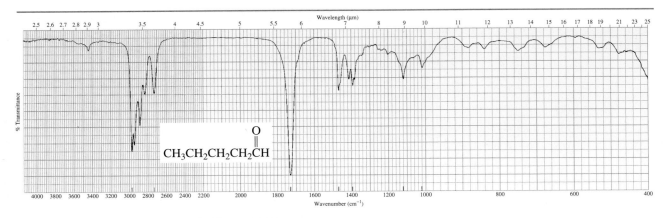

▲ **Figure 12.22**
The IR spectrum of pentanal.

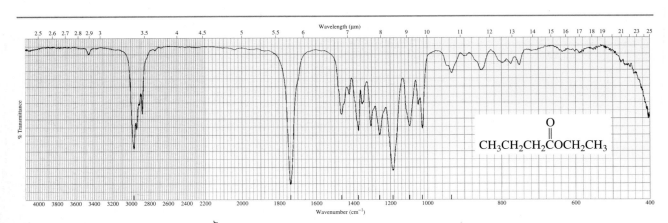

▲ **Figure 12.23**
The IR spectrum of ethyl butanoate.

PROBLEM 22

For each of the following pairs of compounds, give one absorption band that could be used to distinguish between them.

a. $CH_3CH_2CH_2CH_2CH_3$ and $CH_3CH_2CH_2OCH_2CH_3$

b. $CH_3CH_2CH_2\overset{\overset{\displaystyle O}{\|}}{C}OH$ and $CH_3CH_2CH_2CH_2OH$

c. ⬡   and   ⬡

d. $CH_3CH_2CH_2\overset{\overset{\displaystyle O}{\|}}{C}OCH_3$ and $CH_3CH_2CH_2\overset{\overset{\displaystyle O}{\|}}{C}OH$

e. (hexagon) and (hexagon with CH₃)

f. $CH_3CH_2C\equiv CCH_3$ and $CH_3CH_2C\equiv CH$

Not all vibrations result in absorption bands. In order for a vibration to cause absorption of IR radiation, the dipole moment of the molecule must change when the vibration occurs. In other words, the vibrating bond must be polar. For example, 1-butene has a dipole moment along the carbon–carbon double bond. The dipole moment is equal to the magnitude of the charge on the atoms multiplied by the distance between them (Section 1.3). When the carbon–carbon double bond stretches, the distance between the atoms increases. This in turn increases the dipole moment. The change in dipole moment allows absorption of IR radiation; consequently, an absorption band is observed for the carbon–carbon double-bond stretch.

12.9
INFRARED INACTIVE VIBRATIONS

1-butene   2,3-dimethyl-2-butene   2,3-dimethyl-2-heptene

Because 2,3-dimethyl-2-butene is a symmetrical molecule, it does not have a dipole moment along the carbon–carbon double bond. When the bond stretches, it still has no dipole moment. Since stretching is not accompanied by a change in dipole moment, an absorption band is not observed for the carbon–carbon double-bond stretch. The vibration is *infrared inactive*. 2,3-Dimethyl-2-heptene experiences a very small change in dipole moment when it stretches, so only an extremely weak (if any) absorption band will be detected for the stretching vibration of the carbon–carbon double bond.

PROBLEM 23 ◆

Which of the following compounds has a vibration that is infrared inactive: acetone, CO_2, CO, 1-butyne, 2-butyne, H_2, H_2O, Cl_2, ethene?

We saw in Section 12.9 that the intensity of an absorption band depends on the size of the dipole moment change associated with the vibration; i.e., on the polarity of the vibrating bond. For example, the stretching vibration of an oxygen–hydrogen bond will absorb more energy and its absorption band will be more intense than the stretching vibration of a carbon–hydrogen bond because the oxygen–hydrogen bond is more polar.

The intensity of an absorption band also depends on the number of bonds responsible for the absorption. For example, the absorption band for the carbon–hydrogen stretch will be more intense for a compound such as octyl iodide, which

12.10
INTENSITY OF INFRARED ABSORPTION BANDS

has 17 carbon–hydrogen bonds, than for methyl iodide, which has only three carbon–hydrogen bonds. The concentration of the sample used to obtain an IR spectrum also affects the intensity of the absorption bands. Concentrated samples have greater numbers of absorbing molecules and therefore more intense absorption bands.

12.11 OVERTONE AND COMBINATION BANDS

Overtone and combination bands can make an IR spectrum more difficult to interpret. **Overtone bands** are bands that occur at multiples of the fundamental absorption frequency. For example, a compound that shows a fundamental absorption band at 720 cm^{-1} can show a first overtone at 1440 cm^{-1} (2×720) and a second overtone at 2160 cm^{-1} (3×720). Notice the absorption band at ~3400 cm^{-1} in Figure 12.21; it is an overtone of the carbonyl absorption band at ~1700 cm^{-1}.

Combination bands arise when two fundamental vibrations, absorbing energy at $\tilde{\nu}_1$ and $\tilde{\nu}_2$, absorb energy simultaneously. The combination band will appear at $\tilde{\nu}_1 + \tilde{\nu}_2$. Care must be taken not to misinterpret an overtone or combination band as being a result of the fundamental vibration of a functional group. Fortunately, overtone and combination bands are very weak.

Weak overtone and combination bands in the 2000–1650 cm^{-1} region are useful in determining the number and relative positions of the substituents on a benzene ring: the overtone bands are different for monosubstituted and disubstituted benzenes, and they are different for 1,2-, 1,3-, and 1,4-disubstituted benzenes (Figures 12.24 and 12.25).

PROBLEM 24 ◆

A compound has strong fundamental bands at 680, 1420, and 2910 cm^{-1}.

a. At what wavenumbers are the first overtones?

b. At what wavenumbers are the combination bands?

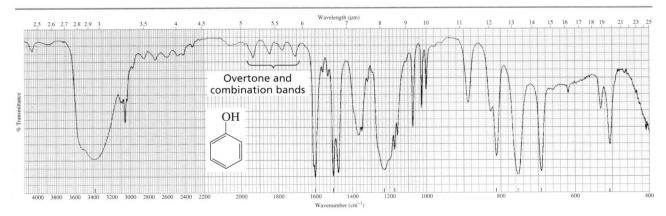

▲ **Figure 12.24**
The IR spectrum of phenol.

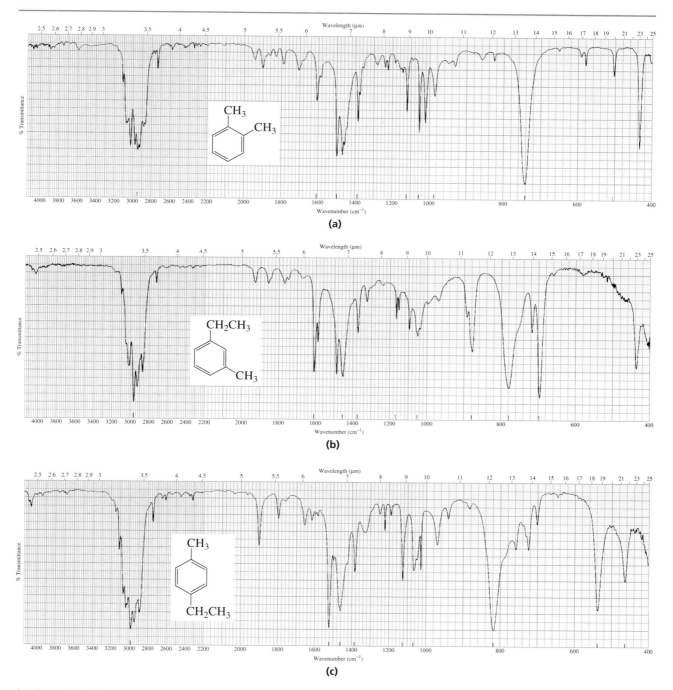

▲ **Figure 12.25**
The IR spectra of three dialkyl-substituted benzenes: (a) 1,2-dimethylbenzene, (b) 1-ethyl-3-methylbenzene, and 1-ethyl-4-methylbenzene.

PROBLEM 25◆

A dibromobenzene gives the IR spectrum shown in Figure 12.26. Identify the dibromobenzene.

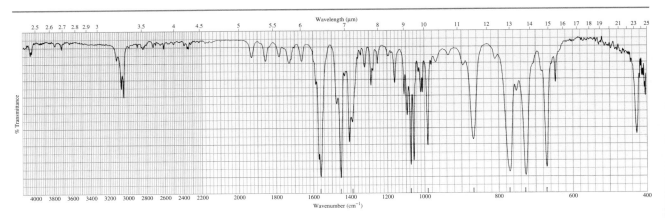

▲ **Figure 12.26**
The IR spectrum for Problem 25 .

PROBLEMS

26. For each of the following pairs of compounds, give one absorption band that could be used to distinguish between them.

    a. $CH_3CH_2CH_2\overset{\displaystyle O}{\overset{\|}{C}}H$ and $CH_3CH_2CH_2\overset{\displaystyle O}{\overset{\|}{C}}CH_3$

    b. (cyclohexane with CH₃) and $CH_3CH_2CH_2CH_2CH_2CH_2CH_3$

    c. $CH_3CH_2CH_2CH{=}CH_2$ and $CH_3CH_2CH_2CH{=}\overset{\displaystyle CH_3}{\overset{|}{C}}CH_3$

d. $CH_3CH_2\overset{\overset{\displaystyle O}{\|}}{C}NH_2$ and $CH_3CH_2\overset{\overset{\displaystyle O}{\|}}{C}OCH_3$

e. 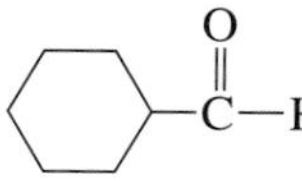 (cyclohexyl)$-\overset{\overset{\displaystyle O}{\|}}{C}-H$ and (phenyl)$-\overset{\overset{\displaystyle O}{\|}}{C}-H$

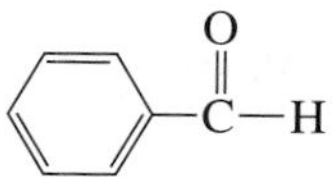

f. $CH_3CH_2CH_2CH_2OH$ and $CH_3CH_2CH_2OCH_3$

g. *cis*-2-butene and *trans*-2-butene

h. $CH_3CH_2CH_2\overset{\overset{\displaystyle O}{\|}}{C}OCH_3$ and $CH_3CH_2CH_2\overset{\overset{\displaystyle O}{\|}}{C}CH_3$

i. 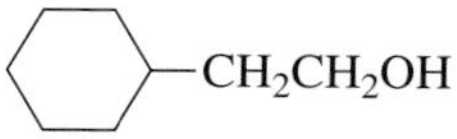 (cyclohexyl)$-CH_2CH_2OH$ and (cyclohexyl)$-\underset{\underset{\displaystyle OH}{|}}{C}HCH_3$

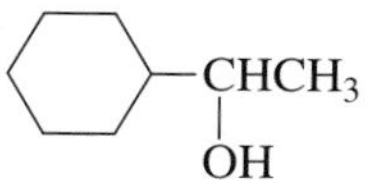

j. $CH_3CH_2CH\!=\!CHCH_3$ and $CH_3CH_2C\!\equiv\!CCH_3$

27. Which peak would be more intense in the mass spectrum of the following compounds, the peak at $m/z = 57$ or the peak at $m/z = 71$?
   a. 3-methylpentane
   b. 2-methylpentane

28. List three factors that influence the intensity of an IR absorption band.

29. How could you determine by IR spectroscopy that the following reaction had occurred?

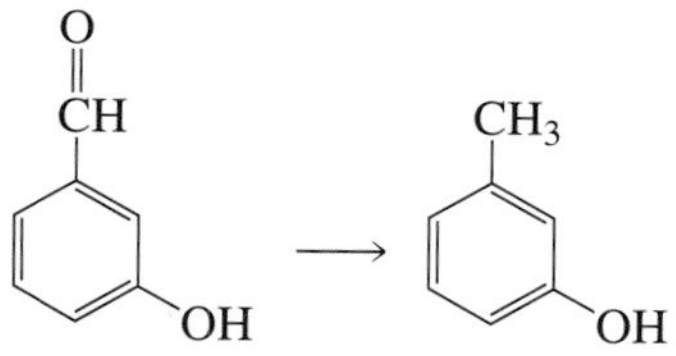

30. What identifying characteristic would be present in the mass spectrum of a compound containing two bromine atoms?

31. Assuming that the force constant is approximately the same for carbon–carbon, carbon–nitrogen, and carbon–oxygen single bonds, predict the relative positions of their stretching vibrations.

32. A mass spectrum shows significant peaks at $m/z = 87, 115, 140,$ and 143. Which compound is responsible for that mass spectrum:
   4,7-dimethyl-1-octanol, 2,6-dimethyl-4-octanol, or 2,2,4-trimethyl-4-heptanol?

33. How could you use IR spectroscopy to distinguish between 1,5-hexadiene and 2,4-hexadiene?

34. How could you determine the relative strengths of the double bonds in the following compounds?

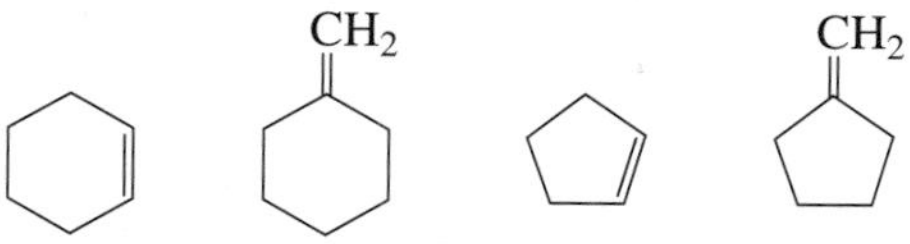

35. For each of the IR spectra in Figures 12.27 to 12.29, four compounds are shown. In each case, indicate which of the four compounds is responsible for the spectrum.

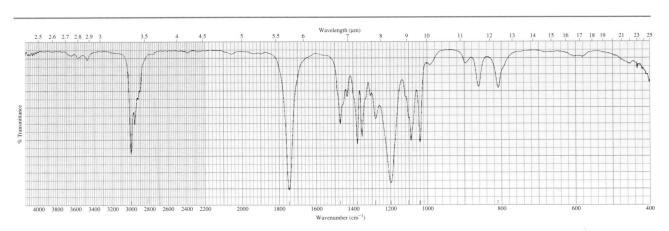

a. $CH_3CH_2CH_2C\equiv CCH_3$ $CH_3CH_2CH_2CH_2OH$ $CH_3CH_2CH_2\overset{O}{\overset{\|}{C}}OH$

$CH_3CH_2CH_2CH_2C\equiv CH$

b. $CH_3CH_2\overset{O}{\overset{\|}{C}}OH$ $CH_3CH_2\overset{O}{\overset{\|}{C}}OCH_2CH_3$ $CH_3CH_2\overset{O}{\overset{\|}{C}}H$ $CH_3CH_2\overset{O}{\overset{\|}{C}}CH_3$

c. $C(CH_3)_3$ CH_2CH_2Br $CH=CH_2$ CH_2OH

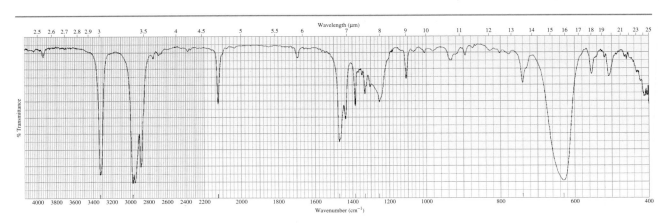

▲ **Figure 12.27**
The IR spectrum for Problem 35a.

▲ **Figure 12.28**
The IR spectrum for Problem 35b.

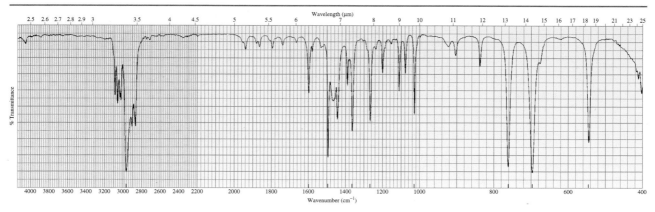

▲ **Figure 12.29**
The IR spectrum for Problem 35c.

36. What hydrocarbons will have a molecular ion peak at $m/z = 112$?

37. In the following boxes, list the types of bonds and the approximate wavenumber where each type of bond is expected to show an absorption.

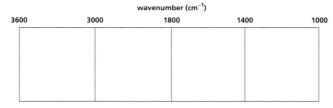

wavenumber (cm^{-1})

3600	3000	1800	1400	1000

38. What peaks in the mass spectra would you use to distinguish between 4-methyl-2-pentanone and 2-methyl-3-pentanone?

39. How could you use IR spectroscopy to distinguish among 1-hexyne, 2-hexyne, and 3-hexyne?

40. For each of the IR spectra in Figures 12.30 to 12.32, indicate which of the five given compounds is responsible for the spectrum.

a. $CH_3CH_2CH{=}CH_2$ $CH_3CH_2CH_2CH_2OH$ $CH_2{=}CHCH_2CH_2OH$ $CH_3CH_2CH_2OCH_3$ $CH_3CH_2CH_2\overset{\displaystyle O}{\overset{\|}{C}}OH$

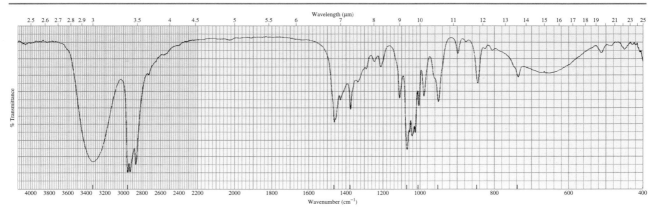

▲ **Figure 12.30**
The IR spectrum for Problem 40a.

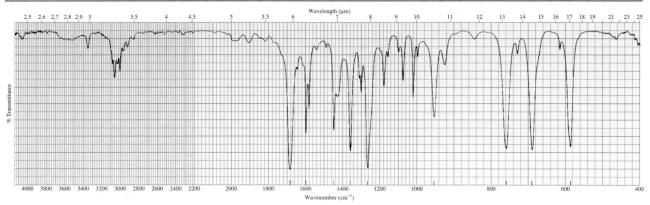

b.

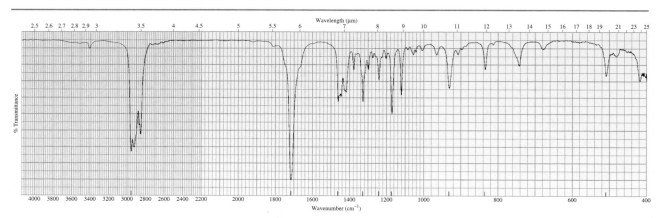

▲ **Figure 12.31**
The IR spectrum for Problem 40b.

c.

▲ **Figure 12.32**
The IR spectrum for Problem 40c.

41. Each of the IR spectra shown in Figure 12.33 is the spectrum of one of the following compounds. Identify the spectra.

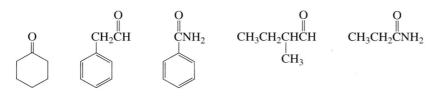

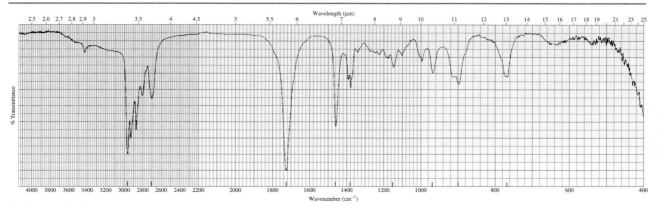

▲ **Figure 12.33a**
The IR spectra for Problem 41.

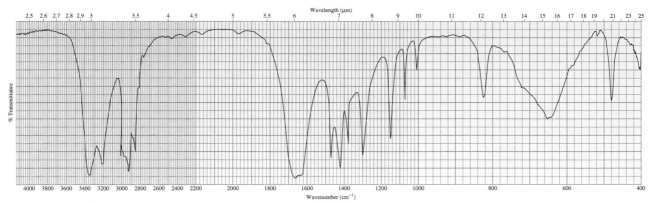

▲ **Figure 12.33b**

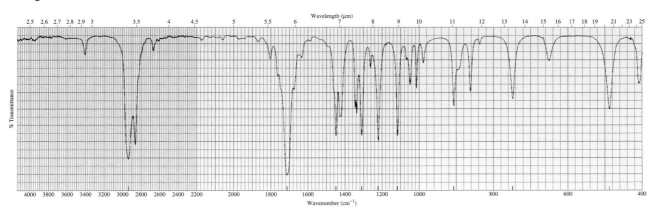

▲ **Figure 12.33c**

42. Predict the major characteristic IR absorption bands that would be given by each of the following compounds.

a. $CH_2{=}CHCH_2\overset{\displaystyle O}{\overset{\displaystyle \|}{C}}H$

b.

c. $CH_3CH_2\overset{\displaystyle O}{\overset{\displaystyle \|}{C}}CH_2CH_2NH_2$

d.

e.

f.

43. The IR spectrum of a compound with molecular formula C_5H_8O was obtained in CCl_4 and is shown below in Figure 12.34. Identify the compound.

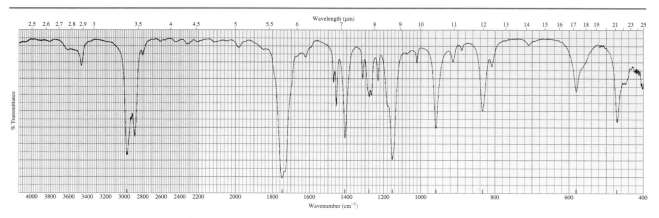

▲ **Figure 12.34**
The IR spectrum for Problem 43.

44. The IR spectrum shown in Figure 12.35 is the spectrum of one of the following compounds. Identify the compound.

CH_2OH　　CH_2CH_3　　$\overset{\displaystyle O}{\overset{\displaystyle \|}{C}}H$　　$CH_2\overset{\displaystyle O}{\overset{\displaystyle \|}{C}}CH_3$　　CH_2CH_3

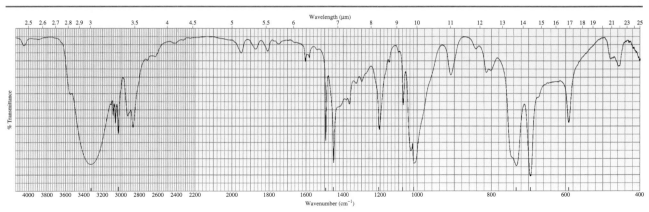

▲ **Figure 12.35**
The IR spectrum for Problem 44.

---

45. The IR spectrum shown in Figure 12.36 is the spectrum of one of the following compounds. Identify the compound.

$$CH_3CH_2CH_2\overset{\overset{\displaystyle O}{\|}}{C}CH_3$$

$$CH_3CH_2CH_2\overset{\overset{\displaystyle O}{\|}}{C}H$$

$$CH_2\overset{\overset{\displaystyle O}{\|}}{C}CH_3$$

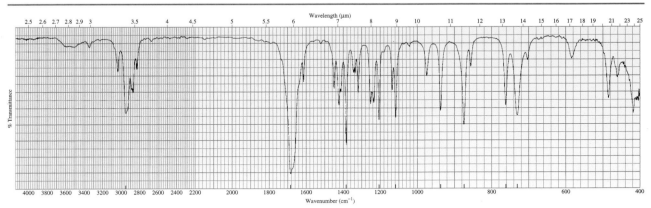

▲ **Figure 12.36**
The IR spectrum for Problem 45.

---

46. Determine the molecular formula of a saturated acyclic hydrocarbon with an M peak at $m/z = 100$ with a relative intensity of 27.32% and an M + 1 peak with a relative intensity of 2.10%.

47. Calculate the approximate wavenumber at which a C=C stretch will occur given that the force constant for the carbon–carbon double bond is 10×10^5 g s^{-2}.

48. Determine the structure of each of the following unknown compounds based on its mass spectrum and IR spectrum.

a.

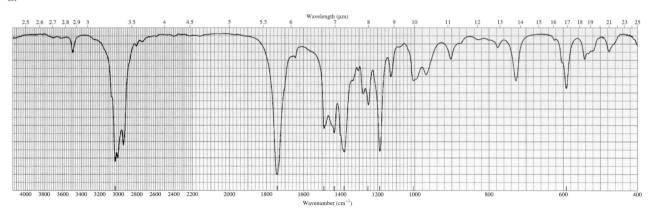

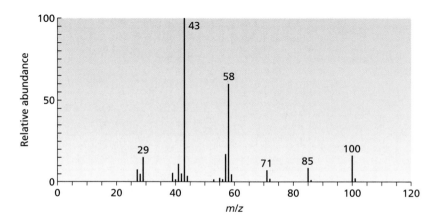

b.

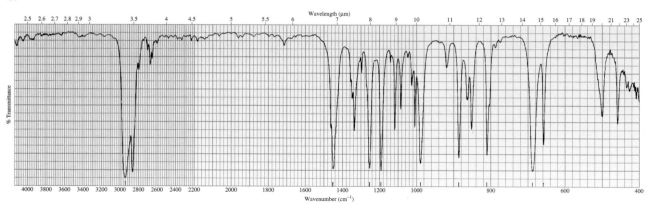

b. (cont'd)

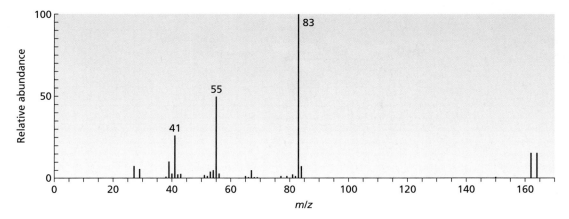

c.

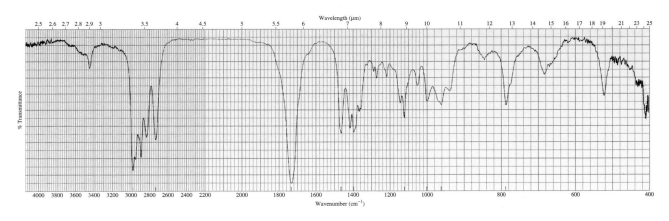

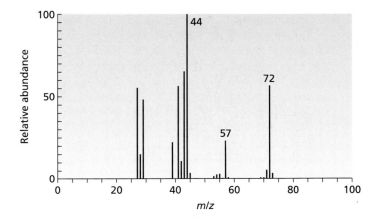

49. Given that the force constants are similar for carbon–hydrogen and carbon–carbon bonds, explain why the stretching vibration of a carbon–hydrogen bond occurs at a greater wave number.

13

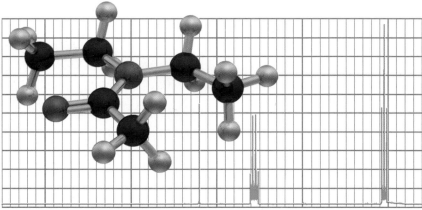

N,N-diethylethanamide

NMR SPECTROSCOPY AND ULTRAVIOLET/ VISIBLE SPECTROSCOPY

Identification of compounds is an important part of organic chemistry. For example, chemists who study natural products must determine the structure of a naturally occurring compound before they can either design a synthesis to produce the compound in greater quantities than nature can provide or design and synthesize related compounds with modified properties. And, after any compound has been synthesized, it must be identified to confirm its structure. In Chapter 12, you were introduced to two instrumental techniques used to identify organic compounds: mass spectrometry and infrared spectroscopy. Mass spectrometry gives the molecular weight and the molecular formula of a compound as well as information about its structure; IR spectroscopy tells what functional groups are present in the compound.

This chapter deals with two additional instrumental techniques that chemists use for structure determination: nuclear magnetic resonance (NMR) spectroscopy and ultraviolet/visible (UV/Vis) spectroscopy. **NMR spectroscopy** helps to identify the carbon–hydrogen framework of an organic compound. **UV/Vis spectroscopy** gives information about compounds with conjugated π electrons. Like IR spectroscopy, NMR and UV/Vis spectroscopy involve the absorption of electromagnetic radiation. The electromagnetic radiation used for NMR spectroscopy is lower in energy than that used for IR spectroscopy, and the electromagnetic radiation used for UV/Vis spectroscopy is higher in energy than that used for IR spectroscopy (Figure 12.11).

13.1 INTRODUCTION TO NMR SPECTROSCOPY

We have seen that electrons are charged, spinning bodies with two allowed spin states, $+\frac{1}{2}$ and $-\frac{1}{2}$. Certain nuclei also have spin with allowed spin states of $+\frac{1}{2}$ and $-\frac{1}{2}$. Examples of such nuclei are ^{1}H, ^{13}C, ^{19}F, and ^{31}P.

NMR spectroscopy was developed by physical chemists in the late 1940s to study nuclei with different spin states. In 1951, chemists realized that NMR spectroscopy could also be used to determine the structures of organic compounds. Since hydrogen nuclei (protons) were the first nuclei studied by *nuclear magnetic resonance,* the acronym "NMR" is generally assumed to mean **^{1}H NMR** (*proton magnetic resonance.*) Spectrometers were later developed for **^{13}C NMR,** ^{15}N NMR, ^{19}F NMR, ^{31}P NMR, and other magnetic nuclei.

A nucleus with spin behaves like a tiny bar magnet. In the absence of an external magnetic field, the individual nuclei in a compound are randomly oriented. However, when an external magnetic field is applied to the nuclei (Figure 13.1), nuclei with spin $+\frac{1}{2}$ orient themselves so that their own tiny magnetic fields are in the same direction as the applied magnetic field (*with the field*). This is similar to a bar magnet or a compass needle orienting itself with respect to Earth's magnetic field. Nuclei with spin $-\frac{1}{2}$ orient themselves so that their magnetic fields are opposite to the direction of the applied magnetic field (*against the field*).

The nuclei aligned with the field (the α-**spin state**) are lower in energy than those aligned against the field (the β-**spin state**). The difference in the number of nuclei in the α-spin and β-spin states is not great: for every million protons (hydrogen nuclei), there are only some 10 to 20 more in the lower-energy state than in the higher-energy state at room temperature. This difference, however, is sufficient to form the basis of NMR spectroscopy.

If electromagnetic radiation of the appropriate energy is applied to nuclei that have been oriented by a magnetic field, a nucleus in the α-spin state will absorb the radiation, flipping its spin and entering the β-spin state. This absorption is detected, resulting in a signal in the NMR spectrum. The energy absorbed in this transition equals the energy difference (ΔE) between the α- and β-spin states. The molecule then returns to the α-spin state, releasing energy as heat. The term "nuclear magnetic resonance" comes from the fact that the nuclei are *in resonance* with the electromagnetic radiation. In other words, they are flipping back and forth between the α-spin and β-spin states in response to the electromagnetic radiation.

The energy difference between the α-spin and β-spin states depends on the strength of the externally applied magnetic field (H_0). The greater the strength of the applied magnetic field, the greater the difference in energy (ΔE) (Figure 13.2).

The following equation describes the relationship between the energy difference of the α-spin and β-spin states and the strength of the applied magnetic field:

$$\Delta E = h\nu = h\frac{\gamma}{2\pi}H_0$$

where ΔE is the energy difference between the α-spin and β-spin states; h is Planck's constant; ν is frequency; H_0 is the strength of the applied magnetic field, measured in tesla (T); and γ is the gyromagnetic ratio, in radians (rad) $\text{T}^{-1}\,\text{s}^{-1}$.

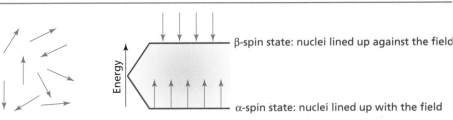

no magnetic field   **a magnetic field (↑) has been applied with the direction shown**

◀ **Figure 13.1**
In the absence of an applied magnetic field, the spins of the nuclei are randomly oriented. In the presence of an applied magnetic field, the spins of the nuclei line up with or against the field.

Figure 13.2 ▶
The greater the strength of the applied magnetic field, the greater the difference in energy between the α- and β-spin states.

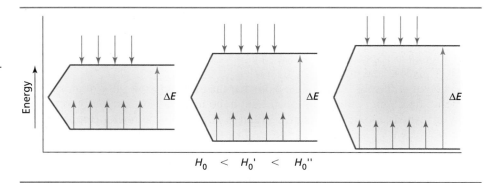

$$H_0 \; < \; H_0' \; < \; H_0''$$

Until recently, the gauss (G) was the unit in which magnetic field strength was commonly measured (1 T = 10^4 G). **Nikola Tesla (1856–1943)** *was born in Croatia, the son of a clergyman. He emigrated to the United States in 1884, becoming a citizen in 1891. He was a proponent of alternating current and bitterly fought Edison, who promoted direct current. Although he did not win his dispute with Marconi over which of them invented the radio, Tesla is given credit for developing the motor for the refrigerator. His other inventions include the Tesla coil, a type of transformer.*

Earth's magnetic field is 5×10^{-5} T, measured at the equator. Its maximum surface magnetic field is 7×10^{-5} T, measured at the south magnetic pole.

The **gyromagnetic ratio** (γ) is the ratio of the magnetic moment of a rotating charged particle to its angular momentum. The size of the gyromagnetic ratio depends on the particular kind of nucleus. In the case of the proton, it is 26.75×10^7 rad T^{-1} s^{-1}. The following calculation shows that, if an NMR spectrometer is equipped with a magnet with a magnetic field of 1.4092 T, the spectrometer will require an energy source with a frequency of 60×10^6 Hz [60 MHz (megahertz), or 60 million cycles per second (cps)].

$$\nu = \frac{\gamma}{2\pi} H_0$$

$$= \frac{26.75 \times 10^7 \, \text{rad}}{2(3.1416) \, \text{rad}} T^{-1} \, s^{-1} \times 1.4092 \, T$$

$$= 60 \times 10^6 \, s^{-1}$$

$$= 60 \times 10^6 \, \text{Hz} = 60 \, \text{MHz}$$

If the spectrometer has a more powerful magnet, the energy source must have a higher frequency. For example, a magnetic field of 2.3486 T requires an energy source with a frequency of 100 MHz. Radiation of 60 MHz or 100 MHz is in the radiofrequency (rf) region of the electromagnetic spectrum, so it is sometimes referred to as **rf radiation.**

The most common NMR spectrometers operate at frequencies of 60, 100, 200, 300, 360, and 500 MHz. The **operating frequency** of a particular spectrometer depends on the strength of the built-in magnet. Some NMR spectrometers have an operating frequency as high as 1000 MHz. The greater the operating frequency of the instrument—and the stronger the magnet—the better the resolution of the NMR spectrum (Section 13.11).

Because each kind of nucleus has its own gyromagnetic ratio, different energy sources are required to bring different kinds of nuclei into resonance in a spectrometer with a given operating frequency. For example, an NMR spectrometer with a magnet requiring an energy source of 60 MHz to flip the spin of an ^{1}H nucleus requires an energy source of 15 MHz to flip the spin of a ^{13}C nucleus. Many NMR spectrometers are equipped with radiation sources that can be tuned to different frequencies so they can be used to obtain NMR spectra of different kinds of nuclei (^{1}H, ^{13}C, ^{15}N, ^{19}F, ^{31}P, etc.).

We have just seen that a proton[1] in an organic compound absorbs energy when the energy difference between the α-spin and β-spin states (ΔE) equals the energy defined by the operating frequency of the spectrometer, and ΔE is determined by the strength of the **applied magnetic field** (H_0). If all the protons in an organic compound had exactly the same environment, they would all require the same applied magnetic field to cause ΔE to correspond to the constant energy source of the spectrometer. If this were the case, all NMR spectra would consist of one absorption signal. Such a spectrum would tell us nothing about the structure of the compound except that it contains protons.

Protons, however, are surrounded by electrons, so the protons exist in an environment that partly shields them from the applied magnetic field. Fortunately for chemists, the **shielding** varies for different protons within a compound. All protons require the same effective magnetic field to achieve ΔE. The **effective magnetic field** is the magnetic field that the proton "senses." Protons that are more shielded from the applied magnetic field require a stronger applied magnetic field in order to "sense" a given effective magnetic field. Protons that are less shielded from the applied magnetic field require a weaker applied magnetic field in order to "sense" the same effective magnetic field (Figure 13.3).

In NMR spectroscopy, the frequency is held constant and the applied magnetic field is slowly varied, which changes the energy separation between the spin states. (This is technically easier than holding the applied magnetic field constant and varying the frequency.) When the applied magnetic field has just the right strength to cause the energy separation between the two spin states for a given set of protons to match the operating frequency of the spectrometer, the protons absorb energy from the rf source (Figure 13.4).

In order to take an NMR spectrum, 5 to 50 mg of a compound is dissolved in about 0.5 mL of solvent, and the solution is put into a long, thin glass tube, which is placed between the poles of a magnet. (The solvents used in NMR are discussed in Section 13.10.) Spinning the sample tube about its long axis averages the molecules with respect to their position in the magnetic field; this greatly increases the

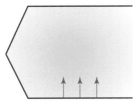

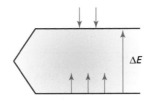

These protons are deshielded, in comparison with those in (a), from the applied magnetic field, so ΔE is too big. They require a smaller applied magnetic field in order to "sense" the effective magnetic field that corresponds to the correct ΔE.	ΔE is constant and is defined by the operating frequency of the spectrophotometer. A proton will absorb energy when the effective magnetic field corresponds to ΔE.	These protons are shielded, in comparison with those in (a), from the applied magnetic field, so ΔE is too small. They require a greater applied magnetic field in order to "sense" the effective magnetic field that corresponds to the correct ΔE.
(c)	**(a)**	**(b)**

▲ **Figure 13.3**
Some protons are shielded from the applied magnetic field and therefore require a stronger magnetic field in order to come into resonance. Deshielded protons require a weaker magnetic field in order to come into resonance.

[1]The terms "proton" and "hydrogen" are both used to describe covalently bound hydrogen in discussions of NMR spectroscopy.

Figure 13.4 ▶
Shielded nuclei come into resonance at greater values of
H_0 than deshielded nuclei.

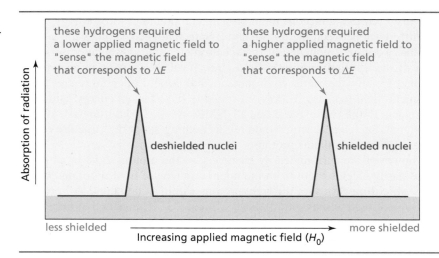

these hydrogens required
a lower applied magnetic field to
"sense" the magnetic field
that corresponds to ΔE

these hydrogens required
a higher applied magnetic field to
"sense" the magnetic field
that corresponds to ΔE

deshielded nuclei

shielded nuclei

Absorption of radiation

less shielded

more shielded

Increasing applied magnetic field (H_0)

*The 1991 Nobel Prize in chem-
istry was awarded to* **Richard
R. Ernst** *for two important
contributions: FT–NMR spectroscopy, and an NMR tomography method that forms the
basis of MRI. Ernst was born
in 1933, received a Ph.D. from
the Swiss Federal Institute of
Technology [Eidgenössische
Technische Hochschule (ETH)]
in Zurich, and became a
research scientist at Varian
Associates in Palo Alto,
California. In 1968 he returned
to ETH, where he is a professor of chemistry.*

resolution of the spectrum. The sample is irradiated with rf radiation from a coil. In
continuous-wave spectrometers, the strength of the magnetic field is varied continuously by a magnet controller, and the recorder registers the signal when the
nucleus flips its spin (Figure 13.5). It takes several minutes to scan from low magnetic field to high magnetic field to obtain an NMR spectrum by this technique.

In more modern instruments, called pulsed Fourier transform (FT) spectrometers, an rf pulse of short duration excites all the nuclei simultaneously, and all the
signals are collected at the same time with a computer. Many hundreds of individual runs are collected (each can be done in less than a second), and the signals
are averaged. The data are mathematically converted (a Fourier transform) to a
spectrum that looks like a spectrum obtained in a continuous-wave spectrometer
with one important difference: **Fourier transform NMR (FT–NMR)** spectra show
almost no electronic background noise, so the signals stand out clearly. Compare
Figure 13.29(a) (a continuous-wave spectrum) with Figure 13.29(b) (an FT–NMR
spectrum). Because background noise is random, it averages close to zero when it
is averaged in a Fourier transform. FT–NMR is much faster and more sensitive than

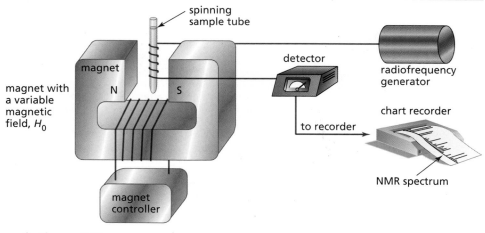

spinning
sample tube

magnet

detector

radiofrequency
generator

magnet with
a variable
magnetic
field, H_0

N   S

chart recorder

to recorder

NMR spectrum

magnet
controller

▲ **Figure 13.5**
Schematic of an NMR spectrometer.

continuous-wave NMR. Consequently, FT–NMR spectra can be obtained with less than 5 mg of compound. The NMR spectra in this book are FT–NMR spectra that were taken on a spectrometer with an operating frequency of 300 MHz.

PROBLEM 1◆

a. Calculate the magnetic field (in tesla) required to flip an ^{1}H nucleus in an NMR spectrometer that operates at 360 MHz.

b. What size magnetic field is required when a 500-MHz instrument is used?

c. What can you conclude from these calculations?

PROBLEM 2◆

What frequency (in MHz) is required to cause a proton to absorb energy when it is exposed to a magnetic field of 1 tesla?

Protons in the same environment are called **chemically equivalent protons.** For example, the three methyl protons in 1-bromopropane are chemically equivalent because of rotation about the carbon–carbon bonds. Each set of chemically equivalent protons in a compound gives rise to a signal in the ^{1}H NMR spectrum of that compound. For example, 1-bromopropane has three sets of chemically equivalent protons, so it has three signals in its ^{1}H NMR spectrum. Absorption signals are also called **resonances.** (This use of the word "resonance" has nothing to do with the resonance associated with delocalized electrons.) The chemically equivalent protons in the following compounds are designated by the same letter.

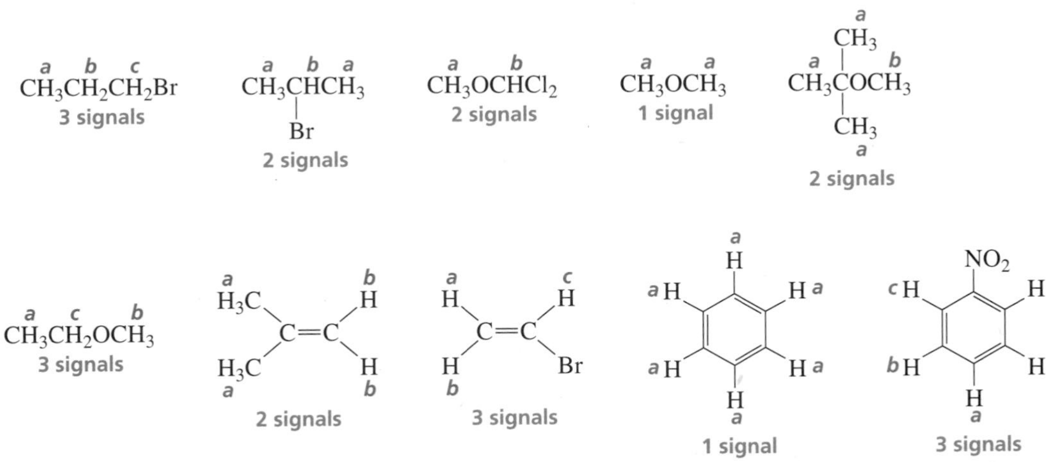

13.2
THE NUMBER OF SIGNALS IN THE ^{1}H NMR SPECTRUM

From the number of signals in an ^{1}H NMR spectrum, you can tell how many kinds of chemically nonequivalent protons the compound has. (Sometimes the signals in an ^{1}H NMR spectrum are not sufficiently separated and overlap each other. When this happens, one sees fewer signals than anticipated.)

The ^{1}H NMR spectrum of chlorocyclobutane has five signals. Even though they are bonded to the same carbon, the H$_a$ and H$_b$ protons are not equivalent because they are not in the same environment: H$_a$ is trans to Cl and H$_b$ is cis to Cl. Similarly, the H$_c$ and H$_d$ protons are not equivalent.

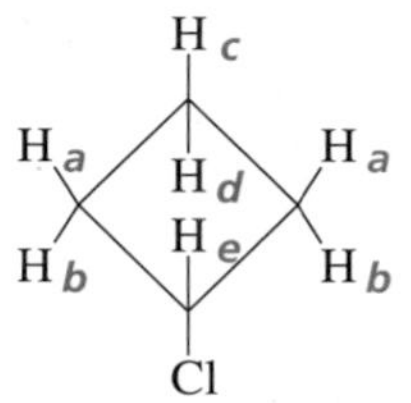

chlorocyclobutane
the NMR spectrum has 5 signals
H_a and H_b are not equivalent
H_c and H_d are not equivalent

PROBLEM 3 ◆

How many signals would you expect to see in the NMR spectrum of each of the following compounds?

a. $CH_3CH_2CH_2\overset{\overset{\text{O}}{\|}}{C}CH_3$

i.

b. $CH_2{=}CH\overset{\overset{\text{O}}{\|}}{C}H$

j.

c. $CH_3CH_2\underset{\underset{\text{Cl}}{|}}{C}HCH_2CH_3$

k.

d. $CH_2{=}CHCl$

l.

e. $CH_3CH_2CH_2CH_3$

m.

f. $CH_3\underset{\underset{\text{CH}_3}{|}}{C}HCH_2\underset{\underset{\text{CH}_3}{|}}{C}HCH_3$

n.

g. $BrCH_2CH_2Br$

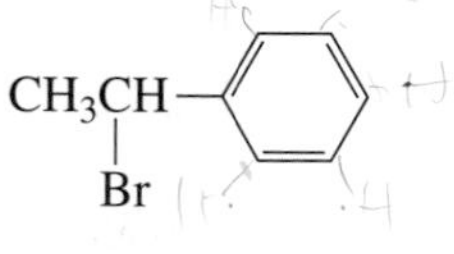

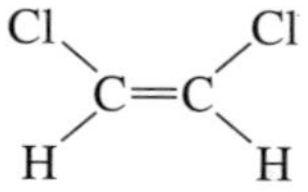

o.

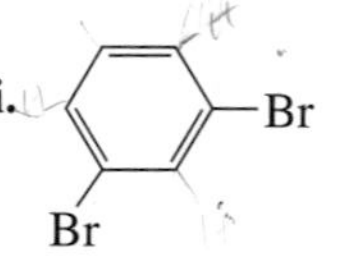

h. $CH_3\underset{\underset{\text{Br}}{|}}{C}H$—

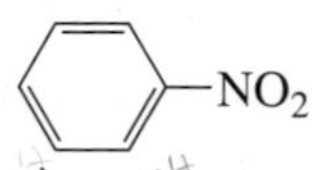

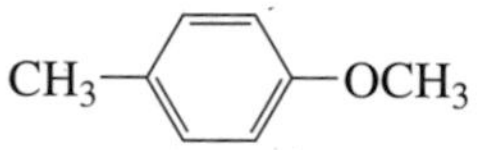

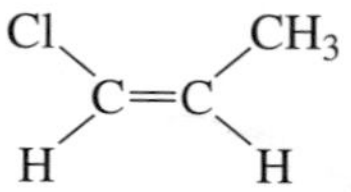

PROBLEM 4

How could you distinguish between the NMR spectra of the following compounds?

a. $CH_3OCH_2OCH_3$

b. CH_3OCH_3

c. $CH_3OCH_2\underset{\underset{\text{CH}_3}{|}}{\overset{\overset{\text{CH}_3}{|}}{C}}CH_2OCH_3$

PROBLEM 5

There are three isomeric dichlorocyclopropanes. The NMR spectrum of isomer 1 has
one signal, that of isomer 2 has two signals, and that of isomer 3 has three signals.
Draw the structures of isomers 1, 2, and 3.

We have seen that all protons absorb energy at the same *effective magnetic field* but
not at the same *applied magnetic field*. This is because some protons are shielded
more than others from the applied magnetic field and therefore require a greater
applied field in order to "sense" the same effective magnetic field (Section 13.1).
What causes shielding? A nucleus is embedded in a cloud of electrons that shield
the nucleus from the applied magnetic field. These spinning electrons induce a
local magnetic field that opposes (subtracts from) the applied magnetic field. The
effective magnetic field is what the nuclei "sense" through the surrounding electronic environment. In other words, to achieve a given effective magnetic field
($H_{effective}$), the greater the induced local magnetic field (H_{local}) is, the greater the
applied magnetic field ($H_{applied}$) must be.

$$H_{effective} = H_{applied} - H_{local}$$

This means that the greater the electron density of the environment in which the
proton is located, the more the proton will be shielded from the magnetic field and
the greater the applied field will have to be to overcome the effects of the local
magnetic field. This type of shielding is known as **diamagnetic shielding.**

It will be easier to understand this concept if we look at an ^{1}H NMR spectrum.
The ^{1}H NMR spectrum of 1-bromo-2,2-dimethylpropane (Figure 13.6) has two signals because the compound has two different kinds of protons. (The third signal, at
$\delta = 0$, is a reference signal that will be discussed later.) The methylene protons are
in a less electron-dense environment than the methyl protons because the methylene protons are closer to the electron-withdrawing bromine. Because the methylene protons are in a less electron-dense environment, they are less shielded from

13.3
THE POSITION OF THE
^{1}H NMR SIGNALS

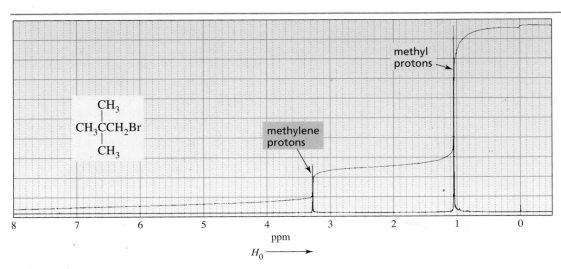

▲ **Figure 13.6**
^{1}H NMR spectrum of 1-bromo-2,2-dimethylpropane.

the applied magnetic field. The signal for these protons therefore occurs at a lower magnetic field than the signal for the more shielded methyl protons. *Notice that the right-hand side of an NMR spectrum is the high magnetic field (more shielded) side, where protons with electron-dense environments show a signal. The left-hand side is the low magnetic field (less shielded) side, where protons with less electron-dense environments show a signal.*

PROBLEM 6 ◆

Which set of protons in each of the following compounds is the least shielded?

Which set of protons in each compound is the most shielded?

                                                  O

                                                  ∥

a. $CH_3CH_2CH_2Cl$ **b.** $CH_3CH_2COCH_3$ **c.** $CH_3CHCHBr$

                                                        Br Br

---

Because it is difficult to measure the exact value of the applied magnetic field at which an absorption occurs, a small amount of a **reference compound** is added to the sample tube containing the compound whose NMR spectrum is to be taken. Measurements can then be made with respect to the signal of the reference compound. The most commonly used reference compound is tetramethylsilane (TMS).

$$CH_3-\underset{\underset{CH_3}{|}}{\overset{\overset{CH_3}{|}}{Si}}-CH_3$$

tetramethylsilane
TMS

The methyl protons of TMS are in a more electron-dense environment than are most protons in organic molecules because silicon is less electronegative than carbon (electronegativity of silicon = 1.8; electronegativity of carbon = 2.5). Consequently, the signal for the methyl protons of TMS is farther upfield than most other signals. (Upfield is toward the right in the NMR spectrum; downfield is toward the left.) The locations of the signals in an NMR spectrum are defined according to how far they are from the reference signal. (TMS is a highly volatile compound, so—after the NMR spectrum is taken—it can be removed from the sample by evaporation.)

The position in an NMR spectrum where a signal occurs is called the **chemical shift,** because the signal has been shifted upfield or downfield depending on the environment in which the proton resides: upfield for protons with electron-dense environments, and downfield for protons with less electron-dense environments. The most common scale for chemical shifts is the δ (delta) scale. The TMS signal is used to define the zero position on the δ scale. The ^{1}H NMR spectrum for 1-bromo-2,2-dimethylpropane shows that the chemical shift of the methyl protons is δ 1.05 and the chemical shift of the methylene protons is δ 3.28. *Notice that signals occurring at high magnetic field strengths (upfield) have small δ values, whereas those occurring at low magnetic field strengths (downfield) have large δ values.*

Electron withdrawal causes NMR signals to appear downfield (at greater δ values).

Compare the chemical shifts of the methylene protons immediately adjacent to the halogen in each of the alkyl halides below. The position of the signal depends on the electronegativity of the halogen. The signal for the methylene protons adjacent to fluorine (the most electronegative of the halogens) is the farthest downfield, while the signal for the methylene protons adjacent to iodine (the least electronegative of the halogens) is the farthest upfield.

$CH_3CH_2CH_2CH_2CH_2F$ $CH_3CH_2CH_2CH_2CH_2Cl$ $CH_3CH_2CH_2CH_2CH_2Br$ $CH_3CH_2CH_2CH_2CH_2I$

δ 4.50   δ 3.50   δ 3.40   δ 3.20

The chemical shift indicates how far the signal is from the TMS peak. It is determined by measuring the distance from the TMS peak (in hertz) and dividing by the operating frequency of the instrument (in megahertz). Because the units are Hz/MHz, a chemical shift has units of parts per million (ppm) of the operating frequency. Thus a chemical shift can be reported as occurring at δ 1.05 or at 1.05 ppm. Most chemical shifts fall in the range 0 to 12 ppm.

$$\delta = \text{chemical shift (ppm)} = \frac{\text{distance downfield from TMS (Hz)}}{\text{operating frequency of the spectrometer (MHz)}}$$

Notice that the distance from the TMS peak is not measured in magnetic field units (tesla) but in frequency units (hertz). A frequency scale can be used to denote the magnetic field strength because the two are proportional (Section 13.1). The advantage of the δ scale is that the chemical shift of a given nucleus is *independent of the operating frequency of the NMR spectrometer*. The chemical shift of the methyl protons of 1-bromo-2,2-dimethylpropane occurs at 63 Hz in a 60-MHz instrument and at 378 Hz in a 360-MHz instrument. Both, however, occur at δ 1.05 (63/60 = 1.05; 378/360 = 1.05).

PROBLEM 7◆

A signal has been reported to occur at 120 Hz downfield from TMS in an NMR spectrometer with a 60-MHz operating frequency.

a. What is its chemical shift?

b. What would its chemical shift be in an instrument operating at 100 MHz?

c. How many hertz downfield from TMS would the signal be in a 100-MHz spectrometer?

PROBLEM 8◆

a. If two signals differ by 1.5 ppm, how do they differ in hertz in a 60-MHz spectrometer?

b. If two signals differ by 1.5 ppm in a 60-MHz spectrometer, how do they differ in ppm in a 100-MHz spectrometer?

PROBLEM 9

Where would you expect to find the ^1H NMR signal for $(CH_3)_2Mg$ relative to the TMS peak? (*Hint:* see Table 11.3.)

The two signals in the ^1H NMR spectrum of 1-bromo-2,2-dimethylpropane are not the same size (Figure 13.6), because the area under each signal is proportional to the number of protons that cause the absorption. The area under the signal

13.4
INTEGRATION OF THE NMR SIGNALS

occurring farther upfield is larger, because the signal is caused by nine methyl protons while the smaller downfield signal results from two methylene protons.

From your calculus course, you probably remember that the precise area under a curve can be determined by integration. An ^{1}H NMR spectrometer is equipped with an integrator that integrates each signal, and the result of this process is displayed on the spectrum. It looks like a series of steps. The height of each step is proportional to the number of protons giving rise to the signal. By measuring the heights of the steps in Figure 13.6 (you can use a ruler or you can count spaces on the spectrum), you see that the ratio of the integrals is 1:4.5 (going from left to right across the spectrum). Thus you know that there are 4.5 times as many protons causing absorption at δ 1.05 as those causing absorption at δ 3.28.

The integration tells us the relative number of protons that give rise to each signal, not the absolute number. For example, integration could not distinguish between 1,1-dichloroethane and 1,2-dichloro-2-methylpropane because both compounds would show integral ratios of 1:3.

$$CH_3-\overset{\displaystyle |}{\underset{\displaystyle Cl}{CH}}-Cl \qquad\qquad CH_3-\overset{\displaystyle CH_3}{\underset{\displaystyle Cl}{\overset{|}{\underset{|}{C}}}}-CH_2Cl$$

1,1-dichloroethane
ratio of protons = 1 : 3

1,2-dichloro-2-methylpropane
ratio of protons = 1 : 3

PROBLEM 10

How could you distinguish among the NMR spectra of the following compounds?

$$CH_3-\overset{\displaystyle CH_3}{\underset{\displaystyle CH_3}{\overset{|}{\underset{|}{C}}}}-CH_2Br \qquad CH_3-\overset{\displaystyle CH_3}{\underset{\displaystyle Br}{\overset{|}{\underset{|}{C}}}}-CH_2Br \qquad CH_3-\overset{\displaystyle CH_2Br}{\underset{\displaystyle CH_2Br}{\overset{|}{\underset{|}{C}}}}-CH_2Br$$

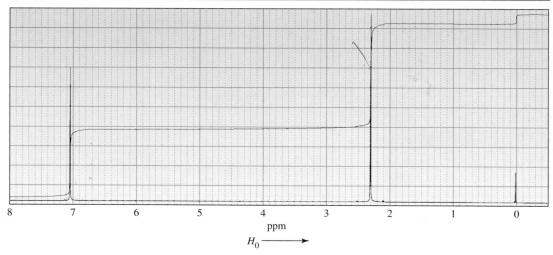

8   7   6   5   4   3   2   1   0

ppm

$H_0 \longrightarrow$

▲ **Figure 13.7**
^{1}H NMR spectrum for Problem 12.

PROBLEM 11 / SOLVED

a. Calculate the ratios of the different kinds of protons in a compound with an integral ratio of $4:18.4:6$ (going from left to right across the spectrum).

b. Give the structure of a compound that would give these relative integrals in the observed order.

SOLUTION

a. Divide by the smallest number:

$$\frac{4}{4} = 1 \qquad\qquad \frac{18.4}{4} = 4.6 \qquad\qquad \frac{6}{4} = 1.5$$

Multiply by a number that will cause all the numbers to be close to whole numbers.

$$1 \times 2 = 2 \qquad\qquad 4.6 \times 2 = 9.2 \qquad\qquad 1.5 \times 2 = 3$$

Round to whole numbers, because there can be only whole numbers of protons. (The measured integrals contain experimental error and thus the heights of the integration steps are approximate.) The ratio of the different kinds of protons is then $2:9:3$.

b. $$\text{CH}_3\text{CH}_2\overset{\displaystyle O}{\overset{\displaystyle \|}{\text{C}}}\overset{\displaystyle CH_3}{\underset{\displaystyle CH_3}{\text{OC}}}\text{CH}_3$$

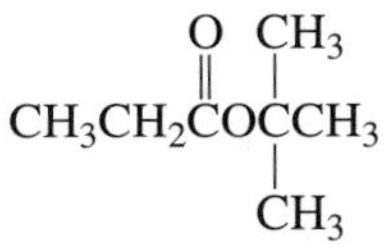

The methylene protons are the least shielded, because they are bonded to a carbon that is adjacent to a carbon bonded to two oxygens. The *tert*-butyl methyl groups are in the middle, because they are bonded to a carbon that is adjacent to a carbon bonded to one oxygen. The methyl group is the most shielded, because it is farthest away from the oxygens.

PROBLEM 12 ◆

The ^{1}H NMR spectrum shown in Figure 13.7 is the spectrum of one of the compounds shown below. Which compound is responsible for this spectrum?

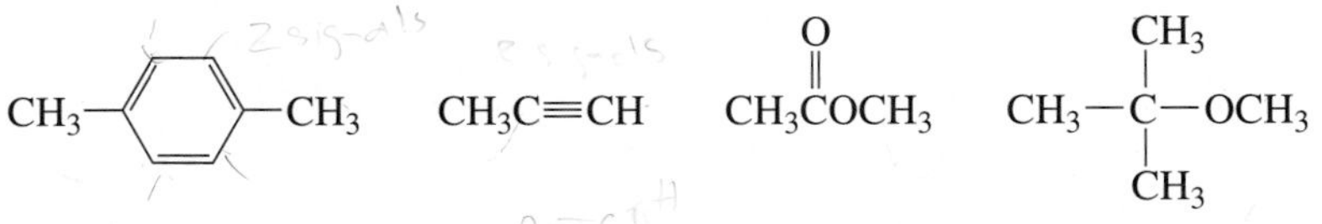

$$\text{CH}_3-\!\!\!\bigcirc\!\!\!-\text{CH}_3 \qquad \text{CH}_3\text{C}\!\equiv\!\text{CH} \qquad \text{CH}_3\overset{\displaystyle O}{\overset{\displaystyle \|}{\text{C}}}\text{OCH}_3 \qquad \text{CH}_3-\overset{\displaystyle CH_3}{\underset{\displaystyle CH_3}{\overset{\displaystyle |}{\underset{\displaystyle |}{\text{C}}}}}-\text{OCH}_3$$

Approximate values of chemical shifts for different kinds of protons are shown in Table 13.1. A more extensive compilation of chemical shifts is given in Appendix V.

Notice that the chemical shift of methyl protons is farther upfield (0.9 ppm) than the chemical shift of methylene protons (1.3 ppm) in the same environment, and that the chemical shift of methylene protons is farther upfield than the chemical shift of a methine proton (1.4 ppm). (When an sp^3 carbon is bonded to only one hydrogen, the hydrogen is called a **methine hydrogen**.) For example, the ^{1}H NMR spectrum of butanone shows three signals. The *a* proton signal of butanone appears farthest upfield because the protons are farthest from the electron-withdrawing

**13.5
CHARACTERISTIC
VALUES OF
CHEMICAL SHIFTS**

TABLE 13.1   Approximate Values of Chemical Shifts for ^{1}H NMR^a

Type of proton	Approximate chemical shift (δ)	Type of proton	Approximate chemical shift (δ)
$(CH_3)_4Si$	0	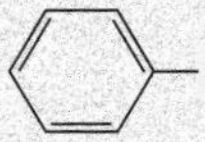 ⬡—H	6.5–8
—CH_3	0.9	$\overset{\displaystyle O}{\overset{\|}{-C}}-H$	9.7–10
—CH_2—	1.3	$I-\overset{\|}{\underset{\|}{C}}-H$	2–4
$-\overset{\|}{\underset{\|}{C}}H-$	1.4	$Br-\overset{\|}{\underset{\|}{C}}-H$	2.5–4
$-C{=}C-CH_3$	1.7	$Cl-\overset{\|}{\underset{\|}{C}}-H$	3–4
$-\overset{\displaystyle O}{\overset{\|}{C}}-CH_3$	2.1	$F-\overset{\|}{\underset{\|}{C}}-H$	4–4.5
⬡—CH_3	2.3		
—$C{\equiv}C-H$	2.4	RNH_2	variable, 1.5–4
$R-O-CH_3$	3.3	ROH	variable, 2–5
$R-C{=}CH_2$ with R below	4.7	$ArOH$	variable, 4–7
$R-C{=}C-H$ with R, R below	5.3	$-\overset{\displaystyle O}{\overset{\|}{C}}-OH$	variable, 10–12

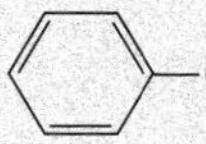

^{a}The values are approximate because they are affected by neighboring substituents.

carbonyl group. (In illustrations of NMR spectra, the set of protons responsible for the signal farthest upfield will be labeled *a*, the next set downfield will be labeled *b*, the next set *c*, etc.) The *b* and *c* protons are the same distance from the carbonyl group, but the signal for the *c* protons appears farther downfield because methylene protons appear farther downfield than methyl protons in the same environment.

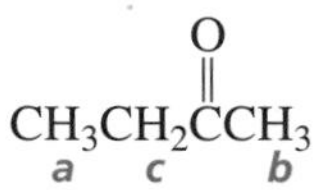

$$\underset{\substack{a \quad\quad c \quad\quad b}}{CH_3CH_2\overset{\displaystyle O}{\overset{\|}{C}}CH_3}$$

butanone

$$\underset{\substack{\quad\quad\quad a \\ \quad\quad\quad CH_3}}{\overset{\substack{b \quad\quad c \ a}}{CH_3OCHCH_3}}$$

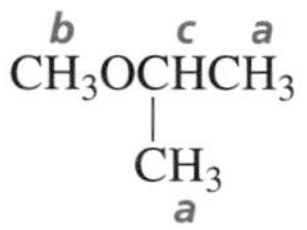

2-methoxypropane

The signal for the *a* protons of 2-methoxypropane appears farthest upfield in the ^{1}H NMR spectrum for this compound, because these protons are farthest from the electron-withdrawing oxygen. The *b* and *c* protons are the same distance from the oxygen, but the signal for the *c* proton appears farthest downfield, because, in the same environment, a methine proton signal appears farther downfield than a methyl proton signal.

PROBLEM 13◆

In each of the following pairs of compounds, which of the underlined protons has the greater chemical shift (is farther downfield)?

a. $CH_3CH_2C\underline{H}_2Cl$ or $CH_3CH_2C\underline{H}CH_3$
$\qquad\qquad\qquad\qquad\qquad\qquad\quad |$
$\qquad\qquad\qquad\qquad\qquad\qquad\quad Cl$

b. $CH_3C\underline{H}_2CHCH_3$ or $CH_3CH_2CHC\underline{H}_3$
$\qquad\qquad\quad |\qquad\qquad\qquad\qquad\qquad |$
$\qquad\qquad\quad Cl\qquad\qquad\qquad\qquad\quad Cl$

c. $CH_3CH_2C\underline{H}_2Cl$ or $CH_3CH_2C\underline{H}_2Br$

d. $CH_3CH_2CH{=}C\underline{H}_2$ or $CH_3C\underline{H}_2CH{=}CH_2$

e. $CH_3CH_2C\underline{H}$ or $CH_3CH_2COC\underline{H}_3$
$\qquad\quad\overset{\displaystyle O}{\overset{\displaystyle \|}{}}\qquad\qquad\qquad\overset{\displaystyle O}{\overset{\displaystyle \|}{}}$

f. $C\underline{H}_3CHOCH_3$ or $CH_3CHOC\underline{H}_3$
$\qquad\;\;\overset{\displaystyle CH_3}{\overset{\displaystyle |}{}}\qquad\qquad\quad\overset{\displaystyle CH_3}{\overset{\displaystyle |}{}}$

g. $CH_3C\underline{H}CCH_2CH_3$ or $CH_3CHCC\underline{H}_2CH_3$
$\qquad\quad\;\;\overset{}{\underset{\displaystyle CH_3}{\underset{\displaystyle |}{}}}\overset{\displaystyle O}{\overset{\displaystyle \|}{}}\qquad\qquad\quad\overset{}{\underset{\displaystyle CH_3}{\underset{\displaystyle |}{}}}\overset{\displaystyle O}{\overset{\displaystyle \|}{}}$

h. $CH_3C\underline{H}CHBr$ or $CH_3CHC\underline{H}Br$
$\qquad\quad\;\;|\;\;|\qquad\qquad\qquad\;\;|\;\;|$
$\qquad\quad\;\;Br\;Br\qquad\qquad\qquad Br\;Br$

PROBLEM 14◆

Without referring to Table 13.1, label the protons in the following compounds. The proton that gives the farthest upfield signal should be labeled *a*, the next *b*, etc.

a. $CH_3CH_2CH_2Cl$

b. $CH_3CHCHCH_3$
$\qquad\quad\overset{\displaystyle CH_3}{\overset{\displaystyle |}{}}$
$\qquad\qquad\;\;|$
$\qquad\qquad\;\;Cl$

c. $CH_3CH_2CH_2COCH_3$
$\qquad\qquad\qquad\overset{\displaystyle O}{\overset{\displaystyle \|}{}}$

d. $CH_3CHCHBr$
$\qquad\quad\;\;|\;\;|$
$\qquad\quad\;\;Br\;Br$

e. $CH_3CHCH_2OCH_3$
$\qquad\quad\;\;\overset{}{\underset{\displaystyle CH_3}{\underset{\displaystyle |}{}}}$

f. CH_3CH_2CH
$\qquad\qquad\quad\overset{\displaystyle O}{\overset{\displaystyle \|}{}}$

g. $CH_3CH_2CH_2CCH_3$
$\qquad\qquad\qquad\;\overset{\displaystyle O}{\overset{\displaystyle \|}{}}$

h. $ClCH_2CH_2CH_2Cl$

i. $CH_3CH_2CHCH_2CH_3$
$\qquad\qquad\quad\;\;|$
$\qquad\qquad\quad\;\;OCH_3$

j. $CH_3CH_2CH_2OCHCH_3$
$\qquad\qquad\qquad\qquad\quad|$
$\qquad\qquad\qquad\qquad\quad CH_3$

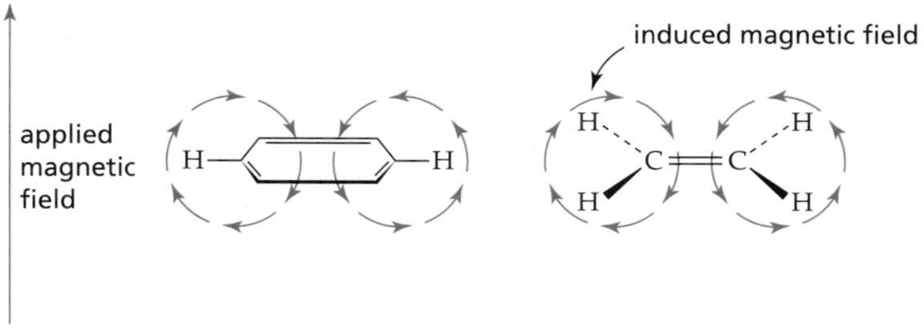

▲ **Figure 13.8**
The magnetic field induced by the π electrons of a benzene ring and the magnetic field induced by the π electrons of an alkene, in the area of space where the protons are located, are in the same direction as the applied magnetic field.

The chemical shifts of hydrogens bonded to sp^2 hybridized carbons do not occur where one would expect them based on electronegativity. The signals for the hydrogens bonded to sp^2 carbons appear farther downfield are (less shielded) than one would predict: 4.7 ppm for a hydrogen bonded to a terminal sp^2 carbon; 5.3 ppm for a hydrogen bonded to an internal sp^2 carbon; and 6.5 to 8.0 ppm for a hydrogen of a benzene ring.

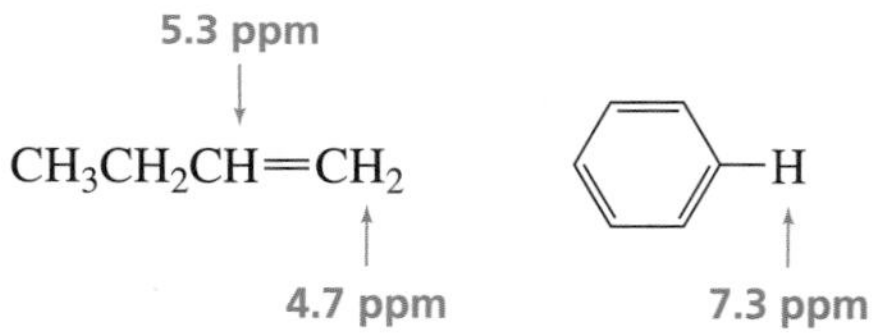

These unusual chemical shifts are due to **magnetic anisotropy.** *Anisotropic* is Greek for "different in different directions." Magnetic anisotropy means different magnetic properties at different points in space. Because π electrons are less tightly held by nuclei than σ electrons, they are more free to move in response to a magnetic field. When a magnetic field is applied to a compound with π electrons, the π electrons move in a circular path. This induced electron motion causes an induced magnetic field. How this induced magnetic field affects the chemical shift depends on the direction of the induced magnetic field relative to the direction of the applied magnetic field.

The magnetic field induced by the π electrons of a benzene ring, in the region where benzene's protons are located, is oriented in the same direction as the applied field (Figure 13.8). The magnetic field induced by the π electrons of an alkene, in the region where the protons bonded to the sp^2 carbons of the alkene are located, is also oriented in the same direction as the applied field. Thus, in both cases, a smaller applied field is required for resonance than would be required if the π electrons had not induced a magnetic field.

PROBLEM 15

[18]-Annulene shows two signals in the NMR, one at 9.25 ppm and the other very far upfield at −2.88 ppm. What hydrogens are responsible for each of the signals?

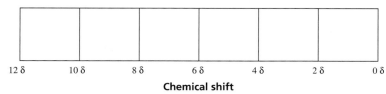

PROBLEM 16

Six boxes are shown that divide the NMR spectrum into six areas. Fill in each box with the kinds of protons that you would expect to show a signal in that region. (If you can remember the kinds of protons that are in each box, you will be able to recognize the kinds of protons that are in a compound from a quick look at its NMR spectrum.)

12 δ	10 δ	8 δ	6 δ	4 δ	2 δ 0 δ

Chemical shift

The signals in the ^{1}H NMR spectrum of 1,1-dichloroethane (Figure 13.9) look different from the signals in the ^{1}H NMR spectrum of 1-bromo-2,2-dimethylpropane (Figure 13.6). The signals in the latter are both **singlets,** but the signal for the methyl protons of 1,1-dichloroethane (the peak farthest upfield) is split into two peaks (a **doublet**) and the signal for the methine proton is split into four peaks (a **quartet**). (Magnifications of the doublet and quartet are shown in insets in Figure 13.9.)

13.6
SPLITTING OF
THE SIGNALS

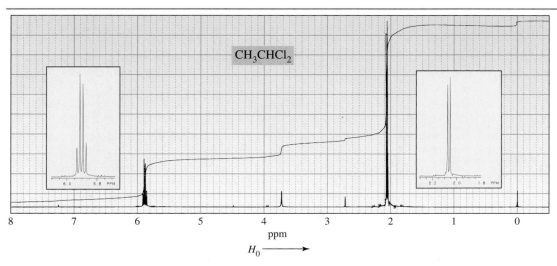

▲ **Figure 13.9**
^{1}H NMR spectrum of 1,1-dichloroethane.

The splitting of a signal is described by the **$N + 1$ rule,** where N is the number of equivalent protons bonded to an *adjacent* carbon. By equivalent protons we mean the protons bonded to an adjacent carbon that are equivalent to each other but not equivalent to the proton giving rise to the signal. For example, the carbon adjacent to the methyl group in 1,1-dichloroethane is bonded to one proton, so the signal for the methyl protons is split into a doublet ($N + 1 = 1 + 1 = 2$). The carbon adjacent to the carbon bonded to the methine proton is bonded to three equivalent protons, so the signal for the methine proton is split into a quartet ($N + 1 = 3 + 1 = 4$). Because the methine proton and the methyl protons split each other's signal, the two sets of protons are called **coupled protons.** *Be sure to notice that it is not the number of protons giving rise to a signal that determines the splitting of the signal; rather, it is the number of protons bonded to the adjacent carbon that determines the splitting.* For example, the signal for the *a* protons in the following compound will be split into three peaks (a **triplet**) because the adjacent carbon is bonded to two hydrogens. The signal for the *b* protons will appear as a quartet because the adjacent carbon is bonded to three hydrogens, and the signal for the *c* protons will be a singlet.

$$\underset{a \quad\quad b \quad\quad\quad c}{CH_3CH_2\overset{\overset{\textstyle O}{\textstyle \|}}{C}OCH_3}$$

The splitting of signals is caused by **spin–spin coupling.** In Section 13.1 we learned that, when a nucleus is placed in a magnetic field, the magnetic moment of the nucleus lines up either with the field or against it. The magnitude of the applied magnetic field required for the methyl protons of 1,1-dichloroethane to come into resonance is affected by the magnetic moment of the methine proton. If the magnetic moment of the methine proton is lined up with the applied magnetic field, the magnetic moment of the methine proton will be added to the applied magnetic field, so a slightly lower applied magnetic field will induce the methyl protons to produce a signal. In contrast, if the magnetic moment of the methine proton is lined up against the applied magnetic field, the magnetic moment of the methine proton will be subtracted from the applied magnetic field and a slightly higher applied magnetic field will be required to induce resonance (Figure 13.10). So the signal for the methyl protons is split into two

Figure 13.10 ▶
The signal for the methyl protons of 1,1-dichloroethane is split into a doublet by the methine proton.

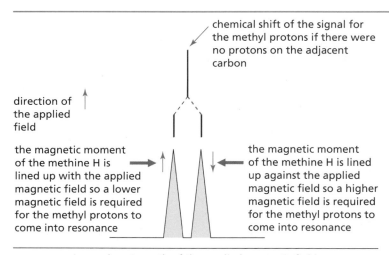

chemical shift of the signal for the methyl protons if there were no protons on the adjacent carbon

direction of the applied field

the magnetic moment of the methine H is lined up with the applied magnetic field so a lower magnetic field is required for the methyl protons to come into resonance

the magnetic moment of the methine H is lined up against the applied magnetic field so a higher magnetic field is required for the methyl protons to come into resonance

Increasing strength of the applied magnetic field

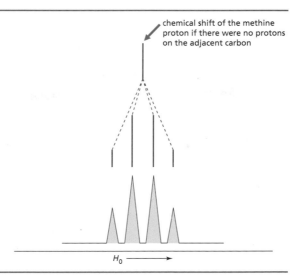

chemical shift of the methine
proton if there were no protons
on the adjacent carbon

H_0 ⟶

◀ **Figure 13.11**
The signal for the methine
proton of 1,1-dicloroethane is
split into a quartet by the
methyl protons.

peaks, one corresponding to the lower applied magnetic field and one corresponding to the higher applied magnetic field. Since about half of the methine protons are lined up with the applied magnetic field and half lined up against it, the two peaks of the *doublet* each have approximately the same area.

Similarly, the magnitude of the magnetic field required for the methine proton to give rise to a signal is affected by the magnetic moments of the three protons bonded to the adjacent carbon. The magnetic moments of the three methyl protons can all line up with the applied magnetic field; two can line up with the field and one against it; one can line up with it and two against it; or all three can line up against it. Since the magnetic field the proton senses is affected in four different ways, the signal for the methine proton is a *quartet* (Figure 13.11). The number of peaks in the signal is called the **multiplicity** of the signal (Table 13.2).

A quartet has relative peak areas of 1:3:3:1 (Table 13.2). There is only one way to align the magnetic moments of three protons so they are all with the field, and only one way to align them if they are all against the field. But there are three ways to align the magnetic moments of three protons so that two are lined up with the

TABLE 13.2 Multiplicity of the Signal and Relative Intensities of the Peaks in the Signal		
Number of equivalent protons causing splitting	Multiplicity of the signal	Relative peak intensities
0	singlet	1
1	doublet	1:1
2	triplet	1:2:1
3	quartet	1:3:3:1
4	quintet	1:4:6:4:1
5	sextet	1:5:10:10:5:1
6	septet	1:6:15:20:15:6:1

Figure 13.12 ▶
There is only one way to align three protons so that their magnetic moments are lined up either all with or all against the applied magnetic field. There are three ways to align three protons so that two have their magnetic moments lined up in one direction and the third in the opposite direction.

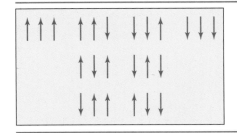

field and one is lined up against the field; the first two can be with it and the third against it; the first and third with it and the second against it; or the second and third with it and the first against it (Figure 13.12). There are also three ways to align the magnetic moments so that two are lined up against the field and one is lined up with the field.

The relative peak intensities shown in Table 13.2 can be determined by figuring out how many ways the given number of protons can be aligned. The relative intensities obey the mathematical mnemonic known as Pascal's triangle. (You may remember this from your math classes.) According to Pascal, each number in the triangle is the sum of the two numbers closest to it in the row above.

Blaise Pascal (1623–1662)

was born in France. At age 16 he published a book on geometry, and at 19 invented a calculating machine. He propounded the modern theory of probability, developed the principle underlying the hydraulic press, and showed that atmospheric pressure decreases as altitude increases. In 1644, he narrowly escaped death when his carriage horses bolted. That scare caused him to devote the rest of his life to meditation and religious writings.

PROBLEM 17

Using a diagram like the one in Figure 13.12, predict

a. the relative intensities of the peaks in a triplet.

b. the relative intensities of the peaks in a quintet.

PROBLEM 18

One of the spectra in Figure 13.13 is due to 1-chloropropane and the other to 1-iodopropane. Which is which?

Protons normally do not split each other if they are separated by more than three σ bonds. If, however, they are separated by four bonds and one of the bonds is a double or triple bond, small splitting is usually observable. Coupling through four or more bonds is called **long-range coupling.**

H_a and H_b will split each other because they are separated by 3 σ bonds

H_a and H_b will not split each other because they are separated by 4 σ bonds

H_a and H_b will split each other because they are separated by 4 bonds including one double bond

The ^{1}H NMR spectrum of bromomethane shows one singlet. The three methyl protons are chemically equivalent, and splitting is not observed between chemically equivalent protons. The four protons in 1,2-dichloroethane are chemically equivalent. Consequently, the ^{1}H NMR spectrum of 1,2-dichloroethane shows one singlet.

CH_3Br $ClCH_2CH_2Cl$
bromomethane   1,2-dichloroethane
both compounds have an NMR spectrum that shows one singlet because equivalent protons do not split each other

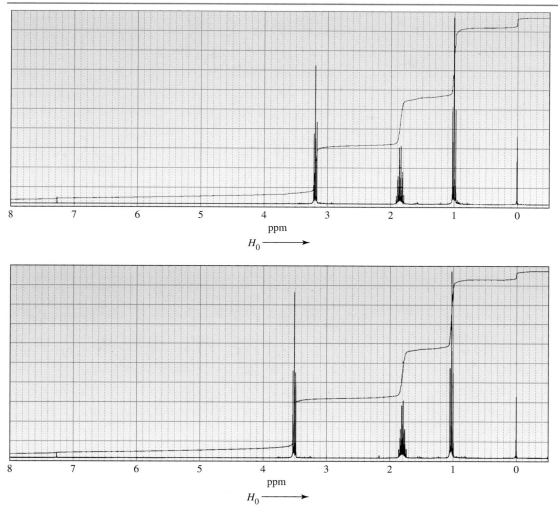

▲ **Figure 13.13**
^{1}H NMR spectra for Problem 18.

There are two signals in the ^{1}H NMR spectrum of 1,3-dibromopropane (Figure 13.14). The signal for the H_b protons is split into a triplet by the two hydrogens on the adjacent carbon. The H_a protons have two adjacent carbons that are bonded to protons. The protons on one adjacent carbon are equivalent to the protons on the other adjacent carbon. Because the two sets of protons are equivalent, the $N + 1$ rule is applied to both sets at the same time. So the signal for the H_a protons is split into a quintet ($4 + 1 = 5$). The fact that two methylene groups contribute to the H_b signal is indicated by the integration, which shows twice as many protons giving rise to the H_b signal as to the H_a signal.

The ^{1}H NMR spectrum of isopropyl butanoate shows five signals (Figure 13.15). The signal for the H_a protons is split into a triplet by the H_c protons. The signal for the H_b protons is split into a doublet by the H_e proton. The signal for the H_d protons is split into a triplet by the H_c protons, and the signal for the H_e proton is split into a septet by the H_b protons. The signal for the H_c protons is split by both the H_a and H_d protons. Because the H_a and H_d protons are not equivalent, the $N +$ 1 rule has to be applied separately to each set: the signal for the H_c protons will be

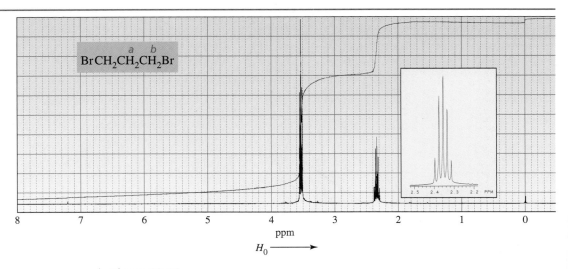

▲ **Figure 13.14**
^{1}H NMR spectrum of 1,3-dibromopropane.

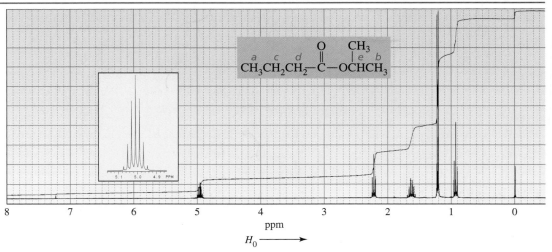

▲ **Figure 13.15**
^{1}H NMR spectrum of isopropyl butanoate.

split into a quartet by the H_a protons, and each of these four peaks will be split into a triplet by the H_d protons. Thus the signal for the H_c protons is a **multiplet** (a signal with more than seven peaks). The reason we do not see 12 peaks is that some of the peaks overlap.

PROBLEM 19◆

Indicate the number of signals and the multiplicity of each signal in the NMR spectrum of each of the following compounds.

a. $ICH_2CH_2CH_2Br$ **b.** $ClCH_2CH_2CH_2Cl$ **c.** ICH_2CH_2CHBr
$\overset{|}{Br}$

The ¹H NMR spectrum of 3-bromo-1-propene shows four signals (Figure 13.16). Although the H_b and H_c protons are bonded to the same carbon, they are not chemically equivalent, so each produces a separate signal. The signal for the H_a protons is split into a doublet by the H_d proton. Notice that the H_a protons are more shielded than the protons bonded to the sp^2 carbons (Appendix V). The signal for the H_d proton is a multiplet because it is split by the H_a, H_b, and H_c protons.

Because the H_b and H_c protons are not equivalent, the signal for the H_b proton can be split by the H_c proton and the signal for the H_c proton can be split by the H_b proton. This means that the signal for the H_b proton can be split into a doublet by the H_d proton and that each of the resulting peaks can be split into a doublet by the H_c proton. Therefore, the signal for the H_b proton should be what is called a **doublet of doublets.** The signal for the H_c proton should also be a doublet of doublets. However, the mutual splitting of the signals of two nonidentical protons bonded to the same carbon, known as **geminal coupling,** is often too small to be observed when the two protons are bonded to an sp^2 carbon. Therefore, the signals for the H_b and H_c protons each appears as a doublet rather than as a doublet of doublets.

There are five sets of chemically equivalent protons in ethylbenzene. We see the expected triplet for the H_a protons and the quartet for the H_b protons (Figure 13.17). (Notice that this is a characteristic pattern for an ethyl group.) We expect the signal for the H_c proton to be a doublet and the signal for the H_e proton to be a triplet. Because the H_c and H_e protons are not equivalent, they must be considered separately when determining the splitting of the signal for the H_d proton. We expect the signal for the H_d proton to be split into a doublet by the H_c proton and each peak of the doublet to be split into another doublet by the H_e proton, forming a doublet of doublets. However, we do not see three distinct signals for the H_c, H_d, and H_e protons in Figure 13.17; we see overlapping signals. Apparently the electronic effect of an ethyl substituent is not sufficiently different from that of a hydrogen to cause a difference in the environments of the H_c, H_d, and H_e protons that is large enough to allow them to appear as separate signals.

Notice that the benzene ring protons in Figure 13.17 resonate at 7.1 to 7.3 ppm. It is easy to recognize an aromatic ring in an ¹H NMR spectrum because aromatic

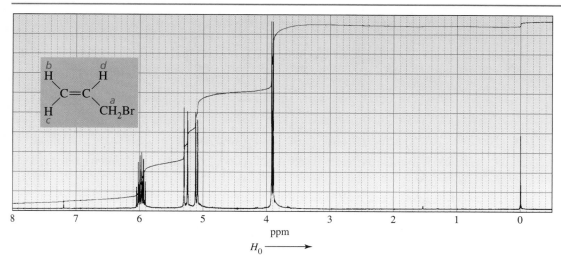

▲ **Figure 13.16**
¹H NMR spectrum of 3-bromo-1-propene.

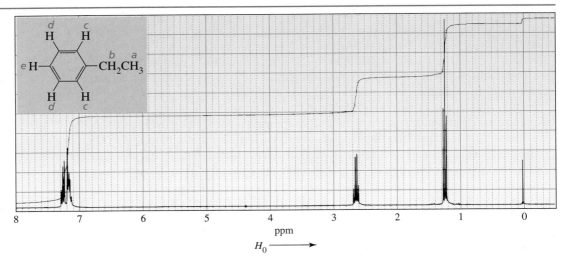

▲ **Figure 13.17**
^{1}H NMR spectrum of ethylbenzene.

protons give signals between 6.5 and 8.0 ppm, a region in which other kinds of protons usually do not resonate.

The H_a, H_b, and H_c protons of nitrobenzene (Figure 13.18) show three distinct signals, and the multiplicity of each signal is what we predicted for the signals for the benzene ring protons in ethylbenzene. The nitro group is sufficiently electron withdrawing to cause the H_a, H_b, and H_c protons to be in sufficiently different environments that their signals do not overlap.

Although the information in Table 13.2 suggests that splitting patterns should be symmetrical, this is often not the case. An interesting effect, known as **leaning,** is illustrated in Figure 13.19. The outside peaks of the quartet have different heights—as do the inside peaks. The outside peaks of the triplet also have different heights. If an arrow is drawn over the tops of the peaks with its head on the high end (Figure 13.19), it will point toward the signal given by the protons that cause the splitting. The closer together the chemical shifts of the mutually split sets of protons are, the more obvious the leaning becomes.

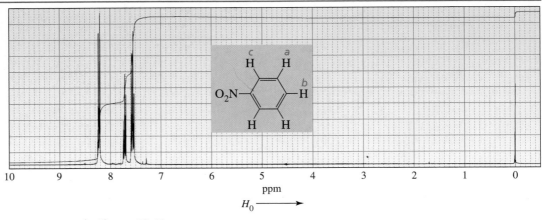

▲ **Figure 13.18**
^{1}H NMR spectrum of nitrobenzene.

PROBLEM 20

Identify each compound from its molecular formula and its ^1H NMR spectrum.

a. C_9H_{12}

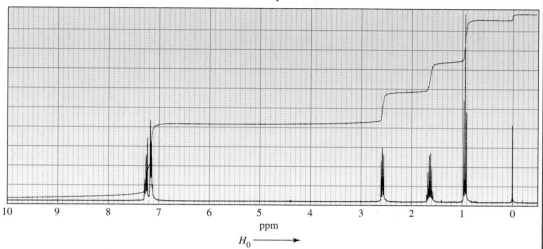

b. $C_5H_{10}O$

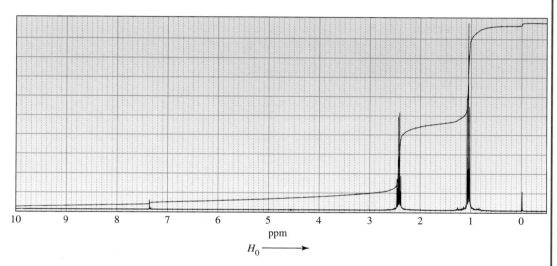

c. $C_9H_{10}O_2$

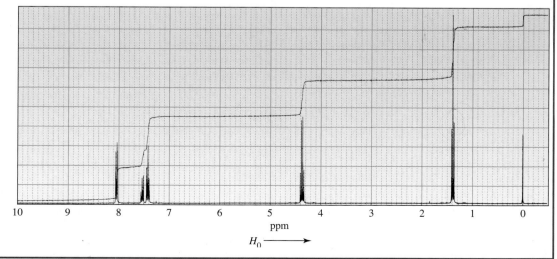

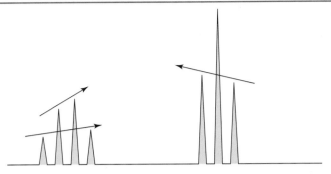

▲ **Figure 13.19**
"Leaning" in the NMR spectrum of ethylbenzene. The arrow drawn over the tops of
the outside peaks of the triplet for the methyl protons points toward the signal for the
methylene protons because these are the protons responsible for the methyl protons
being split into a triplet. The arrows drawn over the tops of the outside and inside
peaks in the quartet for the methylene protons point toward the signal for the methyl
protons because these are the protons responsible for the methylene protons being
split into a quartet.

PROBLEM 21

How would the ^{1}H NMR spectra for the four compounds with molecular formula
$C_3H_6Br_2$ differ?

PROBLEM 22

Predict the splitting pattern that you would expect for the signals given by each of the
compounds in Problem 3.

PROBLEM 23

How could ^{1}H NMR be used to prove that the addition of HBr to propene follows
Markovnikov's rule?

PROBLEM 24 ◆

Identify the following compounds. (Relative integrals are stated going from left to
right across the spectrum.)

a. The ^{1}H NMR spectrum of a compound with molecular formula $C_4H_{10}O_2$ has two
singlets with an area ratio of $2:3$.

b. The ^{1}H NMR spectrum of a compound with molecular formula $C_6H_{10}O_2$ has two
singlets with an area ratio of $2:3$.

c. The ^{1}H NMR spectrum of a compound with molecular formula $C_{10}H_{14}$ has two
singlets with an area ratio of $5:9$.

PROBLEM 25

Describe the NMR spectrum that you would expect for each of the following compounds, using relative chemical shifts rather than absolute chemical shifts.

a. $BrCH_2CH_2CH_2Br$

c. $CH_3CH_2OCH_2CH_3$

b. $CH_3OCH_2CH_2CH_2Br$

d. $CH_3CH_2OCH_2Cl$

e. O=⟨cyclohexanedione⟩=O

f. $CH_3\overset{\overset{\displaystyle CH_3}{|}}{C}CH_2CH_3$
$\qquad\underset{|}{Br}$

g. $CH_3\overset{\overset{\displaystyle O}{||}}{C}CH_2\overset{\overset{\displaystyle O}{||}}{C}OCH_3$

h. $CH_3\overset{}{C}HCHCl_2$
$\qquad\underset{|}{Cl}$

i. ⟨oxetane ring structure⟩

j. $CH_3\overset{\overset{\displaystyle CH_3}{|}}{C}HCH_2\overset{\overset{\displaystyle O}{||}}{C}H$

k. $\underset{H}{\overset{H}{\diagdown}}C{=}C\underset{\diagup Cl}{\overset{Cl}{\diagdown}}$

l. $\underset{H}{\overset{Cl}{\diagdown}}C{=}C\underset{\diagdown Cl}{\overset{H}{\diagup}}$

m. $\underset{H}{\overset{H}{\diagdown}}C{=}C\underset{\diagdown Cl}{\overset{H}{\diagup}}$

n. $CH_3OCH_2CH_2CH_2OCH_3$

13.7 COUPLING CONSTANTS

The distance (in hertz) between two adjacent peaks of a split NMR signal is called the **coupling constant** (denoted by J). The coupling constant for H_a being split by H_b is denoted by J_{ab}. We have seen that because the methine proton and the methyl protons of 1,1-dichloroethane split each other's signals, they are called *coupled protons*. The signals of coupled protons have the same coupling constant; in other words, $J_{ab} = J_{ba}$ (Figure 13.20). Coupling constants are useful in analyzing complex NMR spectra because protons on adjacent carbons can be identified by the identical coupling constants of their signals.

The magnitude of a coupling constant is independent of the operating frequency of the spectrometer: the same coupling constant is obtained from a 60-MHz instrument as from a 100-MHz instrument. The magnitude of a coupling constant depends on the number and type of bonds that connect the coupled protons. Characteristic coupling constants are shown in Table 13.3; they range from 0 to 15 Hz. Because coupling occurs through bonds, J is largest for protons on the same carbon and falls off rapidly as the number of bonds between the coupled protons increases.

◀ **Figure 13.20**
The H_a and H_b protons of 1,1-dichloroethane are coupled protons, so their signals have the same coupling constant, $J_{ab} = J_{ba}$.

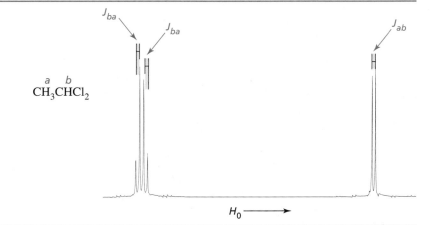

TABLE 13.3 Approximate Values of Coupling Constants		
Approximate value of J_{ab} (Hz)	**Approximate value of J_{ab} (Hz)**	
$\begin{array}{c} H_a \\ \mid \\ -C- \\ \mid \\ H_b \end{array}$ 12	$\begin{array}{c} H_a \\ \diagdown \\ C=C \\ \diagdown \\ H_b \end{array}$	2
$\begin{array}{c} H_a \;\; H_b \\ \mid \;\; \mid \\ -C-C- \\ \mid \;\; \mid \end{array}$ 7	$\begin{array}{c} H_a \\ \diagdown \\ C=C \\ \diagdown \\ H_b \end{array}$	15 (trans)
$\begin{array}{c} H_a \;\;\;\; H_b \\ \mid \;\;\;\; \mid \\ -C-C-C- \\ \mid \;\; \mid \;\; \mid \end{array}$ 0	$\begin{array}{c} H_a \;\;\;\; H_b \\ \diagdown \;\; \diagup \\ C=C \\ \diagup \;\; \diagdown \end{array}$	10 (cis)
	$\begin{array}{c} H_a \\ \diagdown \\ C=C \\ \diagup \;\; C \\ \;\;\;\; H_b \end{array}$	1

The coupling constant for two nonequivalent hydrogens on the same sp^2 carbon is quite small, but it is large for hydrogens bonded to adjacent sp^2 carbons. Apparently the interaction between the hydrogens is strongly affected by the intervening π electrons. In compounds with π bonds, we again see that J decreases as the number of σ bonds between the coupled protons increases.

The splitting pattern obtained when a signal is split by more than one set of protons can best be understood by using a splitting diagram. In a **splitting diagram**, the NMR peaks are shown as vertical lines and the effect of each of the splittings is shown one at a time. For example, a splitting diagram is shown in Figure 13.21 for splitting of the signal for the H_c proton of 1,1,2-trichloro-3-methylbutane into a doublet of doublets by the H_b and H_d protons.

Notice the difference between a quartet and a doublet of doublets. Both have four peaks. A quartet results from splitting by three equivalent adjacent protons; it has relative peak intensities of $1:3:3:1$, and the individual peaks are equally spaced. A doublet of doublets results from splitting by two nonequivalent adjacent protons; it has relative peak intensities of $1:1:1:1$, and the individual peaks are not necessarily equally spaced (Figure 13.21).

a quartet
relative intensities: $1:3:3:1$

a doublet of doublets
relative intensities: $1:1:1:1$

The signal for the H_b protons of propyl bromide is split into a quartet by the H_a protons, and each of the resulting four peaks is split into a triplet by the H_c protons (Figure 13.22). How many of the 12 peaks are actually seen depends on the relative magnitudes of the two coupling constants, J_{ba} and J_{bc}. For example, Figure 13.22 shows that 12 peaks are visible when J_{ba} is much greater than J_{bc}, nine peaks are visible when $J_{ba} = 2\,J_{bc}$, and only six peaks are visible when $J_{ba} = J_{bc}$. When peaks overlap, their intensities add together.

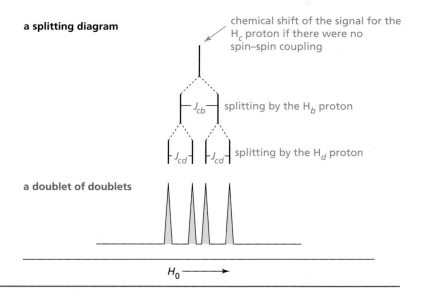

◀ **Figure 13.21**
A splitting diagram for a doublet of doublets.

Cl Cl
   a   b | |
CH₃CHCHCHCl
       | c   d
      CH₃

1,1,2-trichloro-3-methylbutane

a splitting diagram

chemical shift of the signal for the H$_c$ proton if there were no spin–spin coupling

J_{cb} splitting by the H$_b$ proton

J_{cd} J_{cd} splitting by the H$_d$ proton

a doublet of doublets

H_0 ⟶

We would expect the signal for the H$_a$ protons of 1-chloro-3-iodopropane to be split into nine peaks, because the signal would be split into a triplet by the H$_b$ protons and each of the resulting peaks into a triplet by the H$_c$ protons. The signal, however, is a quintet (Figure 13.23).

CH₃CH₂CH₂Br
  a   b   c
propyl bromide

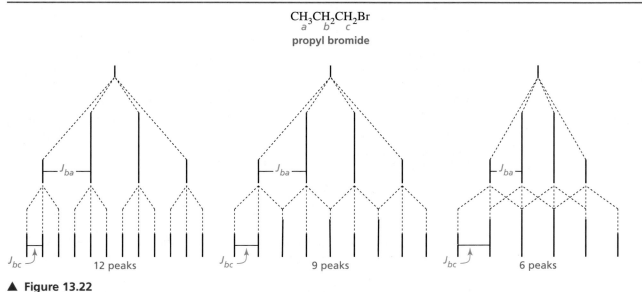

J_{ba} J_{ba} J_{ba}

J_{bc} 12 peaks J_{bc} 9 peaks J_{bc} 6 peaks

▲ **Figure 13.22**
A splitting diagram for a quartet of triplets: the number of peaks actually observed when a signal is split by two sets of protons depends on the relative magnitudes of the two coupling constants.

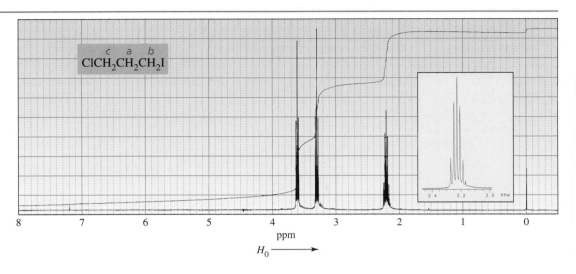

▲ **Figure 13.23**
^{1}H NMR spectrum of 1-chloro-3-iodopropane.

Finding that the signal for the H_a protons of 1-chloro-3-iodopropane is a quintet indicates that J_{ab} and J_{ac} have about the same value. The splitting diagram shows that a quintet results if $J_{ab} = J_{ac}$.

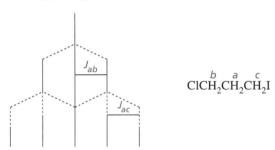

From this we can conclude that, when two different sets of protons split a signal, the multiplicity of the signal should be determined using the $N + 1$ rule separately for each set of hydrogens when the coupling constants for the two sets are different. When, however, the coupling constants are similar, the multiplicity of a signal can be determined by treating both sets of hydrogens as if they were equivalent, applying the $N + 1$ rule to both sets as a whole.

PROBLEM 26 / SOLVED

The two hydrogens of a methylene group adjacent to a chirality center are not equivalent hydrogens because they are in different environments owing to the proximity of the chirality center. Therefore, the $N + 1$ rule must be applied to them separately when determining the multiplicity of the signal for the adjacent methyl hydrogens. We expect the signal to be a doublet of doublets. The signal, however, is a triplet. Explain why it is a triplet rather than a doublet of doublets, using a splitting diagram to explain your answer.

$$CH_3\overset{*}{C}HCH_2CH_3$$
$$|$$
$$Br$$

nonequivalent hydrogens

SOLUTION   The observation of a triplet means that the $N + 1$ rule did not have to be applied to the two protons separately but could have been applied to the two protons as a set. This means that the coupling constant for splitting of the methyl signal by one of the methylene hydrogens is about the same as the coupling constant for splitting by the other methylene hydrogen.

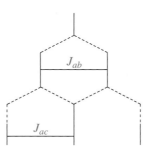

Coupling constants can be used to distinguish between the ^{1}H NMR spectra of cis and trans isomers because the coupling constant of *trans*-protons is significantly greater than the coupling constant of *cis*-protons (Figure 13.24). This is because the coupling constant is dependent on the dihedral angle between the two C—H bonds in the H—C—C—H unit; the coupling constant is greatest when the angle between the two C—H bonds is 180° (trans) and smaller when it is 0° (cis). Notice also the difference in the coupling constants for the H_b and H_c protons in 3-bromo-1-propene (Figure 13.16).

The dependence of the coupling constant on the angle between the two C—H bonds is called the Karplus relationship. **Martin Karplus** *was born in 1930. He received a B.A. from Harvard University and a Ph.D. from The California Institute of Technology. He is currently a professor of chemistry at Harvard University.*

P R O B L E M   2 7

Draw a splitting diagram for H_b where

a. $J_{ba} = 12$ Hz and $J_{bc} = 6$ Hz.   **b.** $J_{ba} = 12$ Hz and $J_{bc} = 12$ Hz.

$$-\overset{|}{\underset{H_a}{C}}-\overset{|}{\underset{H_b}{C}}-\overset{|}{\underset{H_c}{C}}-$$

H_b H_a H_b H_a

$J_{ba} = 14$ Hz $J_{ab} = 14$ Hz $J_{ba} = 9$ Hz $J_{ab} = 9$ Hz

$$\underset{Cl}{\overset{b}{\underset{}{H}}}\diagdown\underset{}{C}=\underset{}{C}\overset{\overset{c}{COOH}}{\diagup_{\underset{a}{H}}}$$

trans-**3-chloropropenoic acid**

$$\underset{Cl}{\overset{b}{\underset{}{H}}}\diagdown\underset{}{C}=\underset{}{C}\overset{\overset{a}{H}}{\diagup_{\underset{c}{COOH}}}$$

cis-**3-chloropropenoic acid**

◀ **Figure 13.24**
The doublets observed for the H_a and H_b protons in the ^{1}H NMR spectra of *trans*-3-chloropropenoic acid and *cis*-3-chloropropenoic acid. The coupling constant for trans protons (14 Hz) is greater than the coupling constant for cis protons (9 Hz).

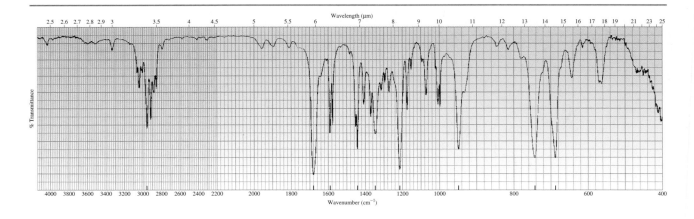

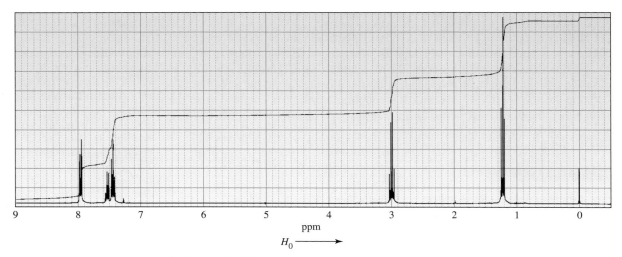

▲ **Figure 13.25**
Spectra for the Problem-Solving Strategy.

Let's now summarize the kind of information that can be obtained from an ^1H NMR spectrum.

1. The number of signals indicates the number of different kinds of protons that are in the compound.
2. The position of a signal indicates the electron density around the proton(s) responsible for the signal.
3. The multiplicity of the signal tells the number of protons bonded to adjacent carbons.
4. The integration of the signal tells the relative number of protons responsible for the signal.

PROBLEM-SOLVING STRATEGY

Identify the compound that gives the IR and ^1H NMR spectra in Figure 13.25.

One way to approach this kind of problem is to identify whatever structural features you can from the IR spectrum and then use the information from the ^1H NMR spectrum to expand on that knowledge. From the IR spectrum we learn that the compound is a ketone (a carbonyl group at $\sim1700\ cm^{-1}$ and the absence of bands that would indicate an aldehyde, carboxylic acid, ester, or amide), contains a benzene ring (>3000 cm^{-1}, ~1600 cm^{-1}, and ~1580 cm^{-1}), and has hydrogens bonded to sp^3 carbons (<3000 cm^{-1}). In the NMR spectrum, the triplet at ~1.2 ppm and quartet at ~3.0 ppm indicate the presence of an ethyl group that is attached to an electron-withdrawing group. From this information we can conclude that the compound is the following ketone. The integration ratio $(5:2:3)$ confirms this answer.

$$\text{C}_6\text{H}_5-\overset{\displaystyle O}{\overset{\|}{\text{C}}}\text{CH}_2\text{CH}_3$$

Now continue on to do Problem 28.

PROBLEM 28◆

Identify the compound with molecular formula $C_8H_{10}O$ that gives the IR and ^{1}H NMR spectra in Figure 13.26.

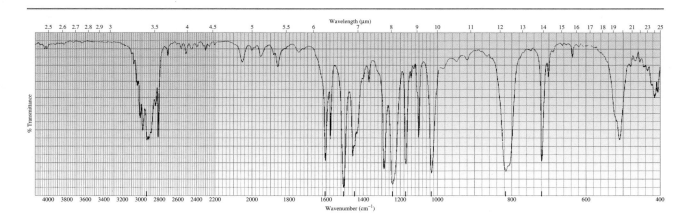

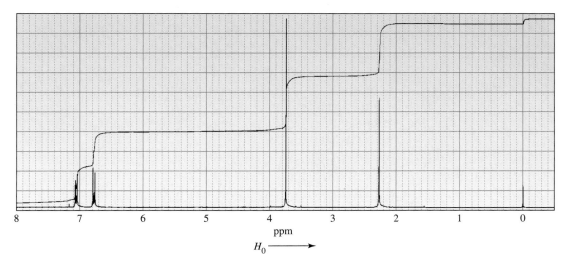

▲ **Figure 13.26**
Spectra for Problem 28.

13.8 TIME DEPENDENCE OF NMR SPECTROSCOPY

Although cyclohexane has two kinds of protons (axial protons and equatorial protons), the ^{1}H NMR spectrum for cyclohexane shows only one unsplit signal. There is only one signal because the interconversion of the chair conformers of cyclohexane occurs too rapidly at room temperature to be detected by an NMR spectrometer. Because axial protons in one chair are equatorial protons in the other chair, all the protons in cyclohexane have the same average environment on the NMR time scale. So the NMR spectrum shows one singlet.

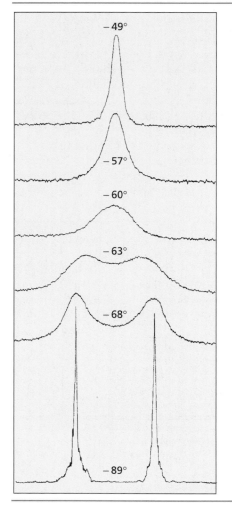

The rate of chair–chair interconversion is temperature-dependent: the lower the temperature, the slower the rate. Cyclohexane-d_{11} has 11 deuteriums, which means that it has only one hydrogen. ^{1}H NMR spectra of cyclohexane-d_{11} taken at various temperatures are shown in Figure 13.27. Cyclohexane with only one hydrogen was

Figure 13.27 ▶
^{1}H NMR spectra of cyclohexane-d_{11} at various temperatures. (Reproduced with permission from F. A. Bovey, *Nuclear Magnetic Resonance Spectroscopy*, Academic Press, New York, 1988.)

used for this experiment to prevent spin–spin coupling, which would have complicated the spectrum. Deuterium signals are not detectable, and splitting by a deuterium on the same or on an adjacent carbon is not normally detectable at the
operating frequency of an ^{1}H NMR spectrometer.

At room temperature, the ^{1}H NMR spectrum of cyclohexane-d_{11} shows one
signal, which is an average for the axial proton of one chair and the equatorial
proton of the other chair. As the temperature decreases, the signal becomes broader
and eventually splits into two signals, which are equidistant from the original
signal. At $-89\ ^\circ$C, two sharp singlets are observed, because, at this temperature,
the rate of chair–chair interconversion has decreased sufficiently to allow the two
kinds of protons to be individually detected on the NMR time scale.

The chemical shift of a proton bonded to an oxygen or a nitrogen depends on the
degree to which the proton is hydrogen bonded. For example, the chemical shift of
the OH proton of an alcohol ranges from 2.0 to 5.5 ppm, the chemical shift of the
OH proton of a carboxylic acid from 10 to 12 ppm, and the chemical shift of a
proton bonded to a nitrogen from 1.5 to 4.0 ppm; the greater the extent of hydrogen
bonding, the greater the chemical shift (the more downfield the signal).

The ^{1}H NMR spectrum of pure, dry ethanol is shown in Figure 13.28(a), and the
^{1}H NMR spectrum of ethanol with a trace amount of acid (or base) is shown in
Figure 13.28(b). The spectrum shown in Figure 13.28(a) is what we would predict
from what we have learned so far. The signal for the proton bonded to oxygen is farthest downfield and is split into a triplet by the neighboring methylene protons; the
signal for the methylene protons is split into a multiplet by the combined effects of
the methyl protons and the OH proton. The spectrum shown in Figure 13.28(b) is
the type of spectrum that is most often obtained for alcohols. The signal for the
proton bonded to oxygen is not split, and this proton does not split the signal for
adjacent protons. So the signal for the OH proton is a singlet and the signal for the
methylene protons is a quartet because it is split only by the methyl protons.

The difference in the two spectra arises from the fact that protons bonded to
oxygen or nitrogen undergo **chemical exchange,** which means that they are transferred from one molecule to another. In a sample of dry alcohol, the rate of chemical exchange is slower than can be detected on the NMR time scale. This causes
the spectrum to look identical to one that would be obtained if chemical exchange
did not occur. If the alcohol is contaminated with acid, base, or water, chemical
exchange becomes very rapid. When chemical exchange is rapid, the spectrum
records only an average of all possible environments. So a rapidly exchanging
proton is recorded as a singlet and its effect on adjacent protons is averaged, which
means that it does not cause splitting. This is called **spin-decoupling:** The protons
undergoing rapid chemical exchange are decoupled from the rest of the molecule.

13.9
PROTONS BONDED TO OXYGEN AND NITROGEN

mechanism for acid-catalyzed proton exchange

The signal for an OH or an NH proton is often easy to spot in an ^{1}H NMR spectrum because it frequently is somewhat broader than other signals (see the signal at
δ 4.9 in Figure 13.29b). The broadening occurs because the rate of proton exchange
is generally not slow enough to result in a cleanly split signal, as in Figure 13.28(a),
or fast enough for a cleanly averaged signal as in Figure 13.28(b).

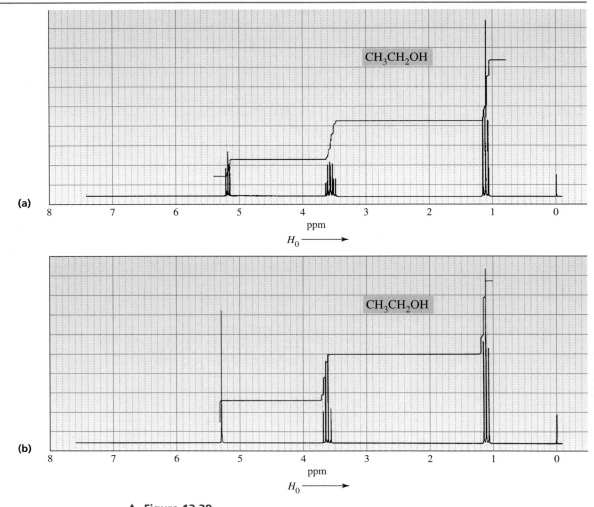

(a)

(b)

▲ **Figure 13.28**
(a) ^{1}H NMR spectrum of pure ethanol. (b) ^{1}H NMR spectrum of ethanol containing a trace amount of acid.

PROBLEM 29◆

Which would show the signal for the OH proton at a greater chemical shift, the ^{1}H NMR spectrum of pure ethanol or the ^{1}H NMR spectrum of ethanol dissolved in CCl_4?

PROBLEM 30

Propose a mechanism for base-catalyzed proton exchange.

13.10 USE OF DEUTERIUM IN ^{1}H NMR SPECTROSCOPY

Because deuterium signals are not seen in ^{1}H NMR, substituting a deuterium for a hydrogen is a technique used to identify signals and to simplify ^{1}H NMR spectra.

If, after obtaining an ^{1}H NMR spectrum of an alcohol, a few drops of D_2O are added to the sample and the spectrum is taken again, the OH signal in the first spectrum can be identified. It will be the signal that disappears or becomes less intense in the second spectrum because of the proton exchange process just discussed.

If the ^{1}H NMR spectrum of $CH_3CH_2OCH_3$ were compared with the ^{1}H NMR spectrum of $CH_3CD_2OCH_3$, the peak farthest downfield in the first spectrum would be absent in the second spectrum, indicating that this signal is attributable to the methylene group. This can be a helpful technique for analysis of complicated ^{1}H NMR spectra.

The sample used to obtain an ^{1}H NMR spectrum is made by dissolving the compound in an appropriate solvent (Section 13.1). Solvents with protons cannot be used, because the signal for the protons in the solvent could cover up the signals for all other protons. Therefore, deuterated solvents such as $CDCl_3$ and D_2O are commonly used in NMR spectroscopy.

The ^{1}H NMR spectrum of 2-*sec*-butylphenol taken on a 60-MHz NMR spectrometer is shown in Figure 13.29(a), and the ^{1}H NMR spectrum of the same compound taken on a 300-MHz instrument is shown in Figure 13.29(b). Why is the resolution of the second spectrum so much better?

13.11
HIGH-RESOLUTION ^{1}H NMR SPECTROSCOPY

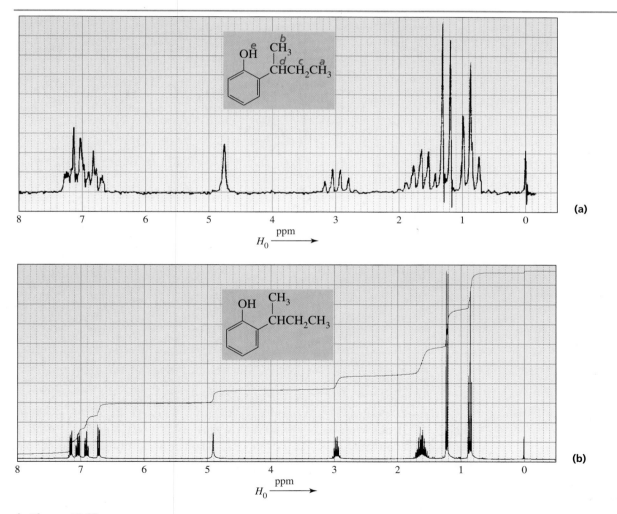

▲ **Figure 13.29**
(a) 60-MHz ^{1}H NMR spectrum of 2-*sec*-butylphenol. (b) 300-MHz ^{1}H NMR spectrum of 2-*sec*-butylphenol.

Figure 13.30 ▶
The splitting pattern of an
ethyl group as a function of
the $\Delta v/J$ ratio. (Reproduced
with permission of J. Houser.)

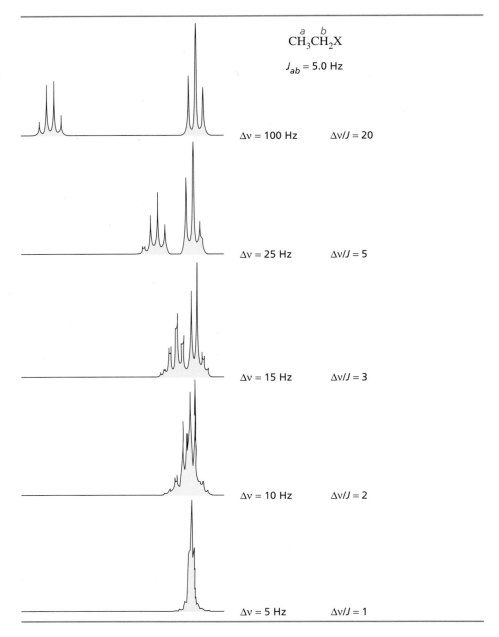

To observe separate signals with "clean" splitting patterns, the difference in the chemical shifts of two adjacent protons (Δv) must be at least ten times the value of the coupling constant (J). As $\Delta v/J$ decreases, the outer peaks of the signals become less intense and the inner peaks become more intense (Figure 13.30).

The difference in the chemical shifts of the H_a and H_c protons of 2-*sec*-butylphenol is 0.8 ppm, which corresponds to 48 Hz in a 60-MHz instrument and 240 Hz in a 300-MHz instrument. Coupling constants are independent of the operating frequency of the NMR spectrometer, so J_{ac} is 7 Hz whether the spectrum is taken on a 60-MHz or a 300-MHz instrument. Only in the latter case is the difference in chemical shift (in hertz) more than ten times the value of the coupling constant (in hertz). ^{1}H NMR spectra taken in ^{1}H NMR spectrometers with high operating frequency are called **high-resolution ^{1}H NMR spectra.**

Since high-resolution spectra are easier to interpret, one might expect that all ^{1}H NMR spectra would be run in spectrometers with high operating frequencies. The reason they are not is one of economics. The higher the operating frequency of the ^{1}H NMR spectrometer, the more expensive it is and the more difficult it is to maintain in good operating condition. Typically, then, only spectra that are difficult to interpret are run in spectrometers with high operating frequencies.

PROBLEM 31

You have seen that splitting of signals is not observed between chemically equivalent protons (Section 13.6). Yet, according to Table 13.3, protons on the same carbon have coupling constants of 12 Hz. If they are able to split each other, why is the splitting not observed? (*Hint:* Calculate $\Delta \nu / J$).

The number of signals in a ^{13}C NMR spectrum tells you how many different kinds of carbons a compound has—just as the number of signals in an ^{1}H NMR spectrum tells you how many different kinds of hydrogens a compound has. The principles behind ^{1}H NMR and ^{13}C NMR spectroscopy are essentially the same. There are, however, some differences that make ^{13}C NMR easier to interpret.

The development of ^{13}C NMR spectroscopy as a routine analytical instrument was not possible until small, powerful computers were developed that could carry out a Fourier transform (Section 13.1). ^{13}C NMR requires Fourier transform techniques because the signals obtained from a single ^{13}C NMR spectrum are too weak to be distinguished from background electronic noise. However, FT–^{13}C NMR spectra can be obtained rapidly, so a large number of spectra can be recorded and stored in a computer and the signals can be averaged. Since noise is random, it averages close to zero, and the ^{13}C signals add up and stand out when hundreds of spectra are averaged. Without Fourier transform, it could take days to record the number of spectra required for a ^{13}C NMR spectrum.

The weakness of the individual ^{13}C signals is due to the fact that the isotope of carbon (^{13}C) that gives rise to ^{13}C NMR signals constitutes only 1.11% of carbon atoms (Section 12.3). (The most abundant isotope of carbon, ^{12}C, has no nuclear spin and therefore cannot produce an NMR signal.) This means that the intensities of the signals in ^{13}C NMR compared with those in ^{1}H NMR are reduced by a factor of approximately 100. Because the gyromagnetic ratio of ^{13}C is only one-fourth that of ^{1}H, the signal intensities in ^{13}C NMR are reduced even further.

One advantage of ^{13}C NMR spectroscopy is that the chemical shifts range over about 220 ppm, compared with about 12 ppm for ^{1}H NMR (Table 13.1). This means that signals are less likely to overlap. The ^{13}C NMR chemical shifts of different kinds of carbons are shown in Table 13.4. TMS is also used as the reference compound in ^{13}C NMR. A disadvantage of ^{13}C NMR spectroscopy is that, unless special techniques are used, the area under a ^{13}C NMR signal is not proportional to the number of atoms giving rise to the signal. This means that ^{13}C NMR spectra cannot be routinely analyzed by integration.

The ^{13}C NMR spectrum of 2-butanol is shown in Figure 13.31. 2-Butanol has carbons in four different chemical environments, so there are four signals in the spectrum. The relative positions of the signals depend on the same factors that determine the relative positions of the proton signals in ^{1}H NMR. Carbons in

13.12
^{13}C NMR SPECTROSCOPY

TABLE 13.4 Approximate Values of Chemical Shifts for ^{13}C NMR

Type of carbon	Approximate chemical shift (δ)	Type of carbon	Approximate chemical shift (δ)
$(CH_3)_4Si$	0	C—I	0–40
—CH_3	8–35	C—Br	25–65
—CH_2—	15–50	C—Cl	35–80
—CH— C	20–60	C—O	50–80
C—C—C C	30–40	C=O	170–210
$\equiv$C	65–85	(aromatic)C	110–170
=C	100–150		

electron-dense environments produce upfield signals, and carbons close to electron-withdrawing groups produce downfield signals. This means that the signals for the carbons of 2-butanol are in the same relative order that we would expect for the signals of the protons on those carbons in the ^1H NMR spectrum. The carbon of the methyl group farthest away from the electron-withdrawing OH group gives the farthest upfield signal. As you move downfield, the other methyl carbon comes next, followed by the methylene carbon, and the carbon attached to the OH group gives the farthest downfield signal.

The signals are not normally split by neighboring carbons, because there is little likelihood of an adjacent carbon being a ^{13}C. The probability of two ^{13}C carbons being next to each other is 1.11% $\times$ 1.11% (about 1 in 10,000). (Since ^{12}C does not have a magnetic moment, it cannot affect the position of a signal.)

The signals in a ^{13}C NMR spectrum can be split by nearby hydrogens. However, this splitting is not usually observed, because the spectra are recorded using spin-

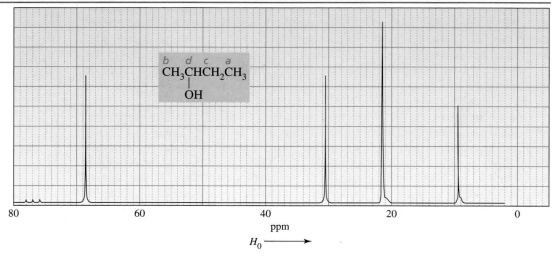

▲ **Figure 13.31**
Proton-decoupled ^{13}C NMR spectrum of 2-butanol.

decoupling. In other words, the carbon–proton interactions are decoupled. Thus all the signals are singlets in an ordinary ^{13}C NMR spectrum (Figure 13.31).

If the spectrometer is run in a *spin-coupled* mode, the signals show spin–spin coupling. The splitting is not caused by adjacent carbons but by protons bonded to the carbon that produces the signal. Protons on other carbons do not cause splitting. The multiplicity of the signal is determined by the $N + 1$ rule. The **spin-coupled** ^{13}**C NMR spectrum** of 2-butanol is shown in Figure 13.32. (The triplet at 78 ppm is produced by the solvent.) The signals for the methyl carbons are each split into a quartet because each methyl carbon is bonded to three protons $(3 + 1 = 4)$. The signal for the methylene carbon is split into a triplet $(2 + 1 = 3)$, and the signal for the carbon bonded to the OH group is split into a doublet $(1 + 1 = 2)$.

PROBLEM 32◆

Answer the following questions for each of the compounds.

  a. How many signals would you expect to see in the ^{13}C NMR spectrum?

  b. Which signal would you expect to be farthest upfield?

    1. $CH_3CH_2CH_2Br$　　　　　　　　**6.** $(CH_3)_2C{=}CH_2$

    2. $CH_3CH_2OCH_3$　　　　　　　　　**7.** $CH_2{=}CHBr$

    3. $CH_3\underset{\underset{\displaystyle Br}{|}}{CH}CH_3$　　　　　　　　　**8.** $CH_3\underset{\underset{\displaystyle CH_3}{|}}{CH}\overset{\overset{\displaystyle O}{\|}}{C}H$

    4. $CH_3\underset{\underset{\displaystyle CH_3}{|}}{\overset{\overset{\displaystyle CH_3}{|}}{C}}OCH_3$　　　　　　　**9.** ⬡—Cl

    5. $CH_3CH_2\overset{\overset{\displaystyle O}{\|}}{C}OCH_3$　　　　　**10.** $CH_3\overset{\overset{\displaystyle O}{\|}}{C}CH_2CH_2\overset{\overset{\displaystyle O}{\|}}{C}CH_3$

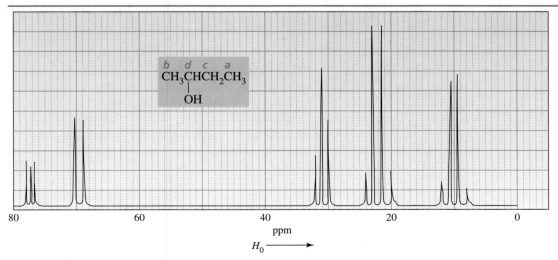

▲ **Figure 13.32**
Spin-coupled ^{13}C NMR spectrum of 2-butanol.

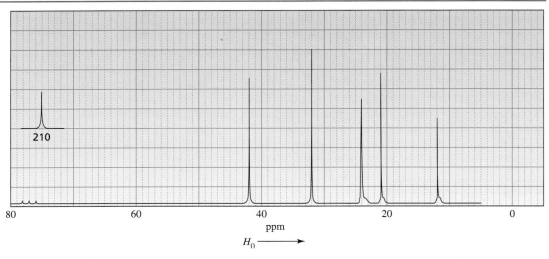

▲ **Figure 13.33**
The ^{13}C NMR spectrum for Problem 35.

PROBLEM 33

Describe the spin-coupled ^{13}C NMR spectrum for compounds 1 to 5 in Problem 32, showing relative values of chemical shifts (not absolute values).

PROBLEM 34

How can 1,2-, 1,3-, and 1,4-dinitrobenzene be distinguished by

a. ^{1}H NMR spectroscopy?　　　　　**b.** ^{13}C NMR spectroscopy?

PROBLEM 35 ◆

The ^{13}C NMR spectrum for a compound with molecular formula $C_{11}H_{22}O$ is shown in Figure 13.33. Identify the compound.

**13.13
DEPT ^{13}C NMR
SPECTRA**

A new technique known as DEPT ^{13}C NMR (distortionless enhancement by polarization transfer) has been developed to distinguish among CH_3, CH_2, and CH groups. It is now much more widely used than **spin-coupling** to determine the number of hydrogens attached to a carbon.

The DEPT ^{13}C NMR spectrum of citronellal is shown in Figure 13.34. A **DEPT ^{13}C NMR spectrum** is actually four spectra of the same compound. The bottommost spectrum shows signals *for all carbons that are covalently bonded to hydrogens.* The next-to-the-bottom spectrum is run under conditions that cause only signals resulting from CH carbons to appear. The third spectrum is run under conditions that cause signals produced by CH_2 carbons to appear, and the top spectrum shows signals produced by CH_3 carbons. Information from the upper three spectra allows you to determine whether a signal in a ^{13}C NMR spectrum is produced by a CH_3, CH_2, or CH group.

Notice that a DEPT ^{13}C NMR spectrum does not show a signal for a carbon that is not attached to a hydrogen. For example, the ^{13}C NMR spectrum of 2-butanone

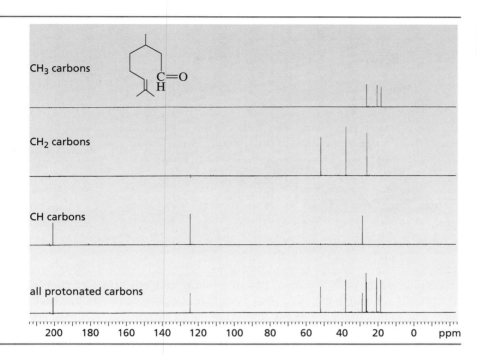

will show four signals because it has four nonequivalent carbons, but the DEPT ^{13}C NMR spectrum of 2-butanone will show only three signals because the carbonyl carbon will not produce a signal since it is not bonded to a hydrogen.

13.14 TWO-DIMENSIONAL NMR SPECTROSCOPY

Complex molecules such as proteins and nucleic acids are difficult to analyze by NMR because their signals overlap. Such compounds are now being analyzed by two-dimensional (2-D) NMR. **2-D NMR** techniques also allow scientists to determine the structures of molecules in solution. This is particularly important for biological molecules whose properties depend on how they fold in water. More recently, 3-D and 4-D NMR spectroscopy have been developed and can be used to determine the structures of very complex molecules. A thorough discussion of 2-D NMR is beyond the scope of this book, but the following discussion will give you a brief introduction to this increasingly important spectrophotometric technique.

The ^{1}H NMR and ^{13}C NMR spectra discussed in the preceding sections have one frequency axis and one intensity axis; 2-D NMR spectra have two frequency axes and one intensity axis. The most common 2-D spectra are ^{1}H–^{1}H shift correlations; they identify protons that are coupled to each other (i.e., that split each other). This is known as ^{1}H–^{1}H shift correlated spectroscopy, which is known by the acronym COSY.

A portion of the **COSY spectrum** of ethyl vinyl ether is shown in Figure 13.35(a); it looks like a mountain range viewed from the air. These "mountain-like" spectra (known as *stack plots*) are not the spectra actually used to identify a compound. Instead, the compound is identified using a contour plot (Figure 13.35(b)), where each mountain in Figure 13.35(a) is represented by a large dot (as if its top had been cut off). The two mountains shown in Figure 13.35(a) correspond to the two shaded dots in Figure 13.35(b).

In the contour plot, the usual one-dimensional ^{1}H NMR spectrum is plotted on both the x- and y-axes. To analyze the spectrum, a diagonal is drawn through the dots. The dots that are not on the diagonal (A, B, C) are called *cross peaks*. The cross

Figure 13.35 ▶
(a) COSY spectrum of ethyl
vinyl ether (stack plot). (b)
COSY spectrum of ethyl vinyl
ether (contour plot). The dots
in (b) that represent the
mountains in (a) are shaded.

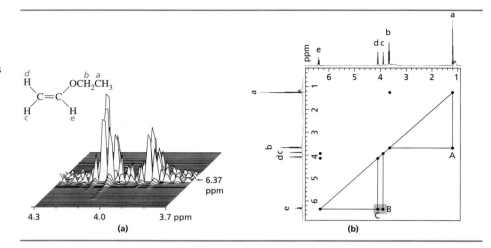

Figure 13.35 ▶
(a) COSY spectrum of ethyl
vinyl ether (stack plot). (b)
COSY spectrum of ethyl vinyl
ether (contour plot). The dots
in (b) that represent the
mountains in (a) are shaded.

peaks indicate the pairs of protons that are splitting each other (that are coupled). For example, if we start at the cross peak labeled A and draw a straight line back to the diagonal parallel to the y-axis, we hit the dot on the diagonal at ~1.1 ppm produced by the H_a protons. If we then go back to A and draw a straight line back to the diagonal parallel to the x-axis, we hit the dot on the diagonal at ~3.8 ppm produced by the H_b protons. This means that the H_a and H_b protons are coupled. If we then go to the cross peak labeled B and draw two perpendicular lines back to the diagonal, we see that the H_c and H_e protons are coupled; the cross peak labeled C indicates that the H_d and H_e protons are coupled. Notice that we used only cross peaks below the diagonal. The cross peaks above the diagonal give the same information.

The COSY spectrum of 1-nitropropane is shown in Figure 13.36. Cross peak A shows that the H_a protons are coupled to the H_b protons, cross peak B shows that the H_b protons are coupled to the H_c protons. Notice that the two triangles in Figure 13.36 have a common vertex since the H_b protons are coupled with both the H_a and H_c protons.

PROBLEM 36

Identify pairs of coupled protons in 2-methyl-3-pentanone using the COSY spectrum in Figure 13.37.

Figure 13.36 ▶
COSY spectrum of
1-nitropropane.

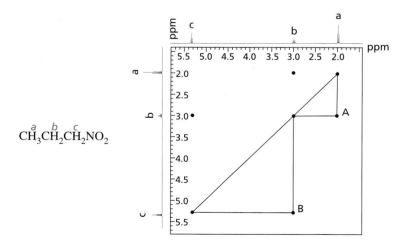

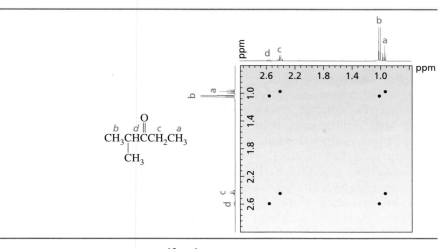

2-D NMR spectra that show ^{13}C–^1H shift correlations are called HETCOR (from heteronuclear correlation) spectra. **HETCOR spectra** indicate coupling between protons and the carbon to which they are attached.

The HETCOR spectrum of 2-methyl-3-pentanone is shown in Figure 13.38. The ^{13}C NMR spectrum is shown on the *x*-axis and the ^1H NMR spectrum is shown on the *y*-axis. Cross peak A indicates that the hydrogens that show a signal at ~0.9 ppm in the ^1H NMR spectrum are bonded to the carbon that shows a signal at ~8 ppm in the ^{13}C NMR spectrum. Cross peak C shows that the hydrogens that show a signal at ~2.5 ppm are bonded to the carbon that shows a signal at ~34 ppm.

Clearly, 2-D NMR techniques are not necessary for assigning the signals in the NMR spectrum of a simple compound such as 2-methyl-3-pentanone. However, in the case of many complicated molecules, signal assignment can be done only with the aid of 2-D NMR.

Spectra showing ^{13}C–^{13}C shift correlations (called 2-D ^{13}C INADEQUATE spectra) identify directly bonded carbons. There also are spectra in which chemical shifts are plotted on one of the frequency axes and coupling constants on the other. Other 2-D spectra involve the nuclear Overhauser effect (NOESY for very large molecules; ROESY for mid-sized molecules).[2] They are used to define through-space interactions for stereochemical studies.

Albert Warner Overhauser was born in San Diego in 1925. He has been a professor of chemistry at Cornell University and Purdue University. From 1958 to 1973, he was at the Physical Science Laboratory of Ford Motor Company.

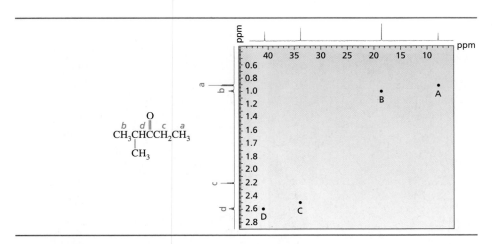

◀ **Figure 13.38**
HETCOR spectrum of
2-methyl-3-pentanone.

[2]NOESY and ROESY are the acronyms for Nuclear Overhauser Effect Spectroscopy and Rotation-frame Overhauser Effect Spectroscopy.

13.15 MAGNETIC RESONANCE IMAGING

NMR has become an important tool in medical diagnosis. When it was first introduced into clinical practice in 1981, considerable attention was given to the selection of an appropriate name. Because some members of the public associate nuclear processes with harmful radiation, the "N" is dropped from NMR when this technique is used in medicine. It is called **magnetic resonance imaging (MRI),** and the spectrometer is called an **MRI scanner.**

An MRI scanner consists of a magnet sufficiently large to accommodate an entire patient, along with additional coils for exciting the nuclei, for modifying the magnetic field, and for receiving signals. Different tissues yield different signals. The signals are separated by Fourier transform analysis into components. Each component can be attributed to a specific site of origin in the patient. This allows a cross-sectional image of the patient's body to be constructed.

Most of the signals in an MRI scan originate from the hydrogens of water molecules, because these hydrogens are far more abundant in tissues than are the hydrogens of organic compounds. The difference in the way water is bound in different tissues is what produces much of the signal variation among organs, as well as the signal variation between healthy and diseased tissue (Figure 13.39).

MRI has become an important diagnostic tool for several reasons. First of all, an MRI scan can be obtained without exposing the patient to the ionizing radiation of X-rays. In addition, an MRI image may be obtained in any plane essentially independent of the patient's position. This allows optimum visualization of the anatomy of interest. By contrast, the plane of an X-ray image is defined by the physical position of the patient. Likewise, the plane of a CT (computed tomography) scan is defined by the position of the patient within the machine and is usually perpendicular to the long axis of the body. CT scans in other planes may be obtained only if the patient is a skilled contortionist. Recent technology has allowed CT scans to be obtained independent of the patient's position, but the scans are of very low resolution.

Unlike X-rays or CT scans, MRI "sees through" hard tissues and provides images of soft tissues. This means that MRI scans can provide much more information than images obtained using other techniques.

For example, MRI can provide detailed images of blood vessels. Flowing fluids such as blood respond differently than do stationary tissues to excitation in an MRI scanner and normally do not produce a signal. However, the data may be processed

Figure 13.39 ▶
(a) MRI of a normal brain. The pituitary is highlighted (pink). (b) MRI of an axial section through the brain showing a tumor (purple) surrounded by damaged, fluid-filled tissue (red).

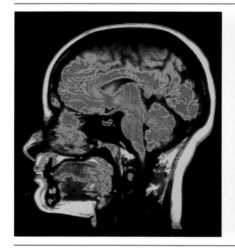
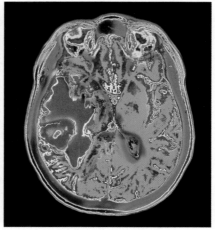

to eliminate signals from motionless structures instead, thereby showing signals only from flowing fluids. This technique, which is just beginning to be used clinically, will eventually replace more invasive methods of examining the vascular tree. NMR spectroscopy using ^{23}Na, ^{39}K, ^{31}P, and ^{15}N is not yet in routine clinical use, but is being used widely in clinical research and is of particular interest because of the importance of these atoms in cellular metabolism and cell physiology.

13.16 ULTRAVIOLET AND VISIBLE SPECTROSCOPY

We learned in Section 1.6 that orbitals are conserved. Consequently, if two atomic orbitals are combined, two molecular orbitals result: a bonding molecular orbital and an antibonding molecular orbital.

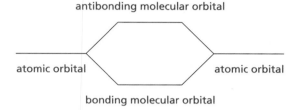

The normal electronic configuration of a molecule is known as its **ground state.** When a molecule absorbs light of a particular wavelength, an electron can be promoted from its **highest occupied molecular orbital (HOMO)** to its **lowest unoccupied molecular orbital (LUMO).** Such a transition is called an **electronic transition.** The molecule is then in an **excited state.** The relative energies of the ground and excited states are shown in Figure 13.40.

Ultraviolet and visible light have sufficient energy to cause only the two electronic transitions shown in Figure 13.40. The one with the lowest energy is excitation of a nonbonding electron (n) into a π antibonding molecular orbital. This is called an $n \longrightarrow \pi^*$ transition (stated as "n to pi star"). The other electronic transition is excitation of an electron from a π bonding molecular orbital into a π antibonding molecular orbital, known as a $\pi \longrightarrow \pi^*$ transition (stated as "pi to pi star").

The absorption of energy as a result of an electronic transition causes an absorption band in the UV/Vis spectrum. This means that only compounds with π electrons or nonbonding electrons can produce UV/Vis spectra. Ultraviolet and visible spectroscopy are based on the same theory: if a molecule absorbs ultraviolet light, a UV spectrum is obtained; if it absorbs visible light, a visible spectrum is obtained.

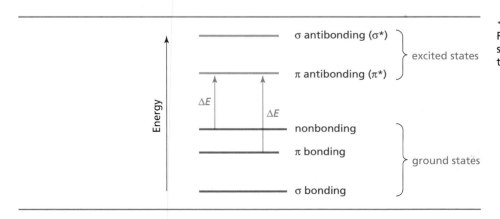

◀ **Figure 13.40**
Relative energies of ground state and excited state electrons.

Figure 13.41 ▶
The UV spectrum of acetone.

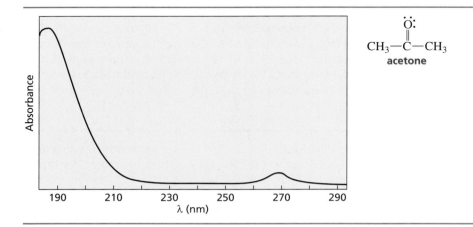

Ultraviolet light is electromagnetic radiation with wavelengths ranging from 180 to 400 nm (nanometers). **Visible light** has wavelengths ranging from 400 to 780 nm. [One nanometer is 10^{-9} m or 10 Å. In older literature, millimicron (mμ) is used instead of nanometer, but interconversion is easy because 1 mμ = 1 nm.] We have seen (Section 12.6) that wavelength (λ) is related to energy of the radiation by the following equation.

$$E = \frac{hc}{\lambda}$$

A compound such as acetone has both π electrons and nonbonding electrons. The UV spectrum of acetone is shown in Figure 13.41. There are two absorption bands, one for the $\pi \longrightarrow \pi^*$ transition and one for the $n \longrightarrow \pi^*$ transition. The $\lambda_{\mathbf{max}}$ (stated as "lambda max") of an absorption band is the wavelength at which the greatest absorbtion occurs. The $\pi \longrightarrow \pi^*$ transition has a λ_{max} at 187 nm, and the $n \longrightarrow \pi^*$ transition has a λ_{max} at 270 nm. We know that the $\pi \longrightarrow \pi^*$ transition corresponds to the λ_{max} at the shorter wavelength because this transition requires more energy than does the $n \longrightarrow \pi^*$ transition (Figure 13.40).

The shorter the wavelength, the greater its energy.

When a UV spectrum is taken under normal conditions, absorption bands below about 210 nm cannot be detected because oxygen absorbs in that region and covers up all other absorption bands. To be able to see the absorption band at 187 nm, the spectrum of acetone must be obtained in a spectrophotometer purged with nitrogen or by using vacuum UV techniques.

PROBLEM 37 ◆

a. Which has greater energy, UV light or visible light?

b. Which has greater energy, visible light or IR radiation?

ULTRAVIOLET LIGHT AND SUNSCREENS

Exposure to ultraviolet light causes specialized cells in the skin to produce a black pigment known as melanin, which causes the skin to look tanned. Melanin is the body's natural protection, absorbing potentially harmful UV light. If more UV light reaches the skin than the melanin can absorb, the light will burn the skin and can cause photochemical reactions that can result in skin cancer (Section 20.6). UV-A is the lowest-energy UV light (315 to 400 nm) and does the least biological damage. Fortunately, most of the more dangerous, higher-energy UV light, UV-B (290 to 315 nm) and UV-C (180 to 290 nm), is filtered out by the ozone layer in the stratosphere. This is why there is such great concern about the diminishing ozone layer (Section 8.9).

Applying a sunscreen to skin can protect it against UV light. Some sunscreens contain an inorganic reagent such as zinc oxide that reflects the light as it reaches the skin. Others contain a compound that absorbs UV light. PABA was the first commercially available UV-absorbing sunscreen; it absorbs UV-B light. However, it is not very soluble in oily skin lotions. More nonpolar compounds such as Padimate O are now commonly used. Giv Tan F absorbs both UV-B and UV-A light, so it gives better protection. Recent research has shown that sunscreens that absorb only UV-B light do not give adequate protection against skin cancer; both UV-A and UV-B protection are needed.

para-aminobenzoic acid
PABA

(2-ethylhexyl) 4-(dimethylamino) benzoate
Padimate O

(2-ethylhexyl) 3-(4-methoxyphenyl) -2-propenoate
Giv Tan F

The amount of protection provided by a particular sunscreen is indicated by its SPF (sun protection factor).

$$SPF = \frac{\text{UV energy required to produce minimal sunburn on protected skin}}{\text{UV energy required to produce minimal sunburn on unprotected skin}}$$

Wilhelm Beer and Johann Lambert independently proposed that absorbance depends on the amount of absorbing species in the path of the light as it passes through the solution. In other words, absorbance depends on both the concentration of the sample and the length of the light path through the sample. The relationship among absorbance, concentration, and length of the light path is known as the **Beer–Lambert law.**

the Beer–Lambert law

$$A = cl\epsilon$$

$$A = \text{absorbance} = \log\frac{I_0}{I}$$

I_0 = intensity of the radiation entering the sample
I = intensity of the radiation emerging from the sample
c = concentration of the sample in moles/liter
l = length of the light path through the sample in centimeters
ϵ = molar absorptivity

The **molar absorptivity** (formerly called the extinction coefficient) is a constant that is characteristic of a compound at a particular wavelength. It is the absorbance that would be observed when using a 1.00 M solution in a cell with a 1.00 cm path length. The molar absorptivity of acetone, for example, is 900 at 187 nm and 15 at 270 nm. The solvent in which the sample is dissolved when the spectrum is taken is reported because molar absorptivity is not exactly the same in all solvents. So the

Wilhelm Beer (1797–1850) *was born in Germany. He was a banker whose hobby was astronomy. He was the first to make a map of the darker and lighter areas of Mars.*

Johann Heinrich Lambert (1728–1777), *a German-born mathematician, was the first to make accurate measurements of light intensities and to introduce hyperbolic functions into trigonometry.*

Although the equation that relates absorbance, concentration, and light path bears the names of Beer and Lambert, it is believed that **Pierre Bouguer,** *a French mathematician, first formulated this relationship in 1729.*

UV spectrum of acetone in hexane would be reported as λ_{max} 187 nm ($\epsilon = 900$, hexane); λ_{max} 270 nm ($\epsilon = 15$, hexane). Because absorbance is proportional to concentration, the concentration of a solution can be determined if the absorbance and molar absorptivity at a particular wavelength are known.

The two absorption bands of acetone in Figure 13.41 are very different in size because of the difference in molar absorptivity at the two wavelengths. Small molar absorptivities are characteristic of $n \longrightarrow \pi^*$ transitions. Such transitions therefore can be difficult to detect. Consequently, $\pi \longrightarrow \pi^*$ transitions are much more useful in UV/Vis spectroscopy.

In order to obtain a UV or visible spectrum, the solution is placed in a cell. Most cells have light paths of 1 cm. Either glass or quartz cells can be used for visible spectra, but quartz cells must be used for UV spectra because glass absorbs UV light.

The $n \longrightarrow \pi^*$ transition for methyl vinyl ketone is at 324 nm, and the $\pi \longrightarrow \pi^*$ transition is at 219 nm.

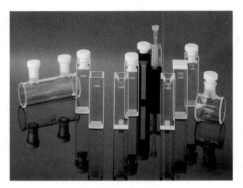

Both λ_{max}'s of methyl vinyl ketone are at longer wavelengths than the corresponding λ_{max}'s of acetone because methyl vinyl ketone has two conjugated double bonds. Because conjugation raises the energy of the HOMO and lowers the energy of the LUMO, electronic transitions are easier for conjugated systems than for nonconjugated systems (Figure 13.42). The more conjugated double bonds there are in a compound, the less energy is required for the electronic transition and therefore the longer the wavelength at which the transition occurs.

The λ_{max} of the $\pi \longrightarrow \pi^*$ transition for several conjugated dienes are shown in Table 13.5. Notice that both the λ_{max} and the molar absorptivity increase as the

Figure 13.42 ▶
Conjugation raises the energy of the HOMO and lowers the energy of the LUMO.

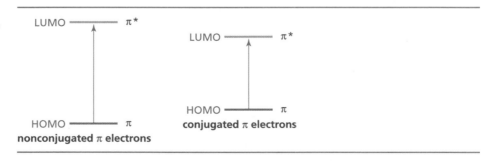

TABLE 13.5   Values of λ_{max} and ϵ for Ethylene and Conjugated Dienes

Compound	λ_{max} (nm)	ϵ
$H_2C{=}CH_2$	165	15,000
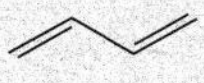	217	21,000
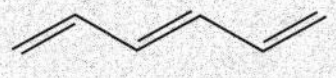	256	50,000
	290	85,000
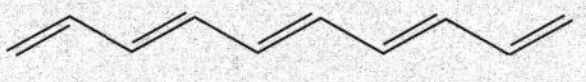	334	125,000
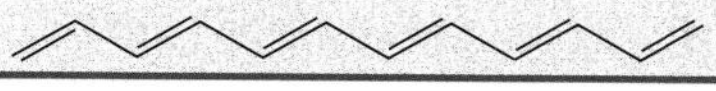	364	138,000

number of conjugated double bonds increases. We see that the λ_{max} of a compound can be used to predict the number of conjugated double bonds in the compound.

If a compound has enough double bonds, it will absorb visible light ($\lambda_{max} > 400$ nm) and the compound will be colored. β-Carotene, a precursor of vitamin A, is an orange substance found in carrots, apricots, and sweet potatoes. Lycopene is red and is found in tomatoes, watermelon, and pink grapefruit.

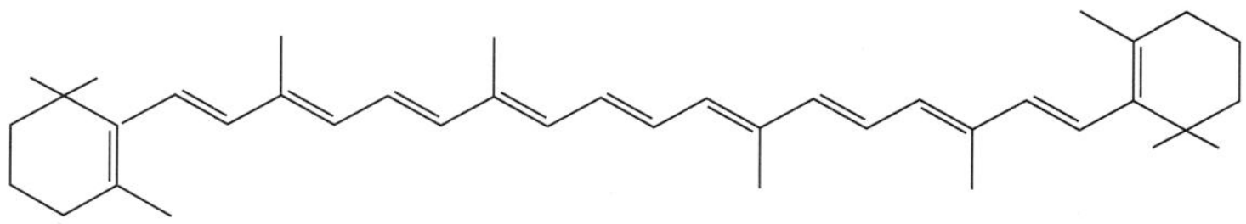

β-carotene
λ_{max} = 455 nm

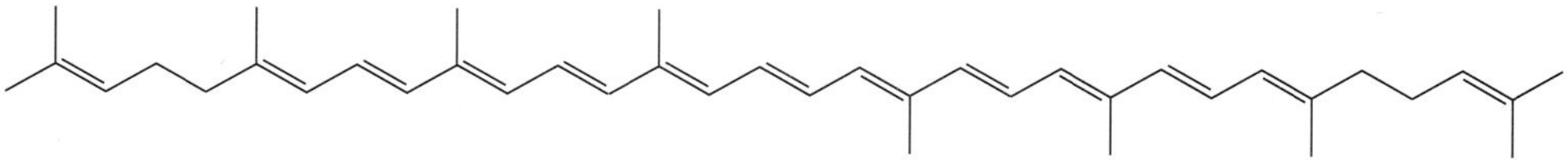

lycopene
λ_{max} = 474 nm

A **chromophore** is that part of a molecule that is responsible for a UV or visible spectrum. The carbonyl group is the chromophore of acetone. The following compounds all have the same chromophore, so they all have approximately the same λ_{max}.

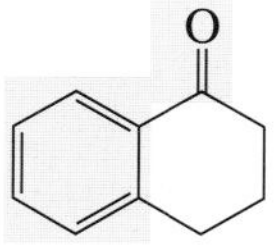

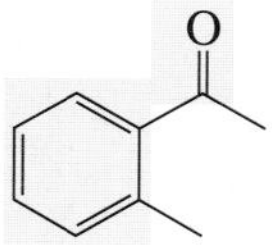

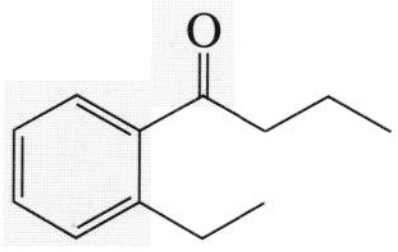

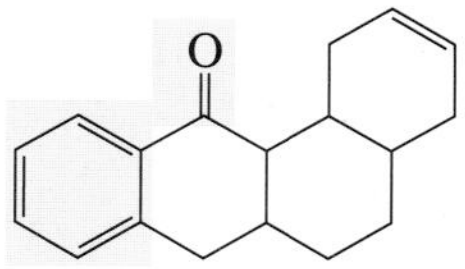

An **auxochrome** is a substituent that, when attached to a chromophore, alters the λ_{max} and the intensity of the absorption; OH and NH_2 groups are examples of auxochromes. The lone pair (nonbonding) electrons on oxygen and nitrogen are available for interaction with the π electron cloud of the benzene ring, which increases the λ_{max}. Because the anilinium ion does not have an auxochrome, its λ_{max} is similar to that of benzene.

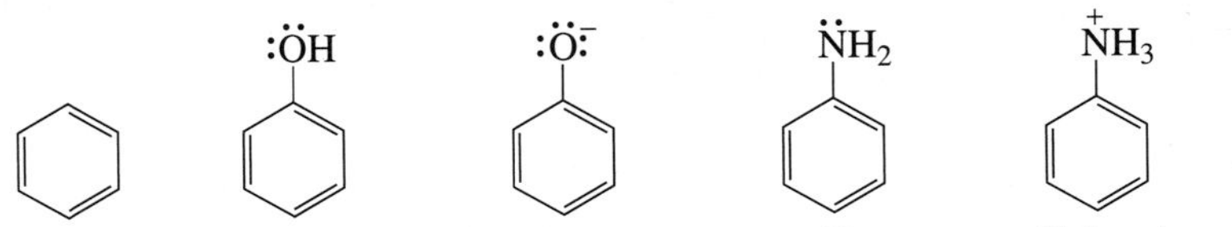

benzene	phenol	phenolate ion	aniline	anilinium ion
255 nm	270 nm	287 nm	280 nm	254 nm

:ÖH   :Ö⁻   N̈H₂   ⁺N̈H₃

Removal of a proton from phenol (forming the phenolate ion) further increases the λ_{max} because an additional pair of nonbonding electrons is available for interaction with the ring. Protonating aniline (forming the anilinium ion) decreases its λ_{max} because the nonbonding electron pair is no longer available to interact with the benzene ring. Because wavelengths of red light are longer than those of blue light, a shift to a longer wavelength is called a **red shift,** and a shift to a shorter wavelength is called a **blue shift.** Deprotonation of phenol results in a red shift, while protonation of aniline produces a blue shift.

red shift ⟶

⟵ blue shift

200 m   400 m

The intensity of the absorption depends on the probability of interaction between the light and the electron system: the greater the dipole moment of the compound, the greater the probability of interaction. For example, resonance structures of aniline indicate how the auxochrome increases the dipole moment of the molecule and therefore the molar absorptivity of the compound.

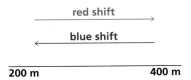

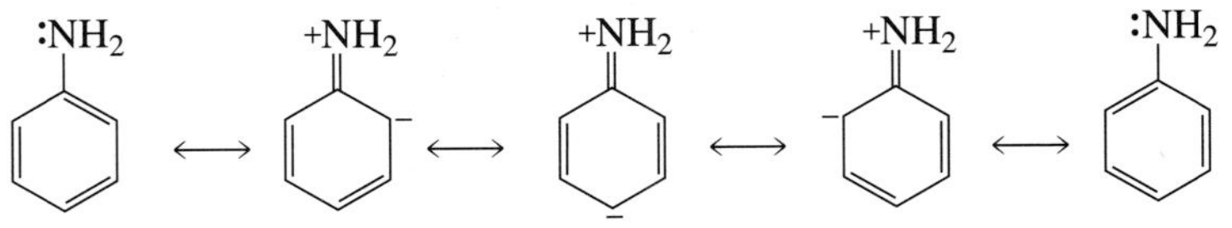

:NH₂   +NH₂   +NH₂   +NH₂   :NH₂

PROBLEM 38◆

Rank the compounds in each group in order of decreasing λ_{max}.

a.

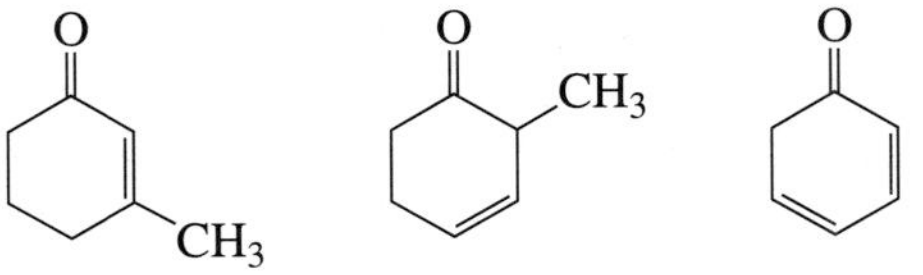

b.

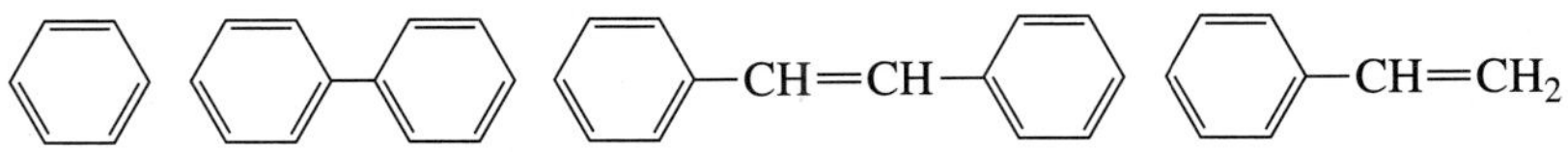

PROBLEM 39◆

A solution of 4-methyl-3-penten-2-one in ethanol gives an absorbance of 0.52 at 236 nm in a cell with a 1 cm light path. Its molar absorptivity in ethanol at that wavelength is 12,600. What is the concentration of the compound?

The λ_{max} of the $\pi \longrightarrow \pi^*$ transition for compounds with four or fewer conjugated double bonds can be calculated by using a simple set of rules known as the **Woodward–Fieser rules.** The following base numbers are used for the calculations: $\lambda_{max} = 217$ nm for a conjugated diene, 210 nm for a conjugated aldehyde, and 215 nm for a conjugated ketone.

$$CH_2{=}CH{-}CH{=}CH_2$$
$$\lambda_{max} = \textbf{217 nm}$$

$\lambda_{max} = \textbf{210 nm}$

$\lambda_{max} = \textbf{215 nm}$

To the base number is added:

1. 30 for each extra conjugated double bond
2. 5 each time a conjugated double bond is an exocyclic double bond
3. 36 for each conjugated double bond that is frozen in the *s*-cis conformation
4. 5 for each alkyl group or halogen bonded to the conjugated system of a polyene
5. 10 for an α-substituent of a conjugated aldehyde or ketone
6. 12 for a β-substituent of a conjugated aldehyde or ketone

an exocyclic double bond   **an endocyclic double bond**   **double bonds frozen in the
s-cis conformation**

The rules enabling us to calculate λ_{max} can be best understood by looking at a few examples.

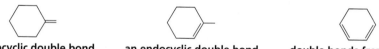

alkyl substituents

$$CH_2{=}\overset{\displaystyle CH_3}{\underset{}{C}}{-}CH{=}\overset{\displaystyle CH_3}{\underset{}{C}}{-}CH_3$$

base:	217
3 alkyl substituents:	15

calculated λ_{max} 232 nm
observed λ_{max} 232 nm

alkyl substituents

base:	217
additional double bond:	30
s-cis conformation:	36
2 alkyl substituents:	10

calculated λ_{max} 293 nm
observed λ_{max} 293 nm

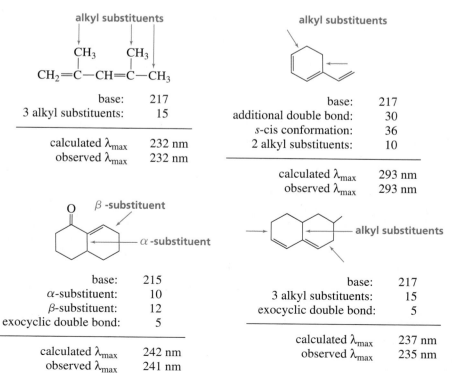

β-substituent
α-substituent

base:	215
α-substituent:	10
β-substituent:	12
exocyclic double bond:	5

calculated λ_{max} 242 nm
observed λ_{max} 241 nm

alkyl substituents

base:	217
3 alkyl substituents:	15
exocyclic double bond:	5

calculated λ_{max} 237 nm
observed λ_{max} 235 nm

Notice that in the last two examples, one of the double bonds is endocyclic to one ring but exocyclic to the other.

PROBLEM 40

Why do alkyl groups and halogens increase the λ_{max}?

PROBLEM 41 ◆

Predict the λ_{max} of each of the following compounds.

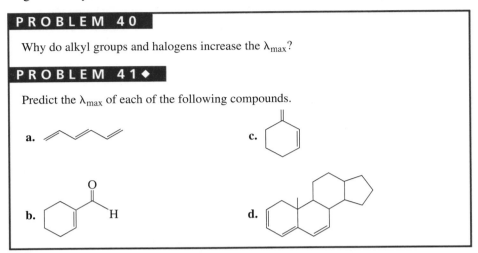

a.

b.

c.

d.

13.18
THE VISIBLE
SPECTRUM
AND COLOR

White light is a mixture of all possible wavelengths of visible light. If any color is removed from white light, the remaining light appears colored. So, if a compound absorbs any visible light, it will appear colored. Its color depends on the color of the light transmitted to the eye. In other words, it depends on the color produced from the wavelengths of light that are not absorbed.

The relationship between the wavelengths of the light absorbed and the color observed is shown in Table 13.6. Notice that two absorption bands are necessary to produce a green color. Most colored compounds have fairly broad absorption bands; vivid colors have narrow absorption bands. The human eye is able to distinguish more than a million different shades of color.

Azobenzenes (benzene rings connected by a nitrogen–nitrogen double bond) have an extended conjugated system that causes them to absorb light from the visible region of the spectrum. Some substituted azobenzenes are used commercially as dyes. Varying the extent of conjugation and the substituents attached to the conjugated system provides a large number of different colors. Notice that the only difference between butter yellow and methyl orange is an SO_3^- Na^+ group. Methyl

TABLE 13.6 Dependence of the Color Observed on the Wavelength of Light Absorbed	
Wavelengths absorbed (nm)	**Observed color**
380–460	yellow
380–500	orange
440–560	red
480–610	purple
540–650	blue
380–420 and 610–700	green

orange is a commonly used acid–base indicator. When margarine was first produced, it was colored with butter yellow to make it look more like butter. (White
margarine would not have been very appetizing.) This dye was no longer used
when it was found to be carcinogenic. β-Carotene is currently used to color margarine. There is some evidence that β-carotene is an anticancer agent.

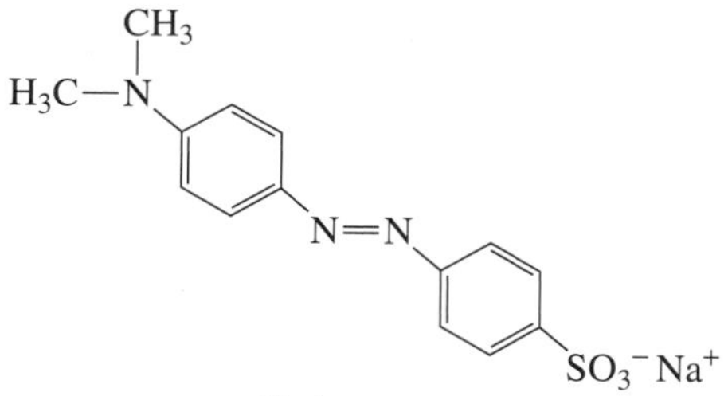

methyl orange
an azobenzene

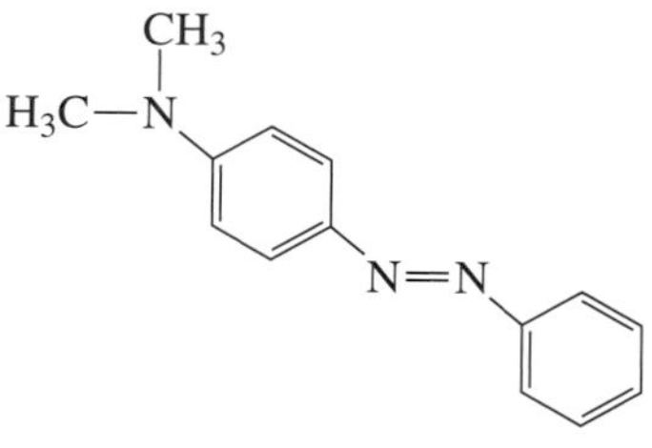

butter yellow
an azobenzene

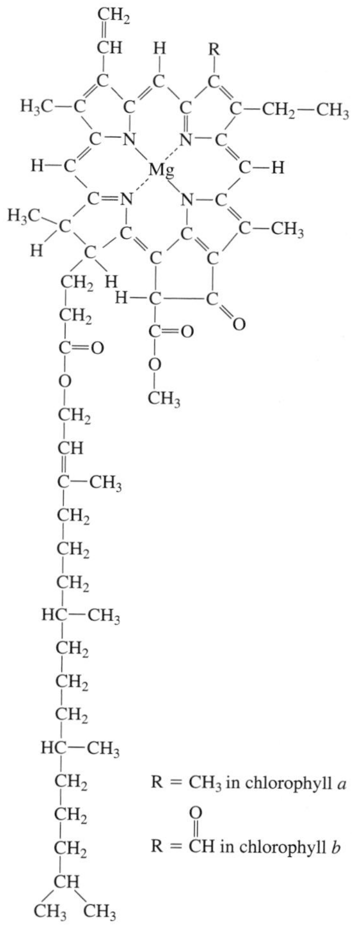

Chlorophyll *a* and *b* are the
pigments that make plants
look green. This highly conjugated compound absorbs
nongreen light. Therefore,
plants reflect green light.

PROBLEM 42

a. At pH = 7, one of the following ions is purple and the other is blue. Which is
which?

b. What would be the difference in the colors of the compounds at pH = 3?

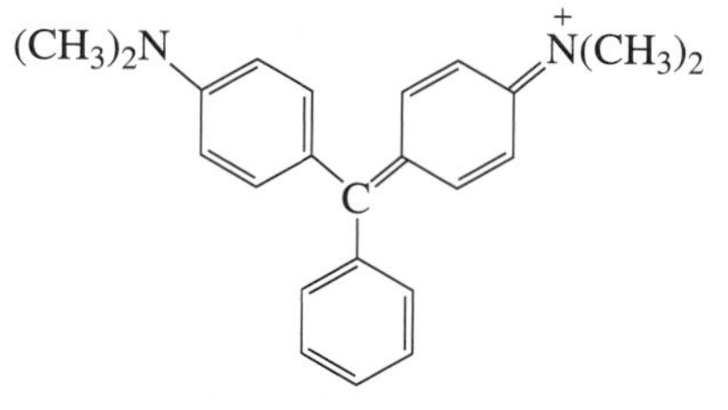

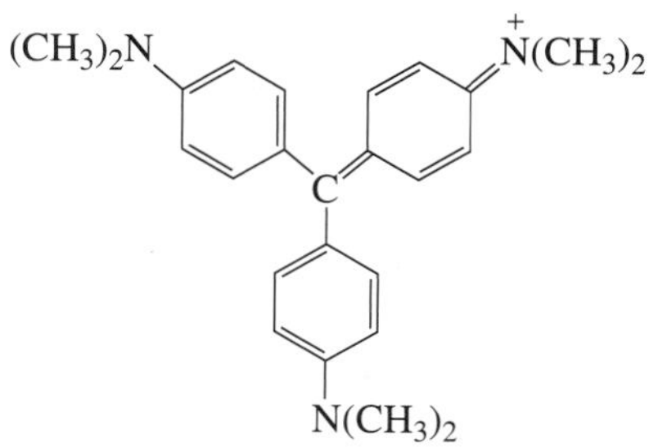

UV/Vis spectroscopy is not nearly as useful as IR or NMR spectroscopy for structure determination. UV/Vis spectroscopy, however, has other uses that make it a
very powerful analytical tool.

Reaction rates are commonly measured using UV/Vis spectroscopy. The rate of
any reaction can be measured using UV/Vis spectroscopy as long as one of the reactants or one of the products absorbs UV or visible light at a wavelength at which the
other reactants and products have little or no absorbance. For example, the anion of
nitroethane has a λ_{max} at 240 nm, but neither H_2O nor the reactants show any significant absorbance at that wavelength. In order to measure the rate at which hydroxide
ion removes a proton from nitroethane (that is, the rate at which the nitroethane
anion is formed), the UV spectrophotometer is adjusted to measure absorbance as a
function of time instead of absorbance as a function of wavelength. Nitroethane is
added to a quartz cell containing a basic solution, and the rate of the reaction is determined by monitoring the increase in absorbance at 240 nm (Figure 13.43).

13.19
USES OF UV/Vis
SPECTROSCOPY

Figure 13.43 ▶
The rate of proton removal from nitroethane is determined by monitoring the increase in absorbance at 240 nm.

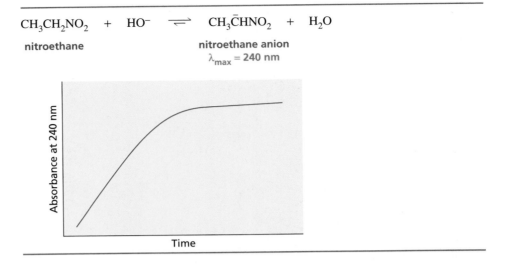

$$CH_3CH_2NO_2 \ + \ HO^- \ \rightleftharpoons \ CH_3\bar{C}HNO_2 \ + \ H_2O$$

nitroethane   nitroethane anion
                                       $\lambda_{max} = 240$ nm

The enzyme lactate dehydrogenase catalyzes the reduction of pyruvate by NADH to form lactate. NADH is the only species in the reaction mixture that absorbs light at 340 nm. So the rate of the reaction can be determined by monitoring the decrease in absorbance at 340 nm (Figure 13.44).

> **PROBLEM 43**
>
> Describe how one could determine the rate of the alcohol dehydrogenase catalyzed oxidation of ethanol by NAD$^+$ (Section 17.11).

The pK_a of a compound can be determined by UV/Vis spectroscopy if either the acidic form or the basic form of the compound absorbs UV or visible light. For example, the phenolate ion has a λ_{max} at 287 nm. If the absorbance at 287 nm is determined as a function of pH, the pK_a of phenol can be ascertained by determining the pH at which exactly one-half the increase in absorbance has occurred

Figure 13.44 ▶
The rate of reduction of pyruvate by NADH is measured by monitoring the decrease in absorbance at 340 nm.

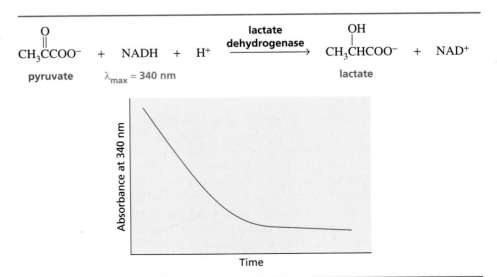

pyruvate $\lambda_{max} = 340$ nm   lactate

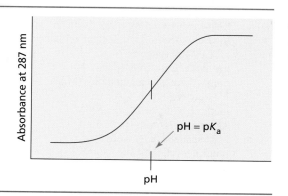

◀ **Figure 13.45**
The absorbance of an aqueous solution of phenol as a function of pH.

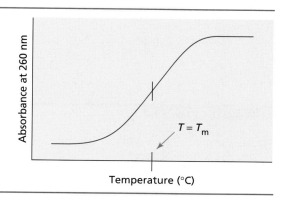

◀ **Figure 13.46**
The absorbance of a solution of DNA as a function of temperature.

(Figure 13.45). At this pH, half of the phenol has been converted into phenolate ion, so this pH is the pK_a of the compound. From the Henderson–Hasselbalch equation, we know that the pK_a of the compound is the pH at which half the compound exists in its acidic form and half exists in its basic form ([HA] = [A$^-$]; Section 1.19).

UV spectroscopy can be used to estimate the nucleotide composition of DNA. The two strands of DNA are held together by both A–T base pairs and G–C base pairs (Section 24.4). When DNA is heated, the two strands break apart. Single-stranded DNA has a greater molar absorptivity at 260 nm compared with double-stranded DNA. The melting temperature (T_m) of DNA is the midpoint of an absorbance versus temperature curve (Figure 13.46). The T_m of double-stranded DNA increases with increasing content of G–C base pairs, because G–C base pairs are held together by three hydrogen bonds whereas A–T base pairs are held together by only two hydrogen bonds. Therefore, the T_m can be used to estimate the G–C content. These are just a few examples of the many uses of UV/Vis spectroscopy.

KEY TERMS

applied magnetic field (page 525)
auxochrome (page 571)
Beer–Lambert law (page 569)
blue shift (page 572)
chemical exchange (page 555)

chemically equivalent protons
(page 527)
chemical shift (page 530)
chromophore (page 571)
COSY spectrum (page 563)

coupled protons (page 538)
coupling constant (page 547)
DEPT ^{13}C NMR spectrum (page 562)
diamagnetic shielding (page 529)
doublet (page 537)

PROBLEMS

44. How many signals would you expect for each of the following compounds in its:
   a. ^{1}H NMR spectrum?
   b. ^{13}C NMR spectrum?

1. CH_3—⟨benzene ring⟩—$OCHCH_3$ with CH_3 below

2. ⟨benzene ring⟩—$\overset{O}{\overset{\|}{C}}$—$OCH_2CH_3$

3. ⟨six-membered ring lactone⟩=O with O

4. ⟨four-membered ring with O⟩ (oxetane)

5. ⟨cyclopropane ring with Cl⟩

6. ⟨structure with CH3 groups⟩

45. 4-Methyl-3-penten-2-one has two absorption bands in its UV spectrum, one at 236 nm and one at 314 nm.

$$CH_3\overset{O}{\overset{\|}{C}}CH=\overset{CH_3}{\overset{|}{C}}CH_3$$
4-methyl-3-penten-2-one

   a. Why are there two absorption bands?
   b. Which band shows the greater absorbance?

46. Draw a splitting diagram for the H_b proton and indicate its multiplicity if:
   a. $J_{ba} = J_{bc}$ **b.** $J_{ba} = 2 J_{bc}$

$$H_a-\overset{H_a}{\underset{H_a}{\overset{|}{\underset{|}{C}}}}-\overset{X}{\underset{H_b}{\overset{|}{\underset{|}{C}}}}-\overset{X}{\underset{H_c}{\overset{|}{\underset{|}{C}}}}-X$$

47. Label each set of chemically equivalent protons, using *a* for the set that will be farthest upfield in the ^{1}H NMR spectrum, *b* for the next, etc. Indicate the multiplicity of each signal.

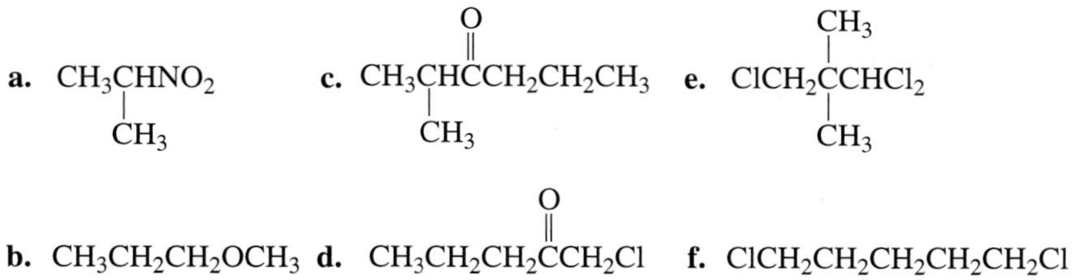

a. CH_3CHNO_2
       |
       CH_3

c. $CH_3CHCCH_2CH_2CH_3$ (with O double bonded and CH_3 substituent)

e. $ClCH_2CCHCl_2$ (with CH_3 substituents)

b. $CH_3CH_2CH_2OCH_3$

d. $CH_3CH_2CH_2CCH_2Cl$ (with O double bonded)

f. $ClCH_2CH_2CH_2CH_2CH_2Cl$

48. **a.** Methyl orange (its structure is given in Section 13.18) is an acid–base indicator. In solutions of pH < 4 it is yellow, and in solutions of pH > 4 it is red. Account for the change in color.
    b. Phenolphthalein is also an indicator, but it exhibits a much more dramatic color change. In solutions of pH < 8.5 it is colorless, and in solutions of pH > 8.5 it is deep red–purple. Account for the change in color.

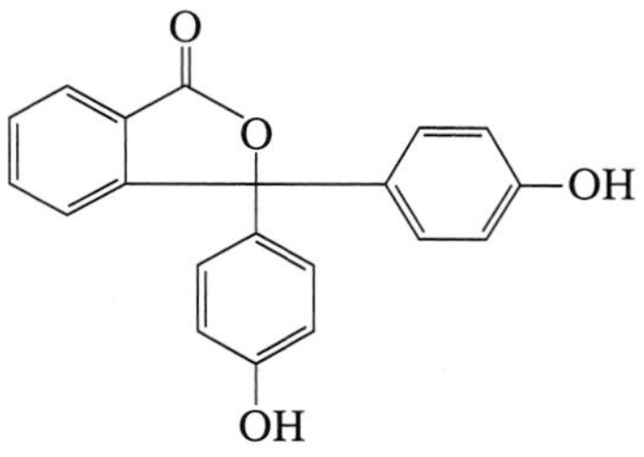

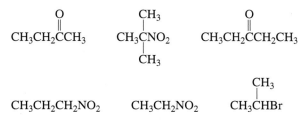

phenolphthalein

49. Match each of the ^{1}H NMR spectra shown below with one of the following compounds.

$CH_3CH_2CCH_3$ (with O double bonded) CH_3CNO_2 (with CH_3 substituents) $CH_3CH_2CCH_2CH_3$ (with O double bonded)

$CH_3CH_2CH_2NO_2$ $CH_3CH_2NO_2$ CH_3CHBr (with CH_3 substituent)

a.

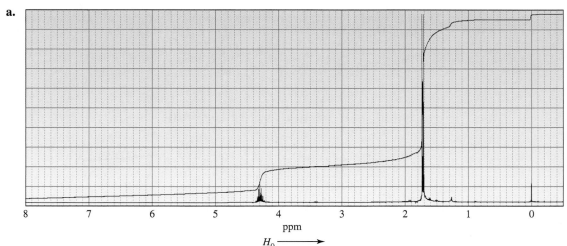

ppm

$H_0 \longrightarrow$

b.

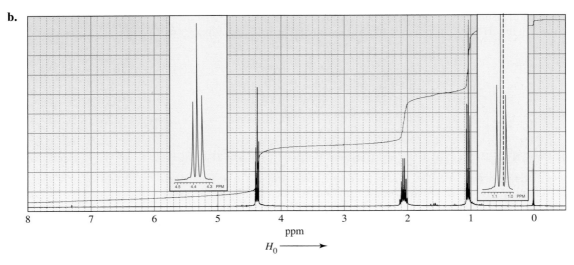

c.

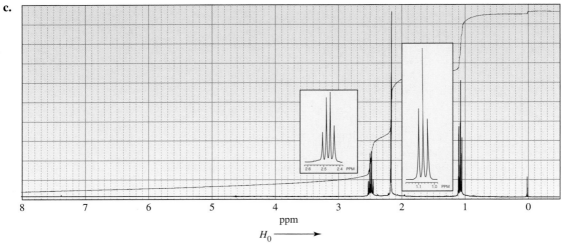

50. Determine the ratios of the magnetically nonequivalent hydrogens in a compound if the steps of the integration curves measure 40.5, 27, 13, and 118 mm, going from left to right across the spectrum. Give the structure of a compound whose ¹H NMR spectrum would show these relative integrals in the observed order.

51. How could you distinguish between the compounds in each of the following pairs using UV spectroscopy?

a. 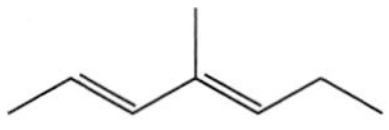 [structure] and [structure]

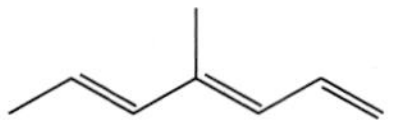

b. CH₂=CH—C(=O)—CH₃ and CH₃—C(CH₃)=CH—C(=O)—CH₃

c. [structure] and [structure]

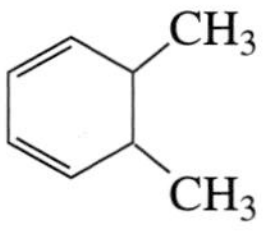

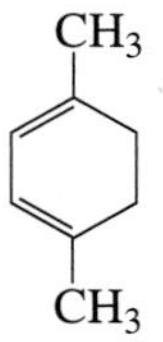

d. $CH_2=CHCH=CHCH=CH_2$ and $CH_2=CHCH=CHCCH_3$ (with C=O)

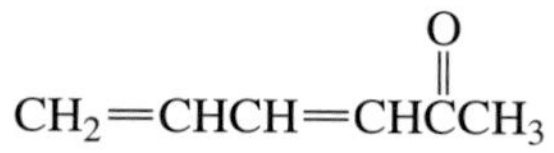

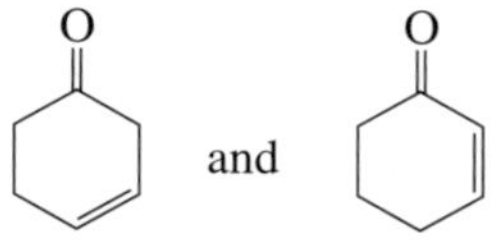

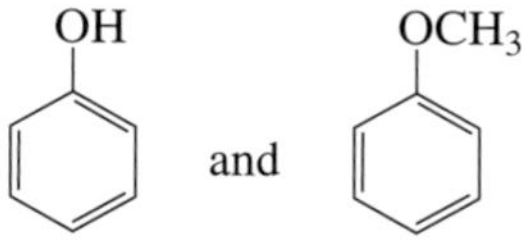

e. and

f. (OH) and (OCH$_3$)

g. and

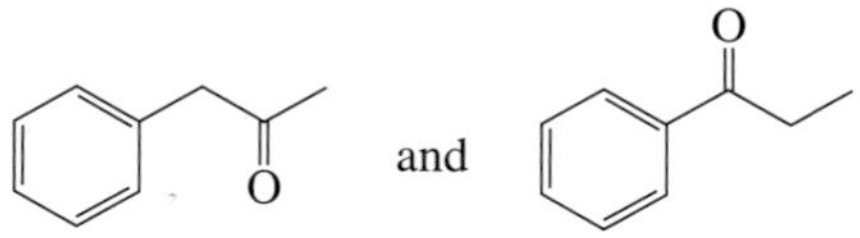

52. How could you distinguish between the compounds in each of the following pairs by ^1H NMR?

a. $CH_3CH_2CH_2OCH_3$ and $CH_3CH_2OCH_2CH_3$

b. $BrCH_2CH_2CH_2Br$ and $BrCH_2CH_2CH_2NO_2$

c. $CH_3\underset{\overset{|}{CH_3}}{CH}-\underset{\overset{|}{CH_3}}{CH}CH_3$ and $CH_3\underset{\overset{|}{CH_3}}{\overset{\overset{CH_3}{|}}{C}}CH_2CH_3$

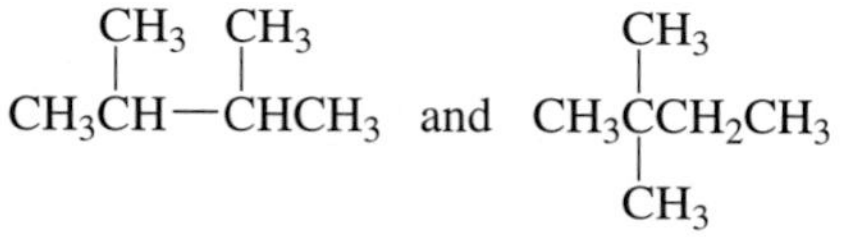

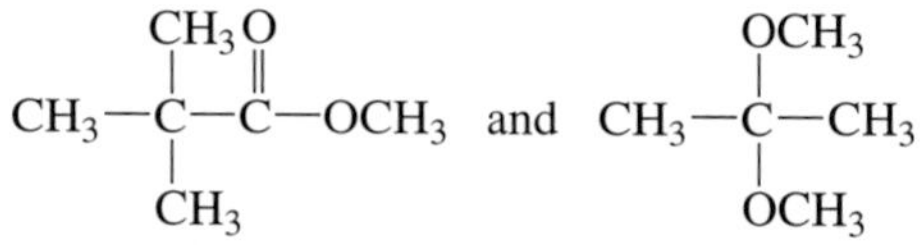

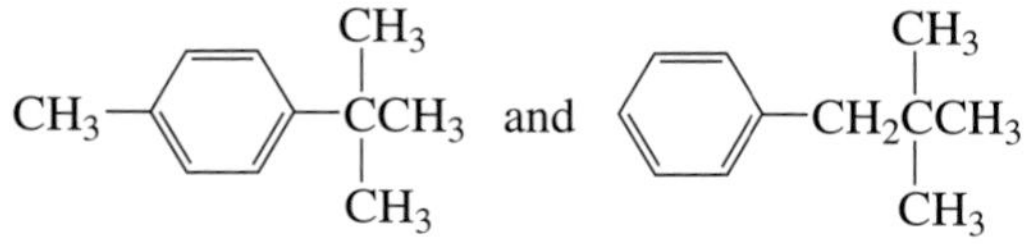

d. $CH_3-\underset{\overset{|}{CH_3}}{C}-\overset{\overset{O}{\|}}{C}-OCH_3$ and $CH_3-\underset{\overset{|}{OCH_3}}{\overset{\overset{OCH_3}{|}}{C}}-CH_3$

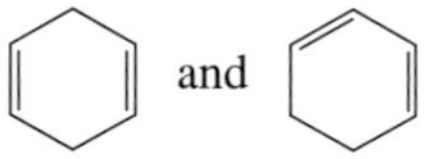

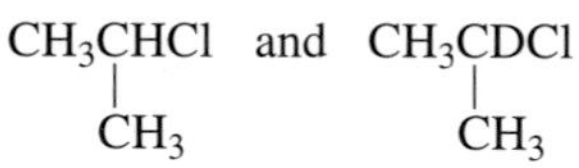

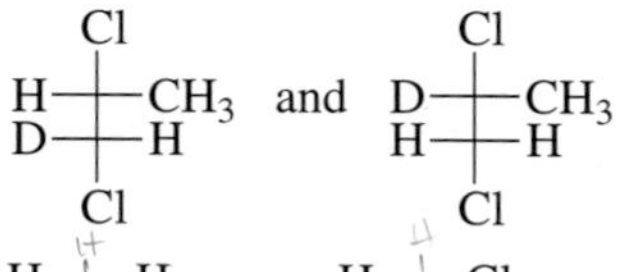

e. $CH_3-$$-\underset{\overset{|}{CH_3}}{\overset{\overset{CH_3}{|}}{C}}CH_3$ and $-CH_2\underset{\overset{|}{CH_3}}{\overset{\overset{CH_3}{|}}{C}}CH_3$

f. and

g. CH_3CHCl and CH_3CDCl
   $\quad\ \ \underset{CH_3}{|}$ $\quad\ \ \underset{CH_3}{|}$

h. $\begin{matrix} Cl \\ H{-}{-}CH_3 \\ D{-}{-}H \\ Cl \end{matrix}$ and $\begin{matrix} Cl \\ D{-}{-}CH_3 \\ H{-}{-}H \\ Cl \end{matrix}$

i. and

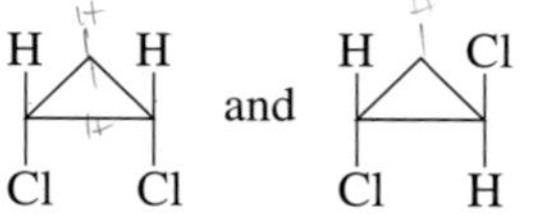

53. Answer the following:
   a. What is the relationship between chemical shift in ppm and operating frequency?
   b. What is the relationship between chemical shift in hertz and operating frequency?
   c. What is the relationship between coupling constant and operating frequency?
   d. How does the operating frequency in NMR spectroscopy compare with the operating frequency in IR and UV/Vis spectroscopy?

54. The ^{1}H NMR spectra of three isomers with molecular formula C_4H_9Br are given below. Tell which isomer produces each spectrum.

a.

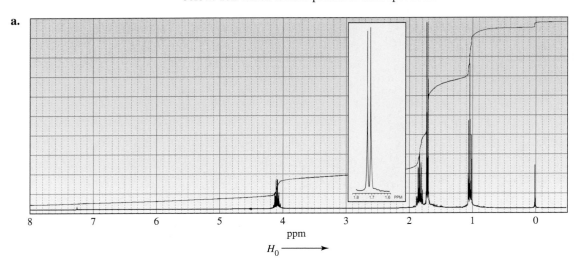

b.

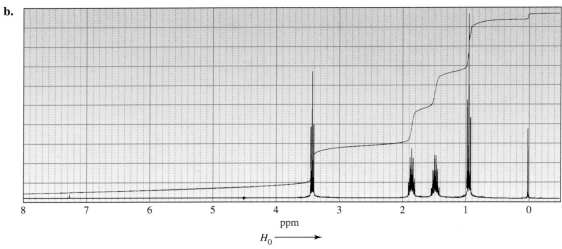

c.

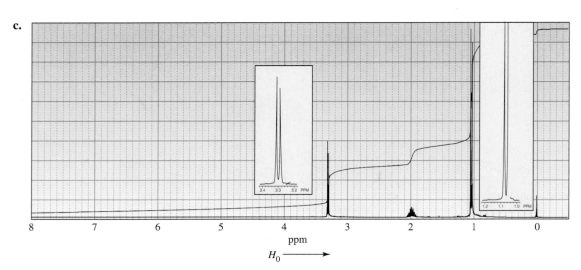

55. Identify each of the following compounds from the ^{1}H NMR data and molecular formula. The number of hydrogens responsible for each signal is shown in parentheses.

   a. $C_4H_8Br_2$ δ 1.97 (6) singlet
                      δ 3.89 (2) singlet
   b. C_8H_9Br δ 2.01 (3) doublet
                      δ 5.14 (1) quartet
                      δ 7.35 (5) broad singlet
   c. $C_5H_{10}O_2$ δ 1.15 (3) triplet
                      δ 1.25 (3) triplet
                      δ 2.33 (2) quartet
                      δ 4.13 (2) quartet

56. Compound A, with molecular formula C_4H_9Cl, shows two signals in its ^{13}C NMR spectrum. Compound B, an isomer of compound A, shows four signals and in the spin-coupled mode the signal farthest downfield is a doublet. Identify the compounds.

57. The ^{1}H NMR spectra of three isomers with molecular formula $C_7H_{14}O$ are given below. Tell which isomer produces each spectrum.

a.

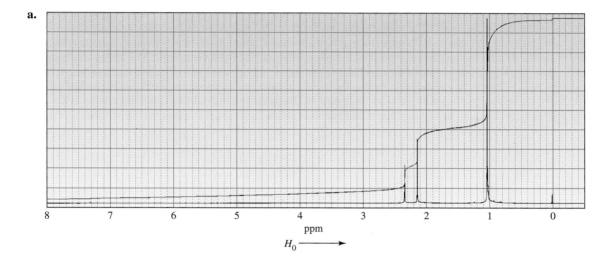

b.

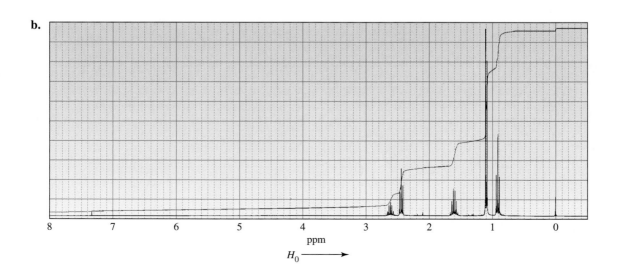

c.

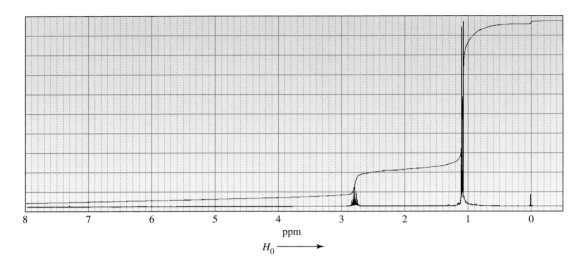

58. Would it be better to use ^1H NMR or ^{13}C NMR to distinguish among 1-butene, *cis*-2-butene, and 2-methylpropene? Explain your answer.

59. Many credit card slips do not have carbon paper. Nevertheless, when you sign the slip, an imprint of your signature is made on the bottom copy. The carbonless paper contains tiny capsules that are filled with the colorless compound shown below. When you press on the paper, the capsules burst and the colorless contents come into contact with the acid-treated paper, forming a highly colored compound. What is the structure of the colored compound?

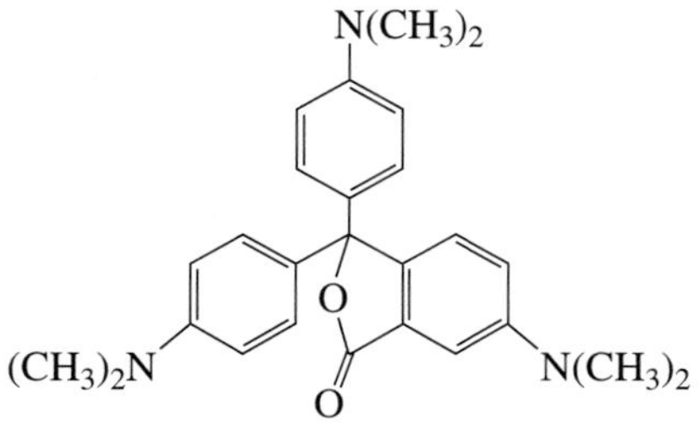

60. Determine the structure of each unknown compound based on its molecular formula and its IR and ^1H NMR spectra.

a. C_3H_6O

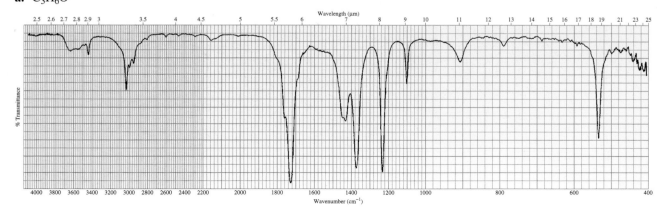

a. continued

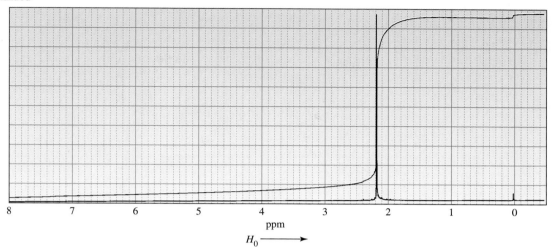

b. C₅H₁₂O

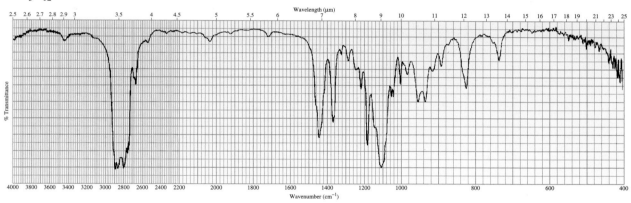

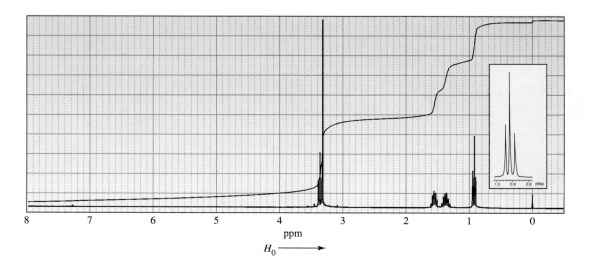

c. $C_6H_{12}O_2$

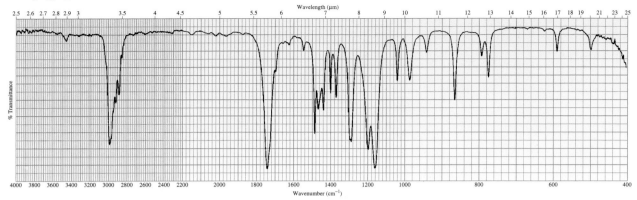

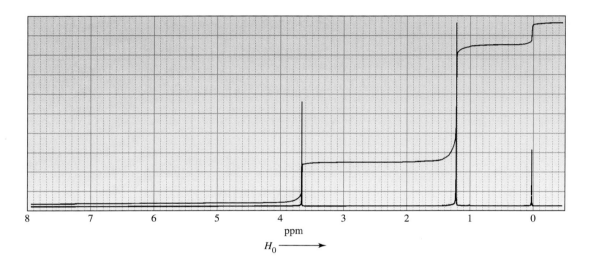

d. $C_4H_7ClO_2$

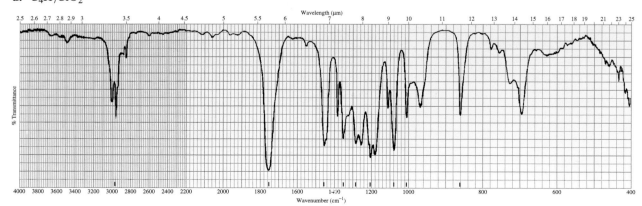

d. continued

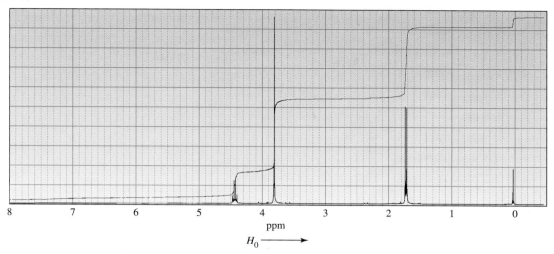

e. $C_4H_8O_2$

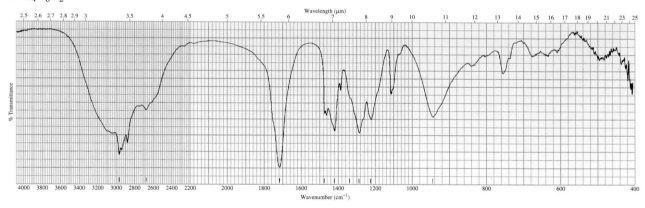

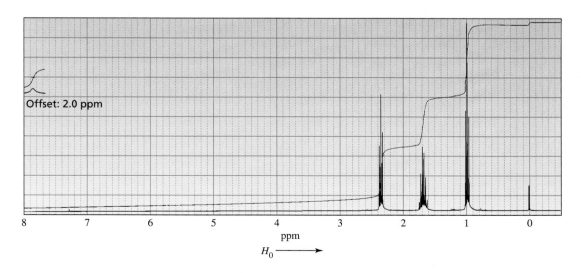

61. There are four esters with molecular formula $C_4H_8O_2$. How could you distinguish among them by ^1H NMR?

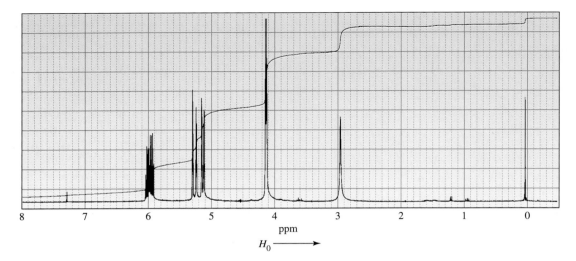

62. The ^{1}H NMR spectrum of 2-propen-1-ol is shown above. Indicate the protons in the molecule that give rise to each of the signals in the spectrum.

63. The ^{1}H NMR spectra for two compounds with molecular formula $C_{11}H_{16}$ are shown below. Identify the compounds.

a.

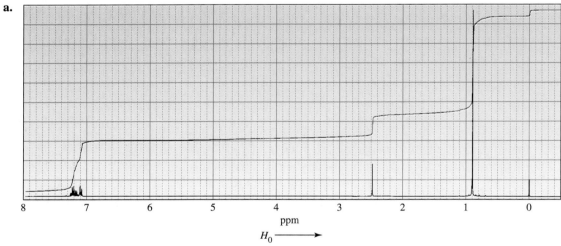

b.

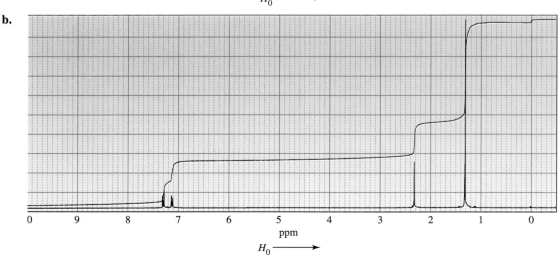

64. Draw a splitting diagram for the H_b proton if $J_{bc} = 10$ and $J_{ba} = 5$.

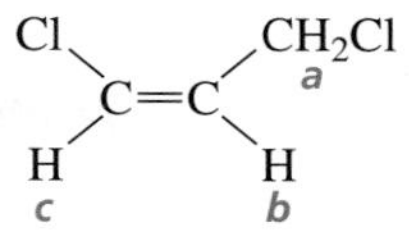

Cl—C=C with CH₂Cl (a), H (c), H (b) substituents

65. Sketch the following spectra that would be obtained for 2-chloroethanol.
   a. The ^1H NMR spectrum for a dry sample of the alcohol
   b. The ^1H NMR spectrum for a sample of the alcohol that contains a trace amount of acid
   c. The ^{13}C NMR spectrum
   d. The spin-coupled ^{13}C NMR spectrum
   e. The four parts of a DEPT ^{13}C NMR spectrum

66. Identify each compound from its molecular formula and its ^1H NMR spectrum.

   a. C_8H_8

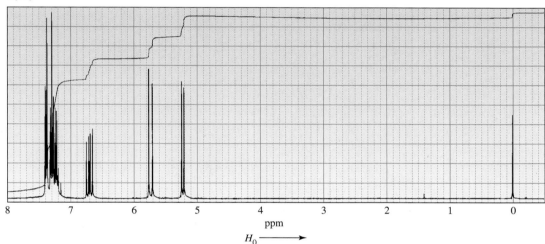

   b. $C_6H_{12}O$

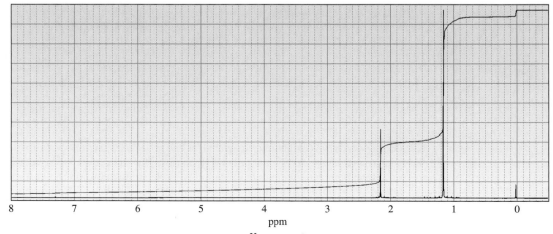

c. $C_9H_{18}O$

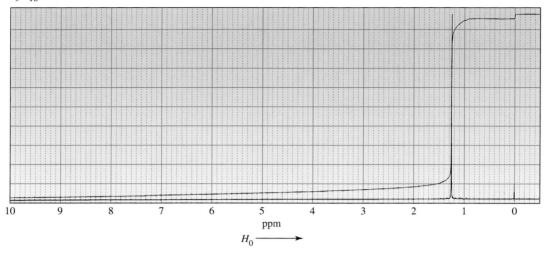

d. C_4H_8O

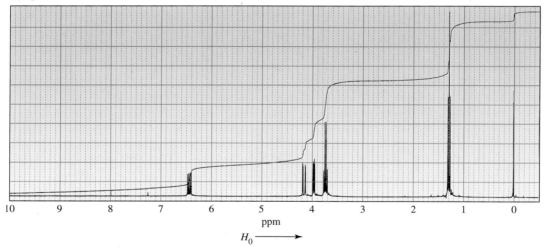

67. Dr. N. M. Arr was called in to help analyze the ^1H NMR spectrum of a mixture of compounds known to contain only C, H, and Br. The mixture showed two singlets, one at 1.8 ppm and the other at 2.7 ppm, with relative integrals of $1:6$, respectively. Dr. Arr determined that the spectrum was that of a mixture of bromomethane and 2-bromo-2-methylpropane. What was the ratio of bromomethane to 2-bromo-2-methylpropane in the mixture?

68. Calculate the amount of energy (in calories) required to flip a ^1H nucleus in an NMR spectrometer that operates at 60 MHz.

69. The ^1H NMR spectra for four compounds with molecular formula $C_6H_{12}O_2$ are shown below. Identify the compounds.

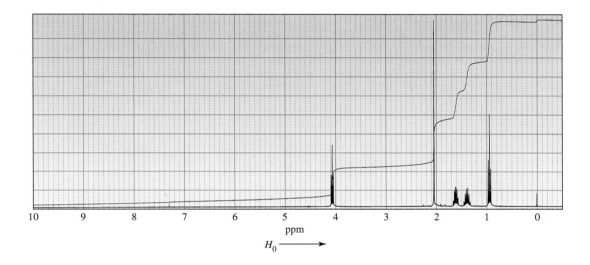

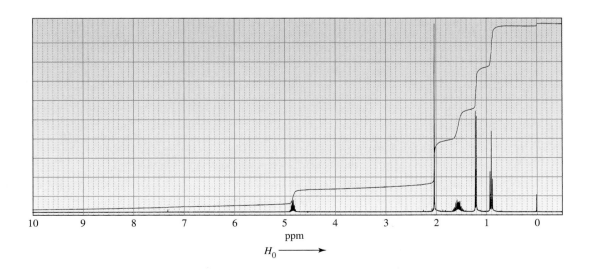

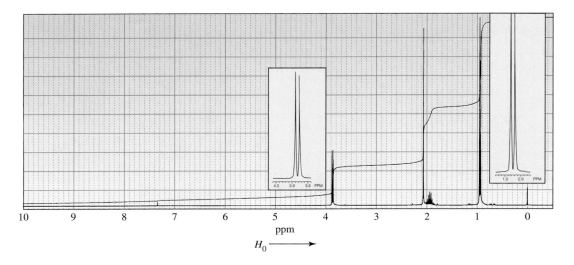

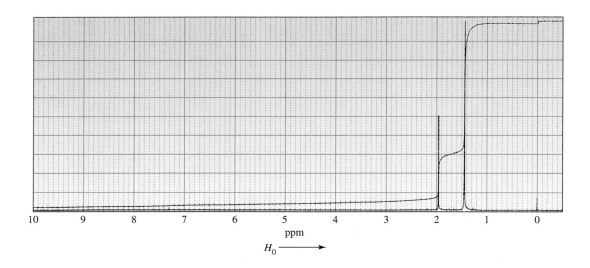

70. A solution of ethanol has been contaminated with benzene. Benzene has a molar absorptivity of 230 at 260 nm, and ethanol shows no absorbance at 260 nm. How could you determine the concentration of benzene in the solution?

71. When compound A ($C_5H_{12}O$) is treated with HBr, it forms compound B ($C_5H_{11}Br$). The ^1H NMR spectrum of compound A has one singlet (1), two doublets (3, 6), and two multiplets (both 1). (The relative areas are indicated in parentheses.) The NMR spectrum of compound B has a singlet (6), a triplet (3), and a quartet (2). Identify compounds A and B.

72. Determine the structure of each unknown compound based on its molecular formula and its IR and ^1H NMR spectra.

a. $C_6H_{12}O$

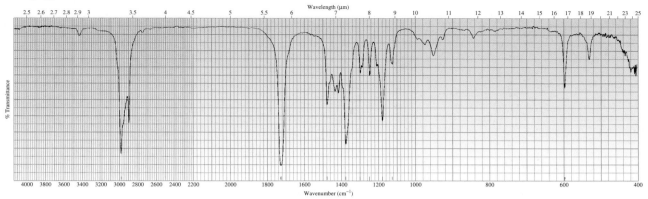

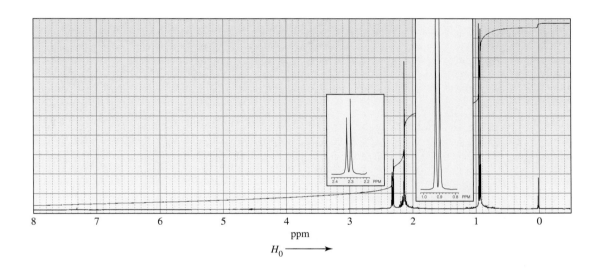

b. $C_6H_{14}O$

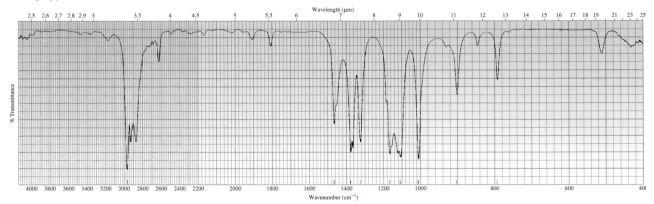

b. continued

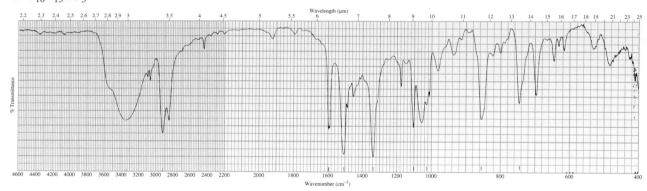

c. $C_{10}H_{13}NO_3$

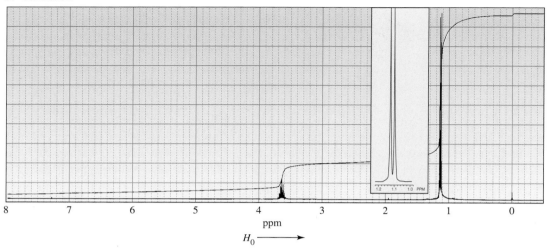

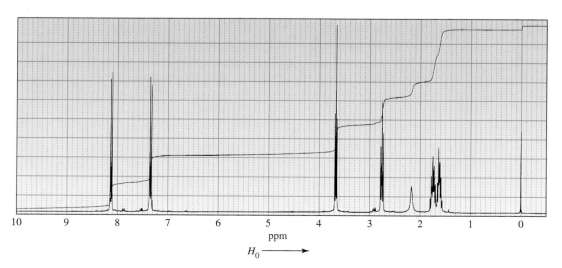

d. $C_{11}H_{14}O_2$

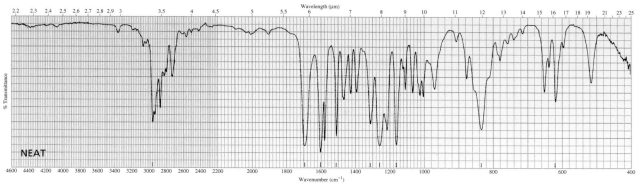

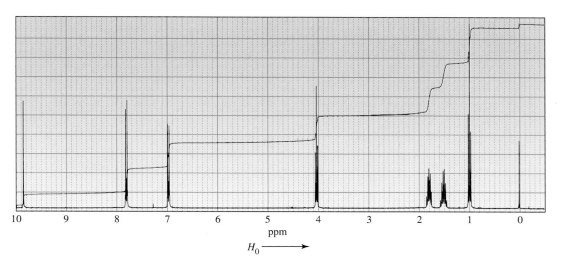

73. Identify the compound with molecular formula $C_6H_{10}O$ that is responsible for the following DEPT ^{13}C NMR spectrum.

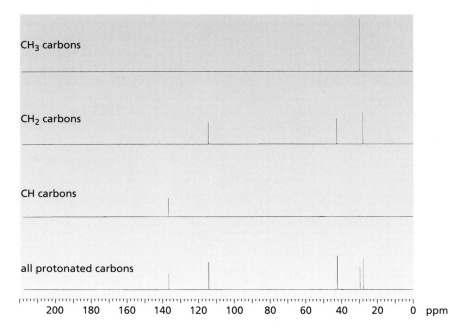

74. Determine the structure of each unknown compound based on its mass, IR, and ^{1}H NMR spectra.

a.

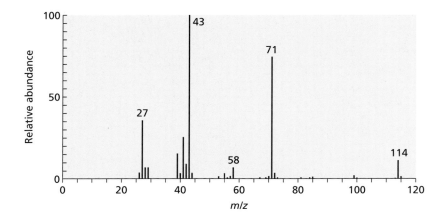

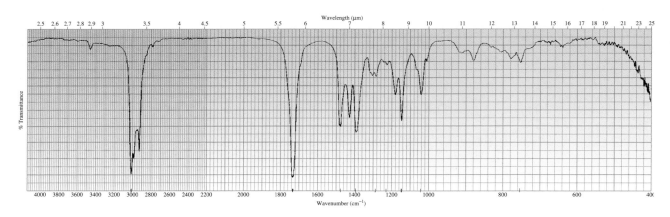

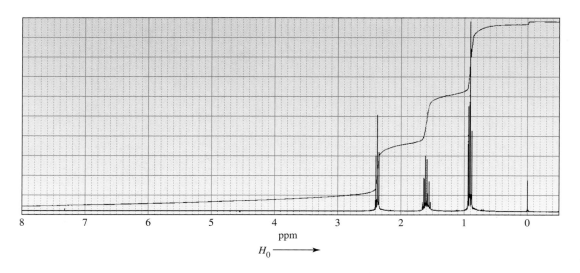

b.

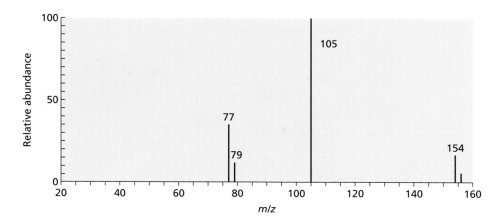

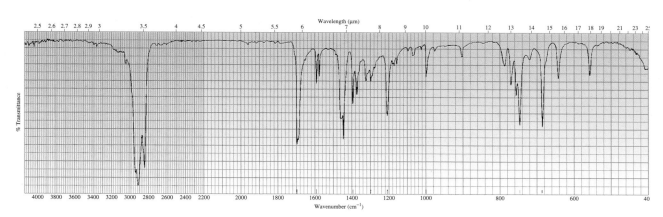

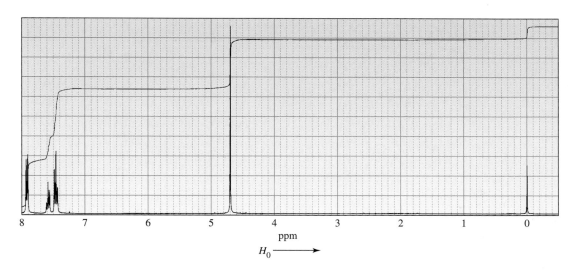

75. Identify the compound with molecular formula $C_7H_{14}O$ that gives the spin-coupled ^{13}C NMR spectrum shown below.

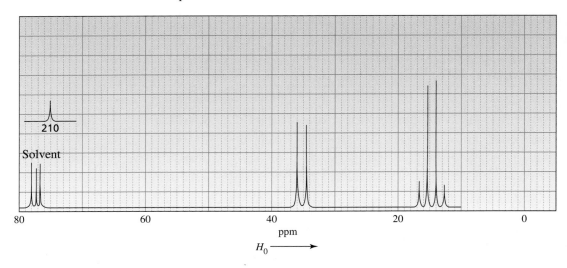

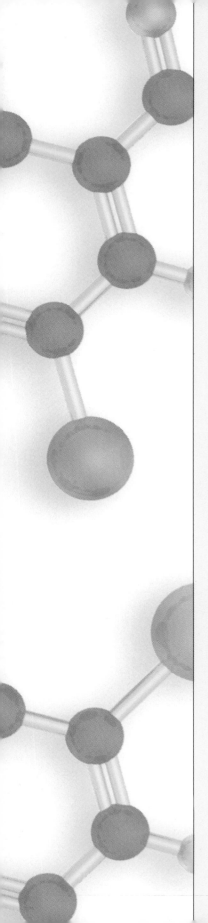

PART V

AROMATIC COMPOUNDS

Chapter 14   Reactions of Benzene and Substituted Benzenes

In Chapter 6, you learned about resonance and aromaticity, and you discovered that benzene was a particularly stable compound because of its aromaticity. In **Chapter 14** you will study the reactions that benzene undergoes. You will see that, although benzene has many of the same structural features that alkenes and dienes have (they all have carbon–carbon π bonds and therefore they are all nucleophiles), benzene's aromaticity causes it to undergo reactions that are quite different from the reactions that alkenes and dienes undergo.

You will see that if the benzene ring has a substituent, the nature of the substituent will affect both the reactivity of the ring and the placement of any incoming substituent. You will then have the opportunity to design syntheses of compounds that contain benzene rings. The reactions of aromatic compounds in which one of the ring atoms is an atom other than a carbon (such compounds are called heterocyclic compounds) are discussed in Chapter 26.

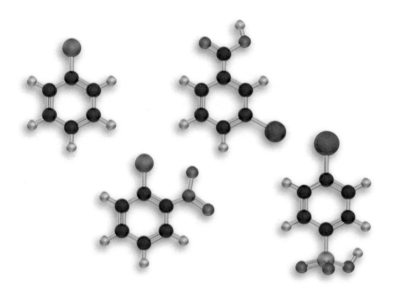

14

REACTIONS OF BENZENE AND SUBSTITUTED BENZENES

chlorobenzene, *meta*-bromobenzoic acid,
ortho-chloronitrobenzene, *para*-iodobenzenesulfonic acid

The compound we know as benzene was first isolated in 1825 by Michael Faraday. He extracted the compound from a gas obtained from whale oil that was being used to illuminate buildings in London. Because of its origin, chemists suggested that it should be called "pheno" from the Greek word *phainein* ("to shine").

In 1834, Eilhardt Mitscherlich correctly determined the molecular formula of benzene (C_6H_6) and decided to call it benzin because of its relationship to benzoic acid, a known substituted form of the compound. Later its name was changed to benzene.

Compounds like benzene, with relatively few hydrogens compared with the number of carbons, are typically found in oils produced by trees and other plants. Early chemists called such compounds **aromatic compounds** because of their pleasing fragrances. In this way, they were distinguished from **aliphatic compounds,** which have higher hydrogen-to-carbon ratios and are obtained from the chemical degradation of fats. The chemical meaning of the word "aromatic" now signifies certain kinds of chemical structures. A compound must meet certain criteria to be classified as aromatic, which we discussed in Section 6.11. Benzene is an aromatic compound based on these criteria.

Since the isolation of benzene from whale oil, many naturally occurring, substituted benzenes have been found. A few examples that show physiological activity in humans are presented here.

Michael Faraday (1791–1867) *was born in England, one of ten children of a blacksmith. At the age of 14 he was apprenticed to a bookbinder and educated himself by reading the books that he bound. In 1833 he became a professor of chemistry at the Royal Institution. He is best known for his work on electricity.*

adrenaline
epinephrine

mescaline
active agent of the peyote cactus

ephedrine
a bronchodilator

chloramphenicol
an antibiotic used against rickettsial infections such as Rocky Mountain spotted fever; particularly effective against typhoid fever

PEYOTE CULTS

For several centuries, a peyote cult existed among the Aztecs in Mexico, later spreading to many Native North American tribes. By 1880, a religion that combined Christian beliefs with the ancient Native American use of the peyote cactus had developed in the southwestern United States, primarily among Native Americans. The followers of this religion believed that the cactus is divinely endowed to shape each person's life. Currently, the only people in the United States legally permitted to use peyote are members of the Native American Church, and they are allowed to use it only in their religious rites.

In addition to naturally occurring substituted benzenes, there are also many synthetic substituted benzenes. The diet drug "fen/phen" is a mixture of two synthetic substituted benzenes, fenfluramine and phentermine. Agent Orange, a defoliant widely used in the 1960s during the Vietnam War, is also a mixture of two synthetic substituted benzenes, 2,4-D and 2,4,5-T. The compound TCDD is a contaminant formed during the manufacture of Agent Orange, and has been implicated as the causative agent of the various symptoms suffered by those exposed to the defoliant during the war.

Because of the known physiological activities of adrenaline (epinephrine) and mescaline, chemists have synthesized compounds with similar structures. One such compound is amphetamine, a central nervous system stimulant. Amphetamine and a close relative, methamphetamine, are used clinically as appetite suppressants. Methamphetamine is the street drug known as "speed" because of its rapid and intense psychological effects. Two other synthetic substituted benzenes, BHA and BHT, are preservatives (discussed in Section 8.8) found in a wide variety of packaged foods. These compounds represent just a few of the many substituted benzenes that have been synthesized for commercial use by the chemical and pharmaceutical industries.

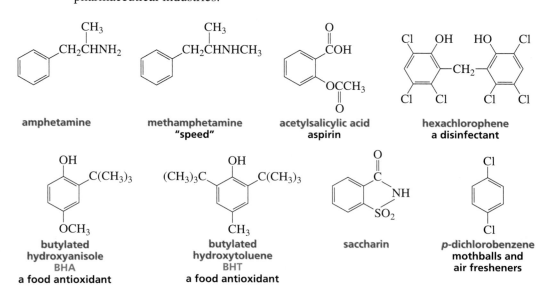

We discussed the structure of benzene in Chapter 6, but we have not yet discussed the reactions it undergoes. This chapter deals primarily with the reactions of benzene and substituted benzenes. You should recall that the structure of benzene can be represented by two contributing resonance structures and that, as a result of its aromaticity, benzene is a particularly stable compound. The physical properties of several substituted benzenes are given in Appendix I.

For a compound to be aromatic, it must be cyclic and planar and it must have an uninterrupted cloud of delocalized π electrons. The cloud must contain an odd number of pairs of π electrons.

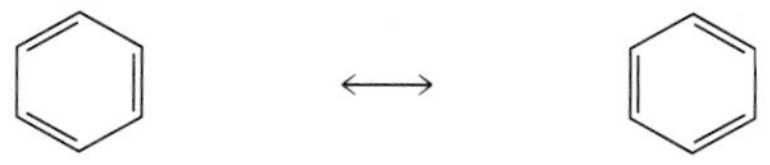

resonance contributor
of benzene

⟷

resonance contributor
of benzene

resonance hybrid
of benzene

Monosubstituted Benzenes

Some monosubstituted benzenes are named simply by stating the name of the substituent followed by the word "benzene."

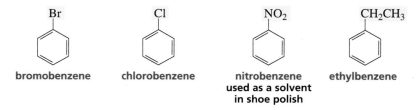

bromobenzene   chlorobenzene   nitrobenzene
**used as a solvent
in shoe polish**   ethylbenzene

Some monosubstituted benzenes have names that incorporate the name of the substituent. Unfortunately, these names must be memorized.

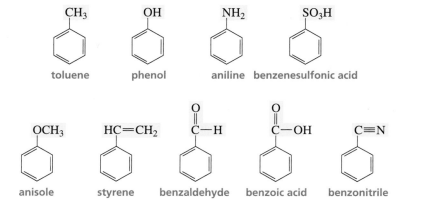

toluene   phenol   aniline   benzenesulfonic acid

anisole   styrene   benzaldehyde   benzoic acid   benzonitrile

With the exception of toluene, benzene rings with an alkyl substituent are named as substituted benzenes.

Eilhardt Mitscherlich (1794–1863) *was born in Germany. He studied Oriental languages at the University of Heidelberg and the Sorbonne, where he concentrated on Farsi, hoping that Napoleon would include him in a delegation he intended to send to Persia. That ambition ended with Napoleon's defeat. Mitscherlich returned to Germany to study science, simultaneously receiving a doctorate in Persian studies. He became a professor of chemistry at the University of Berlin.*

CH₃CHCH₃   CH₃CHCH₂CH₃   CH₃ĊCH₃ (CH₃)

isopropylbenzene   *sec*-**butylbenzene**   *tert*-**butylbenzene**
cumene

If the alkyl substituent cannot be unambiguously named, the compound must be named as a substituted alkane. As a substituent, the benzene ring itself is called a **phenyl group,** and a benzene ring with a methylene group is a **benzyl group.** The phenyl group gets its name from "pheno," the name that had been rejected for benzene.

a phenyl group   a benzyl group

CH₃CHCH₂CH₂CH₃   CH₂Cl

2-phenylpentane   **chloromethylbenzene**   **diphenyl ether**   **dibenzyl ether**
benzyl chloride

An aryl group (Ar) is the general term for either a phenyl group or a substituted phenyl group, just as an alkyl group (R) is the general term for a group derived from an alkane. In other words, ArOH could be used to designate any of the following phenols.

OH   OH (Br)   OH (CH₂CH₃)   OH (OCH₃)

Disubstituted Benzenes

Benzene rings with two substituents must be named to indicate the relative locations of the substituents. Their relative positions can be indicated by numbers or by the prefixes *ortho, meta,* and *para.* Adjacent substituents are called *ortho,* substituents separated by one carbon are called *meta,* and substituents located opposite one another are designated *para.* Often only their abbreviations (*o, m, p*) are used in naming compounds.

Br (Br)   Br (Br)   Br (Br)

1,2-dibromobenzene   **1,3-dibromobenzene**   **1,4-dibromobenzene**
ortho-dibromobenzene   *meta*-dibromobenzene   *para*-dibromobenzene
o-dibromobenzene   *m*-dibromobenzene   *p*-dibromobenzene

If the two substituents are different, they are listed in alphabetical order. The ring is numbered so that the smallest numbers are used to name the compound and the first stated substituent is given the 1-position. If one of the substituents can be incorporated into a name, that name is used and the incorporated substituent is given the 1-position.

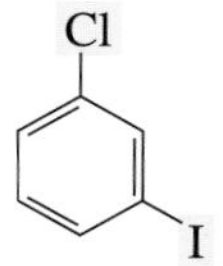

1-chloro-3-iodobenzene
meta-chloroiodobenzene
not
1-iodo-3-chlorobenzene
meta-iodochlorobenzene

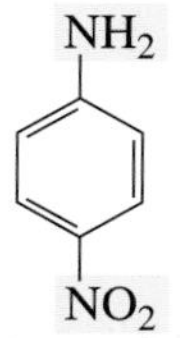

4-nitroaniline
para-nitroaniline
not
para-aminonitrobenzene

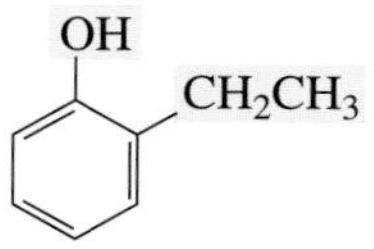

2-ethylphenol
ortho-ethylphenol
not
ortho-ethylhydroxybenzene

A few disubstituted benzenes have names that incorporate both substituents.

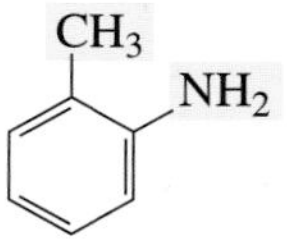

ortho-toluidine

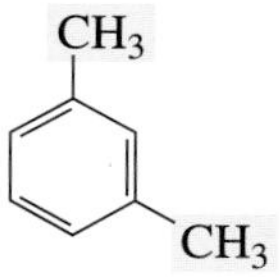

meta-xylene

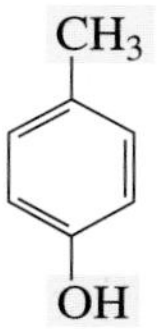

para-cresol
**used as a wood preservative
until prohibited for
environmental reasons**

PROBLEM 1

Give the structures of:

a. *para*-toluidine   **b.** *meta*-cresol   **c.** *para*-xylene

---

Polysubstituted Benzenes

If the benzene ring has more than two substituents, the substituents are numbered so the lowest possible numbers are used. The substituents are listed in alphabetical order with their appropriate numbers.

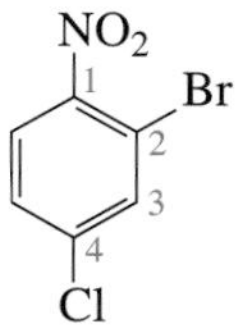

2-bromo-4-chloro-1-nitrobenzene 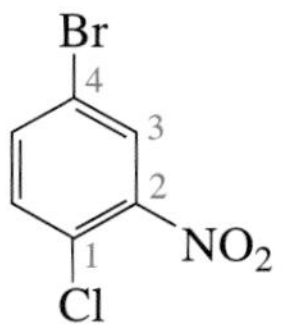   **4-bromo-1-chloro-2-nitrobenzene** 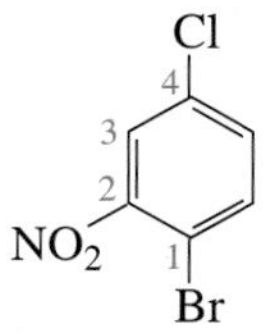   **1-bromo-4-chloro-2-nitrobenzene**

As with disubstituted benzenes, if one of the substituents can be incorporated into a name, that name is used and the incorporated substituent is given the 1-position.

PROBLEM 2 ◆

Draw the structure of each of the following compounds.

a. *m*-dichlorobenzene

b. *p*-bromophenol

c. *o*-nitroaniline

d. 2-bromo-4-iodo-1-nitrobenzene

e. 2-phenylhexane

f. 3-benzylpentane

g. *m*-chlorotoluene

h. 2,5-dinitrobenzaldehyde

i. *o*-xylene

j. *m*-chlorobenzonitrile

PROBLEM 3 ◆

Correct the following incorrect names:

a. 2,4,6-tribromobenzene

b. 3-hydroxynitrobenzene

c. *para*-methylbromobenzene

d. 1,6-dichlorobenzene

14.2 REACTIVITY CONSIDERATIONS

We have seen (Section 6.2) that benzene is a planar molecule with a doughnut-shaped cloud of π electrons both above and below the plane of the ring (Figure 14.1). As a consequence of these π electron clouds, benzene is a nucleophile and therefore will be attracted to an electrophile (Y^+). When an electrophile attaches itself to a benzene ring, a carbocation intermediate is formed.

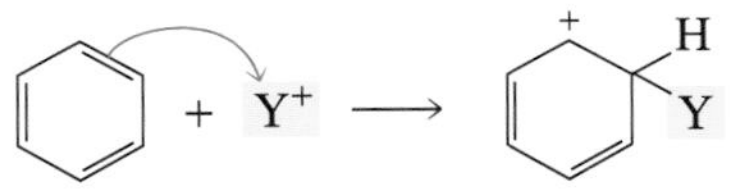

carbocation intermediate

This is reminiscent of the first step in an electrophilic addition reaction of an alkene, where a nucleophilic alkene reacts with an electrophile, forming a carbocation intermediate (Section 3.6). In alkene reactions, we saw that the carbocation subsequently reacts with a nucleophile (Z^-) to form an addition product.

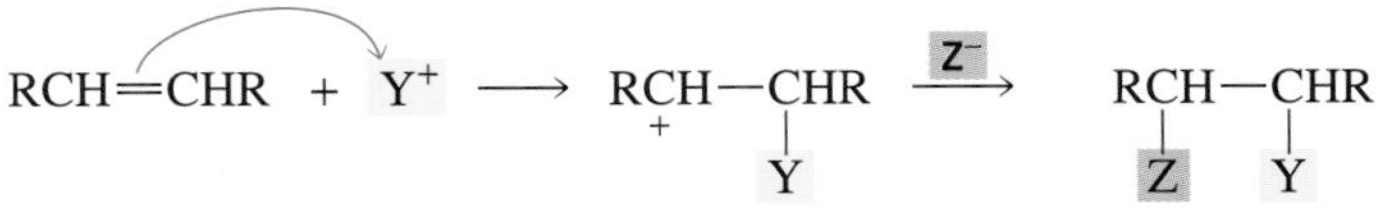

$$RCH{=}CHR + Y^+ \longrightarrow RCH{-}CHR \xrightarrow{Z^-} RCH{-}CHR$$

carbocation intermediate — **product of electrophilic addition**

▲ **Figure 14.1**
Overlap of *p* orbitals results in a cloud of high π electron density above and below the plane of the benzene ring.

If the carbocation intermediate formed from the reaction of benzene with an electrophile were to react similarly with a nucleophile, the product would not be aromatic. Because there is a great deal of stabilization associated with an aromatic ring, the carbocation, instead of reacting with a nucleophile, loses a proton from the site of electrophilic attack. In this way, the aromaticity of the benzene ring is restored. The end result is that a hydrogen attached to the benzene ring is replaced by an electrophile.

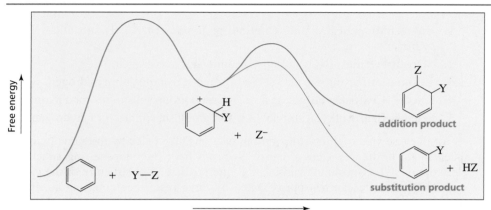

▲ **Figure 14.2**
Reaction coordinate diagrams for electrophilic substitution of benzene and for electrophilic addition to benzene.

The reaction of benzene to form a substituted benzene has a $\Delta G°$ close to zero (Figure 14.2). The reaction of benzene to form the much less stable nonaromatic addition product would have been a highly endergonic reaction. Consequently, aromatic compounds such as benzene and substituted benzenes undergo *electrophilic substitution reactions* (an electrophile substitutes for a hydrogen) rather than *electrophilic addition reactions,* which are characteristic of alkenes.

PROBLEM 4

If electrophilic addition to benzene is an endergonic reaction, how can electrophilic addition to an alkene be an exergonic reaction?

In an **electrophilic aromatic substitution reaction,** an electrophile substitutes for a hydrogen of an aromatic compound.

an electrophilic aromatic substitution reaction

**14.3
GENERAL
MECHANISM FOR
ELECTROPHILIC
AROMATIC
SUBSTITUTION
REACTIONS**

In an electrophilic aromatic substitution reaction, an electrophile (Y^+) is put on the ring and an H^+ comes off the ring.

The five most common electrophilic aromatic substitution reactions are:

1. **Halogenation:** a bromine, chlorine, or iodine (Br, Cl, I) substitutes for a hydrogen.
2. **Nitration:** a nitro (NO_2) group substitutes for a hydrogen.
3. **Sulfonation:** a sulfonic acid (SO_3H) group substitutes for a hydrogen.
4. **Friedel–Crafts acylation:** an acyl ($RC{=}O$) group substitutes for a hydrogen.
5. **Friedel–Crafts alkylation:** an alkyl (R) group substitutes for a hydrogen.

All of these electrophilic substitution reactions take place by the same two-step mechanism. In the first step, a pair of electrons from the benzene ring attacks an electrophile (Y^+), forming a carbocation intermediate. The structure of the carbocation intermediate can be approximated by three resonance contributors. In the second step of the reaction, a proton comes off (it is pulled off by a base in the reaction mixture), and the electrons that held the proton move into the ring to reestablish its aromaticity.

general mechanism for electrophilic aromatic substitution

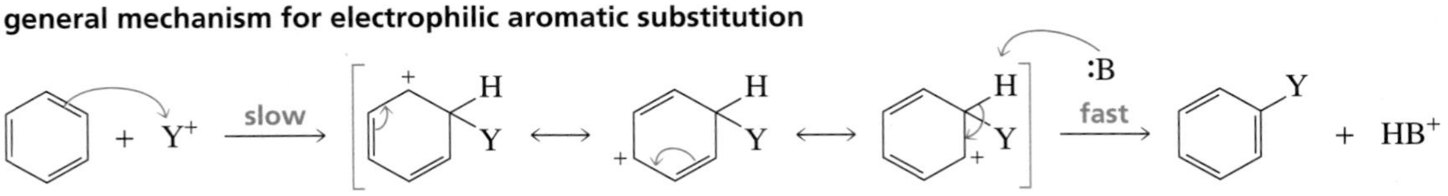

Without looking at the reaction coordinate diagram (Figure 14.2) to see which transition state has the higher energy, we can predict that the first step is the rate-determining or slow step of the reaction, because the aromaticity of the benzene ring is lost in this step. The second step is a relatively fast step, because this step restores the stability-enhancing aromaticity.

We will look at each of these five electrophilic aromatic substitution reactions individually. As you study them, notice that they differ only in how the electrophile (Y^+) that is needed to start the reaction is generated. Once the electrophile has been formed, all five reactions follow this two-step mechanism for electrophilic aromatic substitution.

14.4
HALOGENATION OF BENZENE

bromination

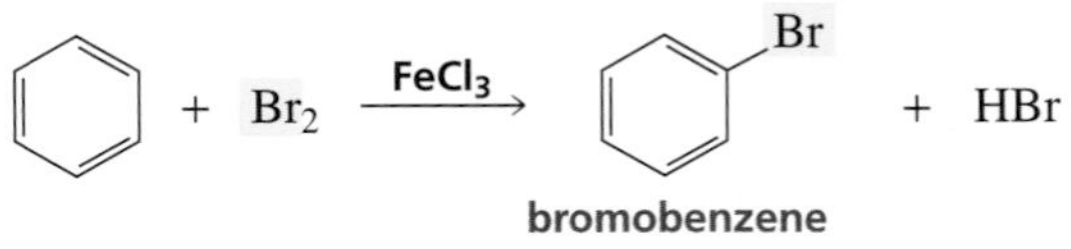

bromobenzene

chlorination

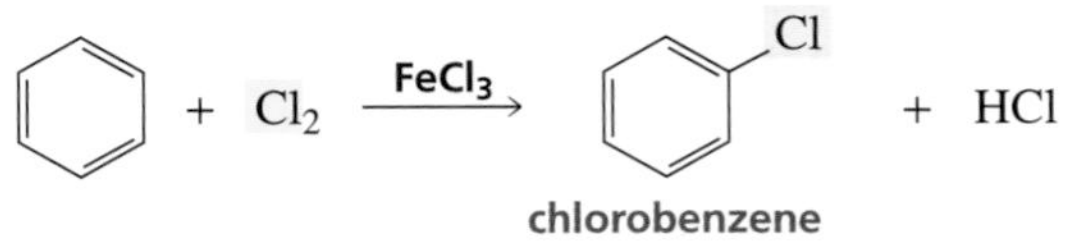

chlorobenzene

iodination

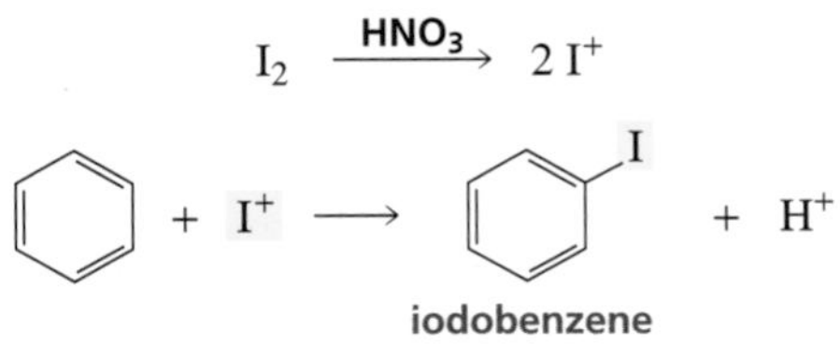

$$I_2 \xrightarrow{\text{HNO}_3} 2\ I^+$$

iodobenzene

A Lewis acid such as ferric bromide ($FeCl_3$ or $FeBr_3$) is required as a catalyst for the bromination of benzene. Recall that a *Lewis acid* is a compound that accepts a share in a pair of electrons (Section 1.20). In the first step of the reaction, bromine donates a lone pair of electrons to the Lewis acid. This weakens the Br—Br bond, thereby providing the electrophile necessary for electrophilic aromatic substitution. (For the sake of simplicity the electrophile is shown as Br^+. In actuality Br^+ is complexed with $^-FeCl_3Br$.)

mechanism for bromination

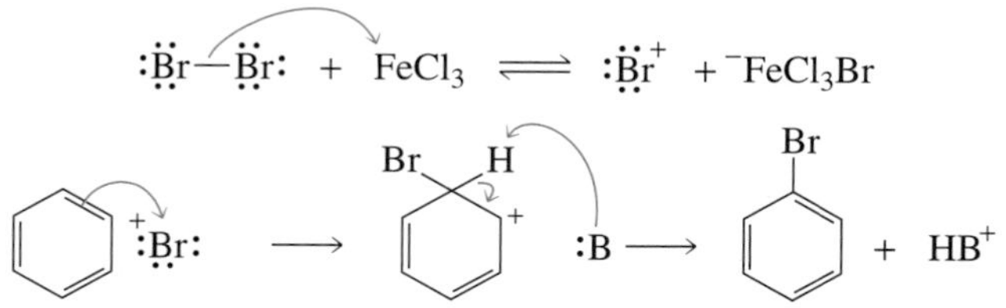

For brevity, only one of the three resonance contributors of the carbocation intermediate is shown (Section 14.3). In the last step of the reaction, a base from the reaction mixture ($^-FeCl_3Br$, Br^-, or solvent) removes the proton from the carbocation intermediate.

Since $FeBr_3$ and $FeCl_3$ readily react with moisture in the air during handling, and since the hydrated Lewis acid is inactive as a catalyst, hydration is prevented by generating the catalyst *in situ* (in the reaction mixture) by adding iron filings and bromine to the reaction mixture.

$$2\ Fe\ +\ 3\ Br_2\ \longrightarrow\ 2\ FeBr_3$$

PROBLEM 5

a. Why is hydrated $FeBr_3$ inactive as a Lewis acid catalyst?

b. Why does generating the catalyst *in situ* prevent hydration?

Chlorination of benzene occurs by the same mechanism as bromination.

mechanism for chlorination

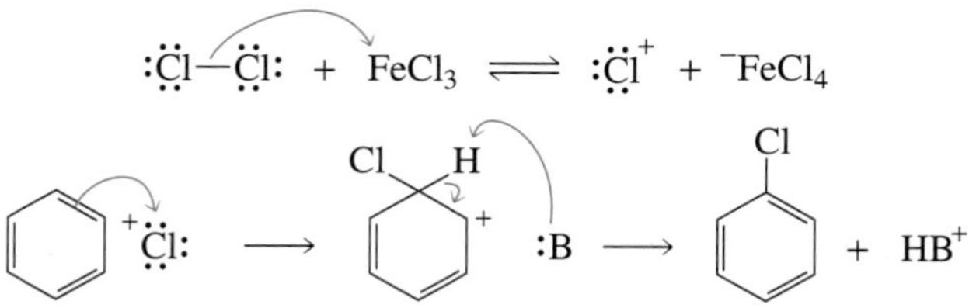

Unlike halogenation of benzene, halogenation of an alkene does not require a Lewis acid catalyst (Section 3.15). This is because an alkene is a stronger nucleophile than benzene and, as a result, the Br—Br or Cl—Cl bond does not have to be broken to form a stronger electrophile.

Electrophilic iodine (I^+) is obtained by oxidizing I_2 with an oxidizing agent such as nitric acid.

mechanism for iodination

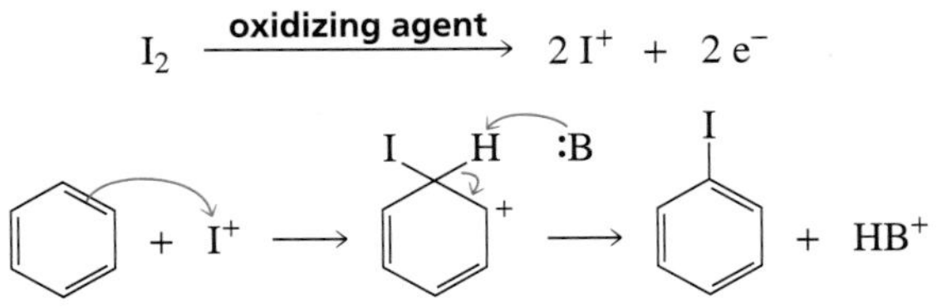

$$I_2 \xrightarrow{\text{oxidizing agent}} 2\,I^+ + 2\,e^-$$

14.5 NITRATION OF BENZENE

nitration

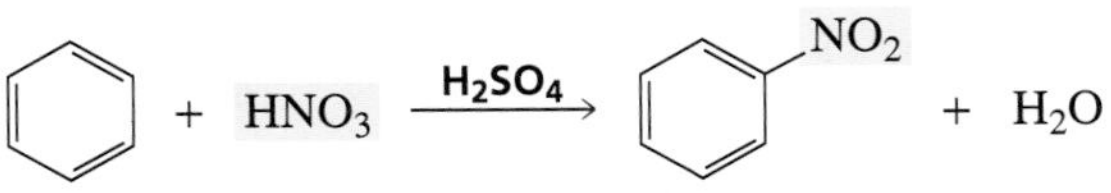

nitrobenzene

Nitration of benzene requires sulfuric acid as a catalyst. Sulfuric acid protonates nitric acid. Loss of water from protonated nitric acid forms a nitronium ion, the electrophile required for nitration. Any base (B = H_2O, HSO_4^-) present in the reaction mixture can remove the proton in the second step of the aromatic substitution reaction.

mechanism for nitration

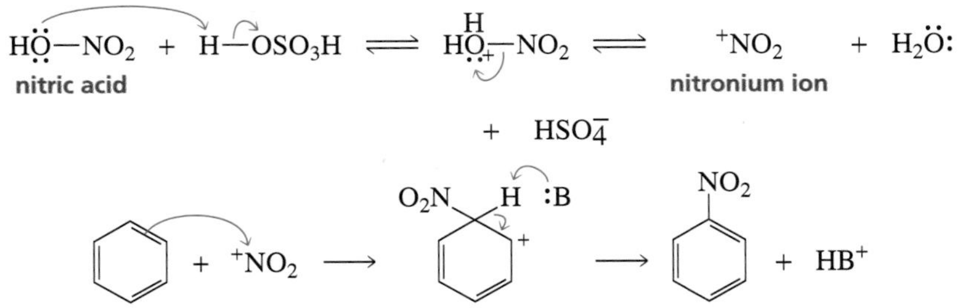

nitric acid

nitronium ion

$$+ \quad HSO_4^-$$

14.6 SULFONATION OF BENZENE

sulfonation

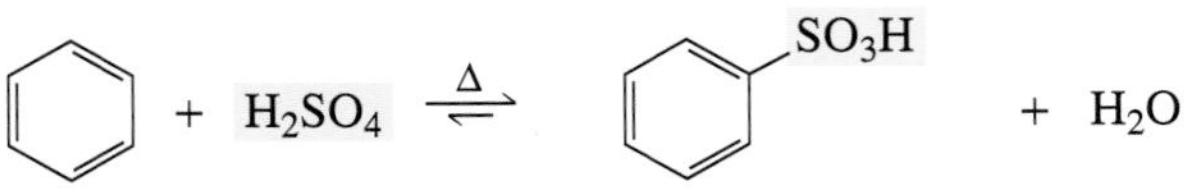

benzenesulfonic acid

Fuming sulfuric acid (a solution of SO_3 in sulfuric acid) or concentrated sulfuric acid is used to sulfonate aromatic rings. As the following mechanism shows, a substantial amount of electrophilic sulfur trioxide (SO_3) is formed when concentrated sulfuric acid is heated, as a result of protonation of sulfuric acid and subsequent loss of water and a proton. Take a minute to note the similarities in the mechanisms for generation of the $^+SO_3H$ electrophile for sulfonation and the $^+NO_2$ electrophile for nitration.

mechanism for sulfonation

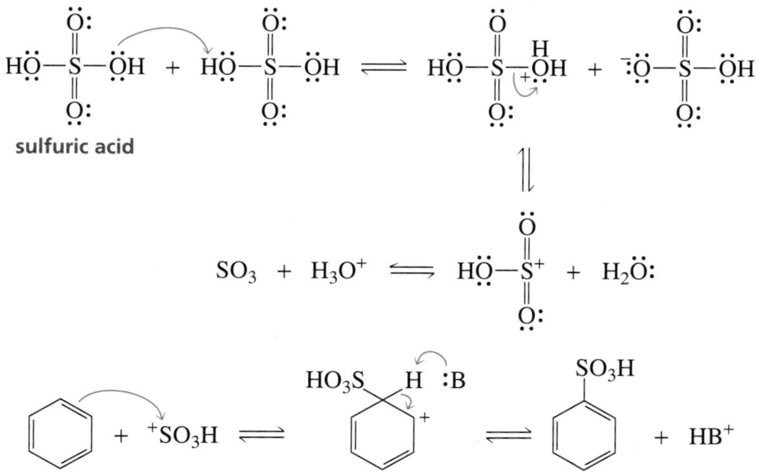

$$SO_3 + H_3O^+ \rightleftharpoons HO-S^+ \ \| \ O + H_2O:$$

Sulfonic acids are strong acids because of the three electron-withdrawing oxygen atoms and the resonance stabilization of the sulfonate ion.

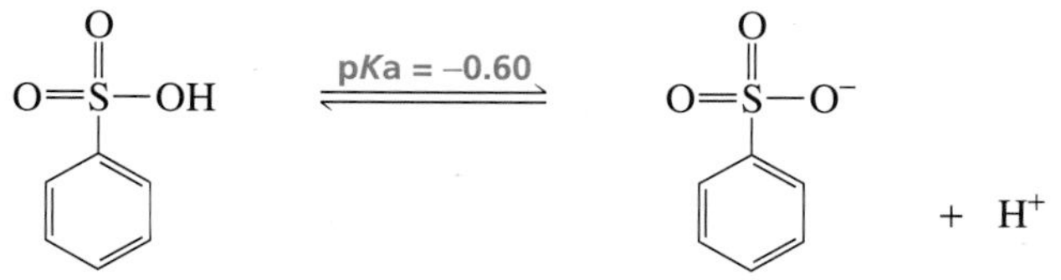

benzenesulfonic acid   benzenesulfonate ion

Sulfonation of benzene is a reversible reaction. If benzenesulfonic acid is heated in dilute acid, the reaction proceeds in the reverse direction.

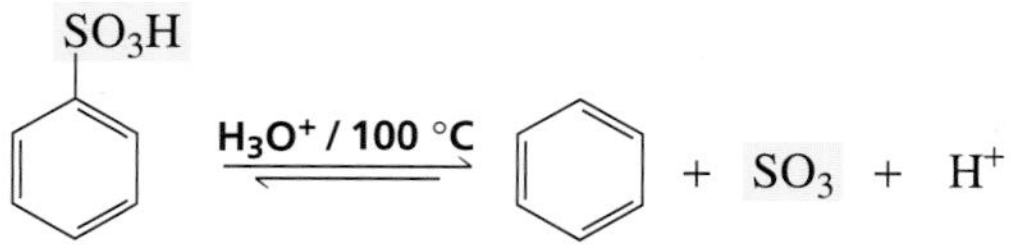

The **principle of microscopic reversibility** applies to all reactions. It states that the mechanism of the reaction in the reverse direction must retrace each step of the mechanism of the reaction in the forward direction in microscopic detail. This means that both reactions must have the same intermediates and that the rate-determining "energy hill" is the same in both directions. For example, sulfonation is described by the reaction coordinate diagram in Figure 14.3 going from left to right. Therefore, desulfonation is described by the same reaction coordinate diagram going from right to left. In sulfonation, the rate-determining step is nucleophilic attack of benzene on the $^+SO_3H$ ion. In desulfonation, the rate-limiting step is loss of the $^+SO_3H$ ion from the benzene ring. An example of the usefulness of desulfonation to synthetic chemists is given in Problem 23.

mechanism for desulfonation

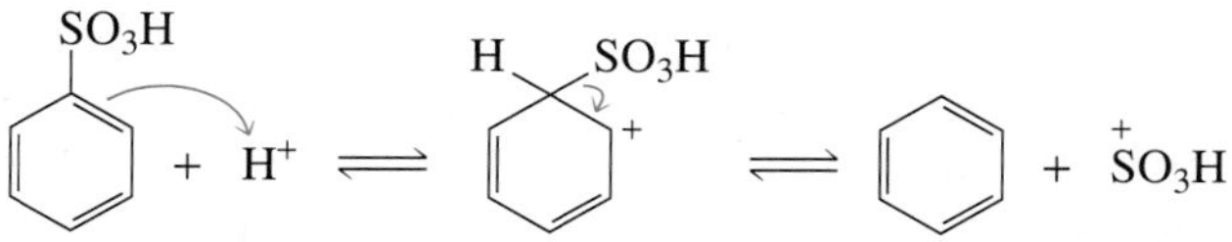

Figure 14.3 ▶
Reaction coordinate diagram
for the sulfonation of
benzene (left to right) and
for the desulfonation of
benzenesulfonic acid
(right to left).

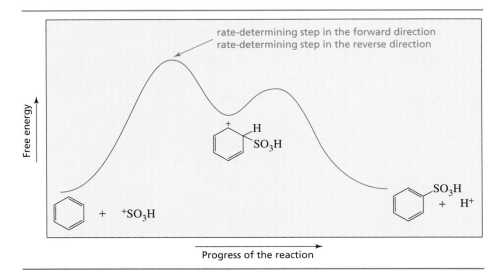

PROBLEM 6

Sulfonation of benzene and desulfonation of benzenesulfonic acid are both two-step
reactions. The rate-determining step for sulfonation is the step with the smaller rate constant. The rate-determining step for desulfonation is the step with the larger rate constant.
Explain how the step with the larger rate constant can be the rate-determining step.

14.7 FRIEDEL–CRAFTS ACYLATION OF BENZENE

Two electrophilic substitution reactions bear the names of chemists Charles Friedel
and James Crafts. *Friedel–Crafts acylation* places an acyl group on a benzene ring,
and *Friedel–Crafts alkylation* places an alkyl group on a benzene ring.

$$\underset{\textbf{an acyl group}}{R-\overset{\displaystyle O}{\overset{\|}{C}}-} \qquad \underset{\textbf{an alkyl group}}{R-}$$

Either an acyl halide or an acid anhydride can be used for acylation.

Friedel–Crafts acylation

Charles Friedel (1832–1899)
*was born in Strasbourg,
France. He was a professor of
chemistry and director of
research at the Sorbonne. At
one point his interest in miner-
alogy led him to attempt to
make synthetic diamonds. He
met Crafts when they both
were doing research at L'Ecole
de Médicine in Paris. They col-
laborated scientifically for
most of their lives, discovering
the Friedel–Crafts reactions in
Friedel's laboratory in 1877.*

The acylium ion is the required electrophile in the Friedel–Crafts acylation reaction. This ion is formed by the reaction of an acyl chloride or an acid anhydride
with $AlCl_3$, a Lewis acid.

mechanism for Friedel–Crafts acylation

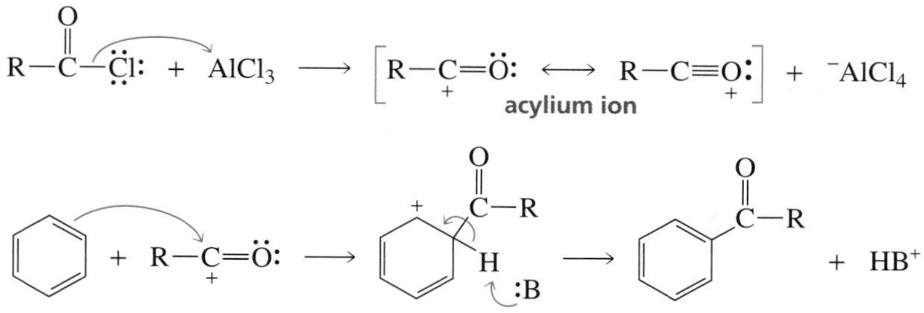

Charles Friedel

PROBLEM 7

Show the mechanism for the generation of the acylium ion if an anhydride instead of an acyl chloride is used in a Friedel–Crafts acylation reaction.

Because the product of a Friedel–Crafts acylation reaction contains a carbonyl group that can complex with $AlCl_3$, Friedel–Crafts acylation reactions must be carried out with more than one equivalent of $AlCl_3$. When the reaction is over, water is added to the reaction mixture to liberate the product from the complex by hydrolyzing the aluminum salts.

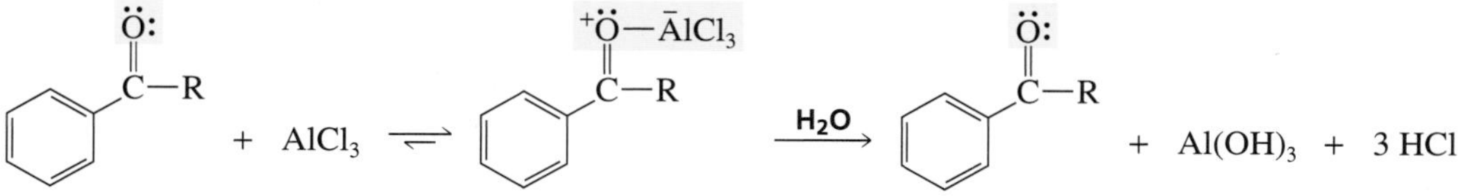

$$+ \; AlCl_3 \;\rightleftharpoons\; \xrightarrow{H_2O} \; + \; Al(OH)_3 \; + \; 3 \, HCl$$

Friedel–Crafts alkylation

$$+ \; RCl \; \xrightarrow{AlCl_3} \; + \; HCl$$

14.8
FRIEDEL–CRAFTS ALKYLATION OF BENZENE

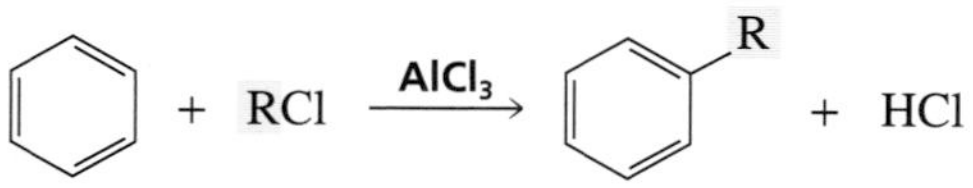

In the first step in a Friedel–Crafts alkylation reaction, a carbocation is formed from the reaction of an alkyl halide with a Lewis acid ($AlCl_3$). Alkyl fluorides, chlorides, bromides, and iodides can all be used. Vinyl halides and aryl halides cannot be used, because their carbocations are too unstable to be formed (Section 9.8).

mechanism for Friedel–Crafts alkylation

$$R-\ddot{C}l\colon + \; AlCl_3 \;\longrightarrow\; R^+ \; + \; {}^-AlCl_4$$

$$+ \; R^+ \;\longrightarrow\; \;\longrightarrow\; + \; HB^+$$

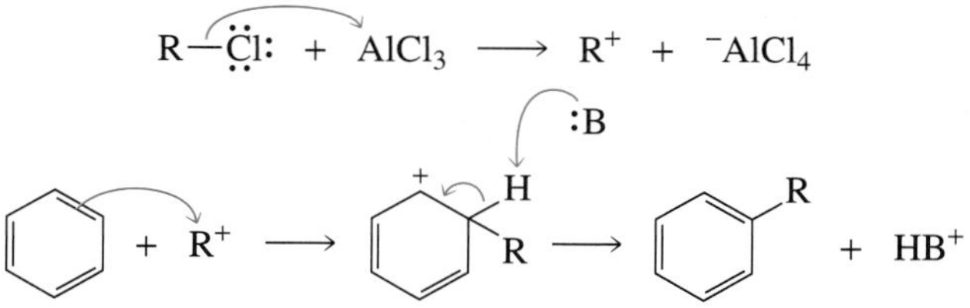

Because an alkylbenzene is more reactive than benzene (Section 14.11), a large excess of benzene is used in Friedel-Crafts alkylation reactions to ensure that the electrophile is more likely to encounter a molecule of benzene than a molecule of alkylated benzene.

As we have seen previously, a carbocation will rearrange if rearrangement leads to a more stable carbocation (Section 3.14). If the carbocation rearranges, the major product of a Friedel–Crafts alkylation reaction will be the product with the rearranged alkyl group on the benzene ring. The relative amounts of rearranged and unrearranged product depend on the increase in carbocation stability achieved as a result of the rearrangement.

In the reaction of benzene with 1-chlorobutane, where rearrangement converts a primary carbocation into a secondary carbocation, 65% of the product is the rearranged product.

$$\text{benzene} + CH_3CH_2CH_2CH_2Cl \xrightarrow[0\ ^\circ C]{AlCl_3}$$

1-chlorobutane

2-phenylbutane
rearranged alkyl
substituent
65%

1-phenylbutane
unrearranged alkyl
substituent
35%

$$CH_3CH_2\overset{+}{C}HCH_2 \xrightarrow{\textbf{1,2-hydride shift}} CH_3CH_2\overset{+}{C}HCH_3$$
  $\overset{|\curvearrowright}{H}$

a primary carbocation   **a secondary carbocation**

In the reaction of benzene with 1-chloro-2,2-dimethylpropane, the extra stability achieved by rearrangement is greater (rearrangement converts a primary carbocation into a tertiary carbocation) than in the preceding reaction. Here, 100% of the product has the rearranged alkyl substituent.

$$\text{benzene} + CH_3\overset{\overset{CH_3}{|}}{\underset{\underset{CH_3}{|}}{C}}CH_2Cl \xrightarrow{AlCl_3}$$

1-chloro-2,2-dimethylpropane

2-methyl-2-phenylbutane
rearranged alkyl
substituent
100%

2,2-dimethyl-1-phenylpropane
unrearranged alkyl
substituent
0%

$$CH_3\overset{\overset{CH_3}{|}}{\underset{\underset{CH_3}{|}}{\overset{+}{C}}}CH_2 \xrightarrow{\textbf{1,2-methyl shift}} CH_3\overset{\overset{CH_3}{|}}{C}CH_2CH_3$$

a primary carbocation   **a tertiary carbocation**

In addition to reacting with carbocations generated from alkyl halides, benzene also reacts with carbocations generated from alkenes.

$$\text{benzene} + CH_3CH=CHCH_3 \xrightarrow{HF}$$

sec-butylbenzene

James Mason Crafts

mechanism of the reaction

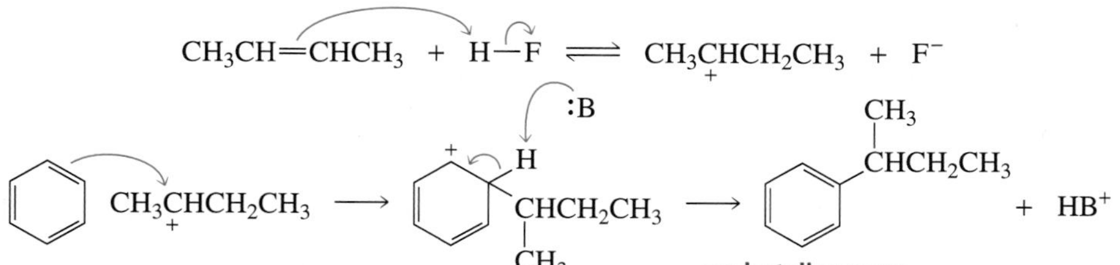

$$CH_3CH{=}CHCH_3 \ + \ H{-}F \ \rightleftharpoons \ CH_3\overset{+}{C}HCH_2CH_3 \ + \ F^-$$

:B

$$CH_3\overset{+}{C}HCH_2CH_3 \longrightarrow \quad \text{(benzenonium intermediate)} \longrightarrow \quad \text{(sec-butylbenzene)} \ + \ HB^+$$

sec-butylbenzene

INCIPIENT PRIMARY CARBOCATIONS

For simplicity, we have shown the formation of a primary carbocation in the preceding two examples. However, we learned in Section 9.4 that primary carbocations are too unstable to be formed. We must conclude, then, that a true primary carbocation is never formed in a Friedel–Crafts alkylation reaction. The carbocation remains complexed with the catalyst; it is an "incipient" carbocation. The fact that carbocation rearrangement occurs means that the incipient carbocation has sufficient carbocation character to permit the rearrangement.

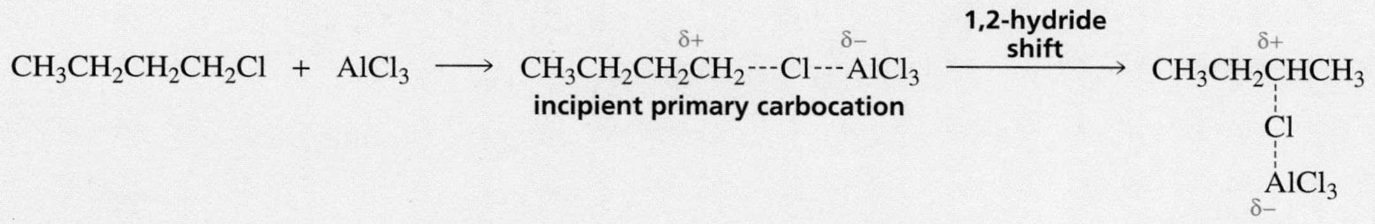

$$CH_3CH_2CH_2CH_2Cl \ + \ AlCl_3 \longrightarrow \ CH_3CH_2CH_2\overset{\delta+}{CH_2}{-}{-}{-}\overset{\delta-}{Cl}{-}{-}{-}AlCl_3 \ \xrightarrow{\text{1,2-hydride shift}} \ CH_3CH_2\overset{\delta+}{C}HCH_3$$

incipient primary carbocation

with Cl and AlCl_3 (δ−)

PROBLEM 8 ◆

What would be the major product of Friedel–Crafts alkylation of benzene using each of the following alkyl halides?

a. CH_3CH_2Cl

b. $CH_3CH_2CH_2Cl$

c. $CH_3CH_2CH(Cl)CH_3$

d. $(CH_3)_3CCH_2Cl$

e. $(CH_3)_2CHCH_2Cl$

f. $CH_2{=}CHCH_2Cl$

It is not possible using a Friedel–Crafts alkylation reaction, to obtain a good yield of an alkylbenzene containing a straight-chain alkyl group. This is because the incipient primary carbocation will rearrange to a more stable carbocation.

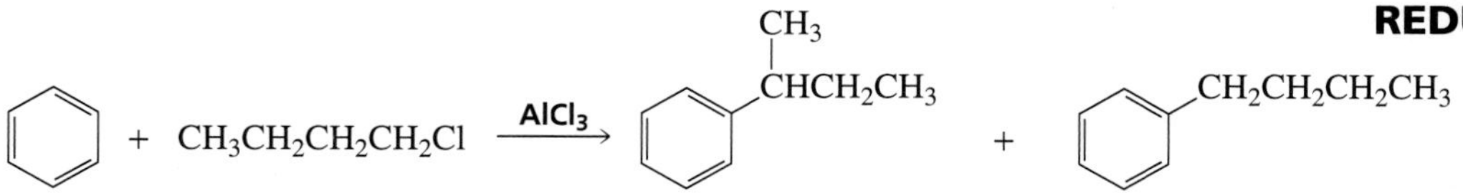

**14.9
ALKYLATION
OF BENZENE BY
ACYLATION–
REDUCTION**

$$\text{(benzene)} \ + \ CH_3CH_2CH_2CH_2Cl \ \xrightarrow{AlCl_3} \ \text{(sec-butylbenzene)} \ + \ \text{(n-butylbenzene)}$$

major product   **minor product**

Because acylium ions do not rearrange, straight-chain alkyl groups can be placed on benzene rings using a Friedel–Crafts acylation reaction, followed by reduction of the carbonyl group to a methylene group.

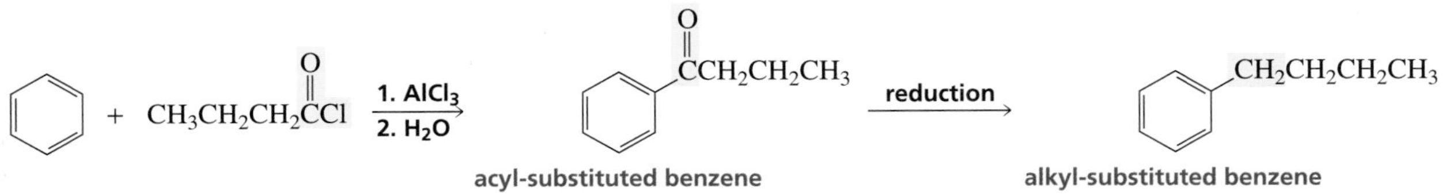

Another advantage of preparing alkylbenzenes by acylation–reduction rather than by direct alkylation is that, while alkylbenzenes are more reactive than benzene, acylbenzenes are less reactive than benzene (Section 14.11). Therefore, alkylbenzenes can be prepared without having to use a large excess of benzene (Section 14.8).

There are two methods commonly used to reduce the carbonyl group of an aldehyde or a ketone to a methylene group. The **Clemmensen reduction** uses an acidic solution of zinc dissolved in mercury as the reducing reagent. The **Wolff–Kishner reduction** employs hydrazine (H_2NNH_2) under basic conditions. The mechanism of the Wolff–Kishner reaction is shown in Section 16.7.

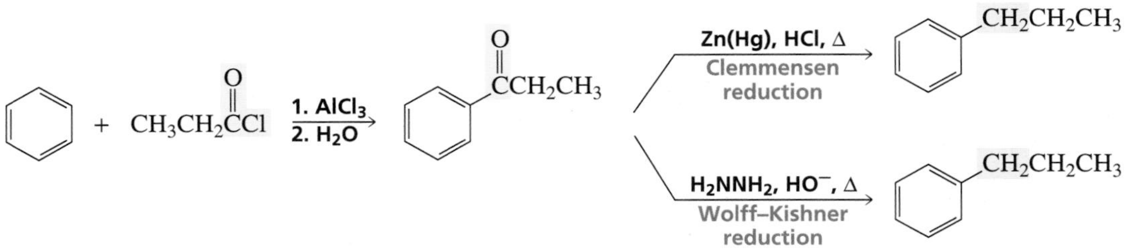

E. C. Clemmensen (1876–1941) *was born in Denmark and received a Ph.D. from the University of Copenhagen. He was a scientist at Clemmensen Corp. in Newark, New York.*

Ludwig Wolff (1857–1919) *was born in Germany. He received a Ph.D. from the University of Strasbourg. He was a professor at the University of Jena in Germany.*

At this point, you may wonder why it is necessary to have two different ways to carry out the same reaction. The reason is that there may be another functional group in the molecule that you do not want to react with the reagents you are using to carry out the desired reaction. For example, heating the following compound with HCl as required by the Clemmensen reduction would cause the alcohol to undergo substitution (Section 11.1). Under the basic conditions of the Wolff–Kishner reduction, the alcohol group would remain unchanged.

N. M. Kishner (1867–1935) *was born in Moscow. He received a Ph.D. from the University of Moscow under the direction of Prof. Markovnikov. He was a professor at the University of Tomsk and later at the University of Moscow.*

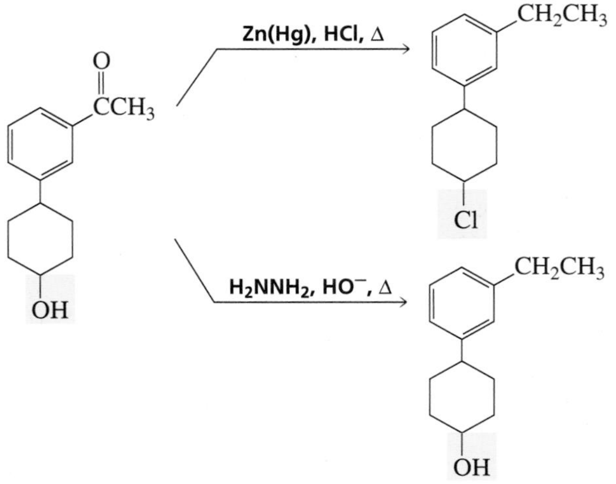

The quaternary ammonium ion would undergo a Hofmann elimination reaction under the basic conditions required for the Wolff–Kishner reduction (Section 11.11), but it would be inert to the conditions of the Clemmensen reduction.

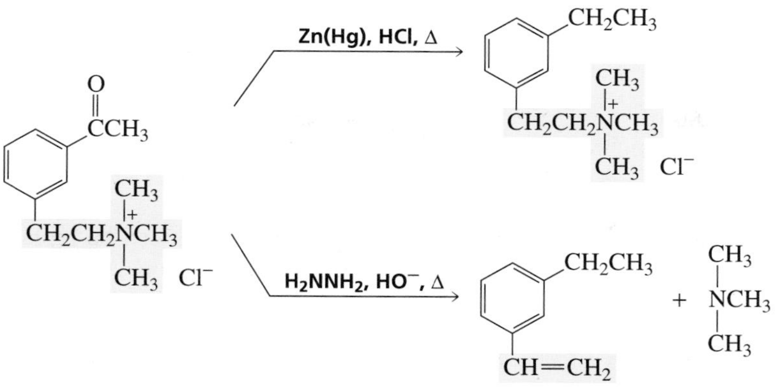

PROBLEM 9

Describe how the following compounds could be prepared from benzene.

a. 2-phenylpentane

b. 1-phenylpentane

14.10 REACTIONS OF SUBSTITUENTS ON BENZENE

Benzenes with substituents other than those listed in Section 14.3 can be prepared by first synthesizing one of the substituted benzenes listed in that section and then chemically changing the substituent. For example, a nitro substituent can be reduced to an amino substituent. Either catalytic hydrogenation or tin (or sometimes iron) and hydrochloric acid can be used to carry out the reduction. If acidic conditions are employed in the reduction, the product is in its acidic form (anilinium ion). When the reaction is over, base can be added to convert the product into its basic form (aniline).

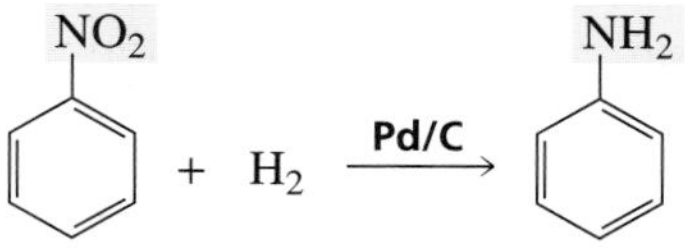

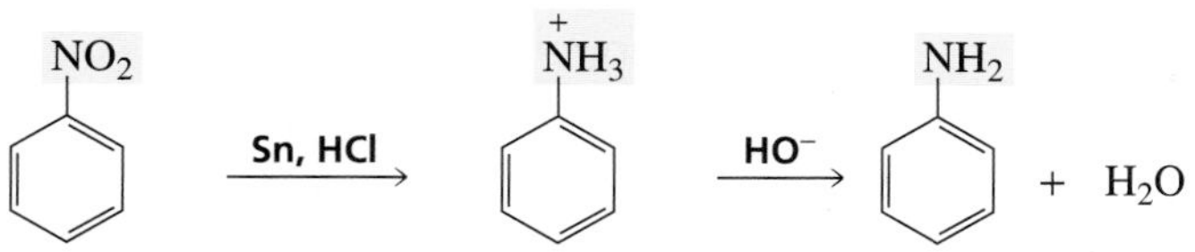

nitrobenzene   protonated aniline / anilinium ion   aniline

Conversely, aniline can be oxidized to nitrobenzene using trifluoroperoxyacetic acid as the oxidizing agent. (trifluoroperoxyacetic acid is a *peroxy acid*—a carboxylic acid with an additional oxygen; Section 17.8.)

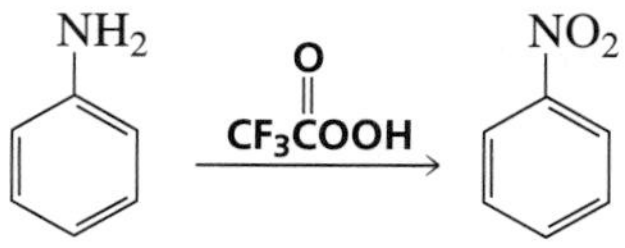

A bromine will selectively substitute for a benzylic hydrogen in a radical substitution reaction. (NBS stands for *N*-bromosuccinimide; Section 8.5.)

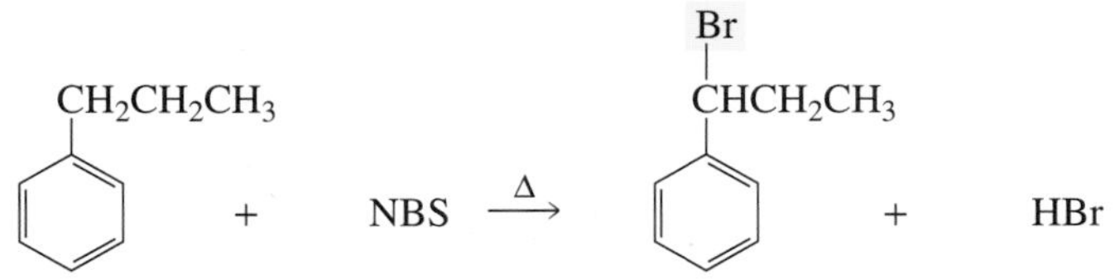

propylbenzene   +   NBS $\xrightarrow{\Delta}$ 1-bromo-1-phenylpropane   +   HBr

Once a halogen has been placed in the benzylic position, it can be replaced by a variety of nucleophiles by means of an S_N2 or an S_N1 reaction. A wide variety of compounds can be prepared this way.

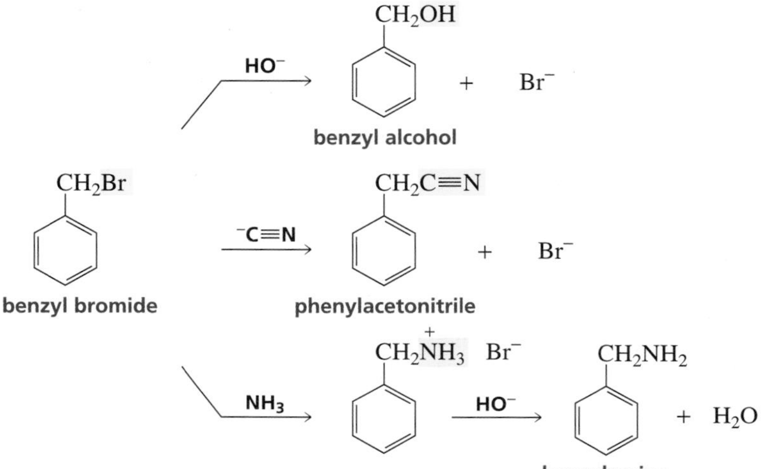

The benzene ring itself is not readily oxidized. But an alkyl group bonded to a benzene ring can be oxidized to a carboxylic acid group. Commonly used oxidizing agents are acidic solutions of sodium dichromate or potassium permanganate.

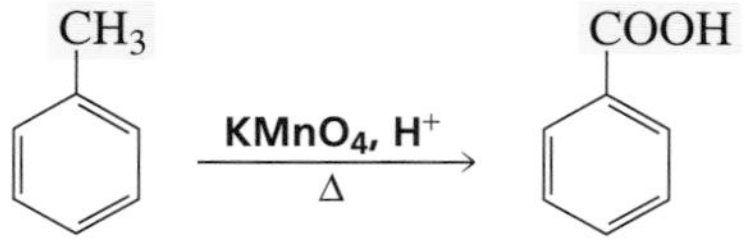

toluene   benzoic acid

The first step in the mechanism is removal of a hydrogen from the benzylic carbon. This means that, if the alkyl group does not have a hydrogen bonded to the benzylic carbon, the oxidation reaction will not occur.

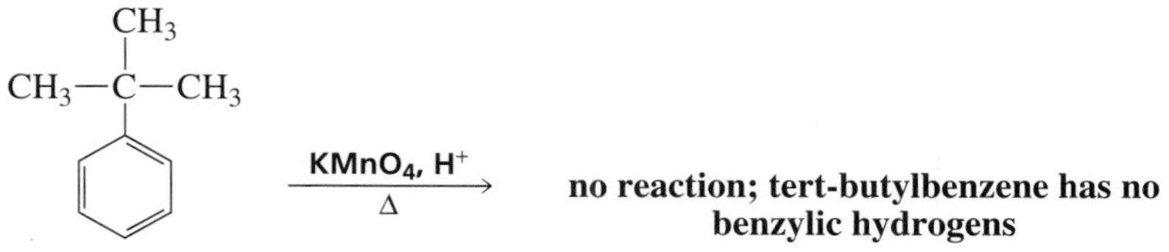

tert-butylbenzene $\xrightarrow[\Delta]{\text{KMnO}_4,\ \text{H}^+}$ **no reaction; tert-butylbenzene has no benzylic hydrogens**

Regardless of the length of the alkyl group, it will be oxidized to a COOH group, provided that it has a benzylic hydrogen.

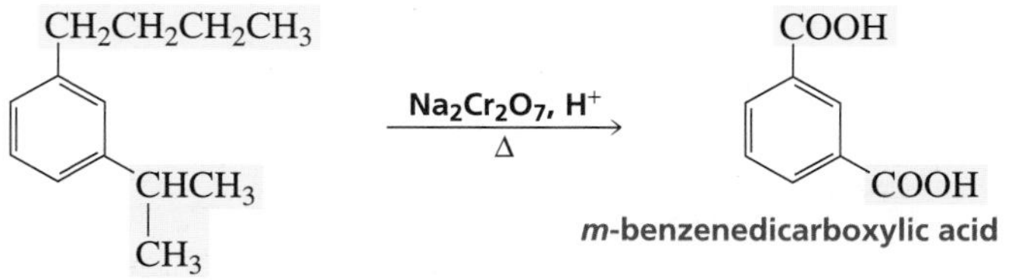

m-butylisopropylbenzene

m-benzenedicarboxylic acid

Under the vigorous conditions employed for the oxidation of alkyl substituents, benzylic alcohols are oxidized to benzoic acid.

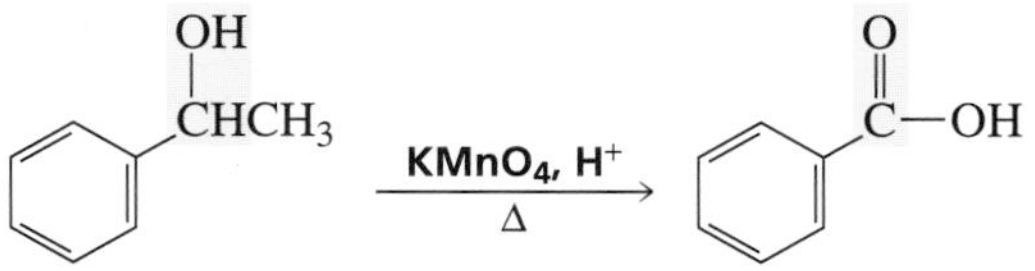

1-phenylethanol   benzoic acid

If, however, a mild oxidizing agent is used (MnO_2), benzylic alcohols are oxidized to aldehydes or ketones.

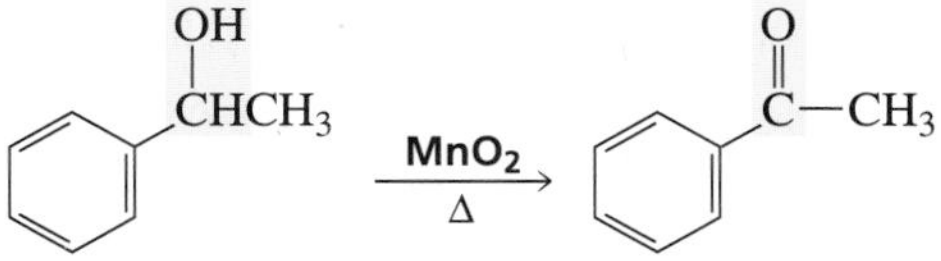

1-phenylethanol   acetophenone

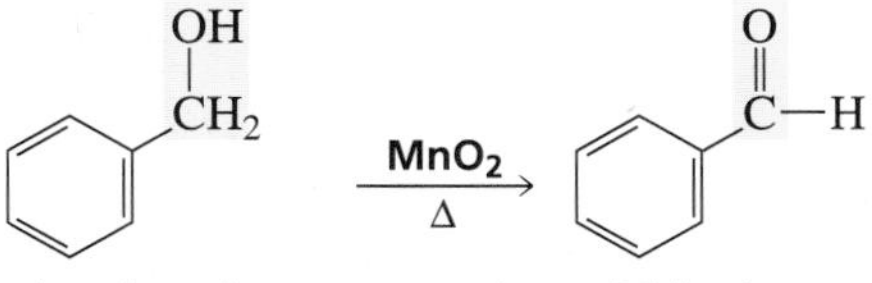

phenylmethanol   benzaldehyde
benzyl alcohol

Substituents with double and triple bonds can undergo catalytic hydrogenation (Section 3.19).

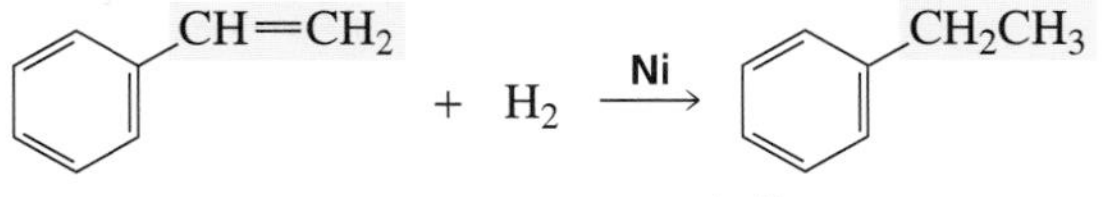

styrene   ethylbenzene

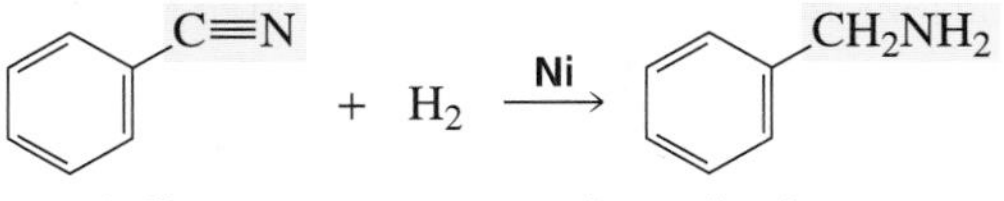

benzonitrile   benzylamine

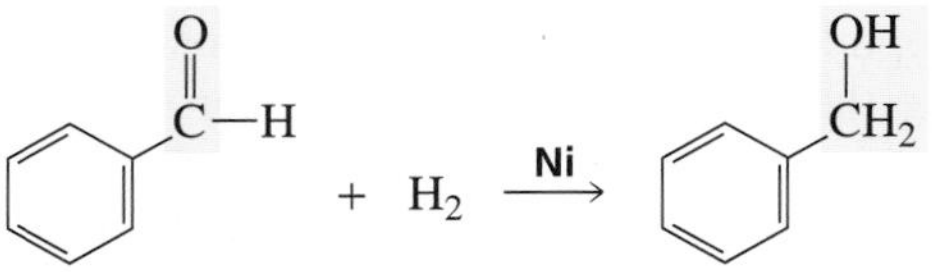

benzaldehyde   benzyl alcohol

Because of the stability of its aromatic ring, benzene undergoes catalytic hydrogenation only under extreme conditions.

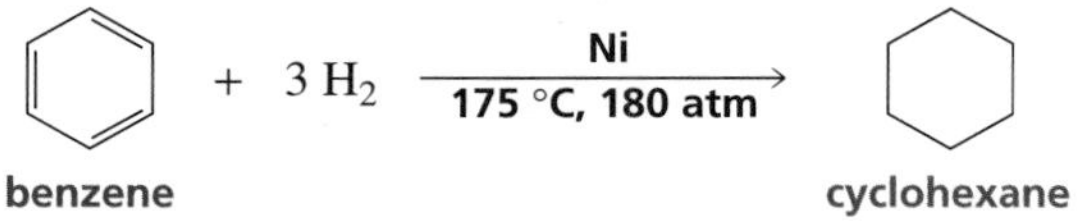

benzene + 3 H$_2$ $\xrightarrow[\text{175 °C, 180 atm}]{\text{Ni}}$ cyclohexane

PROBLEM 10◆

Give the product of each of the following reactions.

a. 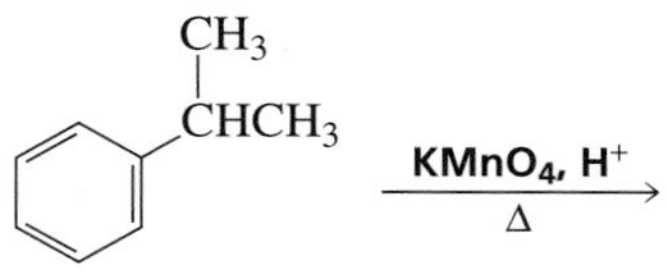 $\xrightarrow[\Delta]{\text{KMnO}_4,\ \text{H}^+}$

c. 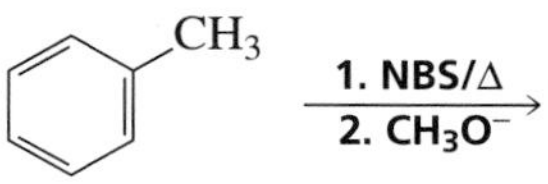 $\xrightarrow{\begin{array}{l}\text{1. NBS/}\Delta\\ \text{2. CH}_3\text{O}^-\end{array}}$

b. 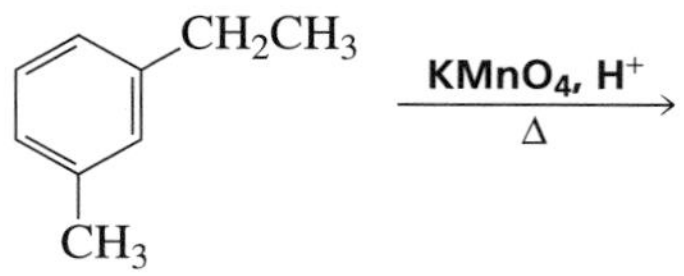 $\xrightarrow[\Delta]{\text{KMnO}_4,\ \text{H}^+}$

d. 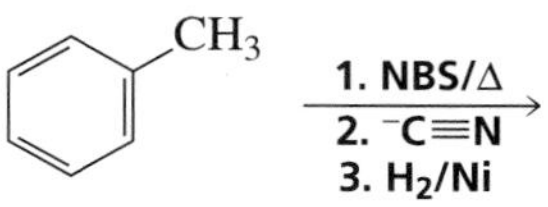 $\xrightarrow{\begin{array}{l}\text{1. NBS/}\Delta\\ \text{2. }^-\text{C}\equiv\text{N}\\ \text{3. H}_2\text{/Ni}\end{array}}$

PROBLEM 11

Show how the following compounds could be prepared from benzene.

a. benzoic acid

b. styrene

c. 2-phenylethanol

d. 1-bromo-2-phenylethane

e. aniline

f. *N,N,N*-trimethylanilinium iodide

A FEW WORDS OF WARNING

Although benzene is widely used in chemical synthesis and is frequently used as a solvent, it is toxic. Its major toxic effect is on the central nervous system and on bone marrow. Chronic exposure to benzene causes leukemia and aplastic anemia. In addition, a higher-than-average incidence of leukemia has been found in industrial workers with long-term exposure to as little as 1 ppm benzene in the atmosphere. Toluene has replaced benzene as a solvent because, although it is also a central nervous system depressant, it does not produce leukemia or aplastic anemia. "Glue sniffers" seek the narcotic central nervous system effects of solvents such as toluene. This can be a highly dangerous activity.

14.11 THE EFFECT OF SUBSTITUENTS ON REACTIVITY

Is a substituted benzene more reactive or less reactive than benzene itself in an electrophilic substitution reaction? The answer depends on the substituent. Some substituents make the ring more reactive and some make the ring less reactive toward electrophilic substitution.

The rate-determining step of an electrophilic aromatic substitution reaction is attack of the nucleophilic aromatic ring on the positively charged electrophile. Therefore, increasing the electron density of the benzene ring increases its nucleophilicity. The greater its nucleophilicity, the more readily it will attack the electrophile and the faster the electrophilic substitution reaction will take place. *Therefore, substituents that are capable of donating electrons into the ring will increase the rate of electrophilic substitution, while substituents that withdraw electrons from the ring will decrease the rate of electrophilic substitution.*

relative rates of electrophilic substitution

X donates electrons
into the benzene ring

Y withdraws electrons
from the benzene ring

Electron-donating substituents increase the reactivity of the benzene ring toward electrophilic substitution.

There are two ways in which substituents can donate electrons into a benzene ring: inductive electron donation and electron donation by resonance. There are also two ways by which substituents can withdraw electrons from a benzene ring: inductive electron withdrawal and electron withdrawal by resonance.

Electron-withdrawing substituents decrease the reactivity of the benzene ring toward electrophilic substitution.

Inductive Electron Donation and Withdrawal

If a substituent bonded to a benzene ring is *less electron-withdrawing than a hydrogen,* the substituent will be less effective than a hydrogen at withdrawing the σ electrons (that attach it to the ring) from the ring. Such a substituent, compared with a hydrogen, donates electrons inductively into the ring. In other words, the electrons in the σ bond that attaches such a substituent to the benzene ring move toward the ring more readily than do the electrons in the σ bond that attaches a hydrogen to the ring. Donation of electrons through a σ bond is called **inductive electron donation.** Alkyl substituents (such as CH_3) donate electrons inductively compared with a hydrogen.

substituent donates electrons inductively
(compared with a hydrogen)

substituent withdraws electrons inductively
(compared with a hydrogen)

Notice that it is an alkyl *group* whose electron donating ability is being compared with that of hydrogen. A carbon atom is actually *more* electronegative than hydrogen, but an alkyl group is *more* electron-donating than hydrogen (Section 3.10).

If a substituent is *more electron-withdrawing than a hydrogen,* it will withdraw the σ electrons away from the benzene ring more strongly than will a hydrogen. Withdrawal of electrons through a σ bond is called **inductive electron withdrawal.** The $^+NH_3$ group is an example of a substituent that withdraws electrons inductively, because the atom attached to the benzene ring is more electronegative than a hydrogen.

Resonance Electron Donation and Withdrawal

If a substituent has a pair of nonbonded electrons on the atom directly attached to the benzene ring, these electrons can be delocalized into the ring through p orbital overlap. Such substituents are said to donate electrons by resonance. Substituents such as NH_2, OH, OR, and Cl **donate electrons by resonance.** These substituents also withdraw electrons inductively, because the atom attached to the benzene ring is more electronegative than a hydrogen.

donation of electrons into a benzene ring by resonance

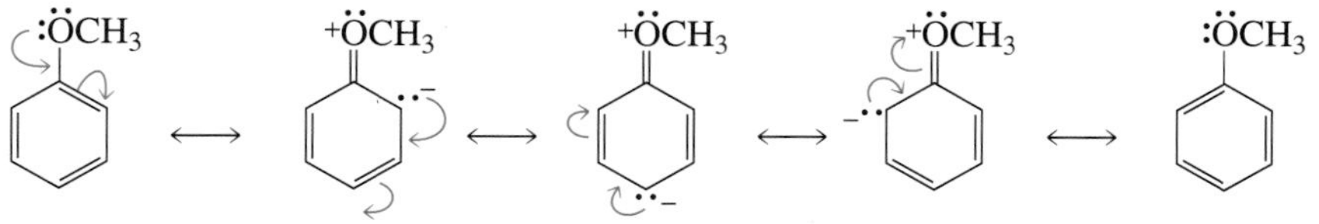

If a substituent is attached to the benzene ring by an atom that is doubly or triply bonded to a more electronegative atom, the π electrons of the ring can be delocalized onto the substituent. Such substituents are said to withdraw electrons by resonance. Substituents such as C=O, C≡N, NO_2, and SO_3H **withdraw electrons by resonance.** These substituents also withdraw electrons inductively, because the atom attached to the benzene ring is more electronegative than a hydrogen.

withdrawal of electrons from a benzene ring by resonance

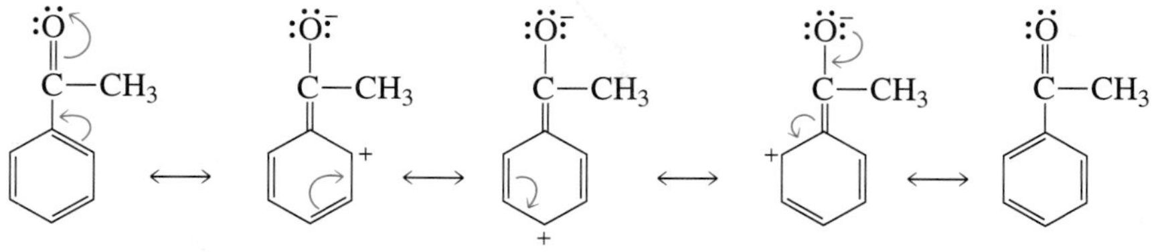

PROBLEM 12 ◆

For each of the following substituents, indicate whether it donates electrons inductively, withdraws electrons inductively, donates electrons by resonance, or withdraws electrons by resonance. Inductive effects should be compared with a hydrogen. (Remember that many substituents can be characterized in more than one way.)

a. Br

b. CH_2CH_3

c. $\overset{\overset{\textstyle O}{\|}}{C}CH_3$

d. $NHCH_3$

e. OCH_3

f. $\overset{+}{N}(CH_3)_3$

Relative Reactivity of Substituted Benzenes

The substituents shown in Table 14.1 are listed according to how they affect the reactivity of the benzene ring toward electrophilic substitution. *The* **activating substituents** *make the benzene ring more reactive toward electrophilic substitution; the* **deactivating substituents** *make the benzene ring less reactive toward electrophilic substitution.* Recall that activating substituents donate electrons into the ring and deactivating substituents withdraw electrons from the ring.

All the *strongly activating substituents* donate electrons into the ring by resonance and withdraw electrons from the ring inductively. The fact that they have been found experimentally to be strong activators indicates that electron donation into the ring by resonance is much greater than electron withdrawal from the ring by the inductive effect.

strongly activating substituents

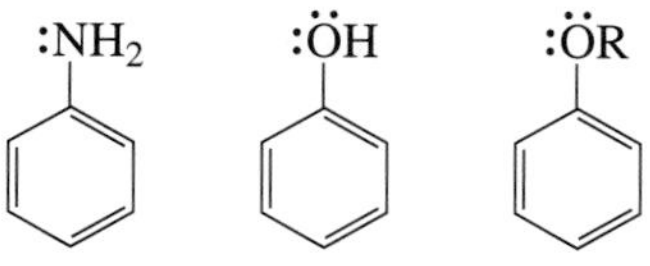

TABLE 14.1   The Effects of Substituents on the Reactivity of a Benzene Ring Toward Electrophilic Substitution

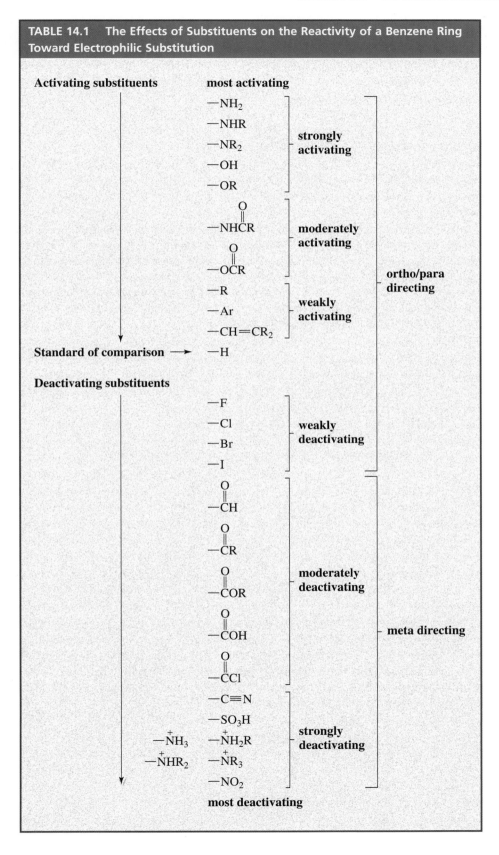

Activating substituents

most activating

—NH$_2$

—NHR

—NR$_2$ } strongly activating

—OH

—OR

—NHCR (with C=O) } moderately activating

—OCR (with C=O)

—R

—Ar } weakly activating

—CH=CR$_2$

Standard of comparison ⟶   —H

Deactivating substituents

—F

—Cl

—Br } weakly deactivating

—I

—CH (with C=O)

—CR (with C=O)

—COR (with C=O) } moderately deactivating

—COH (with C=O)

—CCl (with C=O)

—C≡N

—SO$_3$H

—$\overset{+}{N}$H$_3$ —$\overset{+}{N}$H$_2$R } strongly deactivating

—$\overset{+}{N}$HR$_2$ —$\overset{+}{N}$R$_3$

—NO$_2$

most deactivating

ortho/para directing

meta directing

The *moderately activating substituents* also donate electrons into the ring by resonance and withdraw electrons from the ring inductively. However, they donate electrons into the ring by resonance less effectively than do the strongly activating substituents.

moderately activating substituents

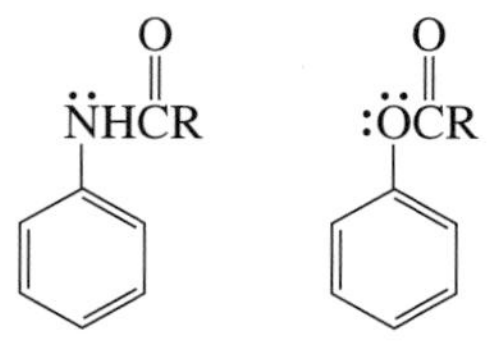

These substituents are less effective at resonance electron donation because, unlike the strongly activating substituents that donate electrons by resonance only *into* the ring, these substituents can donate electrons by resonance in two competing directions—*into* the ring and *away* from the ring. The fact that these substituents are activators indicates that, despite their diminished resonance electron donation into the ring, overall they donate electrons by resonance more strongly than they withdraw electrons inductively.

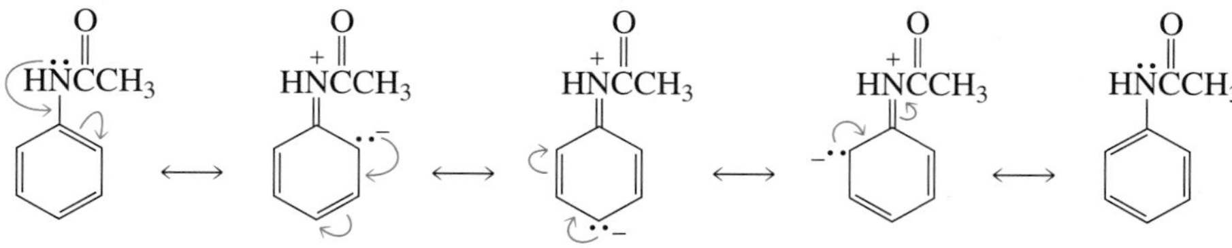

substituent donates electrons by resonance into the benzene ring

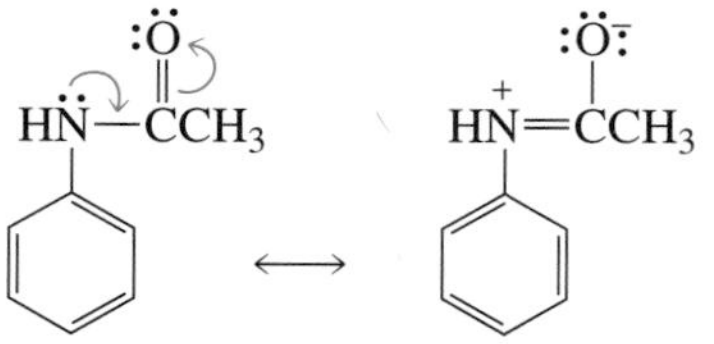

substituent donates electrons by resonance away from the benzene ring

Alkyl, aryl, and CH=CHR groups are *weakly activating substituents*. Two of the three groups (aryl and CH=CHR) can donate electrons into the ring by resonance and can withdraw electrons from the ring by resonance. The fact that they are weak activators indicates that they are slightly more electron-donating than they are electron-withdrawing. An alkyl substituent is a weak electron donor relative to a hydrogen because of hyperconjugation (Section 3.10).

weakly activating substituents

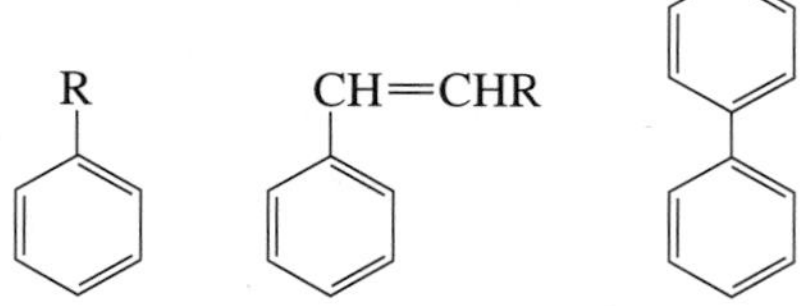

The halogens are *weakly deactivating substituents.* They donate electrons into the ring by resonance and withdraw electrons from the ring inductively. Because the halogens have been found experimentally to be deactivators, we can conclude that they withdraw electrons inductively more strongly than they donate electrons by resonance.

weakly deactivating substituents

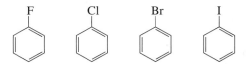

The *moderately deactivating substituents* all have carbonyl groups directly attached to the benzene ring. Carbonyl groups withdraw electrons both inductively and by resonance.

moderately deactivating substituents

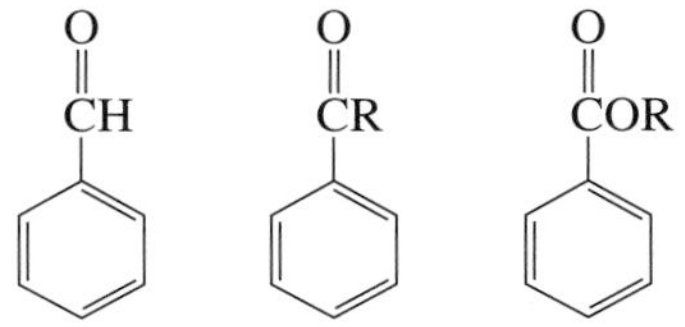

The *strongly deactivating substituents* are powerful electron withdrawers. With the exception of $^+NHR_2$, $^+NH_2R$, $^+NR_3$, and $^+NH_3$, they withdraw electrons both inductively and by resonance. The ammonium ions have no resonance effect, but the positive charge on the nitrogen atom causes them to strongly withdraw electrons inductively.

strongly deactivating substituents

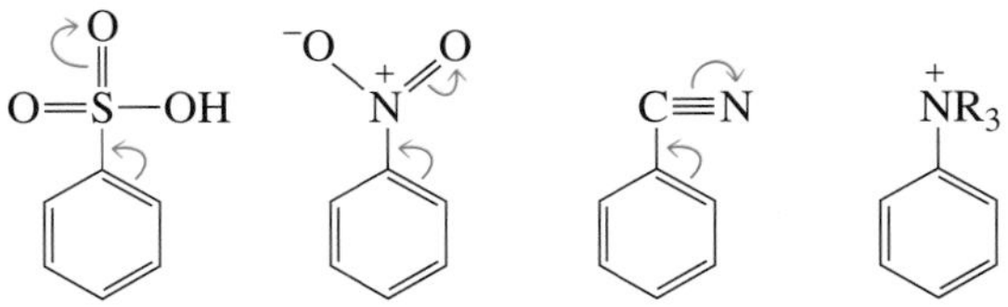

PROBLEM 13 ◆

List each of the following sets of compounds in order of decreasing reactivity toward electrophilic substitution.

a. benzene, phenol, toluene, nitrobenzene, bromobenzene

b. dichloromethylbenzene, difluoromethylbenzene, toluene, chloromethylbenzene

---

PROBLEM 14 / SOLVED

Explain why the halogens have the relative order of reactivity shown in Table 14.1.

SOLUTION   Table 14.1 shows that fluorine is the least deactivating of the halogen substituents and iodine is the most deactivating. We know that fluorine is the most electronegative of the halogens, which means that it is best at withdrawing electrons inductively. Fluorine is also best at donating electrons by resonance because its $2p$ orbital, compared with the $3p$ orbital of chlorine, the $4p$ orbital of bromine, or the $5p$ orbital of iodine, can better overlap with the $2p$ orbital of carbon. So the fluorine substituent is best at withdrawing electrons inductively and best at donating electrons by resonance. The fact that, of the halogens, it is the weakest deactivator means that electron donation by resonance is the more important factor in determining the relative order of reactivity of halosubstituted benzenes.

14.12
THE EFFECT OF SUBSTITUENTS ON ORIENTATION

When a substituted benzene undergoes an electrophilic substitution reaction, where does the new substituent attach? Is the product of the reaction the ortho isomer, the meta isomer, or the para isomer?

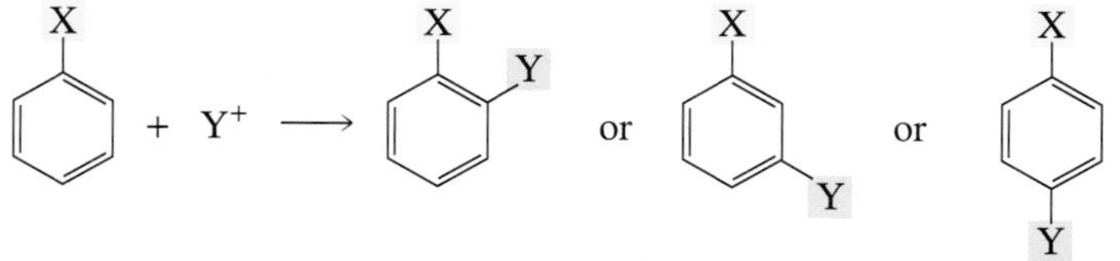

ortho isomer   meta isomer   para isomer

The substituent already bonded to benzene determines the location of the new substituent. The substituents can be divided into two groups: a substituent is either an **ortho- and para-directing substituent** or a **meta-directing substituent.**

1. All the activating substituents and weakly deactivating substituents (the halogens) direct an incoming electrophile to the ortho and para positions.

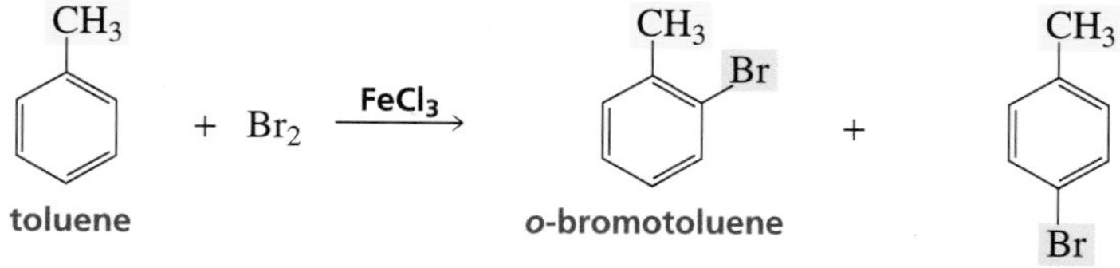

toluene   o-bromotoluene   p-bromotoluene

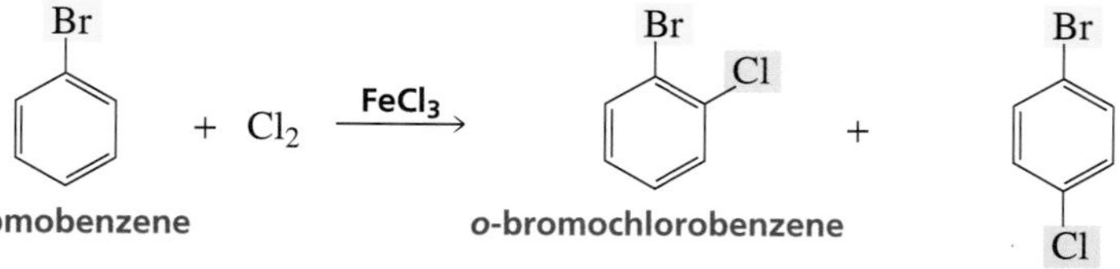

bromobenzene   o-bromochlorobenzene   p-bromochlorobenzene

2. All moderately deactivating and strongly deactivating substituents direct an incoming substituent to the meta position.

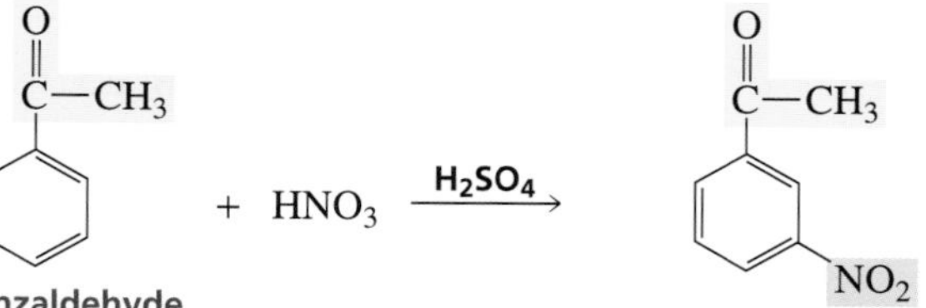

benzaldehyde   m-nitrobenzaldehyde

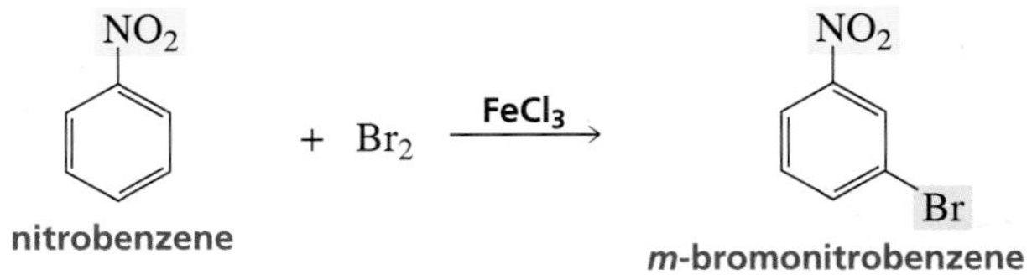

nitrobenzene   + Br$_2$ $\xrightarrow{\text{FeCl}_3}$ m-bromonitrobenzene

All activating substituents are ortho/para directors.

The weakly deactivating halogens are also ortho/para directors.

All moderately deactivating and strongly deactivating substituents are meta directors.

It is easy to determine whether a substituent is an ortho/para director or a meta director without resorting to memorization. All ortho/para directors have a lone pair of electrons on the atom directly attached to the ring (with the exception of alkyl, aryl, and CH=CHR groups). All meta directors have a positive charge or a partial positive charge on the atom attached to the ring. Take a few minutes to examine the substituents listed in Table 14.1 to convince yourself that this is true.

To understand why a substituent directs an incoming electrophile to a particular position, we must look at the stability of the carbocation intermediate formed in the rate-determining step. When a substituted benzene undergoes an electrophilic substitution reaction, three different carbocation intermediates can be formed. Which one is actually formed depends on the relative stabilities of the carbocations (Figure 14.4). Comparing the relative stabilities of these three carbocations allows us to determine the preferred pathway of the reaction, because the more stable the carbocation, the less energy is required to make it (Section 3.11).

If a substituent donates electrons inductively (a methyl group, for example), the indicated resonance contributors in Figure 14.4 are the most stable. This is because the substituent is attached directly to the positively charged carbon, which it can stabilize by inductive electron donation. These relatively stable resonance contributors are obtained only when the substituent is directed to an ortho or para position. Therefore, the most stable carbocation is obtained by directing the incoming group to the ortho and para positions. This means that any substituent that donates electrons inductively is an ortho/para director.

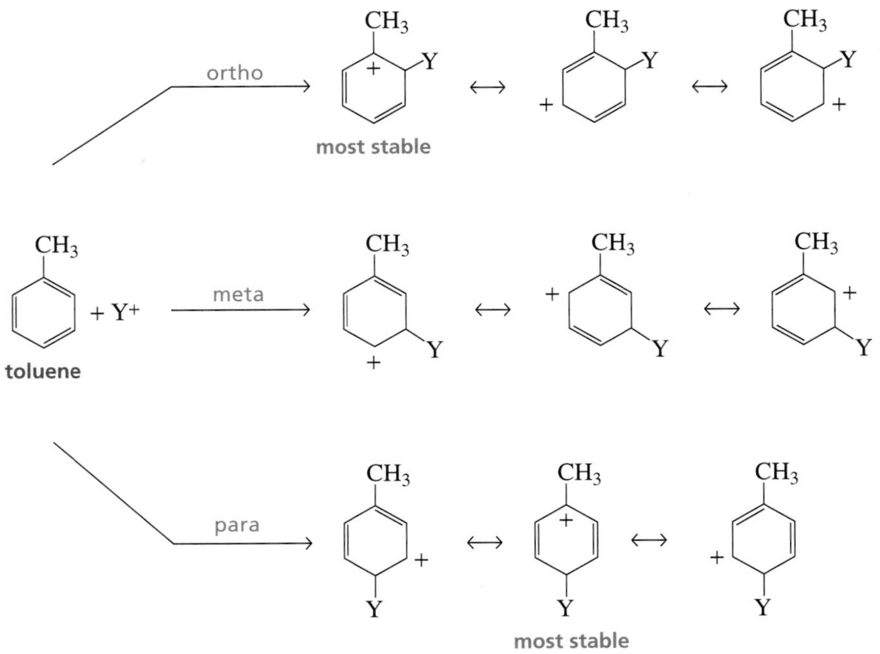

◀ **Figure 14.4**
The structures of the carbocation intermediates formed from the reaction of an electrophile with toluene at the ortho, meta, and para positions.

Figure 14.5 ▶
The structures of the carbocation intermediates formed from the reaction of an electrophile with anisole at the ortho, meta, and para positions.

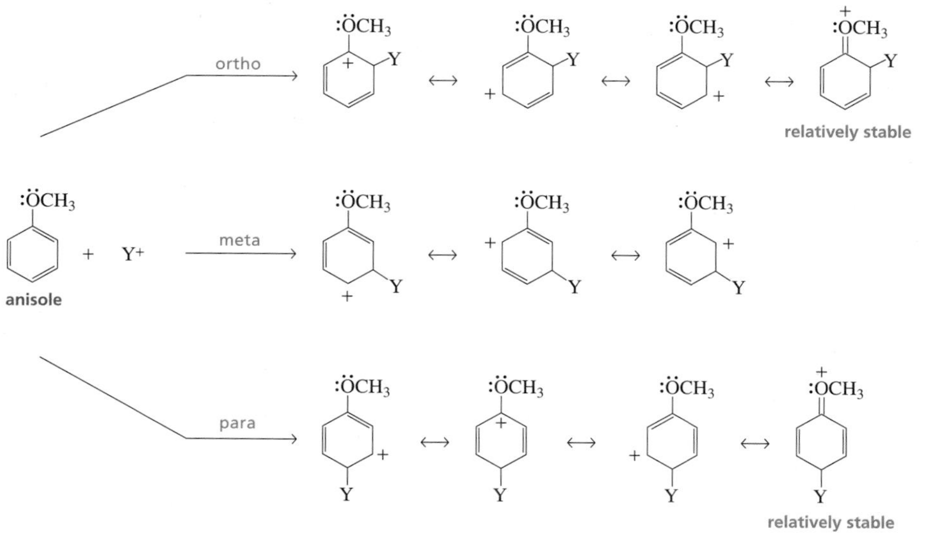

If a substituent donates electrons by resonance, the carbocations formed by putting the incoming electrophile on the ortho and para positions have a fourth resonance contributor as a result of resonance electron donation by the substituent (Figure 14.5). This is an especially stable resonance contributor because it is the only one whose atoms (except for hydrogen) all have complete octets. Therefore, all groups that donate electrons by resonance are ortho/para directors.

Substituents with a positive charge or a partial positive charge on the atom attached to the benzene ring withdraw electrons from the ring. All of these substituents withdraw electrons inductively and most withdraw electrons by resonance as well. For all such substituents, the indicated resonance contributors in Figure 14.6 are the least stable. This is because the substituent is in position to further

Figure 14.6 ▶
The structures of the carbocation intermediates formed from the reaction of an electrophile with protonated aniline at the ortho, meta, and para positions.

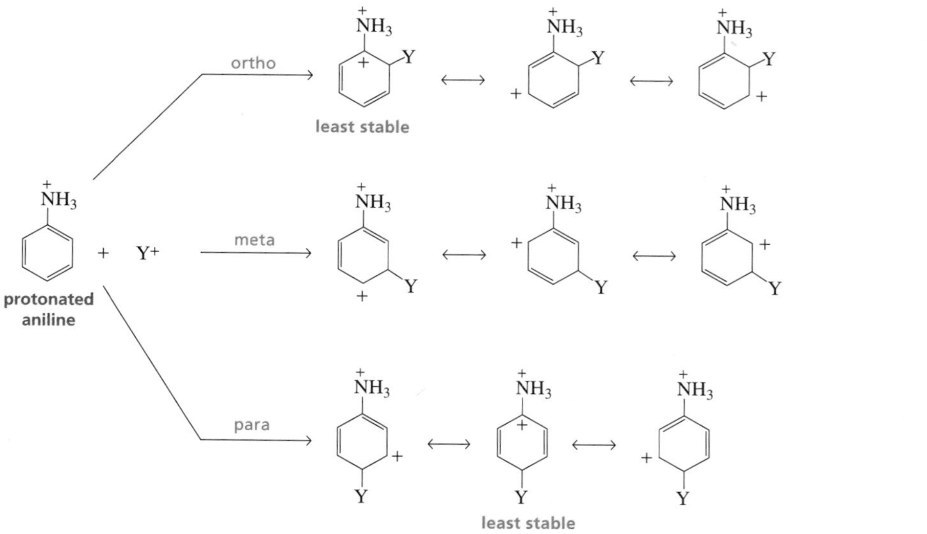

increase the size of the positive charge on the carbon by inductive electron withdrawal. So the most stable carbocation occurs when the incoming electrophile is directed to the meta position.

Notice that the three possible carbocation intermediates in Figure 14.4 and the three in Figure 14.6 are exactly the same. The only difference in the two figures is that the resonance contributors with the substituent directly attached to the positively charged carbon are the most stable resonance contributors when the substituent donates electrons and the least stable resonance contributors when the substituent withdraws electrons.

The only deactivating substituents that are ortho/para directors are the halogens, which are the weakest of the deactivators. We have seen that they are deactivators because they inductively withdraw electrons from the ring more strongly than they donate electrons by resonance. All the ring positions are deactivated by inductive electron withdrawal, but this deactivation is partially made up for at the ortho and para positions by electron donation by resonance. Because the meta position cannot benefit from resonance electron donation, it is the most deactivated position. In other words, the halogens are ortho/para directors.

The directing effects of substituents in electrophilic aromatic substitution reactions can be summarized as follows:

- All activating substituents are ortho/para directors.
- The weakly deactivating halogens are ortho/para directors.
- All other deactivating substituents are meta directors.

PROBLEM 15

a. Draw the contributing resonance structures for nitrobenzene.

b. Why is it a meta director?

PROBLEM 16

a. Draw the contributing resonance structures for chlorobenzene.

b. Why is it an ortho/para director?

PROBLEM 17◆

What product(s) would result from nitration of each of the following compounds?

a. propylbenzene

b. benzenesulfonic acid

c. bromobenzene

d. benzaldehyde

e. cyclohexylbenzene

f. benzonitrile

PROBLEM 18◆

Are the following substituents ortho/para directors or meta directors?

a. CF_3 b. $N=O$ c. NO_2

We have seen that electron-withdrawing substituents increase the acidity of a compound (Sections 1.18 and 6.10). The ability of a substituent to either withdraw electrons from a benzene ring or donate electrons into a benzene ring is reflected in the pK_a values of substituted phenols, benzoic acids, and protonated anilines. For

**14.13
THE EFFECT OF
SUBSTITUENTS
ON pK_a**

example, the pK_a of phenol in H$_2$O at 25 °C is 9.95. The pK_a of *para*-chlorophenol is lower (9.38) because of the electron-withdrawing chloro substituent, while the pK_a of *para*-methyl phenol is higher (10.19) because of the electron-donating methyl substituent.

OH	OH	OH	OH	OH	OH
OCH$_3$	CH$_3$		Cl	HC$=$O	NO$_2$
pK_a = 10.20	pK_a = 10.19	pK_a = 9.95	pK_a = 9.38	pK_a = 7.66	pK_a = 7.14

The more deactivating (electron-withdrawing) the substituent, the more it increases the acidity of a COOH, OH, or $^+$NH$_3$ group attached to a benzene ring.

Take a minute to compare the effect of a substituent on the pK_a of phenol with its effect on the reactivity of a benzene ring toward electrophilic substitution. Notice that the more strongly electron-withdrawing the substituent, the lower the pK_a of the phenol and the less reactive the ring towards electrophilic substitution.

A similar substituent effect on pK_a is observed for substituted benzoic acids and substituted protonated anilines.

COOH	COOH	COOH	COOH	COOH	COOH
OCH$_3$	CH$_3$		Br	CH$_3$C$=$O	NO$_2$
pK_a = 4.47	pK_a = 4.34	pK_a = 4.20	pK_a = 4.00	pK_a = 3.70	pK_a = 3.44

$^+$NH$_3$	$^+$NH$_3$	$^+$NH$_3$	$^+$NH$_3$	$^+$NH$_3$	$^+$NH$_3$
OCH$_3$	CH$_3$		Br	HC$=$O	NO$_2$
pK_a = 5.29	pK_a = 5.07	pK_a = 4.58	pK_a = 3.91	pK_a = 1.76	pK_a = 0.98

PROBLEM-SOLVING STRATEGY

The *para*-nitroanilinium ion is 3.60 pK_a units more acidic than the anilinium ion, but *para*-nitrobenzoic acid is only 0.76 pK_a unit more acidic than benzoic acid. Explain why the nitro substituent causes a large change in pK_a in one case and a small change in pK_a in the other.

Don't expect to be able to solve this kind of problem simply by reading it. First pay close attention to exactly what the question is asking. You are being asked why the addition of a particular substituent has a greater effect on the acidity of one compound than it does on the acidity of another compound.

Now let's recall what causes compounds to have different acidities. We have seen that the acidity of a compound depends on how well the electrons that are left behind when the proton is removed can be accommodated (Sections 1.18 and 6.10).

Next, draw structures of the compounds in question. Check to see how well the substituent that has changed the acidity of a given compound can accommodate the electrons that are left behind when the proton is removed.

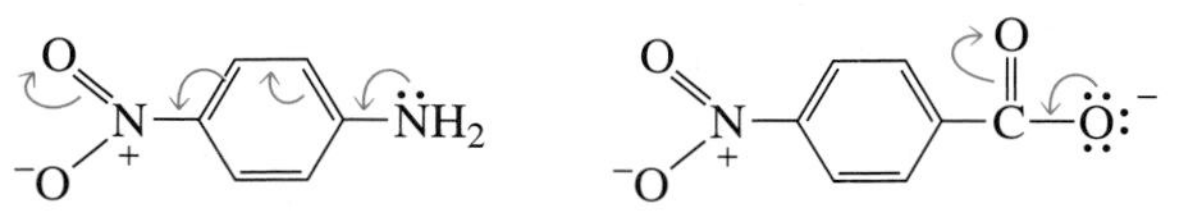

The structures allow us to answer the question. When a proton is lost from the *para*-nitroanilinium ion, the electrons that are left behind can be delocalized onto the nitro substituent. When a proton is lost from *para*-nitrobenzoic acid, the electrons that are left behind cannot be delocalized onto the nitro substituent. Therefore, the addition of a nitro substituent will have a greater affect on the acidity of an anilinium ion than on benzoic acid. Now continue on to do Problem 19.

PROBLEM 19

p-Nitrophenol has a pK_a of 7.14, while the pK_a of *m*-nitrophenol is 8.39. Explain.

PROBLEM 20 ◆

Which of the compounds in each of the following pairs is more acidic?

a. CH_3COH or $ClCH_2COH$ (both with C=O)

b. O_2NCH_2COH or $O_2NCH_2CH_2COH$ (both with C=O)

c. CH_3CH_2COH or $H_3\overset{+}{N}CH_2COH$ (both with C=O)

d. $HOCCH_2COH$ or $^-OCCH_2COH$ (both with two C=O)

e. CH_3CH_2COH or CH_3COH (both with C=O)

f.

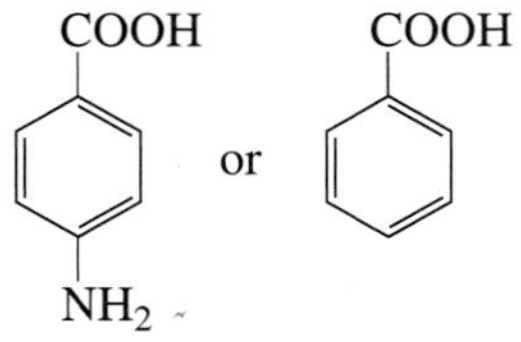

or

g. FCH_2COH or $ClCH_2COH$ (both with C=O)

h.

or

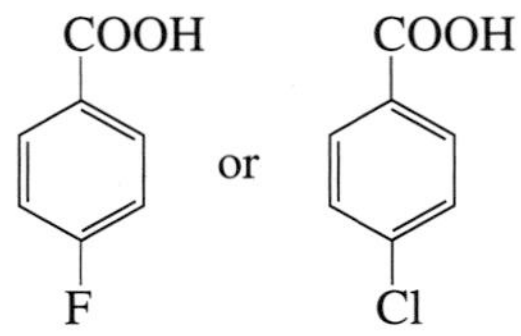

14.14
THE ORTHO/PARA RATIO

When a benzene ring with an ortho/para-directing substituent undergoes an electrophilic substitution reaction, what percentage of the product is the ortho isomer and what percentage is the para isomer? In terms of probability, one would expect more of the ortho product because there are two ortho positions available to the incoming electrophile and only one para position. The ortho position, however, is sterically hindered and the para position is not. Consequently, the para isomer will be formed preferentially if either the substituent on the ring or the incoming electrophile is large. The following nitration reactions illustrate the decrease in the ortho/para ratio with an increase in size of the directing alkyl substituent.

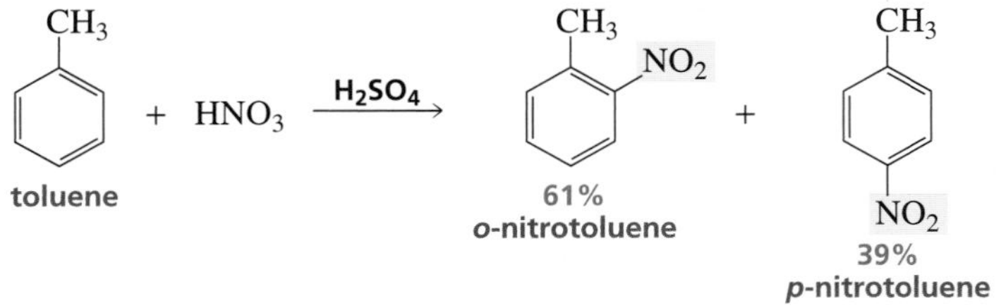

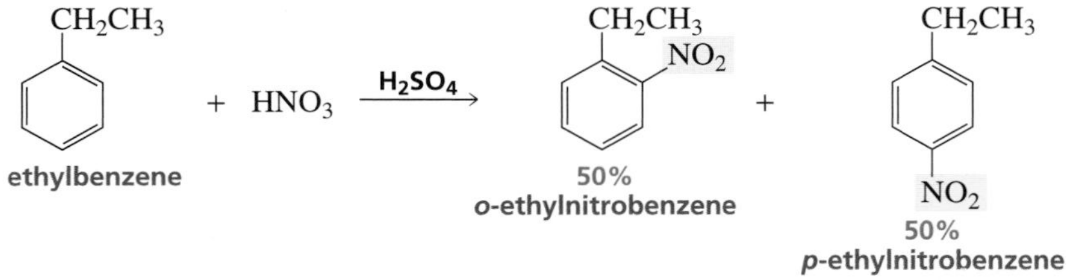

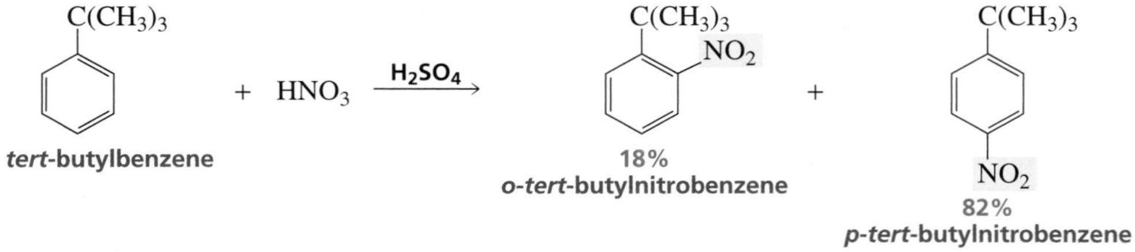

Fortunately, the difference in the physical properties of the ortho- and para-substituted isomers is sufficient to allow them to be easily separated. Consequently, electrophilic substitution reactions that lead to both ortho and para isomers are useful in synthetic schemes even though they lead to a mixture of products.

14.15
ADDITIONAL CONSIDERATIONS REGARDING SUBSTITUENT EFFECTS

It is important to know whether a substituent is activating or deactivating in order to determine the conditions to use when carrying out a reaction. For example, sulfonation of benzene requires fuming sulfuric acid, but sulfonation of toluene can be done with concentrated sulfuric acid because the activating methyl group allows the reaction to proceed under less rigorous conditions.

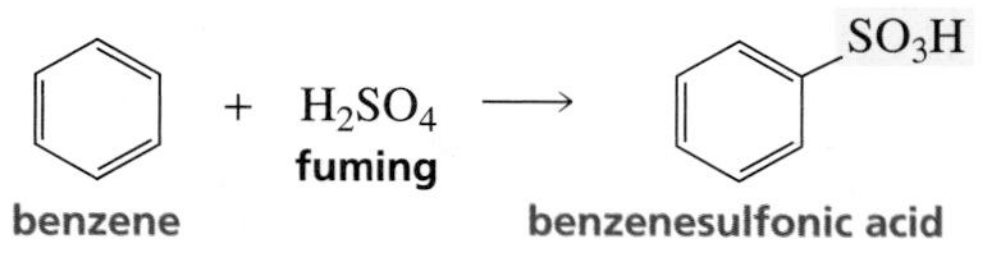

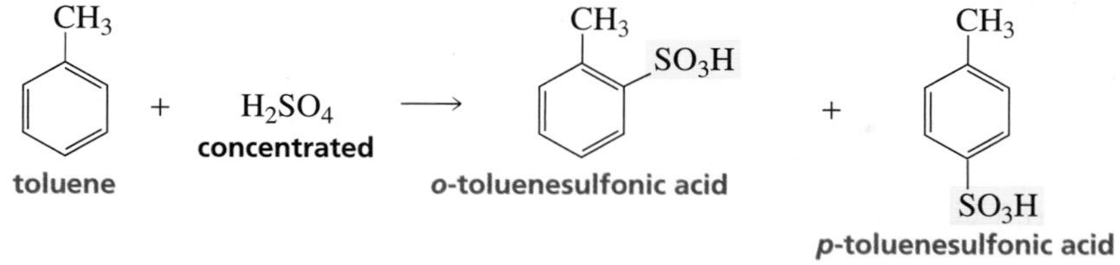

The methoxy and hydroxy substituents are so strongly activating that bromination is done without the Lewis acid ($FeCl_3$) catalyst.

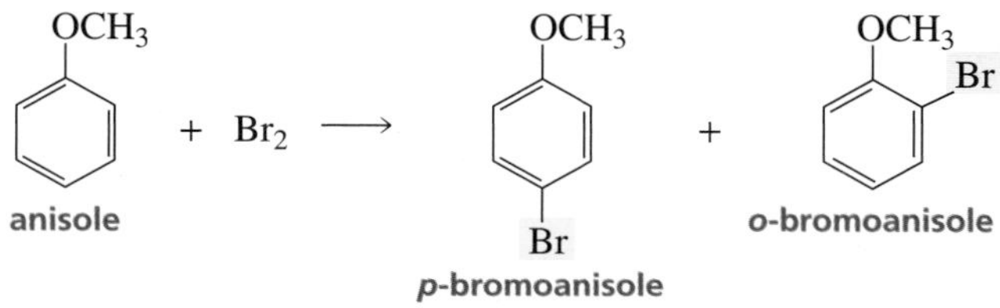

If a catalyst and excess bromine are used, the tribromide is obtained.

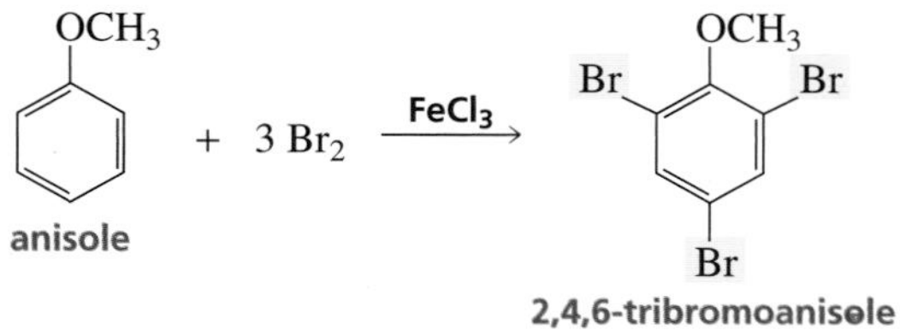

Aniline cannot be nitrated, because nitric acid is an oxidizing agent and primary amines are susceptible to oxidation. (Nitric acid and aniline can be an explosive combination.) However, a nitro-substituted aniline can be prepared if aniline is first converted to an amide (Section 15.6). The acetyl group is a protecting group. It protects the primary amine from oxidation. After nitration, the protecting group can be removed by hydrolysis (Section 15.13). Because of the bulkiness of the directing group, only the para isomer is formed.

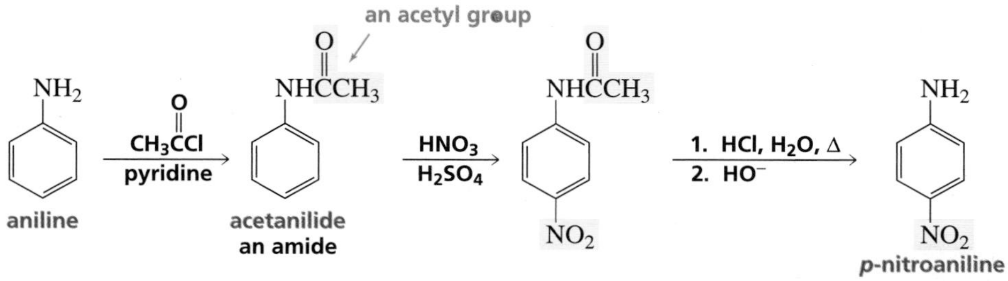

Tertiary aromatic amines can be nitrated. Since the tertiary amino group is a strong activator, nitration is carried out using nitric acid in acetic acid, milder conditions than nitric acid in sulfuric acid. About twice as much para isomer is formed as ortho isomer.

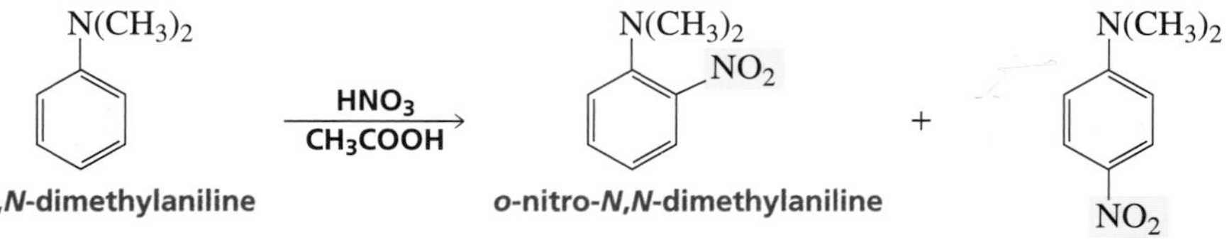

N,N-dimethylaniline → (HNO₃ / CH₃COOH) → *o*-nitro-*N,N*-dimethylaniline + *p*-nitro-*N,N*-dimethylaniline

---

PROBLEM 21

When *N,N*-dimethylaniline is nitrated, some *meta*-nitro-*N,N*-dimethylaniline is formed. Why does the meta isomer form? (*Hint:* The pK_a of *N,N*-dimethylaniline is 5.07, and more meta isomer is formed if the reaction is carried out at pH = 3.5 than if it is carried out at pH = 4.5.)

Of the five reactions listed in Section 14.3, the two Friedel–Crafts reactions are the most sluggish. Therefore, if there is a meta director on the ring (remember that meta directors are all moderate or strong deactivators), the ring will be too unreactive to undergo either Friedel–Crafts acylation or Friedel–Crafts alkylation.

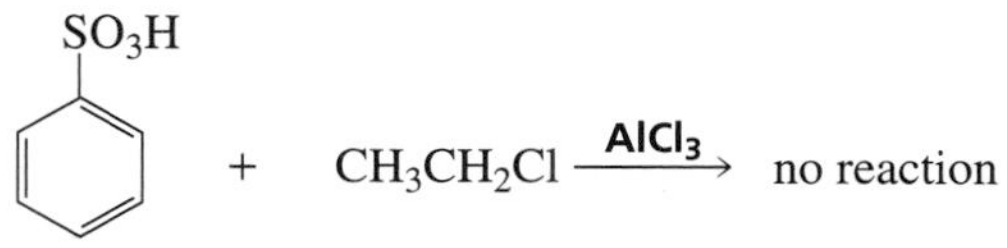

benzenesulfonic acid + CH_3CH_2Cl → (AlCl₃) → no reaction

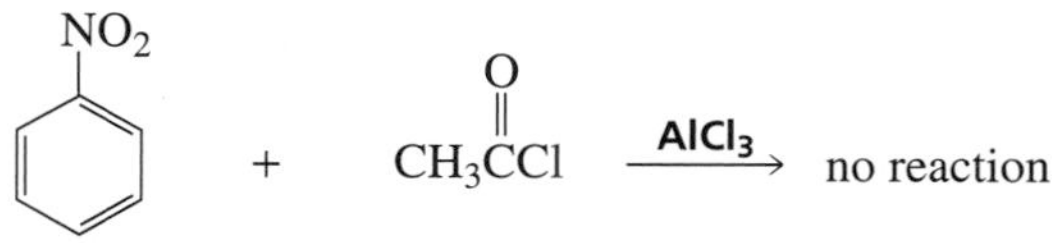

nitrobenzene + CH_3CCl (O) → (AlCl₃) → no reaction

Aniline and *N*-substituted anilines cannot undergo Friedel–Crafts reactions because the lone pair electrons on the amino group will complex with the Lewis acid ($AlCl_3$) needed to catalyze the reaction. Complexation converts the NH_2 substituent into a deactivating meta director, and we have just seen that Friedel–Crafts reactions do not occur on benzene rings containing meta-directing substituents.

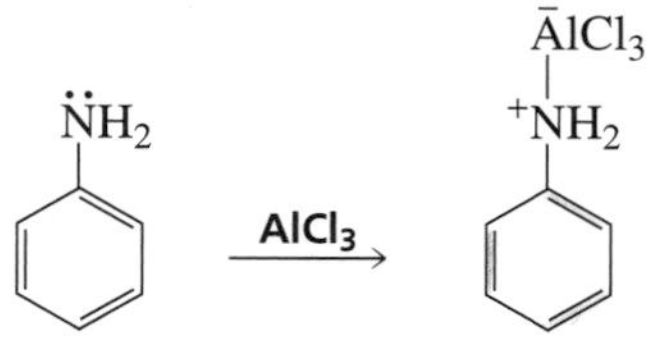

aniline → (AlCl₃) → substituent is a meta director

PROBLEM 22 ◆

Give the products, if any, of each of the following reactions.

a. benzonitrile + methyl chloride + AlCl₃   **c.** benzoic acid + CH₃CH₂Cl + AlCl₃

b. aniline + 3 Br₂   **d.** benzene + 2 CH₃Cl + AlCl₃

As the number of reactions we know increases, we have more reactions to choose from in designing a synthesis. For example, we can now design two very different routes for the synthesis of 2-phenylethanol. Which route we choose depends on how easy it is to do the required reactions and on the expected yield of the target molecule. For example, the first route shown for the synthesis of 2-phenylethanol is the better procedure. The second route has more steps, it requires excess benzene since the alkylated product is more reactive than benzene, it involves a radical reaction that can produce unwanted side products, the yield of the elimination reaction is not high because some substitution product is formed as well, and hydroboration–oxidation is not an easy reaction to carry out.

**14.16
DESIGNING A
SYNTHESIS III:
SYNTHESIS OF
MONOSUBSTITUTED
AND DISUBSTITUTED
BENZENES**

2-phenylethanol

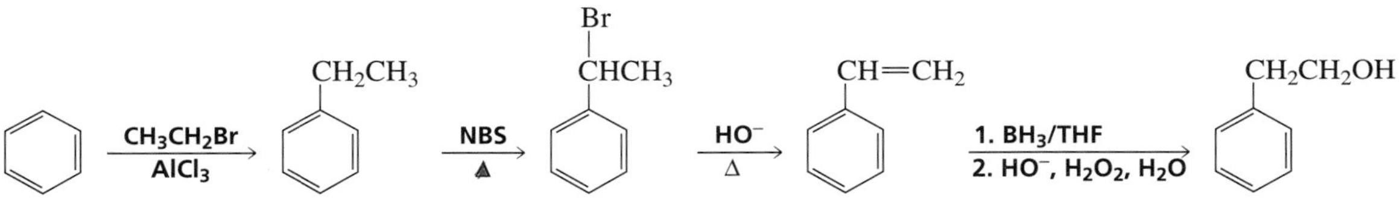

excess   **2-phenylethanol**

In designing the synthesis of disubstituted benzenes, it is important to pay attention to the order in which the substituents are placed on the ring. For example, if you want to synthesize *meta*-bromobenzenesulfonic acid, the sulfonic acid group has to be placed on the ring first because that group will direct the bromo substituent into the desired meta position.

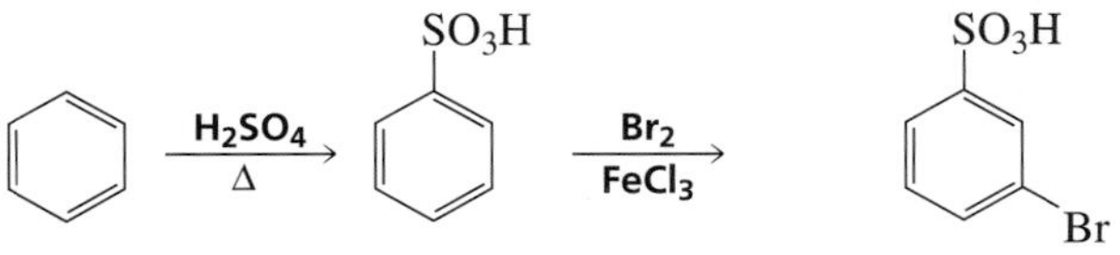

***m*-bromobenzenesulfonic acid**

On the other hand, if the desired product is *para*-bromobenzenesulfonic acid, the order of the two reactions must be reversed to take advantage of the fact that the bromo group is an ortho/para director.

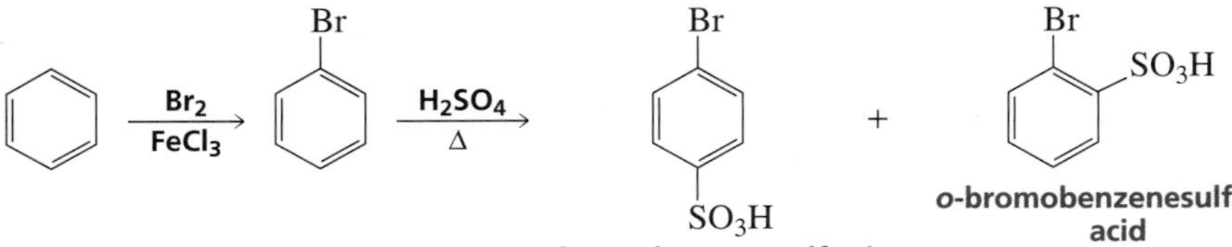

***p*-bromobenzenesulfonic
acid**   ***o*-bromobenzenesulfonic
acid**

Both substituents in *meta*-nitroacetophenone are meta directors. However, the Friedel–Crafts acylation reaction must be carried out first because nitrobenzene is too deactivated to undergo a Friedel–Crafts reaction (Section 14.15).

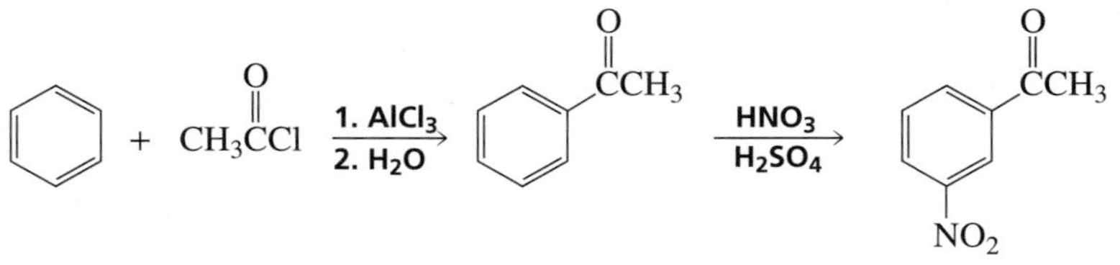

m-nitroacetophenone

It is also important to determine the point in a reaction sequence at which a substituent should be chemically modified. In the synthesis of *para*-chlorobenzoic acid from toluene, the methyl group is oxidized after it directs the chloro substituent to the para position (*ortho*-chlorobenzoic acid is also formed in this reaction).

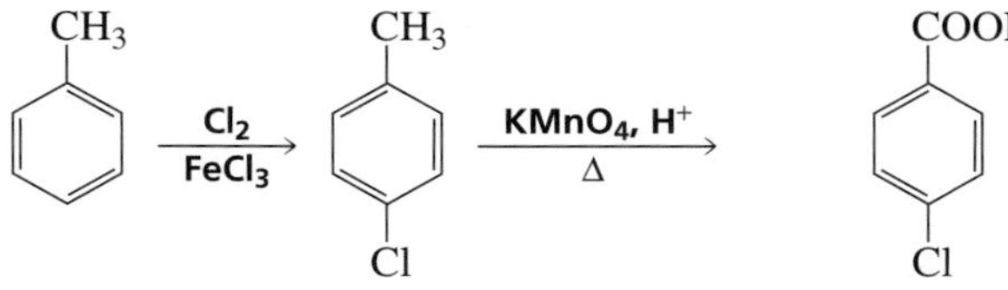

para-chlorobenzoic acid

In the synthesis of *meta*-chlorobenzoic acid, the methyl group is oxidized before chlorination because a meta director is needed to obtain the desired product.

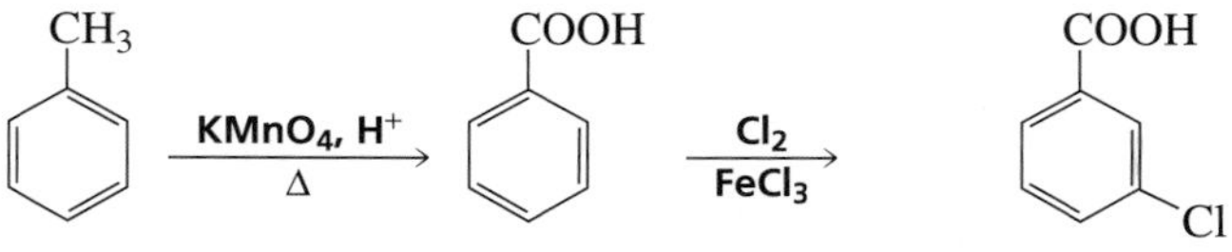

meta-chlorobenzoic acid

In the following synthesis of *para*-propylbenzenesulfonic acid, the type of reaction employed, the order in which the substituents are put on the benzene ring, and the point at which a substituent is chemically modified all must be considered. The straight-chain propyl substituent must be put on the ring by a Friedel–Crafts acylation reaction because of the carbocation rearrangement that would occur with a Friedel–Crafts alkylation reaction. The Friedel–Crafts acylation must be carried out before sulfonation, since acylation could not be carried out on a ring with a strongly deactivating sulfonic acid substituent. The sulfonic acid group must be put on the ring after the carbonyl group is reduced to a methylene group so it will be directed primarily to the para position by the alkyl group.

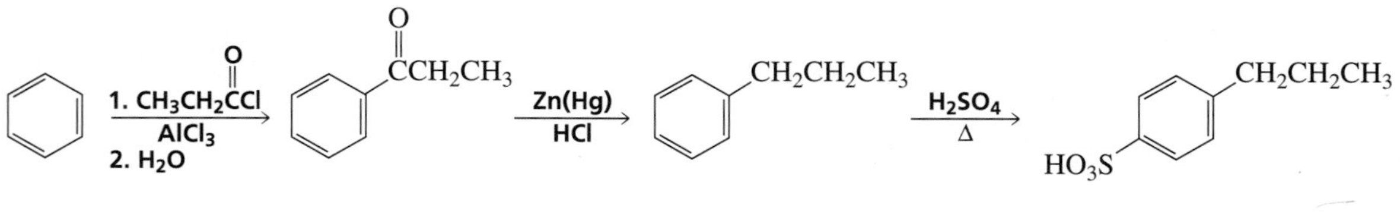

para-propylbenzenesulfonic acid

PROBLEM 23 / SOLVED

When phenol is treated with Br_2, a mixture of monobromo, dibromo, and tribromophenols is obtained. Design a synthesis that would convert phenol primarily to:

a. *ortho*-bromophenol   **b.** *para*-bromophenol

SOLUTION TO 23a The bulky sulfonic acid group will add preferentially to the para position. Both the OH and SO_3H groups will direct bromine to the position ortho to the OH group. Heating in dilute acid removes the sulfonic acid group.

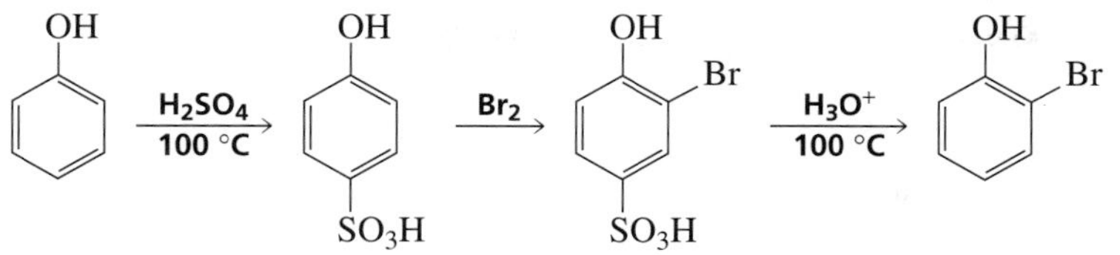

SOLUTION TO 23b An acetyl group is put on phenol (Section 14.15). The bulkiness of the acetyl group causes the incoming bromine to be directed primarily to the para position. The acetyl group is removed by heating in aqueous acid.

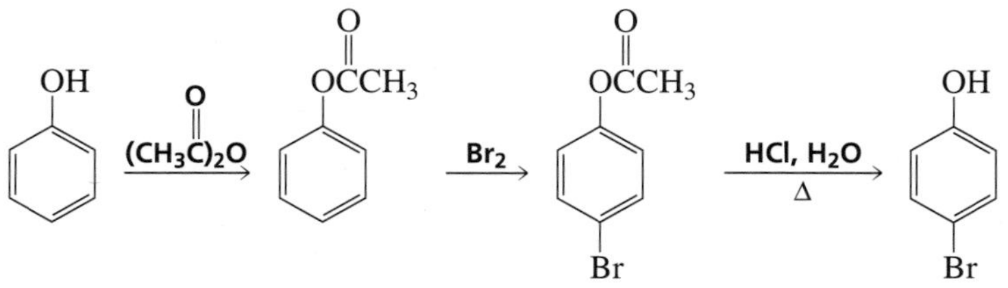

PROBLEM 24

Show how each of the following compounds could be synthesized from benzene.

a. *p*-chloroaniline
b. *m*-nitrobenzoic acid
c. *p*-nitrobenzoic acid
d. *m*- bromopropylbenzene
e. *o*-bromopropylbenzene
f. 1-phenyl-2-propanol
g. *o*-nitrophenol
h. *m*-chloroaniline
i. 2-phenylpropene

14.17 SYNTHESIS OF TRISUBSTITUTED BENZENES

The directing effects of both substituents must be considered in carrying out reactions on disubstituted benzenes. If both substituents direct to the same position, the product of the reaction is easily predicted.

the methyl and nitro substituents direct the incoming substituent to the positions indicated by the arrows

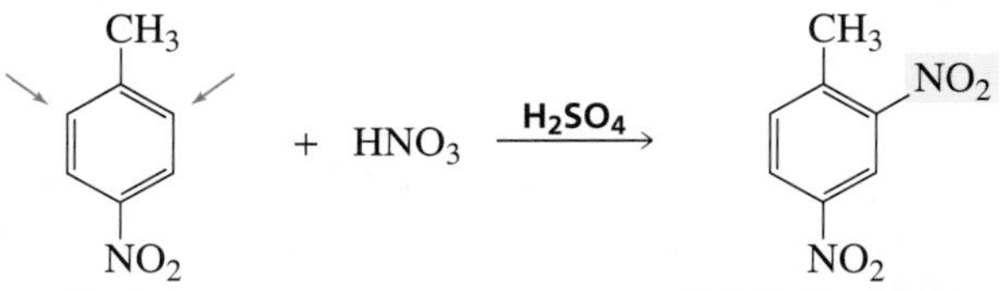

p-nitrotoluene   2,4-dinitrotoluene

Notice that three positions are activated in the following reaction, but the new substituent ends up on only two of the three positions. Steric hindrance makes the position between the substituents less accessible.

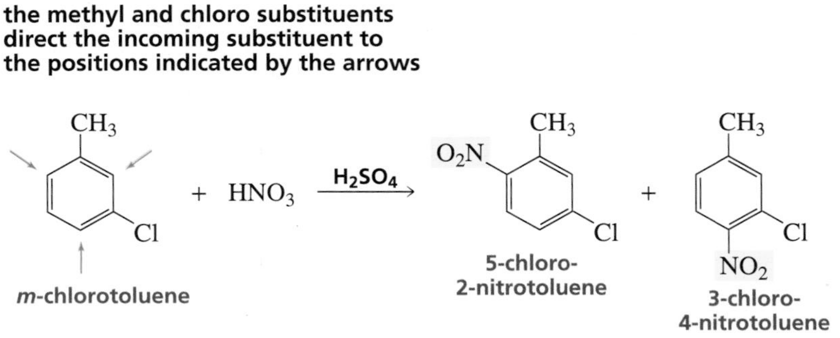

the methyl and chloro substituents direct the incoming substituent to the positions indicated by the arrows

m-chlorotoluene   + HNO₃ →(H₂SO₄)→   5-chloro-2-nitrotoluene   +   3-chloro-4-nitrotoluene

If the two substituents direct to different positions, a strongly activating substituent will win out over a weakly activating substituent or a deactivating substituent.

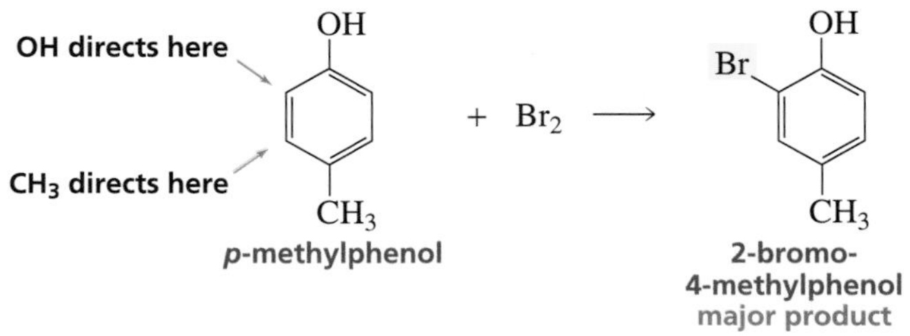

OH directs here

CH₃ directs here

p-methylphenol   + Br₂ →   2-bromo-4-methylphenol
major product

If the two substituents have similar activating properties, neither will win and a mixture of products will be obtained.

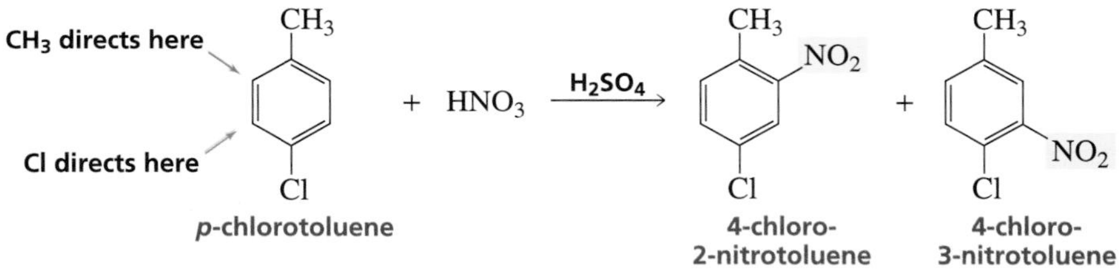

CH₃ directs here

Cl directs here

p-chlorotoluene   + HNO₃ →(H₂SO₄)→   4-chloro-2-nitrotoluene   +   4-chloro-3-nitrotoluene

---

PROBLEM 25

Give the major products of each of the following reactions.

a. nitration of *p*-fluoroanisole

b. chlorination of *o*-benzenedicarboxylic acid

c. bromination of *p*-chlorobenzoic acid

**14.18
SYNTHESIS OF
SUBSTITUTED
BENZENES USING
DIAZONIUM SALTS**

So far we have learned how to place a limited number of different substituents on a benzene ring—those substituents listed in Section 14.3, and those that can be obtained from these substituents by chemical conversion (Section 14.10). However, the kinds of substituents that can be placed on benzene rings can be greatly expanded by the use of aryl **diazonium salts.**

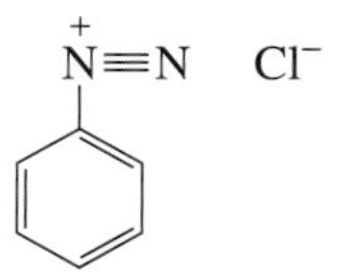

an aryl diazonium salt

The drive to form a molecule of stable nitrogen gas (N_2) causes the leaving group of a diazonium ion to be easily displaced by a wide variety of nucleophiles. The mechanism by which a nucleophile displaces the diazonium group depends on the nucleophile; some displacements involve phenyl cations, others involve radicals.

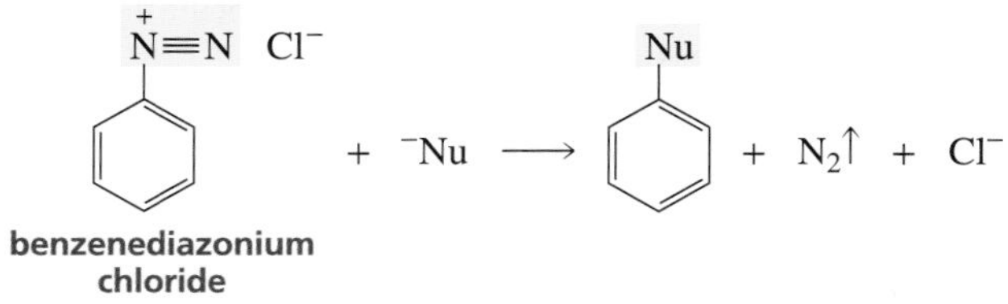

A primary amine can be converted into a diazonium salt by treatment with nitrous acid (HNO_2). Nitrous acid is unstable, so it is formed *in situ* using an aqueous solution of sodium nitrite and HCl or HBr; N_2 is such a good leaving group that the diazonium salt must be synthesized at 0 °C and used immediately. [The mechanism for the conversion of a primary amino group (NH_2) to a diazonium group ($^+N{\equiv}N$) is shown in Section 14.20.]

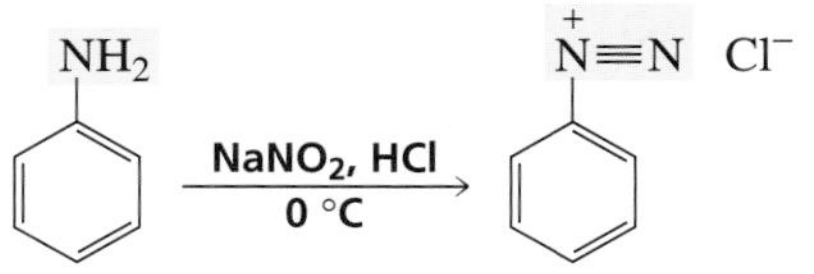

Nucleophiles such as $^-C{\equiv}N$, Cl^-, and Br^- will replace the diazonium group if the appropriate cuprous salt is added to the solution containing the diazonium salt. The reaction of an aryl diazonium salt with a cuprous salt is known as a **Sandmeyer reaction.**

Sandmeyer reactions

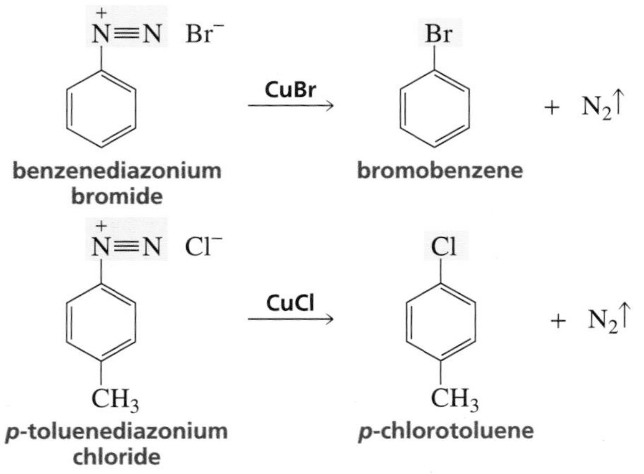

benzenediazonium bromide   **bromobenzene**

p-toluenediazonium chloride   **p-chlorotoluene**

Traugott Sandmeyer (1854–1922) *was born in Switzerland. He received a Ph.D. from the University of Heidelberg and discovered the reaction that bears his name in 1884. He was a scientist at Geigy Co. in Basel, Switzerland.*

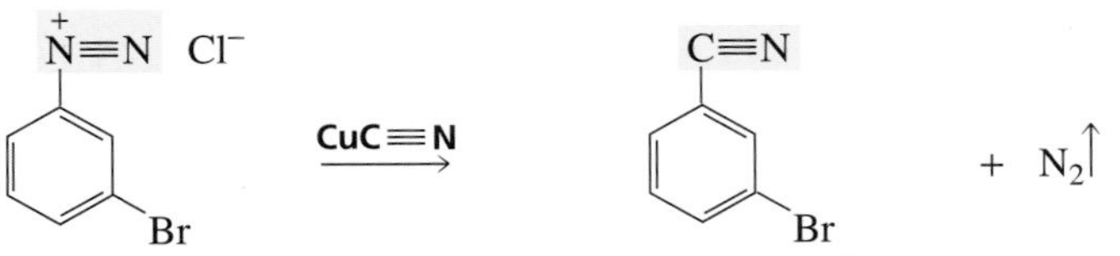

m-bromobenzenediazonium chloride   *m*-bromobenzonitrile

KCl and KBr cannot be used in place of CuCl and CuBr in Sandmeyer reactions; the cuprous salts are required. This indicates that the cuprous ion is catalyzing the Sandmeyer reaction, most likely by serving as a radical initiator.

Although chloro and bromo substituents can be placed directly on a benzene ring by halogenation, the Sandmeyer reaction can be a useful alternative. For example, if you wanted to make *para*-chloroethylbenzene, chlorination of ethylbenzene would lead to a mixture of the ortho and para isomers.

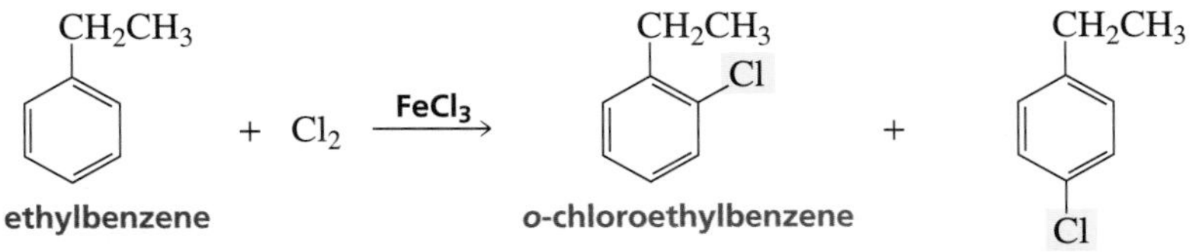

ethylbenzene   o-chloroethylbenzene   p-chloroethylbenzene

If, however, you started with *para*-ethylaniline and used a Sandmeyer reaction for chlorination, only the desired para product would be formed.

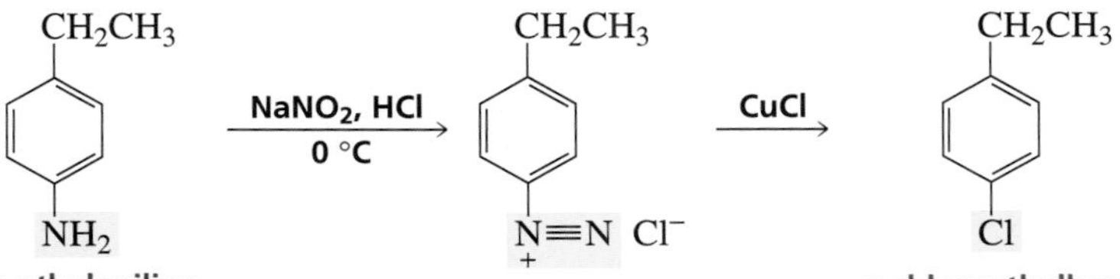

p-ethylaniline   *p*-chloroethylbenzene

An iodo substituent will replace the diazonium group if potassium iodide is added to the solution containing the diazonium ion.

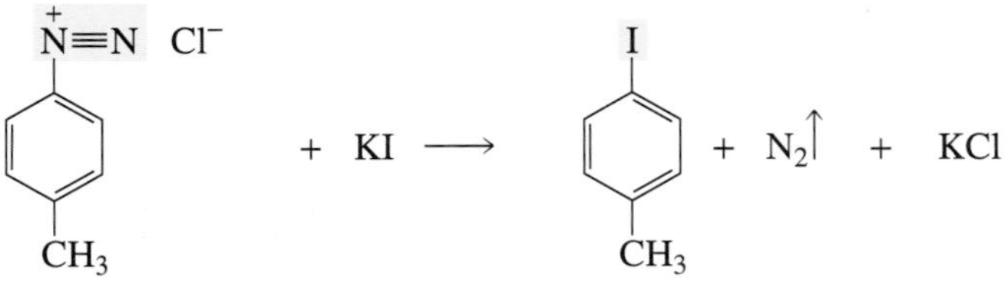

p-toluenediazonium chloride   *p*-iodotoluene

Fluoro substitution occurs if the diazonium salt is heated with fluoroboric acid (HBF_4). This reaction is known as the **Schiemann reaction.**

Günther Schiemann (1899–1969) *was born in Germany. He was a professor of chemistry at the Technologische Hochschule in Hanover.*

Schiemann reaction

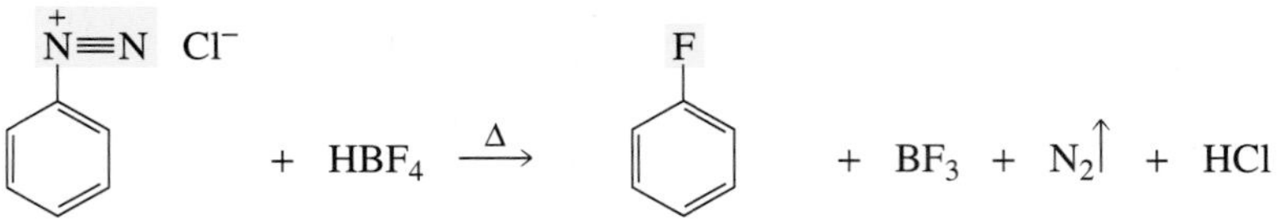

fluorobenzene

If the aqueous solution in which the benzenediazonium salt has been synthesized is allowed to warm up to room temperature, a hydroxy group will replace the diazonium group. This is a convenient way to synthesize phenols.

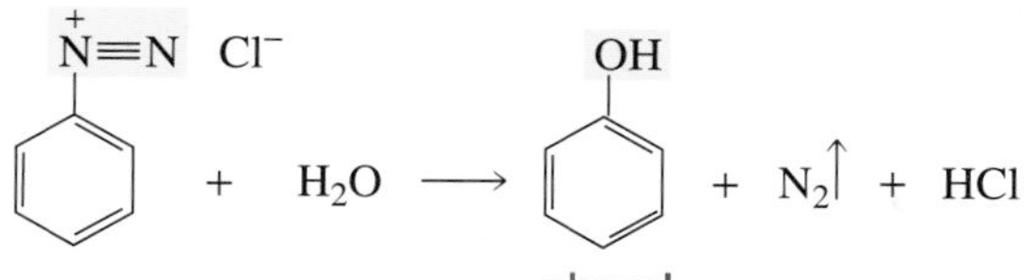

phenol

A hydrogen will replace a diazonium group if the diazonium ion is treated with hypophosphorous acid (H_3PO_2). This is a useful reaction if an amino group is needed for directing purposes and subsequently must be removed. It would be difficult to envision how 1,3,5-tribromobenzene could be synthesized without this reaction.

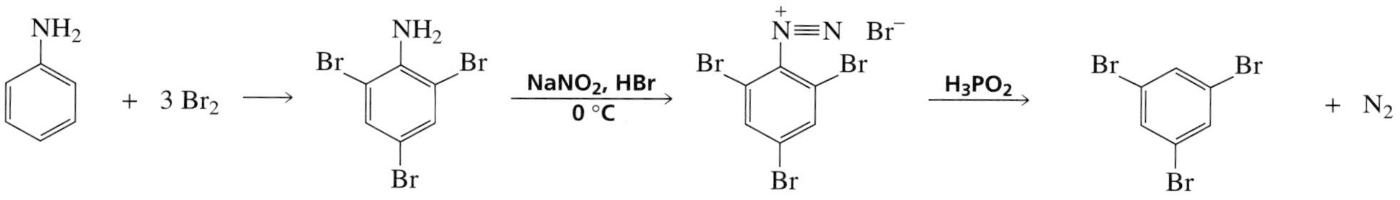

1,3,5-tribromobenzene

PROBLEM 26

Write the sequence of steps involved in the conversion of benzene into benzenediazonium chloride.

PROBLEM 27◆

In an attempt to synthesize benzonitrile, a student allowed a freshly prepared solution of benzenediazonium chloride to warm up to room temperature before adding CuCN.

a. Was the synthesis successful?

b. What was the major product of the reaction?

PROBLEM 28 / SOLVED

Show how the following compounds could be synthesized from benzene.

a. *m*-dibromobenzene

b. *p*-methylbenzonitrile

c. *m*-bromophenol

d. *o*-chlorophenol

e. *m*-nitrotoluene

SOLUTION TO 28a   A bromo substituent is an ortho/para director, so halogenation cannot be used to introduce both bromo substituents of *m*-dibromobenzene. If we recall that a bromo substituent can be placed on a benzene ring with a Sandmeyer reaction and that the bromo substituent in a Sandmeyer reaction replaces what originally was a meta-directing nitro substituent, we have a route to the synthesis of the target compound.

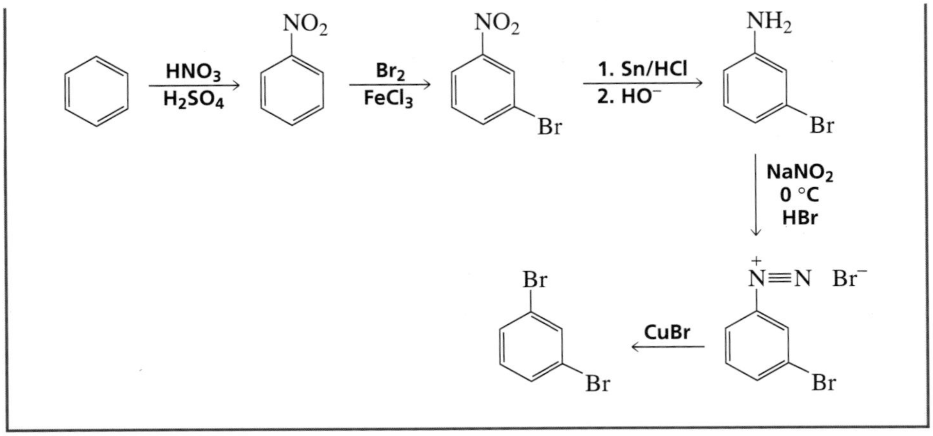

14.19
THE DIAZONIUM ION AS AN ELECTROPHILE

In addition to their use in nucleophilic displacement reactions to place substituents on a benzene ring, aryl diazonium ions can also be used as electrophiles in electrophilic aromatic substitution reactions. Because an aryl diazonium ion is unstable at room temperature, it can be used as an electrophile only in electrophilic substitution reactions that can be carried out well below room temperature. In other words, only highly activated benzene rings (phenols, anilines, and N-alkylanilines) can undergo electrophilic substitution reactions with aryl diazonium ion electrophiles. The product of the reaction is an *azobenzene*. The —N=N— linkage is called an **azo linkage.** Because the electrophile is so large, substitution takes place preferentially at the less sterically hindered para position.

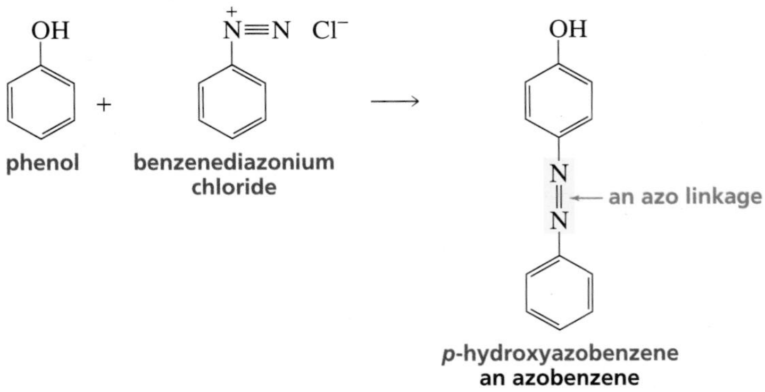

phenol   benzenediazonium chloride

p-hydroxyazobenzene
an azobenzene

If, however, the para position is blocked, substitution will occur at an ortho position.

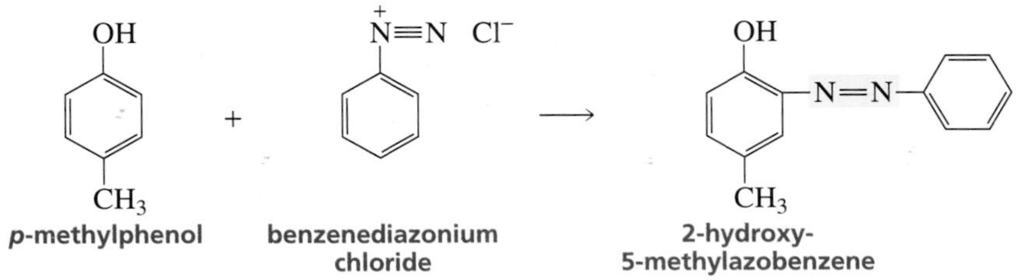

p-methylphenol   benzenediazonium chloride

2-hydroxy-5-methylazobenzene

The mechanism for electrophilic substitution using a diazonium ion electrophile is the same as the mechanism for electrophilic substitution using any other electrophile.

mechanism for electrophilic substitution using an aryl diazonium ion electrophile

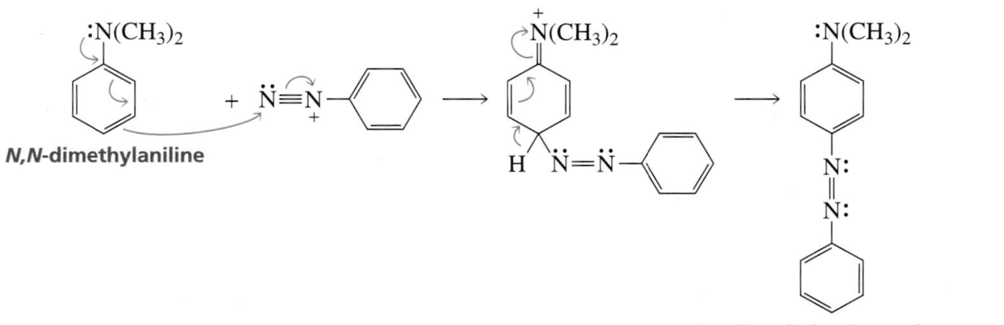

N,N-dimethylaniline

p-N,N-dimethylaminoazobenzene

PROBLEM 29

In the mechanism for electrophilic substitution using a diazonium ion as the electrophile, why does nucleophilic attack occur on the terminal nitrogen atom of the diazonium ion rather than on the nitrogen atom bonded to the benzene ring?

Azo compounds, like alkenes, can exist in cis and trans forms. Because of steric strain, the trans isomer is considerably more stable than the cis isomer (Section 3.19).

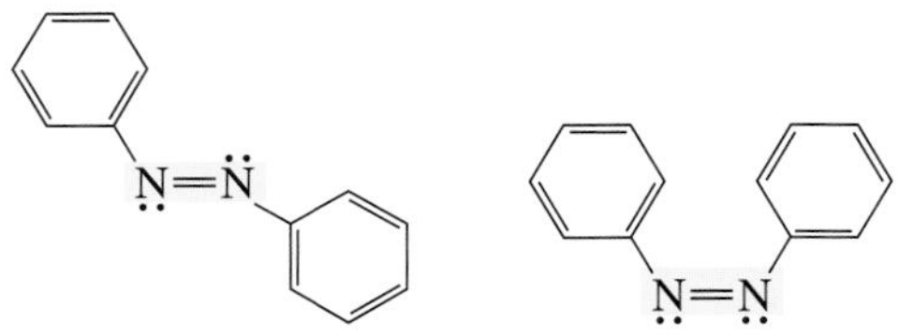

trans-azobenzene   *cis*-azobenzene

We have seen that azobenzenes are colored compounds because of their extended conjugation. We have also learned that they are used commercially as dyes (Section 13.18).

PROBLEM 30

Give the structure of the activated benzene ring and the diazonium ion used in the synthesis of:

a. butter yellow   **b.** methyl orange

(The structures of these compounds can be found in Section 13.18.)

We have seen that the reaction of a primary amine with nitrous acid produces a diazonium salt. Both aryl amines and alkyl amines undergo this reaction, and both follow the same mechanism.

$$\text{aryl}-NH_2 \xrightarrow[\text{0 °C}]{\text{NaNO}_2,\ \text{HCl}} \text{aryl}-\overset{+}{N}\equiv N \ \ Cl^-$$

$$\text{alkyl}-NH_2 \xrightarrow[\text{0 °C}]{\text{NaNO}_2,\ \text{HCl}} \text{alkyl}-\overset{+}{N}\equiv N \ \ Cl^-$$

14.20
MECHANISM FOR THE REACTION OF AMINES WITH NITROUS ACID

The first step in the reaction is formation of nitrous acid from sodium nitrite and HCl. Loss of water from protonated nitrous acid generates the nitrosonium ion, which is the reactive species in the reaction of amines with nitrous acid.

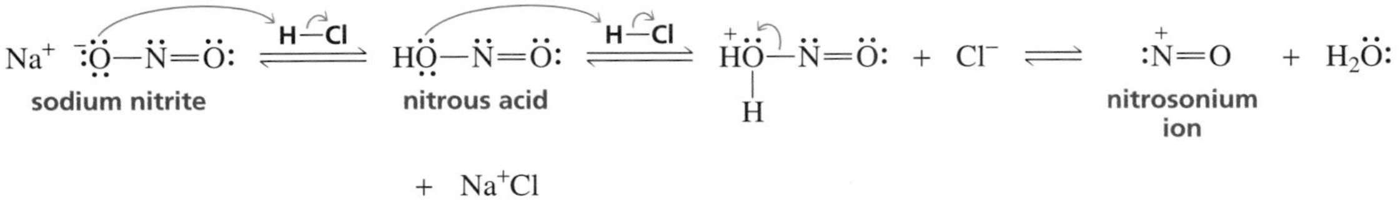

$$+ \quad Na^+Cl$$

The nitrogen atom of the amine shares its lone pair electrons with the nitrosonium ion. Loss of a proton from nitrogen forms a **nitrosamine** (also called an **N-nitroso compound** since a nitroso substituent is bonded to a nitrogen). Delocalization of nitrogen's lone pair electrons and protonation of oxygen form a protonated *N*-hydroxyazo compound. This compound is in equilibrium with its nonprotonated form, which can be reprotonated on nitrogen (reverse reaction) or protonated on oxygen (forward reaction). Elimination of water results in formation of a diazonium ion.

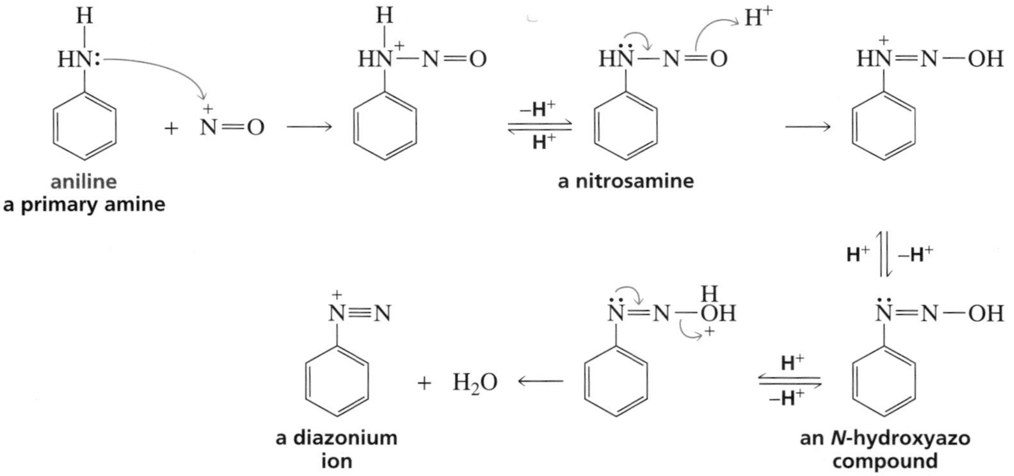

Notice how proton transfer steps are indicated: $-H^+$ means loss of a proton; H^+ means gain of a proton.

Remember that reactions in which aryl diazonium ions are involved must be carried out at 0 °C because of their instability at higher temperature. Alkyl diazonium ions are even less stable. They lose molecular N_2 as they are formed, reacting with whatever nucleophiles are present in the reaction mixture by both $S_N1/E1$ and $S_N2/E2$ mechanisms. Because of the mixture of products obtained, alkyl diazonium ions are of limited synthetic use.

PROBLEM 31

What products would be formed from the reaction of isopropylamine with sodium nitrite and HCl?

PROBLEM 32

Diazomethane can be used to convert a carboxylic acid into a methyl ester. Propose a mechanism for this reaction.

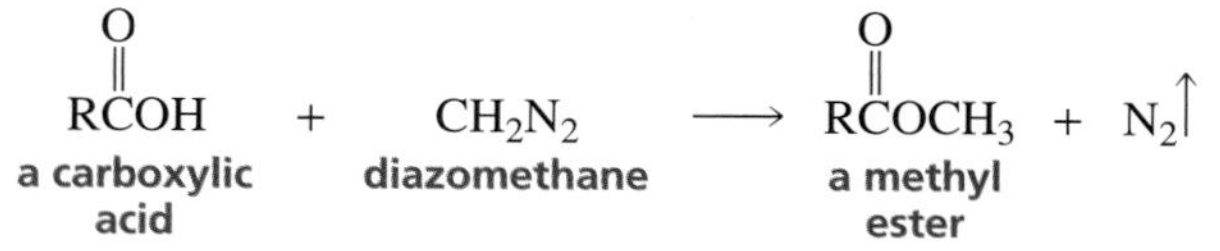

$$\text{RCOH} \ + \ \text{CH}_2\text{N}_2 \ \longrightarrow \ \text{RCOCH}_3 \ + \ \text{N}_2\uparrow$$

a carboxylic   diazomethane   a methyl
acid   ester

Secondary aryl and alkyl amines react with the nitrosonium ion generated from nitrous acid to form nitrosamines. The mechanism of the reaction is similar to that for the reaction of a primary amine with a nitrosonium ion except that the reaction stops at the nitrosamine stage. The reaction stops at this point because a secondary amine, unlike a primary amine, does not have the second proton that must be lost in order to generate the diazonium ion.

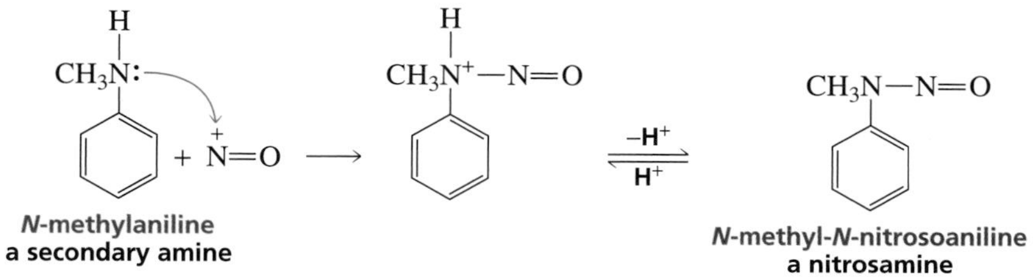

N-methylaniline
a secondary amine

N-methyl-N-nitrosoaniline
a nitrosamine

The product obtained from the reaction of a tertiary amine with a nitrosonium ion cannot be stabilized by loss of a proton. A tertiary aryl amine, therefore, undergoes an electrophilic aromatic substitution reaction with a nitrosonium ion. The product of the reaction is primarily the para isomer because the bulky dimethylamino group blocks approach of the nitrosonium ion to the ortho position.

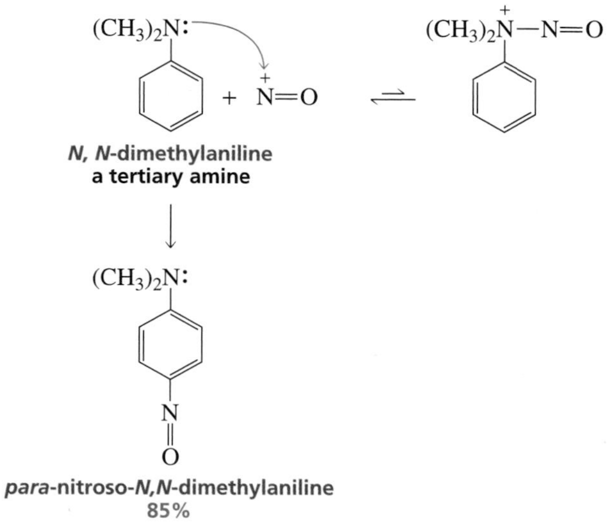

N, N-dimethylaniline
a tertiary amine

para-nitroso-N,N-dimethylaniline
85%

NITROSAMINES AND CANCER

A 1962 outbreak of food poisoning in sheep in Norway was traced to their ingestion of nitrite-treated fish meal. This incident immediately raised concerns about human consumption of nitrite-treated foods. Sodium nitrite is used as a food preservative; it can react with naturally occurring secondary amines present in food to produce nitrosamines, which are known to be carcinogenic. Smoked fish, cured meats, and beer have all been found to contain nitrosamines. Nitrosamines have also been found in cheese, which is not surprising, because some cheeses are preserved with nitrite, and cheese is rich in secondary amines. In the United States, consumer groups asked the Food and Drug Administration to ban the use of sodium nitrite as a preservative. This request was vigorously opposed by the meatpacking industry. Despite extensive investigations, it has not been determined whether or not the small amounts of nitrosamines present in our food pose a hazard to our health. Until this question can be answered, it will be hard to avoid sodium nitrite in our diet. It is worrisome to note that Japan, with one of the highest gastric cancer rates, also has the highest average ingestion of sodium nitrite. The good news is that, in recent years, there has been a considerable reduction in the concentration of nitrosamines present in bacon as a result of the addition of ascorbic acid (a nitrosamine inhibitor) to the curing mixture, and improvements in the malting process have led to a reduction in the level of nitrosamines found in commercial beer. Dietary sodium nitrite does have a redeeming feature: there is some evidence that it protects against a type of food poisoning called botulism.

14.21 NUCLEOPHILIC AROMATIC SUBSTITUTION REACTIONS

Benzene does not react with nucleophiles under standard reaction conditions for two reasons: (1) the π electron clouds above and below the plane of the ring repel the approach of a nucleophile, and (2) the leaving group would be the very basic hydride ion. If, however, the benzene ring has (1) one or more substituents that strongly withdraw electrons from the ring by resonance, and (2) a good leaving group (such as a halogen), **nucleophilic aromatic substitution** reactions can occur without using extreme conditions. These electron-withdrawing groups must be positioned ortho or para to the leaving group. The greater the number of electron-withdrawing substituents, the easier it will be to carry out the nucleophilic aromatic substitution reaction. Notice the different conditions under which the following reactions occur.

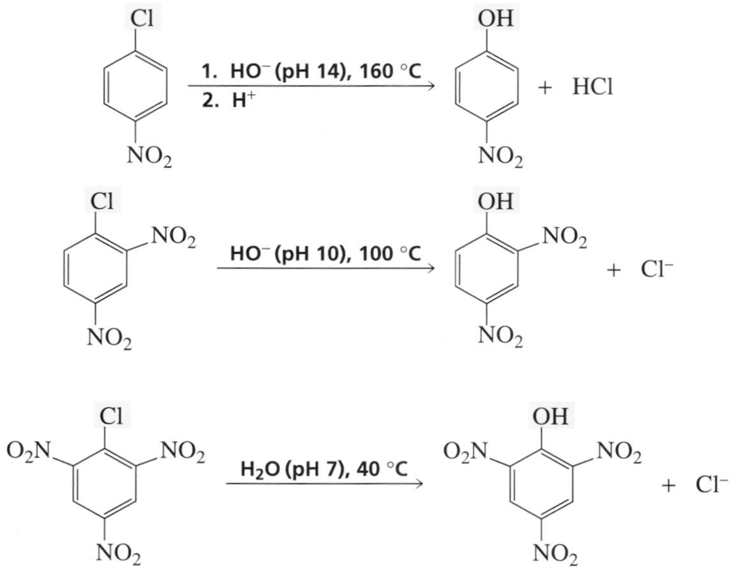

Notice also that the strong electron-withdrawing substituents that activate the benzene ring toward nucleophilic aromatic substitution reactions are the same substituents that deactivate the ring toward electrophilic aromatic substitution. In other words, making the ring less electron rich makes it easier for a nucleophile to approach, but makes it more difficult for an electrophile to approach. Thus, any substituent that deactivates the ring toward electrophilic substitution activates the ring toward nucleophilic substitution, and vice versa.

Nucleophilic aromatic substitution takes place by a two-step reaction known as an **S$_N$Ar reaction** (substitution nucleophilic aromatic). In the first step, the nucleophile attacks the carbon bearing the leaving group from a trajectory nearly perpendicular to the aromatic ring. (Recall from Section 9.8 that leaving groups cannot be displaced from sp^2 carbon atoms by backside attack.) Nucleophilic attack results in formation of a carbanion intermediate that can be approximated by four resonance contributors. In the second step of the reaction, the leaving group departs, reestablishing the aromaticity of the ring.

Electron-withdrawing substituents increase the reactivity of the benzene ring toward nucleophilic substitution and decrease the reactivity of the benzene ring toward electrophilic substitution.

general mechanism for nucleophilic aromatic substitution

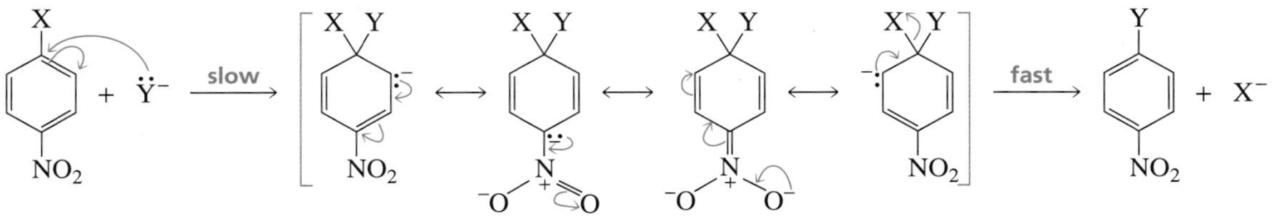

The substituent that is replaced must be a weaker base than the incoming nucleophile so that it, rather than the incoming group, is eliminated from the intermediate.

The electron-withdrawing substituent must be ortho or para to the site of nucleophilic attack because the electrons of the attacking nucleophile can be delocalized onto the substituent only if the substituent is in one of those positions.

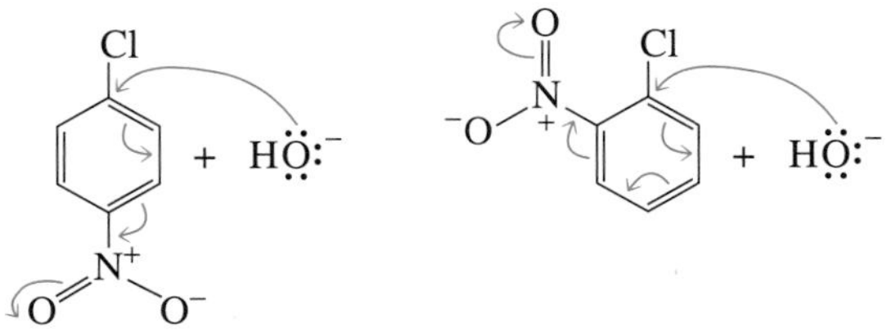

PROBLEM 33

Draw resonance structures for the carbanion that would be formed if *meta*-chloronitrobenzene could react with hydroxide ion. Why does it not react?

A variety of substituents can be placed on a benzene ring by means of nucleophilic aromatic substitution reactions. The only requirement is that the incoming substituent be a stronger base than the substituent that is being replaced.

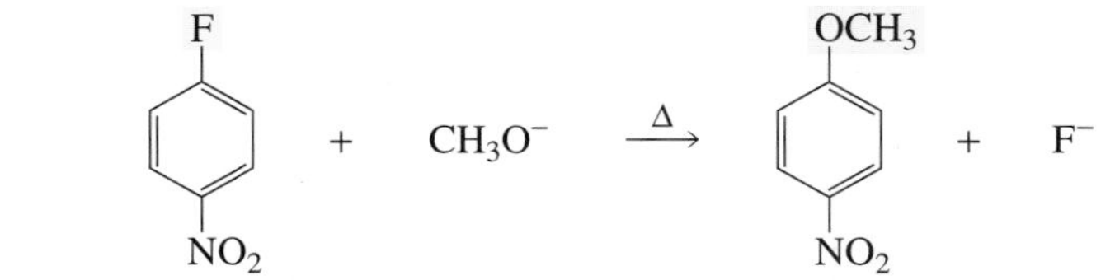

p-fluoronitrobenzene *p*-nitroanisole

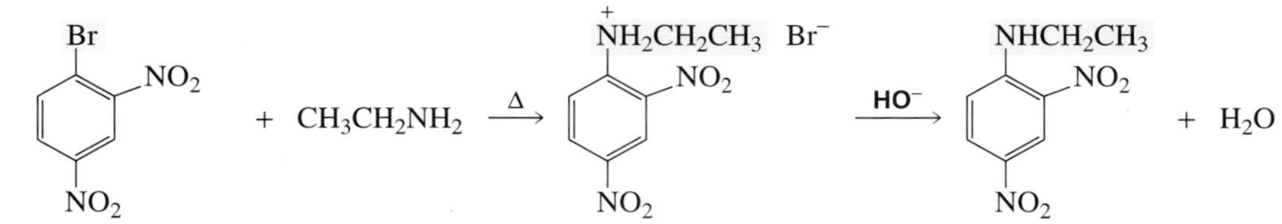

1-bromo-2,4-dinitrobenzene *N*-ethyl-2,4-dinitro-aniline

PROBLEM 34◆

a. List the following compounds in order of decreasing reactivity toward nucleophilic aromatic substitution.

chlorobenzene 1-chloro-2,4-dinitrobenzene *p*-chloronitrobenzene

b. List the same compounds in order of decreasing reactivity toward electrophilic aromatic substitution.

PROBLEM 35◆

p-Fluoronitrobenzene is more reactive than *p*-chloronitrobenzene toward hydroxide ion. What does this tell you about the rate-determining step for nucleophilic aromatic substitution?

PROBLEM 36

Show how each of the following compounds could be synthesized from benzene.

a. *o*-nitroaniline

b. *p*-nitro-*N*-methylaniline

c. *p*-bromoanisole

14.22 BENZYNE

An aryl halide can undergo a nucleophilic substitution reaction in the presence of a very strong base such as $^{-}NH_2$. There are two surprising features about this reaction: the aryl halide does not have to contain an electron-withdrawing group, and the incoming substituent does not always end up on the carbon vacated by the leaving group. For example, when chlorobenzene—that has the carbon to which the chlorine is attached isotopically labeled with ^{14}C—is treated with amide ion in liquid ammonia, aniline is obtained as a product. Half of the product has the amino group attached to the isotopically labeled carbon (denoted by the asterisk) as expected, but the other half has the amino group attached to the carbon adjacent to the labeled carbon.

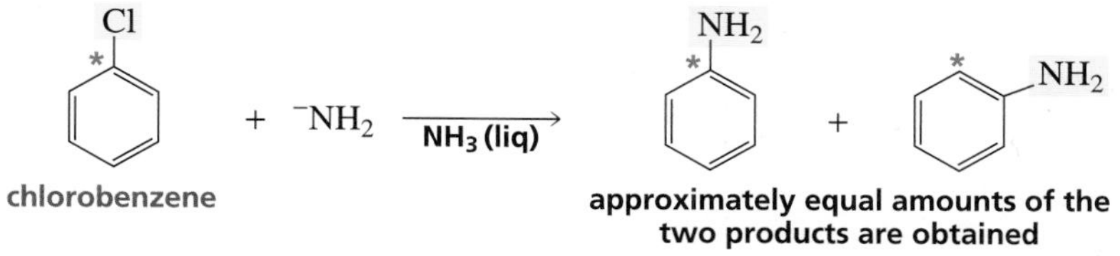

These are the only products formed. Anilines with the amino group two or three carbons removed from the labeled carbon are not formed.

The fact that the two products are formed in approximately equal amounts indicates that the reaction takes place by a mechanism that forms an intermediate in which the two carbons to which the amino group is attached in the product are equivalent. The mechanism that accounts for the experimental observations involves formation of a **benzyne intermediate.** Benzyne has an extra π bond between two adjacent carbon atoms of benzene. In the first step of the mechanism, the strong base (pK_a of $NH_3 = 36$) removes a proton from the position ortho to the halogen. The resulting anion eliminates the halide ion, thereby forming benzyne.

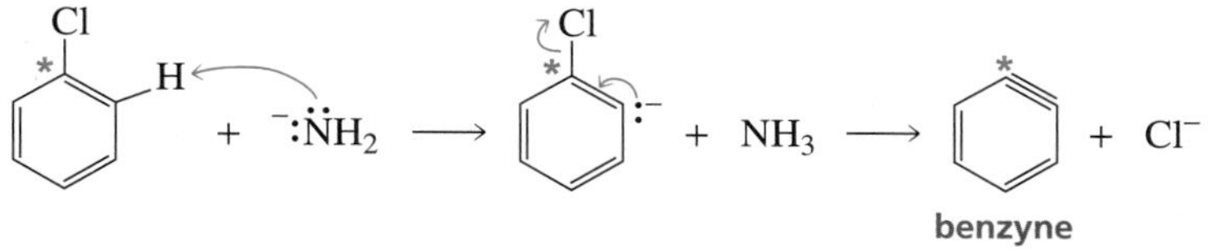

The incoming nucleophile can attack either of the carbons of the "triple bond" of benzyne. Protonation of the resulting anion forms the substitution product. The overall reaction is an elimination–addition reaction: benzyne is formed in an elimination reaction and immediately undergoes an addition reaction.

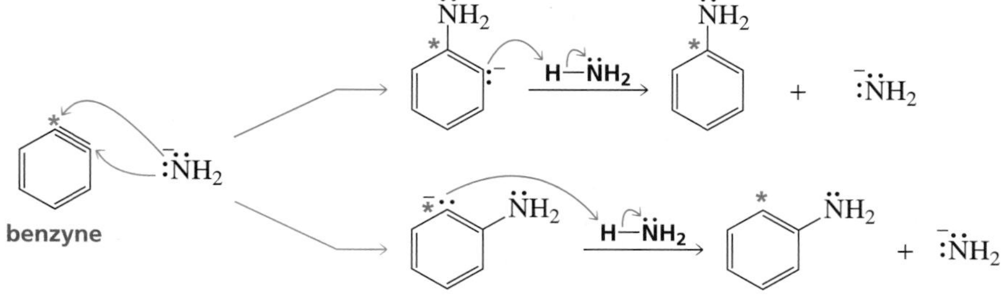

Benzyne is an extremely reactive species. In Section 5.3 we saw that, in a molecule with a triple bond, the sp hybridized carbons and the atoms attached to these carbons ($C—C\equiv C—C$) are linear since the bond angles are 180°. A linear structure cannot be incorporated into a six-membered ring, so the $C—C\equiv C—C$ system is distorted: the orginal π bond is unchanged, but the orbitals that form the new π

Martin D. Kamen *first isolated* ^{14}C. *It immediately became the most useful of all the isotopes in chemical and biochemical research. Kamen was born in Toronto in 1913. He received a B.S. and a Ph.D. from the University of Chicago, becoming a U.S. citizen in 1938. He was a professor at the University of California, Berkeley, at Washington University, and at Brandeis University. He was one of the founding professors at the University of California, San Diego. In later years he became a member of the faculty at the University of Southern California. He received the Fermi medal in 1996.*

Figure 14.7 ▶
Orbital pictures of the second
π bond in (a) a normal triple
bond and (b) the distorted
"triple bond" in benzyne.

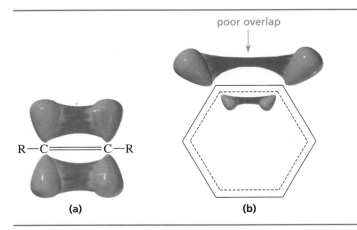

bond are not parallel to each other (Figure 14.7). Because of the distortion, they cannot overlap as well as the orbitals that form a normal π bond, leading to a weaker and more reactive π bond. This means that benzyne can be thought of as a highly distorted and therefore very reactive alkyne.

Since the following reaction forms a benzyne intermediate, two products are formed. Substitution at the carbon that was attached to the leaving group is called **direct substitution.** Substitution at the adjacent carbon is called **cine substitution** (cine comes from *kinesis,* which is Greek for "movement"). *o*-Toluidine is the direct substitution product; *m*-toluidine is the cine substitution product.

o-bromotoluene + NaNH₂ $\xrightarrow{\text{NH}_3 \text{ (liq)}}$ **o-toluidine** direct substitution product + **m-toluidine** cine substitution product

PROBLEM 37

Give the products that would be obtained from reaction of the following compounds with sodium amide in liquid ammonia.

a.

b.

c.

PROBLEM 38

Although benzyne is too unstable to be isolated, evidence for its formation can be obtained by a trapping experiment. When furan is added to a reaction that forms a benzyne intermediate, it traps the benzyne intermediate by forming a Diels–Alder adduct. Give the structure of the Diels–Alder adduct.

Naphthalene, like benzene, is an aromatic hydrocarbon. The resonance hybrid of naphthalene has three contributing resonance structures.

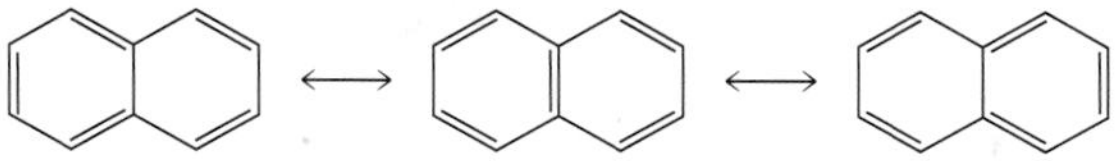

resonance contributors of naphthalene

14.23
ELECTROPHILIC
SUBSTITUTION
REACTIONS OF
NAPHTHALENE
AND SUBSTITUTED
NAPHTHALENES

Like benzene, naphthalene undergoes electrophilic aromatic substitution reactions. Substitution occurs preferentially at the 1-position. In common nomenclature, the 1-position is called the α-position, and the 2-position is called the β-position.

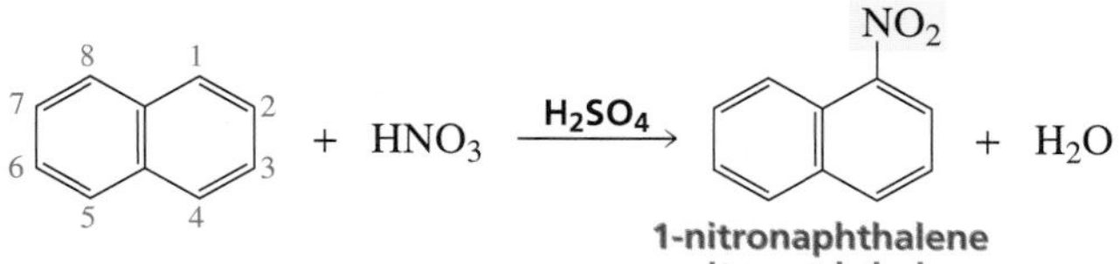

1-nitronaphthalene
α-nitronaphthalene

   Naphthalene is more reactive than benzene toward electrophilic substitution because the carbocation formed in the rate-determining step is more stable and therefore easier to form than the carbocation formed from benzene. Because of naphthalene's increased reactivity, a Lewis acid is not needed for bromination or chlorination.

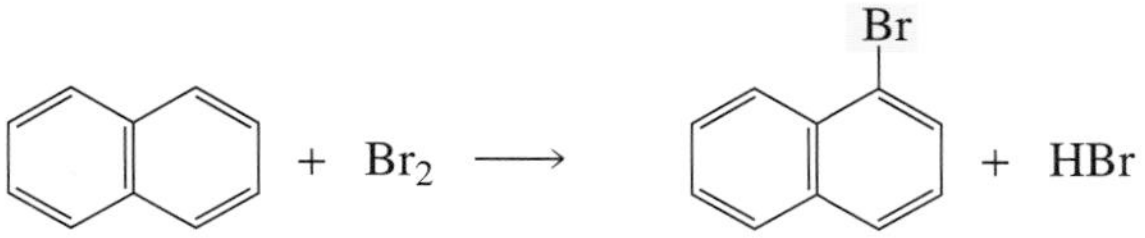

1-bromonaphthalene

PROBLEM 39

From the contributing resonance structures, predict whether the carbon–carbon bonds in naphthalene are all of the same length, as they are in benzene.

PROBLEM 40

Draw the contributing resonance structures for the carbocation intermediates obtained from electrophilic substitution at the 1-position and at the 2-position of naphthalene. Explain why substitution at the 1-position is preferred.

   Sulfonation of naphthalene does not always lead to substitution at the 1-position. If the reaction is carried out at 80 °C, a temperature at which sulfonation is not reversible, substitution occurs at the 1-position.

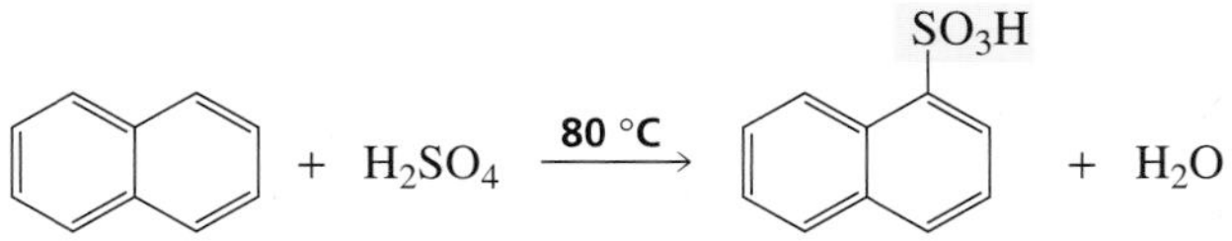

**naphthalene-
1-sulfonic acid**

But if the reaction is carried out under conditions where sulfonation is readily reversible (160 °C), substitution occurs predominantly at the 2-position.

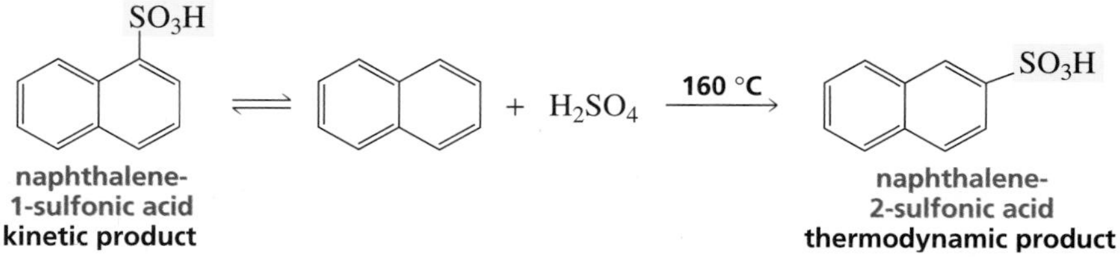

naphthalene-
1-sulfonic acid
kinetic product

naphthalene-
2-sulfonic acid
thermodynamic product

This is another example of kinetic versus thermodynamic control. The 1-substituted product is the kinetic product because it is easier to form. Therefore, it is the predominant product under mild reaction conditions. The 2-substituted product is the thermodynamic product because it is more stable. It is the predominant product under conditions (higher temperatures) that cause the reaction to be readily reversible (Section 7.7).

The 1-substituted product is easier to form because the carbocation leading to its formation is more stable. The 2-substituted product is more stable because there is more room for the sulfonic acid group at the 2-position. In the 1-substituted product, the bulky sulfonic acid group is too close to the hydrogen at the 8- position.

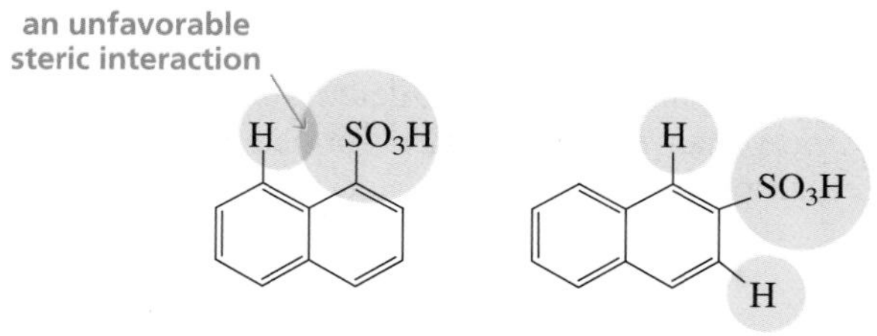

an unfavorable
steric interaction

In the case of substituted naphthalenes, the nature of the substituent determines which ring will undergo electrophilic substitution. If the substituent is deactivating, the electrophile will attack the 1-position of the ring without the substituent because that ring is more reactive than the ring containing the deactivating substituent.

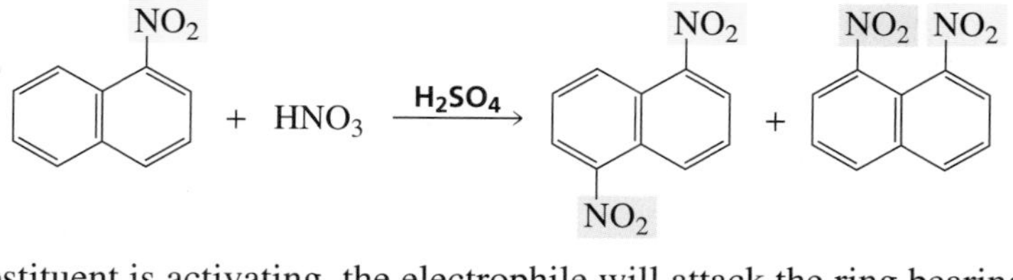

If the substituent is activating, the electrophile will attack the ring bearing the substituent. The incoming electrophile will be directed to a 1-position that is ortho or para to the substituent because all activating substituents are ortho/para directors (Section 14.12).

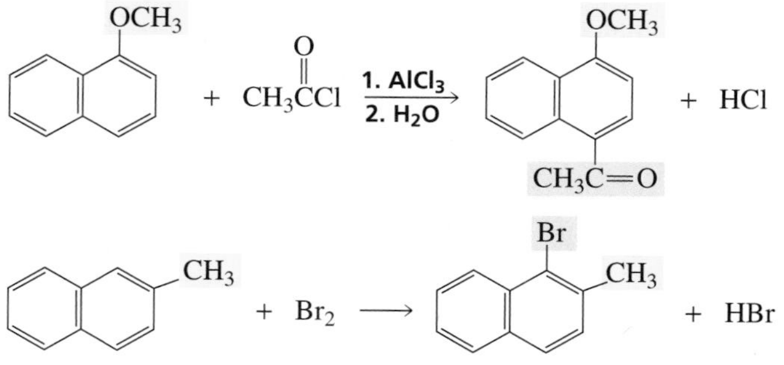

PROBLEM 41◆

Give the products that would be obtained from the reaction of the following compounds with Cl_2.

a.

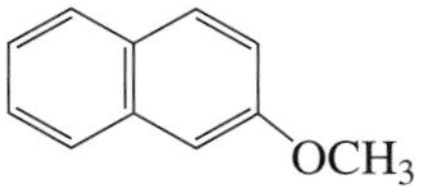

OCH₃

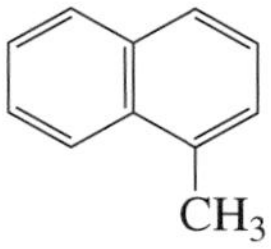

b.

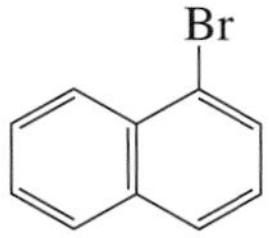

Br

c.

CH₃

d.

NO₂

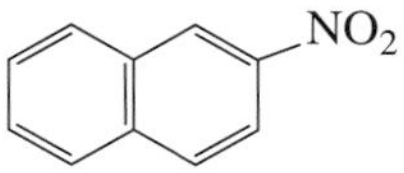

14.24 POLYCYCLIC BENZENOID HYDROCARBONS

Polycyclic benzenoid hydrocarbons are compounds that contain two or more fused benzene rings. **Fused rings** share two adjacent carbons: naphthalene has two fused rings; anthracene and phenanthrene have three fused rings; and tetracene, triphenylene, pyrene, and chrysene have four fused rings. There are many polycyclic benzenoid hydrocarbons with more than four fused rings.

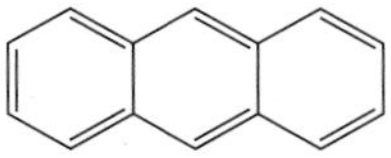

anthracene 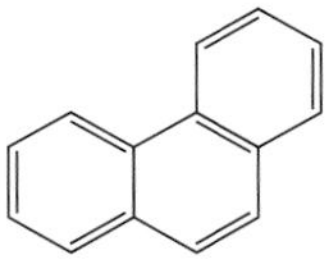   phenanthrene 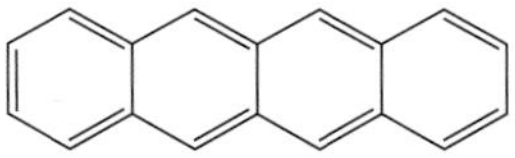   tetracene

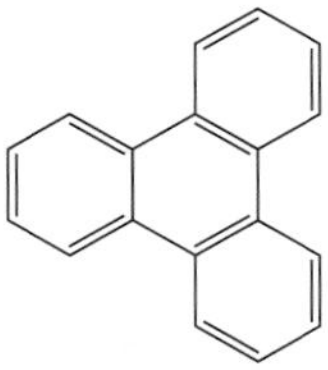

triphenylene 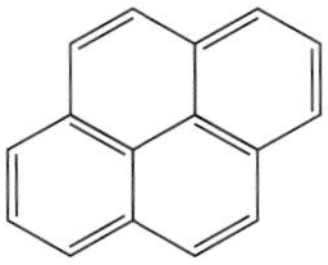   pyrene 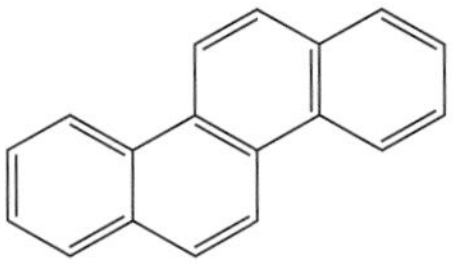   chrysene

   Like benzene and naphthalene, all of the larger polycyclic benzenoid hydrocarbons undergo electrophilic substitution reactions. Some of these compounds are well-known carcinogens. The chemical reactions responsible for causing cancer and how you can predict which compounds are carcinogenic were discussed in Section 11.7.

SUMMARY OF REACTIONS

1. Electrophilic aromatic substitution reactions
   a. Halogenation (Section 14.4)

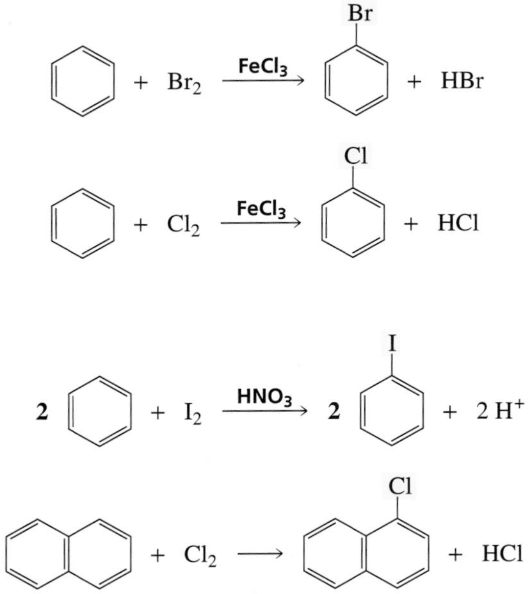

b. Nitration, sulfonation, and desulfonation (Sections 14.5 and 14.6)

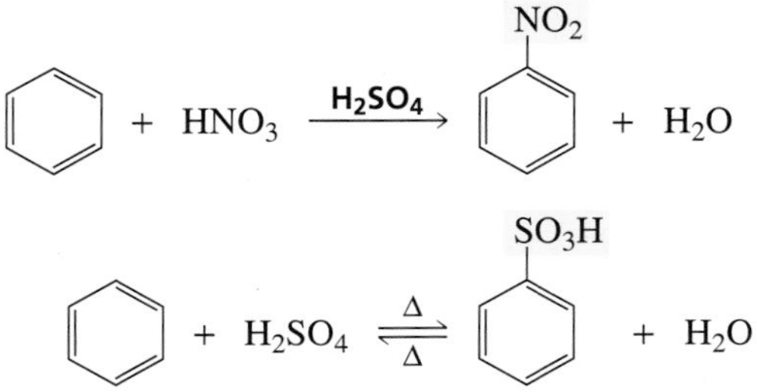

c. Friedel–Crafts acylation and alkylation (Sections 14.7 and 14.8)

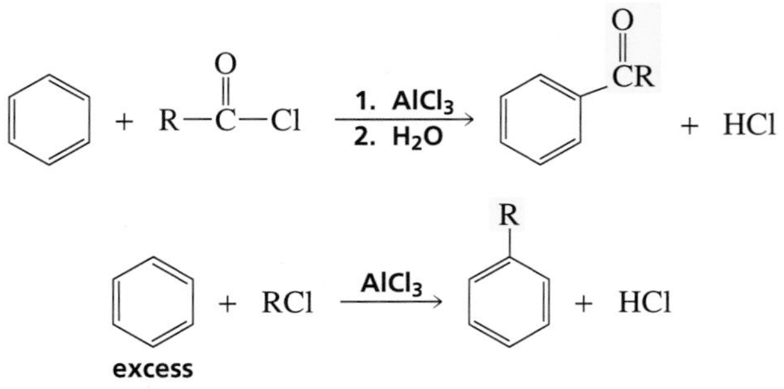

excess

d. Formation of an azobenzene (Section 14.19)

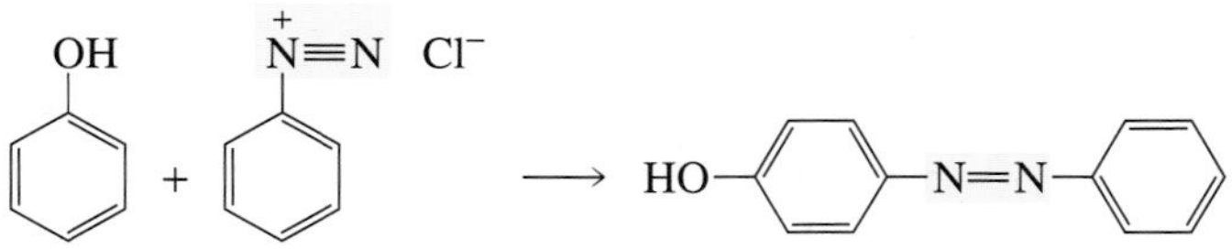

2. Reactions of substituents on a benzene ring (Sections 14.9 and 14.10)

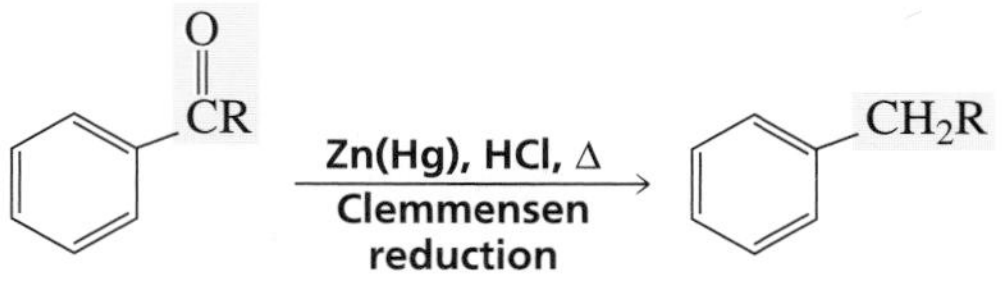

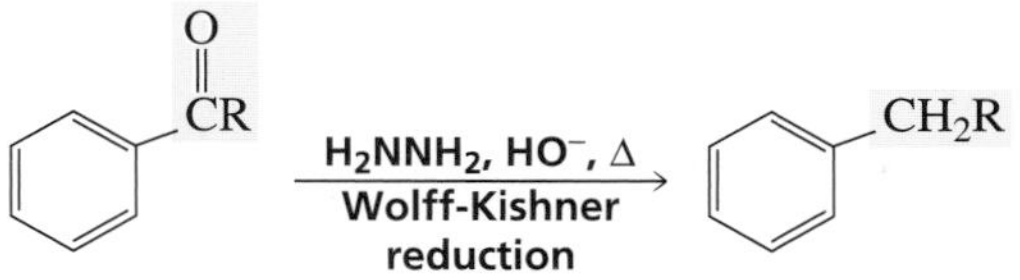

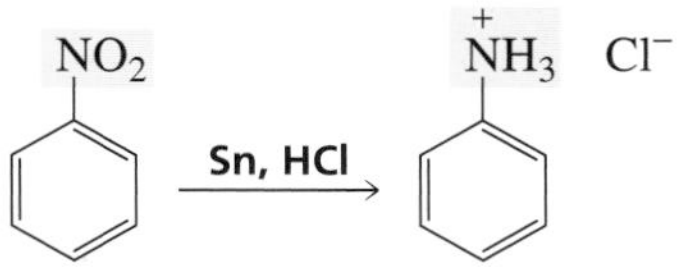

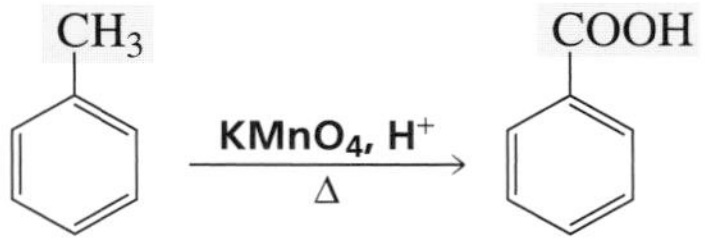

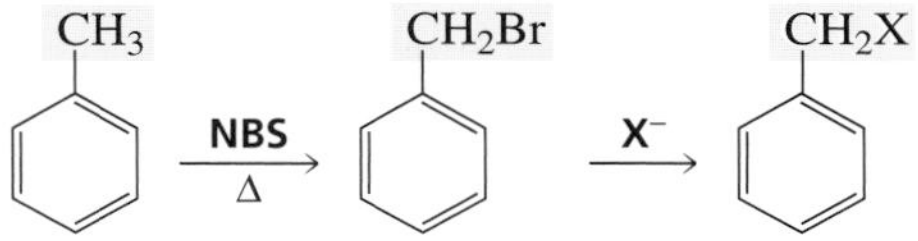

Clemmensen reduction: ArC(=O)R + Zn(Hg), HCl, Δ → ArCH₂R

Wolff-Kishner reduction: ArC(=O)R + H₂NNH₂, HO⁻, Δ → ArCH₂R

NO_2 (benzene) $\xrightarrow{\text{Sn, HCl}}$ $\overset{+}{N}H_3$ Cl⁻ (benzene)

CH_3 (benzene) $\xrightarrow[\Delta]{\text{KMnO}_4,\ \text{H}^+}$ COOH (benzene)

CH_3 (benzene) $\xrightarrow[\Delta]{\text{NBS}}$ CH_2Br (benzene) $\xrightarrow{\text{X}^-}$ CH_2X (benzene)

3. Reaction of amines with nitrous acid (Section 14.20)

primary amine: NH_2 (benzene) $\xrightarrow[0\,°C]{\text{NaNO}_2,\ \text{HCl}}$ $\overset{+}{N}\equiv N$ Cl⁻ (benzene)

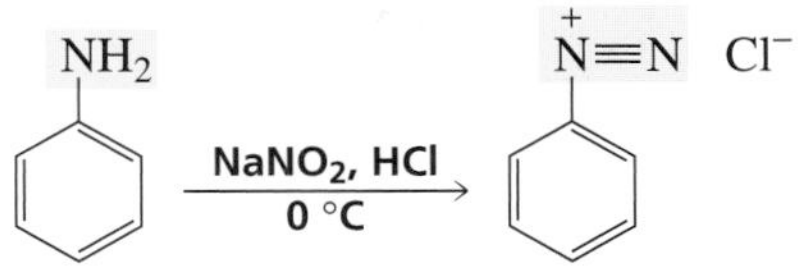

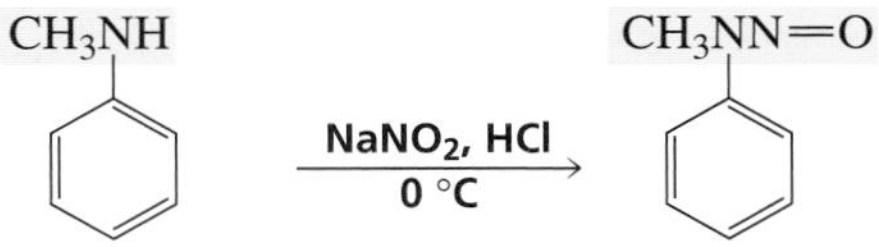

secondary amine: CH_3NH (benzene) $\xrightarrow[0\,°C]{\text{NaNO}_2,\ \text{HCl}}$ $CH_3NN=O$ (benzene)

tertiary amine: $(CH_3)_2N$ (benzene) $\xrightarrow[0\,°C]{\text{NaNO}_2,\ \text{HCl}}$ $(CH_3)_2N$ (benzene) with N=O para

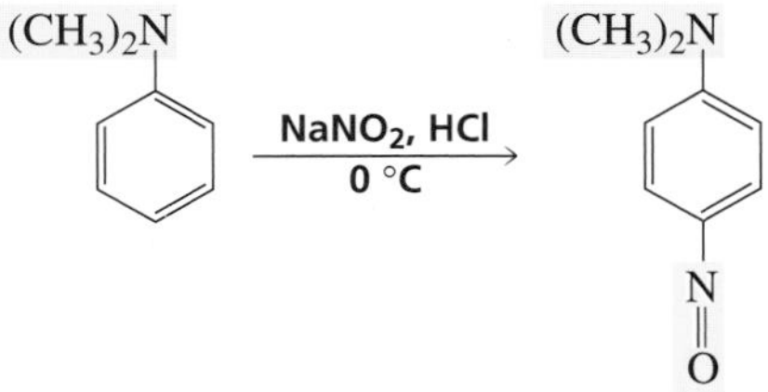

4. Replacement of a diazonium group (Section 14.18)

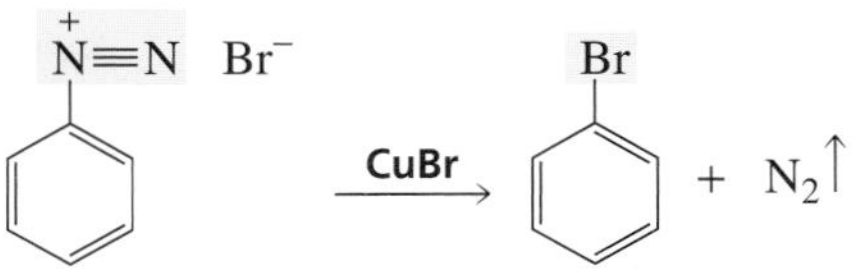

$\overset{+}{N}\equiv N$ Br⁻ (benzene) $\xrightarrow{\text{CuBr}}$ Br (benzene) + $N_2\uparrow$

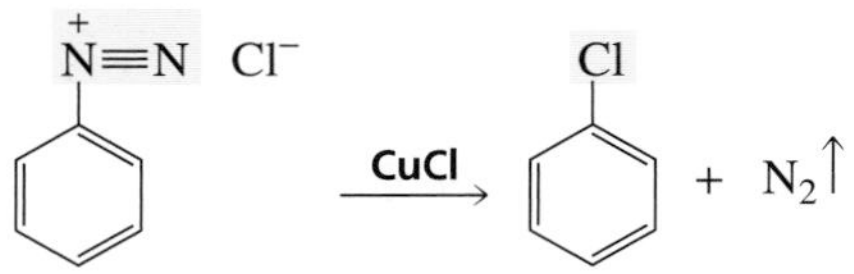

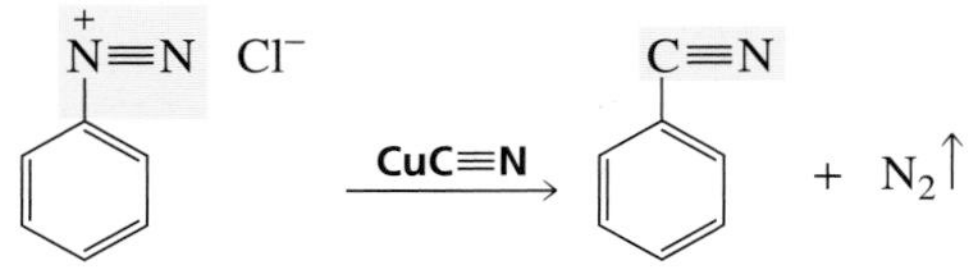

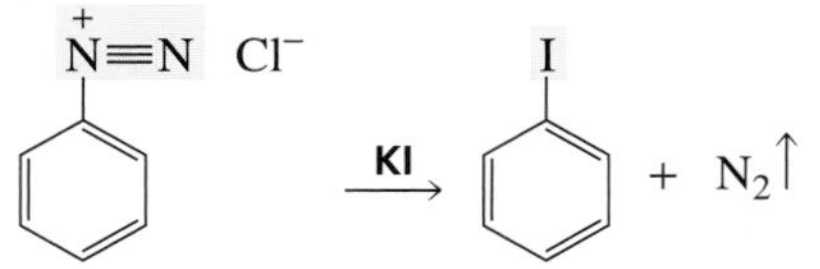

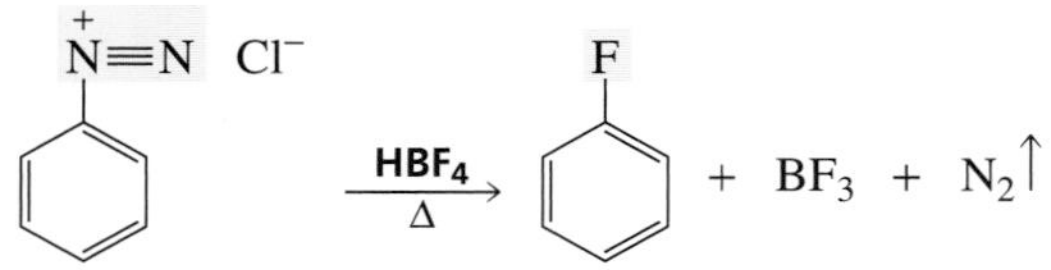

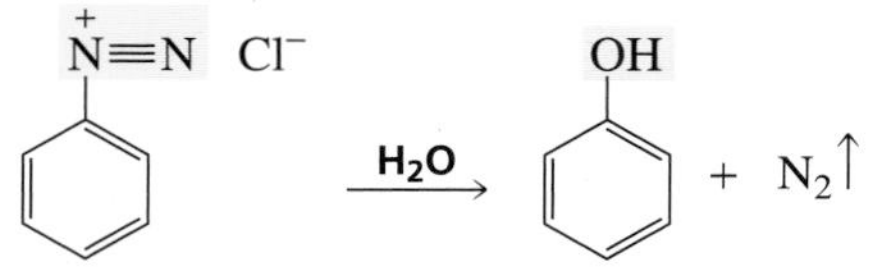

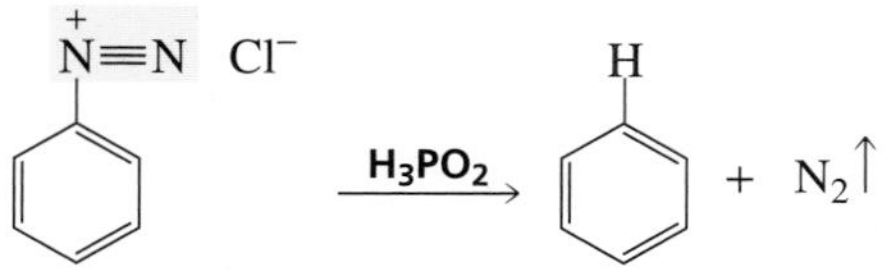

$$\underset{}{\text{C}_6\text{H}_5\text{N}\equiv\text{N}^+ \ \text{Cl}^-} \quad \xrightarrow{\text{CuCl}} \quad \text{C}_6\text{H}_5\text{Cl} \ + \ \text{N}_2\uparrow$$

$$\underset{}{\text{C}_6\text{H}_5\text{N}\equiv\text{N}^+ \ \text{Cl}^-} \quad \xrightarrow{\text{CuC}\equiv\text{N}} \quad \text{C}_6\text{H}_5\text{C}\equiv\text{N} \ + \ \text{N}_2\uparrow$$

$$\underset{}{\text{C}_6\text{H}_5\text{N}\equiv\text{N}^+ \ \text{Cl}^-} \quad \xrightarrow{\text{KI}} \quad \text{C}_6\text{H}_5\text{I} \ + \ \text{N}_2\uparrow$$

$$\underset{}{\text{C}_6\text{H}_5\text{N}\equiv\text{N}^+ \ \text{Cl}^-} \quad \xrightarrow[\Delta]{\text{HBF}_4} \quad \text{C}_6\text{H}_5\text{F} \ + \ \text{BF}_3 \ + \ \text{N}_2\uparrow$$

$$\underset{}{\text{C}_6\text{H}_5\text{N}\equiv\text{N}^+ \ \text{Cl}^-} \quad \xrightarrow{\text{H}_2\text{O}} \quad \text{C}_6\text{H}_5\text{OH} \ + \ \text{N}_2\uparrow$$

$$\underset{}{\text{C}_6\text{H}_5\text{N}\equiv\text{N}^+ \ \text{Cl}^-} \quad \xrightarrow{\text{H}_3\text{PO}_2} \quad \text{C}_6\text{H}_6 \ + \ \text{N}_2\uparrow$$

5. Nucleophilic aromatic substitution reactions (Sections 14.21 and 14.22)

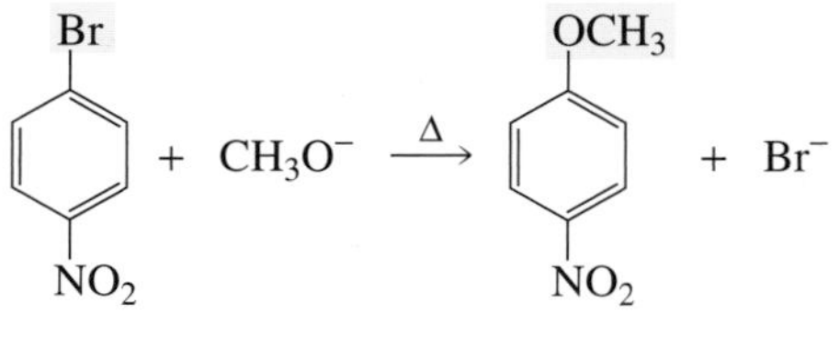

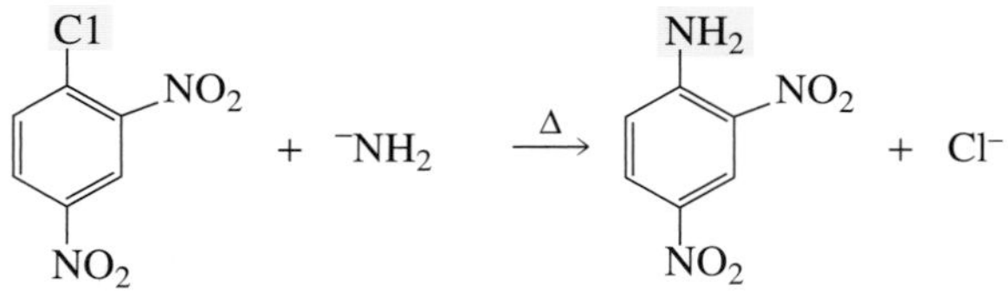

$$\text{4-Br-C}_6\text{H}_4\text{-NO}_2 \ + \ \text{CH}_3\text{O}^- \ \xrightarrow{\Delta} \ \text{4-CH}_3\text{O-C}_6\text{H}_4\text{-NO}_2 \ + \ \text{Br}^-$$

$$\text{2,4-(NO}_2)_2\text{-C}_6\text{H}_3\text{-Cl} \ + \ {}^-\text{NH}_2 \ \xrightarrow{\Delta} \ \text{2,4-(NO}_2)_2\text{-C}_6\text{H}_3\text{-NH}_2 \ + \ \text{Cl}^-$$

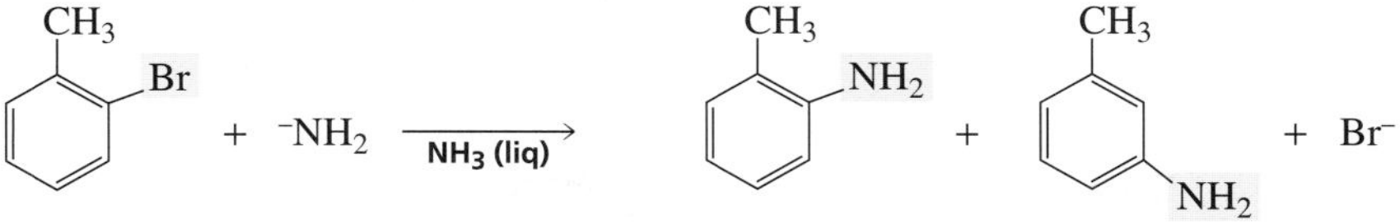

CH₃ / Br + ⁻NH₂ →(NH₃ (liq)) → **direct substitution product** + **cine substitution product** + Br⁻

KEY TERMS

activating substituent (page 622)
aliphatic compound (page 601)
aromatic compound (page 601)
azo linkage (page 642)
benzyl group (page 604)
benzyne intermediate (page 649)
cine substitution (page 650)
Clemmensen reduction (page 616)
deactivating substituent (page 622)
diazonium salt (page 638)
direct substitution (page 650)
electrophilic aromatic substitution
   reaction (page 607)
Friedel–Crafts acylation (page 608)

Friedel–Crafts alkylation (page 608)
fused rings (page 653)
halogenation (page 608)
inductive electron donation (page 621)
inductive electron withdrawal
   (page 621)
meta-directing substituent (page 626)
nitration (page 608)
nitrosamine (page 644)
N-nitroso compound (page 644)
nucleophilic aromatic substitution
   (page 646)
ortho-and para-directing substituent
   (page 626)

phenyl group (page 604)
principle of microscopic reversibility
   (page 611)
resonance electron donation
   (page 621)
resonance electron withdrawal
   (page 622)
Sandmeyer reaction
   (page 639)
Schiemann reaction (page 640)
S$_N$Ar reaction (page 647)
sulfonation (page 608)
Wolff–Kishner reduction
   (page 616)

PROBLEMS

42. Draw the structure of each of the following compounds.
   a. *m*-ethylphenol
   b. *p*-nitrobenzenesulfonic acid
   c. (*E*)-2-phenyl-2-pentene
   d. *o*-bromoaniline
   e. 2-chloroanthracene
   f. *m*-chlorostyrene
   g. *o*-nitroanisole
   h. 2,4-dichlorotoluene

43. Name the following compounds.

a. 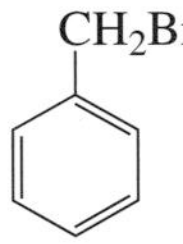 CH₂Br

b.  CHBr₂

c. 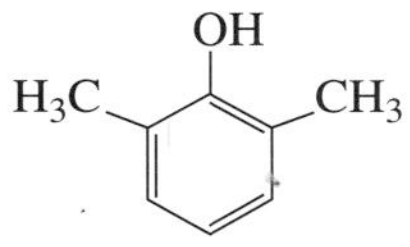 H₃C — OH — CH₃

d.  COOH, Br

e. 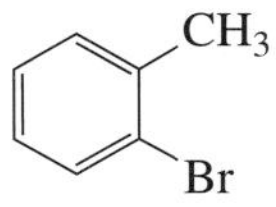 CH₃, Br

f. 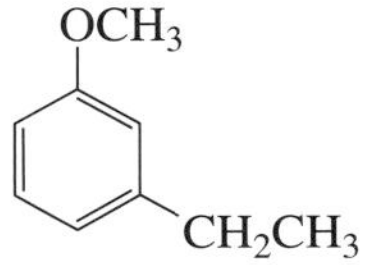 OCH₃, CH₂CH₃

g.

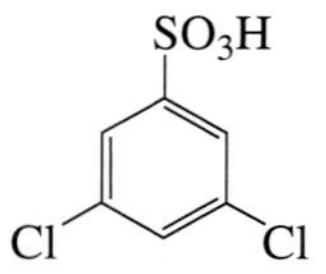

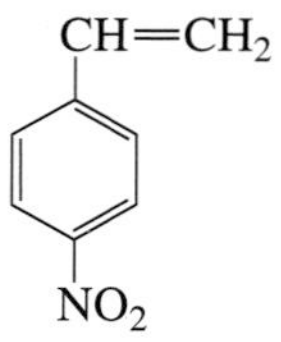

h.

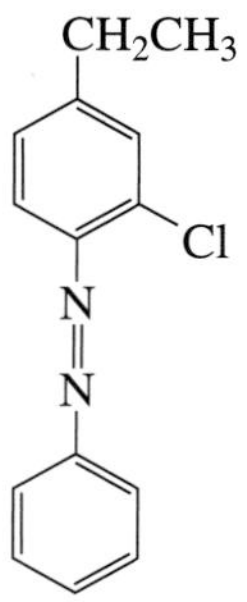

i.

j.

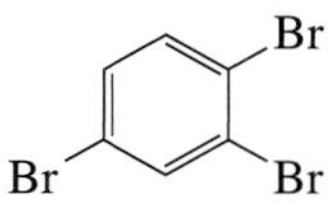

k. 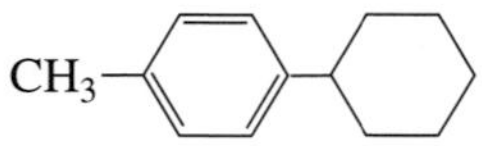 CH_3-

44. Provide the necessary reagents next to the arrows.

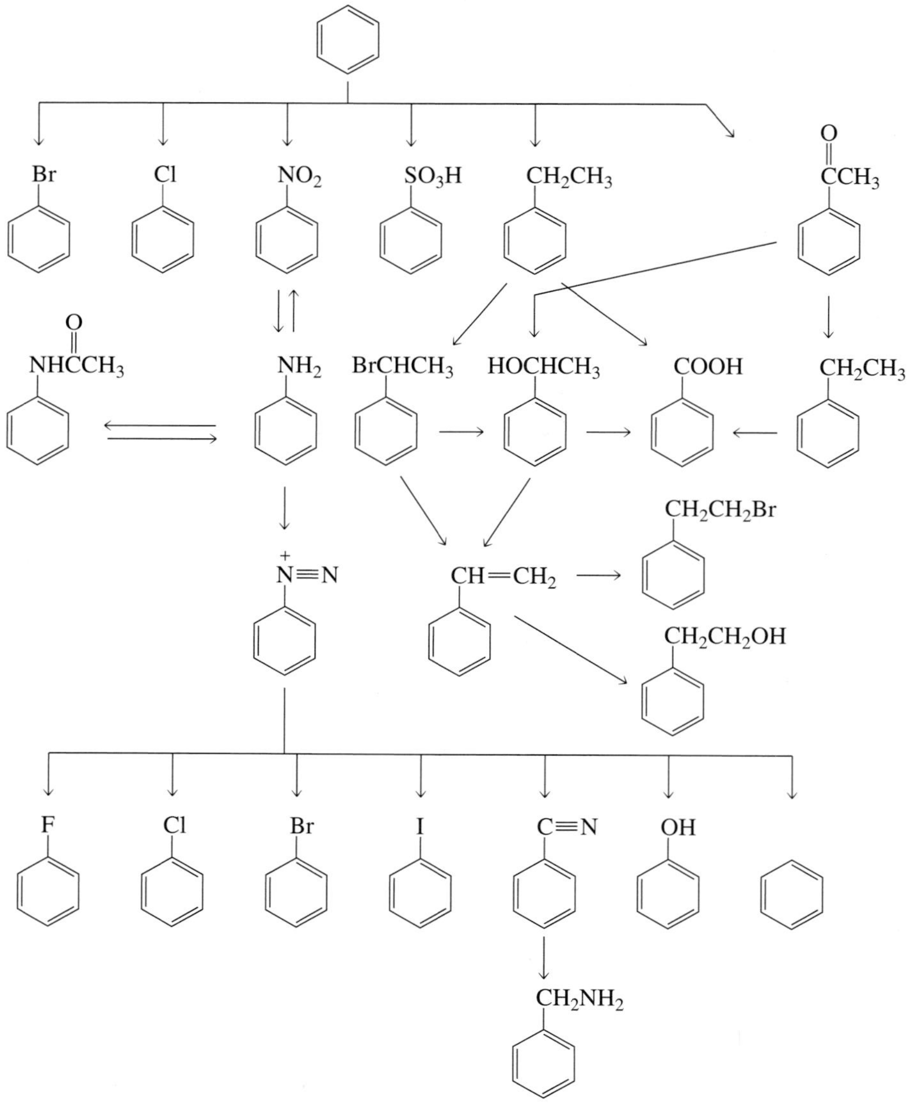

45. The pK_a values of a few ortho-, meta-, and para-substituted benzoic acids are shown below. The relative pK_a values depend on the substituent. For chloro-substituted benzoic acids, the ortho isomer is the most acidic and the para isomer is the least acidic; for nitro-substituted benzoic acids, the ortho isomer is the most acidic and the meta isomer is the least acidic; and for amino-substituted benzoic acids, the meta isomer is the most acidic and the ortho isomer is the least acidic. Explain these differences.

Cl: ortho > meta > para

NO_2: ortho > para > meta

NH_2: meta > para > ortho

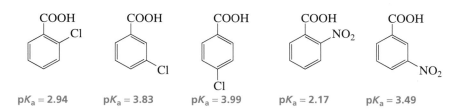

COOH Cl	COOH Cl	COOH Cl	COOH NO₂	COOH NO₂
pK_a = 2.94	pK_a = 3.83	pK_a = 3.99	pK_a = 2.17	pK_a = 3.49

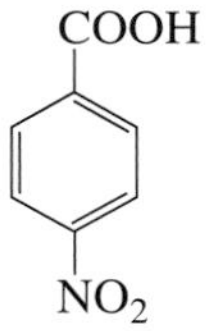

COOH NO₂	COOH NH₂	COOH NH₂	COOH NH₂
pK_a = 3.44	pK_a = 4.95	pK_a = 4.73	pK_a = 4.89

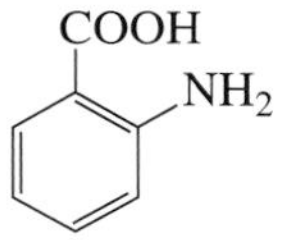

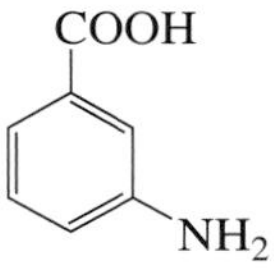

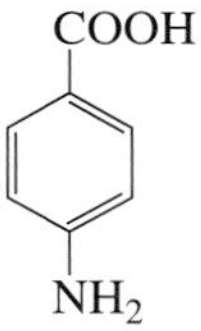

46. Give the product(s) of each of the following reactions.
   a. benzoic acid + HNO_3/H_2SO_4
   b. isopropylbenzene + cyclohexene + HF
   c. naphthalene + acetyl chloride + $AlCl_3$
   d. o-methylaniline + benzenediazonium chloride
   e. cyclohexyl phenyl ether + $Br_2/FeCl_3$
   f. phenol + H_2SO_4 + Δ
   g. ethylbenzene + $Br_2/FeCl_3$
   h. m-xylene + $KMnO_4$ + H^+ + Δ

47. Show how the following compounds could be synthesized from benzene.
   a. m-chlorobenzenesulfonic acid
   b. m-chloroethylbenzene
   c. benzyl alcohol
   d. m-bromobenzonitrile
   e. p-nitrophenol
   f. benzyl methyl ether
   g. p-benzylchlorobenzene
   h. 1-phenylpentane
   i. m-hydroxybenzoic acid
   j. m-bromobenzoic acid
   k. o-chloroanisole
   l. p-nitroaniline
   m. m-bromoiodobenzene
   n. p-dideuteriobenzene

48. For each group of substituted benzenes shown below, indicate:
   a. the one that would be the most reactive in an electrophilic substitution reaction.
   b. the one that would be the least reactive in an electrophilic substitution reaction.
   c. the one that would yield the highest percentage of meta product.

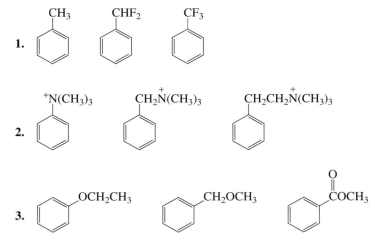

49. Arrange the following groups of compounds in order of decreasing reactivity toward electrophilic substitution.
   a. benzene, ethylbenzene, chlorobenzene, nitrobenzene, anisole
   b. 1-chloro-2,4-dinitrobenzene, 2,4-dinitrophenol, 2,4-dinitrotoluene
   c. toluene, *p*-cresol, benzene, *p*-xylene
   d. benzene, benzoic acid, phenol, propylbenzene
   e. *p*-nitrotoluene, 2-chloro-4-nitrotoluene, 2,4-dinitrotoluene, *p*-chlorotoluene
   f. bromobenzene, chlorobenzene, fluorobenzene, iodobenzene

50. Give the products of the following reactions.

a. 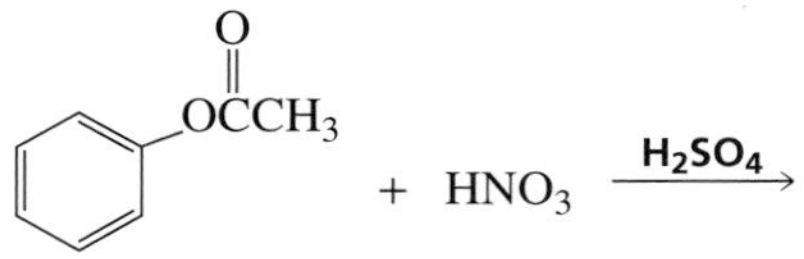
$$\text{(phenyl acetate)} \quad + \quad HNO_3 \xrightarrow{\ H_2SO_4\ }$$

b. 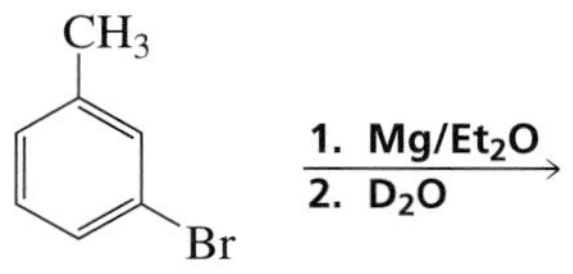
$$\text{(3-bromotoluene)} \xrightarrow[\text{2. } D_2O]{\text{1. } Mg/Et_2O}$$

c.
$$\text{(anisole)} \quad + \quad \text{(succinic anhydride)} \xrightarrow{\ AlCl_3\ }$$

d.
$$\text{(1-phenylpiperidine)} \quad + \quad Br_2 \longrightarrow$$

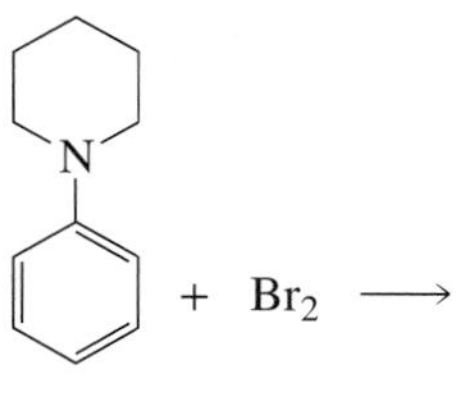

e. 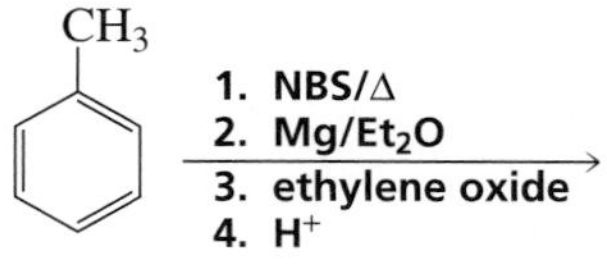
$$\text{(toluene)} \xrightarrow[\substack{\text{2. } Mg/Et_2O \\ \text{3. ethylene oxide} \\ \text{4. } H^+}]{\text{1. } NBS/\Delta}$$

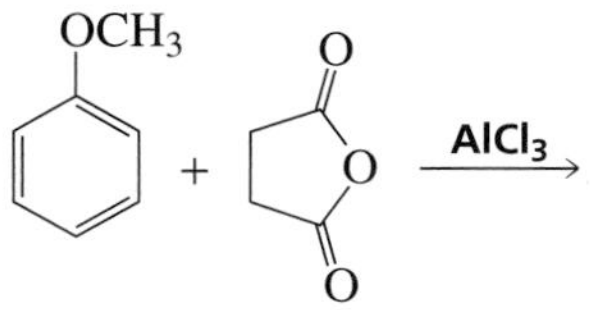

f. 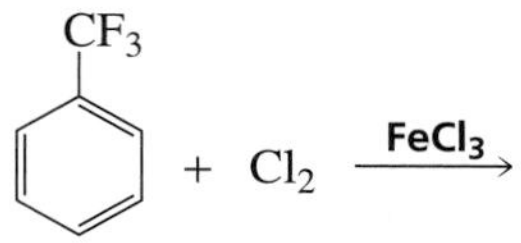 (benzene ring with CF$_3$) + Cl$_2$ $\xrightarrow{\text{FeCl}_3}$

51. For each of the statements in Column I, choose a substituent from Column II that fits the description for the compound at the right.

Column I
a. X donates electrons inductively but does not donate or withdraw electrons by resonance.
b. X withdraws electrons inductively and withdraws electrons by resonance.
c. X deactivates the ring and directs ortho/para.
d. X withdraws electrons inductively, donates electrons by resonance, and activates the ring.
e. X withdraws electrons inductively but does not donate or withdraw electrons by resonance.

Column II
OH

Br

$^+$NH$_3$
CH$_2$CH$_3$

NO$_2$

(benzene ring with X substituent)

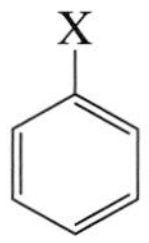

52. For each of the following compounds, indicate the ring carbon that would be nitrated if the compound were treated with HNO$_3$/H$_2$SO$_4$.

a. 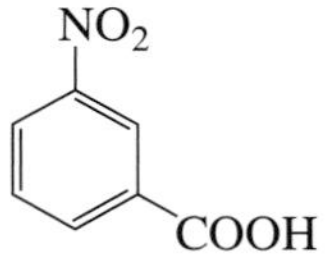 (benzene ring with NO$_2$ and COOH, meta)

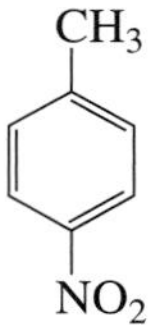

b. 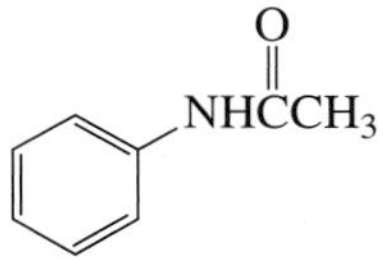 (benzene ring with NHCCH$_3$, amide C=O)

c. (benzene ring with OCH$_3$ and Cl, para)

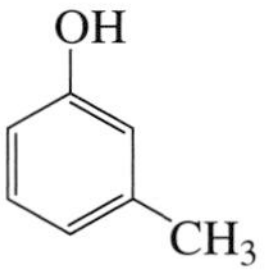

d. 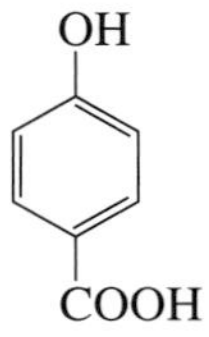 (benzene ring with OH and COOH, para)

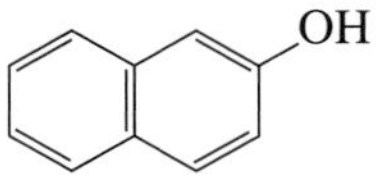

e. 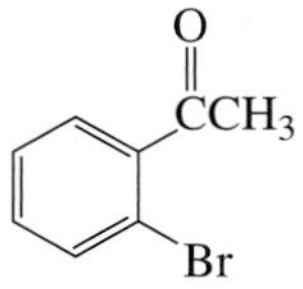 (benzene ring with CCH$_3$ ketone C=O and Br, ortho)

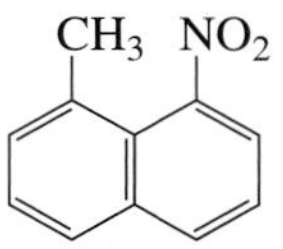

f. 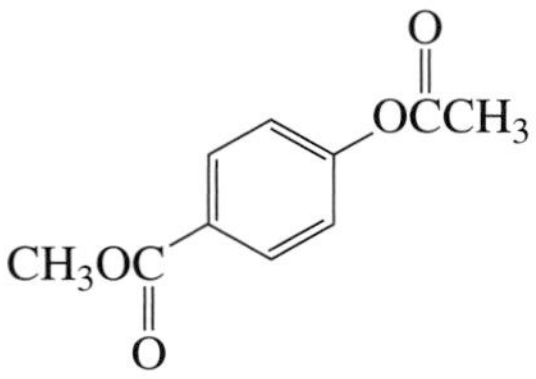 (benzene ring with OCCH$_3$ ester and CH$_3$OC ester; CH$_3$OC=O)

g. (benzene ring with CH$_3$ and NO$_2$, para)

h. (benzene ring with Cl and Cl, meta)

i. (benzene ring with OH and CH$_3$, meta)

j. (naphthalene with OH)

k. (naphthalene with CH$_3$ and NO$_2$)

l. 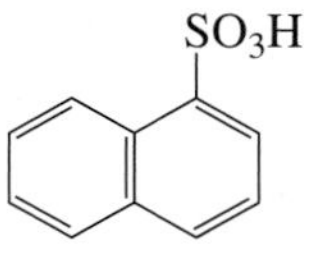 (naphthalene with SO$_3$H)

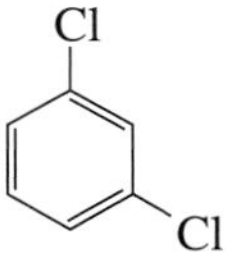

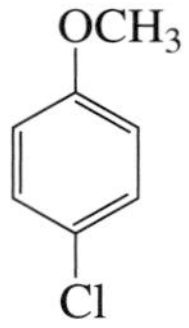

53. Give the product(s) obtained from the reaction of each of the following compounds with $Br_2/FeCl_3$.

a.

b.

c. H_3C

d.

54. Give the product of the reaction of excess benzene with each of the following reagents.
   a. isobutyl chloride + $AlCl_3$
   b. propene + HF
   c. neopentyl chloride + $AlCl_3$
   d. dichloromethane + $AlCl_3$

55. Which would react more rapidly with Cl_2 + $FeCl_3$, m-xylene or p-xylene? Explain.

56. Benzene underwent a Friedel–Crafts acylation reaction followed by a Clemmensen reduction. The ^1H NMR spectrum of the product is shown below. What acyl chloride was used in the Friedel–Crafts acylation reaction?

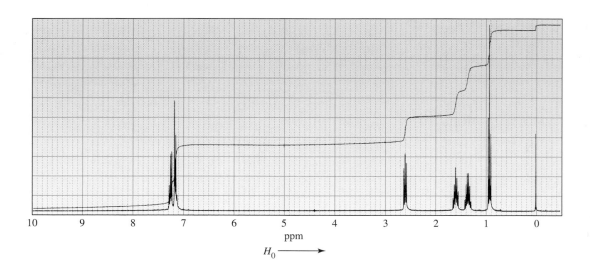

57. What products would be obtained from the reaction of the following compounds with $KMnO_4 + H^+ + \Delta$?

a. 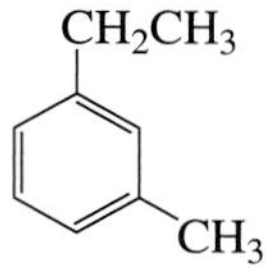 CH₂CH₃ / CH₃ (benzene ring)

c. 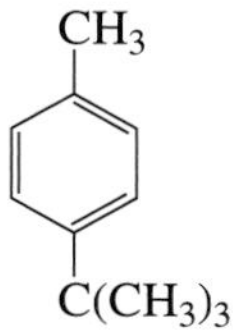 CH₃ / C(CH₃)₃ (benzene ring)

b. 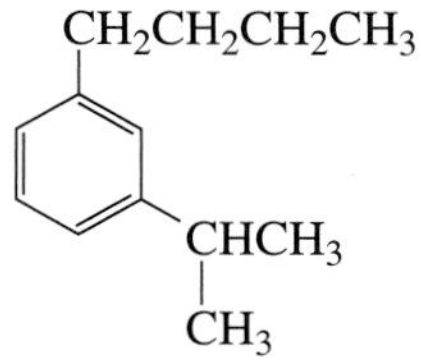 CH₂CH₂CH₂CH₃ / CHCH₃ / CH₃ (benzene ring)

58. Propose a mechanism for each of the following reactions.

a. 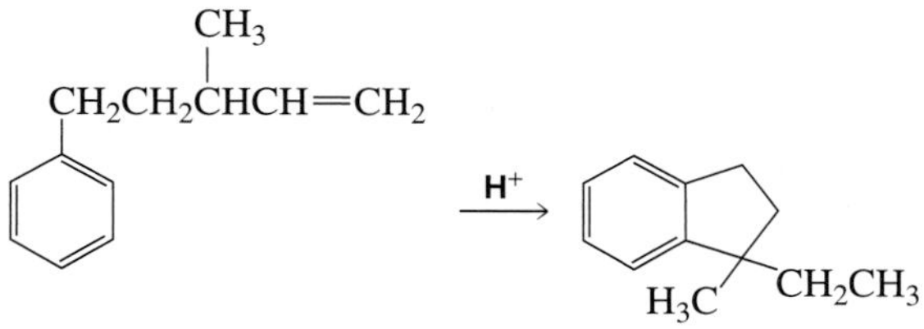 CH₃ | CH₂CH₂CHCH=CH₂ (benzene), H⁺ → indane with H₃C, CH₂CH₃

b. 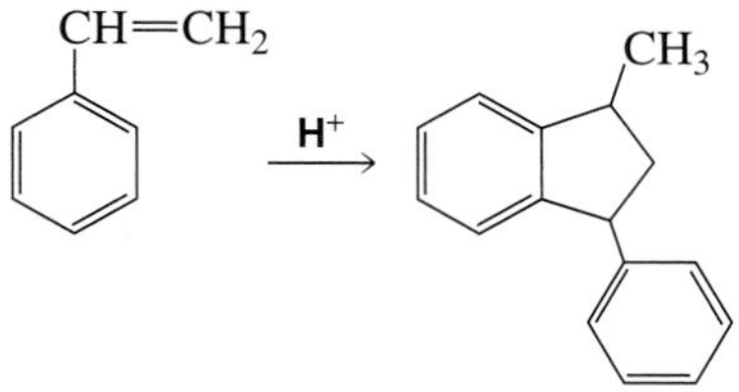 CH=CH₂ (benzene), H⁺ → indane with CH₃ and phenyl

59. A student had prepared three ethyl-substituted benzaldehydes, X, Y, and Z, but had neglected to label them. The premed student at the next bench said that they could be identified by brominating a sample of each and determining how many bromo-substituted products were formed. Is the premed student's advice sound?

60. Explain, using resonance contributors, why a phenyl group is an ortho/para director.

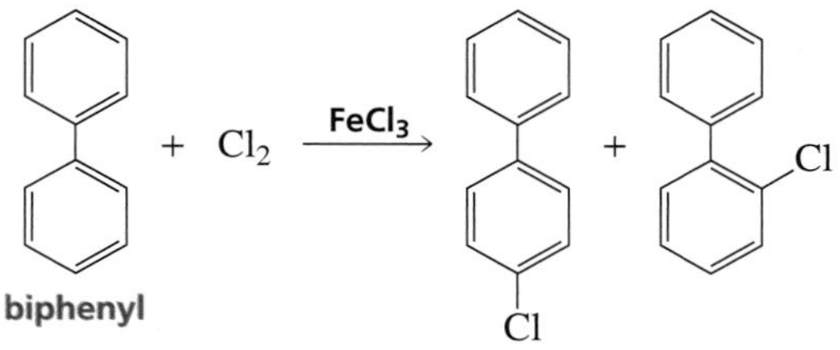

biphenyl + Cl_2 →(FeCl₃) products + (with Cl substituents)

61. Give two synthetic routes for the preparation of p-methoxyaniline using benzene as the starting material.

62. Using resonance contributors, answer the following.
  a. Which is a more stable carbocation intermediate?

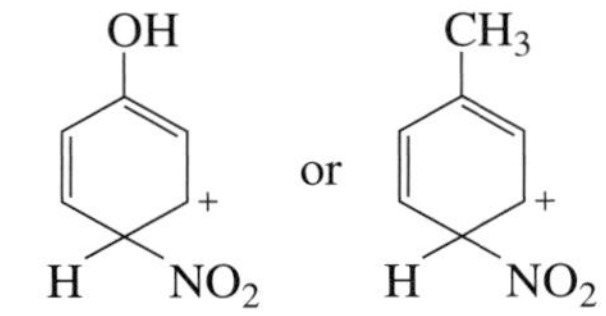

or

  b. Which is a more stable carbanion intermediate?

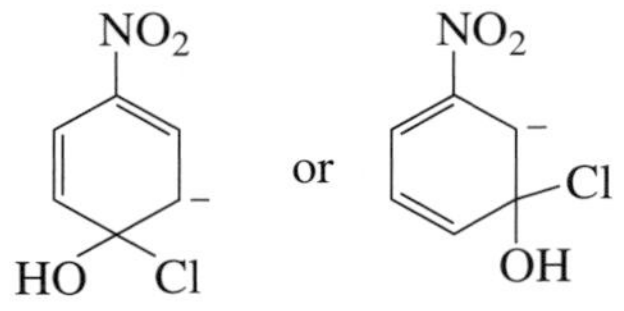

or

63. Show how the following compounds could be prepared from benzene.

  a.

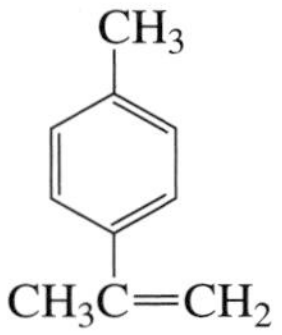

  c.

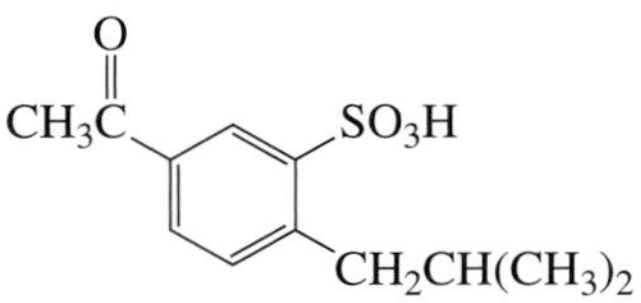

  b.

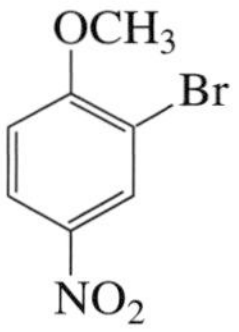

64. An OH substituent and an NH_2 substituent both strongly donate electrons into a benzene ring by resonance. However, while phenol is a good substrate for electrophilic substitution reactions, aniline is usually a poor substrate. Explain. How can the problem with aniline be circumvented?

65. How could you distinguish among the infrared spectra of the following compounds?

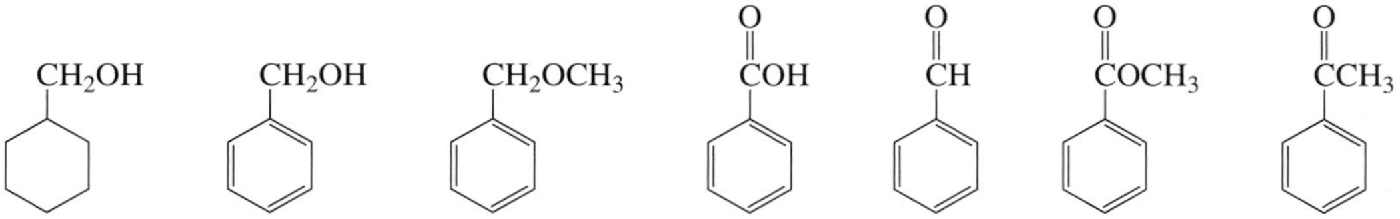

66. The following tertiary alkyl bromides undergo an S_N1 reaction in aqueous acetone to form the corresponding tertiary alcohols. List the alkyl bromides in order of decreasing reactivity.

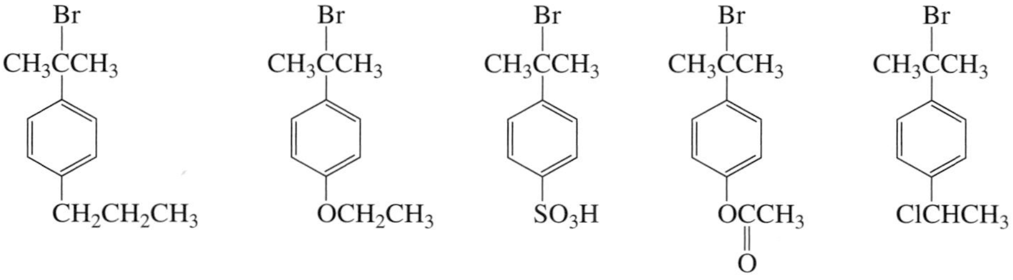

67. a. Explain why the following reaction leads to the products that are shown.

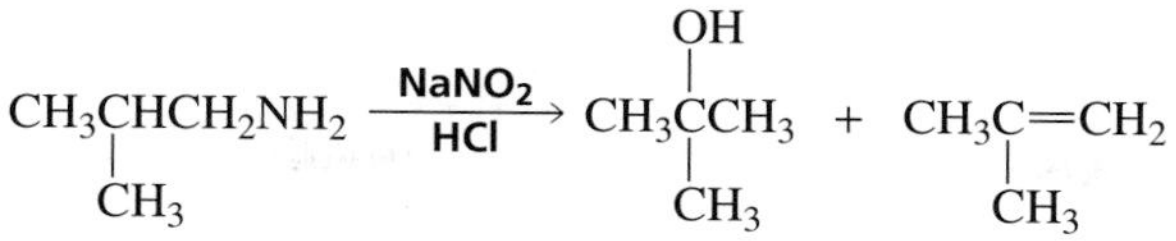

$$CH_3CHCH_2NH_2 \xrightarrow[\text{HCl}]{\text{NaNO}_2} \underset{\underset{CH_3}{|}}{CH_3}\overset{\overset{OH}{|}}{C}CH_3 \ + \ CH_3C\!\!=\!\!CH_2$$
$$\underset{CH_3}{|} \qquad\qquad\qquad \underset{CH_3}{|}$$

b. What product would be obtained from the following reaction?

$$CH_3\!-\!\overset{\overset{OH}{|}}{C}\!-\!\overset{\overset{NH_2}{|}}{C}\!-\!CH_3 \xrightarrow[\text{HCl}]{\text{NaNO}_2}$$
$$\underset{CH_3}{|}\ \ \underset{CH_3}{|}$$

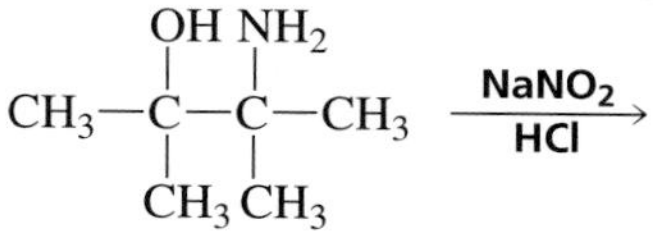

68. Describe how mescaline could be synthesized from benzene.

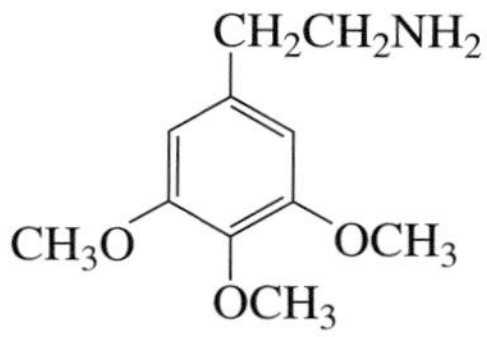

mescaline

69. Hydroxide ion catalyzes the reaction of piperidine with 2,4-dinitroanisole but has no effect on the reaction of piperidine with 1-chloro-2,4-dinitrobenzene. Explain why hydroxide ion acts as a catalyst in one reaction but has no effect in the other reaction. (*J. Am. Chem. Soc.*, 1965, *87*, 5209.)

70. Benzene can be partially reduced to 1,4-cyclohexadiene by an alkali metal (Na, Li, or K) in liquid ammonia and a low molecular weight alcohol. This reaction is called the Birch reduction. Propose a mechanism for this reaction. (*Hint:* see Section 5.8.)

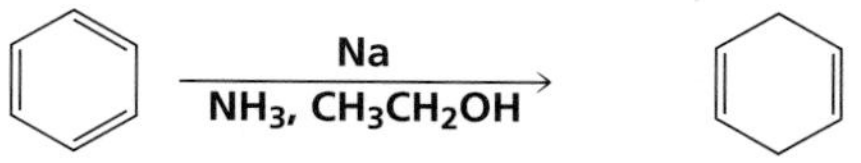

1,4-cyclohexadiene

71. One reason that has been suggested to explain why the Birch reduction forms only 1,4-cyclohexadiene is the *principle of least motion,* which states that the reaction that (all else being equal) involves the least change in atomic positions or electronic configuration is favored. How does this account for the observation that no 1,3-cyclohexadiene is obtained from a Birch reduction?

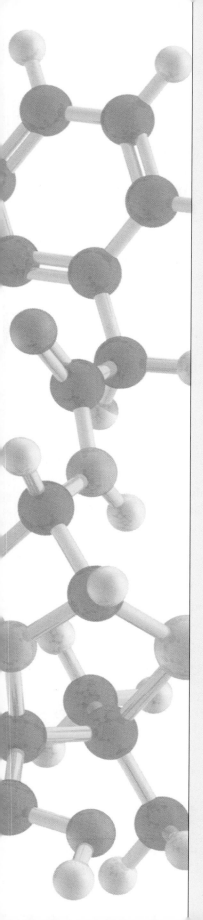

PART VI

CARBONYL COMPOUNDS

The chapters in this section focus on the reactions of carbonyl compounds. Carbonyl compounds can be placed in one of two groups: those that contain a group that can be replaced by another group (Class I), and those that do not contain a group that can be replaced by another group (Class II).

Chapter 15 discusses the reactions of Class I carbonyl compounds with relatively poor nucleophiles. Carboxylic acids are the Class I carbonyl compounds most commonly available both in the laboratory and in biological systems. Because carboxylic acids are not very reactive, they must be activated to undergo reactions at reasonable rates. This chapter describes the techniques chemists use to activate carboxylic acids and compares these techniques with the activation techniques used by biological systems.

The first part of **Chapter 16** compares the reactivities of Class I and Class II carbonyl compounds by discussing the reactivity of each class with good nucleophiles. The chapter goes on to discuss the reactions of Class II carbonyl compounds with the poorer nucleophiles whose reactions with Class I carbonyl compounds you studied in Chapter 15.

Chapter 17 covers oxidation–reduction reactions of organic compounds. You have seen many of these reactions in other chapters, but here they are discussed as a group, which gives you the opportunity to compare and contrast them. First you will study reduction reactions, where you will see that all reduction reactions follow one of three mechanisms. Then you will study oxidation reactions. Many carbonyl compounds have two sites of reactivity, the carbonyl group and the α-carbon. While Chapters 15 and 16 discuss the reactions that carbonyl compounds undergo that take place at the carbonyl group, **Chapter 18** examines the reactions that carbonyl compounds undergo that take place at the α-carbon.

15

CARBONYL COMPOUNDS I:
Reactions of Carboxylic Acids and Their Derivatives with Oxygen and Nitrogen Nucleophiles

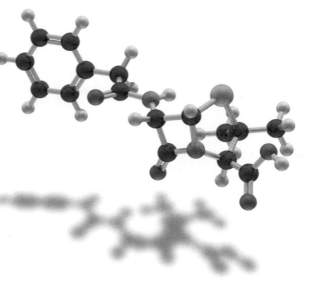

penicillin G

an acyl chloride

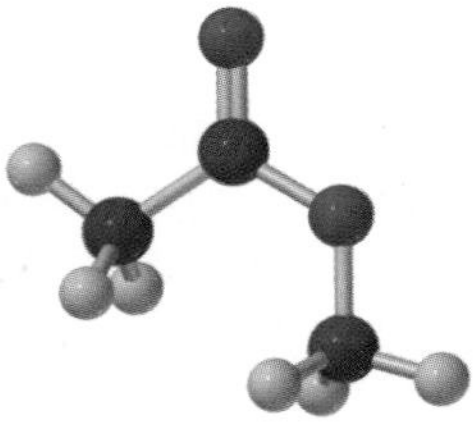

an ester

a carboxylic acid

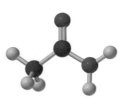

an amide

Compounds containing carbonyl groups abound in nature. Many play important roles in biological processes. Hormones, vitamins, amino acids, drugs, and flavorings are just a few of the carbonyl compounds that affect us daily. Recall that a **carbonyl group** is a carbon doubly bonded to an oxygen. An **acyl group** consists of a carbonyl group attached to an alkyl group or to an aryl group.

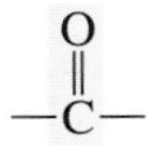

a carbonyl group   acyl groups

Carbonyl compounds can be divided into two classes. One class includes compounds in which the acyl group is attached to an atom or group that can be replaced by another group. Carboxylic acids, acyl halides, acid anhydrides, esters, and amides belong to this class. All of these compounds contain a group (—OH, —Cl, —Br, —O(CO)R, —OR, —NH$_2$, —NHR, or —NR$_2$) that can be replaced by a nucleophile.

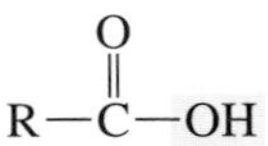

a carboxylic acid 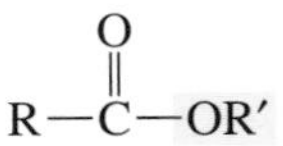   an ester 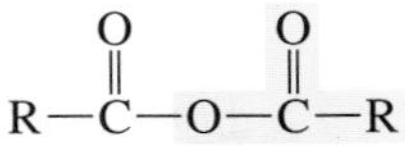   an acid anhydride

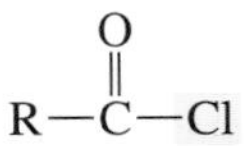

an acyl chloride 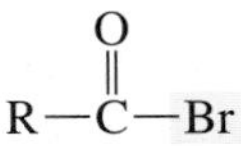   an acyl bromide   amides
acyl halides

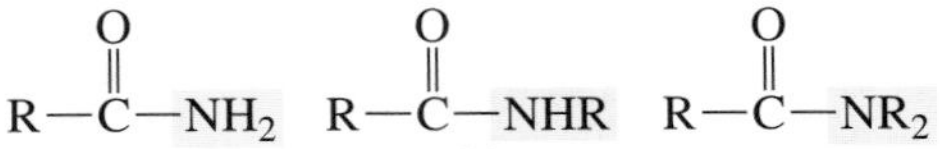

The second class consists of carbonyl compounds in which the acyl group is attached to a group that cannot be replaced by a nucleophile. Aldehydes and ketones belong to this class.

$$
\underset{\text{an aldehyde}}{R-\overset{\overset{\displaystyle O}{\|}}{C}-H}
\qquad
\underset{\text{a ketone}}{R-\overset{\overset{\displaystyle O}{\|}}{C}-R}
$$

In Chapter 9 we saw that the ability of a group to leave (that is, to be replaced by another group) depends on its basicity: *the weaker the basicity of a group, the better its leaving ability.* This is because weak bases do not share their electrons as well as do strong bases (Section 9.3).

> **The weaker the base, the better it is as a leaving group.**

The pK_a values of the conjugate acids of the leaving groups of the various carbonyl compounds are listed in Table 15.1. Notice that the acyl groups of the carbonyl compounds in the first class are attached to weaker bases than are the acyl groups of the carbonyl compounds in the second class. (Remember that the lower the pK_a, the stronger the acid and the weaker its conjugate base.) The H of an aldehyde and the alkyl or aryl (R or Ar) group of a ketone are too basic to be replaced by another group.

In this chapter we will discuss the chemistry of carbonyl compounds in the first class. Because these compounds have an acyl group attached to a group that can be replaced by a nucleophile, we will see that they undergo substitution reactions.

TABLE 15.1   The pK_a Values of the Conjugate Acids of the Leaving Groups of Carbonyl Compounds

Carbonyl compound	Leaving group	Conjugate acid of the leaving group	pK_a
Class I			
$R-\overset{O}{\overset{\|}{C}}-Br$	Br^-	HBr	-9
$R-\overset{O}{\overset{\|}{C}}-Cl$	Cl^-	HCl	-7
$R-\overset{O}{\overset{\|}{C}}-O-\overset{O}{\overset{\|}{C}}-R$	^-OCR (with $\overset{O}{\overset{\|}{}}$)	$RCOH$ (with $\overset{O}{\overset{\|}{}}$)	$\sim 3\text{–}5$
$R-\overset{O}{\overset{\|}{C}}-OR'$	^-OR	ROH	$\sim 15\text{–}16$
$R-\overset{O}{\overset{\|}{C}}-OH$	^-OH	H_2O	15.7
$R-\overset{O}{\overset{\|}{C}}-NH_2$	$^-NH_2$	NH_3	36
Class II			
$R-\overset{O}{\overset{\|}{C}}-R$	R^-	RH	~ 50
$R-\overset{O}{\overset{\|}{C}}-H$	H^-	H_2	very large

The chemistry of aldehydes and ketones will be considered in the next chapter. Because these compounds have an acyl group that is attached to a group that cannot be replaced by a nucleophile, we can correctly predict that these compounds will not undergo substitution reactions.

15.1 NOMENCLATURE

Carboxylic Acids

In the IUPAC system of nomenclature, a **carboxylic acid** is named by replacing the "e" ending of the alkane name with "oic acid." For example, a one-carbon alkane is methane; a one-carbon carboxylic acid is methanoic acid. Long-chain carboxylic acids are called **fatty acids**.

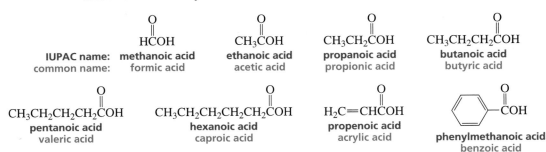

HCOH	CH₃COH	CH₃CH₂COH	CH₃CH₂CH₂COH
IUPAC name: methanoic acid	**ethanoic acid**	**propanoic acid**	**butanoic acid**
common name: formic acid	acetic acid	propionic acid	butyric acid

CH₃CH₂CH₂CH₂COH	CH₃CH₂CH₂CH₂CH₂COH	H₂C=CHCOH	⬡—COH
pentanoic acid	**hexanoic acid**	**propenoic acid**	**phenylmethanoic acid**
valeric acid	caproic acid	acrylic acid	benzoic acid

Carboxylic acids containing six or fewer carbons are frequently called by their common names. These names were chosen by early chemists to describe some feature of the compound, usually its origin. For example, formic acid is found in ants, bees, and other stinging insects. Its name comes from *formica,* which is Latin for "ant." Acetic acid—contained in vinegar—got its name from *acetum,* the Latin word for "vinegar." Propionic acid is the smallest acid that shows some of the characteristics of the larger fatty acids. Its name comes from the Greek words *pro* ("the first") and *pion* ("fat"). Butyric acid is found in rancid butter, and the Latin word for "butter" is *butyrum.* Caproic acid is found in goat's milk, and if you have the occasion to smell both a goat and caproic acid, you will find that they have similar odors. *Caper* is the Latin word for "goat."

When IUPAC nomenclature is used, the position of a substituent is designated by a number. The carbonyl carbon is always the C-1 carbon of a carboxylic acid. In common nomenclature, the position of a substituent is designated by a lower case Greek letter and the carbonyl carbon is not given a designation. The carbon adjacent to the carbonyl carbon is the α-**carbon.**

α = alpha
β = beta
γ = gamma
δ = delta
ε = epsilon

$$\underset{6 \quad 5 \quad 4 \quad 3 \quad 2 \quad 1}{CH_3CH_2CH_2CH_2CH_2COH}$$
IUPAC nomenclature

$$\underset{\epsilon \quad \delta \quad \gamma \quad \beta \quad \alpha}{CH_3CH_2CH_2CH_2CH_2COH}$$
common nomenclature

Take a careful look at the following examples to make certain that you understand the difference between IUPAC and common nomenclature.

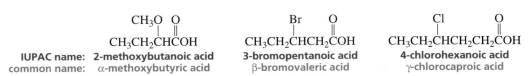

CH₃CH₂CHCOH	CH₃CH₂CHCH₂COH	CH₃CH₂CHCH₂CH₂COH
IUPAC name: 2-methoxybutanoic acid	**3-bromopentanoic acid**	**4-chlorohexanoic acid**
common name: α-methoxybutyric acid	β-bromovaleric acid	γ-chlorocaproic acid

The functional group of a carboxylic acid is called a **carboxyl group.**

O
‖
—C—OH   —COOH   —CO₂H
a carboxyl group   **carboxyl groups are frequently**
shown in abbreviated forms

Carboxylic acids in which a carboxyl group is attached to a cyclic compound can be named by adding "carboxylic acid" to the name of the cyclic compound.

cyclohexylmethanoic acid   *trans*-**3-methylcyclopentanecarboxylic**   **1,2,4-benzenetricarboxylic**
or   **acid**   **acid**
cyclohexanecarboxylic acid

Acyl Halides

The most common **acyl halides** are acyl chlorides and acyl bromides. Acyl halides are named by using the acid name and replacing "ic acid" with "yl chloride (bromide)." For acids ending with "carboxylic acid," "carboxylic acid" is replaced with "carbonyl chloride (bromide)."

	O ‖ CH₃CCl	CH₃ O ‖ CH₃CH₂CHCH₂CBr	O ‖ CCl (cyclopentyl)
IUPAC name:	ethanoyl chloride	3-methylpentanoyl bromide	cyclopentylmethanoyl chloride
common name:	acetyl chloride	β-methylvaleryl bromide	cyclopentanecarbonyl chloride

Acid Anhydrides

Loss of water from two molecules of a carboxylic acid results in an **acid anhydride.** Anhydride means "without water."

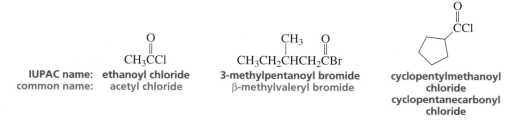

O   O   O   O
‖   ‖   ‖   ‖
R—C—O—H   H—O—C—R   ⟶   R—C—O—C—R + H₂O
                                 an acid anhydride

If the two carboxylic acid molecules forming the acid anhydride are the same, the anhydride is a **symmetrical anhydride.** If the two carboxylic acid molecules are different, the anhydride is a **mixed anhydride.** Symmetrical anhydrides are named by using the acid name and replacing "acid" with "anhydride." Mixed anhydrides are named by stating the names of both acids in alphabetical order, followed by "anhydride."

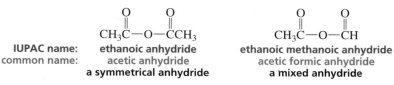

	O O ‖ ‖ CH₃C—O—CCH₃	O O ‖ ‖ CH₃C—O—CH
IUPAC name:	ethanoic anhydride	ethanoic methanoic anhydride
common name:	acetic anhydride	acetic formic anhydride
	a symmetrical anhydride	**a mixed anhydride**

Esters

In naming an **ester,** the name of the group attached to the **carboxyl oxygen** is stated first. This is followed by the name of the acid with "ic acid" replaced by "ate."

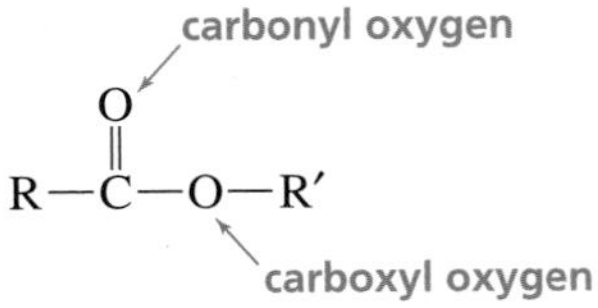

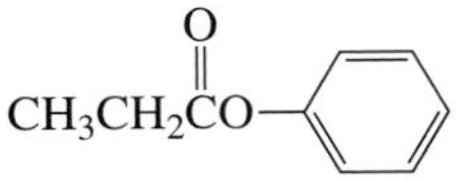

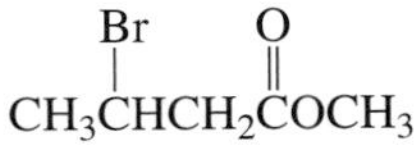

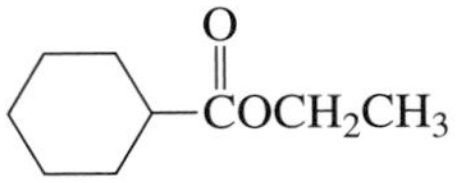

IUPAC name:	ethyl ethanoate
common name:	ethyl acetate

phenyl propanoate
phenyl propionate

methyl 3-bromobutanoate
methyl β-bromobutyrate

ethyl cyclohexylmethanoate
ethyl cyclohexanecarboxylate

Salts of carboxylic acids are named in the same way. The cation is named first, followed by the name of the acid, again with "ic acid" replaced by "ate."

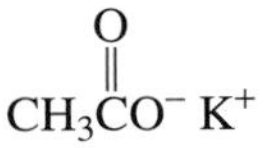

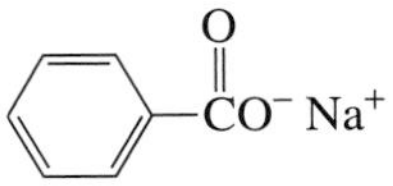

IUPAC name:	sodium methanoate
common name:	sodium formate

potassium ethanoate
potassium acetate

sodium phenylmethanoate
sodium benzoate

Cyclic esters are called **lactones.** The length of the carbon chain is designated by the common name of the carboxylic acid, and a Greek letter is used to indicate the carbon to which the carboxyl oxygen is attached. Thus, four-membered ring lactones are β-lactones (the carboxyl oxygen is on the β-carbon atom), five-membered ring lactones are γ-lactones, and six-membered ring lactones are δ-lactones.

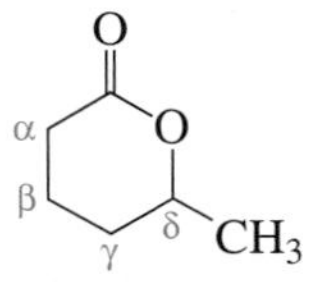

δ-caprolactone 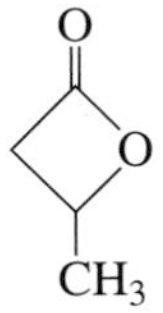   β-butyrolactone 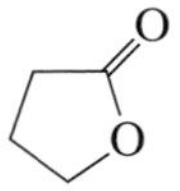   γ-butyrolactone 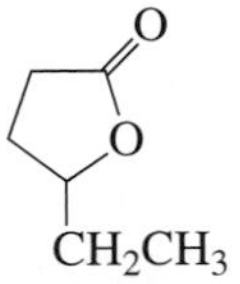   γ-caprolactone

PROBLEM 1

The word "lactone" comes from lactic acid, a three-carbon carboxylic acid with an OH group on the α-carbon. Ironically, lactic acid cannot form a lactone. Explain why not.

Amides

Amides are classified as primary, secondary, or tertiary, depending on the number of alkyl groups bonded to the nitrogen atom. Primary amides have no alkyl groups bonded to nitrogen, secondary amides have one, and tertiary amides have two.

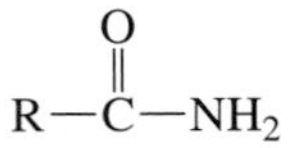

a primary amide 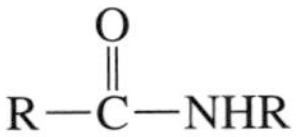   a secondary amide 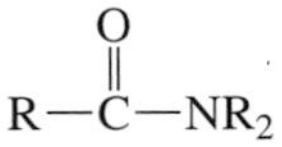   a tertiary amide

Primary amides are named by using the acid name, replacing "oic acid" or "ic acid" with "amide." For acids ending with "carboxylic acid," "ylic acid" is replaced with "amide."

$CH_3\overset{O}{\overset{\|}{C}}NH_2$	$ClCH_2CH_2CH_2\overset{O}{\overset{\|}{C}}NH_2$		

IUPAC name: ethanamide
common name: acetamide

4-chlorobutanamide
γ-chlorobutyramide

phenylmethanamide
benzamide

cis-2-ethylcyclohexanecarboxamide

If there is a substituent bonded to the nitrogen, the name of the substituent is stated first, followed by the name of the amide. The name of each substituent is preceded by a capital *N* (in italic) to indicate that the substituent is bonded to a nitrogen.

$CH_3CH_2\overset{O}{\overset{\|}{C}}NH-$

N-cyclohexylpropanamide

$CH_3CH_2CH_2CH_2\overset{O}{\overset{\|}{C}}-\overset{CH_3}{\underset{}{N}}CH_2CH_3$

N-ethyl-N-methylpentanamide

$CH_3CH_2CH_2\overset{O}{\overset{\|}{C}}-\overset{CH_2CH_3}{\underset{}{N}}CH_2CH_3$

N,N-diethylbutanamide

Cyclic amides are known as **lactams.** Their nomenclature is similar to that of lactones: the length of the carbon chain is indicated by the common name of the carboxylic acid, and a Greek letter is used to indicate the carbon to which the nitrogen is attached.

δ-valerolactam

γ-valerolactam

Nitriles

Nitriles are compounds that contain a C≡N functional group. They are considered to be carboxylic acid derivatives because, like all Class I carbonyl compounds, they react with water to form carboxylic acids (Section 15.15). In IUPAC nomenclature, nitriles are named by adding "nitrile" to the alkane name. Notice that the triply bonded carbon of the nitrile group is included in the number of carbons in the longest continuous chain. In common nomenclature, nitriles are named by replacing "ic acid" with "onitrile," or they are named as alkyl cyanides.

$CH_3C\equiv N$		$CH_3\overset{CH_3}{\underset{\|}{CH}}CH_2CH_2CH_2C\equiv N$	$CH_2=CHC\equiv N$

IUPAC name: ethanenitrile
common name: acetonitrile
                      methyl cyanide

phenylmethanenitrile
benzonitrile
phenyl cyanide

5-methylhexanenitrile

isohexyl cyanide

propenenitrile
acrylonitrile

PROBLEM 2◆

Name the following compounds.

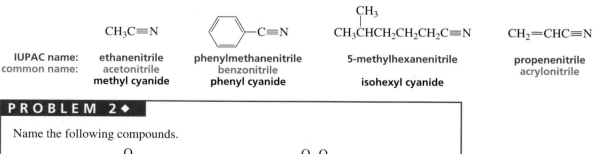

a. $CH_3CH_2CH_2CH_2\overset{O}{\overset{\|}{C}}Cl$

b. (lactam with NH)

c. $CH_3CH_2\overset{O}{\overset{\|}{C}}O\overset{O}{\overset{\|}{C}}CH_3$

d. $CH_3CH_2CH_2\overset{O}{\overset{\|}{C}}OCH_2\overset{CH_3}{\underset{\|}{C}}HCH_3$

e. 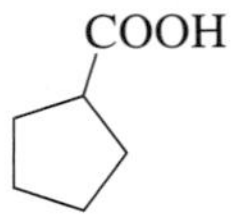 [cyclopentane with COOH]

g. $CH_3CH_2CH_2C\equiv N$

f. $CH_3CH_2CH_2CH_2CH_2\overset{O}{\overset{\|}{C}}N(CH_3)_2$

h. $CH_3CH_2CH_2\overset{O}{\overset{\|}{C}}O^-\ K^+$

15.2
THE STRUCTURE OF CARBOXYLIC ACIDS AND CARBOXYLIC ACID DERIVATIVES

Acyl halides, acid anhydrides, esters, and amides are all called **carboxylic acid derivatives** because they differ from a carboxylic acid only in the nature of the group that has replaced the OH group of the carboxylic acid. The **carbonyl carbon** in carboxylic acids and carboxylic acid derivatives is sp^2 hybridized. It uses its three sp^2 orbitals to form σ bonds to the carbonyl oxygen, the α-carbon, and a substituent (labeled Y in the following schematic). The three atoms attached to the carbonyl carbon are in the same plane, and the bond angles are each approximately 120°.

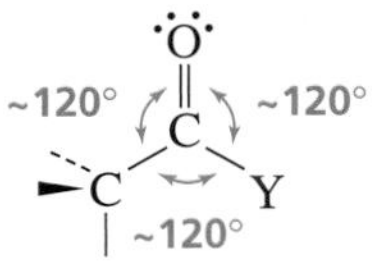

[schematic of carbonyl with ~120° angles]

The **carbonyl oxygen** is also sp^2 hybridized. One of its sp^2 orbitals forms a σ bond with the carbonyl carbon, and each of the other two sp^2 orbitals contains a pair of nonbonding electrons. The remaining p orbital of the carbonyl oxygen overlaps with the remaining p orbital of the carbonyl carbon to form a π bond (Figure 15.1).

Esters, carboxylic acids, and amides have two major contributing resonance structures.

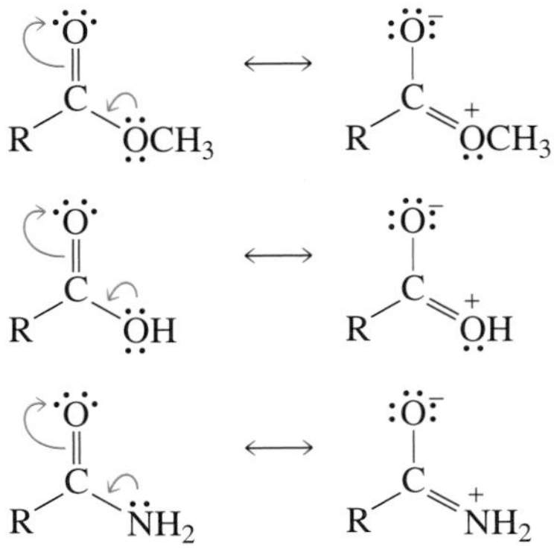

[resonance structures for ester, carboxylic acid, amide]

The resonance contributor on the right is more important for the amide than it is for the ester or the carboxylic acid, because nitrogen is better than oxygen at sharing its electrons. Basicity is a measure of how well a group shares its electrons, and $^-NH_2$ is a stronger base than ^-OH (which has about the same basicity as $^-OCH_3$) (Table 15.1).

PROBLEM 3

Which is longer, the carbon–oxygen single bond in a carboxylic acid or the carbon–oxygen bond in an alcohol? Why?

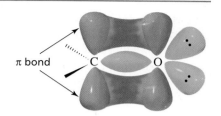

◀ **Figure 15.1**
Bonding in a carbonyl compound.

π bond

PROBLEM 4◆

There are three carbon–oxygen bonds in methyl acetate.

a. What are their relative lengths?

b. What are the relative infrared (IR) stretching frequencies of these bonds?

PROBLEM 5◆

An ester has two oxygen atoms. Which is more basic, the carbonyl oxygen or the carboxyl oxygen?

15.3 REACTIVITY CONSIDERATIONS

The greater electronegativity of oxygen compared with carbon causes the carbonyl group to be polar. Because the carbonyl carbon has a partial positive charge, we can predict that it will be attacked by nucleophiles.

$$\underset{\delta+}{\overset{\displaystyle\overset{\delta-}{O}}{\underset{\|}{R-C-Y}}}$$

When a nucleophile attacks the carbonyl group of a carboxylic acid derivative, the carbon–oxygen π bond breaks. The resulting product is called a **tetrahedral intermediate.** "Tetrahedral" comes from the fact that the trigonal (sp^2) carbon in the reactant has become a tetrahedral (sp^3) carbon in the intermediate.

$$R-\overset{\overset{\ddot{O}:\ sp^2}{\|}}{C}-Y \ +\ Z\!:^{-} \ \underset{k_{-1}}{\overset{k_1}{\rightleftharpoons}}\ R-\underset{\underset{Z}{|}}{\overset{\overset{sp^3\ :\ddot{O}:^{-}}{|}}{C}}-Y \ \underset{k_{-2}}{\overset{k_2}{\rightleftharpoons}}\ R-\overset{\overset{\ddot{O}:\ sp^2}{\|}}{C}-Z \ +\ Y\!:^{-}$$

a tetrahedral intermediate

The tetrahedral intermediate formed when a nucleophile attacks the carbonyl carbon of a carboxylic acid derivative is not stable and cannot be isolated. It is not a final product; it is an intermediate formed on the way to the final product. A pair of nonbonding electrons on the oxygen reforms the π bond, and either Y (k_2) or Z (k_{-1}) is eliminated with its bonding electrons.

Whether Y or Z is eliminated depends on their relative basicities. The weaker base is preferentially eliminated. This is another example of the principle that we saw in Section 9.3: *the weaker the base, the better it is as a leaving group.* This is because a weak base does not share its electrons as well as a strong base does. If it does not share its electrons as well, it forms a weaker bond—one that is easier to break. If Z is a much weaker base than Y, Z will be eliminated. In such a case, $k_{-1} \gg k_2$, and the reaction can be written:

$$R-\overset{\overset{\displaystyle \ddot{O}:}{\|}}{C}-Y \; + \; Z\overset{..}{\vphantom{.}} \; \underset{k_{-1}}{\overset{k_1}{\rightleftharpoons}} \; R-\overset{\overset{\displaystyle :\ddot{O}:^-}{|}}{\underset{\underset{\displaystyle Z}{|}}{C}}-Y$$

In this case, we see that no reaction occurs. The nucleophile attacks the carbonyl carbon, but the tetrahedral intermediate eliminates the nucleophile and reforms the reactant.

On the other hand, if Y is a much weaker base than Z, Y will be eliminated and a new product will be formed. In this case, $k_2 \gg k_{-1}$, and the reaction can be written:

$$R-\overset{\overset{\displaystyle \ddot{O}:}{\|}}{C}-Y \; + \; Z\overset{..}{\vphantom{.}} \; \overset{k_1}{\longrightarrow} \; R-\overset{\overset{\displaystyle :\ddot{O}:^-}{|}}{\underset{\underset{\displaystyle Z}{|}}{C}}-Y \; \overset{k_2}{\longrightarrow} \; R-\overset{\overset{\displaystyle \ddot{O}:}{\|}}{C}-Z \; + \; Y\overset{..}{\vphantom{.}}$$

This is called a **nucleophilic acyl substitution reaction** because the substituent that was attached to the acyl group in the reactant (Y) has been replaced by a nucleophile (Z).

If the basicity of Y and Z are similar, the value of k_{-1} will be similar to the value of k_2. Therefore, some molecules of the tetrahedral intermediate will eliminate Y and others will eliminate Z. When the reaction is over, there will be some reactant present and some product present. The relative amounts of each will depend on the relative basicities of Y and Z (that is, the relative values of k_2 and k_{-1}).

$$R-\overset{\overset{\displaystyle \ddot{O}:}{\|}}{C}-Y \; + \; Z\overset{..}{\vphantom{.}} \; \underset{k_{-1}}{\overset{k_1}{\rightleftharpoons}} \; R-\overset{\overset{\displaystyle :\ddot{O}:^-}{|}}{\underset{\underset{\displaystyle Z}{|}}{C}}-Y \; \underset{k_{-2}}{\overset{k_2}{\rightleftharpoons}} \; R-\overset{\overset{\displaystyle O}{\|}}{C}-Z \; + \; Y\overset{..}{\vphantom{.}}$$

These three cases are illustrated by the reaction coordinate diagrams in Figure 15.2.

1. If the incoming nucleophile is a weaker base than the group attached to the acyl group in the starting material (Figure 15.2a), the easier pathway is for the tetrahedral intermediate (TI) to expel the newly added group and reform the starting materials.

2. If the incoming nucleophile is a stronger base than the group attached to the acyl group in the starting material (Figure 15.2b), the easier pathway is for the tetrahedral intermediate to expel the group that was attached to the acyl group and form a substitution product.

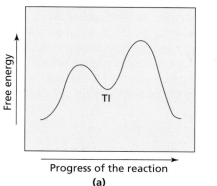
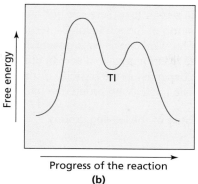
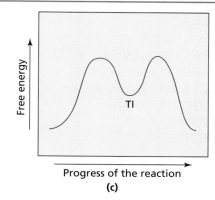

▲ **Figure 15.2**
Reaction coordinate diagrams for nucleophilic acyl substitution reactions in which
(a) the nucleophile is a weaker base than the group attached to the acyl group in the
starting material,
(b) the nucleophile is a stronger base than the group attached to the acyl group in the
starting material, and
(c) the nucleophile and the group attached to the acyl group in the starting material
have similar basicities.

3. If the incoming nucleophile and the group attached to the acyl group in the starting material have similar basicities (Figure 15.2c), the tetrahedral intermediate can expel either group with similar ease. A mixture of starting material and substitution product will result.

We can make the following general statement: *a carboxylic acid derivative will undergo a nucleophilic acyl substitution reaction provided that the incoming nucleophile is not a much weaker base than the substituent attached to the acyl group in the reactant.*

For a carboxylic acid derivative to undergo a nucleophilic acyl substitution reaction, the incoming nucleophile must not be a much weaker base than the group that is to be replaced.

PROBLEM 6◆

Using the pK_a values given in Table 15.1, predict the products of the following reactions.

a. CH$_3$—C(=O)—OCH$_3$ + NaCl $\longrightarrow$

b. CH$_3$—C(=O)—Cl + CH$_3$—C(=O)—O⁻Na⁺ $\longrightarrow$

c. CH$_3$—C(=O)—O—C(=O)—CH$_3$ + NaCl $\longrightarrow$

PROBLEM 7◆

Is the following statement true or false?

As long as the incoming nucleophile is a stronger base than the group attached to the acyl group in the reactant, formation of the tetrahedral intermediate is the rate-limiting step of a nucleophilic acyl substitution reaction.

15.4
RELATIVE REACTIVITIES OF CARBOXYLIC ACIDS, ACYL HALIDES, ACID ANHYDRIDES, ESTERS, AND AMIDES

We have just seen that there are two steps in a nucleophilic acyl substitution reaction: formation of the tetrahedral intermediate and collapse of the tetrahedral intermediate. The weaker the base attached to the acyl group, the easier it is for *both steps* of the reaction to take place. In other words, the reactivity of a carboxylic acid derivative depends on the basicity of the leaving group attached to the acyl group: the less basic the leaving group, the more reactive the carboxylic acid derivative.

relative basicities of the leaving groups

$$ Cl^- \; < \; {}^-OCR \; < \; {}^-OR \; \sim \; {}^-OH \; < \; {}^-NH_2 $$

(with O double-bonded to C in ^-OCR)

relative reactivities of carboxylic acid derivatives

$$ \underset{\text{acyl chloride}}{R-\overset{O}{\overset{\|}{C}}-Cl} \; > \; \underset{\text{acid anhydride}}{R-\overset{O}{\overset{\|}{C}}-O-\overset{O}{\overset{\|}{C}}-R} \; > \; \underset{\text{ester}}{R-\overset{O}{\overset{\|}{C}}-OR'} \; \sim \; \underset{\text{carboxylic acid}}{R-\overset{O}{\overset{\|}{C}}-OH} \; > \; \underset{\text{amide}}{R-\overset{O}{\overset{\|}{C}}-NH_2} $$

increasing reactivity

How does having a weak base attached to the acyl group make the first step of the nucleophilic substitution reaction easier? A weak base makes the carbonyl compound less stable and therefore more reactive. This is because a weak base does not share its electrons as well as a strong base does; thus, the weaker the basicity of Y, the smaller the contribution from the resonance contributor with a positive charge on Y (Section 15.2). Consequently, the weaker the basicity of Y, the less the resonance stabilization of the molecule.

resonance contributors of a carboxylic acid or carboxylic acid derivative

In addition, the weaker the basicity of Y, the more it withdraws electrons inductively from the carbonyl carbon, which increases the carbonyl carbon's susceptibility to nucleophilic attack.

inductive electron withdrawal increases the electrophilicity of the carbonyl carbon

How does having a weak base attached to the acyl group make the second step of the nucleophilic acyl substitution reaction easier? When the tetrahedral intermediate collapses, it eliminates the leaving group, and weak bases are easier to eliminate.

the weaker the base, the easier it is to eliminate

Since the incoming nucleophile in a nucleophilic acyl substitution reaction must be a stronger base than the base that is already there (Section 15.3), a carboxylic acid derivative can be converted into a less reactive carboxylic acid derivative, but not into one that is more reactive. For example, an acyl chloride can be converted into an anhydride, because a carboxylate ion is a stronger base than a chloride ion.

$$
\underset{\substack{\text{an acyl chloride}}}{R-\overset{\overset{\displaystyle O}{\|}}{C}-Cl} \;+\; \underset{\substack{\text{carboxylate ion}}}{R-\overset{\overset{\displaystyle O}{\|}}{C}-O^-} \;\longrightarrow\; \underset{\substack{\text{an anhydride}}}{R-\overset{\overset{\displaystyle O}{\|}}{C}-O-\overset{\overset{\displaystyle O}{\|}}{C}-R} \;+\; Cl^-
$$

An anhydride, however, cannot be converted into an acyl chloride, because a chloride ion is a weaker base than a carboxylate ion.

$$
\underset{\substack{\text{an anhydride}}}{R-\overset{\overset{\displaystyle O}{\|}}{C}-O-\overset{\overset{\displaystyle O}{\|}}{C}-R} \;+\; Cl^- \;\longrightarrow\; \text{no reaction}
$$

Because of their high reactivity, acyl halides and acid anhydrides are not found in nature. Carboxylic acids, being less reactive, are found widely in nature. For example, (+)-lactic acid is the compound responsible for the burning sensation felt in muscles during anaerobic exercise, and it is also found in sour milk. Spinach and other leafy green vegetables are rich in oxalic acid. Succinic acid and citric acid are important intermediates in the Krebs cycle (Appendix VI), a series of reactions that results in the oxidation of acetyl CoA (which is formed from pyruvic acid) to CO_2 in biological systems. Citrus fruits are rich in citric acid. The concentration is greatest in lemons, less in grapefruits, and still less in oranges. (S)-(−)-Malic acid is responsible for the sharp taste of unripe apples and pears. As the fruit ripens, the amount of malic acid in the fruit decreases and the amount of sugar increases. This is important for propagation of the plant. Animals will not eat the fruit until it becomes ripe—at which time its seeds are mature enough to germinate when they are scattered about. Prostaglandins are locally acting hormones that have several different physiological functions (Section 15.8), such as stimulating inflammation, causing hypertension, and producing pain and swelling.

Esters are also commonly found in nature. Many of the fragrances of flowers and fruits are due to esters.

$$CH_3\overset{\text{O}}{\overset{\|}{C}}OCH_2\text{—}\bigcirc$$

benzyl acetate
jasmine

$$CH_3\overset{\text{O}}{\overset{\|}{C}}OCH_2CH_2\overset{CH_3}{\overset{|}{C}}HCH_3$$

isopentyl acetate
banana

$$CH_3CH_2CH_2\overset{\text{O}}{\overset{\|}{C}}OCH_3$$

methyl butyrate
apple

Carboxylic acids with an amino group on the α-carbon are commonly called **amino acids.** Amino acids are linked together by amide bonds to form peptides and proteins (Section 20.7). Caffeine is another example of a naturally occurring amide; it is found in cocoa and coffee beans. Penicillin G, a compound with two amide bonds (one is in a β-lactam ring), was first isolated from a mold in 1928 by Sir Alexander Fleming, a Scottish bacteriologist.

$$\overset{+}{H_3N}\overset{\text{O}}{\overset{\|}{C}}HC\overset{\text{O}}{\overset{\|}{C}}O^-$$
$$\underset{R}{|}$$

an amino acid

$$-NH\underset{\underset{R}{|}}{CH}\overset{\text{O}}{\overset{\|}{C}}-NH\underset{\underset{R}{|}}{CH}\overset{\text{O}}{\overset{\|}{C}}-$$

general structure for a peptide or a protein

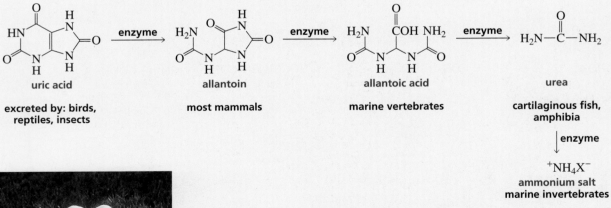

caffeine

piperine
the major component of black pepper

penicillin G

DALMATIANS: DON'T TRY TO FOOL MOTHER NATURE

When amino acids are metabolized, the excess nitrogen is concentrated into uric acid—a compound with five amide bonds—which is excreted by birds, reptiles, and insects. A series of enzyme-catalyzed reactions degrades uric acid to ammonium ion. The extent to which this degradation process occurs depends on the species. Most mammals excrete excess nitrogen as allantoin. Excess nitrogen in aquatic animals is excreted as allantoic acid, as urea, or as ammonium salts.

uric acid

excreted by: birds, reptiles, insects

→ enzyme →

allantoin

most mammals

→ enzyme →

allantoic acid

marine vertebrates

→ enzyme →

$$H_2N\overset{\text{O}}{\overset{\|}{\text{—}C\text{—}}}NH_2$$

urea

cartilaginous fish, amphibia

↓ enzyme

$$^+NH_4X^-$$
ammonium salt
marine invertebrates

Dalmatians, unlike most mammals, excrete high levels of uric acid, because Dalmatian breeders select dogs that have black spots with no white hairs. The gene that determines the presence of white hairs is linked to the gene that converts uric acid to allantoin. Dalmatians, therefore, are known to be susceptible to gout (painful deposits of uric acid in joints).

THE DISCOVERY OF PENICILLIN

Sir Alexander Fleming (1881–1955) was born in Scotland. He was a professor of bacteriology at University College, London. The story is told that one day Fleming was about to throw away a culture of staphylococcal bacteria that had been contaminated by a rare strain of the mold *Penicillium notatum*. He noticed that the bacteria had disappeared wherever there was a particle of mold. This suggested to him that the mold must have produced an antibacterial substance. He isolated an active extract of the mold, which was found 10 years later to be a mixture of seven to nine related compounds. The most reproducible of these was penicillin G (Section 15.13). Although Fleming is generally given credit for the discovery of penicillin, there is clear evidence that the germicidal activity of the mold was recognized in the nineteenth century by Lord Joseph Lister (1827–1912), the English physician renowned for the introduction of aseptic surgery.

All carboxylic acid derivatives undergo nucleophilic acyl substitution reactions by the same mechanism. If the nucleophile is negatively charged, the mechanism discussed in Section 15.3 is followed.

15.5 GENERAL MECHANISM FOR NUCLEOPHILIC ACYL SUBSTITUTION REACTIONS

$$R-\overset{\overset{\displaystyle \cdot\cdot}{O}}{\underset{}{C}}-Y \; + \; H\overset{\cdot\cdot}{\underset{\cdot\cdot}{O}}{}^{-} \; \rightleftharpoons \; R-\overset{:\overset{\cdot\cdot}{O}:^{-}}{\underset{:OH}{C}}-Y \; \longrightarrow \; R-\overset{\overset{\displaystyle :O}{\parallel}}{C}-\overset{\cdot\cdot}{O}H \; + \; Y^{-}$$

If the nucleophile is neutral, the mechanism has an additional step. A proton is lost from the tetrahedral intermediate formed in the first step, resulting in a tetrahedral intermediate equivalent to the one formed by negatively charged nucleophiles.

All carboxylic acid derivatives react by the same mechanism.

$$R-\overset{\overset{\displaystyle \cdot\cdot}{O}}{\underset{}{C}}-Y \; + \; H_2\overset{\cdot\cdot}{O}: \; \rightleftharpoons \; R-\overset{:\overset{\cdot\cdot}{O}:^{-}}{\underset{\overset{+:OH}{\underset{H}{}}}{C}}-Y \; \underset{H^+}{\overset{-H^+}{\rightleftharpoons}} \; R-\overset{:\overset{\cdot\cdot}{O}:^{-}}{\underset{:OH}{C}}-Y \; \rightleftharpoons \; R-\overset{\overset{\displaystyle :O}{\parallel}}{C}-\overset{\cdot\cdot}{O}H \; + \; Y^{-}$$

PROTON-TRANSFER STEPS

Any source of protons present in solution (H_3O^+, HCl, H_2O) can serve as a proton donor. Therefore, we will not specify the proton donor in proton-transfer steps. Instead, we will show a protonation step by placing H^+ over the arrow and the reverse deprotonation step by placing $-H^+$ under the arrow.

$$CH_3-\overset{\overset{\displaystyle O}{\parallel}}{C}-OCH_3 \; \underset{-H^+}{\overset{H^+}{\rightleftharpoons}} \; CH_3-\overset{\overset{\displaystyle \overset{+}{O}H}{\parallel}}{C}-OCH_3$$

Similarly, any base present in solution (HO^-, H_2O) can remove a proton. Therefore, we will not specify the base in proton-transfer steps. We will show a deprotonation step using $-H^+$ over the arrow and the reverse protonation step using H^+ under the arrow.

$$CH_3-\overset{\overset{\displaystyle OH}{|}}{\underset{\overset{+OH}{\underset{H}{}}}{C}}-OCH_3 \; \underset{H^+}{\overset{-H^+}{\rightleftharpoons}} \; CH_3-\overset{\overset{\displaystyle OH}{|}}{\underset{OH}{C}}-OCH_3$$

In the preceding section, you learned that you will be able to determine the outcome of all the reactions of carboxylic acids and carboxylic acid derivatives discussed in this chapter simply by looking at the two leaving groups in the tetrahedral intermediate and remembering that the weaker base is the one that is eliminated.

1. If the new group in the tetrahedral intermediate is a weaker base than the group that was already there, the new group will be eliminated and no reaction will take place.
2. If the new group in the tetrahedral intermediate is a stronger base than the group that was already there, the original group will be eliminated and the overall result will be a nucleophilic acyl substitution reaction.
3. If the two groups in the tetrahedral intermediate have similar basicities, either of the groups can be eliminated and an equilibrium mixture of reactant and product will be formed.

As you read the remaining sections of this chapter, remember that you are seeing specific examples of these general principles. It is important to remember that *all the reactions follow the same mechanism.*

15.6 REACTIONS OF ACYL HALIDES

Acyl halides react with carboxylate ions to form anhydrides, with alcohols to form esters, with water to form carboxylic acids, and with amines to form amides, because in each case, the incoming nucleophile is a stronger base than the departing halide ion (Table 15.1). Notice that both alcohols and phenols can be used to prepare esters.

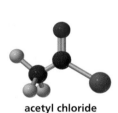

acetyl chloride

$$CH_3\overset{O}{\overset{\|}{C}}-Cl \ + \ CH_3\overset{O}{\overset{\|}{C}}O^- \longrightarrow CH_3\overset{O}{\overset{\|}{C}}-O-\overset{O}{\overset{\|}{C}}CH_3 \ + \ Cl^-$$

acetyl chloride   acetic anhydride

benzoyl chloride + CH_3OH ⟶ methyl benzoate + HCl

$$CH_3CH_2\overset{O}{\overset{\|}{C}}-Cl \ + \ \text{(phenol)}-OH \longrightarrow CH_3CH_2\overset{O}{\overset{\|}{C}}-O-\text{(phenyl)} \ + \ HCl$$

propionyl chloride   phenyl propionate

$$CH_3CH_2CH_2\overset{O}{\overset{\|}{C}}-Cl \ + \ H_2O \longrightarrow CH_3CH_2CH_2\overset{O}{\overset{\|}{C}}-OH \ + \ HCl$$

butyryl chloride   butyric acid

phenylethanoyl chloride + 2 CH_3NH_2 ⟶ *N*-methylphenylethanamide + $CH_3\overset{+}{N}H_3\ Cl^-$

All the reactions follow the general mechanism described in Section 15.5. In the conversion of an acyl chloride into an acid anhydride, the nucleophilic carboxylate ion attacks the carbonyl carbon of the acyl chloride. Because the resulting tetrahedral intermediate is unstable, the electrons on oxygen immediately reform the double bond, eliminating chloride ion because it is a weaker base than the carboxylate ion. The final product is an anhydride.

mechanism for the conversion of an acyl chloride into an acid anhydride

$$CH_3-\overset{\overset{\displaystyle ::O:}{\|}}{C}-\ddot{\underset{..}{C}}l: \;+\; CH_3-\overset{\overset{\displaystyle O:}{\|}}{C}-\ddot{\underset{..}{O}}: \;\rightleftharpoons\; CH_3-\overset{\overset{\displaystyle :O:^-}{|}}{\underset{\underset{\displaystyle CH_3}{\overset{|}{C=\ddot{O}:}}}{\underset{|}{\overset{|}{\underset{:O:}{C}}}}-\ddot{\underset{..}{C}}l: \;\longrightarrow\; CH_3-\overset{\overset{\displaystyle O:}{\|}}{C}-\ddot{\underset{..}{O}}-\overset{\overset{\displaystyle O:}{\|}}{C}-CH_3 \;+\; :\ddot{\underset{..}{C}}l:^-$$

In the conversion of an acyl chloride into an ester, the nucleophilic alcohol attacks the carbonyl carbon of the acyl chloride. Because the protonated alcohol group is a strong acid (Section 1.17), the tetrahedral intermediate loses a proton. Chloride ion is eliminated from the deprotonated tetrahedral intermediate because chloride ion is a weaker base than methoxide ion.

mechanism for the conversion of an acyl chloride into an ester

(mechanism: Ph–C(=O)–Cl + CH₃ÖH ⇌ Ph–C(O⁻)(⁺ÖCH₃H)–Cl $\xrightarrow[H^+]{-H^+}$ Ph–C(O⁻)(ÖCH₃)–Cl ⇌ Ph–C(=O)–ÖCH₃ + :Cl⁻)

The reaction of an acyl chloride with ammonia or with a primary or secondary amine produces an amide and HCl. The acid generated in the reaction will protonate unreacted ammonia or unreacted amine and, because the protonated amines are not nucleophiles, they cannot react with the acyl chloride. The reaction, therefore, must be carried out with twice as much ammonia or amine as acyl chloride or else there will not be enough amine to react with all of the acyl halide.

$$CH_3\overset{\overset{\displaystyle O}{\|}}{C}-Cl \;+\; NH_3 \;\longrightarrow\; CH_3\overset{\overset{\displaystyle O}{\|}}{C}-NH_2 \;+\; HCl \;\xrightarrow{NH_3}\; \overset{+}{N}H_4\,Cl^-$$

$$CH_3CH_2\overset{\overset{\displaystyle O}{\|}}{C}-Cl \;+\; CH_3NH_2 \;\longrightarrow\; CH_3CH_2\overset{\overset{\displaystyle O}{\|}}{C}-NHCH_3 \;+\; HCl \;\xrightarrow{CH_3NH_2}\; CH_3\overset{+}{N}H_3\,Cl^-$$

$$Ph-\overset{\overset{\displaystyle O}{\|}}{C}-Cl \;+\; 2\,CH_3\underset{\underset{\displaystyle CH_3}{|}}{N}H \;\longrightarrow\; Ph-\overset{\overset{\displaystyle O}{\|}}{C}-\underset{\underset{\displaystyle CH_3}{|}}{N}CH_3 \;+\; CH_3\overset{+}{N}H_2\,Cl^-$$

Because tertiary amines cannot form amides, an equivalent of a tertiary amine such as triethylamine or pyridine can be used instead of excess amine.

$$CH_3\overset{\overset{\displaystyle O}{\|}}{C}-Cl \;+\; CH_3NH_2 \;\xrightarrow{\text{pyridine}}\; CH_3\overset{\overset{\displaystyle O}{\|}}{C}-NHCH_3 \;+\; \text{(pyridinium)}\;\overset{+}{N}H\;Cl^-$$

PROBLEM 8 / SOLVED

a. Two amides are obtained from the reaction of acetyl chloride with a mixture of ethylamine and propylamine. Identify the amides.

b. Only one amide is obtained from the reaction of acetyl chloride with a mixture of ethylamine and pyridine. Why is only one amide obtained?

SOLUTION TO 8a   Either of the amines can react with acetyl chloride, so both *N*-ethylacetamide and *N*-propylacetamide are formed.

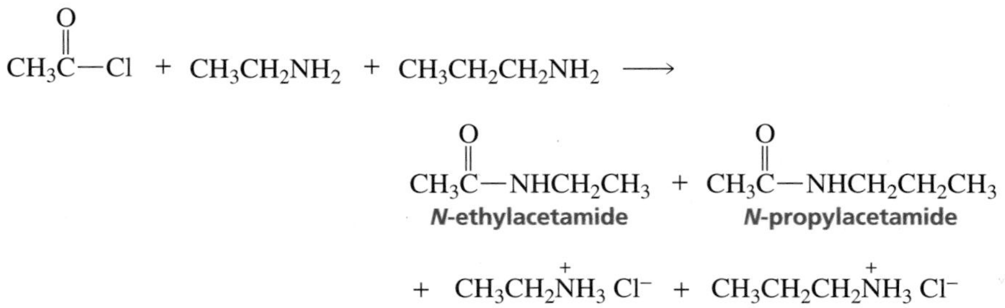

$$CH_3\overset{\displaystyle O}{\overset{\|}{C}}-Cl \ + \ CH_3CH_2NH_2 \ + \ CH_3CH_2CH_2NH_2 \ \longrightarrow$$

$$CH_3\overset{\displaystyle O}{\overset{\|}{C}}-NHCH_2CH_3 \ + \ CH_3\overset{\displaystyle O}{\overset{\|}{C}}-NHCH_2CH_2CH_3$$
N-ethylacetamide   **N-propylacetamide**

$$+ \ CH_3CH_2\overset{+}{N}H_3 \ Cl^- \ + \ CH_3CH_2CH_2\overset{+}{N}H_3 \ Cl^-$$

SOLUTION TO 8b   Initially, two amides can be formed. However, one of the amides is very reactive because it has a positively charged nitrogen atom (an excellent leaving group) since the amine that formed the amide is a tertiary amine. Therefore, it will react immediately with unreacted ethylamine. Consequently, *N*-ethylacetamide is the only amide obtained from the reaction.

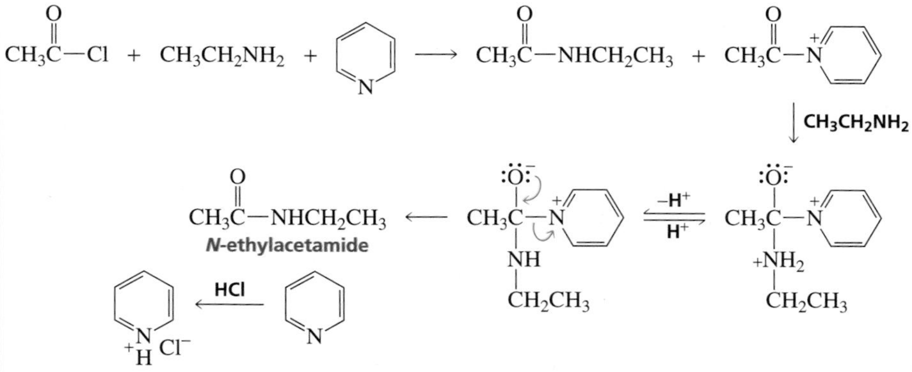

PROBLEM 9

Although excess amine is necessary in the reaction of an acyl chloride with an amine, it is not necessary to use excess alcohol in the reaction of an acyl chloride with an alcohol. Explain.

PROBLEM 10

Give the mechanism for:

a. the reaction of acetyl chloride with water to form acetic acid.

b. the reaction of acetyl bromide with methyl amine to form *N*-methylacetamide.

PROBLEM 11◆

Starting with acetyl chloride, what nucleophile would you use to make each of the following?

a. CH₃COCH₂CH₂CH₃
$$CH_3\overset{O}{\overset{\|}{C}}OCH_2CH_2CH_3$$

b. $CH_3\overset{O}{\overset{\|}{C}}NHCH_2CH_3$

c. $CH_3\overset{O}{\overset{\|}{C}}N(CH_3)_2$

d. $CH_3\overset{O}{\overset{\|}{C}}O\overset{O}{\overset{\|}{C}}CH_3$

e. $CH_3\overset{O}{\overset{\|}{C}}O\text{—}\!\!\bigcirc\!\!\text{—}NO_2$

f. $CH_3\overset{O}{\overset{\|}{C}}OH$

Acid anhydrides do not react with sodium chloride or with sodium bromide, because the incoming halide ion is a weaker base than the departing carboxylate ion (Table 15.1).

$$CH_3\overset{O}{\overset{\|}{C}}\text{—}O\text{—}\overset{O}{\overset{\|}{C}}CH_3 \;+\; Cl^- \longrightarrow \text{ no reaction}$$

Since the incoming halide ion is the weaker base, the halide ion would be the substituent eliminated from the tetrahedral intermediate.

$$CH_3\overset{\ddot{O}:}{\overset{\|}{C}}\text{—}O\text{—}\overset{O}{\overset{\|}{C}}CH_3 \;+\; :\ddot{C}\mathrm{l}:^- \;\rightleftharpoons\; CH_3\overset{:\ddot{O}:^-}{\underset{\underset{Cl}{|}}{\overset{|}{C}}}\text{—}O\text{—}\overset{O}{\overset{\|}{C}}CH_3$$

An acid anhydride reacts with an alcohol to form an ester and a carboxylic acid, with water to form two equivalents of a carboxylic acid, and with an amine to form an amide and a carboxylic acid. In each case, the incoming nucleophile is a stronger base than the departing carboxylate ion. Two equivalents of an amine, or one equivalent of an amine plus one equivalent of a tertiary amine such as pyridine, must be employed in the reaction of an amine with an anhydride so that sufficient amine is present to react with the acid produced in the reaction.

15.7 REACTIONS OF ACID ANHYDRIDES

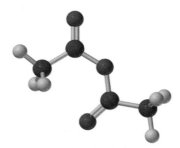

acetic anhydride

$$CH_3\overset{O}{\overset{\|}{C}}\text{—}O\text{—}\overset{O}{\overset{\|}{C}}CH_3 \;+\; CH_3CH_2OH \longrightarrow CH_3\overset{O}{\overset{\|}{C}}\text{—}OCH_2CH_3 \;+\; CH_3\overset{O}{\overset{\|}{C}}OH$$

acetic anhydride · · · · · · · · · · · **ethyl acetate** · · · · **acetic acid**

$$\bigcirc\!\!\overset{O}{\overset{\|}{C}}\text{—}O\text{—}\overset{O}{\overset{\|}{C}}\!\!\bigcirc \;+\; H_2O \longrightarrow 2\;\bigcirc\!\!\overset{O}{\overset{\|}{C}}\text{—}OH$$

benzoic anhydride · · · · · · · · · · · **benzoic acid**

$$CH_3CH_2\overset{O}{\overset{\|}{C}}\text{—}O\text{—}\overset{O}{\overset{\|}{C}}CH_2CH_3 \;+\; 2\,CH_3NH_2 \longrightarrow CH_3CH_2\overset{O}{\overset{\|}{C}}\text{—}NHCH_3 \;+\; CH_3CH_2\overset{O}{\overset{\|}{C}}\text{—}O^-\;H_3\overset{+}{N}CH_3$$

propionic anhydride · · · · · · · · · · · **N-methylpropionamide**

All the reactions follow the general mechanism described in Section 15.5. For example, compare the mechanism for conversion of an acid anhydride into an ester (below) with the mechanism for conversion of an acyl chloride into an ester (Section 15.6).

mechanism for the conversion of an acid anhydride into an ester (and a carboxylic acid)

$$\text{CH}_3\text{C}-\ddot{\text{O}}-\text{CCH}_3 + \text{CH}_3\ddot{\text{O}}\text{H} \rightleftharpoons \text{CH}_3\text{C}-\ddot{\text{O}}-\text{CCH}_3 \underset{\text{H}^+}{\overset{-\text{H}^+}{\rightleftharpoons}} \text{CH}_3\text{C}-\ddot{\text{O}}-\text{CCH}_3$$

$$\downarrow$$

$$\text{CH}_3\text{C}-\ddot{\text{O}}\text{CH}_3 + {}^{-}\ddot{\text{O}}-\text{CCH}_3$$

PROBLEM 12

a. Propose a mechanism for the reaction of acetic anhydride with water.

b. How does this mechanism differ from the mechanism for the reaction of acetic anhydride with an alcohol?

PROBLEM 13

We have seen that acid anhydrides react with alcohols, water, and amines. In which one of these three reactions does the tetrahedral intermediate not have to lose a proton before it eliminates the carboxylate ion? Explain.

15.8
REACTIONS OF ESTERS

methyl acetate

Esters do not react with halide ions or with carboxylate ions because these nucleophiles are much weaker bases than the RO⁻ leaving group of the ester (Table 15.1). Esters react with water to form carboxylic acids and with alcohols to form different esters. A reaction with water is called **hydrolysis,** and a reaction with an alcohol is called **alcoholysis.** The hydrolysis or alcoholysis of an ester is a very slow reaction. Therefore, when these reactions are carried out in the laboratory, they are always catalyzed. Both hydrolysis and alcoholysis of an ester can be catalyzed by acids (H⁺). The rate of hydrolysis can also be increased by hydroxide ion (HO⁻), and the rate of alcoholysis can be increased by alkoxide ion (RO⁻) (Sections 15.9 and 15.10).

$$\underset{\text{methyl acetate}}{\text{CH}_3\text{C}-\text{OCH}_3} + \text{H}_2\text{O} \underset{}{\overset{\text{H}^+}{\rightleftharpoons}} \underset{\text{acetic acid}}{\text{CH}_3\text{C}-\text{OH}} + \text{CH}_3\text{OH}$$

$$\underset{\text{methyl benzoate}}{\bigcirc\!\!-\!\text{C}-\text{OCH}_3} + \text{CH}_3\text{CH}_2\text{OH} \overset{\text{H}^+}{\rightleftharpoons} \underset{\text{ethyl benzoate}}{\bigcirc\!\!-\!\text{C}-\text{OCH}_2\text{CH}_3} + \text{CH}_3\text{OH}$$

Esters also react with amines to form amides. A reaction with an amine is called **aminolysis.** Notice that the aminolysis of an ester requires only one equivalent of amine, unlike the aminolysis of an acyl halide or an acid anhydride, which we saw requires two equivalents (Sections 15.6 and 15.7). This is because the leaving group of an ester (RO^-) is more basic than the amine nucleophile (RNH_2), so the alkoxide ion—rather than unreacted amine—picks up the proton produced in the reaction.

$$CH_3CH_2\overset{\overset{\displaystyle O}{\|}}{C}-OCH_2CH_3 \;+\; CH_3NH_2 \;\longrightarrow\; CH_3CH_2\overset{\overset{\displaystyle O}{\|}}{C}-NHCH_3 \;+\; CH_3CH_2OH$$

ethyl propionate   *N*-methylpropionamide

The tetrahedral intermediate formed when an ester is hydrolyzed has both an RO^- leaving group and an HO^- leaving group. Because both leaving groups have approximately the same basicity (Table 15.1), they are equally likely to be eliminated.

eliminates HO^- to give the ester

eliminates CH_3O^- to give the carboxylic acid

Notice that when CH_3O^- is eliminated, the final products are a carboxylate ion and methanol. If only one species is protonated, it will be the more basic one (CH_3O^- is more basic than $RCOO^-$).

Since the two leaving groups in the tetrahedral intermediate have approximately the same basicity, when the reaction comes to equilibrium about half of the ester molecules will have been converted to carboxylic acid and the other half will remain as ester. The equilibrium can be driven to the right if the reaction is carried out in excess water (Le Châtelier's principle).

$$CH_3\overset{\overset{\displaystyle O}{\|}}{C}-OCH_3 \;+\; H_2O \;\overset{H^+}{\rightleftharpoons}\; CH_3\overset{\overset{\displaystyle O}{\|}}{C}-OH \;+\; CH_3OH$$

excess

An ester reacts with an alcohol to form a new ester and a new alcohol. This is called a **transesterification reaction.** Because both alcohols have approximately the same leaving tendency in the tetrahedral intermediate, an excess of the reactant alcohol is necessary to drive the equilibrium to the right. Or, if the boiling point of the product alcohol is significantly lower than the boiling points of the other reaction components, the reaction can be driven to the right by distilling off the product alcohol as it is formed. The rate of transesterification can be increased by H^+ or by the conjugate base (RO^-) of the reactant alcohol (Sections 15.9 and 15.10).

$$CH_3\overset{\displaystyle O}{\overset{\|}{C}}-OCH_3 \ + \ CH_3CH_2CH_2OH \ \underset{}{\overset{H^+}{\rightleftharpoons}} \ CH_3\overset{\displaystyle O}{\overset{\|}{C}}-OCH_2CH_2CH_3 \ + \ CH_3OH$$

methyl acetate　　**propyl alcohol**　　　　　　**propyl acetate**　　　**methyl alcohol**
　　　　　　　　　　excess

Because phenols are stronger acids than alcohols (Section 6.10), phenolate ions (ArO⁻) are weaker bases than alkoxide ions (RO⁻). This means that phenyl esters are more reactive than alkyl esters.

$CH_3\overset{\displaystyle O}{\overset{\|}{C}}-O\!\!-\!\!\bigcirc$　is more reactive than　$CH_3\overset{\displaystyle O}{\overset{\|}{C}}-OCH_3$

phenyl acetate　　　　　　　　　　　　　　**methyl acetate**

$\bigcirc\!\!-\!\!OH$　　　　　　　　　　　　　　　　CH_3OH

$pK_a = 10.0$　　　　　　　　　　　　　　　$pK_a = 15.5$

The reaction of an ester with an amine is also a slow reaction. However, unlike the reaction of an ester with water or an alcohol, the rate of the reaction of an ester with an amine cannot be increased by H⁺ or by HO⁻ or RO⁻ (Problem 19). Aminolysis of an ester can be driven to completion by using excess amine or by distilling off the alcohol as it is formed.

$$CH_3\overset{\displaystyle O}{\overset{\|}{C}}-OCH_3 \ + \ CH_3(CH_2)_4NH_2 \ \longrightarrow \ CH_3\overset{\displaystyle O}{\overset{\|}{C}}-NH(CH_2)_4CH_3 \ + \ CH_3OH$$
　　　　　　　　　　　excess

ASPIRIN

A transesterification reaction that blocks prostaglandin synthesis is responsible for the activity of aspirin (acetylsalicylic acid) as an antiinflammatory agent. Prostaglandins have several different biological functions, one of which is to stimulate inflammation. The enzyme prostaglandin synthase catalyzes the conversion of arachidonic acid into PGH₂, a precursor of prostaglandins and the related thromboxanes (Section 23.5).

arachidonic acid $\xrightarrow{\text{\textbf{prostaglandin synthase}}}$ PGH₂ $\underset{\searrow}{\overset{\nearrow}{}}$ prostaglandins / thromboxanes .

Prostaglandin synthase is composed of two enzymes. One of the enzymes, cyclooxygenase, has a serine hydroxyl group that is necessary for enzyme activity. In the presence of aspirin, the serine hydroxyl group participates in a transesterification reaction and becomes acetylated. This inactivates the enzyme. Prostaglandin therefore cannot be synthesized, and inflammation is suppressed.

serine hydroxyl group

$CH_3\overset{\displaystyle O}{\overset{\|}{C}}-O\!\!-\!\!\bigcirc$ + $HOCH_2$— $\xrightarrow{\text{\textbf{transesterification}}}$ $CH_3\overset{\displaystyle O}{\overset{\|}{C}}-OCH_2$— + $HO\!\!-\!\!\bigcirc$
　　　　⁻OOC　　　　　　　　　　　　　　　　　　　　　　　　　　⁻OOC

acetylsalicyclic acid　　**enzyme**　　　　　　　　**acetylated enzyme**
aspirin　　　　　　　　active　　　　　　　　　　　　inactive

Thromboxanes stimulate platelet aggregation. Because aspirin inhibits the formation of PGH_2, it inhibits thromboxane production and therefore platelet aggregation. Presumably, this is why low levels of aspirin have been reported to reduce the incidence of strokes and heart attacks.

PROBLEM 14

Give the mechanism for:

a. the noncatalyzed hydrolysis of methyl propionate.

b. the aminolysis of phenyl formate using methylamine.

PROBLEM 15 / SOLVED

a. List the following esters in order of decreasing reactivity toward hydrolysis.

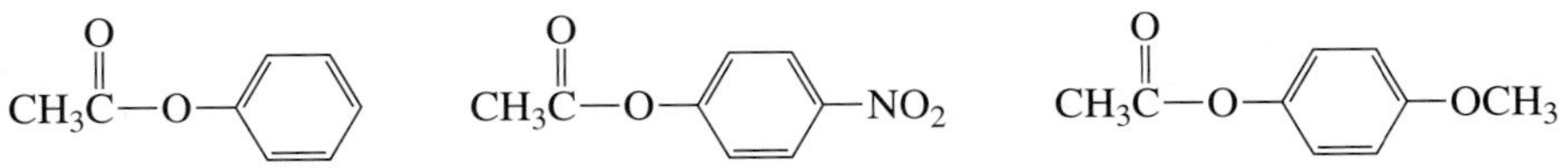

b. How would the rate of hydrolysis of the *para*-methyl-phenyl ester compare with the rate of hydrolysis of these three esters?

SOLUTION TO 15a   Because the nitro group withdraws electrons from the benzene ring and the methoxy group donates electrons into the ring, the nitro-substituted ester will be the most susceptible and the methoxy-substituted ester the least susceptible of the three esters to nucleophilic attack. Because electron withdrawal increases acidity and electron donation decreases acidity, *para*-nitrophenol is a stronger acid than phenol, which is a stronger acid than *para*-methoxyphenol. Therefore, the *para*-nitrophenolate ion is the weakest base and the best leaving group, while the *para*-methoxyphenolate ion is the strongest base and the worst leaving group of the three. As a result, both relatively slow steps of the hydrolysis reaction are fastest for the nitrosubstituted ester and slowest for the methoxy-substituted ester.

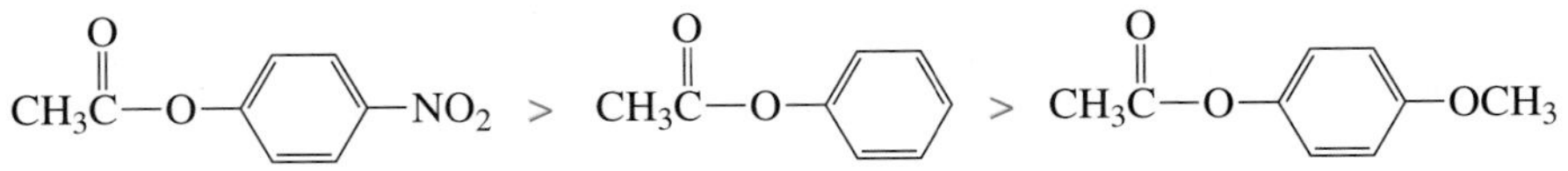

SOLUTION TO 15b   The methyl substituent donates electrons into the benzene ring but does so to a lesser extent than the methoxy substituent. Therefore the rate of hydrolysis of the methyl-substituted ester is slower than the rate of hydrolysis of the unsubstituted ester but faster than the rate of hydrolysis of the methoxy-substituted ester.

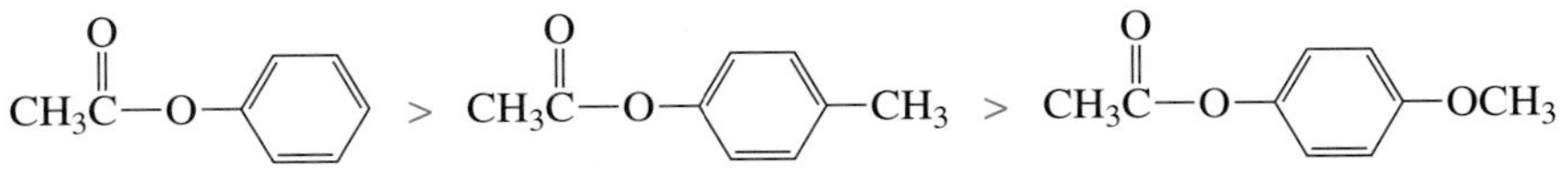

PROBLEM 16 ◆

State three factors that contribute to the fact that the noncatalyzed hydrolysis of an ester is a very slow reaction.

**15.9
ACID-CATALYZED
ESTER HYDROLYSIS**

We have seen that esters undergo nucleophilic substitution reactions very slowly because their leaving groups are quite basic. The rate of hydrolysis of an ester can be increased by carrying out the reaction in the presence of a catalyst. Either H^+ or HO^- can be used to increase the rate of the reaction. Notice that in an acid-catalyzed reaction all organic reactants, intermediates, and products are positively charged or neutral; *there are no negatively charged organic reactants, intermediates, or products in acidic solutions.* Also notice that in a reaction in which HO^- is used to increase the rate of the reaction, all organic reactants, intermediates, and products are negatively charged or neutral; *there are no positively charged organic reactants, intermediates, or products in basic solutions.*

The first step in the mechanism for acid-catalyzed ester hydrolysis is protonation of the carbonyl oxygen by the acid.

$$\underset{\text{O}}{\overset{\text{O}}{\text{CH}_3\overset{\|}{\text{C}}-\text{OCH}_3}} \quad \underset{-\text{H}^+}{\overset{\text{H}^+}{\rightleftharpoons}} \quad \underset{\overset{+}{\text{OH}}}{\text{CH}_3\overset{\|}{\text{C}}-\text{OCH}_3}$$

The carbonyl oxygen is protonated because it is the atom with the greatest electron density, as shown by the contributing resonance structures.

$$\text{CH}_3-\text{C}-\ddot{\text{O}}\text{CH}_3 \quad \longleftrightarrow \quad \text{CH}_3-\text{C}=\overset{+}{\ddot{\text{O}}}\text{CH}_3$$

contributing resonance structures of an ester

In the second step of the mechanism, the nucleophile (H_2O) attacks the protonated carbonyl group. The resulting protonated tetrahedral intermediate (tetrahedral intermediate I) is in equilibrium with its nonprotonated form. Either the OH or the OR group of the nonprotonated intermediate can be protonated. Because the OH and OR groups have approximately the same basicity, both tetrahedral intermediate I (OH is protonated) and tetrahedral intermediate II (OR is protonated) are formed. [From Section 1.19 we know that the relative amounts of the protonated intermediates (I and II) compared with the nonprotonated intermediate depend on the pH of the solution and the pK_a values of the protonated intermediates.] When tetrahedral intermediate I collapses, it eliminates H_2O in preference to CH_3O^- because H_2O is a weaker base, thereby reforming the ester. When tetrahedral intermediate II collapses, it eliminates CH_3OH rather than HO^- because CH_3OH is a weaker base, thereby forming the carboxylic acid.

mechanism for acid-catalyzed ester hydrolysis

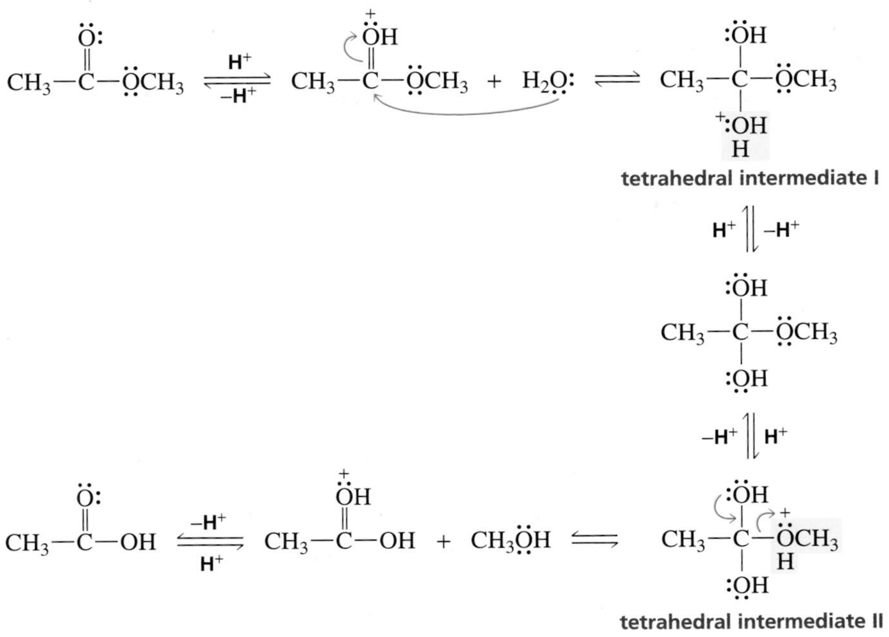

tetrahedral intermediate I

tetrahedral intermediate II

Because H_2O and CH_3OH have approximately the same basicity, it will be equally easy for tetrahedral intermediate I to collapse to reform the ester as it will for tetrahedral intermediate II to collapse to form the carboxylic acid. Consequently, when the reaction has reached equilibrium, both ester and carboxylic acid will be obtained.

both ester and carboxylic acid will be obtained when the reaction has reached equilibrium

Excess water will force the equilibrium to the right.

The mechanism for the acid-catalyzed reaction of a carboxylic acid and an alcohol to form an ester and water is the exact reverse of the mechanism for the acid-catalyzed hydrolysis of an ester to form a carboxylic acid and an alcohol. If the ester is the desired product, the reaction should be carried out under conditions that will drive the equilibrium to the left—using excess alcohol or removing water as it is formed.

PROBLEM 17

Referring to the mechanism for the acid-catalyzed hydrolysis of methyl acetate, propose a mechanism for the acid-catalyzed reaction of acetic acid and methanol to form methyl acetate.

How does H^+ increase the rate of ester hydrolysis? For a catalyst to increase the rate of a reaction, it must increase the rate of the slow step of the reaction. Changing the rate of a fast step will not affect the rate of the overall reaction. Hydrolysis of an ester involves two relatively slow steps: formation of a tetrahedral intermediate and collapse of a tetrahedral intermediate. H^+ increases the rates of both slow steps. (Proton transfer to or from an electronegative atom such as oxygen or nitrogen is a fast step.)

H^+ increases the rate of formation of a tetrahedral intermediate by protonating the carbonyl oxygen. Protonated carbonyl groups are more susceptible than nonprotonated carbonyl groups to nucleophilic attack, because a positively charged oxygen is more electron withdrawing than an uncharged oxygen. This serves to increase the partial positive charge on the carbonyl carbon, which increases its attractiveness to nucleophiles.

more susceptible to attack by a nucleophile   less susceptible to attack by a nucleophile

protonation of the carbonyl oxygen increases the susceptibility of the carbonyl carbon to nucleophilic attack

H^+ increases the rate of collapse of a tetrahedral intermediate by decreasing the basicity of the leaving group, thereby making it easier to eliminate. In the acid-catalyzed hydrolysis of an ester, the leaving group is ROH, a weaker base than the leaving group (RO^-) in the non-acid-catalyzed reaction.

The reaction of an ester with an alcohol is also catalyzed by H^+. The mechanism for transesterification is identical to the mechanism for hydrolysis. The only difference is the nucleophile (ROH versus H_2O). The leaving groups in the tetrahedral intermediate formed in transesterification have approximately the same basicity. Consequently, for the reaction to produce more of the desired product, an excess of the reactant alcohol will have to be used.

methyl benzoate   ethyl alcohol   ethyl benzoate   methyl alcohol
                     excess

PROBLEM 18

Propose a mechanism for the acid-catalyzed transesterification reaction of methyl acetate with ethanol.

Instead of increasing the rate of hydrolysis of an ester by carrying out the reaction in an acidic solution, we can increase the rate by carrying out the reaction in a basic solution. Hydroxide ion increases the rates of both slow steps of the reaction.

Hydroxide ion increases the rate of formation of the tetrahedral intermediate because hydroxide ion is a better nucleophile than H_2O and so it more readily attacks the carbonyl carbon. Hydroxide ion increases the rate of collapse of the tetrahedral intermediate because in a basic solution a smaller fraction of the negatively charged tetrahedral intermediate becomes protonated. The driving force of the negatively charged oxygen is necessary to expel the very basic leaving group (RO^-).

15.10
HYDROXIDE-ION-PROMOTED ESTER HYDROLYSIS

mechanism for hydroxide-ion-promoted hydrolysis of an ester

The final product of this reaction is a carboxylate ion rather than a carboxylic acid because the carboxylic acid is in a basic solution. Because carboxylate ions are negatively charged, they are not easily attacked by nucleophiles. Therefore, the hydroxide-ion-promoted hydrolysis of an ester is not a reversible reaction.

This reaction is called a hydroxide-ion-promoted reaction rather than a base-catalyzed reaction because hydroxide increases the rate of the first step of the reaction by being a better nucleophile than water—not by being a stronger base than water, and because hydroxide ion is consumed in the overall reaction, it is actually a reagent rather than a catalyst. Therefore, it is more accurate to call the reaction a hydroxide-ion-*promoted* reaction than a hydroxide-ion-*catalyzed* reaction; we will use the former term from now on.

Only hydrolysis reactions can be promoted by hydroxide ion. Reactions of carboxylic acid derivatives with alcohols or with amines cannot be promoted by hydroxide ion because one function of hydroxide ion is to provide a strong nucleophile for the first step of the reaction. If the nucleophile is water, nucleophilic attack by hydroxide ion instead of water gives the same product. If, however, the nucleophile is an alcohol or an amine, nucleophilic attack by hydroxide ion would form a product different from the product that would be formed from nucleophilic attack by an alcohol or amine.

Reactions in which the nucleophile is an alcohol can be catalyzed by the conjugate base of the alcohol.

PROBLEM 19

a. What species other than H^+ can be used to increase the rate of a transesterification reaction that converts methyl acetate to propyl acetate?

b. The rate of aminolysis of an ester cannot be increased by H^+, HO^-, or RO^-. Explain.

We have said that nucleophilic acyl substitution reactions take place by a mechanism that involves the formation of a tetrahedral intermediate and its subsequent collapse. The tetrahedral intermediate, however, is too unstable to be isolated. How then do we know that it is formed? How do we know that the reaction doesn't take place by a one-step direct displacement mechanism (similar to the mechanism of an S_N2 reaction) that does not form a tetrahedral intermediate?

$$\underset{H}{\overset{O}{\underset{|}{\overset{||}{HO}}}}\underset{R}{\overset{\delta+}{\underset{|}{---}}}\overset{\delta+}{\underset{H}{C}}---OR}$$

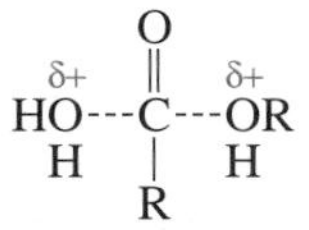

**transition state for a hypothetical one-step
direct displacement mechanism**

**Myron L. Bender
(1924–1988)** *was born in St. Louis. He was a professor of chemistry at the Illinois Institute of Technology and at Northwestern University.*

To answer this question, Myron Bender investigated the hydroxide-ion-promoted hydrolysis of ethyl benzoate, with the carbonyl oxygen of ethyl benzoate labeled with ^{18}O. When he isolated unreacted ethyl benzoate, he found that some of the ester was no longer labeled. This proved that a tetrahedral intermediate had been formed during the course of the reaction. If you take a few minutes to examine the mechanism, you will see that if the reaction had taken place by a direct displacement mechanism, all of the isolated ester would have been labeled.

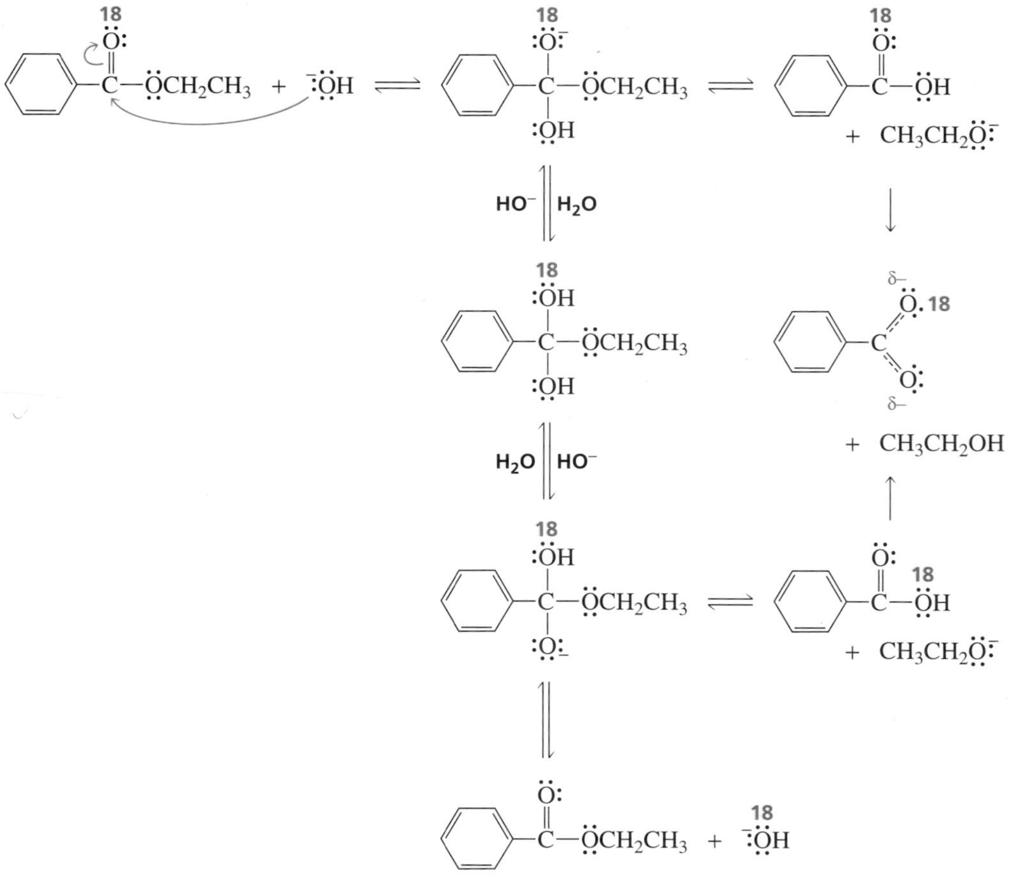

unlabeled ester

PROBLEM 20◆

D. N. Kursanov, a Russian chemist, proved that the bond broken in the hydroxide-ion-promoted hydrolysis of an ester is the acyl C—O bond rather than the alkyl C—O bond by studying the reaction of the following ester with HO^-/H_2O.

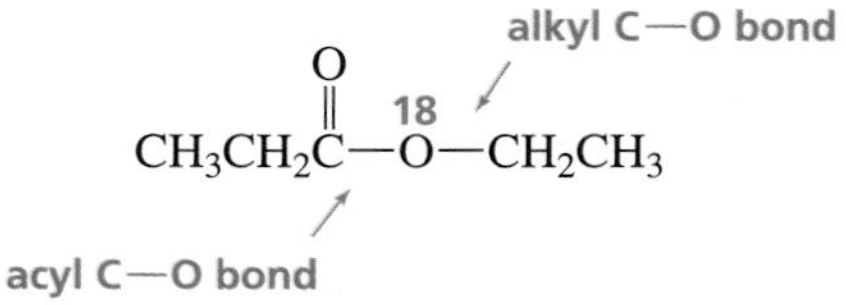

alkyl C—O bond

$$CH_3CH_2\overset{O}{\overset{\|}{C}}\!-\!\!\overset{18}{O}\!-\!CH_2CH_3$$

acyl C—O bond

a. Which of the products contained the ^{18}O label?

b. What product would have contained the ^{18}O label if the alkyl C—O bond had broken?

PROBLEM 21 / SOLVED

Early chemists could envision several possible mechanisms for hydroxide-ion-promoted ester hydrolysis:

1. a nucleophilic acyl substitution mechanism

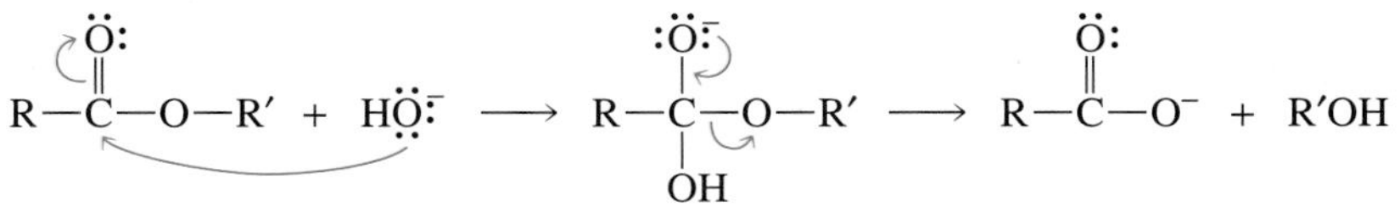

$$R\!-\!\overset{\ddot{O}:}{\overset{\|}{C}}\!-\!O\!-\!R' + H\ddot{O}^- \longrightarrow R\!-\!\overset{:\ddot{O}^-}{\underset{OH}{\overset{|}{C}}}\!-\!O\!-\!R' \longrightarrow R\!-\!\overset{\ddot{O}:}{\overset{\|}{C}}\!-\!O^- + R'OH$$

2. an S_N2 mechanism

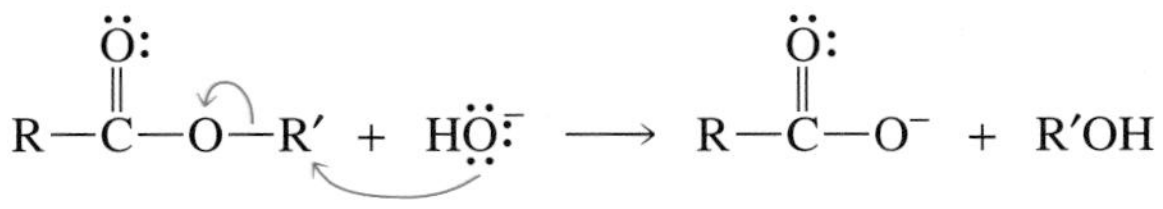

$$R\!-\!\overset{\ddot{O}:}{\overset{\|}{C}}\!-\!O\!-\!R' + H\ddot{O}^- \longrightarrow R\!-\!\overset{\ddot{O}:}{\overset{\|}{C}}\!-\!O^- + R'OH$$

3. an S_N1 mechanism

$$R\!-\!\overset{\ddot{O}:}{\overset{\|}{C}}\!-\!O\!-\!R' \longrightarrow R\!-\!\overset{\ddot{O}:}{\overset{\|}{C}}\!-\!O^- + R'^+ \xrightarrow{H\ddot{O}^-} R'OH$$

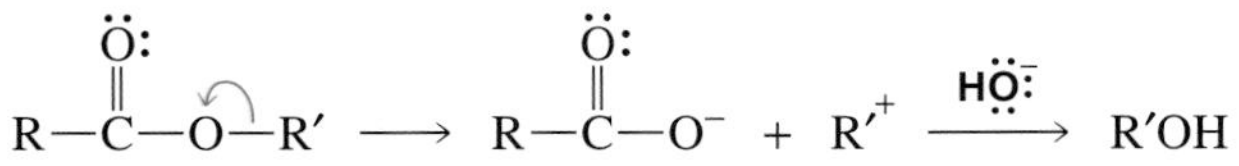

Devise an experiment that would distinguish among these three mechanisms.

SOLUTION   Start with a single stereoisomer of an alcohol that has its OH group bonded to a chirality center, and determine the specific rotation of the alcohol. Convert the alcohol into an ester using a method that does not break any bonds to the chirality center. Hydrolyze the ester. Isolate the alcohol obtained from hydrolysis and determine its specific rotation.

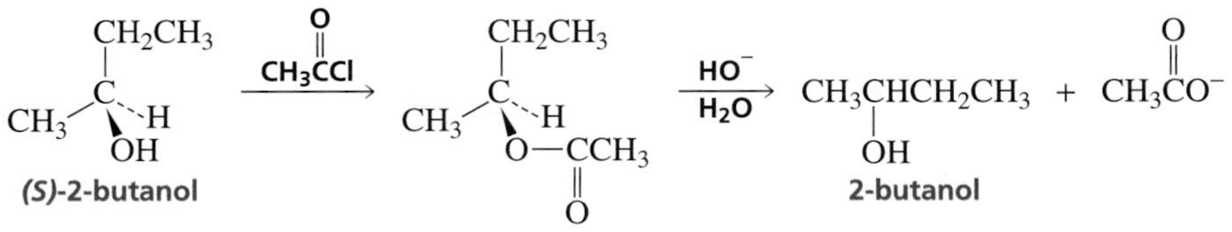

(S)-2-butanol $\xrightarrow{CH_3CCl}$ ester $\xrightarrow[H_2O]{HO^-}$ $CH_3CHCH_2CH_3$ + CH_3CO^- 2-butanol

If the reaction is a nucleophilic acyl substitution reaction, the product alcohol will have the same specific rotation as the reactant alcohol, because no bonds to the chirality center are broken.

If the reaction is an S_N2 reaction, the product alcohol and the reactant alcohol will have opposite specific rotations, because the mechanism requires backside attack on the chirality center.

If the reaction is an S_N1 reaction, the product alcohol will have a small specific rotation, because the mechanism requires carbocation formation, which leads to partial racemization of the alcohol.

15.11 REACTIONS OF CARBOXYLIC ACIDS

The leaving group of a carboxylic acid (HO^-) has approximately the same basicity as the leaving group of an ester (RO^-). Therefore, carboxylic acids have approximately the same reactivity as esters. Like esters, carboxylic acids do not react with halide ions or with carboxylate ions.

A carboxylic acid reacts with an alcohol to form an ester. It is a very slow reaction, so it is always carried out in the presence of an acid catalyst (Section 15.9). Since the tetrahedral intermediate formed in this reaction has two groups of approximately the same basicity (HO^- and RO^-), the reaction must be carried out with excess alcohol to drive it toward products. Because Emil Fischer (Section 4.4) was the one who discovered that an ester could be prepared by treating a carboxylic acid with excess alcohol in the presence of an acid catalyst, the reaction is known as the **Fischer esterification reaction.**

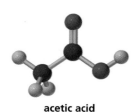
acetic acid

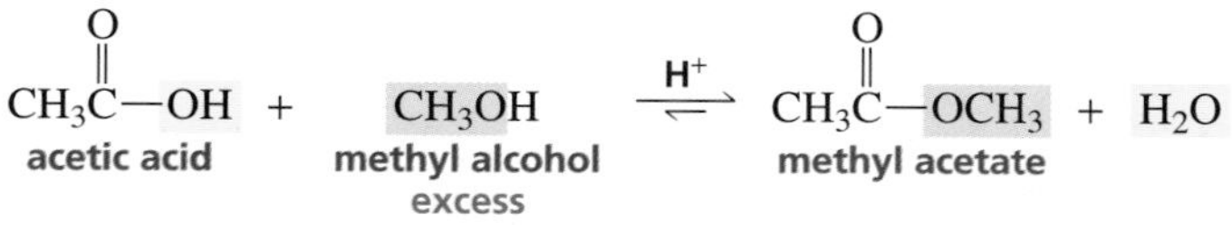

$$CH_3\overset{O}{\overset{\|}{C}}-OH + CH_3OH \underset{}{\overset{H^+}{\rightleftharpoons}} CH_3\overset{O}{\overset{\|}{C}}-OCH_3 + H_2O$$
acetic acid   methyl alcohol   methyl acetate
             excess

A carboxylic acid is an acid (it donates a proton), and an amine is a base (it accepts a proton). Because a carboxylic acid has a lower pK_a than a protonated amine, the carboxylic acid immediately donates a proton to the amine when the two compounds are mixed. The protonated amine is not a nucleophile, so it cannot attack the carbonyl carbon. Therefore the ammonium carboxylate salt is the final product of the reaction. When, however, the salt is heated at a higher temperature, it loses water and forms the amide.

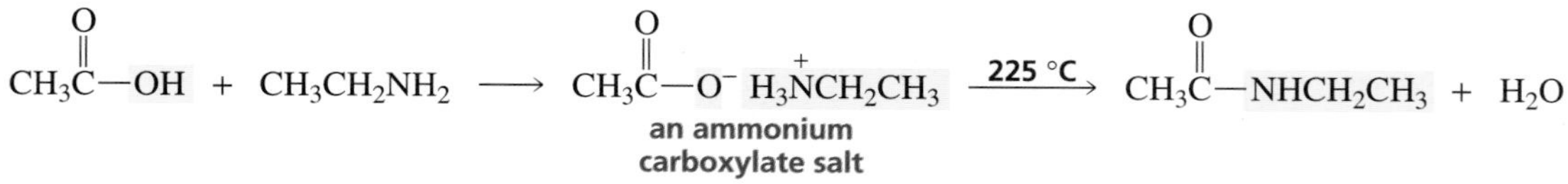

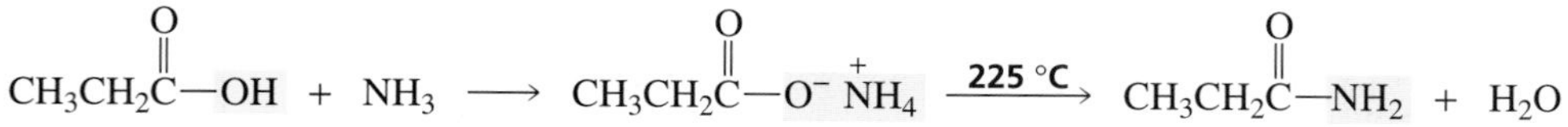

$$CH_3\overset{O}{\overset{\|}{C}}-OH + CH_3CH_2NH_2 \longrightarrow CH_3\overset{O}{\overset{\|}{C}}-O^- \overset{+}{H_3N}CH_2CH_3 \xrightarrow{225\ °C} CH_3\overset{O}{\overset{\|}{C}}-NHCH_2CH_3 + H_2O$$
an ammonium carboxylate salt

$$CH_3CH_2\overset{O}{\overset{\|}{C}}-OH + NH_3 \longrightarrow CH_3CH_2\overset{O}{\overset{\|}{C}}-O^- \overset{+}{NH_4} \xrightarrow{225\ °C} CH_3CH_2\overset{O}{\overset{\|}{C}}-NH_2 + H_2O$$

PROBLEM-SOLVING STRATEGY

Propose a mechanism for the following reaction.

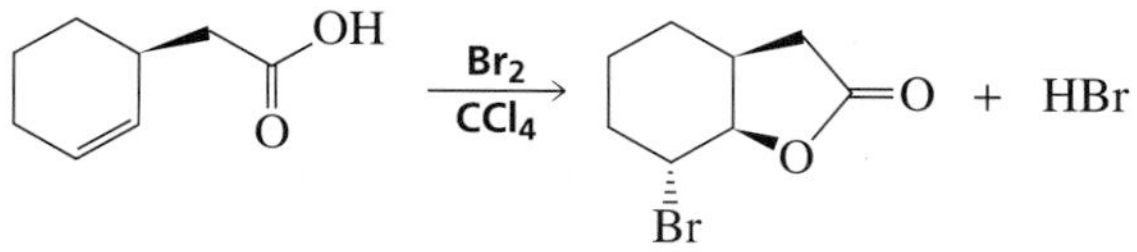

When asked to propose a mechanism, look carefully at the reagents to determine the first step of the mechanism. One of the given reagents has two functional groups, a carboxylic acid and an alkene. The other reagent, Br_2, does not react with carboxylic acids but does react with alkenes, forming a bromonium ion. One side of the alkene is sterically hindered by the carboxylic acid group. Br_2, therefore, will add to the other side of the alkene. We know that the second step of an alkene addition reaction is attack by a nucleophile. Of the two nucleophiles present, the carbonyl oxygen is a better nucleophile and is located closer to the bromonium ion; it attacks the carbon bonded to the bromine on its back side, resulting in a compound with the observed configuration. Loss of a proton gives the final product of the reaction.

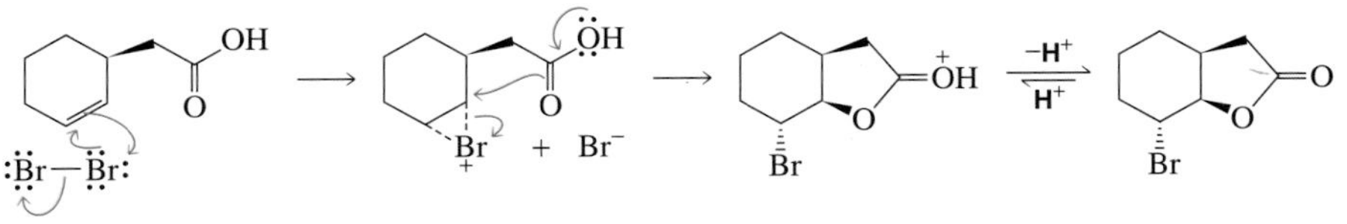

Now continue on to do Problem 22.

PROBLEM 22

Propose a mechanism for the following reaction.

$$CH_2{=}CHCH_2CH_2CH{=}CCH_3 \ \ + \ \ CH_3COH \ \ \xrightarrow{\ H_2SO_4\ }$$

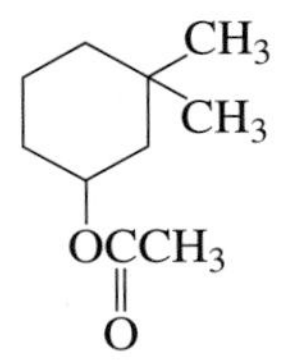

with CH_3 on the central carbon; product is a cyclohexane ring with two CH_3 groups and an $OCCH_3$ ($=O$) substituent.

Amides do not react with halide ions, carboxylate ions, alcohols, or water, because in each case the incoming nucleophile is a weaker base than the leaving group of the amide (Table 15.1).

$$CH_3C{-}NHCH_2CH_2CH_3 \ \ + \ \ Cl^- \ \longrightarrow \ \text{no reaction}$$
N-propylacetamide

$$CH_3CH_2C{-}N(CH_3)_2 \ \ + \ \ CH_3CO^- \ \longrightarrow \ \text{no reaction}$$
N,N-dimethylpropionamide

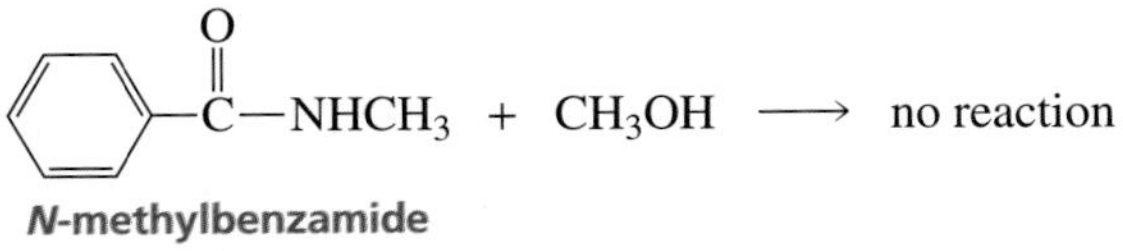

C₆H₅—C—NHCH₃ + CH₃OH ⟶ no reaction
N-methylbenzamide

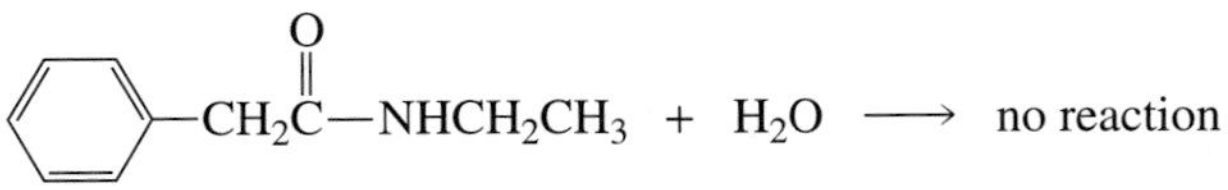

C₆H₅—CH₂C—NHCH₂CH₃ + H₂O ⟶ no reaction
N-ethylphenylacetamide

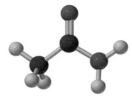

acetamide

Amides, however, can react with water if the reaction mixture is heated in the presence of an acid or hydroxide ion. The reason for this will be explained in Section 15.13.

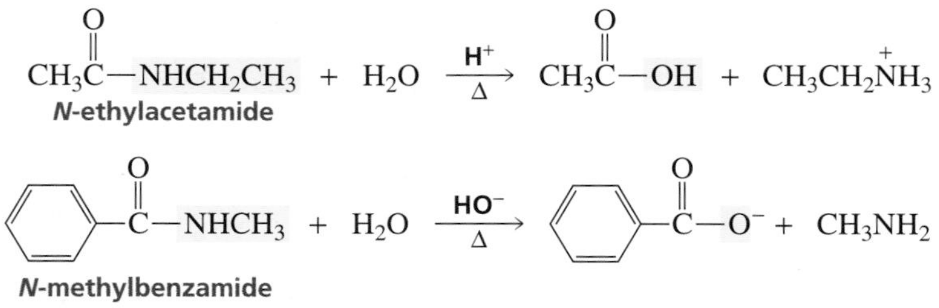

$$CH_3\overset{\displaystyle O}{\overset{\|}{C}}-NHCH_2CH_3 + H_2O \xrightarrow[\Delta]{H^+} CH_3\overset{\displaystyle O}{\overset{\|}{C}}-OH + CH_3CH_2\overset{+}{N}H_3$$

N-ethylacetamide

$$\underset{\text{N-methylbenzamide}}{\text{⬡}}-\overset{\displaystyle O}{\overset{\|}{C}}-NHCH_3 + H_2O \xrightarrow[\Delta]{HO^-} \text{⬡}-\overset{\displaystyle O}{\overset{\|}{C}}-O^- + CH_3NH_2$$

A primary amide can be dehydrated to a nitrile. Dehydration reagents commonly employed for this purpose are P_2O_5, $POCl_3$, or $SOCl_2$.

$$CH_3CH_2\overset{\displaystyle O}{\overset{\|}{C}}-NH_2 \xrightarrow[85\,°C]{P_2O_5} CH_3CH_2C\equiv N + H_2O$$

15.13
HYDROLYSIS OF AMIDES: ACID-CATALYZED HYDROLYSIS AND HYDROXIDE-ION-PROMOTED HYDROLYSIS

An amide cannot be hydrolyzed unless the reaction mixture is heated in the presence of an acid or hydroxide ion.

When the reaction is carried out under acidic conditions, the acid protonates the carbonyl oxygen, increasing the susceptibility of the carbonyl carbon to nucleophilic attack. Nucleophilic attack by water on the carbonyl carbon leads to tetrahedral intermediate I, which is in equilibrium with its nonprotonated form. Reprotonation can occur either on oxygen to reform tetrahedral intermediate I or on nitrogen to form tetrahedral intermediate II. Protonation on nitrogen is favored, because the NH_2 group is a stronger base than the OH group. Tetrahedral intermediate II has two leaving groups, HO^- and NH_3. Because NH_3 is a weaker base, it is eliminated, forming the carboxylic acid as the final product. (Because the reaction is carried out in an acidic solution, NH_3 will be protonated after it is eliminated from the tetrahedral intermediate.)

mechanism for acid-catalyzed hydrolysis of an amide

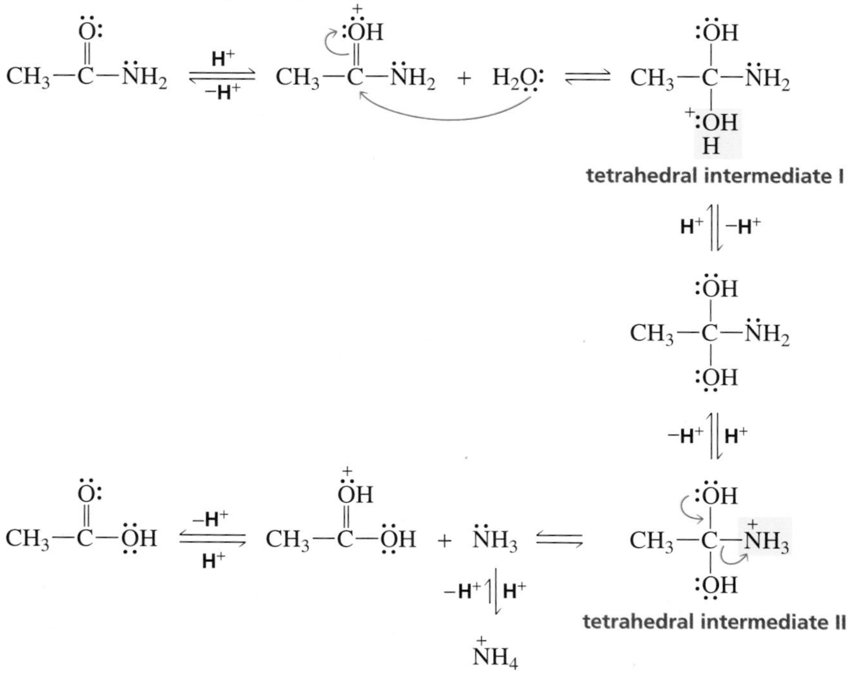

tetrahedral intermediate I

tetrahedral intermediate II

Let's take a minute to see why hydrolysis of an amide does not occur without a catalyst. In the non-acid-catalyzed reaction, the amide is not protonated. Therefore, water, a very poor nucleophile, must attack a neutral amide that is much less susceptible to nucleophilic attack than is a protonated amide. Consequently, the tetrahedral intermediate is formed very slowly. In addition, the NH_2 group of the tetrahedral intermediate is not protonated in the non-acid-catalyzed reaction. The two leaving groups of the tetrahedral intermediate are HO^- and $^-NH_2$. Because HO^- is the weaker base, it is more easily eliminated, which reforms the amide.

tetrahedral intermediate   tetrahedral intermediate
in a noncatalyzed reaction   in an acid-catalyzed reaction

The addition of acid increases the rate of formation of the tetrahedral intermediate by protonating the amide. The acid also changes the relative leaving abilities of the two groups. In an acidic solution, the two leaving groups are HO^- and NH_3. Because NH_3 is the weaker base, it is eliminated and hydrolysis results.

In the hydroxide-ion-promoted hydrolysis of an amide, HO^- rather than water is the nucleophile. Because HO^- is a better nucleophile than water, it is better at forming the tetrahedral intermediate. The two leaving groups of the tetrahedral intermediate are HO^- and $^-NH_2$. Because HO^- is the weaker base, it is more easily eliminated, but occasionally an $^-NH_2$ is ejected. When this happens, the carboxylic acid that is formed immediately loses a proton. Since this step is irreversible (the carboxylate ion cannot be attacked by nucleophiles), it disturbs the equilibrium and drives the reaction toward products (Section 9.3). Because hydroxide ion is consumed in the reaction, it is a reagent, not a catalyst.

mechanism for hydroxide-ion-promoted hydrolysis of an amide

In strongly basic solutions, the hydrolysis reaction is second-order in hydroxide ion. In other words, two equivalents of hydroxide ion participate in the reaction. The second equivalent of hydroxide ion removes a proton from the OH group of the initially formed tetrahedral intermediate. When this tetrahedral intermediate collapses, the possible leaving groups are O^{2-} or $^-NH_2$. Since $^-NH_2$ is a weaker base than O^{2-}, $^-NH_2$ is eliminated and the hydrolysis reaction occurs. Elimination of the strongly basic $^-NH_2$ group is helped by the driving force of the two negatively charged oxygens in the tetrahedral intermediate. Notice that the second equivalent of hydroxide ion is a catalyst because it is regenerated.

PENICILLIN AND DRUG RESISTANCE

Penicillin contains a strained β-lactam ring. The strain in the four-membered ring causes the amide to be much more susceptible to nucleophilic acyl substitution reactions than are nonstrained amides. It is thought that the antibiotic activity of penicillin results from its ability to acylate an OH group of an enzyme that is involved in the synthesis of bacterial cell walls. This inactivates the enzyme, and actively growing bacteria die because they are unable to produce functional cell walls. Penicillin has no effect on mammalian cells because mammalian cells are not enclosed by cell walls.

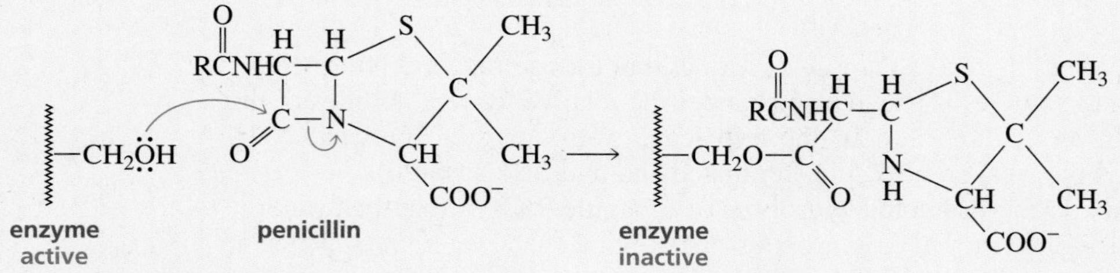

Bacteria that are resistant to penicillin secrete a β-lactamase, an enzyme that catalyzes the hydrolysis of the β-lactam ring of penicillin. The ring-opened product has no antibacterial activity. To minimize hydrolysis during storage, penicillins are refrigerated.

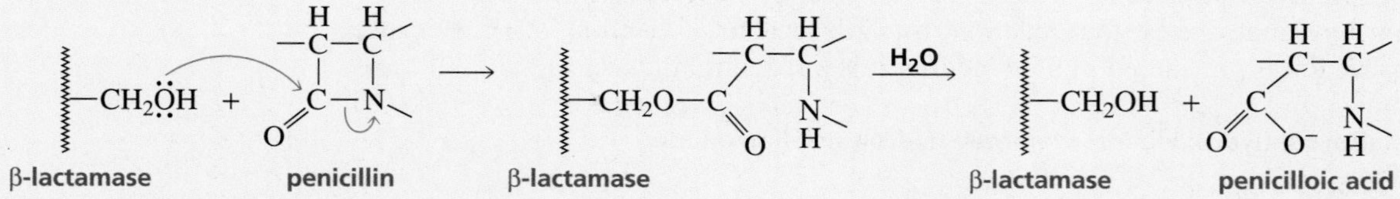

PENICILLINS IN CLINICAL USE

More than ten different penicillins are currently in clinical use. Some of these penicillins are shown here. In addition to their structural differences, the penicillins differ in the organisms against which they are most effective. They also differ in their resistance to β-lactamase. For example, ampicillin, a synthetic penicillin, is clinically effective against bacteria that are resistant to penicillin G. Almost 19% of the population is allergic to penicillin G, a naturally occurring penicillin.

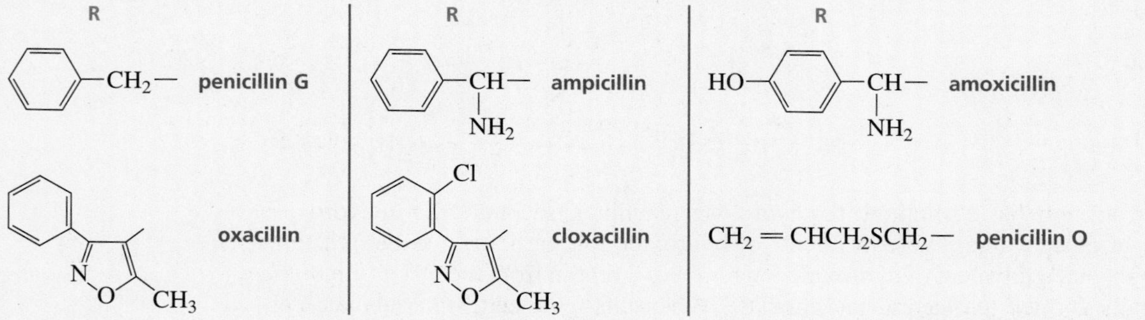

PROBLEM 23◆

List the following amides in order of decreasing reactivity toward hydroxide-ion-promoted hydrolysis.

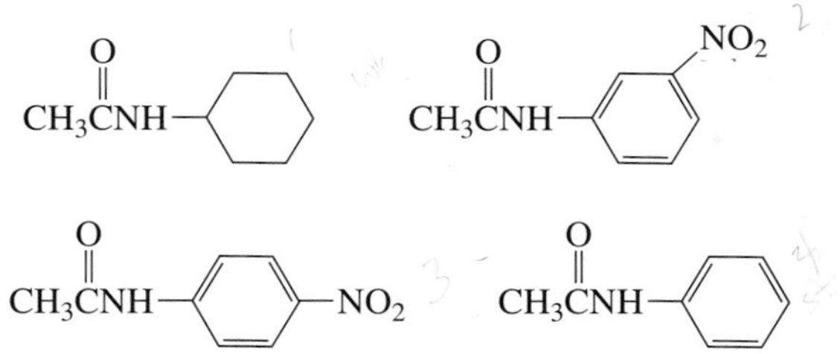

A reaction used for the synthesis of primary amines involves the hydrolysis of an **imide**—a compound with two acyl groups bonded to a nitrogen. The **Gabriel synthesis** converts alkyl halides into primary amines.

$$RCH_2Br \xrightarrow{\text{Gabriel synthesis}} RCH_2NH_2$$

15.14
THE GABRIEL SYNTHESIS OF PRIMARY AMINES

In the first step of the reaction, a base removes a proton from the nitrogen of phthalimide. The resulting nucleophile reacts with an alkyl halide. Because this is an S_N2 reaction, it works best with primary alkyl halides. Hydrolysis of the *N*-substituted phthalimide can be catalyzed by acid or promoted by hydroxide ion. Acid-catalyzed hydrolysis of the product gives the primary ammonium ion and phthalic acid. Neutralization of the primary ammonium ion with base gives the primary amine. The alkyl group of the primary amine is identical to the alkyl group of the alkyl halide.

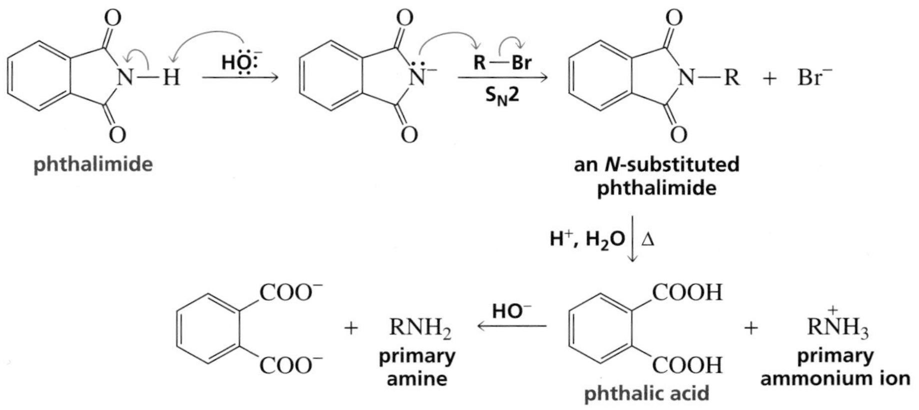

Because there is only one hydrogen bonded to the nitrogen of phthalimide, only one alkyl group can be placed on the nitrogen. This means that the Gabriel synthesis can be used only for the preparation of primary amines.

PROBLEM 24◆

What alkyl bromide would you use in the Gabriel synthesis in order to prepare each of the following amines?

a. pentylamine

b. benzylamine

c. isohexylamine

PROBLEM 25

Primary amines can also be prepared by the reaction of an alkyl halide with azide ion followed by catalytic hydrogenation. What advantage do this method and the Gabriel synthesis have over synthesis of a primary amine using an alkyl halide and ammonia?

$$CH_3CH_2CH_2Br \xrightarrow{{}^-N_3} CH_3CH_2CH_2N{=}\overset{+}{N}{=}\overset{-}{N} \xrightarrow[Pt]{H_2} CH_3CH_2CH_2NH_2 \; + \; N_2$$

15.15 HYDROLYSIS OF NITRILES

Nitriles are even harder to hydrolyze than amides. They are slowly hydrolyzed to carboxylic acids when heated with water and either H^+ or HO^-.

$$CH_3CH_2C{\equiv}N \; + \; H_2O \xrightarrow[\Delta]{H^+} CH_3CH_2\overset{\overset{\displaystyle O}{\|}}{C}OH \; + \; \overset{+}{N}H_4$$

$$CH_3CH_2C{\equiv}N \; + \; H_2O \xrightarrow[\Delta]{HO^-} CH_3CH_2\overset{\overset{\displaystyle O}{\|}}{C}O^- \; + \; NH_3$$

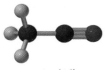

acetonitrile

In the first step of the acid-catalyzed hydrolysis of a nitrile, the acid protonates the nitrogen of the cyano group, making it easier for water to attack the carbon of the cyano group in the next step. Attack on the cyano group by water is analogous to attack on a carbonyl group by water. Because nitrogen is a stronger base than oxygen, oxygen loses a proton and nitrogen gains a proton, resulting in a product whose other resonance contributor is a protonated amide. Since an amide is easier to hydrolyze than a nitrile, the amide is immediately hydrolyzed to a carboxylic acid following the acid-catalyzed mechanism shown in Section 15.13.

mechanism for acid-catalyzed hydrolysis of a nitrile

$$R{-}C{\equiv}N: \underset{-H^+}{\overset{H^+}{\rightleftharpoons}} R{-}C{\equiv}\overset{+}{N}H \; + \; H_2\overset{..}{\underset{..}{O}}: \;\rightleftharpoons\; R{-}\underset{\underset{H}{\overset{+}{\underset{..}{O}H}}}{\overset{..}{C}}{=}\overset{..}{N}H \underset{H^+}{\overset{-H^+}{\rightleftharpoons}} R{-}\underset{\overset{..}{\underset{..}{O}H}}{\overset{..}{C}}{=}\overset{..}{N}H$$

$$R{-}\underset{\overset{\displaystyle \|}{\underset{..}{O}:}}{C}{-}OH \xleftarrow[\text{(several steps)}]{H_2O} R{-}\underset{\overset{\displaystyle \|}{+\overset{..}{O}H}}{C}{-}\overset{..}{N}H_2 \;\longleftrightarrow\; R{-}\underset{\overset{..}{\underset{..}{O}H}}{C}{=}\overset{+}{N}H_2$$

a carboxylic acid   a protonated amide

The first step in the hydroxide-ion-promoted hydrolysis of a nitrile is attack of hydroxide ion on the carbon of the cyano group. Nitrogen gains a proton from water, oxygen loses a proton, and the product rearranges to an amide. The amide is further hydrolyzed to a carboxylate ion by means of the hydroxide-ion-promoted mechanism shown in Section 15.13.

mechanism for hydroxide-ion-promoted hydrolysis of a nitrile

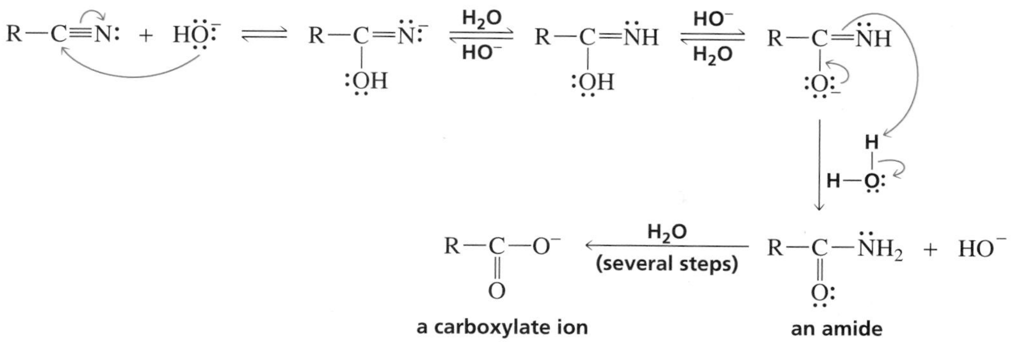

$$R-C{\equiv}N: \ + \ H\ddot{O}{:}^- \ \rightleftharpoons \ R-\underset{:OH}{C}{=}\ddot{N}{:}^- \ \underset{HO^-}{\overset{H_2O}{\rightleftharpoons}} \ R-\underset{:OH}{C}{=}\ddot{N}H \ \underset{H_2O}{\overset{HO^-}{\rightleftharpoons}} \ R-\underset{:\ddot{O}{:}}{C}{=}\ddot{N}H$$

$$\underset{H-\ddot{O}{:}}{\overset{H}{\mid}}$$

$$\underset{\underset{O}{\overset{\parallel}{C}}-O^-}{R-C}-O^- \ \underset{(several\ steps)}{\overset{H_2O}{\longleftarrow}} \ R-\underset{\ddot{O}{:}}{\overset{\cdots}{C}}-\ddot{N}H_2 \ + \ HO^-$$

a carboxylate ion   an amide

Because nitriles can be prepared from the reaction of an alkyl halide with cyanide ion (Section 9.3), we now know how to convert an alkyl halide into a carboxylic acid. Notice that the carboxylic acid has one more carbon than the alkyl halide.

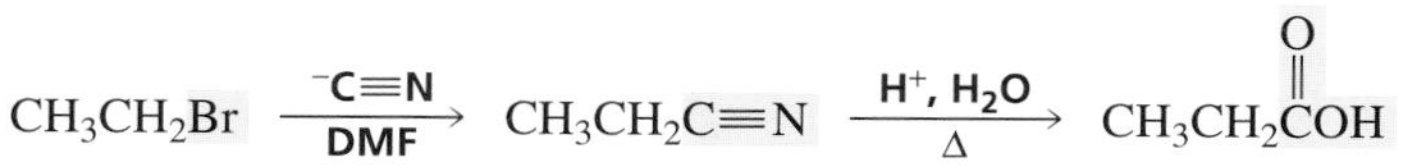

$$CH_3CH_2Br \ \xrightarrow[DMF]{^-C{\equiv}N} \ CH_3CH_2C{\equiv}N \ \xrightarrow[\Delta]{H^+,\ H_2O} \ CH_3CH_2\overset{\overset{O}{\parallel}}{C}OH$$

PROBLEM 26 ◆

What alkyl halides form the following carboxylic acids after reaction with sodium cyanide and heating of the product in an acidic aqueous solution?

a. butyric acid   **c.** cyclohexane carboxylic acid

b. isovaleric acid   **d.** succinic acid (Table 15.2)

Fats and **oils** are triesters of glycerol. Glycerol contains three alcohol groups and therefore can form three ester groups. When the ester groups are hydrolyzed in a basic solution, glycerol and the salts of carboxylic acids are formed. The carboxylic acids that are bonded to glycerol in fats and oils have long, unbranched R groups (Section 23.3). Such carboxylic acids are called *fatty acids*.

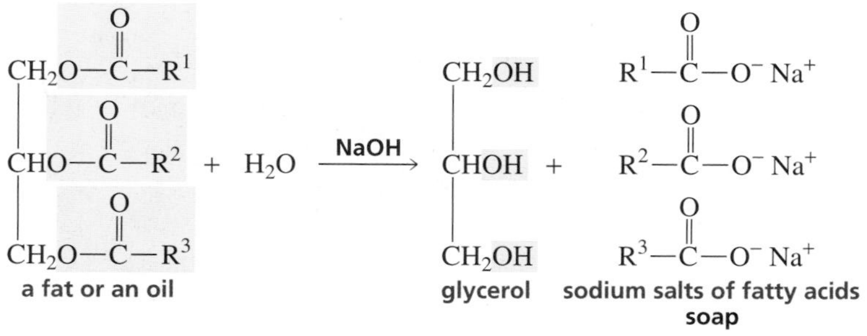

15.16 SOAPS, DETERGENTS, AND MICELLES

$$\begin{matrix} CH_2O-\overset{\overset{O}{\parallel}}{C}-R^1 \\ \\ CHO-\overset{\overset{O}{\parallel}}{C}-R^2 \ + \ H_2O \\ \\ CH_2O-\overset{\overset{O}{\parallel}}{C}-R^3 \end{matrix} \quad \xrightarrow{NaOH} \quad \begin{matrix} CH_2OH \\ \\ CHOH \ + \\ \\ CH_2OH \end{matrix} \quad \begin{matrix} R^1-\overset{\overset{O}{\parallel}}{C}-O^-\ Na^+ \\ \\ R^2-\overset{\overset{O}{\parallel}}{C}-O^-\ Na^+ \\ \\ R^3-\overset{\overset{O}{\parallel}}{C}-O^-\ Na^+ \end{matrix}$$

a fat or an oil   glycerol   sodium salts of fatty acids
                                                 soap

Soaps are sodium or potassium salts of fatty acids. Three of the most common soaps are shown below. Hydrolysis of an ester in a basic solution is called **saponification,** because soaps are obtained when fats or oils are hydrolyzed under basic conditions.

$$CH_3(CH_2)_{16}\overset{\displaystyle O}{\overset{\displaystyle \|}{C}}O^-\ Na^+$$
sodium stearate

$$CH_3(CH_2)_7CH\!=\!CH(CH_2)_7\overset{\displaystyle O}{\overset{\displaystyle \|}{C}}O^-\ Na^+$$
sodium oleate

$$CH_3(CH_2)_4CH\!=\!CHCH_2CH\!=\!CH(CH_2)_7\overset{\displaystyle O}{\overset{\displaystyle \|}{C}}O^-\ Na^+$$
sodium linoleate

PROBLEM 27 ◆

An oil obtained from coconuts is unusual in that all three fatty acid components are identical. The molecular formula of the oil is $C_{45}H_{86}O_6$. What is the molecular formula of the acid obtained when the oil is saponified?

Long-chain carboxylate ions do not exist as individual ions in aqueous solution; instead, they arrange themselves in spherical clusters called **micelles.** Each micelle contains 50 to 100 long-chain carboxylate ions (Figure 15.3). The polar carboxylate end of each ion is on the outside of the micelle because of its attraction for water. The nonpolar end is in the interior of the micelle to minimize its contact with water. The attractive forces of hydrocarbon chains for each other in water are called **hydrophobic interactions** (Section 20.14). The components of water (H_2O, HO^-, H_3O^+) are collectively known as *lyate species.* Hydrophobic interactions result from "fearing lyate species."

Because the surface of the micelle is negatively charged, the individual micelles repel each other instead of clustering to form larger aggregates. The cleaning ability of soap results from the fact that nonpolar oil molecules, which carry dirt, dissolve in the nonpolar interior of the micelle and are carried away with the soap during the rinse.

Soap lowers the surface tension of water, which is why soap solutions feel slippery. Lowering the surface tension enables soap to penetrate the fibers of a fabric,

Figure 15.3 ▶
In aqueous solution, soap forms micelles with the carboxylate groups on the surface and the nonpolar tails in the interior.

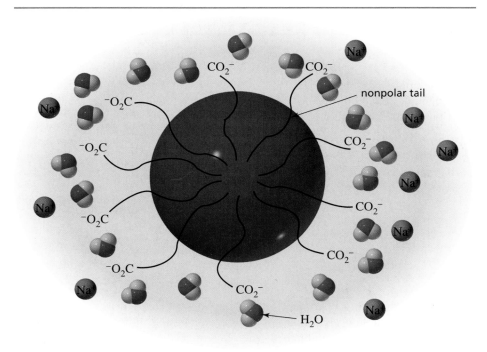

thus enhancing its cleaning ability. Compounds that lower the surface tension of water all have a polar head group and a long-chain nonpolar tail. They are called *surfactants*.

MAKING SOAP

For thousands of years, soap was prepared by heating animal fat with wood ashes. Wood ashes contain potassium carbonate, which makes the solution basic. The modern commercial method of making soap involves boiling fats or oils in aqueous sodium hydroxide and adding sodium chloride to precipitate the soap. The soap is then dried and pressed into bars. Perfume can be added for scented soaps; dyes can be added for colored soaps; sand can be added for scouring soaps; and air can be blown into the soap to make it float in water.

Making soap in the 19th century.

As water flows over and around rocks on its way to the water tap, it leaches out calcium and magnesium ions. The concentration of calcium and magnesium ions in the water is described by its "hardness." Hard water contains high concentrations of these ions; soft water contains few if any calcium and magnesium ions. While micelles with sodium and potassium cations are dispersed in water, micelles with calcium and magnesium cations form aggregates. In hard water, therefore, soaps form a precipitate that we know as "bathtub ring" or "soap scum."

The fact that soaps form scum in hard water led to a search for synthetic materials that would have the cleansing properties of soap but would not form scum when they encountered calcium and magnesium ions. The synthetic "soaps" that were developed are known as detergents. **Detergents** are salts of sulfonic acids. Calcium and magnesium sulfonate salts do not form aggregates. Detergent comes from the Latin *detergere,* which means "to wipe off." After the initial introduction of detergents into the marketplace, it was found to be important that the alkyl substituent of a detergent be a straight-chain alkyl group. Straight-chain alkyl groups are biodegradable, while branched-chain alkyl groups are not. Nonbiodegradable detergents become pollutants in rivers and lakes.

a sulfonic acid

an anionic surfactant
a detergent

a cationic surfactant
a germicide

Because the polar head of a soap or a detergent is negatively charged, it is called an anionic surfactant. Cationic surfactants are used widely as fabric softeners and germicides.

15.17 SYNTHESIS OF CARBOXYLIC ACID DERIVATIVES

Of the various classes of carbonyl compounds discussed in this chapter (acyl halides, acid anhydrides, esters, carboxylic acids, and amides), carboxylic acids are the most commonly available both in the laboratory and in biological systems. This means that carboxylic acids are the reagents most likely to be available when a chemist or a cell needs to synthesize a carboxylic acid derivative. However, we have seen that carboxylic acids are relatively unreactive toward nucleophilic acyl substitution reactions because the OH group of a carboxylic acid is a strong base and therefore a poor leaving group. In neutral solutions (physiological pH = 7.3), a carboxylic acid is even more resistant to nucleophilic acyl substitution reactions because it exists predominantly in its negatively charged basic form, in which case it is resistant to nucleophilic attack and the leaving group would be the strongly basic O^{2-}. Therefore, both organic chemists and living organisms need a way to activate carboxylic acids so they can readily undergo nucleophilic acyl substitution reactions.

Activation of Carboxylic Acids for Nucleophilic Acyl Substitution Reactions in the Laboratory

Because acyl halides are the most reactive of the carboxylic acid derivatives, the easiest way to synthesize any other carboxylic acid derivative is to add the appropriate nucleophile to an acyl halide. Therefore, organic chemists activate carboxylic acids by converting them into acyl halides.

A carboxylic acid can be converted into an acyl chloride by heating it either with thionyl chloride ($SOCl_2$) or with phosphorus trichloride (PCl_3). Acyl bromides can be synthesized by using phosphorus tribromide (PBr_3).

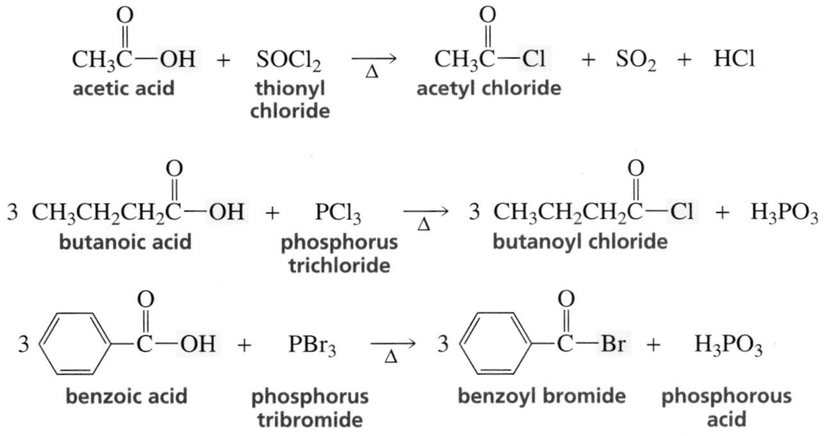

All of these reagents first convert the OH group of the carboxylic acid into a better leaving group than the halide ion. Therefore, when the halide ion subsequently attacks the carbonyl carbon and forms a tetrahedral intermediate, the halide ion is not the group that is eliminated.

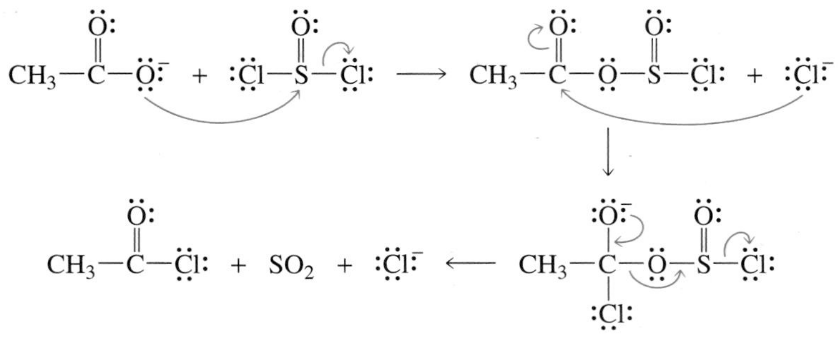

Notice that the reagents that cause the OH group of a carboxylic acid to be replaced by a halogen are the same reagents that cause the OH group of an alcohol to be replaced by a halogen (Section 11.2). In other words, alcohols and carboxylic acids have OH groups that can be activated to make the compound more reactive.

Once the acyl halide has been prepared, a wide variety of carboxylic acid derivatives can be synthesized by adding the appropriate nucleophile (Section 15.6). Because acyl halides are so reactive, they are generally prepared just prior to reaction with the nucleophile.

$$RC\overset{O}{\|}-Cl + RCO^- \longrightarrow RC\overset{O}{\|}-O-\overset{O}{\|}CR + Cl^-$$
an anhydride

$$RC\overset{O}{\|}-Cl + ROH \longrightarrow RC\overset{O}{\|}-OR + HCl$$
an ester

$$RC\overset{O}{\|}-Cl + 2\,RNH_2 \longrightarrow RC\overset{O}{\|}-NHR + R\overset{+}{N}H_3\,Cl^-$$
an amide

Carboxylic acids can also be activated for nucleophilic acyl substitution reactions by being converted into anhydrides. Treatment of a carboxylic acid with a strong dehydrating agent such as P_2O_5 yields an anhydride.

$$2\,C_6H_5C\overset{O}{\|}-OH \xrightarrow{P_2O_5} C_6H_5\overset{O}{\|}C-O-\overset{O}{\|}CC_6H_5 + H_2O$$

Carboxylic acids and carboxylic acid derivatives can also be prepared by methods other than nucleophilic acyl substitution reactions. A summary of the methods used to synthesize these compounds is given in Appendix IV.

Activation of Carboxylate Ions for Nucleophilic Acyl Substitution Reactions in Biological Systems

The synthesis of compounds by biological organisms is called **biosynthesis.** Acyl halides and acid anhydrides are too reactive to be used as reagents by biological organisms. Cells live in a predominantly aqueous environment, and acyl halides and acid anhydrides are rapidly hydrolyzed in water. So living organisms must activate carboxylic acids in a different way.

Carboxylic acids are activated in cells by being converted to acyl phosphates, acyl pyrophosphates, acyl adenylates, or thioesters. An **acyl phosphate** is a mixed anhydride of a carboxylic acid and phosphoric acid; an **acyl pyrophosphate** is a mixed anhydride of a carboxylic acid and pyrophosphoric acid; an **acyl adenylate** is a mixed anhydride of a carboxylic acid and adenosine monophosphate (AMP). Because these mixed anhydrides are negatively charged, they are not readily approached by nucleophiles. One of the functions of enzymes that catalyze biological nucleophilic acyl substitution reactions is to neutralize the negative charges on the mixed anhydride (Section 24.5). A **thioester** is an ester in which the leaving group is a thiol (RSH) instead of an alcohol (ROH). The structure of adenosine triphosphate (ATP) is shown with "Ad" in place of the adenosine ring.

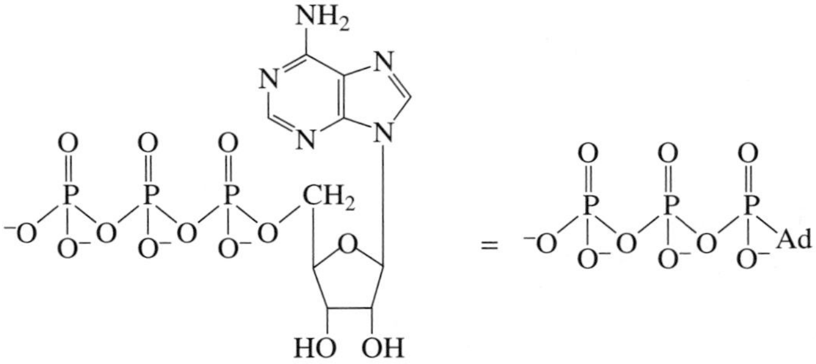

adenosine triphosphate
ATP

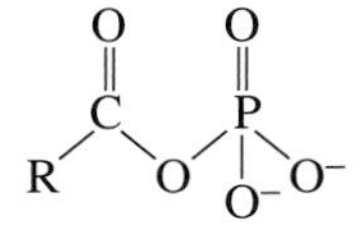

an acyl phosphate 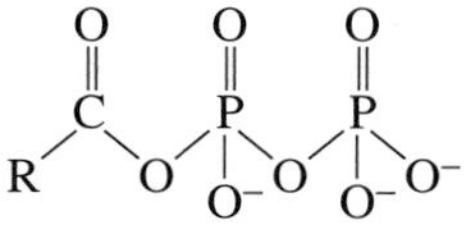   an acyl pyrophosphate 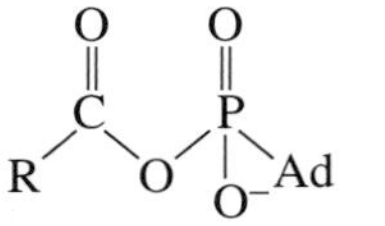   an acyl adenylate   a thioester

"activated" carboxylic acids

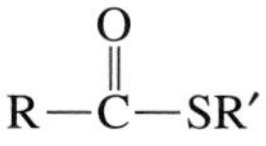

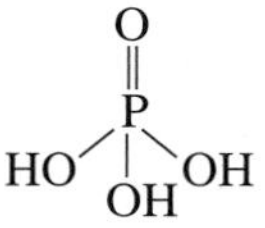

phosphoric acid 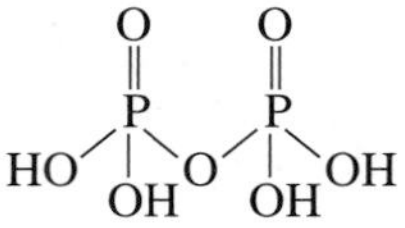   pyrophosphoric acid

Acyl phosphates are formed by nucleophilic attack of a carboxylate ion on the γ-phosphorus (the terminal phosphorus) of ATP. Notice that the reaction is a one-step reaction, very much like an S_N2 reaction. An intermediate is not formed because the π bond does not break. This reaction and the following reactions will be discussed in greater detail in Sections 24.3 and 24.4.

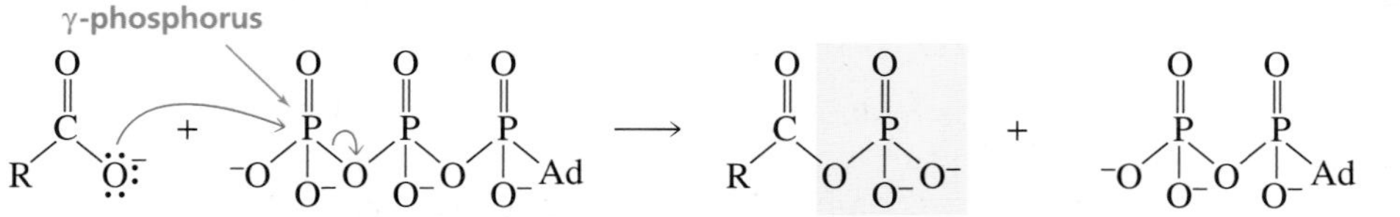

γ-phosphorus

adenosine triphosphate   an acyl phosphate   adenosine diphosphate
ATP   ADP

Acyl pyrophosphates are formed by nucleophilic attack of a carboxylate ion on the β-phosphorus of ATP.

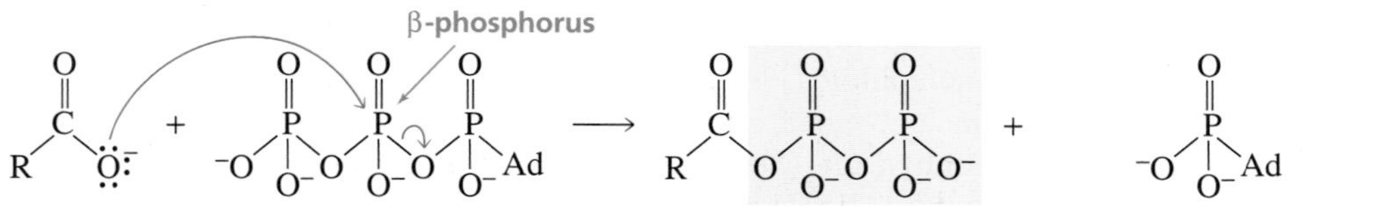

β-phosphorus

adenosine triphosphate   an acyl pyrophosphate   adenosine monophosphate
ATP   AMP

Acyl adenylates are formed by nucleophilic attack of a carboxylate anion on the α-phosphorus of ATP. Acyl phosphates, acyl pyrophosphates, and acyl adenylates will be discussed in Chapter 24.

α-phosphorus

adenosine triphosphate
ATP   an acyl adenylate   pyrophosphate

USING ATP FOR BIOSYNTHESIS

An example of the biosynthesis of a carboxylic acid derivative is the biosynthesis of *N*-carbamoyl-aspartate from aspartate, bicarbonate, and ammonia. *N*-Carbamoylaspartate is a compound needed for the synthesis of pyrimidines. Substituted pyrimidines are found in DNA and RNA (Section 24.5). DNA is the genetic material that makes each living organism unique. The overall reaction requires the activation of both carboxyl oxygens of bicarbonate so that two amide bonds can be formed. Each of the carboxyl oxygens is activated by forming an acyl phosphate. The biosynthesis is catalyzed by enzymes.

ATP

N-carbamoylaspartate

aspartate

Acyl phosphates, acyl pyrophosphates, and acyl adenylates are used only in enzyme-catalyzed reactions where water can be excluded from the active site of the enzyme. Otherwise hydrolysis of the mixed anhydride would compete with the desired nucleophilic acyl substitution reaction.

Thioesters are the most common activated carboxylic acids in the cell. What is remarkable about thioesters is that, although they hydrolyze at about the same rate as oxygen–esters, they are much more reactive than oxygen–esters toward attack by nitrogen and carbon nucleophiles. This allows a thioester to float around in the aqueous environment of the cell without being hydrolyzed, waiting to be a substrate in a nucleophilic acyl substitution reaction.

The thiol used in biological systems for the formation of thioesters is coenzyme A (the A stands for acetylation). The compound is written as CoASH to emphasize that the thiol group is the reactive part of the molecule.

Coenzyme A was discovered by **Fritz A. Lipmann (1899– 1986).** *He also was the first to recognize its importance in intermediary metabolism. Lipmann was born in Germany. To escape the Nazis, he moved to Denmark in 1932 and to the United States in 1939, becoming a United States citizen in 1944. For his work on CoA, he received the Nobel Prize in physiology and medicine in 1953, sharing it with Hans Krebs (Section 3.20).*

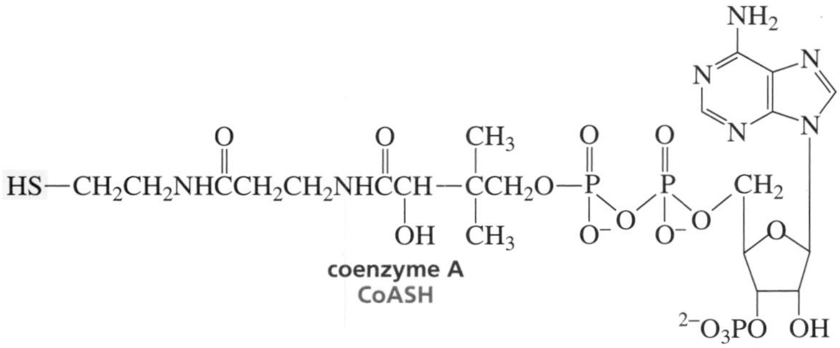

coenzyme A
CoASH

Acetyl-CoA is formed by an enzyme-catalyzed reaction between acetate ion and ATP to activate the acetyl group. The activated compound then reacts with CoASH to form the thioester.

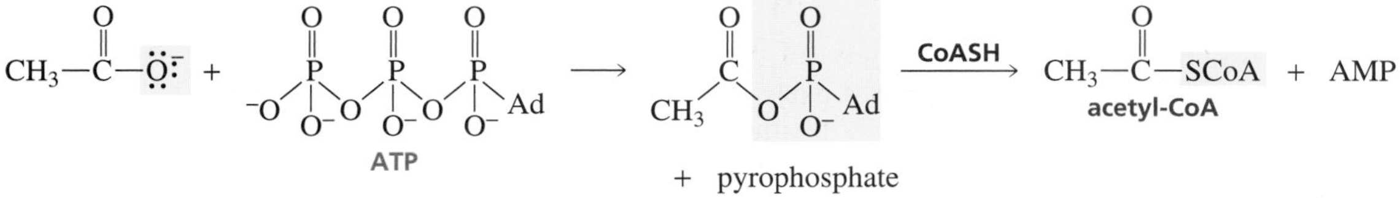

Acetyl-CoA is used by the cell to synthesize needed carboxylic acid derivatives. Acetylcholine is one such derivative. Acetylcholine is a **neurotransmitter,** a compound that transmits nerve impulses across the synapses between nerve cells.

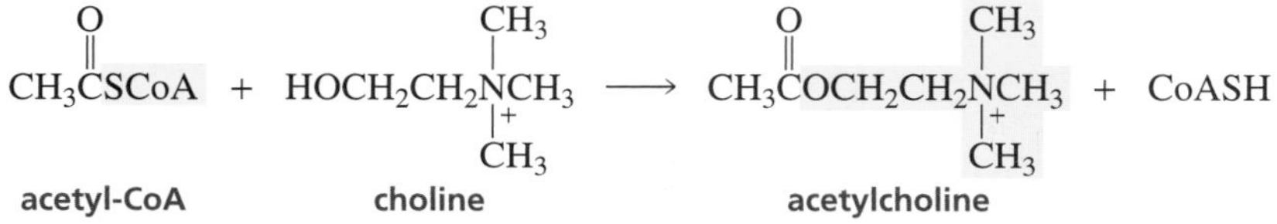

NERVE IMPULSES, PARALYSIS, AND INSECTICIDES

After the nerve impulse is transmitted between cells, acetylcholine must be rapidly hydrolyzed so that the receiving cell can get ready to receive another impulse.

Acetylcholinesterase, the enzyme that catalyzes this hydrolysis, possesses a CH_2OH group that is necessary for its catalytic activity. Diisopropyl fluorophosphate (DFP), a military nerve gas, inhibits the enzyme by reacting with the CH_2OH group. When the enzyme is inhibited, paralysis occurs because the nerve impulses cannot be transmitted properly. DFP's LD_{50} (the lethal dose for 50% of the test animals) is only 0.5 mg/kg of body weight.

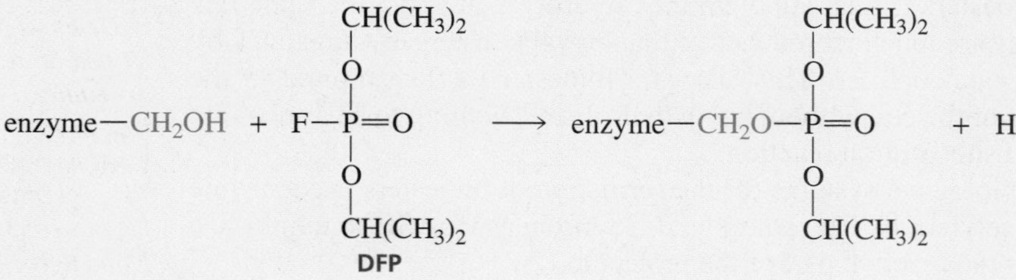

DFP

Malathion and parathion, compounds related to DFP, are used as insecticides. The LD_{50} of malathion is 2800 mg/kg. Parathion is more toxic, with an LD_{50} of 2 mg/kg.

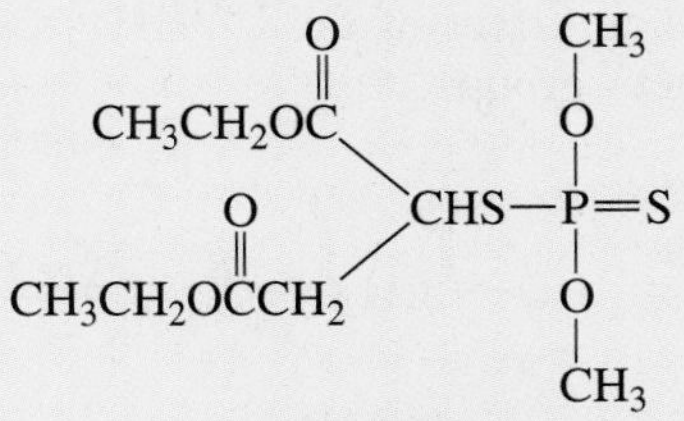

malathion 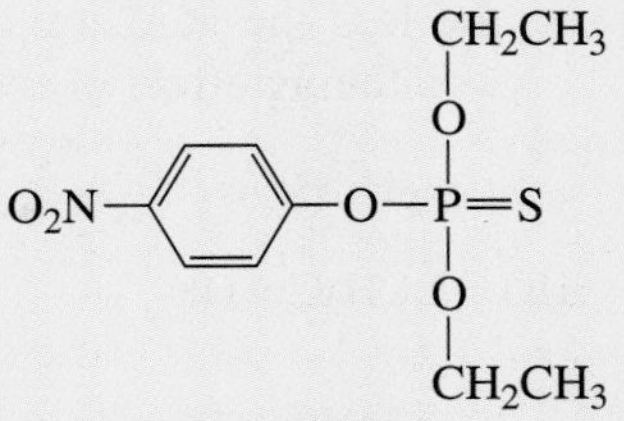   parathion

The acid properties of carboxylic acids have been discussed previously (Sections 1.18 and 6.10). The structures of some common dicarboxylic acids and their pK_a values are shown in Table 15.2. Although the two carboxyl groups of a dicarboxylic acid are identical, the two pK_a values of a dicarboxylic acid are different, because the protons are lost one at a time and therefore leave from different species: the first proton is lost from a neutral molecule; the second proton is lost from a negatively charged ion. A COOH group withdraws electrons (more strongly than an H) and therefore increases the acidity of the other COOH group; in contrast, a COO^-

15.18
DICARBOXYLIC ACID DERIVATIVES

TABLE 15.2   Structures, Names, and pK_a Values of Some Simple Dicarboxylic Acids

Dicarboxylic Acid	Common name	pK_{a1}	pK_{a2}
HOCOH	carbonic acid	3.58	6.35
HOC—COH	oxalic acid	1.27	4.27
$HOCCH_2COH$	malonic acid	2.86	5.70
$HOCCH_2CH_2COH$	succinic acid	4.21	5.64
$HOCCH_2CH_2CH_2COH$	glutaric acid	4.34	5.27
$HOCCH_2CH_2CH_2CH_2COH$	adipic acid	4.41	5.28
phthalic acid structure	phthalic acid	2.95	5.41

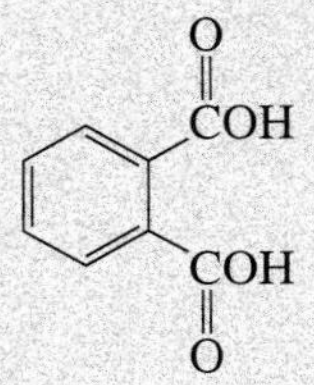

group donates electrons (compared with an H) and so decreases the acidity of the second COOH group. From the pK_a values of the dicarboxylic acids given in Table 15.2, it is apparent that the greater the separation between the two carboxyl groups, the less the acid-strengthening effect of the COOH group and the less the acid-weakening effect of the COO⁻ group.

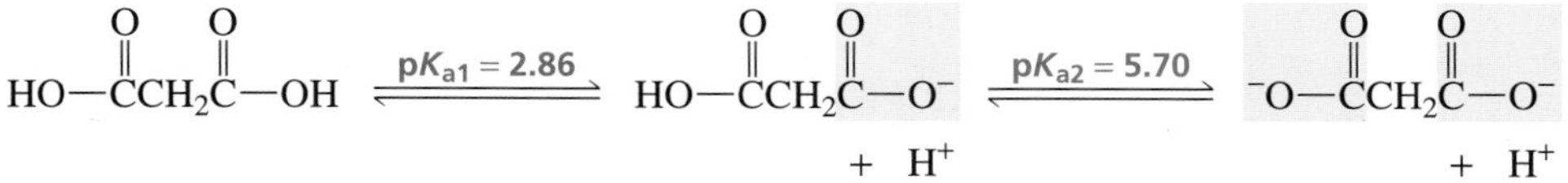

Dicarboxylic acids readily lose water when heated if they can form a cyclic anhydride with a five- or a six-membered ring.

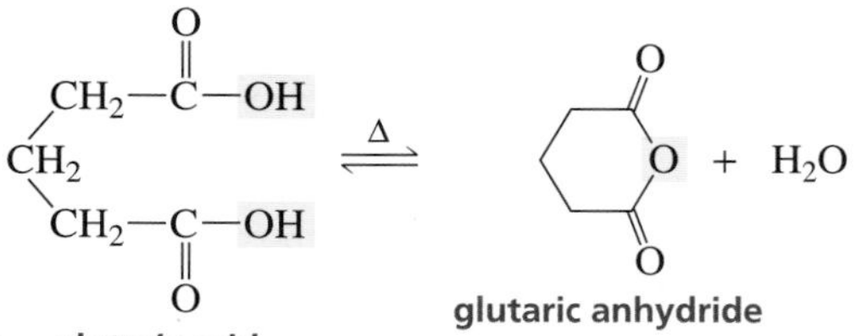

glutaric acid / glutaric anhydride

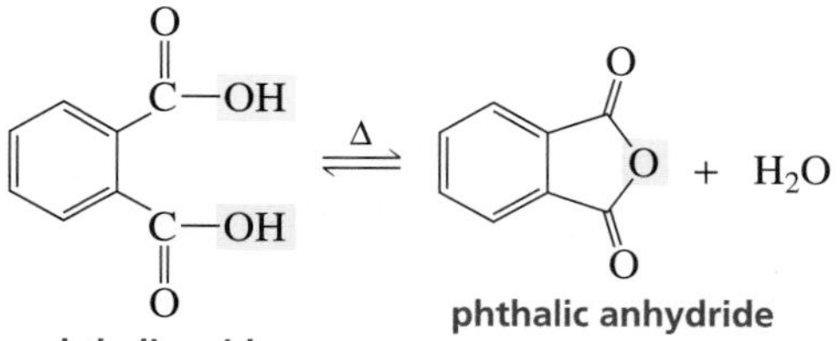

phthalic acid / phthalic anhydride

Cyclic anhydrides are more easily prepared if the dicarboxylic acid is heated in the presence of acetyl chloride or acetic anhydride.

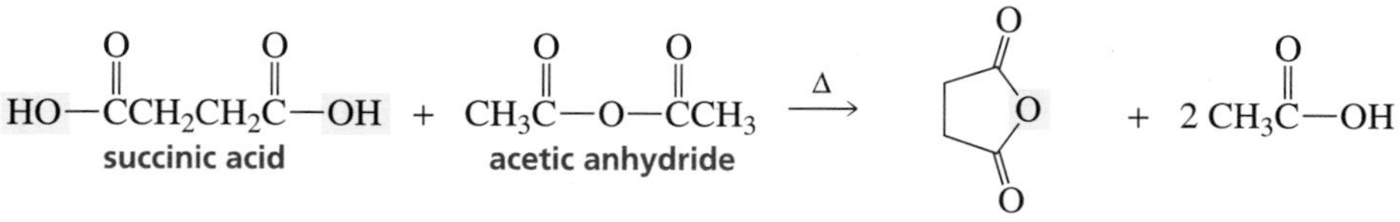

succinic acid + acetic anhydride → succinic anhydride + 2 CH₃C—OH

PROBLEM 28

a. Propose a mechanism for the formation of succinic anhydride in the presence of acetic anhydride.

b. How does acetic anhydride help in the formation of succinic anhydride?

Carbonic acid—a compound with two OH groups bonded to the carbonyl carbon—is unstable. It readily breaks down to CO_2 and H_2O. The reaction is reversible, so carbonic acid is formed when CO_2 is bubbled into water (Section 1.19).

$$HO-\overset{O}{\overset{\|}{C}}-OH \rightleftharpoons CO_2 + H_2O$$

carbonic acid

We have seen that the OH group of a carboxylic acid can be substituted to give a variety of carboxylic acid derivatives. Similarly, the OH groups of carbonic acid can be replaced by other groups.

$$\underset{\textbf{phosgene}}{\overset{\overset{\displaystyle O}{\|}}{Cl-C-Cl}} \quad \underset{\textbf{dimethyl carbonate}}{\overset{\overset{\displaystyle O}{\|}}{CH_3O-C-OCH_3}} \quad \underset{\textbf{urea}}{\overset{\overset{\displaystyle O}{\|}}{H_2N-C-NH_2}} \quad \underset{\textbf{carbamic acid}}{\overset{\overset{\displaystyle O}{\|}}{H_2N-C-OH}} \quad \underset{\textbf{methyl carbamate}}{\overset{\overset{\displaystyle O}{\|}}{H_2N-C-OCH_3}}$$

PROBLEM 29 ◆

What products would you expect to obtain from the following reactions?

a. phosgene + excess diethylamine

b. malonic acid + 2 acetyl chloride

c. methyl carbamate + methylamine

d. urea + water

e. urea + water + H^+

f. β-ethylglutaric acid + acetyl chloride + Δ

SUMMARY OF REACTIONS

1. Reactions of acyl halides (Section 15.6)

$$\overset{\overset{\displaystyle O}{\|}}{RCCl} + \overset{\overset{\displaystyle O}{\|}}{CH_3CO^-} \longrightarrow \overset{\overset{\displaystyle O\ \ O}{\|\ \ \|}}{RCOCCH_3} + Cl^-$$

$$\overset{\overset{\displaystyle O}{\|}}{RCCl} + CH_3OH \longrightarrow \overset{\overset{\displaystyle O}{\|}}{RCOCH_3} + H^+ + Cl^-$$

$$\overset{\overset{\displaystyle O}{\|}}{RCCl} + H_2O \longrightarrow \overset{\overset{\displaystyle O}{\|}}{RCOH} + H^+ + Cl^-$$

$$\overset{\overset{\displaystyle O}{\|}}{RCCl} + 2\,CH_3NH_2 \longrightarrow \overset{\overset{\displaystyle O}{\|}}{RCNHCH_3} + CH_3\overset{+}{N}H_3\,Cl^-$$

2. Reactions of acid anhydrides (Section 15.7)

$$\overset{\overset{\displaystyle O\ \ O}{\|\ \ \|}}{RCOCR} + CH_3OH \longrightarrow \overset{\overset{\displaystyle O}{\|}}{RCOCH_3} + \overset{\overset{\displaystyle O}{\|}}{RCOH}$$

$$\overset{\overset{\displaystyle O\ \ O}{\|\ \ \|}}{RCOCR} + H_2O \longrightarrow 2\,\overset{\overset{\displaystyle O}{\|}}{RCOH}$$

$$\overset{\overset{\displaystyle O\ \ O}{\|\ \ \|}}{RCOCR} + 2\,CH_3NH_2 \longrightarrow \overset{\overset{\displaystyle O}{\|}}{RCNHCH_3} + \overset{\overset{\displaystyle O}{\|}}{RCO^-}\,H_3\overset{+}{N}CH_3$$

3. Reactions of esters (Sections 15.8–15.10)

$$\underset{\text{RCOR}}{O} + CH_3OH \overset{H^+}{\rightleftharpoons} \underset{\text{RCOCH}_3}{O} + ROH$$

$$\underset{\text{RCOR}}{O} + H_2O \overset{H^+}{\rightleftharpoons} \underset{\text{RCOH}}{O} + ROH$$

$$\underset{\text{RCOR}}{O} + H_2O \overset{HO^-}{\longrightarrow} \underset{\text{RCO}^-}{O} + ROH$$

$$\underset{\text{RCOR}}{O} + CH_3NH_2 \longrightarrow \underset{\text{RCNHCH}_3}{O} + ROH$$

4. Reactions of carboxylic acids (Section 15.11)

$$\underset{\text{RCOH}}{O} + CH_3OH \overset{H^+}{\rightleftharpoons} \underset{\text{RCOCH}_3}{O} + H_2O$$

$$\underset{\text{RCOH}}{O} + CH_3NH_2 \longrightarrow \underset{\text{RCO}^- \overset{+}{H_3NCH_3}}{O} \overset{225\,°C}{\longrightarrow} \underset{\text{RCNHCH}_3}{O} + H_2O$$

5. Reactions of amides (Sections 15.12 and 15.13)

$$\underset{\text{RCNH}_2}{O} + H_2O \overset{H^+}{\underset{\Delta}{\longrightarrow}} \underset{\text{RCOH}}{O} + \overset{+}{NH_4}$$

$$\underset{\text{RCNH}_2}{O} + H_2O \overset{HO^-}{\underset{\Delta}{\longrightarrow}} \underset{\text{RCO}^-}{O} + NH_3$$

$$\underset{\text{RCNH}_2}{O} \overset{P_2O_5}{\underset{\Delta}{\longrightarrow}} RC\equiv N + H_2O$$

6. Gabriel synthesis of primary amines (Section 15.14)

$$RCH_2Br \overset{\text{1. phthalimide, HO}^-}{\underset{\text{3. HO}^-}{\overset{\text{2. H}^+, \text{H}_2\text{O}, \Delta}{\longrightarrow}}} RCH_2NH_2$$

7. Reactions of nitriles (Section 15.15)

$$RC\equiv N + H_2O \overset{H^+}{\underset{\Delta}{\longrightarrow}} \underset{\text{RCOH}}{O} + \overset{+}{NH_4}$$

$$RC\equiv N + H_2O \overset{HO^-}{\underset{\Delta}{\longrightarrow}} \underset{\text{RCO}^-}{O} + NH_3$$

8. Activation of carboxylic acids (Section 15.17)

$$\underset{\substack{\text{O}\\\parallel}}{\text{RCOH}} + \text{SOCl}_2 \xrightarrow{\Delta} \underset{\substack{\text{O}\\\parallel}}{\text{RCCl}} + \text{SO}_2 + \text{HCl}$$

$$\underset{\substack{\text{O}\\\parallel}}{\text{RCOH}} + \text{PBr}_3 \xrightarrow{\Delta} \underset{\substack{\text{O}\\\parallel}}{\text{RCBr}} + \text{H}_3\text{PO}_3$$

$$\underset{\substack{\text{O}\\\parallel}}{\text{RCOH}} + \text{PCl}_3 \xrightarrow{\Delta} \underset{\substack{\text{O}\\\parallel}}{\text{RCCl}} + \text{H}_3\text{PO}_3$$

$$2\,\underset{\substack{\text{O}\\\parallel}}{\text{RCOH}} \xrightarrow{\text{P}_2\text{O}_5} \underset{\substack{\text{O}\quad\text{O}\\\parallel\quad\parallel}}{\text{RCOCR}} + \text{H}_2\text{O}$$

9. Dehydration of dicarboxylic acids (Section 15.18)

$$\underset{\substack{\text{O}\quad\quad\text{O}\\\parallel\quad\quad\parallel}}{\text{HOCCH}_2\text{CH}_2\text{COH}} \;\rightleftharpoons^{\Delta}\; \text{[succinic anhydride]} + \text{H}_2\text{O}$$

KEY TERMS

acid anhydride (page 671)
acyl adenylate (page 707)
acyl group (page 668)
acyl halide (page 671)
acyl phosphate (page 707)
acyl pyrophosphate (page 707)
alcoholysis (page 686)
amide (page 672)
amino acid (page 680)
aminolysis (page 686)
biosynthesis (page 707)
α-carbon (page 670)
carbonyl carbon (page 674)
carbonyl compound (page 668)
carbonyl group (page 668)
carbonyl oxygen (page 674)

carboxyl group (page 670)
carboxylic acid (page 670)
carboxylic acid derivative
   (page 674)
carboxyl oxygen (page 672)
detergent (page 705)
ester (page 672)
fat (page 703)
fatty acid (page 670)
Fischer esterification reaction
   (page 696)
Gabriel synthesis (page 701)
hydrolysis (page 686)
hydrophobic interactions
   (page 704)
imide (page 701)

lactam (page 673)
lactone (page 672)
micelle (page 704)
mixed anhydride (page 671)
neurotransmitter (page 710)
nitrile (page 673)
nucleophilic acyl substitution reaction
   (page 676)
oil (page 703)
saponification (page 703)
soap (page 703)
symmetrical anhydride (page 671)
tetrahedral intermediate (page 675)
thioester (page 707)
transesterification reaction
   (page 687)

PROBLEMS

30. Give a structure for each of the following.
   a. *N,N*-dimethylhexanamide
   b. 3,3-dimethylhexanamide
   c. cyclohexanecarbonyl chloride
   d. propanenitrile
   e. propionyl bromide
   f. sodium acetate
   g. benzoic anhydride
   h. β-valerolactone
   i. 3-methylbutanenitrile
   j. cycloheptanecarboxylic acid

31. Name the following compounds.

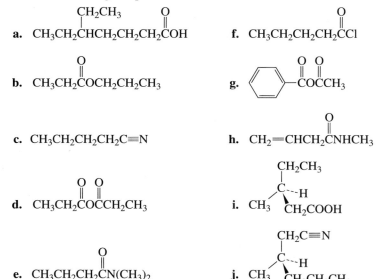

a. CH₃CH₂CHCH₂CH₂CH₂COH (with CH₂CH₃ substituent and O double bond)

f. CH₃CH₂CH₂CH₂CCl (with O double bond)

b. CH₃CH₂COCH₂CH₂CH₃ (with O double bond)

g. (benzene ring)—COCCH₃ (with two O double bonds)

c. CH₃CH₂CH₂CH₂C≡N

h. CH₂=CHCH₂CNHCH₃ (with O double bond)

d. CH₃CH₂COCCH₂CH₃ (with two O double bonds)

i. CH₃ C--H with CH₂CH₃ and CH₂COOH

e. CH₃CH₂CH₂CN(CH₃)₂ (with O double bond)

j. CH₃ C--H with CH₂C≡N and CH₂CH₂CH₃

32. What products would be formed from the reaction of benzoyl chloride with the following reagents?
a. sodium acetate
b. water
c. dimethylamine
d. aqueous HCl
e. aqueous NaOH
f. cyclohexanol
g. benzylamine
h. 4-chlorophenol
i. isopropyl alcohol
j. aniline

33. a. List the following esters in order of decreasing reactivity in the first step of a nucleophilic acyl substitution reaction (formation of the tetrahedral intermediate).
b. List the following esters in order of decreasing reactivity in the second step of a nucleophilic acyl substitution reaction (collapse of the tetrahedral intermediate).

CH₃CO—(benzene)   CH₃CO—(cyclohexane)   CH₃CO—(benzene)—CH₃   CH₃CO—(benzene)—Cl

34. a. Which compound would you expect to have a higher dipole moment: methyl acetate or butanone?
b. Which would you expect to have a higher boiling point?

CH₃COCH₃   CH₃CCH₂CH₃
methyl acetate   butanone

35. Using a carboxylic acid as a starting material, what would be the best method to synthesize the following esters?
a. ethyl phenylethanoate (odor of honey)   b. isobutyl propionate (odor of rum)

36. How could you use nuclear magnetic resonance (¹H NMR) spectroscopy to distinguish among the following esters?

CH₃COCH₂CH₃   HCOCH₂CH₂CH₃

CH₃CH₂COCH₃   HCOCHCH₃ (with O and CH₃)

37. If propionyl chloride is added to one equivalent of methylamine, only a 50% yield of *N*-methylpropanamide is obtained. If, however, the acyl chloride is added to two equivalents of methylamine, the yield of *N*-methylpropanamide is almost 100%. Explain these observations.

38. a. When a carboxylic acid is dissolved in isotopically labeled water, the label is incorporated into both oxygens of the acid. Propose a mechanism to account for this.

$$CH_3-\overset{\overset{\displaystyle O}{\|}}{C}-OH \;+\; H_2\overset{18}{O} \;\rightleftharpoons\; CH_3-\overset{\overset{\displaystyle \overset{18}{O}}{\|}}{C}-\overset{18}{O}H \;+\; H_2O$$

b. If a carboxylic acid is dissolved in isotopically labeled methanol and an acid catalyst is added, where will the label reside in the product?

39. What reagents would you use to convert methyl propanoate into the following compounds?
   a. isopropyl propanoate
   b. sodium propanoate
   c. *N*-ethylpropanamide
   d. propanoic acid

40. Aspartame is 160 times sweeter than sucrose. It is the sweetener used in the commercial products known as NutraSweet and Equal. What products would be obtained if aspartame were hydrolyzed completely in an aqueous solution of HCl?

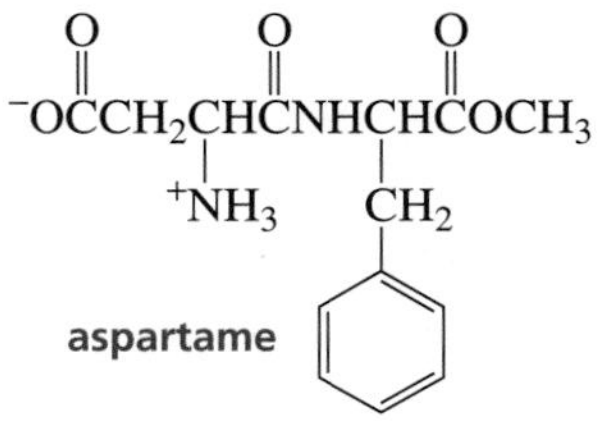

$$^-OCCH_2CHCNHCHCOCH_3$$
$$^+NH_3 \quad CH_2$$

aspartame

41. a. Which of the following reactions will not give the carbonyl product shown?
   b. Which of the reactions that do not occur can be made to occur if an acid catalyst is added to the reaction mixture?

1. $CH_3COH + CH_3CO^- \longrightarrow CH_3COCCH_3$

2. $CH_3CCl + CH_3CO^- \longrightarrow CH_3COCCH_3$ ✓

3. $CH_3CNH_2 + Cl^- \longrightarrow CH_3CCl$ ✗

4. $CH_3COH + CH_3NH_2 \longrightarrow CH_3CNHCH_3$ ✗

5. $CH_3COCH_3 + CH_3NH_2 \longrightarrow CH_3CNHCH_3$ ✓

6. $CH_3COCH_3 + Cl^- \longrightarrow CH_3CCl$ ✗

7. $\underset{\text{O}}{\overset{\text{O}}{CH_3\overset{\parallel}{C}NHCH_3}}$ + $\underset{\text{O}}{\overset{\text{O}}{CH_3\overset{\parallel}{C}O^-}}$ ⟶ $\underset{\text{O O}}{CH_3\overset{\parallel}{C}O\overset{\parallel}{C}CH_3}$ ✗

8. $CH_3\overset{\text{O}}{\overset{\parallel}{C}}Cl$ + H_2O ⟶ $CH_3\overset{\text{O}}{\overset{\parallel}{C}}OH$

9. $CH_3\overset{\text{O}}{\overset{\parallel}{C}}NHCH_3$ + H_2O ⟶ $CH_3\overset{\text{O}}{\overset{\parallel}{C}}OH$ ✗

42. The tertiary amine shown below catalyzes transesterification. Propose a mechanism to show how it catalyzes the reaction.

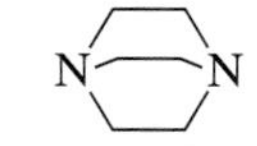

1,4-diazabicyclo[2.2.2]octane
DABCO

43. Identify the major and minor products of the following reaction.

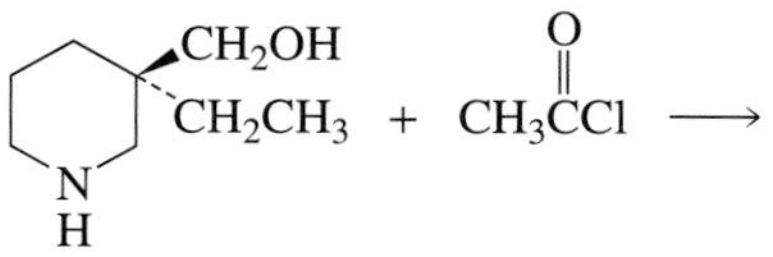

44. Two products (A and B) are obtained from the reaction of 1-bromobutane with NH_3: A reacts with acetyl chloride to form C, and B reacts with acetyl chloride to form D. The IR spectra of C and D are shown below. Identify A, B, C, and D.

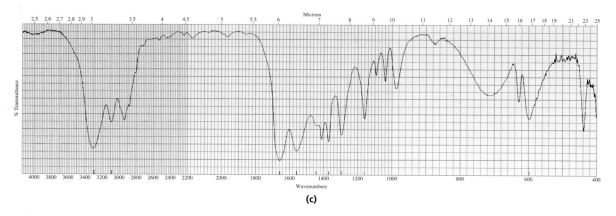

(c)

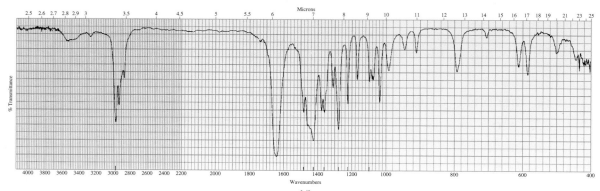

(d)

45. a. Phosgene (COCl$_2$) was used as a poison gas in World War I. Give the product that would be formed from the reaction of phosgene with the following reagents.
    1. one equivalent of methanol
    2. excess methanol
    3. excess propylamine
    4. one equivalent of ethanol followed by one equivalent of methylamine
  b. What is the mode of action of phosgene as a poison gas?

46. When Ethyl Ester treated butanedioic acid with thionyl chloride, she was surprised to find that the product she obtained was an anhydride rather than an acyl chloride. Propose a mechanism to explain why she obtained an anhydride.

47. Give the products of the following reactions.

a. CH$_3$CCl + KF $\longrightarrow$

f. 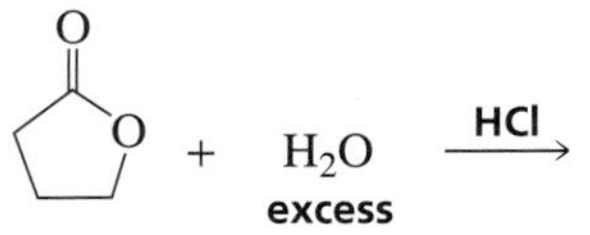 + H$_2$O (excess) $\xrightarrow{\text{HCl}}$

b. 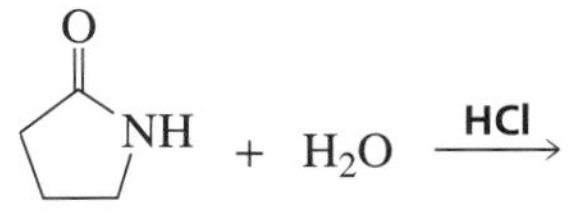 + H$_2$O $\xrightarrow{\text{HCl}}$

g. CH$_3$CCH$_2$OCCH$_3$ + CH$_3$OH (excess) $\xrightarrow{\text{CH}_3\text{O}^-}$

c. $\xrightarrow[\text{2. 2 CH}_3\text{NH}_2]{\text{1. SOCl}_2}$

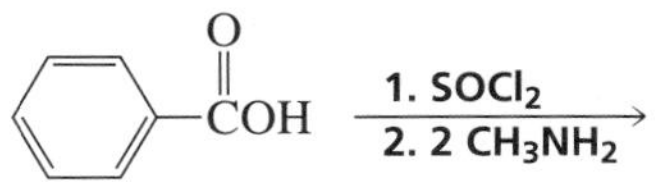

h. 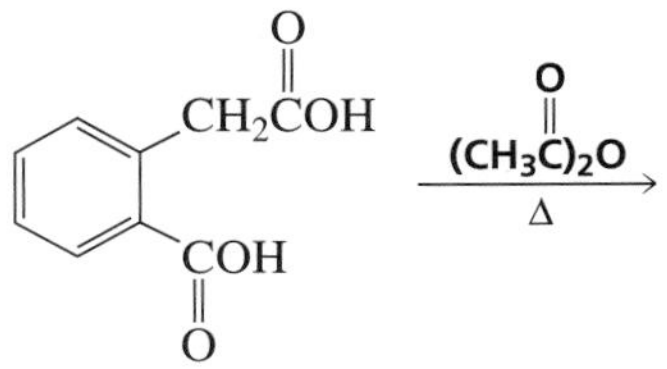 $\xrightarrow[\Delta]{(\text{CH}_3\text{C})_2\text{O}}$

d. 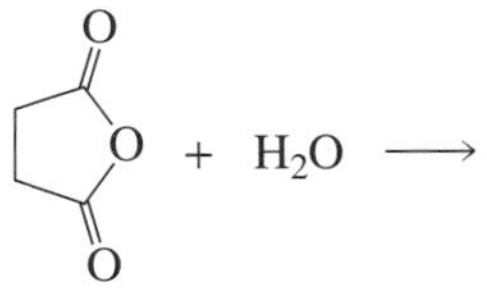 + H$_2$O $\longrightarrow$

i. 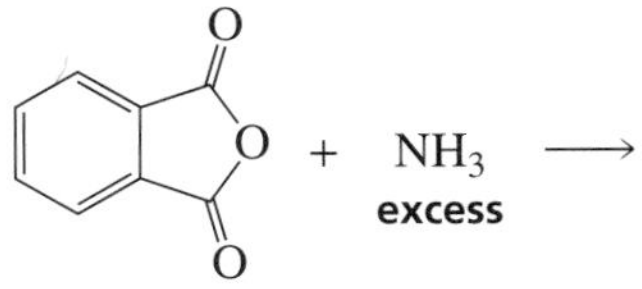 + NH$_3$ (excess) $\longrightarrow$

e. 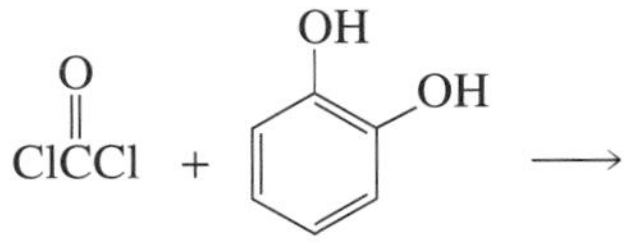 ClCCl + $\longrightarrow$

j. 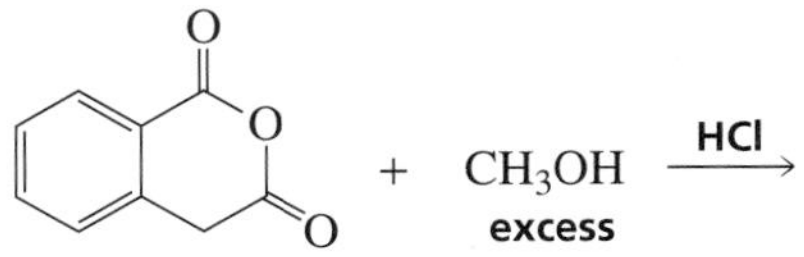 + CH$_3$OH (excess) $\xrightarrow{\text{HCl}}$

48. Compound A (with molecular formula C$_4$H$_6$Cl$_2$O), when treated with an equivalent of methanol, forms the compound whose ^{1}H NMR spectrum is shown below. Identify compound A.

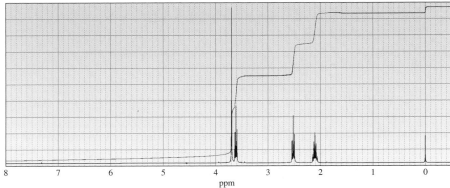

49. a. Identify the two products obtained from the following reaction.

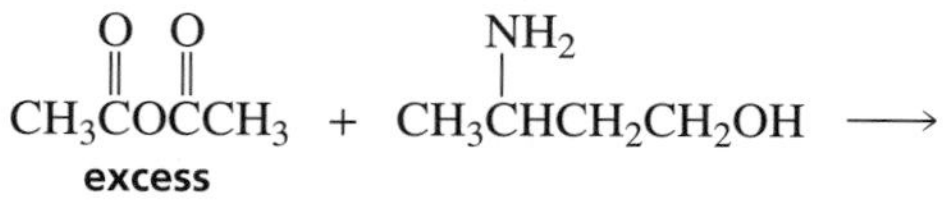

$$\underset{\textbf{excess}}{CH_3\overset{\overset{O}{\|}}{C}\overset{\overset{O}{\|}}{O}CCH_3} + CH_3\overset{\overset{NH_2}{|}}{C}HCH_2CH_2OH \longrightarrow$$

  b. Eddie Amine carried out the reaction but stopped it before it was half over and isolated the major product. He was surprised to find that the product he isolated was neither of the products obtained when the reaction was allowed to go to completion. What product did he isolate?

50. An aqueous solution of a primary or secondary amine reacts with an acyl chloride to form an amide as the major product. However, if the amine is tertiary, an amide is not formed. What product is formed? Explain.

51. a. Ann Hydride did not obtain any ester when she added 2,4,6-trimethylbenzoic acid to an acidic solution of methanol. Why? (*Hint:* Build models.)

  b. Would she have encountered the same problem if she had tried to synthesize the methyl ester of *p*-methylbenzoic acid in the same way?

  c. How could she have prepared the methyl ester of 2,4,6-trimethylbenzoic acid? (*Hint:* See Section 14.20.)

52. List the following compounds in order of decreasing frequency of the carbon–oxygen double bond stretch.

$$CH_3\overset{\overset{O}{\|}}{C}OCH_3 \qquad CH_3\overset{\overset{O}{\|}}{C}Cl \qquad CH_3\overset{\overset{O}{\|}}{C}H \qquad CH_3\overset{\overset{O}{\|}}{C}NH_2$$

53. a. If the equilibrium constant for the reaction of acetic acid and ethanol to form ethyl acetate is 4.02, what will be the concentration of ethyl acetate at equilibrium if the reaction is carried out with equal amounts of acetic acid and ethanol?

  b. What will be the concentration of ethyl acetate at equilibrium if the reaction is carried out with 10 times more ethanol than acetic acid? (*Hint:* Recall the Quadratic Equation.)

$$\text{For:} \quad ax^2 + bx + c = 0$$

$$x = \frac{-b \pm (b^2 - 4ac)^{1/2}}{2a}$$

  c. What will be the concentration of ethyl acetate at equilibrium if the reaction is carried out with 100 times more ethanol than acetic acid?

54. The NMR spectra for two esters with molecular formula $C_8H_8O_2$ are given on the next page. If each of the esters is added to an aqueous solution with a pH of 10, which of the esters will be hydrolyzed more completely when the hydrolysis reaction has reached completion?

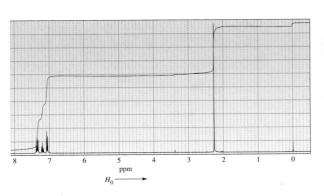

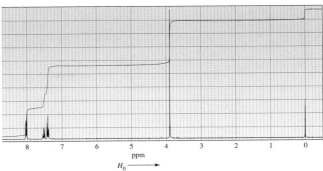

55. A study of the kinetics of the base-promoted hydrolysis of an amide has shown that the reaction is second-order in hydroxide ion. What does this tell you about the mechanism?

56. Show how the following compounds could be prepared from the given starting materials. You can use any necessary organic or inorganic reagents.

a. $CH_3CH_2\overset{\overset{\displaystyle O}{\|}}{C}NH_2 \longrightarrow CH_3CH_2\overset{\overset{\displaystyle O}{\|}}{C}Cl$

b. $CH_3CH_2CH_2CH_2OH \longrightarrow CH_3CH_2CH_2CH_2\overset{\overset{\displaystyle O}{\|}}{C}OH$

c. (benzene ring with CH$_3$) $\longrightarrow$ (benzene ring with CH$_2\overset{\overset{\displaystyle O}{\|}}{C}OH$)

d. (benzene ring with CH$_3$) $\longrightarrow$ (benzene ring with $\overset{\overset{\displaystyle O}{\|}}{C}NHCH_3$)

e. (benzene ring with NH$_2$) $\longrightarrow$ (benzene ring with NH$_2$ and $CH_3\overset{\underset{\displaystyle O}{\|}}{C}$)

f. $CH_3(CH_2)_{10}\overset{\overset{\displaystyle O}{\|}}{C}OH \longrightarrow CH_3(CH_2)_{11}$—(benzene ring)—$SO_3^- Na^+$

a detergent

57. Is the acid-catalyzed hydrolysis of acetamide a reversible or an irreversible reaction? Explain.

58. The reaction of a nitrile with an alcohol in the presence of a strong acid forms a secondary amide. This reaction is known as the Ritter reaction. The Ritter reaction does not work with primary alcohols.

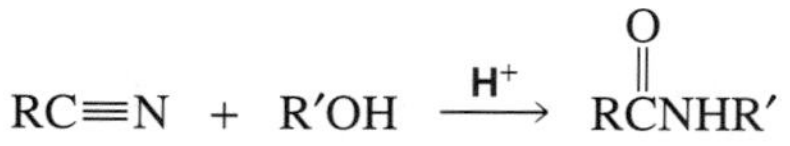

$$RC \equiv N \ + \ R'OH \ \xrightarrow{H^+} \ \overset{\overset{O}{\parallel}}{R\text{C}}NHR'$$

the Ritter reaction

a. Propose a mechanism for the Ritter reaction.

b. Why does the Ritter reaction not work with primary alcohols?

c. How does the Ritter reaction differ from the acid-catalyzed hydrolysis of a nitrile to form a primary amide?

59. The intermediate shown below is formed during the hydroxide-ion-promoted hydrolysis of the ester group. Propose a mechanism for the reaction (*J. Am. Chem. Soc.* 1988, 110, 8157).

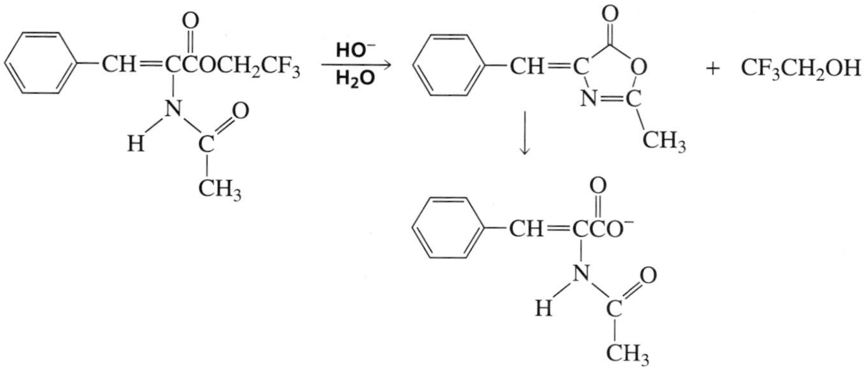

60. What product would you expect to obtain from each of the following reactions?

a. 
$$\underset{OH}{\overset{\text{OH}}{CH_3CH_2CHCH_2CH_2CH_2COH}} \xrightarrow{H^+}$$

b.

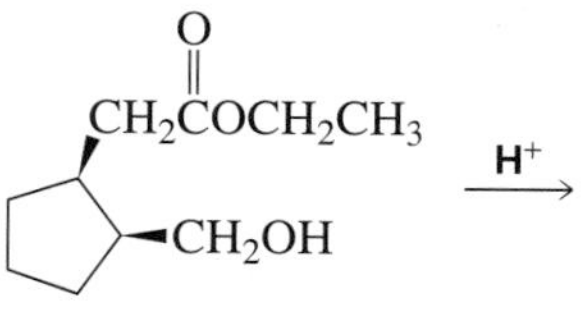

c.

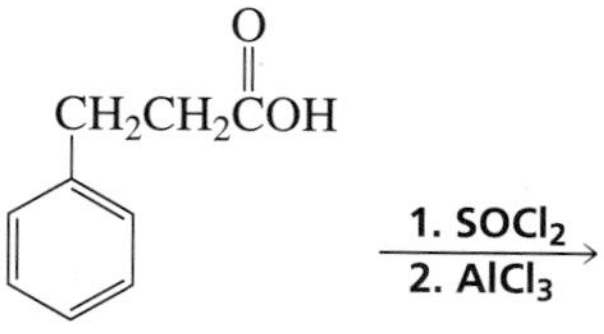

61. a. How could aspirin be synthesized starting with benzene?

b. Ibuprofen is the active ingredient in pain relievers such as Advil, Motrin, and Nuprin. How could ibuprofen be synthesized starting with benzene?

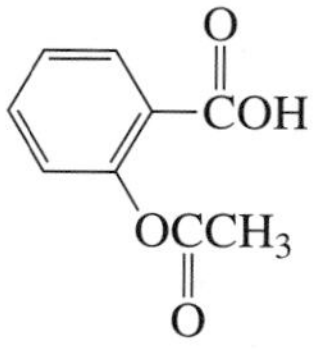

aspirin

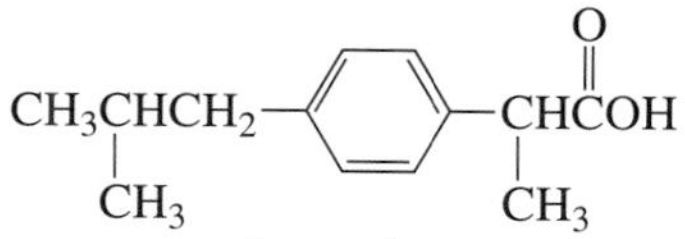

ibuprofen

62. The compound shown below has been found to be an inhibitor of a β-lactamase. The enzyme can be reactivated by hydroxylamine (NH$_2$OH). Propose a mechanism to account for the inhibition and for the reactivation (*Science* 1989, 246, 917).

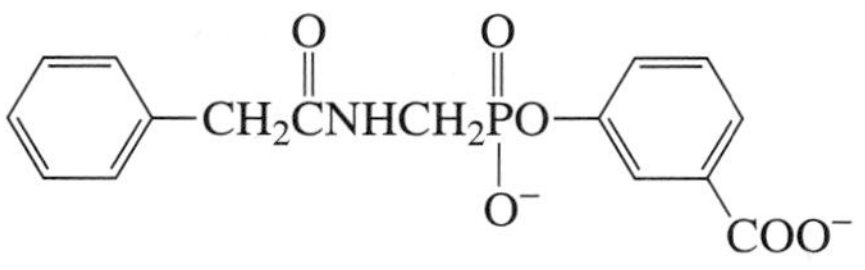

63. For each of the following reactions, propose a mechanism that will account for the formation of the product.

a.

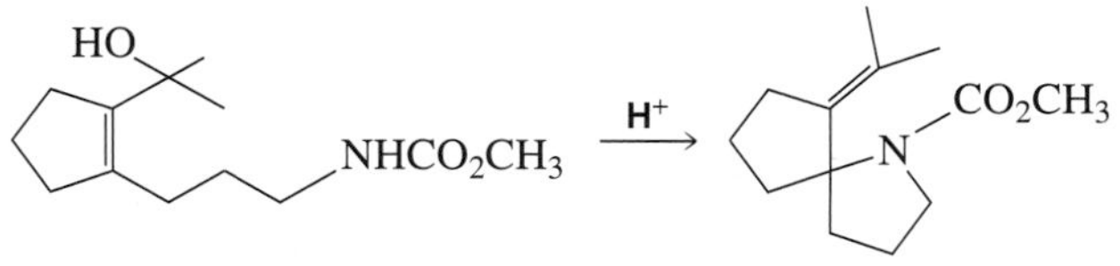

b.

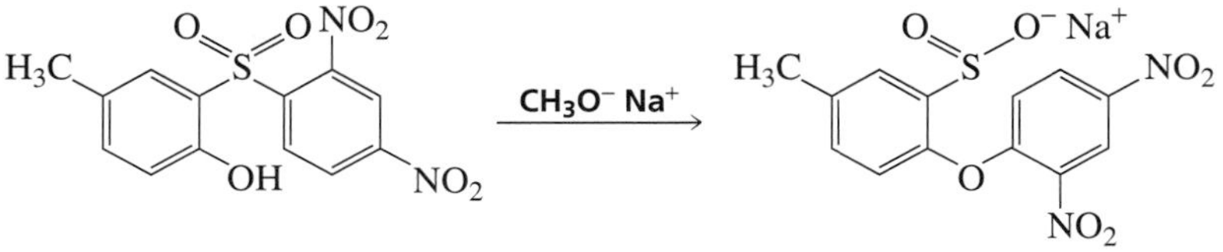

64. Catalytic antibodies were first introduced in 1986. They catalyze a reaction by binding to the transition state, thereby stabilizing it. This lowers the energy of activation and therefore allows the reaction to go faster. The synthesis of the antibody is carried out in the presence of a transition state analog. A transition state analog is a stable molecule that structurally resembles the transition state. This causes generation of an antibody that will recognize and bind to the transition state. For example, the following transition state analog has been used to generate a catalytic antibody that catalyzes the hydrolysis of the structurally similar ester.

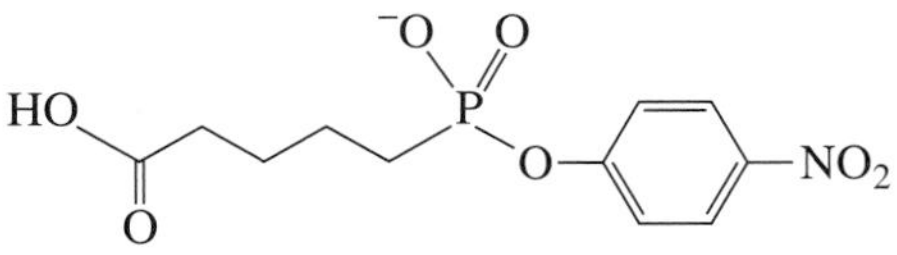

transition state analog

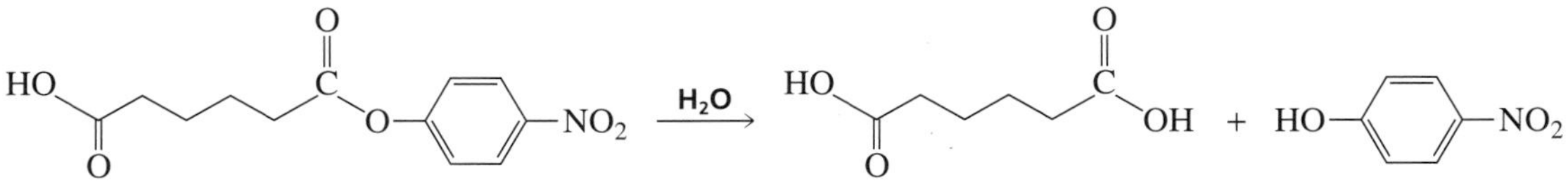

a. Draw the transition state for the hydrolysis reaction.
b. The following transition state analog is used to generate a catalytic antibody for the catalysis of ester hydrolysis. Give the structure of an ester whose rate of hydrolysis would be increased by this catalytic antibody.

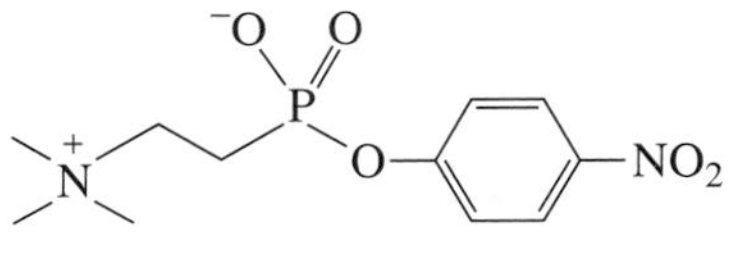

c. Design a transition state analog that would catalyze amide hydrolysis at the amide group indicated by the arrow.

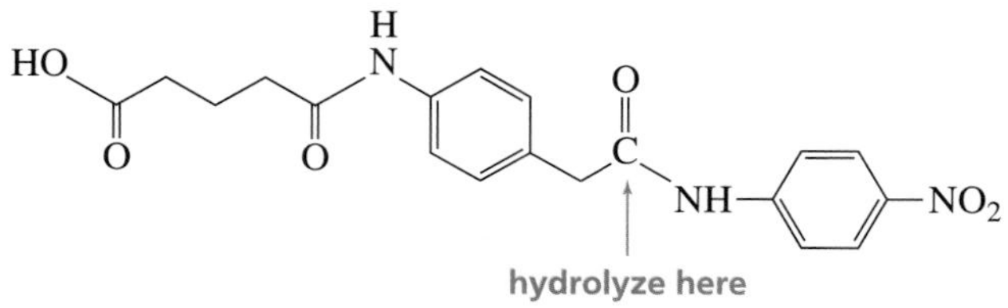

hydrolyze here

65. Information about the mechanism of reaction of a series of substituted benzenes can be obtained by plotting the logarithm of the observed rate constant obtained at a particular pH versus the Hammett substituent constant (σ) for the particular substituent. The Hammett substituent constant for hydrogen is 0. Electron-donating substituents have negative Hammett substituent constants, and electron-withdrawing substituents have positive Hammett substituent constants. The more strongly electron donating the substituent, the greater the negative value of its substituent constant; the more strongly electron withdrawing the substituent, the greater the positive value of its substituent constant. The slope of a plot of the log of the rate constant versus σ is called the ρ (rho) value. The ρ value for the hydroxide-ion-promoted hydrolysis of a series of *meta*- and *para*-substituted ethyl benzoates is $+2.46$; the ρ value for amide formation from the reaction of a series of *meta*- and *para*-substituted anilines with benzoyl chloride is -2.78.

a. Why does one set of experiments give a positive ρ value while the other set of experiments gives a negative ρ value?

b. Why do you think that ortho-substituted compounds were not included in the above-mentioned experiment?

c. What would you predict the sign of the ρ value to be for the ionization of a series of *meta*- and *para*-substituted benzoic acids?

66. Saccharin, an artificial sweetener, is about 300 times sweeter than sucrose. Describe how saccharin could be prepared using benzene as the starting material.

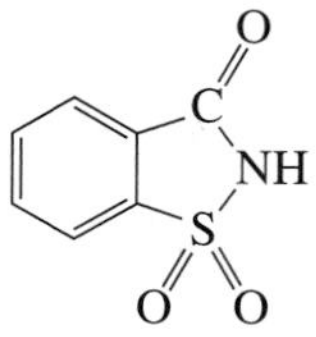

saccharin

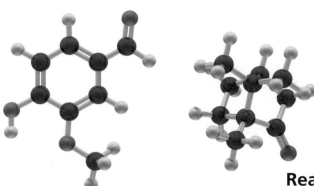

vanillin, camphor

CARBONYL COMPOUNDS II:
Reactions of Carbonyl Compounds with Carbon and Hydrogen Nucleophiles; Reactions of Aldehydes and Ketones with Oxygen and Nitrogen Nucleophiles; Reactions of α,β-Unsaturated Carbonyl Compounds

Aldehydes and ketones possess a carbonyl group and therefore are carbonyl compounds. The carbonyl group of the simplest aldehyde, formaldehyde, is bonded to two hydrogens. In all other **aldehydes,** the carbonyl group is bonded to a *hydrogen* and to an *alkyl* (or an *aryl*) *group*. The carbonyl group of a **ketone** is bonded to *two alkyl* (or *aryl*) *groups*. Aldehydes and ketones differ from the carbonyl compounds discussed in Chapter 15 in that they do not have a group that can be replaced by a nucleophile: carbanions (R^-) and hydride ions (H^-) are too basic to be displaced by nucleophiles. The physical properties of aldehydes and ketones are given in Appendix I. A summary of the methods used to prepare aldehydes and ketones can be found in Appendix IV.

| formaldehyde | an aldehyde | a ketone | formaldehyde | acetaldehyde | acetone |

Many compounds found in nature have aldehyde or ketone functional groups. Aldehydes have pungent odors, whereas the odors of ketones tend to be sweet. Vanilla and cinnamon flavorings are examples of naturally occurring aldehydes. The ketones carvone and camphor are responsible for the characteristic sweet odors of spearmint leaves, caraway seeds, and the camphor tree.

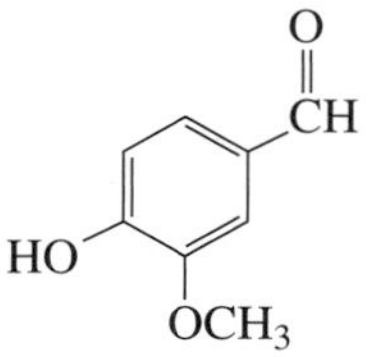

vanillin
vanilla flavoring

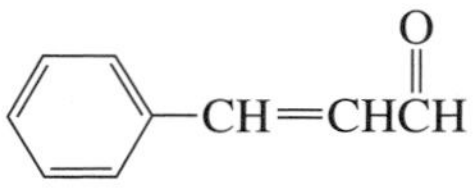

cinnamaldehyde
cinnamon flavoring

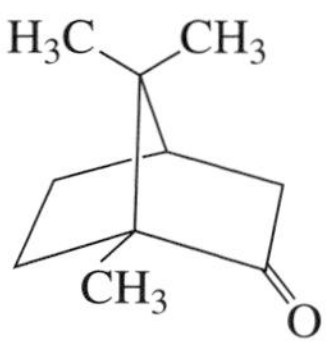

camphor

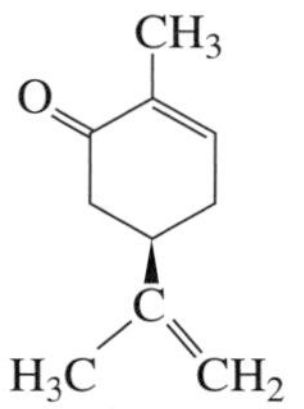

(R)-(−)-carvone
spearmint oil

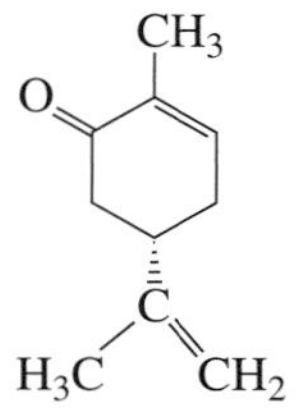

(S)-(+)-carvone
caraway seed oil

In ketosis, a pathological condition that can occur in people with diabetes, the body produces more acetoacetate than can be metabolized. The excess acetoacetate breaks down to acetone (a ketone) and CO_2. This condition can be recognized by the sweet smell of acetone on the patient's breath.

$$\underset{\text{acetoacetate}}{CH_3\overset{O}{\overset{\|}{C}}CH_2\overset{O}{\overset{\|}{C}}O^-} \longrightarrow \underset{\text{acetone}}{CH_3\overset{O}{\overset{\|}{C}}CH_3} + CO_2$$

Two ketones of biological importance illustrate how a small difference in structure can be responsible for a large difference in biological activity: progesterone is a female sex hormone synthesized primarily in the ovaries, and testosterone is a male sex hormone synthesized primarily in the testes.

progesterone
a female sex hormone

testosterone
a male sex hormone

16.1 NOMENCLATURE

Aldehydes

The IUPAC name of an aldehyde is obtained by removing the "e" from the alkane name and adding "al." The position of the carbonyl carbon does not have to be designated since it is always at the end of the parent hydrocarbon and therefore is always at the number 1 position.

The common name of an aldehyde is the same as the common name of the corresponding carboxylic acid, except that "aldehyde" is substituted for "ic acid" (or "oic acid"). When common names are used, the position of a substituent is designated by a lowercase Greek letter. The carbonyl carbon is not given a designation; the carbon adjacent to the carbonyl carbon is the α-carbon.

	$\overset{O}{\overset{\|}{HCH}}$	$CH_3\overset{O}{\overset{\|}{CH}}$	$CH_3\overset{Br}{\underset{}{\overset{\|}{CH}}}\overset{O}{\overset{\|}{CH}}$
IUPAC name:	methanal	ethanal	2-bromopropanal
common name:	formaldehyde	acetaldehyde	α-bromopropionaldehyde

	$CH_3\overset{Cl}{\underset{}{\overset{\|}{CH}}}CH_2\overset{O}{\overset{\|}{CH}}$	$CH_3\overset{CH_3}{\underset{}{\overset{\|}{CH}}}CH_2\overset{O}{\overset{\|}{CH}}$	$\overset{O}{\overset{\|}{HC}}CH_2CH_2CH_2CH_2\overset{O}{\overset{\|}{CH}}$
IUPAC name:	3-chlorobutanal	3-methylbutanal	hexanedial
common name:	β-chlorobutyraldehyde	isovaleraldehyde	

If the aldehyde group is attached to a ring, the aldehyde is named by adding "carbaldehyde" to the name of the cyclic compound.

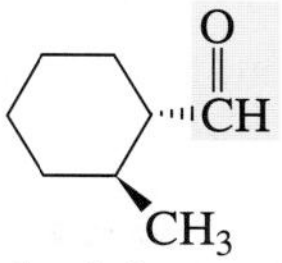

IUPAC name: *trans*-2-methylcyclohexanecarbaldehyde   benzenecarbaldehyde
or
phenylmethanal
benzaldehyde

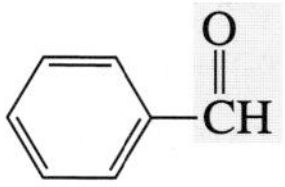

common name:

If a second functional group of higher naming priority is present, the aldehyde group can be designated by the prefix "formyl."

$$CH_3CH_2CHCH_2CH_2COCH_2CH_3$$
$$HC\!=\!O$$
ethyl 4-formylhexanoate

If the compound has both an alkene and an aldehyde functional group, the alkene is cited first, with the "e" ending omitted to avoid two successive vowels (Section 7.1).

$$CH_3CH\!=\!CHCH_2CH$$
3-pentenal

Ketones

Ketones are named in the IUPAC system by removing the "e" from the alkane name and adding "one." The chain is numbered in the direction that gives the carbonyl group the smaller number. In the case of cyclic ketones, a number is not necessary because the carbonyl group is assumed to be at the number 1 position. Frequently, derived names are used for ketones. In derived names, the substituents attached to the carbonyl group are cited in alphabetical order followed by "ketone."

	CH_3CCH_3	$CH_3CH_2CCH_2CH_2CH_3$	$CH_3CHCH_2CH_2CH_2CCH_3$
IUPAC name:	propanone	3-hexanone	6-methyl-2-heptanone
common name:	acetone		
derived name:	dimethyl ketone	ethyl propyl ketone	isohexyl methyl ketone

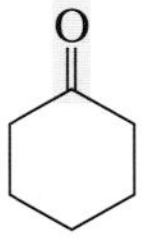

		$CH_3C\!-\!CCH_3$	$CH_3CCH_2CCH_3$	$CH_3CH\!=\!CHCH_2CCH_3$
IUPAC name:	cyclohexanone	butanedione	2,4-pentanedione	4-hexen-2-one
common name:			acetylacetone	

PROBLEM 1

Why is a number not used to designate the position of the functional groups in propanone and butanedione?

Only a few ketones have common names. The smallest ketone, propanone, is usually referred to by its common name, acetone. Acetone is a common laboratory solvent. Common names are also used for some phenyl-substituted ketones; the number of carbons (other than those of the phenyl group) is indicated by the common name of the corresponding carboxylic acid substituting "ophenone" for "ic acid."

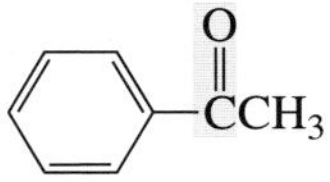

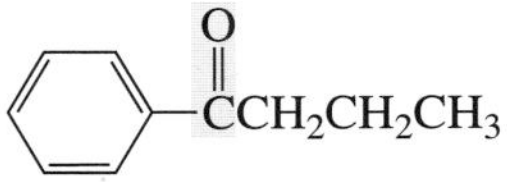

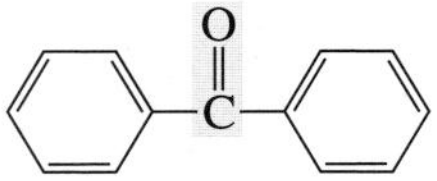

common name: **acetophenone**	**butyrophenone**	**benzophenone**
derived name: **methyl phenyl ketone**	**phenyl propyl ketone**	**diphenyl ketone**

If the ketone has a second functional group of higher naming priority, the ketone group can be indicated by its substituent name, which is "oxo." As with all substituent names, it is cited at the beginning of the name of the compound.

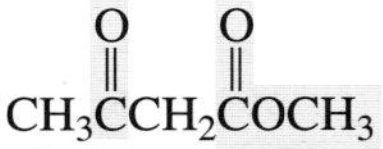

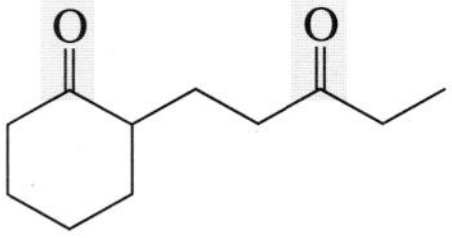

$CH_3CCH_2CH_2CH$	$CH_3CCH_2COCH_3$	
IUPAC name: **4-oxopentanal**	**methyl 3-oxobutanoate**	**2-(3-oxopentyl)-cyclohexanone**

BUTANEDIONE: AN UNPLEASANT COMPOUND

Fresh perspiration is odorless. Certain bacteria that are always present on our skin produce lactic acid. The acidic environment that these bacteria create allows other bacteria to break down components of perspiration, producing compounds with the unappealing odors we associate with armpits and sweaty feet. One such compound is butanedione.

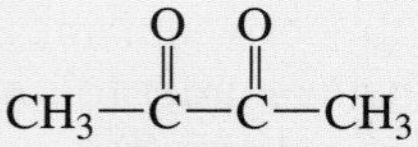

$$CH_3-\overset{O}{\overset{||}{C}}-\overset{O}{\overset{||}{C}}-CH_3$$
butanedione

PROBLEM 2◆

Give two names for each of the following compounds.

a. 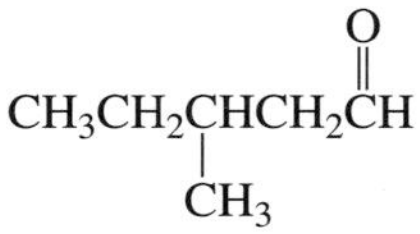 $CH_3CH_2\underset{\underset{CH_3}{|}}{CH}CH_2\overset{O}{\overset{||}{C}}H$

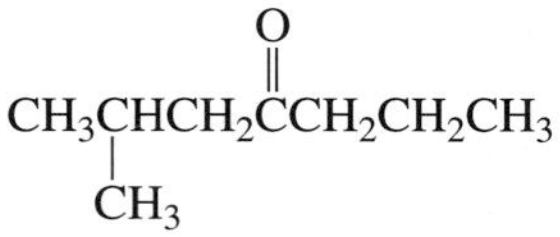

d. 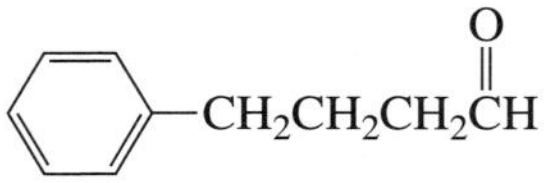 benzene$-CH_2CH_2CH_2\overset{O}{\overset{||}{C}}H$

b. $CH_3CH_2CH_2\overset{O}{\overset{||}{C}}CH_2CH_2CH_3$

e. 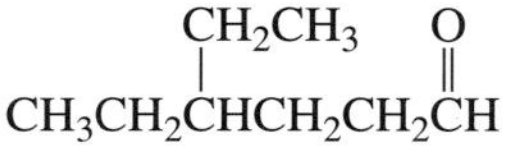 $CH_3CH_2\underset{\underset{CH_2CH_3}{|}}{CH}CH_2CH_2\overset{O}{\overset{||}{C}}H$

c. $CH_3\underset{\underset{CH_3}{|}}{CH}CH_2\overset{O}{\overset{||}{C}}CH_2CH_2CH_3$

f.  $CH_2{=}CH\overset{O}{\overset{||}{C}}CH_2CH_2CH_2CH_3$

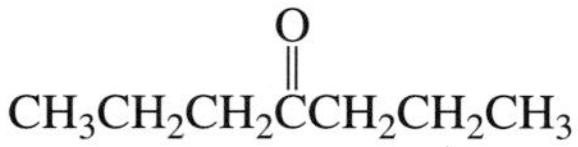

We have seen that the carbonyl group is polar since oxygen, being more electronegative than carbon, has a greater share of the electrons of the double bond. Because of the partial positive charge on the carbonyl carbon, carbonyl compounds are attacked by nucleophiles.

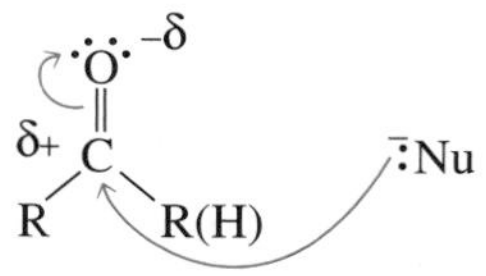

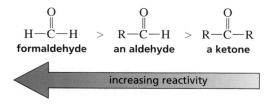

Aldehydes are more reactive toward nucleophilic attack than are ketones. Because alkyl groups are electron-donating compared to a hydrogen, an aldehyde has a greater partial positive charge on its carbonyl carbon, making it more susceptible to attack by a nucleophile.

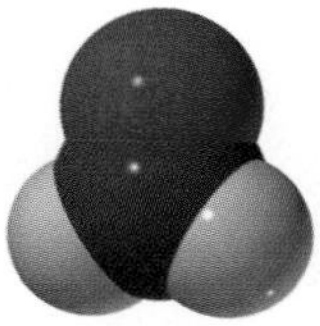

relative reactivities

formaldehyde

$$H-\underset{\underset{\text{formaldehyde}}{}}{\overset{\overset{O}{\|}}{C}}-H \;>\; R-\underset{\underset{\text{an aldehyde}}{}}{\overset{\overset{O}{\|}}{C}}-H \;>\; R-\underset{\underset{\text{a ketone}}{}}{\overset{\overset{O}{\|}}{C}}-R$$

⬅ increasing reactivity

Steric factors also contribute to the greater reactivity of an aldehyde compared with a ketone. Because the hydrogen attached to the acyl group of an aldehyde is smaller than the alkyl group attached to the acyl group of a ketone, the carbonyl carbon of an aldehyde is more accessible to the nucleophile than is the carbonyl carbon of a ketone. For the same reason, ketones with small alkyl groups bonded to the carbonyl carbon are more reactive than ketones with large alkyl groups.

acetaldehyde

relative reactivities

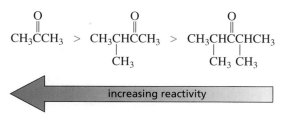

acetone

⬅ increasing reactivity

PROBLEM 3◆

Which ketone is more reactive?

a. 2-heptanone or 4-heptanone

b. *p*-nitroacetophenone or *p*-methoxyacetophenone

How does the reactivity of an aldehyde or a ketone toward nucleophiles compare with the reactivity of the carbonyl compounds whose reactions we studied in Chapter 15? They are right in the middle. Aldehydes and ketones are less reactive than acyl halides and acid anhydrides but are more reactive than esters, carboxylic acids, and amides.

relative reactivities of carbonyl compounds

acyl halide > acid anhydride > aldehyde > ketone > ester ~ carboxylic acid > amide

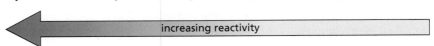

increasing reactivity

Because the carbonyl compounds discussed in Chapter 15 have a pair of non-bonding electrons on an atom attached to the carbonyl group, these compounds have a resonance contributor that aldehydes and ketones do not have. This resonance contributor places some of the positive charge on the atom adjacent to the carbonyl carbon. We have seen that the reactivity of these carbonyl compounds is related to the basicity of Y^-. The weaker the basicity of Y^-, the more reactive the carbonyl group, because weak bases are better able to withdraw electrons inductively from the carbonyl carbon and are less apt to donate electrons by resonance to the carbonyl carbon (Section 15.4).

So aldehydes and ketones are not as reactive as carbonyl compounds in which Y^- is a very weak base (acyl halides and acid anhydrides), but are more reactive than carbonyl compounds in which Y^- is a relatively strong base (carboxylic acids, esters, and amides).

16.3 REACTIVITY CONSIDERATIONS

In Section 15.4 we saw that the acyl group of a carboxylic acid or of a carboxylic acid derivative is attached to a group that can be replaced by another group. These compounds therefore react with nucleophiles to form substitution products.

$$R-\underset{\underset{O}{\|}}{C}-Y \ + \ Z^- \ \longrightarrow \ R-\underset{\underset{O}{\|}}{C}-Z \ + \ Y^-$$

product of nucleophilic substitution

In contrast, the acyl group of an aldehyde or a ketone is attached to a group (H or R) that is too basic (H^- or R^-) to be replaced by another group. Therefore, aldehydes and ketones react with nucleophiles to form addition compounds. Thus, aldehydes and ketones undergo **nucleophilic addition reactions,** while carboxylic acid derivatives undergo **nucleophilic acyl substitution reactions.**

$$R-\underset{\underset{O}{\|}}{C}-R \ + \ HZ \ \longrightarrow \ R-\underset{\underset{Z}{|}}{\overset{\overset{OH}{|}}{C}}-R$$

product of nucleophilic addition

If the nucleophile that adds to the aldehyde or ketone is a good nucleophile, it will readily attack the carbonyl carbon. The resulting addition product can be protonated either by the solvent or by added acid.

$$R-\overset{\overset{\displaystyle O}{\|}}{C}-R \ + \ Z\overset{..}{:} \ \longrightarrow \ R-\overset{\overset{\displaystyle O^-}{|}}{\underset{\underset{\displaystyle Z}{|}}{C}}-R \ \underset{-H^+}{\overset{H^+}{\rightleftharpoons}} \ R-\overset{\overset{\displaystyle OH}{|}}{\underset{\underset{\displaystyle Z}{|}}{C}}-R$$

A poor nucleophile requires an acid catalyst to make the nucleophilic addition reaction occur at a reasonable rate. The acid protonates the carbonyl oxygen, which increases the susceptibility of the carbonyl carbon to nucleophilic attack (Section 15.9).

$$R-\overset{\overset{\displaystyle O}{\|}}{C}-R \ \underset{-H^+}{\overset{H^+}{\rightleftharpoons}} \ R-\overset{\overset{\displaystyle \overset{+}{O}H}{\|}}{C}-R \ + \ Z\overset{..}{:} \ \longrightarrow \ R-\overset{\overset{\displaystyle OH}{|}}{\underset{\underset{\displaystyle Z}{|}}{C}}-R$$

If the attacking atom of the nucleophile has a pair of nonbonding electrons in the addition product, water will be eliminated from the addition product. This is called a **nucleophilic addition–elimination reaction.**

$$R-\overset{\overset{\displaystyle OH}{|}}{\underset{\underset{\displaystyle :Z}{|}}{C}}-R \ \underset{-H^+}{\overset{H^+}{\rightleftharpoons}} \ R-\overset{\overset{\displaystyle \overset{\displaystyle H}{+OH}}{|}}{\underset{\underset{\displaystyle :Z}{|}}{C}}-R \ \rightleftharpoons \ R-\overset{\overset{\displaystyle OH}{\|}}{\underset{\underset{\displaystyle Z^+}{}}{C}}-R \ + \ H_2O$$

16.4 ADDITION OF CARBON NUCLEOPHILES

Few reactions in organic chemistry result in formation of new carbon–carbon bonds. Consequently, those reactions that do are very important to synthetic organic chemists when they need to synthesize larger organic molecules from smaller molecules. The addition of a carbon nucleophile to a carbonyl compound is an example of a reaction that forms a new carbon–carbon bond and, therefore, increases the number of carbon atoms in the starting material.

Addition of Grignard Reagents

Addition of a Grignard reagent to a carbonyl compound is a versatile reaction that leads to formation of a new carbon–carbon bond. Because both the structure of the carbonyl compound and the structure of the Grignard reagent can be varied, the reaction can produce compounds with a variety of structures. We saw in Section 11.8 that a Grignard reagent can be prepared by adding an alkyl halide to magnesium shavings in diethyl ether. We also saw that a Grignard reagent reacts as if it were a carbanion.

$$CH_3CH_2Br \ \xrightarrow[\text{ether}]{\text{Mg}} \ CH_3CH_2MgBr$$

$$CH_3CH_2MgBr \quad \text{reacts as if it were} \quad CH_3\overset{-}{C}H_2 \ \overset{+}{M}gBr$$

Attack of a Grignard reagent on a carbonyl carbon forms an **oxyanion** (an ion with a negatively charged oxygen) that is complexed with magnesium ion. Addition of water or dilute acid breaks up the complex. When a Grignard reagent reacts with an aldehyde, the addition product formed is a secondary alcohol.

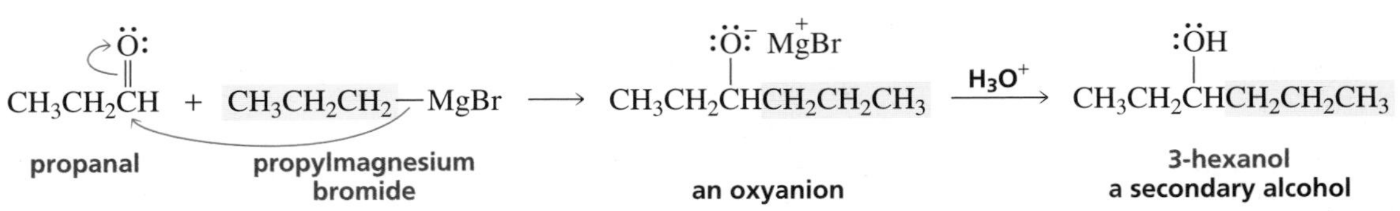

$$CH_3CH_2\overset{\displaystyle\overset{\displaystyle :\!\ddot{O}:}{\|}}{C}H + CH_3CH_2CH_2\!\!-\!\!MgBr \longrightarrow CH_3CH_2\overset{\displaystyle :\!\ddot{O}^-\ \overset{+}{M}gBr}{\underset{\displaystyle |}{C}}HCH_2CH_2CH_3 \overset{H_3O^+}{\longrightarrow} CH_3CH_2\overset{\displaystyle :\!\ddot{O}H}{\underset{\displaystyle |}{C}}HCH_2CH_2CH_3$$

propanal	**propylmagnesium bromide**	**an oxyanion**	**3-hexanol a secondary alcohol**

When a Grignard reagent reacts with a ketone, a tertiary alcohol is formed.

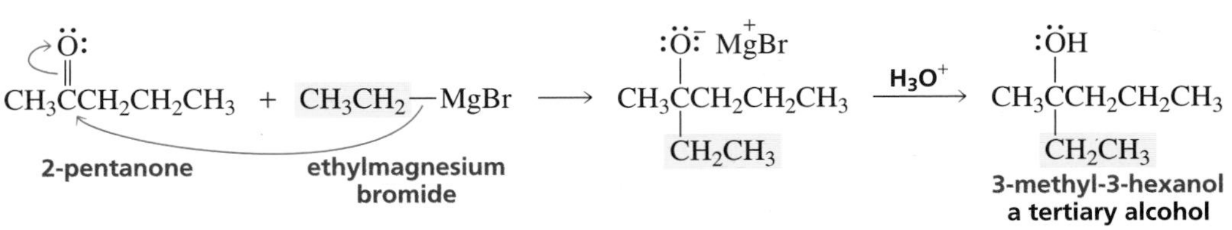

$$CH_3\overset{\displaystyle\overset{\displaystyle :\!\ddot{O}:}{\|}}{C}CH_2CH_2CH_3 + CH_3CH_2\!\!-\!\!MgBr \longrightarrow CH_3\overset{\displaystyle :\!\ddot{O}^-\ \overset{+}{M}gBr}{\underset{\displaystyle |}{\underset{\displaystyle CH_2CH_3}{C}}}CH_2CH_2CH_3 \overset{H_3O^+}{\longrightarrow} CH_3\overset{\displaystyle :\!\ddot{O}H}{\underset{\displaystyle |}{\underset{\displaystyle CH_2CH_3}{C}}}CH_2CH_2CH_3$$

2-pentanone	**ethylmagnesium bromide**		**3-methyl-3-hexanol a tertiary alcohol**

In the following reactions, numbers are used with the reagents to indicate that the acid is not added until reaction with the Grignard reagent is complete.

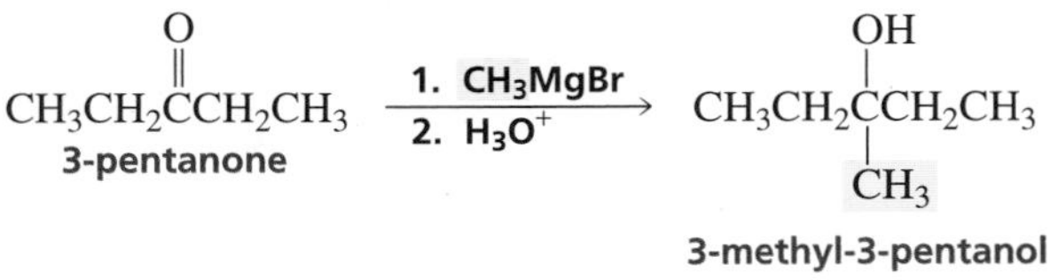

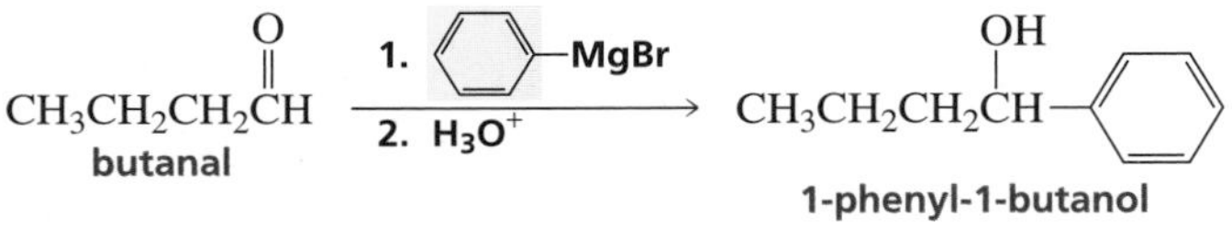

$$CH_3CH_2\overset{\displaystyle\overset{\displaystyle O}{\|}}{C}CH_2CH_3 \quad\overset{\textbf{1. CH}_3\textbf{MgBr}}{\underset{\textbf{2. H}_3\textbf{O}^+}{\longrightarrow}}\quad CH_3CH_2\overset{\displaystyle OH}{\underset{\displaystyle \underset{\displaystyle CH_3}{|}}{\underset{|}{C}}}CH_2CH_3$$

3-pentanone

3-methyl-3-pentanol

$$CH_3CH_2CH_2\overset{\displaystyle\overset{\displaystyle O}{\|}}{C}H \quad\overset{\textbf{1.}\ \text{◯}\!\!-\!\!\textbf{MgBr}}{\underset{\textbf{2. H}_3\textbf{O}^+}{\longrightarrow}}\quad CH_3CH_2CH_2\overset{\displaystyle OH}{\underset{\displaystyle |}{C}}H\!\!-\!\!\text{◯}$$

butanal

1-phenyl-1-butanol

A Grignard reagent can react with carbon dioxide. The product of the reaction is a carboxylic acid with one more carbon atom than the Grignard reagent.

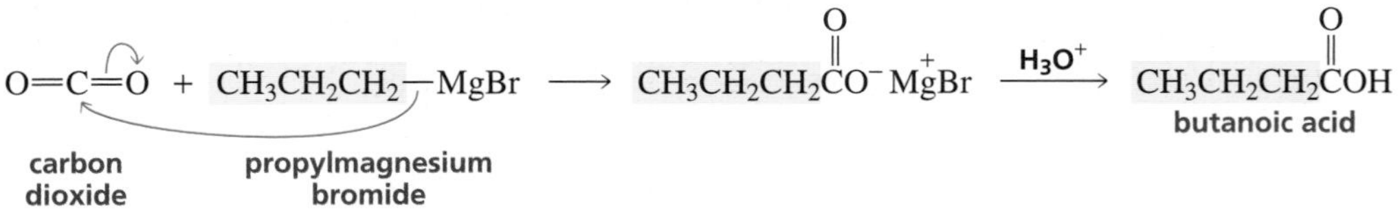

$$O\!\!=\!\!C\!\!=\!\!O + CH_3CH_2CH_2\!\!-\!\!MgBr \longrightarrow CH_3CH_2CH_2\overset{\displaystyle\overset{\displaystyle O}{\|}}{C}O^-\ \overset{+}{M}gBr \overset{H_3O^+}{\longrightarrow} CH_3CH_2CH_2\overset{\displaystyle\overset{\displaystyle O}{\|}}{C}OH$$

carbon dioxide	**propylmagnesium bromide**		**butanoic acid**

PROBLEM 4◆

a. How many isomers are obtained from the reaction of 2-pentanone with ethylmagnesium bromide followed by H_3O^+? Name the isomers.

b. How can a primary alcohol be prepared from a carbonyl compound and a Grignard reagent?

PROBLEM 5◆

We saw that 3-methyl-3-hexanol can be synthesized from the reaction of 2-pentanone with ethylmagnesium bromide. Give two other combinations of ketone and Grignard reagent that could be used to prepare the same tertiary alcohol.

Addition of Acetylide Anions

We have seen that a terminal alkyne can be converted into an acetylide anion by a strong base (Section 5.9).

$$CH_3C{\equiv}CH \xrightarrow[\text{NH}_3]{\text{NaNH}_2} CH_3C{\equiv}C{:}^-$$

An acetylide anion is another example of a carbon nucleophile that adds to aldehydes and ketones. When the reaction is over, acid is added to the reaction mixture to protonate the oxyanion.

$$CH_3CH_2\overset{\overset{\displaystyle :\ddot{O}:}{\|}}{C}H + CH_3C{\equiv}C{:}^- \longrightarrow CH_3CH_2\overset{\overset{\displaystyle :\ddot{O}:^-}{|}}{C}HC{\equiv}CCH_3 \xrightarrow{\text{H}^+} CH_3CH_2\overset{\overset{\displaystyle :\ddot{O}H}{|}}{C}HC{\equiv}CCH_3$$

Addition of Hydrogen Cyanide

Hydrogen cyanide adds to aldehydes and ketones to form **cyanohydrins.** This reaction increases the number of carbon atoms in the reactant by one. In the first step of the reaction, the cyanide ion attacks the carbonyl carbon. The oxyanion then accepts a proton from an undissociated molecule of hydrogen cyanide.

$$CH_3\overset{\overset{\displaystyle :\ddot{O}:}{\|}}{C}CH_3 + {:}C{\equiv}N \rightleftharpoons CH_3\overset{\overset{\displaystyle :\ddot{O}:^-}{|}}{\underset{\underset{\displaystyle CH_3}{|}}{C}}C{\equiv}N \xrightarrow{\overset{\displaystyle H-C{\equiv}N}{}} CH_3\overset{\overset{\displaystyle :\ddot{O}H}{|}}{\underset{\underset{\displaystyle CH_3}{|}}{C}}C{\equiv}N + {:}C{\equiv}N$$

acetone   acetone cyanohydrin

Because hydrogen cyanide is a toxic gas, the best way to carry out this reaction is to generate hydrogen cyanide during the reaction by adding HCl to a mixture of the aldehyde or ketone and excess sodium cyanide. Excess sodium cyanide is used in order to make sure that some cyanide ion is available to act as a nucleophile.

Compared with other carbon nucleophiles, cyanide ion is a relatively weak base (pK_a of HC$\equiv$N is 9.14; pK_a of HC$\equiv$CH is 25; pK_a of CH_3CH_3 is ~50), which means that the cyano group is the most easily eliminated from the addition product. Cyanohydrins, however, are stable because the neutral OH group does not have sufficient driving force to eliminate the cyano group. But if the OH group loses its proton, the cyano group will be eliminated. Therefore, in basic solutions, a cyanohydrin is converted back to the carbonyl compound.

cyclohexanone
cyanohydrin

The addition of hydrogen cyanide to aldehydes and ketones is a synthetically useful reaction because of subsequent reactions that can be carried out on the cyanohydrin. For example, acid-catalyzed hydrolysis of a cyanohydrin forms an α-hydroxy carboxylic acid (Section 15.15).

$$CH_3CH_2\overset{\overset{\displaystyle OH}{|}}{\underset{\underset{\displaystyle CH_2CH_3}{|}}{C}}{-}C{\equiv}N \xrightarrow[\Delta]{\text{H}^+,\ \text{H}_2\text{O}} CH_3CH_2\overset{\overset{\displaystyle OH}{|}}{\underset{\underset{\displaystyle CH_2CH_3}{|}}{C}}{-}\overset{\overset{\displaystyle O}{\|}}{C}OH$$

a cyanohydrin   an α-hydroxy carboxylic acid

The catalytic addition of hydrogen to a cyanohydrin produces a primary amine with an OH group on the β-carbon.

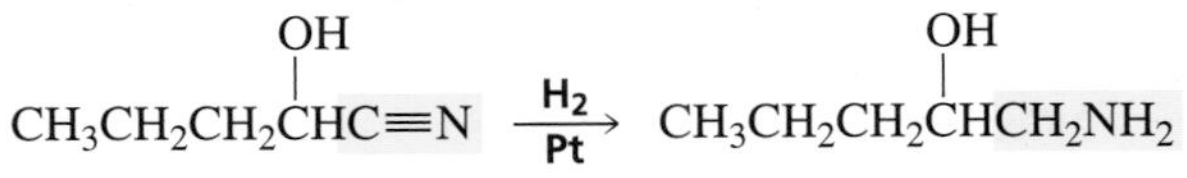

$$\underset{\text{OH}}{\text{CH}_3\text{CH}_2\text{CH}_2\overset{|}{\text{CH}}\text{C}\equiv\text{N}} \xrightarrow[\text{Pt}]{\text{H}_2} \underset{\text{OH}}{\text{CH}_3\text{CH}_2\text{CH}_2\overset{|}{\text{CH}}\text{CH}_2\text{NH}_2}$$

PROBLEM 6 ◆

Describe two ways to convert 1-bromobutane into pentanoic acid.

PROBLEM 7 ◆

Can a cyanohydrin be prepared by treating a ketone with sodium cyanide?

PROBLEM 8

Although aldehydes and ketones react with a weak acid such as hydrogen cyanide in the presence of $^-\text{C}\equiv\text{N}$, they do not react with strong acids such as HCl or H_2SO_4 in the presence of Cl^- or HSO_4^-. Explain.

PROBLEM 9 / SOLVED

How can the following compounds be prepared, starting with a carbonyl compound with one fewer carbon atom than the desired product?

a. $HOCH_2CH_2NH_2$

b. 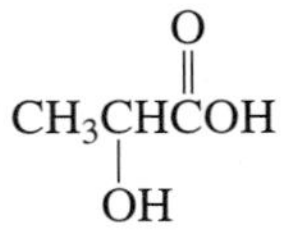 $\underset{\overset{|}{\text{OH}}}{\text{CH}_3\overset{\overset{\displaystyle O}{\|}}{\text{CH}}\text{COH}}$

SOLUTION TO 9a   The starting material for the synthesis of the two-carbon compound must be formaldehyde, the only carbonyl compound with one carbon. Addition of hydrogen cyanide followed by addition of H_2 to the triple bond of the cyanohydrin forms the desired compound.

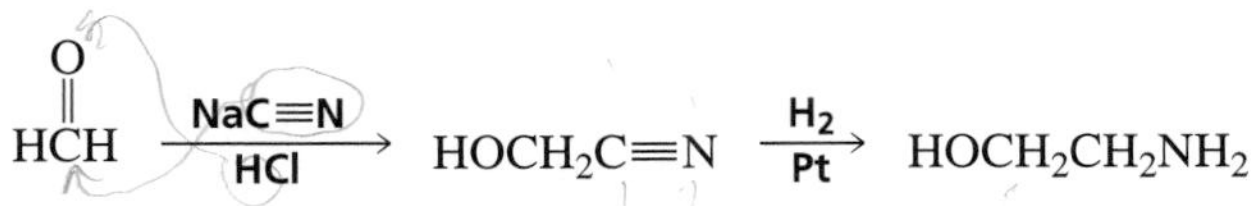

$$\underset{\text{HCH}}{\overset{\overset{\displaystyle O}{\|}}{}} \xrightarrow[\text{HCl}]{\text{NaC}\equiv\text{N}} HOCH_2\text{C}\equiv\text{N} \xrightarrow[\text{Pt}]{\text{H}_2} HOCH_2CH_2NH_2$$

SOLUTION TO 9b   The starting material for the synthesis of the three-carbon α-hydroxy carboxylic acid must be ethanal. Addition of hydrogen cyanide followed by hydrolysis of the cyanohydrin forms the target compound.

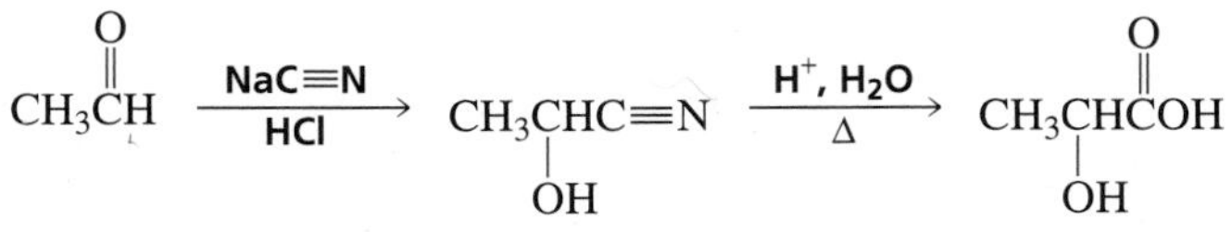

$$\underset{\text{CH}_3\text{CH}}{\overset{\overset{\displaystyle O}{\|}}{}} \xrightarrow[\text{HCl}]{\text{NaC}\equiv\text{N}} \underset{\overset{|}{\text{OH}}}{\text{CH}_3\text{CHC}\equiv\text{N}} \xrightarrow[\Delta]{\text{H}^+, \text{H}_2\text{O}} \underset{\overset{|}{\text{OH}}}{\text{CH}_3\overset{\overset{\displaystyle O}{\|}}{\text{CH}}\text{COH}}$$

Addition of hydride ion to an aldehyde or ketone forms an oxyanion that can subsequently be protonated by adding water or a weak acid to the reaction mixture. The overall reaction adds H_2 to the carbonyl double bond. The addition of hydrogen to an organic compound is a **reduction reaction** (Section 17.1).

**16.5
ADDITION OF
HYDRIDE ION**

$$
\underset{\substack{\text{O}\\ \| \\ \text{R}-\text{C}-\text{R}}}{} + \; :\text{H}^- \longrightarrow \underset{\substack{\text{O}^-\\ | \\ \text{R}-\text{C}-\text{R}\\ | \\ \text{H}}}{} \xrightarrow{\text{H}^+} \underset{\substack{\text{OH}\\ | \\ \text{R}-\text{C}-\text{R}\\ | \\ \text{H}}}{}
$$

Aldehydes and ketones are reduced using sodium borohydride ($NaBH_4$) as the source of hydride ion. Aldehydes are reduced to primary alcohols, and ketones are reduced to secondary alcohols. Notice that the source of protons (water or a weak acid) is not added to the reaction mixture until reaction with the hydride donor is complete.

$$
\underset{\substack{\text{butanal}\\ \text{an aldehyde}}}{\underset{\substack{\text{O}\\ \|}}{\text{CH}_3\text{CH}_2\text{CH}_2\text{CH}}} \xrightarrow[\text{2. H}_2\text{O}]{\text{1. NaBH}_4} \underset{\substack{\text{1-butanol}\\ \text{a primary alcohol}}}{\text{CH}_3\text{CH}_2\text{CH}_2\text{CH}_2\text{OH}}
$$

$$
\underset{\substack{\text{2-pentanone}\\ \text{a ketone}}}{\underset{\substack{\text{O}\\ \|}}{\text{CH}_3\text{CH}_2\text{CH}_2\text{CCH}_3}} \xrightarrow[\text{2. H}_2\text{O}]{\text{1. NaBH}_4} \underset{\substack{\text{2-pentanol}\\ \text{a secondary alcohol}}}{\underset{\substack{\text{OH}\\ |}}{\text{CH}_3\text{CH}_2\text{CH}_2\text{CHCH}_3}}
$$

PROBLEM 10◆

What alcohols are obtained from reduction of the following compounds with sodium borohydride?

a. 2-methylpropanal   **c.** benzaldehyde

b. cyclohexanone   **d.** acetophenone

Reaction with Grignard Reagents

**16.6
REACTIONS OF
CARBONYL
COMPOUNDS THAT
HAVE LEAVING
GROUPS WITH
GRIGNARD
REAGENTS AND
HYDRIDE ION
DONORS**

In addition to reacting with aldehydes and ketones, Grignard reagents and hydride ion donors also react with carbonyl compounds that have leaving groups (the Class I carbonyl compounds discussed in Chapter 15).

The reaction of a carbonyl compound that has a leaving group with a Grignard reagent involves two successive reactions with the nucleophile. For example, when an ester reacts with a Grignard reagent, the first reaction is a nucleophilic acyl substitution reaction because an ester, unlike an aldehyde or a ketone, has a group that can be replaced by the Grignard reagent. The product of the reaction is a ketone. The reaction does not stop at the ketone stage, because ketones are more reactive than esters toward nucleophilic attack. Reaction of the ketone with a second molecule of the Grignard reagent forms a tertiary alcohol. Because the tertiary alcohol is formed as a result of two successive reactions with a Grignard reagent, the alcohol has two identical groups on the tertiary carbon.

mechanism for the reaction of an ester with a Grignard reagent

$$CH_3CH_2COCH_3 + CH_3{-}MgBr \longrightarrow CH_3CH_2\overset{\overset{\displaystyle :\ddot{O}:^-\;MgBr^+}{|}}{\underset{\underset{\displaystyle CH_3}{|}}{C}}{-}OCH_3 \longrightarrow CH_3CH_2\overset{\overset{\displaystyle :\ddot{O}:}{\|}}{C}CH_3 + CH_3O^-$$

an ester   a ketone

$$\downarrow CH_3{-}MgBr$$

$$CH_3CH_2\overset{\overset{\displaystyle :\ddot{O}H}{|}}{\underset{\underset{\displaystyle CH_3}{|}}{C}}CH_3 \xleftarrow{H_3O^+} CH_3CH_2\overset{\overset{\displaystyle :\ddot{O}:^-\;MgBr^+}{|}}{\underset{\underset{\displaystyle CH_3}{|}}{C}}CH_3$$

a tertiary alcohol

Tertiary alcohols are also formed from the reaction of two equivalents of a Grignard reagent with an acyl halide.

$$CH_3CH_2CH_2\overset{\overset{\displaystyle O}{\|}}{C}Cl \xrightarrow[\text{2. } H_3O^+]{\text{1. 2 } CH_3CH_2MgBr} CH_3CH_2CH_2\overset{\overset{\displaystyle OH}{|}}{\underset{\underset{\displaystyle CH_2CH_3}{|}}{C}}CH_2CH_3$$

butyryl chloride

3-ethyl-3-hexanol

In theory, we should be able to stop this reaction at the ketone stage, because a ketone is less reactive than an acyl halide. However, because the Grignard reagent is itself so reactive, it can be prevented from reacting with the ketone only under very carefully controlled conditions. There are better ways to synthesize ketones (Appendix IV).

PROBLEM 11 ◆

What product would be obtained from the reaction of one equivalent of a carboxylic acid with one equivalent of a Grignard reagent?

PROBLEM 12 / SOLVED

a. Which of the following tertiary alcohols cannot be prepared by the reaction of an ester with excess Grignard reagent?

b. For those alcohols that can be prepared by the reaction of an ester with excess Grignard reagent, what ester and what Grignard reagent should be used?

(1) $CH_3\overset{\overset{\displaystyle OH}{|}}{\underset{\underset{\displaystyle CH_3}{|}}{C}}CH_3$ (3) $CH_3CH_2\overset{\overset{\displaystyle OH}{|}}{\underset{\underset{\displaystyle CH_3}{|}}{C}}CH_2CH_2CH_3$ (5) $CH_3\overset{\overset{\displaystyle OH}{|}}{\underset{\underset{\displaystyle CH_2CH_3}{|}}{C}}CH_2CH_2CH_2CH_3$

(2) $CH_3\overset{\overset{\displaystyle OH}{|}}{\underset{\underset{\displaystyle CH_3}{|}}{C}}CH_2CH_3$ (4) $CH_3CH_2\overset{\overset{\displaystyle OH}{|}}{\underset{\underset{\displaystyle CH_3}{|}}{C}}CH_2CH_3$ (6) $\displaystyle \overset{\overset{\displaystyle OH}{|}}{\underset{\underset{\displaystyle CH_3}{|}}{C}}$ (diphenyl)

SOLUTION TO 12a A tertiary alcohol is obtained from the reaction of an ester with two equivalents of a Grignard reagent. Therefore, tertiary alcohols prepared in this way must have two identical substituents on the carbon to which the OH is bonded, because both substituents come from the Grignard reagent. Alcohols (3) and (5) cannot be prepared in this way because they do not have two identical substituents.

> **SOLUTION TO 12b(2)**   Methyl propanoate and excess methylmagnesium bromide.

Reaction with a Hydride Ion Donor

The reaction of a carbonyl compound that has a leaving group with a hydride ion donor—like the reaction with a Grignard reagent—involves two successive reactions with the nucleophile. Sodium borohydride is not a sufficiently strong hydride donor to react with the less reactive esters and carboxylic acids (compared with aldehydes and ketones), so esters and carboxylic acids must be reduced using lithium aluminum hydride ($LiAlH_4$).

Although both sodium borohydride and lithium aluminum hydride are sources of hydride ion, sodium borohydride, because it is less reactive, is safer and easier to use. Lithium aluminum hydride must be used in a dry, aprotic solvent because it reacts violently with protic solvents. Depending on the reactant, $NaBH_4$ may be chosen for its ease of safe handling, or $LiAlH_4$ may be chosen for its greater reactivity.

The reaction of an ester with $LiAlH_4$ produces two alcohols, one corresponding to the acyl portion of the ester and one corresponding to the alkyl portion.

$$\underset{\substack{\text{methyl propanoate}\\\text{an ester}}}{CH_3CH_2\overset{\displaystyle O}{\overset{\|}{C}}OCH_3} \xrightarrow[\text{2. } H_2O]{\text{1. } LiAlH_4} \underset{\text{1-propanol}}{CH_3CH_2CH_2OH} + \underset{\text{methanol}}{CH_3OH}$$

When an ester reacts with hydride ion, the first reaction is a nucleophilic acyl substitution reaction, because an ester has a group that can be substituted by hydride ion. The product of this reaction is an aldehyde. The aldehyde then undergoes a nucleophilic addition reaction with a second equivalent of hydride ion, forming a primary alcohol. The reaction cannot be stopped at the aldehyde stage, because an aldehyde is more reactive than an ester toward nucleophilic attack.

mechanism for the reaction of an ester with hydride ion

$$\underset{\text{an ester}}{CH_3CH_2\overset{\displaystyle \ddot{O}:}{\overset{\|}{C}}OCH_3} + H{-}\bar{A}lH_3 \longrightarrow CH_3CH_2\overset{\displaystyle :\ddot{O}:^-}{\underset{\displaystyle H}{\overset{|}{C}}}{-}OCH_3 \longrightarrow \underset{\text{an aldehyde}}{CH_3CH_2\overset{\displaystyle \ddot{O}:}{\overset{\|}{C}}H} + CH_3O^-$$

$$\Big\downarrow H{-}\bar{A}lH_3$$

$$\underset{\text{an alcohol}}{CH_3CH_2CH_2OH} \xleftarrow{H_3O^+} CH_3CH_2\overset{\displaystyle :\ddot{O}:^-}{\underset{\displaystyle H}{\overset{|}{C}}H}$$

If diisobutylaluminum hydride (DIBAH) is used as the hydride donor, the reaction can be stopped after the addition of one equivalent of hydride ion. This reaction allows esters to be converted into aldehydes.

$$\underset{\text{methyl pentanoate}}{CH_3CH_2CH_2CH_2\overset{\displaystyle O}{\overset{\|}{C}}OCH_3} \xrightarrow[\text{2. } H_2O]{\text{1. DIBAH}} \underset{\text{pentanal}}{CH_3CH_2CH_2CH_2\overset{\displaystyle O}{\overset{\|}{C}}H}$$

The reaction of a carboxylic acid with LiAlH₄ forms a single primary alcohol.

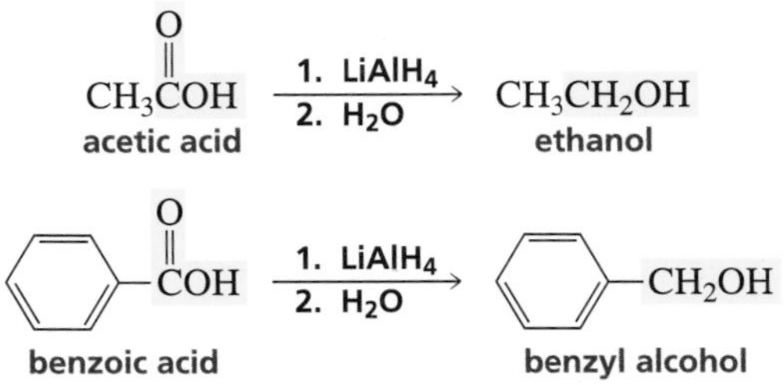

$$
\begin{array}{ccc}
\underset{\text{acetic acid}}{CH_3\overset{\displaystyle O}{\overset{\|}{C}}OH} & \xrightarrow[\text{2. H}_2\text{O}]{\text{1. LiAlH}_4} & \underset{\text{ethanol}}{CH_3CH_2OH}
\end{array}
$$

$$
\begin{array}{ccc}
\underset{\text{benzoic acid}}{C_6H_5\overset{\displaystyle O}{\overset{\|}{C}}OH} & \xrightarrow[\text{2. H}_2\text{O}]{\text{1. LiAlH}_4} & \underset{\text{benzyl alcohol}}{C_6H_5-CH_2OH}
\end{array}
$$

In the first step of the reaction, hydride ion reacts with the acidic hydrogen of the carboxylic acid, forming H₂. Then, as in the reduction of an ester by LiAlH₄, two successive additions of hydride ion take place, with an aldehyde being formed as an intermediate. Notice that ⁻OR is the leaving group in the reduction of an ester, while ⁻OAlH₂ is the leaving group in the reduction of a carboxylic acid.

mechanism for the reaction of a carboxylic acid with hydride ion

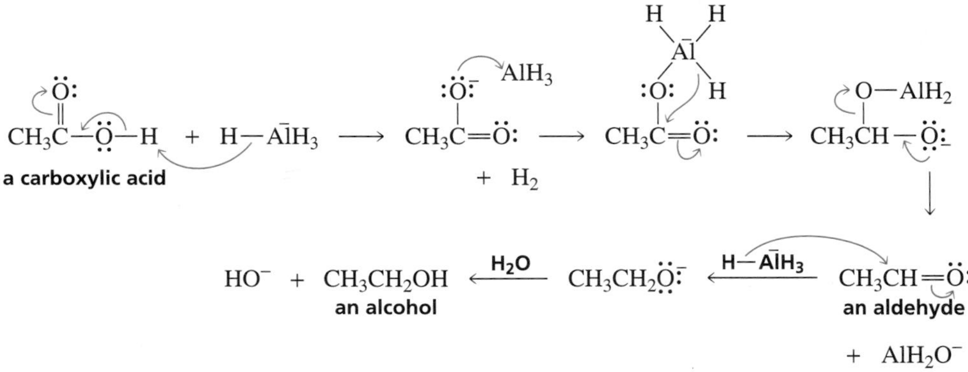

Acyl chlorides, like esters and carboxylic acids, undergo two successive additions of hydride ion when treated with LiAlH₄.

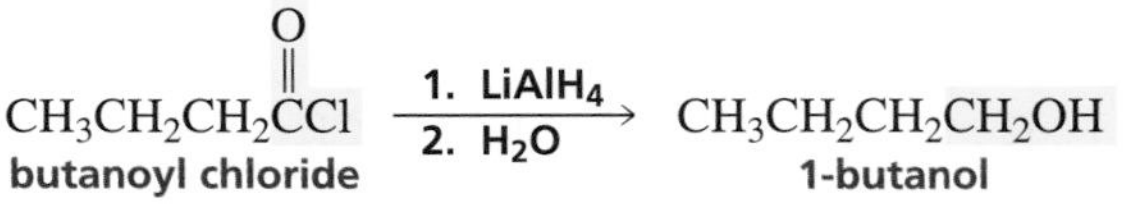

$$
\begin{array}{ccc}
\underset{\text{butanoyl chloride}}{CH_3CH_2CH_2\overset{\displaystyle O}{\overset{\|}{C}}Cl} & \xrightarrow[\text{2. H}_2\text{O}]{\text{1. LiAlH}_4} & \underset{\text{1-butanol}}{CH_3CH_2CH_2CH_2OH}
\end{array}
$$

If a poorer and sterically bulky hydride donor, such as lithium tri-*tert*-butoxyaluminum hydride, is used at low temperatures, the reaction can be stopped after one addition of hydride ion. In this way, an acyl halide can be converted into an aldehyde.

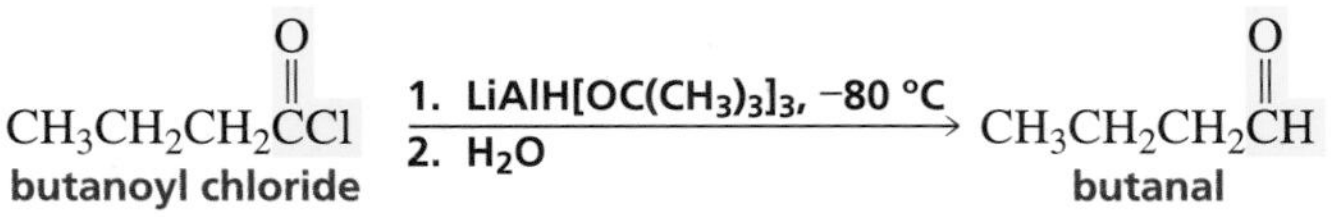

$$
\begin{array}{ccc}
\underset{\text{butanoyl chloride}}{CH_3CH_2CH_2\overset{\displaystyle O}{\overset{\|}{C}}Cl} & \xrightarrow[\text{2. H}_2\text{O}]{\text{1. LiAlH[OC(CH}_3)_3]_3,\ -80\ °C} & \underset{\text{butanal}}{CH_3CH_2CH_2\overset{\displaystyle O}{\overset{\|}{C}}H}
\end{array}
$$

Amides also undergo two successive additions of hydride ion when they react with LiAlH₄. The product of the reaction is an amine. Primary, secondary, and tertiary amines can be formed, depending on the number of substituents on the nitrogen of the amide. Overall, the reaction converts a carbonyl group into a methylene group.

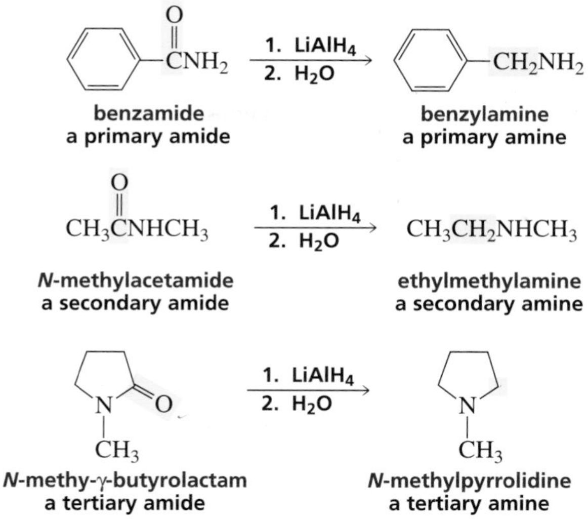

benzamide
a primary amide

benzylamine
a primary amine

CH_3CNHCH_3

N-methylacetamide
a secondary amide

$CH_3CH_2NHCH_3$

ethylmethylamine
a secondary amine

N-methy-γ-butyrolactam
a tertiary amide

N-methylpyrrolidine
a tertiary amine

The mechanism of the reaction shows why the product of the reaction is an amine.

mechanism for the reaction of a primary or secondary amide with hydride ion

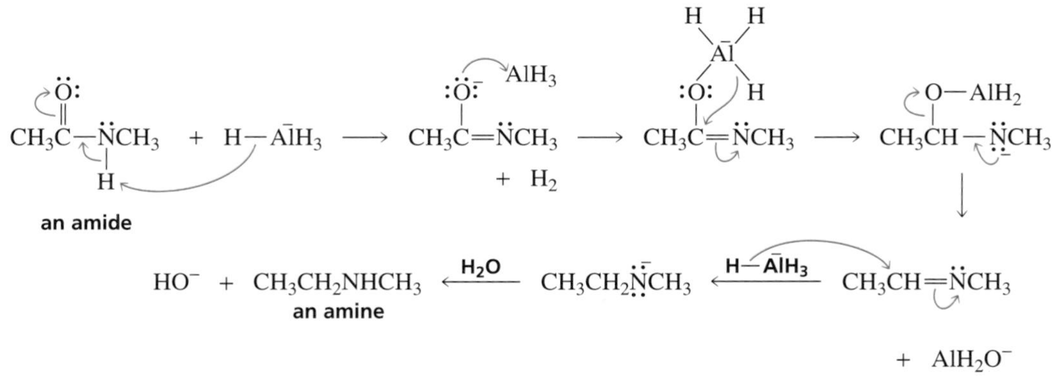

an amide

$+ H_2$

$HO^- + CH_3CH_2NHCH_3 \xleftarrow{H_2O} CH_3CH_2\ddot{N}CH_3 \xleftarrow{H-\bar{A}lH_3} CH_3CH=\ddot{N}CH_3$

an amine

$+ AlH_2O^-$

mechanism for the reaction of a tertiary amide with hydride ion

$CH_3CH_2NCH_3 + AlH_2O^-$
$\quad\quad\quad |$
$\quad\quad\quad CH_3$

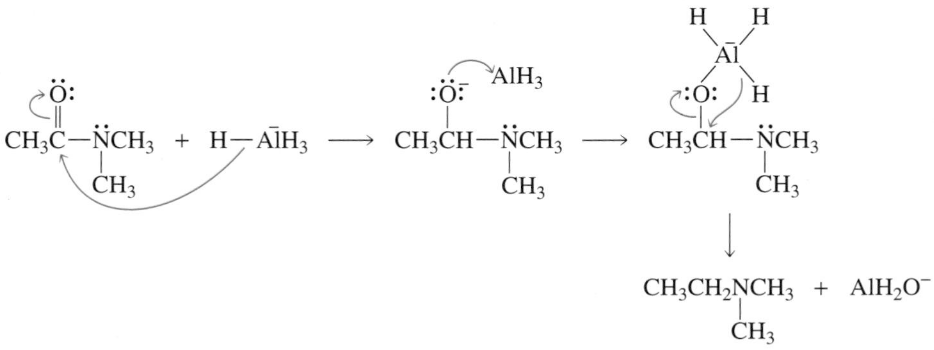

PROBLEM 13◆

What amides would you treat with LiAlH₄ in order to prepare the following amines?

a. benzylmethylamine   **b.** ethylamine

c. diethylamine　　　　　　　　　　　　d. triethylamine

PROBLEM 14◆

Starting with *N*-benzylbenzamide, how would you make the following?

a. dibenzylamine　　　　　　　　**c.** benzaldehyde

b. benzoic acid　　　　　　　　　**d.** benzyl alcohol

**16.7
ADDITION
OF NITROGEN
NUCLEOPHILES**

Addition of Ammonia, Primary Amines, and Other Ammonia Derivatives

Aldehydes and ketones react with primary amines ($R\text{-}NH_2$) and with other ammonia derivatives ($Z\text{-}NH_2$) to form imines. An **imine** is a compound with a carbon–nitrogen double bond.

$$\begin{matrix} R \\ \diagdown \\ C{=}O \\ \diagup \\ R \end{matrix} \; + \quad H_2NZ \quad \rightleftharpoons \quad \begin{matrix} R \\ \diagdown \\ C{=}NZ \\ \diagup \\ R \end{matrix} \; + \; H_2O$$

aldehyde　　　　a primary amine　　　　an imine
or　　　　　　　　or
ketone　　　　other derivative
　　　　　　　of ammonia

The orbital structure of an imine (Figure 16.1) is similar to the orbital structure of a carbonyl group (Figure 15.1). The imine nitrogen is sp^2 hybridized. One of its sp^2 orbitals forms a σ bond with the imine carbon, one forms a σ bond with a substituent, and the third contains a pair of nonbonding electrons. The p orbital of nitrogen and the p orbital of carbon overlap to form a π bond.

Compounds such as hydroxylamine (NH_2OH), hydrazine (NH_2NH_2), semicarbazide ($NH_2NHCONH_2$), and primary amines are all derivatives of ammonia because each has a substituent in place of one of the hydrogens of ammonia (NH_3).

The imine obtained from the reaction of a carbonyl compound and a primary amine is called a **Schiff base;** the imine obtained from reaction with hydroxylamine is called an **oxime;** the imine obtained from reaction with hydrazine is called a **hydrazone;** and the imine obtained from reaction with semicarbazide is called a **semicarbazone.**

Figure 16.1 ▶
Bonding in an imine.

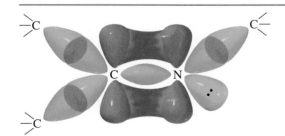

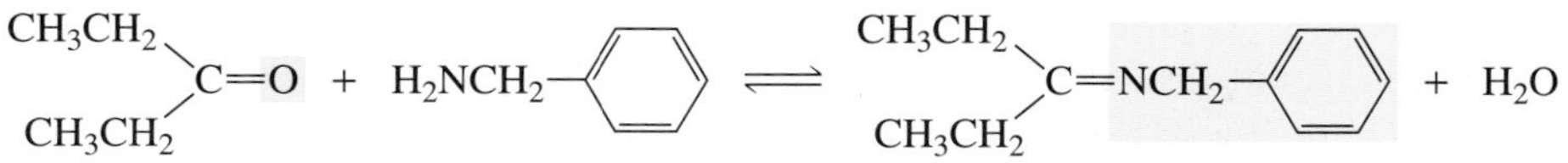

$$\underset{CH_3CH_2}{\overset{CH_3CH_2}{>}}C{=}O \;+\; H_2NCH_2{-}\langle\bigcirc\rangle \;\rightleftharpoons\; \underset{CH_3CH_2}{\overset{CH_3CH_2}{>}}C{=}NCH_2{-}\langle\bigcirc\rangle \;+\; H_2O$$

a primary amine   a Schiff base

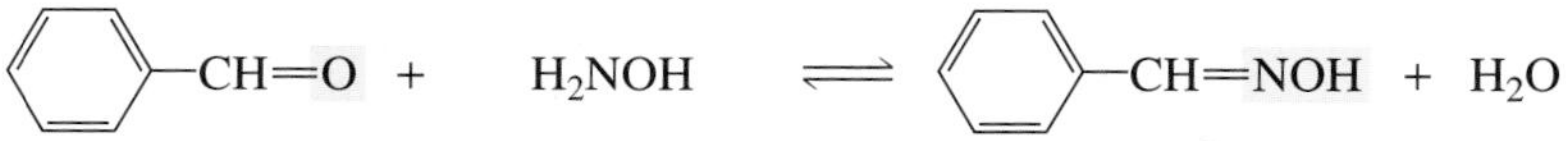

$$\langle\bigcirc\rangle{-}CH{=}O \;+\; H_2NOH \;\rightleftharpoons\; \langle\bigcirc\rangle{-}CH{=}NOH \;+\; H_2O$$

hydroxylamine   an oxime

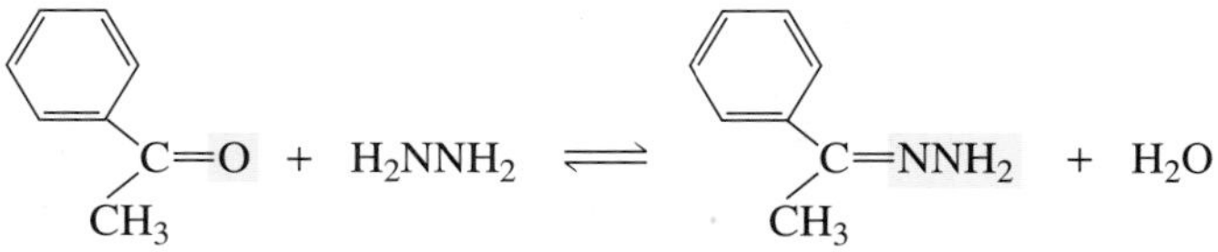

$$\underset{CH_3}{\overset{\langle\bigcirc\rangle}{>}}C{=}O \;+\; H_2NNH_2 \;\rightleftharpoons\; \underset{CH_3}{\overset{\langle\bigcirc\rangle}{>}}C{=}NNH_2 \;+\; H_2O$$

hydrazine   a hydrazone

$$\langle\bigcirc\rangle{=}O \;+\; H_2NNH\overset{O}{\overset{\|}{C}}NH_2 \;\rightleftharpoons\; \langle\bigcirc\rangle{=}NNH\overset{O}{\overset{\|}{C}}NH_2 \;+\; H_2O$$

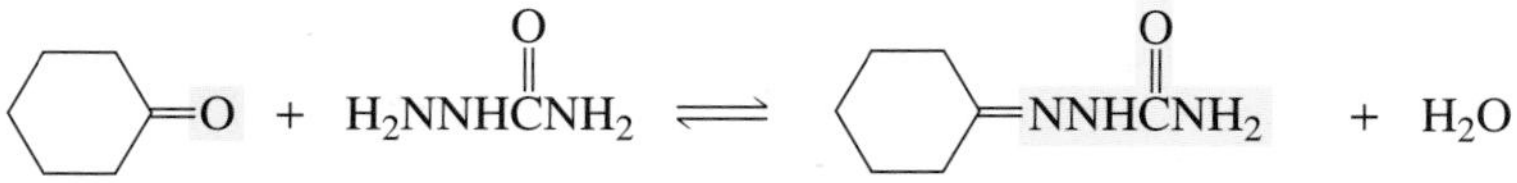

semicarbazide   a semicarbazone

Phenyl-substituted hydrazines react with aldehydes and ketones to form **phenyl-hydrazones.**

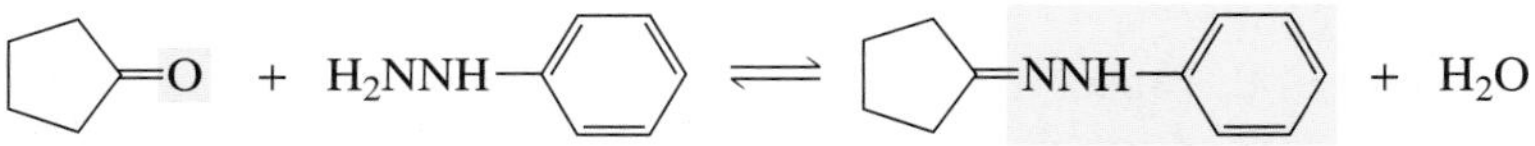

$$\langle\bigcirc\rangle{=}O \;+\; H_2NNH{-}\langle\bigcirc\rangle \;\rightleftharpoons\; \langle\bigcirc\rangle{=}NNH{-}\langle\bigcirc\rangle \;+\; H_2O$$

phenylhydrazine   a phenylhydrazone

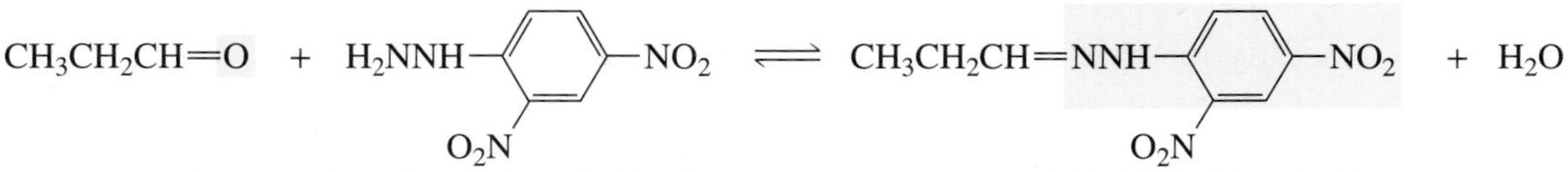

$$CH_3CH_2CH{=}O \;+\; H_2NNH{-}\langle\bigcirc\rangle{-}NO_2 \;\rightleftharpoons\; CH_3CH_2CH{=}NNH{-}\langle\bigcirc\rangle{-}NO_2 \;+\; H_2O$$

O_2N   O_2N
2,4-dinitrophenylhydrazine   a 2,4-dinitrophenylhydrazone

In the first step of the mechanism for imine formation, the amine attacks the carbonyl carbon. Gain of a proton by the oxyanion and loss of a proton by the ammonium ion forms a neutral tetrahedral intermediate. The neutral tetrahedral intermediate is in equilibrium with two protonated forms. Protonation can take place on either the nitrogen or the oxygen atom. Loss of water from the oxygen-protonated intermediate forms a protonated imine that loses a proton to yield the imine.

mechanism for imine formation

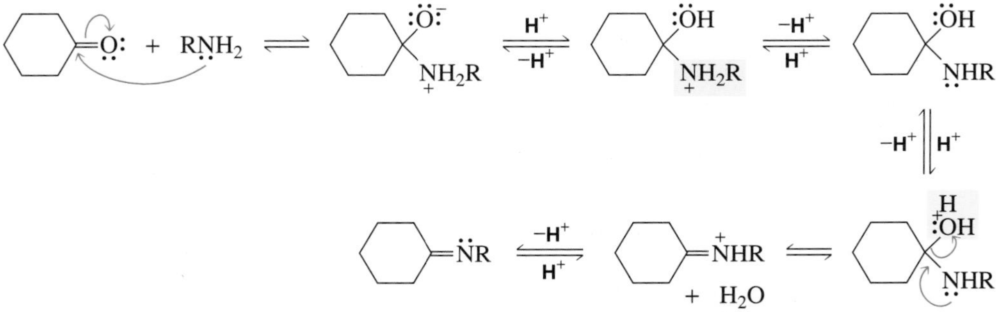

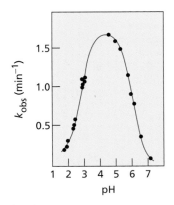

▲ **Figure 16.2**
Dependence of the rate of the reaction of acetone with hydroxylamine on the pH of the reaction mixture.

Because nitrogen is more basic than oxygen, the equilibrium favors the nitrogen-protonated tetrahedral intermediate. The equilibrium can be forced toward the imine by removing water as it is formed or by precipitation of the imine. The formation of all imines (Schiff bases, semicarbazones, oximes, hydrazones, and phenylhydrazones) follows the same mechanism.

The addition of a nitrogen nucleophile to an aldehyde or a ketone is, unlike the addition of a carbon or hydrogen nucleophile, more than a simple addition reaction. The addition of a carbon or hydrogen nucleophile to an aldehyde or a ketone forms a stable tetrahedral compound. But because the nitrogen atom in the tetrahedral compound formed when a nitrogen nucleophile adds to an aldehyde or a ketone still has a nonbonding pair of electrons, the tetrahedral compound is not stable. The nonbonding electrons on nitrogen cause water to be eliminated. Loss of a proton from the resulting protonated imine results in a stable imine. Overall, the reaction is a nucleophilic addition–elimination reaction: nucleophilic addition of amine, followed by elimination of water. Aldehydes and ketones undergo *nucleophilic addition reactions* with carbon and hydrogen nucleophiles, and undergo *nucleophilic addition–elimination* reactions with nitrogen nucleophiles.

The pH at which imine formation is carried out must be carefully controlled. There must be sufficient acid present to protonate the tetrahedral intermediate so that H_2O rather than the much more basic HO^- is the leaving group. However, if too much acid is present, it protonates the reactant amine. Protonated amines are not nucleophiles, so they cannot react with carbonyl groups.

A plot of the observed rate constant for the reaction of acetone with hydroxylamine as a function of the pH of the reaction mixture is shown in Figure 16.2. This type of plot is called a **pH–rate profile.** The pH–rate profile in Figure 16.2 is a bell-shaped curve with the maximum rate occurring at about pH 4.5. As the acidity increases below pH 4.5, the rate of the reaction decreases because more and more of the amine becomes protonated. As a result, less and less of the amine is present in the nucleophilic nonprotonated form. As the acidity decreases above pH 4.5, the rate decreases because less and less of the tetrahedral intermediate is present in the reactive protonated form.

Imine formation is a reversible reaction. In aqueous acidic solutions, imines are hydrolyzed back to the carbonyl compound and amine. In an acidic solution, the amine is protonated and is unable to participate in the reverse reaction. Imine formation and hydrolysis are important reactions in biological systems (Sections 18.20, 21.8, and 22.6).

$$\text{C}_6\text{H}_5\text{—CH}=\text{NCH}_2\text{CH}_3 + H_2O \xrightarrow{\text{H}^+} \text{C}_6\text{H}_5\text{—CH}=\text{O} + \text{CH}_3\text{CH}_2\overset{+}{\text{N}}\text{H}_3$$

Imine hydrolysis is a necessary step in the conversion of a nitrile to a ketone. Reaction of a nitrile with a Grignard reagent forms an imine that is hydrolyzed to a ketone.

$$\text{CH}_3\text{CH}_2\text{C}\equiv\text{N:} + \text{R—MgBr} \longrightarrow \text{CH}_3\text{CH}_2\overset{\overset{\bar{\text{N}}:\ \overset{+}{\text{MgBr}}}{\|}}{\text{C}}\text{—R} \xrightarrow{\text{H}^+} \text{CH}_3\text{CH}_2\overset{\overset{^+\text{NH}_2}{\|}}{\text{C}}\text{—R} \xrightarrow[\text{H}_2\text{O}]{\text{H}^+} \text{CH}_3\text{CH}_2\overset{\overset{\text{O}}{\|}}{\text{C}}\text{—R} + \overset{+}{\text{N}}\text{H}_4$$

NONSPECTROPHOTOMETRIC IDENTIFICATION OF ALDEHYDES AND KETONES

Before spectrophotometric techniques were available, unknown aldehydes and ketones were identified by preparing imine derivatives. For example, let's assume you have an unknown ketone whose boiling point you have determined to be 140 °C. This allows you to narrow the possibilities to five ketones (A to E, below) based on their boiling points. (Ketones boiling at 139 °C and 141 °C cannot be excluded unless your thermometer is calibrated perfectly and your laboratory technique is sensational.) Adding 2,4-dinitro-phenylhydrazine to a sample of the unknown ketone produces crystals that melt at 102 °C. You can now narrow the choice to two ketones, B and D. Preparing the oxime of the unknown ketone will not help in the identification because the oximes of B and D have similar melting points, but preparing the semicarbazone will allow you to identify the ketone. Finding that the semicarbazone of the unknown ketone has a melting point of 112 °C establishes that the unknown ketone is D.

Ketone	bp (°C)	2,4-Dinitrophenylhydrazone mp (°C)	Oxime mp (°C)	Semicarbazone mp (°C)
A	140	94	57	98
B	140	102	68	123
C	139	121	79	121
D	140	101	69	112
E	141	90	61	101

PROBLEM 15

Semicarbazide has two NH_2 groups, but only one of them forms an imine. Explain.

Addition of Secondary Amines

The reaction of an aldehyde or a ketone with a secondary amine produces an enamine (pronounced "ene-amine"). An **enamine** is an α,β-unsaturated tertiary amine (that is, it has a double bond in the α,β-position relative to the nitrogen atom). Notice that the double bond is in the part of the molecule that came from the ketone. The name "enamine" comes from "ene" + "amine," with the "e" omitted in order to avoid two successive vowels.

cyclopentanone · diethylamine

N,N-diethyl-1-cyclopentenamine
an enamine

$+ H_2O$

cyclohexanone   pyrrolidine

N-(1-cyclohexenyl)-pyrrolidine
an enamine

$+ H_2O$

The mechanism for enamine formation is exactly the same as the mechanism for imine formation except for the last step of the reaction. When a primary amine reacts with an aldehyde or a ketone, the protonated imine formed loses a proton from nitrogen, forming a neutral imine. However, when the amine is secondary, the positively charged nitrogen is not bonded to a hydrogen. In order to obtain a stable neutral molecule, a proton must be removed from the α-carbon of the compound that was the carbonyl compound. An enamine results.

mechanism for enamine formation

an enamine

this intermediate cannot lose
a proton from N, so it loses
a proton from an α-carbon

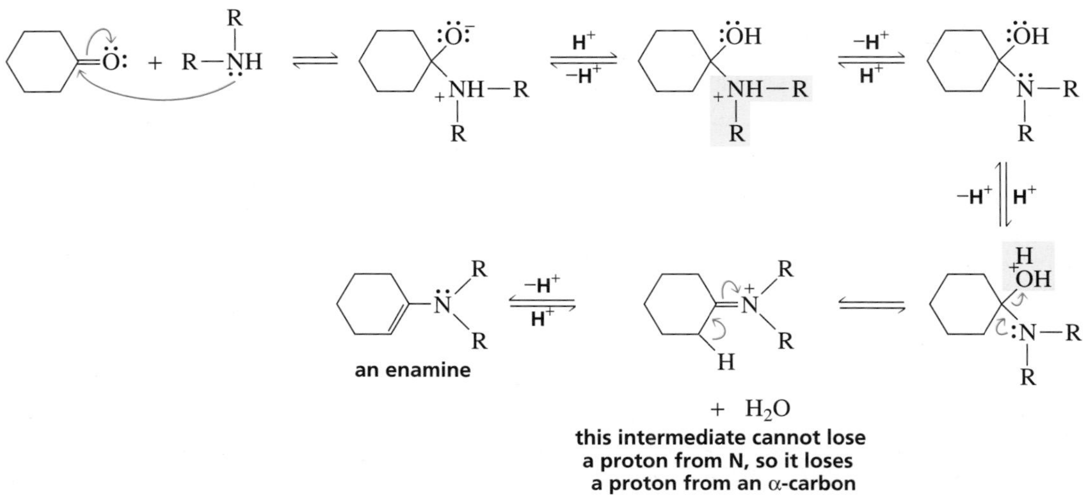

In aqueous acidic solutions, an enamine is hydrolyzed back to the carbonyl compound and secondary amine.

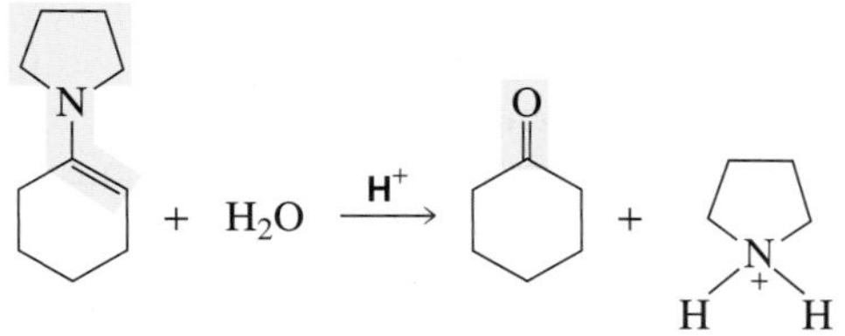

PROBLEM 16

Give the mechanism for:

a. the acid-catalyzed hydrolysis of an imine to a carbonyl compound and a primary amine.

b. the acid-catalyzed hydrolysis of an enamine to a carbonyl compound and a secondary amine.

c. How do these mechanisms differ?

PROBLEM 17◆

Give the products of the following reactions.

a. cyclopentanone + ethylamine

b. cyclopentanone + diethylamine

c. acetophenone + hexylamine

d. acetophenone + cyclohexylamine

Addition of Ammonia: Reductive Amination

The imine formed from the reaction of an aldehyde or a ketone with ammonia is not stable. There must be a substituent other than a hydrogen bonded to the nitrogen of an imine in order to stabilize it. The unstable imine, however, is a useful intermediate. For example, if the reaction with ammonia is carried out in the presence of H_2 and an appropriate metal catalyst, H_2 will add to the C=N bond as it is formed, forming a primary amine.

$$\begin{array}{c} CH_3CH_2 \\ \diagdown \\ CH_3CH_2 \diagup \end{array} C{=}O \ + \ NH_3 \ \xrightarrow{} \ \left[\begin{array}{c} CH_3CH_2 \\ \diagdown \\ CH_3CH_2 \diagup \end{array} C{=}NH \right] \ \xrightarrow[\textbf{Raney Ni}]{H_2} \ \begin{array}{c} CH_3CH_2 \\ \diagdown \\ CH_3CH_2 \diagup \end{array} CHNH_2$$

excess   unstable

Similarly, secondary and tertiary amines can be prepared from imines and enamines by reduction of the imine or enamine. Commonly used reducing agents are hydrogen and a metal catalyst, or sodium cyanoborohydride. The reaction of an aldehyde or a ketone with ammonia (or with an amine) in the presence of a reducing agent is called **reductive amination.**

$$\bigcirc{-}CH{=}O \ + \ CH_3CH_2NH_2 \ \longrightarrow \ \bigcirc{-}CH{=}NCH_2CH_3 \ \xrightarrow[\textbf{Raney Ni}]{H_2} \ \bigcirc{-}CH_2NHCH_2CH_3$$

$$\bigcirc{=}O \ + \ CH_3\underset{\underset{CH_3}{|}}{N}H \ \longrightarrow \ \bigcirc{-}N\overset{CH_3}{\underset{CH_3}{\diagdown}} \ \xrightarrow{\textbf{NaBH}_3\textbf{CN}} \ \bigcirc{-}N\overset{CH_3}{\underset{CH_3}{\diagdown}}$$

PROBLEM 18◆

Excess ammonia must be used when a primary amine is synthesized by reductive amination. What product will be obtained if the reaction is carried out with an excess of the carbonyl compound instead?

The Wolff–Kishner Reduction

In Section 14.9 we saw that when a ketone or an aldehyde is heated in a basic solution of hydrazine, the carbonyl group is converted into a methylene group. This process is called **deoxygenation** because an oxygen is removed from the reactant. The reaction is known as the *Wolff–Kishner reduction.*

$$\bigcirc{-}\overset{\overset{O}{\|}}{C}CH_3 \ \xrightarrow[\textbf{HO}^-, \, \Delta]{\textbf{NH}_2\textbf{NH}_2} \ \bigcirc{-}CH_2CH_3$$

Initially, the ketone reacts with hydrazine to form a hydrazone. Hydroxide ion and heat differentiate the Wolff–Kishner reduction from ordinary hydrazone formation. After the hydrazone is formed, hydroxide ion removes a proton from the NH_2 group. Heat is required because these protons are not easily removed. Resonance places some of the negative charge on carbon, which abstracts a proton from water. The last two steps are repeated to form the deoxygenated product and nitrogen gas.

mechanism for the Wolff–Kishner reduction

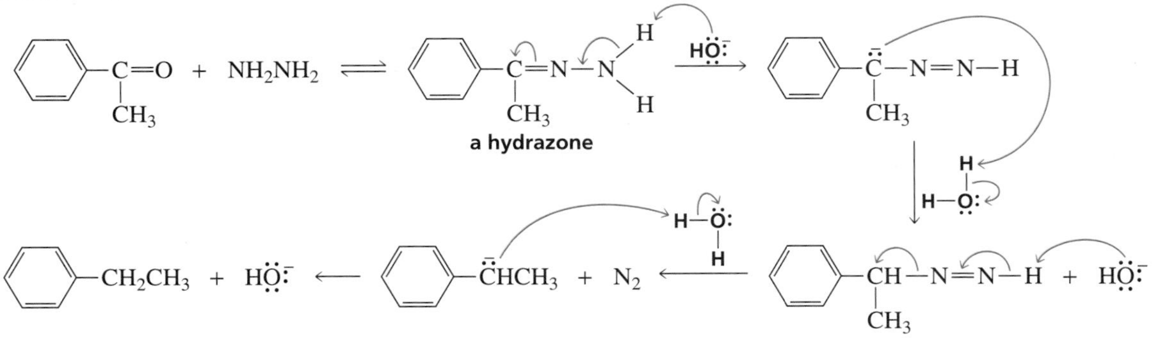

16.8
ADDITION
OF OXYGEN
NUCLEOPHILES

Addition of Water

Water adds to an aldehyde or to a ketone to form a hydrate. A **hydrate** is a molecule with two OH groups on the same carbon. Hydrates also are called ***gem*-diols** (*gem* comes from *geminus,* Latin for "twin").

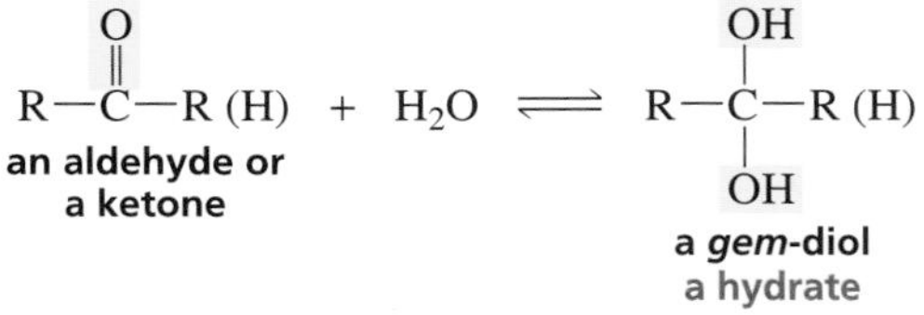

The mechanism for the addition reaction involves nucleophilic attack by water on the carbonyl carbon. Loss of a proton from the **oxonium ion** (a positively charged oxygen) and gain of a proton by the oxyanion forms the diol.

mechanism for hydrate formation

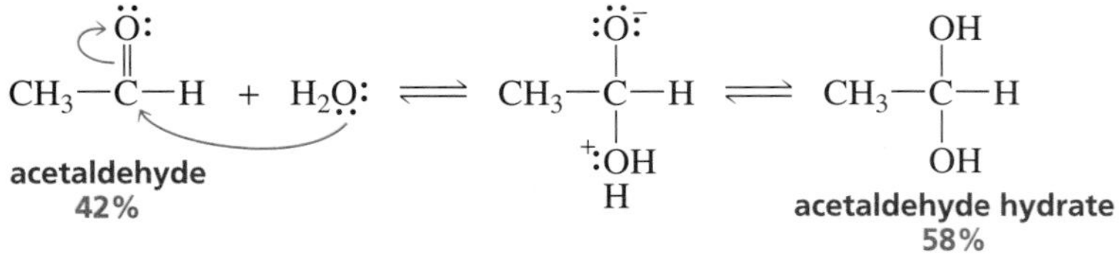

Water is a poor nucleophile and therefore adds relatively slowly to a carbonyl group, but the rate of the reaction can be increased by an acid catalyst. Notice that a catalyst has no effect on the position of the equilibrium. The catalyst affects the *rate* at which an aldehyde or a ketone is converted to a hydrate; it has no effect on the *amount* of aldehyde or ketone converted to hydrate (Section 21.1).

mechanism for acid-catalyzed hydrate formation

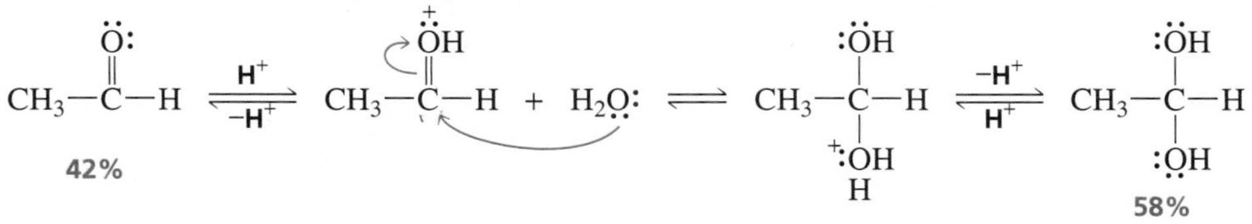

PROBLEM 19

Hydration of an aldehyde can also be catalyzed by hydroxide ion. Propose a mechanism for hydroxide-ion-catalyzed hydration.

The extent to which an aldehyde or a ketone is hydrated in an aqueous solution depends on the aldehyde or ketone. For example, only 0.2% of acetone is hydrated at equilibrium, but 99.9% of formaldehyde is hydrated. Why is there such a great difference?

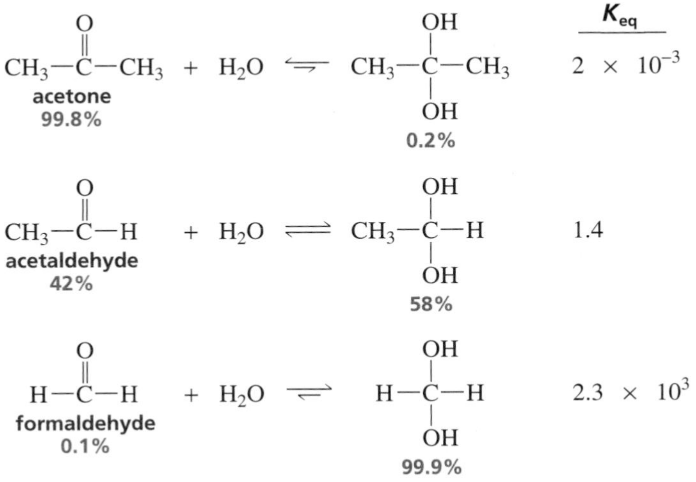

$$
\begin{array}{ccc}
\underset{\substack{\text{acetone}\\\text{99.8\%}}}{\text{CH}_3-\overset{\text{O}}{\overset{\|}{\text{C}}}-\text{CH}_3} + \text{H}_2\text{O} & \rightleftharpoons & \underset{\substack{\text{OH}\\0.2\%}}{\text{CH}_3-\overset{\text{OH}}{\underset{|}{\text{C}}}-\text{CH}_3} \quad 2\times10^{-3}
\end{array}
$$

K_{eq}

$$
\underset{\substack{\text{acetaldehyde}\\42\%}}{\text{CH}_3-\overset{\text{O}}{\overset{\|}{\text{C}}}-\text{H}} + \text{H}_2\text{O} \rightleftharpoons \underset{\substack{\text{OH}\\58\%}}{\text{CH}_3-\overset{\text{OH}}{\underset{|}{\text{C}}}-\text{H}} \quad 1.4
$$

$$
\underset{\substack{\text{formaldehyde}\\0.1\%}}{\text{H}-\overset{\text{O}}{\overset{\|}{\text{C}}}-\text{H}} + \text{H}_2\text{O} \rightleftharpoons \underset{\substack{\text{OH}\\99.9\%}}{\text{H}-\overset{\text{OH}}{\underset{|}{\text{C}}}-\text{H}} \quad 2.3\times10^3
$$

The equilibrium constant for hydrate formation depends on the relative stabilities of the carbonyl compound and the hydrate. We have seen that electron-donating alkyl groups make the carbonyl compound more stable (less reactive), so acetone is more stable than formaldehyde (Section 16.2). However, alkyl groups have the opposite effect on the stability of the hydrate; they make the hydrate less stable, so the hydrate of acetone is less stable than the hydrate of formaldehyde.

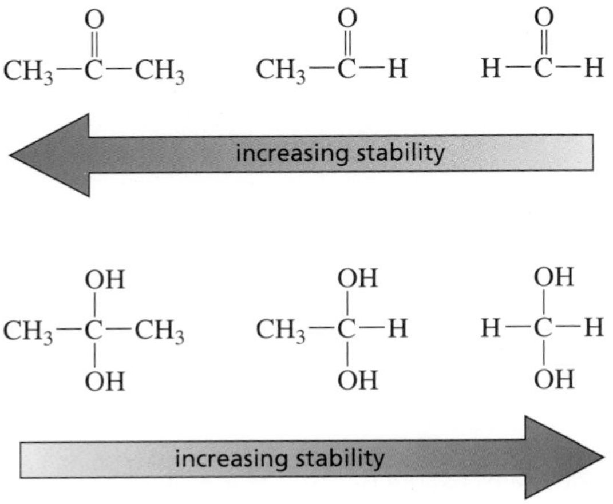

increasing stability

increasing stability

Steric interactions between the alkyl groups are responsible for the decreased stability of the hydrate. The electron clouds of the alkyl substituents do not interfere with each other in the carbonyl compound, since the bond angle between them is 120°. However, the bond angle in the hydrate is reduced to 109.5° so the alkyl groups are closer to each other. Hydrates of aldehydes or ketones are generally too unstable to be isolated and exist only in an aqueous solution.

The reaction coordinate diagrams in Figure 16.3 illustrate how the opposing effects of substituents on the stabilities of the carbonyl compound and the hydrate can have a dramatic effect on the equilibrium constant for hydrate formation.

PRESERVING BIOLOGICAL SPECIMENS

A 37% solution of formaldehyde in water is known as formalin. It was commonly used to preserve biological specimens. Because formaldehyde is an eye and skin irritant, it has been replaced in most biology laboratories by other preservatives. A mixture of 2 to 5% phenol in ethanol with added antimicrobial agents is frequently used.

If the amount of hydrate formed from the reaction of water with a ketone is too small to detect, how do we know that the reaction has even occurred? We can prove that it occurs by treating the ketone with ^{18}O-labeled water and isolating the ketone after the equilibrium has been established. Finding that the label is incorporated into the ketone proves that the reaction has occurred.

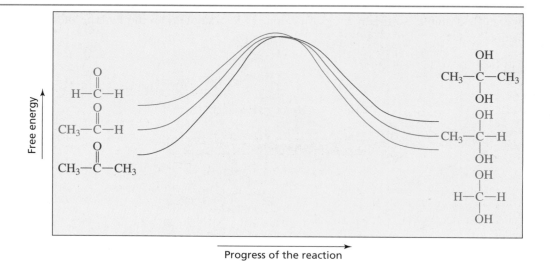

▲ Figure 16.3
Reaction coordinate diagrams for the reaction of acetone, acetaldehyde, and formaldehyde with water.

$$CH_3-\overset{\overset{\displaystyle \ddot{O}:}{\|}}{C}-CH_3 \underset{-H^+}{\overset{H^+}{\rightleftharpoons}} CH_3-\overset{\overset{\displaystyle \overset{+}{\ddot{O}H}}{|}}{C}-CH_3 + H_2\ddot{O}: \rightleftharpoons CH_3-\underset{\underset{\displaystyle H}{\overset{18}{\underset{+}{:}}\ddot{O}H}}{\overset{\overset{\displaystyle :\ddot{O}H}{|}}{C}}-CH_3 \underset{H^+}{\overset{-H^+}{\rightleftharpoons}} CH_3-\underset{\overset{18}{:}\ddot{O}H}{\overset{\overset{\displaystyle :\ddot{O}H}{|}}{C}}-CH_3$$

$$-H^+ \updownarrow H^+$$

$$CH_3-\overset{\overset{\displaystyle \overset{18}{\ddot{O}:}}{\|}}{C}-CH_3 \underset{H^+}{\overset{-H^+}{\rightleftharpoons}} CH_3-\overset{\overset{\displaystyle \overset{18}{\overset{+}{\ddot{O}H}}}{\|}}{C}-CH_3 + H_2O \rightleftharpoons CH_3-\underset{\underset{\displaystyle 18}{:\ddot{O}H}}{\overset{\overset{\displaystyle \overset{H}{\underset{+}{:}\ddot{O}H}}{|}}{C}}-CH_3$$

PROBLEM 20

Trichloroacetaldehyde has such a large equilibrium constant for reaction with water that the reaction is essentially irreversible. Consequently, the product, chloral hydrate is one of the few hydrates that can be isolated. Chloral hydrate is a sedative and is commonly known—in detective novels at least—as a "Mickey Finn." Explain the favorable equilibrium constant.

$$Cl_3C-\overset{\overset{\displaystyle O}{\|}}{C}-H + H_2O \longrightarrow Cl_3C-\underset{\underset{\displaystyle OH}{|}}{\overset{\overset{\displaystyle OH}{|}}{C}}-H$$

trichloroacetaldehyde   **chloral hydrate**

PROBLEM 21◆

Which of the following ketones has the largest equilibrium constant for addition of water?

$$CH_3O-\underset{}{\bigcirc}-\overset{\overset{\displaystyle O}{\|}}{C}-\underset{}{\bigcirc}-OCH_3 \qquad \underset{}{\bigcirc}-\overset{\overset{\displaystyle O}{\|}}{C}-\underset{}{\bigcirc}$$

$$O_2N-\underset{}{\bigcirc}-\overset{\overset{\displaystyle O}{\|}}{C}-\underset{}{\bigcirc}-NO_2$$

Addition of Alcohol

The product of the addition of one equivalent of an alcohol to an aldehyde is called a **hemiacetal.** The product formed when a second equivalent of alcohol is added is called an **acetal.** When the carbonyl compound is a ketone instead of an aldehyde, the addition products are called **hemiketal** and **ketal.**

$$CH_3-\overset{\overset{\displaystyle O}{\|}}{C}-H + CH_3OH \rightleftharpoons CH_3-\underset{\underset{\displaystyle OCH_3}{|}}{\overset{\overset{\displaystyle OH}{|}}{C}}-H \overset{CH_3OH}{\rightleftharpoons} CH_3-\underset{\underset{\displaystyle OCH_3}{|}}{\overset{\overset{\displaystyle OCH_3}{|}}{C}}-H$$

an aldehyde   **a hemiacetal**   **an acetal**

$$CH_3-\overset{\overset{O}{\|}}{C}-CH_3 \ + \ CH_3OH \ \rightleftharpoons \ CH_3-\overset{\overset{OH}{|}}{\underset{\underset{OCH_3}{|}}{C}}-CH_3 \ \xrightleftharpoons{CH_3OH} \ CH_3-\overset{\overset{OCH_3}{|}}{\underset{\underset{OCH_3}{|}}{C}}-CH_3$$

a ketone   a hemiketal   a ketal

Hemi is the Greek prefix for "half." When one equivalent of alcohol has added to an aldehyde or a ketone, the compound is halfway to the final acetal or ketal, which contains two equivalents of alcohol.

Acetal (or ketal) formation requires an acid catalyst. The acid protonates the carbonyl oxygen, making it easier for the alcohol to attack the carbonyl carbon. Loss of a proton from the protonated tetrahedral intermediate gives the hemiacetal (or hemiketal). Because the reaction is carried out in an acidic solution, the hemiacetal (or hemiketal) is in equilibrium with its protonated form. The two oxygen atoms of the hemiacetal (or hemiketal) are equally basic, so either one can be protonated. Loss of water from the tetrahedral intermediate with a protonated OH group forms a compound that is very reactive because of its positively charged oxygen. Nucleophilic attack on this compound by a second molecule of alcohol results in the acetal (or ketal).

mechanism for acid-catalyzed acetal or ketal formation

High yields of acetals and ketals are obtained if the water eliminated from the hemiacetal (or hemiketal) is removed as it is formed. The formation of acetals and ketals requires an acid catalyst because the tetrahedral intermediate readily eliminates water (the leaving group in the presence of an acid catalyst). It is difficult for the tetrahedral intermediate to eliminate the basic ⁻OH group (the leaving group in the absence of an acid catalyst).

Because acetal (or ketal) formation is reversible, an acetal or ketal can be transformed back to the aldehyde or ketone in an acidic aqueous solution.

$$CH_3CH_2-\overset{\overset{OCH_2CH_3}{|}}{\underset{\underset{OCH_2CH_3}{|}}{C}}-CH_3 \ + \ \underset{\text{excess}}{H_2O} \ \xrightleftharpoons{H^+} \ CH_3CH_2-\overset{\overset{O}{\|}}{C}-CH_3 \ + \ 2\ CH_3CH_2OH$$

The mechanism for acetal (or ketal) formation is similar to the mechanism for imine formation. After the nucleophile (an alcohol in the case of acetal and ketal formation; an amine in the case of imine formation) has added to the carbonyl compound, water is eliminated from the tetrahedral intermediate. In imine formation, elimination of water is followed by loss of a proton from nitrogen to form a neutral imine. In acetal formation, the only way a neutral compound can be obtained after water is eliminated is by a second addition of alcohol. So imine formation is a *nucleophilic addition–elimination reaction,* while acetal (or ketal) formation is a **nucleophilic addition–elimination—nucleophilic addition reaction.**

imine formation

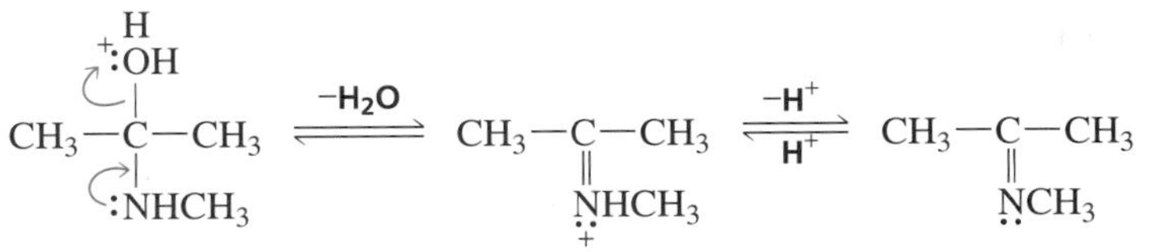

ketal formation

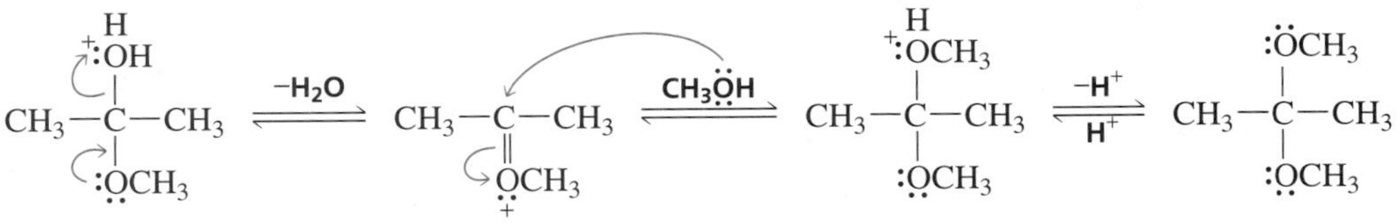

Hydrate formation also involves nucleophilic addition followed by elimination of water. Because the adding nucleophile is water, elimination of water gives back the original aldehyde or ketone. Notice that elimination of water occurs only with nucleophiles that, in the tetrahedral intermediate, have a pair of nonbonding electrons.

PROBLEM-SOLVING STRATEGY

Acetals and ketals are hydrolyzed back to the aldehyde or ketone in acidic aqueous solutions, but they are stable in basic aqueous solutions. Explain.

The easiest way to approach this kind of question is to write out the structures and/or mechanism that describe what the question is asking. Once the mechanism is written, the answer should become apparent. In an acidic solution, the acid protonates an oxygen of the acetal. This creates a weak base (CH_3OH) that can be expelled by the other CH_3O group. Once the group is expelled, water can attack the reactive intermediate and you are then on your way back to the ketone (or aldehyde).

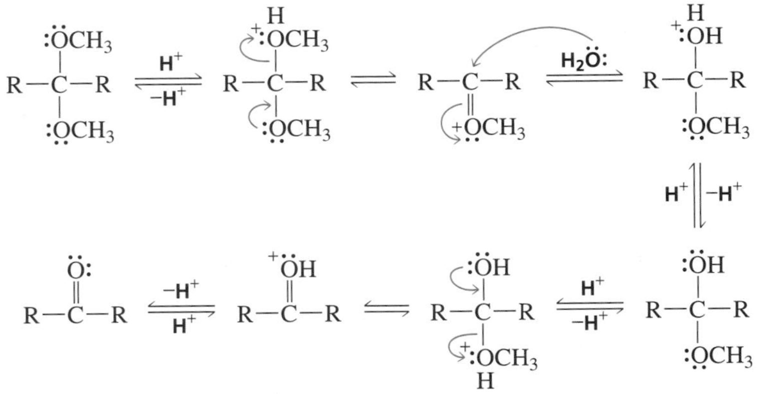

In contrast, the CH_3O group cannot be protonated in a basic solution. Therefore, the group that would have to be eliminated to reform the ketone (or aldehyde) would be the very basic CH_3O^- group. This group is too basic to be eliminated by the other CH_3O group, which has little driving force because of the positive charge that would be placed on the oxygen atom if elimination were to occur.

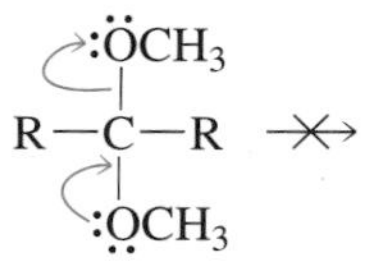

$$R-\underset{\underset{\ddot{\underset{..}{O}}CH_3}{|}}{\overset{\overset{.\ddot{O}CH_3}{|}}{C}}-R \quad \xtimes$$

Now continue on to solve Problem 22.

PROBLEM 22

a. Would you expect hemiacetals to be stable in basic solutions? Explain.

b. Acetal formation must be catalyzed by an acid. It cannot be catalyzed by CH_3O^-. Explain.

c. Can hydrate formation be catalyzed by HO^- as well as by H^+? Explain.

16.9 PROTECTING GROUPS

Ketones (or aldehydes) react with 1,2-diols to form five-membered ring ketals (or acetals) and with 1,3-diols to form six-membered ring ketals (or acetals). The mechanism is the same as that given previously for acetal formation except that, instead of reacting with two separate molecules of alcohol, the carbonyl compound reacts with the two alcohol groups of the diol.

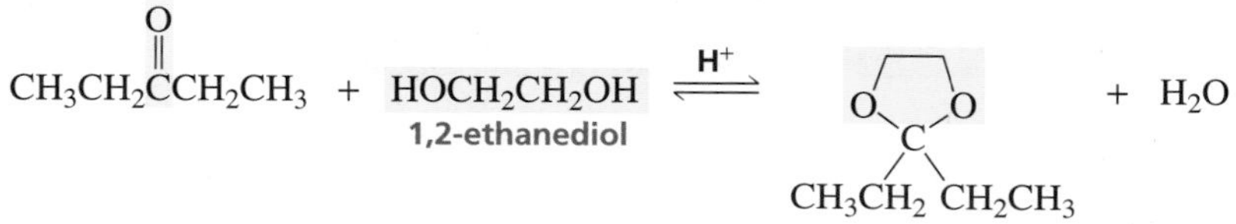

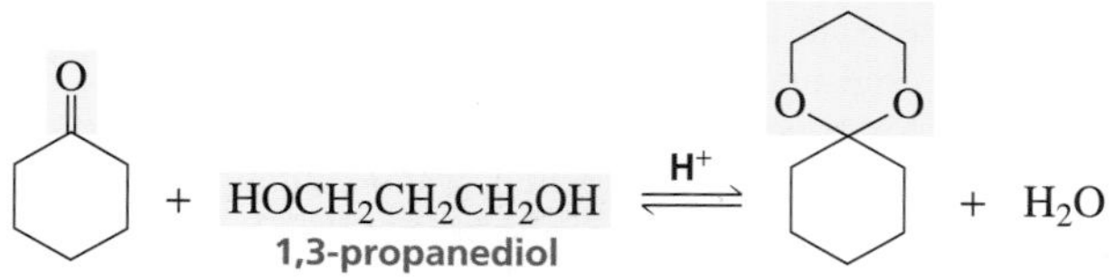

$$CH_3CH_2\overset{\overset{O}{\|}}{C}CH_2CH_3 \; + \; HOCH_2CH_2OH \; \underset{}{\overset{H^+}{\rightleftharpoons}} \; \text{(ring)} \; + \; H_2O$$

1,2-ethanediol

(cyclic ketal: $CH_3CH_2 \quad CH_2CH_3$)

$$\text{(cyclohexanone)} \; + \; HOCH_2CH_2CH_2OH \; \overset{H^+}{\rightleftharpoons} \; \text{(spiro ketal)} \; + \; H_2O$$

1,3-propanediol

If a compound has more than one functional group that will react with a given reagent and you want only one of them to react, it is necessary to protect the second functional group from the reagent. A **protecting group** protects a functional group from a synthetic operation that it would otherwise not survive.

If you have ever painted a room with a spray gun, you may have taped over the things that you do not want to paint, such as baseboards and window frames. After you spray the room and remove the tape, the walls are painted but the areas that have been covered with tape are free of paint. In a similar way, 1,2-diols and 1,3-diols are used to protect the carbonyl group of aldehydes and ketones. The diol is like the tape. For example, in the synthesis of the following hydroxyketone from the keto ester, a problem arises because the keto group is more reactive than the ester group toward nucleophilic attack by a Grignard reagent.

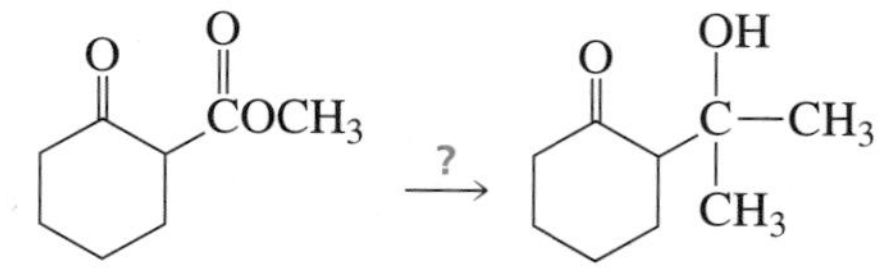

$$\text{(keto ester)} \; \overset{?}{\longrightarrow} \; \text{(hydroxyketone)}$$

If the keto group is converted to a ketal, only the ester group will react with the Grignard reagent. The protecting group can be removed by acid-catalyzed hydrolysis after the Grignard reagent has reacted with the ester. It is critical that the conditions used to remove a protecting group do not affect other groups in the molecule.

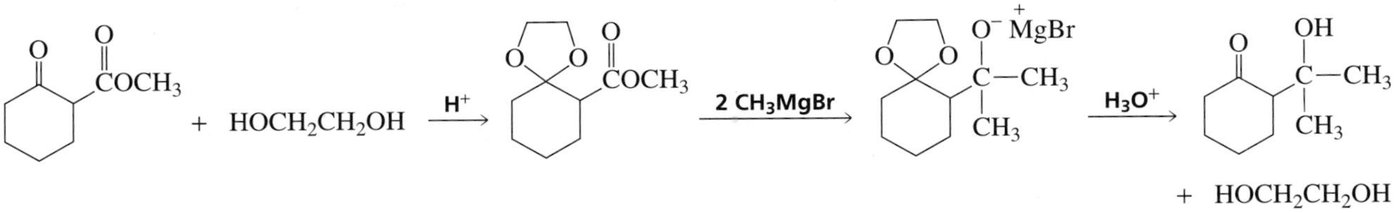

In the following reaction, the aldehyde carbonyl group reacts with the diol because aldehydes are more reactive than ketones. The ketone then can be reduced without simultaneously reducing the aldehyde. After the ketone is reduced, the protecting group can be removed by acid-catalyzed hydrolysis.

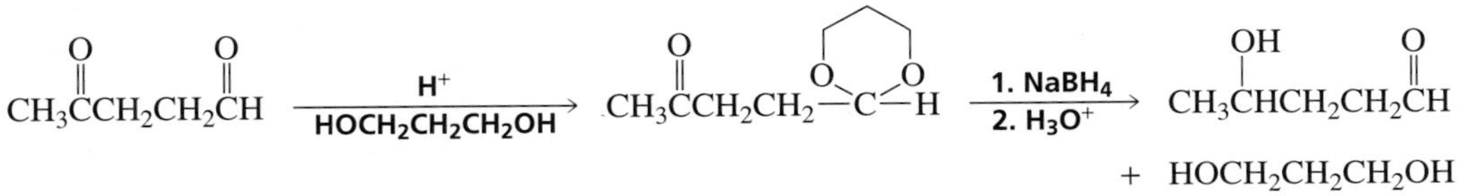

PROBLEM 23

Propose a mechanism for the following reaction.

Aldehydes and ketones react with thiols to form thioacetals and thioketals. The mechanism for addition of a thiol is exactly the same as the mechanism for addition of an alcohol. Recall that thiols are sulfur analogs of alcohols (Section 11.10).

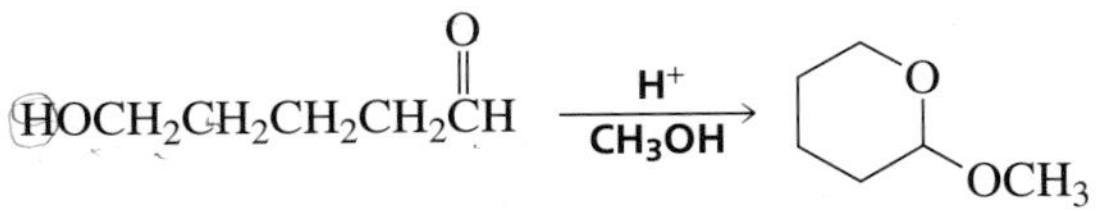

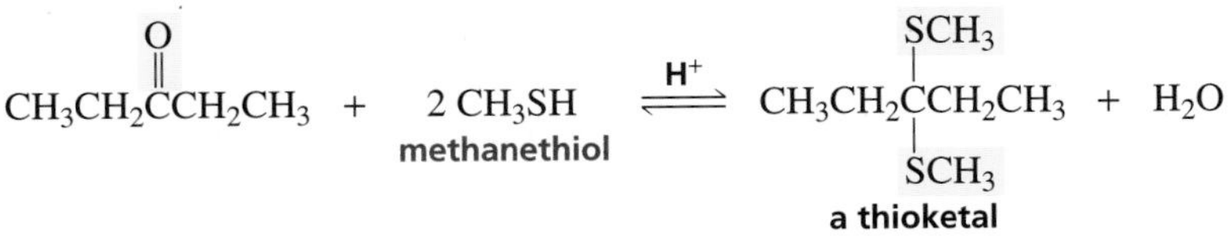

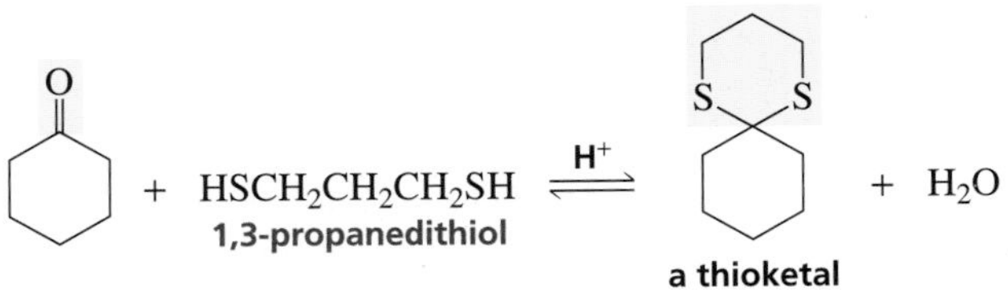

16.10
ADDITION OF SULFUR NUCLEOPHILES

Thioacetal (or thioketal) formation is a synthetically useful reaction because a thioacetal or thioketal is desulfurized when it reacts with H_2 and Raney nickel. Desulfuration replaces the C—S bonds with C—H bonds.

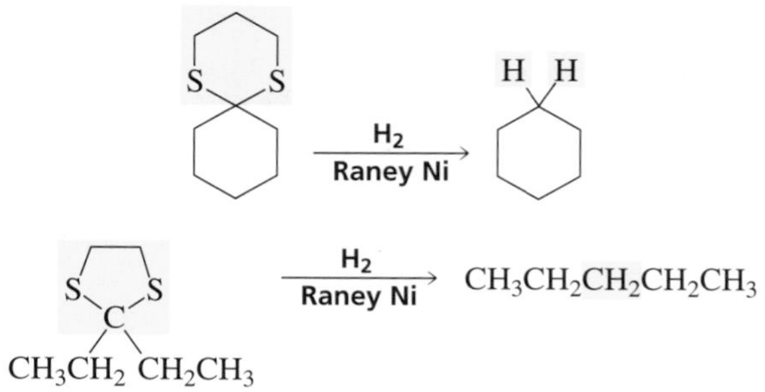

Thioketal formation followed by desulfuration provides us with a third method we can use to convert the carbonyl group of a ketone into a methylene group. We have already seen the other two methods, the Clemmensen reduction and the Wolff–Kishner reduction (Sections 14.9 and 16.7).

16.11
THE WITTIG
REACTION

An aldehyde or a ketone reacts with a phosphonium ylide (pronounced "ill-id") to form an alkene. An **ylide** is a compound that has opposite charges on adjacent, covalently bonded atoms with complete octets. The ylide can also be written in the doubly bonded form because phosphorus can have more than eight valence electrons.

$$(C_6H_5)_3\overset{+}{P}-\overset{-}{C}H_2 \quad \longleftrightarrow \quad (C_6H_5)_3P=CH_2$$
a phosphonium ylide

This reaction is called the **Wittig reaction.** The overall reaction amounts to interchanging the doubly-bonded oxygen of the carbonyl compound and the doubly-bonded carbon group of the phosphonium ylide.

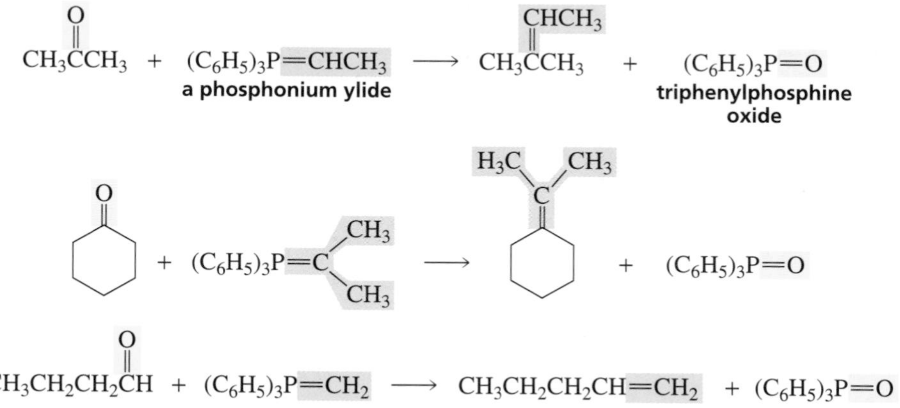

Georg Friedrich Karl Wittig (1897–1987) *was born in Germany. He was a professor of chemistry at the University of Heidelberg, where he studied phosphorus-containing organic compounds. He received the Nobel Prize in chemistry in 1979, sharing it with H. C. Brown (Section 3.17).*

In the first step of the Wittig reaction, the nucleophilic carbon of the ylide attacks the carbonyl carbon of the aldehyde or ketone. Nucleophilic attack results in formation of a dipolar intermediate known as a *betaine* (pronounced either "beta-ene" or "be-tane"). Bond formation between oxygen and phosphorus forms a second intermediate called an *oxaphosphatane*. Elimination of triphenylphosphine oxide from the oxaphosphatane forms the alkene product.

mechanism for the Wittig reaction

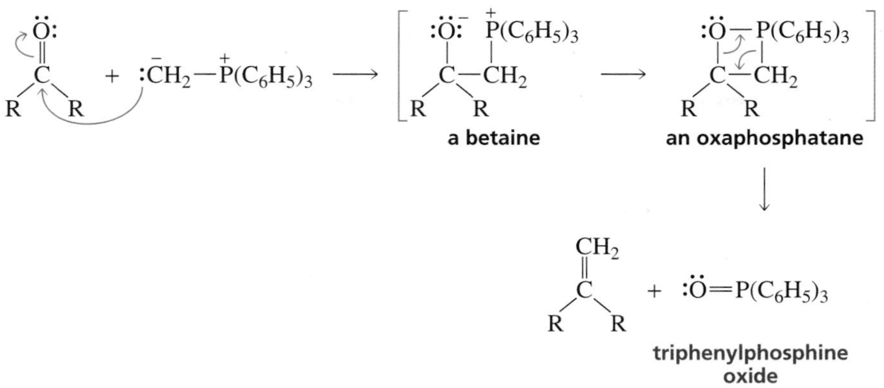

a betaine   an oxaphosphatane

$$CH_2$$
C   + $:\ddot{O}=P(C_6H_5)_3$
R   R

**triphenylphosphine
oxide**

The Wittig reaction is another example of the addition of a carbon nucleophile to an aldehyde or ketone. It is unlike the addition of the carbon nucleophiles discussed in Section 16.4 in that the nucleophilic addition reaction is followed by an elimination reaction.

The phosphonium ylide needed for a particular synthesis is obtained by nucleophilic attack by triphenylphosphine on an alkyl halide with the appropriate number of carbon atoms. A proton on the carbon adjacent to the positively charged phosphorus atom is sufficiently acidic ($pK_a = 35$) to be removed by a strong base such as butyllithium (Section 11.8).

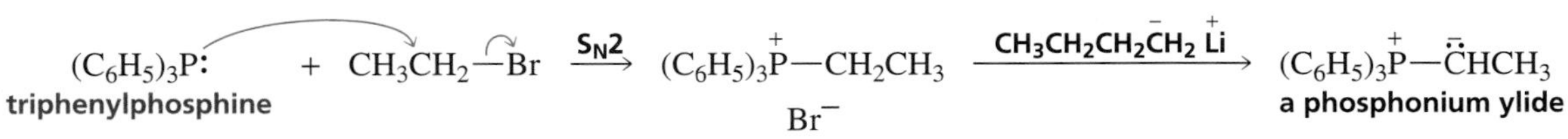

$(C_6H_5)_3P:$ + CH_3CH_2–Br $\xrightarrow{S_N2}$ $(C_6H_5)_3\overset{+}{P}$–CH_2CH_3 $\xrightarrow{CH_3CH_2CH_2\overset{-}{C}H_2\ \overset{+}{Li}}$ $(C_6H_5)_3\overset{+}{P}$–$\overset{..}{\overset{-}{C}}HCH_3$

triphenylphosphine Br^- **a phosphonium ylide**

An advantage of using a Wittig reaction rather than an E2 elimination reaction to prepare an alkene is that a Wittig reaction is completely regioselective: the product has the double bond only in one location. An E2 reaction, in contrast, can lead to the formation of more than one alkene.

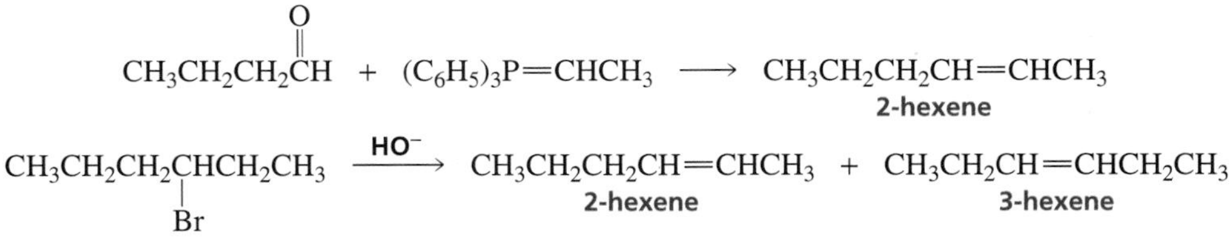

$$CH_3CH_2CH_2\overset{\overset{\displaystyle O}{\|}}{C}H + (C_6H_5)_3P=CHCH_3 \longrightarrow CH_3CH_2CH_2CH=CHCH_3$$
2-hexene

$$CH_3CH_2CH_2\underset{\underset{\displaystyle Br}{|}}{C}HCH_2CH_3 \xrightarrow{HO^-} CH_3CH_2CH_2CH=CHCH_3 + CH_3CH_2CH=CHCH_2CH_3$$
2-hexene   **3-hexene**

The Wittig reaction is also stereoselective: if the alkene product can exist as stereoisomers, more of one isomer is obtained than of the other.

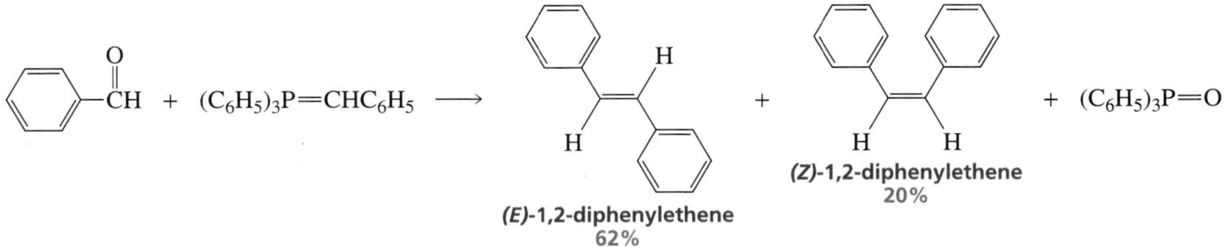

(E)-1,2-diphenylethene
62%

(Z)-1,2-diphenylethene
20%

β-CAROTENE

β-Carotene is found in yellow-orange fruits and vegetables such as apricots, mangos, carrots, and sweet potatoes. The synthesis of β-carotene from vitamin A is an important example of the use of the Wittig reaction in industry. β-Carotene is used in the food industry to color margarine. Many people take β-carotene as a dietary supplement because there is some evidence that high levels of β-carotene are associated with a low incidence of cancer. More recent evidence suggests that β-carotene taken in pill form does not have the cancer-preventing effects of β-carotene obtained from vegetables.

vitamin A aldehyde

β-carotene

PROBLEM 24

a. What carbonyl compound and what phosphonium ylide are required for the synthesis of the following alkenes?

b. What alkyl halide is required to prepare each of the phosphonium ylides?

$$(1) \quad CH_3CH_2CH_2CH{=}\overset{\underset{\displaystyle CH_3}{|}}{C}CH_3$$

$$(2) \quad \text{⬡}{=}CHCH_2CH_3$$

$$(3) \quad \text{⬡}{-}CH{=}CH_2$$

$$(4) \quad (C_6H_5)_2C{=}CHCH_3$$

**16.12
STEREOCHEMISTRY
OF NUCLEOPHILIC
ADDITION
REACTIONS: *re* AND
si FACES**

A carbonyl carbon bonded to two different substituents is a **prochiral carbonyl carbon** because it will become a chirality center if it is attacked by a group unlike either of the groups already bonded to it. The addition product will be a pair of enantiomers.

a prochiral carbonyl carbon

$$\underset{X \qquad Y}{\overset{O}{\overset{\|}{C}}} \quad \xrightarrow{\text{HZ}} \quad X{-}\!\!\!\overset{OH}{\underset{Z}{|}}\!\!\!{-}Y \quad + \quad X{-}\!\!\!\overset{Z}{\underset{OH}{|}}\!\!\!{-}Y$$

a pair of enantiomers

The carbonyl carbon and the three atoms attached to it define a plane. The nucleophile can approach either side of the plane. One side of the carbonyl compound is called the *re* face (pronounced "ree"), and the other side is called the *si* face (pronounced "sigh"); *re* is for *rectus* and *si* is for *sinister* (similar to *R* and *S*). To distinguish between the *re* and *si* faces, the three groups attached to the carbonyl carbon are assigned priorities using the Cahn–Ingold–Prelog system of priorities that are used in *E, Z* and *R, S* nomenclature (Sections 3.5 and 4.5). The *re* face is the one closest to the observer when decreasing priorities (1 > 2 > 3) are in a clockwise direction, and the *si* face is the opposite face (or the one closest to the observer when decreasing priorities are in a counterclockwise direction).

Attack by a nucleophile on the *re* face forms one enantiomer, while attack on the *si* face forms the other enantiomer. For example, attack by hydride ion on the *re* face of butanone forms *(S)*-2-butanol; attack on the *si* face forms *(R)*-2-butanol. (Use models if you cannot see how the three-dimensional structure is turned into a Fischer projection.)

Whether attack on the *re* face gives the *R* or *S* enantiomer depends on the priority of the attacking nucleophile compared with the priorities of the groups attached to the carbonyl carbon. For example, we saw that attack by hydride ion on the *re* face of butanone forms *(S)*-2-butanol. In contrast, attack by a methyl Grignard reagent on the *re* face of propanal forms *(R)*-2-butanol.

Because the carbonyl carbon and the three atoms attached to it define a plane, the *re* and *si* faces have an equal probability of being attacked. Consequently, an addition reaction forms equal amounts of the two enantiomers.

ENZYME-CATALYZED CARBONYL ADDITIONS

In an enzyme-catalyzed addition to a carbonyl compound, only one of the enantiomers is formed. The enzyme can block one face of the carbonyl compound so that it cannot be attacked, or it can position the nucleophile so that it can attack the carbonyl group from only one side of the molecule.

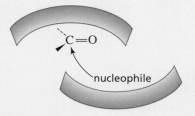

PROBLEM 25◆

Which enantiomer is formed from attack by a methyl Grignard reagent on the *re* face of the following carbonyl compounds?

a. propiophenone   **b.** benzaldehyde   **c.** 2-pentanone   **d.** 3-hexanone

**16.13
DESIGNING A
SYNTHESIS IV:
THE SYNTHESIS OF
CYCLIC COMPOUNDS**

Most of the reactions that we have studied have two reactants. A reactive group in one reactant reacts with a reactive group in the second reactant. This is called an *intermolecular reaction* (a reaction between two molecules). Cyclic structures are formed when the reactive groups are in the same molecule. This is called an *intramolecular reaction* (a reaction within a molecule). We have seen that intramolecular reactions are particularly favorable if the reaction forms a compound with a five- or a six-membered ring (Section 9.4).

In designing the synthesis of a cyclic compound, we must examine the target molecule to determine what kinds of reactive groups will be necessary for a successful synthesis. For example, we know that an ester is formed from the acid-catalyzed reaction of a carboxylic acid with an alcohol. Therefore, a cyclic ester (lactone) can be prepared from a reactant that has both a carboxylic acid and an alcohol group in the same molecule. The size of the lactone ring will be determined by the number of carbon atoms between the carboxylic acid and alcohol groups.

$$HOCH_2CH_2CH_2CH_2\overset{O}{\overset{\|}{C}}OH \xrightarrow{H^+} \text{← a lactone}$$
4 carbon atoms

$$HOCH_2CH_2CH_2\overset{O}{\overset{\|}{C}}OH \xrightarrow{H^+}$$
3 carbon atoms

A compound with a ketone group attached to a benzene ring can be prepared using a Friedel-Crafts acylation reaction. Therefore, a cyclic ketone will result if a Lewis acid (AlCl₃) is added to a compound that contains both a benzene ring and an acyl chloride group.

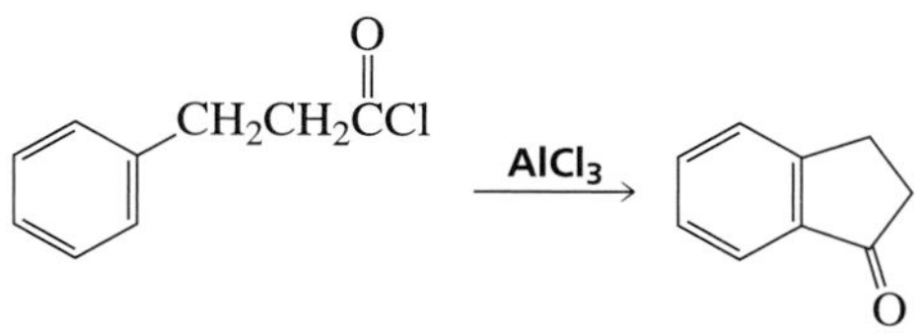

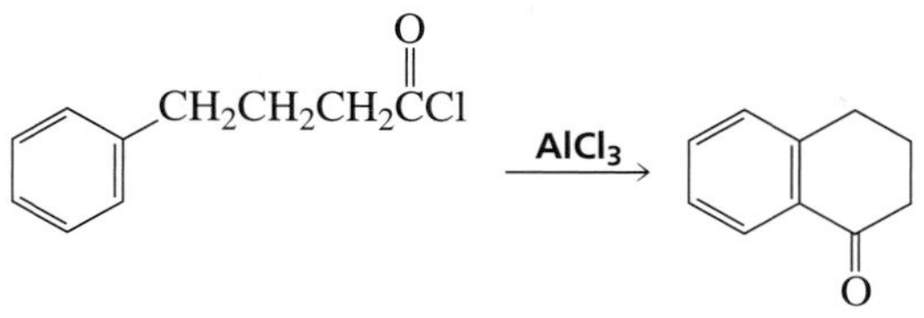

A cyclic secondary amine can be prepared by reductive amination using a compound that contains ketone and primary amine groups that are separated by the appropriate number of carbon atoms.

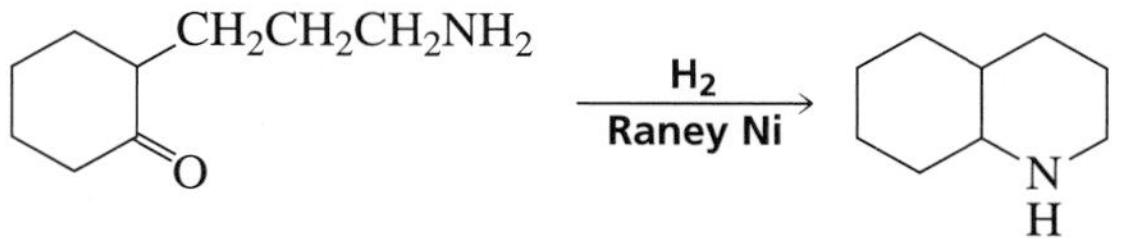

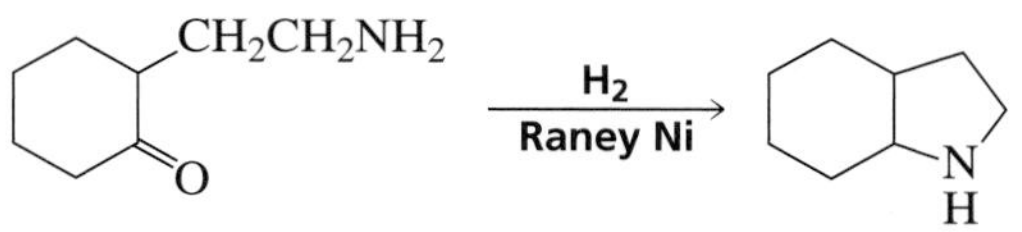

Probably the most obvious method of preparing a cyclic ether is by an intramolecular Williamson ether synthesis.

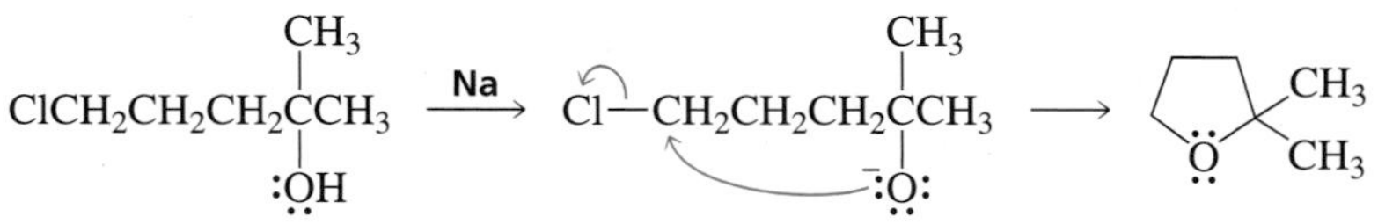

However, if the compound with the alkyl halide and alcohol groups is not available, the ether could be prepared from the reaction of a Grignard reagent with a compound that has an alkyl halide group and a ketone group.

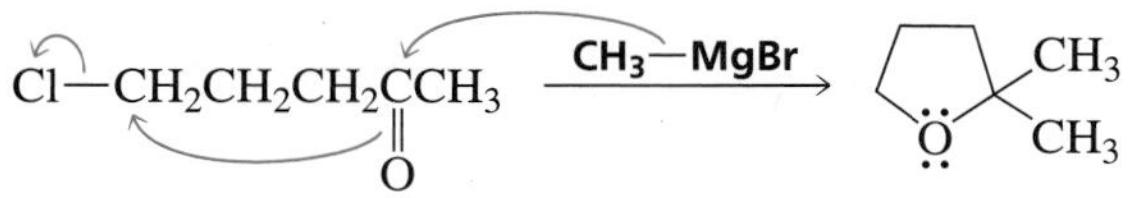

PROBLEM 26

Design a synthesis for each of the following compounds using an intramolecular reaction.

a.

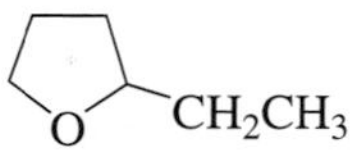

b.

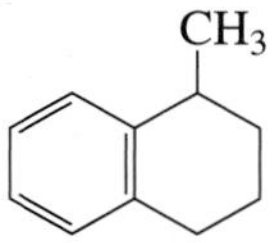

c.

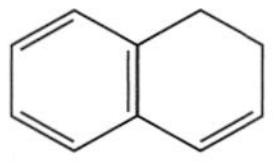

d.

e.

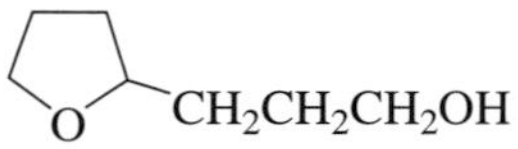

CH₂CH₂CH₂OH

f.

OH

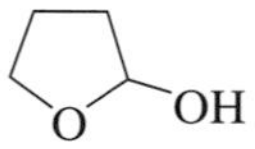

**16.14
NUCLEOPHILIC
ADDITION TO
α,β-UNSATURATED
ALDEHYDES AND
KETONES: DIRECT
ADDITION VERSUS
CONJUGATE
ADDITION**

The contributing resonance structures for an α,β-unsaturated carbonyl compound show that the molecule has two electrophilic sites—the carbonyl carbon and the β-carbon.

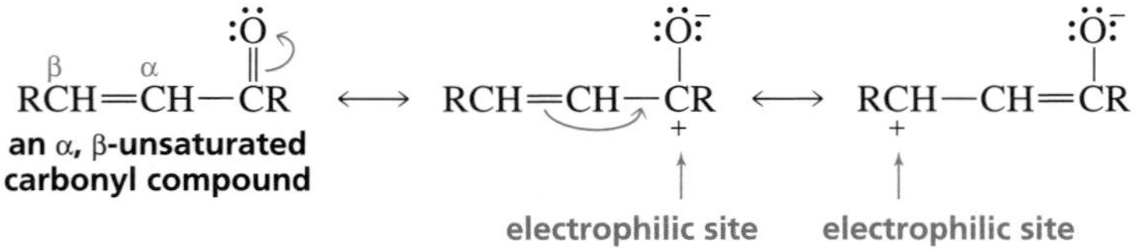

This means that if an aldehyde or a ketone has a double bond in the α,β-position, a nucleophile can add either to the carbonyl carbon or to the β-carbon. Nucleophilic addition to the carbonyl carbon is called **direct addition** or **carbonyl addition.**

direct addition

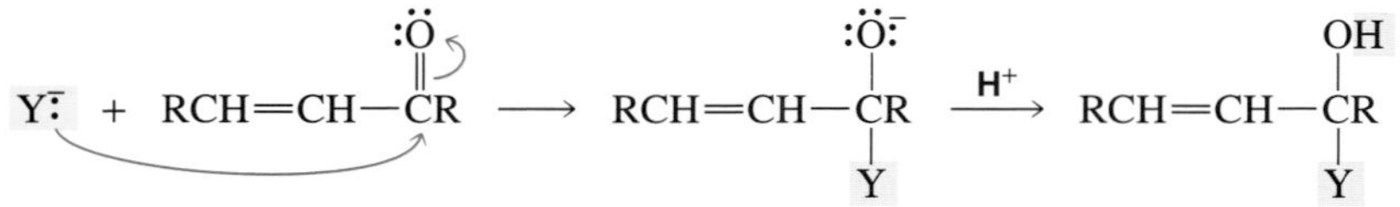

Nucleophilic addition to the β-carbon is called **conjugate addition** because addition occurs to a double bond conjugated with the carbonyl group. The initially formed product of conjugate addition is an enol that immediately tautomerizes to a ketone or aldehyde (Section 5.6). The overall reaction amounts to addition to the carbon–carbon double bond, with the nucleophile adding to the β-carbon and a proton from the reaction mixture adding to the α-carbon.

conjugate addition

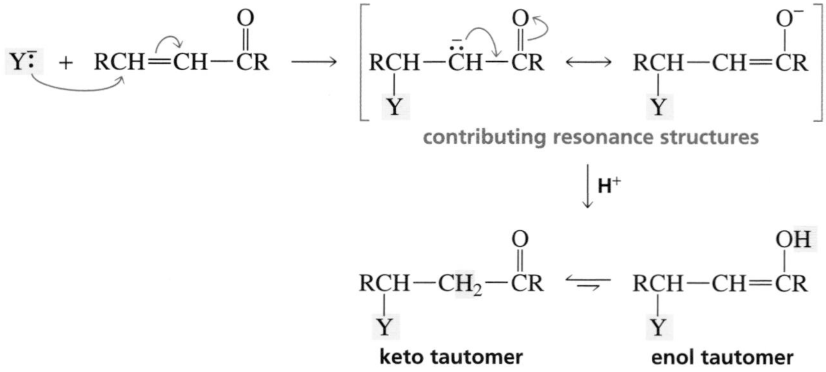

contributing resonance structures

keto tautomer   enol tautomer

Whether the product obtained from nucleophilic addition to an α,β-unsaturated carbonyl compound is the direct addition product or the conjugate addition product depends on the nature of the nucleophile. Nucleophiles that are strong bases, such as organolithium reagents and hydride ion, tend to form direct addition products.

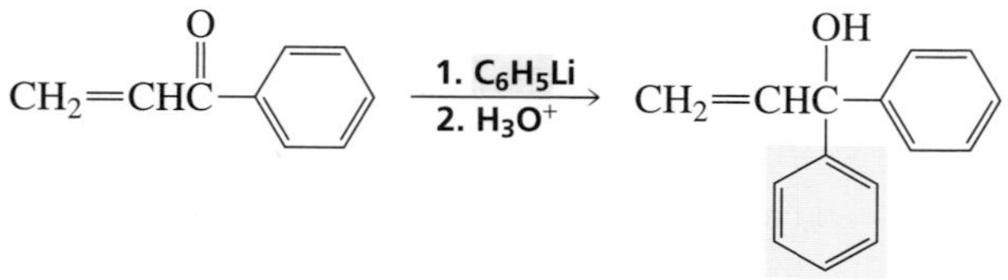

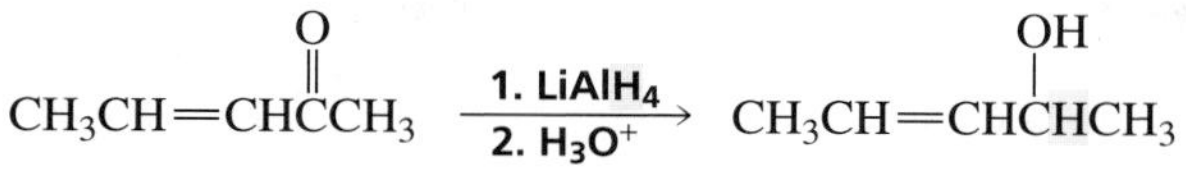

$$CH_2=CHC\!\!-\!\!\langle\text{phenyl}\rangle \xrightarrow[\text{2. } H_3O^+]{\text{1. } C_6H_5Li} CH_2=CHC\!\!-\!\!\langle\text{phenyl}\rangle$$ (OH, phenyl)

$$CH_3CH=CHCCH_3 \xrightarrow[\text{2. } H_3O^+]{\text{1. LiAlH}_4} CH_3CH=CHCHCH_3$$ (OH)

Nucleophiles that are relatively weak bases, such as cyanide ion, amines, thiols, and halide ions, usually form conjugate addition products.

$$\langle\text{cyclohexenone}\rangle + \ ^-C\equiv N \xrightarrow{HC\equiv N} \langle\text{cyclohexanone with } C\equiv N\rangle$$

$$CH_2=CHCCH_3 + (CH_3CH_2)_2NH \longrightarrow (CH_3CH_2)_2NCH_2CH_2CCH_3$$

$$\langle\text{phenyl}\rangle\!-\!CH=CHCH + CH_3SH \longrightarrow \langle\text{phenyl}\rangle\!-\!CHCH_2CH$$ (SCH_3)

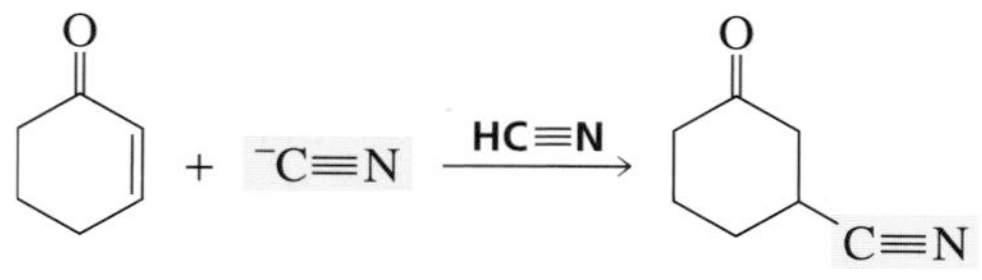

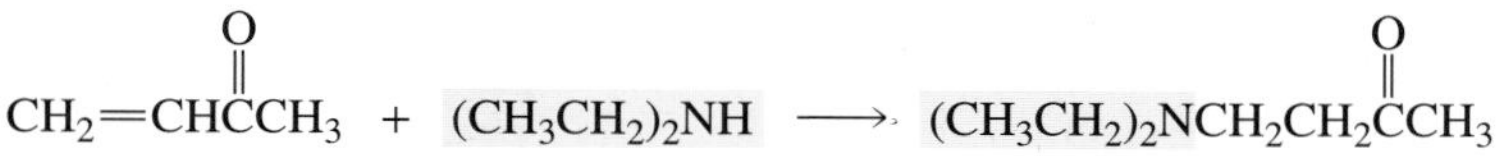

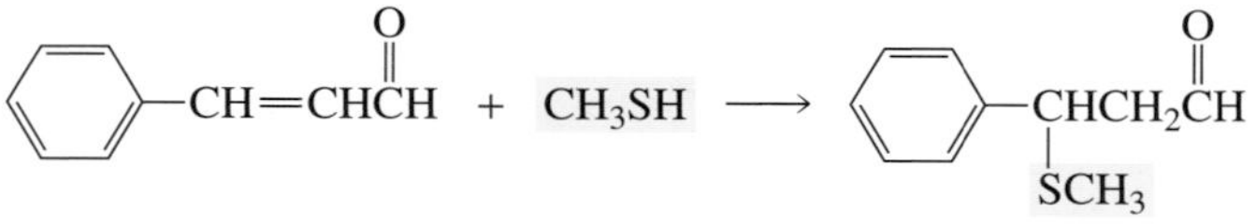

Why do strong bases form direct addition products and weak bases form conjugate addition products? Direct addition occurs more rapidly than conjugate addition. Therefore, the direct addition product is formed first and, if addition is *irreversible,* it will be the final product of the reaction. If direct addition is *reversible,* the initially formed direct addition product can collapse back to the starting material, allowing conjugate addition to occur. Since conjugate addition is irreversible, the conjugate addition product will accumulate and eventually become the major product of the reaction.

Direct addition will be irreversible if the nucleophile is a strong base, because strong bases are poor leaving groups and cannot be eliminated from the direct addition product. Direct addition will be reversible if the nucleophile is a weak base and, therefore, can be eliminated from the direct addition product. When direct addition is reversible, the conjugate addition product will be the major product of the reaction even though it is formed more slowly.

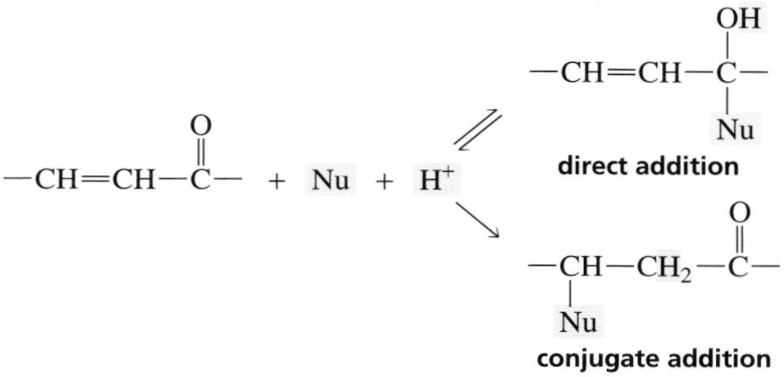

$$-CH=CH-C- + Nu + H^+ \nearrow \searrow$$

$$-CH=CH-\underset{Nu}{\overset{OH}{C}}-$$
direct addition

$$-\underset{Nu}{CH}-CH_2-C-$$
conjugate addition

Because Grignard reagents are strong bases, they react with α,β-unsaturated aldehydes and most α,β-unsaturated ketones to form direct addition products.

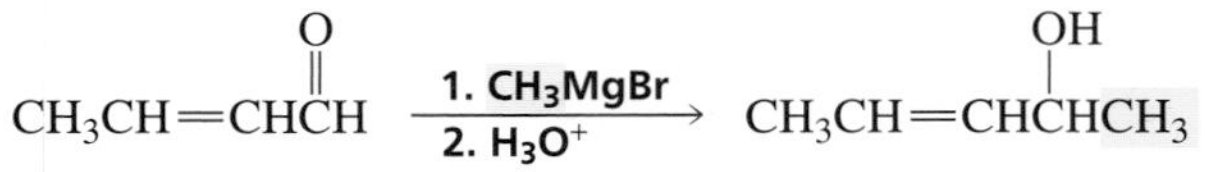

$$\underset{\text{CH}_3\text{CH}=\text{CHCH}}{\overset{O}{\|}} \xrightarrow[\text{2. H}_3\text{O}^+]{\text{1. CH}_3\text{MgBr}} \underset{\text{CH}_3\text{CH}=\text{CHCHCH}_3}{\overset{\text{OH}}{|}}$$

If, however, the carbonyl carbon is bonded to a bulky group, the conjugate addition product is formed preferentially. Apparently, the bulky group sufficiently slows the rate of direct addition to allow competition from conjugate addition.

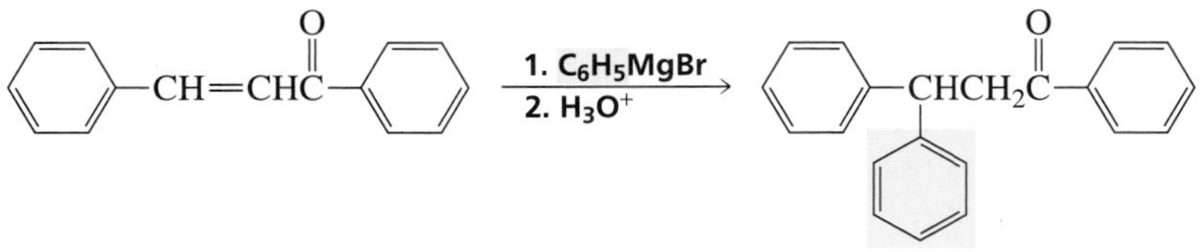

Only conjugate addition occurs when lithium dialkylcuprates react with unsaturated carbonyl compounds. This means that Grignard reagents should be used when you want to add an alkyl group to the carbonyl carbon, while lithium dialkylcuprates should be used when you want to add an alkyl group to the β-carbon.

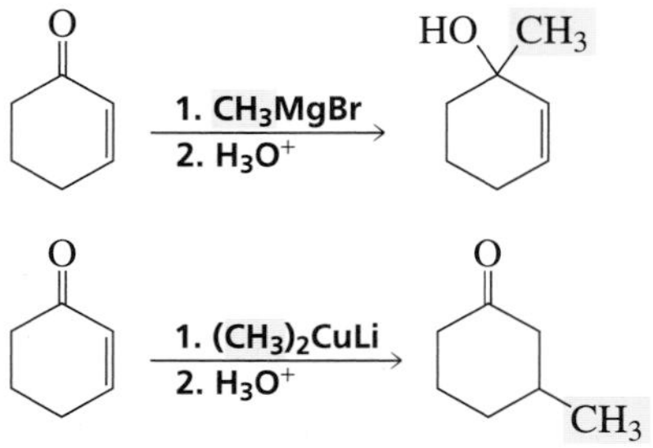

Although only direct addition occurs with a strong reducing agent such as lithium aluminum hydride, both direct and conjugate addition occur when the reducing agent is sodium borohydride, a weaker reducing agent. Cyclohexanone, the product of conjugate addition, is not isolated because it immediately undergoes a second reduction to form cyclohexanol.

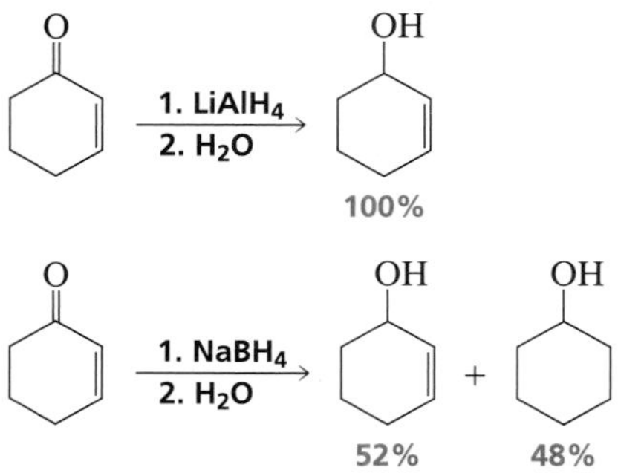

PROBLEM 27◆

Give the major product of each of the following reactions.

a. 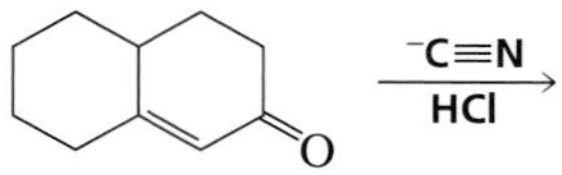 $\xrightarrow[\text{HCl}]{^-\text{C}{\equiv}\text{N}}$

b. 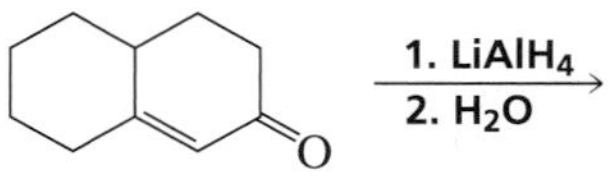 $\xrightarrow[\text{2. H}_2\text{O}]{\text{1. LiAlH}_4}$

$$
\textbf{c.} \quad CH_3\overset{\overset{\displaystyle CH_3}{|}}{C}=CH\overset{\overset{\displaystyle O}{||}}{C}CH_3 \quad \xrightarrow[\textbf{2. H}_3\textbf{O}^+]{\textbf{1. CH}_3\textbf{MgBr}}
$$

$$
\textbf{d.} \quad CH_3\overset{\overset{\displaystyle CH_3}{|}}{C}=CH\overset{\overset{\displaystyle O}{||}}{C}CH_3 \quad \xrightarrow[\textbf{2. H}_3\textbf{O}^+]{\textbf{1. (CH}_3)_2\textbf{CuLi}}
$$

α,β-Unsaturated carboxylic acids and carboxylic acid derivatives, like α,β-unsaturated aldehydes and ketones, have two electrophilic sites for nucleophilic attack. As in the case of α,β-unsaturated aldehydes and ketones, nucleophiles that are weak bases form conjugate addition products. Nucleophiles that are strong bases form nucleophilic acyl substitution products. Notice that nucleophiles that are strong bases form *nucleophilic acyl substitution products* rather than *direct addition products* due to the presence of a group in the reactant that can be replaced by a nucleophile.

16.15 NUCLEOPHILIC ADDITION TO α,β-UNSATURATED CARBOXYLIC ACIDS AND CARBOXYLIC ACID DERIVATIVES

$$
CH_2=CH\overset{\overset{\displaystyle O}{||}}{C}OCH_3 \; + \; HBr \; \longrightarrow \; BrCH_2CH_2\overset{\overset{\displaystyle O}{||}}{C}OCH_3
$$

conjugate addition product

$$
CH_3CH=CH\overset{\overset{\displaystyle O}{||}}{C}OCH_3 \; + \; HO^- \; \xrightarrow{\;H_2O\;} \; CH_3CH=CH\overset{\overset{\displaystyle O}{||}}{C}O^- \; + \; CH_3OH
$$

nucleophilic substitution product

CROSS-CONJUGATION VERSUS LINEAR CONJUGATION

The double bond in the α,β-position of an aldehyde or a ketone increases the stability of the carbonyl compound, because conjugated double bonds are more stable than isolated double bonds. However, a double bond in the α,β-position of a carboxylic acid or an ester has only a small effect on the stability of the compound, because the conjugated system is not linear as it is in an aldehyde or a ketone. The conjugated system in a carboxylic acid or an ester is not linear because the carbon–oxygen double bond is in conjugation with both the carbon–carbon double bond and the nonbonded electrons of the oxygen attached to the carbonyl carbon. This is called **cross-conjugation.** Molecules are significantly more stabilized by **linear conjugation** than by cross-conjugation.

linear conjugation   **cross-conjugation**

16.16
ENZYME-CATALYZED ADDITIONS TO α,β-UNSATURATED CARBONYL COMPOUNDS

Several reactions in biological systems involve addition to α,β-unsaturated carbonyl compounds. Because the nucleophiles present in living organisms—such as H_2O and NH_3—tend to be weak bases, conjugate addition products are obtained from these enzyme-catalyzed addition reactions. The following are examples of conjugate addition reactions that occur in biological systems.

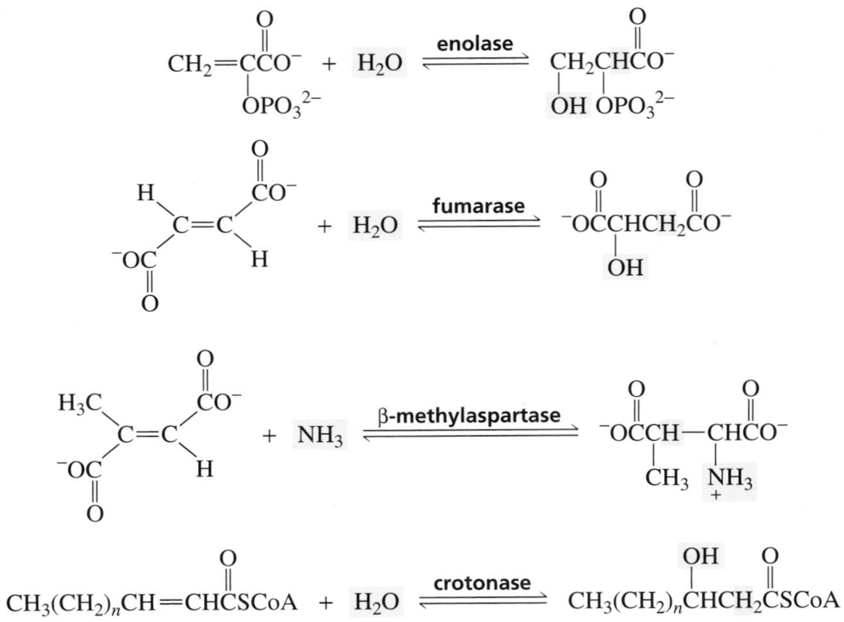

$$CH_2{=}CCO^- + H_2O \underset{\text{enolase}}{\rightleftharpoons} CH_2CHCO^-$$
(with OPO_3^{2-} and $OH\ OPO_3^{2-}$ substituents)

$$\text{fumarate} + H_2O \underset{\text{fumarase}}{\rightleftharpoons} {}^-OCCHCH_2CO^-$$

$$\text{+ } NH_3 \underset{\beta\text{-methylaspartase}}{\rightleftharpoons} {}^-OCCH{-}CHCO^-$$

$$CH_3(CH_2)_nCH{=}CHCSCoA + H_2O \underset{\text{crotonase}}{\rightleftharpoons} CH_3(CH_2)_nCHCH_2CSCoA$$

ENZYME-CATALYZED CIS–TRANS INTERCONVERSION

Enzymes that catalyze the interconversion of cis and trans isomers are called cis–trans isomerases. These isomerases are all known to contain thiol (SH) groups. Thiols are weak bases and therefore add to the β-carbon of an α,β-unsaturated carbonyl compound (conjugate addition). The resulting carbon–carbon single bond rotates before the enol is able to tautomerize to the ketone. When tautomerization occurs, the thiol is eliminated. Rotation results in cis–trans interconversion.

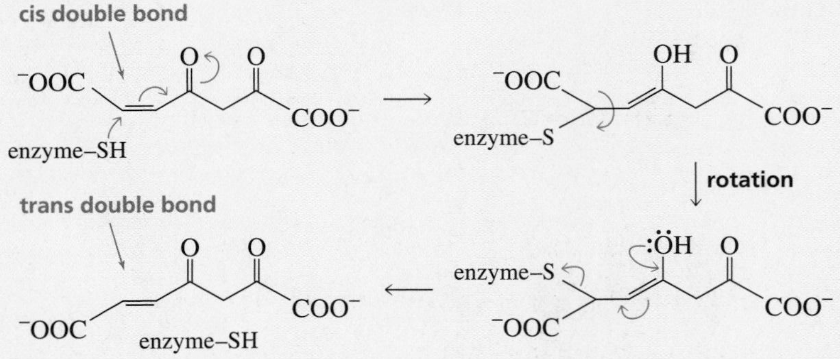

SUMMARY OF REACTIONS

1. Reaction of *carbonyl compounds* with Grignard reagents (Sections 16.4 and 16.6)

a. Reaction of an *aldehyde* with a Grignard reagent forms a secondary alcohol:

$$R-\overset{O}{\overset{\|}{C}}-H \xrightarrow[\text{2. H}_3\text{O}^+]{\text{1. CH}_3\text{MgBr}} R-\overset{OH}{\underset{CH_3}{\overset{|}{\underset{|}{C}}}}-H$$

b. Reaction of a *ketone* with a Grignard reagent forms a tertiary alcohol:

$$R-\overset{O}{\overset{\|}{C}}-R' \xrightarrow[\text{2. H}_3\text{O}^+]{\text{1. CH}_3\text{MgBr}} R-\overset{OH}{\underset{CH_3}{\overset{|}{\underset{|}{C}}}}-R'$$

c. Reaction of an *ester* with a Grignard reagent forms a tertiary alcohol with two identical substituents:

$$R-\overset{O}{\overset{\|}{C}}-OR' \xrightarrow[\text{2. H}_3\text{O}^+]{\text{1. 2 CH}_3\text{MgBr}} R-\overset{OH}{\underset{CH_3}{\overset{|}{\underset{|}{C}}}}-CH_3$$

d. Reaction of an *acyl chloride* with a Grignard reagent forms a tertiary alcohol with two identical substituents:

$$R-\overset{O}{\overset{\|}{C}}-Cl \xrightarrow[\text{2. H}_3\text{O}^+]{\text{1. 2 CH}_3\text{MgBr}} R-\overset{OH}{\underset{CH_3}{\overset{|}{\underset{|}{C}}}}-CH_3$$

e. Reaction of CO_2 with a Grignard reagent forms a carboxylic acid:

$$O{=}C{=}O \xrightarrow[\text{2. H}_3\text{O}^+]{\text{1. CH}_3\text{MgBr}} CH_3-\overset{O}{\overset{\|}{C}}-OH$$

2. Reactions of *aldehydes* and *ketones* with other carbon nucleophiles (Section 16.4)

$$R-\overset{O}{\overset{\|}{C}}-R \xrightarrow[\text{2. H}_3\text{O}^+]{\text{1. RC}{\equiv}\text{C}^-} R-\overset{OH}{\underset{R}{\overset{|}{\underset{|}{C}}}}-C{\equiv}CR$$

$$R-\overset{O}{\overset{\|}{C}}-R \xrightarrow[\text{HCl}]{^-\text{C}{\equiv}\text{N}} R-\overset{OH}{\underset{R}{\overset{|}{\underset{|}{C}}}}-C{\equiv}N$$

3. Reactions of *carbonyl compounds* with hydride ion donors (Sections 16.5 and 16.6)
    a. Reaction of an *aldehyde* with sodium borohydride forms a primary alcohol:

$$R-\overset{O}{\overset{\|}{C}}-H \xrightarrow[\text{2. H}_2\text{O}]{\text{1. NaBH}_4} RCH_2OH$$

b. Reaction of a *ketone* with sodium borohydride forms a secondary alcohol:

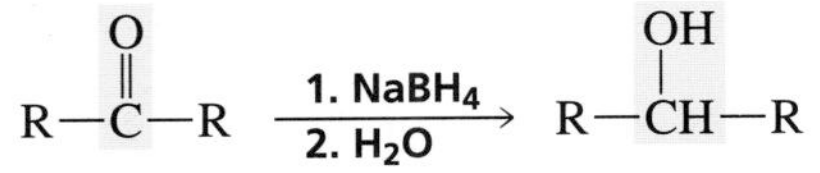

$$R-\overset{\overset{\displaystyle O}{\|}}{C}-R \xrightarrow[\text{2. H}_2\text{O}]{\text{1. NaBH}_4} R-\overset{\overset{\displaystyle OH}{|}}{C}H-R$$

c. Reaction of an *ester* with lithium aluminum hydride forms primary alcohols:

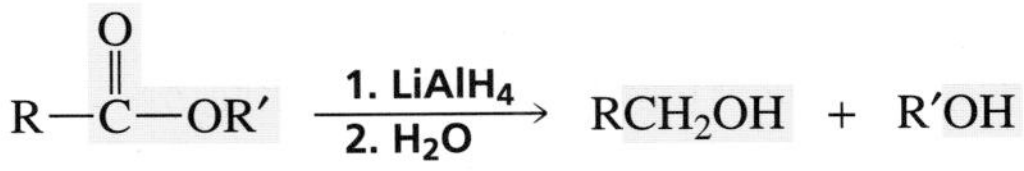

$$R-\overset{\overset{\displaystyle O}{\|}}{C}-OR' \xrightarrow[\text{2. H}_2\text{O}]{\text{1. LiAlH}_4} RCH_2OH + R'OH$$

d. Reaction of an ester with diisobutylaluminum hydride forms an aldehyde:

$$R-\overset{\overset{\displaystyle O}{\|}}{C}-OR' \xrightarrow[\text{2. H}_2\text{O}]{\text{1. DIBAH}} R-\overset{\overset{\displaystyle O}{\|}}{C}-H$$

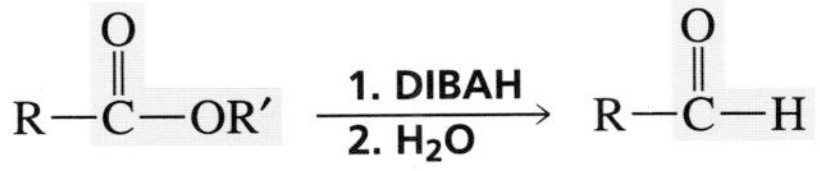

e. Reaction of a *carboxylic acid* with lithium aluminum hydride forms a primary alcohol:

$$R-\overset{\overset{\displaystyle O}{\|}}{C}-OH \xrightarrow[\text{2. H}_2\text{O}]{\text{1. LiAlH}_4} R-CH_2-OH$$

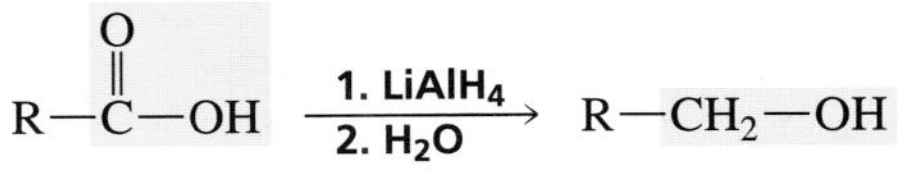

f. Reaction of an *acyl chloride* with lithium aluminum hydride forms a primary alcohol:

$$R-\overset{\overset{\displaystyle O}{\|}}{C}-Cl \xrightarrow[\text{2. H}_2\text{O}]{\text{1. LiAlH}_4} R-CH_2-OH$$

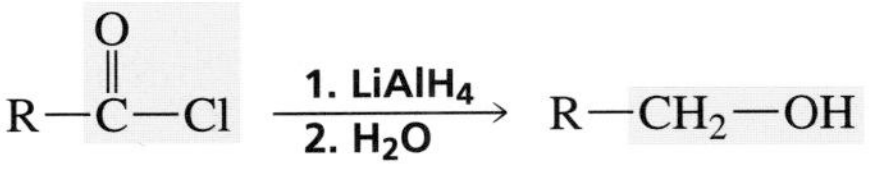

g. Reaction of an *acyl chloride* with lithium tri-*tert*-butoxyaluminum hydride forms an aldehyde:

$$R-\overset{\overset{\displaystyle O}{\|}}{C}-Cl \xrightarrow[\text{2. H}_2\text{O}]{\text{1. LiAlH[OC(CH}_3)_3]_3} R-\overset{\overset{\displaystyle O}{\|}}{C}-H$$

h. Reaction of an *amide* with lithium aluminum hydride forms an amine:

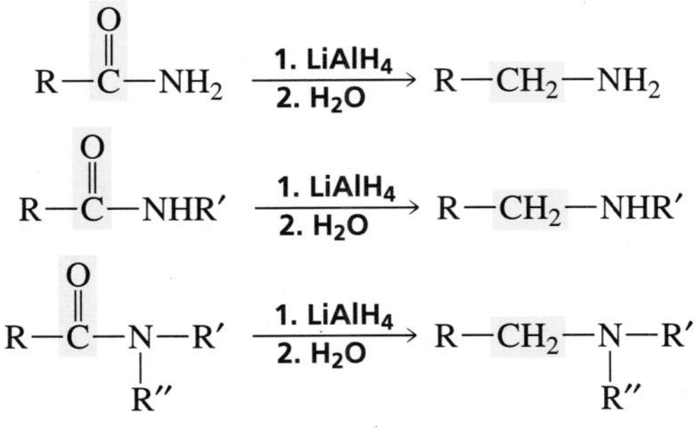

$$R-\overset{\overset{\displaystyle O}{\|}}{C}-NH_2 \xrightarrow[\text{2. H}_2\text{O}]{\text{1. LiAlH}_4} R-CH_2-NH_2$$

$$R-\overset{\overset{\displaystyle O}{\|}}{C}-NHR' \xrightarrow[\text{2. H}_2\text{O}]{\text{1. LiAlH}_4} R-CH_2-NHR'$$

$$R-\overset{\overset{\displaystyle O}{\|}}{C}-\overset{\overset{\displaystyle |}{N}}{\underset{\underset{\displaystyle R''}{|}}{}}-R' \xrightarrow[\text{2. H}_2\text{O}]{\text{1. LiAlH}_4} R-CH_2-\underset{\underset{\displaystyle R''}{|}}{N}-R'$$

4. Reactions of *aldehydes* and *ketones* with amines (Section 16.7)
   a. Reaction with a *primary amine* forms an imine:

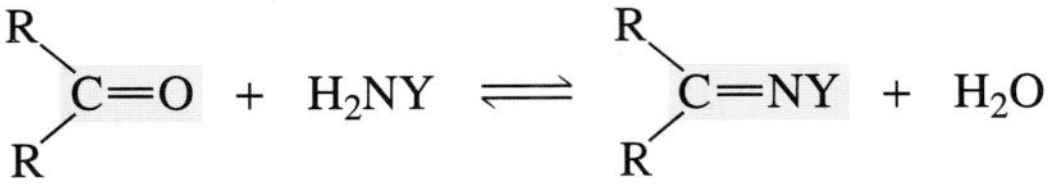

$$\overset{\displaystyle R}{\underset{\displaystyle R}{>}}C=O + H_2NY \rightleftharpoons \overset{\displaystyle R}{\underset{\displaystyle R}{>}}C=NY + H_2O$$

When Y = R, the product is a Schiff base; Y can also be OH, NH_2, NHC_6H_5, $NHC_6H_3(NO_2)_2$, $NHCONH_2$.

b. Reaction with a *secondary amine* forms an enamine:

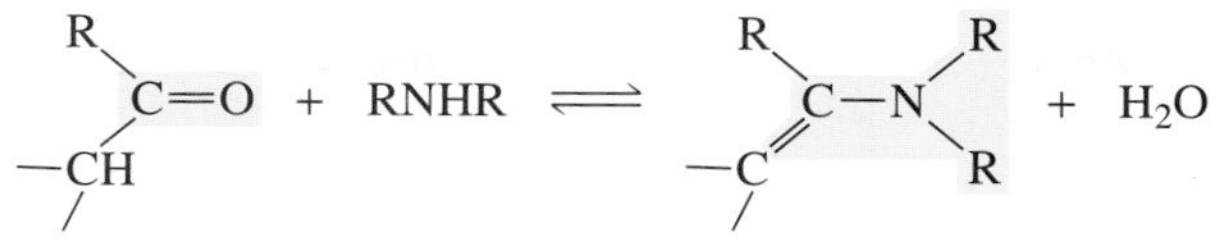

$$\underset{-CH}{\overset{R}{\diagdown}}C=O \ + \ RNHR \ \rightleftharpoons \ \underset{-C}{\overset{R}{\diagdown}}C-\underset{R}{\overset{R}{N}} \ + \ H_2O$$

c. Reaction with *ammonia* plus H_2/metal catalyst or sodium cyanoborohydride forms a primary amine:

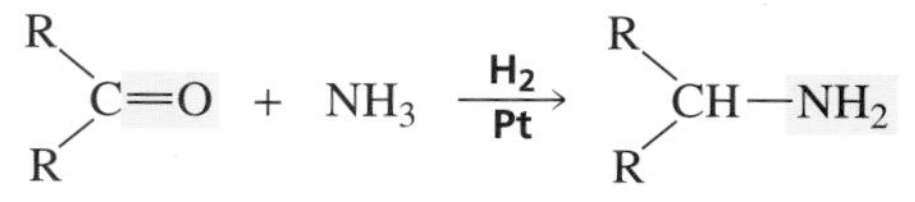

$$\underset{R}{\overset{R}{\diagdown}}C=O \ + \ NH_3 \ \xrightarrow[\text{Pt}]{H_2} \ \underset{R}{\overset{R}{\diagdown}}CH-NH_2$$

d. Reaction with a *primary amine* plus H_2/metal catalyst or sodium cyanoborohydride forms a secondary amine:

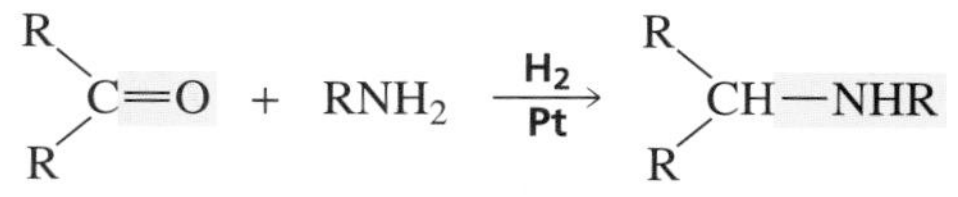

$$\underset{R}{\overset{R}{\diagdown}}C=O \ + \ RNH_2 \ \xrightarrow[\text{Pt}]{H_2} \ \underset{R}{\overset{R}{\diagdown}}CH-NHR$$

e. Reaction with a *secondary amine* plus H_2/metal catalyst or sodium cyanoborohydride forms a tertiary amine:

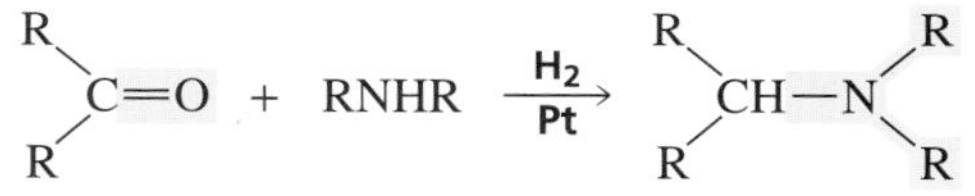

$$\underset{R}{\overset{R}{\diagdown}}C=O \ + \ RNHR \ \xrightarrow[\text{Pt}]{H_2} \ \underset{R}{\overset{R}{\diagdown}}CH-\underset{R}{\overset{R}{N}}$$

f. The Wolff–Kishner reduction converts a carbonyl group to a methylene group:

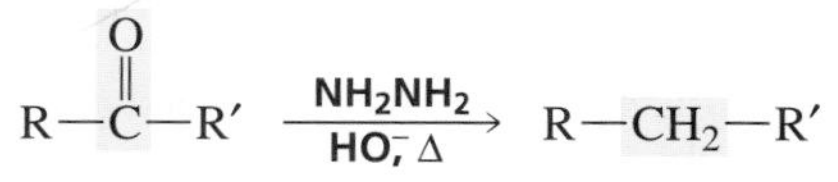

$$R-\overset{\overset{\displaystyle O}{\|}}{C}-R' \ \xrightarrow[\text{HO}^-, \, \Delta]{NH_2NH_2} \ R-CH_2-R'$$

5. Reaction of an *aldehyde* or a *ketone* with water forms a hydrate (Section 16.8)

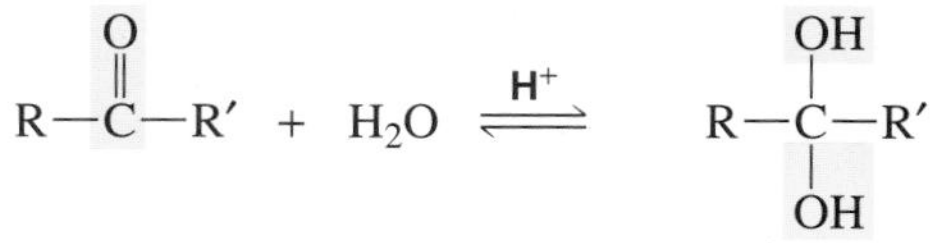

$$R-\overset{\overset{\displaystyle O}{\|}}{C}-R' \ + \ H_2O \ \xrightleftharpoons[\ ]{H^+} \ R-\underset{\underset{\displaystyle OH}{|}}{\overset{\overset{\displaystyle OH}{|}}{C}}-R'$$

6. Reaction of an *aldehyde* or a *ketone* with an alcohol forms an acetal or a ketal (Section 16.8)

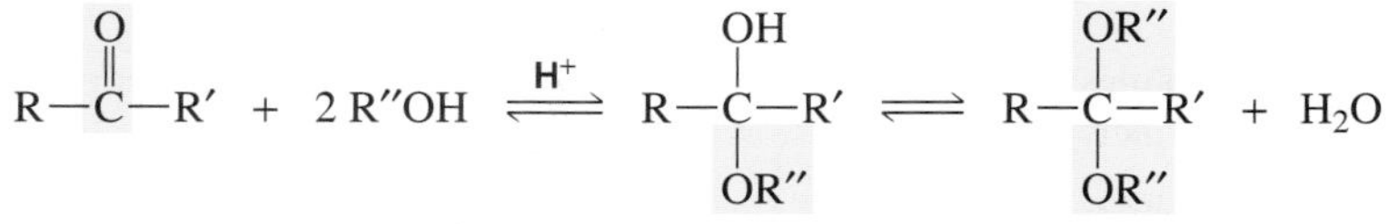

$$R-\overset{\overset{\displaystyle O}{\|}}{C}-R' \ + \ 2\,R''OH \ \xrightleftharpoons[\ ]{H^+} \ R-\underset{\underset{\displaystyle OR''}{|}}{\overset{\overset{\displaystyle OH}{|}}{C}}-R' \ \rightleftharpoons \ R-\underset{\underset{\displaystyle OR''}{|}}{\overset{\overset{\displaystyle OR''}{|}}{C}}-R' \ + \ H_2O$$

7. Reaction of an *aldehyde* or a *ketone* with a thiol forms a thioacetal or a thioketal (Section 16.10)

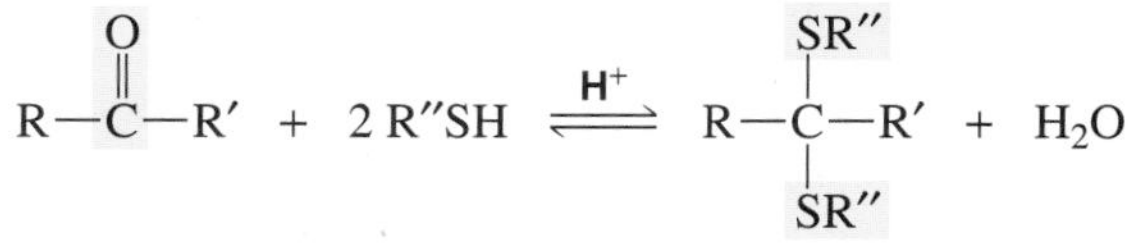

$$R-\overset{\overset{\displaystyle O}{\|}}{C}-R' \ + \ 2\,R''SH \ \xrightleftharpoons[\ ]{H^+} \ R-\underset{\underset{\displaystyle SR''}{|}}{\overset{\overset{\displaystyle SR''}{|}}{C}}-R' \ + \ H_2O$$

8. Desulfuration of *thioacetals* and *thioketals* forms alkanes (Section 16.10)

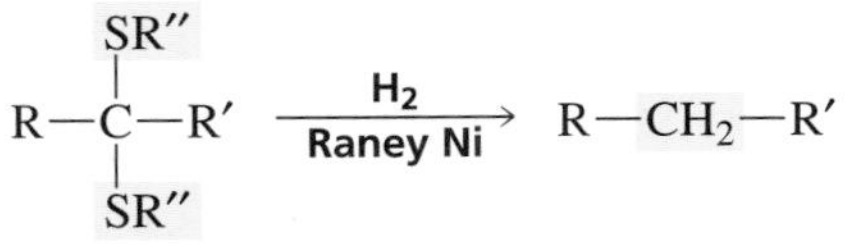

$$R-\underset{\underset{SR''}{|}}{\overset{\overset{SR''}{|}}{C}}-R' \xrightarrow[\textbf{Raney Ni}]{\textbf{H}_2} R-CH_2-R'$$

9. Reaction of an *aldehyde* or a *ketone* with a phosphonium ylide forms an alkene (Section 16.11)

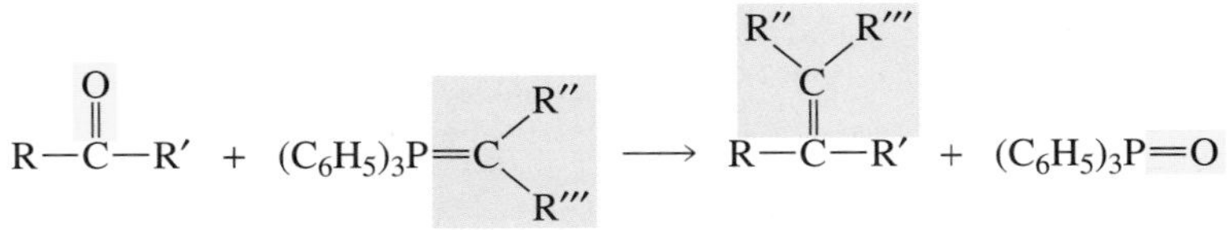

$$R-\overset{\overset{O}{\|}}{C}-R' + (C_6H_5)_3P=C\overset{R''}{\underset{R'''}{<}} \longrightarrow R-\overset{\overset{R''\quad R'''}{C}}{C}-R' + (C_6H_5)_3P=O$$

10. Reactions of α,β-*unsaturated aldehydes* and *ketones* with nucleophiles (Section 16.14)

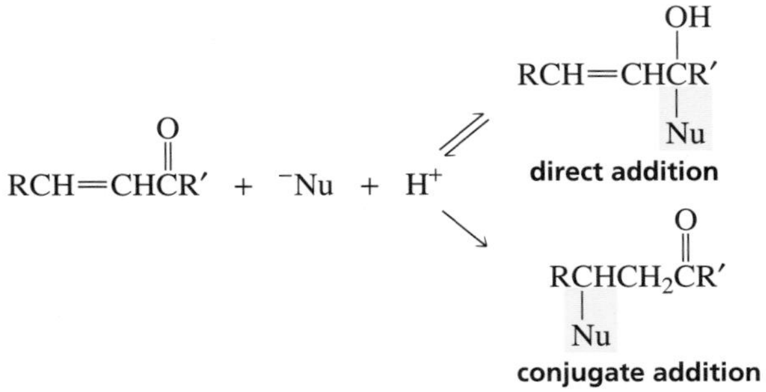

$$RCH=CHCR' + {}^-Nu + H^+ \nearrow \atop \searrow$$

$$\underset{\textbf{direct addition}}{RCH=CH\underset{\underset{Nu}{|}}{\overset{\overset{OH}{|}}{C}}R'}$$

$$\underset{\textbf{conjugate addition}}{R\underset{\underset{Nu}{|}}{CH}CH_2\overset{\overset{O}{\|}}{C}R'}$$

Nucleophiles that are strong bases (RLi, RMgBr, H⁻) form direct addition products.

Nucleophiles that are weak bases (⁻CN, RNH₂, RSH, Br⁻, R₂CuLi) form conjugate addition products.

11. Reactions of α,β-*unsaturated carboxylic acids* and *carboxylic acid derivatives* with nucleophiles (Section 16.15)

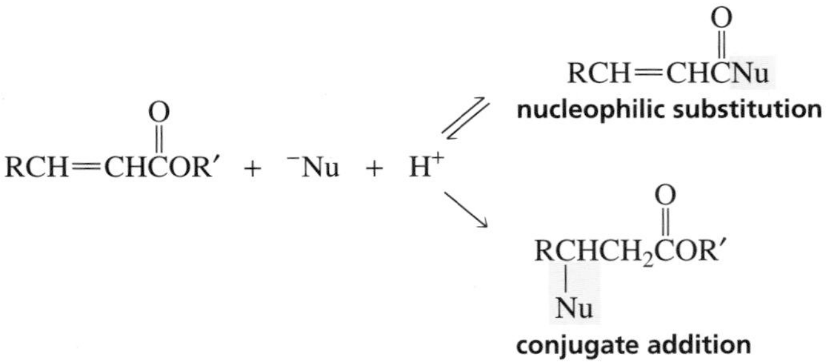

$$RCH=CHCOR' + {}^-Nu + H^+ \nearrow \atop \searrow$$

$$\underset{\textbf{nucleophilic substitution}}{RCH=CH\overset{\overset{O}{\|}}{C}Nu}$$

$$\underset{\textbf{conjugate addition}}{R\underset{\underset{Nu}{|}}{CH}CH_2\overset{\overset{O}{\|}}{C}OR'}$$

Nucleophiles that are strong bases (RLi, RMgBr, H⁻) form nucleophilic substitution products.

Nucleophiles that are weak bases (⁻CN, RNH₂, RSH, Br⁻, R₂CuLi) form conjugate addition products.

KEY TERMS

enamine (page 743)
hemiacetal (page 749)
hemiketal (page 749)
hydrate (page 746)
hydrazone (page 740)
imine (page 740)
ketal (page 749)
ketone (page 725)
linear conjugation (page 763)
nucleophilic acyl substitution reaction
  (page 730)

nucleophilic addition–elimination—
  nucleophilic addition reaction
  (page 751)
nucleophilic addition–elimination
  reaction (page 731)
nucleophilic addition reaction
  (page 730)
oxime (page 740)
oxonium ion (page 746)
oxyanion (page 731)
phenylhydrazone (page 741)

pH–rate profile (page 742)
prochiral carbonyl carbon
  (page 756)
protecting group (page 752)
reduction reaction (page 735)
reductive amination
  (page 745)
Schiff base (page 740)
semicarbazone (page 740)
Wittig reaction (page 754)
ylide (page 754)

PROBLEMS

28. Give a structure for each of the following.
  a. isobutyraldehyde
  b. 4-hexenal
  c. diisopentyl ketone
  d. 3-methylcyclohexanone
  e. 2,4-pentanedione
  f. 4-bromo-3-heptanone
  g. γ-bromocaproaldehyde
  h. 2-ethylcyclopentanecarbaldehyde
  i. 4-methyl-5-oxohexanal
  j. *meta*-formylbenzaldehyde

29. Give the products of each of the following reactions.

a. $CH_3CH_2\overset{\displaystyle O}{\overset{\|}{C}}H$ + CH_3CH_2OH (excess) $\xrightarrow{H^+}$

b. 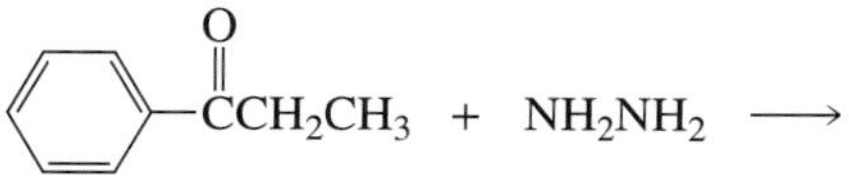 Ph–$\overset{\displaystyle O}{\overset{\|}{C}}CH_2CH_3$ + NH_2NH_2 ⟶

c. Ph–$\overset{\displaystyle O}{\overset{\|}{C}}CH_2CH_3$ + NH_2NH_2 $\xrightarrow[\Delta]{HO^-}$

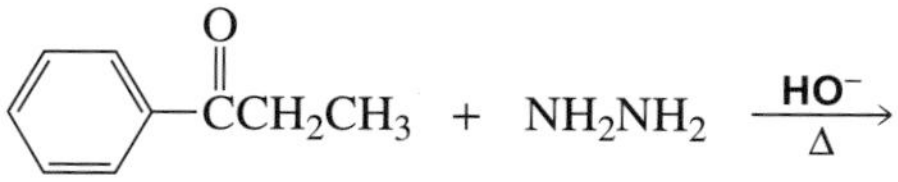

d. $CH_3CH_2\overset{\displaystyle O}{\overset{\|}{C}}CH_3$ $\xrightarrow[\text{2. H}_2\text{O}]{\text{1. NaBH}_4}$

e. $CH_3CH_2\overset{\displaystyle O}{\overset{\|}{C}}CH_2CH_3$ + $NaC{\equiv}N$ (excess) $\xrightarrow{HCl}$

f. $CH_3CH_2CH_2\overset{\displaystyle O}{\overset{\|}{C}}OCH_2CH_3$ $\xrightarrow[\text{2. H}_2\text{O}]{\text{1. LiAlH}_4}$

g. $CH_3CH_2CH_2\overset{\displaystyle O}{\overset{\|}{C}}CH_3$ + $HOCH_2CH_2OH$ $\xrightarrow{H^+}$

h. 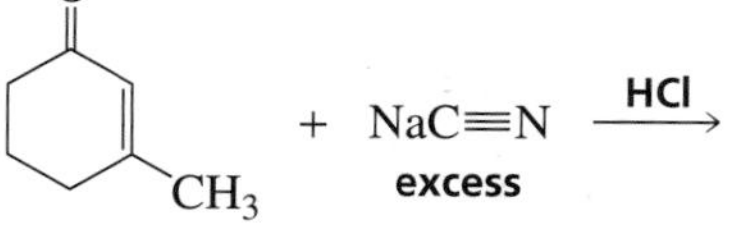 + $NaC{\equiv}N$ (excess) $\xrightarrow{HCl}$

i. 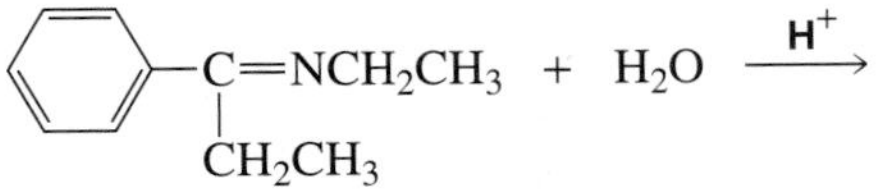 (phenyl)—C=NCH$_2$CH$_3$ + H$_2$O $\xrightarrow{\text{H}^+}$
    $\underset{\text{CH}_2\text{CH}_3}{|}$

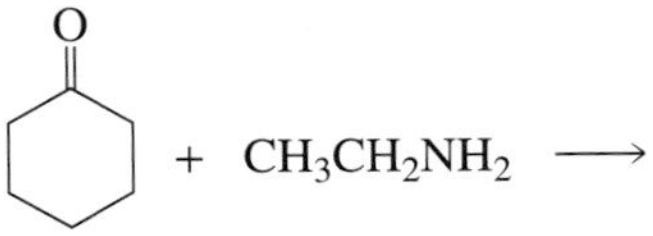

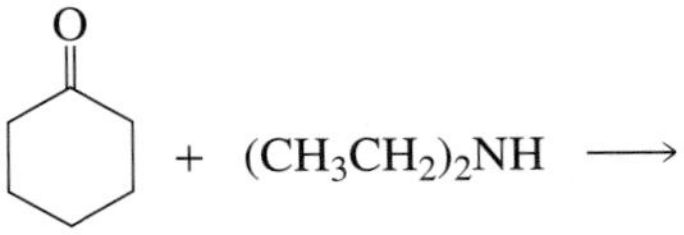

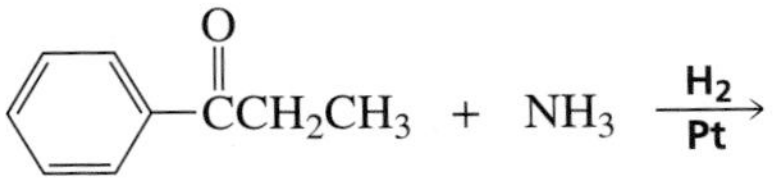

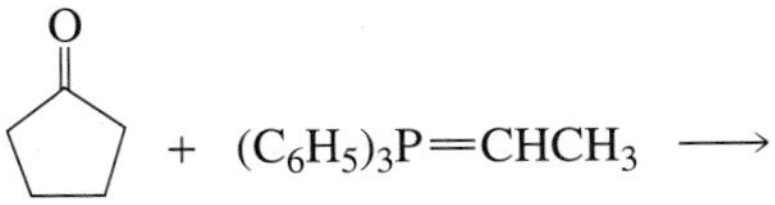

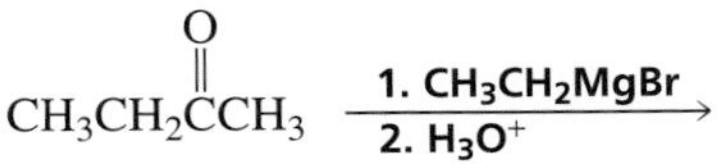

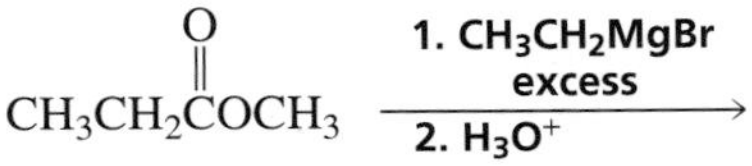

j. (cyclohexanone) + CH$_3$CH$_2$NH$_2$ $\longrightarrow$

k. (cyclohexanone) + (CH$_3$CH$_2$)$_2$NH $\longrightarrow$

l. (phenyl)—$\overset{\text{O}}{\overset{||}{\text{C}}}CH_2CH_3$ + NH$_3$ $\xrightarrow{\text{H}_2}{\text{Pt}}$

m. (cyclopentanone) + (C$_6$H$_5$)$_3$P=CHCH$_3$ $\longrightarrow$

n. CH$_3$CH$_2$$\overset{\text{O}}{\overset{||}{\text{C}}}CH_3$ $\xrightarrow[\text{2. H}_3\text{O}^+]{\text{1. CH}_3\text{CH}_2\text{MgBr}}$

o. CH$_3$CH$_2$$\overset{\text{O}}{\overset{||}{\text{C}}}OCH_3$ $\xrightarrow[\text{2. H}_3\text{O}^+]{\overset{\text{1. CH}_3\text{CH}_2\text{MgBr}}{\text{excess}}}$

p. CH$_3$$\overset{\overset{\text{CH}_3}{|}}{\text{C}}$=CH$\overset{\text{O}}{\overset{||}{\text{C}}}CH_3$ + HBr $\longrightarrow$

q. (pyrrolidinone) $\xrightarrow[\text{2. H}_2\text{O}]{\text{1. LiAlH}_4}$

30. List the following compounds in order of decreasing reactivity toward nucleophilic attack.

CH$_3$CH$_2$CH$\overset{\text{O}}{\overset{||}{\text{C}}}CH_2CH_3$ 　　CH$_3$CH$_2$$\overset{\text{O}}{\overset{||}{\text{C}}}$H 　　CH$_3CH_2CH\overset{\text{OCH}_3}{\overset{||}{\text{C}}}CH_2CH_3$
$\underset{\text{CH}_3}{|}$　　　　　　　　　　　　　　　　　　　　$\underset{\text{CH}_3}{|}$ $\underset{\text{CH}_3}{|}$

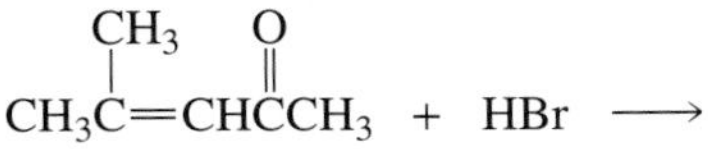

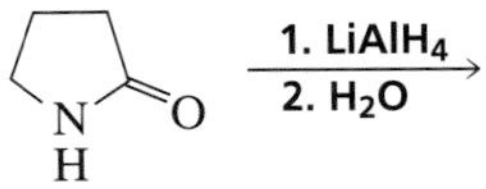

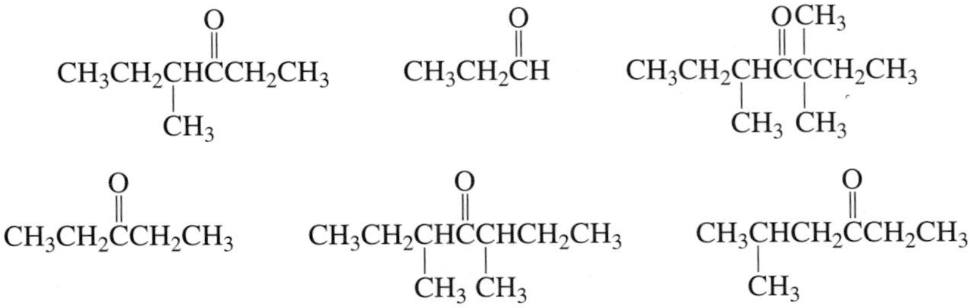

CH$_3$CH$_2$$\overset{\text{O}}{\overset{||}{\text{C}}}CH_2CH_3$ 　　CH$_3$CH$_2$CH$\overset{\text{O}}{\overset{||}{\text{C}}}$CHCH$_2CH_3$ 　　CH$_3$CHCH$_2$$\overset{\text{O}}{\overset{||}{\text{C}}}CH_2CH_3$
　　　　　　　　　　　　　$\underset{\text{CH}_3}{|}$ $\underset{\text{CH}_3}{|}$　　　　　　$\underset{\text{CH}_3}{|}$

31. Using cyclohexanone as the starting material, describe how each of the following compounds could be synthesized.

a. 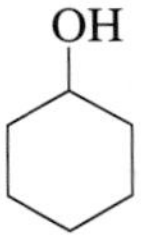 (cyclohexanol with OH)

b.  (cyclohexene)

c.
Br

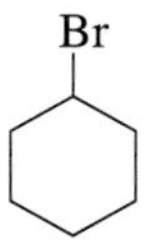

g. 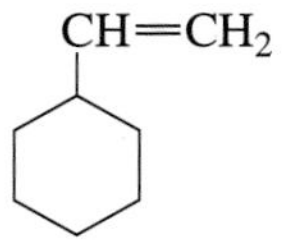
CH=CH₂

d. 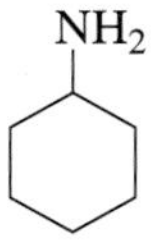
NH₂

h. 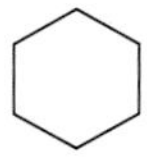 ⬡ (show two methods)

e. 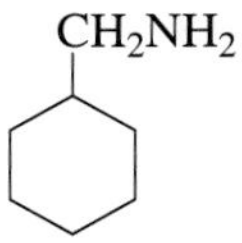
CH₂NH₂

i. 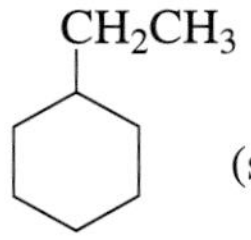 ⬡ (show two methods)
CH₂CH₃

f. 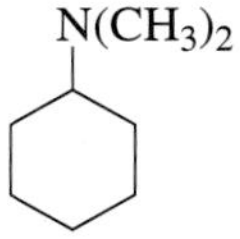
N(CH₃)₂

32. List the following compounds in order of decreasing K_{eq} for hydrate formation.

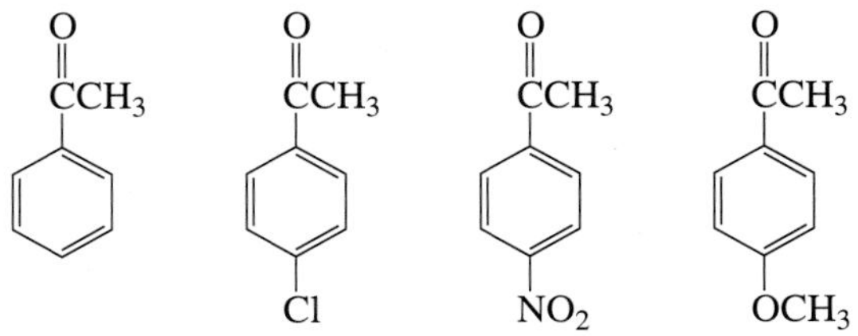

$$\overset{O}{\underset{}{\overset{\|}{C}}}CH_3 \qquad \overset{O}{\underset{}{\overset{\|}{C}}}CH_3 \qquad \overset{O}{\underset{}{\overset{\|}{C}}}CH_3 \qquad \overset{O}{\underset{}{\overset{\|}{C}}}CH_3$$

Cl   NO₂   OCH₃

33. Fill in the boxes.

a. 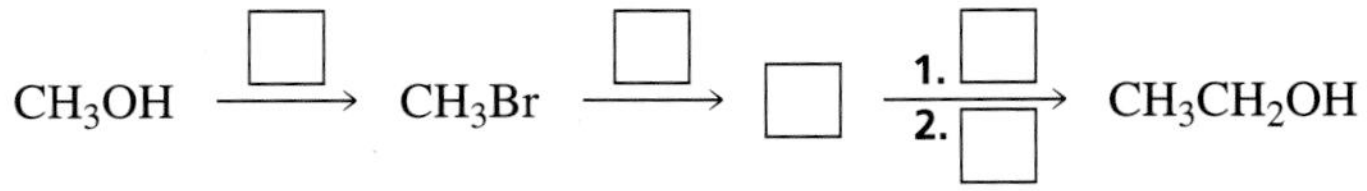 CH₃OH →☐→ CH₃Br →☐→ ☐ $\xrightarrow[2.\,☐]{1.\,☐}$ CH₃CH₂OH

b. 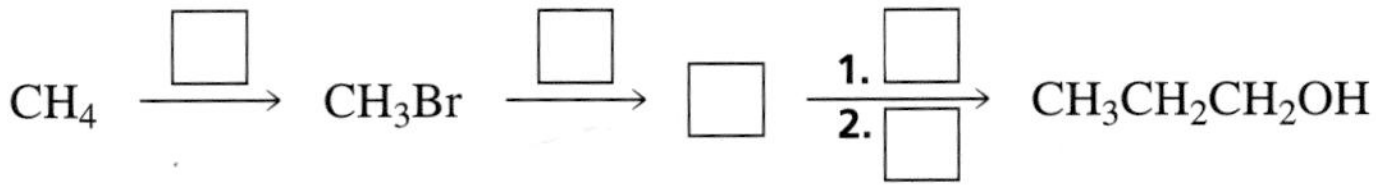 CH₄ →☐→ CH₃Br →☐→ ☐ $\xrightarrow[2.\,☐]{1.\,☐}$ CH₃CH₂CH₂OH

c.

34. Thiols can be prepared from the reaction of thiourea with an alkyl halide followed by hydroxide-ion-promoted hydrolysis and protonation of the resulting products.

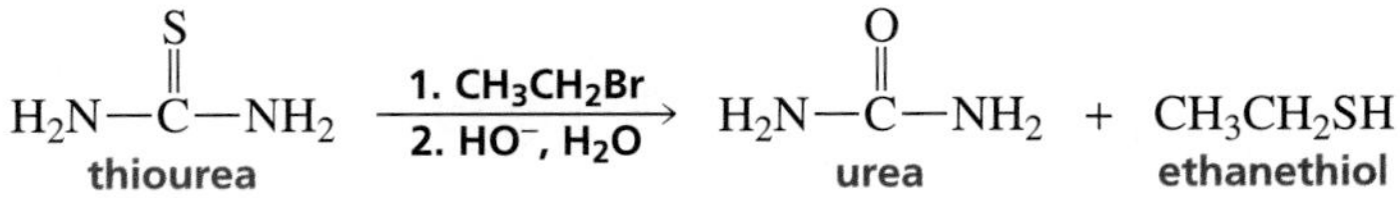

$$\underset{\text{thiourea}}{H_2N-\overset{S}{\overset{\|}{C}}-NH_2} \xrightarrow[\text{2. HO}^-,\,H_2O]{\text{1. CH}_3\text{CH}_2\text{Br}} \underset{\text{urea}}{H_2N-\overset{O}{\overset{\|}{C}}-NH_2} + \underset{\text{ethanethiol}}{CH_3CH_2SH}$$

a. Propose a mechanism for the reaction.
b. What thiol would be formed if the alkyl halide employed were pentyl bromide?

35. Imines can exist as stereoisomers. The isomers are named by the *E, Z* system of nomenclature. (The unshared pair of electrons has the lowest priority.)

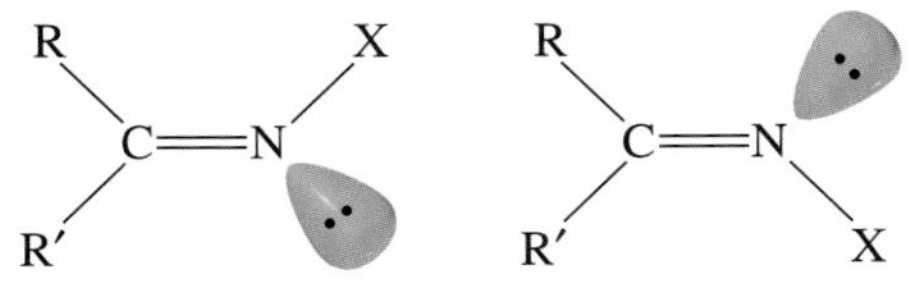

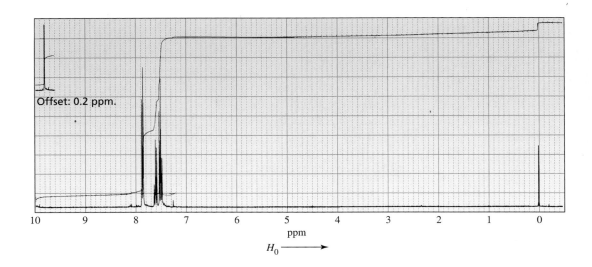

Give a structure for each of the following:
   a. *(E)*-benzaldehyde semicarbazone
   b. *(Z)*-propiophenone oxime
   c. cyclohexanone 2,4-dinitrophenylhydrazone

36. The only organic compound obtained when compound Z undergoes the following sequence of reactions gives the ^{1}H NMR spectrum shown below. Identify compound Z.

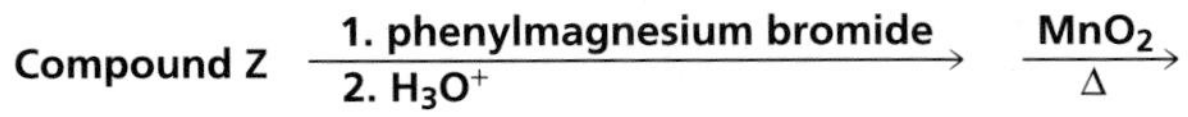

$$\text{Compound Z} \xrightarrow[\text{2. } H_3O^+]{\text{1. phenylmagnesium bromide}} \xrightarrow[\Delta]{MnO_2}$$

Offset: 0.2 ppm.

10   9   8   7   6   5   4   3   2   1   0
ppm

$H_0 \longrightarrow$

37. Propose a mechanism for each of the following reactions.

a. [structure] $\xrightarrow{H_3O^+}$ [structure with $CH_2CH_2CH_2\overset{+}{N}H_3$]

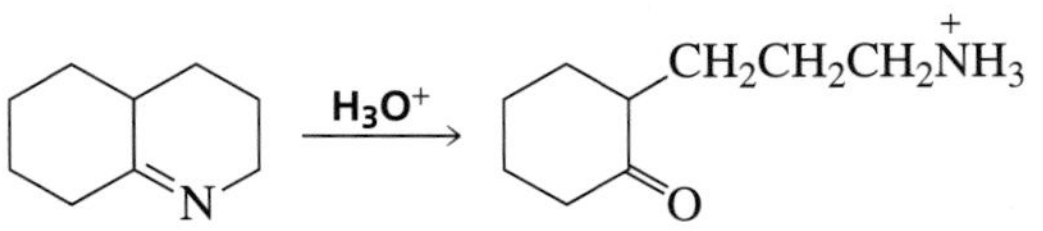

b. [structure with OCH_3] $\xrightarrow{H_3O^+}$ [cyclohexanone structure]

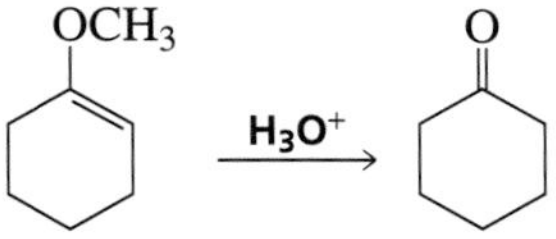

c. [structure] + CH_3CH_2OH $\xrightarrow{H^+}$ [structure with OCH_2CH_3]

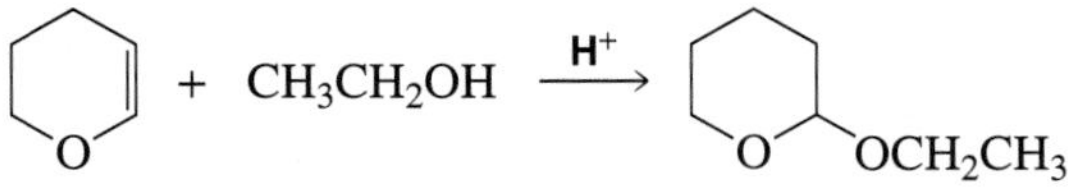

38. How many signals would the product of the following reaction show in
   a. its ^{1}H NMR spectrum?   **b.** its ^{13}C NMR spectrum?

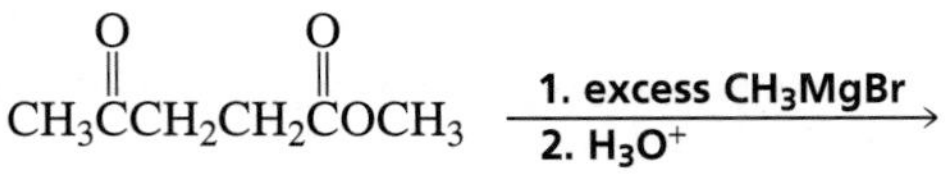

$$CH_3\overset{O}{\overset{\|}{C}}CH_2CH_2\overset{O}{\overset{\|}{C}}OCH_3 \xrightarrow[\text{2. } H_3O^+]{\text{1. excess } CH_3MgBr}$$

39. Each of the following reactions can result in the formation of more than one stereoisomer. Show the stereoisomers.

a.

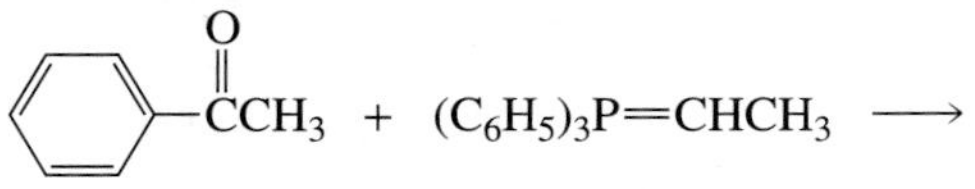

$$\text{(phenyl)}-\overset{\overset{\text{O}}{\|}}{\text{C}}\text{CH}_3 \ + \ (C_6H_5)_3P=CHCH_3 \longrightarrow$$

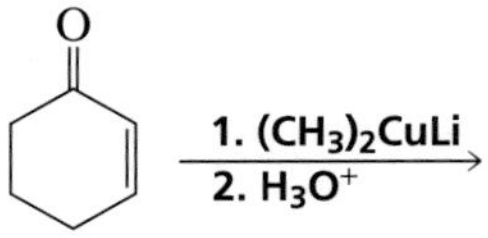

b.

$$\xrightarrow[\text{2. H}_3\text{O}^+]{\text{1. (CH}_3)_2\text{CuLi}}$$

c.

$$\xrightarrow[\text{2. H}_3\text{O}^+]{\text{1. CH}_3\text{MgBr}}$$

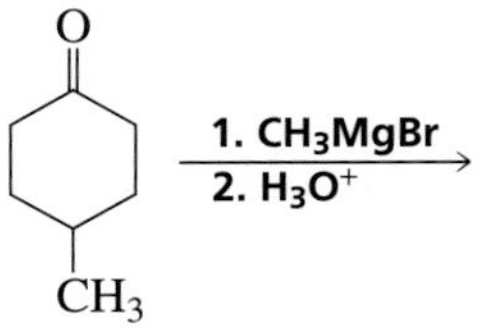

d.

$$\text{(phenyl)}-\overset{\overset{\text{O}}{\|}}{\text{C}}\text{CH}_2\text{CH}_3 \ +$$

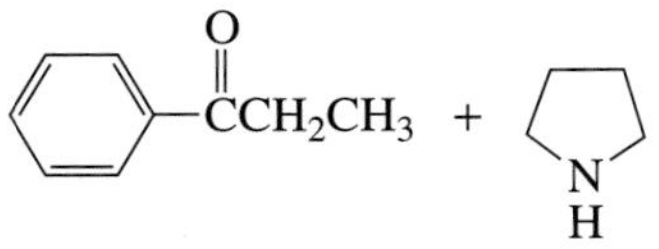

e. $CH_3CH_2\overset{\overset{\text{O}}{\|}}{\text{C}}CH_2CH_2CH_2CH_3$ $\xrightarrow[\text{2. H}_2\text{O}]{\text{1. LiAlH}_4}$

40. List three different sets of reagents (a carbonyl compound and a Grignard reagent) that could be used to prepare each of the following tertiary alcohols.

a. 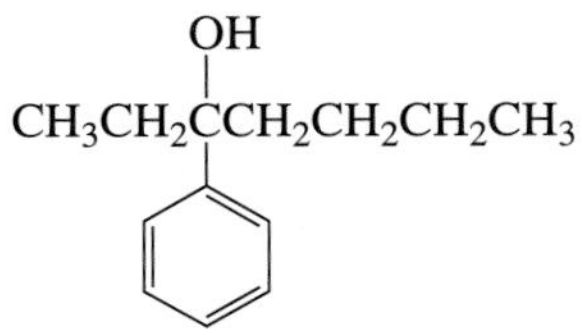
$$\underset{\underset{\text{(phenyl)}}{|}}{CH_3CH_2\overset{\overset{\text{OH}}{|}}{C}CH_2CH_2CH_2CH_3}$$

b.
$$CH_3CH_2\overset{\overset{\text{OH}}{|}}{\underset{\underset{CH_2CH_3}{|}}{C}}CH_2CH_2CH_3$$

41. Give the product of the reaction of 3-methyl-2-cyclohexenone with each of the following reagents.
a. CH_3MgBr followed by H_3O^+ **d.** $(CH_3)_2NH$
b. excess NaCN, HCl   **e.** $(CH_3CH_2)_2CuLi$
c. H_2, Pd   **f.** CH_3CH_2SH

42. Norlutin and Enovid are ketones used clinically to suppress ovulation. Consequently, they are used as contraceptives. For which of these compounds would you expect the infrared carbonyl absorption (C=O stretch) to be at a higher frequency? Explain.

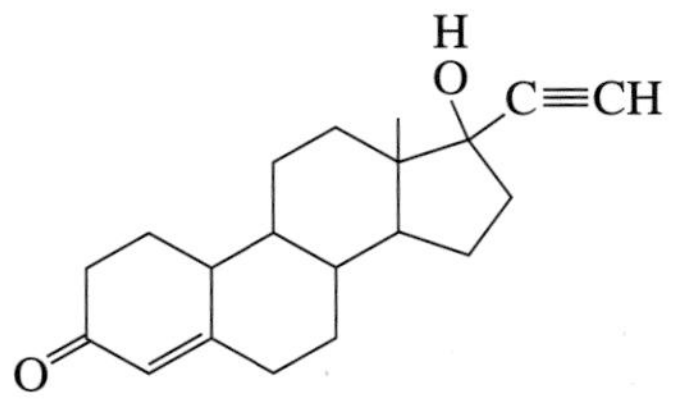

Norlutin 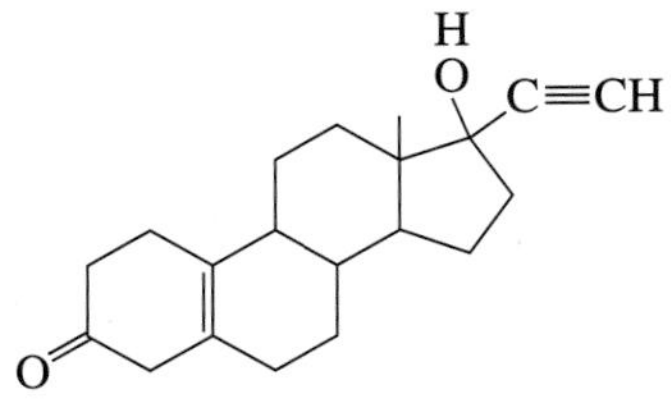   Enovid

43. Unlike a phosphonium ylide that reacts with an aldehyde or ketone to form an alkene, a sulfonium ylide reacts with an aldehyde or ketone to form an epoxide. Explain why one ylide forms an alkene while the other forms an epoxide.

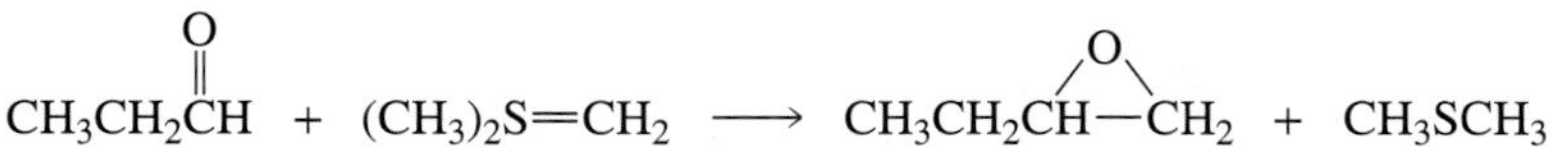

$$CH_3CH_2CH(=O) + (CH_3)_2S=CH_2 \longrightarrow CH_3CH_2CH\underset{O}{-}CH_2 + CH_3SCH_3$$

44. Indicate how the following compounds could be prepared from the given starting materials.

a.

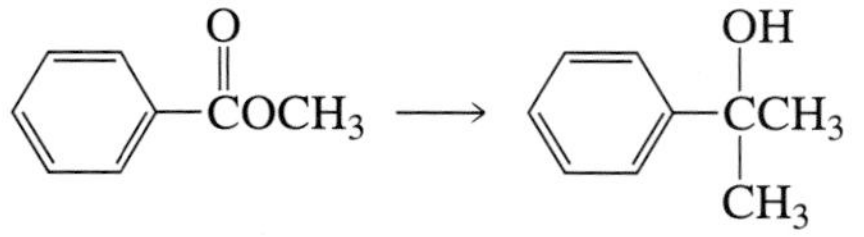

b.

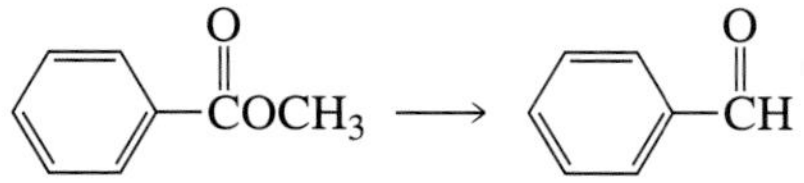

c.

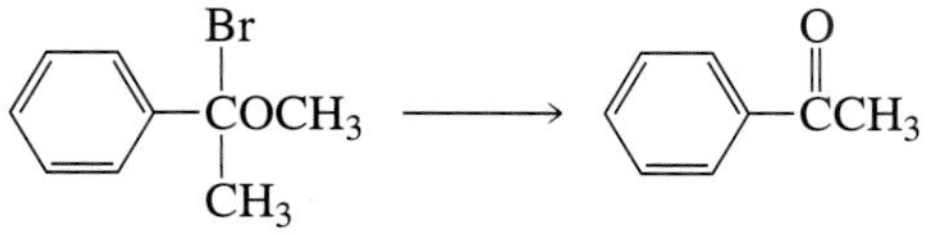

d. $CH_3CH_2CH_2CH_2Br \longrightarrow CH_3CH_2CH_2CH_2\overset{O}{\overset{\|}{C}}OH$

e.

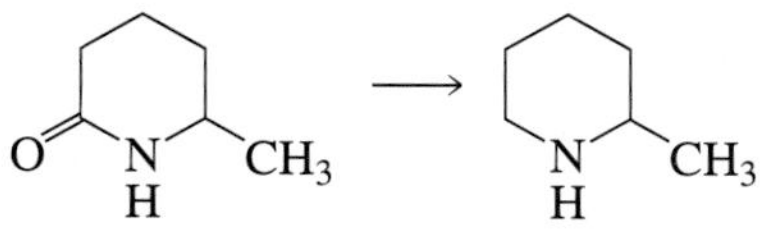

f.

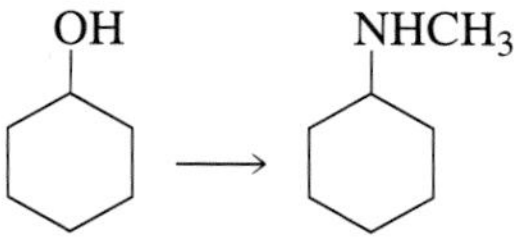

45. Propose a reasonable mechanism for each of the following reactions.

a. 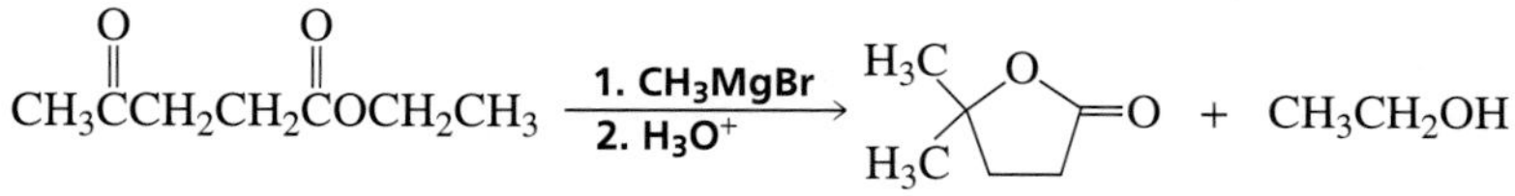 $CH_3\overset{O}{\overset{\|}{C}}CH_2CH_2\overset{O}{\overset{\|}{C}}OCH_2CH_3 \xrightarrow[\text{2. H}_3\text{O}^+]{\text{1. CH}_3\text{MgBr}}$ $+ CH_3CH_2OH$

b.

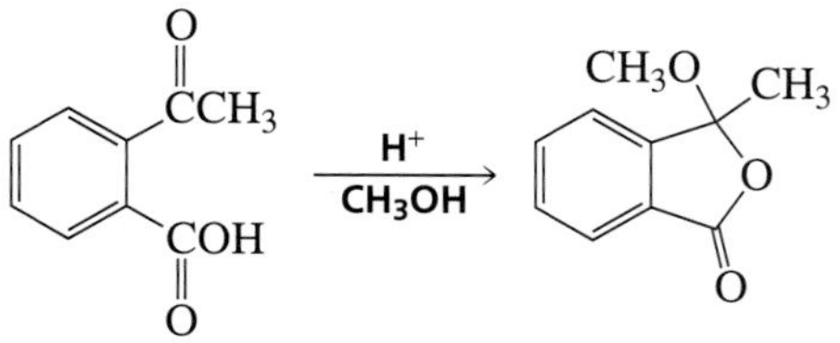

c.

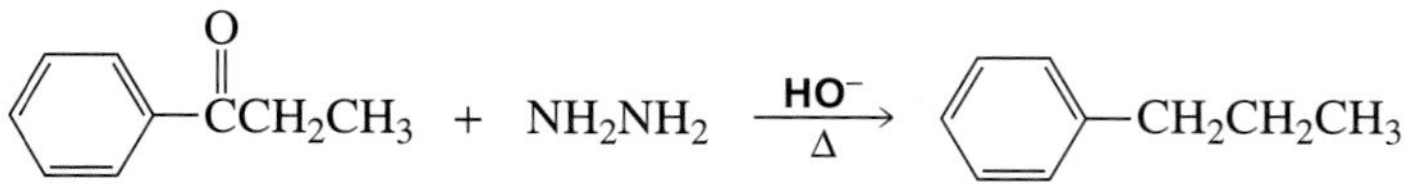

d.

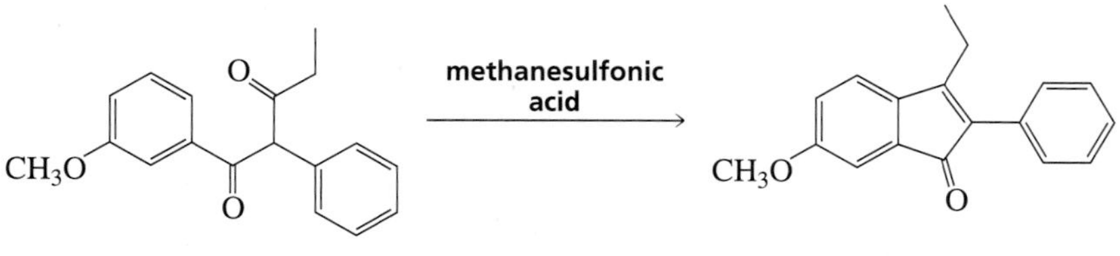

46. a. In aqueous solution, glucose exists in equilibrium with two six-membered ring compounds. Give the structures of these compounds.

b. Which of the six-membered ring compounds will be present in greater amount?

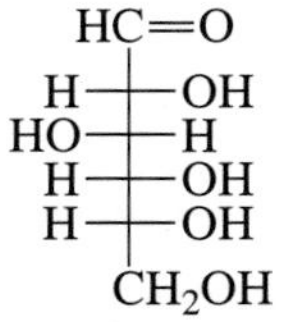

$$HC=O$$
H——OH
HO——H
H——OH
H——OH
$$CH_2OH$$
D-glucose

47. Acetaldehyde, in the presence of an acid catalyst, forms a trimer known as paraldehyde. Paraldehyde induces sleep when it is administered to animals in large doses, and consequently it is used as a sedative or as a hypnotic. Propose a mechanism for the formation of paraldehyde.

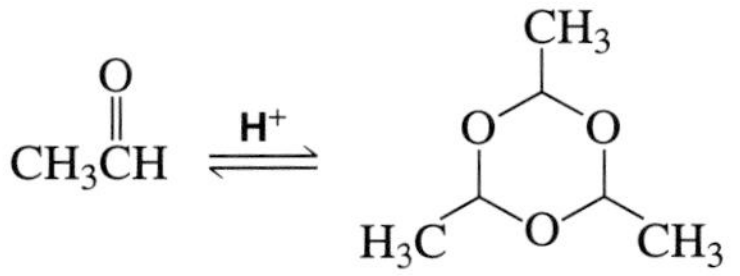

$$CH_3CH \overset{O}{\underset{}{\parallel}} \quad \overset{H^+}{\rightleftharpoons} \quad$$ **paraldehyde**

48. The addition of hydrogen cyanide to benzaldehyde forms a compound called mandelonitrile. *(R)*-Mandelonitrile is formed from the hydrolysis of amygdalin, a compound found in the pits of peaches and apricots. Amygdalin is the principal constituent of Laetrile, a compound that was once highly touted as a treatment for cancer. It was subsequently found to be ineffective. Is *(R)*-mandelonitrile formed by attack of cyanide ion on the *re* face or the *si* face of benzaldehyde?

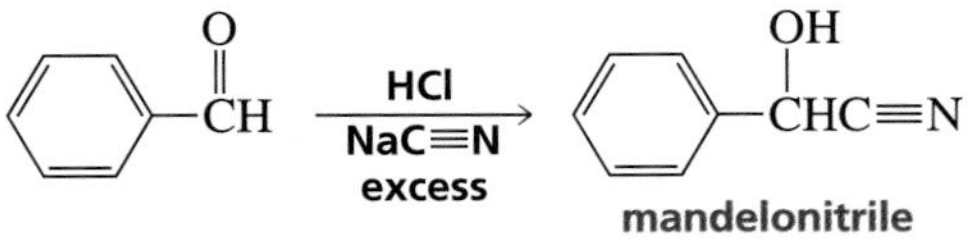

$$\xrightarrow[\text{excess}]{\begin{array}{c}HCl\\ NaC\equiv N\end{array}}$$ **mandelonitrile**

49. What carbonyl compound and what phosphonium ylide are required in order to synthesize the following compounds?

a. 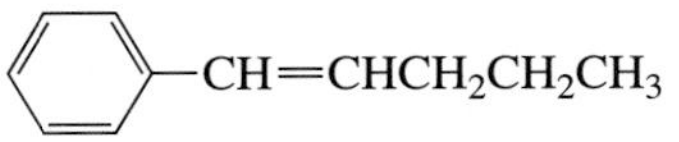 —CH=CHCH$_2$CH$_2$CH$_3$ **c.** 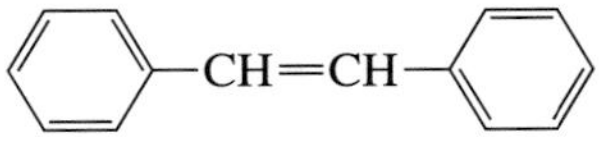 —CH=CH—

b. 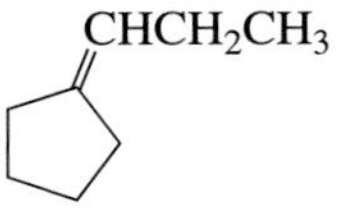 $\overset{CHCH_2CH_3}{}$ **d.** 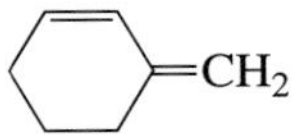 =CH$_2$

50. Identify compounds A and B.

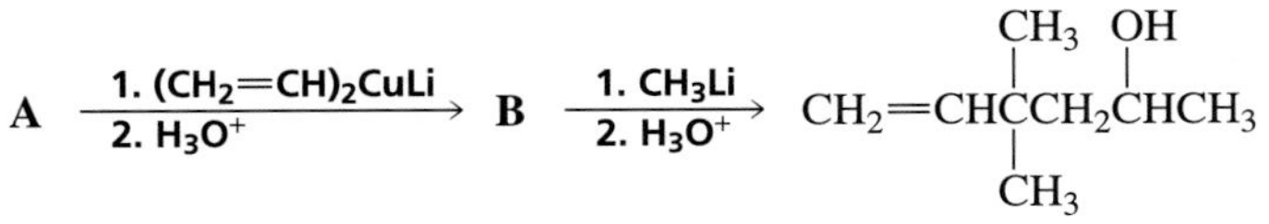

$$A \xrightarrow[\text{2. H}_3\text{O}^+]{\text{1. (CH}_2=\text{CH)}_2\text{CuLi}} B \xrightarrow[\text{2. H}_3\text{O}^+]{\text{1. CH}_3\text{Li}} CH_2=CHCCH_2CHCH_3$$

with CH$_3$, OH substituents and CH$_3$.

51. 4-Methyl-3-pentenoic acid is more stable than 4-methyl-2-pentenoic acid even though the latter has conjugated double bonds. Explain.

52. When a cyclic ketone reacts with diazomethane, the next larger cyclic ketone is formed. This is called a ring expansion reaction. Provide a mechanism for this reaction.

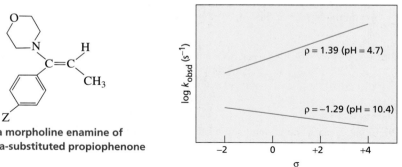

cyclohexanone   diazomethane   cycloheptanone

53. The pK_a values of oxaloacetic acid are 2.22 and 3.98.

$$HOCCCH_2COH$$

oxaloacetic acid

The amount of hydrate present in an aqueous solution of oxaloacetic acid is dependent on the pH of the solution (*J. Am. Chem. Soc.* 1983, *105*, 4982): 95% at pH = 0; 81% at pH = 1.3; 35% at pH = 3.1; 13% at pH = 4.7; 6% at pH = 6.7; 6% at pH = 12.7. Explain this pH dependence.

54. In order to solve this problem you must read the description of the Hammett σ, ρ treatment given in Chapter 15, Problem 65. When the rate constants for hydrolysis of several morpholine enamines of para-substituted propiophenones are determined at pH = 4.7, the ρ value is positive; but when the rates of hydrolysis are determined at pH = 10.4, the ρ value is negative.
   a. What is the rate-determining step of the hydrolysis reaction when it is carried out in basic solution?
   b. What is the rate-determining step of the reaction when it is carried out in acidic solution? (*J. Am. Chem. Soc.* 1970, *92*, 4261).

a morpholine enamine of
para-substituted propiophenone

$\rho = 1.39$ (pH = 4.7)

$\rho = -1.29$ (pH = 10.4)

log k_{obsd} (s^{-1})

σ

55. Propose a mechanism for each of the following reactions.

a. $\xrightarrow{\text{H}_3\text{O}^+}$ $(HOCH_2CH_2CH_2)_2CHCH$

b. $+ CH_3CH_2SH \longrightarrow$

c. $CH_3CH=CHCCH_3 +$ $\xrightarrow{\text{H}_3\text{O}^+}$

NAD$^+$

OXIDATION–REDUCTION REACTIONS

An important category of organic reactions involves the transfer of electrons from one molecule to another. Organic chemists use these reactions—called **oxidation–reduction reactions** or **redox reactions**—to synthesize a large variety of compounds. Redox reactions are also important in biological systems because energy is produced by such reactions.

In an oxidation–reduction reaction, one molecule loses electrons and one molecule gains electrons. The molecule that loses electrons is oxidized, and the molecule that gains electrons is reduced. One simple way to remember this is the phrase "LEO the lion says GER": *Loss of Electrons is Oxidation, Gain of Electrons is Reduction.*

The following is an example of an oxidation–reduction reaction involving inorganic reagents.

$$Cu^+ \ + \ Fe^{3+} \ \longrightarrow \ Cu^{2+} \ + \ Fe^{2+}$$

Here Cu^+ loses an electron, which means that it is oxidized; the other reactant, Fe^{3+}, is the oxidizing agent. Fe^{3+} gains an electron, which means that it is reduced; Cu^+ is the reducing agent. Notice that the oxidizing agent is reduced and the reducing agent is oxidized. Oxidation is always coupled with reduction. In other words, a compound cannot gain electrons (be reduced) unless another compound in the reaction simultaneously loses electrons (is oxidized).

It is easy to tell whether an organic compound is being oxidized or reduced. Simply look at the change in the structure of the compound. If the number of carbon–hydrogen bonds increases, the compound is being reduced. If the number of carbon–hydrogen bonds decreases, or if the number of carbon–oxygen, carbon–nitrogen, or carbon–halogen bonds increases, the compound is being oxidized. We will now take a look at some examples.

In each of the following reactions, the product has more carbon–hydrogen bonds than the reactant. The alkene, aldehyde, and ketone, therefore, are being reduced.

Hydrogen, sodium borohydride, and hydrazine are the reducing agents.

$$\text{RCH=CHR} \xrightarrow[\text{Pt}]{\text{H}_2} \text{RCH}_2\text{CH}_2\text{R}$$

an alkene

Reduction increases the number of carbon–hydrogen bonds.

$$\underset{\text{an aldehyde}}{\overset{\displaystyle O}{\underset{\|}{\text{RCH}}}} \xrightarrow[\text{2. H}_2\text{O}]{\text{1. NaBH}_4} \text{RCH}_2\text{OH}$$

$$\underset{\text{a ketone}}{\overset{\displaystyle O}{\underset{\|}{\text{RCR}}}} \xrightarrow[\text{HO}^-]{\text{H}_2\text{NNH}_2} \text{RCH}_2\text{R}$$

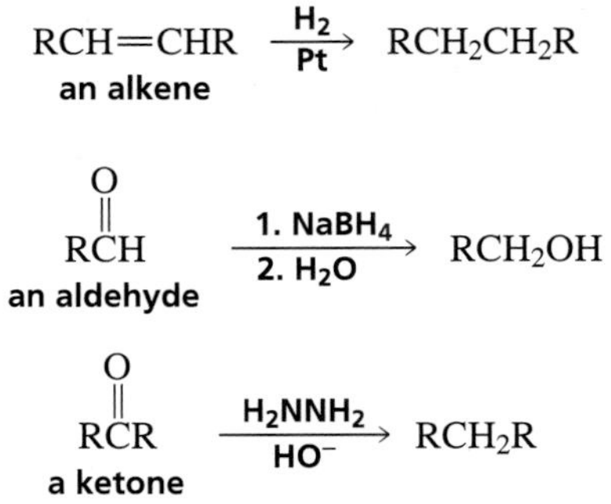

In the next group of reactions, the number of carbon–bromine bonds increases in the first reaction, and the number of carbon–hydrogen bonds decreases (and the number of carbon–oxygen bonds increases) in the next two reactions. This means that the alkene, the aldehyde, and the alcohol have been oxidized. Bromine and sodium dichromate ($\text{Na}_2\text{Cr}_2\text{O}_7$) are the oxidizing agents. In the last reaction, notice that the increase in the number of carbon–oxygen bonds results from a carbon–oxygen single bond becoming a carbon–oxygen double bond.

$$\underset{\text{an alkene}}{\text{RCH=CHR}} \xrightarrow[\text{CCl}_4]{\text{Br}_2} \overset{\text{Br Br}}{\overset{|\quad|}{\text{RCHCHR}}}$$

Oxidation decreases the number of carbon–hydrogen bonds, or increases the number of C–O, C–N, or C–X (X = halogen) bonds.

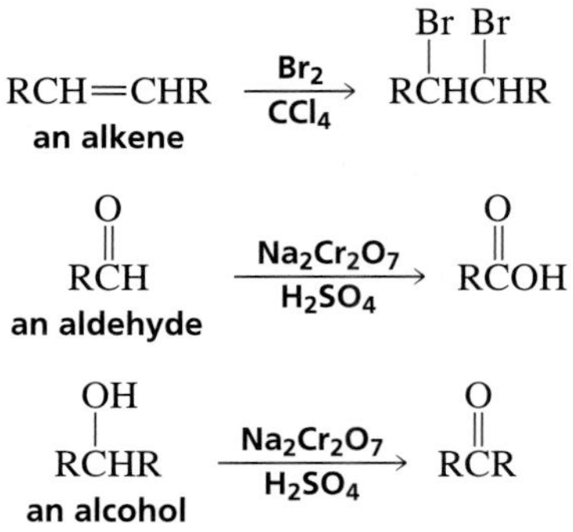

$$\underset{\text{an aldehyde}}{\overset{\displaystyle O}{\underset{\|}{\text{RCH}}}} \xrightarrow[\text{H}_2\text{SO}_4]{\text{Na}_2\text{Cr}_2\text{O}_7} \overset{\displaystyle O}{\underset{\|}{\text{RCOH}}}$$

$$\underset{\text{an alcohol}}{\overset{\text{OH}}{\underset{|}{\text{RCHR}}}} \xrightarrow[\text{H}_2\text{SO}_4]{\text{Na}_2\text{Cr}_2\text{O}_7} \overset{\displaystyle O}{\underset{\|}{\text{RCR}}}$$

If water is added to an alkene, the product has one more carbon–hydrogen bond than the reactant, but it also has one more carbon–oxygen bond. So one carbon is reduced and another is oxidized. The two cancel each other as far as the overall molecule is concerned. Thus the overall reaction is neither an oxidation nor a reduction.

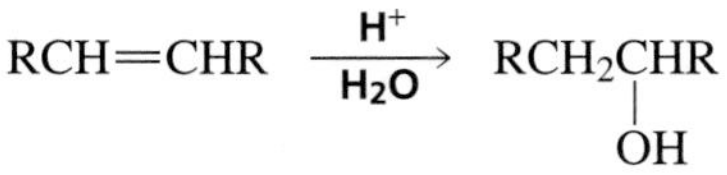

$$\text{RCH=CHR} \xrightarrow[\text{H}_2\text{O}]{\text{H}^+} \underset{\overset{|}{\text{OH}}}{\text{RCH}_2\text{CHR}}$$

Many oxidizing reagents and many reducing reagents are available to organic chemists. This chapter will highlight only a small fraction of the available reagents. The ones selected are some of the more common reagents that illustrate the types of transformations caused by oxidation and reduction.

PROBLEM 1◆

Indicate whether each of the following reactions is an oxidation reaction, a reduction reaction, or neither.

a. $CH_3\overset{\displaystyle O}{\overset{\|}{C}}Cl \xrightarrow[\substack{\textbf{deactivated}\\\textbf{Pd}}]{H_2} CH_3\overset{\displaystyle O}{\overset{\|}{C}}H$ *Red*

b. $RCH=CHR \xrightarrow{HBr} RCH_2\overset{\displaystyle Br}{\overset{|}{C}}HR$ ∿

c. 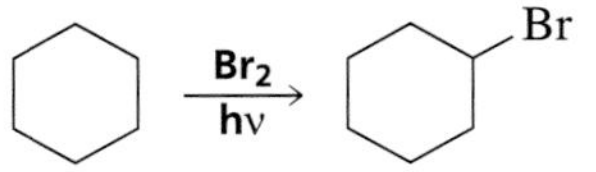 $\xrightarrow[hv]{Br_2}$ *ot*

d. $CH_3CH_2OH \xrightarrow[H_2SO_4]{KMnO_4} CH_3\overset{\displaystyle O}{\overset{\|}{C}}OH$ *ox.*

e. $CH_3C\equiv N \xrightarrow{\substack{H_2\\Pt}} CH_3CH_2NH_2$ *Red*

f. $CH_3CH_2CH_2Br \xrightarrow{HO^-} CH_3CH_2CH_2OH$ *neith*

An organic compound is reduced by adding hydrogen (H_2) to it. We can think of H_2 as being composed of *two hydrogen atoms,* of *two electrons and two protons,* or of *a hydride ion and a proton.* In the following sections we will see that these three ways to describe H_2 correspond to the three kinds of mechanisms by which H_2 is added to an organic compound.

17.1 REDUCTION REACTIONS

components of H:H

H· ·H	$^-$ H$^+$ · $^-$ H$^+$	H:$^-$ H$^+$
two hydrogen atoms	**two electrons and two protons**	**a hydride ion and a proton**

Reduction by Addition of Two Hydrogen Atoms

We have already seen that hydrogen can be added to carbon–carbon double bonds and carbon–carbon triple bonds in the presence of a metal catalyst (Sections 3.19 and 5.8). These reactions are called **catalytic hydrogenations.** Since there are more carbon–hydrogen bonds in the products than in the reactants, catalytic hydrogenations are examples of reduction reactions. Both alkenes and alkynes are reduced to alkanes.

$$CH_3CH_2CH=CH_2 + H_2 \xrightarrow{\textbf{Pt, Pd, or Ni}} CH_3CH_2CH_2CH_3$$
$$\textbf{1-butene} \qquad\qquad\qquad\qquad\qquad \textbf{butane}$$

$$CH_3CH_2CH_2C\equiv CH + 2 H_2 \xrightarrow{\textbf{Pt, Pd, or Ni}} CH_3CH_2CH_2CH_2CH_3$$
$$\textbf{1-pentyne} \qquad\qquad\qquad\qquad\qquad \textbf{pentane}$$

In a catalytic hydrogenation, the hydrogen–hydrogen bond breaks homolytically (Section 3.19). This means that reduction occurs as a result of the addition of two hydrogen atoms to the organic molecule.

We have seen that the catalytic hydrogenation of an alkyne can be stopped at a *cis*-alkene if a deactivated catalyst is used (Section 5.8).

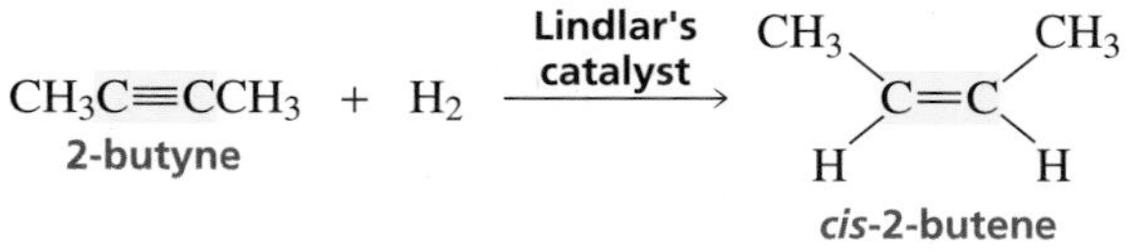

$$CH_3C{\equiv}CCH_3 + H_2 \xrightarrow[\text{catalyst}]{\text{Lindlar's}} \underset{\text{\textit{cis}-2-butene}}{\overset{\displaystyle CH_3 \qquad CH_3}{\underset{\displaystyle H \qquad\quad H}{C{=}C}}}$$

2-butyne

Only the alkene substituent is reduced in the following reaction. The very stable benzene ring can be reduced only under special conditions.

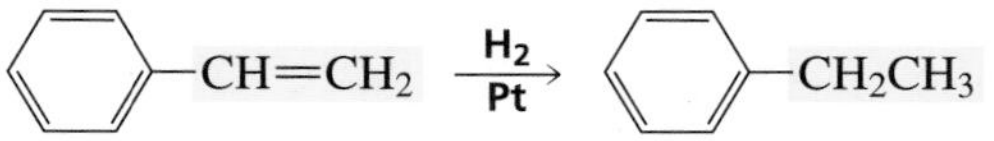

$$\text{⟨benzene⟩-CH{=}CH}_2 \xrightarrow[\text{Pt}]{H_2} \text{⟨benzene⟩-CH}_2\text{CH}_3$$

Catalytic hydrogenation can also be used to reduce carbon–nitrogen double bonds and carbon–nitrogen triple bonds. The reaction products are amines (Section 16.7).

$$CH_3CH_2CH{=}NCH_3 + H_2 \xrightarrow{Pt} CH_3CH_2CH_2NHCH_3$$
methylpropylamine

$$CH_3CH_2CH_2C{\equiv}N + H_2 \xrightarrow{Pt} CH_3CH_2CH_2CH_2NH_2$$
butylamine

Murray Raney (1885–1966) *was born in Tennessee. He received a B.A. from the University of Tennessee. He worked at the Gilman Paint and Varnish Co. from 1925 to 1950, at which point he left to found the Raney Catalyst Co.*

The carbon–oxygen double bonds of aldehydes and ketones can be reduced by catalytic hydrogenation. The metal catalysts most commonly used for the addition of hydrogen to a carbonyl group are Raney nickel (finely dispersed nickel with preadsorbed hydrogen), PtO_2, and Pd/C (Section 3.19). Ketones are reduced to secondary alcohols, and aldehydes are reduced to primary alcohols.

$$\underset{\text{a ketone}}{CH_3CH_2\overset{\displaystyle O}{\overset{\|}{C}}CH_3} \xrightarrow[\text{Raney Ni}]{H_2} \underset{\text{a secondary alcohol}}{CH_3CH_2\overset{\displaystyle OH}{\overset{|}{C}HCH_3}}$$

$$\underset{\text{an aldehyde}}{CH_3CH_2CH_2\overset{\displaystyle O}{\overset{\|}{C}}H} \xrightarrow[\text{Pd/C}]{H_2} \underset{\text{a primary alcohol}}{CH_3CH_2CH_2CH_2OH}$$

The carbon–oxygen double bond of an acyl chloride is reduced by catalytic hydrogenation to an aldehyde, which is further reduced to a primary alcohol.

$$\underset{\text{an acyl chloride}}{CH_3CH_2\overset{\displaystyle O}{\overset{\|}{C}}Cl} \xrightarrow[\text{Pd/C}]{H_2} \left[\underset{\text{an aldehyde}}{CH_3CH_2\overset{\displaystyle O}{\overset{\|}{C}}H}\right] \longrightarrow \underset{\text{a primary alcohol}}{CH_3CH_2CH_2OH}$$

Karl W. Rosenmund *was born in Berlin in 1884. He was a professor of chemistry at the University of Kiel.*

The reaction can be stopped at the aldehyde using a deactivated palladium catalyst. This reaction is known as the **Rosenmund reduction.** [This catalyst is similar to the deactivated palladium catalyst that causes the reduction of an alkyne to stop at a *cis*-alkene (Section 5.8).]

$$\underset{\text{an acyl chloride}}{CH_3CH_2\overset{\displaystyle O}{\overset{\|}{C}}Cl} \xrightarrow[\substack{\text{deactivated}\\\text{Pd}}]{H_2} \underset{\text{an aldehyde}}{CH_3CH_2\overset{\displaystyle O}{\overset{\|}{C}}H}$$

The carbon–oxygen double bonds of carboxylic acids, esters, and amides are harder to reduce than the carbon–oxygen double bonds of aldehydes and ketones. These less reactive carbonyl groups cannot be reduced by catalytic hydrogenation (except under extreme conditions). They can, however, be reduced by another method, discussed next. The relative ease of reduction of various functional groups by catalytic hydrogenation is shown in Table 17.1

$$\underset{\text{a carboxylic acid}}{CH_3CH_2\overset{\displaystyle O}{\overset{\|}{C}}OH} \xrightarrow[\text{Raney Ni}]{H_2} \text{no reaction}$$

$$\underset{\text{an ester}}{CH_3CH_2\overset{\displaystyle O}{\overset{\|}{C}}OCH_3} \xrightarrow[\text{Raney Ni}]{H_2} \text{no reaction}$$

$$\underset{\text{an amide}}{CH_3CH_2\overset{\displaystyle O}{\overset{\|}{C}}NHCH_3} \xrightarrow[\text{Raney Ni}]{H_2} \text{no reaction}$$

PROBLEM 2◆

Give the products of the following reactions.

a. $CH_3CH_2CH_2CH_2\overset{\displaystyle O}{\overset{\|}{C}}H \xrightarrow[\text{Raney Ni}]{H_2}$

b. $CH_3CH_2CH_2C{\equiv}N \xrightarrow[\text{Pt}]{H_2}$

c. $CH_3CH_2CH_2C{\equiv}CCH_3 \xrightarrow[\substack{\text{Lindlar's}\\\text{catalyst}}]{H_2}$

d. 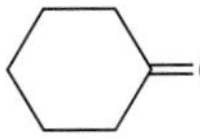 $\langle\text{cyclohexyl}\rangle{=}O \xrightarrow[\text{Raney Ni}]{H_2}$

e. $CH_3\overset{\displaystyle O}{\overset{\|}{C}}OCH_3 \xrightarrow[\text{Raney Ni}]{H_2}$

f. $CH_3\overset{\displaystyle O}{\overset{\|}{C}}Cl \xrightarrow[\text{Raney Ni}]{H_2}$

g. $CH_3\overset{\displaystyle O}{\overset{\|}{C}}Cl \xrightarrow[\substack{\text{deactivated}\\\text{Pd}}]{H_2}$

h. 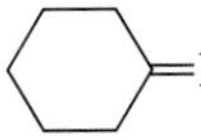 $\langle\text{cyclohexyl}\rangle{=}NCH_3 \xrightarrow[\text{Pt}]{H_2}$

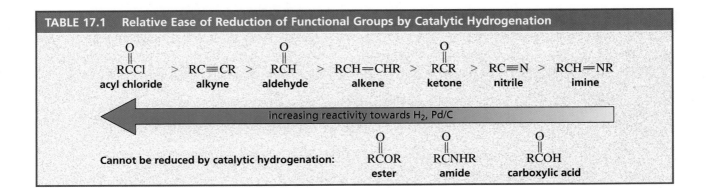

TABLE 17.1	Relative Ease of Reduction of Functional Groups by Catalytic Hydrogenation

$\overset{\displaystyle O}{\overset{\|}{R\overset{}{C}Cl}}$	>	$RC{\equiv}CR$	>	$\overset{\displaystyle O}{\overset{\|}{R\overset{}{C}H}}$	>	$RCH{=}CHR$	>	$\overset{\displaystyle O}{\overset{\|}{R\overset{}{C}R}}$	>	$RC{\equiv}N$	>	$RCH{=}NR$
acyl chloride		alkyne		aldehyde		alkene		ketone		nitrile		imine

increasing reactivity towards H_2, Pd/C

Cannot be reduced by catalytic hydrogenation: $\overset{\displaystyle O}{\overset{\|}{RCOR}}$ $\overset{\displaystyle O}{\overset{\|}{RCNHR}}$ $\overset{\displaystyle O}{\overset{\|}{RCOH}}$

ester   amide   carboxylic acid

Reduction by Addition of an Electron, a Proton, an Electron, a Proton

When a compound is reduced using sodium in liquid ammonia, sodium donates an electron to the compound and ammonia donates a proton. This sequence is repeated, so the overall reduction adds two electrons and two protons to the compound. Such a reaction is called a **dissolving metal reduction.** An alkyne is reduced by sodium in liquid ammonia to a *trans*-alkene. The mechanism for this reaction is shown in Section 5.8.

$$CH_3C{\equiv}CCH_3 \xrightarrow[\text{NH}_3\text{(liq)}]{\text{Na or Li}}$$

2-butyne

trans-2-butene

Sodium in liquid ammonia cannot reduce a carbon–carbon double bond. This makes it a useful reagent for reducing a triple bond in a compound in which there is also a double bond.

$$CH_3C{=}CHCH_2C{\equiv}CCH_3 \xrightarrow[\text{NH}_3\text{(liq)}]{\text{Na or Li}} CH_3C{=}CHCH_2$$

Reduction by Addition of a Hydride Ion and a Proton

Carbon–oxygen double bonds are easily reduced by sodium borohydride ($NaBH_4$) and/or lithium aluminum hydride ($LiAlH_4$). Sodium borohydride and lithium aluminum hydride are sources of hydride ion (H^-), which is the actual reducing agent in these reductions. Hydride ion adds to the carbonyl carbon, and the oxyanion that is formed is subsequently protonated by water. In other words, the carbonyl group is reduced by adding an H^- followed by an H^+. The mechanism for reduction by these reagents is discussed in Sections 16.5 and 16.6.

Aldehydes and ketones are reduced by sodium borohydride.

$$CH_3CH_2CH_2CH \xrightarrow[\text{2. H}_2\text{O}]{\text{1. NaBH}_4} CH_3CH_2CH_2CH_2OH$$

an aldehyde — a primary alcohol

$$CH_3CH_2CH_2CCH_3 \xrightarrow[\text{2. H}_2\text{O}]{\text{1. NaBH}_4} CH_3CH_2CH_2CHCH_3$$

a ketone — a secondary alcohol

Remember that the numbers in front of the reagents along an arrow indicate that the second reagent is not added until reaction with the first reagent is complete.

Lithium aluminum hydride, because of its more polar metal hydrogen bonds, is a stronger reducing agent than sodium borohydride. Only lithium aluminum hydride is a strong enough reducing agent to reduce carboxylic acids, esters, and amides.

$$CH_3CH_2CH_2COH \xrightarrow[\text{2. H}_2\text{O}]{\text{1. LiAlH}_4} CH_3CH_2CH_2CH_2OH + H_2O$$

a carboxylic acid — a primary alcohol

$$CH_3CH_2COCH_3 \xrightarrow[\text{2. H}_2\text{O}]{\text{1. LiAlH}_4} CH_3CH_2CH_2OH + CH_3OH$$

an ester — a primary alcohol

The carbonyl group of an amide is reduced to a methylene group (CH_2) by lithium aluminum hydride. Primary, secondary, and tertiary amines are formed, depending on the number of substituents bonded to the nitrogen of the amide.

$$CH_3CH_2CH_2\overset{\overset{\textstyle O}{\|}}{C}NH_2 \quad \xrightarrow[\text{2. H}_2\text{O}]{\text{1. LiAlH}_4} \quad CH_3CH_2CH_2CH_2NH_2$$
a primary amine

$$CH_3CH_2CH_2\overset{\overset{\textstyle O}{\|}}{C}NHCH_3 \quad \xrightarrow[\text{2. H}_2\text{O}]{\text{1. LiAlH}_4} \quad CH_3CH_2CH_2CH_2NHCH_3$$
a secondary amine

$$CH_3CH_2CH_2\overset{\overset{\textstyle OCH_3}{\|\,|}}{C}NCH_3 \quad \xrightarrow[\text{2. H}_2\text{O}]{\text{1. LiAlH}_4} \quad CH_3CH_2CH_2CH_2\overset{\overset{\textstyle CH_3}{|}}{N}CH_3$$
a tertiary amine

Lithium aluminum hydride also reduces aldehydes and ketones. But $NaBH_4$ is safer and easier to use, so $LiAlH_4$ is generally used only for compounds that cannot be reduced by the milder reagent.

Because sodium borohydride cannot reduce an ester, it can be used to reduce selectively an aldehyde or a ketone group in a compound that also contains an ester group.

$$CH_3\overset{\overset{\textstyle O}{\|}}{C}CH_2CH_2\overset{\overset{\textstyle O}{\|}}{C}OCH_3 \quad \xrightarrow[\text{2. H}_2\text{O}]{\text{1. NaBH}_4} \quad CH_3\overset{\overset{\textstyle OH}{|}}{C}HCH_2CH_2\overset{\overset{\textstyle O}{\|}}{C}OCH_3$$

We have seen that sterically bulky hydride donors can be used to deliver only one equivalent of hydride ion. For example, diisobutylaluminum hydride (DIBAH) reduces an ester to an aldehyde, and lithium tri-*tert*-butoxyaluminum hydride reduces an acyl chloride to an aldehyde (Section 16.6).

$$CH_3CH_2CH_2\overset{\overset{\textstyle O}{\|}}{C}OCH_3 \quad \xrightarrow[\text{2. H}_2\text{O}]{\text{1. DIBAH, }-80\,°\text{C}} \quad CH_3CH_2CH_2\overset{\overset{\textstyle O}{\|}}{C}H$$
an ester   an aldehyde

$$CH_3CH_2CH_2CH_2\overset{\overset{\textstyle O}{\|}}{C}Cl \quad \xrightarrow[\text{2. H}_2\text{O}]{\text{1. LiAl[OC(CH}_3)_3]_3\text{H, }-80\,°\text{C}} \quad CH_3CH_2CH_2CH_2\overset{\overset{\textstyle O}{\|}}{C}H$$
an acyl chloride   an aldehyde

The multiply-bonded carbon atoms of alkenes and alkynes do not possess a partial positive charge and therefore will not react with reagents that reduce compounds by donating a hydride ion.

$$CH_3CH_2CH{=}CH_2 \quad \xrightarrow{\text{NaBH}_4} \quad \text{no reduction reaction}$$

$$CH_3CH_2C{\equiv}CH \quad \xrightarrow{\text{NaBH}_4} \quad \text{no reduction reaction}$$

Because sodium borohydride cannot reduce carbon–carbon double bonds, a carbonyl group in a compound that also has an alkene functional group can be selectively reduced. [If the double bonds are conjugated, some product in which both functional groups are reduced is also formed (Section 16.14).]

TABLE 17.2   Relative Ease of Reduction of Functional Groups by Addition of Hydride Ion

$$
\underset{\text{acyl chloride}}{\overset{\overset{\displaystyle O}{\parallel}}{RCCl}} \;>\; \underset{\text{aldehyde}}{\overset{\overset{\displaystyle O}{\parallel}}{RCH}} \;>\; \underset{\text{ketone}}{\overset{\overset{\displaystyle O}{\parallel}}{RCR}} \;>\; \underset{\text{imine}}{RCH{=}NR} \;>\; \underset{\text{ester}}{\overset{\overset{\displaystyle O}{\parallel}}{RCOR}} \;>\; \underset{\text{carboxylic acid}}{\overset{\overset{\displaystyle O}{\parallel}}{RCOH}} \;>\; \underset{\text{amide}}{\overset{\overset{\displaystyle O}{\parallel}}{RCNHR}} \;>\; \underset{\text{nitrile}}{RC{\equiv}N}
$$

⟵ increasing reactivity toward H⁻

Cannot be reduced by hydride ion: $\underset{\text{alkene}}{RCH{=}CHR}$ $\underset{\text{alkyne}}{RC{\equiv}CR}$

$$
\underset{}{CH_3CH{=}CHCH_2\overset{\overset{\displaystyle O}{\parallel}}{C}CH_3} \quad \xrightarrow[\text{2. H}_2\text{O}]{\text{1. NaBH}_4} \quad CH_3CH{=}CHCH_2\overset{\overset{\displaystyle OH}{|}}{C}HCH_3
$$

The relative ease of reduction of various functional groups by the addition of hydride ion is shown in Table 17.2.

PROBLEM 3

Terminal alkynes cannot be reduced by LiAlH₄ or by Na in liquid NH₃. Explain.

PROBLEM 4 ◆

Give the products of the following reactions.

a. 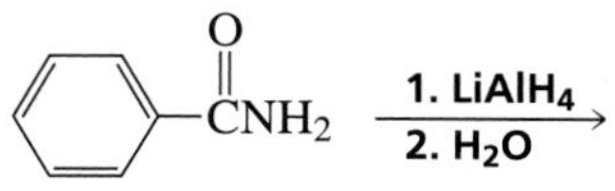 (phenyl)$\overset{\overset{\displaystyle O}{\parallel}}{C}NH_2$ $\xrightarrow[\text{2. H}_2\text{O}]{\text{1. LiAlH}_4}$

d. (cyclohexyl)$\overset{\overset{\displaystyle O}{\parallel}}{C}OCH_2CH_3$ $\xrightarrow[\text{2. H}_2\text{O}]{\text{1. LiAlH}_4}$

b. (phenyl)$\overset{\overset{\displaystyle O}{\parallel}}{C}OH$ $\xrightarrow[\text{2. H}_2\text{O}]{\text{1. LiAlH}_4}$

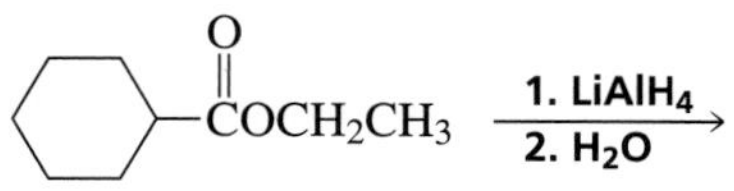

e. $CH_3CH_2\overset{\overset{\displaystyle O}{\parallel}}{C}NHCH_2CH_3$ $\xrightarrow[\text{2. H}_2\text{O}]{\text{1. LiAlH}_4}$

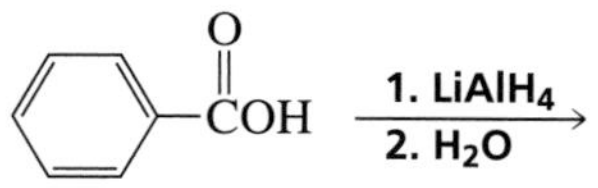

c. $CH_3CH_2\overset{\overset{\displaystyle O}{\parallel}}{C}CH_2CH_3$ $\xrightarrow[\text{2. H}_2\text{O}]{\text{1. NaBH}_4}$

f. $CH_3CH_2CH_2\overset{\overset{\displaystyle O}{\parallel}}{C}OH$ $\xrightarrow[\text{2. H}_2\text{O}]{\text{1. LiAlH}_4}$

PROBLEM 5

Can carbon–nitrogen double and triple bonds be reduced by lithium aluminum hydride? If not, why not? If so, why?

PROBLEM 6

Give the products of the following reactions. Assume that excess reducing agent is used in **d.**

a. 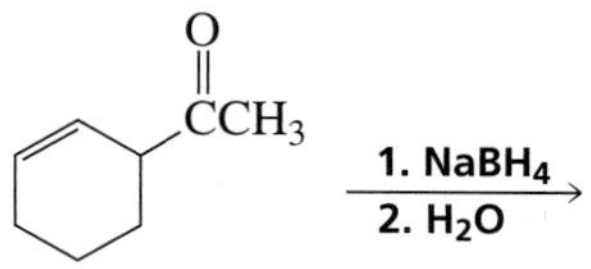 (cyclohexenyl)$\overset{\overset{\displaystyle O}{\parallel}}{C}CH_3$ $\xrightarrow[\text{2. H}_2\text{O}]{\text{1. NaBH}_4}$

b. 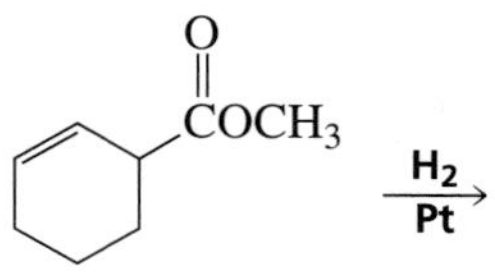 (cyclohexenyl)$\overset{\overset{\displaystyle O}{\parallel}}{C}OCH_3$ $\xrightarrow[\text{Pt}]{\text{H}_2}$

c. (structure: cyclohexanone with CH₂COCH₃ substituent) → 1. NaBH₄ 2. H₂O

d. (structure: cyclohexanone with CH₂COCH₃ substituent) → 1. LiAlH₄ 2. H₂O

17.2 OXIDATION OF ALCOHOLS

Oxidation is the reverse of **reduction.** For example, a ketone can be reduced to a secondary alcohol. The reverse reaction is the oxidation of a secondary alcohol to a ketone.

$$\text{ketone} \underset{\text{oxidation}}{\overset{\text{reduction}}{\rightleftarrows}} \text{secondary alcohol}$$

The reagents most commonly used for the oxidation of alcohols are acidic solutions of chromic anhydride (CrO_3), sodium chromate (Na_2CrO_4), sodium dichromate ($Na_2Cr_2O_7$), or potassium permanganate ($KMnO_4$). These reactions are easily recognized as oxidation reactions because the number of carbon–hydrogen bonds in the reactant decreases (and the number of carbon–oxygen bonds increases).

$$CH_3CH_2\overset{OH}{\underset{|}{C}}HCH_3 \xrightarrow[H_2SO_4]{CrO_3} CH_3CH_2\overset{O}{\overset{||}{C}}CH_3$$

(cyclohexanol) $\xrightarrow[H_2SO_4]{Na_2Cr_2O_7}$ (cyclohexanone)

(cyclopentyl)$-\overset{OH}{\underset{|}{C}}HCH_2CH_3 \xrightarrow[H_2SO_4]{KMnO_4}$ (cyclopentyl)$-\overset{O}{\overset{||}{C}}CH_2CH_3$

secondary alcohols   **ketones**

Primary alcohols are initially oxidized to aldehydes by these reagents. The reaction, however, does not stop at the aldehyde. The aldehyde is further oxidized to a carboxylic acid.

$$CH_3CH_2CH_2CH_2OH \xrightarrow[H_2SO_4]{Na_2Cr_2O_7} CH_3CH_2CH_2\overset{O}{\overset{||}{C}}H \xrightarrow[\text{oxidation}]{\text{further}} CH_3CH_2CH_2\overset{O}{\overset{||}{C}}OH$$

a primary alcohol   **an aldehyde**   **a carboxylic acid**

The aldehyde can be isolated if a milder chromic acid oxidizing agent (such as pyridinium dichromate) is used in aqueous solution and the aldehyde is distilled off as it is formed—before it can be oxidized. Only aldehydes with boiling points less than 100 °C can be prepared in this way, because the boiling point of the aldehyde to be distilled off must be lower than that of the aqueous mixture.

Chromic acid oxidizes an alcohol by forming a chromate ester. (Chromic acid is formed *in situ* when sulfuric acid protonates the chromate ion.) The carbonyl compound is formed when the chromate ester undergoes an E2 elimination.

$$CH_3CH_2\ddot{O}H + HO-\underset{\underset{O}{||}}{\overset{\overset{O}{||}}{Cr}}-OH \longrightarrow CH_3CH-\underset{\underset{H}{|}}{\ddot{O}}-\underset{\overset{||}{O}}{\overset{O}{||}}{Cr}-OH \xrightarrow{\text{E2}} CH_3CH{=}O + \underset{HO\quad OH}{\overset{O}{||}}{Cr} + H_2O$$

chromic acid

$H_2\ddot{O}{:}$

a chromate ester

The oxidation of a primary alcohol can be easily stopped at the aldehyde if *pyridinium chlorochromate* (PCC) is used as the oxidizing agent and the reaction is carried out in an anhydrous solvent such as dichloromethane.

$$CH_3CH_2CH_2CH_2OH \xrightarrow[\text{dry CH}_2\text{Cl}_2]{\text{PCC}} CH_3CH_2CH_2\overset{\overset{O}{||}}{C}H$$

a primary alcohol   an aldehyde

Notice that oxidation of either a primary or a secondary alcohol involves removal of a hydrogen from the carbon to which the oxygen is attached. In a tertiary alcohol, the carbon bearing the OH group has no hydrogen, so the OH group of a tertiary alcohol cannot be oxidized to a carbonyl group.

THE ROLE OF HYDRATES IN THE OXIDATION OF PRIMARY ALCOHOLS

When a primary alcohol is oxidized to a carboxylic acid, the alcohol is initially oxidized to an aldehyde. The aldehyde is in equilibrium with its hydrate, since the reaction is carried out in an acidic aqueous solution. It is the hydrate that is subsequently oxidized to a carboxylic acid.

$$CH_3CH_2OH \xrightarrow[\text{H}_2\text{SO}_4]{\text{Na}_2\text{Cr}_2\text{O}_7} CH_3\overset{\overset{O}{||}}{C}H \underset{\text{H}_2\text{O}}{\overset{\text{H}^+}{\rightleftharpoons}} CH_3\underset{\underset{OH}{|}}{\overset{\overset{OH}{|}}{C}}H \xrightarrow[\text{H}_2\text{SO}_4]{\text{Na}_2\text{Cr}_2\text{O}_7} CH_3\overset{\overset{O}{||}}{C}OH$$

The oxidation reaction can be stopped at the aldehyde if the reaction is carried out with pyridinium chlorochromate (PCC), because PCC is used in an anhydrous solvent. If water is not present, the hydrate cannot be formed.

BLOOD ALCOHOL CONTENT

Because blood passes through the arteries in the lungs, an equilibrium is established between the alcohol in one's blood and the alcohol in one's breath. So if the concentration of one is known, the concentration of the other can be estimated. The test that law enforcement agencies use to approximate a person's blood alcohol level is based on the oxidation of breath ethanol by sodium dichromate. A sealed glass tube is used, which contains the oxidizing agent impregnated onto an inert material. The ends of the tube are broken off. One end of the tube is attached to a mouthpiece and the other to a balloon-type bag. The person undergoing the test blows into the mouthpiece until the bag is filled with air.

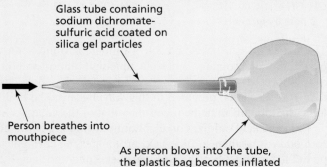

Glass tube containing sodium dichromate-sulfuric acid coated on silica gel particles

Person breathes into mouthpiece

As person blows into the tube, the plastic bag becomes inflated

Any ethanol in the breath is oxidized as it passes through the column. When ethanol is oxidized, the red–orange oxidizing agent ($Cr_2O_7^{2-}$) is reduced to green chromic ion. The greater the concentration of alcohol in the breath, the farther the green color spreads through the tube.

$$CH_3CH_2OH \;+\; \underset{\textbf{red–orange}}{Cr_2O_7^{2-}} \;\xrightarrow{H^+}\; \underset{O}{\overset{\displaystyle\|}{CH_3C}}OH \;+\; \underset{\textbf{green}}{Cr^{3+}}$$

If the person fails this test—determined by the extent to which the green color spreads through the tube—a more accurate Breathalyzer test is administered. This test also depends on the oxidation of breath ethanol by sodium dichromate, but it provides more accurate results. Here, an accurate volume of breath is bubbled through an acidic solution of sodium dichromate. The concentration of chromic ion is measured precisely using a spectrophotometer.

PROBLEM 7◆

Give the product formed from the reaction of each of the following alcohols with an acidic solution of sodium dichromate.

a. 3-pentanol

b. 1-pentanol

c. 2-methyl-2-pentanol

d. 2,4-hexanediol

PROBLEM 8 / SOLVED

How could the following ketone be prepared from the given starting material?

$$CH_3CH_2CH_2CH_3 \longrightarrow CH_3\overset{\displaystyle\overset{O}{\|}}{C}CH_2CH_3$$

SOLUTION Because the starting material is an alkane, we know that the first reaction must be a radical halogenation, because this is the only reaction that an alkane undergoes. Bromination, as a result of the greater selectivity of the bromine radical, will lead to a greater yield of the desired 2-halosubstituted compound than would chlorination. Reaction of 2-bromobutane with HO^- leads to a secondary alcohol, which, when oxidized, forms the target compound.

$$CH_3CH_2CH_2CH_3 \xrightarrow[h\nu]{Br_2} CH_3\overset{\overset{Br}{|}}{C}HCH_2CH_3 \xrightarrow{HO^-} CH_3\overset{\overset{OH}{|}}{C}HCH_2CH_3 \xrightarrow[H_2SO_4]{Na_2CrO_4} $$

$$CH_3\overset{\overset{O}{\|}}{C}CH_2CH_3$$

Aldehydes are oxidized to carboxylic acids. Since aldehydes are generally easier to oxidize than primary alcohols, any of the reagents described in the preceding section for oxidizing primary alcohols to carboxylic acids can be used to oxidize aldehydes to carboxylic acids.

17.3 OXIDATION OF ALDEHYDES AND KETONES

$$CH_3CH_2\overset{\overset{O}{\|}}{C}H \xrightarrow[\text{H}_2\text{SO}_4]{\text{Na}_2\text{CrO}_4} CH_3CH_2\overset{\overset{O}{\|}}{C}OH$$

aldehydes   carboxylic acids

**Bernhard Tollens
(1841–1918)** *was born in
Germany. He was a professor
of chemistry at the University
of Göttingen.*

Silver oxide is a mild oxidizing agent. A dilute solution of silver oxide in ammonia (*Tollens' reagent*) will oxidize an aldehyde, but it is too weak an oxidizing agent to oxidize an alcohol. An advantage to using Tollens' reagent to oxidize an aldehyde is that the reaction occurs under basic conditions. Therefore, you do not have to worry about harming other functional groups in the molecule that may react in an acidic solution.

$$CH_3CH_2\overset{\overset{O}{\|}}{C}H \xrightarrow[\text{2. H}_3\text{O}^+]{\text{1. Ag}_2\text{O, NH}_3} CH_3CH_2\overset{\overset{O}{\|}}{C}OH + \underset{\substack{\text{metallic}\\\text{silver}}}{\text{Ag}}$$

The oxidizing agent in Tollens' reagent is Ag^+, which is reduced to metallic silver. This forms the basis of the **Tollens test.** If Tollens' reagent is added to a small amount of an aldehyde in a test tube, the inside of the test tube becomes coated with a shiny mirror of metallic silver. If a mirror is not formed when Tollens' reagent is added to a compound, we can conclude that the compound does not have an aldehyde functional group.

Ketones do not react with most of the reagents used to oxidize aldehydes. However, both aldehydes and ketones can be oxidized by a peroxyacid. Aldehydes are oxidized to carboxylic acids and ketones are oxidized to esters. A **peroxyacid** (also called a percarboxylic acid or an acyl hydroperoxide) contains one more oxygen than a carboxylic acid, and it is this oxygen that is inserted between the carbonyl carbon and the H of an aldehyde or the R of a ketone. This oxidation is called a **Baeyer–Villiger oxidation.**

Johann Friedrich Wilhelm
Adolf von Baeyer.

$$\underset{\text{an aldehyde}}{CH_3CH_2CH_2\overset{\overset{O}{\|}}{C}H} + \underset{\text{a peroxyacid}}{R\overset{\overset{O}{\|}}{C}OOH} \longrightarrow \underset{\text{a carboxylic acid}}{CH_3CH_2CH_2\overset{\overset{O}{\|}}{C}OH} + R\overset{\overset{O}{\|}}{C}OH$$

$$\underset{\text{a ketone}}{CH_3CH_2\overset{\overset{O}{\|}}{C}CH_2CH_3} + \underset{\text{a peroxyacid}}{R\overset{\overset{O}{\|}}{C}OOH} \longrightarrow \underset{\text{an ester}}{CH_3CH_2\overset{\overset{O}{\|}}{C}OCH_2CH_3} + R\overset{\overset{O}{\|}}{C}OH$$

**Johann Friedrich Wilhelm
Adolf von Baeyer
(1835–1917)** *started his study
of chemistry under Bunsen and
Kekulé at the University of
Heidelberg and received a
Ph.D. from the University of
Berlin, studying under
Hofmann. Victor Villiger was
Baeyer's student. They first
published their finding of the
Baeyer–Villiger oxidation in*
Chemische Berichte *in 1899
(see also Section 2.12).*

If the ketone is not symmetrical, on which side of the carbonyl carbon is the oxygen inserted? For example, does the oxidation of cyclohexyl methyl ketone lead to methyl cyclohexanecarboxylate or to cyclohexyl acetate?

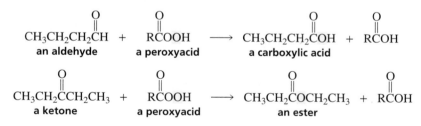

cyclohexyl methyl
ketone

methyl
cyclohexanecarboxylate

cyclohexyl
acetate

We can answer this question by looking at the mechanism of the reaction. The ketone and the peroxyacid react to form an unstable tetrahedral intermediate. The oxygen–oxygen bond in the intermediate is very weak. As it breaks heterolytically, one of the groups bonded to the carbonyl group of the ketone migrates to the oxygen. This is similar to the 1,2-shifts that occur when carbocations rearrange.

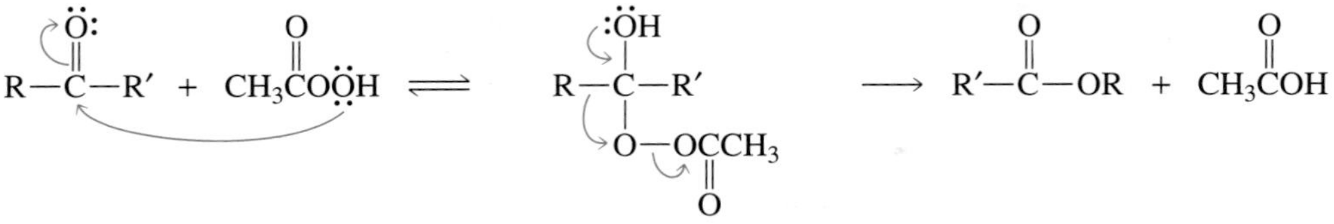

an unstable intermediate

Several studies have shown the following order of group migration tendencies:

relative migration tendencies

H > *tert*-alkyl > *sec*-alkyl = phenyl > primary alkyl > methyl

increasing tendency to migrate

So the product of the Baeyer–Villiger oxidation of cyclohexyl methyl ketone will be cyclohexyl acetate because a secondary alkyl group (the cyclohexyl group) is more likely to migrate than is a methyl group. Because H has the greatest tendency to migrate, aldehydes are always oxidized to carboxylic acids.

PROBLEM 9

Give the products of the following reactions.

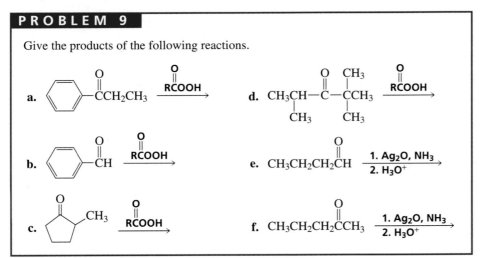

An alkene can be oxidized to a 1,2-diol either by potassium permanganate ($KMnO_4$) in a cold basic solution or by osmium tetroxide (OsO_4). It is important that the solution of potassium permanganate be basic and that the oxidation be carried out at room temperature or below. If the solution is heated or if it is acidic, the diol will be further oxidized (Section 17.6). A diol is also called a **glycol**. Since the OH groups are on adjacent carbons, 1,2-diols are also known as **vicinal diols** or **vicinal glycols**.

**17.4
HYDROXYLATION
OF ALKENES**

$$CH_3CH=CHCH_3 \xrightarrow[\text{cold}]{KMnO_4, HO^-, H_2O} \underset{\text{a vicinal diol}}{CH_3\overset{OH}{\underset{|}{C}}H\overset{OH}{\underset{|}{C}}HCH_3}$$

$$CH_3CH_2CH=CH_2 \quad \xrightarrow[\text{2. NaHSO}_3,\ H_2O]{\text{1. OsO}_4} \quad CH_3CH_2\overset{\overset{\displaystyle OH}{|}}{C}HCH_2OH$$

a vicinal diol

Each of these reactions forms a cyclic intermediate. The reactions occur because manganese and osmium are in a highly positive oxidation state ($+7$ and $+8$, respectively) and therefore attract electrons. Formation of the cyclic intermediate is a syn addition because both oxygens are delivered to the same side of the double bond. Therefore, the oxidation of a cycloalkene is a stereoselective reaction; only the *cis*-diol is obtained.

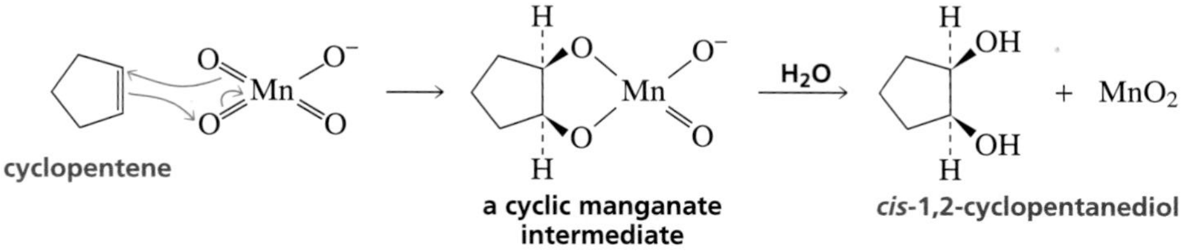

cyclopentene → a cyclic manganate intermediate → $\xrightarrow{H_2O}$ *cis*-1,2-cyclopentanediol + MnO_2

Notice that aqueous sodium bisulfite ($NaHSO_3$) is used to cleave the less reactive cyclic osmate.

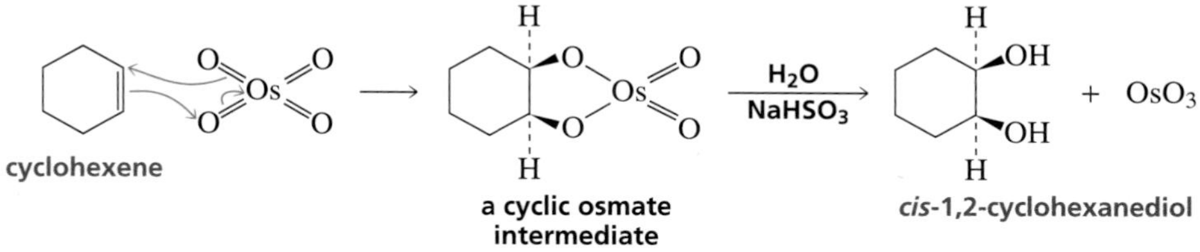

cyclohexene → a cyclic osmate intermediate → $\xrightarrow[\text{NaHSO}_3]{H_2O}$ *cis*-1,2-cyclohexanediol + OsO_3

Because the cyclic osmate has less tendency to undergo side reactions, higher yields of the diol are obtained when osmium tetroxide is used. Osmium tetroxide, however, is considerably more expensive than potassium permanganate and is also toxic. Typically it is used only for small-scale laboratory preparations; potassium permanganate is preferred for large-scale preparations. To minimize the cost of an osmium tetroxide oxidation, a technique has been developed that uses a catalytic amount of OsO_4 together with an equivalent of a tertiary amine oxide (Section 11.11), which oxidizes the reduced oxidizing reagent back to OsO_4.

PROBLEM 10◆

Give the products that would be formed from the reaction of each of the following alkenes with OsO_4 followed by aqueous $NaHSO_3$.

a. $CH_3\overset{\overset{\displaystyle }{|}}{C}=CHCH_2CH_3$
    $\overset{\displaystyle |}{CH_3}$

b. 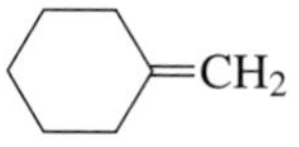 ⬡=CH_2

PROBLEM 11

What stereoisomers would be formed from the reaction of each of the following alkenes with OsO_4 followed by hydrolysis?

a. *trans*-2-butene

b. *cis*-2-butene

c. *cis*-2-pentene

d. *trans*-2-pentene

17.5 OXIDATIVE CLEAVAGE OF 1,2-DIOLS

1,2-Diols are oxidized to ketones and/or aldehydes by periodic acid (HIO_4). Periodic acid reacts with the diol to form a cyclic intermediate. The reaction takes place because iodine is in a highly positive oxidation state (+7). Consequently, it readily accepts electrons. When the intermediate breaks down, the bond between the two carbons bonded to the OH groups breaks. If the carbon that is bonded to an OH group is also bonded to two R groups, the product will be a ketone. If it is bonded to an R and an H, the product will be an aldehyde. Because this oxidation reaction cuts the reactant into two pieces, it is called **oxidative cleavage.**

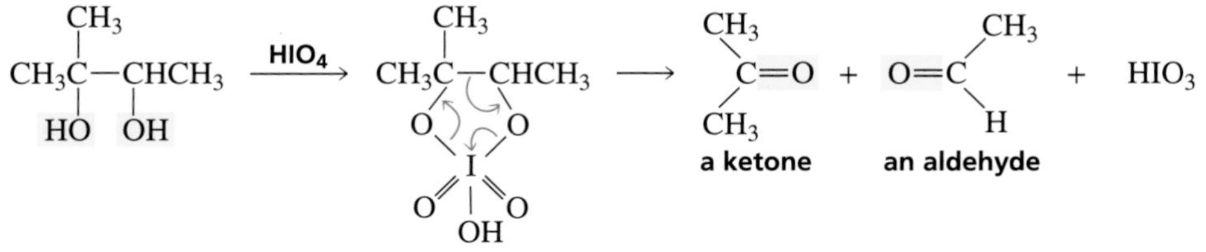

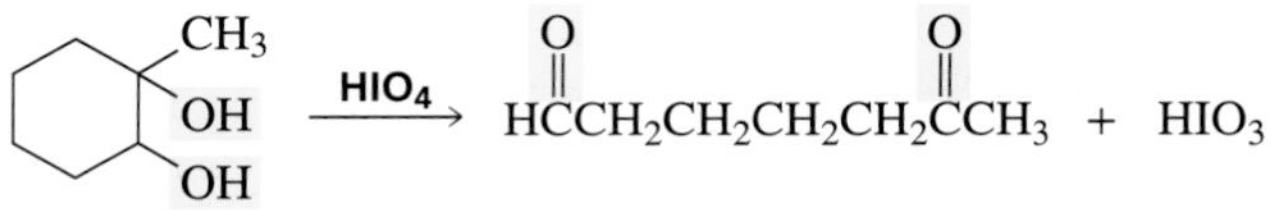

PROBLEM-SOLVING STRATEGY

Of the five compounds shown, only D cannot be cleaved by periodic acid. Explain.

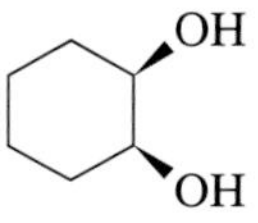
A

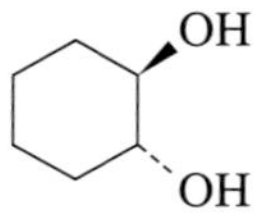
B

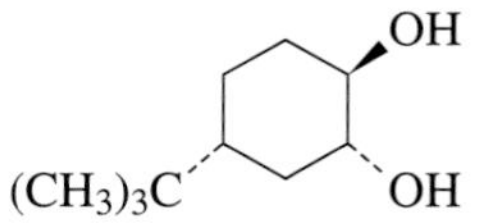
C

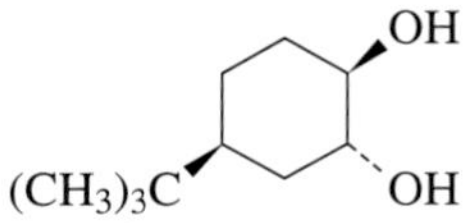
D

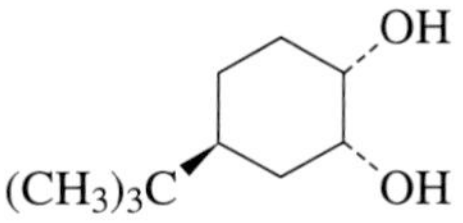
E

In solving this kind of problem, first consider what kinds of compounds undergo the reaction and any stereochemical requirements of the reaction. We know that periodic acid cleaves 1,2-diols. Because the reaction forms a cyclic intermediate, the two OH groups of the diol must be positioned so they can form the intermediate. Since the reactants are all cyclic, we should look at where the OH groups are in relation to each other in a cyclic molecule: they can both be equatorial, they can both be axial, or one can be equatorial and the other axial.

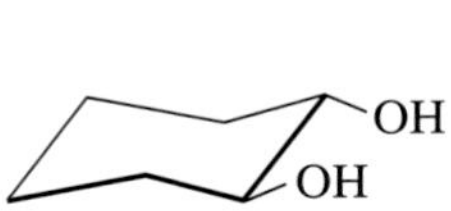
both equatorial
trans

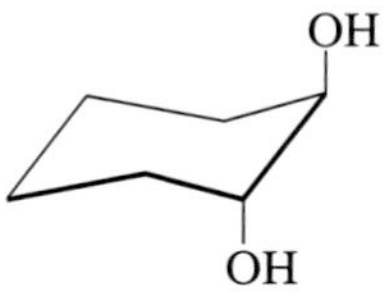
both axial
trans

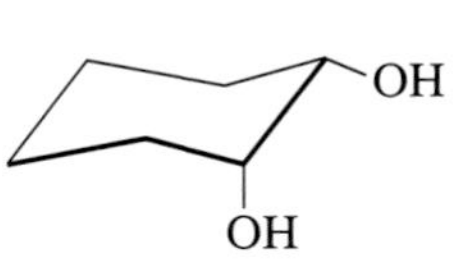
one equatorial, one axial
cis

Since both *cis*-1,2-diols (**A** and **E**) are cleaved, we know that the cyclic intermediate can be formed when one OH group is equatorial and the other is axial. A *trans*-1,2-diol has both OH groups equatorial or both OH groups axial (Section 4.16). Since two of the *trans*-1,2-diols can be cleaved (**B** and **C**) and one cannot (**D**), we can conclude from the structures above that the one that cannot be cleaved must have both OH groups in axial positions because these clearly are too far from each other to be able to form a cyclic intermediate. Now we should draw the most stable conformation for **B**, **C**, and **D** to see why only **D** has both OH groups in axial positions.

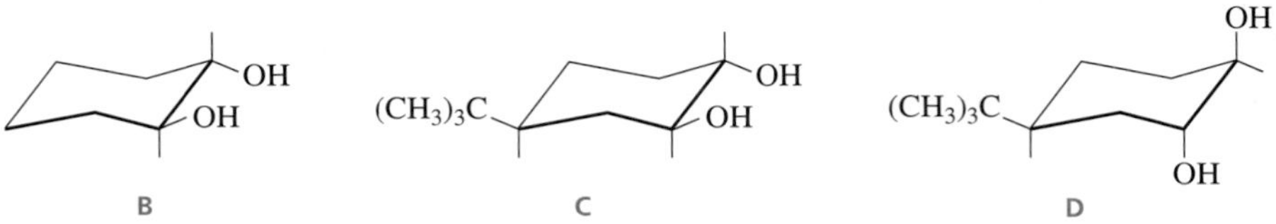

The most stable conformation for **B** is with both OH groups in equatorial positions. The steric requirements of the bulky *tert*-butyl group force it into an equatorial position. This causes both OH groups in compound **C** to be in equatorial positions and both OH groups in compound **D** to be in axial positions. So **D** cannot be cleaved by periodic acid.

Now continue on to do Problem 12.

PROBLEM 12◆

Which of each pair of diols is cleaved more rapidly by periodic acid?

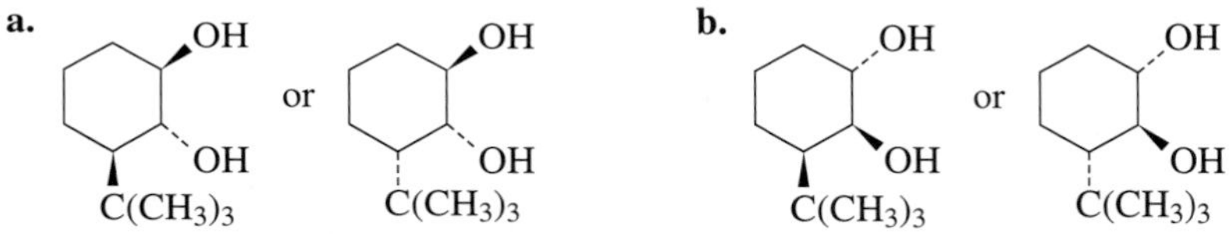

PROBLEM 13◆

An alkene is treated with OsO_4 followed by aqueous $NaHSO_3$. When the resulting diol is treated with HIO_4, the only product obtained is an unsubstituted cyclic ketone with molecular formula $C_6H_{10}O$. Give the structure of the alkene.

17.6 OXIDATIVE CLEAVAGE OF ALKENES: OZONOLYSIS

We have seen that alkenes can be oxidized to 1,2-diols, and that 1,2-diols can be further oxidized to aldehydes and ketones (Sections 17.4 and 17.5). Alternatively, alkenes can be directly oxidized to aldehydes and ketones by ozone (O_3). When an alkene is treated with ozone at low temperatures, both the σ bond and the π bond of the double bond break, and the carbons that were doubly bonded to each other find themselves doubly bonded to oxygens instead. This oxidation reaction is known as **ozonolysis.**

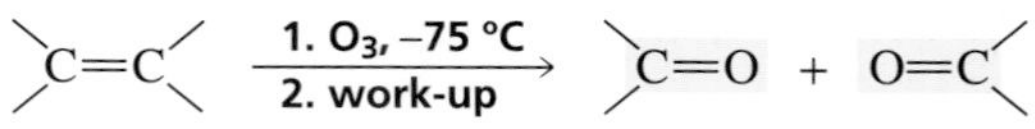

Ozone is produced by passing an electric discharge through oxygen gas. Its structure can be represented by the following resonance contributors.

contributing resonance structures of ozone

Ozone and the alkene undergo a concerted cycloaddition reaction; the oxygen atoms that add to the two sp^2 carbons do so in a single step. Addition of ozone to the alkene should remind you of the electrophilic addition reactions of alkenes in Chapter 3. An electrophile adds to one of the sp^2 carbons, and a nucleophile adds to the other. The electrophile is the oxygen at one end of the ozone molecule, and the nucleophile is the oxygen at the other end. The product of ozone addition to an alkene is an unstable **molozonide.** (The name "molozonide" indicates that one mole of ozone has added to the alkene.) The molozonide rearranges to a more stable **ozonide,** by a mechanism that we will not discuss here.

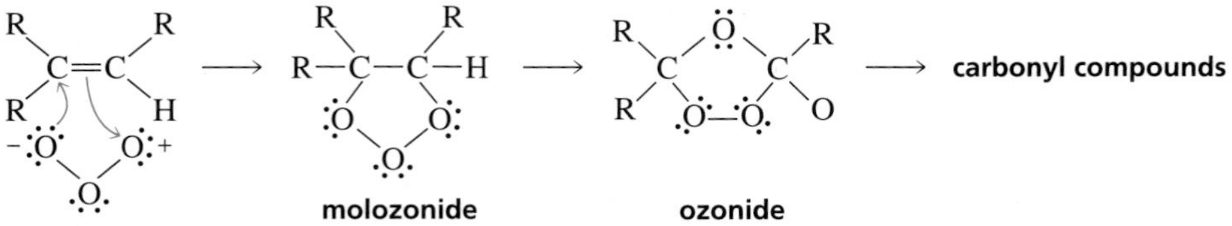

molozonide   ozonide   carbonyl compounds

Because some of the by-products formed in ozonolysis are explosive, ozonides are seldom isolated. The ozonide is easily cleaved to form carbonyl compounds.

If the ozonide is cleaved in the presence of a reducing agent such as zinc or dimethyl sulfide, the products will be ketones and/or aldehydes. (The product will be a ketone if the sp^2 carbon of the alkene is bonded to two carbon-containing substituents. It will be an aldehyde if at least one of the substituents bonded to the sp^2 carbon is a hydrogen.) The reducing agent prevents aldehydes from being oxidized to carboxylic acids. Cleaving the ozonide in the presence of zinc or dimethyl sulfide is referred to as "working up the ozonide under reducing conditions."

If the ozonide is cleaved in the presence of an oxidizing agent such as hydrogen peroxide (H_2O_2), the products will be ketones and/or carboxylic acids. Cleavage in the presence of H_2O_2 is referred to as "working up the ozonide under oxidizing conditions." Under oxidizing conditions, carboxylic acids are formed instead of aldehydes.

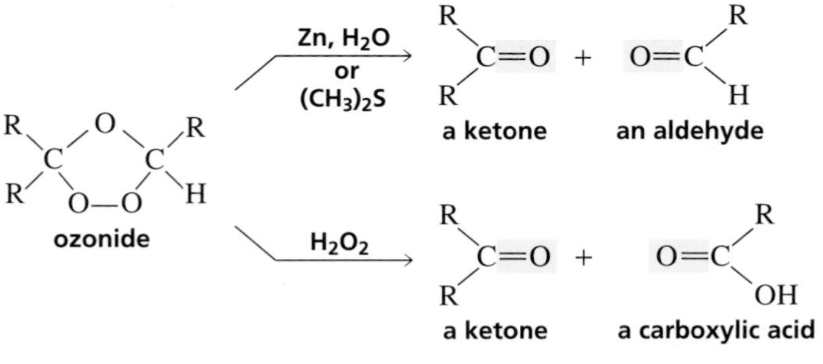

The following reactions are examples of the oxidative cleavage of alkenes by ozonolysis.

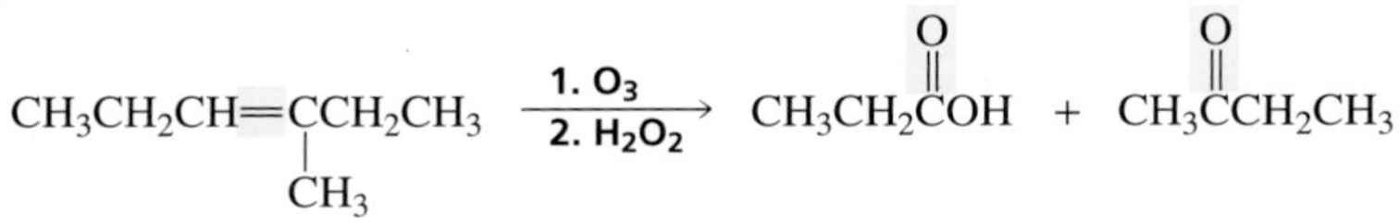

$$CH_3CH_2CH=CHCH_2CH_3 \xrightarrow[\text{2. (CH}_3)_2S]{\text{1. O}_3} 2\ CH_3CH_2\overset{\overset{\displaystyle O}{\|}}{C}H$$

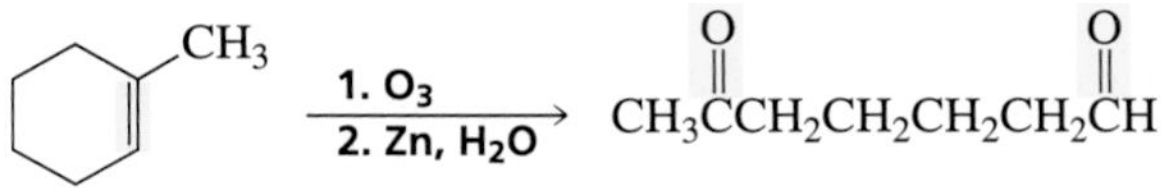

$$\xrightarrow[\text{2. Zn, H}_2O]{\text{1. O}_3} CH_3\overset{\overset{\displaystyle O}{\|}}{C}CH_2CH_2CH_2CH_2\overset{\overset{\displaystyle O}{\|}}{C}H$$

The one-carbon fragment obtained from the reaction of terminal alkenes with ozone will be oxidized to formaldehyde if the ozonide is worked up under reducing conditions and to CO_2 if it is worked up under oxidizing conditions.

$$CH_3CH_2CH_2CH=CH_2 \xrightarrow[\text{2. Zn, H}_2O]{\text{1. O}_3} CH_3CH_2CH_2\overset{\overset{\displaystyle O}{\|}}{C}H + H\overset{\overset{\displaystyle O}{\|}}{C}H$$

$$CH_3CH_2CH_2CH=CH_2 \xrightarrow[\text{2. H}_2O_2]{\text{1. O}_3} CH_3CH_2CH_2\overset{\overset{\displaystyle O}{\|}}{C}OH + CO_2$$

Only the side chain double bond will be oxidized in the following reaction because the stable benzene ring can be oxidized only under prolonged exposure to ozone.

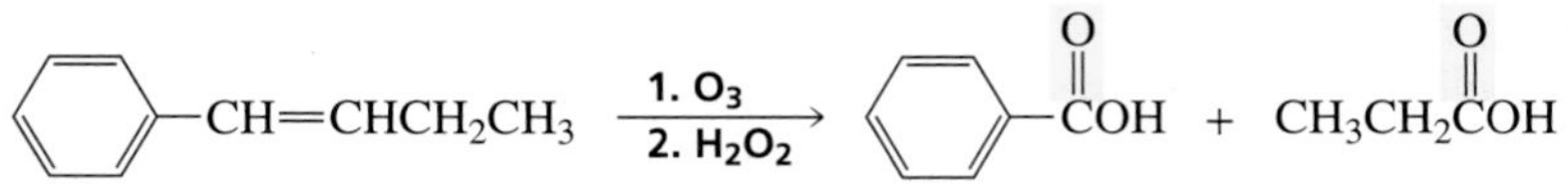

Notice that the products obtained from ozonolysis of an alkene under reducing conditions are the same as the products obtained from oxidation of the alkene to a 1,2-diol with osmium tetroxide (or with a basic solution of potassium permanganate) followed by oxidation of the diol with periodic acid.

PROBLEM 14

Give an example of an alkene that will form the same ozonolysis products regardless of whether the ozonide is worked up under reducing conditions (Zn, H_2O) or under oxidizing conditions (H_2O_2).

PROBLEM 15

Give the products that you would expect to obtain when the following compounds are treated with ozone followed by:

a. Zn, H_2O

b. H_2O_2

(1) $CH_3CH_2CH_2\overset{\overset{\displaystyle CH_3}{|}}{C}=CHCH_3$

(2) $CH_2=CHCH_2CH_2CH_2CH_3$

(3)

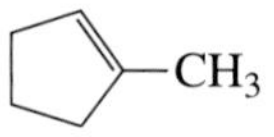

(4)

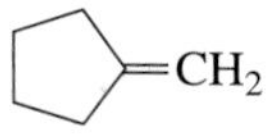

(5) $CH_3CH_2CH_2CH=CHCH_2CH_2CH_3$

(6)

Ozonolysis can be used to determine the structure of an unknown alkene. If you know what carbonyl compounds are formed by ozonolysis, you can mentally work backward to deduce the structure of the alkene. For example, if ozonolysis of an alkene followed by a work-up under reducing conditions forms acetone and butanal as products, you can conclude that the alkene was 2-methyl-2-hexene.

$$
\underset{\substack{\text{acetone}}}{\overset{\overset{\displaystyle O}{\|}}{CH_3CCH_3}} + \underset{\substack{\text{butanal}}}{\overset{\overset{\displaystyle O}{\|}}{CH_3CH_2CH_2CH}} \Longrightarrow \underset{\substack{\text{2-methyl-2-hexene}}}{\overset{\overset{\displaystyle CH_3}{|}}{CH_3C}=CHCH_2CH_2CH_3}
$$

ozonolysis products   alkene that underwent
ozonolysis

In Section 17.4 we saw that alkenes are oxidized to 1,2-diols by a basic solution of potassium permanganate at room temperature or below. If, however, the basic reaction mixture is heated or if the potassium permanganate solution is acidic, the reaction will not stop at the diol. Instead, the alkene will be cleaved. The reaction products are ketones and carboxylic acids. If the reaction is carried out under basic conditions, the carboxylic acids formed will be in their basic forms ($RCOO^-$); if the reaction is carried out under acidic conditions, the carboxylic acids formed will be in their acidic forms ($RCOOH$). Terminal alkenes form CO_2 as a product. The products formed by cleavage of an alkene with potassium permanganate are identical to the products formed by ozonolysis followed by a work-up under oxidizing conditions. However, the yield of the carbonyl compounds formed by cleavage with permanganate is lower than the yield obtained by cleavage with ozone.

$$
\underset{\substack{}}{\overset{\overset{\displaystyle CH_3}{|}}{CH_3CH_2C}=CHCH_3} \xrightarrow[\Delta]{\textbf{KMnO}_4\text{, HO}^-} \overset{\overset{\displaystyle O}{\|}}{CH_3CH_2CCH_3} + \overset{\overset{\displaystyle O}{\|}}{CH_3CO^-}
$$

$$
CH_3CH_2CH=CH_2 \xrightarrow[H^+]{\textbf{KMnO}_4} \overset{\overset{\displaystyle O}{\|}}{CH_3CH_2COH} + CO_2
$$

$$
\bigcirc{=}CH_2 \xrightarrow[\Delta]{\textbf{KMnO}_4\text{, HO}^-} \bigcirc{=}O + CO_2
$$

The various methods used to oxidize an alkene that were discussed in Sections 17.4-17.6 are summarized in Table 17.3.

PROBLEM 16◆

a. What alkene would give only acetone as an ozonolysis product?

b. What alkenes would give only butanal as an ozonolysis product?

PROBLEM 17◆

What aspect of the structure of the alkene does ozonolysis not tell you?

TABLE 17.3   **Summary of the Methods Used to Oxidize an Alkene**

$$\underset{\substack{\text{CH}_3\\|}}{\text{CH}_3\text{C}}=\text{CHCH}_3 \xrightarrow[\text{2. Zn, H}_2\text{O}]{\text{1. O}_3} \underset{\substack{\text{O}\\||}}{\text{CH}_3\text{CCH}_3} + \underset{\substack{\text{O}\\||}}{\text{CH}_3\text{CH}}$$

$$\xrightarrow[\text{2. H}_2\text{O}_2]{\text{1. O}_3} \underset{\substack{\text{O}\\||}}{\text{CH}_3\text{CCH}_3} + \underset{\substack{\text{O}\\||}}{\text{CH}_3\text{COH}}$$

$$\xrightarrow[\text{H}^+]{\text{KMnO}_4} \underset{\substack{\text{O}\\||}}{\text{CH}_3\text{CCH}_3} + \underset{\substack{\text{O}\\||}}{\text{CH}_3\text{COH}}$$

$$\xrightarrow[\Delta]{\text{KMnO}_4,\ \text{HO}^-} \underset{\substack{\text{O}\\||}}{\text{CH}_3\text{CCH}_3} + \underset{\substack{\text{O}\\||}}{\text{CH}_3\text{CO}^-}$$

$$\xrightarrow[\text{2. H}_2\text{O}]{\text{1. KMnO}_4,\ \text{HO}^-\ \text{cold}} \underset{\substack{\text{CH}_3\\|\\\quad\ \ |\\\text{OH OH}}}{\text{CH}_3\text{C}-\text{CHCH}_3} \xrightarrow{\text{HIO}_4} \underset{\substack{\text{O}\\||}}{\text{CH}_3\text{CCH}_3} + \underset{\substack{\text{O}\\||}}{\text{CH}_3\text{CH}}$$

$$\xrightarrow[\text{2. NaHSO}_3,\ \text{H}_2\text{O}]{\text{1. OsO}_4} \underset{\substack{\text{CH}_3\\|\\\quad\ \ |\\\text{OH OH}}}{\text{CH}_3\text{C}-\text{CHCH}_3} \xrightarrow{\text{HIO}_4} \underset{\substack{\text{O}\\||}}{\text{CH}_3\text{CCH}_3} + \underset{\substack{\text{O}\\||}}{\text{CH}_3\text{CH}}$$

PROBLEM 18 / SOLVED

The following products were obtained from ozonolysis of a diene followed by a work-up under oxidizing conditions. Give the structure of the diene.

$$\underset{\substack{\text{O}\\||}}{\text{HOCCH}_2\text{CH}_2\text{CH}_2}\underset{\substack{\text{O}\\||}}{\text{COH}} + \text{CO}_2 + \underset{\substack{\text{O}\\||}}{\text{CH}_3\text{CH}_2\text{COH}}$$

SOLUTION   Because the first compound is a five-carbon dicarboxylic acid, the diene must contain five carbons flanked by two double bonds.

$$\underset{\substack{\text{O}\\||}}{\text{HOCCH}_2\text{CH}_2\text{CH}_2}\underset{\substack{\text{O}\\||}}{\text{COH}} \Longrightarrow =\text{CHCH}_2\text{CH}_2\text{CH}_2\text{CH}=$$

5-carbon dicarboxylic acid   **5 carbons flanked by double bonds**

The other products obtained from ozonolysis are carbon dioxide (one carbon atom) and propanoic acid (three carbon atoms). Therefore, one carbon has to be added to one end of the diene and three carbons have to be added to the other end.

$$\text{CH}_2=\text{CHCH}_2\text{CH}_2\text{CH}_2\text{CH}=\text{CHCH}_2\text{CH}_3$$

**17.7
OXIDATIVE
CLEAVAGE
OF ALKYNES**

Alkynes are oxidized by the same reagents that oxidize alkenes. Alkynes are oxidized to diketones by a basic solution of KMnO_4 at room temperature, and are cleaved by ozonolysis to carboxylic acids. Ozonolysis requires neither oxidative nor reductive work-up; ozonolysis is followed only by hydrolysis. Carbon dioxide is obtained from the CH group of a terminal alkyne.

$$CH_3C{\equiv}CCH_2CH_3 \xrightarrow[HO^-]{KMnO_4} CH_3\overset{\overset{O}{\|}}{C}-\overset{\overset{O}{\|}}{C}CH_2CH_3$$
2-pentyne

$$CH_3C{\equiv}CCH_2CH_3 \xrightarrow[\text{2. } H_2O]{\text{1. } O_3} CH_3\overset{\overset{O}{\|}}{C}OH + CH_3CH_2\overset{\overset{O}{\|}}{C}OH$$
2-pentyne

$$CH_3CH_2CH_2C{\equiv}CH \xrightarrow[\text{2. } H_2O]{\text{1. } O_3} CH_3CH_2CH_2\overset{\overset{O}{\|}}{C}OH + CO_2$$
1-pentyne

PROBLEM 19◆

Give the structure of the alkyne that gives each of the following products upon ozonolysis followed by treatment with H_2O_2.

a.

COOH

(cyclohexane ring with COOH) $+ CO_2$

b. $HO\overset{\overset{O}{\|}}{C}CH_2CH_2CH_2\overset{\overset{O}{\|}}{C}OH + 2\ CH_3CH_2\overset{\overset{O}{\|}}{C}OH$

An alkene can be oxidized to an epoxide by a peroxyacid. The overall reaction amounts to the transfer of an oxygen atom from the oxidizing agent to the alkene.

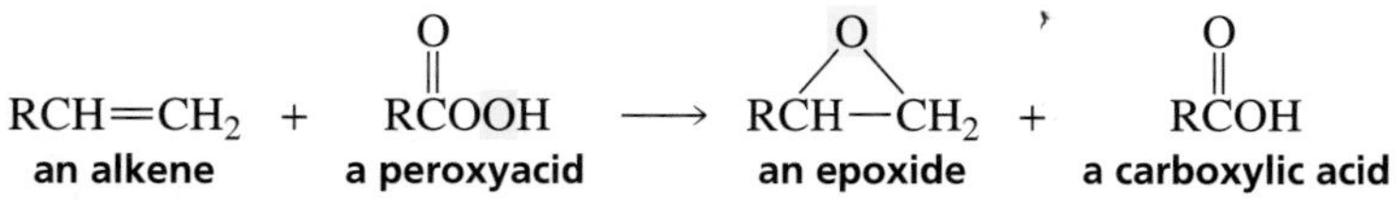

17.8
OXIDATION OF ALKENES WITH PEROXYACIDS

$$RCH{=}CH_2 + R\overset{\overset{O}{\|}}{C}OOH \longrightarrow R\overset{O}{\overset{/\backslash}{CH-CH_2}} + R'\overset{\overset{O}{\|}}{C}OH$$
an alkene   **a peroxyacid**   **an epoxide**   **a carboxylic acid**

The oxygen–oxygen single bond (a peroxide bond) of the peroxyacid is weak and easily broken.

a weak peroxide bond

$$R-\overset{\overset{O}{\|}}{C}-O-O-H$$

The oxygen atom of the OH group of the peroxyacid accepts a pair of electrons from the π bond of the alkene, causing the weak oxygen–oxygen single bond to break. The electrons from the oxygen–oxygen bond are delocalized onto the carbonyl group. The electrons left behind as the oxygen–hydrogen bond breaks add to the carbon of the alkene that becomes electron deficient when the π bond breaks. Notice that **epoxidation** of an alkene is a concerted reaction: all of the bond-forming and bond-breaking processes take place in a single step.

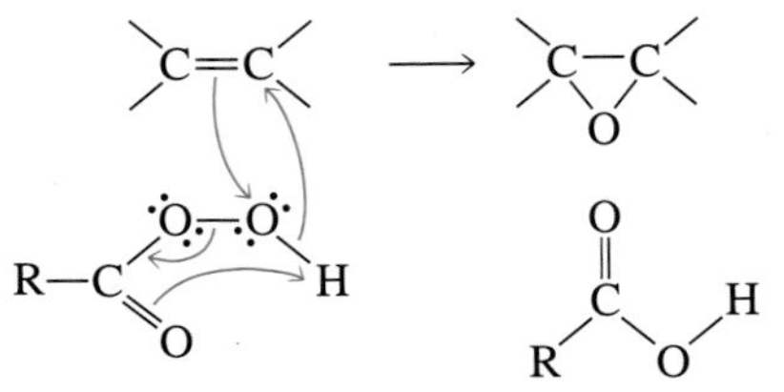

The mechanism for addition of oxygen to a double bond to form an epoxide is exactly the same as the mechanism for addition of bromine to a double bond to form a cyclic bromonium ion (Section 3.15). In one case the electrophile is oxygen, and in the other it is bromine. So the reaction of an alkene with a peroxyacid, like the reaction of an alkene with Br_2, is another example of an electrophilic addition reaction.

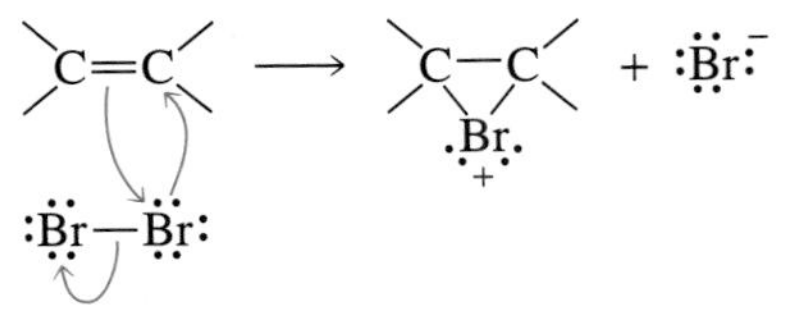

STABLE PEROXYACIDS

For a long time, the most commonly used peroxyacid was *meta*-chloroperoxybenzoic acid (MCPBA), because it is an unusually stable peroxyacid and therefore relatively easily handled. Chemists have found that magnesium monoperoxyphthalate (MMPP) is even more stable, so it is currently the most widely used reagent for epoxidation.

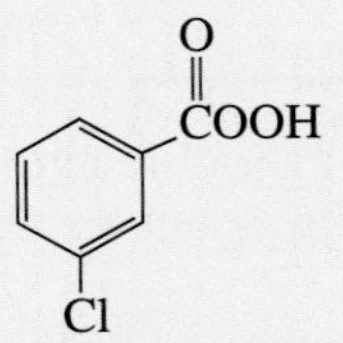

meta-chloroperoxybenzoic acid
MCPBA

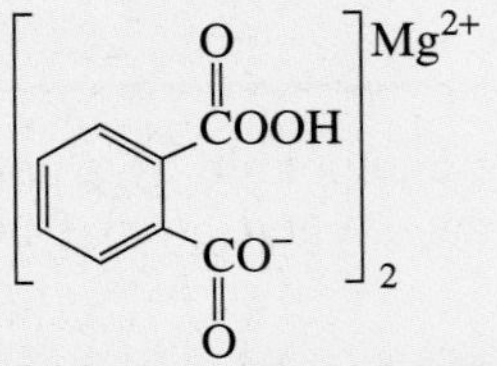

magnesium monoperoxyphthalate
MMPP

Since epoxidation results from the reaction of an electrophilic oxygen atom with a nucleophilic alkene, increasing the electron density of the π bond of the alkene increases the rate of epoxidation. We know that alkyl substituents increase the electron density of the π bond (Section 3.19). Therefore, if a diene is epoxidized with only enough peroxyacid to react with one of the double bonds, the most substituted double bond will be the bond that is epoxidized.

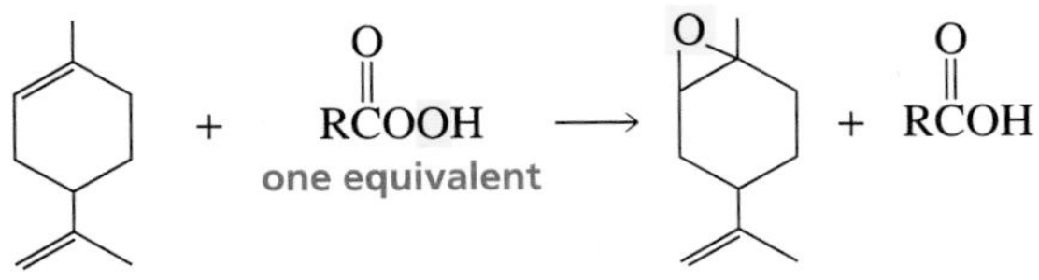

limonene

one equivalent

The addition of oxygen to an alkene is a stereospecific reaction. If the reactant has the cis configuration, the epoxide will also have the cis configuration. Similarly, a *trans*-alkene forms a *trans*-epoxide.

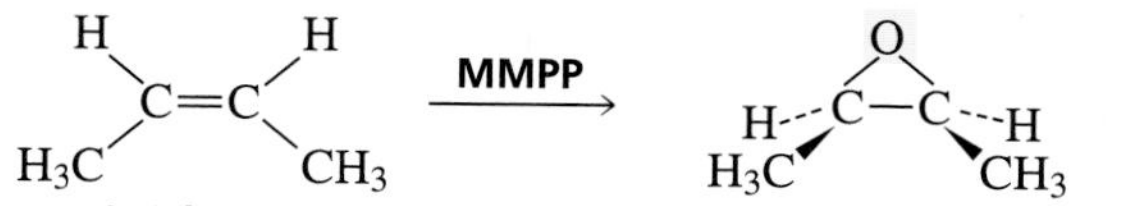

H, H (C=C) H₃C, CH₃ →(MMPP) O (H--C—C--H) H₃C, CH₃
cis-2-butene → *cis*-2,3-dimethyloxirane

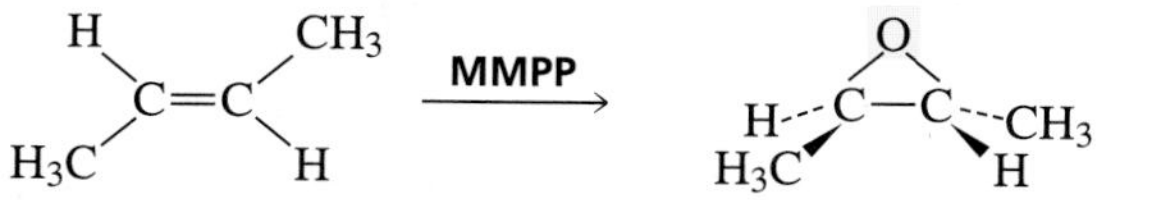

H, CH₃ (C=C) H₃C, H →(MMPP) O (H--C—C--CH₃) H₃C, H
trans-2-butene → *trans*-2,3-dimethyloxirane

PROBLEM 20◆

What alkene would you treat with MMPP in order to obtain each of the following epoxides?

a. 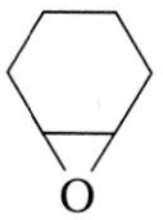 (bicyclic epoxide with O)

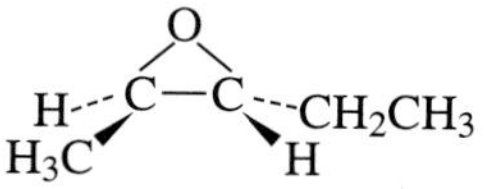

b.  H₂C—CHCH₂CH₃ (with O epoxide)

c. O (H--C—C--CH₂CH₃) H₃C, H

d. 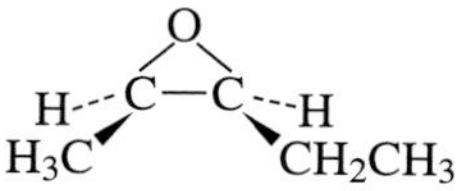 O (H--C—C--H) H₃C, CH₂CH₃

PROBLEM 21

Give the major product of the reaction of each of the following compounds with one equivalent of MMPP. Indicate the configuration of the product.

a. 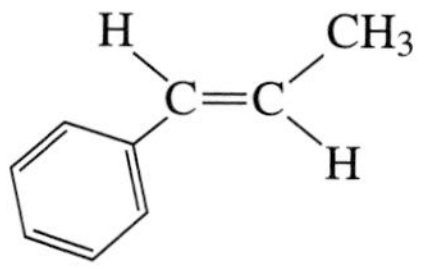 H, CH₃ (C=C) with phenyl, H

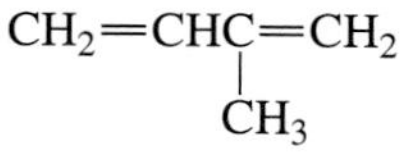

b. 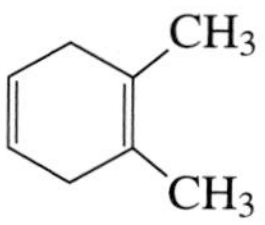 (cyclohexene ring with CH₃ and CH₃)

c. CH₂=CHC=CH₂, CH₃

d. 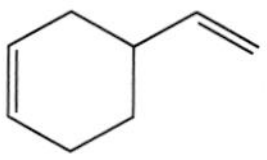 (cyclohexene with vinyl substituent)

PROBLEM 22

Why is an epoxide a relatively stable product while a bromonium ion is a reactive intermediate?

A *para*-benzenediol such as hydroquinone is easily oxidized to *para*-benzoquinone. Although a wide variety of oxidizing agents can be used, Fremy's salt (dipotassium nitrosodisulfonate) is the preferred oxidizing agent. The quinone can be easily reduced back to hydroquinone.

17.9
OXIDATION OF HYDROQUINONES/ REDUCTION OF QUINONES

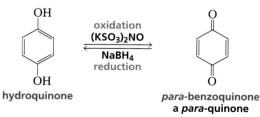

OH ⬡ OH
hydroquinone

oxidation (KSO₃)₂NO ⇌ NaBH₄ reduction

O ⬡ O
para-benzoquinone
a *para*-quinone

Similarly, *ortho*-benzenediols are oxidized to *ortho*-quinones.

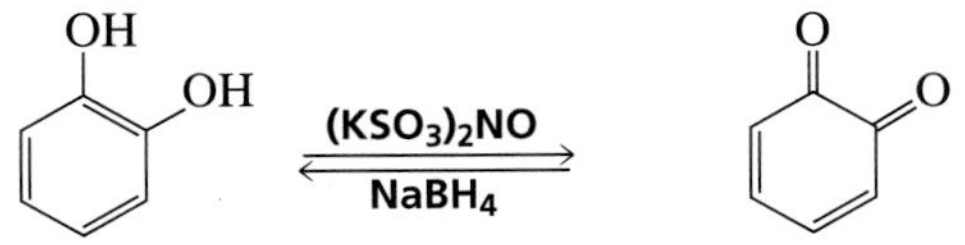

ortho-benzoquinone
an ortho-quinone

The oxidation–reduction reaction takes place by a two-step mechanism with H·
being lost in each of the steps. Overall, therefore, the reaction involves loss of H_2.
In Section 8.8 we saw that the ability of a phenol to lose a hydrogen atom is why
phenols are used as radical scavengers.

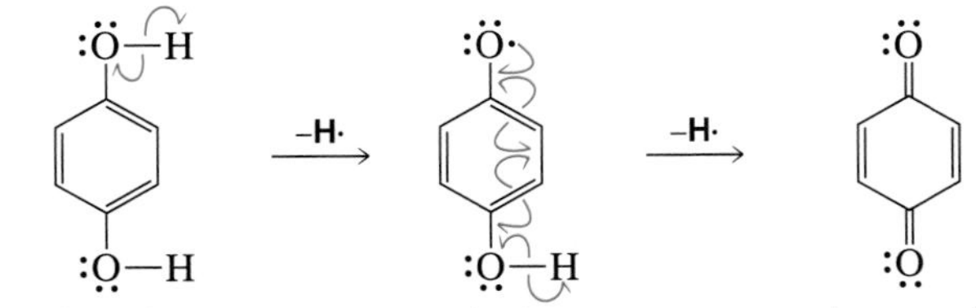

hydroquinone   semiquinone   *para*-benzoquinone

THE CHEMISTRY OF PHOTOGRAPHY

Black-and-white photography depends on the fact that hydroquinone is easily oxidized. Photographic
film is covered by an emulsion of silver bromide. When light hits the film, the silver bromide is sensitized. Sensitized silver bromide is a better oxidizing agent than silver bromide that has not been
exposed to light. When the exposed film is put into a solution of hydroquinone (a common photographic devel-
oper), the hydroquinone is oxidized to quinone by sensitized silver ion and the silver ion is reduced to silver
metal, which remains in the emulsion. The exposed film is "fixed" by washing away unsensitized silver bromide with $Na_2S_2O_3/H_2O$. Black silver deposits are left in regions where light has struck the film. This is the
black, opaque part of a photographic negative.

COENZYME Q

Coenzyme Q (CoQ) is a quinone found in the mitochondria of cells. It is also called ubiquinone
because it is ubiquitous (found everywhere). Its function is to carry electrons in the electron-transport
chain. The oxidized form of CoQ accepts a pair of electrons from NADH or $FADH_2$ (biological
reducing agents, Section 17.11) and thereby becomes reduced. The reduced form of CoQ then transfers the electrons to cytochrome b.

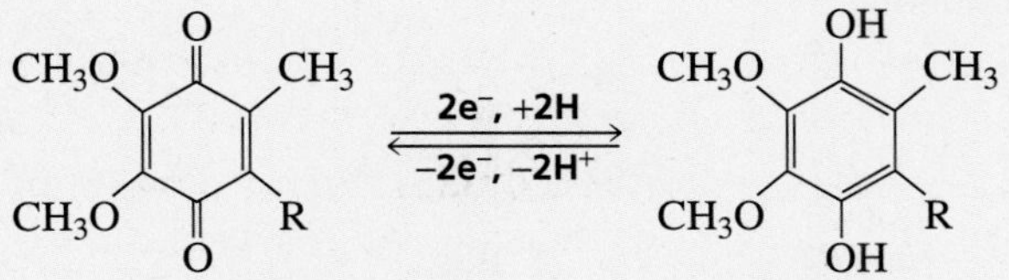

Coenzyme Q
oxidized form   **Coenzyme Q**
reduced form

A knowledge of oxidation–reduction reactions greatly expands the kinds of synthetic transformations we can carry out. For example, in Chapter 11 we learned how to convert an alkyl halide into an alcohol that contains two additional carbon atoms.

$$CH_3CH_2CH_2Br \xrightarrow[Et_2O]{Mg} CH_3CH_2CH_2MgBr \xrightarrow[2.\ H^+]{1.\ \triangle\!\!\!-\!\!\!O} CH_3CH_2CH_2CH_2CH_2OH$$

17.10
DESIGNING A SYNTHESIS V: FUNCTIONAL GROUP INTERCONVERSION

Now we can convert an alkyl halide into an alcohol that contains only one additional carbon atom, and we can do it in more than one way.

$$CH_3CH_2CH_2Br \xrightarrow[Et_2O]{Mg} CH_3CH_2CH_2MgBr$$

1. $H_2C{=}O$ 2. H^+

$$CH_3CH_2CH_2\overset{O}{\overset{\|}{C}}OH \xrightarrow[2.\ H_2O]{1.\ LiAlH_4} CH_3CH_2CH_2CH_2OH$$

$$CH_3CH_2CH_2Br \xrightarrow{{}^-C{\equiv}N} CH_3CH_2CH_2C{\equiv}N \xrightarrow[H_2O,\ H^+]{\triangle}$$

We can also convert an alkyl halide into an amine that contains one additional carbon atom.

$$CH_3CH_2CH_2Br \xrightarrow{{}^-C{\equiv}N} CH_3CH_2CH_2C{\equiv}N \xrightarrow[Pt]{H_2} CH_3CH_2CH_2CH_2NH_2$$

Or, we can convert a primary alkyl halide into an amine that contains one fewer carbon atoms.

$$CH_3\underset{CH_3}{CH}CH_2Br \xrightarrow{tert\text{-}BuO^-} CH_3\underset{CH_3}{C}{=}CH_2 \xrightarrow[2.\ Zn,\ H_2O]{1.\ O_3} CH_3\underset{CH_3}{C}{=}O \xrightarrow[H_2,\ Pt]{NH_3} CH_3\underset{CH_3}{CH}NH_2$$

We can convert a ketone into a compound that has substituents on adjacent carbons.

Or, we can convert a compound with substituents on adjacent carbons atoms into a ketone, into an aldehyde, or into an alcohol.

Converting one functional group into another is called **functional group interconversion.** Our knowledge of oxidation–reduction reactions has greatly expanded our ability to carry out functional group interconversions. For example, an aldehyde can be converted into a primary alcohol, an alkene, a secondary alcohol, a ketone, a carboxylic acid, an acyl chloride, an ester, or an amide.

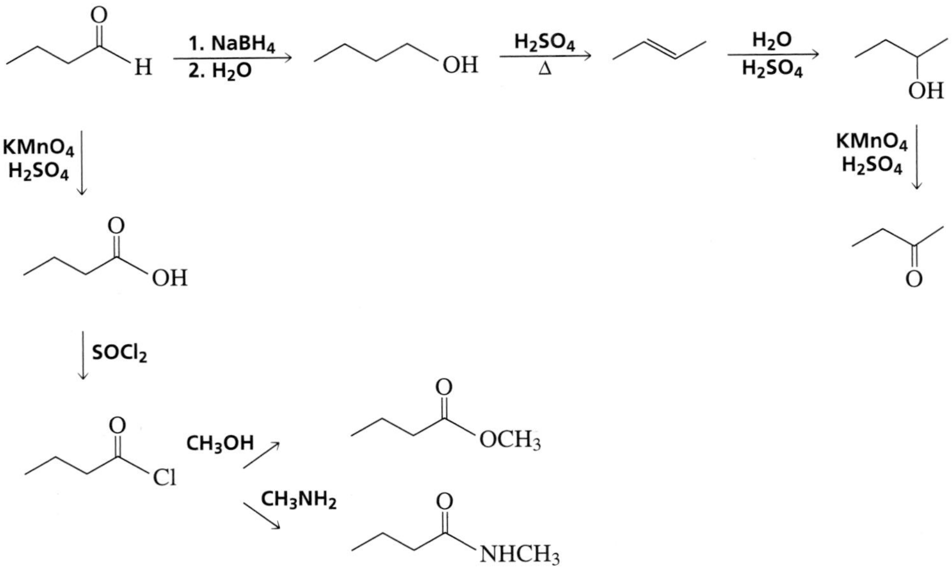

A ketone can be converted into an ester or an alcohol.

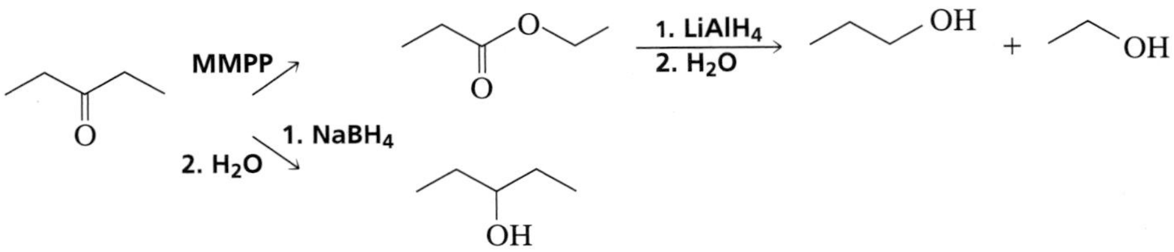

As the number of reactions you learn increases, not only will the number of functional group interconversions you can do increase, but you will also find that you have more than one route available when you design a synthesis. The route you actually decide to use will depend on the availability of the starting materials, the cost of the starting materials, and the ease with which the reactions in the synthetic pathway can be carried out.

PROBLEM 23

How many different functional groups can you use to synthesize a primary alcohol? (*Hint:* See Figure 28.1.)

PROBLEM 24◆

Add reagents over the arrows.

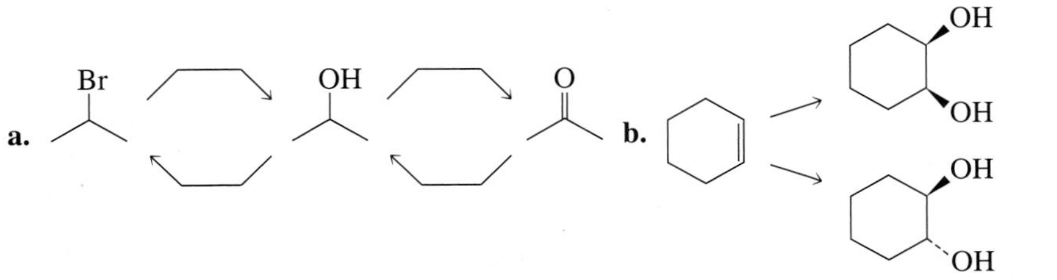

PROBLEM 25

Show how each of the following compounds could be synthesized using the given starting material.

a. ⬡–CH₂Br ⟶ ⬡–CH₂–C(=O)–OH

b. ⬡–CH₂Br ⟶ ⬡–CH₂–C(=O)–CH₃

c. ⬡–CH₂Br ⟶ cyclohexanone (⬡=O)

d. ⬡–CH₂Br ⟶ ⬡–C(=O)–OH

Oxidation and reduction reactions are important reactions in living systems. An example of an oxidation reaction that takes place in animal cells is the oxidation of ethanol to acetaldehyde, a reaction catalyzed by the enzyme alcohol dehydrogenase. Ingestion of a moderate amount of ethanol lowers inhibitions and causes a lightheaded feeling, but the physiological effects of acetaldehyde are not as pleasant. Acetaldehyde is responsible for the feeling known as a hangover. (In Section 22.4, we will see how vitamin B_1 can help to cure a hangover.)

17.11 BIOLOGICAL OXIDATION–REDUCTION REACTIONS

$$CH_3CH_2OH + NAD^+ \xrightarrow{\substack{\text{alcohol} \\ \text{dehydrogenase}}} \underset{\text{acetaldehyde}}{CH_3\overset{\displaystyle O}{\overset{\|}{C}}H} + NADH + H^+$$

ethanol

TREATING ALCOHOLICS WITH ANTABUSE

Disulfiram, most commonly known by one of its trade names, Antabuse, is used to treat alcoholics. It causes violently unpleasant effects when consumed with alcohol even when the alcohol is consumed a day or two after the drug is taken.

$$\underset{CH_3CH_2}{\overset{CH_3CH_2}{>}}N-\overset{\overset{\displaystyle S}{\|}}{C}-S-S-\overset{\overset{\displaystyle S}{\|}}{C}-N\underset{CH_2CH_3}{\overset{CH_2CH_3}{<}}$$

Antabuse

Antabuse inhibits aldehyde dehydrogenase, the enzyme responsible for oxidizing acetaldehyde to acetic acid, resulting in a buildup of acetaldehyde. It is the acetaldehyde that is responsible for the unpleasant physiological effects—intense flushing, nausea, dizziness, sweating, throbbing headaches, decreased blood pressure, and ultimately shock. Consequently, Antabuse should be taken only under strict medical supervision. In many people, aldehyde dehydrogenase is a nonfunctional enzyme. Their symptoms in response to ingestion of alcohol are nearly the same as those of individuals who are medicated with Antabuse.

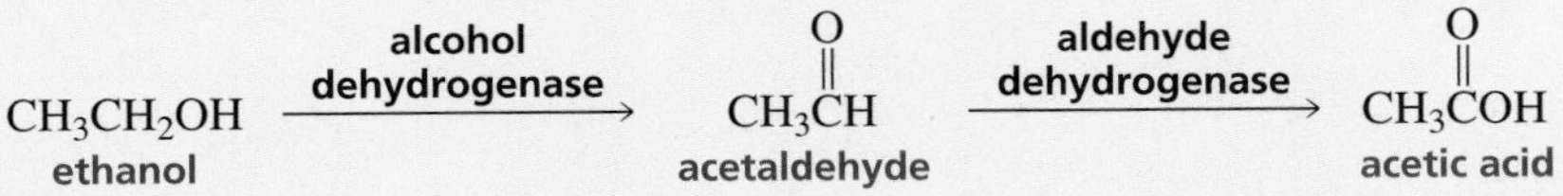

$$CH_3CH_2OH \xrightarrow[\text{dehydrogenase}]{\text{alcohol}} CH_3\overset{\displaystyle O}{\overset{\displaystyle \|}{C}}H \xrightarrow[\text{dehydrogenase}]{\text{aldehyde}} CH_3\overset{\displaystyle O}{\overset{\displaystyle \|}{C}}OH$$

ethanol   acetaldehyde   acetic acid

Ethanol cannot be oxidized to acetaldehyde unless an oxidizing agent is present. Oxidizing agents used by organic chemists, such as chromate and permanganate salts, are not present in living systems. NAD^+ (nicotinamide adenine dinucleotide) is the most common oxidizing agent available in living systems and is used by cells to oxidize alcohols to aldehydes. Notice that NAD^+ is written with a positive charge to reflect the positive charge on the nitrogen atom of the pyridine ring.

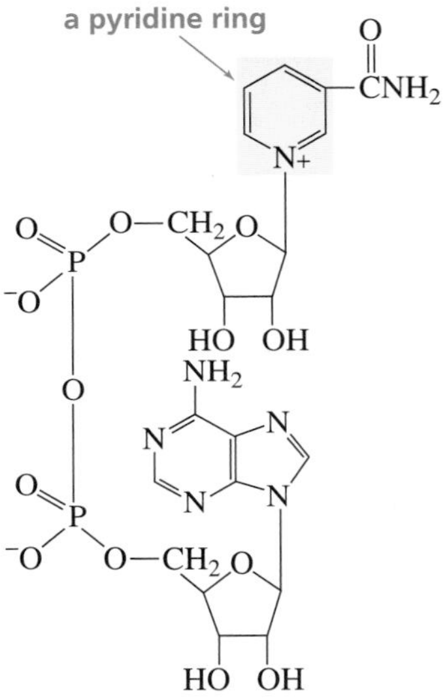

nicotinamide adenine dinucleotide
NAD^+

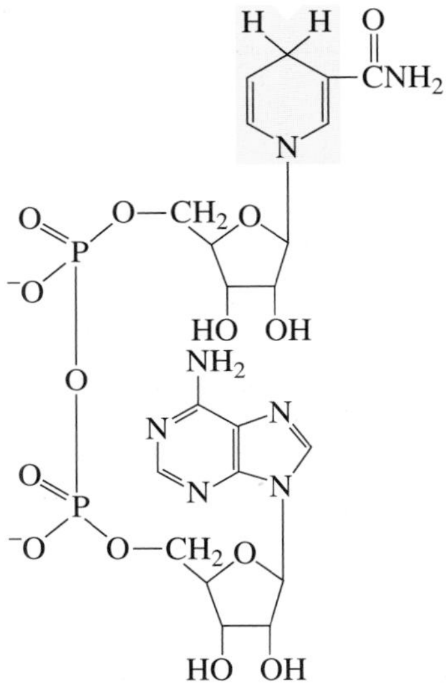

reduced nicotinamide adenine dinucleotide
NADH

NAD^+ is reduced to NADH when it oxidizes a compound. NADH is used by the cell as a reducing agent. When NADH reduces a compound, it is oxidized back to NAD^+, which can then be used for another oxidation. Although NAD^+ and NADH are complicated-looking molecules, it is important to notice that the structural changes that occur when they act as oxidizing and reducing agents take place on a relatively small part of the molecule. The rest of the molecule is used to bind it to the proper site on the enzyme that catalyzes the reaction.

NAD^+ oxidizes a compound by accepting a hydride ion from it. In this way, the number of carbon–hydrogen bonds in the compound is decreased (the compound is oxidized) and the number of carbon–hydrogen bonds in NAD^+ is increased (NAD^+ is reduced). The contributing resonance structures for NAD^+ show why it is able to accept a hydride ion: the carbon atoms at positions 2, 4, and 6 of the pyridine ring have a partial positive charge. The hydride ion removed from the substrate is always donated to the 4-position of the oxidizing agent.

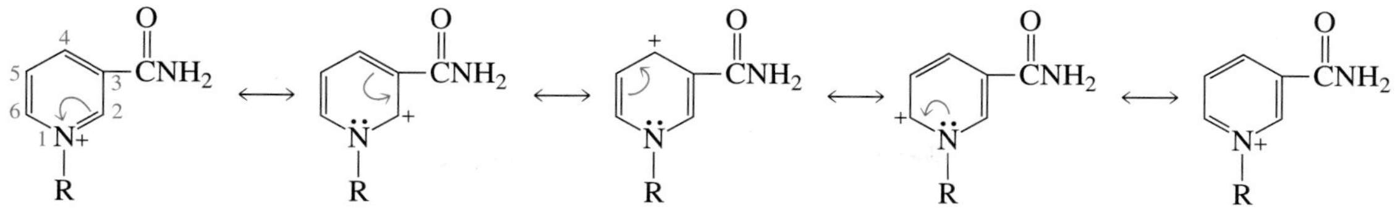

contributing resonance structures

We saw in Section 4.14 that, because enzymes are chiral, alcohol dehydrogenase can tell the difference between the two hydrogens bonded to the α-carbon of the alcohol. It removes only the pro-*R* hydrogen.

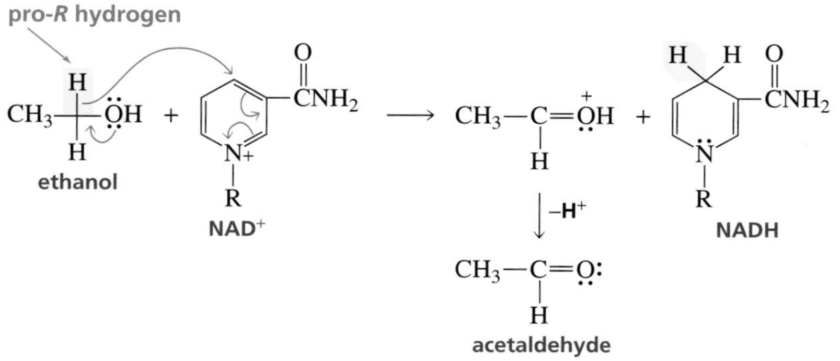

NADH reduces a compound by donating a hydride ion from the 4-position of its pyridine ring: NADH, $NaBH_4$, and $LiAlH_4$ all act as reducing agents in the same way: they donate a hydride ion.

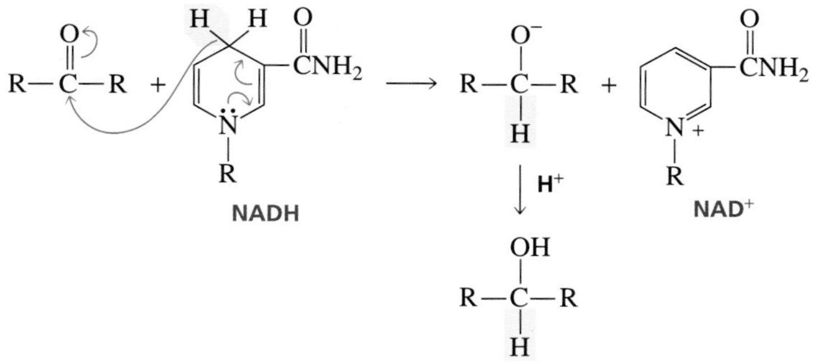

AN UNUSUAL ANTIDOTE

Alcohol dehydrogenase, the enzyme that catalyzes the oxidation of ethanol to acetaldehyde, catalyzes the oxidation of other alcohols as well. For example, it catalyzes the oxidation of methanol to formaldehyde.

$$\underset{\text{methanol}}{CH_3OH} + NAD^+ \xrightarrow{\text{alcohol dehydrogenase}} \underset{\text{formaldehyde}}{HC(=O)H} + NADH$$

Methanol itself is not harmful, but ingestion of methanol can be fatal because formaldehyde is extremely toxic. The treatment for methanol ingestion consists of making the patient drink substantial quantities of ethanol. Alcohol dehydrogenase has 25 times the affinity for ethanol that it has for methanol. If the patient drinks ethanol, the enzyme is kept busy oxidizing the ethanol. The patient can stop drinking when sufficient time has elapsed to allow the unmetabolized methanol to pass through the body.

FETAL ALCOHOL SYNDROME

The damage done to a human fetus when the mother drinks alcohol during her pregnancy is known as fetal alcohol syndrome. It has been shown that the harmful effects are attributable to acetaldehyde formed from the oxidation of ethanol, which crosses the placenta and accumulates in the liver of the fetus.

SUMMARY OF REACTIONS

1. Catalytic hydrogenation of double and triple bonds (Section 17.1)

$$RCH{=}CHR + H_2 \xrightarrow{\text{Pt, Pd, or Ni}} RCH_2CH_2R$$

$$RC{\equiv}CR + H_2 \xrightarrow{\text{Pt, Pd, or Ni}} RCH_2CH_2R$$

$$RCH{=}NR + H_2 \xrightarrow{\text{Pt, Pd, or Ni}} RCH_2NHR$$

$$RC{\equiv}N + H_2 \xrightarrow{\text{Pt, Pd, or Ni}} RCH_2NH_2$$

$$\underset{\displaystyle RCH}{\overset{\displaystyle O}{\|}} + H_2 \xrightarrow{\text{Pt, Pd, or Ni}} RCH_2OH$$

$$\underset{\displaystyle RCR}{\overset{\displaystyle O}{\|}} + H_2 \xrightarrow{\text{Pt, Pd, or Ni}} \underset{\displaystyle RCHR}{\overset{\displaystyle OH}{|}}$$

$$\underset{\displaystyle RCCl}{\overset{\displaystyle O}{\|}} + H_2 \xrightarrow{\text{Pt, Pd, or Ni}} RCH_2OH$$

$$\underset{\displaystyle RCCl}{\overset{\displaystyle O}{\|}} + H_2 \xrightarrow{\begin{array}{c}\text{deactivated}\\\text{Pd}\end{array}} \underset{\displaystyle RCH}{\overset{\displaystyle O}{\|}}$$

2. Reduction of alkynes to alkenes (Section 17.1)

$$RC{\equiv}CR \xrightarrow[\substack{\text{Lindlar's}\\\text{catalyst}}]{H_2} \begin{array}{c}H\quad\ \ H\\ \diagdown\ \diagup \\ C{=}C\\ \diagup\quad\diagdown\\ R\qquad R\end{array}$$

$$RC{\equiv}CR \xrightarrow[\substack{NH_3\ (liq)}]{\text{Na or Li}} \begin{array}{c}H\qquad R\\ \diagdown\ \diagup \\ C{=}C\\ \diagup\quad\diagdown\\ R\qquad H\end{array}$$

3. Reduction of carbonyl compounds with reagents that donate hydride ion (Section 17.1)

$$\underset{\displaystyle RCH}{\overset{\displaystyle O}{\|}} \xrightarrow[\text{2. } H_2O]{\text{1. NaBH}_4} RCH_2OH$$

$$\underset{\displaystyle RCR}{\overset{\displaystyle O}{\|}} \xrightarrow[\text{2. } H_2O]{\text{1. NaBH}_4} \underset{\displaystyle RCHR}{\overset{\displaystyle OH}{|}}$$

$$\underset{\displaystyle RCOH}{\overset{\displaystyle O}{\|}} \xrightarrow[\text{2. } H_2O]{\text{1. LiAlH}_4} RCH_2OH$$

$$\overset{\overset{\displaystyle O}{\|}}{RCOR'} \xrightarrow[\text{2. H}_2\text{O}]{\text{1. LiAlH}_4} RCH_2OH \ + \ R'OH$$

$$\overset{\overset{\displaystyle O}{\|}}{RCNHR'} \xrightarrow[\text{2. H}_2\text{O}]{\text{1. LiAlH}_4} RCH_2NHR'$$

$$\overset{\overset{\displaystyle O}{\|}}{RCOR'} \xrightarrow[\text{2. H}_2\text{O}]{\text{1. DIBAH, }-80°} \overset{\overset{\displaystyle O}{\|}}{RCH} \ + \ R'OH$$

$$\overset{\overset{\displaystyle O}{\|}}{RCCl} \xrightarrow[\text{2. H}_2\text{O}]{\text{1. LiAl[OC(CH}_3)_3]_3\text{H, }-80°} \overset{\overset{\displaystyle O}{\|}}{RCH}$$

4. Oxidation of alcohols (Section 17.2)

$$\textbf{primary alcohols} \quad RCH_2OH \xrightarrow[\text{H}_2\text{SO}_4]{\text{KMnO}_4} \overset{\overset{\displaystyle O}{\|}}{RCH} \xrightarrow[\text{oxidation}]{\text{further}} \overset{\overset{\displaystyle O}{\|}}{RCOH}$$

$$RCH_2OH \xrightarrow[\text{dry CH}_2\text{Cl}_2]{\text{PCC}} \overset{\overset{\displaystyle O}{\|}}{RCH}$$

$$\textbf{secondary alcohols} \quad \overset{\overset{\displaystyle OH}{|}}{RCHR} \xrightarrow[\text{H}_2\text{SO}_4]{\text{KMnO}_4} \overset{\overset{\displaystyle O}{\|}}{RCR}$$

5. Oxidation of aldehydes and ketones (Section 17.3)

$$\textbf{aldehydes} \quad \overset{\overset{\displaystyle O}{\|}}{RCH} \xrightarrow[\text{H}_2\text{SO}_4]{\text{Na}_2\text{Cr}_2\text{O}_7} \overset{\overset{\displaystyle O}{\|}}{RCOH}$$

$$\overset{\overset{\displaystyle O}{\|}}{RCH} \xrightarrow[\text{2. H}_3\text{O}^+]{\text{1. Ag}_2\text{O, NH}_3} \overset{\overset{\displaystyle O}{\|}}{RCOH} \ + \ \underset{\substack{\text{metallic}\\\text{silver}}}{\text{Ag}}$$

$$\overset{\overset{\displaystyle O}{\|}}{RCH} \xrightarrow{\overset{\displaystyle O}{\overset{\|}{R'COOH}}} \overset{\overset{\displaystyle O}{\|}}{RCOH} \ + \ \overset{\overset{\displaystyle O}{\|}}{R'COH}$$

$$\textbf{ketones} \quad \overset{\overset{\displaystyle O}{\|}}{RCR} \xrightarrow{\overset{\displaystyle O}{\overset{\|}{R'COOH}}} \overset{\overset{\displaystyle O}{\|}}{RCOR} \ + \ \overset{\overset{\displaystyle O}{\|}}{R'COH}$$

6. Oxidation of alkenes (Sections 17.4, 17.6, 17.8)

$$\overset{\overset{\displaystyle R}{|}}{RC}{=}CHR' \xrightarrow[\text{2. Zn, H}_2\text{O}]{\text{1. O}_3} \overset{\overset{\displaystyle O}{\|}}{RCR} \ + \ \overset{\overset{\displaystyle O}{\|}}{R'CH}$$

$$\xrightarrow[\text{2. H}_2\text{O}_2]{\text{1. O}_3} \overset{\overset{\displaystyle O}{\|}}{RCR} \ + \ \overset{\overset{\displaystyle O}{\|}}{R'COH}$$

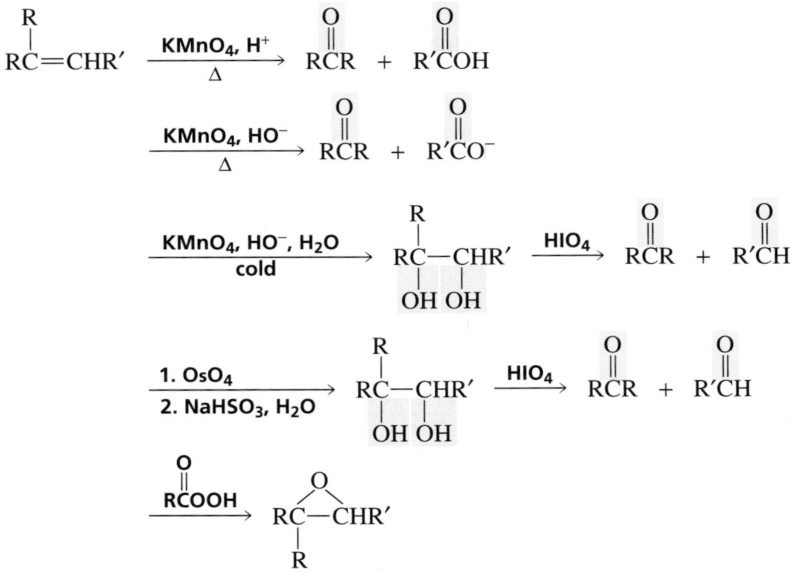

$$\underset{\substack{|\\R}}{RC}\!=\!CHR' \xrightarrow[\Delta]{\textbf{KMnO}_4,\ \textbf{H}^+} \underset{\substack{\|\\O}}{RCR} + \underset{\substack{\|\\O}}{R'COH}$$

$$\xrightarrow[\Delta]{\textbf{KMnO}_4,\ \textbf{HO}^-} \underset{\substack{\|\\O}}{RCR} + \underset{\substack{\|\\O}}{R'CO^-}$$

$$\xrightarrow[\textbf{cold}]{\textbf{KMnO}_4,\ \textbf{HO}^-,\ \textbf{H}_2\textbf{O}} \underset{\substack{|\\OH\ OH}}{\overset{R}{RC\!-\!CHR'}} \xrightarrow{\textbf{HIO}_4} \underset{\substack{\|\\O}}{RCR} + \underset{\substack{\|\\O}}{R'CH}$$

$$\xrightarrow[\textbf{2. NaHSO}_3,\ \textbf{H}_2\textbf{O}]{\textbf{1. OsO}_4} \underset{\substack{|\\OH\ OH}}{\overset{R}{RC\!-\!CHR'}} \xrightarrow{\textbf{HIO}_4} \underset{\substack{\|\\O}}{RCR} + \underset{\substack{\|\\O}}{R'CH}$$

$$\xrightarrow{\overset{\textbf{O}}{\overset{\|}{\textbf{RCOOH}}}} \underset{\substack{|\\R}}{\overset{O}{RC\!-\!CHR'}}$$

7. Oxidation of 1,2-diols (Section 17.5)

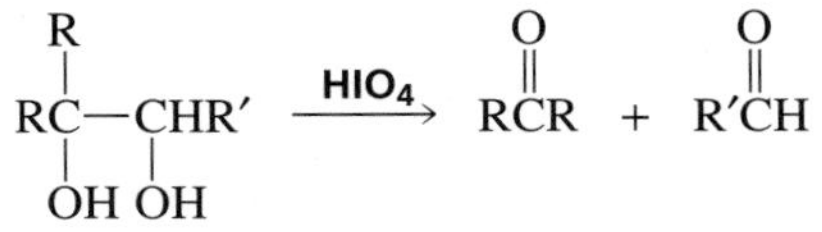

$$\underset{\substack{|\\OH\ OH}}{\overset{R}{RC\!-\!CHR'}} \xrightarrow{\textbf{HIO}_4} \underset{\substack{\|\\O}}{RCR} + \underset{\substack{\|\\O}}{R'CH}$$

8. Oxidation of alkynes (Section 17.7)

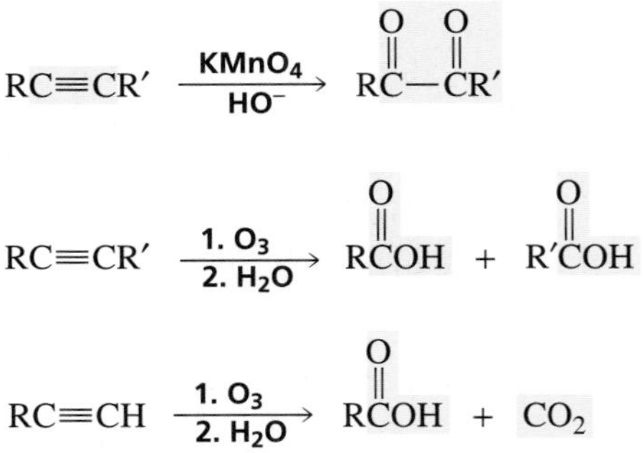

$$RC\!\equiv\!CR' \xrightarrow[\textbf{HO}^-]{\textbf{KMnO}_4} \underset{\substack{\|\quad\|\\O\quad O}}{RC\!-\!CR'}$$

$$RC\!\equiv\!CR' \xrightarrow[\textbf{2. H}_2\textbf{O}]{\textbf{1. O}_3} \underset{\substack{\|\\O}}{RCOH} + \underset{\substack{\|\\O}}{R'COH}$$

$$RC\!\equiv\!CH \xrightarrow[\textbf{2. H}_2\textbf{O}]{\textbf{1. O}_3} \underset{\substack{\|\\O}}{RCOH} + CO_2$$

9. Oxidation of hydroquinones/reduction of quinones (Section 17.9)

Baeyer–Villiger oxidation (page 788)
catalytic hydrogenation (page 779)

dissolving metal reduction (page 782)
epoxidation (page 797)

functional group interconversion
(page 801)

glycol (page 789)
molozonide (page 793)
oxidation (page 785)
oxidation–reduction reaction
   (page 777)

oxidative cleavage (page 791)
ozonide (page 793)
ozonolysis (page 792)
peroxyacid (page 788)
redox reaction (page 777)

reduction (page 785)
Rosenmund reduction (page 780)
Tollens test (page 788)
vicinal diol (page 789)
vicinal glycol (page 789)

PROBLEMS

26. Fill in the blank with "oxidized" or "reduced."
   a. Secondary alcohols are _____ to ketones.
   b. Acyl halides are _____ to aldehydes.
   c. Aldehydes are _____ to primary alcohols.
   d. Alkenes are _____ to aldehydes and/or ketones.
   e. Aldehydes are _____ to carboxylic acids.
   f. Alkenes are _____ to 1,2-diols.
   g. Alkenes are _____ to alkanes.

27. Give the products of the following reactions. Indicate whether each reaction is an oxidation or a reduction.

a. $CH_3CH_2CH_2CH_2CH_2OH$ $\xrightarrow[H_2SO_4]{Na_2Cr_2O_7}$

j. [benzaldehyde structure] $\xrightarrow[Raney\ Ni]{H_2}$

b. 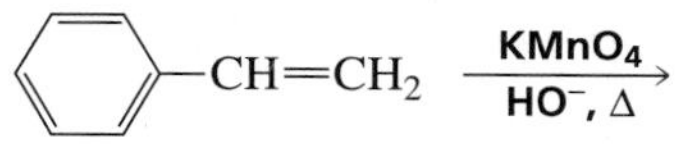 [phenyl]$-CH=CH_2$ $\xrightarrow[HO^-,\ \Delta]{KMnO_4}$

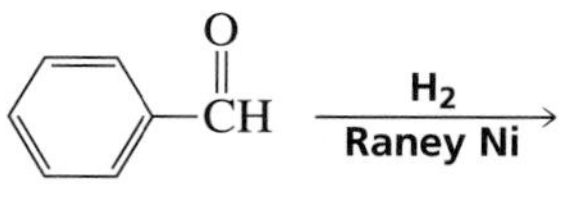

k. $CH_3CH_2CH_2C\equiv CCH_3$ $\xrightarrow[NH_3(liq)]{Na}$

c. $CH_3CH_2CH_2\overset{O}{\overset{\|}{C}}Cl$ $\xrightarrow[Pt]{H_2}$

l. $CH_3CH_2CH_2C\equiv CCH_3$ $\xrightarrow[2.\ H_2O]{1.\ O_3}$

d. $CH_3CH_2C\equiv CH$ $\xrightarrow{\begin{array}{l}1.\ \text{disiamylborane}\\2.\ H_2O_2,\ HO^-,\ H_2O\\3.\ LiAlH_4\\4.\ H_2O\end{array}}$

m. [phenyl]$-CH=CHCH_3$ $\xrightarrow[Pt]{H_2}$

e. $CH_3CH_2CH=CHCH_2CH_3$ $\xrightarrow[2.\ Zn,\ H_2O]{1.\ O_3}$

n. [cyclopentane]$=CH_2$ $\xrightarrow[2.\ (CH_3)_2S]{1.\ O_3}$

f. $CH_3CH_2CH_2\overset{O}{\overset{\|}{C}}NHCH_3$ $\xrightarrow[2.\ H_2O]{1.\ LiAlH_4}$

o. [cyclohexene] $\xrightarrow{\begin{array}{l}1.\ MMPP\\2.\ CH_3MgBr\\3.\ H_3O^+\end{array}}$

g. [benzaldehyde structure] $\xrightarrow{\overset{O}{\overset{\|}{R}COOH}}$

p. [cyclohexene] $\xrightarrow[HO^-,\ cold]{KMnO_4}$

h. [phenyl]$-\overset{O}{\overset{\|}{C}}OCHCH_3$ with CH_3 branch $\xrightarrow[2.\ H_2O]{1.\ LiAlH_4}$

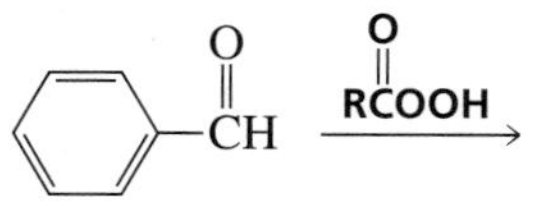

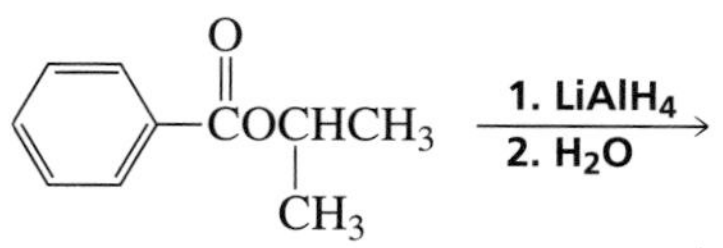

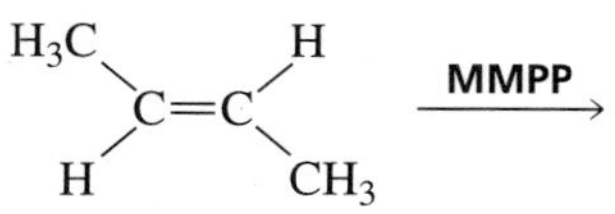

q. [cyclohexene] $\xrightarrow[HO^-,\ \Delta]{KMnO_4}$

i. $\underset{H}{\overset{H_3C}{>}}C=C\underset{CH_3}{\overset{H}{<}}$ $\xrightarrow{MMPP}$

r. [cyclohexene] $\xrightarrow[2.\ H_2O_2]{1.\ O_3}$

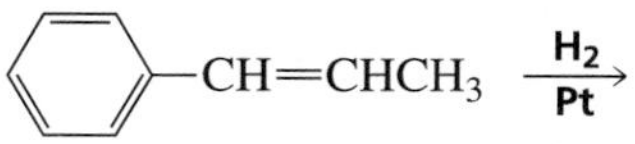

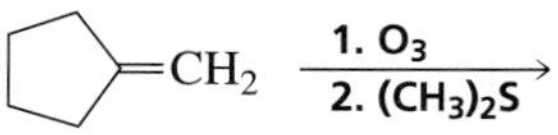

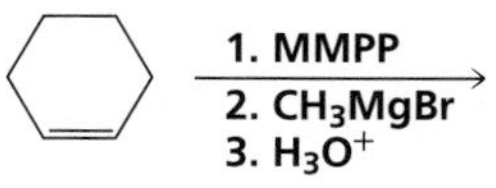

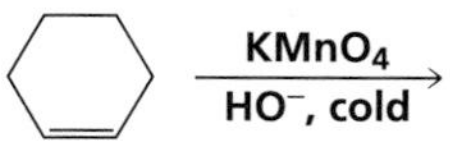

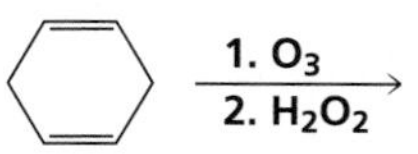

28. How could each of the following compounds be converted to $CH_3CH_2CH_2\overset{\overset{\displaystyle O}{\|}}{C}OH$?

a. $CH_3CH_2CH_2\overset{\overset{\displaystyle O}{\|}}{C}H$

b. $CH_3CH_2CH_2CH_2OH$

c. $CH_3CH_2CH_2CH_2Br$

d. $CH_3CH_2CH{=}CH_2$

29. Identify the alkene that would give each of the following products upon ozonolysis followed by treatment with hydrogen peroxide.

a. $CH_3CH_2CH_2\overset{\overset{\displaystyle O}{\|}}{C}OH \ + \ CH_3\overset{\overset{\displaystyle O}{\|}}{C}CH_3$

b. 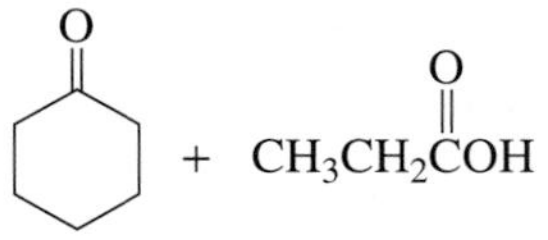 $+ \ CH_3CH_2\overset{\overset{\displaystyle O}{\|}}{C}OH$

c. $CH_3\overset{\overset{\displaystyle O}{\|}}{C}CH_2CH_2CH_2CH_2\overset{\overset{\displaystyle O}{\|}}{C}CH_2CH_3$

d. 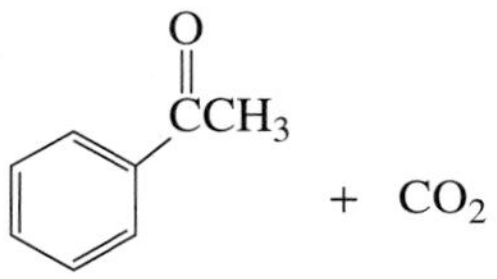 $+ \ CO_2$

e.

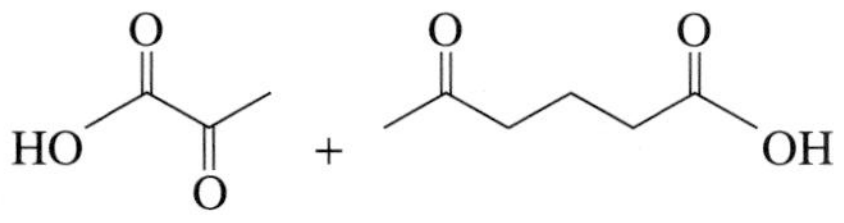

30. Fill in each box with the appropriate reagent.

a.

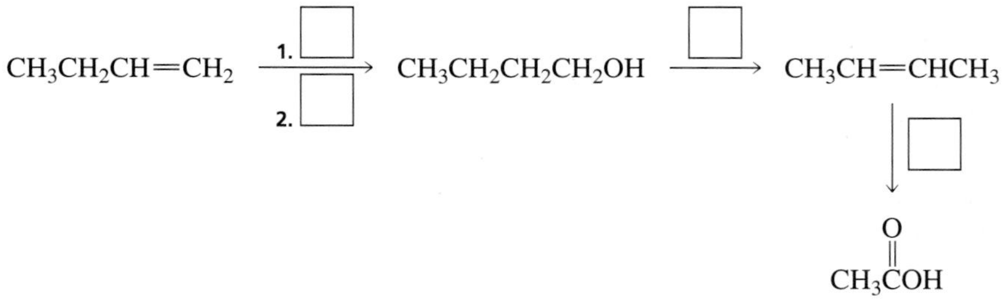

$CH_3CH_2CH{=}CH_2 \ \xrightarrow[\text{2.}\ \Box]{\text{1.}\ \Box} \ CH_3CH_2CH_2CH_2OH \ \xrightarrow{\Box} \ CH_3CH{=}CHCH_3$

$\Big\downarrow \Box$

$CH_3\overset{\overset{\displaystyle O}{\|}}{C}OH$

b.

$CH_3CH_2Br \ \xrightarrow{\Box} \ \Box \ \xrightarrow[\text{2.}\ \Box]{\text{1.}\ \Box} \ CH_3CH_2CH_2CH_2OH \ \xrightarrow{\Box} \ CH_3CH_2CH_2\overset{\overset{\displaystyle O}{\|}}{C}H$

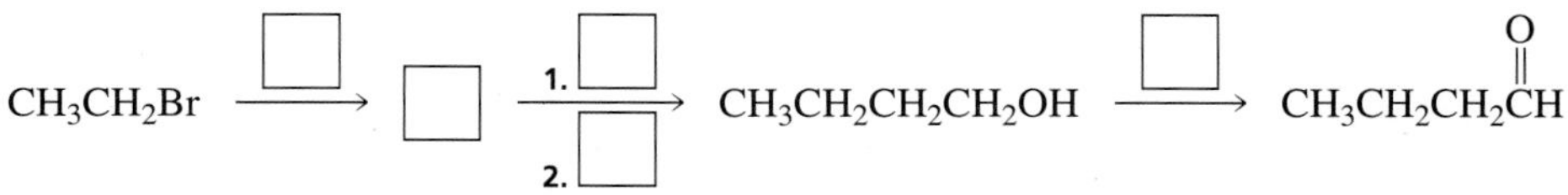

c.

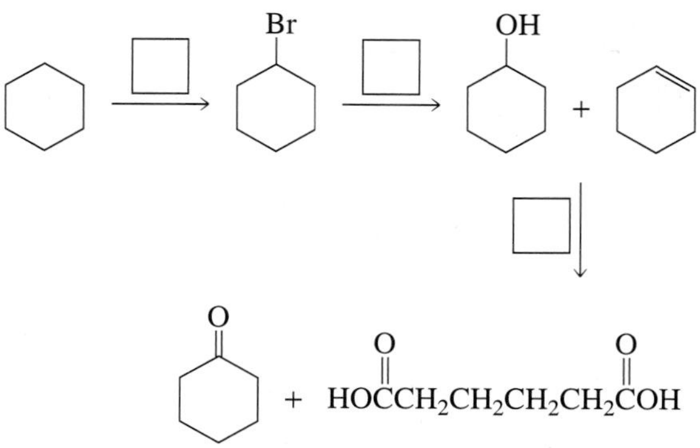

31. Describe how 1-butyne can be converted into each of the following compounds.

a.

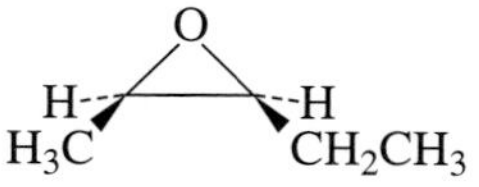

b.

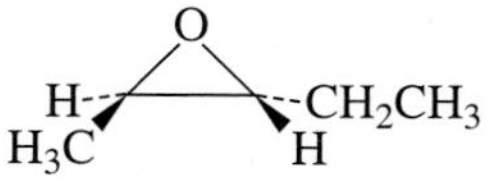

32. Show how each of the following compounds can be prepared from the given starting material.

a.

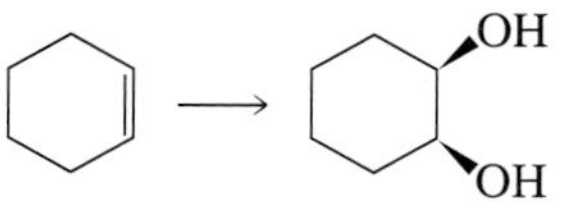

b.

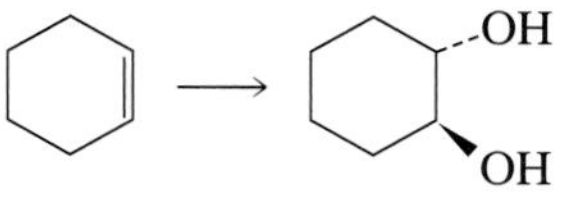

c.

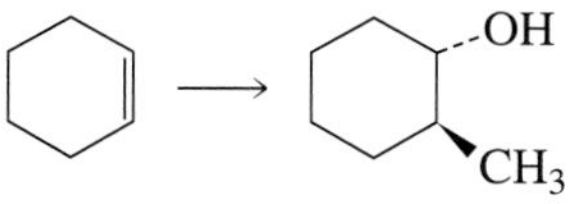

d.

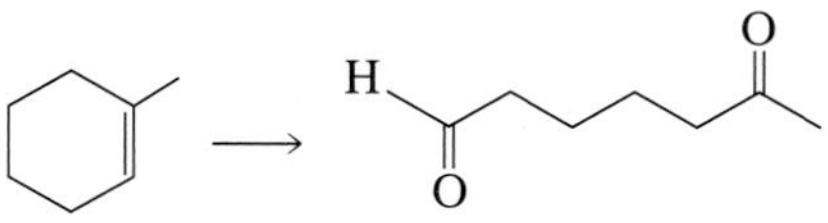

e.

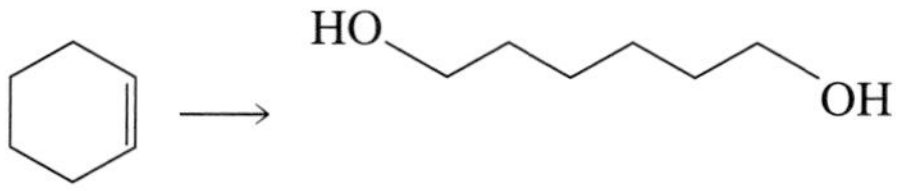

f.

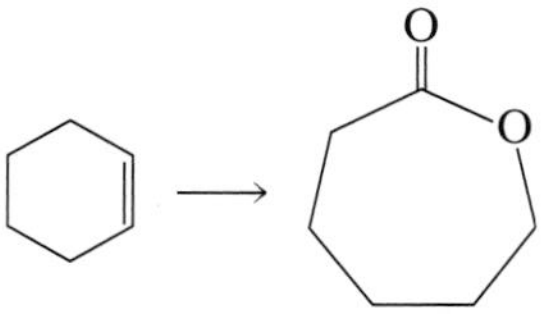

g.

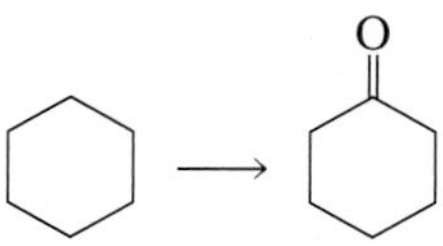

h.

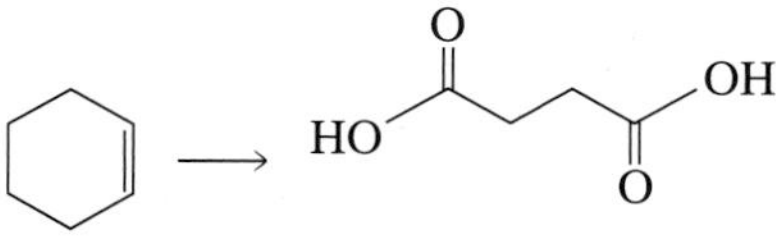

33. The ^{1}H NMR spectrum shown below is that of the product obtained when an unknown alkene reacts with ozone and the ozonolysis product is worked up under oxidizing conditions. Identify the alkene.

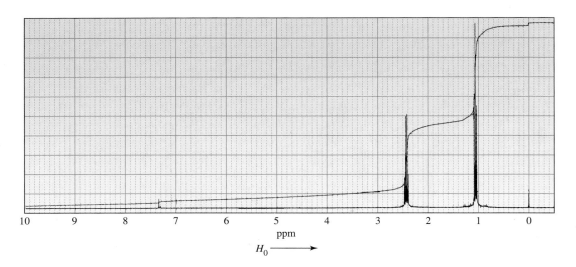

34. The rate of oxidation of 2-propanol by chromic acid is six times greater than the rate of oxidation of 2-deuterio-2-propanol by the same reagent. Explain.

35. Show how each of the following compounds could be prepared using the given starting material.

a. $CH_3CH_2\overset{\displaystyle O}{\overset{\displaystyle \|}{C}}H \longrightarrow CH_3CH_2\overset{\displaystyle O}{\overset{\displaystyle \|}{C}}OCH_2CH_2CH_3$

b. $CH_3CH_2CH_2CH_2OH \longrightarrow CH_3CH_2CH_2\overset{\displaystyle O}{\overset{\displaystyle \|}{C}}CH_2CH_3$

c.

d. $\longrightarrow HO\overset{\displaystyle O}{\overset{\displaystyle \|}{C}}CH_2CH_2CH_2CH_2\overset{\displaystyle O}{\overset{\displaystyle \|}{C}}OH$

e. $\longrightarrow HO\overset{\displaystyle O}{\overset{\displaystyle \|}{C}}CH_2CH_2CH_2CH_2\overset{\displaystyle O}{\overset{\displaystyle \|}{C}}CH_3$

36. It is known that HIO_4 cleaves A much more rapidly than it cleaves B. What does this tell you about the mechanism of the reaction?

A   B

37. Show how cyclohexylacetylene can be converted into each of the following compounds.

a. 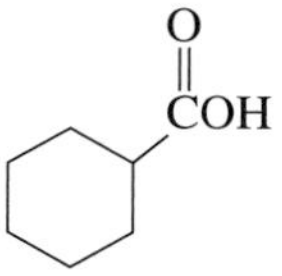 (cyclohexyl)—COH with =O

b. 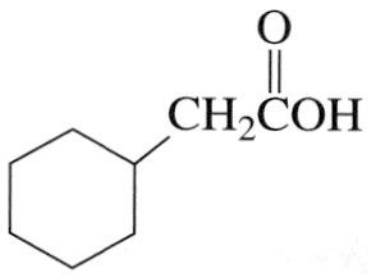 (cyclohexyl)—CH₂COH with =O

38. Catalytic hydrogenation of 0.5 g of a hydrocarbon at 25 °C consumed about 200 mL of H_2 under 1 atm of pressure. Reaction of the hydrocarbon with ozone followed by treatment with hydrogen peroxide gave one product, which was found to be a four-carbon carboxylic acid. Identify the hydrocarbon.

39. Tom Thumbs was asked to prepare the compounds listed below from the given starting materials. The reagents that he chose to use for each synthesis are shown.
   a. Which of his syntheses were successful?
   b. What products did he obtain from the other syntheses?
   c. In his unsuccessful syntheses, what reagents should he have used to obtain the desired product?

1.  $CH_3CH_2\overset{\underset{|}{CH_3}}{C}=CHCH_3 \xrightarrow[H_2SO_4]{KMnO_4} CH_3CH_2\overset{\underset{|}{CH_3}}{\underset{\underset{OH}{|}}{C}}-\overset{}{\underset{\underset{OH}{|}}{C}HCH_3}$

2. $CH_3CH_2\overset{\overset{O}{||}}{C}OCH_3 \xrightarrow[2.\ H_2O]{1.\ NaBH_4} CH_3CH_2CH_2OH + CH_3OH$

3. [cyclohexene with H₃C and CH₃ substituents] $\xrightarrow[2.\ HO^-]{1.\ MMPP}$ [cyclohexane with OH OH, H₃C CH₃ substituents]

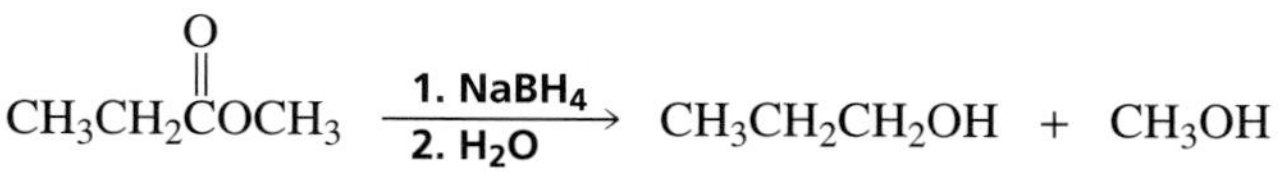

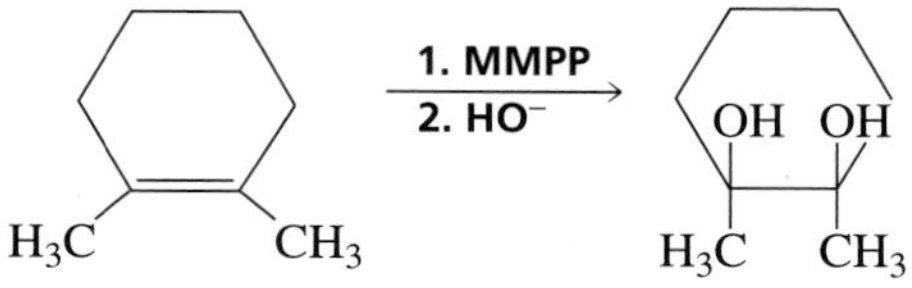

4. $CH_3CH=CH\overset{\overset{O}{||}}{C}Cl \xrightarrow[Pd/C]{excess\ H_2} CH_3CH_2CH_2CH_2OH$

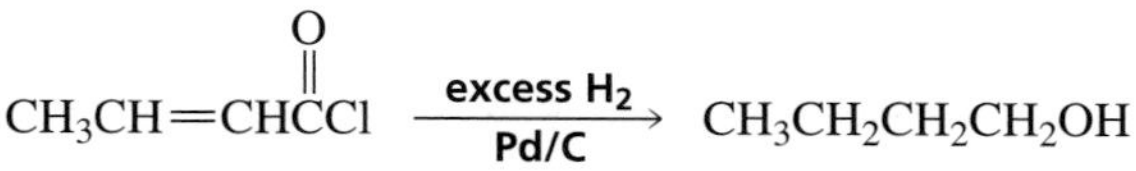

40. Diane Diol had worked for several days to prepare the compounds shown below. She carefully labeled them and went to lunch. To her horror, she found when she returned that the labels had fallen off the bottles and onto the floor. Gladys Glycol, the student at the next bench, told her that the diols could be easily distinguished by two experiments. All Diane had to do was determine which ones were optically active and how many products were obtained when each was treated with periodic acid. Diane did what Gladys suggested and found the following:

   1. Compounds A, E, and F are optically active, and B, C, and D are optically inactive.

   2. One product is obtained from the reaction of A, B, and D with periodic acid.

   3. Two products are obtained from the reaction of F with periodic acid.

   4. C and E do not react with periodic acid.

Will Diane be able to distinguish between the six diols and label them A to F with only this information? Label the structures.

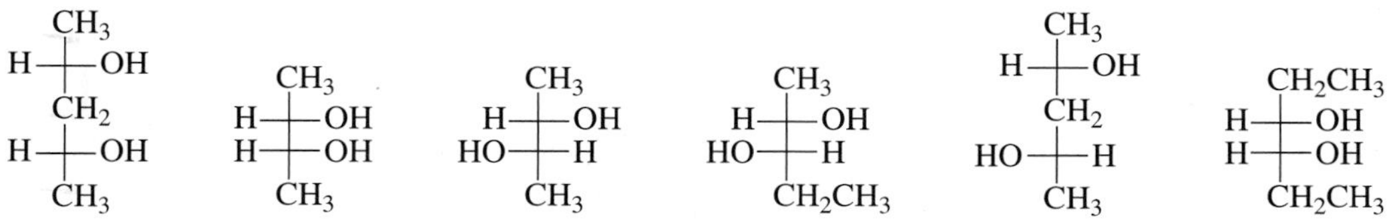

[Six Fischer projection structures:]

Structure 1: CH₃ / H—OH / CH₂ / H—OH / CH₃

Structure 2: CH₃ / H—OH / H—OH / CH₃

Structure 3: CH₃ / H—OH / HO—H / CH₃

Structure 4: CH₃ / H—OH / HO—H / CH₂CH₃

Structure 5: CH₃ / H—OH / CH₂ / HO—H / CH₃

Structure 6: CH₂CH₃ / H—OH / H—OH / CH₂CH₃

41. Show how propyl propionate could be prepared using allyl alcohol as the only source of carbon.

42. Compound A has a molecular formula of $C_5H_{12}O$ and is oxidized by an acidic solution of sodium dichromate to give compound B, which has a molecular formula of $C_5H_{10}O$. When compound A is heated with H_2SO_4, C and D are obtained. Considerably more D is obtained than C. Compound C reacts with O_3, followed by treatment with H_2O_2 to give two products: CO_2 and compound E, with molecular formula C_4H_8O. The same treatment of compound D also gives two products: compound F, with molecular formula C_3H_6O, and compound G, with molecular formula $C_2H_4O_2$. Give the structures of compounds A to G.

43. Primary amines can be synthesized by catalytic hydrogenation of the alkyl azide formed from the reaction of an alkyl halide with azide ion ($^-N_3$), a good nucleophile and a weak base. Why is this a better method of primary amine synthesis than the reaction of an alkyl halide with sodium amide ($Na^+{}^-NH_2$)?

$$RCH_2Br \;+\; \overset{-}{N}=\overset{+}{N}=\overset{-}{N} \longrightarrow RCH_2\overset{+}{N}=\overset{+}{N}=\overset{-}{N} \xrightarrow[\text{Pt}]{H_2} RCH_2NH_2 \;+\; N_2$$
$$\text{alkyl halide} \qquad \text{azide ion} \qquad\qquad \text{alkyl azide} \qquad\qquad \text{primary amine}$$
$$+\; Br^-$$

44. Show how the following compounds could be prepared using only the indicated carbon-containing reagents.

a. $\underset{\displaystyle\text{O}}{CH_3\overset{\displaystyle\text{O}}{\overset{\|}{C}}CH_3}$ from $CH_3\overset{\displaystyle CH_3}{\overset{|}{C}H}CH_3$

b. $CH_3CH=\overset{\displaystyle CH_3}{\overset{|}{C}}CH_3$ using propane as the only source of carbon

c. $CH_3\overset{\displaystyle O}{\overset{\|}{C}}\overset{\displaystyle CH_3}{\overset{|}{C}H}CH_3$ from propane and any molecule with two carbon atoms

d. $CH_3CH_2\overset{\displaystyle O}{\overset{\|}{C}}H$ from two molecules of ethane

45. A primary alcohol can be oxidized only as far as the aldehyde stage if the alcohol is first treated with tosyl chloride (TsCl) and the resulting tosylate is allowed to react with dimethyl sulfoxide (DMSO). Provide a mechanism for reaction of the tosylate with DMSO. (*Hint:* See Section 17.5.)

$$CH_3CH_2CH_2CH_2OH \xrightarrow[\text{pyridine}]{\text{TsCl}} CH_3CH_2CH_2CH_2OTs \xrightarrow{\text{DMSO}} CH_3CH_2CH_2\overset{\displaystyle O}{\overset{\|}{C}}H$$

46. Identify the alkene that gives each of the following products upon ozonolysis followed by treatment with dimethyl sulfide.

a.

b.

47. Propose a mechanism for the following enzyme-catalyzed reaction. (*Hint:* Notice that addition of the OH and Br groups does not follow Markovnikov's rule.)

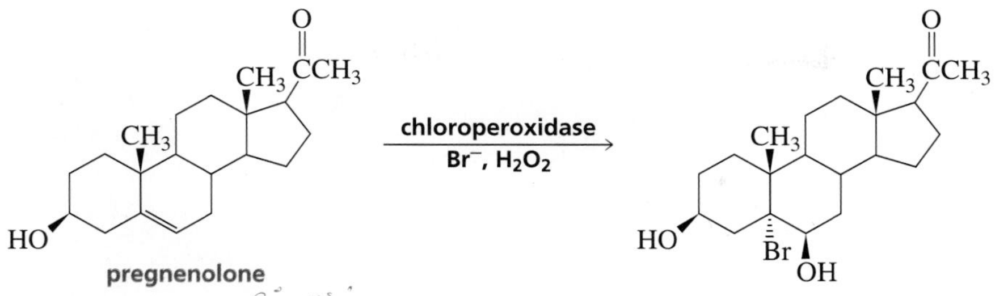

pregnenolone

48. Terpineol ($C_{10}H_{18}O$) is an optically active compound with one chirality center. It is used as an antiseptic. Reaction of terpineol with H_2/Pt forms an optically inactive compound ($C_{10}H_{20}O$). Heating the reduced compound in acid followed by ozonolysis and work-up under reducing conditions produces the compounds shown below. What is the structure of terpineol?

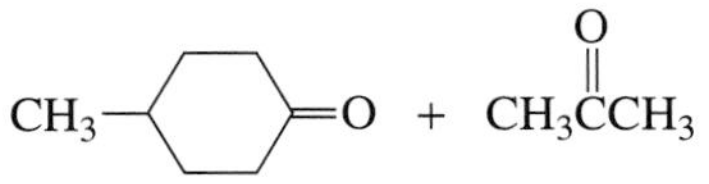

18

CARBONYL COMPOUNDS III: REACTIONS AT THE α-CARBON

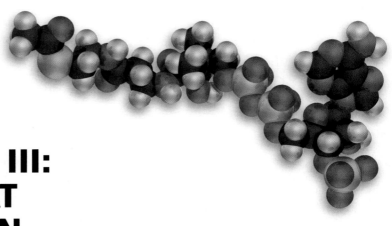

acetyl-CoA

W hen we studied the reactions of carbonyl compounds in Chapters 15 and 16, we saw that their site of reactivity is the partially positively charged carbonyl carbon that is attacked by nucleophiles.

Aldehydes, ketones, esters, and amides have a second site of reactivity. A hydrogen bonded to *a carbon adjacent to a carbonyl carbon* is sufficiently acidic to be removed by a strong base. This carbon is called an α-**carbon.** A hydrogen bonded to an α-carbon is called an α-**hydrogen.**

In the next section, we will find out why a hydrogen bonded to a carbon adjacent to a carbonyl carbon is acidic. We then will look at reactions resulting from this acidity. At the end of this chapter, we will see that a proton is not the only substituent that can be removed from an α-carbon: a carboxyl group bonded to a carbon adjacent to a carbonyl carbon can be removed as CO_2. Finally, we will look at synthetic schemes that take advantage of the ability to remove protons and carboxyl groups from α-carbons.

Hydrogen and carbon have similar electronegativities, which means that the electrons binding them together are shared almost equally by the two atoms. Consequently, a hydrogen bonded to a carbon is usually not acidic. This is particularly true for hydrogens bonded to sp^3 hybridized carbons because these carbons are the most similar to hydrogen in electronegativity (Section 5.9). The high pK_a of ethane (50) is evidence of the low acidity of hydrogens bonded to sp^3 hybridized carbons.

$$CH_3CH_3 \overset{\diagdown}{}$$
$$pK_a = 50$$

A hydrogen bonded to an sp^3 hybridized carbon that is adjacent to a carbonyl carbon is considerably more acidic than hydrogens bonded to other sp^3 hybridized carbons. For example, the pK_a of a hydrogen bonded to the α-carbon of an aldehyde or a ketone ranges from 16 to 20, and the pK_a of a hydrogen bonded to the α-carbon of an ester is about 25 (Table 18.1). A compound that contains a carbon bonded to a relatively acidic hydrogen is called a **carbon acid.**

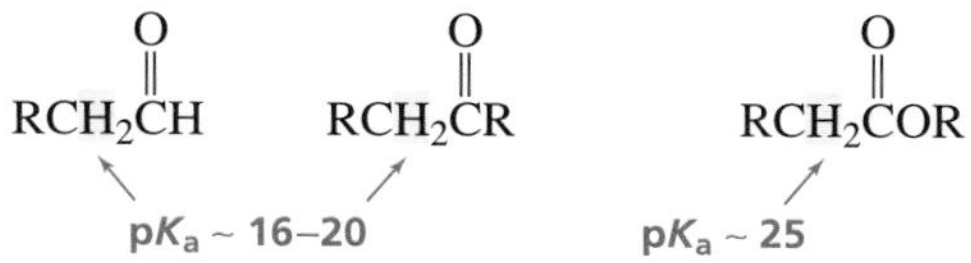

Why is a hydrogen bonded to a carbon adjacent to a carbonyl carbon so much more acidic than hydrogens bonded to other sp^3 hybridized carbons? It is because the base formed when a proton is removed from the α-carbon is more stable than

18.1
THE ACIDITY OF A HYDROGEN BONDED TO A CARBON ADJACENT TO A CARBONYL CARBON

TABLE 18.1	The pK_a Values of Some Carbon Acids		
	pK_a		pK_a
$\underset{\text{H}}{CH_2CN(CH_3)_2}$ (C=O)	30	$\underset{\text{H}}{N{\equiv}CCHC{\equiv}N}$	11.8
$\underset{\text{H}}{CH_2COCH_2CH_3}$ (C=O)	25	$\underset{\text{H}}{CH_3CCHCOCH_2CH_3}$ (C=O, C=O)	10.7
$\underset{\text{H}}{CH_2C{\equiv}N}$	25	$\underset{\text{H}}{C_6H_5CCHCCH_3}$ (C=O, C=O)	9.4
$\underset{\text{H}}{CH_2CCH_3}$ (C=O)	20	$\underset{\text{H}}{CH_3CCHCCH_3}$ (C=O, C=O)	8.9
$\underset{\text{H}}{CH_2CH}$ (C=O)	17	$\underset{\text{H}}{CH_3CCHCH}$ (C=O, C=O)	5.9
$\underset{\text{H}}{CH_3CHNO_2}$	8.6	$\underset{\text{H}}{O_2NCHNO_2}$	3.6

the base formed when a proton is removed from other sp^3 hybridized carbons, and the more stable the base, the easier it is to generate by removing a proton.

What accounts for the greater stability of the α-carbanion? When a proton is removed from ethane, the electrons left behind reside on a carbon. Because carbon is not an electronegative atom, a carbanion is unstable and therefore difficult to form. So the pK_a of its conjugate acid is very high.

$$CH_3CH_3 \;\rightleftharpoons\; CH_3\ddot{C}H_2^- + H^+$$

When a proton is removed from a carbon adjacent to a carbonyl carbon, two factors combine to increase the stability of the base that is formed. First, the electrons left behind when the proton is removed are delocalized, and electron delocalization increases the stability of a compound (Section 6.6). More importantly, the electrons are delocalized onto an oxygen, an atom that is better able to accommodate the electrons because oxygen is more electronegative than carbon.

contributing resonance structures

PROBLEM 1

The pK_a of propene is 42, which is greater than the pK_a of a carbonyl compound but less than the pK_a of an alkane. Explain.

Why are the α-hydrogens of aldehydes and ketones ($pK_a = 16-20$) more acidic than the α-hydrogens of esters ($pK_a = 25$)? The electrons left behind when an α-hydrogen is removed from an ester are not as readily delocalized onto the carbonyl oxygen as are the electrons left behind when an α-hydrogen is removed from an aldehyde or a ketone. That is because a pair of nonbonding electrons on the oxygen of the OR group of the ester can also be delocalized onto the carbonyl oxygen, so the two pairs of electrons compete for delocalization onto oxygen.

contributing resonance structures

Hydrogens bonded to the α-carbon of nitroalkanes, nitriles, and tertiary amides are also more acidic than the hydrogens of ethane (Table 18.1), because, in each case, the electrons left behind when the proton is removed can be delocalized onto an atom that is more electronegative than carbon.

$CH_3CH_2NO_2$	$CH_3CH_2C{\equiv}N$	$CH_3\overset{\overset{\displaystyle O}{\|}}{C}N(CH_3)_2$
nitroethane	propanenitrile	*N,N*-dimethylacetamide
$pK_a = 8.6$	$pK_a = 26$	$pK_a = 30$

If the α-carbon is between two carbonyl groups, the acidity of an α-hydrogen is even greater (Table 18.1). For example, an α-hydrogen of ethyl acetoacetate, a compound with an α-carbon between a ketone carbonyl group and an ester carbonyl group, has a pK_a of 10.7. An α-hydrogen of acetylacetone, a compound with an α-carbon between two ketone carbonyl groups, has a pK_a of 8.9. Ethyl aceto-acetate is classified as a **β-keto ester** because the ester has a carbonyl group at the β-position. Acetylacetone is a **β-diketone.**

$$CH_3-\overset{\overset{\displaystyle O}{\|}}{C}-CH_2-\overset{\overset{\displaystyle O}{\|}}{C}-OCH_2CH_3 \qquad CH_3-\overset{\overset{\displaystyle O}{\|}}{C}-CH_2-\overset{\overset{\displaystyle O}{\|}}{C}-CH_3$$

ethyl 3-oxobutyrate
ethyl acetoacetate
a β-keto ester
pK_a = 10.7

2,4-pentanedione
acetylacetone
a β-diketone
pK_a = 8.9

The increased acidity of α-hydrogens bonded to carbons flanked by two car-bonyl groups is due to the fact that the electrons left behind when the proton is removed can be delocalized onto two oxygen atoms. β-Diketones have lower pK_a values than β-keto esters because electrons are more readily delocalized onto ketone carbonyl groups than they are onto ester carbonyl groups.

$$CH_3-\overset{\overset{\displaystyle :\ddot{O}:}{\|}}{C}-CH=\overset{\overset{\displaystyle :\ddot{O}:^-}{|}}{C}-CH_3 \longleftrightarrow CH_3-\overset{\overset{\displaystyle :\ddot{O}:}{\|}}{C}-\underset{\cdot\cdot}{C}H-\overset{\overset{\displaystyle :\ddot{O}:}{\|}}{C}-CH_3 \longleftrightarrow CH_3-\overset{\overset{\displaystyle :\ddot{O}:^-}{|}}{C}=CH-\overset{\overset{\displaystyle :\ddot{O}:}{\|}}{C}-CH_3$$

contributing resonance structures for acetylacetone anion

PROBLEM 2◆

Give an example of:

a. a β-keto nitrile.

b. a β-diester.

PROBLEM 3◆

List the compounds in each of the following groups in order of decreasing acidity.

a. $CH_2{=}CH_2$ 3 CH_3CH_3 4 $CH_3\overset{\overset{\displaystyle O}{\|}}{C}H$ 1 $HC{\equiv}CH$ 2

b. $CH_3\overset{\overset{\displaystyle O}{\|}}{C}CH_2\overset{\overset{\displaystyle O}{\|}}{C}CH_3$ 1 $CH_3O\overset{\overset{\displaystyle O}{\|}}{C}CH_2\overset{\overset{\displaystyle O}{\|}}{C}OCH_3$ 3 $CH_3\overset{\overset{\displaystyle O}{\|}}{C}CH_2\overset{\overset{\displaystyle O}{\|}}{C}OCH_3$ 2 $CH_3\overset{\overset{\displaystyle O}{\|}}{C}CH_3$ 4

c. [ring structure with NCH$_3$] 3   [ring structure with O] 2   [ring structure] 1

A ketone exists in equilibrium with its enol tautomer (Section 5.6). The two consti-tutional isomers are **tautomers** because they differ in the location of a double bond and a hydrogen.

18.2
KETO–ENOL
TAUTOMERISM

$$RCH_2-\overset{\displaystyle O}{\overset{\displaystyle \|}{C}}-R \rightleftharpoons RCH=\overset{\displaystyle OH}{\overset{\displaystyle |}{C}}-R$$

keto tautomer **enol tautomer**

For most ketones, the enol tautomer is much less stable than the keto tautomer. For example, an aqueous solution of acetone exists as an equilibrium mixture of more than 99.9% keto tautomer and less than 0.1% enol tautomer.

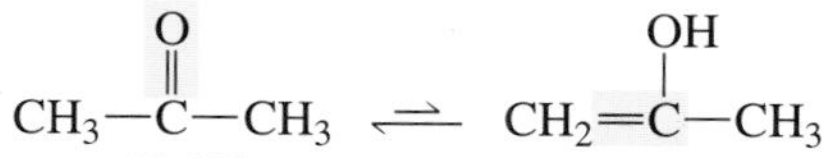

$$CH_3-\overset{\displaystyle O}{\overset{\displaystyle \|}{C}}-CH_3 \rightleftharpoons CH_2=\overset{\displaystyle OH}{\overset{\displaystyle |}{C}}-CH_3$$

> 99.9%   < 0.1%
keto tautomer   **enol tautomer**

The fraction of enol tautomer is considerably greater for a β-diketone because its enol tautomer is stabilized by hydrogen bonding and by conjugation of the carbon–carbon double bond with the second carbonyl group.

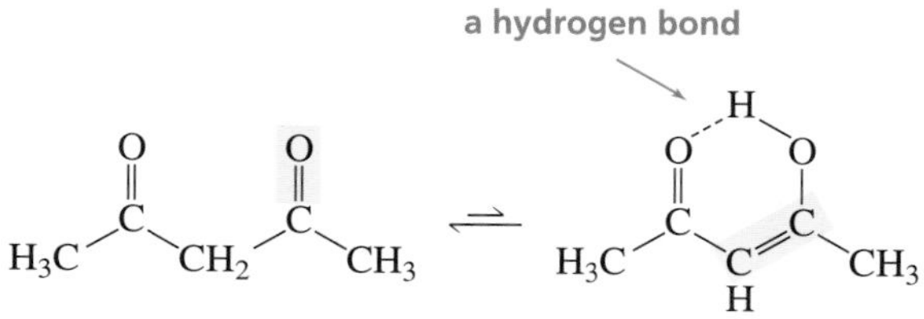

a hydrogen bond

85%   15%
keto tautomer   **enol tautomer**

Phenol is unusual in that its enol tautomer is more stable than its keto tautomer, because the enol tautomer is aromatic but the keto tautomer is not.

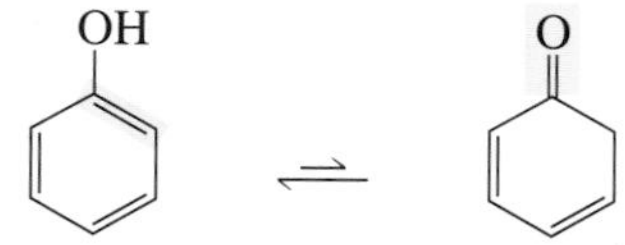

enol tautomer   **keto tautomer**

PROBLEM 4

While 15% of acetylacetone exists as the enol tautomer in water, 92% is the enol tautomer in hexane. Explain why this is so.

Now that we know that a hydrogen on a carbon adjacent to a carbonyl carbon is acidic, we can understand the mechanism for the interconversion of keto and enol tautomers first encountered in Chapter 5. **Keto–enol interconversion** is also called **keto–enol tautomerization** or **enolization.** The interconversion of the tautomers can be catalyzed by either acids or bases.

In a basic solution, hydroxide ion removes a proton from the α-carbon of the keto tautomer and the electrons are delocalized onto oxygen, forming an enolate ion. Protonation on oxygen forms the enol tautomer; protonation on carbon reforms the keto tautomer.

base-catalyzed keto–enol interconversion

$$RCH\!-\!\underset{\underset{H}{|}}{\overset{\overset{\ddot{O}:}{||}}{C}}\!-\!R \;\rightleftharpoons\; RCH\!=\!\overset{:\ddot{O}:^{-}}{\underset{}{C}}\!-\!R \;\rightleftharpoons\; RCH\!=\!\overset{:\ddot{O}H}{\underset{}{C}}\!-\!R \;+\; HO^{-}$$

keto tautomer   an enolate ion   enol tautomer

HÖ:⁻

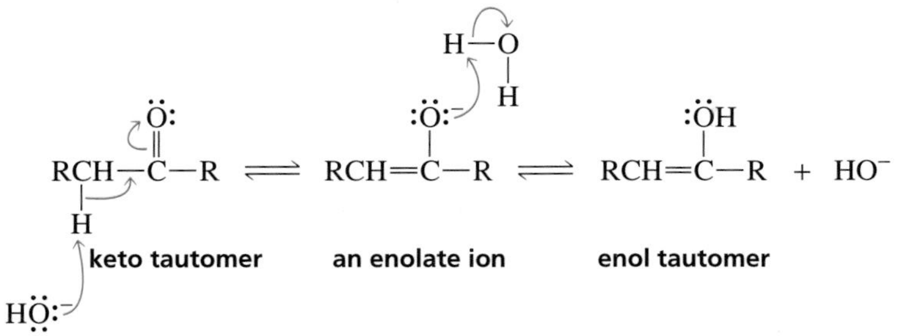

In an acidic solution, the carbonyl oxygen of the keto tautomer is protonated and water removes a proton from the α-carbon, forming the enol.

acid-catalyzed keto–enol interconversion

$$RCH_2\!-\!\overset{\overset{\ddot{O}:}{||}}{C}\!-\!R \;\underset{-H^{+}}{\overset{H^{+}}{\rightleftharpoons}}\; RCH\!-\!\underset{\underset{H}{|}}{\overset{\overset{\overset{+}{\ddot{O}H}}{||}}{C}}\!-\!R \;\rightleftharpoons\; RCH\!=\!\overset{:\ddot{O}H}{\underset{}{C}}\!-\!R \;+\; H_3O^{+}$$

keto tautomer   enol tautomer

H₂Ö:

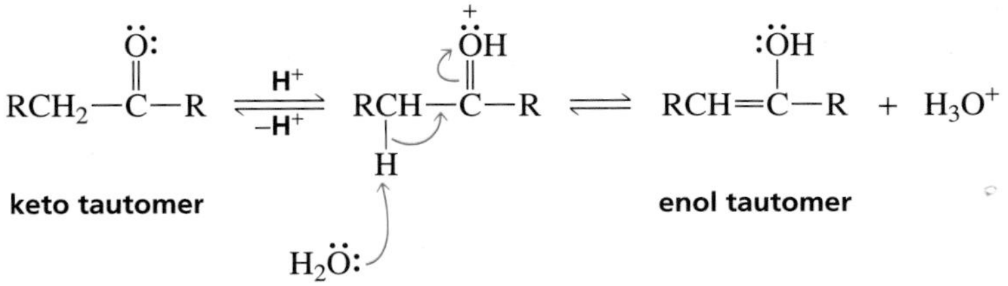

Notice that the steps are reversed in the base- and acid-catalyzed reactions. In the base-catalyzed reaction, the first step is removal of an α-proton and the second step is protonation of the oxygen. In the acid-catalyzed reaction, the first step is protonation of the oxygen and the second step is removal of an α-proton.

PROBLEM 5◆

Draw the enol tautomers for each of the following compounds. For those compounds that have more than one enol tautomer, indicate which is more stable.

a. 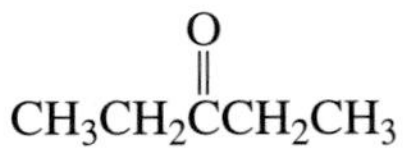 $CH_3CH_2\overset{\overset{O}{||}}{C}CH_2CH_3$

b. (phenyl)–$\overset{\overset{O}{||}}{C}CH_3$

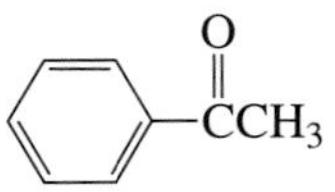

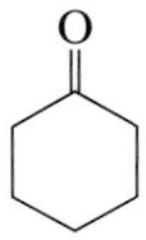

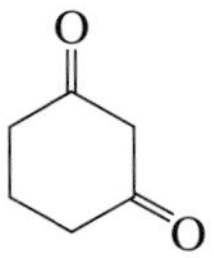

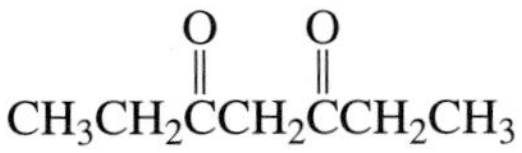

c. (cyclohexanone)

d. (cyclohexane-1,3-dione)

e. $CH_3CH_2\overset{\overset{O}{||}}{C}CH_2\overset{\overset{O}{||}}{C}CH_2CH_3$

f. (phenyl)–$CH_2\overset{\overset{O}{||}}{C}CH_3$

The anion formed when a base removes a hydrogen from a carbon adjacent to a carbonyl carbon is a good nucleophile. Because the α-carbanion and the enolate ion are resonance contributors, the anion has two sites that can react with an electrophile—the electron-rich carbon and the electron-rich oxygen. The anion is an

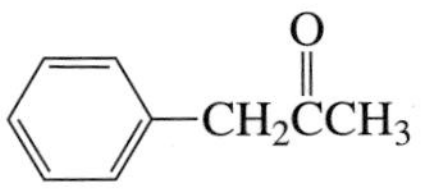

18.3 REACTIVITY CONSIDERATIONS

example of an ambident nucleophile (*ambi* is Latin for "both"; *dent* is Latin for "teeth"). An **ambident nucleophile** is a nucleophile with two nucleophilic sites ("two teeth").

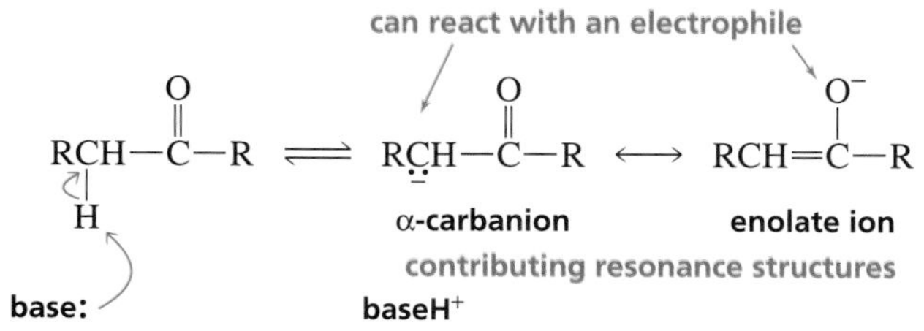

Which nucleophilic site (C or O) will react with an electrophile depends on the electrophile and on the reaction conditions. Protonation occurs preferentially on oxygen, because the resonance hybrid more closely resembles the enolate since it is the more stable resonance contributor (the negative charge is on a more electronegative atom). However, when the electrophile is something other than a proton, carbon is more likely to be the nucleophile.

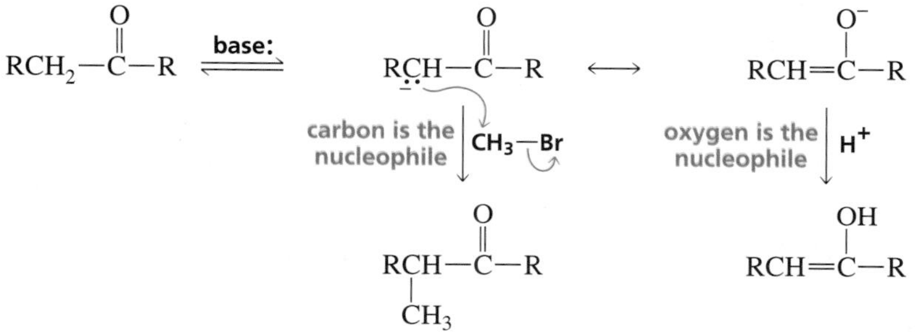

Most of the reactions in this chapter follow the same mechanism. A base removes an α-proton, forming an anion that reacts with an electrophile. The overall reaction is an α-**substitution reaction.** As various α-substitution reactions are discussed, notice that they differ only in the nature of the base and the electrophile. (Although the α-carbanion is in resonance with the enolate ion, we will frequently show only the α-carbanion in the mechanisms, and we will refer to the resonance hybrid as the enolate.)

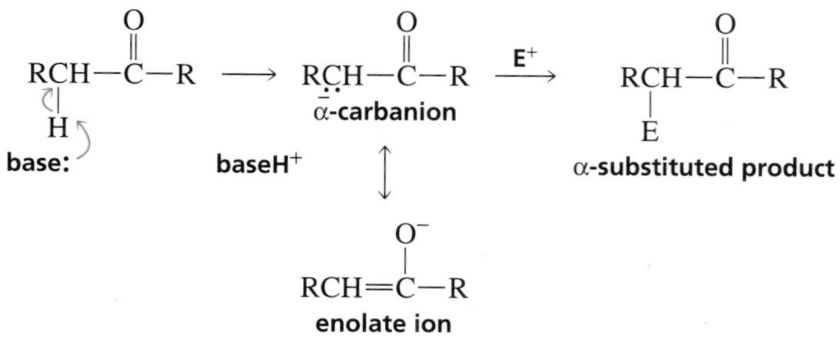

An α-substitution reaction can also take place by an acid-catalyzed pathway. Protonation of the carbonyl oxygen followed by removal of an α-proton forms an enol. Nucleophilic attack of the enol on the electrophile and loss of a proton from the carbonyl oxygen gives the α-substituted product.

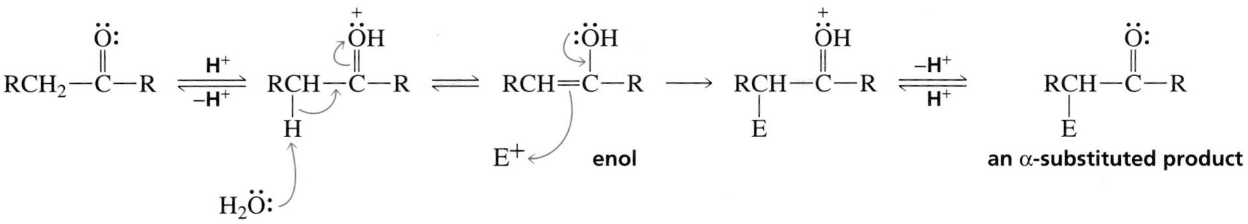

Notice the similarity between keto–enol interconversion and α-substitution. Actually, keto–enol interconversion is an α-substitution reaction with H^+ serving as the electrophile.

PROBLEM 6

Explain why the aldehyde hydrogen is not exchanged for deuterium.

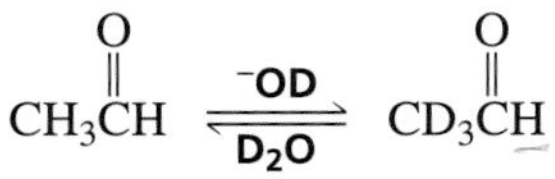

If Br_2, Cl_2, or I_2 is added to an acidic solution of an aldehyde or a ketone, a halogen will replace one of the α-hydrogens of the carbonyl compound.

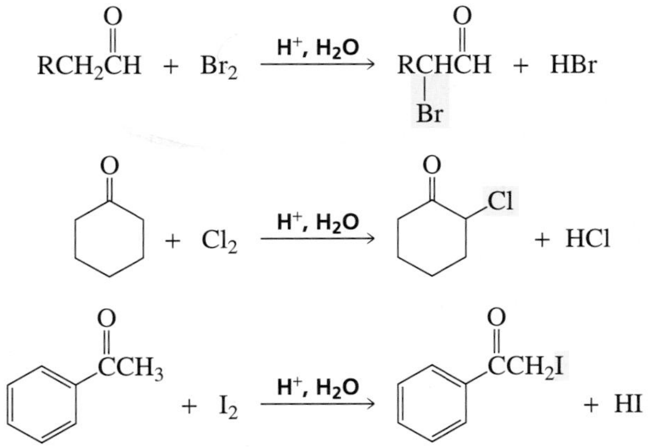

18.4
HALOGENATION OF THE α-CARBON OF ALDEHYDES AND KETONES

In the first step of this acid-catalyzed reaction, the carbonyl oxygen is protonated. Water is the base that removes a proton from the α-carbon, forming an enol that attacks an electrophilic bromine. The carbonyl oxygen loses a proton, because the pH of the solution is greater than the pK_a of the protonated α-halogenated ketone (Section 1.19).

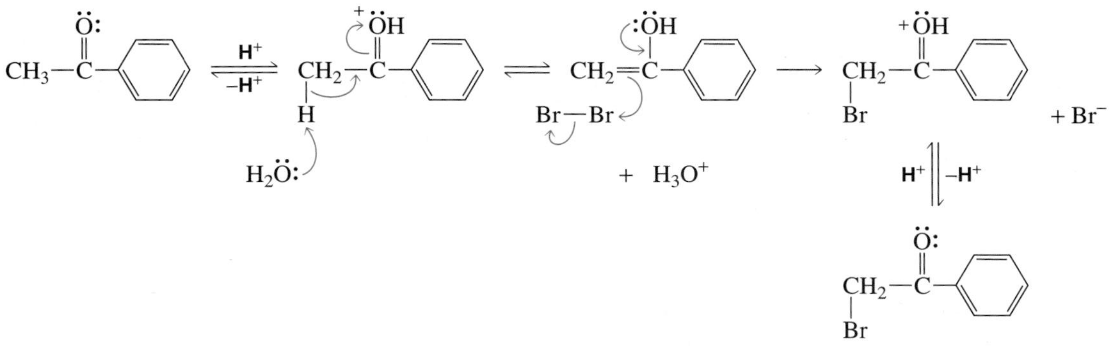

PROBLEM 7

How would you prepare the following compounds from the given starting materials?

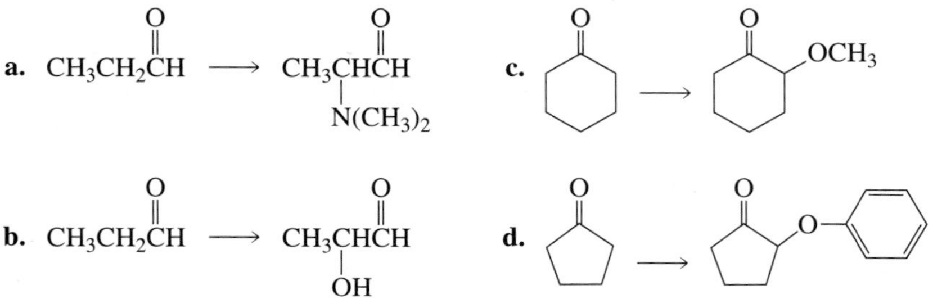

a. $CH_3CH_2CH(=O) \longrightarrow CH_3CHCH(=O)$ with $N(CH_3)_2$

b. $CH_3CH_2CH(=O) \longrightarrow CH_3CHCH(=O)$ with OH

c. (cyclohexanone) $\longrightarrow$ (cyclohexanone with OCH_3 at α-carbon)

d. (cyclopentanone) $\longrightarrow$ (cyclopentanone with O-phenyl at α-carbon)

If excess Br_2, Cl_2, or I_2 is added to a basic solution rather than to an acidic solution of an aldehyde or a ketone, the halogen will replace all the α-hydrogens.

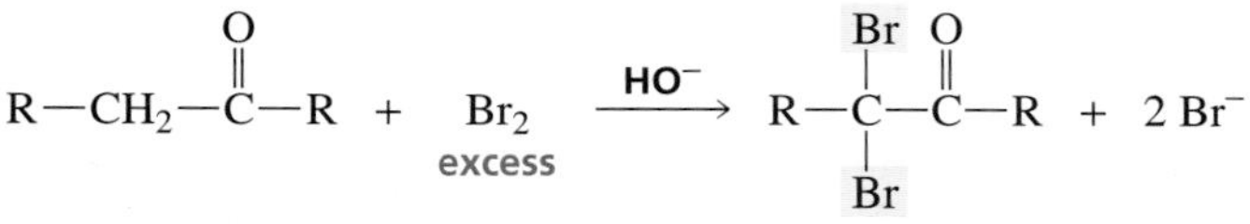

The first step in the mechanism for this base-promoted reaction is removal of a proton from the α-carbon by hydroxide ion. The second step is nucleophilic attack of the enolate on Br_2. These two steps are repeated until all the α-hydrogens are replaced by bromine.

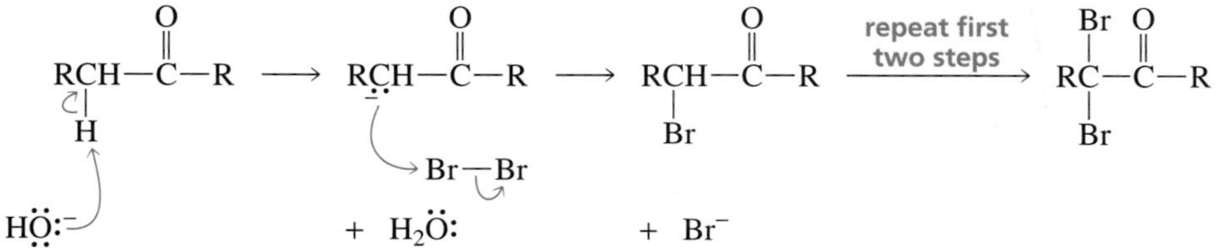

Carl Magnus von Hell (1849–1926) *was born in Germany. He studied with H. Fehling at the University of Stuttgart and with Richard Erlenmeyer (1825–1909) at the University of Munich. In 1883, he succeeded Erlenmeyer. Von Hell synthesized $C_{60}H_{122}$, which was the largest known alkane at the time. He reported the HVZ reaction in 1881, and the reaction was independently confirmed by both Volhard and Zelinski in 1887.*

In the presence of excess base and excess halogen, a methyl ketone will be converted into a carboxylic acid. After the trihalo-substituted ketone is formed, hydroxide ion attacks the carbonyl carbon. Since the trihalomethyl anion is a weaker base than hydroxide ion (the pK_a of CHI_3 is 14, the pK_a of H_2O is 15.7), the trihalomethyl group will be the group eliminated from the tetrahedral intermediate. The reaction is called a haloform reaction because one of the products is haloform—$CHCl_3$ (chloroform), $CHBr_3$ (bromoform), or CHI_3 (iodoform).

the haloform reaction

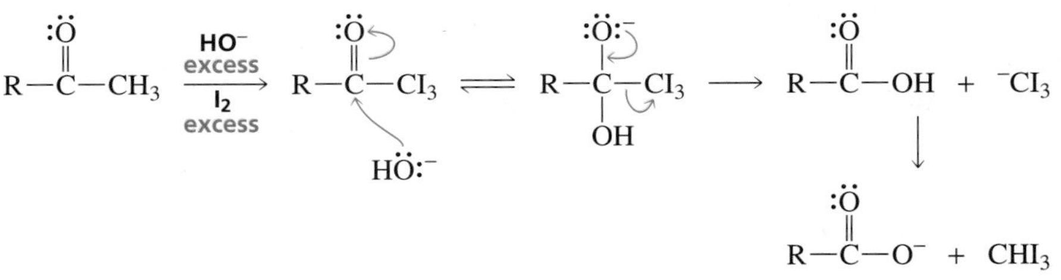

Jacob Volhard (1834–1910) *was also born in Germany. Brilliant but lacking direction, he was sent by his parents to England to be with August Hofmann (Section 10.11), a family friend. After working with Hofmann, he became a professor of chemistry—first at the University of Munich, then at the University of Erlangen, and later at the University of Halle. He was the first to synthesize sarcosine and creatine.*

PROBLEM 8 ◆

Base-promoted bromination of a given ketone takes place at the same rate as base-promoted chlorination. What does this tell you about the mechanism of the reaction?

Carboxylic acids do not undergo substitution reactions at the α-carbon because the α-hydrogens of carboxylic acids are even less acidic than the α-hydrogens of esters. However, if a carboxylic acid is treated with Br_2 and PBr_3, bromination at the α-carbon occurs. (Red phosphorus can be used in place of PBr_3 because P and Br_2 react to form PBr_3.) This reaction is known as the **Hell–Volhard–Zelinski reaction** or, more simply, as the **HVZ reaction.** We will see when we look at the mechanism of the reaction that α-substitution occurs because an acyl bromide, rather than a carboxylic acid, undergoes substitution.

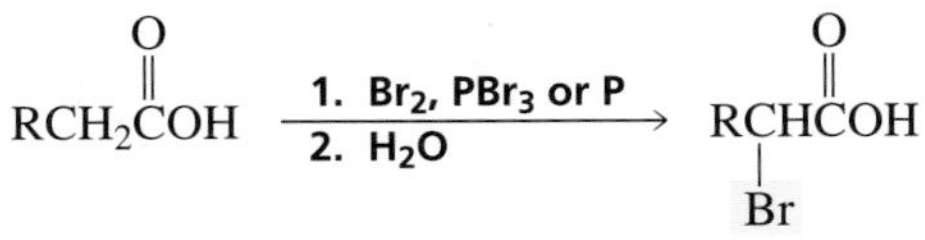

18.5
HALOGENATION OF THE α-CARBON OF CARBOXYLIC ACIDS: THE HELL–VOLHARD–ZELINSKI REACTION

The first step in the HVZ reaction is conversion of the carboxylic acid into an acyl bromide. (In Section 11.2 we saw that PBr_3 replaces an OH with a Br.) The acyl bromide is in equilibrium with its enol. Bromination of the enol followed by loss of a proton gives the α-brominated acyl bromide, which is hydrolyzed to the α-brominated carboxylic acid.

mechanism for the Hell–Volhard–Zelinski reaction

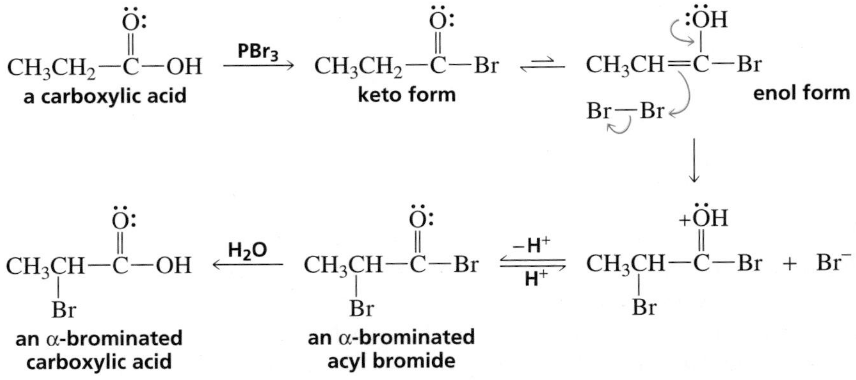

Nikolai Dimitrievich Zelinski (1861–1953) *was born in Moldavia. He was a professor of chemistry at the University of Moscow. In 1911 he left the university to protest the firing of the entire administration by the Ministry of Education. He went to St. Petersburg, where he directed the laboratory of the Ministry of Finances. In 1917, after the revolution, he returned to the University of Moscow.*

18.6
SYNTHESIS OF
α,β-UNSATURATED
CARBONYL
COMPOUNDS

Once a bromine has been introduced into the α-position of a carbonyl compound, an α,β-unsaturated carbonyl compound can be prepared by means of an elimination reaction. The yield of this reaction is not very high because of the competing substitution reaction (as well as a competing aldol addition reaction, Section 18.10).

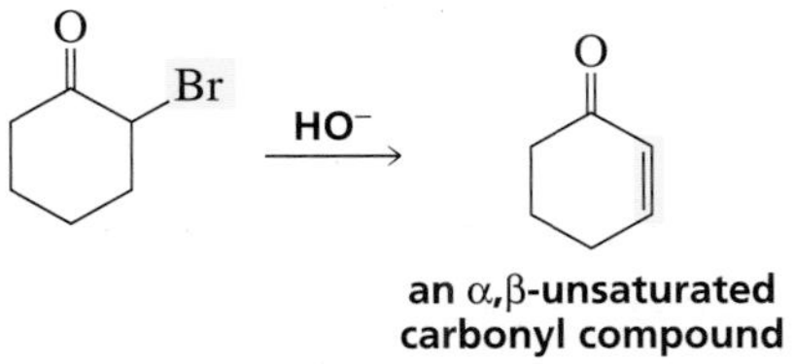

an α,β-unsaturated
carbonyl compound

Much better yields of α,β-unsaturated carbonyl compounds can be obtained if the carbonyl compound is selenenylated rather than brominated. Subsequent elimination of the selenenyl group forms the α,β-unsaturated carbonyl compound.

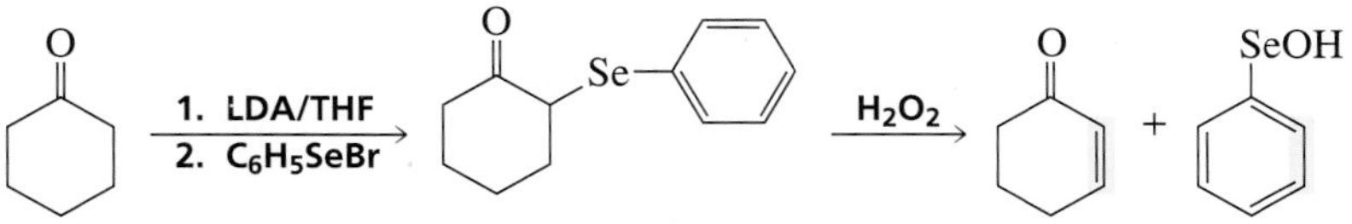

In the first step of the **selenenylation reaction,** an α-hydrogen is removed by lithium diisopropylamide (LDA). Benzeneselenenyl bromide (C_6H_5SeBr) is then added to the solution. In the following steps:

- the benzeneselenenyl bromide is attacked by the enolate, displacing bromide ion.
- the α-phenylseleno ketone is oxidized to a selenoxide by hydrogen peroxide. [This is similar to the oxidation of an amine to an amine oxide (Section 11.11).]
- the selenoxide forms the α,β-unsaturated ketone by an intramolecular elimination reaction. [This is similar to a Cope reaction (Section 11.11).]

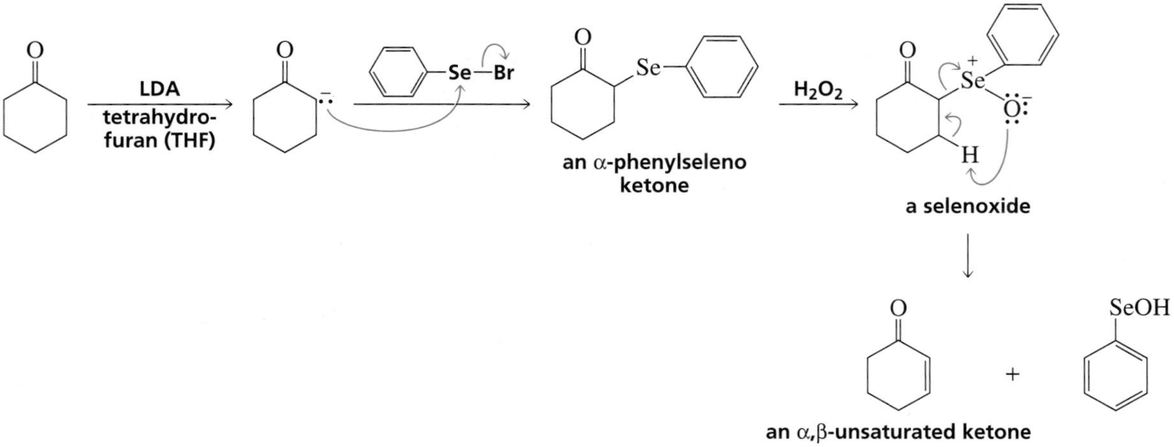

an α-phenylseleno
ketone

a selenoxide

an α,β-unsaturated ketone

In addition to making α,β-unsaturated ketones, selenenylation also can be used to prepare α,β-unsaturated esters and α,β-unsaturated nitriles in good yield. The yield of α,β-unsaturated aldehydes, however, is poor because of the tendency for aldehydes to undergo aldol addition reactions (Section 18.10).

$$CH_3CH_2CH_2\overset{\overset{\displaystyle O}{\|}}{C}OCH_3 \quad \xrightarrow[\begin{array}{l}\textbf{1. LDA/THF}\\\textbf{2. C}_6\textbf{H}_5\textbf{SeBr}\\\textbf{3. H}_2\textbf{O}_2\end{array}]{} \quad CH_3CH=CHCOCH_3$$

an α,β-unsaturated ester

$$\text{(phenyl)}-CH_2CH_2C\equiv N \quad \xrightarrow[\begin{array}{l}\textbf{1. LDA/THF}\\\textbf{2. C}_6\textbf{H}_5\textbf{SeBr}\\\textbf{3. H}_2\textbf{O}_2\end{array}]{} \quad \text{(phenyl)}-CH=CHC\equiv N$$

an α,β-unsaturated nitrile

The amount of carbonyl compound converted to enolate depends on the pK_a of the carbonyl compound and the base used to remove the α-proton. For example, when hydroxide ion (the pK_a of its conjugate acid is 15.7) is used to remove an α-proton from cyclohexanone ($pK_a = 17$), only a small amount of the carbonyl compound is converted into enolate because hydroxide ion is a weaker base than the base being formed. When lithium diisopropylamide (LDA) is used to remove the α-proton [the pK_a of its conjugate acid (DIA, or diisopropylamine) is about 35], essentially all the carbonyl compound is converted to enolate because LDA is a much stronger base than the base being formed (Section 1.17).

**18.7
USING LITHIUM
DIISOPROPYLAMIDE
(LDA) TO FORM
AN ENOLATE**

$$\text{(cyclohexanone)} + HO^- \;\rightleftharpoons\; \text{(cyclohexenolate)} + H_2O$$

$pK_a = 17$ < 0.1% $pK_a = 15.7$

$$\text{(cyclohexanone)} + LDA \;\longrightarrow\; \text{(cyclohexenolate)} + DIA$$

$pK_a = 17$ ~100% $pK_a = 35$

Therefore, LDA is the base of choice for those reactions that require the carbonyl compound to be completely converted to enolate before it reacts with an electrophile. For example, when we use LDA as the base in the selenenylation of a ketone such as cyclohexanone, only monoselenenylation occurs because all the cyclohexanone has been converted to enolate before benzeneselenenyl bromide is added to the solution. If we were to use HO^- as the base, only a small amount of cyclohexanone would be immediately selenenylated. A proton could subsequently be removed either from cyclohexanone or from the α-phenylseleno ketone, resulting in a mixture of mono- and diselenenylated products.

A problem with using a nitrogen base to remove a proton is that such a base can also react as a nucleophile and attack the carbonyl carbon (Section 16.7). However, the two bulky alkyl substituents bonded to the nitrogen of LDA make it difficult for the nitrogen to get close enough to the carbonyl carbon to attack it. Consequently, LDA is a strong base but a poor nucleophile; therefore, its rate of removal of an α-hydrogen is much greater than its rate of attack on a carbonyl carbon. LDA is easily prepared by adding butyllithium to diisopropylamine in THF at $-78\ °C$.

$$\underset{\textbf{diisopropylamine}}{\overset{\overset{\displaystyle CH_3\;\; CH_3}{|\quad\quad|}}{CH_3CHNHCHCH_3}} + \underset{\textbf{butyllithium}}{CH_3CH_2CH_2\overset{-}{C}H_2\overset{+}{Li}} \xrightarrow[\textbf{-78 °C}]{\textbf{THF}} \underset{\textbf{lithium diisopropylamide}}{\overset{\overset{\displaystyle CH_3\;\; CH_3}{|\quad\quad|}}{\underset{\underset{\displaystyle Li^+}{}}{CH_3CHNCHCH_3}}} + CH_3CH_2CH_2CH_3$$

PROBLEM 9

How could the following compounds be prepared from a carbonyl compound with no carbon–carbon double bonds?

a. $CH_3CH{=}CHCCH_2CH_2CH_3$ (with O double bond on the C)

b. 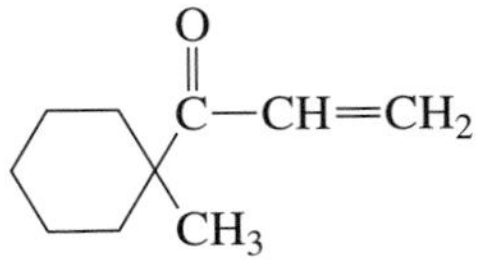 (cyclohexane ring with) $\overset{O}{\overset{\|}{C}}{-}CH{=}CH_2$ and CH_3

18.8
ALKYLATION OF THE α-CARBON OF CARBONYL COMPOUNDS

Alkylation of the α-carbon of a carbonyl compound is an important reaction because it gives us another way to form a carbon–carbon bond. Alkylation is carried out by first removing a proton from the α-carbon with a strong base such as LDA and then adding the appropriate alkyl halide.

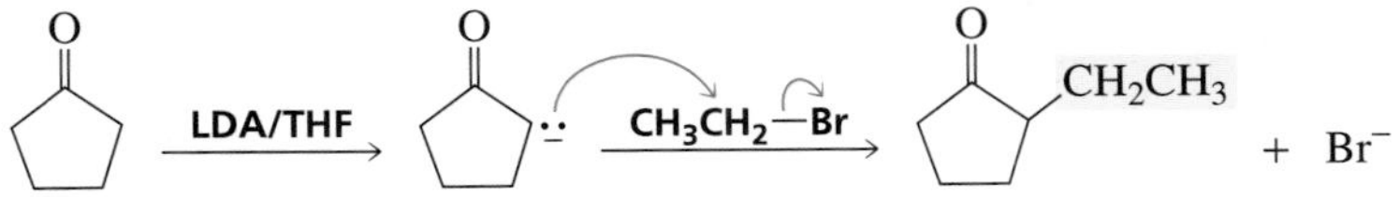

Ketones, esters, and nitriles can be alkylated at the α-carbon in this way. Aldehydes give poor yields of α-alkylated products.

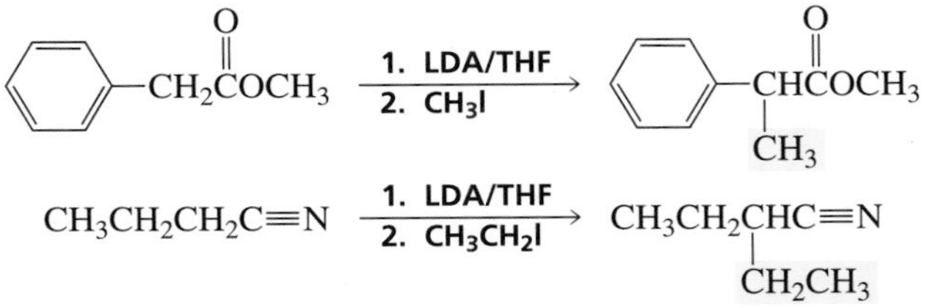

Alkylation of unsymmetrical ketones can form two different products because both α-carbons can be alkylated. For example, methylation of 2-methylcyclohexanone can form both 2,6-dimethylcyclohexanone and 2,2-dimethylcyclohexanone. The relative amounts of the two products depend on the reaction conditions: the kinetic product is the major product under mild conditions, and the thermodynamic product is the major product under more vigorous conditions (Section 7.7).

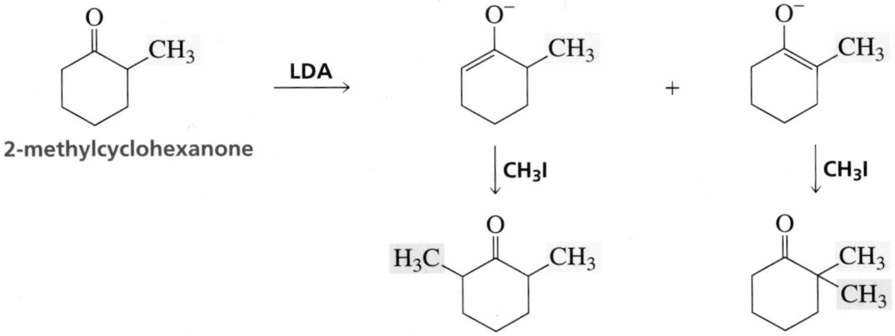

2-methylcyclohexanone

2,6-dimethylcyclohexanone
kinetic product

2,2-dimethylcyclohexanone
thermodynamic product

2,6-Dimethylcyclohexanone is the kinetic product because the α-proton that is removed to make the enolate intermediate that leads to this product is more accessible and slightly more acidic. So 2,6-dimethylcyclohexanone is the major product if the reaction is carried out at −78 °C.

2,2-Dimethylcyclohexanone is the thermodynamic product because its enolate intermediate has the more substituted double bond, making it the more stable intermediate. (Substitution increases enolate stability for the same reason that substitution increases alkene stability, Section 3.19). Therefore, 2,2-dimethylcyclohexanone is the major product if the reaction is carried out under conditions that make enolate formation reversible (high temperature and a protic solvent so that the kinetic enolate can be reprotonated).

The less substituted α-carbon can be alkylated—without having to control the conditions to make certain that the reaction does not become reversible—by first making the *N,N*-dimethylhydrazone of the ketone.

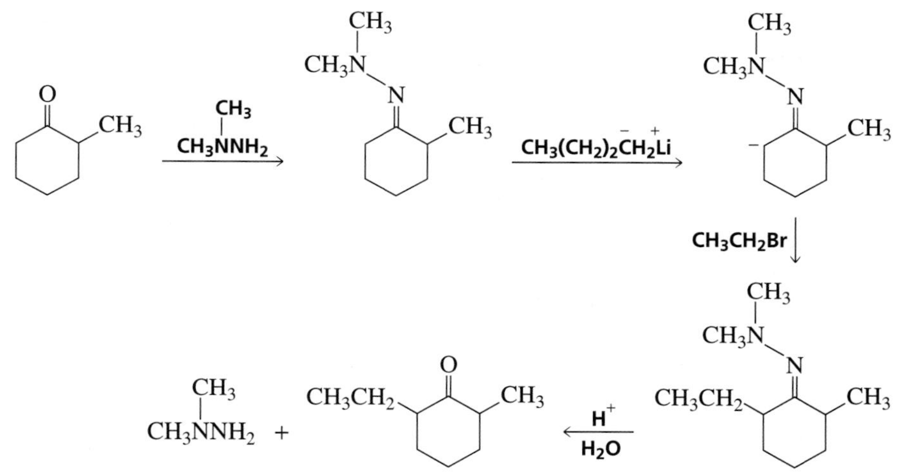

The *N,N*-dimethylhydrazone will form so that the dimethylamino group is pointing away from the more substituted α-carbon. The nitrogen of the dimethylamino group then directs the base to the less substituted carbon by coordinating with the lithium ion. Butyllithium is the base commonly employed in this reaction. Hydrolysis of the hydrazone reforms the ketone.

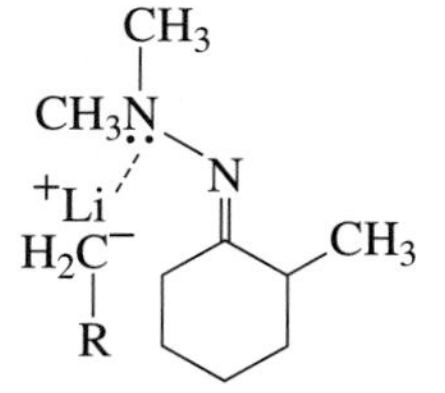

What compound is formed when cyclohexanone is shaken with NaOD in D₂O for several hours?

Alkylation of an α-carbon works best if the alkyl halide used in the reaction is a primary alkyl halide and does not work at all if it is a tertiary alkyl halide. Explain.

Herman Kolbe (1818–1884)
and **Rudolph Schmitt**
(1830–1898) *were born in*
Germany. Kolbe was a pro-
fessor at the universities of
Marburg and Leipzig. Schmitt
received a Ph.D. from the
University of Marburg and was
a professor at the University of
Dresden. Kolbe discovered
how to prepare aspirin in
1859. Schmitt modified the
synthesis in 1885, making it
available in large quantities at
a low price.

PROBLEM 12◆

How could each of the following compounds be prepared from the given material?

a.

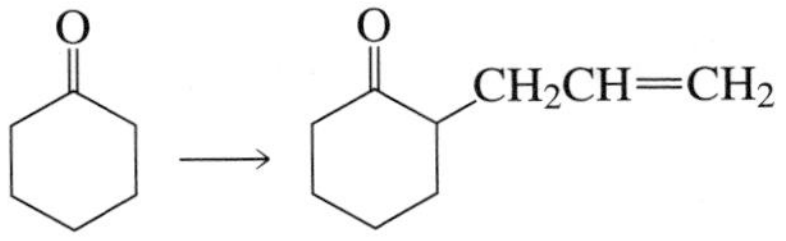

b. 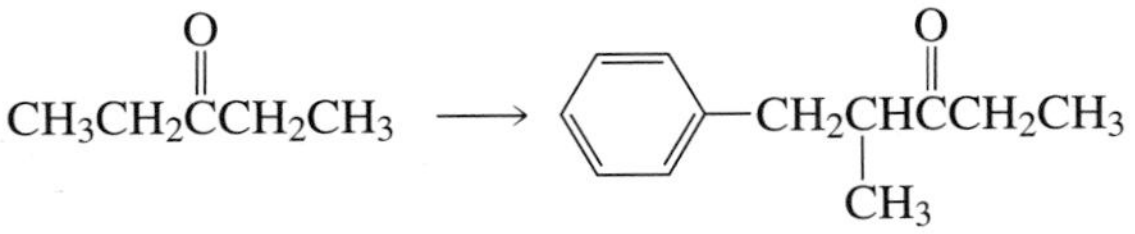 $CH_3CH_2CCH_2CH_3 \longrightarrow$

We have seen that when an aldehyde or a ketone reacts with a secondary amine, an enamine is formed (Section 16.7).

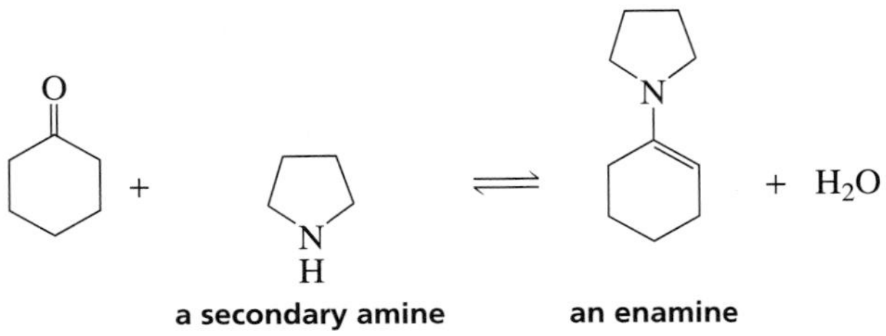

a secondary amine   **an enamine**

Enamines react with electrophiles in the same way that enolates do.

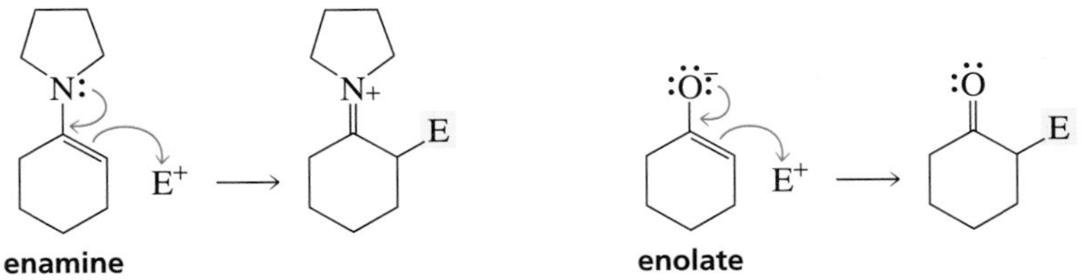

enamine   **enolate**

This means that electrophiles can be added to the α-carbon of an aldehyde or a ketone by first converting the carbonyl compound to an enamine by treating the carbonyl compound with a secondary amine (Section 16.7), adding the electrophile, and then hydrolyzing the imine back to the ketone.

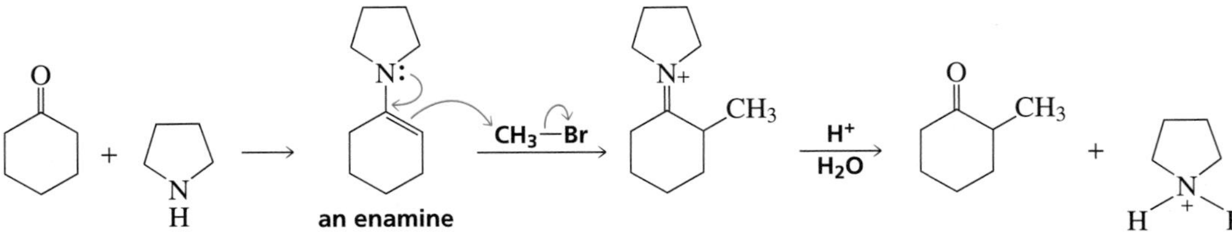

an enamine

Because the alkylation step is an S_N2 reaction, only primary alkyl halides or methyl halides should be used.

An advantage of using an enamine intermediate to alkylate an aldehyde or a ketone is that only the monoalkylated product is formed.

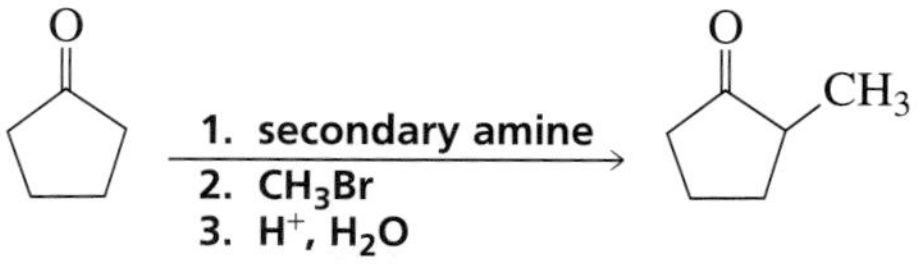

When a carbonyl compound is alkylated directly, dialkylated and O-alkylated products can also be formed.

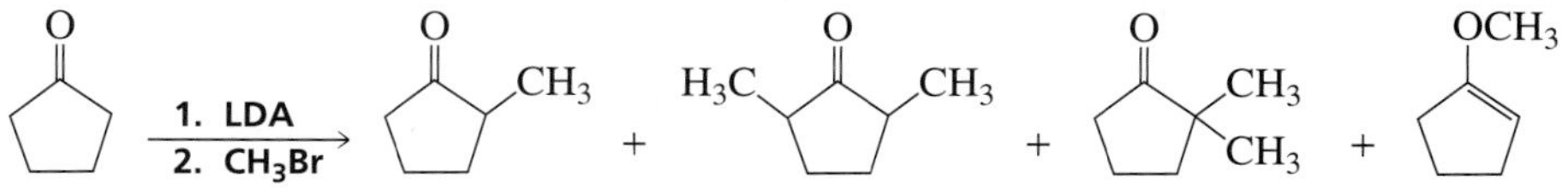

THE SYNTHESIS OF ASPIRIN

The first step in the industrial synthesis of aspirin is known as the **Kolbe–Schmitt carboxylation reaction.** The phenolate ion reacts with carbon dioxide under pressure to form *o*-hydroxybenzoic acid, also known as salicylic acid. Acetylation of salicylic acid forms acetylsalicylic acid (aspirin).

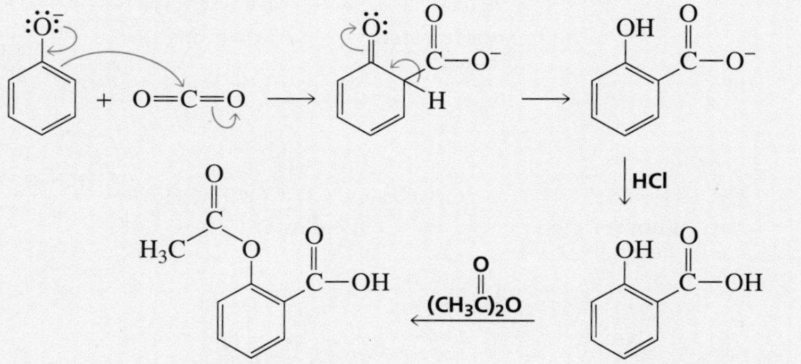

acetylsalicylic acid
aspirin

salicylic acid
o-hydroxybenzoic acid

During World War I, the American subsidiary of Bayer Co. bought as much phenol as they could from the international market with the knowledge that they could eventually convert it all into aspirin. This left little phenol available for other countries to purchase for the synthesis of 2,4,6-trinitrophenol, a common explosive.

PROBLEM 13

Describe how the following compounds could be prepared using an enamine intermediate.

a. 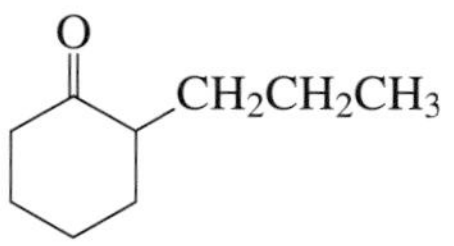 $CH_2CH_2CH_3$

b. 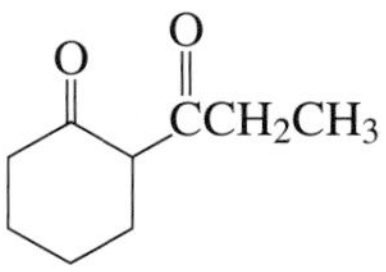 CCH_2CH_3

In Section 16.14 we saw that nucleophiles react with α,β-unsaturated carbonyl compounds, forming either direct addition products or conjugate addition products. Nucleophiles that are strong bases, unless they are bulky, add preferentially to the

**18.9
THE MICHAEL
REACTION**

carbonyl carbon, forming direct addition products. Weak bases and bulky strong bases add preferentially to the β-carbon, forming conjugate addition products.

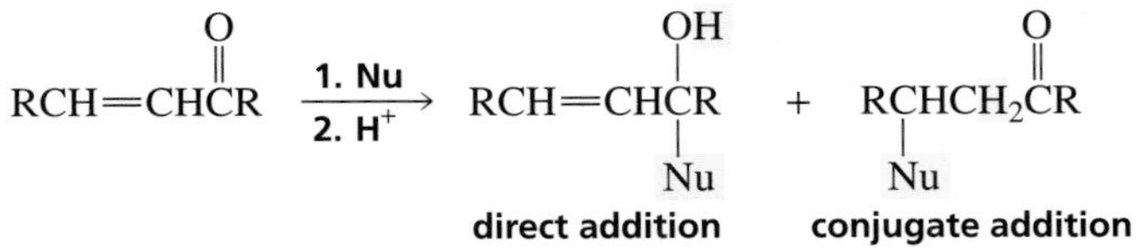

$$RCH\!=\!CHCR \xrightarrow[\text{2. H}^+]{\text{1. Nu}} RCH\!=\!CHCR + RCHCH_2CR$$

direct addition   conjugate addition

Arthur Michael (1853–1942)
was born in Buffalo, New York. He studied at the University of Heidelberg, at the University of Berlin, and at L'École de Médicine in Paris. He was a professor of chemistry at Tufts and Harvard universities, retiring from Harvard when he was 83.

Many different nucleophiles can add to α,β-unsaturated carbonyl compounds (Section 16.14). When the nucleophile is an enolate, the addition reaction has a special name: it is called a **Michael reaction.** The enolates that work best in Michael reactions are enolates that are flanked by two electron-withdrawing groups: enolates of β-diketones, β-diesters, β-keto esters, and β-keto nitriles. Because these enolates are relatively weak bases and also can be quite bulky, addition occurs at the β-carbon. Notice that a wide variety of α,β-unsaturated compounds undergo Michael reactions.

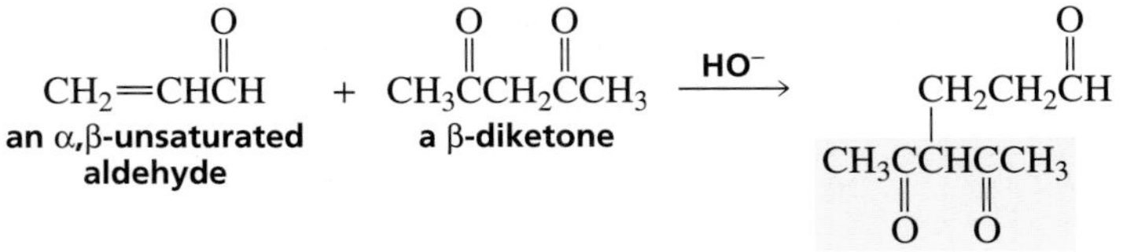

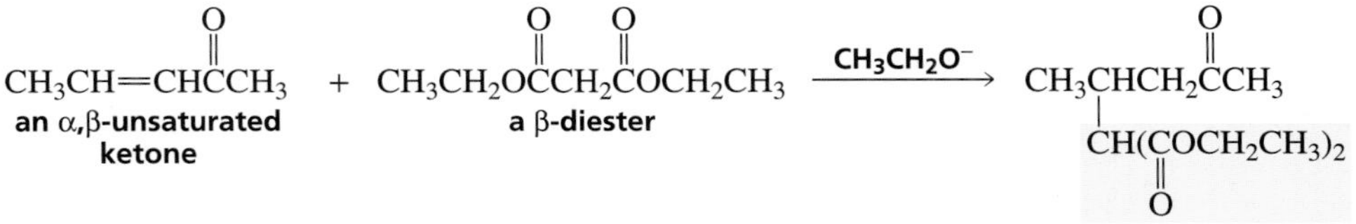

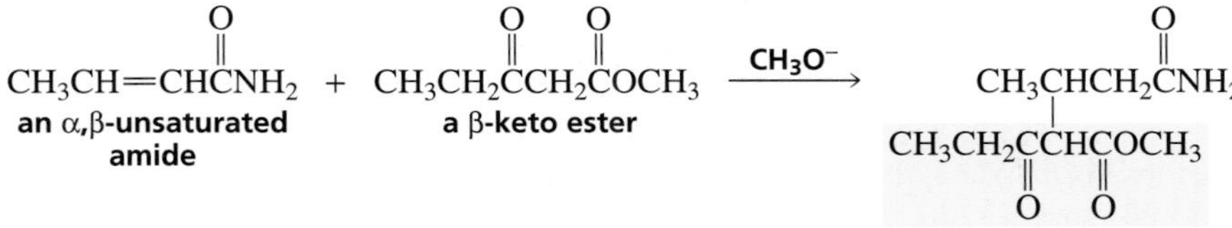

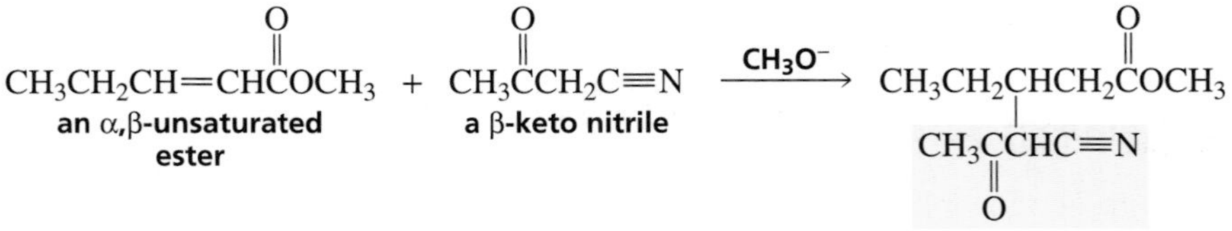

$$CH_2\!=\!CHCH + CH_3CCH_2CCH_3 \xrightarrow{HO^-}$$

an α,β-unsaturated aldehyde   a β-diketone

$$CH_3CH\!=\!CHCCH_3 + CH_3CH_2OCCH_2COCH_2CH_3 \xrightarrow{CH_3CH_2O^-} $$

an α,β-unsaturated ketone   a β-diester

$$CH_3CH\!=\!CHCNH_2 + CH_3CH_2CCH_2COCH_3 \xrightarrow{CH_3O^-}$$

an α,β-unsaturated amide   a β-keto ester

$$CH_3CH_2CH\!=\!CHCOCH_3 + CH_3CCH_2C\!\equiv\!N \xrightarrow{CH_3O^-}$$

an α,β-unsaturated ester   a β-keto nitrile

All of these reactions take place by the same mechanism. A base removes a proton from the α-carbon of the carbon acid. The enolate adds to the β-carbon of an α,β-unsaturated carbonyl compound, and the α-carbon obtains a proton from the solvent.

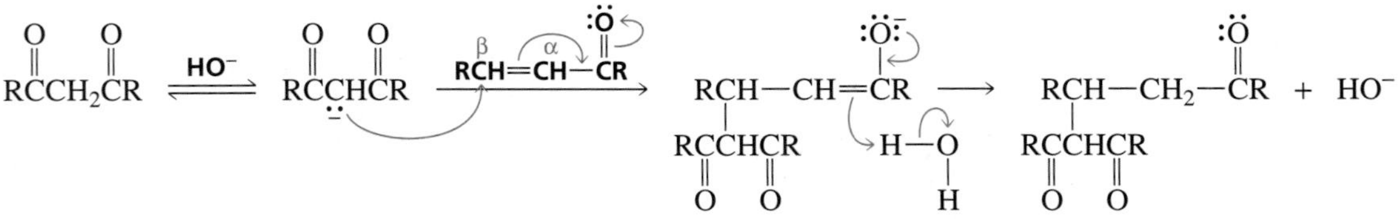

Notice that if either of the reactants in a Michael reaction has an ester group, the base used to remove the α-proton is the same as the leaving group of the ester. This is done because the base, in addition to being able to remove an α-proton, can react as a nucleophile and attack the carbonyl group of the ester. If the nucleophile is identical to the leaving group of the ester, nucleophilic attack on the carbonyl group will not change the reactant (no transesterification reaction will occur).

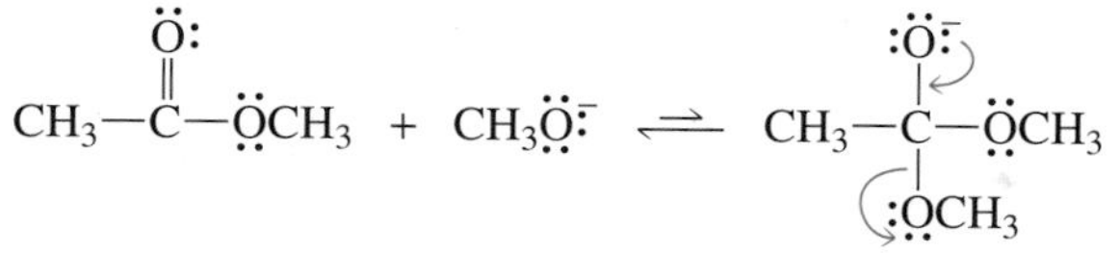

Gilbert Stork was born in Belgium in 1921. He received a Ph.D. from the University of Wisconsin. He was a professor of chemistry at Harvard University and has been a professor at Columbia University since 1953.

Enamines can be used in place of enolates in Michael reactions. When an enamine is used as a nucleophile in a Michael reaction, the reaction is called a **Stork enamine reaction.** Gilbert Stork was the first to show how enamines could be used to alkylate aldehydes and ketones.

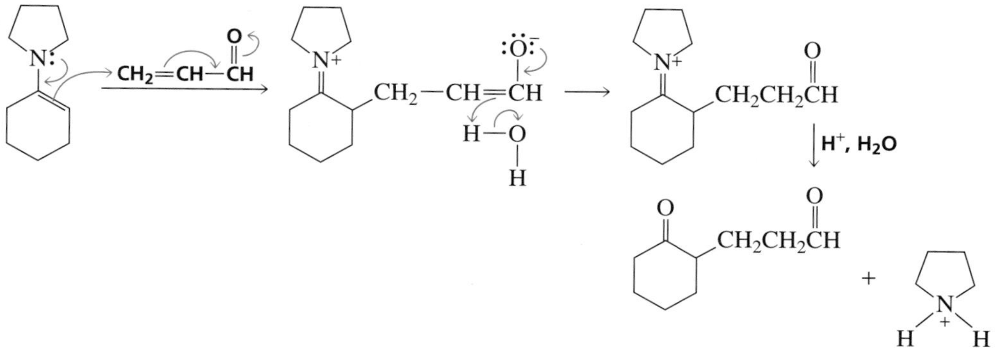

PROBLEM 14 ◆

What reagents would you use to prepare the following compounds?

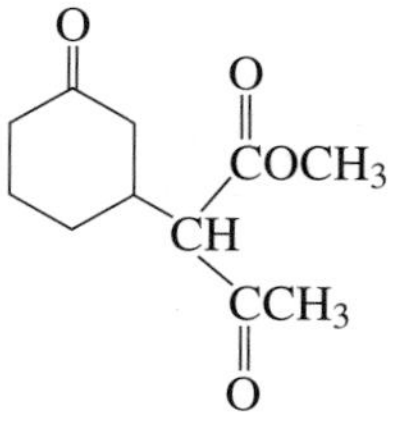

In Chapter 16 we saw that aldehydes and ketones are electrophiles and therefore react with nucleophiles. In the preceding sections we have seen that, when a proton is removed from the α-carbon of an aldehyde or a ketone, the resulting anion is a nucleophile and therefore reacts with electrophiles. An **aldol addition** is a reaction in which both of these activities are observed: one molecule of a carbonyl compound—after a proton is removed from an α-carbon—reacts as a nucleophile and attacks the electrophilic carbonyl carbon of a second molecule of the carbonyl compound.

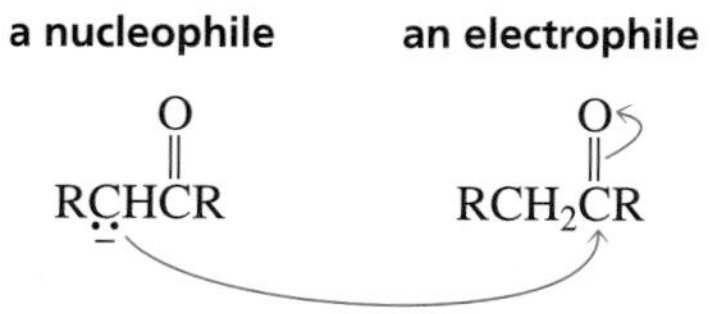

**18.10
THE ALDOL
ADDITION**

An aldol addition is a reaction between two molecules of aldehyde or two molecules of ketone. When the reactant is an aldehyde, the addition product is a β-hydroxyaldehyde, which is why the reaction is called an aldol ("ald" for aldehyde, "ol" for alcohol) addition. When the reactant is a ketone, the addition product is a β-hydroxyketone. Because the addition reaction is reversible, good yields of the addition product are obtained only if it is removed from the solution as it is formed.

aldol additions

a β-hydroxyaldehyde

a β-hydroxyketone

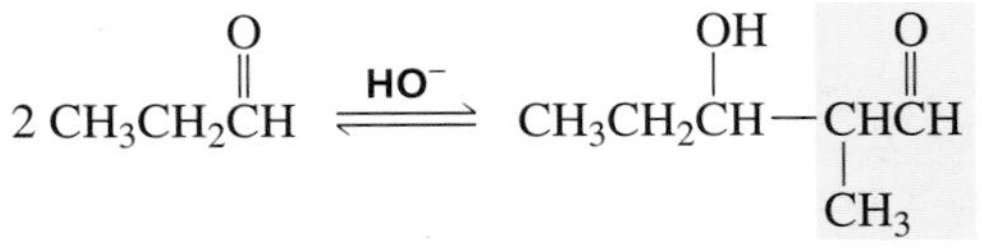

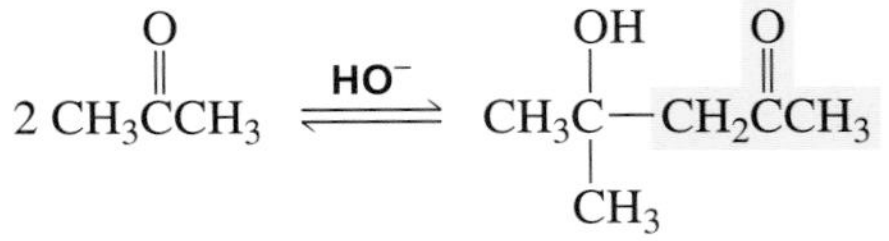

In the first step of an aldol addition, a base removes an α-proton from the carbonyl compound, creating an enolate. The enolate adds to the carbonyl carbon of a second molecule of carbonyl compound, and the resulting negatively charged oxygen is protonated by the solvent.

mechanism for the aldol addition

a β-hydroxyaldehyde

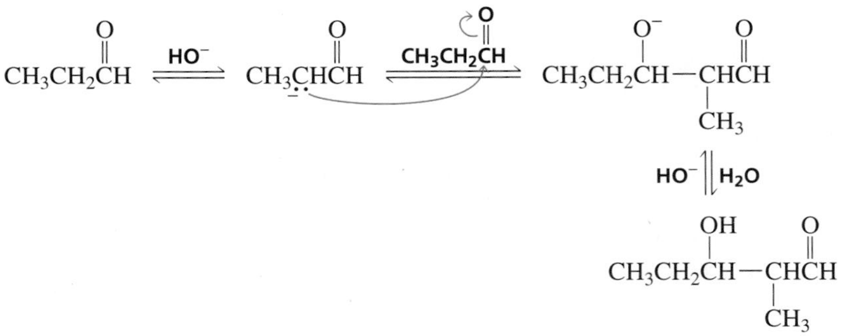

Ketones are less susceptible than aldehydes to attack by nucleophiles, so aldol additions occur more slowly with ketones. The relatively high reactivity of aldehydes in aldol addition reactions is what causes them to give low yields of α-selenenylation or α-alkylation products (Sections 18.6 and 18.8).

a β-hydroxyketone

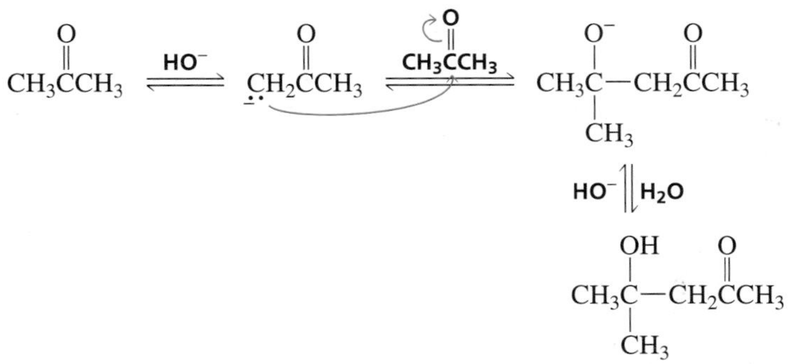

Notice that the product of an aldol addition has twice as many carbons as the reacting aldehyde or ketone.

PROBLEM 15◆

Show the aldol addition product for each of the following compounds.

a. $CH_3CH_2CH_2CH_2\overset{\displaystyle O}{\overset{\|}{C}}H$

c. $CH_3CH_2\overset{\displaystyle O}{\overset{\|}{C}}CH_2CH_3$

b. $CH_3\underset{\underset{\displaystyle CH_3}{|}}{C}HCH_2CH_2\overset{\displaystyle O}{\overset{\|}{C}}H$

d.

PROBLEM 16

An aldol addition can be catalyzed by acids as well as by bases. Propose a mechanism for the acid-catalyzed aldol addition of propanal.

We have seen that alcohols are dehydrated when they are heated with acid (Section 11.4). The β-hydroxyaldehyde and β-hydroxyketone products of aldol addition reactions are easier to dehydrate than other alcohols because the double bond formed as the result of dehydration is conjugated with a carbonyl group. Conjugation increases the stability of the product (Section 7.3) and therefore makes it easier to form. The product is called an enone: "ene" for the double bond, "one" for the carbonyl group.

18.11
DEHYDRATION OF ALDOL ADDITION PRODUCTS: FORMATION OF α,β-UNSATURATED ALDEHYDES AND KETONES

$$2\ CH_3CH_2\overset{\displaystyle O}{\overset{\|}{C}}H \ \underset{}{\overset{HO^-}{\rightleftharpoons}}\ CH_3CH_2\underset{\underset{\displaystyle CH_3}{|}}{C}H\!-\!CH\overset{\displaystyle O}{\overset{\|}{C}}H\ \overset{H^+}{\underset{\Delta}{\longrightarrow}}\ CH_3CH_2CH\!=\!\underset{\underset{\displaystyle CH_3}{|}}{C}\overset{\displaystyle O}{\overset{\|}{C}}H\ +\ H_2O$$

a β-hydroxyaldehyde   an α,β-unsaturated aldehyde

If the product of an aldol addition is dehydrated, the overall reaction is called an **aldol condensation.** A **condensation reaction** is a reaction that combines two molecules while removing a small molecule (usually water or an alcohol).

β-Hydroxyaldehydes and β-hydroxyketones can also be dehydrated under basic conditions. So heating the aldol addition product in either acid or base leads to dehydration.

$$2\ CH_3\overset{\displaystyle O}{\overset{\|}{C}}CH_3 \ \underset{}{\overset{HO^-}{\rightleftharpoons}}\ CH_3\underset{\underset{\displaystyle CH_3}{|}}{C}\!-\!CH_2\overset{\displaystyle O}{\overset{\|}{C}}CH_3\ \overset{HO^-}{\underset{\Delta}{\longrightarrow}}\ CH_3\underset{\underset{\displaystyle CH_3}{|}}{C}\!=\!CH\overset{\displaystyle O}{\overset{\|}{C}}CH_3\ +\ H_2O$$

a β-hydroxyketone   an α,β-unsaturated ketone

Dehydration sometimes occurs under the conditions in which the aldol addition is carried out, without additional heating. In such cases, the β-hydroxycarbonyl compound is an intermediate and the enone is the final product of the reaction. For example, the β-hydroxyketone formed from the aldol addition of acetophenone loses water as soon as it is formed, because the double bond formed by loss of water is conjugated not only with the carbonyl group but also with the benzene ring. This makes the product a very stable compound and therefore relatively easy to form.

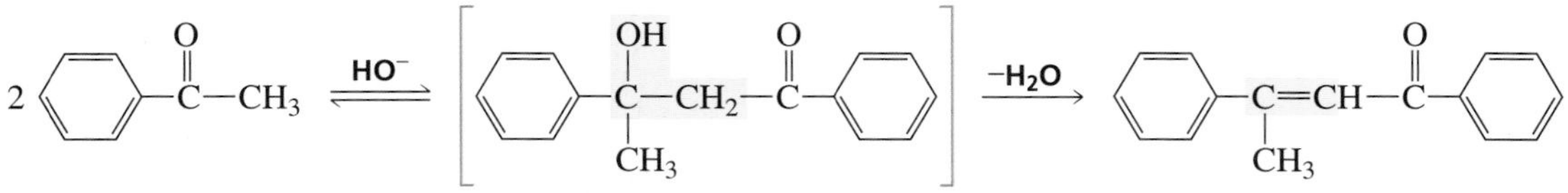

PROBLEM 17 / SOLVED

How would you prepare the following compounds using a starting material containing no more than three carbons?

a. 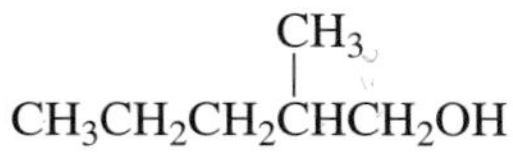 $CH_3CH_2CH_2\overset{\overset{\displaystyle CH_3}{|}}{C}HCH_2OH$

b. $CH_3CH_2CH_2\overset{\overset{\displaystyle O}{||}}{C}OH$

c. $CH_3\overset{\overset{\displaystyle}{|}}{C}HCH_2\overset{\overset{\displaystyle O}{||}}{C}CH_3$
   $\quad\;\; |$
   $\quad\; CH_3$

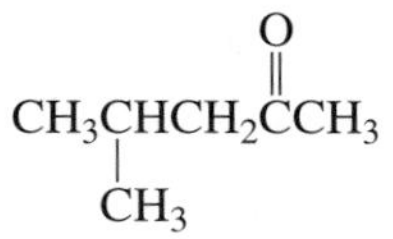

SOLUTION TO 17b Because the target compound is a carboxylic acid that has four carbons, the starting material should be a two-carbon aldehyde. Dehydration of the addition product forms an α,β-unsaturated aldehyde. Catalytic hydrogenation forms an aldehyde, but some of the α,β-unsaturated aldehyde might be reduced to an alcohol. Both the aldehyde and the alcohol are oxidized to the target compound.

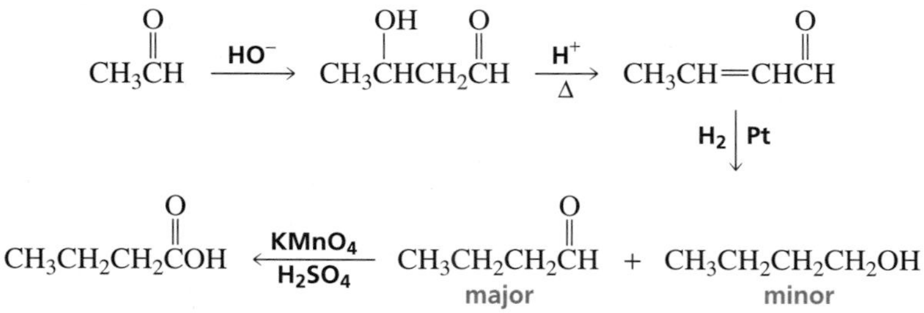

18.12
THE MIXED ALDOL
ADDITION

If two different carbonyl compounds are used in an aldol addition, four products can be formed, because each carbonyl compound can react with itself as well as with the other carbonyl compound. In the following example, both carbonyl compound A and carbonyl compound B can lose a proton from an α-carbon to form nucleophiles A⁻ and B⁻; A⁻ can react with either A or B, and B⁻ can react with either A or B. This is called a **mixed aldol addition** or a **crossed aldol addition.** The four products have similar physical properties, making them difficult to separate. Consequently, a mixed aldol addition that forms four products is not a useful reaction.

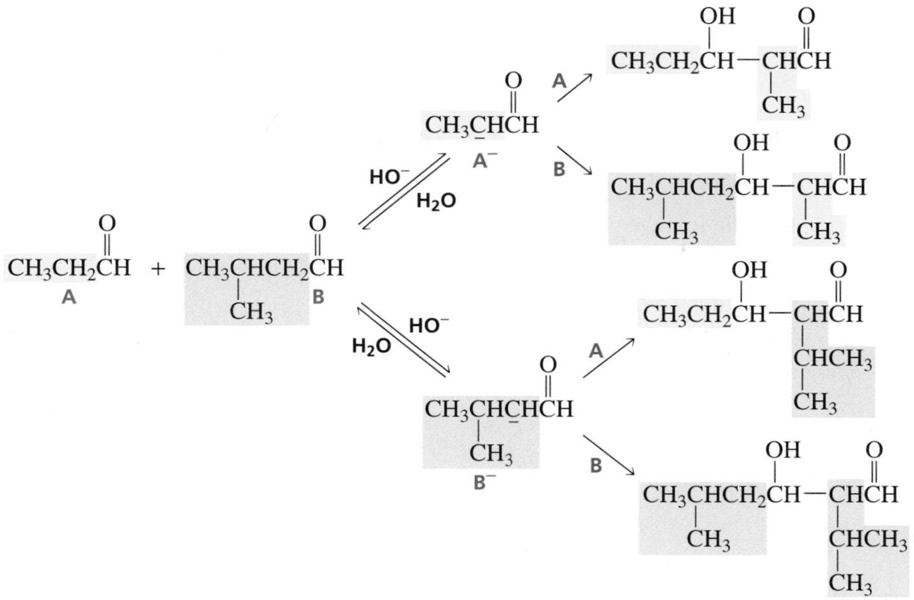

Under certain conditions, a mixed aldol addition can lead primarily to one product. If one of the carbonyl compounds does not have any α-hydrogens, it cannot form an enolate. This reduces the number of possible products from four to two. A greater amount of one of the two products will be formed if the compound without α-hydrogens is always present in excess. The enolate will be more likely to attack the carbonyl compound *without* α-hydrogens because there is more of it in solution.

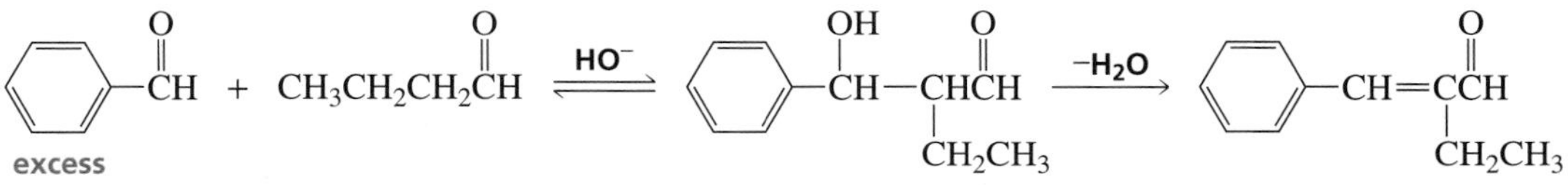

A common method used to obtain a single aldol product is to use LDA to remove the α-proton when creating the enolate. Because LDA is a strong base (Section 18.7), all the carbonyl compound will be converted into an enolate so there will be no carbonyl compound left with which to react. Addition will occur when the second carbonyl compound is added slowly to the reaction mixture.

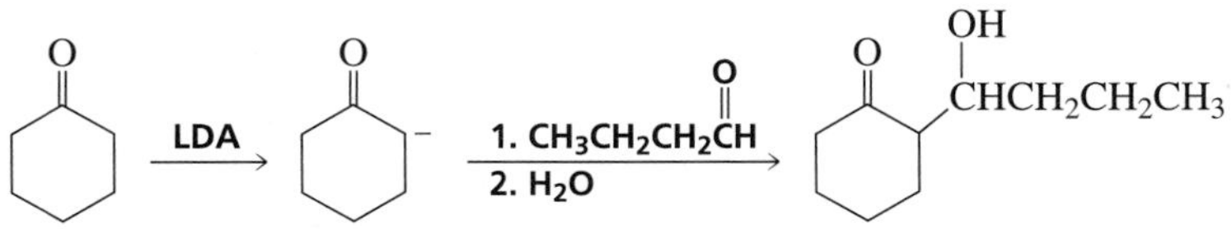

PROBLEM 18

Give the products obtained from mixed aldol additions of the following compounds.

a. $CH_3CH_2CH_2\overset{O}{\overset{\|}{C}}H$ + $CH_3CH_2CH_2CH_2\overset{O}{\overset{\|}{C}}H$

b. $CH_3\overset{O}{\overset{\|}{C}}CH_3$ + $CH_3CH_2\overset{O}{\overset{\|}{C}}CH_2CH_3$

c. [cyclohexanone structure] + CH$_3$CH$_2$CH (with C=O)   **d.** HCH + CH$_3$CH$_2$CH (both with C=O)
                                                                    excess

PROBLEM 19

Describe how the following compounds could be prepared using an aldol addition in the first step of the synthesis.

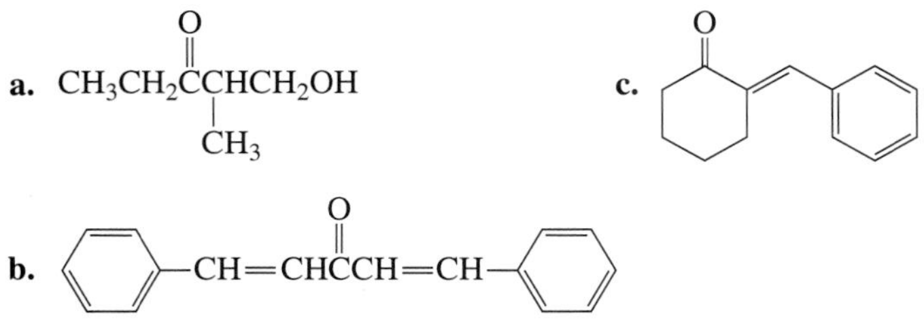

a. CH$_3$CH$_2$CCHCH$_2$OH (with C=O and CH$_3$ branch)   **c.** [cyclohexanone with benzylidene substituent]

b. [phenyl]—CH=CHCCH=CH—[phenyl] (with C=O)

18.13
THE CLAISEN
CONDENSATION

When two molecules of ester undergo a condensation reaction, the reaction is called a **Claisen condensation.** The product of a Claisen condensation is a β-keto ester.

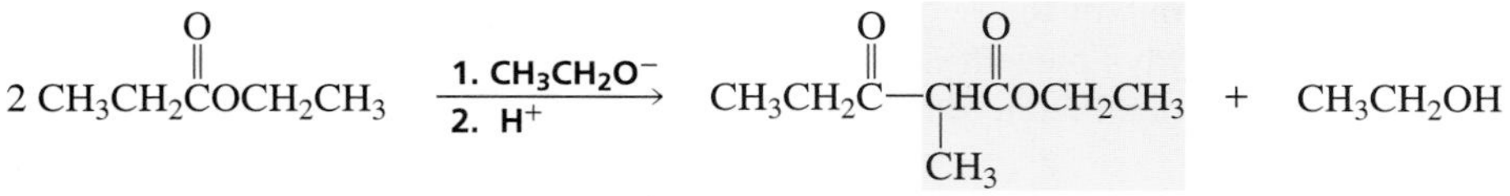

$$2\ CH_3CH_2COCH_2CH_3 \xrightarrow[\text{2. H}^+]{\text{1. CH}_3\text{CH}_2\text{O}^-} CH_3CH_2C-CHCOCH_2CH_3 + CH_3CH_2OH$$

(with CH$_3$ branch, two C=O groups)

a β-keto ester

As in the aldol addition, one molecule of carbonyl compound is converted into an enolate by having an α-proton removed by a strong base. The enolate attacks the carbonyl carbon of a second molecule of carbonyl compound. The base employed corresponds to the leaving group of the ester in order to prevent a transesterification reaction from occurring (Section 18.9).

After nucleophilic attack, the aldol addition and the Claisen condensation differ. In the Claisen condensation, the negatively charged oxygen reforms the carbon–oxygen π bond and eliminates the OR group.

mechanism for the Claisen condensation

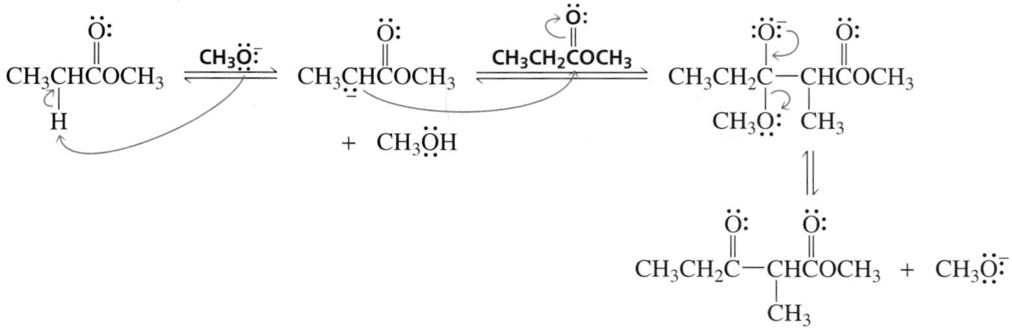

Let's compare this with the aldol addition, where the negatively charged oxygen picks up a proton from the solvent. The difference in the two reactions arises from the fact that with esters, the carbon to which the negatively charged oxygen is bonded is also bonded to a group weakly basic enough to be eliminated. With aldehydes or ketones, the carbon to which the negatively charged oxygen is bonded is not bonded to a group that can be eliminated. Thus the Claisen condensation is a substitution reaction, whereas the aldol reaction is an addition reaction.

Ludwig Claisen (1851–1930) *was born in Germany and received a Ph.D. from the University of Bonn, studying under Kekulé. He was a professor of chemistry at the University of Bonn, Owens College (Manchester, England), the University of Munich, the University of Aachen, the University of Kiel, and the University of Berlin.*

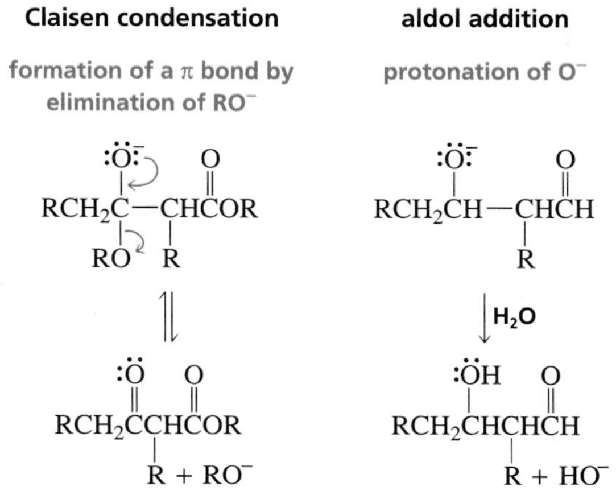

Claisen condensation	**aldol addition**
formation of a π bond by elimination of RO⁻	protonation of O⁻

$$RCH_2\overset{:\overset{..}{O}:^-}{\underset{\underset{RO}{|}\ \underset{R}{|}}{C}}-\overset{O}{\overset{||}{C}}HCOR \qquad RCH_2\overset{:\overset{..}{O}:^-}{\underset{\underset{R}{|}}{C}}H-\overset{O}{\overset{||}{C}}HCH$$

$$\Updownarrow \qquad\qquad \downarrow H_2O$$

$$RCH_2\overset{:\overset{..}{O}}{\overset{||}{C}}\overset{O}{\overset{||}{C}}HCOR \qquad RCH_2\overset{:\overset{..}{O}H}{\underset{\underset{R}{|}}{C}}HCHCH$$

$$R + RO^- \qquad\qquad R + HO^-$$

The elimination step in the Claisen condensation is a reversible reaction since the alkoxide ion can readily attack the keto group of the β-keto ester. The reaction can be driven to completion by removing a proton from the β-keto ester. This is easy to do because, since the central α-carbon of the β-keto ester is flanked by two carbonyl groups, its hydrogens are more acidic than the α-hydrogens of the reacting ester. Consequently, a successful Claisen condensation requires an ester with two α-hydrogens and an equivalent amount of base rather than a catalytic amount of base. When the reaction is over, addition of acid to the reaction mixture reprotonates the β-keto ester anion.

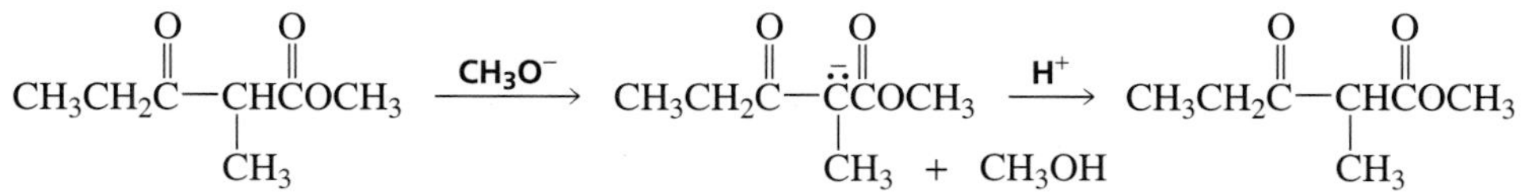

$$CH_3CH_2\overset{O}{\overset{||}{C}}-\overset{O}{\underset{\underset{CH_3}{|}}{\overset{||}{C}}}HCOCH_3 \xrightarrow{CH_3O^-} CH_3CH_2\overset{O}{\overset{||}{C}}-\overset{O}{\overset{||}{\underset{\underset{CH_3 \ + \ CH_3OH}{|}}{C}}}COCH_3 \xrightarrow{H^+} CH_3CH_2\overset{O}{\overset{||}{C}}-\overset{O}{\underset{\underset{CH_3}{|}}{\overset{||}{C}}}HCOCH_3$$

PROBLEM 20 ◆

Give the products of the following reactions.

a.  $CH_3CH_2CH_2\overset{O}{\overset{||}{C}}OCH_3 \quad \xrightarrow[\text{2. H}^+]{\text{1. CH}_3O^-}$

b. 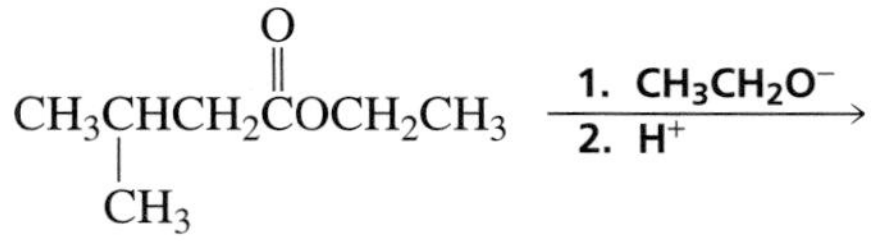 $CH_3\underset{\underset{CH_3}{|}}{C}HCH_2\overset{O}{\overset{||}{C}}OCH_2CH_3 \quad \xrightarrow[\text{2. H}^+]{\text{1. CH}_3CH_2O^-}$

PROBLEM 21

Explain why a Claisen condensation product is not obtained from an ester that has only one α-hydrogen.

18.14
THE MIXED CLAISEN
CONDENSATION

A **mixed Claisen condensation** is a condensation between two different esters. Like a mixed aldol reaction, a mixed Claisen condensation is a useful reaction only if it is carried out under conditions that foster the formation of primarily one product. Otherwise, the result is a mixture of products that are difficult to separate. Primarily one product will be formed if one of the esters has no α-hydrogens and the other ester is added slowly so that the ester without α-hydrogens is always in excess.

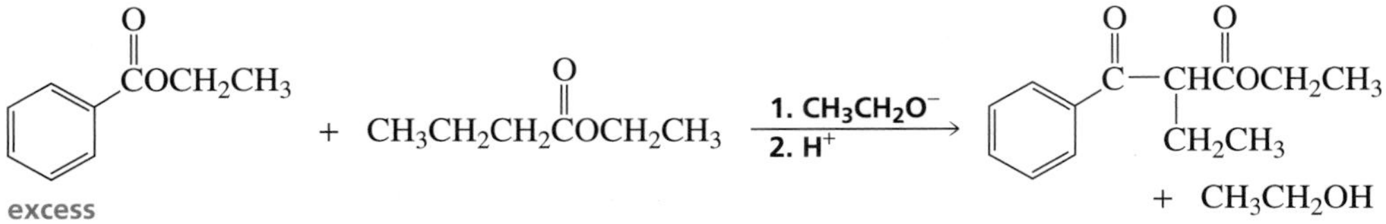

A reaction similar to a mixed Claisen condensation is the condensation of a ketone and an ester. Because the α-hydrogens of a ketone are more acidic than those of an ester, primarily one product is formed if the ketone is added slowly to a mixture of the base and excess ester. The product is a β-diketone.

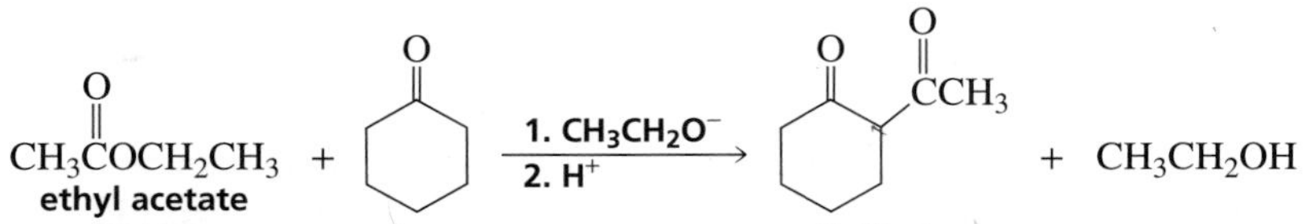

A β-keto aldehyde is formed when a ketone condenses with a formate ester.

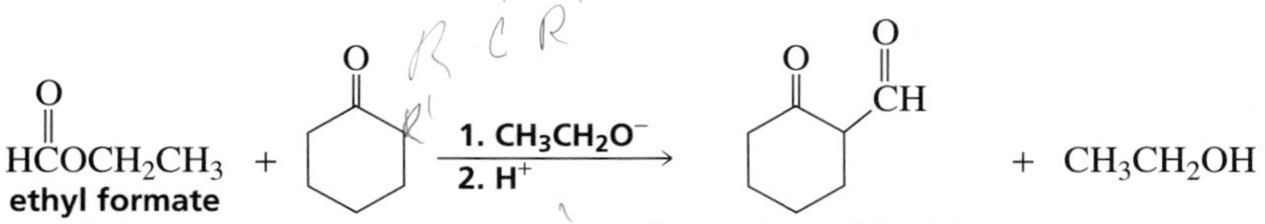

A β-keto ester is formed when a ketone condenses with diethyl carbonate.

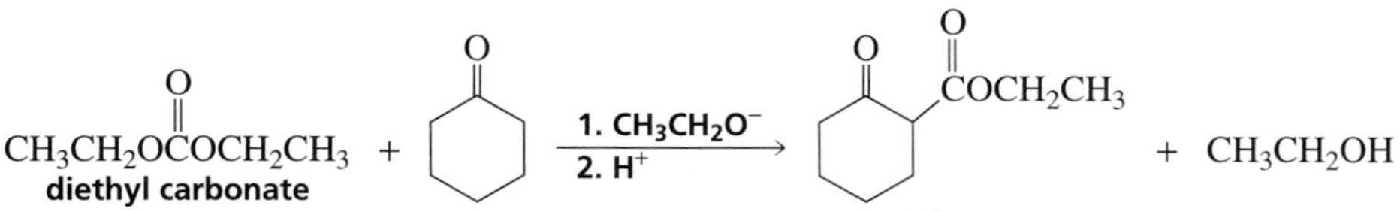

PROBLEM 22◆

Give the product of each of the following reactions.

a. 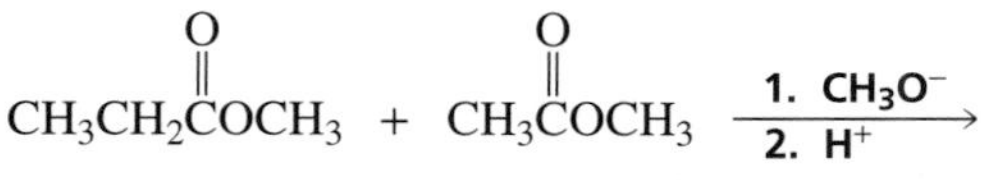 CH₃CH₂COCH₃ + CH₃COCH₃ →(1. CH₃O⁻ 2. H⁺)

b. 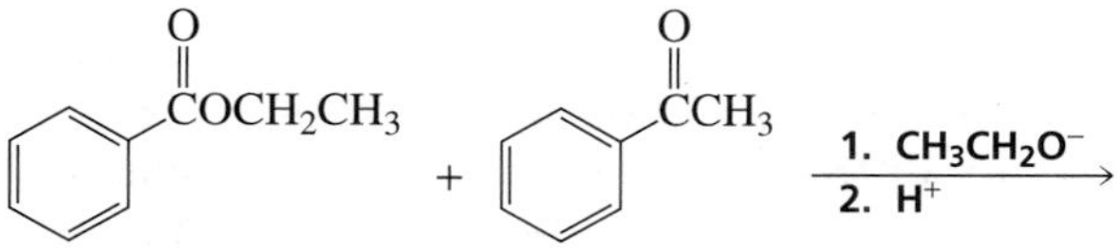 (phenyl)COCH₂CH₃ (excess) + (phenyl)CCH₃ →(1. CH₃CH₂O⁻ 2. H⁺)

c. $\overset{\overset{\text{O}}{\|}}{\text{HCOCH}_3}$ + $\text{CH}_3\text{CH}_2\text{CH}_2\overset{\overset{\text{O}}{\|}}{\text{COCH}_3}$ $\xrightarrow[\text{2. H}^+]{\text{1. CH}_3\text{O}^-}$
excess

d. $\text{CH}_3\text{CH}_2\text{OCOCH}_2\text{CH}_3$ + $\text{CH}_3\text{CH}_2\overset{\overset{\text{O}}{\|}}{\text{COCH}_2\text{CH}_3}$ $\xrightarrow[\text{2. H}^+]{\text{1. CH}_3\text{CH}_2\text{O}^-}$
excess

We have seen that, if a compound has two functional groups that can react with each other, an intramolecular reaction readily occurs if the reaction leads to the formation of a five- or six-membered ring (Section 9.4). Consequently, if base is added to a compound that contains two carbonyl groups, an intramolecular reaction occurs if a product with a five- or six-membered ring can be formed. A compound with two ester groups undergoes an intramolecular Claisen condensation; a compound with two aldehyde or ketone groups undergoes an intramolecular aldol addition.

18.15 INTRAMOLECULAR CONDENSATION AND ADDITION REACTIONS

Intramolecular Claisen Condensations

The addition of base to a 1,6-diester causes the diester to undergo an intramolecular Claisen condensation, thereby forming a five-membered ring β-keto ester. An intramolecular Claisen condensation is called a **Dieckmann condensation.**

Walter Dieckmann
(1869–1925) *was born in Germany. He received a Ph.D. from the University of Munich and became a professor of chemistry at that university*

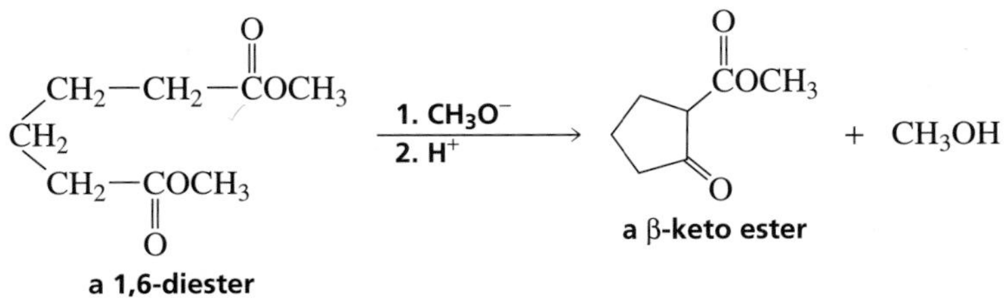

a 1,6-diester → a β-keto ester + CH₃OH

A six-membered ring β-keto ester is formed from a Dieckmann condensation of a 1,7-diester.

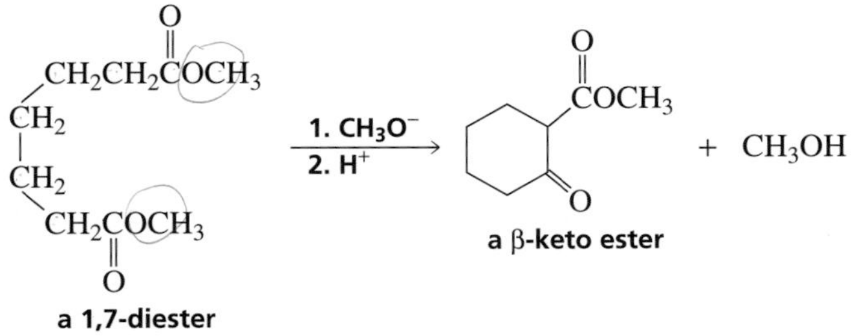

a 1,7-diester → a β-keto ester + CH₃OH

The mechanism of the Dieckmann condensation is the same as the mechanism of the Claisen condensation. The only difference in the two reactions is that the attacking enolate and the carbonyl group undergoing nucleophilic attack are in different molecules in the Claisen condensation but are in the same molecule in the Dieckmann condensation. The Dieckmann condensation, like the Claisen condensation, is driven to completion by carrying out the reaction with enough base to remove a proton from the α-carbon of the β-keto ester product. When the reaction is over, acid is added to reprotonate the condensation product.

mechanism for the Dieckmann condensation

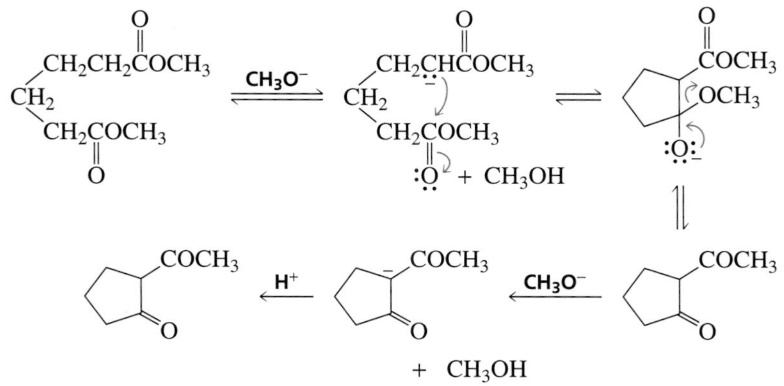

PROBLEM 23

Give the mechanism for the base-catalyzed formation of a cyclic β-keto ester from a 1,7-diester.

Intramolecular Aldol Additions

Because a 1,4-diketone has two different sets of α-hydrogens, two different intramolecular addition products can potentially form: one product would contain a five-membered ring and the other a three-membered ring. The greater stability of five- and six-membered rings causes them to be formed preferentially (Section 2.12). The five-membered ring product is the only product formed from the intramolecular aldol addition of a 1,4-diketone.

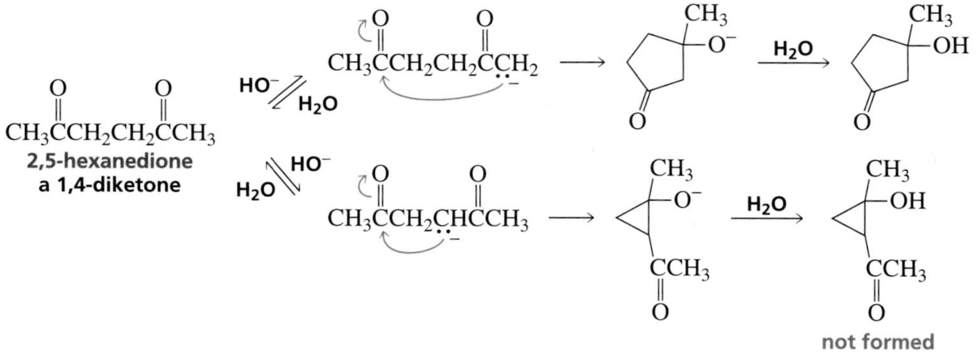

An intramolecular aldol addition of a 1,6-diketone can lead to either a seven- or a five-membered ring product. Again, the five-membered ring product is the only product of the reaction.

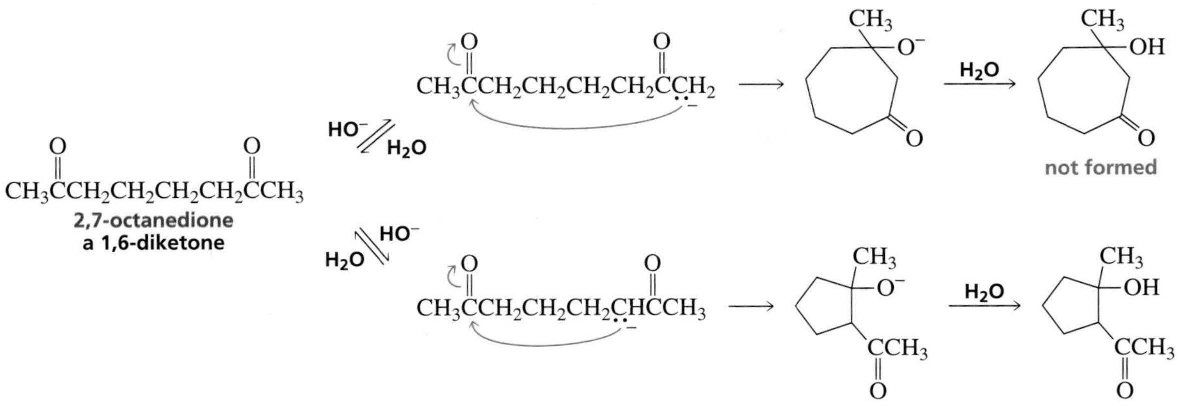

1,5-Diketones and 1,7-diketones undergo intramolecular aldol additions to form six-membered ring products.

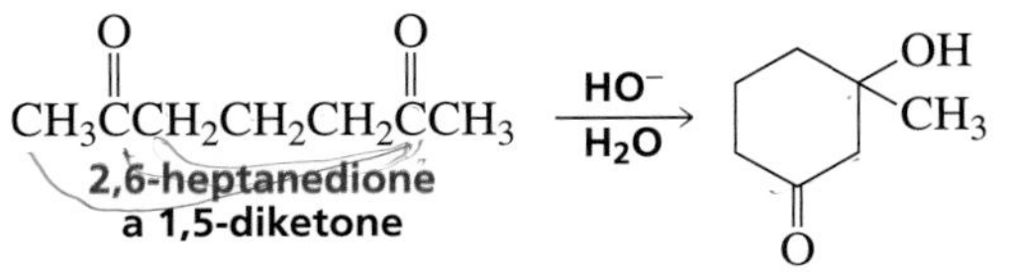

PROBLEM 24 ◆

If the preference for formation of a six-membered ring were not so great, what other cyclic product would be formed from the intramolecular aldol addition of:

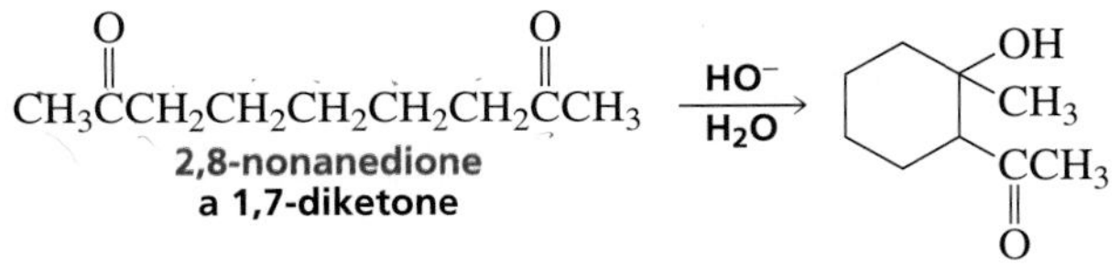

a. 2,6-heptanedione?   **b.** 2,8-nonanedione?

PROBLEM 25

Can 2,4-pentanedione undergo an intramolecular aldol addition? If so, why? If not, why not?

PROBLEM 26 / SOLVED

The ketoaldehyde shown below forms three products when treated with a base. Identify the products. Which is the major product when the reaction is carried out under conditions in which enolate formation is reversible?

$$CH_3\overset{O}{\overset{\|}{C}}CH_2CH_2CH_2CH_2\overset{O}{\overset{\|}{C}}H$$

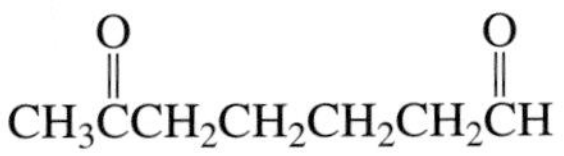

SOLUTION   Three products can form, because there are three different sets of α-hydrogens. More B and C are formed than A because a five-membered ring is formed in preference to a seven-membered ring. When the reaction is carried out under conditions in which enolate formation is reversible, more B is formed than C because the double bond of the enolate leading to B is more substituted and therefore that enolate is more stable.

$$CH_3\overset{O}{\overset{\|}{C}}CH_2CH_2CH_2CH_2\overset{O}{\overset{\|}{C}}H$$

$$H_2O \,\|\, HO^-$$

$$\bar{C}H_2\overset{O}{\overset{\|}{C}}CH_2CH_2CH_2CH_2\overset{O}{\overset{\|}{C}}H \qquad CH_3\overset{O}{\overset{\|}{C}}\bar{C}HCH_2CH_2CH_2\overset{O}{\overset{\|}{C}}H \qquad CH_3\overset{O}{\overset{\|}{C}}CH_2CH_2CH_2\bar{C}H\overset{O}{\overset{\|}{C}}H$$

$$\updownarrow \qquad\qquad\qquad\qquad \updownarrow \qquad\qquad\qquad\qquad \updownarrow$$

$$CH_2{=}\overset{O^-}{\overset{\|}{C}}CH_2CH_2CH_2CH_2\overset{O}{\overset{\|}{C}}H \qquad CH_3\overset{O^-}{\overset{\|}{C}}{=}CHCH_2CH_2CH_2\overset{O}{\overset{\|}{C}}H \qquad CH_3\overset{O}{\overset{\|}{C}}CH_2CH_2CH_2CH{=}\overset{O^-}{\overset{\|}{C}}H$$

$$\downarrow \qquad\qquad\qquad\qquad \downarrow \qquad\qquad\qquad\qquad \downarrow$$

A   B   C

PROBLEM 27◆

Give the product of the reaction of each of the following compounds with base.

a. [structure: cyclohexanone with CH$_2$CH$_2$CCH$_3$ (O) substituent]

c. $H\overset{O}{\overset{\|}{C}}CH_2CH_2CH_2CH_2CH_2\overset{O}{\overset{\|}{C}}H$

b. [structure: bicyclic decalindione with two O groups]

d. [structure: cyclohexanone with CH$_2$CH$_2$CH$_2$CCH$_3$ (O) substituent]

**Sir Robert Robinson
(1886–1975)** *was born in
England, the son of a manufacturer. After receiving a Ph.D.
from the University of
Manchester, he joined the faculty at the University of Sydney
in Australia. He returned to
England three years later,
becoming a professor of chemistry at Oxford in 1929.
Knighted in 1939, he received
the 1947 Nobel Prize in chemistry for his work on alkaloids.*

The Robinson Annulation

Reactions that form carbon–carbon bonds are important to synthetic chemists. Without such reactions, large organic molecules could not be prepared from smaller ones. We have seen that Michael reactions and aldol additions are examples of reactions that form carbon–carbon bonds. The **Robinson annulation** is a reaction that puts these two carbon–carbon bond forming reactions together, providing a route to the synthesis of many complicated organic molecules. Annulation comes from *annulus,* Latin for "ring." An **annulation reaction** is a ring-forming reaction.

The first stage of a Robinson annulation is a Michael reaction. The second stage is an intramolecular aldol addition. Notice that a Robinson annulation results in a product that contains a cyclohexenone ring.

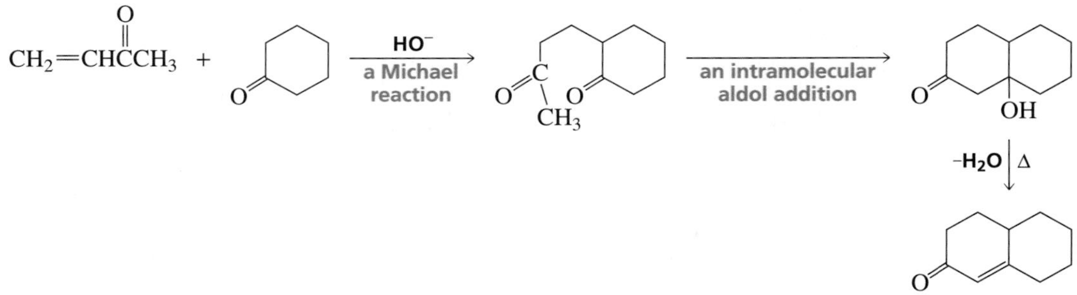

PROBLEM-SOLVING STRATEGY

Propose a synthesis for each of the following compounds using a Robinson annulation.

a.

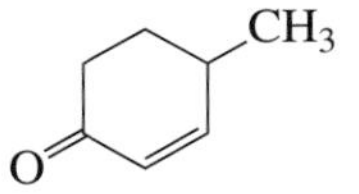

b.

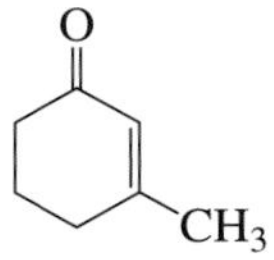

By analyzing the Robinson annulation, we will be able to determine which part of the molecule comes from which reactant. This allows us to choose appropriate reactants for any other Robinson annulation. Analysis shows that the keto group of the cyclohexenone comes from the α,β-unsaturated carbonyl compound, and the double bond results from attack of the enolate of the α,β-unsaturated carbonyl compound on the carbonyl group of the other reactant. Thus we can arrive at the appropriate reactants by cutting through the double bond and between the β- and γ-carbons on the other side of the carbonyl group.

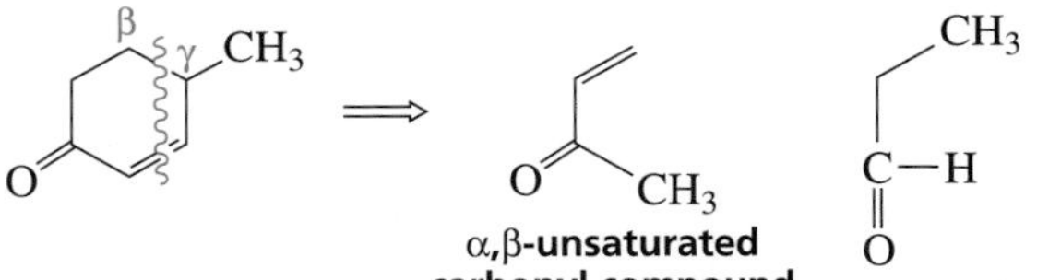

α,β-**unsaturated carbonyl compound**

Therefore, the required reactants for **a** are:

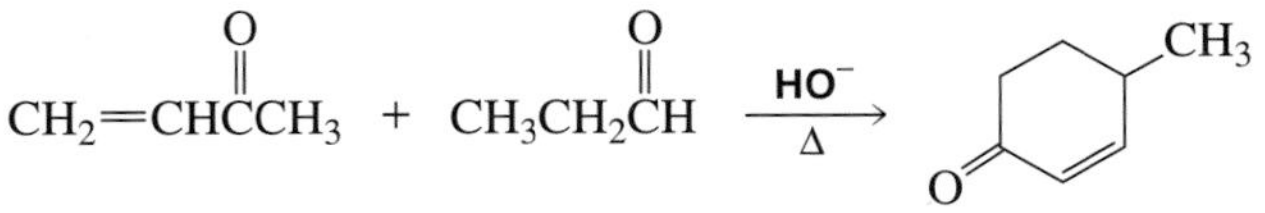

By cutting through **b,** we can determine the required reactants for its synthesis:

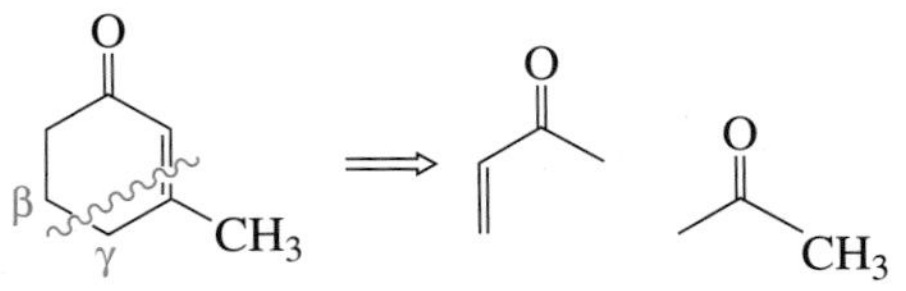

Therefore, the required reactants for **b** are:

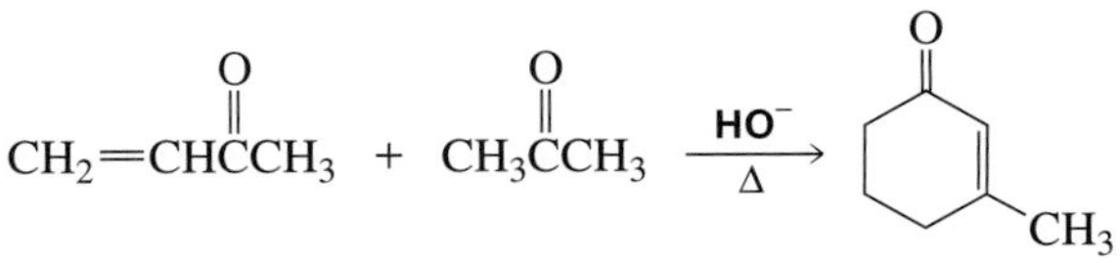

Now continue on to solve Problem 28.

PROBLEM 28

Propose a synthesis for each of the following compounds using a Robinson annulation.

a.

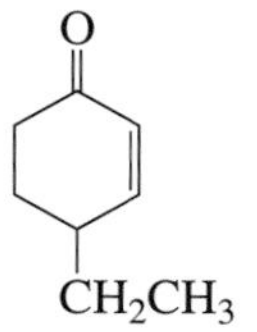

b.

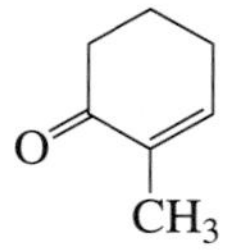

c.

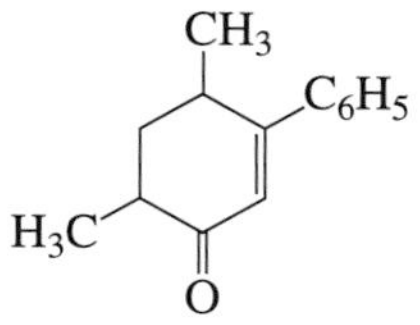

d.

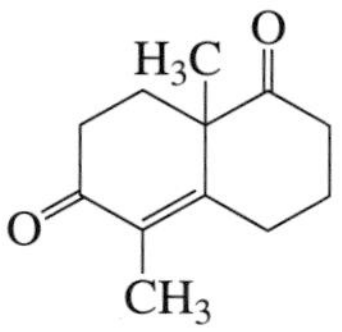

18.16
DECARBOXYLATION OF 3-OXOCARBOX-YLIC ACIDS

Carboxylate ions do not lose CO_2 for the same reason that ethane does not lose a proton. If a carboxylate ion were to lose CO_2 or if ethane were to lose a proton, the leaving group would be a carbanion. Carbanions are very strong bases and therefore very poor leaving groups.

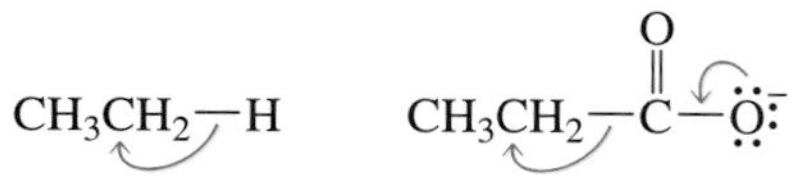

$$CH_3CH_2{-}H \qquad CH_3CH_2{-}\overset{\overset{\displaystyle O}{\|}}{C}{-}\ddot{\underset{..}{O}}{:}^-$$

If, however, the CO_2 group is bonded to a carbon adjacent to a carbonyl carbon, the CO_2 group can be removed because the electrons left behind can be delocalized onto the carbonyl oxygen. Consequently, 3-oxocarboxylate ions lose CO_2 when they are heated. Loss of CO_2 from a molecule is called **decarboxylation.**

removing CO₂ from an α-carbon

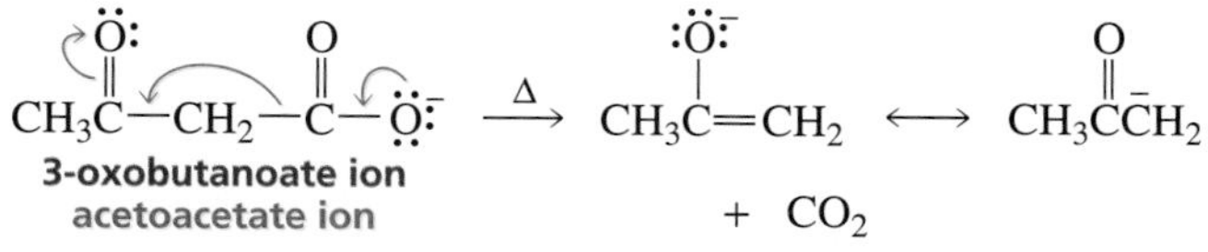

$$CH_3\overset{\overset{\displaystyle \ddot{O}:}{\|}}{C}{-}CH_2{-}\overset{\overset{\displaystyle O}{\|}}{C}{-}\ddot{\underset{..}{O}}{:}^- \overset{\Delta}{\longrightarrow} CH_3\overset{\overset{\displaystyle :\ddot{O}:^-}{|}}{C}{=}CH_2 \longleftrightarrow CH_3\overset{\overset{\displaystyle O}{\|}}{C}\overset{-}{C}H_2$$

3-oxobutanoate ion
acetoacetate ion

$+\ CO_2$

Notice the similarity between removal of CO_2 from a 3-oxocarboxylate ion and removal of a proton from an α-carbon. In both reactions, a substituent (CO_2 in one case; H^+ in the other) is removed from an α-carbon and its bonding electrons are delocalized onto an oxygen.

removing a proton from an α-carbon

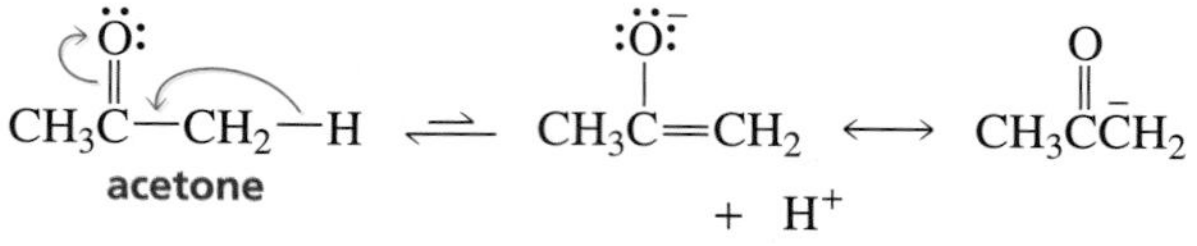

$$CH_3\overset{\overset{\displaystyle \ddot{O}:}{\|}}{C}{-}CH_2{-}H \rightleftharpoons CH_3\overset{\overset{\displaystyle :\ddot{O}:^-}{|}}{C}{=}CH_2 \longleftrightarrow CH_3\overset{\overset{\displaystyle O}{\|}}{C}\overset{-}{C}H_2$$

acetone

$+\ H^+$

Decarboxylation is even easier if the reaction is carried out under acidic conditions. Decarboxylation is catalyzed by an intramolecular transfer of a proton from the carboxyl group to the carbonyl oxygen. The enol that is formed immediately tautomerizes to a ketone.

$$CH_3\overset{O}{\underset{}{C}}-CH_2-\overset{O}{\underset{}{C}}=O \xrightarrow{\Delta} CH_3\overset{OH}{\underset{}{C}}=CH_2 \xrightleftharpoons[\text{tautomerization}]{} CH_3\overset{O}{\underset{}{C}}CH_3$$

3-oxobutanoic acid
acetoacetic acid
a β-keto acid

$+ CO_2$

Earlier in this chapter, we saw that it is harder to remove a proton from an α-carbon if the electrons are delocalized onto the carbonyl group of an ester rather than onto the carbonyl group of a ketone (Section 18.1). For the same reason, a higher temperature is required to decarboxylate a β-dicarboxylic acid such as malonic acid than to decarboxylate a β-keto acid.

$$HO\overset{O}{\underset{}{C}}-CH_2-\overset{O}{\underset{}{C}}=O \xrightarrow{135\ °C} HO\overset{OH}{\underset{}{C}}=CH_2 \xrightleftharpoons[\text{tautomerization}]{} HO\overset{O}{\underset{}{C}}CH_3$$

malonic acid

$+ CO_2$

So 3-oxocarboxylic acids as well as 3-oxocarboxylate ions lose CO_2 when they are heated.

$$CH_3CH_2CH_2\overset{O}{\underset{}{C}}CH_2\overset{O}{\underset{}{C}}OH \xrightarrow{\Delta} CH_3CH_2CH_2\overset{O}{\underset{}{C}}CH_3 \ + \ CO_2$$

3-oxohexanoic acid   **2-pentanone**

2-oxocyclohexane-
carboxylic acid   **cyclohexanone**

$$HO\overset{O}{\underset{}{C}}\underset{\underset{CH_3}{|}}{CH}\overset{O}{\underset{}{C}}OH \xrightarrow{\Delta} CH_3CH_2\overset{O}{\underset{}{C}}OH \ + \ CO_2$$

α-methylmalonic acid   **propionic acid**

Heinz and Clare Hunsdiecker *were both born in Germany—Heinz in Cologne in 1904 and Clare in Kiel in 1903. They both received Ph.D.'s from the University of Cologne and spent their careers working in a private laboratory in that city.*

THE HUNSDIECKER REACTION

Heinz and Clare Hunsdiecker found that a carboxylic acid can be decarboxylated if a heavy metal salt of the carboxylic acid is heated with bromine or iodine. The product is an alkyl halide with one less carbon than the starting carboxylic acid. The heavy metal can be silver ion, mercuric ion, or lead(IV). The reaction is now known as the **Hunsdiecker reaction.**

$$CH_3CH_2CH_2CH_2\overset{O}{\underset{}{C}}OH \xrightarrow[\text{2. Br}_2,\ \Delta]{\text{1. Ag}_2O} CH_3CH_2CH_2CH_2Br \ + \ CO_2 \ + \ AgBr$$

pentanoic acid   **1-bromobutane**

The mechanism of the reaction involves the initial formation of a hypobromite as a result of precipitation of AgBr. A radical reaction is initiated by homolytic cleavage of the oxygen–bromine bond of the hypobromite. The carboxyl radical loses CO_2, and the alkyl radical thus formed abstracts a bromine radical from the hypobromite to propagate the reaction.

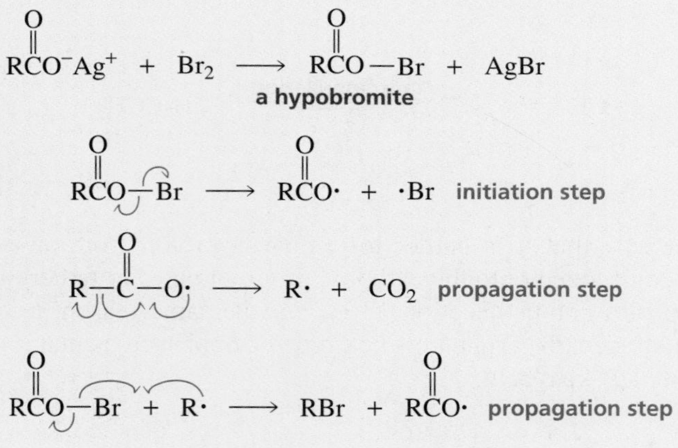

$$RCO^-Ag^+ + Br_2 \longrightarrow RCO-Br + AgBr$$

a hypobromite

$$RCO-Br \longrightarrow RCO\cdot + \cdot Br \quad \text{initiation step}$$

$$R-C-O\cdot \longrightarrow R\cdot + CO_2 \quad \text{propagation step}$$

$$RCO-Br + R\cdot \longrightarrow RBr + RCO\cdot \quad \text{propagation step}$$

PROBLEM 29 ◆

Which of the following compounds would be expected to decarboxylate when heated?

a.

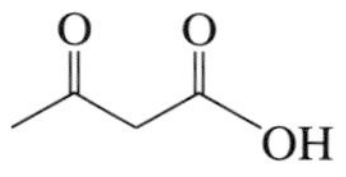

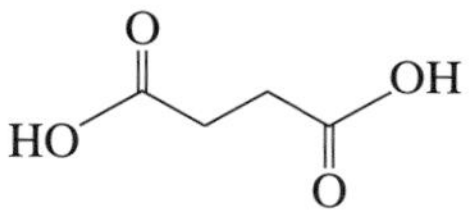

b.

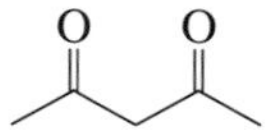

c. HO OH

d. 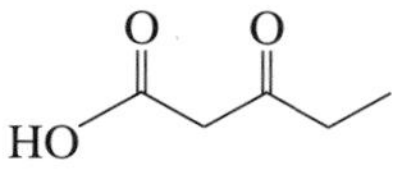 HO

18.17

THE MALONIC ESTER SYNTHESIS: SYNTHESIS OF CARBOXYLIC ACIDS

A combination of two of the reactions discussed in this chapter (alkylation of an α-carbon and decarboxylation of a β-dicarboxylic acid) can be used to prepare carboxylic acids of any desired chain length. The procedure is called the **malonic ester synthesis** because the starting material for the synthesis is the diethyl ester of malonic acid. The first two carbons of the carboxylic acid come from malonic ester, and the rest of the carboxylic acid comes from the alkyl halide used in the second step of the reaction.

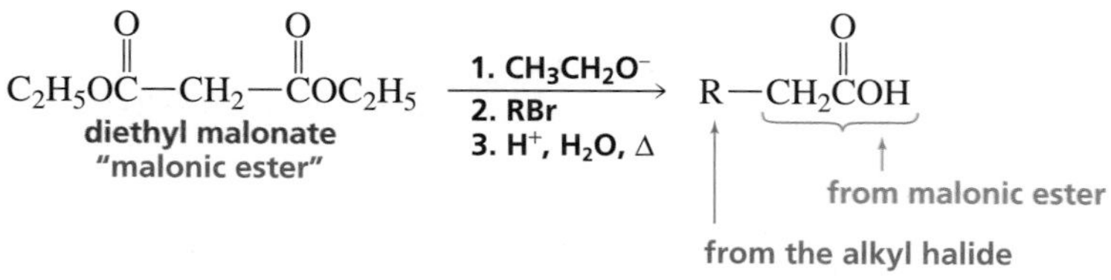

$$C_2H_5OC-CH_2-COC_2H_5 \xrightarrow[\substack{2.\ RBr \\ 3.\ H^+,\ H_2O,\ \Delta}]{1.\ CH_3CH_2O^-} R-CH_2COH$$

diethyl malonate
"malonic ester"

from malonic ester

from the alkyl halide

The first part of the malonic ester synthesis involves alkylation of the α-carbon of the diester (Section 18.8). A base readily removes an α-proton because it is bonded to a carbon flanked by two ester groups ($pK_a = 13$). The resulting α-carbanion reacts with an alkyl halide (a nucleophilic substitution reaction), forming an α-substituted malonic ester. In a heated acidic aqueous solution, the α-substituted malonic ester is hydrolyzed to an α-substituted malonic acid, which upon further heating loses CO_2, forming a carboxylic acid with two more carbons than the alkyl halide.

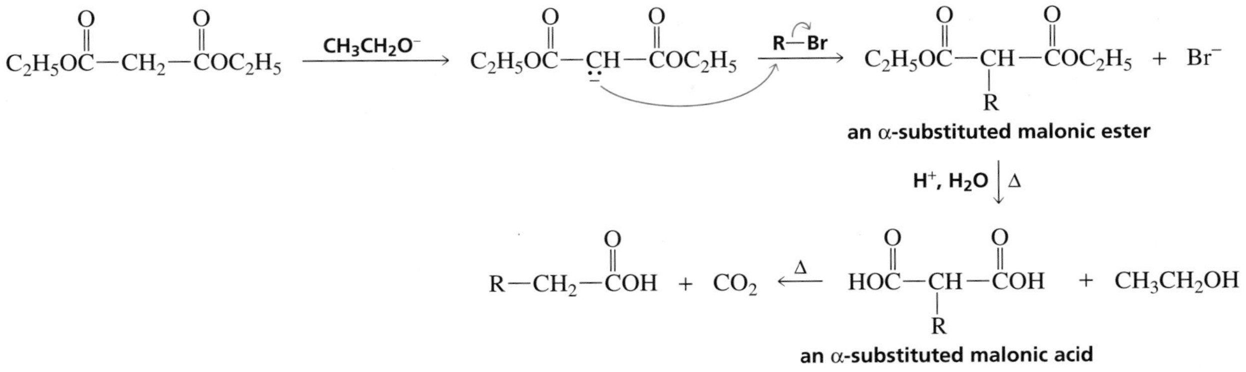

an α-substituted malonic ester

H^+, H_2O | Δ

an α-substituted malonic acid

Carboxylic acids with two substituents bonded to the α-carbon can be prepared by doing two successive α-carbon alkylations.

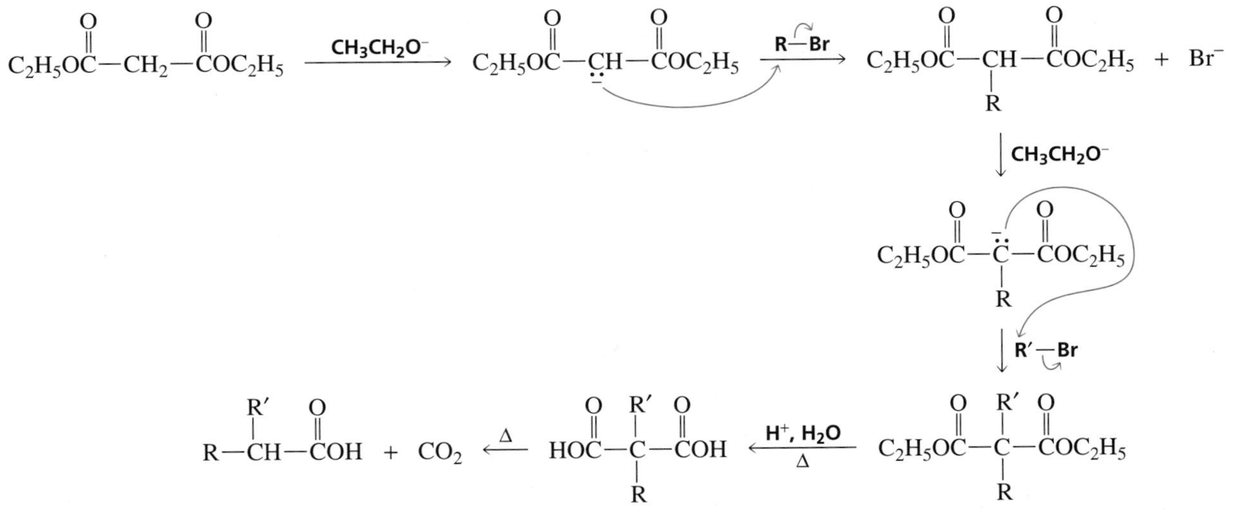

PROBLEM 30◆

What alkyl bromides should be used in the malonic ester synthesis of the following carboxylic acids?

a. propanoic acid

b. 2-methylpropanoic acid

c. 3-phenylpropanoic acid

d. pentanoic acid

e. 4-methylpentanoic acid

f. 2-methylpentanoic acid

PROBLEM 31

Explain why the following acids cannot be prepared by the malonic ester synthesis.

a.

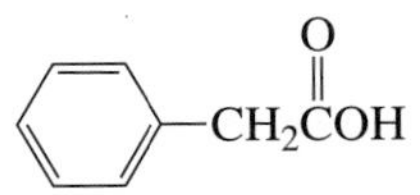

b. 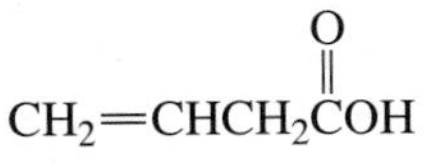 $CH_2{=}CHCH_2COH$ (with =O above)

$$\underset{\substack{\displaystyle CH_3 \\ \displaystyle |}}{\overset{\substack{\displaystyle CH_3 \quad O \\ \displaystyle | \quad \; \|}}{}} $$

c. CH_3CCH_2COH

PROBLEM 32◆

a. What carboxylic acid would be formed if the malonic ester synthesis were carried out with one equivalent of malonic ester, one equivalent of 1,5-dibromopentane, and two equivalents of base?

b. What carboxylic acid would be formed if the malonic ester synthesis were carried out with two equivalents of malonic ester, one equivalent of 1,5-dibromopentane, and two equivalents of base?

18.18
THE ACETOACETIC ESTER SYNTHESIS: SYNTHESIS OF METHYL KETONES

The only difference between the acetoacetic ester synthesis and the malonic ester synthesis is the use of acetoacetic ester rather than malonic ester as the starting material. Because of the difference in starting material, the product of the **acetoacetic ester synthesis** is a methyl ketone rather than a carboxylic acid. The carbonyl group and the carbon atoms on either side of it come from acetoacetic ester, and the rest of the ketone comes from the alkyl halide used in the second step of the reaction.

$$\underset{\substack{\text{ethyl acetoacetate} \\ \text{"acetoacetic ester"}}}{CH_3\overset{O}{\overset{\|}{C}}-CH_2-\overset{O}{\overset{\|}{C}}OC_2H_5} \quad \xrightarrow[\substack{\text{2. RBr} \\ \text{3. H}^+, \text{H}_2\text{O}, \Delta}]{\text{1. CH}_3\text{CH}_2\text{O}^-} \quad R-CH_2\overset{O}{\overset{\|}{C}}CH_3$$

from the alkyl halide | from acetoacetic ester

The mechanisms for the acetoacetic ester synthesis and the malonic ester synthesis are identical. The last step in the acetoacetic ester synthesis is decarboxylation of a substituted acetoacetic acid rather than a substituted malonic acid.

$$CH_3\overset{O}{\overset{\|}{C}}-CH_2-\overset{O}{\overset{\|}{C}}OC_2H_5 \xrightarrow{CH_3CH_2O^-} CH_3\overset{O}{\overset{\|}{C}}-\overset{..}{C}H-\overset{O}{\overset{\|}{C}}OC_2H_5 \xrightarrow{R-Br} \underset{\substack{| \\ R}}{CH_3\overset{O}{\overset{\|}{C}}-CH-\overset{O}{\overset{\|}{C}}OC_2H_5} + Br^-$$

$$H^+, H_2O \Big| \Delta$$

$$CH_3\overset{O}{\overset{\|}{C}}-CH_2-R + CO_2 \xleftarrow{\Delta} \underset{\substack{| \\ R}}{CH_3\overset{O}{\overset{\|}{C}}-CH-\overset{O}{\overset{\|}{C}}OH} + CH_3CH_2OH$$

PROBLEM 33◆

What alkyl bromide should be used in the acetoacetic ester synthesis of each of the following methyl ketones?

a. 2-pentanone   b. 2-octanone   c. 4-phenyl-2-butanone

When you are planning the synthesis of a compound that requires the formation of a new carbon–carbon bond, first locate the new bond that must be made. For example, in the synthesis of the β-diketone shown below, the new bond is the one that makes the second five-membered ring.

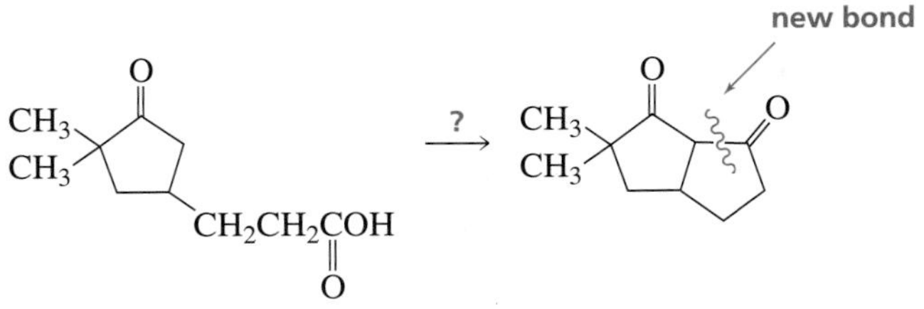

18.19
DESIGNING A SYNTHESIS VI: MAKING NEW CARBON–CARBON BONDS

Next, determine which atom at the end of the bond should be the nucleophile and which should be the electrophile. Because you know that a carbonyl carbon is an electrophile, it is easy to choose between the two possibilities.

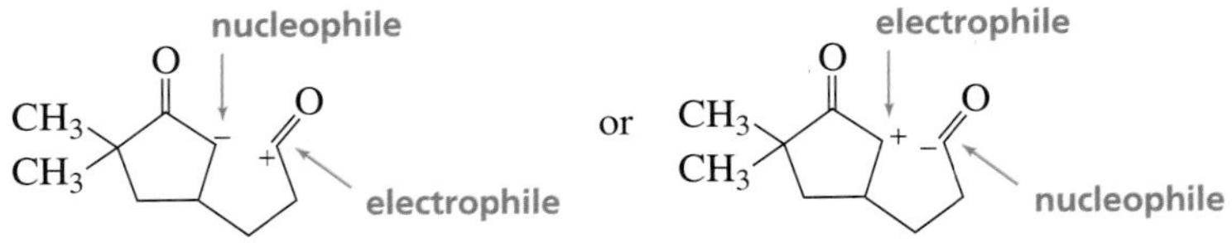

Now you need to determine what compound you could use that would give you the desired electrophilic and nucleophilic sites. If the starting material is indicated, use that compound as a clue to arrive at the desired compound. For example, an ester carbonyl group would be a good electrophile for this synthesis because it has a group that would be eliminated. Further, since the α-hydrogens of the ketone are more acidic than the α-hydrogens of the ester, it would be easy to obtain the desired nucleophile. The ester can be easily prepared from the carboxylic acid starting material.

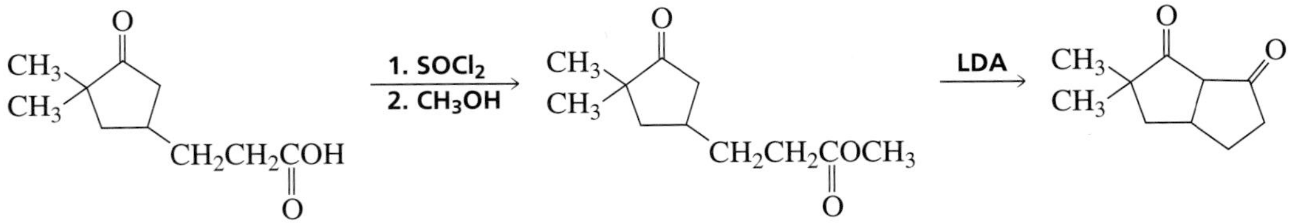

In the following synthesis, two new carbon–carbon bonds must be formed.

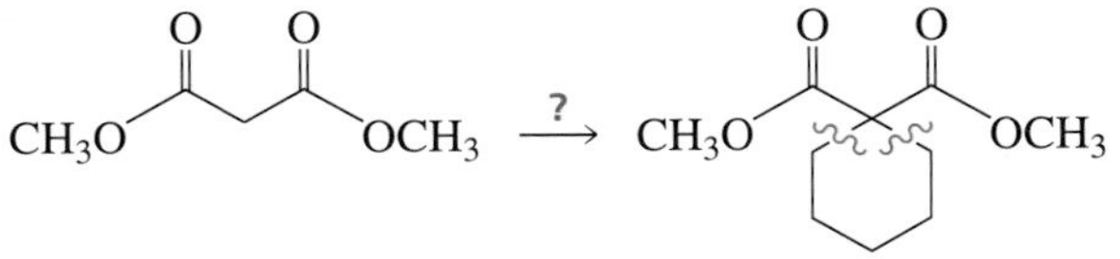

After determining the electrophilic and nucleophilic sites, you can see that two successive alkylations of a diester of malonic acid, using 1,5-dibromopentane as the alkyl halide, will produce the desired compound.

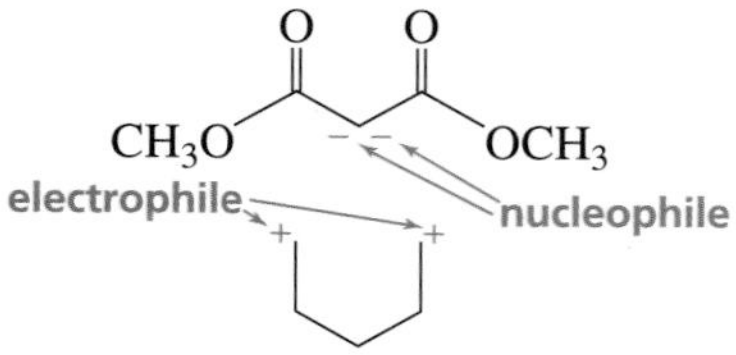

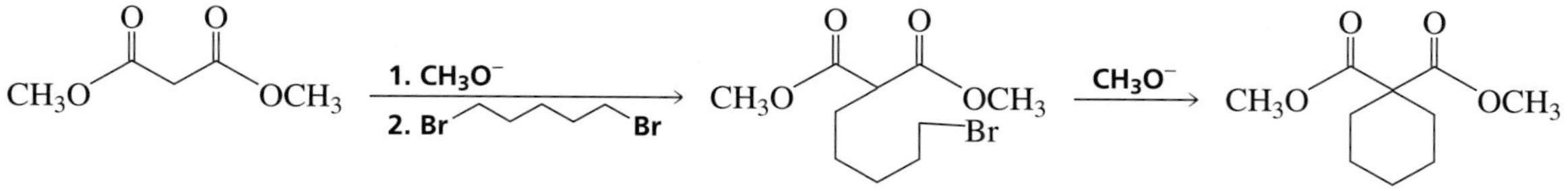

In planning the following synthesis, the diester that is given as the starting material suggests that we should use a Dieckmann condensation to obtain the cyclic compound.

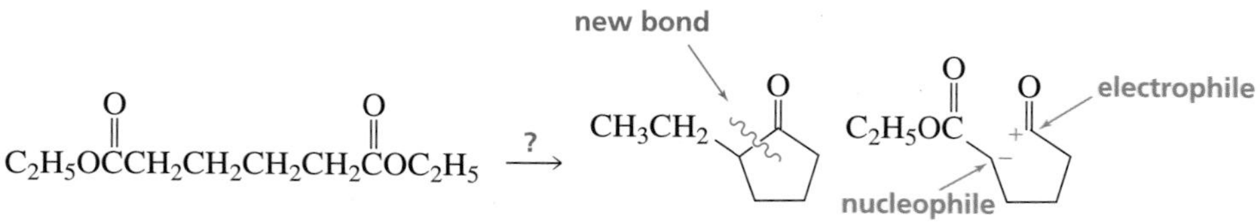

After the cyclopentanone ring is formed from a Dieckmann condensation of the starting material, alkylation of the α-carbon followed by hydrolysis of the β-keto ester and decarboxylation forms the desired product.

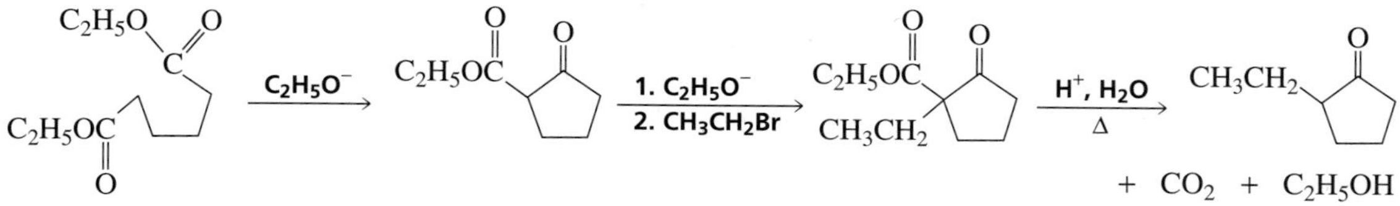

$+ CO_2 + C_2H_5OH$

PROBLEM 34

Design a synthesis for each of the following compounds.

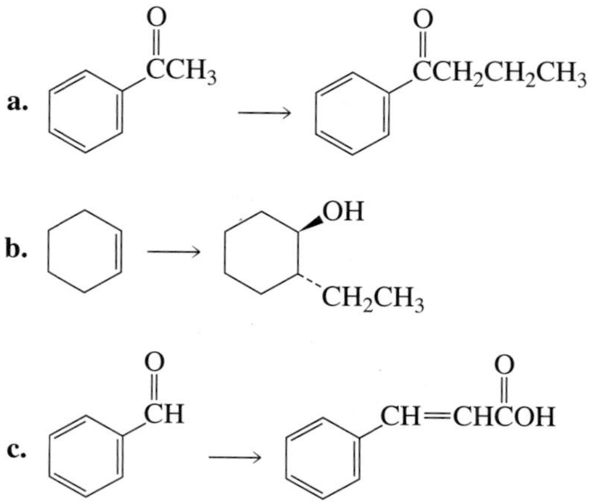

a.

b.

c.

d. 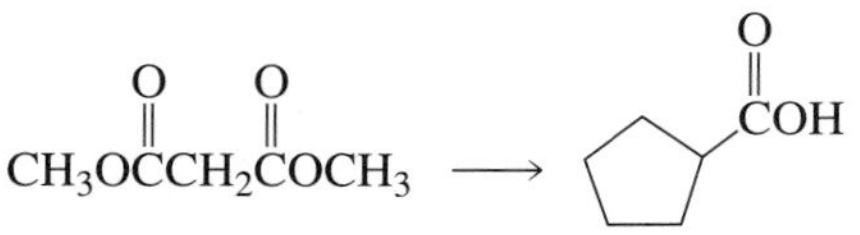 $CH_3OCCH_2COCH_3 \longrightarrow$

(with structures of diketone and cyclopentane-COOH product shown)

A Biological Aldol Addition

D-glucose, the most abundant sugar found in nature, is synthesized by biological systems from two molecules of pyruvate. This is called **gluconeogenesis.** The breakdown of D-glucose into two molecules of pyruvate (the reverse of gluconeogenesis) is called **glycolysis.**

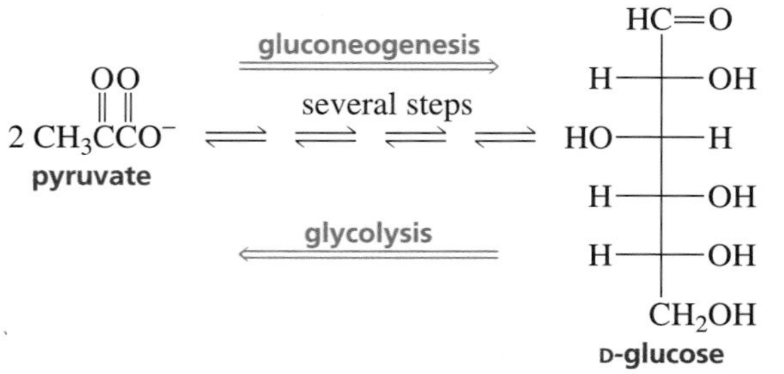

(Scheme showing: $2\ CH_3CCO^-$ **pyruvate** ⇌ via **gluconeogenesis / several steps** and **glycolysis** to **D-glucose** structure)

Because D-glucose has twice as many carbons as does pyruvate, it is not surprising that one of the steps in the biosynthesis of D-glucose is an aldol addition. In this step, the enzyme aldolase catalyzes the aldol addition between a molecule of dihydroxyacetone phosphate and a molecule of D-glyceraldehyde-3-phosphate. The product is fructose-1,6-diphosphate, which is subsequently converted to D-glucose.

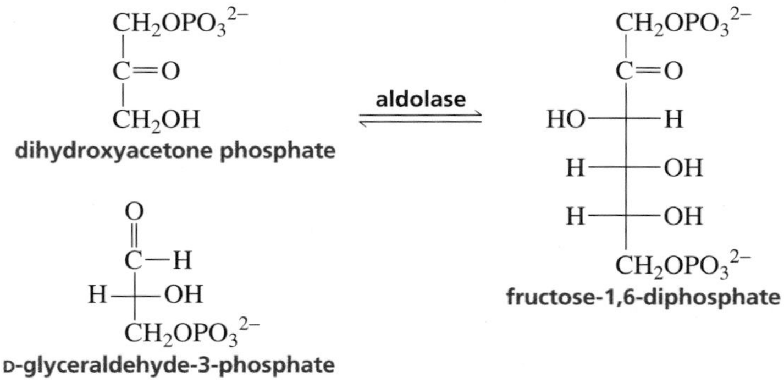

(Structures: **dihydroxyacetone phosphate** + **D-glyceraldehyde-3-phosphate** ⇌ **aldolase** → **fructose-1,6-diphosphate**)

PROBLEM 35

Propose a mechanism for the formation of fructose-1,6-diphosphate from dihydroxyacetone phosphate and glyceraldehyde-3-phosphate using HO^- as the catalyst. (The mechanism by which aldolase catalyzes the reaction will be discussed in Section 21.8.)

A Biological Aldol Condensation

Collagen is the most abundant protein in mammals, amounting to about one-fourth of the total protein. It is the major fibrous component of bone, teeth, skin, cartilage, and tendons. It also is responsible for holding groups of cells together in discrete units. Individual molecules of collagen are called tropocollagen. Tropocollagen can be isolated only from tissues of young animals. As animals age, the individual molecules become cross-linked. This is why meat from older animals is tougher than meat from younger ones. Collagen cross-linking is an example of an aldol condensation.

Before collagen molecules can cross-link, the primary amino groups of the lysine residues of collagen must be converted to aldehyde groups. The enzyme that catalyzes this reaction is called lysyl oxidase. An aldol condensation between two aldehyde residues results in a cross-linked protein.

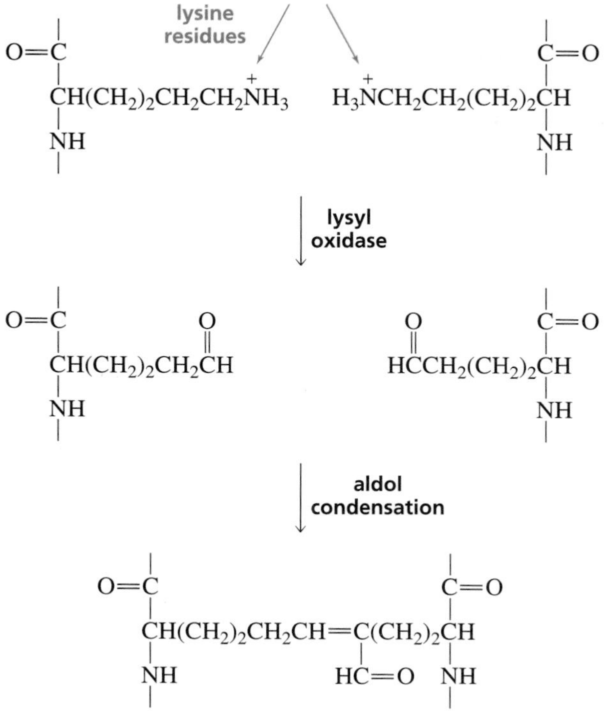

cross-linked collagen

A Biological Claisen Condensation

Fatty acids are long-chain, unbranched carboxylic acids. Most naturally occurring fatty acids contain an even number of carbons because they are synthesized from acetate, an ion with two carbons.

In Section 15.17 we saw that carboxylic acids can be activated in biological systems by being converted to thioesters of coenzyme A.

$$\underset{\text{acetate}}{CH_3\overset{O}{\overset{\|}{C}}{-}O^-} + \underset{\text{coenzyme A}}{CoASH} + ATP \longrightarrow \underset{\text{acetyl-CoA}}{CH_3\overset{O}{\overset{\|}{C}}{-}SCoA} + AMP + \underset{\text{pyrophosphate}}{PP_i}$$

One of the necessary reactants for fatty acid synthesis is malonyl-CoA, which is obtained by carboxylation of acetyl-CoA. A possible mechanism for this reaction will be discussed in Section 22.5.

$$CH_3\overset{O}{\overset{\|}{C}}-SCoA + HCO_3^- \longrightarrow {}^-O-\overset{O}{\overset{\|}{C}}-CH_2\overset{O}{\overset{\|}{C}}SCoA$$

acetyl-CoA   **malonyl-CoA**

Before fatty acid synthesis can occur, however, the acyl groups of acetyl-CoA and malonyl-CoA are transferred to other thiols by means of a transesterification reaction.

$$CH_3\overset{O}{\overset{\|}{C}}-SCoA + RSH \longrightarrow CH_3\overset{O}{\overset{\|}{C}}-SR + CoASH$$

$${}^-O-\overset{O}{\overset{\|}{C}}-CH_2\overset{O}{\overset{\|}{C}}-SCoA + RSH \longrightarrow {}^-O-\overset{O}{\overset{\|}{C}}-CH_2\overset{O}{\overset{\|}{C}}-SR + CoASH$$

The first step in the biosynthesis of a fatty acid is a Claisen condensation between a molecule of acetyl thioester and a molecule of malonyl thioester. We have seen that in the laboratory the nucleophile needed for a Claisen condensation is obtained by using a strong base to remove an α-proton. A strong base is not available for biological reactions because they take place at neutral pH. So the required nucleophile is generated by removing CO_2 from the α-carbon of the malonylthioester. Loss of CO_2 also serves to drive the condensation reaction to completion (Le Châtelier's principle; Section 9.3). The product of the condensation reaction undergoes a reduction, a dehydration, and a second reduction to give a four-carbon thioester.

$$CH_3\overset{O}{\overset{\|}{C}}-SR \quad :\overset{..}{O}-\overset{O}{\overset{\|}{C}}-CH_2\overset{O}{\overset{\|}{C}}-SR \longrightarrow CH_3\overset{:\overset{..}{O}:}{\overset{\|}{C}}-CH_2\overset{O}{\overset{\|}{C}}-SR \longrightarrow CH_3\overset{O}{\overset{\|}{C}}-CH_2\overset{O}{\overset{\|}{C}}-SR$$

$$SR + CO_2 \qquad\qquad \big\downarrow \text{reduction}$$

$$CH_3CH_2CH_2\overset{O}{\overset{\|}{C}}SR \xleftarrow{\text{reduction}} CH_3CH=CH\overset{O}{\overset{\|}{C}}-SR \xleftarrow{\text{dehydration}} CH_3\overset{OH}{\overset{|}{C}H}-CH_2\overset{O}{\overset{\|}{C}}-SR$$

The four-carbon thioester undergoes a Claisen condensation with another molecule of malonyl thioester. Again, the product of the condensation reaction undergoes a reduction, a dehydration, and a second reduction to form a six-carbon thioester. This sequence of reactions is repeated, and each time two more carbons are added to the chain. This mechanism explains why most naturally occurring fatty acids contain an even number of carbons.

$$CH_3CH_2CH_2\overset{O}{\overset{\|}{C}}-SR + {}^-O-\overset{O}{\overset{\|}{C}}-CH_2\overset{O}{\overset{\|}{C}}-SR \xrightarrow[\text{condensation}]{\text{Claisen}} CH_3CH_2CH_2\overset{O}{\overset{\|}{C}}-CH_2\overset{O}{\overset{\|}{C}}-SR$$

$$\big\downarrow \begin{array}{l} \text{1. reduction} \\ \text{2. dehydration} \\ \text{3. reduction} \end{array}$$

$$CH_3CH_2CH_2CH_2CH_2\overset{O}{\overset{\|}{C}}-SR$$

Once a thioester with the appropriate number of carbons is obtained, it undergoes a transesterification reaction with glycerol in order to form fats, oils, and phospholipids (Section 23.3).

PROBLEM 36 ◆

Palmitic acid (hexadecanoic acid) is a 16-carbon fatty acid. How many moles of malonyl-CoA are required for the synthesis of one mole of palmitic acid?

A Biological Decarboxylation

An example of a decarboxylation of a 3-oxocarboxylic acid that occurs in biological systems is the decarboxylation of acetoacetate. Acetoacetate decarboxylase, the enzyme that catalyzes the reaction, first forms an imine with acetoacetate. The imine is protonated under physiological conditions and therefore readily accepts the pair of electrons left behind when the substrate loses CO_2. Decarboxylation leads to the formation of an enamine. Hydrolysis of the enamine produces the decarboxylated product (acetone) and regenerates the enzyme (Section 16.7).

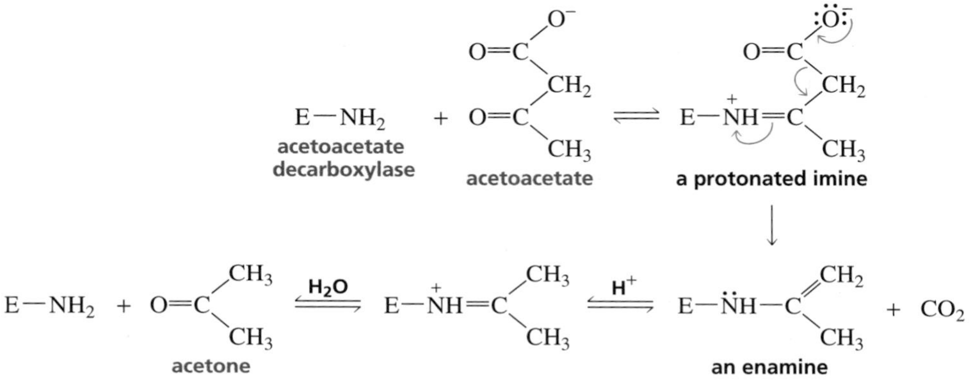

PROBLEM 37

When the enzymatic decarboxylation of acetoacetate is carried out in $H_2^{18}O$, the acetone that is formed contains ^{18}O. What does this tell you about the mechanism of the reaction?

SUMMARY OF REACTIONS

1. Halogenation of the α-carbon of aldehydes and ketones (Section 18.4)

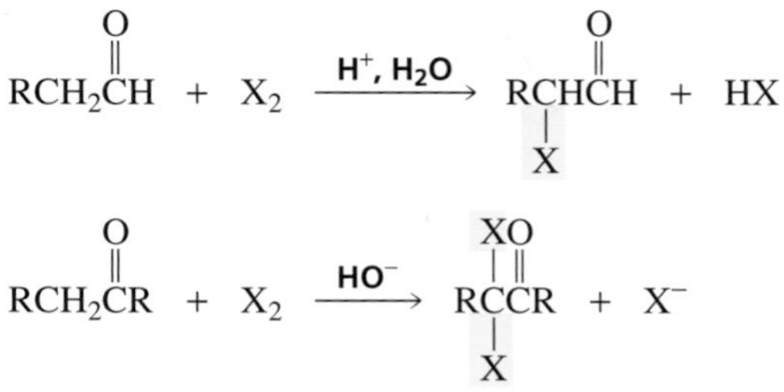

$$RCH_2\overset{O}{\overset{\|}{C}}H + X_2 \xrightarrow{H^+, H_2O} RCH\overset{O}{\overset{\|}{C}}H + HX$$
$$\underset{X}{|}$$

$$RCH_2\overset{O}{\overset{\|}{C}}R + X_2 \xrightarrow{HO^-} R\overset{XO}{\overset{|\,\|}{C}}CR + X^-$$
$$\underset{X}{|}$$

$$X_2 = Cl_2, Br_2, I_2$$

2. Halogenation of the α-carbon of carboxylic acids (Section 18.5)

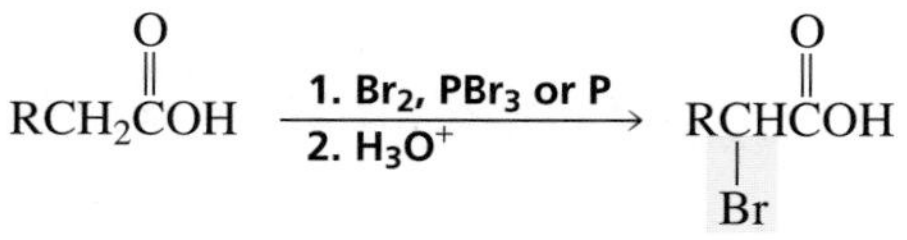

$$RCH_2\overset{\overset{\displaystyle O}{\|}}{C}OH \xrightarrow[\text{2. H}_3\text{O}^+]{\text{1. Br}_2, \text{PBr}_3 \text{ or P}} R\overset{\overset{\displaystyle O}{\|}}{\underset{\underset{\displaystyle Br}{|}}{C}H}COH$$

3. Synthesis of α,β-unsaturated carbonyl compounds (Section 18.6)

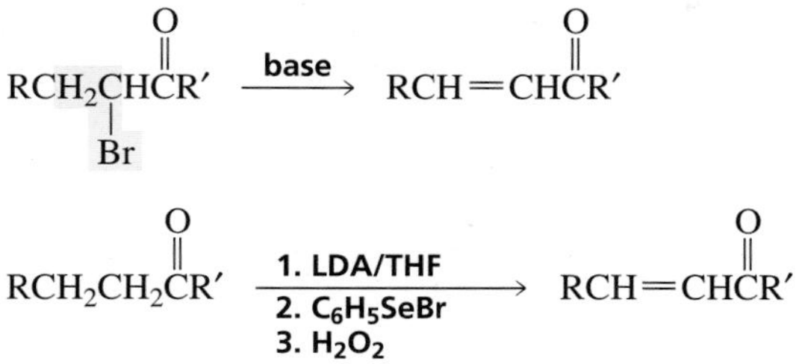

$$RCH_2\overset{\overset{\displaystyle O}{\|}}{\underset{\underset{\displaystyle Br}{|}}{C}H}CR' \xrightarrow{\text{base}} RCH{=}CH\overset{\overset{\displaystyle O}{\|}}{C}R'$$

$$RCH_2CH_2\overset{\overset{\displaystyle O}{\|}}{C}R' \xrightarrow[\substack{\text{2. C}_6\text{H}_5\text{SeBr} \\ \text{3. H}_2\text{O}_2}]{\text{1. LDA/THF}} RCH{=}CH\overset{\overset{\displaystyle O}{\|}}{C}R'$$

4. Alkylation of the α-carbon of carbonyl compounds (Section 18.8)

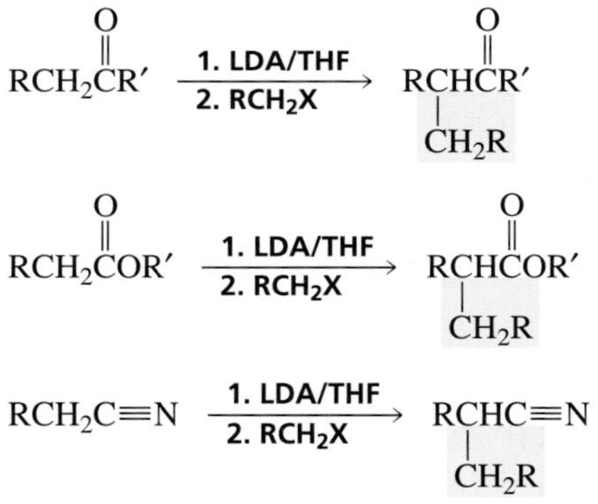

$$RCH_2\overset{\overset{\displaystyle O}{\|}}{C}R' \xrightarrow[\text{2. RCH}_2\text{X}]{\text{1. LDA/THF}} R\overset{\overset{\displaystyle O}{\|}}{\underset{\underset{\displaystyle CH_2R}{|}}{C}H}CR'$$

$$RCH_2\overset{\overset{\displaystyle O}{\|}}{C}OR' \xrightarrow[\text{2. RCH}_2\text{X}]{\text{1. LDA/THF}} R\overset{\overset{\displaystyle O}{\|}}{\underset{\underset{\displaystyle CH_2R}{|}}{C}H}COR'$$

$$RCH_2C{\equiv}N \xrightarrow[\text{2. RCH}_2\text{X}]{\text{1. LDA/THF}} R\underset{\underset{\displaystyle CH_2R}{|}}{C}HC{\equiv}N$$

$$X = Cl, Br, I$$

5. Alkylation and acylation of the α-carbon of aldehydes and ketones by means of an enamine intermediate (Section 18.8)

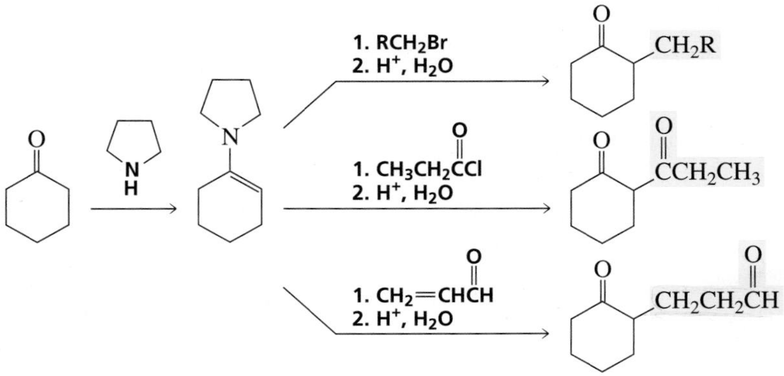

6. Michael reaction: attack of an enolate on an α,β-unsaturated carbonyl compound (Section 18.9)

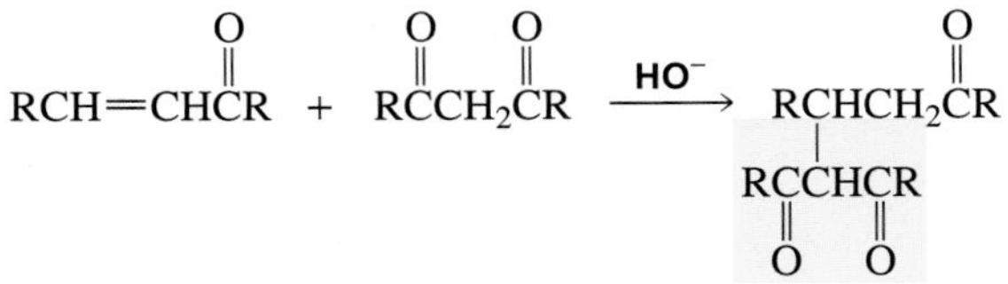

$$RCH=CHCR \;+\; RCCH_2CR \;\xrightarrow{HO^-}\; RCHCH_2CR$$

7. Aldol addition of two aldehydes, two ketones, or an aldehyde and a ketone (Sections 18.10 and 18.12)

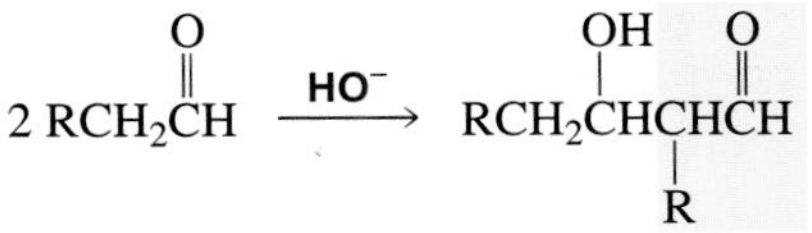

$$2\,RCH_2CH \;\xrightarrow{HO^-}\; RCH_2CHCHCH$$

8. Aldol condensation: dehydration of the product of an aldol addition (Section 18.11)

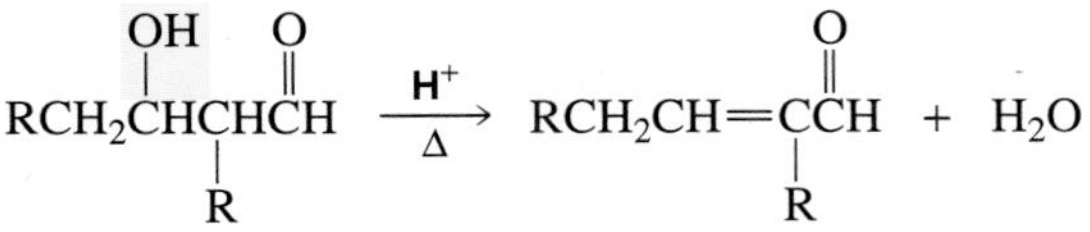

$$RCH_2CHCHCH \;\xrightarrow[\Delta]{H^+}\; RCH_2CH=CCH \;+\; H_2O$$

9. Claisen condensation of two esters (Sections 18.13 and 18.14)

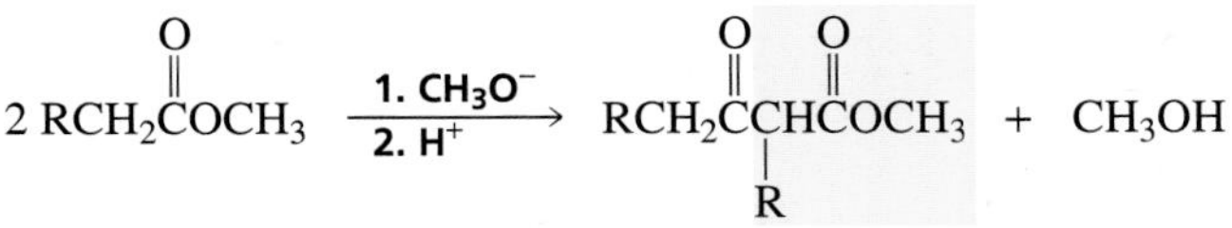

$$2\,RCH_2COCH_3 \;\xrightarrow[\text{2. }H^+]{\text{1. } CH_3O^-}\; RCH_2CCHCOCH_3 \;+\; CH_3OH$$

10. Condensation of a ketone and an ester (Section 18.14)

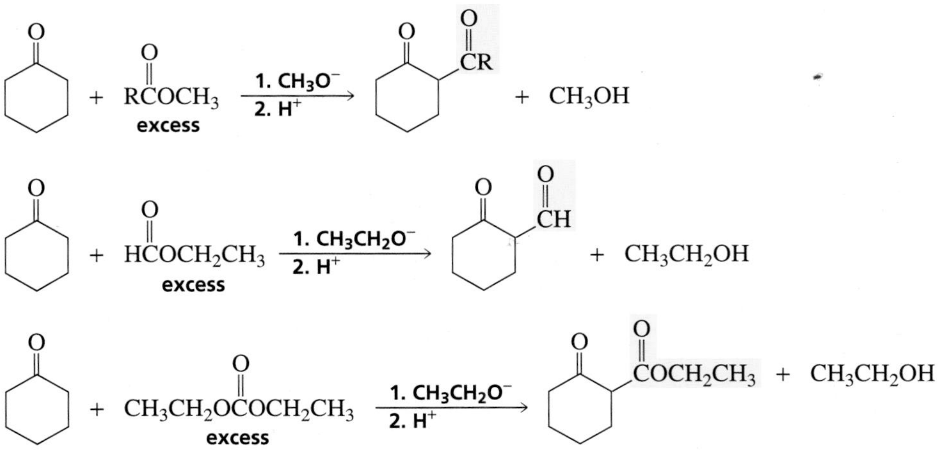

11. Robinson annulation (Section 18.15)

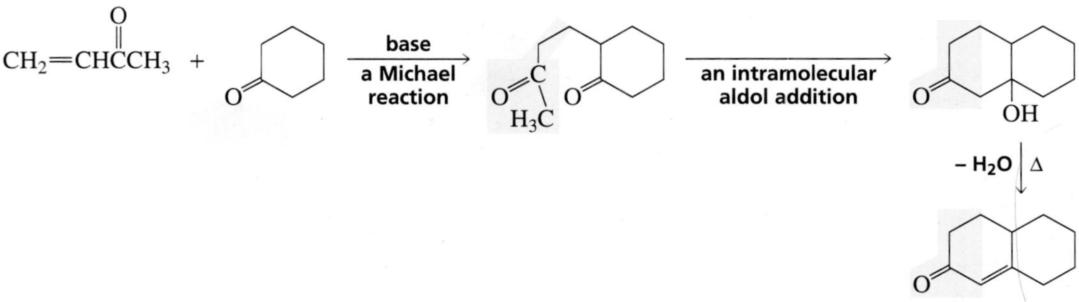

12. Decarboxylation of 3-oxocarboxylic acids (Section 18.16)

$$\underset{\text{RCCH}_2\text{COH}}{\overset{\text{O} \quad \text{O}}{\parallel \quad \parallel}} \xrightarrow{\Delta} \underset{\text{RCCH}_3}{\overset{\text{O}}{\parallel}} + CO_2$$

13. Malonic ester synthesis: preparation of carboxylic acids (Section 18.17)

$$\underset{\text{C}_2\text{H}_5\text{OCCH}_2\text{COC}_2\text{H}_5}{\overset{\text{O} \quad \text{O}}{\parallel \quad \parallel}} \xrightarrow[\substack{\text{2. RBr} \\ \text{3. H}^+, \text{H}_2\text{O}, \Delta}]{\text{1. CH}_3\text{CH}_2\text{O}^-} \underset{\text{RCH}_2\text{COH}}{\overset{\text{O}}{\parallel}} + CO_2$$

14. Acetoacetic ester synthesis: preparation of methyl ketones (Section 18.18)

$$\underset{\text{CH}_3\text{CCH}_2\text{COC}_2\text{H}_5}{\overset{\text{O} \quad \text{O}}{\parallel \quad \parallel}} \xrightarrow[\substack{\text{2. RBr} \\ \text{3. H}^+, \text{H}_2\text{O}, \Delta}]{\text{1. CH}_3\text{CH}_2\text{O}^-} \underset{\text{RCH}_2\text{CCH}_3}{\overset{\text{O}}{\parallel}} + CO_2$$

KEY TERMS

acetoacetic ester synthesis (page 850)
aldol addition (page 833)
aldol condensation (page 835)
ambident nucleophile (page 822)
annulation reaction (page 844)
α-carbon (page 816)
carbon acid (page 817)
Claisen condensation (page 838)
condensation reaction (page 835)
crossed aldol addition (page 836)
decarboxylation (page 846)

Dieckmann condensation (page 841)
β-diketone (page 819)
enolization (page 820)
gluconeogenesis (page 853)
glycolysis (page 853)
Hell–Volhard–Zelinski (HVZ)
    reaction (page 825)
Hunsdiecker reaction (page 847)
α-hydrogen (page 816)
keto–enol interconversion
    (page 820)
keto–enol tautomerization (page 820)

β-keto ester (page 819)
Kolbe–Schmitt carboxylation reaction
    (page 831)
malonic ester synthesis (page 848)
Michael reaction (page 832)
mixed aldol addition (page 836)
mixed Claisen condensation (page 840)
Robinson annulation (page 844)
selenenylation reaction (page 826)
Stork enamine reaction (page 833)
α-substitution reaction (page 822)
tautomers (page 819)

PROBLEMS

38. Give a structure for each of the following:
   a. ethyl acetoacetate
   b. α-methylmalonic acid
   c. a β-keto ester
   d. benzeneselenenyl bromide
   e. the enol tautomer of cyclopentanone
   f. the carboxylic acid obtained from the
      malonic ester synthesis when the alkyl
      halide is propyl bromide

39. Give the products of the following reactions.
   a. diethyl heptanedioate: (1) sodium ethoxide; (2) H$^+$, H$_2$O
   b. pentanoic acid + PBr$_3$ + Br$_2$ followed by hyrolysis
   c. acetone + ethyl acetate: (1) sodium ethoxide; (2) H$^+$, H$_2$O
   d. diethyl 2-ethylhexanedioate: (1) sodium ethoxide; (2) H$^+$, H$_2$O
   e. methyl cyclohexanecarboxylate: (1) LDA in THF; (2) C$_6$H$_5$SeBr; (3) H$_2$O$_2$
   f. diethyl malonate: (1) sodium ethoxide; (2) isobutyl bromide; (3) H$^+$, H$_2$O + Δ
   g. acetophenone + diethyl carbonate: (1) sodium ethoxide; (2) H$^+$
   h. 1,3-cyclohexanedione + allyl bromide + sodium hydroxide
   i. dibenzyl ketone + methyl vinyl ketone + excess sodium hydroxide

    j. cyclopentanone: (1) pyrrolidine + Δ; (2) ethyl bromide; (3) H$^+$, H$_2$O
    k. γ-butyrolactone + LDA in THF followed by methyl iodide
    l. 2,7-octanedione + sodium hydroxide
    m. cyclohexanone + NaOD in D$_2$O
    n. diethyl 1,2-benzenedicarboxylate + ethyl acetate: (1) excess sodium ethoxide; (2) H$^+$

40. **a.** When (*R*)-2-methyl-1-phenyl-1-butanone is dissolved in an acidic or basic aqueous solution, a racemic mixture of 2-methyl-1-phenyl-1-butanone is formed. Explain.
    b. Give an example of another ketone that would undergo acid- or base-catalyzed racemization.

41. A β,γ-unsaturated carbonyl compound rearranges to a more stable conjugated α,β-unsaturated compound in the presence of either acid or base.
    a. Propose a mechanism for the base-catalyzed rearrangement.
    b. Propose a mechanism for the acid-catalyzed rearrangement.

                    H$^+$ or HO$^-$

a β,γ-unsaturated   an α,β-unsaturated
carbonyl compound   carbonyl compound

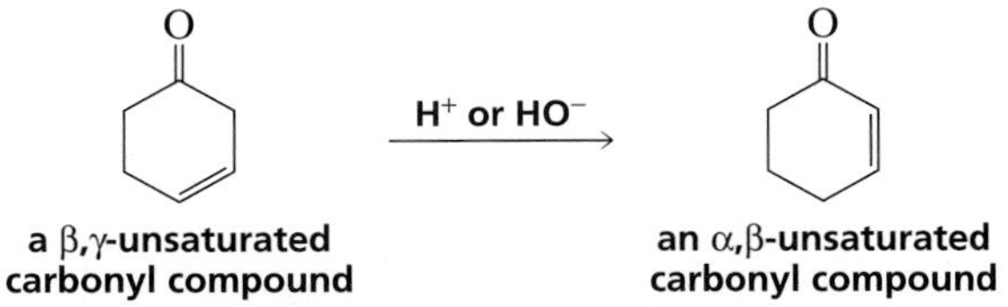

42. There are other condensation reactions similar to the aldol and Claisen condensations.
    a. The Perkin condensation is the condensation of an aromatic aldehyde and acetic anhydride. Give the product obtained from the following Perkin condensation.

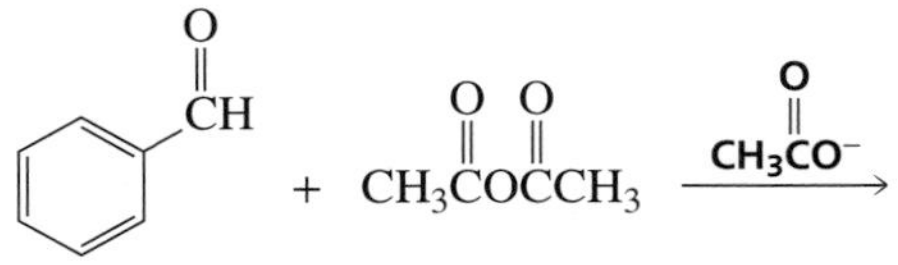

    b. What compound would result if water were added to the product of the Perkin condensation?
    c. The Knoevenagel condensation is the condensation of an aldehyde or a ketone that has no α-hydrogens and a compound such as diethyl malonate that has an α-carbon flanked by two electron-withdrawing groups. Give the product obtained from the following Knoevenagel condensation.

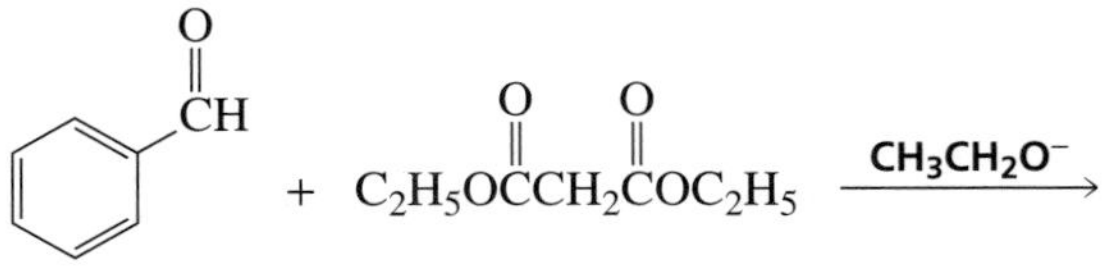

    d. What product would be obtained if the product of the Knoevenagel condensation were heated in an aqueous acidic solution?

43. The Reformatsky reaction is an addition reaction in which an organozinc reagent is used, instead of a Grignard reagent, to attack the carbonyl group of an aldehyde or a ketone. Because the organozinc reagent is less reactive than a Grignard reagent, a second nucleophilic addition to the ester group does not occur. The organozinc reagent is prepared by treating an α-bromo ester with zinc.

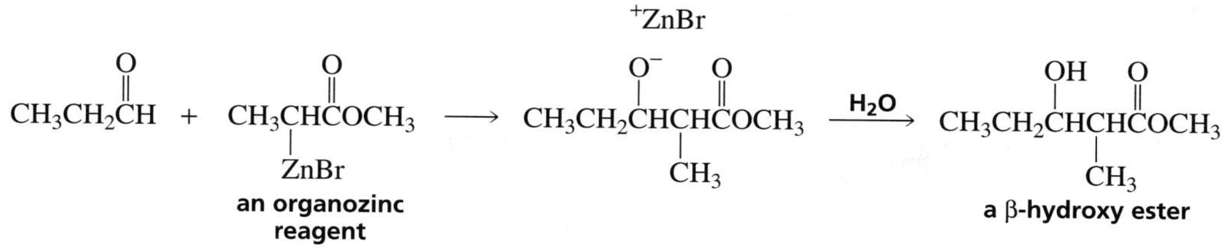

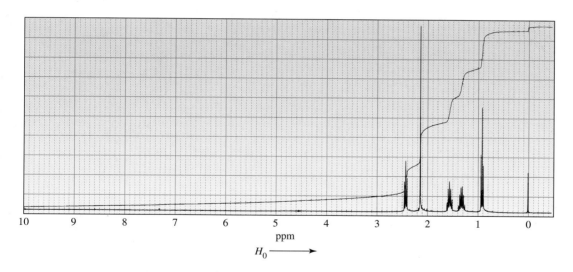

$$\underset{\substack{\text{O} \\ \text{CH}_3\text{CH}_2\overset{\|}{\text{CH}}}}{} + \underset{\substack{\text{O} \\ \text{CH}_3\overset{\|}{\text{CHCOCH}_3} \\ | \\ \text{ZnBr}}}{} \longrightarrow \underset{\substack{\text{O}^- \quad \text{O} \\ \text{CH}_3\text{CH}_2\overset{}{\text{CHCHCOCH}_3} \\ | \\ \text{CH}_3}}{} \xrightarrow{\text{H}_2\text{O}} \underset{\substack{\text{OH} \quad \text{O} \\ \text{CH}_3\text{CH}_2\overset{}{\text{CHCHCOCH}_3} \\ | \\ \text{CH}_3}}{}$$

an organozinc
reagent

a β-hydroxy ester

Describe how each of the following compounds could be prepared using a
Reformatsky reaction.

a. 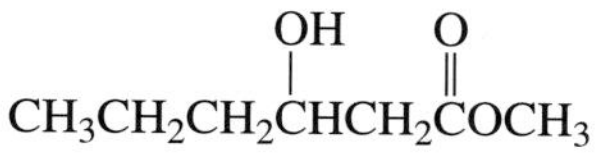 $\underset{\substack{\text{OH} \quad \text{O} \\ \text{CH}_3\text{CH}_2\text{CH}_2\text{CHCH}_2\text{COCH}_3}}{}$

c. 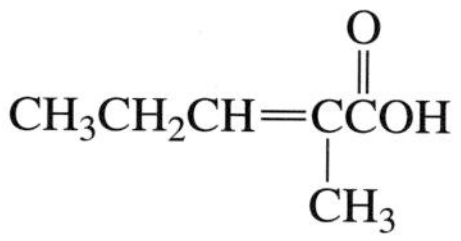 $\underset{\substack{\text{O} \\ \text{CH}_3\text{CH}_2\text{CH}{=}\text{CCOH} \\ | \\ \text{CH}_3}}{}$

b. 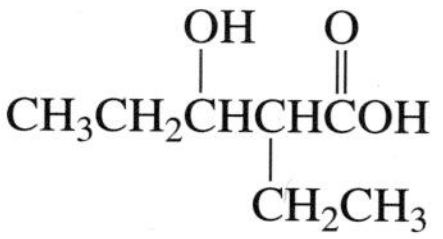 $\underset{\substack{\text{OH} \quad \text{O} \\ \text{CH}_3\text{CH}_2\text{CHCHCOH} \\ | \\ \text{CH}_2\text{CH}_3}}{}$

d. 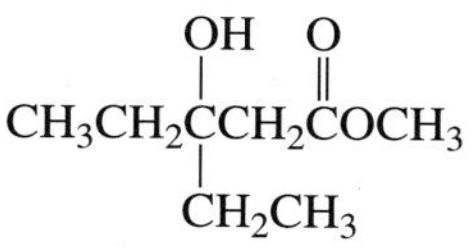 $\underset{\substack{\text{OH} \quad \text{O} \\ \text{CH}_3\text{CH}_2\text{CCH}_2\text{COCH}_3 \\ | \\ \text{CH}_2\text{CH}_3}}{}$

44. The ketone whose ^{1}H NMR spectrum is shown below was obtained as the product of
an acetoacetic ester synthesis. What alkyl halide was used in the synthesis?

[NMR spectrum with x-axis labeled from 10 to 0 ppm, $H_0 \longrightarrow$]

45. Indicate how the following compounds could be synthesized from cyclohexanone and
any other necessary reagents.

a. 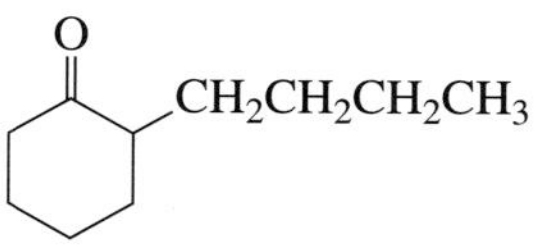 [cyclohexanone with CH₂CH₂CH₂CH₃ substituent]

c. 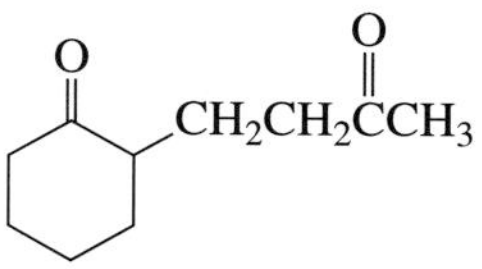 [cyclohexanone with CH₂CH₂CCH₃ (ketone) substituent]

b. 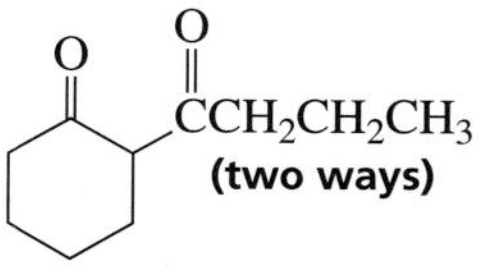 [cyclohexanone with CCH₂CH₂CH₃ (ketone) substituent]
(two ways)

d. 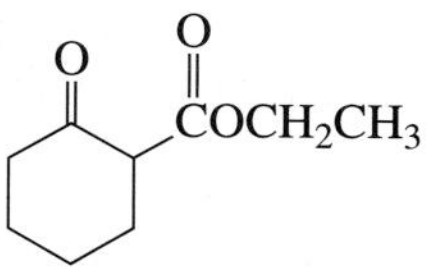 [cyclohexanone with COCH₂CH₃ (ester) substituent] e.

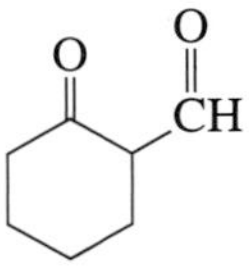

f.

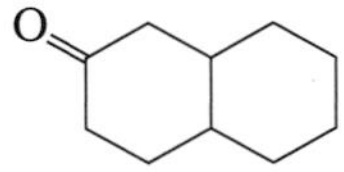

46. The iodoform test is specific for a methyl ketone. When excess I_2 is added to a basic solution of the ketone, a yellow precipitate forms if the ketone is a methyl ketone. The precipitate is iodoform (CHI_3).

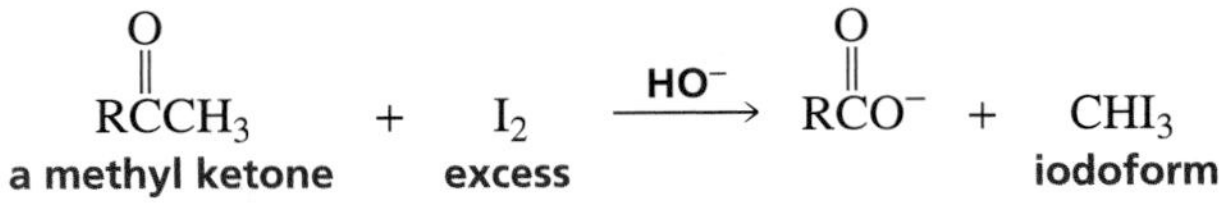

$$\underset{\text{a methyl ketone}}{RCCH_3} \quad + \quad \underset{\text{excess}}{I_2} \quad \xrightarrow{HO^-} \quad RCO^- \quad + \quad \underset{\text{iodoform}}{CHI_3}$$

a. Propose a mechanism for this reaction.
b. Will 2-butanol give a positive iodoform test?

47. A compound with a molecular formula of C_6H_{10} has two peaks in its NMR spectrum, both of which are singlets (ratio, 9:1). It reacts with an acidic aqueous solution containing mercuric sulfate to form a compound that also has an NMR spectrum that shows two singlets (ratio, 3:1) and that gives a positive iodoform test. Identify this compound.

48. Indicate how each of the following compounds could be synthesized from the given starting material and any other necessary reagents.

a. 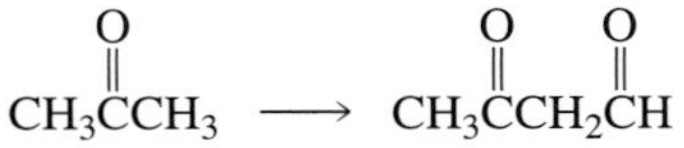 $CH_3CCH_3 \longrightarrow CH_3CCH_2CH$

b. 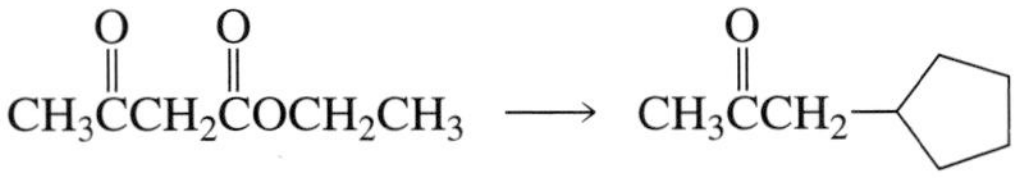 $CH_3CCH_2COCH_2CH_3 \longrightarrow CH_3CCH_2-$

c. 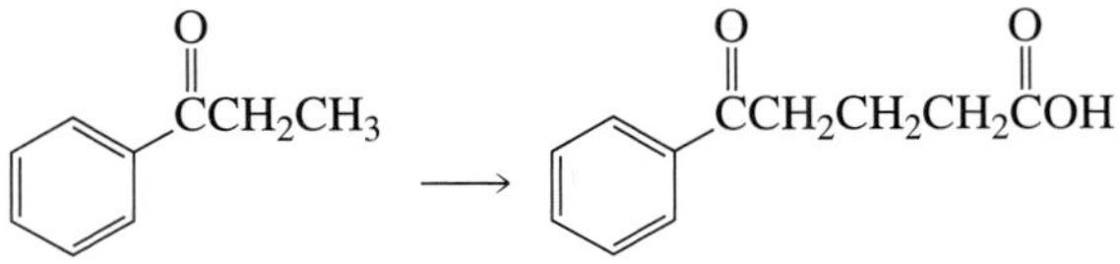 $\longrightarrow$

d. 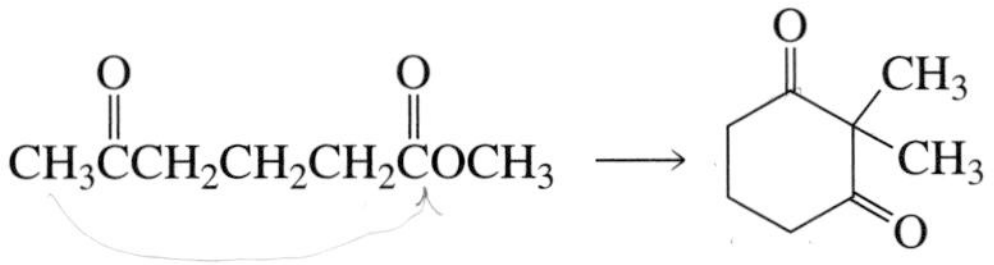 $CH_3CCH_2CH_2CH_2COCH_3 \longrightarrow$

e. 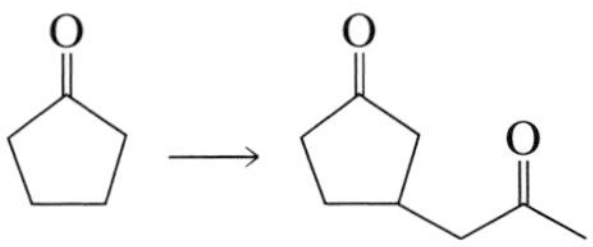 $\longrightarrow$

f.  $CH_3CH_2OCCH_2CH_2CH_2CH_2COCH_2CH_3 \longrightarrow$

49. Bupropion hydrochloride is an antidepressant marketed under the trade name Wellbutrin. Propose a synthesis of bupropion hydrochloride starting with benzene.

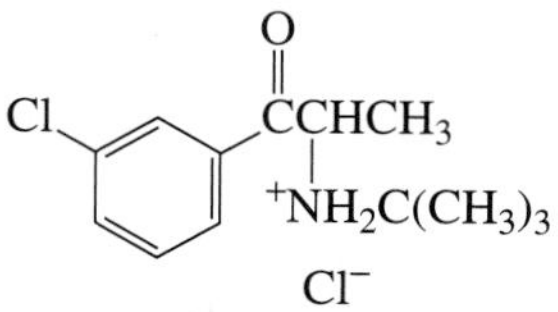

bupropion hydrochloride

50. What reagents would be required to carry out the following transformations?

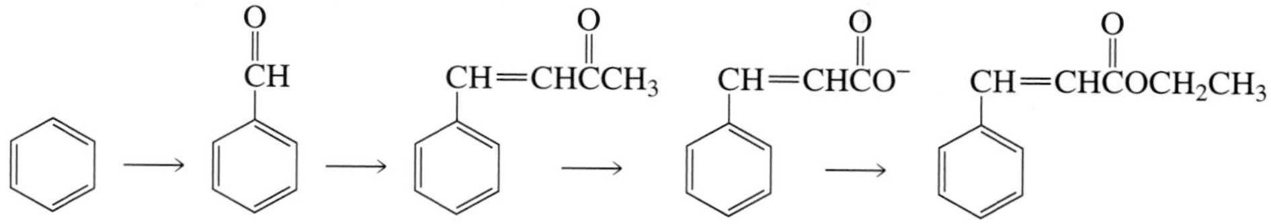

51. **a.** If the biosynthesis of palmitic acid, a 16-carbon fatty acid, were carried out with CD_3COSR and nondeuterated malonyl thioester, how many deuteriums would be incorporated into palmitic acid?

 b. If the biosynthesis of palmitic acid were carried out with $^-OOCCD_2COSR$ and nondeuterated acetyl thioester, how many deuteriums would be incorporated into palmitic acid?

52. Give the products of the following reactions.

a. 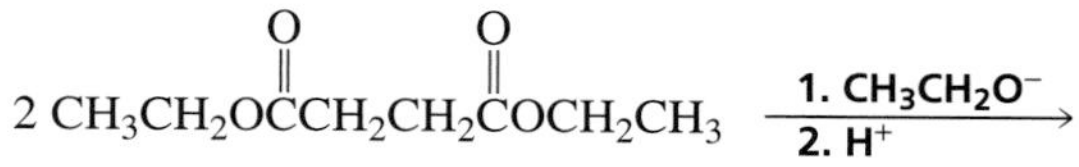 2 $CH_3CH_2OCCH_2CH_2COCH_2CH_3$ $\xrightarrow[\text{2. H}^+]{\text{1. CH}_3\text{CH}_2\text{O}^-}$

b.

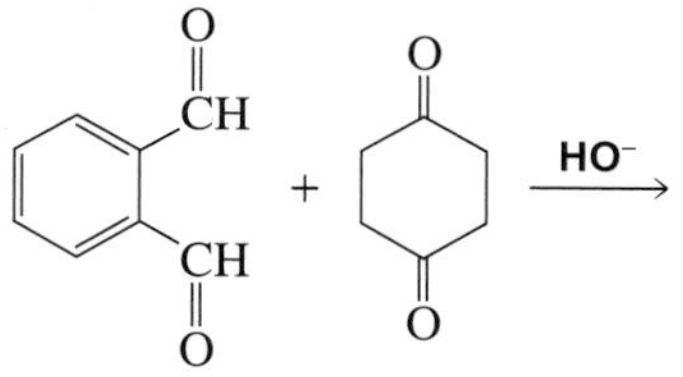

53. **a.** Show how the amino acid alanine can be synthesized from propanoic acid.

 b. Show how the amino acid glycine can be synthesized from phthalimide and diethyl 2-bromomalonate.

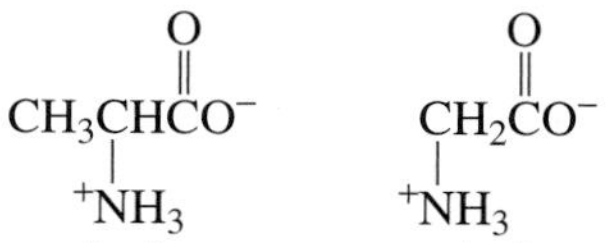

54. Cindy Synthon tried to prepare the following compounds using aldol condensations. Which of the compounds was she successful in synthesizing? Explain why the other syntheses were not successful.

a. 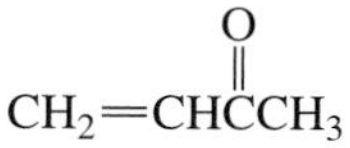 $CH_2=CHCCH_3$

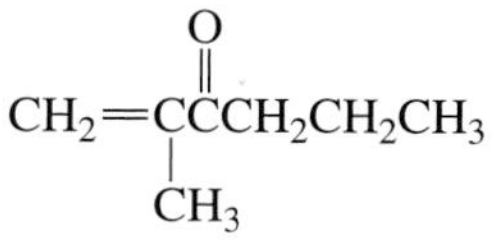

b. 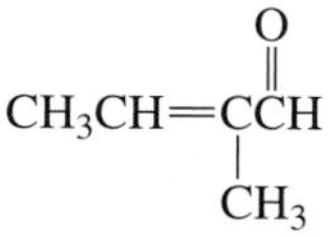 $CH_3CH=CCH$ with CH_3

c. $CH_2=CCH_2CH_2CH_3$ with CH_3

d. 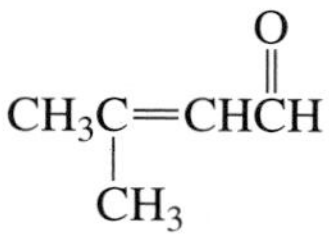 $CH_3C=CHCH$ with CH_3

e. 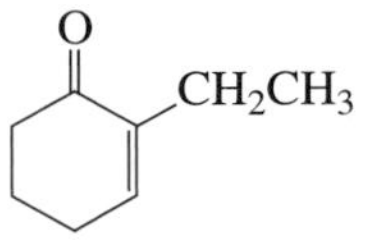 (cyclohexenone with CH_2CH_3 substituent)

h. 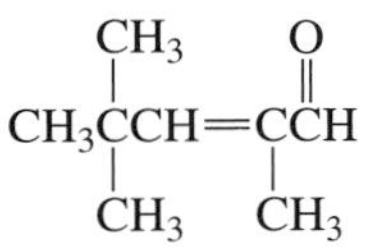 $CH_3\overset{\displaystyle CH_3}{\underset{\displaystyle CH_3}{C}}CH=\overset{\displaystyle O}{C}CH$

f. 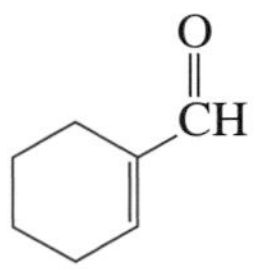 (cyclohexene-carbaldehyde)

i. 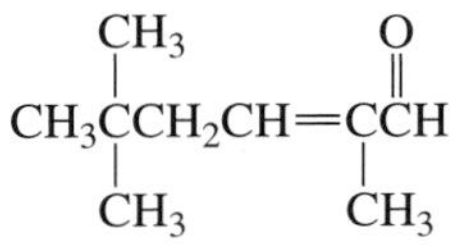 $CH_3\overset{\displaystyle CH_3}{C}CH_2CH=\overset{\displaystyle O}{C}CH$ with CH_3

g. 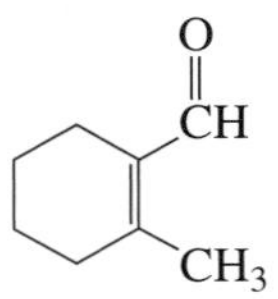 (cyclohexene with CH=O and CH_3 substituents)

55. Explain why the product obtained depends on the number of equivalents of base used in the reaction.

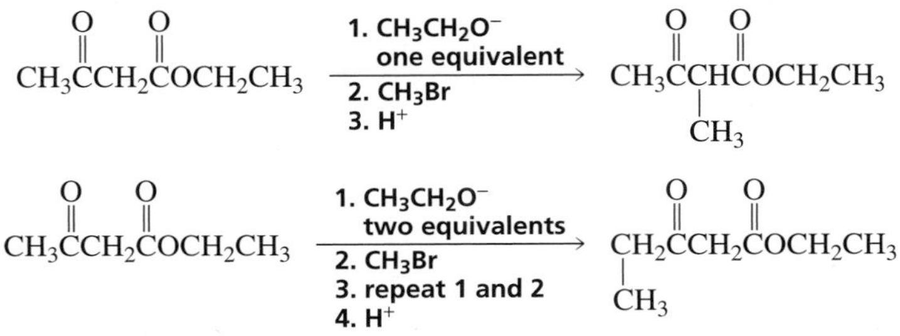

$$CH_3\overset{O}{C}CH_2\overset{O}{C}OCH_2CH_3 \quad \xrightarrow[\substack{\text{2. CH}_3\text{Br} \\ \text{3. H}^+}]{\substack{\text{1. CH}_3\text{CH}_2\text{O}^- \\ \text{one equivalent}}} \quad CH_3\overset{O}{C}\overset{CH_3}{CH}\overset{O}{C}OCH_2CH_3$$

$$CH_3\overset{O}{C}CH_2\overset{O}{C}OCH_2CH_3 \quad \xrightarrow[\substack{\text{2. CH}_3\text{Br} \\ \text{3. repeat 1 and 2} \\ \text{4. H}^+}]{\substack{\text{1. CH}_3\text{CH}_2\text{O}^- \\ \text{two equivalents}}} \quad CH_2\overset{O}{C}CH_2\overset{O}{C}OCH_2CH_3 \text{ with } CH_3$$

56. Amobarbital is a sedative marketed under the trade name Amytal. Propose a synthesis of amobarbital starting with diethyl malonate.

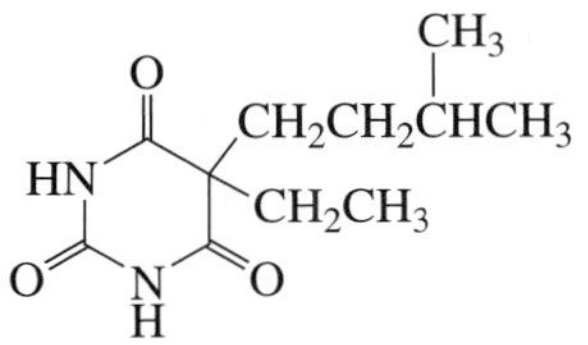

(structure of amobarbital: a barbituric acid ring with substituents $CH_2CH_2CHCH_3$ / CH_3 and CH_2CH_3)

amobarbital

57. Fill in the boxes.

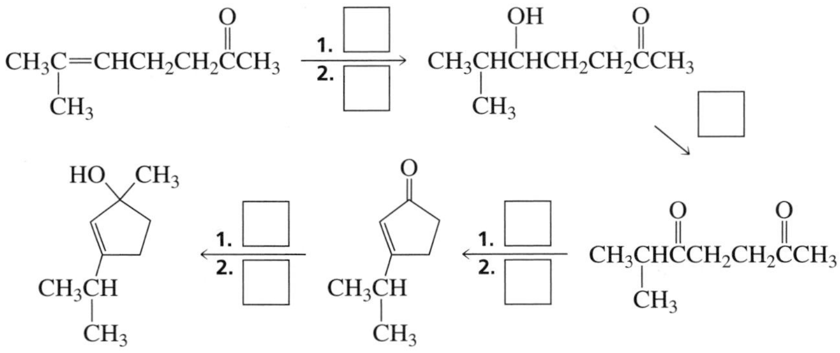

$$CH_3\underset{CH_3}{C}=CHCH_2CH_2\overset{O}{C}CH_3 \quad \xrightarrow[\text{2.}]{\text{1.}\ \square} \quad CH_3\underset{CH_3}{CH}\overset{OH}{CH}CH_2CH_2\overset{O}{C}CH_3$$

(cyclopentene structure with HO and CH_3, CH_3CH/CH_3 substituents) $\xleftarrow[\text{2.}]{\text{1.}\ \square}$ (cyclopentenone with CH_3CH/CH_3) $\xleftarrow[\text{2.}]{\text{1.}\ \square}$ $CH_3\underset{CH_3}{CH}\overset{O}{C}CH_2CH_2\overset{O}{C}CH_3$

58. Ninhydrin reacts with an amino acid in basic solution to form a purple-colored compound. Propose a mechanism to account for formation of the colored compound.

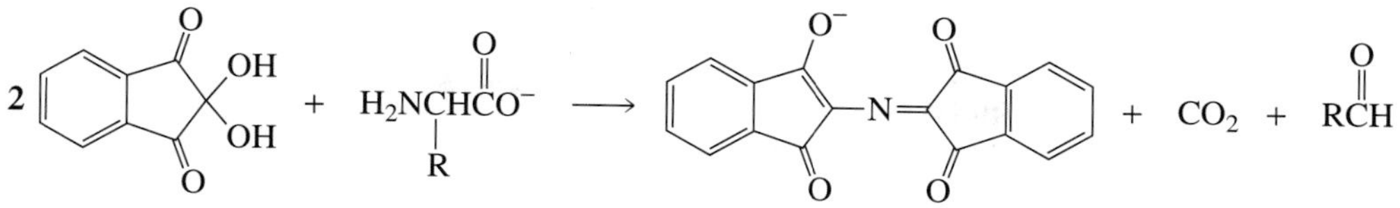

an amino acid   purple-colored
                                  compound

59. A carboxylic acid is formed when an α-haloketone reacts with hydroxide ion. This reaction is called a Favorskii reaction. Propose a mechanism for the following Favorskii reaction. (*Hint:* In the first step, HO⁻ removes a proton from the α-carbon not bonded to Br; a three-membered ring is formed in the second step; HO⁻ is a nucleophile in the third step.)

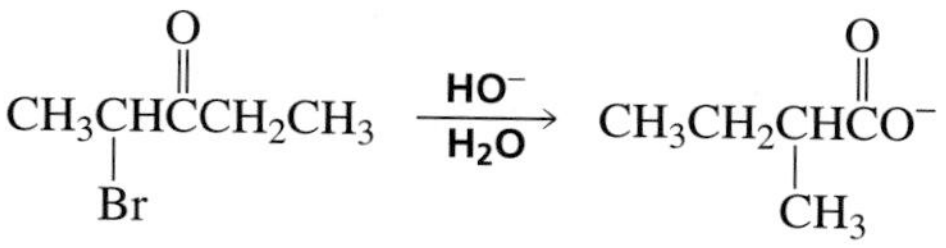

60. Give the products of the following reactions. (*Hint:* See Problem 59.)

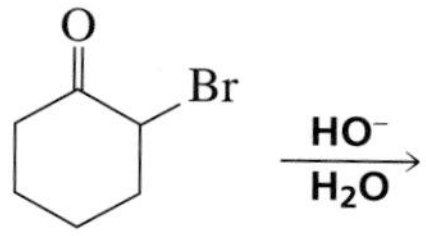

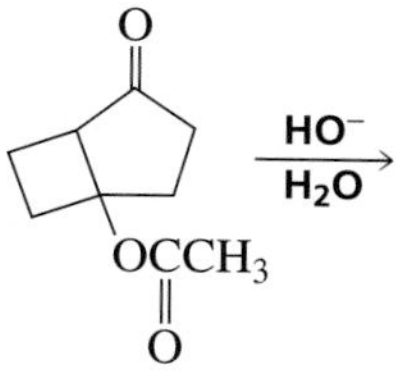

61. When an aldehyde with no α-hydrogens reacts with concentrated sodium hydroxide, half of the aldehyde is converted to a carboxylic acid and the other half is converted to an alcohol. In other words, half of the reactant is oxidized and the other half is reduced. This reaction is known as the Cannizzaro reaction. Propose a reasonable mechanism for the following Cannizzaro reaction.

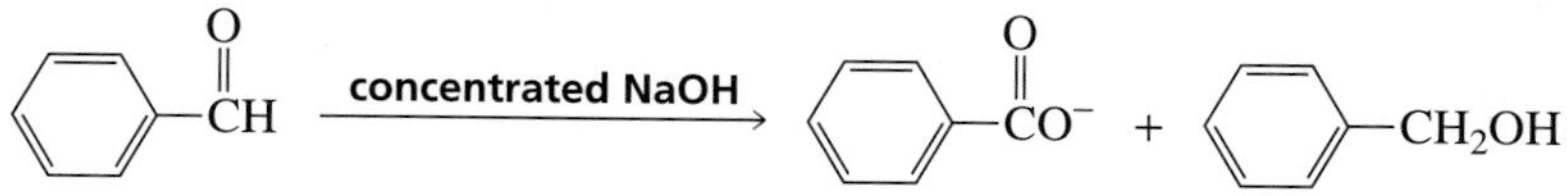

62. Propose a reasonable mechanism for each of the following reactions.

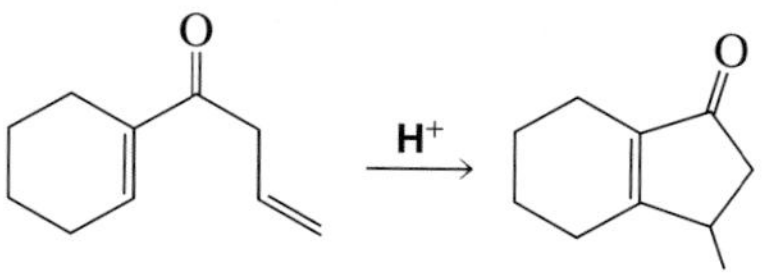

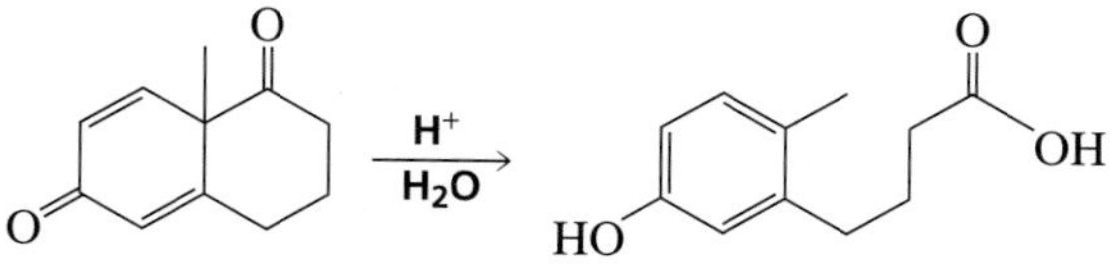

63. Show how the following compounds could be synthesized. The only carbon-containing reagents that are available to you for each synthesis are the ones shown.

a. 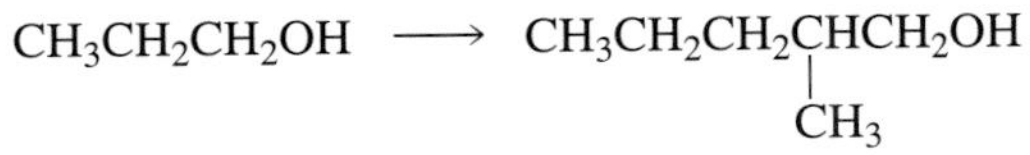 $CH_3CH_2CH_2OH \longrightarrow CH_3CH_2CH_2CHCH_2OH$
                                                             |
                                                            CH_3

b. 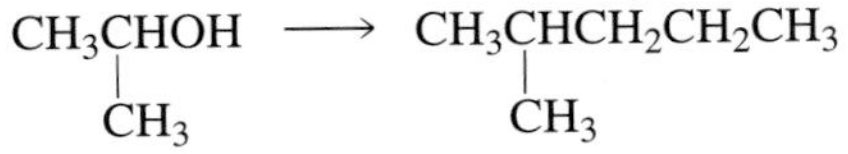 $CH_3CHOH \longrightarrow CH_3CHCH_2CH_2CH_3$
          | |
         CH_3 CH_3

c. 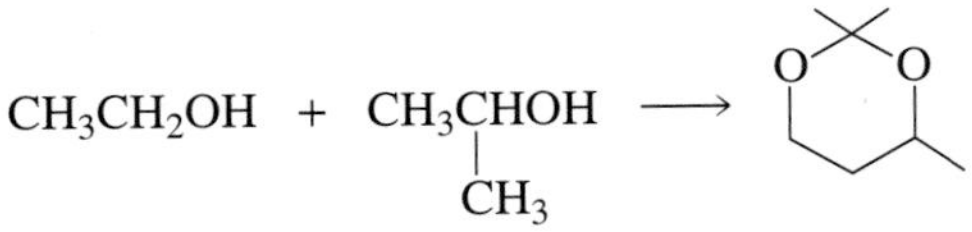 $CH_3CH_2OH + CH_3CHOH \longrightarrow$
                              |
                             CH_3

64. The following compound can be prepared in high yield using a mixed aldol condensation. What starting materials are required?

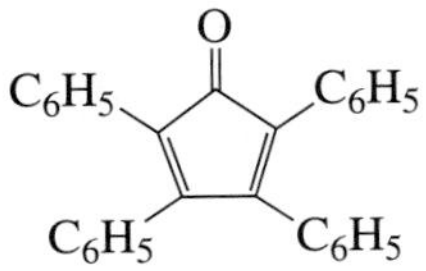

65. Propose a reasonable mechanism for each of the following reactions.

a. 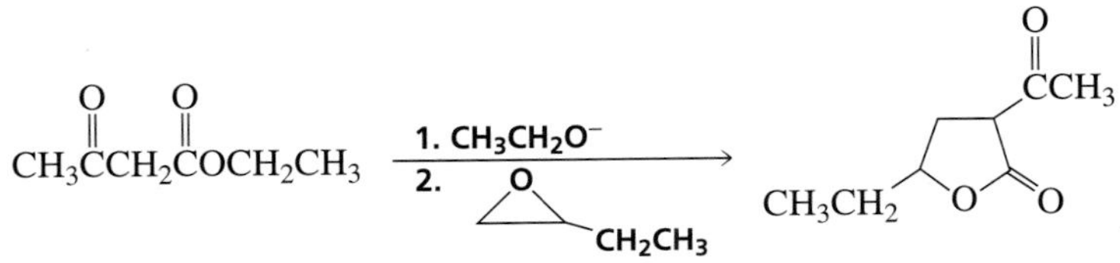 $CH_3CCH_2COCH_2CH_3$ $\xrightarrow[\text{2.}]{\text{1. } CH_3CH_2O^-}$

b.

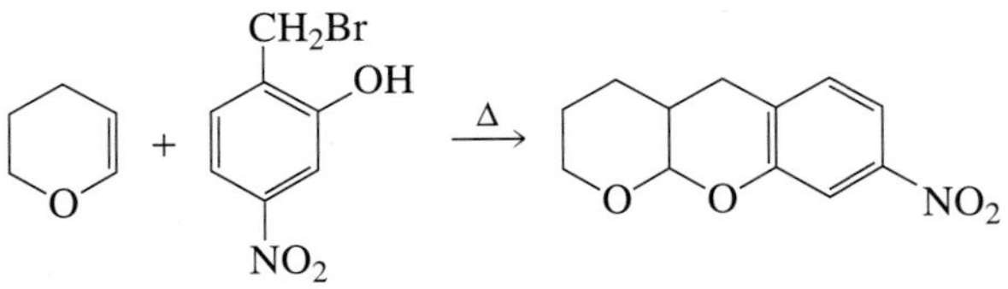

66. Orsellinic acid, a common constituent of lichens, is synthesized from the condensation of acetyl thioester and malonyl thioester. If the lichen were grown on a medium containing acetate that was radioactively labeled with ^{14}C at the carbonyl carbon, which carbons would be labeled in orsellinic acid?

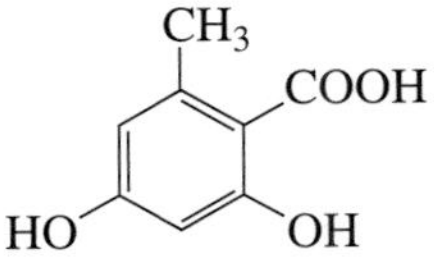

orsellinic acid

67. Tyramine is an alkaloid found in mistletoe, ripe cheese, and putrefied animal tissue. Dopamine is a neurotransmitter involved in the regulation of the central nervous system.

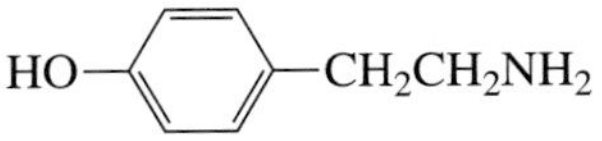

$HO-\langle\rangle-CH_2CH_2NH_2$ $HO-\langle\rangle-CH_2CH_2NH_2$

tyramine 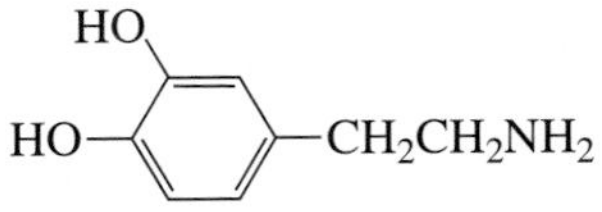   dopamine

a. Give two ways to prepare β-phenylethylamine from β-phenylethyl chloride.
b. How can β-phenylethylamine be prepared from benzyl chloride?
c. How can β-phenylethylamine be prepared from benzaldehyde?
d. How can tyramine be prepared from β-phenylethylamine?
e. How can dopamine be prepared from tyramine?

68. A Robinson annulation is used in the commercial synthesis of estrone, a steroid hormone. Propose a mechanism for the reaction, using the given starting materials.

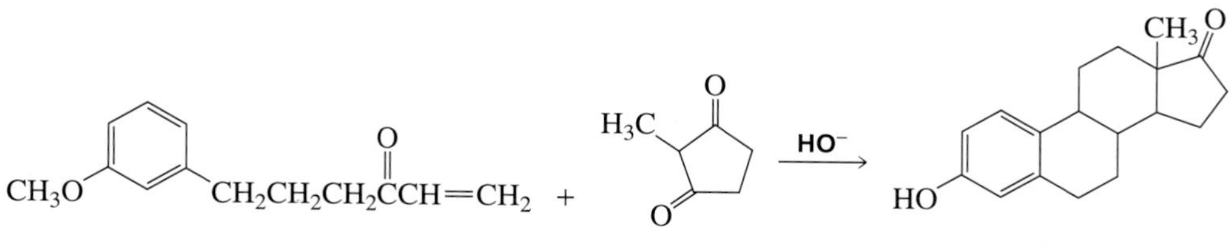

estrone

PART VII

BIOORGANIC COMPOUNDS

Chapters 19 to 24 discuss the chemistry of organic compounds that are found in biological systems. Many of these compounds may be larger than the organic compounds you have seen up to this point and they may have more than one functional group, but the reaction principles are essentially the same as those of the compounds you have been studying. These chapters will give you the opportunity to review and to apply your knowledge of organic reactions to compounds found in the biological world.

Chapter 19 introduces you to the chemistry of carbohydrates.

Chapter 20 discusses the chemistry of amino acids, peptides, and proteins.

Chapter 21 first describes the various ways that organic reactions are catalyzed, and then shows you how enzymes use these same principles in their catalysis of reactions that occur in biological systems.

Chapter 22 describes the chemistry of the coenzymes—organic compounds that some enzymes need to catalyze biological reactions. Coenzymes play a variety of chemical roles: some function as oxidizing and reducing agents, some allow electrons to be delocalized, some activate groups for further reaction, and some provide good nucleophiles or strong bases needed for reactions. Because coenzymes are derived from vitamins, you will see why vitamins are necessary for many of the organic reactions that occur in biological systems.

The chemistry of lipids is discussed in **Chapter 23.** Lipids are water-insoluble compounds found in animals and plants.

Chapter 24 discusses the chemistry of nucleosides, nucleotides, and nucleic acids (RNA and DNA).

19

CARBOHYDRATES

α-D-glucose, β-D-glucose

Carbohydrates are the first group of bioorganic compounds we will study. **Bioorganic compounds** are organic compounds that are found in biological systems. The structures of most bioorganic compounds are more complicated than the structures of many of the organic compounds you are used to seeing, but do not let the complicated-looking structures fool you into thinking their chemistry is equally complicated. Bioorganic compounds follow the same principles of structure and reactivity as the organic molecules we have discussed so far. One reason bioorganic compounds have complicated structures is because the compounds must be able to recognize each other, and much of their structure is for this purpose—a concept known as **molecular recognition.**

Carbohydrates are the most abundant class of compounds in the biological world, making up more than 50% of the dry weight of Earth's biomass. Carbohydrates are important constituents of all living organisms, and have a variety of different functions. Some are important structural components of cells, and some act as recognition sites on cell surfaces. Others serve as a major source of metabolic energy. For example, the leaves, stems, and roots of plants contain carbohydrates that plants use both for their own metabolic needs and for the metabolic needs of the animals that eat them.

The first carbohydrates investigated had molecular formulas that made them appear to be hydrates of carbon, $C_n(H_2O)_n$; hence the name. Later structural studies revealed that these compounds were not hydrates because they did not contain intact water molecules. However, the term "carbohydrate" persisted. It now refers either to polyhydroxy aldehydes such as D-glucose and polyhydroxy ketones such as D-fructose, or to compounds such as sucrose that can be hydrolyzed to polyhydroxy aldehydes or polyhydroxy ketones. The chemical structures of carbohydrates are commonly represented by wedge-and-dash structures or by Fischer projections. Notice that both D-glucose and D-fructose have molecular formulas $[C_6(H_2O)_6]$ that make them appear to be hydrates of carbon.

wedge-and-dash structure | Fischer projection

D-glucose
a polyhydroxy aldehyde

wedge-and-dash structure | Fischer projection

D-fructose
a polyhydroxy ketone

D-glucose

D-fructose

The most abundant carbohydrate is D-glucose. Cells of organisms oxidize D-glucose in the first of a series of processes that provide energy to the cells. When animals have more D-glucose than they need for energy, they convert the excess D-glucose into a polymer called glycogen. When an animal needs energy, glycogen is broken down into individual D-glucose molecules. Plants convert excess D-glucose into a polymer known as starch. Cellulose is another polymer of D-glucose. It is the major structural component of plants. Chitin, a carbohydrate similar to cellulose, makes up the exoskeletons of crustaceans, insects and other arthropods, and the structural material of fungi.

Animals obtain glucose by eating plants or by eating food containing glucose. Plants obtain glucose by a process known as **photosynthesis.** During photosynthesis, plants take up water through their roots and use carbon dioxide from the air to synthesize glucose and oxygen. Because photosynthesis is the opposite of the process used by organisms to obtain energy, plants require energy to carry out the process of photosynthesis. They acquire this energy from sunlight. Chlorophyll molecules in green plants capture the light energy used in photosynthesis. Notice that photosynthesis uses the CO_2 that animals exhale as waste and generates the O_2 that animals inhale. Nearly all the oxygen in Earth's atmosphere has been released by photosynthetic processes.

$$\underset{\text{glucose}}{C_6H_{12}O_6} + 6\,O_2 \underset{\text{photosynthesis}}{\overset{\text{oxidation}}{\rightleftharpoons}} 6\,CO_2 + 6\,H_2O + \text{energy}$$

The terms "carbohydrate," "saccharide," and "sugar" are often used interchangeably. Saccharide comes from the word for "sugar" in several early languages (*sarkara* in Sanskrit, *sakcharon* in Greek, and *saccharum* in Latin).

There are two classes of carbohydrates—simple carbohydrates and complex carbohydrates. **Simple carbohydrates** are **monosaccharides** (single sugars). **Complex carbohydrates** contain two or more sugar subunits linked together: **disaccharides** have two linked sugar subunits; **oligosaccharides** have three to ten sugar subunits (*oligos* is Greek for "few"); and **polysaccharides** have more than ten sugar subunits linked together. Disaccharides, oligosaccharides, and polysaccharides can be broken down to monosaccharide subunits by hydrolysis.

**19.1
CLASSIFICATION OF CARBOHYDRATES**

$$-M-M-M-M-M-M-M-M-M- \xrightarrow{\text{hydrolysis}} \quad M$$

polysaccharide   monosaccharide

A *monosaccharide* can be a polyhydroxy aldehyde or a polyhydroxy ketone. Polyhydroxy aldehydes are known as **aldoses** ("ald" is for aldehyde; "ose" is the suffix for a sugar), while polyhydroxy ketones are called **ketoses.** Monosaccharides are also classified according to the number of carbons they contain: monosaccharides with three carbons are **trioses;** those with four carbons are **tetroses;** those with five carbons are **pentoses;** and those with six and seven carbons are **hexoses** and **heptoses,** respectively. A six-carbon polyhydroxy aldehyde such as D-glucose is an aldohexose; a six-carbon polyhydroxy ketone such as D-fructose is a ketohexose.

PROBLEM 1◆

Classify the following monosaccharides.

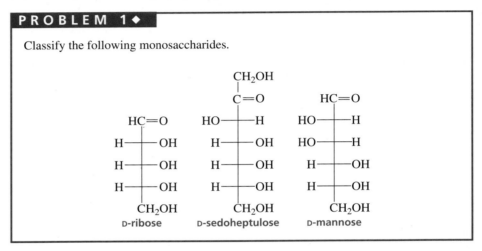

19.2
THE D AND L
NOTATION

The smallest aldose and the only one whose name does not end in "ose" is an aldotriose. Chemists and biochemists invariably refer to this compound by its common name, glyceraldehyde.

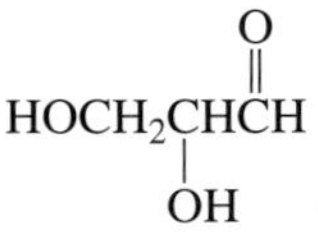

$$HOCH_2\overset{\overset{\displaystyle O}{\|}}{C}H\overset{\|}{C}H$$
$$|$$
$$OH$$

glyceraldehyde

Recall that in Fischer projections, horizontal bonds point toward the viewer while vertical bonds point away from the viewer (Section 4.4).

Because glyceraldehyde has a chirality center, it can exist as a pair of enantiomers. The Fischer projections of the enantiomers of glyceraldehyde are shown below.

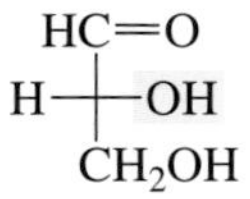

(*R*)-(+)-glyceraldehyde 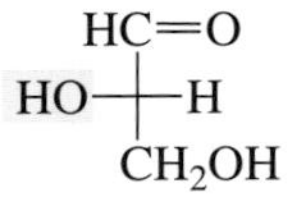   (*S*)-(−)-glyceraldehyde

Emil Fischer and his colleagues studied carbohydrates in the late nineteenth century, when techniques for determining the configurations of compounds were not available. Fischer arbitrarily assigned the configuration shown above on the left to the dextrorotatory isomer of glyceraldehyde that we call D-glyceraldehyde. He turned out to be correct: D-glyceraldehyde is (*R*)-(+)-glyceraldehyde, and L-glyceraldehyde is (*S*)-(−)-glyceraldehyde (Section 4.13).

$$
\begin{array}{cc}
\text{HC}\!=\!\text{O} & \text{HC}\!=\!\text{O} \\
\text{H}\!-\!\!\text{OH} & \text{HO}\!-\!\!\text{H} \\
\text{CH}_2\text{OH} & \text{CH}_2\text{OH} \\
\text{D-glyceraldehyde} & \text{L-glyceraldehyde}
\end{array}
$$

The D and L notations are used to describe the configuration of carbohydrates and amino acids (Section 20.2), so it is important to learn what D and L signify. In Fischer projections of monosaccharides, the carbonyl group is always placed on top (in the case of aldoses) or as close to the top as possible (in the case of ketoses). From its structure, you can see that galactose has four chirality centers (C-2, C-3, C-4, and C-5). If the OH group attached to the bottom-most chirality center (the second from the bottom carbon) is on the right, the sugar is a D-sugar. The mirror image of the D-sugar is an L-sugar. Almost all sugars found in nature are D-sugars.

Recall that a carbon with four different groups attached to it is a chirality center.

$$
\begin{array}{cc}
\text{HC}\!=\!\text{O} & \text{HC}\!=\!\text{O} \\
\text{H}\!-\!\!\text{OH} & \text{HO}\!-\!\!\text{H} \\
\text{HO}\!-\!\!\text{H} & \text{H}\!-\!\!\text{OH} \\
\text{HO}\!-\!\!\text{H} & \text{H}\!-\!\!\text{OH} \\
\text{H}\!-\!\!\text{OH} & \text{HO}\!-\!\!\text{H} \\
\text{CH}_2\text{OH} & \text{CH}_2\text{OH} \\
\text{D-galactose} & \text{L-galactose}
\end{array}
$$

the OH group is on the right

mirror image of D-galactose

Notice that D and L are like *R* and *S*; they indicate the configuration of a chirality center, but they do not indicate whether the compound rotates polarized light to the right (+) or to the left (−). For example, D-glyceraldehyde is dextrorotatory but D-lactic acid is levorotatory.

$$
\begin{array}{cc}
\text{HC}\!=\!\text{O} & \text{COOH} \\
\text{H}\!-\!\!\text{OH} & \text{H}\!-\!\!\text{OH} \\
\text{CH}_2\text{OH} & \text{CH}_3 \\
\text{D-(+)-glyceraldehyde} & \text{D-(–)-lactic acid}
\end{array}
$$

The common name of the monosaccharide together with the D or L designation completely defines its structure, since the relative configurations of all the chirality centers are implicit in the common name.

PROBLEM 2

Draw Fischer projections for L-glucose and L-fructose.

PROBLEM 3 ◆

Indicate whether each of the following is D-glyceraldehyde or L-glyceraldehyde (Section 4.5).

$$
\begin{array}{ccc}
\text{HC}\!=\!\text{O} & \text{H} & \text{CH}_2\text{OH} \\
\textbf{a.}\ \ \text{HOCH}_2\!-\!\!\text{OH} & \textbf{b.}\ \ \text{HO}\!-\!\!\text{CH}_2\text{OH} & \textbf{c.}\ \ \text{HO}\!-\!\!\text{H} \\
\text{H} & \text{HC}\!=\!\text{O} & \text{HC}\!=\!\text{O}
\end{array}
$$

19.3
CONFIGURATIONS
OF THE ALDOSES

Aldotetroses have two chirality centers and therefore four stereoisomers. Two of the stereoisomers are D-sugars and two are L-sugars. The names of the erythro and threo pairs of enantiomers described in Section 4.8 originated from the names of the aldotetroses, erythrose and threose.

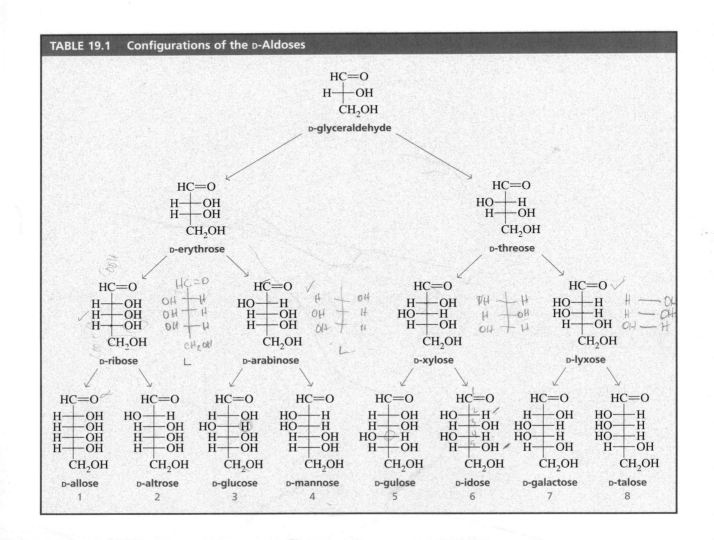

Aldopentoses have three chirality centers and therefore eight stereoisomers (four pairs of enantiomers), while aldohexoses have four chirality centers and 16 stereoisomers (eight pairs of enantiomers). The four D-aldopentoses and the eight D-aldohexoses are shown in Table 19.1.

Two diastereomers that differ in configuration at only one chirality center are called **epimers.** For example, D-ribose and D-arabinose are C-2 epimers (they differ in configuration only at C-2), and D-idose and D-talose are C-3 epimers.

Diastereomers are configurational isomers that are not enantiomers.

$$
\begin{array}{c}
\text{HC}=\text{O} \\
\text{H} \!-\!\!\!-\! \text{OH} \\
\text{H} \!-\!\!\!-\! \text{OH} \\
\text{H} \!-\!\!\!-\! \text{OH} \\
\text{CH}_2\text{OH}
\end{array}
\qquad
\begin{array}{c}
\text{HC}=\text{O} \\
\text{HO} \!-\!\!\!-\! \text{H} \\
\text{H} \!-\!\!\!-\! \text{OH} \\
\text{H} \!-\!\!\!-\! \text{OH} \\
\text{CH}_2\text{OH}
\end{array}
\qquad
\begin{array}{c}
\text{HC}=\text{O} \\
\text{HO} \!-\!\!\!-\! \text{H} \\
\text{H} \!-\!\!\!-\! \text{OH} \\
\text{HO} \!-\!\!\!-\! \text{H} \\
\text{H} \!-\!\!\!-\! \text{OH} \\
\text{CH}_2\text{OH}
\end{array}
\qquad
\begin{array}{c}
\text{HC}=\text{O} \\
\text{HO} \!-\!\!\!-\! \text{H} \\
\text{HO} \!-\!\!\!-\! \text{H} \\
\text{HO} \!-\!\!\!-\! \text{H} \\
\text{H} \!-\!\!\!-\! \text{OH} \\
\text{CH}_2\text{OH}
\end{array}
$$

D-ribose   D-arabinose   D-idose   D-talose

C-2 epimers   **C-3 epimers**

D-Glucose, D-mannose, and D-galactose are the most common aldohexoses in biological systems. An easy way to learn their structures is to memorize the structure of D-glucose and then remember that D-mannose is a C-2 epimer of D-glucose and D-galactose is a C-4 epimer of D-glucose. Sugars such as D-glucose and D-galactose are also diastereomers because they are configurational isomers that are not enantiomers (Section 4.8). An epimer is a particular kind of diastereomer.

PROBLEM 4 ◆

a. Are D-erythrose and L-erythrose a pair of enantiomers or a pair of diastereomers?

b. Are L-erythrose and L-threose a pair of enantiomers or a pair of diastereomers?

PROBLEM 5 ◆

a. Which sugar is the C-3 epimer of D-xylose?

b. Which sugar is the C-5 epimer of D-allose?

c. Which sugar is the C-4 epimer of L-gulose?

PROBLEM 6 ◆

Name the following compounds using IUPAC rules and indicating the configuration (R or S) of each of the chirality centers.

a. D-glucose   **b.** L-glucose

Naturally occurring ketoses have the ketone group in the 2-position. The configurations of the D-2-ketoses are shown in Table 19.2. A 2-ketose has one fewer chirality center than does an aldose with the same number of carbon atoms. Therefore, a 2-ketohexose has only half as many stereoisomers as an aldohexose.

19.4
CONFIGURATIONS OF THE KETOSES

PROBLEM 7 ◆

Which sugar is the C-3 epimer of D-fructose?

PROBLEM 8 ◆

How many stereoisomers are possible for:

a. a 2-ketoheptose?

b. an aldoheptose?

c. a ketotriose?

TABLE 19.2   Configurations of the D-Ketoses

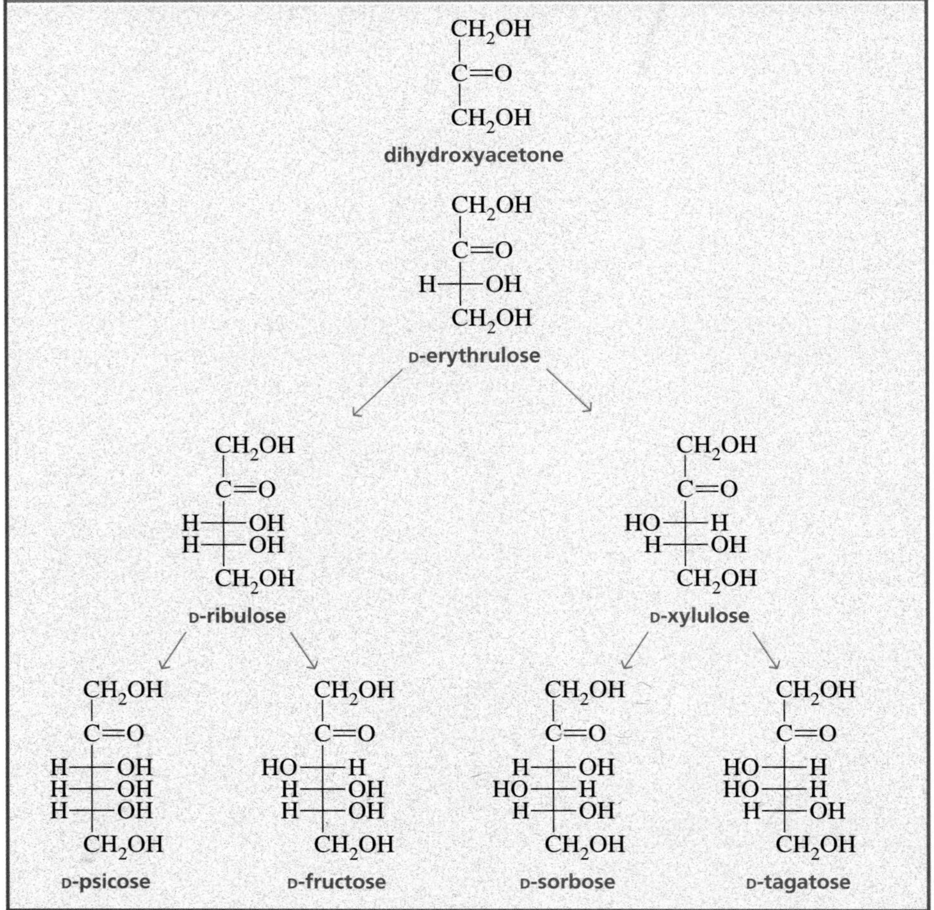

$$
\begin{array}{c}
CH_2OH \\
| \\
C{=}O \\
| \\
CH_2OH
\end{array}
$$

dihydroxyacetone

$$
\begin{array}{c}
CH_2OH \\
| \\
C{=}O \\
H\!-\!\!-\!OH \\
CH_2OH
\end{array}
$$

D-erythrulose

$$
\begin{array}{c}
CH_2OH \\
| \\
C{=}O \\
H\!-\!\!-\!OH \\
H\!-\!\!-\!OH \\
CH_2OH
\end{array}
\qquad
\begin{array}{c}
CH_2OH \\
| \\
C{=}O \\
HO\!-\!\!-\!H \\
H\!-\!\!-\!OH \\
CH_2OH
\end{array}
$$

D-ribulose   **D-xylulose**

$$
\begin{array}{c}
CH_2OH \\
| \\
C{=}O \\
H\!-\!\!-\!OH \\
H\!-\!\!-\!OH \\
H\!-\!\!-\!OH \\
CH_2OH
\end{array}
\quad
\begin{array}{c}
CH_2OH \\
| \\
C{=}O \\
HO\!-\!\!-\!H \\
H\!-\!\!-\!OH \\
H\!-\!\!-\!OH \\
CH_2OH
\end{array}
\quad
\begin{array}{c}
CH_2OH \\
| \\
C{=}O \\
H\!-\!\!-\!OH \\
HO\!-\!\!-\!H \\
H\!-\!\!-\!OH \\
CH_2OH
\end{array}
\quad
\begin{array}{c}
CH_2OH \\
| \\
C{=}O \\
HO\!-\!\!-\!H \\
HO\!-\!\!-\!H \\
H\!-\!\!-\!OH \\
CH_2OH
\end{array}
$$

D-psicose   **D-fructose**   **D-sorbose**   **D-tagatose**

19.5 REDOX REACTIONS OF MONO-SACCHARIDES

Because monosaccharides contain alcohol functional groups and either aldehyde or ketone functional groups, the reactions of monosaccharides are an extension of what you have already learned about the reactions of alcohols, aldehydes, and ketones. For example, an aldehyde group in a monosaccharide can be oxidized or reduced and can react with nucleophiles to form imines, hemiacetals, and acetals. When you read the sections that deal with the reactions of monosaccharides, you will find cross-references to the sections where we discussed the same reactivity for simple organic molecules. As you study, refer back to these sections, it will make learning about carbohydrates a lot easier as well as give you a good review of chemistry you already learned.

Reduction

The carbonyl group of aldoses and ketoses can be reduced by the usual carbonyl-group reducing agents (H_2 + Pd/C, $NaBH_4$). The product of the reduction is a polyalcohol, known as an **alditol.** Reduction of an aldose forms only one alditol; reduction of a ketose forms two alditols because the reaction creates a new chirality center in the product. D-Mannitol, the alditol formed from the reduction of D-mannose, is found in mushrooms, olives, and onions. The reduction of D-fructose forms D-mannitol and D-glucitol, the C-2 epimer of D-mannitol.

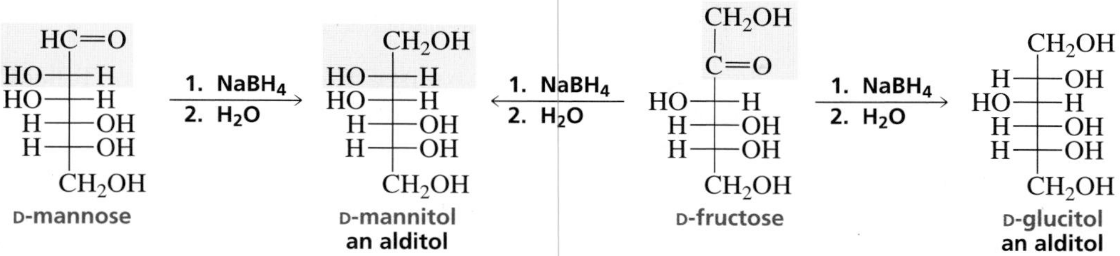

HC=O | HO—H | HO—H | H—OH | H—OH | CH₂OH — **D-mannose**

→ 1. NaBH₄ 2. H₂O →

CH₂OH | HO—H | HO—H | H—OH | H—OH | CH₂OH — **D-mannitol** **an alditol**

← 1. NaBH₄ 2. H₂O ←

CH₂OH | C=O | HO—H | H—OH | H—OH | CH₂OH — **D-fructose**

→ 1. NaBH₄ 2. H₂O →

CH₂OH | H—OH | HO—H | H—OH | H—OH | CH₂OH — **D-glucitol** **an alditol**

D-Glucitol is also obtained from the reduction of either D-glucose or L-gulose.

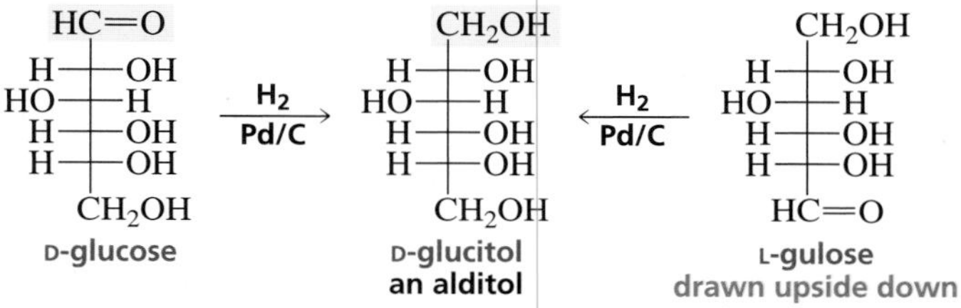

HC=O | H—OH | HO—H | H—OH | H—OH | CH₂OH — **D-glucose**

→ H₂ Pd/C →

CH₂OH | H—OH | HO—H | H—OH | H—OH | CH₂OH — **D-glucitol** **an alditol**

← H₂ Pd/C ←

CH₂OH | H—OH | HO—H | H—OH | H—OH | HC=O — **L-gulose** drawn upside down

---

PROBLEM 9◆

What product is obtained from the reduction of:

a. D-idose?

b. D-sorbose?

PROBLEM 10◆

a. What other monosaccharide is reduced only to the alditol obtained from the reduction of:

  1. D-talose?

  2. D-galactose?

b. What monosaccharide is reduced to two alditols, one of which is the alditol obtained from the reduction of:

  1. D-talose?

  2. D-mannose?

Oxidation

Aldoses and ketoses can be distinguished by observing what happens to the color of an aqueous solution of bromine when it is added to the sugar. Br_2 is a mild oxidizing agent and easily oxidizes the aldehyde group; it is not a sufficiently strong oxidizing agent to oxidize ketones or alcohols. Consequently, if a small amount of an aqueous solution of Br_2 is added to an unknown monosaccharide, the reddish-brown color of Br_2 will disappear if the monosaccharide is an aldose but will persist if the monosaccharide is a ketose. The product of the oxidation reaction is an **aldonic acid.**

$$
\begin{array}{c}
\text{HC=O} \\
\text{H}\!-\!\text{OH} \\
\text{HO}\!-\!\text{H} \\
\text{H}\!-\!\text{OH} \\
\text{H}\!-\!\text{OH} \\
\text{CH}_2\text{OH}
\end{array}
\;+\; \text{Br}_2 \;\xrightarrow{\;\text{H}_2\text{O}\;}\;
\begin{array}{c}
\text{COOH} \\
\text{H}\!-\!\text{OH} \\
\text{HO}\!-\!\text{H} \\
\text{H}\!-\!\text{OH} \\
\text{H}\!-\!\text{OH} \\
\text{CH}_2\text{OH}
\end{array}
\;+\; 2\,\text{Br}^-
$$

D-glucose   red   D-gluconic acid   colorless
an aldonic acid

Because both aldoses and ketoses are oxidized to aldonic acids by Tollens' reagent, this reagent cannot be used to distinguish between aldoses and ketoses. Why are ketoses oxidized by Tollens' reagent when only aldehydes can be oxidized by this reagent? It is because the reaction is carried out under basic conditions, and in a basic solution ketoses are converted into aldoses by enolization (Section 18.2). For example, the ketose D-fructose is in equilibrium with its enol. But the enol of D-fructose is also the enol of D-glucose as well as the enol of D-mannose. Therefore, when the enol reketonizes, all three carbonyl compounds are formed.

D-fructose (a ketose) $\xrightarrow[\text{HO}^-,\,\text{H}_2\text{O}]{\text{HO}^-,\,\text{H}_2\text{O}}$ enol of D-fructose / enol of D-glucose / enol of D-mannose $\xrightarrow[\text{HO}^-,\,\text{H}_2\text{O}]{\text{HO}^-,\,\text{H}_2\text{O}}$ D-glucose (an aldose) + D-mannose (an aldose)

PROBLEM 11

Show the mechanism for the base-catalyzed conversion of D-fructose into D-glucose and D-mannose.

PROBLEM 12 ◆

When D-tagatose is added to a basic aqueous solution, an equilibrium mixture of three monosaccharides is obtained. What are these monosaccharides?

If an oxidizing agent stronger than those discussed above is used (say, HNO$_3$), one or more of the alcohol groups can be oxidized in addition to the aldehyde group. The primary alcohol is the one most easily oxidized. The product obtained when both the aldehyde and the primary alcohol groups are oxidized is called an aldaric acid. (In an aldonic acid, one end is oxidized; in an **aldaric acid,** both ends are oxidized.)

D-glucose $\xrightarrow[\Delta]{\text{HNO}_3}$ D-glucaric acid (an aldaric acid)

PROBLEM 13 ◆

a. Name an aldohexose other than D-glucose that is oxidized to D-glucaric acid by nitric acid.

b. What is another name for D-glucaric acid?

c. Name another pair of aldohexoses that are oxidized to identical aldaric acids.

The tendency of monosaccharides to form syrups that do not crystallize made the purification and isolation of monosaccharides difficult. Fischer found that when phenylhydrazine is added to an aldose or a ketose, a yellow crystalline solid that is insoluble in water is formed. He called this derivative an **osazone** ("ose" for sugar; "azone" for hydrazone). Osazones are easily isolated and purified and were once used extensively to identify monosaccharides.

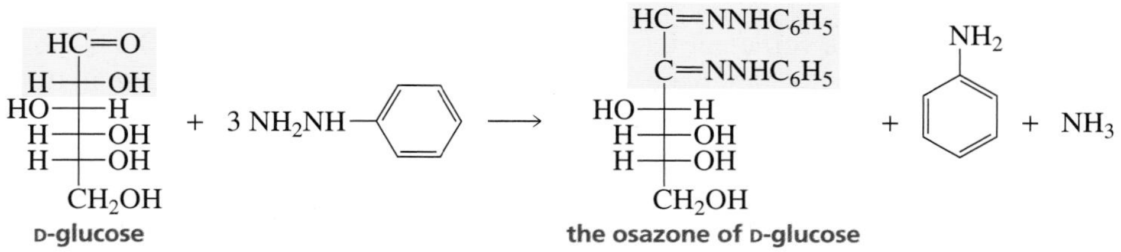

**19.6
OSAZONE
FORMATION**

D-glucose + 3 NH₂NH–C₆H₅ → the osazone of D-glucose + aniline + NH₃

Aldehydes and ketones react with one equivalent of phenylhydrazine, forming phenylhydrazones (Section 16.7). Aldoses and ketoses, in contrast, react with three equivalents of phenylhydrazine. The first equivalent forms an imine with the aldehyde or ketone group as expected. The second equivalent oxidizes the alcohol group adjacent to the imine to a carbonyl group. This newly formed carbonyl group reacts with the third equivalent of phenylhydrazine. The reaction stops at this point. Oxidation of additional alcohol groups does not occur no matter how much extra phenylhydrazine is present. The equivalent of phenylhydrazine that functions as an oxidizing agent is reduced to aniline and ammonia.

Because the configuration of the number 2 carbon is destroyed during osazone formation, C-2 epimers form identical osazones. For example, D-idose and D-gulose are C-2 epimers, so each forms the same osazone.

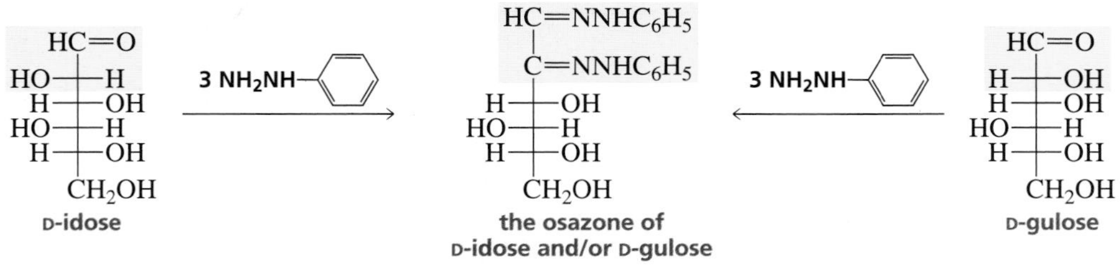

D-idose → the osazone of D-idose and/or D-gulose ← D-gulose

The number 1 and number 2 carbons of ketoses react with phenylhydrazine. Consequently, D-fructose, D-glucose, and D-mannose all form the same osazone.

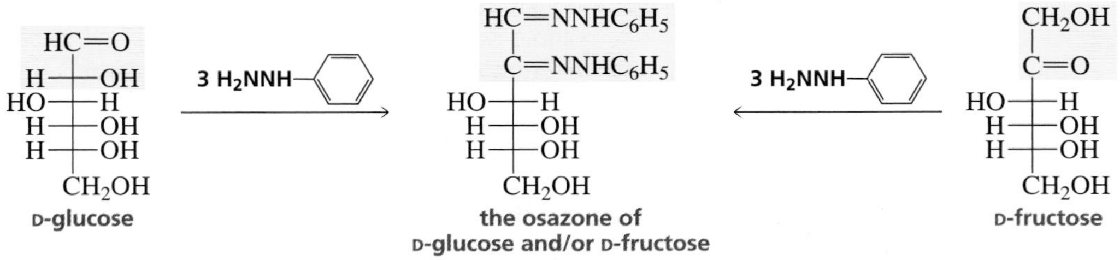

D-glucose → the osazone of D-glucose and/or D-fructose ← D-fructose

MEASURING THE BLOOD GLUCOSE LEVELS OF DIABETICS

Glucose reacts with an NH_2 group of hemoglobin to form an imine by means of the same mechanism as the first step in osazone formation. The imine undergoes an irreversible rearrangement to form a more stable α-aminoketone known as hemoglobin-A_{Ic}.

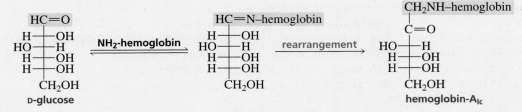

Diabetes results when the body does not produce sufficient insulin, or when the insulin it produces does not properly stimulate its target cells. Since insulin is the hormone that maintains the proper level of glucose in the blood, diabetics have increased blood glucose levels. Because the amount of hemoglobin-A_{Ic} formed is proportional to the concentration of glucose in the blood, diabetics have a higher concentration of hemoglobin-A_{Ic} than do nondiabetics. Measuring the hemoglobin-A_{Ic} level is a way to determine whether the blood glucose level of a diabetic is being adequately controlled. Cataracts, a common complication in diabetics, are caused by the reaction of glucose with the NH_2 group of proteins in the lens of the eye. It is thought that the arterial rigidity common in old age may be attributable to a similar reaction of glucose with the NH_2 group of long-lived proteins.

PROBLEM 14 ◆

Name a ketose and another aldose that form the same osazone as:

a. D-ribose

c. L-idose

b. D-altrose

d. D-galactose

PROBLEM 15 ◆

What pairs of epimers form the same osazone as D-sorbose?

19.7 CHAIN ELONGATION: THE KILIANI–FISCHER SYNTHESIS

Heinrich Kiliani (1855–1945) *was born in Germany. He received a Ph.D. from the University of Munich, studying under Professor Emil Erlenmeyer. Kiliani became a professor of chemistry at the University of Freiburg.*

The carbon chain of an aldose can be increased by one carbon by means of the **Kiliani–Fischer synthesis.** In other words, tetroses can be converted into pentoses, and pentoses can be converted into hexoses.

In the first step of the synthesis (the Kiliani portion), hydrogen cyanide is added to the aldose. Addition of cyanide ion to the carbonyl group creates a new chirality center. Consequently, two cyanohydrins that differ only in configuration at C-2 are formed. The configurations of the other chirality centers do not change, because no bond to any of the chirality centers is broken during the course of the reaction. Kiliani went on to hydrolyze the cyanohydrins to aldonic acids; Fischer had previously developed a method to convert aldonic acids to aldoses. This reaction sequence was used for many years, but the method currently used to convert the cyanohydrins to aldoses was developed by Serianni and Barker in 1979; it is referred to as the modified Kiliani–Fischer synthesis. Serianni and Barker reduced the cyanohydrins to imines using a deactivated palladium (on barium sulfate) catalyst so that the imines would not be further reduced to amines. The imines could then be hydrolyzed to aldoses.

the modified Kiliani–Fischer synthesis

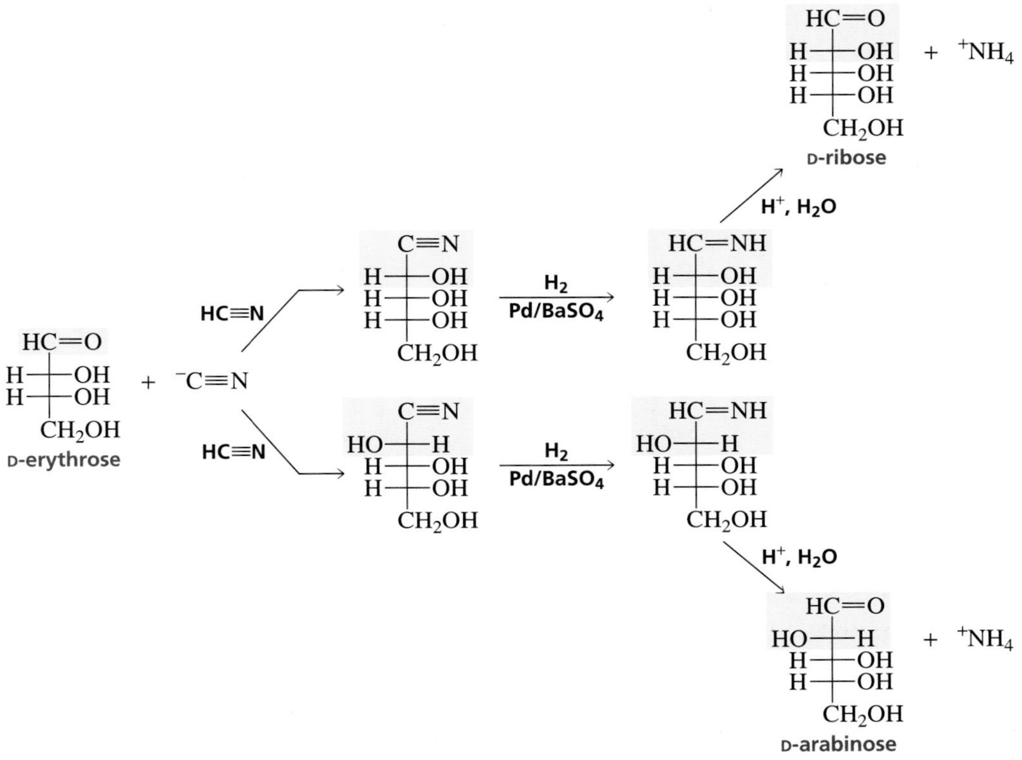

Notice that the synthesis leads to a pair of C-2 epimers, since the first step of the reaction converts the carbonyl carbon in the starting material to a chirality center. Therefore the OH on C-2 in the product can be on the right or on the left in the Fischer projection. The two epimers are not obtained in equal amounts, because the first step of the reaction produces a pair of diastereomers and diastereomers are always formed in unequal amounts (Section 4.19).

PROBLEM 16◆

What monosaccharides would be formed in a Kiliani–Fischer synthesis starting with:

a. D-xylose?   **b.** L-threose?

19.8
CHAIN SHORTENING: THE RUFF DEGRADATION

The **Ruff degradation** is the opposite of the Kiliani–Fischer synthesis. The degradation shortens an aldose chain by one carbon: hexoses are converted into pentoses, and pentoses are converted into tetroses. In the Ruff degradation, the calcium salt of an aldonic acid is oxidized with hydrogen peroxide. Ferric ion catalyzes the reaction. The oxidation reaction cleaves the bond between C-1 and C-2, forming CO_2 and an aldehyde. It is known that the reaction involves the formation of radicals, but the precise mechanism is not well understood.

Otto Ruff (1871–1939) was born in Germany. He received a Ph.D. from the University of Berlin. He was a professor of chemistry at the University of Danzig and later at the University of Breslau.

the Ruff degradation

$$
\begin{array}{c}
\text{COO}^- \ (\text{Ca}^{2+})_{1/2} \\
\text{H}-\text{OH} \\
\text{HO}-\text{H} \\
\text{H}-\text{OH} \\
\text{H}-\text{OH} \\
\text{CH}_2\text{OH}
\end{array}
\; + \; \text{H}_2\text{O}_2 \;\xrightarrow{\text{Fe}^{3+}}\;
\begin{array}{c}
\text{HC}=\text{O} \\
\text{HO}-\text{H} \\
\text{H}-\text{OH} \\
\text{H}-\text{OH} \\
\text{CH}_2\text{OH}
\end{array}
\; + \; \text{CO}_2
$$

calcium D-gluconate   D-arabinose

The calcium salt of the aldonic acid necessary for the Ruff degradation is easily obtained by oxidizing an aldose with an aqueous solution of bromine and then adding calcium hydroxide to the reaction mixture.

$$
\begin{array}{c}
\text{HC}=\text{O} \\
\text{H}-\text{OH} \\
\text{HO}-\text{H} \\
\text{H}-\text{OH} \\
\text{H}-\text{OH} \\
\text{CH}_2\text{OH}
\end{array}
\;\xrightarrow[\text{2. Ca(OH)}_2]{\text{1. Br}_2, \text{H}_2\text{O}}\;
\begin{array}{c}
\text{COO}^- \ (\text{Ca}^{2+})_{1/2} \\
\text{H}-\text{OH} \\
\text{HO}-\text{H} \\
\text{H}-\text{OH} \\
\text{H}-\text{OH} \\
\text{CH}_2\text{OH}
\end{array}
$$

D-glucose   calcium D-gluconate

PROBLEM 17◆

Which two monosaccharides can be degraded to:

a. D-arabinose?

b. D-glyceraldehyde?

c. L-ribose?

19.9 STEREOCHEMISTRY OF GLUCOSE: THE FISCHER PROOF

In 1891, Emil Fischer determined the stereochemistry of glucose using one of the most brilliant examples of reasoning in the history of chemistry. He set out to determine the stereochemistry of (+)-glucose because it was the most common monosaccharide found in nature.

Fischer knew that (+)-glucose was an aldohexose. But 16 different structures can be written for an aldohexose. Which of them represents the structure of (+)-glucose? Since the 16 stereoisomers of the aldohexoses are actually eight pairs of enantiomers, Fischer considered only the eight stereoisomers that had the C-5 OH group on the right in the Fischer projection (the stereoisomers that we now call the D-sugars). One of these would be one enantiomer of glucose, and its mirror image would be the other enantiomer. It was not possible to determine whether (+)-glucose was D-glucose or L-glucose until 1951 (Section 4.13). The D-sugars are shown in Table 19.1, numbered 1 to 8. These are the steps that Fischer followed:

1. When the Kiliani–Fischer synthesis is done on the sugar known as (−)-arabinose, the two sugars obtained are the sugars known as (+)-glucose and (+)-mannose. This means that (+)-glucose and (+)-mannose are C-2 epimers. In other words, they have the same configuration at C-3, C-4, and C-5. Consequently, (+)-glucose and (+)-mannose have to be one of the following pairs: sugars 1 and 2, 3 and 4, 5 and 6, or 7 and 8 (from Table 19.1).

2. (+)-Glucose and (+)-mannose are both oxidized by nitric acid to optically active aldaric acids. The aldaric acids of sugars 1 and 7 would not be

optically active because they have a plane of symmetry. Excluding sugars 1 and 7 means that (+)-glucose and (+)-mannose must be sugars 3 and 4 or 5 and 6.

3. Because (+)-glucose and (+)-mannose are the products obtained when the Kiliani–Fischer synthesis is carried out on (−)-arabinose, there are only two possibilities for the structure of (−)-arabinose: if (+)-glucose and (+)-mannose are sugars 3 and 4, (−)-arabinose has the structure shown on the left; but if (+)-glucose and (+)-mannose are sugars 5 and 6, (−)-arabinose has the structure shown on the right.

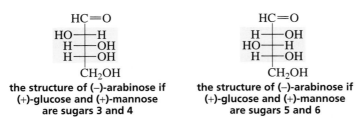

the structure of (–)-arabinose if (+)-glucose and (+)-mannose are sugars 3 and 4 — the structure of (–)-arabinose if (+)-glucose and (+)-mannose are sugars 5 and 6

When (−)-arabinose is oxidized with nitric acid, the aldaric acid obtained is optically active. This means that the aldaric acid does not have a plane of symmetry. Therefore, (−)-arabinose must have the structure shown on the left, because the aldaric acid of the sugar on the right has a plane of symmetry. Thus, (+)-glucose and (+)-mannose are represented by sugars 3 and 4.

4. The last step in the Fischer proof was to determine whether (+)-glucose is sugar 3 or 4. In order to answer this question, Fischer had to develop a chemical method to interchange the aldehyde and primary alcohol groups of an aldohexose. When he chemically interchanged the aldehyde and primary alcohol groups of the sugar known as (+)-glucose, he obtained an aldohexose that was different from (+)-glucose. When he chemically interchanged the aldehyde and primary alcohol groups of (+)-mannose, he still had (+)-mannose. Therefore, he concluded that (+)-glucose is sugar 3 because reversing the aldehyde and alcohol groups of sugar 3 leads to a different sugar (L-gulose).

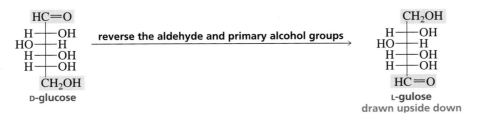

D-glucose — reverse the aldehyde and primary alcohol groups → L-gulose drawn upside down

If (+)-glucose is sugar 3, (+)-mannose must be sugar 4. As predicted, when the aldehyde and primary alcohol groups of sugar 4 are reversed, the same sugar is obtained.

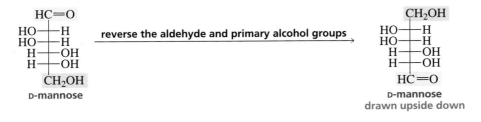

D-mannose — reverse the aldehyde and primary alcohol groups → D-mannose drawn upside down

Using similar reasoning, Fischer went on to determine the stereochemistry of 14 of the 16 aldohexoses. He received the Nobel Prize in chemistry in 1902 for this achievement. His original guess that (+)-glucose is a D-sugar was later shown to be correct, so all of his structures are correct. If he had been wrong and (+)-glucose had been an L-sugar, his contribution to the stereochemistry of aldoses would still have had the same significance, but all his stereochemical assignments would have been reversed.

GLUCOSE/DEXTROSE

André Dumas first used the term glucose (in 1838) to refer to the sweet compound that comes from honey and grapes. Later, Kekulé (Section 6.1) decided that it should be called dextrose because it was dextrorotatory. When Fischer studied the sugar, he called it glucose, and chemists have called it glucose ever since (although dextrose is often found on food labels).

19.10 CYCLIC STRUCTURE OF MONO- SACCHARIDES: HEMIACETAL FORMATION

D-Glucose exists in three different forms. There is the open-chain form of D-glucose that we have been discussing and there are two cyclic forms: α-D-glucose and β-D-glucose. The fact that the two cyclic forms are different is shown by their physical properties: α-D-glucose melts at 146 °C, while β-D-glucose melts at 150 °C; α-D-glucose has a specific rotation of +112.2°, and β-D-glucose has a specific rotation of +18.7°.

How can D-glucose exist in a cyclic form? In Section 16.8, we saw that an aldehyde reacts with an equivalent of an alcohol to form a hemiacetal. A monosaccharide such as D-glucose has an aldehyde group and several alcohol groups. The alcohol group bonded to C-5 of D-glucose reacts intramolecularly with the aldehyde group, forming a six-membered ring hemiacetal. Why are there two different cyclic forms? Two different hemiacetals are formed, because the carbonyl carbon of the open-chain sugar becomes a chirality center in the hemiacetal. This new chirality center can have the OH group on the right (α-D-glucose) or on the left (β-D-glucose). Formation of the cyclic hemiacetal takes place by the same mechanism that occurs in hemiacetal formation between individual aldehyde and alcohol molecules (Section 16.8).

Jean Baptiste André Dumas (1800–1884) *was born in France. Apprenticed to an apothecary, he left to study chemistry in Switzerland. He became a professor of chemistry at the Athenaeum and was the first French chemist to teach laboratory courses. In 1848, he left science for a political career. He became a senator, master of the French mint, and mayor of Paris.*

α-D-Glucose and β-D-glucose are anomers. **Anomers** are two sugars that differ in configuration only at the carbon that is the carbonyl carbon in the open-chain form. This is known as the **anomeric carbon.** The anomeric carbon is the only carbon bonded to two oxygens. The prefixes α and β denote the configuration of the anomeric carbon. Anomers, like epimers, are a particular kind of diastereomers.

In an aqueous solution, the open-chain compound is in equilibrium with the two hemiacetals. Formation of five- and six-membered ring cyclic hemiacetals, unlike acyclic hemiacetals, proceeds nearly to completion; very little of the compound exists in the open-chain form (about 0.02%). At equilibrium, there is almost twice as much β-D-glucose (64%) as α-D-glucose (36%). The sugar still undergoes the reactions discussed in previous sections (oxidation, reduction, and osazone formation), because the reagents can react with the small amount of open-chain aldehyde that is present. As the aldehyde reacts, the equilibrium shifts to form more aldehyde, which can then undergo reaction. So, eventually, all the glucose molecules react by means of the open-chain aldehyde.

When crystals of pure α-D-glucose are dissolved in water, the specific rotation gradually changes from +112.2° to +52.7°. When crystals of pure β-D-glucose are dissolved in water, the specific rotation gradually changes from +18.7° to +52.7°. This change in rotation occurs because, in water, the hemiacetal opens to form the aldehyde and, when the aldehyde recyclizes, both α-D-glucose and β-D-glucose are formed. Eventually, the three forms of glucose reach equilibrium concentrations. The specific rotation of the equilibrium mixture is +52.7°, which is why the same specific rotation results whether the crystals originally dissolved in water are α-D-glucose or β-D-glucose. A slow change in optical rotation to an equilibrium value is known as **mutarotation.**

If an aldose can form a five- or a six-membered ring, it will exist predominantly as a cyclic hemiacetal in solution. Whether a five- or a six-membered ring is formed depends on their relative stabilities. Six-membered ring sugars are known as **pyranoses,** and five-membered ring sugars are called **furanoses.** These names come from pyran and furan, the names of the cyclic ethers shown below. Consequently, α-D-glucose is also known as α-D-glucopyranose. The prefixes α and β indicate that the sugar is in the cyclic form; "pyranose" indicates that the sugar is in a six-membered ring.

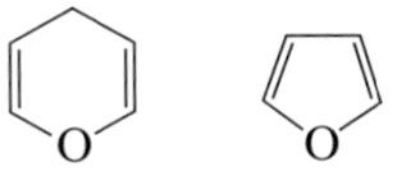

pyran   furan

PROBLEM 18

4-Hydroxy- and 5-hydroxyaldehydes exist primarily in the cyclic hemiacetal form. Give the structure of the cyclic hemiacetal formed by each of the following.

a. 4-hydroxybutanal   c. 4-hydroxypentanal

b. 5-hydroxypentanal   d. 4-hydroxyheptanal

Using Fischer projections to show the structure of a cyclic sugar is unsatisfactory because of the way that the C—O—C bond is represented. A somewhat more satisfactory representation is given by a **Haworth projection.** In a Haworth projection of a D-pyranose, the six-membered ring is represented as being flat and is viewed edge on. The ring oxygen is always placed in the back right-hand corner of the ring, with the anomeric carbon (C-1) on the right-hand side and the primary alcohol group drawn *up* from the back left-hand corner (C-5). Groups on the *right* in a Fischer projection are *down* in a Haworth projection, while groups on the *left* in a Fischer projection are *up* in a Haworth projection.

The Haworth projection of a D-furanose is viewed on edge with the ring oxygen away from the viewer. The anomeric carbon is on the right-hand side of the molecule and the primary alcohol group is drawn *up* from the back left-hand corner.

Sir Walter Norman Haworth (1883–1950) *was born in England. He received a Ph.D. in Germany from the University of Göttingen and later was a professor of chemistry at the universities of Durham and Birmingham in Britain. He was the first to synthesize vitamin C and was the one who named it ascorbic acid. During World War II, he worked on the atomic bomb project. He was knighted in 1947.*

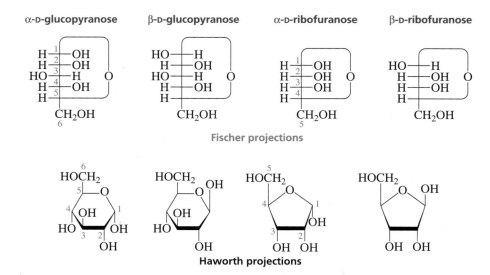

α-D-**glucopyranose** β-D-**glucopyranose** α-D-**ribofuranose** β-D-**ribofuranose**

Fischer projections

Haworth projections

Ketoses also exist predominantly in cyclic forms. D-Fructose forms a five-membered ring hemiketal as a consequence of the C-5 OH group reacting with the ketone group (Section 16.8). If the new chirality center has the OH group on the right, the compound is α-D-fructose; if the OH group is on the left, the compound is β-D-fructose. These sugars can also be called α-D-fructofuranose and β-D-fructofuranose. Notice that in fructose the anomeric carbon is the C-2 carbon. D-Fructose can also form a six-membered ring by using the C-6 OH group. The pyranose form predominates in the monosaccharide, while the furanose form predominates when the sugar is in a disaccharide.

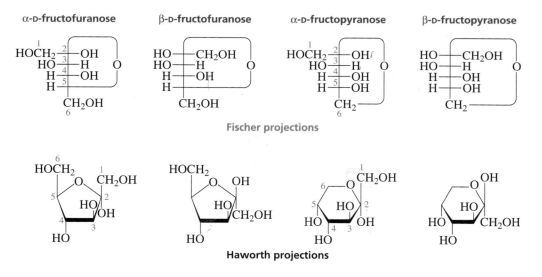

α-D-**fructofuranose** β-D-**fructofuranose** α-D-**fructopyranose** β-D-**fructopyranose**

Fischer projections

Haworth projections

Haworth projections are useful because they allow us to see easily whether the OH groups on the ring are cis or trans to each other. Because five-membered rings are close to planar, furanoses are well represented by Haworth projections. However, Haworth projections are structurally misleading for pyranoses because a six-membered ring is not flat but exists preferentially in a chair conformation (Section 2.13).

PROBLEM 19

Draw the following sugars using Haworth projections.

a. β-D-galactose

b. α-D-tagatose

c. α-L-glucose

PROBLEM 20◆

D-Glucose most often exists as a pyranose, but it can also exist as a furanose. Draw the Haworth projection formula of α-D-glucofuranose.

Drawing glucose in its chair conformation shows why it is the most common aldo-hexose in nature. To convert a Haworth projection into a chair conformation, start by placing the ring oxygen at the back right-hand corner and the primary alcohol group in the equatorial position. The primary alcohol group is the largest of all the substituents, and large substituents are more stable in the equatorial position because there is less steric strain in that position (Section 2.14). Because the OH group bonded to C-4 is trans to the primary alcohol group (this is easily seen in the Haworth projection), the C-4 OH group is also in the equatorial position (1,2-diequatorial substituents are trans to one another, Section 4.16). The C-3 OH group is trans to the C-4 OH group, so the C-3 OH group is also in the equatorial position. As you move around the ring, you will find that all the OH substituents in β-D-glucose are in equatorial positions. The axial positions are all occupied by hydro-gens, which require little space and therefore experience little steric strain. No other aldohexose exists in such a strain-free conformation. This means that glucose is the most stable of all the aldohexoses, so it is not surprising that it is the most prevalent aldohexose in nature.

19.11
STABILITY
OF GLUCOSE

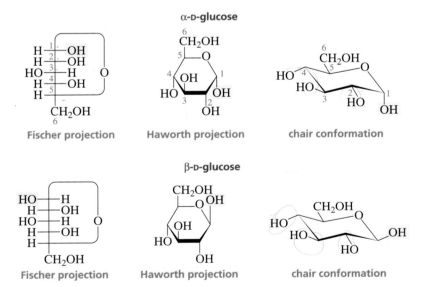

The α-position is to the right in a Fischer projec-tion; it is down in a Haworth projection; and it is axial in a chair confor-mation.

The β-position is to the left in a Fischer projection; it is up in a Haworth pro-jection; and it is equatorial in a chair conformation.

Why is there more β-D-glucose than α-D-glucose in an aqueous solution at equilibrium? The OH group bonded to the anomeric carbon is in the equatorial position in β-D-glucose, but it is in the axial position in α-D-glucose. Therefore, β-D-glucose is more stable than α-D-glucose, so β-D-glucose predominates at equilibrium in an aqueous solution.

If you remember that all the OH groups in β-D-glucose are in equatorial positions, it is easy to draw the chair conformation of any other pyranose. For example, if you needed to draw α-D-galactose, you would put all the OH groups in equatorial positions except the OH groups at C-4 (since galactose is a C-4 epimer of glucose) and at C-1 (since it is the α-anomer); these two OH groups would be in axial positions.

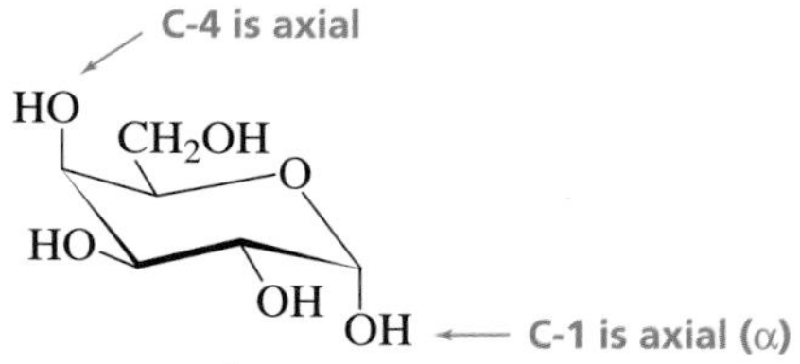

α-D-**galactose**

To draw an L-pyranose, draw the D-pyranose first, and then draw its mirror image. For example, to draw β-L-gulose, first draw β-D-gulose (gulose differs from glucose at C-3 and C-4, so the OH groups at these positions are in the axial position). Then draw the mirror image of β-D-gulose to get β-L-gulose.

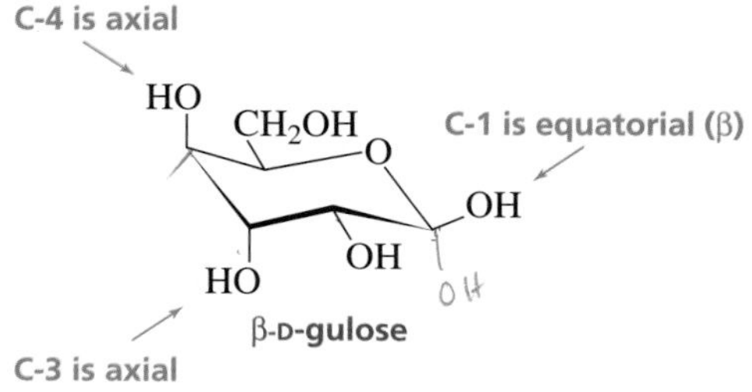

β-D-**gulose** 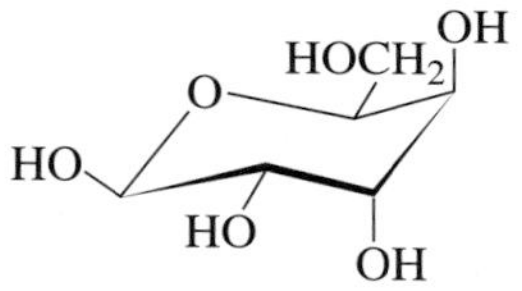   β-L-**gulose**

PROBLEM 21◆

Which OH groups are in the axial position in:

a. β-D-mannose?

b. β-D-idose?

c. α-D-allose?

**19.12
FORMATION
OF GLYCOSIDES**

In Section 16.8 we saw that, after an aldehyde reacts with an equivalent of an alcohol to form a hemiacetal, the hemiacetal reacts with a second equivalent of alcohol to form an acetal. Similarly, the cyclic hemiacetal (or hemiketal) formed by a monosaccharide can react with an alcohol to form an acetal (or ketal). The acetal (or ketal) of a sugar is called a **glycoside**, and the bond between the anomeric carbon and the alkoxy oxygen is called a **glycosidic bond.** Glycosides are named by replacing the "ose" ending of the sugar's name with "oside": the glycoside of glucose is a glucoside, the glycoside of galactose is a galactoside, etc. If the pyranose or furanose name is used, the acetal is called a **pyranoside** or a **furanoside.**

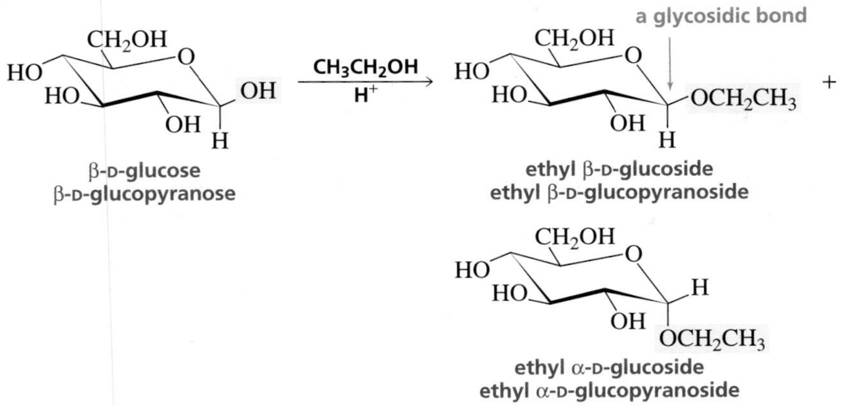
a glycosidic bond
CH3CH2OH
H+
β-D-glucose
β-D-glucopyranose
ethyl β-D-glucoside
ethyl β-D-glucopyranoside
ethyl α-D-glucoside
ethyl α-D-glucopyranoside

Notice that the reaction of a single anomer leads to the formation of both the α- and β-glycosides. The mechanism of the reaction shows why both glycosides are formed. The OH group bonded to the anomeric carbon becomes protonated in the acidic solution, and a nonbonding pair of electrons on the ring oxygen helps expel a molecule of water. The anomeric carbon in the resulting oxonium ion is sp^2 hybridized, causing that part of the molecule to be planar. When the alcohol comes in from the top of the plane, the β-glycoside is formed; when the alcohol comes in from the bottom of the plane, the α-glycoside is formed. Notice that the mechanism is exactly the same as the mechanism for acetal formation (Section 16.8).

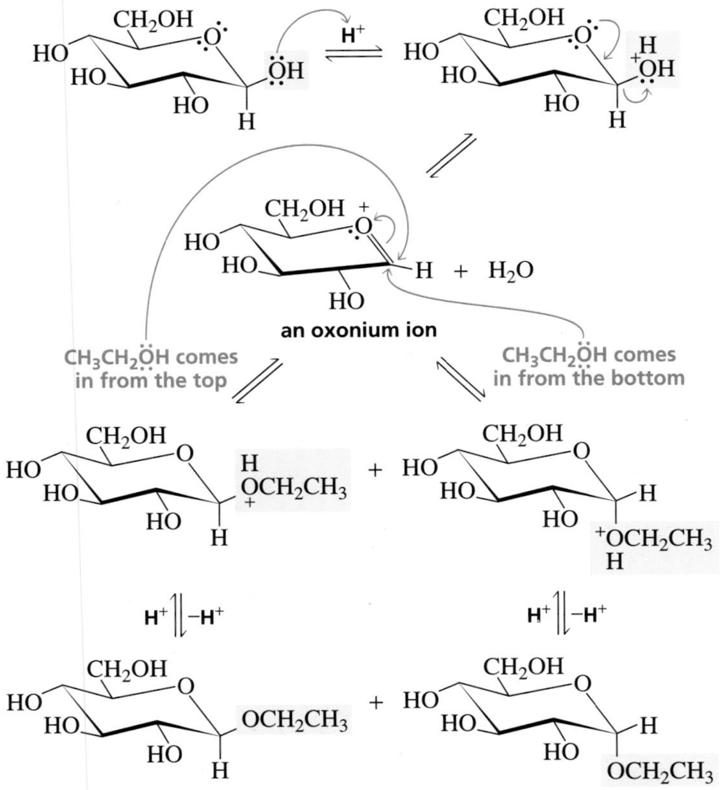
an oxonium ion
CH3CH2ÖH comes in from the top
CH3CH2ÖH comes in from the bottom

Because glycosides are acetals (or ketals), they are not in equilibrium with the open-chain aldehyde (or ketone) in aqueous solution. Because they are not in equilibrium with a compound with a carbonyl group, they cannot be oxidized by reagents such as Ag^+ or Br_2. Glycosides, therefore, are **nonreducing sugars.** (They cannot reduce Ag^+ or Br_2.)

Hemiacetals (or hemiketals) are in equilibrium with the open-chain sugars in aqueous solution. So as long as a sugar has an aldehyde, a ketone, a hemiacetal, or a hemiketal group, it is able to reduce an oxidizing agent and therefore is classified as a **reducing sugar.** Without one of these groups, it is a nonreducing sugar.

A reaction similar to the reaction of a monosaccharide with an alcohol is the reaction of a monosaccharide with an amine in the presence of a trace amount of acid. The product of the reaction is an **N-glycoside;** it has a nitrogen in place of the oxygen at the glycosidic linkage. The subunits of DNA and RNA are β-N-glycosides (Section 24.1).

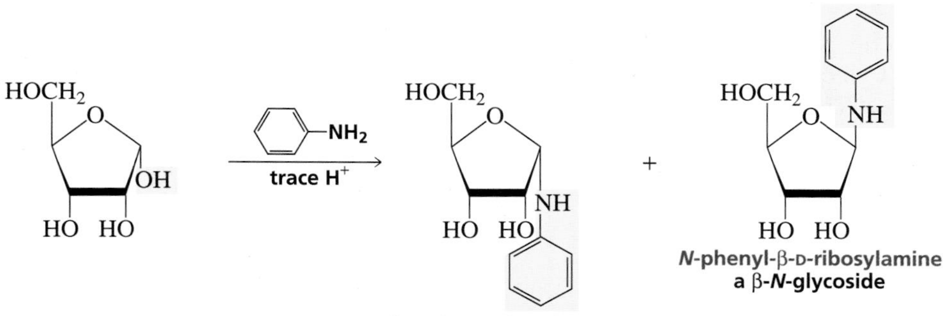

> **A sugar with an aldehyde, ketone, hemiacetal, or hemiketal group is a reducing sugar. A sugar without one of these groups is a nonreducing sugar.**

N-phenyl-α-D-ribosylamine
an α-N-glycoside

N-phenyl-β-D-ribosylamine
a β-N-glycoside

THE ANOMERIC EFFECT

Surprisingly, when the OH substituent bonded to the anomeric carbon is alkylated, it is more stable in the axial position than in the equatorial position. In other words, α-glycosides are more stable than β-glycosides. The preference for the axial position by certain substituents bonded to the anomeric carbon is called the **anomeric effect.** Several hypotheses have been suggested to account for the anomeric effect, but as yet it is not completely understood.

PROBLEM 22 / SOLVED

Name the following compounds and indicate whether each is a reducing sugar or a nonreducing sugar.

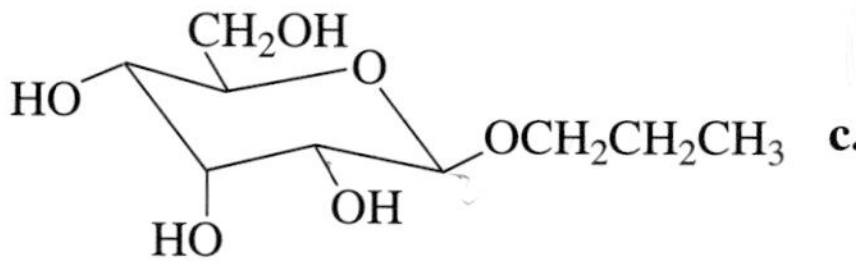

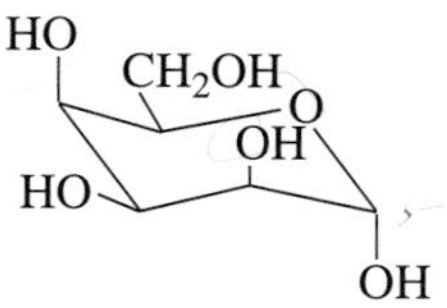

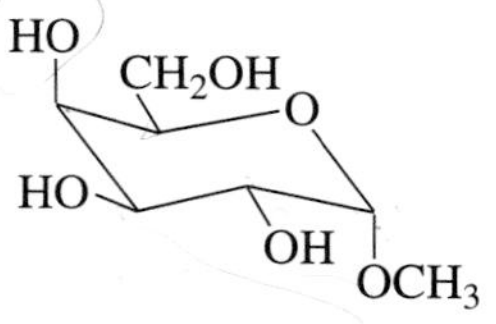

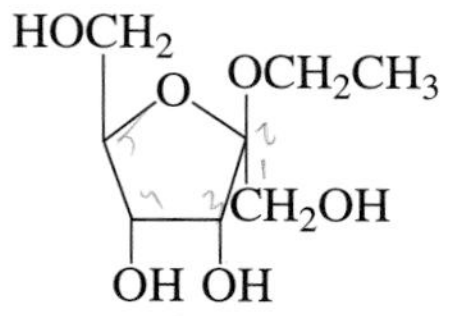

SOLUTION TO 22a   The only OH group in an axial position in **a** is the one at C-3. Therefore, this sugar is the C-3 epimer of glucose, which is allose. The substituent at the anomeric carbon is in the β-position. Thus, the sugar's name is propyl β-D-alloside or propyl β-D-allopyranoside. Because the sugar is an acetal, it is nonreducing sugar.

PROBLEM 23

The oxonium ion formed as an intermediate in glycoside formation can be represented by two contributing resonance structures. Draw the resonance contributors and indicate which is more stable.

PROBLEM 24

Why is only a trace amount of acid used in the formation of an *N*-glycoside?

19.13 DISACCHARIDES

If the hemiacetal group of a monosaccharide uses an alcohol group of another monosaccharide to form an acetal, the glycoside that is formed is a disaccharide. *Disaccharides* are compounds with two monosaccharide subunits hooked together by an acetal linkage. For example, maltose, a disaccharide formed from the hydrolysis of starch, contains two D-glucose subunits hooked together by an **α-1,4'-glycosidic linkage.** The linkage is between C-1 of one sugar subunit and C-4 of the other. The "prime" superscript indicates that C-4 is not in the same ring as C-1. It is an α-1,4'-glycosidic linkage because the oxygen atom involved in the glycosidic linkage is in the α-position. *Remember that the α-position is axial when a sugar is shown in a chair conformation and is down when the sugar is shown in a Haworth projection; the β-position is equatorial when a sugar is shown in a chair conformation and is up when the sugar is shown in a Haworth projection.*

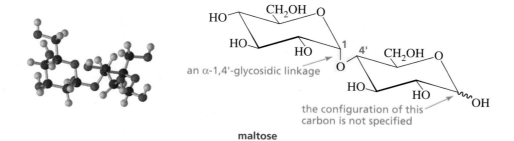

maltose

In the disaccharide above, the configuration of the anomeric carbon not involved in acetal formation (the anomeric carbon of the subunit on the right) is not specified, because maltose can exist in both the α and β forms. In α-maltose, the OH group bonded to this anomeric carbon is in the axial position. In β-maltose, this OH group is in the equatorial position. Because maltose can exist in both α and β forms, mutarotation occurs when crystals of one form are dissolved in a solvent. Maltose is a reducing sugar because the right-hand subunit is a hemiacetal and therefore is in equilibrium with the open-chain aldehyde that is easily oxidized.

Cellobiose, a disaccharide obtained from the hydrolysis of cellulose, also contains two D-glucose subunits. It differs from maltose in that the two glucose subunits are hooked together by a **β-1,4′-glycosidic linkage.** Thus the only difference in the structures of maltose and cellobiose is the configuration of the glycosidic linkage. Like maltose, cellobiose exists in both α and β forms because the OH group bonded to the anomeric carbon not involved in acetal formation can be in either the axial position (α-cellobiose) or the equatorial position (β-cellobiose). Cellobiose is a reducing sugar because the subunit on the right is a hemiacetal.

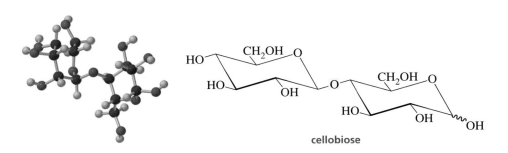

cellobiose

Lactose is a disaccharide found in milk. It constitutes 4.5% of cow's milk by weight and 6.5% of human milk. One of the subunits of lactose is D-galactose, and the other is D-glucose. The D-galactose subunit is an acetal, and the D-glucose subunit is a hemiacetal. The subunits are joined through a β-1,4′-glycosidic linkage. Because one of the subunits is a hemiacetal, lactose is a reducing sugar and undergoes mutarotation.

LACTOSE INTOLERANCE

Lactase is an enzyme that specifically breaks the β-1,4′-glycosidic linkage of lactose. Cats and dogs lose their intestinal lactase when they become adults and therefore are no longer able to digest lactose. Consequently, when they are fed milk or milk products, the undegraded lactose causes digestive problems such as bloating, abdominal pain, and diarrhea. This is because only monosaccharides can pass into the bloodstream, so lactose has to pass undigested into the large intestine. When humans have stomach flu or other intestinal disturbances, they can temporarily lose their lactase, thereby becoming lactose intolerant. Some humans lose their lactase permanently as they mature. This occurs in approximately 10% of the white population of the United States. It is much more common in people whose ancestors came from non-dairy-producing countries. For example, only 3% of Danes but 97% of Thais are lactose-intolerant.

GALACTOSEMIA

After lactose is degraded into glucose and galactose, the galactose has to be converted into glucose before it can be used by cells. Those who do not have the enzyme that converts galactose into glucose have the genetic disease known as galactosemia. Without this enzyme, galactose accumulates in the bloodstream. This accumulation can cause mental retardation in infants and even death. Galactosemia is treated by excluding galactose from the diet.

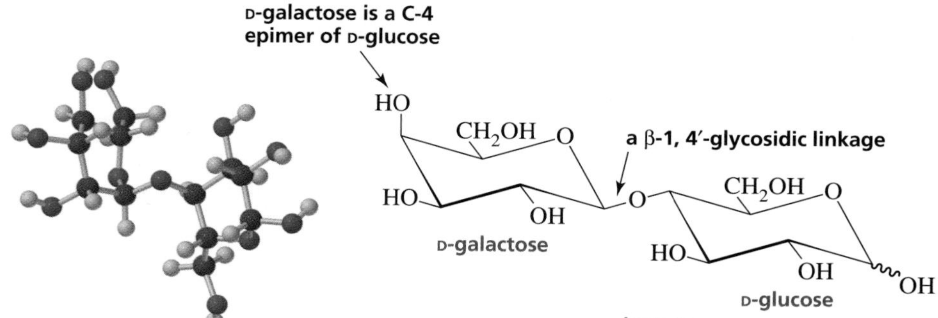

D-galactose is a C-4 epimer of D-glucose

a β-1, 4′-glycosidic linkage

D-galactose

D-glucose

lactose

The most common disaccharide is sucrose (table sugar). Sucrose is obtained from sugar beets and sugar cane. About 90 million tons of sucrose is produced in the world each year. Sucrose consists of a D-glucose subunit and a D-fructose subunit linked by a glycosidic bond between C-1 of glucose (in the α-position) and C-2 of fructose (in the β-position).

Sucrose, unlike the other disaccharides that have been discussed, is not a reducing sugar and does not exhibit mutarotation. This is because the glycosidic bond is between the anomeric carbon of glucose and the anomeric carbon of fructose. Because sucrose does not have a hemiacetal group, it is not in equilibrium with the readily oxidized open-chain aldehyde or ketone form in aqueous solution.

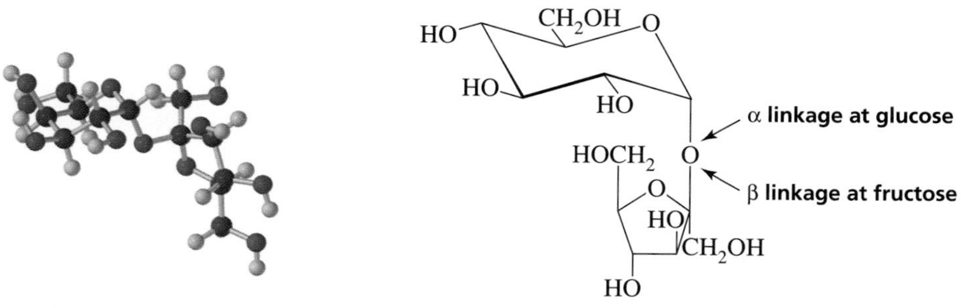

α linkage at glucose

β linkage at fructose

sucrose

Sucrose has a specific rotation of +66.5°. When it is hydrolyzed, the resulting equimolar mixture of glucose and fructose has a specific rotation of −22.0°. Because of the change in the sign of the rotation when sucrose is hydrolyzed, a 1 : 1 mixture of glucose and fructose is called invert sugar. The enzyme that catalyzes the hydrolysis of sucrose is called invertase. Honeybees have this enzyme. The honey they produce is a mixture of sucrose, glucose, and fructose. Because fructose is sweeter than sucrose, invert sugar is sweeter than sucrose. Some foods are advertised as containing fructose instead of glucose, which means they achieve the same level of sweetness with a lower sugar content.

19.14 POLYSACCHARIDES

Polysaccharides contain as few as ten or as many as several thousand monosaccharide units joined together by glycosidic linkages. The molecular weight of the individual polysaccharide chains is variable. The most common polysaccharides are starch and cellulose.

Starch is the major component of flour, potatoes, rice, beans, corn, and peas. It is a mixture of two different polysaccharides: amylose (about 20%) and amylopectin (about 80%). Amylose is a chain of several thousand D-glucose units joined by α-1,4'-glycosidic linkages.

an α-1,4'-glycosidic linkage

3 subunits of amylose

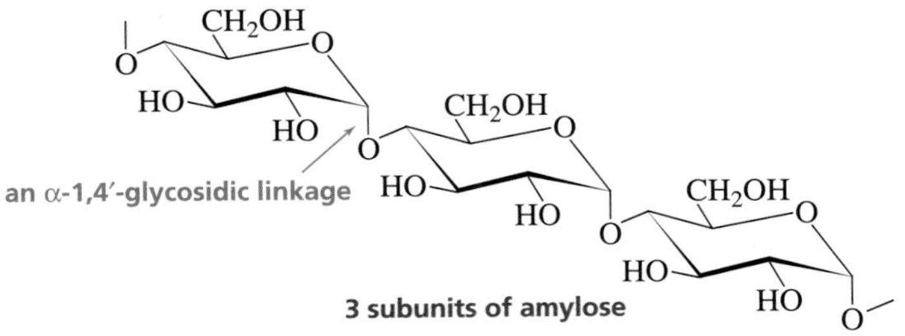

Amylopectin is a branched polysaccharide. Like amylose, it is composed of D-glucose units joined by α-1,4'-glycosidic linkages. Unlike amylose, amylopectin also contains α-1,6'-glycosidic linkages. These linkages create the branches in the polysaccharide. Amylopectin can contain up to 10^6 glucose units, making it one of the largest molecules found in nature.

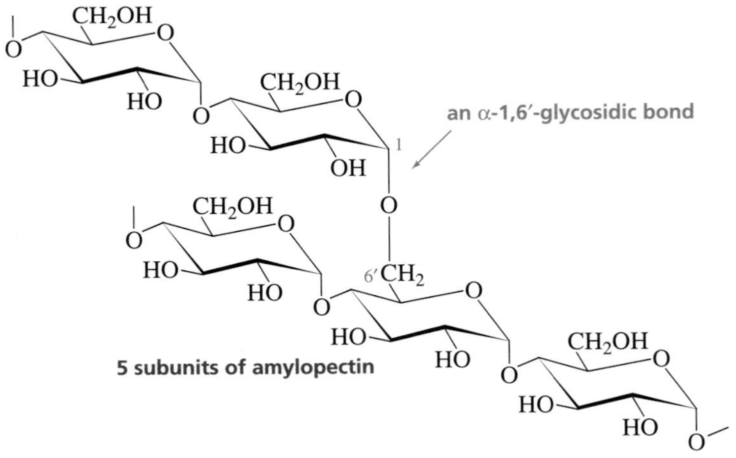

an α-1,6'-glycosidic bond

5 subunits of amylopectin

Animals store their excess glucose in a polysaccharide known as glycogen. Glycogen has a structure similar to that of amylopectin, but glycogen has more branches (Figure 19.1). The branch points in glycogen occur about every ten residues, while those in amylopectin occur about every 25 residues. The high degree of branching in glycogen has important physiological effects. When the body needs energy, many individual glucose units can be simultaneously removed from the ends of many branches.

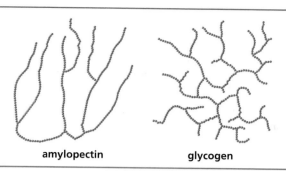

amylopectin   glycogen

WHY THE DENTIST IS RIGHT

Bacteria found in the mouth have an enzyme that converts sucrose into a polysaccharide called dextran. Dextran is made up of glucose units joined mainly through α-1,3′- and α-1,6′-glycosidic linkages. About 10% of dental plaque is composed of dextran. Now you know the chemical reason why your dentist cautions you not to eat candy.

Cellulose is the structural material of higher plants. For example, cotton is composed of about 90% cellulose, and wood is made up of about 50% cellulose. Like amylose, cellulose is composed of an unbranched chain of D-glucose units. However, it differs from amylose in that the glucose units are joined by β-1,4′-glycosidic linkages.

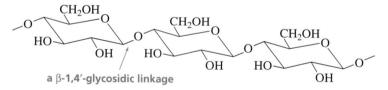

a β-1,4′-glycosidic linkage

3 subunits of cellulose

All mammals have the enzyme (α-glucosidase) that hydrolyzes the α-1,4′-glycosidic linkages that join glucose units, but they do not have the enzyme (β-glucosidase) that hydrolyzes β-1,4′-glycosidic linkages. This means that mammals cannot obtain the glucose they need by eating cellulose. Bacteria that possess β-glucosidase inhabit the digestive tracts of grazing animals. So cows can eat grass and horses can eat hay to meet their nutritional requirements for glucose. Termites also harbor bacteria that break down the cellulose in the wood they eat.

The different glycosidic linkages in starch and cellulose provide these compounds with very different physical properties. As a consequence of its α-linkages, amylose forms a helix that is extensively hydrogen bonded to water molecules (Figure 19.2). As a result, starch is soluble in water.

The β-linkages in cellulose cause the molecules to form intramolecular hydrogen bonds. Consequently, they are less extensively bonded to water and line up in linear arrays (Figure 19.3). The linear molecules form intermolecular hydrogen bonds and, as a result, cellulose is not soluble in water. The strength of these bundles of polymer chains causes cellulose to be an effective structural material.

Figure19.2 ▶
The α-1,4′-glycosidic linkages in amylose cause it to form a left-handed helix.

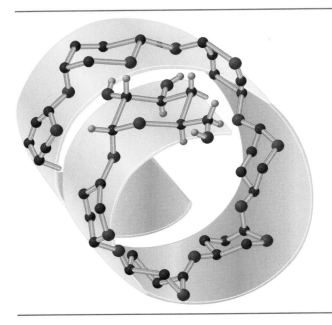

Figure 19.3 ▶
The β-1,4′-glycosidic linkages in cellulose cause it to form intramolecular hydrogen bonds and line up in linear arrays.

SYNTHETIC FIBERS

Cellulose is the raw material for some commercially important products. For example, after being specially treated, cellulose is spun into a fiber known as rayon. The OH groups are esterified with acetic anhydride, resulting in a fabric commonly known as acetate. These cellulose products were the first synthetic fibers. The production of rayon and acetate is now decreasing (20% of all synthetic fibers in 1984; 14% in 1991), partly because their production generates large amounts of polluting waste. Processed cellulose is also used for the production of paper and cellophane.

Chitin is a polysaccharide structurally similar to cellulose. It is the major structural component of the shells of crustaceans (lobsters, crabs, shrimps) and the exoskeletons of insects. It differs from cellulose in that it has an *N*-acetylamino group instead of an OH group at the C-2 position. The β-1,4′-glycosidic linkages give chitin its structural rigidity.

A bright orange crab from Australia.

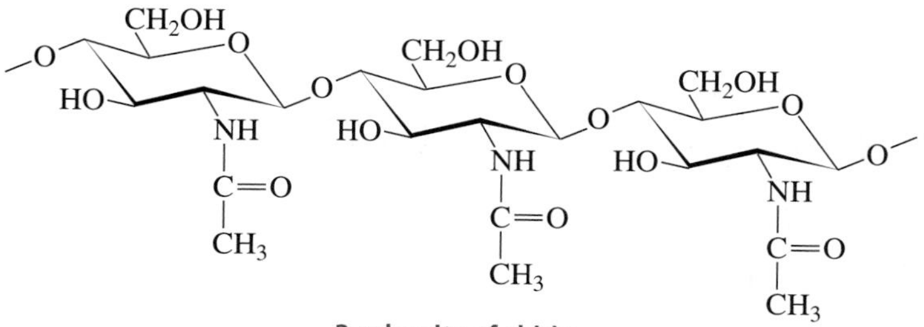

3 subunits of chitin

HEPARIN

Heparin is an anticoagulant that is released to prevent excessive blood clot formation when an injury occurs. It is a polysaccharide made up of glucosamine, glucuronic acid, and iduronic acid subunits. The C-6 OH groups of the glucosamine subunits and the C-2 OH groups of the iduronic acid subunits are sulfonated; some of the amino groups are sulfonated and some are acetylated. Thus heparin is a highly negatively charged molecule. Heparin is found principally in cells that line arterial walls. The longer the polysaccharide chain, the greater the anticoagulant activity. Heparin is widely used clinically as an anticoagulant.

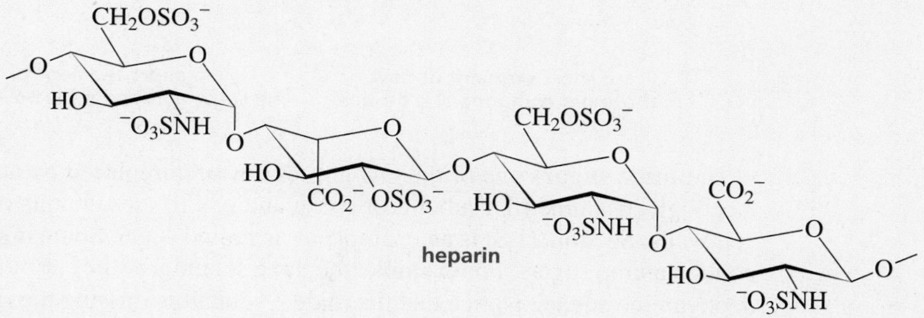

heparin

PROBLEM 25

What is the main structural difference between:

  a. amylose and cellulose?

  b. amylose and amylopectin?

  c. amylopectin and glycogen?

  d. cellulose and chitin?

Deoxy sugars are sugars in which one of the OH groups is replaced by a hydrogen (deoxy means "without oxygen"). 2-Deoxyribose (it is missing the oxygen at the C-2 position) is an important example of a deoxy sugar. Ribose is the sugar component of RNA (ribonucleic acid), while 2′-deoxyribose is the sugar component of DNA (deoxyribonucleic acid). RNA and DNA are *N*-glycosides; their subunits consist of an amine bonded to the β-position of the anomeric carbon of ribose or deoxyribose. The subunits are linked by a phosphate group between C-3 of one sugar and C-5 of the next sugar (Section 24.1).

**19.15
SOME NATURALLY
OCCURRING
PRODUCTS DERIVED
FROM
CARBOHYDRATES**

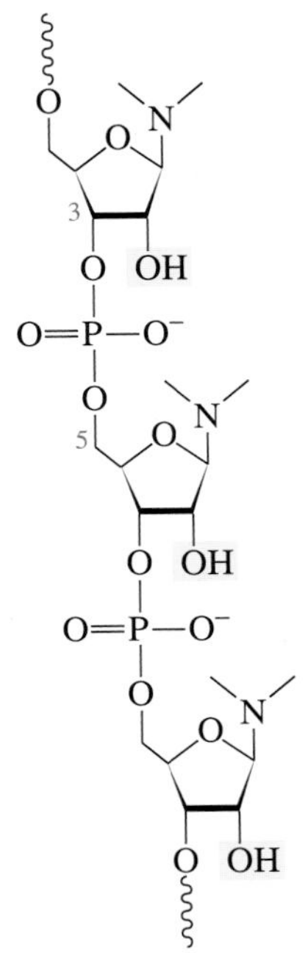

a short segment of RNA
the sugar component is D-ribose

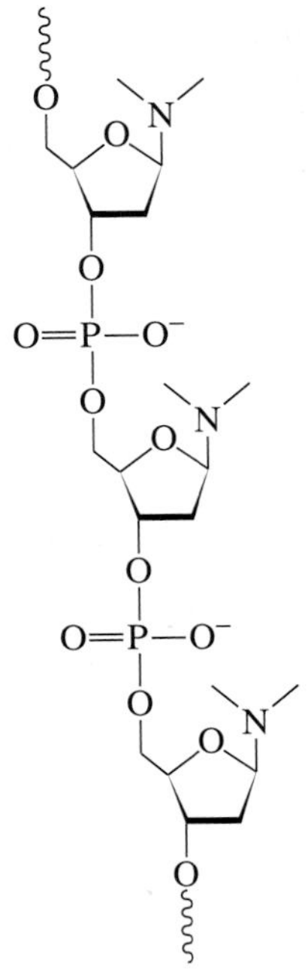

a short segment of DNA
the sugar component is D-2'-deoxyribose

In **amino sugars** one of the OH groups has been replaced by an amino group. *N*-Acetylglucosamine (the subunit of chitin and one of the subunits of certain bacterial cell walls; Section 21.8) is an example of an amino sugar. Some important antibiotics contain amino sugars. For example, the three subunits of the antibiotic gentamicin are deoxyamino sugars. Notice that the middle subunit is missing the ring oxygen.

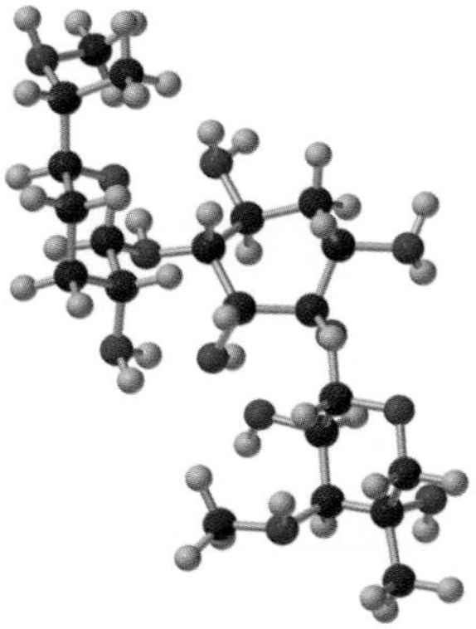

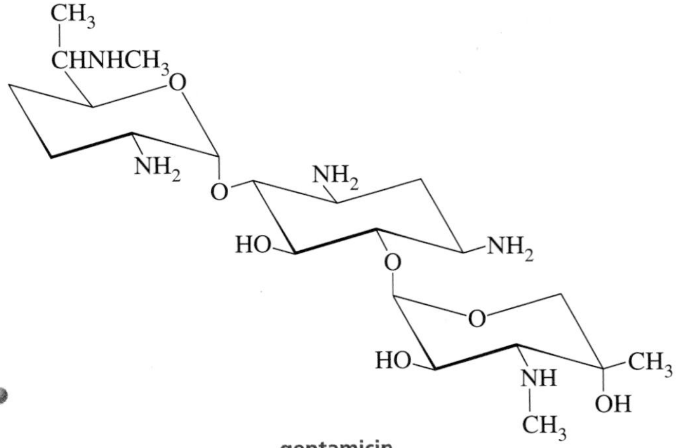

gentamicin
an antibiotic

L-Ascorbic acid (vitamin C) is synthesized in plants and in the livers of most vertebrates. Humans, monkeys, and guinea pigs do not have the enzymes necessary for the biosynthesis of vitamin C, so they must include the vitamin in their diets. The biosynthesis of the vitamin involves enzymatic conversion of D-glucose into L-gulonic acid (reminiscent of the last step in the Fischer proof). L-Gulonic acid is converted into a γ-lactone by the enzyme lactonase, and then an enzyme called oxidase oxidizes the lactone to L-ascorbic acid. The L-configuration of ascorbic acid refers to the configuration at C-5, which was C-2 in D-glucose.

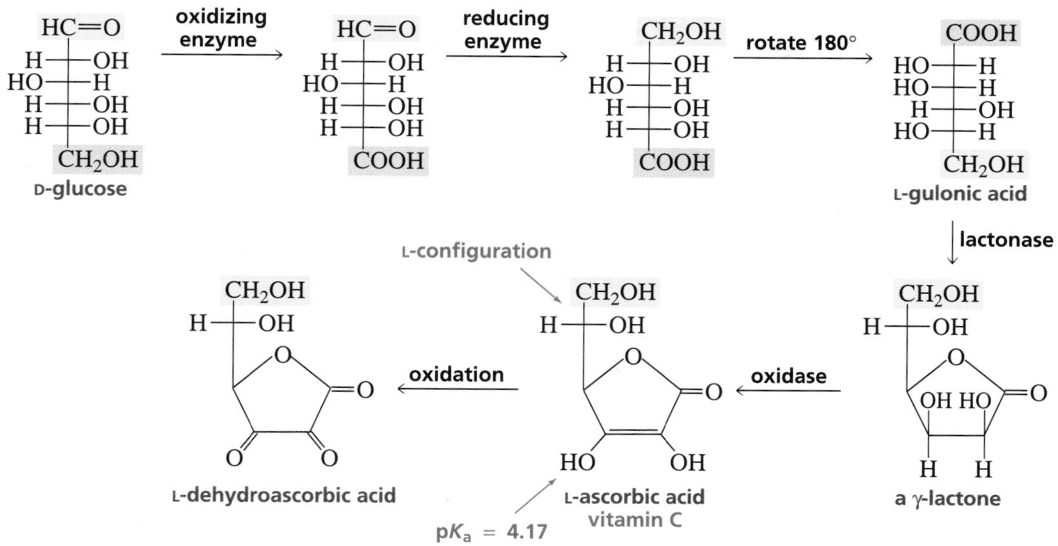

Although L-ascorbic acid does not have a carboxylic acid group, it is an acidic compound because the pK_a of the C-3 OH group is 4.17. L-Ascorbic acid is readily oxidized to L-dehydroascorbic acid, which is also physiologically active. If the lactone ring is opened by hydrolysis, all vitamin C activity is lost. Not much intact vitamin C survives in food that has been thoroughly cooked. And if the food is cooked in water, the water-soluble vitamin is thrown out with the water.

PROBLEM 26

Explain why the C-3 OH group of vitamin C is more acidic than the C-2 OH group.

VITAMIN C

Vitamin C traps radicals formed in aqueous environments (Section 8.8). It is an antioxidant because it prevents oxidation reactions by radicals. Not all the physiological functions of vitamin C are known. It is required for the synthesis of collagen, which is the structural protein of skin, tendons, connective tissue, and bone. If vitamin C is not present in the diet (it is abundant in citrus fruits and tomatoes), lesions appear on the skin, severe bleeding occurs about the gums, in the joints, and under the skin, and wounds heal slowly. The disease caused by a deficiency of vitamin C is known as scurvy. British sailors who shipped out to sea after the late 1700s were required to eat limes to prevent scurvy. This is how they came to be called "limeys." Thus, scurvy was the first disease to be treated by adjusting the diet. Latin for scurvy is *scorbutus;* ascorbic, therefore, means "no scurvy."

The surfaces of many cells contain short polysaccharide chains. These polysaccharides are linked to the cell surface by the reaction of an OH or an NH_2 group of a protein with the anomeric carbon of a cyclic sugar. Proteins that are bonded to polysaccharides are called **glycoproteins.** The percentage of carbohydrate is variable: some glycoproteins contain as little as 1% carbohydrate by weight, whereas others contain as much as 80% carbohydrate.

19.16
CARBOHYDRATES ON CELL SURFACES

$$\text{HO} \overset{\overset{\displaystyle CH_2OH}{|}}{\diagdown} \underset{\text{OH}}{\diagdown} \text{O-protein} \qquad \text{HO} \overset{\overset{\displaystyle CH_2OH}{|}}{\diagdown} \underset{\text{OH}}{\diagdown} \text{NH-protein}$$

Many different types of proteins are glycoproteins. For example, structural proteins such as collagen, proteins found in mucous secretions, immunoglobulins, follicle-stimulating hormone and thyroid-stimulating hormone, interferon (an antiviral protein), and blood plasma proteins are all glycoproteins. One of the functions of the polysaccharide chain is to act as a receptor site on the cell surface in order to transmit signals from hormones and other molecules across the cell membrane into the cell. The carbohydrates on the surfaces of cells also serve as points of attachment for other cells, viruses, and toxins.

Carbohydrates on the surfaces of cells provide a way for cells to recognize one another. The interaction between surface carbohydrates has been found to play a role in inflammatory diseases such as rheumatoid arthritis and septic shock. The fact that several known antibiotics contain amino sugars suggests that they function by recognizing target cells. Carbohydrate interactions also are involved in the regulation of cell growth, so changes in the membrane glycoproteins are thought to be correlated with malignant transformations.

Blood type (A, B, or O) is determined by the nature of the sugar bound to the protein on the outer surface of red blood cells. Each type of blood is associated with a different carbohydrate structure (Figure 19.4). Type AB blood has the carbohydrate structure of both type A and type B.

Antibodies are proteins that are synthesized by the body in response to a foreign substance, called an **antigen.** Interaction with the antibody causes the antigen to precipitate or flags it for destruction by immune-system cells. Blood cannot be transferred from one person to another unless the carbohydrate portions of the donor and acceptor are compatible, otherwise the blood will be considered a foreign substance.

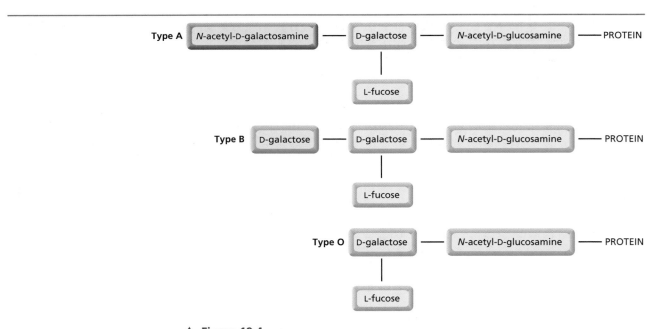

▲ **Figure 19.4**
Blood type is determined by the nature of the sugar on the surfaces of red blood cells.
Fucose is 6-deoxygalactose.

From the nature of the carbohydrates bound to red blood cells, it is apparent why the immune system of type A people recognizes the blood from type B people as being foreign and vice versa. The immune system of people with type A, B, or AB blood does not recognize type O blood as being foreign, since the carbohydrate in type O blood is also a component of types A, B, and AB blood. Thus, anyone can accept type O blood; people with type O blood are called universal donors. Type AB people can accept types A, B, and O blood; people with type AB blood are referred to as universal acceptors.

PROBLEM 27◆

From the nature of the carbohydrate bound to red blood cells, answer the following questions.

a. People with type O blood can donate blood to anyone, but they cannot receive blood from everyone. From whom can they not receive blood? Why?

b. People with type AB blood can receive blood from anyone, but they cannot give blood to everyone. To whom can they not give blood? Why?

For a molecule to taste sweet, it must bind to a receptor on a taste bud cell of the tongue. The binding of this molecule causes a nerve impulse to pass from the taste bud to the brain, where the molecule is interpreted to be sweet. Sugars differ in their degree of "sweetness." The relative sweetness of glucose is 1.00, that of sucrose is 1.45, and that of fructose, the sweetest of all sugars, is 1.65. Developers of synthetic sweeteners must consider several factors in addition to taste, such as toxicity, stability, and cost.

Saccharin, the first synthetic sweetener, was discovered by Ira Remsen and his student Constantine Fahlberg at Johns Hopkins University in 1878. Fahlberg was studying the oxidation of ortho-substituted toluenes in Remsen's laboratory when he found that one of his newly synthesized compounds had an extremely sweet taste. As strange as it may seem today, at one time it was common for chemists to taste a compound in order to characterize it. He called this compound saccharin, and it was eventually found to be about 300 times sweeter than glucose. Notice that, in spite of its name, saccharin is not a saccharide.

19.17 SYNTHETIC SWEETENERS

Ira Remsen (1846–1927) *was born in New York. After receiving an M.D. from Columbia University, he decided to become a chemist. He earned a Ph.D. in Germany, and then returned to the United States in 1872 to accept a faculty position at Williams College. In 1876, he became a professor of chemistry at the newly established Johns Hopkins University. He later became the second president of Johns Hopkins.*

saccharin   dulcin   sodium cyclamate

aspartame

Because it has no caloric value, saccharin became an important substitute for sucrose when it became commercially available in 1885. Its use was spurred by the fact that the chief nutritional problem in the West was (and still is) overconsumption of sugar and its consequences—obesity, heart disease, and dental decay.

Saccharin is also important to diabetics, who must limit their consumption of sucrose and glucose. Although careful toxicity studies had not been done when saccharin first became available to the public (our current concern with toxicity is a fairly recent development), the extensive studies done since then have shown saccharin to be a safe sugar substitute. In 1912, saccharin was temporarily banned in the United States, not because of any concern about its toxicity, but rather because of a concern that people would miss out on the nutritional benefits of sugar.

Dulcin was the second synthetic sweetener to be discovered (in 1884). Even though it did not have the bitter, metallic aftertaste associated with saccharin, it never achieved much popularity. It was taken off the market in 1951 in response to questions about its toxicity.

Sodium cyclamate became a widely used non-nutritive sweetener in the 1950s but was banned in the United States some 20 years later in response to two studies that appeared to show that large amounts of sodium cyclamate cause liver cancer in mice.

Aspartame, the methyl ester of a dipeptide, was approved by the U.S. Food and Drug Administration (FDA) in 1981. It is about 200 times sweeter than sucrose and is sold under the trade name NutraSweet (Section 20.8). Because NutraSweet contains phenylalanine, it should not be used by people with the genetic disease known as PKU (Section 22.6).

The fact that these four synthetic sweeteners have such different structures and that their structures are very different from those of monosaccharides indicates that the sensation of sweetness is not induced by a single molecular shape.

SUMMARY OF REACTIONS

1. Reduction (Section 19.5)

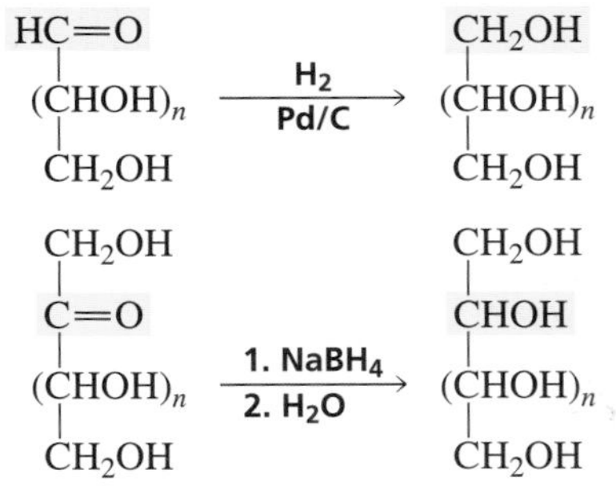

2. Oxidation (Section 19.5)

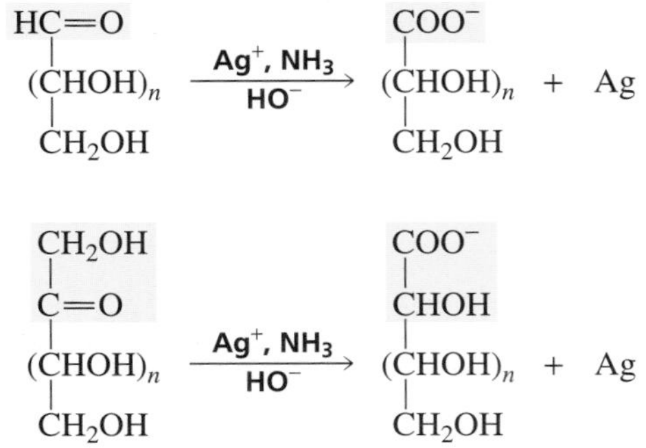

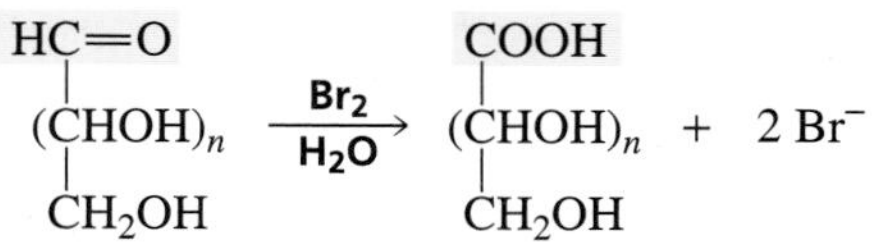

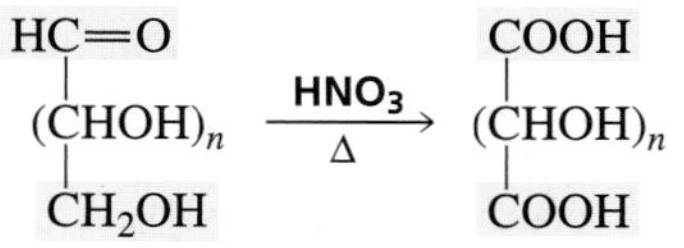

$$\begin{array}{c} \text{HC=O} \\ | \\ \text{(CHOH)}_n \\ | \\ \text{CH}_2\text{OH} \end{array} \xrightarrow[\text{H}_2\text{O}]{\text{Br}_2} \begin{array}{c} \text{COOH} \\ | \\ \text{(CHOH)}_n \\ | \\ \text{CH}_2\text{OH} \end{array} + 2\text{ Br}^-$$

$$\begin{array}{c} \text{HC=O} \\ | \\ \text{(CHOH)}_n \\ | \\ \text{CH}_2\text{OH} \end{array} \xrightarrow[\Delta]{\text{HNO}_3} \begin{array}{c} \text{COOH} \\ | \\ \text{(CHOH)}_n \\ | \\ \text{COOH} \end{array}$$

3. Enolization (Section 19.5)

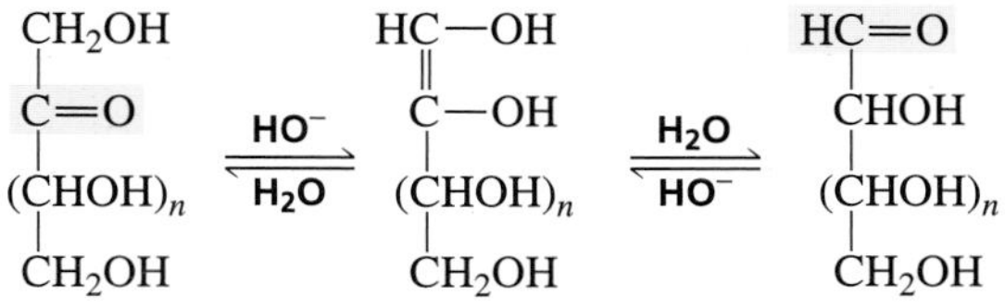

$$\begin{array}{c} \text{CH}_2\text{OH} \\ | \\ \text{C=O} \\ | \\ \text{(CHOH)}_n \\ | \\ \text{CH}_2\text{OH} \end{array} \underset[\text{H}_2\text{O}]{\overset{\text{HO}^-}{\rightleftharpoons}} \begin{array}{c} \text{HC—OH} \\ \| \\ \text{C—OH} \\ | \\ \text{(CHOH)}_n \\ | \\ \text{CH}_2\text{OH} \end{array} \underset[\text{HO}^-]{\overset{\text{H}_2\text{O}}{\rightleftharpoons}} \begin{array}{c} \text{HC=O} \\ | \\ \text{CHOH} \\ | \\ \text{(CHOH)}_n \\ | \\ \text{CH}_2\text{OH} \end{array}$$

4. Osazone formation (Section 19.6)

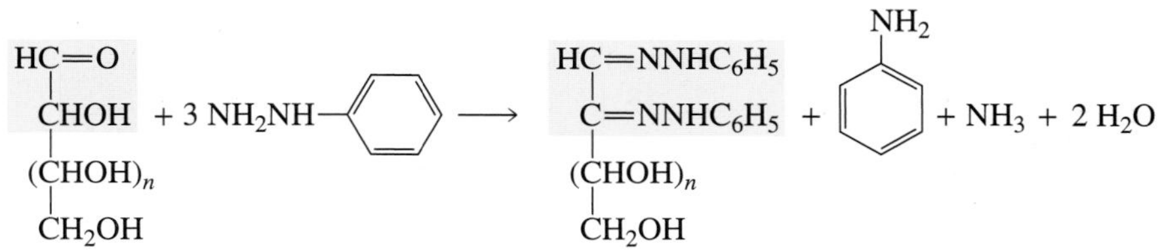

$$\begin{array}{c} \text{HC=O} \\ | \\ \text{CHOH} \\ | \\ \text{(CHOH)}_n \\ | \\ \text{CH}_2\text{OH} \end{array} + 3\text{ NH}_2\text{NH}\text{—}\langle\text{C}_6\text{H}_5\rangle \longrightarrow \begin{array}{c} \text{HC=NNHC}_6\text{H}_5 \\ | \\ \text{C=NNHC}_6\text{H}_5 \\ | \\ \text{(CHOH)}_n \\ | \\ \text{CH}_2\text{OH} \end{array} + \langle\text{C}_6\text{H}_5\text{NH}_2\rangle + \text{NH}_3 + 2\text{ H}_2\text{O}$$

5. Chain elongation (Section 19.7)

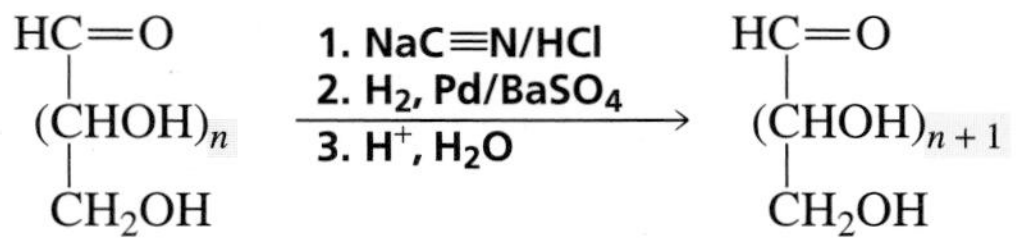

$$\begin{array}{c} \text{HC=O} \\ | \\ \text{(CHOH)}_n \\ | \\ \text{CH}_2\text{OH} \end{array} \xrightarrow[\substack{\text{2. H}_2\text{, Pd/BaSO}_4 \\ \text{3. H}^+\text{, H}_2\text{O}}]{\text{1. NaC}\equiv\text{N/HCl}} \begin{array}{c} \text{HC=O} \\ | \\ \text{(CHOH)}_{n+1} \\ | \\ \text{CH}_2\text{OH} \end{array}$$

6. Chain shortening (Section 19.8)

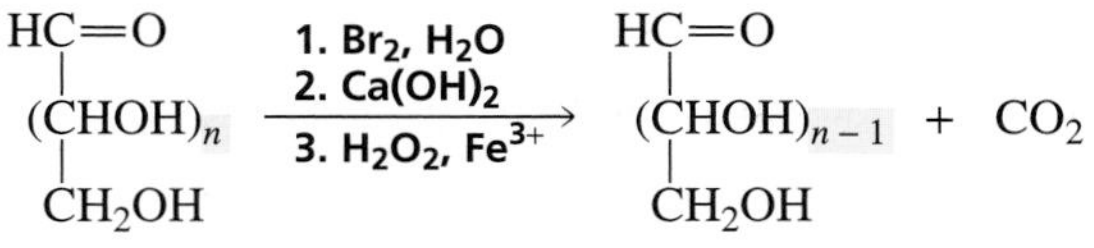

$$\begin{array}{c} \text{HC=O} \\ | \\ \text{(CHOH)}_n \\ | \\ \text{CH}_2\text{OH} \end{array} \xrightarrow[\substack{\text{2. Ca(OH)}_2 \\ \text{3. H}_2\text{O}_2\text{, Fe}^{3+}}]{\text{1. Br}_2\text{, H}_2\text{O}} \begin{array}{c} \text{HC=O} \\ | \\ \text{(CHOH)}_{n-1} \\ | \\ \text{CH}_2\text{OH} \end{array} + \text{CO}_2$$

7. Acetal (and ketal) formation (Section 19.12)

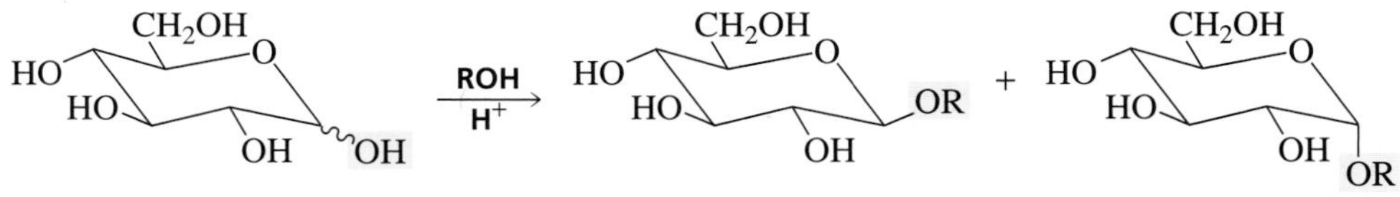

KEY TERMS

aldaric acid (page 878)
alditol (page 877)
aldonic acid (page 877)
aldose (page 872)
amino sugar (page 898)
anomeric carbon (page 884)
anomeric effect (page 890)
anomers (page 884)
antibody (page 900)
antigen (page 900)
bioorganic compound (page 870)
carbohydrate (page 870)
complex carbohydrate (page 871)
deoxy sugar (page 897)
disaccharide (page 871)

epimers (page 874)
furanose (page 885)
furanoside (page 888)
glycoprotein (page 899)
glycoside (page 888)
N-glycoside (page 890)
glycosidic bond (page 888)
α-1,4′-glycosidic linkage (page 891)
β-1,4′-glycosidic linkage (page 892)
Haworth projection (page 885)
heptose (page 872)
hexose (page 872)
ketose (page 872)
Kiliani–Fischer synthesis (page 880)
molecular recognition (page 870)

monosaccharide (page 871)
mutarotation (page 885)
nonreducing sugar (page 890)
oligosaccharide (page 871)
osazone (page 879)
pentose (page 872)
photosynthesis (page 871)
polysaccharide (page 871)
pyranose (page 885)
pyranoside (page 888)
reducing sugar (page 890)
Ruff degradation (page 881)
simple carbohydrate (page 871)
tetrose (page 872)
triose (page 872)

PROBLEMS

28. Give the product(s) that are obtained when D-galactose reacts with:
  a. nitric acid
  b. Tollens' reagent
  c. H_2/Pd/C
  d. three equivalents of phenylhydrazine
  e. Br_2 in water
  f. ethanol + HCl
  g. product of reaction **e** (above) + Ca(OH)$_2$, Fe^{3+}, H_2O_2

29. Identify the following sugars.
  a. An aldopentose that is not D-arabinose forms D-arabinitol when it is reduced with NaBH$_4$.
  b. A sugar forms the same osazone as D-galactose with phenylhydrazine but it is not oxidized by an aqueous solution of Br_2.
  c. A sugar that is not D-altrose forms D-altraric acid when it reacts with nitric acid.
  d. A ketose, when reduced with H_2/Pd/C, forms D-altritol and D-allitol.

30. Pectin is a polysaccharide obtained from fruits. It is used as a jelling agent in making jams and jellies. It can be synthesized by treating amylose with nitric acid. Draw a short segment of pectin.

31. Answer the following questions for the eight aldopentoses.
  a. Which are pairs of enantiomers?
  b. Which give identical osazones?
  c. Which form an optically active compound when oxidized with nitric acid?

32. The reaction of D-ribose with methanol plus HCl forms four products. Give the structures of the products.

33. Determine the structure of D-galactose, using arguments similar to those used by Fischer to prove the structure of D-glucose.

34. Dr. Isent T. Sweet isolated a monosaccharide and determined that it had a molecular weight of 150. Much to his surprise, he found that it was not optically active. What is the structure of the monosaccharide?

35. The NMR spectrum of D-glucose in D$_2$O exhibits two low-field doublets. What is responsible for these doublets?

36. D-Glucuronic acid is found widely in plants and animals. One of its functions is to detoxify poisonous HO-containing compounds by reacting with them in the liver to form glucuronides. Glucuronides are water-soluble and therefore readily excreted. After ingestion of a poison such as turpentine, morphine, or phenol, the glucuronides of these compounds are found in the urine. Give the structure of the glucuronide formed by the reaction of β-D-glucuronic acid and phenol.

β-D-**glucuronic acid**

37. Hyaluronic acid is a component of connective tissue and is the fluid that lubricates the joints. It is an alternating polymer of N-acetyl-D-glucosamine and D-glucuronic acid joined by β-1,3'-glycosidic linkages. Draw a short segment of hyaluronic acid.

38. In order to synthesize D-galactose, Professor Amy Losse went to the stockroom to get some D-lyxose to use as a starting material. She found that the labels had fallen off the bottles containing D-lyxose and D-xylose. How could she have determined which bottle contained D-lyxose?

39. When D-fructose is dissolved in D$_2$O and the solution is made basic, the D-fructose recovered from the solution has an average of 1.7 deuterium atoms attached to carbon per molecule. Show the mechanism that accounts for the incorporation of these deuterium atoms into D-fructose.

40. A D-aldopentose is oxidized by nitric acid to an optically active aldaric acid. A Ruff degradation of the aldopentose leads to a monosaccharide that is oxidized by nitric acid to an optically inactive aldaric acid. Identify the D-aldopentose.

41. How many aldaric acids are obtained from the 16 aldohexoses?

42. Calculate the percentages of α-D-glucose and β-D-glucose present at equilibrium from the specific rotations of α-D-glucose, β-D-glucose, and the equilibrium mixture. Compare your values with those given in Section 19.10.

43. Predict whether D-altrose exists preferentially as a pyranose or a furanose. (*Hint:* The most stable arrangement for a five-membered ring is for all the adjacent substituents to be trans.)

44. In an aqueous solution of D-mannose, there is more α-D-mannose present than β-D-mannose. Explain. (*Hint:* Consider the dipole moments at C-1 and C-2.)

α-D-**mannose**   β-D-**mannose**

45. Propose a mechanism for the rearrangement that converts an α-hydroxyimine into an α-aminoketone (Section 19.6).

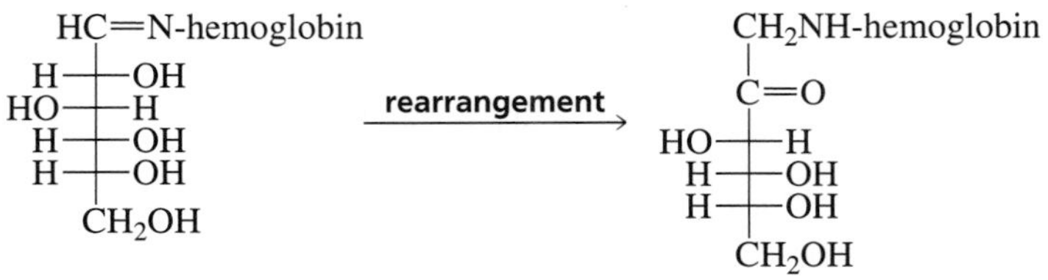

HC=N-hemoglobin

H——OH
HO——H
H——OH
H——OH
CH$_2$OH

$\xrightarrow{\text{rearrangement}}$

CH$_2$NH-hemoglobin

C=O
HO——H
H——OH
H——OH
CH$_2$OH

46. A disaccharide forms a silver mirror with Tollens' reagent and is hydrolyzed by a β-glycosidase. When the disaccharide is treated with excess methyl iodide in the presence of Ag$_2$O (a reaction that converts all the OH groups to OCH$_3$ groups) and then hydrolyzed with water under acidic conditions, two products are formed: 2,3,4-tri-*O*-methylmannose and 2,3,4,6-tetra-*O*-methylgalactose.
  a. Draw the structure of the disaccharide.
  b. What is the function of Ag$_2$O?

47. Devise an experiment to prove that the hemiacetal group in lactose is in the glucose residue rather than in the galactose residue.

48. All of the glucose units in dextran have six-membered rings. When a sample of dextran is treated with methyl iodide and silver oxide, and the product is hydrolyzed under acidic conditions, the products obtained are 2,3,4,6-tetra-*O*-methyl-D-glucose, 2,4,6-tri-*O*-methyl-D-glucose, 2,3,4-tri-*O*-methyl-D-glucose, and 2,4-di-*O*-methyl-D-glucose. Draw a short segment of dextran.

49. When a pyranose is in the chair conformation in which the CH$_2$OH group and the C-1 OH group are both in the axial position, the two groups can react to form an acetal. This is called the anhydro form of the sugar (it has lost water). The anhydro form of D-idose is shown below. In an aqueous solution at 100 °C, a large percentage of D-idose exists in the anhydro form (about 80%). Under the same conditions, only about 0.1% of D-glucose exists in the anhydro form. Explain.

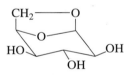
anhydro form of D-idose

50. Oxidation with periodic acid can be used to determine whether a glycoside has the pyranose or the furanose structure. Explain how the products obtained from oxidation of methyl β-D-glucoside can be used to determine whether the glucoside has a five- or a six-membered ring.

51. Devise a method to convert D-glucose into D-allose.

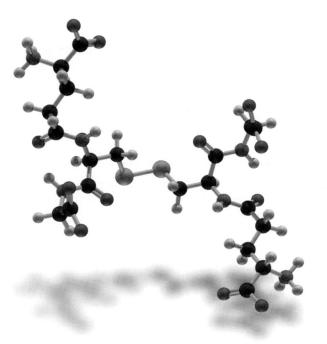

oxidized glutathione

AMINO ACIDS, PEPTIDES, AND PROTEINS

T hree kinds of polymers are prevalent in nature: polysaccharides, proteins, and nucleic acids. We have already studied polysaccharides, which are naturally occurring polymers of sugar subunits (Section 19.14), and nucleic acids will be covered in Chapter 24. We will now look at proteins and the structurally similar but shorter peptides. **Peptides** and **proteins** are polymers of **amino acids** linked together by amide bonds. The monomeric units are called **amino acid residues.**

Amino acid polymers can be composed of any number of monomers. A **dipeptide** contains two amino acid residues, a **tripeptide** contains three amino acid residues, an **oligopeptide** contains three to ten amino acid residues, and a **polypeptide** contains many amino acid residues. Proteins are naturally occurring polypeptides that are made up of 40 to 4000 amino acid residues.

From the structure of an amino acid, we can see that the name is not very precise. The compounds commonly called amino acids are more precisely called α-aminocarboxylic acids.

$$R-\underset{\underset{+NH_3}{|}}{C}H-\overset{\overset{O}{\parallel}}{C}-OH$$

an α-aminocarboxylic acid
an "amino acid"

amide bonds ↓ ↓

$$-NHCH\overset{\overset{O}{\parallel}}{C}-\underset{R'}{NHCH}\overset{\overset{O}{\parallel}}{C}-\underset{R''}{NHCH}\overset{\overset{O}{\parallel}}{C}-$$
$$\underset{R}{}$$

amino acids are linked together by amide bonds

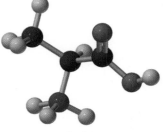

**alanine
an amino acid**

Proteins and peptides serve many functions in biological systems. Some protect biological organisms from their environment and/or impart strength to certain biological structures. Hair, horns, hoofs, feathers, fur, and the tough outer layer of skin are all composed largely of a protein called keratin. Keratin is a **structural protein.** Collagen, another structural protein, is a major component of bones, muscles, and tendons. Some proteins have other protective functions: snake venoms and plant toxins protect their owners from invading species; blood-clotting proteins protect the vascular system when it is injured; and antibodies and protein antibiotics protect us from disease. A group of proteins called **enzymes** catalyze the chemical reactions that occur in living organisms, and some of the hormones that regulate these reactions are peptides. Proteins are also responsible for many physiological functions such as the transport and storage of oxygen in the body and the contraction of muscles.

20.1 CLASSIFICATION AND NOMENCLATURE OF AMINO ACIDS

The structures of the 20 most common naturally occurring amino acids and the frequency with which each occurs in proteins are shown in Table 20.1. Other amino acids occur in nature, but their frequency of occurrence is quite small. All the amino acids except proline contain a primary amino group. Proline contains a secondary amino group incorporated into a five-membered ring. The amino acids differ only in the substituent (R) attached to the α-carbon. The wide variation in these side chains is what gives proteins their great structural diversity and, as a consequence, their great functional diversity.

The amino acids are always called by their common names. Often the name tells you something about the amino acid. For example, glycine got its name because of its sweet taste (*glykos* is Greek for "sweet"), and valine, like valeric acid, has five carbon atoms. Asparagine was first found in asparagus, and tyrosine was isolated from cheese (*tyros* is Greek for "cheese").

Dividing the amino acids into classes makes them easier to learn. The aliphatic amino acids include glycine, the amino acid in which R = H, and four amino acids with alkyl side chains. Alanine is the amino acid with a methyl side chain, and valine has an isopropyl side chain. Can you guess which amino acid—leucine or isoleucine—has an isobutyl side chain? If you gave the obvious answer, you guessed incorrectly. Isoleucine does not have an "iso" group. It is leucine that has an isobutyl substituent; isoleucine has a *sec*-butyl substituent. Each of the amino acids has a three-letter abbreviation (the first three letters of the name in most cases) and a single-letter abbreviation.

Two amino acid side chains contain alcohol groups. Serine is an HO-substituted alanine, and threonine has a branched ethanol substituent. There are also two sulfur-containing amino acids: cysteine is an HS-substituted alanine, and methionine has a 2-methylthioethyl substituent.

There are two acidic amino acids (amino acids with two carboxylic acid groups): aspartic acid, which is a carboxy-substituted alanine; and glutamic acid, which has one more methylene group than aspartic acid. Two amino acids are amides of the acidic amino acids: asparagine, which is the amide of aspartic acid; and glutamine, which is the amide of glutamic acid. Notice that the obvious one-letter abbreviations cannot be used for these four amino acids because A and G are used for alanine and glycine. Aspartic acid and glutamic acid are abbreviated D and E, while asparagine and glutamine are abbreviated N and Q.

There are two basic amino acids (amino acids with two basic nitrogen-containing groups): lysine, which has an ε-amino group; and arginine, which has an δ-guanidino group. The ε and δ can remind you how many methylene groups each amino acid has.

$$\overset{+}{H_3N}-\overset{\varepsilon}{C}H_2\overset{\delta}{C}H_2\overset{\gamma}{C}H_2\overset{\beta}{C}H_2\overset{\alpha}{C}HCO^-$$

an ε-amino group

lysine

$$\overset{+NH_2}{H_2N-C}-NH-\overset{\delta}{C}H_2\overset{\gamma}{C}H_2\overset{\beta}{C}H_2\overset{\alpha}{C}HCO^-$$

a δ-guanidino group

arginine

TABLE 20.1   The 20 Most Common Naturally Occurring Amino Acids
The amino acids are shown in the form that predominates at physiological pH (pH = 7.3).

	Formula	Name	Abbreviations		Average occurrence in proteins
Aliphatic amino acids	$\text{H}-\overset{\displaystyle \overset{\textstyle O}{\|\|}}{\text{C}}\text{HCO}^-$, $\overset{+}{\text{N}}\text{H}_3$	glycine	Gly	G	7.5%
	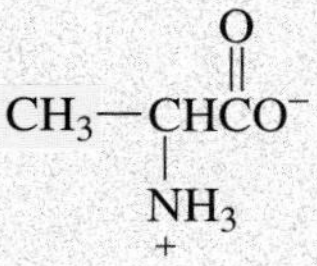 $\text{CH}_3-\text{CHCO}^-$ (C=O), $\overset{+}{\text{N}}\text{H}_3$	alanine	Ala	A	9.0%
	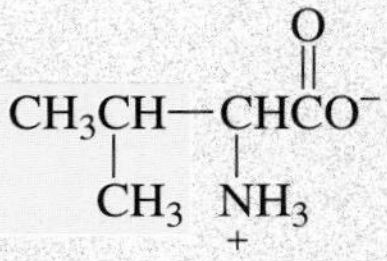 $\text{CH}_3\text{CH}-\text{CHCO}^-$, CH_3 $\overset{+}{\text{N}}\text{H}_3$	valine*	Val	V	6.9%
	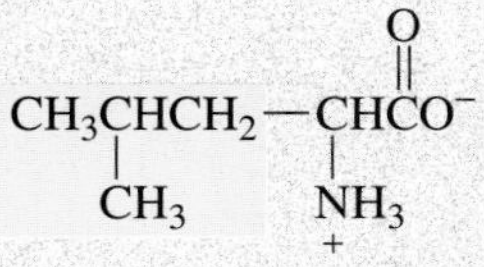 $\text{CH}_3\text{CHCH}_2-\text{CHCO}^-$, CH_3 $\overset{+}{\text{N}}\text{H}_3$	leucine*	Leu	L	7.5%
	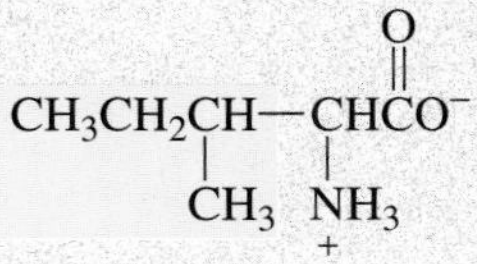 $\text{CH}_3\text{CH}_2\text{CH}-\text{CHCO}^-$, CH_3 $\overset{+}{\text{N}}\text{H}_3$	isoleucine*	Ile	I	4.6%
Hydroxy-containing amino acids	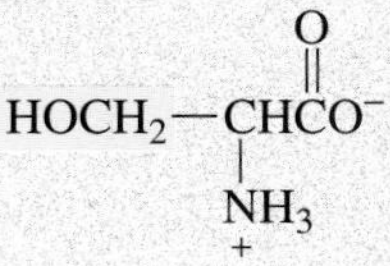 $\text{HOCH}_2-\text{CHCO}^-$, $\overset{+}{\text{N}}\text{H}_3$	serine	Ser	S	7.1%
	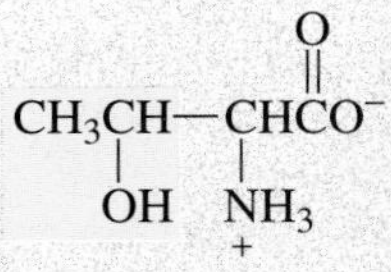 $\text{CH}_3\text{CH}-\text{CHCO}^-$, OH $\overset{+}{\text{N}}\text{H}_3$	threonine*	Thr	T	6.0%
Sulfur-containing amino acids	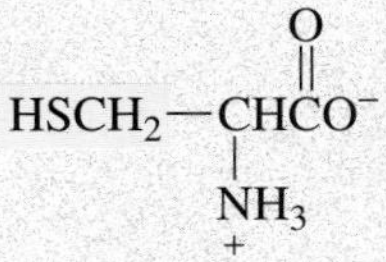 $\text{HSCH}_2-\text{CHCO}^-$, $\overset{+}{\text{N}}\text{H}_3$	cysteine	Cys	C	2.8%

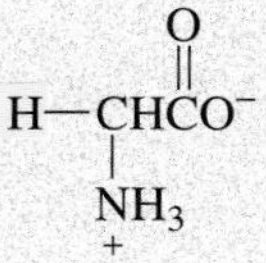

TABLE 20.1
(Continued)

	Formula	Name	Abbreviations		Average occurrence in proteins
	$CH_3SCH_2CH_2-CHCO^-$ (with $\overset{\|}{NH_3^+}$ and C=O)	methionine*	Met	M	1.7%
Acidic amino acids	$^-OCCH_2-CHCO^-$ (with C=O groups and $\overset{\|}{NH_3^+}$)	aspartate (aspartic acid)	Asp	D	5.5%
	$^-OCCH_2CH_2-CHCO^-$ (with C=O groups and $\overset{\|}{NH_3^+}$)	glutamate (glutamic acid)	Glu	E	6.2%
Amides of acidic amino acids	$H_2NCCH_2-CHCO^-$ (with C=O groups and $\overset{\|}{NH_3^+}$)	asparagine	Asn	N	4.4%
	$H_2NCCH_2CH_2-CHCO^-$ (with C=O groups and $\overset{\|}{NH_3^+}$)	glutamine	Gln	Q	3.9%
Basic amino acids	$\overset{+}{H_3}NCH_2CH_2CH_2CH_2-CHCO^-$ (with C=O and $\overset{\|}{NH_3^+}$)	lysine*	Lys	K	7.0%
	$H_2N\overset{\overset{+}{N}H_2}{C}NHCH_2CH_2CH_2-CHCO^-$ (with C=O and $\overset{\|}{NH_3^+}$)	arginine*	Arg	R	4.7%
Benzene-containing amino acids	(phenyl ring)$-CH_2-CHCO^-$ (with C=O and $\overset{\|}{NH_3^+}$)	phenyl-alanine*	Phe	F	3.5%

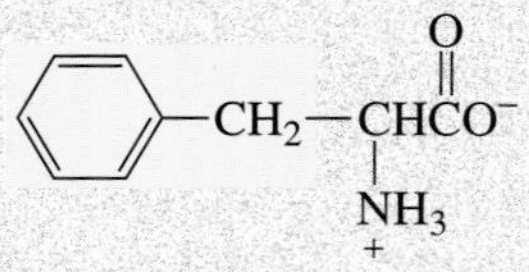

TABLE 20.1
(Continued)

Formula	Name	Abbreviations		Average occurrence in proteins
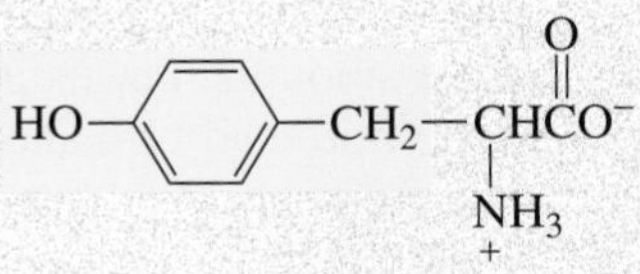	tyrosine	Tyr	Y	3.5%
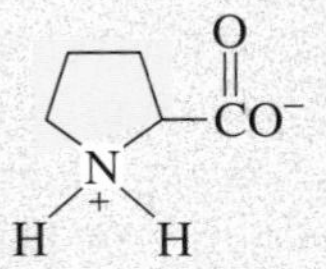	proline	Pro	P	4.6%
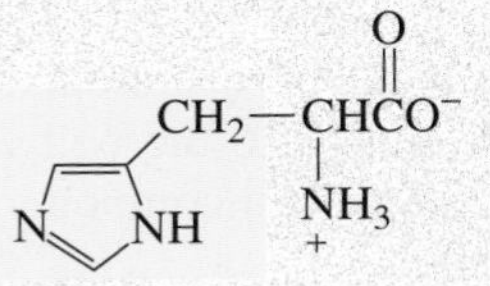	histidine*	His	H	2.1%
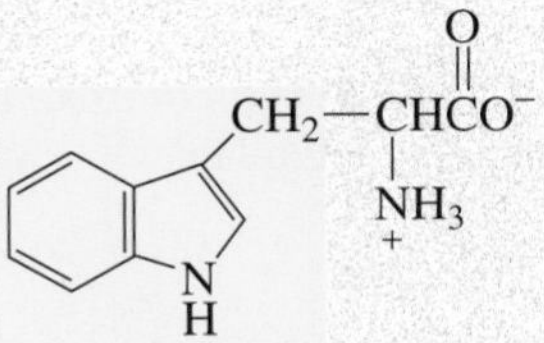	tryptophan*	Trp	W	1.1%

Heterocyclic amino acids

Two amino acids—phenylalanine and tyrosine—contain benzene rings. As its name indicates, phenylalanine is a phenyl-substituted alanine. Tyrosine is phenylalanine with a *para*-hydroxy substituent.

Proline, histidine, and tryptophan are heterocyclic amino acids. Proline has a pyrrolidine ring; it is the amino acid containing a secondary amino group discussed above. Histidine is an imidazole-substituted alanine. Imidazole is an aromatic compound because it is cyclic, planar, and has three pairs of delocalized π electrons. The pK_a of a protonated imidazole ring is 6.0, so the ring will be protonated in acidic solutions and nonprotonated in basic solutions (Section 20.3).

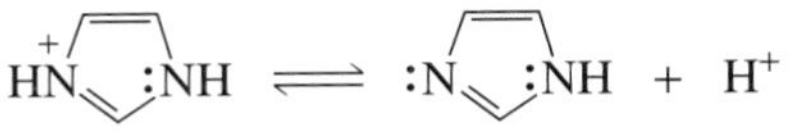

Tryptophan is an indole-substituted alanine (Sections 26.2). Like imidazole, indole is an aromatic compound. Because indole needs the nonbonding pair of electrons on nitrogen for its aromaticity, indole is a very weak base (the pK_a of protonated indole is −2.4). Therefore, the ring nitrogen in tryptophan is never protonated under physiological conditions.

indole

Ten of the amino acids are **essential amino acids.** We humans must obtain these ten amino acids from our diets because we either cannot synthesize them at all or cannot synthesize them in adequate amounts. For example, we must have a dietary source of phenylalanine, but we do not need tyrosine in our diets because we can synthesize the necessary amounts from phenylalanine. The essential amino acids are denoted by red asterisks (*) in Table 20.1. Although humans can synthesize arginine, it is needed for growth in greater amounts than can be synthesized. So arginine is an essential amino acid for children but a nonessential amino acid for adults. Not all proteins contain the same amino acids. Bean protein is deficient in methionine, for example, and wheat protein is deficient in lysine. They are *incomplete* proteins—they contain too little of one or more essential amino acids to support growth. Therefore, a balanced diet must contain proteins from different sources.

Dietary protein is hydrolyzed in the body to individual amino acids. Some of these amino acids are used to synthesize proteins needed by the body, some are broken down further to supply energy to the body, and some are used as starting materials for the synthesis of nonprotein compounds that the body needs, such as adrenaline, thyroxine, and melanin.

PROBLEM 1

a. When the imidazole ring of histidine is protonated, the doubly bonded nitrogen is the nitrogen that accepts the proton. Explain.

b. When the guanidino group of arginine is protonated, the doubly bonded nitrogen is the nitrogen that accepts the proton. Explain.

20.2 CONFIGURATION OF AMINO ACIDS

The α-carbon of all the naturally occurring amino acids except glycine is a chirality center. Therefore, 19 of the 20 amino acids listed in Table 20.1 can exist as enantiomers. The D and L notation used for monosaccharides is also used for amino acids. The D and L isomers of monosaccharides and amino acids are defined the same way. The Fischer projection with the carboxyl group on the top and the R group on the bottom of the vertical axis is a **D-amino acid** if the amino group is on the right of the horizontal axis and is an **L-amino acid** if it is on the left. Unlike monosaccharides, where the D isomer is the one found in nature, most amino acids found in nature have the L configuration. To date, D-amino acid residues have been found only in a few peptide antibiotics and in some small peptides attached to the cell walls of bacteria.

$$
\begin{array}{cc}
\underset{\text{D-glyceraldehyde}}{
\begin{array}{c}
O \\
\parallel \\
C-H \\
H\!-\!\!-\!\!-OH \\
CH_2OH
\end{array}}
&
\underset{\text{L-glyceraldehyde}}{
\begin{array}{c}
O \\
\parallel \\
C-H \\
HO\!-\!\!-\!\!-H \\
CH_2OH
\end{array}}
\end{array}
$$

$$
\begin{array}{cc}
\underset{\text{D-amino acid}}{
\begin{array}{c}
O \\
\parallel \\
C-O^- \\
H\!-\!\!-\!\!-\overset{+}{N}H_3 \\
R
\end{array}}
&
\underset{\text{L-amino acid}}{
\begin{array}{c}
O \\
\parallel \\
C-O^- \\
H_3\overset{+}{N}\!-\!\!-\!\!-H \\
R
\end{array}}
\end{array}
$$

Why D-sugars and L-amino acids? It did not make any difference which isomer was selected to be the one synthesized in nature. But it was important that the same isomer was synthesized by all organisms. For example, if mammals ended up having L-amino acids, it was important for L-amino acids to be the isomers synthesized by the organisms that mammals depend on for food.

AMINO ACIDS AND DISEASE

The Chamorro people of Guam have a high incidence of a syndrome that resembles amyotropic lateral sclerosis (ALS) with elements of Parkinson's disease and dementia. This syndrome developed during World War II when, as a result of food shortages, the tribe ate large quantities of the seeds of *Cycas circinalis*. These seeds contain β-methylamino-L-alanine, an amino acid that binds to glutamate receptors. When monkeys are given β-methylamino-L-alanine, they develop some of the features of this syndrome. There is hope that, by studying the mechanism of action of β-methylamino-L-alanine, we may gain an understanding of how ALS and Parkinson's disease arise.

PROBLEM 2◆

a. Which isomer is D-alanine: (R)-alanine or (S)-alanine?

b. Which isomer is D-aspartic acid: (R)-aspartic acid or (S)-aspartic acid?

c. Can a general statement be made?

PROBLEM 3◆

Which amino acids have more than one chirality center?

Every amino acid has a carboxyl group and an amino group, and each group can exist in an acidic form or a basic form depending on the pH of the solution in which the amino acid is dissolved. The carboxyl groups of the amino acids have pK_a values of approximately 2; the protonated amino groups have pK_a values near 9 (Table 20.2). Therefore, in a very acidic solution (pH near 0), both groups will be in

20.3
ACID–BASE
PROPERTIES OF
AMINO ACIDS

TABLE 20.2 The pK_a Values of the Amino Acids

Amino acid	pK_a α-COOH	pK_a α-NH$_3^+$	pK_a side chain
Alanine	2.34	9.69	—
Arginine	2.17	9.04	12.48
Asparagine	2.02	8.84	—
Aspartic acid	2.09	9.82	3.86
Cysteine	1.71	10.78	8.33
Glutamic acid	2.19	9.67	4.25
Glutamine	2.17	9.13	—
Glycine	2.34	9.60	—
Histidine	1.82	9.17	6.04
Isoleucine	2.36	9.68	—
Leucine	2.36	9.60	—
Lysine	2.18	8.95	10.79
Methionine	2.28	9.21	—
Phenylalanine	1.83	9.13	—
Proline	1.99	10.60	—
Serine	2.21	9.15	—
Threonine	2.63	9.10	—
Tryptophan	2.38	9.39	—
Tyrosine	2.20	9.11	10.07
Valine	2.32	9.62	—

Recall from the Henderson–Hasselbalch equation (Section 1.19) that the acidic form predominates if the pH of the solution is less than the pK_a of the compound, and the basic form predominates if the pH of the solution is greater than the pK_a of the compound.

their acidic forms. At a pH of 7, the pH of the solution is greater than the pK_a of the carboxyl group but less than the pK_a of the protonated amino group. The carboxyl group therefore will be in its basic form; the amino group will be in its acidic form. In a strongly basic solution (pH $\approx$ 11), both groups will be in the basic form.

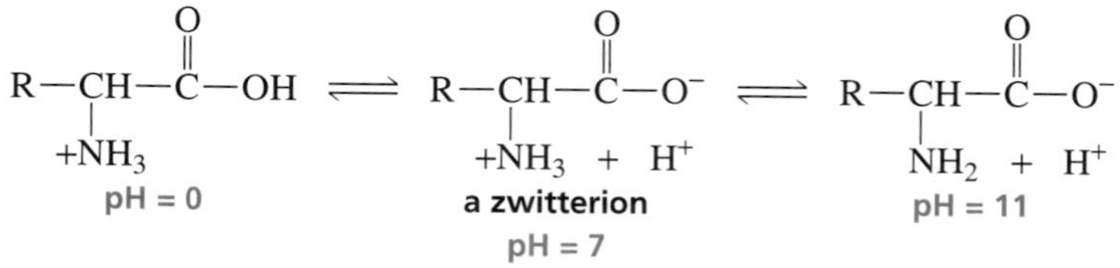

Notice that an amino acid can never exist as an uncharged compound regardless of the pH of the solution. To be uncharged, an amino acid would have to lose a proton from an $^+$NH$_3$ group with a pK_a of about 9 more readily than it would lose a proton from a COOH group with a pK_a of about 2. This clearly is impossible: a weak acid cannot be more acidic than a strong acid. Therefore, at physiological pH (pH = 7.3), an amino acid exists as a dipolar ion. Such a compound is called a **zwitterion.** A zwitterion is a compound that has a negative charge on one atom and a positive charge on a nonadjacent atom. (The name comes from *zwitter,* German for "hermaphrodite" or "hybrid.")

A few amino acids have side chains with ionizable hydrogens (Table 20.2). The protonated imidazole side chain of histidine, for example, has a pK_a of 6.04. Histidine therefore can exist in four different forms, and the form that predominates depends on the pH of the solution.

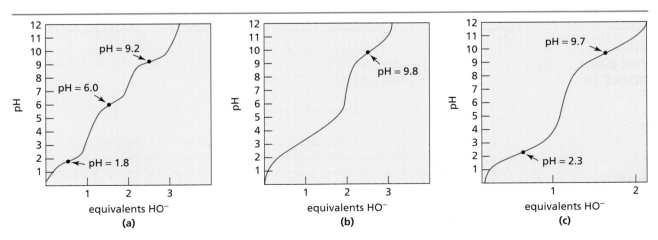

histidine

Some amino acids can be identified by their titration curves. A **titration curve** is a plot of the pH of a solution as a function of added equivalents of hydroxide ion. As hydroxide ion is added to an aqueous solution of the fully protonated form of an amino acid, the pH increases, because hydroxide ion removes protons from the solution by reacting with them to form water. The increase in pH flattens out when HO$^-$ can remove a proton from an ionizable group of the amino acid rather than from the solvent. When all the protons of the ionizable group are removed, the pH again increases. The pH at the midpoint of the flattened-out region, the **inflection point,** is equal to the pK_a of the ionizable group.

The amino acid that gives the titration curve in Figure 20.1a can be positively identified as histidine. No other amino acid has a group with a pK_a near 6. The titration curve in Figure 20.1b shows a broad flattening out requiring two equivalents of HO$^-$. The amino acid, therefore, must have two acidic groups; it could be either aspartic acid or glutamic acid. To distinguish between the two amino acids requires careful analysis of the titration curve. It is actually the titration curve for aspartic acid. The titration curve for alanine is shown in Figure 20.1c.

PROBLEM 4◆

Which other amino acids give a titration curve similar to the one shown in Figure 20.1(c) for alanine?

▲ **Figure 20.1**
Titration curves for a 0.1 M solution of (a) histidine, (b) aspartic acid, and (c) alanine.

PROBLEM 5

The pK_a of a carboxylic acid such as acetic acid is 4.76. Why are the carboxylic acid groups of the amino acids so much more acidic (pK_a values of about 2)?

PROBLEM 6 / SOLVED

Draw the form in which each of the following amino acids predominantly exists at physiological pH (7.3).

   a. aspartic acid   **c.** glutamine   **e.** arginine

   b. histidine   **d.** lysine   **f.** tyrosine

SOLUTION TO 6a   Because the pH is greater than the pK_a of both carboxyl groups, these groups are in their basic forms. Because the pH is less than the pK_a of the protonated amino group, this group is in its acidic form.

$$\overset{\displaystyle O}{\overset{\displaystyle \|}{}}\;\;\;\;\overset{\displaystyle O}{\overset{\displaystyle \|}{}}$$
$$^-OCCH_2\overset{\underset{\displaystyle \overset{|}{\underset{+}{NH_3}}}{}}{C}HCO^-$$

PROBLEM 7 ◆

Draw the form in which glutamic acid predominantly exists at

   a. pH = 0   **b.** pH = 3   **c.** pH = 6   **d.** pH = 11

PROBLEM 8

   a. Why is the pK_a of the glutamic acid side chain greater than the pK_a of the aspartic acid side chain?

   b. Why is the pK_a of the arginine side chain greater than the pK_a of the lysine side chain?

20.4 THE ISOELECTRIC POINT

The **isoelectric point** (pI) of an amino acid is the pH at which it has no net charge. It is the pH at which the amount of negative charge on an amino acid exactly balances the amount of positive charge.

pI **(isoelectric point) = pH at which there is no net charge**

The pI of an amino acid that does not have an ionizable side chain—such as alanine—is midway between its two pK_a values. This occurs because at pH = 2.34, half of the molecules have a negatively charged carboxyl group and half have a neutral carboxyl group, and at pH = 9.69, half of the molecules have a positively charged amino group and half have a neutral amino group. As the pH increases from 2.34, the carboxyl group of more molecules will become negatively charged; as the pH decreases from 9.69, the amino group of more molecules will become positively charged. So, at the average of the two pK_a values, the number of negatively charged groups will equal the number of positively charged groups.

Recall from the Henderson–Hasselbalch equation that when pH = pK_a, half the group is in its acidic form and half is in its basic form (Section 1.19).

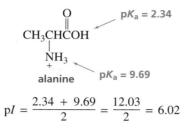

$$pI = \frac{2.34 + 9.69}{2} = \frac{12.03}{2} = 6.02$$

If an amino acid has an ionizable side chain, its pI is the average of the pK_a values of the similarly ionizing groups (positive ionizing to neutral, or neutral ionizing to negative). For example, the pI of lysine is the average of the pK_a values of the two groups that are positively charged in their acidic form and neutral in their basic form; the pI of glutamic acid is the average of the pK_a values of the two groups that are neutral in their acidic form and negatively charged in their basic form.

An amino acid will be positively charged if the pH of the solution is less than its *pI* and will be negatively charged if the pH of the solution is greater than its *pI*.

pKa = 10.79

H₃NCH₂CH₂CH₂CH₂CHCOH   pKₐ = 2.18

NH₃   pKₐ = 8.95

lysine

$$pI = \frac{8.95 + 10.79}{2} = \frac{19.74}{2} = 9.87$$

pKₐ = 4.25

HOCCH₂CH₂CHCOH   pKₐ = 2.19

NH₃   pKₐ = 9.67

glutamic acid

$$pI = \frac{2.19 + 4.25}{2} = \frac{6.44}{2} = 3.22$$

PROBLEM 9

Explain why the pI of lysine is the average of the pK_a values of its two protonated amino groups.

PROBLEM 10◆

Calculate the pI of each of the following amino acids.

a. asparagine   b. arginine   c. serine

PROBLEM 11◆

a. Which amino acid has the smallest pI?

b. Which amino acid has the largest pI?

c. Which amino acid has the greatest amount of negative charge at a pH of 6.20?

d. Which amino acid has a greater amount of negative charge at a pH of 6.20, glycine or methionine?

PROBLEM 12

The pI values of two amino acids, tyrosine and cysteine, cannot be determined by the method described above. Explain.

A mixture of amino acids can be separated by several different techniques. **Electrophoresis** separates amino acids on the basis of their pI values. A few drops of a solution of an amino acid mixture are applied to the middle of a piece of filter

20.5
SEPARATION OF AMINO ACIDS

paper or to a gel. When the paper (or the gel) is placed in a buffered solution between two electrodes and an electric field is applied, an amino acid with a pI greater than the pH of the solution will have an overall positive charge and will migrate toward the cathode (the negative electrode). The farther its pI is from the pH of the buffer, the more positive it will be and the farther it will migrate toward the cathode in a given amount of time. An amino acid with a pI less than the pH of the buffer will have an overall negative charge and will migrate toward the anode (the positive electrode). If two molecules have the same charge, the larger one will move more slowly during electrophoresis because the same charge has to move a greater mass.

How can we detect the separation of the amino acids in the mixture? Amino acids are colorless; however, when they are heated with ninhydrin, they form a colored product. After electrophoretic separation of the amino acids, the filter paper is sprayed with ninhydrin and dried in a warm oven. Most amino acids form a purple product. The number of different kinds of amino acids in the mixture is determined by the number of colored spots on the filter paper (Figure 20.2b). The individual amino acids are identified by comparing the chromatogram with a standard.

The mechanism for formation of the colored product is shown below, omitting the mechanisms for the steps involving dehydration, imine formation, and imine hydrolysis. (These mechanisms are shown in Section 16.7 and 16.8.)

mechanism for the reaction of an amino acid with ninhydrin to form a colored product

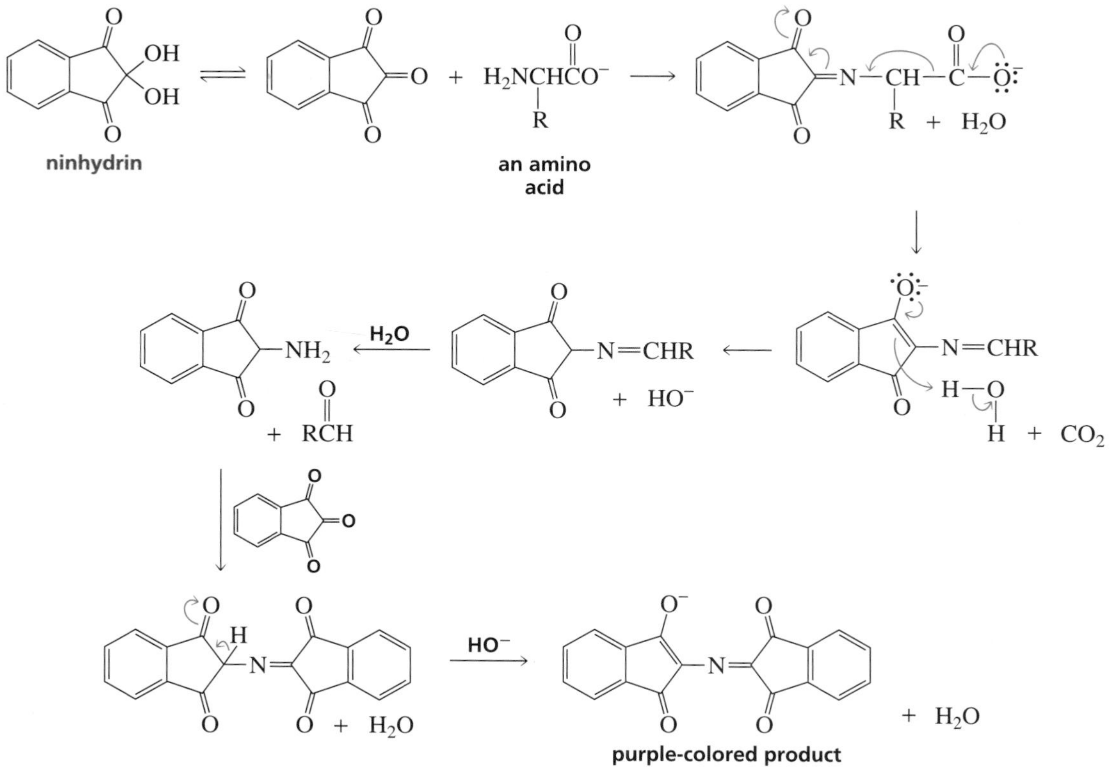

Paper chromatography once played an important role in biochemical analysis because it provided a method to separate amino acids using very simple equipment. Although more modern techniques are now more commonly used, we will describe the principles behind paper chromatography since many of the same principles are employed in modern separation techniques.

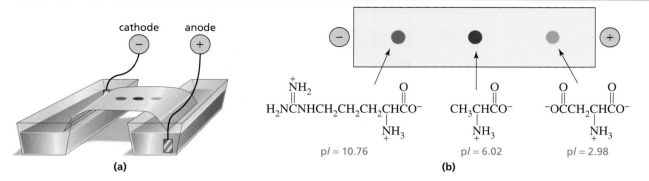

(a)   (b)

▲ **Figure 20.2**
(a) Separation of arginine, alanine, and aspartic acid by electrophoresis at pH = 5.
(b) A chromatogram.

The technique of paper chromatography separates amino acids on the basis of polarity. A few drops of a solution of an amino acid mixture are applied to the bottom of a strip of filter paper. The edge of the paper is placed in a solvent (typically a mixture of water, acetic acid, and butanol). The solvent moves up the paper by capillary action, carrying the amino acids with it. The aqueous component of the solvent binds to the polar cellulose of the paper, and the organic component continues to migrate up the paper. The more polar the amino acid, the sooner it is adsorbed onto the paper. The less polar amino acids remain with the less polar solvent longer and attach to the strip farther up. So when the strip is developed with ninhydrin, the colored spot closest to the origin is the most polar amino acid and the spot farthest away from the origin is the least polar amino acid (Figure 20.3).

The most polar amino acids are those with charged side chains; the next most polar are those with side chains that can form hydrogen bonds; and the least polar are those with hydrocarbon side chains. For amino acids with hydrocarbon side chains, the larger the alkyl group the less polar the amino acid. In other words, leucine is less polar than valine.

Paper chromatography has largely been replaced by **thin-layer chromatography** (TLC). TLC is similar to paper chromatography. However instead of filter paper, thin-layer chromatography uses a glass plate with a coating of solid material. The physical property on which the separation is based depends on the solid material and the solvent chosen for the mobile phase.

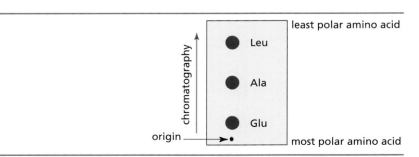

◀ **Figure 20.3**
Separation of glutamic acid, alanine, and leucine by paper chromatography.

PROBLEM 13

A mixture of seven amino acids (glycine, glutamic acid, leucine, lysine, alanine, isoleucine, and aspartic acid) are separated by thin-layer chromatography. When the chromatographic plate is sprayed with ninhydrin and heated, only six spots show up. Explain.

Electrophoresis and thin-layer chromatography are analytical separations—small amounts of amino acids are separated for analysis. Preparative separation—in which larger amounts of amino acids are separated for use in subsequent processes—can be achieved using **ion-exchange chromatography.** This technique uses a column that is packed with an insoluble resin. A solution of a mixture of amino acids is loaded onto the top of the column and eluted with a buffer. The amino acids separate because they flow through the column at different rates (Figure 20.4).

The resin is a chemically inert material with charged side chains. One commonly used resin is a polymer of styrene with negatively charged sulfonic acid groups on some of the benzene rings (Figure 20.5). If a mixture of lysine and glutamic acid in a solution with a pH of 6 were loaded onto the column, glutamic acid would travel down the column rapidly because its negatively charged side chain would be repelled by the negatively charged sulfonic acid groups of the resin. The positively charged side chain of lysine would cause it to be retained on the column. This kind of resin is called a **cation-exchange resin** because it exchanges the Na^+ counterions of the SO_3^- groups for the positively charged species that are added to the

Cations bind most strongly to cation-exchange resins.

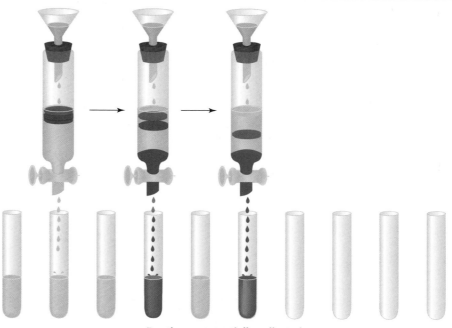

Fractions sequentially collected

▲ **Figure 20.4**
Separation of amino acids by ion-exchange chromatography.

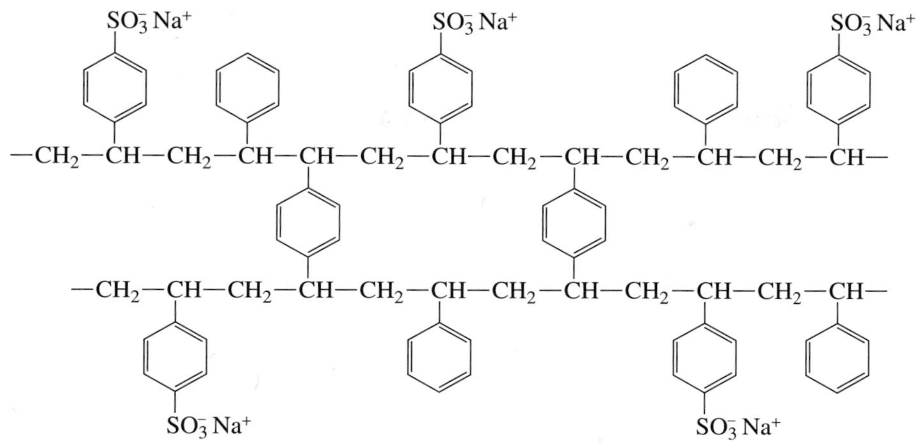

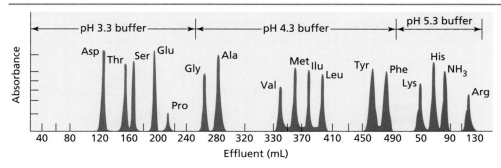

◀ **Figure 20.5**
A section of a cation-exchange resin. This particular resin is called Dowex 50.

column. In addition, the relatively nonpolar nature of the column causes it to retain nonpolar amino acids longer than polar amino acids. Resins with positively charged groups are called **anion-exchange resins** because they impede the flow of anions. A common anion-exchange resin (Dowex 1) has $CH_2N^+(CH_3)_3$ Cl^- groups in place of the SO_3^- Na^+ groups in Figure 20.4.

An **amino acid analyzer** is an instrument that automates ion-exchange chromatography. When a solution of an amino acid mixture passes through the column of an amino acid analyzer containing a cation-exchange resin, the amino acids move through the column at different rates depending on their overall charge. The solution leaving the column is collected in fractions. Fractions are collected often enough that a different amino acid ends up in each fraction (Figure 20.6). If ninhydrin is added to each of the fractions, the concentration of the amino acid in each fraction can be determined by the amount of absorption at 570 nm. This is because the colored compound formed by the reaction of an amino acid with ninhydrin has a λ_{max} of 570 (Section 13.16). In this way, the identity and the relative amount of each amino acid can be determined.

Anions bind most strongly to anion-exchange resins.

▲ **Figure 20.6**
A typical chromatogram obtained from separation of a mixture of amino acids by using an automated amino acid analyzer.

WATER SOFTENERS: EXAMPLES OF CATION-EXCHANGE CHROMATOGRAPHY

Water softeners contain a cation-exchange resin that has been flushed with concentrated sodium chloride. In Section 15.16 we learned that the presence of calcium and magnesium ions in water is what causes it to be "hard." When water passes through the column, the resin binds magnesium and calcium ions more tightly than it binds sodium ions. In this way, the water softener removes magnesium and calcium ions from water, replacing them with sodium ions. The resin must be recharged from time to time. This is done by flushing it with concentrated sodium chloride to replace the bound magnesium and calcium ions with sodium ions.

PROBLEM 14

Why are buffer solutions of increasingly higher pH used to elute the column shown in Figure 20.5?

PROBLEM 15

Explain the order of elution (with a buffer of pH 4) of each of the following pairs of amino acids on a column packed with Dowex 50 (Figure 20.5).

a. aspartic acid before serine

c. valine before leucine

b. glycine before alanine

d. tyrosine before phenylalanine

PROBLEM 16 ◆

In what order would the following amino acids be eluted with a buffer of pH 4 from a column containing an anion-exchange resin?

histidine, serine, aspartic acid, valine

20.6 RESOLUTION OF RACEMIC MIXTURES OF AMINO ACIDS

Chemists do not have to rely on nature to produce amino acids; they can synthesize them in the laboratory using a variety of methods. We will look at a few methods that can be used to prepare amino acids. One of the oldest methods involves replacing an α-hydrogen of a carboxylic acid with a bromine using the Hell–Volhard–Zelinski reaction (Section 18.5). The resulting α-bromocarboxylic acid then undergoes an S_N2 reaction with ammonia to form the amino acid.

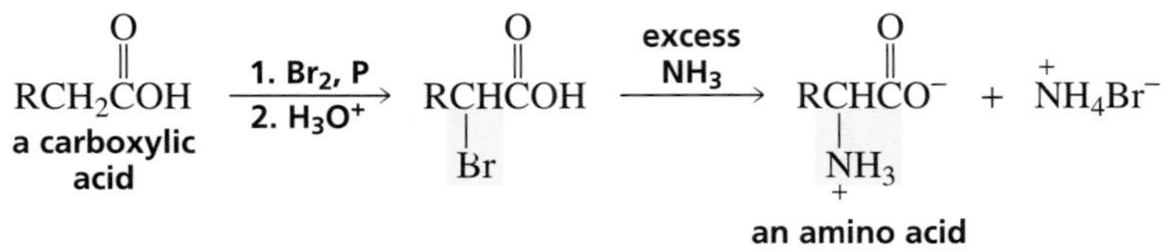

PROBLEM 17

Why is excess ammonia used in the reaction shown above?

When amino acids are synthesized in nature, only the L-enantiomer is formed (Section 4.20). However, when amino acids are synthesized in the laboratory, the product is a racemic mixture—a mixture of D and L enantiomers. If only one isomer is desired, the two enantiomers must be separated. We have seen that a method commonly used to separate enantiomers is to convert them into diastereomers, separate the diastereomers, and convert the separated diastereomers back to enantiomers (Section 4.11). This is a time-consuming technique, and the yields are often low. Fortunately, there is a much easier way to separate a racemic mixture of amino acids.

Because enzymes are chiral, they react at a different rate with two enantiomers (Section 4.20). For example, pig kidney aminoacylase is an enzyme that catalyzes the hydrolysis of N-acetyl-L-amino acids but does not catalyze the hydrolysis of N-acetyl-D-amino acids. Therefore, if the racemic amino acid is N-acetylated and the N-acetylated mixture is hydrolyzed with pig kidney aminoacylase, the products will be the L-amino acid and N-acetyl-D-amino acid, which are easily separated. Because the resolution (separation) of the enantiomers depends on the difference in the rates of reaction of the enzyme with the two N-acetylated compounds, this technique is known as a **kinetic resolution.**

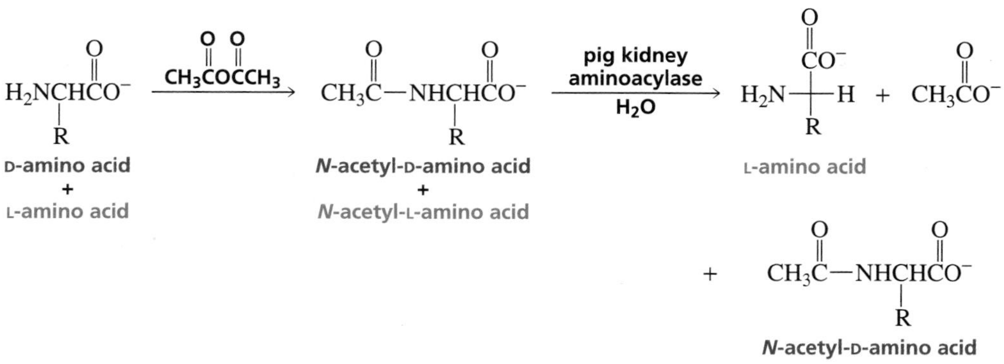

PROBLEM 18

Pig liver esterase is an enzyme that catalyzes the hydrolysis of esters. It hydrolyzes esters of L-amino acids more rapidly than esters of D-amino acids. How can this enzyme be used to separate a racemic mixture of amino acids?

PROBLEM 19 ◆

Amino acids can be synthesized by reductive amination of α-keto acids (Section 16.7).

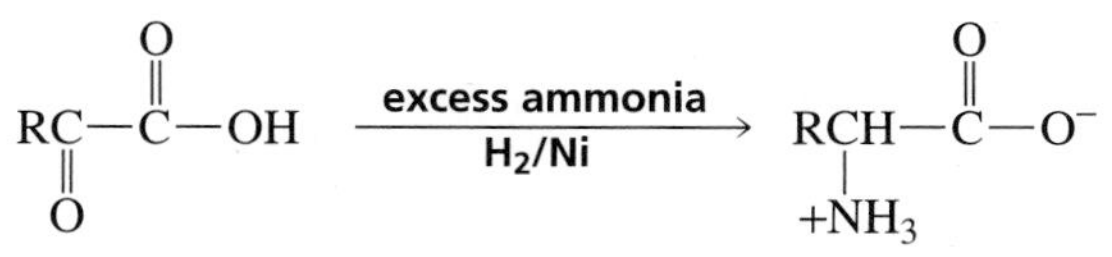

Biological organisms can also convert α-keto acids into amino acids but, since H_2 and metal catalysts are not available to the cell, they do so by a different mechanism (Section 22.6.)

a. What amino acid is obtained from the reductive amination of each of the following metabolic intermediates in the cell?

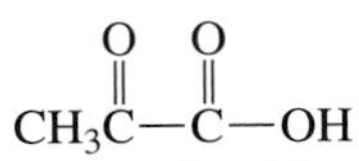

$$CH_3\overset{O}{\underset{}{C}}-\overset{O}{\underset{}{C}}-OH \qquad HO\overset{O}{\underset{}{C}}CH_2-\overset{O}{\underset{}{C}}-\overset{O}{\underset{}{C}}OH \qquad HO\overset{O}{\underset{}{C}}CH_2CH_2-\overset{O}{\underset{}{C}}-\overset{O}{\underset{}{C}}OH$$

pyruvic acid 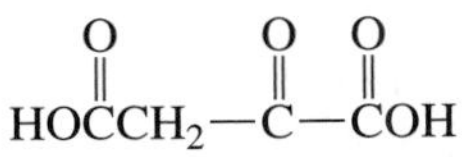   oxaloacetic acid 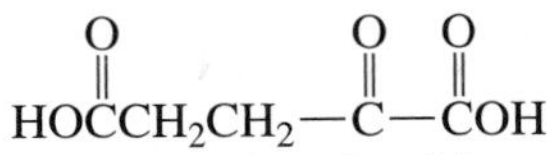   α-ketoglutaric acid

b. What amino acid is obtained from the same metabolic intermediates when it is synthesized in the laboratory?

20.7
PEPTIDE BONDS AND DISULFIDE BONDS

Peptide Bonds

Peptide bonds and disulfide bonds are the only covalent bonds that hold amino acid residues together in a peptide or a protein. The amide bonds that link amino acid residues are called **peptide bonds.** By convention, peptides and proteins are written with the free amino group (the **N-terminal amino acid**) on the left and the free carboxyl group (the **C-terminal amino acid**) on the right. The amino acids are numbered starting with the N-terminal end.

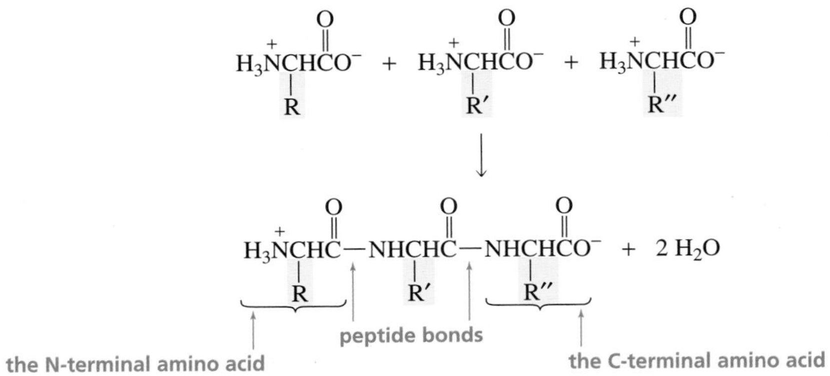

$$\overset{+}{H_3N}\overset{O}{\underset{R}{CHCO^-}} + \overset{+}{H_3N}\overset{O}{\underset{R'}{CHCO^-}} + \overset{+}{H_3N}\overset{O}{\underset{R''}{CHCO^-}}$$

$$\downarrow$$

$$\overset{+}{H_3N}\overset{O}{\underset{R}{CHC}}-NH\overset{O}{\underset{R'}{CHC}}-NH\overset{O}{\underset{R''}{CHCO^-}} + 2\,H_2O$$

peptide bonds

the N-terminal amino acid   the C-terminal amino acid

a tripeptide

When the identities of the amino acids in a peptide are known but their sequence is not known, the amino acids are written separated by commas. When the sequence of amino acids is known, the amino acids are written separated by hyphens. In the pentapeptide shown below, valine is the N-terminal amino acid and histidine is the C-terminal amino acid. The glutamic acid residue is referred to as Glu 4 because it is the fourth amino acid from the N-terminal end. In naming the peptide, adjective names (ending in "yl") are used for all the amino acids except the C-terminal amino acid. Thus, the pentapeptide below is named valylcysteylalanylglutamylhistidine.

Glu, Cys, His, Val, Ala   Val-Cys-Ala-Glu-His

the pentapeptide contains the indicated   the amino acids in the pentapeptide
amino acids, but their sequence is not known   have the indicated sequence

A peptide bond has about 40% double-bond character because of resonance. Steric hindrance causes the trans configuration to be more stable than the cis configuration, so the α-carbons of adjacent amino acids are trans to each other.

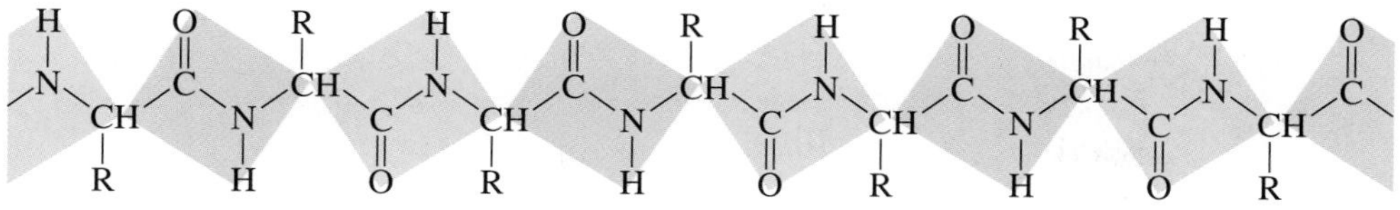

▲ **Figure 20.7**
A segment of a polypeptide chain. The plane defined by each peptide bond is indicated. Notice that the R groups bonded to the α-carbons are on alternate sides of the peptide backbone.

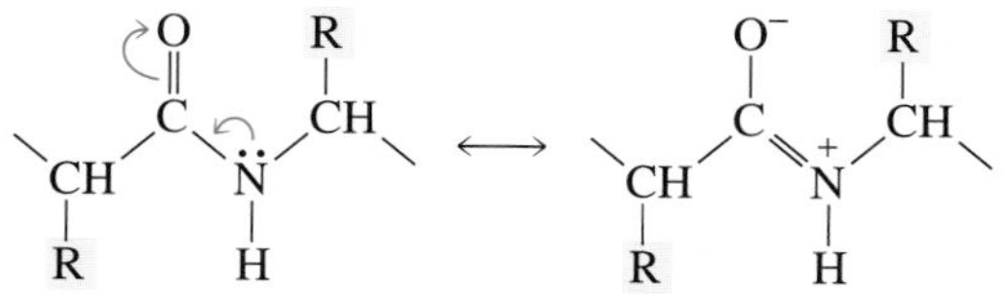

trans configuration

As a consequence of the partial double-bond character of the peptide bond, free rotation is not possible about this bond. The carbon and nitrogen atoms of the peptide bond and the two atoms to which each is attached are held rigidly in a plane (Figure 20.7). This regional planarity affects the way a chain of amino acids can fold, so it has important implications for the three-dimensional shapes of peptides and proteins (Section 20.13).

PROBLEM 20

Draw a peptide bond in a cis configuration.

Disulfide Bonds

When thiols are oxidized under mild conditions, they form disulfides. A **disulfide** is a compound with an S—S bond.

$$2\ R\text{—}SH \xrightarrow{\text{mild oxidation}} RS\text{—}SR$$
$$\text{a thiol} \qquad\qquad\qquad \text{a disulfide}$$

A common oxidizing agent used for this reaction is Br_2 (or I_2) in a basic solution.

mechanism for oxidation of a thiol to a disulfide

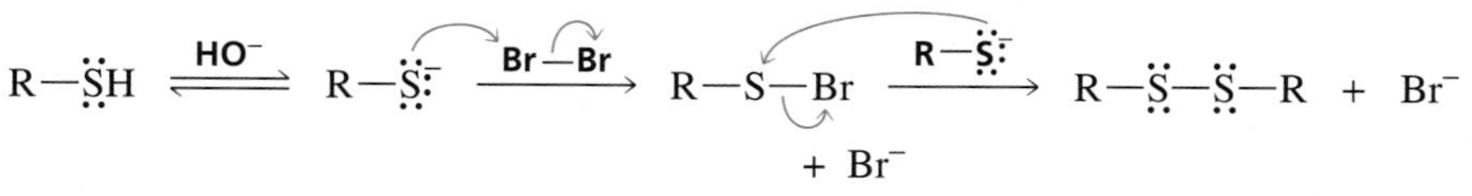

Since thiols are oxidized to disulfides, disulfides can be reduced to thiols.

$$RS\text{—}SR \xrightarrow{\text{reduction}} 2\ R\text{—}SH$$
$$\text{a disulfide} \qquad\qquad \text{a thiol}$$

Cysteine is an amino acid containing a thiol group. Two cysteines therefore can be oxidized to a disulfide. The disulfide is called cystine.

$$2 \; \text{HSCH}_2\overset{\overset{\displaystyle O}{\|}}{\text{CHCO}}{}^- \quad \xrightarrow{\textbf{mild oxidation}} \quad {}^-\overset{\overset{\displaystyle O}{\|}}{\text{OCCHCH}_2}\text{S}-\text{SCH}_2\overset{\overset{\displaystyle O}{\|}}{\text{CHCO}}{}^-$$

$$\underset{\overset{+}{\text{NH}_3}}{} \qquad\qquad\qquad \underset{\overset{+}{\text{NH}_3}}{} \qquad\qquad \underset{\overset{+}{\text{NH}_3}}{}$$

cysteine   cystine

Two cysteine residues in a protein can be oxidized to a disulfide. This is known as a **disulfide bridge.** Disulfide bridges are the only covalent bonds that can form between nonadjacent amino acids. They contribute to the overall shape of a protein by holding the cysteine residues in close proximity (Figure 20.8).

Insulin is a hormone secreted by the pancreas. This hormone controls the level of glucose in the blood by regulating glucose metabolism. Insulin is a polypeptide with two peptide chains. The short chain (the A-chain) contains 21 amino acids, and the long chain (the B-chain) contains 30 amino acids. The two chains are held together by two disulfide bridges. These are **interchain disulfide bridges** (between the A- and B-chains). Insulin also has an **intrachain disulfide bridge** (within the A-chain).

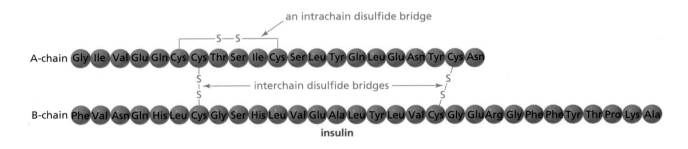

insulin

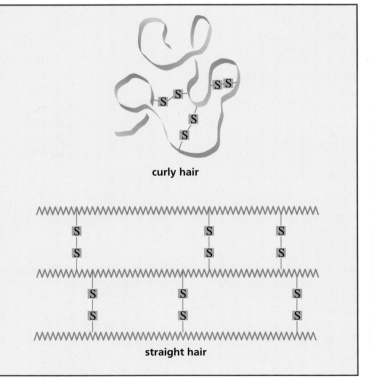

HAIR: STRAIGHT OR CURLY?

Hair is made up of a protein known as keratin. Keratin contains an unusually large number of cysteine residues (about 8%), which give it many disulfide bridges to maintain its three-dimensional structure. People can alter the structure of their hair (if they feel it is either too straight or too curly) by changing the location of these disulfide bridges. This is accomplished by first applying a reducing agent to the hair to reduce all the disulfide bridges in the protein strands. Then the hair is given the desired shape (using curlers to curl it or combing it straight to uncurl it), and an oxidizing agent is applied that forms new disulfide bridges. The new disulfide bridges maintain the hair's new shape. When this treatment is applied to straight hair, it is called a "permanent"; it is called "hair-straightening" when it is applied to curly hair.

curly hair

straight hair

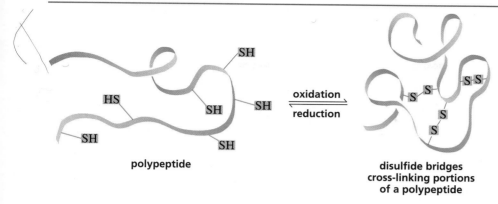

▲ **Figure 20.8**
Disulfide bridges cross-linking portions of a peptide.

P R O B L E M   21◆

a. How many different octapeptides can be made from the 20 naturally occurring amino acids?

b. How many different proteins containing 100 amino acids can be made from the 20 naturally occurring amino acids?

P R O B L E M   22

Which bonds in the backbone of a peptide can rotate freely?

Enkephalins are pentapeptides synthesized by the body to control pain. They decrease the sensitivity to pain by binding to receptors in certain brain cells. Part of their three-dimensional structures must be similar to those of morphine and painkillers such as Darvon and Demerol because they bind to the same receptors.

20.8
SOME INTERESTING PEPTIDES

Tyr-Gly-Gly-Phe-Leu
leucine enkephalin

Tyr-Gly-Gly-Phe-Met
methionine enkephalin

Bradykinin, vasopressin, and oxytocin are peptide hormones. They are all nonapeptides. Bradykinin inhibits the inflammation of tissues. Vasopressin controls blood pressure by regulating the contraction of smooth muscle; it also is an antidiuretic. Oxytocin induces labor in pregnant women and stimulates milk production in nursing mothers. Vasopressin and oxytocin both have an intrachain disulfide bond, and their C-terminal amino acids contain amide rather than carboxyl groups; the C-terminal amide group is indicated by writing "NH_2" after the name of the C-terminal amino acid. Notice that, in spite of their very different physiological

Oxytocin was the first small peptide to be synthesized. This was achieved in 1953 by **Vincent du Vigneaud (1901–1978),** *who later synthesized vasopressin. Du Vigneaud was born in Chicago and was a professor at George Washington School of Medicine and later at Cornell Medical College. For synthesizing these nonapeptides, he received the Nobel Prize in chemistry in 1955.*

effects, vasopressin and oxytocin have similar structures that differ only by two amino acids.

bradykinin   Arg-Pro-Pro-Gly-Phe-Ser-Pro-Phe-Arg

vasopressin   Cys-Tyr-Phe-Gln-Asn-Cys-Pro-Arg-Gly-NH$_2$
                    | |
                    S————————————S

oxytocin   Cys-Tyr-Ile-Gln-Asn-Cys-Pro-Leu-Gly-NH$_2$
                 | |
                 S————————————S

Gramicidin S is an antibiotic produced by a strain of bacteria. It is a cyclic decapeptide. Notice that it contains the amino acids L-ornithine and D-ornithine. Ornithine is not listed in Table 20.1 because it occurs rarely in nature. Ornithine resembles lysine but has one fewer methylene group in its side chain.

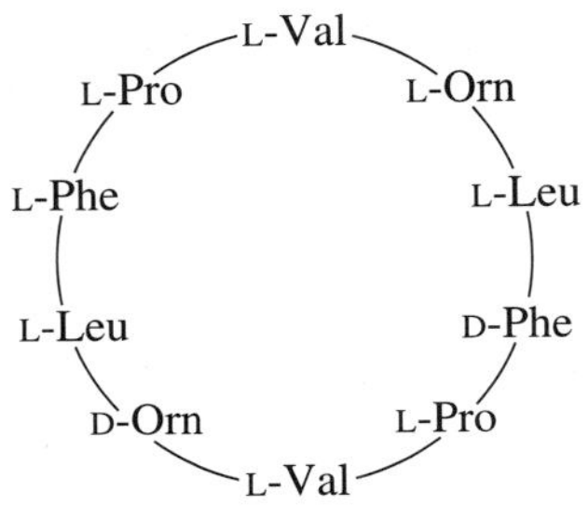

gramicidin S

The synthetic sweetener aspartame, or NutraSweet (Section 19.17), is the methyl ester of a dipeptide of L-aspartic acid and L-phenylalanine. It is about 200 times sweeter than sucrose. The ethyl ester of the same dipeptide is not sweet. If a D-amino acid is substituted for either of the L-amino acids of aspartame, the resulting dipeptide is bitter rather than sweet.

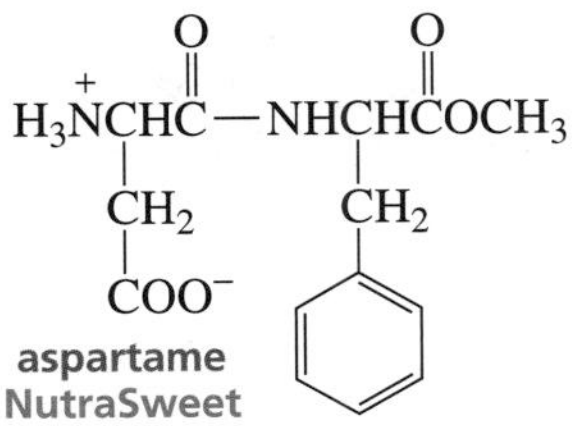

aspartame
NutraSweet

Glutathione is a tripeptide of glutamic acid, cysteine, and glycine. Its function is to destroy harmful oxidizing agents in the body. Oxidizing agents are thought to be responsible for some of the effects of aging and are believed to play a role in cancer. Glutathione removes oxidizing agents by reducing them. Consequently, glutathione is oxidized, forming a disulfide bond between two molecules. An enzyme subsequently reduces the disulfide bond, allowing glutathione to react with more oxidizing agents.

$$2\ \text{H}_3\overset{+}{\text{N}}\text{CHCH}_2\text{CH}_2\overset{\text{O}}{\overset{\|}{\text{C}}}-\text{NHCHC}\overset{\text{O}}{\overset{\|}{}}-\text{NHCH}_2\overset{\text{O}}{\overset{\|}{\text{C}}}\text{O}^-$$

with COO⁻ on first carbon, CH₂—SH branch

glutathione

reducing agent ⇅ **oxidizing agent**

(oxidized glutathione structure with disulfide S—S bridge connecting two glutathione units)

oxidized glutathione

PROBLEM 23

What is unusual about glutathione's structure? (If you can't answer this question, draw the structure you would expect for glutathione and compare your structure with the actual structure.)

Because amino acids have two functional groups, a problem arises when one attempts to make a particular peptide bond. Say, for example, you wanted to make the simple dipeptide Gly-Ala. Heating a mixture of glycine and alanine results in four dipeptides: the one you want and three others (Section 15.11).

20.9 STRATEGY OF PEPTIDE BOND SYNTHESIS: N-PROTECTION AND C-ACTIVATION

$$\text{H}_3\overset{+}{\text{N}}\text{CH}_2\overset{\text{O}}{\overset{\|}{\text{C}}}\text{O}^- + \text{H}_3\overset{+}{\text{N}}\text{CHC}\overset{\text{O}}{\overset{\|}{}}\text{O}^- \xrightarrow{\Delta} \text{H}_3\overset{+}{\text{N}}\text{CH}_2\overset{\text{O}}{\overset{\|}{\text{C}}}-\text{NHCHC}\overset{\text{O}}{\overset{\|}{}}\text{O}^- + \text{H}_3\overset{+}{\text{N}}\text{CHC}\overset{\text{O}}{\overset{\|}{}}-\text{NHCHC}\overset{\text{O}}{\overset{\|}{}}\text{O}^-$$

glycine (CH₃) alanine — Gly-Ala (CH₃) — Ala-Ala (CH₃ CH₃)

$$+\ \text{H}_2\text{O}\ +\ \text{H}_3\overset{+}{\text{N}}\text{CH}_2\overset{\text{O}}{\overset{\|}{\text{C}}}-\text{NHCH}_2\overset{\text{O}}{\overset{\|}{\text{C}}}\text{O}^-\ +\ \text{H}_3\overset{+}{\text{N}}\text{CHC}\overset{\text{O}}{\overset{\|}{}}-\text{NHCH}_2\overset{\text{O}}{\overset{\|}{\text{C}}}\text{O}^-$$

Gly-Gly — Ala-Gly (CH₃)

If the amino group of the amino acid that is to be on the N-terminal end (in this case Gly) is protected, it will not be available to form a peptide bond. If the carboxyl group of this same amino acid is activated before the second amino acid is added, the amino group of the added amino acid (in this case Ala) will react with the activated carboxyl group of glycine in preference to reacting with a nonactivated carboxyl group of another alanine molecule.

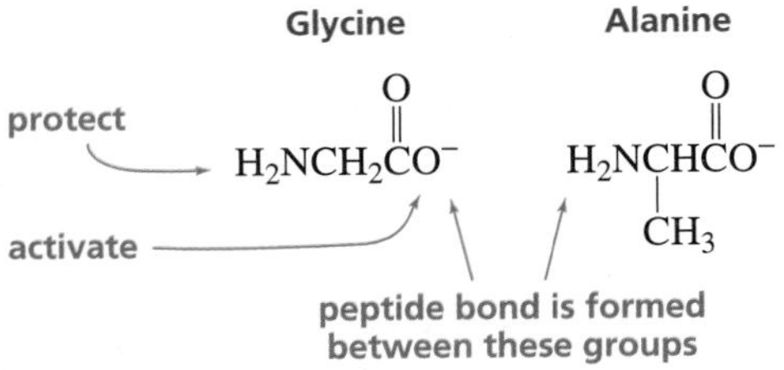

Glycine   Alanine

protect → $H_2NCH_2CO^-$ H_2NCHCO^-

activate —— CH_3

peptide bond is formed
between these groups

The reagent most often used to protect the amino group of an amino acid is di-*tert*-butyl dicarbonate. Its success is due to the ease with which the protecting group can be removed when the need for protection is over (see below). The protecting group is known by the acronym *t*BOC (pronounced tee-boc).

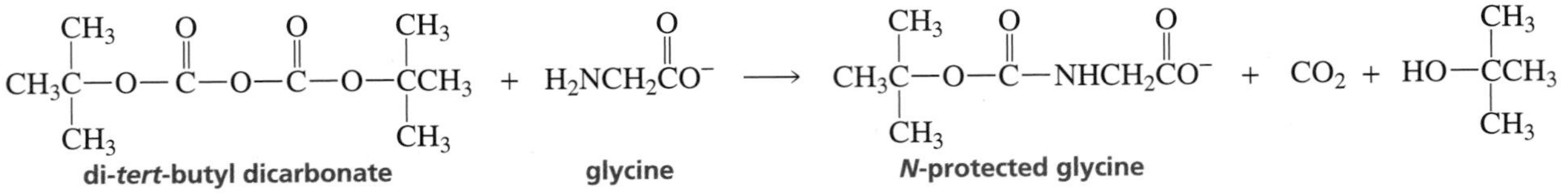

di-*tert*-butyl dicarbonate   glycine   *N*-protected glycine

Carboxylic acids are generally activated by being converted into acyl chlorides (Section 15.17). Acyl chlorides, however, are so reactive that they can readily react with the substituents of some of the amino acids during peptide synthesis, creating unwanted products. The preferred method for activating the carboxyl group of an *N*-protected amino acid is to convert it into an imidate using dicyclohexylcarbodiimide (DCC). (You have probably noticed that biochemists are even more fond of acronyms than are organic chemists.) DCC activates a carboxyl group by putting a good leaving group on the carbonyl carbon.

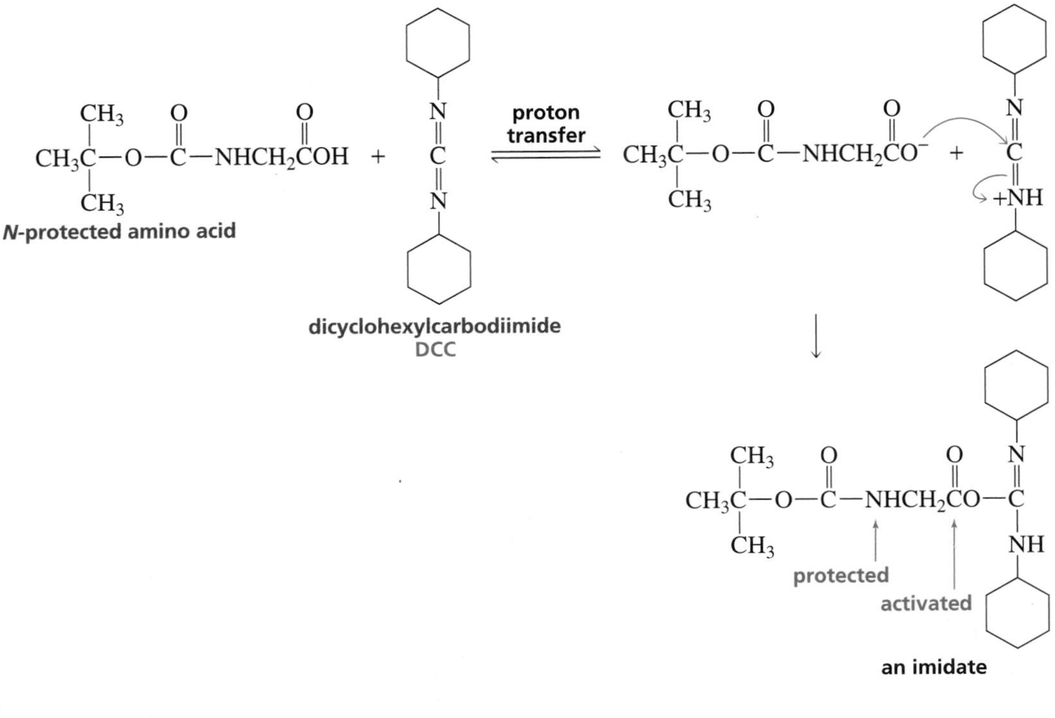

N-protected amino acid

dicyclohexylcarbodiimide
DCC

protected   activated

an imidate

After the amino acid has its N-terminal group protected and its C-terminal group activated, the second amino acid is added to form the new peptide bond. The C—O bond of the tetrahedral intermediate is easily broken because the bonding electrons are delocalized, forming dicyclohexylurea, a stable diamide.

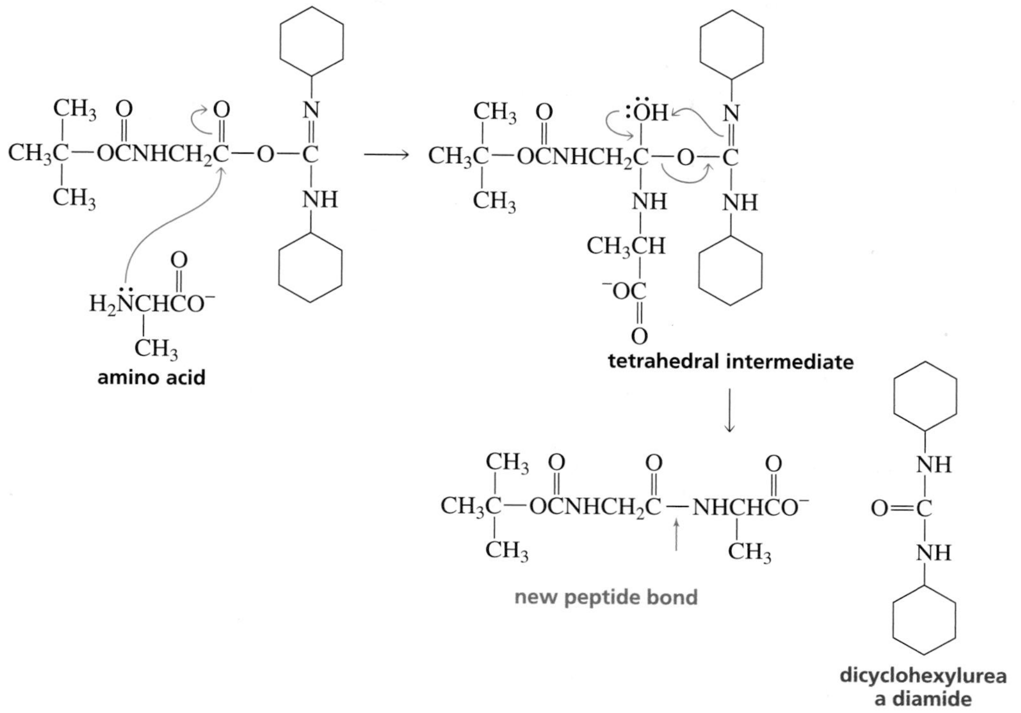

$$CH_3\overset{CH_3}{\underset{CH_3}{C}}-O\overset{O}{C}NHCH_2\overset{O}{C}-O-\overset{N}{\underset{NH}{C}} \longrightarrow CH_3\overset{CH_3}{\underset{CH_3}{C}}-O\overset{O}{C}NHCH_2\overset{O}{C}-O-\overset{N}{\underset{NH}{C}}$$

H_2ṄCHCO⁻ | CH_3
amino acid

CH_3CH | ⁻OC | O
tetrahedral intermediate

$$CH_3\overset{CH_3}{\underset{CH_3}{C}}-O\overset{O}{C}NHCH_2\overset{O}{C}-NHCHCO^- \qquad O=\overset{NH}{\underset{NH}{C}}$$

| CH_3
new peptide bond

dicyclohexylurea
a diamide

Amino acids can be added to the growing C-terminal end by repeating these two steps: activating the carboxyl group of the C-terminal amino acid of the peptide by treating it with DCC and then adding a new amino acid.

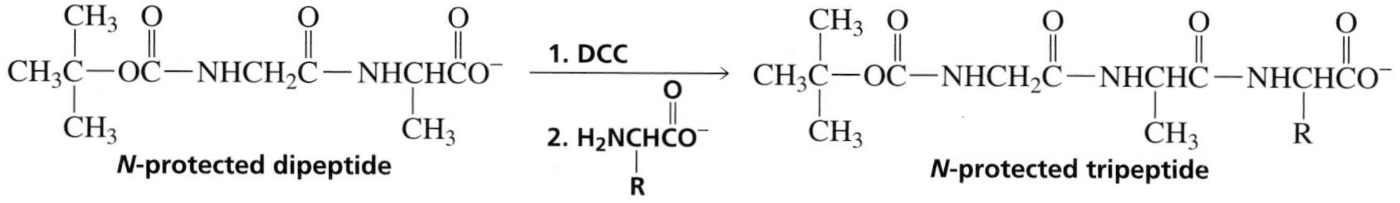

$$CH_3\overset{CH_3}{\underset{CH_3}{C}}-O\overset{O}{C}-NHCH_2\overset{O}{C}-NHCHCO^- \xrightarrow[\text{2. } H_2NCHCO^-]{\text{1. DCC}} CH_3\overset{CH_3}{\underset{CH_3}{C}}-O\overset{O}{C}-NHCH_2\overset{O}{C}-NHCH\overset{O}{C}-NHCHCO^-$$

| CH_3 | CH_3   (2. O, R) | CH_3 | CH_3 | R
N-protected dipeptide   **N-protected tripeptide**

When the desired number of amino acids has been added to the chain, the protecting group on the N-terminal amino acid is removed. *t*BOC is an ideal protecting group because it can be removed by washing with trifluoroacetic acid and methylene chloride, reagents that will not break any other covalent bonds. The protecting group is removed by an elimination reaction, forming isobutylene and carbon dioxide. Because these products are gases, they escape, forcing the reaction to completion.

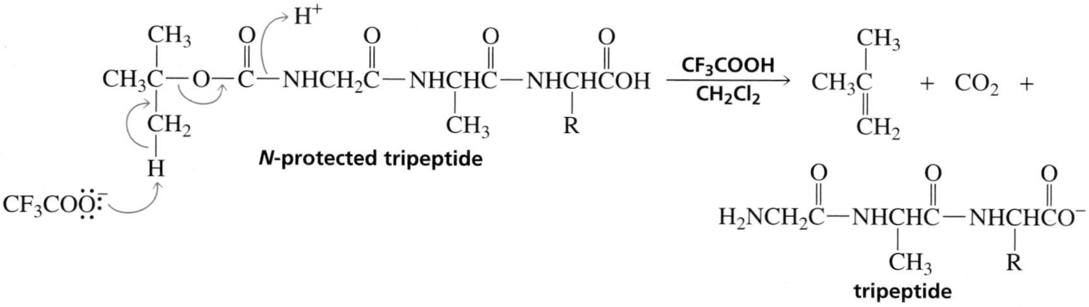

$$CH_3\overset{CH_3}{\underset{CH_2}{C}}-O-\overset{O}{C}-NHCH_2\overset{O}{C}-NHCH\overset{O}{C}-NHCHCOH \xrightarrow[CH_2Cl_2]{CF_3COOH} CH_3\overset{CH_3}{\underset{CH_2}{C}} + CO_2 +$$

H⁺ ; CH_3 ; R
N-protected tripeptide

CF_3COO⁻

$$H_2NCH_2\overset{O}{C}-NHCH\overset{O}{C}-NHCHCO^-$$
| CH_3 | R
tripeptide

Theoretically, one should be able to make as long a peptide as desired using this technique. Reactions do not produce 100% yield, however, and the yield is further decreased during the purification process. After each step of the synthesis, the peptide must be purified to prevent subsequent unwanted reactions with leftover reagents. Assuming that each amino acid can be added to the growing end of the peptide chain in an 80% yield (a relatively high yield, as you can probably appreciate from your own experience in the laboratory), the overall yield of a nonapeptide such as bradykinin would be only 17%. It is clear that large polypeptides could never be synthesized in this way.

Number of amino acids	2	3	4	5	6	7	8	9
Overall yield	80%	64%	51%	41%	33%	26%	21%	17%

PROBLEM 24

The mechanism for the removal of the *tert*-butoxycarbonyl group is shown above as a concerted reaction. The group can also be removed in a stepwise reaction. Write out the steps in the mechanism for the stepwise reaction.

PROBLEM 25

Assume you are trying to synthesize the dipeptide Val-Ser. Compare the product that would be obtained if the carboxyl group of *N*-protected valine were activated with thionyl chloride with the product that would be obtained if the carboxyl group were activated with DCC.

PROBLEM 26

Show the steps in the synthesis of the tetrapeptide Leu-Phe-Lys-Val.

PROBLEM 27 ◆

a. Calculate the overall yield of bradykinin if the yield for the addition of each amino acid to the chain were 70%.

b. What would be the yield of a peptide containing 15 amino acid residues if the yield for incorporation of each were 80%?

20.10 AUTOMATED PEPTIDE SYNTHESIS

In addition to the problem of low overall yields, the method of peptide synthesis described in Section 20.9 is extremely time-consuming because of the purification required at each step of the synthesis. In 1969, Bruce Merrifield described a method that revolutionized the synthesis of peptides because it provided a much faster way to produce peptides in much higher yields. Further, since it is automated, the synthesis requires fewer hours of direct attention. Using this technique, bradykinin was synthesized with an 85% yield in 27 hours. Subsequent refinements in the technique now allow a peptide containing 100 amino acids to be synthesized in 4 days in a reasonable yield.

In the Merrifield method, the C-terminal amino acid is covalently attached to a solid support contained in a column. Each N-terminal blocked amino acid is added one at a time, along with other needed reagents, so the protein is synthesized from the C-terminal end to the N-terminal end. Notice that this is opposite to the way proteins are synthesized in nature (from the N-terminal end to the C-terminal end; Section 24.12). Because it uses a solid support and is automated, this method of protein synthesis is called **automated solid-phase peptide synthesis.**

The solid support to which the C-terminal amino acid is attached is a polystyrene resin similar to the one used in ion-exchange chromatography (Section 20.5) except that the benzene rings have chloromethyl substituents instead of sulfonic acid substituents. Before the C-terminal amino acid is attached to the resin, its amino group is protected with *t*BOC to prevent the amino group from reacting with the resin. The C-terminal amino acid is attached to the resin by means of an S_N2 reaction; its carboxyl group attacks a benzyl carbon of the resin, displacing a chloride ion.

After the C-terminal amino acid is attached to the resin, the protecting group is removed (Section 20.9). The next amino acid, with its amino group protected with *t*BOC and its carboxyl group activated with DCC, is added to the column.

R. Bruce Merrifield

R. Bruce Merrifield was born in 1921 and received a B.S. and a Ph.D. from the University of California, Los Angeles. He is a professor of chemistry at Rockefeller University. Merrifield received the Nobel Prize in chemistry for automated solid-phase peptide synthesis in 1984.

Merrifield automated solid-phase synthesis of a tripeptide

$$CH_3C(CH_3)_2-O-C(=O)-NHCHCOH(R) \xrightarrow{\text{DCC}} CH_3C(CH_3)_2-O-C(=O)-NHCHCO-DCC(R)$$

N-protected amino acid   **N-protected and C-activated amino acid**

$$CH_3C(CH_3)_2-O-C(=O)-NHCHC(=O)-NHCHC(=O)-NHCHCO-CH_2-\langle\!\!\langle\ \rangle\!\!\rangle-\bullet$$

(R, R, R)

$$\downarrow \begin{array}{l} CF_3COOH \\ CH_2Cl_2 \end{array}$$

$$CH_3C(CH_3)(=CH_2) + CO_2 + H_2NCHC(=O)-NHCHC(=O)-NHCHCO-CH_2-\langle\!\!\langle\ \rangle\!\!\rangle-\bullet$$

(R, R, R)

$$\downarrow \text{HF}$$

$$H_3^+NCHC(=O)-NHCHC(=O)-NHCHCOH + HOCH_2-\langle\!\!\langle\ \rangle\!\!\rangle-\bullet$$

(R, R, R)

A huge advantage of the Merrifield method of peptide synthesis is that, after each step of the procedure, the growing peptide can be purified by washing the column with an appropriate solvent. The impurities are washed out of the column because they are not attached to it. Since the peptide is covalently attached to the resin, none of it is lost in the purification step, leading to high yields of purified product.

After the required amino acids have been added one by one, the peptide can be removed from the resin by treatment with HF under mild conditions that do not break the peptide bonds.

Automated solid-phase peptide synthesis is practical only for the synthesis of peptides with fewer than 50 amino acids. However, larger peptides have been successfully synthesized by combining two smaller peptides made by this technique. Although this technique is constantly being improved so that peptides can be made more rapidly and more efficiently, it cannot begin to compare with nature. A bacterial cell is able to synthesize a protein thousands of amino acids long in seconds, and it can simultaneously synthesize thousands of different proteins with no mistakes.

Since the early 1980s, it has been possible to synthesize proteins by genetic engineering techniques. Strands of DNA can be introduced into bacterial cells, which cause the cells to produce large amounts of a desired protein (Section 24.12). For example, mass quantities of human insulin are produced from genetically modified *E. coli*. Genetic engineering techniques also have been useful in synthesizing proteins that differ in one or a few amino acids from the natural protein. Such synthetic proteins have been used, for example, to determine how a single amino acid change affects the properties of a protein (Section 21.8).

PROBLEM 28

Show the steps in the synthesis of the peptide in Problem 26 using Merrifield's method.

20.11 PROTEIN STRUCTURE

Protein molecules are described by several levels of structure. The **primary structure** of a protein is the sequence of amino acids in the chain and the location of all the disulfide bridges. The **secondary structure** describes the regular conformation assumed by segments of the protein's backbone. In other words, the secondary structure describes how the backbone folds. The **tertiary structure** is a description of the three-dimensional structure of the entire polypeptide. If a protein has more than one polypeptide chain, it has quaternary structure. The **quaternary structure** of a protein is the way the individual protein chains are arranged with respect to each other.

Proteins can be divided roughly into two classes. **Fibrous proteins** contain long chains of polypeptides that occur in bundles. These proteins are insoluble in water. All the structural proteins described at the beginning of this chapter are fibrous proteins. **Globular proteins** are soluble in water and tend to have roughly spherical shapes. Essentially all enzymes are globular proteins.

PRIMARY STRUCTURE AND EVOLUTION

When we examine the primary structures of proteins that carry out the same function in different organisms, we can relate the number of amino acid differences between the proteins to the taxonomic differences between the species. For example, cytochrome c, a protein that transfers electrons in biological oxidations, has about 100 amino acid residues. Yeast cytochrome c differs by 48 amino acids from horse cytochrome c, while duck cytochrome c differs by only two amino acids from chicken cytochrome c. Chickens and turkeys have cytochrome c's with identical primary structures. Humans and chimpanzees also have identical cytochrome c's, differing by one amino acid from the cytochrome c of the rhesus monkey.

20.12 DETERMINING THE PRIMARY STRUCTURE OF A PROTEIN

The first step in determining the sequence of amino acids in a peptide or a protein is to reduce any disulfide bridges in the protein. A commonly used reducing agent is 2-mercaptoethanol, which is oxidized to a disulfide. Reaction of the protein thiol groups with iodoacetic acid prevents the disulfide bridges from reforming as a result of oxidation by O_2.

cleaving disulfide bridges

Frederick Sanger

Insulin was the first protein for which the primary sequence was determined. This was done in 1953 by **Frederick Sanger,** *who received the 1958 Nobel Prize in chemistry for this work. Sanger was born in England in 1918 and received a Ph.D. from Cambridge University, where he has been for his entire career. He also received a share of the 1980 Nobel Prize in chemistry (Section 24.14) for being the first to sequence a DNA molecule (with 5375 nucleotide pairs).*

PROBLEM 29

Give the mechanism for the reaction of a cysteine residue with iodoacetic acid.

The next step is to determine the number and kinds of amino acids in the peptide. To do this, a sample of the peptide is dissolved in 6 N HCl and heated at 100 °C for 24 hours. This treatment hydrolyzes all the amide bonds in the protein, including the amide bonds of asparagine and glutamine.

$$\text{protein} \xrightarrow[\substack{100\ °C \\ 24\ h}]{6\ N\ HCl} \text{amino acids}$$

The mixture of amino acids is then passed through an amino acid analyzer to determine the number and kind of each amino acid in the protein (Section 20.5).

Because all the asparagine and glutamine residues have been hydrolyzed to aspartic acid and glutamic acid residues, the number of aspartic acid or glutamic acid residues in the amino acid mixture tells us the number of aspartic acid plus asparagine—or glutamic acid plus glutamine—residues in the original protein. A separate technique must be used to distinguish between aspartic acid and asparagine or between glutamic acid and glutamine in the original protein.

The strongly acidic conditions used for hydrolysis destroy all the tryptophan residues, because the indole ring is unstable in acid (Section 26.2). The tryptophan content of the protein can be determined by hydroxide ion-promoted hydrolysis of the protein. This is not a general method for peptide bond hydrolysis, because the strongly basic conditions destroy several other amino acid residues.

There are several ways to identify the N-terminal amino acid of a protein. One of the most widely used methods is to treat the protein with phenyl isothiocyanate (PITC), more commonly known as **Edman's reagent.** Edman's reagent reacts with the N-terminal amino group and the resulting thiazolinone derivative is cleaved from the protein under mildly acidic conditions. The thiazolinone derivative is extracted into an organic solvent and, in the presence of acid, rearranges to a more stable phenylthiohydantoin (PTH).

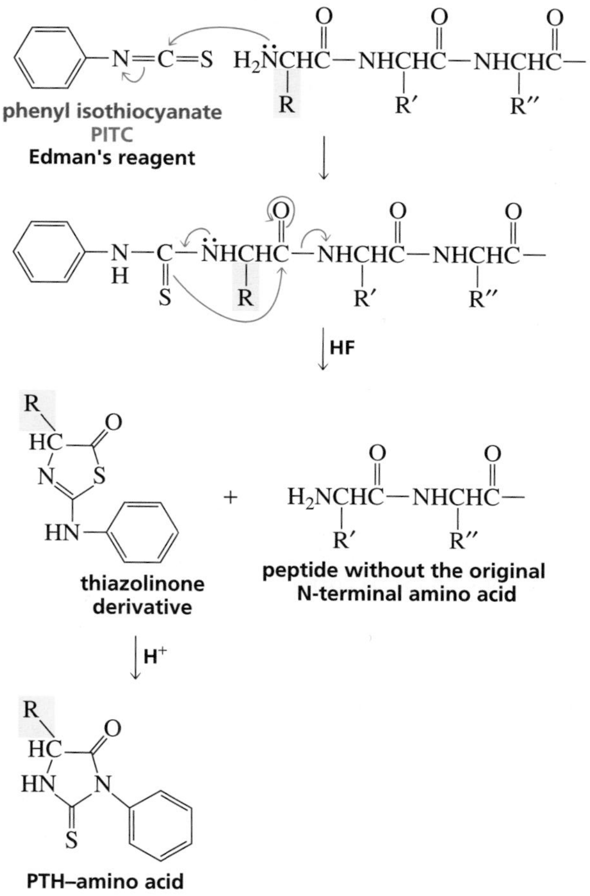

phenyl isothiocyanate
PITC
Edman's reagent

HF

thiazolinone
derivative

peptide without the original
N-terminal amino acid

PTH–amino acid

Since each amino acid has a different substituent (R), each amino acid forms a different PTH. The particular PTH can be identified by chromatography using known standards. Several successive Edman degradations can be carried out on a protein. The entire primary sequence cannot be determined in this way, however, because side products accumulate that interfere with the results. An automated instrument known as a sequenator allows about 50 successive Edman degradations to be carried out on a protein.

The C-terminal amino acid of the protein can be identified by treating the protein with carboxypeptidase A. Carboxypeptidase A cleaves off the C-terminal amino acid as long as it is not arginine or lysine (Section 21.8). Carboxypeptidase B cleaves it off only if it is arginine or lysine. Carboxypeptidases are exopeptidases. An **exopeptidase** is an enzyme that catalyzes the hydrolysis of a peptide bond at the end of a peptide chain.

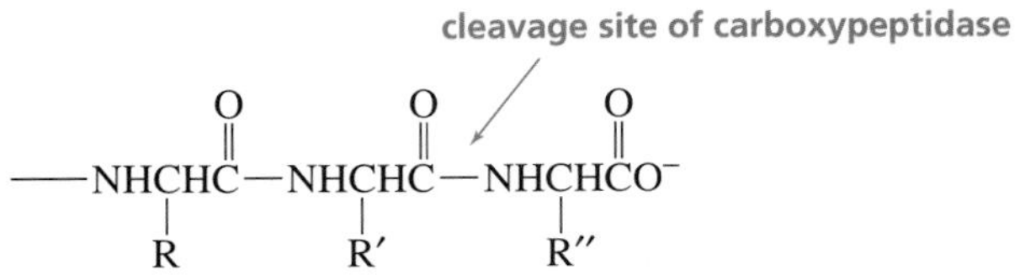

cleavage site of carboxypeptidase

$$\overset{O}{\underset{\underset{R}{|}}{-\!\!-\!\!NHCHC}}\!\!-\!\!\overset{O}{\underset{\underset{R'}{|}}{NHCHC}}\!\!-\!\!\overset{O}{\underset{\underset{R''}{|}}{NHCHCO^-}}$$

Once the N-terminal and C-terminal amino acids have been identified, a sample of the protein is hydrolyzed with dilute acid. This treatment, called **partial hydrolysis,** hydrolyzes only some of the peptide bonds. The resulting fragments are separated, and the amino acid composition of each is determined. The N-terminal and C-terminal amino acids of each peptide can also be determined. The sequence of the original protein can then be determined by lining up the peptides, looking for points of overlap.

PROBLEM 30 ◆

A nonapeptide undergoes partial hydrolysis to give peptides whose amino acid compositions are shown below. Reaction of the intact nonapeptide with Edman's reagent releases PTH-Leu. What is the sequence of the nonapeptide?

a. Pro, Ser   **e.** Glu, Ser, Val, Pro

b. Gly, Glu   **f.** Glu, Pro, Gly

c. Met, Ala, Leu   **g.** Met, Leu

d. Gly, Ala   **h.** His, Val

The protein can also be partially hydrolyzed using endopeptidases. An **endopeptidase** is an enzyme that catalyzes the hydrolysis of a peptide bond that is not at the end of a peptide chain. Trypsin, chymotrypsin, and elastase are endopeptidases that catalyze the hydrolysis of only specific peptide bonds (Table 20.3). Trypsin, for example, catalyzes the hydrolysis of the peptide bond on the C-side of arginine or lysine.

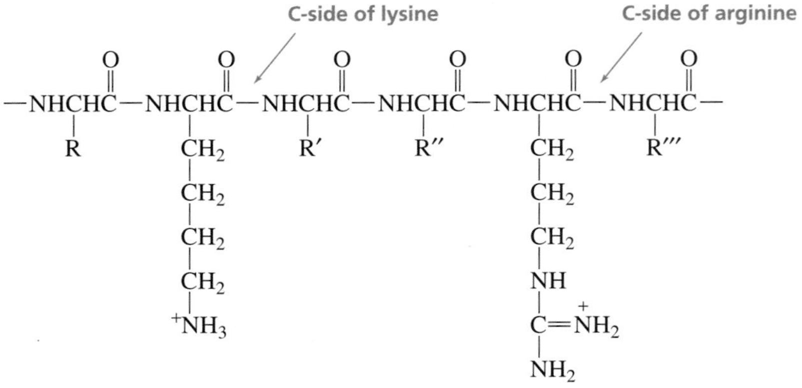

C-side of lysine   C-side of arginine

$$-NHCHC\!-\!NHCHC\!-\!NHCHC\!-\!NHCHC\!-\!NHCHC\!-\!NHCHC-$$

| | | | | | |
|R| CH₂ | R′ | R″ | CH₂ | R‴ |

with side chains:

$$\underset{R}{} \quad \underset{\underset{\underset{\underset{+NH_3}{|}}{CH_2}}{\underset{\underset{CH_2}{|}}{\underset{\underset{CH_2}{|}}{CH_2}}}}{} \quad\quad\quad\quad\quad \underset{\underset{\underset{\underset{\underset{\underset{NH_2}{|}}{C=\overset{+}{N}H_2}}{|}}{NH}}{\underset{\underset{CH_2}{|}}{\underset{\underset{CH_2}{|}}{CH_2}}}}{}$$

So trypsin will catalyze the hydrolysis of three peptide bonds in the following peptide, creating a hexapeptide, a dipeptide, and two tripeptides.

Ala-Lys-Phe-Gly-Asp-Trp-Ser-Arg-Met-Val-Arg-Tyr-Leu-His

↑   ↑   ↑

cleavage by trypsin

Chymotrypsin catalyzes the hydrolysis of the peptide bond on the C-side of amino acids that contain aromatic six-membered rings (Phe, Tyr, Trp).

Ala-Lys-Phe-Gly-Asp-Trp-Ser-Arg-Met-Val-Arg-Tyr-Leu-His

cleavage by chymotrypsin

Elastase catalyzes the hydrolysis of peptide bonds on the C-side of small amino acids (Gly, Ala). Chymotrypsin and elastase are much less specific than trypsin. An explanation for the specificity of these enzymes is given in Section 21.8.

Ala-Lys-Phe-Gly-Asp-Trp-Ser-Arg-Met-Val-Arg-Tyr-Leu-His

cleavage by elastase

None of the exopeptidases or endopeptidases that we have mentioned will catalyze the hydrolysis of an amide bond if proline is at the hydrolysis site. These enzymes recognize the appropriate hydrolysis site by its shape and charge, and proline causes the hydrolysis site to have an unrecognizable three-dimensional shape.

Ala-Lys-Pro   Leu-Phe-Pro   Pro-Phe-Val

trypsin will not cleave   chymotrypsin will not cleave   chymotrypsin will cleave

Cyanogen bromide (BrCN) causes the hydrolysis of the amide bond on the C-side of a methionine residue. Cyanogen bromide is more specific about what peptide bonds it cleaves than are the endopeptidases, so it provides more reliable information about the primary sequence. Because cyanogen bromide is not a protein and therefore does not recognize its substrate by its shape, cyanogen bromide will still cleave the peptide bond if proline is at the cleavage site.

Ala-Lys-Phe-Gly-Lys-Trp-Ser-Arg-Met-Val-Arg-Tyr-Leu-His

cleavage by cyanogen bromide

TABLE 20.3 Specificity of Protein Cleavage	
Reagent	**Specificity**
Chemical reagents	
Edman's reagent	removes the N-terminal amino acid
Cyanogen bromide	hydrolyzes on the C-side of Met
Exopeptidases*	
Carboxypeptidase A	removes the C-terminal amino acid (not Arg or Lys)
Carboxypeptidase B	removes the C-terminal amino acid (only Arg or Lys)
Endopeptidases*	
Trypsin	hydrolyzes on the C-side of Arg and Lys
Chymotrypsin	hydrolyzes on the C-side of amino acids that contain aromatic six-membered rings (Phe, Tyr, Trp)
Elastase	hydrolyzes on the C-side of small amino acids (Gly and Ala)

*Cleavage will not occur if Pro is on either side of the bond to be hydrolyzed

The first step in the mechanism for cyanogen bromide cleavage of a peptide bond is attack by the highly nucleophilic sulfur on cyanogen bromide. Formation of a five-membered ring with departure of the weakly basic leaving group is followed by hydrolysis, which cleaves the protein (Section 16.7). Further hydrolysis can cause the lactone (a cyclic ester) to open to a carboxyl group and an alcohol group (Section 15.8).

mechanism for the cleavage of a peptide bond by cyanogen bromide

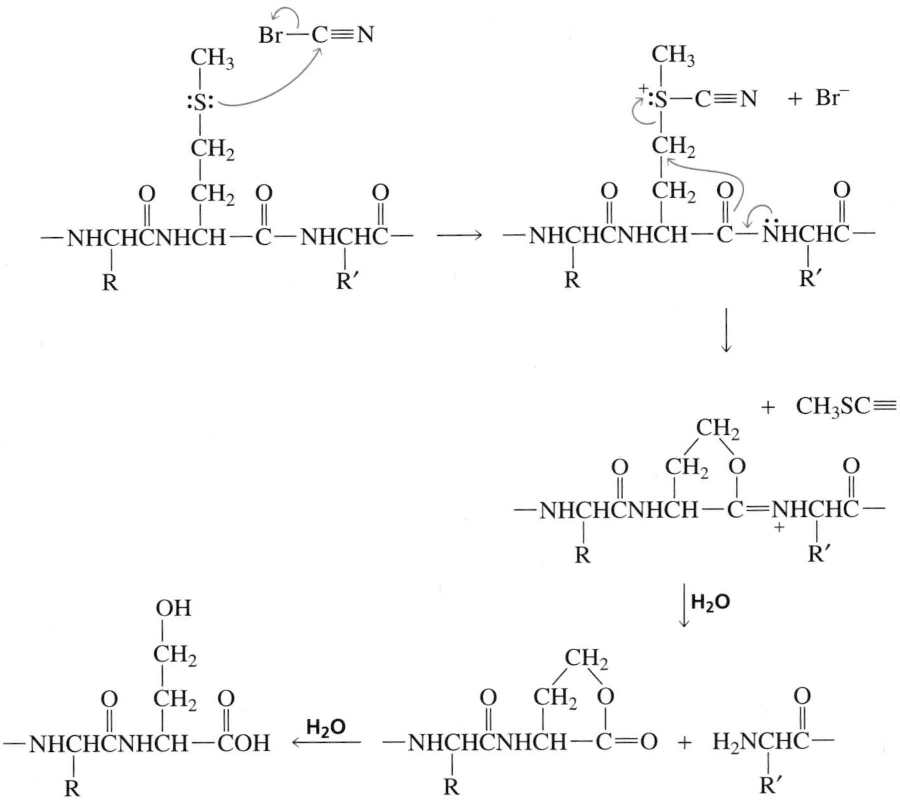

The last step in determining the primary structure of a protein is to figure out the location of any disulfide bonds. This is done by hydrolyzing a sample of the protein that has intact disulfide bonds. From a determination of the amino acids in the cysteine-containing fragments, the locations of the disulfide bonds in the protein can be established (Problem 50).

PROBLEM 31

Why won't cyanogen bromide cleave at cysteine residues?

PROBLEM 32

In determining the primary structure of insulin, what would lead you to conclude that it had more than one polypeptide chain?

PROBLEM 33 / SOLVED

Determine the amino acid sequence of a polypeptide from the following results.

Acid hydrolysis gives Ala, Arg, His, 2 Lys, Leu, 2 Met, Pro, 2 Ser, Thr, Val.

Carboxypeptidase A releases Val.

Edman's reagent releases PTH-Leu.

Cleavage with cyanogen bromide gives three peptides with the following amino acid compositions.

   1. His, Lys, Met, Pro, Ser

   2. Thr, Val

   3. Ala, Arg, Leu, Lys, Met, Ser

Trypsin-catalyzed hydrolysis gives three peptides and a single amino acid.

   1. Arg, Leu, Ser

   2. Met, Pro, Ser, Thr, Val

   3. Lys

   4. Ala, His, Lys, Met

SOLUTION   Acid hydrolysis shows that the polypeptide has 13 amino acids; the N-terminal amino acid is Leu (Edman's reagent), and the C-terminal amino acid is Val (carboxypeptidase A).

Leu __ __ __ __ __ __ __ __ __ __ __ Val

Because cyanogen bromide cleaves on the C-side of Met, any peptide containing Met must have Met as its C-terminal amino acid. The cyanogen bromide data identify the positions of the following amino acids.

        Ala, Arg, Lys, Ser   His, Lys, Pro, Ser

Leu __ __ __ __ Met __ __ __ __ Met Thr Val

Because trypsin cleaves on the C-side of Arg and Lys, any peptide containing Arg or Lys must have that amino acid as its C-terminal amino acid. The trypsin data identifies the positions of the following amino acids.

                                    Pro, Ser

Leu Ser Arg Lys Ala Met His Lys __ __ Met Thr Val

Because trypsin successfully cleaves on the C-side of Lys, Pro cannot be adjacent to Lys.

Leu Ser Arg Lys Ala Met His Lys Ser Pro Met Thr Val

PROBLEM 34 ◆

Determine the primary structure of an octapeptide from the following data.

Acid hydrolysis gives: 2 Arg, Leu, Lys, Met, Phe, Ser, Tyr.

Carboxypeptidase A releases Ser.

Edman's reagent releases Leu.

Cyanogen bromide forms two peptides:

   1. Arg, Phe, Ser

   2. Arg, Leu, Lys, Met, Tyr

Trypsin forms two peptides and two amino acids

1. Arg
2. Ser
3. Arg, Met, Phe
4. Leu, Lys, Tyr

20.13 SECONDARY STRUCTURE OF PROTEINS

Secondary structure describes the conformation of segments of the backbone chain of a peptide or protein. In order to minimize energy, a polypeptide chain tends to fold in a repeating geometric structure. Three factors determine the choice of secondary structure:

- the regional planarity about each peptide bond, which limits the possible conformations of the peptide chain (Section 20.7).
- maximizing the number of peptide groups that engage in hydrogen bonding (hydrogen bonding between the carbonyl oxygen of one amino acid residue and the amide hydrogen of another).

$$\text{C}=\text{O}\text{-----}\text{H}-\text{N}$$

- adequate separation between nearby R groups to avoid steric hindrance and repulsion of like charges.

One type of secondary structure is the **α-helix.** In an α-helix, the backbone of the polypeptide coils around the long axis of the protein molecule (Figure 20.9). The helix is stabilized by hydrogen bonds; each hydrogen attached to an amide nitrogen is hydrogen bonded to a carbonyl oxygen of an amino acid four residues away. The substituents on the α-carbons of the amino acids protrude outward from the helix. Because the amino acids have the L-configuration, the α-helix is a right-handed helix. A right-handed helix rotates in a clockwise direction as it spirals down. Each turn of the helix contains 3.6 amino acid residues, and the repeat distance of the helix is 5.4 Å.

Not all amino acids are able to fit into an α-helix. A proline residue forces a bend in a helix because the bond between the proline nitrogen and the α-carbon cannot rotate to enable it to fit readily into a helix. Two adjacent amino acids that have more than one substituent on a β-carbon (valine, isoleucine, or threonine) cannot fit into a helix because of steric crowding between the R groups. Two adjacent amino acids with like-charged substituents cannot fit into the helix because of electrostatic repulsion between the R groups. The percentage of amino acid residues coiled into an α-helix varies from protein to protein, but overall about 25% of the residues in globular proteins are in α-helices.

The second type of secondary structure is the **β-pleated sheet.** In a β-pleated sheet, the polypeptide backbone is extended in a zigzag structure resembling a series of pleats. A β-pleated sheet is almost fully extended; the average two-residue repeat distance is 7.0 Å The hydrogen bonding in a β-pleated sheet occurs between neighboring peptide chains. The adjacent hydrogen-bonded peptide chains can run in the same direction or in opposite directions. In a **parallel β-pleated sheet,** the

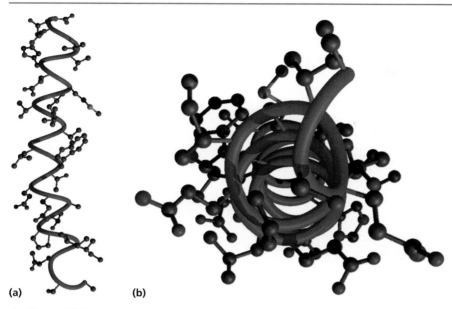

▲ **Figure 20.9**
(a) A segment of a protein in a right-handed α-helix. (b) Looking down the longitudinal axis of an α-helix.

adjacent chains run in the same direction. In an **antiparallel β-pleated sheet,** the adjacent chains run in opposite directions (Figure 20.10).

Because the substituents (R) on the α-carbons of the amino acids on adjacent chains are close to each other, the chains can nestle closely together to maximize hydrogen bonding interactions only if the substituents are small. Silk, a protein with a large number of relatively small amino acids (glycine and alanine), has large segments of β-pleated sheets. The number of side-by-side strands in a β-pleated

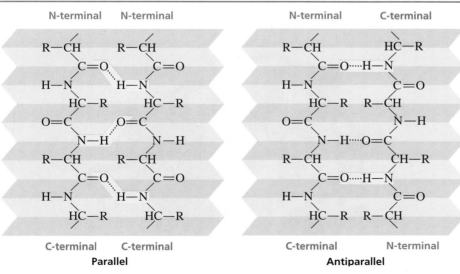

▲ **Figure 20.10**
Segment of a β-pleated sheet drawn to illustrate its pleated character.

sheet ranges from 2 to 15 in a globular protein. The average strand in a β-pleated sheet section of a globular protein contains six amino acid residues.

Wool and the fibrous protein of muscle are examples of proteins with secondary structures that are almost all α-helices. Consequently, these proteins can be stretched. In contrast, the secondary structures of silk and spider webs are predominantly β-pleated sheets. The β-pleated sheet is a fully extended structure, so these proteins cannot be stretched.

Generally less than half of a globular protein is in an α-helix or β-pleated sheet (Figure 20.11). Most of the rest of the protein is still highly ordered but is difficult to describe. These polypeptide fragments are said to be in a **coil conformation** or a **loop conformation.**

β-PEPTIDES: AN ATTEMPT TO IMPROVE ON NATURE

Chemists are currently synthesizing β-peptides, polymers of β-amino acids. These peptides have backbones one carbon longer than the peptides nature synthesizes using α-amino acids. Unlike α-amino acid residues each β-amino acid residue can have two side chains.

$$\overset{+}{H_3N}-\underset{\underset{\mathbf{R}}{|}}{CH}-\overset{\overset{O}{\|}}{C}-O^- \qquad \overset{+}{H_3N}-\underset{\underset{\mathbf{R}}{|}}{CH}-\underset{\underset{\mathbf{R'}}{|}}{CH}-\overset{\overset{O}{\|}}{C}-O^-$$

$$\textbf{α-amino acid} \qquad\qquad\qquad \textbf{β-amino acid}$$

β-Polypeptides, like α-polypeptides, fold into relatively stable helices, causing scientists to wonder if biological activity might be possible. There is hope that β-polypeptides will provide a source of new drugs and catalysts. Surprisingly, the peptide bonds in β-polypeptides are resistant to the enzymes that catalyze the hydrolysis of peptide bonds in α-polypeptides. This resistance to hydrolysis increases the chance that a β-polypeptide drug could be taken orally.

Figure 20.11 ▶
The backbone structure of carboxypeptidase A: α-helical segments are purple; β-pleated sheets are indicated by flat green arrows pointing in the N ⟶ C direction.

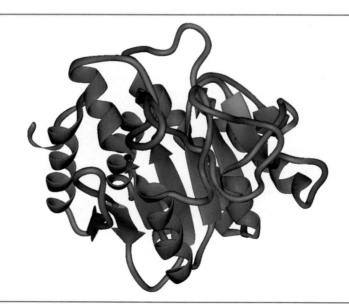

PROBLEM 35◆

How long is an α-helix that contains 74 amino acids? Compare this with the length of a fully extended peptide chain containing the same number of amino acids. (The distance between consecutive amino acids in a fully extended chain is 3.5 Å.)

The *tertiary structure* of a protein is the three-dimensional arrangement of all the atoms in the protein. Proteins fold spontaneously in solution in order to maximize their stability. Every time there is a stabilizing interaction between two atoms, free energy is released; the more free energy that is released (the more negative the $\Delta G°$), the more stable the protein. So a protein tends to fold in a way that maximizes the number of stabilizing interactions (Figure 20.12).

The stabilizing interactions that occur in folding are covalent bonds, hydrogen bonds, electrostatic attractions, and hydrophobic (van der Waals) interactions. The interactions can occur between peptide groups (atoms in the backbone of the protein), between side-chain groups (α-substituents), and between peptide and side-chain groups. Because of the importance of the side-chain groups in determining how a protein folds, the tertiary structure of a protein is determined by its primary structure.

**20.14
TERTIARY
STRUCTURE
OF PROTEINS**

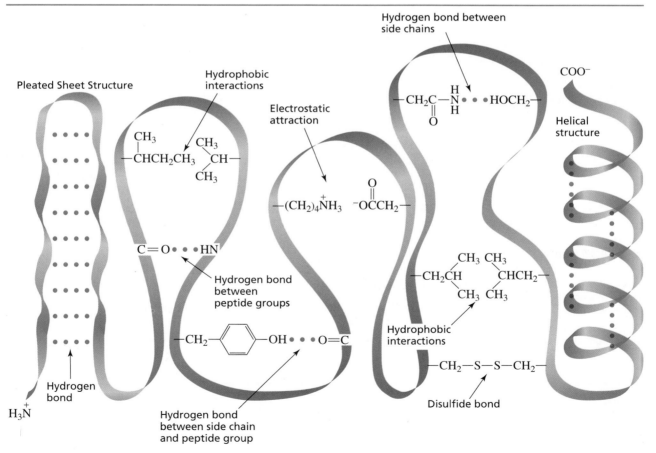

▲ **Figure 20.12**
Stabilizing interactions responsible for the tertiary structure of a protein.

Figure 20.13 ▶
The three-dimensional structure of carboxypeptidase A.

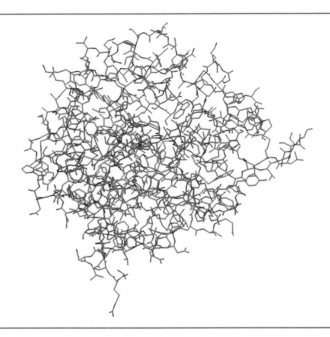

Max Ferdinand Perutz *and* **John Cowdery Kendrew** *were the first to determine the tertiary structure of a protein. Using X-ray diffraction, they determined the tertiary structure of myoglobin (1957) and hemoglobin (1959). For this work they shared the 1962 Nobel Prize in chemistry.*

Max Perutz *was born in Austria in 1914. In 1936, because of the rise of Nazism, he moved to England where he became a professor at Cambridge University. He worked on the three-dimensional structure of hemoglobin and assigned to* **John Kendrew** *the work on myoglobin (a smaller protein). Kendrew was born in England in 1917 and was educated at Cambridge University.*

Disulfide bonds are the only covalent bonds that can form when a protein folds. The other bonding interactions that occur in folding are much weaker but, because there are so many of them (Figure 20.13), they are the most important interactions in determining how a protein folds.

Most proteins exist in aqueous environments. Therefore, they tend to fold in a way that exposes the maximum number of polar groups to the aqueous environment and that buries the nonpolar groups in the interior of the protein, away from water.

The interactions between nonpolar groups are known as **hydrophobic interactions.** Hydrophobic interactions increase the stability of a protein by increasing the entropy of water molecules. Water molecules that surround nonpolar groups are highly structured. When two nonpolar groups come together, the surface area in contact with water decreases, decreasing the amount of structured water. When water loses its structure, there is a gain in entropy. Increasing the entropy decreases the free energy, which increases the stability. (Recall that $\Delta G^\circ = \Delta H^\circ - T\Delta S^\circ$.)

PROBLEM 36

How would a protein that resides in the interior of a membrane fold compared with the water-soluble protein discussed above?

20.15 QUATERNARY STRUCTURE OF PROTEINS

Many proteins have more than one peptide chain. Such proteins are called **oligomers.** The individual chains are called **subunits.** A protein with a single subunit is called a *monomer;* one with two subunits, a *dimer;* one with three subunits, a *trimer;* and one with four subunits, a *tetramer.* Hemoglobin is an example of a tetramer. It has two different kinds of subunits and two of each kind of subunit. The quaternary structure of hemoglobin is shown in Figure 20.14.

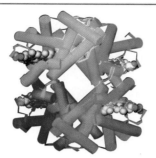

◀ **Figure 20.14**
Computer graphic representation of the quaternary structure of hemoglobin: the
orange and pink subunits are
identical; the green and
purple subunits are identical.
The cylindrical tubes represent the polypeptide chains;
the beads represent the iron-
containing porphyrin rings
(Section 26.4).

The subunits are held together by the same kinds of interactions that hold the individual protein chains in a particular three-dimensional configuration—hydrophobic interactions, hydrogen bonding, and electrostatic attractions. The quaternary structure of a protein is a description of the way the subunits are arranged in space. Some of the possible arrangements of the six subunits of a hexamer are shown below.

possible quaternary structures for a hexamer

PROBLEM 37 ◆

a. Which of the following water-soluble proteins would have the greatest percentage of polar amino acids: a spherical protein, a cigar-shaped protein, or a subunit of a hexamer?

b. Which of these would have the smallest percentage of polar amino acids?

Destroying the highly organized tertiary structure of a protein is called **denaturation.** The totally random conformation of a denatured protein is called a **random coil.** Anything that breaks the bonds responsible for maintaining the three-dimensional shape of the protein will cause the protein to denature (unfold). Since these bonds are weak, proteins are easily denatured.

**20.16
PROTEIN
DENATURATION**

- Changing the pH denatures proteins because it changes the charges on many of the side chains. This disrupts electrostatic attractions and hydrogen bonds.
- Certain reagents such as urea and guanidine hydrochloride denature proteins by forming hydrogen bonds to the protein groups that are stronger than the hydrogen bonds formed between the groups.

- Detergents such as sodium dodecyl sulfate denature proteins by associating with the nonpolar groups of the protein, thus interfering with the normal hydrophobic interactions.
- Organic solvents denature proteins by disrupting hydrophobic interactions.
- Proteins can also be denatured by heat. Heat increases molecular motion, which can disrupt the attractive forces. A well-known example is the change that occurs to the white of an egg when it is heated.

KEY TERMS

amino acid (page 907)
D-amino acid (page 912)
L-amino acid (page 912)
amino acid analyzer (page 921)
amino acid residue (page 907)
anion-exchange resin (page 921)
antiparallel β-pleated sheet (page 943)
automated solid-phase peptide synthesis (page 933)
cation-exchange resin (page 921)
coil conformation (page 944)
denaturation (page 947)
dipeptide (page 907)
disulfide (page 925)
disulfide bridge (page 926)
Edman's reagent (page 937)
electrophoresis (page 917)
endopeptidase (page 938)
enkephalin (page 927)
enzyme (page 908)
essential amino acid (page 912)
exopeptidase (page 937)
fibrous protein (page 935)
globular protein (page 935)
α-helix (page 942)
hydrophobic interactions (page 946)
inflection point (page 915)
interchain disulfide bridge (page 926)
intrachain disulfide bridge (page 926)
ion-exchange chromatography (page 920)
isoelectric point (page 916)
kinetic resolution (page 923)
loop conformation (page 944)
oligomer (page 946)
oligopeptide (page 907)
paper chromatography (page 918)
parallel β-pleated sheet (page 942)
partial hydrolysis (page 938)
peptide (page 907)
peptide bond (page 924)
β-pleated sheet (page 942)
polypeptide (page 907)
primary structure (page 935)
protein (page 907)
quaternary structure (page 935)
random coil (page 947)
secondary structure (page 935)
structural protein (page 907)
subunit (page 946)
C-terminal amino acid (page 924)
N-terminal amino acid (page 924)
tertiary structure (page 935)
thin-layer chromatography (page 919)
titration curve (page 915)
tripeptide (page 907)
zwitterion (page 914)

PROBLEMS

38. Unlike most amines and carboxylic acids, amino acids are insoluble in diethyl ether. Explain.

39. Indicate the peptides that would result from cleavage by the indicated reagent.
  a. His-Lys-Leu-Val-Glu-Pro-Arg-Ala-Gly-Ala by trypsin
  b. Leu-Gly-Ser-Met-Phe-Pro-Tyr-Gly-Val by chymotrypsin
  c. Val-Arg-Gly-Met-Arg-Ala-Ser by carboxypeptidase A
  d. Ser-Phe-Lys-Met-Pro-Ser-Ala-Asp by cyanogen bromide
  e. Arg-Ser-Pro-Lys-Lys-Ser-Glu-Gly by trypsin

40. Identify the amino acid whose titration curve is shown on the following page.

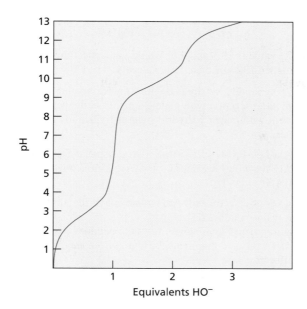

41. Aspartame has a p*I* of 5.9. Draw its most prevalent form at physiological pH.

42. Draw the form of aspartic acid that predominates at
  a. pH 1.0   **b.** pH 2.6   **c.** pH 6.0   **d.** pH 11.0

43. Dr. Kim S. Tree was preparing a manuscript for publication in which she had reported that the p*I* of the tripeptide Lys-Lys-Lys was 10.6. One of her students pointed out that there must be an error in her calculations because the pK_a of the ε-amino group of lysine is 10.8 and the p*I* of the tripeptide has to be greater than any of its individual pK_a values. Was the student correct?

44. A mixture of amino acids that do not separate sufficiently when a single technique is used can often be separated by two-dimensional chromatography. In this technique the mixture of amino acids is applied to a piece of filter paper and separated by chromatographic techniques. The paper is rotated 90°, and the amino acids are further separated by electrophoresis. Identify the spots in the chromatogram obtained from a mixture of Ser, Glu, Leu, His, Met, Thr.

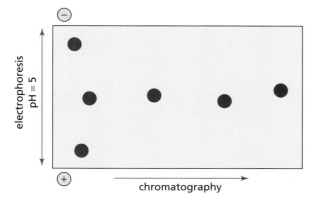

45. Explain the difference in the pK_a values of the carboxyl groups of alanine, serine, and cysteine.

46. Which would be a more effective buffer at physiological pH, a solution of 0.1 M glycylglycylglycylglycine or a solution of 0.2 M glycine?

47. Identify the location and type of charge on the hexapeptide Lys-Ser-Asp-Cys-His-Tyr at
   a. pH 7   **b.** pH 5   **c.** pH 9

48. Glycine has pK_a values of 2.3 and 9.6. Would you expect the pK_a values of glycyl-glycine to be higher or lower than these values?

49. The following polypeptide was treated with 2-mercaptoethanol and then with iodoacetic acid. After reaction with maleic anhydride, the peptide was hydrolyzed by trypsin. Treatment with maleic anhydride causes trypsin to cleave a peptide only at arginine residues.

Gly-Ser-Asp-Ala-Leu-Pro-Gly-Ile-Thr-Ser-Arg-Asp-Val-Ser-Lys-Val-Glu-Tyr-Phe-Glu-Ala-Gly-Arg-Ser-Glu-Phe-Lys-Glu-Pro-Arg-Leu-Tyr-Met-Lys-Val-Glu-Gly-Arg-Pro-Val-Ser-Ala-Gly-Leu-Trp

  a. Why, after a peptide is treated with maleic anhydride, does trypsin no longer cleave it at lysine residues?
  b. How many fragments are obtained from the peptide?
  c. In what order would the fragments be eluted from an anion-exchange column using a buffer of pH 5?

50. Treatment of a polypeptide with 2-mercaptoethanol yields two polypeptides with the following primary sequences.

Val-Met-Tyr-Ala-Cys-Ser-Phe-Ala-Glu-Ser

Ser-Cys-Phe-Lys-Cys-Trp-Lys-Tyr-Cys-Phe-Arg-Cys-Ser

Treatment of the original intact polypeptide with chymotrypsin yields the following peptides.
  a. Ala, Glu, Ser   **d.** Arg, Ser, Cys
  b. 2 Phe, 2 Cys, Ser   **e.** Ser, Phe, 2 Cys, Lys, Ala, Trp
  c. Tyr, Val, Met   **f.** Tyr, Lys
Determine the positions of the disulfide bridges in the original polypeptide.

51. Show how aspartame can be synthesized using DCC.

52. Reaction of a polypeptide with carboxypeptidase A releases Met. It undergoes partial hydrolysis to give the following peptides. What is the sequence of the polypeptide?
  a. Ser, Lys, Trp   **e.** Lys, Ser   **i.** Glu, His
  b. Gly, His, Ala   **f.** Glu, His, Val   **j.** Leu, Lys, Trp
  c. Met, Ala, Gly   **g.** Glu, Val, Ser   **k.** Trp, Leu, Glu
  d. Ser, Lys, Val   **h.** Leu, Glu, Ser   **l.** Ala, Met

53. A mixture of 15 amino acids gave the following fingerprint. Identify the spots. (*Hint #1:* Pro reacts with ninhydrin to form a yellow color; Phe and Tyr form a yellow-green color. *Hint #2:* Count the amino acids and the spots before you start.)

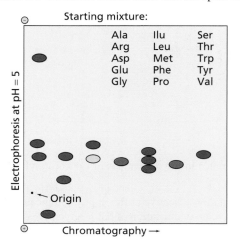

54. Dithiothreitol reacts with disulfide bridges in the same way as does 2-mercaptoethanol. However, with dithiothreitol, the equilibrium lies much more to the right. Explain.

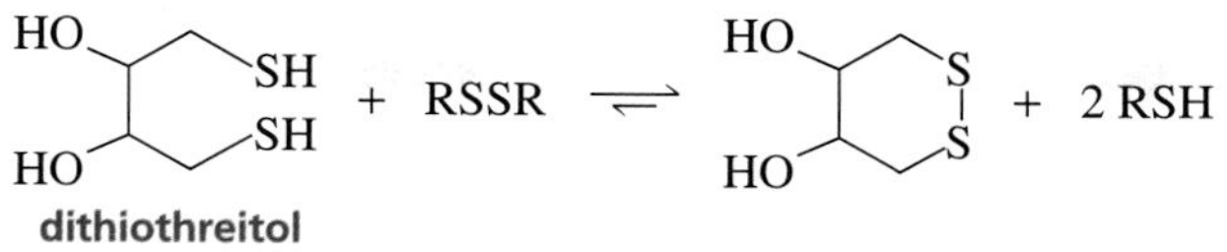

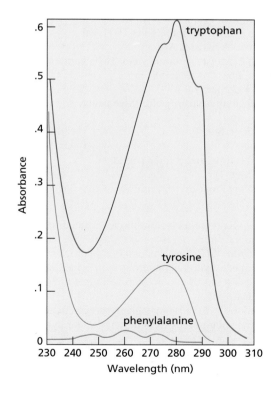

55. Draw the titration curve for the titration of 0.1 M glutamylvalylglutamylarginine with 0.1 N NaOH.

56. α-Amino acids can be prepared by treating an aldehyde with ammonia and hydrogen cyanide, followed by acid-catalyzed hydrolysis.
 a. Give the structures of the two intermediates formed in this reaction.
 b. What amino acid is formed when the aldehyde used is 3-methylbutanal?
 c. What aldehyde would be needed to prepare valine?

57. The UV spectra of tryptophan, tyrosine, and phenylalanine are shown below. Each spectrum is that of a 1×10^{-3} M solution of the amino acid buffered at pH 6.0. Calculate the approximate molar absorptivity of each of the three amino acids at 280 nm.

58. A normal polypeptide and a mutant of the polypeptide were hydrolyzed by an endopeptidase under the same conditions. The normal and mutant differ by one amino acid residue. The fingerprints of the peptides obtained from the normal and mutant polypeptides are shown below. What kind of amino acid substitution occurred as a result of the mutation (i.e., is the substituted amino acid more or less polar than the original amino acid? Is its p*I* lower or higher?)

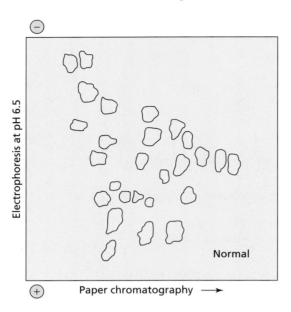

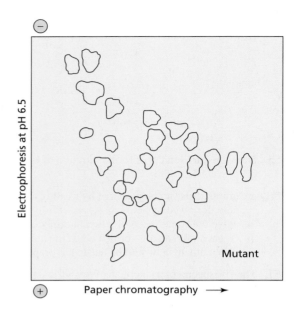

59. Determine the amino acid sequence of a polypeptide from the following results.
  a. Complete hydrolysis of the peptide yields the following amino acids: Ala, Arg, Gly, 2 Lys, Met, Phe, Pro, 2 Ser, Tyr, Val.
  b. Treatment with Edman's reagent gives PTH-Val.
  c. Carboxypeptidase A releases Ala.
  d. Treatment with cyanogen bromide yields two peptides:
    1. Ala, 2 Lys, Phe, Pro, Ser, Tyr
    2. Arg, Gly, Met, Ser, Val
  e. Treatment with chymotrypsin yields three peptides:
    1. 2 Lys, Phe, Pro
    2. Arg, Gly, Met, Ser, Tyr, Val
    3. Ala, Ser
  f. Treatment with trypsin yields three peptides:
    1. Gly, Lys, Met, Tyr
    2. Ala, Lys, Phe, Pro, Ser
    3. Arg, Ser, Val

60. The C-terminal end of a protein extends into the aqueous environment surrounding the protein. The C-terminal amino acids are Gln, Asp, 2 Ser, and three nonpolar amino acids. Assuming that the $\Delta G°$ for formation of a hydrogen bond is -3 kcal/mol and the $\Delta G°$ for removal of a hydrophobic group from water is -4 kcal/mol, calculate the $\Delta G°$ for folding the C-terminal end of the protein into the protein interior under the following conditions:
  a. all the polar groups form intramolecular hydrogen bonds.
  b. all but two of the polar groups form intramolecular hydrogen bonds.

61. Professor Oranga Thorpe wanted to test her hypothesis that the disulfide bridges that form in many proteins do so after the minimum energy conformation of the protein has been achieved. She treated a sample of lysozyme, an enzyme containing four disulfide bridges, with 2-mercaptoethanol and then added urea to denature the enzyme. She slowly removed these reagents so the enzyme could refold and reform the disulfide bridges. The lysozyme she recovered had 80% of its original activity. What would be the percent activity in the recovered enzyme if disulfide bridge formation were entirely random rather than determined by the tertiary structure? Does this experiment support Professor Thorpe's hypothesis?

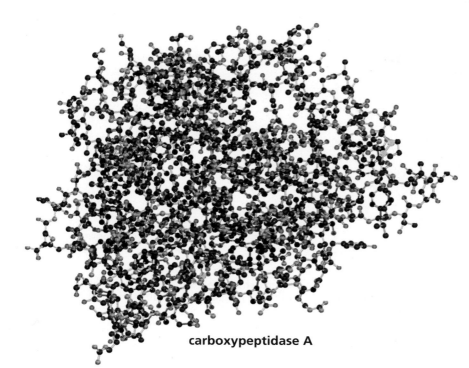

carboxypeptidase A

CATALYSIS

A **catalyst** is a substance that increases the rate of a chemical reaction without itself being consumed or changed in the overall reaction. We have seen that the rate of a chemical reaction depends on the height of the energy hill that must be overcome in the process of converting reactants into products (Section 3.7). The height of the hill is indicated by the free energy of activation ($\Delta G^{\ddagger}$). A catalyst increases the rate of a chemical reaction by decreasing its $\Delta G^{\ddagger}$.

A catalyst can decrease $\Delta G^{\ddagger}$ in one of three ways: (1) the catalyzed and uncatalyzed reactions can follow a similar mechanism, with the catalyst providing a way to make *the reactant less stable* (Figure 21.1a); (2) the catalyzed and uncatalyzed reactions can follow a similar mechanism, with the catalyst providing a way to make *the transition state more stable* (Figure 21.1b); or (3) a catalyst can completely *change the mechanism* of the reaction, providing an alternative pathway with a smaller $\Delta G^{\ddagger}$ than that of the uncatalyzed reaction (Figure 21.2).

As just stated, a catalyst is neither consumed nor changed by the reaction. This does not mean that the catalyst does not participate in the reaction: how could a catalyst make the reaction go faster if it did not participate? It simply means that the catalyst is in the same form when the reaction is over that it was in before the reaction started. Because the catalyst is not used up during the reaction—if it is used up in one step of the reaction, it is regenerated in a subsequent step—only a small amount of the catalyst is needed. Since catalytic molecules can be used again and again, a catalyst is added to a reaction mixture in small, *catalytic* amounts (much less than the number of moles of reactant).

Notice that the stability of the reactants and products is the same in the catalyzed and uncatalyzed reactions. In other words, the catalyst does not change the equilibrium constant of the reaction (ΔG° is the same for the catalyzed and uncatalyzed reactions in Figures 21.1a, 21.1b, and 21.2). Because it does not change the equilibrium constant, it does not change the *amount* of product formed during the reaction. It only changes the *rate* at which the product is formed.

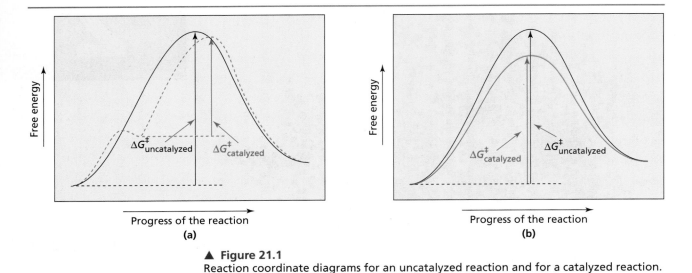

▲ **Figure 21.1**
Reaction coordinate diagrams for an uncatalyzed reaction and for a catalyzed reaction.
(a) The catalyst destabilizes the reactant. (b) The catalyst stabilizes the transition state.

Figure 21.2 ▶
Reaction coordinate diagrams
for an uncatalyzed reaction
and for a catalyzed reaction.
The catalyzed reaction takes
place by an alternative and
energetically more favorable
pathway.

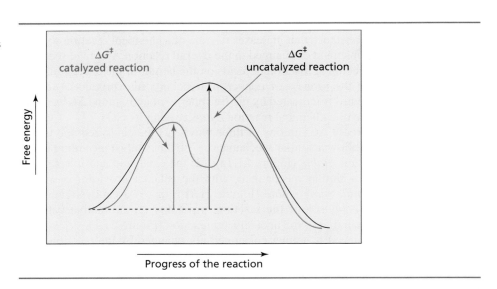

PROBLEM 1 ◆

Which of the following parameters would be different for a reaction carried out in the
presence of a catalyst compared with the same reaction carried out in the absence of
a catalyst?

$$\Delta G°, \Delta H^‡, E_a, \Delta S^‡, \Delta H°, K_{eq}, \Delta G^‡, \Delta S°, k_{rate}$$

There are several ways for a catalyst to provide a more favorable pathway for an organic reaction:

- It can increase the susceptibility of an electrophile to nucleophilic attack.
- It can increase the reactivity of a nucleophile.
- It can increase the leaving ability of a group by converting it into a weaker base.

In this chapter we will look at some of the most common catalysts (nucleophilic catalysts, acid catalysts, base catalysts, and metal-ion catalysts) and the ways in which they provide an energetically more favorable pathway for an organic reaction. We will then see how the same modes of catalysis are used in enzyme-catalyzed reactions.

21.1 CATALYSIS IN ORGANIC REACTIONS

A **nucleophilic catalyst** increases the rate of a reaction by acting as a nucleophile—it forms a covalent bond with one of the reactants. Because a nucleophilic catalyst forms a covalent bond with a reactant, **nucleophilic catalysis** is also known as **covalent catalysis.** A nucleophilic catalyst increases the reaction rate by changing the mechanism of the reaction.

In the following reaction, iodide ion increases the rate of conversion of ethyl chloride into ethyl alcohol by acting as a nucleophilic catalyst.

21.2 NUCLEOPHILIC CATALYSIS

$$CH_3CH_2Cl \ + \ HO^- \ \xrightarrow{I^-} \ CH_3CH_2OH \ + \ Cl^-$$

To understand exactly how iodide ion catalyzes this reaction, we must compare the mechanism of the reaction in the absence of the catalyst with the mechanism of the reaction in the presence of the catalyst. In the absence of iodide ion, ethyl chloride is converted into ethyl alcohol in a one-step S_N2 reaction.

mechanism for the uncatalyzed reaction

$$HO^- \ + \ CH_3CH_2 - Cl \ \longrightarrow \ CH_3CH_2OH \ + \ Cl^-$$

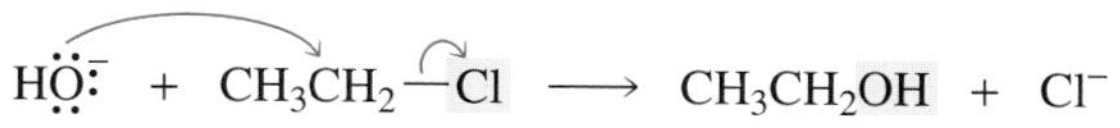

If iodide ion is present in the reaction mixture, the reaction takes place by two successive S_N2 reactions.

mechanism for the iodide-ion-catalyzed reaction

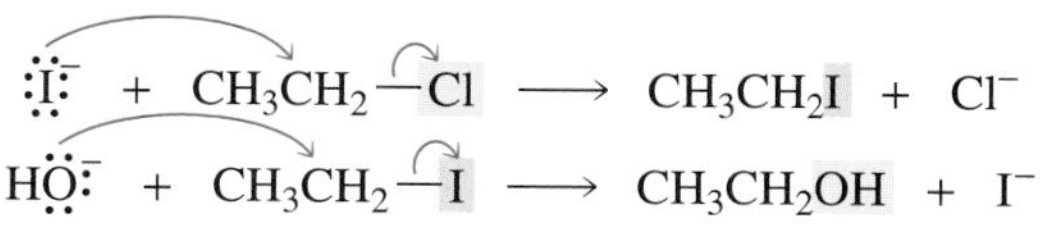

$$I^- \ + \ CH_3CH_2 - Cl \ \longrightarrow \ CH_3CH_2I \ + \ Cl^-$$
$$HO^- \ + \ CH_3CH_2 - I \ \longrightarrow \ CH_3CH_2OH \ + \ I^-$$

The first S_N2 reaction in the catalyzed reaction is faster than the uncatalyzed reaction because iodide ion is a better nucleophile than hydroxide ion, which is the nucleophile in the uncatalyzed reaction. The second S_N2 reaction in the catalyzed reaction is also faster than the uncatalyzed reaction because iodide ion is a weaker base and therefore a better leaving group than chloride ion, the leaving group in the uncatalyzed reaction. Thus, iodide ion increases the rate of formation of ethanol by changing a relatively slow one-step reaction into a reaction with two relatively fast steps (Figure 21.2).

Iodide ion is a nucleophilic catalyst because it reacts as a nucleophile, forming a covalent bond with the reactant. Notice that iodide ion participates in the reaction but comes out of the reaction unchanged.

Another reaction in which a catalyst provides a more favorable pathway by changing the mechanism is the imidazole-catalyzed hydrolysis of phenyl acetate.

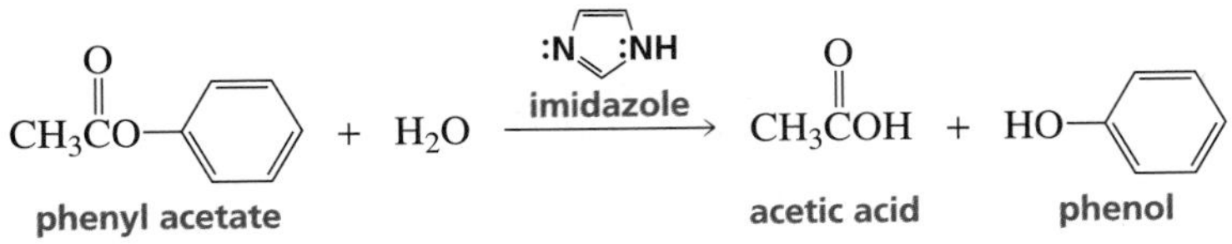

Imidazole is a nucleophilic catalyst. Being a better nucleophile than water, imidazole reacts faster than water with the ester. The acyl imidazole that is formed is particularly reactive because of the positively charged nitrogen. Therefore, it is hydrolyzed much more rapidly than the ester would have been hydrolyzed. Because formation of the acyl imidazole and its subsequent hydrolysis are both faster than ester hydrolysis, imidazole increases the rate of the overall reaction.

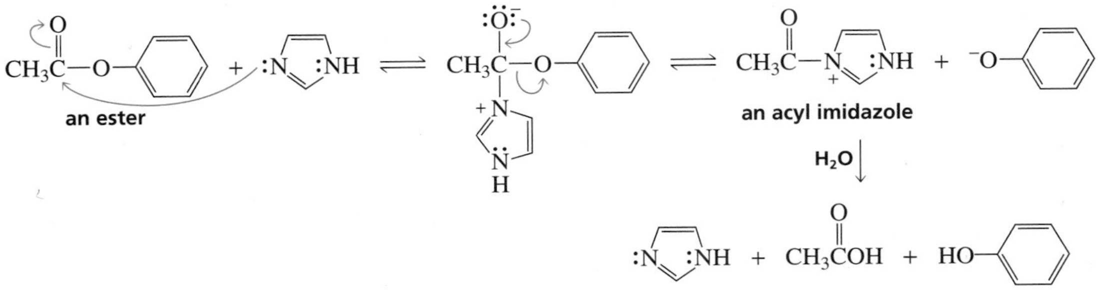

21.3
ACID CATALYSIS

An **acid catalyst** increases the rate of a reaction by donating a proton to a reactant. For example, the rate of hydrolysis of an ester is markedly increased by an acid catalyst (H^+).

$$CH_3COCH_3 + H_2O \xrightleftharpoons{H^+} CH_3COH + CH_3OH$$

A catalyst must increase the rate of a slow step, because increasing the rate of a fast step will not increase the rate of the overall reaction.

How an acid catalyzes the hydrolysis of an ester can be understood by examining the mechanism for the acid-catalyzed reaction. *Donation of a proton to and removal of a proton from an electronegative atom such as oxygen are fast steps.* Therefore, the reaction has two slow steps: formation of the tetrahedral intermediate and collapse of the tetrahedral intermediate. A catalyst must increase the rate of a slow step, because increasing the rate of a fast step will not increase the rate of the overall reaction.

mechanism for the acid-catalyzed hydrolysis of an ester

$$CH_3\overset{\overset{\displaystyle O}{\|}}{C}-OCH_3 \underset{-H^+}{\overset{H^+}{\rightleftharpoons}} CH_3\overset{\overset{\displaystyle \overset{+}{O}H}{\|}}{C}-OCH_3 + H_2\overset{..}{O}: \overset{slow}{\rightleftharpoons} CH_3\overset{\overset{\displaystyle OH}{|}}{\underset{\underset{\displaystyle H}{\overset{+}{O}H}}{C}}-OCH_3$$

$$H^+ \| -H^+$$

$$CH_3\overset{\overset{\displaystyle OH}{|}}{\underset{\underset{\displaystyle OH}{|}}{C}}-OCH_3$$

$$-H^+ \| H^+$$

$$CH_3\overset{\overset{\displaystyle O}{\|}}{C}-OH \underset{H^+}{\overset{-H^+}{\rightleftharpoons}} CH_3\overset{\overset{\displaystyle \overset{+}{O}H}{\|}}{C}-OH + CH_3OH \overset{slow}{\rightleftharpoons} CH_3\overset{\overset{\displaystyle :\overset{..}{O}H}{|}}{\underset{\underset{\displaystyle OH}{|}}{C}}\overset{+}{\underset{H}{O}}CH_3$$

Alfred Bernhard Nobel

The acid increases the rates of both slow steps of this reaction. It increases the rate of formation of the tetrahedral intermediate by protonating the carbonyl oxygen, thereby increasing the reactivity of the carbonyl group. We have seen that a protonated carbonyl group is more electrophilic than an unprotonated carbonyl group. Consequently, the protonated carbonyl group is more susceptible to nucleophilic attack (Section 15.11). Increasing the reactivity of the carbonyl group is an example of making the reactant less stable—i.e., more reactive (Figure 21.1a).

acid-catalyzed first slow step

$$CH_3\overset{\overset{\displaystyle \overset{+}{O}H}{\|}}{C}-OCH_3 + H_2\overset{..}{O}:$$

uncatalyzed first slow step

$$CH_3\overset{\overset{\displaystyle O}{\|}}{C}-OCH_3 + H_2\overset{..}{O}:$$

The acid increases the rate of the second slow step by changing the basicity of the group eliminated when the tetrahedral intermediate collapses. In the presence of an acid, methanol is eliminated; in the absence of an acid, methoxide ion is eliminated. Methanol is a weaker base than methoxide ion, so it is more easily eliminated.

acid-catalyzed second slow step

$$CH_3\overset{\overset{\displaystyle :\overset{..}{O}H}{|}}{\underset{\underset{\displaystyle OH}{|}}{C}}\overset{+}{\underset{H}{O}}CH_3$$

uncatalyzed second slow step

$$CH_3\overset{\overset{\displaystyle :\overset{..}{O}H}{|}}{\underset{\underset{\displaystyle OH}{|}}{C}}OCH_3$$

The mechanism for acid-catalyzed hydrolysis of an ester shows that the reaction can be divided into two distinct phases: formation of a tetrahedral intermediate and collapse of a tetrahedral intermediate. There are three steps in each phase. Notice that the first step in each phase is a protonation step, the second step in each phase involves either breaking a π bond or forming a π bond, and the last step in each phase is a deprotonation step.

Throughout this text you have found biographical sketches that give you some information about those men and women who created the science you are studying. You have seen that many of these people are Nobel Prize winners. The Nobel Prize is considered by many to be the most coveted award that a scientist can receive. The awards were established by **Alfred Bernhard Nobel (1833–1896).** *Nobel was born in Stockholm, Sweden. When he was nine, he moved with his parents to St. Petersburg where his father manufactured the torpedoes and submarine mines he had invented for the Russian government. As a young man Alfred did research on explosives in a factory near Stockholm, where he invented the detonator. In 1864 an explosion in the factory killed his younger brother, causing Alfred to look for ways to make it easier to handle and transport explosives. He established an explosives factory in Germany, where in 1866 he invented dynamite. He also invented blasting gelatin and smokeless powder, and some of his inventions were concerned with artificial leather and silk, and synthetic rubber. Although he was the inventor of explosives used by the military, he was a strong supporter of peace movements.* (continued)

PROBLEM 2

Compare each of the following mechanisms with the mechanism for each phase of the acid-catalyzed hydrolysis of an ester, indicating:

a. similarities.   **b.** differences.

(1) acid-catalyzed formation of a hydrate (Section 16.8)

(2) acid-catalyzed conversion of an aldehyde into a hemiacetal (Section 16.8)

(3) acid-catalyzed conversion of a hemiacetal into an acetal (Section 16.8)

(4) acid-catalyzed hydrolysis of an amide (Section 15.14)

The example of acid catalysis shown on page 957 is known as **specific-acid catalysis.** Both slow steps of the reaction are specific-acid-catalyzed. In specific-acid catalysis, the proton is fully transferred to the reactant *before* the slow step of the reaction begins (Figure 21.3a).

A second kind of acid catalysis is known as general-acid catalysis. In **general-acid catalysis,** the proton is transferred to the reactant *during* the slow step of the reaction (Figure 21.3b). Specific- and general-acid catalysts speed up a reaction in the same way, by donating a proton in order to make bond making and bond breaking easier. The two types of acid catalysts differ in the extent to which the proton is transferred in the transition state of the slow step of the reaction. In a specific-acid catalyzed reaction, the transition state has a fully transferred proton, while in a general-acid-catalyzed reaction, the transition state has a partially transferred proton (Figure 21.3).

In the following pairs of examples, notice the difference in the amount of proton transfer. In specific-acid-catalyzed attack by water on a carbonyl group, the nucleophile attacks a fully protonated carbonyl group. In general-acid-catalyzed attack by water on a carbonyl group, the carbonyl group becomes protonated as the nucleophile attacks it.

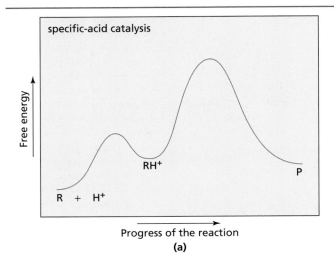

▲ **Figure 21.3**

specific-acid-catalyzed attack by water

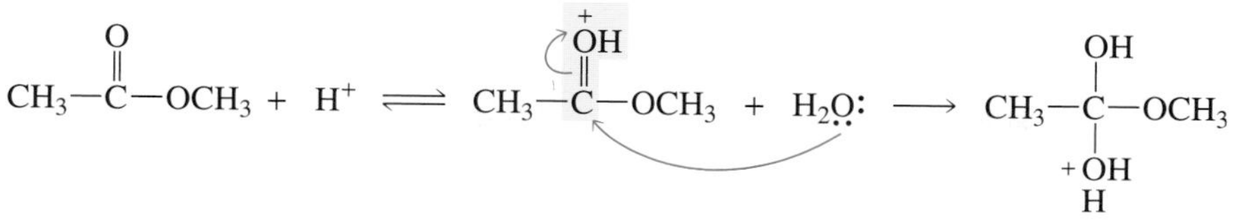

general-acid-catalyzed attack by water

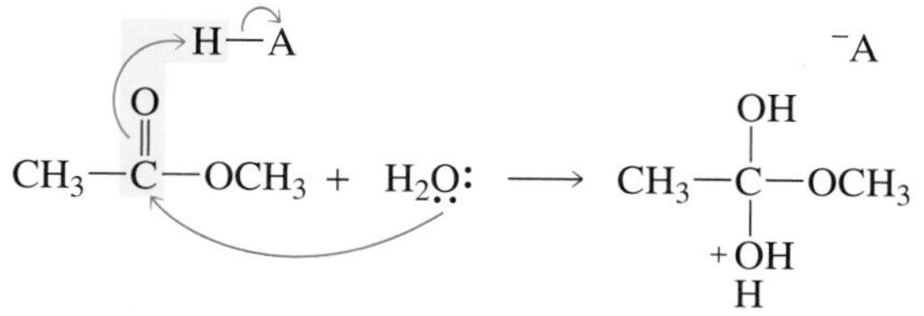

In specific-acid-catalyzed collapse of a tetrahedral intermediate, a fully protonated leaving group is eliminated, while in general-acid-catalyzed collapse of a tetrahedral intermediate, the leaving group picks up a proton as it is eliminated.

specific-acid-catalyzed elimination of the leaving group

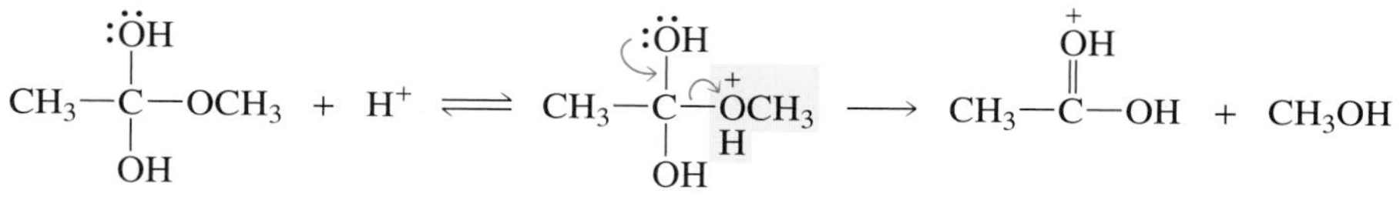

general-acid-catalyzed elimination of the leaving group

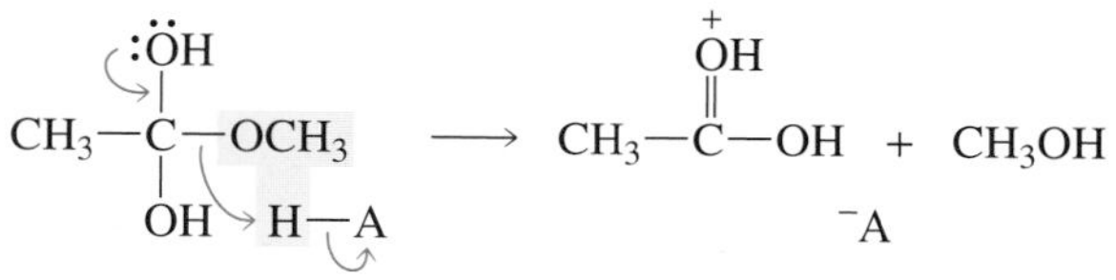

A specific-acid catalyst must be an acid strong enough to protonate the reactant fully before the slow step begins. A general-acid catalyst can be a weaker acid because it only partially transfers a proton in the transition state of the slow step.

Nobel instructed that the prizes were to be awarded by the following groups: the prizes for chemistry and physics by the Swedish Academy of Sciences, for physiology and medicine by the Caroline Institute in Stockholm, for literature by the Swedish Academy, and for peace by a five-person committee elected by the Norwegian Parliament. The deliberations are secret, and appeals cannot be made about the decisions. The first prizes were awarded in 1901. In 1969, the Swedish Central Bank established a prize in economics in Nobel's honor. The recipient of this prize is selected by the Swedish Academy of Sciences. On December 10—the anniversary of Nobel's death—the prizes are awarded in Stockholm, except for the peace prize that is awarded in Oslo.

PROBLEM 3 ◆

The following reaction occurs by a general-acid-catalyzed mechanism. Propose a mechanism for this reaction.

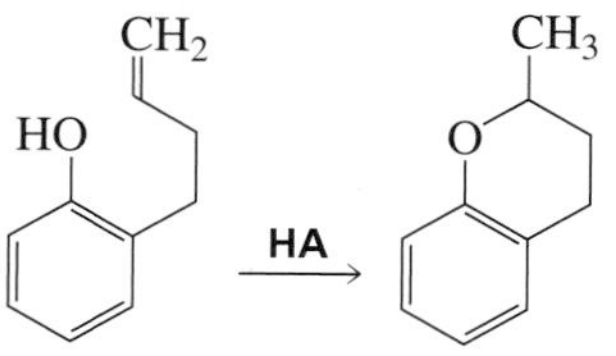

PROBLEM 4 / SOLVED

An alcohol will not react with aziridine unless the reaction is catalyzed by an acid. Why is a catalyst necessary?

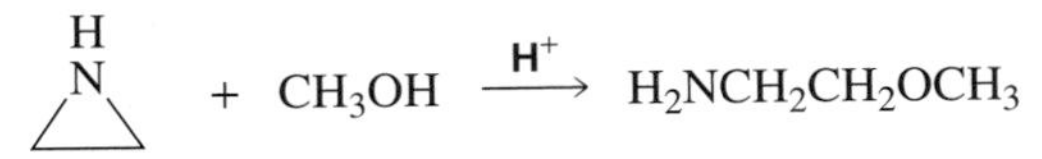

$$\text{aziridine} + CH_3OH \xrightarrow{H^+} H_2NCH_2CH_2OCH_3$$

aziridine

SOLUTION Although relief of ring strain is sufficient by itself to cause a ring-opening reaction of an epoxide, it is not sufficient to cause a ring-opening reaction of an aziridine with its poorer leaving group. (A negatively charged oxygen is a weaker base and therefore a better leaving group than a negatively charged nitrogen.) An acid must protonate the nitrogen, making it a better leaving group, before the reaction can take place.

21.4 BASE CATALYSIS

A **base catalyst** increases the rate of a reaction by removing a proton from a reactant. For example, dehydration of a hydrate in the presence of hydroxide ion is a base-catalyzed reaction. Hydroxide ion (the base) increases the rate of the reaction by removing a proton from the neutral hydrate.

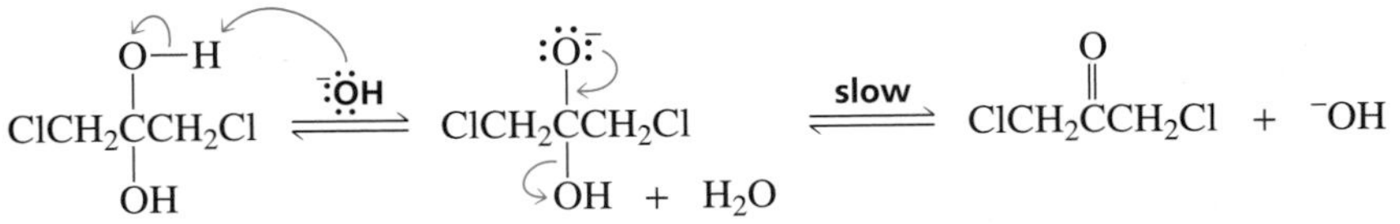

a hydrate

How does the base increase the rate of dehydration? It increases the rate of the reaction by providing a pathway with a more stable transition state. The transition state for elimination of ^-OH from a negatively charged tetrahedral intermediate is more stable than the transition state for elimination of ^-OH from a neutral tetrahedral intermediate. This is because, in the former, a positive charge does not develop on the electronegative oxygen atom.

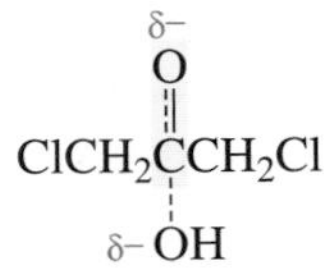

transition state for elimination of $^-$OH from a negatively charged tetrahedral intermediate

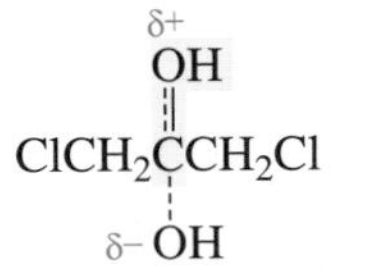

transition state for elimination of $^-$OH from a neutral tetrahedral intermediate

A proton is donated to the reactant in an acid-catalyzed reaction.

The reaction above is an example of specific-base catalysis. In **specific-base catalysis,** the proton is completely removed from the reactant *before* the slow step of the reaction begins. In **general-base catalysis,** the proton is removed from the reactant *during* the slow step of the reaction. General-base-catalyzed dehydration of a hydrate is shown below. Compare the degrees of proton transfer in the slow steps of general-base- and specific-base-catalyzed dehydration.

A proton is removed from the reactant in a base-catalyzed reaction.

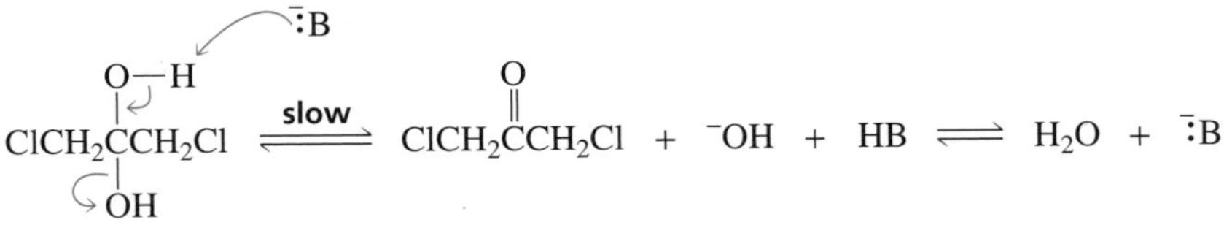

a hydrate

A specific-base catalyst has to be a base strong enough to remove a proton from the reactant completely before the slow step begins. General-base catalysts can be weaker bases because the proton is only partially transferred to the base in the transition state of the slow step. We will see that enzymes catalyze reactions using general-acid and general-base catalytic groups because at physiological pH (pH = 7.3) only a very small concentration of H^+ for specific-acid catalysis or HO^- for specific-base catalysis is available.

PROBLEM 5

The mechanism for hydroxide-ion-promoted ester hydrolysis is shown in Section 15.12. What catalytic role does hydroxide ion play in this mechanism?

PROBLEM 6◆

The following reaction occurs by a mechanism involving general-base catalysis. Propose a mechanism for this reaction.

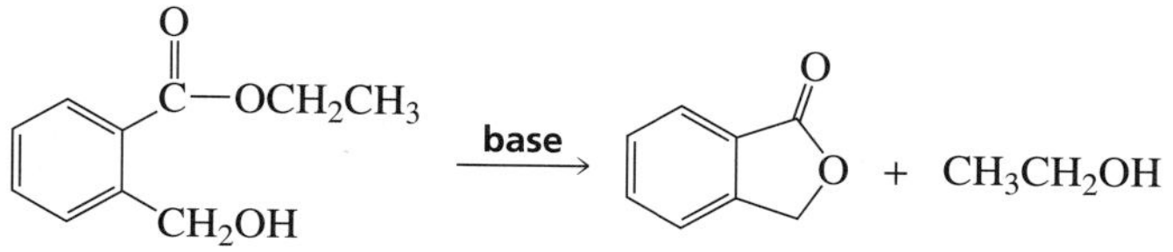

PROBLEM 7

Propose a mechanism for the general-base-catalyzed hydrolysis of an ester.

Metal ions exert their catalytic effect by coordinating (complexing) with atoms that have nonbonding electrons. In other words, metal ions are Lewis acids (Section 1.20). **Metal-ion catalysis** can increase the rate of a reaction in several ways.

- A metal ion can make a reaction center more susceptible to receiving electrons because it can stabilize a developing negative charge on a transition state, as in A in the diagram below.
- A metal ion can make a leaving group a weaker base and therefore a better leaving group, as in B.
- A metal ion can increase the rate of a hydrolysis reaction by forming a complex with water, thereby increasing water's acidity, as in C.

The pK_a of nonmetal-bound water is 15.7; the pK_a of metal-bound water depends on the metal atom (Table 21.1). When metal-bound water loses a proton, metal-bound hydroxide ion is formed. Metal-bound hydroxide ion, while not as good a nucleophile as hydroxide ion, is a better nucleophile than water. Thus, metal-ion complexation increases the reactivity of the attacking nucleophile.

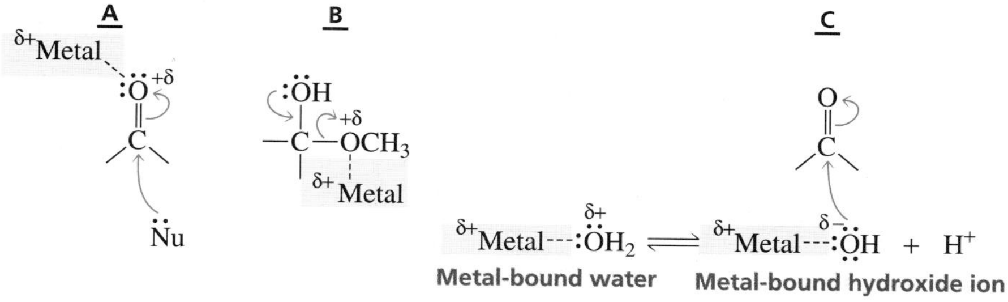

21.5 METAL-ION CATALYSIS

TABLE 21.1 The pK_a of Metal-Bound Water

M^{2+}	pK_a	M^{2+}	pK_a
Ca^{2+}	12.7	Co^{2+}	8.9
Mg^{2+}	11.8	Zn^{2+}	8.7
Cd^{2+}	11.6	Fe^{2+}	7.2
Mn^{2+}	10.6	Cu^{2+}	6.8
Ni^{2+}	9.4	Be^{2+}	5.7

In A and B, the metal ion exerts the same kind of catalytic effect as a proton. In a reaction where the metal ion has the same catalytic effect as a proton (increasing the electrophilicity of a reaction center or decreasing the basicity of a leaving group), the metal ion is often called an **electrophilic catalyst.**

Now we will look at some examples of metal-ion-catalyzed reactions. Co^{3+} catalyzes the condensation of two molecules of glycine ethyl ester to form glycylglycine ethyl ester. The actual catalyst is a cobalt complex, [Co(ethylenediamine)$_2$]$^{3+}$.

$$2\ H_2NCH_2\overset{O}{\overset{\|}{C}}OCH_2CH_3 \xrightarrow{Co^{3+}} H_2NCH_2\overset{O}{\overset{\|}{C}}NHCH_2\overset{O}{\overset{\|}{C}}OCH_2CH_3 + CH_3CH_2OH$$

Coordination of the metal ion with the carbonyl oxygen makes the carbonyl group more susceptible to nucleophilic attack because it stabilizes the developing negative charge on the oxygen in the transition state.

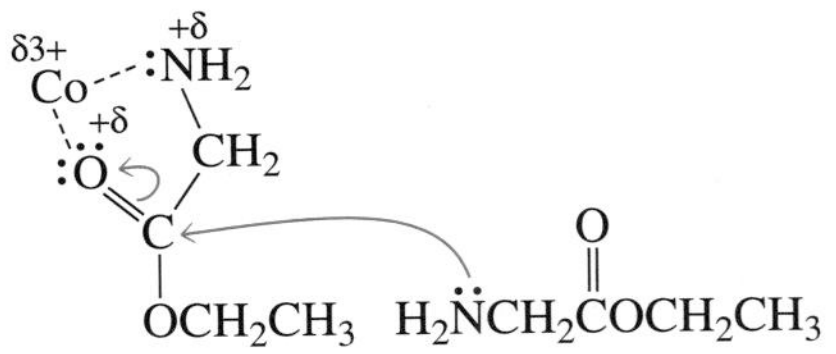

Either Cu^{2+} or Al^{3+} can catalyze the decarboxylation of dimethyloxaloacetate.

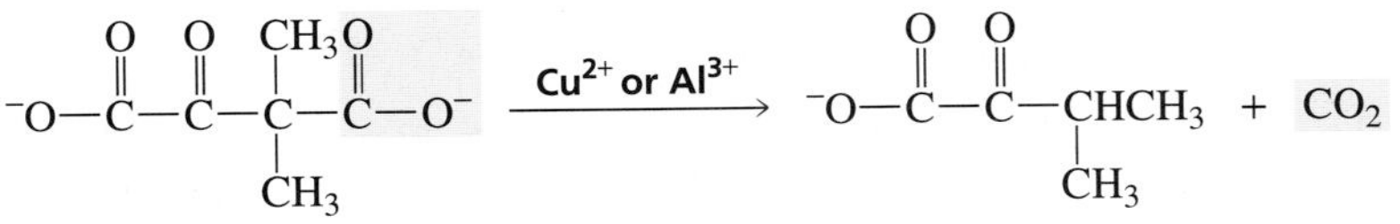

dimethyloxaloacetate

In this reaction, the metal ion complexes with two oxygen atoms of the reactant. Complexation increases the rate of decarboxylation by making the carbonyl group more susceptible to receiving the electrons left behind when CO_2 is eliminated.

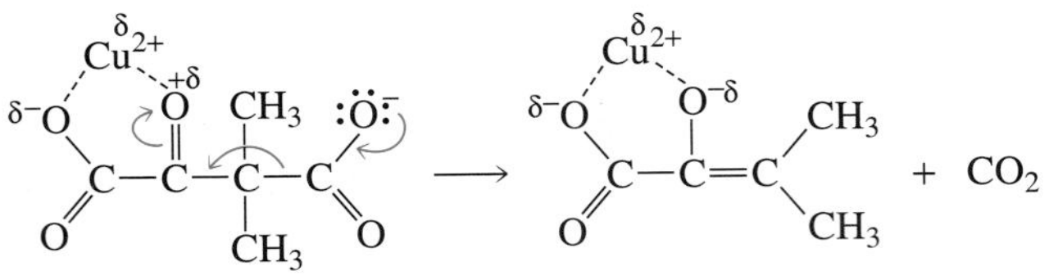

The Zn^{2+}-catalyzed hydrolysis of methyl trifluoroacetate has two slow steps. Zn^{2+} increases the rate of the first slow step by increasing the acidity of water, thereby providing metal-bound hydroxide ion, a better nucleophile than water. It increases the rate of the second slow step by decreasing the basicity of the group that is eliminated from the tetrahedral intermediate.

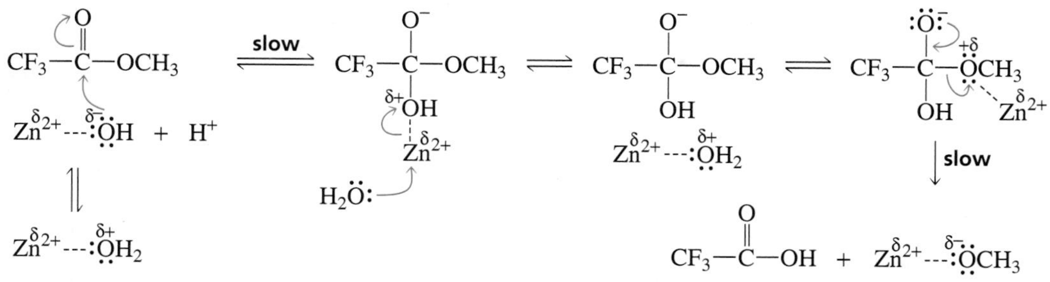

PROBLEM 8◆

The rate constant for the uncatalyzed reaction of two molecules of glycine ethyl ester to form glycylglycine ethyl ester is $0.6\ s^{-1}\ M^{-1}$. In the presence of $[Co(ethylenediamine)_2]^{3+}$, the rate constant is $1.5 \times 10^6\ s^{-1}\ M^{-1}$. What rate enhancement is provided by the catalyst?

PROBLEM 9

Although metal ions increase the rate of decarboxylation of dimethyloxaloacetate, they have no effect on the rate of decarboxylation of either the monoethyl ester of dimethyloxaloacetate or acetoacetate. Explain.

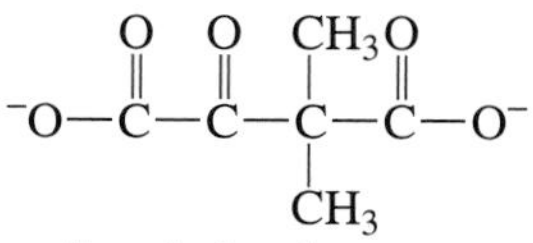

dimethyloxaloacetate

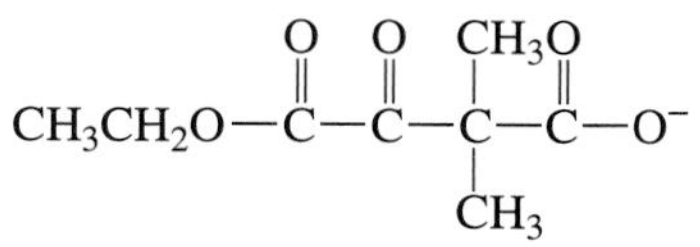

monoethyl ester of dimethyloxaloacetate

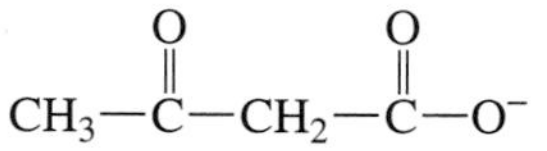

acetoacetate

PROBLEM 10

The hydrolysis of glycinamide is catalyzed by $[Co(ethylenediamine)_2]^{3+}$. Propose a mechanism for this reaction.

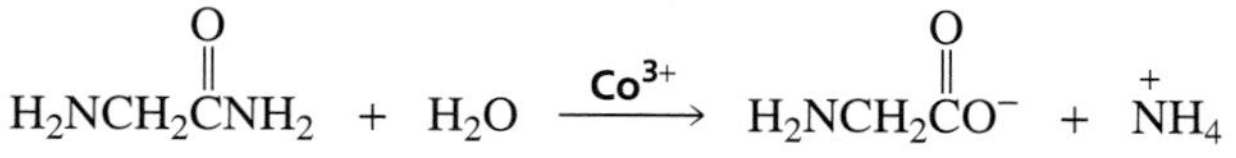

21.6 INTRAMOLECULAR REACTIONS

For a reaction to take place, the reacting groups must come together in solution with the proper orientation. The rate of a chemical reaction is determined by the number of collisions between two molecules with sufficient energy and with the proper orientation in a given time (Section 3.7).

$$\text{rate of a reaction} = \frac{\text{number of collisions}}{\text{per unit of time}} \times \frac{\text{fraction with}}{\text{sufficient energy}} \times \frac{\text{fraction with}}{\text{proper orientation}}$$

Because a catalyst decreases the energy barrier of a reaction, it increases the reaction rate by increasing the fraction of collisions that occur with sufficient energy to overcome the barrier. The rate of a reaction can also be increased by increasing the number of collisions and the fraction of those collisions that take place with the proper orientation. We have seen that an intramolecular reaction that results in the formation of a five- or a six-membered ring occurs more readily than the corresponding intermolecular reaction. This is because an intramolecular reaction has the advantage that the reacting groups are tied together in the same molecule, which gives them a better chance of finding each other than two individual molecules in a solution of the same concentration (Section 9.4). In other words, the number of collisions increases.

If, in addition to being in the same molecule, the reacting groups are juxtaposed in a way that increases the probability of their colliding with each other in the proper orientation, the rate of the reaction is further increased. The relative rates shown in Table 21.2 demonstrate the enormous increase that occurs in the rate of a reaction when the reacting groups are properly juxtaposed.

Rate constants for a series of reactions are generally compared using relative rates because relative rates allow one to see immediately how much faster one reaction is than another. **Relative rates** are obtained by dividing the rate constant for each of the reactions by the rate constant of the slowest reaction in the series. Because an intramolecular reaction is a first-order reaction (units of time^{-1}) and an intermolecular reaction is a second-order reaction (units of $\text{time}^{-1} \text{ M}^{-1}$), the relative rates in Table 21.2 have units of molarity (M) (Section 3.7).

$$\text{relative rate} = \frac{\text{first-order rate constant}}{\text{second-order rate constant}} = \frac{\text{time}^{-1}}{\text{time}^{-1}\text{M}^{-1}} = \text{M}$$

The relative rates shown in Table 21.2 are also called effective molarities. The **effective molarity** is the advantage given to the reaction by having the reacting groups in the same molecule. Effective molarity is the concentration of a reactant that would be required in an *intermolecular* reaction for it to have the same rate as that of the *intramolecular* reaction. In some cases, juxtaposing the reacting groups provides such an enormous increase in rate that the effective molarity is greater than the concentration of the reactant in its solid state.

The first reaction shown in Table 21.2 (reaction A) is an intermolecular reaction between an ester and a carboxylate ion. The second reaction (reaction B) has the same reacting groups in a single molecule. The rate of the intramolecular reaction is 1000 times faster than the rate of the intermolecular reaction.

The reactant in reaction B has four carbon–carbon bonds that are free to rotate, while the reactant in reaction D has only three such bonds. Conformers in which the large groups are rotated away from each other are more stable. However, when these groups are pointed away from each other, they are in an unfavorable conformation for reaction. Because the reactant in reaction D has fewer bonds that are free to rotate, reaction D is faster than reaction B. The relative rate constants for the reactions shown in Table 21.2 are quantitatively related to the calculated probability of forming the conformation where the carboxylate ion is in position to attack the carbonyl carbon.

TABLE 21.2   Relative Rates of an Intermolecular Reaction and Five Intramolecular Reactions

Reaction	Relative rate
A.	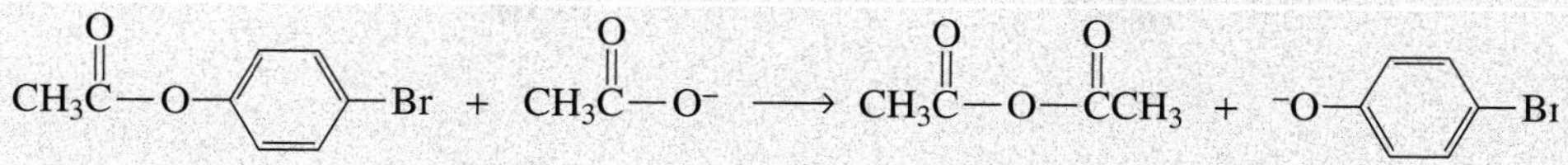 1.0
B.	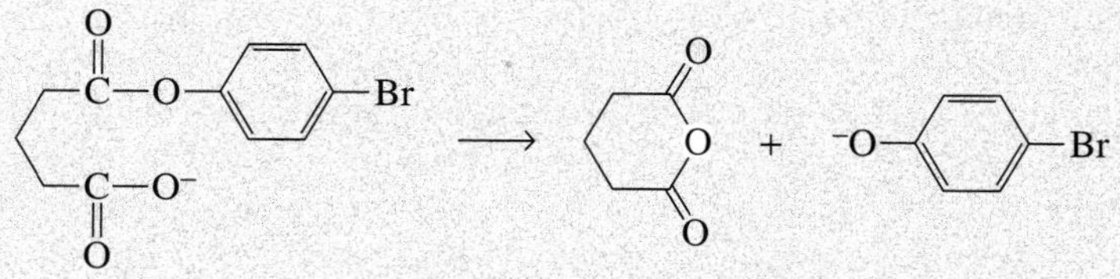 1×10^3 M
C.	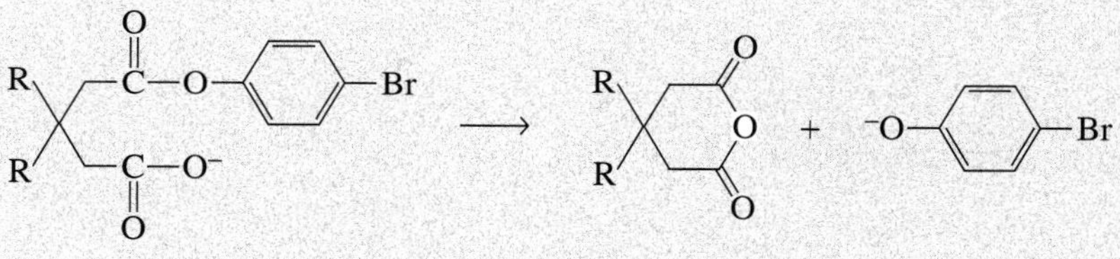 2.3×10^4 M $R = CH_3$ 1.3×10^6 M $R = i\text{-}C_3H_7$
D.	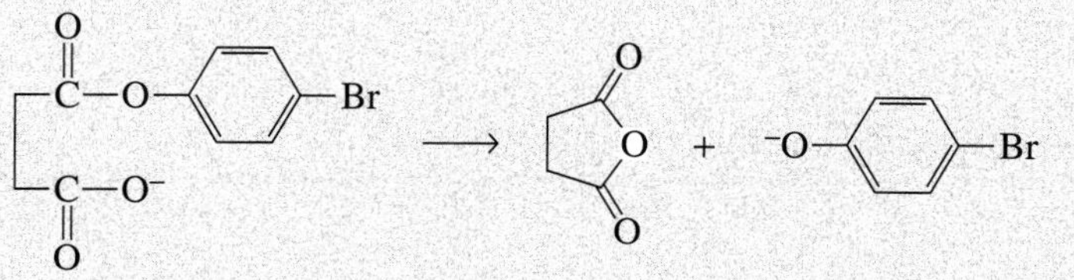 2.2×10^5 M
E.	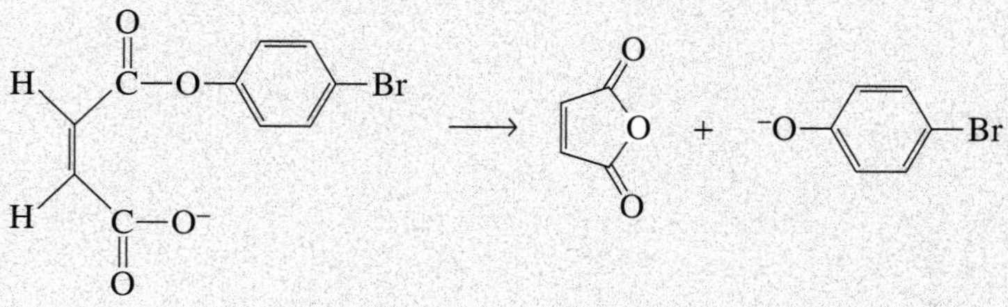 1×10^7 M
F.	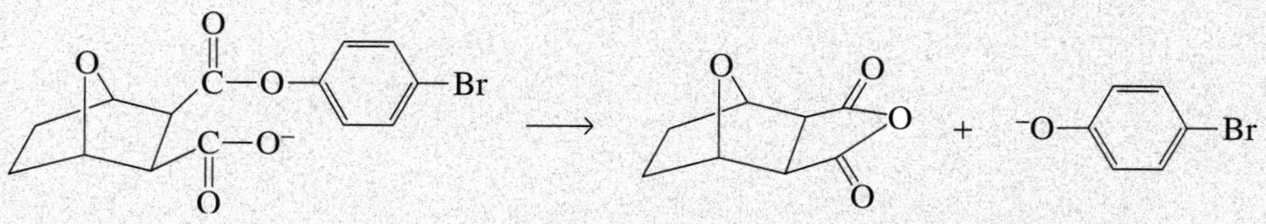 5×10^7 M

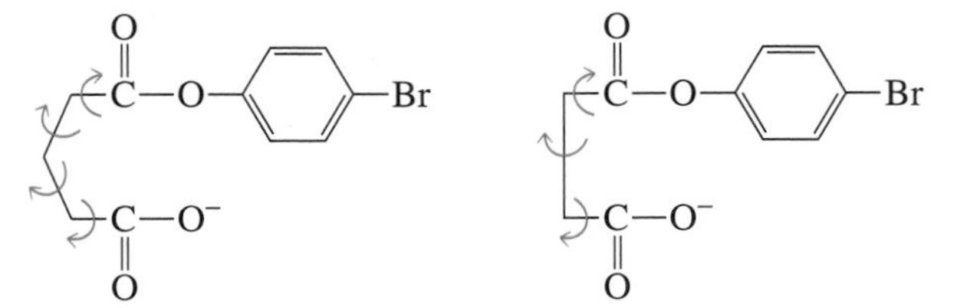

four carbon–carbon bonds   three carbon–carbon bonds
can rotate   can rotate

Reaction C is faster than reaction B because the alkyl substituents of the reactant in C decrease the available volume for rotation of the reactive groups away from each other. Thus, there is a greater probability that the molecule will be in a conformation with the reacting groups positioned for ring closure. This is called the *gem-dialkyl effect,* because the two alkyl substituents are bonded to the same (geminal) carbon. If we compare the rate when the substituents are methyl groups with the rate when the substituents are isopropyl groups, we see that the rate is further increased when the size of the alkyl groups is increased.

The increased rate of reaction E is a result of the double bond that prevents the reacting groups from rotating away from each other. The bicyclic compound in reaction F reacts at an even greater rate, because the reacting groups are frozen in the proper conformation for reaction.

PROBLEM 11◆

The relative rate of reaction of the *cis*-alkene is given in Table 21.2. What would you expect the relative rate of the trans isomer to be?

21.7 INTRAMOLECULAR CATALYSIS

Just as putting two reacting groups in the same molecule increases the rate of a reaction compared with having the groups in separate molecules, putting a reacting group and a catalyst in the same molecule increases the rate of a reaction compared with having them in separate molecules. When a catalyst is part of the reacting molecule, the catalysis that takes place is called **intramolecular catalysis.** Intramolecular nucleophilic catalysis, intramolecular general-acid or general-base catalysis, and intramolecular metal-ion catalysis all occur. Intramolecular catalysis is also known as **anchimeric assistance** (*anchi-meric* means "adjacent parts" in Greek).

Now we will look at some examples of intramolecular catalysis. The intramolecular general-base-catalyzed enolization reaction shown below is much faster than the analogous intermolecular general-base-catalyzed reaction.

intramolecular general-base catalysis

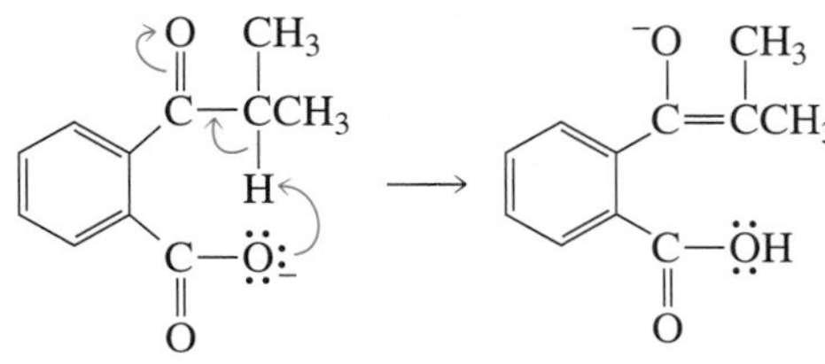

intermolecular general-base catalysis

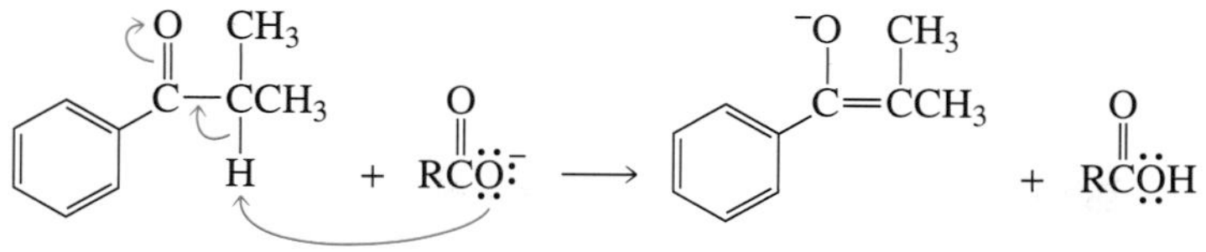

When chlorocyclohexane is added to an aqueous solution of ethanol, an alcohol and an ether are formed. Two products are formed because there are two nucleophiles (H_2O and CH_3CH_2OH) in the solution.

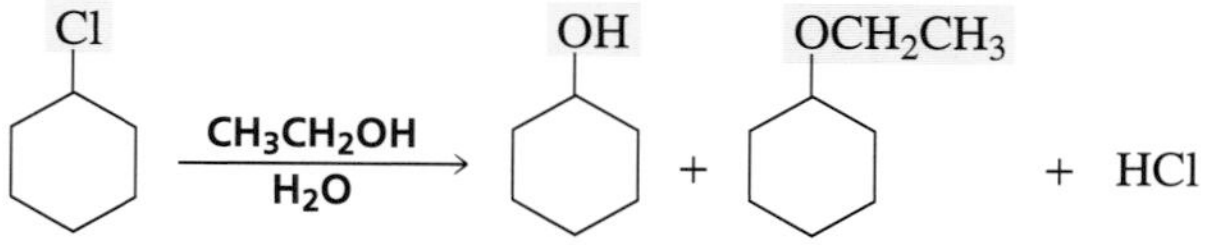

A 2-thio-substituted chlorocyclohexane undergoes the same reaction. However, the rate of the reaction depends on whether the substituent is cis or trans to the chloro substituent. If it is trans, the 2-thio-substituted compound reacts about 70,000 times faster than the unsubstituted compound. But if it is cis, the 2-thio-substituted compound reacts a little more slowly than the unsubstituted compound.

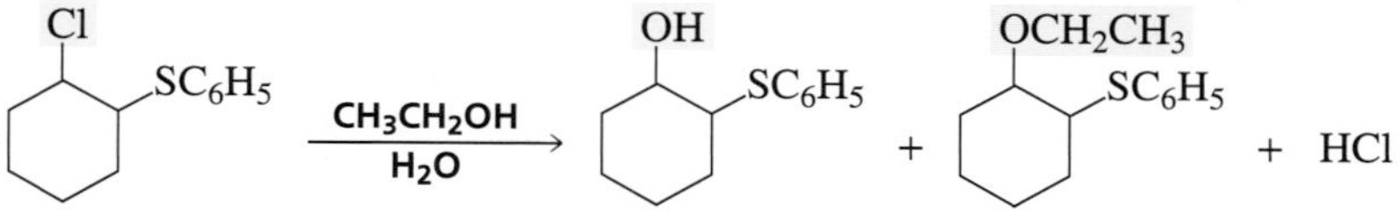

What accounts for the much faster reaction of the *trans*-substituted compound? In this reaction, the thio substituent is an intramolecular nucleophilic catalyst. It displaces the chloro substituent by attacking the back side of the carbon to which the chloro substituent is attached. Back side attack requires both substituents to be in axial positions, which means they must be trans to each other (Section 4.16). Subsequent attack by water or ethanol on the sulfonium ion is rapid, because the positively charged sulfur is an excellent leaving group and the strain in the three-membered ring is released.

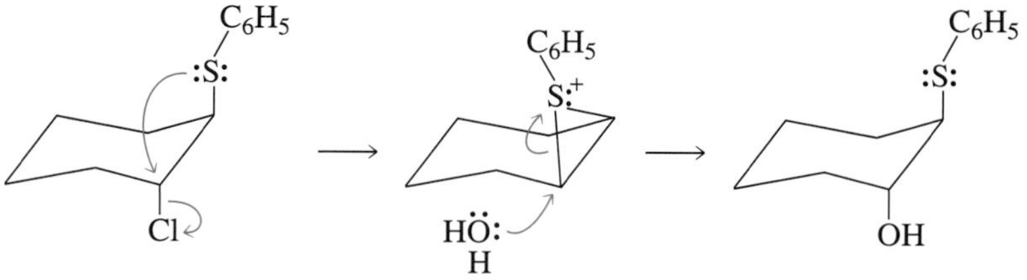

PROBLEM 12◆

a. Why does *cis*-2-thio-substituted chlorocyclohexane react more slowly in aqueous ethanol than does chlorocyclohexane?

b. What does the slower rate of the cis-substituted compound compared with the unsubstituted compound tell you about the mechanism of the reaction of chlorocyclohexane in aqueous ethanol?

The rate of hydrolysis of phenyl acetate is increased about 150-fold at neutral pH by the presence of a carboxylate ion in the ortho position. The *ortho*-carboxyl-substituted ester is commonly known as aspirin.

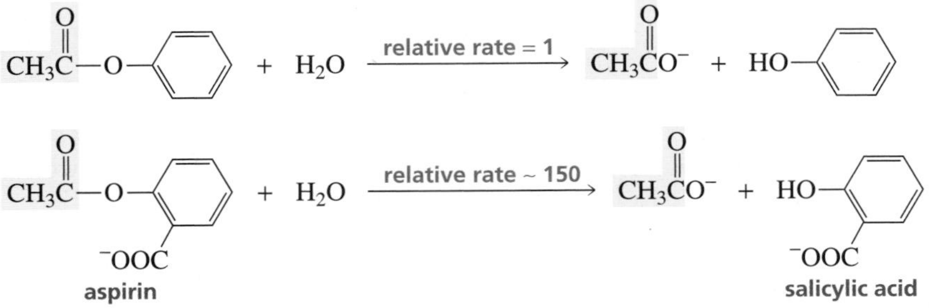

The *ortho*-carboxyl substituent is an intramolecular general-base catalyst. It increases the nucleophilicity of water, thereby increasing the rate of formation of the tetrahedral intermediate.

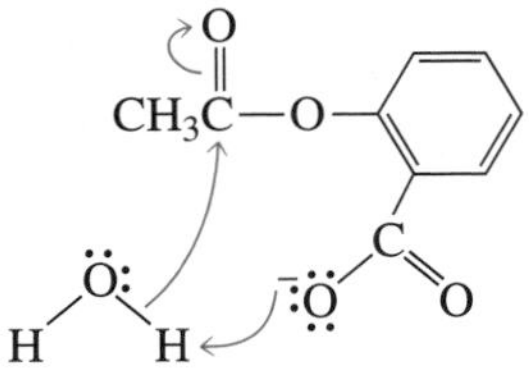

Electron-withdrawing nitro substituents on the benzene ring decrease the basicity of the leaving group, thereby increasing its leaving tendency. Making the group easier to eliminate changes the type of catalysis exerted by the *ortho*-carboxyl substituent. In the case of the dinitro-substituted ester, the *ortho*-carboxyl substituent acts as an intramolecular nucleophilic catalyst. It converts the ester into an anhydride, and an anhydride is more rapidly hydrolyzed than an ester.

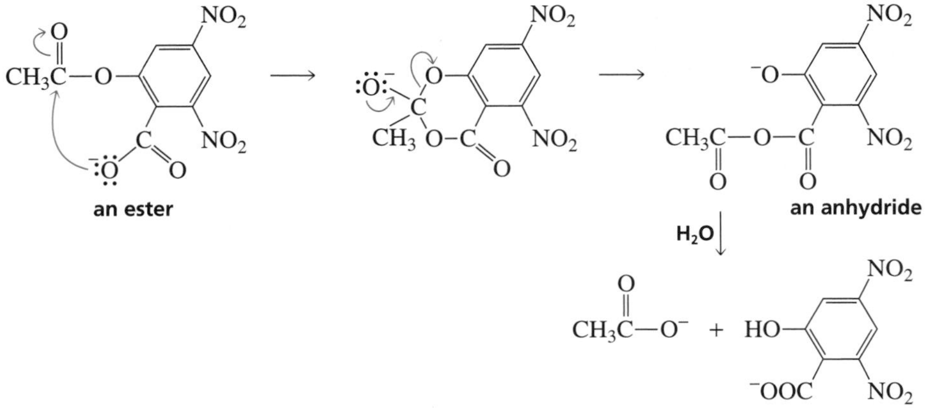

PROBLEM 13 / SOLVED

Explain why the mode of catalysis for the hydrolysis of an *ortho*-carboxyl-substituted phenyl acetate changes from general base to nucleophilic as the leaving tendency of the phenyl group is increased.

SOLUTION The *ortho*-carboxyl substituent is in position to form a tetrahedral intermediate. If the phenoxy group in the tetrahedral intermediate is a poor leaving group, the carboxyl group will be eliminated from the intermediate, reforming the starting material, which will be hydrolyzed by a general-base-catalyzed mechanism (path A). However, if the phenoxy group is a good leaving group, it will be eliminated, forming the anhydride, and the reaction will have occurred by a nucleophilic mechanism (path B).

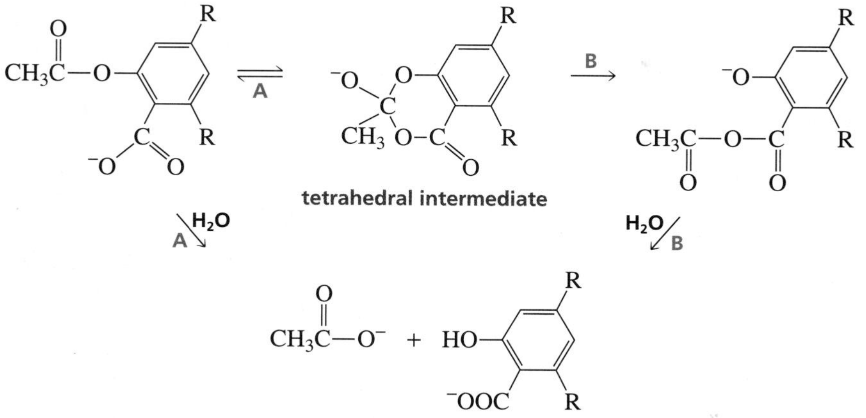

tetrahedral intermediate

PROBLEM 14

Whether the *ortho*-carboxyl substituent acts as an intramolecular general-base catalyst or as an intramolecular nucleophilic catalyst can be determined by carrying out the hydrolysis of aspirin with ^{18}O-labeled water and determining whether ^{18}O is incorporated into salicylic acid. Explain the results that would be obtained with the two types of catalysis.

The following reaction is an example of intramolecular metal-ion catalysis. Hydrolysis of the ester is catalyzed by Ni^{2+} or Co^{2+}.

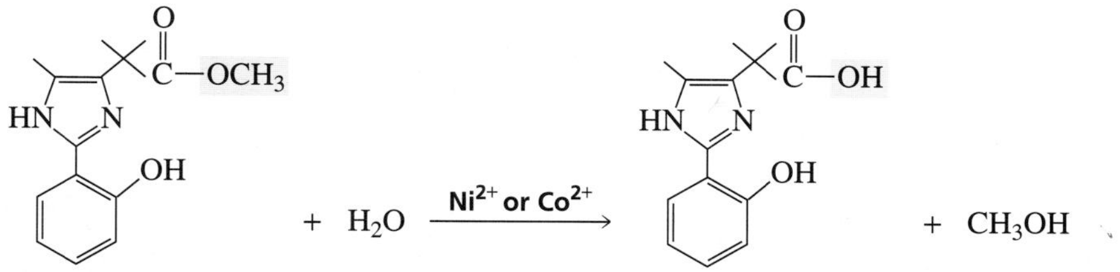

The metal ion complexes with an oxygen and a nitrogen of the reactant as well as with a molecule of water. The metal ion increases the rate of the reaction by positioning the water molecule and making it a stronger acid, which increases its nucleophilicity.

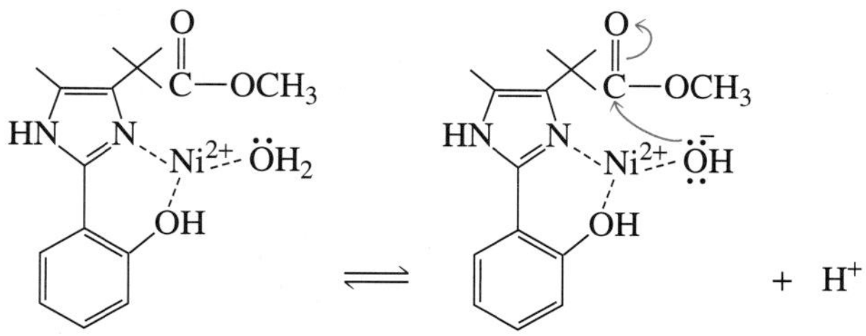

21.8 CATALYSIS IN BIOLOGICAL REACTIONS

Essentially all organic reactions that occur in biological systems require a catalyst. Most biological catalysts are **enzymes.** Enzymes are globular proteins. Each biological reaction is catalyzed by a different enzyme. Enzymes are extraordinarily good catalysts; they can increase the rate of an intermolecular reaction by as much as 10^{16}-fold. This can be compared with rate enhancements achieved by nonbiological catalysts in intermolecular reactions, which are seldom greater than 10^4-fold.

The reactant of an enzyme-catalyzed reaction is called a **substrate.** The enzyme has a pocket or cleft known as an active site. The substrate specifically fits and binds to the **active site,** and all the bond-making and bond-breaking steps of the reaction occur while the substrate is in this site. Enzymes differ from nonbiological catalysts in that they are specific for the reactant whose reaction they catalyze (Section 4.20). Not all enzymes have the same degree of specificity. Some are specific for a single compound and will not tolerate even the slightest variation in structure, while some catalyze the reaction of an entire family of compounds with related structures. The specificity of an enzyme for its substrate is an example of the phenomenon known as **molecular recognition**—one molecule "recognizes" another.

The specificity of an enzyme results from the conformation of the protein and the particular amino acid side chains that make up the active site. For example, a negatively charged side chain of an amino acid residue at the active site of an enzyme can associate with a positively charged group on the substrate; a hydrogen-bond donor on the enzyme can associate with a hydrogen-bond acceptor on the substrate; and hydrophobic groups associate with other hydrophobic groups. The specificity of an enzyme for its substrate is described by the lock-and-key model. In the **lock-and-key model,** the substrate is said to fit the enzyme much as a key fits a lock.

The energy released as a result of binding the substrate to the enzyme can be used to induce a change in the shape of the enzyme, resulting in more precise binding between the substrate and the active site. This change in conformation of the enzyme is known as induced fit. In the **induced-fit model,** the shape of the active site does not become completely complementary to the shape of the substrate until the enzyme has bound the substrate.

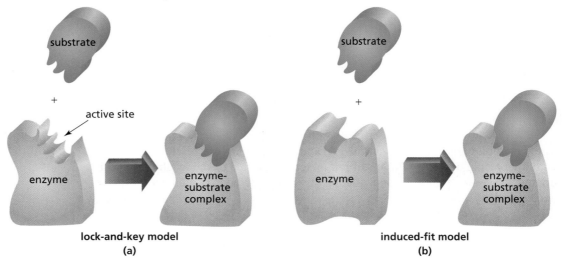

lock-and-key model
(a)

induced-fit model
(b)

An example of induced fit is shown in Figure 21.4. The three-dimensional structure of the enzyme hexokinase is shown before and after binding glucose, its substrate. Notice the change in conformation that occurs upon binding the substrate. Some of the most important factors accounting for the remarkable catalytic ability of enzymes are the following:

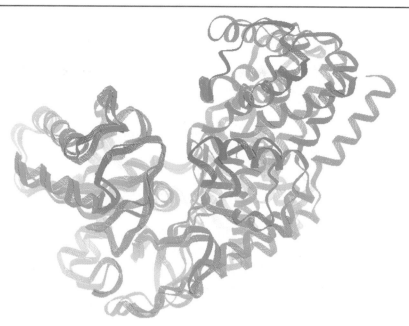

▲ **Figure 21.4**
The structure of hexokinase before binding its substrate is shown in red; the structure of hexokinase after binding its substrate is shown in green.

- All reacting groups are brought together at the active site in the proper position for reaction. This is analogous to the effect that proper positioning of reacting groups has on the rate of intramolecular reactions (Section 21.6).
- Some of the amino acid side chains serve as catalytic groups: these side chains are in the proper position relative to the substrate to act as catalysts. This is analogous to the rate enhancements observed for intramolecular catalysis (Section 21.7).
- The conformational change that an enzyme undergoes after binding a substrate can introduce strain into the substrate, making it more reactive.
- Groups on the enzyme can stabilize an intermediate and, therefore, the transition state leading to the intermediate, by van der Waals and electrostatic interactions and by hydrogen bonding.

As we look at some examples of enzyme-catalyzed reactions, it is important to note that the functional groups on the enzyme side chains are the same functional groups we are used to seeing in simple organic compounds, and the modes of catalysis used by enzymes are exactly the same as the modes of catalysis used in organic reactions. One factor contributing to the remarkable catalytic ability of enzymes is their ability to use several modes of catalysis in the same reaction. Factors other than those listed above can contribute to the increased rate of enzyme-catalyzed reactions, but not all factors are employed by every enzyme. We will consider some of these factors when we discuss individual enzymes. We will now look at the mechanisms for four enzyme-catalyzed reactions.

Mechanism for Carboxypeptidase A

Carboxypeptidase A catalyzes the hydrolysis of the C-terminal peptide bond in peptides and proteins, releasing the C-terminal amino acid (Section 20.13).

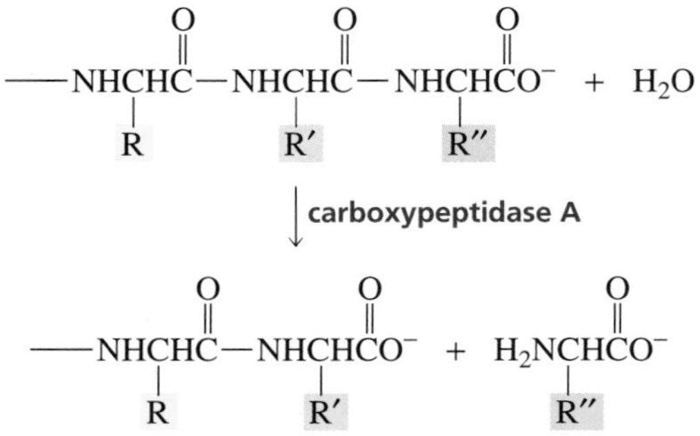

Carboxypeptidase A is a metalloenzyme. A *metalloenzyme* is an enzyme that contains a tightly bound metal ion. The metal ion in carboxypeptidase A is Zn^{2+}. In bovine pancreatic carboxypeptidase A, Zn^{2+} is bound to the enzyme at its active site, by forming a complex with Glu 72, His 196, and His 69 as well as with a water molecule (Figure 21.5). (The source of the enzyme is specified because, although carboxypeptidase A's from different sources follow the same mechanism, they have slightly different primary structures.)

Several groups at the active site of carboxypeptidase A participate in binding the substrate in the optimum position for reaction (Figure 21.5). Arg 145 forms two hydrogen bonds with the C-terminal carboxyl group of the substrate; Tyr 248 is also involved in hydrogen bonding to the carboxyl group. The side chain of the C-terminal amino acid is positioned in a hydrophobic pocket, which is why carboxypeptidase A is not active if the C-terminal amino acid is arginine or lysine. Apparently, the long, positively charged side chains of these amino acid residues cannot fit into the nonpolar pocket.

- When the substrate binds to the active site, Zn^{2+} partially complexes with the oxygen of the carbonyl group that will be hydrolyzed (Figure 21.5). Zn^{2+} polarizes the carbon–oxygen double bond, making the carbonyl carbon more susceptible to nucleophilic attack and stabilizing the negative charge that develops on the oxygen atom in the transition state. Arg 127 also participates in increasing the carbonyl group's electrophilicity and in stabilizing the developing negative charge on the transition state. Zn^{2+} also complexes with water, which increases its acidity, thereby making the nucleophile more like hydroxide ion. Glu 270 functions as a general-base catalyst, further increasing water's nucleophilicity.
- In the second step of the catalytic reaction, Glu 270 functions as a general-acid catalyst, increasing the leaving tendency of the amino group. When the reaction is over, the amino acid and the peptide with one less amino acid residue dissociate from the enzyme and another molecule of substrate binds to the active site. It has been suggested that the unfavorable electrostatic interaction between the negatively charged carboxyl group of the peptide product and the negatively charged Glu 270 residue facilitates release of the product from the enzyme.

PROBLEM 15◆

Which of the following C-terminal peptide bonds would be more readily cleaved by carboxypeptidase A? Explain.

—Ser-Ala-Phe or —Ser-Ala-Asp

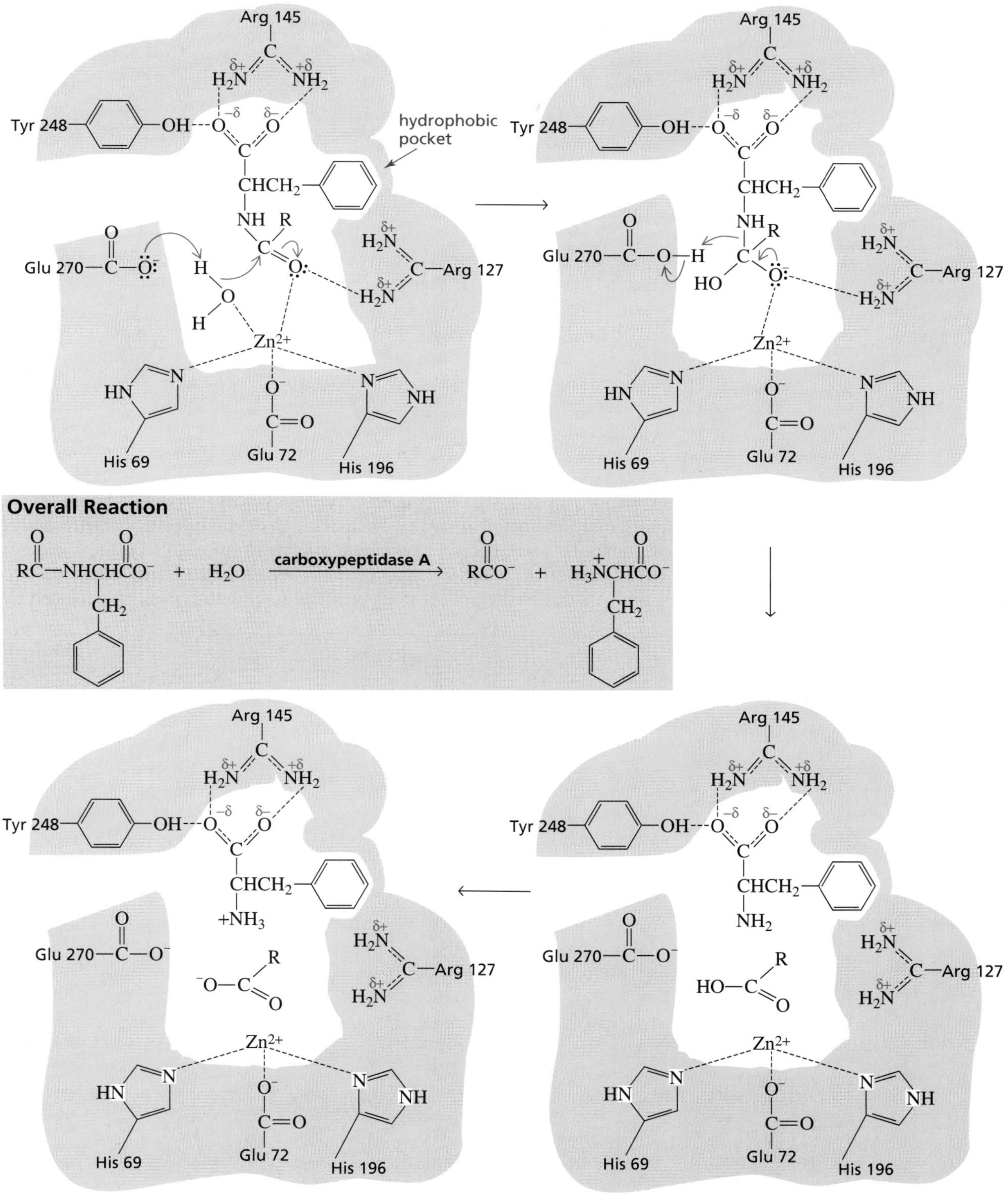

▲ **Figure 21.5**
Proposed mechanism for carboxypeptidase A–catalyzed hydrolysis of a peptide bond.

PROBLEM 16

Carboxypeptidase A has esterase activity as well as peptidase activity. In other words, it can hydrolyze ester bonds as well as peptide bonds. When it hydrolyzes ester bonds, Glu 270 acts as a nucleophilic catalyst instead of a general-base catalyst. Propose a mechanism for carboxypeptidase A–catalyzed hydrolysis of an ester bond.

Mechanism for the Serine Proteases

Trypsin, chymotrypsin, and elastase are members of a large group of endopeptidases known collectively as serine proteases. They are called *proteases* because they catalyze the hydrolysis of protein peptide bonds. They are called *serine proteases* because they all have a serine residue at the active site that participates in the catalysis.

The various serine proteases have similar primary structures, suggesting that they are evolutionarily related. They all have the same three catalytic residues at the active site: an aspartate, a histidine, and a serine. But they have one important difference—the nature of the pocket at the active site that binds the side chain of the amino acid residue undergoing hydrolysis (Figure 21.6). This is what gives the serine proteases their different specificities. The pocket in trypsin is narrow, with a negatively charged aspartic acid and a serine at the bottom of the pocket. The shape and charge of the binding pocket cause it to bind long, positively charged amino acid side chains (Lys and Arg). This is why trypsin hydrolyzes peptide bonds on the C-side of arginine and lysine residues. The pocket in chymotrypsin is narrow and is lined with nonpolar amino acids, causing chymotrypsin to cleave on the C-side of amino acids with flat, nonpolar side chains (Section 20.13). In elastase, two glycines on the sides of the pocket in trypsin and in chymotrypsin are replaced by

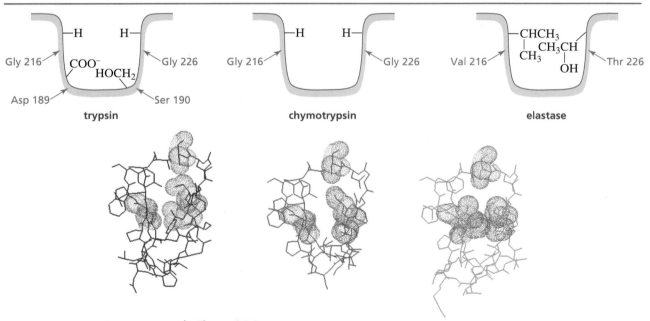

▲ **Figure 21.6**
The binding pockets in trypsin, chymotrypsin, and elastase. The negatively charged aspartic acid is shown in red, the relatively nonpolar amino acids are shown in green. The structures of the binding pockets explain why trypsin binds long positively charged amino acids, chymotrypsin binds flat nonpolar amino acids, and elastase binds only small amino acids.

relatively bulky valine and threonine residues. Consequently, only small amino acids can fit into the pocket. Elastase therefore hydrolyzes peptide bonds on the C-side of small amino acids (glycine and alanine).

The mechanism for bovine chymotrypsin–catalyzed hydrolysis of a peptide bond is shown in Figure 21.7. The other serine proteases follow the same mechanism.

- As a result of binding the flat, nonpolar side chain in the pocket, the amide linkage that is to be hydrolyzed is positioned very close to Ser 195. His 57 functions as a general-base catalyst, increasing the nucleophilicity of serine, which attacks the carbonyl group. This process is helped by Asp 102, which uses its negative charge to stabilize the resulting positive charge on His 57. Formation of the tetrahedral intermediate causes a slight change in the shape of the protein that allows the negatively charged oxygen to slip into a previously unoccupied area of the active site known as the *oxyanion hole*. Once in the oxyanion hole, the negatively charged oxygen can form hydrogen bonds with two peptide groups (Gly 193 and Ser 195), and in this way the tetrahedral intermediate is stabilized.
- In the next step, the tetrahedral intermediate collapses, eliminating the amino group. This is a strongly basic group that cannot be expelled without the participation of His 57, which acts as a general acid catalyst. The product of the second step is an **acyl-enzyme intermediate** because the serine group of the enzyme has been acylated.
- The third step is just like the first step except that water instead of serine is the nucleophile. Water attacks the acylated group, with His 57 functioning as a general-base catalyst to increase the nucleophilicity of water, and Asp 102 stabilizing the positively charged histidine residue.
- In the final step of the reaction, the tetrahedral intermediate collapses, eliminating serine. His 57 functions as a general-acid catalyst in this step, increasing serine's leaving tendency.

The mechanism for chymotrypsin-catalyzed hydrolysis shows the importance of histidine as a catalytic group. Because the pK_a of the imidazole ring of histidine ($pK_a = 6.0$) is close to neutrality, histidine can act both as a general-acid catalyst and as a general-base catalyst (see also Figure 21.11).

Much information about the relationship between the structure of a protein and its function has been determined by **site-specific mutagenesis**—a technique that replaces one amino acid of a protein with another. For example, when Asp 102 of chymotrypsin is replaced with Asn 102, the enzyme's ability to bind the substrate is unchanged, but its ability to catalyze the reaction decreases to less than 0.05% of its value with the native enzyme. Clearly, Asp 102 is involved in the catalytic process; its negative charge stabilizes histidine's positive charge.

side chain of an aspartate (Asp) residue   side chain of an asparagine (Asn) residue

PROBLEM 17◆

Arginine and lysine side chains fit into trypsin's binding pocket. One of these side chains forms a direct hydrogen bond with serine and an indirect (mediated through a water molecule) hydrogen bond with aspartate. The other side chain forms direct hydrogen bonds with both serine and aspartate. Which is which?

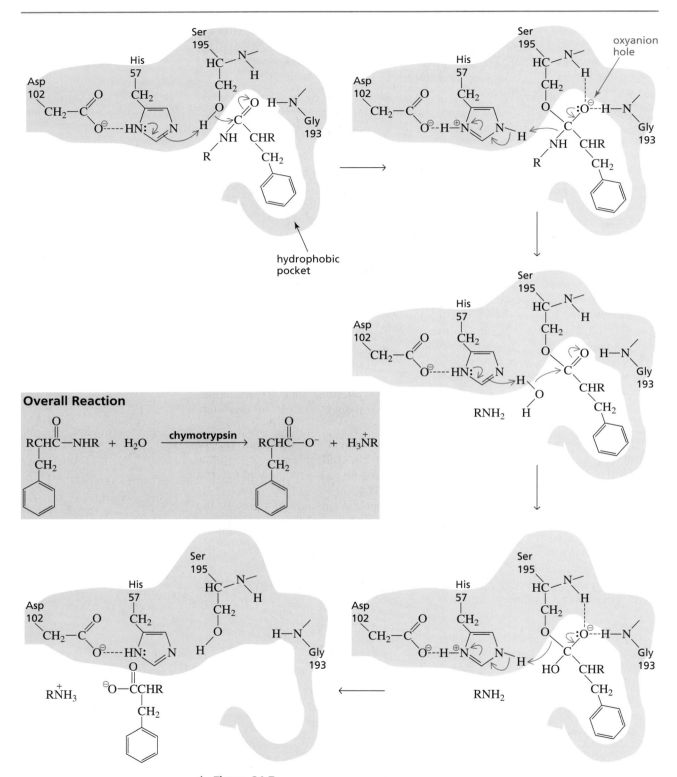

▲ **Figure 21.7**
Proposed mechanism for chymotrypsin-catalyzed hydrolysis of a peptide bond.

PROBLEM 18

Serine proteases do not catalyze hydrolysis if the amino acid at the hydrolysis site is a D-amino acid. For example, trypsin cleaves on the C-side of L-Arg and L-Lys but not on the C-side of D-Arg and D-Lys. Explain.

Mechanism for Lysozyme

Lysozyme is an enzyme that destroys bacterial cell walls. These cell walls are composed of alternating NAM (N-acetylmuramic acid) and NAG (N-acetylglucosamine) units. Lysozyme destroys the cell wall by catalyzing the hydrolysis of the NAM–NAG bond.

The active site of hen egg-white lysozyme binds six sugar residues of the substrate. The many amino acid residues involved in binding the substrate in the correct position in the active site are shown in Figure 21.8. The six sugar residues are labeled A, B, C, D, E, and F. The carboxylic acid substituent of NAM cannot fit into the binding site for C or E. This means that NAM units must be in the sites for B, D, and F. Hydrolysis occurs between D and E.

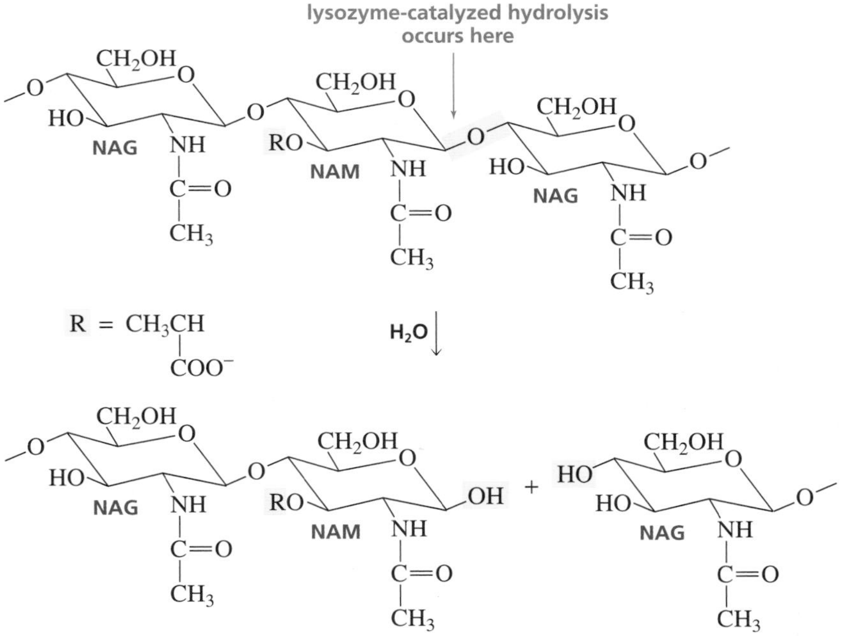

- Lysozyme has two catalytic groups at the active site, Glu 35 and Asp 52 (Figure 21.9). Glu 35 acts as a general-acid catalyst, protonating the leaving group and thereby making it a weaker base and a better leaving group. The developing positive charge on the oxygen atom is stabilized by Asp 52. The stabilization of a charge by an opposite charge is called **electrostatic catalysis.**

- In the second step of the reaction, water attacks the oxonium ion intermediate. The mechanism results in retention of configuration at C-1 of the NAM residue, because the enzyme shields the other side of the oxonium ion intermediate from attack by water.

After lysozyme binds its substrate, the enzyme undergoes a change in shape that distorts the D residue of the substrate from a chair conformation to a half-chair

Figure 21.8 ▶ The amino acids at the active site of lysozyme that are involved in binding the substrate. (Figure copyright by Irving Geis.)

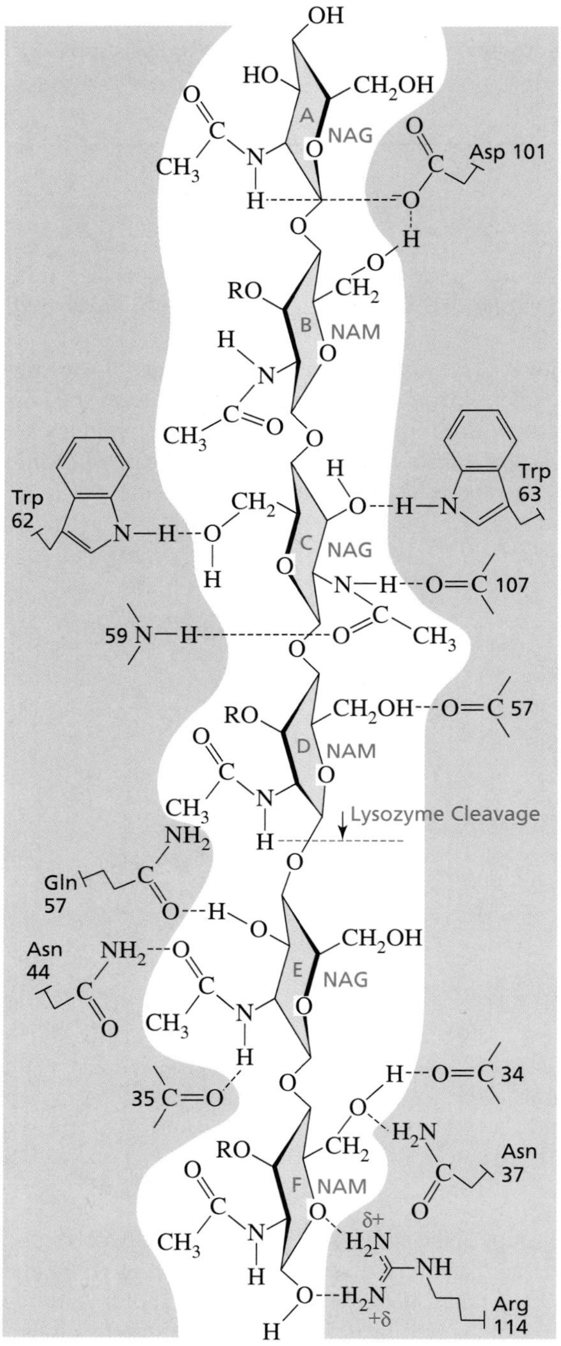

conformation (Section 2.13). This destabilizes the reactant *and* stabilizes the transition state for the first step of the hydrolysis reaction, because the oxonium ion that is formed as the product of this step is in a half-chair conformation (Figure 21.1a and b).

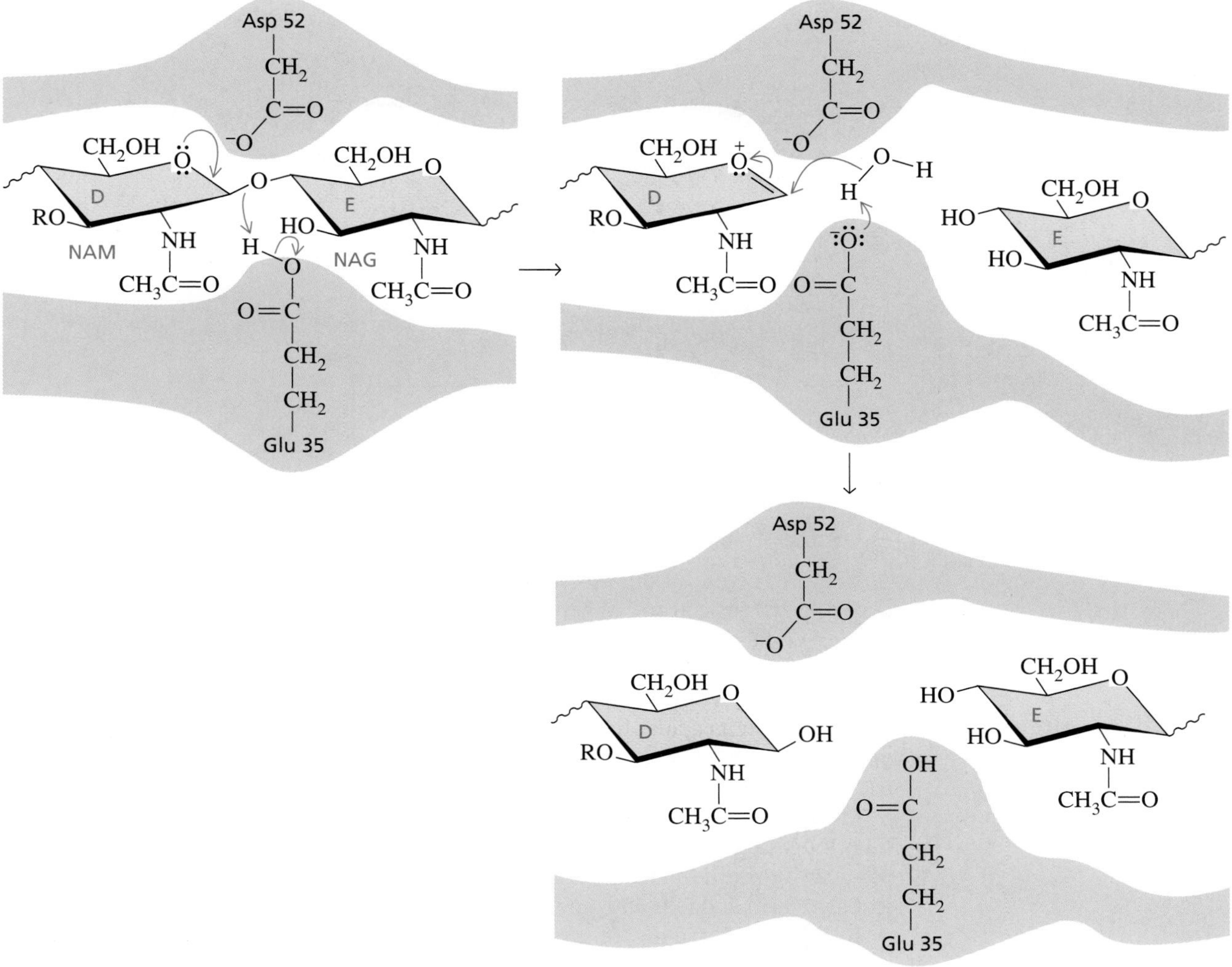

▲ **Figure 21.9**
Proposed mechanism for lysozyme-catalyzed hydrolysis of a cell wall.

PROBLEM 19◆

If H_2O^{18} were used to hydrolyze lysozyme, which ring would contain the label, NAM or NAG?

A plot of the activity of an enzyme as a function of the pH of the reaction mixture is called a **pH-activity profile** or a **pH-rate profile** (Section 16.7). The pH-activity profile for lysozyme is shown in Figure 21.10. It is a bell-shaped curve with the maximum rate occurring at about pH 5.3. The pH at which the enzyme is 50% active is 3.8 on the ascending leg of the profile and 6.7 on the descending leg. These pH values correspond to the pK_a values of the enzyme's catalytic groups. (This is true for all bell-shaped pH-rate profiles provided the pK_a values are at least 2 pK_a units apart. If the difference between them is less than 2 pK_a units, the precise pK_a values of the catalytic groups must be determined in other ways.)

Figure 21.10 ▶
Dependence of lysozyme
activity on the pH of the reaction mixture.

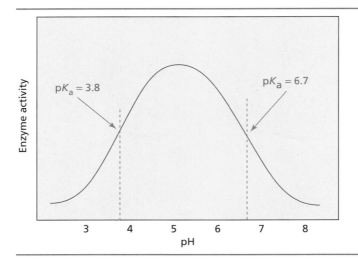

The pK_a given by the ascending leg is the pK_a of a group that is catalytically active in its basic form. When that group is fully protonated, the enzyme is not active. As the pH of the reaction mixture increases, a larger fraction of the group is present in its basic form and, as a result, the enzyme shows increasing activity. Similarly, the pK_a given by the descending leg is the pK_a of a group that is catalytically active in its acidic form. There is maximum catalytic activity when the group is fully protonated, and the activity decreases with increasing pH because a greater fraction of the group lacks a proton.

From the lysozyme mechanism shown in Figure 21.9, we can conclude that Asp 52 is the group with a pK_a of 3.8 and that Glu 35 is the group with a pK_a of 6.7. The pH-activity profile indicates that lysozyme is maximally active when Asp 52 is in its basic form and Glu 35 is in its acidic form.

In Table 20.2, the pK_a of aspartic acid is listed as 3.86, and that of glutamic acid as 4.25. The pK_a of Asp 52 agrees with the pK_a of aspartic acid, but the pK_a of Glu 35 is much greater than the pK_a of glutamic acid. Why is the pK_a of the glutamic acid residue at the active site of the enzyme so much greater than the pK_a given for glutamic acid in Table 20.2? The pK_a values in Table 20.2 were determined in water. In the enzyme, Asp 52 is surrounded by polar groups, which means that its pK_a should be close to the pK_a determined in water, a polar solvent. Glu 35, however, is in a predominantly nonpolar microenvironment. The pK_a of a carboxylic acid is greater in a nonpolar solvent because, in nonpolar solvents, there is less of a tendency to form charged species (Section 9.10).

Part of the catalytic efficiency of lysozyme results from its ability to provide different solvent environments at the active site. This allows one catalytic group to exist in its acidic form at the same surrounding pH at which a second catalytic group exists in its basic form. This property is unique to enzymes. Chemists cannot provide different solvent environments for different parts of nonenzymatic systems.

PROBLEM 20

When cut apples are exposed to oxygen, an enzyme-catalyzed reaction causes them to turn brown. They can be prevented from turning brown by coating them with lemon juice (pH ~ 3.5). Explain.

Mechanism for Aldolase

Glycolysis is the name given to the series of reactions responsible for converting D-glucose into two molecules of pyruvate (Section 18.20 and Appendix VI). Because D-glucose contains six carbons and pyruvate contains three carbons, at some point in the reaction pathway a six-carbon compound must be cleaved into two three-carbon-containing compounds. The enzyme *aldolase* catalyzes this cleavage (Figure 21.11). Aldolase converts D-fructose-1,6-diphosphate into D-glyceraldehyde-3-phosphate and dihydroxyacetone phosphate. The enzyme is called aldolase because the reverse reaction is an aldol addition reaction.

$$
\begin{array}{c}
\text{CH}_2\text{OPO}_3^{2-} \\
| \\
\text{C}=\text{O} \\
| \\
\text{HO}-\text{C}-\text{H} \\
| \\
\text{H}-\text{C}-\text{OH} \\
| \\
\text{H}-\text{C}-\text{OH} \\
| \\
\text{CH}_2\text{OPO}_3^{2-}
\end{array}
\quad \xrightleftharpoons{\text{aldolase}} \quad
\begin{array}{c}
\text{HC}=\text{O} \\
| \\
\text{H}-\text{OH} \\
| \\
\text{CH}_2\text{OPO}_3^{2-}
\end{array}
\; + \;
\begin{array}{c}
\text{CH}_2\text{OPO}_3^{2-} \\
| \\
\text{C}=\text{O} \\
| \\
\text{CH}_2\text{OH}
\end{array}
$$

D-fructose-1,6-diphosphate   D-glyceraldehyde-3-phosphate   dihydroxyacetone phosphate

- In the first step of the aldolase-catalyzed reaction, D-fructose-1,6-diphosphate forms an imine (Schiff base) with a lysine residue at the active site of the enzyme (Section 16.7).
- A cysteine residue functions as a general-base catalyst in the step that cleaves the bond between C-3 and C-4. The molecule of D-glyceraldehyde-3-phosphate formed in this step dissociates from the enzyme.
- The enamine intermediate rearranges to an imine with a histidine residue functioning as a general-acid catalyst.
- Hydrolysis of the imine releases dihydroxyacetone phosphate, the other three-carbon product.

PROBLEM 21

Propose a mechanism for aldolase-catalyzed cleavage of D-fructose-1,6-diphosphate if it did not form an imine with the substrate. What is the advantage of imine formation?

PROBLEM 22

Why must D-glucose-6-phosphate isomerize to D-fructose-6-phosphate before the cleavage reaction with aldolase occurs? (See pages 853 and A-23.)

PROBLEM 23

Aldolase shows no activity if it is incubated with iodoacetic acid before D-fructose-1,6-diphosphate is added to the reaction mixture. What does this tell you about the mechanism of the reaction?

The induced-fit model led chemists to propose that the most precise binding between an enzyme and its substrate occurs when the substrate is in the transition state. Stabilization of the transition state of an enzyme-catalyzed reaction contributes to the catalytic power of the enzyme. Because transition states are only

**21.9
CATALYTIC
ANTIBODIES AND
ARTIFICIAL ENZYMES**

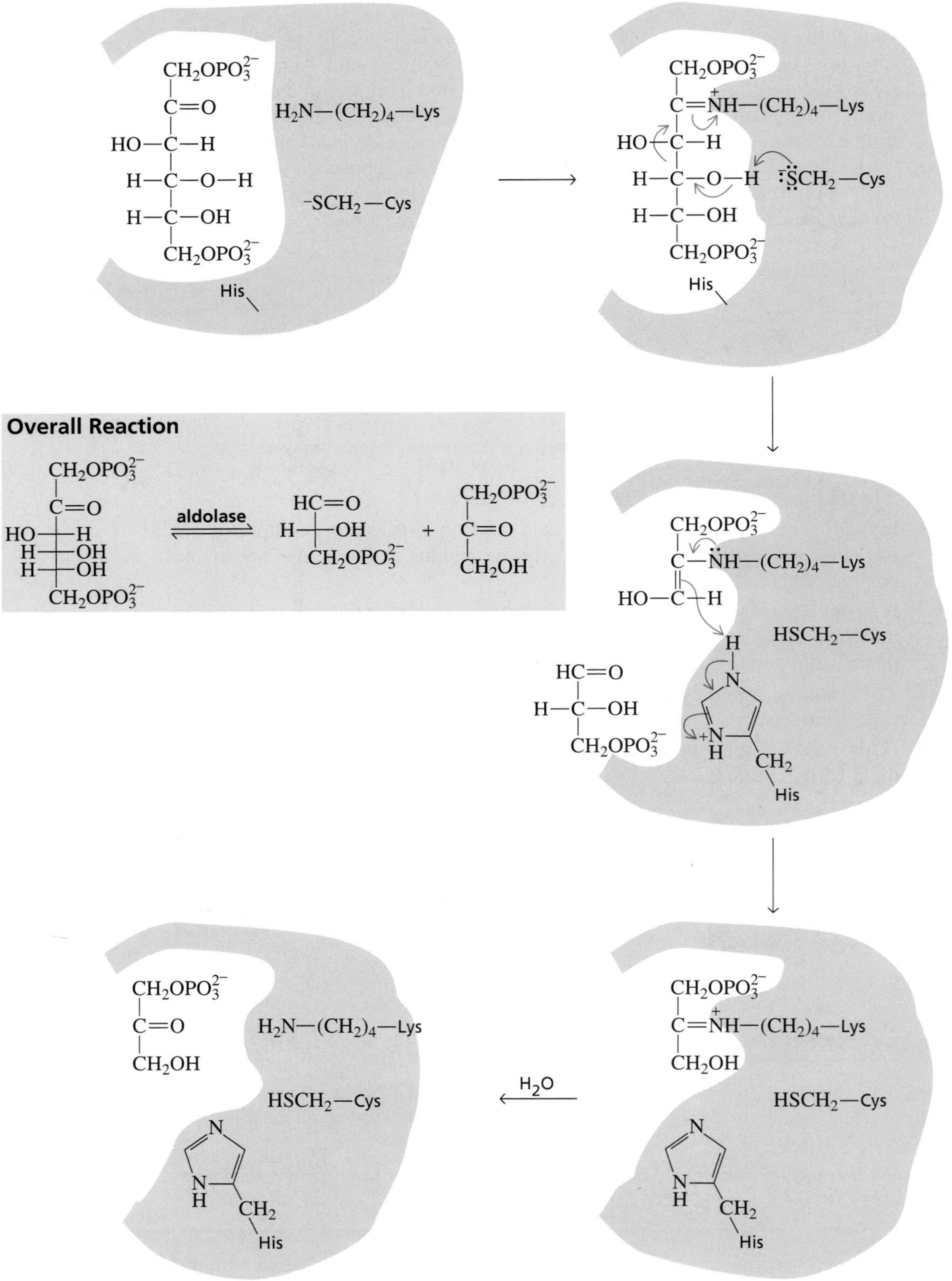

Overall Reaction

▲ Figure 21.11
Proposed mechanism for aldolase-catalyzed cleavage of D-fructose-1,6-diphosphate.

transitory, the binding between an enzyme and a transition state cannot be directly determined. However, the Hammond postulate (Section 3.11) allows us to predict the structure of a transition state. Knowing the approximate structure of a transition state allows stable transition-state analogs to be designed and synthesized. A **transition-state analog** is structurally similar to the transition state of an enzyme-catalyzed reaction.

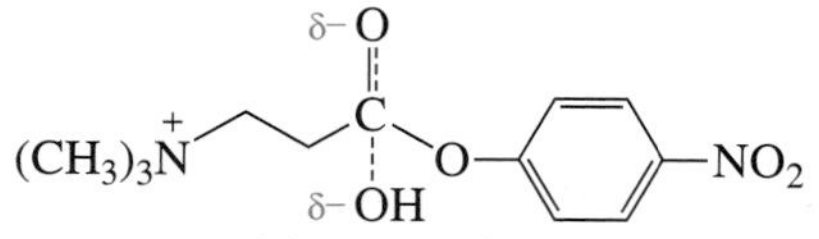

transition state for attack
by HO⁻ on an ester

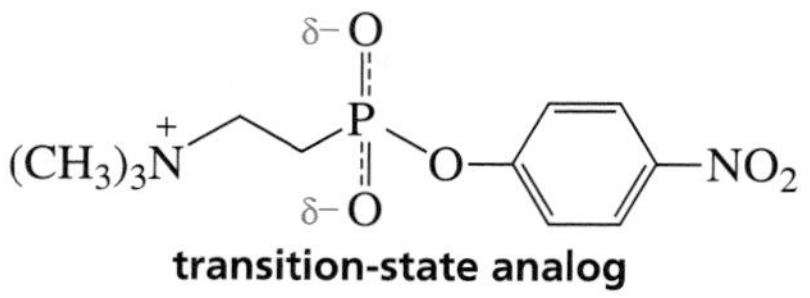

transition-state analog

Chemists have found that enzymes bind a transition-state analog 100 to 1 million times more tightly than they bind the substrate. This is good evidence for strong complementarity between an enzyme and the transition state of the reaction it catalyzes.

Chemists have developed methods for using transition-state analogs as antigens to stimulate cells to synthesize complementary antibodies. These antibodies, therefore, are designed to catalyze reactions with transition states similar to the transition-state analog and are known as **catalytic antibodies** (Chapter 15, Problem 55).

KEY TERMS

acid catalyst (page 956)
active site (page 970)
acyl-enzyme intermediate (page 975)
anchimeric assistance (page 966)
base catalyst (page 960)
catalyst (page 953)
catalytic antibody (page 983)
covalent catalysis (page 955)
effective molarity (page 964)
electrophilic catalyst (page 962)

electrostatic catalysis (page 977)
enzyme (page 970)
gem-dialkyl effect (page 966)
general-acid catalysis (page 958)
general-base catalysis (page 960)
induced-fit model (page 970)
intramolecular catalysis (page 966)
lock-and-key model (page 970)
metal-ion catalysis (page 961)
molecular recognition (page 970)

nucleophilic catalysis (page 955)
nucleophilic catalyst (page 955)
pH-activity profile (page 979)
pH-rate profile (page 979)
relative rate (page 964)
site-specific mutagenesis (page 975)
specific-acid catalysis (page 958)
specific-base catalysis (page 960)
substrate (page 970)
transition-state analog (page 983)

PROBLEMS

24. Which of the following compounds would eliminate HBr more rapidly?

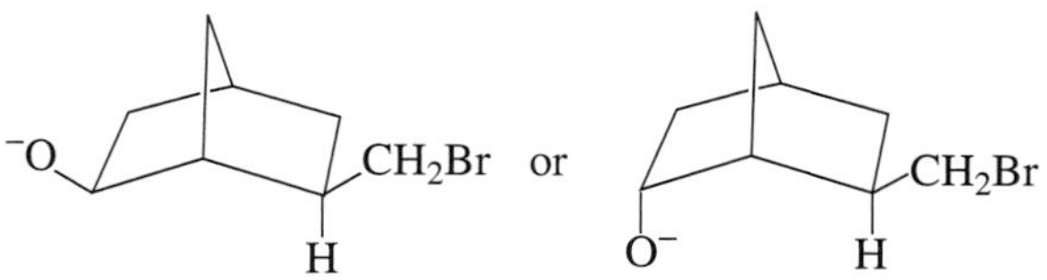

25. For each of the following pairs of compounds, indicate which would form a lactone more rapidly.

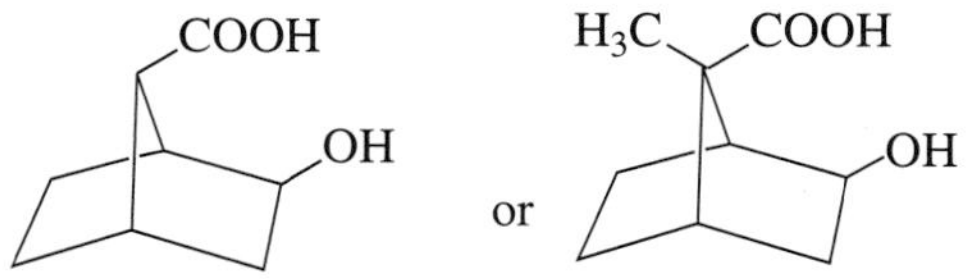

b.

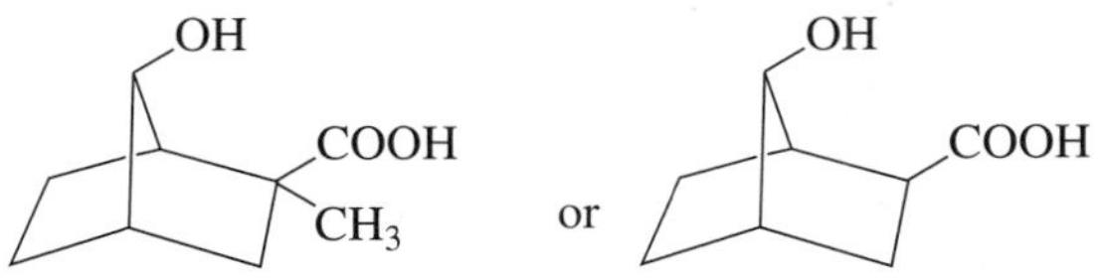

or

26. Which of the following compounds would form an anhydride more rapidly?

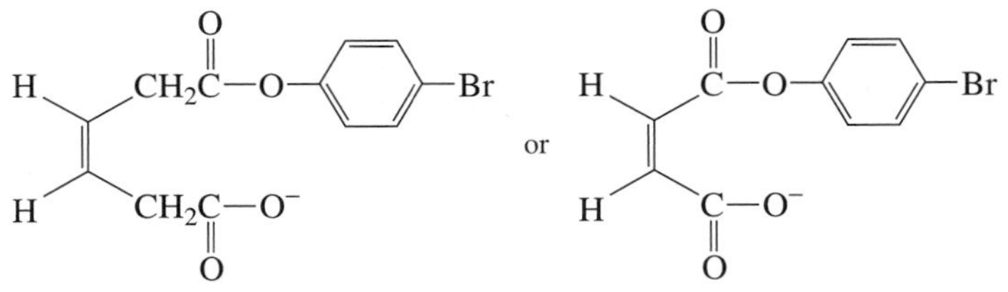

or

27. Which compound has the greatest rate of hydrolysis: benzamide, *o*-carboxybenza-mide, *o*-formylbenzamide, or *o*-hydroxybenzamide?

28. Indicate the type of catalysis occurring in the slow step in each of the following reaction sequences.

a. 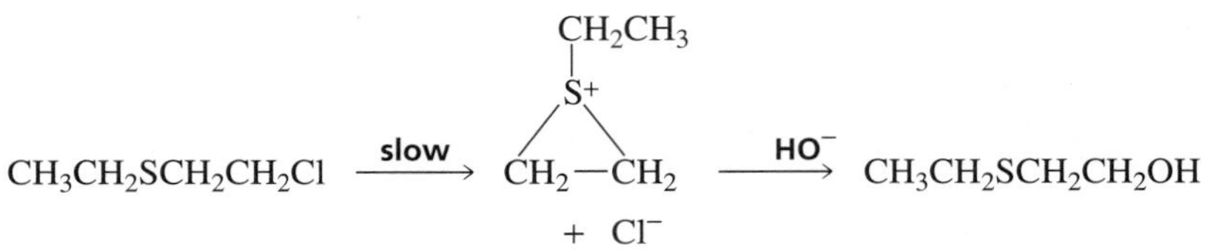 $CH_3CH_2SCH_2CH_2Cl$ $\xrightarrow{\text{slow}}$

$\xrightarrow{HO^-}$ $CH_3CH_2SCH_2CH_2OH$

b.

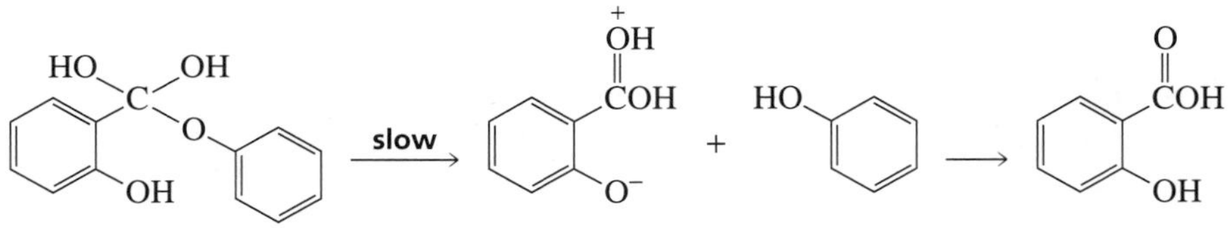

29. The deuterium kinetic isotope effect (k_{H_2O}/k_{D_2O}) for the hydrolysis of aspirin is 2.2. What does this tell you about the kind of catalysis exerted by the *ortho*-carboxyl substituent? (*Hint:* It is easier to break an O—H bond than an O—D bond.)

30. Draw the pH-activity profile for an enzyme with only one catalytic group at the active site. The catalytic group is a general-acid catalyst with a pK_a of 5.6.

31. A Co^{3+} complex catalyzes the hydrolysis of the lactam shown below. Propose a mechanism for the metal-ion catalyzed reaction.

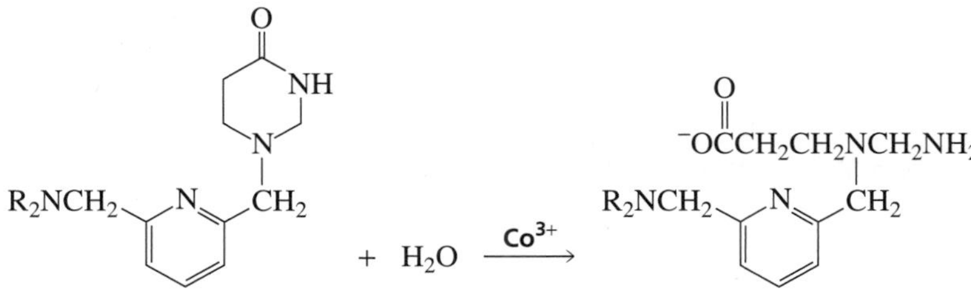

32. There are two kinds of aldolases. Class I aldolases are found in animals and plants, while class II aldolases are found in fungi, algae, and some bacteria. Only class I aldolases form a Schiff base. Class II aldolases have a metal ion (Zn^{2+}) at the active site. The mechanism for the class I aldolases was given in Section 21.8. Propose a mechanism for the class II aldolases.

33. The hydrolysis of the following ester is catalyzed by morpholine, a secondary amine. Propose a mechanism for this reaction. (*Hint:* The pK_a of the conjugate acid of morpholine is 8.0, so it is too weak a base to function as a general-base catalyst in this reaction.)

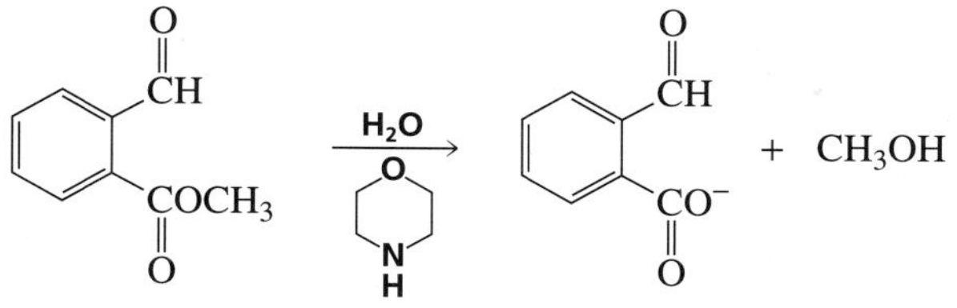

morpholine

34. The enzyme carbonic anhydrase catalyzes the conversion of carbon dioxide into bicarbonate ion (Section 1.19). It is a metalloenzyme with Zn^{2+} coordinated at the active site by three histidine residues. Propose a mechanism for this reaction.

$$CO_2 \ + \ H_2O \ \xrightarrow{\text{carbonic anhydrase}} \ HCO_3^- \ + \ H^+$$

35. At pH = 12, the rate of hydrolysis of ester A is faster than the rate of hydrolysis of ester B. At pH = 8, the relative rates reverse; ester B hydrolyzes faster than ester A. Explain.

A   B

36. Reaction of 2-acetoxycyclohexyl tosylate with acetate ion forms 1,2-cyclohexanediol diacetate. The reaction is stereospecific; the stereoisomers obtained as products depend on the stereoisomer used as a reactant. Explain the following observations.
  a. Both cis reactants form an optically active trans product, but each cis reactant forms a different trans product.
  b. Both trans reactants form the same racemic mixture.
  c. A trans reactant is more reactive than a cis reactant.

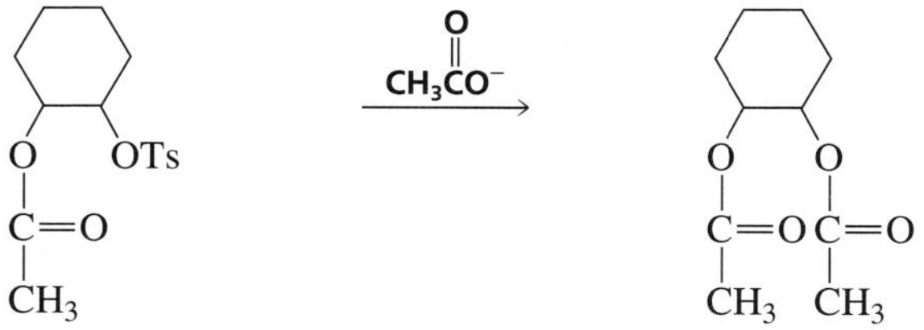

2-acetoxycyclohexyl tosylate   1,2-cyclohexanediol diacetate

37. *Staphylococcus* nuclease is an enzyme that catalyzes the hydrolysis of DNA. The overall hydrolysis reaction is shown below. The reaction is catalyzed by Ca^{2+}, Glu 43, and Arg 87. Propose a mechanism for this reaction.

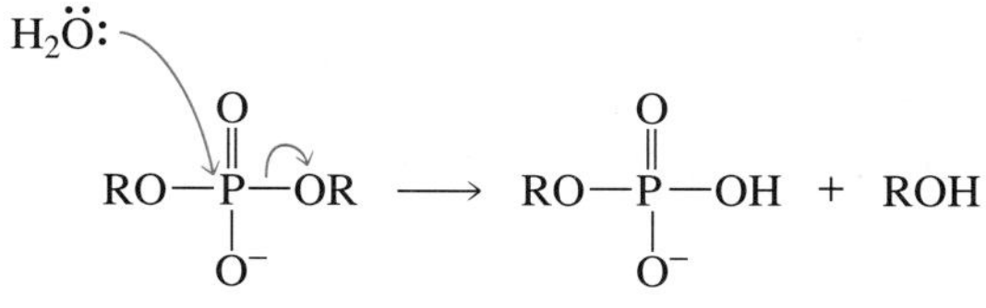

38. Proof for formation of an imine between aldolase and its substrate was obtained using D-fructose-1,6-diphosphate labeled at the C-2 position with ^{14}C as the substrate. NaBH$_4$ was added to the reaction mixture. The radioactive product was isolated from the reaction mixture and hydrolyzed in an acidic solution. A radioactive product was isolated. Give its structure. (*Hint:* NaBH$_4$ reduces an imine linkage.)

39. 3-Amino-2-oxindole catalyzes the decarboxylation of α-keto acids.
   a. Propose a mechanism for the catalyzed reaction.
   b. Would 3-aminoindole be equally effective as a catalyst?

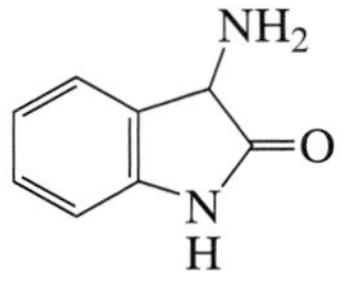

3-amino-2-oxindole

40. **a.** Explain why the alkyl halide shown below reacts much more rapidly than primary alkyl halides, such as butyl chloride and pentyl chloride, with guanine residues.
   b. The alkyl halide can react with two guanine residues in two different chains, thereby cross-linking the chains. Propose a mechanism for the cross-linking reaction.

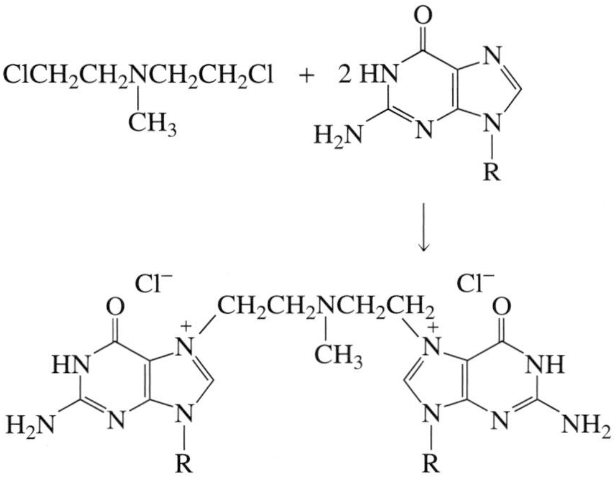

41. Triosephosphate isomerase catalyzes the conversion of dihydroxyacetone phosphate to glyceraldehyde-3-phosphate. The enzyme's catalytic groups are Glu 165 and His 95. In the first step of the reaction, these catalytic groups function as a general-base and a general-acid catalyst, respectively. Propose a mechanism for the reaction.

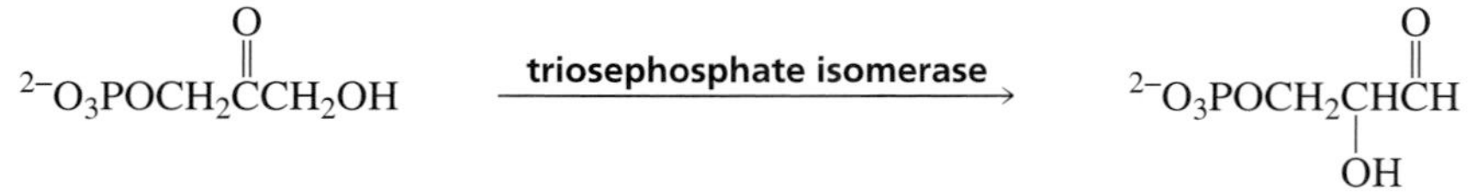

dihydroxyacetone phosphate   **triosephosphate isomerase**   **glyceraldehyde-3-phosphate**

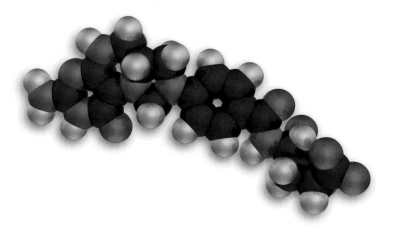

N⁵,N¹⁰-methylenetetrahydrofolate

22

THE ORGANIC MECHANISMS OF THE COENZYMES

Many enzymes cannot catalyze a reaction without the help of a cofactor. **Cofactors** assist enzymes in carrying out a variety of reactions that cannot be carried out solely by the amino acid side chains of the protein. Some cofactors are metal ions; others are organic molecules.

A metal-ion cofactor acts as a Lewis acid in a variety of ways to maximize the catalytic activity of an enzyme. It can coordinate with groups on the enzyme, causing them to align in a geometry that is advantageous for reactivity; it can help bind the substrate to the active site of the enzyme; it can form a coordination complex with the substrate to increase its reactivity; or it can increase the nucleophilicity of water at the active site. An enzyme that has a tightly bound metal ion (Co^{2+}, Cu^{2+}, Fe^{2+}, Mo^{2+}, Zn^{2+}) is known as a **metalloenzyme.** Carboxypeptidase A is an example of an enzyme that has a metal-ion cofactor (Section 21.8); it is a metalloenzyme.

PROBLEM 1

How does the metal ion in carboxypeptidase A increase its catalytic activity?

Cofactors that are organic molecules are called **coenzymes.** Coenzymes are derived from organic compounds that are commonly known as **vitamins.** Table 22.1 lists the vitamins and their biochemically active coenzyme forms.

A vitamin is a substance that is needed in small amounts for normal body function but that the body cannot synthesize. Sir Frederick Hopkins was the first to suggest that diseases such as rickets and scurvy might result from the absence of substances in the diet that are needed only in very small quantities. Because the first such compound recognized to be essential in the diet was an amine, Casimir Funk incorrectly concluded that all such compounds were amines and called them vitamines ("life-amines"). The *e* was later dropped from the name.

Sir Frederick G. Hopkins (1861–1947) *was born in England. His finding that one sample of a protein supported life while another did not led him to conclude that one sample contained a trace amount of a substance essential to life. His postulation later became known as the "vitamin concept," for which he received a share of the 1929 Nobel Prize in medicine. He also originated the concept of "essential amino acids."*

Casimir Funk (1884–1967) *was born in Poland, received his medical degree from the University of Bern, and became a U.S. citizen in 1920. In 1923 he returned to Poland to direct the State Institute of Hygiene but returned to the United States permanently when World War II broke out.*

TABLE 22.1 The Vitamins, Their Coenzymes, and Their Chemical Functions

Vitamin	Coenzyme	Reaction Catalyzed	Human Deficiency Disease
Water-Soluble Vitamins			
niacin (niacinate)	NAD$^+$, NADP$^+$ NADH, NADPH	oxidation reduction	pellagra
riboflavin (vitamin B$_2$)	FAD, FMN FADH$_2$, FMNH$_2$	oxidation reduction	skin inflammation
thiamine (vitamin B$_1$)	thiamine pyrophosphate (TPP)	two-carbon transfer	beriberi
lipoate	lipoate dihydrolipoate	oxidation reduction	—
pantothenate	coenzyme A (CoASH)	acyl transfer	—
biotin (vitamin H)	biotin	carboxylation	—
pyridoxine (vitamin B$_6$)	pyridoxal phosphate (PLP)	decarboxylation transamination racemization C_α—C_β bond cleavage α,β-elimination β-substitution	anemia
vitamin B$_{12}$	coenzyme B$_{12}$	isomerization	pernicious anemia
folate	tetrahydrofolate (THF)	one-carbon transfer	megaloblastic anemia
ascorbic acid (vitamin C)	—	—	scurvy
Water-Insoluble (Lipid-Soluble) Vitamins			
vitamin A	—	—	—
vitamin D	—	—	—
vitamin E	—	—	—
vitamin K	vitamin KH$_2$	carboxylation	—

Casimir Funk

We have seen that enzymes catalyze reactions following the principles of organic chemistry (Section 21.8). Coenzymes also carry out their roles using these same principles. We will see that coenzymes play a variety of chemical roles that the amino acid side chains of enzymes cannot play. Some coenzymes function as oxidizing and reducing agents, some allow electrons to be delocalized, some activate groups for further reaction, and some provide good nucleophiles or strong bases needed for a reaction. Because it would be very inefficient for the body to use a compound only once and then discard it, coenzymes are recycled. Therefore, we will see that after a coenzyme has carried out a reaction, a second reaction converts the coenzyme back to its original form.

An enzyme plus the cofactor it requires to catalyze a reaction is called a **holoenzyme.** An enzyme that has had its cofactor removed is an **apoenzyme.** Holoenzymes are catalytically active; apoenzymes are catalytically inactive because they have lost their cofactors.

Early nutritional studies divided vitamins into two classes: water-soluble vitamins and water-insoluble (lipid-soluble) vitamins (Table 22.1). Vitamins A, D, E, and K are water-insoluble. Vitamin K is the only water-insoluble vitamin currently recognized to function as a coenzyme. Vitamin A is required for proper vision, vitamin D regulates calcium and phosphate metabolism, and vitamin E is an antioxidant. Because they do not function as coenzymes, vitamins A, D, and E are not discussed in this chapter. Vitamins A and E will be discussed in Chapter 23. Vitamin D will be discussed in Section 27.6.

All the water-soluble vitamins function as coenzymes except vitamin C. In spite of its name, vitamin C is actually not a vitamin because most mammals are able to synthesize it (Section 19.15). Humans and guinea pigs cannot synthesize it, so it must be included in their diets. But it still is not a vitamin because it is required in fairly high amounts. We have seen that vitamin C and vitamin E are radical inhibitors and therefore act as antioxidants. Vitamin C traps radicals formed in aqueous environments, while vitamin E traps radicals formed in nonpolar environments (Section 8.8).

It is impossible to overdose on water-soluble vitamins because, since they are water-soluble, the body can readily eliminate any excess. However, you can overdose on water-insoluble vitamins because, since they are not easily eliminated by the body, they can accumulate in cell membranes and other nonpolar components of the body. For example, excess vitamin D causes calcification of soft tissues. The kidneys are particularly susceptible to calcification, and this eventually leads to kidney failure. Vitamin D is formed in the skin as a result of a photochemical reaction caused by the UV rays of the sun (Section 27.6).

VITAMIN B₁

Christiaan Eijkman (1858–1930) was a member of a medical team that was sent to the East Indies to study beriberi in 1886. At that time, all diseases were thought to be caused by microorganisms. When the microorganism that caused beriberi could not be found, the team left the East Indies. Eijkman stayed behind to become the director of a new bacteriological laboratory. In 1896 he accidentally discovered the cause of beriberi when he noticed that chickens being used in the laboratory had developed symptoms characteristic of the disease. He found that the symptoms had developed when a cook had started feeding the chickens rice meant for hospital patients. The symptoms disappeared

Christiaan Eijkman

Sir Frederick Hopkins

when a new cook resumed feeding chicken feed to the chickens. Later it was recognized that thiamine (vitamin B₁) is present in chicken feed but not in polished rice. Eijkman shared the 1929 Nobel Prize in medicine or physiology with Frederick Hopkins.

22.1 OVERALL VIEW OF METABOLISM

The reactions that living organisms carry out to obtain the energy they need and to synthesize the compounds they require are collectively known as **metabolism.** The process of metabolism can be divided into two parts, **catabolism** and **anabolism.** *Catabolic reactions* break down complex nutrient molecules to provide energy and simple precursor molecules for synthesis. *Anabolic reactions* require energy and result in the synthesis of complex biomolecules from simpler precursor molecules.

| **catabolism** | complex molecules $\longrightarrow$ simple molecules + energy |
| **anabolism** | simple molecules + energy $\longrightarrow$ complex molecules |

We will be looking at many different enzyme-catalyzed reactions in this chapter. It is important to remember that almost every reaction that occurs in a living system is catalyzed by an enzyme. The enzyme holds the reactants and any necessary cofactors rigidly in place so that the reacting functional groups and catalytic groups are oriented in a way that will cause a very specific and well-defined chemical reaction. In some cases it may be helpful to see where a particular reaction fits into the overall metabolic scheme. So we will first look at an overall view of catabolism. Catabolism can be divided into four stages.

The *first stage of catabolism* is called digestion. All the reactants required for all life processes ultimately come from our diet. We really are what we eat. In this first stage, fats, carbohydrates, and proteins are hydrolyzed into fatty acids, monosaccharides, and amino acids, respectively (Figure 22.1). In *the second stage of catabolism,* fatty acids, monosaccharides, and amino acids are converted into compounds that can enter the citric acid cycle. These compounds include those that are part of the citric acid cycle (known as citric acid cycle intermediates, Appendix VI), acetyl-CoA (a compound that can enter the cycle), and pyruvate (pyruvate can be converted into acetyl-CoA). Fatty acids are broken down to acetyl-CoA; monosaccharides are broken down to pyruvate; and amino acids are broken down to acetyl-CoA, pyruvate, or citric acid cycle intermediates, depending on the amino acid. The

Figure 22.1 ▶
The four stages of catabolism.

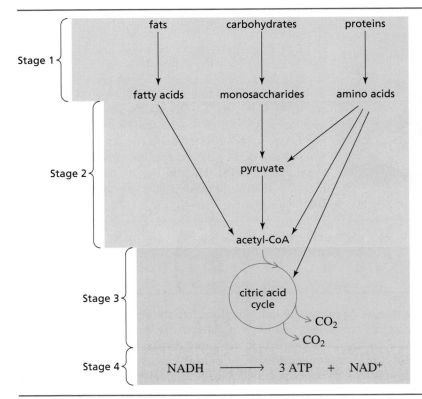

third stage of catabolism is the citric acid cycle. For every acetyl-CoA that enters the cycle, two molecules of CO_2 are formed. In other words, the compounds derived from the first two stages of catabolism are completely oxidized in the citric acid cycle to CO_2.

Metabolic energy is measured in terms of adenosine triphosphate (ATP). How the body uses ATP is described in Sections 24.2 and 24.3. Very little ATP is formed in the first three stages of catabolism. However, in the *fourth stage of catabolism,* every NADH that is formed in the process of carrying out oxidation reactions in the earlier stages is converted into three ATPs in a process known as oxidative phosphorylation. Thus, the bulk of the energy provided by fuel molecules (fats, carbohydrates, and proteins) is obtained in this fourth stage.

Anabolism can be thought of as the reverse of catabolism. Acetyl-CoA, pyruvate, and citric acid cycle intermediates are the starting materials for the synthesis of fatty acids, monosaccharides, and amino acids. These compounds are then used to form fats, carbohydrates, and proteins. The mechanisms that biological systems use to synthesize fats and proteins are discussed in Sections 18.20 and 24.12.

An enzyme that catalyzes an oxidation reaction cannot catalyze that reaction unless a coenzyme is present. Since none of the amino acid side chains can act as an oxidizing agent, the oxidizing agent is the coenzyme. The enzyme's role is to hold the substrate and coenzyme together so the oxidation reaction can take place. The coenzymes most commonly used by enzymes that catalyze oxidation reactions are **nicotinamide adenine dinucleotide (NAD^+)** and **nicotinamide adenine dinucleotide phosphate ($NADP^+$).** (You were introduced to NAD^+ in Section 17.11.)

22.2 PYRIDINE NUCLEOTIDE COENZYMES

$$\text{substrate}_{\text{reduced}} + NAD^+ \xrightleftharpoons{\text{enzyme}} \text{substrate}_{\text{oxidized}} + NADH + H^+$$

$$\text{substrate}_{\text{reduced}} + NADP^+ \xrightleftharpoons{\text{enzyme}} \text{substrate}_{\text{oxidized}} + NADPH + H^+$$

When NAD^+ (or $NADP^+$) oxidizes a substrate, the coenzyme is reduced to NADH (or NADPH). NADH and NADPH are reducing agents. They are used as coenzymes by enzymes that catalyze reduction reactions. Enzymes that catalyze oxidation reactions bind NAD^+ (or $NADP^+$) more tightly than they bind NADH (or NADPH). Enzymes that catalyze reduction reactions bind NADH (or NADPH) more tightly than they bind NAD^+ (or $NADP^+$). When the reaction is over, the relatively loosely bound coenzyme can dissociate from the enzyme.

NAD^+ and $NADP^+$ are oxidizing agents.

NAD^+ and $NADP^+$ are derived from the vitamin commonly known as *niacin.* A deficiency in niacin causes pellagra, a disease that begins with dermatitis and ultimately causes insanity and death. More than 120,000 cases of pellagra were reported in the United States in 1927, mainly among poor people with unvaried diets. A factor present in preparations of vitamin B was known to prevent pellagra, but it was not until 1937 that the factor was identified as nicotinic acid. When bread companies started adding nicotinic acid to their bread, they insisted that its name be changed to niacin because they thought that nicotinic acid sounded too much like nicotine and they did not want their vitamin-enriched bread to be

NADH and NADPH are reducing agents.

associated with a harmful substance. Niacinamide is a nutritionally equivalent form of the vitamin.

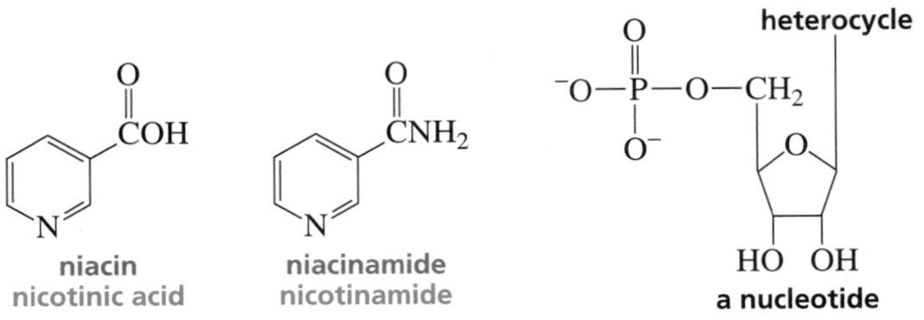

niacin
nicotinic acid

niacinamide
nicotinamide

a nucleotide

NAD^+ is composed of two nucleotides linked together through their phosphate groups. A **nucleotide** consists of a heterocycle attached to produce a β-configuration at C-1 of a phosphorylated ribose. A **heterocycle** is a cyclic compound in which one or more of the ring atoms is an atom other than carbon. The heterocyclic component of one of the nucleotides of NAD^+ is nicotinamide, and the heterocyclic component of the other is adenine. This accounts for the coenzyme's name (nicotinamide adenine dinucleotide). The positive charge in the NAD^+ abbreviation refers to the positively charged nitrogen of the pyridine ring.

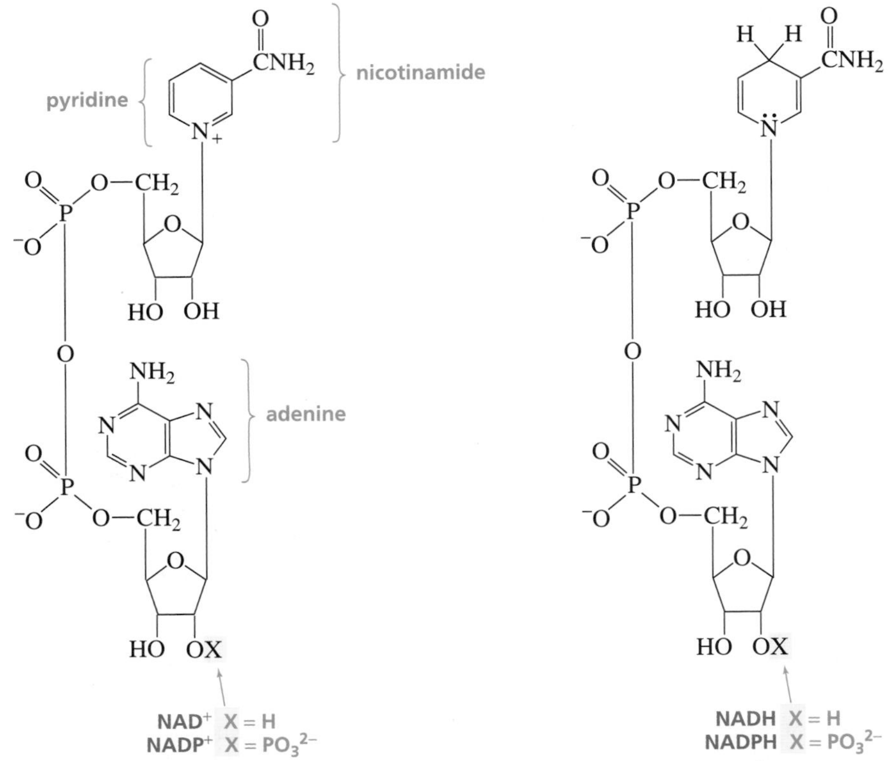

NAD⁺ X = H
NADP⁺ X = PO_3^{2-}

NADH X = H
NADPH X = PO_3^{2-}

The only way in which $NADP^+$ differs structurally from NAD^+ is the phosphate group bonded to the 2'-OH group of the ribose of the adenine nucleotide; this explains the addition of "P" to its name. NAD^+ and NADH are generally used as coenzymes in catabolic reactions, and the phosphorylated derivatives, $NADP^+$ and NADPH, are generally used as coenzymes in anabolic reactions.

The oxidation of the secondary alcohol group of malate to a ketone group is one of the reactions in the sequence of reactions known as the Krebs cycle (a catabolic pathway). NAD^+ is the oxidizing reagent in this reaction. Many enzymes that catalyze oxidation reactions are called **dehydrogenases** (they remove hydrogen).

$$\underset{\text{malate}}{^-OCCH_2CHCO^-} + NAD^+ \overset{\text{malate}}{\underset{\text{dehydrogenase}}{\rightleftharpoons}} \underset{\text{oxaloacetate}}{^-OCCH_2CCO^-} + NADH + H^+$$

Pyruvate is reduced to lactate in a catabolic pathway (NADH is the reducing agent); β-aspartate-semialdehyde is reduced to homoserine in an anabolic pathway (NADPH is the reducing agent).

$$\underset{\text{pyruvate}}{CH_3C-CO^-} + NADH + H^+ \xrightarrow[\substack{\text{a reaction that occurs} \\ \text{in a catabolic pathway}}]{\substack{\text{lactate} \\ \text{dehydrogenase}}} \underset{\text{lactate}}{CH_3CH-CO^-} + NAD^+$$

$$\underset{\substack{\text{β-aspartate-semialdehyde}}}{\underset{^+NH_3}{HCCH_2CHCO^-}} + NADPH + H^+ \xrightarrow[\substack{\text{a reaction that occurs} \\ \text{in an anabolic pathway}}]{\substack{\text{homoserine} \\ \text{dehydrogenase}}} \underset{\substack{^+NH_3 \\ \text{homoserine}}}{HOCH_2CH_2CHCO^-} + NADP^+$$

The differentiation between the coenzymes used in catabolism and anabolism is maintained because the enzymes that catalyze these oxidation–reduction reactions exhibit strong specificity for a particular coenzyme. For example, an enzyme that catalyzes an oxidation reaction can readily tell the difference between NAD^+ and $NADP^+$ and, if it is an enzyme in a catabolic pathway, will bind NAD^+ but not $NADP^+$. In addition, the relative concentrations of the coenzymes in the cell encourage binding of the appropriate coenzyme. For example, because NAD^+ and NADH are catabolic coenzymes and catabolic reactions are most often oxidation reactions, the NAD^+ concentration in the cell is much greater than the NADH concentration (the cell maintains its $[NAD^+]/[NADH]$ ratio near 1000). Because $NADP^+$ and NADPH are anabolic coenzymes and anabolic pathways are predominantly reduction reactions, the concentration of NADPH in the cell is greater than the concentration of $NADP^+$ (the ratio of $[NADP^+]/[NADPH]$ is maintained at about 0.01).

How do these oxidation–reduction reactions take place? All the oxidation/reduction chemistry of the coenzyme takes place on the pyridine ring. The rest of the molecule is important for binding the coenzyme to the proper site on the enzyme. If the reaction is an oxidation, the substrate donates a hydride ion (H^-) to the 4-position of the pyridine ring. A basic group of the enzyme can help the reaction by removing a proton from oxygen.

Glyceraldehyde-3-phosphate dehydrogenase is an example of an enzyme that uses NAD^+ as an oxidizing coenzyme. The enzyme catalyzes the oxidation of the aldehyde group of glyceraldehyde-3-phosphate (GAP) to an anhydride of a carboxylic acid and phosphoric acid.

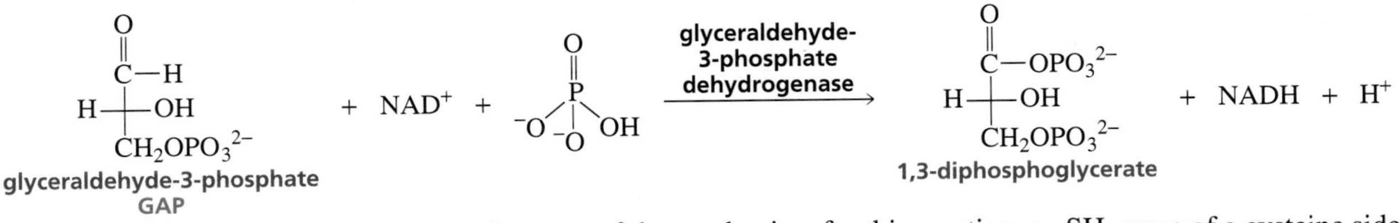

In the first step of the mechanism for this reaction, an SH group of a cysteine side chain at the active site of the enzyme reacts with GAP to form a thiohemiacetal. A basic group on the enzyme increases cysteine's nucleophilicity. The thiohemiacetal transfers a hydride ion to the 4-position of the pyridine ring of an NAD^+ that is bonded to the enzyme at an adjacent site. NADH dissociates from the enzyme, and the enzyme binds a new NAD^+. Phosphate reacts with the thioester, forming the anhydride product and releasing cysteine. Notice that at the end of the reaction the holoenzyme is exactly as it was at the beginning of the reaction, so the catalytic cycle can be repeated. The NADH produced in the reaction is reoxidized to NAD^+ in the fourth stage of catabolism (Figure 22.1).

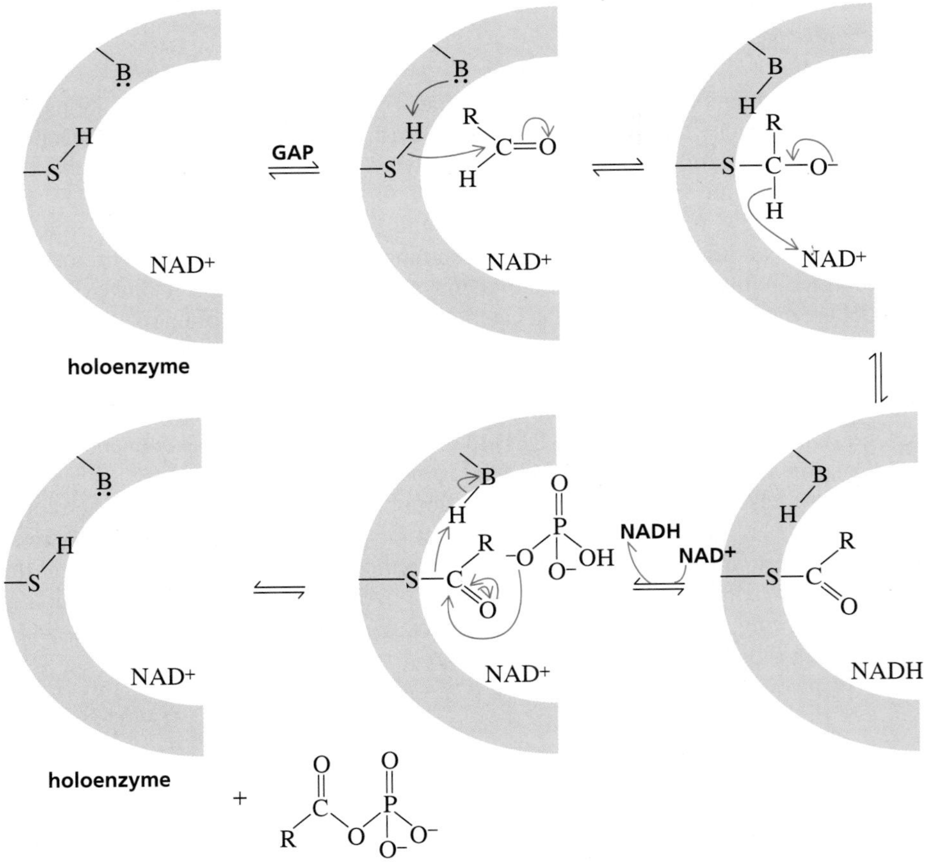

The mechanism for reduction by NADH (or by NADPH) is the reverse of the mechanism for oxidation by NAD^+ (or by $NADP^+$); the dihydropyridine ring donates a hydride ion from its 4-position to the substrate. An acidic group of the enzyme can aid the reaction by donating a proton to the substrate.

Because NADH and NADPH reduce compounds by donating a hydride ion, they can be thought of as biological equivalents of $NaBH_4$ or $LiAlH_4$, which are hydride donors used as reducing reagents in nonbiological reactions (Section 17.1).

Why are biological redox (reducing and oxidizing) reagents so much more complicated than the redox reagents used to carry out the same reaction in the laboratory? NADH is certainly more complicated than $LiAlH_4$, although both reagents reduce compounds by donating a hydride ion. Much of the structural complexity of a coenzyme is for molecular recognition—to allow it to be recognized by the enzyme. *Molecular recognition* allows the enzyme to bind the substrate and the coenzyme in the proper orientation for reaction.

In addition, a redox reagent found in a biological system must be much less reactive than a laboratory redox reagent. A biological reducing agent must also be selective. It cannot reduce just any reducible compound with which it comes into contact. Biological reactions are much more carefully controlled than that. In addition, because a biological reducing agent must be recycled rather than having its oxidized form thrown away (as would be the case for a reducing agent used in a laboratory), the equilibrium constant between the oxidized and reduced forms is generally close to 1. Therefore, biological redox reactions are not highly exergonic; they are equilibrium reactions that are driven in the appropriate direction by reaction products being removed as a result of participation in subsequent reactions.

Because the coenzyme is relatively unreactive compared with nonbiological oxidizing and reducing agents, the reaction between the substrate and the coenzyme does not occur at all, or takes place very slowly, without the enzyme. For example, NADH will reduce an aldehyde or a ketone only in the presence of an enzyme. $NaBH_4$ and $LiAlH_4$ are more reactive hydride donors—much too reactive to exist in the aqueous environment of the cell. Similarly, NAD^+ is a much more selective oxidizing agent than the typical oxidizing agents used in the laboratory. For example, NAD^+ will oxidize an alcohol only in the presence of an enzyme.

As you study the coenzymes in this chapter, don't let the complexity of their structures deter you. Notice that only a small part of the coenzyme is actually involved in the chemical reaction. Notice also that the coenzymes follow the same rules of organic chemistry as do the simple organic molecules with which you are familiar.

We saw in Section 17.11 that an oxidizing enzyme can tell the difference between the two hydrogens on the carbon from which it catalyzes the removal of a hydride ion. For example, alcohol dehydrogenase removes only the *pro-R* hydrogen of ethanol.

Similarly, a reducing enzyme can tell the difference between the two hydrogens at the 4-position of the nicotinamide ring. An enzyme has a specific binding site for the coenzyme and, when it binds the coenzyme, the enzyme blocks one of the coenzyme's sides. If the enzyme blocks the B-side of NADH, the substrate will bind to the A-side and the H_a hydride ion will be transferred to the substrate. If the enzyme blocks the A-side of the coenzyme, the substrate will have to bind to the B-side and the H_b hydride ion will be transferred. Currently, 156 dehydrogenases are known to transfer H_a, and 121 are known to transfer H_b. H_a is the *pro-R* hydrogen of NADH, and H_b is the *pro-S* hydrogen.

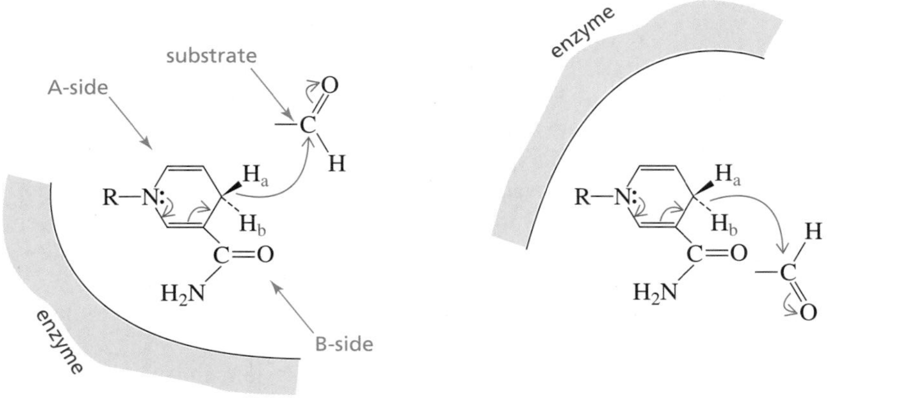

The enzyme blocks the B-side of the coenzyme, so the substrate binds to the A-side.

The enzyme blocks the A-side of the coenzyme, so the substrate binds to the B-side.

22.3
FLAVIN ADENINE DINUCLEOTIDE AND FLAVIN MONONUCLEOTIDE

A *flavoprotein* is an enzyme that contains either **flavin adenine dinucleotide (FAD)** or **flavin mononucleotide (FMN)** as a coenzyme. FAD and FMN, like NAD^+ and $NADP^+$, are coenzymes used in oxidation reactions. As its name indicates, FAD is a dinucleotide in which one of the heterocyclic components is flavin and the other is adenine. FMN contains flavin but not adenine; it is a mononucleotide. (Flavin is a bright yellow compound; *flavus* is Latin for "yellow.") Notice that instead of ribose, the flavin nucleotide has a ribitol group (reduced ribose). Flavin plus ribitol is called *riboflavin*. Riboflavin is also known as vitamin B_2. A vitamin B_2 deficiency causes inflammation of the skin.

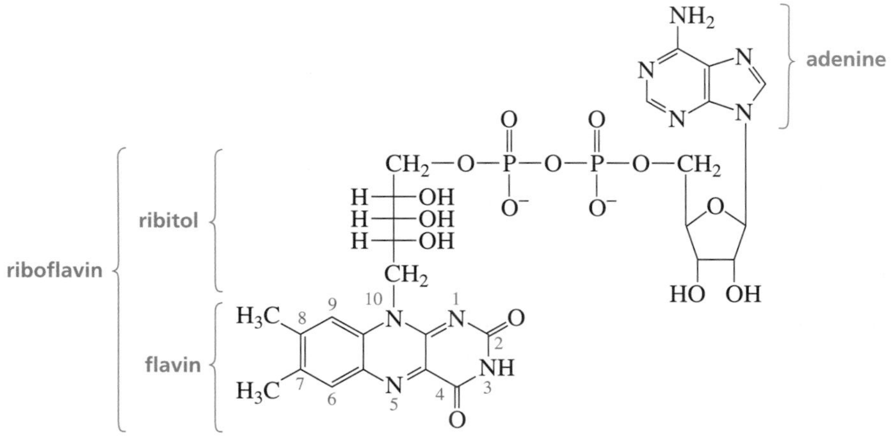

FAD

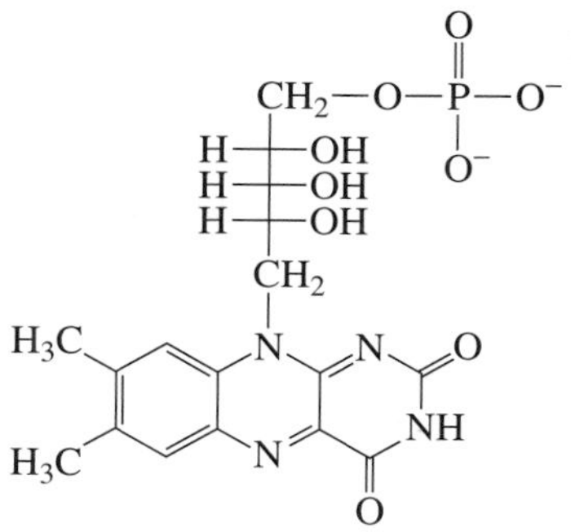

FMN

In most flavoproteins, FAD (or FMN) is bound quite tightly. This tight binding allows the enzyme to control the oxidation potential of the coenzyme. (The more positive the oxidation potential, the stronger the oxidizing agent.) Consequently, some flavoproteins are stronger oxidizing agents than others.

PROBLEM 2◆

FAD is obtained by an enzyme-catalyzed reaction that uses FMN and ATP as substrates. What is the other product of this reaction?

How can we tell which enzymes use FAD (or FMN) rather than NAD^+ (or $NADP^+$) as the oxidizing coenzyme? A rough guideline is that NAD^+ and $NADP^+$ are generally the coenzymes used in enzyme-catalyzed oxidation reactions involving carbonyl compounds (alcohols oxidized to ketones, aldehydes, or carboxylic acids), while FAD and FMN are generally the coenzymes used in other types of oxidations. For example, in the following reactions, FAD oxidizes a dithiol to a disulfide, an amine to an imine, and a saturated alkyl group to an alkene, and FMN oxidizes NADH to NAD^+. This is only an approximate guideline, because FAD is involved in some oxidations that involve carbonyl compounds, while NAD^+ and $NADP^+$ are involved in some oxidations that do not involve carbonyl compounds.

FAD and FMN are oxidizing agents.

$FADH_2$ and $FMNH_2$ are reducing agents.

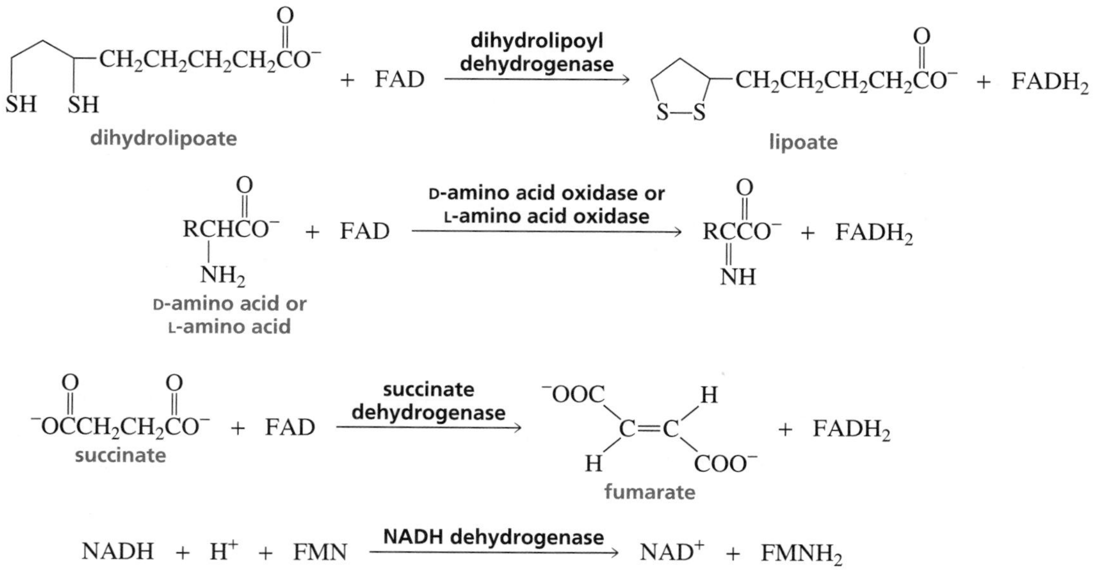

$$NADH + H^+ + FMN \xrightarrow{\text{NADH dehydrogenase}} NAD^+ + FMNH_2$$

When FAD (or FMN) oxidizes a substrate, the coenzyme is reduced to $FADH_2$ (or $FMNH_2$). All the oxidation/reduction chemistry takes place on the flavin ring. Reduction of the flavin ring interferes with the conjugated system, so the reduced coenzymes are less colored than their oxidized forms.

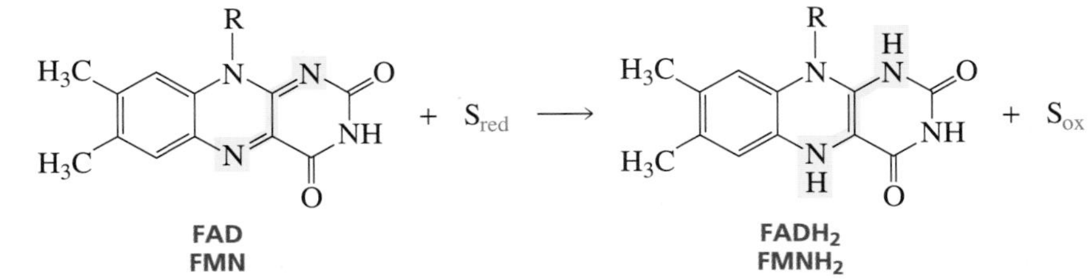

FAD
FMN

$FADH_2$
$FMNH_2$

PROBLEM 3

How many conjugated double bonds are there in:

a. FAD?

b. $FADH_2$?

In the first step of the mechanism for the oxidation of dihydrolipoate to lipoate, the thiolate ion attacks the 4a-position of the flavin ring. This reaction is general-acid-catalyzed; as the thiolate ion attacks the ring, a proton is donated to the N-5 nitrogen. A second attack of a thiolate ion on the sulfur covalently attached to the coenzyme generates the oxidized product and $FADH_2$.

mechanism for dihydrolipoyl dehydrogenase

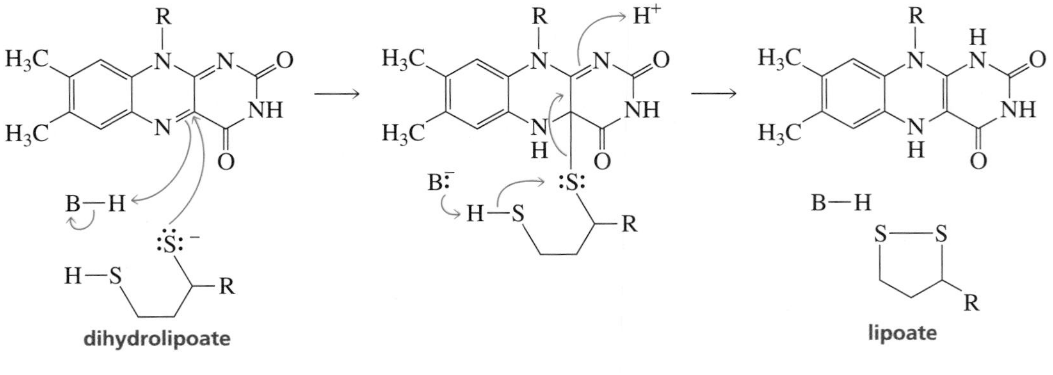

dihydrolipoate

lipoate

In the first step of the flavin-catalyzed oxidation of an amino acid to an imino acid, a basic group at the active site of the enzyme removes a proton from the α-carbon of the amino acid. The carbanion that is formed attacks the N-5 position of the flavin ring. Breakdown of the resulting tetrahedral intermediate gives the oxidized amino acid and the reduced coenzyme ($FADH_2$).

mechanism for D- or L-amino acid oxidase

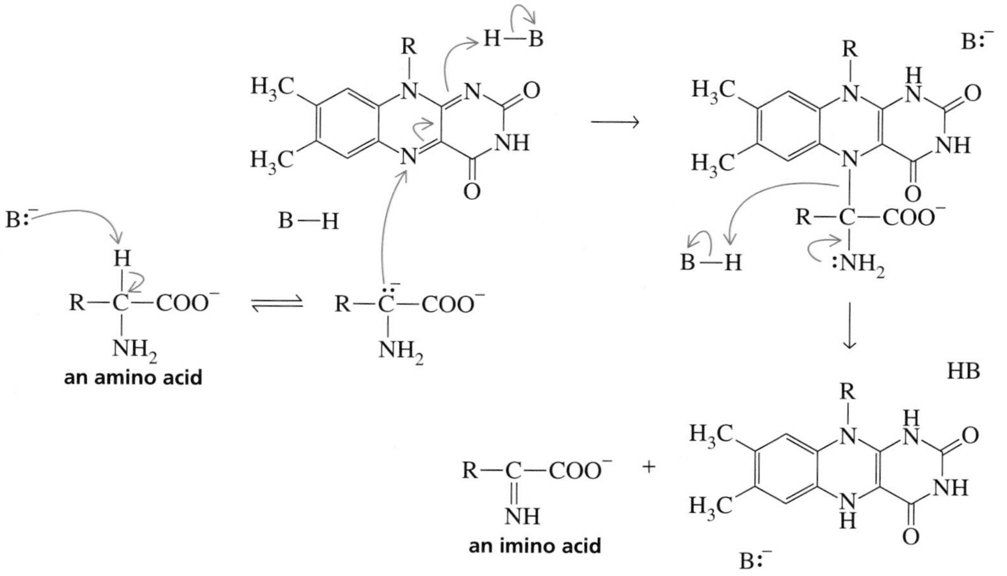

Notice that FAD is a more versatile coenzyme than NAD^+. Unlike NAD^+ which always uses the same mechanism, flavin coenzymes can use several different mechanisms to carry out an oxidation. For example, we have seen that oxidation of dihydrolipoate by FAD involves nucleophilic attack on the C-4a position of the flavin ring, while oxidation of an amino acid involves nucleophilic attack on the N-5 position.

Cells contain very low concentrations of FAD and much higher concentrations of NAD^+. This difference in concentration is responsible for a significant difference in the behavior of enzymes (**E**) that use NAD^+ as an oxidizing agent and those that use FAD: $FADH_2$ (unlike NADH) does not dissociate from the enzyme when the oxidation reaction is over. FAD is tightly bound to its enzyme and remains attached to it after FAD is reduced to $FADH_2$. This means that before the enzyme can enter another round of catalysis the $FADH_2$ bound to the enzyme must be reoxidized to FAD. The oxidizing agent used for this reaction is NAD^+ or O_2. Therefore, an enzyme that uses an oxidizing coenzyme other than NAD^+ may still require NAD^+ to reoxidize the reduced coenzyme. For this reason NAD^+ has been called the common coinage of biological oxidation– reduction reactions.

PROBLEM 4

Propose a mechanism for the reduction of lipoate by $FADH_2$.

PROBLEM 5

The hydrogens of the methyl group bonded to flavin at C-8 are more acidic than those of the methyl group bonded at C-7. Explain why this is so.

PROBLEM 6 / SOLVED

A common way for FAD to become covalently bound to its enzyme is by having a proton removed from the C-8 methyl group and a proton donated to N-1. Then a cysteine or other nucleophilic side chain of the enzyme attacks the methylene at C-8 as a proton is donated to N-5. Describe these events mechanistically.

SOLUTION

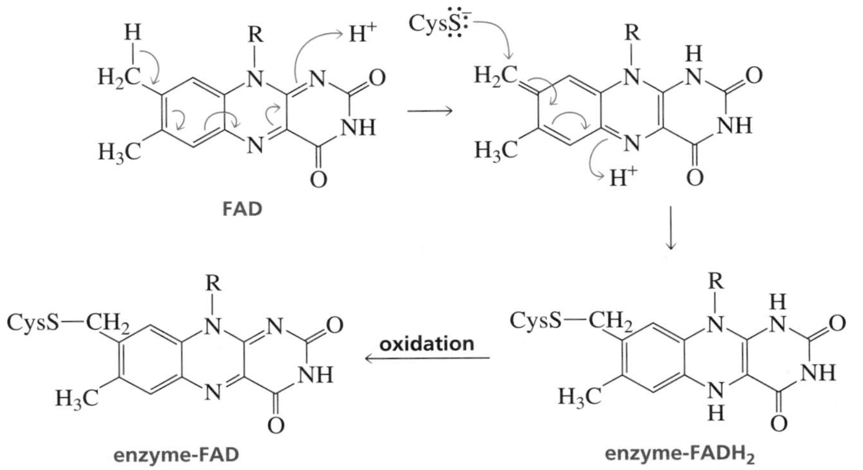

Notice that during the process of attaching FAD to the enzyme, FAD is reduced to $FADH_2$. It is subsequently oxidized back to FAD. Once the coenzyme is attached to the enzyme, it does not come off.

**22.4
THIAMINE
PYROPHOSPHATE:
VITAMIN B₁**

Thiamine was the first of the B vitamins to be identified. Consequently, it became known as vitamin B_1. The absence of thiamine in the diet causes a disease called beriberi, which damages the heart and impairs nerve reflexes and, in extreme cases, causes paralysis. One major dietary source of the vitamin is the outer layer of rice kernels. A deficiency is therefore most likely to occur when highly polished rice is a major component of the diet. A deficiency is also seen in alcoholics who are severely malnourished.

The coenzyme form of the vitamin is **thiamine pyrophosphate (TPP).** TPP is the coenzyme required by enzymes that catalyze the transfer of a two-carbon fragment from one species to another.

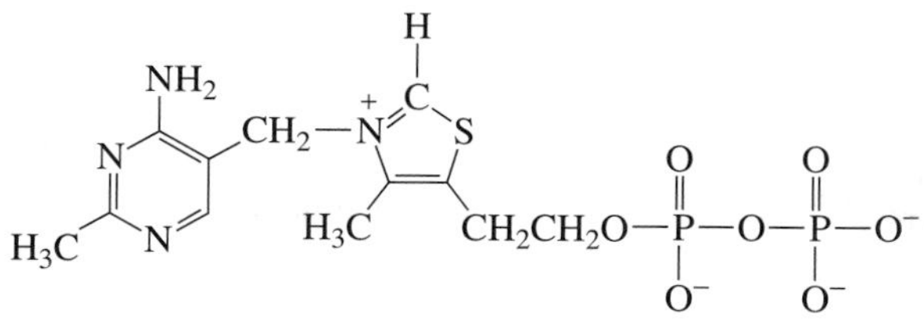

thiamine pyrophosphate
TPP

Pyruvate decarboxylase is an example of an enzyme that requires thiamine pyrophosphate. The enzyme catalyzes the decarboxylation of pyruvate and transfers the resulting two-carbon fragment to a proton, resulting in the formation of acetaldehyde.

Thiamine pyrophosphate (TPP) is required by enzymes that catalyze the transfer of a two-carbon fragment from one species to another.

The fact that an α-keto acid such as pyruvate can be decarboxylated should come as a surprise, because the electrons left behind when CO_2 is removed cannot be delocalized onto the carbonyl oxygen. We will see that the thiazolium ring of the coenzyme provides an "electron sink" for the delocalization of the electrons. **Electron sink** is the name given to a site to which electrons can be delocalized.

The hydrogen bonded to the imine carbon of TPP is relatively acidic (pK_a = 12.7), because the carbanion formed when the proton is removed is stabilized by the adjacent positively charged nitrogen. Loss of the proton creates a good nucleophile that attacks α-carbonyl groups.

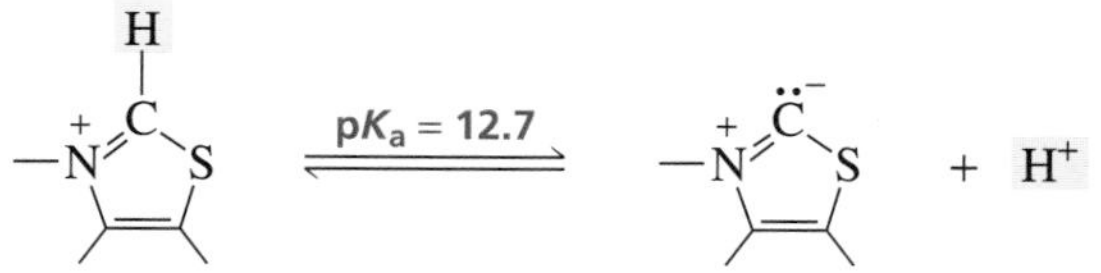

In the first step of the mechanism for pyruvate decarboxylase, the ylide carbanion attacks the ketone group of the α-keto acid. The resulting intermediate can easily undergo decarboxylation because the electrons left behind when CO_2 is removed can be delocalized onto the positively charged nitrogen. The positively charged nitrogen of TPP is a more effective electron sink than the β-keto group of a β-keto acid, a class of compounds that are fairly easily decarboxylated (Section 18.16). The product of decarboxylation is stabilized by resonance; one of the resonance contributors is neutral, and the other has separated charges. We will refer to this compound as a "resonance-stabilized carbanion." Protonation of the resonance-stabilized carbanion and a subsequent elimination reaction result in formation of acetaldehyde and regeneration of the coenzyme.

mechanism for pyruvate decarboxylase

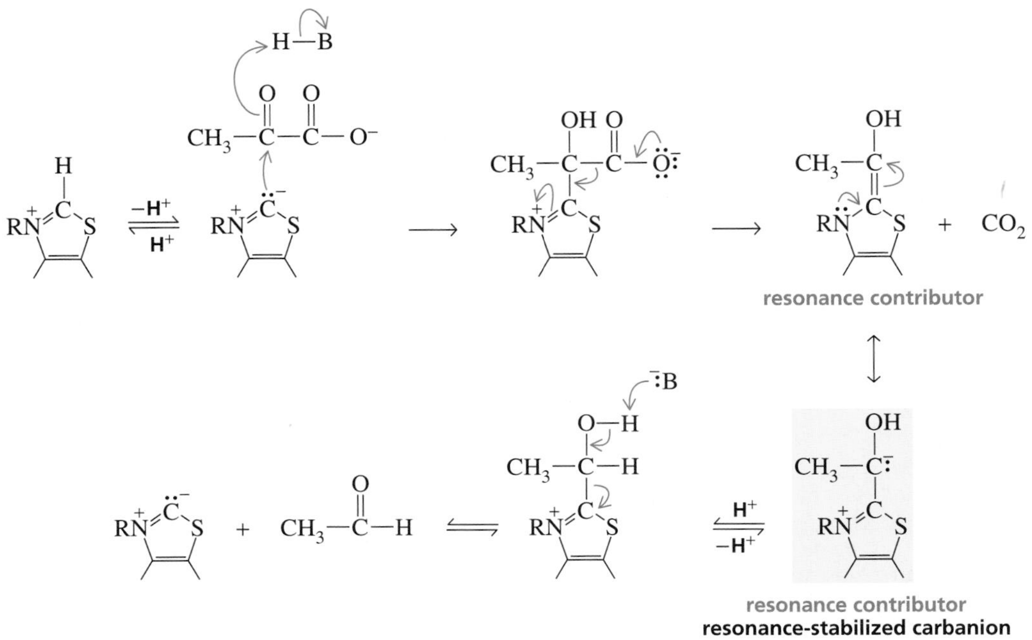

resonance contributor

resonance contributor
resonance-stabilized carbanion

PROBLEM 7

Draw structures that show the similarity between decarboxylation of the pyruvate–TPP intermediate and decarboxylation of a β-keto acid.

PROBLEM 8

Acetolactate synthase is another example of a TPP-requiring enzyme. It also catalyzes the decarboxylation of pyruvate but transfers the resulting two-carbon fragment to another molecule of pyruvate, forming acetolactate. This is the first step in the biosynthesis of the amino acids valine and leucine. Propose a mechanism for acetolactate synthase.

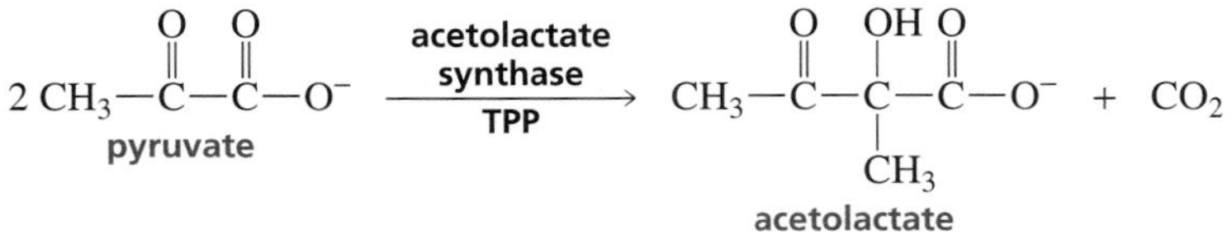

PROBLEM 9

Acetolactate synthase can also transfer the two-carbon fragment from pyruvate to α-ketobutyrate, forming α-aceto-α-hydroxybutyrate. This is the first step in the formation of isoleucine. Propose a mechanism for this reaction.

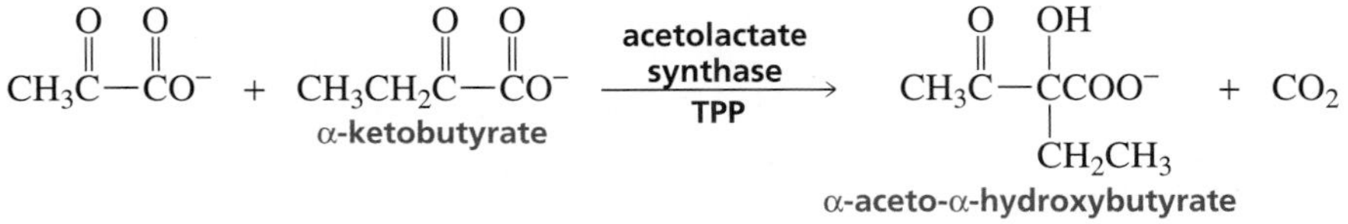

PROBLEM 10

An unfortunate effect of drinking too much alcohol—a hangover—is attributable to the acetaldehyde formed when ethanol is oxidized. There is some evidence that vitamin B$_1$ can cure a hangover. How can the vitamin do this?

Figure 22.1 shows that the compounds obtained from the catabolism of fats, carbohydrates, and amino acids must ultimately enter the citric acid cycle. Entrance into the cycle is required for a compound to be completely metabolized. Thus fats, carbohydrates, and amino acids must be converted into compounds that are part of the citric acid cycle or can enter the cycle. Acetyl-CoA is the only compound capable of entering the citric acid cycle. The final product of carbohydrate metabolism is pyruvate. Pyruvate, therefore, must be converted into acetyl-CoA before it can enter the citric acid cycle.

The pyruvate dehydrogenase system is a group of three enzymes responsible for the conversion of pyruvate to acetyl-CoA. The system requires TPP and four other coenzymes (lipoate, coenzyme A, FAD, and NAD$^+$).

$$\underset{\text{pyruvate}}{CH_3-\overset{\overset{\displaystyle O}{\|}}{C}-\overset{\overset{\displaystyle O}{\|}}{C}-O^-} \quad \xrightarrow[\text{system}]{\text{pyruvate dehydrogenase}} \quad \underset{\text{acetyl-CoA}}{CH_3-\overset{\overset{\displaystyle O}{\|}}{C}-SCoA} + CO_2$$

The first enzyme in the system is a TPP-requiring enzyme that catalyzes the reaction of TPP with pyruvate to form the same resonance-stabilized carbanion formed by pyruvate decarboxylase and by the enzyme in Problems 8 and 9. The second enzyme of the system (E$_2$) requires **lipoate,** a coenzyme that is attached to its enzyme by an amide linkage to a lysine side chain. The disulfide linkage of lipoate is cleaved when it undergoes nucleophilic attack by the carbanion. In the next step, the TPP ylide carbanion is eliminated from the tetrahedral intermediate. **Coenzyme A (CoASH)** reacts with the thioester in a transthioesterification reaction (one thioester is converted into another), substituting coenzyme A for dihydrolipoate. At this point the final reaction product (acetyl-CoA) has been formed. However, before another cycle of catalysis can occur, dihydrolipoate must be oxidized back to lipoate. This is done by the third enzyme (E$_3$), a FAD-requiring enzyme. Oxidation of dihydrolipoate by FAD forms enzyme-bound FADH$_2$. NAD$^+$ then oxidizes FADH$_2$ back to FAD.

mechanism for the pyruvate dehydrogenase system

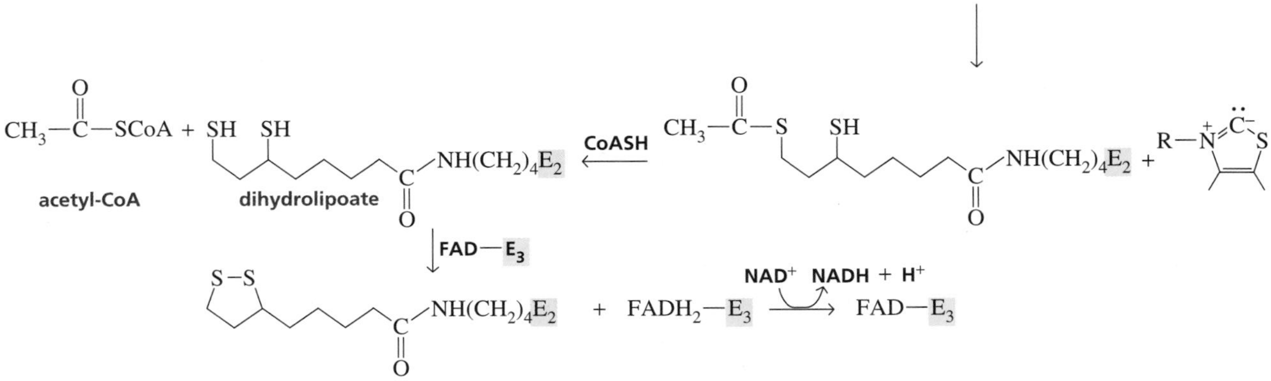

The vitamin precursor of coenzyme A is pantothenate. We have seen that CoASH is used in biological systems to activate a carboxylic acid by converting it into a thioester, which is many times more reactive toward nucleophilic acyl substitution reactions than is a carboxylic acid (Section 15.17).

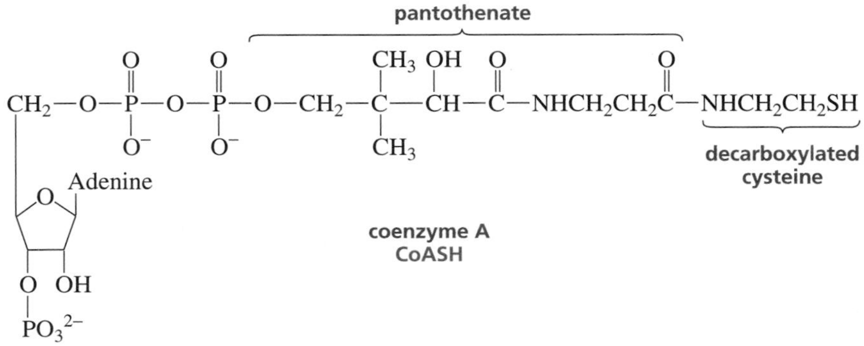

The increased reactivity of thioesters compared with that of carboxylic acids is attributable to two factors. First, the carbonyl carbon of a thioester is more susceptible to nucleophilic attack because there is less contribution to the hybrid from the resonance structure with a positive charge on the leaving group (Y), since there is less overlap between the $3p$ orbital of sulfur and the $2p$ orbital of carbon compared with the amount of overlap between the $2p$ orbital of oxygen and the $2p$ orbital of carbon (Section 15.2). Second, a thiolate ion is a weaker base and therefore a better leaving group than hydroxide ion.

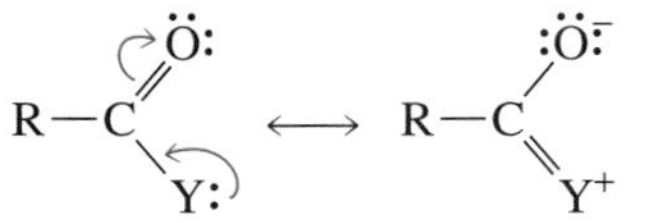

PROBLEM 11 / SOLVED

TPP is a coenzyme for transketolase, the enzyme that catalyzes the conversion of a ketopentose (xylulose-5-P) and an aldopentose (ribose-5-P) into an aldotriose (glyceraldehyde-3-P) and a ketoheptose (sedoheptulose-7-P). Notice that the total number of carbon atoms in the sugars is conserved (5 + 5 = 3 + 7). Propose a mechanism for this reaction.

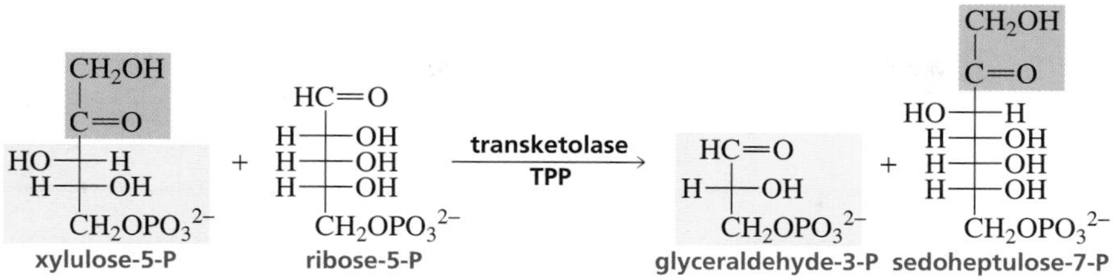

SOLUTION   The reaction shows that a two-carbon fragment is transferred from xylulose-5-P to ribose-5-P. Since TPP transfers two-carbon fragments, we know that TPP must remove the two-carbon fragment that is to be transferred from xylulose-5-P. Thus the reaction must start by TPP attacking the carbonyl group of xylulose-5-P. We can add an acid group to accept the electrons from the carbonyl group and a basic group to aid in the removal of the two-carbon fragment. The two-carbon fragment, when attached to TPP, is a resonance-stabilized carbanion that is added to the carbonyl group of ribose-5-P. Again, an acid group accepts the electrons from the carbonyl group and a basic group aids in the removal of TPP.

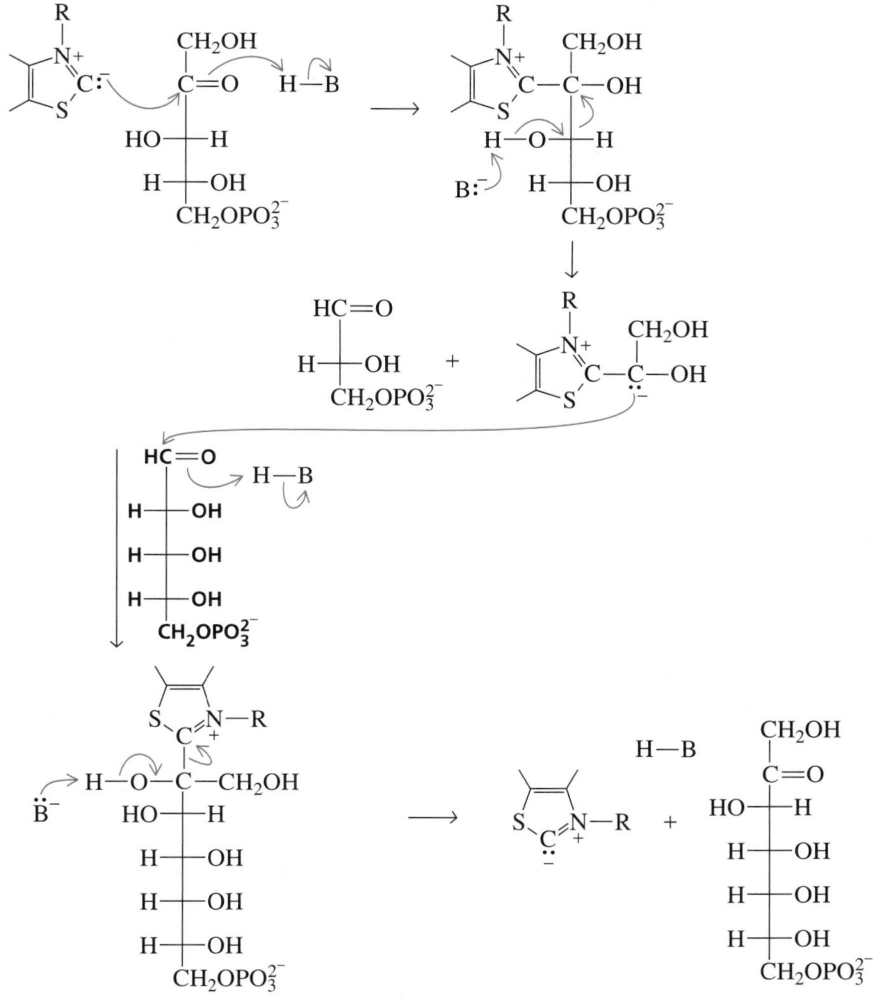

Notice the similar function of TPP in all TPP-requiring enzymes. The coenzyme attacks a carbonyl group of the substrate and allows a bond to that carbonyl group to be broken, because the electrons left behind can be delocalized into the thiazolium ring. The resulting two-carbon fragment is then transferred—to a proton in the case of pyruvate decarboxylase, to coenzyme A (via lipoate) in the pyruvate dehydrogenase system, and to a carbonyl group in Problems 8, 9, and 11.

22.5 BIOTIN

Biotin is an unusual vitamin in that it is not required in the diet because it can be synthesized by bacteria that live in the intestine. Consequently, biotin deficiencies are rare. However, this type of deficiency can be found in people who maintain a diet that is high in raw eggs. Raw eggs can cause a biotin deficiency because egg whites contain a protein (avidin) that tightly binds biotin and thereby prevents it from acting as a coenzyme. When eggs are cooked, avidin is denatured, and the denatured protein does not bind biotin. Biotin is attached to its enzyme through an amide linkage with the amino group of a lysine side chain.

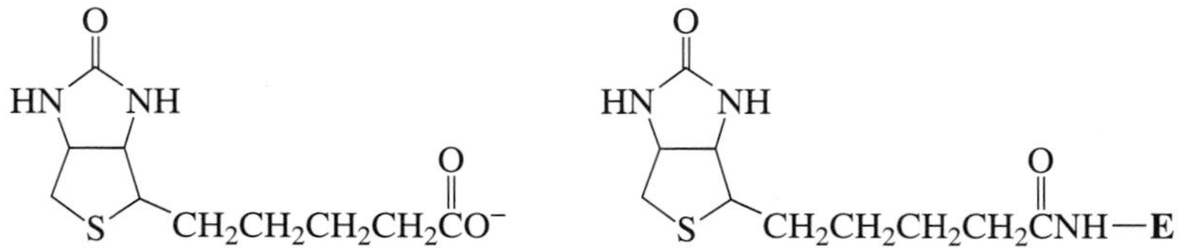

biotin   enzyme-bound biotin

Biotin is required by enzymes that catalyze the carboxylation of a carbon adjacent to a carbonyl group.

Biotin is the coenzyme required by enzymes that catalyze carboxylation of a carbon adjacent to a carbonyl group. Biotin-requiring enzymes use bicarbonate (HCO_3^-) for the source of the carboxyl group that becomes attached to the substrate.

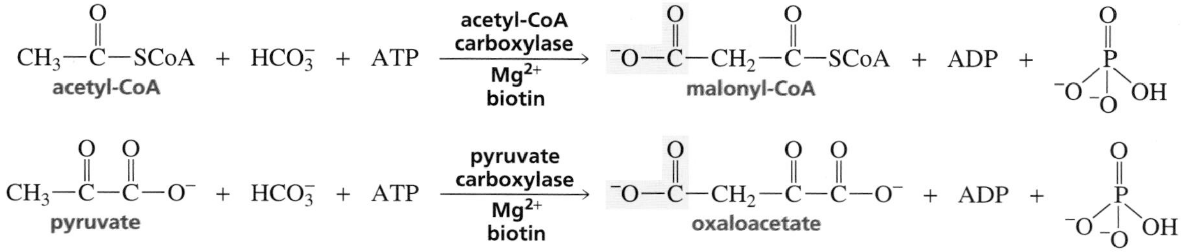

Biotin-requiring enzymes also require Mg^{2+} and ATP. The function of Mg^{2+} is to decrease the overall negative charge on ATP by complexing with two of its negatively charged oxygens. The function of ATP is to increase the reactivity of bicarbonate by giving it a good leaving group (Section 24.2).

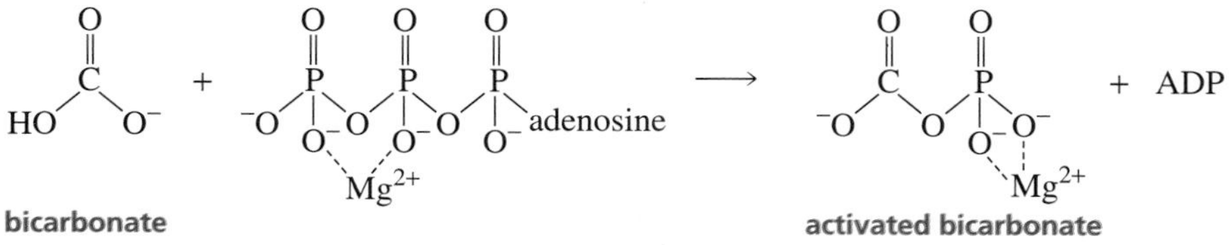

bicarbonate   activated bicarbonate

Nucleophilic attack by biotin on activated bicarbonate results in formation of carboxybiotin. Since the nitrogen of an amide is not nucleophilic, it is likely that the active form of biotin has an "enol-like" structure. Nucleophilic attack by the substrate (in this case, the enol form of acetyl-CoA) on carboxybiotin results in transfer of the carboxyl group from carboxybiotin to the substrate.

"enol-like" structure of enzyme-bound biotin

carboxybiotin

acetyl-CoA

malonyl-CoA

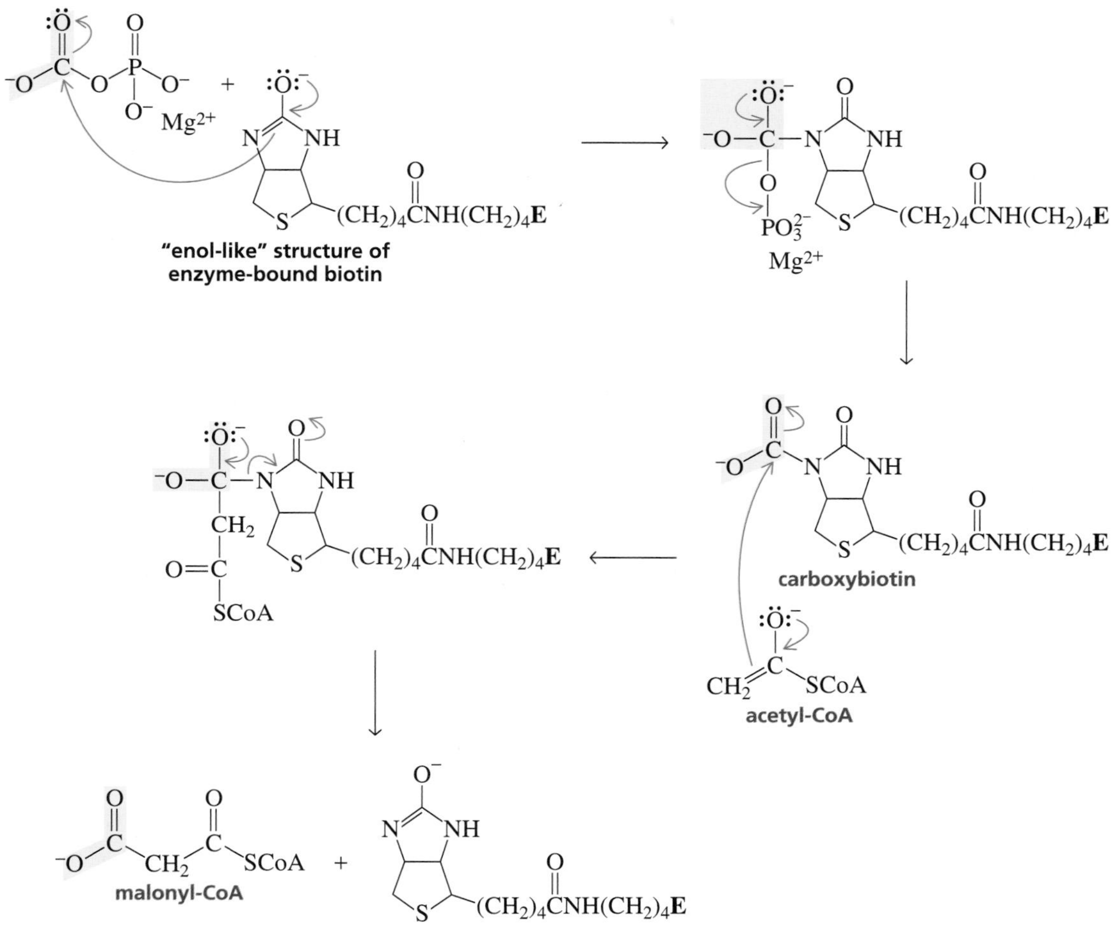

All biotin-requiring enzymes follow the same three steps: activation of bicarbonate by ATP, reaction of activated bicarbonate with biotin to form carboxybiotin, and transfer of the carboxyl group from carboxybiotin to the substrate.

Pyridoxal phosphate (PLP) is the coenzyme required by enzymes that catalyze certain transformations of amino acids. The coenzyme is derived from pyridoxine, the vitamin known as vitamin B$_6$. The name "pyridoxal" indicates that the coenzyme is a pyridine aldehyde. A deficiency in vitamin B$_6$ causes anemia. Severe deficiencies can cause seizures and death.

**22.6
PYRIDOXAL
PHOSPHATE:
VITAMIN B$_6$**

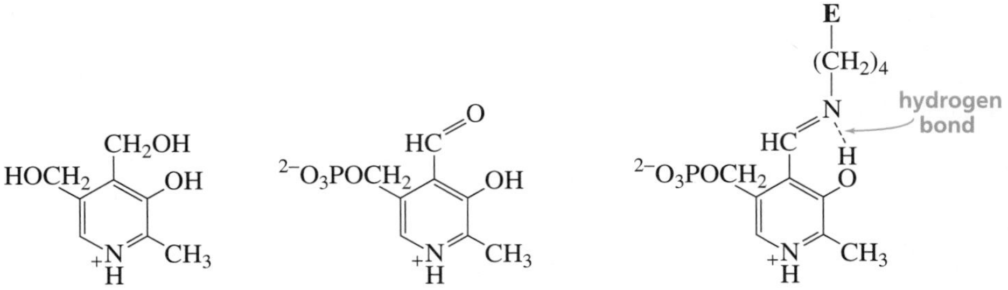

pyridoxine
vitamin B_6

pyridoxal phosphate
PLP

the coenzyme is bound to the enzyme by means of an imine linkage with a lysine residue

There are several different kinds of amino acid transformations that PLP-requiring enzymes catalyze: decarboxylation, transamination, racemization (interconversion of L- and D-amino acids), C_α—C_β bond cleavage, and α,β-elimination are the most common.

decarboxylation

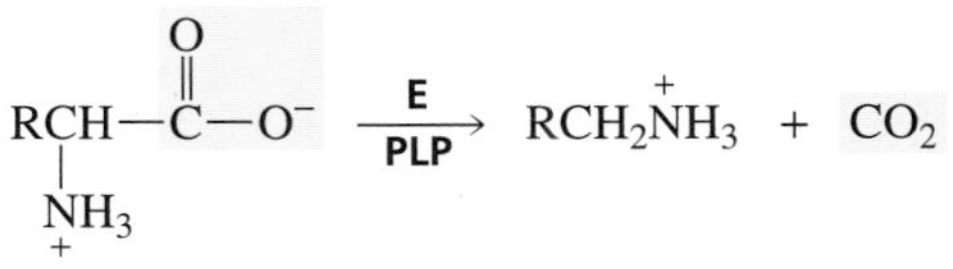

transamination

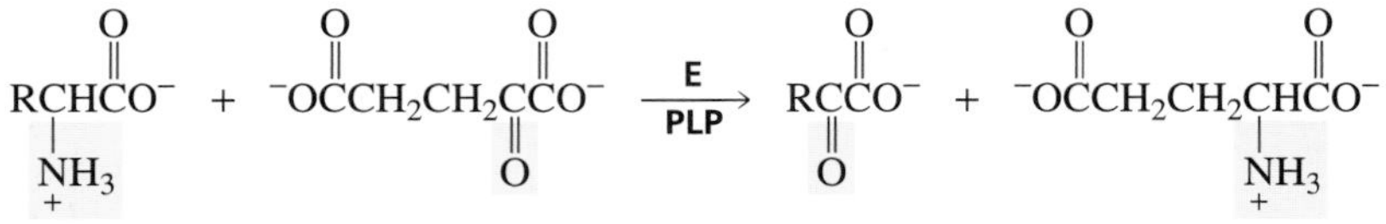

α-ketoglutarate

glutamate

racemization

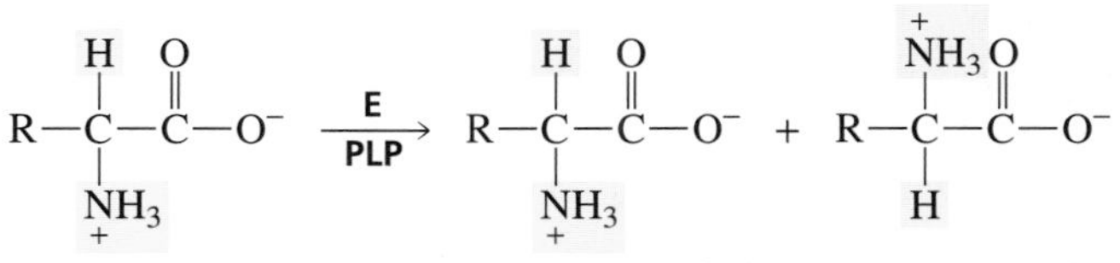

L-amino acid

L-amino acid

D-amino acid

C_α—C_β bond cleavage

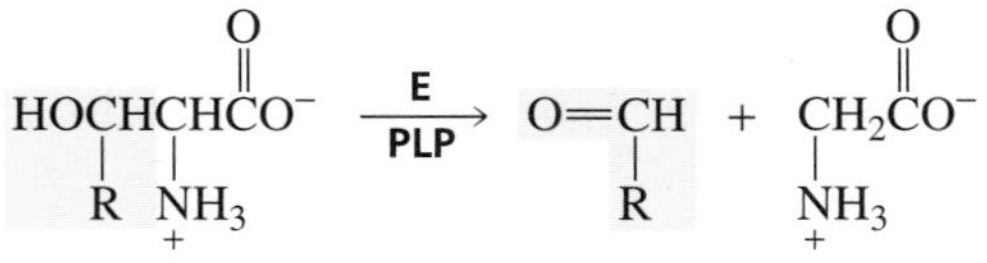

α,β-elimination

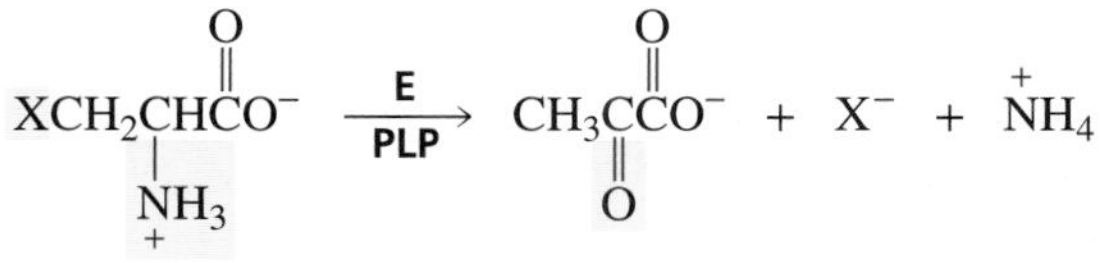

Each of these transformations requires that one of the bonds to the α-carbon of the amino acid substrate be broken in the first step of the reaction. Decarboxylation requires that the bond joining the carboxyl group to the α-carbon be broken; transamination, racemization, and α,β-elimination require that the bond joining the hydrogen to the α-carbon be broken; C$_\alpha$—C$_\beta$ bond cleavage requires that the bond joining the R group to the α-carbon be broken.

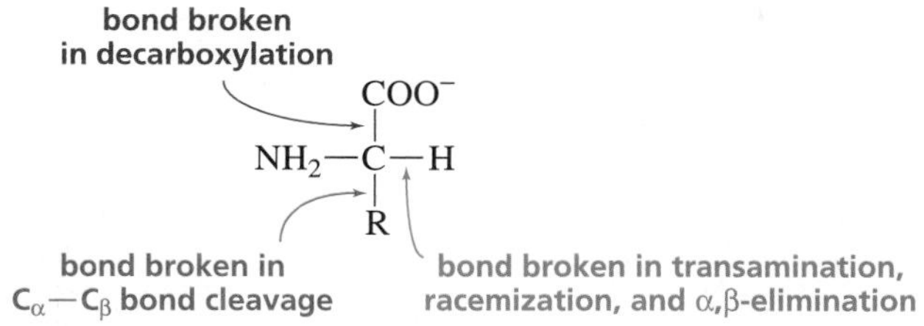

Pyridoxal phosphate (PLP) is required by enzymes that catalyze certain transformations of amino acids.

PLP attaches to its enzyme by forming an imine (a Schiff base) with an amino group of a lysine side chain. The first thing that happens in all PLP-requiring enzymes is a **transimination** reaction: the amino acid substrate reacts with the imine formed by PLP and the enzyme, forming a tetrahedral intermediate. The lysine group of the enzyme is eliminated, resulting in the formation of a new imine between PLP and the amino acid.

transimination

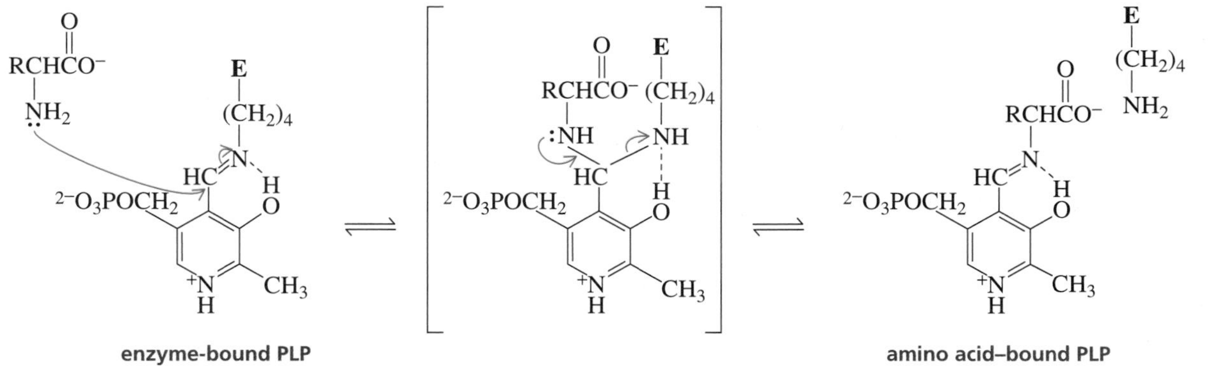

Once the amino acid has formed an imine with PLP, a bond to the α-carbon can be broken because the electrons left behind when the bond breaks can be delocalized onto the positively charged nitrogen of the pyridine ring. In other words, the nitrogen of the pyridine ring is an electron sink. If the OH group is removed from the pyridine ring, the cofactor loses much of its activity. Apparently, the hydrogen bond formed by the OH group helps weaken the bond to the α-carbon.

If the PLP-catalyzed reaction is a decarboxylation, the carboxyl group is removed from the α-carbon of the amino acid. Electron rearrangement and protonation of the α-carbon of the decarboxylated intermediate by protonated lysine or some other acid group reestablishes the aromaticity of the pyridine ring. Transimination with a lysine side chain releases the decarboxylated substrate and regenerates enzyme-bound PLP.

mechanism for PLP-catalyzed decarboxylation of an amino acid

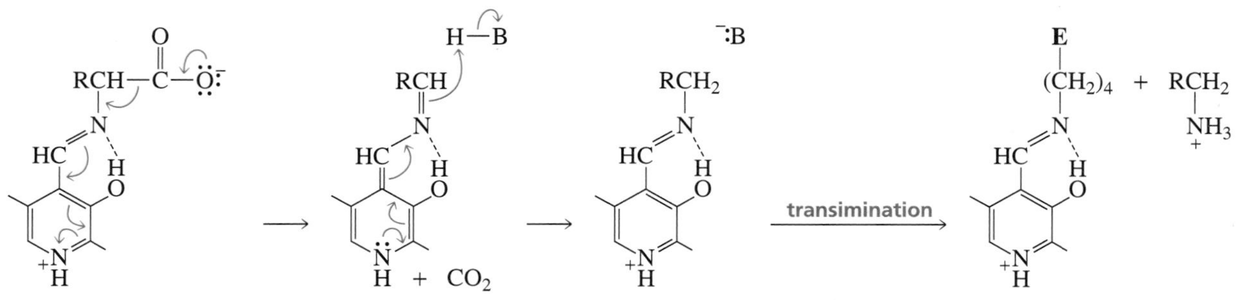

The first reaction in the catabolism of most amino acids is substitution of the amino group of the amino acid by a ketone group. This reaction is known as **transamination** because the amino group removed from the amino acid is not lost but is transferred to the ketone group of α-ketoglutarate, thereby forming glutamate. The enzymes that catalyze transamination are called *aminotransferases*. Transamination allows the amino groups of the various amino acids to be collected into a single amino acid (glutamate) so that excess nitrogen can be easily excreted. (Do not confuse *transamination* with *transimination* discussed above.)

In the first step of transamination, a proton is removed from the α-carbon of the amino acid. Rearrangement of electrons and protonation of the carbon attached to the pyridine ring, followed by hydrolysis of the imine, forms the α-keto acid and pyridoxamine. At this point the amino group has been removed from the amino acid but pyridoxamine has to be converted back to enzyme-bound PLP so another round of catalysis can occur. Pyridoxamine forms an imine with α-ketoglutarate, the second substrate of the reaction. Removal of a proton from the carbon attached to the pyridine ring, followed by rearrangement of electrons and donation of a proton to the α-carbon of the substrate, results in an imine that, when transiminated with a lysine side chain, releases glutamate and reforms enzyme-bound PLP.

mechanism for PLP-dependent transamination of an amino acid

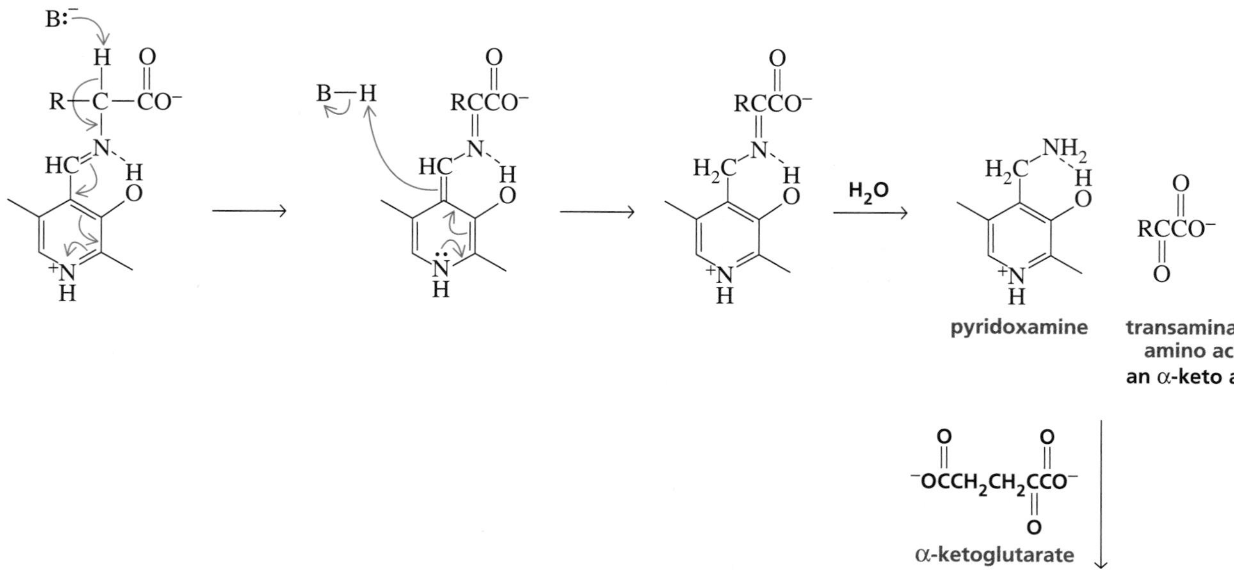

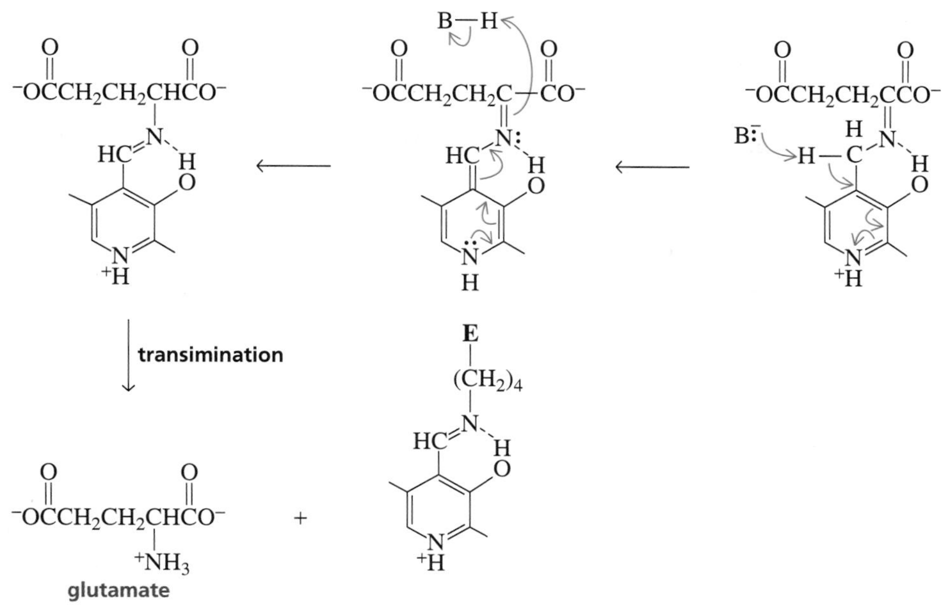

Notice that the proton transfer steps are reversed in the two phases of the reaction. Transfer of the amino group of the amino acid to pyridoxal requires proton removal from the α-carbon and proton donation to the carbon bonded to the pyridine ring. Transfer of the amino group of pyridoxamine to α-ketoglutarate requires proton removal from the carbon bonded to the pyridine ring and proton donation to the α-carbon.

HEART ATTACKS: ASSESSING THE DAMAGE

Damage to heart muscle after myocardial infarction allows aminotransferases and other enzymes to leak from the damaged cells of the heart into the bloodstream. After a heart attack, the severity of damage done to the heart can be determined from the concentrations of alanine aminotransferase and aspartate aminotransferase in the bloodstream.

PHENYLKETONURIA: AN INBORN ERROR OF METABOLISM

Tyrosine is a nonessential amino acid because the body can make it by hydroxylating phenylalanine. About 1 in every 20,000 babies is born without phenylalanine hydroxylase, the enzyme that converts phenylalanine into tyrosine. This genetic disease is called phenylketonuria (PKU).

Tyrosine is transaminated to *para*-hydroxyphenylpyruvate. Without phenylalanine hydroxylase, the level of phenylalanine builds up and, when it reaches a high concentration, it is transaminated to phenylpyruvate. The high level of phenylpyruvate found in urine gave the disease its name.

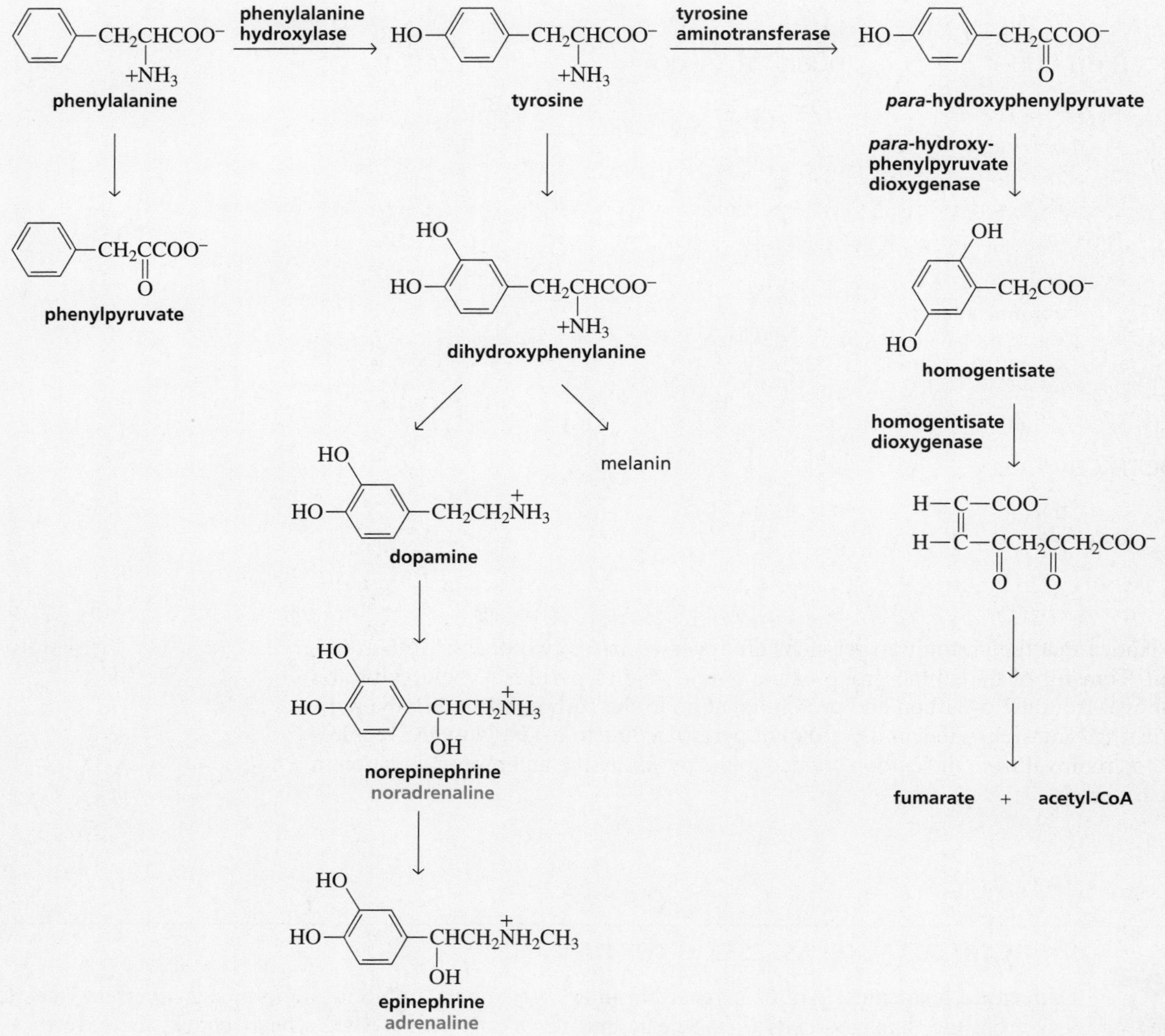

phenylalanine →(**phenylalanine hydroxylase**)→ tyrosine →(**tyrosine aminotransferase**)→ *para*-hydroxyphenylpyruvate

phenylalanine → phenylpyruvate

tyrosine → dihydroxyphenylanine

para-hydroxyphenylpyruvate →(***para*-hydroxy-phenylpyruvate dioxygenase**)→ homogentisate

homogentisate →(**homogentisate dioxygenase**)→ (intermediate) → fumarate + acetyl-CoA

dihydroxyphenylanine → dopamine and melanin

dopamine → norepinephrine / noradrenaline → epinephrine / adrenaline

All babies born in the United States are tested about 3 days after they begin drinking milk for high serum phenylalanine levels, which would indicate a build up of phenylalanine. If the baby is found to lack phenylalanine hydroxylase, he or she is immediately put on a diet low in phenylalanine and high in tyrosine. As long as the phenylalanine level is kept under careful control for the first 5 to 10 years of life, the baby will experience no adverse effects. If, however, the diet is not controlled, the baby will be severely mentally retarded by the time he or she is a few months old. Untreated children have paler skin and fairer hair than other members of the family because, since they don't synthesize tyrosine, their melanin levels are low. Melanin is the compound responsible for pigmentation. Half of untreated phenylketonurics are dead by age twenty. When a woman with PKU becomes pregnant, she must return to the low phenylalanine diet she had as a child, because a high level of phenylalanine can cause abnormal development of the fetus.

Another genetic disease that results from a deficiency of an enzyme in the pathway discussed above is alcaptonuria, which is caused by lack of homogentisate dioxygenase. The only ill effect of this enzyme deficiency is that the urine of those who don't have this enzyme is black as a result of immediate air oxidation of the homogentisate they excrete.

PROBLEM 12◆

Using the pathway for phenylalanine degradation, answer the following questions:

a. What coenzyme and what other organic compound are needed by tyrosine aminotransferase?

b. What compound is used to supply the methyl group needed to convert noradrenaline into adrenaline? (*Hint:* See Section 9.11)

c. What bond in homogentisate is oxidized by homogentisate dioxygenase? (*Hint:* Keto-enol tautomerism occurs after oxidation.)

PROBLEM 13◆

α-Keto acids other than α-ketoglutarate can be used to accept the amino group from pyridoxamine in enzyme-catalyzed transaminations. What amino acids are formed from the following α-keto acids?

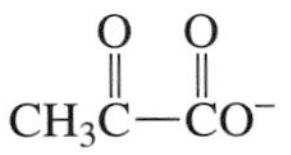

$$CH_3\overset{O}{\underset{}{C}}-\overset{O}{\underset{}{C}}O^-$$
pyruvate

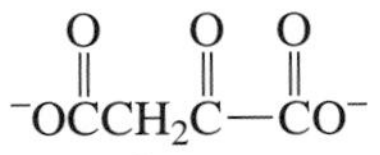

$$^-O\overset{O}{\underset{}{C}}CH_2\overset{O}{\underset{}{C}}-\overset{O}{\underset{}{C}}O^-$$
oxaloacetate

The first step in the PLP-catalyzed racemization of an amino acid is the same as the first step in the PLP-catalyzed transamination of an amino acid: removal of a proton from the α-carbon of the amino acid. In the second step of racemization, reprotonation occurs on the α-carbon. The proton can be donated to the sp^2 hybridized α-carbon from either the top or the bottom of the plane defined by the double bond. Consequently, both D- and L-amino acids are formed. In other words, the L-amino acid is racemized.

mechanism for PLP-catalyzed racemization of an amino acid

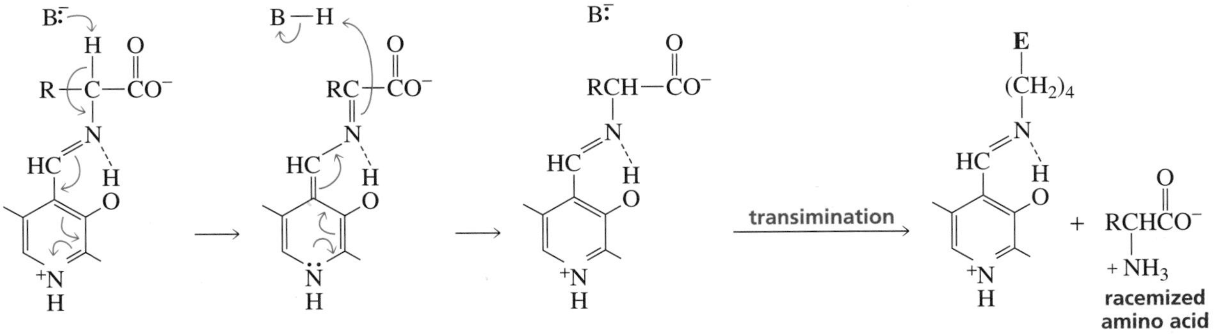

Compare the second step in a PLP-catalyzed transamination with the second step in a PLP-catalyzed racemization. In an enzyme that catalyzes transamination, an acidic group at the active site of the enzyme is in position to donate a proton to the carbon attached to the pyridine ring. The enzyme that catalyzes racemization does not have this acidic group, so the substrate is protonated at the α-carbon. In other words, the coenzyme is carrying out the chemical reaction but the enzyme is determining the course of the reaction.

In the first step of the mechanism for PLP-catalyzed C_α—C_β bond cleavage, a basic group at the active site of the enzyme removes a proton from an OH group bonded to the β-carbon of the amino acid. This causes the C_α—C_β bond to be cleaved. Serine and threonine are the only two amino acids that can serve as substrates for this reaction, because they are the only amino acids that have an OH group bonded to their β-carbon. When serine is the substrate, the product of the cleavage reaction is formaldehyde (R = H); when threonine is the substrate, the product of the cleavage reaction is acetaldehyde (R = CH_3). Electron rearrangement and protonation of the α-carbon of the amino acid followed by transimination with a lysine side chain releases glycine.

mechanism for PLP-catalyzed C_α — C_β bond cleavage

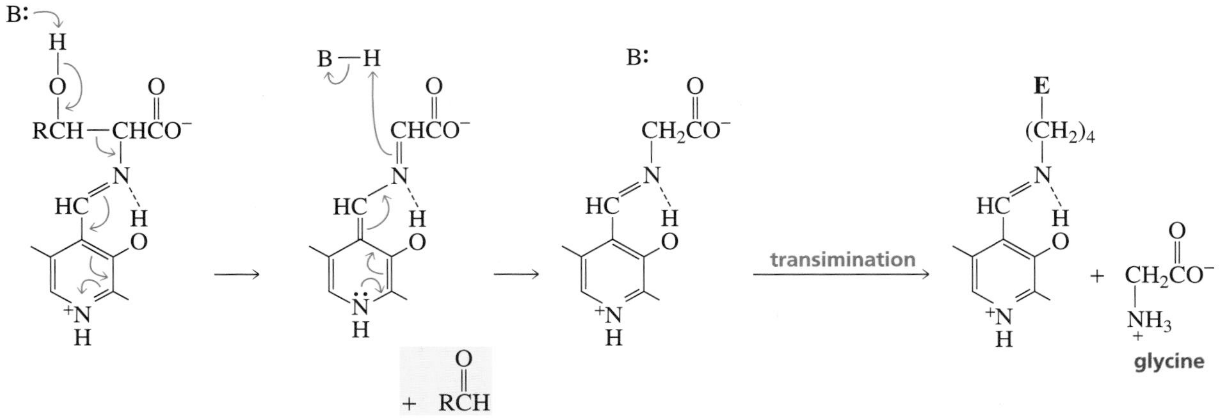

The formaldehyde formed when serine undergoes C_α—C_β bond cleavage never leaves the active site of the enzyme; it is immediately transferred to tetrahydrofolate (Section 22.8).

PROBLEM 14

Propose a mechanism for a PLP-catalyzed α,β-elimination.

If all PLP-requiring enzymes start with the same substrate, how can three different bonds be cleaved in the first step of the reaction? The bond that is cleaved in the first step depends on the conformation of the amino acid that the enzyme binds. Each enzyme binds a particular conformation of the amino acid. There is free rotation about the C_α—N bond of the amino acid, and an enzyme can bind any of the possible conformations about this bond. Which conformation a particular enzyme binds depends on the orientation of the orbitals of the bond that will be broken in the first step of the reaction. The enzyme will bind the conformation in which these orbitals lie parallel to the p orbitals of the conjugated system. In this way the orbital containing the electrons left behind when the bond is broken can overlap with the orbitals of the conjugated system. If this overlap cannot occur, the electrons cannot be delocalized into the conjugated system.

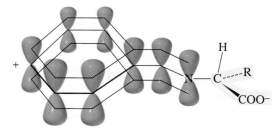

**Schiff base between PLP and the amino acid
before breaking the carbon–hydrogen bond**

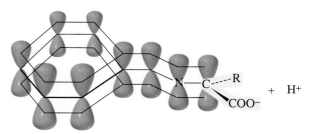

Schiff base between PLP and the amino acid

PROBLEM 15◆

Which of the following compounds is more easily decarboxylated?

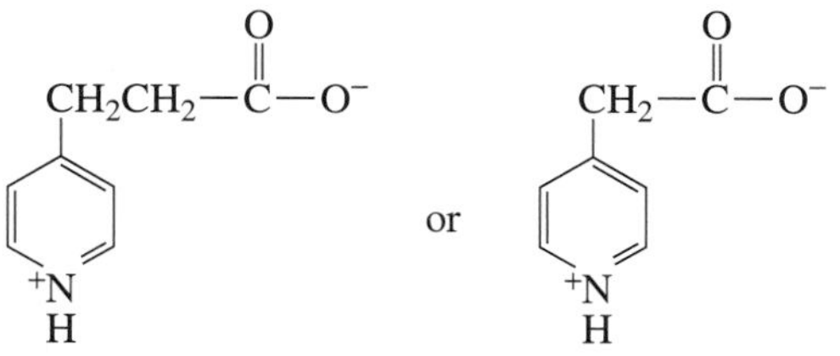

$$CH_2CH_2-\overset{\overset{\displaystyle O}{\|}}{C}-O^-$$

or

$$CH_2-\overset{\overset{\displaystyle O}{\|}}{C}-O^-$$

PROBLEM 16

If a PLP-requiring enzymatic reaction is carried out at a pH where the pyridine nitrogen is not protonated, PLP's ability to catalyze an amino acid transformation is greatly reduced. Account for this observation.

PROBLEM 17

If the OH substituent of pyridoxal phosphate is replaced by an OCH$_3$ substituent, PLP's ability to catalyze an amino acid transformation is greatly reduced. Account for this observation.

Enzymes that catalyze certain rearrangement reactions require **coenzyme B$_{12}$,** a coenzyme derived from vitamin B$_{12}$. The structure of vitamin B$_{12}$ was determined by Dorothy Crowfoot Hodgkin using X-ray crystallography. The vitamin has a cyano group (or HO$^-$ or H$_2$O) coordinated to cobalt (Section 26.4). In coenzyme B$_{12}$, the cyano group is replaced by an adenosyl group.

22.7
COENZYME B$_{12}$:
VITAMIN B$_{12}$

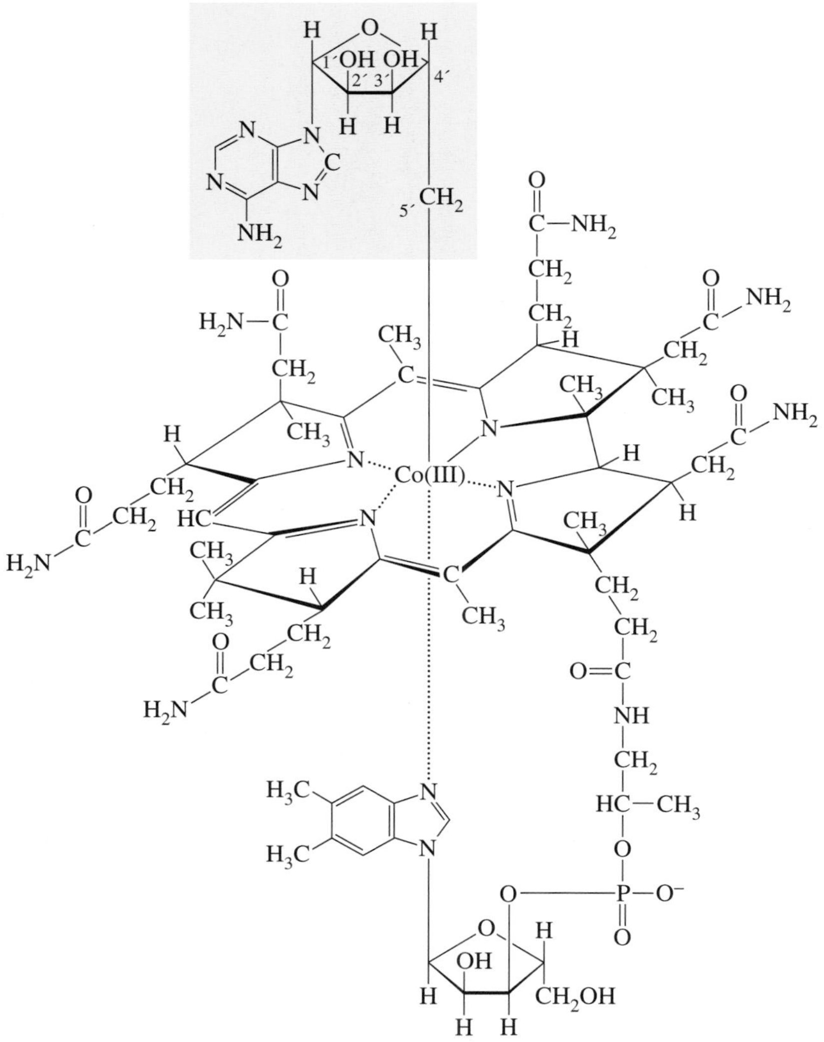

Dorothy Crowfoot Hodgkin

coenzyme B$_{12}$

Dorothy Crowfoot Hodgkin (1910–1994) *was born in Egypt to English parents. She received an undergraduate degree from Somerville College, Oxford and earned a Ph.D. from Cambridge University. She determined the structures of penicillin, insulin, and vitamin B$_{12}$. For her work on vitamin B$_{12}$, she received the 1964 Nobel Prize in chemistry. She was a professor of chemistry at Somerville, where one of her students was former Prime Minister Margaret Thatcher. Hodgkin was a founding member of Pugwash, an organization whose purpose was to further communication between scientists on both sides of the Iron Curtain.*

Animals and plants cannot synthesize vitamin B$_{12}$; in fact, only a few microorganisms can synthesize it. Humans must obtain all their vitamin B$_{12}$ from their diet, particularly from meat. Because vitamin B$_{12}$ is needed in only very small amounts, deficiencies caused by consumption of insufficient amounts of the vitamin are rare but have been found in vegetarians who eat no animal products. A vitamin B$_{12}$ deficiency is most commonly caused by an inability to absorb the vitamin into the intestine. The deficiency causes pernicious anemia. Examples of enzyme-catalyzed reactions that require coenzyme B$_{12}$ are shown below.

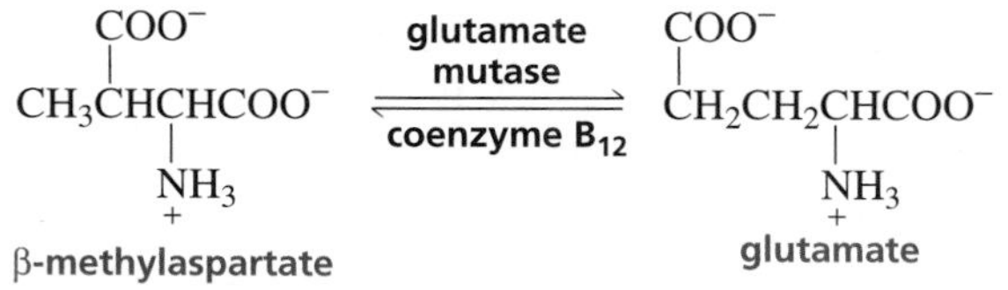

$$\underset{\text{β-methylaspartate}}{CH_3\underset{\underset{NH_3^+}{|}}{CH}CHCOO^-} \underset{\text{coenzyme B}_{12}}{\overset{\text{glutamate mutase}}{\rightleftharpoons}} \underset{\text{glutamate}}{CH_2CH_2\underset{\underset{NH_3^+}{|}}{CH}COO^-}$$

$$\underset{\substack{\text{methylmalonyl-CoA}}}{\underset{\substack{| \\ \text{COO}^-}}{\text{CH}_3\text{CHCSCoA}}\overset{\text{O}}{\parallel}} \quad \underset{\substack{\text{coenzyme B}_{12}}}{\overset{\substack{\text{methylmalonyl-CoA} \\ \text{mutase}}}{\rightleftharpoons}} \quad \underset{\substack{\text{succinyl-CoA}}}{\underset{\substack{| \\ \text{COO}^-}}{\text{CH}_2\text{CH}_2\text{CSCoA}}\overset{\text{O}}{\parallel}}$$

$$\underset{\substack{\text{1,2-propanediol}}}{\underset{\substack{| \\ \text{OH}}}{\text{CH}_3\text{CHCH}_2\text{OH}}} \quad \underset{\substack{\text{coenzyme B}_{12}}}{\overset{\substack{\text{dioldehydrase}}}{\longrightarrow}} \quad \left[\underset{\substack{\text{a hydrate}}}{\underset{\substack{| \\ \text{OH}}}{\text{CH}_3\text{CH}_2\text{CHOH}}}\right] \quad \overset{-\text{H}_2\text{O}}{\longrightarrow} \quad \underset{\substack{\text{propionaldehyde}}}{\text{CH}_3\text{CH}_2\text{CH}}\overset{\text{O}}{\overset{\parallel}{}}$$

In each of these coenzyme B$_{12}$–requiring reactions, a group (Y) bonded to one carbon changes places with a hydrogen bonded to an adjacent carbon.

$$\underset{\substack{\text{Y} \quad \text{H}}}{-\overset{|}{\underset{|}{\text{C}}}-\overset{|}{\underset{|}{\text{C}}}-} \quad \underset{\substack{\text{dependent enzyme}}}{\overset{\substack{\text{a coenzyme B}_{12^-}}}{\rightleftharpoons}} \quad \underset{\substack{\text{H} \quad \text{Y}}}{-\overset{|}{\underset{|}{\text{C}}}-\overset{|}{\underset{|}{\text{C}}}-}$$

For example, glutamate mutase and methylmalonyl-CoA mutase both catalyze a reaction in which the COO$^-$ group bonded to one carbon changes places with a hydrogen of an adjacent methyl group. In the reaction catalyzed by dioldehydrase, an OH group changes places with a methylene hydrogen. The resulting product is a hydrate that loses water to form propionaldehyde.

The chemistry of coenzyme B$_{12}$ takes place at the bond joining the cobalt and the adenosyl group. The currently accepted mechanism for dioldehydrase involves an initial homolytic cleavage of this unusually weak bond (26 kcal/mol compared with 98 kcal/mol for a C—H bond). Breaking the bond forms an adenosyl radical and reduces Co(III) to Co(II). The adenosyl radical abstracts a hydrogen atom from the C-1 carbon of the substrate, thereby becoming 5'-deoxyadenosine. A hydroxyl radical ($\cdot$OH) migrates from C-2 to C-1, creating a radical at C-2. This radical abstracts a hydrogen atom from 5'-deoxyadenosine, forming the rearranged product and reforming the adenosyl radical. The adenosyl radical recombines with Co(II), reforming the coenzyme. The enzyme–coenzyme complex is then ready for another catalytic cycle. The initial product is a hydrate that loses water to form propionaldehyde, the final product of the reaction.

Coenzyme B$_{12}$ is required by enzymes that catalyze the exchange of a hydrogen bonded to one carbon with a group bonded to an adjacent carbon.

the role of adenosylcobalamin in a coenzyme B$_{12}$–dependent enzyme-catalyzed reaction

It is likely that all coenzyme B$_{12}$–requiring enzymes catalyze reactions by means of the same general mechanism. The role of the coenzyme is to provide a way to remove a hydrogen atom from the substrate. Once the hydrogen atom has been removed, an adjacent group can migrate to take its place. The coenzyme then gives back the hydrogen atom, delivering it to the carbon that lost the migrating group.

PROBLEM 18

Ethanolamine ammonia lyase, a coenzyme B_{12}–requiring enzyme, catalyzes the following reaction. Propose a mechanism for this reaction.

$$HOCH_2CH_2NH_2 \longrightarrow CH_3\overset{\overset{\displaystyle O}{\|}}{C}H + NH_3$$

PROBLEM 19 ◆

A fatty acid with an even number of carbon atoms is metabolized to acetyl-CoA, which can enter the citric acid cycle. A fatty acid with an odd number of carbon atoms is metabolized to acetyl-CoA and one equivalent of propionyl-CoA. Two coenzyme-requiring enzymes are needed to convert propionyl-CoA into succinyl-CoA, a citric acid cycle intermediate. Give the two reactions and indicate the required coenzymes.

22.8 TETRAHYDROFOLATE: FOLIC ACID

Tetrahydrofolate (THF) is the coenzyme used by enzymes catalyzing reactions that donate a group containing a single carbon to their substrates. The one-carbon group can be a methyl group (CH_3), a methylene group (CH_2), or a formyl group (HC=O). Tetrahydrofolate results from the reduction of two double bonds of folic acid (folate), its precursor vitamin. Bacteria synthesize folate, but mammals cannot.

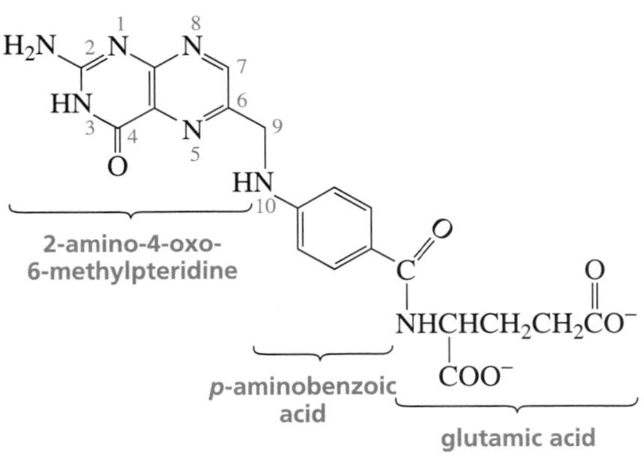

2-amino-4-oxo-6-methylpteridine

p-aminobenzoic acid

glutamic acid

folic acid (folate)

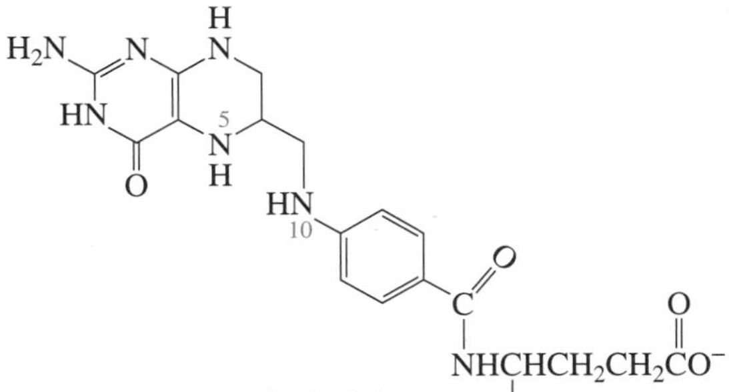

tetrahydrofolate
THF

There are six different THF-coenzymes. N^5-Methyl-THF transfers a methyl group (CH_3), N^5,N^{10}-methylene-THF transfers a methylene group (CH_2), and the others transfer a formyl group (HC=O).

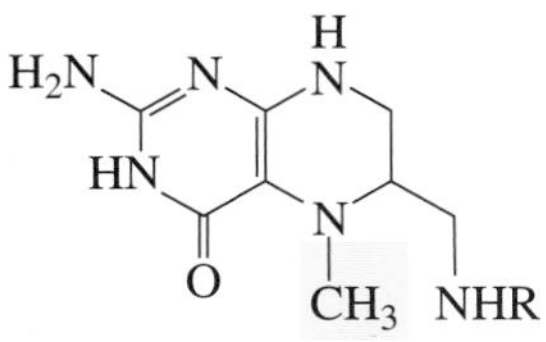

N^5-methyl-THF

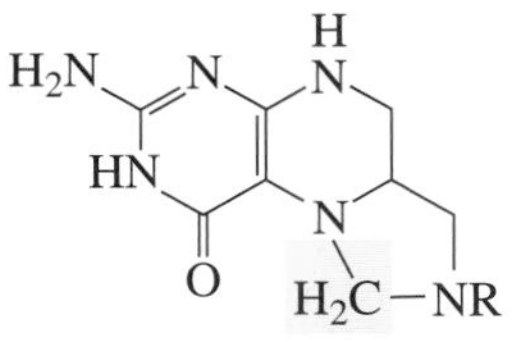

N^5,N^{10}-methylene-THF

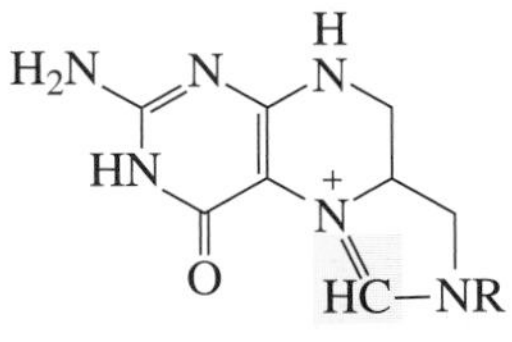

N^5,N^{10}-methenyl-THF

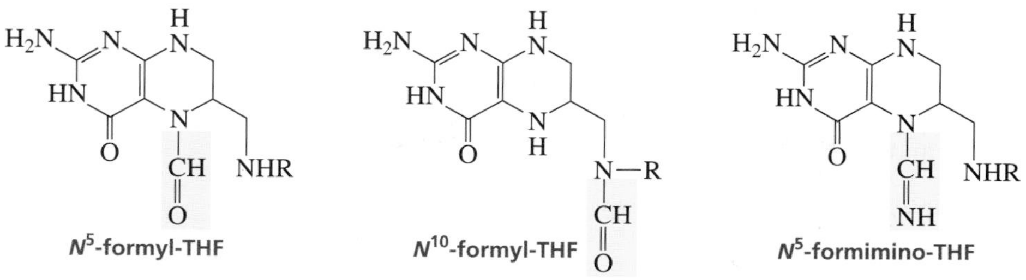

N^5-formyl-THF
N^{10}-formyl-THF
N^5-formimino-THF

Homocysteine methyl transferase and glycinamide ribonucleotide (GAR) transformylase are examples of enzymes that require THF-coenzymes.

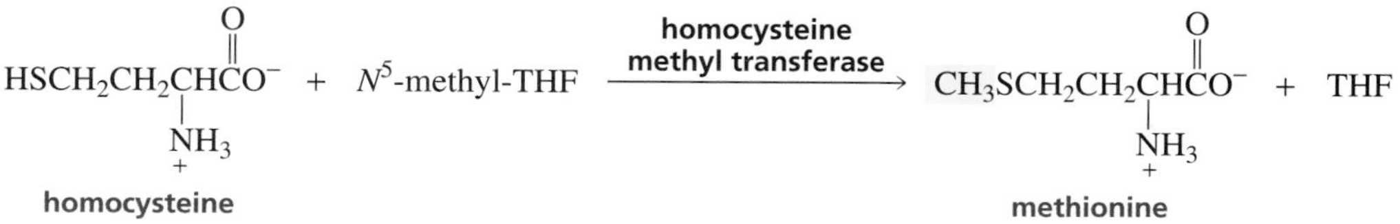

homocysteine
methyl transferase
homocysteine
methionine

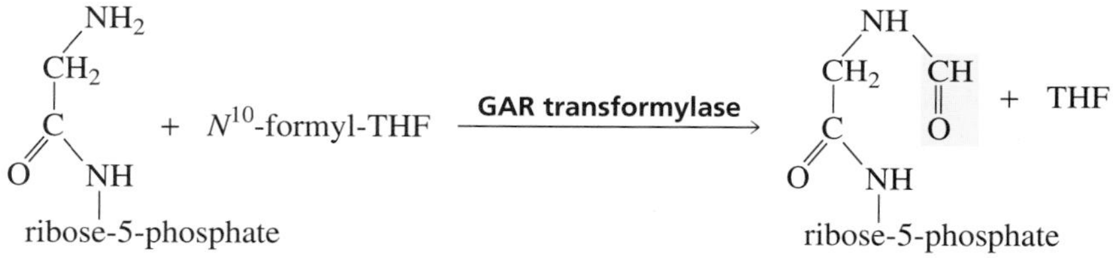

GAR transformylase
ribose-5-phosphate
ribose-5-phosphate

Tetrahydrofolate (THF) is the coenzyme required by enzymes that catalyze the donation of a group containing one carbon to their substrates.

Thymidylate Synthase: the Enzyme That Converts U's into T's

RNA is a polymer of nucleotides in which the heterocyclic components are adenine, guanine, cytosine, and uracil (A, G, C, U). DNA is a polymer of nucleotides in which the heterocyclic components are adenine, guanine, cytosine, and thymine (A, G, C, T). In other words, the heterocyclic bases in RNA and DNA differ in that RNA contains U's, while DNA contains T's (Sections 24.1, and 24.13). The T's used for the biosynthesis of DNA are synthesized from U's by thymidylate synthase, an enzyme that requires N^5,N^{10}-methylene-THF as a coenzyme. Even though the only structural difference between a U and a T is a methyl group, a T is synthesized by first transferring a methylene group to a U.

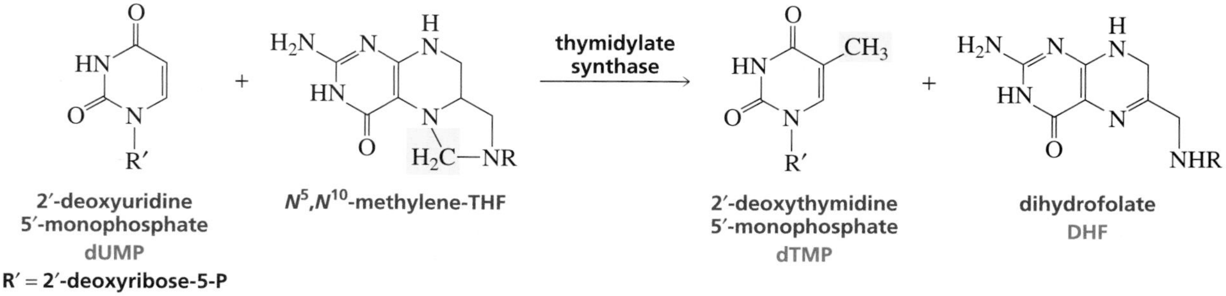

thymidylate
synthase
2′-deoxyuridine
5′-monophosphate
dUMP
R′ = 2′-deoxyribose-5-P
N^5,N^{10}-methylene-THF
2′-deoxythymidine
5′-monophosphate
dTMP
dihydrofolate
DHF

In the first step of the reaction catalyzed by thymidylate synthase, a nucleophilic cysteine group at the active site of the enzyme attacks the β-carbon of uridine (an example of conjugate addition; Section 16.14). A subsequent nucleophilic attack by the α-carbon of uridine on the methylene group of N^5,N^{10}-

methylene-THF forms a covalent bond between uridine and the coenzyme. A proton on the α-carbon of uridine is removed with the help of a basic group at the active site of the enzyme, eliminating the coenzyme. Transfer of a hydride ion from the coenzyme to uridine and elimination of the enzyme results in formation of thymidine and dihydrofolate (DHF).

mechanism for catalysis by thymidylate synthase

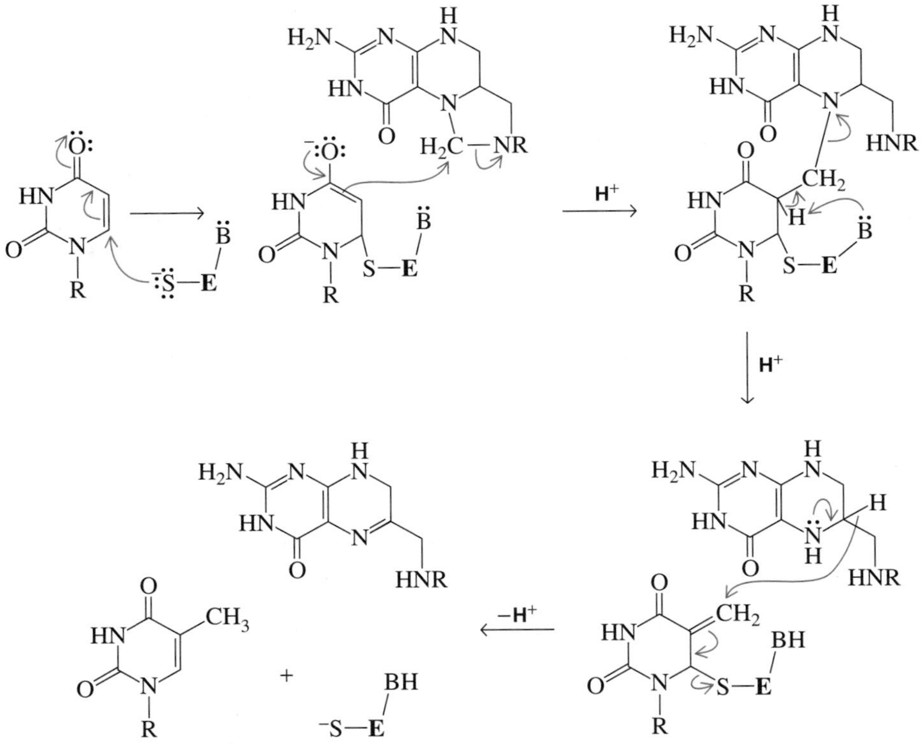

Notice that the coenzyme initially transfers a methylene group to the substrate. The methylene group is subsequently reduced to a methyl group. Because the coenzyme is the reducing agent, it is oxidized. The oxidized coenzyme is dihydro-folate.

When the reaction is over, dihydrofolate must be converted back to N^5,N^{10}-methylene-THF so that it can undergo another catalytic cycle. This reaction occurs in two steps. In the first step, dihydrofolate is reduced to tetrahydrofolate. In the second step, serine hydroxymethyl transferase catalyzes the transfer of the hydroxymethyl group from serine to the coenzyme. Serine hydroxymethyl transferase is the PLP-requiring enzyme that cleaves the C_α—C_β bond of serine to form glycine and formaldehyde (Section 22.6). In other words, the formaldehyde cleaved off serine is immediately transferred to THF to form N^5,N^{10}-methylene-THF—which is fortunate since formaldehyde is cytotoxic (it kills cells).

$$\text{dihydrofolate} + \text{NADPH} + \text{H}^+ \xrightarrow{\substack{\text{dihydrofolate} \\ \text{reductase}}} \text{tetrahydrofolate} + \text{NADP}^+$$

$$\text{tetrahydrofolate} + \text{HOCH}_2\underset{\underset{\text{serine}}{\overset{|}{\text{NH}_2}}}{\text{CHCOO}^-} \xrightarrow[\text{PLP}]{\substack{\text{serine hydroxymethyl} \\ \text{transferase}}} N^5, N^{10}\text{-methylene-THF} + \underset{\underset{\text{glycine}}{\overset{|}{\text{NH}_2}}}{\text{CH}_2\text{COO}^-}$$

Cancer Chemotherapy

Cancer is associated with rapidly growing and proliferating cells. Because cells cannot multiply if they cannot synthesize DNA, several cancer chemotherapeutic agents have been developed based on inhibiting thymidylate synthase and dihydrofolate reductase. If a cell cannot make thymidine, it cannot synthesize DNA. Inhibiting dihydrofolate reductase also prevents the synthesis of thymidine, because cells have a limited amount of tetrahydrofolate. If they cannot convert dihydrofolate back to tetrahydrofolate, they cannot continue to synthesize thymidine.

A common anticancer drug that inhibits thymidylate synthase is 5-fluorouracil. 5-Fluorouracil and uracil react with thymidylate synthase in the same way. However, the fluorine at the 5-position cannot be removed by the base in the second step of the mechanism, since fluorine is too electronegative to come off as F^+. Because fluorine cannot be removed, the reaction stops at this point, leaving the enzyme permanently attached to the substrate. The active site of the enzyme is now blocked with 5-fluorouracil, so thymidine can no longer be synthesized. This means that the synthesis of DNA is also stopped. Unfortunately, most anticancer drugs cannot discriminate between diseased and normal cells. This is why cancer chemotherapy is associated with terrible side effects. However, since cancer cells undergo uncontrolled cell division, they are dividing more rapidly than normal cells and thus are the hardest hit by cancer chemotherapeutic agents.

5-fluorouracil
5-FU

the enzyme has become
irreversibly attached to the substrate

5-Fluorouracil is an example of a **mechanism-based inhibitor;** it inactivates the enzyme by undergoing part of the normal catalytic mechanism. It is also called a **suicide inhibitor** because when the enzyme reacts with it, the enzyme "commits suicide." The example of 5-fluorouracil illustrates the importance of knowing the mechanism for an enzyme-catalyzed reaction. If you know the mechanism, you may be able to design an inhibitor to turn the reaction off at a certain step.

Aminopterin and Methotrexate are anticancer drugs that are inhibitors of dihydrofolate reductase. Because their structures are similar to that of dihydrofolate, they compete with dihydrofolate for binding at the active site of the enzyme. Because they bind 1000 times more tightly to the enzyme than does dihydrofolate, they inhibit the enzyme. These two compounds are examples of **competitive inhibitors.**

Donald D. Woods
(1912–1964) *was born in Ipswich, England and received a B.A. and a Ph.D. from Cambridge University. He worked in Paul Flores's laboratory at the London Hospital Medical College.*

Trimethoprim

Paul B. Flores (1882–1971) *was born in London. He moved his laboratory to Middlesex Hospital Medical School when a bacterial chemistry unit was established there. He was knighted in 1946.*

Aminopterin R = H
Methotrexate R = CH₃

Because these compounds inhibit the synthesis of THF, they interfere with the synthesis of any compound that requires a THF-coenzyme in one of the steps of its synthesis. Thus, not only do they prevent the synthesis of thymidine, they also inhibit the synthesis of adenine and guanine since their synthesis also requires a THF-coenzyme. Adenine and guanine are other heterocyclic compounds needed for the synthesis of DNA. One clinical technique used in cancer chemotherapy calls for the patient to be given a lethal dose of methotrexate and then to "save" him or her by giving N^5-formyl-THF.

Trimethoprim is used as an antibiotic because it binds to bacterial dihydrofolate reductase much more tightly than to mammalian dihydrofolate reductase.

THE FIRST ANTIBACTERIAL DRUGS

Sulfonamides—the sulfa drugs—were introduced clinically in 1934 as the first effective antibacterial drugs. Donald Woods, a British bacteriologist, noticed that sulfanilamide, the most widely used sulfonamide, was structurally similar to *p*-aminobenzoic acid, a compound necessary for bacterial growth. He proposed that sulfanilamide's antibacterial properties were the result of its blocking the normal utilization of *p*-aminobenzoic acid.

a sulfonamide   **sulfanilamide**   ***p*-aminobenzoic acid**

Woods and Paul Flores suggested that sulfanilamide acts by inhibiting the enzyme that incorporates *p*-aminobenzoic acid into folic acid. Because the enzyme cannot tell the difference between sulfanilamide and *p*-aminobenzoic acid, both compounds compete for the active site of the enzyme. Humans do not synthesize folate; they depend on dietary folate. Therefore, they are not affected by the drug.

PROBLEM 20 ◆

What is the source of the methyl group in thymidine?

22.9 VITAMIN K

Vitamin K is required for proper clotting of blood. The letter K comes from *koagulation,* which is German for "clotting." A series of reactions involving six proteins is involved in blood clotting. In order for blood to clot, the blood-clotting proteins must bind Ca^{2+}. Vitamin K is required for proper Ca^{2+} binding. Vitamin K deficiencies are rare, because the vitamin is synthesized by intestinal bacteria. The vitamin is also found in the leaves of green plants. **Vitamin KH$_2$** (the hydroquinone of vitamin K) is the coenzyme form of the vitamin.

Paul Dowd (1936–1996) *was born in Brockton, Massachusetts. He did his undergraduate work at Harvard University and received a Ph.D. from Columbia University. He was a professor of chemistry at Harvard University and, from 1970 to 1996, was a professor of chemistry at the University of Pittsburgh.*

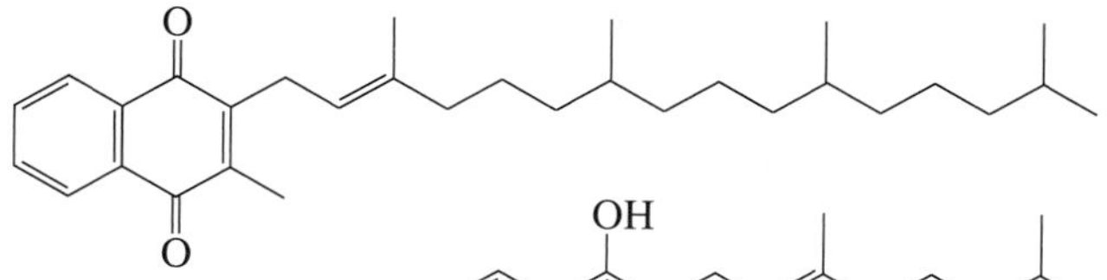

**vitamin K
a quinone**

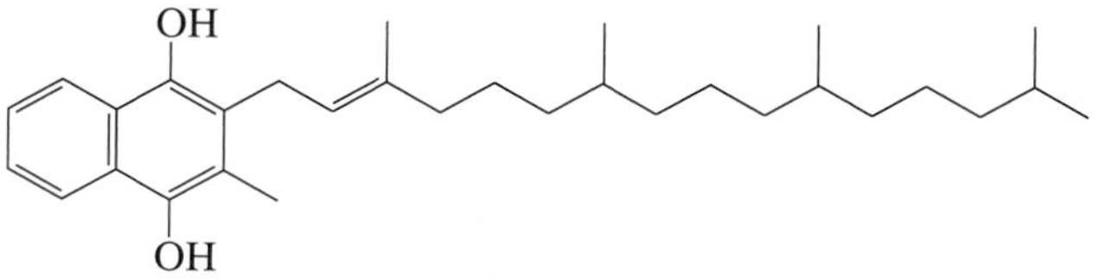

**vitamin KH$_2$
a hydroquinone**

Vitamin KH$_2$ is the coenzyme for the enzyme that catalyzes the carboxylation of the γ-carbon of glutamate side chains in proteins, forming γ-carboxyglutamates. γ-Carboxyglutamates complex Ca^{2+} much more effectively than do glutamates. The proteins involved in blood clotting all have several glutamates near their N-terminal ends. For example, prothrombin has glutamates at positions 7, 8, 15, 17, 20, 21, 26, 27, 30, and 33.

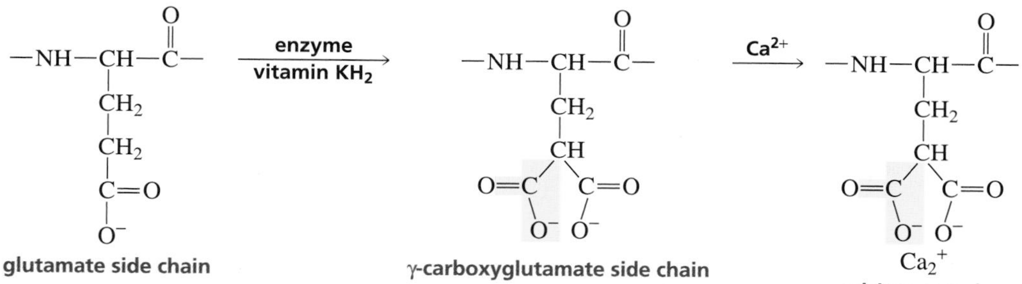

glutamate side chain

γ-carboxyglutamate side chain

calcium complex

The mechanism for the vitamin KH$_2$–catalyzed carboxylation of glutamate has been a puzzle to chemists for quite some time because the γ-proton that must be removed from glutamate before it can attack CO_2 is not very acidic. Therefore, the mechanism must involve the creation of a strong base. The following mechanism has been proposed by Paul Dowd. The vitamin loses a proton from an OH group, and the base that is formed attacks molecular oxygen. A dioxetane is formed that collapses to give a vitamin K base that is strong enough to remove a proton from the γ-carbon of glutamate. The glutamate carbanion attacks CO_2 to form γ-carboxyglutamate, and the protonated vitamin K base loses water, forming vitamin K epoxide.

mechanism for the vitamin KH$_2$–dependent carboxylation of glutamate

a dioxetane

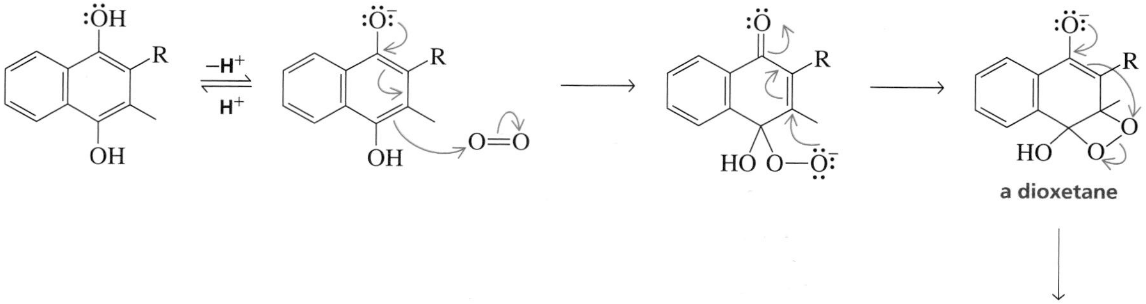

Vitamin KH₂ is required by the enzyme that catalyzes the carboxylation of the γ-carbon of a glutamate side chain in a protein.

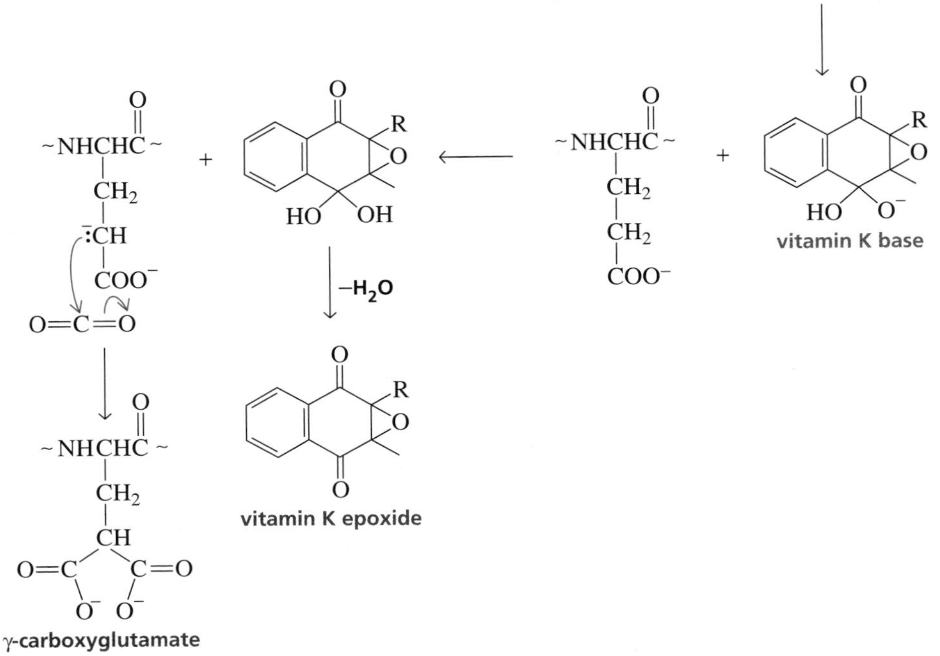

γ-carboxyglutamate

vitamin K base

vitamin K epoxide

Vitamin K epoxide is converted back to vitamin KH₂ by an enzyme that uses dihydrolipoate as a cofactor. The epoxide is first reduced to vitamin K, which is further reduced to vitamin KH₂.

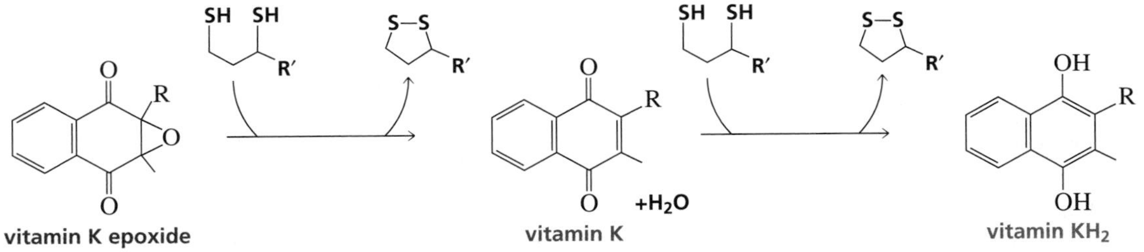

vitamin K epoxide   vitamin K   vitamin KH₂

Warfarin and dicumarol are used clinically as anticoagulants. Warfarin is also a common rat poison, causing death by internal bleeding. These compounds prevent the clotting of blood by inhibiting the enzyme that reduces vitamin K epoxide to vitamin KH₂, thereby preventing carboxylation of glutamate. The enzyme cannot tell the difference between these two compounds and vitamin K epoxide, so the compounds act as *competitive inhibitors*.

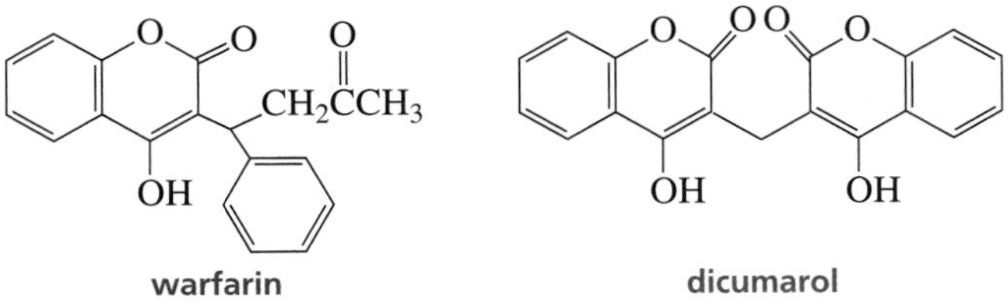

warfarin   dicumarol

Vitamin E has recently been found to be an anticoagulant. Unlike warfarin and dicumarol, which inhibit the enzyme that returns vitamin K epoxide back to vitamin KH₂, vitamin E directly inhibits the enzyme that carboxylates glutamate residues.

TOO MUCH BROCCOLI

An article describing two women with diseases character-ized by abnormal blood clotting reported that they did not improve when they were given warfarin. When questioned about their diets, one woman reported that she ate at least a pound (0.45 kg) of broccoli every day, and the other ate broccoli soup and a broccoli salad every day. When broccoli was removed from their diets, warfarin became effective in preventing the abnormal clot-ting of their blood. Because broccoli is high in vitamin K, these patients had been getting enough dietary vitamin K to compete with the drug, thereby making the drug ineffective.

PROBLEM 21

Thiols such as ethanethiol and propanethiol can be used to reduce vitamin K epoxide back to vitamin KH_2, but they react much more slowly than dihydrolipoate. Explain.

KEY TERMS

anabolism (page 989)
apoenzyme (page 988)
biotin (page 1006)
catabolism (page 989)
coenzyme (page 987)
coenzyme A (CoASH)
    (page 1003)
coenzyme B_{12} (page 1015)
cofactor (page 987)
competitive inhibitor (page 1021)
dehydrogenase (page 993)
electron sink (page 1001)

flavin adenine dinucleotide (FAD)
    (page 996)
flavin mononucleotide (FMN)
    (page 996)
heterocycle (page 992)
holoenzyme (page 988)
lipoate (page 1003)
mechanism-based inhibitor (page 1021)
metabolism (page 989)
metalloenzyme (page 987)
nicotinamide adenine dinucleotide
    (NAD^+) (page 991)

nicotinamide adenine dinucleotide
    phosphate ($NADP^+$) (page 991)
nucleotide (page 992)
pyridoxal phosphate (PLP) (page 1007)
suicide inhibitor (page 1021)
tetrahydrofolate (THF) (page 1018)
thiamine pyrophosphate (TPP)
    (page 1000)
transamination (page 1010)
transimination (page 1009)
vitamin (page 987)
vitamin KH_2 (page 1023)

PROBLEMS

22. Answer the following:
   a. Name six cofactors that act as oxidizing agents.
   b. What are the cofactors that donate one-carbon groups?
   c. What three one-carbon groups are various tetrahydrofolates capable of donating to substrates?
   d. What is the function of FAD in the pyruvate dehydrogenase complex?
   e. What is the function of NAD^+ in the pyruvate dehydrogenase complex?
   f. What is the reaction necessary for proper blood clotting catalyzed by vitamin KH_2?
   g. What coenzymes are used for decarboxylation reactions?
   h. What kinds of substrates do the decarboxylating coenzymes work on?
   i. What coenzymes are used for carboxylation reactions?
   j. What kinds of substrates do the carboxylating coenzymes work on?

23. Name the coenzymes that
    a. allow electrons to be delocalized.
    b. activate groups for further reaction.
    c. provide a good nucleophile.
    d. provide a strong base.

24. For each of the following reactions, give the name of the enzyme that catalyzes the reaction and the name of the required coenzyme.

  a. 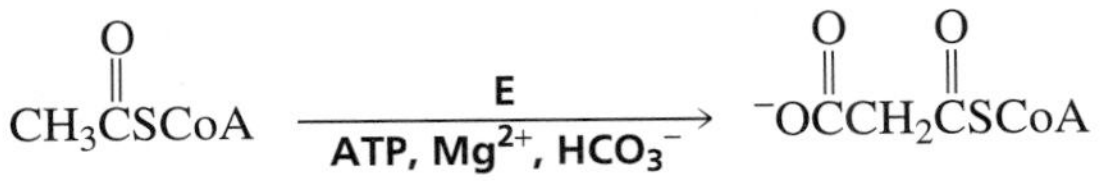 $CH_3\overset{O}{\overset{\|}{C}}SCoA \xrightarrow[\text{ATP, Mg}^{2+},\text{ HCO}_3^-]{\text{E}} {}^-O\overset{O}{\overset{\|}{C}}CH_2\overset{O}{\overset{\|}{C}}SCoA$

  b. (structure: dithiol chain with $(CH_2)_4CO^-$ and two SH groups) $\xrightarrow{\text{E}}$ (disulfide ring with $(CH_2)_4CO^-$, S—S)

  c. ${}^-O\overset{O}{\overset{\|}{C}}\underset{\underset{CH_3}{|}}{C}H\overset{O}{\overset{\|}{C}}SCoA \xrightarrow{\text{E}} {}^-O\overset{O}{\overset{\|}{C}}CH_2CH_2\overset{O}{\overset{\|}{C}}SCoA$

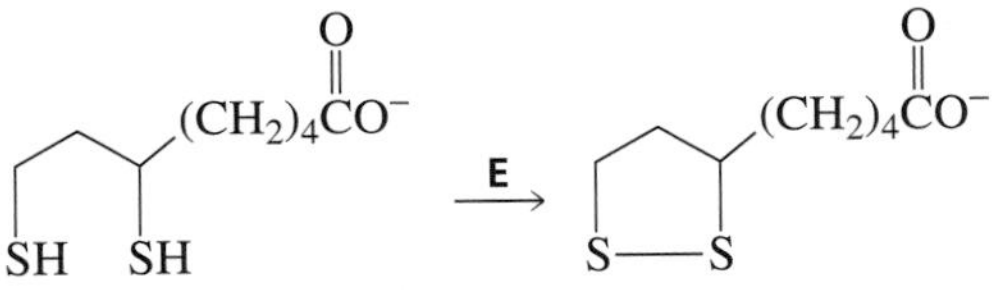

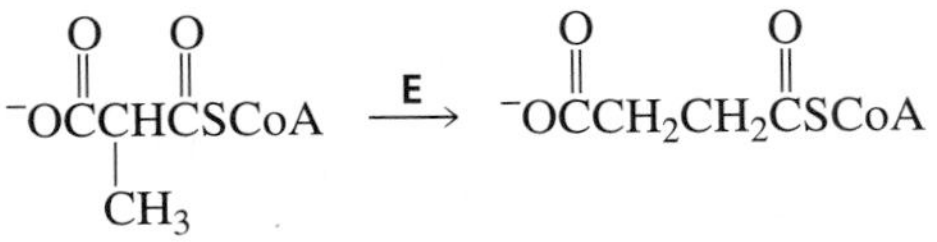

  d. $CH_3\overset{O}{\overset{\|}{C}}{-}\overset{O}{\overset{\|}{C}}O^- \xrightarrow[\text{a catabolic reaction}]{\text{E}} CH_3\underset{\underset{OH}{|}}{C}H\overset{O}{\overset{\|}{C}}O^-$

  e. ${}^-O\overset{O}{\overset{\|}{C}}CH_2\underset{\underset{\overset{NH_3}{+}}{|}}{C}H\overset{O}{\overset{\|}{C}}O^- \xrightarrow{\text{E}} {}^-O\overset{O}{\overset{\|}{C}}CH_2\overset{O}{\overset{\|}{C}}C\overset{O}{\overset{\|}{}}O^-$

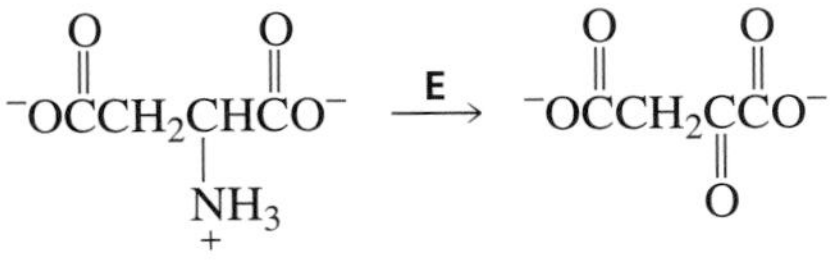

  f. 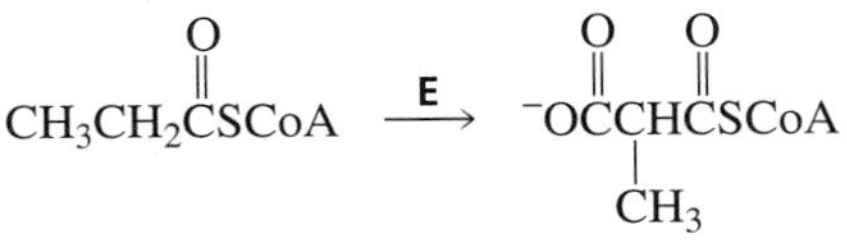 $CH_3CH_2\overset{O}{\overset{\|}{C}}SCoA \xrightarrow{\text{E}} {}^-O\overset{O}{\overset{\|}{C}}\underset{\underset{CH_3}{|}}{C}H\overset{O}{\overset{\|}{C}}SCoA$

25. S-Adenosylmethionine is formed as a result of the reaction between ATP (Section 9.11) and methionine. The other product of the reaction is triphosphate. Propose a mechanism for this reaction.

26. Five coenzymes are required by α-ketoglutarate dehydrogenase, the enzyme in the citric acid cycle that converts α-ketoglutarate to succinyl-CoA.
    a. Identify the coenzymes.
    b. Propose a mechanism for this reaction.

$${}^-O\overset{O}{\overset{\|}{C}}CH_2CH_2\overset{O}{\overset{\|}{C}}{-}\overset{O}{\overset{\|}{C}}O^- \xrightarrow{\text{α-ketoglutarate dehydrogenase}} {}^-O\overset{O}{\overset{\|}{C}}CH_2CH_2\overset{O}{\overset{\|}{C}}SCoA + CO_2$$

α-ketoglutarate   succinyl-CoA

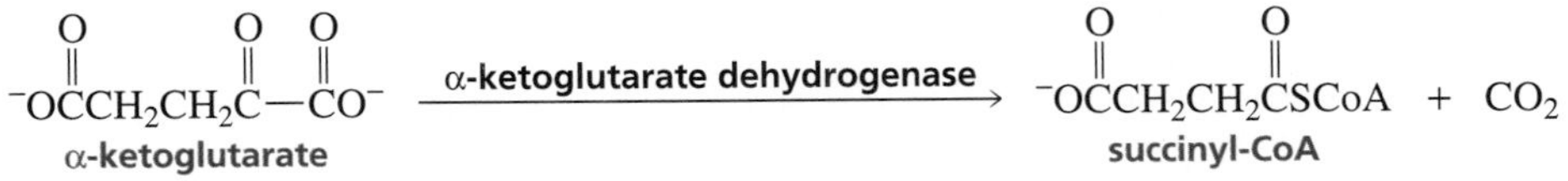

27. Give the products of the following reaction. (*Hint:* Tritium is a hydrogen atom with two neutrons. Although a carbon–tritium bond breaks four times slower than a carbon–hydrogen bond, it is still the first bond in the substrate that breaks.)

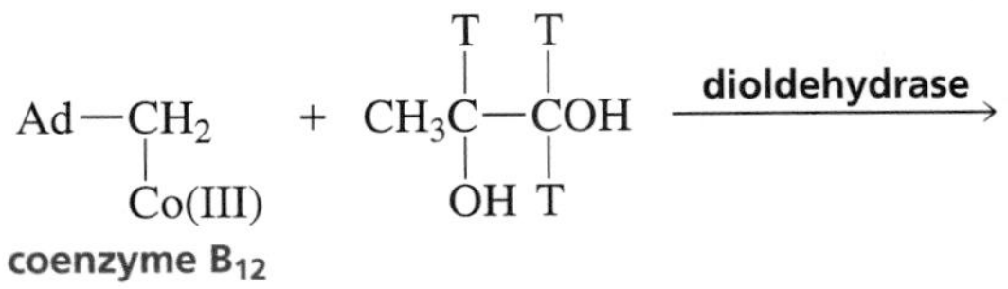

$$Ad{-}\underset{\underset{Co(III)}{|}}{C}H_2 \quad + \quad CH_3\overset{\overset{T\ T}{|\ \ |}}{\underset{\underset{OH\ T}{|\ \ \ |}}{C{-}C}}OH \xrightarrow{\text{dioldehydrase}}$$

coenzyme B$_{12}$

28. Propose a mechanism for methylmalonyl-CoA mutase, the enzyme that converts methylmalonyl-CoA into succinyl-CoA.

29. When transaminated, the three branched-chain amino acids (valine, leucine, and isoleucine) form compounds that have the characteristic odor of maple syrup. An enzyme known as branched-chain α-keto acid dehydrogenase converts these compounds into CoA esters. People who do not have this enzyme have the genetic disease known as maple syrup urine disease—so-called because their urine smells like maple syrup.
   a. Give the structures of the compounds that smell like maple syrup.
   b. Give the structures of the CoA esters.
   c. Branched-chain α-keto acid dehydrogenase has five coenzymes. Identify the coenzymes.
   d. How can this disease be treated?

30. When UMP is dissolved in T_2O (T = tritium; see Problem 27), exchange of T for H occurs at the 5-position. Propose a mechanism for this exchange.

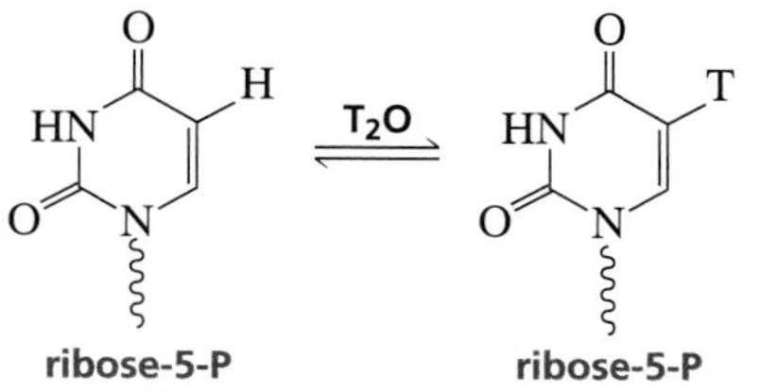

31. Dehydratase is an example of a pyridoxal-requiring enzyme that catalyzes an α,β-elimination reaction. Propose a mechanism for this reaction.

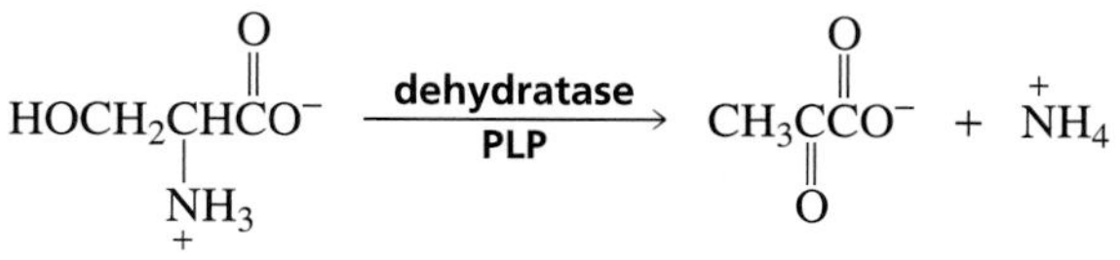

32. In addition to the reactions mentioned in Section 22.6, PLP can also catalyze β-substitution reactions. Propose a mechanism for the following PLP-catalyzed β-substitution.

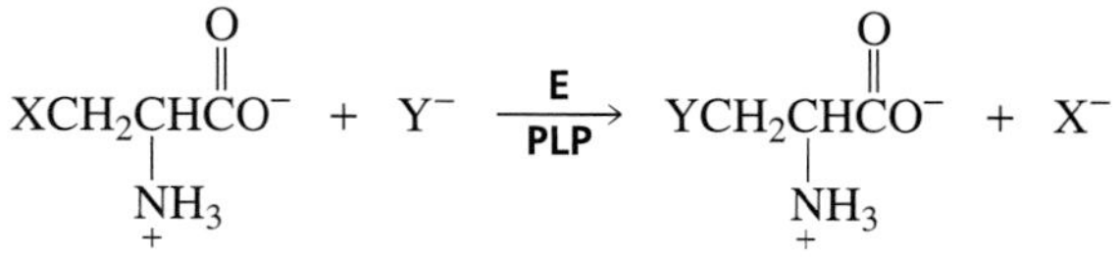

33. PLP can catalyze both α,β-elimination reactions (Problem 31) and β,γ-elimination reactions. Propose a mechanism for the following PLP-catalyzed β,γ-elimination.

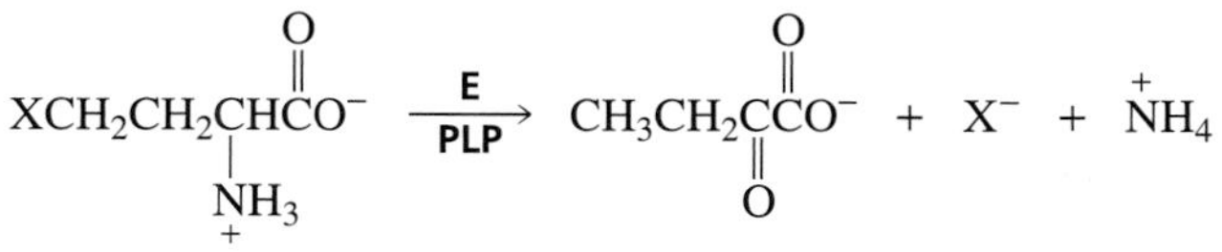

34. The glycine cleavage system is a group of four enzymes that together catalyze the following reaction.

$$\text{glycine} + \text{THF} \xrightarrow{\textbf{glycine cleavage system}} N^5, N^{10}\text{-methylene-THF} + CO_2$$

Use the following information to determine the sequence of reactions involved in the glycine cleavage system.

a. The first enzyme involved in the reaction is a PLP-requiring decarboxylase.

b. The second enzyme is aminomethyltransferase. This enzyme has a lipoate coenzyme.

c. N^5, N^{10}-methylene-THF synthesizing enzyme is the third enzyme. It catalyzes a reaction that forms $^+NH_4$ as one of the products.

d. The fourth enzyme is a FAD-requiring enzyme.

e. The cleavage system also requires NAD^+.

35. Nonenzyme-bound FAD is a stronger oxidizing agent than NAD^+. How then can NAD^+ oxidize the reduced flavoenzyme in the pyruvate dehydrogenase system?

36. $FADH_2$ reduces α,β-unsaturated thioesters to saturated thioesters. The reaction is thought to take place by a mechanism that involves radicals. Propose a mechanism for this reaction.

$$\underset{\text{RCH}=\text{CHCSR}}{\overset{\text{O}}{\|}} + FADH_2 \longrightarrow \underset{\text{RCH}_2\text{CH}_2\text{CSR}}{\overset{\text{O}}{\|}} + FAD$$

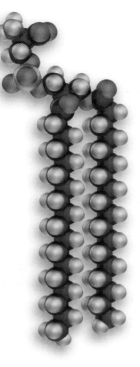

phosphatidylserine

LIPIDS

Lipids are organic compounds found in living organisms, that are soluble in nonpolar organic solvents. Because compounds are classified as lipids as a result of their physical properties rather than their structures, lipids have a variety of structures and functions, as the examples below indicate.

The solubility of lipids in nonpolar organic solvents results from their significant hydrocarbon component. The hydrocarbon portion of the compound is responsible for the "oiliness" or "fattiness" associated with lipids. The word *lipid* comes from the Greek *lipos,* which means "fat."

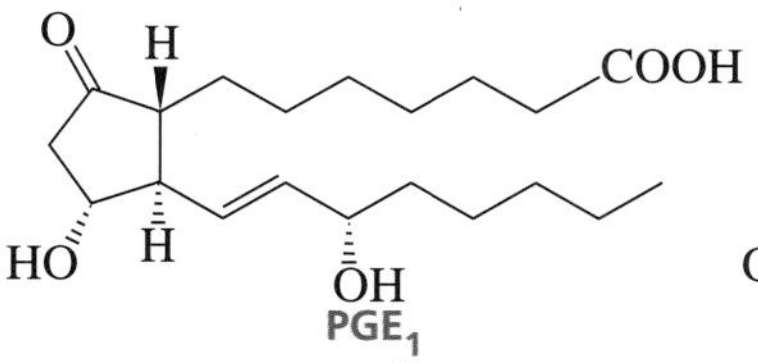

PGE$_1$
a vasodilator

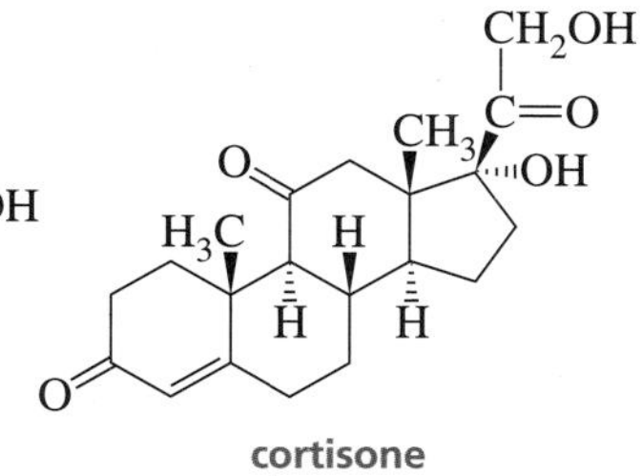

cortisone
a hormone

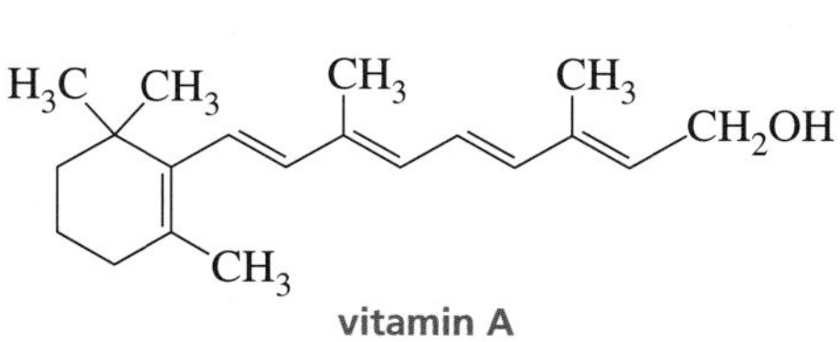

vitamin A
a vitamin

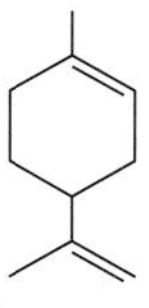

limonene
in orange and lemon oils

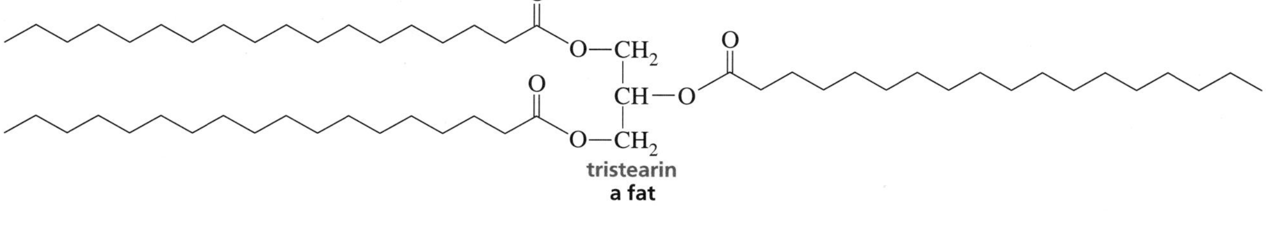

tristearin
a fat

23.1
FATTY ACIDS

Fatty acids are carboxylic acids with long hydrocarbon chains. The fatty acids most frequently found in nature are shown in Table 23.1. Because they are synthesized from acetate, a compound with two-carbon atoms, most naturally occurring fatty acids contain an even number of carbon atoms and are unbranched. The mechanism for their biosynthesis is discussed in Section 18.20. Fatty acids can be saturated ("saturated" with hydrogen, therefore containing no carbon–carbon double bonds) or unsaturated (containing carbon–carbon double bonds). Fatty acids with more than one double bond are called **polyunsaturated fatty acids.** Double bonds in unsaturated fatty acids are never conjugated; they are always separated by a methylene group.

The physical properties of a fatty acid depend on the length of the hydrocarbon chain and the degree of unsaturation. As expected, the melting points of saturated fatty acids increase with increasing molecular weight because of increased intermolecular van der Waals interactions (Section 2.9).

TABLE 23.1 Common Naturally Occurring Fatty Acids

Number of carbons	Common name	IUPAC name		Melting Point °C
Saturated				
12	lauric acid	dodecanoic acid	 COOH	44
14	myristic acid	tetradecanoic acid	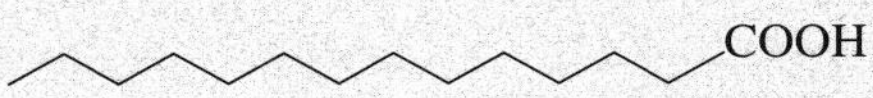 COOH	58
16	palmitic acid	hexadecanoic acid	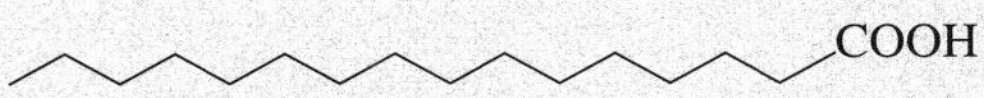 COOH	63
18	stearic acid	octadecanoic acid	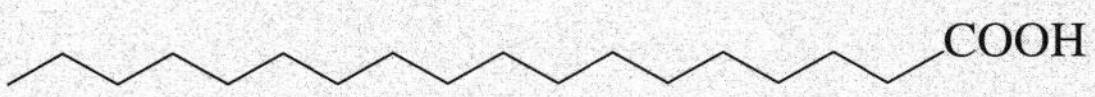 COOH	69
20	arachidic acid	eicosanoic acid	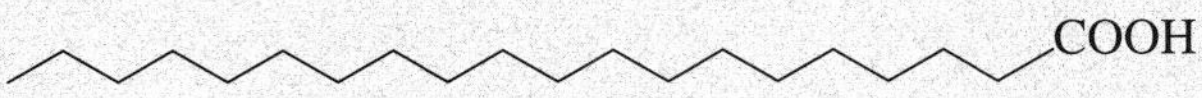 COOH	77
Unsaturated				
16	palmitoleic acid	9-hexadecenoic acid	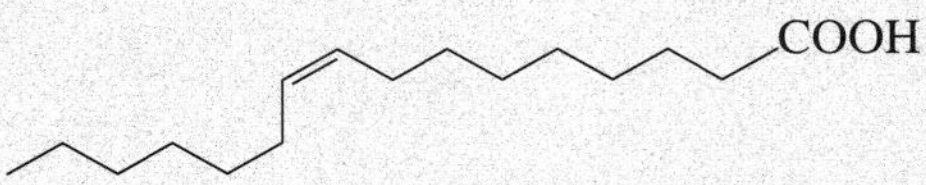 COOH	0
18	oleic acid	9-octadecenoic acid	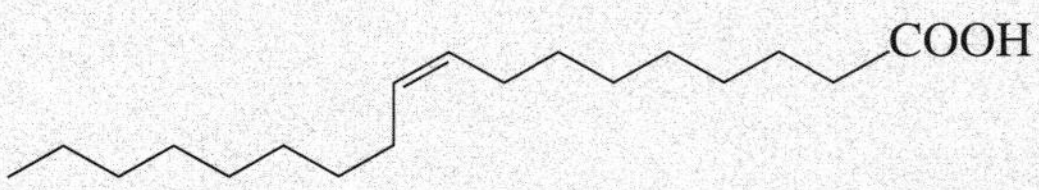 COOH	13
18	linoleic acid	9,12-octadecadienoic acid	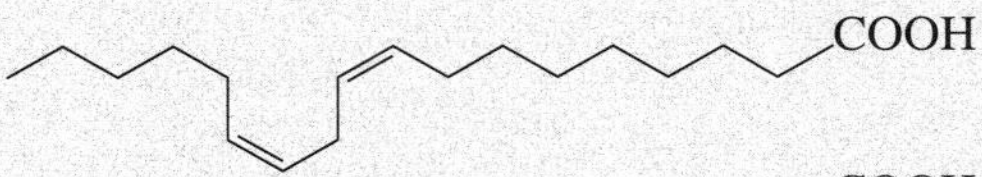 COOH	−5
18	linolenic acid	9,12,15-octadecatrienoic acid	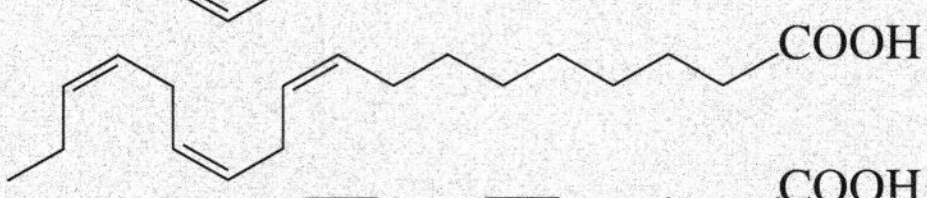 COOH	−11
20	arachidonic acid	5,8,11,14-eicosatetraenoic acid	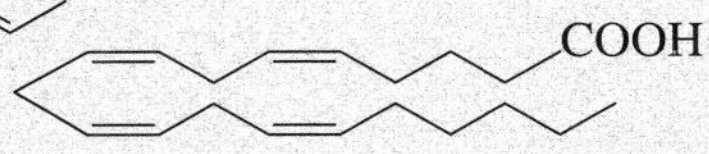 COOH	−50
20	EPA	5,8,11,14,17-eicosapentaeneoic acid	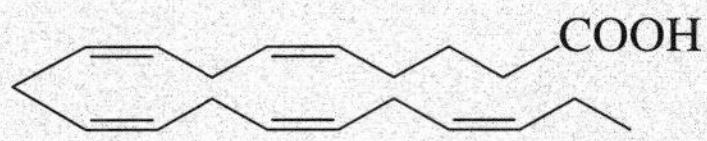 COOH	−50

The double bonds in unsaturated fatty acids generally have the cis configuration. This produces a bend in the molecules, which prevents them from packing together as tightly as fully saturated fatty acids. As a result, unsaturated fatty acids have fewer intermolecular interactions and, therefore, lower melting points than saturated fatty acids with comparable molecular weights. The data in Table 23.2 also show that the melting points of the unsaturated fatty acids decrease as the number of double bonds in the molecule increases.

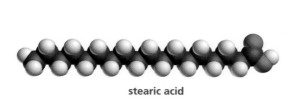

stearic acid

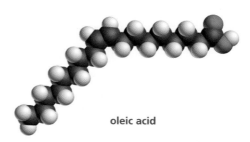

oleic acid

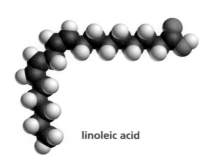

linoleic acid

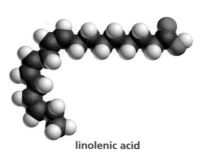

linolenic acid

PROBLEM 1

Explain the difference in the melting points of the following:

a. palmitic acid and stearic acid   **c.** oleic acid and linoleic acid

b. palmitic acid and palmitoleic acid

PROBLEM 2 ◆

What products result from the reaction of arachidonic acid with excess ozone followed by treatment with H_2O_2?

Waxes are esters formed from long-chain carboxylic acids and long-chain alcohols. For example, beeswax, the structural material of beehives, has a 16-carbon carboxylic acid component and a 30-carbon alcohol component. The word *wax* comes from the Old English *weax,* meaning "material of the honeycomb." Carnauba wax

23.2
WAXES

Layers of honeycomb in a beehive.

is a particularly hard wax because of its relatively high molecular weight (a 32-carbon carboxylic acid component and a 34-carbon alcohol component). It is widely used in car waxes and floor polishes.

Waxes are common among living organisms. The feathers of birds are coated with wax to make them water-repellent. Some vertebrates secrete wax in order to keep their hair lubricated and water-repellent. Insects secrete a waterproof, waxy layer on the outside of their exoskeletons. Wax is also found on the surfaces of certain leaves and fruits, where it serves as a protectant against parasites and minimizes evaporation of water.

$$CH_3(CH_2)_{14}\overset{\overset{O}{\|}}{C}O(CH_2)_{29}CH_3$$

a major component of beeswax structural material of beehives

$$CH_3(CH_2)_{30}\overset{\overset{O}{\|}}{C}O(CH_2)_{33}CH_3$$

a major component of carnauba wax coating on the leaves of a Brazilian palm

$$CH_3(CH_2)_{14}\overset{\overset{O}{\|}}{C}O(CH_2)_{15}CH_3$$

a major component of spermaceti wax from the heads of sperm whales

23.3 TRIACYLGLYCEROLS

Triacylglycerols, also called triglycerides, are compounds in which the three OH groups of glycerol are esterified with fatty acids. If the three fatty acid components of a triacylglycerol are the same, the triacylglycerol is called a **simple triacylglycerol. Mixed triacylglycerols,** with two or three different fatty acid components, are more common. Not all triacylglycerol molecules from a single source are necessarily identical. For example, substances such as lard and olive oil are mixtures of several different triacylglycerols (Table 23.2).

TABLE 23.2 Approximate Fatty Acid Compositions of Some Common Fats and Oils

		Saturated fatty acids				Unsaturated fatty acids		
	mp °C	lauric C_{12}	myristic C_{14}	palmitic C_{16}	stearic C_{18}	oleic C_{18}	linoleic C_{18}	linolenic C_{18}
Animal fats								
butter	32	2	11	29	9	27	4	—
lard	30	—	1	28	12	48	6	—
human fat	15	1	3	25	8	46	10	—
whale blubber	24	—	8	12	3	35	10	—
Plant oils								
corn	20	—	1	10	3	50	34	—
cottonseed	−1	—	1	23	1	23	48	—
linseed	−24	—	—	6	3	19	24	47
olive	−6	—	—	7	2	84	5	—
peanut	3	—	—	8	3	56	26	—
safflower	−15	—	—	3	3	19	70	3
sesame	−6	—	—	10	4	45	40	—
soybean	−16	—	—	10	2	29	51	7

$$\begin{array}{ccc}
\text{CH}_2\text{—OH} & \overset{\displaystyle O}{\underset{\displaystyle \parallel}{R^1\text{—C—OH}}} & \overset{\displaystyle O}{\underset{\displaystyle \parallel}{\text{CH}_2\text{—O—C—R}^1}} \\[2mm]
\text{CH—OH} & \overset{\displaystyle O}{\underset{\displaystyle \parallel}{R^2\text{—C—OH}}} & \overset{\displaystyle O}{\underset{\displaystyle \parallel}{\text{CH—O—C—R}^2}} \\[2mm]
\text{CH}_2\text{—OH} & \overset{\displaystyle O}{\underset{\displaystyle \parallel}{R^3\text{—C—OH}}} & \overset{\displaystyle O}{\underset{\displaystyle \parallel}{\text{CH}_2\text{—O—C—R}^3}} \\[1mm]
\textbf{glycerol} & \textbf{fatty acids} & \textbf{a triacylglycerol}
\end{array}$$

a fat or an oil

Triacylglycerols that are solids or semisolids at room temperature are called **fats.** They are usually obtained from animals and are composed largely of triacylglycerols with either saturated fatty acids or fatty acids with only one double bond. The saturated fatty acid tails can pack closely together, causing them to be solids at room temperature. Dietary fat is hydrolyzed in the intestine, regenerating glycerol and fatty acids. In Section 15.16, we saw that hydrolysis of fats under basic conditions forms glycerol and salts of fatty acids that are commonly known as *soap.*

Liquid triacylglycerols are called **oils.** Oils typically come from plant products such as corn, soybeans, olives, and peanuts. They are composed primarily of triacylglycerols with unsaturated fatty acids that cannot pack tightly together. This causes them to be liquids at room temperature. The approximate fatty acid compositions of some common fats and oils are listed in Table 23.2.

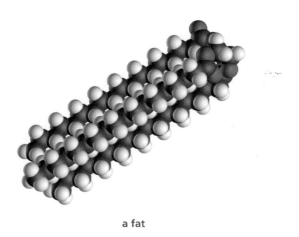

a fat

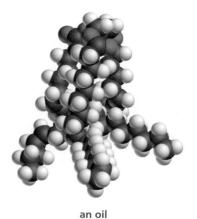

an oil

Some or all of the double bonds of polyunsaturated oils can be reduced by catalytic hydrogenation. Margarine and shortening (for example, Crisco) are prepared by hydrogenating vegetable oils such as soybean oil and safflower oil until they have the desired consistency. The reaction is carefully controlled, because if all the carbon–carbon double bonds were reduced, a hard fat with the consistency of beef tallow would be formed. This process is called "hardening of oils."

$$\text{RCH}{=}\text{CHCH}_2\text{CH}{=}\text{CHCH}_2\text{CH}{=}\text{CH—} \xrightarrow{\overset{\displaystyle \textbf{H}_2}{\textbf{Pt}}} \text{RCH}_2\text{CH}_2\text{CH}_2\text{CH}{=}\text{CHCH}_2\text{CH}_2\text{CH}_2\text{—}$$

Vegetable oils have become popular for food preparation because of studies linking the consumption of saturated fats with heart disease. However, recent studies have shown that unsaturated fats may also be implicated in heart disease. A 20-carbon fatty acid with five double bonds (known as EPA), found in high concentrations in fish oils, is thought to lower the chance of developing certain forms of heart disease.

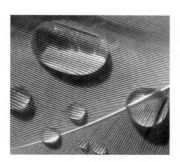

Raindrops on a feather.

OLESTRA: NONFAT WITH FLAVOR

Chemists have been searching for ways to reduce the caloric intake of foods without decreasing flavor. Many people who believe that "no-fat" is synonymous with "no-flavor" can understand this problem. Recently the Federal Food and Drug Administration (Section 29.13) approved the use of Olestra for limited use as a substitute for dietary fat in "snack foods." Procter and Gamble spent 30 years and more than 200 million dollars to develop this compound. Its approval was based on the results of more than 150 studies.

Olestra is a semisynthetic compound. In other words, the compound does not exist in nature but its component parts do. Developing a compound that can be made from units that are a normal part of our diet decreases the potential toxic effects of a new compound. Olestra is made by esterifying all the OH groups of sucrose with fatty acids obtained from cottonseed oil and soybean oil. Therefore its component parts are table sugar and vegetable oil. The ester linkages of Olestra are too hindered to be hydrolyzed by digestive enzymes. So even though it tastes like fat, it has no caloric value since it cannot be digested.

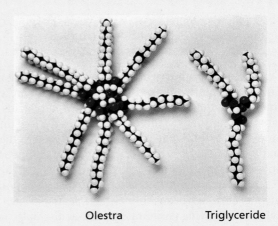

Olestra   Triglyceride

PROBLEM 3

Do all triacylglycerols have the same number of chirality centers?

PROBLEM 4 ◆

Which has a higher melting point, glyceryl tripalmitoleate or glyceryl tripalmitate?

Living organisms store energy in the form of triacylglycerols. A fat provides about six times as much metabolic energy as an equal weight of hydrated glycogen.

Triacylglycerols provide more energy than carbohydrates because they are less oxidized and, since they are nonpolar, they do not bind water. In contrast, two-thirds of the weight of stored glycogen is water.

Animals have a subcutaneous layer of fat cells that serves as both an energy source and an insulator. The fat content of the average man is about 21%, while the fat content of the average woman is about 25%. Humans can store sufficient fat to provide for the body's metabolic needs for 2 to 3 months but can store only enough carbohydrate to provide for the body's metabolic needs for less than 24 hours. Carbohydrates, therefore, are used primarily as a quick, short-term energy source.

Polyunsaturated fats are easily oxidized by O_2. The oxidation takes place by means of a radical chain reaction. In the initiation step, a radical removes a hydrogen from a methylene group that is flanked by two double bonds. This is the most easily removed hydrogen, because the resulting radical is stabilized by resonance with both double bonds. The resulting radical reacts with O_2, forming a peroxy radical with conjugated double bonds. The peroxy radical removes a hydrogen from a methylene group of another molecule of fatty acid, forming an alkyl hydroperoxide. The two propagating steps are repeated over and over.

His diet is high in fish oil.

RCH=CH—CH—CH=CH—   +   X· $\xrightarrow{\text{initiation}}$ RCH=CH—ĊH—CH=CH—   +   HX
　　　　　　|
　　　　　　H
　　　　　　　　　　　　　　　　　　　　　　resonance contributor
　　　　　　　　　　　　　　　　　　　　　　with isolated double bonds

$\updownarrow$

RĊH—CH=CH—CH=CH—
resonance contributor
with conjugated double bonds

·Ö—Ö· | propagation

RCH—CH=CH—CH=CH—
|
:Ö—Ö·

a peroxy radical

RCH=CH—CH₂—CH=CH— | propagation

RCH=CH—ĊH—CH=CH—   +   RCH—CH=CH—CH=CH—
　　　　　　　　　　　　　　　　　　　　|
　　　　　　　　　　　　　　　　　　　:Ö—ÖH

an alkyl hydroperoxide

The reaction of fatty acids with O_2 causes them to become rancid. The unpleasant taste and smell associated with rancidity are the results of further oxidation of the alkyl hydroperoxide to shorter-chain carboxylic acids that have strong odors. The same process produces the odor associated with sour milk.

PROBLEM 5

Draw the resonance contributors for the radical formed when a hydrogen atom is removed from the CH_2 group of a polyunsaturated fatty acid.

WHALES AND ECHOLOCATION

Whales have enormous heads, accounting for 33% of their total weight. They have large deposits of fat in their heads and lower jaws.

This fat is very different from both the whale's normal body fat and its dietary fat. We might expect that the fat in the head would be used as a source of metabolic energy, but this fat is not used for energy. Since major anatomical modifications were necessary to accommodate this fat, it must have some important function for the animal. It is now believed that the fat is used for echolocation. Echolocation is the process of emitting sounds in pulses and gaining information by analyzing the returning echoes. The fat in the whale's head focuses the emitted sound waves in a directional beam, and the echoes are received by the fat organ in the lower jaw. This fat organ transmits the sound to the brain for processing and interpretation, providing the animal with information about the depth of the water, changes in the sea floor, and the position of the coastline. Thus, these fat deposits give whales a unique acoustic sensory system and allow them to compete successfully for survival with sharks—who also have a well-developed sense of sound direction.

Humpback whale in Alaska

23.4 PHOSPHOLIPIDS AND SPHINGOLIPIDS

For biological systems to operate, some parts of organisms must be separated from other parts. On a cellular level, the outside of the cell must be separated from the inside of the cell. "Greasy" lipid **membranes** serve as the barrier. In addition to isolating the contents of the cell, these membranes allow the selective transport of ions and organic molecules into and out of the cell.

Phosphoacylglycerols (also called **phosphoglycerides**) are the major components of cell membranes. They are similar to triacylglycerols except that a terminal OH group of glycerol is esterified with phosphoric acid rather than with a fatty acid, forming a **phosphatidic acid.** Because these are lipids that contain a phosphate group, they are classified as **phospholipids.** The C-2 carbon of glycerol in phosphoacylglycerols has the R configuration.

a phosphatidic acid

Phosphatidic acids are present only in small amounts in membranes. The most common phosphoacylglycerols in membranes have a second phosphate ester linkage. The alcohols most commonly used to form this second ester group are ethanolamine, choline, and serine. Phosphatidylethanolamines are also called

cephalins, and phosphatidylcholines are called **lecithins.** Lecithins are used in food as emulsifying agents. They are added to foods such as mayonnaise to prevent the aqueous and fat components from separating.

$$
\begin{array}{l}
CH_2-O-\overset{\displaystyle O}{\overset{\|}{C}}-R^1 \\[2pt]
\overset{O}{\overset{\|}{}} \\[-2pt]
CH-O-\overset{\displaystyle O}{\overset{\|}{C}}-R^2 \\[2pt]
\overset{O}{\overset{\|}{}} \\[-2pt]
CH_2-O-\overset{\displaystyle }{\underset{O^-}{P}}-OCH_2CH_2\overset{+}{N}H_3
\end{array}
$$

a phosphatidylethanolamine
a cephalin

$$
\begin{array}{l}
CH_2-O-\overset{\displaystyle O}{\overset{\|}{C}}-R^1 \\[2pt]
CH-O-\overset{\displaystyle O}{\overset{\|}{C}}-R^2 \\[2pt]
CH_2-O-\overset{\displaystyle }{\underset{O^-}{P}}-OCH_2CH_2\overset{\overset{CH_3}{|+}}{\underset{CH_3}{N}}CH_3
\end{array}
$$

a phosphatidylcholine
a lecithin

$$
\begin{array}{l}
CH_2-O-\overset{\displaystyle O}{\overset{\|}{C}}-R^1 \\[2pt]
CH-O-\overset{\displaystyle O}{\overset{\|}{C}}-R^2 \\[2pt]
CH_2-O-\overset{\displaystyle }{\underset{O^-}{P}}-OCH_2\overset{}{\underset{\overset{}{\overset{+}{N}H_3}}{C}}HCOO^-
\end{array}
$$

a phosphatidylserine

Phosphoacylglycerols form membranes by arranging themselves in a **lipid bilayer.** The polar heads of the phosphoacylglycerols are on the outside of the bilayer, while the fatty acid chains and cholesterol form the interior of the bilayer (Figure 23.1). A typical bilayer is about 50 Å thick. Compare the bilayer with the micelles formed by soap in aqueous solution (Section 15.16).

The fluidity of a membrane is controlled by the fatty acid components of the phosphoacylglycerols. Long-chain saturated fatty acids decrease membrane fluidity because their hydrocarbon chains can pack closely together. Unsaturated fatty acids increase fluidity. Cholesterol also decreases fluidity (Section 23.9). Only animal membranes contain cholesterol; consequently, animal membranes are more rigid than plant membranes.

The unsaturated fatty acid chains of phosphoacylglycerols are susceptible to reaction with O_2, similar to the reaction described on page 1035 with fats. This oxidation reaction can lead to the degradation of membranes. Vitamin E is an important antioxidant that protects fatty acid chains from this reaction. Vitamin E, also called α-tocopherol, is classified as a lipid since it is soluble in nonpolar organic solvents. Because vitamin E reacts more rapidly than triacylglycerols with oxygen and other radicals, the vitamin prevents biological membranes from reacting with

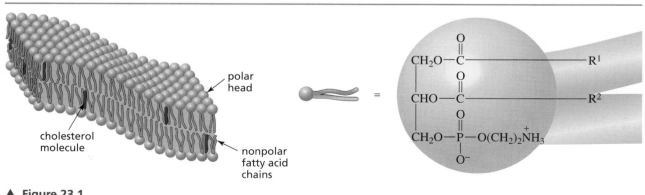

▲ **Figure 23.1**
A lipid bilayer

oxygen (Section 8.8). There are some who believe that this vitamin slows the aging process. Since vitamin E also reacts with oxygen more rapidly than fats do, the vitamin is added to many foods to prevent spoilage.

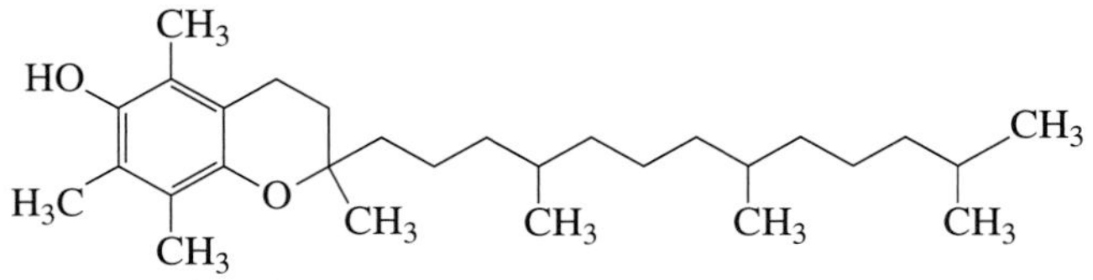

α-tocopherol
vitamin E

PROBLEM 6◆

Membranes contain proteins. Integral membrane proteins extend partly or completely through the membrane, while peripheral membrane proteins are found on the inner or outer surfaces of the membrane. What is the likely difference in the overall amino acid composition of integral and peripheral membrane proteins?

PROBLEM 7◆

A colony of bacteria accustomed to an environment of 25 °C was moved to an environment of 35 °C. The increased temperature increased the fluidity of the bacterial membranes. What could the bacteria do to regain their original membrane fluidity?

Sphingolipids are also found in membranes. They are the major lipid components in the myelin sheaths of nerve fibers. Sphingolipids contain sphingosine instead of glycerol. In sphingolipids, the amino group of sphingosine is bonded to the acyl group of a fatty acid.

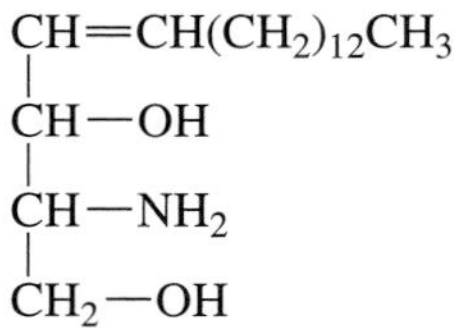

sphingosine

Two of the most common kinds of sphingolipids are **sphingomyelins** and **cerebrosides.** In sphingomyelins, the primary OH group of sphingosine is bonded to a phosphocholine or to a phosphoethanolamine, similar to the bonding in lecithins and cephalins. In cerebrosides, the primary OH group of sphingosine is bonded to a sugar residue through a β-glycosidic linkage. Sphingomyelins are phospholipids because they contain a phosphate group; cerebrosides are not phospholipids.

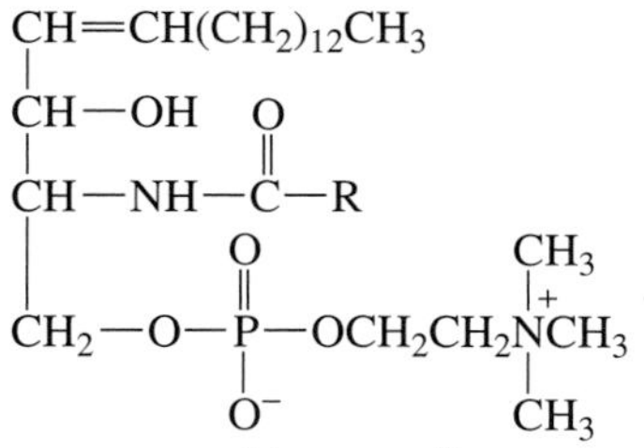

a sphingomyelin

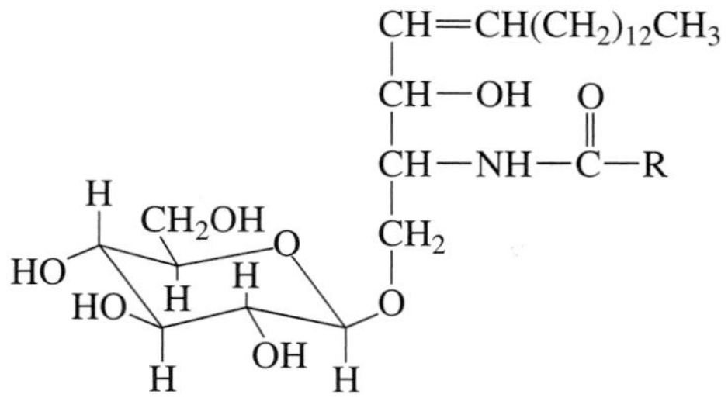

a glucocerebroside

MULTIPLE SCLEROSIS AND THE MYELIN SHEATH

The myelin sheath is a lipid-rich material that is wrapped around the axons of nerve cells. This sheath is composed largely of sphingomyelins and cerebrosides. The sheath's function is to increase the velocity of nerve impulses. Multiple sclerosis is a disease characterized by loss of the myelin sheath and a consequent slowing of nerve impulses and eventual paralysis.

PROBLEM 8

a. Give the structures of three different sphingomyelins.

b. Give the structure of a galactocerebroside.

PROBLEM 9

The membrane phospholipids in animals such as deer and elk have a higher degree of unsaturation in cells closer to the hoof than in cells closer to the body. Explain how this can be important for survival.

Prostaglandins are responsible for regulating a variety of physiological responses such as inflammation, blood pressure, blood clotting, fever, pain, induction of labor, and the sleep/wake cycle. All prostaglandins have a carbon skeleton with 20 carbons: a five-membered ring with a seven-carbon carboxylic acid substituent and an eight-carbon hydrocarbon substituent. The two substituents are trans to each other.

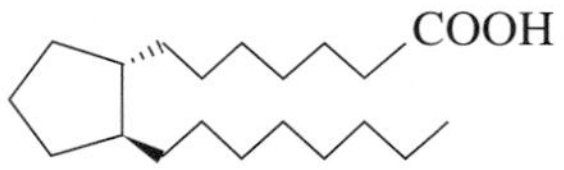

23.5 PROSTAGLANDINS

prostaglandin skeleton

The prostaglandins are named using the format PGX, where X designates the functional groups of the five-membered ring. The five-membered rings of PGA's, PGB's, and PGC's contain a carbonyl group and a double bond; the location of the double bond determines whether a prostaglandin is a PGA, PGB, or PGC. PGD's and PGE's are β-hydroxy ketones, and PGF's are 1,3-diols. A subscript indicates the total number of double bonds in the side chains, and α and β indicate the configuration of the two OH groups in a PGF: α indicates that the OH groups are on the same side of the ring, and β indicates they are on opposite sides.

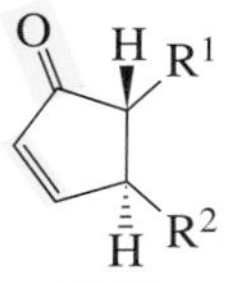

PGA's 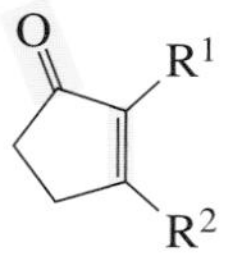   PGB's 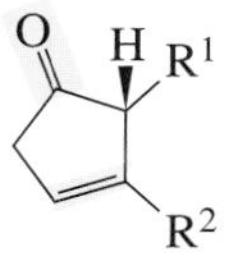   PGC's 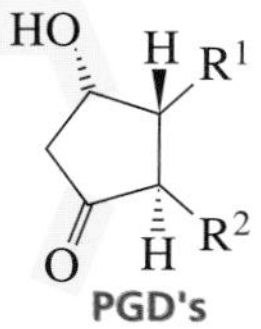   PGD's

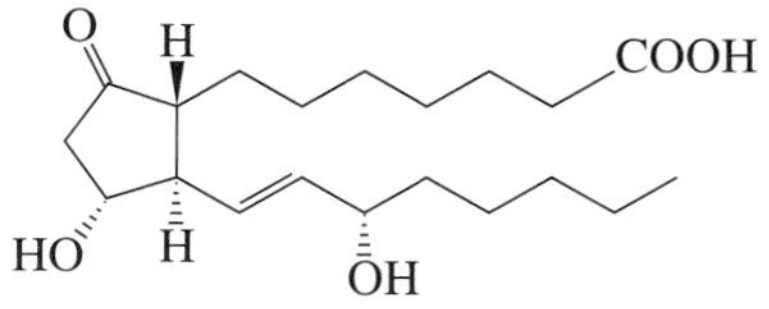

PGE₁   PGE₂

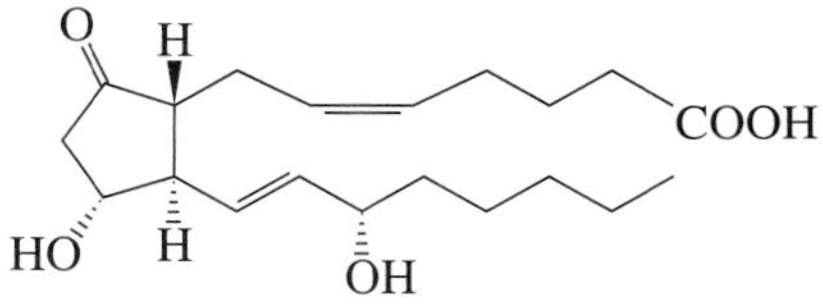

PGF$_{2\alpha}$

Prostaglandins are synthesized from arachidonic acid, a 20-carbon fatty acid with four cis double bonds. Arachidonic acid in turn is synthesized from linoleic acid. Because linoleic acid cannot be synthesized by mammals, it is essential that this fatty acid be included in the diet.

An enzyme called prostaglandin endoperoxide synthase catalyzes the conversion of arachidonic acid to PGH$_2$, the precursor of all prostaglandins. This enzyme has two activities: a cyclooxygenase activity and a hydroperoxidase activity. It uses its cyclooxygenase activity to form the five-membered ring. In the first step of this transformation, a hydrogen atom is removed from a carbon flanked by two double bonds. This hydrogen is removed relatively easily because the resulting radical is stabilized by resonance. The radical reacts with oxygen to form a peroxy radical. Notice that these two steps are the same as the first two steps in the reaction that causes fats to become rancid (Section 23.3). The peroxy radical rearranges and reacts with a second molecule of oxygen. The enzyme then uses its hydroperoxidase activity to convert the OOH group into an OH group, forming PGH$_2$. PGH$_2$ rearranges to form PGE$_2$, a prostaglandin.

In addition to serving as an intermediate for the synthesis of prostaglandins, PGH$_2$ is an intermediate for the synthesis of **thromboxanes** and **prostacyclins.** Thromboxanes constrict blood vessels and stimulate platelet aggregation, the first step in blood clotting. Prostacyclins have the opposite effect. They dilate blood vessels and inhibit platelet aggregation. The levels of these two compounds are carefully controlled to maintain a balance in the blood.

Aspirin (acetylsalicylic acid) inhibits the cyclooxygenase activity of prostaglandin endoperoxide synthase. It does this by transferring an acetyl group to a serine hydroxyl group of the enzyme (Section 15.8). Acetylating the serine residue inhibits the enzyme. Aspirin therefore inhibits the synthesis of prostaglandins and in this way decreases the inflammation produced by these compounds. Aspirin also inhibits the synthesis of thromboxanes and prostacyclins. Overall, this causes a slight decrease in the rate of blood clotting, which is why some doctors recommend one aspirin tablet every other day to reduce the chance of heart attacks caused by clotting in blood vessels.

serine hydroxyl
group

**acetylsalicylic acid
aspirin**   **cyclooxygenase
active enzyme**   **cyclooxygenase
inactive enzyme**

biosynthesis of prostaglandins, thromboxanes, and prostacyclins

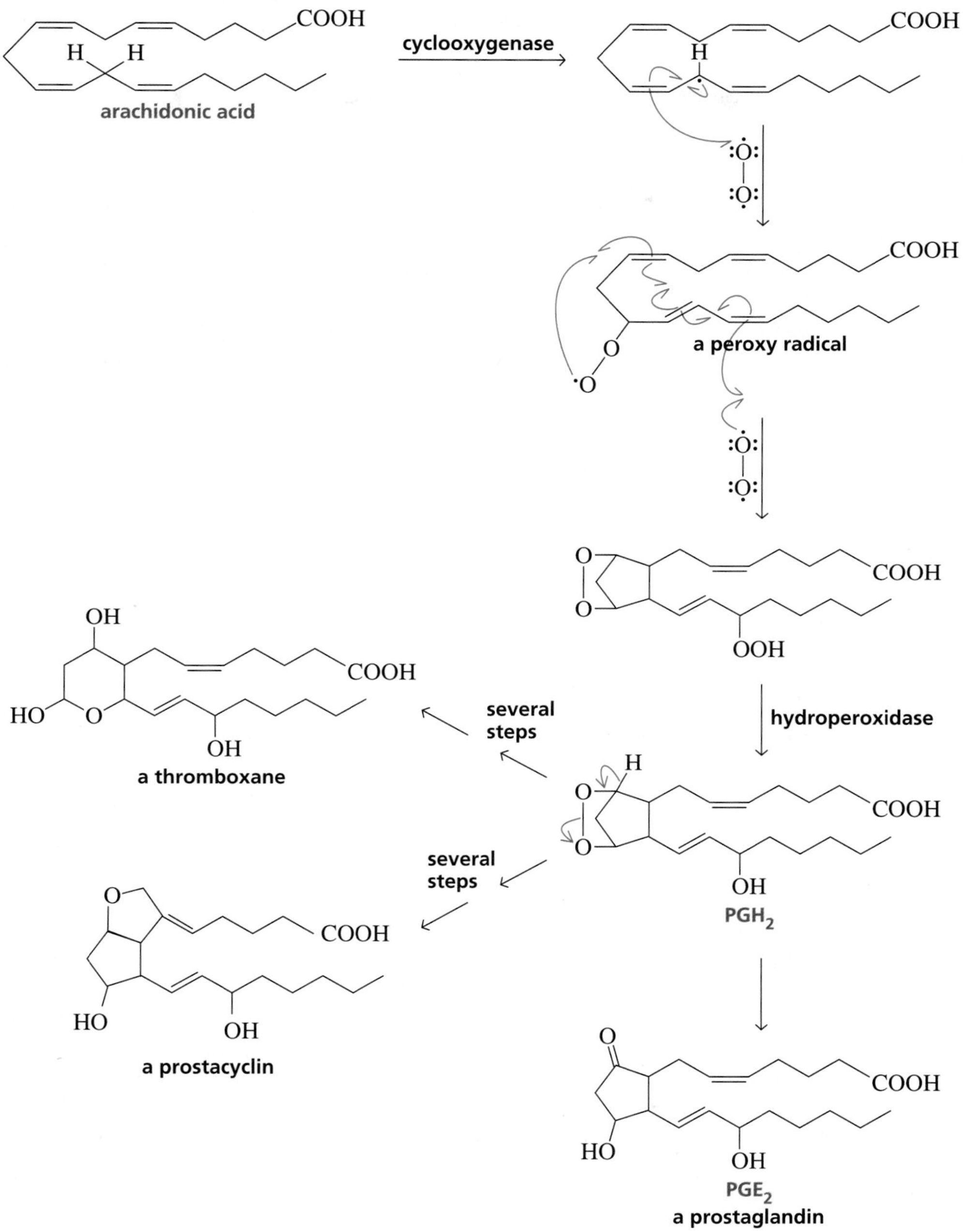

a thromboxane

a prostacyclin

PGH₂

PGE₂
a prostaglandin

   Other anti-inflammatory drugs such as ibuprofen (the active ingredient in Advil, Motrin, and Nuprin) and naproxen (the active ingredient in Aleve) also inhibit the synthesis of prostaglandins. They compete with either arachidonic acid or the peroxy radical for the enzyme's binding site.

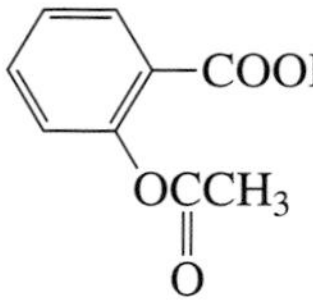

aspirin

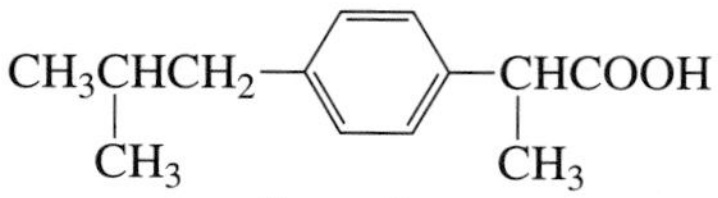

ibuprofen

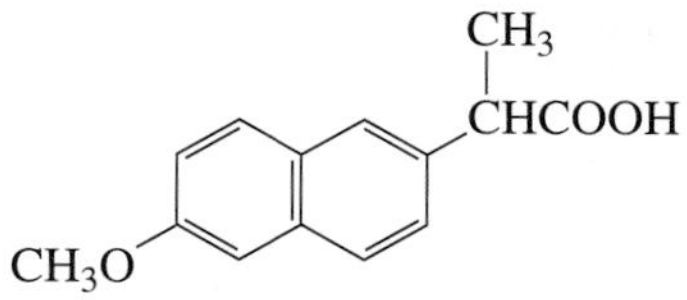

naproxen

Arachidonic acid can also be converted into a leukotriene. **Leukotrienes** contain three conjugated double bonds. Because they induce contraction of the muscle that lines the airways to the lungs, they are implicated in allergic reactions, inflammatory reactions, and heart attacks. Leukotrienes are the compounds that bring on the symptoms of asthma. They are also implicated in anaphylactic shock, a potentially fatal allergic reaction. Scientists are currently trying to design drugs to control the effects of leukotrienes.

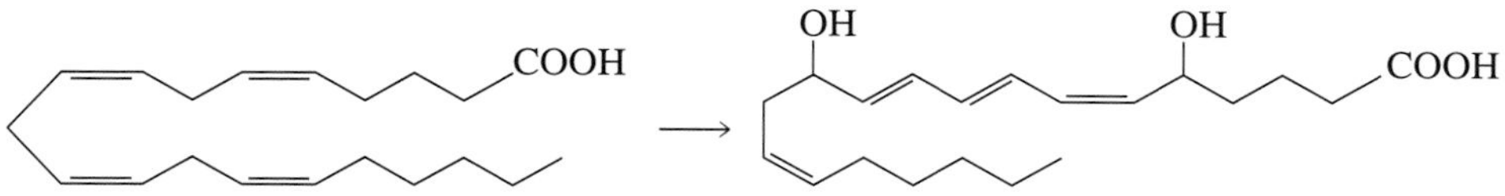

arachidonic acid

a leukotriene

PROBLEM 10

Treating PGA$_2$ with a strong base such as sodium *tert*-butoxide followed by addition of acid converts it to PGC$_2$. Propose a mechanism for this reaction.

23.6

TERPENES

Terpenes are a very diverse class of lipids. More than 20,000 terpenes are known. They can be hydrocarbons or they can contain oxygen and be alcohols, ketones, or aldehydes. Oxygen-containing terpenes are sometimes called **terpenoids.** Some of these compounds have been used as spices, perfumes, and medicines for many thousands of years. After analyzing a large number of terpenes, organic chemists realized that many of them share a common feature: they contain carbon atoms in multiples of 5. These naturally occurring compounds contain 10, 15, 20, 25, 30, and 40 carbon atoms, which suggests that there is a compound with five carbon atoms that serves as a building block for the synthesis of these organic molecules.

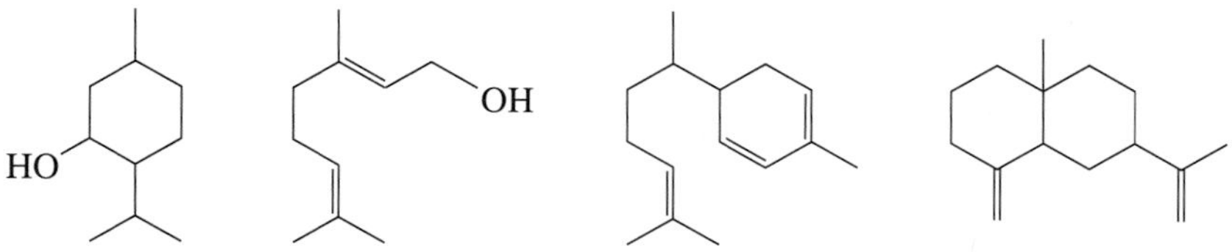

menthol
peppermint oil

geraniol
geranium oil

zingiberene
oil of ginger

β-selinene
oil of celery

Investigation of these compounds showed that their structures could be understood by assuming that they were made by joining together isoprene units in a "head-to-tail" fashion. Isoprene is the common name for 2-methyl-1,3-butadiene, a compound containing five carbon atoms.

$$CH_2=\overset{\overset{\displaystyle CH_3}{|}}{C}-CH=CH_2$$

2-methyl-1,3-butadiene
isoprene

The branched end of isoprene is called the "head," and the unbranched end is called the "tail." This head-to-tail pattern of linkage that is used to form terpenes is known as the **isoprene rule.**

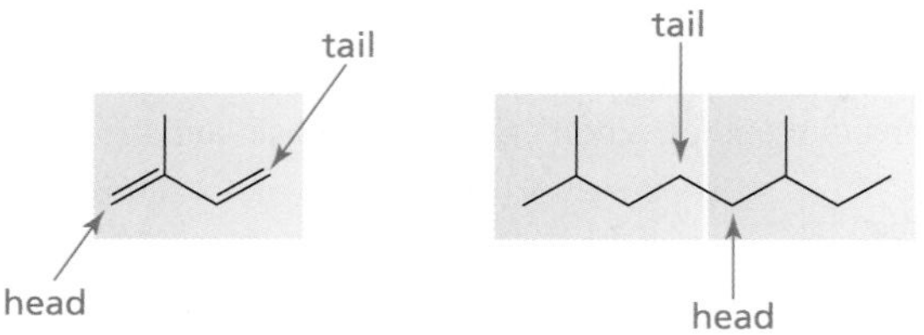

carbon skeleton of two isoprene units with a bond between the tail of one and the head of another

In the case of cyclic compounds, linkage of the head of one isoprene unit to the tail of another is followed by an additional attachment to form the ring. The second attachment is not necessarily head-to-tail but is whatever is necessary to form a stable (five- or six-membered) ring.

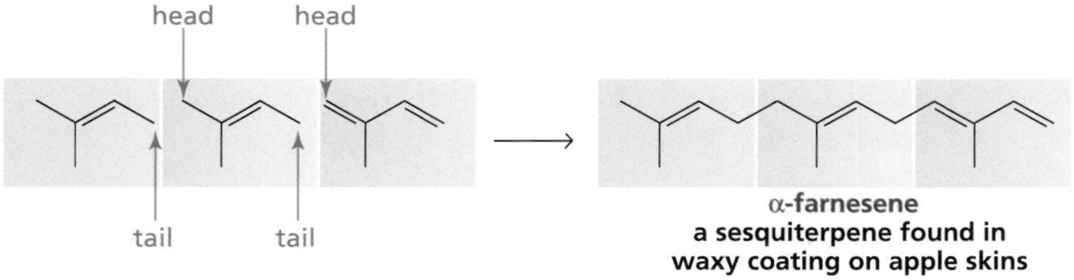

Leopold Ružička

Leopold Stephen Ružička (1887–1976) *was the first to recognize that many organic compounds contain multiples of five carbons. Ružička, a Croatian, attended college in Switzerland and became a Swiss citizen in 1917. He was a professor of chemistry at the University of Utrecht and later at the State Technical College in Zurich. For his work on terpenes, he shared the 1939 Nobel Prize in chemistry with Butenandt (page 1054).*

α-farnesene
a sesquiterpene found in waxy coating on apple skins

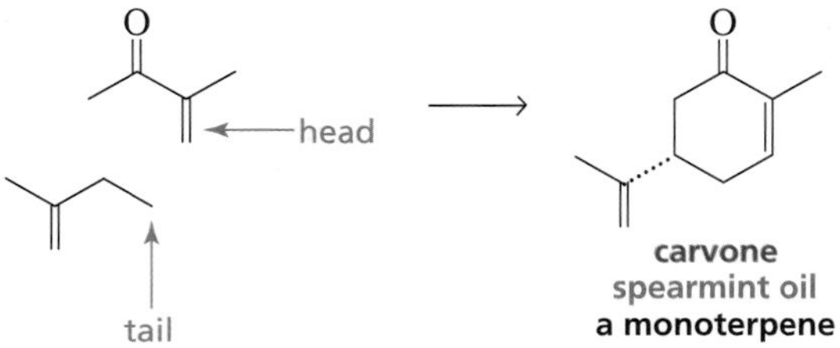

carvone
spearmint oil
a monoterpene

In Section 23.8, we will see that the compound actually used in the biosynthesis of these compounds is not isoprene but isopentenyl pyrophosphate, a compound that has the same carbon skeleton as isoprene. We will also look at the mechanism by which isopentenyl pyrophosphate units are joined together in a head-to-tail fashion.

Terpenes are classified according to the number of carbons they contain (Table 23.3). **Monoterpenes** have ten carbons and are therefore composed of two isoprene

TABLE 23.3 Classification of Terpenes

Carbon atoms	Classification	Carbon atoms	Classification
10	monoterpenes	25	sesterterpenes
15	sesquiterpenes	30	triterpenes
20	diterpenes	40	tetraterpenes

units. Many fragrances and flavorings found in plants are monoterpenes and **sesquiterpenes** (15 carbons). These compounds are known as **essential oils.**

PROBLEM 11 / SOLVED

Mark off the isoprene units in menthol, zingiberene, β-selinene, and squalene.

SOLUTION   For zingiberene:

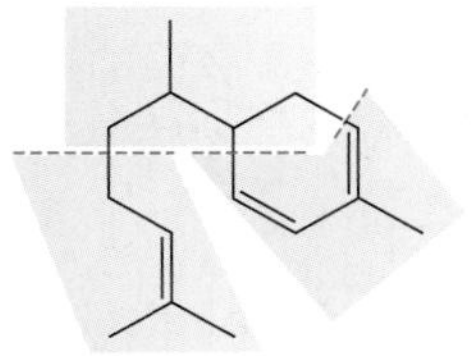

Triterpenes (six isoprene units) and **tetraterpenes** (eight isoprene units) have important biological roles. **Squalene,** a triterpene, is a precursor of steroid molecules (Section 23.9).

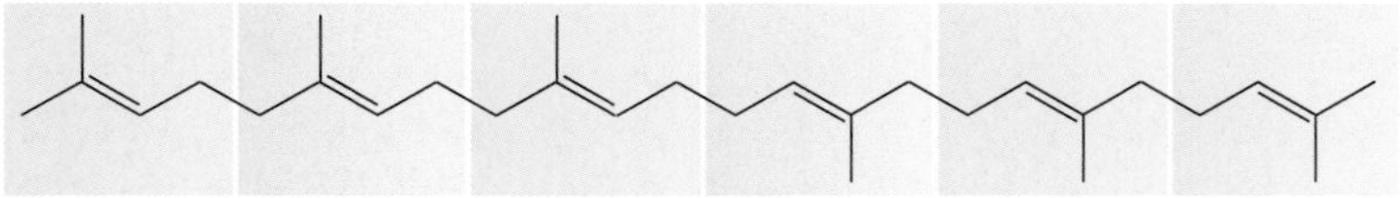

squalene

Carotenoids become visible in autumn when chlorophyll degrades.

Carotenoids are tetraterpenes. Lycopene, the compound responsible for the red coloring of tomatoes and watermelon, and β-carotene, the compound that causes carrots and apricots to be orange, are examples of carotenoids. β-Carotene is also the coloring agent used in margarine. β-Carotene and other colored compounds are found in the leaves of trees, but their characteristic colors are usually obscured by the green color of chlorophyll. In the fall when chlorophyll degrades, the colors become apparent. It is the many conjugated double bonds in lycopene and β-carotene that cause the compounds to be colored (Section 13.18).

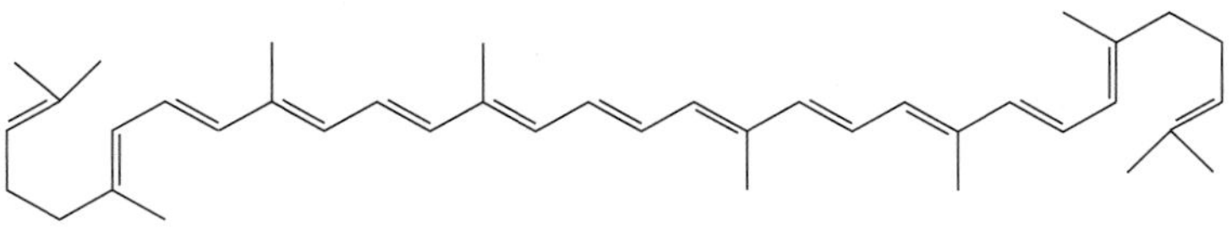

lycopene

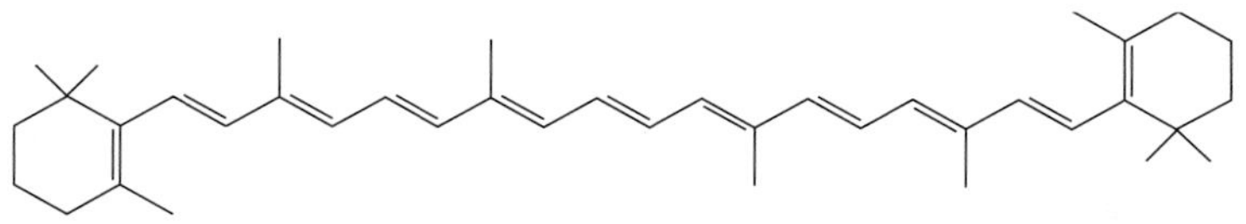

β-carotene

PROBLEM 12◆

One of the linkages in squalene is a tail-to-tail linkage. What does this suggest about how squalene is synthesized in nature? (Locating the position of the tail-to-tail linkage will help you answer this question.)

PROBLEM 13

Mark off the isoprene units in lycopene and β-carotene. Can you detect a similarity in the way in which squalene, lycopene, and β-carotene are biosynthesized?

Four vitamins are lipids: A, D, E, and K (Sections 22.9 and 27.6). Vitamin A is the only water-insoluble vitamin we have not already discussed. β-Carotene, which is cleaved to form two molecules of vitamin A, is the major dietary source of the vitamin. Vitamin A, also called retinol, plays an important role in vision.

**23.7
VITAMIN A**

The retina of the eye contains cone cells and rod cells. The cone cells are responsible for color vision and for vision in bright light; the rod cells are responsible for vision in dim light. In rod cells, vitamin A is oxidized to an aldehyde and the trans double bond at C-11 is isomerized to a cis double bond. The mechanism for the enzyme-catalyzed interconversion of cis and trans double bonds is discussed in Section 16.16. The protein *opsin* uses a lysine side chain (Lys 216) to form an imine with (11Z)-retinal, resulting in a complex known as *rhodopsin*. When rhodopsin absorbs visible light, it isomerizes to the trans isomer. This change in molecular geometry causes an electrical signal to be sent to the brain, where it is perceived as a visual image. The trans isomer is not stable and is hydrolyzed to (11E)-retinal and opsin. This reaction is referred to as *bleaching* of the visual pigment.

the chemistry of vision

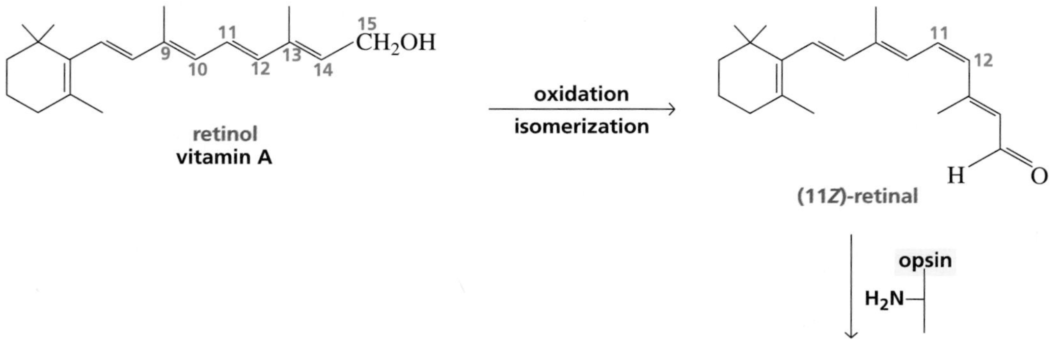

retinol
vitamin A

oxidation
isomerization

(11Z)-retinal

opsin

H_2N—

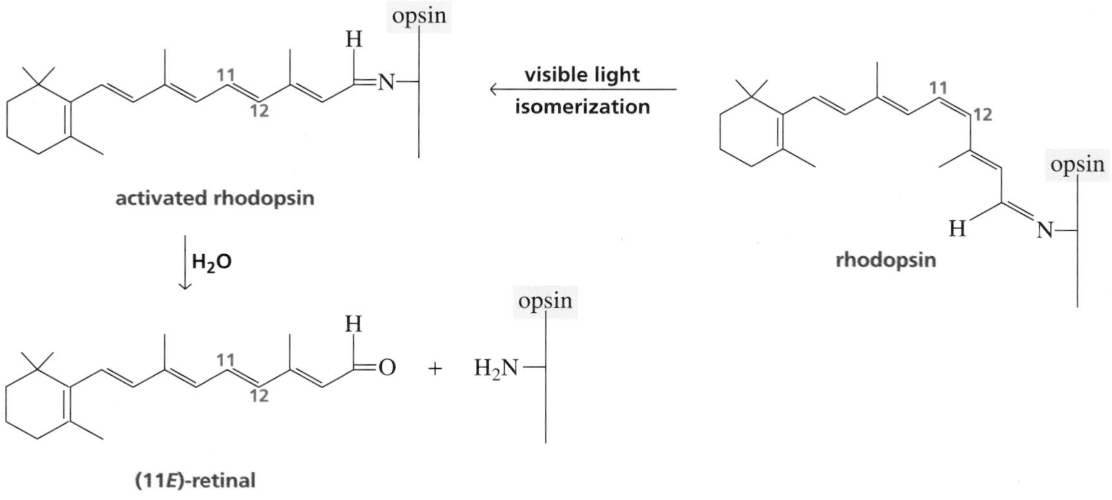

activated rhodopsin

rhodopsin

(11E)-retinal

The details of how the sequence of reactions above creates a visual image are not clearly understood. The fact that a simple change in configuration can be responsible for initiating a process as complicated as vision is remarkable.

23.8 BIOSYNTHESIS OF TERPENES

The five-carbon compound used for the biosynthesis of terpenes is isopentenyl pyrophosphate. Each step in the biosynthesis is catalyzed by a different enzyme. The first step is the same Claisen condensation that occurs in the first step of the biosynthesis of fatty acids, except that the acetyl and malonyl groups remain attached to coenzyme A rather than being transferred to the acyl carrier protein (Section 18.20). The Claisen condensation is followed by an aldol addition with a second molecule of malonyl-CoA. The thioester is reduced with two equivalents of NADPH (Section 22.2). A pyrophosphate group is added by means of two separate phosphorylations with ATP (the mechanism is given below). Decarboxylation and loss of the OH group result in isopentenyl pyrophosphate.

biosynthesis of isopentenyl pyrophosphate

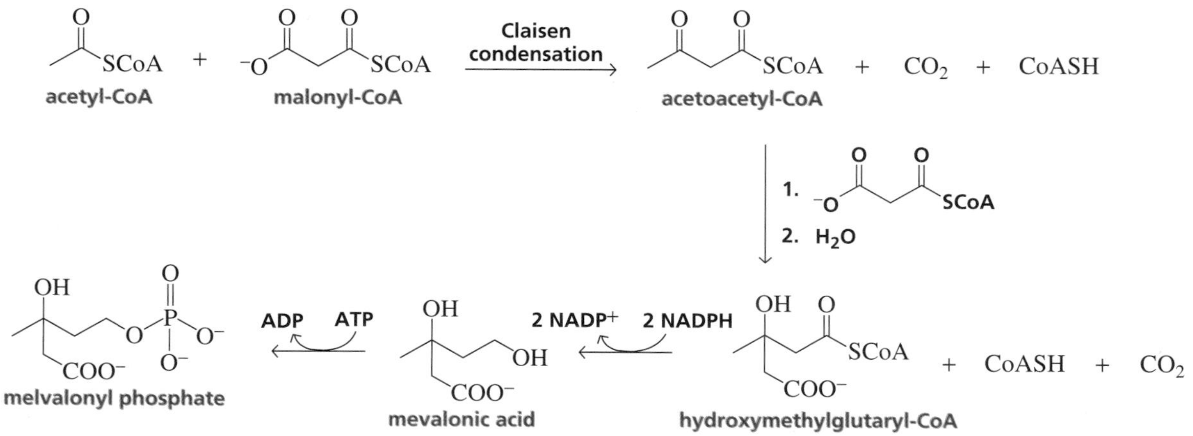

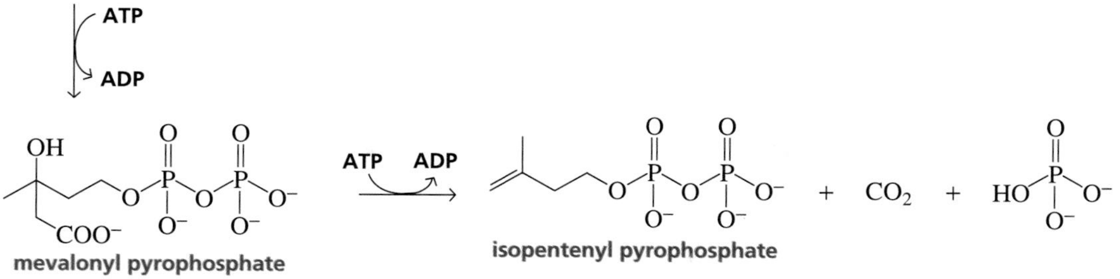

mevalonyl pyrophosphate → isopentenyl pyrophosphate + CO₂ + phosphate

The mechanism for converting mevalonic acid into mevalonyl pyrophosphate is shown below. ATP is an excellent phosphorylating reagent for nucleophiles, because the phosphoanhydride bonds of ATP are easily broken. The reason that phosphoanhydride bonds are so easily broken is discussed in Section 24.4.

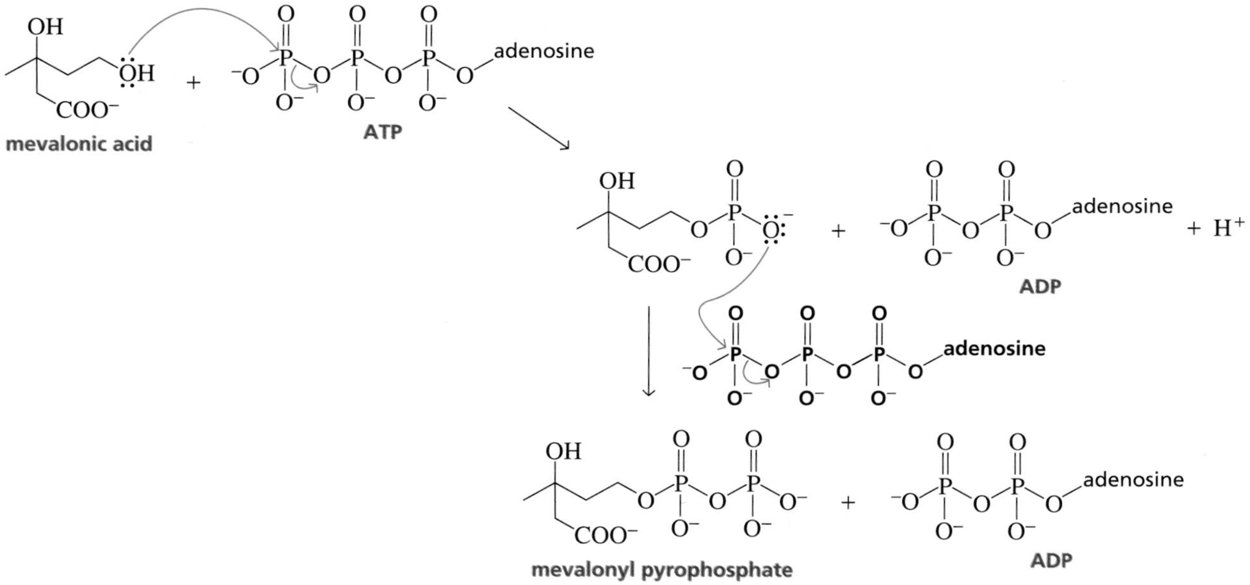

PROBLEM 14 / SOLVED

Give the mechanism for the last step in the biosynthesis of isopentenyl pyrophosphate, showing why ATP is required.

SOLUTION   In the last step of the biosynthesis of isopentenyl pyrophosphate, elimination of CO_2 is accompanied by elimination of an OH group. The OH group is a strong base and therefore a poor leaving group. ATP is used to convert the OH group into a phosphate group, which—because it is a good leaving group—is easily eliminated.

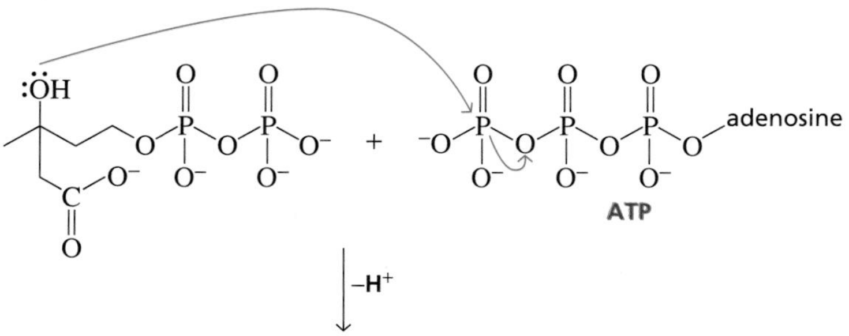

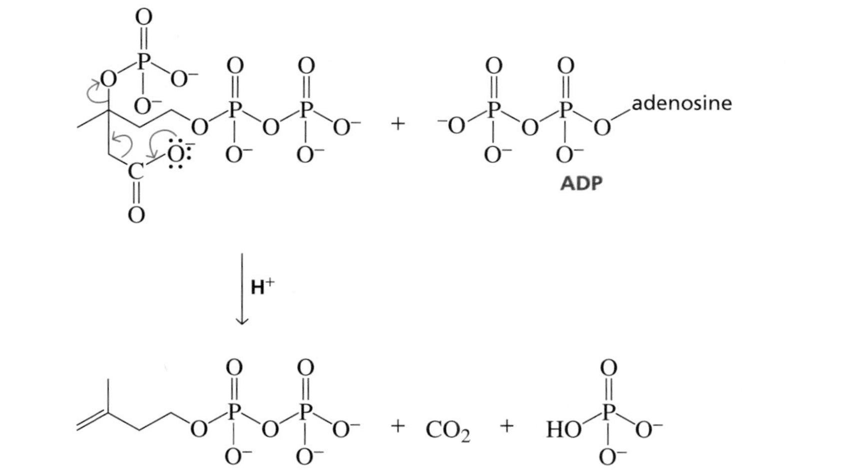

PROBLEM 15

Give the mechanisms for the Claisen condensation and aldol addition that occur in the first two steps of the biosynthesis of isopentenyl pyrophosphate.

An enzyme catalyzes the isomerization of isopentenyl pyrophosphate to dimethylallyl pyrophosphate, so that both compounds are available as starting materials for the biosynthesis of terpenes. The isomerization involves addition of a proton to isopentenyl pyrophosphate following Markovnikov's rule and elimination of a proton from the carbocation intermediate following Zaitsev's rule.

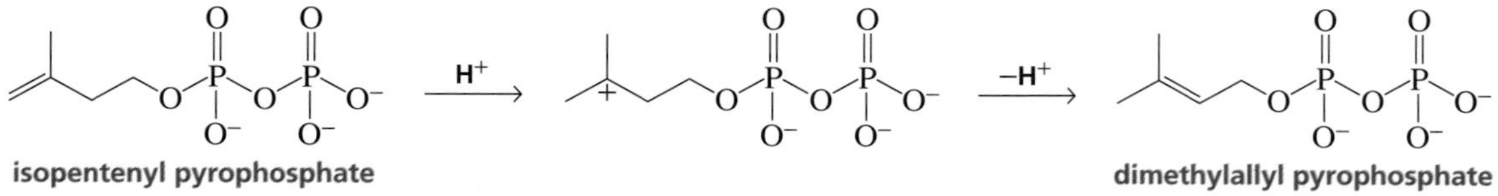

isopentenyl pyrophosphate   dimethylallyl pyrophosphate

The reaction of dimethylallyl pyrophosphate with isopentenyl pyrophosphate forms a ten-carbon compound. In the first step of the reaction, isopentenyl pyrophosphate acts as a nucleophile and displaces a pyrophosphate group from dimethylallyl pyrophosphate. Pyrophosphate is an excellent leaving group; its four OH groups have pK_a's of 0.9, 2.0, 6.6, and 9.4. A proton is removed in the next step, resulting in the formation of geranyl pyrophosphate.

dimethylallyl pyrophosphate   isopentenyl pyrophosphate

geranyl pyrophosphate   pyrophosphate

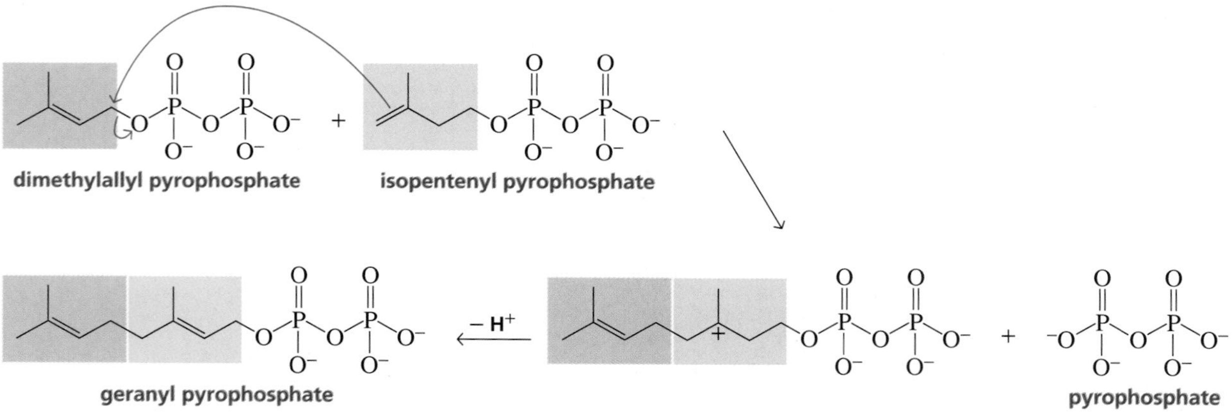

The following scheme shows how some of the many monoterpenes are synthesized from geranyl pyrophosphate.

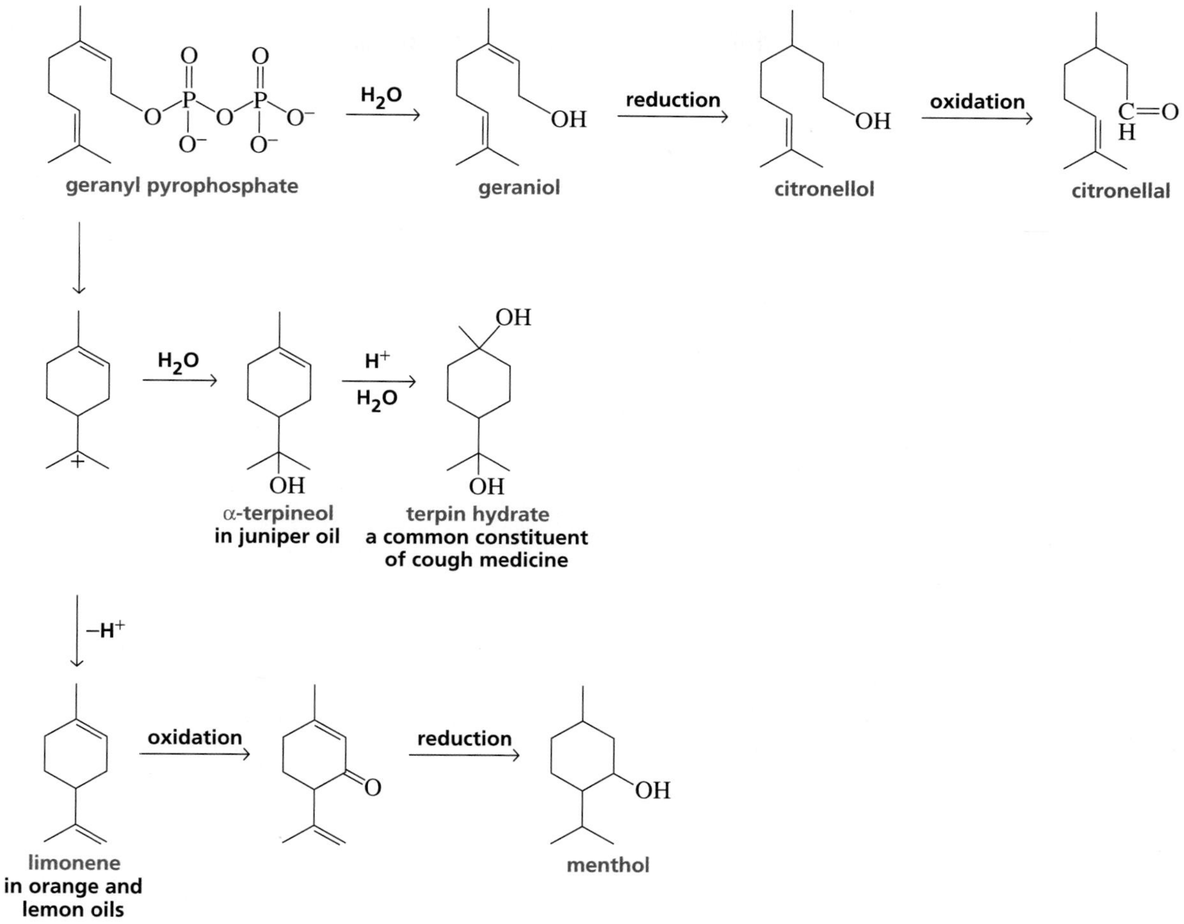

PROBLEM 16

Propose a mechanism for conversion of the *E* isomer of geranyl pyrophosphate to the *Z* isomer.

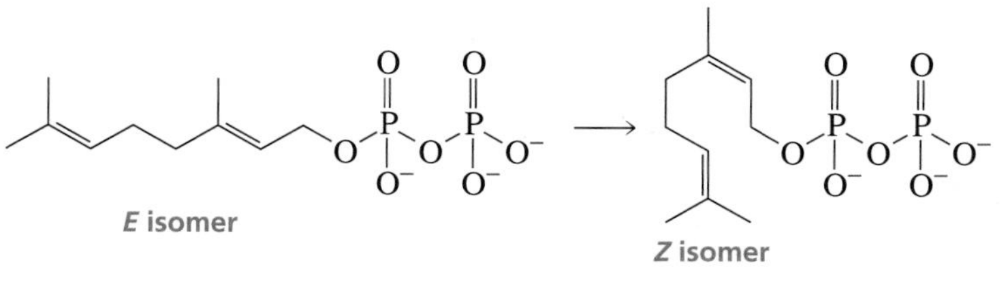

PROBLEM 17

Propose mechanisms for the formation of α-terpineol and limonene from geranyl pyrophosphate.

Geranyl pyrophosphate can react with another molecule of isopentenyl pyrophosphate to form farnesyl pyrophosphate, a 15-carbon compound.

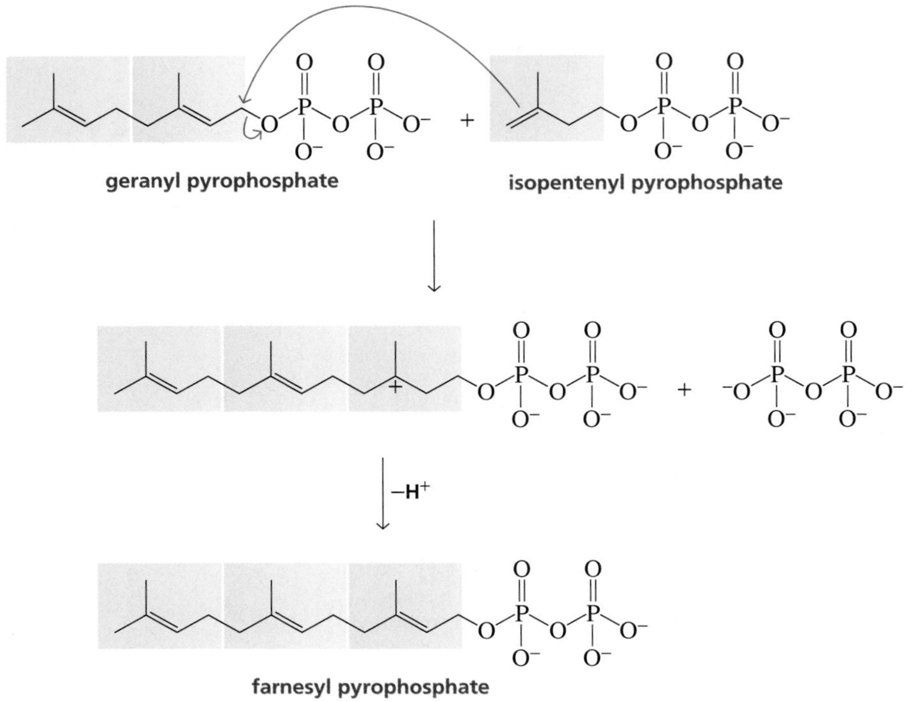

geranyl pyrophosphate   **isopentenyl pyrophosphate**

$-H^+$

farnesyl pyrophosphate

Two molecules of farnesyl pyrophosphate form squalene, a 30-carbon compound. This reaction is catalyzed by the enzyme squalene synthase. Squalene synthase joins the two molecules in a tail-to-tail linkage. Squalene is the precursor of cholesterol, and cholesterol is the precursor of all other steroids.

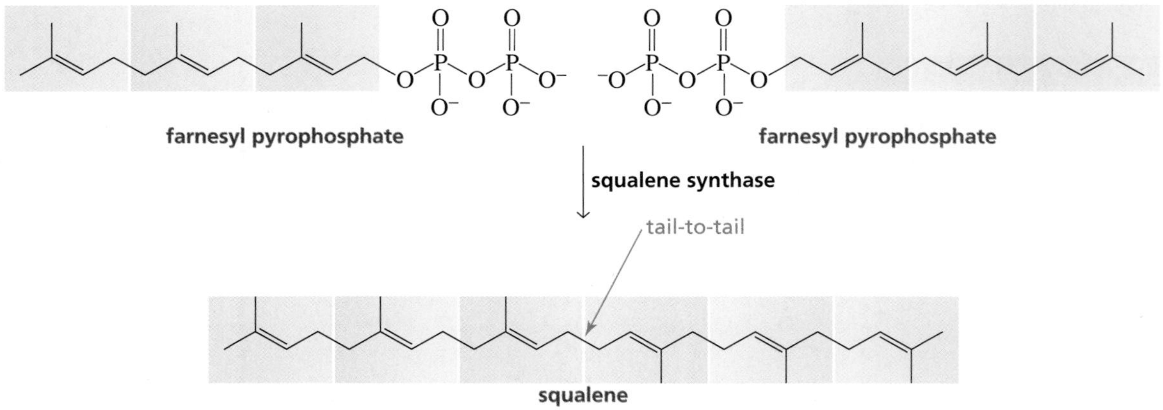

farnesyl pyrophosphate   **farnesyl pyrophosphate**

squalene synthase

tail-to-tail

squalene

Farnesyl pyrophosphate can react with another molecule of isopentenyl pyrophosphate to form geranylgeranyl pyrophosphate, a 20-carbon compound. Two geranylgeranyl pyrophosphates can join to form phytoene, a 40-carbon compound. Phytoene is the precursor of the carotenoid (tetraterpene) pigments of plants.

PROBLEM 18

In aqueous acidic solution, farnesyl pyrophosphate forms the sesquiterpene shown here. Propose a mechanism for this reaction.

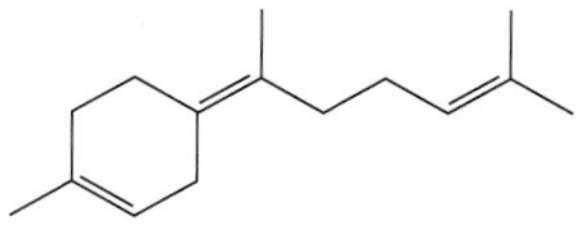

PROBLEM 19 / SOLVED

If squalene were synthesized in a medium containing acetate in which the carbonyl carbon of acetate were radioactively labeled with ^{14}C, which carbons in squalene would be labeled?

SOLUTION   Since malonyl-CoA is prepared from acetyl-CoA, the carbonyl carbon of malonyl-CoA will also be labeled. Examining each step of the mechanism for the biosynthesis of isopentenyl pyrophosphate from acetyl-CoA and malonyl-CoA allows you to determine the location of each of the radioactively labeled carbons in isopentenyl pyrophosphate. The locations of each of the radioactively labeled carbons in geranyl pyrophosphate can be determined from the mechanism for its biosynthesis from isopentenyl pyrophosphate. The locations of the radioactively labeled carbons in farnesyl pyrophosphate can be determined from the mechanism for its biosynthesis from geranyl pyrophosphate. Knowing that squalene is obtained from a tail-to-tail linkage of two farnesyl pyrophosphates gives you the answer.

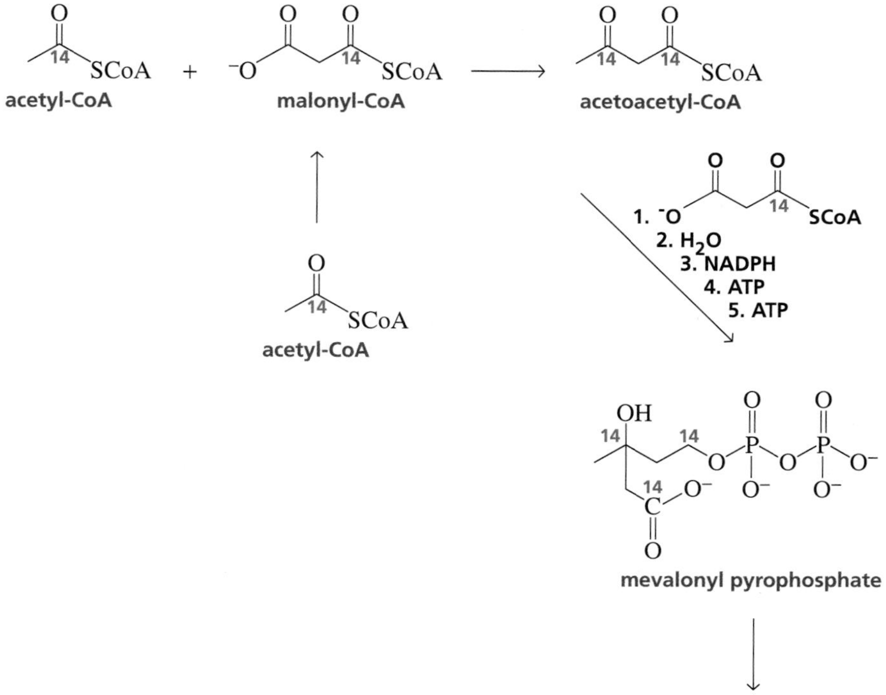

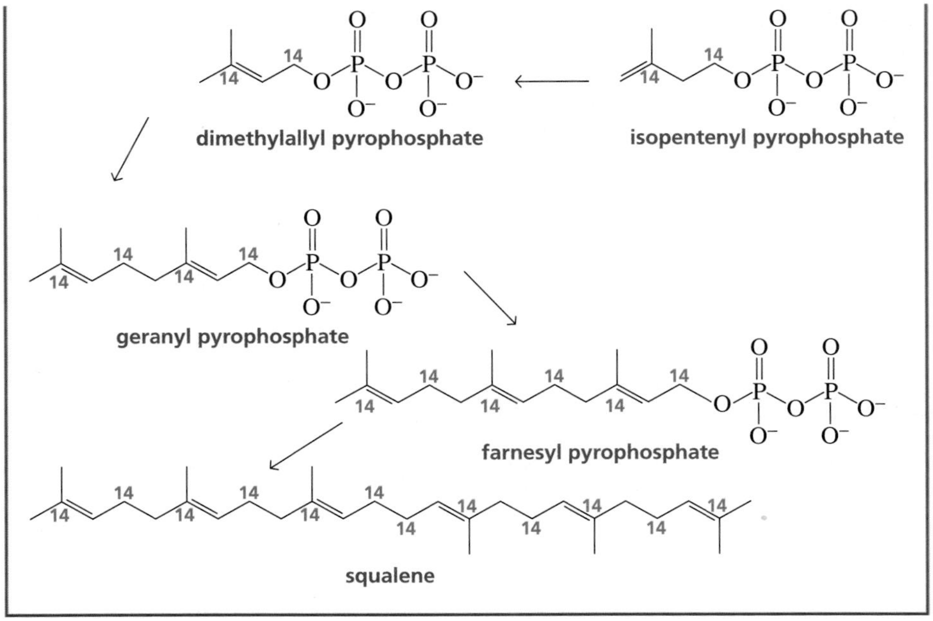

dimethylallyl pyrophosphate ← isopentenyl pyrophosphate

geranyl pyrophosphate

farnesyl pyrophosphate

squalene

23.9
STEROIDS

Hormones are chemical messengers. They are organic compounds that are synthesized in glands and delivered by the bloodstream to target tissues to stimulate or inhibit some process. Many hormones are **steroids.** Because steroids are nonpolar compounds, they are lipids. Their nonpolar character allows them to cross cell membranes so they can leave the cells in which they are synthesized and enter their target cells.

All steroids contain the following tetracyclic ring system. The four rings are designated A, B, C, and D. A, B, and C are six-membered rings; D is a five-membered ring. The carbons in the steroid ring system are numbered as shown.

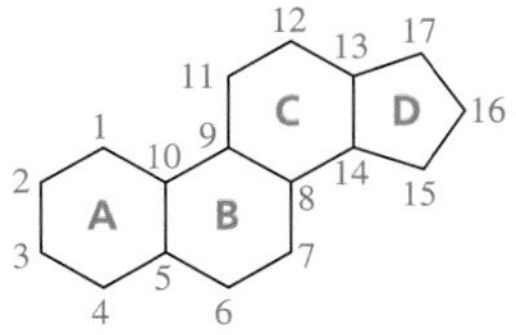

the steroid ring system

We have seen that rings can be **trans fused** or **cis fused** and that trans fused rings are more stable (Section 4.17). In steroids, the B, C, and D rings are all trans fused. In most naturally occurring steroids, the A and B rings are also trans fused.

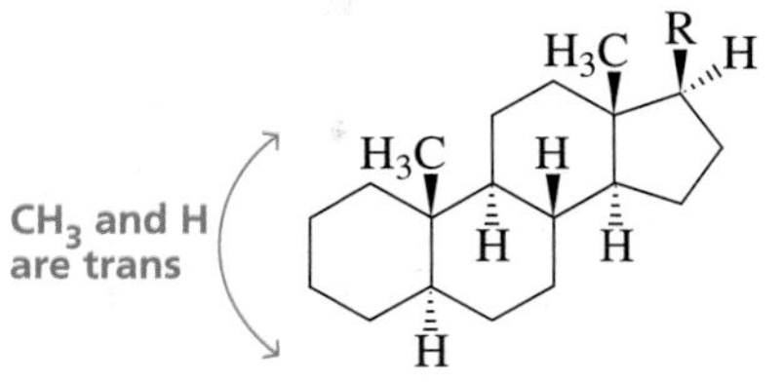

CH$_3$ and H
are trans

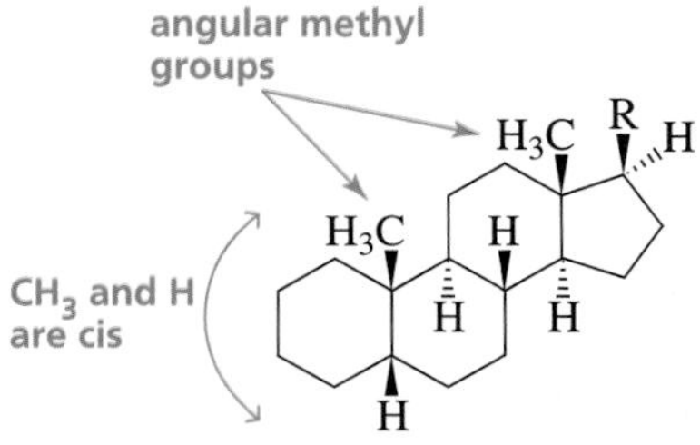

CH$_3$ and H
are cis

angular methyl
groups

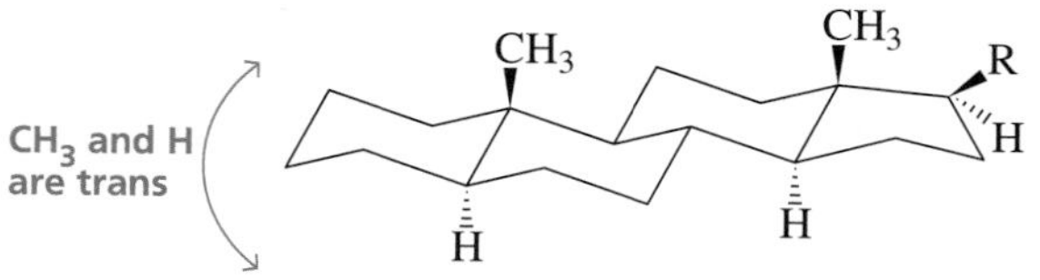

A and B rings are trans fused 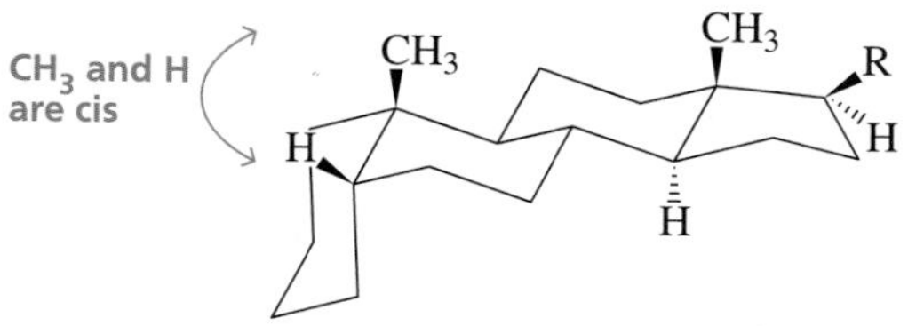   **A and B rings are cis fused**

Many steroids have methyl groups at the 10- and 13-positions. These are called **angular methyl groups.** When steroids are drawn as shown above, both angular methyl groups are always axial and above the plane of the steroid ring system. Substituents on the same side of the steroid ring system as the angular methyl groups are designated β-**substituents** (written with a solid wedge). Those on the opposite side of the plane of the ring system are α-**substituents** (indicated with a dashed wedge).

PROBLEM 20◆

A β-hydrogen at C-5 means that the A and B rings are _____ fused, while an α-hydrogen at C-5 means that the A and B rings are _____ fused.

The most prominent member of the steroid family is **cholesterol.** There are more than a 100 different steroids, and cholesterol is the precursor for all of them. Cholesterol is biosynthesized from squalene, a triterpene (Section 23.6). Cholesterol is an important component of cell membranes (Figure 23.1). Its ring structure makes it more rigid than other membrane lipids. Because cholesterol has eight centers of chirality, 256 stereoisomers are possible, but only one exists in nature (Chapter 4, Problem 38).

cholesterol

The steroid hormones can be divided into five classes: glucocorticoids, mineralocorticoids, androgens, estrogens, and progestins. Glucocorticoids and mineralocorticoids are synthesized in the adrenal cortex and are collectively known as **adrenal cortical steroids.** Glucocorticoids, as their name suggests, are involved in glucose metabolism as well as in the metabolism of proteins and fatty acids. Cortisone is an example of a glucocorticoid. Because of its anti-inflammatory effect, it is used clinically in the treatment of arthritis. Mineralocorticoids cause increased reabsorption of Na^+, Cl^-, and HCO_3^- by the kidneys, leading to an increase in blood pressure. Aldosterone is an example of a mineralocorticoid. All adrenal cortical steroids have an oxygen at C-11.

Two German chemists, **Heinrich Otto Wieland (1877–1957)** *and* **Adolf Windaus (1876–1959),** *each received a Nobel Prize in chemistry (Wieland in 1927 and Windaus in 1928) for work that led to the determination of the structure of cholesterol.*

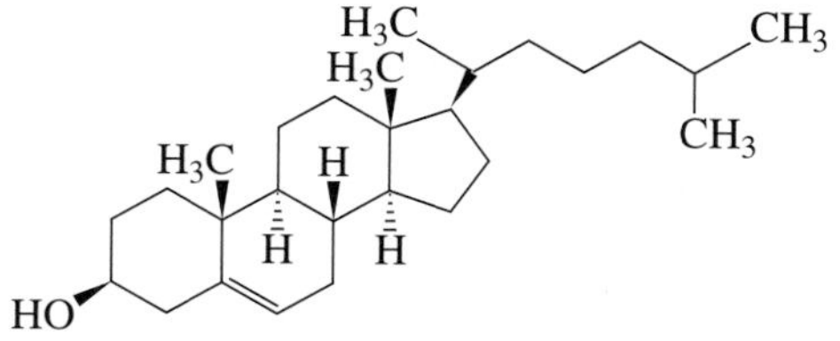

Heinrich Wieland, *the son of a chemist, was a professor at the University of Munich, where he showed that the bile acids were steroids and determined their individual structures. During World War II, he remained in Germany but was openly anti-Nazi.*

Adolph Windaus *originally intended to be a physician, but the experience of working with Emil Fischer for a year changed his mind. He discovered that vitamin D was a steroid, and he was the first to recognize that vitamin B_1 contained sulfur.*

Adolf Friedrich Johann Butenandt (1903–1995) *was born in Germany. He shared the 1939 Nobel Prize in chemistry (with Ružička) for isolating and determining the structures of estrone, androsterone, and progesterone. Forced by the Nazi government to refuse the prize, he accepted it after World War II. He was the director of the Kaiser Wilhelm Institute in Berlin and later was a professor at the Universities of Tübingen and Munich.*

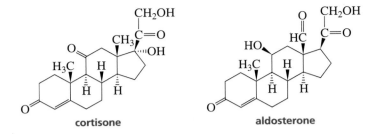

cortisone   aldosterone

PROBLEM 21◆

Is the OH substituent of the A ring of cholesterol an α-substituent or a β-substituent?

PROBLEM 22◆

Aldosterone is in equilibrium with its cyclic hemiacetal. Draw the hemiacetal form of aldosterone.

The male sex hormones, known as **androgens,** are secreted by the testes. They are responsible for the development of male secondary sex characteristics during puberty. They also promote muscle growth. Testosterone and 5α-dihydrotestosterone are androgens.

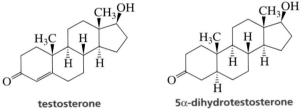

testosterone   5α-dihydrotestosterone

Estradiol and estrone are female sex hormones, known as **estrogens.** They are secreted by the ovaries and are responsible for the development of female secondary sex characteristics. They also regulate the menstrual cycle. Progesterone is the hormone that prepares the lining of the uterus for implantation of an ovum and is essential for the maintenance of pregnancy. It also prevents ovulation during pregnancy. This hormone belongs to the class of hormones known as **progestins.** Progesterone, which is biosynthesized from cholesterol, is a precursor for the glucocorticoids, the mineralocorticoids, and the sex hormones.

Michael Brown and Joseph Goldstein

Michael S. Brown *and* **Joseph Leonard Goldstein** *shared the 1985 Nobel Prize in medicine or physiology for their work on the regulation of cholesterol metabolism and the treatment of disease caused by elevated cholesterol levels in the blood. Brown was born in New York in 1941, and Goldstein in South Carolina in 1940. They are both professors of medicine at the University of Texas Southwestern Medical Center.*

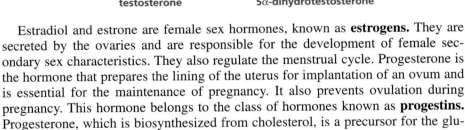

estradiol   estrone   progesterone

Although the various steroid hormones have remarkably different physiological effects, their structures are quite similar. For example, the only difference between testosterone and progesterone is the substituent at C-17, and the only difference between 5α-dihydrotestosterone and estradiol is one carbon and six hydrogens, but

CHOLESTEROL AND HEART DISEASE

Cholesterol is probably the best known lipid because of the correlation between cholesterol and heart disease. It is synthesized in the liver and is found in almost all body tissues. Cholesterol is found in many foods, but we do not require it in our diet because the body can synthesize all we need. A diet high in cholesterol can lead to high levels of cholesterol in the bloodstream. This excess cholesterol can accumulate on the walls of arteries, restricting the flow of blood. This disease of the circulatory system is known as *atherosclerosis* and is a primary cause of heart disease. Cholesterol travels through the bloodstream packaged in particles containing cholesterol, cholesterol esters, phospholipids, and proteins. The particles are classified according to their density. LDL (low-density lipoprotein) particles transport cholesterol from the liver to other tissues. Receptors on the surfaces of cells bind LDL particles, allowing them to be brought into the cell so that the cell can use the cholesterol. HDL (high-density lipoprotein) is a cholesterol scavenger, removing cholesterol from the surfaces of membranes and delivering it back to the liver, where it is converted into bile acids. LDL is the so-called "bad" cholesterol, while HDL is the "good" cholesterol. The more cholesterol we eat, the less the body synthesizes. But this does not mean that the presence of dietary cholesterol has no effect on the total amount of cholesterol in the bloodstream. Unfortunately, dietary cholesterol also inhibits the synthesis of the LDL receptors. So the more cholesterol we eat, the less the body synthesizes but also the less the body can get rid of by bringing it into target cells.

CLINICAL TREATMENT OF HIGH CHOLESTEROL

Compactin and Lovastatin are used clinically to decrease the concentration of serum cholesterol levels. They have been found to decrease the concentration by as much as 60%. They inhibit the enzyme that catalyzes the reduction of hydroxymethylglutaryl-CoA to mevalonic acid (Section 23.8). Decreasing the mevalonic acid concentration decreases the isopentenyl pyrophosphate concentration, so the biosynthesis of all terpenes, including cholesterol, is decreased.

Lovastatin   **Compactin**

The enzyme catalyzing the formation of mevalonic acid cannot tell the difference between its substrate and Compactin or Lovastatin. Therefore, the drugs are competitive inhibitors of the enzyme. Compare the structures of the substrate and the drugs to explain their ability to act as competitive inhibitors.

these two compounds make the difference between being male or female. This illustrates the extreme specificity of biochemical reactions.

PROBLEM 23

The acid component of a cholesterol ester is a fatty acid such as linoleic acid. Draw the structure of a cholesterol ester.

In addition to being the precursor of all other steroids, cholesterol is also the precursor of the **bile acids.** In fact, the word *cholesterol* is derived from the Greek words *chole* meaning "bile" and *stereos* meaning "solid." The bile acids—cholic acid and chenodeoxycholic acid—are synthesized in the liver, stored in the gall-bladder, and secreted into the small intestine where they act as emulsifying agents so that fats and oils can be digested by water-soluble digestive enzymes. Cholesterol is also the precursor of vitamin D (Section 27.6).

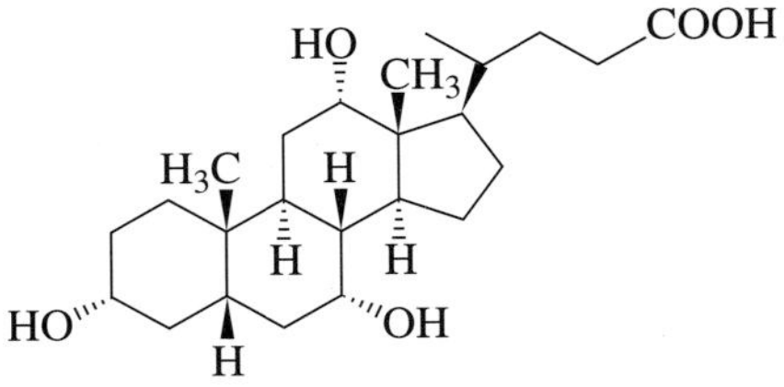

cholic acid 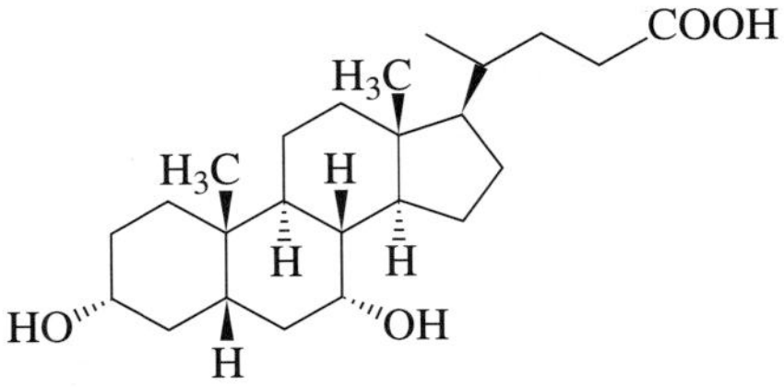   chenodeoxycholic acid

Konrad Bloch *and* **Feodor Lynen** *shared the 1964 Nobel Prize in medicine or physiology. Bloch showed how fatty acids and cholesterol are biosynthesized from acetate. Lynen showed that the two-carbon acetate unit is actually acetyl-CoA, and he determined the structure of coenzyme A.*

PROBLEM 24◆

Are the three OH groups of cholic acid axial or equatorial?

23.10 BIOSYNTHESIS OF CHOLESTEROL

How is cholesterol, the precursor of all the other steroid hormones, biosynthesized? The starting material for the biosynthesis is the triterpene squalene, which must first be converted to lanosterol. Lanosterol is converted to cholesterol in a series of 19 steps.

The first step in the conversion of squalene to lanosterol is epoxidation of the 2,3-double bond of squalene. Acid-catalyzed opening of the epoxide initiates a series of cyclizations resulting in the protosterol cation. Elimination of a C-9 proton from the cation initiates a series of 1,2-hydride and 1,2-methyl shifts, resulting in lanosterol.

biosynthesis of lanosterol and cholesterol

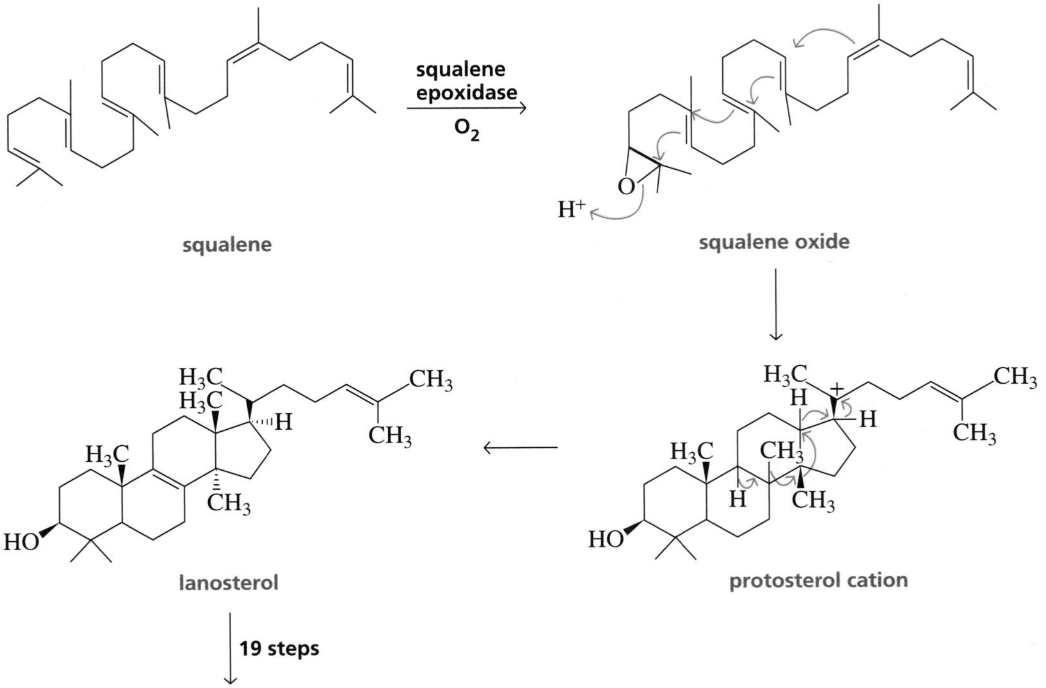

squalene

squalene epoxidase, O_2

squalene oxide

lanosterol

protosterol cation

19 steps

cholesterol

Konrad Emil Bloch *was born in Poland in 1912, studied in Germany, left Nazi Germany for Switzerland in 1934, and came to the United States in 1936, becoming a U.S. citizen in 1944. He received a Ph.D. from Columbia, taught at the University of Chicago, and became a professor of bio-chemistry at Harvard in 1954.*

Converting lanosterol to cholesterol requires removing three methyl groups from lanosterol (in addition to reducing two double bonds and creating a double bond). Removing methyl groups from carbon atoms is not an easy process—many different enzymes are required to carry out the 19 steps. Why does nature bother? Why not just use lanosterol instead of cholesterol? Konrad Bloch found that membranes that contain lanosterol instead of cholesterol are much more permeable. Small molecules are able to pass easily through lanosterol-containing membranes. As each methyl group is removed from lanosterol, the resulting membrane becomes less and less permeable.

Feodor Lynen (1911–1979) *was born in Germany, received a Ph.D. under Heinrich Wieland, and married Wieland's daughter. He was head of the Institute of Cell Chemistry at the University of Munich.*

PROBLEM 25

Draw the individual 1,2-hydride and 1,2-methyl shifts responsible for conversion of the protosterol cation to lanosterol. How many hydride shifts are involved? How many methyl shifts?

The potent physiological effects of steroids led scientists, in their search for new drugs, to synthesize steroids not available in nature and to investigate their physiological effects. Stanozolol and Dianabol are examples of drugs developed in this way. They have the same muscle-building effect as testosterone. Steroids that aid in the development of muscle are called **anabolic steroids.** These drugs are available by prescription and are used to treat people suffering from traumas accompanied by muscle deterioration. The same drugs have been administered to athletes and racehorses to increase their muscle mass. Stanozolol was the drug detected in several athletes in the 1988 Olympics. Anabolic steroids, when taken in relatively high dosages, have been found to cause liver tumors, personality disorders, and testicular atrophy.

23.11
SYNTHETIC STEROIDS

Stanozolol

Dianabol

Oral contraceptives have structures similar to that of progesterone. Norethindrone is better than progesterone in arresting ovulation. RU 486, when taken along with prostaglandins, terminates pregnancy within the first 9 weeks of gestation.

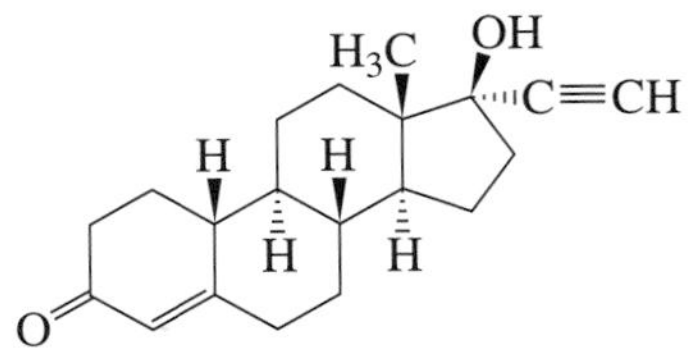

Norethindrone

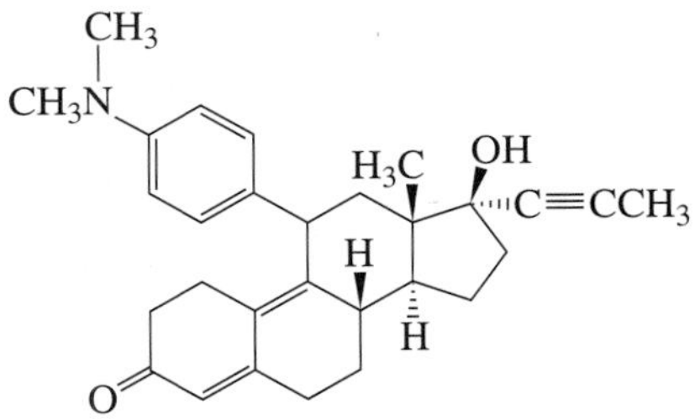

RU 486

KEY TERMS

adrenal cortical steroids (page 1053)
anabolic steroids (page 1057)
androgens (page 1054)
angular methyl group (page 1053)
bile acids (page 1055)
carotenoid (page 1044)
cephalin (page 1037)
cerebroside (page 1038)
cholesterol (page 1053)
cis fused (page 1052)
essential oil (page 1044)
estrogens (page 1054)
fat (page 1033)
fatty acid (page 1030)
hormone (page 1052)
isoprene rule (page 1043)
lecithin (page 1037)
leukotriene (page 1042)
lipid (page 1029)
lipid bilayer (page 1037)
membrane (page 1036)
mixed triacylglycerol (page 1032)
monoterpene (page 1043)
oil (page 1033)
phosphatidic acid (page 1036)
phosphoacylglycerol (page 1036)
phosphoglyceride (page 1036)
phospholipid (page 1036)
polyunsaturated fatty acid (page 1030)
progestins (page 1054)
prostacyclin (page 1040)
prostaglandin (page 1039)
sesquiterpene (page 1044)
simple triacylglycerol (page 1032)
sphingolipid (page 1038)
sphingomyelin (page 1038)
squalene (page 1044)
steroid (page 1052)
α-substituent (page 1053)
β-substituent (page 1053)
terpene (page 1042)
terpenoid (page 1042)
tetraterpene (page 1044)
thromboxane (page 1040)
trans fused (page 1052)
triacylglycerol (page 1032)
triterpene (page 1044)
wax (page 1031)

PROBLEMS

26. An optically active fat, when completely hydrolyzed, yields twice as much stearic acid as palmitic acid. Give the structure of the fat.

27. **a.** How many different triacylglycerols are there in which one of the fatty acid components is lauric acid and two are myristic acid? (Include stereoisomers.)
b. How many different triacylglycerols are there in which one of the fatty acid components is lauric acid, one is myristic acid, and one is palmitic acid? (Include stereoisomers.)

28. Cardiolipins are found in some membranes. Give the products formed when a cardiolipin undergoes complete hydrolysis.

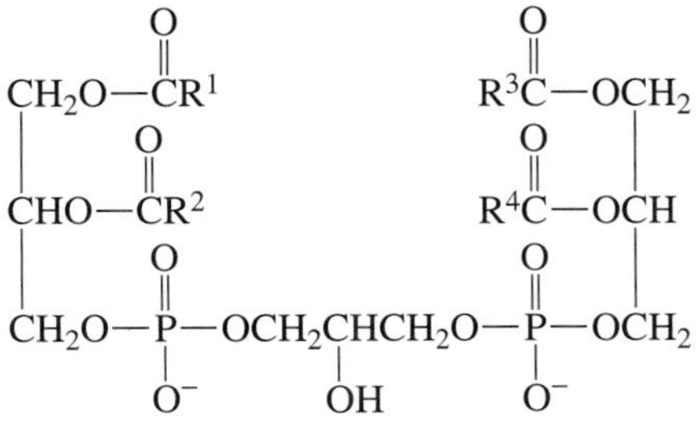

a cardiolipin

29. Nutmeg contains a simple, fully saturated triacylglycerol with a molecular weight of 722. Give its structure.

30. Give the product that would be obtained from the reaction of cholesterol with each of the following reagents. (*Hint:* Because of steric hindrance from the angular methyl groups, the α-face is more susceptible to reagents than the β-face.)
  a. H_2O, H^+
  b. BH_3 in THF followed by H_2O_2 + HO^-
  c. H_2,Pd/C
  d. Br_2 + H_2O
  e. peroxybenzoic acid
  f. the product of part **e** + CH_3O^-

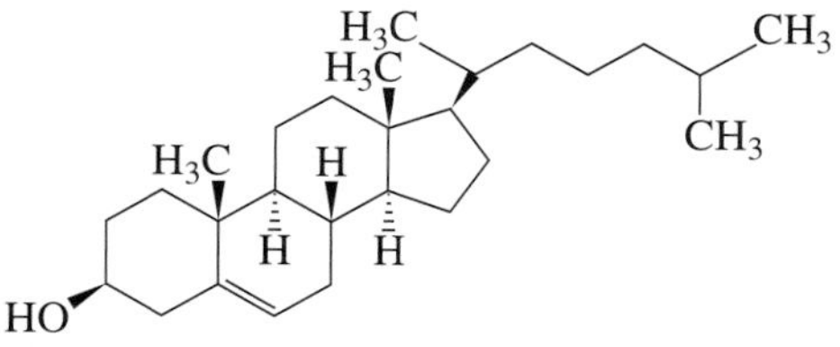

cholesterol

31. Dr. Chole S. Terol synthesized the following samples of mevalonic acid and fed them to a group of lemon trees. Which carbons will be labeled in citronellal, which is isolated from lemon oil, that is produced by trees that were fed
  a. sample A?   **b.** sample B?   **c.** sample C?

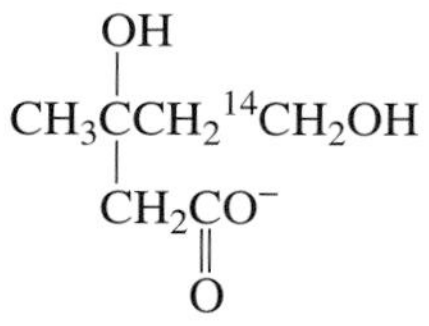

sample A 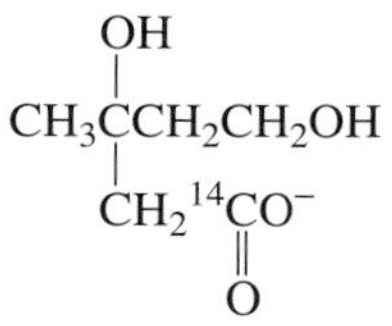   sample B 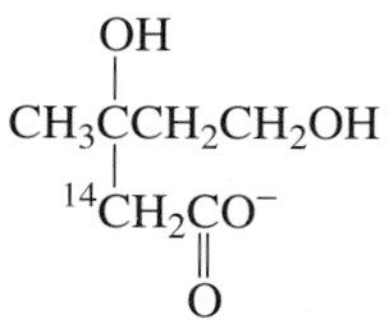   sample C

32. An optically active monoterpene (compound A) with molecular formula $C_{10}H_{18}O$ undergoes catalytic hydrogenation to form an optically inactive compound with molecular formula $C_{10}H_{20}O$ (compound B). When compound B is heated with acid, followed by reaction with O_3 and work-up under reducing conditions (Zn, H_2O), one of the products obtained is 4-methylcyclohexanone. Give possible structures for compound A.

33. If junipers were allowed to grow in a medium containing acetate in which the methyl carbon was radioactively labeled with ^{14}C, which carbons in α-terpineol would be labeled?

34. **a.** Propose a mechanism for the reaction shown below.
  b. What class of terpene is the starting material? Mark off the isoprene units in the starting material.

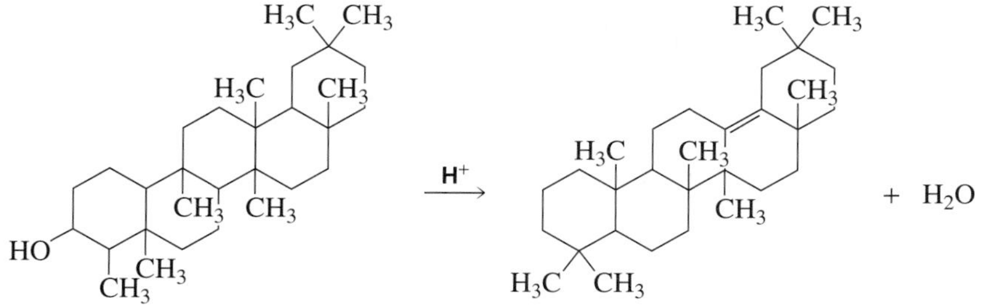

35. 5-Androstene-3,17-dione is isomerized to 4-androstene-3,17-dione by hydroxide ion. Propose a mechanism for this reaction (*J. Am. Chem. Soc.* 1989, *111*, 6419).

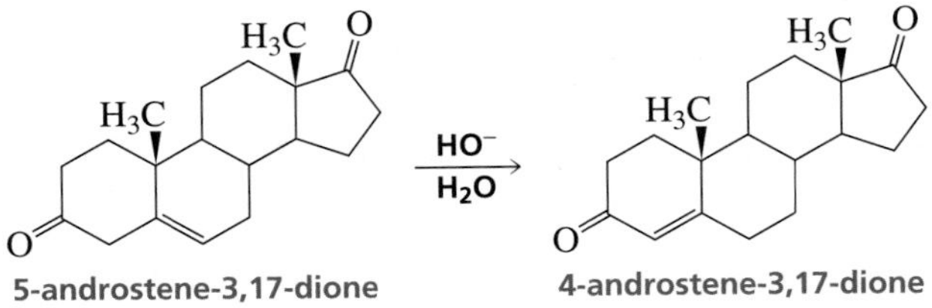

5-androstene-3,17-dione   4-androstene-3,17-dione

36. Both OH groups of one of the steroid diols shown below react with excess ethyl chloroformate, but only one OH group of the other steroid diol reacts under the same conditions. Explain this difference in reactivity.

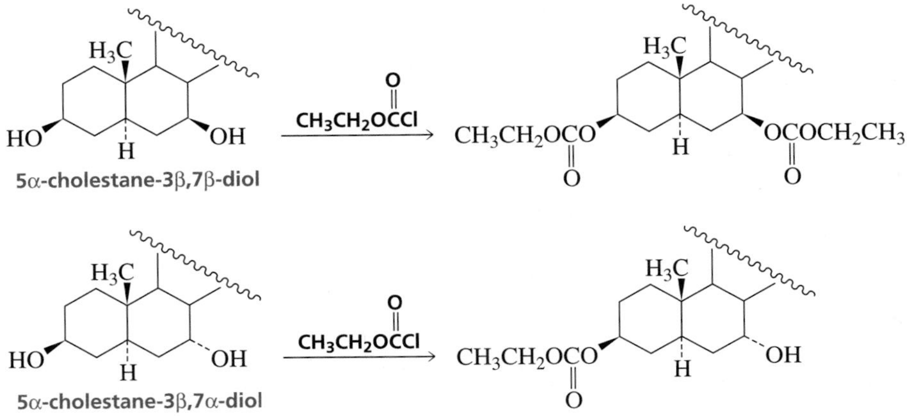

5α-cholestane-3β,7β-diol

5α-cholestane-3β,7α-diol

37. The acid-catalyzed dehydration of an alcohol to a rearranged alkene is known as a *Wagner–Meerwein rearrangement*. Propose a mechanism for the following Wagner–Meerwein rearrangement.

isoborneol   camphene

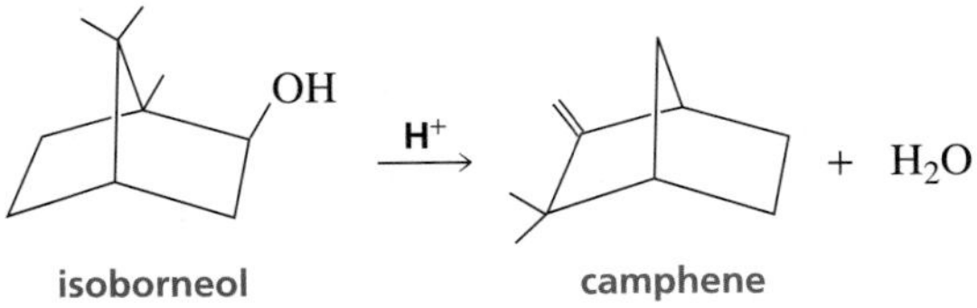

38. Diethylstilbestrol (DES) was given to pregnant women to prevent miscarriage until it was found that the drug caused cancer in both the mothers and their female children. DES has estradiol activity even though it is not a steroid. Draw DES so it is structurally similar to estradiol.

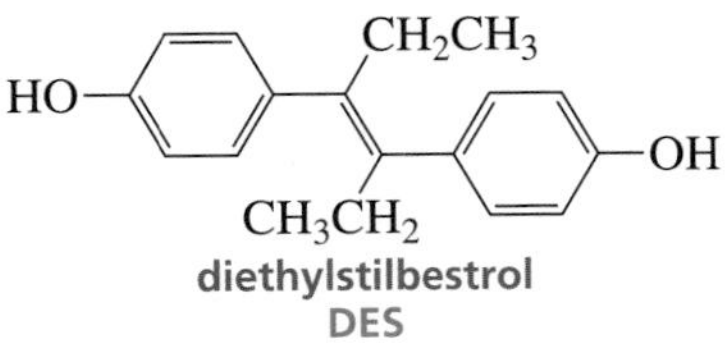

diethylstilbestrol
DES

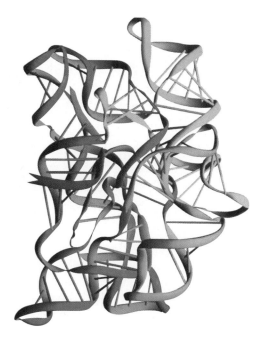

an RNA catalyst

NUCLEOSIDES, NUCLEOTIDES, AND NUCLEIC ACIDS

In previous chapters, we studied two of the three major kinds of biopolymers: polysaccharides and proteins. Now we will look at the third: nucleic acids. There are two types of nucleic acids: **deoxyribonucleic acid (DNA)** and **ribonucleic acid (RNA).** DNA encodes an organism's entire hereditary information and controls the growth and division of cells. In most organisms, the genetic information stored in DNA is transcribed into RNA. This information can then be translated for the synthesis of all the proteins needed for cellular structure and function.

DNA was first isolated in 1869 from the nuclei of white blood cells. Because this material was found in the nucleus and was acidic, it was called *nucleic acid.* Eventually, scientists found that the nuclei of all cells contain DNA, but it wasn't until 1944 that they realized that nucleic acids are the carriers of genetic information. In 1953, James Watson and Francis Crick described the three-dimensional structure of DNA—the famed double helix.

THE STRUCTURE OF DNA: WATSON, CRICK, FRANKLIN, AND WILKINS

James D. Watson was born in Chicago in 1928. He graduated from the University of Chicago at the age of 19 and received a Ph.D. three years later from Indiana University. In 1951, as a postdoctoral fellow at Cambridge University, Watson worked on determining the three-dimensional structure of DNA.

Francis H. C. Crick was born in Northampton, England in 1916. Originally trained as a physicist, Crick was involved in radar research during World War II. After the war, he entered Cambridge University to study for a Ph.D. in chemistry, which he received in 1953. He was a graduate student when he carried out his portion of the work that led to the proposal of the double-helical structure of DNA.

Rosalind Franklin was born in London in 1920. She graduated from Cambridge University and in 1942 quit her graduate studies to accept a position as a research officer in the British Coal Utilization Effort. After the war, she studied X-ray diffraction techniques in Paris. In 1951 she returned to England, accepting a position to develop an X-ray diffraction unit in the biophysics department at King's College. Her X-ray studies showed that DNA was a helix with phosphate groups on the outside of the molecule. Franklin died in 1958 without knowing the significance her work had played in determining the structure of DNA.

Francis Crick and James Watson

Rosalind Franklin

Watson and Crick shared the 1962 Nobel Prize in medicine or physiology with Maurice Wilkins for determining the double helical structure of DNA. Wilkins contributed X-ray studies that confirmed the double-helical structure. Wilkins was born in New Zealand in 1916 and moved to England 6 years later with his parents. During World War II he joined other British scientists who were working with American scientists on the development of the atomic bomb.

24.1 NUCLEOSIDES AND NUCLEOTIDES

The nucleic acids are chains of five-membered-ring sugars linked by phosphate groups (Figure 24.1). The anomeric carbon of each sugar is bonded to a nitrogen of a heterocyclic compound in a β-glycosidic linkage. (Recall that a β-linkage is one in which the substituents at C-1 and C-4 are on the same side of the furanose ring; Section 19.10.) Because the heterocyclic compounds are all basic amines, they are commonly referred to as **bases.** In RNA, the five-membered ring-sugar is D-ribose. In DNA, it is D-2′-deoxyribose (D-ribose without an OH group in the 2′-position).

Phosphoric acid links the sugars in both RNA and DNA. The acid has three dissociable OH groups with pK_a values of 2.1, 7.2, and 12.3. Each of the OH groups can react with an alcohol to form a phosphomonoester, a phosphodiester, or a phosphotriester. In nucleic acids, the phosphate group is a phosphodiester.

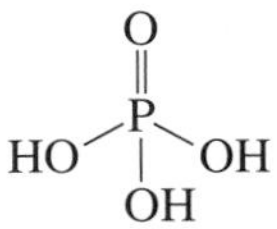

$$
\begin{array}{cccc}
\overset{\displaystyle O}{\underset{\displaystyle OH}{\overset{\|}{HO-P-OH}}} & \overset{\displaystyle O}{\underset{\displaystyle OH}{\overset{\|}{HO-P-OR}}} & \overset{\displaystyle O}{\underset{\displaystyle OH}{\overset{\|}{RO-P-OR}}} & \overset{\displaystyle O}{\underset{\displaystyle OR}{\overset{\|}{RO-P-OR}}} \\[4pt]
\text{phosphoric acid} & \text{a phosphomonoester} & \text{a phosphodiester} & \text{a phosphotriester}
\end{array}
$$

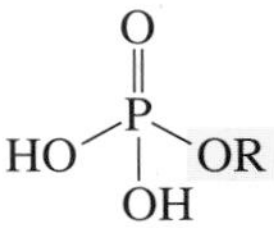

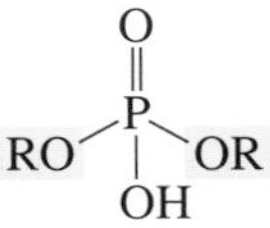

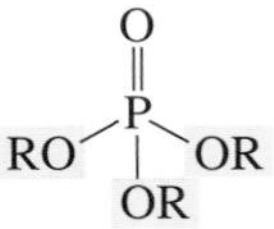

The vast differences in heredity among species and among members of the same species are determined by the sequence of the bases in DNA. Surprisingly, there are only four bases in DNA: two are substituted purines (adenine and guanine), and two are substituted pyrimidines (cytosine and thymine).

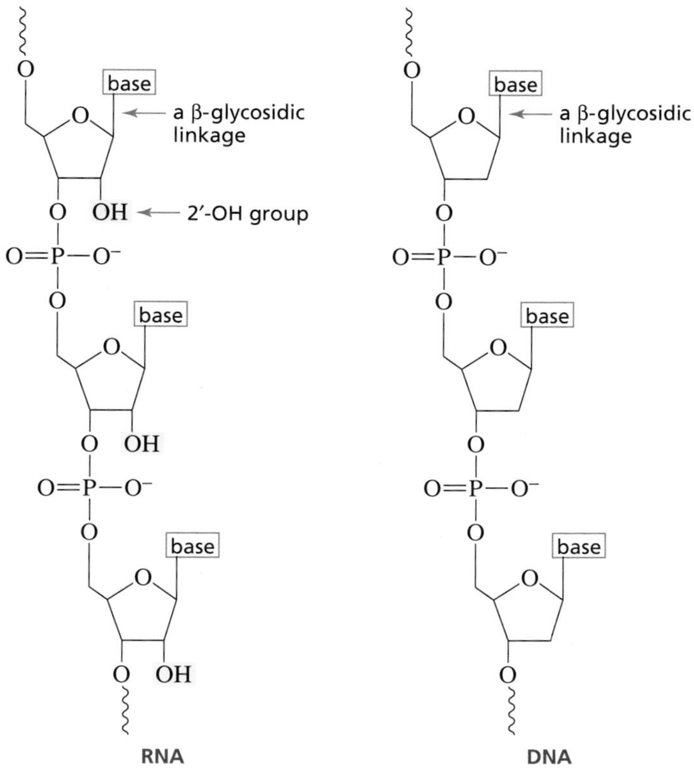

◀ **Figure 24.1**
Nucleic acids consist of a chain of five-membered-ring sugars linked by phosphate groups. Each sugar is bonded to a heterocyclic amine by a β-glycosidic linkage.

Studies that determined the structures of the nucleic acids and paved the way for the discovery of the DNA double helix were carried out by Phoebus Levene and elaborated by Sir Alexander Todd.

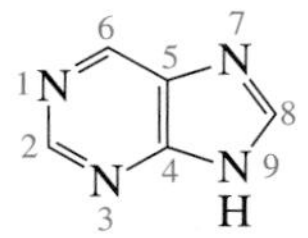

purine

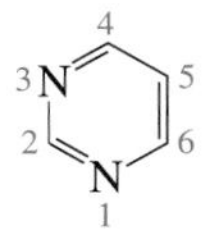

pyrimidine

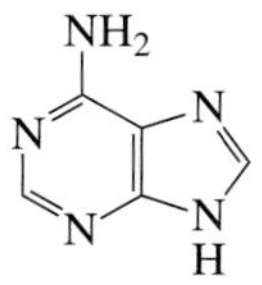

adenine

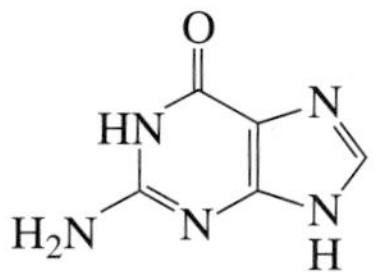

guanine

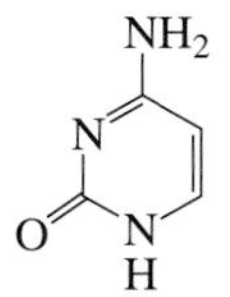

cytosine

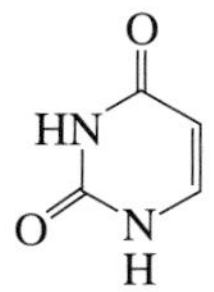

uracil

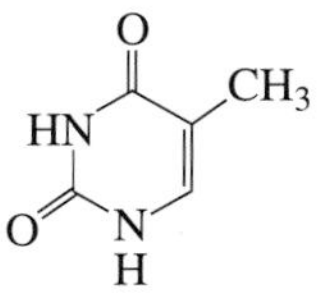

thymine

RNA also contains four bases. Three are the same as those in DNA, but the fourth base in RNA is uracil rather than thymine. Notice that thymine and uracil differ only by a methyl group: thymine is 5-methyluracil. The reason that DNA contains thymine instead of uracil is explained in Section 24.13.

The purines and pyrimidines are bonded to the anomeric carbon of the furanose ring: purines at N-9 and pyrimidines at N-1. A base bonded to D-ribose or to D-2'-deoxyribose is called a **nucleoside.** In a nucleoside, the ring positions of the sugar are indicated by prime numbers to distinguish them from the ring positions of the base. This is why the sugar component of DNA is referred to as 2'-deoxyribose. Notice the difference in the base names and the nucleoside names (adenine versus adenosine,

Phoebus Aaron Theodor Levene (1869–1940) *was born in Russia. When he emigrated to the United States with his family in 1891, his Russian name Fyodor was changed to Phoebus. Because his medical school education had been interrupted, he returned to Russia to complete his studies. When he returned to the United States he took chemistry courses at Columbia University. Deciding to forgo medicine for a career in chemistry, he went to Germany to study under Emil Fischer. He was a professor of chemistry at Rockefeller Institute (now Rockefeller University).*

cytosine versus cytidine, and so forth). Because uracil is found only in RNA, it is shown below attached to D-ribose but not to D-2'-deoxyribose; because thymine is found only in DNA, it is shown attached to D-2'-deoxyribose but not to D-ribose.

nucleosides

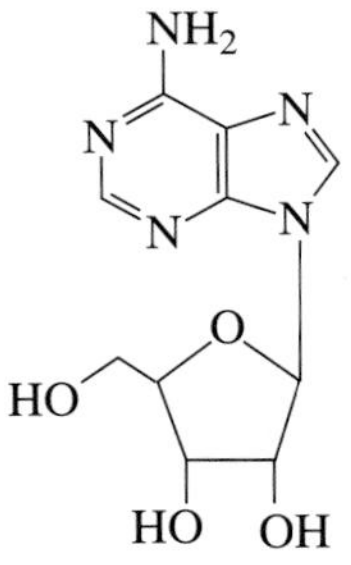

adenosine 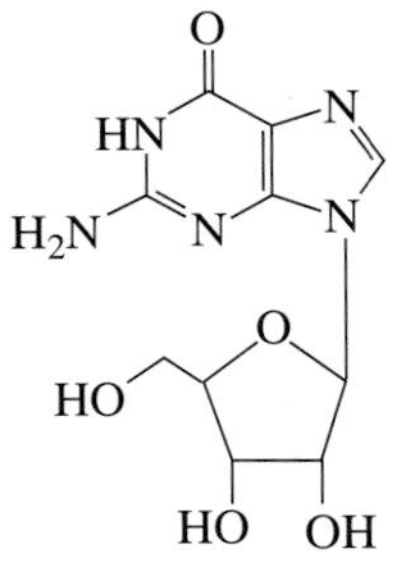   guanosine 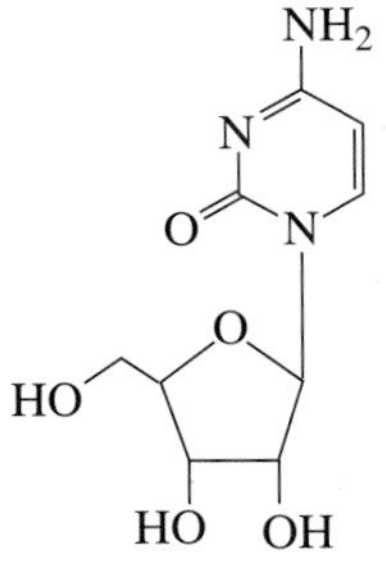   cytidine 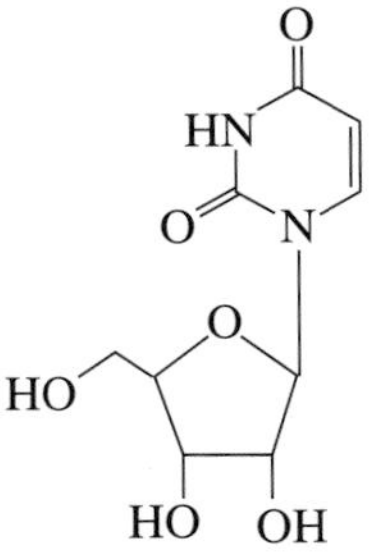   uridine

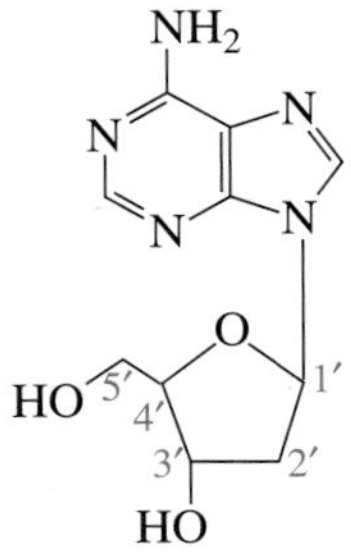

2'-deoxyadenosine 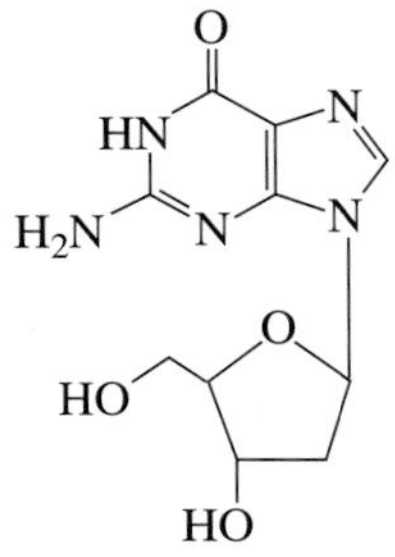   2'-deoxyguanosine 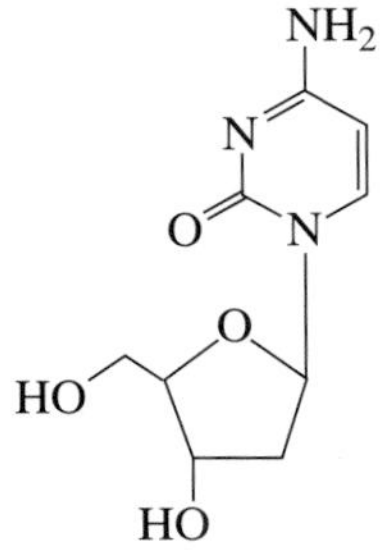   2'-deoxycytidine 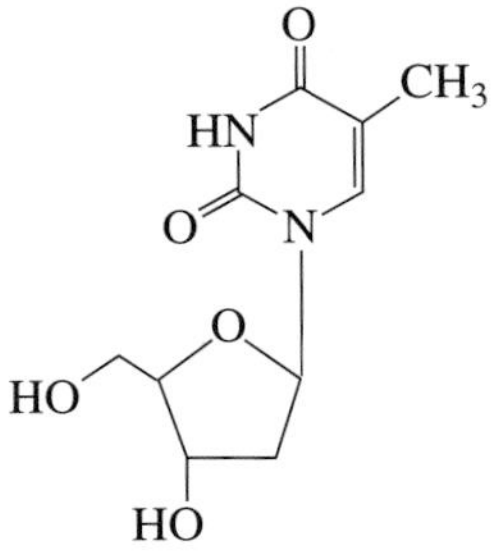   thymidine

PROBLEM 1

In acidic solutions, nucleosides are hydrolyzed to the sugar and heterocyclic base. Propose a mechanism for this hydrolysis.

A **nucleotide** is a nucleoside with one of its OH groups bonded to phosphoric acid by an ester linkage. The phosphate group in a nucleotide is attached to either the 5'- or the 3'-OH group. The nucleotides of RNA (where the sugar is D-ribose) are more precisely called **ribonucleotides,** while the nucleotides of DNA (where the sugar is D-2'-deoxyribose) are called **deoxyribonucleotides.**

nucleoside = base + sugar

**nucleotide =
base + sugar + phosphate**

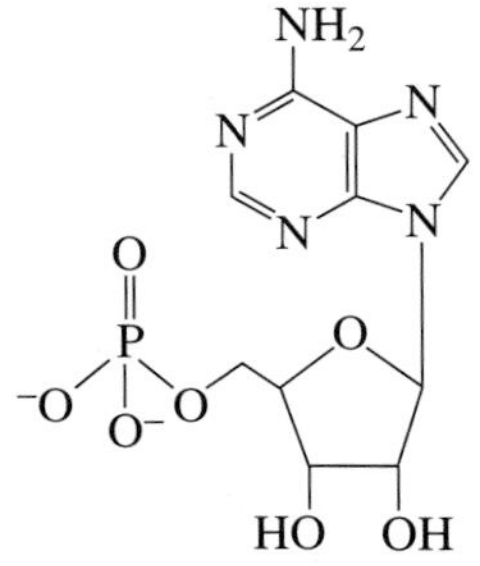

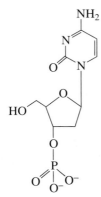

adenosine 5'-monophosphate
a ribonucleotide

2'-deoxycytidine 3'-monophosphate
a deoxyribonucleotide

When phosphoric acid is heated, it loses water, forming an anhydride known as pyrophosphoric acid. Its name comes from *pyros,* the Greek word for "fire." Thus, pyrophosphoric acid is prepared by "fire"—that is, by heating. When pyrophosphoric acid is heated with P_2O_5, triphosphoric acid and higher polyphosphoric acids are formed.

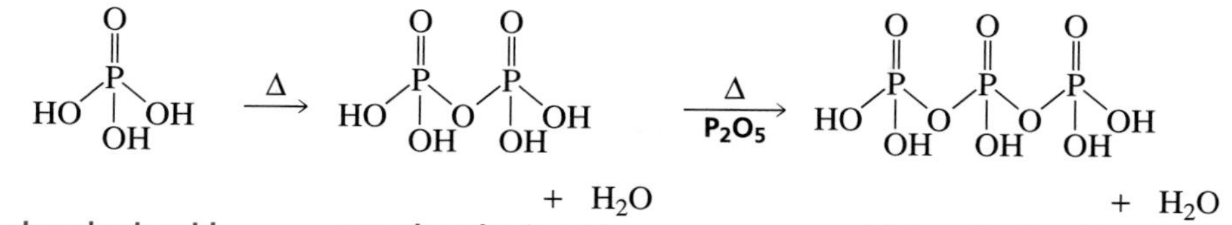

phosphoric acid   pyrophosphoric acid   triphosphoric acid

$+ H_2O$ $+ H_2O$

Alexander R. Todd (1907–1997) *was born in Scotland. He received two Ph.D.'s, one from the University of Frankfurt-am-Main (1931) and one from Oxford University (1933). He was a professor of chemistry at the University of Edinburgh, at the University of Manchester, and from 1944 to 1971 at Cambridge University. He was knighted in 1954 and was made a baron in 1962 (Baron Todd of Trumpington). For his work on nucleotides, he was awarded the 1957 Nobel Prize in chemistry.*

Because phosphoric acid can form an anhydride, nucleotides can exist as monophosphates, diphosphates, and triphosphates. They are named by adding *monophosphate* (or *diphosphate* or *triphosphate*) to the name of the nucleoside (Table 24.1).

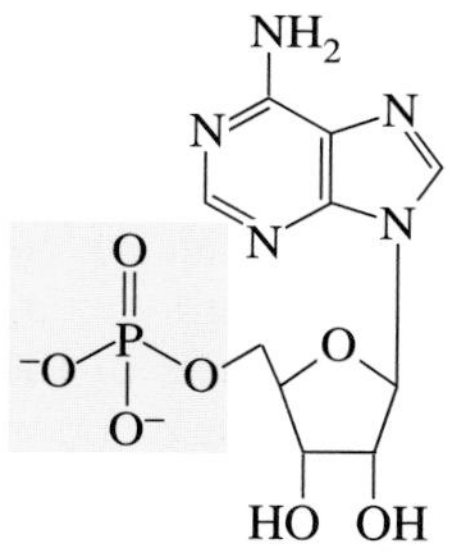

adenosine 5′-monophosphate AMP — adenosine 5′-diphosphate ADP — adenosine 5′-triphosphate ATP

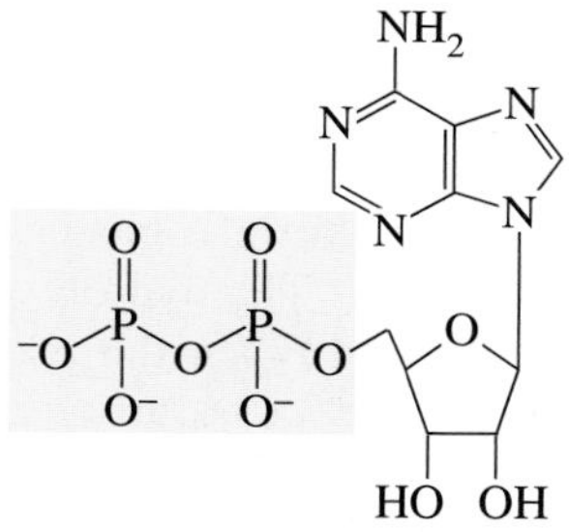

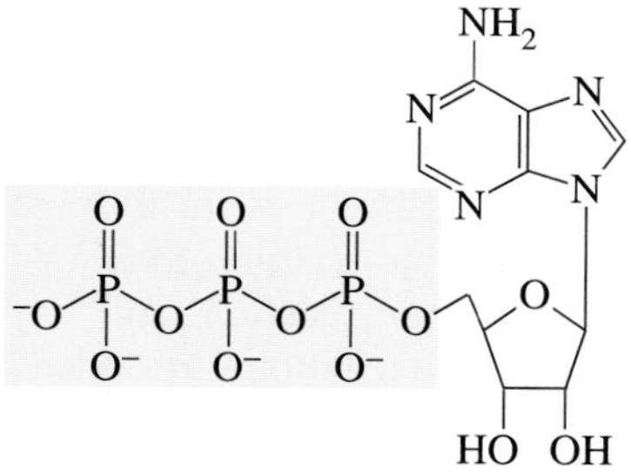

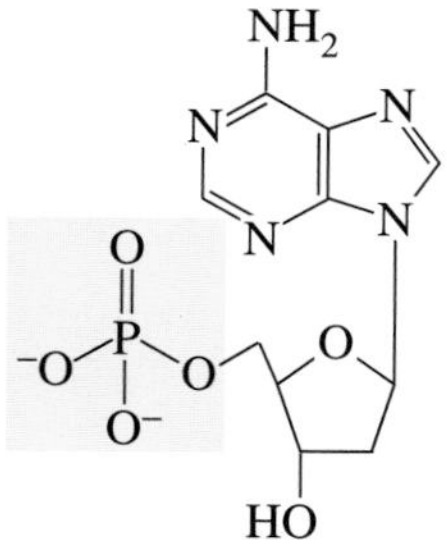

2′-deoxyadenosine 5′-monophosphate dAMP — 2′-deoxyadenosine 5′-diphosphate dADP — 2′-deoxyadenosine 5′-triphosphate dATP

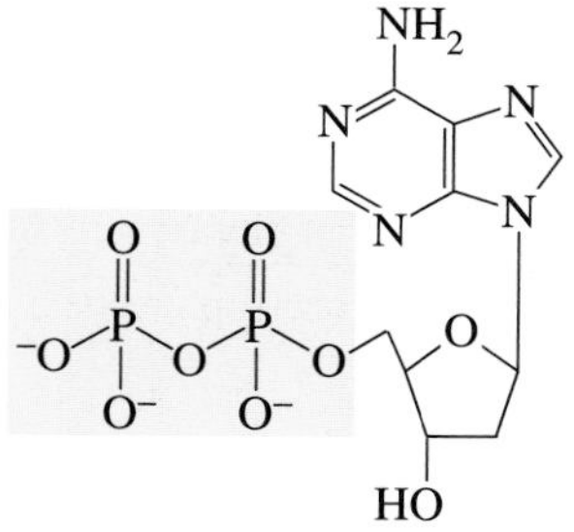

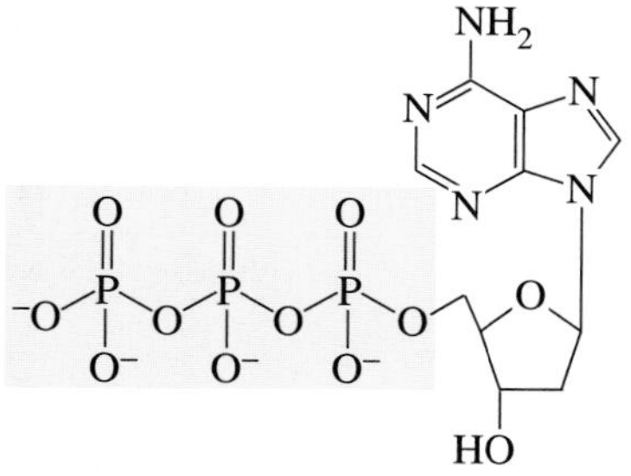

PROBLEM 2◆

Draw a structure for each of the following.

a. dCDP

b. guanosine 5'-triphosphate

c. dUMP

d. UDP

e. dTTP

TABLE 24.1 The Names of the Bases, the Nucleosides, and the Nucleotides				
Base	**Nucleoside**	**Deoxynucleoside**	**Ribonucleotide**	**Deoxyribonucleotide**
adenine	adenosine	2'-deoxyadenosine	adenosine 5'-phosphate	2'-deoxyadenosine 5'-phosphate
guanine	guanosine	2'-deoxyguanosine	guanosine 5'-phosphate	2'-deoxyguanosine 5'-phosphate
cytosine	cytidine	2'-deoxycytidine	cytidine 5'-phosphate	2'-deoxycytidine 5'-phosphate
thymine	—	thymidine	—	thymidine 5'-phosphate
uracil	uridine	—	uridine 5'-phosphate	

24.2
ATP: THE CARRIER OF CHEMICAL ENERGY

Adenosine 5'-triphosphate (ATP) is known as the universal carrier of chemical energy because, as it is commonly stated, "its energy of hydrolysis can be used to convert an endergonic reaction into an exergonic reaction." The importance of ATP to biological reactions is shown by its turnover rate in humans: each day, a person uses an amount of ATP equivalent to his or her body weight. Clearly, an enormous amount of ATP is continually being broken down and synthesized in biological organisms.

ATP's ability to enable reactions to occur that would not be able to occur without it is attributed to the large amount of energy released when ATP is hydrolyzed. This energy can be used to drive an endergonic reaction. For example, the reaction of D-glucose with hydrogen phosphate to form D-glucose-6-phosphate is an endergonic reaction ($\Delta G^{\circ\prime} = +3.3$ kcal/mol).[1] The hydrolysis of ATP is a highly exergonic reaction ($\Delta G^{\circ\prime} = -7.3$ kcal/mol). When the reactants and products of the two reactions are added together (the species occurring on both sides of the reaction arrow cancel) and the ΔG° values of both reactions are added, it is apparent that the energy released from the hydrolysis of ATP is more than enough to drive the phosphorylation of D-glucose ($\Delta G^{\circ\prime} = -4.0$ kcal/mol). Two reactions in which the energy of one is used to drive the other are known as coupled reactions.

$$\Delta G^{\circ\prime}$$

D-glucose + hydrogen phosphate $\longrightarrow$ D-glucose-6-phosphate + H_2O	+ 3.3 kcal/mol
ATP + H_2O $\longrightarrow$ ADP + hydrogen phosphate	− 7.3 kcal/mol
D-glucose + ATP $\longrightarrow$ D-glucose-6-phosphate + ADP	− 4.0 kcal/mol

This nonmechanistic description of ATP's power makes it appear to be some kind of magical source of energy. Let's look at the mechanism of the reaction to see what really happens. The reaction is a simple one-step nucleophilic substitution reaction. The 6-OH group of glucose attacks the terminal phosphate of ATP, breaking a **phosphoanhydride bond** without forming an intermediate. This is called an **in-line displacement mechanism.** Essentially it is an S_N2 reaction with an adenosine pyrophosphate leaving group. Notice that nucleophilic attack breaks

---

[1]The prime in $\Delta G^{\circ\prime}$ indicates that two additional parameters have been added to the ΔG° defined in Section 3.7: the reaction occurs in aqueous solution at pH = 7 and the concentration of water is assumed to be constant.

the phosphoanhydride bond rather than the π bond—unlike nucleophilic attack on a carbonyl group where the π bond is the first bond to break. The reason the phosphoanhydride bond is so easily broken is discussed in Section 24.4.

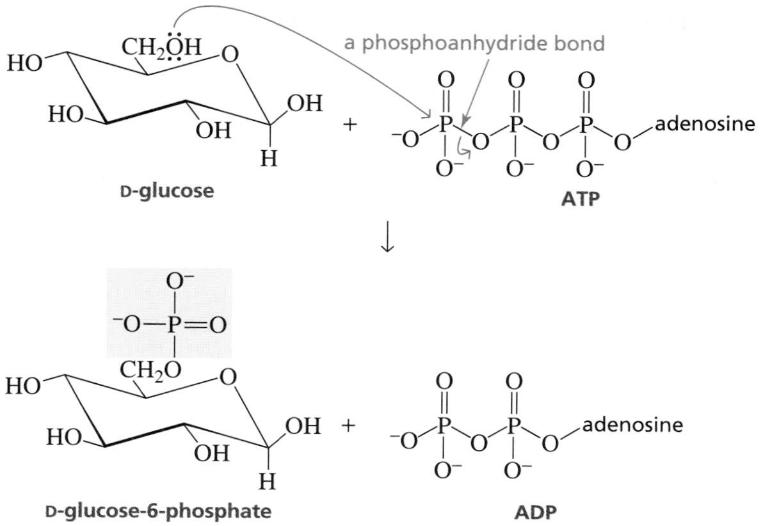

We can understand why the phosphorylation of glucose requires ATP. Without ATP, the 6-OH group of D-glucose would have to displace a very basic $^-$OH group. With ATP, the 6-OH group of D-glucose displaces weakly basic ADP.

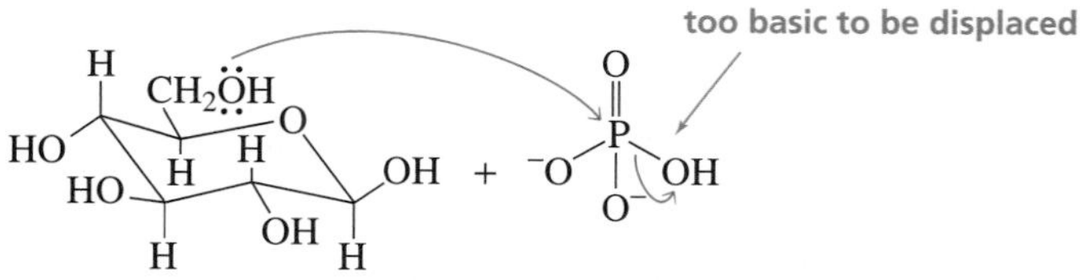

Although the phosphorylation of glucose is described as being driven by the hydrolysis of ATP, you can see from the mechanism that glucose does not react with phosphate and that ATP is not hydrolyzed because it does not react with water. In other words, neither of the coupled reactions actually occurs. The phosphate group of ATP is transferred directly to D-glucose.

The reaction above is an example of a **phosphoryl transfer reaction.** There are many such reactions in biological systems. In all of these reactions, the electrophilic phosphate group is transferred from one nucleophile to another as a result of breaking a phosphoanhydride bond. This example of a phosphoryl transfer reaction demonstrates the actual chemical function of ATP: it provides a reaction pathway involving a good leaving group for a reaction that cannot occur because of a poor leaving group.

PROBLEM 3 / SOLVED

The hydrolysis of phosphoenolpyruvate is so highly exergonic ($\Delta G^{\circ\prime} = -14.8$ kcal/mol) that it can be used to "drive the formation" of ATP from ADP ($\Delta G^{\circ\prime} = +7.3$ kcal/mol). Propose a mechanism for this reaction.

SOLUTION As we saw in the example with ATP, although the overall reaction is hydrolysis of phosphoenolpyruvate, phosphoenolpyruvate does not actually react with

water. Just as ATP "drives the formation" of glucose-1-phosphate by supplying glucose
with a phosphate that has a good leaving group (ADP), phosphoenolpyruvate "drives
the formation" of ATP by supplying ADP with a phosphate that has a good leaving
group (pyruvate).

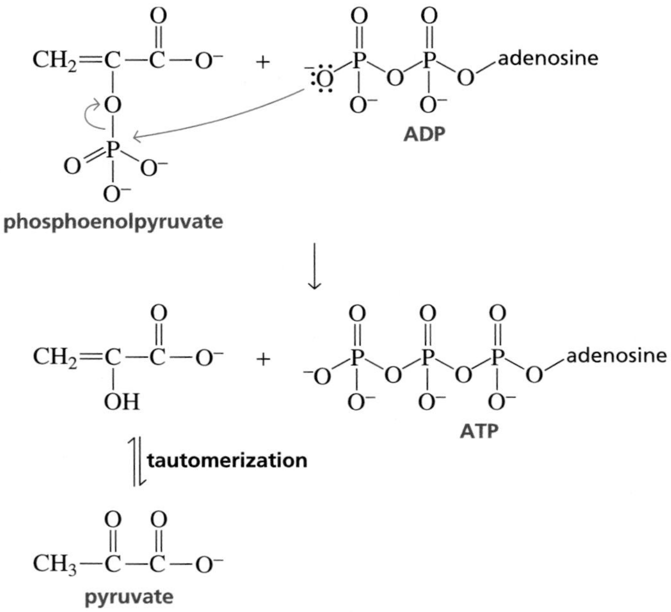

phosphoenolpyruvate

ADP

ATP

tautomerization

pyruvate

PROBLEM 4◆

Several important biomolecules and the $\Delta G^{\circ\prime}$ values for their hydrolysis are listed
below. Which of them hydrolyzes with sufficient energy to drive the formation of ATP?

glycerol-1-phosphate; -2.2 kcal/mol   phosphocreatine; -11.8 kcal/mol

fructose-6-phosphate; -3.8 kcal/mol   glucose-6-phosphate; -3.3 kcal/mol

24.3 THREE MECHANISMS FOR PHOSPHORYL TRANSFER REACTIONS

There are three possible mechanisms for phosphoryl transfer reactions. We will
illustrate them using the following nucleophilic acyl substitution reaction.

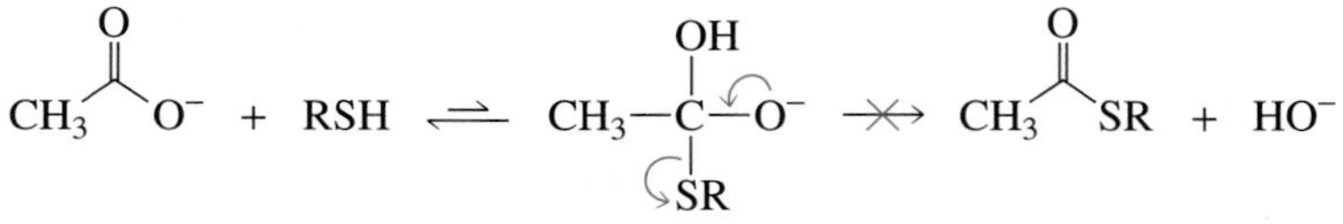

This reaction does not occur without ATP, since the carboxylate ion is resistant to
nucleophilic attack because of its negative charge, and the incoming nucleophile is
a weaker base than the base that would have to be expelled from the tetrahedral
intermediate to form the thioester. Instead, the thiol would be expelled from the
tetrahedral intermediate, reforming the carboxylate ion.

If ATP is added to the reaction mixture, the reaction occurs, because the carboxylate
ion attacks one of the phosphate groups of ATP, breaking a phosphoanhydride bond.
This puts a leaving group on the carboxyl group that can be displaced by the thiol.

There are three possible mechanisms for the reaction of a carboxylate ion with ATP, because each of the three phosphorus atoms of ATP can be attacked by the nucleophilic carboxylate ion. Each mechanism puts a different phosphate leaving group on the carboxyl group. The phosphate group is then displaced by the thiol.

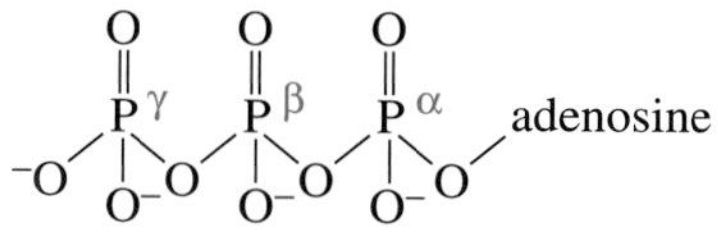

ATP

If the carboxylate ion attacks the γ-phosphorus of ATP, an **acyl phosphate** is formed. The acyl phosphate reacts with the thiol in a nucleophilic acyl substitution reaction, forming the thioester.

nucleophilic attack on the γ-phosphorus

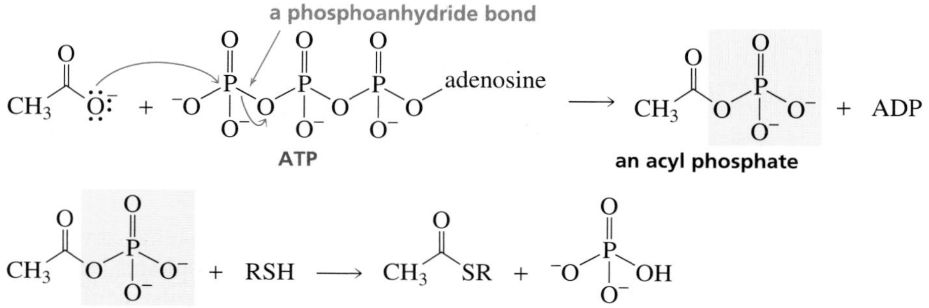

overall reaction

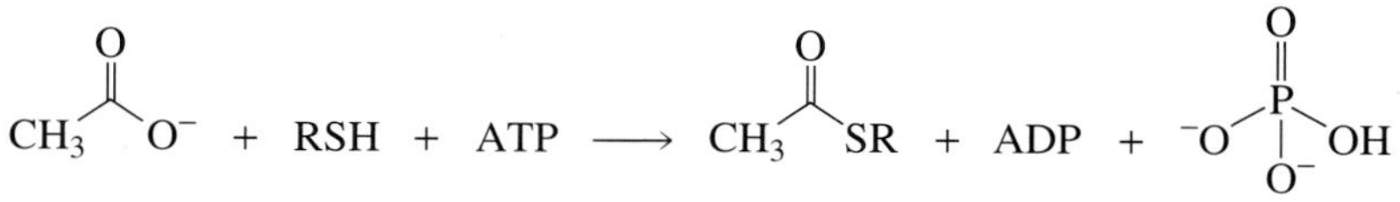

If the carboxylate ion attacks the β-phosphorus of ATP, an **acyl pyrophosphate** is formed. Attack by the thiol on the intermediate forms the thioester, because pyrophosphate is a good leaving group.

nucleophilic attack on the β-phosphorus

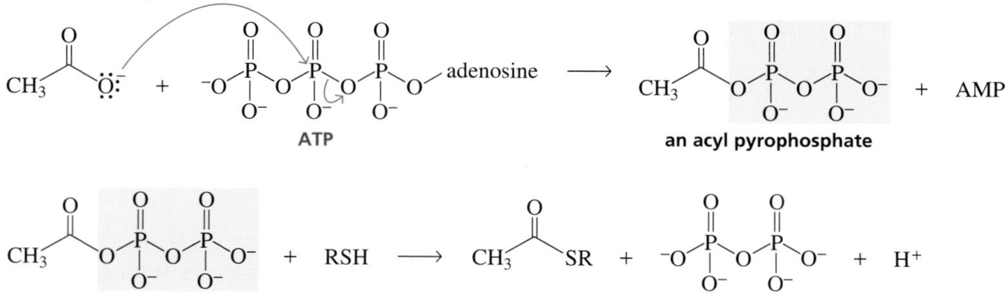

overall reaction

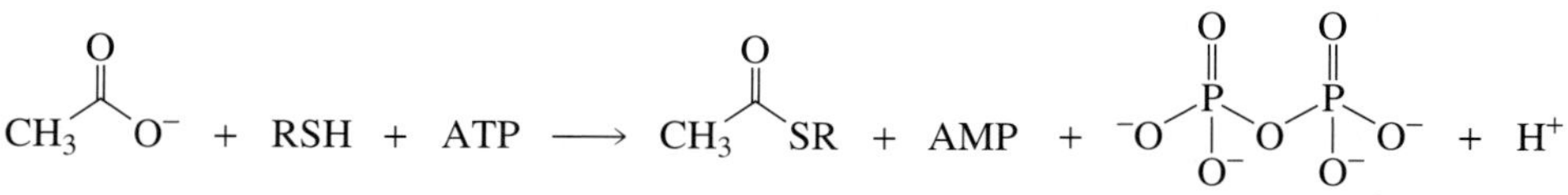

In the third possible mechanism, the carboxylate ion attacks the α-phosphorus of ATP, forming an **acyl adenylate.** Nucleophilic attack of the thiol on the acyl adenylate forms the thioester, because AMP is a good leaving group.

nucleophilic attack on the α-phosphorus

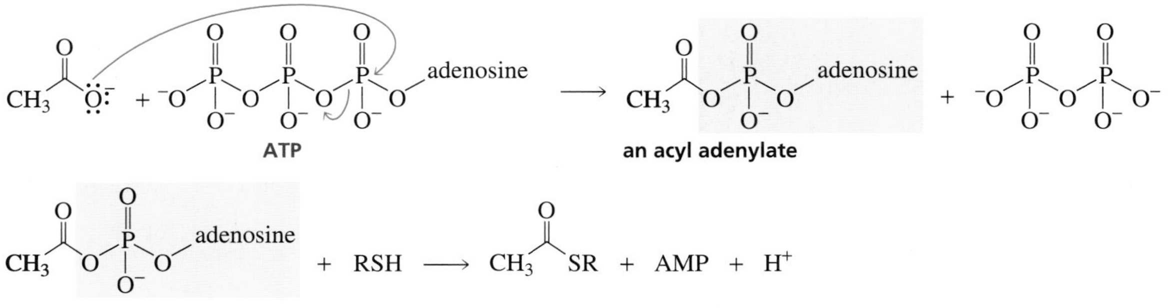

overall reaction

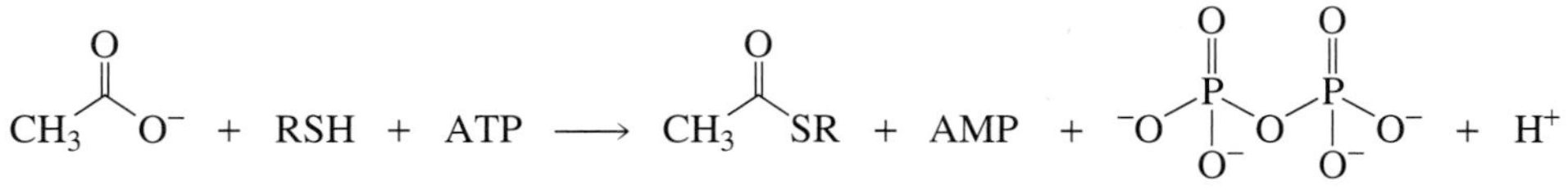

In Section 15.17, we saw that carboxylic acids in biological systems can be activated by being converted into acyl phosphates, acyl pyrophosphates, and acyl adenylates. Formation of these three classes of activated carboxylic acids corresponds to the three mechanisms discussed above for reaction of a nucleophile with ATP. Nucleophilic attack of a carboxylate ion on the γ-phosphorus forms an acyl phosphate, nucleophilic attack on the β-phosphorus forms an acyl pyrophosphate, and nucleophilic attack on the α-phosphorus forms an acyl adenylate.

ATP provides a reaction pathway involving a good leaving group for a reaction that cannot occur because of a poor leaving group.

Many different nucleophiles react with ATP in biological systems. Whether nucleophilic attack occurs on the α-, β-, or γ-phosphorus in any particular reaction depends on the enzyme catalyzing the reaction (Section 24.5). Mechanisms involving nucleophilic attack on the γ-phosphorus form ADP and phosphate as side products, while mechanisms involving nucleophilic attack on the α- or β-phosphorus form AMP and pyrophosphate as side products.

Pyrophosphate is hydrolyzed to two equivalents of phosphate. Consequently, in mechanisms in which pyrophosphate is formed as a product, its subsequent hydrolysis drives the reaction to the right, ensuring its irreversibility.

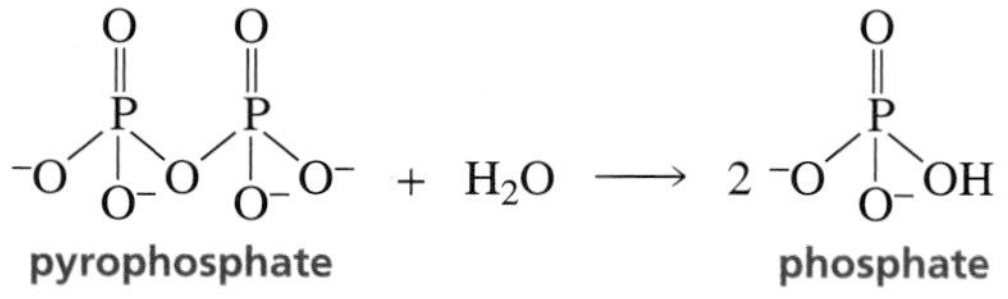

pyrophosphate   phosphate

Therefore, enzyme-catalyzed reactions in which irreversibility is important take place by one of the mechanisms that form pyrophosphate as a product (attack on the α- or β-phosphorus of ATP). For example, the reaction that links nucleotide subunits to form nucleic acids (Section 24.7) and the reaction that binds an amino acid to a tRNA (the first step in translating RNA into a protein; Section 24.11) both involve nucleophilic attack on the α-phosphorus of ATP.

PROBLEM 5

The β-phosphorus of ATP has two phosphoanhydride linkages, but only the one linking the β-phosphorus to the α-phosphorus is broken in phosphoryl transfer reactions. Explain why the one linking the β-phosphorus to the γ-phosphorus is never broken.

Because hydrolysis of a phosphoanhydride bond is a highly exergonic reaction, phosphoanhydride bonds are called "**high-energy bonds.**" The term "high-energy" in this context means that a lot of energy is released when the bond is broken. Do not confuse it with "bond energy," the term chemists use to describe how difficult it is to break a bond. A bond with a high bond energy is hard to break, while a high-energy bond breaks readily.

24.4
THE "HIGH-ENERGY" CHARACTER OF PHOSPHOANHYDRIDE BONDS

   Why is hydrolysis of a phosphoanhydride bond so exergonic? In other words, why is the $\Delta G°'$ value for its hydrolysis large and negative? A large negative $\Delta G°'$ means that the products of the reaction are much more stable than the reactants. Let's look at ATP and its hydrolysis products to see why this is so.

Three factors contribute to the greater stability of ADP and phosphate compared with ATP:

1. At physiological pH (pH = 7.3), ATP has 3.3 negative charges, ADP has 2.8 negative charges, and phosphate has 1.7 negative charges. Because of ATP's greater negative charge, more electrostatic repulsions are present in ATP than in ADP or phosphate. Electrostatic repulsions destabilize a molecule.

2. Negative charges in an aqueous solution are stabilized by solvation. Because the reactant has 3.3 negative charges, while the sum of the negative charges on the products is 4.5 (2.8 + 1.7), there is more solvation in the products than in the reactant.

3. There is greater resonance stabilization in the products than in the reactant. Delocalization of a pair of nonbonding electrons on the oxygen joining the two phosphorus atoms is not very effective because it puts a positive charge on an oxygen that is next to two partially positively charged phosphorus atoms. Delocalization of the nonbonding electrons in either of the products of phosphoanhydride hydrolysis does not have such a disadvantage.

an anhydride ion

   Additionally, because a single molecule is cleaved into two separate molecules, the entropy of the products is greater than the entropy of the reactants. Since

$\Delta G^{\circ\prime} = \Delta H^{\circ\prime} - T\Delta S^{\circ\prime}$, a positive $\Delta S^{\circ\prime}$ makes $\Delta G^{\circ\prime}$ more negative and therefore increases the equilibrium constant (Section 3.7). Similar factors explain the large, negative $\Delta G^{\circ\prime}$ when ATP is hydrolyzed to AMP and pyrophosphate and when pyrophosphate is hydrolyzed to two equivalents of phosphate.

PROBLEM 6 / SOLVED

Do the calculation showing that at pH 7.3:

a. the charge on ATP is -3.3.

b. the charge on ADP is -2.8.

c. the charge on phosphate is -1.6.

ATP has pK_a values of 0.9, 1.5, 2.3, and 7.7; ADP has pK_a values of 0.9, 2.8, and 6.8; and phosphoric acid has pK_a values of 2.1, 7.2, and 12.3.

SOLUTION TO 6a   Because pH 7.3 is much more basic than the pK_a values of the first three ionizations of ATP, we know that these three groups will be entirely in their basic forms at that pH, giving ATP three negative charges. We need to determine what fraction of the group with pK_a 7.7 will be in its basic form at pH 7.3. In other words, we must solve for

$$\frac{\text{concentration in the basic form}}{\text{total concentration}} = \frac{[A^-]}{[A^-] + [HA]}$$

$[A^-]$ = concentration of the basic form

$[HA]$ = concentration of the acidic form

Because this equation has two unknowns, one of the unknowns must be expressed in terms of the other unknown and of known quantities. Using the definition of the acid dissociation constant (K_a), we can define $[HA]$ in terms of $[A^-]$, K_a, and $[H^+]$.

$$K_a = \frac{[A^-][H^+]}{[HA]}$$

$$[HA] = \frac{[A^-][H^+]}{K_a}$$

$$\frac{[A^-]}{[A^-] + [HA]} = \frac{[A^-]}{[A^-] + \dfrac{[A^-][H^+]}{K_a}} = \frac{K_a}{K_a + [H^+]}$$

Now we can calculate the fraction of the group with pK_a 7.7 that will be in the basic form.

$$\frac{K_a}{K_a + [H^+]} = \frac{2.0 \times 10^{-8}}{2.0 \times 10^{-8} + 5.0 \times 10^{-8}} = 0.3$$

total negative charge on ATP = 3.0 + 0.3 = 3.3

**24.5
KINETIC STABILITY
OF ATP IN THE CELL**

Although ATP reacts readily in enzyme-catalyzed reactions, in the absence of an enzyme it reacts quite slowly. For example, carboxylic acid anhydrides hydrolyze in a matter of minutes, but ATP takes several weeks to hydrolyze. This low rate of ATP hydrolysis is important because ATP must exist in the cell until it is needed for an enzyme-catalyzed reaction.

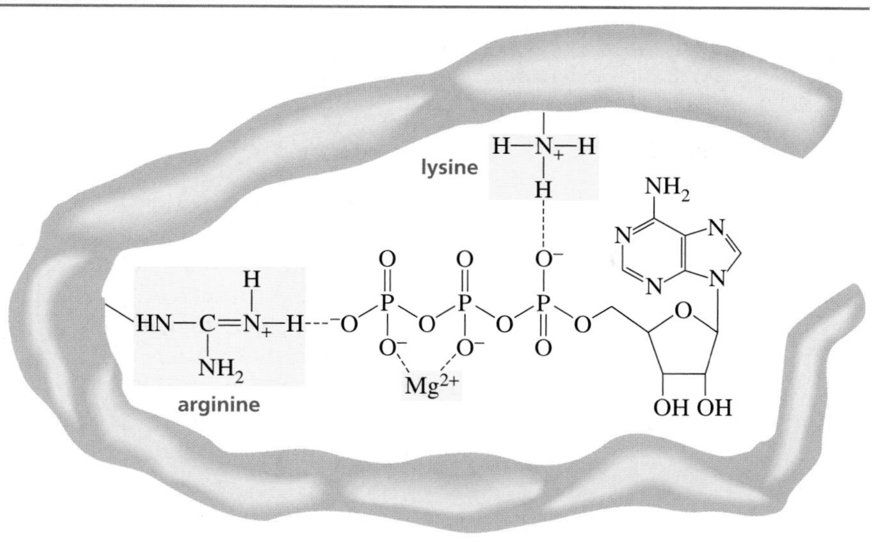

◀ **Figure 24.2**
The interactions among ATP, Mg^{2+}, and arginine and lysine residues at the active site of an enzyme.

The negative charges on ATP are what make it relatively unreactive. These negative charges discourage the approach of nucleophiles. When ATP is bound at an active site of an enzyme, it complexes with magnesium, neutralizing two of its negative charges. (This is why ATP-requiring enzymes use metal ions, Section 22.5.) The other two negative charges can be stabilized by positively charged groups such as arginine or lysine residues at the active site (Figure 24.2). In this form, ATP is readily approached by nucleophiles and therefore reacts rapidly in enzyme-catalyzed nucleophilic substitution reactions, while it reacts very slowly in the absence of the enzyme.

24.6 OTHER IMPORTANT NUCLEOTIDES

ATP is not the only biologically important nucleotide. GTP is used in place of ATP in some phosphoryl transfer reactions. We have also seen the use of dinucleotides as oxidizing agents (NAD^+, $NADP^+$, FAD, FMN) and reducing agents (NADH, NADPH, $FADH_2$, $FMNH_2$).

Another important nucleotide is adenosine 3',5'-phosphate, commonly known as cyclic AMP or cAMP. Cyclic AMP is called a "second messenger" because it serves as a link between several hormones (the first messengers) and certain enzymes that regulate cellular function. Secretion of certain hormones, such as adrenaline, causes the activation of adenylate cyclase, the enzyme responsible for the synthesis of cyclic AMP from ATP. Cyclic AMP then activates an enzyme, generally by phosphorylating it. Cyclic nucleotides are so important in regulating cellular reactions that an entire scientific journal is devoted to these processes.

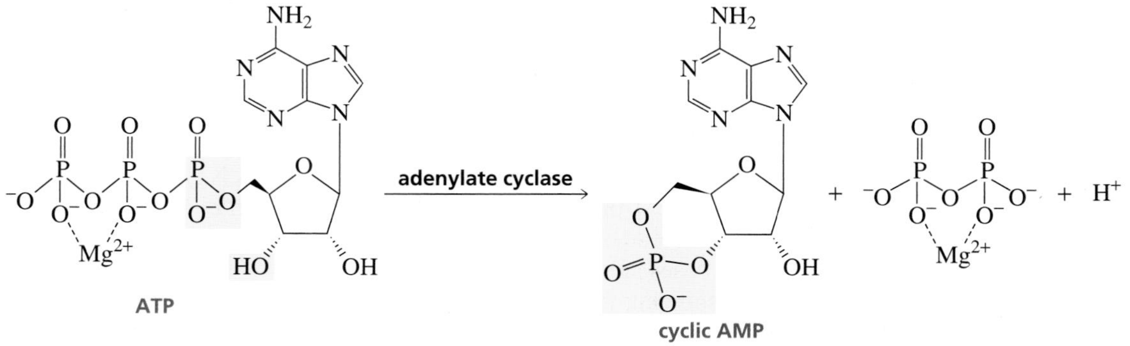

> ### PROBLEM 7
>
> Propose a mechanism for the formation of cyclic AMP from ATP.

24.7
THE NUCLEIC ACIDS

Erwin Chargaff

Erwin Chargaff *was born in Austria in 1905 and received a Ph.D. from the University of Vienna. To escape Hitler, he came to the United States in 1935, becoming a professor at Columbia University College of Physicians and Surgeons. He modified paper chromatography, a technique developed to identify amino acids (Section 20.5), so that it could be used to quantify the different bases in a sample of DNA.*

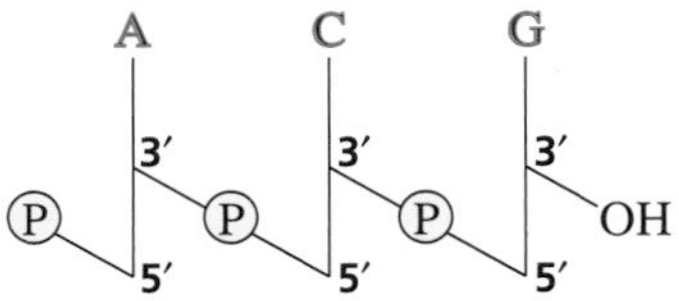

[A] = [T]
[G] = [C]

Nucleic acids are composed of long strands of nucleotide subunits linked by phosphodiester bonds. These linkages join the 3'-OH group of one nucleotide to the 5'-OH group of the next nucleotide (Figure 24.1). A **dinucleotide** contains two nucleotide subunits, an **oligonucleotide** contains three to ten subunits, and a **polynucleotide** contains many subunits. DNA and RNA are polynucleotides. Notice that the nucleotide at one end of the strand has an unlinked 5'-substituent, and the nucleotide at the other end of the strand has an unlinked 3'-substituent.

$$A \qquad C \qquad G$$

a trinucleotide

Nucleotide triphosphates are the starting materials for the biosynthesis of nucleic acids. DNA is synthesized by enzymes called DNA polymerases, and RNA is synthesized by enzymes known as RNA polymerases. The nucleotide strand is formed as a result of nucleophilic attack by a 3'-OH group of one nucleotide triphosphate on the α-phosphorus of another nucleotide triphosphate, breaking a phosphoanhydride bond and eliminating pyrophosphate (Figure 24.3). This means that the growing polymer is synthesized in the 5' ⟶ 3' direction; in other words, new nucleotides are added to the 3'-end. (Notice that the strand grows in the 5' ⟶ 3' direction by attacking from 3' to 5'.) Pyrophosphate is subsequently hydrolyzed, which guarantees the irreversibility of the reaction (Section 24.3). RNA strands are biosynthesized in the same way, using ribonucleotides instead of 2'-deoxyribonucleotides. Like proteins, nucleic acids have both primary and secondary structures. The **primary structure** of a nucleic acid is simply the sequence of bases in the strand.

Watson and Crick concluded that DNA consists of two strands of nucleic acids with the sugar–phosphate backbone on the outside and the bases on the inside. The chains are held together by hydrogen bonds between the bases on one strand and the bases on the other strand. The distance between the two strands is relatively constant, indicating that a purine must pair with a pyrimidine. Pairing of the larger purines would cause the strand to bulge, and the strands would have to cave in to allow the smaller pyrimidines to get close enough to form hydrogen bonds.

Critical to Watson and Crick's proposal for the secondary structure of DNA were experiments carried out by Erwin Chargaff. His research showed that the number of adenines in DNA equals the number of thymines, and the number of guanines equals the number of cytosines. Chargaff also noted that the number of adenines and thymines relative to the number of guanines and cytosines is characteristic of a given species but varies from species to species. For example, in human DNA, 60.4% of the bases are adenines and thymines, while in the bacterium *Sarcina lutea*, 74.2% of the bases are adenines and thymines.

Chargaff's data showing that [adenine] = [thymine] and [guanine] = [cytosine] could be explained if adenine (A) always paired with thymine (T) and guanine (G)

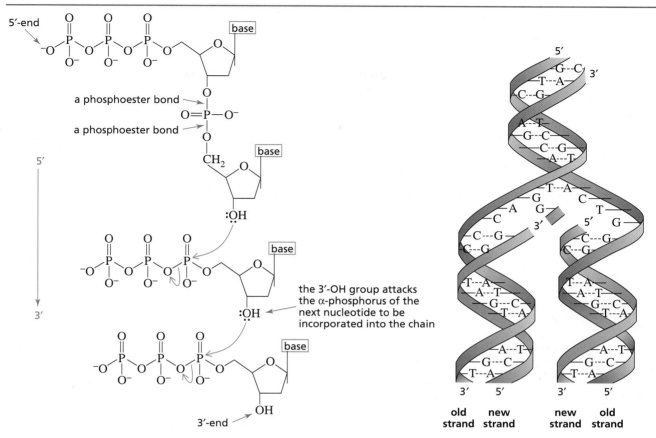

▲ **Figure 24.3**
Addition of nucleotides to a growing strand of DNA. Biosynthesis occurs in a 5' ⟶ 3' direction.

always paired with cytosine (C). This means that the two strands are complementary: where there is an A in one strand, there is a T in the opposing strand, and where there is a G in one strand there is a C in the other strand (Figure 24.4). Thus, if you know the sequence of bases in one strand, you can figure out the sequence of bases in the other strand.

What causes adenine to pair with thymine rather than with cytosine (the other pyrimidine)? The base pairing is dictated by hydrogen bonding. Learning that the bases exist in the keto form allowed Watson to explain the pairing.[2] Adenine forms two hydrogen bonds with thymine but would form no hydrogen bonds with cytosine. Guanine forms three hydrogen bonds with cytosine but would form only one hydrogen bond with thymine (Figure 24.5). Notice that the hydrogen bonds holding the bases together are all about the same length (2.9 ± 0.1 Å).

[2]Watson was having difficulty understanding the base pairing in DNA because he thought the bases existed in the enol form (see Problem 9). When Jerry Donohue, an American crystallographer, informed him that the bases more likely existed in the keto form, Chargaff's data could easily be explained by hydrogen bonding between adenine and thymine and between guanine and cytosine.

A double helix of DNA.

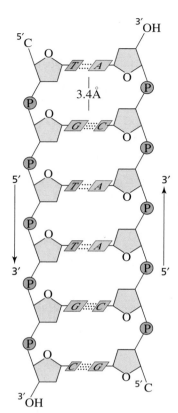

▲ **Figure 24.4**
Complementary base pairing in DNA.

The two DNA strands are antiparallel; they run in opposite directions, with the sugar–phosphate backbone on the outside and the bases on the inside (Figures 24.4 and 24.6). By convention, when base sequences of a portion of a nucleic acid are written, they are written in the 5′ $\longrightarrow$ 3′ direction (the 5′-end is on the left).

ATGAGCCATGTAGCCTAATCGGC

5′-end   3′-end

The DNA strands are not linear but are twisted into a helix around a common axis (Figure 24.7a). The base pairs are planar and parallel to each other on the inside of the helix (Figure 24.7b and c). The secondary structure is therefore known as a **double helix.** The double helix resembles a ladder (the base pairs are the rungs) twisted around an axis running down through its rungs (Figure 24.4 and 24.7c). The sugar–phosphate backbone is wrapped around the bases. The phosphate OH group has a pK_a of about 2, causing it to be in its basic form (negatively charged) at physiological pH. The negatively charged backbone repels nucleophiles, thereby preventing cleavage of the phosphodiester bonds.

Unlike DNA, RNA is easily cleaved, because the 2′-OH group of ribose can act as the nucleophile that cleaves the strand (Figure 24.8). Now we can understand why the 2′-OH group is missing in DNA. To preserve the genetic information, DNA must remain intact throughout the life span of a cell. Cleavage of DNA would have disastrous consequences for the cell and for life itself. RNA, in contrast, is synthesized as it is needed and is degraded when it has served its purpose.

Hydrogen bonding between base pairs is just one of the forces holding the two strands of the DNA double helix together. The bases are planar, aromatic molecules that stack on top of one another. Each pair is slightly rotated with respect to the next pair, like a partially opened hand of cards (Figure 24.7b). There are favorable van der Waals interactions between the mutually induced dipoles of adjacent pairs of bases. These interactions, known as **stacking interactions,** are weak attractive forces, but when added together they contribute significantly to the stability of the

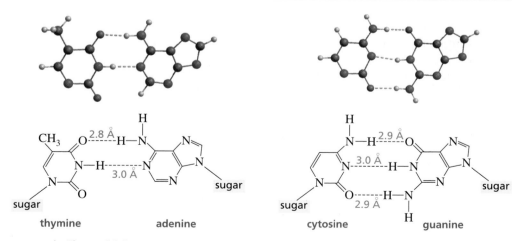

▲ **Figure 24.5**
Base pairing in DNA:
adenine and thymine form two hydrogen bonds;
cytosine and guanine form three hydrogen bonds.

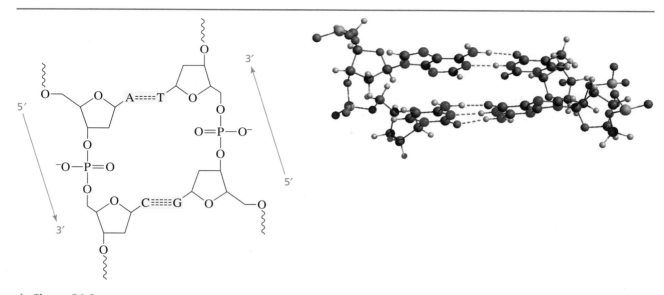

▲ **Figure 24.6**
The sugar–phosphate backbone of DNA is on the outside, and the bases are on the inside, with A's pairing with T's and G's pairing with C's. The two strands are antiparallel; they run in opposite directions.

double helix. Stacking interactions are strongest between two purines and weakest between two pyrimidines.

Bringing the bases together into the inside of the helix also stabilizes the helix because it reduces the surface area of the relatively nonpolar residues exposed to water. This increases the entropy of the surrounding water molecules (Section 20.14).

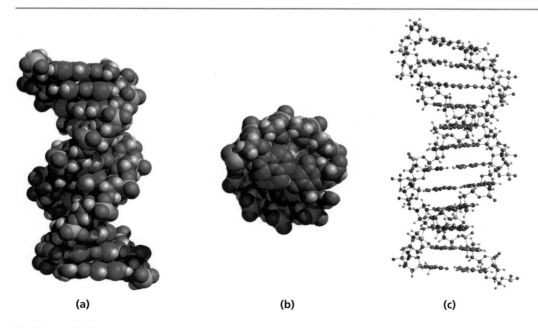

(a)   (b)   (c)

▲ **Figure 24.7**
(a) The DNA double helix. (b) View looking down the long axis of the helix. (c) The bases are planar and parallel on the inside of the helix.

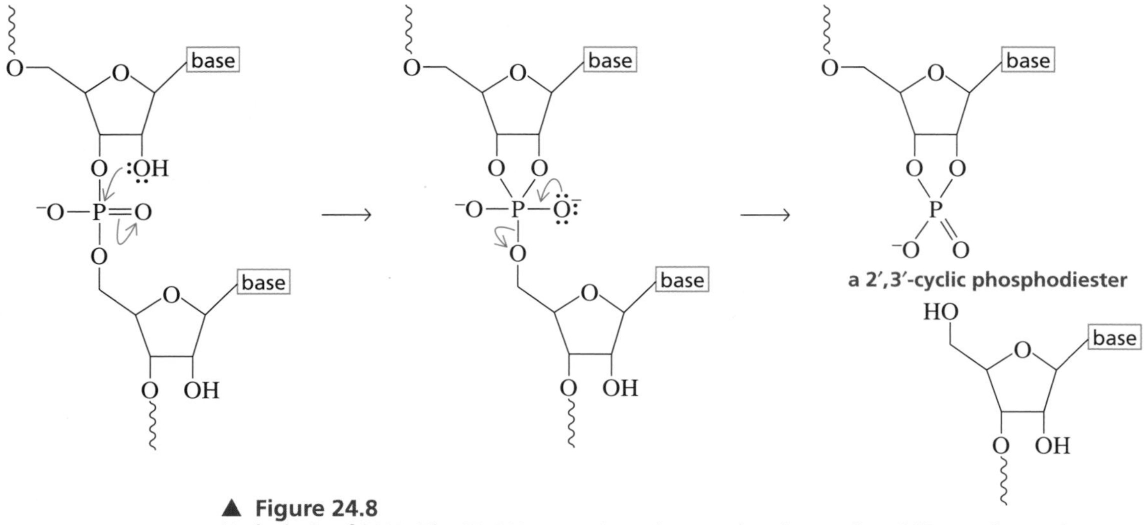

▲ Figure 24.8
Hydrolysis of RNA. The 2'-OH group is an intramolecular nucleophilic catalyst. It has been estimated that RNA is hydrolyzed three billion times faster than DNA.

PROBLEM 8

Indicate whether each functional group of the five heterocyclic bases in nucleic acids can function as a hydrogen bond acceptor (A), a hydrogen bond donor (D), or both (D/A).

PROBLEM 9

Using the D, A, and D/A designations of Problem 8, explain how base pairing would be affected if the bases existed in the enol form.

PROBLEM 10

The 2',3'-cyclic phosphodiester, formed when RNA is hydrolyzed, reacts with water, forming a mixture of nucleotide 2'- and 3'-phosphates. Propose a mechanism for this reaction.

PROBLEM 11◆

If one of the strands of DNA has the following sequence of bases running in the 5' ⟶ 3' direction,

$$5'—G–G–A–C–A–A–T–C–T–G–C—3'$$

a. what is the sequence of bases in the complementary strand?

b. what base is closest to the 5' end in the complementary strand?

24.8 HELICAL FORMS OF DNA

Naturally occurring DNA can exist in three different helical forms (Figure 24.9). The B- and A-helices are both right-handed helices. The B-helix is the predominant form in aqueous solution, while the A-helix is the predominant form in nonpolar solvents. Nearly all the DNA in living organisms is in a B-helix. The Z-helix is a

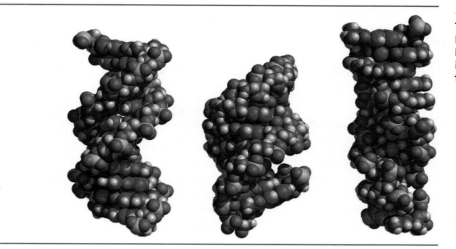

◀ **Figure 24.9**
The three helical forms of
DNA. The helix on the left is a
B-helix, the one in the middle
is an A-helix, and the one on
the right is a Z-helix.

left-handed helix. It occurs in regions where there is a high content of G—C base pairs. The A-helix is shorter (for a given number of base pairs) and about 3% broader than the B-helix, which is shorter and broader than the Z-helix.

Helices are characterized by the number of bases per 360° turn and the distance (the rise) between adjacent base pairs. A-DNA has 11 base pairs per turn and a 2.3Å rise; B-DNA has ten base pairs per turn and a 3.3Å rise; and Z-DNA has 12 base pairs per turn and 2.8Å rise.

If you examine Figure 24.7(a), you will see that there are two kinds of alternating grooves in DNA. In B-DNA, the **major groove** is wider and deeper than the **minor groove.** Cross sections of the double helix show that one side of each base pair faces into the major groove and the other side faces into the minor groove (Figure 24.10).

Proteins and other molecules can bind to the grooves. The hydrogen-bonding properties of the functional groups facing into each groove determine what kind of molecules will bind to the groove. Transcription factors, proteins involved in transcription (Section 24.10), bind to the major groove of DNA. Distamycin, a naturally occurring compound that has been found to have both antibacterial activity and anticancer activity, works by binding to the minor groove of DNA. It binds at regions rich in A's and T's (Section 29.10).

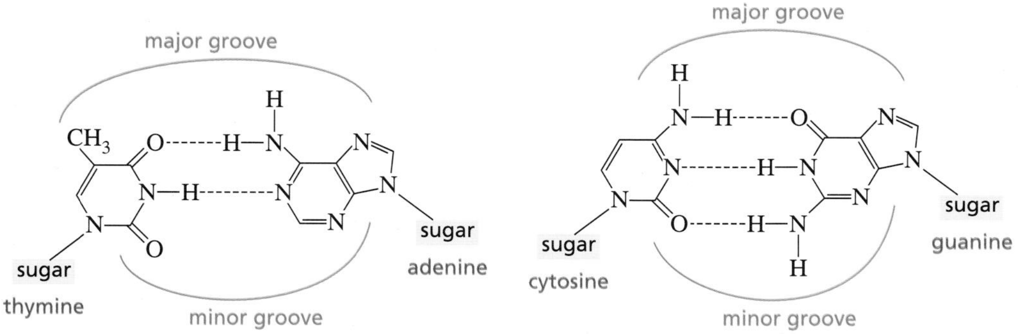

▲ **Figure 24.10**
One side of each base pair faces into the major groove, and the other side faces into the minor groove.

> ### PROBLEM 12◆
>
> Calculate the length of a turn in:
>
> **a.** A-DNA.   **b.** B-DNA.   **c.** Z-DNA.

24.9
BIOSYNTHESIS OF DNA: REPLICATION

Watson and Crick's proposal for the structure of DNA was an exciting development because the structure immediately suggested how DNA works—how it is able to pass on genetic information to succeeding generations. Because the two strands are complementary, both strands carry the same genetic information. Both strands serve as templates for the synthesis of complementary new strands (Figure 24.11). The new (daughter) DNA molecules are identical to the original. Thus they contain all the original genetic information. The synthesis of identical copies of DNA is called **replication.**

All the reactions involved in nucleic acid synthesis are catalyzed by enzymes. Synthesis of DNA takes place in a region of the molecule where the strands have started to separate, called a **replication fork.** Because a nucleic acid can be synthesized only in the 5' ⟶ 3' direction, one strand of DNA is synthesized continuously in a single piece (because it is synthesized in the 5' ⟶ 3' direction). The other strand of DNA is synthesized discontinuously in small pieces. Each piece is synthesized in the 5' ⟶ 3' direction, and the fragments are later joined together by an enzyme called DNA ligase. Each of the two daughter molecules of DNA that result contains one of the original strands plus a newly synthesized strand. This process is called **semiconservative replication.**

The genetic information of a human cell is contained in 23 pairs of chromosomes. Each chromosome is composed of several thousand **genes** (segments of DNA). It has been estimated that the total DNA of a human cell (the **human genome**) contains 2.9 billion base pairs.

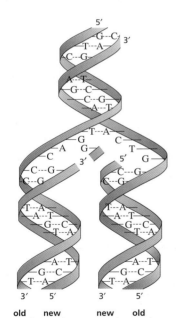

▲ **Figure 24.11**
Replication of DNA.

> ### PROBLEM 13
>
> Using a dark line for parental DNA and wavy lines for DNA synthesized from parental DNA, show what the population of DNA molecules would look like in the fourth generation.

> ### PROBLEM 14◆
>
> Assuming that the human genome, with its 2.9 billion base pairs, is entirely in a B-helix, how long is the DNA in a human cell?

> ### PROBLEM 15
>
> Why does DNA not unravel completely before replication begins?

24.10
BIOSYNTHESIS OF MESSENGER RNA: TRANSCRIPTION

The sequence of DNA bases provides the blueprint for the synthesis of messenger RNA (mRNA). The synthesis of mRNA from a DNA blueprint, called **transcription,** takes place in the nucleus. The newly synthesized mRNA can then leave the nucleus, carrying the genetic information into the cytoplasm (the cell material outside the nucleus) where translation of this information into proteins takes place.

DNA contains sequences of bases known as promoter sites. The **promoter sites** are at the beginning of genes. An enzyme recognizes a promoter site and binds to it, initiating RNA synthesis. The DNA at a promoter site unwinds to give two single strands, exposing the bases. One of the strands is called the **sense strand** or **informational strand.** The complementary strand is called the **template strand** or **antisense strand.** The template strand is read in the 3' $\longrightarrow$ 5' direction so that mRNA can be synthesized in the 5' $\longrightarrow$ 3' direction (Figure 24.12). The bases in the template strand specify the bases that need to be incorporated into mRNA, following the same base pairing found in DNA. For example, each guanine in the template strand specifies the incorporation of a cytosine into mRNA, and each adenine in the template strand specifies the incorporation of a uracil into mRNA. (Recall that in RNA, uracil is used instead of thymine.)

Since both the sense strand and mRNA are complementary to the template strand, the sense strand and mRNA have the same base sequence except that mRNA has a uracil wherever the sense strand has a thymine. Just as there are promoter sites that signal the places to start mRNA synthesis, there are other sites in DNA that signal that no more bases should be added to the growing strand of mRNA, at which point synthesis stops.

Surprisingly, a gene is not necessarily a continuous sequence of bases. Often the bases of a gene are interrupted by bases that appear to have no informational content. A stretch of bases representing a portion of a gene is called an **exon,** while a stretch of bases that contains no genetic information is called an **intron.** The mRNA that is synthesized is complementary to the entire sequence of DNA bases—exons and introns. So after the mRNA is synthesized, but before it leaves the nucleus, the so-called nonsense bases (encoded by the introns in the DNA) are cut out and the informational fragments are spliced together, resulting in a much shorter mRNA molecule. This RNA processing step is known as **RNA splicing.** Scientists have found that only about 2% of DNA contains genetic information, while 98% consists of introns.

It has been suggested that the purpose of introns is to make mRNA more versatile. The originally synthesized long strand of mRNA could be spliced in different ways to create a variety of shorter mRNAs. Each short mRNA specifies the synthesis of a different protein (Section 24.12), so different proteins could be prepared from the same gene.

Severo Ochoa was the first to prepare synthetic strands of RNA by incubating nucleotides in the presence of enzymes that are involved in the biosynthesis of RNA. **Arthur Kornberg** *prepared synthetic strands of DNA in a similar manner. For this work they shared the 1959 Nobel Prize in medicine or physiology.*

Severo Ochoa *was born in Spain in 1905. He graduated from the University of Malaga in 1921 and received an M.D. from the University of Madrid. He spent the next 4 years studying in Germany and England and then joined the faculty at New York University College of Medicine. He became a U.S. citizen in 1956.*

Arthur Kornberg *was born in New York in 1918. He graduated from the College of the City of New York and received an M.D. from the University of Rochester. He is a member of the faculty of the biochemistry department at Stanford University.*

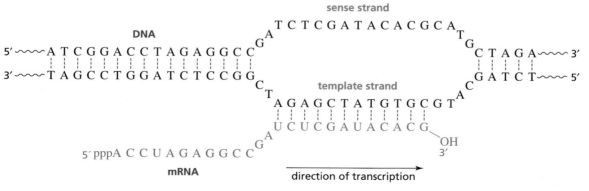

▲ Figure 24.12
Transcription: using DNA as a blueprint to form mRNA.

> ### PROBLEM 16
>
> Why do both thymine and uracil specify the incorporation of adenine?

24.11
RIBOSOMAL RNA
AND TRANSFER RNA

Sidney Altman *and* **Thomas R. Cech** *received the 1989 Nobel Prize in chemistry for their discovery of the catalytic properties of RNA.*

Sidney Altman *was born in Montreal in 1939. He received a B.S. from MIT and a Ph.D. from the University of Colorado, Boulder. He was a postdoctoral fellow in Francis Crick's laboratory at Cambridge University. He is a professor of biology at Yale University.*

Thomas Cech *was born in Chicago in 1947. He received a B.A. from Grinnell College and a Ph.D. from the University of California, Berkeley. He was a postdoctoral fellow at MIT. He is a professor of chemistry at the University of Colorado, Boulder.*

RNA is much shorter than DNA and is generally single-stranded. Although DNA molecules can have billions of base pairs, RNA molecules rarely have more than 10,000 nucleotides. Messenger RNA (mRNA) is one of three kinds of RNA; the other two kinds of RNA are ribosomal RNA (rRNA), a structural component of ribosomes, and transfer RNA (tRNA), the carrier of an amino acid for protein synthesis.

Biosynthesis of proteins takes place on particles known as **ribosomes.** A ribosome is composed of about 40% protein and about 60% rRNA. There is increasing evidence that protein synthesis is catalyzed by rRNA molecules rather than by enzymes. RNA molecules—found in ribosomes—that act as catalysts are known as **ribozymes.** The protein molecules in the ribosome enhance the functioning of the rRNA molecules.

Ribosomes are made up of two subunits. The size of the subunits depends on whether they are found in prokaryotic cells or eukaryotic cells. **Prokaryotic cells** (*pro*, Greek for "before"; *karyon*, Greek for "kernel" or "nut") were the earliest cells. They are unicellular and do not have nuclei. A **eukaryotic cell** (*eu*, Greek for "well") has a thousand to a million times the volume of a prokaryotic cell and is much more complicated. Eukaryotic cells have nuclei and can be unicellular or multicellular. A prokaryotic ribosome is composed of a 50S subunit and a smaller 30S subunit; together they form a 70S ribosome. A eukaryotic ribosome has a 60S subunit and a 40S subunit; together they form an 80S ribosome. The S stands for the **sedimentation constant,** which designates where a given component sediments during centrifugation.[3]

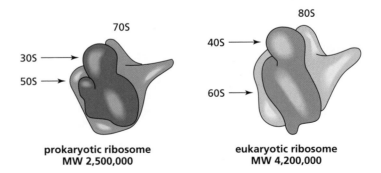

prokaryotic ribosome
MW 2,500,000

eukaryotic ribosome
MW 4,200,000

Transfer RNA (tRNA) is much smaller than mRNA or rRNA. It contains only 70 to 90 nucleotides. The single strand of tRNA is folded into a characteristic cloverleaf structure strung out with three loops and a little bulge next to the right-hand loop (Figure 24.13). There are at least four regions with complementary base pairing. All tRNAs have a CCA sequence at the 3'-end. The three bases at the bottom of the loop directly opposite the 5'- and 3'-ends are called an **anticodon.**

[3]Sedimentation constants are not additive, which is why a 50S and a 30S can equal a 70S.

Each tRNA can carry an amino acid bound as an ester to its terminal 3'-OH group. The amino acid will be inserted into a protein during protein biosynthesis. Each tRNA can carry only one particular amino acid. A tRNA that carries alanine is designated as tRNA^{Ala}. Most amino acids are carried by more than one tRNA.

How does a tRNA molecule become attached to the amino acid? Attachment of the amino acid is catalyzed by an enzyme called aminoacyl-tRNA synthetase. In the first step of the enzyme-catalyzed reaction, the carboxyl group of the amino acid attacks the α-phosphorus of ATP, activating the carboxyl group by forming an acyl adenylate. The pyrophosphate that is eliminated is subsequently hydrolyzed, forcing the phosphoryl transfer reaction to the right (Section 24.3). Then a nucleophilic acyl substitution reaction occurs; the 3'-OH group of tRNA attacks the carbonyl carbon of the amino acid forming a tetrahedral intermediate. The amino acyl tRNA is formed when AMP is eliminated from the tetrahedral intermediate. All the steps take place at the active site of the enzyme.

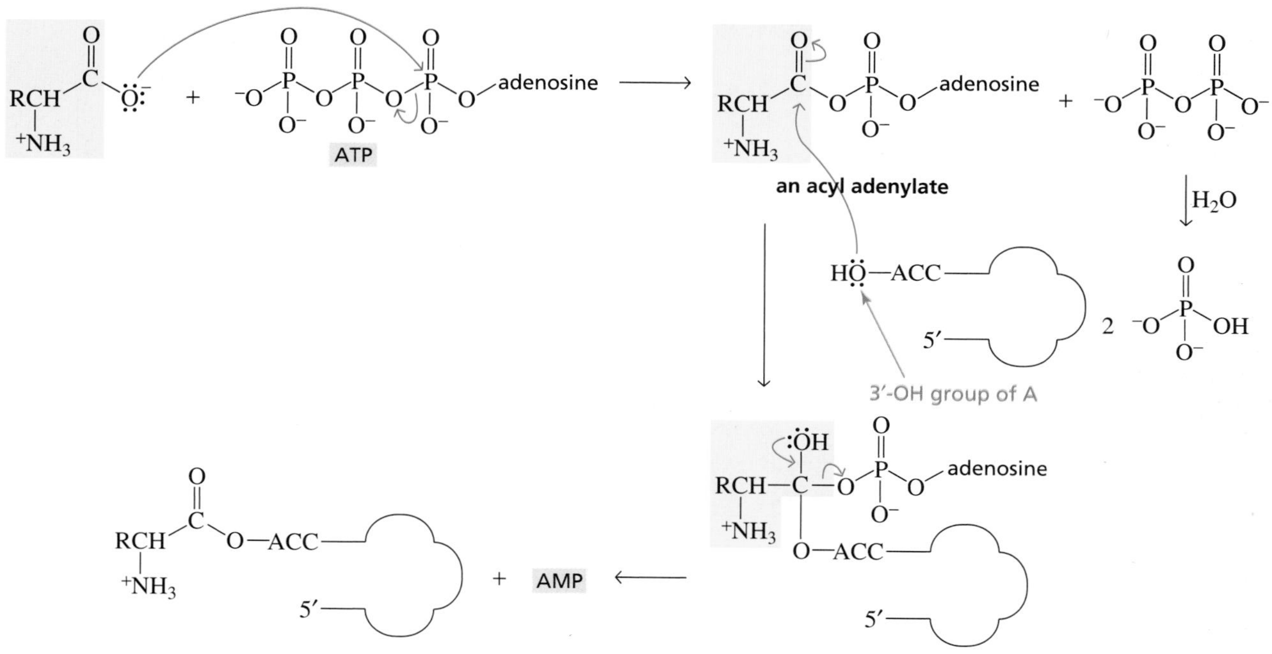

Each amino acid has its own aminoacyl-tRNA synthetase. Each synthetase has two specific binding sites, one for the amino acid and one for the tRNA that carries that amino acid (Figure 24.14). It is critical that the correct amino acid is attached to the tRNA. Otherwise, the correct protein will not be synthesized. Fortunately, the synthetases correct their own mistakes. For example, valine and threonine are approximately the same size: valine has a CH_3 where threonine has an OH. Therefore, both can bind at the amino acid binding site of the aminoacyl-tRNA synthetase for valine, and both can then be activated by reacting with ATP to form an acyl adenylate. The aminoacyl-tRNA synthetase for valine has two adjacent catalytic sites, one for attaching the acyl adenylate to tRNA and one for hydrolyzing the acyl adenylate. The acylation site is hydrophobic, so valine is preferred over threonine for the tRNA acylation reaction. The hydrolytic site is polar, so threonine is preferred over valine for the hydrolysis reaction. Thus, if threonine is activated by the aminoacyl-tRNA synthetase for valine, it will be hydrolyzed rather than transferred to the tRNA.

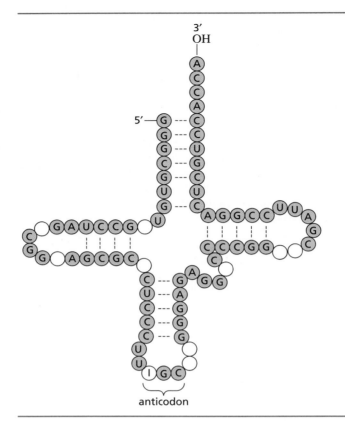

anticodon

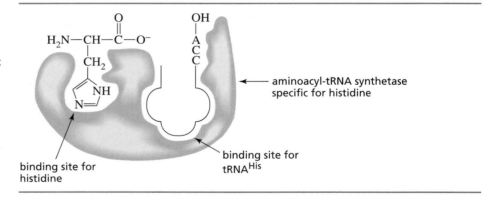

aminoacyl-tRNA synthetase specific for histidine

binding site for tRNA^{His}

binding site for histidine

24.12
BIOSYNTHESIS OF PROTEINS: TRANSLATION

A protein is synthesized from its N-terminal end to its C-terminal end by reading the bases along the mRNA strand in the 5' ⟶ 3' direction. A sequence of three bases, called a **codon,** specifies a particular amino acid that is to be incorporated into a protein. The bases are read consecutively and are never skipped. A codon is written with the 5'-nucleotide on the left. The amino acid specified by each three-base sequence is known as the **genetic code** (Table 24.2). For example, UCA on mRNA codes for the amino acid serine, while CAG codes for glutamine.

Because a codon is a triplet and because there are four different bases to choose from, $4^3 = 64$ different codons are possible. This is many more than are needed to

TABLE 24.2 The Genetic Code

5'-Position		Middle position				3'-Position
		U	C	A	G	
U		Phe	Ser	Tyr	Cys	U
		Phe	Ser	Tyr	Cys	C
		Leu	Ser	STOP	STOP	A
		Leu	Ser	STOP	Trp	G
C		Leu	Pro	His	Arg	U
		Leu	Pro	His	Arg	C
		Leu	Pro	Gln	Arg	A
		Leu	Pro	Gln	Arg	G
A		Ile	Thr	Asn	Ser	U
		Ile	Thr	Asn	Ser	C
		Ile	Thr	Lys	Arg	A
		Met	Thr	Lys	Arg	G
G		Val	Ala	Asp	Gly	U
		Val	Ala	Asp	Gly	C
		Val	Ala	Glu	Gly	A
		Val	Ala	Glu	Gly	G

specify the 20 different amino acids, so all the amino acids—except methionine and tryptophan—have more than one codon. It is not surprising, therefore, that methionine and tryptophan are the least abundant amino acids in proteins. Actually, 61 of the bases specify amino acids, and three bases are stop codons. **Stop codons** say "stop protein synthesis here."

Translation is the process by which the genetic message in DNA that has been passed to mRNA is decoded and used to build proteins. Each of the approximately 100,000 proteins in the human body is synthesized from a different mRNA. Don't confuse transcription and translation—these words are used just as in regular English. Transcription (DNA to RNA) is copying *within the same language* of nucleotides. Translation (RNA to protein) is *changing to another language*—the language of amino acids.

How the information in mRNA is translated into a polypeptide is shown in Figure 24.15. Serine was the last amino acid incorporated into the growing polypeptide chain. It was specified by the AGC codon because the anticodon of the tRNA that carries serine is GCU (3'-UCG-5'). (Remember that a base sequence is read in the 5' to 3' direction, so the sequence of bases in an anticodon must be read from right to left.) The next codon is CUU, signaling for a tRNA with an anticodon of AAG (3'-GAA-5'). That particular tRNA carries leucine. The amino group of leucine reacts, in an enzyme-catalyzed nucleophilic acyl substitution reaction, with the ester on the adjacent tRNA, displacing the tRNA. The next codon (GCC) brings in a tRNA carrying alanine. The amino group of alanine displaces the tRNA that brought in leucine. Subsequent amino acids are brought in one at a time in this way, with the codon in mRNA specifying the amino acid to be incorporated by complementary base pairing with the anticodon of the tRNA carrying that amino acid.

The genetic code was worked out independently by **Marshall Nirenberg** and **Har Gobind Khorana,** *for which they shared the 1968 Nobel Prize in medicine or physiology.* **Robert Holley,** *who worked on the structure of tRNA molecules, also shared that year's prize.*

Marshall Nirenberg *was born in New York in 1927. He received a bachelor's degree from the University of Florida and a Ph.D. from the University of Michigan. He is a scientist at the National Institutes of Health.*

Robert W. Holley (1922–1993) *was born in Illinois and received a bachelor's degree from the University of Illinois and a Ph.D. from Cornell University. During World War II he worked on the synthesis of penicillin at Cornell Medical School. He was a professor at Cornell and later at the University of California, San Diego. He also was a noted sculptor.*

Har Gobind Khorana *was born in India in 1922. He received a bachelor's and a master's degree from the University of Punjab and a Ph.D. from the University of Liverpool. In 1960 he joined the faculty at the University of Wisconsin and subsequently moved to MIT.*

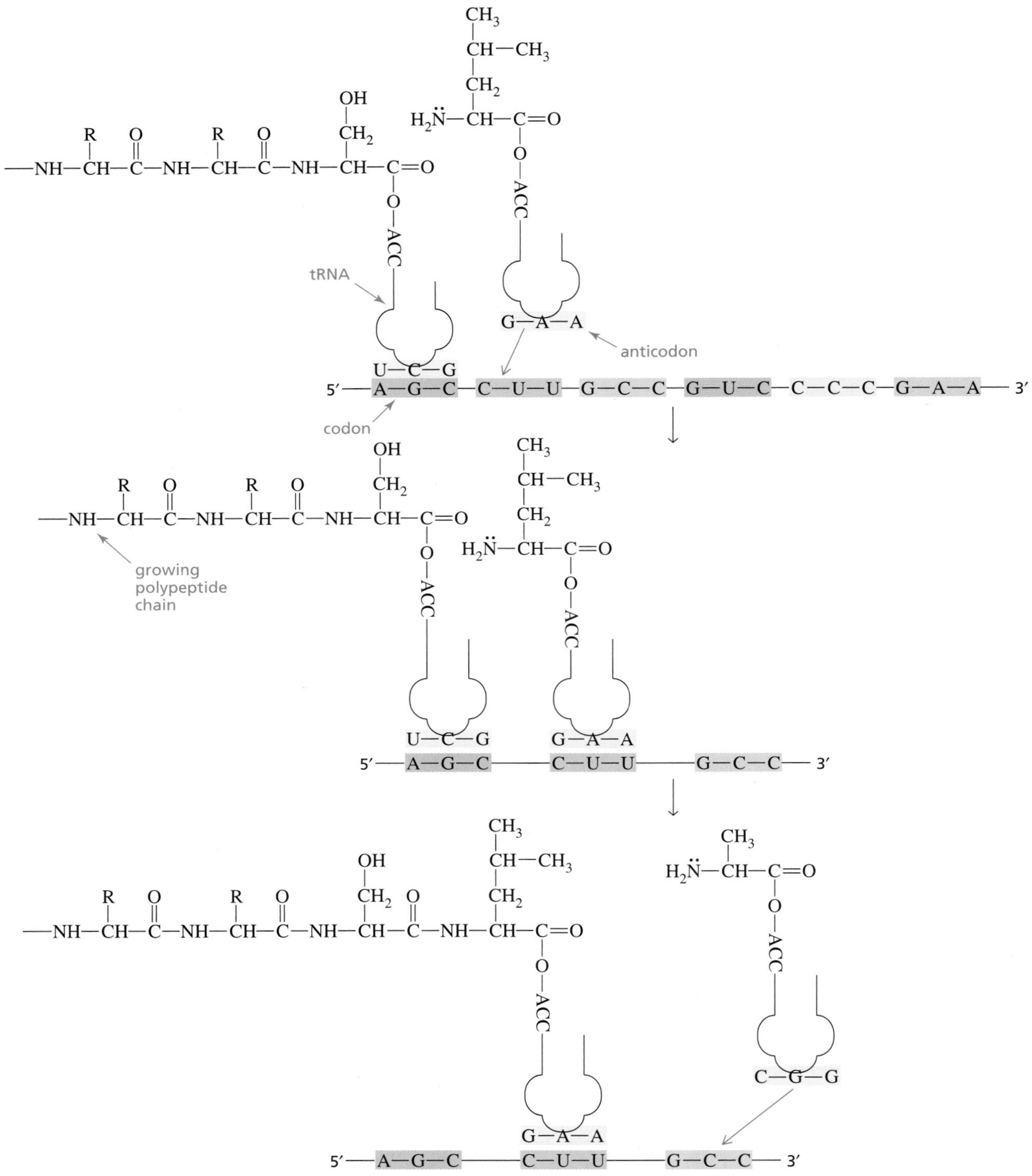

▲ **Figure 24.15**
Translation. The sequence of bases in mRNA determines the sequence of amino acids in a protein.

Protein synthesis takes place on the ribosomes. The smaller subunit of the ribosome (30S) has three binding sites for RNA molecules. It binds the mRNA whose base sequence is to be read, the tRNA carrying the growing peptide chain (at its P site), and the tRNA carrying the next amino acid to be incorporated into the protein (at its A site). The larger ribosome subunit catalyzes peptide bond formation.

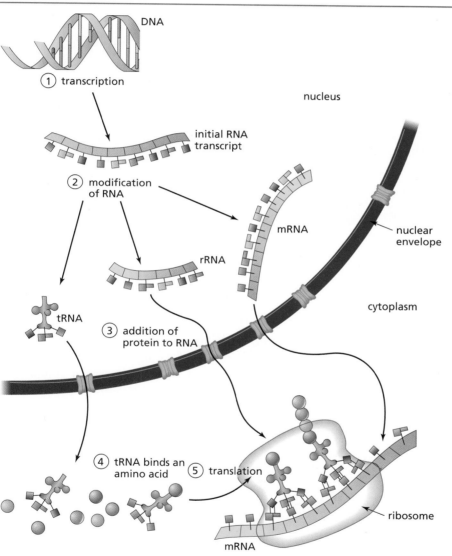

transcription:
DNA ⟶ RNA

translation:
mRNA ⟶ protein

▲ **Figure 24.16**
1. Transcription of DNA occurs in the nucleus. The initial RNA transcript is the precursor of all RNA: tRNA, rRNA, and mRNA. 2. The initially formed RNA generally must be chemically modified before it acquires biological activity. Modification can entail removing nucleotide segments, adding nucleotides to the 5' or 3' ends, or chemically altering certain nucleosides. 3. Proteins are added to rRNA to form ribosomal subunits. tRNA, mRNA, and ribosomal subunits leave the nucleus. 4. Each tRNA binds the appropriate amino acid. 5. tRNA, mRNA, and a ribosome participate in translation of the mRNA information into a protein.

SICKLE CELL ANEMIA

Sickle cell anemia is an example of the damage that can be caused by a change in a single base of DNA. It is a hereditary disease caused by a GAG triplet in the section of DNA that codes for the β-subunit of hemoglobin becoming a GTG. As a consequence, a mRNA codon becomes GUG—causing incorporation of valine—rather than GAG, which would have signaled for incorporation of glutamic acid. The change from a polar to a nonpolar amino acid is sufficient to change the shape of the hemoglobin molecule and reduce its solubility, causing it to precipitate in red blood cells. This stiffens the cells, making it difficult for them to squeeze through a capillary. Blocking capillaries causes severe pain and can be fatal (Problem 58 in Chapter 20).

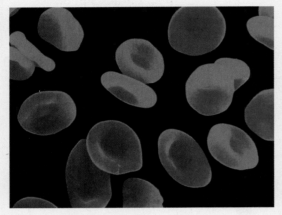

Normal red blood cells.

Sickle red blood cells.

ANTIBIOTICS THAT ACT BY INHIBITING TRANSLATION

Puromycin is a naturally occurring antibiotic. It is one of several antibiotics that act by inhibiting translation. Puromycin mimics the 3'-CCA-aminoacyl portion of a tRNA. If, during the translation process, the enzyme is fooled into transferring the growing peptide chain to the amino group of puromycin rather than to the amino group of the incoming 3'-CCA-aminoacyl tRNA, protein synthesis stops. Because puromycin blocks protein synthesis in eukaryotes as well as in prokaryotes, it is poisonous to humans and therefore is not a clinically useful antibiotic. To be clinically useful, an antibiotic must affect protein synthesis only in prokaryotic cells.

puromycin

clinically useful antibiotic	mode of action
tetracycline	prevents the aminoacyl-tRNA from binding to the ribosome
erythromycin	prevents elongation of the protein by preventing the tRNA from moving from the A site to the P site so a new tRNA can bind to the A site
streptomycin	binds to the 30S ribosome and inhibits the initiation of translation
chloramphenicol	prevents the new peptide bond from being formed

PROBLEM 17 ◆

What is the sequence of the amino acids in the heptapeptide encoded for by the following stretch of m-RNA? Note that methionine is the first amino acid incorporated into the heptapeptide.

5'–G–C–A–U–G–G–A–C–C–C–C–G–U–U–A–U–U–A–A–A–C–A–C–3'

PROBLEM 18 ◆

Four C's occur in a row in the segment of mRNA above. What polypeptide would be formed from the mRNA in Problem 17 if one of the four C's were cut out of the strand?

PROBLEM 19

UAA is a stop codon. Why does the UAA sequence in mRNA in Problem 17 not cause protein synthesis to stop?

PROBLEM 20 ◆

Give the sequences of bases in the sense strand of DNA that resulted in the mRNA in Problem 17.

PROBLEM 21

List the possible codons on mRNA that specify each amino acid in Problem 17 and the anticodon on the tRNA that carries that amino acid.

tRNA^{Ser}: the anticodon is shown in red; the serine binding site is shown in yellow.

In Section 22.8, we saw that dTMP is formed by methylation of dUMP with the coenzyme N^5,N^{10}-methylenetetrahydrofolate (N^5,N^{10}-methylene-THF) supplying the methyl group. Because the incorporation of the methyl group into uracil oxidizes tetrahydrofolate to dihydrofolate, dihydrofolate must be reduced back to tetrahydrofolate to prepare the cofactor for another catalytic cycle. The reducing agent is NADPH. Every NADPH formed in a biological organism can drive the formation of three ATP's, so using an NADPH to reduce dihydrofolate comes at the expense of ATP. This means that the synthesis of thymine is energetically expensive, so there must be a good reason for DNA to contain thymine instead of uracil.

24.13
WHY DNA CONTAINS THYMINE INSTEAD OF URACIL

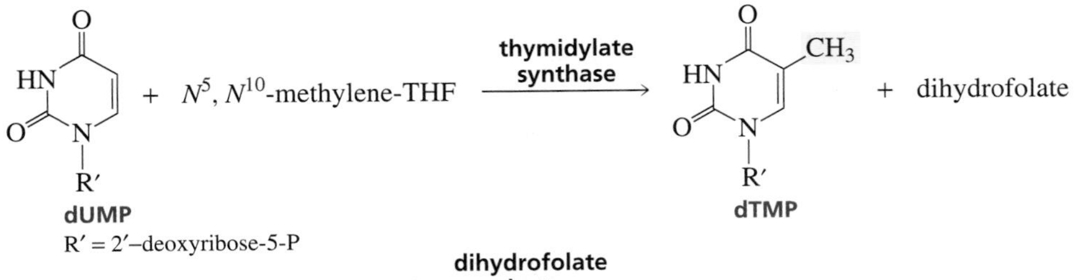

$$\text{dUMP} + N^5, N^{10}\text{-methylene-THF} \xrightarrow{\text{thymidylate synthase}} \text{dTMP} + \text{dihydrofolate}$$

R′ = 2′–deoxyribose-5-P

$$\text{dihydrofolate} + \text{NADPH} \xrightarrow{\text{dihydrofolate reductase}} \text{tetrahydrofolate} + \text{NADP}^+$$

The presence of thymine instead of uracil in DNA prevents potentially lethal mutations. Cytosine can tautomerize to form an imine, which can be hydrolyzed to uracil (Section 16.7). Because the overall reaction results in the removal of ammonia, it is called a **deamination.**

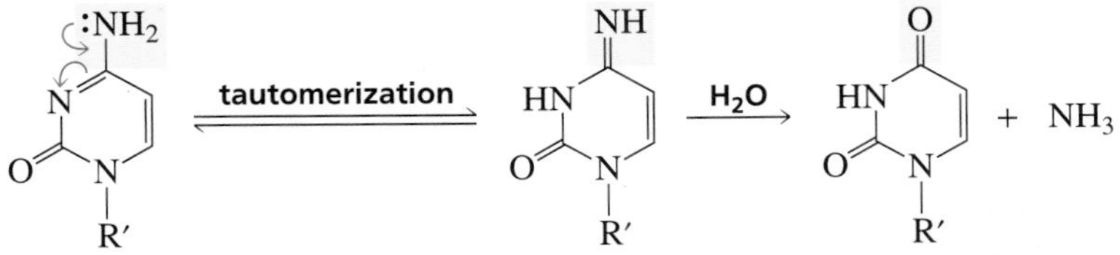

If a cytosine in DNA is deaminated to a uracil, the uracil will specify incorporation of an adenine into the daughter strand during replication instead of the guanine that would have been specified by cytosine. Fortunately, a U in DNA is recognized as a "mistake" by cell enzymes before an incorrect base can be inserted into the daughter strand. These enzymes cut the U out and replace it with a C. If U's were normally found in DNA, the enzymes could not distinguish between a normal U and a U formed by deamination of cytosine. Having T's in place of U's in DNA allows the U's that are found in DNA to be recognized as mistakes.

Unlike DNA, which replicates itself, any mistake in RNA does not survive for long because RNA is constantly being degraded and then resynthesized from the DNA template. Therefore, it is not worth spending the extra energy to incorporate T's into RNA.

PROBLEM 22◆

Adenine can be deaminated to hypoxanthine, and guanine can be deaminated to xanthine. Give structures for hypoxanthine and xanthine.

PROBLEM 23

Explain why thymine cannot be deaminated.

24.14 DETERMINING THE BASE SEQUENCE OF DNA

A current major undertaking in biomedical research is determining the sequence of the 2.9 billion base pairs in the human genome. This is an enormous project: if the rate of base sequencing can be increased to 1 million base pairs a day, it will still take more than 10 years to complete the sequence of the human genome.

DNA molecules are too large to sequence as a unit. So DNA is first cleaved at specific base sequences and the resulting DNA fragments are sequenced. The enzymes that cleave DNA at specific base sequences are called **restriction**

endonucleases. The DNA fragments that result are called **restriction fragments.** Several hundred restriction enzymes are now known. A few examples of restriction enzymes, the base sequence each recognizes, and the point of cleavage in that base sequence are shown below.

restriction enzyme	recognition sequence
A*lu*I	AG|CT TC|GA
F*nu*DI	GG|CC CC|GG
P*st*I	CTGCA|G G|ACGTC

Walter Gilbert *and* **Frederick Sanger** *(Section 20.12) shared half of the 1980 Nobel Prize in chemistry. The other half went to* **Paul Berg,** *who had developed a method of cutting nucleic acids at specific sites and recombining the fragments in new ways, a technique known as recombinant DNA technology.*

The base sequences that most restriction enzymes recognize are *palindromes.* A palindrome is a word or a group of words that reads the same forward and backward. "Toot" and "poor Dan in a droop" are examples of palindromes.[4] A restriction enzyme recognizes a piece of DNA in which the template strand is a palindrome of the sense strand. In other words, the sequence of bases in the template strand (reading from right to left) is identical to the sequence of bases in the sense strand (reading from left to right).

Walter Gilbert *was born in Boston in 1932. He received a master's degree from Harvard and a Ph.D. in physics from Cambridge University. In 1957 he joined the faculty at Harvard, where he became interested in molecular biology.*

PROBLEM 24◆

Which of the following base sequences would most likely be recognized by a restriction endonuclease?

a. ACGCGT **c.** ACGGCA **e.** ACATCGT

b. ACGGGT **d.** ACACGT **f.** CCAACC

Paul Berg *was born in New York in 1926. He received a Ph.D. from Western Reserve University (now Case Western Reserve University). He joined the faculty at Washington University in St. Louis in 1955 and moved to Stanford in 1959.*

The restriction fragments can be sequenced using a chemical cleavage method known as Maxam–Gilbert sequencing, or by a chain-terminator procedure developed by Sanger known as the dideoxy method. In **Maxam–Gilbert sequencing,** each double-stranded restriction fragment is labeled at the 5'-end with a phosphate group containing radioactive phosphorus (^{32}P). The strands are then separated and isolated. Each labeled single-stranded fragment of DNA is treated with reagents that cleave the strand selectively at the 5'- and 3'-sides of a particular base—a cytosine, for example—cutting out that base. The conditions used for cleavage are such that if there is more than one C in the strand, each strand is cleaved at only one of the C's. Each cleavage leads to a labeled fragment and an unlabeled fragment.

^{32}P—AGCACTTAACGGT $\xrightarrow{\text{cleavage at C}}$ ^{32}P—AG + ACTTAACGGT

^{32}P—AGCA + TTAACGGT

^{32}P—AGCACTTAA + GGT

Allan M. Maxam *received a Ph.D. from Harvard University in 1980 and subsequently became a professor at the Dana Farber Cancer Institute at Harvard. Is it a coincidence that his last name is a palindrome?*

This technique is repeated with a reagent that causes specific cleavage at a different base. For example, treating the same strand with a reagent that causes cleavage at the 5'- and 3'-sides of G forms the following fragments.

[4]Some other palindromes are "Mom," "Dad," "Bob," "Lil," "radar," "race car," "a man, a plan, a canal, Panama," "Sex at noon taxes," "He lived as a devil, eh?"

$$^{32}\text{P—AGCACTTAACGGT} \xrightarrow{\text{cleavage at G}} {}^{32}\text{P—A} + \text{CACTTAACGGT}$$

$$^{32}\text{P—AGCACTTAAC} + \text{GT}$$

$$^{32}\text{P—AGCACTTAACG} + \text{T}$$

The fragments obtained from each cleavage experiment are loaded onto separate lanes of a buffered polyacrylamide gel; the fragments obtained from cleavage at A are loaded onto one lane, the fragments obtained from cleavage at C onto another lane, and so on. An electric field is applied across the ends of the gel, causing the negatively charged fragments to travel toward the positively charged electrode (the anode). The smaller fragments fit through the spaces in the gel relatively easily and therefore travel through the gel faster, while the larger fragments pass through the gel more slowly.

After separation, the gel is placed in contact with a photographic plate. Radiation from ^{32}P causes a dark spot to appear on the plate opposite the location of the labeled fragment in the gel. Unlabeled fragments do not show up on the photographic plate. This technique is called **autoradiography,** and the exposed photographic plate is known as an **autoradiograph** (Figure 24.17).

The sequence of bases in the original restriction fragment can be read directly from the autoradiograph. The identity of each base is determined by noting the column where each successive dark spot (larger piece of labeled fragment) appears, starting at the bottom of the gel. The sequence of the fragment of DNA responsible for the autoradiograph in Figure 24.17 is shown on the left-hand side of the figure.

A reagent is not available for specific cleavage at A bases, but there is a procedure for cleavage at both G and A. A dark spot in the A + G column indicates either an A or a G. The absence of a spot at the same level in the G column means that the A + G spot is caused by an A, while the presence of a spot in the G column indicates that the spot is caused by a G. Similarly, the reagent responsible for cleavage

Figure 24.17 ▶
An autoradiograph.

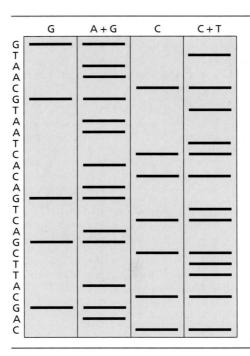

at C cleaves at both C and T under certain conditions. A spot in the C + T column is a C if there is a spot at the corresponding level in the C column, and is a T if the spot is missing in the C column.

Once the sequence of bases in a restriction fragment is determined, the results can be checked by determining the base sequence of the complementary strand. The sequence of the various restriction fragments in DNA can be determined by repeating the entire procedure with a different restriction endonuclease and noting overlapping fragments.

Why are restriction fragments cleaved at specific bases? This happens as a result of a chemical reaction with a specific type of base. For example, dimethyl sulfate alkylates guanine at N-7. This provides an "electron sink" so that **depurination** (elimination of the purine ring) can occur. Addition of piperidine to the reaction mixture forms an imine. Piperidine then acts as a base, removing an α-hydrogen that has been made more acidic by the electron-withdrawing iminium group, and eliminating the phosphate group. Piperidine catalyzes a second elimination reaction because the conjugated imine increases the acidity of the hydrogen at C-4. The overall reaction cuts the DNA chain on either side of guanine.

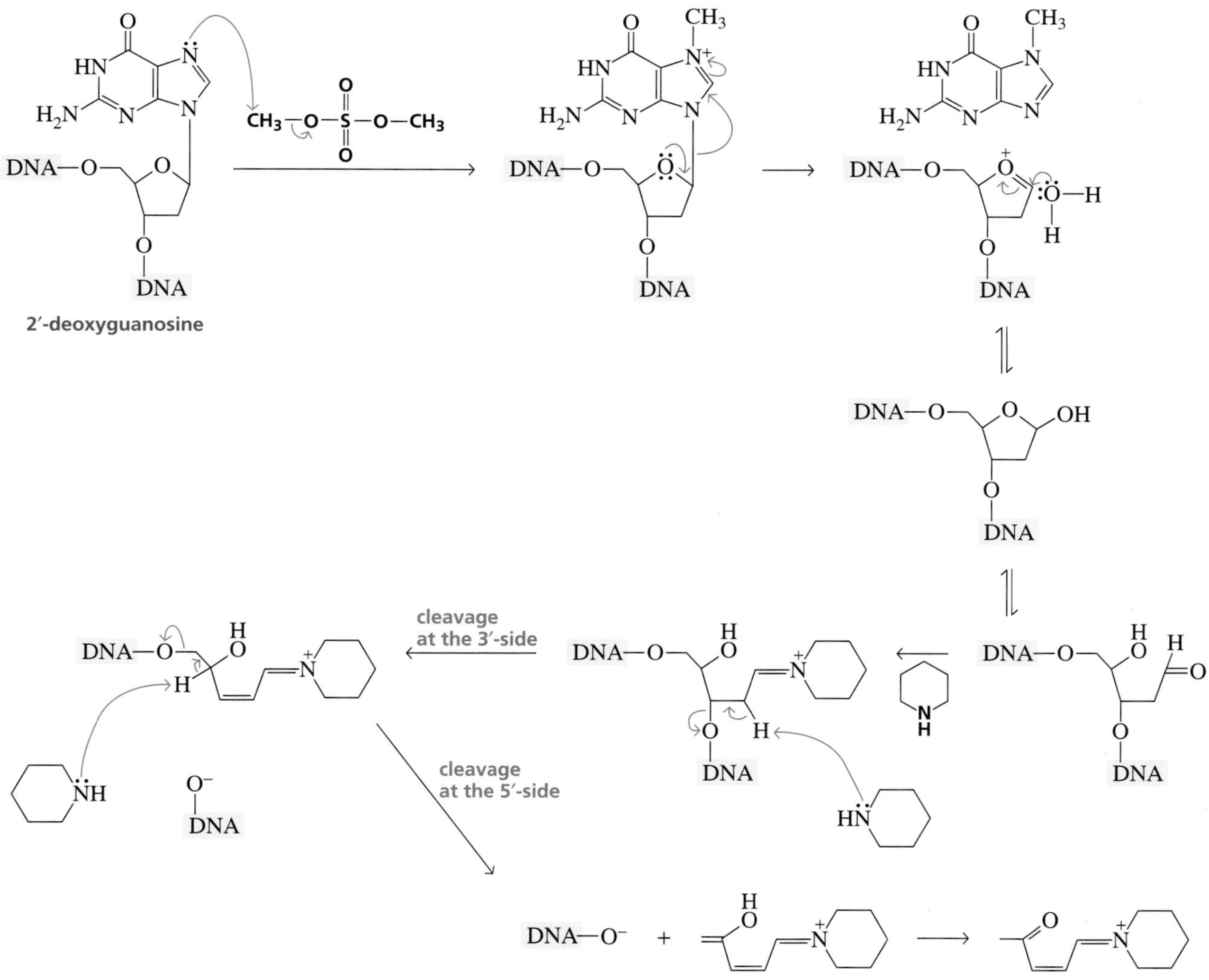

Adenine is methylated by dimethyl sulfate at N-3 rather than at N-7, so DNA is much less apt to be cleaved at A by the treatment discussed above. If DNA is treated with formic acid, both purines (A's and G's) will be cut out of DNA, and subsequent treatment with piperidine will cleave the DNA strand on either side of the deoxyribose that was bonded to G or A.

PROBLEM 25

Why is a purine less apt to be cleaved out of DNA if the purine is methylated at N-3 rather than at N-7?

Pyrimidines (C's and T's) are cut out of restriction fragments by treating the single-stranded DNA with hydrazine, followed by heating with piperidine. If the hydrazine treatment is carried out in the presence of 1.5 M NaCl, only C's are cut out.

Hydrazine attacks the pyrimidine ring at C-6, an example of conjugate addition to an α,β-unsaturated carbonyl compound (Section 16.15). Tautomerization is followed by an intramolecular acyl substitution reaction that opens the pyrimidine ring. The furanose ring is broken in the next step. Reaction with piperidine causes a transimination reaction (Section 22.6) that forms the same compound that is formed when purines are cleaved out of DNA. This compound undergoes the two piperidine-catalyzed elimination reactions shown earlier, with the result that DNA fragments are cut on either side of C's and T's.

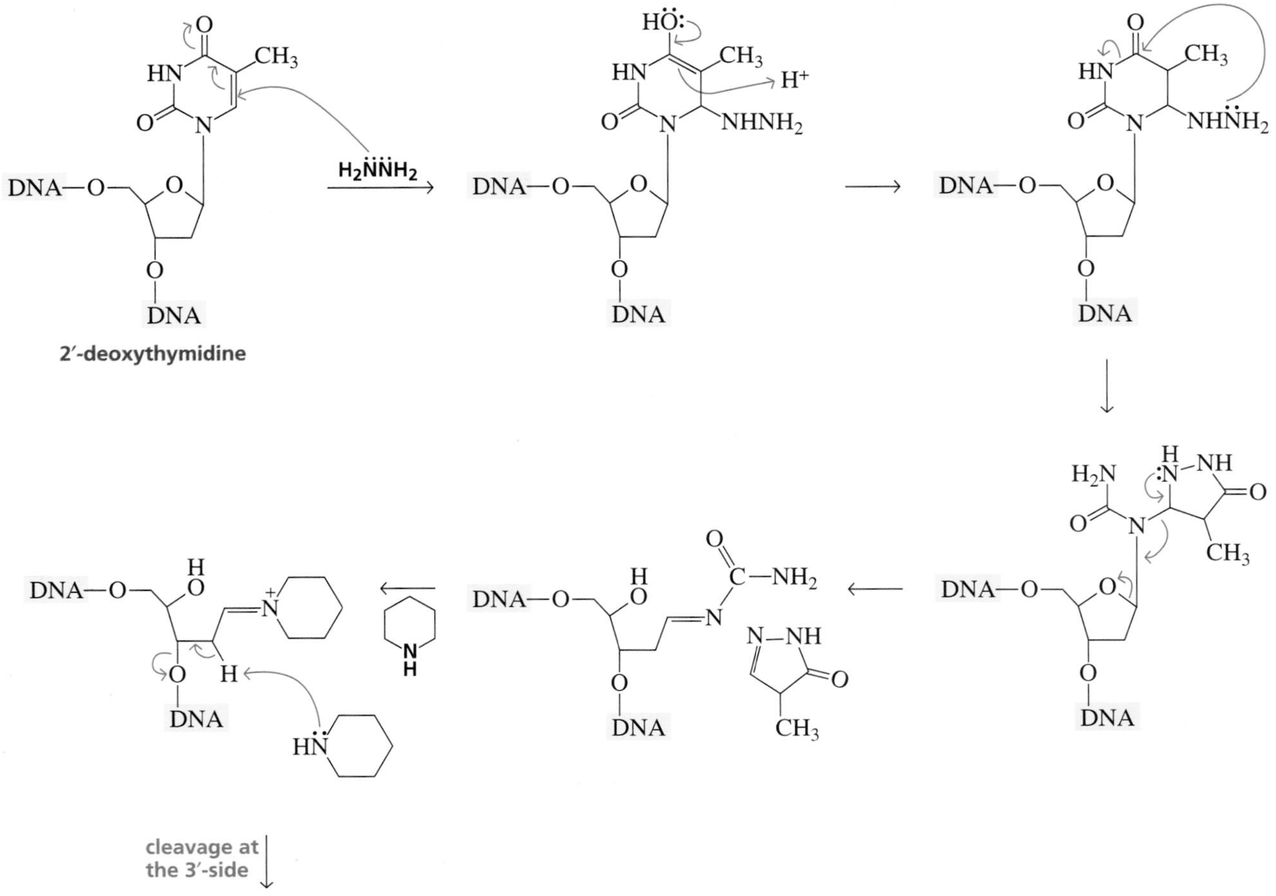

cleavage at the 3'-side

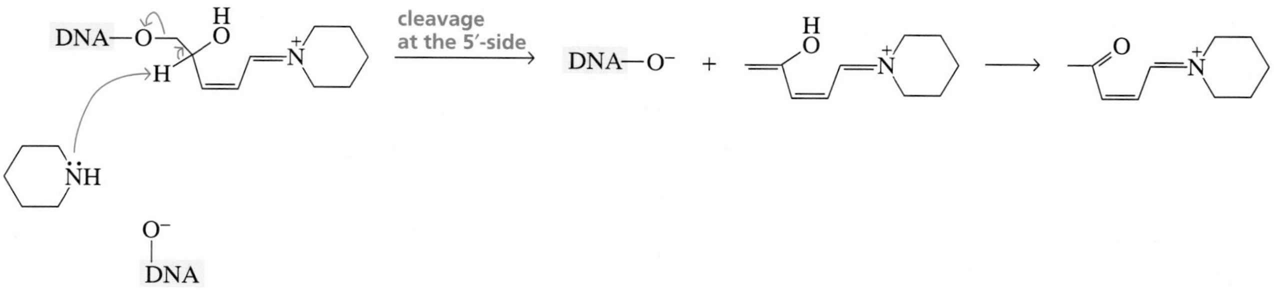

The **dideoxy method** is the preferred method of determining the sequence of bases in DNA. The Maxam–Gilbert chemical cleavage method is primarily used in DNA footprinting—determining where a protein binds to DNA. The protein binding site on DNA can be located because it will not be methylated by dimethyl sulfate, since the protein protects the guanines at the binding site from methylation.

In the dideoxy method, a small piece of DNA, called a primer, labeled at the 5'-end with ^{32}P is added to the restriction fragment whose sequence is to be determined. Next, the four 2'-deoxyribonucleoside triphosphates are added as well as DNA polymerase, the enzyme that adds nucleotides to a strand of DNA. In addition, a small amount of the 2',3'-dideoxynucleoside triphosphate of one of the bases is also added to the reaction mixture. A 2',3'-dideoxynucleoside triphosphate has no OH groups at the 2'- and 3'-positions.

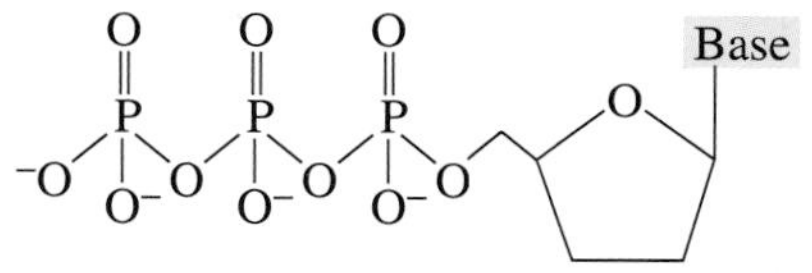

a 2',3'-dideoxynucleoside triphosphate

Nucleotides will be added to the primer by base pairing with the restriction fragment. When the 2',3'-dideoxynucleoside triphosphate is bonded to an adenine, whenever an A is to be added to the growing polymer, synthesis will stop if the 2',3'-analog of dATP is added instead of dATP, because the 2',3'-analog does not have a 3'-OH to which additional nucleotides can be added.

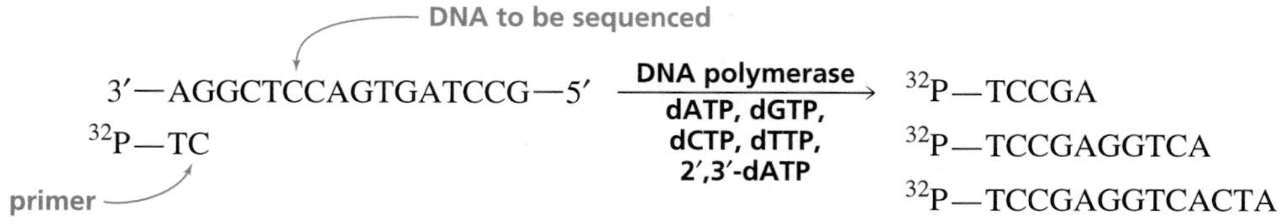

The procedure is repeated three more times using a 2',3'-analog of dGTP, then a 2',3'-analog of dCTP, and then a 2',3'-analog of dTTP. Each of the four sets of chain-terminated fragments is added to one of the four lanes of a buffered polyacrylamide gel, and the base sequence can be determined from the autoradiograph.

DNA FINGERPRINTING

The base sequence of the human genome varies from individual to individual, generally by a single base change every few hundred base pairs. Because some of these changes occur in base sequences recognized by restriction endonucleases, the fragments formed when human DNA reacts with a particular restriction endonuclease vary in size depending on the individual. It is this variation that forms the basis of DNA fingerprinting (also called DNA profiling or DNA typing). This technique is used by forensic chemists to compare DNA samples collected at the scene of a crime with the DNA of the suspected perpetrator. The most powerful technique for DNA identification analyzes restriction fragment length polymorphisms (RFLPs) obtained from regions of DNA in which individual variations are most common. This technique takes 4 to 6 weeks and requires a blood stain about the size of a dime. The chance of identical results from two different persons is thought to be 1 in a million. The second type of DNA profiling uses a polymerase chain reaction (PCR), which amplifies a specific region of DNA and compares differences at that site among individuals. This technique can be done in less than a week and requires only 1% of the amount required for RFLP, but does not discriminate as well among individuals. The chance of identical results from two different people is 1 in 500 to 1 in 2000. DNA fingerprinting is also being used to establish paternity, accounting for about 100,000 DNA profiles a year.

PROBLEM 26

a. Give the labeled fragments that would be formed from the following segment of DNA under the indicated conditions:

$$^{32}P—GCATAGATCGACTGACAACT$$

1. cleavage at G

2. cleavage at G and A

3. cleavage at C

4. cleavage at C and T

b. List the labeled fragments in order of increasing size.

PROBLEM 27

Propose a mechanism for attack of hydrazine on C. How does it differ from the mechanism shown for attack of hydrazine on T?

24.15
LABORATORY SYNTHESIS OF DNA STRANDS

There is a great deal of interest in the synthesis of oligonucleotides with specific base sequences. This would allow scientists to synthesize genes that could be inserted into the DNA of microorganisms, causing the organisms to synthesize a particular protein. Or a synthetic gene could be inserted into the DNA of an organism defective in that gene—a process known as **gene therapy.**

Synthesizing an oligonucleotide with a particular base sequence is an even more challenging task than synthesizing a polypeptide with a specific amino acid sequence. The nucleotides must be strung together in a particular order, and each nucleotide has several groups that must be protected and then deprotected at the proper times. The approach taken was to develop an automated method similar to automated peptide synthesis (Section 20.10). The growing nucleotide is attached to a solid support so it can be purified by flushing the reaction container with a solvent.

One currently used method synthesizes oligonucleotides using phosphoramidite monomers. The 5'-OH group of each phosphoramidite monomer is attached to a *para*-dimethoxytrityl (DMTr) protecting group. The particular group (Pr) used to protect the base depends on the base.

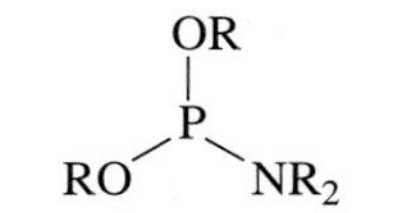

a phosphoramidite

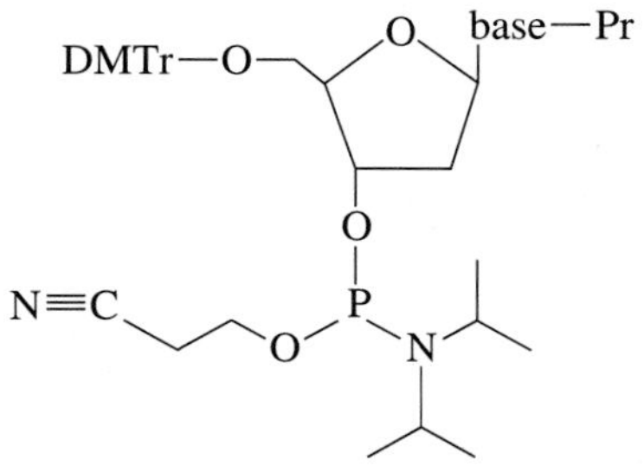

a phosphoramidite monomer
used for
oligonucleotide synthesis

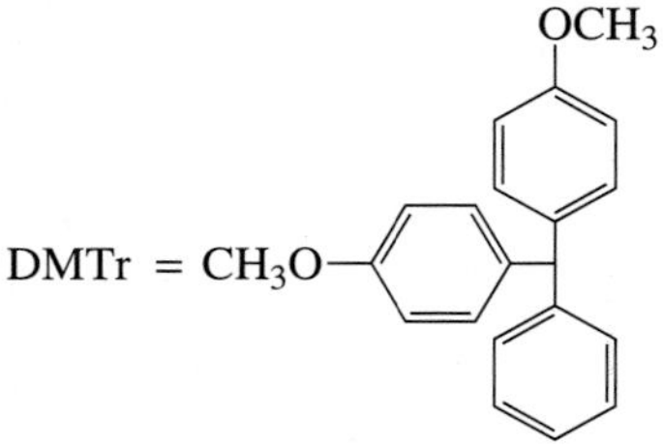

The 3'-nucleoside of the oligonucleotide being synthesized is attached to a controlled-pore glass solid support by an ester bond, and the oligonucleotide is synthesized from the 3'-end. When a monomer is added to the nucleoside attached to the solid support, the only nucleophile in the reaction mixture is the 5'-OH group of the sugar bonded to the solid support. This nucleophile attacks the phosphorus of the phosphoramidite, displacing the amine and forming a phosphite. The amine is too strong a base to be eliminated without being protonated. Protonated tetrazole is the acid used for protonation because it is strong enough to protonate the diisopropylamine leaving group, but not strong enough to remove the DMTr protecting group. The phosphite is oxidized to a phosphate using I_2 or *tert*-butylhydroperoxide as the oxidizing agent. The DMTr protecting group on the 5'-end of the dinucleotide is removed with mild acid. The cycle of (1) monomer addition, (2) oxidation, and (3) deprotection with acid is repeated over and over until a polymer of the desired length and sequence is obtained. Notice that the DNA polymer is synthesized in the $3' \longrightarrow 5'$ direction, which is opposite to the direction ($5' \longrightarrow 3'$) DNA is synthesized in nature.

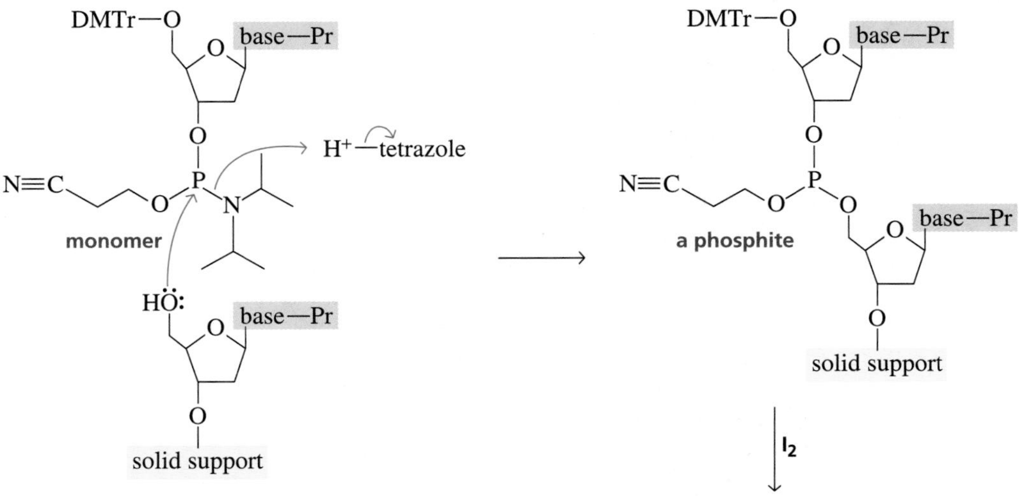

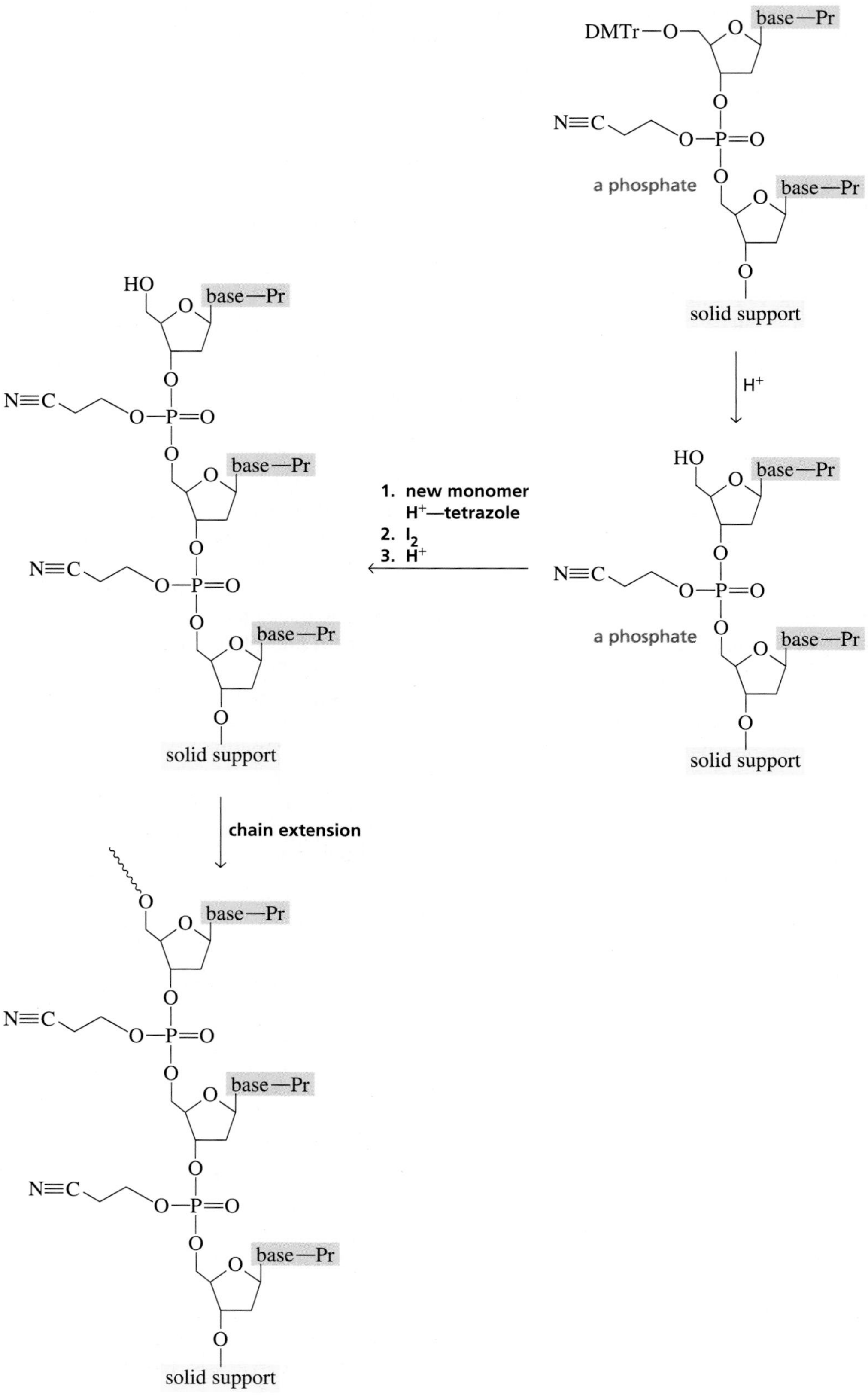

DMTr—O
base—Pr
N≡C
O—P=O
a phosphate
base—Pr
solid support
H+
HO
base—Pr
a phosphate
solid support
1. new monomer
H+—tetrazole
2. I2
3. H+
chain extension
solid support

The NH$_2$ groups of cytosine, adenine, and guanine are nucleophiles and therefore have to be protected to prevent them from reacting with a newly added monomer. The NH$_2$ groups of cytosine and adenine are protected as benzamides, and the NH$_2$ group of guanine is protected as a butyramide. Thymine does not contain any nucleophilic groups, so it does not have to be protected.

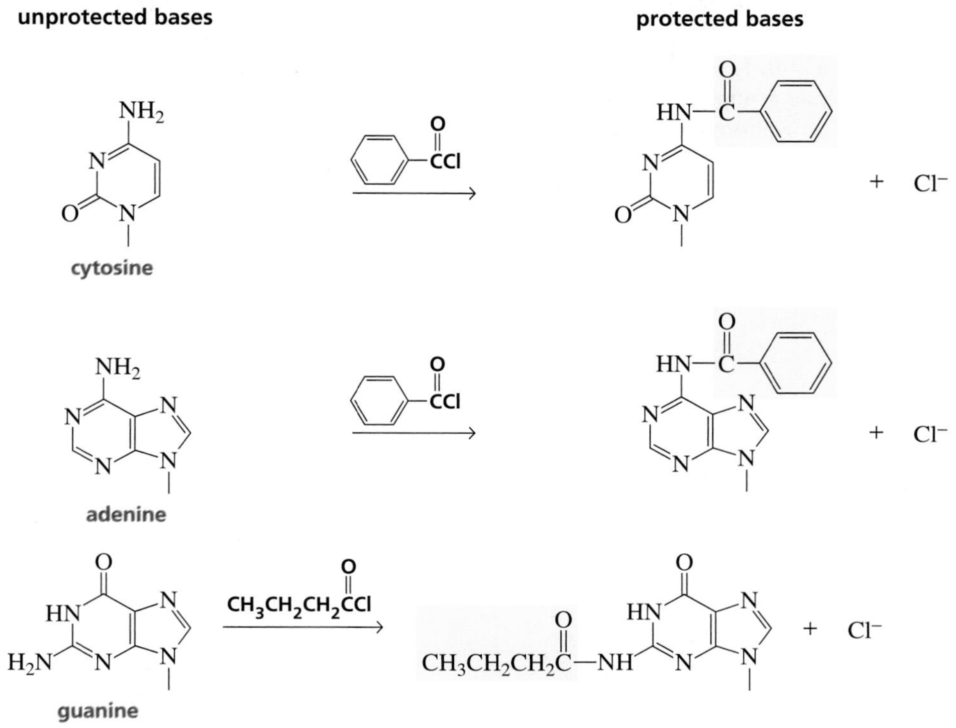

unprotected bases   **protected bases**

After the oligonucleotide is synthesized, the protecting groups on the phosphates and the protecting groups on the bases must be removed, and the oligonucleotide has to be detached from the solid support. This can all be done in a single step using aqueous ammonia. Because a hydrogen bonded to a carbon adjacent to a cyano group is relatively acidic (Section 18.1), NH$_3$ removes the phosphate protecting group.

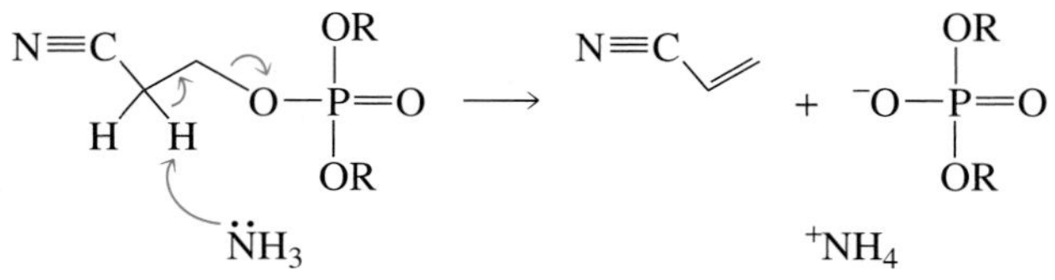

Currently, automated DNA synthesizers can synthesize nucleic acids containing as many as 130 nucleotides in acceptable yields, adding one nucleotide every 2 to 3 minutes. Longer nucleic acids can be prepared by splicing together two or more individually prepared strands. To ensure a good overall yield of oligonucleotide, the addition of each monomer must occur in high yield (>98%). This can be accomplished by using a large excess of monomer. However, the unreacted monomer is wasted, making oligonucleotide synthesis very expensive.

A second method, using H-phosphonate monomers, has an advantage over the phosphoramidite method in that the monomers are easier to handle and phosphate protecting groups are not needed; however, the yields are not as good. The H-phosphonate monomers are activated by being converted into anhydrides as a result of reacting with an acyl chloride. The 5'-OH group of the nucleoside attached to the solid support reacts with the anhydride, forming a dimer. The DMTr protecting group is removed with acid under mild conditions, and a second activated monomer is added. Monomers are added one at a time in this way until the strand is complete. Oxidation with I_2 (or *tert*-butylhydroperoxide) converts the H-phosphonate groups to phosphate groups. The base protecting groups are removed by aqueous ammonia as in the phosphoramidite method.

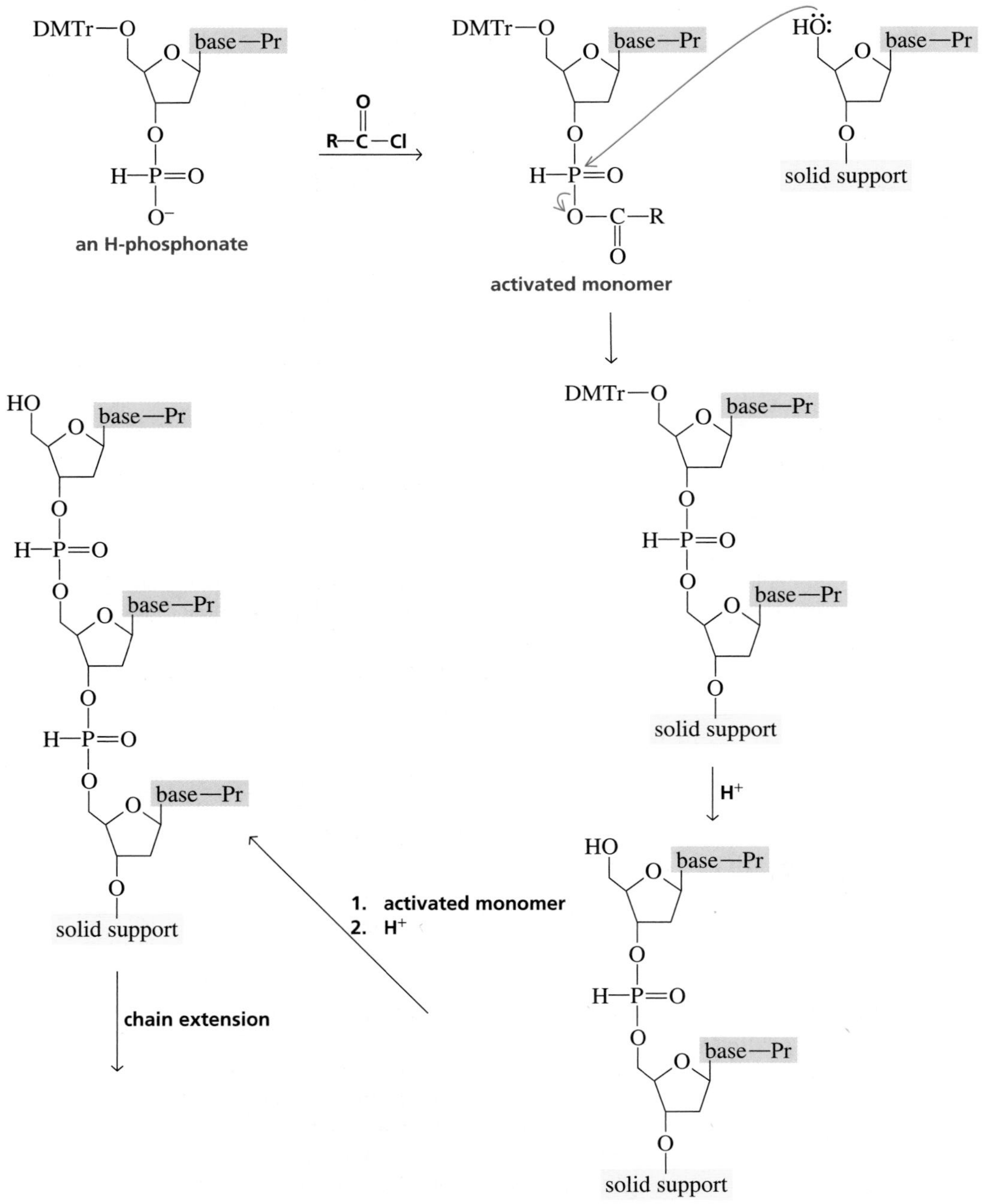

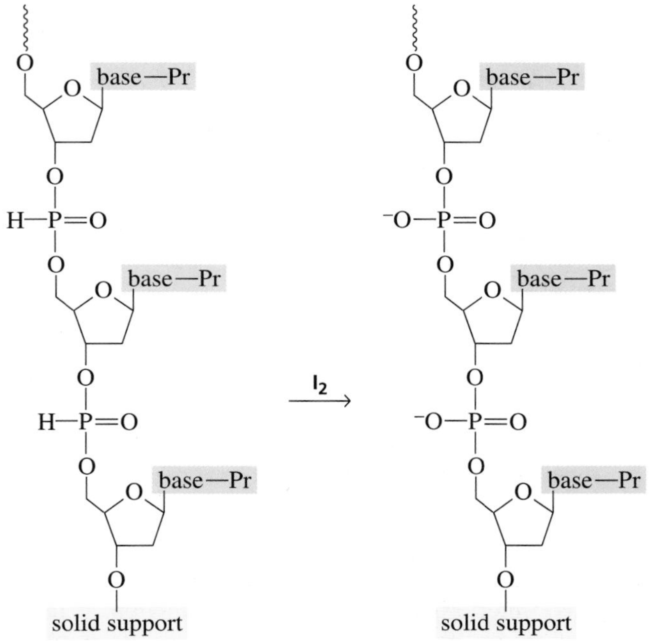

In the phosphoramidite method, notice that an oxidation is done each time a monomer is added, whereas in the H-phosphonate method a single oxidation is carried out after the entire strand has been synthesized.

PROBLEM 28

Propose a mechanism for removal of the DMTr protecting group by treatment with acid.

Certain diseases such as AIDS and herpes are caused by **retroviruses.** The genetic information of a retrovirus is contained in its RNA. The retrovirus uses the sequence of bases in RNA as a template to synthesize DNA. It is called a retrovirus because its genetic information flows from RNA to DNA instead of the more typical flow from DNA to RNA.

Drugs that interfere with the synthesis of DNA by retroviruses have been designed and developed. If the retrovirus cannot synthesize DNA, it cannot take over the genetic machinery of the cell to produce more retroviral RNA and retroviral proteins. Designing drugs with particular structures to achieve specific purposes is called **rational drug design.** AZT is perhaps the best known of the drugs designed to interfere with retroviral DNA synthesis. AZT is taken up by the T lymphocytes, cells that are particularly susceptible to human immunodeficiency virus (HIV), the retrovirus that causes acquired immunodeficiency syndrome (AIDS). Once in the cell, AZT is converted to AZT–triphosphate. The retroviral enzyme (reverse transcriptase) that catalyzes DNA formation from RNA binds AZT–triphosphate more tightly than it binds dTTP. Therefore, AZT rather than T is added to the growing DNA chain. Because AZT does not have a 3'-OH group, no additional nucleotides can be added to the chain, and DNA synthesis comes to a sudden halt. Fortunately, the concentration of AZT required to affect reverse

24.16
RATIONAL DRUG DESIGN

transcription is too low to affect most cellular DNA replication. A newer drug, 2′,3′-dideoxyinosine (DDI), has a similar mechanism of action.

**3′-azido-3′-deoxythymidine
AZT**

**2′,3′-dideoxyinosine
DDI**

Because viruses, bacteria, and cancer cells all require DNA to grow and reproduce, chemists are trying to design compounds that will bind to the DNA of invading organisms, interfering with their reproduction. Chemists are also attempting to design compounds that will bind to specific sequences of human DNA. Such compounds could disrupt the expression of a gene (interfere with its transcription into RNA). For example, there is hope that compounds can be designed that will interfere with the expression of genes that contribute to the development of cancer.

For a compound to target a particular gene, it must be able to recognize a specific sequence of 15 to 20 bases. A sequence that long might occur only once in the human genome, so the compound would be specific for a particular site on DNA. In contrast, if the compound can recognize a sequence of only six bases, statistically that sequence would occur once in every 2000 bases. Such a drug could affect the human genome at more than a million locations. However, since only 10% of the genes are expressed in most cells, a compound that recognizes a specific sequence of 10 to 12 bases may confer a gene-specific effect.

One approach to **site-specific recognition** uses a synthetic strand of oligonucleotides. When a strand of oligonucleotides is added to natural double-stranded

Figure 24.18 ▶
A triple helix. A synthetic strand of oligonucleotides (magenta atoms) is wrapped around double-stranded DNA.

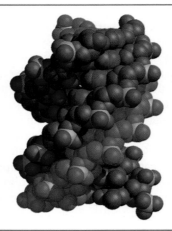

DNA, the strand wraps around the DNA, forming a triple helix (Figure 24.18). The hope is that if the DNA sequence of a particular gene is known, a deoxyribonucleotide can be synthesized that will bind to the gene.

A triple helix is formed through Hoogsteen base pairing between the existing base pair in DNA and the third synthetic strand. In **Hoogsteen base pairing,** a T in the synthetic strand binds to an A of an A–T base pair, while a protonated cytosine in the synthetic strand binds to a G of a G–C base pair (Figure 24.19). Thus, oligonucleotides can be prepared with sequences that will base-pair to the sense strand of DNA at the desired location. Several methods are being investigated that will cut out the piece of targeted DNA after the synthetic strand binds to the double helix.

Hoogsteen base pairing requires that the synthetic strands of oligonucleotides contain only pyrimidines (T's and C's). The targeted site of DNA, therefore, must contain only purines (A's and G's). This limitation has recently been overcome by creating bases that can be used in the synthetic strand that will recognize pyrimidines in double-stranded DNA.

A problem that occurs when synthetic oligonucleotides are used to target DNA is that the synthetic strands are susceptible to enzymes, such as restriction endonucleases, that catalyze the hydrolysis of DNA. Consequently, other polymers are being designed that will recognize specific DNA sequences but will not be enzymatically hydrolyzed. A compound that has shown some promise is a polymer of phosphorothioates. The polymer differs from DNA in that the O^- group bonded to the phosphorous is replaced by S^-. The polymer binds to both DNA and RNA with complementary base pairing, and it also has been found to bind to certain proteins. Polymers, such as oligonucleotide phosphorothioates, that bind to DNA are called **antigene agents;** those that bind to mRNA are called **antisense agents.**

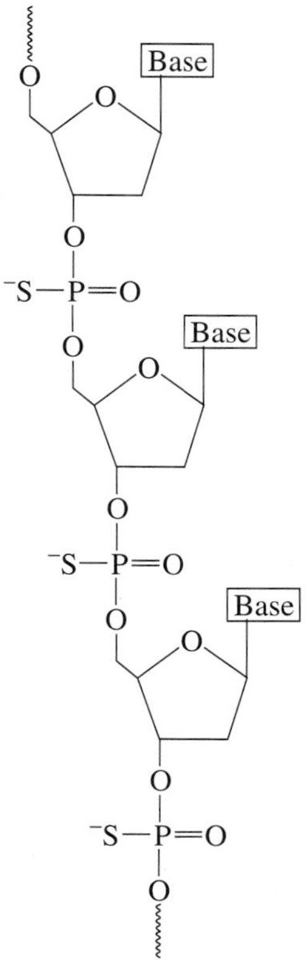

Figure 24.19 ▶
Hoogsteen base pairing:
a T in a synthetic strand of
oligonucleotides binds to
the A of an A–T base pair in
double-stranded DNA;
a protonated C (CH⁺) in the
synthetic strand binds to the
G of a G–C base pair in DNA.

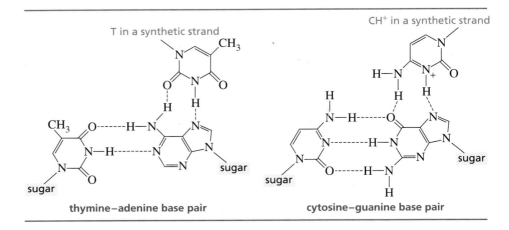

thymine–adenine base pair   cytosine–guanine base pair

Oligonucleotide phosphorothioates consisting of various lengths of deoxycytidine residues recently have been found to be effective in protecting T cells from the HIV virus.

PROBLEM 29

5-Bromouracil, a highly mutagenic compound, is used in cancer chemotherapy. When administered to a patient, it is converted to the triphosphate and incorporated into DNA in place of thymine, which it resembles sterically. Why does it cause mutations? (*Hint:* The bromo substituent increases the stability of the enol tautomer.)

5-bromouracil   **thymine**

Maxam–Gilbert sequencing (page 1091)
minor groove (page 1079)
nucleic acid (page 1074)
nucleoside (page 1063)
nucleotide (page 1064)
oligonucleotide (page 1074)
phosphoanhydride bond (page 1066)
phosphoryl transfer reaction (page 1067)
polynucleotide (page 1074)
primary structure (page 1074)
prokaryotic cell (page 1082)
promoter site (page 1081)
rational drug design (page 1101)
replication (page 1080)
replication fork (page 1080)
restriction endonuclease (page 1090)
restriction fragment (page 1091)
retrovirus (page 1101)
ribonucleic acid (RNA) (page 1061)
ribonucleotide (page 1064)
ribosome (page 1082)
ribozyme (page 1082)
RNA splicing (page 1081)
sedimentation constant (page 1082)
semiconservative replication (page 1080)
sense strand (page 1081)
site-specific recognition (page 1102)
stacking interactions (page 1076)
stop codon (page 1085)
template strand (page 1081)
transcription (page 1080)
translation (page 1085)

PROBLEMS

30. Name the following compounds.

a.

c.

b.

d.

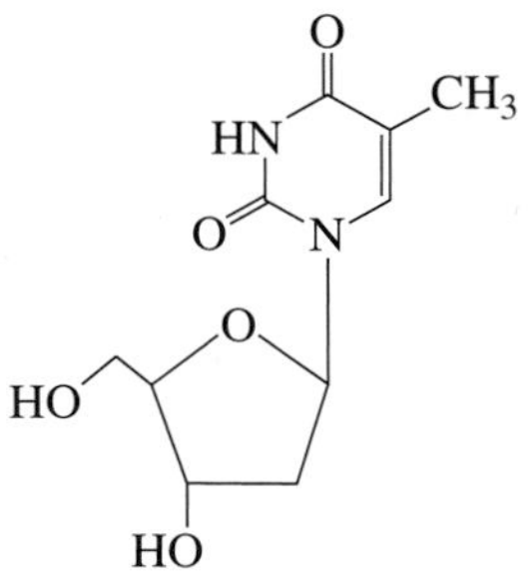

31. What nonapeptide is coded for by the following piece of mRNA?

5'–AAA–GUU–GGC–UAC–CCC–GGA–AUG–GUG–GUC–3'

32. What would be the base sequence of the segment of DNA that is responsible for the biosynthesis of the following hexapeptide?

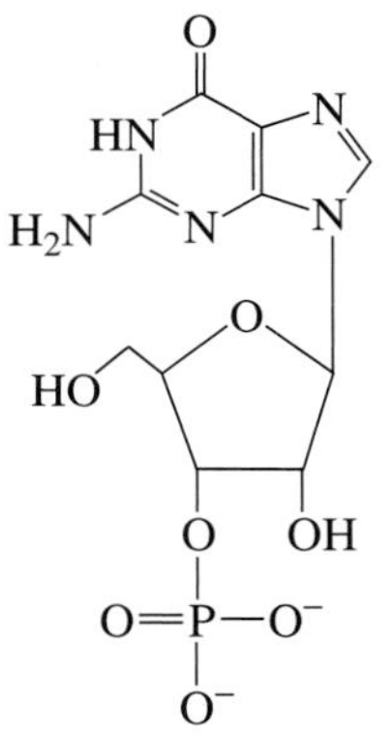

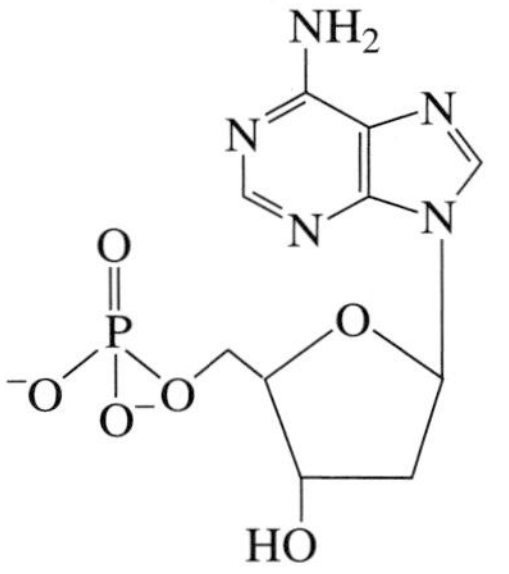

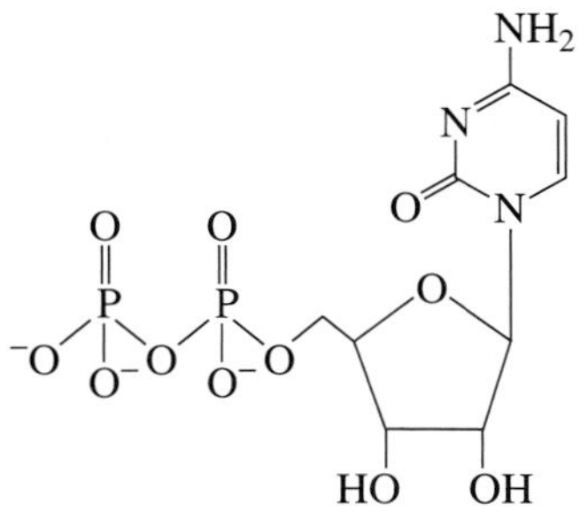

Gly-Ser-Arg-Val-His-Glu

33. Propose a mechanism for the following reaction.

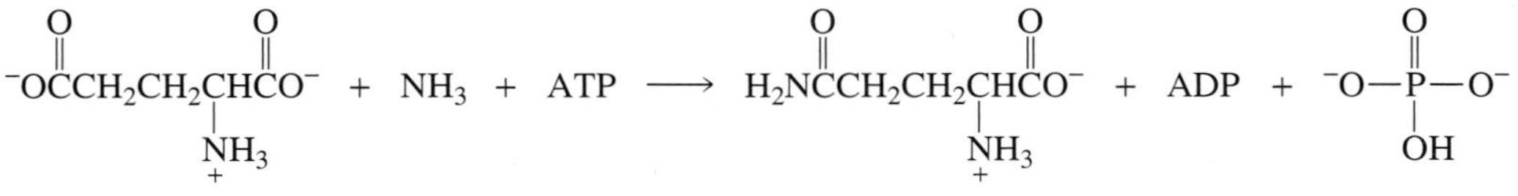

$$
\text{^{-}OCCH}_2\text{CH}_2\text{CHCO}^- + \text{NH}_3 + \text{ATP} \longrightarrow \text{H}_2\text{NCCH}_2\text{CH}_2\text{CHCO}^- + \text{ADP} + \text{^{-}O-P-O}^-
$$

(with $\overset{\text{O}}{\overset{\|}{}}$ groups on the carbonyls, $\overset{+}{\text{NH}_3}$ substituents on both α-carbons, and the phosphate bearing $\overset{\text{O}}{\overset{\|}{}}$ above and OH below)

34. Match the codon with the anticodon.

codon	anticodon
AAA	ACC
GCA	CCU
CUU	UUU
AGG	AGG
CCU	UGA
GGU	AAG
UCA	GUC
GAC	UGC

35. Using the single-letter abbreviations for the amino acids in Table 20.1, write the sequence of amino acids in a polypeptide represented by the letters in your first name, using at least four letters. Do not use any letter twice. (Because not all letters are assigned to amino acids, you might have to use letters in your last name.) Write the sequence of bases in mRNA that would result in the synthesis of that polypeptide. Write the sequence of bases in the sense strand of DNA that would result in formation of the appropriate mRNA.

36. Which of the following pairs of dinucleotides will occur in equal amounts in DNA?
   a. CC and GG
   b. CG and GT
   c. CA and TG
   d. CG and AT
   e. GT and CA
   f. TA and AT

37. Why is the codon a triplet rather than a doublet or a quartet?

38. RNAase, the enzyme that catalyzes the hydrolysis of RNA, has two catalytically active histidine residues at its active site. One of the histidine residues is catalytically active in its acidic form and the other is catalytically active in its basic form. Propose a mechanism for RNAase.

39. The amino acid sequences of peptide fragments obtained from a normal protein and from the same protein synthesized by a defective gene were compared. They were found to differ in only one peptide fragment. The primary sequences of the fragments are shown below.

   Normal: Gln-Tyr-Gly-Thr-Arg-Tyr-Val
   Mutant: Gln-Ser-Glu-Pro-Gly-Thr

   a. What is the defect in DNA?
   b. It was later determined that the normal peptide fragment is an octapeptide with a C-terminal Val-Leu. What is the C-terminal amino acid of the mutant peptide?

40. Whether the mechanism requiring activation of a carboxylate ion by ATP involves attack of the carboxylate ion on the α-phosphorus or the β-phosphorus of ATP cannot be determined from the reaction products because AMP and pyrophosphate are obtained as products in both mechanisms. The mechanisms, however, can be distinguished by a labeling experiment. In such an experiment, the enzyme, the carboxylate ion, ATP, and radioactively labeled pyrophosphate are incubated, and the ATP is isolated. If the isolated ATP is radioactive, the mechanism involves attack on the α-phosphorus. If it is not radioactive, the mechanism involves attack on the β-phosphorus. Explain these conclusions.

41. What would be the results of the experiment in Problem 40 if radioactive AMP were added to the incubation mixture instead of radioactive pyrophosphate?

42. Which cytosine in the following sense strand of DNA could cause the most damage to the organism if it were deaminated?

$$5'\text{——}A\text{—}T\text{—}G\text{—}T\text{—}C\text{—}G\text{—}C\text{—}T\text{—}A\text{—}A\text{—}T\text{—}C\text{——}3'$$

43. Sodium nitrite, a common food preservative (page 645), is capable of causing mutations in an acidic environment by converting cytosines to uracils. Explain how this occurs.

44. The first amino acid incorporated into a polypeptide chain during its biosynthesis is always *N*-formylmethionine. Explain the purpose of the formyl group.

PART VIII

SPECIAL TOPICS IN ORGANIC CHEMISTRY

You have studied polymers synthesized by biological systems—proteins, carbohydrates, and nucleic acids. **Chapter 25** discusses synthetic polymers—polymers synthesized by chemists. These polymers have physical properties that make them useful in everyday life, in applications such as beverage bottles, clothing, food wrap, and compact discs.

Chapter 26 introduces the reactions of heterocyclic compounds—cyclic compounds in which one of the ring atoms is a hetero atom (an atom other than carbon). Many natural products—such as caffeine, nicotine, chlorophyll, and serotonin—and most drugs are heterocyclic compounds.

Chapter 27 discusses pericyclic reactions—reactions that occur as a result of cyclic reorganization of electrons. In this chapter you will learn how to use the conservation of orbital symmetry theory to explain the relationships among reactant, product, and reaction conditions in a pericyclic reaction.

Throughout this text you have designed a wide variety of multistep organic syntheses. In **Chapter 28** you will explore in greater depth how organic chemists approach a synthetic problem. You will follow a few examples of some complicated syntheses so that you can gain an understanding of the power of organic synthesis and some of the ways in which chemists go about solving difficult problems.

Chapter 29 introduces you to medicinal chemistry. Here you will see how many of our commonly used drugs were discovered, and you will learn about some of the techniques used to develop new drugs.

25

SYNTHETIC POLYMERS

Super Glue

No group of synthetic compounds is more important to modern life than synthetic polymers. A **polymer** is a large molecule made by linking together repeating units of small molecules called **monomers.** The process of linking them up is called **polymerization.**

$$n\text{M} \xrightarrow{\text{polymerization}} \text{—M—M—M—M—M—M—M—M—M—}$$

monomer   polymer

ethylene monomers   polyethylene

Unlike small organic molecules that are of interest because of their chemical properties, these giant molecules are interesting because of their physical properties. Some synthetic polymers resemble natural substances, but most are materials that are unavailable in nature and have physical properties that make them useful in everyday life. Such diverse items as photographic film, records, compact discs, rugs, food wrap, artificial joints, Super Glue, toys, beverage bottles, weather stripping, automobile body parts, shoe soles, and condoms are made of synthetic polymers. Humans first relied on natural polymers for clothing, wrapping themselves in animal skins and furs. Later, they learned to spin natural fibers into thread and weave the thread into cloth. Now much of our clothing is made of synthetic polymers (nylon, polyester, polyacrylonitrile). Many people prefer clothing made of natural polymers (cotton, wool, silk), but it has been estimated that, if synthetic polymers were not available, all of the arable land in the United States would have to be used for the production of cotton and wool for clothing.

Polymers can be divided into two broad groups: **biopolymers** and **synthetic polymers.** Biopolymers are synthesized by organisms and are essential for our very existence. Examples of biopolymers are DNA, the storage molecule for genetic information (the molecule that determines whether a fertilized egg becomes a human or a honeybee); RNA and proteins, the molecules by which biochemical transformations are made to occur; and polysaccharides. Their structures and properties are presented in other chapters. In this chapter, we will explore synthetic polymers.

The first plastic—a polymer capable of being molded—was celluloid. Invented in 1856 by Alexander Parke, it was a mixture of nitrocellulose and camphor. Celluloid was used in the manufacture of billiard balls and piano keys, replacing scarce ivory. This discovery provided a reprieve for many elephants but caused some moments of consternation in billiard parlors because nitrocellulose is flammable and explosive. Celluloid was used for motion picture film until it was replaced by cellulose acetate, a less dangerous polymer.

The first synthetic fiber was rayon. In 1865, the French silk industry was threatened by an epidemic that killed many silkworms, highlighting the need for an artificial silk substitute. Louis Chardonnet accidentally discovered the starting material for a synthetic fiber when, while wiping up some spilled nitrocellulose from a table, he noticed long, silklike strands adhering to both the cloth and the table. "Chardonnet silk" was introduced at the Paris Exposition in 1891. It was called rayon because it was so shiny that it appeared to give off rays of light.

The first synthetic rubber was synthesized by German chemists in 1917. Their efforts were in response to a severe shortage of raw materials as a result of blockading during World War I.

Hermann Staudinger was the first to recognize that the various polymers being produced were not disorderly conglomerates of monomers but rather they were made up of continuous chains of monomers joined together. Today, the synthesis of polymers has grown from a process carried out with little chemical understanding to a sophisticated science in which molecules are engineered with predetermined specifications in order to produce new materials that are tailored to fit human needs. New polymers that make new products possible are constantly being designed. Recent examples include Lycra, a fabric with elastic properties, and Dyneema, the strongest fabric commercially available. **Polymer chemistry** is part of the larger discipline of **materials science,** which involves creation of new materials to replace metal, glass, ceramics, wood, cardboard, and paper. Currently there are approximately 30,000 patented polymers in the United States.

Alexander Parke (1813–1890) *was born in Birmingham, England. He called the polymer that he invented "pyroxylin" but was unable to market it.*

The inventor **John Wesley Hyatt (1837–1920)** *was born in New York. Because a New York firm was offering a prize of $10,000 for a substitute for ivory billiard balls, Hyatt improved the synthesis of pyroxylin. He called the improved polymer "celluloid" and patented a method for making billiard balls. Unfortunately, he did not win the prize.*

Louis Marie Hilaire Bernigaud, Comte de Chardonnet (1839–1924) *was born in France. In the early stages of his career, he was an assistant to Louis Pasteur. Since the rayon he initially produced was made from nitrocellulose, it was dangerously flammable. Eventually, chemists learned to remove some of the nitro groups after the fiber was formed, which made the fiber much less flammable but not as strong.*

25.1 GENERAL CLASSES OF SYNTHETIC POLYMERS

Synthetic polymers can be divided into two major classes, depending on their method of preparation. **Chain-growth polymers** (also known as **addition polymers**) are made by the addition of monomers to the end of a growing chain. The end of the chain is reactive because it is a radical, a cation, or an anion. Polystyrene—used for disposable food containers, insulation, "beanbag" pillows (Styrofoam), and toothbrush handles—is an example of a chain-growth polymer.

Hermann Staudinger (1881–1965), *the son of a professor, was born in Germany. He became a professor at the Technical Institute of Karlsruhe and at the University of Freiburg. He received the Nobel Prize in chemistry in 1953 for his contributions to polymer chemistry.*

repeating unit

$CH_2{=}CH$ $CH_2{=}CH$ $CH_2{=}CH$ $\longrightarrow$ $-CH_2-CH{\left[CH_2-CH\right]}_n CH_2-CH-$

styrene

polystyrene
a chain-growth polymer

Step-growth polymers (also called **condensation polymers**) are made by combining two molecules while, in most cases, removing a small molecule (generally water or an alcohol). The reacting molecules have reactive functional groups at both ends. Unlike chain-growth polymer formation, which requires the individual molecules to add to the end of a growing chain, any two reactive molecules can combine in step-growth polymer formation. Dacron is an example of a step-growth polymer.

repeating unit

$CH_3O-C(=O)-\langle\rangle-C(=O)-OCH_3$ + $HOCH_2CH_2OH$ $\xrightarrow{\Delta}$ ${\left[OCH_2CH_2O-C(=O)-\langle\rangle-C(=O)\right]}_n OCH_2CH_2O-$ + $2n\ CH_3OH$

dimethyl terephthalate   ethylene glycol

poly(ethylene terephthalate)
Dacron
a step-growth polymer

Dacron is the most common of the group of polymers known as polyesters (polymers with many ester groups). Polyesters are used for clothing and are responsible for the wrinkle-resistant behavior of many fabrics. Polyester is also used to make the plastic film used for making magnetic recording tape, which is known by the trade name Mylar. This film is tear-resistant and, when processed, has a tensile strength nearly as great as that of steel. The polymer used to make soft drink bottles is also a polyester.

25.2 CHAIN-GROWTH POLYMERS

The monomers used most commonly in chain-growth polymerization are ethylene and substituted ethylenes. In the chemical industry, monosubstituted ethylenes are known as **alpha olefins.** Polymers formed from ethylene or substituted ethylenes are often called **vinyl polymers.** Some of the many vinyl polymers synthesized by chain-growth polymerization are listed in Table 25.1.

When plastics are recycled, the various types have to be separated from one another. To aid in the separation, many states require manufacturers to include a recycling symbol on their products to indicate the type of plastic it is. You are probably familiar with these symbols, which consist of three arrows around one of seven numbers, often found on the bottom of plastic containers. The lower the number in the middle of the symbol, the greater the ease with which the material can be recycled, and the abbreviation below the symbol indicates the type of polymer from which the container is made: 1 for poly(ethylene) terephthlate (PET), 2 for high-density polyethylene (HDPE), 3 for poly(vinyl chloride) (V), 4 for low-density polyethylene (LDPE), 5 for polypropylene (PP), 6 for polystyrene (PS), and 7 for all other plastics.

Chain-growth polymerization proceeds by one of three possible mechanisms: **radical polymerization, cationic polymerization,** or **anionic polymerization.** Each mechanism has three distinct phases: an *initiation step* that starts the polymerization, *propagation steps* that allow the chain to grow, and *termination steps*

TABLE 25.1   Some Important Chain-Growth Polymers and Their Uses

Monomer	Repeating Unit	Polymer Name	Uses
$CH_2=CH_2$	$-CH_2-CH_2-$	polyethylene	film, toys, bottles, plastic bags
$CH_2=CH$ $\quad\ \|$ $\quad\ Cl$	$-CH_2-CH-$ $\qquad\ \|$ $\qquad\ Cl$	poly(vinyl chloride)	"squeeze" bottles, pipe, siding, flooring
$CH_2=CH-CH_3$	$-CH_2-CH-$ $\qquad\ \|$ $\qquad\ CH_3$	polypropylene	molded caps, margarine tubs, indoor/outdoor carpeting, upholstery
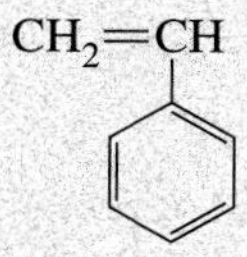 $CH_2=CH$ (phenyl ring)	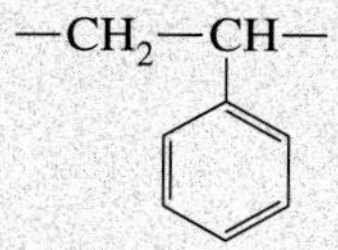 $-CH_2-CH-$ (phenyl ring)	polystyrene	packaging, toys, clear cups, egg cartons, hot drink cups
$CF_2=CF_2$	$-CF_2-CF_2-$	poly(tetrafluoroethylene) Teflon	nonsticking surfaces, liners, cable insulation
$CH_2=CH$ $\quad\ \|$ $\quad\ C\equiv N$	$-CH_2-CH-$ $\qquad\ \|$ $\qquad\ C\equiv N$	poly(acrylonitrile) Orlon, Acrilan	rugs, blankets, yarn, apparel, simulated fur
$CH_2=C-CH_3$ $\qquad\ \|$ $\qquad\ COCH_3$ $\qquad\ \|\|$ $\qquad\ O$	$\qquad\ CH_3$ $\qquad\ \|$ $-CH_2-C-$ $\qquad\ \|$ $\qquad\ COCH_3$ $\qquad\ \|\|$ $\qquad\ O$	poly(methyl methacrylate) Plexiglas, Lucite	lighting fixtures, signs, solar panels, skylights
$CH_2=CH$ $\quad\ \|$ $\quad\ OCCH_3$ $\quad\ \|\|$ $\quad\ O$	$-CH_2-CH-$ $\qquad\ \|$ $\qquad\ OCCH_3$ $\qquad\ \|\|$ $\qquad\ O$	poly(vinyl acetate)	latex paints, adhesives

that stop the growth of the chain. We will see that the choice of mechanism depends on the structure of the monomer and the initiator used to activate the monomer.

Radical Polymerization

For chain-growth polymerization to occur by a radical mechanism, a radical initiator must be added to the monomer to convert some of the monomer molecules into radicals. The initiator breaks homolytically into radicals, and each radical adds to an alkene monomer, converting it into a radical. This radical reacts with another monomer, adding a new subunit that propagates the chain. The radical site is now at the end of the most recent unit added to the end of the chain. This is called the **propagating site.**

chain-initiating steps

$$RO{-}OR \xrightarrow{\Delta} 2\ RO\cdot$$

a radical initiator   radicals

$$RO\cdot\ +\ CH_2{=}\underset{Z}{CH} \longrightarrow ROCH_2\underset{Z}{\overset{\cdot}{C}H}$$

chain-propagating steps

propagating sites

$$ROCH_2\overset{\cdot}{C}H_2\ +\ CH_2{=}\underset{Z}{CH} \longrightarrow ROCH_2\underset{Z}{CH}CH_2\underset{Z}{\overset{\cdot}{C}H}$$

$$ROCH_2CH_2CH_2\overset{\cdot}{C}H_2\ +\ CH_2{=}\underset{Z}{CH} \longrightarrow ROCH_2\underset{Z}{CH}CH_2\underset{Z}{CH}CH_2\underset{Z}{\overset{\cdot}{C}H} \xrightarrow{\text{etc.}}$$

This process is repeated over and over. Hundreds or even thousands of alkene monomers can add, one at a time, to the growing chain. Eventually the chain reaction stops, because the propagating sites are lost. Propagating sites can be destroyed when two chains combine at their propagating sites; when two chains undergo disproportionation, with one chain being oxidized to an alkene and the other being reduced to an alkane (Section 8.2), or when a chain reacts with an impurity that consumes the radical.

methods of terminating the chain

chain combination

$$2\ RO{-}[CH_2\underset{Z}{CH}]_n CH_2\underset{Z}{\overset{\cdot}{C}H} \longrightarrow RO{-}[CH_2\underset{Z}{CH}]_n CH_2\underset{Z}{CH}\underset{Z}{CH}CH_2[\underset{Z}{CH}CH_2]_n OR$$

disproportionation

$$2\ RO{-}[CH_2\underset{Z}{CH}]_n CH_2\underset{Z}{\overset{\cdot}{C}H} \longrightarrow RO{-}[CH_2\underset{Z}{CH}]_n CH{=}\underset{Z}{CH}\ +\ RO{-}[CH_2\underset{Z}{CH}]_n CH_2\underset{Z}{CH}$$

reaction with an impurity

$$2\ RO{-}[CH_2\underset{Z}{CH}]_n CH_2\underset{Z}{\overset{\cdot}{C}H}\ +\ \text{impurity} \longrightarrow RO{-}[CH_2\underset{Z}{CH}]_n CH_2\underset{Z}{CH}{-}\text{impurity}$$

Thus, *radical polymerizations* have chain-initiating, chain-propagating, and chain-terminating steps similar to the radical reactions discussed in Sections 3.18 and 8.2. Alkenes other than ethylene can be used in radical polymerizations.

As long as the polymer has a high molecular weight, the groups at the ends of the polymer are relatively unimportant in determining the physical properties of the polymer and are generally not even specified. It is the rest of the molecule that determines the properties of the polymer.

The molecular weight of the polymer can be controlled by a process known as **chain transfer.** In chain transfer, the growing chain reacts with a molecule XY in a manner that allows X· to terminate the chain, leaving behind Y· to initiate a new chain. XY can be a solvent, a radical initiator, or any molecule with a bond that can be cleaved homolytically.

$$-CH_2[CH_2\underset{Z}{CH}]_n CH_2\underset{Z}{\overset{\cdot}{C}H}\ +\ XY \longrightarrow -CH_2[CH_2\underset{Z}{CH}]_n CH_2\underset{Z}{CH}X\ +\ Y\cdot$$

Chain-growth polymerization of substituted ethylenes exhibits a marked preference for **head-to-tail addition,** where the head of one monomer is attached to the tail of another.

tail   head

$$CH_2{=}CH$$
$$|$$
$$Z$$

$-CH_2CHCH_2CH-$ $-CH_2CHCHCH_2-$ $-CHCH_2CH_2CH-$
$\quad\;|\quad\;\;\;|$ $\quad\;\;\;|\;\;\;|$ $|\qquad\qquad|$
$\quad\;Z\quad\;\;Z$ $\quad\;\;\;Z\;\;\;Z$ $Z\qquad\qquad Z$

head-to-tail   **head-to-head**   **tail-to-tail**

Head-to-tail addition of a substituted ethylene results in a polymer in which every other carbon bears a substituent.

$$CH_2{=}CH \longrightarrow -CH_2CHCH_2CHCH_2CHCH_2CHCH_2CHCH_2CH-$$
$$\qquad\;|\qquad\qquad\qquad\;\;|\quad\;\;\;|\quad\;\;\;|\quad\;\;\;|\quad\;\;\;|\quad\;\;\;|$$
$$\qquad Cl\qquad\qquad\qquad Cl\quad\;Cl\quad\;Cl\quad\;Cl\quad\;Cl\quad\;Cl$$

poly(vinyl chloride)

Head-to-tail addition is preferred for steric reasons, because the propagating site preferentially attacks the less sterically hindered carbon of the alkene. Groups that stabilize adjacent radicals also favor head-to-tail addition. For example, when X is a phenyl substituent, the benzene ring stabilizes the radical by resonance, so the propagating site is the carbon that bears the phenyl substituent.

$$-CH_2CHCH_2\dot{C}H$$

In cases where X is small—which makes steric considerations less important—and is less able to stabilize the growing end of the chain by resonance, some head-to-head addition and some tail-to-tail addition also occur. This has been observed primarily in situations where X = F. Abnormal addition, however, has never been found to constitute more than 10% of the overall chain.

Monomers that most readily undergo chain-growth polymerization by a radical mechanism are those in which the substituent (X) is able to stabilize the growing radical species by resonance. Examples of such monomers are shown in Table 25.2.

Any compound that readily undergoes homolytic cleavage to form radicals that are sufficiently energetic to convert an alkene into a radical can serve as a radical

---

TABLE 25.2 Examples of Alkenes That Undergo Radical Polymerization

$$CH_2{=}CH$$

styrene

$$CH_2{=}CH$$
$$\qquad\quad|$$
$$\qquad\quad OCCH_3$$
$$\qquad\qquad\|$$
$$\qquad\qquad O$$

vinyl acetate

$$CH_2{=}CCH_3$$
$$\qquad\quad|$$
$$\qquad\quad COCH_3$$
$$\qquad\qquad\|$$
$$\qquad\qquad O$$

methyl methacrylate

$$CH_2{=}CH$$
$$\qquad\quad|$$
$$\qquad\quad Cl$$

vinyl chloride

$$CH_2{=}CH$$
$$\qquad\quad|$$
$$\qquad\quad C{\equiv}N$$

acrylonitrile

$$CH_2{=}CH$$
$$\qquad\quad|$$
$$\qquad\quad CH{=}CH_2$$

1,3-butadiene

TABLE 25.3 Some Radical Initiators

$$HO-OH \longrightarrow 2 \cdot OH$$

$$\underset{\underset{CH_3}{|}}{\overset{\overset{CH_3}{|}}{CH_3CO}}-OH \longrightarrow \underset{\underset{CH_3}{|}}{\overset{\overset{CH_3}{|}}{CH_3CO}}\cdot + \cdot OH$$

$$\underset{\overset{||}{O}}{\overset{\overset{O}{||}}{KOSO}}-\underset{\overset{||}{O}}{\overset{\overset{O}{||}}{OSOK}} \longrightarrow 2 \underset{\overset{||}{O}}{\overset{\overset{O}{||}}{KOSO}}\cdot$$

$$\underset{\underset{CH_3}{|}}{\overset{\overset{CH_3}{|}}{CH_3CO}}-\underset{\underset{CH_3}{|}}{\overset{\overset{CH_3}{|}}{OCCH_3}} \longrightarrow 2 \underset{\underset{CH_3}{|}}{\overset{\overset{CH_3}{|}}{CH_3CO}}\cdot$$

$$\text{C}_6\text{H}_5\overset{\overset{O}{||}}{C}O-O\overset{\overset{O}{||}}{C}\text{C}_6\text{H}_5 \longrightarrow 2\ \text{C}_6\text{H}_5\overset{\overset{O}{||}}{C}O\cdot$$

$$\underset{\underset{C\equiv N}{|}}{\overset{\overset{CH_3}{|}}{CH_3CN}}=\underset{\underset{C\equiv N}{|}}{\overset{\overset{CH_3}{|}}{NCCH_3}} \longrightarrow 2 \underset{\underset{C\equiv N}{|}}{\overset{\overset{CH_3}{|}}{CH_3C}}\cdot + N_2$$

initiator for radical polymerization. Several radical initiators are shown in Table 25.3.

A common feature of all the radical initiators is a relatively weak bond that readily undergoes homolytic cleavage. In all but one of the radical initiators shown in Table 25.3, the weak bond is an oxygen–oxygen bond. Two factors enter into the choice of radical initiator for a particular chain-growth polymerization. The first is the desired solubility property of the initiator. For example, potassium persulfate is often used if the initiator needs to be water-soluble, whereas an initiator with several carbons is chosen if it is necessary for the initiator to be soluble in a nonpolar solvent. The second factor is the temperature at which the polymerization reaction is to be carried out. For example, a *tert*-butoxy radical is relatively stable, so an initiator that forms a *tert*-butoxy radical is used for polymerizations carried out at relatively high temperatures.

PROBLEM 1◆

What monomer would you use to form each of the following polymers?

a. $\underset{\overset{|}{Cl}}{-CH_2CH}\underset{\overset{|}{Cl}}{CH_2CH}\underset{\overset{|}{Cl}}{CH_2CH}\underset{\overset{|}{Cl}}{CH_2CH}\underset{\overset{|}{Cl}}{CH_2CH}-$

b. $-CH_2\overset{\overset{\displaystyle CH_3}{|}}{C}CH_2\overset{\overset{\displaystyle CH_3}{|}}{C}CH_2\overset{\overset{\displaystyle CH_3}{|}}{C}CH_2\overset{\overset{\displaystyle CH_3}{|}}{C}CH_2\overset{\overset{\displaystyle CH_3}{|}}{C}CH_2\overset{\overset{\displaystyle CH_3}{|}}{C}-$

$$\underset{\overset{\displaystyle |}{CH_3}}{\overset{\displaystyle C=O}{|}}\ \underset{\overset{\displaystyle |}{CH_3}}{\overset{\displaystyle C=O}{|}}\ \underset{\overset{\displaystyle |}{CH_3}}{\overset{\displaystyle C=O}{|}}\ \underset{\overset{\displaystyle |}{CH_3}}{\overset{\displaystyle C=O}{|}}\ \underset{\overset{\displaystyle |}{CH_3}}{\overset{\displaystyle C=O}{|}}\ \underset{\overset{\displaystyle |}{CH_3}}{\overset{\displaystyle C=O}{|}}$$

c. $-CF_2CF_2CF_2CF_2CF_2CF_2CF_2CF_2CF_2CF_2-$

PROBLEM 2 ◆

Which polymer would be more apt to contain abnormal head-to-head linkages, poly(vinyl chloride) or polystyrene?

PROBLEM 3

Draw a segment of polystyrene that contains abnormal head-to-head and tail-to-tail linkages.

PROBLEM 4

Show the mechanism for the formation of a segment of poly(vinyl chloride) containing three molecules of vinyl chloride and initiated by hydrogen peroxide.

Branching of the Polymer Chain

If the propagating site abstracts a hydrogen atom from a chain, a branch will grow off the chain at that point.

$-CH_2CH_2CH_2\overset{\cdot}{C}H_2$ + $-CH_2CH_2\overset{\overset{\displaystyle H}{|}}{C}HCH_2CH_2CH_2-$

$\downarrow$

$-CH_2CH_2CH_2\overset{\overset{\displaystyle H}{|}}{C}H_2$ + $-CH_2CH_2\overset{\cdot}{C}HCH_2CH_2CH_2-$ $\xrightarrow{CH_2=CH_2}$ $-CH_2CH_2\overset{\overset{\displaystyle \overset{\cdot}{C}H_2}{\overset{\displaystyle |}{CH_2}}}{C}HCH_2CH_2CH_2-$

Abstraction of a hydrogen atom from a carbon near the end of the chain leads to short branches, while abstraction of a hydrogen atom from a carbon near the middle of the chain results in long branches. Short branches are more likely to be formed than are long branches.

chain with short branches

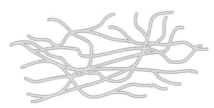

chain with long branches

Branching greatly affects the physical characteristics of the polymer. Linear, unbranched chains can pack together more closely than branched chains can. Consequently, linear polyethylene (known as high-density polyethylene) is a relatively hard plastic, used for the production of such things as artificial hip joints, while branched polyethylene (low-density polyethylene) is a much more flexible polymer, used for trash bags and dry cleaning bags.

PROBLEM 5◆

Polyethylene can be used for the production of beach chairs and beach balls. Which of these items is made from more highly branched polyethylene?

PROBLEM 6

Draw a short segment of branched polystyrene that clearly shows the linkages at the branch point.

Cationic Polymerization

In cationic polymerization, the initiator is an electrophile that adds to the alkene, causing it to become a cation. The initiator most often used in cationic polymerization is a Lewis acid, such as BF_3 or $AlCl_3$. The advantage of such an initiator is that it does not have an accompanying nucleophile that could act as a chain terminator, as would be the case with a proton-donating acid such as HCl. The cation formed in the initiation step reacts with a second monomer, forming a new cation that reacts in turn with a third monomer. As each subsequent monomer adds to the chain, the positively charged propagating site always ends up on the last added unit.

chain-initiating step

$$BF_3 \;+\; CH_2{=}C{\overset{CH_3}{\underset{CH_3}{\big\backslash}}} \longrightarrow F_3\bar{B}{-}CH_2\overset{+}{C}{\overset{CH_3}{\underset{CH_3}{\big\backslash}}}$$

chain-propagating steps   propagating sites

$$F_3\bar{B}{-}CH_2\overset{+}{C}{\overset{CH_3}{\underset{CH_3}{\big\backslash}}} \;+\; CH_2{=}C{\overset{CH_3}{\underset{CH_3}{\big\backslash}}} \longrightarrow F_3\bar{B}{-}CH_2\underset{CH_3}{\overset{CH_3}{C}}CH_2\overset{+}{C}{\overset{CH_3}{\underset{CH_3}{\big\backslash}}}$$

$$F_3\bar{B}{-}CH_2\underset{CH_3}{\overset{CH_3}{C}}CH_2\overset{+}{C}{\overset{CH_3}{\underset{CH_3}{\big\backslash}}} \;+\; CH_2{=}C{\overset{CH_3}{\underset{CH_3}{\big\backslash}}} \longrightarrow F_3\bar{B}{-}CH_2\underset{CH_3}{\overset{CH_3}{C}}CH_2\underset{CH_3}{\overset{CH_3}{C}}CH_2\overset{+}{C}{\overset{CH_3}{\underset{CH_3}{\big\backslash}}}$$

Cationic polymerization can be terminated by loss of a proton or by addition of a nucleophile that reacts with the propagating site. The chain can also be terminated by a chain transfer reaction with the solvent (XY).

chain-terminating steps

$$F_3\bar{B}-CH_2\underset{\underset{CH_3}{|}}{\overset{\overset{CH_3}{|}}{C}}\left[-CH_2\underset{\underset{CH_3}{|}}{\overset{\overset{CH_3}{|}}{C}}\right]_n-CH_2\overset{+}{C}\overset{CH_3}{\underset{CH_3}{\diagdown}} \xrightarrow{-H^+} F_3\bar{B}-CH_2\underset{\underset{CH_3}{|}}{\overset{\overset{CH_3}{|}}{C}}\left[-CH_2\underset{\underset{CH_3}{|}}{\overset{\overset{CH_3}{|}}{C}}\right]_n-CH=C\overset{CH_3}{\underset{CH_3}{\diagdown}}$$

$$F_3\bar{B}-CH_2\underset{\underset{CH_3}{|}}{\overset{\overset{CH_3}{|}}{C}}\left[-CH_2\underset{\underset{CH_3}{|}}{\overset{\overset{CH_3}{|}}{C}}\right]_n-CH_2\overset{+}{C}\overset{CH_3}{\underset{CH_3}{\diagdown}} \xrightarrow{Nu^-} F_3\bar{B}-CH_2\underset{\underset{CH_3}{|}}{\overset{\overset{CH_3}{|}}{C}}\left[-CH_2\underset{\underset{CH_3}{|}}{\overset{\overset{CH_3}{|}}{C}}\right]_n-CH_2\underset{\underset{CH_3}{|}}{\overset{\overset{CH_3}{|}}{C}}-Nu$$

$$F_3\bar{B}-CH_2\underset{\underset{CH_3}{|}}{\overset{\overset{CH_3}{|}}{C}}\left[-CH_2\underset{\underset{CH_3}{|}}{\overset{\overset{CH_3}{|}}{C}}\right]_n-CH_2\overset{+}{C}\overset{CH_3}{\underset{CH_3}{\diagdown}} \xrightarrow{XY} F_3\bar{B}-CH_2\underset{\underset{CH_3}{|}}{\overset{\overset{CH_3}{|}}{C}}\left[-CH_2\underset{\underset{CH_3}{|}}{\overset{\overset{CH_3}{|}}{C}}\right]_n-CH_2\underset{\underset{CH_3}{|}}{\overset{\overset{CH_3}{|}}{C}}-X \; + \; Y^+$$

The carbocation intermediates formed during cationic polymerization, like any other carbocations, can undergo rearrangement by either a 1,2-hydride shift or a 1,2-methyl shift if rearrangement leads to a more stable carbocation (Section 3.14). For example, the polymer formed from cationic polymerization of 3-methyl-1-butene contains both unrearranged and rearranged units. The unrearranged unit is a secondary carbocation, while the rearranged unit is a more stable tertiary carbocation. The extent of rearrangement depends on the reaction temperature.

$$CH_2=CHCHCH_3 \longrightarrow$$
$$\qquad\qquad\; \underset{CH_3}{|}$$

3-methyl-1-butene

$$-CH_2CH_2\underset{\underset{CH_3}{|}}{\overset{\overset{CH_3}{|}}{C}}CH_2\underset{\underset{\underset{\underset{CH_3}{|}}{CH_3}}{|}}{\overset{\overset{CH_3}{|}}{CH}}CH_2\underset{\underset{\underset{CH_3}{|}}{CH_3}}{\overset{\overset{CH_3}{|}}{CH}}CH_2CH_2\underset{\underset{CH_3}{|}}{\overset{\overset{CH_3}{|}}{C}}-$$

$$-CH_2-\overset{+}{\underset{\underset{\underset{CH_3}{|}}{CHCH_3}}{CH}} \qquad\qquad -CH_2CH_2\overset{+}{\underset{\underset{CH_3}{|}}{\overset{\overset{CH_3}{|}}{C}}}$$

unrearranged propagating site   **propagating site rearranged as a result of a 1,2-hydride shift**

Monomers that are best able to undergo polymerization by a cationic mechanism are those with electron-donating substituents that can stabilize the positive charge at the propagating site. Examples of monomers that undergo cationic polymerization are shown in Table 25.4.

TABLE 25.4   Examples of Alkenes That Undergo Cationic Polymerization

$$CH_2=\underset{\underset{CH_3}{|}}{CH} \qquad\quad CH_2=\underset{\underset{CH_3}{|}}{CCH_3} \qquad\quad CH_2=\underset{\underset{\underset{O}{\overset{\|}{C}}}{O}\underset{}{CCH_3}}{CH} \qquad\quad CH_2=CH$$

propylene   **isobutylene**   **vinyl acetate**   **styrene**

PROBLEM 7◆

List the following groups of monomers in order of decreasing ability to undergo cationic polymerization.

a. $CH_2=CH$ $CH_2=CH$ $CH_2=CH$

(benzene ring with NO_2)   (benzene ring with CH_3)   (benzene ring with OCH_3)

b. $CH_2=CHCH_3$ $CH_2=CHOCCH_3$ (with =O) $CH_2=CHCOCH_3$ (with =O)

c. $CH_2=CH$ $CH_2=CCH_3$

(benzene ring)   (benzene ring)

Anionic Polymerization

In anionic polymerization, the initiator is a nucleophile that reacts with the alkene to form a propagating site that is an anion. Nucleophilic attack on an alkene is not an easy reaction, because alkenes are themselves nucleophiles. Therefore, the initiator must be a very good nucleophile such as sodium amide or butyllithium, and the alkene must contain an electron-withdrawing substituent to decrease its nucleophilicity. Some alkenes that undergo polymerization by an anionic mechanism are shown in Table 25.5.

chain-initiating step

$$\overset{..}{B}u \ Li^+ \ + \ CH_2=CH \longrightarrow Bu-CH_2\overset{..}{C}H$$
$$\quad\quad\quad\quad\quad\quad\quad | \quad\quad\quad\quad\quad\quad\quad |$$
$$\quad\quad\quad\quad\quad\quad\quad C\equiv N \quad\quad\quad\quad\quad\quad C\equiv N$$

chain-propagating steps

propagating sites

$$Bu-CH_2\overset{..}{C}H \ + \ CH_2=CH \longrightarrow Bu-CH_2CH-CH_2\overset{..}{C}H$$
$$\quad | \quad\quad\quad\quad\quad\quad | \quad\quad\quad\quad\quad | \quad\quad\quad |$$
$$\quad C\equiv N \quad\quad\quad\quad C\equiv N \quad\quad\quad C\equiv N \quad C\equiv N$$

$$Bu-CH_2CH-CH_2\overset{..}{C}H \ + \ CH_2=CH \longrightarrow Bu-CH_2CH-CH_2CH-CH_2\overset{..}{C}H$$
$$\quad | \quad\quad\quad | \quad\quad\quad\quad\quad | \quad\quad\quad\quad | \quad\quad\quad | \quad\quad\quad |$$
$$\quad C\equiv N \quad C\equiv N \quad\quad\quad C\equiv N \quad\quad C\equiv N \quad C\equiv N \quad C\equiv N$$

The chain can be terminated by a chain transfer reaction with the solvent or by reaction with an impurity in the reaction mixture. If the solvent cannot donate a proton to terminate the chain, and if all impurities that can react with a carbanion are rigorously excluded, chain propagation will continue until all the monomer has been consumed. At this point the propagating site will still be active, so the polymerization reaction will continue if more monomer is added to the system. Such nonterminated chains are called **living polymers** since the chains remain active

TABLE 25.5 Examples of Alkenes That Undergo Anionic Polymerization

CH$_2$=CH	CH$_2$=CH	CH$_2$=CCH$_3$	CH$_2$=CH
Cl	C≡N	COCH$_3$	
vinyl chloride	acrylonitrile	O	
		methyl methacrylate	styrene

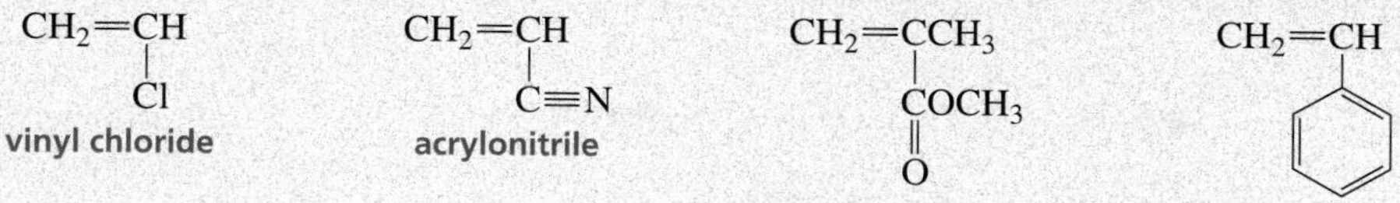

until they are "killed." Living polymers usually result from anionic polymerization because the chains cannot be terminated by proton loss from the polymer as they can in cationic polymerization, or by disproportionation or radical recombination as they can in radical polymerization.

Super Glue is a polymer of methyl α-cyanoacrylate. Because the monomer has two electron-withdrawing groups, it requires only a moderate nucleophile to initiate anionic polymerization. An OH group of cellulose or a nucleophilic group of a protein can act as an initiator. You may well have experienced this reaction if you have ever spilled a drop of Super Glue on your fingers. A nucleophilic group of the protein on the surface of the skin initiates the polymerization reaction with the result that two fingers can become firmly glued together. The ability to form covalent bonds with groups on the surfaces of objects to be glued together is what gives Super Glue its amazing strength.

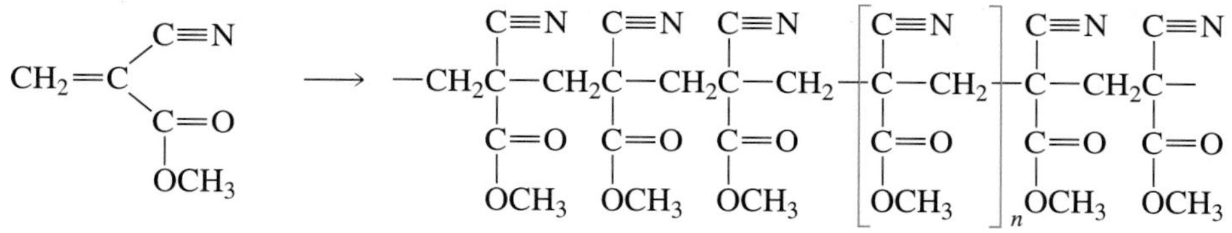

methyl α-cyanoacrylate → Super Glue

PROBLEM 8 ◆

List the following groups of monomers in order of decreasing ability to undergo anionic polymerization.

a. CH$_2$=CH   CH$_2$=CH   CH$_2$=CH

with NO$_2$, CH$_3$, OCH$_3$ substituents on benzene rings

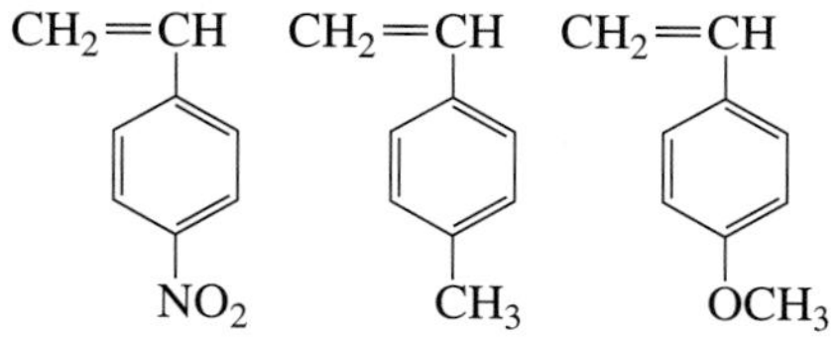

b. CH$_2$=CHCH$_3$ CH$_2$=CHCl   CH$_2$=CHC≡N

We have seen that the substituent on the alkene determines the best mechanism for chain-growth polymerization. Alkenes with substituents that can stabilize radicals readily undergo radical polymerization; alkenes with electron-donating substituents that can stabilize cations undergo cationic polymerization; and alkenes with electron-withdrawing substituents that can stabilize anions undergo anionic polymerizations. Some alkenes can readily undergo polymerization by more than one mechanism. For example, styrene can undergo polymerization by radical, cationic, and anionic mechanisms because the phenyl group can stabilize benzylic radicals, benzylic cations, and benzylic anions. The particular mechanism followed for the polymerization of styrene depends on the nature of the initiator chosen to start the polymerization reaction.

Although ethylene and substituted ethylenes are the monomers most commonly used for chain-growth polymerization reactions, other compounds can polymerize as well. For example, epoxides undergo polymerization reactions. If the initiator is a nucleophile such as HO⁻ or RO⁻, polymerization occurs by an anionic mechanism.

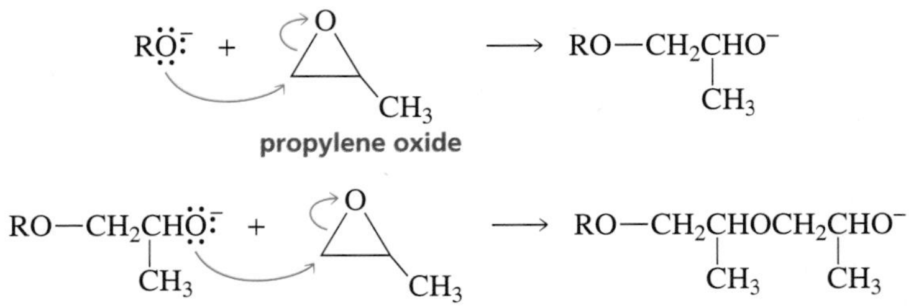

If the initiator molecule is a Lewis acid or a proton-donating acid, polymerization occurs by a cationic mechanism. Polymerization reactions that involve ring-opening reactions, such as the polymerization of propylene oxide, are called **ring-opening polymerizations.**

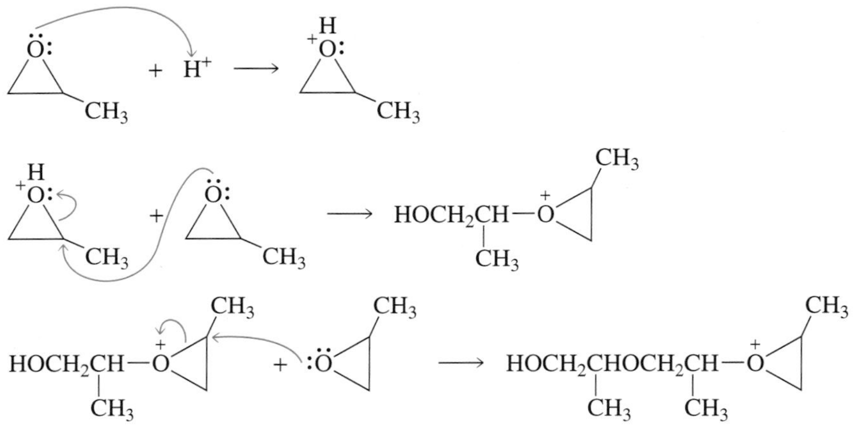

PROBLEM 9

When propylene oxide undergoes anionic polymerization, nucleophilic attack occurs at the less substituted carbon of the epoxide; however, when it undergoes cationic polymerization, nucleophilic attack occurs at the more substituted carbon. Explain why this is so.

PROBLEM 10

Describe the polymerization of 2,2-dimethyloxirane

a. by an anionic mechanism.

b. by a cationic mechanism.

PROBLEM 11◆

What monomer and what initiator would you use to synthesize each of the following polymers?

a. 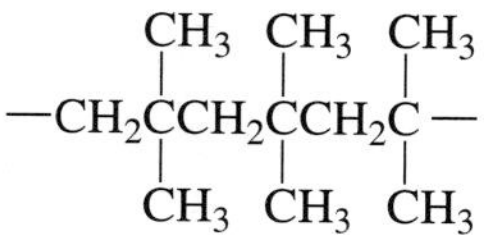 $-CH_2\overset{\overset{\displaystyle CH_3}{|}}{\underset{\underset{\displaystyle CH_3}{|}}{C}}CH_2\overset{\overset{\displaystyle CH_3}{|}}{\underset{\underset{\displaystyle CH_3}{|}}{C}}CH_2\overset{\overset{\displaystyle CH_3}{|}}{\underset{\underset{\displaystyle CH_3}{|}}{C}}-$

c. $-CH_2CH_2OCH_2CH_2O-$

b. 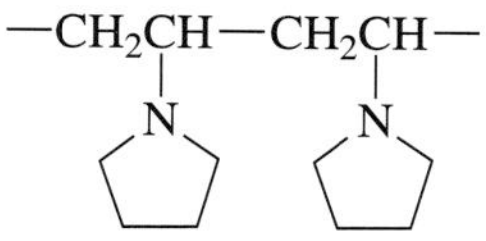 $-CH_2CH-CH_2CH-$
$\qquad\quad$ N $\qquad\quad$ N

d. 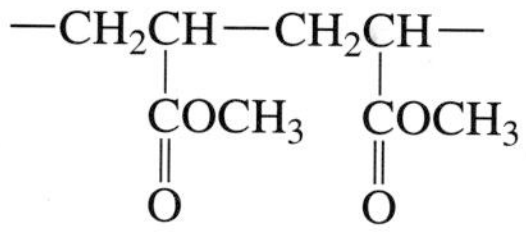 $-CH_2CH-CH_2CH-$
$\qquad\qquad\underset{\underset{\displaystyle O}{\|}}{C}OCH_3 \quad \underset{\underset{\displaystyle O}{\|}}{C}OCH_3$

PROBLEM 12 ◆

Draw a short segment of the polymer formed from cationic polymerization of 3,3-dimethyloxacyclobutane.

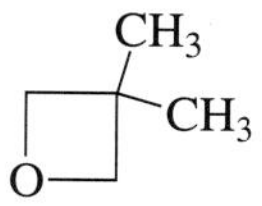

CH₃
CH₃
O

3,3-dimethyloxacyclobutane

Polymers formed from substituted ethylenes can exist in three possible configurations: isotactic, syndiotactic, and atactic. An **isotactic polymer** has all the substituents on the same side of the fully extended carbon chain. *Iso* and *taxis* are Greek for "the same" and "order," respectively. In a **syndiotactic polymer** (*syndio* means "alternating"), the substituents regularly alternate on both sides of the carbon chain. The substituents in an **atactic polymer** are randomly oriented.

25.3 STEREOCHEMISTRY OF POLYMERIZATION. ZIEGLER–NATTA CATALYSTS

isotactic configuration (same order)

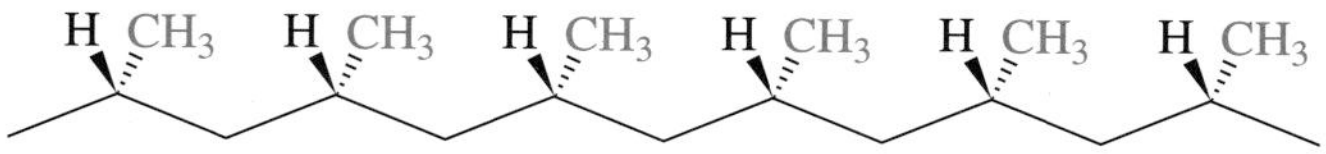

syndiotactic configuration (alternating order)

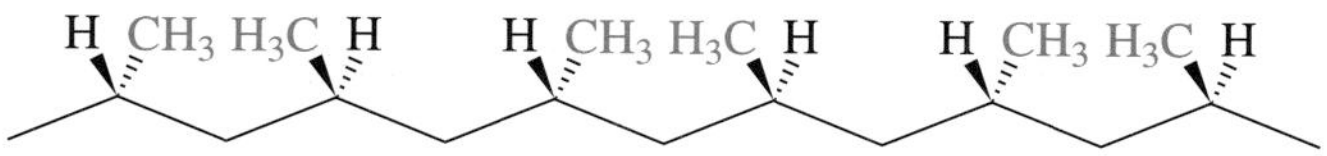

The configuration of the polymer affects its physical properties. Polymers in the isotactic or syndiotactic configuration are more likely to be crystalline solids, because positioning the substituents in a regular order allows for a more regular packing arrangement. Polymers in the atactic configuration are more disordered and cannot pack together as well. These polymers are softer and less rigid.

The structure of the polymer depends on the mechanism by which polymerization occurs. In general, radical polymerization leads primarily to branched polymers in the atactic configuration. Cationic polymerization produces polymers with a considerable fraction of the chains in the isotactic or syndiotactic configuration.

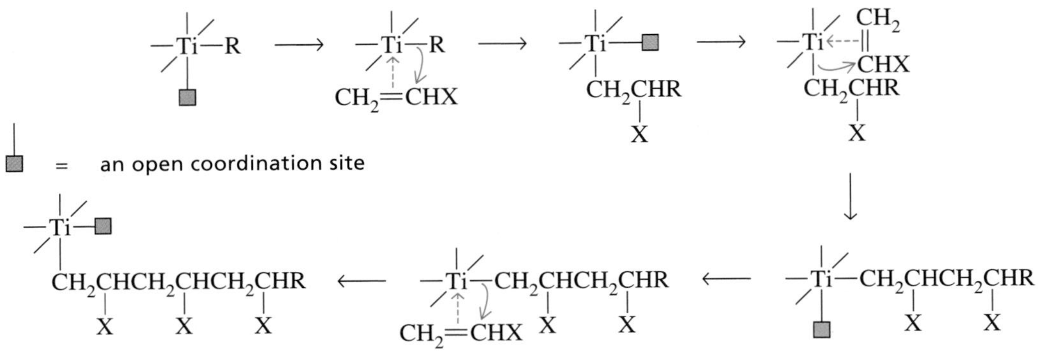

▲ **Figure 25.1**
The mechanism of the Ziegler–Natta-catalyzed polymerization of a substituted ethylene. The propagation steps involve insertion of a monomer into the titanium–carbon bond.

Karl Ziegler (1898–1973), *the son of a minister, was born in Germany. He was a professor at the University of Frankfurt and then at the University of Heidelberg.*

Giulio Natta (1903–1979) *was the son of an Italian judge. He was a professor at the Polytechnic Institute in Milan, where he became the director of the Industrial Chemistry Research Center. Ziegler and Natta did not work together, but each independently developed the catalyst system used in polymerization. They shared the 1963 Nobel Prize in chemistry.*

Anionic polymerization produces polymers with the most stereoregularity. The percentage of chains in the isotactic or syndiotactic configuration increases as the polymerization temperature decreases.

In 1953, Karl Ziegler and Giulio Natta found that the structure of a polymer could be controlled if the growing end of the chain and the incoming monomer were coordinated with an aluminum–titanium initiator. These initiators are now called **Ziegler–Natta catalysts.** Long, unbranched polymers with either the isotactic or the syndiotactic configuration can be prepared using Ziegler–Natta catalysts. Whether the chain is isotactic or syndiotactic depends on the particular Ziegler–Natta catalyst used. These catalysts revolutionized the field of polymer chemistry because they allow the synthesis of stronger and stiffer polymers that have greater resistance to cracking and heat. High-density polyethylene is prepared using a Ziegler–Natta process.

The mechanism of the Ziegler–Natta-catalyzed polymerization of a substituted ethylene is shown in Figure 25.1. Notice that the monomer forms a π complex with the metal at an open coordination site. After coordination, the alkene is inserted into the titanium–carbon bond.

Polyacetylene is another polymer prepared by a Ziegler–Natta process. The conjugated double bonds in polyacetylene cause it to be able to conduct electricity down its backbone after several electrons are removed from or added to the backbone; it is a **conducting polymer.**

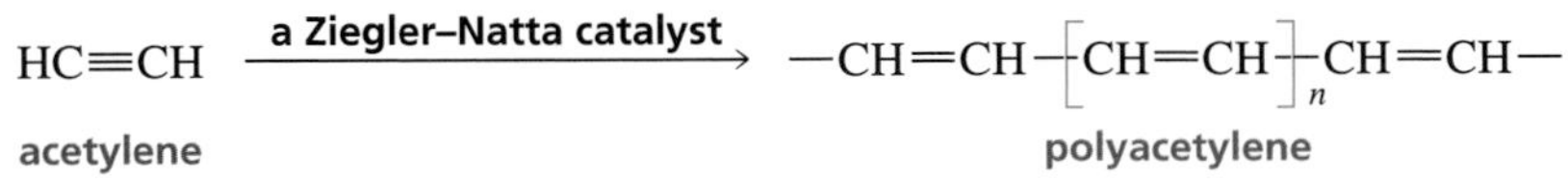

$$HC{\equiv}CH \xrightarrow{\text{a Ziegler–Natta catalyst}} -CH{=}CH{-}\!\!\left[CH{=}CH\right]_n\!\!-CH{=}CH-$$

acetylene   polyacetylene

25.4
POLYMERIZATION OF DIENES. THE MANUFACTURE OF RUBBER

When the bark of a rubber tree is cut, a sticky white liquid oozes out. This is the same liquid that is found inside the stalks of dandelions and milkweed. The sticky material is latex, a suspension of rubber particles in water. Natural rubber is a polymer of 2-methyl-1,3-butadiene (isoprene) (Section 23.6). On average, a molecule of rubber contains 5000 isoprene units. All the double bonds in rubber are cis.

Rubber is a waterproof material because it consists of a tangle of hydrocarbon chains that have no affinity for water.

isoprene units → *cis*-**poly(2-methyl-1,3-butadiene)**
natural rubber

Gutta-percha is a naturally occurring isomer of rubber in which all the double bonds are trans. Like rubber, gutta-percha is exuded by certain trees, but it is much less common. It is harder and more brittle than rubber. It is the filling material that dentists use in root canals.

Latex being collected from a rubber tree.

PROBLEM 13

Draw a short segment of gutta-percha.

By mimicking nature, scientists have learned to make synthetic rubber with properties tailored to meet human needs. These materials are called synthetic rubbers because they have some of the properties of natural rubber, including the fact that they are waterproof and elastic. But they have some improved properties as well: they are more flexible, tougher, and more durable than natural rubber.

Synthetic rubbers have been made by polymerizing dienes other than isoprene. One synthetic rubber is a polymer of 1,3-butadiene in which all the double bonds are cis. Polymerization is carried out in the presence of a Ziegler–Natta catalyst so that the configuration of the double bonds in the polymer can be controlled.

1,3-butadiene monomers → a Ziegler–Natta catalyst → *cis*-**poly(1,3-butadiene)**
a synthetic rubber

Neoprene is a synthetic rubber made by polymerizing 2-chloro-1,3-butadiene (chloroprene) in the presence of a Ziegler–Natta catalyst that causes all the double bonds in the polymer to have the trans configuration. Neoprene is used to make shoe soles, tires, hoses, and coated fabrics.

$$CH_2=CCH=CH_2$$
2-chloro-1,3-butadiene
chloroprene
a Ziegler–Natta catalyst → **neoprene**

A problem common to both natural and synthetic rubber is that the polymers are very soft and sticky. They can be hardened by a process known as *vulcanization.* Charles Goodyear discovered this process when, while trying to improve the properties of rubber, he accidentally spilled a mixture of rubber and sulfur on a hot

Charles Goodyear (1800–1860), the son of an inventor of farm implements, was born in Connecticut. He patented the process of vulcanization in 1844. However, because the process was so simple that it could be easily copied, he spent many years contesting infringements on his patent. In 1852, with Daniel Webster as his lawyer, he obtained the right to the patent.

Figure 25.2 ►
The rigidity of rubber is increased by cross-linking the polymer chains with disulfide bonds. When rubber is stretched, the randomly coiled chains straighten out and orient themselves along the direction of the stretch.

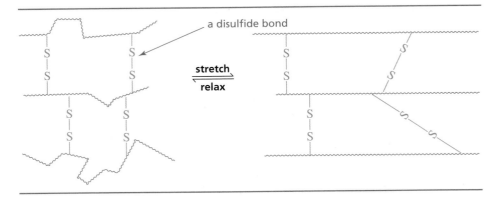

stove. To his surprise, the mixture became hard but flexible. He called the heating of rubber with sulfur **vulcanization,** after Vulcan, the Roman god of fire.

Heating rubber with sulfur causes **cross-linking** of separate polymer chains through disulfide bonds (Figure 25.2). Rather than the individual chains just being entangled together, the vulcanized chains are locked together in one giant molecule. Because the polymer has double bonds, the chains have bends and kinks that prevent them from forming a tightly packed crystalline polymer. When rubber is stretched, the chains straighten out along the direction of the pull. Cross-linking prevents the polymer from being torn when it is stretched, and the cross-links provide a reference framework for the material to return to when the stretching force is removed.

The physical properties of rubber can be controlled by regulating the amount of sulfur used in the vulcanization process. Rubber made with 1 to 3% sulfur is soft and stretchy and is used to make rubber bands. Rubber made with 3 to 10% sulfur is more rigid, and is used in the manufacture of tires. Goodyear's name can be found on many of the tires sold today. The story of rubber is an example of a scientist's taking a natural material and finding ways to improve its useful properties.

PROBLEM 14

The polymer formed from a diene such as 1,3-butadiene contains vinyl branches. Propose a mechanism to account for the formation of these branches.

$$CH_2=CHCH=CH_2 \longrightarrow -CH_2CH=CHCH_2CHCH_2CH_2CH=CHCH_2-$$
$$\overset{|}{C}H=CH_2$$

**25.5
COPOLYMERS**

The polymers we have discussed so far are formed from only one type of monomer and are called **homopolymers.** Often two (or more) different monomers are used to form a polymer. The resulting product is called a **copolymer.** Increasing the number of different monomers used to form the copolymer dramatically increases

TABLE 25.6   Some Examples of Copolymers and Their Uses

Monomers	Copolymer Name	Uses
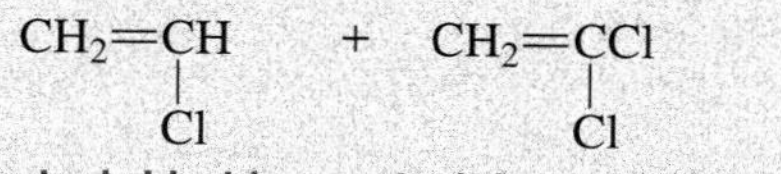 $CH_2{=}CH$ (Cl) vinyl chloride + $CH_2{=}CCl$ (Cl) vinylidene chloride	Saran	film for wrapping food
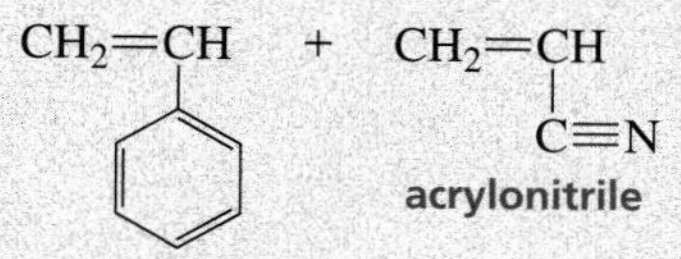 $CH_2{=}CH$ styrene + $CH_2{=}CH$ ($C{\equiv}N$) acrylonitrile	SAN	dishwasher-safe objects, vacuum cleaner parts
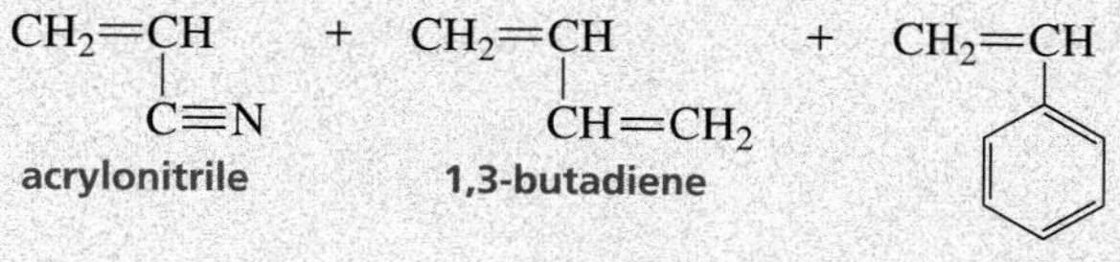 $CH_2{=}CH$ ($C{\equiv}N$) acrylonitrile + $CH_2{=}CH$ ($CH{=}CH_2$) 1,3-butadiene + $CH_2{=}CH$ styrene	ABS	bumpers, crash helmets, telephones, luggage
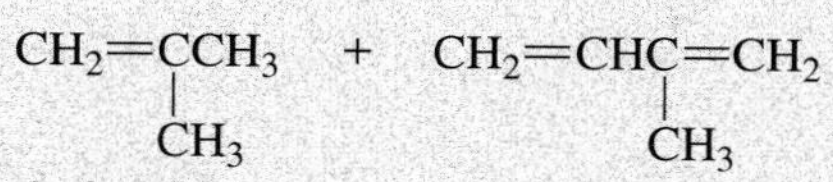 $CH_2{=}CCH_3$ (CH_3) isobutylene + $CH_2{=}CHC{=}CH_2$ (CH_3) isoprene	butyl rubber	inner tubes, balls, inflatable sporting goods

the number of different copolymers that can be formed. Even if only two kinds of monomers are used, copolymers with very different properties can be prepared by varying the amounts of each monomer. Both chain-growth polymers and step-growth polymers can be copolymers. Many of the synthetic polymers used today are copolymers. Table 25.6 shows some common copolymers and the monomers of which they are composed.

There are four types of copolymers. In an **alternating copolymer,** the two monomers alternate. In a **block copolymer,** there are blocks of each kind of monomer. In a **random copolymer,** the distribution of monomers is random. A **graft copolymer** contains branches derived from one monomer grafted onto a backbone derived from another monomer. These structural differences extend the range of physical properties available to the scientist designing the copolymer.

an alternating copolymer   ABABABABABABABABABABABABA

a block copolymer   AAAAABBBBBAAAAABBBBBAAA

a random copolymer   AABABABBBABAABBBABABBAAAB

a graft copolymer   AAAAAAAAAAAAAAAAAAAAAAAAA
B B B
B B B
B B B
B B B
B B B
B B B

25.6
STEP-GROWTH
POLYMERS

Step-growth polymers are formed by the intermolecular reaction of bifunctional molecules (molecules with two functional groups). When the functional groups react, in most cases a small molecule such as H_2O, alcohol, or HCl is lost. This is why these polymers are also known as *condensation polymers*. There are two types of step-growth polymers. One type is formed by the reaction of a single monomer that possesses two different functional groups, A and B. Functional group A of one monomer reacts with functional group B of another monomer. Nylon 6 is an example of this type of step-growth polymer. The carboxylic acid group of one monomer reacts with the amino group of another monomer, resulting in the formation of amide groups. Structurally, this reaction is similar to the polymerization of α-amino acids to form proteins (Section 20.9).

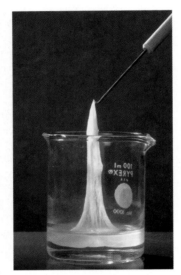

Nylon is pulled from a beaker of adipoyl chloride and hexamethylenediamine.

$$\overset{+}{H_3N}(CH_2)_5\overset{O}{\overset{\|}{C}}O^- \xrightarrow[-H_2O]{\Delta} -NH(CH_2)_5\overset{O}{\overset{\|}{C}}\left[NH(CH_2)_5\overset{O}{\overset{\|}{C}}\right]_n NH(CH_2)_5\overset{O}{\overset{\|}{C}}-$$

6-aminohexanoic acid   **nylon 6**
                                                           a polyamide

Nylon is the common name of a synthetic **polyamide.** This particular nylon is called nylon 6 because it is formed from the polymerization of a compound that contains six carbons, 6-aminohexanoic acid. The starting material for nylon 6 is ε-caprolactam. The lactam is opened by hydrolysis.

$$\text{ε-caprolactam} \xrightarrow[\Delta]{H^+, H_2O} \overset{+}{H_3N}CH_2CH_2CH_2CH_2CH_2\overset{O}{\overset{\|}{C}}OH$$

ε-caprolactam   **ε-aminocaproic acid**
                                                 6-aminohexanoic acid

The other type of step-growth polymer is formed by the reaction of two different bifunctional monomers. One monomer contains two A functional groups, and the other monomer contains two B functional groups. An example of this type of step-growth polymer is nylon 66, which is formed when adipic acid reacts with hexamethylenediamine. Nylon 66 is so named because it is a polyamide formed from a six-carbon diacid and a six-carbon diamine.

$$HO\overset{O}{\overset{\|}{C}}(CH_2)_4\overset{O}{\overset{\|}{C}}OH + H_2N(CH_2)_6NH_2 \xrightarrow[-H_2O]{\Delta} -\overset{O}{\overset{\|}{C}}(CH_2)_4\overset{O}{\overset{\|}{C}}\left[NH(CH_2)_6NH\overset{O}{\overset{\|}{C}}(CH_2)_4\overset{O}{\overset{\|}{C}}\right]_n NH(CH_2)_6NH-$$

adipic acid   **hexamethylenediamine**   **nylon 66**

Nylon first found wide use in textiles and carpets. Because it is resistant to stress, it is now used in many other applications. It is used for mountaineering ropes, for tire cords, for fishing line, and as a substitute for metal in bearings and gears. The extended applications of nylon precipitated a search for new "super fibers" with super strength and super heat resistance.

PROBLEM 15◆

a. Draw a short segment of nylon 4.

b. From what lactam is it synthesized?

c. Give the structure of a short segment of nylon 44.

PROBLEM 16

Give the equation that explains what will happen if a scientist working in the laboratory spills sulfuric acid on her nylon (nylon 66) stockings.

One new "super fiber" is Kevlar, a polymer of 1,4-benzenedicarboxylic acid and 1,4-diaminobenzene. Incorporation of ring structures into polymers has been found to result in polymers with great physical strength. Aromatic polyamides are known as **aramides.** Kevlar is an aramide with a tensile strength greater than that of steel. Army helmets are made of Kevlar. Kevlar is also used for lightweight bulletproof vests and high-performance skis. Because it is stable at very high temperatures, it is used in protective clothing worn by firefighters.

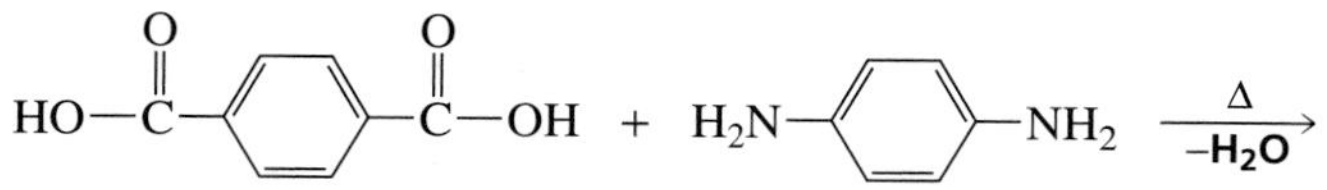

1,4-benzenedicarboxylic acid   **1,4-diaminobenzene**

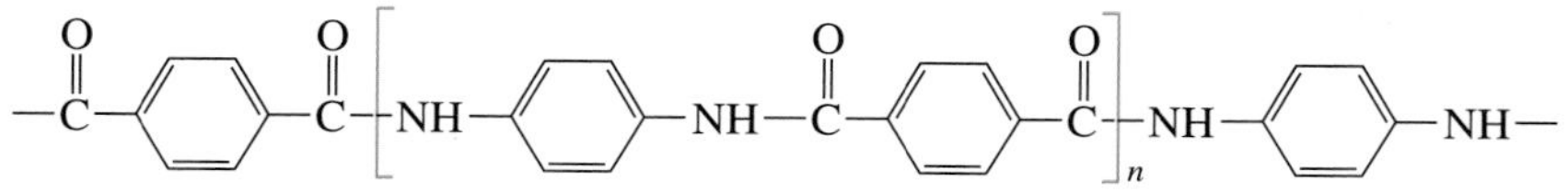

Kevlar
an aramide

Kevlar owes is strength to the way in which the individual polymer chains interact with each other. The chains are hydrogen bonded, which causes them to form a sheet structure.

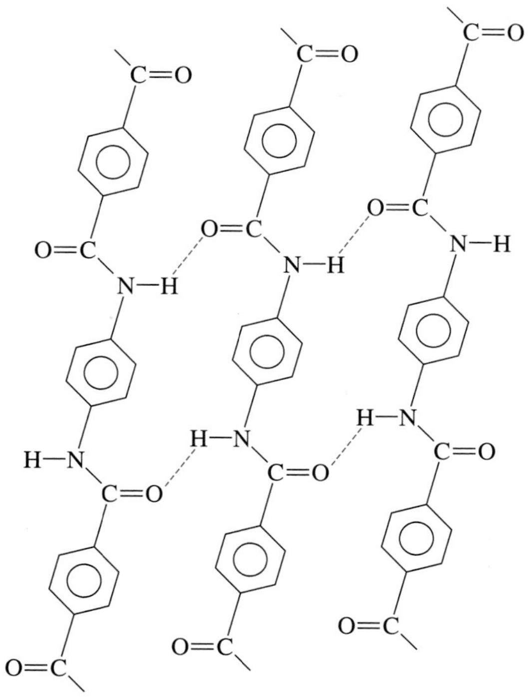

Nylon was first synthesized in 1931 by **Wallace Carothers (1896–1937).** *He was born in Iowa and received a Ph.D. from the University of Illinois. He taught there and at Harvard before being hired by Du Pont to head their program in basic science. The use of nylon by the public was delayed until after World War II, because all nylon produced during the war was used by the military. Carothers died unaware of the era of synthetic fibers that was to occur after the war.*

Polyesters are step-growth polymers in which the monomer units are joined together by ester groups. They have found wide commercial use as fibers, plastics, and coatings. The most common polyester is known by the trade name Dacron and is made by the transesterification of dimethyl terephthalate with ethylene glycol. High resilience, durability, and moisture resistance are the properties of this polymer that contribute to its "wash-and-wear" characteristics.

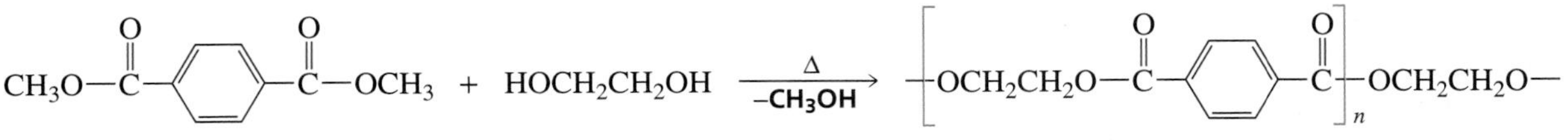

dimethyl terephthalate + ethylene glycol → poly(ethylene) terephthalate / Dacron / a polyester

Kodel polyester is formed by the transesterification of dimethyl terephthalate with 1,4-bis(hydroxymethyl)cyclohexane. The stiff polyester chain causes the fiber to have a harsh feel that can be overcome by blending it with wool or cotton.

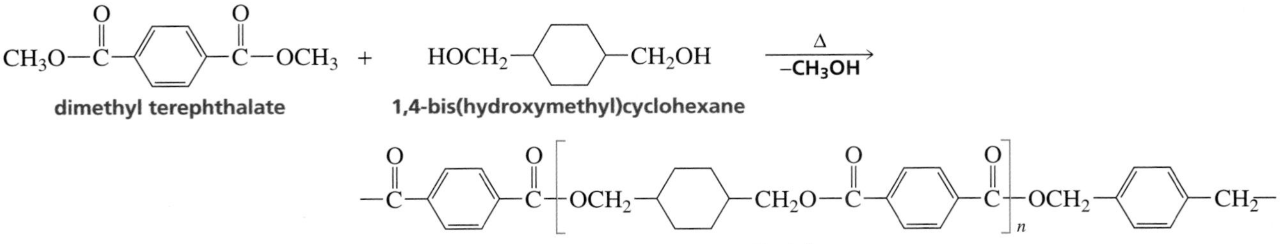

dimethyl terephthalate + 1,4-bis(hydroxymethyl)cyclohexane

Kodel

PROBLEM 17

What happens to polyester slacks if aqueous NaOH is spilled on them?

Polyesters with two ester groups bonded to the same carbon are known as **polycarbonates.** Lexan is produced by the reaction of phosgene with bisphenol A. Lexan is a strong and transparent polymer used for bulletproof windows and traffic-light lenses. In recent years, polycarbonates have become important polymers in the automobile industry as well as in the manufacture of compact discs.

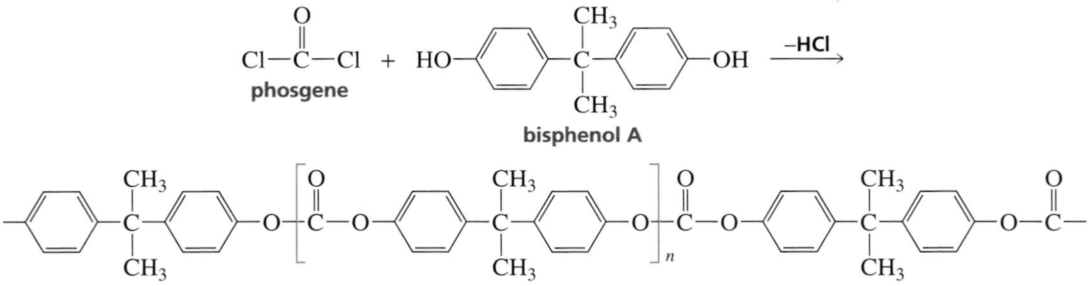

phosgene + bisphenol A

Lexan
a polycarbonate

Epoxy resins are the strongest adhesives known. They can adhere to almost any kind of surface, and they are resistant to solvents and to extremes of temperature. When an epoxy cement is used, a low-molecular-weight *prepolymer* (the most common is a polymer of bisphenol A and epichlorohydrin) is mixed with a *hard-ener*. The two compounds react, forming a cross-linked polymer.

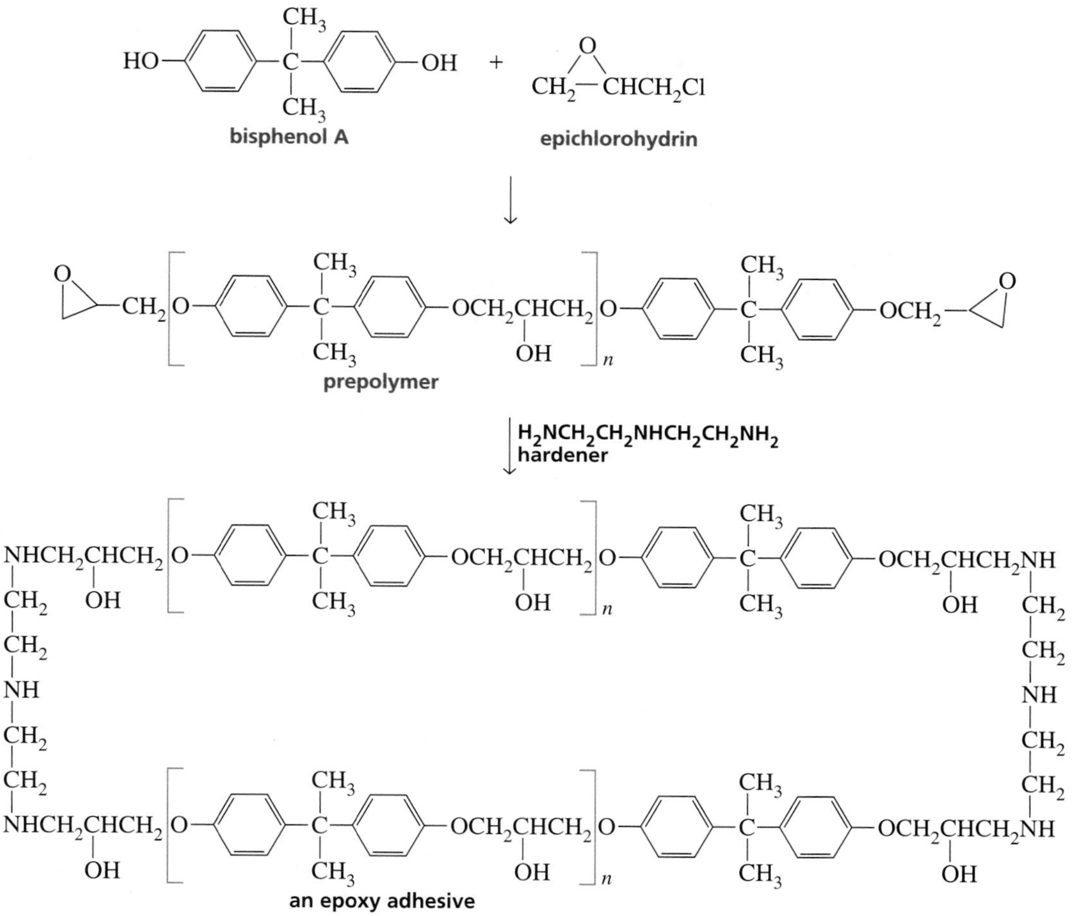

bisphenol A   epichlorohydrin

prepolymer

H$_2$NCH$_2$CH$_2$NHCH$_2$CH$_2$NH$_2$
hardener

an epoxy adhesive

PROBLEM 18

a. Propose a mechanism for the formation of the prepolymer shown above.

b. Propose a mechanism for the reaction of the prepolymer with the hardener.

A **urethane** is a compound that is both an ester and an amide. Urethanes can be prepared by treating an isocyanate with an alcohol.

$$RN{=}C{=}O \;+\; ROH \;\longrightarrow\; RNH{-}\overset{\displaystyle O}{\overset{\|}{C}}{-}OR$$

an isocyanate   an alcohol   a urethane

Figure 25.3 ▶
Progress of a step-growth
polymerization.

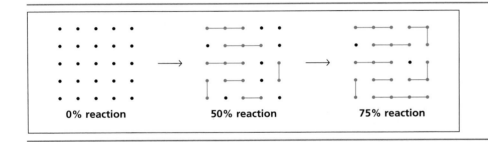

| 0% reaction | 50% reaction | 75% reaction |

Polyurethanes are polymers that contain urethane groups. One of the most common polyurethanes is prepared by the polymerization of toluene-2,6-diisocyanate and ethylene glycol. If the reaction is carried out in the presence of a blowing agent, the product is a polyurethane foam. Blowing agents are gases such as nitrogen or carbon dioxide, or low-boiling liquids such as chlorofluorocarbons (Section 8.9) that vaporize on heating. Polyurethane foams are used for furniture stuffing, carpet backings, and insulation. Notice that polyurethanes prepared from diisocyanates and diols are the only step-growth polymers that we have seen in which a small molecule is not lost during polymerization.

toluene-2,6-diisocyanate   + HOCH$_2$CH$_2$OH ⟶
ethylene glycol

a polyurethane

One of the most important uses of polyurethanes is in fabrics with elastic properties, such as spandex (Lycra). These materials are block copolymers in which some of the polymer segments are polyurethanes, some are polyesters, and some are polyamides. The blocks of polyurethane are soft, amorphous segments that become crystalline on stretching. When the tension is released, they revert to the amorphous state.

The formation of step-growth polymers, unlike the formation of chain-growth polymers, does not involve chain reactions. Any two monomers (or short chains) can react. The progress of a typical step-growth polymerization is shown schematically in Figure 25.3. When the reaction is 50% complete (10 bonds have formed between 20 monomers), the reaction products are primarily dimers and trimers. Even at 75% completion, no long chains have been formed. This means that if step-growth polymerization is to lead to long-chain polymers, very high yields must be achieved.

**25.7
PHYSICAL
PROPERTIES
OF POLYMERS**

The individual chains of a polymer such as polyethylene are held together by van der Waals forces. Because these forces operate only at close distances, they are strongest if the polymer chains can line up in an ordered, closely packed array. The regions of the polymer in which the chains are highly ordered with respect to one

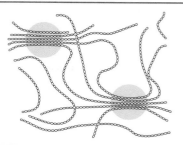

◀ **Figure 25.4**
The circled regions are crystallites, regions where the polymer chains are highly ordered, similar to the ordering found in crystals. Between the circles, the polymer chains are randomly oriented, and these regions are noncrystalline.

another are called **crystallites** (Figure 25.4). Between the crystallites are amorphous, noncrystalline regions in which the chains are randomly oriented. The crystallinity of the polymer (extent of ordering) can be increased by stretching as the molten polymer is drawn through small holes. The more crystalline the polymer is, the denser, harder, and more resistant to heat it is (Table 25.7). If the polymer chains possess substituents [as does poly(methyl methacrylate), for example] or have branches that prevent them from packing closely together, the density of the polymer will be reduced.

TABLE 25.7 Properties of Polyethylene as a Function of Crystallinity					
Crystallinity (%)	55	62	70	77	85
Density (gm/cm^3)	0.92	0.93	0.94	0.95	0.96
Melting point (°C)	109	116	125	130	133

Plastics can be classified according to the physical properties imparted to them by the way in which their individual chains are arranged. **Thermoplastic polymers** are polymers that have both ordered crystalline regions and amorphous, noncrystalline regions. Such polymers are hard at room temperature, but when they are heated, the individual chains can slip past one another and the polymer becomes soft enough to be molded. Thermoplastic polymers are the plastics we encounter most often in our daily lives—in combs, toys, light switch plates, and telephone casings, for example. They are the plastics that are relatively easily cracked.

Very strong and rigid materials can be obtained if the polymer chains are cross-linked. The greater the degree of cross-linking that exists, the more rigid the polymer. Such cross-linked polymers are known as **thermosetting polymers.** After they are hardened, they cannot be remelted by heating. Cross-linking reduces the mobility of the polymer chains, causing them to be relatively brittle materials. Because thermosetting polymers do not have the wide range of properties characteristic of thermoplastic polymers, they are less widely used.

Melmac, a highly cross-linked thermosetting polymer of melamine and formaldehyde, is a hard and moisture-resistant material. Because it is a colorless polymer, it can be made into materials with pastel colors. It is used to make lightweight dishes and counter surfaces.

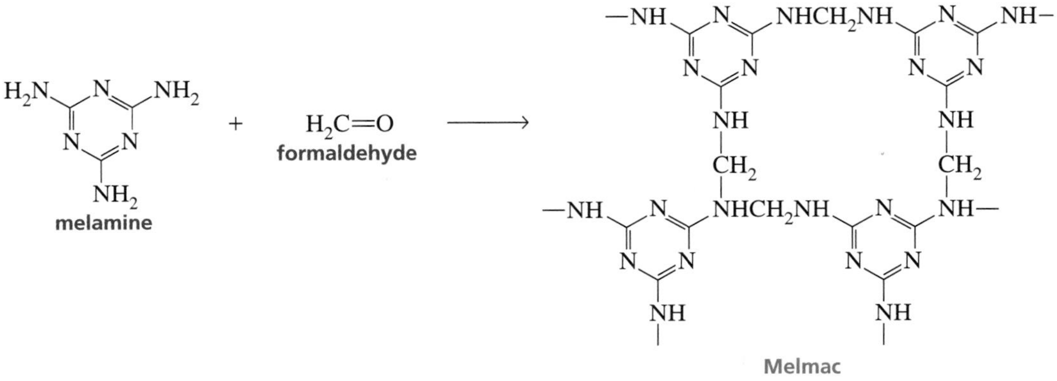

melamine + formaldehyde → Melmac

DESIGNING A POLYMER

A polymer that is used for making dental impressions must be soft enough to mold around the teeth, but later must become hard enough to maintain a fixed shape. The polymer commonly used for dental impressions contains three-membered aziridine rings which react to cross-link the chains. Because arizidine rings are not very reactive (Section 26.1), cross-linking occurs relatively slowly, so most of the hardening of the polymer does not occur until the polymer is removed from the patient's mouth.

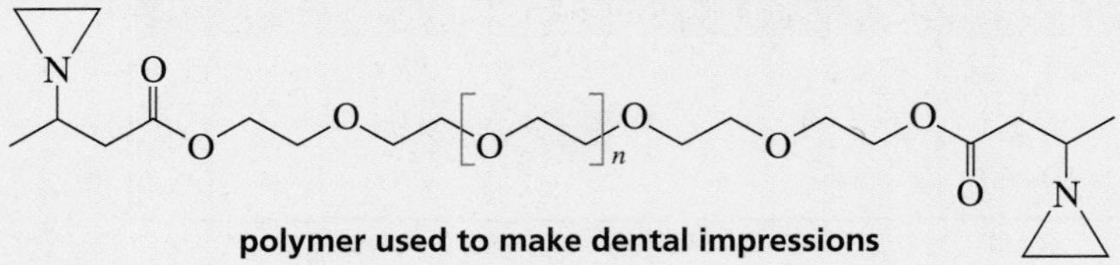

polymer used to make dental impressions

Leo Hendrik Baekeland (1863–1944) *discovered Bakelite while looking for a substitute for shellac in his home laboratory. He was born in Belgium and became a professor of chemistry at the University of Bruges. A fellowship brought him to the United States in 1889, and he decided to stay. His hobby was photography, and he invented photographic paper that could be developed under artificial light, which he sold to Eastman-Kodak.*

PROBLEM 19

Propose a mechanism for the formation of Melmac.

PROBLEM 20

Bakelite was the first of the thermosetting polymers. It is a highly cross-linked polymer formed from the acid-catalyzed polymerization of phenol and formaldehyde. It is a much darker polymer than Melmac; therefore, the color range of the products made from it is limited. Propose a structure for Bakelite.

A plasticizer can be added to a polymer to make it more flexible. A **plasticizer** is an organic compound that dissolves in the polymer and allows the polymer chains to slide past one another. Dibutyl phthalate is a commonly used plasticizer. It is added to poly(vinyl chloride)—normally a brittle polymer—to make such products as vinyl raincoats, shower curtains, and garden hoses.

$$\underset{\substack{\text{dibutyl phthalate}\\\text{a plasticizer}}}{}$$

O
‖
C—OCH₂CH₂CH₂CH₃

C—OCH₂CH₂CH₂CH₃
‖
O

dibutyl phthalate
a plasticizer

An important criterion to consider in choosing a plasticizer is its permanence. Permanence refers to how well the plasticizer remains in the polymer. The "new car smell" appreciated by car owners is the odor of the plasticizer that has vaporized from the vinyl upholstery. When a significant amount of the plasticizer has evaporated, the upholstery becomes brittle and cracks. Phthalates with higher molecular weights and lower vapor pressures than those of dibutyl phthalate are now commonly used for car interiors.

An **elastomer** is a plastic that stretches and then reverts to its original shape. It is a randomly oriented amorphous polymer. It must have some cross-linking so the chains do not slip over one another. When elastomers are stretched, the random chains stretch out, but there are insufficient van der Waals forces to maintain them in that configuration. When the stretching force is removed, they go back to their random shapes. Rubber is an elastomer.

Polymers that are stronger than steel or that conduct electricity almost as well as copper can be made by taking the polymer chains obtained by conventional polymerization, stretching them out, and putting them back together in a parallel fashion (Figure 25.5). Such polymers are called **oriented polymers.** Converting conventional polymers into oriented polymers has been compared to "uncooking" spaghetti. The conventional polymer is disordered cooked spaghetti, while the oriented polymer is ordered raw spaghetti.

Dyneema, the strongest commercially available fabric, is an oriented polyethylene polymer. Its molecular weight is 100 times greater than that of high-density polyethylene. It is lighter than Kevlar and at least 40% stronger. A rope made of Dyneema can lift almost 119,000 pounds, while a steel rope of similar size fails before the weight reaches 13,000 pounds. That a chain of carbon atoms can be stretched and properly oriented to produce a material stronger than steel is somewhat astounding. Dyneema is being used to make full-face crash helmets, protective fencing suits, and hang gliders.

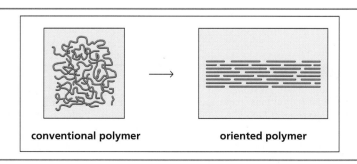

conventional polymer   oriented polymer

◀ **Figure 25.5**
The creation of an oriented polymer.

PROBLEM 21

If a small amount of glycerol is added to the reaction mixture of toluene-2,6-diisocyanate and ethylene glycol during the synthesis of polyurethane foam, a much stiffer foam is obtained. Explain.

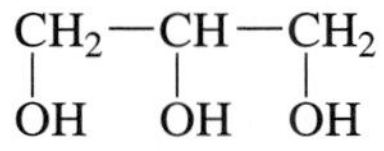

$$CH_2-CH-CH_2$$
$$\quad|\qquad|\qquad|$$
$$OH\quad OH\quad OH$$

glycerol

25.8 BIODEGRADABLE POLYMERS

Biodegradable polymers are polymers that can be broken into small segments by enzyme-catalyzed reactions, using enzymes produced by microorganisms. Because the carbon–carbon bonds of chain-growth polymers are inert to enzyme-catalyzed reactions, which makes them nonbiodegradable, bonds that can be broken by enzymes are inserted into the polymer. Therefore, when they are buried as waste, enzymes present in the ground can degrade the polymer—it is now a biodegradable polymer. One method involves inserting hydrolyzable ester groups into the polymer. For example, if the acetal shown below is added to an alkene undergoing radical polymerization, ester groups will be inserted into the polymer. These "weak links" are susceptible to enzyme-catalyzed hydrolysis.

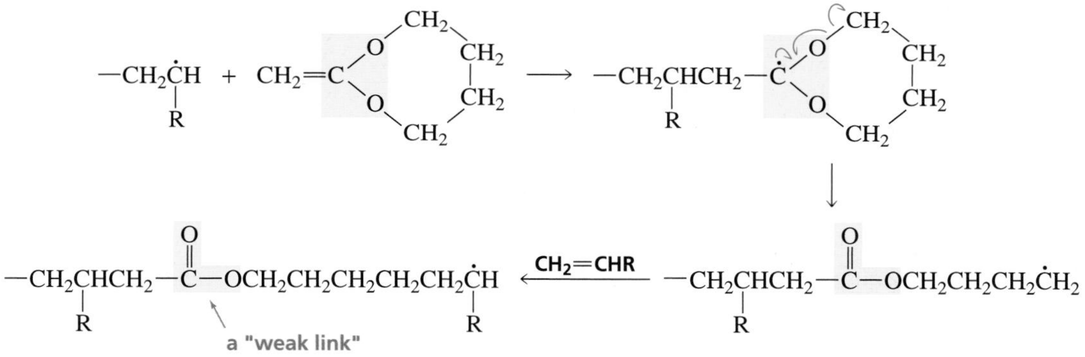

Polymer chemistry has evolved into a multibillion-dollar industry. More than 2.5×10^{13} kilograms of synthetic polymers are produced in the United States each year, and we can expect many more new materials to be developed by scientists in the years to come.

KEY TERMS

addition polymer (page 1111)
alpha olefin (page 1112)
alternating copolymer (page 1127)
anionic polymerization (page 1112)
aramide (page 1129)
atactic polymer (page 1123)

biodegradable polymer (page 1136)
biopolymer (page 1111)
block copolymer (page 1127)
cationic polymerization (page 1112)
chain-growth polymer (page 1111)
chain transfer (page 1114)

condensation polymer (page 1112)
conducting polymer (page 1124)
copolymer (page 1126)
cross-linking (page 1126)
crystallites (page 1133)
elastomer (page 1135)

PROBLEMS

22. Draw short segments of the polymers obtained from the following monomers. Indicate whether the polymerization is a chain-growth or a step-growth polymerization.

a. $CH_2{=}CHF$

b. $CH_2{=}CHCO_2H$

c. $HO(CH_2)_5\overset{\overset{\displaystyle O}{\|}}{C}OH$

d. $Cl\overset{\overset{\displaystyle O}{\|}}{C}(CH_2)_5\overset{\overset{\displaystyle O}{\|}}{C}Cl \;+\; H_2N(CH_2)_5NH_2$

e. 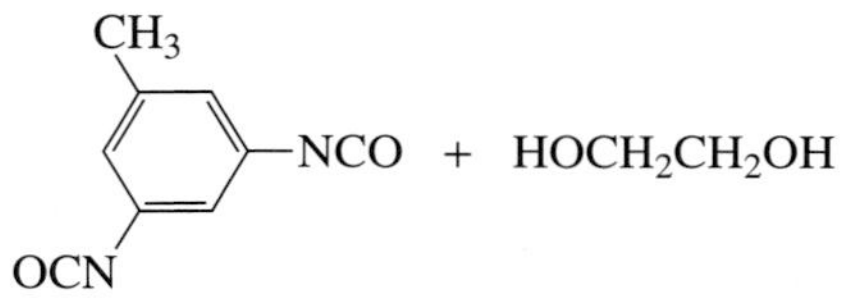 (aromatic ring with CH_3, $-NCO$, and OCN substituents) $+\; HOCH_2CH_2OH$

23. Draw the repeating unit of the step-growth polymer that will be formed from each of the following reactions.

a. 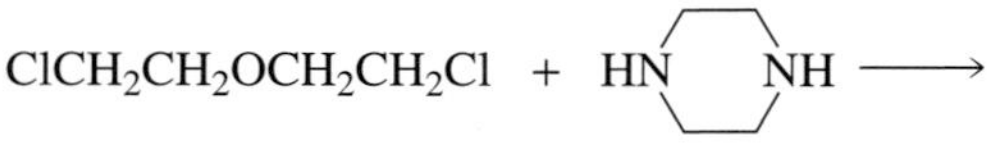 $ClCH_2CH_2OCH_2CH_2Cl \;+\; HN{\Large\bigcirc}NH \longrightarrow$

b. 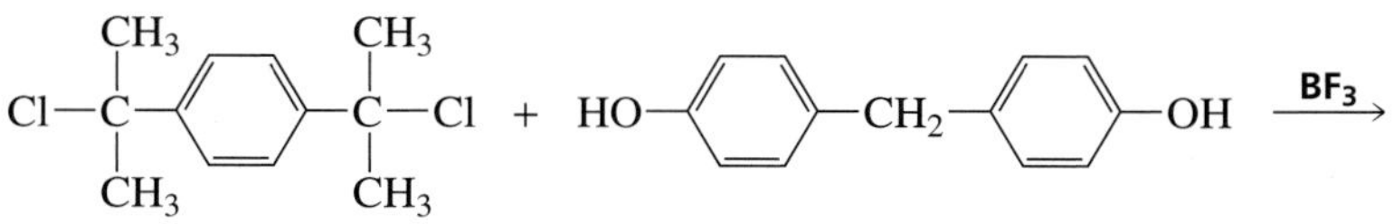 $Cl-\underset{\underset{\displaystyle CH_3}{|}}{\overset{\overset{\displaystyle CH_3}{|}}{C}}-{\large\bigcirc}-\underset{\underset{\displaystyle CH_3}{|}}{\overset{\overset{\displaystyle CH_3}{|}}{C}}-Cl \;+\; HO-{\large\bigcirc}-CH_2-{\large\bigcirc}-OH \xrightarrow{BF_3}$

c. $H_2N-{\large\bigcirc}-OCH_2CH_2CH_2O-{\large\bigcirc}-NH_2 \;+\; H\overset{\overset{\displaystyle O}{\|}}{C}-\overset{\overset{\displaystyle O}{\|}}{C}H \longrightarrow$

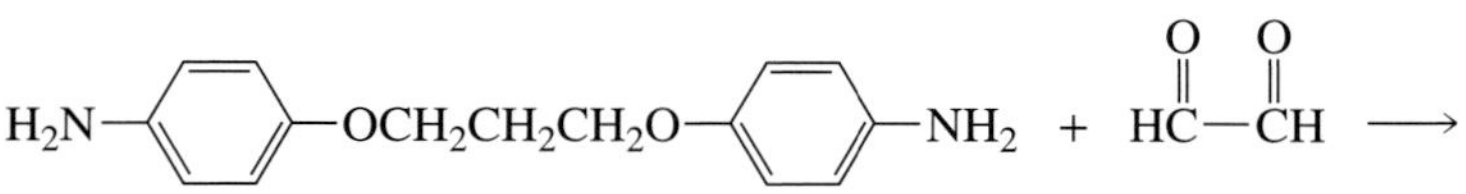

d. 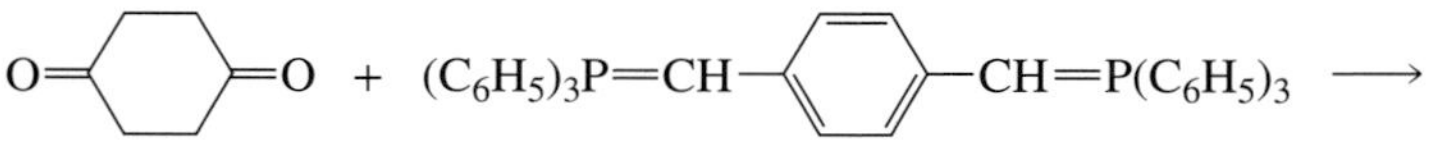 $O{=}{\large\bigcirc}{=}O \;+\; (C_6H_5)_3P{=}CH-{\large\bigcirc}-CH{=}P(C_6H_5)_3 \longrightarrow$

24. Give the structure of the monomer or monomers used to synthesize the following polymers. Indicate whether each polymer is a chain-growth polymer or a step-growth polymer.

a. $-CH_2\underset{\underset{\displaystyle CH_2CH_3}{|}}{CH}-$

b. $-CH_2\underset{\underset{\displaystyle CH_3}{|}}{CHO}-$

c. 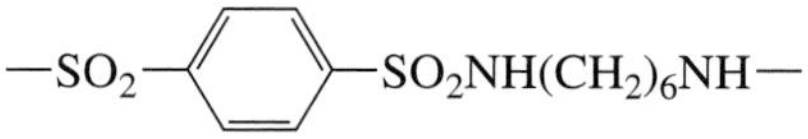 $-SO_2-\!\!\!\!\bigcirc\!\!\!\!-SO_2NH(CH_2)_6NH-$

d. 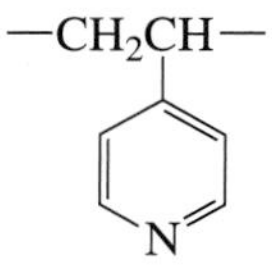 $-CH_2CH-$

(pyridine ring with N)

e. $-CH_2\overset{\underset{|}{CH_3}}{C}=CHCH_2-$

f. $-CH_2CH_2CH_2CH_2\overset{\overset{O}{\parallel}}{C}O-$

g. 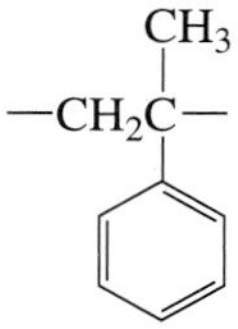 $-CH_2\overset{\underset{|}{CH_3}}{\underset{|}{C}}-$

(with phenyl group)

h. 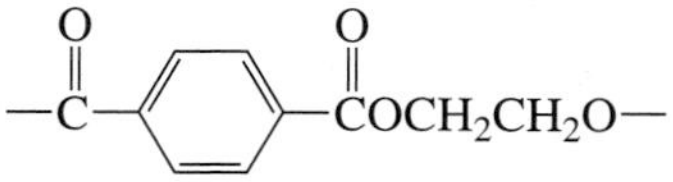 $-\overset{\overset{O}{\parallel}}{C}-\!\!\!\!\bigcirc\!\!\!\!-\overset{\overset{O}{\parallel}}{C}OCH_2CH_2O-$

25. The configuration of a polymer of isobutylene is not isotactic, syndiotactic, or atactic. Explain.

26. Draw short segments of the polymers obtained from the following compounds under the given reaction conditions.

a. 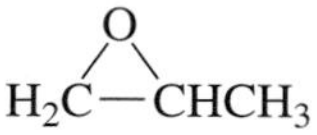 $H_2\overset{\overset{O}{\triangle}}{C}-CHCH_3 \xrightarrow{CH_3O^-}$

b. $CH_2=CHCl \xrightarrow{CH_3CH_2CH_2CH_2Li}$

c. 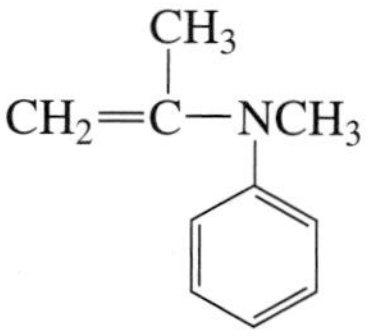 $CH_2=\overset{\underset{|}{CH_3}}{C}-NCH_3 \xrightarrow{\text{peroxide}}$

(with phenyl group)

d. 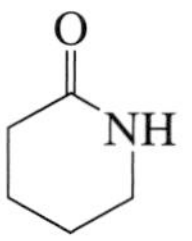 (ring structure with NH and O) $\xrightarrow[\Delta]{H^+, H_2O}$

e. $CH_2=\overset{\underset{|}{CH_3}}{C}-\overset{\underset{|}{CH_3}}{C}=CH_2 \xrightarrow{\text{a Ziegler–Natta catalyst}}$

f. $CH_2=CHCH_2CH_2CH_3 \xrightarrow{BF_3}$

27. Quiana is a synthetic fabric that feels very much like silk.
  a. What monomers are used to synthesize Quiana?
  b. Is Quiana a nylon or a polyester?

$$-NH-\langle\bigcirc\rangle-CH_2-\langle\bigcirc\rangle-NH-\overset{\overset{\displaystyle O}{\|}}{C}-(CH_2)_6-\overset{\overset{\displaystyle O}{\|}}{C}-NH-\langle\bigcirc\rangle-CH_2-\langle\bigcirc\rangle-NH-$$

Quiana

28. When 3,3-dimethyl-1-butene undergoes cationic polymerization, a random copolymer is obtained. Explain.

$$CH_2=CH-\underset{\underset{\displaystyle CH_3}{|}}{\overset{\overset{\displaystyle CH_3}{|}}{C}}-CH_3 \longrightarrow -CH_2-CH-CH_2-CH-\overset{\overset{\displaystyle CH_3}{|}}{\underset{\underset{\displaystyle CH_3}{|}}{C}}-CH_2-CH-\overset{\overset{\displaystyle CH_3}{|}}{\underset{\underset{\displaystyle CH_3}{|}}{C}}-CH_2-CH-$$

(with substituents CH₃CCH₃, CH₃, CH₃, CH₃ CH₃, CH₃CCH₃)

29. Polly Propylene starts two polymerization reactions. One flask contains a monomer that polymerizes by a chain-growth mechanism, and the other flask contains a monomer that polymerizes by a step-growth mechanism. When the reactions are terminated and the contents of the flasks analyzed, one flask contains a high-molecular-weight polymer and some monomer but very little material of intermediate molecular weight. The other flask contains mainly material of intermediate molecular weight and very little monomer or high-molecular-weight material. Which flask is which? Explain.

30. Poly(vinyl alcohol) is a polymer used to make fibers and adhesives. It is synthesized by hydrolysis or alcoholysis of the polymer obtained from polymerization of vinyl acetate.
   a. Why is poly(vinyl alcohol) not prepared by polymerizing vinyl alcohol?
   b. Is poly(vinyl acetate) a polyester?

$$-CH_2-CH-CH_2-CH-CH_2-CH-\ \xrightarrow[\Delta]{\ CH_3OH\ }\ -CH_2-CH-CH_2-CH-CH_2-CH-$$

(left substituents: OCCH₃ with =O, three times — **poly(vinyl acetate)**; right substituents: OH, OH, OH — **poly(vinyl alcohol)**)

31. Five different repeating units are found in the polymer obtained by cationic polymerization of 4-methyl-1-pentene. Identify these repeating units.

32. If a peroxide is added to styrene, the polymer known as polystyrene is formed. If a small amount of 1,4-divinylbenzene is added to the reaction mixture, a stronger and more rigid polymer is formed. Give the structure of a short section of this more rigid polymer.

$$CH_2=CH-\langle\bigcirc\rangle-CH=CH_2$$

1,4-divinylbenzene

33. A particularly strong and rigid polyester used for electronic parts is marketed under the trade name Glyptal. Glyptal is a polymer of terephthalic acid and glycerol. Draw a segment of the polymer and explain why it is so strong.

34. Give the structure of the polymer obtained from anionic polymerization of β-propiolactone.

35. Which monomer would give a greater yield of polymer, 5-hydroxypentanoic acid or 6-hydroxyhexanoic acid?

36. When rubber balls and other objects made of rubber are exposed to the air for long periods of time, they turn brittle and crack. This does not happen to objects made of polyethylene. Explain.

37. Vinyl raincoats become brittle as they get old even if they are not exposed to air or to any pollutants. Explain.

38. The polymer shown below is synthesized by hydroxide-ion-promoted hydrolysis of an alternating copolymer of *para*-nitrophenyl methacrylate and acrylic acid.
   a. Propose a mechanism for the formation of the alternating copolymer.
   b. Explain why hydrolysis of the copolymer to form the polymer occurs much more rapidly than hydrolysis of *para*-nitrophenyl acetate.

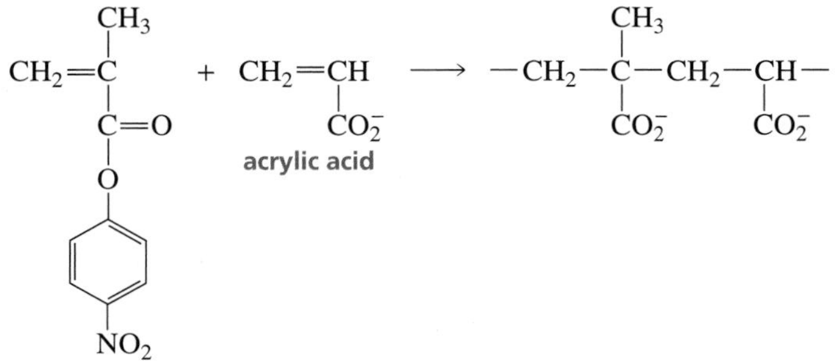

para-nitrophenyl
methacrylate

acrylic acid

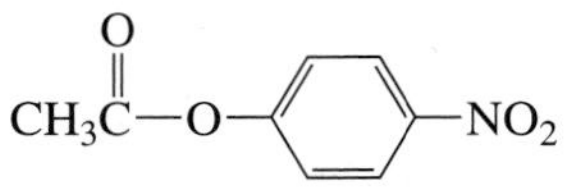

para-nitrophenyl acetate

39. An alternating copolymer of styrene and vinyl acetate can be turned into a graft copolymer by hydrolyzing it and then adding ethylene oxide. Draw the structure of the graft copolymer.

40. How could "head-to-head" poly(vinyl bromide) be synthesized?

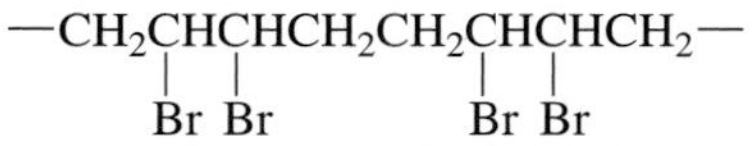

$$-CH_2CHCHCH_2CH_2CHCHCH_2-$$
$$\quad\quad |\;\; |\quad\quad\quad\; |\;\; |$$
$$\quad\quad Br\; Br\quad\quad\; Br\; Br$$

"head-to-head" poly(vinyl bromide)

HETEROCYCLIC COMPOUNDS

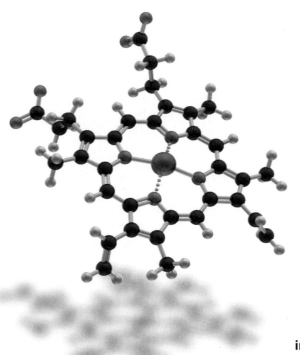

**iron protoporphyrin IX
heme**

Heterocyclic compounds, or heterocycles, are cyclic compounds in which one or more of the atoms of the ring are hetero atoms. A **hetero atom** is an atom other than carbon. The name comes from the Greek word *heteros,* which means "different." A variety of atoms, such as N, O, S, Se, P, Si, B, and As, can be incorporated into ring structures; in this chapter, however, we will consider only the most prevalent heterocyclic compounds—the ones that contain the hetero atoms N, O, and S.

Heterocycles make up an exceedingly important class of compounds—more than half of all organic compounds are heterocycles. Almost all the compounds we know as drugs, most vitamins (Chapter 22), and many other natural products are heterocycles.

A **natural product** is a compound synthesized in nature. **Alkaloids** are natural products containing one or more nitrogen hetero atoms that are found in the leaves, bark, roots, or seeds of plants. Examples include caffeine (found in tea leaves, coffee beans, and cola nuts) and nicotine (found in tobacco leaves). Morphine is an alkaloid obtained from opium, the juice derived from a species of poppy. Morphine is 50 times stronger than aspirin as an analgesic, but it is addictive and suppresses respiration. Heroin is a synthetic compound that is made by acetylating morphine.

Other examples of heterocycles include Valium, a synthetic tranquilizer, and serotonin, a neurotransmitter. Serotonin is responsible for, among other things, the feeling of having had enough to eat. When food is ingested, brain neurons are signaled to release serotonin. A widely used diet drug (actually a combination of two drugs, phentermine and fenfluramine) popularly known as fen/phen works by causing brain neurons to release extra serotonin (Chapter 14, page 602). After finding that 30% of those who took fenfluramine had abnormal echocardiograms because of heart valve problems, the Food and Drug Administration asked the manufacturer of the diet drugs to withdraw their products. There is also some evidence that faulty metabolism of serotonin plays a role in manic depressive psychoses.

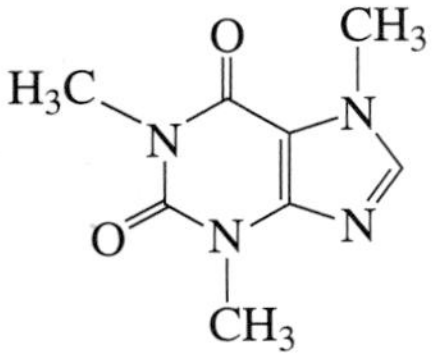

caffeine 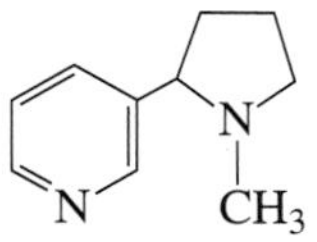   **nicotine** 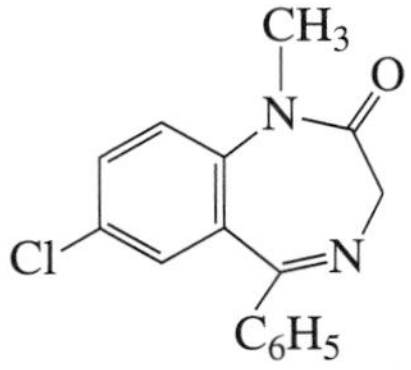   **Valium** 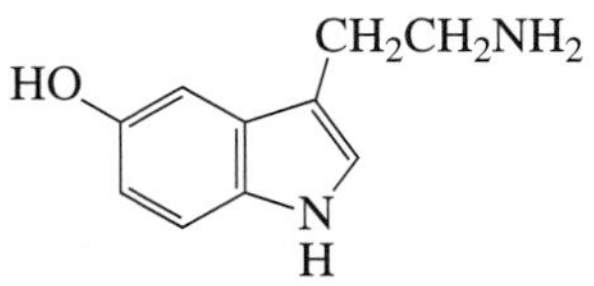   **serotonin**

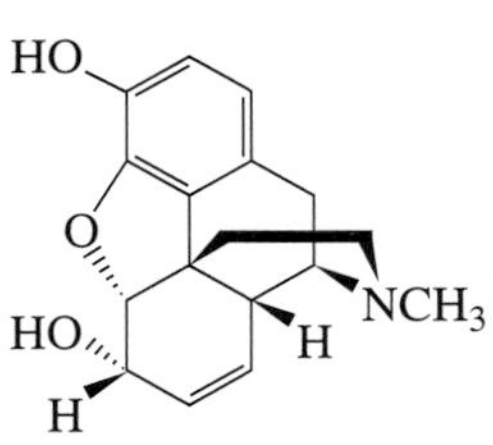

morphine 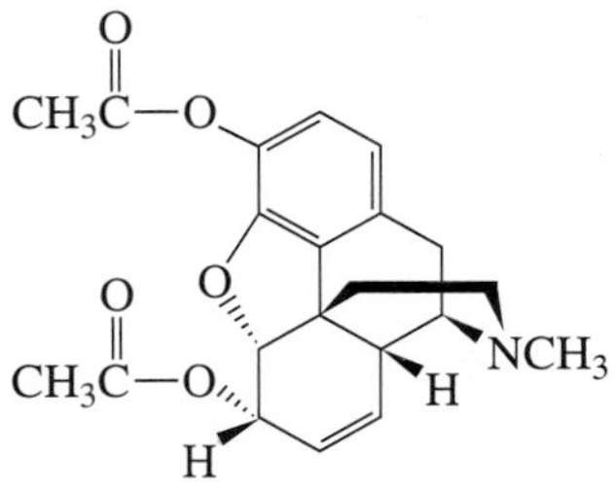   **heroin**

26.1
SATURATED
HETEROCYCLES

Saturated heterocycles do not contain double bonds. A saturated heterocycle is easily named as a cycloalkane using a prefix to denote the hetero atom ("aza" for nitrogen, "oxa" for oxygen, "thia" for sulphur). There are, however, other acceptable names; some of the more commonly used names are shown here.

oxacyclopropane **azacyclopropane** **thiacyclopropane** **oxacyclobutane** **thiacyclobutane** **oxacyclopentane**
oxirane **aziridine** **thiirane**
ethylene oxide ethyleneimine tetrahydrofuran

Heterocyclic rings are numbered so that the hetero atoms have the lowest possible numbers.

3-methylazacyclopentane **2-methylazacyclohexane** **N-ethylazacyclopentane** **1,3-dithiacyclohexane**
3-methylpyrrolidine 2-methylpiperidine N-ethylpyrrolidine 1,3-dithiane

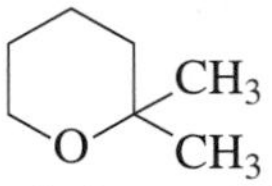

2,2-dimethyloxacyclohexane
2,2-dimethyltetrahydropyran

PROBLEM 1◆

Name the following compounds.

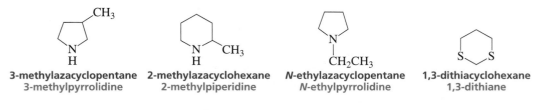

a. 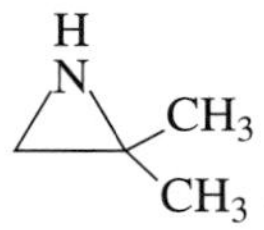   b. 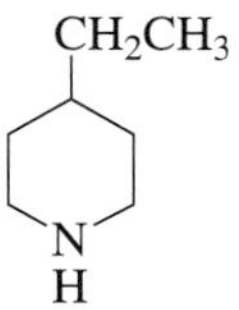   c.

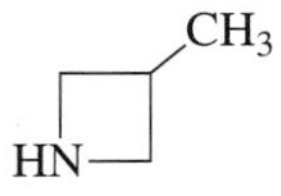

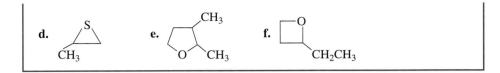

d.   e.   f.

Saturated heterocycles containing five or more atoms have physical and chemical properties typical of acyclic compounds that contain the same hetero atom. For example, tetrahydrofuran, tetrahydropyran, and 1,4-dioxane are typical ethers. Since ethers have little chemical reactivity, these compounds are commonly used as solvents (Section 11.5).

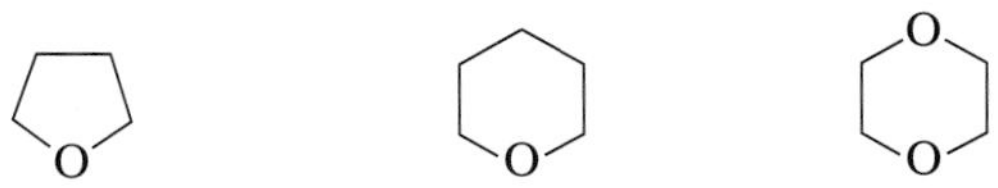

tetrahydrofuran   **tetrahydropyran**   **1,4-dioxane**

Pyrrolidine, piperidine, and morpholine are typical secondary amines, while *N*-methylpyrrolidine and quinuclidine are typical tertiary amines. The conjugate acids of these amines have pK_a values expected for ammonium ions.

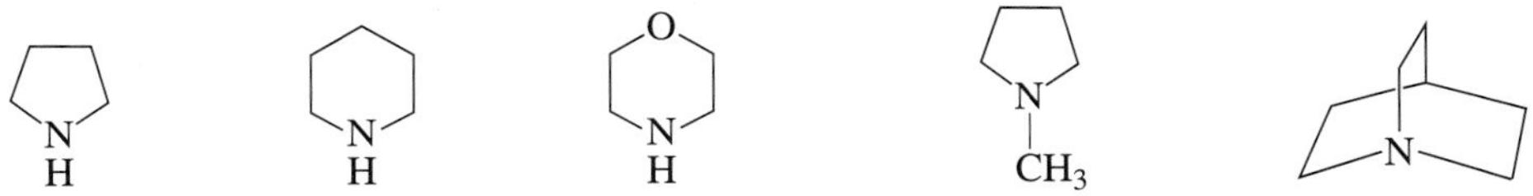

pyrrolidine   **piperidine**   **morpholine**   ***N*-methylpyrrolidine**   **quinuclidine**
pK_a = 11.27   pK_a = 11.12   pK_a = 9.28   pK_a = 10.32   pK_a = 11.38

the pK_a values are for the conjugate acids of the structures shown

Because of its three-membered ring, the aziridinium ion has a considerably lower pK_a than that of a typical secondary ammonium ion. The internal bond angle is smaller than usual, causing the external bond angles to be somewhat larger than usual. The larger external bond angles cause the orbitals nitrogen uses when it overlaps with the hydrogen to have more *s* character than those of a typical sp^3 hybridized nitrogen (the larger the bond angle, the more *s* character; Section 1.14). This makes nitrogen more electronegative, which lowers the pK_a.

external bond angles
are greater than normal

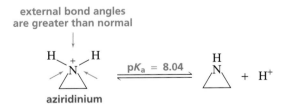

pK_a = 8.04   + H$^+$

aziridinium

PROBLEM 2

Why is the pK_a of the conjugate acid of morpholine significantly lower than the pK_a of the conjugate acid of piperidine?

PROBLEM 3 ◆

a. Draw the structure of 3-quinuclidinone.

b. What is the approximate pK_a of its conjugate acid?

c. Which has a lower pK_a, the conjugate acid of 3-bromoquinuclidine or the conjugate acid of 3-chloroquinuclidine?

Compounds with saturated five- and six-membered rings undergo the same reactions as their open-chain analogs undergo.

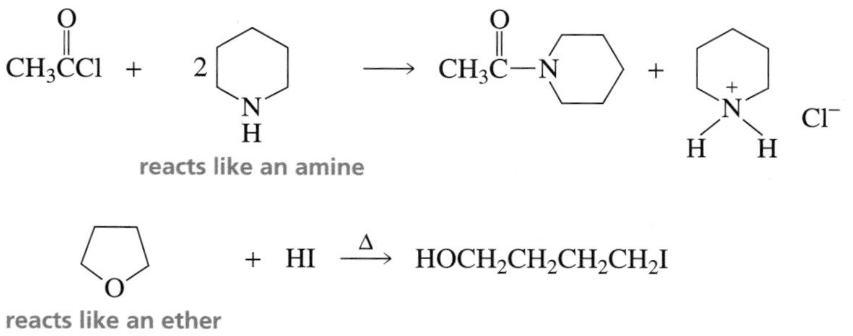
reacts like an amine
reacts like an ether

PROBLEM 4◆

Give the product of each of the following reactions.

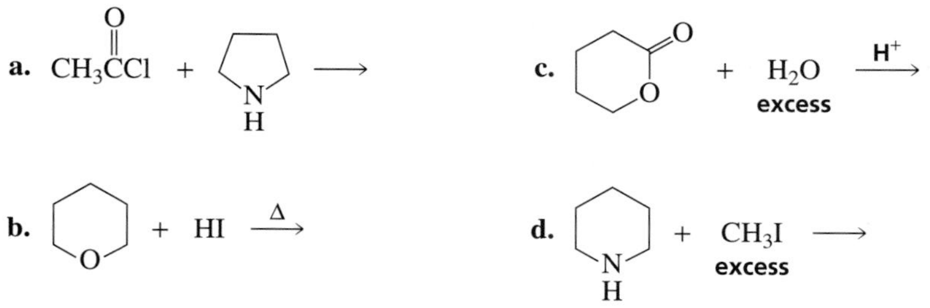
excess
excess

We have already studied the reactions of cyclopropanes, epoxides, and arene oxides and have seen that three-membered rings do not have the same chemical reactivity as their open-chain analogs (Sections 8.7, 11.6, and 11.7). Three-membered rings readily undergo ring-opening reactions because this releases the angle strain and torsional strain inherent in their structures.

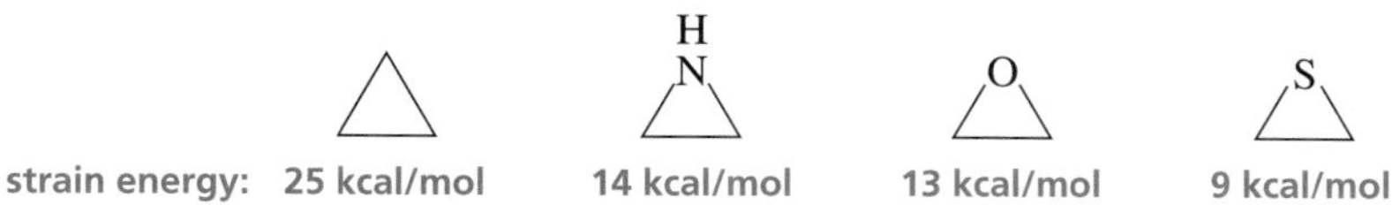
strain energy: 25 kcal/mol
14 kcal/mol
13 kcal/mol
9 kcal/mol

When three-membered-ring heterocycles react under acidic conditions, the hetero atom is protonated. Because the ring begins to break before it encounters the nucleophile, it breaks in the direction that puts the partial positive charge on the more substituted carbon (Section 11.6).

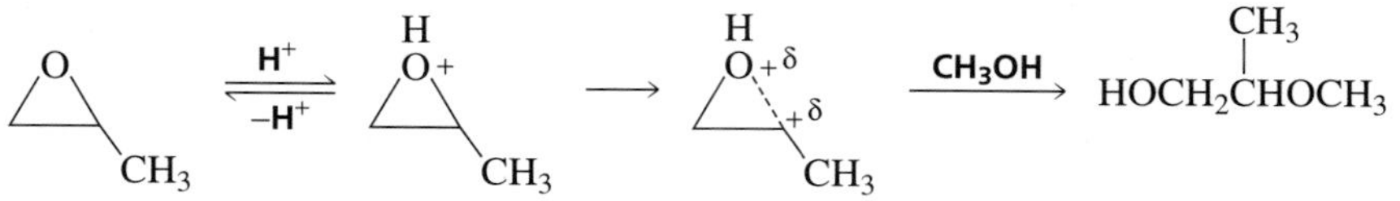
CH3OH
HOCH2CHOCH3

Therefore, under acidic conditions, nucleophilic attack occurs on the more substituted carbon.

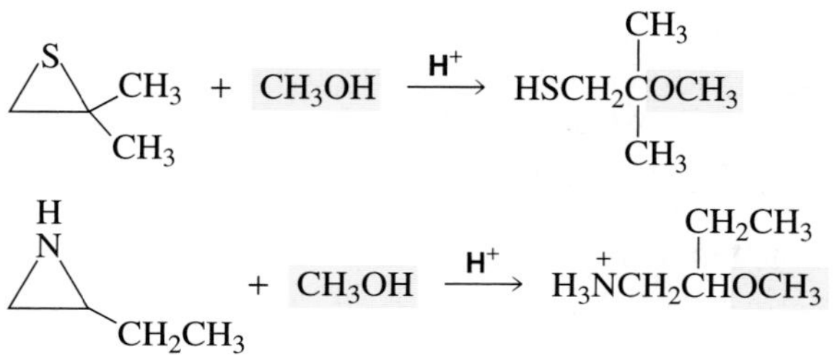

$$\text{(thiirane with CH}_3\text{, CH}_3\text{)} + CH_3OH \xrightarrow{H^+} HSCH_2\underset{\underset{CH_3}{|}}{\overset{\overset{CH_3}{|}}{C}}OCH_3$$

$$\text{(aziridine with CH}_2CH_3\text{)} + CH_3OH \xrightarrow{H^+} \overset{+}{H_3}NCH_2\underset{}{\overset{\overset{CH_2CH_3}{|}}{C}}HOCH_3$$

When, however, the ring-opening reaction is carried out under conditions in which the hetero atom does not become protonated, the three-membered ring does not begin to open until it is attacked by the nucleophile. Therefore, the nucleophile attacks the less substituted carbon since it is more accessible (Section 11.6).

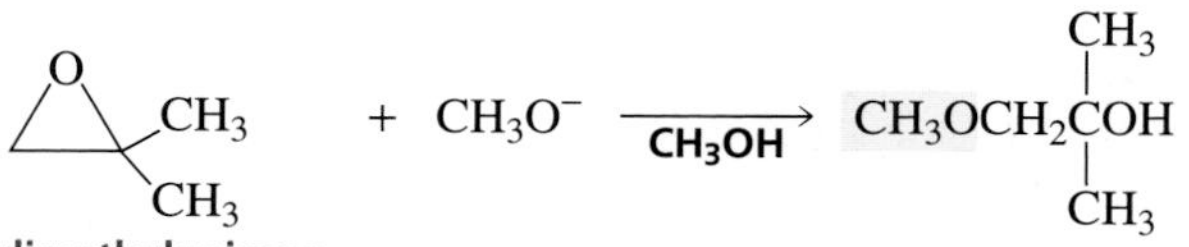

$$\text{(2,2-dimethyloxirane)} + CH_3O^- \xrightarrow[CH_3OH]{} CH_3OCH_2\underset{\underset{CH_3}{|}}{\overset{\overset{CH_3}{|}}{C}}OH$$

2,2-dimethyloxirane

$$\text{(2-ethylthiirane)} \xrightarrow[\text{2. H}^+, \text{H}_2\text{O}]{\text{1. CH}_3\text{CH}_2\text{MgBr}} CH_3CH_2CH_2\underset{}{\overset{\overset{SH}{|}}{C}}HCH_2CH_3$$

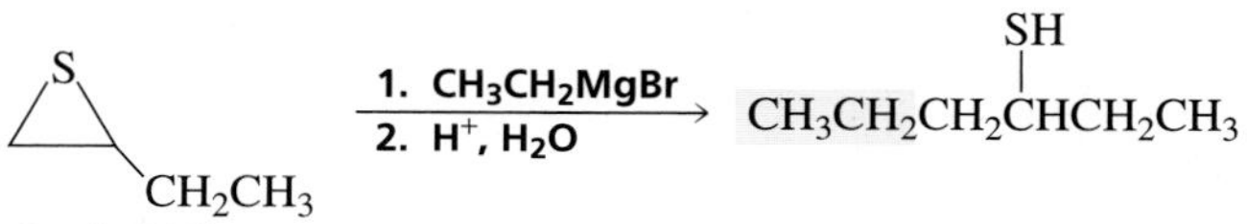

2-ethylthiirane

Aziridines do not undergo ring-opening reactions under conditions in which the hetero atom is not protonated. Apparently, relief of strain is not sufficient to make up for the highly basic (poor) leaving group.

Four-membered rings are not as reactive as three-membered rings. They readily undergo acid-catalyzed ring-opening reactions, but rigorous conditions are required for ring-opening reactions under conditions in which the hetero atom is not protonated. We have seen that penicillin's antibiotic activity is attributable to its easily opened four-membered ring (Section 15.13).

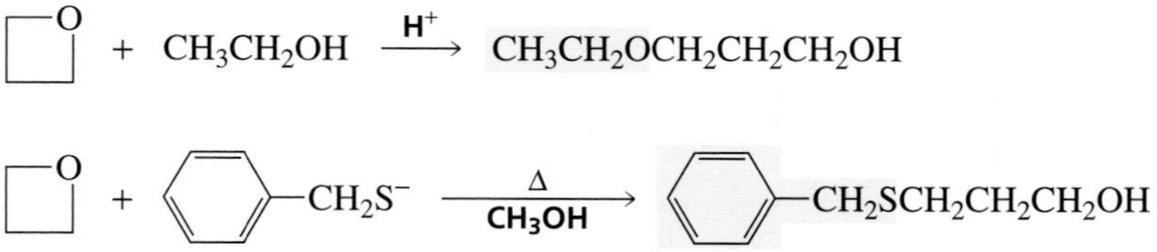

$$\text{(oxetane)} + CH_3CH_2OH \xrightarrow{H^+} CH_3CH_2OCH_2CH_2CH_2OH$$

$$\text{(oxetane)} + \text{(C}_6\text{H}_5)-CH_2S^- \xrightarrow[CH_3OH]{\Delta} \text{(C}_6\text{H}_5)-CH_2SCH_2CH_2CH_2OH$$

PROBLEM 5 ◆

Give the major product of each of the following reactions.

a. 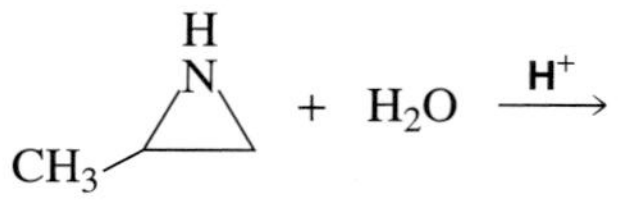 (methylaziridine) $+ H_2O \xrightarrow{H^+}$

b. 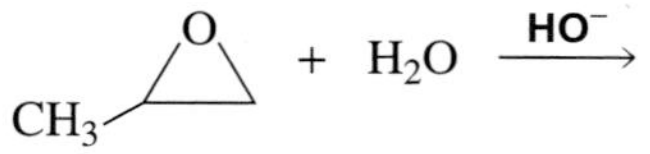 (methyloxirane) $+ H_2O \xrightarrow{HO^-}$

26.2
UNSATURATED FIVE-MEMBERED-RING HETEROCYCLES

Pyrrole, Furan, and Thiophene

Pyrrole, furan, and **thiophene** are five-membered-ring heterocycles. Each of these compounds has three pairs of delocalized π electrons. Two of the pairs are shown as π bonds, and one pair is shown as a pair of nonbonding electrons on the hetero atom. Furan and thiophene have a second pair of nonbonding electrons that are not part of the π cloud. These electrons are in an sp^2 orbital perpendicular to the p orbitals. Since pyrrole, furan, and thiophene are cyclic, planar molecules with three pairs of delocalized π electrons, each fulfills the criteria for aromaticity (Section 6.11).

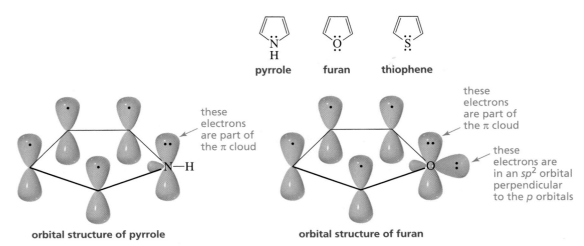

Pyrrole is an extremely weak base because nitrogen's nonbonding pair of electrons is needed for pyrrole's aromaticity. When the nitrogen is protonated, pyrrole's aromaticity is destroyed. In other words, the drive to become aromatic causes the conjugate acid of pyrrole to lose a proton readily—that is, to be a very strong acid ($pK_a = -3.8$).

The resonance contributors of pyrrole show that nitrogen donates its nonbonding electrons into the five-membered ring.

Pyrrolidine has a dipole moment of 1.57 D because of the electron-withdrawing nitrogen atom. Pyrrole has a slightly larger dipole moment (1.80 D), but it is in the opposite direction. This indicates that nitrogen's ability to donate electrons into the ring by resonance more than makes up for its inductive electron withdrawal.

In Section 6.6 we saw that the more stable and more nearly equivalent the resonance contributors, the greater the resonance energy. The resonance energies of pyrrole, furan, and thiophene are not as great as the resonance energies of benzene and the cyclopentadienyl anion, compounds for which the resonance contributors are all equivalent. Thiophene, with the least electronegative hetero atom, has the greatest resonance energy of the three five-membered heterocycles, and furan, with the most electronegative hetero atom, has the smallest resonance energy.

relative resonance energies of some aromatic compounds

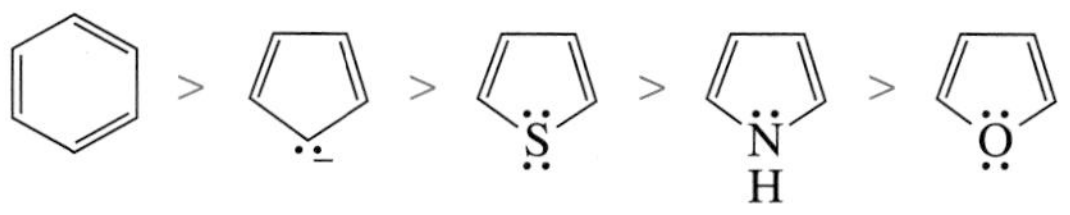

Because pyrrole, furan and thiophene are aromatic, they undergo electrophilic aromatic substitution reactions. All three compounds are more reactive than benzene toward electrophilic substitution. The reason for this greater reactivity is apparent from the resonance hybrid for pyrrole, which shows that there is increased electron density on each of the carbons as a result of delocalization of the non-bonding pair of electrons into the π cloud. Because the rate-limiting step of an electrophilic substitution reaction is attachment of the electrophile to the aromatic compound, increasing the electron density of the ring carbons increases the rate of electrophilic substitution.

Pyrrole, furan, and thiophene are more reactive than benzene toward electrophilic substitution.

mechanism for electrophilic aromatic substitution

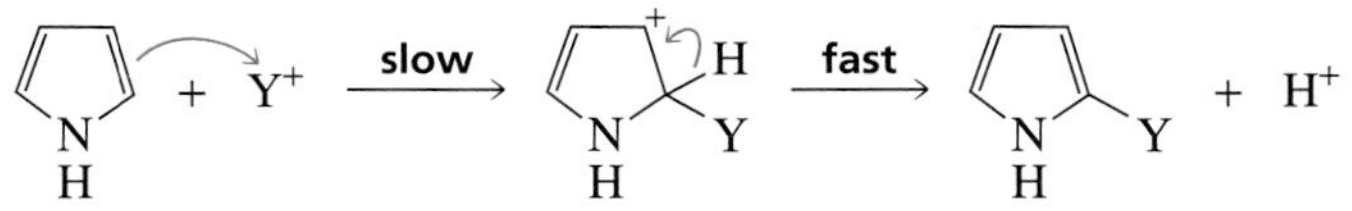

Pyrrole, furan, and thiophene undergo electrophilic substitution at C-2.

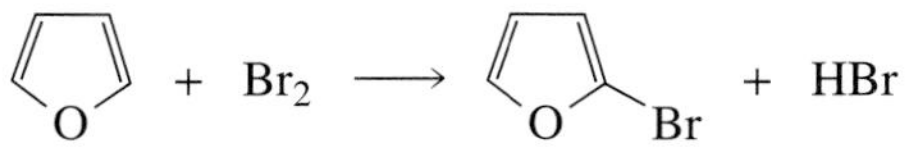

2-bromofuran

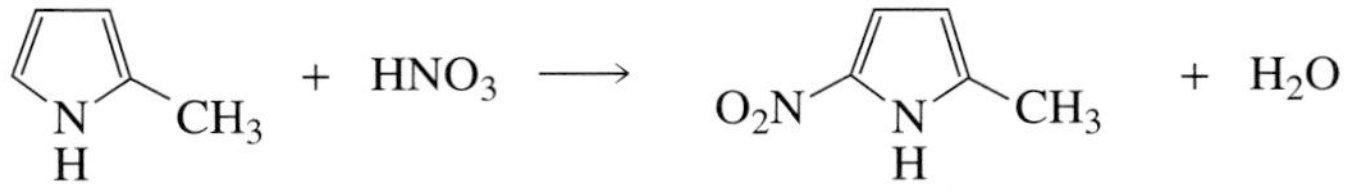

2-methyl-5-nitropyrrole

Substitution occurs preferentially at C-2 because the intermediate obtained by putting a substituent at this position is more stable than the intermediate obtained by putting a substituent at C-3. Both intermediates have a resonance contributor in which all the atoms (except H) have complete octets. The intermediate resulting from C-2 substitution of pyrrole has two additional resonance contributors; they both have a positive charge on a relatively stable secondary allylic carbon. The intermediate resulting from C-3 substitution has only one additional resonance contributor, and it has a positive charge on a secondary carbon, which is less stable than a resonance contributor with a positive charge on a secondary allylic carbon (Figure 26.1). If both positions adjacent to the hetero atom are occupied, electrophilic substitution will take place at C-3.

Pyrrole, furan, and thiophene undergo electrophilic substitution at C-2.

Figure 26.1 ▶
Structures of the intermediates that can be formed from the reaction of an electrophile with pyrrole at C-2 and C-3.

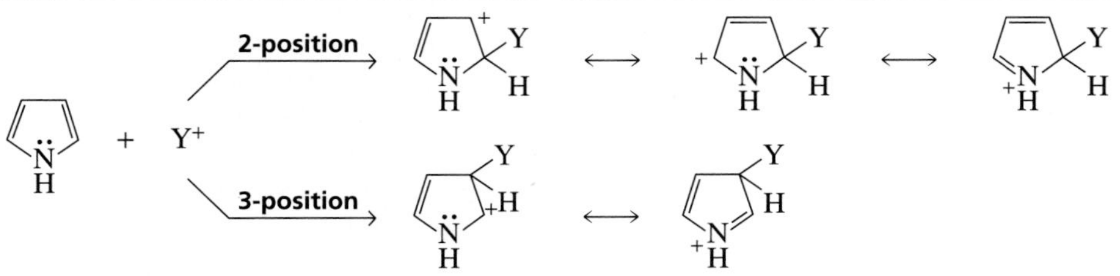

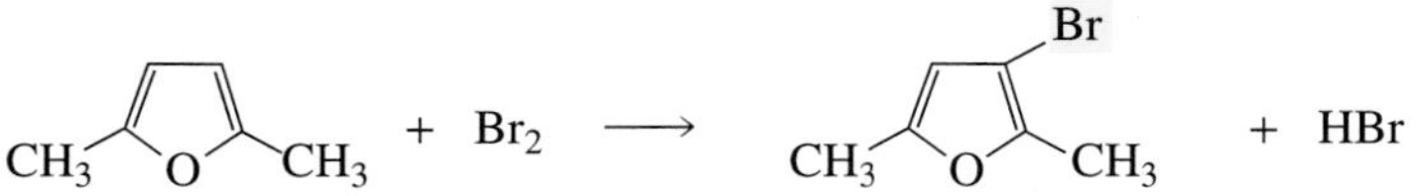

3-bromo-2,5-dimethylfuran

Furan is not as reactive as pyrrole in electrophilic substitution reactions. This is because oxygen, as a result of its greater electronegativity, is not as effective as nitrogen in donating its nonbonding pair of electrons into the ring. Thiophene is not as reactive as furan toward electrophilic substitution, because thiophene's nonbonding electrons are in a 3*p* orbital, which overlaps less effectively than the 2*p* orbitals of nitrogen or oxygen with the 2*p* orbital of carbon. More of the electron density, therefore, is maintained in the vicinity of the sulfur.

relative reactivity toward electrophilic aromatic substitution

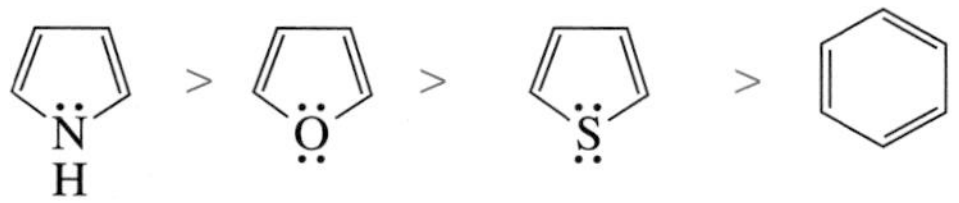

pyrrole furan thiophene benzene

The relative reactivities of the five-membered-ring heterocycles are reflected in the Lewis acid required to catalyze a Friedel–Crafts acylation reaction (Section 14.7). Benzene requires $AlCl_3$, a relatively strong Lewis acid. Because thiophene is more reactive than benzene, it can undergo a Friedel–Crafts reaction using $SnCl_4$, a weaker Lewis acid. An even weaker Lewis acid, BF_3, can be used when the substrate is furan. Pyrrole is so reactive that an anhydride is used instead of a more reactive acyl chloride, and no catalyst is necessary.

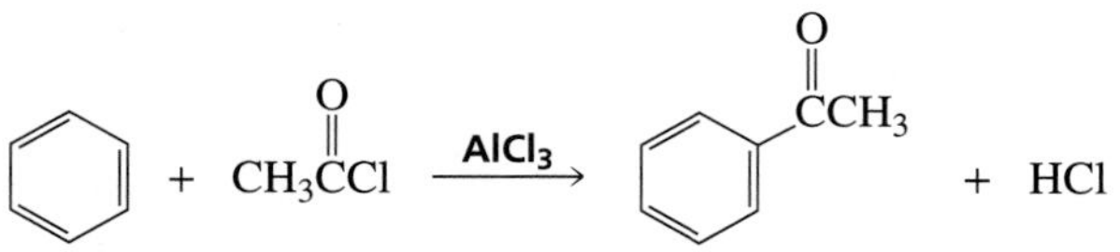

phenylethanone

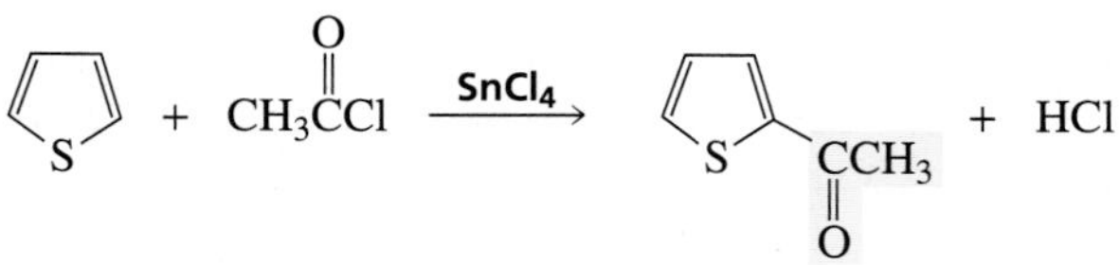

2-acetylthiophene

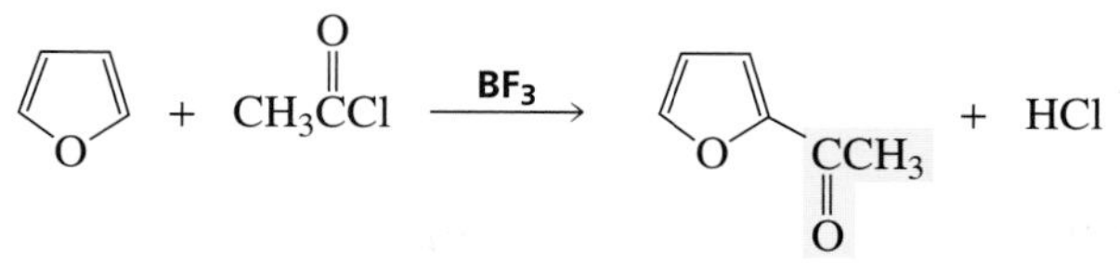

2-acetylfuran

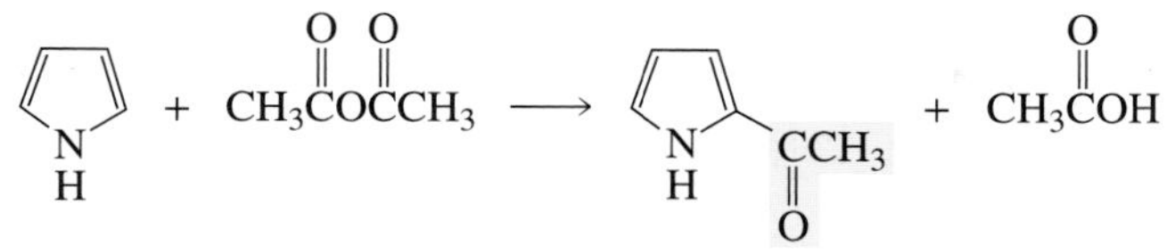

2-acetylpyrrole

The resonance hybrid of pyrrole indicates that there is a partial positive charge on the nitrogen. This means that, when pyrrole is protonated, it is not protonated on nitrogen; it is protonated on C-2, because the proton is an electrophile and, like other electrophiles, attaches to the position most vulnerable to electrophilic attack.

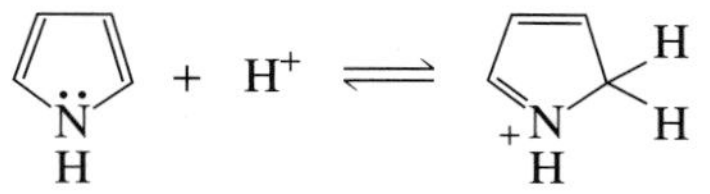

$pK_a = -3.8$

Pyrrole is unstable in strongly acidic solutions because it polymerizes readily.

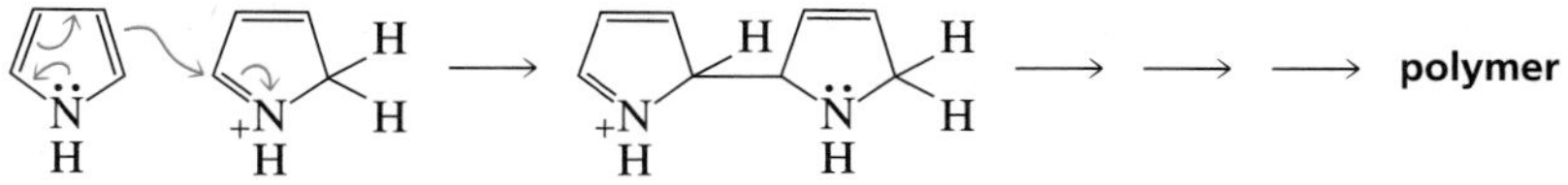

Neutral pyrrole is more acidic than saturated neutral amines because the nitrogen in pyrrole is sp^2 hybridized and thus is more electronegative than the sp^3 nitrogen of a saturated amine. In addition, because the nitrogen in pyrrole donates electrons into the ring, it has a partial positive charge, which further increases its electron withdrawing ability (Table 26.1).

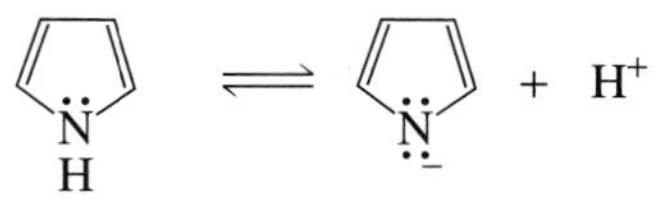

$pK_a = \sim17$

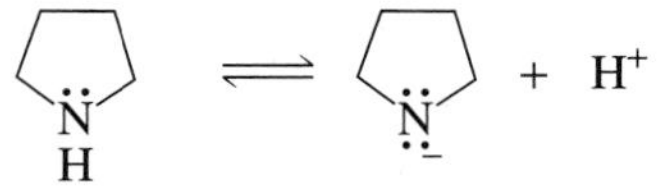

$pK_a = \sim36$

PROBLEM 6

When pyrrole is added to a dilute solution of D_2SO_4 in D_2O, 2-deuteriopyrrole is formed. Propose a mechanism to account for the formation of this compound.

PROBLEM 7

Use resonance contributors to explain why pyrrole is protonated on C-2 rather than on nitrogen.

TABLE 26.1 Summary of the pK_a Values of Several Nitrogen Heterocycles

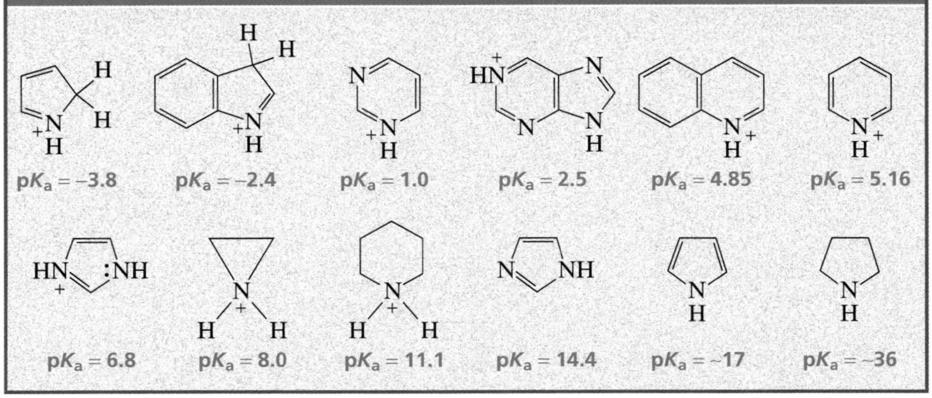

pK_a = –3.8   pK_a = –2.4   pK_a = 1.0   pK_a = 2.5   pK_a = 4.85   pK_a = 5.16

pK_a = 6.8   pK_a = 8.0   pK_a = 11.1   pK_a = 14.4   pK_a = ~17   pK_a = ~36

PROBLEM 8

In spite of the fact that nitrogen is considerably more electronegative than carbon, pyrrole (pK_a = 17) is less acidic than cyclopentadiene (pK_a = 15). Explain.

Indole, Benzofuran, and Benzothiophene

Indole, benzofuran, and benzothiophene contain a five-membered aromatic ring fused with a benzene ring. The rings are numbered in a way that gives the hetero atom the lowest possible number. Indole, benzofuran, and benzothiophene are aromatic because they are cyclic, planar, molecules, and each has five pairs of delocalized π electrons (Section 6.11). Indole, like pyrrole, needs nitrogen's nonbonding pair of electrons for its aromaticity. The conjugate acid of indole is therefore a strong acid (pK_a = −2.4).

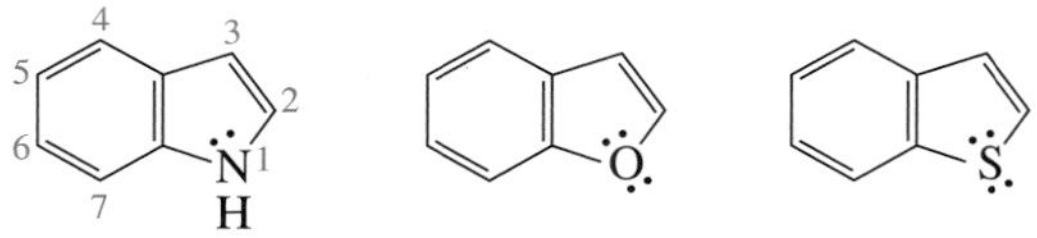

indole   benzofuran   benzothiophene

Electrophilic aromatic substitution in these compounds takes place on the five-membered ring because it is more reactive toward electrophilic aromatic substitution than the benzene ring. Indole undergoes electrophilic aromatic substitution primarily at C-3, benzofuran primarily at C-2, and benzothiophene about equally well at C-2 and C-3.

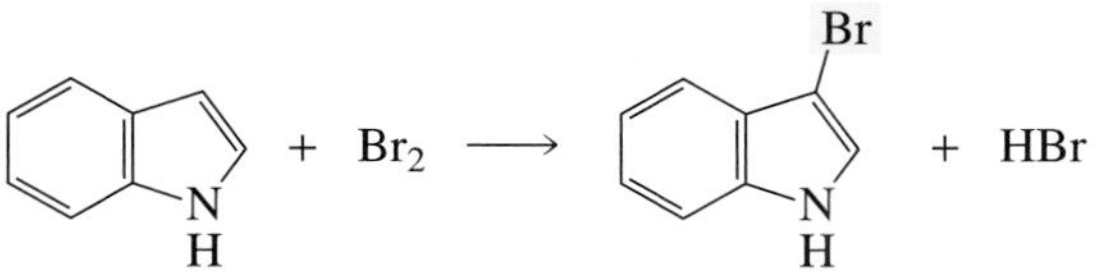

indole   + Br$_2$ →   3-bromoindole   + HBr

benzofuran   + CH$_3$COCCH$_3$ →   2-acetylbenzofuran   + CH$_3$COH

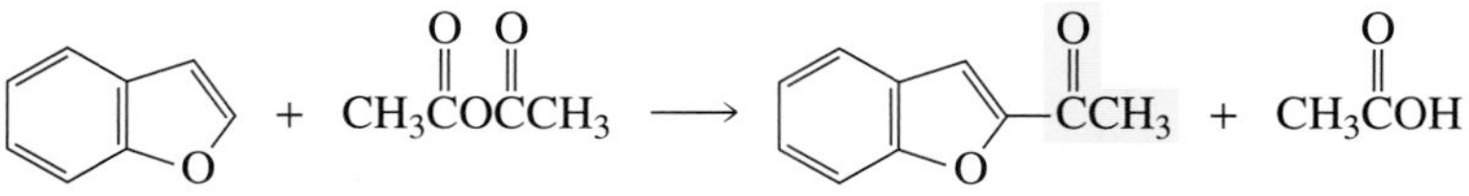

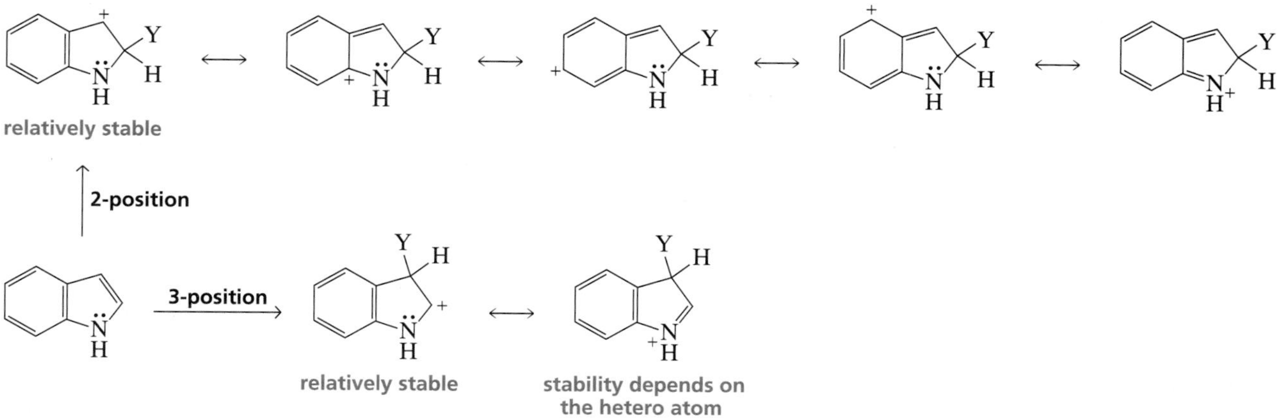

▲ **Figure 26.2**
Structures of the intermediates that can be formed from the reaction of an electrophile with indole at C-2 and C-3.

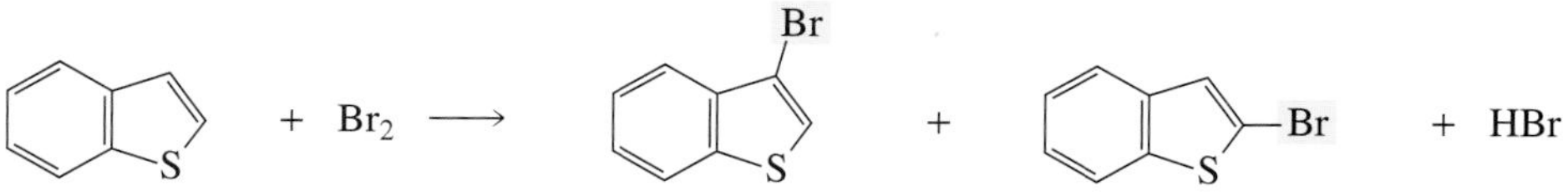

benzothiophene   3-bromobenzothiophene   2-bromobenzothiophene

   The difference in orientation can be explained by comparing the stabilities of the intermediates obtained as a result of electrophilic attack at the C-2 and C-3 positions (Figure 26.2). Substitution at C-2 gives only one relatively stable resonance contributor, because the six-π-electron system of the benzene ring is disturbed in the other four resonance contributors. Substitution at C-3 also gives one relatively stable resonance contributor. The stability of the second resonance contributor depends on the nature of the hetero atom. In the case of indole, it is more stable than the four contributors in which the aromaticity of the benzene ring has been lost, so indole undergoes electrophilic substitution primarily at C-3. In the case of benzofuran, the second resonance contributor is less stable than the four contributors without an intact benzene ring, because oxygen, being more electronegative than nitrogen, is less able to tolerate a positive charge. Benzofuran therefore undergoes electrophilic aromatic substitution primarily at C-2. Benzothiophene undergoes electrophilic aromatic substitution equally well at C-2 and C-3, indicating that the two intermediates have similar stabilities.

Pyridine

When one of the carbons of a benzene ring is replaced by a nitrogen, the resulting compound is **pyridine**.

**26.3
UNSATURATED SIX-
MEMBERED-RING
HETEROCYCLES**

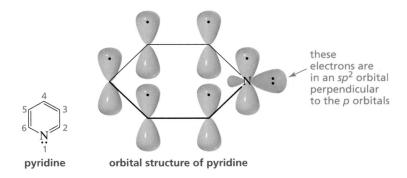

pyridine   orbital structure of pyridine

Pyridine is a tertiary amine and undergoes reactions characteristic of tertiary amines. For example, it undergoes S_N2 reactions with alkyl halides (Section 9.3), and it reacts with hydrogen peroxide to form an *N*-oxide (Section 11.11).

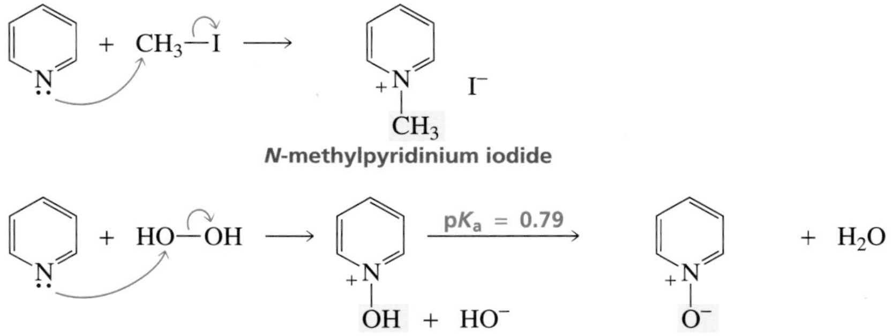

N-methylpyridinium iodide

pyridine-*N*-oxide

PROBLEM 9 / SOLVED

Will an amide be formed from the reaction of an acyl chloride with an aqueous solution of pyridine? If so, why? If not, why not?

SOLUTION   An amide will not be formed, because a positively charged nitrogen is an excellent leaving group. The final product of the reaction will therefore be a carboxylic acid. (If the final pH of the solution is greater than the pK_a of the carboxylic acid, the carboxylic acid will be predominantly in its basic form.)

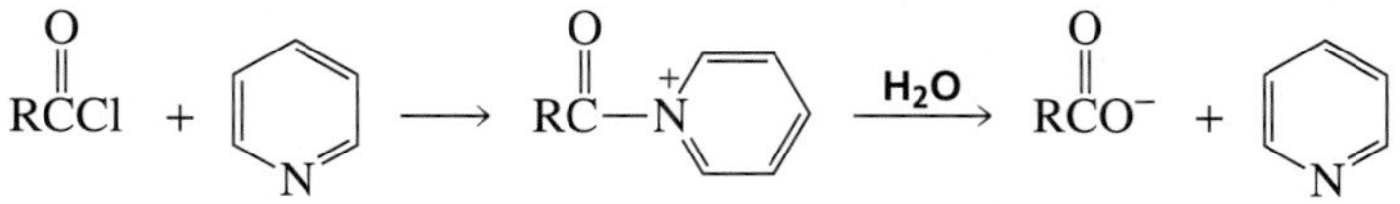

The pK_a of the pyridinium ion is lower than that of a typical ammonium ion, because the acidic hydrogen of a pyridinium ion is attached to an sp^2 hybridized nitrogen, which is more electronegative than an sp^3 hybridized nitrogen (Section 5.9).

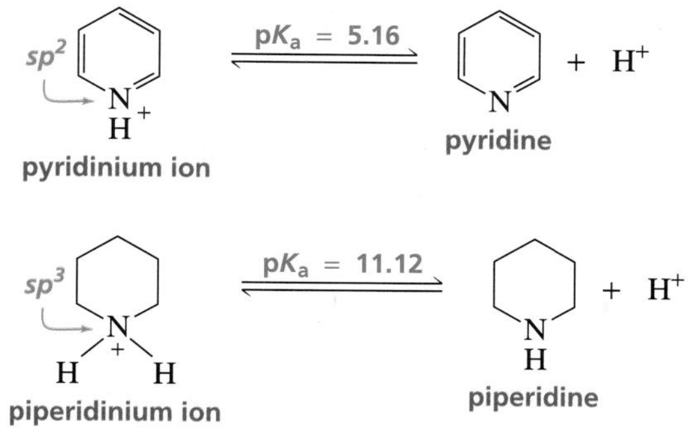

pyridinium ion $pK_a = 5.16$ pyridine $+ H^+$

piperidinium ion $pK_a = 11.12$ piperidine $+ H^+$

Pyridine is aromatic. Like benzene, it has two neutral resonance contributors. Because of the electron-withdrawing nitrogen, it also has three charged resonance contributors that benzene does not have.

resonance contributors of pyridine

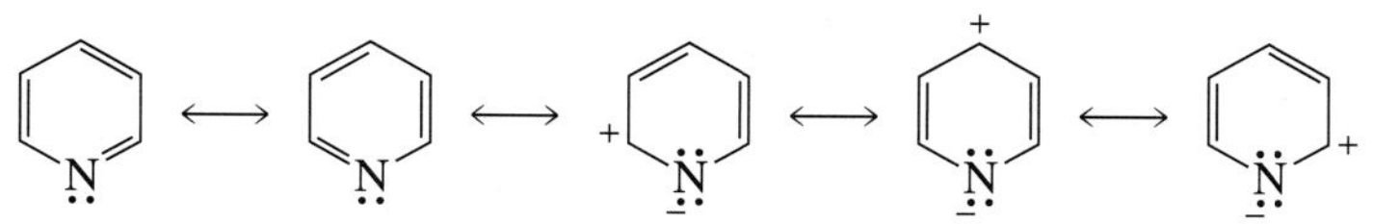

The dipole moment of pyridine is 1.57 D. As the resonance contributors indicate, the electron-withdrawing nitrogen is the negative end of the dipole.

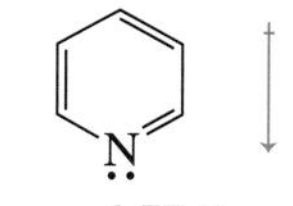

$\mu = 1.57 \text{ D}$

Because it is aromatic, pyridine, like benzene, undergoes electrophilic aromatic substitution reactions.

mechanism for electrophilic aromatic substitution

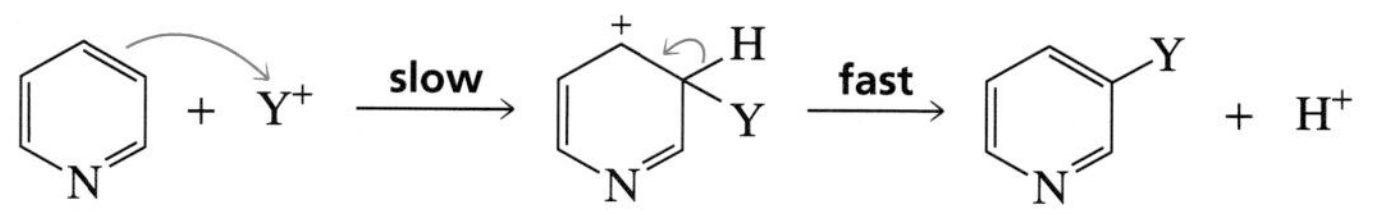

Pyridine's electron-withdrawing nitrogen causes the ring carbons to have significantly less electron density than the ring carbons of benzene. Pyridine is therefore less reactive than benzene toward electrophilic aromatic substitution. It is even less reactive than nitrobenzene. (Recall that an electron-withdrawing nitro group strongly deactivates a benzene ring toward electrophilic aromatic substitution; Section 14.11).

relative reactivity toward electrophilic aromatic substitution

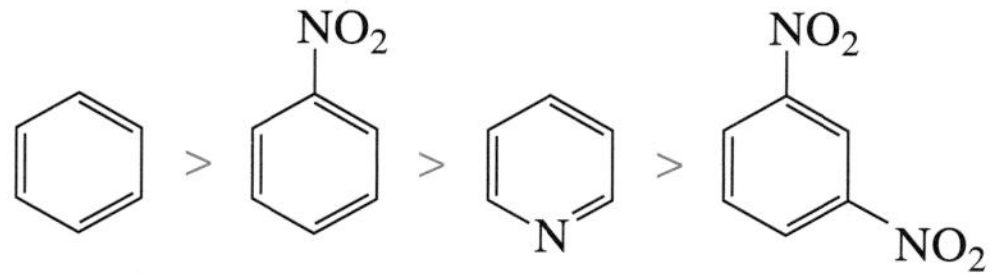

Pyridine undergoes electrophilic aromatic substitution reactions only under vigorous conditions, and the yields of these reactions are often quite low. If the nitrogen becomes protonated under the reaction conditions, the reactivity is further decreased because a positively charged nitrogen is more electron-withdrawing than a neutral nitrogen.

3-bromopyridine
30%

pyridine-3-sulfonic acid
71%

Pyridine undergoes electrophilic substitution at C-3.

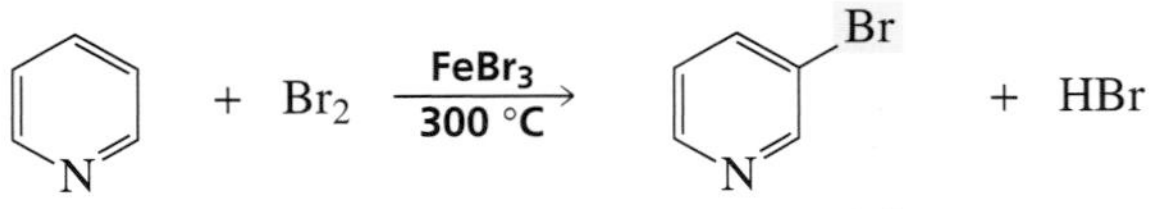

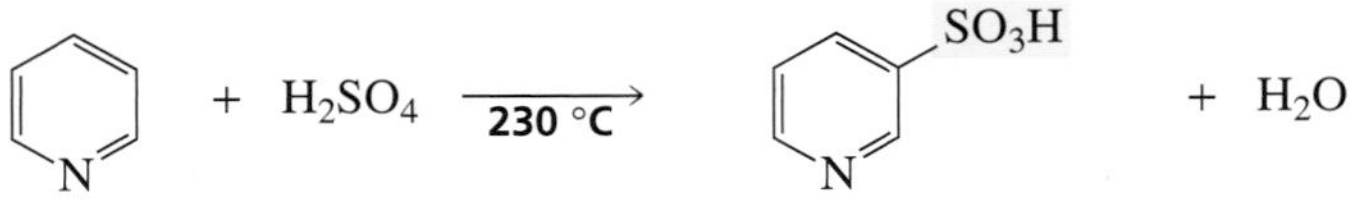

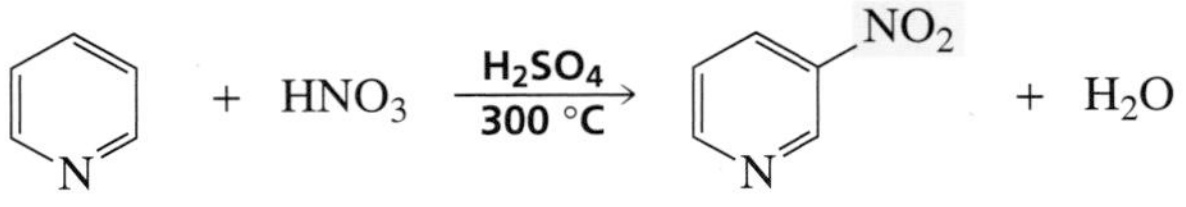

3-nitropyridine
22%

Electrophilic aromatic substitution of pyridine takes place at C-3, because the most stable intermediate is obtained by placing an electrophilic substituent at this position (Figure 26.3). When the substituent is placed at C-2 or C-4, one of the resulting resonance contributors is particularly unstable because its nitrogen atom has an incomplete octet as well as a positive charge.

We have seen that highly deactivated benzene rings do not undergo Friedel–Crafts alkylation reactions; therefore pyridine, whose reactivity is similar to that of a highly deactivated benzene, does not undergo these reactions.

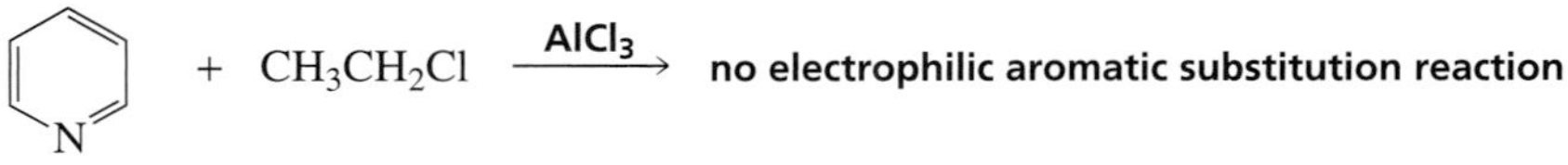

PROBLEM 10◆

Give the product of the following reaction.

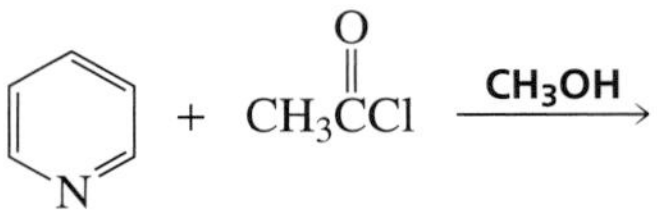

Pyridine is more reactive than benzene toward nucleophilic aromatic substitution for the same reason it is less reactive than benzene toward electrophilic aromatic substitution.

mechanism for nucleophilic aromatic substitution

Pyridine is less reactive than benzene toward electrophilic aromatic substitution and more reactive than benzene toward nucleophilic aromatic substitution.

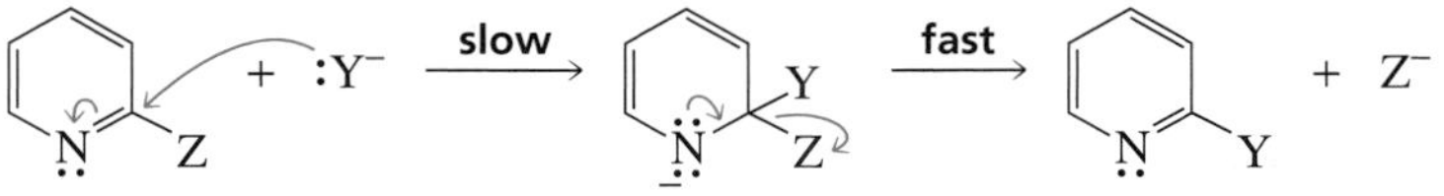

Nucleophilic aromatic substitution of pyridine takes place at C-2 and C-4, because attack at these positions leads to the most stable intermediate. Only when nucleophilic attack occurs at these positions is a resonance contributor obtained that has

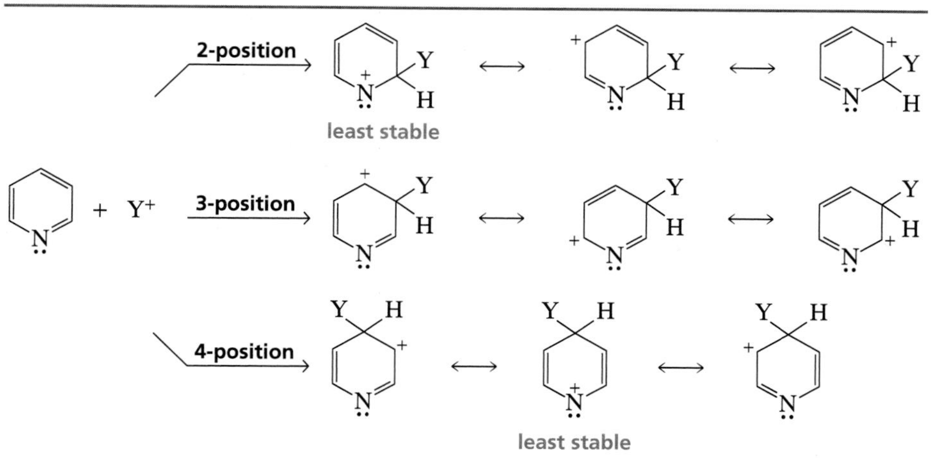

◀ **Figure 26.3**
Structures of the intermediates that can be formed from the reaction of an electrophile with pyridine.

the greatest electron density on nitrogen, the most electronegative of the ring atoms (Figure 26.4).

An example of a nucleophilic aromatic substitution reaction is the reaction of pyridine with sodium amide in liquid ammonia. This is known as the **Chichibabin reaction.** The major product of the Chichibabin reaction is the 2-substituted product, because nucleophilic attack at both the 2- and 6-positions leads to that product.

Alexei E. Chichibabin (1871–1945) *was born in Russia. He received a Ph.D. from the University of St. Petersburg and was a professor of chemistry at the University of Moscow and at the University of Paris.*

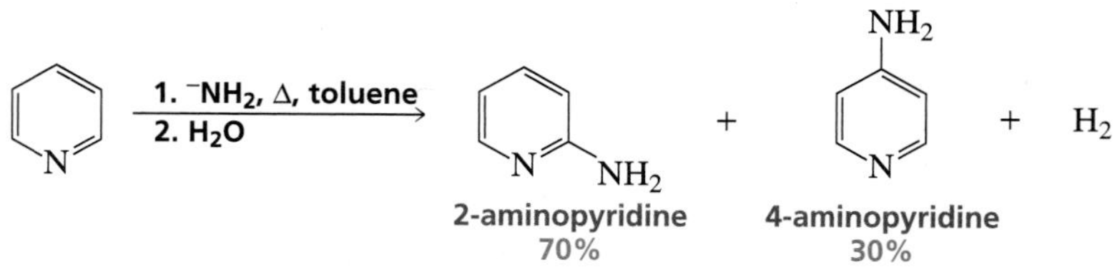

We have previously seen that a hydride ion is normally too strong a base to serve as a leaving group. It is a leaving group in the Chichibabin reaction because, after it is eliminated, it can remove a proton from the NH_2 group, thereby forming H_2. Evolution of hydrogen gas forces the reaction to completion (Le Châtelier's principle; Section 9.3).

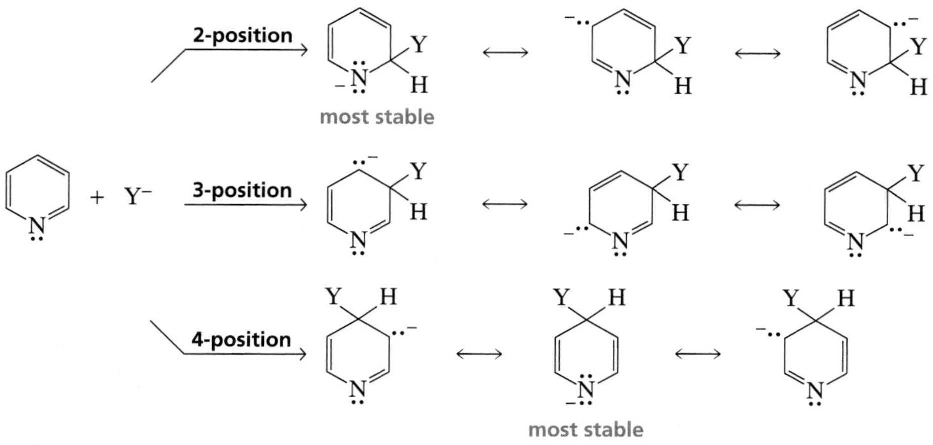

◀ **Figure 26.4**
Structures of the intermediates that can be formed from the reaction of a nucleophile with pyridine.

mechanism for the Chichibabin reaction

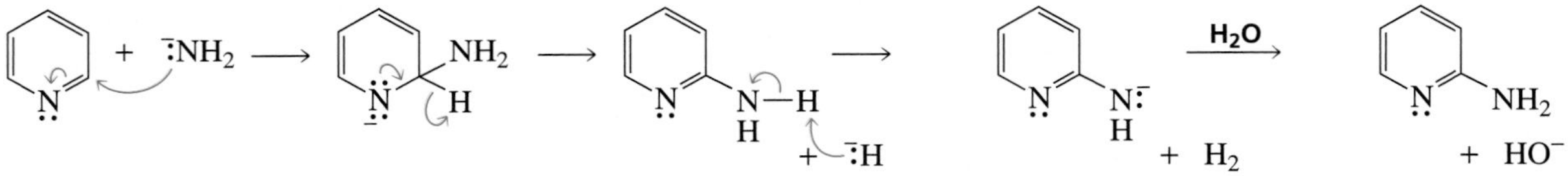

Pyridines can be alkylated with organolithium reagents in a reaction similar to the Chichibabin reaction (PhLi = phenyllithium).

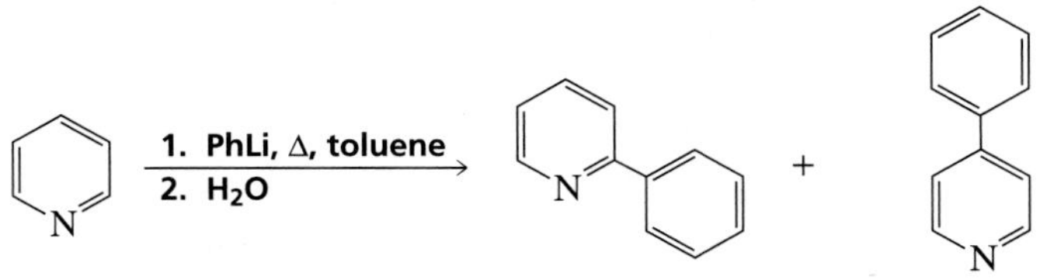

2-phenylpyridine   **4-phenylpyridine**

Pyridine undergoes nucleophilic substitution at C-2 and C-4.

If the leaving groups at C-2 and C-4 are different, the incoming nucleophile will preferentially substitute for the weaker base (the better leaving group).

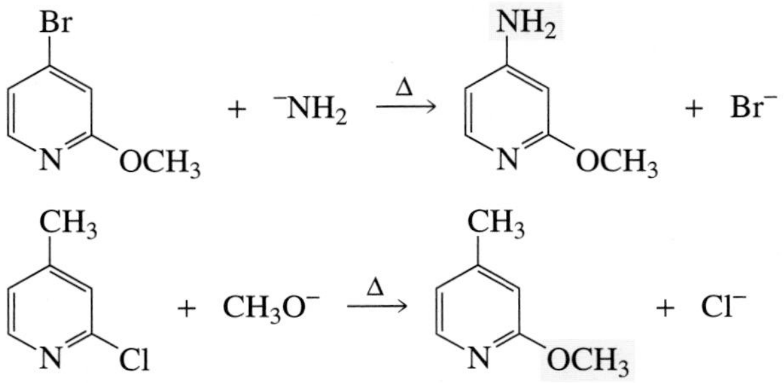

PROBLEM 11

Compare the mechanisms for the following reactions.

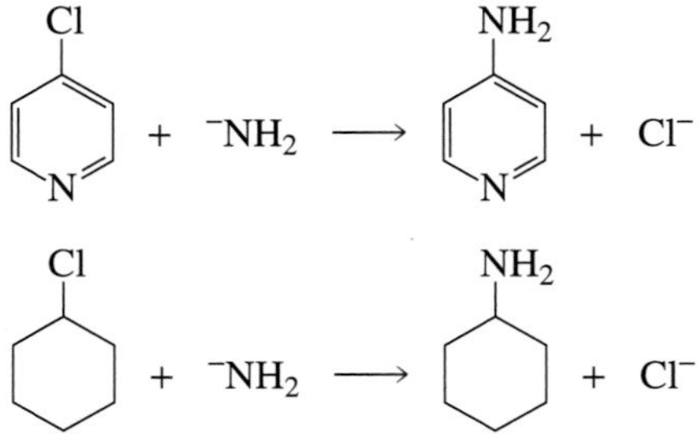

PROBLEM 12

a. Propose a mechanism for the following reaction.

b. What other product is formed?

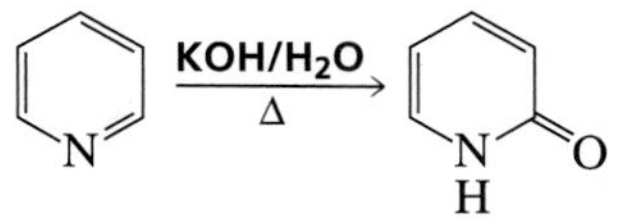

Substituted pyridines undergo many of the side-chain reactions that substituted benzenes undergo.

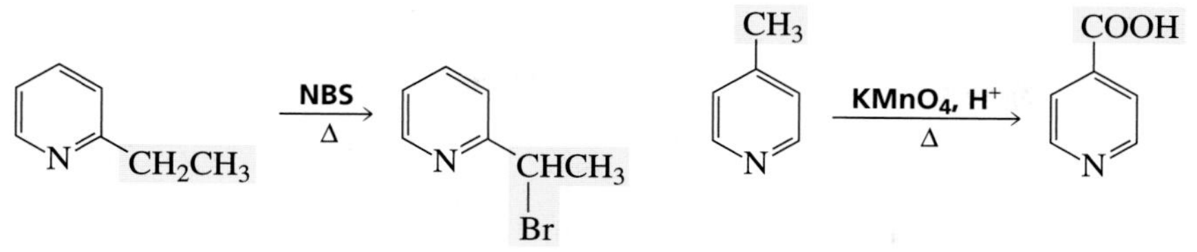

When 2- or 4-aminopyridine is diazotized, α-pyridone or γ-pyridone is formed. We have seen the mechanism for conversion of an amino group into a diazonium group (Section 14.20). Apparently the diazonium salt reacts immediately with water to form a hydroxypyridine (Section 14.18). The keto form of a hydroxypyridine is more stable than the enol form.

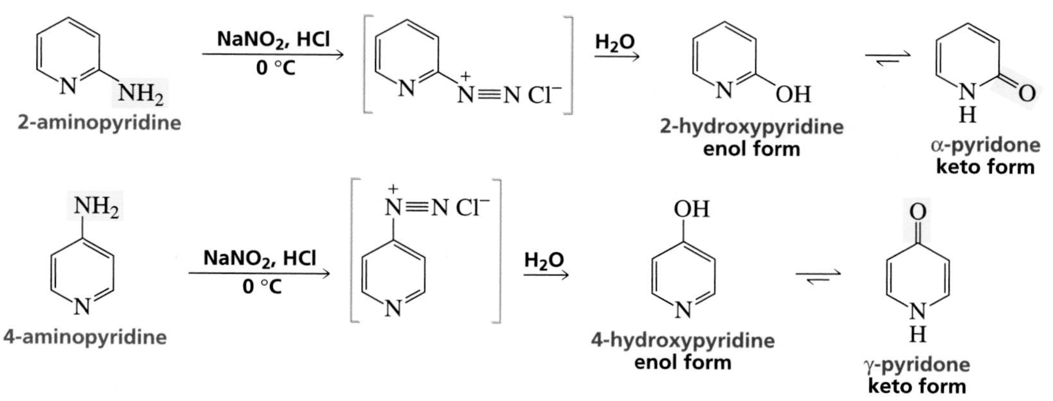

The electron-withdrawing nitrogen causes the α-hydrogens of alkyl groups attached to the 2- and 4-positions of the pyridine ring to have about the same acidity as the α-hydrogens of ketones (Section 18.1).

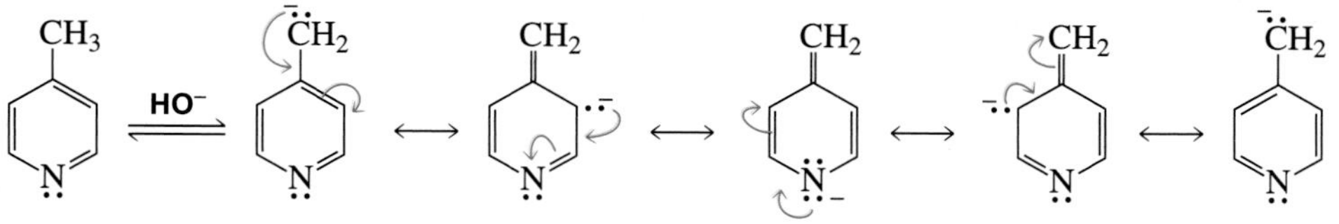

Consequently, they can be removed by base, and the resulting carbanions can react as nucleophiles.

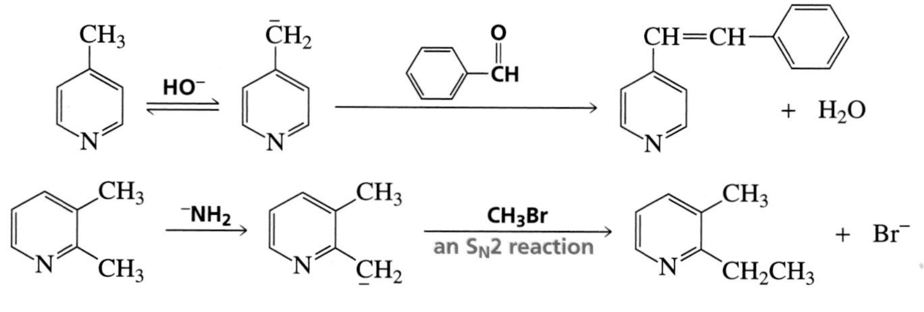

Pyridine-*N*-oxide is considerably more reactive toward electrophilic aromatic substitution than pyridine because of resonance electron donation by the oxygen. It undergoes electrophilic aromatic substitution primarily at C-4. (Recall that the $^+NO_2$ electrophile is generated from the reaction of HNO_3 with H_2SO_4; Section 14.5.) Because the N-oxide can be converted to pyridine by treatment with PCl_3, electrophilic aromatic substitution of pyridine-N-oxide is a synthetically useful reaction because it provides a way to introduce an electrophile at the 4-position of pyridine.

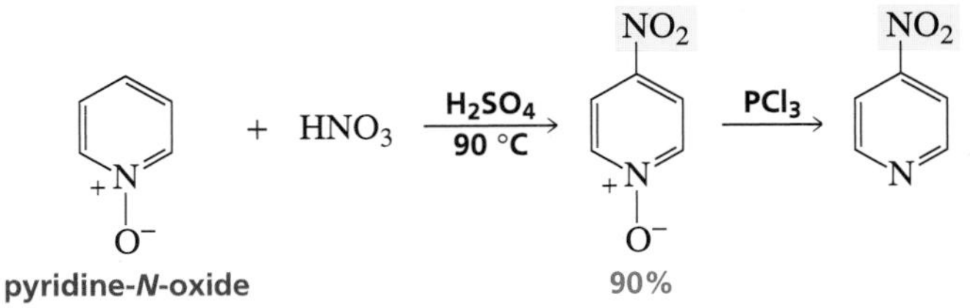

pyridine-N-oxide   90%

PROBLEM 13

Draw contributing resonance structures to show why pyridine-N-oxide is more reactive than pyridine toward electrophilic aromatic substitution and why it undergoes electrophilic aromatic substitution primarily at C-4.

PROBLEM 14

Show how the following compounds could be prepared from pyridine.

a.

c.

b.

d.

PROBLEM 15◆

Rank the following compounds in order of decreasing ease of removing a proton from a methyl group.

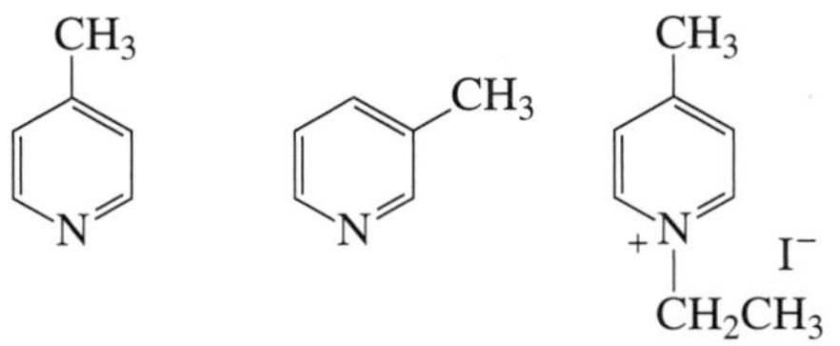

Quinoline and Isoquinoline

Since quinoline and isoquinoline have both a benzene ring and a pyridine ring, they are known as *benzopyridines*. Like pyridine, they are aromatic compounds. The pK_a values of their conjugate acids are similar to the pK_a of the conjugate acid of pyridine. (In order for the carbons in quinoline and isoquinoline to have the same numbers, the rule that the hetero atom has the lowest possible number is violated for isoquinoline, and the nitrogen is assigned the 2-position.)

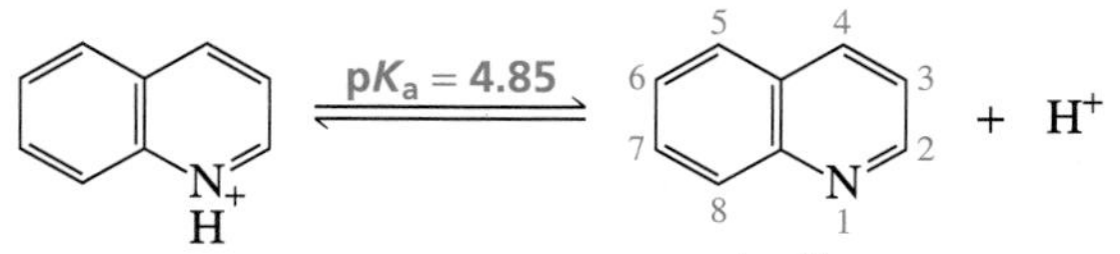

quinoline

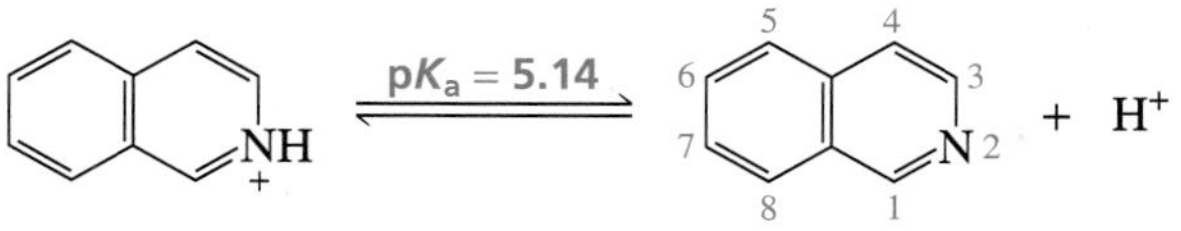

isoquinoline

Quinoline and isoquinoline undergo *electrophilic aromatic substitution* on the benzene ring, because a benzene ring is more reactive than a pyridine ring toward electrophilic aromatic substitution. Substitution takes place primarily at C-5 and C-8, which is reminiscent of naphthalene's preference for electrophilic aromatic substitution at the α-position (Section 14.23).

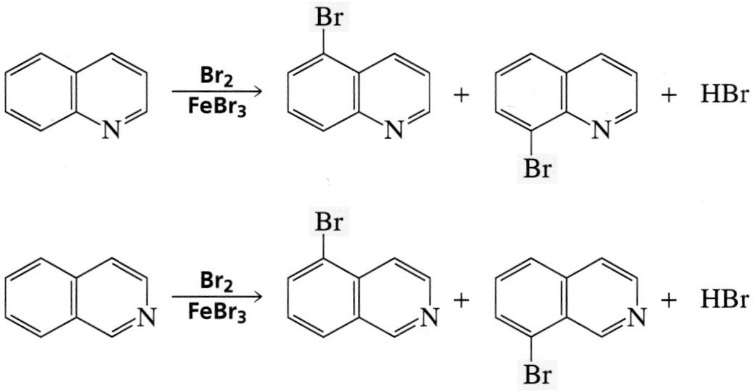

Because a pyridine ring is more reactive than a benzene ring toward nucleophilic aromatic substitution, *nucleophilic aromatic substitution* occurs on the pyridine ring. As expected, quinoline undergoes nucleophilic aromatic substitution at C-2 and C-4.

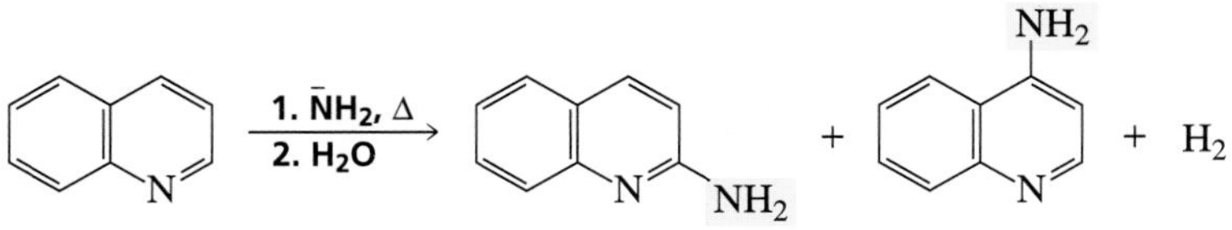

Isoquinoline undergoes nucleophilic aromatic substitution only at C-1.

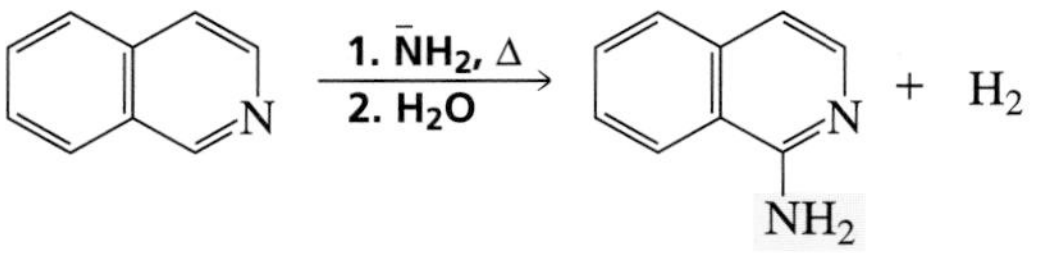

PROBLEM 16◆

Give the products of the following reactions.

a. 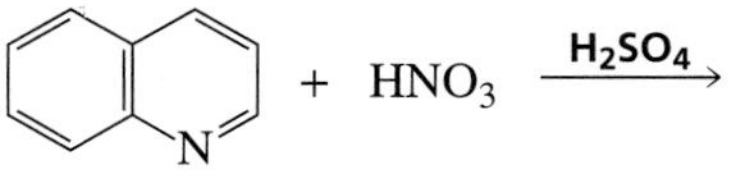 + HNO_3 $\xrightarrow{H_2SO_4}$ b. 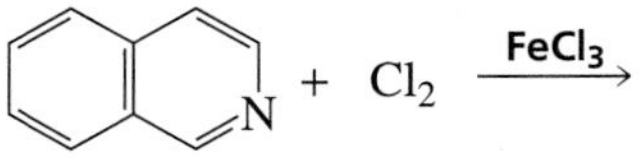 + Cl_2 $\xrightarrow{FeCl_3}$

c. 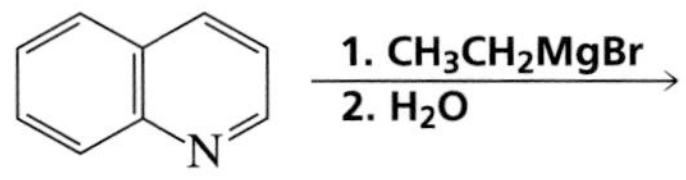 $\xrightarrow[\text{2. H}_2\text{O}]{\text{1. CH}_3\text{CH}_2\text{MgBr}}$ **d.** 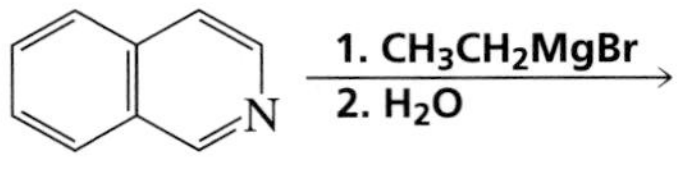 $\xrightarrow[\text{2. H}_2\text{O}]{\text{1. CH}_3\text{CH}_2\text{MgBr}}$

26.4 BIOLOGICALLY IMPORTANT HETEROCYCLES

Proteins are naturally occurring polymers of α-amino acids (Chapter 20). Three of the 20 naturally occurring amino acids contain heterocyclic systems. Proline contains a pyrrolidine ring, tryptophan contains an indole ring, and histidine contains an imidazole ring.

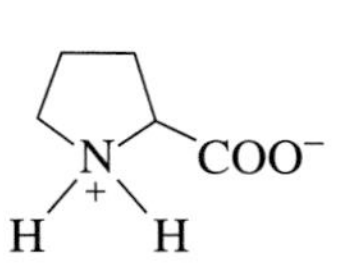

proline

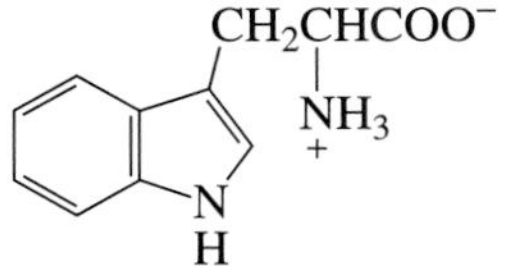

tryptophan

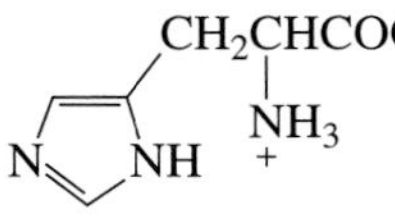

histidine

We have already seen that a saturated pyrrolidine ring behaves like a typical secondary amine (Section 26.1), and we have studied the chemistry of indole (Section 26.2). Imidazole, the heterocyclic ring of histidine, is the first heterocyclic compound we have encountered that has two hetero atoms.

Imidazole

Imidazole is an aromatic compound because it is cyclic, planar, and has three pairs of delocalized π electrons. The nonbonding electrons on N-1 are part of the π cloud because they are in a p orbital, while the nonbonding electrons on N-3 are in an sp^2 orbital, perpendicular to the p orbitals. The resonance energy of imidazole is 14 kcal/mol, significantly less than the resonance energy of benzene (36 kcal/mol).

resonance contributors of imidazole

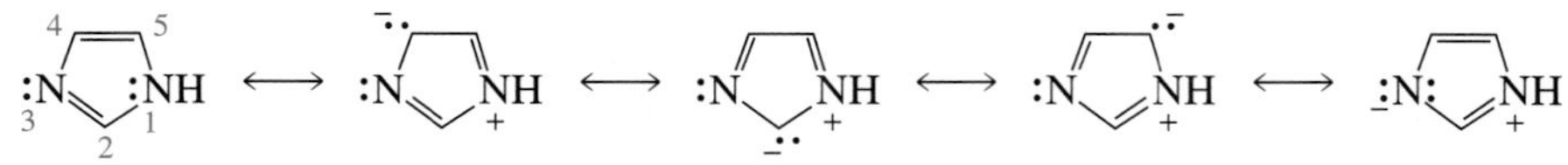

Because imidazole has a pair of nonbonding electrons that are not involved in resonance, it can be protonated; when imidazole is protonated, the proton attaches to the nonbonding electrons that are in an sp^2 orbital, because protonating the nonbonding electrons in a p orbital would destroy the compound's aromaticity. Because the conjugate acid of imidazole has a pK_a of 6.8, it exists in both the protonated and unprotonated forms at physiological pH (pH = 7.3). We have seen that this is one of the reasons why histidine, the imidazole-containing amino acid, is an important component of many enzymes (Section 21.8).

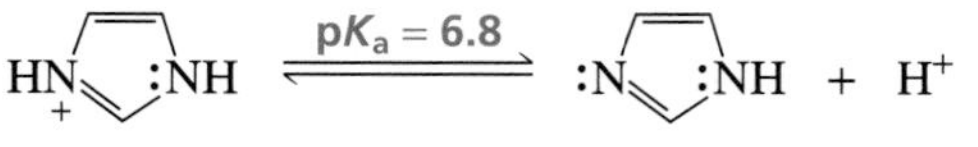

Because of the presence of the second ring nitrogen, neutral imidazole is a stronger acid ($pK_a = 14.4$) than neutral pyrrole ($pK_a = 17$).

$$:N \diagup NH \quad \xrightarrow{pK_a = 14.4} \quad :N \diagup N: \overline{} \quad + \quad H^+$$

Notice that both protonated imidazole and the imidazole ion have two equivalent resonance contributors. This means that C-4 and C-5 become equivalent when imidazole is either protonated or deprotonated.

protonated imidazole   **imidazole ion**

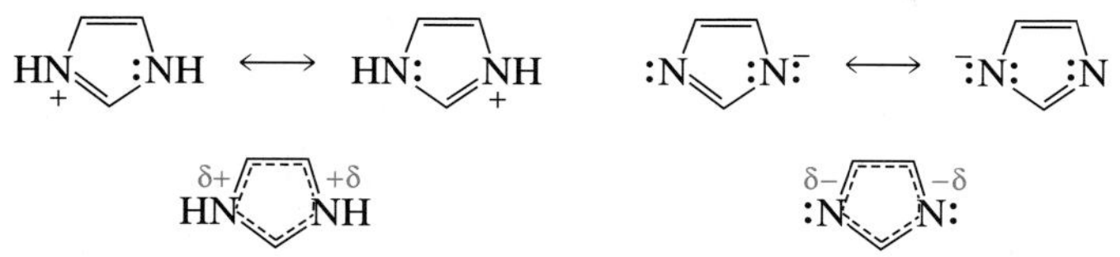

$$HN \diagup :NH \longleftrightarrow HN: \diagup NH \qquad :N \diagup N: \overline{} \longleftrightarrow \overline{} :N: \diagup N$$

$$\overset{\delta+}{HN} \diagup \overset{+\delta}{NH} \qquad\qquad \overset{\delta-}{:N} \diagup \overset{-\delta}{N:}$$

resonance hybrid   resonance hybrid

PROBLEM 17

a. Give the major product of the following reaction.

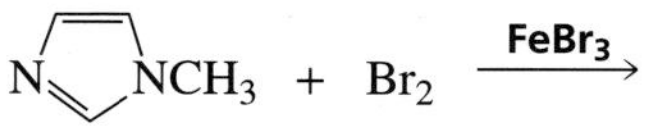

$$N \diagup NCH_3 \quad + \quad Br_2 \quad \xrightarrow{FeBr_3}$$

b. List imidazole, pyrrole, and benzene in order of decreasing reactivity toward electrophilic aromatic substitution.

PROBLEM 18

Imidazole boils at 257 °C, while *N*-methylimidazole boils at 199 °C. Explain this difference in boiling points.

PROBLEM 19◆

What percent of imidazole will be protonated at physiological pH (pH = 7.3)?

Purine and Pyrimidine

Nucleic acids (DNA and RNA) contain substituted **purines** and substituted **pyrimidines** (Section 24.1). Unsubstituted purine and pyrimidine are not found in nature. Notice that hydroxypurines and hydroxypyrimidines are more stable in the keto form. We have seen that this preference for the keto form is important in DNA base pairing (Section 24.7).

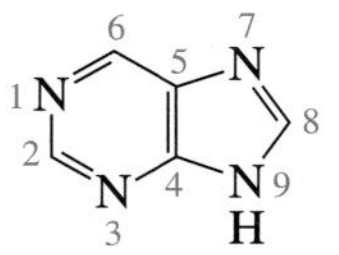

purine 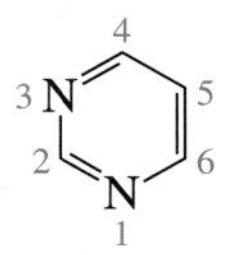   pyrimidine

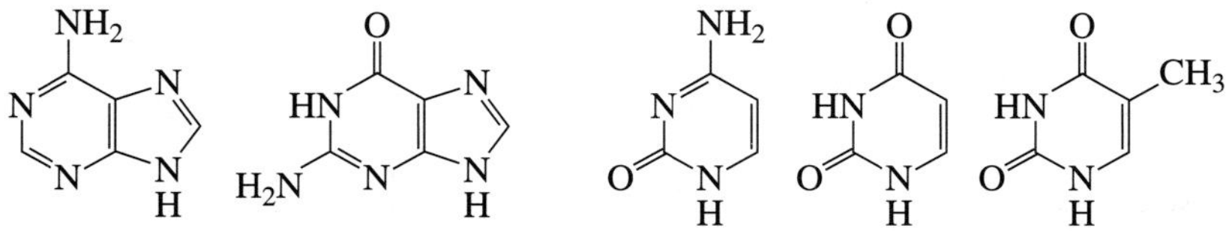

adenine   guanine   cytosine   uracil   thymine

Substituted purines and pyrimidines were some of the first compounds to be studied by organic chemists. Uric acid was isolated in 1776, and alloxan in 1818.

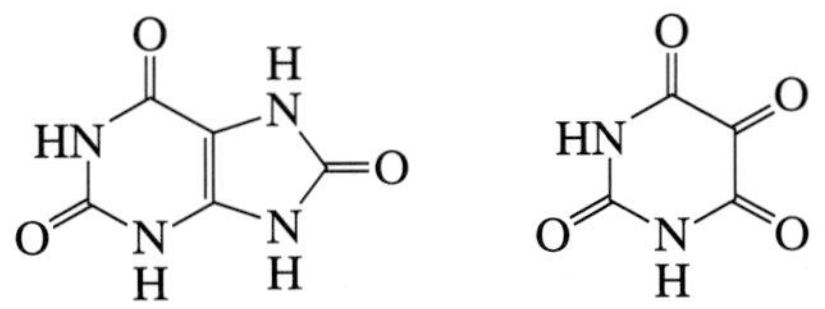

uric acid   alloxan

Because of the presence of the second nitrogen, pyrimidine is a much weaker base than pyridine (the conjugate acid is a stronger acid).

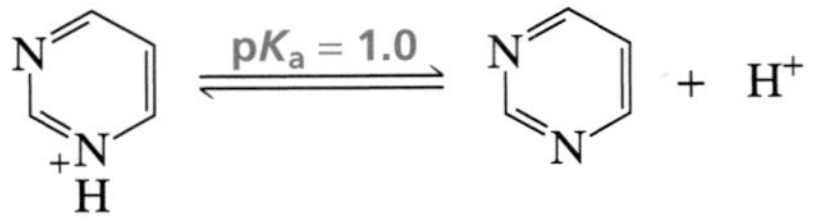

The pK_a values of the N-1 hydrogen in uracil, thymine, and cytosine are 9.5, 9.8, and 12.1, respectively. The relative acidity of this hydrogen means that substituents can easily be placed on N-1.

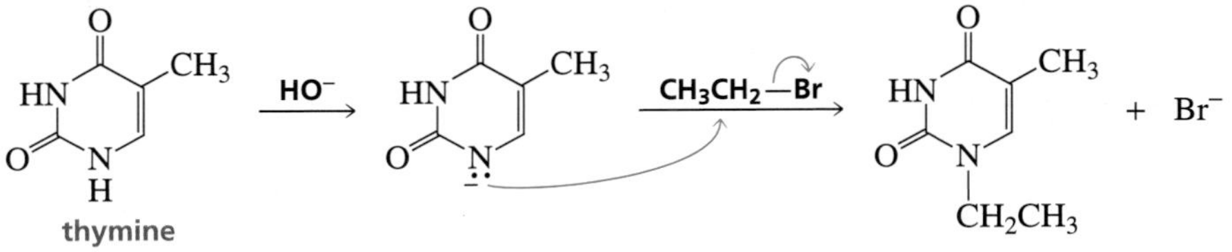

thymine

Because it has two electron-withdrawing hetero atoms, pyrimidine is considerably less reactive than pyridine toward electrophilic aromatic substitution. It can undergo electrophilic aromatic substitution only if there are strongly electron-donating substituents on the ring. Cytosine and uracil undergo electrophilic aromatic substitution reactions because of their electron-donating amino and/or "hydroxy" substituents. The most activated position of unsubstituted pyrimidine toward electrophilic aromatic substitution is C-5 (it corresponds to C-3 of pyridine), and the electron-donating substituents of cytosine and uracil are located so that they further activate this position. 5-Bromouracil is used as an anticancer agent (Section 24.16).

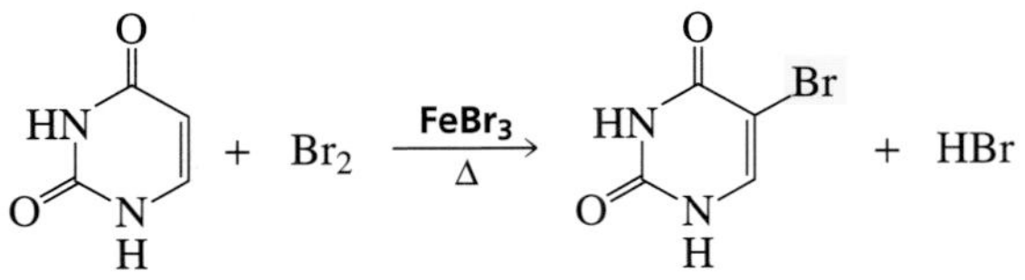

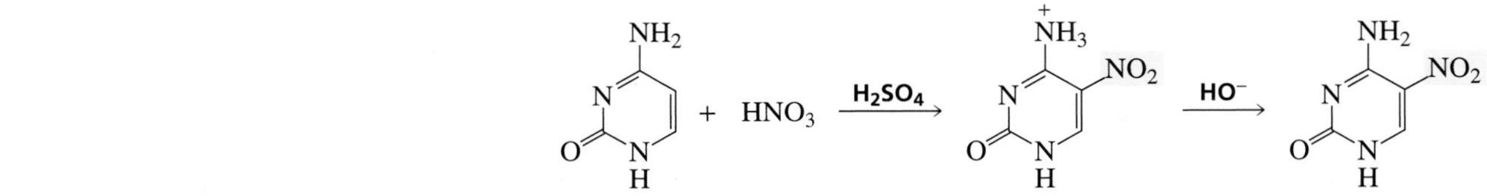

uracil   5-bromouracil

cytosine   5-nitrocytosine

Pyrimidine is more reactive than pyridine toward nucleophilic aromatic substitution for the same reason it is less reactive than pyridine toward electrophilic aromatic substitution.

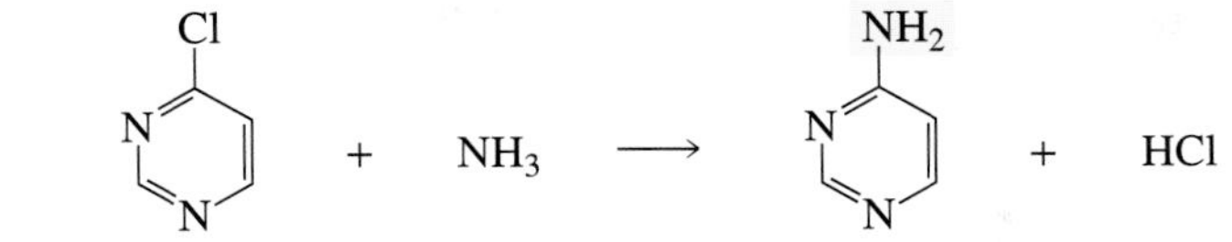

4-chloropyrimidine + NH₃ ⟶ 4-aminopyrimidine + HCl

Purine consists of a pyrimidine ring fused to an imidazole ring, so it has the properties of both ring systems. The electron-donating imidazole ring makes the protonated pyrimidine ring less acidic than unsubstituted protonated pyrimidine ($pK_a = 1.0$). The electron-withdrawing pyrimidine ring makes the hydrogen on N-9 ($pK_a = 8.9$) more acidic than the corresponding N-1 hydrogen of imidazole ($pK_a = 14.4$).

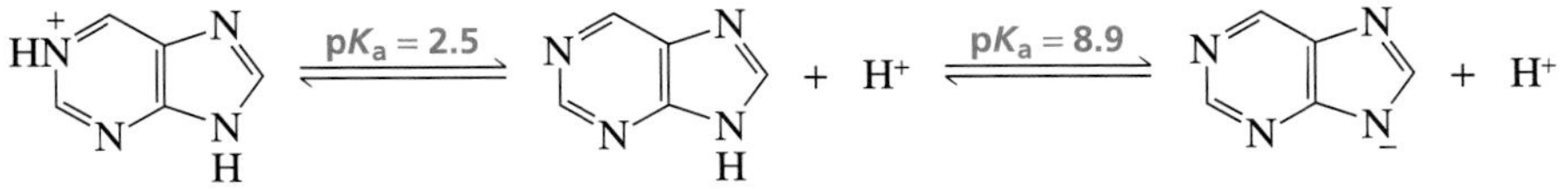

Adenine can be deaminated to give hypoxanthine, and guanine can be deaminated to give xanthine. **Deamination** means loss of ammonia as a result of hydrolysis. Hypoxanthine and xanthine are both enzymatically oxidized to uric acid. Humans form about 600 mg of uric acid each day as a result of the metabolism of no-longer-needed nucleic acids. The uric acid is oxidized to allantoin, which is excreted (Section 15.4).

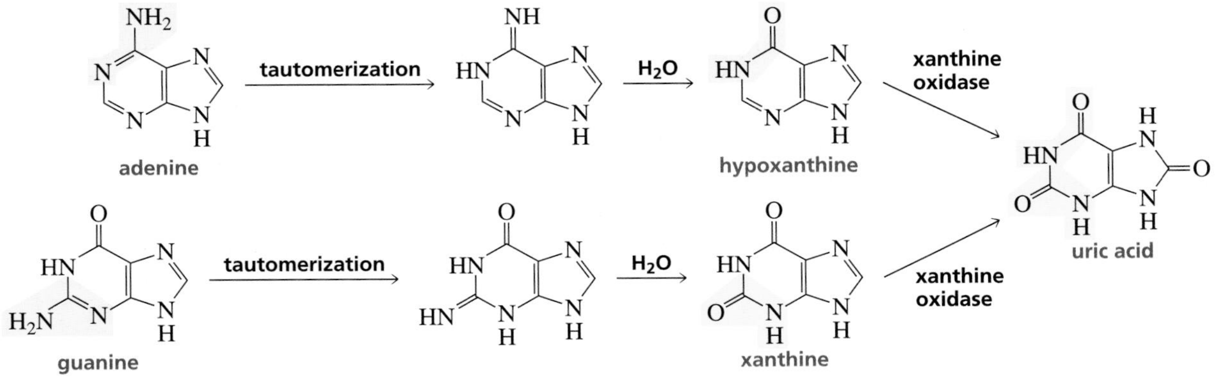

Porphyrin

Substituted porphyrins are important, naturally occurring heterocyclic compounds. A **porphyrin ring system** consists of four pyrrole rings joined by one-carbon bridges. Heme, which is found in hemoglobin and myoglobin, contains an iron atom (Fe^{2+}) ligated by the four nitrogens of a porphyrin ring system. **Ligation** is the sharing of nonbonding electrons with a metal ion. The porphyrin ring system of heme is known as **protoporphyrin IX.** The ring system plus the iron atom is called **iron protoporphyrin IX.** In hemoglobin or myoglobin, the iron is also ligated to a histidine of the protein component (globin), and its sixth ligand is oxygen or carbon dioxide.

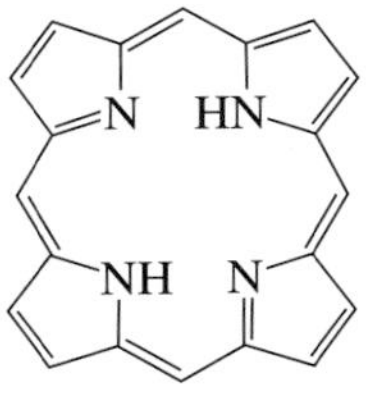

a porphyrin ring system

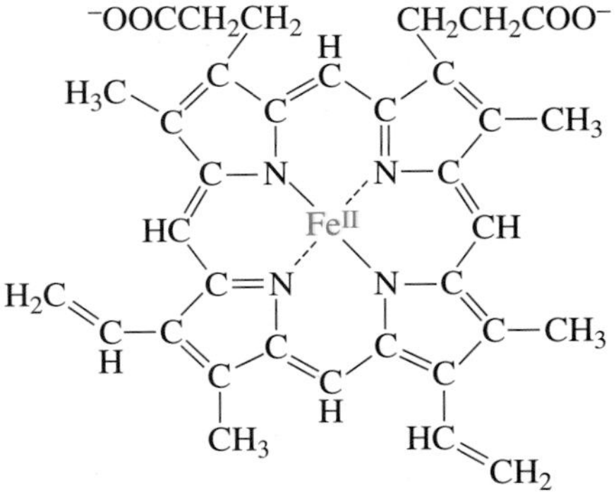

iron protoporphyrin IX
heme

Hemoglobin is responsible for transporting oxygen to cells and carbon dioxide away from cells, while myoglobin is responsible for oxygen storage in cells. Carbon monoxide binds more tightly than does oxygen to Fe^{2+}. Consequently, breathing carbon monoxide can be fatal because it prevents the transport of oxygen.

The extensive, conjugated system of porphyrin gives blood its characteristic red color. Its high molar absorptivity (about 160,000) allows concentrations as low as 1×10^{-8} M to be detected by UV spectroscopy (Section 13.16).

The biosynthesis of porphyrin involves the formation of porphobilinogen from two molecules of δ-aminolevulinic acid. The precise mechanism for this biosynthesis is not known. A possible mechanism starts with the formation of a Schiff base between the enzyme that catalyzes the reaction and one of the molecules of δ-aminolevulinic acid. An aldol condensation occurs between the Schiff base and a free molecule of δ-aminolevulinic acid. Nucleophilic attack by the amino group on the imine closes the ring. The enzyme is then eliminated, and removal of a proton creates the aromatic ring.

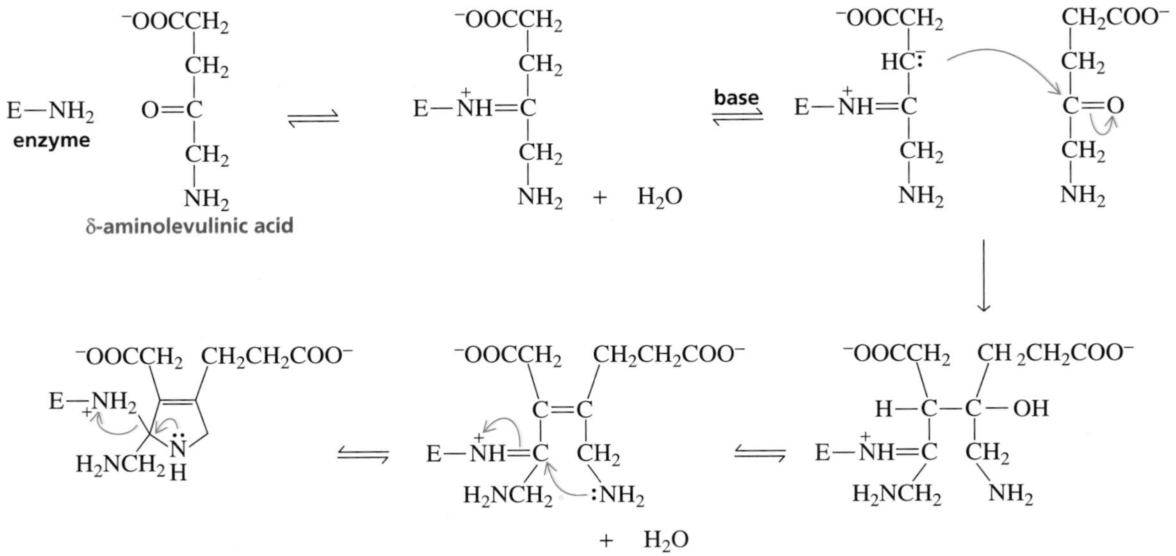

δ-aminolevulinic acid

enzyme

base

+ H₂O

+ H₂O

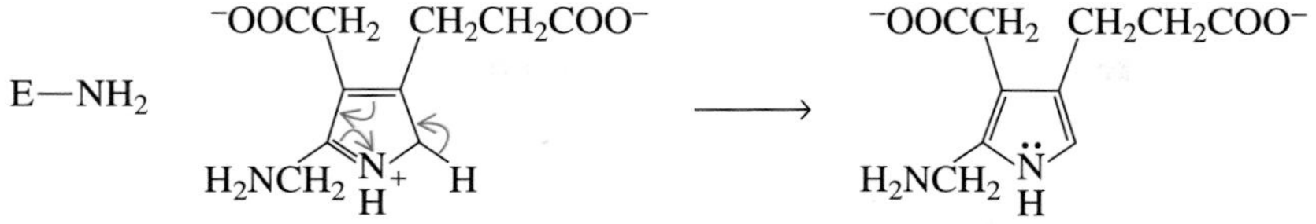

porphobilinogen

Four porphobilinogen molecules react to form porphyrin.

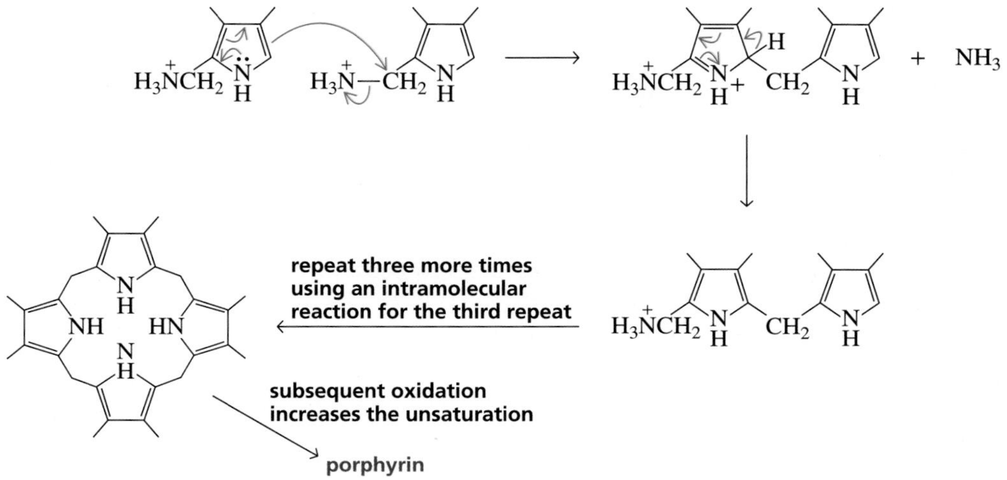

+ NH_3

**repeat three more times
using an intramolecular
reaction for the third repeat**

**subsequent oxidation
increases the unsaturation**

porphyrin

The ring system in chlorophyll *a,* the substance that causes plants to be green, is similar to porphyrin but contains a cyclopentanone ring, and one of its pyrrole rings is partially reduced. The metal atom in chlorophyll *a* is magnesium (Mg^{2+}).

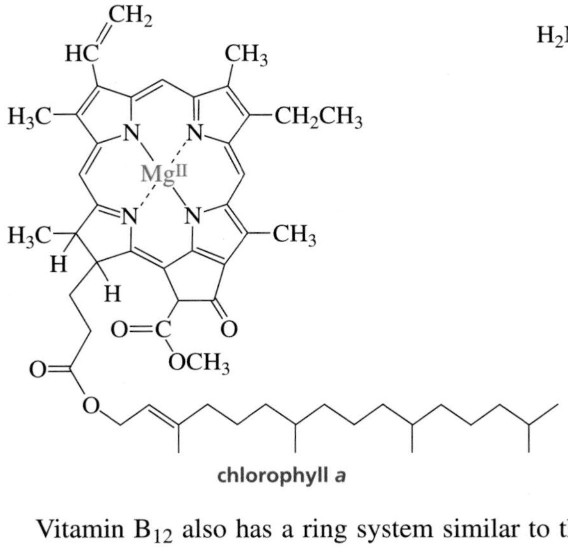

chlorophyll *a*

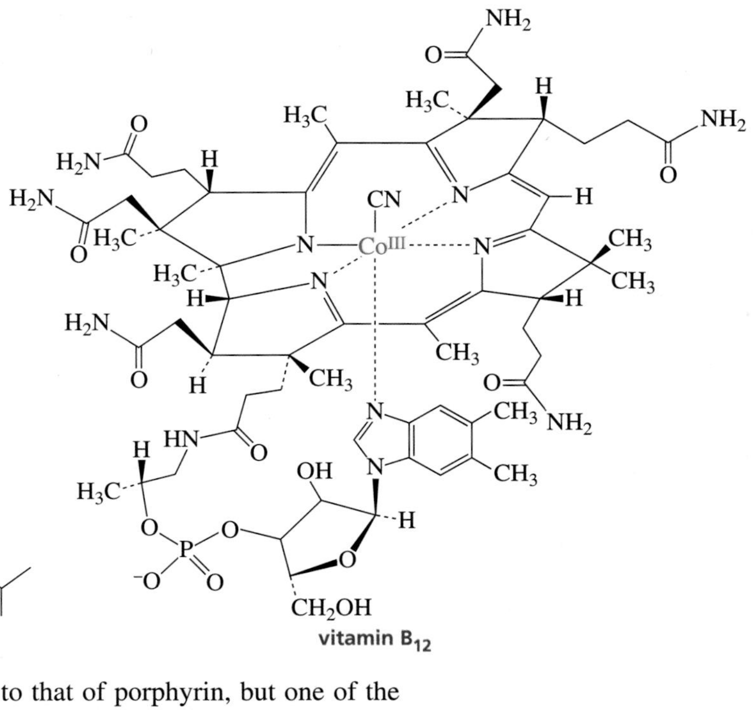

vitamin B$_{12}$

Vitamin B_{12} also has a ring system similar to that of porphyrin, but one of the methine bridges is missing. This is known as a **corrin ring system.** The metal atom in vitamin B_{12} is cobalt (Co^{3+}). The chemistry of this vitamin is discussed in Section 22.7.

PROBLEM 20◆

Is porphyrin aromatic?

PROBLEM 21

Show how the last two porphobilinogen molecules are incorporated into the porphyrin ring.

PORPHYRIN, BILIRUBIN, AND JAUNDICE

The average human turns over about 6 g of hemoglobin each day. The protein portion (globin) and the iron are reutilized, but the porphyrin ring is broken down. First it is reduced to biliverdin, a green compound, which is subsequently reduced to bilirubin, a yellow compound. If more bilirubin is formed than can be excreted by the liver, bilirubin accumulates in the blood. When the concentration of bilirubin in the blood reaches a certain level, it diffuses into the tissues, causing them to become yellow. This condition is known as jaundice.

SUMMARY OF REACTIONS

1. Ring-opening reactions (Section 26.1)

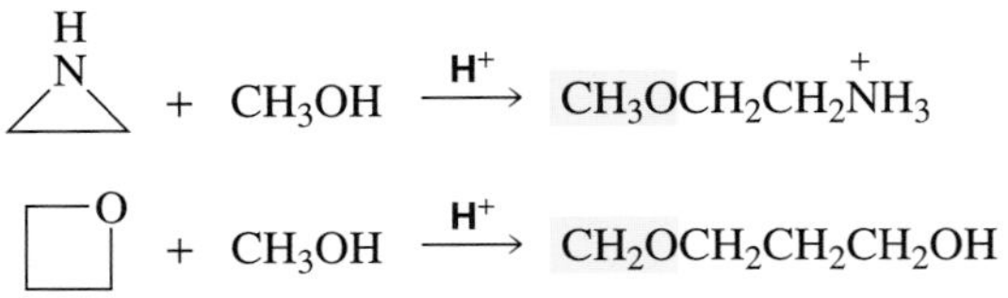

$$+ CH_3OH \xrightarrow{H^+} CH_3OCH_2CH_2\overset{+}{N}H_3$$

$$+ CH_3OH \xrightarrow{H^+} CH_2OCH_2CH_2CH_2OH$$

2. Electrophilic aromatic substitution reactions
   a. pyrrole, furan, and thiophene (Section 26.2)

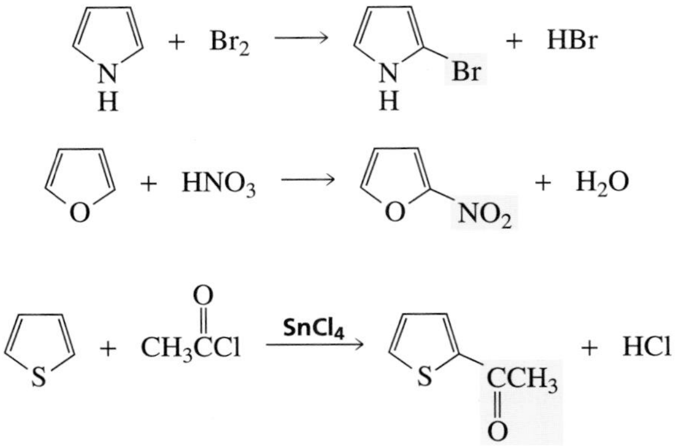

$$+ Br_2 \longrightarrow \quad + HBr$$

$$+ HNO_3 \longrightarrow \quad + H_2O$$

$$+ CH_3\overset{O}{\underset{\|}{C}}Cl \xrightarrow{SnCl_4} \quad + HCl$$

   b. pyridine and pyridine-*N*-oxide (Section 26.3)

c. indole, benzofuran, and benzothiophene (Section 26.2)

d. quinoline and isoquinoline (Section 26.3)

3. Nucleophilic substitution reactions
    a. pyridine and pyrimidine (Sections 26.3 and 26.4)

    b. quinoline and isoquinoline (Section 26.3)

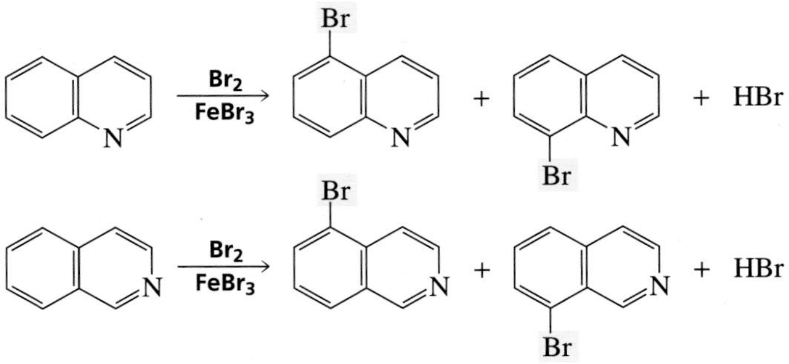

KEY TERMS

alkaloid (page 1141)
aziridine (page 1145)
Chichibabin reaction (page 1155)
corrin ring system (page 1165)
deamination (page 1163)
furan (page 1146)
hetero atom (page 1141)

heterocycle (page 1141)
heterocyclic compound (page 1141)
imidazole (page 1160)
iron protoporphyrin IX (page 1163)
ligation (page 1163)
natural product (page 1141)
porphyrin ring system (page 1163)

protoporphyrin IX (page 1163)
purine (page 1161)
pyrimidine (page 1161)
pyridine (page 1151)
pyrimidine (page 1161)
pyrrole (page 1146)
saturated heterocycles (page 1142)
thiophene (page 1146)

PROBLEMS

22. Name the following compounds.

a.

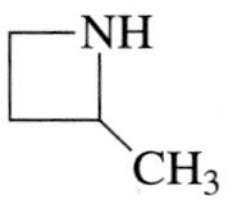

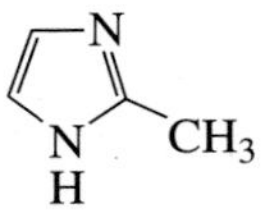

b.

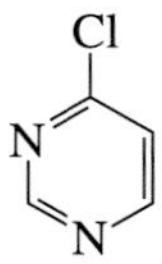

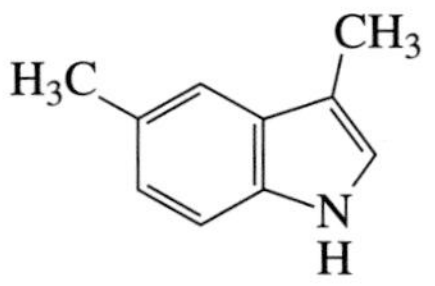

c.

d.

e.

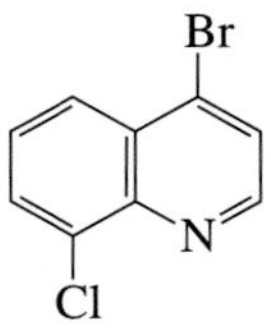

f.

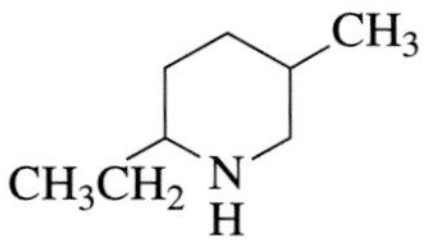

23. Give the product of each of the following reactions.

a. 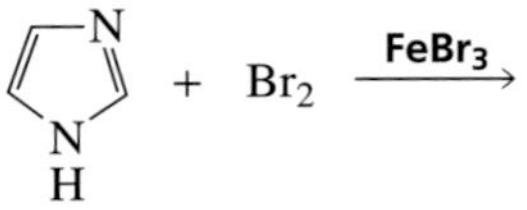 + Br₂ $\xrightarrow{\text{FeBr}_3}$

g. 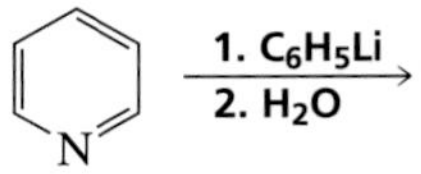 $\xrightarrow[\text{2. H}_2\text{O}]{\text{1. C}_6\text{H}_5\text{Li}}$

b. [structure: 4-bromopyrimidine] + [structure: piperidine] $\longrightarrow$

h. [structure: 2-methylfuran] + CH₃$\overset{\text{O}}{\overset{\|}{\text{C}}}$Cl $\longrightarrow$

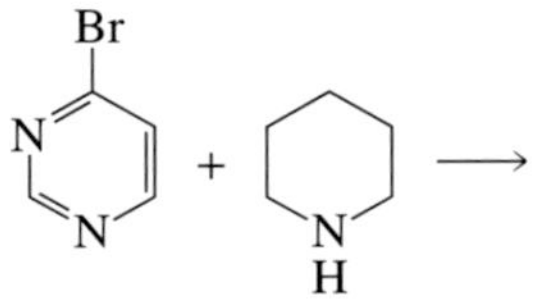

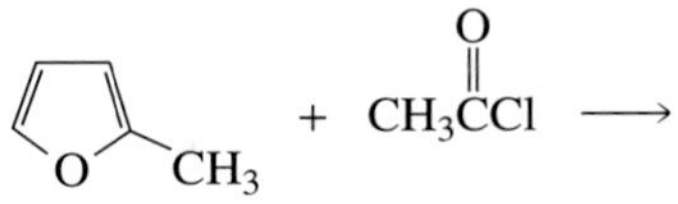

c. [structure: 2-methylazetidine with NH] + CH₃CH₂OH $\xrightarrow{\text{H}^+}$

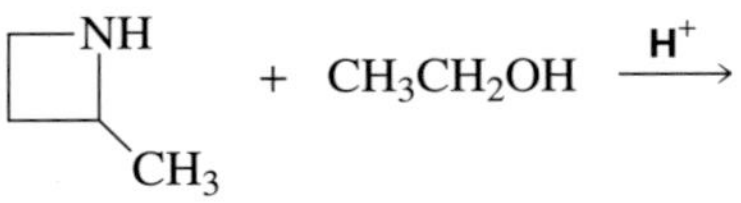

i. 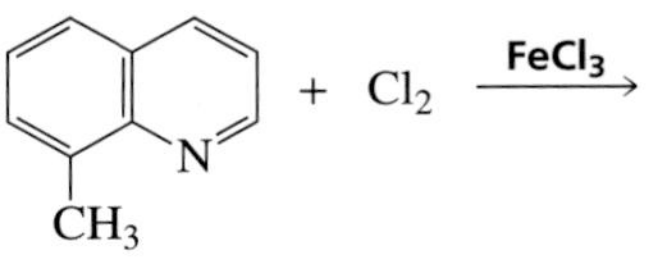 [structure: 8-methylquinoline] + Cl₂ $\xrightarrow{\text{FeCl}_3}$

d. 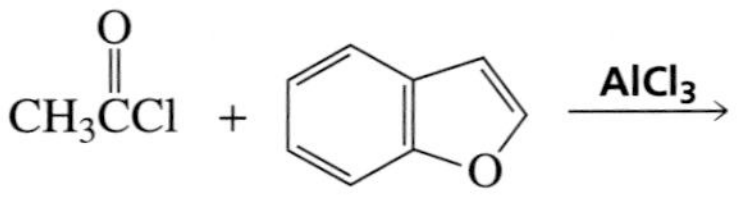 CH₃$\overset{\text{O}}{\overset{\|}{\text{C}}}$Cl + [structure: benzofuran] $\xrightarrow{\text{AlCl}_3}$

j. 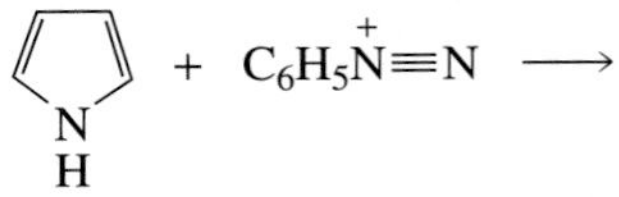 [structure: pyrrole] + C₆H₅$\overset{+}{\text{N}}$≡N $\longrightarrow$

e. 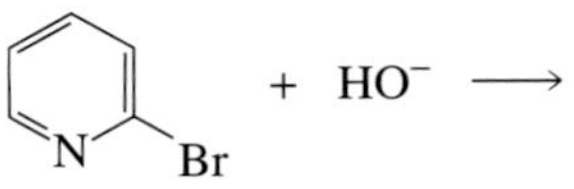 [structure: 2-bromopyridine] + HO⁻ $\longrightarrow$

k. 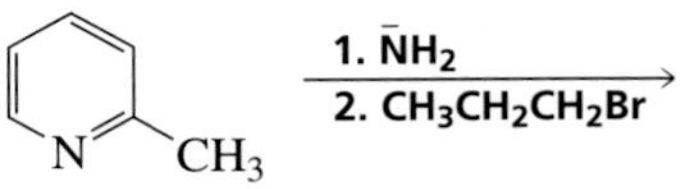 [structure: 2-methylpyridine] $\xrightarrow[\text{2. CH}_3\text{CH}_2\text{CH}_2\text{Br}]{\text{1. }\overset{-}{\text{N}}\text{H}_2}$

f. 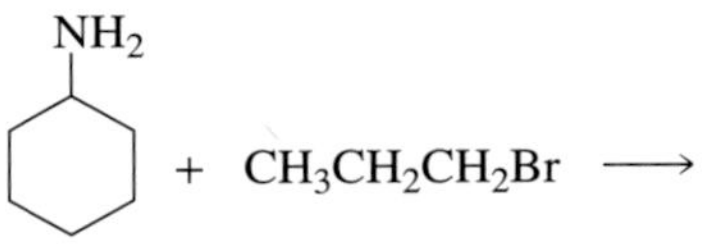 [structure: cyclohexylamine] + CH₃CH₂CH₂Br $\longrightarrow$

l. 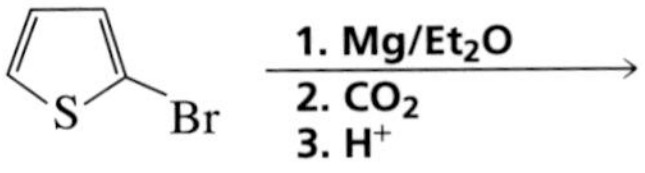 [structure: 2-bromothiophene] $\xrightarrow[\text{3. H}^+]{\substack{\text{1. Mg/Et}_2\text{O} \\ \text{2. CO}_2}}$

24. List the following compounds in order of decreasing acidity.

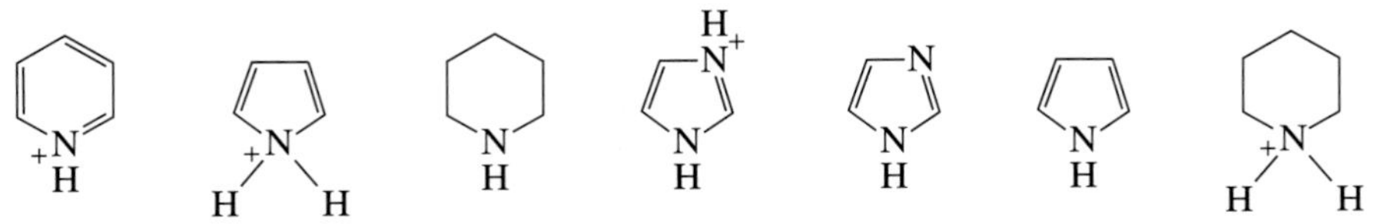

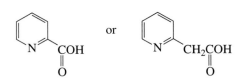

25. Which of the following compounds is easier to decarboxylate?

[structure: pyridine-2-carboxylic acid]   or   [structure: 2-pyridylacetic acid]

26. Rank the following compounds in order of decreasing reactivity in an electrophilic aromatic substitution reaction.

27. One of the following compounds undergoes electrophilic aromatic substitution predominantly at C-3, and one undergoes electrophilic aromatic substitution predominantly at C-4. Which is which?

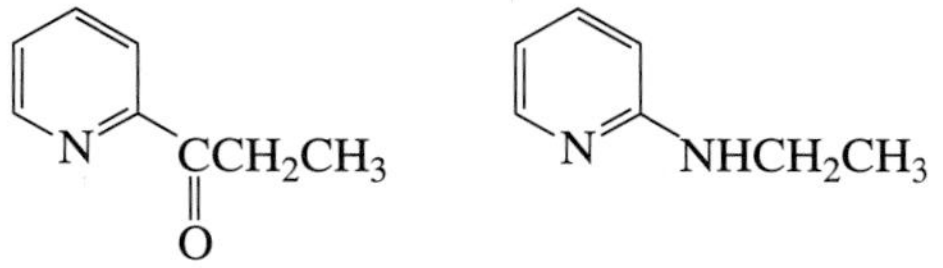

28. Benzene undergoes electrophilic aromatic substitution reactions with aziridines in the presence of a Lewis acid such as $AlCl_3$.
   a. What are the major and minor products of the following reaction?

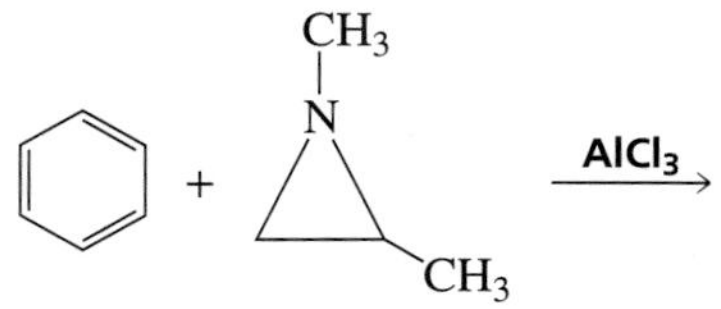

   b. Do epoxides undergo similar reactions?

29. The dipole moments of furan and tetrahydrofuran are in the same direction. One compound has a dipole moment of 0.70 D, and the other, 1.73 D. Which is which?

30. Show how the vitamin niacin can be synthesized from nicotine.

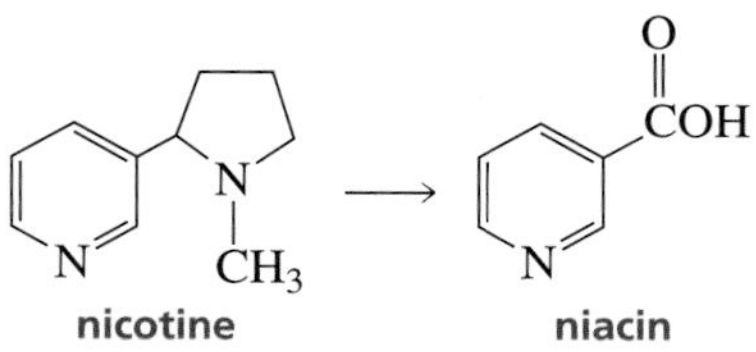

31. Propose a mechanism for the following reaction.

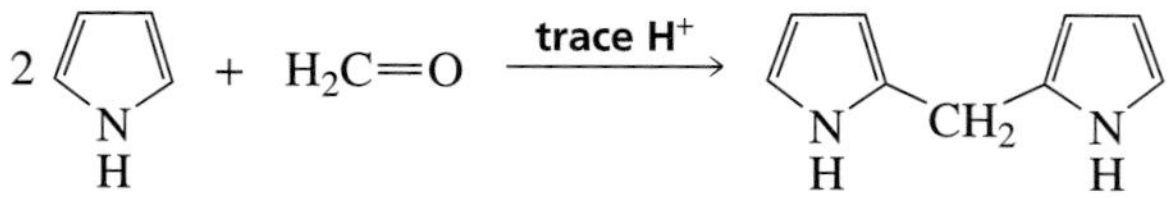

32. The chemical shifts of the C-2 hydrogen in the NMR spectra of pyrrole, pyridine, and pyrrolidine are: δ 2.82, δ 6.42, and δ 8.50. Match each chemical shift with its heterocycle.

33. Explain why protonation of aniline has a dramatic effect on its UV spectrum, while protonation of pyridine has only a small effect on its UV spectrum.

34. Explain why pyrrole is a much stronger acid than ammonia.

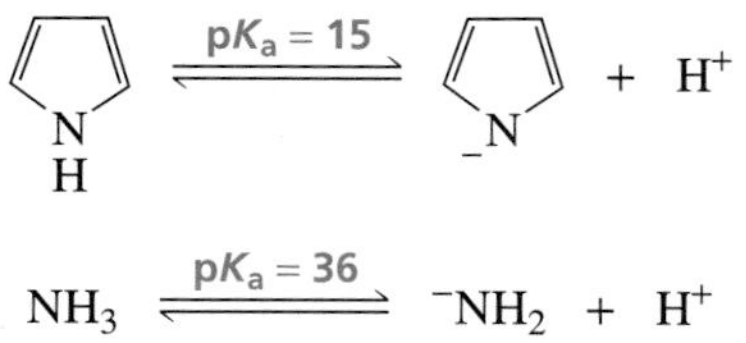

35. Quinolines are commonly synthesized by a method known as the *Skraup synthesis.* This synthesis involves reaction of aniline with glycerol under acidic conditions. Nitrobenzene is added to the reaction mixture to serve as an oxidizing agent. The first step in the synthesis is the dehydration of glycerol to acrolein.

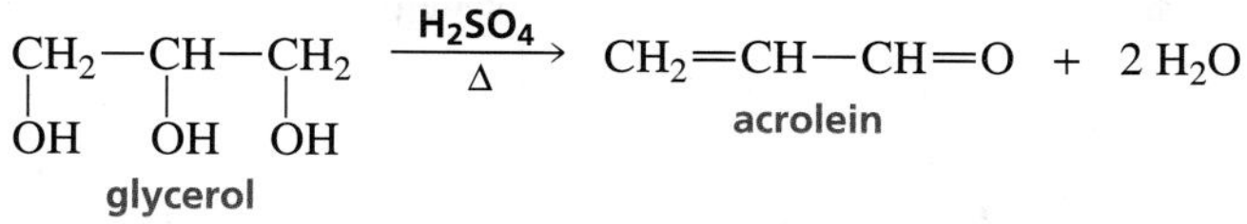

$$CH_2-CH-CH_2 \xrightarrow[\Delta]{H_2SO_4} CH_2=CH-CH=O + 2\ H_2O$$

$$\underset{\text{glycerol}}{\underset{OH\quad OH\quad OH}{|\quad\ \ |\quad\ \ |}}$$

acrolein

a. What product would be obtained if *para*-ethylaniline were used instead of aniline?
b. What product would be obtained if 3-hexen-2-one were used instead of glycerol?
c. What starting materials are needed for the synthesis of 2,7-diethyl-3-methylquinoline?

36. Propose a mechanism for each of the following reactions.

a. 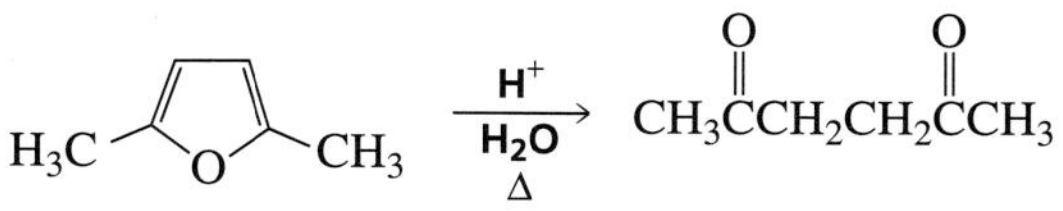 $\xrightarrow[\underset{\Delta}{H_2O}]{H^+}$ $\underset{O}{CH_3}\overset{O}{C}CH_2CH_2\overset{O}{C}CH_3$

b. 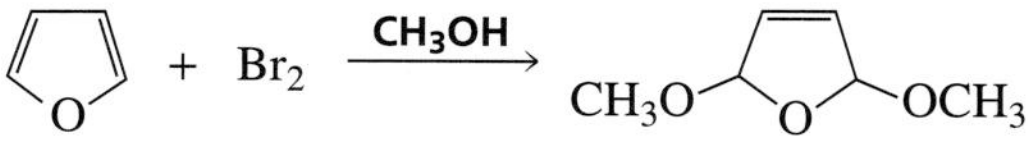 $+\ Br_2 \xrightarrow{CH_3OH}$

37. Give the major product of each of the following reactions.

a. 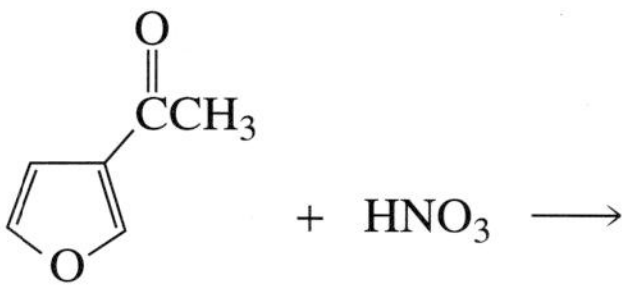 $+\ HNO_3 \longrightarrow$

b. 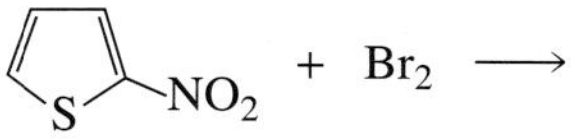 $+\ Br_2 \longrightarrow$

c. 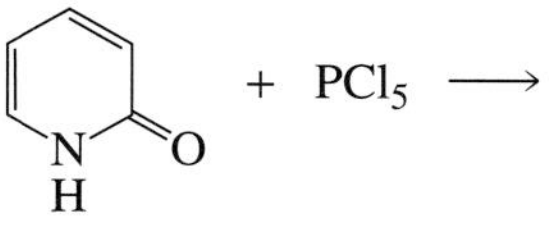 $+\ PCl_5 \longrightarrow$

d. 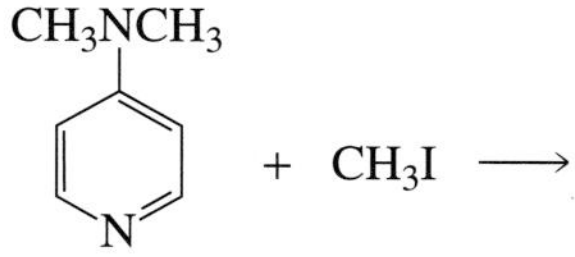 $+\ CH_3I \longrightarrow$

e. 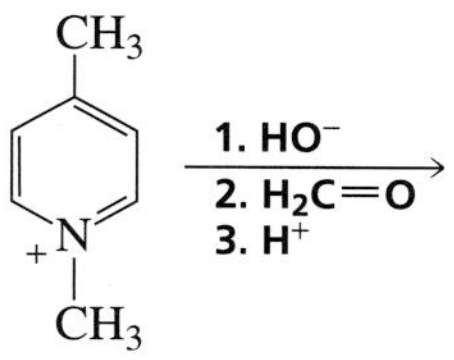 $\xrightarrow[\text{3. H}^+]{\begin{array}{l}\text{1. HO}^-\\\text{2. H}_2\text{C}=\text{O}\\\text{3. H}^+\end{array}}$

f. 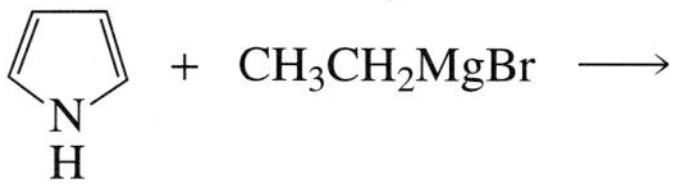 $+\ CH_3CH_2MgBr \longrightarrow$

g. 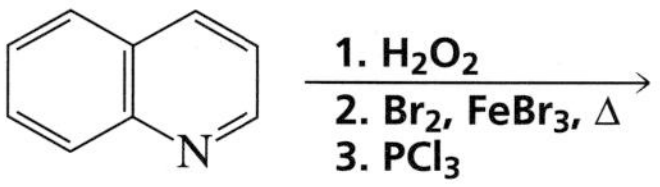 $\xrightarrow[\text{3. PCl}_3]{\begin{array}{l}\text{1. H}_2\text{O}_2\\\text{2. Br}_2\text{, FeBr}_3\text{, }\Delta\\\text{3. PCl}_3\end{array}}$

38. Propose a mechanism for the following reaction.

$+\ CH_3\overset{O}{C}O\overset{O}{C}CH_3 \longrightarrow$ $+\ CH_3\overset{O}{C}O^-$

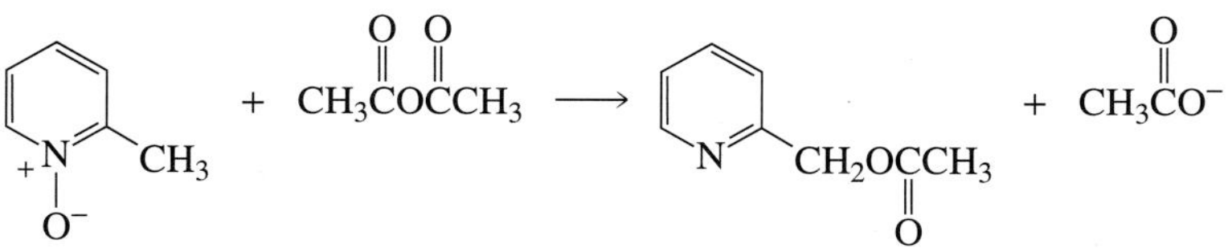

39. Propose a different mechanism than the one shown in Section 26.4 for the biosynthesis of porphobilinogen.

40. Pyrrole reacts with excess *para*-(*N,N*-dimethylamino)benzaldehyde to form a highly colored compound. Give the structure of the colored compound.

41. 2-Phenylindole is prepared from the reaction of acetophenone and phenylhydrazine, a method known as the Fischer indole synthesis. Propose a mechanism for this reaction. (*Hint:* The reactive species is the enamine tautomer of the phenylhydrazone.)

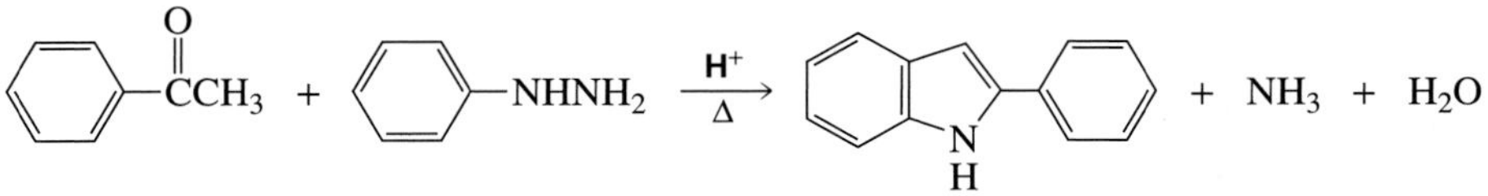

42. What starting materials are required for the synthesis of the following compounds using the Fischer indole synthesis? (*Hint:* See Problem 41.)

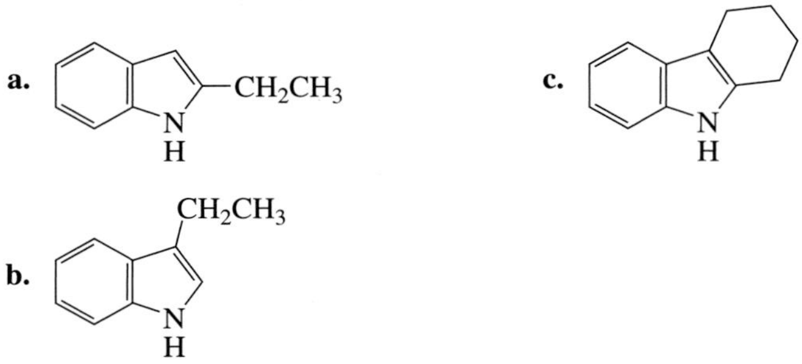

a.

b.

c.

43. Organic chemists work with tetraphenylporphyrins rather than porphyrins because of the greater resistance of the former to air oxidation. Tetraphenylporphyrin can be prepared by the reaction of benzaldehyde with pyrrole. Propose a mechanism for the formation of the ring system shown.

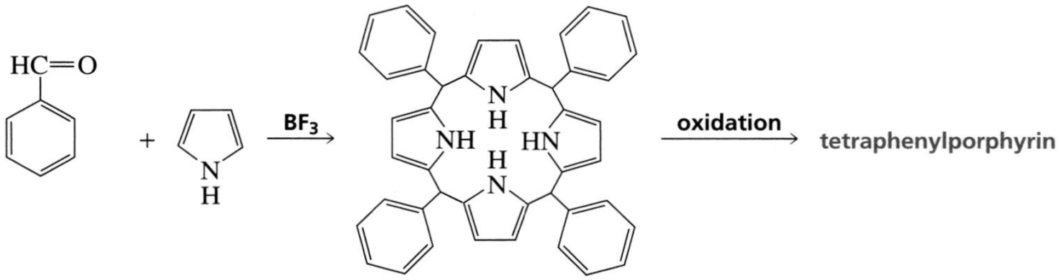

27

PERICYCLIC REACTIONS

vitamin D

Reactions of organic compounds can be divided into three classes: polar reactions, radical reactions, and pericyclic reactions. The most common are polar reactions. A **polar reaction** is a reaction in which a nucleophile reacts with an electrophile. Both electrons in the new bond come from the nucleophile.

a polar reaction

$$H:\ddot{\ddot{O}}:^- \ + \ \overset{\delta+}{CH_3}\!-\!\overset{\delta-}{Br} \ \longrightarrow \ CH_3OH \ + \ Br^-$$

A **radical reaction** is a reaction in which a new bond is formed using one electron from each of the reactants.

a radical reaction

$$CH_3\dot{C}H_2 \ + \ Cl\!-\!Cl \ \longrightarrow \ CH_3CH_2Cl \ + \ \cdot Cl$$

Pericyclic reactions make up the third class of organic reactions. A **pericyclic reaction** occurs as a result of a cyclic reorganization of electrons. In this chapter we will look at the three most common types of pericyclic reactions: electrocyclic reactions, cycloaddition reactions, and sigmatropic rearrangements.

The first class of pericyclic reactions that we will consider are electrocyclic reactions. An **electrocyclic reaction** is an intramolecular reaction in which a new σ (sigma) bond is formed between the ends of a conjugated π (pi) system. This reaction is easy to recognize: the product is a cyclic compound that has one less π bond than the reactant.

**27.1
THREE KINDS
OF PERICYCLIC
REACTIONS**

an electrocyclic reaction

new σ bond

1,3,5-hexatriene   **1,3,-cyclohexadiene**
the product has one less
π bond than the reactant

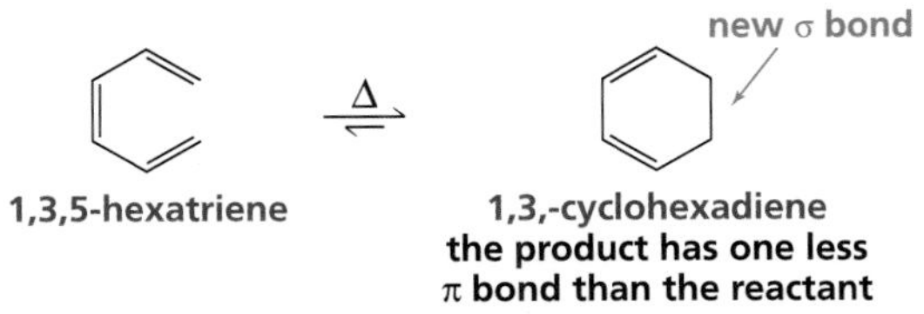

Or, since electrocyclic reactions are reversible, an electrocyclic reaction is a reaction in which a σ bond in a cyclic reactant breaks, forming a conjugated π system that has one more π bond than the cyclic reactant.

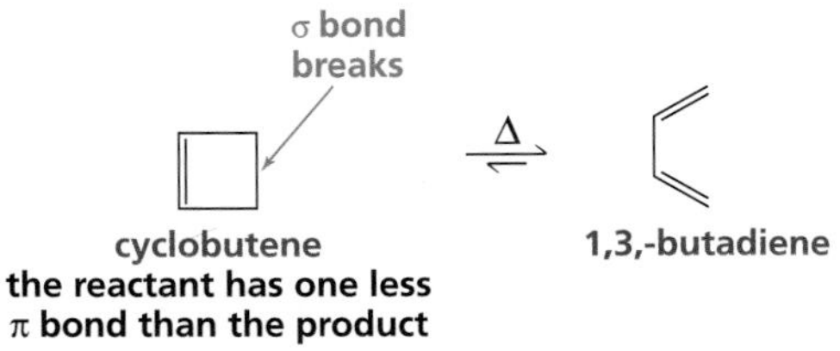

σ bond
breaks

Δ

cyclobutene   **1,3,-butadiene**
the reactant has one less
π bond than the product

In a **cycloaddition reaction,** two different π-bond-containing molecules react to form a cyclic compound. Each of the reactants loses a π bond, and the resulting cyclic product has two new σ bonds. The Diels–Alder reaction is a familiar example of a cycloaddition reaction (Section 7.8).

a cycloaddition reaction

new σ bond

1,3,-butadiene   **ethene**   new σ bond
cyclohexene
the product has two fewer π
bonds than the sum of the
π bonds in the reactants

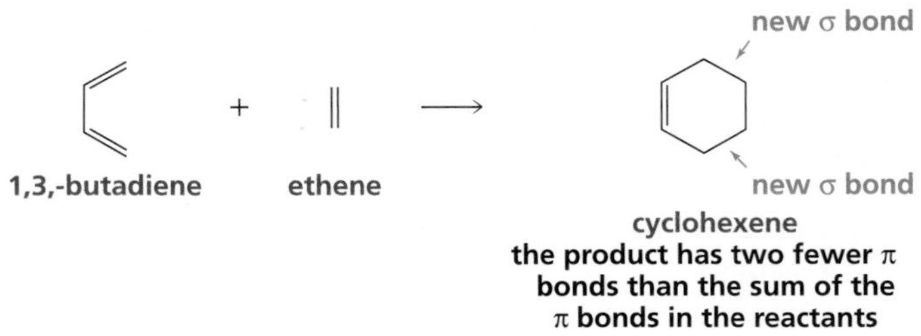

In a **sigmatropic rearrangement,** a σ bond is broken in the reactant, a new σ bond is formed in the product, and the π bonds rearrange. The number of π bonds does not change (the reactant and the product have the same number of π bonds). The σ bond that is broken can be in the middle of the π system or at the end of the π system. The π system consists of the doubly bonded carbons and the carbons immediately adjacent to them.

sigmatropic rearrangements

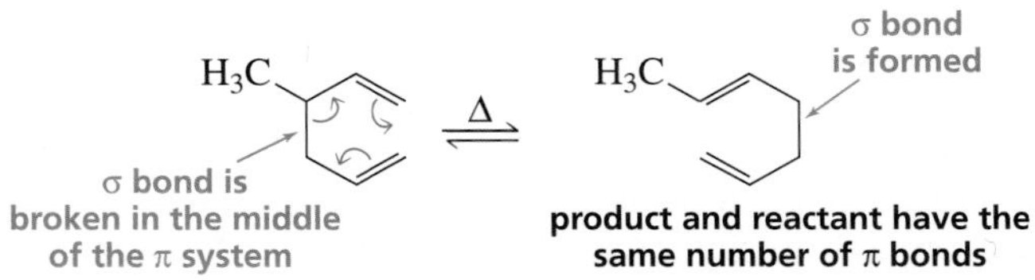

σ bond
is formed

H_3C H_3C

σ bond is
broken in the middle
of the π system

product and reactant have the
same number of π bonds

σ bond is
broken at the end
of the π system

σ bond
is formed

Roald Hoffmann *and* **Kenichi Fukui** *shared the 1981 Nobel Prize in chemistry for the conservation of orbital symmetry theory and the frontier orbital theory.* **R. B. Woodward** *did not receive a share in the prize because he had died 2 years before it was awarded, and Alfred Nobel's will stipulates that the prize cannot be awarded posthumously. Woodward, however, had received the 1965 Nobel Prize in chemistry for his work in organic synthesis (page 1229).*

Notice that electrocyclic reactions and sigmatropic rearrangements occur within a single π system (they are *intra*molecular reactions), while cycloaddition reactions involve the interaction of two different π systems. The three kinds of pericyclic reactions share certain common features.

a. They are all concerted reactions. This means that all the electron reorganization takes place in a single step. Therefore, there is one transition state and no intermediates.

b. Because the reactions are concerted, they are highly stereoselective.

c. The reactions are generally not affected by catalysts or by a change in solvent.

We will see that the configuration of the product that is formed in a pericyclic reaction depends on the following conditions:

a. the configuration of the reactant;

b. the number of double bonds in the reactant;

c. whether the reaction is a thermal reaction or a photochemical reaction.

A **photochemical reaction** is a reaction that takes place when a reactant absorbs light. A **thermal reaction** takes place without the absorption of light. Despite its name, a thermal reaction does not necessarily require more heat than what is available to the reaction at room temperature. Some thermal reactions do require additional heat in order to take place at a reasonable rate, but others readily occur at, or even below, room temperature.

For many years, pericyclic reactions puzzled chemists. Why did some pericyclic reactions take place only under thermal conditions, while others took place only under photochemical conditions, and yet others were successfully carried out under both thermal and photochemical conditions? Another puzzling aspect of pericyclic reactions was the configurations of the products that were formed. After many pericyclic reactions had been investigated, it became apparent that, if a pericyclic reaction could take place under both thermal and photochemical conditions, the configuration of the product formed under one set of conditions was different from the configuration of the product formed under the other set of conditions. For example, if the cis isomer was obtained under thermal conditions, the trans isomer was obtained under photochemical conditions and vice versa.

It took two very talented chemists, each bringing their own expertise to the problem, to solve the puzzling behavior of pericyclic reactions. In 1965, R. B. Woodward, an experimentalist, and Roald Hoffmann, a theorist, developed the **conservation of orbital symmetry theory** to explain the relationship among the structure and configuration of the reactant, the conditions (thermal and/or photochemical) under which the reaction takes place, and the configuration of the product. Since the behavior of pericyclic reactions is so precise, it is not surprising that everything about their behavior can be explained by one simple theory. The difficult part was having the insight to arrive at the theory.

Roald Hoffmann *was born in Poland in 1937 and received a B.S. from Columbia and a Ph.D. from Harvard. When the conservation of orbital symmetry theory was proposed, he and Woodward were both on the faculty at Harvard. Hoffmann is presently a professor of chemistry at Cornell.*

Kenichi Fukui *was born in Japan in 1918. He was a professor at Kyoto Imperial University until 1982, when he became president of Kyoto University of Industrial Arts.*

The conservation of orbital symmetry theory states that, in a concerted reaction, *the molecular orbitals of the reactant evolve smoothly into the molecular orbitals of the product.* This means that the molecular orbitals of the reactant and product must have similar symmetry. The conservation of orbital symmetry theory was based on the **frontier orbital theory** put forth by Kenichi Fukui in 1954. Fukui's theory had been overlooked because of its mathematical complexity and Fukui's failure to apply it to stereoselective reactions.

According to the conservation of orbital symmetry theory, whether or not a compound will undergo a pericyclic reaction under particular conditions and what product will form if the reaction takes place both depend on molecular orbital symmetry. To understand pericyclic reactions, therefore, we must now spend some time looking at molecular orbital theory. We will then be able to understand how the symmetry of a molecular orbital controls both the conditions under which a pericyclic reaction takes place and the configuration of the product that is formed.

PROBLEM 1 ◆

Examine the following pericyclic reactions. For each reaction, indicate whether it is an electrocyclic reaction, a cycloaddition reaction, or a sigmatropic rearrangement.

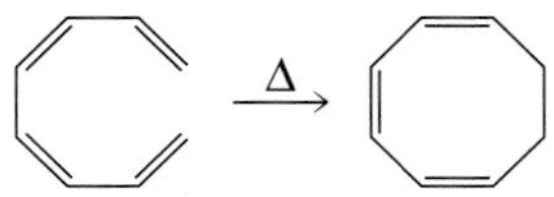

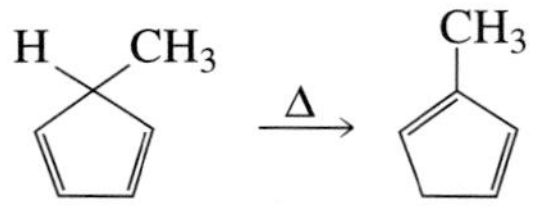

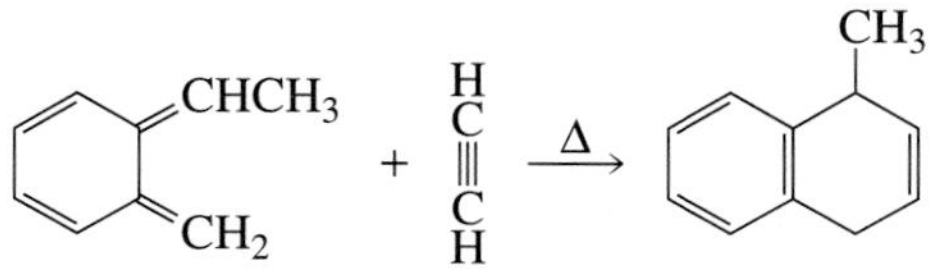

27.2 MOLECULAR ORBITALS AND ORBITAL SYMMETRY

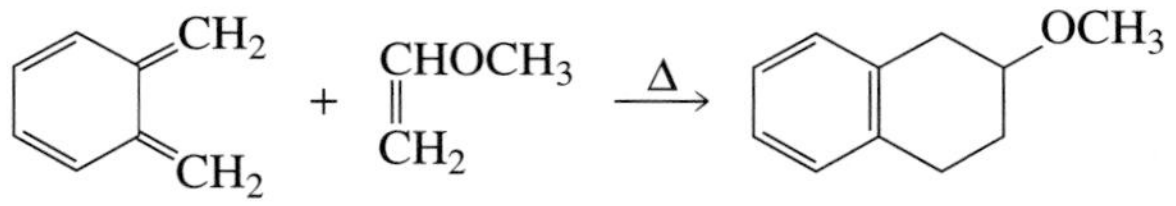

The overlap of *p* orbitals to form π bonds can be described mathematically using quantum mechanics. The result of the mathematical treatment can be described simply in nonmathematical terms using **molecular orbital (MO) theory.** You were introduced to molecular orbital theory in Sections 1.6, 6.15, and 7.3, where you learned the following:

   a. The two lobes of a *p* orbital are of opposite phase. When two in-phase atomic orbitals overlap, a covalent bond is formed. When two out-of-phase atomic orbitals overlap, a node is created between the two nuclei.

   b. Electrons fill molecular orbitals according to the same rules that are used when they fill atomic orbitals: an electron goes into the available molecular orbital with the lowest energy, and only two electrons can occupy a molecular orbital (Section 1.2).

You should take a few minutes to review these sections.

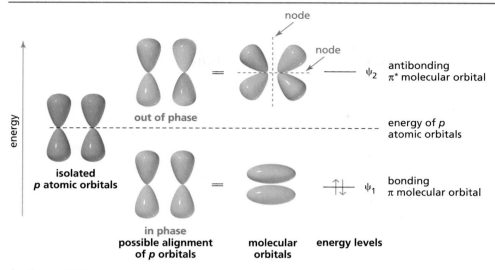

▲ Figure 27.1
Overlap of in-phase *p* atomic orbitals gives a bonding π molecular orbital that is lower in energy than the *p* atomic orbitals. Overlap of out-of-phase *p* atomic orbitals gives an antibonding π molecular orbital that is higher in energy than the *p* atomic orbitals.

Since the π-bonding portion of a molecule is perpendicular to the framework of the σ bonds, the π bonds can be treated independently. Each carbon atom that forms a π bond has a *p* orbital, and the *p* orbitals of the carbon atoms combine to produce a π molecular orbital. Thus a molecular orbital can be described by the **linear combination of atomic orbitals (LCAO).** In a π molecular orbital, each electron that previously occupied a *p* atomic orbital surrounding an individual carbon nucleus now surrounds the entire part of the molecule that is included in the overlapping *p* orbitals.

A molecular orbital description of ethene is shown in Figure 27.1. (To show the different phases of the two lobes of a *p* orbital, one phase is represented by a blue lobe and the other phase by a lavender lobe).[1] Because ethene has one π bond, it has two *p* atomic orbitals that combine to produce two π molecular orbitals. The in-phase overlap of the two *p* atomic orbitals of ethene gives a **bonding π molecular orbital,** designated by ψ_1 (ψ is the Greek letter psi). The bonding molecular orbital is of lower energy than the isolated *p* atomic orbitals. The two *p* atomic orbitals of ethene can also overlap out of phase. Overlapping of out-of-phase orbitals gives an **antibonding π molecular orbital,** ψ_2, which is of higher energy than the *p* atomic orbitals. The bonding molecular orbital results from additive overlap of the atomic orbitals, while the antibonding molecular orbital results from subtractive overlap. In other words, the overlap of in-phase orbitals holds atoms together, while the overlap of out-of-phase orbitals pushes atoms apart. Because electrons reside in the available molecular orbitals with the lowest energy and only two electrons can occupy a molecular orbital, the two π electrons reside in the bonding π molecular orbital. This molecular orbital picture describes all molecules with one π bond.

[1]Because the different phases of the *p* orbital result from the different mathematical signs ($+$ and $-$) of the wave function of the electron, some chemists represent the different phases by a $(+)$ and a $(-)$.

Orbitals are conserved: two atomic orbitals combine to produce two molecular orbitals; four atomic orbitals combine to produce four molecular orbitals; six atomic orbitals combine to produce six molecular orbitals.

Because 1,3-butadiene has two conjugated π bonds, it has four p atomic orbitals (Figure 27.2). Four atomic orbitals can combine linearly in four different ways. Consequently, there are four π molecular orbitals: ψ_1, ψ_2, ψ_3, and ψ_4. Notice that orbitals are conserved: four atomic orbitals combine to produce four molecular orbitals. Half of the molecular orbitals are bonding molecular orbitals (ψ_1 and ψ_2), and the other half are antibonding molecular orbitals (ψ_3 and ψ_4). Because the four π electrons will reside in the available molecular orbitals with the lowest energy, there are two electrons in ψ_1 and two electrons in ψ_2. Remember that although the molecular orbitals are of different energies, they are all valid and they all coexist. This molecular orbital picture describes all molecules with two conjugated π bonds.

In-phase orbitals overlap to give a bonding interaction; out-of-phase orbitals overlap to create a node. Recall that a node is a place in which there is zero probability of finding an electron (Section 1.6). If you examine the overlapping orbitals in Figure 27.2, you will see that, as the energy of the molecular orbital increases, the number of bonding interactions decreases and the number of nodes between the nuclei increases: ψ_1 has three bonding interactions and zero nodes between the nuclei; ψ_2 has two bonding interactions and one node between the nuclei; ψ_3 has one bonding interaction and two nodes between the nuclei; and ψ_4 has zero bonding interactions and three nodes between the nuclei. *Notice that a molecular orbital is bonding if the number of bonding interactions is greater than the number of nodes between the nuclei, and a molecular orbital is antibonding if the number of bonding interactions is fewer than the number of nodes between the nuclei.*

The normal electronic configuration of a molecule is known as its **ground state.** In the ground state of 1,3-butadiene, the *highest occupied molecular orbital (HOMO)* is ψ_2, and the *lowest unoccupied molecular orbital (LUMO)* is ψ_3. If a

Figure 27.2 ▶
Four p atomic orbitals overlap to give the four π molecular orbitals of 1,3-butadiene.

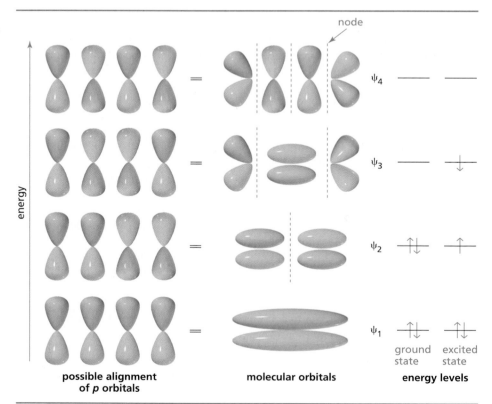

possible alignment of p orbitals

molecular orbitals

energy levels

molecule absorbs light of an appropriate wavelength, the light will promote an electron from its ground-state HOMO to its LUMO (from ψ_2 to ψ_3). The molecule is then in an **excited state.** In the excited state, the HOMO is ψ_3 and the LUMO is ψ_4. In a thermal reaction, the reactant is in its ground state; in a photochemical reaction, the reactant is in an excited state.

Some molecular orbitals are symmetric and some are asymmetric, and they are easy to distinguish. Imagine holding a mirror perpendicular to the plane of the paper along the center of the molecular orbital. If the left half of the molecular orbital is a mirror image of the right half, the molecular orbital is symmetric. If it is not, it is asymmetric. In Figure 27.2, ψ_1 and ψ_3 are **symmetric molecular orbitals,** and ψ_2 and ψ_4 are **asymmetric molecular orbitals.** There is an even easier way to determine whether a molecular orbital is symmetric or asymmetric. If the p orbitals at the ends of the molecular orbital are identical (both have blue lobes on the top and lavender lobes on the bottom), the molecular orbital is symmetric. If the two end p orbitals are not identical, the molecular orbital is asymmetric. Notice that as the molecular orbitals increase in energy, they alternate from being symmetric to being asymmetric. *This means that the ground-state HOMO and the excited-state HOMO always have opposite symmetries.* A molecular orbital description of 1,3,5-hexatriene, a compound with three conjugated double bonds, is shown in Figure 27.3. As a review, examine Figure 27.3 to see:

> **The ground-state HOMO and the excited-state HOMO have opposite symmetries.**

- the distribution of electrons in the ground and excited states;
- that the number of bonding interactions decreases and the number of nodes increases as the molecular orbitals increase in energy;
- that the molecular orbitals alternate between being symmetric and being asymmetric.

A knowledge of all the molecular orbitals in a compound is necessary to fully understand its chemistry. However, a great deal can be learned by looking at only two of the orbitals, the **highest occupied molecular orbital (HOMO)** and the **lowest unoccupied molecular orbital (LUMO).** These two molecular orbitals are known as the **frontier orbitals.** Now we will see that—simply by evaluating one of the frontier molecular orbitals of the reactant(s) in a pericyclic reaction—we can predict the conditions under which the reaction will occur (thermal or photochemical, or both) and the products that will be obtained from it.

PROBLEM 2◆

Answer the following questions for the π molecular orbitals of 1,3,5-hexatriene.

a. Which are the bonding orbitals and which are the antibonding orbitals?

b. Which orbitals are the HOMO and the LUMO in the ground state?

c. Which orbitals are the HOMO and the LUMO in the excited state?

d. Which orbitals are symmetric and which are asymmetric?

e. What is the relationship between HOMO and LUMO and symmetric and asymmetric orbitals?

PROBLEM 3◆

a. How many π molecular orbitals does 1,3,5,7-octatetraene have?

b. What is the designation of its HOMO (ψ_1, ψ_2, etc.)?

c. How many nodes does its highest-energy molecular orbital have between the nuclei?

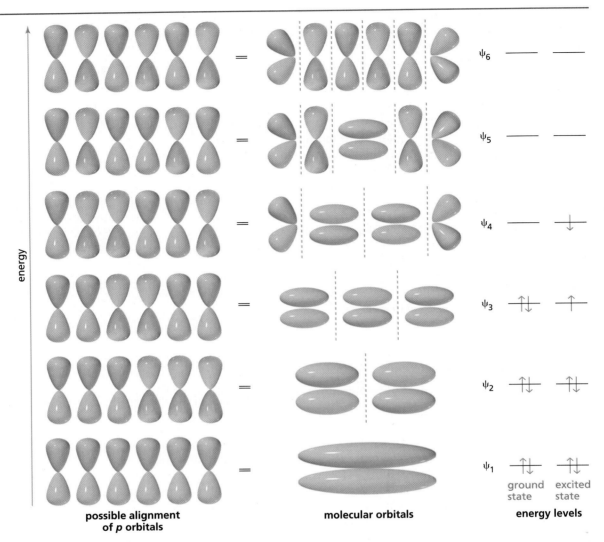

energy

possible alignment
of *p* orbitals

molecular orbitals

ψ_6

ψ_5

ψ_4

ψ_3

ψ_2

ψ_1

ground   excited
state   state

energy levels

▲ Figure 27.3
Six *p* atomic orbitals overlap to give the six π molecular orbitals of 1,3,5-hexatriene.

PROBLEM 4

Give a molecular orbital description for each of the following.

a. 1,3-pentadiene

b. 1,4-pentadiene

c. 1,3,5-heptatriene

d. 1,3,5,8-nonatetraene

27.3 ELECTROCYCLIC REACTIONS

An *electrocyclic reaction* is an intramolecular reaction in which the rearrangement of π electrons leads to a cyclic product; the cyclic product has one less π bond than the reactant. An electrocyclic reaction is completely stereoselective. For example, when (2E,4Z,6E)-octatriene undergoes an electrocyclic reaction under thermal conditions, only the cis product is formed; when (2*E*,4*Z*,6*Z*)-octatriene undergoes an electrocyclic reaction under thermal conditions, only the trans product is formed. Recall that *E* means that the high-priority groups are on opposite sides of the double bond, and *Z* means that the high-priority groups are on the same side of the double bond (Section 3.5).

(2*E*,4*Z*,6*E*)-octatriene   *cis*-**5,6-dimethyl-1,3-cyclohexadiene**

(2*E*,4*Z*,6*Z*)-octatriene   *trans*-**5,6-dimethyl-1,3-cyclohexadiene**

However, when the reactions are carried out under photochemical conditions, the products have the opposite configuration: the compound that forms the cis isomer under thermal conditions forms the trans isomer under photochemical conditions, and the compound that forms the trans isomer under thermal conditions forms the cis isomer under photochemical conditions.

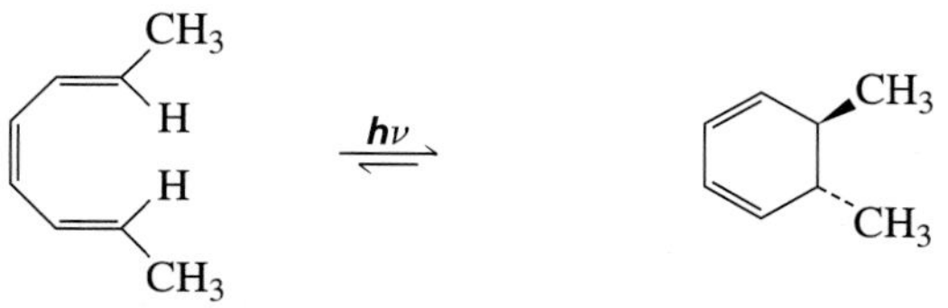

(2*E*,4*Z*,6*E*)-octatriene   *trans*-**5,6-dimethyl-1,3-cyclohexadiene**

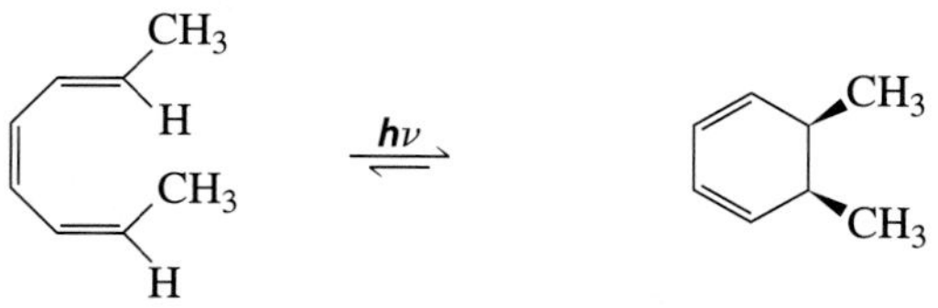

(2*E*,4*Z*,6*Z*)-octatriene   *cis*-**5,6-dimethyl-1,3-cyclohexadiene**

Under thermal conditions, (2*E*,4*Z*)-hexadiene cyclizes to *cis*-3,4-dimethylcy-clobutene, and (2*E*,4*E*)-hexadiene cyclizes to *trans*-3,4-dimethylcyclobutane.

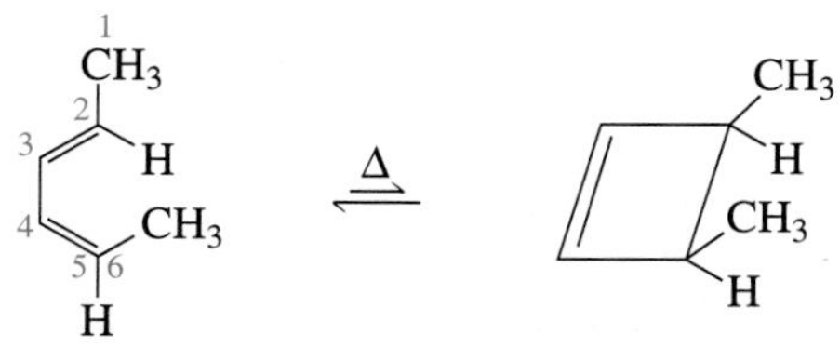

(2*E*,4*Z*)-hexadiene   *cis*-**3,4-dimethylcyclobutene**

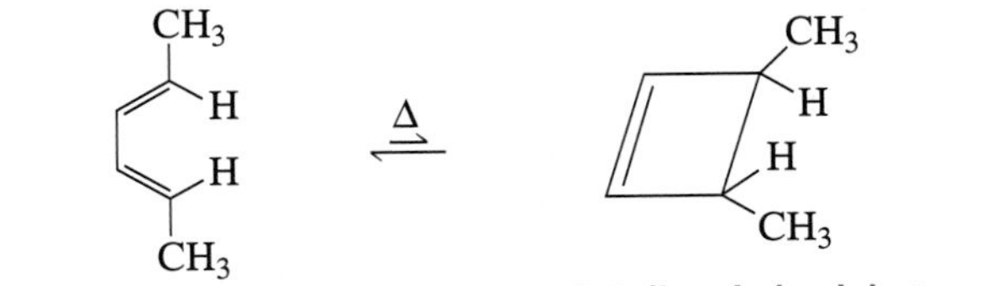

$$(2E,4E)\text{-hexadiene} \quad \xrightarrow{\Delta} \quad trans\text{-3,4-dimethylcyclobutene}$$

(2E,4E)-hexadiene   **trans-3,4-dimethylcyclobutene**

As we saw with the octatrienes, the configuration of the product changes if the reactions are carried out under photochemical conditions: the trans isomer is obtained instead of the cis isomer; the cis isomer is obtained instead of the trans isomer.

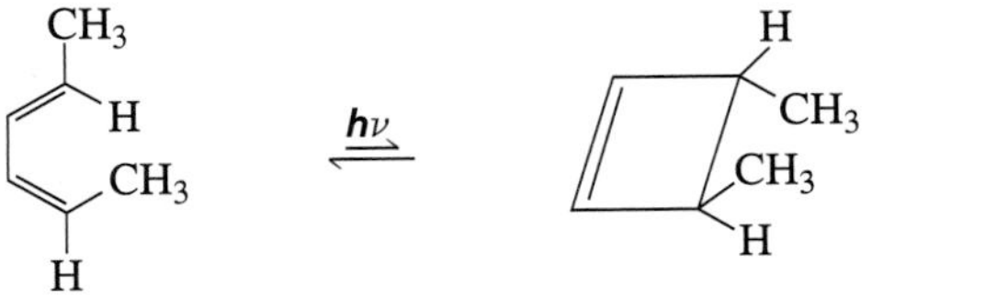

(2E,4Z)-hexadiene $\xrightarrow{h\nu}$ **trans-3,4-dimethylcyclobutene**

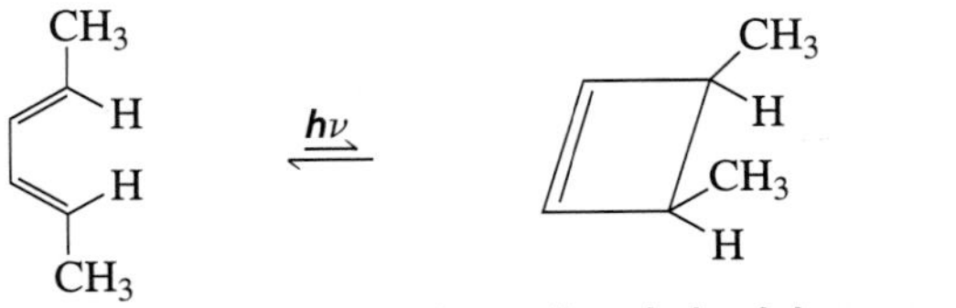

(2E,4E)-hexadiene $\xrightarrow{h\nu}$ **cis-3,4-dimethylcyclobutene**

Electrocyclic reactions are reversible. The cyclic compound is favored for electrocyclic reactions that form six-membered rings, while the open-chain compound is favored for electrocyclic reactions that form four-membered rings because of the angle strain associated with four-membered rings (Section 2.12).

Now we will use what we have learned about molecular orbitals to explain the configuration of the products of the above reactions. We will then be able to predict the configuration of the product of any other electrocyclic reaction.

The product of an electrocyclic reaction is formed as a result of formation of a new σ bond. In order to form this bond, the *p* orbitals at the ends of the conjugated system must rotate so they can overlap head-to-head (and rehybridize to sp^3). Rotation can occur in two ways. If both orbitals rotate in the same direction (both clockwise or both counterclockwise), ring closure is **conrotatory.**

If the orbitals rotate in opposite directions, ring closure is **disrotatory.**

The mode of ring closure depends on the symmetry of the HOMO of the compound undergoing ring closure. Only the symmetry of the HOMO is important in determining the course of the reaction, because it is in this orbital that the highest-energy electrons reside. These are the most loosely held electrons and therefore the ones most easily moved during a reaction.

To form the new σ bond, the orbitals must rotate so that in-phase p orbitals overlap, because in-phase overlap is a bonding interaction. Out-of-phase overlap would be an antibonding interaction. If the HOMO is symmetric (the end orbitals are identical), rotation will have to be disrotatory to achieve in-phase overlap. In other words, disrotatory ring closure is symmetry-allowed, while conrotatory ring closure is symmetry-forbidden.

disrotatory
ring closure

HOMO is symmetric

A **symmetry-allowed pathway** leads to overlap of in-phase orbitals; a **symmetry-forbidden pathway** leads to overlap of out-of-phase orbitals. A symmetry-allowed reaction can take place under relatively mild conditions. If a reaction is symmetry-forbidden, it cannot take place by a concerted pathway. If a symmetry-forbidden reaction takes place at all, it must do so by a nonconcerted mechanism.

If the HOMO is asymmetric, rotation has to be conrotatory in order to achieve in-phase overlap.

A symmetry-allowed pathway requires in-phase orbital overlap.

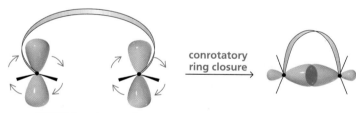

conrotatory
ring closure

HOMO is asymmetric

Now we are ready to learn why the electrocyclic reactions discussed at the beginning of this section formed the indicated products, and why the configuration of the product is changed if the reaction is carried out under photochemical conditions.

The ground-state HOMO (ψ_3) of a compound with three conjugated π bonds, such as (2E,4Z,6E)-octatriene, is symmetric (Figure 27.3). This means that ring closure under *thermal conditions* is disrotatory. In disrotatory ring closure of (2E,4Z,6E)-octatriene, the methyl groups are both pushed up (or down) which results in the *cis* product being formed.

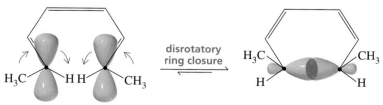

disrotatory
ring closure

(2E,4Z,6E)-octatriene   *cis*-5,6-dimethyl-1,3-cyclohexadiene

CHAPTER 27 Pericyclic Reactions

In disrotatory ring closure of (2E,4Z,6Z)-octatriene, one methyl group is pushed up and the other is pushed down which results in the *trans* product being formed.

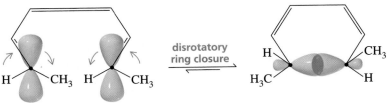

(2E,4Z,6Z)-octatriene   *trans*-5,6-dimethyl-1,3-cyclohexadiene

If the reaction takes place under *photochemical conditions,* we must consider the excited-state HOMO rather than the ground-state HOMO. The excited-state HOMO (ψ_4) of a compound with three π bonds is asymmetric. Therefore, under photochemical conditions, (2E,4Z,6Z)-octatriene undergoes conrotatory ring closure, so both methyl groups are pushed down (or up) and the *cis* product is formed.

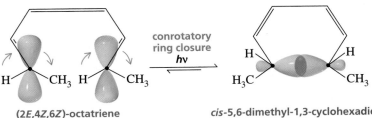

(2E,4Z,6Z)-octatriene   *cis*-5,6-dimethyl-1,3-cyclohexadiene

We have just seen why the configuration of the product formed under photochemical conditions is the exact opposite of the configuration of the product formed under thermal conditions: the HOMO of the ground state is symmetric, leading to disrotatory ring closure, while the HOMO of the excited state is asymmetric, leading to conrotatory ring closure. Thus, the outcome of an electrocyclic reaction is controlled by the symmetry of the HOMO of the compound undergoing ring closure.

Now let's see why ring closure of (2E,4Z)-hexadiene forms *cis*-3,4-dimethylcyclobutene. The compound undergoing ring closure has two conjugated π bonds. The ground-state HOMO of a compound with two conjugated π bonds is asymmetric (Figure 27.2), so ring closure is conrotatory. Conrotatory ring closure of (2E,4Z)-hexadiene leads to the cis product.

The symmetry of the HOMO of the compound undergoing ring closure controls the outcome of an electrocyclic reaction.

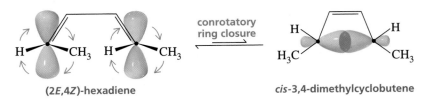

(2E,4Z)-hexadiene   *cis*-3,4-dimethylcyclobutene

Similarly, conrotatory ring closure of (2E,4E)-hexadiene leads to the trans product.

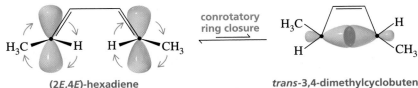

(2E,4E)-hexadiene   *trans*-3,4-dimethylcyclobutene

However, if the reaction is carried out under photochemical conditions, the excited-state HOMO of a compound with two conjugated π bonds is symmetric. (Recall that the ground-state HOMO and the excited-state HOMO have opposite symmetries.) So (2E,4Z)-hexadiene will undergo disrotatory ring closure, resulting in the trans product, while (2E,4E)-hexadiene will undergo disrotatory ring closure and form the cis product.

We have seen that the HOMO of a compound with two conjugated double bonds is asymmetric, while the HOMO of a compound with three conjugated double bonds is symmetric. If we examine molecular orbital diagrams for compounds with four, five, six, and more conjugated double bonds, we can conclude that *the ground-state HOMO of a compound with an even number of conjugated double bonds is asymmetric, while the ground-state HOMO of a compound with an odd number of conjugated double bonds is symmetric.* Therefore, from the number of conjugated double bonds in a compound, we can immediately tell whether ring closure will be conrotatory (an even number of conjugated double bonds) or disrotatory (an odd number of conjugated double bonds) under thermal conditions. However, if the reaction takes place under photochemical conditions, everything is reversed, because the ground-state and excited-state HOMOs have opposite symmetries (if the ground-state HOMO is symmetric, the excited-state HOMO is asymmetric).

We have seen that the stereochemistry of an electrocyclic reaction depends on the mode of ring closure. In turn, the mode of ring closure depends on the number of conjugated π bonds in the reactant and on whether the reaction is carried out under thermal or photochemical conditions. What we have learned about pericyclic reactions can be summarized with a series of **selection rules** (Table 27.1). These are also known as the **Woodward–Hoffmann rules** for electrocyclic reactions.

The rules in Table 27.1 are for determining whether a given electrocyclic reaction is "allowed by orbital symmetry." There are also selection rules to determine whether cycloaddition reactions (Table 27.3) and sigmatropic rearrangements (Table 27.4) are "allowed by orbital symmetry." It can be rather burdensome to memorize these rules (and worrisome if they are forgotten during an exam), but all the rules can be summarized by the word "TE-AC." How to use "TE-AC" is explained in Section 27.7.

> The ground-state HOMO of a compound with an even number of conjugated double bonds is asymmetric.

> The ground-state HOMO of a compound with an odd number of conjugated double bonds is symmetric.

TABLE 27.1 Woodward–Hoffmann Rules for Electrocyclic Reactions		
Number of Conjugated π Bonds	**Reaction Conditions**	**Allowed Mode of Ring Closure**
Even number	Thermal	Conrotatory
	Photochemical	Disrotatory
Odd Number	Thermal	Disrotatory
	Photochemical	Conrotatory

PROBLEM 5

a. For conjugated systems with two, three, four, five, six, and seven conjugated π bonds, construct quick molecular orbitals (just draw the p orbitals at the ends of the conjugated system to show whether the HOMO is symmetric or asymmetric as they are drawn on page 1184).

b. Using these drawings, convince yourself that the summary in Table 27.1 is valid.

c. Using these drawings, convince yourself that the "TE-AC" shortcut method for learning the information in Table 27.1 is valid (Section 27.7).

PROBLEM 6 ◆

a. Under thermal conditions, will ring closure of (2E,4Z,6Z,8E)-decatetraene be conrotatory or disrotatory?

b. Will the product have the cis or the trans configuration?

c. Under photochemical conditions, will ring closure be conrotatory or disrotatory?

d. Will the product have the cis or the trans configuration?

The series of reactions in Figure 27.4 illustrates just how easy it is to determine the mode of ring closure and therefore the product of an electrocyclic reaction. The reactant of the first reaction has three conjugated double bonds and is undergoing ring closure under thermal conditions. Ring closure is therefore disrotatory (Table 27.1). Disrotatory ring closure of this reactant causes the hydrogens to be cis in the ring-closed product. To determine the relative configurations of the hydrogens, draw them in the reactant and then draw arrows showing disrotatory ring closure (Figure 27.4a).

The second step is a ring-opening electrocyclic reaction that takes place under photochemical conditions. Because of the principle of microscopic reversibility (Section 14.6), the orbital symmetry rules used for a ring-closure reaction also apply to the reverse ring-opening reaction. The compound undergoing ring closure (the reverse of the ring-opening reaction) has three conjugated double bonds. Because the reaction occurs under photochemical conditions, ring opening (or ring closure) is conrotatory. (Notice that the number of conjugated double bonds used to determine the mode of ring opening/closure in reversible electrocyclic reactions is the number in the compound undergoing ring closure.) In order for conrotatory rotation to result in a product with cis hydrogens, the hydrogens in the compound undergoing ring closure must point in the same direction (Figure 27.4b). The third step is a thermal ring closure of a compound with three conjugated double bonds, so ring closure is disrotatory. Drawing the hydrogens and the arrows (Figure 27.4c) allows you to determine the configurations of the hydrogens in the ring-closed product.

Notice that, in all these electrocyclic reactions, if the bonds to the substituents in the reactant point in opposite directions (as in Figure 27.4a), the substituents will be cis in the product if ring closure is disrotatory and trans if ring closure is conrotatory. On the other hand, if they point in the same direction (as in Figure 27.4b or 27.4c), they will be trans in the product if ring closure is disrotatory and cis if ring closure is conrotatory (Table 27.2).

Figure 27.4 ▶
Determining the stereochemistry of the product of an electrocyclic reaction.

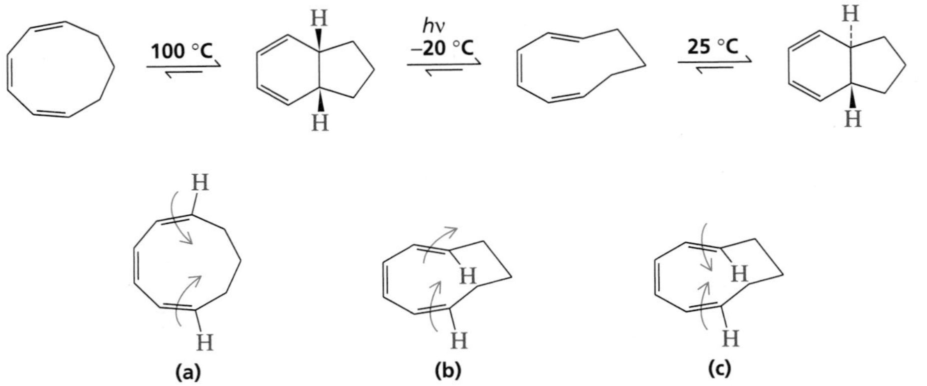

TABLE 27.2   Configuration of the Product of an Electrocyclic Reaction

Substituents in the Reactant	Mode of Ring Closure	Configuration of the Product
Point in opposite directions	Disrotatory	cis
	Conrotatory	trans
Point in the same direction	Disrotatory	trans
	Conrotatory	cis

PROBLEM 7◆

Which of the following are correct? Correct any false statements.

a. A conjugated diene with an even number of double bonds undergoes conrotatory ring closure under thermal conditions.

b. A conjugated diene with asymmetric HOMO undergoes conrotatory ring closure under thermal conditions.

c. A conjugated diene with an odd number of double bonds has a symmetric HOMO.

PROBLEM 8◆

a. Identify the mode of ring closure for each of the following electrocyclic reactions.

b. Are the indicated hydrogens cis or trans?

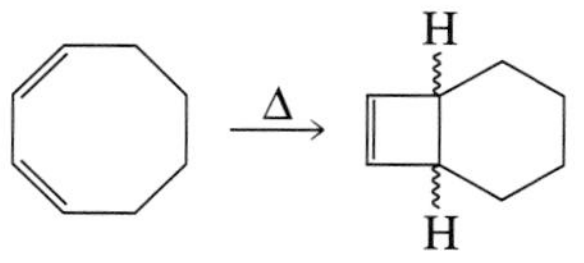

In a *cycloaddition reaction,* two different π-bond-containing molecules react to form a cyclic molecule. The Diels–Alder reaction is one of the best known examples of a cycloaddition reaction (Section 7.8). Cycloaddition reactions are classified according to the number of π electrons that interact in the reaction. The Diels–Alder reaction is a [4 + 2] cycloaddition because one reactant has four interacting π electrons and the other reactant has two interacting π electrons. Only the π electrons participating in electron rearrangement are counted.

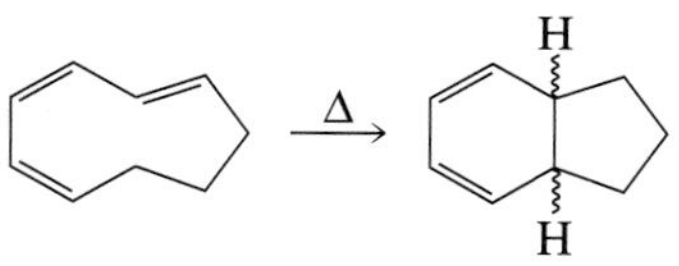

27.4 CYCLOADDITION REACTIONS

[4 + 2] cycloaddition (a Diels–Alder reaction)

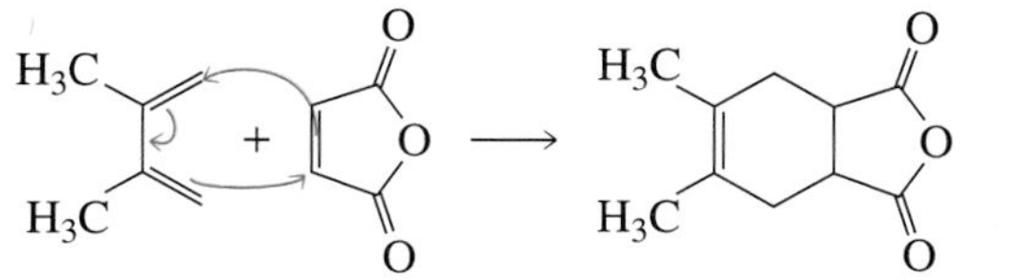

[2 + 2] cycloaddition

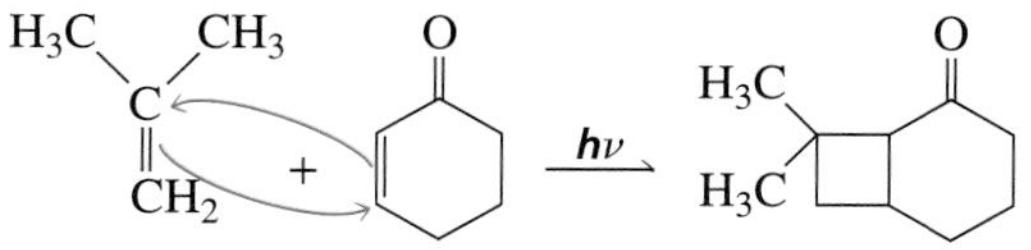

[8 + 2] cycloaddition

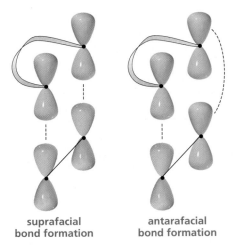

In a cycloaddition reaction, the orbitals of one molecule must overlap with the orbitals of the second molecule. Therefore, the frontier molecular orbitals of both reactants must be evaluated to determine the outcome of the reaction. Since the new σ bonds in the product are formed by donation of electron density from one reactant to the other reactant, we must consider the HOMO of one of the molecules and the LUMO of the other, because only an empty orbital can accept electrons. It does not matter which reacting molecule's HOMO is used. It is required only that we use the HOMO of one and the LUMO of the other.

There are two modes of orbital overlap for the simultaneous formation of two σ bonds: suprafacial and antarafacial. Bond formation is **suprafacial** if both σ bonds form from the same side of the π system. Bond formation is **antarafacial** if the two σ bonds form from opposite sides of the π system. Suprafacial bond formation is similar to syn addition, while antarafacial bond formation resembles anti addition (Section 4.19).

suprafacial bond formation   antarafacial bond formation

A cycloaddition reaction that forms a four-, five-, or six-membered ring must involve suprafacial bond formation. The geometric constraints of these small rings make the antarafacial approach highly unlikely even if it is symmetry-allowed (the overlapping orbitals are in phase). Antarafacial bond formation is more likely in cycloaddition reactions that form larger rings.

Frontier orbital analysis of a [4 + 2] cycloaddition reaction shows that overlap of in-phase orbitals to form the two new σ bonds requires suprafacial orbital overlap (Figure 27.5). This is true whether we use the HOMO of the alkene (a system with one π bond; Figure 27.1) and the LUMO of the diene (a system with two conjugated π bonds; Figure 27.2), or the LUMO of the alkene and the HOMO of the diene. Now we can understand why Diels–Alder reactions occur with relative ease (Section 7.8).

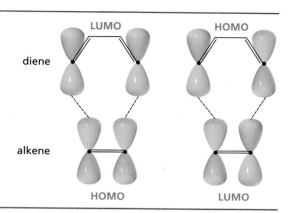

◀ **Figure 27.5**
Frontier molecular orbital analysis of a [4 + 2] cycloaddition reaction. The HOMO of either of the reactants can be used with the LUMO of the other. Both situations result in suprafacial overlap.

A [2 + 2] cycloaddition reaction does not occur under thermal conditions but does take place under photochemical conditions.

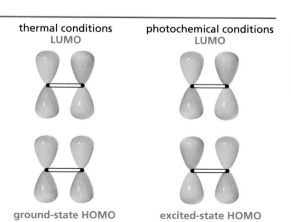

The reason for this is apparent from an examination of the frontier molecular orbitals (Figure 27.6). Under thermal conditions, suprafacial overlap is not symmetry-allowed (the overlapping orbitals are out of phase). Antarafacial overlap is symmetry-allowed but is not possible because of the small size of the ring. Under photochemical conditions, the reaction can take place because the excited-state HOMO has symmetry opposite that of the ground-state HOMO. Therefore, overlap of the excited-state HOMO of one alkene with the LUMO of the second alkene involves symmetry-allowed suprafacial bond formation. Notice that in the photochemical reaction only one of the reactants is in an excited state. Owing to the very short lifetimes of excited states, there is little likelihood that two excited states will

◀ **Figure 27.6**
Frontier molecular orbital analysis of a [2 + 2] cycloaddition reaction under thermal and photochemical conditions.

TABLE 27.3 Woodward–Hoffmann Rules for Cycloaddition Reactions

Sum of the Number of π Bonds in the Reacting Systems of Both Reagents	Reaction Conditions	Allowed Mode of Ring Closure
Even number	Thermal	Antarafacial[a]
	Photochemical	Suprafacial
Odd number	Thermal	Suprafacial
	Photochemical	Antarafacial[a]

[a]Although antarafacial ring closure is symmetry-allowed, it can occur only with large rings.

find one another to interact. The selection rules for cycloaddition reactions are summarized in Table 27.3.

PROBLEM 9

Maleic anhydride reacts rapidly with 1,3-butadiene but does not react at all with ethene under thermal conditions. Explain.

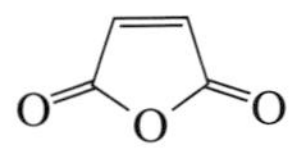

maleic anhydride

PROBLEM 10 / SOLVED

Compare the reaction of cycloheptatrienone with cyclopentadiene and with ethene. Why does cycloheptatrienone use two π electrons in one reaction and four π electrons in the other reaction?

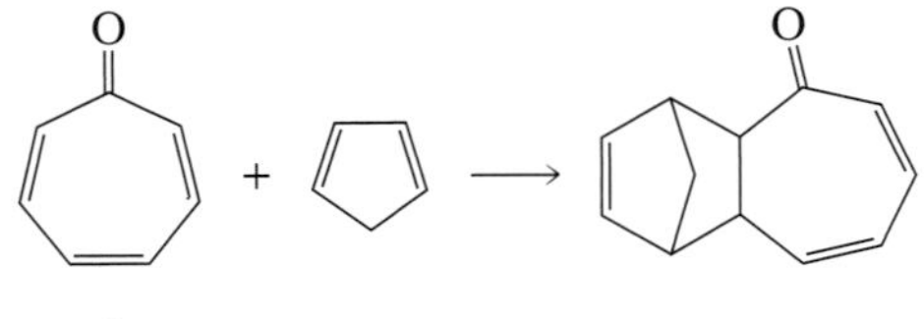

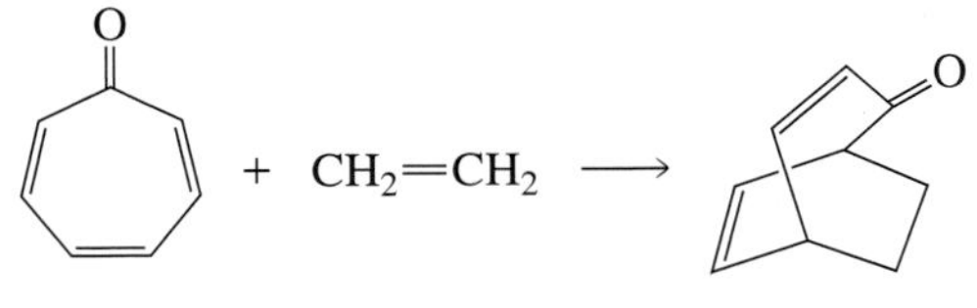

SOLUTION Both reactions are [4 + 2] cycloaddition reactions. When cycloheptatrienone reacts with cyclopentadiene, it uses two of its π electrons because cyclopentadiene is the four-π-electron reactant. When cycloheptatrienone reacts with ethene, it uses four of its π electrons because ethene is the two-π-electron reactant.

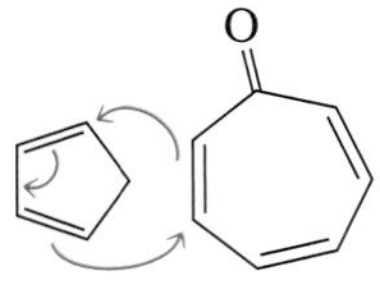

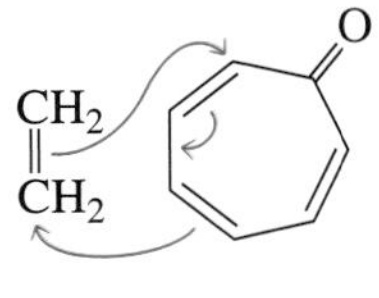

PROBLEM 11◆

Will a concerted reaction take place between 1,3-butadiene and 2-cyclohexenone in the presence of ultraviolet light?

The last class of concerted pericyclic reactions that we will consider is the group of reactions known as *sigmatropic rearrangements*. In a sigmatropic rearrangement, a σ bond in a molecule is broken, a new σ bond is formed, and the π electrons rearrange. The σ bond that breaks is bonded to an allylic carbon. It can be a σ bond between a carbon and a hydrogen, between a carbon and another carbon, or between a carbon and an oxygen, nitrogen, or sulfur. "Sigmatropic" comes from the Greek word *tropos,* which means "change," so sigmatropic means "sigma-change."

The numbering system used to describe a sigmatropic rearrangement differs from any numbering system you have seen previously. First, mentally break the σ bond in the reactant into two pieces, and then look at the new σ bond in the product. Count the number of atoms in each of the fragments that connect the broken σ bond and the new σ bond. The two numbers are put in brackets with the smaller number stated first. For example, in the [2,3] sigmatropic rearrangement shown below, two atoms (N, N) connect the old and new σ bonds in one fragment and three atoms (C, C, C) connect the old and new σ bonds in the other fragment.

**27.5
SIGMATROPIC
REARRANGEMENTS**

a [2,3] sigmatropic rearrangement

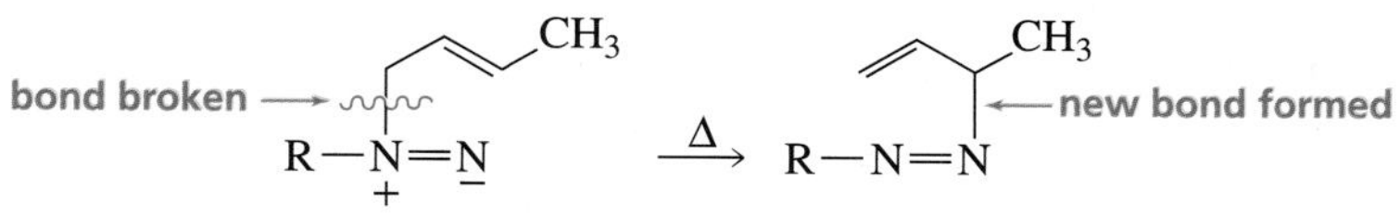

a [1,5] sigmatropic rearrangement

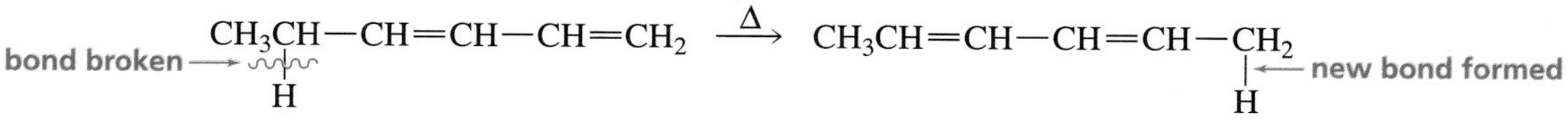

a [1,3] sigmatropic rearrangement

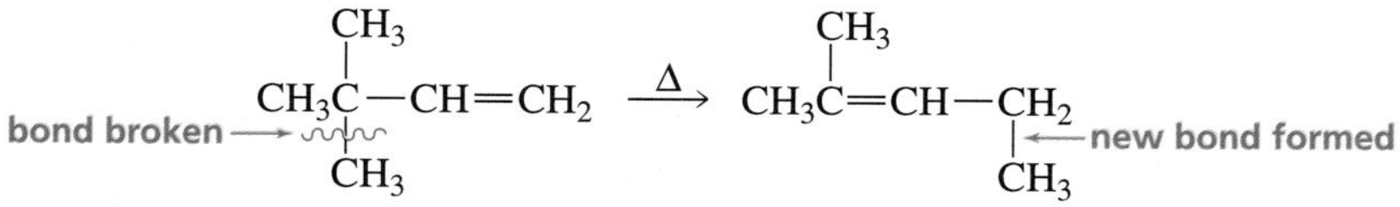

a [3,3] sigmatropic rearrangement

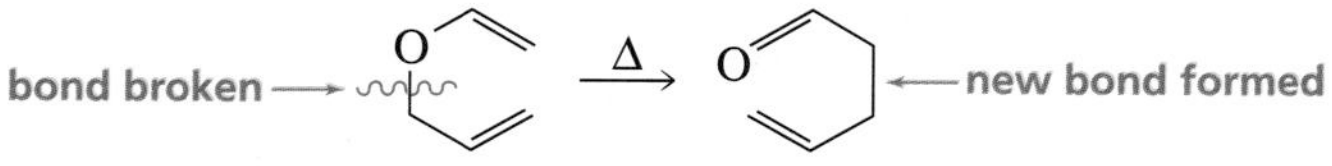

PROBLEM 12◆

a. Name the kind of sigmatropic rearrangement that occurs in each of the following reactions.

b. Using arrows, show the electron rearrangement that takes place in each of the reactions.

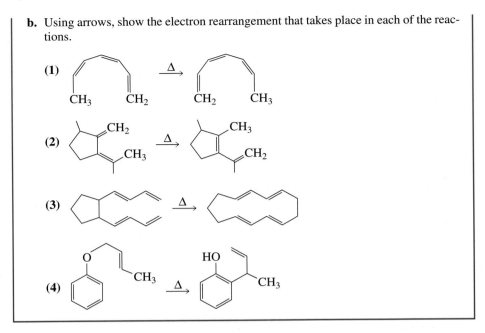

In the transition state of a sigmatropic rearrangement, the group that migrates is partially bonded to the migration source and partially bonded to the migration terminus. There are two possible modes for rearrangement. If the migrating group remains on the same face of the π system, the rearrangement is **suprafacial.** If the migrating group moves to the opposite face of the π system, the rearrangement is **antarafacial.**

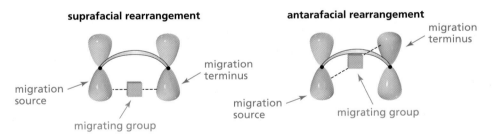

Sigmatropic rearrangements have cyclic transition states. If the transition state has six or fewer atoms in the ring, rearrangement must be suprafacial because of the geometric constraints of small rings.

TABLE 27.4 Woodward–Hoffmann Rules for Sigmatropic Rearrangements		
Number of Pairs of Electrons in the Reacting System	**Reaction Conditions**	**Allowed Mode of Ring Closure**
Even number	Thermal	Antarafacial[a]
	Photochemical	Suprafacial
Odd number	Thermal	Suprafacial
	Photochemical	Antarafacial[a]
[a]Although antarafacial ring closure is symmetry-allowed, it can occur only with large rings.		

A [1,3] sigmatropic rearrangement involves a π bond and a pair of σ electrons, or we can say that it involves two pairs of electrons. A [1,5] sigmatropic rearrangement involves two π bonds and a pair of σ electrons (three pairs of electrons), and a [1,7] sigmatropic rearrangement involves four pairs of electrons. We can use the same symmetry rules for sigmatropic rearrangements that we used for cycloaddition reactions if we substitute "number of pairs of electrons" for "sum of the number of π bonds." (Compare Tables 27.3 and 27.4.) *Recall that the ground-state HOMO of a compound with an even number of conjugated double bonds is asymmetric, while the ground-state HOMO of a compound with an odd number of conjugated double bonds is symmetric.*

A **Cope**[2] **rearrangement** is a [3,3] sigmatropic rearrangement of a 1,5-diene. A **Claisen**[3] **rearrangement** is a [3,3] sigmatropic rearrangement of an allyl vinyl ether. Both rearrangements form six-membered-ring transition states. Therefore, the reactions must be able to take place by a suprafacial pathway. Whether or not a suprafacial pathway is symmetry-allowed depends on the number of pairs of electrons involved in the rearrangement (Table 27.4). Because [3,3] sigmatropic rearrangements involve three pairs of electrons, they occur by a suprafacial pathway under thermal conditions. Therefore, both Cope and Claisen rearrangements readily take place under thermal conditions.

a Cope rearrangement

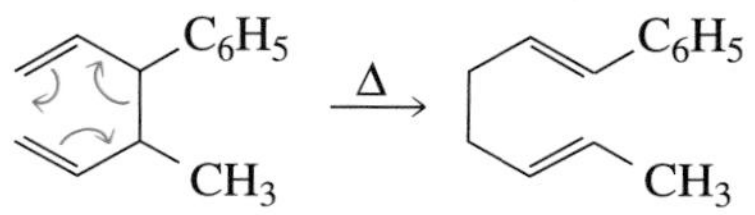

a Claisen rearrangement

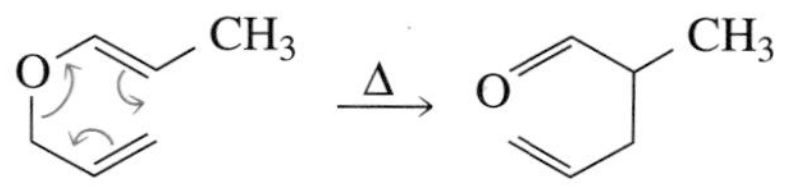

PROBLEM 13◆

a. Give the product of the following reaction.

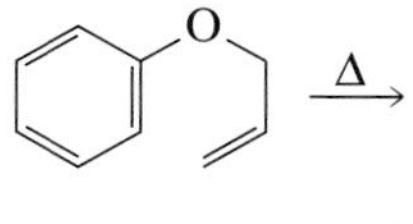

b. If the terminal sp^2 carbon of the substituent bonded to the benzene ring is labeled with ^{14}C, where will the label be in the product?

Migration of Hydrogen

When a hydrogen migrates in a sigmatropic rearrangement, the s orbital of hydrogen is partially bonded to both the migration source and the migration terminus in the transition state. Therefore, a [1,3] sigmatropic migration of hydrogen involves a four-membered-ring transition state. Because two pairs of electrons are involved (the HOMO is asymmetric), the selection rules require an antarafacial

---

[2]Cope also discovered the Cope elimination (page 466).
[3]Claisen also discovered the Claisen condensation (page 840).

rearrangement for a 1,3-hydrogen shift under thermal conditions (Table 27.4). Consequently, 1,3-hydrogen shifts do not occur under thermal conditions, because the four-membered-ring transition state does not allow the required antarafacial rearrangement.

migration of hydrogen

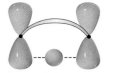 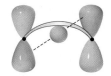

suprafacial rearrangement   **antarafacial rearrangement**

1,3-Hydrogen shifts can take place if the reaction is carried out under photochemical conditions, because under such conditions the HOMO is symmetric which means that hydrogen can migrate by a suprafacial pathway (Table 27.4).

1,3-hydrogen shift

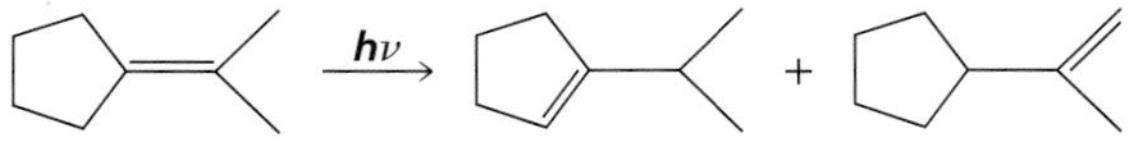

Two products are obtained in the reaction above because two different allyl hydrogens can undergo a 1,3-hydrogen shift.

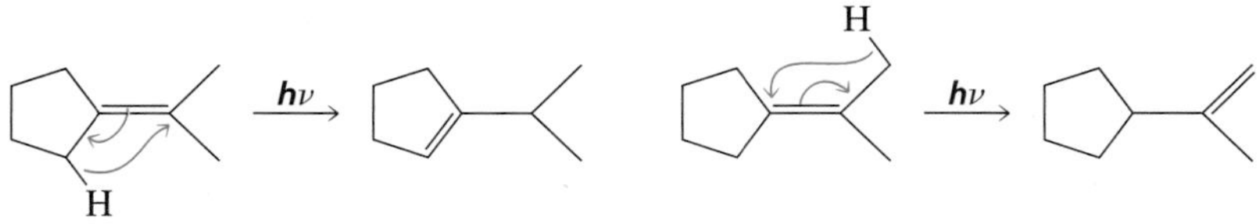

[1,5] Sigmatropic migrations of hydrogen are well known. Because they involve three pairs of electrons, they take place by a suprafacial pathway under thermal conditions.

1,5-hydrogen shifts

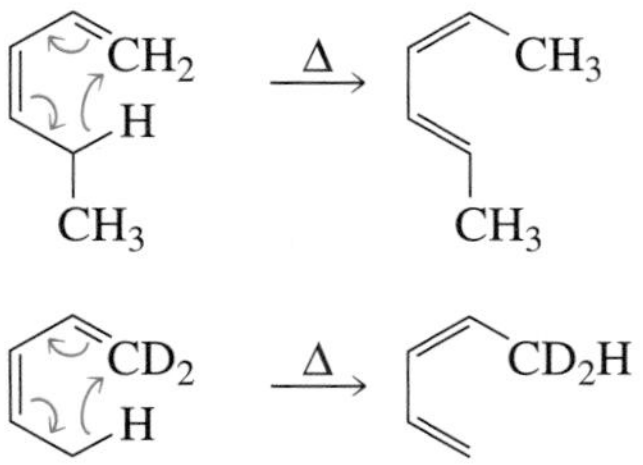

PROBLEM 14

Why was a deuterated compound used in the last example?

PROBLEM 15

Account for the difference in the products obtained under photochemical and thermal conditions.

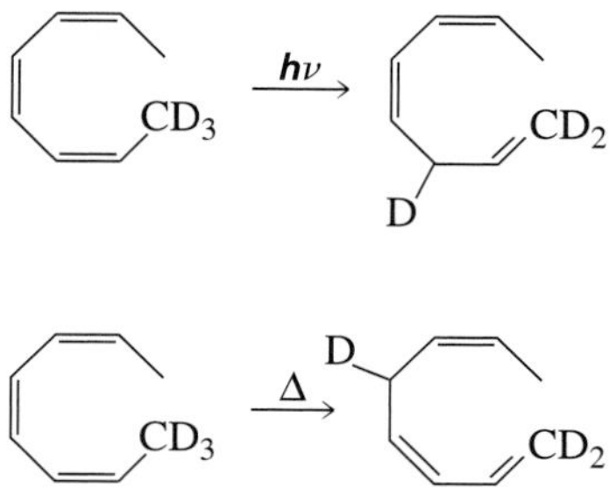

[1,7] Sigmatropic hydrogen migrations involve four pairs of electrons. They can take place under thermal conditions because the eight-membered-ring transition state allows the required antarafacial rearrangement.

1,7-hydrogen shift

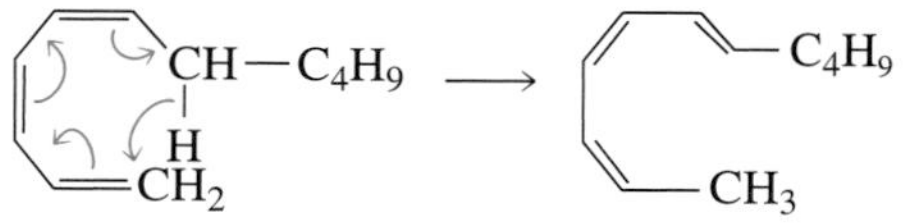

PROBLEM 16 / SOLVED

5-Methylcyclopentadiene rearranges to give a mixture of 5-methylcyclopentadiene, 2-methylcyclopentadiene, and 1-methylcyclopentadiene. Show how these products form.

SOLUTION Notice that both equilibria involve [1,5] sigmatropic rearrangements. Although a hydride moves from one carbon to an adjacent carbon, the rearrangements are not called 1,2-shifts because this would not account for all the atoms involved in the rearranged π electron system.

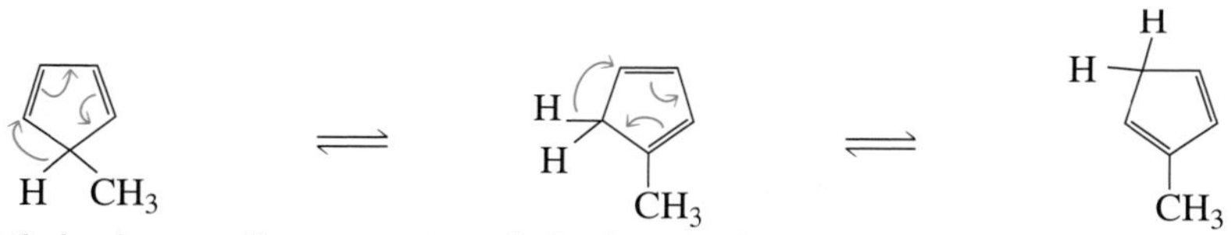

5-methylcyclopentadiene   1-methylcyclopentadiene   2-methylcyclopentadiene

---

Migration of Carbon

Unlike hydrogen, which can migrate in only one way because of its spherical *s* orbital, carbon has two ways to migrate because it has a two-lobed *p* orbital. Carbon can simultaneously interact with the migration source and the migration terminus using one of its lobes.

carbon migrating with one of its lobes interacting

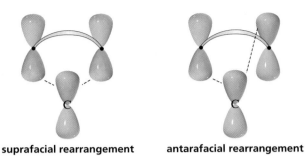

suprafacial rearrangement   **antarafacial rearrangement**

Carbon can also interact with the migration source and the migration terminus using both of its lobes.

carbon migrating with both of its lobes interacting

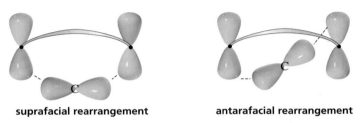

suprafacial rearrangement   **antarafacial rearrangement**

If the reaction requires a suprafacial rearrangement, carbon will migrate using one of its lobes if the HOMO is symmetric, and will migrate using both of its lobes if the HOMO is asymmetric.

When carbon migrates with only one of its lobes interacting with the migration source and migration terminus, the migrating group retains its configuration because bonding is always to the same lobe. When carbon migrates with both of its lobes interacting, bonding in the reactant and bonding in the product involve different lobes. Therefore, migration occurs with inversion of configuration.

The following [1,3] sigmatropic rearrangement has a four-membered-ring transition state that requires a suprafacial pathway. The reacting system has two pairs of electrons, so its HOMO is asymmetric. Therefore, the migrating carbon interacts with the migration source and the migration terminus using both of its lobes, so it undergoes inversion of configuration.

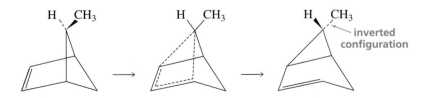

PROBLEM 17

[1,3] Sigmatropic migrations of hydrogen cannot occur under thermal conditions, but [1,3] sigmatropic migrations of carbon can occur under thermal conditions. Explain.

PROBLEM 18◆

a. Will thermal 1,3-migrations of carbon occur with retention or inversion of configuration?

b. Will thermal 1,5-migrations of carbon occur with retention or inversion of configuration?

A Biological Cycloaddition Reaction

Exposure to ultraviolet light is a well-known cause of skin cancer. This is one of the reasons for the current concern about the disappearance of the ozone layer. The ozone layer absorbs ultraviolet radiation high in the atmosphere, protecting organisms on Earth's surface (Section 8.9). One cause of skin cancer is the formation of *thymine dimers*. At any point in DNA where there are two adjacent thymine residues, a [2 + 2] cycloaddition reaction can occur, resulting in the formation of a thymine dimer. Because [2 + 2] cycloaddition reactions take place only under photochemical conditions, the reaction takes place only in the presence of ultraviolet light.

two adjacent thymine residues on DNA   mutation-causing thymine dimer

Thymine dimers can cause cancer because they interfere with the structural integrity of DNA. Any modification of DNA structure can lead to mutations and possibly to cancer.

Fortunately, there are enzymes that repair damaged DNA. When a repair enzyme recognizes a thymine dimer, it removes it and inserts two individual thymine residues. Repair enzymes, however, are not perfect, and some damage always remains uncorrected. People who do not have the repair enzyme (called DNA photolyase) that removes thymine dimers rarely live beyond the age of 20. Fortunately, this genetic defect is rare.

A Biological Reaction Involving an Electrocyclic Reaction and a Sigmatropic Rearrangement

7-Dehydrocholesterol, a steroid found in skin, is converted into vitamin D_3 by two pericyclic reactions. The first is an electrocyclic reaction that opens one of the six-membered rings in the starting material to form provitamin D_3. This reaction occurs under photochemical conditions. The provitamin then undergoes a [1,7] sigmatropic rearrangement to form vitamin D_3. The sigmatropic rearrangement takes place under thermal conditions and is slower than the electrocyclic reaction, so vitamin D_3 continues to be synthesized for several days after exposure to sunlight. Vitamin D_3 is not the active form of the vitamin. The active form requires two successive hydroxylations of vitamin D_3; the first occurs in the liver and the second in the kidney.

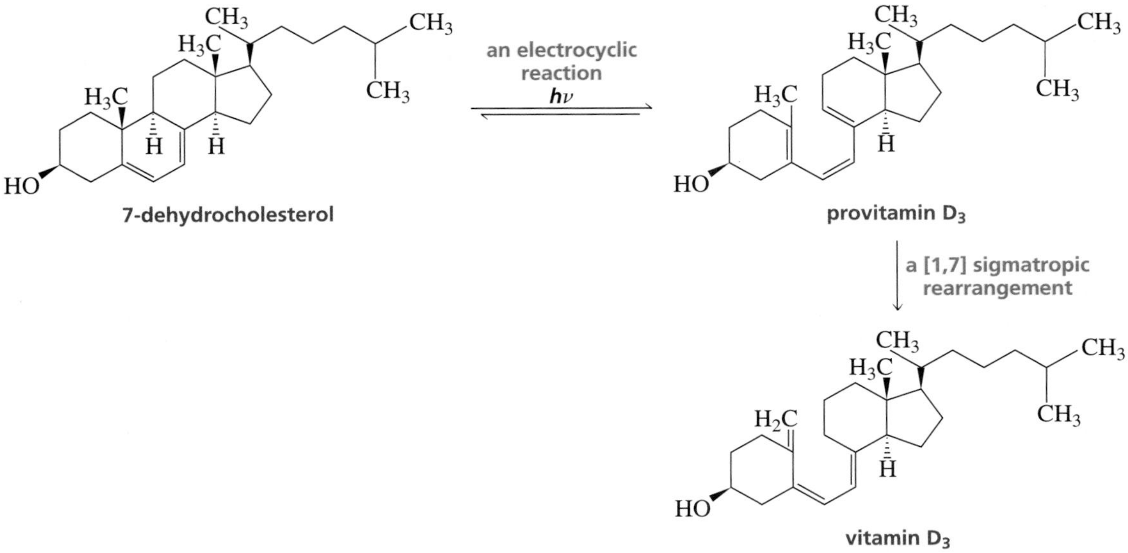

A deficiency in vitamin D, which can be prevented by getting enough sun, causes a disease known as rickets. Rickets is characterized by deformed bones and stunted growth. Too much vitamin D is harmful because it causes the calcification of soft tissues. It is thought that skin pigmentation evolved to protect the skin from the sun's ultraviolet rays in order to prevent the synthesis of too much vitamin D_3. This agrees with the observation of greater skin pigmentation in peoples who are indigenous to countries close to the equator.

PROBLEM 19 ◆

Does the [1,7] sigmatropic rearrangement that converts provitamin D_3 to vitamin D_3 involve suprafacial or antarafacial rearrangement?

PROBLEM 20

Explain why photochemical ring closure of provitamin D_3 to form 7-dehydrocholesterol results in the hydrogen and methyl substituents being trans to one another.

27.7 SUMMARY OF THE SELECTION RULES FOR PERICYCLIC REACTIONS

The selection rules that determine the outcome of electrocyclic reactions, cycloaddition reactions, and sigmatropic rearrangements are summarized in Tables 27.1, 27.3, and 27.4, respectively. This is still a lot to remember. Fortunately, the selection rules for all pericyclic reactions can be summarized in one word: "**TE-AC.**"

- If **TE** describes the reaction (**T**hermal/**E**ven), the outcome is given by **AC** (**A**ntarafacial or **C**onrotatory).
- If one of the letters of **TE** is incorrect (it is not **T**hermal/**E**ven but is **T**hermal/**O**dd or **P**hotochemical/**E**ven), the outcome is not given by **AC** (the outcome is **S**uprafacial or **D**isrotatory).
- If both of the letters of **TE** are incorrect (**P**hotochemical/**O**dd), the outcome is given by **AC** (**A**ntarafacial or **C**onrotatory)–"two negatives make a positive."

KEY TERMS

antarafacial bond formation (page 1188)
antarafacial rearrangement (page 1192)
antibonding molecular orbital (page 1177)
asymmetric molecular orbital (page 1179)
bonding molecular orbital (page 1177)
Claisen rearrangement (page 1193)
conrotatory ring closure (page 1182)
conservation of orbital symmetry theory (page 1175)
Cope rearrangement (page 1193)
cycloaddition reaction (page 1174)
disrotatory ring closure (page 1182)
electrocyclic reaction (page 1173)
excited state (page 1179)
frontier orbital analysis (page 1188)
frontier orbitals (page 1179)
frontier orbital theory (page 1176)
ground state (page 1178)
highest occupied molecular orbital (HOMO) (page 1179)
linear combination of atomic orbitals (LCAO) (page 1177)
lowest unoccupied molecular orbital (LUMO) (page 1179)
molecular orbital (MO) theory (page 1176)
pericyclic reaction (page 1173)
photochemical reaction (page 1175)
polar reaction (page 1173)
radical reaction (page 1173)
selection rules (page 1185)
sigmatropic rearrangement (page 1174)
suprafacial bond formation (page 1188)
suprafacial rearrangement (page 1192)
symmetric molecular orbital (page 1179)
symmetry-allowed pathway (page 1183)
symmetry-forbidden pathway (page 1183)
thermal reaction (page 1175)
Woodward–Hoffmann rules (page 1185)

PROBLEMS

21. Give the product of each of the following reactions.

a. 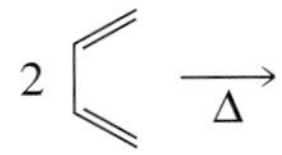

b. 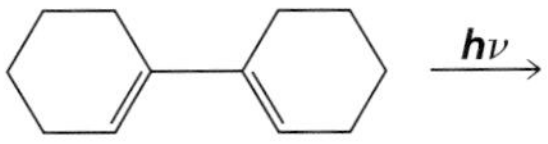

c. 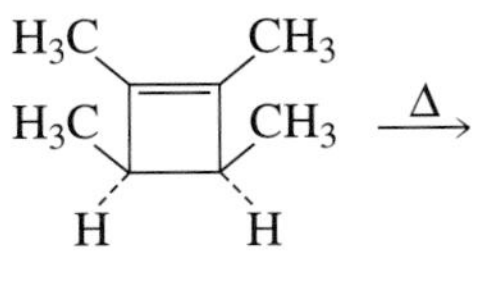

d. 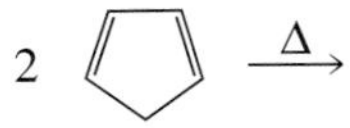

e. 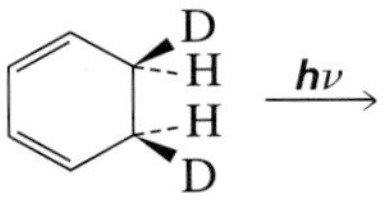

f. 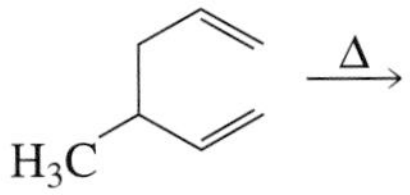

g. 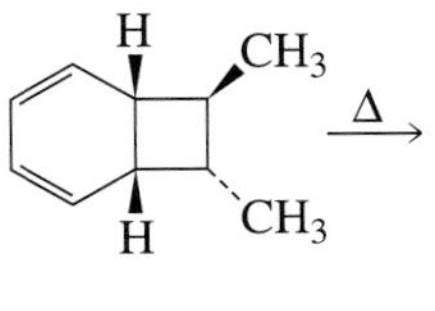

h. 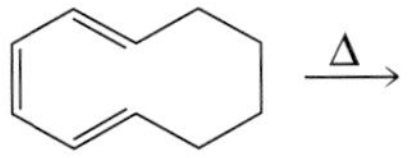

22. Give the product formed when each of the following compounds undergoes an electrocyclic reaction:
a. under thermal conditions
b. under photochemical conditions

1. 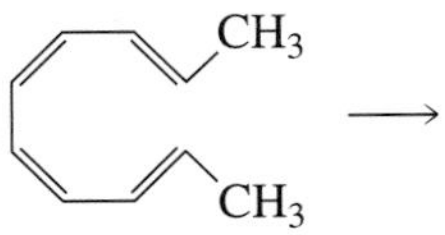

2. 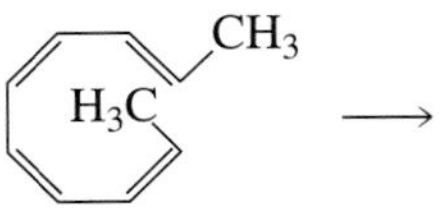

23. Chorismate mutase is an enzyme that catalyzes a pericyclic reaction that forms prephenate, which is subsequently converted into the amino acids phenylalanine and tyrosine. What kind of a pericyclic reaction does chorismate mutase catalyze?

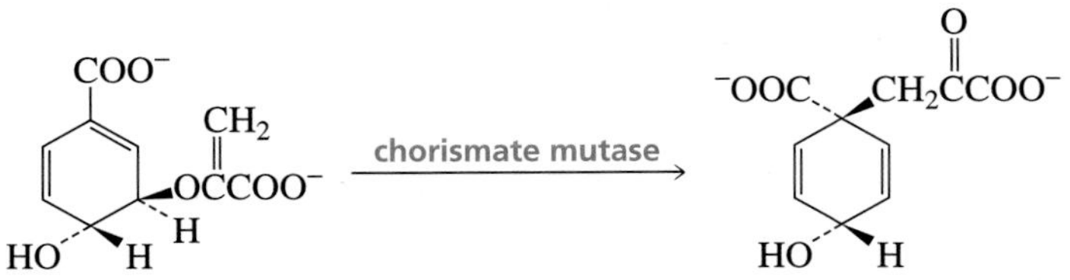

chorismate   chorismate mutase   prephenate

24. Account for the difference in the products.

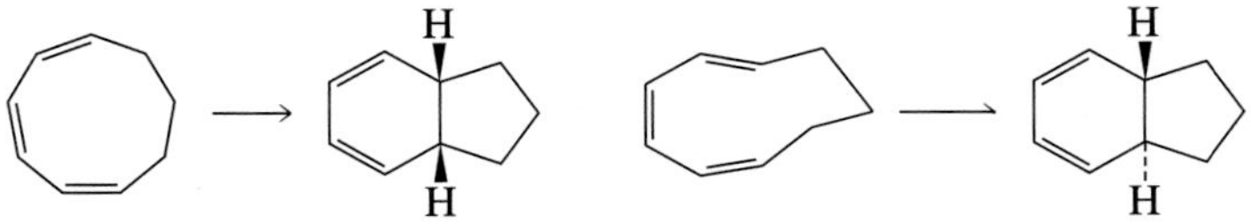

25. Describe how norbornane could be prepared from cyclopentadiene.

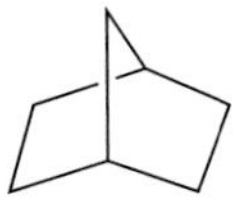

norbornane

26. Give the product of each of the following reactions.

a. $\xrightarrow{h\nu}$

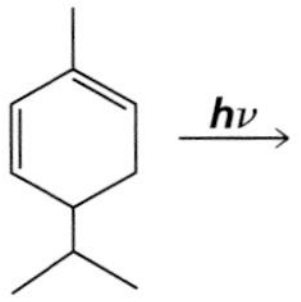

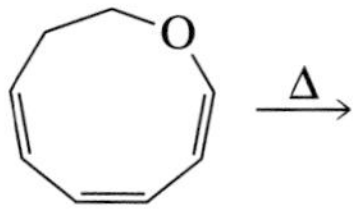

b. $\xrightarrow{\Delta}$

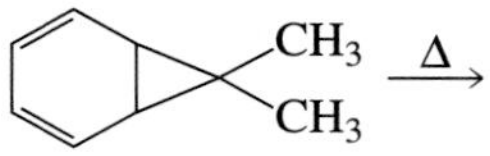

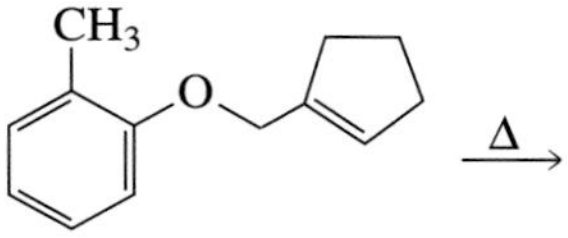

c. $\xrightarrow{\Delta}$

d. $\xrightarrow{\Delta}$

e. $\xrightarrow{\Delta}$

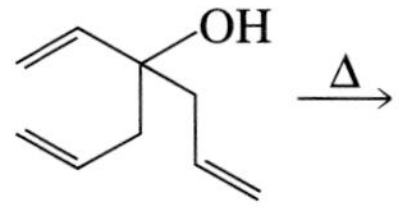

27. Which is the product of the following [1,3] sigmatropic rearrangement, A or B?

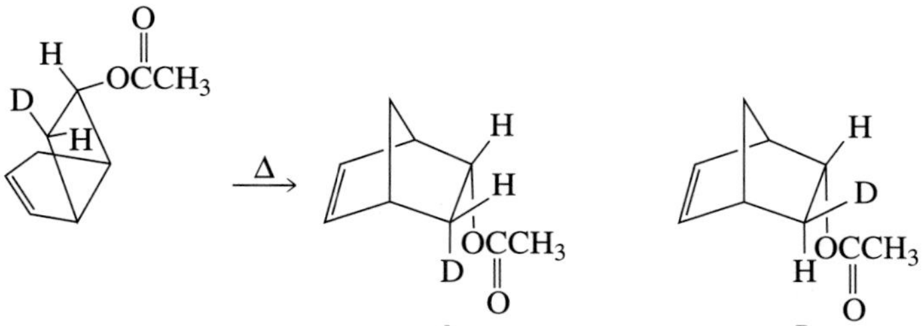

A   B

28. Dewar benzene is a highly strained isomer of benzene. In spite of its thermodynamic instability, it is very stable kinetically. It will rearrange to benzene, but only if heated to a very high temperature. Why is it kinetically stable?

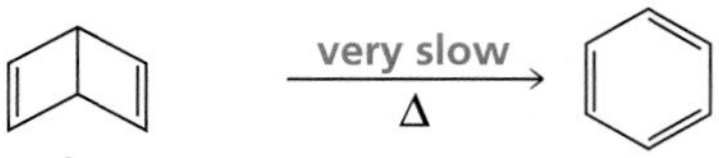

Dewar benzene $\xrightarrow[\Delta]{\text{very slow}}$

29. If the compounds below are heated, one will form one product as a result of a [1,3] sigmatropic rearrangement, and the other will form two products as a result of two different [1,3] sigmatropic rearrangements. Give the products of the reactions.

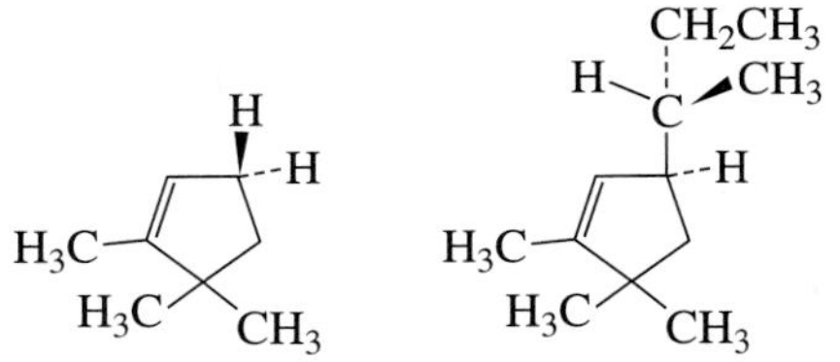

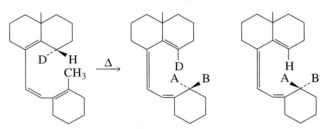

30. When the following compound is heated, a product is formed that shows an infrared absorption band at 1715 cm^{-1}. Give the structure of the product.

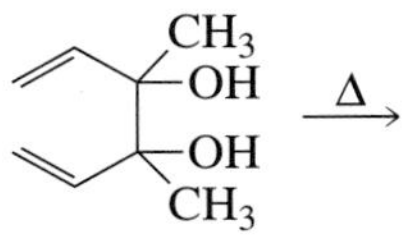

31. Two products are formed in the following [1,7] sigmatropic rearrangement, one due to hydrogen migration and the other to deuterium migration. Show the configuration of the products by replacing A and B with the appropriate substituents (H or D).

32. **a.** Propose a mechanism for the following reaction. (*Hint:* An electrocyclic reaction is followed by a Diels–Alder reaction.)
   b. What would be the reaction product if *trans*-2-butene was used instead of ethene?

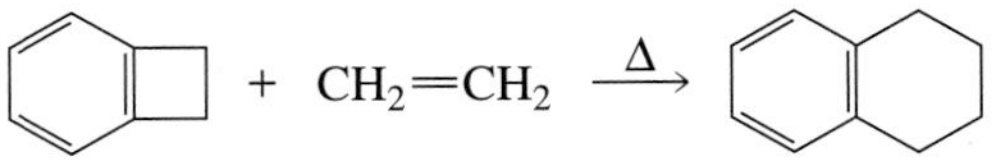

33. Two different products are formed from disrotatory ring closure of (2E,4Z,6Z)-octatriene, but only one product is formed from disrotatory ring closure of (2E,4Z,6E)-octatriene. Explain.

34. Give the product of each of the following sigmatropic rearrangements.

a. [3,3] sigmatropic rearrangement Δ

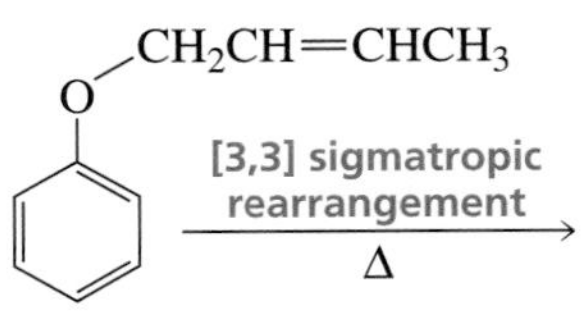

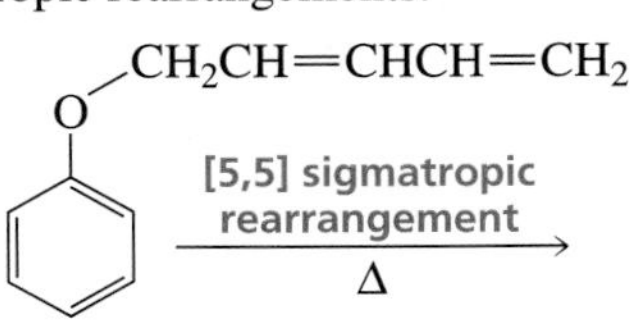

b. [3,3] sigmatropic rearrangement Δ

c. [5,5] sigmatropic rearrangement Δ

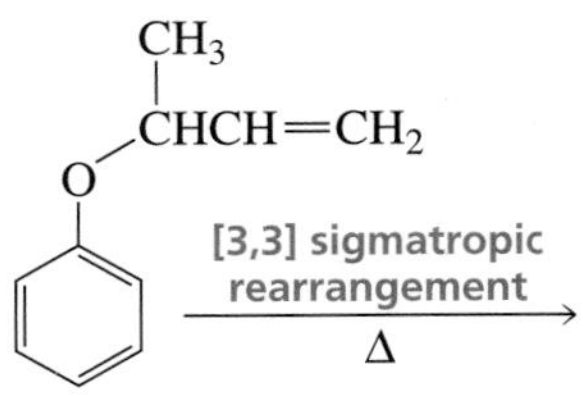

d. [5,5] sigmatropic rearrangement Δ

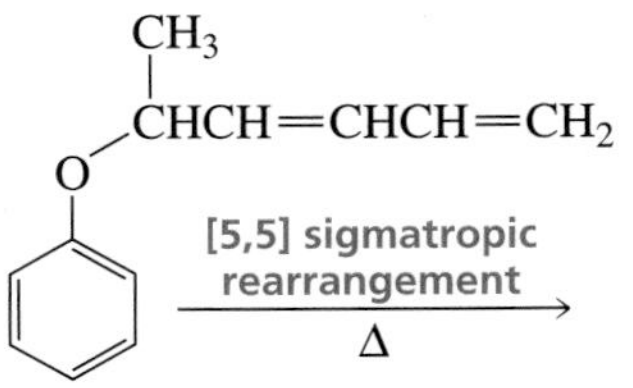

35. *cis*-3,4-Dimethylcyclobutene undergoes thermal ring opening to form the two products shown. One of the products is formed in 99% yield, the other in 1% yield. Which is which?

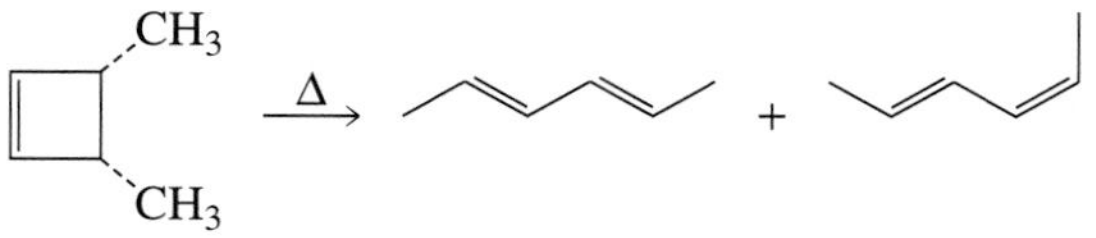

36. If isomer A is heated to about 100 °C, a mixture of isomers A and B is formed. However, there is no trace of isomer C or D. Explain.

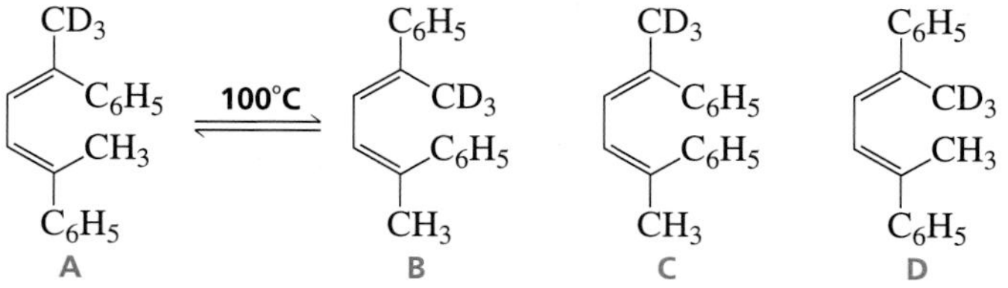

37. Propose a mechanism for the following reaction.

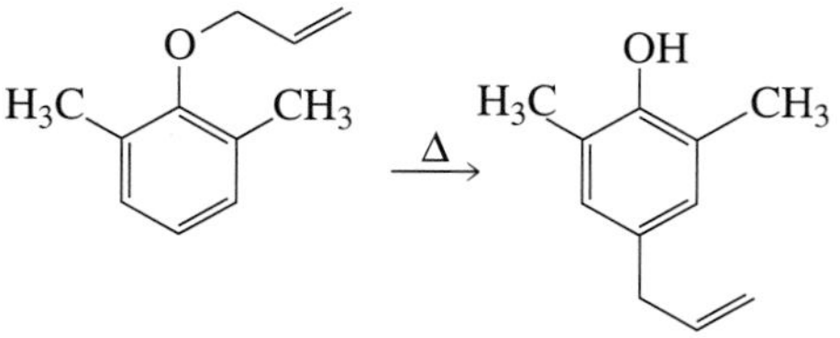

38. Compound A will not undergo a ring-opening reaction under thermal conditions, but compound B will. Explain.

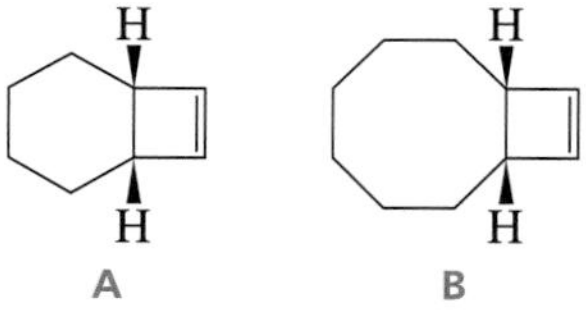

39. Professor Perry C. Lick found that heating any one of the following isomers resulted in scrambling of the deuterium to all three positions on the five-membered ring. Propose a mechanism to account for this.

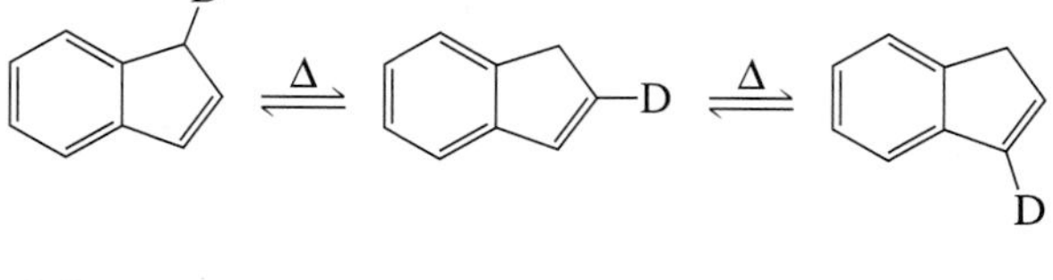

40. How could the following transformation be carried out, using only heat or light?

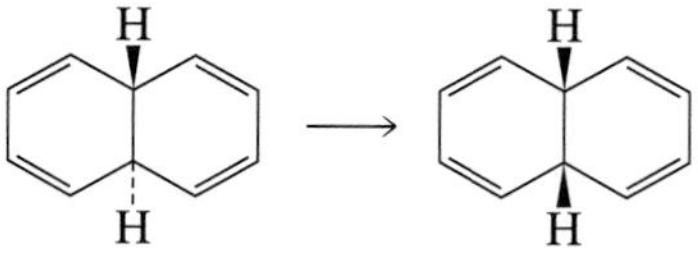

41. Show the steps involved in the following reaction.

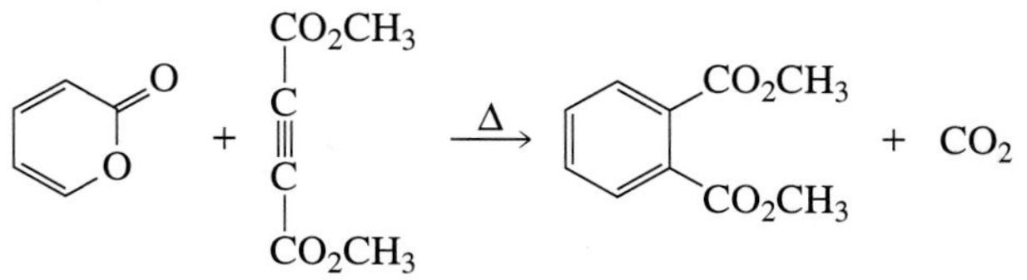

42. Propose a mechanism for the following reaction.

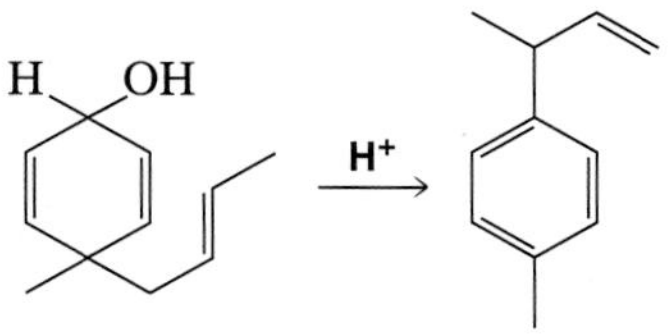

MORE ABOUT MULTISTEP ORGANIC SYNTHESIS

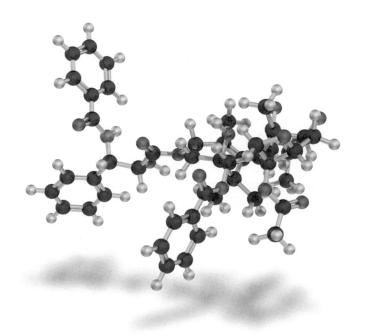

taxol

Organic synthesis is the preparation of organic compounds from other organic compounds. The word "synthesis" comes from the Greek word *synthesis,* which means "a putting together." Because of the ability of carbon atoms to form rings, chains, and multiple bonds, an almost infinite number of organic compounds can be conceived. Chemists synthesize many hundreds of thousands of them each year. Many organic compounds exist only because of the ability of chemists to synthesize them, but many are synthesized in nature. Compounds synthesized in nature are called **natural products.** Two examples are penicillin G and cocaine.

natural products

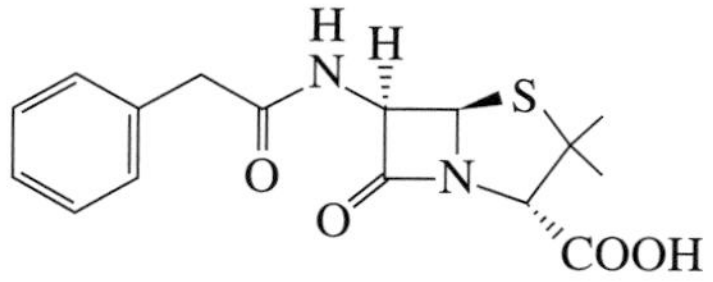

penicillin G 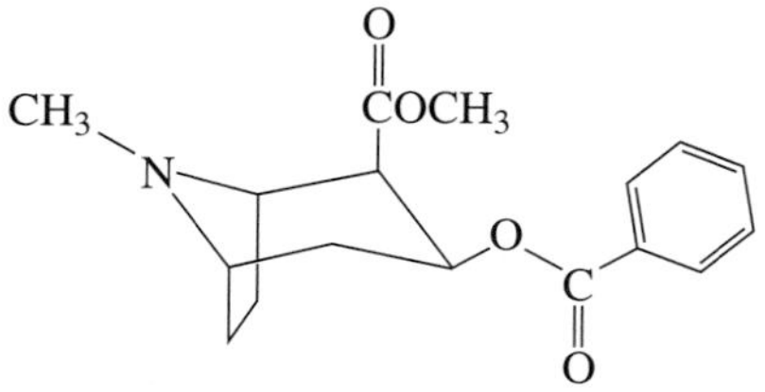   cocaine

You have been introduced to many aspects of organic synthesis (Sections 5.11, 10.11, 14.16, 16.13, 17.10, and 18.19), and you have had the opportunity to design some relatively simple syntheses. In this chapter we will examine in greater depth how organic chemists approach a synthetic problem. We will start by describing some of the basic strategies chemists use when designing syntheses; then we will take a look at how they use these tools to carry out the syntheses of some complicated organic molecules. It is impossible to present all the tools available to the synthetic chemist, but this chapter will give you some idea of the thought processes involved in designing a synthesis and the great diversity of operations available to the synthetic chemist.

Organic chemists synthesize compounds for many reasons: to study their properties, to answer a variety of chemical questions, because they have unusual shapes or other unusual structural features, or because they have useful properties. One reason chemists synthesize natural products is to provide us with greater supplies of these compounds than nature can produce. For example, taxol—a compound that has been successful in treating certain kinds of cancer—is extracted from the bark of the Pacific yew tree. The supply of natural taxol is limited because yew trees are uncommon and grow very slowly, and because stripping the bark kills the tree. Environmentalists' fears are further aggravated by the fact that *Taxus* forests serve as habitats for the spotted owl, which is an endangered species. Once chemists had determined the structure of taxol, efforts were undertaken to synthesize it in order to make it more widely available as an anticancer drug. A successful synthesis has recently been accomplished.

taxol

Once a compound has been synthesized, chemists can study its properties to learn how it works; then they can design and synthesize safer or more potent analogs. For example, chemists have found that the anticancer activity of taxol is substantially reduced if its four ester groups are hydrolyzed. This gives one small clue as to how the molecule functions.

Organic chemists can structurally modify a natural product to see if they can produce a related compound with different properties. For example, cocaine is a natural product found in the leaves of the coca plant. In the nineteenth century, cocaine was widely used as a local anesthetic because it deadens nerve endings quite effectively. Its wide use in medicine meant that legal supplies of the drug were readily available for misuse. An analog of the drug was synthesized that had the pain-deadening effects but lacked the psychological side effects (Section 29.3). This compound was called procaine (Novocain). Substitution of Novocain for cocaine in medical practice caused a decline in the early abuse of cocaine.

procaine
Novocain

Many syntheses, such as the synthesis of an ester from an acyl chloride, can be done in a single step (Section 15.6).

Some syntheses, however, require many steps. The synthesis of adenine from hydrogen cyanide and ammonia is an example of a **multistep synthesis.** The number of steps required to synthesize a particular compound depends on the starting materials that are available.

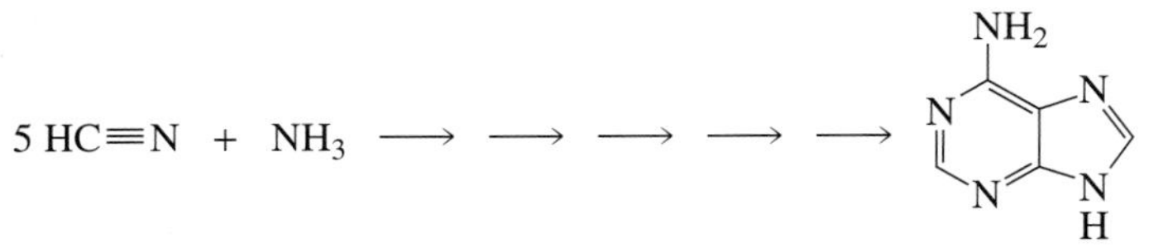

$$5 \, HC \equiv N \; + \; NH_3 \longrightarrow \longrightarrow \longrightarrow \longrightarrow \longrightarrow$$

adenine

The design of a complicated synthesis requires analysis, insight, and intuition coupled with a sound knowledge of the reactivity of organic compounds. The particular synthetic scheme that results depends on the background and training of the individual designing the synthesis, because generally several different approaches can be used. There are often many "correct" answers to a synthesis problem. The only "incorrect" answer is one that fails to produce a pure sample of the desired compound. One method may be superior to others because it leads to a higher yield, requires fewer steps, uses less complicated reactions, or requires cheaper starting materials (Section 5.11).

Frequently, a chemist plans a synthetic route by first constructing a **synthetic tree.** A synthetic tree is an outline of the available routes that lead to the desired product—the **target molecule (TM)**— from available starting materials. Available starting materials are those compounds that can be obtained from chemical suppliers. For example, the synthetic tree shown below indicates that four different compounds (A, E, L, and O) can be converted into the target molecule, and that there is more than one way to produce each of the four from available starting materials.

a synthetic tree

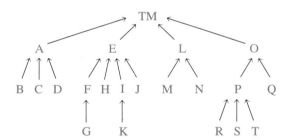

A **linear synthesis** builds a molecule step by step from the starting materials. In a **convergent synthesis,** pieces of the molecule are prepared and then assembled late in the synthesis. A completely convergent synthesis would involve the combination of all the subunits of a molecule in a single step. Because the overall yield of the target molecule decreases as the number of sequential steps that must be performed on the starting material increases, convergent syntheses are preferred over linear syntheses.

a linear synthesis

$$A \longrightarrow B \longrightarrow C \longrightarrow D \longrightarrow E \longrightarrow F \longrightarrow G$$

a convergent synthesis

$$A \longrightarrow B \longrightarrow C + F \longrightarrow G$$
$$\uparrow$$
$$E$$
$$\uparrow$$
$$D$$

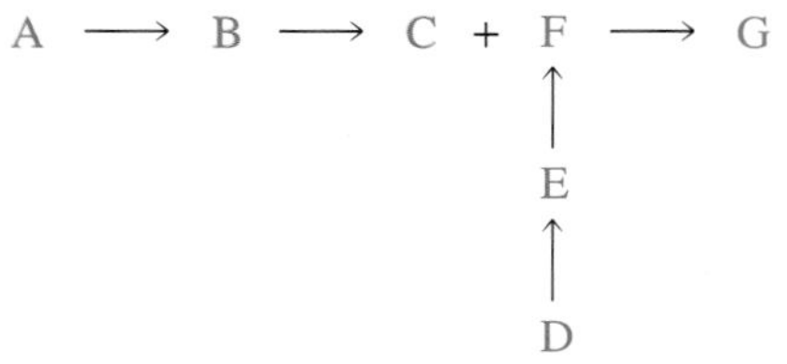

Before starting a synthesis, a chemist carefully examines each of the possible routes to the desired product, aiming for a scheme with the fewest steps, the greatest convergence, the lowest overall cost, the highest yield, and the easiest purification. There are now computer programs that can assist in the design of complicated syntheses. Computer programmers have instructed such programs to use "thought processes" similar to those presented in this chapter.

PROBLEM 1◆

Starting with A, compare the overall yield of G using a six-step linear synthesis and the overall yield of G using the five-step convergent synthesis outlined above if each of the reactions employed gives an 80% yield (a relatively high laboratory yield).

Combinatorial organic synthesis is a technique that uses the concept of mass production to prepare a large number of compounds with closely related structures. This technique looks at an organic compound as a group of building blocks connected together. By systematically changing one building block at a time, a **combinatorial library** (group of structurally related compounds) is obtained. An example of the use of combinatorial organic synthesis by the pharmaceutical industry in an attempt to find new tranquilizers is described in Section 29.11.

SEMISYNTHETIC DRUGS

Chemists have learned to synthesize compounds jointly with nature. For example, in synthesizing the anticancer drug taxol, the yew tree is responsible for the first part of the synthesis. Chemists extract the drug precursor from the needles of the tree, and the precursor is converted to taxol in the laboratory. Thus the precursor is isolated from a renewable resource, while the drug itself could be obtained only by killing the tree.

Sometimes it is the chemist that does the first part of the synthesis, and the synthetic compound is fed to a biological organism which completes the synthesis. Penicillin V is prepared by feeding 2-phenoxyethanol to a culture of *Penicillium*. The mold oxidizes the alcohol to the carboxylic acid it needs for the side chain of penicillin V.

2-phenoxyethanol   *Penicillium* mold   penicillin V

The more methods that are available to form a specific functional group, the easier it is to design a synthesis. It is not difficult to recall the different reactions a particular functional group undergoes, because most of the reactions are related to one another. It is harder to recall all the different ways a functional group can be synthesized, because these reactions can be very different. Therefore, it is a good idea to have a classification scheme that allows easy access to the many available reactions. For this purpose, we have assembled lists of reactions that produce particular functional groups. These lists are found in Appendix IV. The reactions listed there will be familiar to you from earlier chapters.

In designing a synthesis, we often find that the desired product contains more carbons than the available starting material contains. In other words, the carbon

28.1
FUNCTIONAL GROUP INTRODUCTION, REMOVAL, AND INTERCONVERSION

skeleton in the reactant is not the same as the carbon skeleton in the product. Such syntheses require reactions that form new carbon–carbon bonds. Included in Appendix IV is a compilation of reactions that form carbon–carbon bonds.

We have seen that a functional group can be *introduced* into an alkane by first halogenating it and then replacing the halogen with a nucleophile (Sections 8.4 and 9.3).

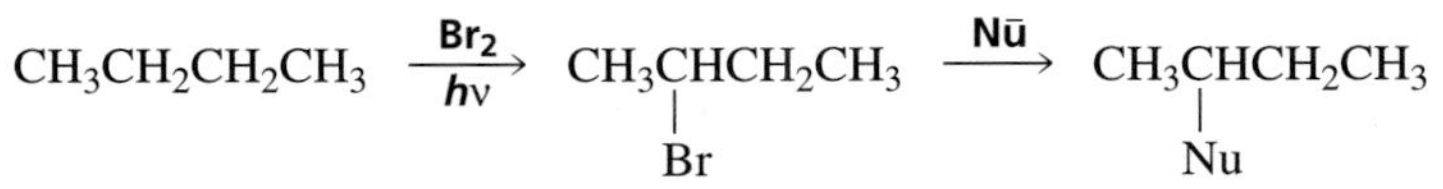

$$CH_3CH_2CH_2CH_3 \xrightarrow[h\nu]{Br_2} CH_3\underset{\underset{Br}{|}}{C}HCH_2CH_3 \xrightarrow{N\bar{u}} CH_3\underset{\underset{Nu}{|}}{C}HCH_2CH_3$$

We have also seen how a functional group can be *removed* from a molecule.

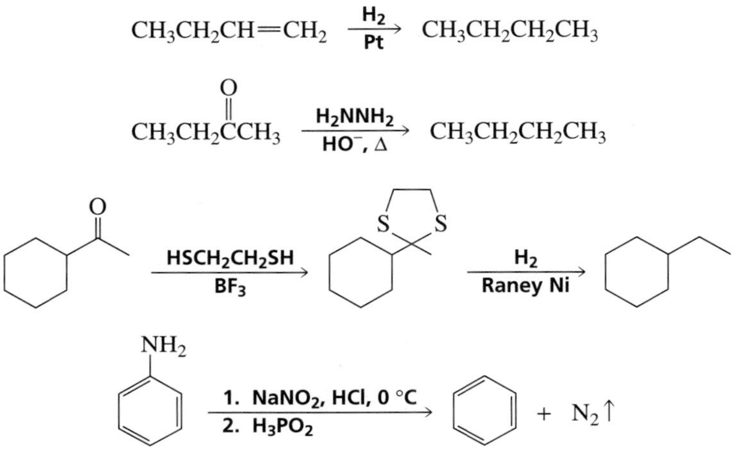

$$CH_3CH_2CH{=}CH_2 \xrightarrow[Pt]{H_2} CH_3CH_2CH_2CH_3$$

$$CH_3CH_2\overset{\overset{O}{\|}}{C}CH_3 \xrightarrow[HO^-,\,\Delta]{H_2NNH_2} CH_3CH_2CH_2CH_3$$

We have explored many ways in which one functional group can be converted into another functional group. Such a reaction is known as a **functional group interconversion (FGI).** For example, a primary alcohol can be formed from a wide variety of other functional groups (Figure 28.1). The method actually used depends on the primary alcohol being synthesized and the available starting materials.

PROBLEM 2 ◆

Which of the functional group interconversions in Figure 28.1 cannot be used for the synthesis of isobutyl alcohol?

PROBLEM 3 ◆

Which of the functional group interconversions in Figure 28.1 change the carbon skeleton of the starting material?

PROBLEM 4

Without looking at Appendix IV, list as many ways to synthesize butanoic acid in one step from another organic compound as you can recall.

PROBLEM 5

Design a synthesis of 1-butanamine using 1-butene as the starting material.

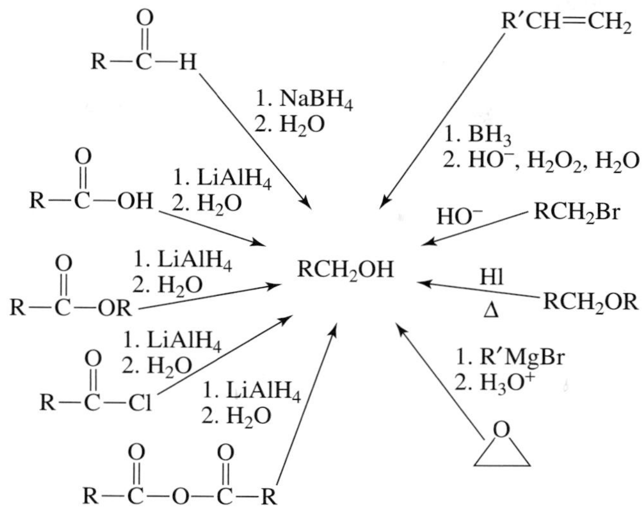

The synthesis of a complicated molecule from simple starting materials is not always obvious. We have seen that it is often easier to work backward from the desired product to the available starting materials (Section 5.11). This process is called **retrosynthesis** or **retrosynthetic analysis.** In a retrosynthetic analysis, the chemist dissects a molecule into smaller and smaller pieces until readily available starting materials are reached.

28.2 RETROSYNTHETIC ANALYSIS. DISCONNECTIONS

retrosynthetic analysis

$$\text{target molecule} \implies Y \implies X \implies W \implies \text{starting materials}$$

There are two kinds of steps in a retrosynthetic analysis: functional group interconversions and disconnections. A **disconnection** involves breaking a bond to carbon to produce two ionic fragments. One fragment is positively charged and one is negatively charged. The fragments of a disconnection are called **synthons.** Synthons are generally not real compounds; they are imaginary constructs. For example, if we consider the retrosynthetic analysis of cyclohexanol, a disconnection gives a hydroxycarbocation and a hydride ion. The hydroxycarbocation and the hydride ion are synthons.

retrosynthetic analysis

open arrow represents a
retrosynthetic operation

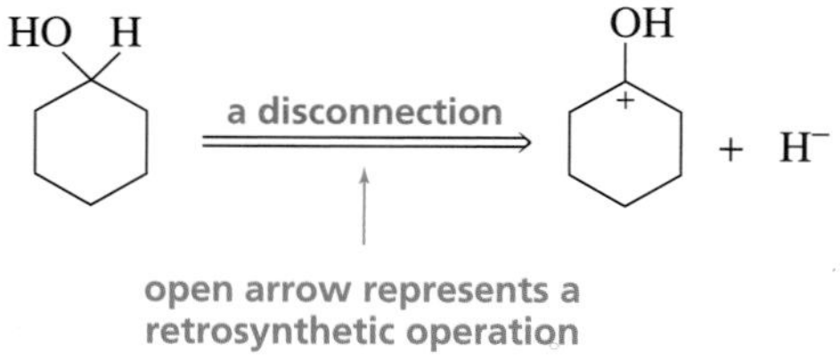

Cyclohexanone is the synthetic equivalent for the hydroxycarbocation, while sodium borohydride is the synthetic equivalent for hydride ion. A **synthetic**

A synthetic equivalent is
the real compound that is
as close as we can get to
an imaginary synthon.

equivalent is the reagent actually used as the source of the synthon. The synthetic equivalents represent the nucleophile and the electrophile that must come together in a reaction. So cyclohexanol, the target molecule, can be prepared by treating cyclohexanone with sodium borohydride.

synthesis

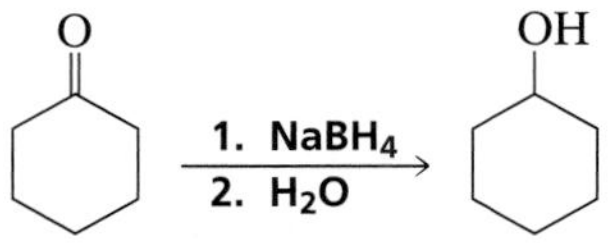

When carrying out a disconnection, one has to decide, after breaking the bond, which fragment gets the positive charge and which gets the negative charge. In the example above, we could have given the positive charge to the hydrogen, and many acids (HCl, HBr, etc.) could have been used as synthetic equivalents for H^+. However, we would have been at a loss to find a synthetic equivalent for an α-hydroxycarbanion. Therefore, when we carry out the disconnection, we assign the positive charge to the carbon and the negative charge to the hydrogen.

Cyclohexanol can also be disconnected by breaking the carbon–oxygen bond instead of the carbon–hydrogen bond, forming a carbocation and hydroxide ion.

retrosynthetic analysis

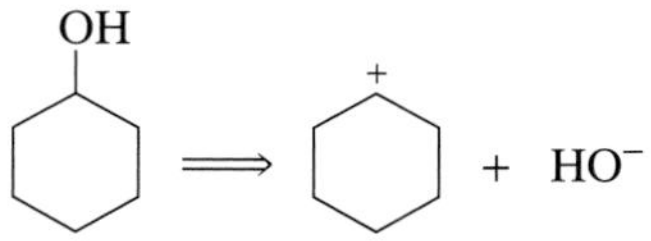

The problem then becomes choosing a synthetic equivalent for the carbocation. A synthetic equivalent for a positively charged synthon needs an electron-withdrawing group at just the right place. Cyclohexyl bromide, with an electron-withdrawing bromine, is a synthetic equivalent for the cyclohexyl carbocation. Cyclohexanol, therefore, can be prepared by treating cyclohexyl bromide with hydroxide ion. This method, however, is not as good as the previous one, because it leads to a lower yield of the target molecule since some of the alkyl halide is converted into an alkene.

synthesis

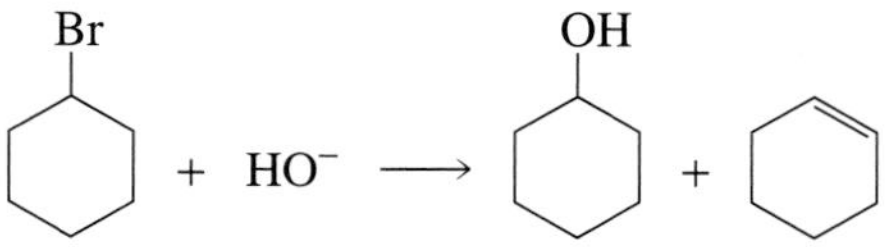

Retrosynthetic analysis shows that 1-methylcyclohexanol can be formed from the reaction of cyclohexanone, the synthetic equivalent for the hydroxycarbocation, and methylmagnesium bromide, the synthetic equivalent for the methyl anion (Section 16.4).

retrosynthetic analysis

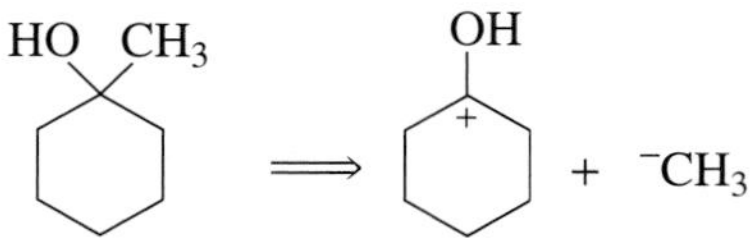

synthesis

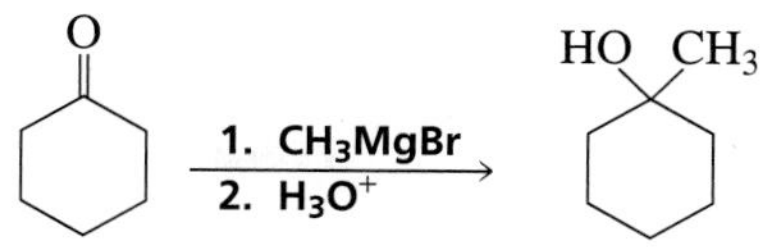

Other disconnections of 1-methylcyclohexanol are possible. For example, one of the ring carbon–carbon bonds could be broken. However, these are not useful disconnections, because the synthetic equivalents of these synthons are not easily prepared. A retrosynthetic step must lead to readily obtainable starting materials. If not, how will the starting materials be obtained to make the disconnected compound?

retrosynthetic analysis

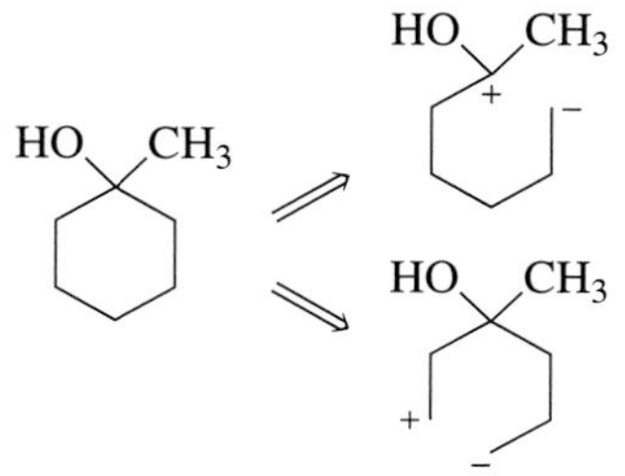

2-Methylcyclohexanone can be disconnected into an α-carbanion and a methyl cation.

retrosynthetic analysis

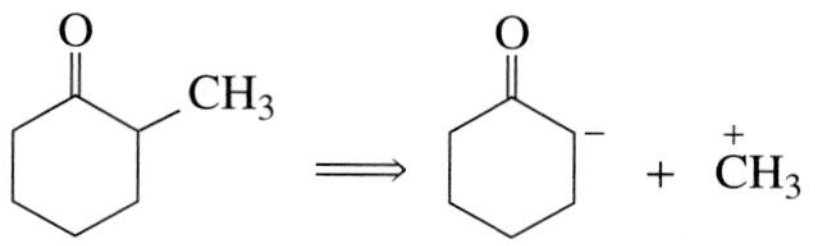

The α-carbanion can be prepared by treating cyclohexanone with LDA (Section 18.7). Methyl iodide is a synthetic equivalent for the methyl cation.

synthesis

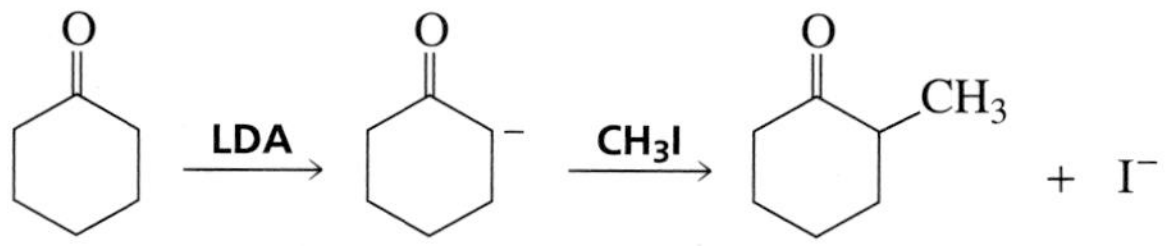

2-Aminocyclohexanone can be disconnected into an electrophilic α-carbocation and a nucleophilic amide anion.

retrosynthetic analysis

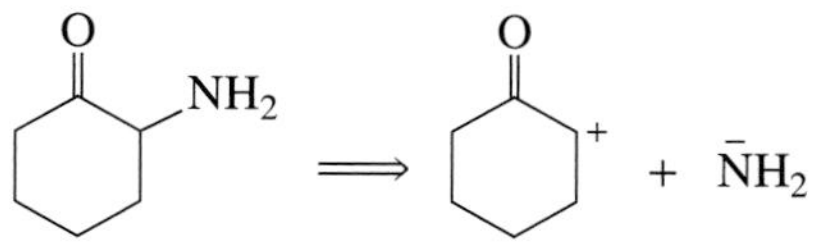

The α-carbon of a ketone is normally nucleophilic, because an α-hydrogen can be removed by a strong base to form an α-carbanion, as was done in the synthesis of 2-methylcyclohexanone. To carry out this synthesis, the α-carbon must be electrophilic. In other words, we must reverse its normal polarity. Reversing the normal polarity is called **umpolung** (German for "polarity reversal"). The α-carbon can be made electrophilic by substituting an electron-withdrawing group for an α-hydrogen. The α-bromo-substituted ketone is easily produced by treating the ketone with bromine under acidic conditions (Section 18.4).

synthesis

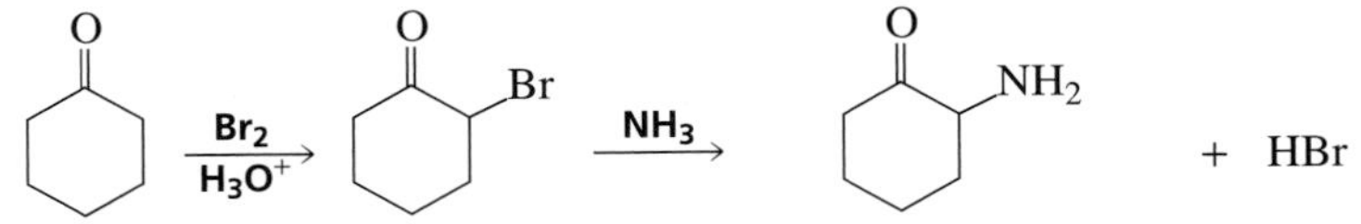

2-aminocyclohexanone

PROBLEM 6

How could you apply the principle of *umpolung* to bromocyclohexane? In other words, how could you convert an ordinarily electrophilic bromocyclohexane into a nucleophile?

PROBLEM 7 ◆

Label each of the following as a normal synthon or an *umpolung* synthon.

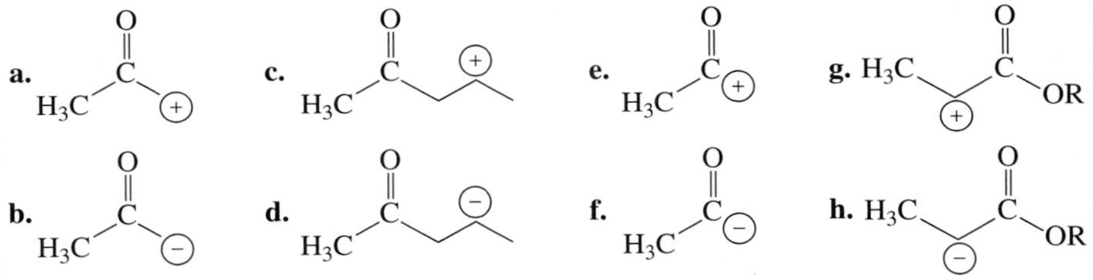

Butanal can be disconnected into either a propyl anion and a formyl cation or a propyl cation and a formyl anion. The formyl cation is the normal synthon; the formyl anion is the *umpolung* synthon. Propylmagnesium bromide is the synthetic equivalent for the propyl anion, and ethyl formate is the synthetic equivalent for the formyl cation. However, once the target molecule (butanal) has been formed, it can react with the Grignard reagent, forming a secondary alcohol. Therefore, the second disconnection may provide a more promising synthesis, since it does not require the use of a Grignard reagent.

retrosynthetic analysis

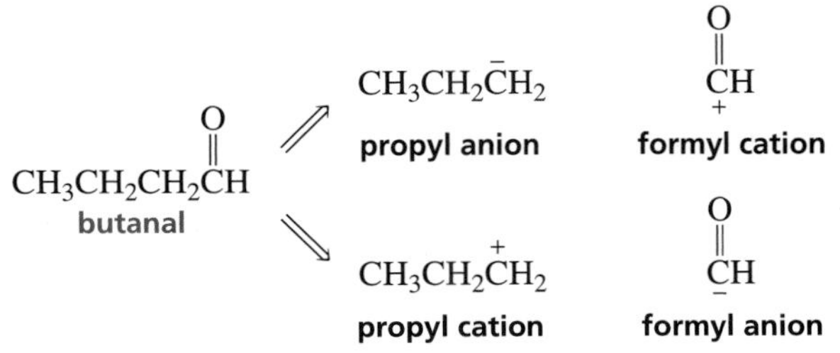

Propyl bromide is a synthetic equivalent for the propyl cation, and dithianyl-lithium is the synthetic equivalent for the formyl anion. Dithianyllithium is easily prepared by treating 1,3-dithiane with a strong base such as butyllithium, because the C-2 hydrogens of 1,3-dithiane are relatively acidic. Reaction of dithianyllithium with an alkyl halide leads to the formation of a thioacetal. The carbon–sulfur bonds are cleaved by mercuric ion in an aqueous acetonitrile solution, forming a hydrate that loses water to form the target molecule.

synthesis

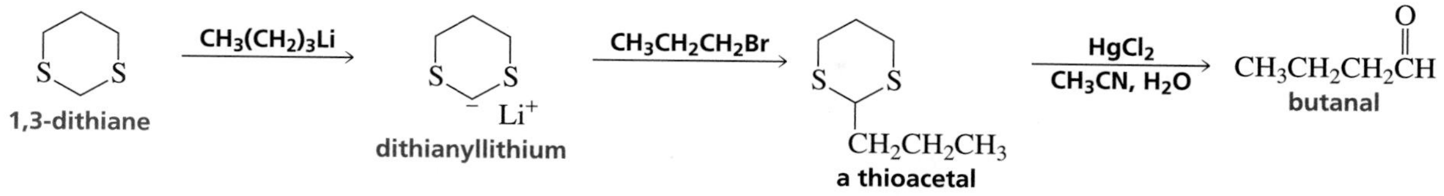

1,3-dithiane → dithianyllithium → a thioacetal → butanal

$$CH_3(CH_2)_3Li$$

$$CH_3CH_2CH_2Br$$

$$\xrightarrow[CH_3CN, H_2O]{HgCl_2}$$

$$CH_3CH_2CH_2\overset{\displaystyle O}{\overset{\|}{C}}H$$

PROBLEM 8 ◆

What change would you make in the preceding synthesis if the desired product were

a. heptanal?   **b.** 2-pentanone?

Because dithianyllithium is a formyl anion synthetic equivalent, it can be used to convert aldehydes and ketones into 1,2-dioxygenated compounds.

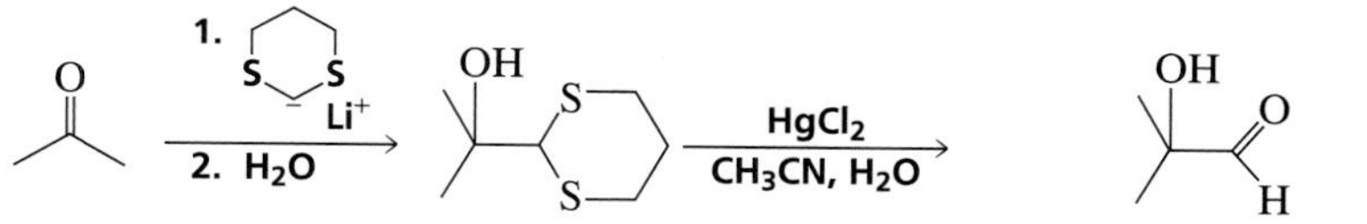

an α-hydroxyaldehyde
a 1,2-dioxygenated compound

1,2-Dioxygenated compounds can also be prepared using cyanide ion rather than formyl anion as the nucleophile. When formyl anion is the nucleophile, the product is an α-hydroxyaldehyde; when cyanide is the nucleophile, the product is an α-hydroxynitrile which can be converted to an α-hydroxycarboxylic acid. Notice that in both cases the number of carbon atoms in the starting material is increased by one.

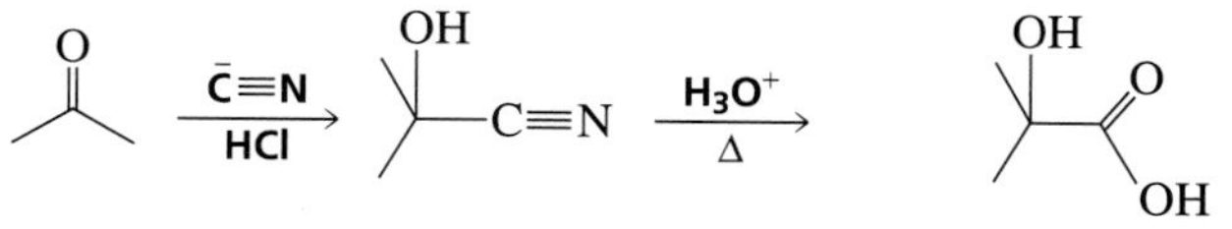

an α-hydroxycarboxylic acid
a 1,2-dioxygenated compound

PROBLEM 9

Using bromocyclohexane as a starting material, how could you synthesize the following compounds?

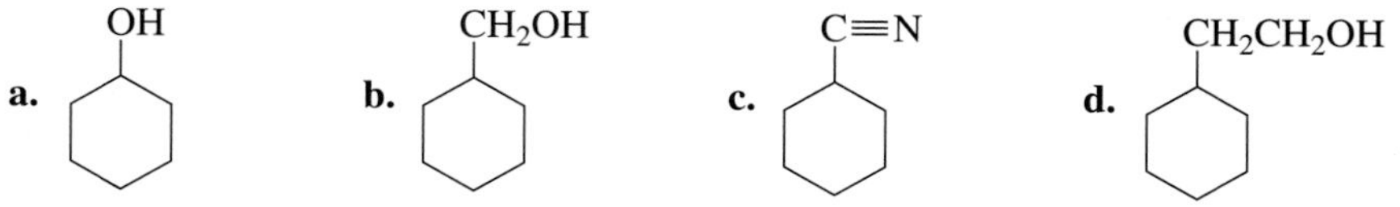

a. OH   **b.** CH$_2$OH   **c.** C≡N   **d.** CH$_2$CH$_2$OH

PROBLEM 10

Give a synthetic equivalent for each of the following.

a. $\overset{+}{C}H_2CH_2CH_3$ **b.** $\bar{C}H_2CH_2CH_3$ **c.** $\overset{+}{C}H_2\overset{\displaystyle O}{\overset{\|}{C}}H$ **d.** $\bar{C}H_2\overset{\displaystyle O}{\overset{\|}{C}}H$

e. $\overset{\overset{\displaystyle O}{\|}}{\underset{+}{C}CH_3}$ g. $\overset{\overset{\displaystyle O}{\|}}{\underset{-}{C}H}$ i. $\overset{+}{C}H_2CH_2OH$

f. $\overset{\overset{\displaystyle O}{\|}}{\underset{-}{C}CH_3}$ h. $\overset{\overset{\displaystyle O}{\|}}{C}CH_2CH_3$ j. $\overset{+}{C}H_2CH_2CH_2OH$

PROBLEM 11 / SOLVED

Describe how the following compounds could be prepared using the given starting materials. Perform retrosynthetic analyses to help you arrive at your answers.

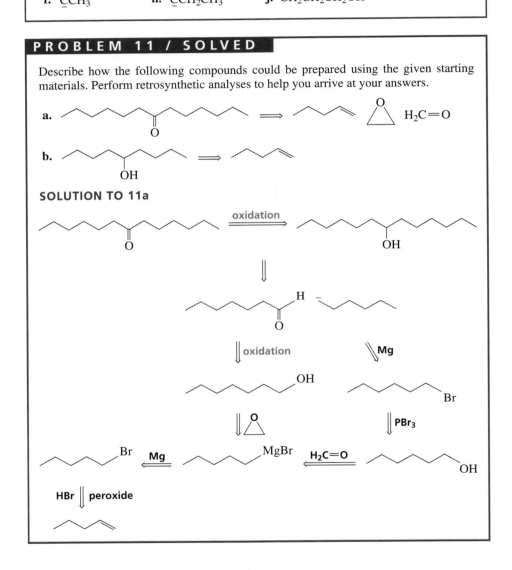

SOLUTION TO 11a

28.3 RETROSYNTHETIC ANALYSIS OF DIOXYGENATED COMPOUNDS

As we have just seen in planning the syntheses of 1,2-dioxygenated compounds, if a compound has two functional groups, the relative locations of the two groups provide valuable hints for designing a synthesis. A 1,3-dioxygenated compound such as 3-hydroxybutanal can be disconnected to give a hydroxycarbocation and an α-carbanion. These are both synthons for the same aldehyde. A 1,3-dioxygenated compound can therefore be synthesized by means of an aldol addition. Acetaldehyde is used as the starting material for the synthesis of 3-hydroxybutanal. Dehydration of the β-hydroxycarbonyl compound results in formation of 2-butenal, an α,β-unsaturated carbonyl compound.

retrosynthetic analysis

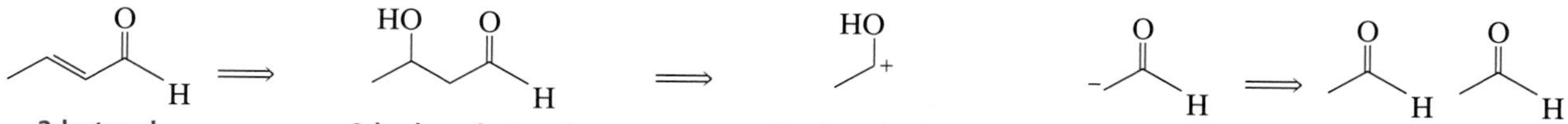

| 2-butenal | 3-hydroxybutanal <br> a 1,3-dioxygenated compound | an α-hydroxycarbocation an α-carbanion | acetaldehyde |

Retrosynthetic analysis shows that 2,5-hexanedione can be used as the starting material for the synthesis of 3-methyl-2-cyclopentenone, an α,β-unsaturated ketone.

retrosynthetic analysis

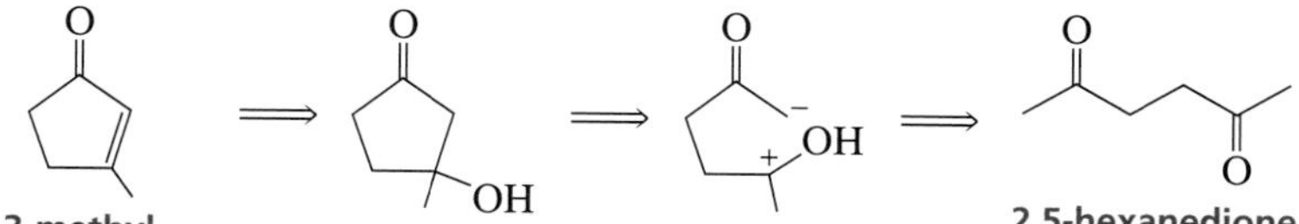

3-methyl-2-cyclopentenone   2,5-hexanedione

PROBLEM 12

Using retrosynthetic analysis, plan a synthesis of the following compound from the given starting materials.

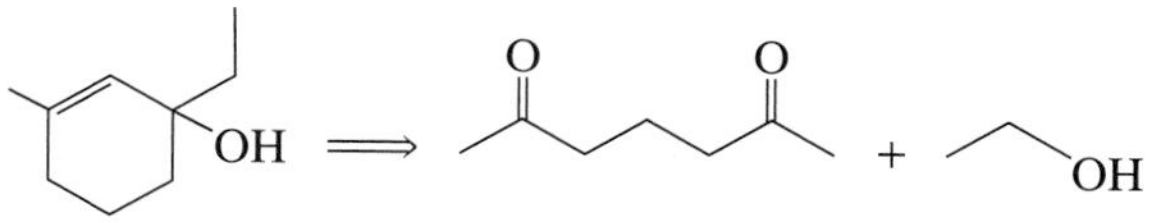

Disconnection of the following 1,4-dioxygenated compound can be done in two different ways. Both disconnections require the application of *umpolung*, because in each case one of the synthons is an α-carbocation. The polarity of an α-carbon can be reversed by putting a bromine in the α-position of the carbonyl compound to make the compound electrophilic at that position. However, generating a negative charge on the α-carbon of the nonbrominated carbonyl compound poses a problem. Bromination increases the acidity of the protons on the brominated carbon. So, if a base is used to create the α-carbanion, the base may remove a proton from the α-carbon of the brominated carbonyl compound rather than from the α-carbon of the nonbrominated carbonyl compound.

retrosynthetic analysis

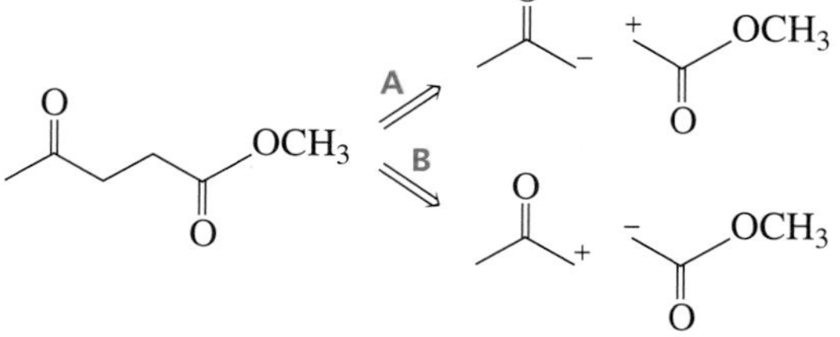

This problem can be solved if an enamine is used as the synthetic equivalent of the α-carbanion (Section 18.8). Because esters cannot form enamines, disconnection A leads to the best starting materials. The enamine can be prepared by treating the carbonyl compound with a secondary amine. After the enamine has reacted with the α-brominated ester, the carbonyl group can be regenerated by hydrolysis under conditions that won't also hydrolyze the ester (Section 16.7).

synthesis

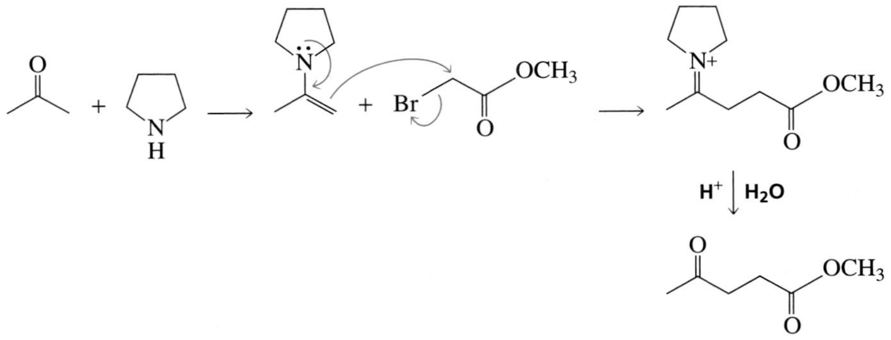

a 1,4-dioxygenated compound

PROBLEM 13

Using a base to generate an α-carbanion can result in the condensation shown below, known as the **Darzens condensation.** Suggest a mechanism for this reaction.

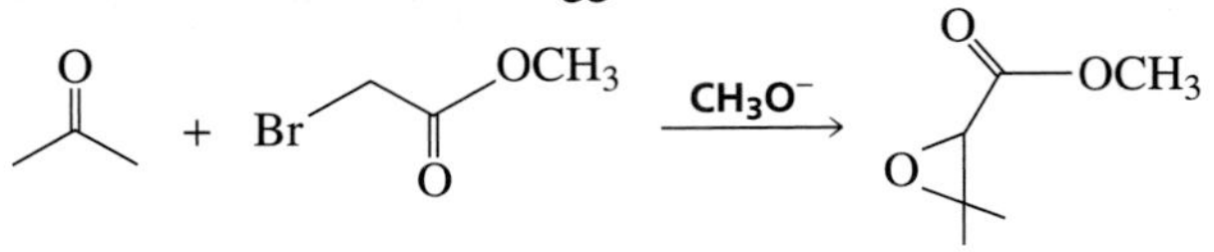

PROBLEM 14

Explain why an α-monobromocarbonyl compound cannot be prepared by treating a carbonyl compound with bromine under basic conditions.

Retrosynthetic analysis indicates that 1,5-dioxygenated compounds can be synthesized by the reaction of an α-carbanion with a carbonyl compound that has a positively charged β-carbon.

retrosynthetic analysis

a 1,5-dioxygenated compound

An α,β-unsaturated carbonyl compound is a synthetic equivalent for the β-carbocation. This suggests that a Michael reaction will form the desired 1,5-dioxygenated product (Section 18.9). The Michael reaction provides an example of how the polar reactivity of a functional group, such as a carbonyl group, can be transmitted through a carbon–carbon double bond. Transmission of reactivity through double bonds is called **vinylogy.**

synthesis

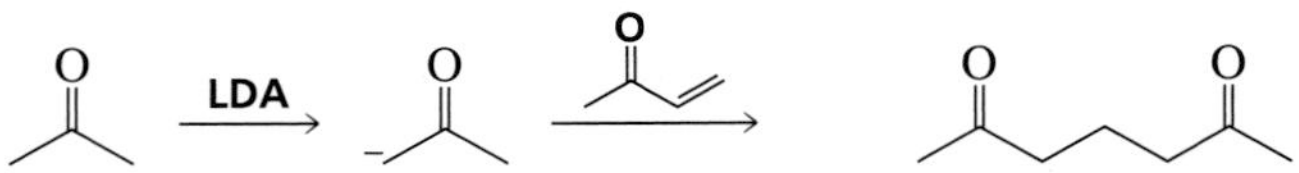

2,6-heptanedione
a 1,5-dioxygenated compound

PROBLEM 15

Describe how the following compound could be synthesized from compounds containing no more than six carbon atoms.

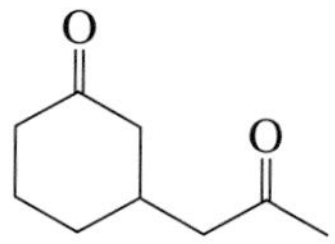

Retrosynthetic analysis shows that a 1,6-dioxygenated compound can be prepared by nucleophilic attack of an α-carbanion on a carbonyl compound with a positively charged γ-carbon.

retrosynthetic analysis

a 1,6-dioxygenated compound

Because there are no good general methods for the synthesis of γ-halosubstituted carbonyl compounds, we must consider another route. Recognizing that a 1,6-dioxygenated compound can be prepared by cleaving a cyclohexene provides an easy route to the target compound, because a compound with a six-membered ring can be readily prepared by means of a Diels–Alder reaction (Section 7.8). The convergence (assembly of pieces) of the Diels–Alder reaction is one of its most useful features.

retrosynthetic analysis

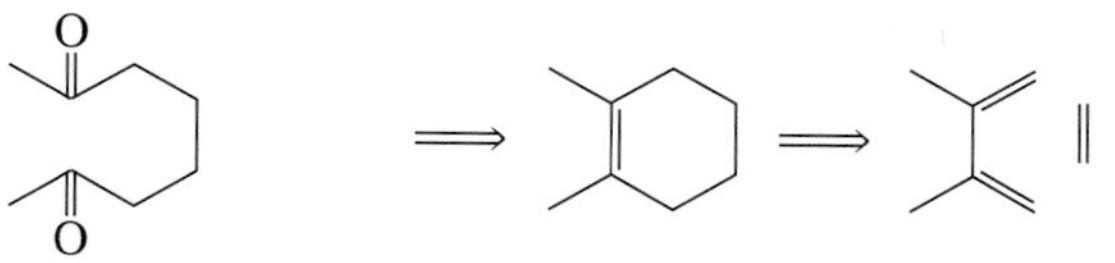

a 1,6-dioxygenated compound

When the target molecule has a six-membered ring, it is quite easy to determine that a Diels–Alder reaction should be employed in the synthetic scheme. However, when the product does not have a six-membered ring, as in the case of the 1,6-dioxygenated compound shown above, it is not immediately obvious that a Diels–Alder reaction should be used.

Another example of a synthesis in which the use of a Diels–Alder reaction is not immediately obvious is the extrusion reaction shown below. In an **extrusion reaction,** a neutral molecule (for example, carbon dioxide, carbon monoxide, nitrogen, or sulfur dioxide) is eliminated. A Diels–Alder reaction leads to a compound that can readily eliminate carbon monoxide in order to form a hexasubstituted benzene ring.

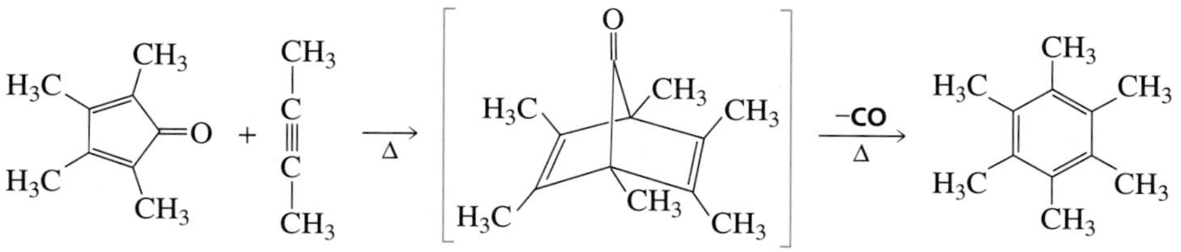

PROBLEM 16

Do a retrosynthetic analysis on each of the following compounds, ending with available starting materials.

a. 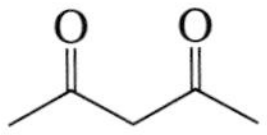 (structure: 2,4-pentanedione)

c. 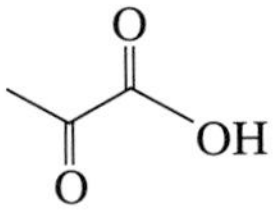 (structure: pyruvate / 2-oxopropanoic acid with OH)

b. 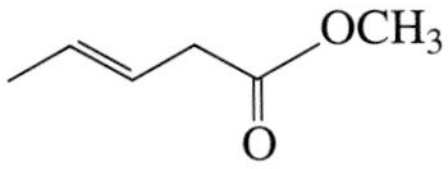 (structure: methyl ester of unsaturated acid, OCH₃)

PROBLEM 17 / SOLVED

How could you synthesize the following compounds from starting materials containing no more than four carbons?

a. 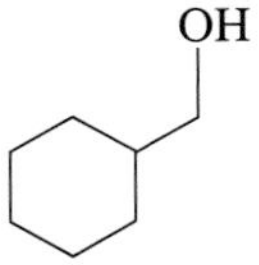 (cyclohexane with CH₂OH)

c. 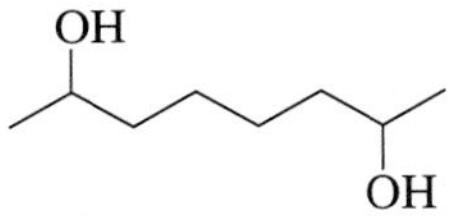 (diol chain with two OH)

b. 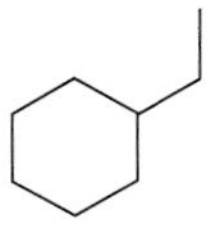 (cyclohexane with ethyl)

d. 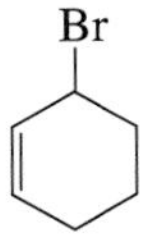 (cyclohexene with Br)

SOLUTION TO 17a   The six-membered ring indicates that the compound can be made by means of a Diels–Alder reaction.

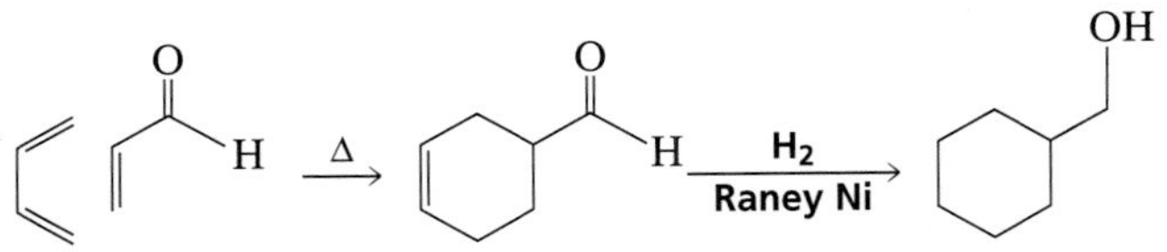

(reaction scheme: diene + acrolein →Δ→ cyclohexene carbaldehyde →H₂ / Raney Ni→ cyclohexyl methanol OH)

We have seen that the Wittig reaction is used to form carbon–carbon double bonds (Section 16.11). This is the most useful procedure for forming carbon–carbon double bonds, because the reaction results in the unambiguous placement of the double bond. Retrosynthetically, the reaction is represented as shown below.

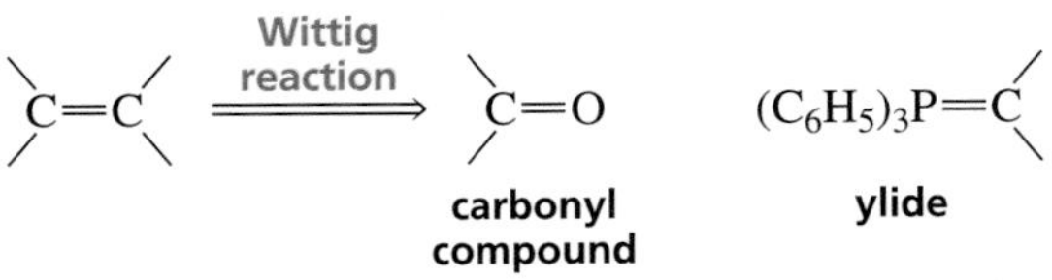

$$\begin{array}{c} \backslash\,/\\ C=C\\ /\,\backslash \end{array} \xrightarrow[\text{reaction}]{\text{Wittig}} \begin{array}{c} \backslash\\ C=O\\ / \end{array} \quad (C_6H_5)_3P=C\begin{array}{c} /\\ \backslash \end{array}$$

carbonyl compound   ylide

When the alkene to be synthesized is not symmetric, you must carefully consider which part of the product should be derived from the carbonyl compound and which part from the ylide. For example, the following enone has two disconnections. Disconnection A is preferred, because the ylide is stabilized by resonance and the ketone is readily available.

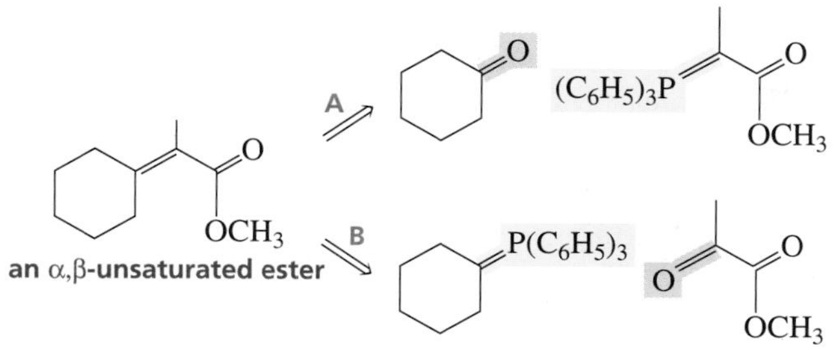

an α,β-unsaturated ester

Catalytic hydrogenation of Wittig products is a useful method for synthesizing complicated alkanes with precisely controlled structures.

PROBLEM 18◆

a. What reagents would you use for the synthesis of 3-ethyl-1-pentene using a Wittig reaction?

b. What advantage does this method of synthesis have over a synthesis using an alkyl halide and acetylide ion?

PROBLEM 19

Propose a method for synthesizing the following compound using starting materials containing no more than six carbons and triphenylphosphine.

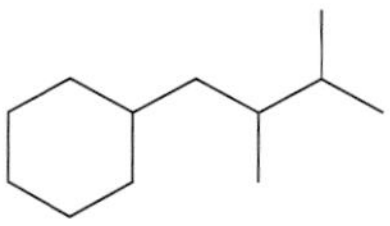

Sometimes when a complicated molecule is broken down into its component parts, the route to its synthesis becomes readily apparent. At first glance, buspirone appears to be difficult to synthesize. However, it can be broken down into readily available starting materials. The recognition of smaller pieces allows for an easy retrosynthetic analysis. With very complicated molecules, this is often the best approach. You can think of this approach as the "divide and conquer" method of analysis. This method leads to convergence, which in turn leads to higher yields.

retrosynthetic analysis

buspirone

A **protecting group** protects a functional group from a synthetic operation that it would otherwise not survive. We have seen that 1,2- and 1,3-diols are used to protect the carbonyl groups of aldehydes and ketones from reaction with nucleophiles

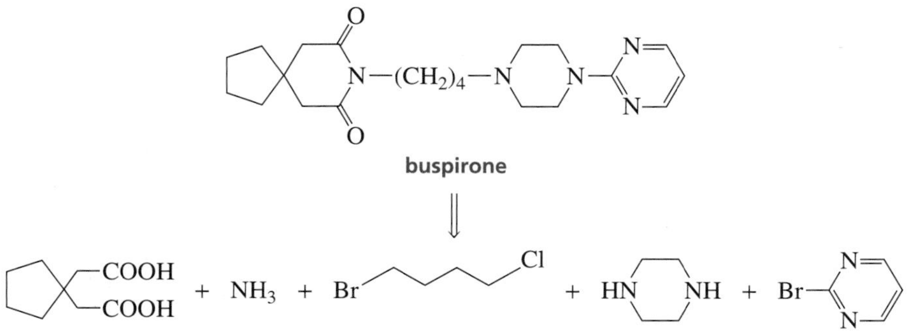

28.4
MORE ABOUT
PROTECTING GROUPS

(Section 16.9). For example, LiAlH$_4$ can reduce both functional groups of a keto ester. The ketone group can be protected from reduction by being converted into an acetal. After the ester has been reduced, the protecting group can easily be removed by acid-catalyzed hydrolysis. It is critical that protecting groups are able to be removed under conditions that will not affect other groups in the molecule.

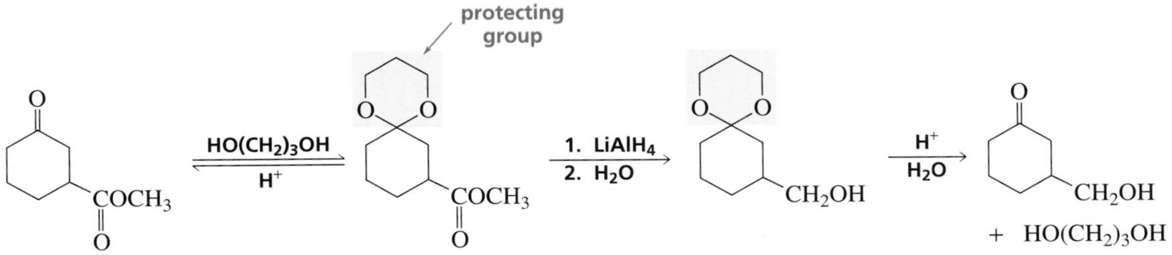

Protecting groups should be used only when absolutely necessary because each time a protecting group is used, it must be attached and then taken off. This adds two linear steps to the synthesis, which decreases the overall yield of the target molecule.

The following disconnection indicates that the desired transformation can be carried out by treating a ketone with a Grignard reagent.

retrosynthetic analysis

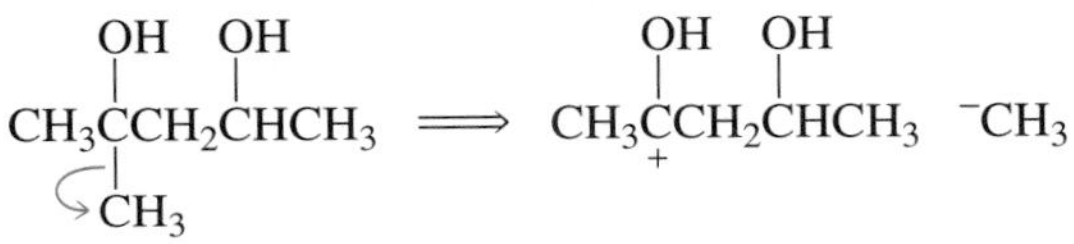

However a problem arises in the synthesis, because the labile proton of the OH group will react with the Grignard reagent (Section 11.9). Therefore, the OH group must be protected in order to make it immune to the Grignard reagent. There are many protecting groups available for alcohols, but the *tert*-butyldimethylsilyl (TBDMS) group has several advantages. It reacts with an alcohol under mild conditions, it is unaffected by acid and Grignard reagents, and it is easily removed by treatment with tetrabutylammonium fluoride. The specificity of fluoride ion for silicon is due to the unusually large energy of the Si—F bond (135 kcal/mol).

synthesis

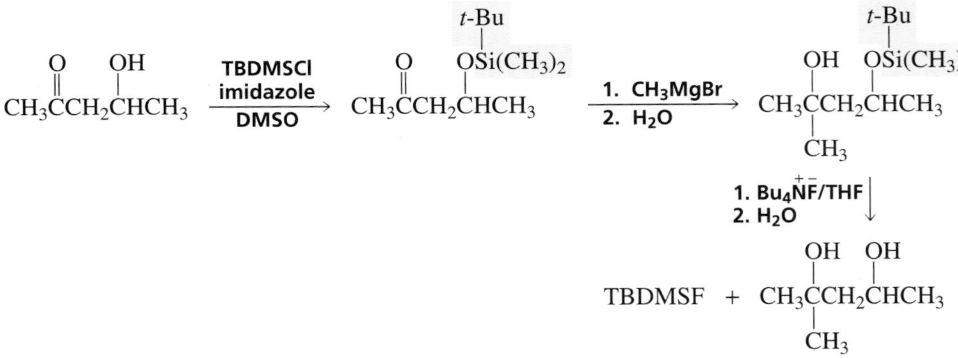

PROBLEM 20◆

An alcohol can be left unprotected if an extra equivalent of the Grignard reagent is used. This is sometimes the method of choice if the Grignard reagent is inexpensive. Explain the fate of the excess Grignard reagent.

A carboxyl group can be protected by being converted into an oxazoline, because oxazolines are not attacked by nucleophiles. The carboxyl group can be regenerated by acid-catalyzed hydrolysis.

$$\text{CH}_3\overset{O}{\overset{||}{C}}\text{CH}_2\text{CH}_2\overset{O}{\overset{||}{C}}\text{OH} + \text{HOCH}_2\underset{\underset{NH_2}{|}}{\overset{\overset{CH_3}{|}}{C}}\text{CH}_3 \xrightarrow{\Delta} \text{CH}_3\overset{O}{\overset{||}{C}}\text{CH}_2\text{CH}_2\text{C}$$

an oxazoline

$$\downarrow \text{CH}_3\text{MgBr}$$

$$\text{HOCH}_2\underset{\underset{NH_2}{|}}{\overset{\overset{CH_3}{|}}{C}}\text{CH}_3 + \text{CH}_3\underset{\underset{CH_3}{|}}{\overset{\overset{OH}{|}}{C}}\text{CH}_2\text{CH}_2\overset{O}{\overset{||}{C}}\text{OH} \underset{H_2O}{\overset{H^+}{\longleftarrow}} \text{CH}_3\underset{\underset{CH_3}{|}}{\overset{\overset{O^-}{|}}{C}}\text{CH}_2\text{CH}_2\text{C}$$

If only the OH group of the carboxyl group needs to be protected, the carboxylic acid can be converted into an ester. When the need for protection is over, hydrolysis will regenerate the carboxylic acid.

$$\text{CH}_3\underset{\underset{OH}{|}}{\text{CH}}\text{CH}_2\overset{O}{\overset{||}{C}}\text{OH} \xrightarrow[\substack{CH_3CH_2OH \\ \textbf{excess}}]{H^+} \text{CH}_3\underset{\underset{OH}{|}}{\text{CH}}\text{CH}_2\overset{O}{\overset{||}{C}}\text{OCH}_2\text{CH}_3 \xrightarrow{\text{SOCl}_2} \text{CH}_3\underset{\underset{Cl}{|}}{\text{CH}}\text{CH}_2\overset{O}{\overset{||}{C}}\text{OCH}_2\text{CH}_3$$

$$\downarrow {}^-\text{C}\equiv\text{N}$$

$$\text{CH}_3\text{CH}_2\text{OH} + \text{CH}_3\underset{\underset{COOH}{|}}{\text{CH}}\text{CH}_2\overset{O}{\overset{||}{C}}\text{OH} \underset{\underset{\Delta}{H_2O}}{\overset{H^+}{\longleftarrow}} \text{CH}_3\underset{\underset{C\equiv N}{|}}{\text{CH}}\text{CH}_2\overset{O}{\overset{||}{C}}\text{OCH}_2\text{CH}_3$$

PROBLEM 21◆

What product would have been formed from the reaction above if the OH group of the carboxylic acid had not been protected?

An amino group can be protected by being converted into an amide with acetyl chloride (Section 14.15). The acetyl group can subsequently be removed by acid-catalyzed hydrolysis.

aniline   acetanilide   *p*-nitroacetanilide   *p*-nitroaniline

An amino group can also be protected with di-*tert*-butyl dicarbonate. Because this group can be removed under mildly acidic conditions, it is used as a protecting group when the compound also contains a group that would be adversely affected by acid hydrolysis.

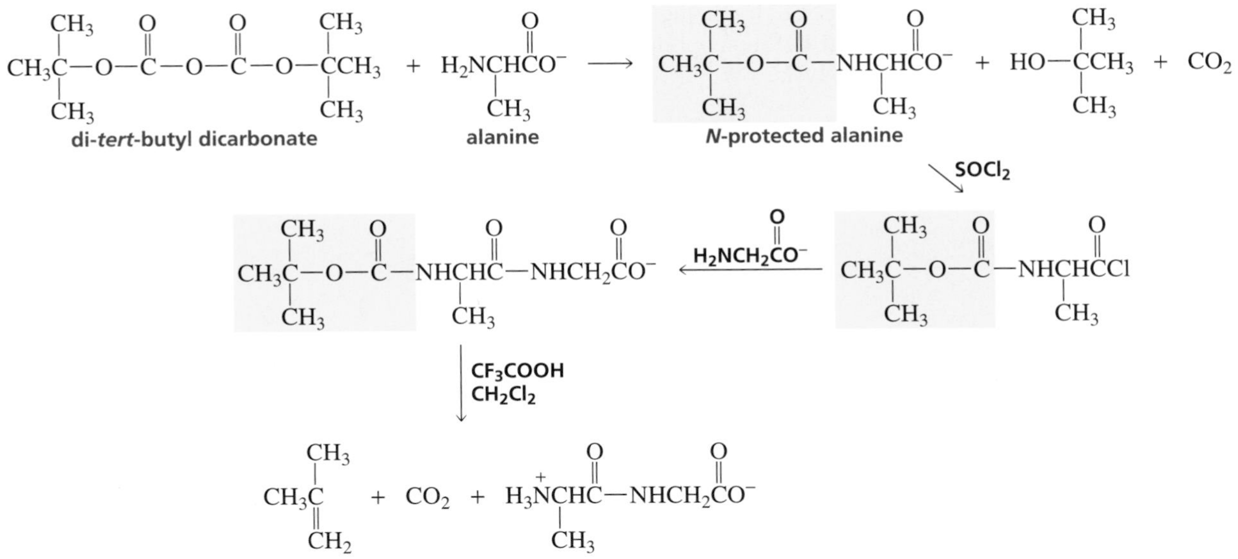

Another advantage the *tert*-butylcarboxy group has as a protecting group is that the products formed when it is removed are gases (isobutylene and carbon dioxide) that force the reaction to completion as they escape (Section 20.9).

PROBLEM 22

Why is di-*tert*-butyl dicarbonate rather than acetyl chloride used as a protecting group in the reaction shown above?

PROBLEM 23 ◆

What products would be formed from the reaction above if alanine's amino group were not protected?

28.5 CONTROL OF STEREOCHEMISTRY

The target molecule of a synthesis may be one of several stereoisomers. The actual number of stereoisomers depends on the number of double bonds and chirality centers in the molecule, because each double bond can exist in the *E* or *Z* configuration (Section 3.5), and each chirality center can have an *R* or *S* configuration (Section 4.5). In addition, if the target molecule has rings with a common bond, the rings can be either trans fused or cis fused (Section 4.17). In designing a synthesis, care must be taken to make sure that each double bond, each chirality center, and each ring fusion in the target molecule has the appropriate configuration. If the stereochemistry of the reactions is not controlled, the resulting mixture of stereoisomers may be difficult or even impossible to separate. Organic chemists must therefore consider the stereochemical outcomes of all reactions in planning a synthesis and use highly stereoselective reactions to achieve the desired configuration.

We have seen examples of stereoselective reactions in previous chapters, with the simplest being the S_N2 reaction. In an S_N2 reaction, the configuration of the chirality center is inverted during the course of the reaction, so two successive S_N2 reactions lead to retention of configuration at a chirality center. S_N1 reactions are much less useful to the synthetic chemist, because they lead to racemic mixtures. Some other examples of stereoselective reactions with which you are familiar are addition of bromine to an alkene (forms anti addition products), addition of H_2 to an alkene (forms syn addition products), and elimination under E2 conditions (forms anti elimination products).

PROBLEM 24

Explain how you could use S_N2 reactions to synthesize either enantiomer of 2-bromo-octane starting with (S)-2-octanol. (*Hint:* See Section 11.3.)

PROBLEM 25

How could the following compounds be prepared from 1-butyne and any other organic reagent with fewer than four carbon atoms? (*Hint:* See Sections 5.8 and 5.10).

a. *cis*-3-hexene

b. *trans*-3-heptene

PROBLEM 26 ◆

How could *trans*-2-butene be converted to *cis*-2-butene? (*Hint:* See Section 10.10.)

Some stereoselective reactions are also enantioselective. An **enantioselective reaction** is a reaction that forms an excess of one enantiomer of a pair. Several methods can be used to synthesize an enantiomerically pure target molecule. We will look at the four most common.

One method employs an enzyme to catalyze the synthetic transformation. Because enzymes are chiral, enzyme-catalyzed reactions can result in the exclusive formation of one enantiomer (Section 4.20). For example, ketones are enzymatically reduced to alcohols by enzymes known as alcohol dehydrogenases. Whether the R or the S enantiomer is formed depends on the particular alcohol dehydrogenase used: alcohol dehydrogenase from the bacterium *Lactobacillus kefir* forms R alcohols, while alcohol dehydrogenases from yeast, horse liver, and the bacterium *Thermoanaerobium brocki* form S alcohols. The alcohol dehydrogenases use NADPH as the actual reducing agent (Section 22.1). The use of an enzyme is not a universally useful method for controlling the configuration of a product, because enzymes require substrates of very specific size and shape (Section 21.8).

(R)-2,2,2-trifluoro-
1-phenyl-1-ethanol

(S)-2,2,2-trifluoro-
1-phenyl-1-ethanol

A second method of preparing a single enantiomer involves the use of a chiral auxiliary. A **chiral auxiliary** is an enantiomerically pure compound that, when attached to a reactant, causes the product with the desired configuration to be formed preferentially. When the reaction is over, the chiral auxiliary is removed. An example of a chiral auxiliary is the chiral hydrazine in the reaction shown on page 1224. The hydrazine reacts with a ketone, forming a chiral hydrazone. The chiral auxiliary blocks one face of the α-carbon from the sterically hindered base (LDA) used to form the α-carbanion (Section 18.7) and also blocks the subsequent approach of the alkyl halide. Consequently, alkylation of the α-carbanion occurs on only one face. When the chiral auxiliary is removed by hydrolysis, only a single enantiomer of the α-alkylated ketone is left behind. In the absence of the auxiliary, the alkylation reaction would form a racemic mixture.

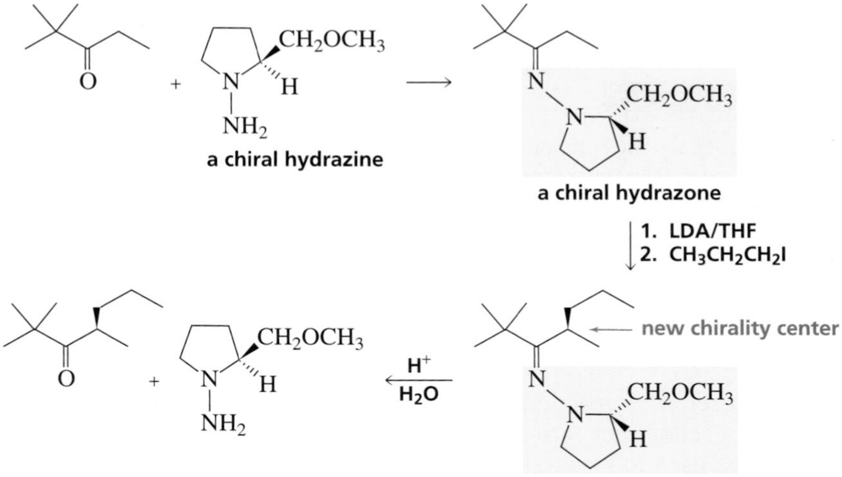

a chiral hydrazine

a chiral hydrazone

1. LDA/THF
2. CH$_3$CH$_2$CH$_2$I

← **new chirality center**

A disadvantage of using a chiral auxiliary is that a molar equivalent of the auxiliary must be used. However, if it is an expensive compound, it can often be recovered and reused.

A third method of controlling the configuration of a target molecule is to use an enantiomerically pure starting material. For example, an enantiomerically pure epoxide of an allyl alcohol can be prepared by treating the alcohol with *tert*-butyl hydroperoxide, titanium isopropoxide, and enantiomerically pure diethyl tartrate (DET). The structure of the epoxide depends on the enantiomer of diethyl tartrate used.

*This third method was developed by **K. Barry Sharpless,** who was born in Philadelphia in 1941. He received a Ph.D. in chemistry from Stanford and served as a professor at MIT, Stanford, and again at MIT, where he developed this procedure. He is currently at Scripps Research Institute, La Jolla, California.*

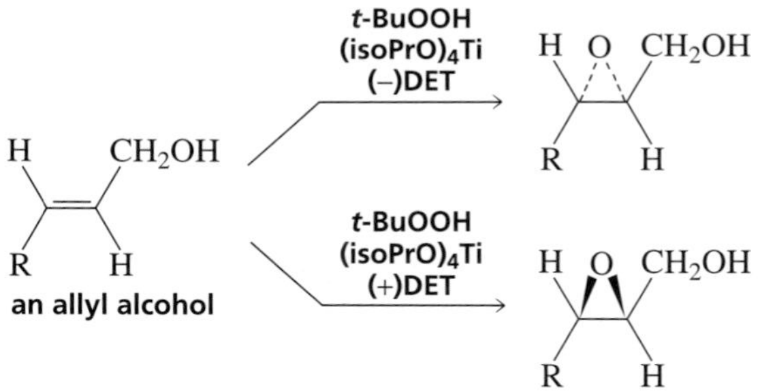

t-BuOOH
(isoPrO)$_4$Ti
(−)DET

t-BuOOH
(isoPrO)$_4$Ti
(+)DET

an allyl alcohol

An enantiomerically pure epoxide is a useful starting material for a wide variety of optically active compounds, because it can easily be converted into a compound with two adjacent chirality centers. This is because epoxides are very susceptible to attack by nucleophiles, and since attack by a nucleophile on an epoxide is a highly stereoselective reaction, only one stereoisomer is formed. In the following example, an allyl alcohol is converted into an enantiomerically pure epoxide, which is used to form an enantiomerically pure diol.

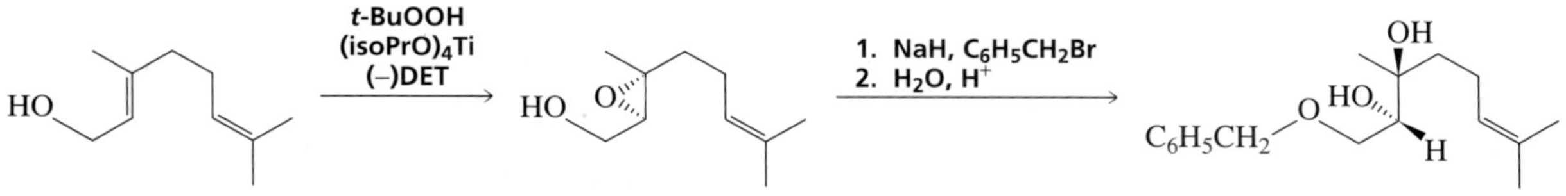

t-BuOOH
(isoPrO)$_4$Ti
(−)DET

1. NaH, C$_6$H$_5$CH$_2$Br
2. H$_2$O, H$^+$

PROBLEM 27

What is the product of the reaction of methylmagnesium bromide with either of the enantiomerically pure epoxides that can be prepared from (*E*)-3-methyl-2-pentene by the method shown above? Assign *R* or *S* configurations to the chirality centers of each product.

PROBLEM 28 ◆

Is the addition of Br_2 to an alkene such as *trans*-2-pentene a stereoselective reaction? Is it a stereospecific reaction? Is it an enantioselective reaction?

A fourth method of controlling stereochemistry is to use a reactant that has a chirality center with the needed configuration. Such reactants are often naturally occurring molecules such as amino acids or carbohydrates. For example, cysteine was used in the synthesis of enantiomerically pure biotin. An advantage to this method is that naturally occurring compounds are often inexpensive, even if they have many chirality centers.

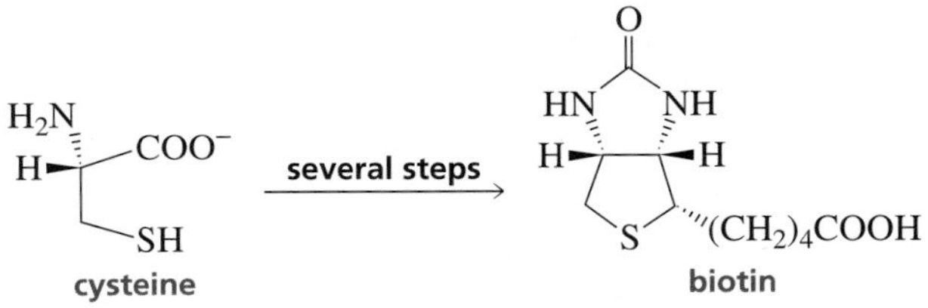

cysteine   several steps   biotin

PROBLEM 29

Show how the following compound could be synthesized from the given starting materials.

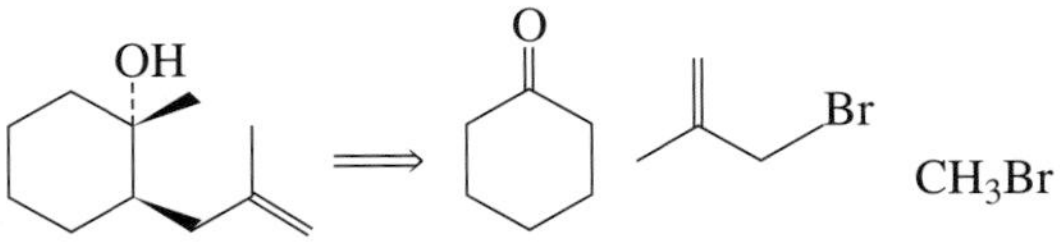

Organic chemists prepare thousands of compounds each year. In this section we will look at several interesting syntheses. First we will look at the syntheses of triptycene and cubane, compounds that were synthesized because of their unusual structures; then we will look at the syntheses of two natural products. These syntheses illustrate the power of organic synthesis and the ability of chemists to come up with clever solutions to difficult problems. They are given here to show you examples of brilliantly designed syntheses. It may have taken the scientists who designed them many months (or even years) of false starts and wrong turns before the successful synthesis was achieved. So don't be intimidated by these syntheses. You certainly would not be expected at this point in your study of organic chemistry to design such syntheses on your own. But you can appreciate them, and you can see how they bring together many of the organic reactions that are familiar to you.

Triptycene was first synthesized in 1942 by P. D. Bartlett. Once it had been synthesized, its cation was found to be considerably less stable than the triphenylmethyl cation that can exist in a planar conformation. This allowed chemists to

28.6
SELECTED EXAMPLES OF SYNTHESES

Paul D. Bartlett *was born in Michigan in 1907. He received a Ph.D. from Harvard University and was a professor of chemistry at Harvard and at Texas Christian University.*

conclude that resonance interactions must have a planar conformation in order to be effective.

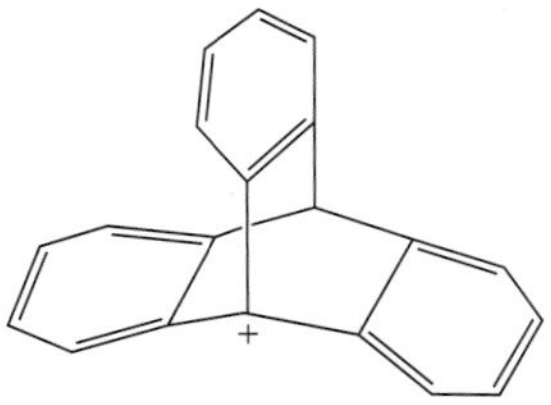

triptycene cation 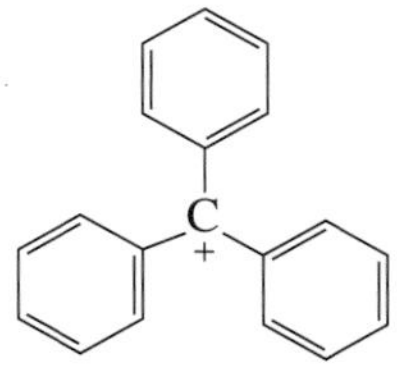   triphenylmethyl cation

The only step that forms a carbon–carbon bond in the synthesis of triptycene is a Diels–Alder cycloaddition of anthracene and 1,4-benzoquinone. All the other steps serve only to convert the newly installed ring into a benzene ring. Treatment of the enedione with hydrochloric acid is an acid-catalyzed enolization of a compound whose enol tautomer is more stable than its keto tautomer. The product is oxidized to a quinone by sodium bromate (Section 17.9). Oxime formation (Section 16.7) followed by reduction with Na_2S forms the diamino-substituted benzene ring. The amino substituents are removed by diazotization followed by treatment with hypophosphorous acid (Section 14.18).

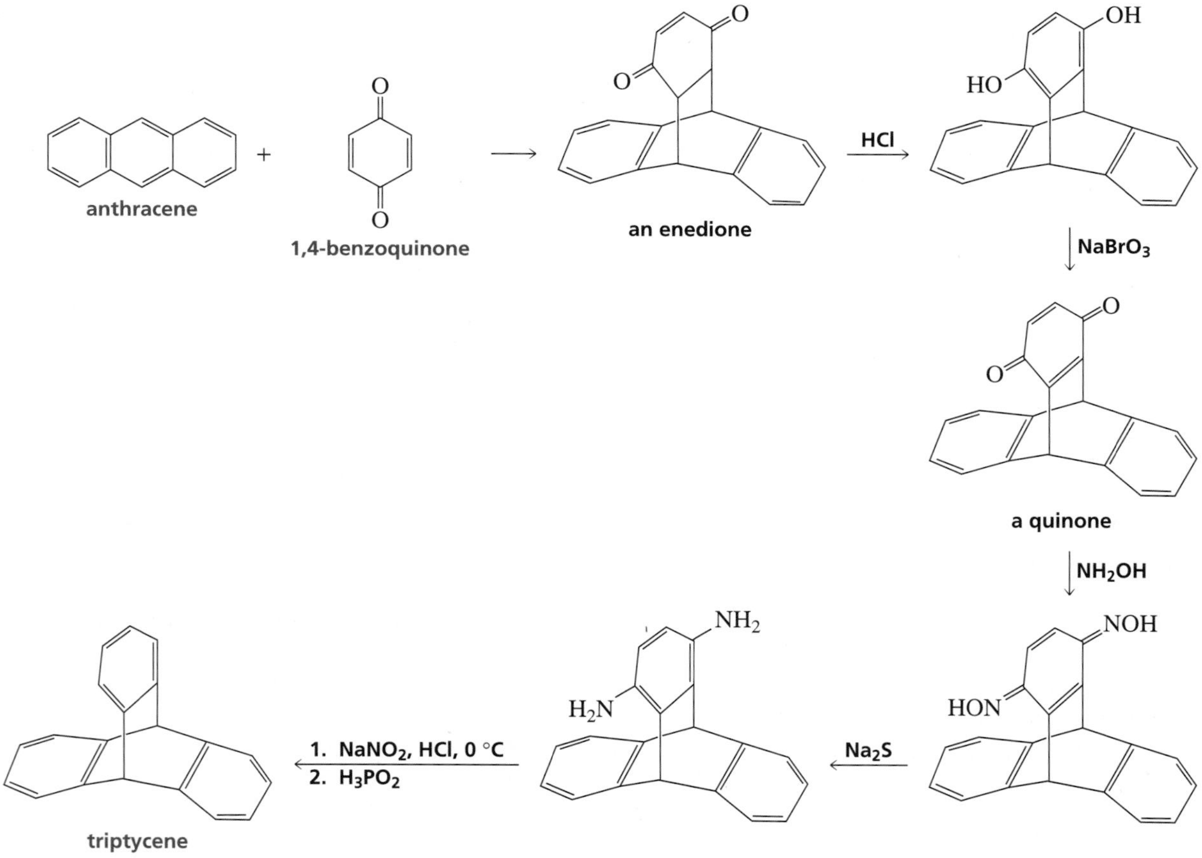

anthracene

1,4-benzoquinone

an enedione

a quinone

triptycene

1. $NaNO_2$, HCl, 0 °C
2. H_3PO_2

Na_2S

NH_2OH

$NaBrO_3$

HCl

PROBLEM 30

Show the mechanism for conversion of the enedione to the diol in the synthetic scheme shown above.

A more convergent synthesis of triptycene has since been developed. It has been prepared in a single step from a Diels–Alder reaction of a diazonium derivative of anthranilic acid with anthracene.

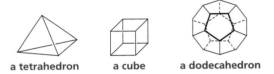

anthranilic acid   anthracene   triptycene

1. NaNO$_2$, HCl, 0 °C
2. HO$^-$

PROBLEM 31

Propose a mechanism for the formation of triptycene from the diazonium salt of anthranilic acid. (*Hint:* A benzyne intermediate is formed; Section 14.22.)

Chemists have synthesized three of the Platonic solids: the tetrahedron, the cube, and the dodecahedron.

a tetrahedron   a cube   a dodecahedron

Plato *was born in Athens in about 428 B.C. Plato and other philosophers of his time believed that world structures could be explained in terms of geometric forms. The Platonic solids are the five "perfect" polyhedra, all faces and angles of which are congruent: the tetrahedron, octahedron, and icosahedron are built of equilateral triangles, the cube is built of squares, and the dodecahedron is built of equilateral pentagons.*

Cubane is the common name for the C_8H_8 hydrocarbon that resembles a cube. The severe strain (angle strain and torsional strain) of this compound was the major problem that had to be overcome in its synthesis. Any approach had to rely on methods that would work with severely strained systems. It was also important to use as few steps as possible, because each step forms a new strained intermediate that has to be stable enough to survive.

Chemists made several attempts to synthesize cubane by two simultaneous [2 + 2] cycloaddition reactions of 1,3-cyclobutadiene under photochemical conditions (Section 27.4), but all these attempts failed. The first successful synthesis of cubane was reported by Philip Eaton in 1964.

The starting material for the synthesis was a dimer of 2-bromocyclopentadienone. In the first step of the synthesis, one of the ketone groups is protected by being converted into an acetal. The first four-membered ring forms as the result of a [2 + 2] cycloaddition reaction under photochemical conditions. The third step is a **Favorskii ring contraction** of an α-bromoketone (Chapter 18, Problems 59 and 60). The carboxyl group is removed after first being converted into a perester. When the perester is heated, the weak oxygen–oxygen bond breaks homolytically. As a result, CO_2 is eliminated, and the resultant cubyl radical abstracts a hydrogen atom from the solvent. The protecting group is removed by hydrolysis, and a second Favorskii ring contraction completes the synthesis of the cube. The remaining carboxyl group is removed by the same technique used to remove the first carboxyl group. This is an example of an iterative synthesis. An **iterative synthesis** is a synthesis in which a reaction sequence is carried out more than once.

Philip Eaton *was born in Brooklyn in 1936. He received an A.B. from Princeton and a Ph.D. from Harvard. He was a professor of chemistry at the University of California, Berkeley, and is now at the University of Chicago.*

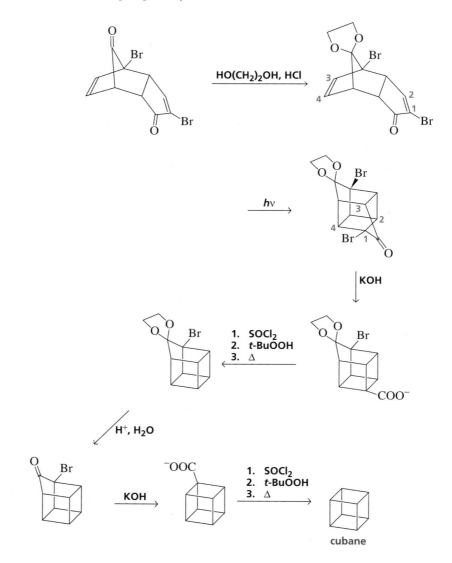

A more convergent synthesis of cubane was later carried out by R. Petit. The first step is a Diels–Alder reaction of a cyclobutadiene equivalent and 2,5-dibromo-1,4-benzoquinone. Cyclobutadiene itself cannot be used as the starting material because its antiaromaticity makes it too unstable. A metal complex of cyclobutadiene is stable enough to be handled easily. The complex is mixed with the dibromosubstituted quinone in the presence of ceric ammonium nitrate, which liberates cyclobutadiene from the metal complex. The second step is a [2 + 2] cycloaddition reaction under photochemical conditions. The third step involves Favorskii ring contractions of the two α-bromoketones. The carboxyl groups are removed by the same method employed by Eaton.

Rowland Petit (1927–1981) *was born in Australia. He earned two Ph.D.'s in chemistry, the first at the University of Adelaide in Australia and the second at the University of London. He was a professor of chemistry at the University of Texas, Austin.*

This synthesis was designed as a result of a retrosynthetic analysis that indicated that a pair of Favorskii ring contractions was the best route to cubane. From that point it was relatively easy to decide to use a quinone and cyclobutadiene to get to the intermediate required for the ring contractions.

Notice that in the syntheses of triptycene and cubane, reactions that remove functional groups are important because in each case the target molecule is a hydrocarbon.

Robert B. Woodward (1917–1979) *was born in Boston and first became acquainted with chemistry in his home laboratory. He entered MIT at the age of 16 and received a Ph.D. in the same year that those who entered MIT with him received their B.A.'s. He went to Harvard as a postdoctoral fellow and remained there for his entire career. Cholesterol, cortisone, strychnine, reserpine (the first tranquilizing drug), chlorophyll, tetracycline, and vitamin B$_{12}$ are just some of the complicated organic molecules that R. B. Woodward synthesized. He received the Nobel Prize in chemistry in 1964. A description of his synthesis of lysergic acid can be found in J. Am. Chem. Soc. **1954**, 76, 5256; **1956**, 78, 3087.*

PROBLEM 35

Propose a mechanism for the Favorskii ring contraction used in the synthesis of cubane.

Thousands of natural products have been synthesized by organic chemists. The two syntheses we will discuss represent examples that use several reactions with which you are familiar. These syntheses should give you an insight into how chemists plan a synthesis based on a sound knowledge of mechanism and structural theory.

Lysergic acid was first synthesized by R. B. Woodward in 1954. The diethylamide of lysergic acid (LSD) is a better-known compound because of its hallucinogenic properties. It is a tribute to Woodward's ability as a synthetic organic chemist that the next synthesis of lysergic acid was not accomplished until 1969.

Lysergic acid is a tetracyclic molecule derived from indole. Because indole is unstable under acidic conditions (Section 26.2), Woodward designed a synthesis that did not form the indole portion of the molecule until the final step. His synthesis began with a dihydroindole carboxylic acid with the nitrogen of the five-membered ring protected by being converted into an amide. This starting material contained two of the required rings. An intramolecular Friedel–Crafts acylation reaction was used to form the third ring. Bromination of the α-carbon followed by S_N2 substitution of the bromine by an amine and removal of the carbonyl protecting group produced a compound that could form the fourth ring by means of an aldol condensation. The carbonyl group was reduced, and the resulting alcohol group was activated and replaced by a cyano group. The cyano group and the protecting amide group were then hydrolyzed. The last step is the reverse of hydrogenation of a double bond and was carried out using a typical hydrogenation catalyst in the absence of hydrogen. The molecule supplied the hydrogen to the catalyst.

PROBLEM 36

Write out all the intermediates formed in Woodward's synthesis of lysergic acid.

PROBLEM 37

Show the retrosynthetic analysis that gets you to the starting material used by Woodward. For each step, indicate whether: carbon–carbon bonds are formed, functional groups are introduced, functional groups are removed, or functional groups are interconverted.

PROBLEM 38

Give the reaction that shows why indole is unstable under acidic conditions.

R. B. Woodward lecturing at Harvard.

Caryophyllene is an oil found in cloves. The most important strategic element in its synthesis was the realization that the nine-membered ring could be formed by fragmenting a six-membered ring fused to a five-membered ring. Then the problem was to design the synthesis of the compound that would undergo fragmentation.[1]

[1] The synthesis was designed and executed by E. J. Corey and his students at Harvard (Chapter 5, page 248). A description of the synthesis is published in *J. Am. Chem. Soc.* **1963**, 85, 362; **1964**, 86, 485.

   The first step in the synthesis is a photochemical [2 + 2] cycloaddition reaction of 2-methylpropene and 2-cyclohexenone. One α-hydrogen of the resulting ketone is substituted with an acyl group, and the second α-hydrogen with a methyl group. An acetylide anion is used to add a three-carbon fragment to the ketone carbonyl group. The triple bond of the alkyne is reduced, the acetal is hydrolyzed, and the resulting aldehyde is oxidized to a carboxylic acid that forms a lactone. A Claisen condensation and hydrolysis of the lactone form the 3-oxocarboxylic acid that is decarboxylated to give the cyclic ketone. Reduction with sodium borohydride gives the diol. Because the secondary alcohol is a more reactive nucleophile than the more sterically hindered tertiary alcohol, the desired tosylate is formed. The stage is now set for the key fragmentation reaction. This reaction involves removal of a proton from the alcohol with base and departure of the tosylate leaving group. Epimerization of the chirality center alpha to the ketone occurs once the ketone has been formed. **Epimerization** is changing the configuration at a chirality center by removing a proton from it and then reprotonating the molecule at the same atom. In the final step of the synthesis, a Wittig reaction is used to form the exo-cyclic double bond.

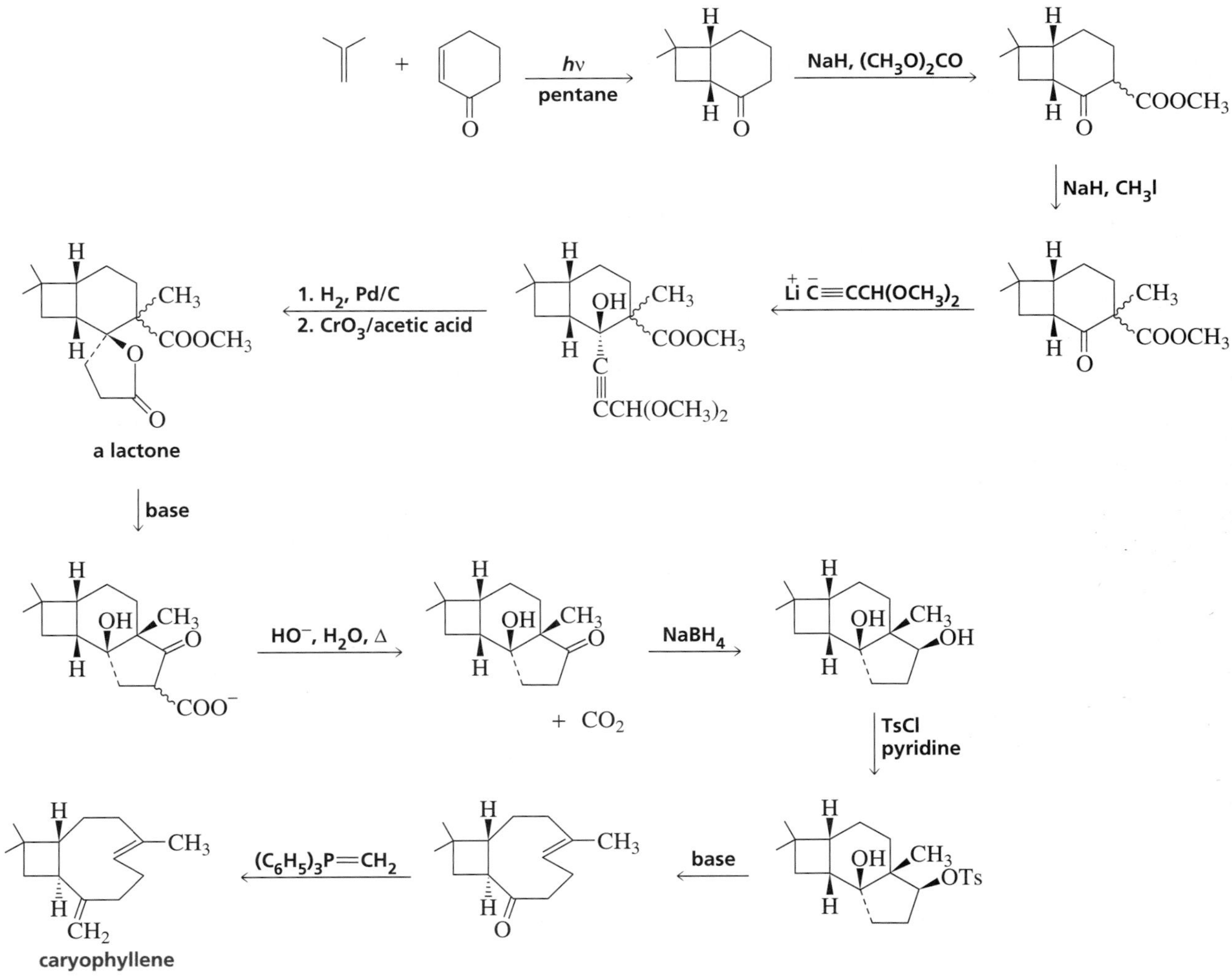

PROBLEM 39◆

Why is the first step in the synthesis of caryophyllene carried out under photochemical conditions? (*Hint:* See Section 27.4.)

PROBLEM 40

Propose a mechanism for the fragmentation reaction in the synthesis of caryophyllene.

PROBLEM 41

Explain why, in the second step in the synthesis of caryophyllene, an α-proton is removed preferentially from one side of the carbonyl group.

You now should have an appreciation of some of the mental processes involved in designing a synthesis. The most important factor in synthesis design is to have good command of organic reactions. The more reactions you know, the better your chances of coming up with a useful reaction. The guiding factor in planning a synthesis is to keep it as simple as possible. The simpler the synthetic plan, the greater the chance it will be successful.

KEY TERMS

chiral auxiliary (page 1223)
combinatorial library (page 1207)
combinatorial organic synthesis
   (page 1207)
convergent synthesis (page 1206)
Darzens condensation (page 1216)
disconnection (page 1209)
enantioselective reaction (page 1223)
epimerization (page 1231)

extrusion reaction (page 1217)
Favorskii ring contraction (page 1227)
functional group interconversion
   (FGI) (page 1208)
iterative synthesis (page 1227)
linear synthesis (page 1206)
multistep synthesis (page 1205)
natural product (page 1204)
organic synthesis (page 1204)

protecting group (page 1219)
retrosynthesis (page 1209)
retrosynthetic analysis (page 1209)
synthetic equivalent (page 1209)
synthetic tree (page 1206)
synthon (page 1209)
target molecule (TM) (page 1206)
umpolung (page 1212)
vinylogy (page 1216)

PROBLEMS

42. Describe how each of the following compounds could be synthesized using starting materials containing no more than five carbon atoms.
    a. 1-hexanol   **d.** 2-hexanol
    b. 1-heptanol   **e.** 2-methyl-2-hexanol
    c. 1-octanol   **f.** 4-methyl-4-heptanol

43. a. Disconnect the following molecule at bond **a** and give a retrosynthetic analysis for its synthesis.
    b. Disconnect the following molecule at bond b and give a retrosynthetic analysis for its synthesis.

$$HO \overset{a}{\diagdown} CH_2 \overset{b}{-} CH = CH_2$$

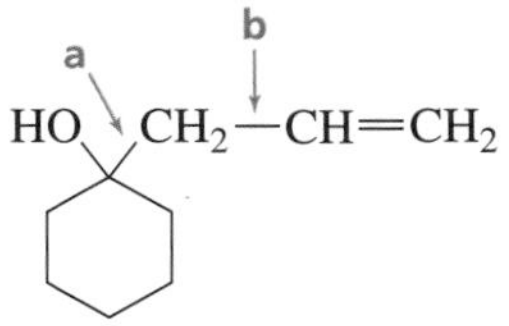

44. Using retrosynthetic analysis, plan a synthesis of the following target molecules using only the given starting materials.

target molecules

a.

b.

c.

d.

starting materials

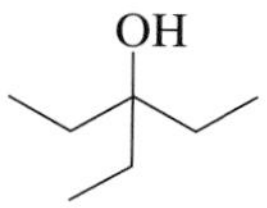

CO_2

CH_3CH_2OH

OH

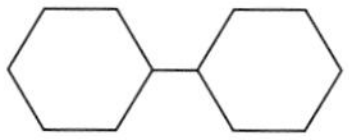

45. What reagents are required to convert 1-methylcyclohexene into the following compounds?

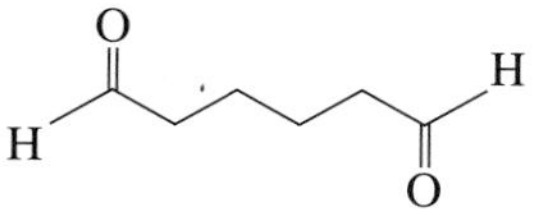

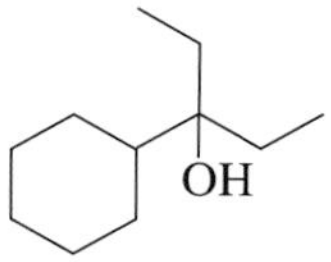

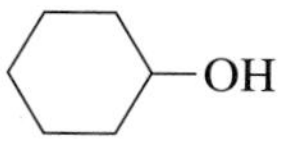

a. 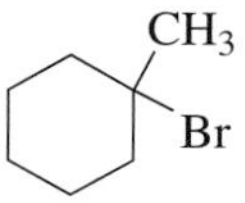 CH_3 / Br

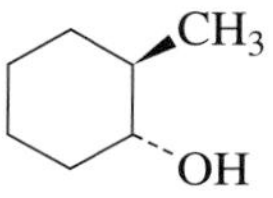

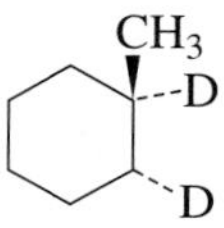

b. 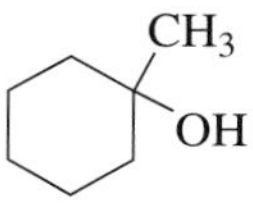 CH_3 / OH

c. CH_3 / OH

d. 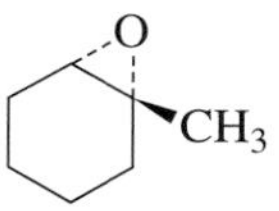 O / CH_3

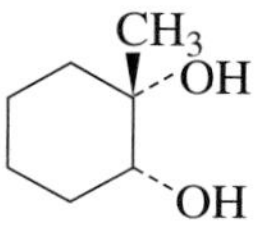

e. CH_3 / OH / OH

f. 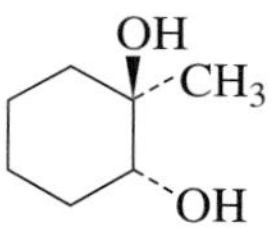 OH / CH_3 / OH

g. CH_3 / D / D

h. 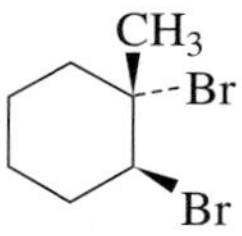 CH_3 / Br / Br

46. Describe a method for the synthesis of each of the following compounds from starting materials containing no more than four carbon atoms.

a. 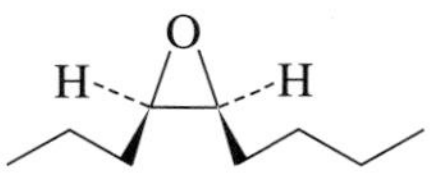 H- O -H

b. 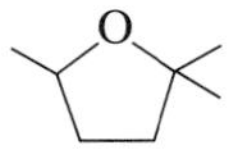 O

47. Sodium borohydride is used to reduce cyclohexanone to cyclohexanol. Why can't sodium hydride be used?

48. Describe two methods for the synthesis of the following compound.

O

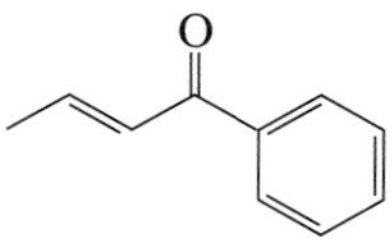

49. Show how each of the following compounds can be prepared from the given starting material. In each case, you will need to use a protecting group.

a. $ClCH_2CH_2\overset{O}{\overset{\|}{C}}H \longrightarrow H_2NCH_2CH_2CH_2\overset{O}{\overset{\|}{C}}H$

b. $CH_3CHCH_2\overset{O}{\overset{\|}{C}}OCH_3 \longrightarrow CH_3CHCH_2\overset{CH_3}{\overset{|}{C}}CH_3$
      | | |
      OH   OH   OH

c.

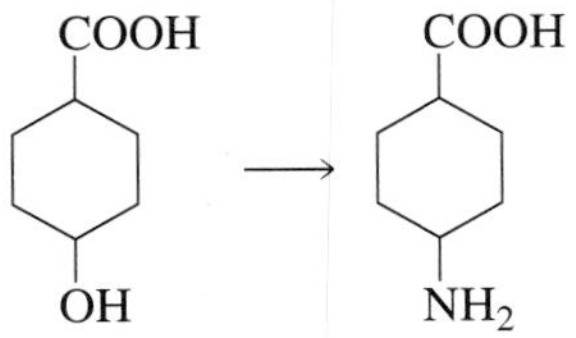

d.

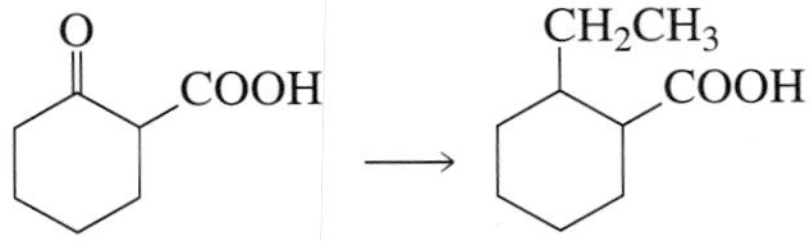

50. Describe how the following compounds could be synthesized from starting materials containing no more than four carbon atoms.

a.

b.

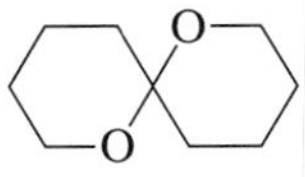

51. Do a retrosynthetic analysis on each of the following compounds, ending with the starting materials.

a. 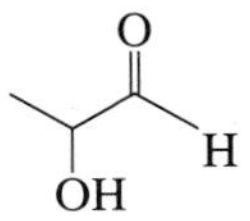   c. 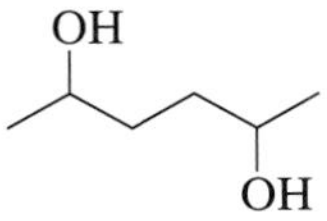   e.

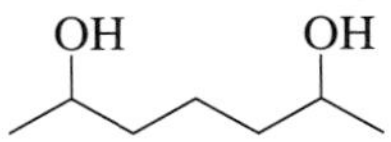

b. 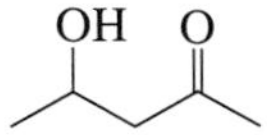   d.    f.

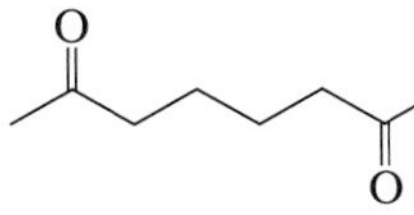

52. Outline syntheses for the following compounds using starting materials containing no more than five carbon atoms.

a. 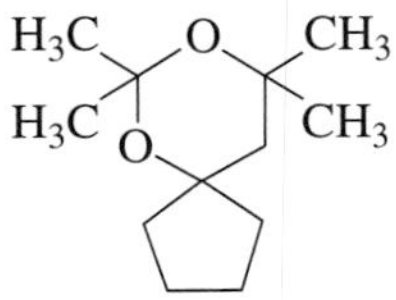   b. 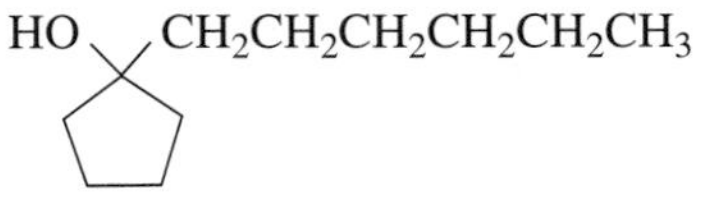 HO⟍ $CH_2CH_2CH_2CH_2CH_2CH_3$

53. Using *para*-benzoquinone (see Problem 55) as one of the starting materials, what diene would you use in order to synthesize the following compounds?

a. 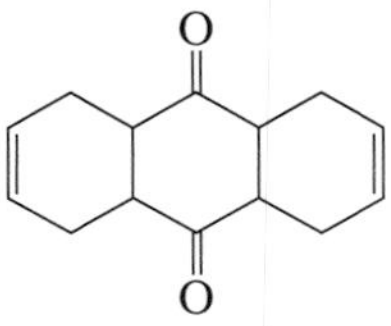   c.

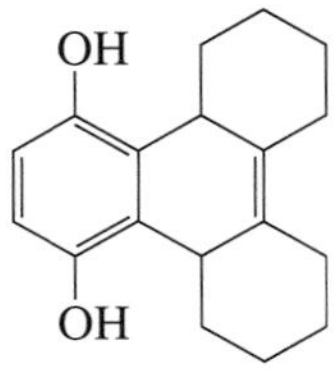

b.

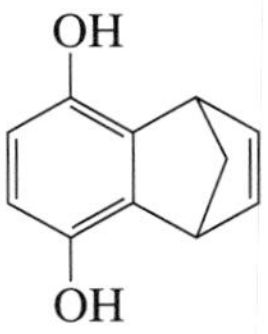

54. Propose a mechanism for each of the following reactions.

a. 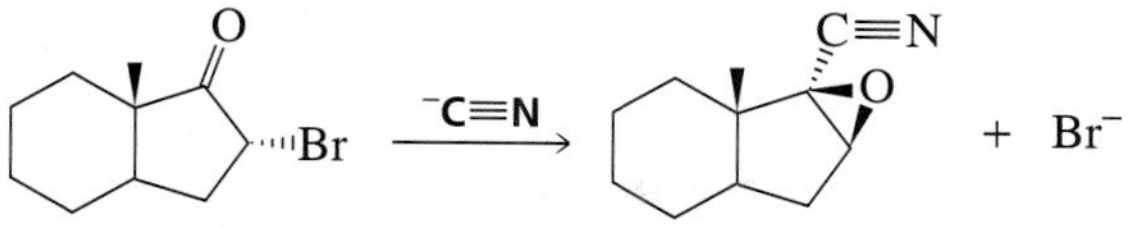 + Br⁻

b. 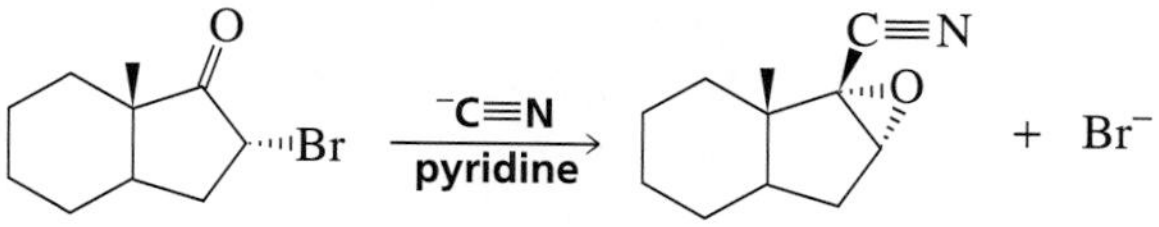 + Br⁻

55. The following *para*-quinone derivative can be used to protect OH and NH_2 groups by converting them into esters and amides, respectively. An advantage to this protecting group is that it can be removed by a mild reducing agent. Consequently, its removal will not hurt groups in the molecule that would be affected by acid- or base-catalyzed hydrolysis. Explain why the protecting group can be removed by reduction. (*Hint:* See Section 17.8.)

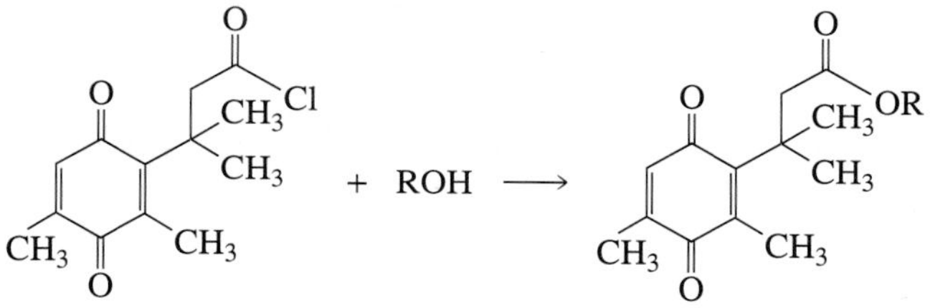

+ ROH ⟶

56. Show how each of the following compounds could be prepared from the given starting material.

a.

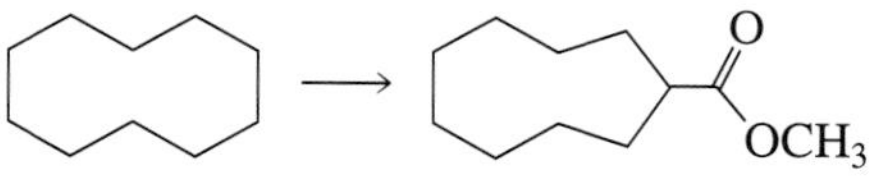

b.

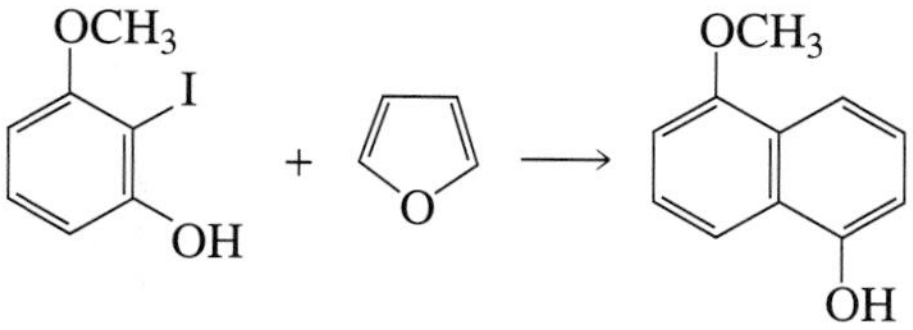

29

THE ORGANIC CHEMISTRY OF DRUGS:
Discovery and Design

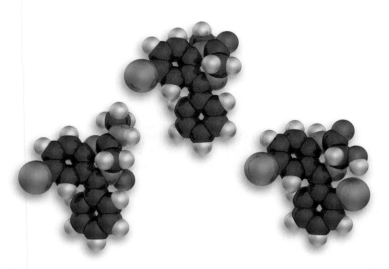

Librium, Valium, and Ativan

A **drug** is a compound that interacts with a biological molecule, triggering a physiological effect. Drugs have been used by humans for thousands of years to alleviate pain and illness. By trial and error, people learned which herbs, berries, roots, and bark could be used for medicinal purposes. However, as late as the beginning of the twentieth century, relatively few truly effective medicinal agents were known. Local anesthetics had just been discovered, and there were only two analgesics to relieve major pain. There were no drugs for the dozens of functional, degenerative, neurological, and psychiatric disorders; no hormone therapies; no vitamins; and—most significantly—no effective drug for the cure of any infectious disease. One reason that families had many children was because some of the children were bound to succumb to childhood diseases. Life spans were generally short: in 1900, the average life expectancy in the United States was 46 years for a male and 48 years for a female. Now there is a drug for treating almost every disease, and this is reflected by current life expectancies: 72 years for a male and 79 years for a female.

The shelves of a typical modern pharmacy contain almost 2000 preparations, most of which contain a single active ingredient, usually an organic compound. The ten most widely prescribed drugs in the United States are listed in Table 29.1. Antibiotics are the most widely prescribed class of drugs in the world. In the developed world, heart drugs are the most widely prescribed class, partly because they are generally taken for the remainder of the patient's life. In the past 10 years prescriptions for psychotropic drugs have decreased as doctors have become more aware of problems associated with addiction, and prescriptions for asthma have increased, reflecting a greater incidence of the disease.

Throughout this book we have encountered many examples of drugs, vitamins, and hormones, and in many cases we have discussed the mechanism by which each compound produces its physiological effect (Table 29.2). Now we will take a look at how drugs are discovered, how they are named, and some of the techniques currently used by scientists in their search for new drugs.

TABLE 29.1   The Most Widely Prescribed Drugs in the United States

Proprietary name	Generic name		Use
Amoxil	amoxicillin	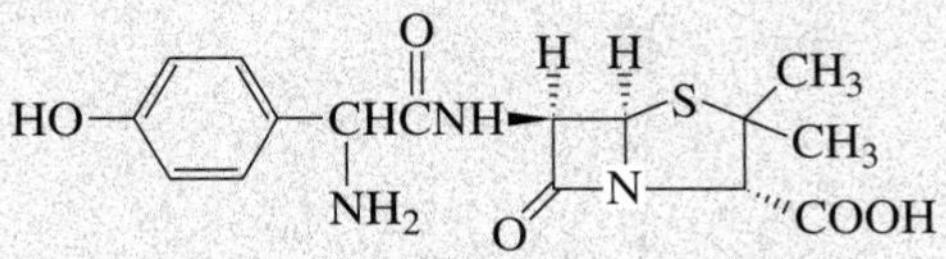	antibiotic
Premarin	conjugated estrogenic hormones	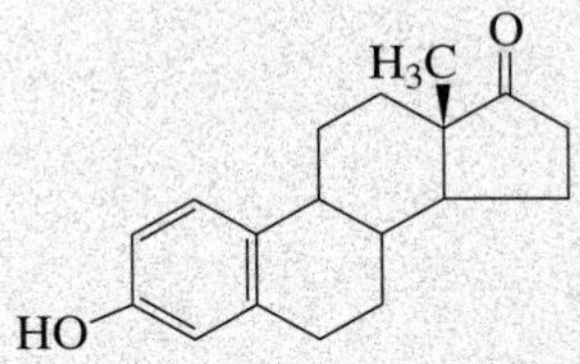	hormone replacement therapy
Zantac	ranitidine	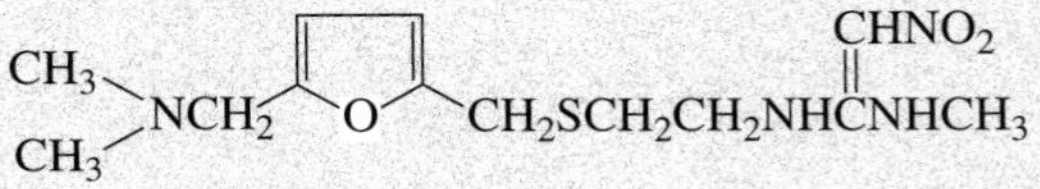	antagonist of the H_2 histamine receptor for the treatment of gastric ulcers
Ortho-N	norethindrone	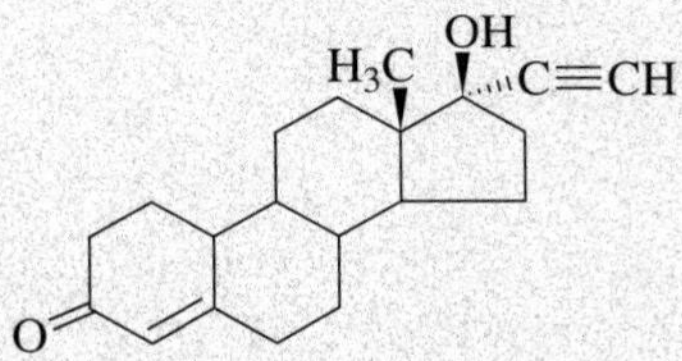	steroid component of oral contraceptives
Hydro-Diuril	hydrochlorothiazide	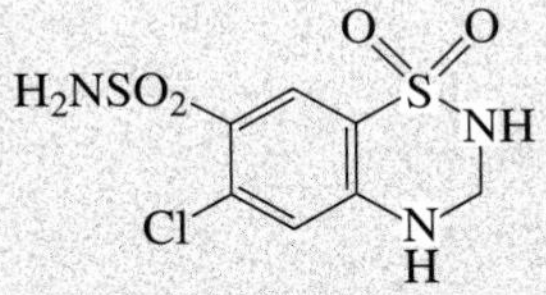	diuretic
Tylenol/codeine	acetaminophen/ codeine	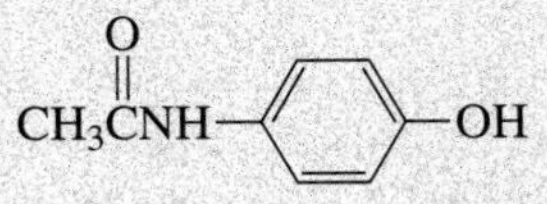 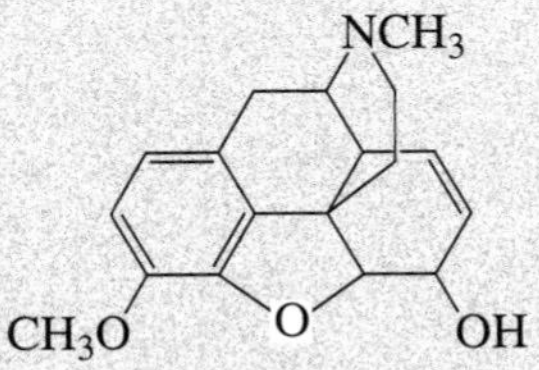	analgesic
Ventolin	salbutamol	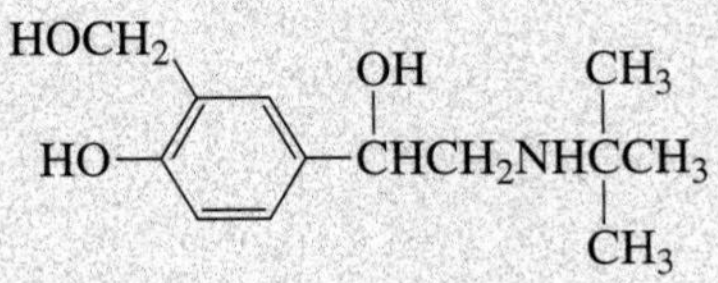	bronchodilator

continues

TABLE 29.1 Continued

Proprietary name	Generic name		Use
Lanoxin	digoxin	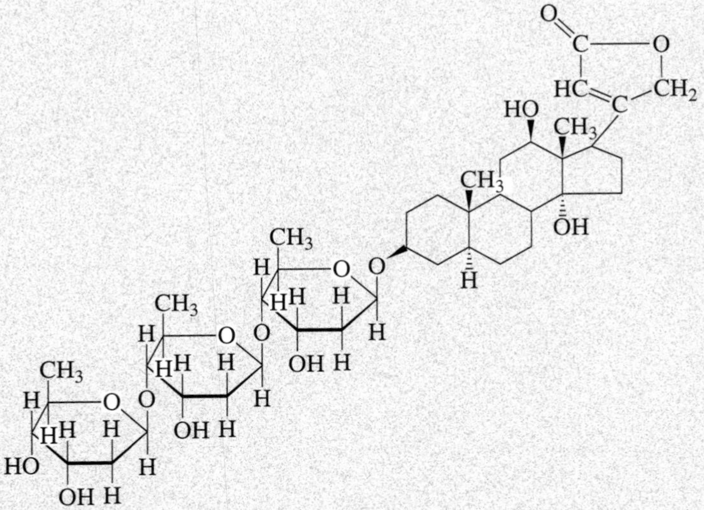	cardiac glycoside for heart failure
V-cillin	penicillin V	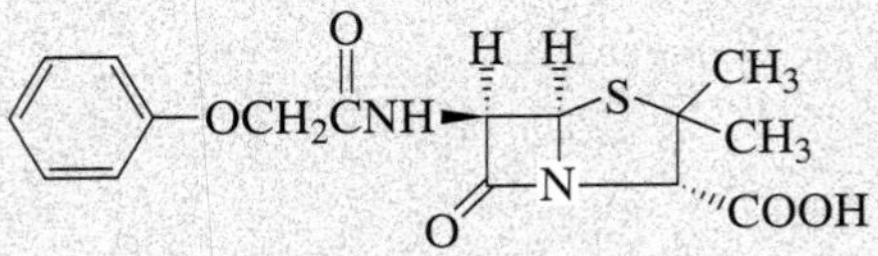	antibiotic
Lasix	furosemide	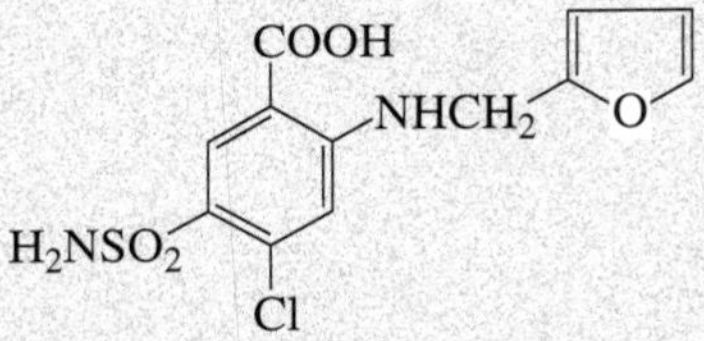	diuretic

29.1 NAMING DRUGS

The most accurate names of drugs are the chemical names that define their structures. However, these names are too long and complicated to appeal to physicians and to the general public. The pharmaceutical company is allowed to choose the proprietary name for a drug it develops. A **proprietary name (trade name or brand name)** identifies a commercial product and distinguishes it from other products. A proprietary name can be used only by the owner of the registered trademark; a **trademark** can be a name, a symbol, or a picture. It is in the best interest of the company to choose a name that is easy to remember and pronounce so, when the patent expires, the public will continue to request the drug by its proprietary name.

Each drug is also given a **generic name** that any pharmaceutical company can use to identify a given product. The pharmaceutical company that develops a drug is allowed to choose the generic name from a list of possible names provided by an independent group. Generic names are usually not as easily remembered as proprietary names, and this gives a competitive advantage to the company holding the registered trademark. Proprietary names must always be capitalized, while generic names are not capitalized.

Drug manufacturers are permitted to patent and retain exclusive rights to the drugs they develop. A patent is valid for 20 years. Once the patent expires, other drug com-

TABLE 29.2 Drugs, Hormones, and Vitamins Discussed in Earlier Chapters

sulfa drugs	the first antibiotics	Section 22.8
tetracycline	broad-spectrum antibiotic	Section 4.8
puromycin	broad-spectrum antibiotic	Section 24.12
nonactin	ionophorous antibiotic	Section 11.9
penicillin	antibiotic	Sections 15.4 and 15.13
gentamicin	antibiotic	Section 19.15
gramicidin S	antibiotic	Section 20.9
aspirin	analgesic, anti-inflammatory agent	Sections 18.8 and 23.5
enkephalins	pain killers	Section 20.9
diethyl ether	anesthetic	Section 11.5
sodium pentothal	sedative hypnotic	Section 11.5
5-fluorouracil	anticancer agent	Section 22.8
methotrexate	anticancer agent	Section 22.8
taxol	anticancer agent	Chapter 28, introduction
thalidomide	sedative with teratogenic side effects	Section 4.11
AZT*	anti-AIDS agent	Section 24.16
warfarin	anticoagulant	Section 22.9
epinephrine	vasoconstrictor, bronchodilator	Section 9.11
vitamins		Chapter 22; Sections 8.8, 16.11, 19.15, 23.4, 23.7, 26.4
hormones		Sections 20.7 and 23.9–23.11

*3′-Azido-3′-deoxythymidine.

panies can market the drug under the generic name or under their own proprietary names, which are called branded generic names. For example, the antibiotic ampicillin is sold as Penbritin by the company that held the original patent. Now that the patent has expired, this drug is sold by other companies as Ampicin, Ampilar, Amplital, Binotal, Nuvapen, Pentrex, Ultrabion, Viccillin, and 30 other branded generic names.

The over-the-counter drugs that line the shelves of drugstores are available without prescription. They are often mixtures containing one or more active ingredients (generic or proprietary drugs) plus sweeteners and inert fillers. For example, the preparations called Advil (Whitehall Laboratories), Motrin (Upjohn), and Nuprin (Bristol-Meyers Squibb) all contain ibuprofen, a mild analgesic and anti-inflammatory drug. Ibuprofen was patented in Britain in 1964 by Boots, Inc., and the U.S. Food and Drug Administration (FDA) approved its use as a nonprescription drug in 1984.

29.2 LEAD COMPOUNDS

The goal of the medicinal chemist is to find compounds that have potent effects on given diseases with minimum side effects. In other words, a drug must react selectively with its target and have a minimal effect on the host cell. A drug must get to the right place in the body, at the right concentration, and at the right time. Therefore, a drug must have the appropriate solubility to allow it to be transported to the target cell. If it is taken orally, it must be insensitive to the acid conditions of the stomach, and it also must resist enzymatic degradation before it reaches its target. Finally, it must eventually be either excreted or degraded to harmless compounds that can be excreted.

Foxglove.

Herbs used by humans since ancient times provided the starting point for the development of our current arsenal of drugs. The active ingredients were isolated from herbs and roots used in traditional medicine. Foxglove furnished digitoxin, a cardiac stimulant. The bark of the cinchona tree yielded quinine for relief from malaria. Willow bark contains salicylates used to control fever and pain. The sticky juice of the oriental opium poppy provided morphine for severe pain and codeine for the control of cough. By 1882 there were more than 50 different herbs commonly used to make medicines. Many of these herbs were grown in the gardens of religious establishments used to treat the sick.

Scientists still search the world for plants and berries and the oceans for flora and fauna that might yield new medicinal compounds. Taxol, a compound isolated from the bark of the Pacific yew tree, is a recently recognized anticancer agent (Chapter 28, introduction).

Once a naturally occurring drug is isolated and its structure determined, it can serve as a prototype in a search for other biologically active compounds. The prototype is called a **lead compound.** Analogs of the lead compound are synthesized in order to find one that might have improved therapeutic properties or fewer side effects. The analog may have a different substituent than the lead compound, a branched chain instead of a straight chain, or a different ring system. Changing the structure of the lead compound is called **molecular modification.**

29.3
MOLECULAR
MODIFICATION

A classic example of molecular modification is the development of synthetic local anesthetics using cocaine as the lead compound. Cocaine comes from the leaves of *Erythroxylon coca,* a bush native to the highlands of the South American Andes. It is an effective local anesthetic, but it produces disturbing effects on the central nervous system (CNS) ranging from initial euphoria to severe depression. By degrading the cocaine molecule step by step (removing the carbomethoxy group and cleaving the seven-membered-ring system), scientists identified the portion of the molecule that carries the local anesthetic activity without the damaging CNS effects. This knowledge gave an improved lead compound, an ester of benzoic acid with the alcohol component of the ester having a terminal tertiary amino group.

Coca leaves.

cocaine
lead compound

improved lead compound

Hundreds of esters were then synthesized, putting substituents on the aromatic ring, changing the alkyl groups bonded to the nitrogen, and changing the length of the connecting alkyl chain. Successful anesthetics obtained as a result of molecular modification were Benzocaine, a topical anesthetic, and procaine, commonly known by the trade name Novocain.

Because the ester group of procaine is hydrolyzed relatively rapidly by serum esterases (enzymes that catalyze ester hydrolysis), the drug has a short lifetime. Therefore compounds with less easily hydrolyzed amide groups were synthesized. In this way lidocaine, one of the most widely used injectable anesthetics, was discovered. The rate at which the drug is hydrolyzed is further decreased by the two *ortho*-methyl substituents, which provide steric hindrance to the vulnerable carbonyl group.

Benzocaine

procaine
Novocain

lidocaine
Xylocaine

Later, physicians recognized that the action of an anesthetic administered *in vivo* (in a living organism) could be lengthened considerably if it was administered along with epinephrine. Because epinephrine is a vasoconstrictor, it reduces the blood supply, allowing the drug to remain at its targeted site for a longer period of time.

In screening the structurally modified compounds for biological activity, scientists were surprised to find that replacing the ester linkage of procaine with an amide linkage led to a compound, procainamide hydrochloride, that had activity as a cardiac depressant in addition to its activity as a local anesthetic. It is currently used clinically as an antiarrhythmic.

procainamide hydrochloride

Morphine is the most widely used analgesic for severe pain. Although scientists have learned how to synthesize morphine, all commercial morphine is obtained from opium, the juice obtained from a species of poppy. Morphine occurs in opium to the extent of 10%. Morphine is the standard by which other painkilling medications are measured. Codeine, which has one-tenth the analgesic activity of morphine, profoundly inhibits the cough reflex. Although 3% of opium is codeine, most commercial codeine is obtained by methylating morphine. Acetylating the two OH groups of morphine forms heroin, a drug that has been banned in most countries because it is widely abused. Because heroin is less polar than morphine, it crosses the blood–brain barrier more rapidly, resulting in a more rapid "high."

morphine

codeine

heroin

Molecular modification of codeine led to dextromethorphan, the active ingredient in most cough medicines. Etorphine was synthesized when scientists realized that analgesic potency was related to the ability of the drug to bind hydrophobically to the opiate receptor (Section 29.6). Etorphine is about 2000 times more potent than morphine, but it is not safe enough for use in humans. It has been used to tranquilize elephants and other large animals. Pentazocine is useful in obstetrics because it dulls the pain of labor but does not depress the respiration of the infant as morphine does.

Paul Ehrlich

Paul Ehrlich (1854–1915) *was a German bacteriologist. He received a medical degree from the University of Leipzig and was a professor at the University of Berlin. In 1892 he developed an effective diphtheria antitoxin. For his work on immunity, he received the 1908 Nobel Prize in medicine or physiology, sharing it with Ilya Ilich Mechnikov.*

Ilya Ilich Mechnikov (1845–1916), *later known as Elie Metchnikoff, was born in the Ukraine. He was the first scientist to recognize that white corpuscles were important in resistance to infection. Metchnikoff succeeded Pasteur as director of the Pasteur Institute.*

dextromethorphan   etorphine   pentazocine

29.4 RANDOM SCREENING

Gerhard Domagk (1895–1964) *was a research scientist at I. G. Farbenindustrie, a German manufacturer of dyes and other chemicals. He carried out studies that showed Prontosil to be an effective antibacterial agent. His daughter, who was dying of a streptococcal infection as a result of having cut her finger, was the first patient to receive and be cured by the drug (1935). Prontosil received wider fame when it was used to save the life of Franklin D. Roosevelt, Jr., son of the U.S. president. Domagk received the Nobel Prize in medicine or physiology in 1939, but Hitler did not allow Germans to accept Nobel Prizes because Karl von Ossietsky, a German who was in a concentration camp, had been awarded the Nobel Prize for peace in 1935. Domagk was eventually able to accept the prize in 1947.*

The lead compound for the development of most drugs is found by random screening of thousands of compounds. A **random screen,** also known as a **blind screen,** is the search for a pharmacologically active compound without any information about what chemical structures might show activity. The first blind screen was carried out by Paul Ehrlich, who was searching for a "magic bullet" against trypanosomes—the microorganisms that cause African sleeping sickness. Having tested more than 900 compounds against trypanosomes, Ehrlich tested some of his compounds against other bacteria. Compound 606 (salvarsan) was found to be dramatically effective against the microorganisms (spirochetes) that cause syphilis.

An important part of random screening is recognizing an effective compound. This requires the development of an assay for the desired biological activity. Some assays can be done *in vitro* (in glass)—for example, searching for a compound that will inhibit a particular enzyme. Others are done *in vivo*—for example, searching for a compound that will save a mouse from a lethal dose of a virus. One problem with *in vivo* assays is that drugs can be metabolized differently by different animals. An effective drug in a mouse may be less effective or useless in a human. Another problem is regulation of the dosages of both the virus and the drug. If the dosage of the virus is too high, it might kill the mouse in spite of the presence of a biologically active compound that could save it. If the dosage of the potential drug is too high, the drug might kill the mouse when a lesser dosage would have saved it.

The knowledge that azo dyes were effective in dyeing wool fibers (animal protein) gave scientists the idea that these compounds might selectively bind to bacterial proteins. Well over 10,000 dyes were screened *in vitro* in antibacterial tests, but none showed any antibiotic activity. Some scientists pointed out that the dyes should be screened *in vivo* since what physicians really needed were antibacterial agents that would cure infections in humans and animals, not in test tubes.

The *in vivo* studies were done in mice that had been infected with a bacterial culture. Now the luck of the investigators improved. Several dyes turned out to counteract gram-positive infections. The least toxic of these, Prontosil (a bright red dye), became the first drug to treat bacterial infections.

Prontosil

The fact that Prontosil was inactive *in vitro* but active *in vivo* should have suggested that the dye was converted to an active compound by the mammalian organism, but this did not occur to the bacteriologists, who were content to have found a useful antibiotic. When scientists at the Pasteur Institute later investigated Prontosil, they noted that mice given the compound did not excrete a red compound. Urine analysis showed that the mice excreted *para*-acetamidobenzenesulfonamide, a colorless compound. Chemists knew that anilines are acetylated *in vivo,* so they prepared the nonacetylated compound (sulfanilamide). When it was tested in mice infected with streptococcus, all the mice were cured, while untreated control mice died. Sulfanilamide was the first of the sulfa drugs.

$$CH_3\overset{\displaystyle O}{\overset{\|}{C}}-NH-\underset{}{\bigcirc}-SO_2NH_2 \qquad H_2N-\underset{}{\bigcirc}-SO_2NH_2$$

para-**acetamidobenzenesulfonamide**   *para*-**aminobenzenesulfonamide**
                                                 sulfanilamide

We have seen that sulfanilamide acts by inhibiting the bacterial enzyme that incorporates *para*-aminobenzoic acid into folic acid (Section 22.8). Thus sulfanilamide is a **bacteriostatic drug,** one that inhibits the further growth of the bacteria, rather than a **bactericidal drug,** one that kills the bacteria. Sulfanilamide's activity as an inhibitor is due to the similar sizes of the sulfonamide and carboxylic acid groups. Many successful drugs have been designed using similar isosteric replacements (Sections 22.8 and 22.9).

Gram-negative bacteria have an outer membrane that covers their cell walls; gram-positive bacteria do not have an outer membrane but tend to have thicker and more rigid cell walls. They can be distinguished by a stain, invented by Hans Christian Gram, that turns gram-negative bacteria pink and gram-positive bacteria purple.

Many drugs have been discovered accidentally. Nitroglycerin, the drug used to relieve the symptoms of angina pectoris, was discovered when workers handling nitroglycerine in the explosives industry experienced severe headaches. Investigation revealed that the headaches were caused by nitroglycerine's ability to produce a marked dilation of blood vessels. The pain associated with an angina attack results from the inability of the blood vessels to supply the heart with adequate blood. The drug relieves the pain by dilating cardiac blood vessels.

29.5
SERENDIPITY IN DRUG DEVELOPMENT

$$\begin{array}{l} CH_2-ONO_2 \\ | \\ CH-ONO_2 \\ | \\ CH_2-ONO_2 \end{array}$$
nitroglycerin

The tranquilizer Librium is another example of a drug that was discovered accidentally. Leo Sternbach synthesized a series of quinazoline 3-oxides, but none of them showed any pharmacological activity. One of the compounds was not submitted for testing because it was not the quinazoline 3-oxide he had set out to synthesize. Two years after the project was abandoned, a laboratory worker came across this compound when cleaning up the lab, and Sternbach decided he might as well submit it for testing before it was thrown away. The compound was shown to have tranquilizing properties and, when its structure was investigated, was found be a benzodiazepine 4-oxide. Methylamine, instead of displacing the chloro substituent to form a quinazoline 3-oxide, had added to the imine carbon of the six-membered ring, causing this ring to open and reclose to a seven-membered ring. The compound was given the trade name Librium when it was put into clinical use in 1960.

Leo H. Sternbach

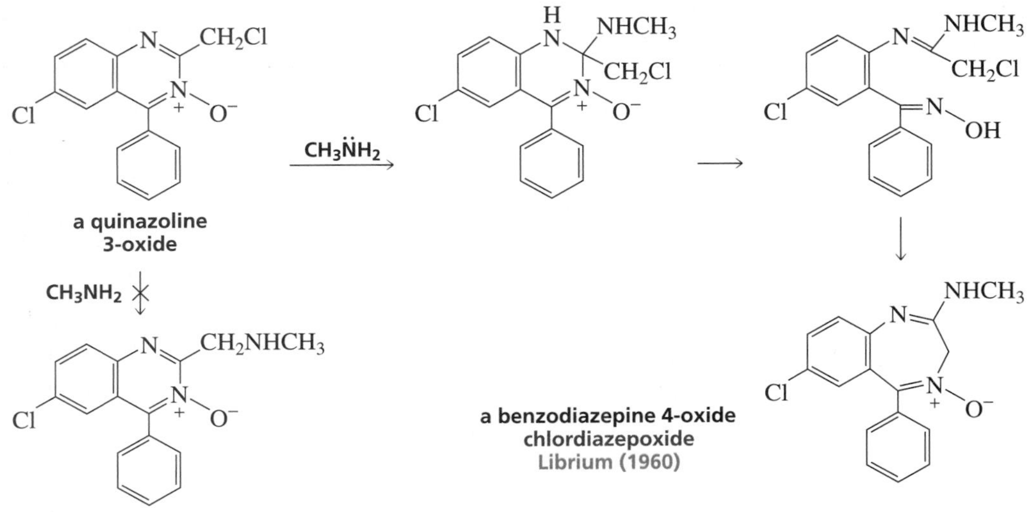

a quinazoline 3-oxide

a benzodiazepine 4-oxide
chlordiazepoxide
Librium (1960)

Librium was structurally modified in an attempt to find other tranquilizers. One successful modification produced Valium, a tranquilizer almost ten times more potent than Librium. Currently there are eight benzodiazepines in clinical use as tranquilizers in the United States and some 15 others abroad. Valium is one of the most widely prescribed medications.

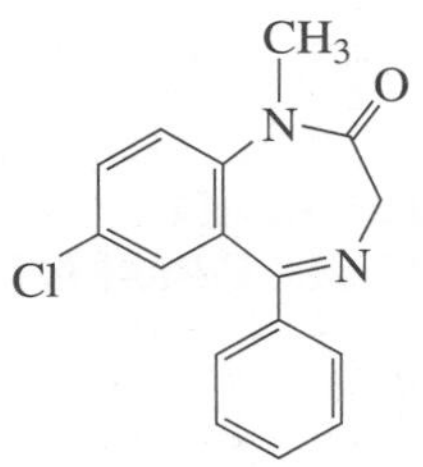

diazepam
Valium (1963)

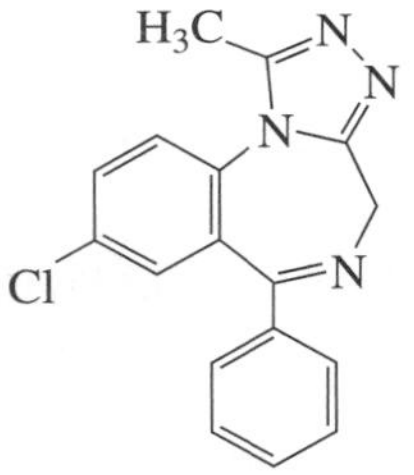

alprazolam
Xanax (1970)

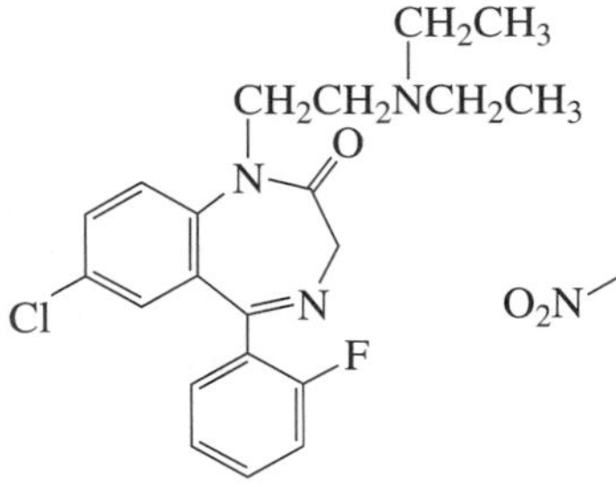

flurazepam
Dalmane (1970)

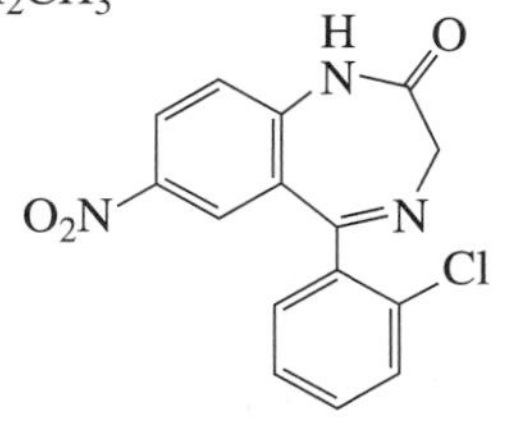

clonazepam
Klonopin (1975)

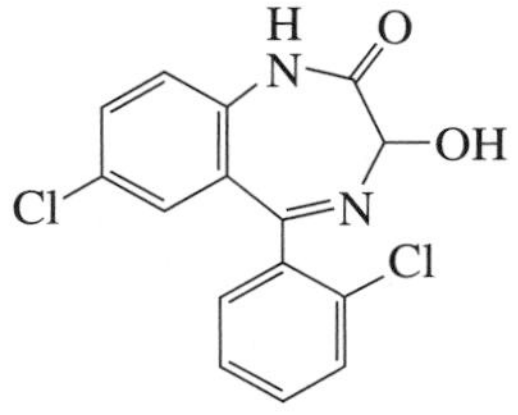

lorazepam
Ativan (1977)

**29.6
RECEPTORS**

Some drugs exert their physiological effects by interacting with a particular cellular binding site called a **receptor.** Drug receptors are generally glycoproteins or lipoproteins. Some receptors are part of cell membranes, while others are found in the cytoplasm (the material outside the nucleus). A drug interacts with its receptor using the same kinds of bonding interactions—hydrogen bonding, electrostatic attractions, van der Waals interactions—that we have encountered in other examples of molecular recognition (Section 21.8). The most important factor in bringing together a drug and a receptor is a snug fit: the greater the affinity of a drug for its binding site, the higher its potential biological activity.

Knowing something about the molecular basis of drug action allows scientists to design and synthesize compounds that might have the desired biological activity. For example, when excess histamine is produced by the body, it causes the symptoms associated with the common cold and allergic responses. This is thought to be the result of the protonated ethylamino group anchoring the histamine molecule to a negatively charged portion of the histamine receptor.

histamine

histamine
receptor

Drugs that have been found to be effective antihistamines bind to the histamine receptor but do not trigger the same response as histamine. Like histamine, these drugs have an amino group that binds to the histamine receptor. The drugs also have bulky groups that keep the histamine molecule from approaching the receptor. Because these compounds interfere with the natural action of histamine, they are called antihistamines.

antihistamines

diphenhydramine
Benadryl

promethazine
Promine

promazine
Talofen

Acetylcholine is a neurohormone that enhances peristalsis, wakefulness, and memory and is essential for nerve transmission. A deficiency of brain cell receptors that bind acetylcholine (cholinergic receptors) contributes to the characteristic loss of memory in Alzheimer's disease. Cholinergic receptors are structurally similar to those that bind histamine. Therefore, antihistamines and cholinergic agents show overlapping activities. For example, the antihistamine diphenhydramine has been used to treat insomnia and to combat motion sickness.

acetylcholine

cholinergic
receptor

Excess histamine production by the body also causes the hypersecretion of stomach acid by the cells of the stomach lining, leading to the development of ulcers. The antihistamines that block the histamine receptors (thereby preventing the allergic responses associated with excess histamine production) have no effect on HCl production. This fact led scientists to conclude that a second kind of histamine receptor triggers the release of acid into the stomach.

Because 4-methylhistamine was found to cause weak inhibition of HCl secretion, it was used as a lead compound. About 500 molecular modifications were performed over a 10-year period before four clinically useful antiulcer agents were found. Two of these are cimetidine and ranitidine. Notice that steric blocking of the receptor site is not a factor in these compounds. Compared with the antihistamines, the effective antiulcer drugs have more polar imidazole rings, longer side chains, and less basic substituents on the side chains.

Leo H. Sternbach *was born in Austria in 1908. In 1918, after World War I and the dissolution of the Austro-Hungarian Empire, Sternbach's father moved to Krakow in the recreated Poland and obtained a concession to open a pharmacy. As a pharmacist's son, Sternbach was accepted at the Jagiellonian University School of Pharmacy, where he received both a Master of Pharmacy and a Ph.D. in chemistry. With discrimination against Jewish scientists growing in Eastern Europe in 1937, Sternbach moved to Switzerland to work with Ružička (p. 1043) at the ETH (The Swiss Federal Institute of Technology). In 1941, Hoffmann–LaRoche brought several scientists out of Europe, and Sternbach was hired as a research chemist at the United States headquarters in Nutley, New Jersey, where he later became director of medicinal chemistry.*

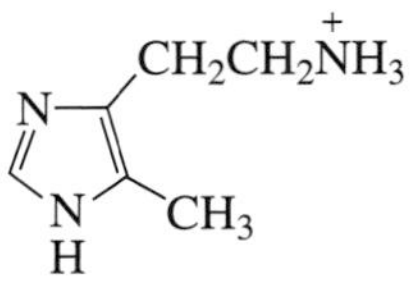

4-methylhistamine

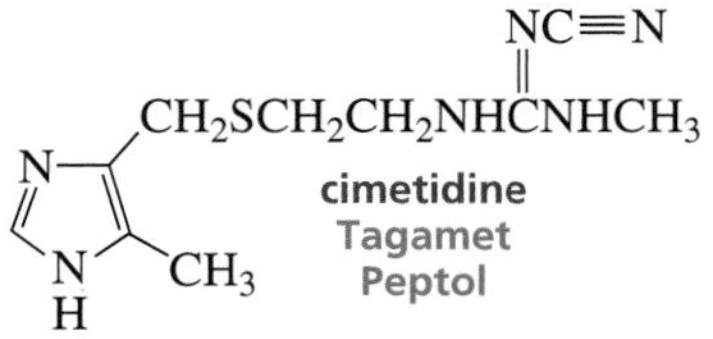

cimetidine
Tagamet
Peptol

ranitidine
Ulcex
Trigger

It is not unusual when screening modified compounds to find a compound with completely different pharmacological activity than the lead compound. For example, a molecular modification of a sulfonamide, an antibiotic, led to tolbutamide, which is a drug with hypoglycemic activity (Section 22.8).

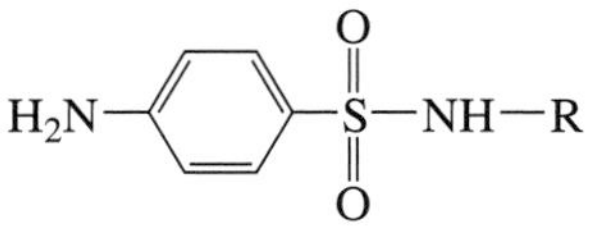

a sulfonamide

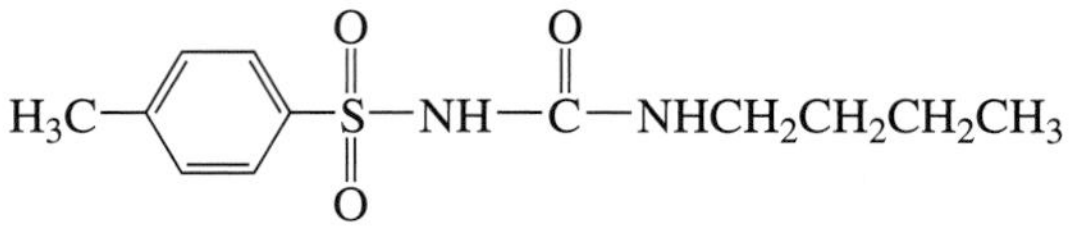

tolbutamide

Molecular modification of promethazine, an antihistamine, led to chlorpromazine. Chlorpromazine did not show any antihistamine activity, but it lowered the body temperature. This drug found clinical use in chest surgery, where patients previously had to be cooled down by wrapping them in cold wet sheets. Because wrapping in cold wet sheets had been an old method of calming psychotic patients, a French psychiatrist tried the drug on some of his patients. He found that chlorpromazine was able to suppress psychotic symptoms to the point that his patients assumed almost normal behavioral characteristics. However, they soon developed uncoordinated involuntary movements. After thousands of molecular modifications, thioridazine was found to have the appropriate calming effect with less problematic side effects. It is now in clinical use as an antipsychotic.

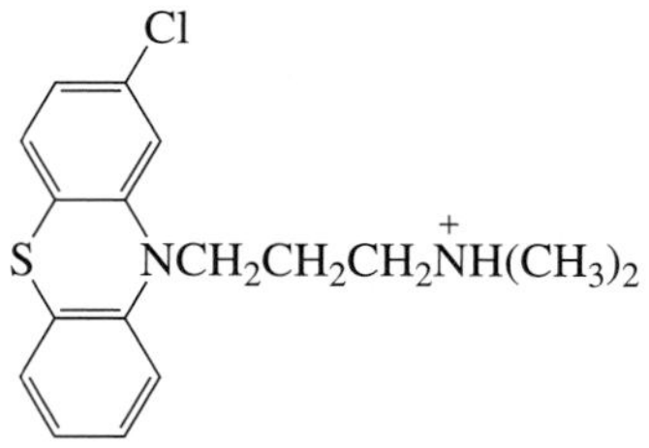

chlorpromazine

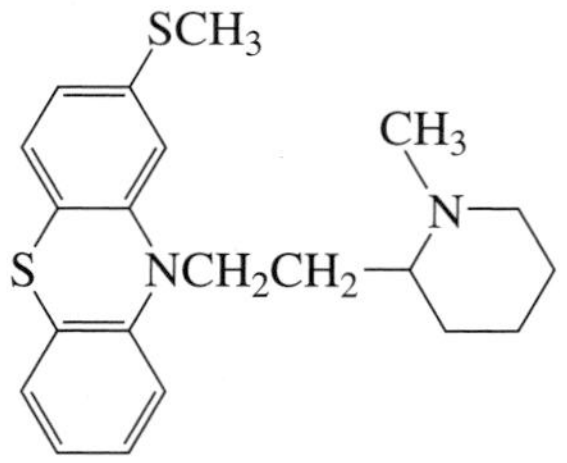

thioridazine
Mellaril

Sometimes a drug initially developed for one purpose is later found to have properties that make it a better drug for a different purpose. Beta-blockers were originally intended to be used to alleviate the pain associated with angina by reducing the amount of work done by the heart. They were later found to have hypotensive (antihypertensive) properties, so now they are used primarily to manage hypertension, a disease very prevalent in the Western world.

In earlier chapters, we saw several examples of drugs that act by inhibiting enzymes (Sections 22.8 and 22.9). One example is penicillin, which destroys bacteria by inhibiting the enzyme that synthesizes bacterial cell walls (Section 15.13).

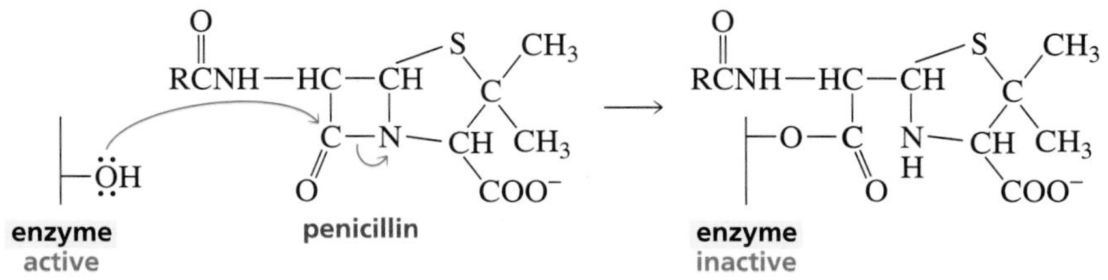

29.7 DRUGS AS ENZYME INHIBITORS

Bacteria develop resistance to penicillin by secreting an enzyme (β-lactamase) that hydrolyzes penicillin before it has an opportunity to interfere with bacterial cell wall synthesis.

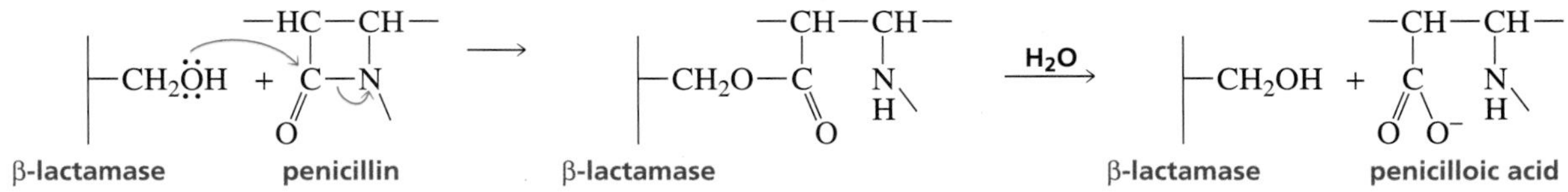

Chemists have developed drugs that inhibit β-lactamase. If such a drug is given to the patient along with penicillin, the antibiotic is not destroyed. This is an example of a drug that has no therapeutic effect itself but acts by protecting a therapeutic drug.

One β-lactamase inhibitor is a sulfone, easily prepared from penicillin by oxidizing the sulfur atom with MMPP (Section 17.8).

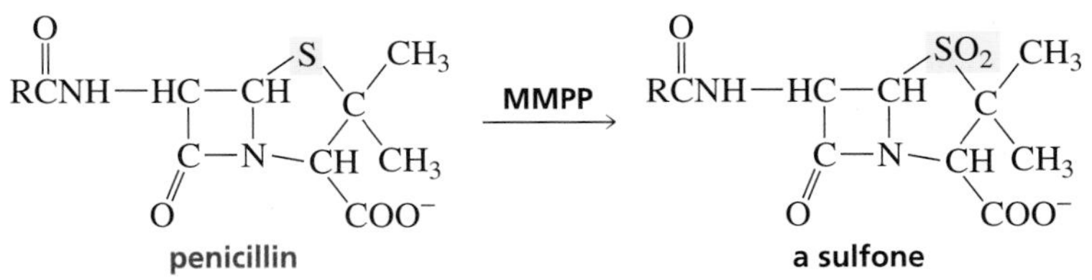

Because the sulfone looks like the original antibiotic, β-lactamase accepts it as a substrate, forming an ester as it does with penicillin. If the ester were then hydrolyzed, β-lactamase would be free to react with penicillin. However, because the substrate is a sulfone, there is an alternative pathway to hydrolysis that results in formation of a stable imine. Because imines are susceptible to nucleophilic attack, an amino group at the active site of β-lactamase can react with the imine, forming a second covalent bond between the enzyme and the sulfone. This inactivates β-lactamase, thereby wiping out penicillin resistance. The sulfone is another example of a mechanism-based **suicide inhibitor** (Section 22.8).

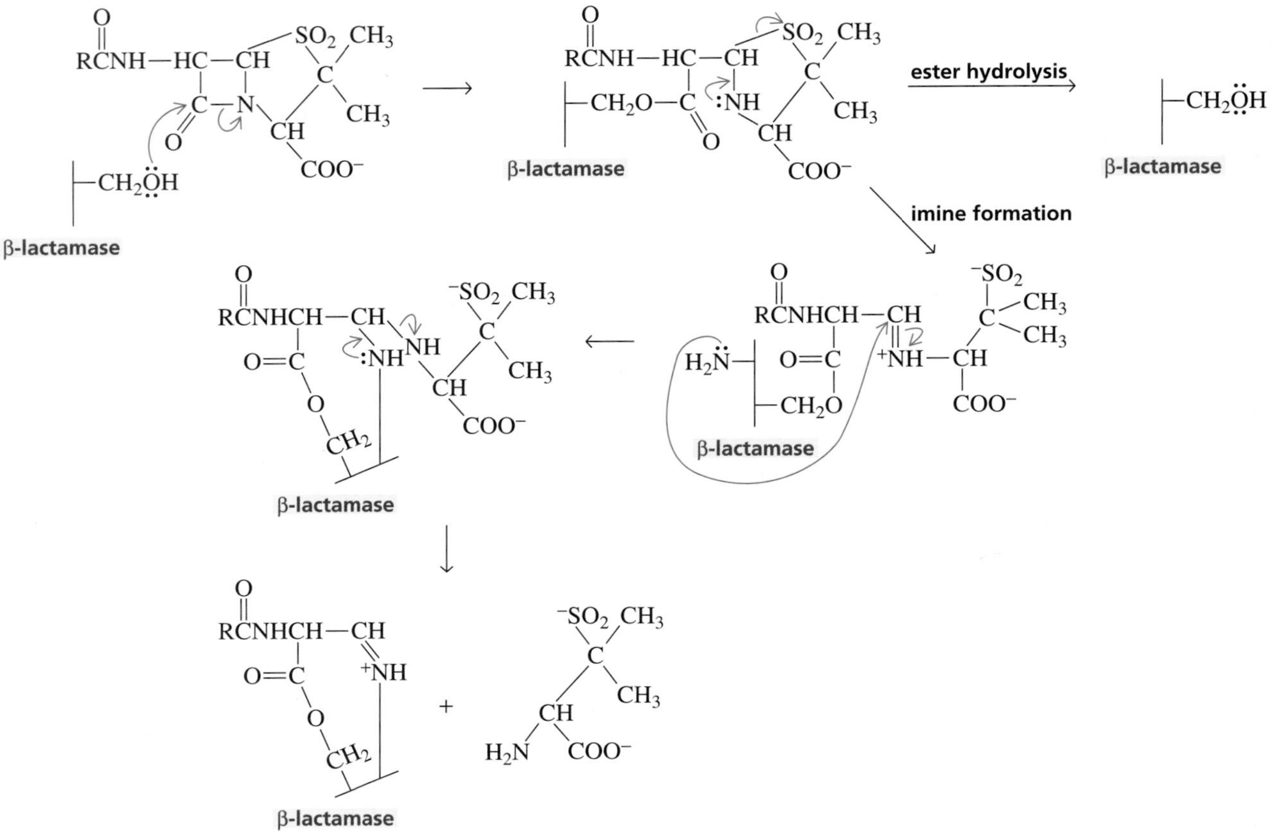

What makes the sulfone such an effective drug is that the reactive imine group does not appear until after the drug has bound to the enzyme that is to be inactivated. In other words, the drug has a specific target. In contrast, if an imine were directly administered to the patient, it would be very nonspecific, reacting with whatever nucleophile it first encountered.

When two drugs are given simultaneously to a patient, their combined effect can be additive, antagonistic, or synergistic. Synergistic means that the effect of two drugs used in combination is greater than the sum of the effects obtained when the drugs are administered individually. Administering penicillin and the sulfone in combination results in **drug synergism.**

Another reason to administer two drugs in combination is that if some bacteria are resistant to one of them, the second drug will minimize the chance that the resistant strain will proliferate. For example, two antimicrobial agents, isoniazid and rifampin, are given in combination to treat tuberculosis.

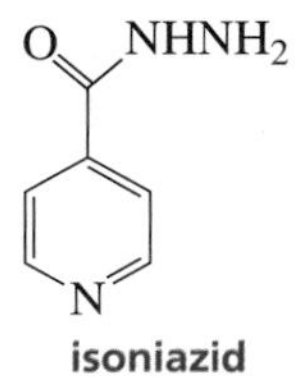

isoniazid

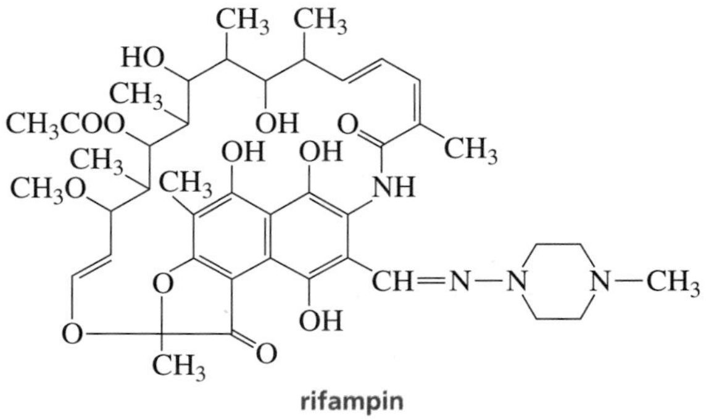

rifampin

Drug resistance is an increasingly important problem in medicinal chemistry. More and more bacterial strains are evolving that show resistance to traditional antibiotics. For example, a strain of tubercle bacillus has appeared in recent years that seems to be resistant to all antibiotics. Typically it takes a bacterial strain 15 to 20 years to evolve resistance to antibiotics. That the fluoroquinolones, the last class of antibiotics to be discovered, were discovered more than 20 years ago is reason for alarm.

The antibiotic activity of the fluoroquinolones results from their ability to inhibit DNA gyrase, an enzyme required for transcription. Fortunately, the bacterial and mammalian forms of the enzyme are sufficiently different that the fluoroquinolones inhibit only the bacterial enzyme.

There are many different fluoroquinolones. They all have fluorine substituents, which increase the lipophilicity of the drug to enable it to penetrate into tissues and cells. If either the carboxyl group or the double bond in the 4-pyridinone ring is removed, all activity is lost. By changing the substituents on the piperazine ring, excretion of the drug can be shifted from the liver to the kidney, which is useful to patients with impaired liver function. The substituents on the piperazine ring also affect the half-life of the drug.

ciprofloxacin
active against gram-negative bacteria

sparfloxacin
**active against gram-negative bacteria
and gram-positive bacteria**

29.8 DESIGNING A SUICIDE SUBSTRATE

It is important for a drug to have a minimum of undesirable side effects. A drug must be administered in sufficient quantity to achieve a therapeutic effect; however, too much of a drug can be lethal. The **therapeutic index** of a drug is the ratio of the lethal dose to the therapeutic dose. The higher the therapeutic index, the greater the margin of safety of the drug.

Penicillin is an effective antibiotic that has a high therapeutic ratio because it interferes with cell wall synthesis—and bacterial cells have cell walls but human cells do not. What else is characteristic about cell walls that could lead to the design of an antibiotic? We know that enzymes and other proteins are polymers of L-amino acids. Cell walls, however, contain both L-amino acids and D-amino acids. Therefore, if the racemization of naturally occurring L-amino acids to D-amino acids could be prevented, D-amino acids would not be available for incorporation into cell walls, and bacterial cell wall synthesis could be stopped.

We have seen that amino acid racemization is catalyzed by an enzyme that requires pyridoxal phosphate as a coenzyme (Section 22.6). What we need, then, is a compound that will inhibit this enzyme. Because the natural substrate for this enzyme is an amino acid, an amino acid analog should be a good potential inhibitor.

The first step in racemization is removal of the α-hydrogen of the amino acid. If the inhibitor has a leaving group on the β-carbon, the electrons left behind when the proton is removed can displace the leaving group instead of being delocalized into the pyridine ring. (Compare the mechanism shown below with that shown for racemization in Section 22.6.) Transimination forms a reactive α,β-unsaturated amino acid that irreversibly reacts with the enzyme. Because the enzyme is now bound to the coenzyme in an amine linkage rather than in an imine linkage, the enzyme can no longer undergo a transimination reaction with the enzyme. It has been irreversibly inactivated. This is another example of an inhibitor that does not become chemically active until it is at the active site of the targeted enzyme.

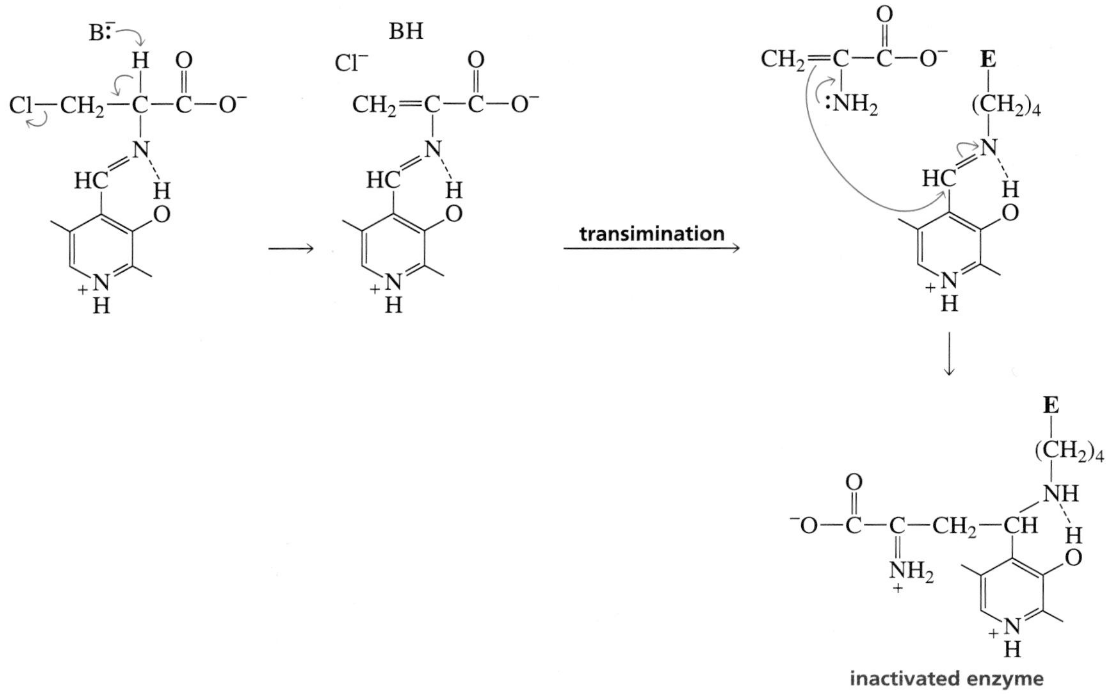

inactivated enzyme

29.9 QUANTITATIVE STRUCTURE–ACTIVITY RELATIONSHIPS (QSAR)

The enormous cost involved in synthesizing and testing thousands of modified compounds in an attempt to find an active drug led scientists to develop a more rational approach to the design of biologically active molecules. They realized that if a physical or chemical property of a series of compounds could be correlated with biological activity, they would know what property of the drug was related to the biological activity. Armed with this knowledge, scientists could design compounds that would have a good chance of exhibiting the desired activity. This would be a great improvement over the random approach to molecular modification that they had traditionally employed.

The first hint that a physical property of a drug could be related to biological activity was made almost 100 years ago when scientists recognized that chloroform ($CHCl_3$), diethyl ether, cyclopropane, and nitrous oxide (N_2O) were all useful general anesthetics. Clearly, the chemical structures of these diverse compounds could not account for their similar pharmacological effects. Therefore, some physical property must explain the similarity of their biological activities.

In the early 1960s, Corwin Hansch postulated that the biological activity of a drug depended on two processes. First, the drug must be able to get from the point where it enters the body to the receptor where it exerts its effect. For example, an anesthetic must be able to cross the aqueous milieu (blood) and penetrate the lipid

Corwin H. Hansch *was born in North Dakota in 1918. He received a B.S. from the University of Illinois and a Ph.D. from New York University. He has been a professor of chemistry at Pomona College since 1946.*

barrier of nerve cell membranes. Secondly, when a drug reaches its receptor, it must properly interact with it.

Chloroform, diethyl ether, cyclopropane, and nitrous oxide were each put into a mixture of 1-octanol (a nonpolar compound) and water. When the amount of drug dissolving in each of the layers was measured, it turned out that they all had a similar **distribution coefficient** (the ratio of the amount of a compound dissolving in each of the solvents). In other words, the distribution coefficient could be related to biological activity. Compounds with greater distribution coefficients (more non-polar compounds) could not cross the aqueous phase. Compounds with lower distribution coefficients could not penetrate the nonpolar cell membrane. This meant that the distribution coefficient of a compound could be used to determine whether or not a compound should be tested *in vivo*. This technique of relating a property of a series of compounds to biological activity is known as a **quantitative structure–activity relationship (QSAR).**

Determining the physical property of a drug cannot take the place of *in vivo* testing because it cannot be predicted by the distribution coefficient alone how the drug will behave once it reaches a suitable receptor. Nevertheless, QSAR provides a way to shortcut extensive molecular modification.

In the following example, QSAR was useful in determining not only the structure of a potentially active drug but also something about the structure of the receptor site. A series of substituted 2,4-diaminopyrimidines used as inhibitors of dihydrofolate reductase (Section 22.8) was investigated.

a 2,4-diaminopyrimidine

The potency of the inhibitors could be described by the following equation, where σ and π are substituent parameters.

$$\text{potency} = 0.80\pi - 7.34\sigma - 8.14$$

The σ parameter is a measure of the electron-donating or electron-withdrawing ability of the substituents R and R'. The negative coefficient of σ indicates that potency is increased by electron donation (Chapter 15, Problem 65). The fact that increasing the basicity of the drug increases its potency suggests that the protonated drug is more active than the nonprotonated drug.

The π parameter is a measure of the hydrophobicity of the substituents. Potency was found to be better related to π when the π value for the more hydrophobic of the two substituents was used rather than the sum of the π values for both substituents. This suggests that the receptor has a hydrophobic pocket that can accommodate one but not both of the substituents.

In addition to solubility and substituent parameters, some of the properties that have been correlated with biological activity are oxidation–reduction potentials, molecular size, interatomic distances between functional groups, degree of ionization, and configuration.

Because the shape of a molecule determines whether it will be recognized by a receptor and therefore whether it will exhibit biological activity, compounds with similar biological activity often have similar structures. Because computers can draw molecular models of compounds on a video display and move them around to

**29.10
MOLECULAR
MODELING**

Figure 29.1 ▶
The antibiotic distamycin A (magenta atoms) bound in the minor groove of DNA.

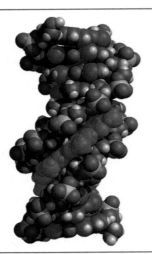

assume different conformations, the use of computer **molecular modeling** is being investigated as an approach to more rational drug design. There are computer programs that allow chemists to scan existing collections of thousands of compounds to find those with appropriate structural and conformational properties.

Any compound that shows promise can be drawn on a computer display along with the three-dimensional image of a receptor site. For example, the binding of distamycin A, a compound with both antibiotic and antiviral activity, to the minor groove of DNA is shown in Figure 29.1.

The fit between the two molecules may suggest modifications that can be made to the compound that will result in more favorable binding. In this way, the selection of compounds to be synthesized for the purpose of screening for biological activity will be more rational and will allow pharmacologically active compounds to be more rapidly discovered. This technique will become more valuable as scientists learn more about receptor sites. However, structure-based drug design considers only the fit between the drug and its receptor; it does not consider the transport of the drug, its distribution properties, or how it is metabolized.

29.11 COMBINATORIAL ORGANIC SYNTHESIS

The need for large collections of compounds that can be screened for biological activity in the constant search for new drugs has led organic chemists to a synthetic strategy that employs the concept of mass production. This strategy, called **combinatorial organic synthesis,** involves the synthesis of a large group of related compounds (known as a library) by covalently connecting sets of building blocks of various structures. For example, if a compound can be synthesized by connecting three different building blocks, and if each set of building blocks contains ten interchangeable compounds, 1000 ($10 \times 10 \times 10$) different compounds can be prepared. This approach clearly mimics nature, which uses amino acids and nucleic acids as building blocks to synthesize an enormous number of different proteins and nucleic acids.

The first requirement in combinatorial synthesis is the availability of an assortment of reactive small molecules to be used as building blocks. Because of the ready availability of amino acids, the use of combinatorial synthesis made its first appearance in the creation of peptide libraries. Peptides, however, have limited use as therapeutic agents because they generally cannot be taken orally and are rapidly metabolized. Currently, organic chemists are attempting to create libraries of small organic molecules.

An example of the approach used in combinatorial synthesis is the creation of a library of benzodiazepines. These compounds can be thought of as originating from three different sets of building blocks: a substituted 2-aminobenzophenone, an amino acid, and an alkylating agent.

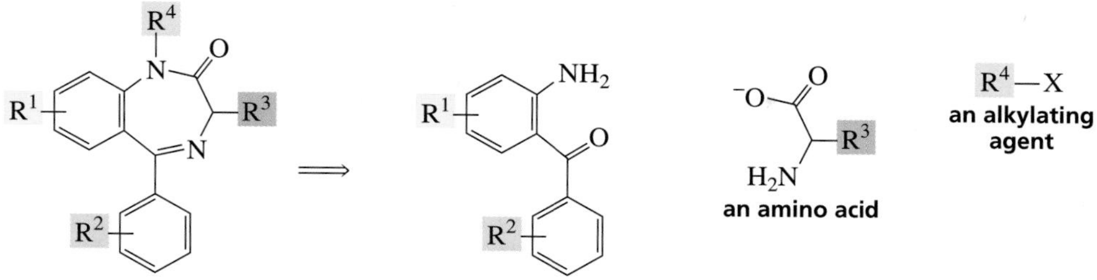

a benzodiazepine   a 2-aminobenzophenone   an amino acid   an alkylating agent

The 2-aminobenzophenone is attached to a solid support in a manner that allows it to be readily removed by acid hydrolysis (Figure 29.2). An *N*-protected amino acyl fluoride is then added. After the amide forms, the protecting group is

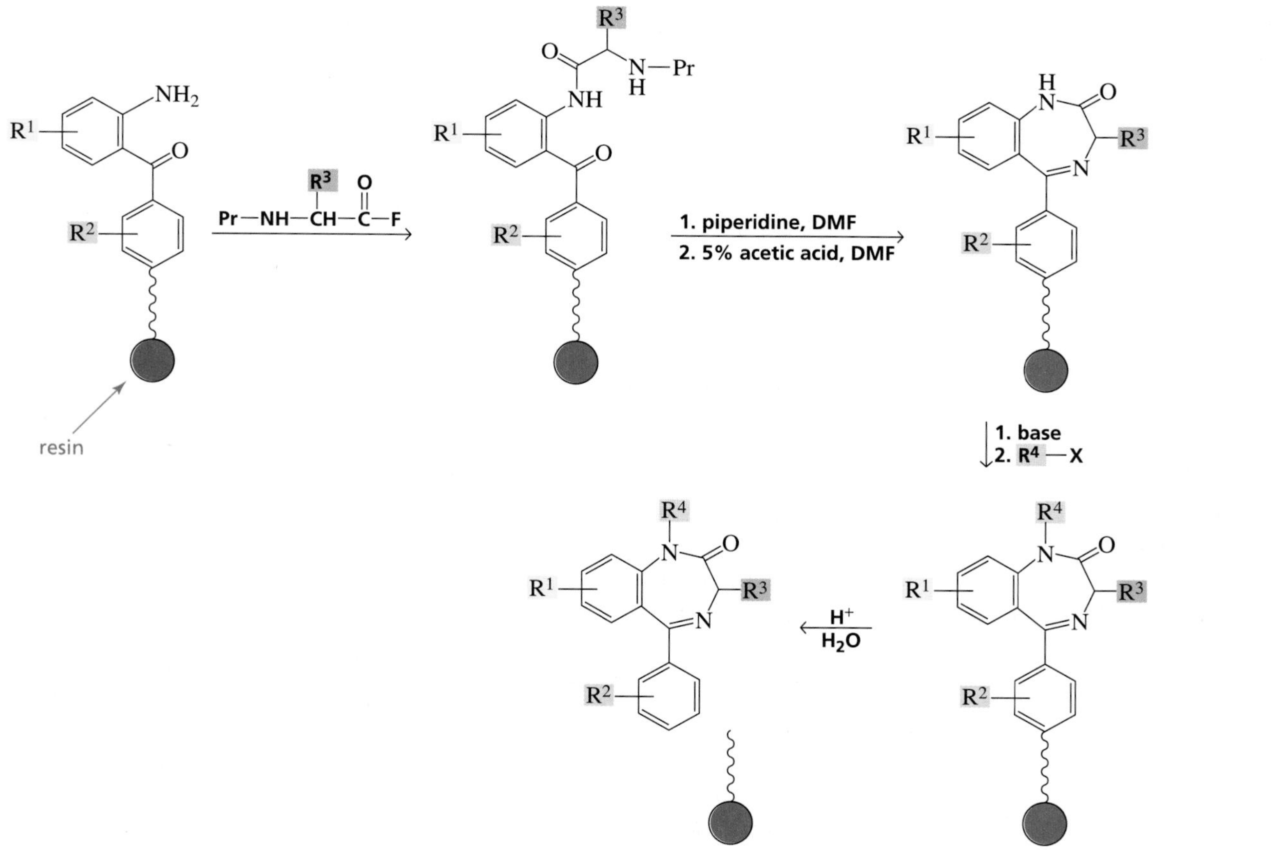

▲ **Figure 29.2**
Combinatorial organic synthesis of benzodiazepines.

removed and the seven-membered ring is formed as a result of imine formation. A base is added to remove the amide hydrogen, forming a nucleophile that reacts with the added alkylating agent. The final product is then removed from the solid support.

The solid support containing the 2-aminobenzophenone can be divided into several portions, and a different amino acid can then be added to each portion. Each ring-closed product can also be divided into several portions, and a different alkylating agent can be added to each portion. In this way, many different benzodiazepines can be prepared.

29.12 ANTIVIRAL DRUGS

Relatively few clinically useful drugs have been developed for viral infections. This slow progress is due to the nature of viruses and the way they replicate. Viruses are smaller than bacteria. They consist of nucleic acid (either DNA or RNA) surrounded by a coat of protein. A virus penetrates a host cell or merely injects its nucleic acid. In either case, its nucleic acid is transcribed and takes over the host's nucleic acids. Viruses have no metabolism of their own; they rely entirely on the enzymes of the host cell.

Most **antiviral drugs** are analogs of nucleosides. They interfere with DNA or RNA synthesis and in this way prevent the virus from replicating.

cytarabine
Cytosar
used against rabies

vidarabine
Vira-A
used against encephalitis

idoxuridine
Herplex
approved for topical ophthalmic use

acyclovir
Aclovir
used against shingles (herpes zoster) and herpes simplex infections

ribavirin
Viramid
a broad-spectrum antiviral agent

29.13 ECONOMICS OF DRUGS. GOVERNMENTAL REGULATIONS

By some estimates, the average cost of launching a new drug is about $230 million. This cost has to be amortized quickly by the manufacturer because the starting date of a patent is the date the drug is first discovered. A patent is good for 20 years from the date of the patent application but, because it takes an average of 12 years to market a drug after its initial discovery, the patent protects the discoverer of the drug for an average of only 8 years. In addition, the average lifetime of a drug is only 15 to 20 years. After that time it is generally replaced by a newer and improved drug.

Why does it cost so much to develop a new drug? First of all, the Food and Drug Administration (FDA) has very high standards that must be met before it approves a drug for a particular use. Before the U.S. government became involved with the regulation of drugs, it was not uncommon for charlatans to dispense useless and even harmful medical preparations. Starting in 1906, Congress passed laws governing the manufacture, distribution, and use of drugs. These laws are amended from time to time to reflect changing situations. The present law requires that all new drugs used by physicians must first be thoroughly tested for effectiveness and safety.

An important factor leading to the high price of many drugs is the low success rate in progressing from the initial concept for a drug to an approved product: only one or two of every hundred drugs tested become lead compounds; out of a hundred structural modifications of a lead compound, only one is worthy of further study; and only 10% of these compounds actually become marketable drugs.

ORPHAN DRUGS

Because of the high cost associated with developing a drug, drug companies are reluctant to carry out research on drugs for rare diseases. Even if a company were to find a drug, there would be no way to recoup its expenditure because of the limited demand. Drugs that no one wants to develop are called **orphan drugs.** In 1983, the U.S. Congress passed the Orphan Drugs Act, which creates public subsidies to fund research and tax credits for up to 50% of the costs of developing and marketing drugs for diseases or conditions that affect fewer than 200,000 people. In addition, the company that develops the drug has 4 years of exclusive marketing rights if the drug is nonpatentable. In the 10 years prior to passage of this act, fewer than ten drugs were developed to treat orphan diseases. Today, they are more than 100, and some 600 others are now in development.

Drugs developed as orphan drugs include AZT (to treat AIDS), taxol (to treat ovarian cancer), exosurf neonatal (to treat respiratory distress syndrome in infants), and opticrom (to treat corneal swelling).

KEY TERMS

antiviral drug (page 1254)
bactericidal drug (page 1243)
bacteriostatic drug (page 1243)
blind screen (page 1242)
brand name (page 1238)
combinatorial organic synthesis
   (page 1252)
distribution coefficient
   (page 1251)

drug (page 1236)
drug resistance (page 1249)
drug synergism (page 1248)
generic name (page 1238)
lead compound (page 1240)
molecular modeling (page 152)
molecular modification
   (page 1240)
orphan drug (page 1255)

proprietary name (page 1238)
quantitative structure–reactivity
   relationship (QSAR) (page 1251)
random screen (page 1242)
receptor (page 1244)
suicide inhibitor (page 1247)
therapeutic index (page 1249)
trademark (page 1238)
trade name (page 1238)

PROBLEMS

1. What is the chemical name of each of the following drugs?
   a. benzocaine   **b.** procaine

2. Based on the compound used as the lead compound for the development of procaine and lidocaine, propose structures for other compounds that you would like to see tested for potential use as anesthetics.

3. Which of the following compounds is more likely to exhibit activity as a tranquilizer?

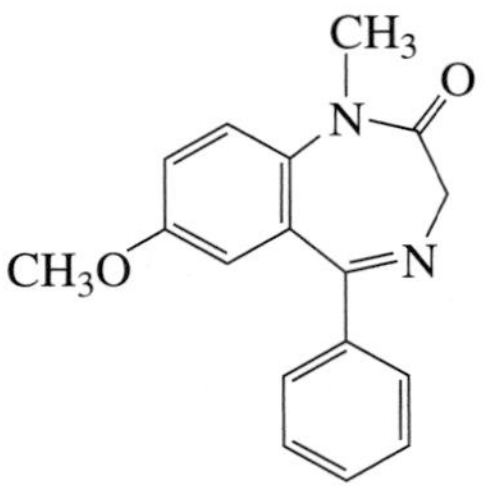

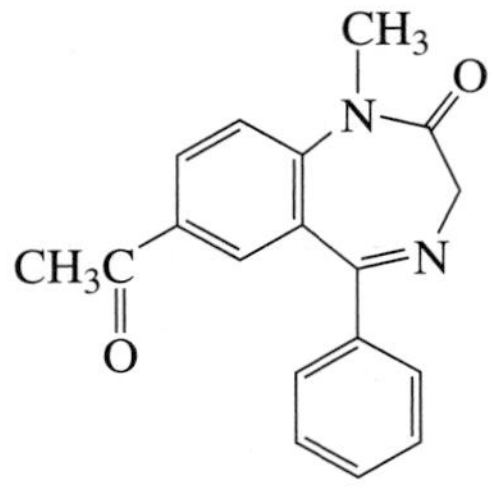

4. Which compound is more likely to be a general anesthetic?

$$CH_3CH_2CH_2OH \quad \text{or} \quad CH_3OCH_2CH_3$$

5. What accounts for the ease of imine formation between β-lactamase and the sulfone antibiotic that counteracts penicillin resistance?

6. For each of the following pairs of compounds, indicate the compound that you would expect to be a more potent inhibitor of dihydrofolate reductase.

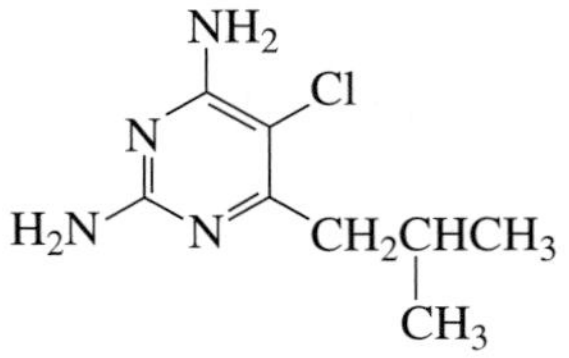

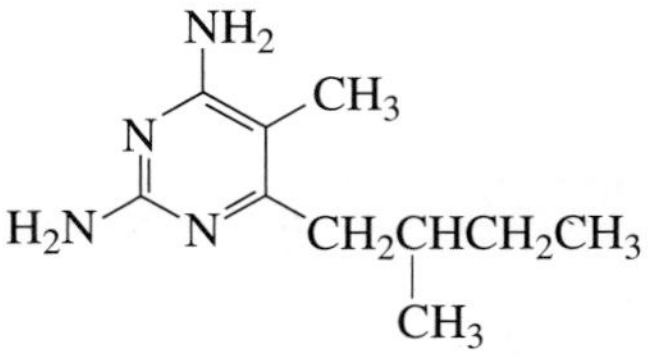

7. The lethal dose of tetrahydrocannabinol in mice is 2.0 g/kg, and the therapeutic dose is 20 mg/kg. The lethal dose of sodium pentothal in mice is 100 mg/kg, and the therapeutic dose is 30 mg/kg. Which is a safer drug?

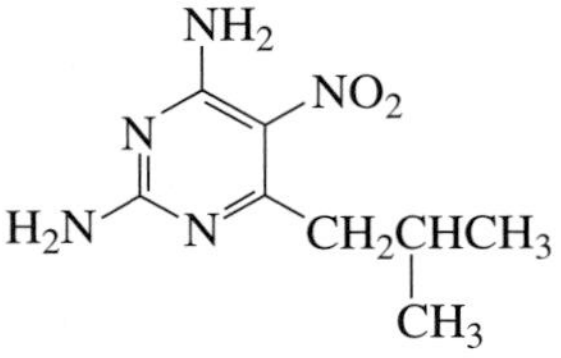

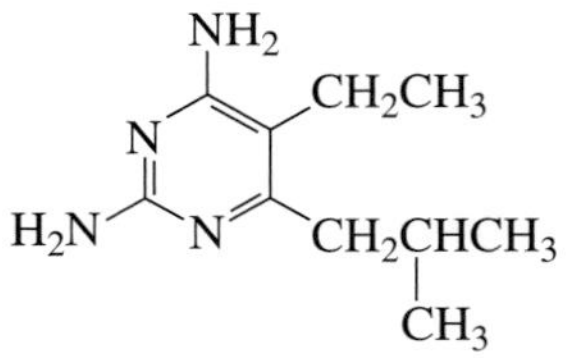

8. The following compound is a suicide inhibitor of the enzyme that catalyzes amino acid racemization. Propose a mechanism that explains how it irreversibly inactivates the enzyme.

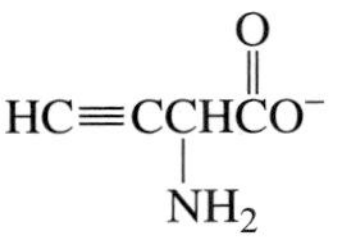

9. Explain how each of the antiviral drugs shown in Section 29.12 differs from the naturally occurring nucleoside that it most closely resembles.

10. Show a mechanism for the formation of a benzodiazepine 4-oxide from the reaction of a quinazoline 3-oxide with methylamine.

APPENDIX I

Physical Properties of Organic Compounds

Physical Properties of Alkenes

Name	Structure	mp (°C)	bp (°C)	Density (g/mL)
Ethene	$CH_2{=}CH_2$	−169	−103.7	
Propene	$CH_2{=}CHCH_3$	−185	−47.4	
1-Butene	$CH_2{=}CHCH_2CH_3$		−6.3	
1-Pentene	$CH_2{=}CH(CH_2)_2CH_3$		30	0.643
1-Hexene	$CH_2{=}CH(CH_2)_3CH_3$	−138	63.5	0.675
1-Heptene	$CH_2{=}CH(CH_2)_4CH_3$	−119	93	0.698
1-Octene	$CH_2{=}CH(CH_2)_5CH_3$	−104	122.5	0.716
1-Nonene	$CH_2{=}CH(CH_2)_6CH_3$		146	0.731
1-Decene	$CH_2{=}CH(CH_2)_7CH_3$	−87	171	0.743
cis-2-Butene	cis-$CH_3CH{=}CHCH_3$	−139	3.7	0.621
trans-2-Butene	trans-$CH_3CH{=}CHCH_3$	−106	0.9	0.604
Methylpropene	$CH_2{=}C(CH_3)_2$	−141	−6.9	0.594
cis-2-Pentene	cis-$CH_3CH{=}CHCH_2CH_3$	−139	3.7	0.621
trans-2-Pentene	trans-$CH_3CH{=}CHCH_2CH_3$	−106	0.9	0.604
Cyclohexene		−104	83	0.810

Physical Properties of Alkynes

Name	Structure	mp (°C)	bp (°C)	Density (g/mL)
Ethyne	$HC{\equiv}CH$	−82	−84.0	
Propyne	$HC{\equiv}CCH_3$	−101.5	−23.2	
1-Butyne	$HC{\equiv}CCH_2CH_3$	−122	8.1	
2-Butyne	$CH_3C{\equiv}CCH_3$	−24	27	0.694
1-Pentyne	$HC{\equiv}C(CH_2)_2CH_3$	−98	39.3	0.695
2-Pentyne	$CH_3C{\equiv}CCH_2CH_3$	−101	55.5	0.714
3-Methyl-1-butyne	$HC{\equiv}CCH(CH_3)_2$		29	0.665
1-Hexyne	$HC{\equiv}C(CH_2)_3CH_3$	−124	71	0.719
2-Hexyne	$CH_3C{\equiv}C(CH_2)_2CH_3$	−92	84	0.730
3-Hexyne	$CH_3CH_2C{\equiv}CCH_2CH_3$	−51	81	0.725
1-Heptyne	$HC{\equiv}C(CH_2)_4CH_3$	−80	100	0.733
1-Octyne	$HC{\equiv}C(CH_2)_5CH_3$	−70	126	0.747
1-Nonyne	$HC{\equiv}C(CH_2)_6CH_3$	−65	151	0.763
1-Decyne	$HC{\equiv}C(CH_2)_7CH_3$	−36	182	0.770

Physical Properties of Alkanes   see page 54

Physical Properties of Cyclic Saturated Alkanes

Name	mp (°C)	bp (°C)	Density (g/mL)
Cyclopropane	−127	33	
Cyclobutane	−80	13	
Cyclopentane	−94	49	0.746
Cyclohexane	6.5	81	0.778
Cycloheptane	−12	118	0.810
Cyclooctane	14	149	0.830
Methylcyclopentane	−142	72	0.749
Methylcyclohexane	−126	100	0.769
cis-1,2-Dimethylcyclopentane	−62	99	0.772
trans-1,2-Dimethylcyclopentane	−120	92	0.750

Physical Properties of Ethers

Name	Structure	mp (°C)	bp (°C)
Dimethyl ether	CH_3OCH_3	−141	−24.8
Diethyl ether	$CH_3CH_2OCH_2CH_3$	−116	34.6
Dipropyl ether	$CH_3(CH_2)_2O(CH_2)_2CH_3$	−122	91
Diisopropyl ether	$(CH_3)_2CHOCH(CH_3)_2$	−60	69
Dibutyl ether	$CH_3(CH_2)_3O(CH_2)_3CH_3$	−95	142
Divinyl ether	$CH_2{=}CHOCH{=}CH_2$		35
Diallyl ether	$CH_2{=}CHCH_2OCH_2CH{=}CH_2$		94
Tetrahydrofuran		−108	66
Dioxane	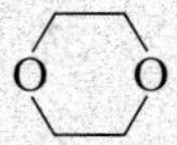	11	101

Physical Properties of Alcohols

Name	Structure	mp (°C)	bp (°C)	Solubility (g/100 g H_2O at 25 °C)
Methanol	CH_3OH	−97.8	64.5	∞
Ethanol	CH_3CH_2OH	−114.7	78.5	∞
1-Propanol	$CH_3(CH_2)_2OH$	−126.5	97.4	∞
1-Butanol	$CH_3(CH_2)_3OH$	−89.5	117.3	7.9
1-Petanol	$CH_3(CH_2)_4OH$	−78.5	138	2.3
1-Hexanol	$CH_3(CH_2)_5OH$	−52	156.5	0.6
1-Heptanol	$CH_3(CH_2)_6OH$	−34	176	0.2
1-Octanol	$CH_3(CH_2)_7OH$	−15	195	0.05
2-Propanol	$CH_3CHOHCH_3$	−89.5	92.5	∞
2-Butanol	$CH_3CHOHCH_2CH_3$	−115	99.5	12.5
2-Methyl-1-propanol	$(CH_3)_2CHCH_2OH$	−108	107.9	10.0
2-Methyl-2-propanol	$(CH_3)_3COH$	25.5	82.2	∞
3-Methyl-1-butanol	$(CH_3)_2CH(CH_2)_2OH$	−117	132	2
2-Methyl-2-butanol	$(CH_3)_2COHCH_2CH_3$	−12	102	12.5
2,2-Dimethyl-1-propanol	$(CH_3)_3CCH_2OH$	53	114	∞
Allyl alcohol	$CH_2{=}CHCH_2OH$	−129	97	∞
Cyclopentanol	C_5H_9OH		140	s. sol.
Cyclohexanol	$C_6H_{11}OH$	24	161.5	s. sol.
Benzyl alcohol	$C_6H_5CH_2OH$	−15	205	4

Physical Properties of Alkyl Halides

Name	bp (°C)			
	Fluoride	Chloride	Bromide	Iodide
Methyl	−78.4	−24.2	3.6	42.4
Ethyl	−37.7	12.3	38.4	72.3
Propyl	−2.5	46.6	71.0	102.5
Isopropyl	−9.4	34.8	59.4	89.5
Butyl	32.5	78.4	101.6	130.5
Isobutyl		68.8	90	120
sec-Butyl		68.3	91.2	120.0
tert-Butyl		50.2	73.1	dec.
Pentyl	62.8	107.8	129.6	157.0
Hexyl		133	154	179
Benzyl		179	201	93

Physical Properties of Amines

Name	Structure	mp (°C)	bp (°C)	Solubility (g/100 g H_2O at 25 °C)
Primary Amines				
Methylamine	CH_3NH_2	−94	−6.3	v. sol.
Ethylamine	$CH_3CH_2NH_2$	−81	16.6	∞
Propylamine	$CH_3(CH_2)_2NH_2$	−83	47.8	∞
Isopropylamine	$(CH_3)_2CHNH_2$	−95.2	32.4	∞
Butylamine	$CH_3(CH_2)_3NH_2$	−49	77.8	v. sol.
Isobutylamine	$(CH_3)_2CHCH_2NH_2$	−85	63	∞
sec-Butylamine	$CH_3CH_2CH(CH_3)NH_2$	−104	63	∞
tert-Butylamine	$(CH_3)_3CNH_2$	−67.5	44.4	∞
Cyclohexylamine	$C_6H_{11}NH_2$	−18	134	s. sol.
Secondary Amines				
Dimethylamine	$(CH_3)_2NH$	−93	7.4	v. sol.
Diethylamine	$(CH_3CH_2)_2NH$	−48	56.3	10.0
Dipropylamine	$(CH_3CH_2CH_2)_2NH$	−63	110	10.0
Dibutylamine	$(CH_3CH_2CH_2CH_2)_2NH$	−60	159	s. sol.
Tertiary Amines				
Trimethylamine	$(CH_3)_3N$	−117	2.9	91
Triethylamine	$(CH_3CH_2)_3N$	−114	89.3	14
Tripropylamine	$(CH_3CH_2CH_2)_3N$	−93	157	s. sol.

Physical Properties of Benzene and Substituted Benzenes

Name	Structure	mp (°C)	bp (°C)	Solubility (g/100 g H_2O at 25 °C)
Aniline	$C_6H_5NH_2$	−6	184	3.7
Benzene	C_6H_6	5.5	80.1	s. sol.
Benzaldehyde	C_6H_5CHO	−26	178	s. sol.
Benzamide	$C_6H_5CONH_2$	132	290	s. sol.
Benzoic acid	C_6H_5COOH	122	250	0.34
Bromobenzene	C_6H_5Br	−30.8	156	insol.
Chlorobenzene	C_6H_5Cl	−45.6	132	insol.
Nitrobenzene	$C_6H_5NO_2$	5.7	210.8	s. sol.
Phenol	C_6H_5OH	43	182	s. sol.
Styrene	$C_6H_5CH{=}CH_2$	−30.6	145.2	insol.
Toluene	$C_6H_5CH_3$	−95	110.6	insol.

Physical Properties of Carboxylic Acids

Name	Structure	mp (°C)	bp (°C)	Solubility (g/100 g H_2O at 25 °C)
Formic acid	HCOOH	8.4	101	∞
Acetic acid	CH_3COOH	16.6	118	∞
Propionic acid	CH_3CH_2COOH	−21	141	∞
Butanoic acid	$CH_3(CH_2)_2COOH$	−5	164	∞
Pentanoic acid	$CH_3(CH_2)_3COOH$	−34	186	4.97
Hexanoic acid	$CH_3(CH_2)_4COOH$	−3	205	0.97
Heptanoic acid	$CH_3(CH_2)_5COOH$	−8	223	0.24
Octanoic acid	$CH_3(CH_2)_6COOH$	17	239	0.068
Nonanoic acid	$CH_3(CH_2)_7COOH$	15	255	0.026
Decanoic acid	$CH_3(CH_2)_8COOH$	32	270	0.015

Physical Properties of Dicarboxylic Acids

Name	Structure	mp (°C)	Solubility (g/100 g H_2O at 25 °C)
Oxalic acid	HOOCCOOH	189	s
Malonic acid	$HOOCCH_2COOH$	136	v. sol.
Succinic acid	$HOOC(CH_2)_2COOH$	185	s. sol.
Glutaric acid	$HOOC(CH_2)_3COOH$	98	v.sol.
Adipic acid	$HOOC(CH_2)_4COOH$	151	s. sol.
Pimelic acid	$HOOC(CH_2)_5COOH$	106	s
Phthalic acid	$1,2\text{-}C_6H_4(COOH)_2$	231	s. sol.
Maleic acid	cis-HOOCCH=CHCOOH	130.5	v. sol.
Fumaric acid	trans-HOOCCH=CHCOOH	302	s. sol.

Physical Properties of Acyl Chlorides and Acid Anhydrides

Name	Structure	mp (°C)	bp (°C)
Acetyl chloride	CH_3COCl	−112	51
Propionyl chloride	CH_3CH_2COCl	−94	80
Butyryl chloride	$CH_3(CH_2)_2COCl$	−89	102
Valeryl chloride	$CH_3(CH_2)_3COCl$	−110	128
Acetic anhydride	$CH_3(CO)O(CO)CH_3$	−73	140
Succinic anhydride	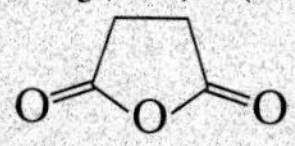		120

Physical Properties of Esters

Name	Structure	mp (°C)	bp (°C)
Methyl formate	$HCOOCH_3$	−99	31.5
Ethyl formate	$HCOOCH_2CH_3$	−80	54
Methyl acetate	CH_3COOCH_3	−98	57.5
Ethyl acetate	$CH_3COOCH_2CH_3$	−84	77
Propyl acetate	$CH_3COO(CH_2)_2CH_3$	−92	102
Methyl propionate	$CH_3CH_2COOCH_3$	−87.5	80
Ethyl propionate	$CH_3CH_2COOCH_2CH_3$	−74	99
Methyl butyrate	$CH_3CH_2CH_2COOCH_3$	−84.8	102.3
Ethyl butyrate	$CH_3CH_2CH_2COOCH_2CH_3$	−93	121

Physical Properties of Amides

Name	Structure	mp (°C)	bp (°C)
Formamide	$HCONH_2$	3	200 d*
Acetamide	CH_3CONH_2	82	221
Propanamide	$CH_3CH_2CONH_2$	79	213
Butanamide	$CH_3(CH_2)_2CONH_2$	116	216
Pentanamide	$CH_3(CH_2)_3CONH_2$	106	232

*d means the substance decomposes.

Physical Properties of Aldehydes

Name	Structure	mp (°C)	bp (°C)	Solubility (g/100 g H_2O at 25 °C)
Formaldehyde	$HCHO$	−92	−21	v. sol.
Acetaldehyde	CH_3CHO	−121	21	∞
Propionaldehyde	CH_3CH_2CHO	−81	49	16
Butyraldehyde	$CH_3(CH_2)_2CHO$	−99	76	7
Pentanal	$CH_3(CH_2)_3CHO$	−92	103	s. sol.
Hexanal	$CH_3(CH_2)_4CHO$	−56	128	s. sol.
Heptanal	$CH_3(CH_2)_5CHO$	−43	153	0.1
Octanal	$CH_3(CH_2)_6CHO$		171	insol.
Nonanal	$CH_3(CH_2)_7CHO$		192	insol.
Decanal	$CH_3(CH_2)_8CHO$	−5	209	insol.
Benzaldehyde	C_6H_5CHO	−26	178	0.3

Physical Properties of Ketones

Name	Structure	mp (°C)	bp (°C)	Solubility (g/100 g H_2O at 25 °C)
Acetone	CH_3COCH_3	−95	56	∞
2-Butanone	$CH_3COCH_2CH_3$	−86	80	25.6
2-Pentanone	$CH_3CO(CH_2)_2CH_3$	−78	102	5.5
2-Hexanone	$CH_3CO(CH_2)_3CH_3$	−57	128	1.6
2-Heptanone	$CH_3CO(CH_2)_4CH_3$	−36	151	0.4
2-Octanone	$CH_3CO(CH_2)_5CH_3$	−16	173	insol.
2-Nonanone	$CH_3CO(CH_2)_6CH_3$	−7	195	insol.
2-Decanone	$CH_3CO(CH_2)_7CH_3$	14	210	insol.
3-Pentanone	$CH_3CH_2COCH_2CH_3$	−40	102	4.8
3-Hexanone	$CH_3CH_2CO(CH_2)_2CH_3$		123	1.5
3-Heptanone	$CH_3CH_2CO(CH_2)_3CH_3$	−39	149	0.3
Acetophenone	$CH_3COC_6H_5$	21	202	insol.
Propiophenone	$CH_3CH_2COC_6H_5$	21	218	insol.

APPENDIX II

pK_a Values

Compound	pK_a	Compound	pK_a	Compound	pK_a
$CH_3C\equiv\overset{+}{N}H$	−10.1	O_2N—⟨⟩—$\overset{+}{N}H_3$	1.0	CH_3—⟨⟩—$\overset{O}{\overset{\|}{C}}OH$	4.3
HI	−10	(pyrimidinium)	1.0		
HBr	−9			CH_3O—⟨⟩—$\overset{O}{\overset{\|}{C}}OH$	4.5
$CH_3\overset{+OH}{\overset{\|}{C}}H$	−8	$Cl_2CH\overset{O}{\overset{\|}{C}}OH$	1.3		
		HSO_4^-	2.0	⟨⟩—$\overset{+}{N}H_3$	4.6
$CH_3\overset{+OH}{\overset{\|}{C}}CH_3$	−7.3	H_3PO_4	2.1		
HCl	−7	(purinium)	2.5	$CH_3\overset{O}{\overset{\|}{C}}OH$	4.8
CH_3SH	−6.8				
$CH_3\overset{+OH}{\overset{\|}{C}}OCH_3$	−6.5	$FCH_2\overset{O}{\overset{\|}{C}}OH$	2.7	(quinolinium)	4.9
$CH_3\overset{+OH}{\overset{\|}{C}}OH$	−6.1	$ClCH_2\overset{O}{\overset{\|}{C}}OH$	2.8	CH_3—⟨⟩—$\overset{+}{N}H_3$	5.1
H_2SO_4	−5	$BrCH_2\overset{O}{\overset{\|}{C}}OH$	2.9		
(pyrrolium)	−3.8	$ICH_2\overset{O}{\overset{\|}{C}}OH$	3.1	(pyridinium)	5.2
$CH_3CH_2\overset{H}{\underset{+}{O}}CH_2CH_3$	−3.6	HF	3.2	CH_3O—⟨⟩—$\overset{+}{N}H_3$	5.3
$CH_3CH_2\overset{H}{\underset{+}{O}}H$	−2.4	HNO_2	3.4	$CH_3\overset{CH_3}{\underset{\|}{C}}=\overset{+}{N}HCH_3$	5.5
$CH_3\overset{H}{\underset{+}{O}}H$	−2.5	O_2N—⟨⟩—$\overset{O}{\overset{\|}{C}}OH$	3.4		
H_3O^+	−1.7			$CH_3\overset{O}{\overset{\|}{C}}CH_2\overset{O}{\overset{\|}{C}}H$	5.9
HNO_3	−1.3	$H\overset{O}{\overset{\|}{C}}OH$	3.8	$HO\overset{+}{N}H_3$	6.0
CH_3SO_3H	−1.2	Br—⟨⟩—$\overset{+}{N}H_3$	3.9	H_2CO_3	6.4
⟨⟩—SO_3H	−0.60	Br—⟨⟩—$\overset{O}{\overset{\|}{C}}OH$	4.0	(imidazolium)	6.8
$CH_3\overset{+OH}{\overset{\|}{C}}NH_2$	0.0			H_2S	7.0
$F_3C\overset{O}{\overset{\|}{C}}OH$	0.2	⟨⟩—$\overset{O}{\overset{\|}{C}}OH$	4.2	O_2N—⟨⟩—OH	7.1
$Cl_3C\overset{O}{\overset{\|}{C}}OH$	0.64			$H_2PO_4^-$	7.2
$\overset{+}{N}$—OH	0.78			⟨⟩—SH	7.8

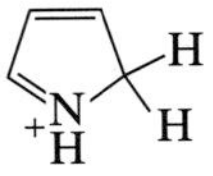

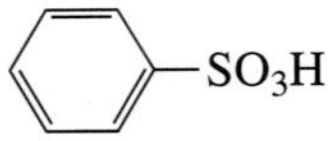

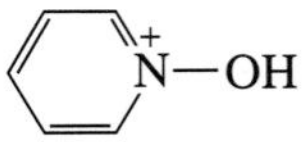

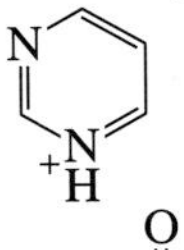

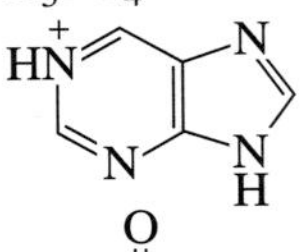

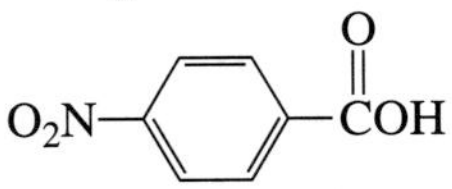

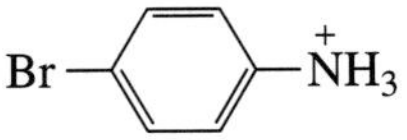

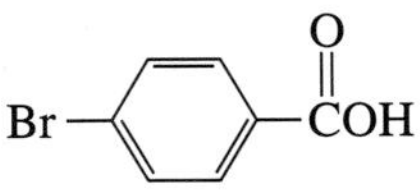

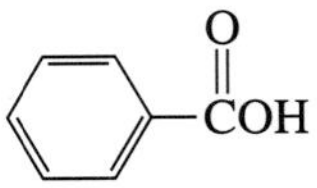

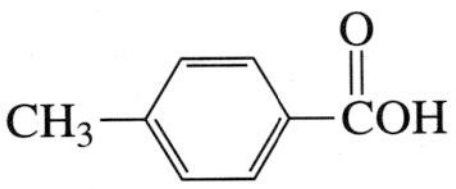

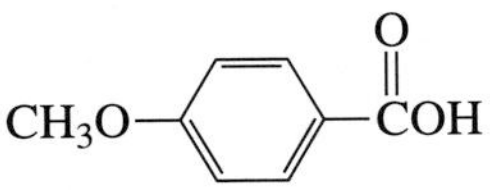

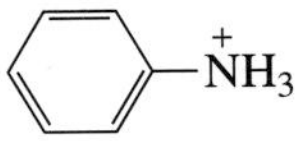

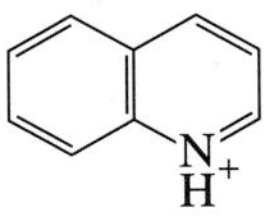

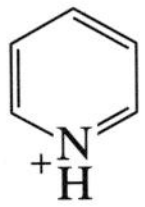

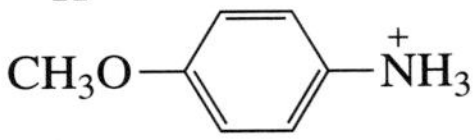

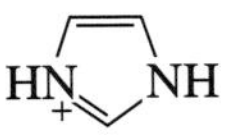

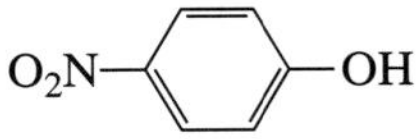

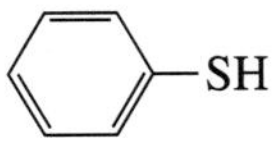

Compound	pKa	Compound	pKa	Compound	pKa
aziridinium ($^+$NH$_2$ ring)	8.0	cyclohexyl–$\overset{+}{N}H_3$	10.7	$CH_3\overset{O}{\overset{\|}{C}}H$	17
$H_2N\overset{+}{N}H_3$	8.1	$(CH_3)_2\overset{+}{N}H_2$	10.7	$(CH_3)_3COH$	18
$CH_3\overset{O}{\overset{\|}{C}}OOH$	8.2	piperidinium	11.1	$CH_3\overset{O}{\overset{\|}{C}}CH_3$	20
$CH_3CH_2NO_2$	8.6	$CH_3CH_2\overset{+}{N}H_3$	11.2	$CH_3\overset{O}{\overset{\|}{C}}OCH_2CH_3$	24.5
$CH_3\overset{O}{\overset{\|}{C}}CH_2\overset{O}{\overset{\|}{C}}CH_3$	8.9	pyrrolidinium	11.3	$HC\equiv CH$	25
purine	8.9	HPO_4^{2-}	12.3	$CH_3C\equiv N$	25
$HC\equiv N$	9.1	CF_3CH_2OH	12.4	$CH_3\overset{O}{\overset{\|}{C}}N(CH_3)_2$	30
morpholinium	9.3	$CH_3CH_2O\overset{O}{\overset{\|}{C}}CH_2\overset{O}{\overset{\|}{C}}OCH_2CH_3$	13.3	NH_3	36
$Cl-C_6H_4-OH$	9.4	$HC\equiv CCH_2OH$	13.5	pyrrolidine	36
$\overset{+}{N}H_4$	9.4	$H_2N\overset{O}{\overset{\|}{C}}NH_2$	13.7	CH_3NH_2	40
$HOCH_2CH_2\overset{+}{N}H_3$	9.5	$CH_3\overset{CH_3}{\overset{\|}{\underset{\|}{\overset{+}{N}}}}CH_2CH_2OH$ (CH_3)	13.9	$CH_3-C_6H_4-CH_3$	41
$H_3\overset{+}{N}CH_2\overset{O}{\overset{\|}{C}}O^-$	9.8	imidazole	14.4	benzene	43
C_6H_5-OH	10.0	CH_3OH	15.5	$CH_2=CHCH_3$	43
$CH_3-C_6H_4-OH$	10.2	H_2O	15.7	$CH_2=CH_2$	44
HCO_3^-	10.2	CH_3CH_2OH	15.9	cyclopropane	46
CH_3NO_2	10.2	$CH_3\overset{O}{\overset{\|}{C}}NH_2$	16	CH_4	50
$H_2N-C_6H_4-OH$	10.3	acetophenone ($C_6H_5\overset{O}{\overset{\|}{C}}CH_3$)	16.0	CH_3CH_3	50
CH_3CH_2SH	10.5	pyrrole	17		
$(CH_3)_3\overset{+}{N}H$	10.6				
$CH_3\overset{O}{\overset{\|}{C}}CH_2\overset{O}{\overset{\|}{C}}OCH_2CH_3$	10.7				
$CH_3\overset{+}{N}H_3$	10.7				

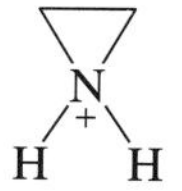

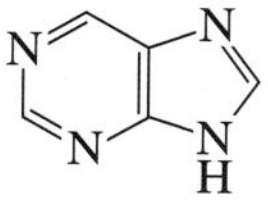

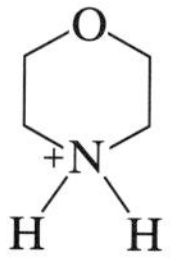

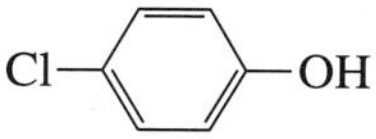

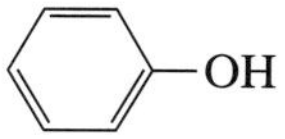

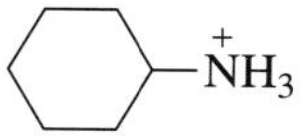

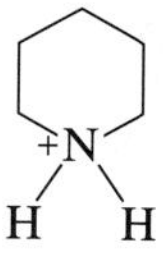

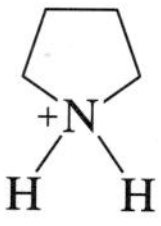

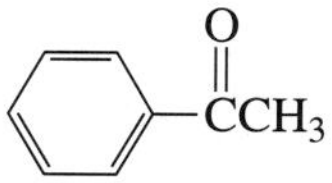

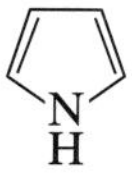

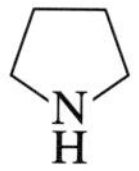

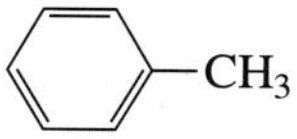

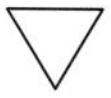

APPENDIX III

Derivations of Rate Laws

A **reaction mechanism** is a detailed analysis of how the chemical bonds (or the electrons) in the reactants rearrange to form the products. The mechanism for a given reaction must obey the observed rate law for the reaction.

A **rate law** tells how the rate of a reaction depends on the concentration of the species involved in the reaction.

First-Order Reaction

$$A \xrightarrow{k_1} \text{products}$$

Change in the concentration of A with respect to time:

$$\frac{-d[A]}{dt} = k_1[A]$$

Let a = the initial concentration of A;
let x = the concentration of A that has reacted up to time t.
Therefore, the concentration of A left at time $t = (a - x)$.
Substitution in the above equation gives:

$$\frac{-d(a - x)}{dt} = k_1(a - x)$$

$$\frac{-da}{dt} + \frac{dx}{dt} = k_1(a - x)$$

$$0 + \frac{dx}{dt} = k_1(a - x)$$

$$\frac{dx}{(a - x)} = k_1\, dt$$

Integration of the above equation gives

$$-\ln(a - x) = k_1 t + \text{constant}$$

At $t = 0$, $x = 0$; therefore,

$$\text{constant} = -\ln a$$

$$-\ln(a - x) = k_1 t - \ln a$$

$$\ln \frac{a}{a - x} = k,\, t$$

$$\ln \frac{a - x}{a} = -k_1 t$$

$$2.303 \log \frac{a - x}{a} = -k_1 t$$

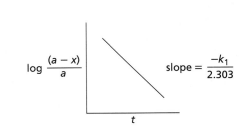

$$\log \frac{(a - x)}{a} \qquad \text{slope} = \frac{-k_1}{2.303}$$

Half-Life of a First-Order Reaction

The **half-life ($t_{1/2}$)** of a reaction is the time it takes for half the reactant to react (or for half the product to form.

$$\ln \frac{a}{(a-x)} = k_1 t$$

At $t_{1/2}$, $x = \dfrac{a}{2}$; therefore,

$$\ln \frac{a}{\left(a - \dfrac{a}{2}\right)} = k_1 t_{1/2}$$

$$\ln \frac{a}{\dfrac{a}{2}} = k_1 t_{1/2}$$

$$\ln 2 = k_1 t_{1/2}$$

$$0.693 = k_1 t_{1/2}$$

$$t_{1/2} = \frac{0.693}{k_1}$$

Notice that the half-life of a first-order reaction is independent of the concentration of the reactant.

Second-Order Reaction

$$A + B \xrightarrow{k_2} \text{products.}$$

Change in the concentration of A with respect to time:

$$\frac{-d[A]}{dt} = k_2[A]\,[B]$$

Let a = the initial concentration of A;
let b = the initial concentration of B;
let x = the concentration of A that has reacted at time t.
Therefore, the concentration of A left at time $t = (a - x)$, and the concentration of B left at time $t = (b - x)$.

Substitution gives:

$$\frac{dx}{dt} = k_2\,(a - x)(b - x)$$

For the case where $a = b$ (this condition can be arranged experimentally):

$$\frac{dx}{dt} = k_2\,(a - x)^2$$

$$\frac{dx}{(a - x)^2} = k_2\,dt$$

Integrating the above equation gives:

$$\frac{1}{(a - x)} = k_2 t + \text{constant}$$

At $t = 0$, $x = 0$; therefore,

$$\text{constant} = \frac{1}{a}$$

$$\frac{1}{(a - x)} - \frac{1}{a} = k_2 t$$

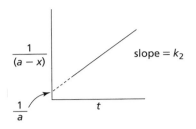

Half-Life of a Second-Order Reaction

$$\frac{1}{(a - x)} - \frac{1}{a} = k_2 t$$

At $t_{1/2}$, $x = \dfrac{a}{2}$; therefore,

$$\frac{1}{a} = k_2 t_{1/2}$$

$$t_{1/2} = \frac{1}{k_2 a}$$

Pseudo-First-Order Reaction

It is easier to determine a first-order rate constant than a second-order rate constant because the kinetic behavior of a first-order reaction is independent of the initial concentration of the reactant. Therefore, a first-order rate constant can be determined without knowing the initial concentration of the reactant. Determination of a second-order rate constant requires not only that the initial concentration of the reactants be known but also that the initial concentrations of the two reactants be identical in order to simplify the kinetic equation.

However, if the concentration of one of the reactants in a second-order reaction is much greater than the concentration of the other, the reaction can be treated as a first-order reaction. Such a reaction is known as a **pseudo-first-order reaction.**

$$\frac{-d\,[A]}{dt} = k_2\,[A]\,[B]$$

If $[B] \gg [A]$,

$$\frac{-d\,[A]}{dt} = k_2'\,[A]$$

The rate constant obtained for a pseudo-first-order reaction (k_2') includes the concentration of B, but k_2 can be determined by carrying out the reaction at several different concentrations of B and determining the slope of a plot of the observed rate versus [B].

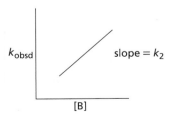

APPENDIX IV

Summary of Methods Used to Synthesize a Particular Functional Group

SYNTHESIS OF ACETALS
1. Acid-catalyzed reaction of an aldehyde or a ketone with 2 equivalents of an alcohol (16.8).

SYNTHESIS OF ACYL CHLORIDES OR ACYL BROMIDES
1. Reaction of a carboxylic acid with $SOCl_2$, PCl_3, or PBr_3 (15.17).

SYNTHESIS OF ALCOHOLS
1. Acid-catalyzed hydration of an alkene (3.13).
2. Oxymercuration–demercuration of an alkene (3.16).
3. Hydroboration–oxidation of an alkene (3.17).
4. Reaction of an alkyl halide with HO^- (9.3).
5. Reaction of a Grignard reagent with an epoxide (11.8).
6. Reduction of an aldehyde, ketone, acyl chloride, anhydride, ester, or carboxylic acid (16.5, 16.6, 17.1).
7. Reaction of a Grignard reagent with an aldehyde, ketone, acyl chloride, or ester (16.4, 16.6).
8. Cleavage of an ether with HI or HBr (11.5).

SYNTHESIS OF ALDEHYDES
1. Hydroboration–oxidation of a terminal alkyne with disiamylborane followed by $H_2O_2 + HO^-$ (5.7).
2. Oxidation of a primary alcohol with pyridinium chlorochromate (17.2).
3. Rosenmund reduction: catalytic hydrogenation of an acyl chloride (17.1).
4. Reaction of an acyl chloride with lithium aluminum tri(*tert*-butoxy)hydride (17.1).
5. Cleavage of a 1,2-diol with periodic acid (17.5).
6. Ozonolysis of an alkene followed by work-up under reducing conditions (17.6).
7. Reaction of dithianyllithium with an alkyl halide followed by treatment with an aqueous solution of mercuric ion (28.2).

SYNTHESIS OF ALKANES
1. Catalytic hydrogenation of an alkene or an alkyne (3.19, 5.8).
2. Reaction of a Grignard reagent with a source of protons (11.8).
3. Wolff–Kishner or Clemmensen reduction of an aldehyde or a ketone (14.9, 16.7).
4. Reduction of a thioacetal with H_2 and Raney nickel (16.10).
5. Reaction of a Gilman reagent with an alkyl halide (11.8).
6. Preparation of a cyclopropane by the reaction of an alkene with a carbene (3.17).

SYNTHESIS OF ALKENES
1. Elimination of hydrogen halide from an alkyl halide (10.1, 10.2, 10.3).
2. Acid-catalyzed dehydration of an alcohol (11.4).
3. Hofmann elimination reaction: elimination of a proton and a tertiary amine from a quaternary ammonium hydroxide (11.11).
4. Exhaustive methylation of an amine followed by a Hofmann elimination reaction (11.11).
5. Hydrogenation of an alkyne with Lindlar's catalyst to form a *cis*-alkene (5.8).
6. Reduction of an alkyne with Na (or Li) and liquid ammonia to form a *trans*-alkene (5.8).
7. Formation of a cyclic alkene using a Diels–Alder reaction (7.8, 27.4).
8. Wittig reaction: reaction of an aldehyde or a ketone with a phosphonium ylide (16.11).

SYNTHESIS OF ALKYL HALIDES
1. Addition of hydrogen halide (HX) to an alkene (3.9).
2. Addition of HBr + peroxide (3.18).
3. Addition of halogen to an alkene (3.15).
4. Addition of hydrogen halide or a halogen to an alkyne (5.5).
5. Radical halogenation of an alkane, an alkene, or an alkyl benzene (8.2, 8.5).
6. Reaction of an alcohol with hydrogen halide, $SOCl_2$, PCl_3, or PBr_3 (11.1, 11.2).
7. Reaction of a sulfonate ester with halide ion (11.3).
8. Cleavage of an ether with HI or HBr (11.5).
9. Halogenation of an α-carbon of an aldehyde, a ketone, or a carboxylic acid (18.4, 18.5).
10. Hunsdiecker reaction: reaction of a carboxylic acid with Br_2 and Ag_2O (18.16).

SYNTHESIS OF ALKYNES
1. Elimination of hydrogen halide from a vinyl halide (10.10).
2. Two successive eliminations of hydrogen halide from a vicinal dihalide or a geminal dihalide (10.10).
3. Reaction of an acetylide ion (formed by removing a proton from a terminal alkyne) with an alkyl halide (5.10).

SYNTHESIS OF AMIDES
1. Reaction of an acyl chloride, an acid anhydride, or an ester with ammonia or with an amine (15.6, 15.7, 15.8).
2. Reaction of a carboxylic acid with ammonia or with an amine and heat (15.11).
3. Reaction of a carboxylic acid and an amine with dicyclohexylcarbodiimide (20.9, 20.10).
4. Reaction of a nitrile with a secondary or tertiary alcohol (page 721).

SYNTHESIS OF AMINES
1. Reaction of an alkyl halide with NH_3, RNH_2, or R_2NH (9.3).
2. Reaction of an alkyl halide with azide ion followed by reduction of the alkyl azide (15.14).
3. Reduction of an imine, a nitrile, or an amide (17.1).
4. Reductive amination of an aldehyde or a ketone (16.7).
5. Gabriel synthesis of primary amines: reaction of a primary alkyl halide with potassium phthalimide (15.14).
6. Reduction of a nitro compound (14.10).

SYNTHESIS OF AMINO ACIDS
1. Hell–Volhard–Zelinski reaction: halogenation of a carboxylic acid followed by treatment with excess NH_3 (18.5).
2. Reductive amination of an α-keto acid (20.6).

SYNTHESIS OF ACID ANHYDRIDES
1. Reaction of an acyl halide with a carboxylate ion (15.6).
2. Preparation of a cyclic anhydride by heating a dicarboxylic acid (15.18).

SYNTHESIS OF CARBOXYLIC ACIDS
1. Oxidation of a primary alcohol (17.2).
2. Oxidation of an aldehyde (17.3).
3. Ozonolysis of a monosubstituted alkene or a 1,2-disubstituted alkene followed by work-up under oxidizing conditions (17.6).
4. Ozonolysis of an alkyne (17.7).
5. Oxidation of an alkyl benzene (14.10).
6. Hydrolysis of an acyl halide, acid anhydride, ester, amide, or nitrile (15.6, 15.7, 15.8, 15.12, 15.15).
7. Haloform reaction: reaction of a methyl ketone with Br_2 (or Cl_2 or I_2) + HO^- (18.4).
8. Reaction of a Grignard reagent with CO_2 (16.4).
9. Malonic ester synthesis (18.17).

SYNTHESIS OF CYANOHYDRINS
1. Reaction of an aldehyde or a ketone with sodium cyanide and HCl (16.4).

SYNTHESIS OF 1,2-DIOLS
1. Reaction of an epoxide with water (11.6).
2. Reaction of an alkene with osmium tetroxide or potassium permanganate (17.4).

SYNTHESIS OF DISULFIDES
1. Milk oxidation of a thiol (20.7).

SYNTHESIS OF ENAMINES
1. Reaction of an aldehyde or a ketone with a secondary amine (16.7).

SYNTHESIS OF EPOXIDES
1. Reaction of an alkene with a peroxyacid (17.8).
2. Reaction of a halohydrin with hydroxide ion (p. 476).
3. Reaction of an aldehyde or a ketone with a sulfonium ylide (p. 773).

SYNTHESIS OF ESTERS
1. Reaction of an acyl halide or an acid anhydride with an alcohol (15.6, 15.7).
2. Acid-catalyzed reaction of an ester or a carboxylic acid with an alcohol (15.8, 15.11).
3. Reaction of an alkyl halide or a sulfonate ester with a carboxylate ion (9.3, 11.3).
4. Oxidation of a ketone (17.3).
5. Preparation of a methyl ester by the reaction of a carboxylate ion with diazomethane (14.20).

SYNTHESIS OF ETHERS
1. Acid-catalyzed addition of an alcohol to an alkene (3.13).
2. Alkoxymercuration–demercuration of an alkene (3.16).
3. Williamson ether synthesis: reaction of an alkoxide ion with an alkyl halide (10.9).
4. Formation of symmetrical ethers by heating an acidic solution of a primary alcohol (11.4).

SYNTHESIS OF HALOHYDRINS
1. Reaction of an alkene with Br_2 (or Cl_2) and H_2O (3.15).
2. Reaction of an epoxide with a hydrogen halide (11.6).

SYNTHESIS OF IMINES
1. Reaction of an aldehyde or a ketone with a primary amine (16.7).

SYNTHESIS OF KETONES
1. Mercuric acid-catalyzed hydration of an alkyne (5.6).
2. Hydroboration–oxidation of an alkyne (5.7).
3. Oxidation of a secondary alcohol (17.2).
4. Cleavage of a 1,2-diol with periodic acid (17.5).
5. Ozonolysis of an alkene (17.6).
6. Friedel–Crafts acylation of an aromatic ring (14.7).
7. Preparation of a methyl ketone by the acetoacetic ester synthesis (18.18).
8. Reaction of a Gilman reagent with an acyl chloride (11.8).
9. Preparation of a cyclic ketone by the reaction of the next size smaller cyclic ketone with diazomethane (p. 764).

SYNTHESIS OF α,β-UNSATURATED KETONES
1. Elimination from an α-haloketone (18.6).
2. Selenenylation of a ketone followed by oxidative elimination (18.6).

SYNTHESIS OF NITRILES
1. Reaction of an alkyl halide with cyanide ion (9.3).

SYNTHESIS OF SUBSTITUTED BENZENES
1. Halogenation with Br_2 or Cl_2 and a Lewis acid (14.4).

2. Nitration with HNO_3 + H_2SO_4 (14.5).
3. Sulfonation: reaction with H_2SO_4 (14.6).
4. Friedel–Crafts acylation (14.7).
5. Friedel–Crafts alkylation (14.8).
6. Sandmeyer reaction: reaction of a benzenediazonium salt with a CuBr, CuCl, or CuCN (14.18).
7. Formation of a phenol by reaction of a benzenediazonium salt with water (14.18).
8. Formation of an aniline by reaction of a benzyne intermediate with NH_3 (14.22).

SYNTHESIS OF SULFIDES
1. Reaction of a thiol with an alkyl halide (9.3, 11.3).
2. Catalytic hydrogenation of a disulfide (20.7).

SYNTHESIS OF THIOLS
1. Reaction of an alkyl halide with hydrogen sulfide (9.3).

SUMMARY OF METHODS THAT CAN BE USED TO FORM CARBON–CARBON BONDS
1. Reaction of an acetylide ion with an alkyl halide or a sulfonate ester (5.10, 9.3, 11.3).
2. Diels–Alder reaction (7.8, 27.4).
3. Reaction of a Grignard reagent with an epoxide (11.8).
4. Friedel–Crafts alkylation and acylation (14.7, 14.8, 14.9).
5. Reaction of cyanide ion with an alkyl halide or a sulfonate ester (9.3, 11.3).
6. Reaction of cyanide ion with an aldehyde or a ketone (16.4).
7. Reaction of a Grignard reagent with an aldehyde, a ketone, an ester, an amide, or CO_2 (16.4, 16.6).
8. Reaction of an alkene with a carbene (3.17).
9. Reaction of a lithium dialkylcuprate with an α,β-unsaturated ketone or an α,β-unsaturated aldehyde (16.14).
10. Aldol addition (18.10, 18.11, 18.12).
11. Claisen condensation (18.13, 18.14).
12. Malonic ester synthesis and acetoacetic ester synthesis (18.17, 18.18).
13. Michael addition reaction (18.9).
14. Alkylation of an enamine (18.8).
15. Alkylation of an α-carbon (18.8).

APPENDIX V

Spectroscopy Tables

Common Fragment Ions*

m/z	Ion	m/z	Ion
14	CH_2	46	NO_2
15	CH_3	47	CH_2SH, CH_3S
16	O	48	$CH_3S + H$
17	OH	49	CH_2Cl
18	H_2O, NH_4	51	CHF_2
19	F, H_3O	53	C_4H_5
26	$C{\equiv}N$	54	$CH_2CH_2C{\equiv}N$
27	C_2H_3	55	C_4H_7, $CH_2{=}CHC{=}O$
28	C_2H_4, CO, N_2, $CH{=}NH$	56	C_4H_8
29	C_2H_5, CHO	57	C_4H_9, $C_2H_5C{=}O$
30	CH_2NH_2, NO		
31	CH_2OH, OCH_3		$\overset{\text{O}}{\overset{\|}{}}$
32	O_2 (air)	58	$CH_3CCH_2\ +\ H$, $C_2H_5CHNH_2$, $(CH_3)_2NCH_2$,
33	SH, CH_2F		$C_2H_5NHCH_2$, C_2H_2S
34	H_2S		
35	Cl		$\overset{\text{O}}{\overset{\|}{}}$
36	HCl	59	$(CH_3)_2COH$, $CH_2OC_2H_5$, $COCH_3$,
39	C_3H_3		$CH_2C{=}O\ +\ H$, CH_3OCHCH_3,
40	$CH_2C{\equiv}N$		$\qquad\underset{NH_2}{\|}$
41	C_3H_5, $CH_2C{\equiv}N + H$, C_2H_2NH		
42	C_3H_6		CH_3CHCH_2OH
43	C_3H_7, $CH_3C{=}O$, C_2H_5N	60	$CH_2COOH + H$, CH_2ONO
44	$CH_2CH{=}O + H$, CH_3CHNH_2, CO_2, $NH_2C{=}O$, $(CH_3)_2N$		
45	CH_3CHOH, CH_2CH_2OH, CH_2OCH_3, COOH, $CH_3CHO + H$		

*All of these ions have a single positive charge.

MASS SPECTROMETRY

Common Fragment Lost

Molecular Ion Minus	Fragment Lost	Molecular Ion Minus	Fragment Lost
1	H	43	C_3H_7, $CH_3\overset{O}{\overset{\|}{C}}$, $CH_2{=}CHO$, $CH_3 + CH_2{=}CH_2$, $HCNO$
15	CH_3		
17	HO		
18	H_2O	44	$CH_2{=}CHOH$, CO_2, N_2O, $CONH_2$, $NHCH_2CH_3$
19	F		
20	HF	45	CH_3CHOH, CH_3CH_2O, CO_2H, $CH_3CH_2NH_2$
26	$CH{\equiv}CH$, $C{\equiv}N$		
27	$CH_2{=}CH$, $HC{\equiv}N$	46	$H_2O + CH_2{=}CH_2$, CH_3CH_2OH, NO_2
28	$CH_2{=}CH_2$, CO, (HCN + H)	47	CH_3S
29	CH_3CH_2, CHO	48	CH_3SH, SO, O_3
30	NH_2CH_2, CH_2O, NO	49	CH_2Cl
31	OCH_3, CH_2OH, CH_3NH_2	51	CHF_2
32	CH_3OH, S	52	C_4H_4, C_2N_2
33	HS, (CH_3 and H_2O)	53	C_4H_5
34	H_2S	54	$CH_2{=}CHCH{=}CH_2$
35	Cl	55	$CH_2{=}CHCHCH_3$
36	HCl, $2\ H_2O$	56	$CH_2{=}CHCH_2CH_3$, $CH_3CH{=}CHCH_3$
37	HCl + H	57	C_4H_9
38	C_3H_2, C_2N, F_2	58	NCS, NO + CO, CH_3COCH_3
39	C_3H_3, HC_2N	59	$CH_3\overset{O}{\overset{\|}{O}}C$, $CH_3\overset{O}{\overset{\|}{C}}NH_2$
40	$CH_3C{\equiv}CH$		
41	$CH_2{=}CHCH_2$	60	C_3H_7OH
42	$CH_2{=}CHCH_3$, $CH_2{=}C{=}O$, $\overset{\ \ \ \ CH_2}{CH_2{-}CH_2}$, NCO		

^{1}H NMR CHEMICAL SHIFTS

Legend: $|\ X = CH_3$ $\substack{\circ\\\circ}\ X = CH_2-$ $\substack{\bullet\\\bullet}\ X = \overset{|}{C}H-$

Scale (ppm): 5   4   3   2   1   0

Group	Approximate shift markers
RCH_2-X	$\bullet\bullet \approx 2.1$; $\substack{\circ\\\circ} \approx 1.3$; $\| \approx 0.9$
$RCH=CH-X$	$\bullet\bullet \approx 2.6$; $\substack{\circ\\\circ} \approx 2.1$
$RC\equiv C-X$	$\bullet\bullet \approx 2.7$; $\substack{\circ\\\circ} \approx 2.2$; $\| \approx 1.8$
C_6H_5-X (phenyl)	$\bullet\bullet \approx 2.9$; $\substack{\circ\\\circ} \approx 2.7$; $\| \approx 2.3$
$F-X$	$\bullet\bullet \approx 4.6$; $\substack{\circ\\\circ} \approx 4.4$; $\| \approx 4.3$
$Cl-X$	$\bullet\bullet \approx 4.0$; $\substack{\circ\\\circ} \approx 3.5$; $\| \approx 3.0$
$Br-X$	$\bullet\bullet \approx 4.1$; $\substack{\circ\\\circ} \approx 3.4$; $\| \approx 2.7$
$I-X$	$\bullet\bullet \approx 4.2$; $\substack{\circ\\\circ} \approx 3.2$; $\| \approx 2.2$
$HO-X$	$\bullet\bullet \approx 3.9$; $\substack{\circ\\\circ} \approx 3.6$
$RO-X$	$\bullet\bullet \approx 3.3$; $\substack{\circ\\\circ} \approx 3.2$; $\| \approx 3.0$
C_6H_5-O-X	$\bullet\bullet \approx 4.3$; $\substack{\circ\\\circ} \approx 3.9$; $\| \approx 3.7$
$R-\overset{O}{\underset{\|}{C}}-O-X$	$\bullet\bullet \approx 4.7$; $\substack{\circ\\\circ} \approx 4.1$; $\| \approx 3.5$
$C_6H_5-\overset{O}{\underset{\|}{C}}-O-X$	$\bullet\bullet \approx 5.0$; $\substack{\circ\\\circ} \approx 4.2$; $\| \approx 3.6$
$H-\overset{O}{\underset{\|}{C}}-X$	$\bullet\bullet \approx 2.8$; $\substack{\circ\\\circ} \approx 2.7$
$R-\overset{O}{\underset{\|}{C}}-X$	$\bullet\bullet \approx 2.7$; $\substack{\circ\\\circ} \approx 2.4$; $\| \approx 2.1$
$C_6H_5-\overset{O}{\underset{\|}{C}}-X$	$\bullet\bullet \approx 3.3$; $\substack{\circ\\\circ} \approx 3.0$
$HO-\overset{O}{\underset{\|}{C}}-X$	$\bullet\bullet \approx 2.8$; $\substack{\circ\\\circ} \approx 2.5$; $\| \approx 2.1$
$RO-\overset{O}{\underset{\|}{C}}-X$	$\bullet\bullet \approx 2.8$; $\substack{\circ\\\circ} \approx 2.5$
$R_2N-\overset{O}{\underset{\|}{C}}-X$	$\bullet\bullet \approx 2.6$; $\substack{\circ\\\circ} \approx 2.4$
$N\equiv C-X$	$\bullet\bullet \approx 3.2$; $\substack{\circ\\\circ} \approx 2.7$; $\| \approx 2.0$
H_2N-X	$\bullet\bullet \approx 3.3$; $\substack{\circ\\\circ} \approx 2.9$; $\| \approx 2.7$
R_2N-X	$\bullet\bullet \approx 3.2$; $\substack{\circ\\\circ} \approx 2.5$; $\| \approx 2.3$
$C_6H_5-\overset{|}{\underset{R}{N}}-X$	$\bullet\bullet \approx 3.7$; $\substack{\circ\\\circ} \approx 3.3$
$R_3\overset{+}{N}-X$	$\bullet\bullet \approx 3.7$; $\substack{\circ\\\circ} \approx 3.3$; $\| \approx 3.2$
$R-\overset{O}{\underset{\|}{C}}-NH-X$	$\bullet\bullet \approx 3.9$; $\substack{\circ\\\circ} \approx 3.4$; $\| \approx 3.0$
O_2N-X	$\bullet\bullet \approx 4.4$; $\substack{\circ\\\circ} \approx 4.3$; $\| \approx 4.2$

Characteristic infrared group frequencies (s = strong, m = medium, w = weak). (Courtesy of N. B. Colthup, Stamford Research Laboratories, American Cynamid Company, and the editor of the journal of the Optical Society.) Overtone bands are marked 2v.

Characteristic infrared group frequency correlation chart.

Axis labels (cm⁻¹): 4000, 3500, 3000, 2500, 2000, 1800, 1600, 1400, 1200, 1000, 800, 600, 400

Axis labels (μm): 2.50, 2.75, 3.00, 3.25, 3.50, 3.75, 4.00, 4.5, 5.0, 5.5, 6.0, 6.5, 7.0, 7.5, 8.0, 9.0, 10, 11, 12, 13, 14, 15, 20, 25

ALKANE GROUPS
- CH_3—C methyl
- CH_3—(C=O)
- —CH_2— methylene
- —CH_2—(C=O), —CH_2—(C≡N)
- ≡CH
- ethyl
- n-propyl
- iso-propyl
- tertiary butyl
- —CH_2—CH_2—

ALKENE
- vinyl —CH=CH_2
- (trans)
- (cis)
- C=CH_2
- C=CH—

ALKYNE
- —C≡C—H
- —C≡C—

AROMATIC
- monosubstituted benzene
- ortho disubstituted
- meta
- para
- vicinal trisubstituted
- unsymmetrical
- symmetrical
- α-naphthalenes
- β-naphthalenes
- (conj.)

ETHERS
- aliphatic ethers . . . CH_2—O—CH_2
- aromatic ethers . . . φ—O—CH_2

ALCOHOLS
- (free)
- (sharp)
- (bonded)
- (broad)
- primary alcohols . . . RCH_2—OH
- secondary . . . R_2CH—OH
- tertiary . . . R_3C—OH
- aromatic . . . φ—OH
- (unbonding lowers)

ACIDS
- carboxylic acids . . . COOH
- ionized carboxyl (salts, zwitterions, etc.) . . . C=O
- (absent in monomer)

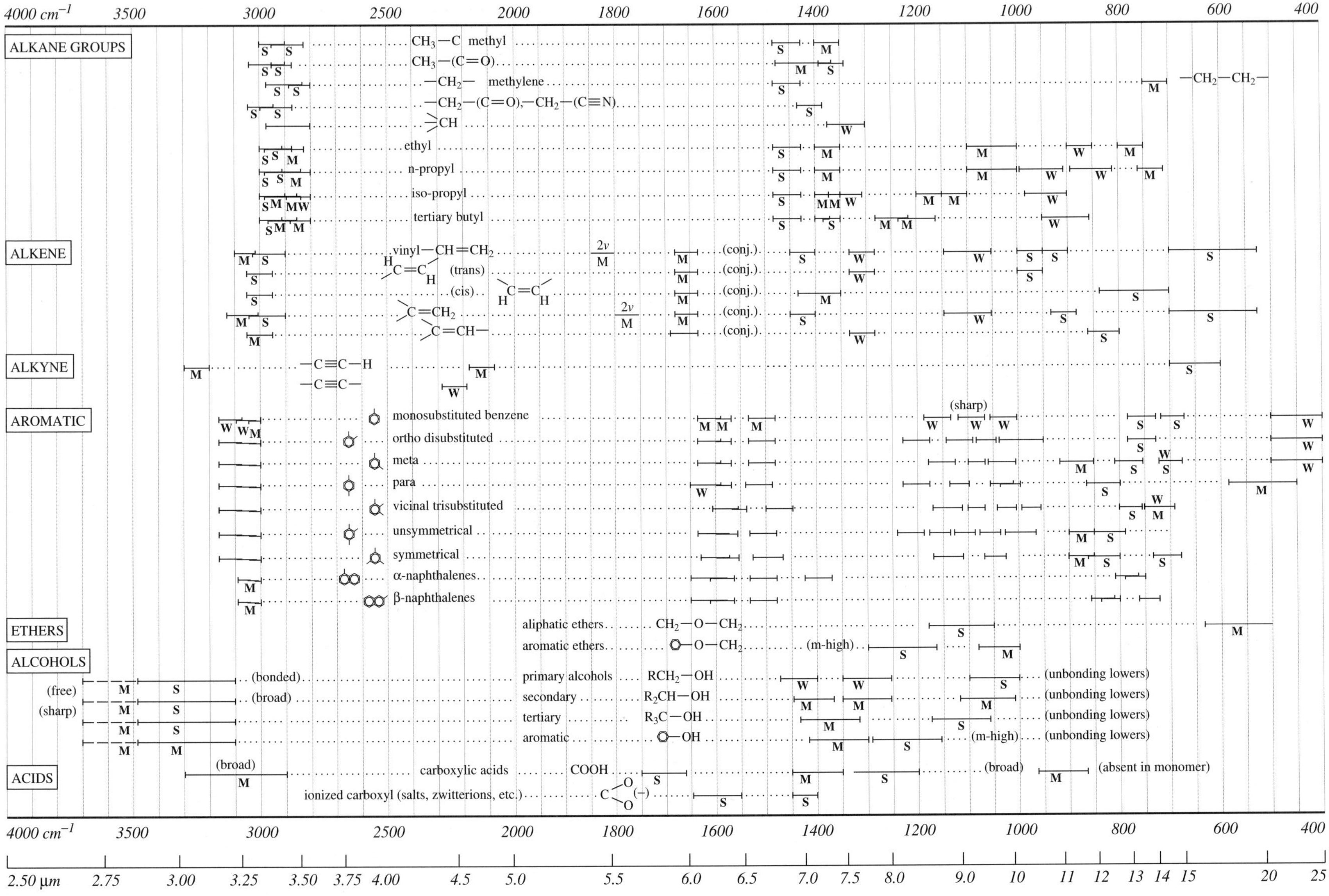

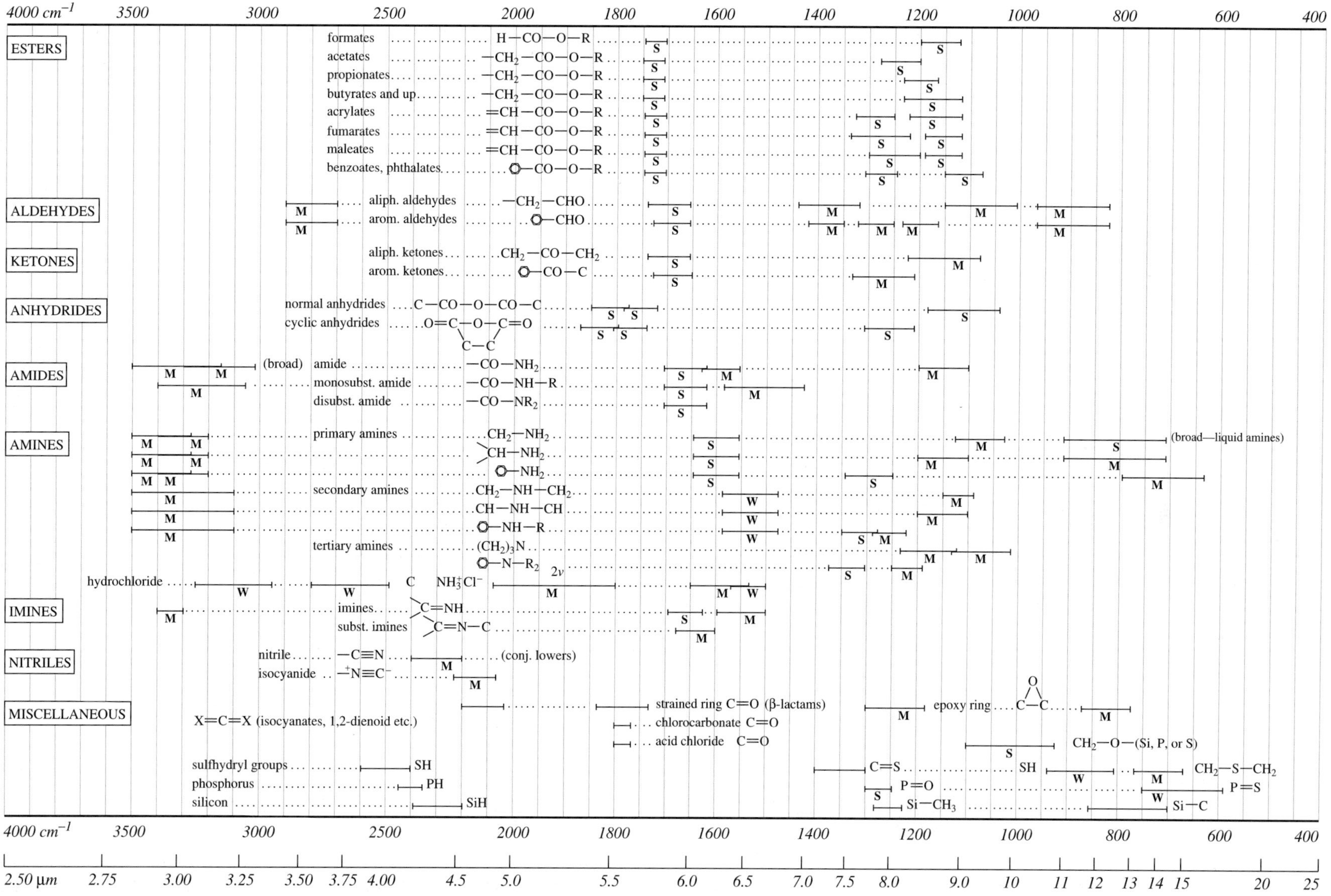
4000 cm⁻¹ 3500 3000 2500 2000 1800 1600 1400 1200 1000 800 600 400
ESTERS
formates ... H—CO—O—R
acetates ... —CH2—CO—O—R
propionates ... —CH2—CO—O—R
butyrates and up ... —CH2—CO—O—R
acrylates ... =CH—CO—O—R
fumarates ... =CH—CO—O—R
maleates ... =CH—CO—O—R
benzoates, phthalates ... —CO—O—R
ALDEHYDES
aliph. aldehydes ... —CH2—CHO
arom. aldehydes ... —CHO
KETONES
aliph. ketones ... CH2—CO—CH2
arom. ketones ... —CO—C
ANHYDRIDES
normal anhydrides ... C—CO—O—CO—C
cyclic anhydrides ... O=C—O—C=O
C—C
AMIDES
(broad) amide ... —CO—NH2
monosubst. amide ... —CO—NH—R
disubst. amide ... —CO—NR2
AMINES
primary amines ... CH2—NH2
CH—NH2
—NH2
secondary amines ... CH2—NH—CH2
CH—NH—CH
—NH—R
tertiary amines ... (CH2)3N
—N—R2
2ν
(broad—liquid amines)
hydrochloride
C NH3+Cl−
IMINES
imines ... C=NH
subst. imines ... C=N—C
NITRILES
nitrile ... —C≡N ... (conj. lowers)
isocyanide ... —+N≡C−
MISCELLANEOUS
X=C=X (isocyanates, 1,2-dienoid etc.)
strained ring C=O (β-lactams)
chlorocarbonate C=O
acid chloride C=O
epoxy ring ... C—C
O
CH2—O—(Si, P, or S)
sulfhydryl groups ... SH
phosphorus ... PH
silicon ... SiH
C=S ... SH
P=O
Si—CH3
CH2—S—CH2
P=S
Si—C
S M W
4000 cm⁻¹ 3500 3000 2500 2000 1800 1600 1400 1200 1000 800 600 400
2.50 μm 2.75 3.00 3.25 3.50 3.75 4.00 4.5 5.0 5.5 6.0 6.5 7.0 7.5 8.0 9.0 10 11 12 13 14 15 20 25

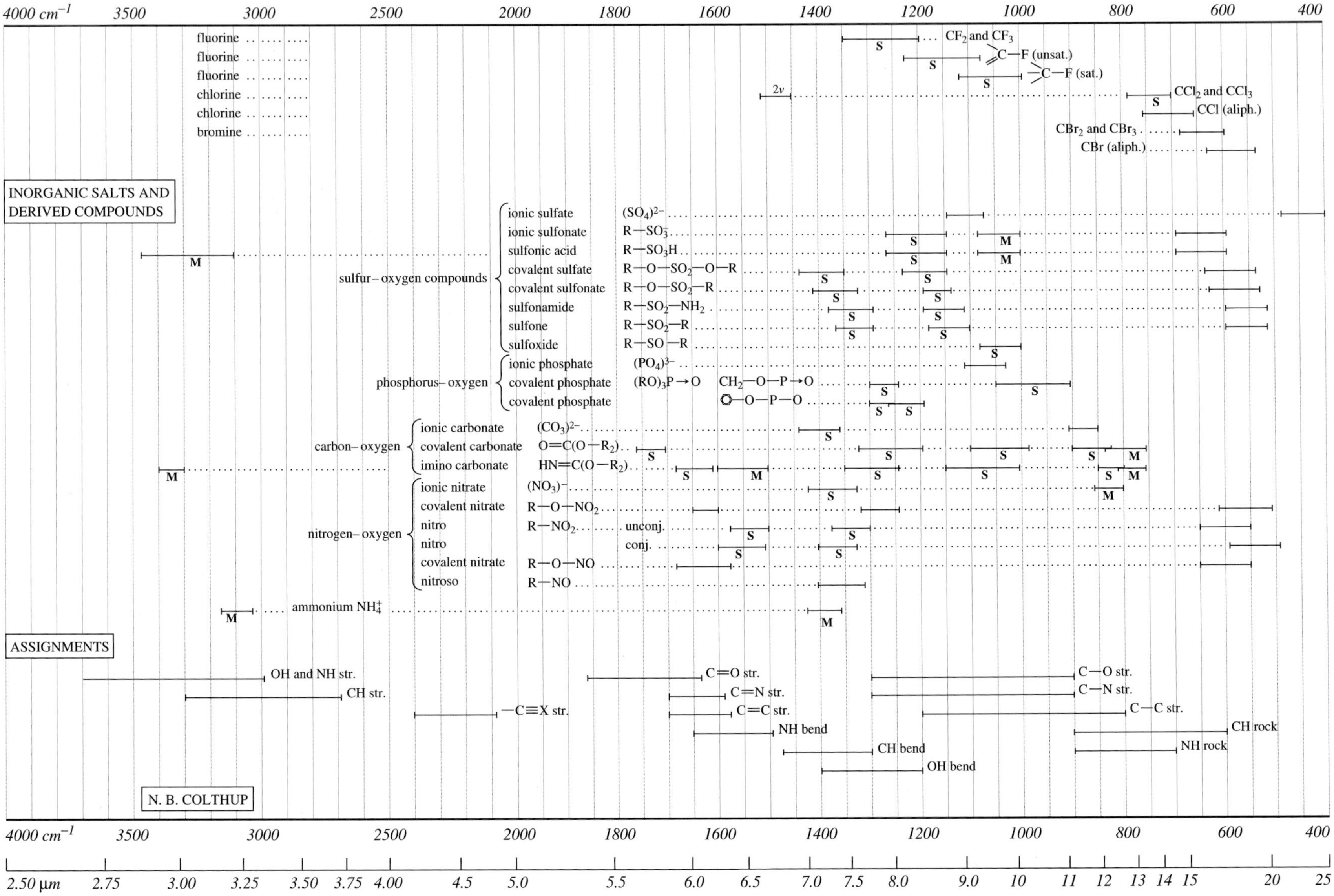

Appendix V (*Continued*)

INORGANIC SALTS AND
DERIVED COMPOUNDS

fluorine
fluorine
fluorine
chlorine
chlorine
bromine

CF₂ and CF₃
>C=C—F (unsat.)
—C—F (sat.)
CCl₂ and CCl₃
CCl (aliph.)
CBr₂ and CBr₃
CBr (aliph.)

2ν

sulfur–oxygen compounds

(SO₄)²⁻   ionic sulfate
R—SO₃⁻   ionic sulfonate
R—SO₃H.   sulfonic acid
R—O—SO₂—O—R   covalent sulfate
R—O—SO₂—R   covalent sulfonate
R—SO₂—NH₂.   sulfonamide
R—SO₂—R   sulfone
R—SO—R   sulfoxide

phosphorus–oxygen

(PO₄)³⁻   ionic phosphate
(RO)₃P→O   covalent phosphate
CH₂—O—P→O   covalent phosphate
O—O—P—O

carbon–oxygen

(CO₃)²⁻   ionic carbonate
O=C(O—R₂).   covalent carbonate
HN=C(O—R₂)   imino carbonate

nitrogen–oxygen

(NO₃)⁻   ionic nitrate
R—O—NO₂.   covalent nitrate
R—NO₂.   nitro   unconj.
                nitro   conj.
R—O—NO   covalent nitrate
R—NO   nitroso

ammonium NH₄⁺

ASSIGNMENTS

OH and NH str.
CH str.

—C≡X str.

C=O str.
C=N str.
C=C str.

NH bend
CH bend

C—O str.
C—N str.
C—C str.

OH bend

NH rock
CH rock

N. B. COLTHUP

S M S M

25   20   15 14 13 12 11 10 9.0 8.0 7.5 7.0 6.5 6.0 5.5 5.0 4.5 4.00 3.75 3.50 3.25 3.00 2.75 2.50 μm

400   600   800   1000 1200 1400 1600 1800 2000 2500 3000 3500 4000 cm⁻¹

APPENDIX VI

Glycolysis and the Krebs Cycle

Glycolysis: the series of enzyme-catalyzed reactions responsible for the conversion of 1 mol of D-glucose into 2 mol of pyruvate and/or lactate.

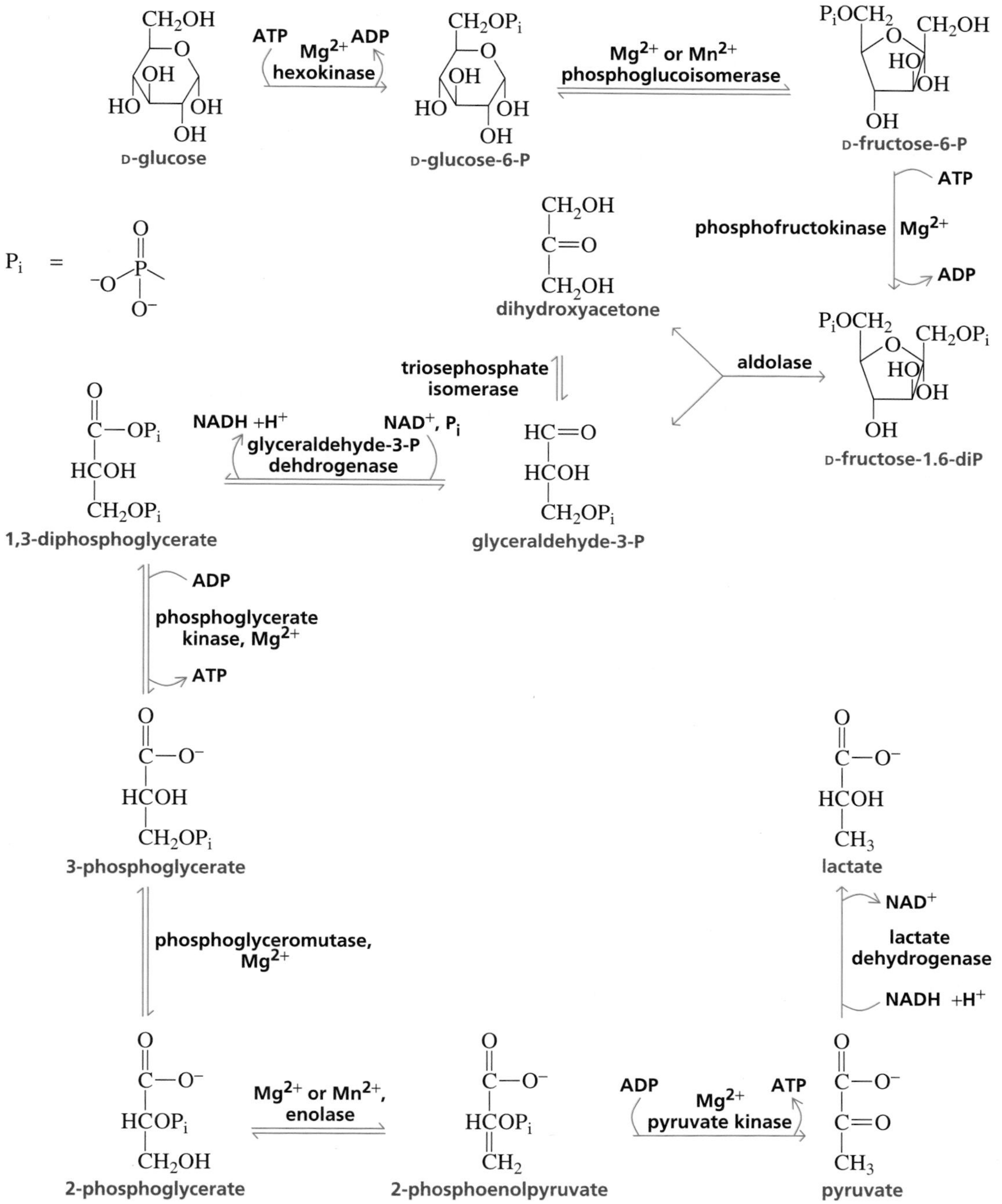

The Krebs cycle (also known as the **citric acid cycle**): The series of enzyme-catalyzed reactions responsible for the complete oxidation of pyruvate and acetyl-CoA to CO_2.

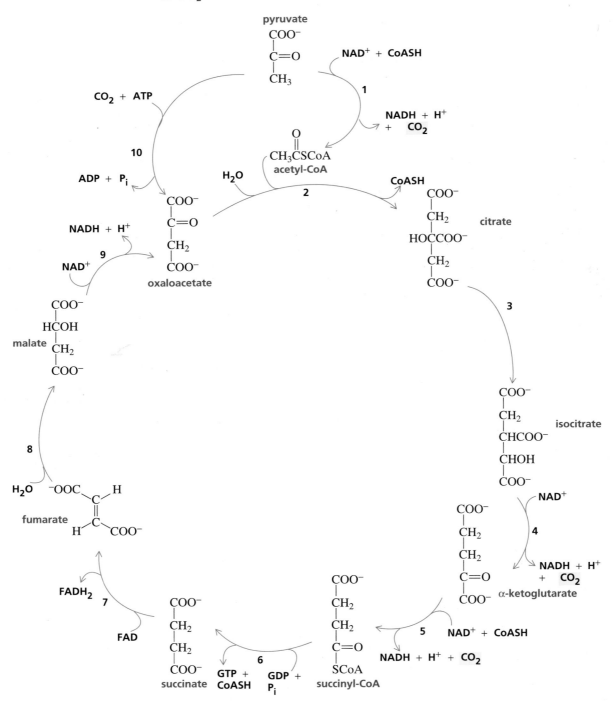

1	pyruvate dehydrogenase	5	α-ketogutarate dehydrogenase	8	fumarase
2	citrate synthase	6	succinyl-CoA synthetase	9	malate dehydrogenase
3	aconitase	7	succinate dehydrogenase	10	pyruvate carboxylase
4	isocitrate dehydrogenase				

ANSWERS TO SELECTED PROBLEMS

CHAPTER 1

1-1. a. 16, 17, and 18 **b.** 15.9994 amu **1-2.** $4s$
1-3. Cl $1s^2 2s^2 2p^6 3s^2 3p^5$ Br $1s^2 2s^2 2p^6 3s^2 3p^6 4s^2 3d^{10} 4p^5$
I $1s^2 2s^2 2p^6 3s^2 3p^6 4s^2 3d^{10} 4p^6 5s^2 4d^{10} 5p^5$ **1-4. b, c, e, f, g, h**

1-5.
 a. $\overset{\delta+}{H_3C}-\overset{\delta-}{Cl}$
 d. $\overset{\delta+}{H_3C}-\overset{\delta-}{OH}$

 b. $\overset{\delta-}{F}-\overset{\delta+}{Br}$
 e. $\overset{\delta-}{HO}-\overset{\delta+}{Br}$

 c. $\overset{\delta+}{H_3C}-\overset{\delta-}{NH_2}$
 f. $\overset{\delta-}{H_3C}-\overset{\delta+}{MgBr}$

1-7. H:C̈:O:C̈:H H:C̈:C̈:O:H (with H's around carbons)

1-9. a. relative lengths: $Br_2 > Cl_2$ relative strengths: $Cl_2 > Br_2$
b. relative lengths: HBr > HCl > HF relative strengths: HF > HCl > HBr
1-10. a. one **b.** two **1-11.** carbon–carbon bonds: sp^3–sp^3 overlap
carbon–hydrogen bonds: sp^3–s overlap **1-12.** greater than 104.5° and
less than 109.5° **1-13.** $^-CH_3$ and H_3O^+ **1-16. a, d, e, g, h**
1-17. a. (1) $^+NH_4$ (2) HCl (3) H_2O (4) H_3O^+ **b.** (1) $^-NH_2$
(2) Br^- (3) NO_3^- (4) HO^- **1-19. a.** 5.2 **b.** 3.4×10^{-3}
1-20. 8.16×10^{-8} **1-22. a.** CH_3COO^- **b.** $^-NH_2$ **c.** H_2O
1-24. a. 2.0×10^5 **b.** 3.2×10^{-7} **c.** 5.0×10^{-7} **1-25. a.** CH_3OCH_2OH
b. $CH_3CH_2CH_2\overset{+}{O}H_2$ **c.** $CH_3CH_2OCH_2OH$

d. $CH_3CH_2\overset{O}{\overset{\|}{C}}OH$

1-26. $CH_3\underset{F}{\overset{}{C}}HOH > CH_2\underset{F}{\overset{}{C}}H_2OH > CH_2\underset{Cl}{\overset{}{C}}H_2OH > CH_3CH_2OH$

1-28. a. F^- **b.** I^- **1-29. a.** oxygen **b.** H_2S **c.** CH_3SH **1-31. a.** 10.4
b. 2.7 **c.** 6.9 **d.** 7.3 **e.** 9.3 **1-32. a.** (1) neutral (2) neutral (3) 1/2
neutral and 1/2 charged (4) charged (5) charged (6) charged
(7) charged **b.** (1) charged (2) charged (3) charged (4) charged
(5) 1/2 charged and 1/2 neutral (6) neutral (7) neutral **1-33. a.** (1) 4.9
(2) 10.7 **b.** (1) 6.9 (2) 8.7 **1-34. a.** CH_3COO^- **b.** $CH_3CH_2\overset{+}{N}H_3$ **c.** H_2O
d. CH_3CH_2OH **e.** CH_3CH_2OH **f.** $\overset{+}{N}H_4$ **g.** $HC\equiv N$ **h.** NO_2^- **i.** NO_3^- **j.** Br^-

CHAPTER 2

2-1.
 a. $CH_3\underset{CH_3}{\overset{}{C}}HOH$
 d. $CH_3\underset{CH_3}{\overset{CH_3}{C}}CH_2Cl$

 b. $CH_3\underset{CH_3}{\overset{}{C}}HCH_2CH_2F$
 e. $CH_3\underset{CH_3}{\overset{CH_3}{C}}NH_2$

 c. $CH_3CH_2\underset{CH_3}{\overset{}{C}}HI$
 f. $CH_3\underset{CH_3}{\overset{}{C}}HCH_2CH_2CH_2CH_2CH_2Br$

2-5. a. 2,2,4-trimethylhexane **b.** 2,2-dimethylbutane **c.** 2,5-dimethylhep-
tane **d.** 3,3-diethylhexane **e.** 3,3-diethyl-4-methyl-5-propyloctane
f. 3-methyl-4-propylheptane **g.** 5-ethyl-4,4-dimethyloctane **h.** 4-isopropyl-
octane **2-7. a.** 1-ethyl-2-methylcyclopentane **b.** ethylcyclobutane
c. 4-ethyl-1,2-dimethylcyclohexane **d.** 3,6-dimethyldecane **e.** 2-cyclo-
propylpentane **f.** 1-ethyl-3-isobutylcyclohexane **g.** 5-isopropylnonane
h. 1-*sec*-butyl-4-isopropylcyclohexane **2-9. a.** *sec*-butyl chloride
2-chlorobutane secondary **b.** isoheptyl chloride 1-chloro-5-methyl-
hexane primary **c.** cyclohexyl bromide bromocyclohexane secondary
d. isopropyl fluoride 2-fluoropropane secondary
2-10. a. $CH_3CH_2CH_2CH_2CH_3$ pentane

b. $CH_3\underset{CH_3}{\overset{CH_3}{C}}CH_3$ dimethylpropane

c. $CH_3\underset{CH_3}{\overset{}{C}}HCH_2CH_3$ methylbutane

2-12. a. 1-hexanol primary **b.** 5-methyl-3-hexanol secondary
c. 5-chloro-2-methyl-2-pentanol tertiary **d.** 2-methyl-4-octanol
secondary **e.** 4-chloro-3-ethylcyclohexanol secondary

2-13. $CH_3\underset{OH}{\overset{CH_3}{C}}CH_2CH_2CH_3$ $CH_3CH_2\underset{OH}{\overset{CH_3}{C}}CH_2CH_3$ $CH_3\underset{OH}{\overset{CH_3}{C}}-\underset{CH_3}{\overset{}{C}}HCH_3$
 2-methyl-2-pentanol   3-methyl-3-pentanol   2,3-dimethyl-2-butanol

2-14. a. hexylamine 1-hexanamine **b.** butylpropylamine *N*-propyl-1-
butanamine **c.** *sec*-butylisobutylamine *N*-isobutyl-2-butanamine
d. diethylpropylamine *N,N*-diethyl-1-propanamine **e.** cyclohexylamine
cyclohexanamine **2-16. a.** 6-methyl-1-heptanamine isooctylamine
primary **b.** 3-methyl-*N*-propyl-1-butanamine isopentylpropylamine
secondary **c.** *N*-ethyl-*N*-methylethanamine diethylmethylamine tertiary
d. 2,5-dimethylcyclohexanamine no common name primary
2-17. a. 104.5° **b.** 107.3° **c.** 104.5° **d.** 109.5° **2-18. a.** 1, 4, 5
b. 1, 2, 4, 5, 6

2-21. HO—⟨OH⟩—OH > ⟨OH⟩—OH > ⟨OH⟩ > ⟨NH_2⟩ >

⟨——⟩ > ⟨—⟩

2-24. ethanol

2-25.
 a. $CH_3\underset{CH_3}{\overset{CH_3}{C}}CH_3$

 b. $CH_3CH_2CH_2CH_2CH_3$

2-28. a. 135° **b.** 140°

CHAPTER 3

3-2. a. 3 **b.** 4 **c.** 1 **d.** 3

3-3.
 a. $CH_3CH=CH_2$ △
 b. $CH_3C\equiv CH$ $CH_2=C=CH_2$ △

 c. $HC\equiv CCH_2CH_3$ ☐ △
 $CH_3C\equiv CCH_3$
 $CH_2=CHCH=CH_2$ △ △
 $CH_2=C=CHCH_3$

3-4.
 a. ⟨cyclopentene with CH₃/CH₃⟩
 c. ⟨cyclohexane with CH=CH_2⟩

 b. $CH_3C=\underset{CH_3}{\overset{CH_3}{C}}CH_2CH_2CH_2Br$
 d. $CH_2=CHCH_2OH$

3-5. a. 4-methyl-2-pentene **b.** 2-chloro-3,4-dimethyl-3-hexene **c.** 1-bro-
mocyclopentene **d.** 1-bromo-4-methyl-3-hexene **e.** 1,5-dimethylcyclo-
hexene **f.** 1-butoxy-1-propene **3-6. a.** 1 and 3

b. **(1)** cis   trans

(3) cis   trans

3-8. nucleophiles: H^- CH_3O^- $CH_3C\equiv CH$ NH_3 electrophiles: $AlCl_3$ $CH_3\overset{+}{C}HCH_3$ **3-14. a.** $A + B \rightleftharpoons C$ **b.** $A + B \rightleftharpoons C$
3-15. a. $\Delta G^\circ = -15$ $K_{eq} = 6.2 \times 10^{10}$ **b.** $\Delta G^\circ = -16$ $K_{eq} = 1.9 \times 10^8$
c. the greater the temperature, the more negative the ΔG° **d.** the greater the temperature, the smaller the K_{eq} **3-16. a.** -16 kcal/mol **b.** -34 kcal/mol
c. exothermic **d.** exergonic **3-17. a** and **c** **3-19.** by decreasing; by increasing **3-20. a.** 1×10^{-7} M s^{-1} **b.** decrease the rate **c.** none
3-21. second step

3-23. **a.** $CH_3CH_2\overset{CH_3}{\underset{+}{C}}CH_3 > CH_3CH_2\overset{+}{C}HCH_3 > CH_3\overset{+}{C}HCH_2CH_2$

b. $CH_3CH_2CH_2\overset{+}{C}H_2 > CH_3CHCH_2\overset{+}{\underset{Cl}{C}}H_2 > CH_3CH_2\overset{+}{\underset{Cl}{C}}HCH_2$

3-24. a. none **b.** ethyl cation **3-25.** methylpropene

3-26. **a.** $CH_3CH_2\underset{Br}{C}HCH_3$ **c.**

b. $CH_3CH_2\overset{CH_3}{\underset{Br}{C}}CH_3$ **d.** $CH_3CH_2\underset{Br}{C}HCH_3$

3-27. **a.** $CH_2=\overset{CH_3}{\underset{}{C}}CH_3$

b.

c.

d.

3-28.

3-29. **a.** $CH_3CH_2CH_2\underset{OH}{C}HCH_3$ **b.**

c. $CH_3CH_2CH_2CH_2\underset{OH}{C}HCH_3$ and $CH_3CH_2CH_2\underset{OH}{C}HCH_2CH_3$

d.

3-32. **a.** $+ CH_3OH \xrightarrow{H^+}$

b. $CH_3\overset{CH_3}{\underset{}{C}}=CH_2 + CH_3OH \xrightarrow{H^+} CH_3\overset{CH_3}{\underset{OCH_3}{C}}CH_3$

c. $CH_3CH=CHCH_3 + CH_3CH_2OH \xrightarrow{H^+} CH_3\underset{OCH_2CH_3}{C}HCH_2CH_3$

d. $CH_3CH=CHCH_3 + H_2O \xrightarrow{H^+} CH_3\underset{OH}{C}HCH_2CH_3$

3-38. $CH_3CH_2\underset{Cl}{C}HCH_2I$

3-39. **a.** $CH_2\underset{Br}{C}H\underset{Br}{C}H_2CH_3$ **c.** $CH_2\underset{Br}{C}H\underset{OCH_2CH_3}{C}H_2CH_3$

b. $CH_2\underset{Br}{C}H\underset{OH}{C}H_2CH_3$ **d.** $CH_2\underset{Br}{C}H\underset{OCH_3}{C}H_2CH_3$

3-41. 2/3 mole

3-42. **a.** $CH_3\underset{OH}{C}HCH\overset{CH_3}{\underset{}{}}CH_3$ **b.**

3-44. A

CHAPTER 4

4-1. **a.** $CH_3CH_2CH_2OH$ $CH_3\underset{CH_3}{C}HOH$ $CH_3CH_2OCH_3$

b. 7

4-2. **a.**

b.

c.

d. cis-1,3-dibromocyclobutane   trans-1,3-dibromocyclobutane

4-3. a, c, and **f**   **4-4. a, c,** and **f**

4-5.

4-6. a. P, F, J, L, G, R, Q, N, Z
b. T, M, O, A, U, V, C, D, E, H, I, X

4-7. **a.** **(1)**

(2)

(3)

b. **(1)**

(2)

(3) $(CH_3)_2CH-\overset{OH}{\underset{H}{C}}-CH_3$ $CH_3-\overset{OH}{\underset{H}{C}}-CH(CH_3)_2$

4-8. a. R **b.** R **c.** S **d.** S **4-11. a.** S **b.** R **c.** S **d.** S **4-12.** $+6.7°$
4-13. a. $-24°$ **b.** $0°$ **4-14. a.** $+79°$ **b.** $0°$ **c.** $-79°$ **4-15.** 58%
4-17. a. enantiomers **b.** identical **c.** diastereomers
4-20. a.

(structures)

b. (structures)

c. (structures)

d. $BrCH_2CH_2-\overset{Br}{\underset{H}{C}}-CH_2CH_3$ $CH_3CH_2-\overset{Br}{\underset{H}{C}}-CH_2CH_2Br$

4-21. (structures)

4-24. b, d, and **f** **4-28. a.** 4
b. $\cdots\overset{\overset{R}{|}}{C}-\overset{\overset{S}{|}}{C}\cdots$

4-31. R **4-32. a.** R **b.** R **c.** S **4-33. a.** enantiotopic **b.** homotopic
c. diastereotopic **d.** homotopic **e.** diastereotopic **f.** homotopic
4-36. a. trans **b.** cis **c.** trans **4-38. a.** 8 **b.** $2^8 = 256$ **4-39. a.** no **b.** no

4-40. a. (structures)
b. (structures)
c. (structures) **d.** (structures)
e. (structures)
f. (structures)

4-50. a. *(S)*-malate + *(R)*-malate **b.** *(S)*-malate + *(R)*-malate

CHAPTER 5

5-1. $C_{14}H_{20}$
5-3. a. 5-bromo-2-pentyne **b.** 6-bromo-2-chloro-4-octyne **c.** 3-ethyl-1-hexyne **d.** 1-methoxy-2-pentyne **5-5.** internal alkyne
5-6. If the less stable reactant has the more stable transition state, or if the difference in the stabilities of the reactants is greater than the difference in the stabilities of the transition states.

5-7. a. $CH_2{=}CCH_2CH_3$ with Br **d.** $HC{=}CCH_2CH_3$ Br Br
b. $CH_3CCH_2CH_3$ Br Br **e.** $ClCHCCH_2CH_3$ Cl Cl
c. $BrCH{=}CHCH_2CH_3$

5-8. a. (structure)
b. (structures)

5-9. $CH_3CH_2CCH_2CH_2CH_2CH_3$ and $CH_3CH_2CH_2CCH_2CH_2CH_3$

5-10. a. $CH_3{\equiv}CH$ **b.** $CH_3CH_2C{\equiv}CCH_2CH_3$
c. $HC{\equiv}C{-}$(cyclohexane)

5-11. a. $CH_2{=}CCH_3$ (OH)
b. $CH_3CH{=}CCH_2CH_2CH_3$ and $CH_3CH_2C{=}CHCH_2CH_3$ (OH)
c. $CH_2{=}C$(cyclohexane) and CH_3C(cyclohexane) (OH)

5-12. a. (1) $CH_3CH_2CCH_3$ **(2)** $CH_3CH_2CH_2CH$
b. (1) $CH_3CH_2CCH_3$ **(2)** $CH_3CH_2CCH_3$
c. (1) $CH_3CH_2CH_2CCH_3$ and $CH_3CH_2CCH_2CH_3$
(2) $CH_3CH_2CH_2CCH_3$ and $CH_3CH_2CCH_2CH_3$

5-13. acetylene

5-14. a. $CH_3CH_2CH_2C{\equiv}CH$ or $CH_3CH_2C{\equiv}CCH_3$ $\xrightarrow{H_2}{Pt}$
b. $CH_3C{\equiv}CCH_3$ $\xrightarrow{H_2}{\text{Lindlar's catalyst}}$
c. $CH_3CH_2C{\equiv}CCH_3$ $\xrightarrow{Na}{NH_3}$
d. $CH_3CH_2CH_2CH_2C{\equiv}CH$ $\xrightarrow{H_2}{\text{Lindlar's catalyst}}$ or $\xrightarrow{Na}{NH_3}$

5-15. 25
5-16. a. $CH_3\overset{+}{C}H_2$ **b.** $H_2C{=}\overset{+}{C}H$

CHAPTER 6

6-1. a. (1) 3 **(2)** 2 **b. (1)** 5 **(2)** 5
6-3. a. All are the same length. **b.** 2/3 of a negative charge
6-4. a. (2), (3), (5), and (6)
b.
(2) (resonance structures)
(3) (resonance structures)
(5) (resonance structures)

$CH_3CH=CH-CH=CH-\overset{+}{C}H_2$ $CH_3\overset{+}{C}H-CH=CH-CH=CH_2$

(6) $CH_3CH=CH-\overset{+}{C}H-CH=CH_2$

6-6. **a.** $CH_3N\overset{+}{H}CH_2$ **b.**

6-8. **a.** $CH_3CH=CHOH$ **b.** $CH_3CH=CHOH$
6-9. **a.** ethylamine **b.** ethoxide ion **c.** ethoxide ion
6-10.

6-12. **a.** 5 **b.** It will be aromatic if it is cyclic, planar, and has an inter-rupted cloud of π electrons. **6-13.** b, e, f **6-17.** Cyclopentadiene has the lower pK_a. **6-19.** **a.** Cyclopropane has the lower pK_a. **b.** 3-bromocyclo-propene **6-20.** a, c, g

CHAPTER 7

7-1 **a.** 1,5-cyclooctadiene **b.** 1,6-dimethyl-1,3-cyclohexadiene **c.** 1-hepten-4-yne **d.** 3-butyn-1-ol **e.** 4-methyl-1,4-hexadiene **f.** 1,3,5-hepta-triene **g.** 2,4-dimethyl-4-hexen-1-ol **h.** 5-vinyl-5-octen-1-yne **7-3.** ψ_3 has three nodes, ψ_4 has four nodes

7-4. **a.** **c.** $CH_2=CHCH_2CH_2CHCH_2Br$

b. $CH_2=CHCH_2CH_2\overset{CH_3}{\underset{Br}{C}}CH_3$ **d.** $HC\equiv CCH_2CHCH_2Cl$ with Cl

7-6. **a.** $CH_3CHCHCH=CHCH_3$ + $CH_3CHCH=CHCHCH_3$ (Cl Cl)

b. $CH_3CH_2\overset{CH_3}{\underset{CH_3}{C}}-C=CHCH_3$ + $CH_3CH_2\overset{CH_3}{\underset{CH_3}{C}}=\overset{Br}{C}CHCH_3$

c. $CH_3CHCH=CCH_2CH_3$ + $CH_3CH=CHCHCH_2CH_3$ (Br CH₃) (CH₃)

d.

7-7. **a.** 3 **b.** 4-bromo-2-pentene

7-9. **a.** **b.** **c.** **d.**

7-10.

7-11. a and d

7-13. **a.** + $\overset{CHC\equiv N}{\underset{CH_2}{}}$ **c.** +

b. + **d.** +

7-14. **a.** exo-2-chlorobicyclo[2.2.1]heptane

b. 3-methylbicyclo[4.4.0]decane **c.** bicyclo[2.2.2]octane
d. cis-6,7-dimethylbicyclo[3.2.0]heptane

CHAPTER 8

8-4. **a.** $CH_2=CHCH_3$ + $CH_3CH_2CH_3$

b. $CH_2=\overset{CH_3}{\underset{}{C}}CH_3$ + $CH_3\overset{CH_3}{\underset{}{C}}HCH_3$

8-5. **a.** 3 **b.** 3 **c.** 1 **d.** 5 **e.** 2 **f.** 1 **g.** 5 **h.** 4
8-6.
a. $CH_3CH_2CH_2CH_2CH_2Cl$ $CH_3CH_2CH_2\overset{Cl}{\underset{}{C}}HCH_3$ $CH_3CH_2\overset{Cl}{\underset{}{C}}HCH_2CH_3$

21% 53% 26%

b. $ClCH_2\overset{CH_3}{\underset{}{C}}HCH_2CH_2\overset{CH_3}{\underset{}{C}}HCH_3$ $CH_3\overset{CH_3}{\underset{Cl}{C}}CH_2CH_2\overset{CH_3}{\underset{}{C}}HCH_3$ $CH_3\overset{CH_3}{\underset{}{C}}HCHCH_2\overset{CH_3}{\underset{Cl}{C}}HCH_3$

32% 27% 41%

g. $ClCH_2\overset{CH_3}{\underset{}{C}}HCH_2CH_2CH_3$ $CH_3\overset{CH_3}{\underset{Cl}{C}}CH_2CH_2CH_3$ $CH_3\overset{CH_3}{\underset{Cl}{C}}HCHCH_2CH_3$

21% 17% 26%

$CH_3\overset{CH_3}{\underset{Cl}{C}}HCH_2CHCH_3$ $CH_3\overset{CH_3}{\underset{}{C}}HCH_2CH_2CH_2Cl$

26% 10%

8-7. 5 times easier **8-9.** 99.4% for bromination
8-10.
a. $CH_3CH_2CH_2CH_2CH_2Br$ $CH_3CH_2\overset{Br}{\underset{}{C}}HCH_2CH_3$ $CH_3CH_2CH_2\overset{Br}{\underset{}{C}}HCH_3$

1% 33% 66%

b. $BrCH_2\overset{CH_3}{\underset{}{C}}HCH_2CH_2\overset{CH_3}{\underset{}{C}}HCH_3$ $CH_3\overset{CH_3}{\underset{Br}{C}}CH_2CH_2\overset{CH_3}{\underset{}{C}}HCH_3$ $CH_3\overset{CH_3}{\underset{}{C}}HCHCH_2\overset{CH_3}{\underset{Br}{C}}HCH_3$

0.3% 90.4% 9.3%

g. $BrCH_2\overset{CH_3}{\underset{}{C}}HCH_2CH_2CH_3$ $CH_3\overset{CH_3}{\underset{Br}{C}}CH_2CH_2CH_3$ $CH_3\overset{CH_3}{\underset{Br}{C}}HCHCH_2CH_3$

0.3% 83.6% 8.5%

$CH_3\overset{CH_3}{\underset{Br}{C}}HCH_2CHCH_3$ $CH_3\overset{CH_3}{\underset{}{C}}HCH_2CH_2CH_2Br$

8.5% 0.2%

8-11. **a.** chlorination **b.** bromination **c.** no preference **8-12.** 1-bromo-2-butene

8-16. **a.** **(1)** + + **(3)**

(2) **(4)**

b. **(1)**

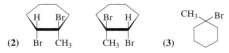

(2) **(3)**

(4)

CHAPTER 9

9-1. decrease

9-2. $CH_3CH_2CH_2CH_2CH_2Cl$ > $CH_3\overset{\underset{\displaystyle CH_3}{|}}{CH}CH_2CH_2Cl$ > $CH_3CH_2\overset{\underset{\displaystyle CH_3}{|}}{CH}CH_2Cl$ > $CH_3CH_2\overset{\overset{\displaystyle CH_3}{|}}{\underset{\underset{\displaystyle CH_3}{|}}{C}}Cl$

9-3. a. 2-butanol **b.** (S)-2-butanol **c.** (R)-3-hexanol **d.** 3-pentanol

9-4. a. RO^- **b.** RS^- **c.** RO^-

9-5. CH_3S^- > HO^- > ⟨benzene⟩$-O^-$ > $CH_3\overset{\overset{\displaystyle O}{\|}}{C}O^-$ > CH_3OH

9-7. a. $CH_3CH_2OCH_3$ **b.** $CH_3CH_2NH_2$ **c.** $CH_3CH_2\overset{+}{N}(CH_3)_3Br^-$

9-9. a. $CH_3CH_2Br + HO^-$

b. $CH_3\overset{\underset{\displaystyle CH_3}{|}}{CH}CH_2Br + HO^-$

c. $CH_3CH_2Cl + CH_3S^-$ **d.** $CH_3Br + I^-$ **e.** $BrCH_2CH_2CH_2CH_2NH_2$

f. Br⌒⌒⌒OH

9-10. $CH_3\overset{\overset{\displaystyle CH_3}{|}}{\underset{\underset{\displaystyle CH_3}{|}}{C}}Br$ > $CH_3\overset{\underset{\displaystyle CH_3}{|}}{CH}Br$ > $CH_3CH_2CH_2Br$ > CH_3Br

9-11. $CH_3CH_2\overset{\overset{\displaystyle CH_3}{|}}{\underset{\underset{\displaystyle Br}{|}}{C}}CH_2CH_3$ > $CH_3\overset{\underset{\displaystyle Br}{|}}{CH}CH_2CH_2CH_3$ > $CH_3\overset{\underset{\displaystyle Cl}{|}}{CH}CH_2CH_2CH_3$ > $ClCH_2CH_2CH_2CH_2CH_3$

9-12. a, b, c

9-13. $CH_2=CH\overset{\underset{\displaystyle O}{|}}{C}CH_3$ $CH_2CH=\overset{\underset{\displaystyle O}{|}}{C}CH_3$

$\overset{|}{C}=O$ $\overset{|}{C}=O$

$\overset{|}{CH_3}$ $\overset{|}{CH_3}$

kinetic product **thermodynamic product**

9-14. the trans isomer

9-16. a and b

9-17. a. $CH_3CH_2CH_2I$ **b.** CH_3OCH_2Cl

c. $CH_3CH_2\overset{\underset{\displaystyle CH_3}{|}}{CH}Br$ **f.** ⟨benzene⟩$-CH_2Br$

d. $CH_3CH_2\overset{\underset{\displaystyle CH_3}{|}}{CH}CH_2Br$ **g.** $CH_3CH=CHCH\overset{\underset{\displaystyle Br}{|}}{}CH_3$

e. ⟨benzene⟩$-CH_2CH_2Br$

9-18. a. $CH_3CH_2CH_2I$ **b.** CH_3OCH_2Cl **c.** equally reactive

d. $CH_3CH_2CH_2\overset{\underset{\displaystyle CH_3}{|}}{CH}Br$ **f.** ⟨benzene⟩$-CH_2Br$

e. ⟨benzene⟩$-CH_2\overset{\underset{\displaystyle Br}{|}}{CH}CH_3$ **g.** $CH_3CH=CHCH\overset{\underset{\displaystyle Br}{|}}{}CH_3$

9-22. a. aprotic **b.** aprotic **c.** protic **d.** aprotic

9-23. a. decrease **b.** decrease **c.** increase

9-24. a. $CH_3Br + HO^- \longrightarrow CH_3OH + Br^-$

b. $CH_3I + HO^- \longrightarrow CH_3OH + I^-$

c. $CH_3Br + NH_3 \longrightarrow CH_3\overset{+}{N}H_3 + Br^-$

d. $CH_3Br + HO^- \xrightarrow{\text{DMSO}} CH_3OH + Br^-$

e. $CH_3Br + NH_3 \xrightarrow{\text{Et}_2\text{O}} CH_3\overset{+}{N}H_3 + Br^-$

9-26. dimethylsulfoxide **9-27.** HO^- in water.

CHAPTER 10

10-2. a. $CH_3CH=CHCH_3$ **b.** $CH_2=CHCH_2CH_3$

c. $CH_3\overset{\overset{\displaystyle CH_3}{|}}{C}=CHCH_2CH_3$ **e.** ⟨cyclohexene⟩

d. $CH_3\overset{\underset{\displaystyle CH_3}{|}}{CH}CH=CHCH_3$ **f.** $CH_3CH=CHCH=CH_2$

10-3. a. $CH_3CH_2\overset{\underset{\displaystyle Br}{|}}{CH}CH_2CH_3$ **c.** $CH_3\overset{\underset{\displaystyle Br}{|}}{CH}\overset{\overset{\displaystyle CH_3}{|}}{CH}CH_2CH_3$

b. ⟨benzene⟩$-CH_2\overset{\underset{\displaystyle Br}{|}}{CH}CH_2CH_3$ **d.** ⟨cycloheptene with Br⟩

10-4. $CH_3\overset{\overset{\displaystyle CH_3}{|}}{C}=\overset{\underset{\displaystyle CH_3}{|}}{C}CH_2CH_3$ > $CH_3\overset{\overset{\displaystyle CH_3}{|}}{CH}\overset{\underset{\displaystyle CH_3}{|}}{C}=CHCH_3$ > $CH_3\overset{\underset{\displaystyle CH_3}{|}}{CH}\overset{\overset{\displaystyle CH_2}{\|}}{C}CH_2CH_3$

10-7. a. E2 $CH_3CH=CHCH_3$ **b. E1** $CH_3CH=CHCH_3$

c. E1 $CH_3\overset{\overset{\displaystyle CH_3}{|}}{C}=CH_2$ **e. E1** $CH_3\overset{\overset{\displaystyle CH_3}{|}}{C}=CHCH_3$

d. E2 $CH_3\overset{\overset{\displaystyle CH_3}{|}}{C}=CH_2$ **f. E2** $CH_3\overset{\underset{\displaystyle CH_3}{|}}{CH}\overset{\overset{\displaystyle CH_3}{|}}{C}=CH_2$

10-8. a. 96% **b.** 1.2%

10-10. a. ⟨structure: C=C with CH₃ and H, two phenyl groups⟩ **b.** ⟨structure: C=C with two phenyl, H and CH₃⟩

10-11. a. ⟨CH_3CH_2 / CH_3 C=C H / H⟩ + ⟨CH_3CH_2 / H C=C H / CH_3⟩

b. $CH_3CH_2CH=\overset{\overset{\displaystyle CH_3}{|}}{C}CH_3$

c. ⟨CH_3CH_2 / H_3C C=C CH_2CH_3 / H⟩ + ⟨CH_3CH_2 / H_3C C=C H / CH_2CH_3⟩

⟨$CH_3CH_2CH_2$ / H_3C C=C CH_3 / H⟩ + ⟨$CH_3CH_2CH_2$ / H_3C C=C H / CH_3⟩

d. ⟨1-methylcyclohexene⟩

10-14. ~1 **10-15.** It will increase. **10-16. a. (1)** no reaction **(2)** no reaction **(3)** substitution and elimination **(4)** substitution and elimination **b. (1)** primarily substitution **(2)** substitution and elimination **(3)** substitution and elimination **(4)** elimination

10-18. a. $CH_3CH{=}CH_2$ **b.** $CH_3CH_2CH{=}CH_2$

10-19.

a. $CH_3CH_2\underset{\underset{CH_3}{|}}{C}HOH \xrightarrow{Na} CH_3CH_2\underset{\underset{CH_3}{|}}{C}HO^- \xrightarrow{CH_3CH_2CH_2Br} CH_3CH_2\underset{\underset{CH_3}{|}}{C}HOCH_2CH_2CH_3$

b. $CH_3\underset{\underset{CH_3}{|}}{\overset{\overset{CH_3}{|}}{C}}OH \xrightarrow{Na} CH_3\underset{\underset{CH_3}{|}}{\overset{\overset{CH_3}{|}}{C}}O^- \xrightarrow{\text{—Br}} \text{cyclohexyl—}O\underset{\underset{CH_3}{|}}{\overset{\overset{CH_3}{|}}{C}}CH_3$

c. or

$CH_3CH_2CH_2CH_2CH_2OH \xrightarrow{Na} CH_3CH_2CH_2CH_2CH_2O^- \xrightarrow{CH_3CH_2Br} CH_3CH_2CH_2CH_2CH_2OCH_2CH_3$

$CH_3CH_2OH \xrightarrow{Na} CH_3CH_2O^- \xrightarrow{CH_3CH_2CH_2CH_2Br}$

d. $\text{—OH} \xrightarrow{Na} \text{—O}^- \xrightarrow{\text{—CH}_2Br} \text{—CH}_2O\text{—}$

10-20. $CH_3\underset{\underset{CH_3}{|}}{\overset{\overset{CH_3}{|}}{C}}{-}OH + CH_3\underset{\underset{CH_3}{|}}{\overset{\overset{CH_3}{|}}{C}}{-}OCH_2CH_3 + CH_3\underset{\underset{CH_3}{|}}{\overset{\overset{CH_3}{|}}{C}}{=}CH_2$

CHAPTER 11

11-2. **a.** $CH_3CH_2\underset{\underset{Br}{|}}{C}HCH_3$ **c.** $CH_3\underset{\underset{Br}{|}}{\overset{\overset{CH_3}{|}}{C}}{-}\underset{\underset{CH_3}{|}}{C}HCH_3$

b. cyclopentane with CH_3 and Cl

d. cyclohexane with CH_3, CH_2CH_3, Cl

11-3. **a.** $CH_3{-}\overset{\overset{D}{|}}{\underset{\underset{C{\equiv}N}{}}{C}}{\cdots}H$ **b.** $CH_3{-}\overset{\overset{D}{|}}{\underset{\underset{H}{}}{C}}{\cdots}C{\equiv}N$

11-6. cyclopentane with Br and CH_3

11-9. **a.** $\underset{CH_3CH_2}{\overset{CH_3}{}}C{=}C\underset{CH_3}{\overset{CH_3}{}}$ **d.** cyclohexene

b. $\underset{CH_3CH_2CH_2}{\overset{CH_3}{}}C{=}C\underset{CH_3}{\overset{CH_3}{}}$ **e.** cyclohexene with CH_3

c. cyclopentane${-}CH_2CH_3$ or cyclopentane${=}CHCH_3$

f. $\underset{H}{\overset{CH_3CH_2}{}}C{=}C\underset{H}{\overset{CH_3}{}} + \underset{H}{\overset{CH_3CH_2}{}}C{=}C\underset{CH_3}{\overset{H}{}}$

11-11. **a.** $CH_3CH_2\underset{\underset{CH_3}{|}}{C}HCH_2OH + CH_3CH_2I$

b. $CH_3CH_2CH_2OH + CH_3CH_2\underset{\underset{I}{|}}{\overset{\overset{CH_3}{|}}{C}}CH_3$

c. $CH_3CH_2\overset{\overset{O}{\|}}{C}H + CH_3CH_2I$

d. $\text{—CH}_2I + HO\text{—}$

11-12. **a.** epoxide${-}CH_2CH_2CH_3$ **b.** bicyclic epoxide

c. $\underset{H_3C}{\overset{O}{\diagup\diagdown}}\underset{CH_3}{\overset{}{}}$ (with H_3C, CH_3) **d.** $\underset{H_3C}{\overset{O}{\diagup\diagdown}}\underset{CH_2CH_3}{\overset{}{}}$ (with H_3C)

11-13. **a.** $HOCH_2\underset{\underset{OCH_3}{|}}{\overset{\overset{CH_3}{|}}{C}}CH_3$ **c.** $CH_3\underset{\underset{H_3C\ OCH_3}{}}{\overset{\overset{HO\ CH_3}{}}{C}}{-}CCH_3$

b. $CH_3OCH_2\underset{\underset{CH_3}{|}}{\overset{\overset{OH}{|}}{C}}CH_3$ **d.** $CH_3\underset{\underset{H_3C\ OCH_3}{}}{\overset{\overset{HO\ CH_3}{}}{C}}{-}CCH_3$

11-14. noncyclic ether

11-16. benzene with OH and CH_3 or   benzene with D, OH, CH_3

11-17. The products are the same. **11-20. a.** $CH_3CH_2CH_2CH_2CH_2OH$

b. benzene${-}CH_2CH_2CH_2OH$ **c.** cyclohexane${-}CH_2CH_2OH$

11-22. They all occur. **11-23.** $(CH_3)_4Si$ **11-27.** $CH_3CH_2CH_2OH$

11-28. $CH_3CH{=}CH_2 + \underset{\underset{CH_3}{|}}{N}CH_2CH_2CH_3$ (with CH_3)

11-29. a. $CH_3CH{=}CH_2 + \underset{\underset{CH_3}{|}}{N}CH_3$ (with CH_3) **c.** cyclohexane$={CH_2} + N(CH_3)_3$

b. $CH_2{=}CHCH\underset{\underset{CH_3}{|}}{\overset{\overset{CH_3}{|}}{C}}H_2CH_2NCH_3$ (with CH_3) **d.** $CH_3NCH_2\underset{\underset{CH_3}{|}}{C}HCH_2CH{=}CH_2$ (with CH_3)

11-32. "carbanion-like"

11-33. a. $CH_3\underset{\underset{OH}{|}}{N} + CH_2{=}CHCH_3$ (with CH_3) **c.** $CH_2{=}CH_2 + \underset{\underset{OH\ CH_3}{}}{N}CH_2\underset{\underset{CH_3}{|}}{C}H$ (with CH_3)

$CH_3\underset{\underset{OH}{|}}{N} + CH_2{=}CHCH_3$ (with CH_3)

b. benzene with $\overset{\overset{OH}{|}}{N}$

d. $CH_2{=}CHCH_2CH_2CH_2CH_2\underset{\underset{OH}{|}}{N}CH_3$

CHAPTER 12

12-2. $m/z = 57$ **12-5.** 8 **12-6.** C_6H_{14} **12-8. a.** 2-methoxy-2-methyl-propane **b.** 2-methoxybutane **c.** 1-methoxybutane

12-9. $CH_2{=}\overset{+}{O}H$

12-10. a. 2-pentanone **b.** 3-pentanone **12-12. a.** 3-pentanol **b.** 2-pentanol **12-13. a.** $2000\ cm^{-1}$ **b.** $8\ \mu m$ **c.** $2\ \mu m$ **12-14. a.** $2500\ cm^{-1}$ **b.** $50\ \mu m$ **12-15. a.** (1) $C{=}O$ (2) C-H stretch (3) $\equiv$C-H
b. (1) C-O (2) C-C **12-16. a.** carbon–nitrogen stretch of an amide **b.** carbon–oxygen stretch of phenol **c.** carbon–oxygen double bond stretch of a ketone **d.** carbon–oxygen stretch

12-17. $H{-}\overset{\overset{O}{\|}}{C}{-}H > CH_3{-}\overset{\overset{O}{\|}}{C}{-}H > CH_3{-}\overset{\overset{O}{\|}}{C}{-}CH_3$

12-18. an ether **12-19.** ethanol dissolved in carbon disulfide **12-20.** sp^3 hybridized carbon
12-23. CO_2, 2-butyne, H_2, Cl_2, ethene **12-24. a.** $1360\ cm^{-1}$, $2840\ cm^{-1}$, $5820\ cm^{-1}$ **b.** $2100\ cm^{-1}$, $3590\ cm^{-1}$, $4330\ cm^{-1}$
12.25 *meta*-dibromobenzene

CHAPTER 13

13-1. **a.** 8.46 T **b.** 11.75 T **c.** The greater the operating frequency of the instrument, the more powerful the magnet required to operate it.

13-2. 43 Mhz **13-3.** **a.** 4 **b.** 4 **c.** 3 **d.** 3 **e.** 2 **f.** 3 **g.** 1 **h.** 5 **i.** 3 **j.** 3 **k.** 3 **l.** 4 **m.** 2 **n.** 1 **o.** 3

13-6. **a.** $CH_3CH_2CH_2Cl$ — most (left), least (right)

c. $CH_3CHCHBr$ with Br, Br — most (left), least (right)

b. $CH_3CH_2COCH_3$ (with O) — most (left), least (right)

13-7. **a.** 2 ppm **b.** 2 ppm **c.** 200 Hz **13-8.** **a.** 90 Hz **b.** 1.5 ppm

13-12. CH_3—⟨benzene⟩—CH_3

13-13. **a.** $CH_3CH_2CHCH_3$ with Cl

b. $CH_3CH_2CHCH_3$ with Cl

c. CH_3CH_2CHCl

d. $CH_3CH_2CH=CH_2$

e. CH_3CH_2CH (with O)

f. CH_3CHOCH_3 with CH_3

g. $CH_3CHCCH_2CH_3$ (with O, CH_3)

h. $CH_3CHCHBr$ with Br, Br

13-14. **a.** $\overset{a}{CH_3}\overset{b}{CH_2}\overset{c}{CH_2}Cl$

b. $\overset{a}{CH_3}\overset{d}{CH}\overset{b}{CHCH_3}$ with CH_3 (a), $\overset{c}{Cl}$

c. $\overset{a}{CH_3}\overset{b}{CH_2}\overset{c}{CH_2}\overset{d}{COCH_3}$ (with O)

d. $\overset{a}{CH_3}\overset{b}{CH}\overset{c}{CHBr}$ with Br, Br

e. $\overset{a}{CH_3}\overset{b}{CH}\overset{d}{CH_2}\overset{c}{OCH_3}$ with CH_3; **3 signals**

f. $\overset{a}{CH_3}\overset{b}{CH_2}\overset{c}{CH_2}$ (with O)

g. $\overset{a}{CH_3}\overset{b}{CH_2}\overset{d}{CH_2}\overset{c}{CCH_3}$ (with O)

h. $ClCH_2CH_2CH_2Cl$

i. $\overset{a}{CH_3}\overset{b}{CH_2}\overset{c}{CH}\overset{d}{CH_2}CH_3$ with $\overset{c}{OCH_3}$

j. $\overset{a}{CH_3}\overset{c}{CH_2}\overset{d}{CH_2}\overset{}{OCHCH_3}$ with $\overset{e}{CH_3}$, $\overset{b}{}$; **3 signals**

13-19. **a.** $ICH_2CH_2CH_2Cl$ — triplet, triplet, multiplet; **2 signals**

c. $ICH_2CH_2CHBr_2$ — triplet, triplet, multiplet

b. $ClCH_2CH_2CH_2Cl$ — triplet, quintet

13-24. **a.** $CH_3OCH_2CH_2OCH_3$

b. $CH_3CCH_2CH_2CCH_3$ (with two O)

$CH_3OCH_2C≡CCH_2OCH_3$

⟨1,4-dioxane ring with two CH_3⟩

c. ⟨benzene⟩—$C(CH_3)_3$ (with CH_3, CH_3, CH_3)

13-28. CH_3O—⟨benzene⟩—CH_3

13-29. pure ethanol **13-32.** **a.** (1) 3 (2) 3 (3) 2 (4) 3 (5) 4 (6) 3 (7) 2 (8) 3 (9) 4 (10) 3

b. (1) $CH_3CH_2CH_2Br$

(2) $CH_3CH_2OCH_3$

(3) CH_3CHCH_3 with Br

(7) $\underset{H}{\overset{H}{C}}=\underset{H}{\overset{Br}{C}}$

(8) CH_3CHCH (with O) with CH_3

(4) CH_3COCH_3 with CH_3, CH_3

(5) $CH_3CH_2COCH_3$ (with O)

(6) $H_3C-C=C\overset{H}{\underset{H}{}}$ with H_3C

(9) ⟨benzene with Cl and H's⟩

(10) $CH_3CCH_2CH_2CCH_3$ (with two O)

13-35. $CH_3CH_2CH_2CH_2CH_2CCH_2CH_2CH_2CH_2CH_3$ (with O)

13-37. **a.** UV light **b.** visible light

13-38. **a.** ⟨cyclohexenone⟩ > ⟨cyclohexenone with CH_3⟩ > ⟨cyclohexenone with CH_3⟩

b. ⟨Ph—CH=CH—Ph⟩ > ⟨biphenyl⟩ > ⟨Ph—CH=CH₂⟩ > ⟨benzene⟩

13-39. 4.1×10^{-5} M **13-41.** **a.** 247 nm **b.** 232 nm **c.** 232 nm **d.** 303 nm

CHAPTER 14

14-2. **a.** ⟨benzene with two Cl (1,3)⟩

b. ⟨benzene with OH and Br (para)⟩

c. ⟨benzene with NH_2 and NO_2 (ortho)⟩

d. ⟨benzene with NO_2, Br, I⟩

e. $CH_3CHCH_2CH_2CH_2CH_3$ attached to ⟨benzene⟩

f. $CH_3CH_2CHCH_2CH_3$ with CH_2 attached to ⟨benzene⟩

g. ⟨benzene with CH_3 and Cl (meta)⟩

h. ⟨benzene with HC=O, NO_2, O_2N⟩

i. (o-xylene: CH₃, CH₃) **j.** (3-methylphenol: CH₃, OH)

14-3. **a.** 1,3,5-tribromobenzene **b.** 3-nitrophenol **c.** *para*-bromotoluene **d.** 1,2-dichlorobenzene

14-8. **a.** (ethylbenzene CH₂CH₃) **d.** (CH₃CH₂CH₃ sec)
b. (CH₃CHCH₃) **e.** (CH₃CCH₂C)
c. (CH₃CH₂CHCH₃) **f.** (CH₂CH=CH₂)

14-10. **a.** (COOH) **c.** (CH₂OCH₃)
b. (m-COOH, COOH) **d.** (CH₂CH₂NH₂)

14-12. **a.** donates electrons by resonance and withdraws electrons inductively **b.** donates electrons inductively **c.** withdraws electrons by resonance and withdraws electrons inductively **d.** donates electrons by resonance and withdraws electrons inductively **e.** donates electrons by resonance and withdraws electrons inductively **f.** withdraws electrons inductively **14-13.** **a.** phenol > toluene > benzene > bromobenzene > nitrobenzene **b.** toluene > chloromethylbenzene > dichloromethylbenzene > difluoromethylbenzene

14-17. **a.** (CH₂CH₂CH₃ para-NO₂ + CH₂CH₂CH₃ ortho-NO₂)
b. (SO₃H, m-NO₂)
c. (Br para-NO₂ + Br ortho-NO₂)
e. (cyclohexyl para-NO₂ + cyclohexyl ortho-NO₂)
f. (C≡N, m-NO₂)
d. (CH, O, m-NO₂)

14-18. They are all meta directors.

14-20. **a.** ClCH₂COH (O) **b.** O₂NCH₂COH (O) **c.** H₃N⁺CH₂COH (O)

d. HOCCH₂COH (O O) **f.** (COOH benzene) **h.** (COOH para-Cl)

e. CH₃COH (O) **g.** FCH₂COH (O)

14-22. **a.** no reaction **c.** no reaction
b. (NH₂, 2,4,6-tribromo: Br, Br, Br)
d. (p-xylene: CH₃, CH₃ + o-xylene: CH₃, CH₃)

14-27. **a.** no **b.** phenol **14-34.** **a.** 1-chloro-2,4-dinitrobenzene > p-chloronitrobenzene > chlorobenzene **b.** chlorobenzene > p-chloronitrobenzene > 1-chloro-2,4-dinitrobenzene **14-35.** Attack of the nucleophile on the aromatic ring is the rate-determining step of the reaction.

14-41. **a.** (naphthalene: OCH₃, Cl)
b. (naphthalene Cl, Br + naphthalene Br, Cl) **c.** (naphthalene Cl, CH₃)
d. (naphthalene Cl, NO₂ + naphthalene NO₂, Cl)

CHAPTER 15

15-2. **a.** pentanoyl chloride   valeryl chloride **b.** γ-butyrolactam **c.** ethanoic propanoic anhydride   acetic propionic anhydride **d.** isobutyl butanoate   isobutyl butyrate **e.** cyclopentanecarboxylic acid **f.** *N,N*-dimethylhexanamide **g.** butanenitrile   propyl cyanide **h.** potassium butanoate **15-4.** The shortest bond has the highest frequency

CH₃—C(O)—O—CH₃   CH₃—C(O)—O—CH₃

a. 1 = longest   **b.** 1 = highest frequency
3 = shortest   3 = lowest frequency

15-5. the carbonyl oxygen **15-6.** **a.** no reaction
b. CH₃—C(O)—O—C(O)—CH₃
c. no reaction **15-7.** true

15-11. **a.** CH₃CH₂CH₂OH **b.** CH₃CH₂NH₂ **c.** (CH₃)₂NH
d. CH₃CO⁻ (O) **e.** HO—⟨⟩—NO₂ **f.** H₂O

15-16. **(1)** The carbonyl group of an ester is not very reactive. **(2)** Water is not a very good nucleophile **(3)** ⁻OCH₃ is a strong base and a poor leaving group. **15-20.** **a.** ethanol **b.** propanoic acid

15-23.

CH₃CNH—⟨⟩—NO₂ (O) > CH₃CNH—⟨NO₂⟩ (O) > CH₃CNH—⟨⟩ (O) > CH₃CNH—⟨cyclohexyl⟩ (O)

15-24. **a.** pentyl bromide **b.** benzyl bromide **c.** isohexyl bromide

15-26. **a.** CH₃CH₂CH₂Br

b. CH_3CHCH_2Br **c.** ⬡—Br   **d.** $BrCH_2CH_2Br$
       |
     CH_3

15-27. $C_{14}H_{28}O_2$

15-29. a. $(CH_3CH_2)_2NCN(CH_2CH_3)_2$ **d.** no reaction
             $\overset{O}{\|}$

b. $CH_3\overset{O}{\overset{\|}{C}}CH_2\overset{O}{\overset{\|}{C}}CH_3$ **e.** $HO\overset{O}{\overset{\|}{C}}OH \rightleftharpoons CO_2 + H_2O$

c. $H_2N\overset{O}{\overset{\|}{C}}NHCH_3$ **f.** CH_3CH_2⬠ (cyclic anhydride)

CHAPTER 16

16-2. a. 3-methylpentanal, β-methylvaleraldehyde **b.** 4-heptanone, dipropyl ketone **c.** 2-methyl-4-heptanone, isobutyl propyl ketone **d.** 4-phenylbutanal, γ-phenylbutyraldehyde **e.** 4-ethylhexanal, γ-ethyl-caproaldehyde **f.** 1-hepten-3-one, butyl vinyl ketone

16-3. a. 2-heptanone **b.** *para*-nitroacetophenone

16-4. a. two **b.** (R)-3-methyl-3-hexanol and (S)-3-methyl-3-hexanol

16-5. $CH_3\overset{O}{\overset{\|}{C}}CH_2CH_3 + CH_3CH_2CH_2MgBr$
    $CH_3CH_2\overset{O}{\overset{\|}{C}}CH_2CH_2CH_3 + CH_3MgBr$

16-6.

$CH_3CH_2CH_2CH_2Br \xrightarrow{NaC\equiv N} CH_3CH_2CH_2CH_2C\equiv N \xrightarrow[\Delta]{H^+ \ H_2O} CH_3CH_2CH_2CH_2\overset{O}{\overset{\|}{C}}OH$

$CH_3CH_2CH_2CH_2Br \xrightarrow[Et_2O]{Mg} CH_3CH_2CH_2CH_2MgBr \xrightarrow[2.\ H^+,\ H_2O]{1.\ CO_2} CH_3CH_2CH_2CH_2\overset{O}{\overset{\|}{C}}OH$

16-7. No, an acid must be present.

16-10. a. CH_3CHCH_2OH **c.** ⬡—CH_2OH
        |
     CH_3

b. ⬡—OH   **d.** ⬡—$CHCH_3$
                  |
                 OH

16-11. an alkane and a carboxylate ion

16-13. a. ⬡—$\overset{O}{\overset{\|}{C}}NHCH_3$ **c.** $CH_3\overset{O}{\overset{\|}{C}}NHCH_2CH_3$

b. $CH_3\overset{O}{\overset{\|}{C}}NH_2$ **d.** $CH_3\overset{O}{\overset{\|}{C}}N\overset{CH_2CH_3}{\underset{CH_2CH_3}{<}}$

16-14. a. (1) $LiAlH_4$ (2) H_2O **b.** H^+, H_2O, Δ **c.** (1) H^+, H_2O, Δ (2) $SOCl_2$ (3) $LiAlH(OC(CH_3)_3)_3$, −80 °C (4) H_3O^+ **d.** (1) H^+, H_2O, Δ (2) $LiAlH_4$ (3) H_3O^+

16-17. a. ⬠=$NCH_2CH_3 + H_2O$

b. ⬠-$N\overset{CH_2CH_3}{\underset{CH_2CH_3}{<}} + H_2O$

c. ⬡—$\overset{}{\underset{CH_3}{C}}$=$N(CH_2)_5CH_3 + H_2O$

d. ⬡—$\overset{}{\underset{CH_3}{C}}$=$N$—⬡$ + H_2O$

16-18. a tertiary amine
16-21. the ketone with the nitro substituents

16-25. a. ⬡—$\overset{OH}{\underset{CH_3}{\overset{|}{C}}}CH_2CH_3$ **b.** $CH_3\overset{OH}{\underset{H}{\overset{|}{C}}}$—⬡
        S   *R*

c. $CH_3CH_2\overset{OH}{\underset{CH_3}{\overset{|}{C}}}CH_2CH_2CH_3$
compound does not have a chirality center

d. $CH_3CH_2CH_2\overset{OH}{\underset{H}{\overset{|}{C}}}CH_2CH_3$
                  S

16-27. a. (decalin ketone with C≡N)   **c.** CH_3C=$CHCHCH_3$
                                | |
                              CH_3 CH_3

b. (octalin with OH)   **d.** $CH_3\overset{CH_3}{\underset{CH_3}{\overset{|}{C}}}$—$CH_2\overset{O}{\overset{\|}{C}}CH_3$

CHAPTER 17

17-1. a. reduction **b.** neither **c.** oxidation **d.** reduction **e.** reduction **f.** neither

17-2. a. $CH_3CH_2CH_2CH_2CH_2OH$ **b.** $CH_3CH_2CH_2CH_2NH_2$

c. $\overset{CH_3CH_2CH_2}{\underset{H}{>}}C$=$C\overset{CH_3}{\underset{H}{<}}$ **f.** CH_3CH_2OH

d. ⬡—OH   **g.** $CH_3\overset{O}{\overset{\|}{C}}H$

e. no reaction   **h.** ⬡—$NHCH_3$

17-4. a. ⬡—CH_2NH_2 **d.** ⬡—CH_2OH + CH_3CH_2OH

b. ⬡—CH_2OH **e.** $CH_3CH_2CH_2NHCH_2CH_3$

c. $CH_3CH_2\overset{OH}{\overset{|}{C}}HCH_2CH_3$ **f.** $CH_3CH_2CH_2CH_2OH$

17-7. a. $CH_3CH_2\overset{O}{\overset{\|}{C}}CH_2CH_3$ **c.** no reaction

b. $CH_3CH_2CH_2CH_2\overset{O}{\overset{\|}{C}}OH$ **d.** $CH_3\overset{O}{\overset{\|}{C}}CH_2\overset{O}{\overset{\|}{C}}CH_2CH_3$

17-10. a. $CH_3\overset{CH_3}{\overset{|}{C}}$—$CHCH_2CH_3$ **b.** (cyclohexane)$\overset{CH_2OH}{\underset{OH}{<}}$
             | |
            OH OH

17-12. a. (cyclohexane diol with C(CH₃)₃)   **b.** (cyclohexane diol with C(CH₃)₃)

17-13. (bicyclohexylidene)

17-16. a. $\overset{CH_3}{\underset{CH_3}{>}}C$=$C\overset{CH_3}{\underset{CH_3}{<}}$
2,3-dimethyl-2-butene

b.

$CH_3CH_2CH_2$ and $CH_2CH_2CH_3$ groups on $C=C$ with H's — *cis*-4-octene ; *trans*-4-octene

17-17. Whether the compound has the cis or the trans configuration.

17-19. **a.** cyclohexyl$-C\equiv CH$

b. $CH_3CH_2C\equiv CCH_2CH_2CH_2C\equiv CCH_2CH_3$

17-20. **a.** cyclohexene **b.** 1-butene **c.** *trans*-2-pentene **d.** *cis*-2-butene

17-24. **a.**

(isopropyl bromide) $\xrightarrow{HO^-}$ (propene) $\xrightarrow{}$ (isopropyl alcohol, OH) $\xrightarrow{KMnO_4/H_2SO_4}$ (acetone)

$\xrightarrow{HBr/\Delta}$; $\xrightarrow[2.\ H_2O]{1.\ NaBH_4}$

b.

cyclohexene $\xrightarrow[\text{or}]{\substack{1.\ OsO_4 \\ 2.\ H_2O,\ NaHSO_3}}$ *cis*-cyclohexane-1,2-diol (OH, OH)

$\xrightarrow{\substack{1.\ KMnO_4,\ HO^- \\ 2.\ H_2O}}$

$\xrightarrow[2.\ HO^-]{1.\ RCOOH}$ *trans*-cyclohexane-1,2-diol (OH, OH)

CHAPTER 18

18-2. **a.** $CH_3\overset{O}{\overset{\|}{C}}CH_2C\equiv N$ — a β-keto nitrile

b. $CH_3O\overset{O}{\overset{\|}{C}}CH_2\overset{O}{\overset{\|}{C}}OCH_3$ — a β-diester

18-3. **a.** $CH_3\overset{O}{\overset{\|}{C}}H$ > $HC\equiv CH$ > $CH_2=CH_2$ > CH_3CH_3

b.

$CH_3\overset{O}{\overset{\|}{C}}CH_2\overset{O}{\overset{\|}{C}}CH_3$ > $CH_3\overset{O}{\overset{\|}{C}}CH_2\overset{O}{\overset{\|}{C}}OCH_3$ > $CH_3O\overset{O}{\overset{\|}{C}}CH_2\overset{O}{\overset{\|}{C}}OCH_3$ > $CH_3\overset{O}{\overset{\|}{C}}CH_3$

c. cyclohexanone > tetrahydropyran-2-one > N-methyl-piperidin-2-one (NCH$_3$)

18-5. **a.** $CH_3CH=\overset{OH}{\overset{|}{C}}CH_2CH_3$

c. 1-cyclohexenol (OH on cyclohexene)

b. phenyl$-\overset{OH}{\overset{|}{C}}=CH_2$

d. 3-hydroxycyclohex-2-enone (OH, =O)

e. $CH_3CH_2\overset{OH}{\overset{|}{C}}=CH\overset{O}{\overset{\|}{C}}CH_2CH_3$

f. phenyl$-CH=\overset{OH}{\overset{|}{C}}CH_3$ and phenyl$-CH_2\overset{OH}{\overset{|}{C}}=CH_2$
more stable

18-8. The rate-determining step must be removal of the proton from the α-carbon of the ketone

18-10. cyclohexanone with D substituents at α-carbons (D, D, D, D)

18-12. **a.** cyclohexanone $\xrightarrow[2.\ ICH_2CH=CH_2]{1.\ LDA/THF}$ cyclohexanone with $CH_2CH=CH_2$ at α-position

b. $CH_3CH_2\overset{O}{\overset{\|}{C}}CH_2CH_3$ $\xrightarrow[2.\ \text{benzyl}-CH_2Br]{1.\ LDA/THF}$ benzyl$-CH_2\overset{O}{\overset{\|}{C}}H\overset{O}{\overset{\|}{C}}CH_2CH_3$ with CH_3

18-14. **a.** cyclohex-2-enone ; $CH_3\overset{O}{\overset{\|}{C}}CH_2\overset{O}{\overset{\|}{C}}CH_3$; CH_3O^-

b. $CH_3\overset{O}{\overset{\|}{C}}CH=CH_2$; $CH_3CH_2O\overset{O}{\overset{\|}{C}}CH_2\overset{O}{\overset{\|}{C}}OCH_2CH_3$; $CH_3CH_2O^-$

18-15. **a.** $CH_3CH_2CH_2CH_2\overset{OH}{\overset{|}{C}}H\overset{O}{\overset{\|}{C}}CH$ with $CH_2CH_2CH_3$ branch

c. $CH_3CH_2\overset{OH}{\underset{CH_2CH_3}{\overset{|}{C}}}-\overset{CH_3}{\overset{|}{C}}H\overset{O}{\overset{\|}{C}}CH_2CH_3$

b. $CH_3\overset{OH}{\overset{|}{C}}HCH_2CH_2\overset{O}{\overset{\|}{C}}H\overset{}{C}CH$ with CH_3, CH_2CHCH_3, CH_3 branches

d. 1-hydroxy-bicyclohexyl with ketone (OH; =O)

18-20. **a.** $CH_3CH_2CH_2\overset{O}{\overset{\|}{C}}\overset{O}{\overset{\|}{C}}HCOCH_3$ with CH_2CH_3 branch

b. $CH_3\overset{}{C}HCH_2\overset{O}{\overset{\|}{C}}\overset{O}{\overset{\|}{C}}HCOCH_2CH_3$ with CH_3, $CHCH_3$, CH_3 branches

18-22.

a. $CH_3CH_2\overset{O}{\overset{\|}{C}}\overset{O}{\overset{\|}{C}}HCOCH_3$ (CH_3) ; $CH_3\overset{O}{\overset{\|}{C}}CH_2\overset{O}{\overset{\|}{C}}OCH_3$; $CH_3CH_2\overset{O}{\overset{\|}{C}}CH_2\overset{O}{\overset{\|}{C}}OCH_3$; $CH_3\overset{O}{\overset{\|}{C}}CH\overset{O}{\overset{\|}{C}}OCH_3$ (CH_3)

b. phenyl$\overset{O}{\overset{\|}{C}}CH_2\overset{O}{\overset{\|}{C}}$phenyl

c. $H\overset{O}{\overset{\|}{C}}\overset{}{C}H\overset{O}{\overset{\|}{C}}OCH_3$ with CH_2CH_3

d. $CH_3CH_2O\overset{O}{\overset{\|}{C}}\overset{}{C}HCOCH_2CH_3$ with CH_3

18-24. **a.** cyclobutane with OH, CH$_3$, and $\overset{O}{\overset{\|}{C}}CH_3$ groups

b. cyclooctanone with OH and CH$_3$

18-27. **a.** decalin ring system with OH and ketone (=O)

c. cyclohexane with OH and $\overset{O}{\overset{\|}{C}}H$

b. fused bicyclic with HO and ketone (=O)

d. decalin with OH and $\overset{O}{\overset{\|}{C}}CH_3$

18-29. **a** and **d**   **b** doesn't have a carboxyl group.   **18-30.** **a.** methyl bromide **b.** methyl bromide (twice) **c.** benzyl bromide **d.** propyl bromide **e.** isobutyl bromide **f.** propyl bromide and methyl bromide

18-32. **a.** cyclohexyl$-\overset{O}{\overset{\|}{C}}OH$ **b.** $HO\overset{O}{\overset{\|}{C}}CH_2CH_2CH_2CH_2CH_2CH_2CH_2\overset{O}{\overset{\|}{C}}OH$

18-33. **a.** ethyl bromide **b.** pentyl bromide **c.** benzyl bromide   **18-36.** 7

CHAPTER 19

19-1. D-Ribose is an aldopentose.   D-Sedoheptulose is a ketoheptose.
D-Mannose is an aldohexose.

19-3.

L-glyceraldehyde   L-glyceraldehyde   D-glyceraldehyde

19-4. a. a pair of enantiomers **b.** a pair of diastereomers
19-5. a. D-ribose **b.** L-talose **c.** L-allose   **19-6. a.** (2R,3S,4R,5R)-
2,3,4,5,6-pentahydroxyhexanal **b.** (2S,3R,4S,5S)-2,3,4,5,6-pentahydroxy-
hexanal   **19-7.** D-psicose   **19-8. a.** A ketoheptose has four chirality
centers (2^4 = 16 stereoisomers). **b.** An aldoheptose has five chirality centers (2^5 = 32 stereoisomers). **c.** A ketotriose has no chirality centers; therefore, it has no stereoisomers.   **19-9. a.** D-iditol **b.** D-iditol and D-gulitol
19-10. a. (1) D-altrose (2) L-galactose **b.** (1) D-tagatose (2) D-fructose
19-12. D-tagatose, D-galactose, and D-talose   **19-13. a.** L-gulose
b. L-gularic acid **c.** D-allose and L-allose, D-altrose and D-talose, L-altrose
and L-talose, D-galactose and L-galactose   **19-14. a.** D-arabinose and
D-ribulose **b.** D-allose and D-psicose **c.** L-gulose and L-sorbose **d.** D-talose
and D-tagatose   **19-15.** D-gulose and D-idose   **19-16.** a. D-gulose and
D-idose **b.** L-xylose and L-lyxose   **19-17. a.** D-glucose and D-mannose
b. D-erythrose and D-threose **c.** L-allose and L-altrose

19-20.
19-21. a. the OH group at C-2 **b.** the OH group at C-2, C-3, and C-4
c. the OH group at C-3 and C-1   **19-27. a.** They cannot receive type A, B,
or AB blood because these blood types have sugar components that type O
blood does not have. **b.** They cannot give blood to those with type A, B, or
O blood because AB blood has sugar components that these other blood
types do not have.

CHAPTER 20

20-2. a. (R)-alanine **b.** (R)-aspartic acid **c.** The α-carbons of all the D-
amino acids except cysteine have the R configuration.   **20-3.** Ile, Thr
20-4. Gly, Ile, Leu, Val

20-7.

a. **b.** **c.** **d.**

20-10. a. 5.43 **b.** 10.76 **c.** 5.68
20-11. a. Asp **b.** Arg **c.** Asp **d.** Met
20-16. His > Val > Ser > Asp
20-19. a. L-Ala, L-Asp, L-Glu **b.** L-Ala and D-Ala, L-Asp and D-Asp, L-Glu
and D-Glu
20-21. a. 2^8 = 25,600,000,000 **b.** 20^{100}
20-27. a. 5.8% **b.** 4.4%
20-30. Leu-Met-Ala-Gly-Glu-Pro-Ser-Val-His
20-34. Leu-Tyr-Lys-Arg-Met-Phe-Arg-Ser
20-35. 110 Å in an α-helix and 260 Å in a straight chain
20-37. a. cigar-shaped protein **b.** subunit of a hexamer

CHAPTER 21

21-1. $\Delta H^{\ddagger}$, E_a, $\Delta S^{\ddagger}$, $\Delta G^{\ddagger}$, k_{rate}

21-3.

21-6.

21-8. 2.5×10^6 **21-11.** close to 1   **21-12. a.** The 2-thio-substituted
chlorocyclohexane reacts more slowly because the stability of the carbocation that is formed in the slow first step is decreased by the electron-withdrawing 2-thio-substituent. **b.** The fact that 2-thio-substituted
chlorocyclohexane reacts more slowly than chlorocyclohexane tells you
that the reaction takes place by an S_N1 mechanism.   **21-15.** -Ser-Ala-Phe
because the phenyl substituent of phenylalanine would be more attracted to
the hydrophobic pocket of the enzyme than would the negatively charged
substituent of aspartic acid.   **21-17.** Arginine forms a direct hydrogen
bond; lysine forms an indirect hydrogen bond.   **21-19.** NAM

CHAPTER 22

22-2. pyrophosphate   **22-12. a.** pyridoxal phosphate and α-ketoglutarate
b. S-adenoslymethionine

c.

22-13. alanine and aspartic acid

22-15.

22-19.

22-20. the methylene group of N^5, N^{10}-methylene-THF

CHAPTER 23

23-2. hexanoic acid, 3 malonic acid, and glutaric acid   **23-4.** glyceryl
tripalmitate   **23-6.** Integral proteins will have a higher percentage of nonpolar amino acids.   **23-7.** The bacteria could synthesize phosphoacylglycerols with more saturated fatty acids.   **23-12.** The two halves are
synthesized in a head-to-tail fashion and then joined together in a tail-to-tail
linkage.   **23-20.** cis; trans   **23-21.** a β-substituent

23-22.

23-24. Two are axial substituents and one is an equatorial substituent.

CHAPTER 24

24-2. a.

dCDP

b.

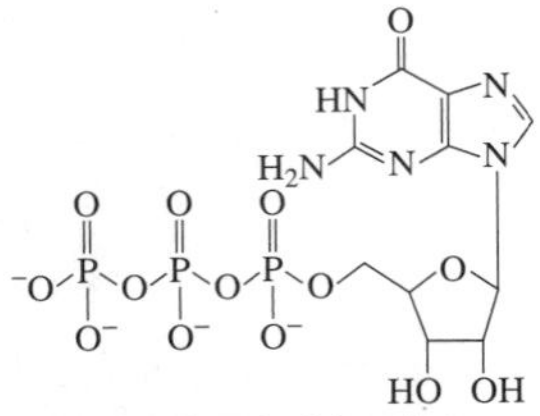

guanosine 5′-triphosphate
GTP

c.

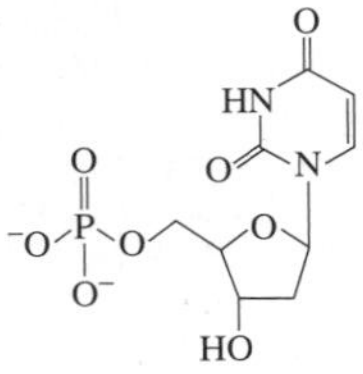

dUMP

d.

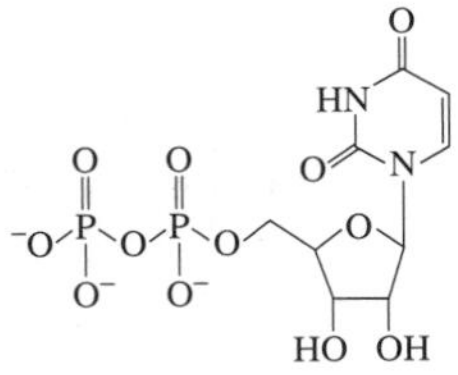

UDP

e.

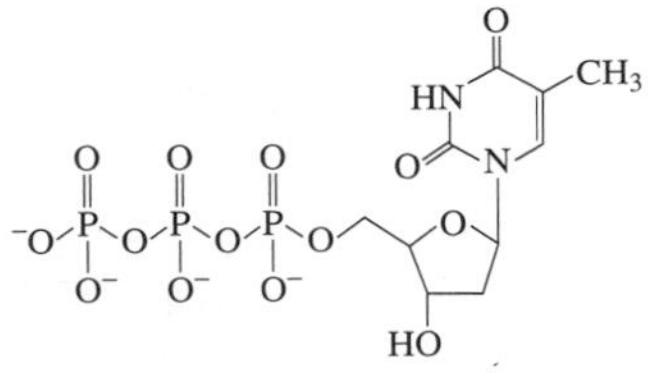

dTTP

24-4. phosphocreatine
24-11. a. 3′——C—C—T—G—T—T—A—G—A—C—G——5′
b. guanine   **24-12. a.** 23 Å **b.** 30 Å **c.** 31 Å   **24-14.** 10^{10} Å
24-17. Met-Asp-Pro-Val-Ile-Lys-His   **24-18.** Met-Asp-Pro-Leu-Leu-Asn
24-20. 5′–G-C-A-T-G-G-A-C-C-C-C-G-T-T-A-T-T-A-A-A-A-C-A-C–3′

24-22. a.

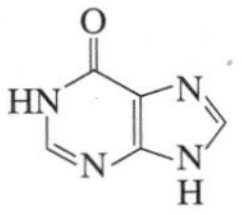

hypoxanthine 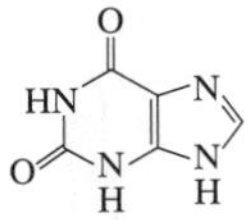   xanthine

24-24. a

CHAPTER 25

25-1. a. $CH_2{=}CHCl$

b.

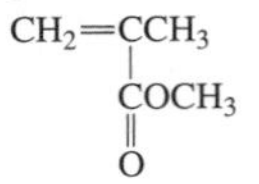

$CH_2{=}CCH_3$

$COCH_3$

c. $CF_2{=}CF_2$
25-2. poly(vinyl chloride)
25-5. beach balls

25-7. a.

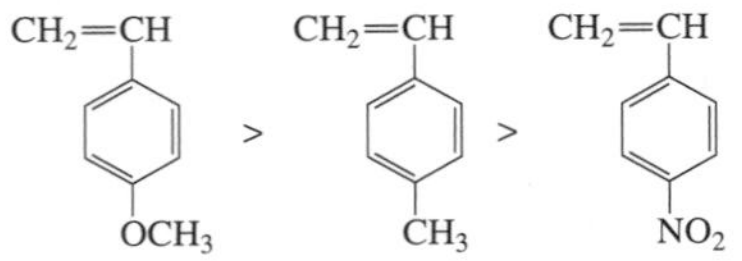

b.

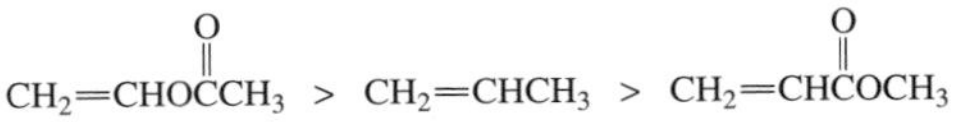

$CH_2{=}CHOCCH_3 > CH_2{=}CHCH_3 > CH_2{=}CHCOCH_3$

c.

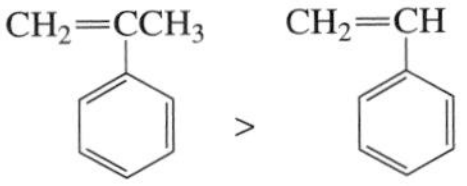

25-8. a.

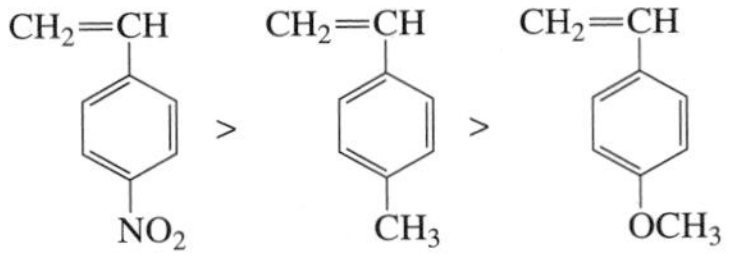

b. $CH_2{=}CHC{\equiv}N > CH_2{=}CHCl > CH_2{=}CHCH_3$

25-11. a. 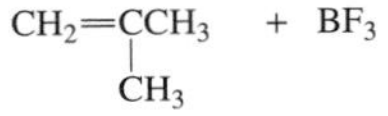 $CH_2{=}CCH_3$ $+$ BF_3 **c.** 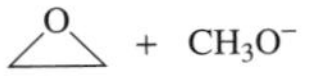 $+$ CH_3O^-
               $|$
               CH_3

b. 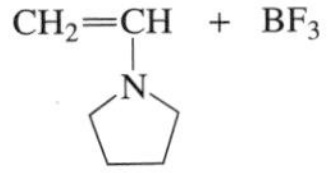 $+$ BF_3 **d.** 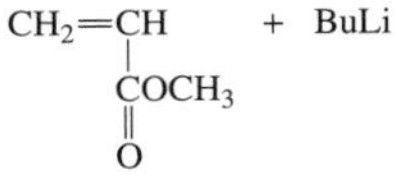 $+$ BuLi

25-12.

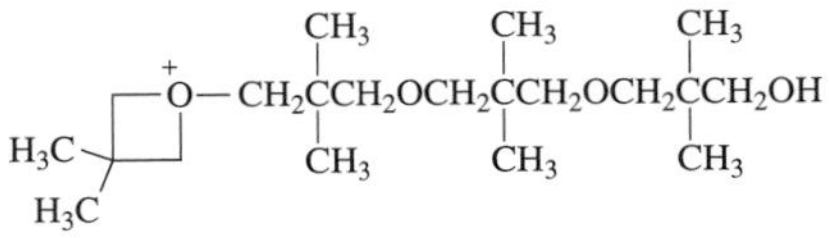

25-15. a.

b.

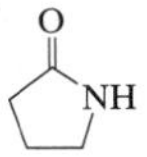

c.

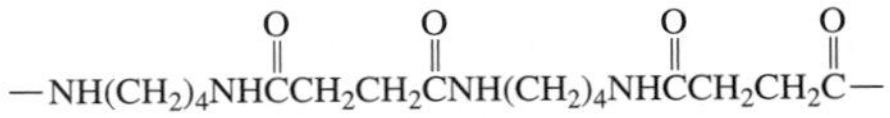

CHAPTER 26

26-1. a. 2,2-dimethylaziridine **b.** 4-ethylpiperidine **c.** 3-methylazacyclobutane **d.** 2-methylthiacyclopropane **e.** 2,3-dimethyltetrahydrofuran **f.** 2-ethyloxacyclobutane

26-3. a.

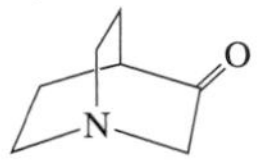

b. pH $\sim$ 8 **c.** 3-chloroquiniculidine

26-4. a.

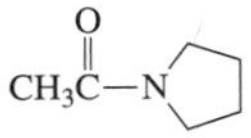

c. 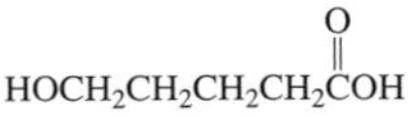 $HOCH_2CH_2CH_2CH_2COH$

b. $HOCH_2CH_2CH_2CH_2CH_2I$

d.

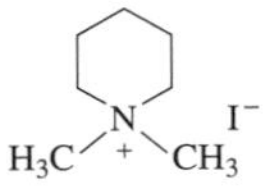

26-5. a. 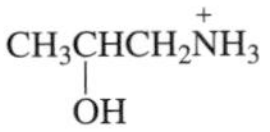 $CH_3\overset{+}{C}HCH_2NH_3$ **b.** 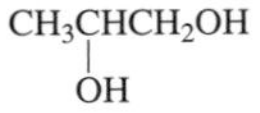 CH_3CHCH_2OH
                $|$ $|$
                OH OH

26-10. CH_3COCH_3

26-15.

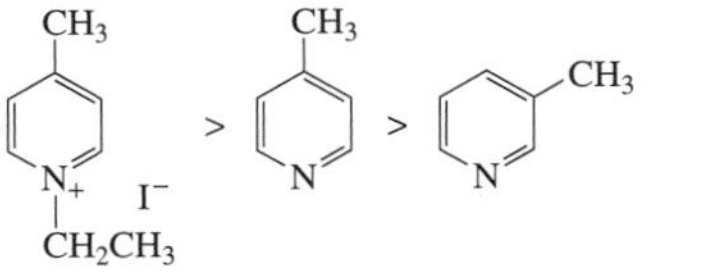

26-16. a.

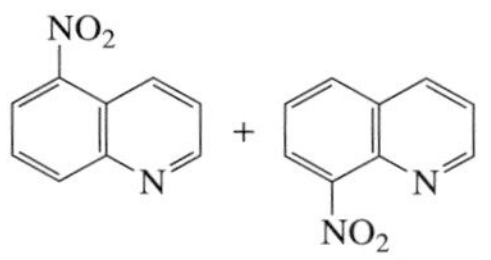

b.

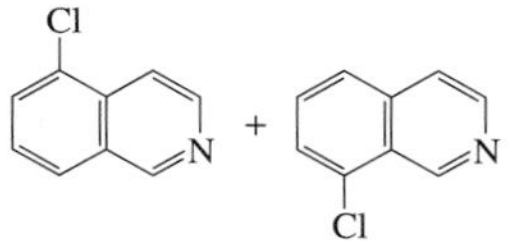

c.

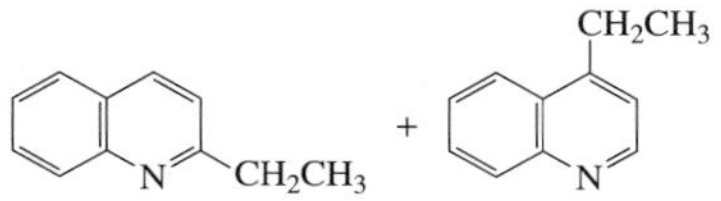

d.

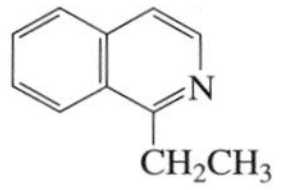

26-19. 24% **26-20.** yes

CHAPTER 27

27-1. a. electrocyclic reaction **b.** sigmatropic rearrangement **c.** cycloaddition reaction **d.** cycloaddition reaction **27-2. a.** bonding orbitals = ψ_1, ψ_2, ψ_3 antibonding orbitals = ψ_4, ψ_5, ψ_6 **b.** ground state HOMO = ψ_3 ground state LUMO = ψ_4 **c.** excited state HOMO = ψ_4 excited state LUMO = ψ_5 **d.** symmetric orbitals = ψ_1, ψ_3, ψ_5 asymmetric orbitals = ψ_2, ψ_4,ψ_6 **e.** The HOMO and LUMO have opposite symmetries. **27-3. a.** 8 **b.** ψ_4 **c.** 8 **27-6. a.** conrotatory **b.** trans **c.** disrotatory **d.** cis **27-7. a.** correct **b.** correct **c.** correct **27-8. (1) a.** conrotatory **b.** trans **(2) a.** disrotatory **b.** cis **27-11.** yes **27-12. a. (1)** [1,7] sigmatropic rearrangement **(2)** [1,5] sigmatropic rearrangement **(3)** [5,5] sigmatropic rearrangement **(4)** [3,3] sigmatropic rearrangement

b. (1)

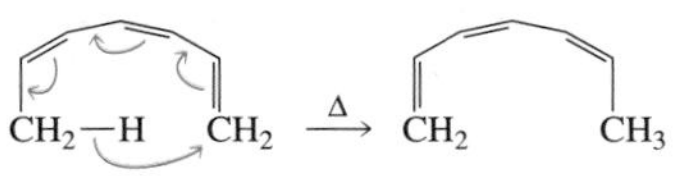

(2)

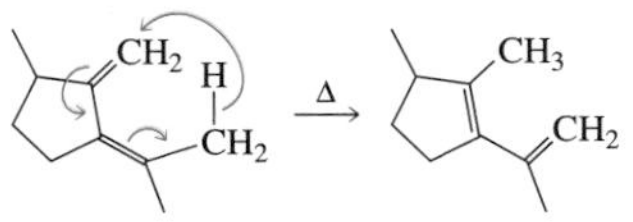

(3)

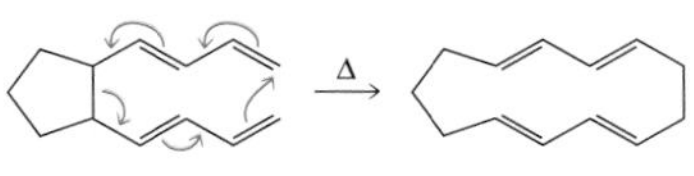

(4)

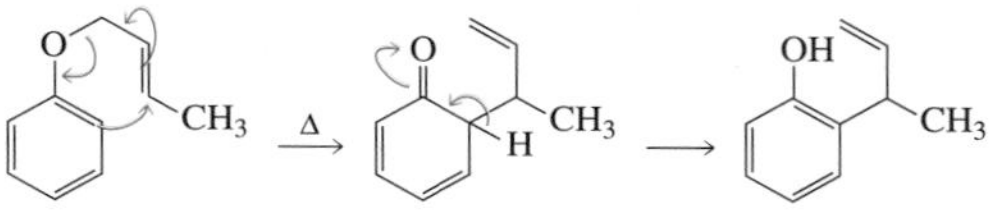

27-13.

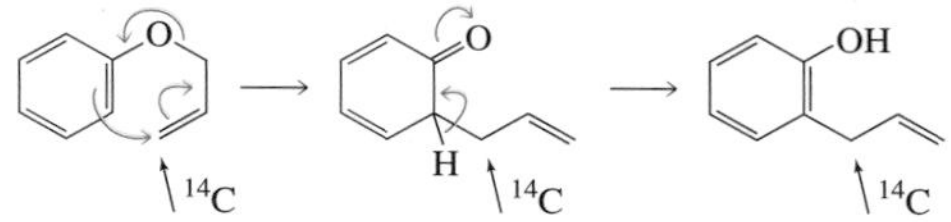

27-18. a. inversion **b.** retention **27-19. a.** antarafacial

CHAPTER 28

28-1. The linear synthesis results in a 21% yield; the convergent synthesis results in a 41% yield. **28-2.** the reaction of a Grignard reagent with ethylene oxide **28-3.** the reaction of a Grignard reagent with ethylene oxide **28-7. a.** *umpolung* **b.** normal **c.** normal **d.** *umpolung* **e.** normal **f.** *umpolung* **g.** *umpolung* **h.** normal **28-8. a.** Use 1-bromohexane instead of 1-bromopropane. **b.** Use 2-methyl-1,3-dithiane instead of 1,3-dithiane.

28-18. a. $CH_3CH_2CH(CH_2CH_3)CH{=}O$ + $CH_2{=}P(C_6H_5)_3$

b. Substitution will not compete with elimination. **28-20.** It will become an alkane as a result of protonation by the OH group.

28-21. $CH_3CH(COOH)CH_2C({=}O)COOH$

28-23. Both Gly-Gly and Gly-Ala would be formed.
28-26.

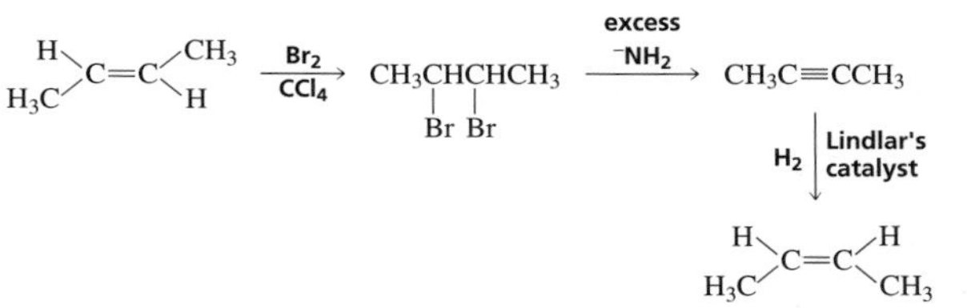

28-28. It is a stereoselective reaction; it is a stereospecific reaction; it is not an enantioselective reaction. **28-39.** Only under photochemical conditions can the required suprafacial mode of ring closure occur.

GLOSSARY

absolute configuration the three-dimensional structure of a chiral compound. The configuration is designated by *R* or *S*.

absorption band a peak in a spectrum that occurs as a result of absorption of energy.

acetal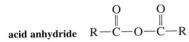

acetoacetic ester synthesis synthesis of a methyl ketone using ethyl acetoacetate as the starting material.

achiral (optically inactive) an achiral molecule has a superimposable mirror image.

acid a substance that donates a proton

acid anhydride
$$R-\overset{\overset{\displaystyle O}{\|}}{C}-O-\overset{\overset{\displaystyle O}{\|}}{C}-R$$

acid–base reaction a reaction in which an acid donates a proton to a base.

acid catalyst a catalyst that increases the rate of a reaction by donating a proton.

acid-catalyzed reaction a reaction catalyzed by an acid.

acid dissociation constant a measure of the degree to which an acid dissociates.

activating substituent a substituent that increases the reactivity of an aromatic ring. Electron-donating substituents activate aromatic rings toward electrophilic attack, and electron-withdrawing substituents activate aromatic rings toward nucleophilic attack.

active site a pocket or cleft in an enzyme where the substrate is bound.

acyclic noncyclic.

acyl adenylate a carboxylic acid derivative with AMP as the leaving group.

acyl–enzyme intermediate an amino acid residue of an enzyme has been acetylated.

acyl group a carbonyl group bonded to an alkyl group or to an aryl group.

acyl halide
$$R-\overset{\overset{\displaystyle O}{\|}}{C}-Cl$$

acyl phosphate a carboxylic acid derivative with a phosphate leaving group.

acyl pyrophosphate a carboxylic acid derivative with a pyrophosphate leaving group.

1,2-addition (direct addition) addition to the 1- and 2-positions of a conjugated system.

1,4-addition (conjugate addition) addition to the 1- and 4-positions of a conjugated system.

addition polymer (chain-growth polymer) made by adding monomers to the growing end of a chain.

addition reaction a reaction in which atoms or groups are added to the reactant.

adrenal cortical steroids glucocorticoids and mineralocorticoids.

alcohol a compound with an OH group in place of one of the hydrogens of an alkane (ROH).

alcoholysis reaction with alcohol

aldaric acid a dicarboxylic acid with an OH group bonded to each carbon. Obtained by oxidizing the aldehyde and primary alcohol groups of an aldose.

aldehyde

alditol a compound with an OH group bonded to each carbon. Obtained by reducing an aldose or a ketose.

aldol addition a reaction between two molecules of an aldehyde (or two molecules of a ketone) that connects the α-carbon of one with the carbonyl carbon of the other.

aldol condensation an aldol addition followed by elimination of water.

aldonic acid a carboxylic acid with an OH group bonded to each carbon. Obtained by oxidizing the aldehyde group of an aldose.

aldose a polyhydroxyaldehyde.

aliphatic compound a nonaromatic organic compound.

alkaloid a natural product with one or more nitrogen hetero atoms found in the leaves, bark, or seeds of plants.

alkane a hydrocarbon that contains only single bonds.

alkene a hydrocarbon that contains a double bond.

alkoxymercuration addition of alcohol using a mercuric salt of a carboxylic acid as a catalyst.

alkylation reaction a reaction that adds an alkyl group to a reactant.

alkyl halide a compound with a halogen in place of one of the hydrogens of an alkane.

alkyl substituent (alkyl group) formed by removing a hydrogen from an alkane.

alkyl tosylate an ester of *para*-toluenesulfonic acid.

alkyne a hydrocarbon that contains a triple bond.

allene a compound with two adjacent double bonds.

allyl group $CH_2=CHCH_2-$

allylic carbon an sp^3 carbon adjacent to a vinyl carbon.

allylic cation a compound with a positive charge on an allylic carbon.

alpha olefin a monosubstituted olefin.

alternating copolymer a copolymer in which two monomers alternate.

ambident nucleophile a nucleophile with two nucleophilic sites.

amide
$$R-\overset{\overset{\displaystyle O}{\|}}{C}-NH_2, \quad R-\overset{\overset{\displaystyle O}{\|}}{C}-NHR, \quad R-\overset{\overset{\displaystyle O}{\|}}{C}-NR_2$$

amine a compound with a nitrogen in place of one of the hydrogens of an alkane (RNH_2, R_2NH, R_3N).

amine inversion a compound containing an sp^3 hybridized nitrogen with a nonbonding pair of electrons rapidly turns inside out.

amino acid an α-aminocarboxylic acid. Naturally occurring amino acids have the L configuration.

amino acid analyzer an instrument that automates the ion-exchange separation of amino acids.

amino acid residue a monomeric unit of a peptide or protein.

aminolysis reaction with an amine.

amino sugar a sugar in which one of the OH groups is replaced by an NH_2 group.

amphoteric compound a compound that can behave either as an acid or as a base.

anabolic steroids steroids that aid in the development of muscle.

anabolism reactions living organisms carry out in order to synthesize complex molecules from simple precursor molecules.

anchimeric assistance (intramolecular catalysis) catalysis in which the catalyst that facilitates the reaction is part of the molecule undergoing reaction.

androgens male sex hormones

angle strain the strain introduced into a molecule as a result of its bond angles being distorted from their ideal values.

angular methyl group a methyl substituent at the 10- or 13-position of a steroid ring system.

anion-exchange resin a positively charged resin used in ion-exchange chromatography.

anionic polymerization chain-growth polymerization in which the initiator is a nucleophile; the propagation site therefore is an anion.

annulation reaction a ring-forming reaction.

annulene a monocyclic hydrocarbon with alternating double and single bonds.

anomeric carbon the carbon in a cyclic sugar that is the carbonyl carbon in the open-chain form.

anomeric effect the preference for the axial position shown by certain substituents bonded to the anomeric carbon of a six-membered ring sugar.

anomers two cyclic sugars that differ in configuration only at the carbon that is the carbonyl carbon in the open-chain form.

antarafacial bond formation formation of two σ bonds from opposites sides of the π system.

antarafacial rearrangement rearrangement in which the migrating group moves to the opposite face of the π system.

anti addition an addition reaction in which the two added substituents add to opposite sides of the molecule.

antiaromatic a cyclic and planar compound with an uninterrupted ring of p orbital-bearing atoms containing an even number of pairs of π electrons.

antibiotic a compound that interferes with the growth of a microorganism.

antibodies compounds that recognize foreign particles in the body.

antibonding molecular orbital a molecular orbital that results when two atomic orbitals with opposite signs interact. Electrons in an antibonding orbital decrease bond strength.

anticodon the three bases at the bottom of the middle loop in a tRNA.

anti conformer the most stable of the staggered conformers.

anti elimination an elimination reaction in which the two substituents eliminated are removed from opposite sides of the molecule.

antigene agent a polymer designed to bind to DNA at a particular site.

antigens compounds that can generate a response from the immune system.

anti-periplanar parallel substituents on opposite sides of a molecule.

antisense agent a polymer designed to bind to mRNA at a particular site.

antisense strand (template strand) the strand in DNA that is read during transcription.

antiviral drug a drug that interferes with DNA or RNA synthesis in order to prevent a virus from replicating.

apoenzyme an enzyme without its cofactor.

applied magnetic field the externally applied magnetic field.

aprotic solvent a solvent that does not have a hydrogen bonded to an oxygen or to a nitrogen.

aramide an aromatic polyamide.

arene oxide an aromatic compound that has had one of its double bonds converted to an epoxide.

aromatic a cyclic and planar compound with an uninterrupted ring of p orbital-bearing atoms containing an odd number of pairs of π electrons.

Arrhenius equation relates the rate constant of a reaction to the energy of activation and to the temperature at which the reaction is carried out ($k + Ae^{-E_a/RT}$).

aryl group general term for a phenyl or substituted phenyl group.

asymmetrical ether an ether with two different substituents bonded to the oxygen.

asymmetric molecular orbital a molecular orbital in which the left half is not a mirror of the right half.

atactic polymer a polymer in which the substituents are randomly oriented on the extended carbon chain.

atomic number tells how many protons (or electrons) the neutral atom has.

atomic orbital an orbital associated with an atom.

atomic weight the average mass of the atoms in the naturally occurring element.

aufbau principle states that an electron will always go into the available orbital with the lowest energy.

automated solid-phase peptide synthesis an automated technique that synthesizes a peptide while its C-terminal amino acid is attached to a solid support.

autoradiograph the exposed photographic plate obtained in autoradiography

autoradiography a technique used to determine the base sequence of DNA.

auxochrome a substituent that, when attached to a chromophore, alters the λ_{max} and intensity of absorption of UV/Vis radiation.

axial bond a bond of the chair form of cyclohexane that is perpendicular to the plane in which the chair is drawn (an up–down bond).

aziridine a three-membered ring compound in which one of the ring atoms is a nitrogen.

azo linkage a —N=N— bond.

back side attack nucleophilic attack on the side of the carbon opposite to the side bonded to the leaving group.

bactericidal drug a drug that kills bacteria.

bacteriostatic drug a drug that inhibits the further growth of bacteria.

Baeyer–Villiger oxidation oxidation of aldehydes or ketones with H_2O_2 to form carboxylic acids or esters, respectively.

banana bond the σ bonds in small rings that are weaker as a result of overlapping at an angle rather than overlapping head-on.

base^1 a substance that accepts a proton.

base^2 a heterocyclic compound (a purine or a pyrimidine) in DNA and RNA.

base catalyst a catalyst that increases the rate of a reaction by removing a proton.

base peak the peak with the greatest abundance in a mass spectrum.

basicity describes the tendency of a compound to share its electrons with a proton.

Beer–Lambert law relationship among the absorbance of UV/Vis light, the concentration of the sample, the length of the light path, and the molar absorptivity.

bending vibration a vibration that does not occur along the line of the bond.

benzoyl group a benzene ring bonded to a carbonyl group.

benzylic carbon an sp^3 hybridized carbon bonded to a benzene ring.

benzylic cation a compound with a positive charge on a benzylic carbon.

benzyl group

benzyne intermediate a compound with a triple bond in place of one of the double bonds of benzene.

bicyclic compound a compound that contains two rings that share at least one carbon.

bifunctional molecule a molecule with two functional groups.

bile acids steroids that act as emulsifying agents so that water-insoluble compounds can be digested.

bimolecular reaction (second-order reaction) a reaction whose rate is dependent on the concentration of two reactants.

biochemistry (biological chemistry) the chemistry of biological systems.

biodegradable polymer a polymer that can be broken into small segments by an enzyme-catalyzed reaction.

bioorganic compound an organic compound that is found in biological systems.

biopolymer a polymer that is synthesized in nature.

biosynthesis synthesis in a biological system.

biotin the coenzyme required by enzymes that catalyze carboxylation of a carbon adjacent to an ester or a keto group.

Birch reduction the partial reduction of benzene to 1,4-cyclohexadiene.

blind screen (random screen)   the search for a pharmacologically active compound without any information about what chemical structures might show activity.

block copolymer   a copolymer in which there are blocks of each kind of monomer.

blue shift   a shift to a shorter wavelength.

boat conformation   the conformation of cyclohexane that roughly resembles a boat.

boiling point   the temperature at which a liquid vaporizes.

bonding molecular orbital   a molecular orbital that results when two atomic orbitals with the same sign interact. Electrons in a bonding orbital increase bond strength.

bond length   the internuclear distance between two atoms at minimum energy (maximum stability).

brand name (proprietary name, trade name)   identifies a commercial product and distinguishes it from other products. It can be used only by the owner of the registered trademark.

bridged bicyclic compound   a bicyclic compound in which rings share two nonadjacent carbons.

Brønsted acid   a substance that donates a proton.

Brønsted base   a substance that accepts a proton.

buffer   an acid and its conjugate base.

C-terminal amino acid   the terminal amino acid of a peptide (or protein) that has a free carboxyl group.

carbanion   a compound containing a negatively charged carbon.

carbene   a species with a carbon that has a nonbonded pair of electrons and an empty orbital ($H_2C:$).

carbocation   a compound containing a positively charged carbon.

carbocation rearrangement   the rearrangement of a carbocation to a more stable carbocation.

carbohydrate   a sugar, a saccharide. Naturally occurring carbohydrates have the D configuration.

α-carbon   a carbon adjacent to a carbonyl carbon.

carbon acid   a compound that contains a carbon that is bonded to a relatively acidic hydrogen.

carbonyl addition (direct addition)   nucleophilic addition to the carbonyl carbon.

carbonyl carbon   the carbon of a carbonyl group.

carbonyl compound   a compound that contains a carbonyl group.

carbonyl group   a carbon doubly bonded to an oxygen.

carbonyl oxygen   the oxygen of a carbonyl group.

carboxyl group   COOH

carboxylic acid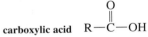

carboxylic acid derivative   a compound that is hydrolyzed to a carboxylic acid.

carboxyl oxygen   the singly bonded oxygen of a carboxlic acid or an ester.

carotenoid   a class of compound (a tetraterpene) responsible for the red and orange colors of fruits, vegetables, and fall leaves.

catabolism   reactions living organisms carry out in order to break down complex molecules into simple molecules and energy.

catalyst   a species that increases the rate at which a reaction occurs without being consumed in the reaction. Because it does not change the equilibrium constant of the reaction, it does not change the amount of product that is formed.

catalytic antibody   a compound that facilitates a reaction by forcing the conformation of the substrate in the direction of the transition state.

catalytic hydrogenation   the addition of hydrogen to a double or a triple bond with the aid of a metal catalyst.

cation-exchange resin   a negatively charged resin used in ion-exchange chromatography.

cationic polymerization   chain-growth polymerization where the initiator is an electrophile; the propagation site therefore is a cation.

cephalin   a phosphoacylglycerol in which the second OH group of phosphate has formed an ester with ethanolamine.

cerebroside   a sphingolipid in which the terminal OH group of sphingosine is bonded to a sugar residue.

chain-growth polymer (addition polymer)   made by adding monomers to the growing end of a chain.

chain transfer   a growing polymer chain reacts with a molecule XY in a manner that allows X to terminate the chain, leaving behind Y to initiate a new chain.

chair conformation   the conformation of cyclohexane that roughly resembles a chair. It is the most stable conformation of cyclohexane.

chemical exchange   the transfer of a proton from one molecule to another.

chemically equivalent protons   protons with the same connectivity relationship to the rest of the molecule.

chemical shift   the location of a signal in an NMR spectrum. It is measured downfield from a reference compound (most often TMS).

Chichibabin reaction   the reaction of pyridine with sodium amide to form 2-aminopyridine (and some 4-aminopyridine).

chiral (optically active)   a chiral molecule has a nonsuperimposable mirror image.

chiral auxiliary   an enantiomerically pure compound that, when attached to a reactant, causes a product with particular configuration to be formed.

chirality center   an atom bonded to four different groups.

cholesterol   a steroid that is the precursor of all other steroids.

chromatography   a separation technique in which the mixture to be separated is dissolved in a solvent and the solvent passed through a column packed with an adsorbent stationary phase.

chromophore   the part of a molecule responsible for a UV or visible spectrum.

cine substitution   substitution at the carbon adjacent to the carbon that was bonded to the leaving group.

cis fused   two cyclohexane rings fused together such that, if the second ring were considered to be two substituents of the first ring, one substituent would be in an axial position and the other would be in an equatorial position.

cis isomer   the isomer with identical substituents on the same side of the double bond.

cis-trans isomers   geometric isomers.

Claisen condensation   a reaction between two molecules of an ester that connects the α-carbon of one with the carbonyl carbon of the other and eliminates an alkoxide ion.

Claisen rearrangement   a [3,3] sigmatropic rearrangement of an allyl vinyl ether.

α-cleavage   homolytic cleavage of an alpha substituent.

Clemmensen reduction   a reaction that reduces the carbonyl group of a ketone to a methylene group using Zn(Hg)/HCl.

codon   a sequence of three bases in mRNA that specifies the amino acid to be incorporated into a protein.

coenzyme   a cofactor that is an organic molecule.

coenzyme A   a thiol used by biological organisms to form thioesters.

coenzyme B$_{12}$ the coenzyme required by enzymes that catalyze certain rearrangement reactions.

cofactor   an organic molecule or a metal ion that certain enzymes need in order to catalyze a reaction.

coil (loop) conformation   that part of a protein that is highly ordered but not in an α-helix or a β-sheet.

combination band   occurs at the sum of two fundamental absorption frequencies ($v_1 + v_2$).

combinatorial library   a group of structurally related compounds.

combinatorial organic synthesis   the synthesis of a library of compounds by covalently connecting sets of building blocks of varying structure.

common intermediate   an intermediate that two compounds have in common.

common name   nonsystematic nomenclature.

competitive inhibitor   a compound that inhibits an enzyme by competing with the substrate for binding at the active site.

complete racemization   formation of a pair of enantiomers in equal amounts.

complex carbohydrate   contains two or more sugar molecules linked together.

concerted reaction   a reaction in which all the bond-making and bond-breaking processes occur in one step.

condensation polymer (step-growth polymer)   made by combining two molecules while removing a small molecule (usually water or an alcohol).

condensation reaction   a reaction combining two molecules while removing a small molecule (usually water or an alcohol).

conducting polymer   a polymer that can conduct electricity down its backbone.

configuration   the three-dimensional structure of a chiral compound. The configuration is designated by *R* or *S*.

configurational isomers   stereoisomers that cannot interconvert unless a covalent bond is broken. Cis-trans isomers and optical isomers are configurational isomers.

conformation   the three-dimensional shape of a molecule at a given instant.

conformational analysis   the investigation of the various conformations of a compound and their relative stabilities.

conformers (conformational isomers)   different conformations of a molecule.

conjugate acid   a compound accepts a proton to form its conjugate acid.

conjugate addition   1,4-addition.

conjugate base   a compound loses a proton to form its conjugate base.

conjugated double bonds   double bonds separated by one single bond.

conrotatory ring closure   achieves head-to-head overlap of *p* orbitals by rotating the orbitals in the same direction.

conservation of orbital symmetry theory   a theory that explains the relationship between the structure and configuration of the reactant, the conditions under which a pericyclic reaction takes place, and the configuration of the product.

constitutional isomers (structural isomers)   molecules that have the same molecular formula but differ in the way the atoms are connected.

contributing resonance structure   a structure with localized electrons that approximates the true structure of a compound with delocalized electrons.

convergent synthesis   a synthesis in which pieces of the target compound are individually prepared and then assembled.

Cope elimination reaction   elimination of a proton and a hydroxyl amine from an amine oxide.

Cope rearrangement   a [3,3] sigmatropic rearrangement of a 1,5-diene.

copolymer   a polymer formed using two or more different monomers.

corrin ring system   a porphyrin ring system without one of the methine bridges.

COSY spectrum   A 2-D NMR spectrum that shows coupling between sets of protons.

coupled protons   protons that split each other. Coupled protons have the same coupling constant.

coupling constant   the distance (in hertz) between two adjacent peaks of a split NMR signal.

coupling reaction   a reaction that joins two alkyl groups.

covalent bond   a bond created as a result of sharing electrons.

covalent catalysis (nucleophilic catalysis)   catalysis that occurs as a result of a nucleophile forming a covalent bond with one of the reactants.

Cram's rule   the rule used to determine the major product of a carbonyl addition reaction in a compound with a chirality center adjacent to the carbonyl group.

cross-conjugation   nonlinear conjugation.

crossed (mixed) aldol addition   an aldol addition in which two different carbonyl compounds are used.

cross-linking   connecting polymer chains by intermolecular bond formation.

crown ether   a cyclic molecule that contains several ether linkages.

crown–guest complex   the complex formed when a crown ether binds a substrate.

cryptand   a three-dimensional polycyclic compound that binds a substrate by encompassing it.

cryptate   the complex formed when a cryptand binds a substrate.

crystallites   regions of a polymer in which the chains are highly ordered.

cumulated double bonds   double bonds that are adjacent to one other.

Curtius rearrangement   conversion of an acyl chloride into a primary amine using $^-N_3$.

cyanohydrin
$$\begin{array}{c} OH \\ | \\ RCR(H) \\ | \\ C\equiv N \end{array}$$

cycloaddition reaction   a reaction in which two π-bond-containing molecules react to form a cyclic compound.

[4 + 2] cycloaddition reaction   a cycloaddition reaction in which 4 π electrons come from one reactant and 2 π electrons come from the other reactant.

cycloalkane   an alkane with its carbon chain arranged in a closed ring.

deactivating substituent   a substituent that decreases the reactivity of an aromatic ring. Electron-withdrawing substituents deactivate aromatic rings toward electrophilic attack, and electron-donating substituents deactivate aromatic rings toward nucleophilic attack.

deamination   loss of ammonia.

decarboxylation   loss of carbon dioxide.

degenerate orbitals   orbitals that have the same energy.

dehydration   loss of water.

dehydrogenase   an enzyme that carries out an oxidation reaction by removing hydrogen from the substrate.

dehydrohalogenation   elimination of a proton and a halide ion.

delocalization energy (resonance energy)   the extra stability a compound achieves as a result of having delocalized electrons.

delocalized electrons   electrons that are shared by more than two atoms.

denaturation   destruction of the highly organized tertiary structure of a protein.

deoxygenation   removal of an oxygen from a reactant.

deoxyribonucleic acid (DNA)   a polymer of deoxyribonucleotides.

deoxyribonucleotide   a nucleotide in which the sugar component is D-2-deoxyribose.

deoxy sugar   a sugar in which one of the OH groups has been replaced by an H.

DEPT ^{13}C NMR spectrum   a series of four spectra that distinguishes among —CH₃, —CH₂, and —CH groups.

depurination   elimination of a purine ring.

detergent   a salt of a sulfonic acid.

deuterium kinetic isotope effect   ratio of the rate constant obtained for a compound containing hydrogen and the rate constant obtained for an identical compound in which one or more of the hydrogens have been replaced by deuterium.

dextrorotatory   the enantiomer that rotates polarized light in a clockwise direction.

diastereomer   a configurational isomer that is not an enantiomer.

diastereotopic hydrogens   two hydrogens bonded to a carbon that when replaced in turn with a deuterium results in a pair of diastereomers.

1,3-diaxial interaction   the interaction between an axial substituent and the other two axial substituents on the same side of the cyclohexane ring.

diazonium ion $Ar\overset{+}{N}\equiv N$ or $R\overset{+}{N}\equiv N$

diazonium salt   a diazonium ion and an anion $Ar\overset{+}{N}\equiv N\ X^-$).

Dieckmann condensation   an intramolecular Claisen condensation.

dielectric constant   a measure of how well a solvent can insulate opposite charges from one another.

Diels–Alder reaction   a [4 + 2] cycloaddition reaction.

diene   a hydrocarbon with two double bonds.

dienophile   an alkene that reacts with a diene in a Diels–Alder reaction.

β-diketone   a ketone with a second carbonyl group at the β-position.

dimer   a molecule formed by the joining together of two identical molecules.

dinucleotide   two nucleotides linked by phosphodiester bonds.

dipeptide   two amino acids linked by an amide bond.

dipole–dipole interaction   an interaction between the dipole of one molecule and the dipole of another.

dipole moment   (μ) a measure of the separation of charge in a bond or in a molecule.

direct addition   1,2-addition.

direct displacement mechanism   a reaction in which nucleophile displaces the leaving group in a single step.

direct substitution   substitution at the carbon that was bonded to the leaving group.

disaccharide   a compound containing two sugar molecules linked together.

disconnection   breaking a bond to carbon to give a simpler species.

disproportionation   transfer of a hydrogen atom by a radical to another radical with the result that an alkane and an alkene are formed.

disrotatory ring closure   achieves head-to-head overlap of p orbitals by rotating the orbitals in opposite directions.

dissociation energy   the amount of energy required to break a bond / the amount of energy released when a bond is formed.

dissolving metal reduction   a reduction using sodium or lithium metal dissolved in liquid ammonia.

distribution coefficient   the ratio of the amount of a compound dissolving in each of two solvents.

disulfide bridge   a disulfide (—S—S—) bond in a peptide or protein.

DNA (deoxyribonucleic acid)   a polymer of deoxyribonucleotides.

double bond   a sigma bond and a pi bond.

doublet   an NMR signal split into two peaks.

doublet of doublets   an NMR signal split into four peaks of approximately equal height. Caused by splitting a signal into a doublet by one hydrogen and into another doublet by another (nonequivalent) hydrogen.

drug   a compound that reacts with a biological molecule, triggering a physiological effect.

drug resistance   resistance to a particular drug.

drug synergism   when the effect of two drugs used in combination is greater than the sum of the effects obtained when administered individually.

eclipsed conformation   a conformation in which the bonds on adjacent carbons are parallel to each other as viewed looking down the carbon–carbon bond.

E conformation   the conformation of a carboxylic acid or carboxylic acid derivative in which the carbonyl oxygen and the substituent bonded to the carboxyl oxygen or nitrogen are on opposite sides of the single bond.

Edman's reagent   phenyl isothiocyanate. A reagent used to determine the N-terminal amino acid of a polypeptide.

effective magnetic field   the magnetic field that a proton "senses" through the surrounding cloud of electrons.

effective molarity   the concentration of the reagent that would be required in an intermolecular reaction for it to have the same rate as an intramolecular reaction.

E isomer   the isomer with the high-priority groups on opposite sides of the double bond.

elastomer   a polymer that can stretch and then revert back to its original shape.

electrocyclic reaction   a reaction in which a π bond in the reactant is lost so that a cyclic compound with a new σ bond can be formed.

electromagnetic radiation   radiant energy that displays wave properties.

electron affinity   the energy given off when an atom acquires an electron.

electronegative   describes an element that readily acquires an electron.

electronegativity   tendency of an atom to pull electrons toward itself.

electronic transition   promotion of an electron from its HOMO to its LUMO.

electron sink   site to which electrons can be delocalized.

electrophile   an electron-deficient atom or molecule.

electrophilic addition reaction   an addition reaction in which the first species that adds to the reactant is an electrophile.

electrophilic aromatic substitution   a reaction in which an electrophile substitutes for a hydrogen of an aromatic ring.

electrophilic catalysis   catalysis in which the species that facilitates the reaction is an electrophile.

electrophoresis   a technique that separates amino acids on the basis of their pI values.

electropositive   describes an element that readily loses an electron.

electrostatic attraction   attractive force between opposite charges.

electrostatic catalysis   stabilization of a charge by an opposite charge.

elemental analysis   a determination of the relative proportions of the elements present in a compound.

α-elimination   removal of two atoms or groups from the same carbon.

β-elimination   removal of two atoms or groups from adjacent carbons.

E1 reaction   a first-order elimination reaction.

E2 reaction   a second-order elimination reaction.

elimination reaction   a reaction that involves the elimination of atoms (or molecules) from the reactant.

empirical formula   the relative numbers of the different kinds of atoms in a molecule.

enamine   an α,β-unsaturated tertiary amine.

enantiomerically pure   containing only one enantiomer.

enantiomeric excess (optical purity)   how much excess of one enantiomer is present in a mixture of a pair of enantiomers.

enantiomers   nonsuperimposable mirror-image molecules.

enantioselective reaction   a reaction that forms an excess of one enantiomer.

enantiotopic hydrogens   two hydrogens bonded to a carbon that is bonded to two other groups that are nonidentical.

endergonic reaction   a reaction with a positive $\Delta G°$.

endo   a substituent is endo if it is closer to the longer or more unsaturated bridge.

endopeptidase   an enzyme that hydrolyzes a peptide bond that is not at the end of a peptide chain.

endothermic reaction   a reaction with a positive $\Delta H°$.

enkephalins   pentapeptides synthesized by the body to control pain.

enolization   keto–enol interconversion.

enthalpy   the heat given off ($-\Delta H°$) or the heat absorbed ($+\Delta H°$) during the course of a reaction.

entropy   a measure of the freedom of motion in a system.

enzyme   a protein that is a catalyst.

epimerization   changing the configuration at a chirality center by removing a proton from it and then reprotonating the molecule at the same site.

epimers   monosaccharides that differ in configuration at only one carbon.

epoxidation   formation of an epoxide.

epoxide (oxirane)   an ether in which the oxygen is incorporated into a three-membered ring.

epoxy resin   formed by mixing a low-molecular-weight prepolymer with a compound that forms a cross-linked polymer.

equatorial bond   a bond of the chair form of cyclohexane that juts out from the ring in approximately the same plane that contains the chair.

equilibrium constant   the ratio of products to reactants at equilibrium or the ratio of the rate constants for the forward and reverse reactions.

equilibrium control   thermodynamic control.

erythro enantiomers   the pair of enantiomers with similar groups on the same side when drawn in a Fischer projection.

essential amino acid   an amino acid that humans must obtain from their diet because they cannot synthesize it or cannot synthesize it in adequate amounts.

essential oils fragrances and flavorings isolated from plants that do not leave residues when they evaporate. Most are terpenes.

ester   R—C(=O)—OR

estrogens female sex hormones.

ether a compound containing an oxygen bonded to two carbons (ROR).

eukaryote a unicellular or multicellular body whose cell or cells contain a nucleus.

excited-state electronic configuration the electronic configuration that results when an electron in the ground-state electronic configuration has been moved to a higher-energy orbital.

exergonic reaction a reaction with a negative $\Delta G°$.

exhaustive methylation reaction of an amine with excess methyl iodide resulting in the formation of a quaternary ammonium iodide.

exo a substituent is exo if it closer to the shorter or more saturated bridge.

exon a stretch of bases in DNA that are a portion of a gene.

exopeptidase an enzyme that hydrolyzes a peptide bond at the end of a peptide chain.

exothermic reaction a reaction with a negative $\Delta H°$.

experimental energy of activation ($E_a = \Delta H^{\ddagger} - RT$) a measure of the approximate energy barrier to a reaction. (It is approximate because it does not contain an entropy component).

extrusion reaction a reaction in which a neutral molecule (for example, CO_2, CO, or N_2) is eliminated from a molecule.

fat a triester of glycerol that exists as a solid at room temperature.

fatty acid a long-chain carboxylic acid.

fibrous protein a water-insoluble protein in which the polypeptide chains are arranged in bundles.

fingerprint region the right-hand third of an IR spectrum where the absorption bands are characteristic of the compound as a whole.

first-order rate constant the rate constant of a first-order reaction.

first-order reaction (unimolecular reaction) a reaction whose rate is dependent on the concentration of one reactant.

Fischer esterification reaction the reaction of a carboxylic acid with excess alcohol in the presence of an acid catalyst to form an ester.

Fischer projection a method of representing the spatial arrangement of groups bonded to a chirality center. The chirality center is the point of intersection of two perpendicular lines; the horizontal lines represent bonds that project out of the plane of the paper toward the viewer, and the vertical lines represent bonds that point back from the plane of the paper away from the viewer.

flagpole hydrogens (transannular hydrogens) the two hydrogens in the boat conformation of cyclohexane that are closest to each other.

flavin adenine dinucleotide (FAD) a coenzyme required in certain oxidation reactions. It is reduced to $FADH_2$, which can act as a reducing agent in another reaction.

flavin mononucleotide (FMN) a coenzyme required in certain oxidation reactions. It is reduced to $FMNH_2$, which can act as a reducing agent in another reaction.

formal charge the number of valence electrons − (the number of nonbonding electrons + 1/2 the number of bonding electrons).

Fourier transform NMR a technique in which all the nuclei are excited simultaneously by an rf pulse, their relaxation monitored, and the data mathematically converted to a spectrum.

free energy of activation ($\Delta G^{\ddagger}$) the true energy barrier to a reaction.

free induction decay relaxation of excited nuclei.

frequency the velocity of a wave divided by its wavelength (in units of cycles/s).

Friedel–Crafts acylation an electrophilic substitution reaction that puts an acyl group on a benzene ring.

Friedel–Crafts alkylation an electrophilic substitution reaction that puts an alkyl group on a benzene ring.

frontier orbital analysis determining the outcome of a pericyclic reaction using frontier orbitals.

frontier orbitals the HOMO and the LUMO.

frontier orbital theory a theory that, like the conservation of orbital symmetry theory, explains the relationships, among reactant, product, and reaction conditions in a pericyclic reaction.

functional group the center of reactivity in a molecule.

functional group interconversion conversion of one functional group into another functional group.

functional group region the left-hand two-thirds of an IR spectrum where most functional groups show absorption bands.

furanose a five-membered ring sugar.

furanoside a five-membered ring glycoside.

fused bicyclic compound a bicyclic compound in which the rings share two adjacent carbons.

Gabriel synthesis conversion of an alkyl halide into a primary amine using phthalimide as a starting material.

gauche X and Y are gauche to each other.

gauche conformer a staggered conformer in which the largest substituents are gauche to each other.

gauche interaction the interaction between two atoms or groups that are gauche to each other.

***gem*-dialkyl effect** two alkyl groups on a carbon whose effect is to increase the probability that the molecule will be in the proper conformation for ring closure.

***gem*-diol** a compound with two OH groups on the same carbon.

geminal coupling the mutual splitting of two nonidentical protons bonded to the same carbon.

geminal dihalide a compound with two halogen atoms bonded to the same carbon.

gene a segment of DNA.

general-acid catalysis catalysis in which a proton is transferred to the reactant during the slow step of the reaction.

general-base catalysis catalysis in which a proton is removed from the reactant during the slow step of the reaction.

generic name a name for a drug that anyone can use.

gene therapy a technique that inserts a synthetic gene into the DNA of an organism defective in that gene.

genetic code the amino acid specified by each three-base sequence of mRNA.

geometric isomers cis-trans (or *E-Z*) isomers.

Gibbs standard free energy change the difference between the free energy content of the products and the free energy content of the reactants at equilibrium under standard conditions (1 M, 25 °C, 1 atm).

Gilman reagent a dialkylcopper reagent used to replace a halogen with an alkyl group.

globular protein a water-soluble protein that tends to have a roughly spherical shape.

gluconeogenesis the synthesis of D-glucose from pyruvate.

glycol a compound containing two or more OH groups.

glycolysis (glycolytic cycle) the sequence of reactions that convert D-glucose into two molecules of pyruvate.

glycoprotein a protein that is covalently bonded to a polysaccharide.

glycoside the acetal of a sugar.

glycosidic bond the bond between the anomeric carbon and the alcohol in a glycoside.

α–1,4′-glycosidic linkage a glycosidic linkage between the C-1 oxygen of one sugar and the C-4 of a second sugar with the oxygen atom of the glycosidic linkage in the axial position.

β–1,4′-glycosidic linkage a glycosidic linkage between the C-1 oxygen of one sugar and the C-4 of a second sugar with the oxygen atom of the glycosidic linkage in the equatorial position.

graft copolymer a copolymer that contains branches of a polymer of one monomer grafted onto the backbone of a polymer made from another monomer.

Grignard reagent   the compound that results when magnesium is inserted between the carbon and halogen of an alkyl halide (RMgBr, RMgCl).

ground-state electronic configuration   a description of which orbitals the electrons of an atom or molecule occupy when they are all in their lowest-energy orbitals.

half-chair conformation   the least stable conformation of cyclohexane.

haloform reaction   the reaction of a halogen and HO^- with a methyl ketone.

halogenation   reaction with halogen (Br_2, Cl_2).

halohydrin   an organic molecule that contains a halogen and an OH group on adjacent carbons.

Hammond postulate   states that the transition state will be more similar in structure to the species (reactants or products) that it is closer to energetically.

Haworth projection   a way to show the structure of a sugar; the five- and six-membered rings are represented as being flat.

head-to-tail addition   the head of one molecule is added to the tail of another molecule.

heat of combustion   the amount of heat given off when a carbon-containing compound reacts completely with O_2 to form CO_2 and H_2O.

heat of formation   the heat given off or consumed when a compound is formed from its elements under standard conditions.

heat of hydrogenation   the heat ($\Delta H°$) released in a hydrogenation reaction.

Heisenberg uncertainty principle   states that both the precise location and the momentum of an atomic particle cannot be simultaneously determined.

α-helix   the backbone of a polypeptide coiled in a right-handed spiral with hydrogen bonding occurring within the helix.

Hell–Volhard–Zelinski (HVZ) reaction   heating a carboxylic acid with Br_2 + P in order to convert it into an α-bromocarboxylic acid.

hemiacetal
$$\begin{array}{c} OH \\ | \\ RCH \\ | \\ OR \end{array}$$

hemiketal
$$\begin{array}{c} OH \\ | \\ RCR \\ | \\ OR \end{array}$$

Henderson–Hasselbalch equation $pK_a = pH + \log [HA]/[A^-]$

heptose   a monosaccharide with 7 carbons.

HETCOR spectrum   A 2-D NMR spectrum that shows coupling between protons and the carbons to which they are attached.

hetero atom   an atom other than carbon.

heterocyclic compound (heterocycle)   a cyclic compound in which one or more of the atoms of the ring are hetero atoms.

heterogeneous catalyst   a catalyst that is insoluble in the reaction mixture.

heterolytic bond cleavage (heterolysis)   breaking a bond with the result that both bonding electrons stay with one of the atoms.

hexose   a monosaccharide with six carbons.

high-energy bond   a bond that releases a great deal of energy when it is broken.

highest occupied molecular orbital (HOMO)   the molecular orbital of highest energy that contains an electron.

high-resolution NMR spectroscopy   NMR spectroscopy that uses a spectrometer with a high operating frequency.

Hofmann degradation   exhaustive methylation of an amine, followed by reaction with Ag_2O, followed by heating to achieve a Hofmann elimination reaction.

Hofmann elimination (anti-Zaitsev elimination)   a hydrogen is removed from the β-carbon bonded to the most hydrogens.

Hofmann elimination reaction   elimination of a proton and a tertiary amine from a quaternary ammonium hydroxide.

Hofmann rearrangement   conversion of an amide into an amine using Br_2/HO^-.

holoenzyme   an enzyme plus its cofactor.

homogeneous catalyst   a catalyst that is soluble in the reaction mixture.

homolog   a member of a homologous series.

homologous series   a family of compounds in which each member differs from the next by one methylene group.

homolytic bond cleavage (homolysis)   breaking a bond with the result that each of the atoms gets one of the bonding electrons.

homopolymer   a polymer that contains only one kind of monomer.

homotopic hydrogens   two hydrogens bonded to a carbon that is bonded to two other groups that are identical.

Hoogsteen base pairing   the pairing between a base in a synthetic strand of DNA with a base pair in double-stranded DNA.

Hooke's law   an equation that describes the motion of a vibrating string.

hormone   an organic compound synthesized in a gland and delivered by the bloodstream to its target tissue.

Hückel's rule   states that for a compound to be aromatic its cloud of electrons must contain ($4n + 2$) π electrons, where n is an integer. This is the same as saying it must contain an odd number of pairs of π electrons.

human genome   the total DNA of a human cell.

Hund's rule   states that when there are degenerate orbitals, an electron will occupy an empty orbital before it will pair up with another electron.

Hunsdiecker reaction   conversion of a carboxylic acid into an alkyl halide by heating a heavy metal salt of the carboxylic acid with bromine or iodine.

hybridized orbital   an orbital formed by mixing (hybridizing) orbitals.

hydrate (*gem*-diol)
$$\begin{array}{c} OH \\ | \\ RCR(H) \\ | \\ OH \end{array}$$

hydrated   water has been added to a compound.

hydration   addition of water to a compound.

hydrazone $R_2C=NNH_2$

hydride ion   a negatively charged hydrogen.

1,2-hydride shift   the movement of a hydride ion from one carbon to an adjacent carbon.

hydroboration–oxidation   the addition of borane to an alkene or an alkyne followed by reaction with hydrogen peroxide and hydroxide ion.

hydrocarbon   a compound that contains only carbon and hydrogen.

α-hydrogen   a hydrogen bonded to the carbon adjacent to a carbonyl carbon.

hydrogenation   addition of hydrogen.

hydrogen bond   an unusually strong dipole–dipole attraction (5 kcal/mol) between a hydrogen bonded to O, N, or F and the nonbonding electrons of an O, N, or F of another molecule.

hydrogen ion (a proton)   a positively charged hydrogen.

hydrolysis   reaction with water.

hydrophobic interactions   interactions between nonpolar groups. They increase stability by decreasing the amount of structured water (increasing entropy).

hyperconjugation   delocalization of electrons by overlap of carbon–hydrogen or carbon–carbon σ bonds with an empty p orbital.

imine $R_2C=NR$

inclusion compound   a compound that specifically binds a metal ion or an organic molecule.

induced dipole–induced dipole interaction   an interaction between a temporary dipole in one molecule and the dipole the temporary dipole induces in another molecule.

induced-fit model   a model that describes the specificity of an enzyme for its substrate: the shape of the active site does not become completely complementary to the shape of the substrate until after the enzyme has bound the substrate.

inductive electron donation   donation of electrons through a σ bond.

inductive electron withdrawal   withdrawal of electrons through a σ bond.

inflection point   the midpoint of the flattened-out region of a titration curve.

informational strand (sense strand)   the strand in DNA that is not read during transcription; it has the same sequence of bases as the synthesized mRNA strand (with a U, T difference).

infrared radiation   electromagnetic radiation familiar to us as heat.

infrared spectroscopy   uses infrared energy to provide a knowledge of the functional groups in a compound.

infrared (IR) spectrum   a plot of relative absorption versus wave number (or wavelength) of infrared radiation.

initiation step   the step in which radicals are created, or the step in which the radical needed for the first propagation step is created.

in-line displacement mechanism   nucleophilic attack on a phosphorus concerted with breaking a phosphoanhydride bond.

interchain disulfide bridge   a disulfide bridge between two cysteine residues in different peptide chains.

intermediate   a species formed during a reaction that is not the final product of the reaction.

intermolecular reaction   a reaction that takes place between two molecules.

internal alkyne   an alkyne with the triple bond not at the end of the carbon chain.

intimate ion pair   the covalent bond that joined the cation and the anion has broken, but the cation and anion are still next to each other.

intrachain disulfide bridge   a disulfide bridge between two cysteine residues in the same peptide chain.

intramolecular catalysis (anchimeric assistance)   catalysis in which the catalyst that facilitates the reaction is part of the molecule undergoing reaction.

intramolecular reaction   a reaction that takes place within a molecule.

intron   a stretch of bases in DNA that contain no genetic information.

inversion of configuration   turning the carbon inside-out like an umbrella in a windstorm so that the resulting product has a configuration opposite to that of the reactant.

iodoform test   addition of I_2/HO^- to a methyl ketone forms a yellow precipitate.

ion–dipole interaction   the interaction between an ion and the dipole of a molecule.

ion-exchange chromatography   a technique that uses a column packed with an insoluble resin to separate compounds on the basis of their charges and polarities.

ionic bond   a bond formed through the attraction of two ions of opposite charges.

ionization energy   the energy required to remove an electron from an atom.

ionophore   a compound that binds metal ions tightly.

iron protoporphyrin IX   the porphyrin ring system of heme plus an iron atom.

isoelectric point (pI)   the pH at which there is no net charge on an amino acid.

isolated double bonds   double bonds separated by more than one single bond.

isomers   nonidentical compounds with the same molecular formula.

isoprene rule   head-to-tail linkage of isoprene units.

isopropyl split   a split in the IR absorption band attributable to a methyl group. It is characteristic of an isopropyl group.

isotactic polymer   a polymer in which all the substituents are on the same side of the fully extended carbon chain.

isotopes   atoms with the same number of protons but different numbers of neutrons.

iterative synthesis   a synthesis in which a reaction sequence is carried out more than once.

IUPAC nomenclature   systematic nomenclature.

Kekulé structure   a model that represents the bonds between atoms as lines.

ketal $\begin{matrix} OR \\ | \\ RCR \\ | \\ OR \end{matrix}$

keto–enol tautomerism (keto–enol interconversion)   interconversion of keto and enol tautomers.

keto–enol tautomers   a ketone and its isomeric α,β-unsaturated alcohol.

β-keto ester   an ester with a second carbonyl group at the β-position.

ketone $\begin{matrix} O \\ \| \\ RCR \end{matrix}$

ketose   a polyhydroxyketone.

Kiliani–Fischer synthesis   a method used to increase the number of carbons in an aldose by one, resulting in the formation of a pair of C-2 epimers.

kinetic control   when a reaction is under kinetic control, the relative amounts of the products depend on the rates at which they are formed.

kinetic isotope effect   a comparison of the rate of reaction of a compound with the rate of reaction of an identical compound in which one of the atoms has been replaced by an isotope.

kinetic product   the product that is formed the fastest.

kinetic resolution   separation of enantiomers based on the difference in their rate of reaction with an enzyme.

kinetics   the field of chemistry that deals with the rates of chemical reactions.

kinetic stability   chemical reactivity, indicated by $\Delta G^{\ddagger}$. If $\Delta G^{\ddagger}$ is large, the compound is kinetically stable (not very reactive). If $\Delta G^{\ddagger}$ is small, the compound is kinetically unstable (very reactive).

Kolbe–Schmitt carboxylation reaction   a reaction that uses CO_2 to carboxylate phenol.

Krebs cycle (citric acid cycle, tricarboxylic acid cycle, TCA cycle)   a series of reactions that converts the acetyl group of acetyl-CoA into two molecules of CO_2.

lactam   a cyclic amide.

lactone   a cyclic ester.

lead compound   the prototype in a search for other biologically active compounds.

leaning   a line drawn over the outside peaks of an NMR signal points in the direction of the signal given by the protons that cause the splitting.

leaving group   the group that is displaced in a nucleophilic substitution reaction.

Le Châtelier's principle   states that if an equilibrium is disturbed, the components of the equilibrium will adjust in a way that will offset the disturbance.

lecithin   a phosphoacylglycerol in which the second OH group of phosphate has formed an ester with choline.

levorotatory   the enantiomer that rotates polarized light in a counterclockwise direction.

Lewis acid   a substance that accepts an electron pair.

Lewis base   a substance that donates an electron pair.

Lewis structure   a model that represents the bonds between atoms as lines or dots and the valence electrons as dots.

ligation   sharing of nonbonding electrons with a metal.

linear combination of atomic orbitals (LCAO)   the combination of p atomic orbitals to produce a molecular orbital.

linear conjugation   the atoms in the conjugated system are in a linear arrangement.

linear synthesis   a synthesis that builds a molecule step by step from starting materials.

lipid   a water-insoluble compound found in a living system.

lipid bilayer   two layers of phosphoacylglycerols arranged so that their polar heads are on the outside and their nonpolar fatty acid chains are on the inside.

lipoate   a coenzyme required in certain oxidation reactions.

living polymer a nonterminated chain-growth polymer that remains active. This means that the polymerization reaction can continue on addition of more monomer.

λ_{max} the wavelength at which there is maximum UV/Vis absorbance.

localized electrons electrons that are restricted to a particular locality.

lock-and-key model a model that describes the specificity of an enzyme for its substrate: the substrate fits the enzyme like a key fits a lock.

London forces Induced dipole–induced dipole interactions.

lone pair electrons (nonbonding electrons) valence electrons not used in bonding.

long-range coupling splitting of a proton by a proton more than three σ bonds away.

loop (coil) conformation that part of a protein that is highly ordered but not in an σ-helix or β-sheet.

lowest unoccupied molecular orbital (LUMO) the molecular orbital of lowest energy that does not contain an electron.

Lucas test a test that determines whether an alcohol is primary, secondary, or tertiary.

magnetic anisotropy the term used to describe the greater freedom of a π electron cloud to move in response to a magnetic field as a consequence of its greater polarizability compared with σ electrons.

magnetic resonance imaging (MRI) NMR used in medicine. The difference in the way water is bound in different tissues produces the signal variation between organs as well as between healthy and diseased states.

magnetogyric ratio a property (measured in rad $T^{-1}s^{-1}$) that depends on the magnetic properties of a particular kind of nucleus.

major groove the wider and deeper of the two alternating grooves in DNA.

malonic ester synthesis synthesis of a carboxylic acid using diethyl malonate as the starting material.

Markovnikov's rule the actual rule is: "when a hydrogen halide adds to an asymmetrical alkene, the addition occurs such that the halogen attaches itself to the sp^2 carbon of the alkene bearing the lowest number of hydrogen atoms." A more useful version is: the electrophile adds to the sp^2 carbon that is bonded to the greater number of hydrogens.

mass number the number of protons plus the number of neutrons in an atom.

mass spectrometry provides a knowledge of the molecular weight, molecular formula, and certain structural features of a compound.

mass spectrum a plot of the relative abundance of the positively charged fragments produced in a mass spectrometer versus their m/z values.

materials science the science of creating new materials to be used in place of known materials such as metal, glass, wood, cardboard, and paper.

Maxam–Gilbert sequencing a technique used to sequence restriction fragments.

McLafferty rearrangement rearrangement of the molecular ion of a ketone. The bond between the α- and β-carbons breaks, and a γ-hydrogen migrates to the oxygen.

mechanism-based inhibitor (suicide inhibitor) a compound that inactivates an enzyme by undergoing part of its normal catalytic mechanism.

mechanism of a reaction a description of the step-by-step process by which reactants are changed into products.

melting point the temperature at which a solid becomes a liquid.

membrane the material that surrounds the cell in order to isolate its contents.

mercaptan (thiol) the sulfur analog of an alcohol (RSH).

meso compound a compound that contains chirality centers and a plane of symmetry.

metabolism reactions living organisms carry out in order to obtain the energy they need and to synthesize the compounds they require.

meta-directing substituent a substituent that directs an incoming substituent meta to an existing substituent.

metal-activated enzyme an enzyme that has a loosely bound metal ion.

metal-ion catalysis catalysis in which the species that facilitates the reaction is a metal ion.

metalloenzyme an enzyme that has a tightly bound metal ion.

methine hydrogen a tertiary hydrogen.

methylene group a CH_2 group.

1,2-methyl shift the movement of a methyl group with its bonding electrons from one carbon to an adjacent carbon.

micelle a spherical aggregation of molecules each with a long hydrophobic tail and a polar head, arranged so that the polar head points to the outside of the sphere.

Michael reaction the addition of an α-carbanion to the β-carbon of an α,β-unsaturated carbonyl compound.

minor groove the narrower and more shallow of the two alternating grooves in DNA.

mixed (crossed) aldol addition an aldol addition in which two different carbonyl compounds are used.

mixed anhydride an acid anhydride with two different R groups.

$$\underset{R-C-O-C-R'}{\overset{\displaystyle O \qquad\quad O}{\overset{\displaystyle \| \qquad\quad \|}{}}}$$

mixed Claisen condensation a Claisen condensation in which two different esters are used.

mixed triacylglycerol a triacylglycerol in which the fatty acid components are different.

molar absorptivity the absorbance obtained from a 1.00-M solution in a cell with a 1.00-cm light path.

molecular ion (parent ion) peak in the mass spectrum with the greatest m/z.

molecular modeling computer-assisted design of a compound with a structure similar to that of a compound with the desired activity.

molecular modification changing the structure of a lead compound.

molecular orbital an orbital associated with a molecule.

molecular orbital theory describes a model in which the electrons occupy orbitals as they do in atoms but with the orbitals extending over the entire molecule.

molecular recognition the recognition of one molecule by another as a result of specific interactions; for example, the specificity of an enzyme for its substrate.

molozonide an unstable intermediate containing a five-membered ring with three oxygens in a row that is formed from the reaction of an alkene with ozone.

monomer a repeating unit in a polymer.

monosaccharide (simple carbohydrate) a single sugar molecule.

monoterpene a terpene that contains ten carbons.

MRI scanner an NMR spectrometer used in medicine for whole-body NMR.

multiplet an NMR signal split into more than seven peaks.

multiplicity the number of peaks in an NMR signal.

multistep synthesis preparation of a compound by a route that requires several steps.

mutarotation a slow change in optical rotation to an equilibrium value.

$N + 1$ rule an ^1H NMR signal for a hydrogen with N equivalent hydrogens bonded to an adjacent carbon is split into $N + 1$ peaks. A ^{13}C NMR signal for a carbon bonded to N hydrogens is split into $N + 1$ peaks.

N-glycoside a glycoside with a nitrogen instead of an oxygen at the glycosidic linkage.

N-terminal acid the terminal amino acid of a peptide (or protein) that has a free amino group.

natural abundance atomic weight the average mass of the atoms in the naturally occurring element.

natural product a product synthesized in nature.

neurotransmitter a compound that transmits nerve impulses.

nicotinamide adenine dinucleotide (NAD^+) a coenzyme required in certain oxidation reactions. It is reduced to NADH, which can act as a reducing agent in another reaction.

nicotinamide adenine dinucleotide phosphate ($NADP^+$) a coenzyme required in certain oxidation reactions. It is reduced to NADPH, which can act as a reducing agent in another reaction.

NIH shift   the 1,2-hydride shift of a carbocation (obtained from an arene oxide) that leads to an enone.

nitration   substitution of a nitro group (NO_2) for a hydrogen of a benzene ring.

nitrile   a compound that contains a carbon–nitrogen triple bond ($RC\equiv N$).

nitrosamine (*N*-nitroso compound) $R_2NN{=}O$.

NMR spectroscopy   the absorption of electromagnetic radiation to determine the structural features of an organic compound. In the case of ^1H NMR spectroscopy, it determines the carbon–hydrogen framework.

node   that part of an orbital in which there is zero probability of finding an electron.

nominal mass   mass rounded to the nearest whole number.

nonbonding electrons (lone pair electrons)   valence electrons not used in bonding.

nonbonding molecular orbital   the *p* orbitals are too far apart to overlap significantly, so the molecular orbital that results neither favors nor disfavors bonding.

nonpolar covalent bond   a bond formed between two atoms that share the bonding electrons equally.

nonreducing sugar   a sugar that cannot be oxidized by reagents such as Ag^+ and Cu^+. Nonreducing sugars are not in equilibrium with the open-chain aldose or ketose.

normal alkane (straight-chain alkane)   an alkane in which the carbons form a continuous chain with no branches.

nucleic acid   the two kinds of nucleic acid are DNA and RNA.

nucleophile   an electron-rich atom or molecule.

nucleophilic acyl substitution reaction   a reaction in which a group bonded to an acyl or aryl group is substituted by another group.

nucleophilic addition–elimination–nucleophilic addition reaction   a nucleophilic addition reaction that is followed by an elimination reaction that is followed by a nucleophilic addition reaction. Acetal formation is an example: an alcohol adds to the carbonyl carbon, water is eliminated, and a second molecule of alcohol adds to the dehydrated product.

nucleophilic addition–elimination reaction   a nucleophilic addition reaction that is followed by an elimination reaction. Imine formation is an example: an amine adds to the carbonyl carbon, and water is eliminated.

nucleophilic addition reaction   a reaction that involves the addition of a nucleophile to a reagent.

nucleophilic aromatic substitution   a reaction in which a nucleophile substitutes for a substituent of an aromatic ring.

nucleophilic catalysis (covalent catalysis)   catalysis that occurs as a result of a nucleophile forming a covalent bond with one of the reactants.

nucleophilic catalyst   a catalyst that increases the rate of a reaction by acting as a nucleophile.

nucleophilicity   a measure of how readily an atom or molecule with a pair of nonbonding electrons attacks an atom.

nucleophilic substitution reaction   a reaction in which a nucleophile substitutes for an atom or group.

nucleoside   a heterocyclic base (a purine or a pyrimidine) bonded to the anomeric carbon of a sugar (D-ribose or D-2-deoxyribose).

nucleotide   a heterocycle attached in the β-position to a phosphorylated ribose.

observed rotation   the amount of rotation observed in a polarimeter.

octet rule   states that an atom will give up, accept, or share electrons in order to achieve a filled shell. Because a filled second shell contains eight electrons, this is known as the octet rule.

off-resonance decoupling   the mode in ^{13}C NMR spectroscopy in which spin–spin splitting occurs between carbons and the hydrogens attached to them.

oil   a triester of glycerol that exists as a liquid at room temperature.

olefin   an alkene.

oligomer   a protein with more than one peptide chain.

oligonucleotide   three to ten nucleotides linked by phosphodiester bonds.

oligopeptide   three to ten amino acids linked by amide bonds.

oligosaccharide   three to ten sugar molecules linked by glycosidic bonds.

open-chain compound   an acyclic compound.

operating frequency   the frequency at which an NMR spectrometer operates.

optical isomers   stereoisomers that contain chirality centers.

optically active (chiral)   rotates the plane of polarized light.

optically inactive (achiral)   does not rotate the plane of polarized light.

optical purity   how much excess of one enantiomer is present in a mixture of a pair of enantiomers.

orbital   the volume of space around the nucleus in which an electron is most likely to be found.

orbital hybridization   mixing of orbitals.

organic compound   a compound that contains carbon.

organic synthesis   preparation of organic compounds from other organic compounds.

organometallic compound   a compound containing a carbon–metal bond.

oriented polymer   a polymer obtained by stretching out polymer chains and putting them back together in a parallel fashion.

orphan drugs   drugs for diseases or conditions that affect fewer than 200,000 people.

ortho/para-directing substituent   a substituent that directs an incoming substituent ortho and para to an existing substituent.

osazone   the product obtained by treating an aldose or a ketose with excess phenylhydrazine. An osazone contains two imine bonds.

overtone band   occurs at a multiple of the fundamental absorption frequency ($2v_1$, $3v_1$).

oxidation   loss of electrons by an atom or molecule.

oxidation–reduction reaction (redox reaction)   a reaction that involves the transfer of electrons from one species to another.

oxidative cleavage   an oxidation reaction that cuts the reactant into two or more pieces.

oxime $R_2C{=}NOH$

oxirane (epoxide)   an ether in which the oxygen is incorporated into a three-membered ring.

oxonium ion   a compound with a positively charged oxygen.

oxyanion   a compound with a negatively charged oxygen.

oxymercuration   addition of water using a mercuric salt of a carboxylic acid as a catalyst.

ozonide   the five-membered ring compound formed as a result of rearrangement of a molozonide.

ozonolysis   reaction of a carbon–carbon double or triple bond with ozone.

packing   the fitting of individual molecules into a frozen crystal lattice.

paraffin   an alkane.

parent hydrocarbon   the longest continuous carbon chain in a molecule.

parent ion (molecular ion)   peak in the mass spectrum with the greatest *m/z*.

partial hydrolysis   a technique that hydrolyzes only some of the peptide bonds in a polypeptide.

partial racemization   formation of a pair of enantiomers in unequal amounts.

Pauli exclusion principle   states that no more than two electrons can occupy an orbital and that the two electrons must have opposite spin.

pentose   a monosaccharide with five carbons.

peptide   polymer of amino acids linked together by amide bonds. A peptide contains fewer amino acid residues than a protein.

peptide bond   the amide bond that links the amino acids in a peptide or protein.

peptide nucleic acid (PNA)   a polymer containing both an amino acid and a base designed to bind to specific residues on DNA or mRNA.

pericyclic reaction   a concerted reaction that takes place as the result of a cyclic rearrangement of electrons.

peroxy acid   a carboxylic acid with an OOH group instead of an OH group.

perspective formula   a method of representing the spatial arrangement of groups bonded to a chirality center. Two bonds are drawn in the plane of the paper; a solid wedge is used to depict a bond that projects out of the plane of the paper toward the viewer, and a dashed line is used to represent a bond that projects back from the plane of the paper away from the viewer.

pH   the pH scale is used to describe the acidity of a solution (pH = $-\log [\text{H}^+]$).

pH-activity profile   a plot of the activity of an enzyme as a function of the pH of the reaction mixture.

phase transfer catalysis   catalysis of a reaction by providing a way to bring a polar reagent into a nonpolar phase so that the reaction between a polar and a nonpolar compound can occur.

phase transfer catalyst   a compound that carries a polar reagent into a nonpolar phase.

$$\text{phenone} \quad \text{C}_6\text{H}_5\overset{\overset{\text{O}}{\|}}{\text{C}}\text{R}$$

phenyl group   C$_6$H$_5$—

phenylhydrazone   R$_2$C=NNHC$_6$H$_5$

pheromone   a compound secreted by an animal that stimulates a physiological or behavioral response from a member of the same species.

phosphatidic acid   a phosphoacylglycerol in which only one of the OH groups of phosphate is in an ester linkage.

phosphoacylglycerol (phosphoglyceride)   formed when two OH groups of glycerol form esters with fatty acids and the terminal OH group forms a phosphate ester.

phosphoanhydride bond   the bond holding two phosphoric acid molecules together.

phospholipid   a lipid that contains a phosphate group.

phosphoryl transfer reaction   the transfer of a phosphate group from one compound to another.

photochemical reaction   a reaction that takes place when a reactant absorbs light.

photosynthesis   the synthesis of glucose and O$_2$ from CO$_2$ and H$_2$O.

pH-rate profile   a plot of the observed rate for a reaction as a function of the pH of the reaction mixture.

pi (π) bond   a bond formed as a result of side-to-side overlap of p orbitals.

pi-complex   a complex formed between an electrophile and a triple bond.

pinacol rearrangement   rearrangement of a vicinal diol.

pK_a   describes the tendency of a compound to lose a proton (pK_a = $-\log K_a$, where K_a is the acid dissociation constant).

plane of symmetry   an imaginary plane that bisects a molecule into a pair of mirror images.

plasticizer   an organic molecule that dissolves in a polymer and allows the polymer chains to slide by each other.

β-pleated sheet   the backbone of a polypeptide is extended in a zigzag structure with hydrogen bonding between neighboring chains.

polar covalent bond   a bond formed by the unequal sharing of electrons.

polarimeter   an instrument that measures the rotation of polarized light.

polarizability   an indication of the ease with which the electron cloud of an atom can be distorted.

polarized light   light that oscillates only in one plane.

polar reaction   the reaction between a nucleophile and an electrophile.

polyamide   a polymer in which the monomers are amides.

polycarbonate   a step-growth polymer in which the dicarboxylic acid is carbonic acid.

polyene   a compound that has several double bonds.

polyester   a polymer in which the monomers are esters.

polymer   a large molecule made by linking monomers together.

polymer chemistry   the field of chemistry that deals with synthetic polymers; part of the larger discipline known as materials science.

polymerization   the process of linking up monomers to form a polymer.

polynucleotide   many nucleotides linked by phosphodiester bonds.

polypeptide   many amino acids linked by amide bonds.

polysaccharide   a compound containing ten or more sugar molecules linked together.

polyunsaturated fatty acid   a fatty acid with more than one double bond.

polyurethane   a polymer in which the monomers are urethanes.

porphyrin ring system   consists of four pyrrole rings joined by one-carbon bridges.

primary alcohol   an alcohol in which the OH group is bonded to a primary carbon.

primary alkyl halide   an alkyl halide in which the halogen is bonded to a primary carbon.

primary amine   an amine with one alkyl group bonded to the nitrogen.

primary carbocation   a carbocation with the positive charge on a primary carbon.

primary carbon   a carbon bonded to only one other carbon.

primary hydrogen   a hydrogen bonded to a primary carbon.

primary radical   a radical with the unpaired electron on a primary carbon.

primary structure (of a nucleic acid)   the sequence of bases in the nucleic acid.

primary structure (of a protein)   the sequence of amino acids in the protein.

principle of microscopic reversibility   states that the mechanism for a reaction in the forward direction has the same intermediates and the same rate-determining step as the mechanism for the reaction in the reverse direction.

prochirality center   a carbon bonded to two hydrogens that will become a chirality center if one of the hydrogens is replaced by deuterium.

prokaryote   a unicellular body without a nucleus.

promoter site   a short sequence of bases at the beginning of a gene.

propagating site   the reactive end of a chain-growth polymer.

propagation step   in the first of a pair of propagation steps, a radical (or an electrophile or a nucleophile) reacts to produce another radical (or an electrophile or a nucleophile) that reacts in the second propagation step to produce the radical (or the electrophile or the nucleophile) that was the reactant in the first propagation step.

proprietary name (trade name, brand name)   identifies a commercial product and distinguishes it from other products.

pro-R-hydrogen   replacing this hydrogen with deuterium creates a chirality center with the R configuration.

pro-S-hydrogen   replacing this hydrogen with deuterium creates a chirality center with the S configuration.

prostacyclin   a lipid, derived from arachidonic acid, that dilates blood vessels and inhibits platelet aggregation.

prochiral carbonyl carbon   a carbonyl carbon that will become a chirality center if it is attacked by a group unlike any of the groups already bonded to it.

prosthetic group   a tightly bound coenzyme.

protecting group   a reagent that protects a functional group from a synthetic operation that it would otherwise not survive.

protein   polymer of amino acids linked by amide bonds.

protic solvent   a solvent that has a hydrogen bonded to an oxygen or a nitrogen.

proton   a positively charged hydrogen; a positively charged particle in an atomic nucleus.

proton-decoupled ^{13}C NMR spectrum   a ^{13}C NMR spectrum in which all the signals appear as singlets because there is no coupling between the ^{13}C nucleus and its bonded hydrogens.

proton transfer reaction   a reaction in which a proton is transferred from an acid to a base.

protoporphyrin-IX   the porphyrin ring system of heme.

pseudo first-order reaction   a second-order reaction in which the concentration of one of the reactants is much greater than the other, which allows the reaction to be treated as a first-order reaction.

pyranose   a six-membered ring sugar.

pyranoside   a six-membered ring glycoside.

pyridoxal-phosphate   the coenzyme required by enzymes that catalyze certain transformations of amino acids.

quantitative structure–activity relationship (QSAR)   the relation between a particular property of a series of compounds and their biological activity.

quantum numbers   numbers arising from the quantum mechanical treatment of an atom that describe the properties of the electrons in the atom.

quartet   an NMR signal split into four peaks.

quaternary ammonium ion   an ion containing a nitrogen bonded to four alkyl groups (R_4N^+).

quaternary ammonium salt   a quaternary ammonium ion and an anion ($R_4N^+X^-$).

quaternary structure   a description of the way the individual polypeptide chains of a protein are arranged with respect to each other.

racemic mixture (racemate, racemic modification)   a mixture of equal amounts of a pair of enantiomers.

radical   an atom or molecule with an unpaired electron.

radical addition reaction   an addition reaction in which the first species that adds is a radical.

radical anion   a species with a negative charge and an unpaired electron.

radical cation   a species with a positive charge and an unpaired electron.

radical chain reaction   a reaction in which radicals are formed and react in repeating propagating steps.

radical inhibitor   a compound that traps radicals.

radical initiator   a compound that creates radicals.

radical polymerization   chain-growth polymerization in which the initiator is a radical; the propagation site therefore is a radical.

radical reaction   a reaction in which a new bond is formed using one electron from one reagent and one electron from another reagent.

radical substitution reaction   a substitution reaction that has a radical intermediate.

random coil   the conformation of a totally denatured protein.

random copolymer   a copolymer with a random distribution of monomers.

random screen (blind screen)   the search for a pharmacologically active compound without any information about what chemical structures might show activity.

rate constant   a measure of how easy it is to reach the transition state of a reaction (to get over the energy barrier to the reaction).

rate-determining step (rate-limiting step)   the step in a reaction that has the transition state with the highest energy.

rational drug design   designing drugs with a particular structure to achieve a specific purpose.

_R_ configuration   after assigning relative priorities to the four groups bonded to a chirality center, if the lowest-priority group is on a vertical axis in a Fischer projection (or pointing away from the viewer in a perspective formula), an arrow drawn from the highest-priority group to the next-highest-priority group goes in a clockwise direction.

reaction coordinate diagram   describes the energy changes that take place during the course of a reaction.

reactivity–selectivity principle   states that the greater the reactivity of a species, the less selective it will be.

receptor site   the site at which a drug binds in order to exert its physiological effect.

redox reaction (oxidation–reduction reaction)   a reaction that involves the transfer of electrons from one species to another.

red shift   a shift to a longer wavelength.

reducing sugar   a sugar that can be oxidized by reagents such as Ag^+ or Br_2. Reducing sugars are in equilibrium with the open-chain aldose or ketose.

reduction   gain of electrons by an atom or molecule.

reductive amination   the reaction of an aldehyde or a ketone with ammonia or with a primary amine in the presence of a reducing agent (H_2/Raney Ni).

reference compound   a compound added to the sample whose NMR spectrum is to be taken. The positions of the signals in the NMR spectrum are measured from the position of the signal given by the reference compound.

regioselective reaction   a reaction that leads to the preferential formation of one constitutional isomer over another.

relative configuration   the configuration of a compound relative to the configuration of another compound.

relative rate   obtained by dividing the actual rate constant by the rate constant of the slowest reaction in the group being compared.

replication   the synthesis of identical copies of DNA.

replication fork   the position on DNA at which replication begins.

resolution of a racemic mixture   separation of a racemic mixture into the individual enantiomers.

resonance   a compound with delocalized electrons is said to have resonance.

resonance contributor (resonance structure)   a structure with localized electrons that approximates the true structure of a compound with delocalized electrons.

resonance electron donation   donation of electrons through _p_ orbital overlap with neighboring π bonds.

resonance electron withdrawal   withdrawal of electrons through _p_ orbital overlap with neighboring π bonds.

resonance energy (delocalization energy)   the extra stability associated with a compound as a result of its having delocalized electrons.

resonance hybrid   the actual structure of a compound with delocalized electrons; it is represented by two or more structures with localized electrons.

resonances   NMR absorption signals.

restriction endonuclease   an enzyme that cleaves DNA at a specific base sequence.

restriction fragment   a fragment that is formed when DNA is cleaved by a restriction endonuclease.

retrosynthesis (retrosynthetic analysis)   working backward (on paper) from the target molecule to available starting materials.

retrovirus   a virus whose genetic information is stored in its RNA.

rf radiation   radiation in the radiofrequency region of the electromagnetic spectrum.

ribonucleic acid (RNA)   a polymer of ribonucleotides.

ribonucleotide   a nucleotide in which the sugar component is D-ribose.

ribosome   a particle composed of about 40% protein and 60% RNA on which protein biosynthesis takes place.

ribozyme   an RNA molecule that acts as a catalyst.

ring current   the movement of π electrons around a benzene ring.

ring-expansion rearrangement   rearrangement of a carbocation in which the positively charged carbon is bonded to a cyclic compound and as a result of rearrangement the size of the ring increases by one carbon.

ring-flip (chair–chair interconversion)   the conversion of the chair conformer of cyclohexane into the other chair conformer. Bonds that are axial in one chair conformer are equatorial in the other chair conformer.

ring-opening polymerization   a chain-growth polymerization that involves opening the ring of the monomer.

Ritter reaction   reaction of a nitrile with a secondary or tertiary alcohol to form a secondary amide.

RNA (ribonucleic acid)   a polymer of ribonucleotides.

RNA splicing   the step in RNA processing that cuts out nonsense bases and splices informational pieces together.

Robinson annulation   a Michael reaction followed by an intramolecular aldol condensation.

Rosenmund reduction   reduction of an acyl chloride to an aldehyde using H_2 and a deactivated palladium catalyst.

Ruff degradation   a method used to shorten an aldose by one carbon.

Sandmeyer reaction   the reaction of an aryl diazonium salt with a cuprous salt.

saponification   hydrolysis of a fat under basic conditions.

saturated hydrocarbon   a hydrocarbon that is completely saturated with hydrogen (contains no double or triple bonds).

Schiemann reaction   the reaction of an aryl diazonium salt with HBF_4.

Schiff base $R_2C=NR$

s-cis-conformation   the conformation in which two double bonds are on the same side of a single bond.

S configuration   after assigning relative priorities to the four groups bonded to a chirality center, if the lowest-priority group is on a vertical axis in a Fischer projection (or pointing away from the viewer in a perspective formula), an arrow drawn from the highest-priority group to the next-highest-priority group goes in a counterclockwise direction.

secondary alcohol   an alcohol in which the OH group is bonded to a secondary carbon.

secondary alkyl halide   an alkyl halide in which the halogen is bonded to a secondary carbon.

secondary amine   an amine with two alkyl groups bonded to the nitrogen.

secondary carbocation   a carbocation with the positive charge on a secondary carbon.

secondary carbon   a carbon bonded to two other carbons.

secondary hydrogen   a hydrogen bonded to a secondary carbon.

secondary radical   a radical with the unpaired electron on a secondary carbon.

secondary structure   a description of the conformation of the backbone of a protein.

second-order rate constant   the rate constant of a second-order reaction.

second-order reaction (bimolecular reaction)   a reaction whose rate is dependent on the concentration of two reactants.

sedimentation constant   designates where a species sediments in an ultracentrifuge.

selection rules   the rules that determine the outcome of a pericyclic reaction.

selenenylation reaction   conversion of an α-bromoketone into an α,β-unsaturated ketone via formation of an α-phenylseleno ketone.

semicarbazone $R_2C=NNHCNH_2$ (with O double-bonded to C)

semiconservative replication   the mode of replication that results in a daughter molecule of DNA having one of the original DNA strands plus a newly synthesized strand.

sense strand (informational strand)   the strand in DNA that is not read during transcription; it has the same sequence of bases as the synthesized mRNA strand (with a U, T difference).

separated charges   a positive and a negative charge that can be neutralized by the movement of electrons.

sesquiterpene   a terpene that contains 15 carbons.

shielding   caused by electron donation to the environment of a proton. The electrons shield the proton from the full effect of the applied magnetic field. The more a proton is shielded, the farther to the right its signal appears in an NMR spectrum.

sigma (σ) bond   a bond with a symmetrical distribution of electrons.

sigmatropic rearrangement   a reaction in which a σ bond is broken in the reactant, a new σ bond is formed in the product, and the π bonds rearrange.

Simmons–Smith reaction   formation of a cyclopropane using CH_2I_2 + Zn(Cu).

simple carbohydrate (monosaccharide)   a single sugar molecule.

simple triacylglycerol   a triacylglycerol in which the fatty acid components are the same.

single bond   a σ bond.

singlet   an unsplit NMR signal.

site-specific mutagenesis   a technique that substitutes one amino acid of a protein for another.

site-specific recognition   recognition of a particular site on DNA.

skeletal structure   shows the carbon–carbon bonds as lines and does not show the carbon–hydrogen bonds.

S_N1 reaction   a unimolecular nucleophilic substitution reaction.

S_N2 reaction   a bimolecular nucleophilic substitution reaction.

S_NAr reaction   a nucleophilic aromatic substitution reaction.

soap   a sodium or potassium salt of a fatty acid.

solid-phase synthesis   a technique in which one end of the compound being synthesized is covalently attached to a solid support.

solvation   the interaction between a solvent and another molecule (or ion).

solvent-separated ion pair   the cation and anion are separated by a solvent molecule.

solvolysis   reaction with the solvent.

specific-acid catalysis   catalysis in which the proton is fully transferred to the reactant before the slow step of the reaction.

specific-base catalysis   catalysis in which the proton is completely removed from the reactant before the slow step of the reaction.

specific rotation   the amount of rotation that will be caused by a compound with a concentration of 1.0 g/mL in a sample tube 1.0 dm long.

spectroscopy   study of the interaction of matter and electromagnetic radiation.

sphingolipid   a lipid that contains sphingosine.

sphingomyelin   a sphingolipid in which the terminal OH group of sphingosine is bonded to phosphocholine or phosphoethanolamine.

spin-coupled ^{13}C NMR spectrum   a ^{13}C NMR spectrum in which each signal for a carbon is split by the hydrogens bonded to that carbon.

spin-coupling   the atom that gives rise to an NMR signal is coupled to the rest of the molecule.

spin-decoupling   the atom that gives rise to an NMR signal is decoupled from the rest of the molecule.

spin–spin coupling   the splitting of a signal in an NMR spectrum described by the N + 1 rule.

α-spin state   nuclei in this spin state have their magnetic moments oriented in the same direction as the applied magnetic field.

β-spin state   nuclei in this spin state have their magnetic moments oriented opposite to the direction of the applied magnetic field.

spirocyclic compound   a bicyclic compound in which the rings share one carbon.

splitting diagram   a diagram that describes the splitting of a set of protons.

squalene   a triterpene that is a precursor of steroid molecules.

stacking interactions   van der Waals interactions between the mutually induced dipoles of adjacent pairs of bases in DNA.

staggered conformation   a conformation in which the bonds on one carbon bisect the bond angle on the adjacent carbon when viewed looking down the carbon–carbon bond.

step-growth polymer (condensation polymer)   made by combining two molecules while removing a small molecule (usually water or an alcohol).

stereochemistry   the field of chemistry that deals with the structures of molecules in three dimensions.

stereoelectronic effects   the combination of steric effects and electronic effects.

stereogenic center (chirality center)   a carbon bonded to four different substituents.

stereoisomers   isomers that differ in the way the atoms are arranged in space.

stereoselective reaction   a reaction that leads to the preferential formation of one stereoisomer over another.

stereospecific reaction   a reaction in which the reactant can exist as stereoisomers and each stereoisomeric reactant leads to a different stereoisomeric product.

steric effects effects due to the fact that groups occupy a certain volume of space.

steric hindrance refers to bulky groups at the site of a reaction that make it difficult for the reactants to approach each other.

steric strain (van der Waals strain or van der Waals repulsion) the repulsion between the electron cloud of an atom or group of atoms and the electron cloud of another atom or group of atoms.

steroid a class of compounds that contains a steroid ring system.

stop codon a codon that says "stop protein synthesis here."

Stork enamine reaction uses an enamine as a nucleophile in a Michael reaction.

straight-chain alkane an alkane in which the carbons form a continuous chain with no branches.

***s*-trans-conformation** a conformation in which two double bonds are on opposite sides of a single bond.

Strecker synthesis a method used to synthesize an amino acid: an aldehyde reacts with NH_3, forming an imine that is attacked by cyanide ion. Hydrolysis of the product gives an amino acid.

stretching frequency the frequency at which a stretching vibration occurs.

stretching vibration a vibration occurring along the line of the bond.

structural isomers (constitutional isomers) molecules that have the same molecular formula but differ in the way the atoms are connected.

structural protein a protein that gives strength to a biological structure.

α-substituent a substituent on the opposite side of a steroid ring system as the angular methyl groups.

β-substituent a substituent on the same side of a steroid ring system as the angular methyl groups.

α-substitution reaction a reaction that puts a substituent on an α-carbon in place of an α-hydrogen.

substrate the reactant of an enzyme-catalyzed reaction.

subunit an individual chain of an oligomer.

suicide inhibitor (mechanism-based inhibitor) a compound that inactivates an enzyme by undergoing part of its normal catalytic mechanism.

sulfide (thioether) the sulfur analog of an ether (RSR).

sulfonate ester the ester of a sulfonic acid (RSO_2OR).

sulfonation substitution of a hydrogen of a benzene ring by a sulfonic acid group ($-SO_3H$).

suprafacial bond formation formation of two σ bonds from the same side of the π system.

suprafacial rearrangement rearrangement in which the migrating group remains on the same face of the π system.

symmetrical anhydride an acid anhydride with identical R groups.

$$R-\overset{\overset{O}{\|}}{C}-O-\overset{\overset{O}{\|}}{C}-R$$

symmetrical ether an ether with two identical substituents bonded to the oxygen.

symmetric molecular orbital a molecular orbital in which the left half is a mirror image of the right half.

symmetry-allowed pathway a pathway that leads to overlap of in-phase orbitals.

symmetry-forbidden pathway a pathway that leads to overlap of out-of-phase orbitals.

syn addition an addition reaction in which the two added substituents add to the same side of the molecule.

syndiotactic polymer a polymer in which the substituents regularly alternate on both sides of the fully extended carbon chain.

syn elimination an elimination reaction in which the two substituents eliminated are removed from the same side of the molecule.

syn-periplanar parallel substituents on the same side of a molecule.

synthetic equivalent the reagent actually used as the source of a synthon.

synthetic polymer a polymer that is not synthesized in nature.

synthetic tree an outline of the available routes to get to the desired

product from available starting materials.

synthon a fragment of a disconnection.

systematic nomenclature IUPAC nomenclature.

target molecule the desired end product of a synthesis.

tautomerism interconversion of tautomers.

tautomers isomers that differ in the location of a double bond and a hydrogen.

template strand (antisense strand) the strand in DNA that is read during transcription.

terminal alkyne an alkyne with the triple bond at the end of the carbon chain.

termination step two radicals combine to produce a molecule in which all the electrons are paired.

terpene a lipid, isolated from a plant, that contains carbon atoms in multiples of 5.

terpenoid a terpene that contains oxygen.

tertiary alcohol an alcohol in which the OH group is bonded to a tertiary carbon.

tertiary alkyl halide an alkyl halide in which the halogen is bonded to a tertiary carbon.

tertiary amine an amine with three alkyl groups bonded to the nitrogen.

tertiary carbocation a carbocation with the positive charge on a tertiary carbon.

tertiary carbon a carbon bonded to three other carbons.

tertiary hydrogen a hydrogen bonded to a tertiary carbon.

tertiary radical a radical with the unpaired electron on a tertiary carbon.

tertiary structure a description of the three-dimensional arrangement of all the atoms in a protein.

tetraene a hydrocarbon with four double bonds.

tetrahedral bond angle the bond angle (109.5°) formed by adjacent bonds of an sp^3 hybridized carbon.

tetrahedral carbon an sp^3 hybridized carbon; a carbon that forms covalent bonds using four sp^3 hybridized orbitals.

tetrahedral intermediate the intermediate formed in a nucleophilic acyl substitution reaction.

tetrahydrofolate (THF) the coenzyme required by enzymes that catalyze reactions that donate a group containing a single carbon to their substrates.

tetraterpene a terpene that contains 40 carbons.

tetrose a monosaccharide with four carbons.

therapeutic index the ratio of the lethal dose of a drug to the therapeutic dose.

thermal cracking using heat to break a molecule apart.

thermal reaction a reaction that takes place without the reactant having to absorb light.

thermodynamic control when a reaction is under thermodynamic control, the relative amounts of the products depend on their stabilities.

thermodynamic product the most stable product.

thermodynamics the field of chemistry that describes the properties of a system at equilibrium.

thermodynamic stability is indicated by $\Delta G°$. If $\Delta G°$ is negative, the products are more stable than the reactants. If $\Delta G°$ is positive, the reactants are more stable than the products.

thermoplastic polymer a polymer that has both ordered crystalline regions and amorphous noncrystalline regions.

thermosetting polymer cross-linked polymers that, after they are hardened, cannot be remelted by heating.

thiamine pyrophosphate (TPP) the coenzyme required by enzymes that catalyze a reaction that transfers a two-carbon fragment to a substrate.

thiirane a three-membered ring compound in which one of the ring atoms is a sulfur.

thin-layer chromatography a technique that separates compounds on the basis of their polarity.

thioester the sulfur analog of an ester. $R-\overset{\overset{O}{\|}}{C}-SR$

thioether (sulfide)   the sulfur analog of an ether (RSR).

thiol (mercaptan)   the sulfur analog of an alcohol (RSH).

threo enantiomers   the pair of enantiomers with similar groups on opposite sides when drawn in a Fischer projection.

titration curve   a plot of pH versus added equivalents of hydroxide ion.

Tollens test   an aldehyde can be identified by observation of the formation of a silver mirror in the presence of Tollens' reagent (Ag_2O/NH_3).

torsional strain   the repulsion felt by the bonding electrons of one substituent as they pass close to the bonding electrons of another substituent.

trademark   a registered name, symbol, or picture.

trade name (proprietary name, brand name)   identifies a commercial product and distinguishes it from other products.

transamination   a reaction in which an amino group is transferred from one compound to another.

transannular hydrogens (flagpole hydrogens)   the two hydrogens in the boat conformation of cyclohexane that are closest to each other.

transcription   the synthesis of mRNA from a DNA blueprint.

transesterification reaction   the reaction of an ester with an alcohol to form a different ester.

trans fused   two cyclohexane rings fused together such that if the second ring were considered to be two substituents of the first ring, both substituents would be in equatorial positions.

transimination   the reaction of a primary amine with an imine to form a new imine and a primary amine derived from the original imine.

trans isomer   the isomer with identical substituents on opposite sides of the double bond.

transition state   the highest point on a hill in a reaction coordinate diagram. In the transition state, bonds in the reactant that will break are partially broken and bonds in the product that will form are partially formed.

transition state analog   a compound that is structurally similar to the transition state of an enzyme-catalyzed reaction.

translation   the synthesis of a protein from an mRNA blueprint.

transmetallation   metal exchange.

triacylglycerol   the compound formed when the three OH groups of glycerol are esterified with fatty acids.

triene   a hydrocarbon with three double bonds.

trigonal planar carbon   an sp^2 hybridized carbon.

triose   a monosaccharide with three carbons.

tripeptide   three amino acids linked by amide bonds.

triple bond   a σ bond plus two π bonds.

triplet   an NMR signal split into three peaks.

triterpene   a terpene that contains 30 carbons.

twist-boat (skew-boat) conformation   a conformation of cyclohexane.

ultraviolet light   electromagnetic radiation with wavelengths ranging from 180 to 400 nm.

umpolung   reversing the normal polarity of a functional group.

unimolecular reaction (first-order reaction)   a reaction whose rate is dependent on the concentration of one reactant.

unsaturated hydrocarbon   a hydrocarbon that contains one or more double or triple bonds.

urethane   a compound with a carbonyl group that is both an amide and an ester.

UV/Vis spectroscopy   the absorption of electromagnetic radiation to determine information about conjugated systems.

valence electron   an electron in an unfilled shell.

valence shell electron pair repulsion (VSEPR) model   combines the concept of atomic orbitals with the concept of shared electron pairs and the minimization of electron pair repulsion.

van der Waals forces (London forces)   induced dipole–induced dipole interactions.

van der Waals radius   a measure of the effective size of an atom or group. A repulsive force occurs (van der Waals repulsion) if two atoms approach each other at a distance less than the sum of their van der Waals radii.

vector sum   takes into account both the magnitudes and the directions of the bond dipoles.

vicinal dihalide   a compound with halogens bonded to adjacent carbons.

vicinal diol (vicinal glycol)   a compound with OH groups bonded to adjacent carbons.

vinyl group $CH_2{=}CH{-}$

vinylic carbon   a carbon in a carbon–carbon double bond.

vinylic cation   a compound with a positive charge on a vinylic carbon.

vinylic radical   a compound with an unpaired electron on a vinylic carbon.

vinylogy   transmission of reactivity through double bonds.

vinyl polymer   a polymer in which the monomers are ethylene or a substituted ethylene.

visible light   electromagnetic radiation with wavelengths ranging from 400 to 780 nm.

vitamin   a substance needed in small amounts for normal body function that the body cannot synthesize or cannot synthesize in adequate amounts.

vitamin KH_2   the coenzyme required by the enzyme that catalyzes the carboxylation of glutamate side chains.

vulcanization   increasing the flexibility of rubber by heating it with sulfur.

wave equation   an equation that describes the behavior of each electron in an atom or a molecule.

wave functions   a series of solutions of a wave equation.

wavelength   distance from any point on one wave to the corresponding point on the next wave (in units of μm).

wavenumber   the number of waves in 1 cm.

wax   an ester formed from a long-chain carboxylic acid and a long-chain alcohol.

wedge-and-dash structure   a method of representing the spatial arrangement of groups bonded to a chirality center. Wedges are used to represent bonds that point out of the plane of the paper toward the viewer, and dashed lines are used to represent bonds that point back from the plane of the paper away from the viewer.

Williamson ether synthesis   formation of an ether from the reaction of an alkoxide ion with an alkyl halide.

Wittig reaction   the reaction of an aldehyde or a ketone with a phosphonium ylide, resulting in formation of an alkene.

Wolff–Kishner reduction   a reaction that reduces the carbonyl group of a ketone to a methylene group using NH_2NH_2/HO^-.

Woodward–Fieser rules   allow the calculation of the λ_{max} of the $\pi \longrightarrow \pi^*$ transition for compounds with four or fewer conjugated double bonds.

Woodward–Hoffmann rules   a series of selection rules for pericyclic reactions.

ylide   a compound with opposite charges on adjacent, covalently bonded atoms with complete octets.

Zaitsev's rule   the more stable alkene product is obtained by removing a proton from the β-carbon that is bonded to the fewest hydrogens.

Z conformation   the conformation of a carboxylic acid or carboxylic acid derivative in which the carbonyl oxygen and the substituent bonded to the carboxyl oxygen or nitrogen are on the same side of the single bond.

Ziegler–Natta catalyst   an aluminum–titanium initiator that controls the stereochemistry of a polymer.

Z isomer   the isomer with the high-priority groups on the same side of the double bond.

zwitterion   a compound with a negative charge and a positive charge on nonadjacent atoms.

PHOTO AND ART CREDITS

Chapter 1

CO: Kenneth Eward/Biografx 1997. Page 6: UPI Corbis-Bettmann. Page 6: Institute for Advanced Study Archives. Page 9: Paul Silverman, Fundamental Photographs. Page 18: AP/Wide World Photos. Page 23: Joe McNally, Sygma. Page 45: Richard Megna, Fundamental Photographs.

Chapter 2

CO: Kenneth Eward/Biografx 1997. Page 81: AP/Wide World Photos.

Chapter 3

CO: Kenneth Eward/Biografx 1997. Page: 103: Hans Reinhard/Okapia., Photo Researchers, Inc. Page 121: Sidney Harris. Page 128: Hulton Getty, Tony Stone Images. Page: 138: University of Pennsylvania, Van Pelt Library.

Chapter 4

CO: Kenneth Eward/Biografx 1997. Page 174: Genentech, Inc. Page 180: Brown Brothers. Page 187: Diane Schiumo, Fundamental Photographs. Page 200: Archive Photos. Page 201: David Parker, Photo Researchers, Inc.

Chapter 5

CO: Kenneth Eward/Biografx 1997. Page 256: Archive Photos.

Chapter 6

CO: Kenneth Eward/Biografx 1997. Page 264: Corbis-Bettmann. Page 284: Ken Eward, Photo Researchers, Inc.

Chapter 7

CO: Kenneth Eward/Biografx 1997. Page 317: AP/Wide World Photos. Page 317: AP/Wide World Photos.

Chapter 8

CO: Kenneth Eward/Biografx 1997. Page 350: NASA/Science Photo Library, Photo Researchers, Inc. Page 351: Phillippe Plailly/Science Photo Library, Photo Researchers, Inc.

Chapter 9

CO: Kenneth Eward/Biografx 1997. Page 358: Mike Severns, Tom Stack & Associates. Page 362: University of Pennsylvania, Van Pelt Library. Page 357: Mike Severns, Tom Stack & Associates.

Chapter 10

CO: Kenneth Eward/Biografx 1997. Page 414: Dr. Paula Bruice. Page 414: Dr. Paula Bruice. Page 414: Dr. Paula Bruice. Page 414: Dr. Paula Bruice. Page 420: University of Pennsylvania, Van Pelt Library.

Chapter 11

CO: Kenneth Eward/Biografx 1997. Page 445: Corbis-Bettmann. Page 454: Archive Photos. Page 463: University of Pennsylvania, Van Pelt Library.

Chapter 12

CO: Kenneth Eward/Biografx 1997. Page 501: Dr. Jeremy Burgess/Science Photo Library, Photo Researchers, Inc.

Chapter 13

CO: Kenneth Eward/Biografx 1997. Page 566: Scott Camazine, Photo Researchers, Inc. Page 566: Simon Fraser/Science Photo Library, Photo Researchers, Inc. Page 570: The Perkin-Elmer Corporation.

Chapter 14

CO: Kenneth Eward/Biografx 1997. Page 613: University of Pennsylvania, Van Pelt Library. Page 614: University of Pennsylvania, Van Pelt Library.

Chapter 15

CO: Kenneth Eward/Biografx 1997. Page 680: Kathleen Campbell, Gamma-Liaison, Inc. Page 705: Corbis-Bettmann.

Chapter 16

CO: Kenneth Eward/Biografx 1997. Page 756: Grant Heilman Photography, Inc.

Chapter 17

CO: Kenneth Eward/Biografx 1997. Page 788: German Information Center.

Chapter 19

CO: Kenneth Eward/Biografx 1997. Page 892: Dr. Paula Bruice. Page 892: Dr. Paula Bruice. Page 896: Stanley Breeden, DRK Photo.

Chapter 20

CO: Kenneth Eward/Biografx 1997. Page 932: D. Goldberg, Sygma. Page 936: AP/Wide World Photos. Page 947: Photo Researchers, Inc.

Chapter 21

CO: Kenneth Eward/Biografx 1997. Page 957: Corbis-Bettmann. Page 971: Dr. Paula Bruice. Page: 974: Dr. Paula Bruice.

Chapter 22

CO: Kenneth Eward/Biografx 1997. Page 988: AP/Wide World Photos. Page 989: University of Pennsylvania, Van Pelt Library. Page 989: Corbis-Bettmann. Page 1016: UPI Corbis-Bettmann, Corbis-Bettmann. Page 1025: Eric L. Heyer, Grant Heilman Photography, Inc.

Chapter 23

CO: Kenneth Eward/Biografx 1997. Page 1032: Scott Camazine, Photo Researchers, Inc. Page 1033: D. Cavagnaro, David Cavagnaro. Page 1034: Procter & Gamble. Page 1035: Tom Tietz, Tony Stone Images. Page 1036: Kim Heacox, Tony Stone Images. Page 1040: Hulton Getty, Tony Stone Images. Page 1043: Corbis-Bettmann. Page 1044: Jerry Alexander, Tony Stone Images. Page 1054: University of Texas at Dallas.

Chapter 24

CO: Kenneth Eward/Biografx 1997. Page 1062: National Portrait Gallery, London. Page 1062: A. Barrington Brown, Photo Researchers, Inc. Page 1074: Photo Researchers, Inc. Page 1085: AP/Wide World Photos. Page 1088: Dr. Gopal Murti/Science Photo Library, Custom Medical Stock Photo, Inc. Page 1088: OMIKRON, Photo Researchers, Inc. Page 1089: Ken Eward/Science Source, Photo Researchers, Inc. Page 1061: Institute of Molecular and Cellular Biology.

Chapter 25

CO: Kenneth Eward/Biografx 1997. Page 1125: Billy Hustace, Tony Stone Images. Page 1128: Stephen Frisch, Stock Boston.

Chapter 26

CO: Kenneth Eward/Biografx 1997. Page 1164: John Gerlach, DRK Photo.

Chapter 28

CO: Kenneth Eward/Biografx 1997. Page 1230: Harvard University Archives.

Chapter 29

CO: Kenneth Eward/Biografx 1997. Page 1240: Bill Ivy, Tony Stone Images. Page 1240: Gregory G. Dimijian, Photo Researchers, Inc. Page 1241: AP/Wide World Photos. Page 1243: Hoffmann-La Roche, Inc.

Table of Contents

INDEX

Puromycin, 1088
PVC (poly(vinyl chloride)), 256,
    1115, 1134
Pyranoses, 885
Pyranoside, 888
Pyrene, 653
Pyridine, 285, 434, 441, 1151–58
    electrophilic aromatic substitution
        reactions of, 1153–54
    nucleophilic aromatic substitution
        reactions of, 1154–58
    properties of, 1153
    S_N2 reactions of, 1152
Pyridine nucleotide, 991–96
Pyridine-N-oxide, 1152, 1157–58
Pyridine ring, 992-93
Pyridinium chlorochromate (PCC),
    786
Pyridone, 1157
Pyridoxal phosphate (PLP), 1007–15
Pyridoxamine, 1010, 1011
Pyridoxine (vitamin B$_6$), 988, 1007,
    1008
    coenzyme derived from, 1007–15
Pyrimidine, 1062, 1063, 1103,
    1161–63
    cut out of restriction fragments,
        1094
Pyrolysis (cracking), 329
Pyrophosphate, 1070
Pyrophosphoric acid, 1065
Pyrrole, 285, 286, 1146–50
Pyrrolidine, 743–44, 1143, 1146
Pyruvate, 991
    conversion to acetyl-CoA, 1003
    decarboxylation of, 1001
    rate of reduction by NADH, 576
    reduction to lactate, 993
Pyruvate decarboxylase, 1001
Pyruvate dehydrogenase system, 1003
Pyruvic acid, 679

Quantitative structure-activity rela-
    tionship (QSAR), 1250–51
Quantum mechanics, 5
Quartet, 537, 538, 539
    doublet of doublets vs., 548
    of triplets, 549
Quaternary ammonium compounds,
    463–67
    Cope elimination, 466–67
    Hofmann elimination, 463–66
Quaternary ammonium ion, 463
Quaternary ammonium salts, 71, 468
Quaternary structure of proteins, 935,
    946–47
Quinazoline 3-oxides, 1243
Quinoline, 247, 1158–60
Quinones, 799–800
Quintet, 549–50
Quinuclidine, 1143

R, S system of nomenclature, 181–85,
    196–200
Racemic mixtures, 188, 201, 922–24
Racemization
    of amino acids, 1249–50
    complete, 378
    partial, 378
    PLP-catalyzed, 1008, 1009, 1013
Radiation
    electromagnetic, 494–96
    infrared, 495, 498
    rf, 524

Radical(s), 13
    addition of, 155–60
    alkyl, 156, 157
    defined, 156
    ethyl, 334
    fluorine, 340
    iodine, 340
    methyl, 29
    reactivity of, 331
    resonance structures for, 270
    stabilities of, 157, 278, 335, 341
    stratospheric ozone and, 349–51
    vinylic, 247–48
Radical anion, 247
Radical cation, 481–82
Radical chain reaction, 158, 331, 332
Radical formation, reactivity differ-
    ences for, 337–41
Radical inhibitors, 159, 348
Radical initiators, 1115–16
Radical intermediate, 155–58,
    219–20, 242, 247, 344–45
Radical polymerization, 1112,
    1113–17, 1123
Radical reactions, 347–49, 1173
Radical substitution, 330–41
    of benzylic and allylic hydrogens,
        341–43
    stereochemistry of, 344–45
Radio waves, 495
Radius, van der Waals, 87
Rancidity, 1035
Random (blind) screening, 1242–43
Random coil, 946, 947
Random copolymer, 1127
Raney, Murray, 780
Ranitidine (Zantac), 1237, 1245
Rate constant (k), 126–29
Rate-determining step, 130, 337–38
Rate law, 359–60
Rate-limiting step, 130
Rate of formation, 126–27
Rate of reaction. See under
    Reaction(s)
Rational drug design, 1101–4
Rayon, 896, 1111
R configuration, 181–85
Reaction(s). See also specific classes
    of reactions, e.g. Addition reac-
    tions; Elimination reaction(s);
    Enzyme-catalyzed reactions;
    Pericyclic reactions; Substitution
    reactions
    acid-catalyzed, 141
    of arene oxides, 450–52
    Cannizzaro, 856
    Chichibabin, 1155–56
    of compounds with a chirality
        center, 203–4, 337–39, 436
    concerted, 152, 1175
    Diels-Alder, 316–22, 1187, 1217
    direct displacement, 360–61
    of enantiomers, 203
    enantioselective, 1223
    endergonic, 118, 1066
    endothermic, 121
    of epoxides (oxiranes), 447–50
    exergonic, 118, 1066, 1071
    exothermic, 121
    extrusion, 1217
    first-order, 125–26, 372
    Fischer esterification, 696
    Hell-Volhard-Zelinski (HVZ), 825,
        922

Hunsdiecker, 847–48
    intermolecular, 370–72, 758, 964,
        965
    intramolecular, 370–72, 758,
        841–46, 963–66
    mechanism of, 114–16
    Michael, 831–33, 1216
    phosphoryl transfer, 1067, 1068–71
    polar, 1173
    of quaternary ammonium com-
        pounds, 463–67
        Cope elimination, 466–67
        Hofmann elimination, 463–66
    radical, 158, 331, 332–33, 341–45,
        347–49, 1173
    rate of, 125, 126
        catalyst and, 141, 953
        solvent and, 387–91
        UV/Vis spectroscopy to measure,
            575
    Reformatsky, 860
    regioselective, 137, 215, 216
    Sandmeyer, 639–40
    SNAr, 647
    Schiemann, 640–41
    second-order, 126, 359
    selenenylation, 826–27
    stereochemistry of, 215–28
        alkene addition reactions,
            216–27
        enzyme-catalyzed reactions,
            227–28
    stereoselective, 215, 216, 1222–23
    stereospecific, 216
    synthesis and, 164–65
    thermodynamic vs. kinetic control
        of, 312–16
    Wittig, 754–56, 1218–19
Reaction coordinate diagrams, 117,
    124
Reactivity-selectivity principle,
    337–41
Reagent(s)
    Edman's, 937, 939
    Gilman, 458
    Grignard. See Grignard reagents
    methylating, 391–93
    organolithium, 1156
    for oxidation-reduction reactions,
        778, 995
    for protein cleavage, 938–40
    Tollens', 788, 878
Rearrangement
    of carbocations, 143–46,
        376–77
    Claisen, 1193
    Cope, 1193
    McLafferty, 492
    ring-expansion, 439–40
    sigmatropic, 1174–75, 1191–97
    suprafacial, 1192, 1194, 1196
    Wagner-Meerwein, 1060
Receptors, 204
    drug, 1244–46
Recrystallization, fractional, 480
Red algae, 358
Red blood cells, normal vs. sickle,
    1088
Redox reactions. See Oxidation-
    reduction reactions
Red shift, 572
Reducing agents, 935, 991
Reducing sugars, 890
Reduction. See also Oxidation-reduc-

tion reactions
    Clemmensen, 616
    of monosaccharides, 876–77
    Rosenmund, 780
    Wolff-Kishner, 616, 745–46
Reductive amination, 745
re face, 756–58
Reference compound, 530–31
Reformatsky reaction, 860
Regioselective reaction, 137, 215, 216
Regioselectivity of E2 reaction, 399,
    409
Regnault, Henri Victor, 119
Relative configuration, 205–7
Relative rates, 964, 965
Remsen, Ira, 901
Replication fork, 1080
Replication of DNA, 1080
Resins
    epoxy, 1130–31
    in ion-exchange chromatography,
        920, 921
Resistance, drug, 700, 1249
Resolution of racemic mixtures, 201,
    922–24
Resonance, 265–66, 291
    diene stability and, 306
    electron donation and withdrawal
        by, 621–22
Resonance contributors, 265–66
    drawing, 266–71
Resonance (delocalization) energy,
    274–76, 280, 281, 1147
Resonance hybrid, 265–66, 272–74
Resonances (absorption signals), 527
Resonance stabilization, 1071
Resonance-stabilized carbanion, 1001
Resonance structures of allylic radi-
    cals, 343
Restriction endonucleases, 1090–91,
    1103
Restriction fragment length polymor-
    phisms (RFLPs), 1096
Restriction fragments, 1091–95
(Z)-11-Retinal, 301
Retinol (vitamin A), 756, 988, 989,
    1045–46
Retrosynthetic analysis, 252–55,
    424–25, 1209–19
    of dioxygenated compounds,
        1214–19
    disconnections, 1209–14
Retroviruses, 1101
rf radiation, 524
Rhodopsin, 111, 1045
Ribavin, 1255
Ribitol, 996
Riboflavin (vitamin B2), 988, 996
Ribonucleic acid (RNA), 1061, 1090,
    1111
    bases in, 1063, 1064
    biosynthesis of, 348, 1074,
        1080–82
    heterocyclic bases in, 1019
    hydrolysis of, 1076, 1078
    messenger (mRNA), 1080–82,
        1084, 1085–87
    nucleotides of, 1064
    ribosomal (rRNA), 1082
    sugar component of, 897, 898
    transfer (tRNA), 1082–84, 1085,
        1087
Ribonucleotides, 1064
D-Ribose, 874–75, 897, 1062

Periodic Table of the Elements

Main groups

Main groups

Transition metals

1A																		8A
1 **H** 1.00794	2A											3A	4A	5A	6A	7A		2 **He** 4.00260
3 **Li** 6.941	4 **Be** 9.01218											5 **B** 10.81	6 **C** 12.011	7 **N** 14.0067	8 **O** 15.9994	9 **F** 18.998403		10 **Ne** 20.1797
11 **Na** 22.98977	12 **Mg** 24.305	3B	4B	5B	6B	7B	8B	8B	8B	1B	2B	13 **Al** 26.98154	14 **Si** 28.0855	15 **P** 30.97376	16 **S** 32.066	17 **Cl** 35.453		18 **Ar** 39.948
19 **K** 39.0983	20 **Ca** 40.078	21 **Sc** 44.9559	22 **Ti** 47.88	23 **V** 50.9415	24 **Cr** 51.996	25 **Mn** 54.9380	26 **Fe** 55.847	27 **Co** 58.9332	28 **Ni** 58.69	29 **Cu** 63.546	30 **Zn** 65.39	31 **Ga** 69.72	32 **Ge** 72.61	33 **As** 74.9216	34 **Se** 78.96	35 **Br** 79.904		36 **Kr** 83.80
37 **Rb** 85.4678	38 **Sr** 87.62	39 **Y** 88.9059	40 **Zr** 91.224	41 **Nb** 92.9064	42 **Mo** 95.94	43 **Tc** (98)	44 **Ru** 101.07	45 **Rh** 102.9055	46 **Pd** 106.42	47 **Ag** 107.8682	48 **Cd** 112.41	49 **In** 114.82	50 **Sn** 118.710	51 **Sb** 121.757	52 **Te** 127.60	53 **I** 126.9045		54 **Xe** 131.29
55 **Cs** 132.9054	56 **Ba** 137.33	57 ***La** 138.9055	72 **Hf** 178.49	73 **Ta** 180.9479	74 **W** 183.85	75 **Re** 186.207	76 **Os** 190.2	77 **Ir** 192.22	78 **Pt** 195.08	79 **Au** 196.9665	80 **Hg** 200.59	81 **Tl** 204.383	82 **Pb** 207.2	83 **Bi** 208.9804	84 **Po** (209)	85 **At** (210)		86 **Rn** (222)
87 **Fr** (223)	88 **Ra** 226.0254	89 **†Ac** 227.0278	104 **Rf** (261)	105 **Db** (262)	106 **Sg** (263)	107 **Bh** (262)	108 **Hs** (265)	109 **Mt** (266)	110 (269)	111 (272)	112 (277)							

*Lanthanide series	58 **Ce** 140.12	59 **Pr** 140.9077	60 **Nd** 144.24	61 **Pm** (145)	62 **Sm** 150.36	63 **Eu** 151.96	64 **Gd** 157.25	65 **Tb** 158.9254	66 **Dy** 162.50	67 **Ho** 164.9304	68 **Er** 167.26	69 **Tm** 168.9342	70 **Yb** 173.04	71 **Lu** 174.967
†Actinide series	90 **Th** 232.0381	91 **Pa** 231.0359	92 **U** 238.0289	93 **Np** 237.048	94 **Pu** (244)	95 **Am** (243)	96 **Cm** (247)	97 **Bk** (247)	98 **Cf** (251)	99 **Es** (252)	100 **Fm** (257)	101 **Md** (258)	102 **No** (259)	103 **Lr** (260)

Masses in parentheses apply to mass number of the longest-lived isotope.

s block elements

p block elements

d block elements

f block elements

Common Functional Groups

Alkane RCH_3

Alkene
$$\text{C}=\text{C} \qquad \text{C}=\text{CH}_2$$
internal   terminal

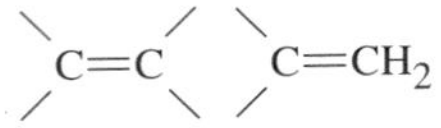

Alkyne $RC\equiv CR$ $RC\equiv CH$
internal   terminal

Nitrile $RC\equiv N$

Ether $R-O-R$

Thiol RCH_2-SH

Sulfide $R-S-R$

Disulfide $R-S-S-R$

Epoxide

Aniline

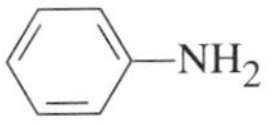

Phenol

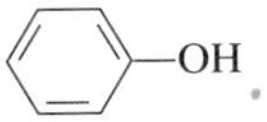

Carboxylic acid
$$R-\overset{\overset{\displaystyle O}{\|}}{C}-OH$$

Acyl chloride
$$R-\overset{\overset{\displaystyle O}{\|}}{C}-Cl$$

Acid anhydride
$$R-\overset{\overset{\displaystyle O}{\|}}{C}-O-\overset{\overset{\displaystyle O}{\|}}{C}-R$$

Ester
$$R-\overset{\overset{\displaystyle O}{\|}}{C}-OR$$

Aldehyde
$$R-\overset{\overset{\displaystyle O}{\|}}{C}-H$$

Ketone
$$R-\overset{\overset{\displaystyle O}{\|}}{C}-R$$

	primary	secondary	tertiary
Alkyl halide	RCH_2-X X = F, Cl, Br, I	$R-\overset{\overset{\displaystyle R}{\|}}{C}H-X$	$R-\overset{\overset{\displaystyle R}{\|}}{\underset{\underset{\displaystyle R}{\|}}{C}}-X$
Alcohol	RCH_2-OH	$R-\overset{\overset{\displaystyle R}{\|}}{C}H-OH$	$R-\overset{\overset{\displaystyle R}{\|}}{\underset{\underset{\displaystyle R}{\|}}{C}}-OH$
Amine	$R-NH_2$	$R-\overset{\overset{\displaystyle R}{\|}}{N}H$	$R-\overset{\overset{\displaystyle R}{\|}}{\underset{\underset{\displaystyle R}{\|}}{N}}$
Amide	$R-\overset{\overset{\displaystyle O}{\|}}{C}-NH_2$	$R-\overset{\overset{\displaystyle O}{\|}}{C}-NHR$	$R-\overset{\overset{\displaystyle O}{\|}}{C}-NR_2$